D0488184

LIBRARY

Total Automotive Technology

FOURTH EDITION

Anthony E. Schwaller

We Encourage Professionalism

Through Technical Certification

THOMSON
DELMAR LEARNING
Australia Canada Mexico Singapore Spain United Kingdom United States

Total Automotive Technology, 4th Edition
Anthony E. Schwaller

Vice President, Technology and Trades SBU:
Alar Elken

Editorial Director:
Sandy Clark

Acquisitions Editor:
David Boelio

Development Editor:
Matthew Thouin

Marketing Director:
Dave Garza

Channel Manager:
Bill Lawrensen

Marketing Coordinator:
Mark Pierro

Production Director:
Mary Ellen Black

Production Coordinator:
Dawn Jacobson

Project Editor:
Ruth Fisher

Art/Design Specialist:
Cheri Plasse

Editorial Assistant:
Kevin Rivenberg

Printed in the United States of America
3 4 5 XX 06

For more information contact
Delmar Learning
Executive Woods
5 Maxwell Drive, PO Box 8007
Clifton Park, NY 12065-8007
Or find us on the World Wide Web at
www.delmarlearning.com

Library of Congress Cataloging-in-Publication Data:
Schwaller, Anthony E.
Total automotive technology / Anthony Schwaller.—4th ed.
p. cm.
Includes bibliographical references and index.
ISBN 1-4018-2476-5
1. Automobiles, Design and construction. 2. Automobiles—Maintenance and repair. I. Title.
TL240.S354 2004
629.28'72—dc22
2004045974

ISBN: 1-4018-2476-5

Contents

SECTION 7 VEHICLE SUSPENSION AND CONTROL SYSTEMS 808

SECTION 8 VEHICLE ACCESSORY SYSTEMS 935

CHAPTER 51
Heating and Ventilation Systems 966

CHAPTER 52
Cruise Control Systems . . 987

CHAPTER 53
Auxiliary and Electrical Systems 1005

Preface

INTRODUCTION

The importance of the automobile in our society is without question. As a prime mover of people, the automobile contributes daily to our economic and social systems. This basic form of transportation gives people the freedom and choice to travel farther from their homes to pursue careers, indulge in leisure activities, and act as consumers of a wide variety of goods and services.

Increased use of the automobile over the years has resulted in a continual evolution in design to achieve faster, more efficient, cleaner, more economical, more reliable, and safer vehicles. This pattern of change continues and is accelerating. Each year innovations resulting from advancing technology and computers appear on new automobile models. The modern automobile is vastly different from the automobile of years ago. Significant changes continue to occur. Automobiles are now designed to produce fewer emissions and less pollution. Today, electronic and computer monitoring systems provide a means of checking and controlling emissions, while still improving fuel efficiency. In addition, automotive computers are now used to monitor and control engine combustion, ignition timing and spark, braking systems, transmissions, suspension, and many other systems. Such monitoring in engine design and vehicle configuration has helped to improve fuel economy, ease of driving, vehicle safety, and control of the vehicle.

The increasing sophistication of electronic and computer control systems has led to wider applications in automobiles, resulting in more efficient and safer operation. The implementation of on-board computers to monitor and control performance and enhanced mechanical systems also results in improved reliability. Each new year will bring further and more innovative changes in automobile design. Future developments in communications, electronics, materials, manufacturing processes, and energy will have significant effects on the automobile.

SCOPE OF THE TEXT

The automobile is a composite of many complex systems. These systems require routine diagnosis, maintenance, and service, and at times may require more extensive service. The automotive technician faces the challenge of understanding each of the systems found in the automobile and the interrelationships of these systems. This understanding rests on a knowledge of basic physical principles. Another challenge the technician faces is the need to stay current with changes as each new model year appears.

Total Automotive Technology is a basic text designed to help students achieve the necessary understanding of automotive principles. On this foundation the student will build skills through actual lab and shop work. All of the following essential information is thoroughly explained for each automotive system: scientific principles, theory of operation, safety considerations, diagnosis, troubleshooting, and service. Computers are prominent in most automobile systems. Thus, the use of computer controls has been included throughout the text.

TOTAL AUTOMOTIVE TECHNOLOGY TEXT SUPPLEMENTS

The text is supported by several well-designed supplements that will assist the instructor in presenting the content. In addition, the student will find that the supplements provide many additional opportunities for learning and for hands-on application of theory to actual shop situations. The supplements consist of a *Student Workbook,* an *Instructor's Guide,* and an e.resource™ CD-ROM.

- The ***Student Workbook*** provides numerous questions to help reinforce student learning. The questions consist of multiple choice, matching, and identification questions, in addition to crossword puzzles and ASE-type multiple choice questions. The workbook also provides ASE/NATEF-correlated worksheets covering typical service procedures. Each worksheet includes a statement of the task objective, appropriate references to the related text chapter, equipment and materials needed, safety precautions, step-by-step procedures, and questions relating to the service procedures, as required.
- The ***Instructor's Guide*** contains the correct answers for the review questions in the core text and the activities in the Student Workbook.
- The ***e.resource***™ CD-ROM contains several components to assist the instructor with the presentation of the material:
  1. ***PowerPoint*** presentations for every chapter of the text, which may be modified to suit the instructor's lecture style.

2. An ***Image Library*** containing hundreds of images from the text, which may be used to customize and visually enhance the PowerPoint presentations.
3. ***Computerized Test Bank*** in ExamView containing an average of 35 questions per chapter, offering an easy way to generate tests, quizzes, and homework assignments; answers are provided.
4. Eight ***ASE-style Practice Exams,*** each covering one of the main eight ASE certification areas, to help prepare students for certification.
5. Over 150 ***Electronic Lesson Plans*** based on concrete learning and performance objectives offer instructors a structured learning approach, with correlations to relevant ASE task lists, chapter number in the text, and topics for classroom discussion.

The combination of *Total Automotive Technology* and the supplements forms a comprehensive, unequaled teaching/learning package. The text and supplements supply the basic automotive theory and general diagnosis and service procedures.

ABOUT THE AUTHOR

Anthony Schwaller has been involved with automotive technology for many years, beginning as an automotive service technician. He began his teaching career as a technical trainer for General Motors in Detroit, Michigan. After leaving Detroit, he taught automotive technology at Eastern Illinois University, Charleston, Illinois, and St. Cloud University, St. Cloud, Minnesota, where he is currently serving both as a professor and administrator. The author received his B.S. and M.S. from the University of Wisconsin-Stout and his Ph.D. from Indiana State University. He has authored seven other textbooks, over forty articles, and has presented more than forty-five papers at various conferences in the field of technology.

Acknowledgments

Many organizations and people contributed to the development of this text. Among the companies that provided valuable technical information, photographs, and illustrations are:

AE Piston Products, Inc.

Allied Aftermarket Division

American Honda Motor Co., Inc.

American Isuzu Motors Inc.

American Petroleum Institute

Americam Society for Testing and Materials

API

ASE (National Institute for Automotive Service Excellence)

ASTM International

Austin Rover Group Limited

BMW NA, LLC

Breton Publishers

Broadhead-Garrett

Champion Spark Plug Company

Chilton Book Company

Clayton Industries

Cooper Tire & Rubber Company

CR Industries

CRC Industries, Inc.

CTS Corporation, Electromechanical Group

Cummins Inc.

DaimlerChrysler Corporation

Dana Corporation

Davis Publications, Inc.

DCA Educational Products, Inc.

Eaton Corporation

Echlin Corporation

EIS Brake Parts, Division Standard Motor Products, Inc.

Federal-Mogul Corporation

Ferodo America

Firestone Tire and Rubber Company

First Brands Corporation (Formerly Union Carbide Corp.)

Ford Motor Company

General Fire Extinguisher Corporation

General Motors Corporation
- Buick Motor Division
- Chevrolet-Pontiac-Canada Group
- Delco-Moraine
- Delco-Remy
- General Motors Product Service Training
- General Motors Proving Grounds
- Harrison Radiator Division
- Oldsmobile Division
- Rochester Products
- United Delco

Goodyear Tire and Rubber Company

Hastings Manufacturing

Honeywell International Inc.

Hunter Engineering

Ignition Manufacturer's Institute

Ingersoll-Rand Power Tool Division

Jasper Engines and Transmissions

K.D. Tools

Lisle Corporation

Loctite Corporation

Lucas Electrical, Parts and Service

Masco Industries

Mazda Motor Corporation

Mazda Motor of America, Inc.

Mitchell Manuals

Mitsubishi Motors North America, Inc.

MOTOR Information Systems

Nissan North America, Inc.

NILFISK of America, Inc.

NTN Bearing Corporation of America

OTC Division of Sealed Power Corp.

Peugeot Motors of America, Inc.
Phillips 66 Company
Plumley Companies, Inc.
Robertshaw Controls Company
Sachs North America
SAE
School Products Co., Inc.
Sellstrom Manufacturing Co.
Sioux Tools, Inc.
Snap-On Tools
Society of Automotive Engineers, Inc.
SPX Corporation
Stanadyne, Diesel Systems Division
Sun Electric Corporation
Thermadyne
Tire Industry Safety Council
Toyota Motor Sales, USA, Inc.
Tracer Products
Tune-Up Manufacturers Institute
United Technologies (Formerly American Bosch)
US EPA
Volkswagen of America, Inc.
Volvo Cars of North America
Western Emergency Equipment
Worldwatch Institute

The author would like to thank Margaret Magnarelli for her participation in the development of the success stories.

A number of instructors reviewed the text and provided suggestions for improvements. Their assistance is appreciated.

David Foster
Austin Community College
Austin, TX

Earl J. Friedell
Dekalb Technical College
Clarkston, GA

John Linden
Midas International
Pittsburgh, PA

Glenn Marcucio
HFM Career and Technical Center
Johnstown, NY

James Meyer
Wheeling High School
Wheeling, IL

Michael M. Millet
Salt Lake Community College
West Valley City, UT

Stan Moczulski
Career Institute of Technology
Easton, PA

Craig Schoenberger
Garden City Community College
Garden City, KS

Mike Setzer
Sheridan Technical Center
Hollywood, FL

DEDICATION

This book is dedicated to several people. First I would like to thank my wife, Renee, and my sons, Matthew and Joshua, and their families for their continued support and understanding during the writing of this textbook. In addition, I would like to thank my parents, Omer and Garnett Schwaller, for the drive and motivation they instilled in me as a child. Without these ingredients, the writing of this textbook would not have been possible.

Special thanks should also go to the employees of Dondelinger Incorporated, Mills Automotive Group, and NAPA Auto Parts Stores in Brainerd, Minnesota, for their help in providing parts for many of the photographs in the textbook. In addition, I would like to thank Kattie Fuchs and Michelle Normand for their help in preparing the manuscript.

Features of the Text

The content is divided into eight sections and 53 chapters. Each chapter includes a number of learning aids to help students in their study of automotive mechanics.

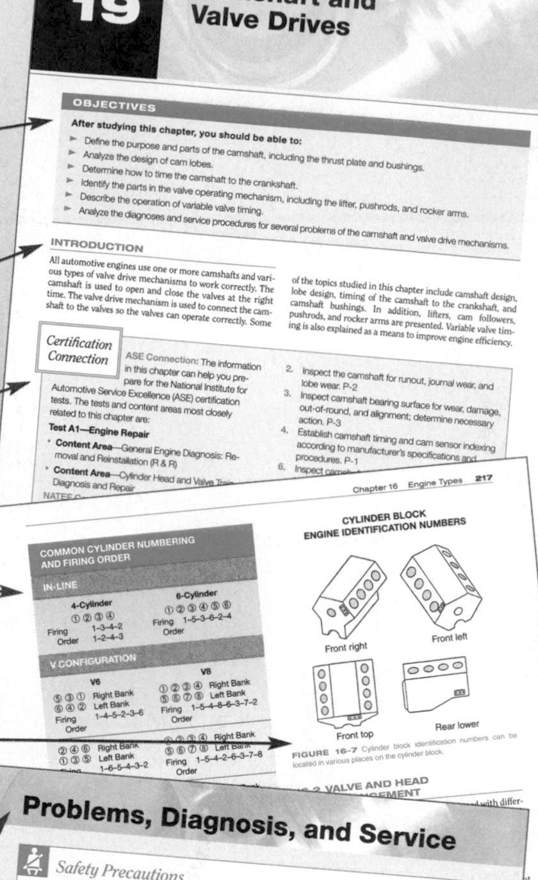
CHAPTER 19

Camshaft and Valve Drives

OBJECTIVES

After studying this chapter, you should be able to:

- Define the purpose and parts of the camshaft, including the thrust plate and bushings.
- Analyze the design of cam lobes.
- Determine how to time the camshaft to the crankshaft.
- Identify the parts in the valve operating mechanism, including the lifter, pushrods, and rocker arms.
- Describe the operation of variable valve timing.
- Analyze the diagnoses and service procedures for several problems of the camshaft and valve drive mechanisms.

INTRODUCTION

All automotive engines use one or more camshafts and various types of valve drive mechanisms to work correctly. The camshaft is used to open and close the valves at the right time. The valve drive mechanism is used to connect the camshaft to the valves so the valves can operate correctly. Some of the topics studied in this chapter include camshaft design, lobe design, timing of the camshaft to the crankshaft, and camshaft bushings. In addition, lifters, cam followers, pushrods, and rocker arms are presented. Variable valve timing is also explained as a means to improve engine efficiency.

Certification Connection

ASE Connection: The information in this chapter can help you prepare for the National Institute for Automotive Service Excellence (ASE) certification tests. The tests and content areas most closely related to this chapter are:

Test A1—Engine Repair

- Content Area—General Engine Diagnosis: Removal and Reinstallation (R & R)
- Content Area—Cylinder Head and Valve Train Diagnosis and Repair

2. Inspect the camshaft for runout, journal wear, and lobe wear. P-2
3. Inspect camshaft bearing surface for wear, damage, out-of-round, and alignment; determine necessary action. P-3
4. Establish camshaft timing and cam sensor indexing according to manufacturer's specifications and procedures. P-1

Chapter 16 Engine Types 217

COMMON CYLINDER NUMBERING AND FIRING ORDER

CYLINDER BLOCK ENGINE IDENTIFICATION NUMBERS

FIGURE 16-7 Cylinder block identification numbers can be located in various places on the cylinder block.

OBJECTIVES

State the expected learning outcomes that will take place as a result of studying the chapter.

INTRODUCTION

Provides a statement of the intent of the chapter.

CERTIFICATION CONNECTION

Identifies the ASE/NATEF tasks that parallel the content of each chapter.

TABLES

Summarize important points, measurements, statistics, and troubleshooting.

FIGURES

Numerous line drawings and photographs illustrate concepts and show current equipment, components, and systems.

PROBLEMS, DIAGNOSIS, AND SERVICE

Problems are stated, then diagnosis and service procedures are suggested to effectively correct the problem.

SAFETY PRECAUTIONS

For most Problems/Diagnosis/Service sections, specific precautions are listed.

PARTS LOCATOR PHOTO NUMBERS

Reference specific photos in the Parts Locator section which provide a real world illustration of the text discussion.

CAR CLINICS

Describe a common problem with an automobile, followed by a tip for diagnosing the problem or the solution to the problem (two or three per chapter).

SERVICE MANUAL CONNECTION

Describes various specifications and procedures that can be found in service manuals pertaining to chapter content.

CHAPTER SUMMARY

Highlights the important concepts covered in the chapter; also serves as a ready reference.

TERMS TO KNOW

Highlights vocabulary to be learned; each term is highlighted and defined in the chapter. Definitions are also provided in the Glossary at the back of the text.

REVIEW QUESTIONS

Reinforce and test students' comprehension of content. Four types of questions are provided: standard multiple choice, ASE-style multiple choice, essay, and short answer.

APPLIED ACADEMICS

Shows how automotive technology is interdisciplinary and describes how current automotive technology is linked to other areas of study including:

- Research/development/design
- Performance and testing
- Safety
- The future
- Manufacturing
- The environment
- Service and education
- History
- Science and mathematics

OTHER IMPORTANT FEATURES OF THE TEXT INCLUDE:

- An in-depth, easy reference full-color Table of Contents
- A full color Parts Locator photo layout showing major automotive systems, subsystems, and components
- Appendix of Abbreviations—a list of abbreviations commonly used in automotive technology, including abbreviations relating to newer technologies, such as computer control
- Appendix of Automotive Web sites—a list of Web sites for automotive-related companies and the products, information, or services they provide
- Glossary of definitions in English and Spanish of all terms introduced in the chapters

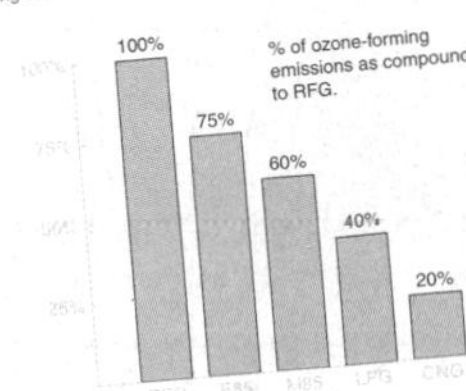

Parts Locator

Today's automobiles are a complex combination of many parts and systems. Often these parts and systems are very difficult to locate or interface with other parts, making correct troubleshooting, diagnosis, and service difficult for the service technician. For example, computers are interfaced with brakes, fuel systems are combined with electronics, ignition systems are combined with various sensors, and most accessory systems require mechnical as well as electrical parts, sensors, and relays.

This section in the textbook is called the Parts Locator. The Parts Locator will help you locate many of the vehicle parts that are found on a typical automobile by showing their location. As you read through each chapter, a reference number will be made to the Parts Locator. Match this reference number to the number under the photo in the Parts Locator to identify the exact photo discussed in that section of this textbook. By viewing the correct photo, you will get a good idea of the location and appearance of the part you are studying.

1. Keyway, Chapter 5

2. Overhead Camshaft, Chapter 12

3. Timing Belt, Chapter 16

4. Oil Passageway, Chapter 17

5. Crankshaft Seals, Chapter 17

6. Valve Seats (positioning), Chapter 18

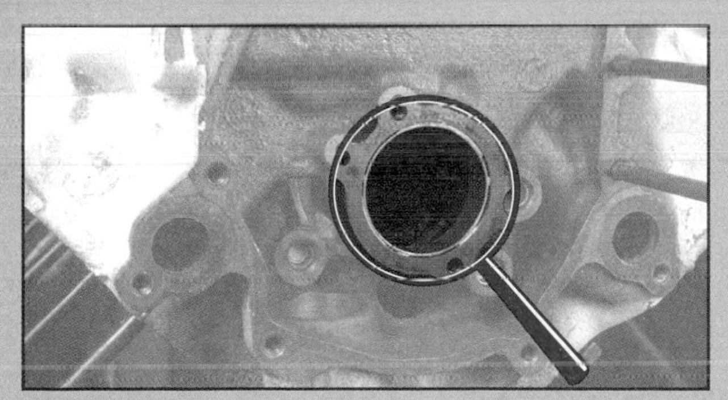

7. Camshaft Bushings, Chapter 19

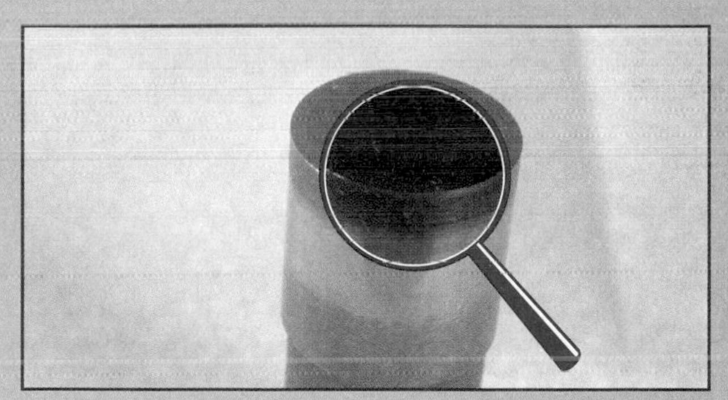

8. Worn Lifters, Chapter 19

9. ECII Oils, Chapter 20

10. Oil Pump, Chapter 21

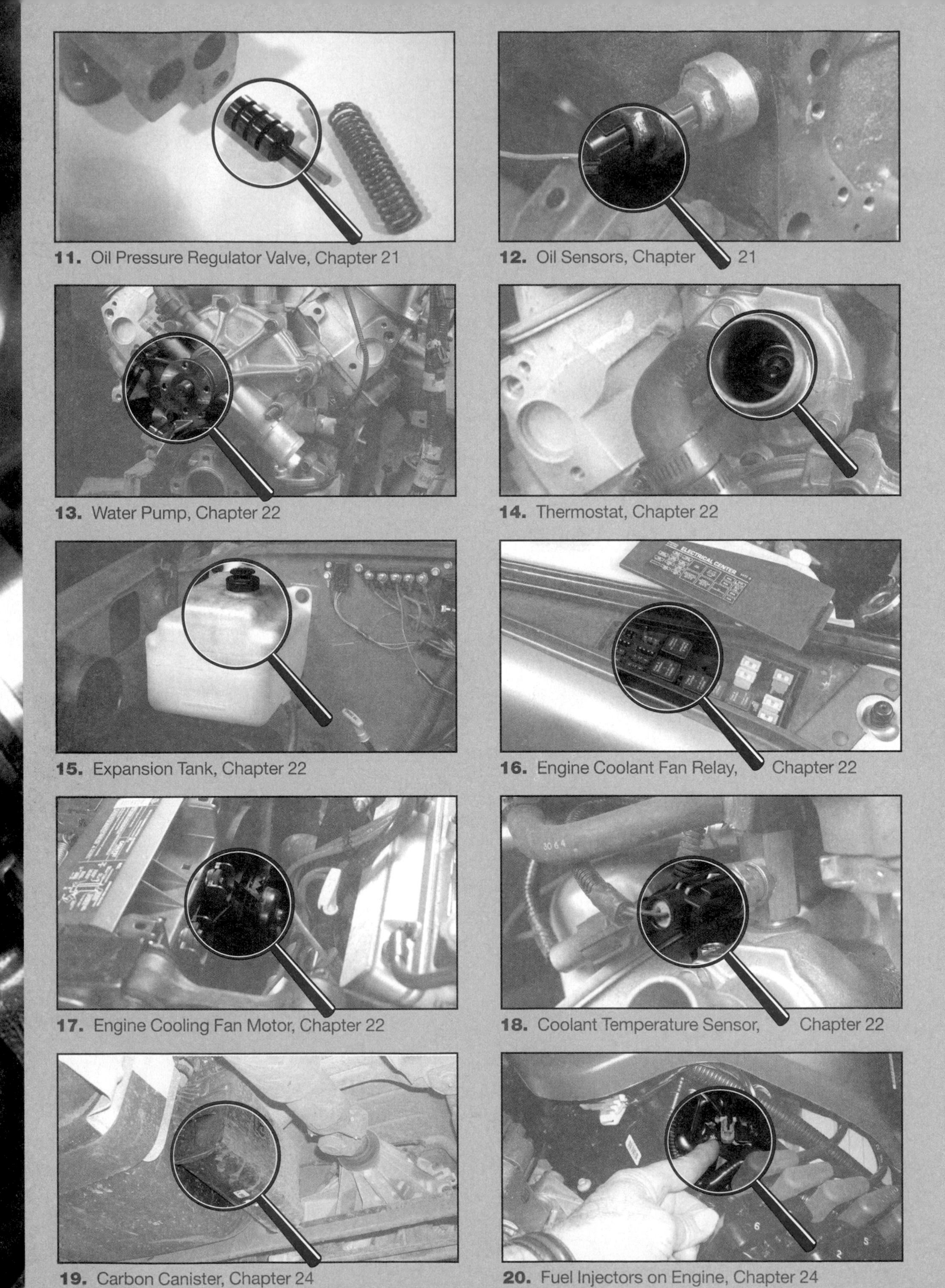

11. Oil Pressure Regulator Valve, Chapter 21

12. Oil Sensors, Chapter 21

13. Water Pump, Chapter 22

14. Thermostat, Chapter 22

15. Expansion Tank, Chapter 22

16. Engine Coolant Fan Relay, Chapter 22

17. Engine Cooling Fan Motor, Chapter 22

18. Coolant Temperature Sensor, Chapter 22

19. Carbon Canister, Chapter 24

20. Fuel Injectors on Engine, Chapter 24

21. Fuel Tank Inlet, Chapter 24

22. Fuel Metering Unit, Chapter 24

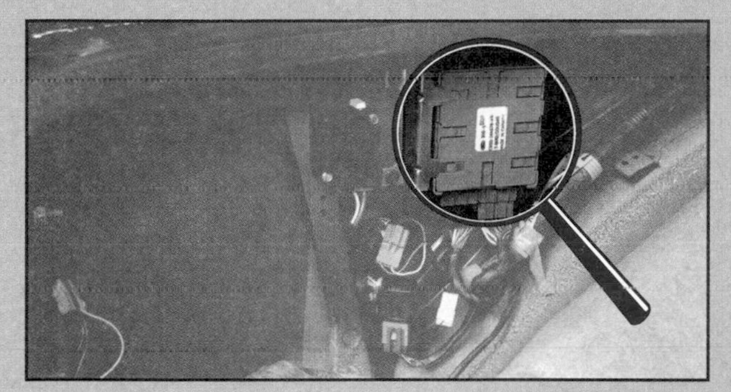

23. Fuel System Inertia Switch, Chapter 24

24. Fuel Pressure Regulator, Chapter 24

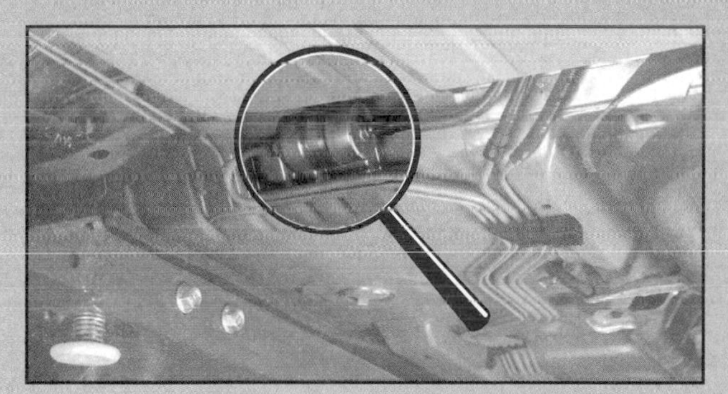

25. Fuel Filters, Chapter 24

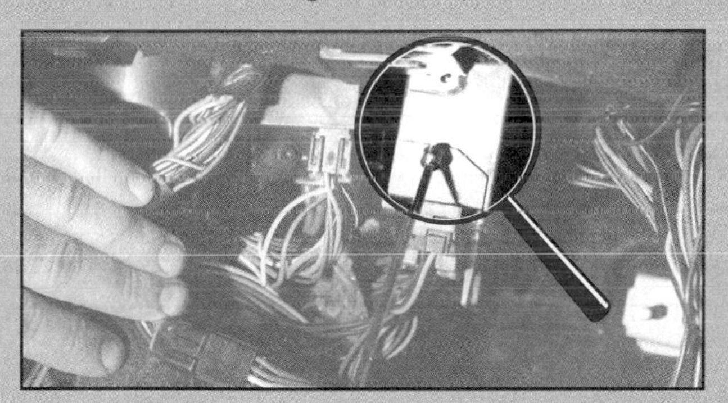

26. Computer, Chapter 25

27. Electronic Carburetor Solenoid, Chapter 25

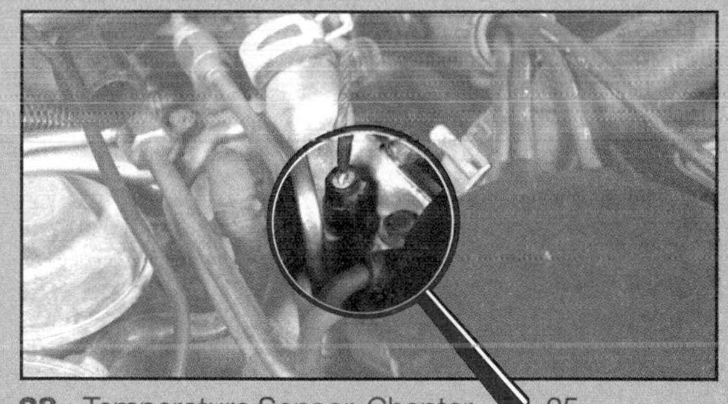

28. Temperature Sensor, Chapter 25

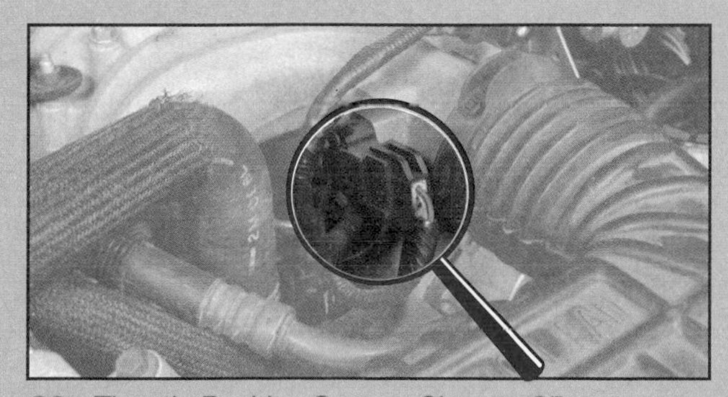

29. Throttle Position Sensor, Chapter 25

30. Engine Speed Sensor, Chapter 25

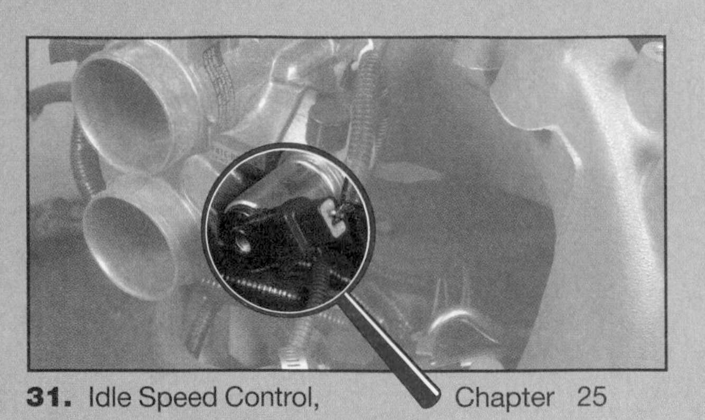

31. Idle Speed Control, Chapter 25

32. Throttle Body Fuel Injector, Chapter 26

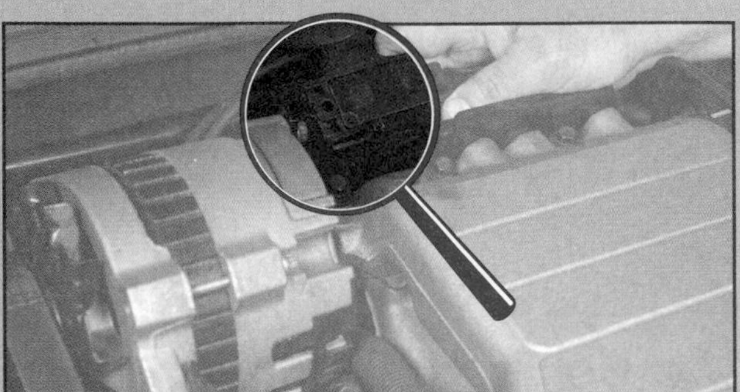

33. Manifold Absolute Pressure (MAP) Sensor, Chapter 26

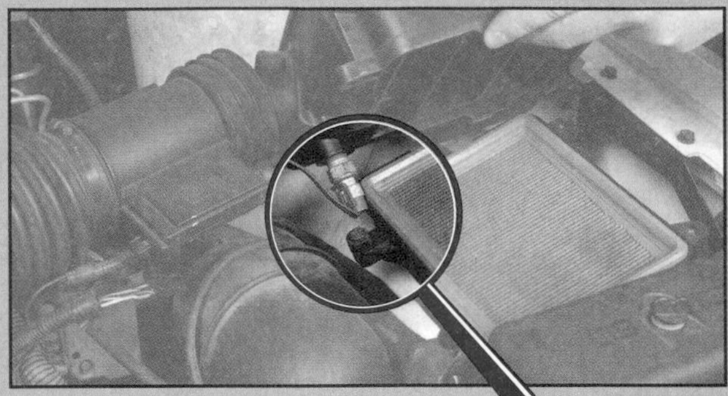

34. Manifold Absolute Temperature (MAT) Sensor, Chapter 26

35. Port Fuel Injectors, Chapter 26

36. Fuel Rails, Chapter 26

37. Throttle Blade, Chapter 26

38. Mass Airflow (MAF) Sensor, Chapter 26

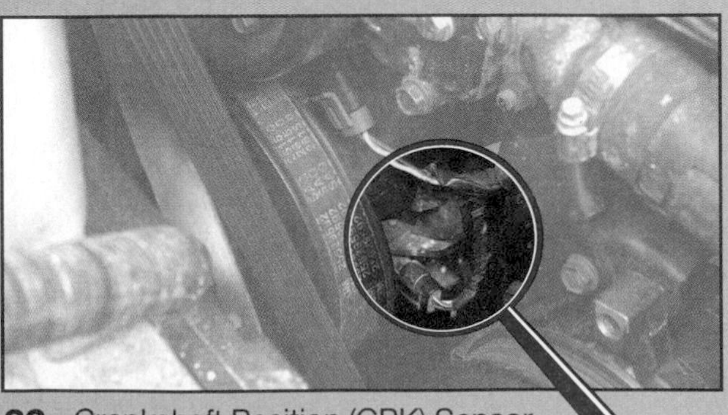

39. Crankshaft Position (CPK) Sensor, Chapter 26

40. Air Cleaner Location, Chapter 27

41. Air Cleaner, Chapter 27

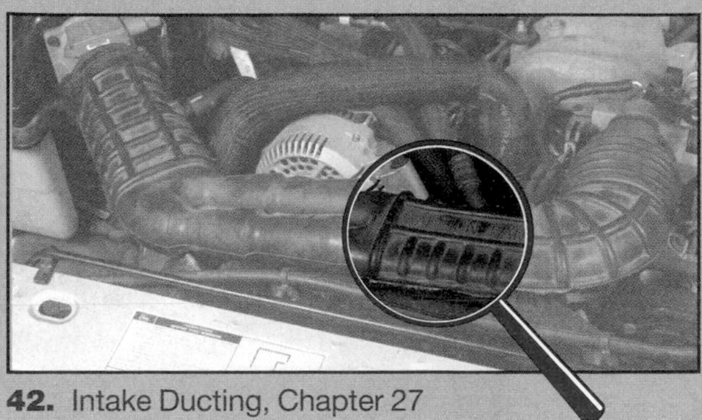

42. Intake Ducting, Chapter 27

43. Intake Manifold, Chapter 27

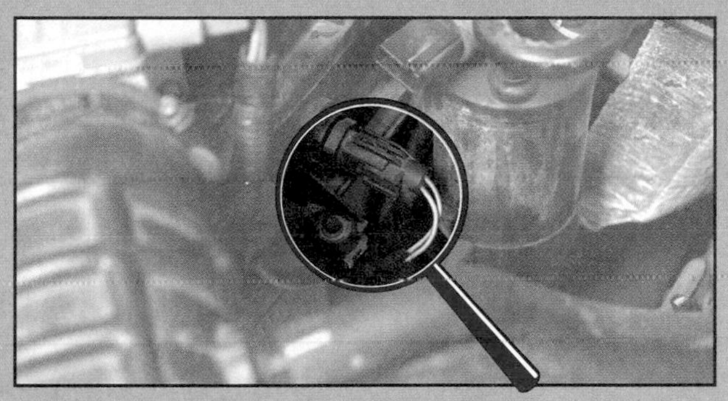

44. Intake Air Temperature (IAT) Sensor, Chapter 27

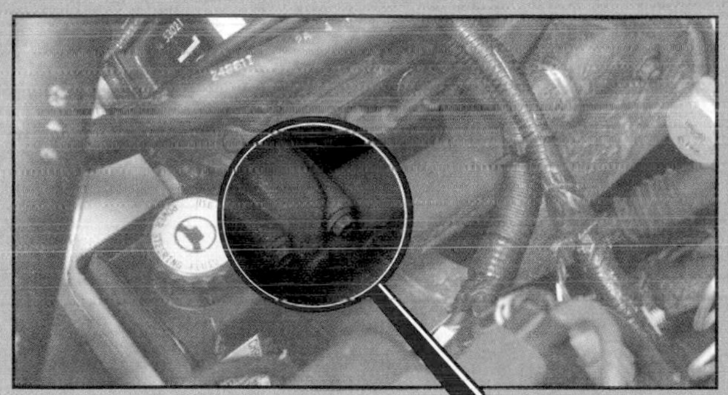

45. Exhaust Manifolds, Chapter 27

46. Exhaust Resonator, Chapter 27

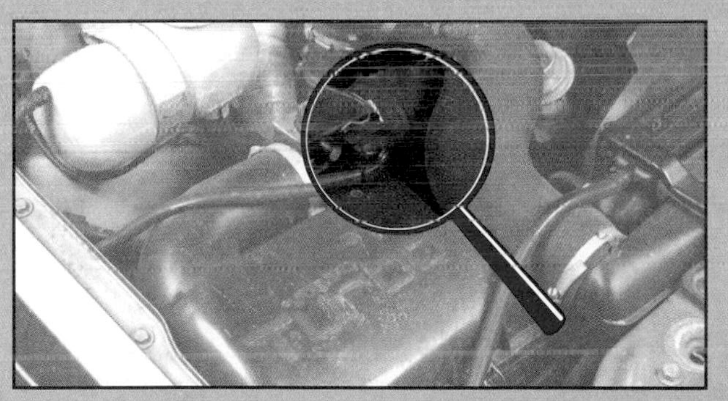

47. Turbocharger Ducting, Chapter 28

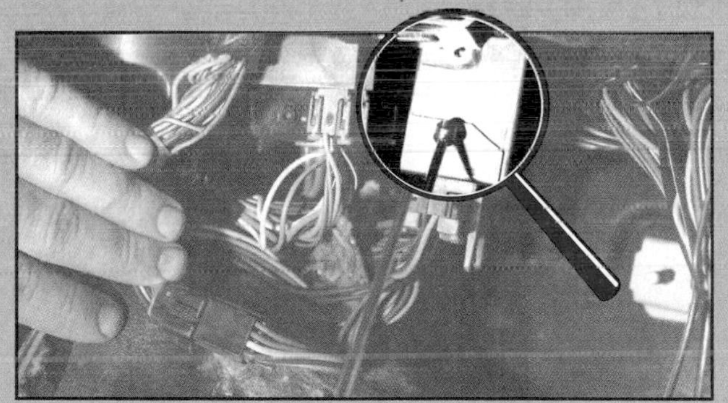

48. Computer in Vehicle, Chapter 30

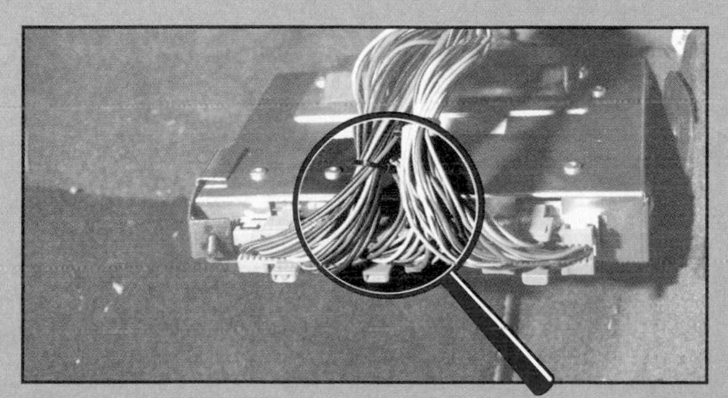

49. Computer in Vehicle, Chapter 30

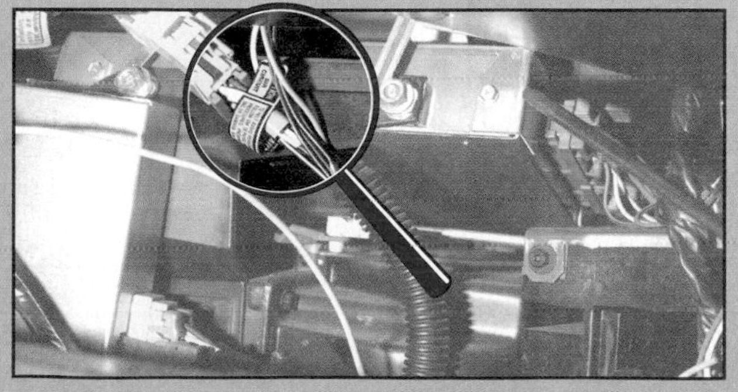

50. Computer in Vehicle, Chapter 30

51. Computer in Vehicle, Chapter 30

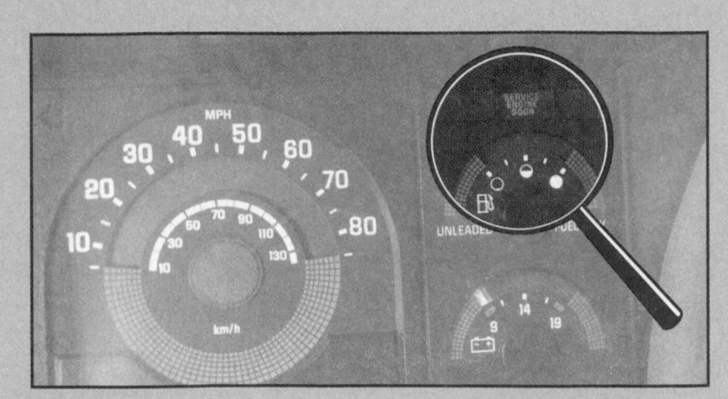

52. Service Engine Soon Light on Dashboard, Chapter 30

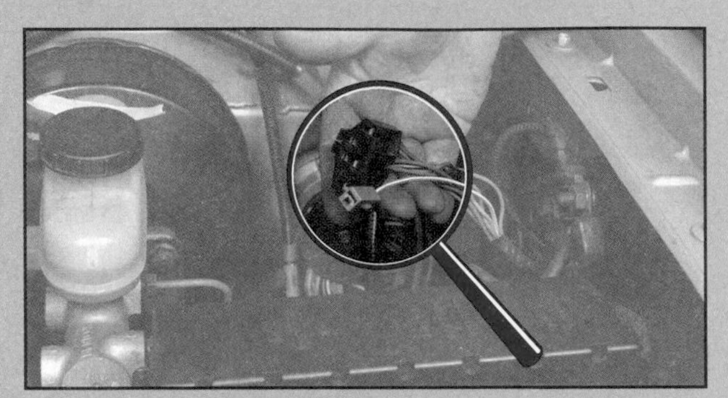

53. DLC Connector, Chapter 30

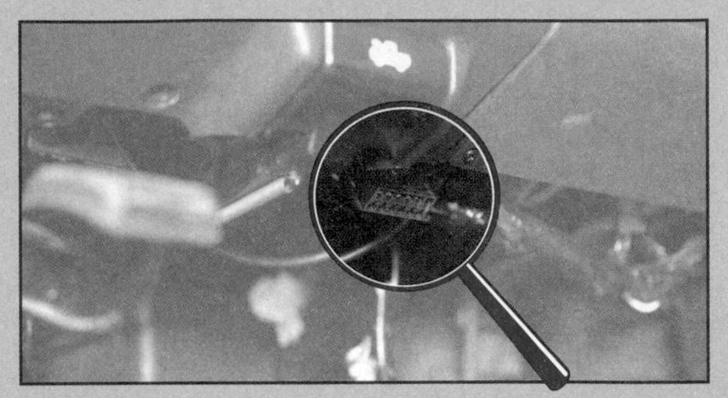

54. DLC Connector, Chapter 30

55. Sight Glass on Battery, Chapter 31

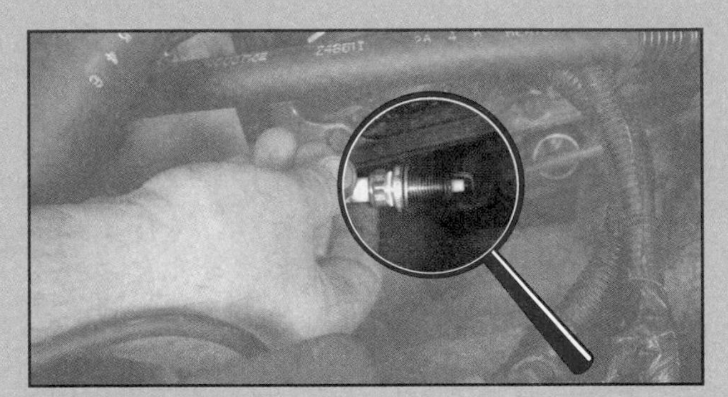

56. Spark Plug Tip, Chapter 32

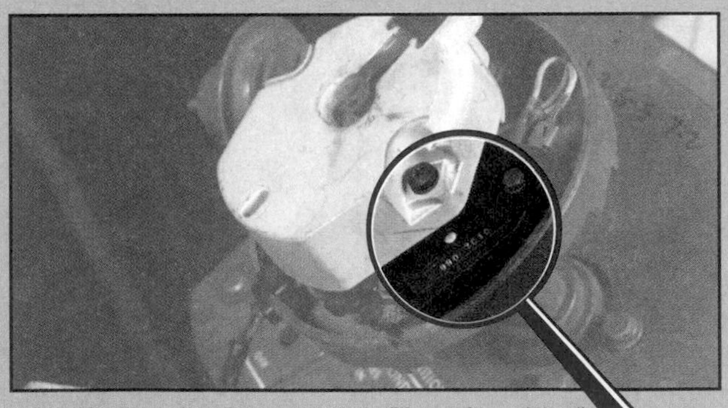

57. Electronic Control Unit (Transistorized), Chapter 33

58. Distributorless Ignition System Coil, Chapter 33

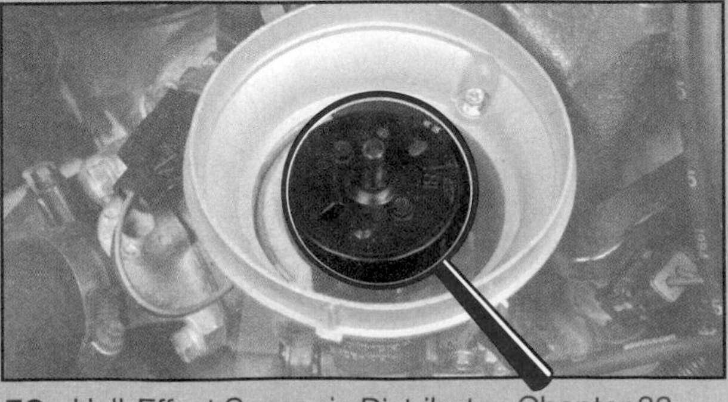

59. Hall-Effect Sensor in Distributor, Chapter 33

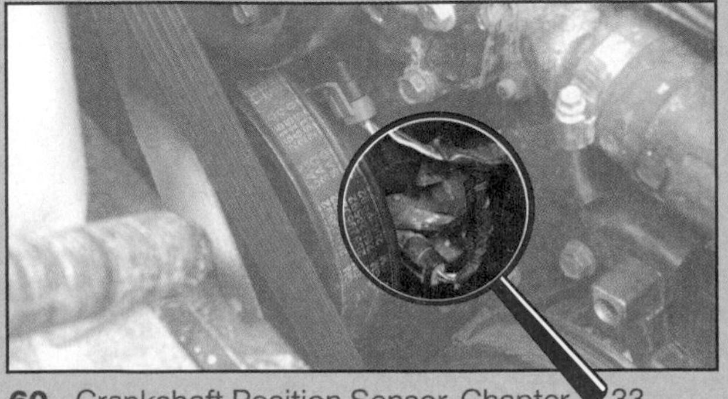

60. Crankshaft Position Sensor, Chapter 33

61. Detonation Sensor, Chapter 33

62. Camshaft Position Sensor, Chapter 33

63. Profile Ignition Pickup, Chapter 33

64. Ignition Coil Pack for a DIS Ignition System, Chapter 33

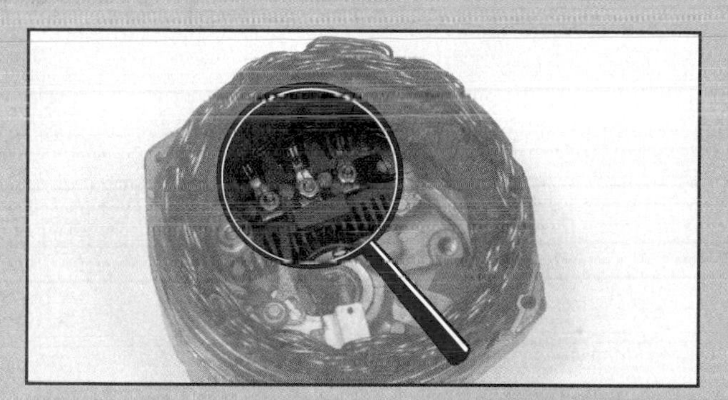

65. Charging System Rectifier, Chapter 34

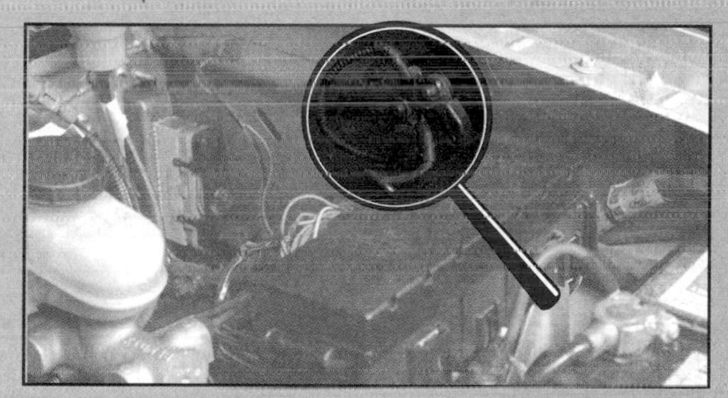

66. Starting System Solenoid, Chapter 35

67. Neutral Position Switch, Chapter 35

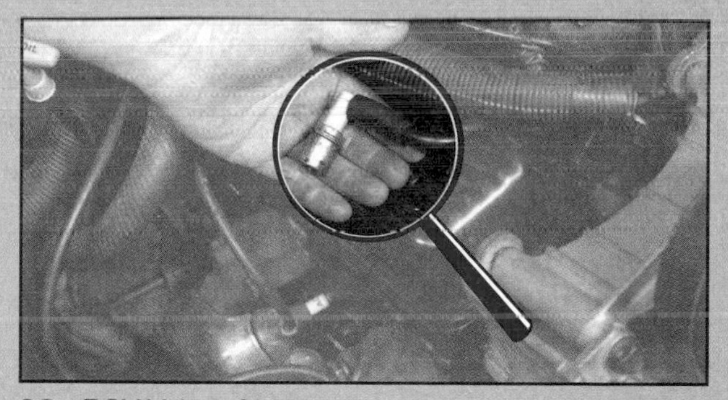

68. PCV Valve, Chapter 37

69. Carbon Canister, Chapter 37

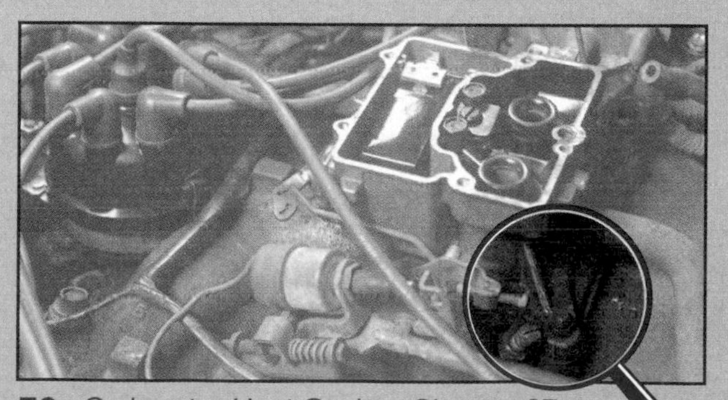

70. Carburetor Heat Gasket, Chapter 37

71. EGR Valve, Chapter 37

72. Air Injection Pump, Chapter 37

73. Air Change Temperature Sensor, Chapter 38

74. Mass Airflow Indicator, Chapter 38

75. Vehicle Speed Sensor, Chapter 38

76. Hall-Effect Crankshaft Position Sensor, Chapter 38

77. Cooling Fan Relay, Chapter 38

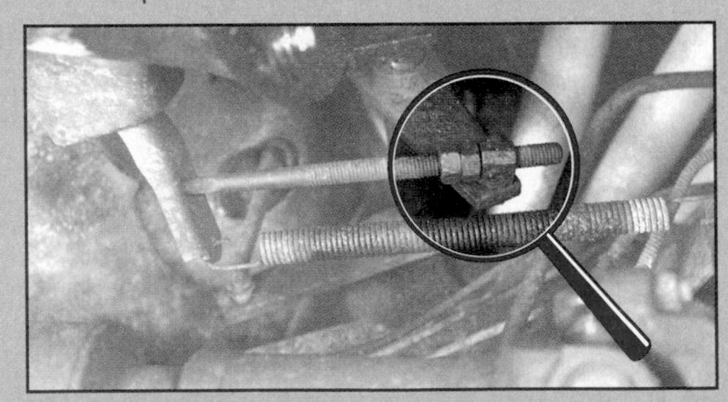

78. Clutch Linkage, Chapter 39

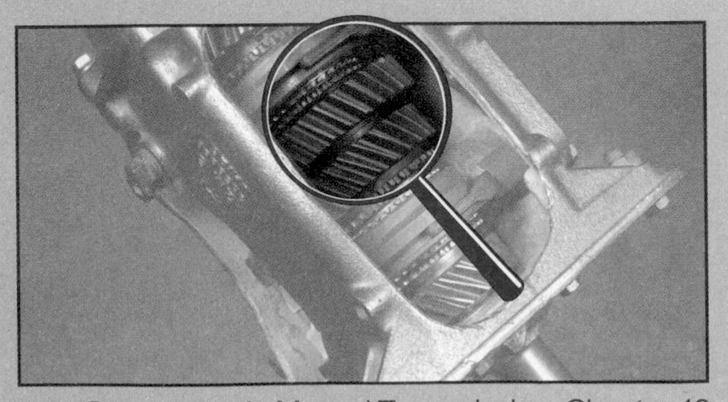

79. Syncronizer in Manual Transmission, Chapter 40

80. Transaxle, Chapter 40

81. Shifting Mechanism on Manual Transmission, Chapter 40

82. Planetary Gear System, Chapter 41

83. Automatic Transmission Valve Body, Chapter 41

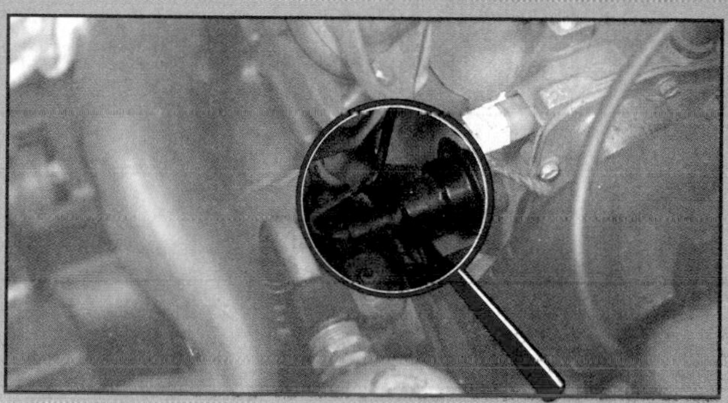

84. Vacuum Modulator on Automatic Transmission, Chapter 41

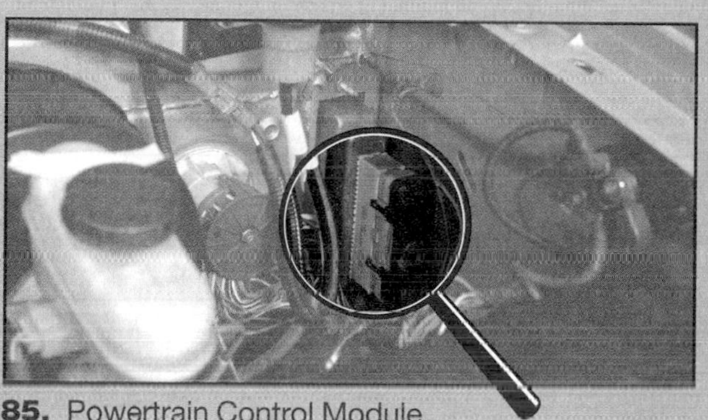

85. Powertrain Control Module, Chapter 42

86. Vehicle Speed Sensor on Electronic Transmission, Chapter 42

87. Transmission Input Speed Sensor, Chapter 42

88. Pressure Switch Assembly on Electronic Transmission, Chapter 42

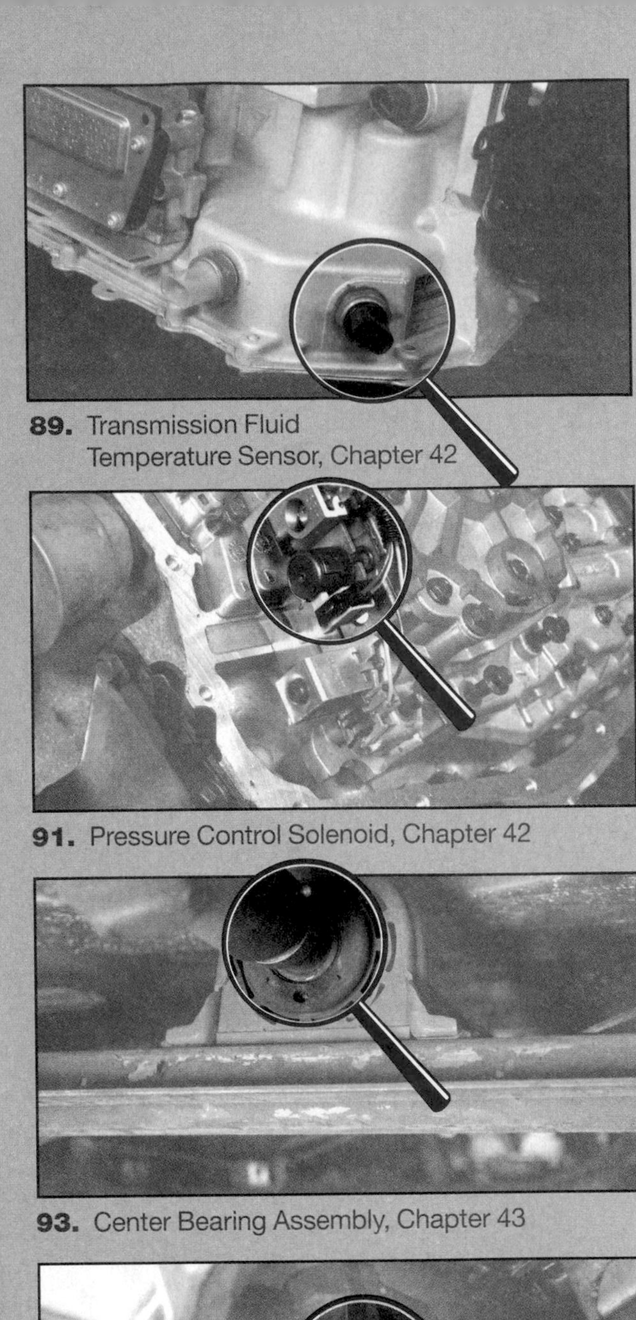

89. Transmission Fluid Temperature Sensor, Chapter 42

90. Shift Solenoids A and B, Chapter 42

91. Pressure Control Solenoid, Chapter 42

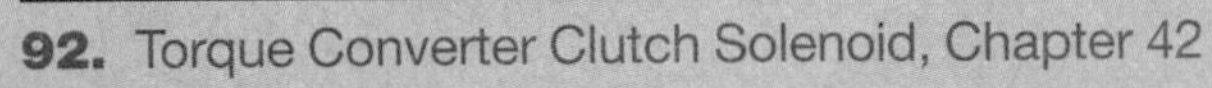

92. Torque Converter Clutch Solenoid, Chapter 42

93. Center Bearing Assembly, Chapter 43

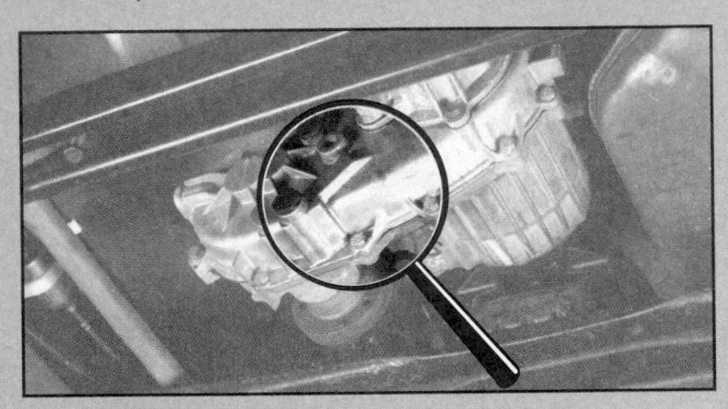

94. 4WD Transfer Case, Chapter 44

95. Viscous Coupling for 4WD Vehicles, Chapter 44

96. Front Axle Disconnect, Chapter 44

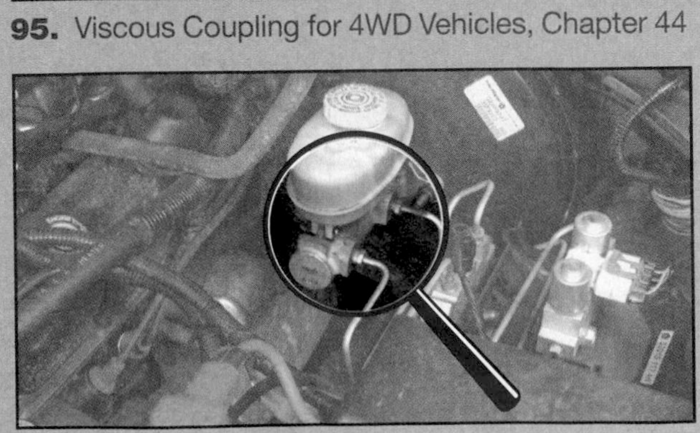

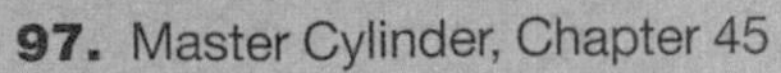

97. Master Cylinder, Chapter 45

98. Drum Brake Wheel Cylinder, Chapter 45

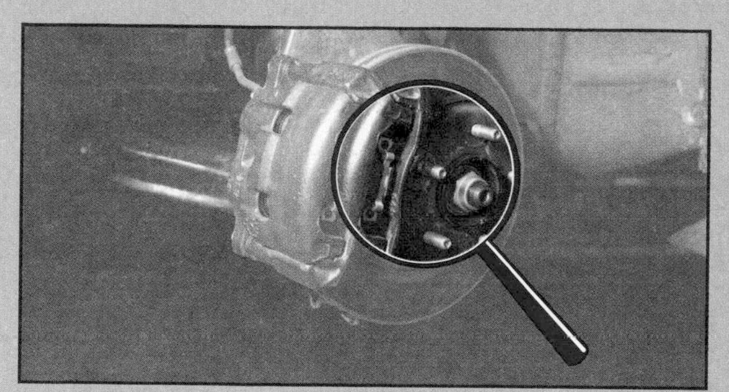

99. Disc Brake Assembly, Chapter 45

100. Brake Light Switch, Chapter 45

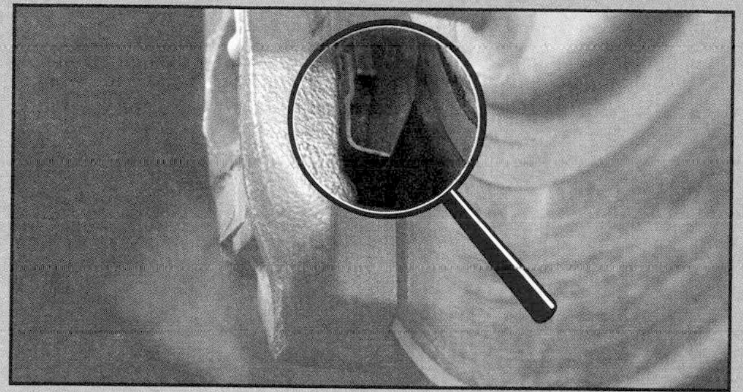

101. Brake Pad Wear Sensor, Chapter 45

102. ABS Wheel Speed Sensor, Chapter 46

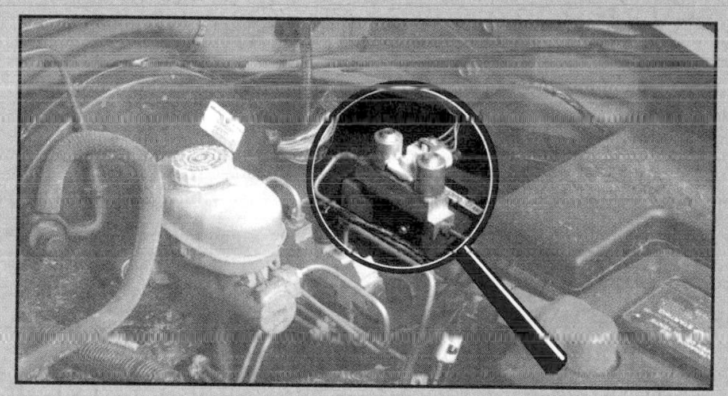

103. Pressure Modular Assembly, Chapter 46

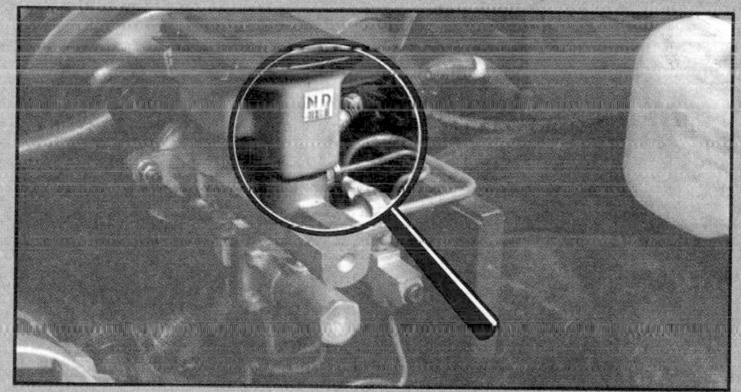

104. Electronic Brake Control Module (EBCM), Chapter 46

105. Shock Absorber, Chapter 47

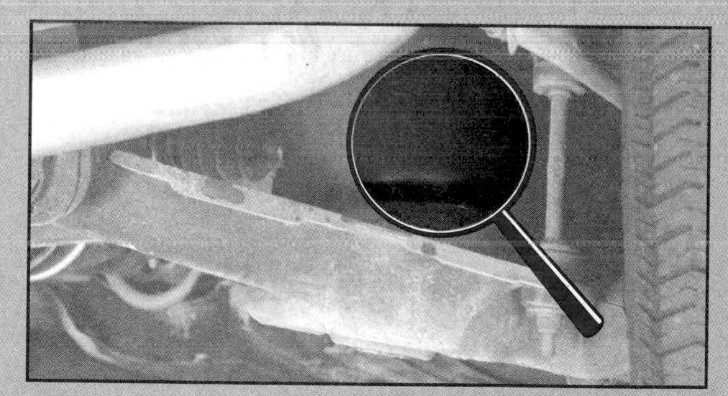

106. Air Suspension Springs, Chapter 47

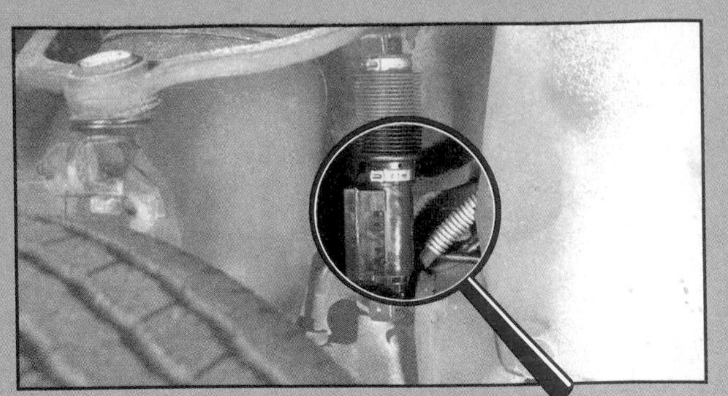

107. Front Height Sensors, Chapter 47

108. Electronic Suspension Control Module, Chapter 47

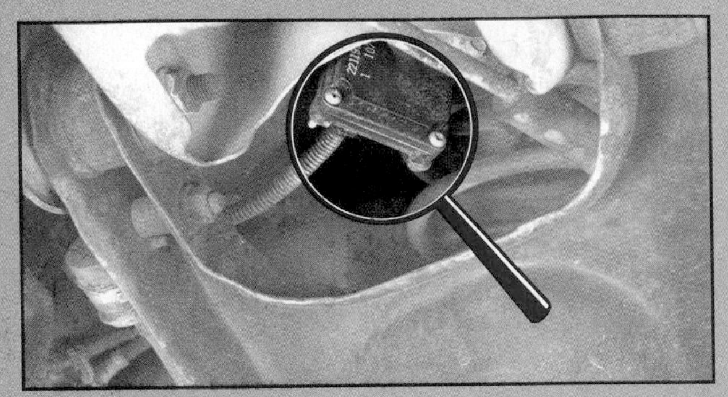

109. Rear Height Sensor, Chapter 47

110. Power Steering Pump and Assembly, Chapter 48

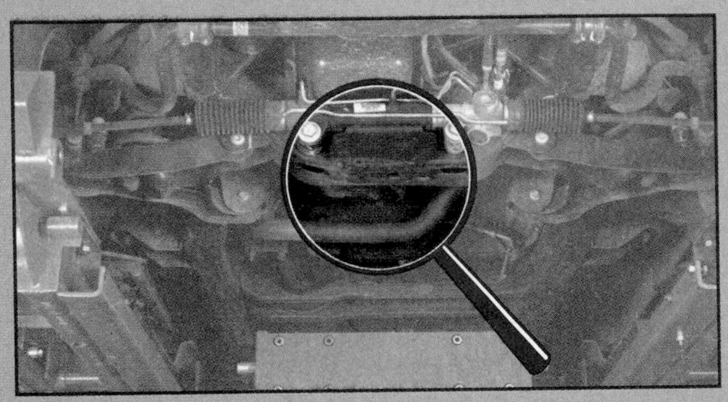

111. Rack-and-Pinion Steering System, Chapter 48

112. Tire Placard, Chapter 49

113. Evaporator Sensor, Chapter 50

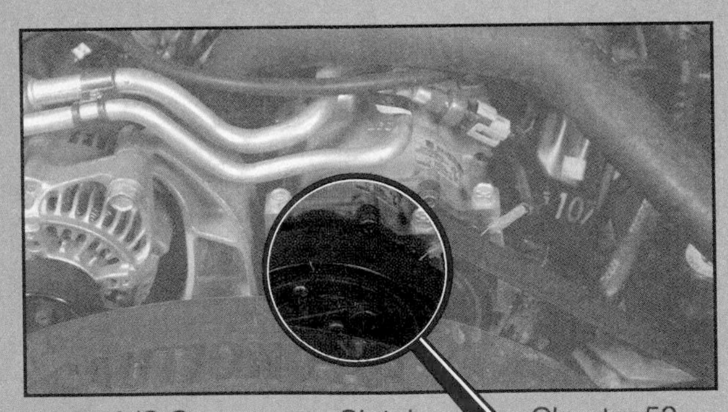

114. A/C Compressor Clutch, Chapter 50

115. A/C High-Pressure Cutout Switch, Chapter 50

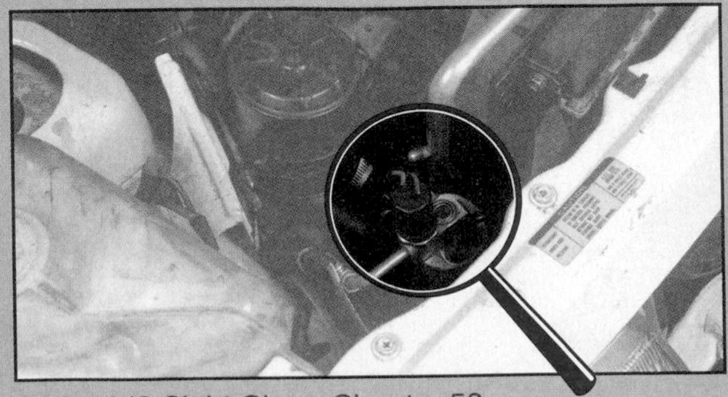

116. A/C Sight Glass, Chapter 50

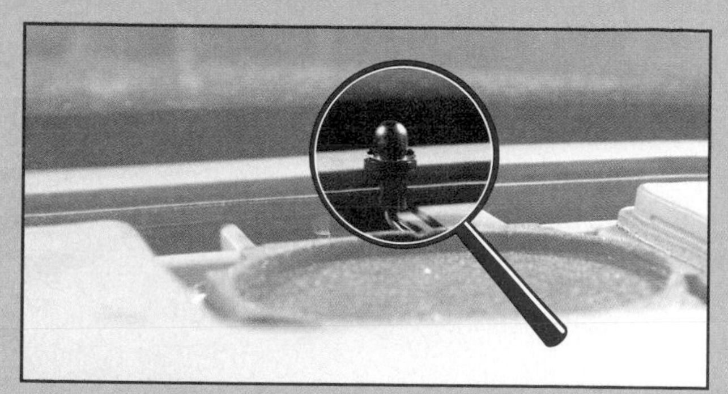

117. A/C Solar Sensor System, Chapter 50

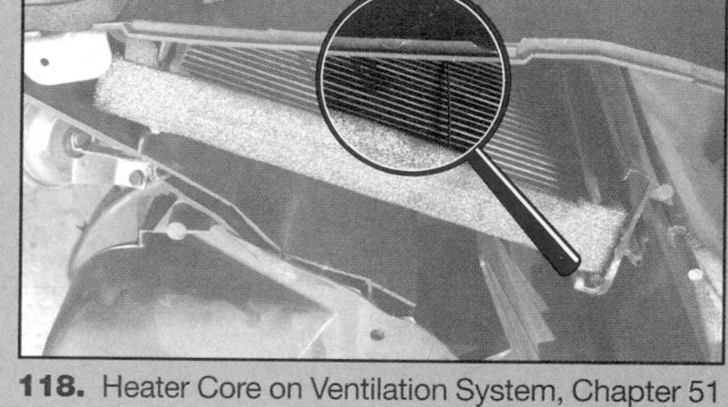

118. Heater Core on Ventilation System, Chapter 51

119. Electrically Controlled Ducts, Chapter 51

120. Ambient Sensor for Ventilation System, Chapter 51

121. A/C Pressure Switch, Chapter 51

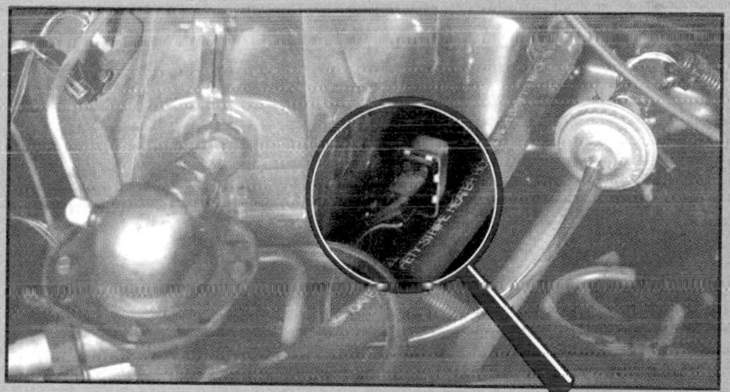

122. Electronic Blower Relay, Chapter 51

123. Cruise Control Transducer, Chapter 52

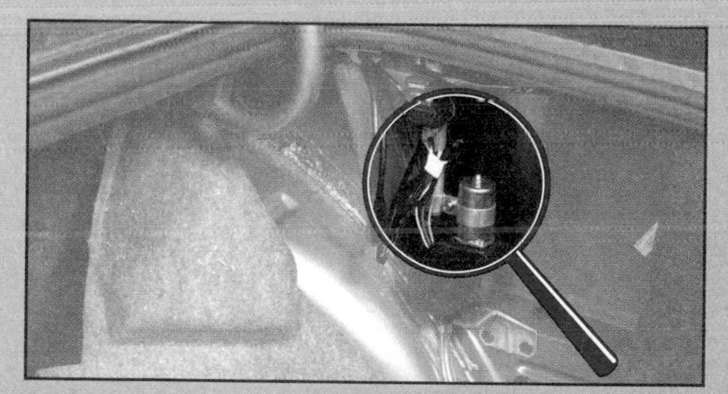

124. Power Antenna, Chapter 53

Working in the Automotive Shop

SECTION 1

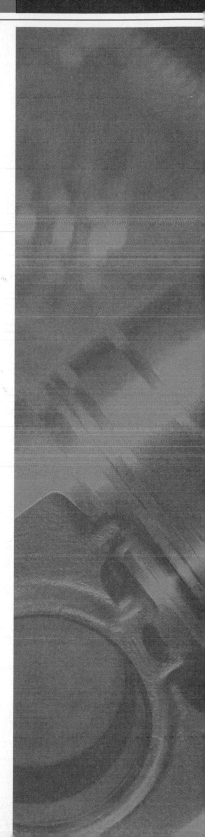

Aron Alexander, 20

Hourly Technician

Young Chevrolet Oldsmobile Cadillac

Owosso, MI

How did you decide to go into this field?

I have a 23-year-old car that leaked oil like crazy; so when I was 16, I decided to rebuild the engine. It took me a whole year, but it really set me on my path in the automotive industry.

What kind of education do you have?

I took autoshop for two years in high school, and was even a student aid assisting my teacher by senior year. Now I'm working on my degree in Automotive Technology from Baker County of Owosso.

Is it necessary to have a college degree?

No, but it helps because the technology is always changing. Plus you can't get ASE-certified without at least one year of training.

What areas are you certified in?

I'm state certified in electrical systems, suspension and steering, engine repair and brakes. And I'm working to get ASE-certified.

How long have you been working at Young?

About a year.

What's a typical day like for you?

I started out doing mostly oil changes. Now I spend most of my day in the alignment rack, or working on steering suspension. I work hourly, so I come in whenever I'm not in class.

What skills are necessary for your job?

Hand–eye coordination definitely! And math—all the adjustments for alignments are in fractions and even in engines you're dealing with tolerances.

What area would you like to be working in?

My heart's really into heavy engine repair. I think my bosses are planning on working me into that.

What happens when you graduate?

I'll stay here, but I'll become flat rate, meaning I'll make a salary rather than hourly wages. I'm a little scared of that because they may quote a job as an hour's worth of work—and if it takes me an hour and a half I'll lose a lot of money!

Introduction to the Automobile Industry

CHAPTER 1

OBJECTIVES

After studying this chapter, you should be able to:

- Identify the importance and influence of the automobile within our society.
- Categorize the many organizations within the automotive industry.
- Describe the effects the automobile has on the environment.

INTRODUCTION

The automobile has become one of the most important technological innovations in our society. People rely on personal vehicles for travel more each day. This chapter is designed to introduce you to the automobile industry and the automobile's effects on the environment.

1.1 AUTOMOBILES AND SOCIETY

AUTOMOBILE USE

People today do not have to be told how important the automobile is to their lives. The automobile (including cars, SUVs, pickups, and vans) is used to get people to work, deliver food and other commodities to stores, and to move people, services, and products throughout the country. Our society is still benefiting from a revolution that began over 100 years ago, called the automotive revolution. The mass production of automobiles affected American social history in the twentieth century more than any other invention. Today, the automobile is so significant that it consumes more than half the total energy used by all types of transportation combined. This proportion is illustrated in ***Figure 1–1***. The energy consumption of automobiles (urban and intercity combined) accounts for 52% of all the energy used by the entire transportation sector of society.

Nowhere is the importance of the automobile greater than in the United States. There are more cars, SUVs, pickups, and vans per person in the United States than anywhere else in the world. This type of transportation in the United States accounts for more than 38% of all the passenger vehicles on the road worldwide. And, as shown in ***Figure 1–2***, there is about one vehicle for every two people in the United States compared, for instance, to Mexico, where there is only one vehicle for every twenty-one people.

The United States remains by far the most mobile society in the world. ***Figure 1–3*** illustrates how many miles Americans drive each year. For example, 30% of the people drive between 20,000 and 30,000 miles per year.

Today, cities and suburbs are being designed on the assumption that the use of the automobile will continue to expand. Millions of dollars are spent to build our society around the automobile. For example, highways have been built to support the millions of vehicles that travel on them. Families can move to the suburbs or the country because

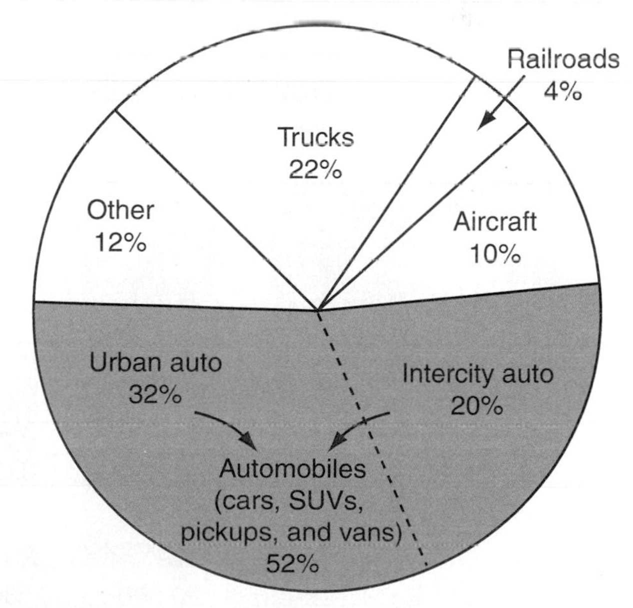

FIGURE 1–1 More than 52% of all energy used by the transportation sector is consumed by automobiles (including cars, SUVs, pickups, and vans).

COUNTRY	PEOPLE FOR EACH VEHICLE
United States	2.0
Japan	5.0
Germany	3.0
France	3.0
Italy	3.0
United Kingdom	4.0
Canada	3.0
Former USSR	32.0
Brazil	15.0
Spain	5.0
Australia	3.0
Netherlands	3.0
Mexico	21.0
Argentina	8.0
Belgium	3.0

FIGURE 1-2 The United States has more vehicles than any other country and averages one car for every two people.

they have transportation to get them to workplaces in the city. Banks and restaurants are designed with drive-through facilities. In fact, our entire society has been designed around the automobile.

ECONOMIC IMPACT

Each working day more than 100,000 automobiles roll off assembly lines around the world. Now the world's largest manufacturing industry, automotive manufacturing has strongly influenced the economic and social evolution of modern technological societies. For example, the following industries have grown because of the development of the automobile:

1. Petroleum refining
2. Road construction and maintenance
3. Motor vehicle manufacturing
4. Parts manufacturing and distribution
5. Automobile sales and servicing
6. Passenger transportation
7. Insurance companies
8. Support companies such as plastics, steel, electronics, rubber, glass, and fabric manufacturers, and many others

The United States leads other countries in how much of the gross domestic product is spent on automobile transportation. Between 10% and 13% of the gross domestic product in the United States has been spent on automobile transportation. It is estimated that 30 million people around the world depend on the automotive industry for their jobs. Close to half of these are in the United States. In fact, 22% of the U.S. workforce is employed in the automotive sector. A recent study by Hertz Corporation indicated that Americans spend 15% of their personal income on automobile transportation.

To get a sense of the importance of the automobile, look at the many types and styles of vehicles that automotive manufacturers are capable of designing. ***Figure 1–4*** lists examples of some of the variations of private passenger vehicles that one can purchase. Of course the type of vehicle purchased depends on the needs of each consumer.

1.2 THE AUTOMOTIVE INDUSTRY

AUTOMOTIVE MANUFACTURERS

There are many manufacturers that currently produce automobiles in the United States. Some of the more well-

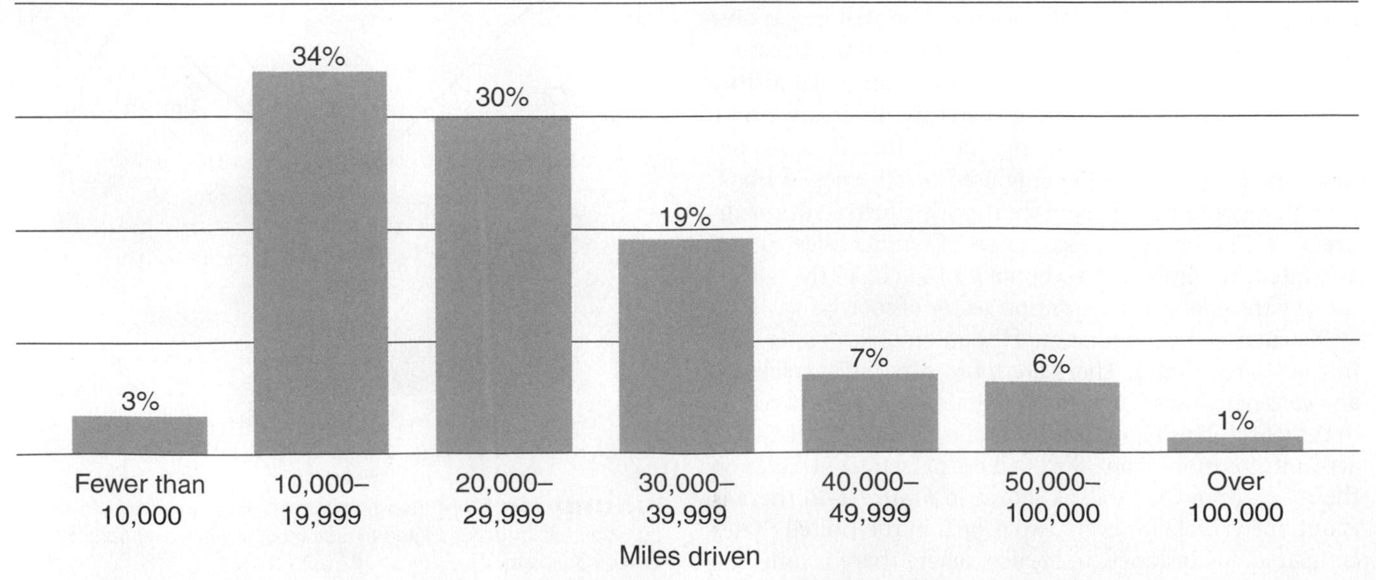

FIGURE 1-3 This chart shows the percentages of Americans driving a certain number of miles each year.

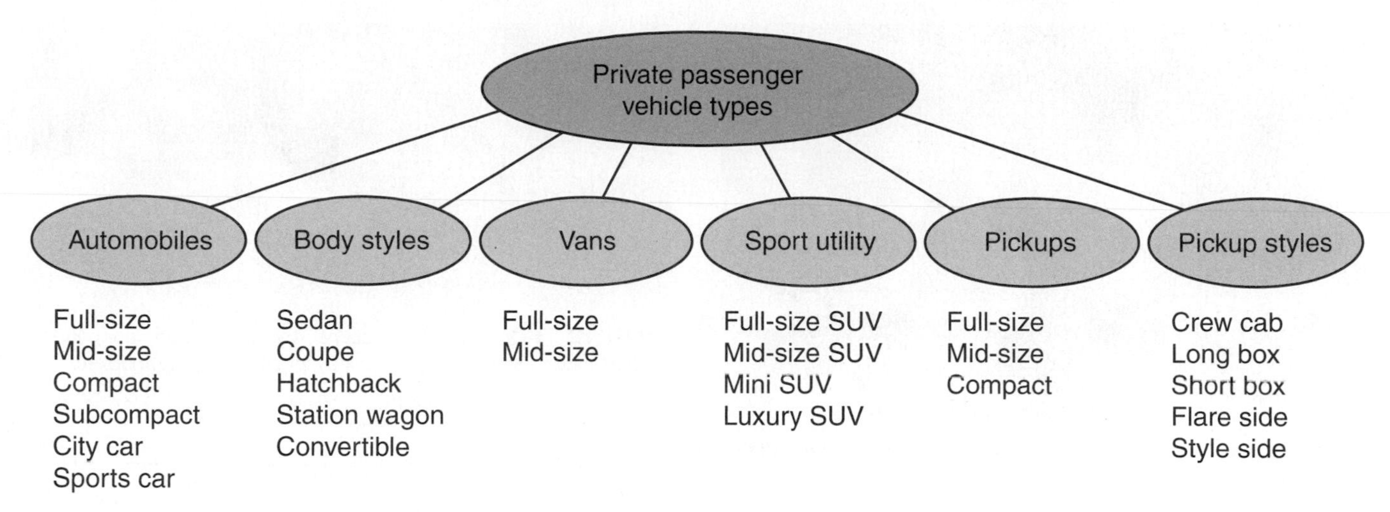

FIGURE 1–4 A variety of types and styles of vehicles have been designed to meet the many needs of Americans.

known automotive manufacturers include General Motors, DaimlerChrysler, Ford Motor Company, Honda, Hyundai, Toyota, Mazda, and Nissan. ***Figure 1–5*** lists some additional automotive companies that manufacture or sell vehicles within the United States.

Many of the major vehicle manufacturers are divided into divisions that manufacture different vehicles. For example, within General Motors, there are the Pontiac division, Cadillac division, Chevrolet division, Buick division, Oldsmobile division, and so on. Within DaimlerChrysler, there are the Jeep Eagle division, Dodge division, and so on. Within Ford Motor Company there are the Ford division and the Mercury/Lincoln division.

Many manufacturers in the United States are combining their products with those of foreign companies. For example, one manufacturer may design and build the basic vehicle in the United States and then purchase the engine or transmission from a foreign manufacturer. On the other hand, some U.S. manufacturers will purchase the engineering from a foreign manufacturer and then build the vehicle in the United States. Many of these manufacturing schemes have the potential to improve foreign relations and manufacturing efficiency, as well as to reduce costs. With the amount of international trade today, it is important to investigate exactly how the vehicle was manufactured and what companies participated in the process. For example, depending upon the vehicle, certain components such as transmissions, computers, etc. are made in other countries. In fact, most vehicles have parts from many countries.

VEHICLE	EXPORTING COUNTRY
Subaru	Japan
Isuzu	Japan
Mitsubishi	Japan
Suzuki	Japan
Volvo	Sweden
Volkswagen	Germany
Mercedes-Benz	Germany
BMW	Germany
Audi	Germany

FIGURE 1–5 This table lists additional automotive companies that distribute vehicles in the United States. Some of these companies have manufacturing facilities in the United States.

Thousands of people are needed to staff many suborganizations and departments in the automotive manufacturing industry. First, the automobile must be designed. Many of the electrical and mechanical engineers employed by automotive manufacturers work primarily on the design of new vehicles. After the design process, a prototype, or a working model, of the vehicle is developed. At this stage, the designs are tested to determine their engineering reliability and economic feasibility. If the prototype passes all the tests, the new vehicle is readied for manufacturing.

All the parts that will go into the automobile must be made first at many supporting plants and companies throughout the United States and certain foreign countries. After the parts are manufactured, the vehicle is assembled. The total manufacturing process is very complex and involves thousands of people.

REGIONAL OFFICES AND DISTRIBUTORSHIPS

Another level of organization in the automotive industry is the **regional office**, sometimes referred to as the **distributorship**. These offices are branches of the main automobile manufacturer. Regional offices are concerned with the selling and service procedures of the company and product. People who work in the regional offices are employed by the main car manufacturer. They are actually considered the link between the car manufacturer and the dealerships where cars are sold and serviced. Regional offices are geographically located throughout the United States.

DEALERSHIPS

After the vehicle has been manufactured, it is sent to the dealer. There are about 25,000 dealers in the United States.

FIGURE 1–6 Car dealerships throughout the United States sell and service automobiles.

FIGURE 1–7 Gasoline stations are considered independent service shops and often have facilities for servicing automobiles.

These vary in size from two or three people up to eighty people working for the **dealership**. The dealership is called a **franchised dealer**. This term means that the dealership has a contract with the main car manufacturer to sell and service its vehicles. Dealerships can be privately owned. People who own dealerships are not employed by the car manufacturing company, but they have a contract to sell and service the manufacturer's products.

The dealership is the main link between the car manufacturing company, through its regional offices, and the customer. All sales and service provided by dealerships are controlled by the policies of the car manufacturing company. Warranty problems are also taken care of through dealerships (***Figure 1–6***).

FLEET SERVICE AND MAINTENANCE ORGANIZATIONS

In certain cases, large **fleet service** companies offer transportation for goods, services, and products throughout our society. These fleet offices have so many vehicles that they usually have their own service and maintenance organizations. For example, car rental companies have fleet service available for all their vehicles. Usually, the service and maintenance procedures are done on a schedule. Having scheduled service periods for the vehicles is referred to as **preventive maintenance**.

INDEPENDENT SERVICE

There are an estimated 190,000 **independent service** repair and maintenance garages within the United States. These vary from small shops with two or three employees to larger shops with thirty or more employees. Typically, car owners will have their vehicles serviced at a dealership until the warranty has expired. Then, depending on the person's preference, future service may be performed by an independent service garage. Cars that have had two to three owners often use the services of independent garages. ***Figure 1–7*** shows an example of an independent service organization. This company not only sells gasoline and other products but also has a service center to complete performance vehicle repairs such as tire replacements, starter and alternator service, air-conditioning service, and so on.

SUPPORTING SPECIALTY SHOPS

Because the automobile has become so technologically complex in the past few years, **specialty shops** have been developed. Rather than trying to understand all of a vehicle's systems and components, a mechanic may specialize in one, two, or even three areas. For example, various transmission and muffler/brake shops have been developed to handle special service needs. Other specialty shops may include alternator, starter, tire, or body shops. The advantage of the specialty shop is that the service technician has a great deal of technical knowledge about a specific area within the automobile. Consequently, prices for services and repairs may be somewhat lower while still maintaining high-quality work.

PARTS DISTRIBUTION

In order for all service, repair, and maintenance shops to continue operating, replacement parts must be readily available. Parts must be distributed throughout the United States for all cars that are driven. **Parts distribution** means parts are sold from parts warehouses to independent parts dealers. Some parts dealers operate on a national level, while some are strictly local. Examples of parts dealers are NAPA (***Figure 1–8***), Automotive Parts and Accessories, Champion Auto Stores, and Crown Auto. Obviously, there are many more.

All of these parts stores are set up as retailers and may be independently owned. Most parts stores today use a computer inventory system. Because of the many parts that must be stocked, a computerized inventory makes the parts business much easier to operate profitably. Parts may also be purchased through the Internet or accessed via a specific online service with certain parts companies, such as NAPA.

Other stores also make automotive parts available. For example, many department stores and hardware stores carry a line of high-volume parts such as oil, air, and fuel filters, batteries, and so on (***Figure 1–9***).

FIGURE 1–8 NAPA is just one of many parts dealers that supply service shops with quality parts.

FIGURE 1–9 Some department stores also carry a line of commonly used automotive parts.

1.3 AUTOMOBILES AND THE ENVIRONMENT

DEFINING AIR POLLUTION

Pollution is defined as the contamination of the environment by harmful products. If **contaminants** exist in large enough numbers, they can harm plants, animals, and humans. There are many forms of pollution such as chemical, thermal, radiation, water, noise, and air pollution. There are also many sources of pollution. Industries, power plants, home heating devices, gasoline and diesel engines, and refuse disposal plants all produce pollution. Pollution is also produced by natural causes such as volcanoes and brush fires.

Air pollution is a by-product of any combustion process. It is produced when coal, oil, natural gas, or other material is burned. Since automobiles use both diesel fuel and gasoline, they have become one of our society's greatest sources of pollution. Although air pollution from automobiles has been reduced drastically in recent years, automotive manufacturers continually seek ways to reduce pollution even further.

SMOG

Smog is the result of having too much pollution suspended in the air. Smog is defined as a fog made heavier and darker by the addition of smoke and chemicals to the air. Both the smoke and chemicals are floating in the air. Automobiles contribute to smog.

PHOTOCHEMICAL SMOG

One type of smog is called **photochemical smog**. The word *photochemical* means that sunlight mixes with the chemicals in the air (***Figure 1–10***). This combination causes photochemical smog. Photochemical smog is more dangerous to plants, animals, and humans than plain smog. For example, when photochemical smog is breathed into the body, the chemicals and acids cause coughing and irritate the nose, throat, and lungs. This type of pollution has been linked to such health problems as asthma, skin damage, cancer of internal organs, emphysema, and heart and circulatory problems.

TEMPERATURE INVERSION

Under normal weather patterns, warm air near the ground tends to rise and cool. When this happens, the smog and pollution in the air **dissipate** and are reduced. During a temperature inversion, however, the warm air becomes trapped and cannot rise. A temperature inversion happens when a layer of warm air acts as a cap or lid to the air near the ground or surface of a city. In this case, the warm air prevents the smog and air pollutants from rising and dissipating into the atmosphere. The inversion layer is usually within 1,000 feet of the surface of the ground. Temperature inversions are common in cities that are located in valleys (***Figure 1–11***). Los Angeles and Denver are examples where temperature inversions usually occur.

UNITS USED TO MEASURE AIR POLLUTION

In order to control air pollution from automobiles, measurements are taken to determine how much pollution is actually produced. Several units have been used to measure pollution. One common unit is a measurement in grams of

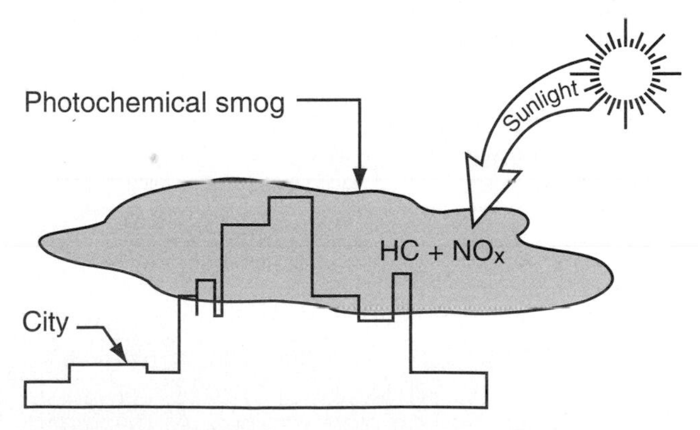

FIGURE 1–10 When sunlight mixes with hydrocarbons (HC) and nitrogen oxides (NO_x), photochemical smog is produced.

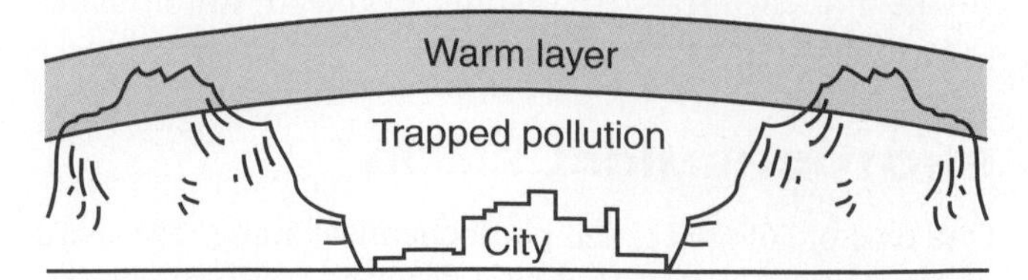

FIGURE 1–11 A temperature inversion occurs when a warm layer of air traps the air around a city. The temperature inversion causes pollutants to stay trapped in the air above the city.

pollution per mile of operation. This unit is abbreviated as g/mi. For example, a vehicle driving down the road may produce 1.5 grams of a certain type of pollution per mile of operation.

Another unit used to measure the amount of pollution is parts per million. It is abbreviated as ppm. This unit represents the number of parts of pollution per million parts of air being put into the atmosphere. This unit is also referred to as micrograms per cubic meter ($\mu g/m^3$). Both of these units are used as guidelines to control the exact amount of pollution that a vehicle is allowed to emit into the air.

A third method of referring to the amount of pollution is to use percentages. For example, the exhaust of a gasoline engine has about 2% to 3% carbon monoxide. This means that about 2% to 3% of the exhaust volume is carbon monoxide.

Other units also are used for pollution readings. These include grams per hour, grams per ton mile, and grams per horsepower. The reason there are so many units is that many manufacturers like to make comparisons between different engines and different applications. These units make it easier to make comparisons.

COMPARING AUTOMOBILES TO OTHER TRANSPORTATION

One way to study the effects of the automobile on the environment is to compare the **miles-per-gallon-per-person** from one form of transportation to another. The automobile

SHARE OF TRANSPORTATION

	PERCENTAGE	MILES PER-GALLON-PER-PERSON
Automobile	80–83%	20–40
Train	4–5%	40–60
Bus	2–5%	100–150

FIGURE 1–12 This chart compares the automobile to two other types of transportation in terms of miles-per-gallon-per-person.

often compared to other forms of transportation in terms of environmental damage.

Miles-per-gallon-per-person is a unit that indicates how much energy is being used to move one person a certain distance. For example, the chart in ***Figure 1–12*** compares three types of transportation: automobile, train, and bus. Two numbers are also shown: the percentage of the share of transportation and the miles-per-gallon-per-person. The percentage of the share of transportation is a measurement of the number of trips taken for each form of transportation. Miles-per-gallon-per-person is calculated by multiplying the average miles-per-gallon times the average number of people in a car. Today, the average number of occupants in a car is about 1.2 people. If 1.2 is multiplied by average gasoline mileage—25 miles per gallon—the result is 30-miles-per-gallon-per-person. It is obvious that trains and buses typically carry many more people. However, these vehicles typically average about 0.5 to 7 miles-per-gallon-per-person.

After studying this chart, the following conclusion can be made: The automobile is the vehicle most used for transporting people, but it uses the greatest amount of fuel and thus produces the most pollution per person. Within the United States, people look at the convenience of the automobile and the price of gasoline. As one can imagine, the

Car Clinic: COMPUTERS AND VEHICLE EMISSIONS

CUSTOMER CONCERN:
Pollution and Computers

A customer asks why computers are needed to manage engine performance. What relationship does the computer have to reducing pollution and vehicle emissions?

SOLUTION: Computers are able to react and make decisions much faster and more precisely than components used on older engines. All manufacturers have now added computers for combustion and pollution control. By using computers to control combusion, the manufacturers can reduce exhaust emissions to such a low level that common exhaust gas analyzers cannot measure them. In addition, many of the mechanical pollution control components used several years ago are no longer needed and are being replaced by computer controls. The result is that computers help to reduce vehicle emissions, thus contributing to a cleaner environment.

automobile is generally more convenient than taking the train or bus. In addition, gasoline prices over the past 30 years have been significantly below the increases in inflation, so the automobile has become the major choice for transportation within our society.

OTHER TYPES OF POLLUTION

The automobile also produces other forms of pollution that can damage our environment. For example, many of the fluids used in the automobile are considered toxic chemicals. Fluids such as battery acid, used engine oil, antifreeze, air-conditioning refrigerant, transmission fluids, and differential lubricants are all considered dangerous to our environment. In addition, many chemicals that are used to service vehicles are considered hazardous. Fluids such as carburetor and fuel injector cleaner, brake cleaners, and so on fall into this category. Finally, as entire vehicles are disposed of, the body, frame, tires, and other parts are also considered a form of pollution.

Today, batteries are being recycled and used engine oils, transmission lubricants, and freon are being collected and in many cases are being recycled. For example, it is important to note that stores that sell lubricants and antifreeze often tell customers where the used oil or antifreeze can be disposed of properly. Also, when new batteries are purchased, many stores take the old discarded batteries and send them to a recycling center. Similarly, tires are being recycled for a variety of purposes. Overall, it is important to be a responsible person when it comes to dealing with the many types of pollutants that an automobile produces. Because there are so many vehicles on the road today, and eventually 99% of these vehicles will be totally discarded, it is important to continue recycling efforts into the future.

SUMMARY

The following statements will help to summarize this chapter:

- The automobile is one of the most influential technological advances of our time.
- The United States has an average of two people per car.
- The automobile has generated an industry composed of many offices and service establishments.
- Regional and independent service garages have been established to service automobiles.
- Fleet service and maintenance shops help to service automobiles.
- Specialty shops and parts distribution shops are needed to keep automobiles serviced correctly.
- Many types and styles of private passenger vehicles are available to consumers.
- Because the automobile is the most used form of transportation, it also consumes the greatest amount of energy and thus creates the most pollution compared to trains and buses.
- The automobile also produces other pollutants such as acid from batteries, antifreeze, used engine oil, used transmission fluid, freon from air-conditioning systems, and tires.
- Recycling becomes an important part of reducing pollutants from the automobile.

TERMS TO KNOW

Can you explain each of the following terms? Review the chapter until you can use each term correctly.

Contaminants
Dealership
Dissipate
Fleet service
Franchised dealer
Independent service
Miles-per-gallon-per-person
Parts distribution
Photochemical smog
Pollution
Preventive maintenance
Regional office or distributorship
Smog
Specialty shops

REVIEW QUESTIONS

Multiple Choice

1. What percentage of energy in the transportation sector is used by the automobile?
   a. 15%
   b. 52%
   c. 95%
   d. 45%
   e. 22%

2. Approximately what percent of all cars purchased are bought from foreign suppliers?
   a. 1–4%
   b. 25–30%
   c. 82–93%
   d. 50–52%
   e. 71–76%

3. Which of the following offices are often considered a franchised office?
   a. Dealerships
   b. Independent service garages
   c. Fleet operations
   d. Supporting specialty shops
   e. All of the above

4. What is the average number of people per car in the United States?
   a. 2
   b. 3
   c. 9
   d. 10
   e. 11

5. What percentage of the U.S. workforce is employed in the automobile sector of society?
   a. 22–25%
   b. 35–45%
   c. 50–60%
   d. 65–70%
   e. None of the above

6. An automotive shop that works only on mufflers and brakes would be classified as a (an):
   a. Independent service shop
   b. Supporting specialty shop
   c. Dealership
   d. Company specialty shop
   e. None of the above

7. Photochemical smog is created when ________ mixes with NO_x and HC emissions.
   a. CO
   b. Particulates
   c. Sunlight
   d. Smog
   e. None of the above

8. Which of the following are considered acceptable units for measuring emissions from the automobile?
   a. Parts per million
   b. Grams per mile
   c. Percentages
   d. All of the above
   e. None of the above

9. Which of the following are considered pollutants from an automobile?
   a. Antifreeze
   b. Used engine oil
   c. Transmission oil (used)
   d. Air-conditioning refrigerant
   e. All of the above

The following questions are similar in format to ASE (Automotive Service Excellence) test questions.

10. *Technician A* says that only a small percentage of people, about 2%, work in the automotive sector of society. *Technician B* says that about 75% of people work in the automotive sector of society. Who is correct?
    a. A only c. Both A and B
    b. B only d. Neither A nor B

11. *Technician A* says that the only pollution from an automobile is the exhaust. *Technician B* says that refrigerant, used oil, and tires are considered part of the pollution produced from an automobile. Who is correct?
    a. A only c. Both A and B
    b. B only d. Neither A nor B

12. *Technician A* says that battery acid is not considered a pollutant from an automobile when a battery is disposed of. *Technician B* says that used transmission oil is considered a form of pollution from the automobile. Who is correct?
    a. A only c. Both A and B
    b. B only d. Neither A nor B

Essay

13. What are some of the organizations within the automotive industry?

14. Identify four sources of pollution produced from an automobile.

15. Define the term *photochemical smog* and state why it is considered a problem.

Short Answer

16. Most Americans drive their cars between ________ miles and ________ miles a year.

17. Automotive manufacturers today in the United States are using products purchased from ________.

18. When a blanket of warm air traps air pollution above a city it is called a ________ ________.

Applied Academics

TITLE: Technical Service Data

ACADEMIC SKILLS AREA: Language Arts

NATEF Automobile Technician Tasks:

The technician adapts a reading strategy for all written materials—customer's notes, service manuals, shop manuals, technical bulletins, and computer/data feed read-outs—that will help identify the solution to an engine repair problem.

CONCEPT:

In the automobile service area, volumes of technical information are available. Technical data can be found in service manuals, technical bulletins, owner's manuals, and computer/data information directly from the manufacturer. It is important that the service technician know how to search for and find the correct information on any vehicle. The service technician must know how to read these documents correctly and then use the right information to correct any problem found in the automobile.

APPLICATION:

All vehicles have an owner's manual. Can you find the owner's manual for a car that you or a friend own? Once you have the owner's manual, locate the vehicle specifications section. Identify the type of specifications that are given and then determine the type of spark plugs, the number of quarts of oil needed for an oil change, the type of oil that should be used, and the suggested time for oil change periods.

CHAPTER 2

Working in the Automobile Industry

OBJECTIVES

After studying this chapter, you should be able to:

- Identify the careers in a car dealership.
- List careers outside of the dealership.
- Define the many educational opportunities in the automobile industry.
- Evaluate automobile technician certification programs.

INTRODUCTION

Working in the automotive industry can be very rewarding and satisfying. There are many careers, from car manufacturing, to sales, to service. In addition, the automobile industry has developed excellent educational and technical certification programs. This chapter is designed to introduce you to numerous automotive careers, as well as to education and certification opportunities within the automotive industry.

2.1 JOBS IN A DEALERSHIP

Numerous careers and jobs are currently available in the automobile industry. Many of these careers are related to servicing vehicles. For example, service managers and service technicians deal with the automobile in a service function. However, many other careers also are supported by the automobile industry. These include marketing and sales positions, parts distribution positions, distributorships, and company sales, service representatives, and training specialists.

Automotive careers are typically available in a variety of working areas, including:

- Dealerships for selling and servicing
- Independent garages for servicing
- Service stations for general service
- Tire and battery dealers (*see **Figure 2–1***)
- Specialty shops that handle wheel alignment, transmissions, body repair, and tune-up work such as Midas and Goodyear.
- Service shops owned by large stores such as Sears, Fleet Farm, etc.
- Fleet repair shops such as truck, bus, and automobile fleets
- Parts stores and parts distribution centers
- Recycling and salvage yard operations

SERVICE TECHNICIAN

One of the most important careers in the automobile industry is that of the **service technician.** This person is educated to diagnose, service, and competently repair any problem on the automobile. Usually these skills come from a sound foundation in auto mechanics, experience, and continuous upgrading and training on new technologies. A study done by *Motor Age* asked the question "Which profession do you

FIGURE 2–1 Tire and battery dealers offer many careers in the automobile industry.

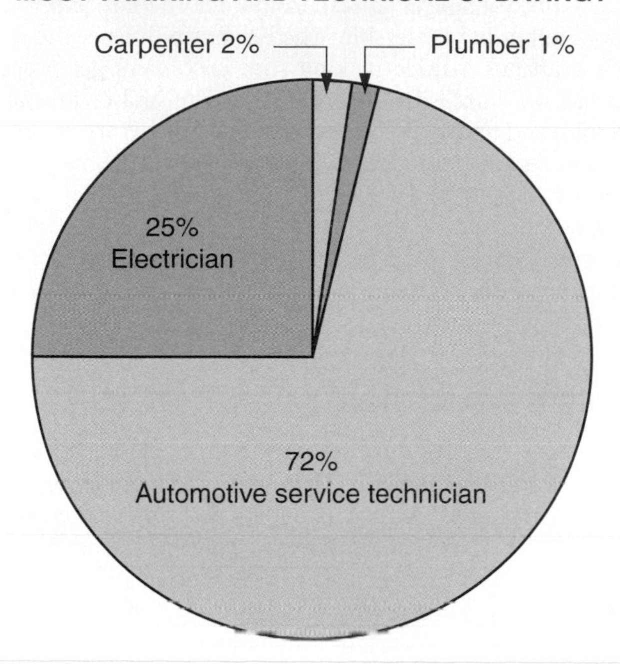

FIGURE 2–2 Seventy-two percent of the people surveyed by *Motor Age* indicated that the position of automotive service technician requires the most training and technical updating.

think takes the most training and technical updating?" ***Figure 2–2*** shows the results.

Service technicians must be able to solve many problems associated with the automobile. Customers often have a complaint but they may not know exactly what is wrong. The service technician must combine the customer's information, past experience, and educational training to determine exactly what the problem is. The service technician must then repair the problem correctly. In order to do this, the service technician must possess both good technical and good people skills.

LIGHT REPAIR TECHNICIAN

In many dealerships and service organizations, there is a person called the light repair technician who helps to do general automotive repair. Often this person helps to disassemble and reassemble automobile components, to clean parts, and to get new parts that are needed. This person helps the service technician do his or her job better. The light repair technician generally needs to understand the total automobile and its mechanical operation.

TRANSMISSION/TRANSAXLE TECHNICIAN

Today's cars are complex and sophisticated. This is also true of the transmission and transaxle components on a car. Today's cars have many types of transmissions. In fact, many electronic transmissions are being used. The transmission and transaxle technician must be able to diagnose transmission and transaxle problems, disassemble, service, and reassemble the parts. This type of service technician must have a sound knowledge of the mechanical, fluid, and electrical components on an automobile.

FRONT-END TECHNICIAN

Automobiles today also have very complex steering systems and front-end suspension systems. Because there is so much to know, dealership and service shops will often have one person who works specifically on the front end. The front-end technician must understand front-end steering geometry, must be able to troubleshoot problems associated with steering and suspension, and must be able to correctly service and repair associated problems. This type of service technician must have a strong mechanical background, be able to understand troubleshooting guides, and be able to align automobiles using the correct alignment equipment (***Figure 2–3***).

AIR-CONDITIONING TECHNICIAN

Air-conditioning systems are now being controlled electronically as well as mechanically. A great deal of knowledge is necessary to service air-conditioning systems correctly. The air-conditioning technician must have a sound knowledge of air conditioning, must be able to diagnose problems, and then be able to correctly service these problems effectively. This type of service technician must have a good mechanical background. In addition, this person needs excellent electrical and fluid, or hydraulic, knowledge to aid in troubleshooting.

BRAKE TECHNICIAN

Although braking systems may seem simple, there are many problems that can develop with brakes. Also, many of today's cars use antilock braking systems. Today's brake

FIGURE 2–3 Technicians who work on the front end of the vehicle are called front-end technicians.

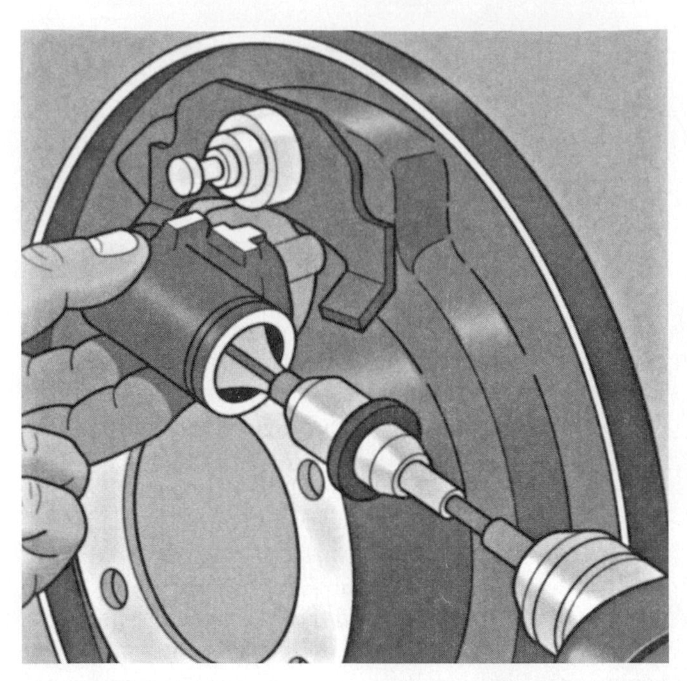

FIGURE 2–4 A technician who works only on brakes is called a brake technician.

technician needs to possess strong mechanical skills and knowledge of fluid and electrical systems. These skills are necessary to correctly troubleshoot and service all of the braking systems used on cars. This type of service technician must also be able to make correct adjustments on braking systems for proper operation (***Figure 2–4***).

OTHER TYPES OF SPECIALISTS

There are many other types of specialists in a typical automotive dealership. Because of the complexity of vehicle systems, many specialized jobs are necessary. For example:

- Engine repair specialist
- Automatic transmission/transaxle specialist
- Manual drivetrain and axles specialist
- Suspension and steering specialist
- Electrical and electronic systems specialist
- Engine performance specialist

SHOP SUPERVISOR

In all automotive shops, there is usually a person called the **shop supervisor**. The shop supervisor is responsible for organizing the work schedules of the service technicians as well as making sure the service function runs smoothly. This person must also be able to work with people, including the customers and the mechanics. In addition, the shop supervisor must have a good deal of technical experience with the automobile, because at times the shop supervisor is expected to work on complex technical problems with the service technicians.

SERVICE MANAGER

The **service manager** is responsible for the entire service operation of the dealership. The responsibilities of the service manager include making sure customers get proper service, working with the shop supervisor and technicians to train and update them, and carrying out factory policies, warranties, and so on. The service manager must have good human relation skills (getting along with people) and a sound knowledge of the automobile and its technical components. Service managers must also have good writing skills to prepare customer orders and obtain customer information for work orders.

PARTS MANAGER

If a service organization is to operate smoothly, parts for the vehicles must be available. The **parts manager** is responsible for making sure the customers' parts are immediately available. This person is then responsible for ordering, stocking, inventorying, and selling high-quality replacement parts and accessories. The parts manager must also have available a network of distributors and suppliers who can get needed parts quickly.

MARKETING AND SALES

There are many positions in the area of automotive marketing and sales. People who sell automobiles usually have to have a broad background of automotive expertise. The salesperson must understand the basic components of the vehicle as well as how these components interrelate with each other. This person must also be interested in dealing directly with people in the selling and marketing of the automobile (***Figure 2–5***).

FIGURE 2–5 Marketing and selling new and used automobiles creates many careers in the automotive industry.

2.2 JOBS OUTSIDE THE DEALERSHIP

COMPANY REPRESENTATIVES—SALES AND SERVICE

Besides careers in the sales and service areas of dealerships and distributors, there are a great many careers working directly for the automotive manufacturers. For example, all major automotive manufacturers have **sales and service representatives**. Many times, these people are the link between the automotive manufacturer and the sales and service dealers throughout the country.

The service representative's responsibilities include working as a link between the dealerships and the specific company, training technicians on new technologies, working with warranty problems, and generally acting as technical experts in complex service problems. Because of these responsibilities, service representatives need to have a detailed exposure to the automobile, including all of the components used.

Sales representatives are the selling link between the dealer or distributor and the specific manufacturer. The sales representatives work with sales and marketing problems that need company attention. Because of these responsibilities, sales representatives must have a broad knowledge of the automotive industry as well as an understanding of the total automobile and their company's vehicles.

SUPPORTING CAREERS

Because the automobile industry is so large, there are many **supporting careers** available to those who are interested in automobiles. Such careers include:

- Claim adjusters—working with vehicles and owners that have been in accidents
- Vocational/technical instructors teaching automotive technology
- Auto body repair technicians—repairing vehicles (***Figure 2–6***)
- Frame and alignment repair—repairing the front-end steering systems and straightening frames after accidents
- Specialty shops—repairing specific components such as tires, fuel injection, muffler and exhaust systems, transmissions, radiators, and so on
- **Parts specialist**—working directly with people who want to purchase automotive parts
- Custom body repair—altering the basic body of the vehicle to include trim, outer vehicle design (van conversions), and general customizing
- Automotive cleaning/detailing—cleaning, polishing, and general maintenance of the vehicle body and interior
- Sports and performance shops—providing parts, basic vehicle design, and construction for various racing activities
- Automotive used and rebuilt parts—providing various used and rebuilt parts such as transmissions, fuel pumps, generators, starter motors, suspension parts, and so on
- Rental shops—providing rental vehicles for consumer transportation purposes (Some of the major rental companies include National Rental, Avis Rental, and Hertz Rent a Car. In addition, various companies also rent special vehicles such as limousines and vans.)
- Automotive loan and leasing—providing the necessary money, programs, papers, and payment schedules for either purchasing or leasing a vehicle
- Service station attendant—providing basic service on most automobiles including filling the fuel tank, cleaning windows, and changing light bulbs—generally an entry-level position
- Street rod shops—providing technical expertise, service, design, and parts for street rods and other classic or vintage vehicles that are being rebuilt or being redesigned
- Alternative fuel conversion shops—providing conversion systems for liquid petroleum gas (LPG) and compressed natural gas (CNG) fuel systems to make vehicles produce less pollution and, in effect, clean up our environment

FIGURE 2–6 There are many careers available in the auto body repair industry.

As you can see, the automobile industry is very large and is capable of supporting thousands of careers for those who are interested and have a good understanding of the automobile.

2.3 EDUCATION IN THE AUTOMOTIVE INDUSTRY

The automotive industry is dynamic and changing. New technology is developed and produced each year. There are many avenues for learning about and keeping current with developments in the automotive industry. Many **work-based learning** programs offer students opportunities to learn about the automobile and its many complex systems. There are internship programs, apprenticeships, on-the-job training, and many more opportunities for students to upgrade their education within the automobile industry. The following sections explain some of the more common work-based learning programs.

INTERNSHIPS

Internships are programs that many vocational and technical schools use to allow students to work in the industry while still going to school. Often internships are provided to allow students to get credit for working after school or during the summer at an automotive dealership or other automotive organization. The benefits of an internship are numerous. First, students usually get paid to work in the automotive industry. Second, students learn firsthand the technical skills and knowledge necessary to develop a career in the automotive industry. Third, students learn how important it is to deal directly with people on a day-to-day basis concerning automotive problems.

APPRENTICESHIPS

An apprenticeship is an educational program that also allows students to work in the automotive industry. However, apprenticeships usually occur after students have finished school. Apprenticeship programs allow students to break in to the automotive career over a period of time. This gives the apprentice time to learn the technology, develop the necessary skills, and improve on the interpersonal skills necessary for the job.

COOPERATIVE SCHOOLS

Cooperative schools are designed to allow students to go to school and work at the same time. Co-op education has well-structured automotive academic programs that tie directly to working in the automotive industry. Often students go to school in the morning and work in the afternoon. With a co-op educational program, students learn automotive principles and then immediately use the principles in the automotive industry.

ON-THE-JOB TRAINING

On-the-job training, often called OJT, can be a less structured form of job-based work experience in the automotive industry. Often students "learn by doing" or "just pick it up" from other experienced automotive technicians. In some automotive dealerships and service organizations, OJT is more structured. In this case an experienced service technician, also called a mentor, works with the new employee on a one-to-one basis. This type of job-based work experience occurs at the workplace, makes use of training programs, and requires active involvement by the service technician.

CORRESPONDENCE COURSES

Another type of education in the automotive industry is correspondence courses. Often technical schools provide all of the information necessary to complete a typical automotive course at home. In this case, the student studies the content, takes tests, and completes various assignments to measure the level of academic achievement about a certain part of an automobile. The tests are then sent to the school and graded. In some schools, these correspondence courses are called "self-paced" courses. The advantage of using these courses is that the student can learn about a specific automotive topic at home and still get credit from the technical school for completing the requirements. However, one disadvantage of correspondence courses is that there is no actual hands-on laboratory experience with the automobile.

FOUR-YEAR DEGREE PROGRAMS

Various colleges and universities offer programs in automotive technology as well. In this case, the information about the automobile is often tied or connected to other areas of study. The exact area may depend on the career direction of the individual. Following are just four examples of some programs automobile technology may be tied to:

1. Service and Marketing—Students may go into automotive marketing and service.
2. Education—Students may become automotive teachers in vocational, technical, and secondary schools.
3. Engineering—Students may work in the automotive engineering field.
4. Management—Students may work as managers in the automotive industry.

THE INTERNET AND WORLD WIDE WEB

A great deal of information on automobiles can also be found on the World Wide Web (www). If you have a computer that is connected to the Internet, you can access many topics concerning automotive technology, automotive performance, automotive consumer information, and so on. In order to access the Internet or Web, you will need to have an electronic address or URL. It would be a good idea to keep a list of all URL addresses that you see or hear about concerning automotive information. (See Appendix B for a list of common Web sites.) This way, you will have more information instantly at your fingertips.

The Internet can be a valuable resource for learning more about the automotive industry. Almost any type of information about automobiles can be found on the Internet. For example, some Web sites deal with car pricing and car reviews, while other Web sites deal with car insurance policies and pricing. Search engines can help you find answers to many technical questions that arise about automobiles. Examples of Web sites with search engines are Yahoo, Google, and msn. In addition, most companies and dealership have Web sites that can answer many of the common questions about specific vehicles. There are also numerous Web sites where you can find new or used parts for street rods, customized vehicles, and vintage vehicles.

SOCIETY OF AUTOMOTIVE ENGINEERING (SAE)

Another excellent resource for learning about automobiles is the Society of Automotive Engineering. The SAE is a worldwide organization involved in issues that deal with land, space, air, and marine transportation. Some of the automotive services and components of SAE include:

- Memberships for those interested in automobiles
- Technical committees that deal with specific areas of the automobile
- Professional development activities for automobile enthusiasts
- Public awareness programs about the automobile
- SAE Foundation awards for students in automotive education programs
- Technical papers on new automobile designs

For more information about the services of SAE, visit http://www.sae.org.

THE SERVICE TECHNICIANS SOCIETY (STS)

The Service Technicians Society (STS) is a professional society dedicated to advancing the knowledge, skills, and image of service technicians. STS is affiliated with the Society of Automotive Engineers. The mission of STS is to:

- Advance the skills and education of service technicians
- Encourage high ethics and performance
- Inspire professionalism and excellence in the mobility service industry
- Disseminate mobility service technology information
- Foster communication and cooperation among service technicians and other professionals
- Serve the public need for environmentally responsible, safe, and efficient mobility systems

STS offers publications and books for service technicians from industry leaders such as Bosch and Jendham at special prices. STS also offers technical and training information on its Web site.

AUTOMOTIVE YOUTH EDUCATIONAL SYSTEM (AYES)

The Automotive Youth Educational System, known as AYES, is an organization that has developed a partnership among automotive manufacturers, local dealers, and selected high schools. The organization's goal is to encourage good students with a sound mechanical aptitude to pursue careers in the fields of automotive service technology or collision repair and refinishing. The intent is to prepare students for entry-level positions in the automotive field. The industries that support AYES include Audi, BMW, DaimlerChrysler, General Motors, Honda, Hyundia, Mercedes-Benz, Mitsubishi, Nissan, Subaru, Toyota, and Volkswagen. Schools that are AYES-certified have an automotive curriculum parallel to the National Institute for Automotive Service Excellent (ASE). For more information about AYES and how to become an AYES school partner, visit http://www.ayes.org.

VOCATIONAL/TECHNICAL SCHOOLS

Most vocational schools and technical colleges offer updated courses, seminars, and workshops on many of the new technologies being developed by the automotive industry.

MANUFACTURING SCHOOLS

Each of the major automobile manufacturers has developed factory-supported training programs and centers for keeping current. These training centers offer special courses on particular engines, vehicles, types of fuel injection, and so on. These schools normally have a service training instructor available from the manufacturer. People who attend these service schools include service technicians and vocational and technical college instructors.

AUTOMOTIVE TEXTBOOKS

Various publishers now offer a complete line of automotive service books for both consumers and automotive technicians. Specialty books are also available on most any vehicle or system within the vehicle. Many of these books can be purchased in local bookstores and chain stores.

2.4 AUTOMOTIVE TECHNICIAN CERTIFICATION

PURPOSE OF CERTIFICATION

Because of the complexity of vehicle design and service on automobiles today, a national certification program has been established. Certification is accomplished through an organization called the National Institute for Automotive Service Excellence, commonly called **ASE**. ASE is a nonprofit corporation dedicated to improving the quality of automotive service and repair throughout the nation. ASE has a board of directors that helps govern its operation. The automotive industry, educational community, government, and consumer groups are all represented on the board of directors.

The primary function of ASE is to test and certify automobile and heavy-duty truck technicians and body repairers and painters. ASE also encourages and assists in the development of effective automotive training programs. Throughout this book there is a great deal of information to help you study for becoming ASE-certified in the automobile technician area.

TYPES OF TESTS

The ASE certification program consists of a series of written tests given twice a year. The tests are designed to measure diagnostic and repair knowledge and skills in eight technical areas. When an automotive technician passes one or more tests and has completed two years of related work experience, the technician is certified by ASE. When the technician has passed all eight of the tests, he or she is then considered an ASE-Certified Master Auto Technician (***Figure 2–7***).

FIGURE 2–7 When service technicians become ASE-certified, they have passed all eight tests and are considered master auto technicians. This ASE patch is worn by the master auto technician.

Figure 2–8 shows some of the technical components in which ASE test questions are written. Each test has a specific number of questions. Note however, that as the technology of the automobile continues to change, additional topics and additional questions may be added. The layout of each of the eight tests is shown below.

Engine Repair Test (A1) This test has a total of eighty questions and is subdivided into the following areas:

1. General Engine Diagnosis
2. Cylinder Head and Valve Train Diagnosis and Repair
3. Engine Block Diagnosis and Repair
4. Lubrication and Cooling Systems Diagnosis and Repair
5. Ignition System Diagnosis and Repair
6. Fuel and Exhaust System Diagnosis and Repair
7. Battery and Starting System Diagnosis and Repair

Automatic Transmissions / Transaxles Test (A2) This test has a total of fifty questions and is subdivided into the following areas:

1. General Transmission/Transaxle Diagnosis
2. Transmission/Transaxle Maintenance and Adjustment
3. In-Vehicle Transmission/Transaxle Repair
4. Off-Vehicle Transmission/Transaxle Repair

Manual Drivetrains and Axles Test (A3) This test has a total of forty questions and is subdivided into the following areas:

1. Clutch Diagnosis and Repair
2. Transmission Diagnosis and Repair
3. Transaxle Diagnosis and Repair
4. Drive (Half) Shaft and Universal Joint Diagnosis and Repair
5. Rear Axle Diagnosis and Repair
6. Four-Wheel Drive Component Diagnosis and Repair

Suspension and Steering Test (A4) This test has a total of forty questions and is subdivided into the following areas:

1. Steering Systems Diagnosis and Repair
2. Suspension System Diagnosis and Repair
3. Wheel Alignment Diagnosis, Adjustment, and Repair
4. Wheel and Tire Diagnosis and Repair

Brakes Test (A5) This test has a total of fifty-five questions and is subdivided into the following areas:

1. Hydraulic System Diagnosis and Repair
2. Drum Brake Diagnosis and Repair
3. Disc Brake Diagnosis and Repair
4. Power Assist Units Diagnosis and Repair
5. Miscellaneous Diagnosis and Repair
6. Antilock Brake System Diagnosis and Repair

Electrical/Electronic Systems Test (A6) This test has a total of fifty questions and is subdivided into the following areas:

1. General Electrical/Electronic System Diagnosis
2. Battery Diagnosis and Service
3. Starting System Diagnosis and Repair

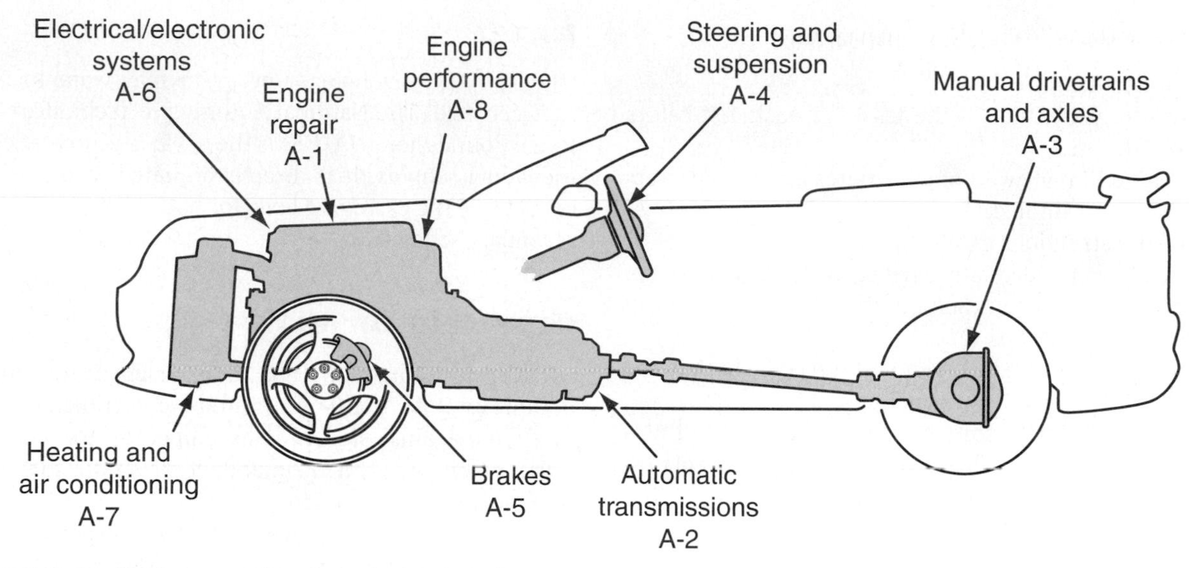

FIGURE 2-8 ASE tests are written for various technical areas on the automobile.

4. Charging System Diagnosis and Repair
5. Lighting Systems Diagnosis and Repair
6. Gauges, Warning Devices, and Drive Information Systems Diagnosis and Repair
7. Horn and Wiper/Washer Diagnosis and Repair
8. Accessories Diagnosis and Repair

Heating and Air Conditioning Test (A7) This test has a total of fifty questions and is subdivided into the following areas:

1. A/C System Diagnosis and Repair
2. Refrigeration System Component Diagnosis and Repair
3. Heating and Engine Cooling System Diagnosis and Repair
4. Operating System and Related Controls Diagnosis and Repair
5. Refrigerant Recovery, Recycling, and Handling

Engine Performance Test (A8) This test has a total of seventy questions and is subdivided into the following areas:

1. General Engine Diagnosis
2. Ignition System Diagnosis and Repair
3. Fuel, Air Induction, and Exhaust System Diagnosis and Repair
4. Emissions Control Systems Diagnosis and Repair
5. Computerized Engine Controls Diagnosis and Repair
6. Engine-Related Service
7. Engine Electrical Systems Diagnosis and Repair

There is also a Recertification Test available every five years. However, it contains about half as many questions in each content area.

If you are interested in becoming ASE-certified, your instructor should have a series of tasks available for each test. Use these tasks to help you study and determine if you can pass each test. You can also write to ASE requesting more information. The address is:

National Institute for Automotive Service Excellence
101 Blue Seal Drive, S.E., Suite 101
Leesburg, VA 20175
http://www.asecert.org

TAKING TESTS

There are several types of test questions used in this book as well as on the ASE certification tests. The most common is the multiple choice test question. At the end of each chapter in this textbook there are sample test questions for you to study and become familiar with.

Generally there are two types of multiple choice test. One type asks a question and then gives several choices for the answer. You are probably most familiar with this type of question. The second type of multiple choice question is the one used on ASE exams. This type gives two statements about a certain subject. You must decide which statement is correct. The following is a typical example:

Technician A says that the engine is used to produce the power to move the vehicle. *Technician B* says that the power to push a car forward comes from coal. Who is correct?

a. A only c. Both A and B
b. B only d. Neither A nor B

When taking this type of test, you must determine which, if any, statements are true. For example, ask yourself if Technician A's statement is correct. Of course you know it is correct. Then ask yourself if Technician B is correct. This statement is not correct. So the answer for this question is *a. A only*. Often this type of comparative question is used to measure knowledge and information about diagnosis and repair of an automobile.

Remember the following tips when taking a test:

- Be confident.
- Make sure you study all the tasks for each test before taking the test.
- Make sure to review all of your notes and the chapters that you have studied.
- Read all instructions carefully.
- Make sure you understand what each question is asking.
- Look for key words in each question that may help you arrive at the right answer.
- Answer questions that you know, then come back to the ones that you are not sure about.
- It is usually best to start by taking one ASE test at a time so as not to overload yourself. Many students try to take too many tests at one time.

LEARNING HOW TO STUDY

Some people have no difficulty when studying, others have more difficulty. The following suggestions may help you to maximize your study time:

- Find a quiet place to study where there are no distractions. (Never study in front of a TV.)
- Look at all the diagrams and photos in the book, making sure you understand each.
- Studying with another person often helps you understand difficult principles.
- Read all necessary information first and develop an outline of the content.
- Make sure you know what part of the automobile you are studying. Make sure you also see the big picture of what you are studying.
- Look over the summary comments in each chapter.
- Make sure you have answered each of the questions at the end of each chapter and know the answers to each question.
- Try to identify the key points in each section of the book.
- Study in small sections. Don't try to learn everything the night before the test.
- Study when you feel the most energetic. There are times in each day when you have much more energy than at other times. Be aware of these times and study when you are most alert.
- If you can't remember what you just studied, stop, take a break and get yourself reenergized. If you get too tired, stop studying immediately.
- Always take clear and neat notes when in class. Use colored pencils to highlight key points of a lecture or demonstration.

Each student has his or her own technique for studying; however, these tips may improve your studying habits. Using these and other good study habits will increase your chance of passing tests.

NATEF

Often an entire school or program may want to become ASE-certified. The National Automotive Technician Educational Foundation (NATEF) is the organization that certifies individual schools. It is directly affiliated with ASE and is considered the certifying body for both secondary and postsecondary schools.

PURPOSE OF NATEF

NATEF was founded in 1983 as an independent, nonprofit organization to evaluate automotive technician training programs against standards developed by the automotive industry and recommend qualifying programs for certification by ASE.

The ASE Board of Trustees is the body responsible for the Automobile Technician Training Certification Program. Through NATEF, ASE grants certification to programs that comply with the evaluation procedure, meet established standards, and adhere to the policies that have been established. NATEF is solely a certification program and is not associated with the accreditation role of other agencies.

The NATEF process has resulted in certified automotive training programs in all fifty states at the secondary and postsecondary levels. The purpose of the certification program is to improve the quality of training offered in the secondary schools. NATEF does not endorse specific curricular materials and does not provide instruction. It does, however, set standards for the content of automotive instruction.

Program standards are developed based on ASE task lists and are designed to bring training programs to a level at which participants are properly trained for entry-level positions in the automotive industry. Once a training program meets the established standards and goes through the certification process, it can be certified as a program that actually teaches technicians to today's industrial standards. The eight automobile areas that may be certified include:

1. Brakes
2. Electrical/Electronic Systems
3. Engine Performance
4. Suspension and Steering
5. Automatic Transmission and Transaxle
6. Engine Repair
7. Heating and Air Conditioning
8. Manual Drivetrain and Axles

Four areas are required for minimum certification or recertification. These include Brakes, Electrical/Electronic Systems, Engine Performance, and Suspension and Steering.

NATEF TASKS

A national committee was recently assembled to review and write the standards used in the automobile certification program. As part of the NATEF tests, a series of tasks have been developed (see ***Figure 2–9***). There are several levels or priorities (P1, P2, and P3) of tasks that are associated with each of the eight ASE test areas. Throughout this textbook, these NATEF tasks have been correlated to each of the appropriate chapters as part of the Certification Connection boxes.

APPLIED ACADEMICS

In addition to its specified tasks, a series of related academic skills have been developed by NATEF. These skills are categorized as Applied Academics and fall into the following three areas:

1. Language Arts and Related Academic Skills
2. Mathematics-Related Academic Skills
3. Science-Related Academic Skills

Each of these skills has been cross-referenced with the eight ASE test areas. For examples of these skills, refer to the Applied Academics section at the end of each chapter in this textbook. For more information on NATEF, its costs, procedures for certification, and so on, visit NATEF's Web site at www.natef.org.

FIGURE 2–9 NATEF is the organization that certifies automotive programs in high schools.

SUMMARY

The following statements will help to summarize this chapter:

- Numerous careers and educational opportunities are available within the automobile industry.
- Some of the major careers in the automobile industry include service technician, shop supervisor, service manager, parts manager, service writer, marketing and sales representative, service company representative, and many additional supporting careers.
- Depending on the specific position and career, a sound knowledge of automobiles is very important to being successful.
- Both technical skills and knowledge as well as good interpersonal skills are needed to be successful in the automobile industry.
- Some of the more important educational opportunities in the automobile industry include co-op programs, internships, apprenticeships, on-the-job training, and manufacturer's schools.
- There are various resources that one can draw on to learn more about the automobile, including the Society of Automotive Engineering, vocational schools, colleges and universities, and the Internet.
- The automobile industry keeps its standards high through the efforts of the National Institute for Automotive Service Excellence, known as ASE.
- ASE provides certification through a series of eight tests designed to evaluate a service technician's auto mobile knowledge.
- An automotive service technician can be certified in any one of the eight test areas with a passing grade, along with two years of work experience.
- NATEF is an organization that certifies secondary and postsecondary automotive technician programs in accordance with the eight ASE test areas.
- NATEF also provides tasks associated with each ASE test as well as a list of Applied Academics skills in language arts, mathematics, and science.

TERMS TO KNOW

Can you explain each of the following terms? Review the chapter until you can use each term correctly.

ASE
Parts manager
Parts specialist
Sales representative
Service manager
Service representative
Service technician
Shop supervisor
Supporting careers
Work-based learning

REVIEW QUESTIONS

Multiple Choice

1. Which of the following careers would require the most technical training and background about the total automobile?
   a. Parts manager
   b. Sales representative
   c. Service technician
   d. Race car driver
   e. All of the above

2. Which automotive specialist position cleans parts and does general automotive repair?
   a. Transmission/transaxle technician
   b. Front-end technician
   c. Light repair technician
   d. Air-conditioning technician
   e. Lubrication technician

3. The transmission/transaxle technician needs knowledge in which of the following areas?
   a. Mechanical
   b. Fluid
   c. Electrical
   d. All of the above
   e. None of the above

4. Which of the following is not considered an automotive specialist career?
   a. Engine repair specialist
   b. Brake specialist
   c. Suspension and steering specialist
   d. Wheel drum specialist
   e. Engine performance specialist

5. Which of the following is considered a supporting career in the automotive industry?
   a. Claims adjuster
   b. Frame and alignment repair technician
   c. Auto body technician
   d. All of the above
   e. None of the above

6. Which educational system allows students to learn the automobile while they are on the job?
   a. Cooperative schools
   b. On-the-job training
   c. Manufacturing schools
   d. Publishers
   e. Vocational/technical schools

7. How many tests must be passed to become an ASE-Certified Master Automobile Technician?
   a. Two
   b. Five
   c. Eight
   d. Ten
   e. Fifteen

8. Which of the following should not be done when studying for an automotive test?
   a. Study all at once, in large blocks of time and content, even if you are tired.
   b. Study with another person.
   c. Try to identify key points of a lecture.
   d. Read the summary of each chapter.
   e. Find a quiet place to study.

The following questions are similar in format to ASE (Automotive Service Excellence) test questions.

9. *Technician A* says one career in the automotive industry is that of the service technician. *Technician B* says that one career in the automotive industry is that of parts washer. Who is correct?
   a. A only
   b. B only
   c. Both A and B
   d. Neither A nor B

10. *Technician A* says that when studying for a test, always study in a quiet place. *Technician B* says that when studying for a test always study when you feel energized. Who is correct?
    a. A only
    b. B only
    c. Both A and B
    d. Neither A nor B

11. *Technician A* says that there are only six ASE tests and a total of twenty-five questions. *Technician B* says that the ASE test only deals with engine controls. Who is correct?
    a. A only
    b. B only
    c. Both A and B
    d. Neither A nor B

Essay

12. Identify at least five careers in the automotive field.
13. Which person is in charge of the entire service operation of a dealership? Describe what that person does.
14. What are some of the supporting careers in the automotive industry?
15. Describe five ways to improve your study habits.
16. Define a correspondence course in the automotive field.
17. Describe the purpose of NATEF.
18. What are the three areas that NATEF designates as Applied Academics?

Short Answer

19. The person who works directly with people who want to purchase parts is called the ________.

20. Tests that are designed to measure diagnostic and repair knowledge are part of the ________ certification program.

21. When studying for an ASE test it is always important to study in ________ sections.

22. A program that allows students to work in the automotive industry during the summer and get credit for work in school is called a/an ________ program.

Applied Academics

TITLE: Automobiles per Population

ACADEMIC SKILLS AREA: Language Arts

NATEF Automobile Technician Tasks:

The technician adapts a style for speaking and writing that will yield effective communication based on the purpose of the interaction.

CONCEPT:

The automobile is a very significant part of our society. In fact, approximately 25% of all jobs in the United States are related directly or indirectly to the manufacturing, sales, and service of automobiles. To understand how many automobiles are used in our society, consider the statistics. The statistics in the chart show the number of automobiles per 1,000 people in selected countries around the world. For example, in the United States, there are about 500 automobiles for every 1,000 people—the highest percentage of automobiles per person in the world. Although these numbers change regularly, they do show the dependence we, as a society, have on the automobile. Asia (excluding Japan) has only about 3 to 5 automobiles per 1,000 people. If you lived in such a developing country, who do you think would be the three people in every thousand to get an automobile?

REGION	AUTOMOBILES PER 1,000 POPULATION
United States	500–550
Canada	400–450
Western Europe	240–270
Oceania and South Africa	160–180
Japan	160–175
Eastern Europe	60–70
Latin America	40–50
Former USSR	10–15
Africa (excluding South Africa)	7–10
Asia (excluding Japan)	3–5

APPLICATION:

Using the information in the chart as a starting point, how would you communicate to fellow students:

- the importance of the automobile in our society?
- the need for the service technician to service the vehicles in our society?
- the reliance that our society has on the automobile?

Introduction to the Automobile

CHAPTER 3

OBJECTIVES

After studying this chapter, you should be able to:

- Describe the basic design of the automobile, including the body, frame, engine, drive lines, running gear, and suspension system.
- Identify the basic technical systems used on an automobile engine.

INTRODUCTION

There are many technical systems used on an automobile engine. In order to understand these technical systems, it is important to know the automobile's basic design. This chapter introduces the basic vehicle design and identifies basic engine systems.

3.1 VEHICLE DESIGN

In order to study the automobile, it is important to review the basic parts of the vehicle (***Figure 3–1***). The vehicle can be subdivided into several major categories: the body and frame, the engine or power source, the drive lines and running gear, and the suspension system.

BODY AND FRAME

The **body and frame** section of the automobile is the basic foundation of the vehicle. All other components and systems are attached to the body and frame. The frame supports the car body, engine, powertrain and wheels, and the drive lines and running gear. ***Figure 3–2*** shows the body and frame of a typical vehicle.

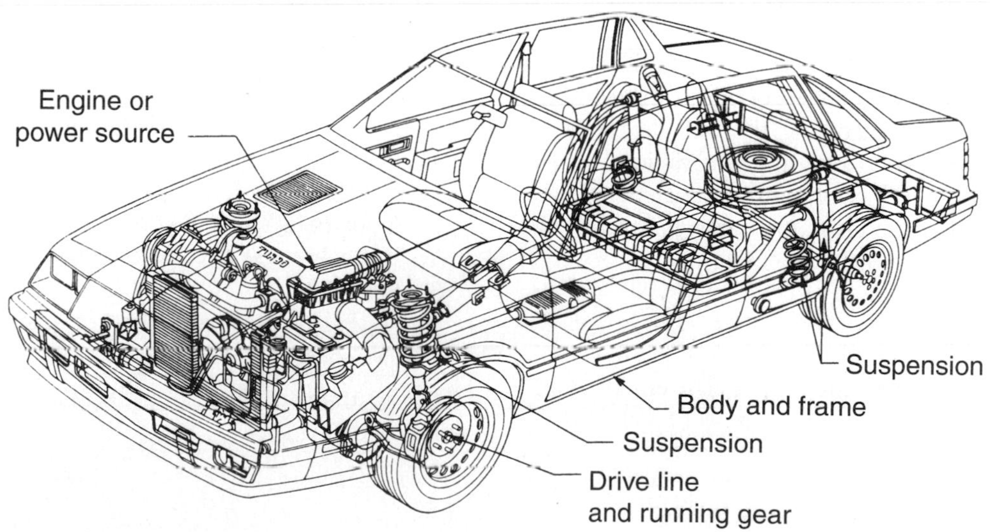

FIGURE 3–1 The basic parts of the automobile include the body and frame, running gear, transmission, suspension, engine, and powertrain. (Courtesy of DaimlerChrysler Corporation)

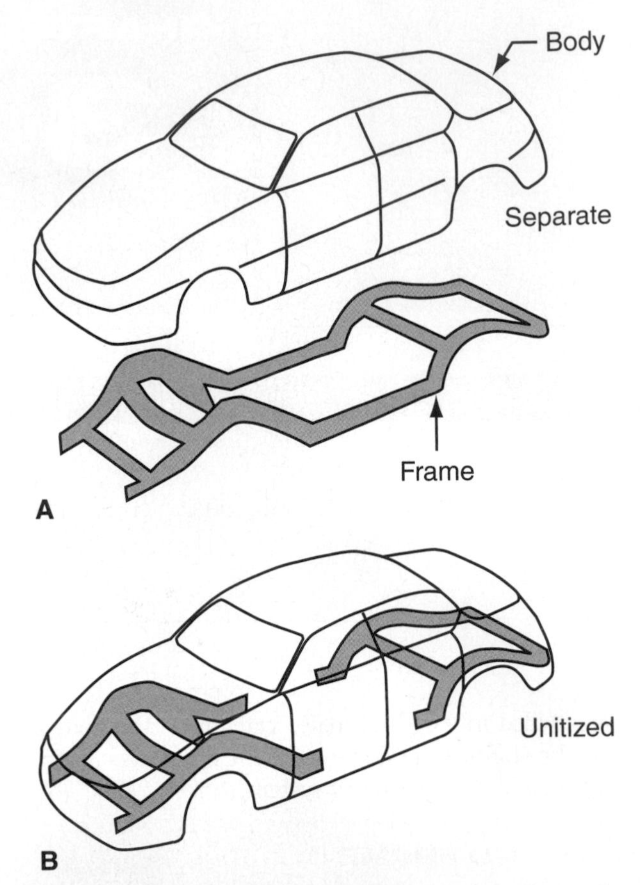

FIGURE 3-2 Two common types of body and frame configurations used in today's automobiles include (A) separate body and frame, and (B) unitized body.

There are several types of body and frame configurations. The separate body and frame construction has been used for the longest time. This type is illustrated in ***Figure 3–2(A)***. Another type of construction is called the unitized body. This type of vehicle is designed with the frame and body in one unit as shown in ***Figure 3–2(B)***. The unitized body is used in many vehicles today.

There are other frame designs. The space frame is made of formed sheet steel. With this type of frame, the vehicle is completely drivable without using the outside plastic or steel panels for support and strength. Some vehicles use the ladder frame design. This type of frame is constructed similar to a ladder so that the cross members connect to the sides of the frame.

BASIC ENGINE

The **engine** is used to power the vehicle. The engine is also called the power source or motor. The word **motor** is defined as that which imparts motion. So a motor can be any device that produces power. However, the power source in the automobile is usually referred to as the engine.

Most automobiles use the gasoline engine as a power source. However, other power sources are being tested and introduced every year. For example, the diesel engine is also being used as a power source in some vehicles today. In addition, some automotive engineers predict the use of gas turbines, Stirling engines, electric batteries, and fuel cells for future power sources in the automobile.

Car Clinic: TROUBLESHOOTING

CUSTOMER CONCERN: Troubleshooting Questions to Ask the Customer and Yourself

When a problem develops in a vehicle, many times the technician will replace parts until the cause of the problem is found. This process costs money and time. What would be a good procedure to identify a problem before trying to fix a car?

SOLUTION: The first step in trying to find the problem in a car is to ask questions. This method will help narrow down the location of the problem to one of the major systems in the vehicle or engine. Some of the more important questions to ask the customer or yourself include:

1. When was the trouble first noticed?
2. Did the problem develop quickly or over a period of time?
3. Was the complaint or problem recorded at an earlier time?
4. Was the problem noticed at all speeds and loads or only at certain speeds and loads?
5. What type of vehicle and engine is the problem associated with?
6. Did any unusual noise develop from the problem?
7. What was the temperature of the engine? Was it above or below the design temperature?
8. Has any work been done on the car recently? Did the problem start before or after the work was done?
9. Is the oil consumption normal?
10. How does the engine respond to acceleration or deceleration?
11. Is there excessive blue (oil) or black (rich) smoke from the exhaust?
12. Does the engine start easily?
13. Does the engine surge or hunt at idle, high idle, or full load?
14. How many miles are on the engine or vehicle?
15. What are the normal driving conditions of the vehicle?

The answers to these questions will help guide the technician and troubleshooter to a logical solution to many of the problems that are found in the automobile.

Today, another power source called the **hybrid** power system, is being studied and used in some vehicles. Several manufacturers are designing and selling hybrid vehicles. The word *hybrid* means that two different power sources are used to power the vehicle. Hybrids use both batteries as well as a small gasoline or diesel engine to power the vehicle. A sophisticated electronic controller is used to determine which of three options will power the vehicle:

1. the gasoline or diesel engine
2. the batteries
3. both the engine and batteries

In the next few years, it will be important for the service technician to learn about hybrid vehicles in order to provide service and effective troubleshooting. More information is presented about hybrid vehicles in Chapter 14, Other Power Sources.

Most automobiles use the reciprocating piston or Otto cycle engine (***Figure 3–3***). However, certain car manufacturers also offer rotary design engines as an optional power source. ***Figure 3–4*** shows the rotary engine used on certain vehicles. Although it looks similar to the reciprocating piston engine, internally there are many differences. The design of the rotary engine is discussed in Chapter 14.

The engine is typically located in the front of the vehicle. But some vehicles have rear engines. In addition, certain manufacturers have developed engines that are placed in the middle of the body and frame.

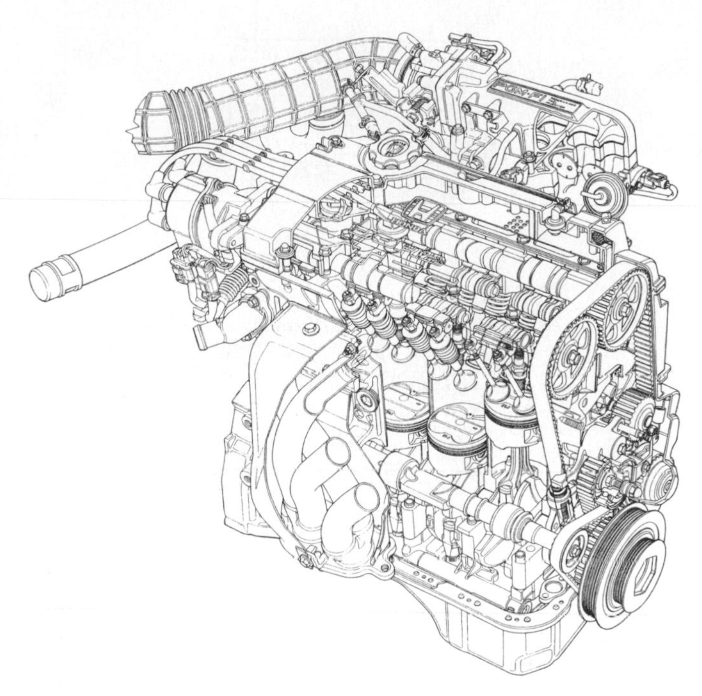

FIGURE 3–3 Most automobiles use reciprocating piston engines. (Courtesy of American Honda Motor Co., Inc.)

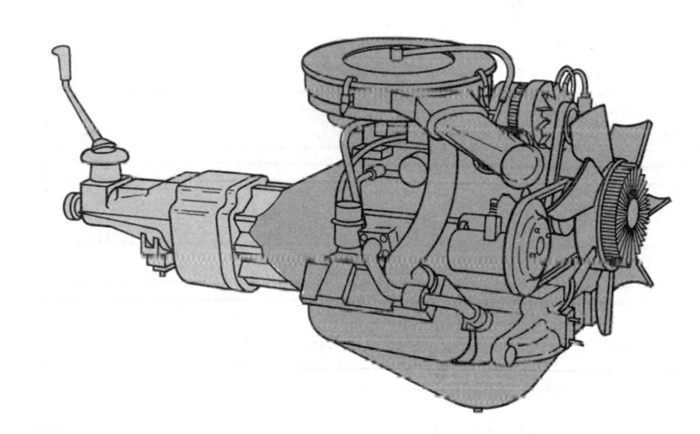

FIGURE 3–4 Some car manufacturers also offer the rotary engine as an alternative power source.

DRIVE LINES AND RUNNING GEAR

The **drive lines** are those components that transmit the power from the engine to the wheels. This action propels the vehicle in a forward or reverse direction. As shown in ***Figure 3–5***, the drive lines include components such as the transmission, drive shafts, differential, and rear axles. Each of these components is discussed in an individual chapter in this text.

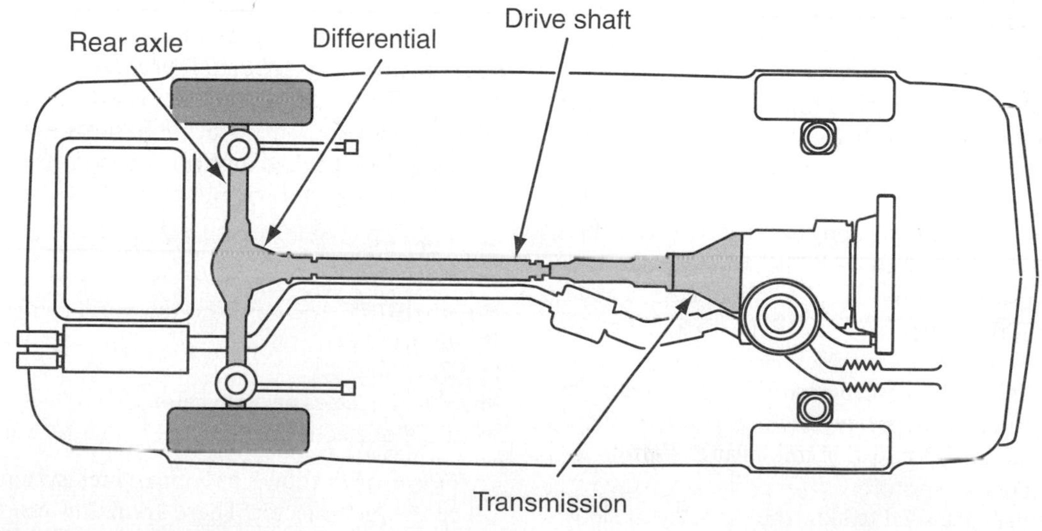

FIGURE 3–5 Many vehicles use the rear-wheel drive line system.

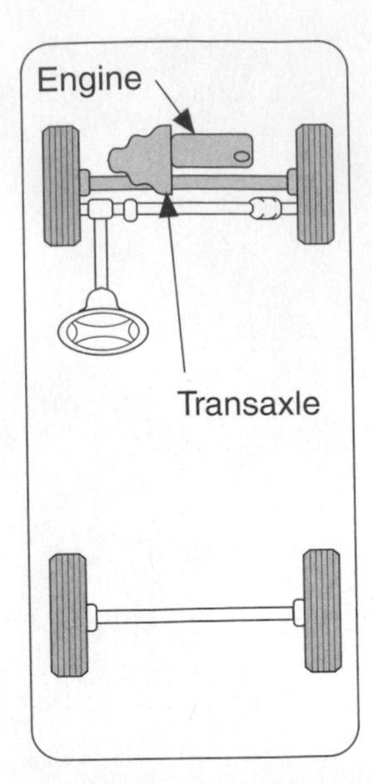

FIGURE 3–6 Many automotive manufacturers place the engine in the front of the vehicle and use front-wheel drive.

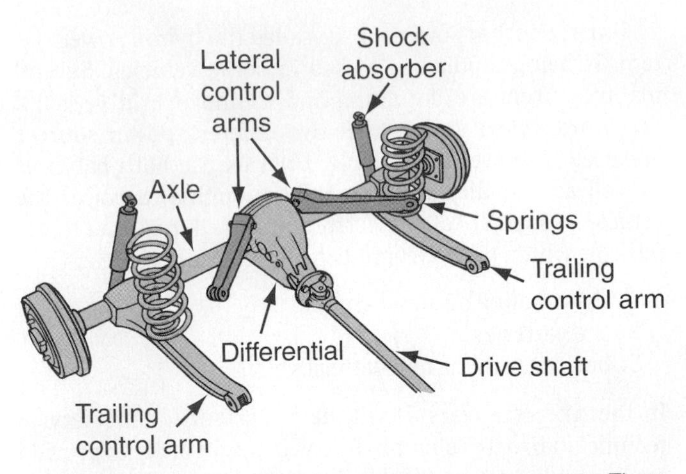

FIGURE 3–7 All vehicles have suspension components. These typically include the shock absorbers, springs, control arms, and axles used to support the total vehicle.

There are two methods in which the drive lines can be designed: the rear-wheel drive lines system and the front-wheel drive lines system. The rear-wheel drive lines system is shown in ***Figure 3–5***. In this system, the engine is in the front of the body and frame. The power is then transmitted through the drive shaft to the rear of the vehicle for propulsion.

The front-wheel drive lines system is used on most vehicles today (***Figure 3–6***). In this system the engine is in the front of the vehicle. The drive lines and running gear are also in the front. Both systems have advantages and disadvantages and are equally reliable in their operation.

The **running gear** consists of components on the automobile that are used to control the vehicle. The running gear is defined as the braking systems, the wheels and tires, and the steering systems. There is a chapter devoted to each of these systems in this text.

SUSPENSION SYSTEM

The **suspension system** on the automobile includes such components as the springs, shock absorbers, MacPherson struts, torsion bars, axles, and connecting linkages. These components are designed to support the body and frame, the engine, and the drive lines on the road. Without these systems, the comfort and ease of driving would be reduced. ***Figure 3–7*** illustrates some of the components that are used on a typical rear-suspension system with a differential used for rear-wheel drive.

The springs and shock absorbers are used to support the axles of the vehicle. The two types of springs commonly used are the leaf spring and the coil spring. ***Figure 3–7*** shows one form of coil spring.

Shock absorbers are used to slow the upward and downward movement of the vehicle. This action occurs when the car goes over a rough road. Shock absorbers are covered in detail in a later chapter.

The axles and control arms (lateral, trailing) are those components that connect the springs, axle, and shock absorbers to the vehicle frame and to the wheels.

Not all rear-suspension systems are the same. Rear-suspension systems are also designed for front-wheel drive vehicles. ***Figure 3–8*** shows a rear-suspension system for a front-wheel-drive vehicle. It uses strut rods and stabilizer bars as well.

3.2 ENGINE DESIGN AND SYSTEMS

All engines are designed to operate with several interconnected technical systems. Generally, there are seven technical systems that make an engine run correctly. Each of these is briefly defined here and will be addressed in more detail in later chapters.

FUEL SYSTEM

The **fuel system** is designed to mix air and fuel in the engine for combustion. This air and fuel mixing should produce an efficient combustion process. If the fuel system is not operating correctly, the engine will run very rough, lack power, or perhaps not run at all. In today's engines, many of the components found in the fuel system are computer controlled. Familiar components in the fuel system include:

- Fuel
- Fuel tank
- Fuel lines
- Fuel pump
- Fuel filter
- Fuel injectors or carburetor
- PCM or electronic controller

Figure 3–9 shows a typical fuel system component called the fuel injector. There are many designs of the fuel injection system, which mixes the air and fuel at the correct

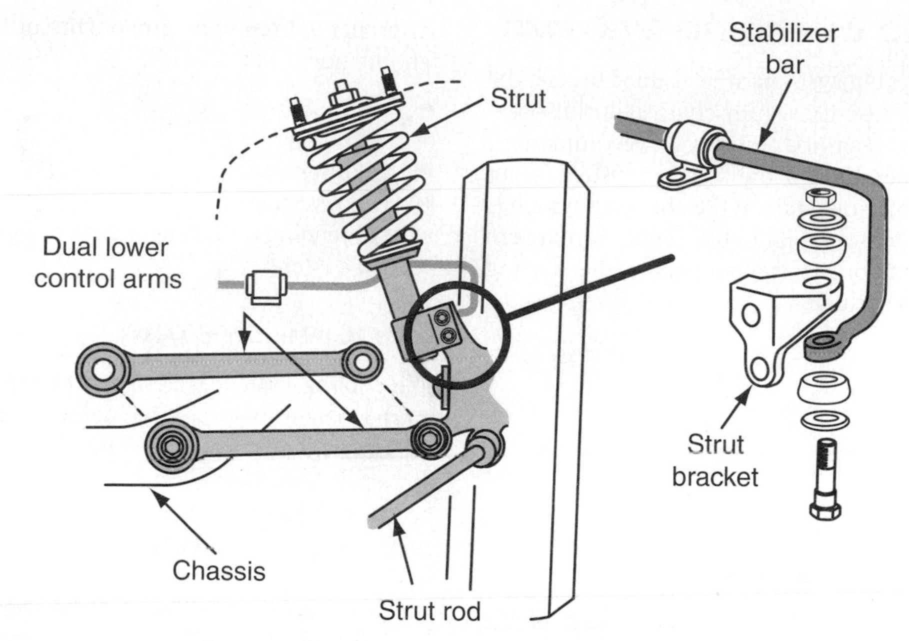

FIGURE 3–8 Rear-suspension systems for front-wheel-drive vehicles do not have a differential.

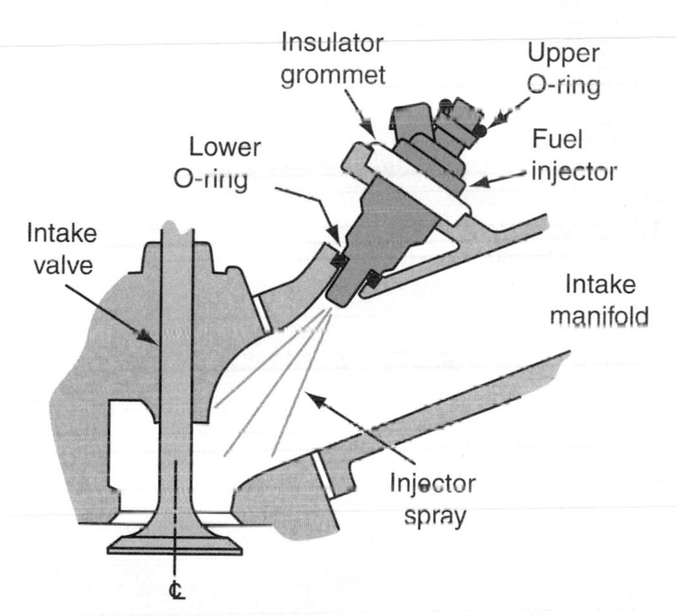

FIGURE 3–9 The fuel injectors are considered part of the fuel system.

and most efficient ratio. Fuel injection is discussed in a later chapter.

IGNITION SYSTEM

The **ignition system** is designed to ignite the air and fuel that have been mixed in the fuel system. In order to do this, a very high voltage is needed to produce a spark within the combustion chamber. Without a properly operating ignition system, there would be no spark to ignite the air and fuel mixture and the engine would not run correctly, or it might not run at all. Each year ignition systems are becoming more and more computerized. Today's ignition systems are totally computer controlled for improved combustion. Some of the components in typical ignition systems include:

- Battery
- Distributor
- Ignition module
- Ignition coil
- Ignition wires
- Spark plugs (***Figure 3–10***)

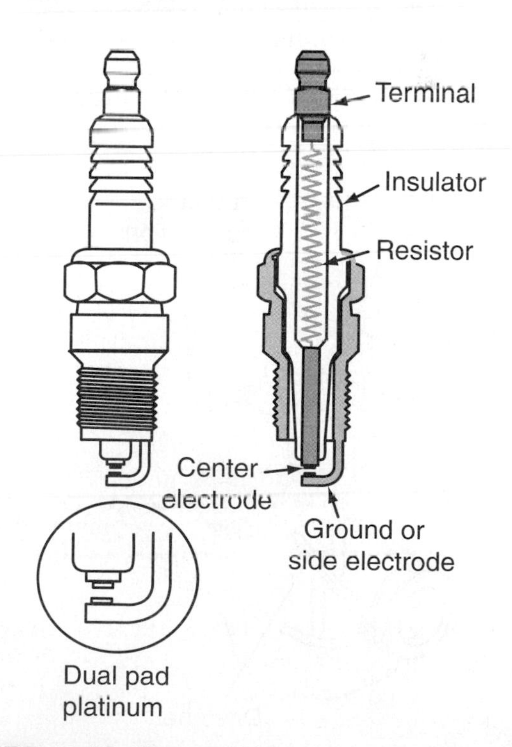

FIGURE 3–10 When servicing spark plugs, the technician is working on the ignition system.

STARTING AND CHARGING SYSTEMS

The **starting and charging systems** are designed to start the engine and then keep the battery fully charged during operation. The starting and charging systems are very important for correct engine operation. If the starting system is not working, the engine may not start. If the charging system is inoperative, then the battery will not have enough charge to start the engine. Some of the components in the starting and charging systems include:

- Battery
- Ignition switch
- Wiring and connectors
- Solenoid
- Starter motor
- Alternator
- Voltage regulator
- Alternator belt and tension adjustment

The alternator shown in ***Figure 3–11*** is part of the charging system. Its purpose is to take the mechanical energy of the engine and convert it to electrical energy to be stored in the battery.

LUBRICATION SYSTEM

The **lubrication system** is designed to keep all of the engine parts lubricated so that friction is reduced. Without lubrication inside the engine, the moving parts that are continuously rubbing against each other would heat up. If they heat up enough, they may be damaged by the heat and metal-to-metal contact. The lubrication system is critical for correct engine operation. Without proper lubrication, the engine would seize up and stop running. The technician who works on a vehicle's lubrication system must understand the types and characteristics of engine oils and how to dispose of used oil properly. Common components of the lubrication system include:

- Oil
- Oil pump
- Oil filter
- Oil passageways
- Oil sending unit, gauge, and wiring/connectors
- Oil cooler

COOLING SYSTEM

The **cooling system** is designed to keep the engine at the most efficient operating temperature. It is important that the engine run at as high a temperature as possible for correct operation. This is usually around 200 degrees Fahrenheit. Consequently, the engine's cooling system is designed to remove just the right amount of heat. If the cooling system is faulty, the engine often overheats and causes internal damage. On the other hand, if the engine is too cool, it is less efficient, gas mileage decreases, and there may be a lack of warm air for heating the passenger compartment. Some of the components in a cooling system include:

- Antifreeze
- Coolant pump
- Thermostat
- Radiator and cap (***Figure 3–12***)
- Radiator fan and clutch
- Radiator hoses
- Radiator overflow tank
- Coolant sensors/wiring/connectors

AIR INTAKE AND EXHAUST SYSTEMS

The **air intake and exhaust systems** are designed to get clean air into the engine and remove the dirty exhaust gases

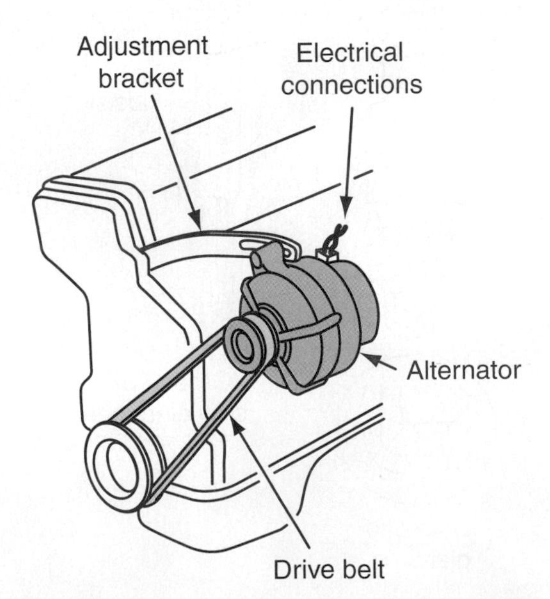

FIGURE 3–11 The alternator and associated parts are considered components of the charging system.

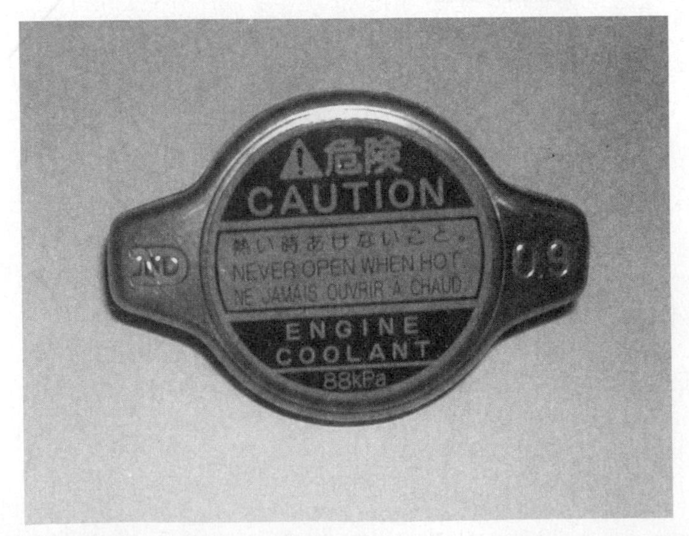

FIGURE 3–12 The radiator cap is considered part of the cooling system.

CUSTOMER CONCERN:
How to Start a Cold Engine
It seems that the best way to start and run a cold engine is to start it up and then let it idle until reaching operating temperature. But it always takes a long time and a lot of gasoline to get the car to operating temperature. Is this the correct way to start a cold engine?

SOLUTION: The best way to start a cold engine is to start the engine and wait about a minute at idle. This will give the oil time to get to all parts of the engine. Then slowly start out, going through all the gears. An engine under a slight to moderate load will reach operating temperature much quicker than an engine at idle.

from the combustion chamber with minimum pollution. Today's engines use sophisticated duct work and exhaust piping to accomplish this. If the air intake and exhaust systems are not working correctly, there may be a lack of clean air for combustion and/or increased pollution from the exhaust system. Today's vehicles use computers to determine how much air is entering the engine and to constantly check the exhaust for the amount of oxygen. Components found in the air intake and exhaust systems include:

Intake system
- Air intake ducting (***Figure 3–13***)
- Air filter
- Intake air sensor, wiring/connectors
- Manifold air pressure sensor
- PCM (computer)
- Turbocharger (if used)

Exhaust system
- Turbocharger, if used
- Exhaust manifold
- Oxygen sensor
- Muffler
- Catalytic converter
- Exhaust pipe
- PCM (computer)

FIGURE 3–13 The intake manifold shown here is part of the air and exhaust system.

POLLUTION CONTROL SYSTEM

The **pollution control system** is designed to reduce various emissions from the engine. Some of the more common pollutants include carbon monoxide (CO), nitrogen oxide (NO_x), and hydrocarbons (HC). Without properly operating pollution control systems, there would be much more pollution both in the air and on the ground. Components found in the pollution control system include:

- Carbon canisters (***Figure 3–14***)
- EGR (exhaust gas recirculator) valves
- PCV (positive crankcase ventilation) valves
- Catalytic converters
- Fuel tanks
- Diverter valves
- Vapor-liquid separators
- Computer controls/wiring/connectors
- Heated air intakes
- Vacuum modulators

When troubleshooting a vehicle problem, remember to identify which technical system you are working with. If this can be done, the process of troubleshooting and diagnosing an engine problem will become much easier. This is because the service technician can then focus only on one system rather than being confused by all of the systems.

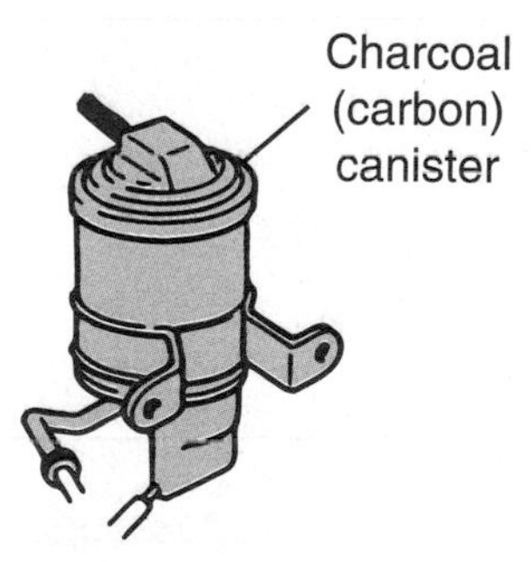

FIGURE 3–14 This charcoal (carbon) canister is considered part of the pollution control system.

SUMMARY

The following statements will help to summarize this chapter:

- The automobile can be subdivided into several major components, including the body and frame, the engine or power source, the drive lines and running gear, and the suspension system.
- The body and frame form the foundation of the automobile.
- The engine is the power source of the vehicle.
- The drive lines and running gear transmit engine power to the wheels.
- The suspension system helps to connect the body and frame to the engine and drive lines.
- An automobile engine is best approached by studying its technical systems.
- The technical systems of the engine include the fuel system, the ignition system, the starting and charging system, the lubrication system, the cooling system, the air intake and exhaust system, and the pollution control system.

TERMS TO KNOW

Can you explain each of the following terms? Review the chapter until you can use each term correctly.

Air intake and exhaust systems
Body and frame
Cooling system
Drive lines
Engine
Fuel system
Hybrid
Ignition system
Lubrication system
Motor
Pollution control system
Running gear
Starting and charging systems
Suspension system

REVIEW QUESTIONS

Multiple Choice

1. Which of the following is considered the mechanical foundation of the automobile?
a. Drive lines
b. Suspension systems
c. Body and frame
d. All of the above
e. None of the above

2. Which type of body and frame is all in one unit?
a. Separate
b. Unitized
c. Stub frame
d. Nose frame
e. Stabilized

3. Which of the following types of power sources are currently used?
a. Diesel power sources
b. Rotary power sources
c. Piston engine (gasoline) power sources
d. All of the above
e. None of the above

4. What technical system on the engine is used to mix the air and fuel correctly?
a. Lubricating
b. Cooling
c. Fuel
d. Drive line
e. Starting

5. The components on the vehicle used to transmit power are referred to as:
a. Running gear
b. Drive lines
c. Suspension systems
d. Shock absorbers
e. Transfer gear

6. Springs and shock absorbers are part of which system?
a. Drive lines
b. Running gear
c. Suspension system
d. Transfer gear
e. Brake system

7. Which of the following technical systems of the engine is used to provide the spark to burn the air and fuel mixture?
a. Lubricating system
b. Cooling system
c. Ignition system
d. Starting system
e. Suspension system

8. Which of the following are used to slow the upward and downward movement of the vehicle?
   a. Springs
   b. Axles
   c. Shock absorbers
   d. Differential
   e. None of the above

The following questions are similar in format to ASE (Automotive Service Excellence) test questions.

9. *Technician A* says that the springs and shock absorbers are part of the suspension system. *Technician B* says springs and shock absorbers are not used on cars. Who is correct?
   a. A only c. Both A and B
   b. B only d. Neither A nor B

10. *Technician A* says the parts used to transmit power to propel the vehicle forward are called the running gear. *Technician B* says the parts used to transmit power to propel the vehicle forward are only used during starting of the vehicle. Who is correct?
    a. A only c. Both A and B
    b. B only d. Neither A nor B

11. *Technician A* says that the ignition system components help to produce the spark for combustion. *Technician B* says that the alternator charges the battery when the engine is running. Who is correct?
    a. A only c. Both A and B
    b. B only d. Neither A nor B

12. *Technician A* says that if the lubrication system fails, the engine may completely seize up. *Technician B* says that the carbon canister is part of the pollution control system. Who is correct?
    a. A only c. Both A and B
    b. B only d. Neither A nor B

Essay

13. Describe the difference between the suspension system and the drive line of an automobile.

14. Describe why it is important to study the engine components as technical systems.

15. Describe the purpose of the cooling system.

16. Identify three components that are used on pollution control systems.

Short Answer

17. The ________ system is designed to reduce emissions from the engine exhaust.

18. The alternator on the charging system is used to keep the ________ fully charged.

19. The thermostat is considered part of the ________ technical system.

20. A solenoid is considered part of the ________ technical system.

Applied Academics

TITLE: The Automobile and Passenger Miles

ACADEMIC SKILLS AREA: Mathematics

NATEF Automobile Technician Tasks:

The service technician can convert variables presented orally into a mathematical form that provides for an algebraic solution.

CONCEPT:

Many scientists compare the automobile to other forms of transportation by comparing "passenger miles." The term *passenger miles* is defined as the average miles per gallon of fuel the vehicle uses multiplied by the average number of people in the vehicle. For example, if a car gets 25 miles per gallon and there are four people in a car pool vehicle, then the passenger miles for this vehicle is (25 × 4) or about 100 passenger miles. When compared to other forms of transportation, the automobile is less efficient. Various modes of transportation are shown on the left side of the following chart. The maximum capacity is provided for each type of transportation. Then the average vehicle mileage is shown. When passenger miles are calculated for each mode, the bicycle is the most efficient; a snowmobile is the least efficient. The energy consumption is also shown for each mode and measured in Btus per passenger mile.

MODE OF TRANSPORT	MAXIMUM CAPACITY	VEHICLE MILEAGE	PASSENGER MILES/ GALLON	ENERGY CONSUMPTION BTU/PASS./MILE
Bicycle	1	1,560	1,560	80
Walking	1	470	470	260
Intercity bus	45	5	225	550
Subway train (10 cars)	600	0.2	120	1,400
Automobile	4	25	100	1,500
Motorcycle	1	60	60	2,060
Snowmobile	1	20	20	5,000

APPLICATION:

In the United States, the average number of people riding in a car at any one time is about 1.2 people. In fact, most people drive to work alone. After reading the information in the chart, calculate the passenger miles of vehicles that get an average of 27 miles per gallon and carry an average of 1.2 people. Can you determine how many Btus per passenger-mile are needed when such a vehicle is used?

Safety in the Automotive Shop

CHAPTER 4

OBJECTIVES

After studying this chapter, you should be able to:

- Identify the hazardous waste produced from automobiles.
- Read a Material Safety Data Sheet.
- Define and illustrate common safety equipment used in the automobile service area.
- List safety rules used in any automobile service area.
- Define OSHA and describe its major functions.
- Develop proper attitudes concerning safety in the automobile service area.
- List the possible danger areas for common chemicals and accidents in the automobile service area.

INTRODUCTION

Most of the service and maintenance work done on automobiles is completed in the automotive shop. Because of the complexity of the automobile, many tools, instruments, and machines are used for service. In addition, there are usually many people in the service area. Complex tools, machines, and instruments, coupled with many people, make the automotive shop a likely place for accidents to happen. In addition, the automobile uses explosive and flammable fuel that can be dangerous. Safety in the automobile shop has become an extremely important aspect in the study of automotive technology.

Safety has become such an important part of our society and industry that the federal government established the Occupational Safety and Health Act of 1970. Known as OSHA, this act makes safety and health on the job a matter of law for 4 million American businesses. It also applies to automotive service shops.

OSHA provides several things. It establishes standards and regulations for safety. It improves unsafe and unhealthful working conditions. It also assists in establishing plans for safe working conditions.

Safety in the automotive shop must also take into account the hazardous waste associated with automobiles. Right-to-Know Laws and Material Safety Data Sheets are used to identify many of these hazardous wastes.

4.1 SAFETY IN THE WORKPLACE

TYPES OF HAZARDS IN THE AUTOMOTIVE SHOP

Since the 1970s, a great deal of research has been done on the various environmental factors related to safety in the workplace. These factors can cause sickness, impaired health, significant discomfort to workers, and even death in some cases. Environmental hazards can be classified as chemical, physical, ergonomic, and biological. Within the automotive area, chemical, physical, and ergonomic hazards are the ones most often found.

Chemical hazards arise from high concentrations of airborne mists, vapors, gases, or solids in the form of dust or fumes. Chemically hazardous substances include cleaning solvents, gasoline, asbestos, and antifreeze.

Hazardous wastes are substances, such as an industrial by-product, that are potentially damaging to the environment and harmful to the health and well-being of humans and other living organisms (***Figure 4–1***).

Physical hazards arise from excessive levels of noise, vibration, temperature, and pressure. There are also cutting and crushing hazards in the automotive shop. Examples include stamping, grinding, exhaust heat, coolant-system pressures, and electrical shock.

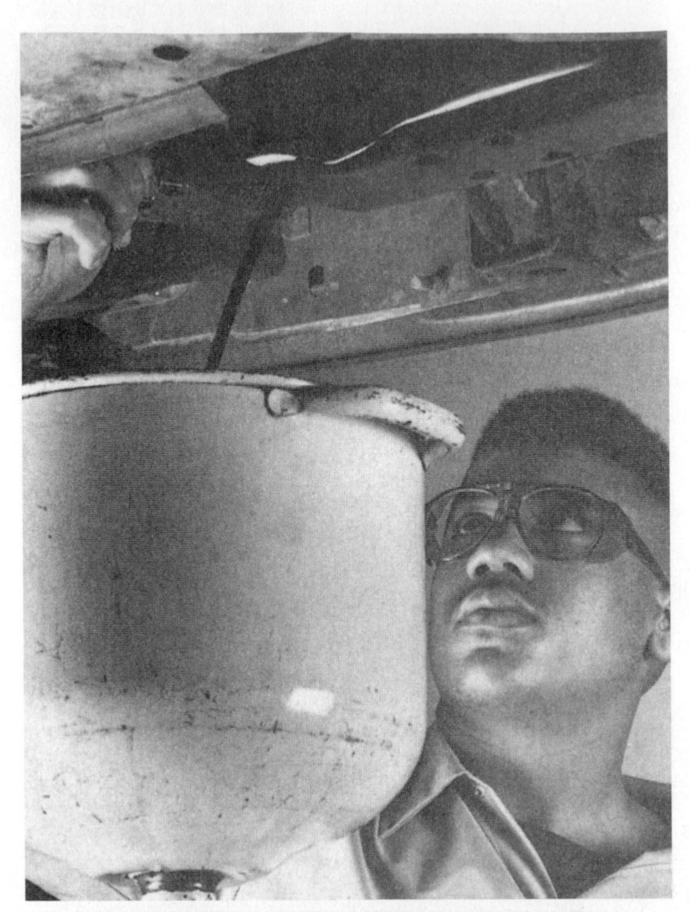

FIGURE 4–1 Used oil from an engine is considered a hazardous waste product.

Ergonomics is the study of human characteristics for the appropriate design of the living and working environment. **Ergonomic hazards** are defined as conditions that relate to the position or proper function of one's body or to motion. Ergonomic hazards include poorly designed tools or work areas, improper lifting or reaching, poor lighting, and so on.

SAFETY AND RESPONSIBILITY

Safety in the workplace is everyone's responsibility. The word *responsibility* is defined as something for which we have a duty or obligation. Technicians in the automotive shop have a duty or an obligation to be as safe as possible.

In every automotive shop there is a great potential for accidents. Accidents happen because people are not careful about what they are doing. Some accidents happen because automotive technicians try to take shortcuts instead of following the correct service procedures. For example, when jacking up a car, the correct procedure should always be followed. This takes time. If a shortcut is taken (for instance, the jacks have not been positioned correctly), the car may not be stable on the jack and may fall. This could cause a serious accident. We have an obligation to ourselves and the other employees in an automotive shop to take time and follow the correct procedures.

Other accidents happen when there are dangerous conditions in the shop. For example, grease or oil often falls on the service area floor making it slippery. This is a dangerous condition. Such conditions should be corrected before an accident happens. We have a duty to make sure there are no dangerous conditions in the automotive shop. If we all try to be more responsible about safety, we can help reduce the potential for accidents in the automotive shop.

OCCUPATIONAL SAFETY AND HEALTH ACT

Safety in the workplace is very important. About one of every four workers today is exposed to known health and safety hazards on the job. Thus, a need exists for a program to monitor, control, and educate the work force about safety. In 1970 the Occupational Safety and Health Administration (**OSHA**) was created by the federal government. This administration provided for the creation of the Occupational Safety and Health Act. The mission of this Act was to:

- Ensure safe and healthful working conditions for working men and women.
- Allow for enforcement of the standards developed under the Act.
- Assist and encourage states in their efforts to assure safe and healthful working conditions by providing research, information, education, and training in the field of occupational safety and health.

Safety standards have been established that are consistent across the country. It is the responsibility of employers to provide a place of employment that is free from all recognized hazards and that will be inspected by government agents knowledgeable in the law of working conditions.

Because of the nature of the automotive industry, especially in the area of automotive service and repair, all safety and health issues have been established and are now controlled by OSHA.

RIGHT-TO-KNOW LAWS

Today, all workplaces across the United States are required to provide safe working conditions for their employees. An important part of this requirement are the Right-to-Know Laws. These laws protect employees from hazardous wastes and other safety hazards. For example, when you start working in an automotive shop, the employer has the responsibility to tell you what safety hazards exist, especially in regard to hazardous materials. Federal and state Right-to-Know Laws protects the employee in three major areas:

1. Your employer must provide training to you about your rights, the nature of hazardous materials in the shop, how hazardous material is labeled, the posting of safety data sheets, the protective equipment required, and the procedures that should be followed in case of accidental spills of hazardous materials.

2. Your employer must label all hazardous materials indicating their health, fire, and reactivity hazards. The labels must be clearly understood by the users. In addition, a list of all hazardous materials used in the work area must be posted where it can be read.
3. Your employer must maintain documentation at your shop that includes proof of training and records of accidents or spills involving hazardous materials or wastes.

4.2 HAZARDOUS MATERIALS AND WASTES

TYPES OF HAZARDOUS MATERIALS

There are many types of hazardous materials and wastes produced in the automotive shop. The Environmental Protection Agency (EPA) helps to define hazardous materials and wastes, which can be both solids or liquids. To be considered hazardous, a material must fall into one of the following four categories:

1. Ignitability—The liquid flash point (temperature at which a liquid will ignite) of a hazardous material is below 140 degrees Fahrenheit. A hazardous solid will self-ignite (spontaneously ignite) due to heat generated by a reaction of two different materials.
2. Corrosivity—A hazardous material burns the skin or dissolves metals.
3. Reactivity—A hazardous material reacts violently with water or other substances, or releases dangerous gases when exposed to low pH (acid) solutions, or generates toxic vapors, fumes, mists, or flammable gases.
4. EP Toxicity—A hazardous material leaches (dissolves or removes) any of eight listed heavy metals in concentrations greater than 100 times the concentration found in standard drinking water.

HAZARDOUS WASTES IN THE AUTOMOTIVE SHOP

The automotive shop where you work may have one or more hazardous waste materials that need to be disposed of. A more complete EPA list of hazardous wastes can be found in the Code of Federal Regulations. Remember that a material is generally not considered a hazardous waste until after it has been used and needs to be disposed of.

There are many types of hazardous waste found in the automotive shop. Some of the more important ones that must be disposed of include:

- Engine oil
- Gasoline
- Diesel fuel
- Solvents
- Transmission fluids
- Transaxle fluids
- Differential fluids
- Engine coolant (ethylene glycol)

FIGURE 4-2 The asbestos dust particles in a brake drum are considered hazardous waste. This vacuum system helps to remove asbestos dust from the wheel area.

- Air-conditioning refrigerants
- Batteries
- Sulfuric acid in batteries
- Tires
- Belts
- Cleaning fluids and chemicals
- Paints
- Brake fluids
- Asbestos brake linings (***Figure 4–2***)
- Greases
- Transmission clutch plates

4.3 MATERIAL SAFETY DATA SHEET (MSDS)

PURPOSE OF MSDS

As part of the Right-to-Know Law, employers must have **MSDS** (Material Safety Data Sheets) available. Material Safety Data Sheets provide critical information about the hazardous materials found in a typical automotive shop. These sheets list all of the known dangers of specific chemicals. They also list first-aid procedures for skin or respiratory contact.

READING A MATERIAL SAFETY DATA SHEET

Figure 4–3 shows a typical Material Safety Data Sheet. Your employer has the responsibility to make sure there is one sheet for each hazardous material that exists in your automotive shop. A typical MSDS shows several items:

1. The product name is generally shown near the top of the sheet.
2. The ingredients are shown for the specific chemical or product.
3. The physical data is shown for the specific chemical or product. Depending on the specific substance, specific gravity, boiling point, freezing point, evaporation data, percent volatility, and other physical data may be shown.
4. The fire and explosion data, such as flash point, is shown.
5. The reactivity and stability data is shown for the chemical or waste.
6. The protection information (proper handling and first aid) is shown.

YOUR RESPONSIBILITY WITH MSDS

The chemical makeup of each person's body is different. Some people react to hazardous materials differently than others. For example, some people may get a rash when they touch cleaning solvents. Other people may get sick when they smell gasoline or diesel fuel. Still others may get dizzy and nauseated when they smell ink markers used in the classroom. It is your personal responsibility to know what hazardous materials exist in an automotive shop, and what ordinary materials you may be sensitive to. Use the four following guidelines when you work in the automotive shop:

1. Know where the MSDS binder is located in your automotive shop.
2. Read each MSDS to know what chemicals and hazardous materials are present in your shop.

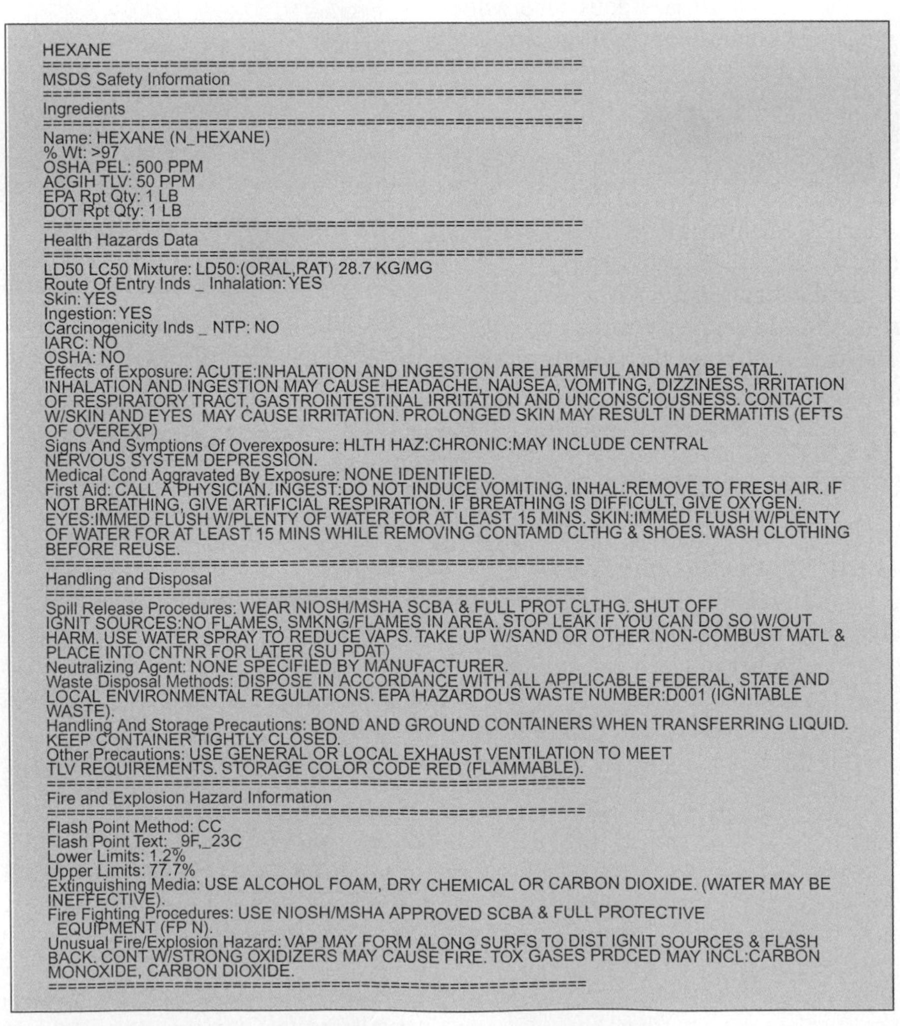

```
HEXANE
============================================================
MSDS Safety Information
============================================================
Ingredients
============================================================
Name: HEXANE (N_HEXANE)
% Wt: >97
OSHA PEL: 500 PPM
ACGIH TLV: 50 PPM
EPA Rpt Qty: 1 LB
DOT Rpt Qty: 1 LB
============================================================
Health Hazards Data
============================================================
LD50 LC50 Mixture: LD50:(ORAL,RAT) 28.7 KG/MG
Route Of Entry Inds _ Inhalation: YES
Skin: YES
Ingestion: YES
Carcinogenicity Inds _ NTP: NO
IARC: NO
OSHA: NO
Effects of Exposure: ACUTE:INHALATION AND INGESTION ARE HARMFUL AND MAY BE FATAL.
INHALATION AND INGESTION MAY CAUSE HEADACHE, NAUSEA, VOMITING, DIZZINESS, IRRITATION
OF RESPIRATORY TRACT, GASTROINTESTINAL IRRITATION AND UNCONSCIOUSNESS. CONTACT
W/SKIN AND EYES  MAY CAUSE IRRITATION. PROLONGED SKIN MAY RESULT IN DERMATITIS (EFTS
OF OVEREXP)
Signs And Symptions Of Overexposure: HLTH HAZ:CHRONIC:MAY INCLUDE CENTRAL
NERVOUS SYSTEM DEPRESSION.
Medical Cond Aggravated By Exposure: NONE IDENTIFIED.
First Aid: CALL A PHYSICIAN. INGEST:DO NOT INDUCE VOMITING. INHAL:REMOVE TO FRESH AIR. IF
NOT BREATHING, GIVE ARTIFICIAL RESPIRATION. IF BREATHING IS DIFFICULT, GIVE OXYGEN.
EYES:IMMED FLUSH W/PLENTY OF WATER FOR AT LEAST 15 MINS. SKIN:IMMED FLUSH W/PLENTY
OF WATER FOR AT LEAST 15 MINS WHILE REMOVING CONTAMD CLTHG & SHOES. WASH CLOTHING
BEFORE REUSE.
============================================================
Handling and Disposal
============================================================
Spill Release Procedures: WEAR NIOSH/MSHA SCBA & FULL PROT CLTHG. SHUT OFF
IGNIT SOURCES:NO FLAMES, SMKNG/FLAMES IN AREA. STOP LEAK IF YOU CAN DO SO W/OUT
HARM. USE WATER SPRAY TO REDUCE VAPS. TAKE UP W/SAND OR OTHER NON-COMBUST MATL &
PLACE INTO CNTNR FOR LATER (SU PDAT)
Neutralizing Agent: NONE SPECIFIED BY MANUFACTURER.
Waste Disposal Methods: DISPOSE IN ACCORDANCE WITH ALL APPLICABLE FEDERAL, STATE AND
LOCAL ENVIRONMENTAL REGULATIONS. EPA HAZARDOUS WASTE NUMBER:D001 (IGNITABLE
WASTE).
Handling And Storage Precautions: BOND AND GROUND CONTAINERS WHEN TRANSFERRING LIQUID.
KEEP CONTAINER TIGHTLY CLOSED.
Other Precautions: USE GENERAL OR LOCAL EXHAUST VENTILATION TO MEET
TLV REQUIREMENTS. STORAGE COLOR CODE RED (FLAMMABLE).
============================================================
Fire and Explosion Hazard Information
============================================================
Flash Point Method: CC
Flash Point Text: _9F,_23C
Lower Limits: 1.2%
Upper Limits: 77.7%
Extinguishing Media: USE ALCOHOL FOAM, DRY CHEMICAL OR CARBON DIOXIDE. (WATER MAY BE
INEFFECTIVE).
Fire Fighting Procedures: USE NIOSH/MSHA APPROVED SCBA & FULL PROTECTIVE
  EQUIPMENT (FP N).
Unusual Fire/Explosion Hazard: VAP MAY FORM ALONG SURFS TO DIST IGNIT SOURCES & FLASH
BACK. CONT W/STRONG OXIDIZERS MAY CAUSE FIRE. TOX GASES PRDCED MAY INCL:CARBON
MONOXIDE, CARBON DIOXIDE.
============================================================
```

FIGURE 4–3 This Material Safety Data Sheet (MSDS) shows the characteristics of a specific hazardous waste material. (Courtesy of CRC Industries, Inc.)

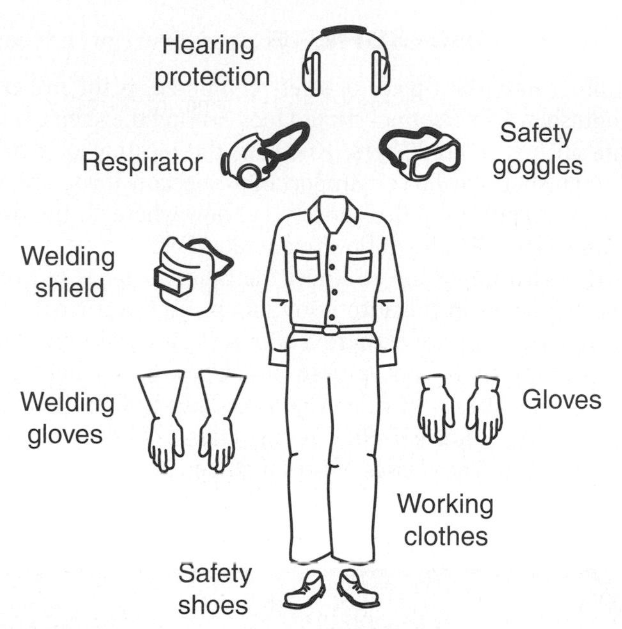

FIGURE 4–4 There are many pieces of safety equipment that the service technician uses in the automotive shop.

3. Be aware of how you feel around the shop. Make sure you can identify how various chemicals and hazardous materials may affect you and your health.
4. Make sure you carefully read the protection information for each chemical and hazardous material just in case you need the information quickly. Remember, you may need the information for yourself or for another person.

4.4 SAFETY EQUIPMENT

To ensure safe working habits in the automotive shop, it is important for all employees to know the location and proper use of safety equipment. Several important types of safety equipment are available. ***Figure 4–4*** shows some of the safety equipment used by service technicians. These include hearing protection, safety goggles, respirators, gloves, safety shoes, working clothes, and welding shields. Other safety equipment in the automotive shop includes fire extinguishers, airtight containers, gasoline containers, and first-aid boxes.

SAFETY GLASSES

One of the most important safety precautions is to have all shop personnel wear **safety glasses**. Many service technicians have been permanently blinded because they thought safety glasses were not important. Safety glasses are now required in all training centers and in all service repair centers.

There are many types of safety glasses. An important rule to remember is that all safety glasses should have safety glass and some sort of side protection. ***Figure 4–5*** illustrates a common type of approved safety glasses.

For some procedures, the entire face should be protected. ***Figure 4–6*** shows a full-face shield. This type of shield may be needed when grinding or cleaning carbon from valves.

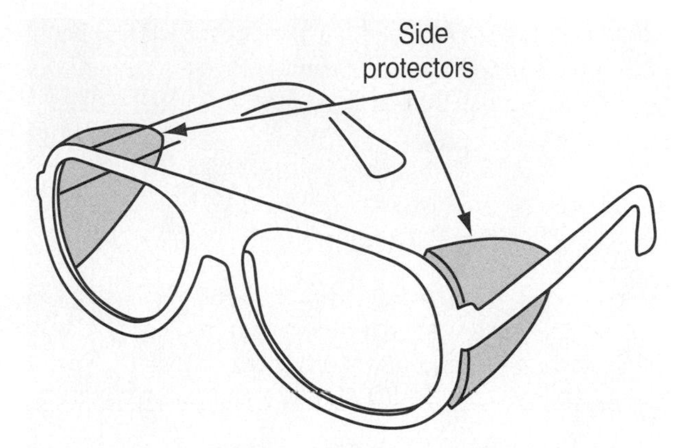

FIGURE 4–5 There are many approved safety glasses used today, such as the one shown. Select a brand that has safety glass and side protectors, and be sure the glasses are comfortable.

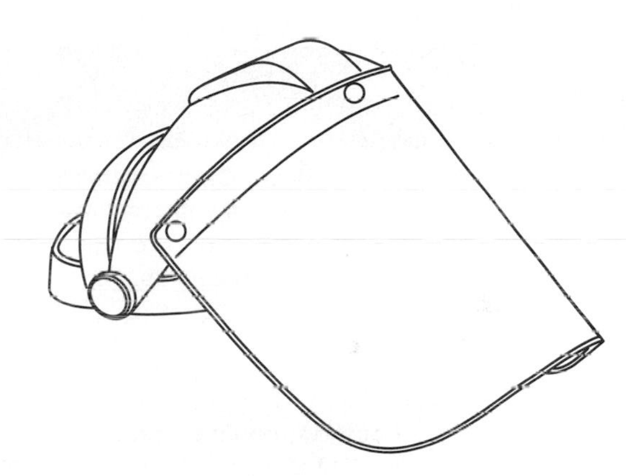

FIGURE 4–6 A full-face shield should be used when grinding or cleaning carbon from valves.

When purchasing safety glasses, always remember to buy a pair that feels comfortable. If glasses are not comfortable, people have a tendency to either remove them or wear them on the top of the head. Both situations leave the eyes totally unprotected.

EYEWASH FOUNTAINS

Various types of damage to the eyes are possible in the automotive shop. Some of the more common eye hazards include:

- A blow from a blunt or sharp object
- Foreign bodies in the eye
- Thermal burns
- Irradiation burns (from too much light)
- Chemical burns (often from battery explosions)

Safety glasses can help protect the eyes from these hazards, but if a chemical does get into the eye it must be washed out to prevent a chemical burn. One of the most effective devices used to wash chemicals from the eyes is an **eyewash fountain**, (***Figure 4–7***). Before you begin working in the automotive shop, be sure to note the location of the eyewash fountain and know how to turn on the water in case of emergency.

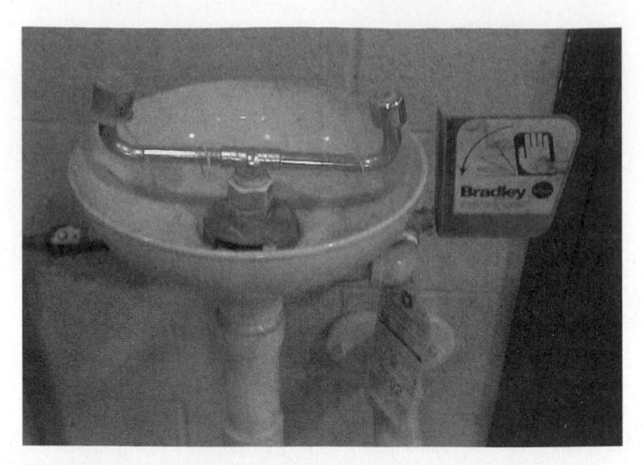

FIGURE 4–7 Eyewash fountains are used to wash and thoroughly clean the eyes, especially if toxic chemicals come in contact with the eyes.

FIRE EXTINGUISHERS

Another important piece of safety equipment is the fire extinguisher. All personnel should look around the shop to locate all fire extinguishers. Knowing the location of a fire extinguisher may be very important, especially if you are in a hurry to put out a fire. Once you know where all the fire extinguishers are, learn how to operate them.

It is also important to know the different types of fires that can occur in the automotive shop and the correct fire extinguishers to use if there is a fire. Only certain types of fire extinguishers can be used on certain fires. There are four major classes of fires: Class A, Class B, Class C, and Class D fires (***Figure 4–8***). Fires are classified by the type of fuel involved. The classes of fires include:

	CLASS OF FIRE	TYPICAL FUEL INVOLVED	TYPE OF EXTINGUISHER
Class A Fire (green)	**For Ordinary Combustibles** Put out a class A fire by lowering its temperature or by coating the burning combustibles.	Wood Paper Cloth Rubber Plastics Rubbish Upholstery	Water*[1] Foam* Multipurpose dry chemical[4]
Class B Fire (red)	**For Flammable Liquids** Put out a class B fire by smothering it. Use an extinguisher that gives a blanketing, flame-interrupting effect; cover whole flaming liquid surface.	Gasoline Oil Grease Paint Lighter fluid	Foam* Carbon dioxide[5] Halogenated agent[6] Standard dry chemical[2] Purple K dry chemical[3] Multipurpose dry chemical[4]
Class C Fire (blue)	**For Electrical Equipment** Put out a class C fire by shutting off power as quickly as possible and by always using a nonconducting extinguishing agent to prevent electric shock.	Motors Appliances Wiring Fuse boxes Switchboards	Carbon dioxide[5] Halogenated agent[6] Standard dry chemical[2] Purple K dry chemical[3] Multipurpose dry chemical[4]
Class D Fire (yellow)	**For Combustible Metals** Put out a class D fire of metal chips, turnings, or shavings by smothering or coating with a specially designed extinguishing agent.	Aluminum Magnesium Potassium Sodium Titanium Zirconium	Dry powder extinguishers and agents only

*Cartridge-operated water, foam, and soda-acid types of extinguishers are no longer manufactured. These extinguishers should be removed from service when they become due for their next hydrostatic pressure test.

Notes:

(1) Freezes in low temperature unless treated with antifreeze solution, usually weighs over 20 pounds, and is heavier than any other extinguisher mentioned.

(2) Also called ordinary or regular dry chemical. (sodium bicarbonate)

(3) Has the greatest initial fire-stopping power of the extinguishers mentioned for class B fires. Be sure to clean residue immediately after using the extinguisher so sprayed surfaces will not be damaged. (potassium bicarbonate)

(4) The only extinguishers that fight A, B, and C classes of fires. However, they should not be used on fires in liquefied fat or oil of appreciable depth. Be sure to clean residue immediately after using the extinguisher so sprayed surfaces will not be damaged. (ammonium phosphates)

(5) Use with caution in unventilated, confined spaces.

(6) May cause injury to the operator if the extinguishing agent (a gas) or the gases produced when the agent is applied to a fire is inhaled.

FIGURE 4–8 The four classes of fires are determined by the type of fuel involved. Fire extinguisher canisters are identified by color markings and geometric shapes for each.

FIGURE 4–9 There are many types of fire extinguishers. The automotive shop should be equipped with fire extinguishers that can handle at least class A, B, and C fires.

1. **Type A fires**—Fires from ordinary combustible materials, such as wood, paper, textiles, and clothing. This type of fire usually requires cooling and quenching.
2. **Type B fires**—Fires from flammable liquids, greases, gasoline, oils, paints, and other liquids. This type of fire requires smothering and blanketing.
3. **Type C fires**—Fires started from electrical equipment malfunctions, motors, switches, and wires. This type of fire requires a nonconducting agent to put it out.
4. **Type D fires**—Not as common in the automotive shop, Type D fires occur where combustible metals such as lithium, sodium, potassium, magnesium, titanium, and zirconium are present.

Figure 4–9 shows several types of fire extinguishers. The outside of each fire extinguisher shows the type of fires that particular fire extinguisher can be used to put out. There are typically four types of fire extinguishers. ***Figure 4–10*** shows that there are foam, carbon dioxide, dry chemical, and soda-acid fire extinguishers. The operation of each is also listed.

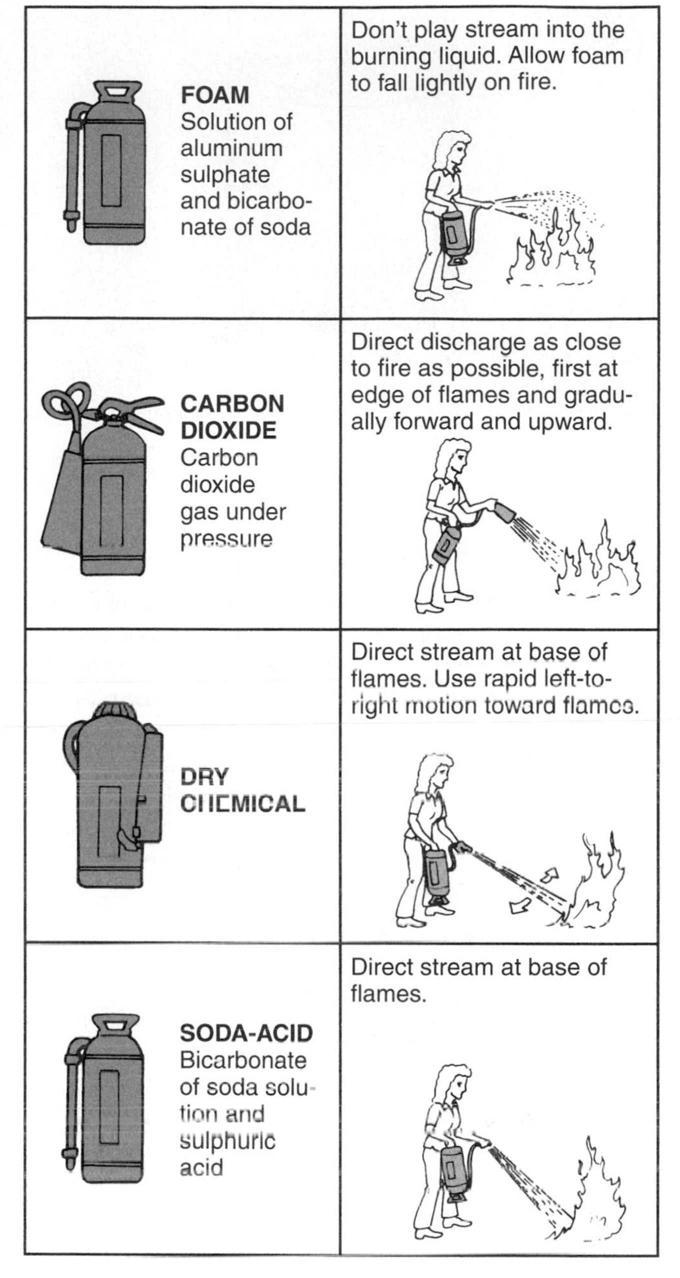

FIGURE 4–10 Each type of fire extinguisher has a slightly different operation and use.

USING A FIRE EXTINGUISHER

Generally, fire extinguishers have directions printed on the canisters. However, the basic rules for use include:

- Do not keep the extinguisher too close to where a fire might occur.
- Mount the fire extinguisher by a door so it will be easily accessible.
- Get as close as possible to the fire without jeopardizing yourself before pulling the trigger. Extinguishers expel their material quickly, generally in 8 to 25 seconds for most small models containing dry chemicals.
- Direct the nozzle at the base of the fire and sweep the extinguisher across the burning area.
- Always keep a door or escape route behind you so that if the fire gets out of control, you can escape easily and quickly.

HALOGEN AND HALON CHEMICALS

For certain types of fires, mostly class B fires, chemicals are used to extinguish the flames. In the past, halogens (a type of chemical) have been used to stop combustion completely. The halogen chemicals help to inhibit chain reactions in the fuel. These chemicals attach to hydrogen, hydroxide, and oxygen molecules to stop the combustion process almost immediately after being applied. However, the gases that result from using halogens have been found to be very toxic. Halon, a different type of chemical, is now being used in place of halogens to help extinguish fires in the same way.

FIGURE 4–11 Approved airtight containers should be used to hold oily rags.

AIRTIGHT CONTAINERS

In the automotive shop there are many oily rags, which can cause a fire by spontaneous combustion. Spontaneous combustion is a chemical reaction. It is produced by the slow generation of heat caused by the oxidation of the oil in the rag. Heat continues to be generated until the ignition temperature is reached. The fuel then begins to burn, causing a fire. Because of this danger, **airtight containers** are used to hold oily rags. ***Figure 4–11*** shows an airtight container for holding waste rags.

GASOLINE CONTAINERS

Gasoline is an extremely explosive and flammable fuel. Because of this danger, it should always be kept in an approved **gasoline container**. The container should always be painted red and should have approved openings for pouring and venting. ***Figure 4–12*** shows a gasoline container approved by OSHA.

Another type of gasoline container is used to store many types of fuel and oil. This container is designed to contain any type of explosion. Any fuel, oil, or other flammable liquid should be stored in the proper small container. The small container should then be stored in **explosion-proof cabinets** (***Figure 4–13***).

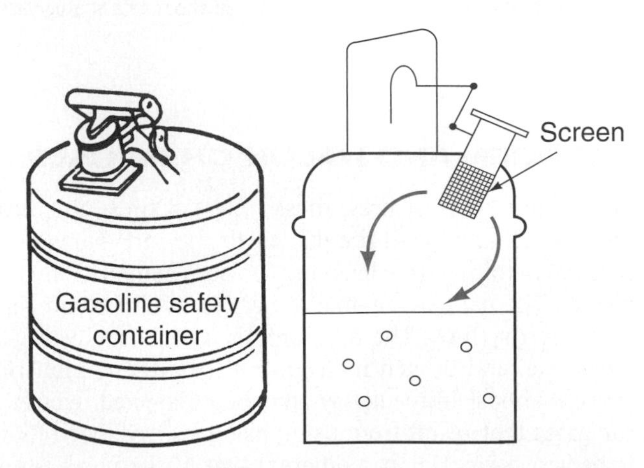

FIGURE 4–12 Gasoline should be kept in an approved gasoline container. Such containers have the correct pouring and venting system.

FIGURE 4–13 All fuel, oil, and other flammable liquids should be stored in an explosion-proof cabinet.

SAFETY AND GASOLINE

Experts suggest that a single gallon of exploding gasoline can equal the force of fourteen sticks of dynamite. A 5-gallon container of gasoline can generate as much heat as 250 pounds of detonating dynamite, enough to completely demolish an automobile or most homes.

It is the expanding vapors of gasoline that are extremely dangerous, even in cold temperatures. In the gasoline tank of an automobile, the vapors are controlled. However, in a portable gasoline container the vapors can possibly move out of the container. Thus, always put a portable gasoline container in a well-vented space, never, for example, in the trunk of a car.

It is recommended that gasoline cans with a capacity of 1 gallon or more should have a flash-arresting screen at the outlet. Such screens prevent external fire or a spark from igniting vapors within the can while the gasoline is being poured. When working with gasoline containers, remember the following suggestions:

1. Leave 1–2 inches of head space in the container to allow for the expansion of gasoline at higher temperatures. If space is not left, as the gasoline expands at higher temperatures it may leak out and cause a serious hazard.
2. Do not store gasoline cans in the home. Keep them in a garage, a well-ventilated shed, or an outbuilding away from the home.

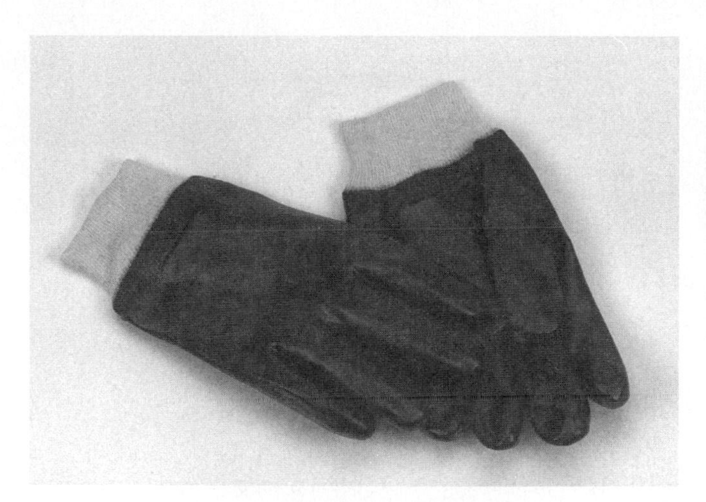

FIGURE 4–14 Always wear protective gloves when working with solvents or hot metal, and when grinding metal.

FIGURE 4–15 Always wear proper clothing when working in the automotive shop.

3. Always secure the gasoline container against upsets when transporting it.
4. Never store a gasoline can with partial fuel for long periods of time. It will emit vapors and thus produce a hazard.
5. Always cap openings for filling, pouring, or venting, except when filling the container or pouring the gasoline.

GLOVES

Many activities in the automotive shop require the use of protective gloves. For example, always use protective gloves when grinding metal, working with caustic cleaning solutions, and when welding or working with hot metals. Also use gloves if there is a possibility of touching another person's blood due to an accident in the shop. This practice may prevent the spread of AIDS.

There are many types of protective gloves used in today's automotive shop. ***Figure 4–14*** shows an example of heavy-duty rubber-coated gloves for heat and solvent protection. In addition, some service technicians use latex gloves when working on electrical circuits and other specialized components.

TYPES OF CLOTHING AND SHOP COATS

The type of clothing that is worn in the automotive service area is important. Always wear clothing that you are willing to get dirty. Constant worry about keeping your clothes clean will most likely affect the quality of your work. However, it is important to customers today that the service technicians are as clean as possible. Customers will not tolerate having their car soiled by a sloppy service technician.

The most common type of clothing that is used in a typical automotive shop is the type of uniform shown in ***Figure 4–15***. Such clothing can help make you more comfortable and productive in the automotive shop. Loose-fitting clothing can get caught on machines or parts of the automobile and cause falls or accidents. This is true of a tie. Ties can get caught in any machine or part that is rotating. When wearing a tie under a shop coat, always make sure the tie is pinned to your shirt or held tightly in place under the coat. Clip-on ties are recommended in shop environments. Watches and jewelry that could get caught on equipment should be avoided.

SHOES

When working in the automobile shop, it is important to not wear sandals or summer-type shoes. When servicing the automobile, heavy parts are often lifted from one spot to another. Heavy objects such as cylinder heads, manifolds, brake drums, and so forth may be dropped accidentally. The person working in the automobile shop should always wear either steel-tipped shoes or shoes that are strong enough to withstand heavy objects to prevent damage or injury to the feet.

FIRST-AID BOXES

All automotive services areas should have several **first-aid boxes**. These boxes, as shown in ***Figure 4–16***, should be located for easy access. Each person who works in the shop should know where all first-aid boxes are located. The first-aid box should always be kept clean and free of blood-borne diseases.

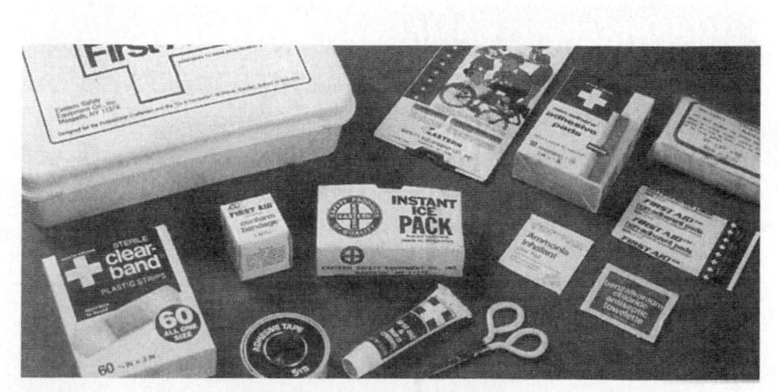

FIGURE 4–16 Every automotive service area should have several first-aid boxes. Each box should include rubber or latex gloves, bandages, and other necessary first-aid equipment.

EAR PROTECTORS

Excessive noise in the automotive shop should be eliminated or reduced whenever possible. In certain cases, noise protection is necessary. For example, if engines are run on dynamometers or put under heavy loads, noise may be above recommended levels.

REPRESENTATIVE SOUND LEVELS		
SOUND LEVEL (dB)	OPERATION OR EQUIPMENT	
150	Jet engine test cell	
145	Threshold of Pain	
130	Pneumatic press (close range)	
	Pneumatic rock drill	
	Riveting steel tank	
125	Pneumatic chipper	
	Pneumatic riveter	
120	Threshold of Discomfort	
	Turbine generator	DANGER ZONE
112	Punch press	
	Sandblasting	
110	Drills, shovels, operating trucks	
	Drop hammer	
105	Circular saw	
	Wire braiders, stranding machine	
	Pin routers	
	Riveting machines	
100	Can-manufacturing plant	
	Portable grinders	
	Ram turret lathes	
	Automatic screw machine	
90	Welding equipment	RISK ZONE
	Weaving mill	
	Milling machine	
	Pneumatic diesel compressor	
	Engine lathes	
	Portable sanders	
85	California freeway traffic (overpass)	
HEARING DAMAGE IF CONTINUED EXPOSURE ABOVE THIS LEVEL		
80	Tabulating machines, electric typewriters	
75	Stenographic room	
70	Electronics assembly plant	
65	Department store	
60	Conversation	SAFE ZONE
35	Quiet home forced air system	
10	Whisper	

FIGURE 4–17 Noise is measured in decibels (db). Different decibel levels are shown. Damage to the ear can result from prolonged exposure to levels of 90 to 100 decibels.

FIGURE 4–18 Ear protectors provide protection from high decibel levels when working in the automotive shop.

Noise is measured in units called **decibels**. Various levels of noise are shown in ***Figure 4–17***. Damage to the ears can result from exposure to constant high levels of noise. It is generally accepted that approximately 90 to 100 decibels can damage the ear.

Because of possible damage, it is recommended that ear protection be used in the automotive shop. Ear protection can be obtained by using commercially available earplugs or earmuffs (***Figure 4–18***).

BREATHING DEVICES

In the automotive shop, there are many times when it is possible to breathe in chemical fumes or other dangerous airborne materials. For example, whenever handling hazardous waste products, painting, or working around dust from brakes, and so on, the service technician should always use protective breathing devices as shown in ***Figure 4–19***.

Other times when you should use breathing devices include:

- When near dust particles from brake systems
- When near paints and varnishes
- When near solvents

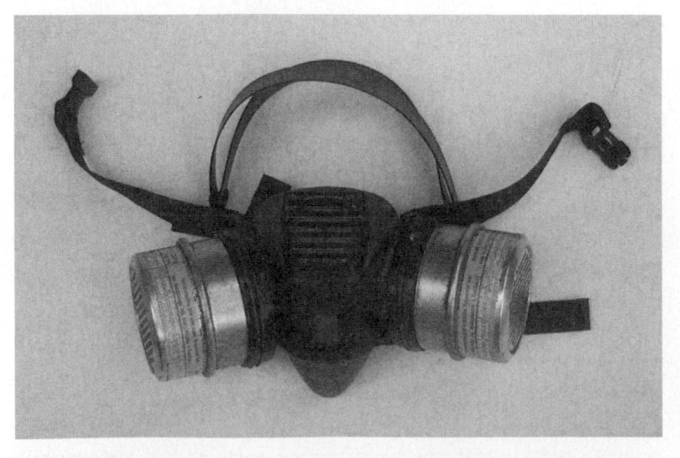

FIGURE 4–19 Always use respiratory breathing protectors when there is a danger of inhaling harmful fumes.

- When near electrical devices that have been overloaded and possibly have begun to burn wires inside of the electrical device
- When near the exhaust of vehicles
- When near clutch plates that have been removed from the vehicle
- When near any electrical fire in a vehicle

4.5 SAFETY RULES IN THE SHOP

There are many safety rules that make working in the automobile shop much easier and safer. These rules concern shop layout, lifting and carrying, good housekeeping, smoking, and awareness of carbon monoxide.

SHOP LAYOUT AND SAFETY

There are many kinds of automotive shops. Each shop will have a different layout. The shop layout determines where the equipment is located, where the cars are worked on, and where special repairs take place. It is important for the service technician or student to become familiar with the total shop layout so that each person can work efficiently. A typical shop layout is shown in ***Figure 4–20***.

In ***Figure 4–20***, the auto bays are used for most service work on the automobile. Parts washers are nearby so that minimum travel is required to wash the needed parts. Tools are located on boards and in special areas shown. Most technicians have their own sets of tools that can be rolled directly up to the vehicle. The service manager's office is located to provide easy access for technicians and customers.

Most automotive shops are designed to have certain types of repair in only one place. For example, shops may have an alignment area, body and fender repair area, painting area, tune-up area, and general repair and maintenance area. Others may include special areas for valve grinding, block boring, and so on. These activities are done in what is called the machining area. Large shops have all of these, while smaller shops may have only some of these areas. It is important for the service person to become completely familiar with the shop layout.

LIFTING AND CARRYING

When working around the automobile, it may be necessary to lift heavy objects. Many **back injuries** have occurred because the service technician did not lift properly. The following procedure, shown in ***Figure 4–21***, should always be used when lifting any heavy object:

1. Consider the size, weight, and shape of the object. Get help if necessary.
2. Set feet solidly, with one foot slightly ahead of the other for stability.
3. Get as close to the object as possible.
4. Keep the back as straight as possible and bend the legs.
5. Grip the object firmly.
6. Straighten the legs to lift the object, bringing the back to a vertical position.
7. Never carry a load you cannot see over or around.
8. Setting down the object requires the reverse procedure.
9. Use mechanical lifts whenever possible.

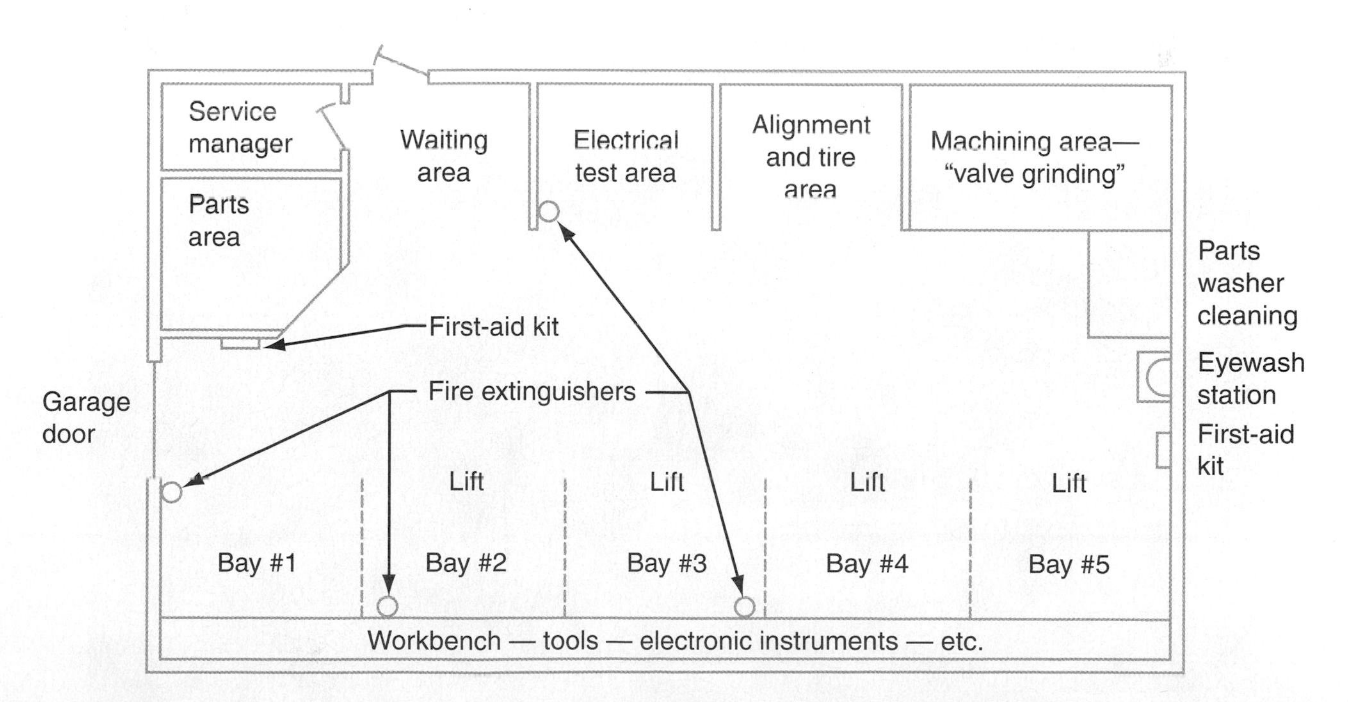

FIGURE 4–20 It is important to be familiar with the automotive service area layout. Improved safety results when employees know the shop layout.

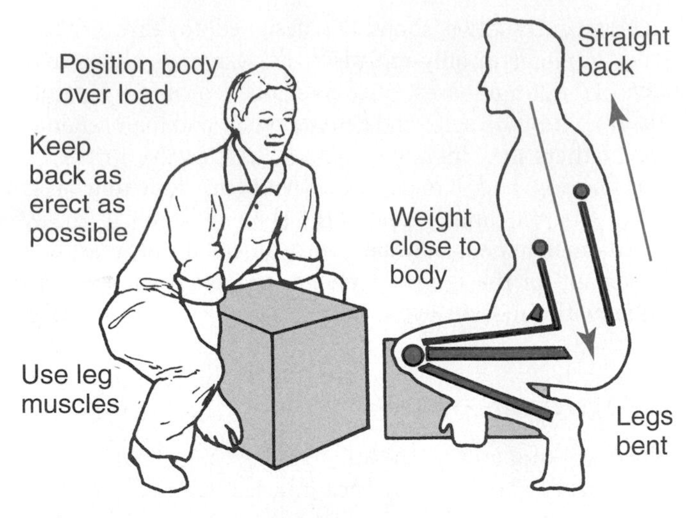

FIGURE 4–21 When lifting heavy objects in the automotive service area, keep a straight back. Be sure to lift with your legs rather than your back.

In addition, safety should always be observed when lifting heavy objects with a chain hoist and car hoist. Make sure the chain hoist is securely connected to the engine before lifting. Bolts should be strong enough to hold the engine and its weight. In addition, when lifting a vehicle on a car hoist, follow the manufacturer's recommended procedure. This usually includes making sure the car is positioned correctly on the hoist as well as making sure the hoist lifts at the correct part of the vehicle.

GOOD HOUSEKEEPING

A clean and orderly shop makes employees respect the equipment and working area. A customer also has more confidence in the work being done if a shop is clean. The following is a list of **housekeeping** questions that should always be answered in the affirmative:

- Is proper light provided?
- Are walls and windows clean?
- Are stairs clean and well lighted?
- Are floor surfaces clean of loose material?
- Are floors free of oil, grease, and so on?
- Are containers provided for refuse?
- Are aisles free of obstructions?
- Are there safe and free passages to fire extinguishers?
- Are tools arranged in proper places? (***Figure 4–22***)
- Are oil rags in the proper container?
- Are tools free of grease and oil?
- Are tools in good working condition?
- Are proper guards provided on all machinery?
- Are benches and seats clean and in good condition?
- Are parts and materials in the proper location?
- Has the creeper (rolling platform for crawling under a car) been put away?

If all shop employees are aware of the importance of these items, the automotive service area will be a safer place to work.

DRUGS AND SMOKING IN THE SHOP

The use of drugs such as tobacco and alcohol during work hours or the night before can seriously affect both a person's performance and safety in the automotive shop. Smoking on the job is especially dangerous. Fuel is often present. It is very important to know when and where to smoke and not to smoke. Awareness of the "smoking" and "no smoking" areas and the **smoking rules** is the responsibility of the automotive service technician as well as the management. Management will help to enforce the smoking rules.

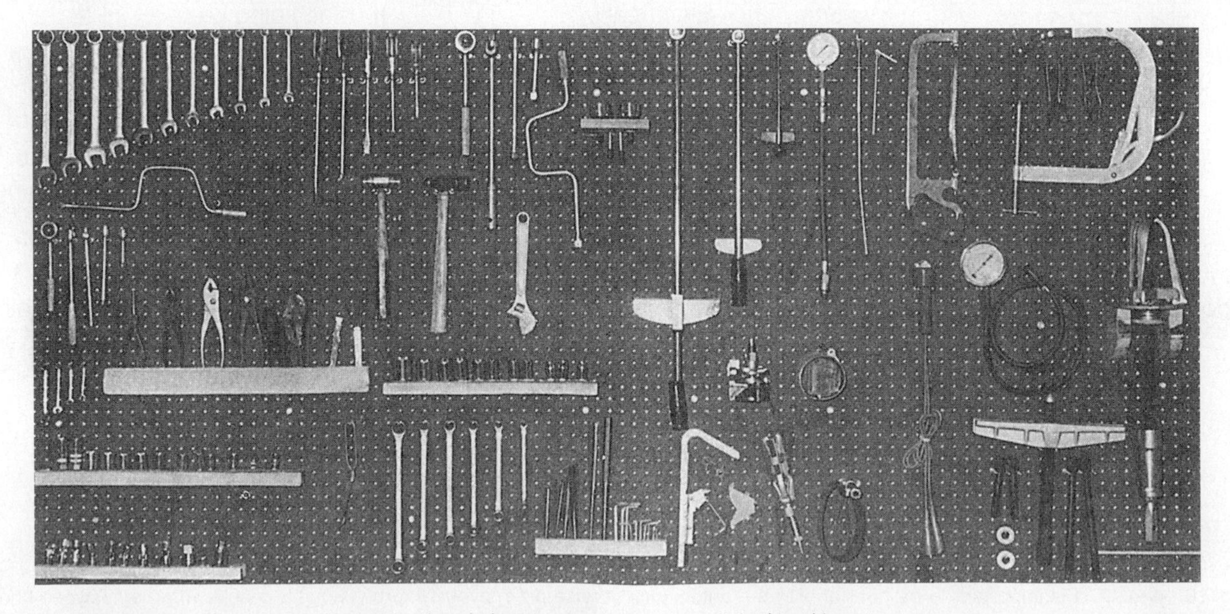

FIGURE 4–22 Tools should be well organized and always returned to their original positions.

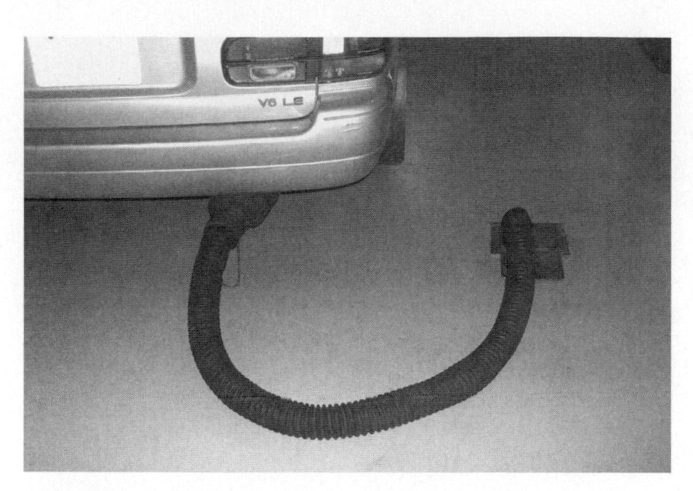

FIGURE 4-23 Remember to use proper exhaust piping when running vehicles in the automotive shop. Carbon monoxide in the exhaust is a colorless, odorless gas that can cause death.

CARBON MONOXIDE

It is important to be familiar with **carbon monoxide** when working with the automobile. Most new cars have exhaust emission controls. However, some carbon monoxide is always given off from the exhaust of running engines. Carbon monoxide is odorless and colorless. You may not even be aware that it is present. If inhaled during normal breathing, carbon monoxide can cause death. Carbon monoxide poisoning is evident from the following symptoms:

- Headache
- Nausea
- Ringing in the ears
- Tiredness
- Fluttering heart

If any of these symptoms appear, it is very important to get fresh air immediately.

To eliminate the possibility of carbon monoxide poisoning, always have good ventilation and make sure that engine exhaust is properly vented out of the shop. Carbon monoxide is eliminated by using the proper exhaust systems, as shown in ***Figure 4–23***.

CARBON MONOXIDE MONITORS

Carbon monoxide is easy to detect in an automotive repair shop by use of various monitors. Some make direct readings of the amount of carbon monoxide in the air. Others have audible alarms that sound when the danger level is reached. Most of these monitors are small and easy to hold in one's hand.

ASBESTOS

One chemical often found in automotive shops is asbestos. **Asbestos** is a term used to describe a number of naturally occurring, fibrous materials. Asbestos fibers are commonly found in various types of brake pads, in clutches, and in other similar materials. (Always clean all brake and clutch dust and powder using an approved vacuum container to avoid getting asbestos into the lungs.) Asbestos has also been used extensively as a building material and as an insulator.

Asbestos has been associated with the development of a variety of diseases. One of the most common types is called mesothelioma, or asbestos-caused cancer. This disease and others, such as asbestosis, cause cancer of the lungs. Asbestos fibers cause a scarring of the lungs. Asbestos inhaled into the lungs can also injure the lining of the lung's air passages. The effects of asbestos may not appear immediately, but may take a month to 45 years after exposure to cause damage. There seems to be no recovery from asbestos damage. In some cases, persons who were exposed to asbestos for only a few days have been afflicted.

DISPOSAL OF SOLVENTS

Many laws and ordinances strictly prohibit pouring chemical solvents into sinks or floor drains that connect with sanitary or storm sewer facilities. All solvents, liquids, fuels, and the like should be disposed of properly by delivery to a waste collection agency. Check your school, city, or state rules and ordinances for such facilities.

ELECTRICAL SAFETY

There are many electrical hazards within the automotive shop. Often electrical tools are used for grinding, drilling, honing, and so on. When using these tools, there is always a possibility of serious shock or electrocution. Poor insulation or defective wiring in tools or frayed electrical cords can cause electrical shock. The severity of the electrical shock is related to the amount and length of the electrical current the victim receives. To reduce the risk of electrical shock from power tools, the following procedures should be observed:

1. Use three-prong grounded plugs or double-insulated equipment. Never cut the grounded prong if the outlet has only two prongs. If an adapter must be used, the pigtail wire (usually green) should be attached to the screw holding the faceplate of the wall socket.
2. Make sure all of the wires to the power tools are not frayed and are in good condition. Buy tools that have the double insulation around the wires.
3. Make sure there is a ground-fault circuit protector. This device operates on the current flowing through a person's body during an accidental line to ground fault. It is attached to the panel circuit breaker or is an integral part of the electrical outlet. This fast-acting circuit breaker will rapidly stop current flow to the tool, and the operator will receive only a modest electrical shock.
4. If an electrical tool such as a drill or hand grinder seems to be shorting out (you might hear electricity arcing inside the tool housing or receive a shock),

immediately shut off the tool. Have an electrician check the tool for shorting or grounding.

5. When working with electrical tools always keep your hands dry, and never stand in water on the floor.

OTHER RULES TO FOLLOW

The following rules should always be observed when working in an automotive service area.

- Make sure all hand tools are in good condition. Using a damaged hand tool or the incorrect tool for the job may result in a severe injury.
- When lifting a car with an air or hydraulic jack, always make sure the jack is centered. When the vehicle is raised, always use safety **jack stands** under the car. Never go under a car without safety jack stands in place. A set of safety jack stands is shown in ***Figure 4–24***.
- Never wear jewelry such as rings, bracelets, necklaces, or watches when working on a car. In addition, if your hair is long, always tie it back. These items can easily catch on moving parts or cause an electrical short and result in serious injury.
- Never use compressed air to remove dirt from your clothing or you may blow dirt in your eyes. Also, never spin bearings with compressed air. If the bearing is damaged, one of the steel balls may come loose and cause serious injury. Damage to the bearing may also result.
- Always be careful where welding sparks are falling. Sparks can cause a fire or explosion if dropped on flammable materials.
- When using any machines, such as hydraulic presses, hoists, drill presses. or special equipment, make absolutely certain that all operational procedures are studied first. Get checked out on the machine from the instructor first.
- When lifting engines out of vehicles always use approved lifting cables or chains, as shown in ***Figure 4–25***.

FIGURE 4-24 Never work under a vehicle without the proper safety jack stands placed securely under the vehicle.

LIFTING DEVICES

Swaged socket

Chains

Wire rope

Clip

Hand-tucked

FIGURE 4-25 Always use approved cables or chains when lifting heavy objects such as an engine block.

Car Clinic: HAZARDOUS MATERIALS

CUSTOMER CONCERN:
Safe Disposal of Asbestos Dust

A customer heard that there is asbestos in vehicle brakes and is worried about possible health effects. What should you tell the customer? What is a safe method to avoid breathing in asbestos dust particles when changing brake shoes?

SOLUTION: The asbestos particles left in the brake drum are carcinogens, or cancer-causing agents. By the time the customer gets the car back, there's virtually no risk. But if inhaled by the service technician, asbestos particles can cause cancer. When removing the brake drum, use an approved vacuum cleaner to remove the dust particles. Then wash all parts in warm soapy water. This process will remove the dust particles, keep the level of asbestos within OSHA standards, and ensure that they are not inhaled. Assure the customer that you follow these high standards.

SUMMARY

The following statements will help to summarize this chapter:

- Safety within the automotive shop is critical to its success.
- Service technicians need to be familiar with all hazardous materials that may be present in the automotive shop, as well as know how to read a Material Safety Data Sheet.
- Always use proper safety equipment in the automotive shop.
- Safety glasses should always be worn in the automotive shop.
- Always know exactly where fire extinguishers are located and how to use them properly.
- Always use an airtight container to store oily rags.
- Make sure you know where all first-aid boxes and eyewash stations are located, so you can find them easily if there is an accident.
- Always wear ear protectors when noise levels are above the recommended decibels.
- Become familiar with the automotive shop layout so that equipment can be easily located and safely used.
- When lifting heavy objects, always lift with the legs and keep the back straight.
- Always maintain a clean and organized automotive workshop to avoid accidents.
- Keep all tools clean and free of oil during and after use.
- Smoke only in designated areas.
- Always be aware of carbon monoxide dangers, and make sure to vent engine exhaust to the outside of the shop.
- Always use the proper jack stand when working under a vehicle.
- Never wear jewelry since it may become caught on moving parts and cause injury.

TERMS TO KNOW

Can you explain each of the following terms? Review the chapter until you can use each term correctly.

Airtight containers
Asbestos
Back injuries
Carbon monoxide
Chemical hazards
Decibels
Ergonomic hazards
Explosion-proof cabinets
Eyewash fountain
First-aid boxes
Gasoline container
Hazardous wastes
Housekeeping
Jack stands
MSDS
OSHA
Physical hazards
Safety glasses
Smoking rules
Type A fires
Type B fires
Type C fires
Type D fires

REVIEW QUESTIONS

Multiple Choice

1. In what year did the federal government pass the Occupational Safety and Health Act?
a. 1925
b. 1945
c. 1970
d. 1980
e. 1985

2. All safety glasses should have:
a. Side protectors
b. Safety glass
c. Steel rims
d. A and B
e. All of the above

3. Which type of fire involves the burning of wood, paper, and clothing?
a. Type A fire
b. Type B fire
c. Type C fire
d. Type D fire
e. Type F fire

4. Airtight containers are used to hold:
a. Wrenches that have grease on them
b. Oily rags
c. Shop parts
d. Gaskets
e. Solvent

5. All gasoline containers should be:
a. Approved by OSHA
b. Handmade
c. Painted black
d. Painted green
e. Never used

6. Which of the following shoes are not acceptable to wear in the automotive shop?
a. Steel-tipped shoes
b. Strong leather shoes
c. Summer sandals
d. Approved work shoes
e. All of the above

7. The unit of measurement for sound is called the:
a. Millimeter
b. Gram
c. Decibel
d. Horsepower
e. Inch

8. What level of sound may damage your ears?
a. 90 to 100 decibels
b. 30 to 40 decibels
c. 10 to 20 decibels
d. 0 to 5 decibels
e. 5 to 9 decibels

9. When lifting heavy objects, always lift with your:
a. Legs
b. Back
c. Neck
d. Arms only
e. Stomach

10. Carbon monoxide is produced from:
a. Transmission oil
b. Gas engine exhaust
c. Radiator coolant
d. Discarded oil
e. None of the above

11. When working in the automotive shop, always:
a. Wear a shop coat
b. Run the vehicle without proper exhaust
c. Wear good clothing
d. Wear white shirts
e. Keep tools greasy

12. Which of the following is (are) important when working in the automotive shop?
a. Know the shop layout
b. Wear a shop coat
c. Wear steel-tipped shoes
d. All of the above
e. None of the above

13. When working in the automotive shop, it is important to:
a. Put oily rags in the proper container
b. Leave tools greasy for later cleanup
c. Let the shop supervisor put tools away
d. Let the customer fix the vehicle
e. None of the above

14. When working in the automotive shop, never:
a. Wear rings or other jewelry
b. Use compressed air to remove dirt from clothing
c. Use tools that are in bad condition
d. All of the above
e. None of the above

15. Three types of hazards found in the automotive shop include:
a. Chemical, physical, and economic
b. Chemical, ergonomic, and political
c. Chemical, ergonomic, and physical
d. Physical, eye damage, and dropping things
e. Carbon monoxide, asbestos, and political

16. Which type of hazard found in the automotive shop deals with how humans interact with the physical world?
a. Physical
b. Ergonomic
c. Chemical
d. Asbestos
e. Carbon monoxide

17. Which of the following is not shown on an MSDS?
a. Name of material
b. Physical data
c. Cost data
d. Protection information
e. Fire and explosion data

18. Which of the following is a type of fire extinguisher?
a. Foam fire extinguisher
b. Carbon dioxide fire extinguisher
c. Dry chemical fire extinguisher
d. All of the above
e. None of the above

The following questions are similar in format to ASE (Automotive Service Excellence) test questions.

19. *Technician A* says it is OK to inhale carbon monoxide and other fumes from the automobile exhaust. *Technician B* says it is OK to inhale carbon monoxide fumes but not carbon dioxide fumes from the automobile exhaust. Who is correct?
a. A only
b. B only
c. Both A and B
d. Neither A nor B

20. *Technician A* says a decibel level of 92 may be damaging to one's ears. *Technician B* says a decibel level of 98 may be damaging to one's ears. Who is correct?
   a. A only
   b. B only
   c. Both A and B
   d. Neither A nor B

21. *Technician A* says that one should use foam-type extinguisher for Type A fires. *Technician B* says that one should use foam-type extinguishers for Type C fires. Who is correct?
   a. A only
   b. B only
   c. Both A and B
   d. Neither A nor B

22. *Technician A* says that asbestos is found in some brake pad materials. *Technician B* says that asbestos is found in some clutch materials. Who is correct?
   a. A only
   b. B only
   c. Both A and B
   d. Neither A nor B

23. *Technician A* says that it is not important to correctly dispose of various chemicals found in the automotive shop. *Technician B* says that cleaning solvents can be disposed of by putting the solvent in a city drain. Who is correct?
   a. A only
   b. B only
   c. Both A and B
   d. Neither A nor B

Essay

24. Discuss some of the safety features of a well-designed shop layout.

25. Describe the correct procedure for lifting heavy objects.

26. Identify at least five housekeeping items that should be observed in the safe automotive shop.

27. State three safety rules to be followed when working in the automotive shop.

Short Answer

28. Hazardous conditions that relate to one's physical body, or motion, are referred to as ________ hazards.

29. The federal program that monitors, controls, and educates the workforce on safety is called ________.

30. If your eyes have been exposed to a toxic chemical, the ________ station should be used.

31. A summary of important health and information on any chemical can be found on a (an) ________.

32. List four types of hazardous waste found in a typical automotive shop.

Applied Academics

TITLE: Safety and the Automobile

ACADEMIC SKILLS AREA: Language Arts

NATEF Automobile Technician Tasks:

Supply clarifying information to customers, associates, parts suppliers, and supervisors.

CONCEPT:

Car manufacturers are very concerned with vehicle crashworthiness and occupant safety. Accident data from various sources throughout the United States have been thoroughly studied. In addition, vehicle collisions are simulated and hundreds of vehicles are demolished each year in carefully planned accident crash tests. In crash tests, dummy occupants, equipped with sensors at many critical points as shown in the photograph, ride remote-controlled vehicles directly into an impact. The crash tests are recorded in slow motion so that the actions of the dummy occupants can be studied. Many improvements have been developed as a result of this type of testing. Designs such as energy-absorbing bumpers and steering columns, side-guard door beams, side door air bags, and injury-mitigating vehicle structures have all been developed based upon information from such tests.

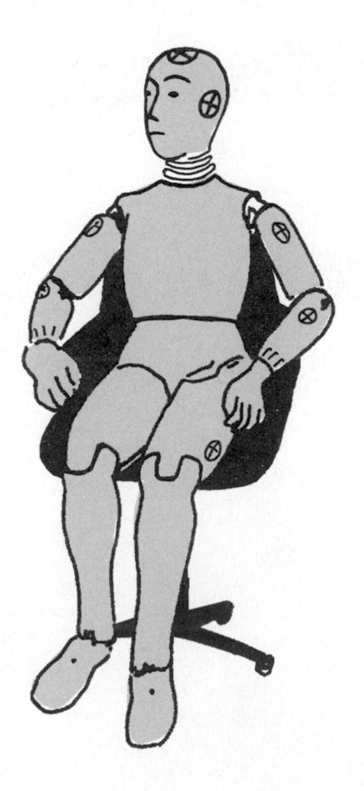

APPLICATION:

Based on the information above and the importance of automobile safety, how would you communicate the importance of using safety belts to a customer? How could you convince another service technician who is not using correct safety procedures in the shop to use the correct safety procedures and equipment?

Fasteners, Gaskets, Seals, and Sealants

CHAPTER 5

OBJECTIVES

After studying this chapter, you should be able to:

- Identify the different types of bolts, washers, splines, keyways, snap rings, and screws used on the automobile.
- Determine how to select torque specifications for different size bolts.
- Examine the purposes and styles of different types of gaskets.
- Describe the purposes and the types of seals, sealants, and gaskets that are used on the automobile.
- Identify various service and diagnostic procedures used when working with fasteners, gaskets, sealants, and seals.

INTRODUCTION

The automobile is a complex combination of many parts. These parts are held together by fasteners such as screws, bolts, and rivets. When parts are placed together near or around oil, different gaskets, sealants, and seals are used to keep the oil and other liquids from leaking. The purpose of this chapter is to examine the many types of fasteners, gaskets, sealants, and seals used on the automobile.

5.1 FASTENERS

Fasteners are those objects that secure or hold together parts of the automobile. Examples include bolts, nuts, washers, snap rings, splines, keyways, rivets, and setscrews. Actually, there are many additional types of fasteners. In this chapter certain fasteners will be discussed and their applications will also be illustrated.

THREADED FASTENERS

One of the most popular types of fastener is the **threaded fastener**. Threaded fasteners include bolts, nuts, screws, and similar items that allow the technician to install or remove parts easily. Examples of common threaded fasteners are shown in ***Figure 5–1***.

TYPES OF THREADED FASTENERS

Figure 5–2 shows a few of the more common types of fasteners used on the automobile. These include bolts, studs, setscrews, cap screws, machine screws, and self-tapping screws.

FIGURE 5–1 Many types of threaded fasteners are used on an automobile. Bolts, nuts, and screws are all threaded fasteners.

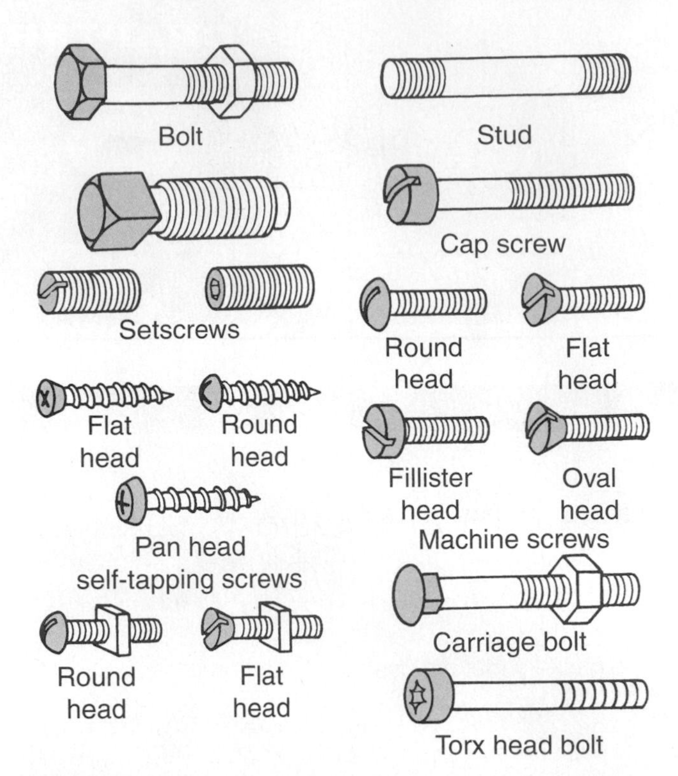

FIGURE 5-2 These are common fasteners used in the automotive industry.

Bolts have a head on one end and threads on the other. Their length is measured from the bottom surface of the head to the end of the threads. Most automotive bolts have a hexagon head.

Studs are rods with threads on both ends. They are used where bolts are not suitable. For example, studs are used on parts that must be removed frequently for service, such as the exhaust manifold. One end of the stud is screwed into a threaded or tapped hole in the exhaust manifold. The other end of the stud passes through the flange on the exhaust pipe. A nut is then used on the projecting end of the stud to hold the parts together.

Setscrews are used to prevent rotary motion between two parts such as a pulley and shaft. Setscrews are either headless or have a square head. As shown in ***Figure 5–2***, they can be turned in or out using either a screwdriver or an Allen wrench.

Cap screws pass through a clearance hole in one member of a part. They then screw into a threaded or tapped hole in another part.

Machine screws are similar to cap screws, but they have a flat point. The threads on a machine screw run the entire length of the stem or shank. Several types of heads are used, including the round, flat, fillister, Torx, and oval heads.

Self-tapping screws are used to fasten sheet-metal parts or to join light metal, wood, or plastic together. These screws form their own threads in the material as they are turned. They are available with different head shapes and points.

Torx head bolts are also used throughout the automotive industry. They are mostly used on the body of the vehicle, such as to secure the headlights. They are also tamper-proof because of their unique design. A star-shaped wrench or screwdriver tip is used to loosen and tighten the bolts.

BOLT IDENTIFICATION

To identify the type of threads on a bolt, bolt terminology must be defined. The bolt has several parts, as shown in ***Figure 5–3***. The **head** is used to **torque**, or tighten, the bolt. A socket fits over the head, which enables the bolts to be tightened. Common USC (U.S. Customary) and metric sizes for bolt heads include those shown in ***Figure 5–4***. The sizes are given in fractions of an inch and in millimeters (metric).

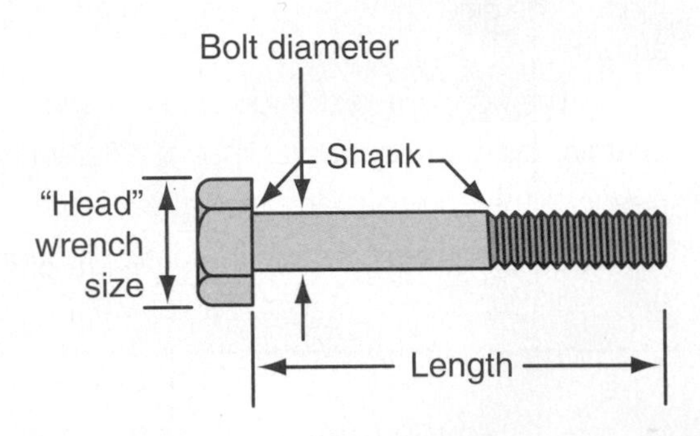

FIGURE 5-3 Bolts are identified by defining the head size, the shank diameter size, the number of threads per inch, and the length.

COMMON ENGLISH (U.S. CUSTOMARY) HEAD SIZES	COMMON METRIC HEAD SIZES
WRENCH SIZE	WRENCH SIZE
3/8"	9 mm
7/16"	10 mm
1/2"	11 mm
9/16"	12 mm
5/8"	13 mm
11/16"	14 mm
3/4"	15 mm
13/16"	16 mm
7/8"	17 mm
15/16"	18 mm
1"	19 mm
1 1/16"	20 mm
1 1/8"	21 mm
1 3/16"	22 mm
1 1/4"	23 mm
1 5/16"	24 mm
1 3/8"	26 mm
1 7/16"	27 mm
1 1/2"	29 mm
	30 mm
	32 mm

FIGURE 5-4 There are many standard bolt head sizes. Both USC and metric sizes are shown.

Some of the USC and metric sockets are very close in size. It is important not to use metric sizes for USC bolts or USC sizes for metric bolts. The bolt heads may be damaged.

The second part of the bolt is called the **shank**. The shank is the distance between the bolt head and the thread. The shank is illustrated in ***Figure 5–3***. Bolts are identified by the shank size or the outer diameter of the threads. Common bolt shank or diameter sizes include 1/4, 5/16, 3/8, 7/16, 1/2, 9/16, and 5/8 inches in diameter.

Another way to identify the size of a bolt is by the number of **threads per inch** as shown in ***Figure 5–5***. This number can be determined by using a ruler and counting the number of threads per inch for each bolt.

The number of threads per inch can also be measured with a **screw-pitch gauge** (***Figure 5–6***). The tool consists of numerous blades with thread-shaped teeth on one edge. The blades are inserted over the thread until one is found that fits the thread exactly. The number stamped on the blade that matches the thread indicates the number of threads per inch of the bolt.

The size of a bolt can also be expressed in terms of its length. A ruler is used to measure the length from the end of the bolt to the bottom surface of the head. Bolts are commonly manufactured in lengths with 1/2-inch increments or sizes.

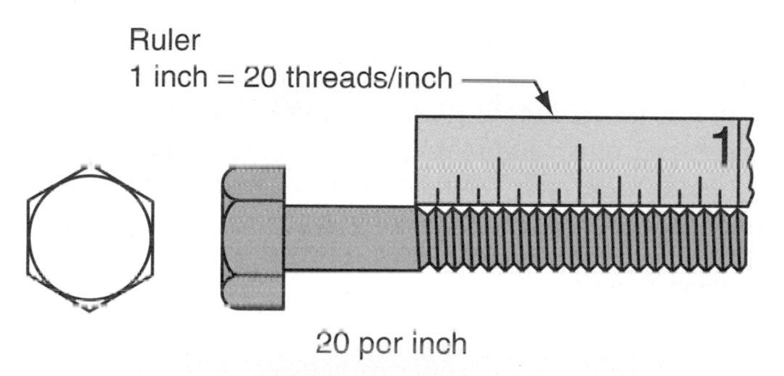

FIGURE 5–5 Threads on bolts can be measured by using a ruler and counting the number of threads per inch.

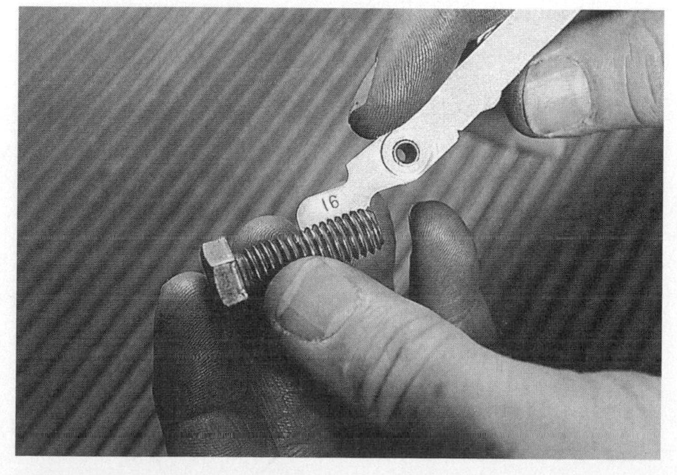

FIGURE 5–6 A screw-pitch gauge is used to determine the exact number of threads per inch.

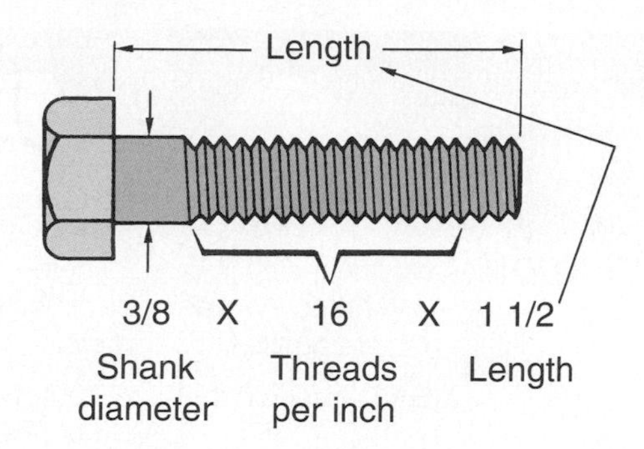

FIGURE 5–7 The shank diameter, the number of threads per inch, and the length are all used to determine the bolt size.

Figure 5–7 shows how the identifying features of a bolt are combined to specify a bolt size. For example, a bolt identified as 3/8 × 16 × 1 1/2 has a shank diameter of 3/8 inch; there are 16 threads per inch; and the bolt is 1 1/2 inches long.

Bolts can have different numbers of threads per inch and still have the same shank size. For example, a 3/8-inch bolt can have either 16 or 24 threads per inch. The greater the number of threads per inch, the finer the thread. The finer the thread, the greater its holding ability.

UNIFIED AND AMERICAN NATIONAL THREAD SIZES

Threads can also be identified by referring to them as either coarse threads or fine threads. Coarse threads identified as NC or UNC (National Coarse or Unified National Coarse) are used for general-purpose work. They are very adaptable for cast iron and soft metals where rapid assembly or disassembly is required. Fine threads identified as NF or UNF (National Fine or Unified National Fine) are used where greater resistance to vibration is required. NF threads are also used where greater strength or holding force is necessary.

BOLT AND NUT HARDNESS AND STRENGTH

Bolts are made from different metals with various degrees of hardness. Softer metal or harder metal can be used to manufacture bolts. Under certain conditions, the standard hardness used in a particular situation is not sufficient. Therefore, bolts are made with different hardnesses and strengths for use in different situations. Bolts are marked with lines on the top of the head to identify **bolt hardness**, as shown in ***Figure 5–8***. The number of lines on the head of the bolt is related to the **tensile strength**. As the number of lines increases, so does the tensile strength. Tensile strength is the amount of pressure, measured in pounds per square inch (psi), that the bolt can withstand before breaking when being pulled apart. The harder or stronger the bolt, the greater its tensile strength.

Nuts also have various grades of hardness. In the automotive shop, always match the hardness of the bolt and the

SAE GRADE MARKINGS	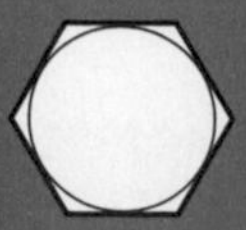				
DEFINITION	No lines—unmarked indeterminate quality SAE Grades 0-1-2	3 lines—common commercial quality Automotive & AN Bolts SAE Grade 5	4 lines—medium commercial quality Automotive & AN Bolts SAE Grade 6	5 lines—rarely used	6 lines—best commercial quality N.A.S. & Aircraft Screws SAE Grade 8
MATERIAL	Low Carbon Steel	Med. Carbon Steel Tempered	Med. Carbon Steel Quenched & Tempered	Med. Carbon Alloy Steel	Med. Carbon Alloy Steel Quenched & Tempered
TENSILE STRENGTH	65,000 psi	120,000 psi	140,000 psi	140,000 psi	150,000 psi

FIGURE 5-8 Bolts are identified by hardness. The more lines shown on the head of the bolt, the stronger the bolt.

hardness of the nut. If this is not done, the threads on the softer component may strip when being torqued to specifications. ***Figure 5–9*** shows an example of the markings and hardness of both customary and metric nuts.

PREVAILING TORQUE FASTENERS

A **prevailing torque nut** is designed to develop an interference between the nut and bolt threads. This interference is most often accomplished by distorting the top of an all-metal nut or by using a nylon patch in the middle of the threads. A nylon insert may also be used as a method of interference between the nut threads and the threads of the tapped hole. ***Figure 5–10*** shows examples of various prevailing torque fasteners used on automobiles.

METRIC THREADS

Metric threads are being used increasingly on all cars manufactured in the United States. This means that metric bolts will also be used more often. Many vehicles have transmissions and other automotive components made in metric sizes. Because metric threads and fasteners are being used more often, it is important that they not be mixed with bolts that have standard threads.

STANDARD NUT STRENGTH MARKING

CUSTOMARY SYSTEM		METRIC SYSTEM	
GRADE	IDENTIFICATION	CLASS	IDENTIFICATION
Hex Nut Grade 5	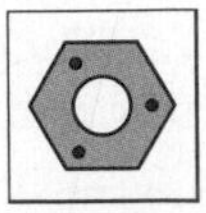 3 Dots	Hex Nut Property Class 9	 9 Arabic 9
Hex Nut Grade 8	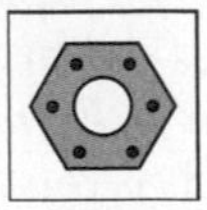 6 Dots	Hex Nut Property Class 10	 10 Arabic 10
Increasing dots represent increasing strength.		Can also have blue finish or paint dab on hex flat. Increasing numbers represent increasing strength.	

FIGURE 5-9 The number of dots show the hardness of a hex nut in the customary system. A metric hex nut has the hardness stamped on the end.

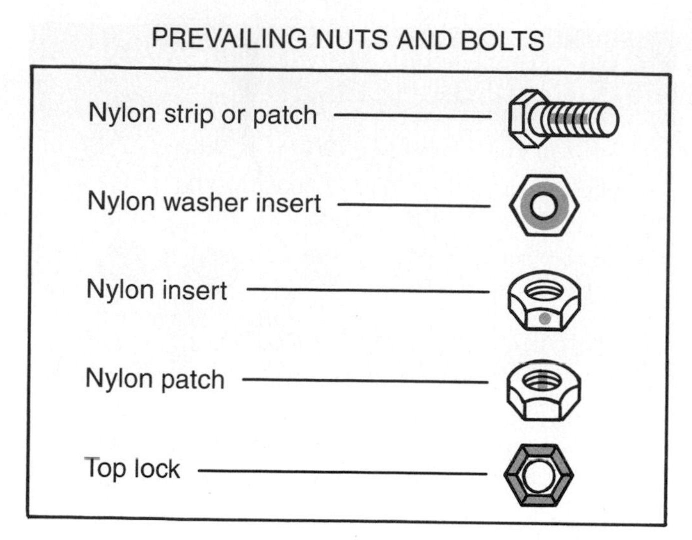

FIGURE 5–10 These are examples of various prevailing threaded fasteners.

For metric bolts, the shank and length are measured in millimeters. To determine the type of thread used, the distance between threads is measured in millimeters. Thus, the bolts are considerably different from inch or USC sizes. ***Figure 5–11*** shows an example of the difference between the USC and metric thread pitch.

The hardness or strength of metric bolts is indicated by using a **property class** number stamped on the head of the bolt as shown in ***Figure 5–12***. The higher the property class number, the harder the bolt. When replacing a bolt, always remember to use the same hardness of bolt as the original application.

DETERMINING TORQUE ON CUSTOMARY BOLTS

All bolts and nuts used on an automobile should be torqued or tightened to specifications. Many repair manuals list common torque specifications. A torque wrench is used to measure the tightening force. The specifications known as **standard bolt and nut torque specifications** are shown in ***Figure 5–13***. The correct torque in pound-inches and/or pound-feet should always be followed when tightening bolts and nuts. (Note that the terms pound-feet and foot-pounds are used interchangeably.) If a bolt is tightened to a value less than that given by the specification, the bolt may vibrate loose. A bolt can also be tightened too much. This may cause the threads to weaken and strip or the bolt to break. In either case, damage to other automobile parts can result. Always torque all bolts and nuts to the manufacturer's specifications. If there are no specifications listed, use the standard torque specifications shown in ***Figure 5–13***.

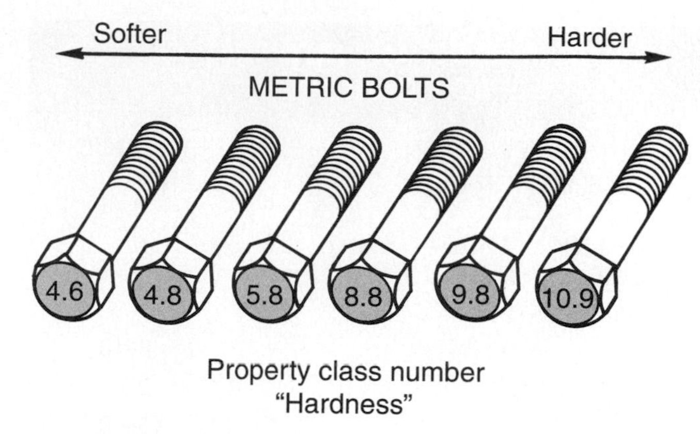

FIGURE 5–12 Metric bolts are rated in hardness by using the property class number.

TORQUE-TO-YIELD BOLTS

Manufacturers also use bolts that are called **torque-to-yield bolts**. Torque-to-yield bolts are those that have been tightened at the manufacturer to a preset yield or stretch point. When a bolt is torqued according to its normal specifications, it is not being stretched beyond its elastic limit. However, it generally is not as tight as it could be. When torque-to-yield bolts are torqued at the manufacturer, they are stretched slightly, so the bolt is generally tighter than a standard bolt. Once stretched, they do not return to their original size, so they should be replaced with new bolts of

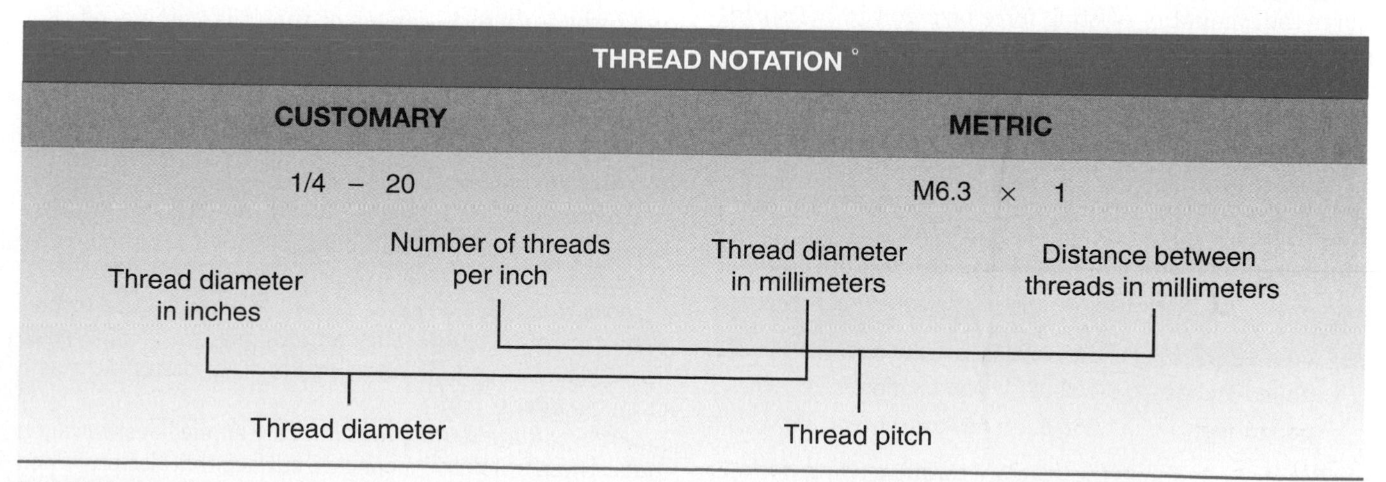

FIGURE 5–11 Customary and metric thread notation.

STANDARD BOLT AND NUT TORQUE SPECIFICATIONS		
SIZE NUT OR BOLT		TORQUE (lb-ft)
SHANK SIZE	THREADS PER INCH	
1/4	20	7–9
1/4	28	8–10
5/16	18	13–17
5/16	24	15–19
3/8	16	30–35
3/8	24	35–39
7/16	14	46–50
7/16	20	57–61
1/2	13	71–75
1/2	20	83–93
9/16	12	90–100
9/16	18	107–117
5/8	11	137–147
5/8	18	168–178
3/4	10	240–250
3/4	16	290–300
7/8	9	410–420
7/8	14	475–485
1	8	580–590
1	14	685–695

FIGURE 5-13 All bolts and nuts have a standard bolt and nut torque specification.

the same hardness. Often these torque-to-yield bolts are used on main or connecting rod bearings and on cylinder head bolts. Usually the automotive manufacturer recommends that these bolts be replaced with new ones or torqued by using a torque angle meter. A torque angle meter measures the amount of twisting force on a bolt by measuring the number of degrees of turning.

DETERMINING TORQUE ON METRIC BOLTS

Metric bolts and nuts are torqued in newton meters (N•m). For reference and a comparison between newton meters and pound-inches or pound-feet use the following formula:

MULTIPLY	BY	TO GET
pound-inches	0.112 98	newton meters
pound-feet	1.355 8	newton meters

Figure 5–14 shows the standard torque readings for various sizes of metric bolts. Note that there are several readings for each bolt. Just as with standard bolts, metric bolts have different grades of hardness and strength. A 5D bolt is the weakest. A 12K bolt is the strongest. The grade of bolt is identified by looking at the head of the bolt. Once the grade of bolt is identified, the correct torque reading in foot-pounds can be determined.

TORQUE SPECIFICATIONS FOR METRIC BOLTS				
Grade of Bolt	5D	8G	10K	12K
Min. Tensile Strength	71,160 PSI	113,800 PSI	142,200 PSI	170,679 PSI
Grade Mark on Head	5D	8G	10K	12K
BOLT DIAMETER	FOOT-POUNDS			
6 mm	5	6	8	10
8 mm	10	16	22	27
10 mm	19	31	40	49
12 mm	34	54	70	86
14 mm	55	89	117	137
16 mm	83	132	175	208
18 mm	111	182	236	283
22 mm	182	284	394	464

FIGURE 5-14 When torquing metric bolts, use these standard torque specifications.

NUTS

Many types and styles of nuts are used on the automobile. ***Figure 5–15*** shows some common types of nuts. The most common type is the hex style.

Hexagon nuts, also called hex nuts, are classified as regular or heavy. They are used on all high-quality work. These nuts are easy to tighten with wrenches in close or tight spaces.

Slotted hexagon nuts, also called castellated nuts, are used where there is danger of the nuts coming off. For example, vibration may cause the nuts to loosen. A cotter pin is used to lock the nut in place.

Jam hexagon nuts are thinner than regular nuts. They are used where height is restricted or as a means of locking the working nut in place.

Square nuts are regular or heavy unfinished nuts. They are used with square-head bolts in rough assembly work such as assembly of body parts.

Lock nuts have a self-contained locking feature to prevent back-off rotation. They are designed with undersized threads and plastic or fiber inserts. This design acts as a gripping force.

Free-running seating lock nuts are applied over hexagon nuts. These nuts have a concave surface that flattens when it contacts the top of the hexagon nut. This causes the

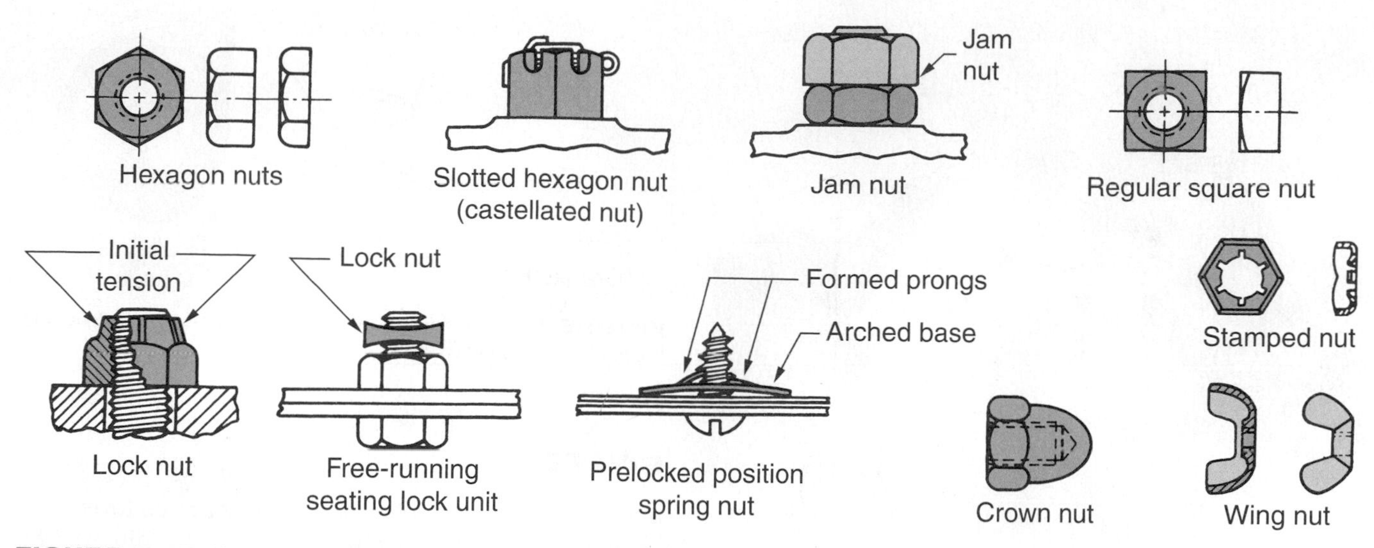

FIGURE 5–15 Many styles of nuts are used on the automobile. Each style has a specific purpose and application.

threads to deflect. The nut binds on the bolt and prevents it from coming loose.

Spring nuts are made of thin spring metal. They are designed with formed prongs as shown in ***Figure 5–15.*** Spring nuts are used in sheet metal construction where high torque is not required.

Crown nuts are used where the ends of the external threaded part must be concealed or hidden or the nut must be attractive. They are often made of stainless steel.

Wing nuts have two arms (or projections) to aid hand tightening and loosening. Typically, these nuts are tightened and loosened frequently. High torque is not a consideration.

WASHERS

Most washers are placed on bolts to lock the bolt and keep it from coming loose. Several types of washers are used for locking bolts in place, as shown in ***Figure 5–16***. Washers are also used for other reasons. For example, copper washers are used to aid in sealing the bolt to a structure. These are called **compression washers.** They may help reduce oil leakage when the threads of the bolt are in or near oil. Certain cars use copper washers on oil pans to eliminate leakage. Flat washers are also used to help spread out the load of tightening the nut to prevent the nut from digging into the material when it is tightened.

When disassembling any automotive component, it is always a good idea to check which type of washer is being used. Then during reassembly, remember to use the same type of washer that was used originally.

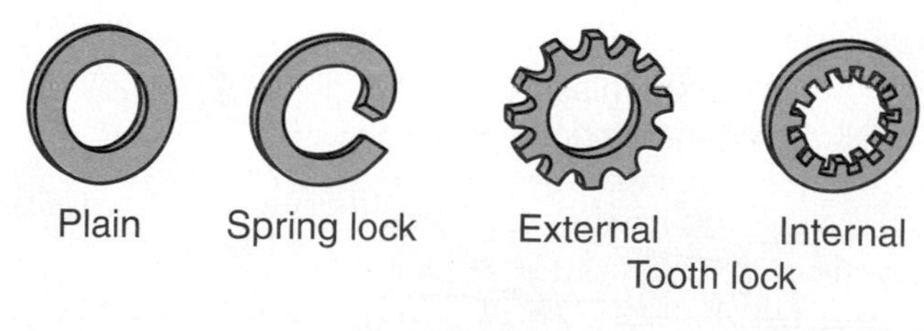

SNAP RINGS

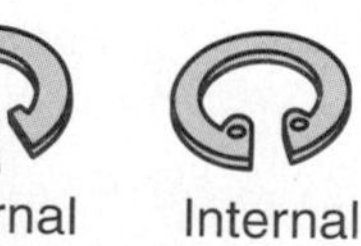

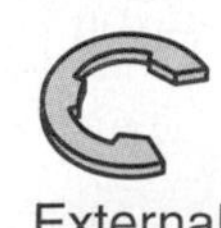

FIGURE 5–16 Washers are typically used to lock bolts to the structure, keeping them from coming loose and preventing damage to softer metal parts.

SNAP RINGS

Snap rings are used to prevent gears and pulleys from sliding off the shaft. There are two types of snap ring: the external and internal snap ring. ***Figure 5–17*** shows snap ring pliers being used to install a snap ring to hold a piston pin in place.

SPLINES

Splines are external or internal teeth cut in a shaft. When a shaft must be inserted into a gear, pulley, or other part, and the part must be able to move on the shaft, a spline is used. The output shaft on the transmission has an external spline. The end of the drive shaft connected to the transmission has an internal spline. This allows the drive shaft to move along the axis when the rear wheels go over a bump in the road (***Figure 5–18***).

KEYWAYS

Keyways are used to lock parts together by fitting a key between a slot on the shaft and a pulley. ***Figure 5–19*** shows a keyway application. Keyways are used to secure the front

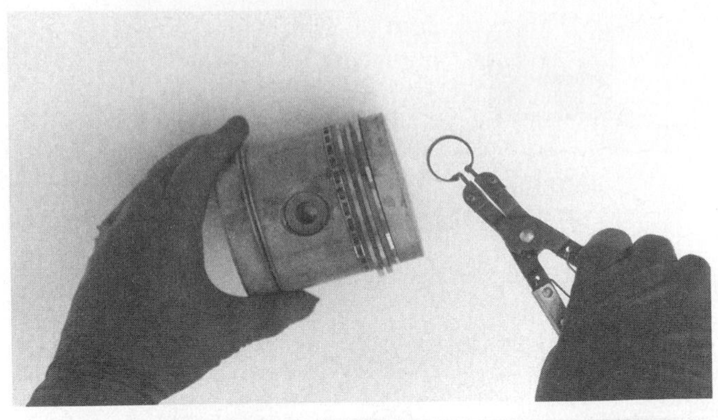

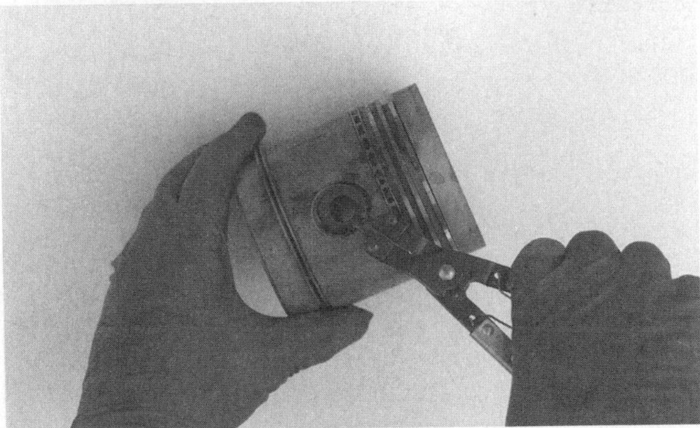

FIGURE 5–17 Snap rings are used to keep gears and pulleys on the shaft or, in this case, hold a piston pin in place.

drive pulley on the engine to the engine crankshaft. The only difference between a keyway and a spline is that, on a spline, the hub or pulley can move parallel or axially along the shaft.

PARTS LOCATOR

Refer to photo #1 in the front of the textbook to see the location of the ***keyway*** *on the front of the crankshaft.*

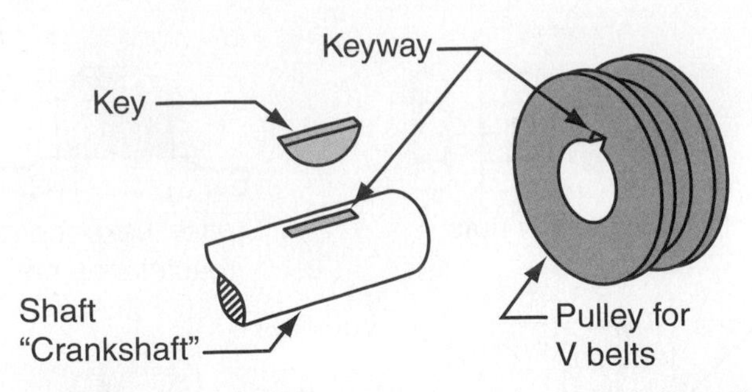

FIGURE 5–19 Keyways are used to hold parts to shafts so that the part and the shaft rotate as one unit.

RIVETS

Another type of automotive fastener is the *rivet*. Rivets look like metal pins. They are often used to fasten thin parts together. ***Figure 5–20*** shows three types of rivets and what the rivet looks like during and after installation. Rivets are installed by first placing the rivet through the two holes. Then a punch or hammer is used to form a head on the end of the rivet. Rivets are most often removed by using a chisel and a hammer to cut off the head of the rivet.

SELF-TAPPING SCREWS

Often it is necessary to join two thin metal plates together. This can be done easily by using self-tapping screws. A self-tapping screw cuts its own threads into the material when the screw is turned into the hole. Self-tapping screws are easy to identify. They usually have small cuts in the end of the screw to help cut the threads. In some cases, the end of the screw looks similar to a drill bit. ***Figure 5–21*** shows several types of self-tapping screws and an example of how one is used.

THREADED INSERTS (HELICOILS)

Threads stripping inside an engine block, cylinder head, or other automotive component is a common problem. Usu-

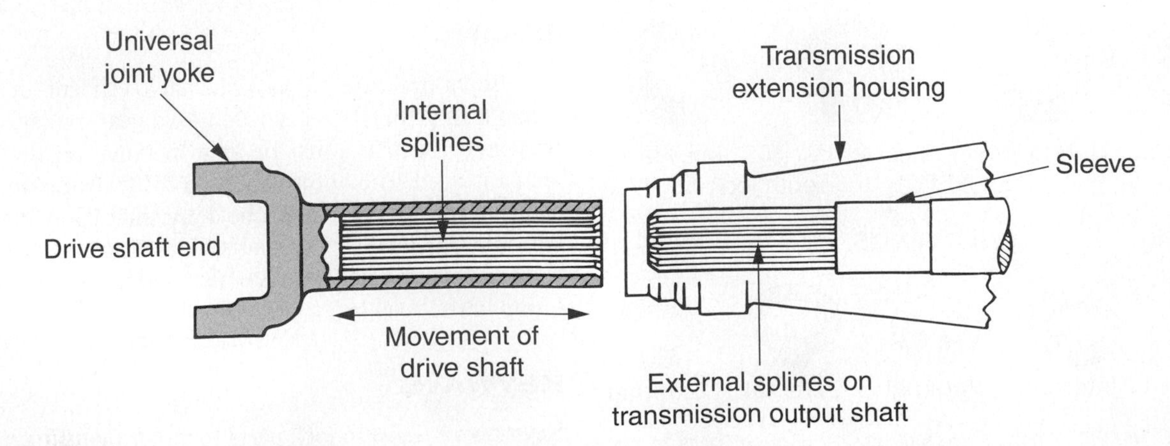

FIGURE 5–18 Both external and internal splines are used to lock rotating shafts together while allowing slide (axial) movement between shafts.

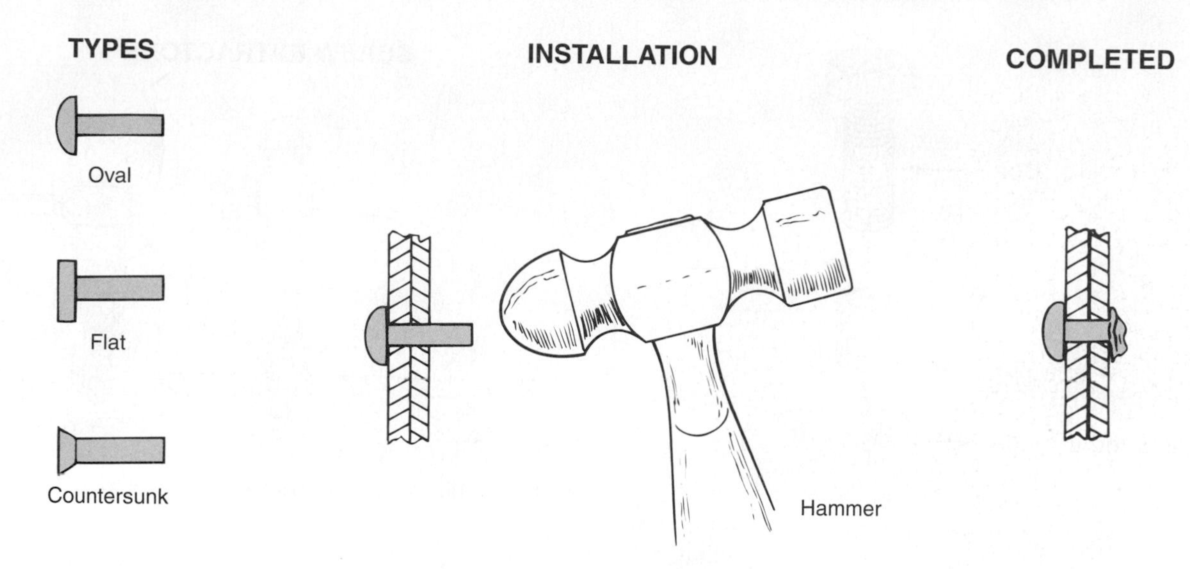

FIGURE 5–20 Various types of rivets are used on the automobile.

SELF-TAPPING SCREWS

For no hole drilled

For drilled holes

Drill

Screwdriver—Hex

Fillister

Round

Flat

Hex

Note cuts in thread

Installation

Drill

Tap

Fasten

FIGURE 5–21 Various types of self-tapping screws are used on the automobile.

ally this problem is caused by too high a torque or by threading the bolt into the hole incorrectly. For example, when a spark plug is threaded back into the cylinder head, it is very easy for the threads to become crossed. If the spark plug is then tightened, the spark plug threads in the cylinder head may be stripped or damaged. A good rule of thumb is to always thread spark plugs by hand two complete revolutions before a wrench is used. If the threads do become stripped, rather than replacing the cylinder head, the threads can be replaced with threaded inserts, or **helicoils.** ***Figure 5–22*** shows a helicoil.

The procedure for replacing damaged threads is as follows:

1. The damaged threads are drilled out to a specific size depending on the helicoil size.
2. The hole is tapped, using the outside diameter of the helicoil.
3. The helicoil is threaded into the new, larger threads.

The inside of the helicoil provides the new threads for the bolt. There are many sizes of helicoils. The tap, drill sizes, and necessary tools are all included in the helicoil kit.

SCREW EXTRACTOR

There are many occasions where bolts are broken off. A broken bolt that is not made of hardened steel may be removed from a hole with a screw extractor. First a hole must be drilled into the broken bolt. Then the correct size screw extractor is placed into the hole. The screw extractor is then turned counterclockwise with a tap wrench. The screw extractor acts much like a corkscrew to remove the broken bolt.

Figure 5–23 shows common screw extractors and how they are used. Figure 5–23A shows a splined screw extractor which is forced into a drilled hole in the broken bolt. Figure 5–23B shows a tapered screw extractor which is also forced

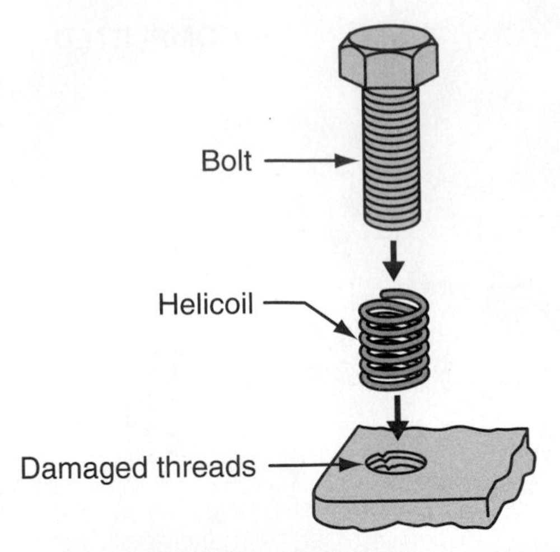

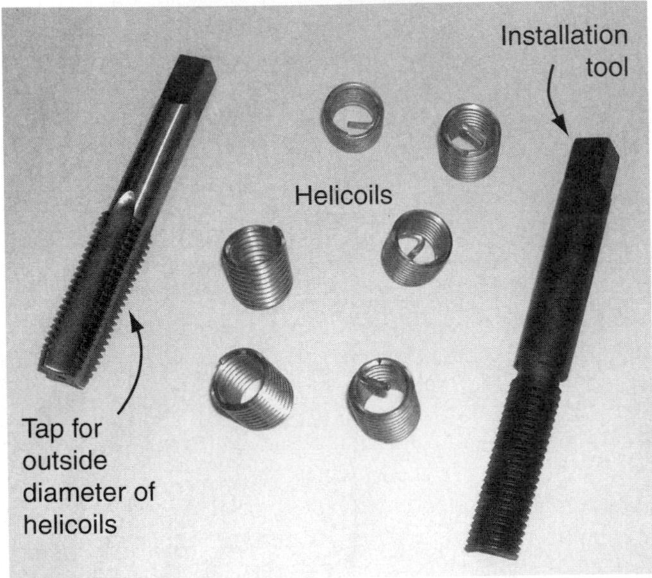

FIGURE 5-22 Helicoils are used to replace damaged internal threads.

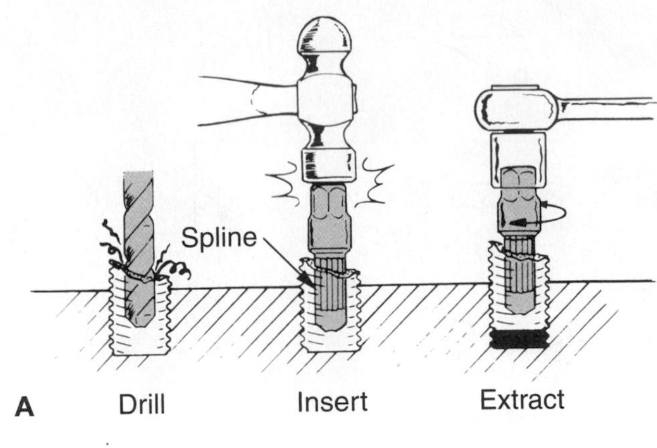

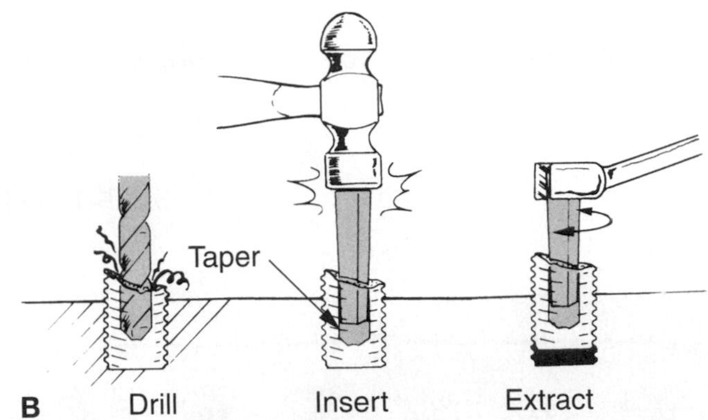

FIGURE 5-23 These screw extractors are used to remove broken bolts.

into a drilled hole in the broken bolt. Both are then unscrewed to remove the broken bolts.

TAPS AND DIES

Both internal and external threads can be repaired using a tap or a die. A tap is a threaded tool used to cut internal threads. A die is a threaded tool used to cut external threads, for example on a bolt. Various sizes of both taps and dies are used. Taps and dies are mounted in special handles, as shown in ***Figure 5–24***.

5.2 GASKETS

PURPOSE OF GASKETS

Gaskets are used on automobiles to prevent leakage of gases, liquids, or greases between two parts bolted together. Examples of gaskets are shown in ***Figure 5–25***. Gaskets are placed between two mating machined surfaces. Bolts or fasteners are then tightened according to standard specifications. Gaskets are used between the cylinder head and the engine block or between the water pump housing and the engine.

Gaskets are designed for particular jobs. Some gaskets may be used in very oily places. Other gaskets are designed to be used on very hot components, while still other gaskets are used for high-pressure applications. For this reason, gaskets are made from different materials. Some of the more common materials used are cork, synthetic rubber, steel, copper, and asbestos.

COMPRESSION GASKETS

Compression gaskets are designed to be squeezed during tightening (***Figure 5–26***). Tightening a compression gasket seals the parts to contain high-pressure gases. Compression gaskets are used most often for head gaskets. As the cylinder head is torqued to specifications, the head gasket is squeezed and forms a good seal for high compression during combustion. When compression gaskets are removed, new compression gaskets should always be used as replacements. If the old gasket is used again, it may not seal as well as originally intended. Some spark plugs also use compression gaskets as washers to seal in the high-compression gases produced during compression and combustion.

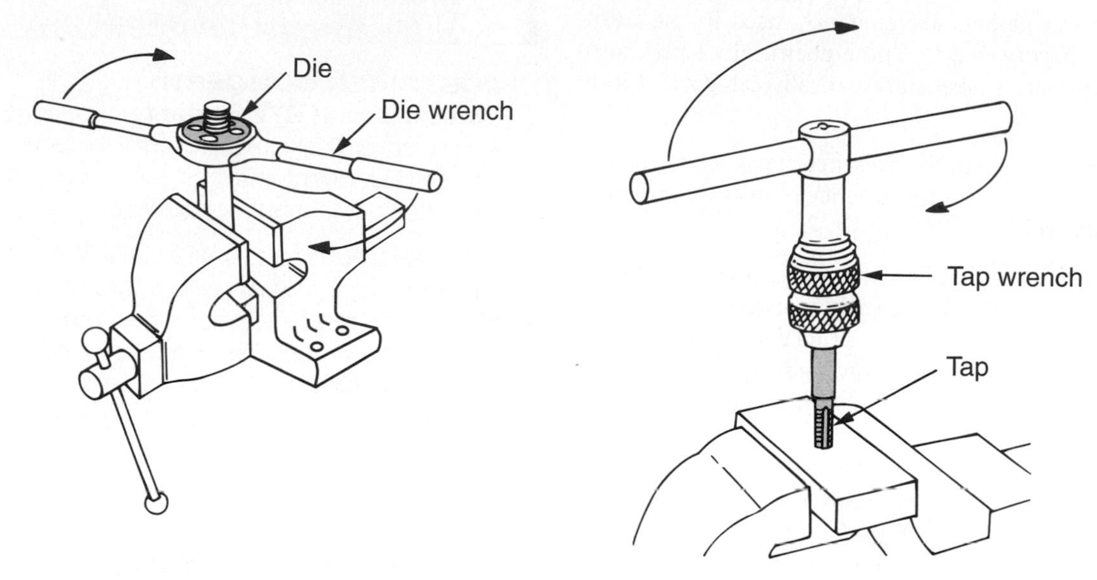

FIGURE 5-24 Taps and dies can be used to repair both internal and external threads.

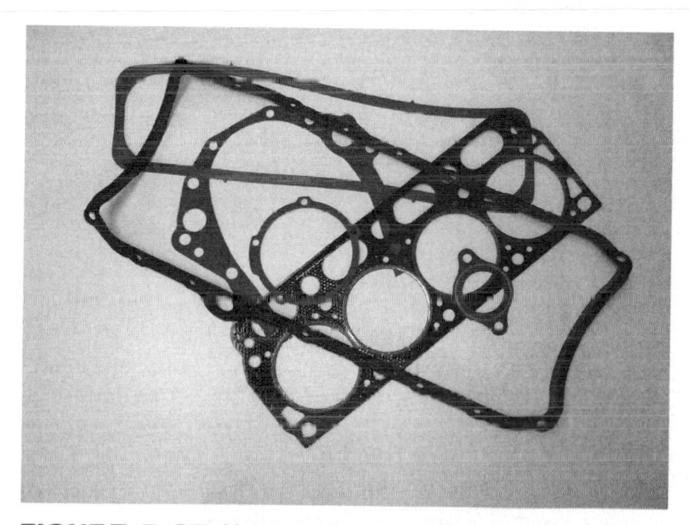

FIGURE 5-25 Many gaskets are used on the automobile engine to seal parts that are bolted together. Gaskets keep grease, gases, and fluids from escaping and dirt from entering the engine.

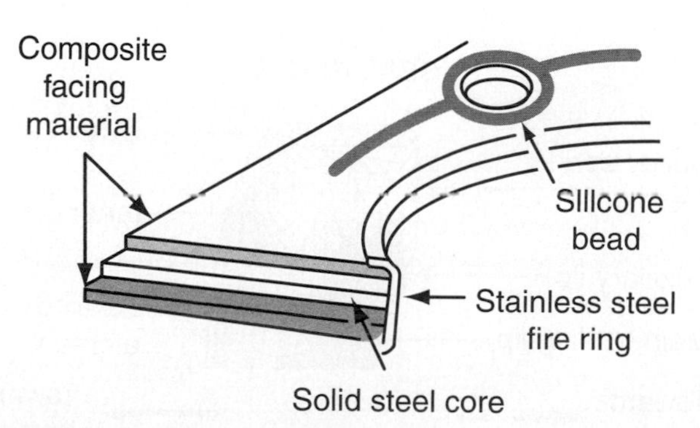

FIGURE 5-26 Compression gaskets are used to seal high pressures produced within the cylinder during the compression and power stroke.

MAKING SPECIAL GASKETS

At times it may be necessary to make your own cork or paper gasket. Rolls of cork and paper gasket material can be purchased from most automotive parts stores. Depending on the size of the gasket needed, many cork or paper gaskets can be made from one roll. Use the following procedures when making cork or paper gaskets:

1. Make sure the thickness of the new gasket is the same as the old gasket.
2. Make sure to completely clean all of the old gasket material from the part. You may need a scraper to clean off all of the old gasket material.
3. If the old gasket is still in its original shape, simply lay the old gasket on the roll of gasket material. Draw the outline of the old gasket onto the new gasket material.
4. Cut out the new gasket as carefully as possible and check to see that it fits correctly.
5. If the old gasket is not in its original shape, you will need to use the part that touches the gasket as a pattern. Using the part, draw the gasket outline as accurately as possible, cut the gasket out of the gasket roll, and check to see that it fits correctly. In this case you may have to try several times to get the gasket to fit correctly.
6. Another way to make a gasket without a pattern is to lay a large piece of gasket material over the part. Then, using a small hammer, gently pound the gasket along the sharp edges of the part to make marks where the gasket will need to be cut. Cut the gasket as needed.

CHEMICAL GASKETS

Several chemical gaskets are currently used in the automotive shop (***Figure 5–27***). These chemical gaskets, also called form-in-place gaskets, are typically categorized into two types:

1. Those that harden in the presence of oxygen (air)
2. Those that harden in the absence of oxygen (air); also called anaerobic

One type of gasket that hardens in the presence of oxygen is a silicon gasket called an **RTV sealant.** (RTV stands for room temperature vulcanizing.) RTV gaskets and sealants dry or cure from the moisture and oxygen in the air. This gasket can be identified by various colors, such as black, red, orange, blue, or some other color. This silicon gasket is used to form a rubberlike gasket on engine parts that have thin flexible flanges (such as a valve cover); see ***Figure 5–28***. Generally, RTV gaskets have a shelf life of about a year. After that time they will not cure properly. Certain types of RTV gaskets are not compatible with the oxygen sensor used on automobiles. Always check the container. It should say "O2 Sensor Safe."

RTV sealant is normally applied in a continuous bead about 1/8 inch in diameter. Make sure to go around all bolt holes and to torque within 10 minutes after application. Also make sure that the RTV gasket is wet at the time the parts are mated.

The second type of chemical gasket is called the **anaerobic sealant.** This gasket or sealant cures to a plasticlike substance in the absence of air. Generally it is applied to very smooth, mating parts. For example, this gasket can be used between a water pump and the engine block. It will cure easily when the two parts are torqued to specifications. Generally, a 1/16-inch diameter bead is placed on the mating surface for correct sealing.

SILICON SEALANTS

CUSTOMER CONCERN:
Correct Use of RTV Sealants

A valve cover gasket was recently replaced with an RTV sealant. Shortly after, the oxygen sensor had to be replaced. What is the problem?

SOLUTION: Some types of RTV silicon sealants may emit vapors when curing. These vapors can damage the oxygen sensor. When using RTV sealants, read the package or tube to make sure the sealant will not damage various components in the engine.

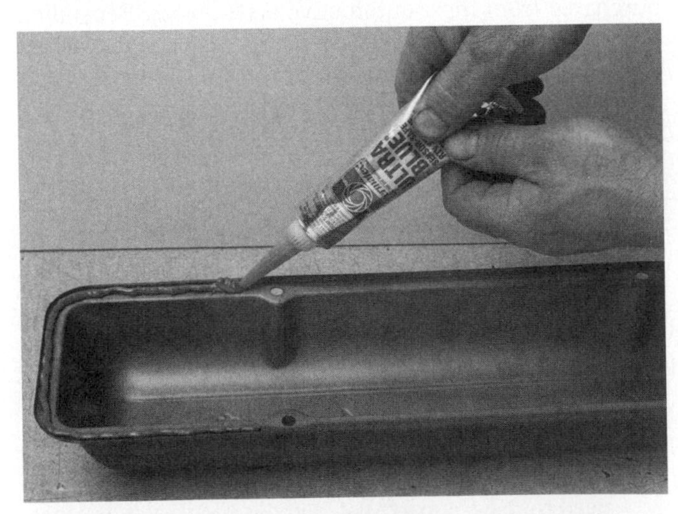

FIGURE 5-27 Silicon and RTV sealants are considered chemical gaskets.

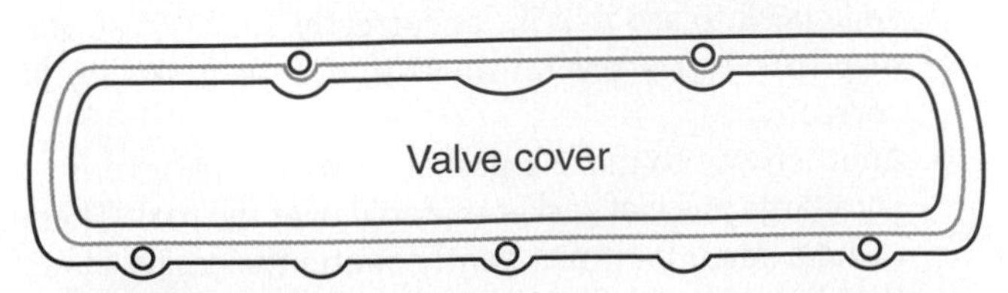

FIGURE 5-28 Apply a 1/8-inch bead of RTV sealant on the valve cover as shown.

5.3 SEALS AND SEALANTS

SEALS

There are many uses for **seals** in the automobile. Seals are placed on rotating shafts to prevent oil, gases, and other fluids from escaping. For example, seals are used on the front and back axles, crankshafts, water pump shafts, and many other locations throughout the vehicle.

Seals are designed to withstand high pressures and to seal fluids in the engine. Seals are made of felt, synthetics, rubber, fiber, or leather. The parts of the common seal are shown in ***Figure 5–29***.

The outer case is usually pressed into the housing that contains the stationary part. The inner case holds the parts of the seal. Seals are sometimes designed with one or two

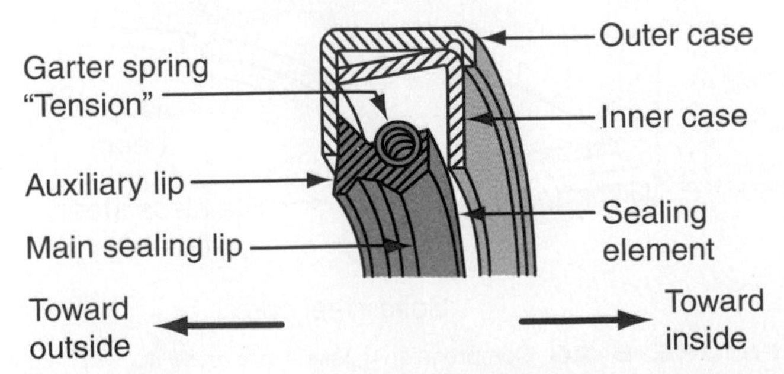

FIGURE 5-29 Seals must be installed so that the main sealing lip points toward the liquid, gas, or pressure to be contained.

lips. The lips are pressed against the rotating part. This pressure causes the sealing action. Springs may be added to the lip tension so that higher pressures can be contained.

Seals should always be installed according to the manufacturer's specifications. In general, the sealing lip should always be placed toward the fluid or pressure being contained.

SEALANTS

Several sealants are used in the automotive industry. **Sealants** are similar to a thick liquid that hardens after being placed on the metal. Some sealants are called form-in-place gaskets. Certain sealants are used to seal between metal surfaces. There are many brands of sealants available, made from many materials. Silicon is one of the most popular sealant materials. Use sealants only where the manufacturer recommends them. If a sealant is not recommended, use the correct gasket for the parts to be sealed.

GASKET SEALANTS

Several types of gasket sealants are also on the market. A **gasket sealant** is a chemical (usually a black, thick liquid) placed on the gasket to (1) help paper and cork gaskets seal better or (2) position gaskets during installation.

ANTISEIZE LUBRICANTS

Often bolts in various engine applications can become "cold-welded" together because the bolt and internal threads are continuously exposed to heating and cooling. This is especially true with aluminum parts. Once the bolts have cold-welded it is almost impossible to remove them. To eliminate this problem, antiseize lubricants are placed on the bolt threads before installation. Antiseize lubricants are most often used on bolts and nuts that are attached to the exhaust system, such as the manifold bolts. Antiseize lubricants are also sometimes used on cylinder head and manifold bolts and spark plugs.

LOCKING THREAD SEALANTS

Locking thread sealants are used on various engine bolts to keep the bolts from loosening. In this case, an anaerobic chemical sealant is used. When placed on bolt threads, the sealant acts as a glue to keep the bolt from coming loose. Typically, two types are used: (1) a blue locking thread sealant is used when the parts may need to be disassembled at a later date; (2) a red locking thread sealant is used when the parts will not need to be disassembled at a later date.

TEFLON TAPE

Some bolts come in contact with a liquid such as oil or antifreeze. These bolts must be sealed to stop the liquid from seeping past the threads. They are usually sealed with Teflon tape as shown in ***Figure 5–30***. Generally, the Teflon tape is wrapped around the threaded part of the bolt before installation. Always remember to wrap the tape in one direction only and make sure no Teflon tape gets into the system.

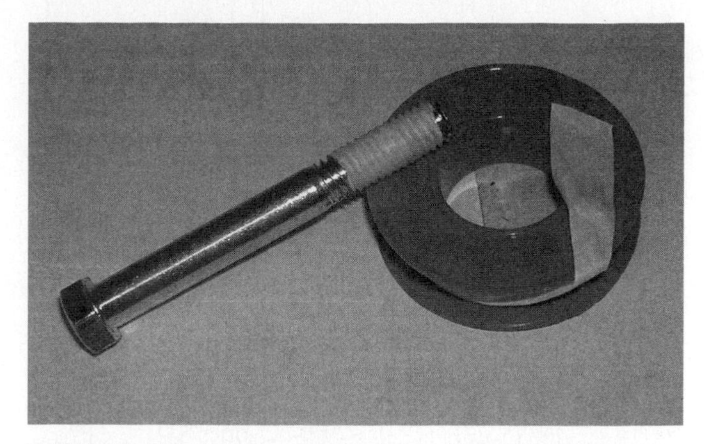

FIGURE 5–30 Teflon tape can be used to seal fluids from seeping past bolt threads.

Car Clinic:
TEFLON TAPE

CUSTOMER CONCERN:
Leaky Bolts Sealed with Teflon Tape

A vehicle was recently overhauled. However, there are several leaks coming from the oil pan. Why are there leaks, and how can they be eliminated?

SOLUTION: Oil will often leak around the threads of a bolt, especially if the bolt threads come into contact with a cavity that has oil in it, as some oil pan bolts do. Many manufacturers (especially of foreign cars) allow the bolt to come into contact with the crankcase area. To eliminate the bolt from leaking oil around the threads, always wrap the bolt threads with Teflon tape. This method will seal the bolt and stop all oil leakage around the bolt threads. Remember to also check for misaligned gaskets or loose oil pan bolts as well.

STATIC AND DYNAMIC SEALS

Seals are classified as either static or dynamic. O-rings are examples of static seals. An **O-ring** is placed between two stationary parts where a fluid passes between the parts. The O-rings shown in ***Figure 5–31*** keep the fluid from leaking out of the two stationary objects.

Dynamic seals provide sealing between a stationary part and a rotating part. A dynamic seal is shown in ***Figure 5–32***. In this case, the fluid is contained by the seals. Note that the lip rides directly on the rotating shaft.

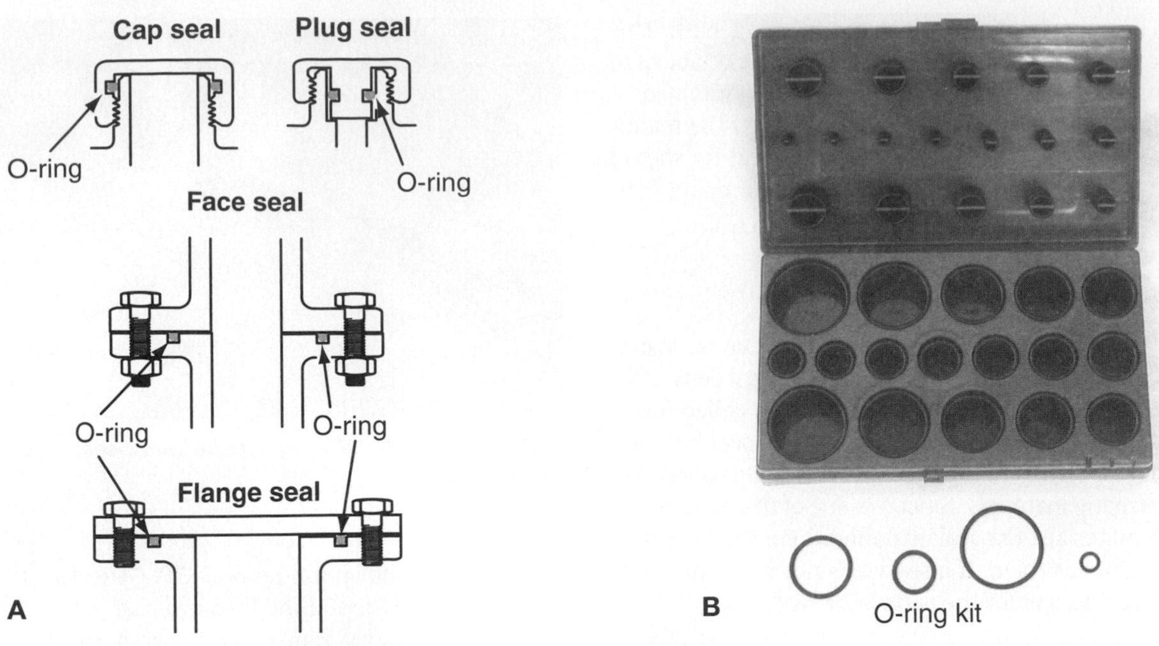

FIGURE 5-31 O-rings are considered static seals. Static seals are placed between two stationary parts to prevent leakage (A). An O-ring kit (B).

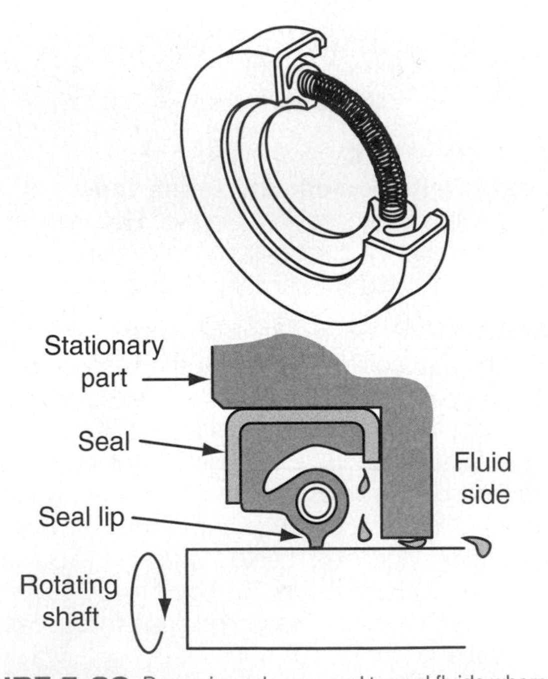

FIGURE 5-32 Dynamic seals are used to seal fluids where one part is stationary and one part is rotating.

Car Clinic:
O-RINGS

CUSTOMER CONCERN:
Bad O-Ring Causes Vehicle to Smoke
A Honda engine has heavy smoking immediately after startup. The smoke lasts about 10 minutes. The rings, valves, and valve guides have been checked and seem OK. What could be the problem?

SOLUTION: This smoking is a common problem in engines that have precombustion chambers. There is a rubber O-ring that seals the chamber where the auxiliary intake valves are located. The O-rings have been known to harden over time. Hardened O-rings do not seal very well, and oil can seep into the combustion chambers. The problem goes away when the engine heats up and the parts expand enough to make a good seal. The solution is to replace the rubber O-rings.

LABYRINTH SEALS

Labyrinth seals are used on high-speed shafts or where the lips of the seal can cause excessive wear on the shaft. A labyrinth seal is shown in ***Figure 5–33***. As liquid is moved toward the sealing area, it contacts the first labyrinth. Centrifugal force causes the fluid to spin outward into the upper groove. If the pressure is great enough to push the fluid farther, the next labyrinth seal will also produce a sealing effect. Because of centrifugal forces, the fluid has difficulty getting past the seal. This action keeps the fluid from leaking out of the area. Labyrinth seals will not work if a fluid touches the seal directly when the shaft is not rotating. There is no contact of the seal to the shaft; therefore, oil will leak out.

Certain foreign car manufacturers use labyrinth seals on the crankshaft to seal oil in the crankcase area. In this case, the seal will not wear on the metal, eliminating possible damage to this area of the crankshaft.

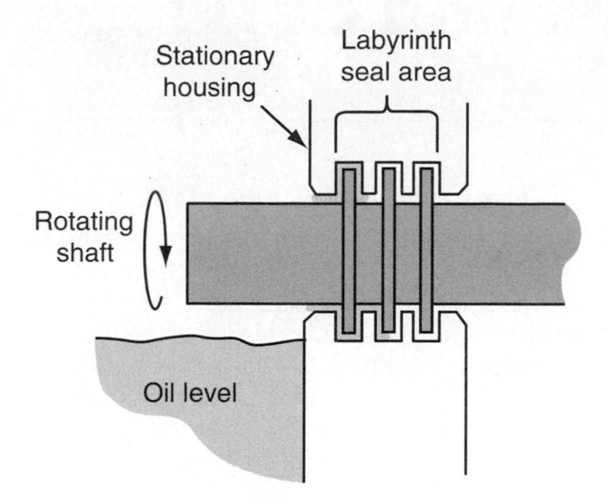

FIGURE 5-33 Labyrinth seals use centrifugal force to spin fluid outward. Because of this action, the fluid cannot pass through to the outside.

ROPE SEALS

On some older vehicles, a **rope seal** was used for sealing on the crankshaft. Although not used anymore, technicians may encounter a rope seal during servicing of older vehicles. This seal is made of two separate parts. One part fits into a groove in the block, and one part fits into a groove in the rear main bearing cap. Rope seals look much like a thin, flexible rope. They are installed by first soaking the seal in oil for several minutes and then lightly pounding it into the grooves using a special tool. Once installed and trimmed to the correct length (***Figure 5–34***) the rope seal takes a more precise shape when the main bearing cap is torqued to its specification.

One advantage of using rope seals is that they can be replaced without removing the crankshaft. With a set of special tools, the old seal can be removed and a new seal installed (***Figure 5–35***).

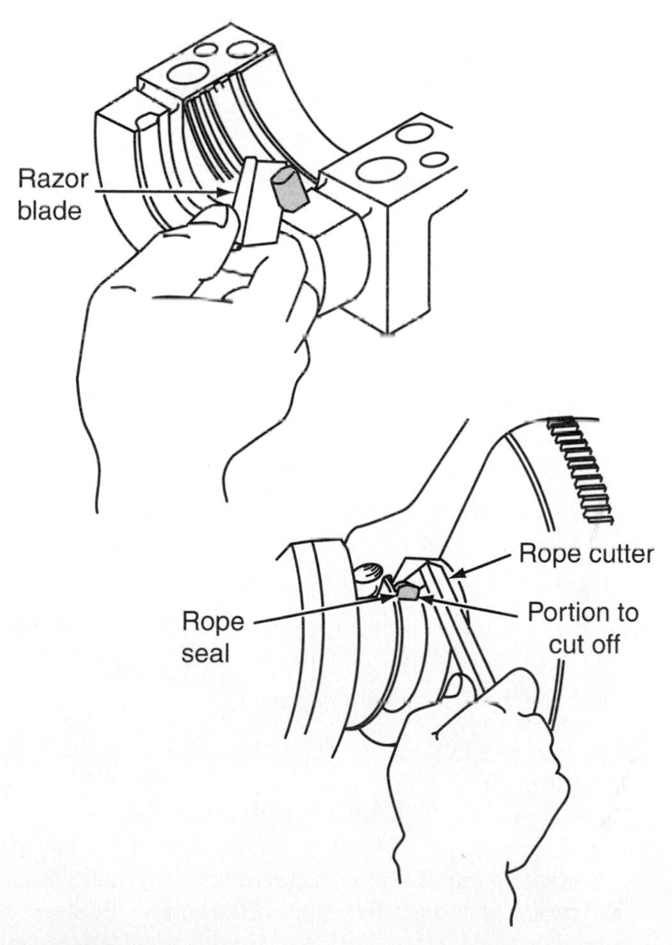

FIGURE 5-34 Rope seals are used on older vehicles and are designed to be replaceable.

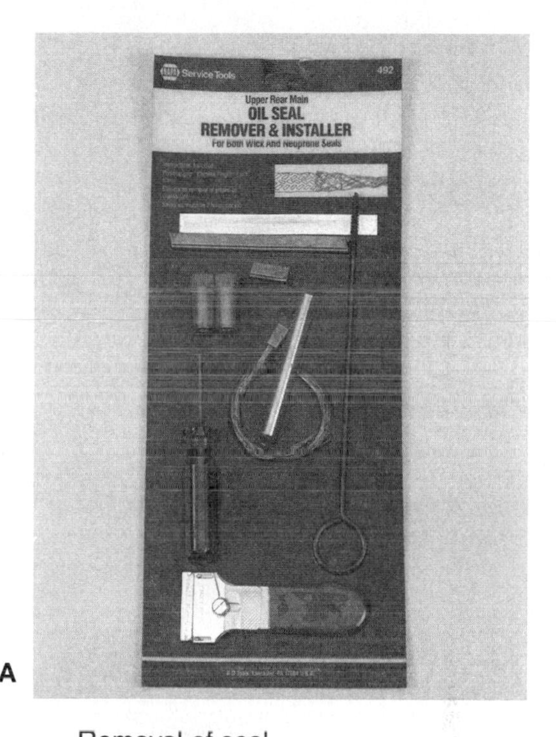

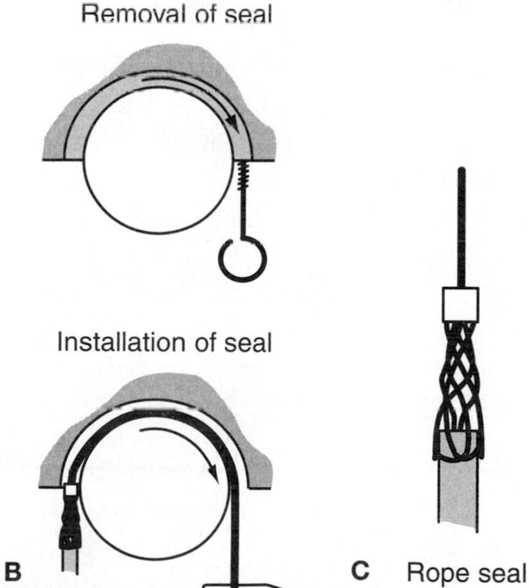

FIGURE 5-35 (A) A tool kit for replacing rope seals. (B) Removal and installation of rope seals with crankshaft in the engine. (C) Rope seal attached to installation tool.

Problems, Diagnosis, and Service

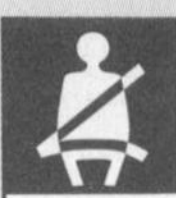

Safety Precautions

1. When tightening bolts and nuts, make sure the wrench is always placed solidly on the bolt head or nut. If the wrench slips off when you are tightening it, you could injure your knuckles and fingers.
2. Always wear safety glasses when tightening or loosening bolts and nuts, especially when using air tools.
3. When using pullers, remember that you are applying very high forces and pressures on the bolts, nuts, and equipment. Always be sure the pullers are securely fastened before beginning.
4. Make sure to immediately clean up any grease spilled on the floor so that you won't slip and injure yourself.
5. Tighten bolts and nuts only to their specified torque. If a bolt breaks, the wrench may slip and injure your hands.
6. When tightening cap and machine screws, always use a screwdriver that is in good condition so it won't slip off the head of the screw.
7. Make sure all tools used in the shop are clean, free of grease, and in good condition.
8. When removing or installing snap rings, always be sure the snap ring pliers have a solid hold on the snap ring. The snap ring may slip off and injure your face or eyes.
9. When placing gaskets on various engine parts, be careful not to cut your hand on the sharp edges of the sealing surface where the gasket is being placed.
10. When tightening bolts, always pull the wrench toward you rather than push it away from you.
11. Be careful not to inhale fumes or odors from chemical gaskets.

PROBLEM: Broken, Loose, or Stripped Bolts

Many types of bolts and nuts are used to fasten components together on the automobile engine. Problems often occur in which the bolts either break, become loose, or are stripped.

DIAGNOSIS

When automotive fasteners break, become loose, or are damaged, check for the following:

1. Make sure the damaged fastener is the correct one recommended by the manufacturer.
2. Make sure the bolt or nut was tightened to the correct torque specification.
3. Make sure you are using the correct size lock-washer.
4. Make sure the length of the bolt is correct. It may be too short for the component being secured or too long for the hole.
5. Look for signs of wear, vibration, or damage near and around the component being checked.

SERVICE

Various service tips are important when working with automotive fasteners. Some include:

1. Always use the standard bolt and nut torque specifications when tightening bolts and nuts.
2. When replacing bolts that are stripped or broken, always use a bolt that has the same hardness.
3. Never mix metric and USC threads.
4. Never use a USC wrench with a metric bolt or nut.
5. Always determine the torque of the bolt by the number of threads per inch or per millimeter and the shank size of the bolt. Never use the head size of the bolt to determine the torque specification.
6. Never use an NC bolt in place of an NF bolt of the same shank size.
7. Always use a lock washer on bolts that could vibrate loose. Make sure the lock washer is the correct size and type recommended by the manufacturer.
8. When removing parts that use a keyway, always store the key in a safe place.
9. Use a helicoil to replace damaged threads.
10. When a bolt breaks use the following procedure to remove the part of the bolt still in the hole:
    a. Soak the bolt in penetrating oil.
    b. Try to remove the broken bolt using a needle nose pliers or vise grip pliers. These tools can be used only if part of the bolt is sticking out of the hole.
    c. If the bolt is too tight or not enough of the bolt is sticking out of the hole, carefully use a small center punch and a hammer, tapping the bolt on the side, to

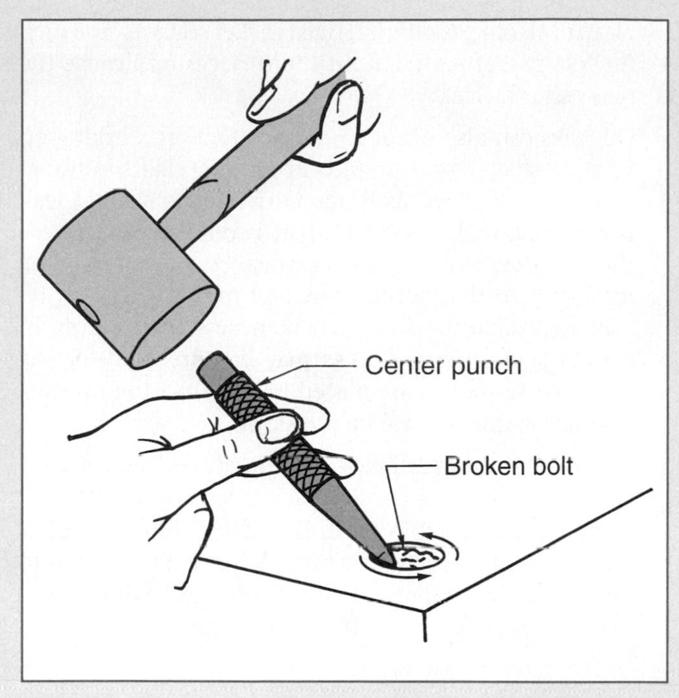

FIGURE 5–36 At times a broken bolt can be removed by using a center punch to loosen the bolt.

try to remove the bolt. ***Figure 5–36*** shows a broken bolt being removed by this method.

d. If the center punch cannot remove the bolt, use a screw extractor as shown in ***Figure 5–37***. First drill a hole in the center of the broken bolt. Carefully tap the correct size screw extractor into the hole. Using a socket and ratchet or equivalent wrench, unscrew the extractor and broken bolt as one piece. Now remove the screw extractor from the broken bolt, using vise grip pliers or other suitable gripping device.

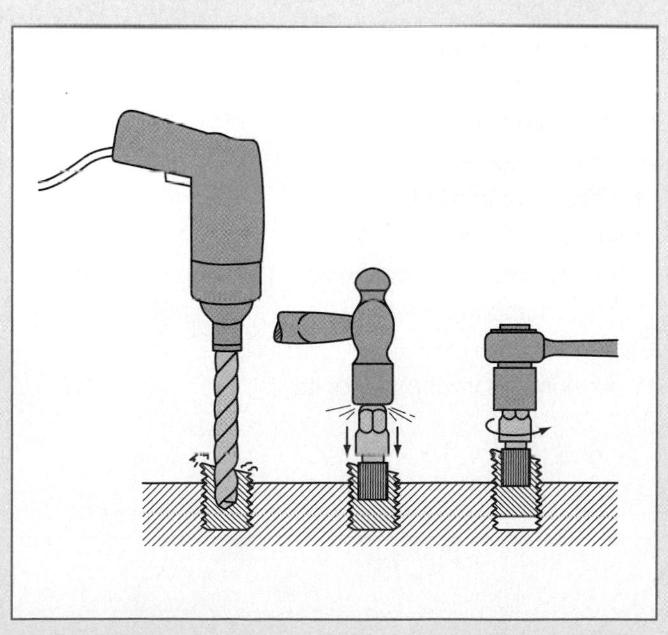

FIGURE 5–37 If a broken bolt cannot be removed easily, a screw extractor may be used.

e. As a last resort for real problem bolts, it is possible to weld a metal tag or nut to a broken bolt and then use a pliers or a wrench on the metal tag or nut to remove the bolt.

PROBLEM: Leaky Gaskets and Seals

Over the life of an engine, many of the gaskets and seals break, shrink, or become brittle. If this happens, oil or other fluids will likely leak from the engine. Components that most often produce leaks include valve covers, fuel pumps, water pumps, timing chain covers, and oil pans.

DIAGNOSIS

Many times it is difficult to locate the component or part that is leaking. To determine the location of the oil leak:

1. Check for spots where the oil has washed dirt away.
2. Check for gaskets that have become brittle.
3. Check for bolts that are loose, causing the gasket to lose its seal.
4. Remember that as the oil is leaking, it flows downward due to gravity and is blown backward by the air as the vehicle moves down the road.

SERVICE

When servicing an engine that has an oil leak or that needs new gaskets or seals, keep the following suggestions in mind:

1. When leaks occur near covers such as valve and timing chain covers, always check the covers for bent sides and flanges. If the flanges are bent slightly, an oil leak may occur that may not be sealed by the gasket. Bend or straighten the cover to eliminate the leak.
2. Sometimes when gaskets are being installed, they slip out of place. To help keep gaskets in place during installation, use a thick grease or gasket sealant. Some service technicians use a small amount of weather stripping as an adhesive to help keep the gasket in place. Also, the gasket can be tied in place using a very thin wire that can easily be removed. Another technique for keeping a gasket in place during installation is shown in ***Figure 5–38***. Use small wires or cut-down paper clips to hold the bolts in place. The gasket is held in place with the bolts. Once the bolts are threaded into the holes, the small clips are removed. The bolts can then be torqued to manufacturer's specifications.
3. Remember that new gaskets must seal on smooth surfaces. Always scrape the machined surfaces and flanges

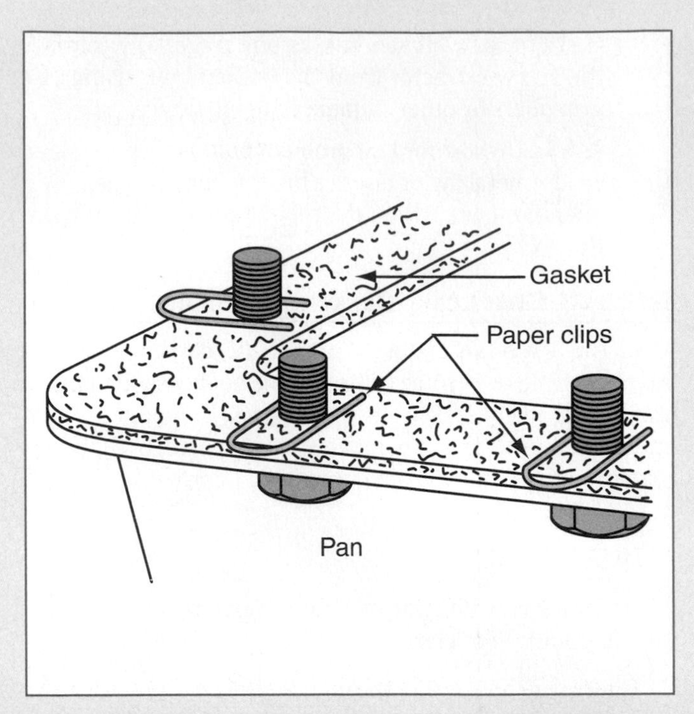

FIGURE 5-38 Gaskets can be held in place using small paper clips cut to hold the bolts in place. The oil pan, gasket, and bolts are shown here.

clean of all old gasket materials, gasket sealants, and dirt. Any old gasket parts left on the surfaces may cause the new gasket to leak.

4. Oil leaks can also occur from seals that are working on rotating shafts, for example, on rotating shafts in transmissions, rear drive shafts, and drive shafts. When a leak is observed, make sure the shaft is not damaged where the seal rides on the metal surface. At times, this area may wear so that even a new seal may not stop an oil leak. Replacing the shaft may be necessary. However, in certain applications, sleeves may be purchased for the shaft. These sleeves are placed on the shaft to provide a new surface for the seal lip to ride on.
5. Never use a compression-type gasket twice. Always use a new one to replace compression-type gaskets. When a compression gasket is used, the material in the gasket (usually a metal piece) is crushed during installation to form-fit to the application. If it is used twice, it may not seal as it was intended by the manufacturer, thus causing leakage.

Service Manual Connection

Fasteners, gaskets, and sealants are all mentioned in many of service manuals. However, most frequently mentioned are fasteners and their torque specifications. Although each size bolt and nut has a standard torque, manufacturers list specific torque readings in the service manuals. To identify the torque specifications for your engine, you will need to know the VIN (vehicle identification number) of the vehicle, the type and year of the vehicle, and the type of engine. Although they may be titled differently, some of the more common torque specifications (not all) found in service manuals are listed to the right. Today, most torque specifications are given in metric newton meters (N•m) or customary pound-feet (lb-ft). Note that these specifications are typical examples. Each vehicle and engine will have different specifications.

Application	N•m	lb-ft
Connecting rod nuts	57	42
Exhaust manifold to cylinder head bolts	34	25
Exhaust system pipes to manifold	34	25
Flywheel to converter	63	46
Flywheel to crankshaft	81	60
Front cover to cylinder block	47	35
Injection pump attaching bolts	34	25
Injection pump fuel inlet line	30	22
Oil pan bolts	14	10
Water pump to front cover bolts	18	13

SUMMARY

The following statements will help to summarize this chapter:

- Many types of fasteners, gaskets, and sealants are used on automobiles.
- Bolts are identified by the shank size, hardness, and number of threads per inch.
- Bolts can be identified as either course or fine thread.
- All bolts are torqued to specifications suggested by the manufacturer.
- Damaged threads can be repaired by using a helicoil, or by using a tap or die to clean up threads.
- Gaskets and seals are used on the automotive vehicle in such places as the engine, differentials, axles, and transmissions.
- The purpose of gaskets and seals is to seal in gases, oil, greases, and other fluids in the vehicle.
- Compression gaskets are used when high-pressure gases, such as those produced by combustion, must be contained.
- Both static and dynamic seals are used on automobiles.
- There are several service and diagnostic procedures used to correctly work with fasteners, gaskets, seals, and sealants.
- Always follow the manufacturers specifications and procedures when working with fasteners, gaskets, seals, and sealants.

TERMS TO KNOW

Can you explain each of the following terms? Review the chapter until you can use each term correctly.

Anaerobic sealant
Bolt hardness
Compression washers
Fasteners
Gaskets
Gasket sealant
Head
Helicoils
Keyways
Labyrinth seals
Metric threads
NC (National Coarse)
NF (National Fine)
O-ring
Prevailing torque nut
Property class
Rope seal
RTV sealant
Screw-pitch gauge
Seals
Sealants
Self-tapping screws
Setscrews
Shank
Snap rings
Splines
Standard bolt and nut torque specifications
Tensile strength
Threaded fastener
Threads per inch
Torque
Torque-to-yield bolts
UNC (Unified National Coarse)
UNF (Unified National Fine)
Wing nuts

REVIEW QUESTIONS

Multiple Choice

1. Which of the following are used to keep a gear on a shaft?
a. Washers
b. Snap rings
c. Gaskets
d. Sealants
e. Rivets

2. Which of the following is used to seal liquid flowing through two stationary parts?
a. Labyrinth seal
b. Dynamic seal
c. O-rings
d. All of the above
e. None of the above

3. In order to replace threads, a ________ is commonly used.
a. Helicoil
b. Stud
c. Snap ring
d. Sealant
e. Teflon tape

4. A gasket used to seal high-pressure gases is called a:
a. Compression gasket
b. Thrust gasket
c. Labyrinth gasket
d. Sealant
e. Rivet

5. Torque for bolts should always be determined by the:
   a. Socket size of the bolt
   b. Length of the bolt
   c. Shank diameter size of the bolt
   d. All of the above
   e. None of the above

6. The greater the number of lines on the bolt head, the:
   a. Softer the bolt
   b. More threads per inch on the bolt
   c. Harder the bolt
   d. Easier it is to turn
   e. Harder it is to turn

7. Which bolt has the greatest ability to hold two objects together?
   a. NC bolt
   b. NF bolt
   c. UNC bolt
   d. All of the above
   e. None of the above

8. The lip of a dynamic seal should always be pointed toward the:
   a. Teflon tape
   b. Fluid to be contained
   c. Stationary part
   d. Air
   e. None of the above

9. Which of the following allows two shafts to be locked together yet slide axially?
   a. Keyways
   b. Rivets
   c. Helicoils
   d. Bearings
   e. None of the above

10. A compression gasket should be used:
    a. Only once
    b. After the gasket has been compressed
    c. To seal oil and low pressures
    d. To seal air only
    e. None of the above

11. Which of the following are used to prevent motion between two rotating parts?
    a. Setscrews
    b. Self-tapping screws
    c. Machine screws
    d. Gaskets
    e. Seals

12. Which of the following are used to join light metal or plastic?
    a. Self-tapping screws
    b. Splines
    c. Setscrews
    d. Seals
    e. Compression gaskets

13. Helicoils are used to:
    a. Replace external threads
    b. Replace internal threads
    c. Lock two shafts together
    d. Seal surfaces together
    e. Tighten bolts correctly

14. A bolt identified as 3/8 × 16 × 1 1/2 means:
    a. 3/8-inch shank size
    b. 16 millimeters long
    c. 1 1/2-inch socket size
    d. All of the above
    e. None of the above

The following questions are similar in format to ASE (Automotive Service Excellence) test questions.

15. *Technician A* says that O-rings are used to form a compression gasket for extremely high pressures. *Technician B* says that O-rings are used to improve the accuracy of torquing a bolt. Who is correct?
    a. A only c. Both A and B
    b. B only d. Neither A nor B

16. A bolt is identified as 3/8 × 16 × 1 1/2. *Technician A* says the bolt has 16 threads per inch. *Technician B* says the bolt is 3/8 inch in shank diameter size. Who is correct?
    a. A only c. Both A and B
    b. B only d. Neither A nor B

17. *Technician A* says the tensile strength of a bolt is identified by the type of chrome on the bolt. *Technician B* says the tensile strength of a bolt is identified by the number of lines on the top of the bolt. Who is correct?
    a. A only c. Both A and B
    b. B only d. Neither A nor B

18. *Technician A* says that metric bolts have the same torque as English or customary bolts. *Technician B* says that metric bolts are not used at all on today's automobiles. Who is correct?
    a. A only c. Both A and B
    b. B only d. Neither A nor B

19. *Technician A* says that RTV sealants should be used in the presence of air, or oxygen. *Technician B* says that RTV sealants should be checked for use with oxygen sensors. Who is correct?
    a. A only c. Both A and B
    b. B only d. Neither A nor B

20. *Technician A* says that prevailing thread fasteners have an interference fit with the mating threads. *Technician B* says that prevailing thread fasteners use nylon inserted into the threads to prevent the fasteners from coming loose. Who is correct?
    a. A only c. Both A and B
    b. B only d. Neither A nor B

Essay

21. Describe how the torque is determined on a bolt.

22. Describe the difference between a spline and a snap ring.

23. What is the purpose of a helicoil?

24. In which direction should the lip of a seal be positioned? Why?

Short Answer

25. A nut that has an interference fit between the nut and bolt threads is called a ________ fastener.

26. Metric bolts are torqued in the unit called ________.

27. A form-in-place gasket is categorized as a ________ type of gasket.

28. When various bolts come into contact with oil or antifreeze, it is a good idea to use ________ to seal the threads.

Applied Academics

TITLE: English and Metric Fasteners

ACADEMIC SKILLS AREA: Language Arts

NATEF Automobile Technician Tasks:

Interpret charts, tables, or graphs to determine the manufacturer's specifications for system operation to identify out-of-tolerance systems and subsystems.

CONCEPT:

Vehicles today use a variety of fasteners, both English and metric. Often it is necessary to locate and use torque specifications for bolts and nuts in the metric system. The accompanying figure shows an example of torque specifications for different classes of metric bolts, in particular, the hexagon head bolt. To find the correct torque, first identify the mark and class of the metric bolt (4T, 5T, 6T, etc.). Then identify the shank diameter of the bolt. From this information the service technician can easily determine the torque in either metric or English units.

	Mark	Class
Hexagon head bolt	Bolt head No. 4—	4T
	5—	5T
	6—	6T
	7—	7T
	8—	8T
	9—	9T
	10—	10T
	11—	11T

Class	Diameter mm	Pitch mm	Specified torque					
			Hexagon head bolt			Hexagon flange bolt		
			N•m	kgf•cm	ft•lbf	N•m	kgf•cm	ft•lbf
4T	6	1	5	55	48 in.•lbf	6	60	52 in.•lbf
	8	1.25	12.5	130	9	14	145	10
	10	1.25	26	260	19	29	290	21
	12	1.25	47	480	35	53	540	39
	14	1.5	74	760	55	84	850	61
	16	1.5	115	1,150	83	—	—	—
5T	6	1	6.5	65	56 in.•lbf	7.5	75	65 in.•lbf
	8	1.25	15.5	160	12	17.5	175	13
	10	1.25	32	330	24	36	360	26
	12	1.25	59	600	43	65	670	48
	14	1.5	91	930	67	100	1,050	76
	16	1.5	140	1,400	101	—	—	—
6T	6	1	8	80	69 in.•lbf	9	90	78 in.•lbf
	8	1.25	19	195	14	21	210	15
	10	1.25	39	400	29	44	440	32
	12	1.25	71	730	53	80	810	59
	14	1.5	110	1,100	80	125	1,250	90
	16	1.5	170	1,750	127	—	—	—

APPLICATION:

From the information presented in the figure, can you determine the metric and English torque readings for a hexagon head metric bolt that has a 6 mark on the bolt head and has a shank diameter of 10 mm? What would happen if the service technician misread the information and used a higher torque reading on the bolt? What might happen if the torque reading was below the specification?

Bearings

CHAPTER 6

OBJECTIVES

After studying this chapter, you should be able to:

- Define the purpose of bearings used on automobiles.
- Identify the parts of a bearing used on the automobile, including the ball, roller, and race.
- Examine the different types of bearings such as the thrust bearing, friction bearing, and sealed bearing.
- Determine how to service common automobile bearings.

INTRODUCTION

Throughout the entire automobile, there are many types and styles of bearings. Bearings are a machine part used to reduce friction between moving surfaces. Bearings are also used to support moving loads. This chapter introduces many of the common types of bearings, explains how they operate, and discusses how to service bearings.

6.1 PURPOSE OF BEARINGS

Bearings are used in the automobile to reduce friction between moving parts and stationary parts. A secondary purpose is to remove the heat produced by unavoidable friction.

Bearings are also used to support moving loads. A good example of a moving load is that of a car moving down the road on the four wheels. The weight of the car is the moving load; this weight must be supported by the bearings on the four axles of the car.

Because there are so many applications for bearings, there are many types. However, most bearings can be classified into two types: **friction bearings** and **antifriction bearings**. Friction bearings have no rotating or moving parts, whereas antifriction bearings use small rollers or steel balls to reduce friction. The two types are shown in ***Figure 6–1***. With friction bearings, rotating or moving parts slide on the stationary part. Antifriction bearings contain balls or rollers to support the rotating part or moving load.

FRICTION BEARINGS

Friction bearings produce more heat and frictional losses (horsepower lost due to friction) in the engine than antifriction bearings. However, this type of bearing usually requires less maintenance and can be replaced easily. When friction bearings are used, the load of the moving part is supported by a layer of oil between the load and the stationary part. The oil molecules act like small ball bearings.

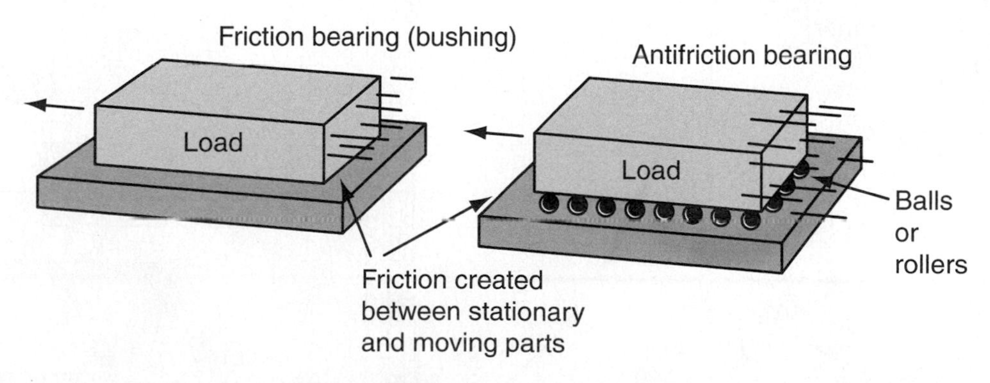

FIGURE 6–1 There are two types of bearings. Friction bearings use oil between the moving and stationary parts. Antifriction bearings have ball or rollers between the moving and stationary parts.

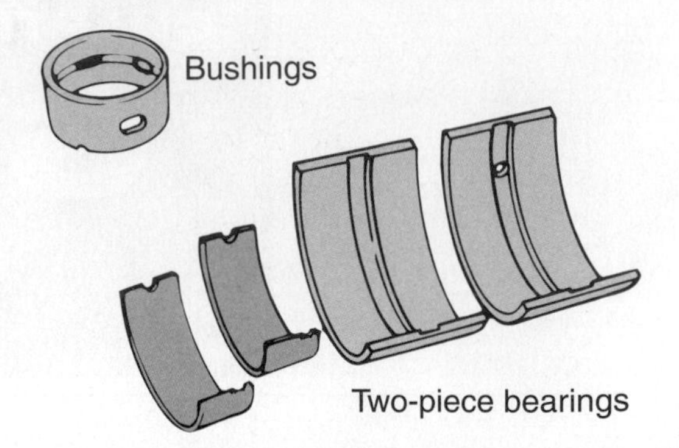

FIGURE 6-2 Bushings are designed as one piece, while bearings are designed with two pieces. (Courtesy of Federal–Mogul Corporation)

There are two types of friction bearings (***Figure 6–2***). One type is a two-piece bearing that is used on such applications as the crankshaft on most gasoline engines. The second type is a one-piece bearing. It is commonly known as a **bushing**.

Bushings are held in place by pressing them into the block or stationary part. In an automobile, bushings are used on the camshaft, some generator and alternator shafts, the starter armature shaft, the transmission, and the distributor shaft.

ANTIFRICTION BEARINGS

Antifriction bearings are also used throughout the automobile in many applications. ***Figure 6–3*** shows the parts of the common ball bearing, which is a type of antifriction bearing. The race is the surface the balls roll on. In operation the balls roll between the inner and outer races to produce minimum friction.

Antifriction bearings rely on rolling friction for operation. This type of friction offers less resistance to rotation, resulting in less frictional loss.

6.2 BEARING LOADS

RADIAL LOADS

Bearings must be able to support various types of loads. Load on a bearing is created by pressure pushing in a certain direction against the bearing. For example, think of a bearing that would be used to allow an axle to turn on one of the rear tires of a car. In this type of application, the weight of the car is pushing down against the bearing. This is called the load of the bearing.

There are several types of loads on a bearing. One type is called the radial load. The word *radial* means moving toward or away from the center of the radius of a circle. In this case, the circle is the shaft of the bearing; so a radial load is one that pushes toward the center of a shaft. ***Figure 6–4*** shows an example of a radial load. In this case, an antifriction bearing is used to support the car's rear axle inside the stationary part. The stationary part is the axle housing, which is attached to the car frame. Note that the load of the car is pushing toward the center of the axle or shaft. The load is perpendicular to the centerline of the rotating shaft. Some common examples of radial loads include axle bearings, water pump bearings, alternator bearings, and air-conditioning pump bearings.

THRUST LOADS

Another type of load on a bearing is called a **thrust load.** A thrust load, also called an **axial load**, is a pressure pushing parallel to the centerline of the rotating shaft. ***Figure 6–5*** shows an example of a thrust load. In this case, the load is parallel to the centerline of the rotating shaft. Note that the bearing is shaped slightly different to allow the thrust load to be absorbed by the bearing. One of the most common applications of **thrust bearings** is on the front axles of a car. As the car goes into a turn, there is a thrust load placed on the

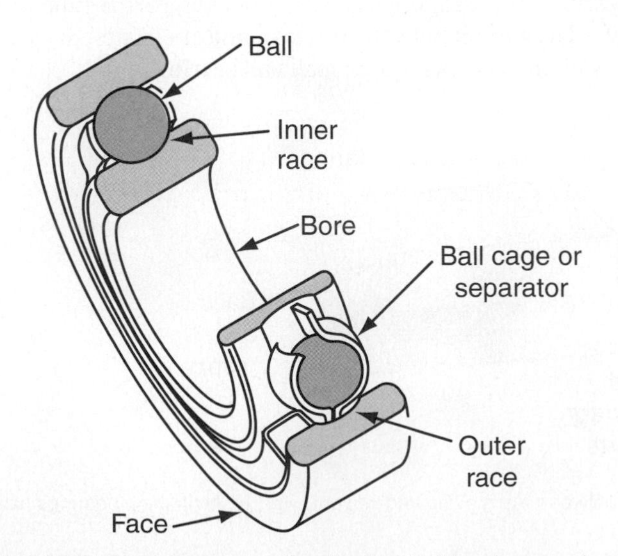

FIGURE 6-3 The parts of a standard ball bearing are shown here.

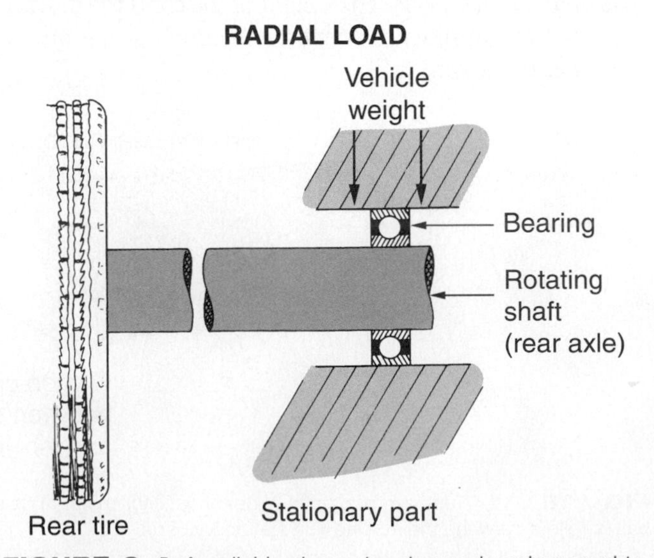

FIGURE 6-4 A radial load on a bearing pushes downward toward the center of the axle.

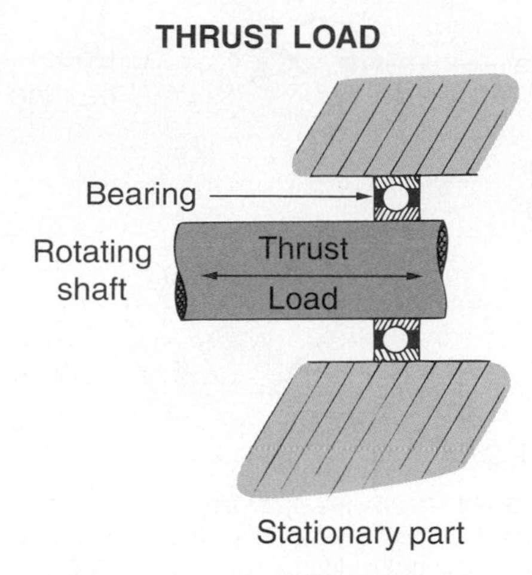

FIGURE 6-5 A thrust load on a bearing is a pressure pushing parallel to the centerline of the rotating shaft.

front wheel bearings of the car. Another common application for a thrust bearing is on a crankshaft. During operation, the crankshaft moves axially back and forth. A thrust bearing (bushing) is located in such a manner as to absorb the crankshaft end thrust.

6.3 BEARING DESIGNS

FRICTION BEARING DESIGNS

Figure 6–6 shows examples of two kinds of friction bearings. These types of bearings can also be called precision insert bearings. They can come in either one circular piece, called the full round, or two separate pieces, called a split. ***Figure 6–7*** shows how these bearings are designed.

Friction bearings have a strong back or shell. This outer layer is generally made of steel. A thin intermediate layer of copper and lead is added on top of the steel back. The inner layer is then placed on top of this material. The inner layer is an antifriction material often called babbitt. **Babbitt** is a very soft material. It is composed mostly of tin and lead. Occasionally silver or aluminum alloy is used. These metals are used because they are very soft. The soft material helps the inner surface of the bearing to take on the shape of the shaft rotating inside it. This material may be only 0.001 inch thick.

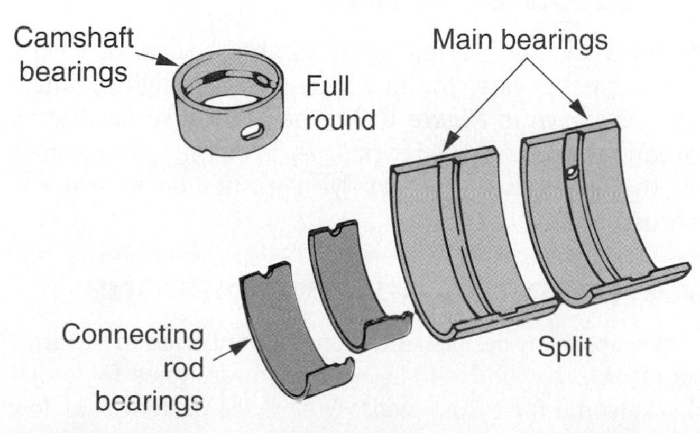

FIGURE 6-6 These friction bearings can be either full, round, or split.

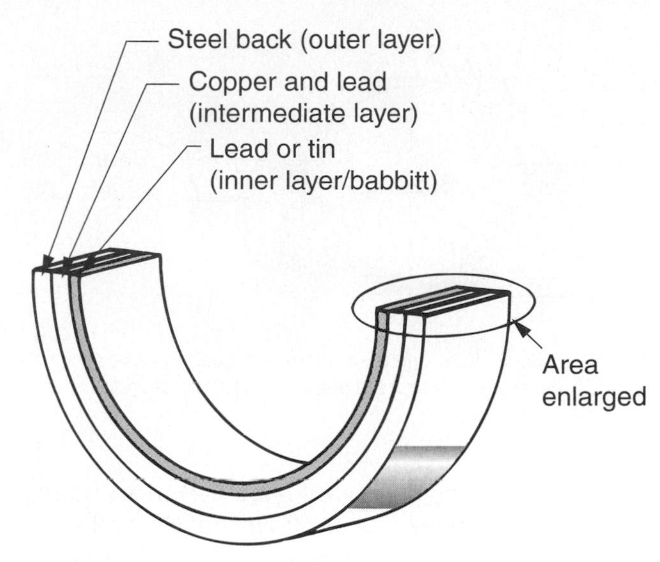

FIGURE 6-7 Friction bearings are made of many materials. The more common materials are shown here.

BEARING SPREAD

Today, many friction bearings and bushings are made with a bearing spread. Bearing spread is shown in ***Figure 6–8***. Note that the top of the bearing half is slightly larger than the diameter of the bore. During installation the bearing is inserted (snapped in place) by placing a slight pressure on the top of the bearing half. The reason for having bearing spread is to keep the bearing in place in the bore during assembly of the engine. If a one-piece bearing or bushing is being installed, remember that its total outside diameter will be slightly larger than the bore. Therefore, bushings or full-round friction bearings must be pressed into the bore using special tools and bearing installers.

BEARING CRUSH

In order to make sure bearings are installed tightly, bearing crush may be designed into the bearing. Bearing crush means that each half of a split bearing is slightly greater

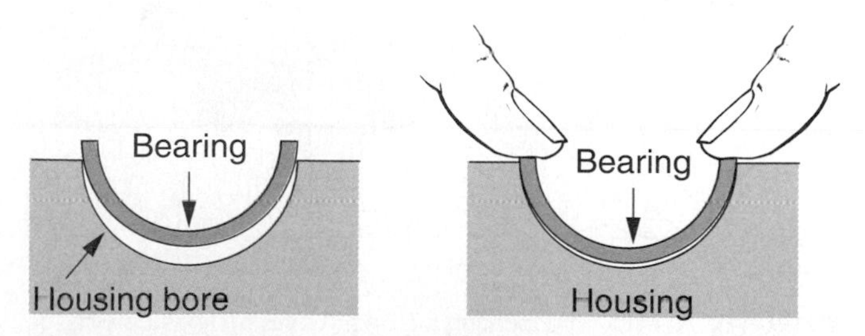

FIGURE 6-8 Bearing spread helps to keep the bearing in place during installation and assembly.

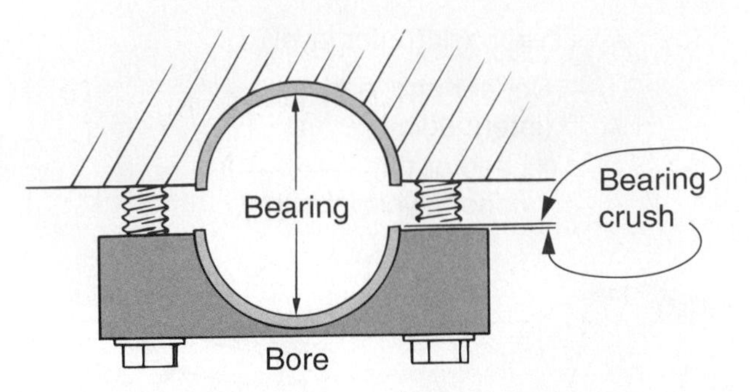

FIGURE 6-9 Bearing crush forces the bearing halves into the bore to produce good contact between the bore and bearings.

than half the bearing diameter. ***Figure 6–9*** shows bearing crush. Note that there is a small amount of the bearing sticking up above the bore on each half. This may be as little as 0.001 inch. When the top half of the bearing is tightened down, the bearing halves are compressed together and the bearing crush forces the bearing tightly into the bore.

BEARING LOCATING LUGS

During operation, when a shaft is spinning inside a bearing, a certain amount of friction is produced. So, there is always a chance that a bearing may also spin. To prevent bearings from spinning in the bearing bore, a small locating lug is designed into the bearing. This locating lug (***Figure 6–10***) fits into a slot in the bearing bore. When installed correctly, the locating lug and slot will prevent the bearing halves from spinning inside of the bore under heavy loads.

BEARING OIL GROOVES AND HOLES

All bearings must be lubricated properly. This is also true of friction bearings. On some friction bearings, there are grooves cut into the bearing surface. These grooves index or line up with an oil hole in the bearing bore. Oil pressure forces lubricating oil through the hole and into the center of the bearing. ***Figure 6–11*** shows a two-piece friction bearing. Note the oil hole on the top half which indexes with the oil groove in the center of each bearing half.

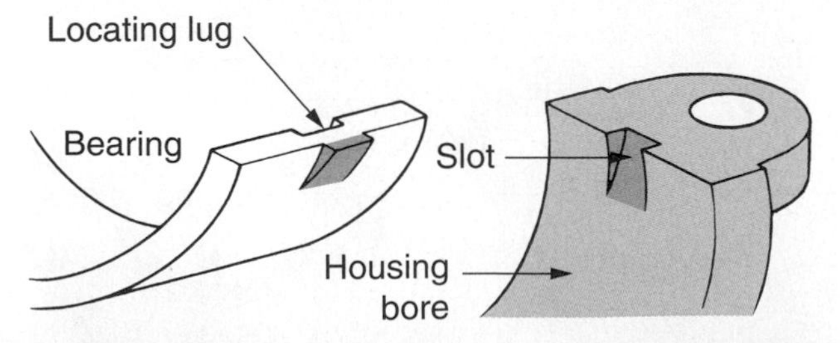

FIGURE 6-10 The locating lug on the bearing lines up with the slot in the housing bore to keep the bearing from spinning during operation.

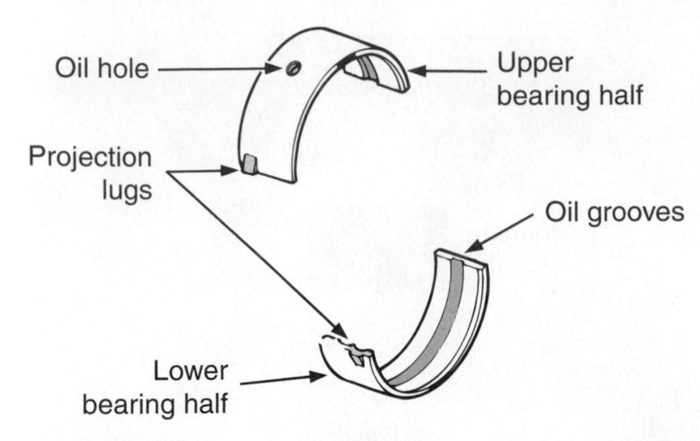

FIGURE 6-11 The oil hole in the upper bearing half lines up with the oil grooves in the inside center of the bearing halves. (Courtesy of American Honda Motor Co., Inc.)

Car Clinic: LOCATING LUGS

CUSTOMER CONCERN:
The Crankshaft Locks Ups during Bearing Installation

During main bearing installation, the service technician noticed that after the second main bearing was tightened down, the crankshaft would no longer turn. Once the bearing cap was loosened up again, the crankshaft turned. What could be the problem?

SOLUTION: Each of the main bearings has a locating lug on the bearing. This lug must be aligned with the slot in the bore. If the locating lug is placed on the opposite side of the bore, when the main bearings are tightened down, the locating lug will be forced into the main bearing journal and cause the crankshaft to lock up. During installation, it is critical that the locating lugs be placed in the slots, otherwise damage to the crankshaft main journals will result.

THRUST SURFACES ON FRICTION BEARINGS

Often a friction bearing needs to absorb axial or thrust loads. In this case, the bearing is shaped slightly different. As shown in ***Figure 6–12***, the bearing surfaces wrap around the sides of the bearing. Then, as the rotating shaft in the center is moved axially, it pushes up against the thrust surfaces.

ANTIFRICTION BEARING DESIGNS

There are many designs and shapes for antifriction bearings used on automobiles today. Some are designed for radial loads, some for thrust loads. Others are designed for low costs, while yet others are designed for higher speed. Some antifriction bearings use rollers, while others use balls.

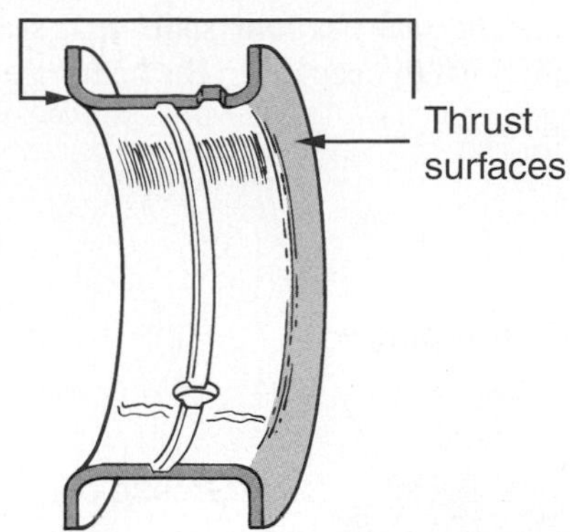

FIGURE 6-12 Thrust surfaces on some friction bearings are used to absorb axial or thrust loads.

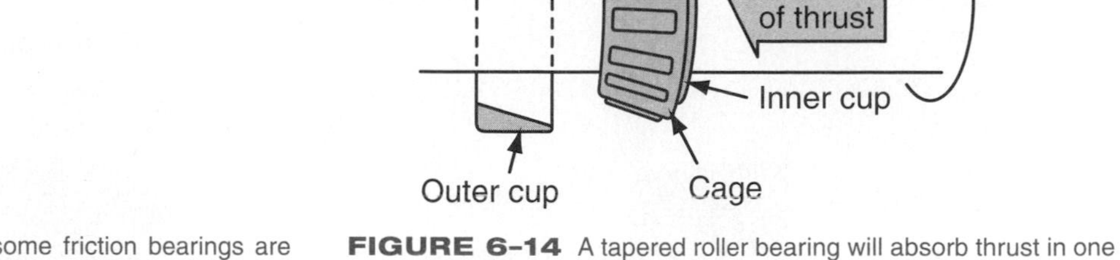

FIGURE 6-14 A tapered roller bearing will absorb thrust in one direction.

Figure 6–13 shows many of the antifriction bearings currently used in the automobile industry.

THRUST AND ANTIFRICTION BEARINGS

Thrust must also be absorbed by antifriction bearings. One common way to absorb thrust is to use tapered roller bearings. *Figure 6–14* shows an example of a tapered roller bearing. In this case, the thrust or axial load is moving toward the left. As the load is applied, the tapered rollers push up against the outer cup absorbing the thrust.

SEALED BEARINGS

At times bearings are used in applications in which they must seal as well as reduce friction. Sealed bearings not only reduce friction but also keep dust, dirt, or other debris

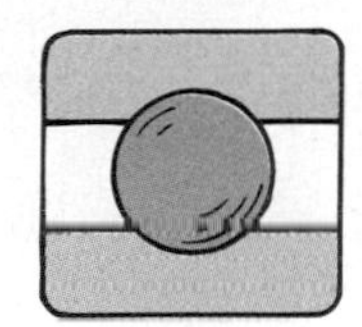

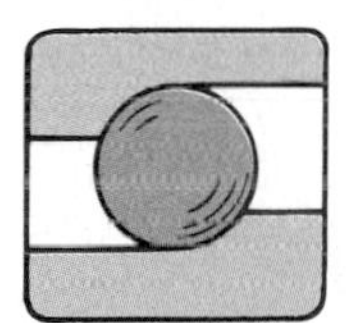

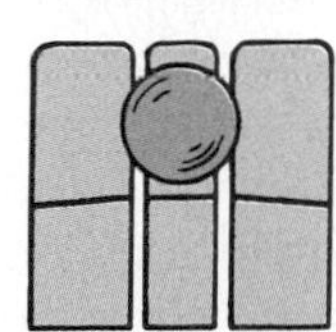

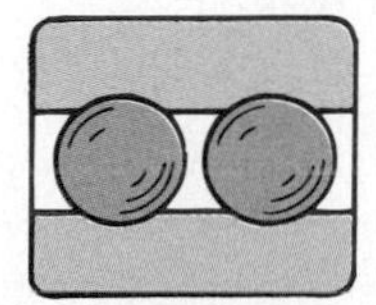

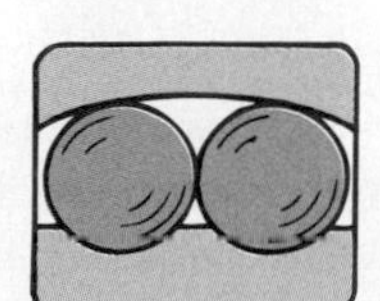

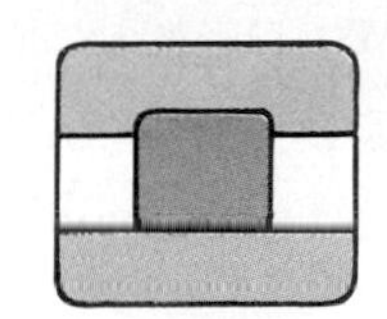

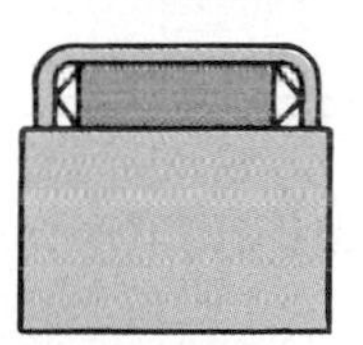

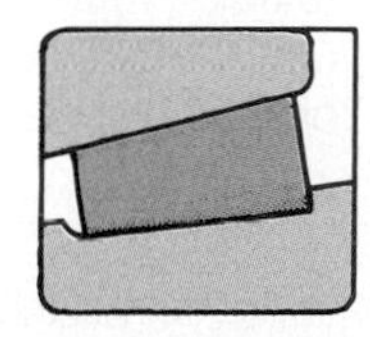

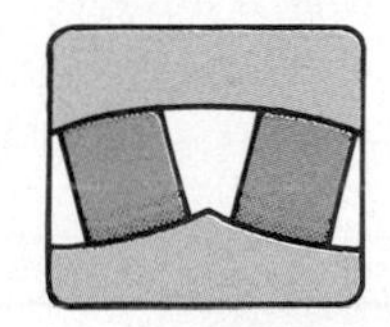

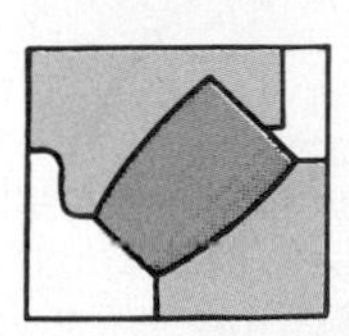

FIGURE 6-13 Both ball and roller bearings are considered antifriction bearings. Many styles and types are used in today's automobiles.

away from the internal parts of the bearing. ***Figure 6–15*** shows a type of sealed bearing in which the seal is housed within the bearing. A plate on the outside of the bearing keeps out dirt. The seal has four spots that seal. The seal also uses a labyrinth to keep oil in the bearing and dirt out of the bearing.

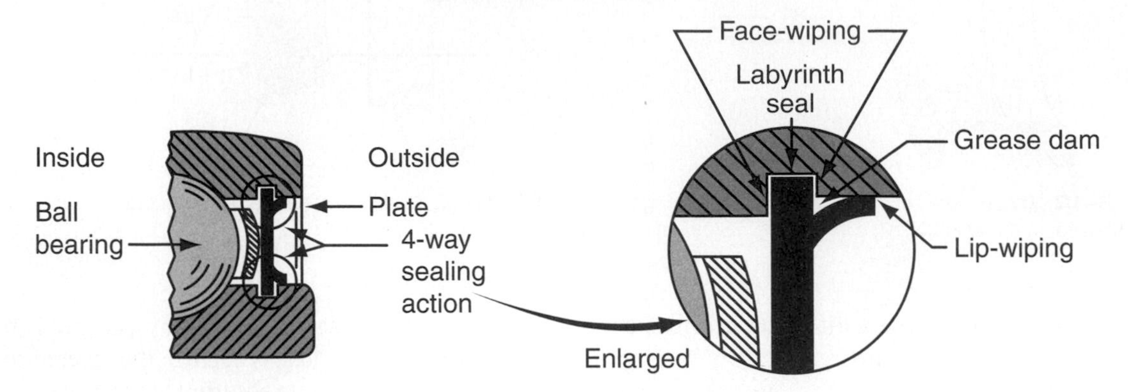

FIGURE 6–15 Seals are used in bearings to keep dirt and water out and lubrication in. In this case, a labyrinth seal design is shown here.

Problems, Diagnosis, and Service

Safety Precautions

1. When replacing bearings, always use the same type and number of bearings that were removed.
2. Always keep dirt away from bearings.
3. Never use an air gun with high pressure to dry off a roller bearing or ball bearing. The roller or ball bearings may fly out under the high speed and injure you.
4. Always use the correct bearing puller or bearing installer when working with bearings. If the correct tool is not used, the bearing may either be damaged or the bearing (friction type) may be installed incorrectly. For example, friction bearings may be installed crooked or misaligned.
5. During installation or removal of bearings, high pressures are often used. Therefore always wear safety glasses when installing or removing bearings.
6. During installation of camshaft bearings, be careful not to pinch your fingers between the bearing and the bore. Also remember that the surfaces and edges of the bearing bores are very sharp.
7. Always use the correct torque when tightening bolts and covers that hold bearings in place.

PROBLEM: Water Pump Bearings

A customer complains that there is a loud, high-pitched sound coming from the front of the engine. The sound occurred suddenly, but now continues all the time.

DIAGNOSIS

The bearings that are used in the water pump are double row ball bearings. ***Figure 6–16*** shows an example of a water pump bearing. The bearing is sealed so that moisture or dirt

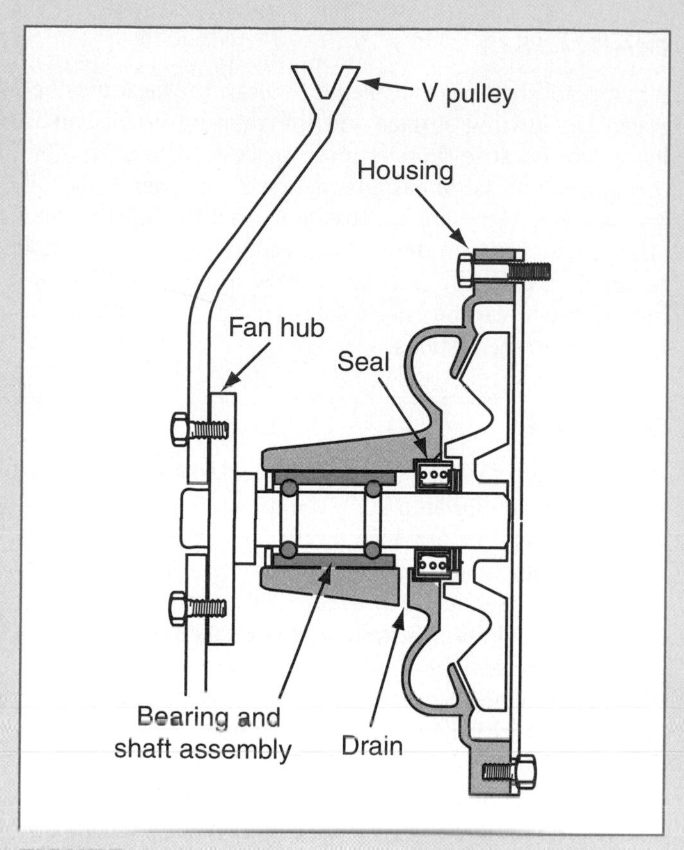

FIGURE 6–16 When evaluating the condition of a water pump bearing, check for minor frictional irregularities as the V pulley is rotated. Also look for side movement on the V pulley.

particles cannot enter the bearing and cause wear. Water pump bearings can be damaged by adjusting the water pump belt too tightly. Tight belt adjustment puts excess loads on the water pump ball bearings and causes them to wear excessively. Damage can also occur when dirt particles and water get behind the bearing seal.

To check if the bearing is worn, first remove the drive belt on the water pump. Then slowly rotate the water pump shaft observing if there is any irregularity in the rotation of the water pump shaft. If there is irregularity or resistance while rotating the shaft, the bearings are bad and the water pump should be replaced.

When a water pump bearing is making loud, high-pitched sounds, the bearing is often severely damaged and the water pump shaft will move easily from side to side. Severely damaged bearings can be checked by grasping the fan or the pulley attached to the main shaft of the water pump, then moving it from side to side. If there is any movement, the water pump bearings are damaged. The water pump should then be replaced.

SERVICE

Following the manufacturer's recommended procedure, remove and replace the water pump by first removing the belts and fan (if necessary) that connect to the water pump. Remove the bolts that hold the water pump to the engine block. Note the location of the gaskets so that the gaskets can be removed, gasket surfaces cleaned, and new gaskets installed. After the water pump is removed, replace it with a new or rebuilt water pump. Note that old water pumps are normally returned to the parts department for rebuilding.

PROBLEM: Bad Bearings

Many types of roller and ball bearings are used in the automobile. For example, the front and rear wheels use bearings. These bearings will often begin to make a soft rumbling noise when they get bad.

DIAGNOSIS

If a bearing is thought to be bad, several checks can be made. These include:

1. Remove all weight and load from the bearing in question.
2. Slowly rotate the wheel or other component that is supported by the bearing.
3. Check to see if there is a rough feeling, coarse sound, or slight vibration as the component is rotated on the bearing. If these conditions are found, the bearing is most likely damaged. If the bearing is good, there should be absolutely no roughness, coarse sound, or any vibration as the component is turned on the rotating shaft.
4. Most bearings are protected from dirt by various types of seals. Seals generally harden with age. Under normal conditions, however, seals should not harden and should remain flexible. A seal can harden because of excessive temperatures near it. Always investigate if you suspect that excessive temperature caused a seal to harden.

SERVICE

Service on bearings can be of several types. If the bearing is determined to be bad or damaged, it must be replaced. Always check the condition of the seals that protect the bearings as well. Make sure the replacement bearing and seal are exact replacements. Other service checks on roller and ball bearings and seals include:

1. Roller and ball bearings should be repacked with grease at regular intervals. Repacking of bearings usually includes the following procedure:
   a. Clean the bearing thoroughly with solvent.
   b. Dry the bearing completely, but never let the bearing sit dry as it can easily rust.
   c. Immediately after drying, put grease into the bearing by using a bearing grease-packing tool. This tool helps to force the grease into the inside of the bearing for correct lubrication. If a bearing grease-packing tool is

not available, place a sufficient amount of grease in one hand. With the bearing in the other hand, force the grease into the bearing. Make sure the grease has completely surrounded the balls or rollers and comes out the outer side.

d. Repeat the previous step on the opposite side of the bearing.

2. **Brinnelling**, small dents in the bearing race, can be caused by improper bearing installation, such as impacts on the outer race. Brinnelling normally causes severe bearing damage.
3. Contamination of bearings usually shows up as small scratches, pitting, and scoring along the raceway. To prevent contamination, keep all parts of the bearing free of dirt, dust, and other particles at all times.
4. Misalignment is another cause of bearing failure. When the balls or rollers of a bearing are running from one side of the race to the other, one race may be misaligned. Misalignment causes uneven load distribution, which causes excess friction and heat buildup. In this case, the bearing was probably installed incorrectly. To prevent misalignment, always follow the manufacturer's recommended procedure for installation.
5. When installing bearings or bushings, always keep the parts well lubricated during assembly and keep them free of all dirt.
6. Some bearings may have the inside race pressed onto the center shaft. If this is the case, always use a hydraulic press and follow the manufacturer's recommendations when removing and installing the bearings.
7. When installing seals, make sure the seal is not cocked or installed at an angle. A cocked seal may leak oil readily and cause dirt to get into the bearings. When a seal has been cocked or replaced incorrectly, it must be removed and replaced with a new seal. Seal installation must be done with the proper tools to avoid damaging the metal that supports the seal.
8. Always place a seal on a rotating shaft so the lip of the seal is toward the fluid being contained or toward the dirt that is being held back.

PROBLEM: Low Engine Oil Pressure

A customer complains about the engine oil pressure dropping over the past year. The vehicle is an older car that has not had the oil changed for many miles.

DIAGNOSIS

When crankshaft bearings begin to wear, the clearance between the bearing surface and the rotating crankshaft is increased. Because there is more space for the oil to flow through, oil pressure has a tendency to be lower. If the oil pressure is lower, there is a possibility that the bearings and other components in the engine need lubrication, may not be able to handle the load, and may wear even more. Bearing clearance can be checked according to the manufacturer's recommendations.

SERVICE

It is important for the service technician to check the wear both on the crankshaft and on the bearings. In most cases, if there is low oil pressure, both the crankshaft journal as well as the bearing have wear. If the wear is out of specifications, the service technician will need to install new bearings. In addition, it may be necessary for the service technician to have the crankshaft journal ground down to a specified dimension so that undersized bearings can be installed. They are called undersized bearings because they fit around the diameter of an undersized crankshaft journal or one that has been machined down. Refer to ***Figure 6–17***.

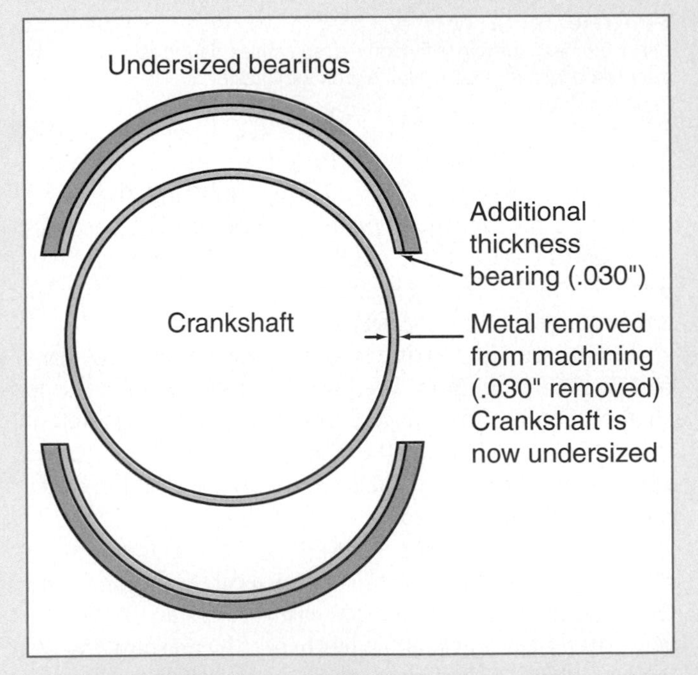

FIGURE 6–17 When a crankshaft has been machined down and its diameter reduced, new undersized bearings are necessary for proper operation.

Service Manual Connection

Bearings are mentioned in many of the service manuals. Among the most common references are the torque specifications associated with bearings throughout the automobile. The following are typical examples of various torque specifications related to bearings.

Application	Torque ft-lb
Camshaft bearing cap bolts	115
Center support bearing	43
Connecting rod bearing cap nuts	22
Main bearing bolt caps	44
Wheel bearing locknut	90
Control arm rear bushing retaining nut	101

SUMMARY

The following statements will help to summarize this chapter:

- Bearings are used to reduce friction between moving and stationary parts.
- There are two types of bearings: friction and antifriction.
- Friction bearings have oil between the rotating shaft and the stationary part.
- There are two types of friction bearings. The one-piece bearing, such as that used on a camshaft, is also called a bushing. Two-piece bearings are used on applications such as the crankshaft of an engine.
- Antifriction bearings use rollers or steel balls between the stationary and rotating shaft to reduce friction.
- Some antifriction bearings are used to absorb end thrust on a rotating shaft.
- Certain antifriction bearings are used to absorb axial loads, such as the weight of a vehicle on a wheel bearing.
- Depending on the application, some bearings are sealed, some have two rows of ball bearings, and some have rollers to absorb thrust.
- There are many service and diagnostic procedures for working with bearings. Typically, the service technician should always be aware of rumbling noises, side movement of a rotating shaft, brinnelling, seal damage, or lack of oil on the bearings.
- Always follow the manufacturer's recommended procedure when servicing bearings.

TERMS TO KNOW

Can you explain each of the following terms? Review the chapter until you can use each term correctly.

Antifriction bearings
Axial load
Babbitt
Bearings
Brinnelling
Bushing
Friction bearings
Thrust bearings
Thrust load

REVIEW QUESTIONS

Multiple Choice

1. Another word for axial load is __________.
a. Force
b. Thrust
c. Bushing
d. End forces
e. Expansion force

2. Which type of bearing uses a set of rollers or balls between the moving and stationary parts?
a. Antifriction
b. Friction
c. Bushing
d. Compression
e. None of the above

3. Which type of bearing produces the most heat and frictional losses in the engine?
   a. Antifriction
   b. Friction
   c. Radial
   d. Roller
   e. Ball bearing

4. Which type of bearing is used on camshafts and crankshafts?
   a. Antifriction bearings
   b. Bushings
   c. Thrust roller bearings
   d. Sealed ball bearings
   e. None of the above

5. Friction bearings are held in place and prevented from spinning with which of the following?
   a. Location lugs
   b. Oil
   c. Bearing crushers
   d. Bearing installers
   e. Torque bolts

6. When a load placed on a bearing is perpendicular to the centerline of the rotating shaft, it is called a (an) ________ load.
   a. Thrust
   b. Axial
   c. Radial
   d. Equal
   e. Parallel

The following questions are similar in format to ASE (Automotive Service Excellence) test questions.

7. *Technician A* says that bearings do not need to be installed with special installers. *Technician B* says that bearings do not need lubrication. Who is correct?
   a. A only
   b. B only
   c. Both A and B
   d. Neither A nor B

8. *Technician A* says that bearing spread keeps the bearing half in place during installation. *Technician B* says that sealed bearings never need to be replaced. Who is correct?
   a. A only
   b. B only
   c. Both A and B
   d. Neither A nor B

9. *Technician A* says that bearings must be designed to accept radial loads. *Technician B* says that thrust loads are parallel to the centerline of the rotating shaft. Who is correct?
   a. A only
   b. B only
   c. Both A and B
   d. Neither A nor B

10. *Technician A* says that antifriction bearings use rollers. *Technician B* says that antifriction bearings use balls. Who is correct?
   a. A only
   b. B only
   c. Both A and B
   d. Neither A nor B

Essay

11. Describe radial loads and compare them to thrust loads.

12. What is the difference between a friction bearing and an antifriction bearing?

13. Define bearing crush and tell why it is important.

14. State where a sealed bearing is normally used.

15. Why are there oil grooves in some friction bearings?

Short Answer

16. A bushing is another name for a ________ bearing.

17. Friction bearings use a soft material on the surface of the bearing called ________.

18. Bearings are used to reduce ________ between two moving parts.

19. Engine camshafts typically will use a ________ friction bearing.

20. When half of a split bearing is slightly greater than half the bearing diameter, it is referred to as bearing ________.

Applied Academics

TITLE: Lubrication of Bearings

ACADEMIC SKILLS AREA: Science

NATEF Automobile Technician Tasks:

Explain to the customer the need for lubrication of adjacent parts to minimize the friction that results from movement at the junction of the parts.

CONCEPT:

Many types of bearings are used in a variety of applications in the automobile. One such bearing is called a spherical bearing. A spherical bearing does not use rollers or balls to separate the moving part from the stationary part. A spherical bearing is designed by machining a metallic sleeve around a spherical ball. This is shown in the accompanying figure. A special material is then injected into the cavity between the two parts. This material, often a polymer resin, has a tendency to bond itself to the sleeve. Together, the sleeve, the resin, and the spherical ball produce an excellent bearing surface. Spherical bearings are often used on applications that produce a significant amount of misalignment between the inner and outer race.

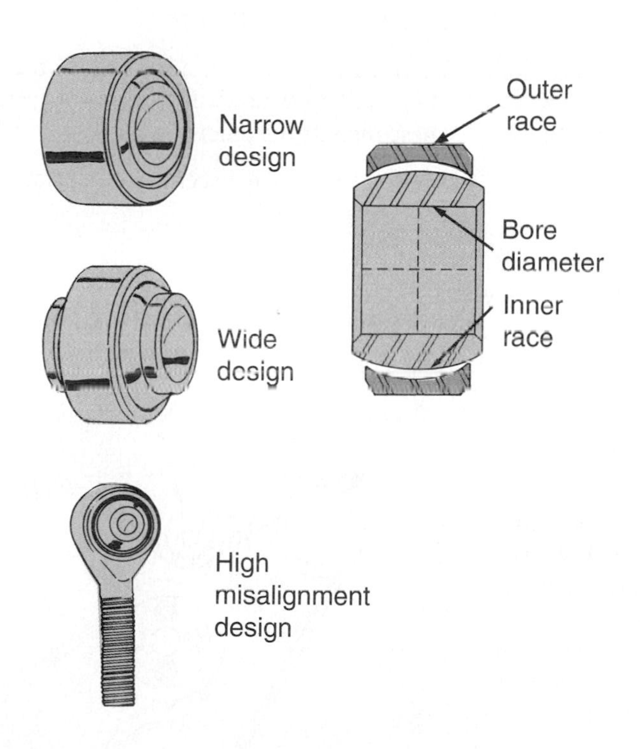

APPLICATION:

Can you think of any applications in which spherical bearings are used on the automobile? What is the difference between a spherical bearing and a ball or roller bearing? Could a spherical bearing be used on a heavy-duty, severe, or dirty application? What is the main advantage of using a spherical bearing?

CHAPTER 7 Automotive Belts, Fittings, and Hoses

OBJECTIVES

After studying this chapter, you should be able to:

- Define the different types of belts used on the automobile.
- Describe how to service automotive belts.
- Identify the various types of fittings used on the automobile.
- Explain how to check and service damaged fittings.
- Identify the various forms of hoses and tubes used on the automobile.
- Describe basic service procedures on automotive hoses and tubes.

INTRODUCTION

On the automobile, especially on the engine, there are many types of belts, fittings, and hoses. Belts are used to connect two rotating shafts together. Various types of fittings are also used to connect tubes and hoses together. This chapter introduces you to the basic types of belts, fittings, and hoses used on today's automobiles. In addition, service information is included that shows how to check and change belts, fittings, hoses, and tubing.

7.1 AUTOMOTIVE BELTS

PURPOSE OF AUTOMOTIVE BELTS

Automotive belts have been used for years to connect one rotating shaft to another. As technology improves, stronger, more flexible belts were developed for new uses. The basic purpose of using an automotive belt is to connect two shafts that are not necessarily next to each other. For example, if two shafts are next to each other and one needs to be driven from the other, gears could be used. But when the shafts are farther apart, belts will be used. In addition, belts are quiet as compared to gears or chains. They are also much less expensive.

The use of automotive belts gives flexibility to where rotating shafts can be placed. On today's automotive engine, belts are used to turn alternators, power steering pumps, water pumps, air-conditioning compressors, and various pollution equipment. ***Figure 7–1*** shows an example of common components that can be driven by belts on the front of an automotive engine.

TYPES OF BELTS

There are several types of belts used on automotive engines. The exact type used depends on several considerations. These include the following:

- The distance between the two rotating shafts
- The speed of the belts
- The load on the belts

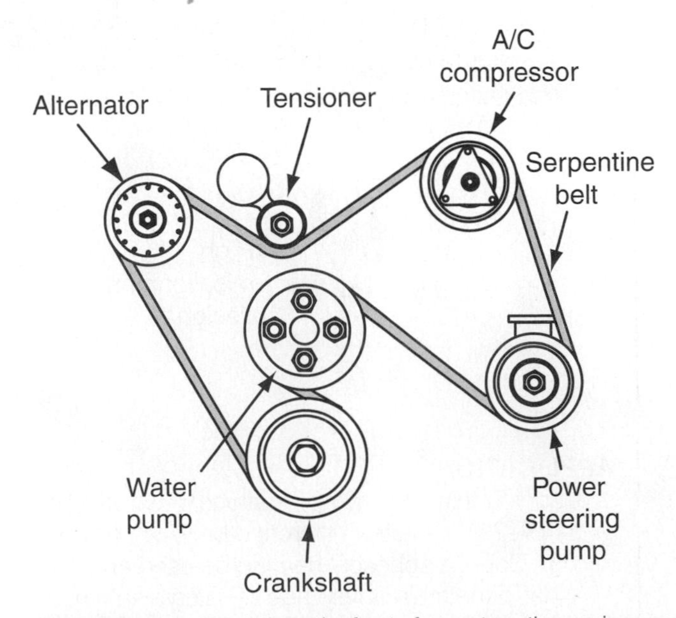

FIGURE 7–1 The belt on the front of an automotive engine can be used to drive many components and accessories.

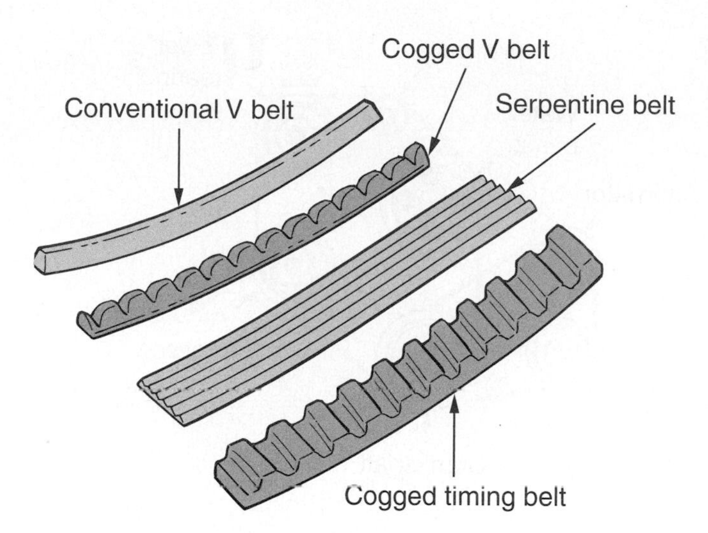

FIGURE 7–2 Various types of belts are used on automotive engines.

- The need for timing or keeping the two shafts in time with each other
- The direction of rotation between the two shafts
- The costs of the application
- The length of service required

Common belts used on the automobile are shown in ***Figure 7–2***. The conventional V belt is used to drive alternators on some cars. The alternator needs to be driven by the crankshaft of the engine. To do this a simple V belt may be used. Note that the V belt will not keep the two rotating shafts timed together. There may be slippage between the two. The V belt is a low-cost and effective way to connect two rotating shafts together.

The cogged V belt has a series of ribs cut into it. These ribs have a tendency to reduce slippage between two rotating shafts during heavier loads. The **serpentine belt** is common today on engines that require several rotating shafts to be turned by a single belt. The serpentine belt is so named because it can be snaked around several rotating shafts. The serpentine belt has several (usually four to seven) angled grooves in the belt surface. The number of grooves is determined by the width and load on the belt. These grooves increase the surface area of the belt so slippage is reduced. In addition, the serpentine belt is wide enough so that pulleys can run on either side of the belt.

BELT SYSTEMS

Each car manufacturer may use a different belt system. ***Figure 7–3*** shows a one-, two-, three-, and four-belt system. Depending on the shafts that must be rotated, a serpentine belt, V belt, or other belt may be used. In each case, the crankshaft is the "drive" component, while the other components are the "driven" components. Note that similar diagrams can usually be found in the service manual or under the hood of the car to help the service technician service belts and pulleys.

Car Clinic: SERPENTINE BELT

CUSTOMER CONCERN:
Alternator Light Comes On

A driver says that while driving home from work, the alternator light came on suddenly. Being only a few blocks from home, the driver continued home and shut the car off. What could be the problem?

SOLUTION: Often the alternator belt is driven from the crankshaft. Sometimes this belt is very hard to see. It is a serpentine-type belt and, if not checked for cracks and other damage, may easily break. The most likely cause of this problem is that the alternator belt broke. Replacing the alternator belt with a new one will easily fix the problem. In this particular case the air-conditioning compressor belt will also have to be removed because it rides on the outer end of the crankshaft pulley. The alternator belt rides on the center of the crankshaft pulley.

BELT CONSTRUCTION

Various materials are used to ensure that automotive belts are strong. All belts are constructed of several materials including rubber, fabric, and rubber-impregnated fabric (rubber and fabric mixed together). Also, some belts are strengthened by using reinforcing fabric or steel cords inside the belt. These materials help to reduce stretch and slippage under heavier loads such as air compressors. ***Figure 7–4*** shows the internal makeup of a typical V belt. Note that there are fiber or metal reinforcement sections, rubber-impregnated sections, and laminated fabric sections.

TIMING BELTS

In addition to the belts just mentioned, many automotive engines also use a **cogged timing belt**. The purpose of the cogged timing belt is to keep the camshaft exactly in time with the crankshaft. This is important so that the valves keep in time with the position of the piston during normal operation. ***Figure 7–5*** shows one example of a timing belt in place on the engine. The cogs in the belt act like a chain, keeping the two rotating shafts from slipping out of time with each other.

SERVICING BELTS

There are many important things to know about servicing belts. For example, when ordering V belts it is very important to replace an existing damaged belt with the correct belt. V belts typically have a certain width and a certain

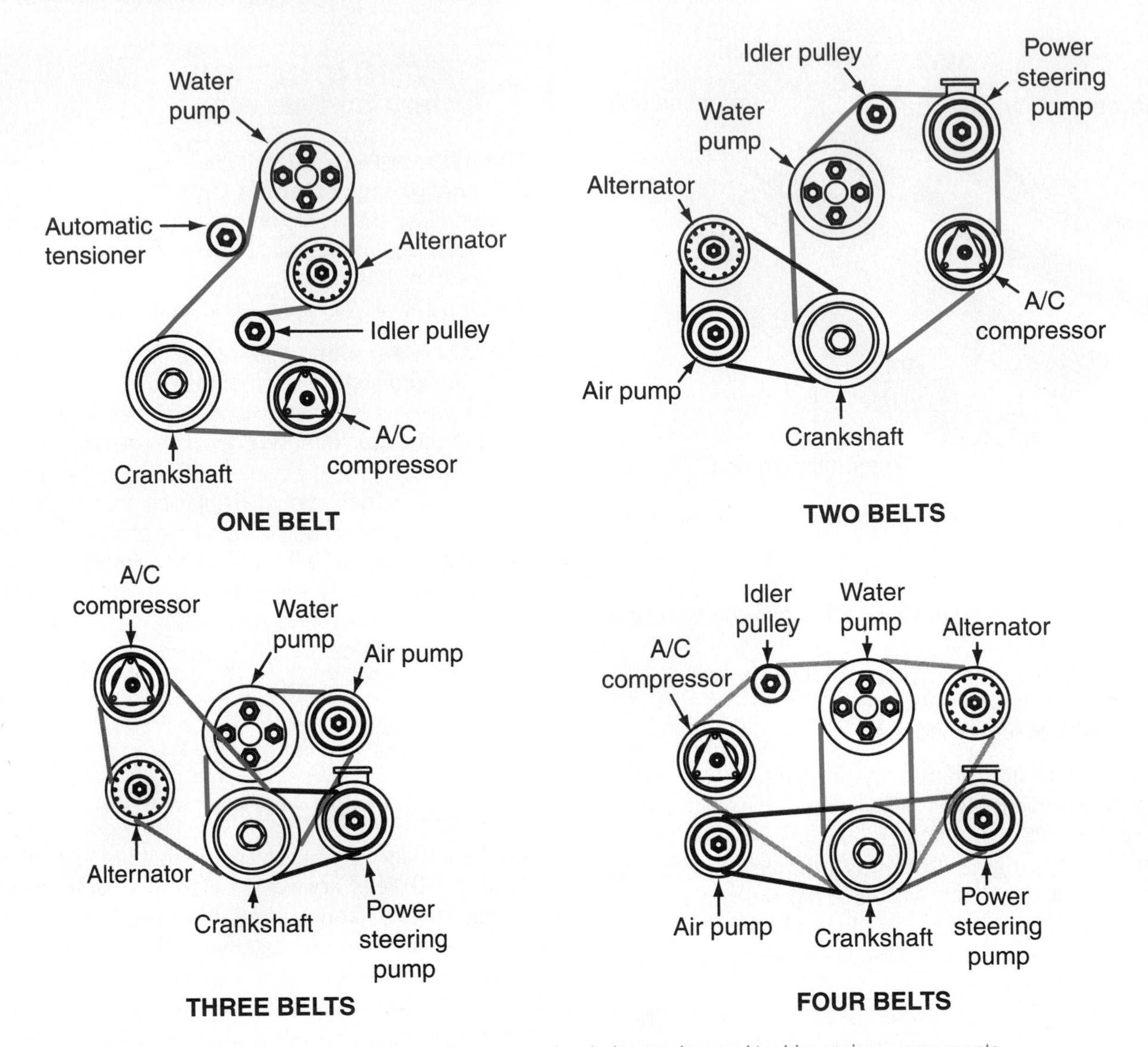

FIGURE 7-3 Depending on the car manufacturer, from one to four belts may be used to drive various components.

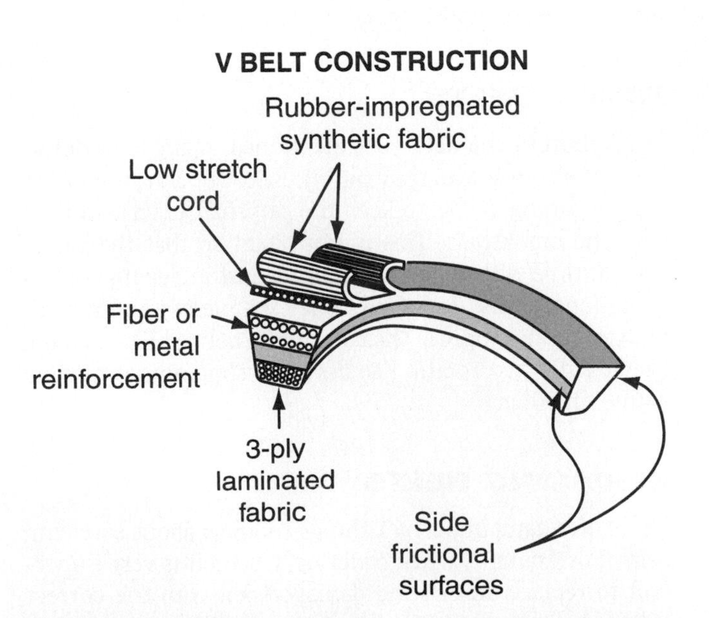

FIGURE 7-4 Belts are constructed with rubber, fabric, and metal cords for strength.

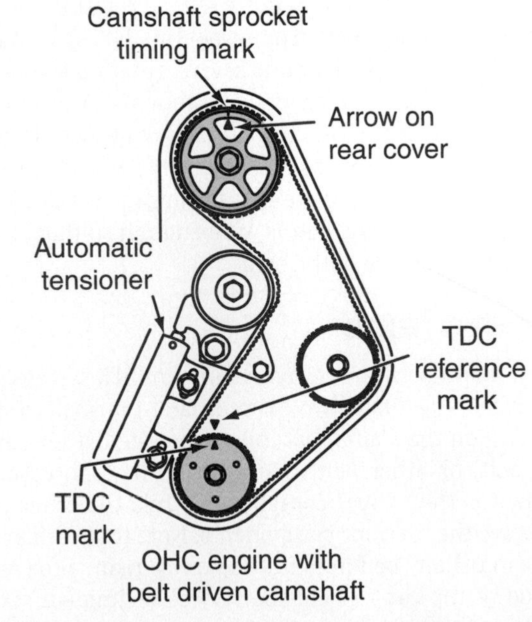

FIGURE 7-5 This cogged timing belt keeps the camshaft and crankshaft in time with each other.

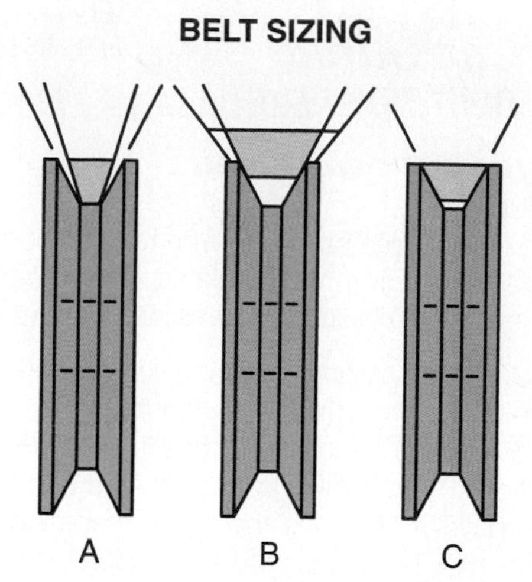

FIGURE 7–6 Always make sure the replacement belt is the correct size recommended by the manufacturer.

pitch or angle of the V. ***Figure 7–6*** shows three examples of V belts fitting into a pulley. Note that the belt is too narrow in diagram A, too wide in diagram B, and the correct size in diagram C.

A V belt may be identified as A, B, C, or D. These letters indicate the width of the belt. A is the narrowest; D is the widest. A belt may also be identified by its length. For example, A35 means a narrow belt that is 35 inches in length. In addition, vehicle manufacturers always provide part numbers for specific belt applications.

When installing serpentine belts always make sure the belt is placed completely on the pulley. It is very easy to have the belt slightly off the pulley as shown in ***Figure 7–7***. In the top drawing, the belt is installed correctly. In the middle drawing, the belt is installed too far in toward the engine. In the bottom drawing, the belt is installed too far out, away from the engine.

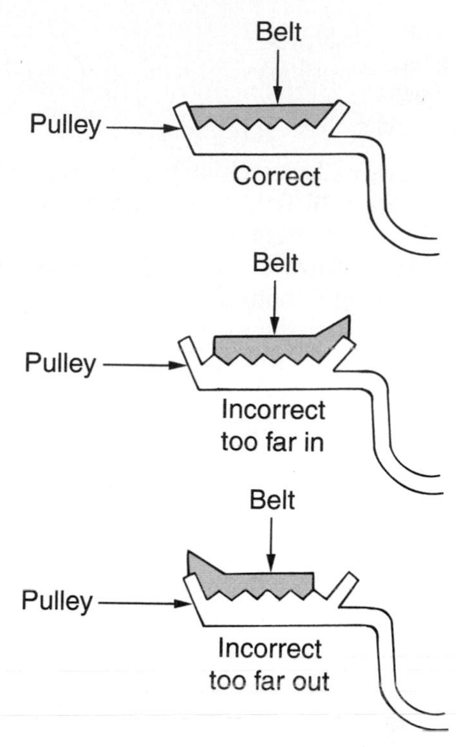

FIGURE 7–7 When installing a serpentine belt, be sure to install it correctly as shown.

It is always important to check the condition of belts on the automobile engine. ***Figure 7–8*** shows typical damage to look for on automobile belts. Problems to look for include the following:

- Cracks in the belt
- Oil soaked into the belt
- Glazing caused by excessive slippage
- Tears or splits
- Peeling

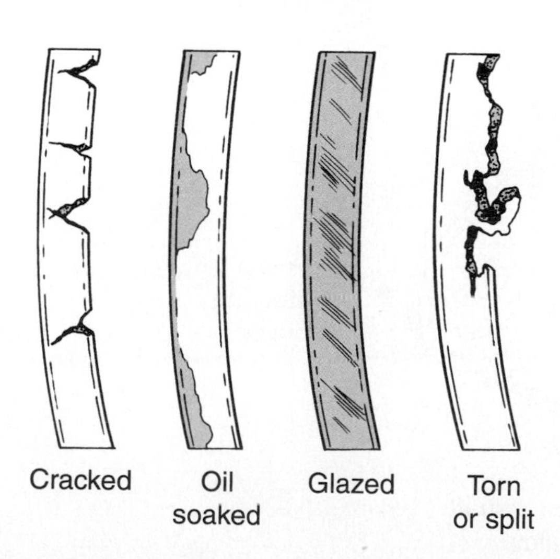

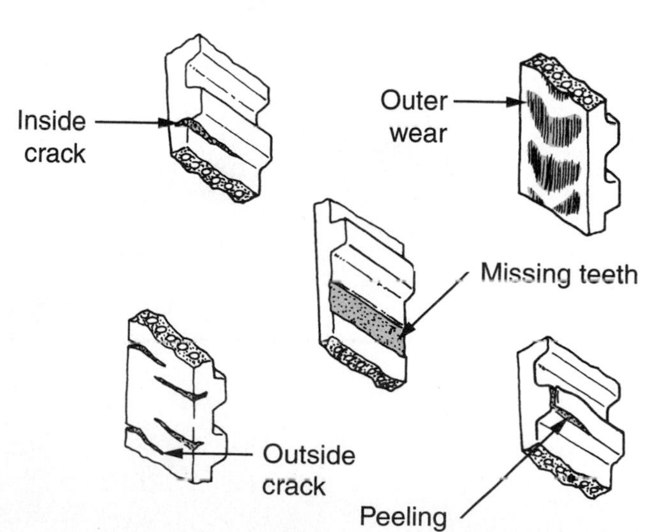

FIGURE 7–8 Always check belts for cracks, peeling, oil, tears, missing teeth, and so on.

- Outer wear
- Missing teeth
- Inside cracks
- Outside cracks

If there are any visible cracks in the belt, it should be replaced immediately. In fact, most manufacturers recommend that if one belt needs replacement, all other belts should be replaced. Often damage to the belt can be better observed if it is held in both hands and then twisted. The twisting helps to expose some of the possible problems with a belt.

ADJUSTING BELTS

Some belts may have a tension device that keeps them at the correct tension. There are many designs of belt-tension devices. In most cases, these tensioners work against the belt under spring pressure. The spring-loaded belt tensioner keeps a constant pressure on the belt during its operation. It is most commonly used on serpentine belts.

Other belts, however, need to be adjusted. It is very important for belts to be adjusted correctly. If a belt is too loose, it is likely to slip. This causes excessive wear on the sides of the belt and operating problems with the driven component. On the other hand, if a belt is too tight, it may damage bearings that are in the alternator, air compressor, water pump, and so on. When a belt is too tight, it puts pressure on the bearings in these components. This causes the bearings to fail prematurely.

Car Clinic: BELT TENSION

CUSTOMER CONCERN:
Belt Squeal
A driver in a Toyota says that when the engine is accelerated, there is a noticeable squeal from the engine area. What could be a possible problem?

SOLUTION: When the engine is accelerated, this puts more load on the belts. If a belt has become loose over time, it may begin to slip only during acceleration. The squeal is caused by the belt slipping on the pulleys. If this is the case, readjustment of the belts may be necessary. Also check the belts for cracks, oil, peeling, and other damage. The problem will most likely be solved by correctly adjusting the tension on all of the belts. Remember that the average life span of a belt is about 5 years or 60,000 miles, depending on the exact application, age of belt, and so on.

Belt tension can be checked several ways. ***Figure 7–9*** shows one simple way to check the tension of a typical belt. Find the longest span in the belt while it is installed on the engine. Then push down as shown in the center of the belt. There should be about 1/2 inch of movement for each foot

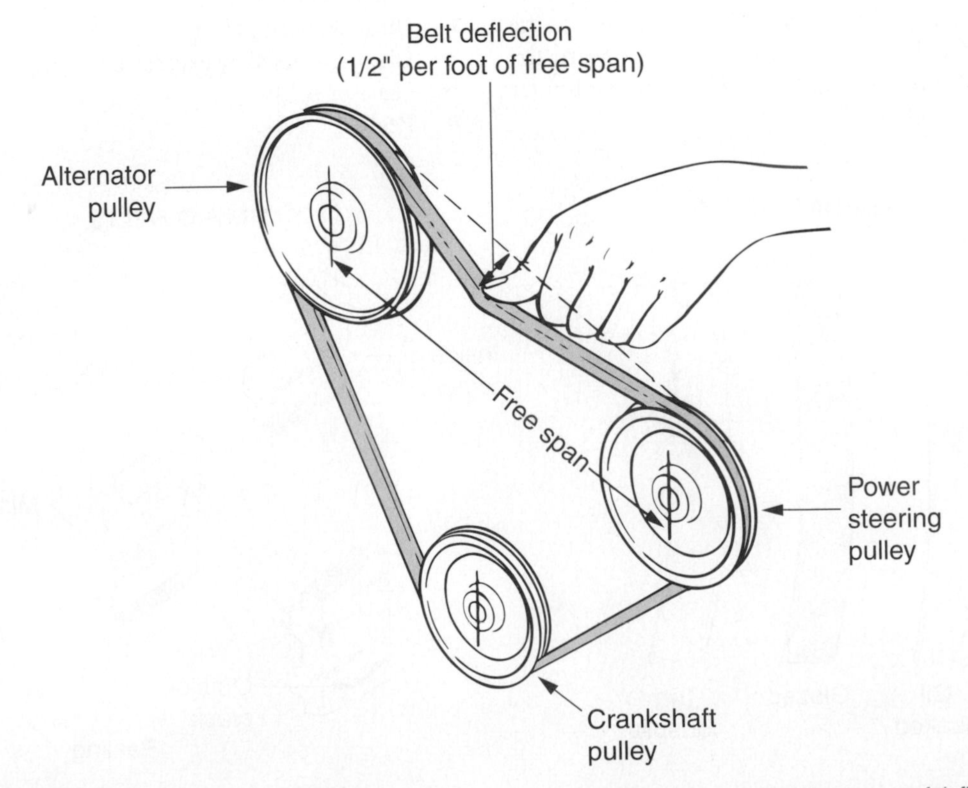

FIGURE 7–9 Belt tension can be checked by pushing in the center of the free span and measuring the amount of deflection.

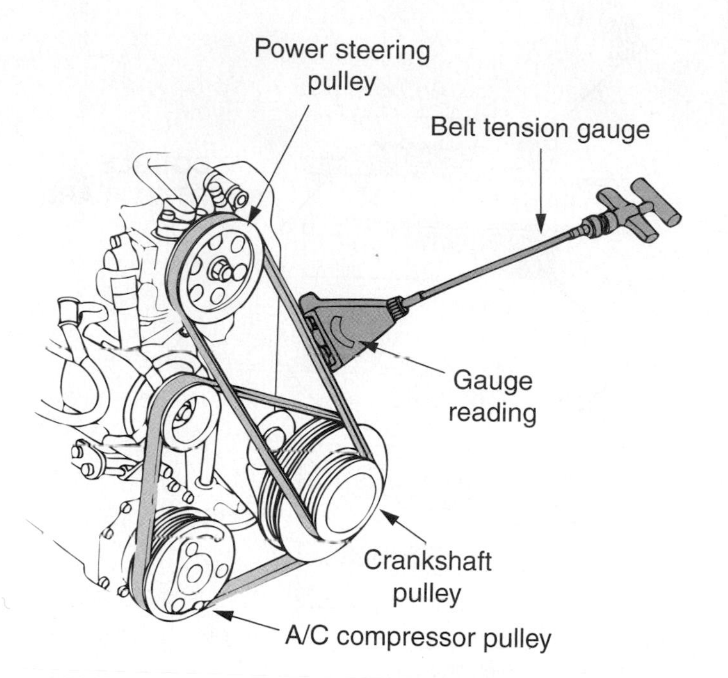

FIGURE 7–10 Belt tension can be checked with a belt tension gauge. (Courtesy of American Honda Motor Co., Inc.)

of free span on the belt. This is just a guideline for adjusting belts. Belt tension adjustment will vary depending on the belt, its type, and the manufacturer. Before adjusting any belt, always refer to the manufacturer's specification in the appropriate service manual.

Belt tension can also be checked using a commercial **belt tension gauge**. ***Figure 7–10*** shows a typical gauge being used to test the belt tension on the power steering belt. When the belt tension tester is placed over the belt, an inside spring pulls the belt inward, causing it to flex. The amount of flex is a measure of its tension and is read directly on the belt tension gauge. When using a belt tension gauge, always compare the gauge readings to the manufacturer's specifications.

7.2 FITTINGS

PURPOSE OF AUTOMOTIVE FITTINGS

As the automobile is carefully analyzed, it is easy to see its numerous **fittings**. Fittings are used to connect many of the cooling, lubrication, fuel, exhaust, brake, power steering, and emission control lines together. There are many types of fluids used in the automobile. Fittings are used to keep oil, coolant, exhaust gases, power steering fluid, brake fluid, refrigerant in air-conditioning systems, and gasoline in the fuel system from leaking. Some fittings are designed for very low pressures. Some fittings are designed to handle very high pressures. For example, a typical brake system has to have fittings that handle very high pressure. On the other hand, cooling systems use lower pressures and so require different types of fittings. ***Figure 7–11*** shows typical fittings used on a typical hydraulic master cylinder brake system.

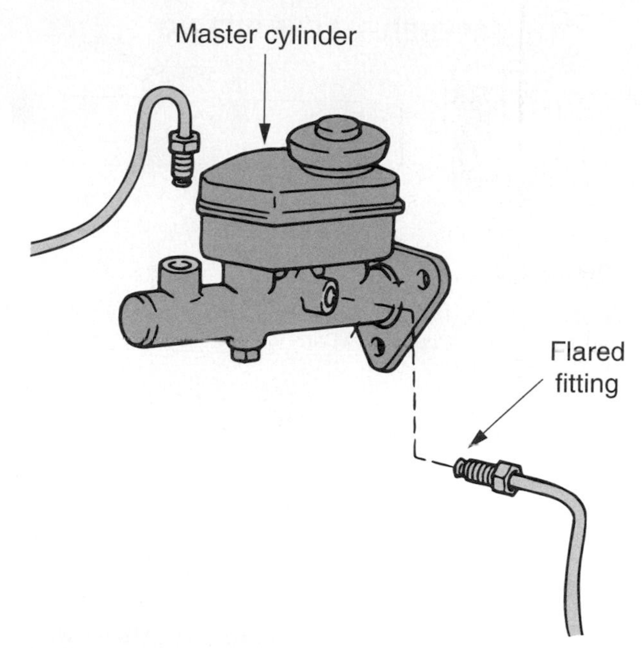

FIGURE 7–11 Flared fittings are used on the master cylinder to seal the brake fluid in the brake lines. (Courtesy of Toyota Motor Sales, USA, Inc. Reprinted with permission.)

TYPES OF FITTINGS

There are many types of fittings used on the automobile. The exact type is determined by several variables, including the following:

- The pressure on the system
- The type of fluid being contained
- The location of the fluid
- The amount of fluid flowing through the fitting
- The amount of vibration and mechanical movement

In most cases, the ends of fittings are identified as being either "male" or "female." The male end of a fitting is the part that fits into the female end. To avoid premature failure or leakage, use only the recommended type and size of fitting for any particular application. ***Figure 7–12*** shows a typical brass compression fitting. A compression fitting is one that seals as the fitting is tightened. The round cone in the center is squeezed between the fitting and the nut when

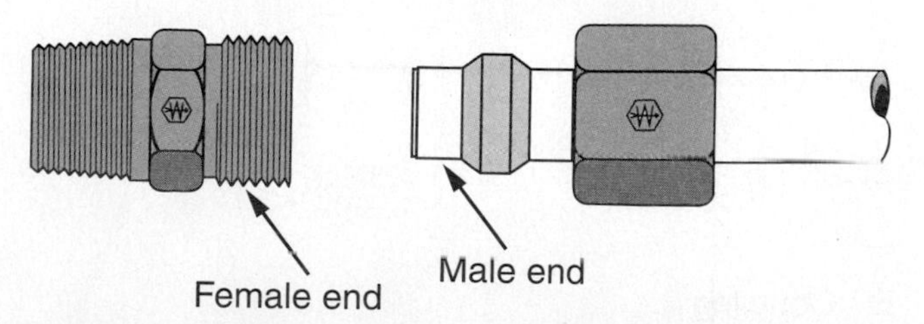

FIGURE 7–12 This compression fitting seals as the fitting is tightened down. (Courtesy of Dana Corporation)

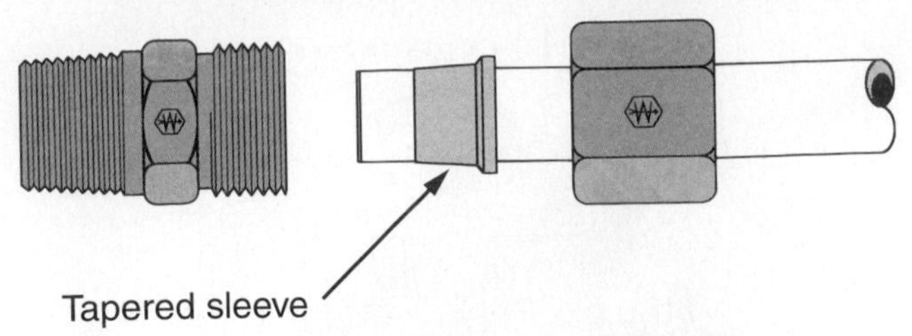

FIGURE 7–13 This tapered sleeve is used to seal the fitting as the nut is tightened. (Courtesy of Dana Corporation)

tightened down. ***Figure 7–13*** shows a similar type of fitting. This one has a tapered sleeve that fits into the female end of the fitting. As the nut is tightened, the tapered sleeve is squeezed into the female end to seal in the fluid.

Other types of brass fittings create a seal by using flares on the end of the steel tubing. ***Figure 7–14*** shows what is called an inverted flare fitting. Note the **flare** on the male end. This flare is pressed into the female end. The nut and the female end must be specially designed to fit the flare. A flaring tool is used to make the flare on the end of the tubing. Also shown is a standard flare fitting. In this case, the flare is fitted directly to the outside of the right side of the coupling.

One of the more popular uses of flare fittings is on brake systems. The fittings used on a typical brake system use flare fittings to help seal in the brake fluid under very high pressures. These flare fittings must be very strong. ***Figure 7–15*** shows two types of flare fittings made to withstand

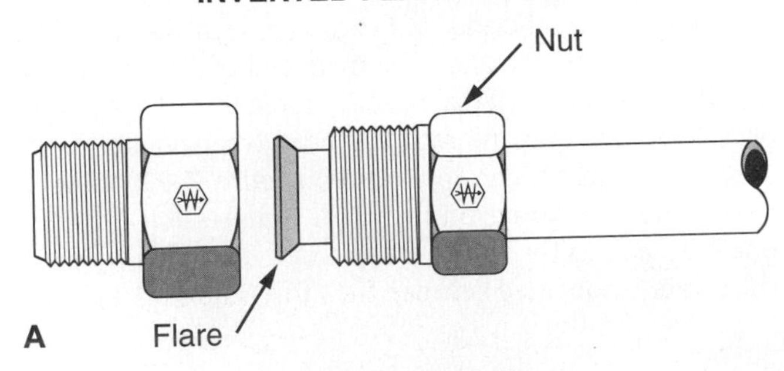

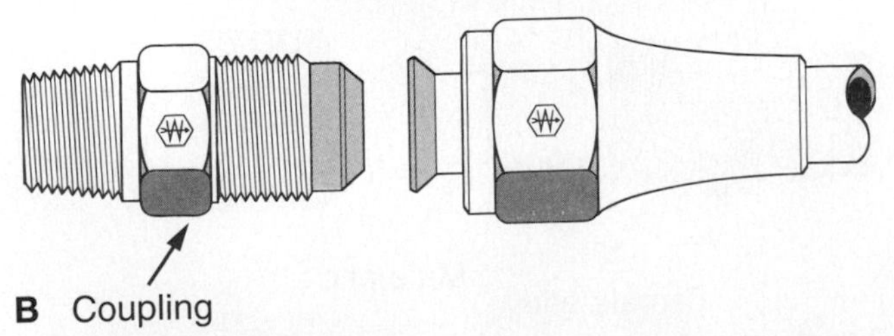

FIGURE 7–14 The flare fitting is used to seal high pressures. (Courtesy of Dana Corporation)

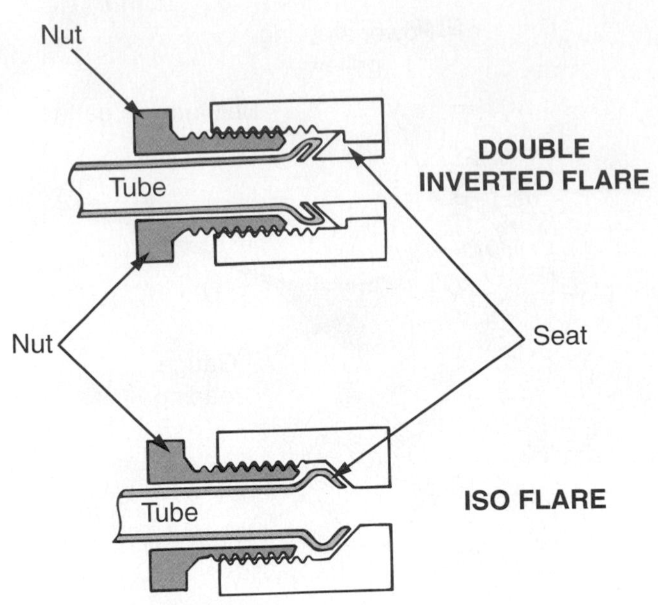

FIGURE 7–15 The double inverted flare and the ISO flare are used to seal very high pressures.

greater pressures. One is called the double inverted flare fitting and the other is called the ISO (International Standards Organization) flare fitting. Both are designed for extremely high pressure. As the nut is tightened the flare is squeezed to aid in sealing.

The fitting shown in ***Figure 7–16*** is called a brass sleeve fitting. In this case, the male end of the tube slips into a mating hole inside the female end. Although not a high-pressure fitting, this type of fitting is used in many automobile applications such as pollution control systems.

The automobile also uses several types of push-on fittings. For low-pressure applications, a simple slip-on fitting may be used. The slip-on type shown in ***Figure 7–17*** may be made of either plastic or brass. There is a small lip on the end of the fitting that slips into the tube. This lip helps to seal the fluid in the line. It also helps to keep the slip-on seal from slipping out of the tube. Other types of push-on fittings use pull tabs, twist fittings, and plastic retainer rings to help lock the tubes ends together.

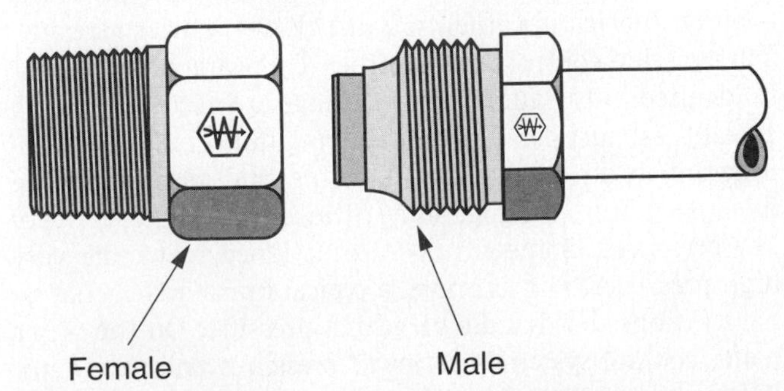

FIGURE 7–16 This is a low-pressure sleeve fitting. (Courtesy of Dana Corporation)

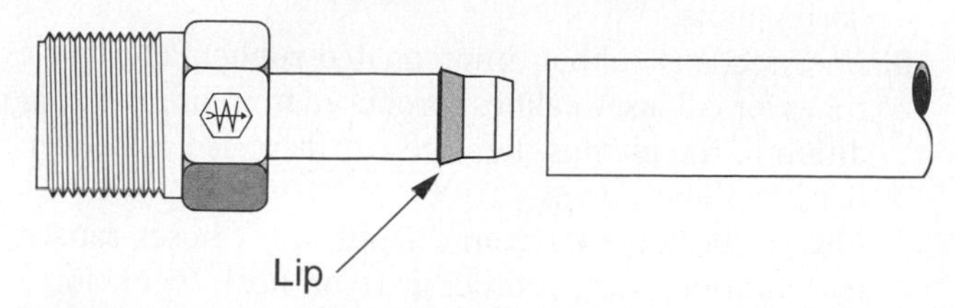

FIGURE 7-17 Slip-on fittings are used for low-pressure applications. (Courtesy of Dana Corporation)

SERVICING FITTINGS

The following are some of the more common service tips for working with different fittings used on the automobile:

1. Always check for damaged threads on brass fittings. Brass is a soft metal and the threads can be easily damaged.
2. Always check the surfaces of flare fittings. At times a small piece of metal may get into the connection and damage the flare end or the compression seal, causing the fitting to leak.
3. Always use the same type fitting that was removed.
4. Remember that fluids such as brake fluid and gasoline can damage certain types of materials. Thus, always follow the manufacturer's recommendations when replacing fittings.
5. Never over-torque a brass fitting as the threads may be easily damaged.
6. When working with fittings, always watch for small sharp metal edges which can cause serious cuts to your hand.
7. When threading a male to a female end, always start threading with your fingers to make sure the threads haven't been crossed. This will ensure that you don't damage the threads when the nut is tightened down with a wrench. You have a much better feel with your fingers than with a wrench. This procedure is especially important when threading brass fittings into steel threads.

7.3 HOSES AND TUBING

PURPOSE OF AUTOMOTIVE HOSES AND TUBES

The automobile uses many hoses and tubes to carry various types of fluids from one point to another. For example, a hose is used to carry coolant from the engine to the radiator. There are many types of hoses and tubes. The exact type used is determined by several variables, including the following:

- Amount of pressure on the system
- Type of fluid being contained
- Location and direction the fluid must be transferred
- Amount of fluid flowing through the hose or tube
- Amount of vibration and mechanical movement
- Temperature in which the hose or tube must operate

USES OF HOSES AND TUBES

Hoses and tubes are used in a variety of locations and for a variety of purposes. Steel tubes are used to carry gasoline from the fuel tank to the engine. In addition, rubber hoses are used to carry fumes from the fuel tank to the carbon canister. Other rubber and plastic hoses are used to carry various pollutants from the engine to pollution control equipment.

Steel and **fabric-reinforced rubber** hoses are used to carry hydraulic fluid in the power steering system. Both steel and fabric-reinforced hoses are used to carry brake fluid from the master cylinder to the individual wheels for braking.

Air-conditioning systems use copper, aluminum, steel, and synthetic-reinforced rubber hoses to carry refrigerant for cooling the passenger compartment. ***Figure 7–18*** shows the internal construction of an air-conditioning hose called a barrier hose.

Because air-conditioning hoses operate under high pressure, special clamps are used to seal the hoses. The hoses are connected to the line fitting with a ferrule. A ferrule is a metal ring placed around the end of the hose. It provides strength and protection from leakage due to high pressure. Because there are several styles of ferrules, always refer to the manufacturer's suggested procedure for replacing air-conditioning high-pressure hoses.

Hoses made of rubber and rubber mixed with fabric are used to carry coolant from the radiator to the engine and back. In addition, smaller rubber hoses are used to carry the coolant to the heater core and back to the engine. ***Figure 7–19*** shows two types of radiator hoses. One is a preformed hose while the other is a universal flexible hose.

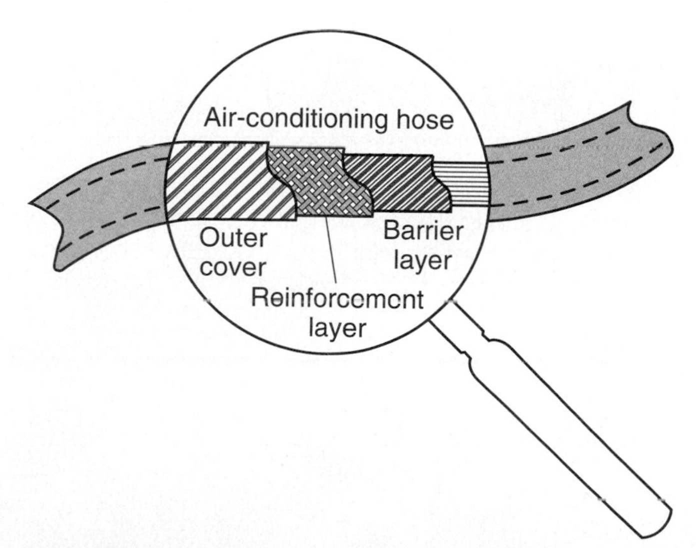

FIGURE 7-18 Air-conditioning hoses are made strong to allow for high pressures and use several layers to increase their strength.

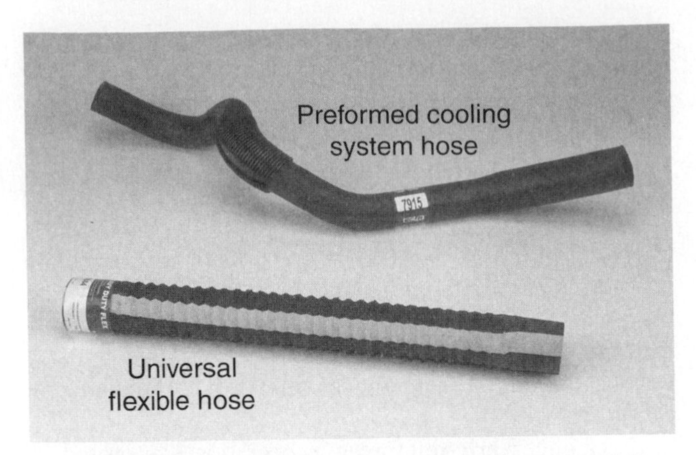

FIGURE 7-19 Cooling system hoses often are preformed. However, a universal flexible hose can be bent to fit various applications.

SERVICING HOSES AND TUBES

There are several service tips for working with hoses and tubes. Some common service techniques include the following:

1. Always be careful not to bend a steel tube too sharply. If bent too sharply, the tube may kink near the bend and cause the fluid to be restricted or leak.
2. Always make sure that rubber hoses are not bent too sharply. Although they may not break or leak, a bend may restrict the fluid flowing through the hoses or tubes.
3. Remember to always use the type of hose or tube recommended by the manufacturer. The hose selected by the manufacturer has been designed for use with the specific fluid, temperature, operating conditions, and so on.
4. Always check the fittings carefully for leaks and for damaged threads and flares that may cause leakage after being installed.
5. When flaring metal tube ends, always use the correct flaring tools.
6. Always check rubber, impregnated-rubber, and fabric tubes for cracks, weakness produced from oil, and condition of the clamps. Replace any damaged or broken hoses with new ones.
7. Always check for wear on the outside of hoses caused by vibration with another part or hose. Look for a shiny spot that may eventually leak when placed under pressure.
8. Always check hoses for soft spots, swelling, hardening, chafing, leaks, and collapsing. If any of these conditions are present, replace the hose.
9. When installing hoses, make sure the hose is not twisted. Some hoses have lines on them to show if the hose is being twisted during installation.

Safety Precautions

When working with automobile belts, fittings, hoses, and tubing, there are several important safety checks to remember.

1. Always wear safety glasses when working with belts, fittings, hoses, and tubing.
2. Always keep your hands away from spinning belts while they are in operation.
3. Remember to watch for sharp edges on metal fittings.
4. Always use the correct type of tubing or hose recommended by the manufacturer.

Service Manual Connection

Belts, fittings, hoses, and tubes are mentioned in many service manuals. The following are typical examples of various specifications related to these automobile components.

Application	Specification
New belt tension A/C and power steering	70 to 110 lb
Used belt tension A/C and power steering	70 to 110 lb
Generator belt tension	100 to 140 lb

SUMMARY

The following statements will help to summarize this chapter:

- V belts, cog belts, and serpentine belts are used on the automobile.
- Belts are used to connect two rotating shafts together.
- Cogged timing belts keep rotating shafts in time with each other.
- Automobile engines use from one to four belts to rotate components, such as the power steering pump, alternator, air-conditioning pump, and so on.
- Belts should be checked for cracks, peeling, missing cog teeth, and glazing.
- The tension on belts can be checked either manually or by using a belt tension tester.
- There are many types of fittings used to seal fluids in tubes and hoses.
- Common fittings include compression fittings, tapered sleeve fittings, flare fittings, and slip-on fittings.
- Systems that use fittings include the brake system, cooling system, lubrication system, power steering system, fuel system, and air-conditioning system.
- Hoses and tubes are used to carry different fluids in the automobile.
- Hoses and tubes are used on the cooling system, power steering system, braking system, fuel system, air-conditioning system, and pollution control system.
- When servicing hoses and tubing, remember to check for leaks, soft spots, cracks, chafing, bends, kinks, and so on.

TERMS TO KNOW

Can you explain each of the following terms? Review the chapter until you can use each term correctly.

Belt tension gauge
Cogged timing belt
Fabric-reinforced rubber
Fittings
Flare
Serpentine belt

REVIEW QUESTIONS

Multiple Choice

1. Which type of belt can be used to turn engine components on both sides of the belt?
a. V belt
b. Serpentine belt
c. Cogged V belt
d. Round belt
e. Width belt

2. Which of the following determines the type of belt needed on an engine?
a. Distance between two rotating shafts
b. Load on the belt
c. Speed of a belt
d. All of the above
e. None of the above

3. Which of the following is not part of a typical belt construction?
a. Fabric
b. Rubber
c. Steel
d. Copper
e. Fabric impregnated with rubber

4. A compression fitting is used when there is:
a. Low-pressure fluid only
b. High-pressure fluid
c. No wrench available
d. No female part needed
e. None of the above

5. The type of fitting used is determined by:
a. Pressure on the system
b. Type of fluid being used
c. Vibration and mechanical movement
d. All of the above
e. None of the above

6. Always check hoses and tubes for:
a. Kinks
b. Soft spots
c. Cracks
d. All of the above
e. None of the above

7. Which of the following automobile systems does not use hoses and tubes?
a. Cooling system
b. Starting system
c. Fuel system
d. Air-conditioning system
e. Power steering system

The following questions are similar in format to ASE (Automotive Service Excellence) test questions.

8. *Technician A* says that the tension on a belt can be checked using a belt tension gauge. *Technician B* says that the tension on a belt can be checked by pushing your finger in the center of the free span and measuring the deflection distance. Who is correct?
a. A only
b. B only
c. Both A and B
d. Neither A nor B

9. *Technician A* says the timing belt on an engine is a V-type belt that should allow for slippage. *Technician B* says the timing belt is considered a serpentine belt. Who is correct?
a. A only
b. B only
c. Both A and B
d. Neither A nor B

10. *Technician A* says that when checking for belt condition, you should always check for cracks in the belt. *Technician B* says that at times a cogged timing belt may be missing teeth. Who is correct?
a. A only
b. B only
c. Both A and B
d. Neither A nor B

11. *Technician A* says it is not necessary to replace fittings with the manufacturer's recommendations. *Technician B* says that when replacing a brass fitting, always start threading using your fingers. Who is correct?
a. A only
b. B only
c. Both A and B
d. Neither A nor B

12. *Technician A* says that brake lines use simple slip-on fittings. *Technician B* says that brake lines use either double flare fittings or ISO fittings. Who is correct?
a. A only
b. B only
c. Both A and B
d. Neither A nor B

13. *Technician A* says that hoses are designed to accommodate several variables, one of which is the pressure in the system. *Technician B* says that hoses are designed for various temperatures. Who is correct?
a. A only
b. B only
c. Both A and B
d. Neither A nor B

Essay

14. Describe the purpose of a serpentine belt.

15. List three things that determine which type of belt will be used in a particular application.

16. Describe one method used to check belt tension.

17. Describe how a compression fitting seals when tightened.

18. Describe the purpose of using hoses on the automobile.

Short Answer

19. Engines can have from one to ________ belts used to drive components on the front.

20. The ________ belt is used to turn many rotating components and can be used on both sides of the belt.

21. To keep the camshaft and crankshaft in time with each other, often a ________ belt is used.

22. A slip-on fitting is used in ________ pressure applications.

23. The end of a fitting that sticks into the nut is called the ________ end.

24. Hoses can be made of metal, fabric, and ________.

25. When installing metal tubes, always check the tubes for ________ in the line.

Applied Academics

TITLE: Efficiency of Belts

ACADEMIC SKILLS AREA: Science

NATEF Automobile Technician Tasks:

Demonstrate an understanding of how cams, pulleys, and levers are used to multiply force or transfer directions of force.

CONCEPT:

The efficiency of a cog belt is very important to conserving energy. Ordinary V belts transmit horsepower at about 94% efficiency under ideal conditions. A cog belt is able to transmit power at about 98% efficiency under the same operating conditions. The difference between these two belts is shown in the accompanying figure. When a V belt bends around a pulley, the flexing of the belt consumes a certain amount of energy due to friction. The wasted energy takes the form of heat, which builds up in the belt and is dissipated into the air. The stiffer the belt, the more energy wasted. The cog belt shown below is made of a combination of molded cogs designed to bend easily around the pulley. A cog belt means less energy wasted, improved fuel mileage, and lower operating costs.

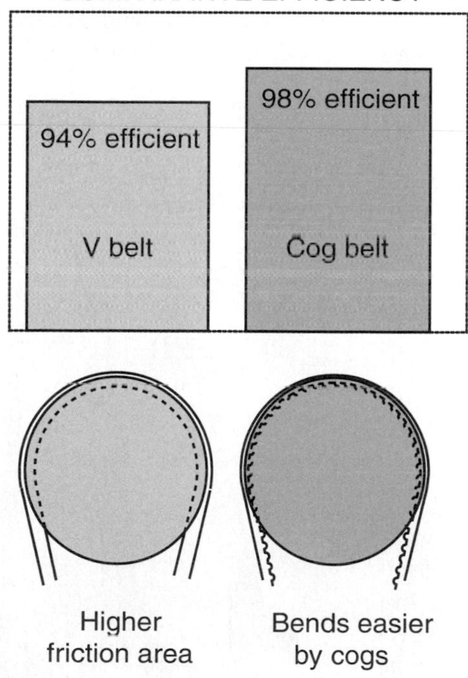

APPLICATION:

What would happen to the belt efficiency if there were more pulleys to go around as compared to only two pulleys? What would happen to the life of the belt if it has more pulleys to go around? Can belts be designed better than the cog belt to improve efficiency?

CHAPTER 8

Common Hand Tools

OBJECTIVES

After studying this chapter, you should be able to:

- Identify common hand tools and their purposes.
- Describe the use of hammers, pliers, screwdrivers, and wrenches.
- Define the use of taps and dies, chisels and punches, hacksaws, and sockets.

INTRODUCTION

In the automotive shop, it is necessary to use many tools. In most shops, each service technician has his or her own set of tools. The tool set may include several hundred tools used to work on the automobile. Tools are used to make service easier for the technician. They are used to remove parts, tighten bolts and nuts to correct specifications, and to assemble parts on the engine or vehicle. These tools maximize the technician's effort by multiplying forces. Today's automotive service technician should be familiar with the use of many tools. Proper tool selection will improve both the quality and efficiency of the service done on the automobile. The purpose of this chapter is to introduce the common tools used on the automobile. Because so many tools are now available, only identification of these tools will be presented.

8.1 COMMON HAND TOOLS

The automotive service technician uses many types of tools. Hand tools are designed to make the technician's work much easier. These tools receive energy from the technician's hand and transform that energy into productive work. Hand tools multiply forces to accomplish work. A person cannot loosen a bolt by hand. However, the bolt can be loosened easily by using the correct wrench. Tools accomplish this work in two ways: (1) by multiplying forces, as when prying a heavy object with a bar, and (2) by concentrating the applied force into a small area, as when using a wrench.

WRENCHES

Many types of wrenches are available. Wrenches are used primarily to turn threaded fasteners with hexagonal heads. Two types of wrench ends are commonly used. They are the open end and box end. ***Figure 8–1*** shows the differences between these wrenches. The box end is closed, the open end

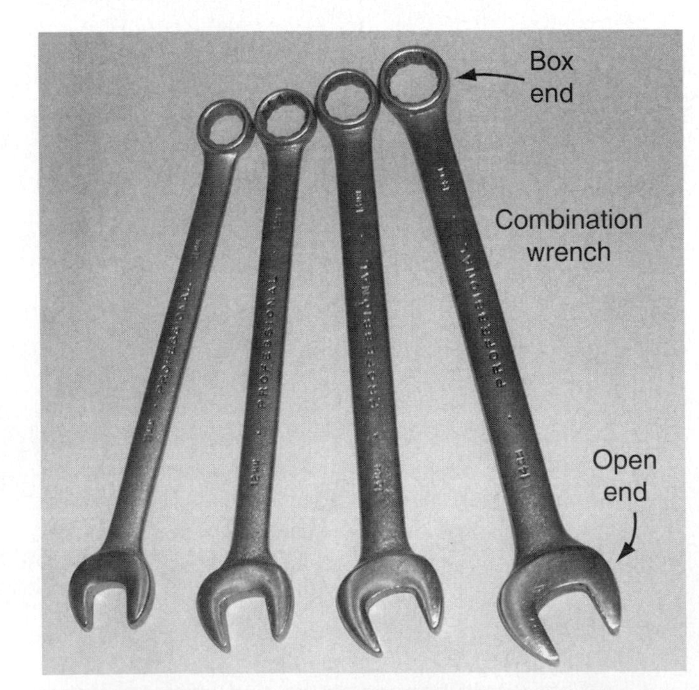

FIGURE 8–1 The most common types of wrench ends are the box end and open end. When one wrench has both types of ends, it is called a combination wrench.

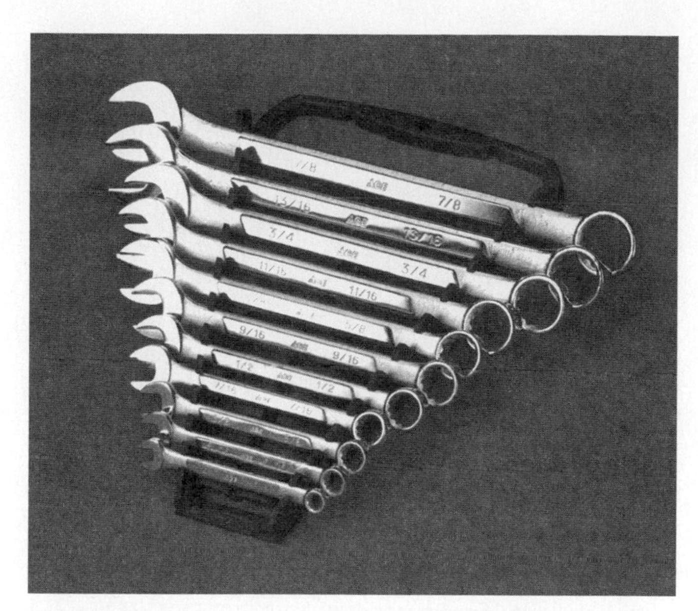

FIGURE 8–2 Wrenches come in a variety of sizes.

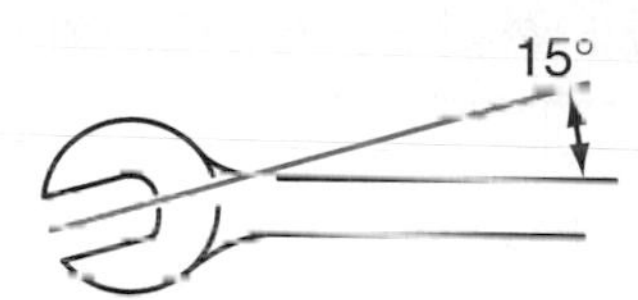

FIGURE 8–3 The open-end wrench has a 15-degree angle to allow for more flexibility.

is open. A combination wrench has both a box end and an open end. ***Figure 8–2*** illustrates many of the different sizes of combination, open-end, and box-end wrenches available today. Wrench sets such as these can be purchased in both English (USC) and metric sizes.

Both the open-end wrench and the box-end wrench have angled ends. ***Figure 8–3*** shows the 15-degree angle on an open-end wrench. Wrench ends are angled so that the wrench can be flipped over in tight areas of movement and still be placed on the head of the bolt or nut. For example, if other parts of the vehicle are in your way, you might have only 25 degrees of rotation available on a bolt. After rotating the bolt 25 degrees, you can flip the wrench over, place it on the bolt, and again rotate 25 degrees. This is shown in ***Figure 8–4***. Angling gives the open-end wrench much more flexibility in its use. ***Figure 8–5*** shows a box-end wrench with a 15-degree angle that allows room for your hands when tightening or loosening bolts or nuts. It should be noted that new wrench designs often are available in many tool stores.

Other styles of wrenches are used on the automobile as well. The adjustable wrench is used when certain size wrenches are not available. This wrench is also called a crescent wrench. It is able to fit bolts and nuts of different sizes by adjusting the size of the jaws. Adjustable wrenches come in different lengths and sizes (***Figure 8–6***).

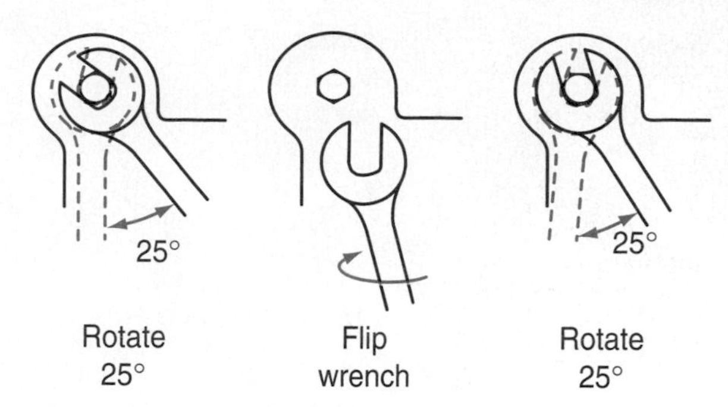

FIGURE 8–4 If there is only a small amount of rotation available, flip the wrench to continue rotating.

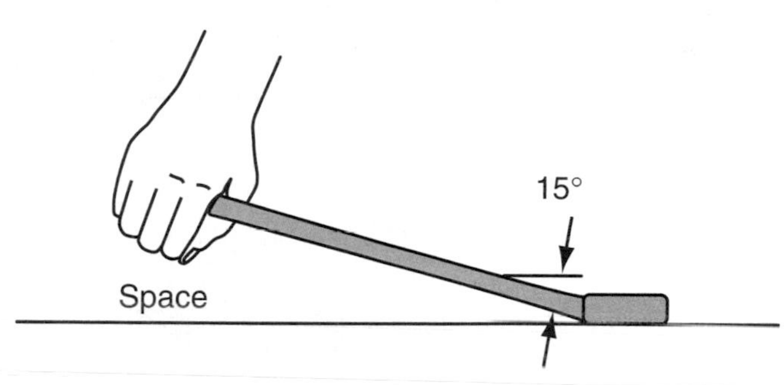

FIGURE 8–5 The box-end wrench has a 15-degree angle to allow room for the service technician's hand.

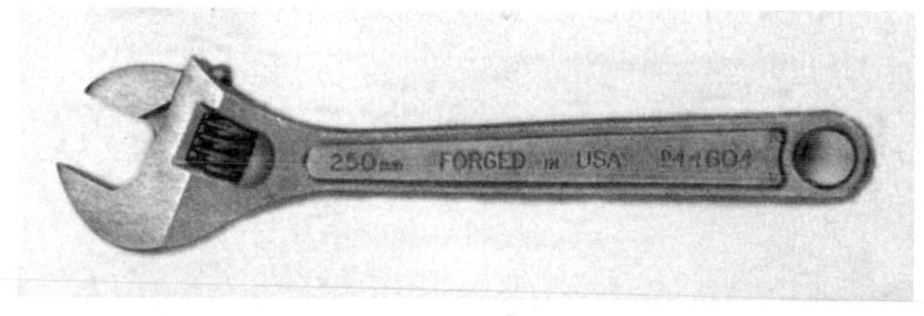

FIGURE 8–6 Adjustable wrenches are used for a variety of service jobs. Always make sure the wrench is adjusted as tightly as possible or the bolt head may be damaged.

SOCKET SETS AND DRIVES

Socket sets, drives, and extensions are used to tighten and loosen bolts and nuts. They are usually faster than open- or box-end wrenches. Socket sets can also loosen bolts and nuts that standard wrenches cannot get at. Socket sets include various sizes of sockets, drives, extensions, and ratchets. They are identified by the drive size. The **drive** on a socket set is the square area that connects the ratchet to the socket. The most common drive sizes are 1/4, 3/8, and 1/2 inch. In some heavier duty applications, 3/4-inch and 1-inch socket drives can also be used.

To get a better idea of the parts of a socket wrench, also called a ratchet, refer to ***Figure 8–7***. Note that there is a reversing lever on the top of this tool. This allows the ratchet to either tighten or loosen a bolt or nut. The drive is on the bottom and is used to match up with the socket.

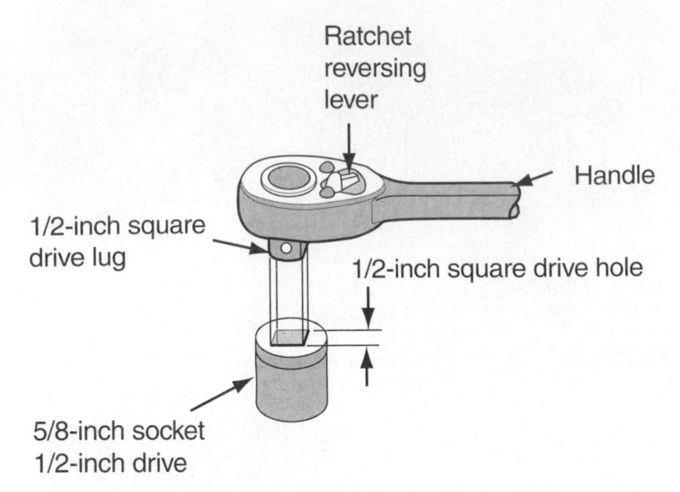

FIGURE 8–7 A ratchet has a reversing lever on its top.

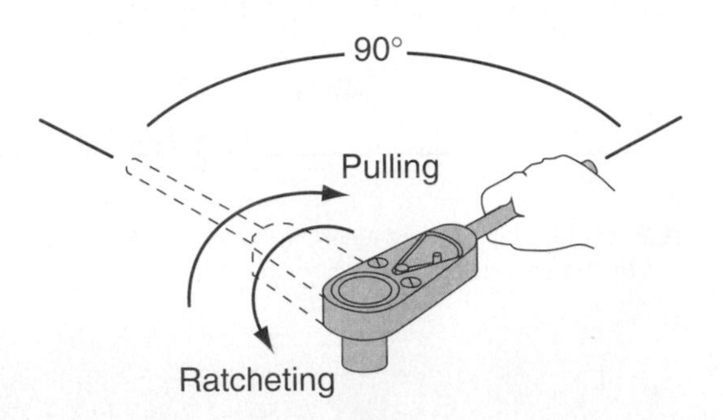

FIGURE 8–8 A ratchet is useful when working in an area where only small rotation spaces are possible.

The ratchet mechanism operation is shown in ***Figure 8–8***. The ratchet allows the wrench to tighten a bolt when a limited rotation is possible. For example, if there is only 90 degrees of rotation available, after the handle is turned 90 degrees, it can then be ratcheted back and rotated another 90 degrees.

Many types of ratchets are available to the service technician. ***Figure 8–9*** shows many of the more common ratchets. This assortment of ratchets allows the service technician to tighten and loosen bolts and nuts in a variety of positions, angles, rotations, and applications.

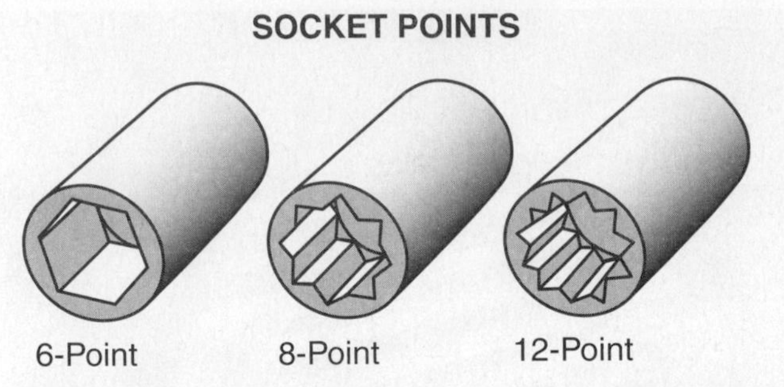

FIGURE 8–10 Sockets are identified by the number of points in the socket. Twelve-point sockets are used for applications in which the degrees of rotation are restricted.

Socket Points Sockets can be purchased with different **socket points**. Either 6-point, 8-point, or 12-point styles are available (***Figure 8–10***). If a bolt is positioned so that only a small amount of rotation is possible with the ratchet, then a 12-point socket should be used. Twelve-point sockets can be repositioned every 30 degrees. Eight-point sockets can be repositioned every 45 degrees, and 6-point sockets can be repositioned every 60 degrees. In addition to these uses, 4-point and 8-point sockets are typically used for square nuts. Six-point and 12-point sockets are used for hex nuts. ***Figure 8–11*** shows a standard socket set with various sockets, drives and extensions, ratchets, and other accessories.

Types of Sockets Sockets are also designed as *deep length* or **deep well**, swivel or universal, and standard. Deep length sockets are used for nuts that are on long studs. ***Figure 8–12*** shows why a deep well socket is handy. When a nut is placed on a long threaded bolt, the deep well socket allows the threaded bolt to fit up into the deep wheel socket. The deep well socket is especially useful when there is not enough room for an open- or box-end wrench to fit on the nut.

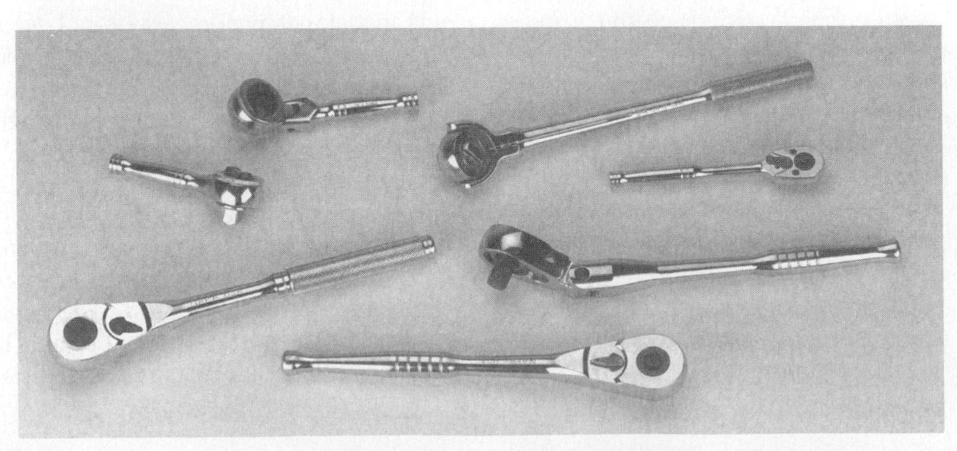

FIGURE 8–9 There are a variety of ratchets available to the service technician for working on automobiles.

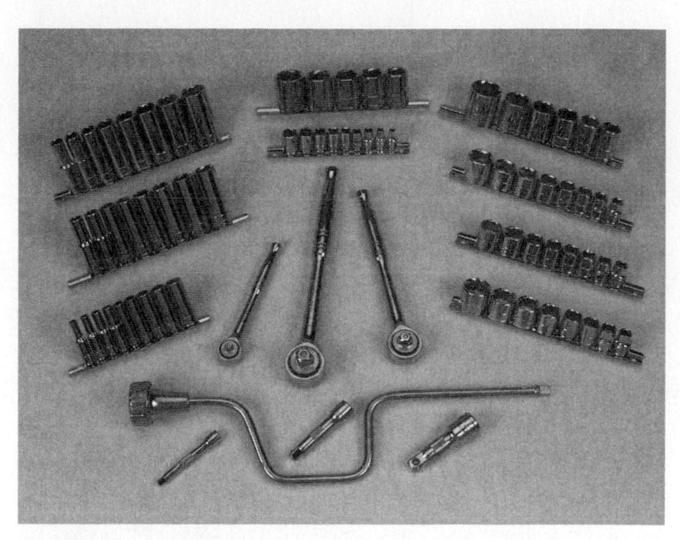

FIGURE 8–11 Socket sets include a variety of sockets, ratchets, drives, and extensions.

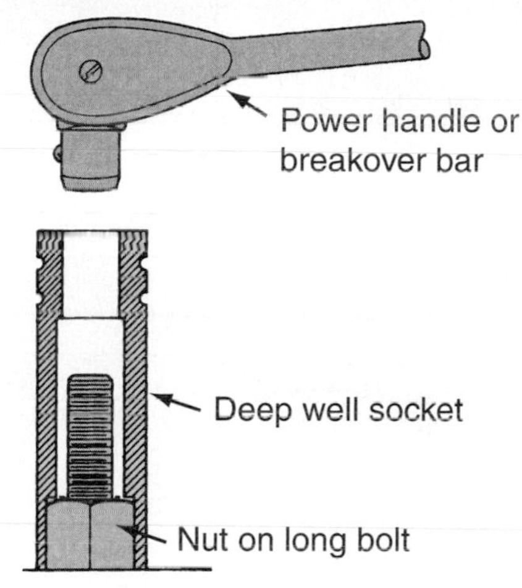

FIGURE 8–12 Deep well sockets are used when the threaded bolt is too long for a standard socket.

Car Clinic: SOCKETS

CUSTOMER CONCERN:
Buying the Correct Type of Sockets for the Job

A service technician seems to always break sockets. The sockets either strip in the middle or crack along the sides when under very heavy loads. What is the problem?

SOLUTION: Sockets are identified as either 6-, 8-, or 12-point sockets. When an 8- or 12-point socket is used on a high-torque application, it may break. For high-impact or high-torque applications, the 6-point socket should always be used. To solve this problem, make sure that only 6-point sockets or high-impact sockets are used for high-torque or high-impact applications.

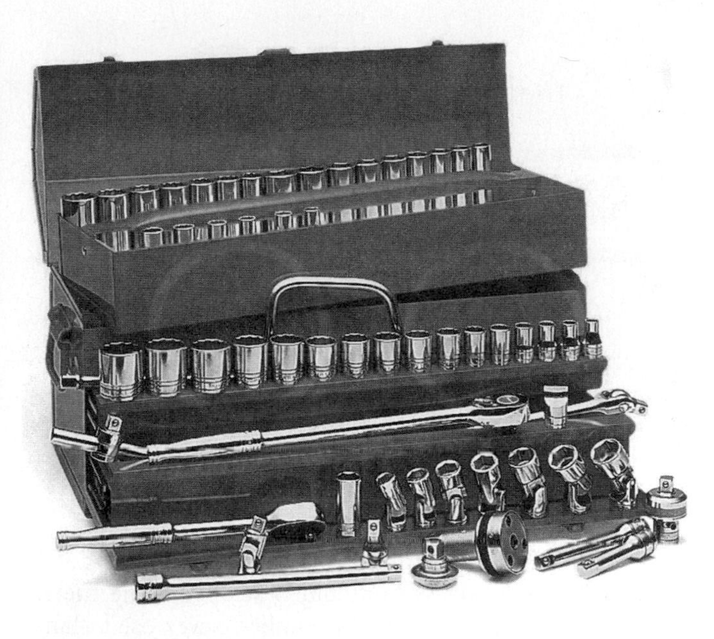

FIGURE 8–13 There are a variety of sockets that can be used in the automotive shop.

Swivel or universal sockets are used for nuts and bolts that cannot be loosened with the ratchet directly above the bolts. In this case, the ratchet and extensions can be rotated from an angle. Standard sockets are used in most applications that simply require tightening and loosening without any restrictions. ***Figure 8–13*** shows several types of sockets that are commonly used.

Impact Sockets In addition to the standard socket sets, there are also heavy-duty sockets used on impact wrenches or wrenches powered by electricity or air. These sockets can be identified by their wall thickness. These sockets should be used for heavy-duty work, such as removing nuts from a wheel and rim. ***Figure 8–14*** shows a set of 6-point-style **impact sockets**.

FIGURE 8–14 Impact sockets are used when high torque is required because they are stronger and capable of more torque than standard sockets.

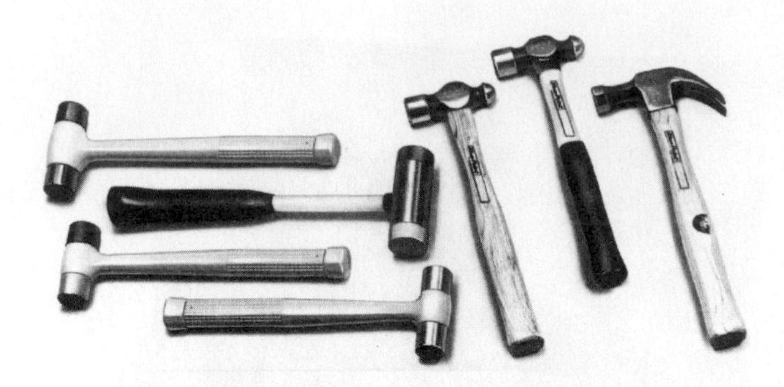

FIGURE 8–15 Many types of hammers are used in the automotive shop. Softer hammers with rubber tips should be used on parts that can be damaged easily.

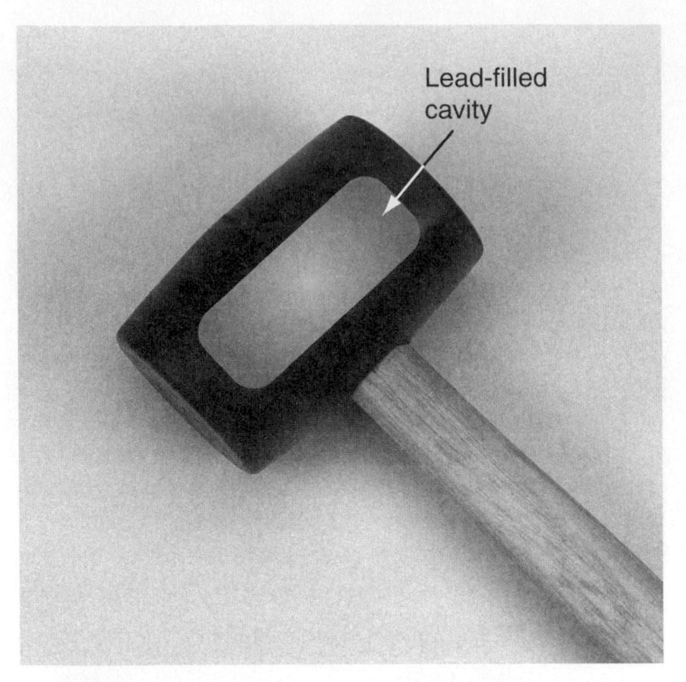

FIGURE 8–16 A lead-filled hammer stops the tool from bouncing when striking an object.

Since impact sockets are used for heavy-duty work and for speed, the socket is built much stronger. The metal around the socket is thicker. Remember, never use a standard socket when an impact socket should be used. The heavy-duty application will often damage the standard socket and may cause injury.

HAMMERS

The automotive shop is not complete unless there are hammers available for various service jobs. There are several types of hammers used in the shop. Correctly using a hammer matched to the job in style, size, and weight helps get the work done faster and safer. The common types of hammers include the ball peen, plastic tip, rubber tip, and bronze tip (***Figure 8–15***).

Hammers also come in different weights. It is important to use the right weight hammer for the job you are doing. The heads of some hammers have a cavity that is filled with lead. This type of hammer does not bounce back upon impact, producing a hitting action like a "dead blow." ***Figure 8–16*** shows an example of a lead-filled hammer.

Some hammers have rubber tips or brass on the ends. These types of hammers should be used when there is a possibility of damaging the part with a harder hammer. Such hammers are often used on gears, crankshafts, camshafts, and so on, to avoid damaging parts.

SCREWDRIVERS

Screwdrivers come in a variety of sizes and shapes. Screwdrivers are identified by the type of tip they have. The two most common screwdrivers are the slotted tip and the Phillips tip. These are shown in ***Figure 8–17***. Always make sure the tip of the screwdriver fits the screw head correctly.

There are other tips on screwdrivers. Several special screwdrivers are used on the automobile. Another type of screwdriver is the Torx®, or star tip, screwdriver. ***Figure 8–18*** shows a Torx® screwdriver. This type of screwdriver is used on headlight assemblies, body molding, mirrors, and

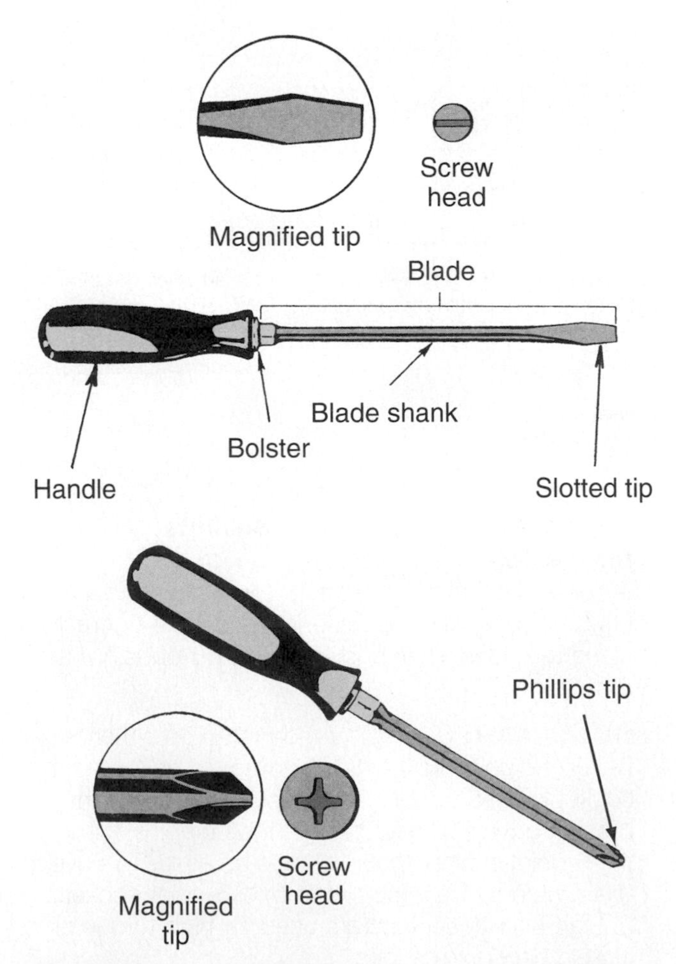

FIGURE 8–17 Of the many types of screwdriver tips, the slotted and the Phillips tips are the most common.

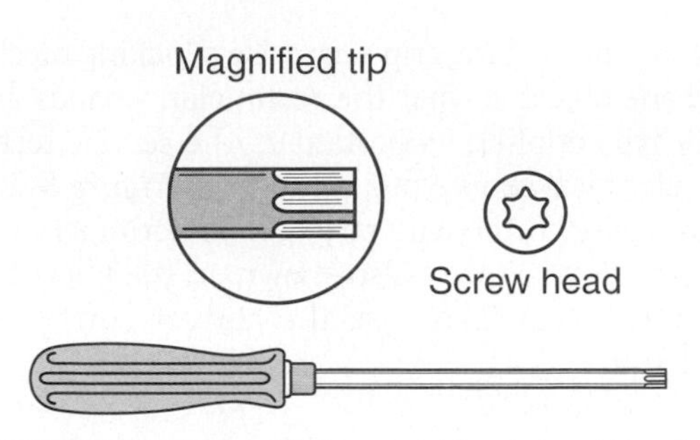

FIGURE 8–18 This Torx® tip screwdriver is used to remove and assemble many automotive body parts.

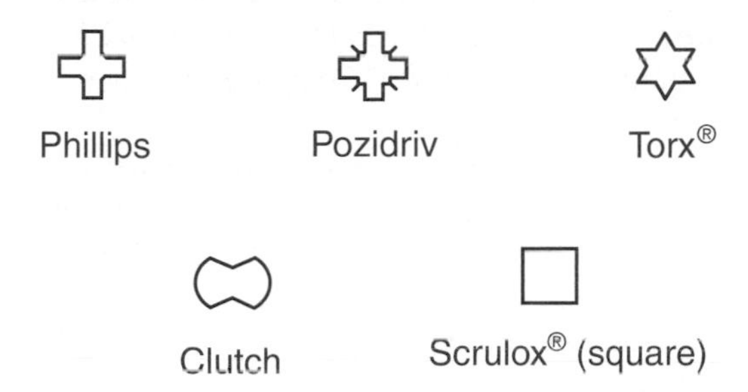

FIGURE 8–19 Various types of screwdriver tips are used in the automotive shop.

other body parts. Some screwdrivers also have a magnetic tip, which helps service technicians pick up small screws that have been dropped into small crevices. In addition, a magnetic tip screwdriver holds the screw on the end of the screwdriver during installation of the screw. This is especially helpful in very tight locations where screws cannot be started by hand.

In the automotive shop, the service technician may find other types of screwdrivers as well. ***Figure 8–19*** shows some of the more common screwdriver tips used to help service the automobile.

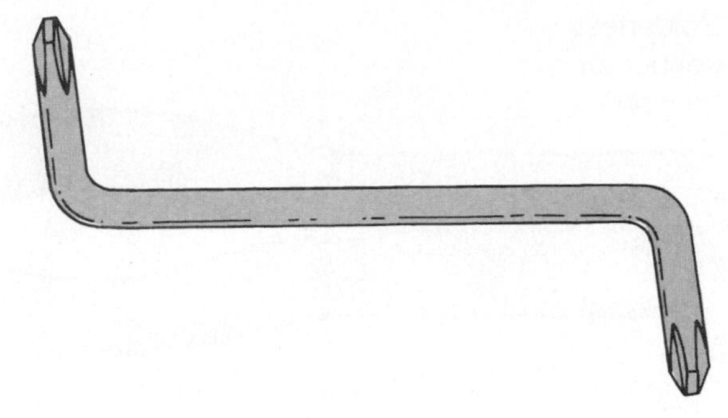

FIGURE 8–20 This offset screwdriver helps the service technician work in tight locations.

Often screws are located in very inconvenient positions and places. To help the service technician, some screwdriver shafts are designed with an offset tip. ***Figure 8–20*** shows a typical offset tip screwdriver.

PLIERS

Pliers (often called dikes) are used to grip objects of different sizes and shapes and for cutting. Because of the various uses of pliers, there are many styles. ***Figure 8–21*** shows pliers commonly used in the automotive shop. Pliers should not be used in place of a wrench. The pliers and the head of the bolt or nut may be damaged.

The different types of pliers give the service technician a variety of options. Some pliers are used for gripping, some for electrical work, some for cutting, and some for special purposes.

The following types of pliers are commonly used in the automotive shop: Slip-joint pliers are used for common gripping applications. Slip-joint pliers have a joint to allow for two different sizes. Needle nose pliers have long, slim jaws used for holding small objects such as pins and electrical

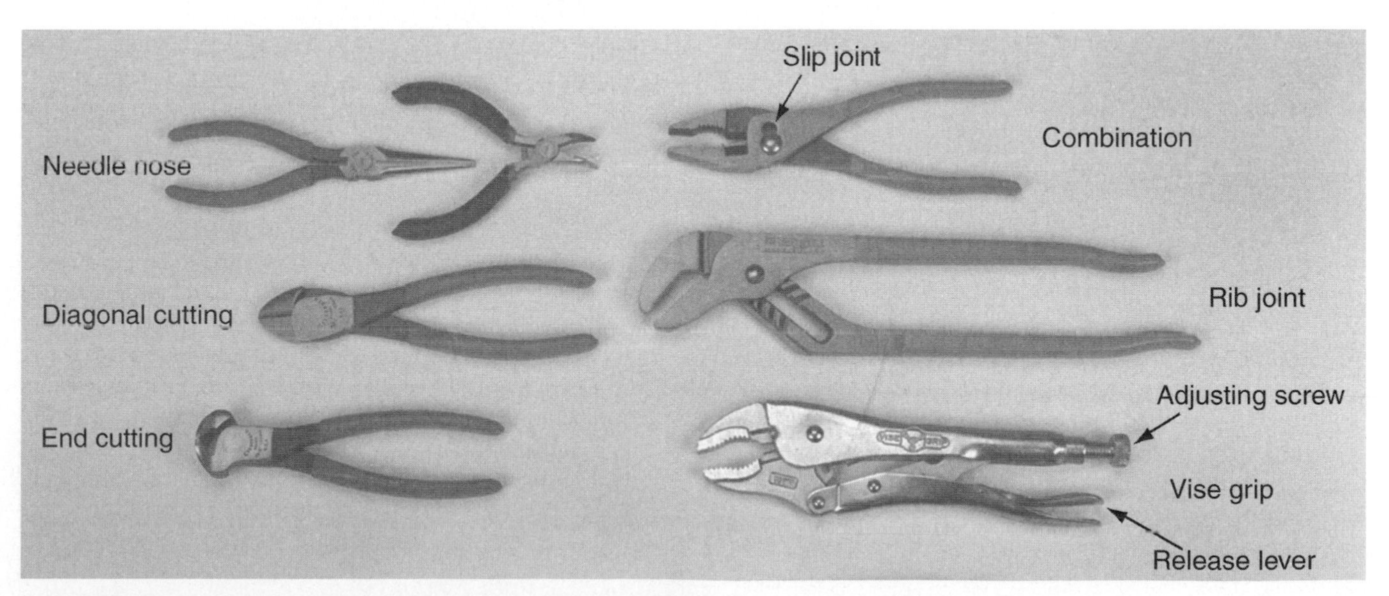

FIGURE 8–21 Many types of pliers are used in today's automotive shops.

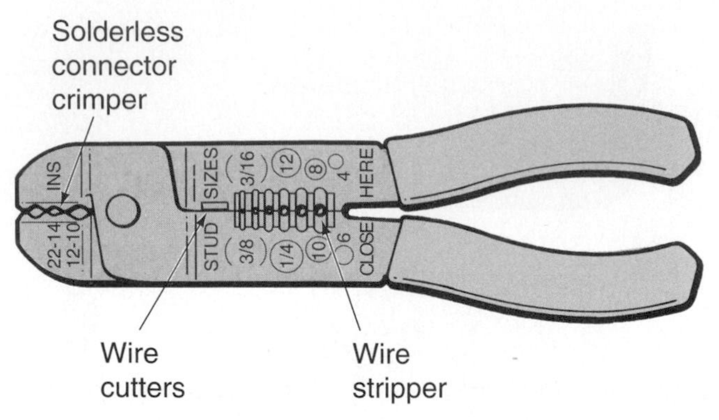

FIGURE 8–22 Electrical pliers cut and strip wire and can be used to crimp electrical connections.

components. Adjustable joint pliers have an adjustable jaw that interlocks in several positions. These pliers have the ability to grip large or small objects depending upon where the jaw is locked. Cutter pliers enable the mechanic to cut wire and other small objects. Retaining ring pliers are designed to install and remove either internal or external retaining rings. Vise-grip pliers are used to grip and hold an object in place. Vise-grip pliers use a locking mechanism to hold the object so that the technician's hands are free. Another type of pliers important to the service technician are the electrical service pliers shown in ***Figure 8–22***. This tool has a wire cutter, wire stripper, and crimper built into the pliers. Wire sizes are also shown on the pliers to make sure service technicians use the correct cutter size and crimper size.

TAPS AND DIES

Threads can be either internal or external. Internal threads are cut with a tap. External threads are formed or cut with a die. ***Figure 8–23*** shows a typical tap and die set for use in the automotive shop. ***Figure 8–24*** shows examples of how a tap and a die fit into a wrench.

A **tap** is a hardened piece of steel with threads on the outside. There are three types of taps. The taper tap has a long taper. It is used to cut threads completely through open holes. It is also used to start threads in blind and partly closed holes. Sometimes this type of tap is called a starter tap.

The plug tap has a shorter taper and is used after the taper tap to provide fuller threads in blind holes.

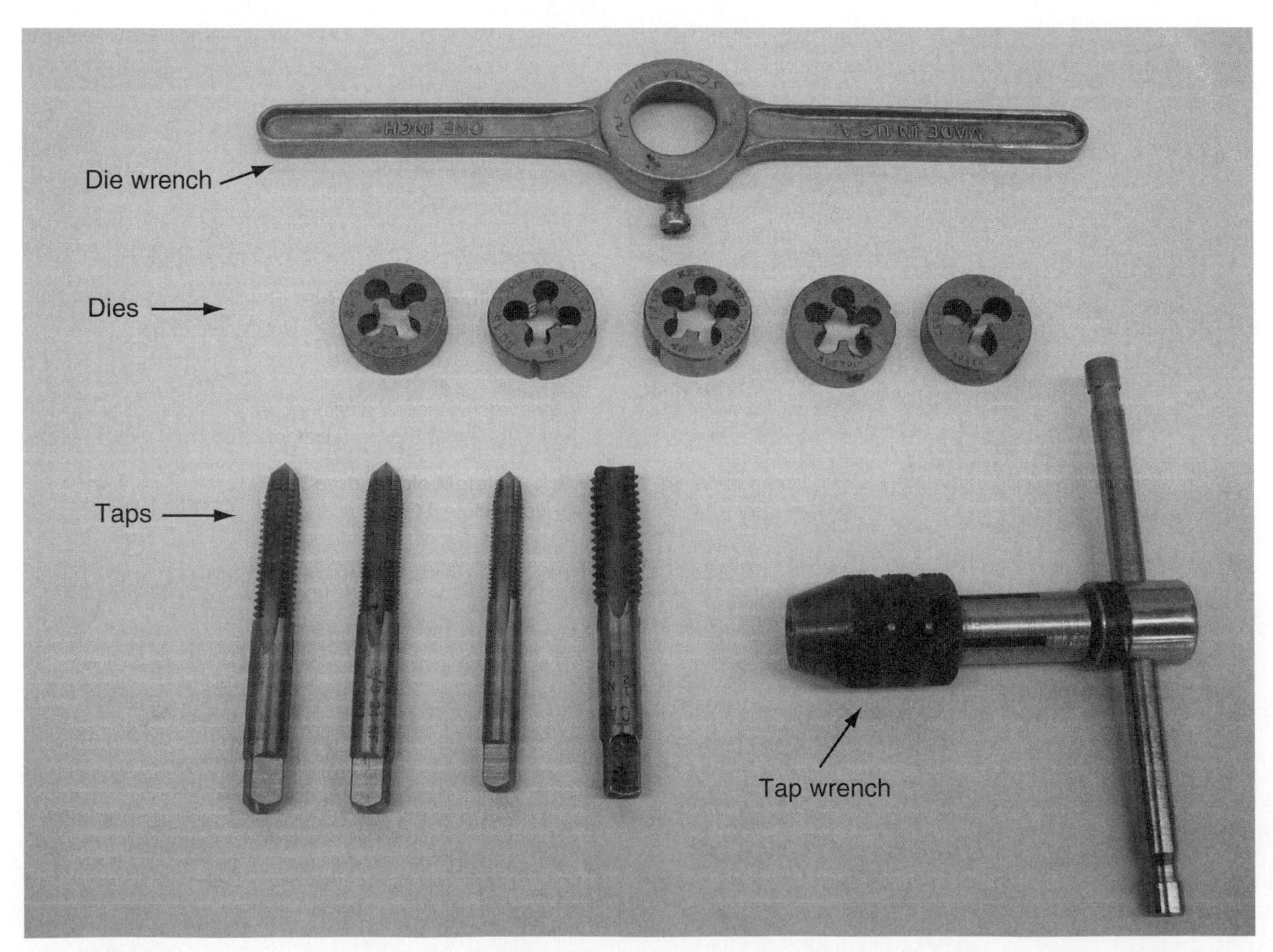

FIGURE 8–23 Taps are used to cut internal threads such as in a cylinder block. Dies are used to form external threads such as on a bolt.

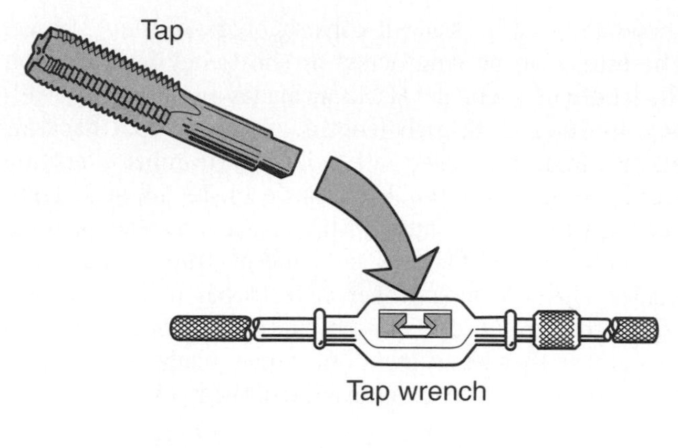

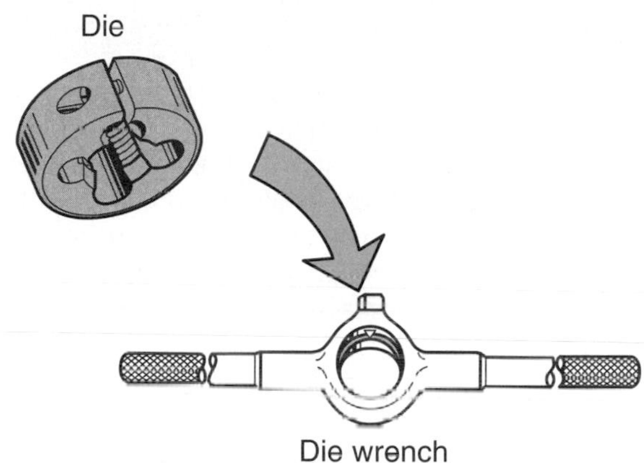

FIGURE 8–24 Both taps and dies fit into wrenches to allow them to be turned easily.

The bottoming tap is used after the taper and plug taps. It is used to cut a full thread to the very bottom of the hole.

A **threading die** is a round, hardened-steel block with a hole containing threads. The threads are slightly tapered on one side to make it easier to start the cutting.

CHISELS AND PUNCHES

Chisels and punches are used in the automotive shop for a variety of jobs, from punching pins through an object to removing a bearing. ***Figure 8–25*** shows a complete set of chisels and punches.

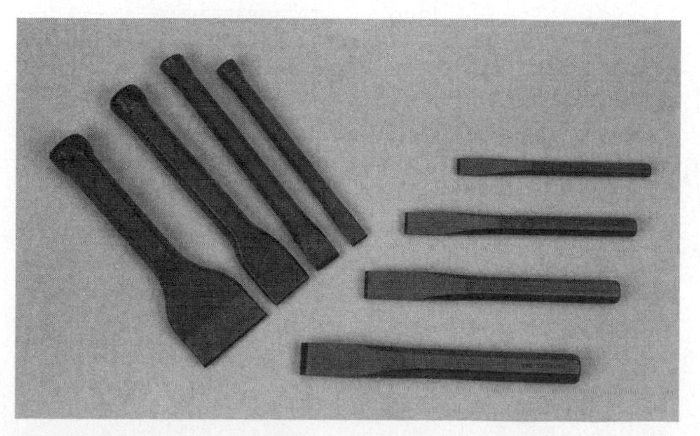

FIGURE 8–25 A quality set of chisels and punches should be available in all automotive shops.

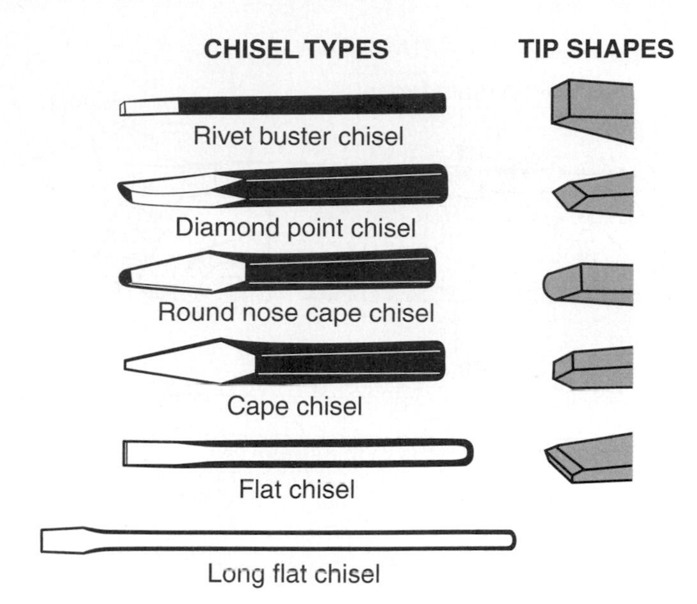

FIGURE 8–26 A variety of chisel tip shapes are available to the service technician.

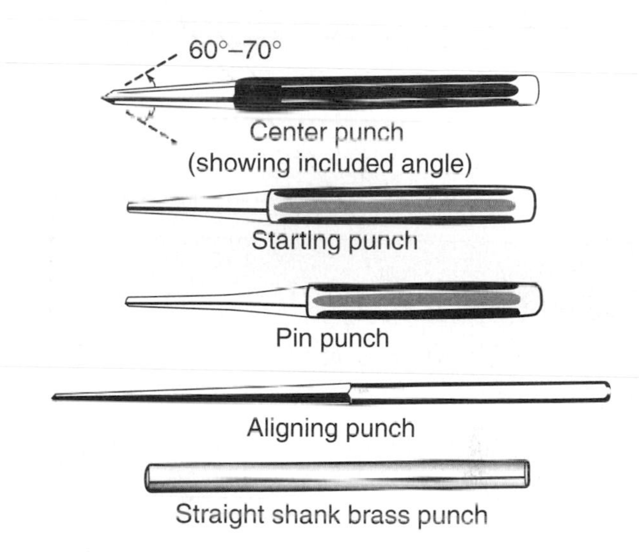

FIGURE 8–27 These punches help the service technician punch out pins and align holes.

There are several types of chisels used in the automotive shop. ***Figure 8–26*** shows different chisel types and tips for specific jobs. A well-equipped automotive shop will have a full set of chisels available to the service technician.

There are also a variety of punches used in the automotive shop (***Figure 8–27***). The center punch can be used to make an indentation in metal before drilling a hole. The starting punch is used to start driving a pin or rivet from a hole. A pin punch is then used to finish driving the pin or rivet from the hole. The aligning punch is used to help align two holes during assembly of many parts.

HACKSAWS

In the automotive shop, it is often necessary to cut various metals. Metals can be cut with a hacksaw. ***Figure 8–28***

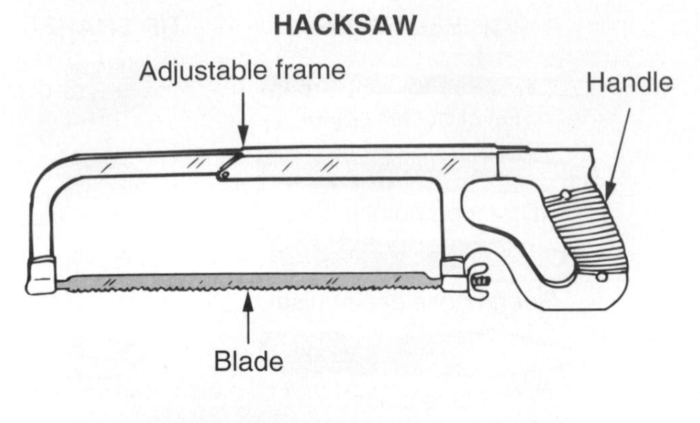

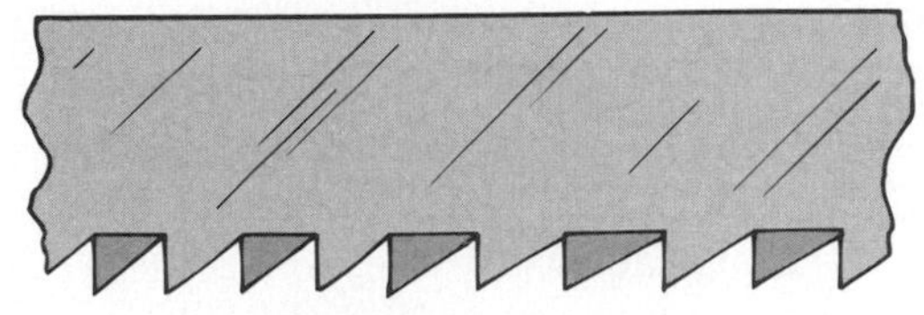

FIGURE 8–28 This hacksaw is used to cut metal (such as a rusted bolt) in the automotive shop.

Safety Precautions

When working with common tools in the automotive shop, keep in mind these important safety suggestions:

1. Always select the right size wrench for each bolt or nut. If a wrench is too large it may slip off the bolt during heavy torque, possibly injuring your hands.
2. Make sure all tools used in the automotive shop are clean and free of grease and oil.
3. Return all tools to their designated toolbox or location on the wall after each use. Tools left on a vehicle may fall into moving parts of the engine. They could then fly out and cause injury.
4. Make sure screwdriver tips are not worn. During high-torque applications, the screwdriver may slip off the screw and cause injury.
5. Occasionally, a tap for cutting internal threads may break. Be careful not to touch the edges of the broken tap, as they are extremely sharp.
6. If you are not familiar with the correct use of a tool, *always* check with your instructor. Never use a tool that you don't know how to use correctly.
7. When using an open-end wrench, make sure the smallest jaw of the wrench is facing in the direction of the turn.

shows a typical hacksaw. It consists of an adjustable frame. The frame can be lengthened or shortened depending on the length of the blade. Hacksaw blades are generally available in 10- and 12-inch lengths. Note also that hacksaw blades can be purchased with a different number of cutting teeth per inch. Common blades have 14, 18, 24, or 32 teeth per inch of blade. Blades with a higher number of teeth per inch are used for finer cuts and on thinner materials. Blades with a lower number of teeth per inch are coarse blades that are used for rough cuts and thicker materials. Remember that when installing a new blade, always keep the teeth pointing toward the front of the hacksaw.

FILES

A file is a hardened steel tool with various types of cutting edges on its face. Files are often used to shape, roughen, or smooth metal in the automotive shop. The parts of a file are shown in ***Figure 8–29***. Generally, a handle is placed on the tang to allow the service technician to hold the file tightly. The face of the file can be made with either a single-cut or double-cut edges. ***Figure 8–30*** shows the different

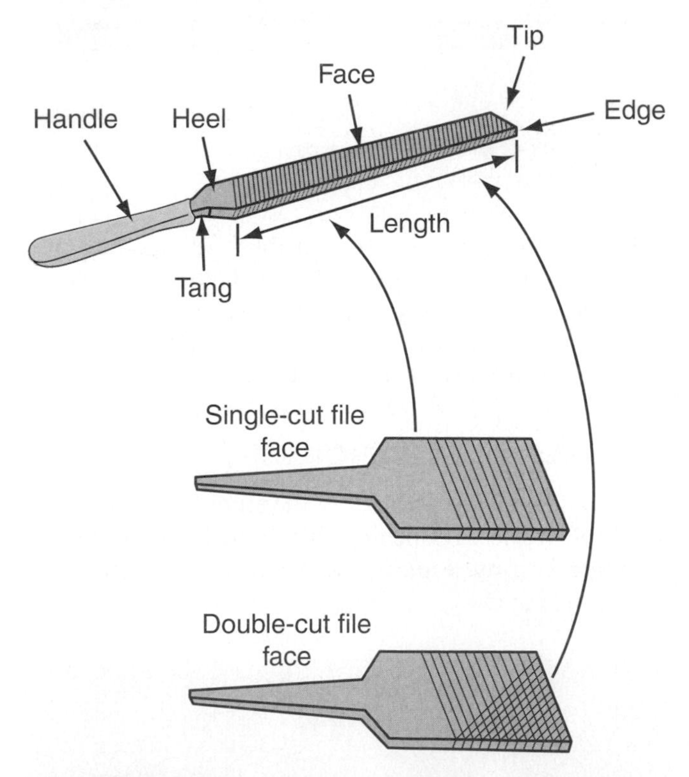

FIGURE 8–29 The parts of the file are shown for reference.

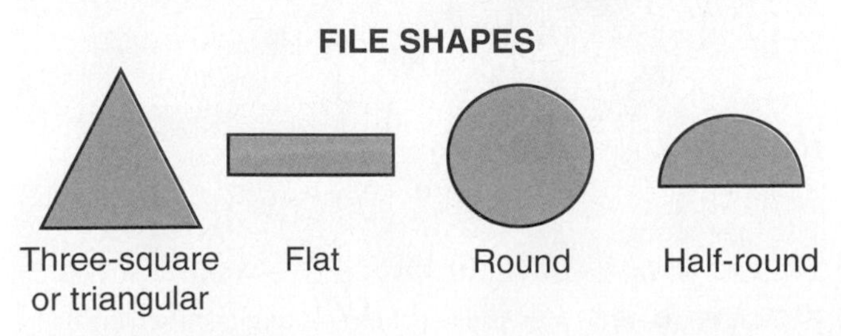

FIGURE 8–30 Files come in a variety of shapes.

file shapes used for various applications in the automotive shop.

There are many other hand tools used in the automotive shop by the skilled service technician. Some of these tools include hex head or Allen wrenches, pipe wrenches, ratchet box wrenches, prybars, and crowfoot wrenches. All of these common hand tools make the automotive service technician's job much easier, safer, and faster.

SUMMARY

The following statements will help to summarize this chapter:

- A variety of common hand tools are used in the automotive shop.
- Types of wrenches include open-end, box-end, or adjustable wrenches.
- Wrenches are manufactured with a 15-degree angle on the end to give the service technician more flexibility when using this tool.
- There are several types of hammers commonly used by service technicians.
- A variety of ratchets and sockets help service technicians tighten and loosen bolts and nuts when only a certain amount of rotation is possible.
- Deep well sockets are used when a nut that is attached to a longer bolt needs to be tightened or loosened.
- Sockets are designed with 6, 8, and 12 points inside the socket.
- There are a variety of screwdrivers and tips used in the automotive shop.
- Pliers are designed in a variety of styles and types for use in cutting and gripping various automotive objects.
- Taps are used to cut internal threads in holes.
- Dies are used to cut external threads on bolts.
- A variety of chisels and punches are designed for use in the automotive shop.
- Hacksaws have a variety of blades with different numbers of teeth per inch, and are used to cut metal.
- Files are used to smooth or shape metal parts and come in a variety of shapes and cuts.

TERMS TO KNOW

Can you explain each of the following terms? Review the chapter until you can use each term correctly.

Deep well socket
Drive
Impact sockets
Pliers
Socket points
Tap
Threading die

REVIEW QUESTIONS

Multiple Choice

1. What type of wrench has the end closed off rather than open?
a. Open-end
b. Adjustable
c. Box-end
d. Closed-end
e. None of the above

2. When there is only a small amount of room for ratchet rotation, which type of socket should be used?
a. 6-point
b. 8-point
c. 12-point
d. 24-point
e. 36-point

3. Which type of socket should be used for high torque when using an air power wrench?
a. Impact socket
b. Standard socket
c. Deep length socket
d. Chrome socket
e. None of the above

4. A ________ is used to cut external threads on a shaft.
a. Tap
b. Die
c. Chisel
d. Wrench
e. Socket

5. Which type of tap is used to start threads?
a. Plug tap
b. Taper tap
c. Bottoming tap
d. Top tap
e. All of the above

6. Which type of screwdriver shaft is bent to get into tight spaces?
a. Torx®
b. Offset
c. Tap and die
d. All of the above
e. None of the above

The following questions are similar in format to ASE (Automotive Service Excellence) test questions.

7. *Technician A* says a tap is used to cut internal threads. *Technician B* says that a die is used to cut external threads on a bolt. Who is correct?
a. A only
b. B only
c. Both A and B
d. Neither A nor B

8. *Technician A* says that Torx® tip screwdrivers are used for air-conditioning compressor disassembly. *Technician B* says that Torx® tip screwdrivers are used for removing body parts such as mirrors and molding. Who is correct?
a. A only
b. B only
c. Both A and B
d. Neither A nor B

9. *Technician A* says it is all right to use standard sockets for heavy-duty applications rather than impact sockets. *Technician B* says that impact sockets are used for light work or light-duty applications. Who is correct?
a. A only
b. B only
c. Both A and B
d. Neither A nor B

Essay

10. What is the difference between 6-, 8-, and 12-point sockets?

11. Describe the different types of taps used to cut threads.

12. What is the difference between a Torx® tip screwdriver and a slotted screwdriver?

13. What is the purpose of using an offset-type screwdriver?

14. What is the purpose of using a 6-point socket rather than a 12-point socket?

Short Answer

15. The ________-end wrench has a closed end.

16. The ________ wrench is also called a crescent wrench.

17. Where the degree of rotation is restricted, a ________-point socket should be used.

18. Pliers that have several interlocking positions are called ________ joint pliers.

19. The ________ tap is also used as a starting tap.

Applied Academics

TITLE: Levers and Wrenches

ACADEMIC SKILLS AREA: Science

NATEF Automobile Technician Tasks:

Explain how levers and pulleys can be used to increase an applied force or distance.

CONCEPT:

There are three types of levers that are used to increase an applied force or distance. These are the first-class, second-class, and third-class levers shown in the accompanying figure. Each lever has an input, an output, and a pivot point for a fulcrum. Depending on the type of lever, either force or distance can be gained. The difference among the three levers is the location of the pivot point, which affects the input and the output forces. All machines today are based on the application of these three levers.

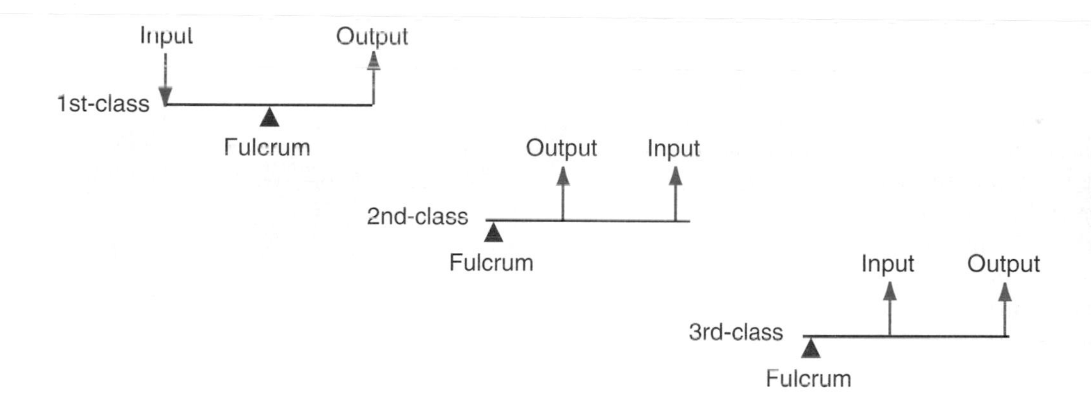

APPLICATION:

Can you think of two examples in the automobile that use levers? What type of lever (or levers) is/are used when using an open-end wrench? Can you determine the type of lever or levers used when a ratchet is rotating a socket? Draw out the input, fulcrum, and output points when using a ratchet.

CHAPTER 9

Measuring Tools and Precision Instruments

OBJECTIVES

After studying this chapter, you should be able to:

- Compare the USC and metric systems of measurement.
- Analyze common measuring tools such as the micrometer and vernier caliper.
- Define the purpose and operation of torque wrenches.
- Analyze the use of feeler gauges, dial indicators, and pressure and vacuum gauges.

INTRODUCTION

In the automotive shop, there are many times when parts, clearances, and specifications must be measured. Measuring tools and instruments help to make the service technician's job much easier and more accurate. In almost any technical system on the automobile, there are specifications, clearances, pressures, and readings to take. The purpose of this chapter is to introduce common measuring tools and precision instruments used on the automobile.

9.1 MEASURING SYSTEMS

Two measuring systems are used in the United States: the United States Customary system, known as USC, and the metric system. In the past, all parts, tools, and measuring instruments made in the United States were designed using the **USC measurements**. However, the United States has been increasingly using the **metric system**. Manufacturers continue to convert to the metric system because many foreign countries use this system. Many American automobile manufacturers now design certain components in metric sizes. Because of these changes, it is important to study and understand both measurement systems.

USC MEASUREMENTS

The measuring system used most often in the United States is the USC. This system measures length, volume, and mass as shown in ***Figure 9–1***.

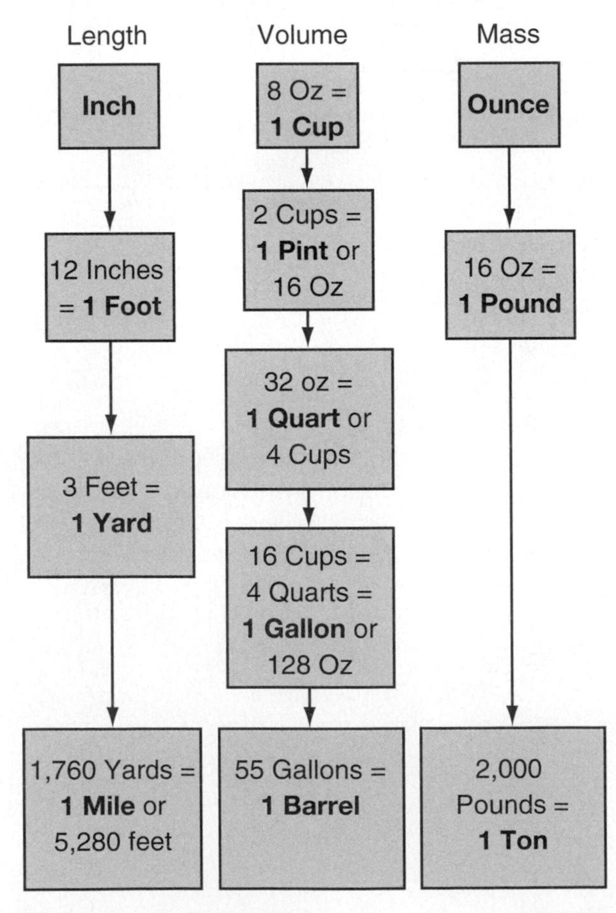

FIGURE 9–1 USC measurements for length, volume, and mass all have specific units of measurements.

COMMON WRENCH SIZES USC	COMMON WRENCH SIZES METRIC
1/4"	7 mm
5/16"	8 mm
3/8"	9 mm
7/16"	10 mm
1/2"	11 mm
	12 mm
9/16"	13 mm
5/8"	14 mm
11/16"	15 mm
3/4"	16 mm
13/16"	17 mm
	18 mm
	19 mm
7/8"	20 mm
15/16"	21 mm
1"	22 mm
	23 mm
	24 mm
	25 mm

FIGURE 9–2 Many length measurements are taken on the automobile. The USC system and the metric system use these common sizes.

Tools in the automobile industry use many length measurements. For example, wrench and socket sizes typically increase in 1/64-, 1/32-, 1/16-, and 1/8-inch sizes (***Figure 9–2***). In addition, many small measurements and clearances are listed in 0.001 of an inch.

METRIC SYSTEM

When you are familiar with it, the metric system is much easier to use than the USC measurements. This is because all of the units use a **base of 10**. This means that all units are multiples of 10. This is not true with USC measurements. The foot has a base of 12 (12 inches in a foot), the yard has a base of 3 (3 feet in a yard), the pound has a base of 16 (16 ounces in a pound), and so on. ***Figure 9–2*** also shows common metric wrench sizes.

MEASURING LENGTH IN METERS

The base unit in the metric system is the **meter**. The distance of a meter is defined as the length equal to 1,650,763.73 wavelengths of krypton in a vacuum. This distance is always constant and can easily be duplicated under laboratory conditions. In relation to the USC measurement, 1 meter is equal to 39.37 inches.

METRIC PREFIXES

Instead of using such phrases as 1/16 of an inch, the metric system uses a set of **prefixes**. These prefixes are shown in ***Figure 9–3***. For example, the prefix *kilo* means 1,000. If a distance is measured as 1,000 meters, it can also be called one kilometer. In terms of small distances, the prefix *milli* means one thousandth of a unit. If the unit is the meter, one millimeter is equal to 1/1,000 of a meter. Each small increment on a meter stick, as shown in ***Figure 9–4***, is a measure of one millimeter. One thousand millimeters make up one complete meter. A decimeter is 1/10 of a meter, while a centimeter is 1/100 of a meter.

MEASURING VOLUME

In the metric system, volume is measured by the **liter**. One-tenth of a meter is called a decimeter. This is about the width of a person's fist. If a cube is made with each edge one decimeter long, the cube will have a volume of one cubic decimeter. A cubic decimeter is equal to a liter (***Figure 9–5***). Note also that the length of one decimeter is equal to 10 centimeters. If one centimeter is cubed, the unit is called a **cubic centimeter**. Both liters and cubic centimeters are used to measure the volume of an engine.

SOME COMMON PREFIXES

NAME	SYMBOL	MEANING	MULTIPLIER
mega	M	one million	1,000,000
kilo	k	one thousand	1,000
hecto	h	one hundred	100
deca	da	ten	10
deci	d	one-tenth of a	0.1
centi	c	one-hundreth of a	0.01
milli	m	one-thousandth of a	0.001
micro	μ	one-millionth of a	0.000 001

FIGURE 9–3 The metric system uses these prefixes to aid in defining measurements.

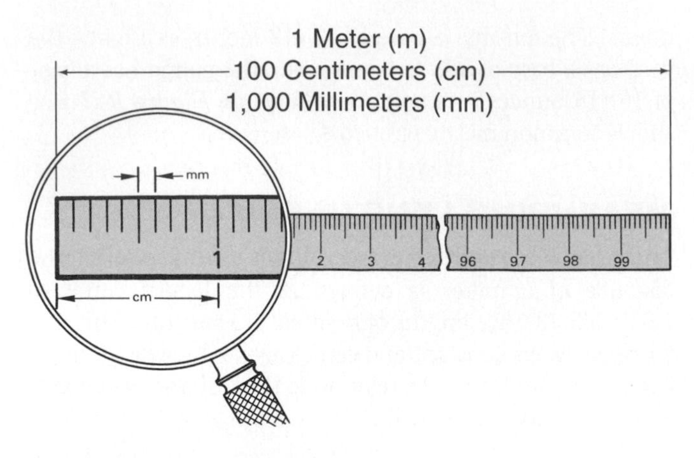

FIGURE 9-4 A meter stick is composed of 1,000 millimeter increments.

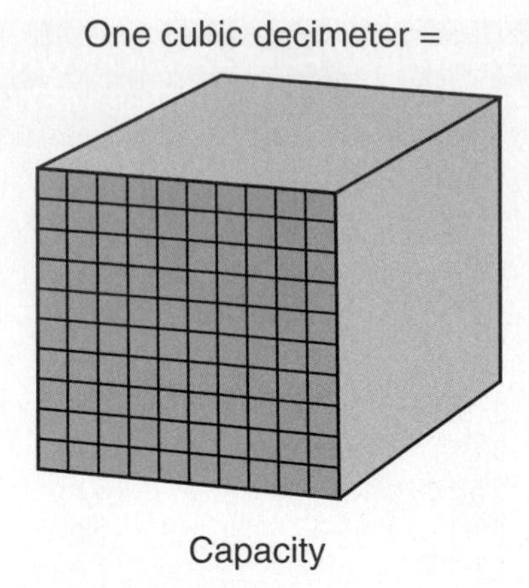

FIGURE 9-5 A cubic decimeter, also called a liter, has 100 cubic centimeters. Engines are sized by indicating the number of cubic centimeters (cc) or liters in the combustion chamber.

MEASURING MASS

The unit used to measure mass in the metric system is called the **gram**. Kilogram is often used because the gram is so small. If one cubic centimeter is filled with water, the mass is one gram. If a liter, 1,000 cubic centimeters, is filled with water, the mass is 1,000 grams, or 1 kilogram.

OTHER METRIC UNITS

Other units are used in the metric system. ***Figure 9–6*** includes a listing of common metric units. The quantity, the unit, and the common symbols are shown. The automotive technician should become familiar with these units.

QUANTITY	UNIT	SYMBOL
Length	millimeter *(one thousandth of a meter)* meter kilometer *(one thousand meters)*	mm m km
Area	square meter hectare *(ten thousand square meters)*	m^2 ha
Volume	cubic centimeter cubic meter milliliter *(one thousandth of a liter)* liter *(one thousandth of a cubic liter)*	cm^3 m^3 ml l
Mass	gram *(one thousandth of a kilogram)* kilogram ton *(one thousand kilograms)*	g kg t
Time	second minute, hour, day, month, year	s min, h, d, mo, yr
Speed	meter per second kilometer per hour	m/s km/h
Power	watt kilowatt *(one thousand watts)*	W kW
Energy	joule kilowatt-hour	J kW•h
Electric potential difference	volt	V
Electrical current	ampere	A
Electrical resistance	ohm	Ω
Frequency	hertz	Hz
Temperature	degree Celsius	°C

FIGURE 9-6 Many units are used in the metric system. Note that each unit has its own symbol.

CONVERSION BETWEEN USC AND METRIC SYSTEMS

At times it is necessary to convert from USC units to metric system units or from metric system units to USC units. The chart shown in ***Figure 9–7*** gives some of the more common **conversion factors**. Specially made slide rules and calculators can be used for making conversions.

Say, for example, a service technician wants to convert 5 inches into millimeters (mm). Using the Length section of this chart, note that there are 25.4 mm in 1 inch. So, if there are 5 inches, then when multiplied by the 25.4 mm/inch the answer is 127 mm in 5 inches (5 in. × 25.4 mm = 127 mm).

As another example, a technician may need to know how many liters are in a 350-cubic inch engine. Referring to the Volume section, note that there are 16.387 cubic centimeters (cm^3) in 1 cubic inch. When the number of cubic inches are multiplied by 16.387, the technician finds there are 5,735 cubic centimeters in 350 cubic inches (350 cubic in. × 16.387 = 5,735). However, remember that there are 1,000 cubic centimeters in 1 liter, so in effect, a 350-cubic inch engine is considered the same as a 5.735 liter engine (5,735 ÷ 1,000 = 5.735).

UNIT	CONVERSION FACTOR
Length	1 inch = 25.4 mm 1 foot = 30.48 cm 1 yard = 0.9144 m 1 mile = 1.609 344 km
Area	1 square inch = 6.4516 cm^2 1 sqare foot = 9.290 304 dm^2 1 square yard = 0.836 127 4 m^2 1 acre = 0.404 685 6 ha 1 square mile = 2.589 988 km^2
Volume	1 cubic inch = 16.387 064 cm^3 1 cubic foot = 28.316 85 dm^3 (or liters) 1 cubic yard = 0.764 555 m^3 1 fluid ounce = 28.413 062 cm^3 1 gallon = 4.546 090 dm^3 (or liters)
Mass	1 ounce (avoirdupois) = 28.349 523 g 1 pound (avoirdupois) = 0.453 592 37 kg 1 ton (short, 2,000 lb) = 907.184 74 kg
Temperature	(5/9) × (number of degrees Fahrenheit − 32) = number of degrees Celsius
Speed	1 mile per hour = 0.447 04 m/s = 1.609 344 km/h
Force	1 pound-force = 4.448 222 N 1 kilogram-force = 9.806 65 N
Pressure	1 pound-force per square inch (psi) = 6.894 757 kPa (kilopascal) 1 inch of mercury (0°C) = 3.386 39 kPa 1 mm of mercury (0°C) = 133.322 Pa 1 standard atmosphere (atm) = 101.325 kPa
Energy, Work	1 British thermal unit (Btu) = 1055.06 J 1 foot-pound force = 1.355 818 J 1 calorie (international) = 4.1868 J 1 kilowatt-hour (kW•h) = 3.6 MJ
Power	1 horsepower (550 ft.-lb/s) = 745.6999 W 1 horsepower (electric) = 746 W

FIGURE 9–7 Some of the more common conversion factors are listed in this chart.

9.2 MEASURING TOOLS

When servicing an automobile, certain dimensions, specifications, and clearances must be measured. These may include torque specifications and various clearances. Measuring tools help the automotive service technician to check these specifications and clearances.

TORQUE WRENCHES

Torque wrenches are used to tighten bolts and nuts to their correct torque specification. Torque specifications were discussed in Chapter 5. Torque wrenches are used to control the amount of tension on a bolt by measuring the amount of twist (torque) developed while tightening the bolt. Torque wrenches are designed to match sockets using 1/4-, 3/8 , and 1/2 inch drives. Other drives are also available but are not as common. They include 3/4-, 1-, and 1 1/2-inch drives for heavy-duty service. Torque wrenches typically have a dial or scale that indicates the amount of torque in pound-inches (lb-in.) or in pound-feet (lb-ft.). ***Figure 9–8*** shows two common torque wrenches used in the automotive shop.

TYPES OF TORQUE WRENCHES

A common type of torque wrench is the adjustable click type. A specified torque value is adjusted on the torque wrench. When that torque level is reached, the wrench clicks so the operator knows that the correct value of torque has been reached. The adjustable click type of torque wrench is very easy to use and also very fast. After setting the wrench, simply torque bolts until you hear a click. The click means you have attained the set amount of torque.

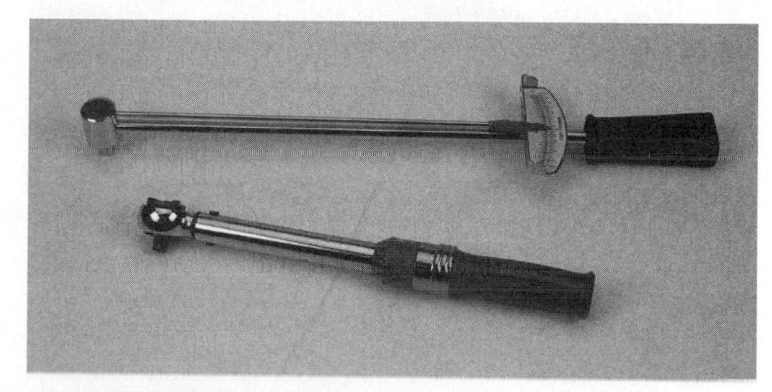

FIGURE 9–8 Torque wrenches are used to tighten bolts and nuts. Several types are available including dial, adjustable, click, and torque drives.

Car Clinic: TORQUE WRENCHES

CUSTOMER CONCERN:
Bolts Keep Breaking
A service technician complains that bolts are too weak and keep breaking when they are tightened down. What is a good method to keep bolts from breaking and to get the right torque on bolts?

SOLUTION: The best way to measure the exact amount of torque on bolts being tightened is to use a torque wrench. Service technicians who have worked with bolts and nuts for many years may think they are putting the right amount of torque on the bolt. However, certain bolts require higher torque, while other bolts require lower torque specifications. Always torque each bolt to its stated torque specifications. This way bolts are not weakened by too much torque.

The dial torque wrench has a dial on top so that as the torque is being applied to the bolt an exact reading can be obtained immediately. ***Figure 9–9*** shows one type of dial used on this type of torque wrench.

T-handle torque wrenches are used with standard ratchet wrenches. A dial is also used to indicate the exact amount of torque being applied. A socket is placed on the bottom of the torque wrench, while the ratchet is placed on top.

The torque driver is used for small torque specifications, where a screwdriver or nut driver would normally be used.

The deflecting beam torque wrench uses a deflecting rod in the center of the wrench to measure the exact value of torque applied to the bolt or nut. ***Figure 9–10*** shows a deflecting beam torque wrench. Note the position of the hands when using this type of torque wrench. One hand is placed on the twisting end of the torque wrench. Its purpose is to place a slight pressure on the end of the wrench so the socket and extension will not slip off the bolt. This slight pressure helps to balance the positioning of the torque wrench. Remember that too much pressure from this hand may produce inaccurate torque readings. The other hand is used for pulling the handle to produce the torque. As the bolt is tightened, the deflecting beam points to the amount of torque being produced on the bolt.

FIGURE 9–9 Several types of dials are used on torque wrenches. Some read in meter-kilograms, some in foot-pounds, and some read both.

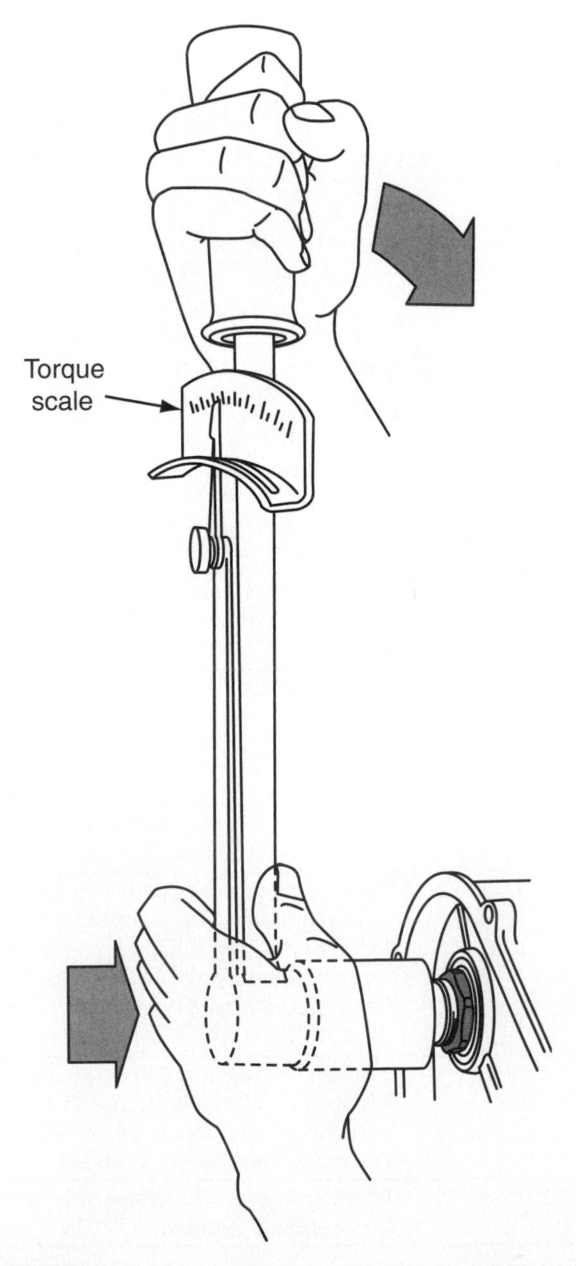

FIGURE 9–10 This is a deflecting beam torque wrench. As the bolt is tightened, the pointer reads the correct torque on the scale.

MICROMETERS

Micrometers are made to measure very small, accurate clearances. A micrometer is capable of measuring length in 0.001 of an inch. There are several types of micrometers. Outside micrometers measure outside dimensions such as shaft diameter, bearing thickness, and shim thickness. Inside micrometers are designed to measure internal dimensions such as engine cylinders and small holes. Inside diameters can also be measured with telescoping gauges. These gauges are used to measure the inside diameter of a bore. They are then measured with an outside micrometer to determine exact bore diameters.

Micrometers come in many sizes and ranges (***Figure 9–11***). Common sizes include 0″–1″, 1″–2″, 2″–3″, 3″–4″, and 4″–5″. In some cases, the micrometer set may include several adapters and extensions. This type of micrometer is often called a ratchet micrometer.

PARTS OF A MICROMETER

Figure 9–12 shows the basic parts of the micrometer. In this case, an outside micrometer is used.

All outside micrometers are the same except for the frame. When larger dimensions are used, the frame is made larger to fit larger dimensions.

READING A MICROMETER

There are several steps to reading a micrometer. The first step is to identify the size of the frame. As mentioned earlier, frame sizes are from 0–1 inch, 1–2 inches, 2–3 inches, 3–4 inches, and so on. Using the ratchet, slowly and gently turn the thimble until there is no clearance between the object being measured and the spindle and frame. At this point, note the highest figure on the barrel that is uncovered by the thimble. This number is the first figure to the right of the decimal point. For example, this number would be 0.200 on the micrometer shown in ***Figure 9–13***.

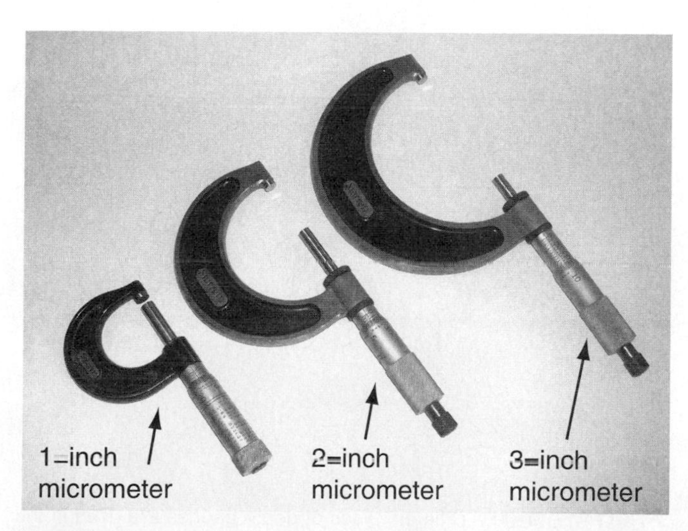

FIGURE 9–11 Micrometers are designed in many sizes.

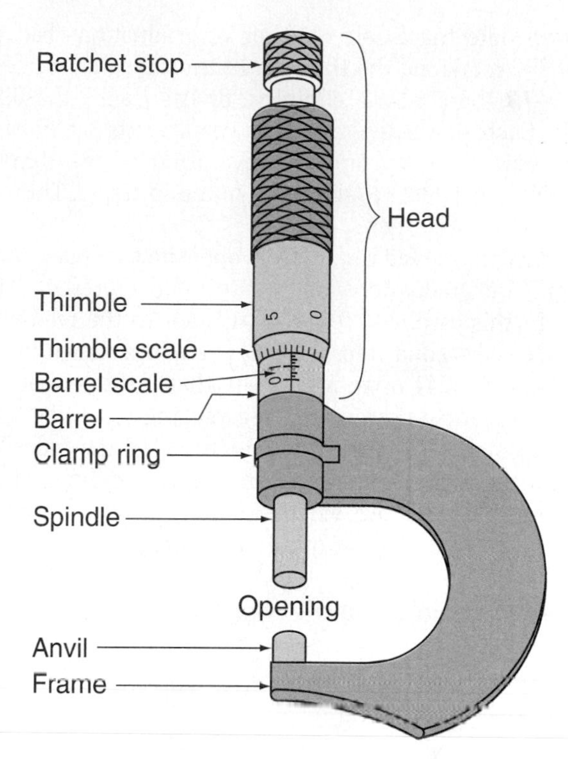

FIGURE 9–12 The outside micrometer has several parts. The object to be measured is placed between the anvil and the spindle. The measurement is taken on the barrel and thimble.

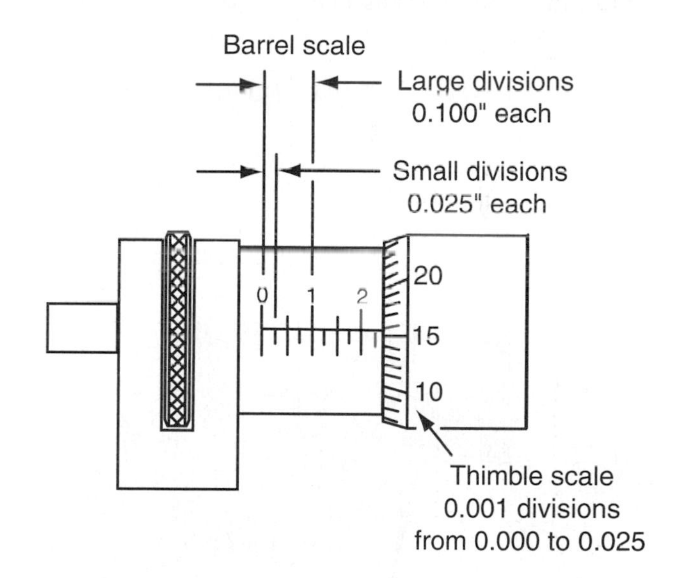

4 steps to read, add together:	Example above:
1. Select frame size	0 -1
2. Large barrel divisions	X 0.100 = 0.200"
3. Small barrel divisions	X 0.025 = 0.025"
4. Thimble divisions	X 0.001 = 0.016"
Reading	0.241"

FIGURE 9–13 A micrometer is read by totaling the number of whole divisions on the barrel scale and adding the thousandths from the thimble scale.

Now, note the whole number of graduations between the 0.200 mark and the thimble. In the case shown in ***Figure 9–13*** there is one complete division after the 0.200 mark. Each of these graduations represents 0.025 of an inch, which is one complete revolution of the thimble. Therefore, add 0.025 to the 0.200 found in step 1. This sum equals 0.225 of an inch.

At this point, read the thimble opposite the index on the barrel. The graduations on the thimble represent 0.001 inch. In this case, add 0.016 of an inch to the reading in the first and second steps. The total reading shown in ***Figure 9–13*** is 0.241 of an inch. Note that if the micrometer were a 2–3 inch micrometer, the reading would be 2.241 inches. When using the metric micrometer, the procedure is the same except the graduations represent different values. ***Figure 9–14*** shows what each graduation represents.

Other styles and types of micrometers read the same as the outside type of micrometer. The only difference is the method in which the micrometer fits on the object to be measured.

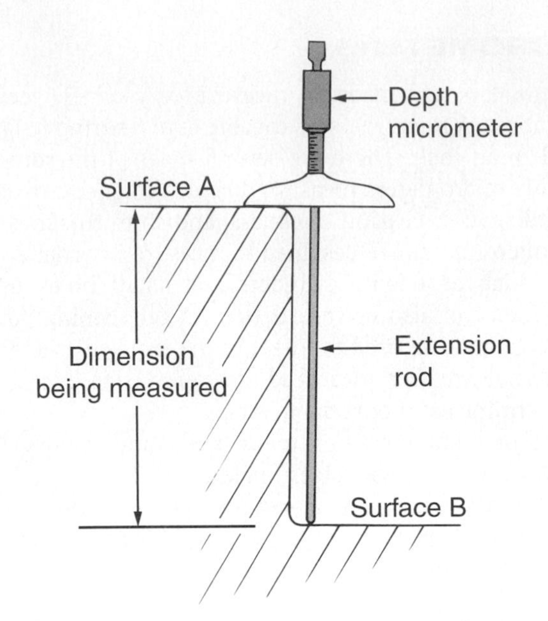

FIGURE 9–15 This depth gauge is read in a similar manner to a standard micrometer.

DEPTH GAUGES

Often it is necessary to measure the difference or dimension between two parallel surfaces. For example, refer to ***Figure 9–15***. In this case, the dimension needed is between surface A and surface B. To measure this dimension accurately a depth gauge can be used. A depth gauge works similar to a micrometer. It has a barrel, a thimble, and a sleeve like an outside micrometer. However, it also has different length extension rods to take the measurement. ***Figure 9–16*** shows two typical depth micrometers. The top one works like a micrometer and the bottom one works like a vernier caliper.

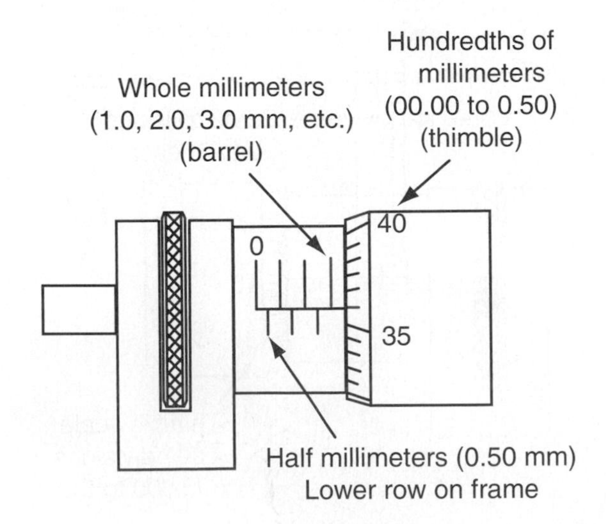

4 steps to read, add together:	Example above
1. Whole mm lines (upper) on barrel	3 = 3.00 mm
2. Half mm line (lower) on barrel	0 = 0.00 mm
3. Lines on thimble	36 = 0.36 mm
Reading	3.36 mm

FIGURE 9–14 The metric micrometer is read in a similar manner to the standard micrometer, though the graduations are read in millimeters rather than inches.

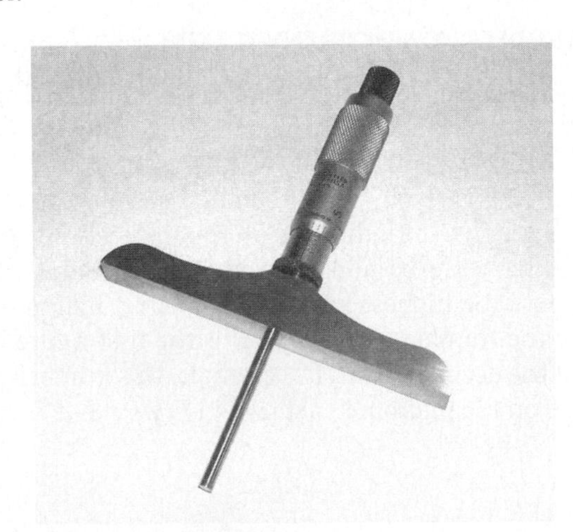

FIGURE 9–16 Different types of depth gauges are used in the automotive shop.

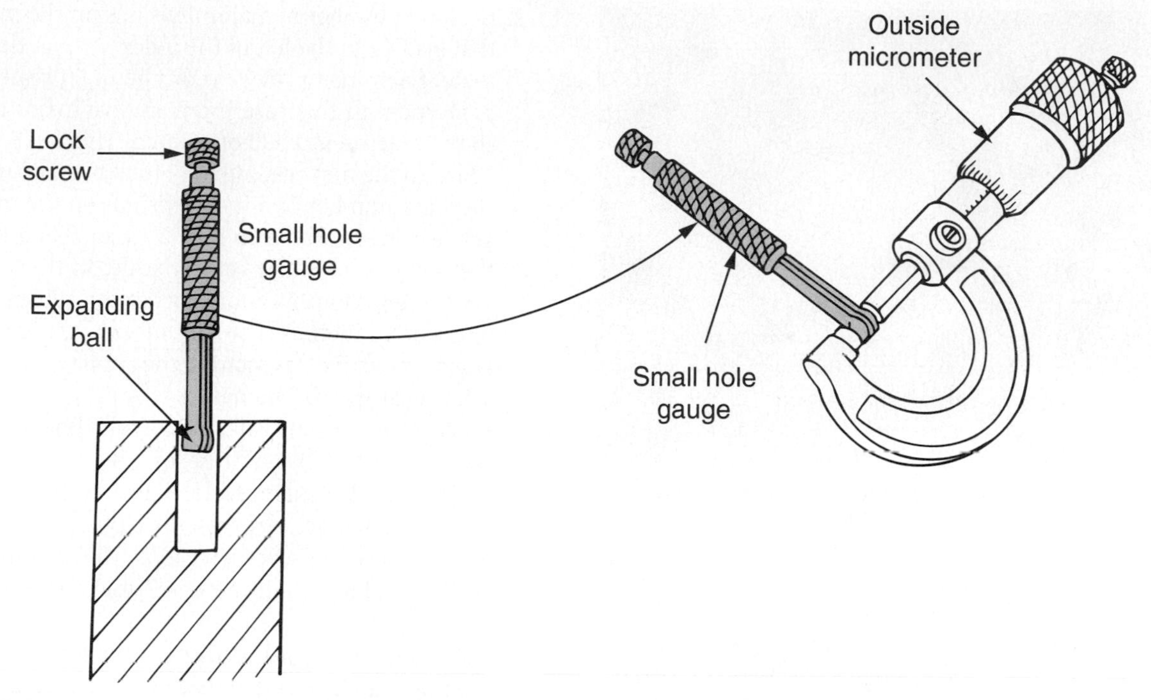

FIGURE 9–17 The small hole gauge is used to measure the inside diameter of a small hole.

SMALL HOLE GAUGES

At times it may be necessary to measure small holes. For example, the hole size of a valve guide may need to be measured. Small holes such as this are measured using a small hole gauge and a micrometer. The small hole gauge is an instrument that is inserted down into a hole. The bore of the small hole gauge is turned to expand a small ball at its end. The small hole gauge is then removed and its diameter measured with a standard outside micrometer (***Figure 9–17***).

TELESCOPING GAUGES

A telescoping gauge is used much the same as a small hole gauge. The telescoping gauge is used to measure slightly larger inside diameters than the small hole gauge. The telescoping gauge has a set of plungers that are spring-loaded. The plungers are placed in a hole and released outward. The plungers are then locked in place by the lock screw. The outside dimension is then measured with a standard outside micrometer. ***Figure 9–18*** shows the use of a telescoping gauge.

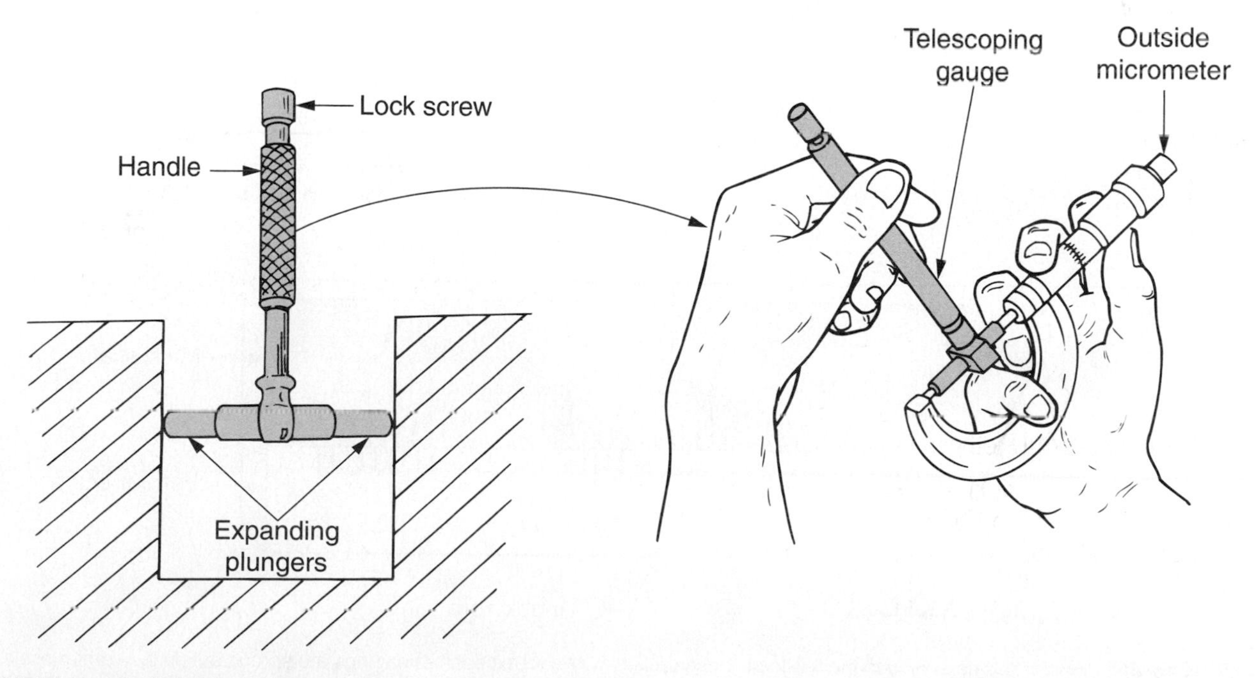

FIGURE 9–18 This telescoping gauge is used along with an outside micrometer to measure hole diameters.

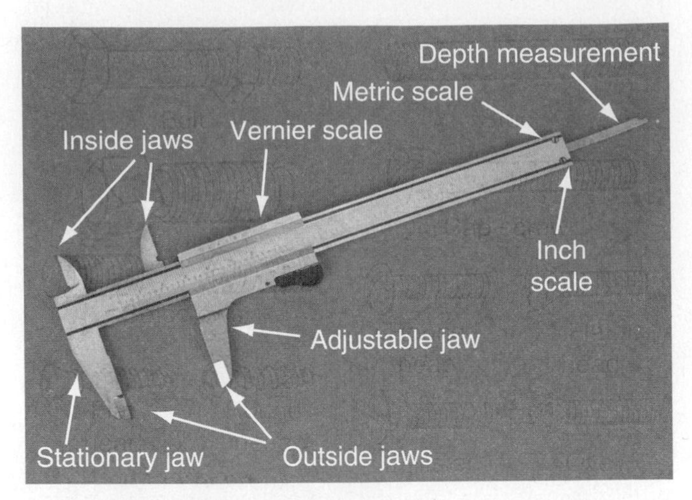

FIGURE 9-19 A vernier caliper is used to measure clearances, both inside and outside, to 0.001 of an inch.

VERNIER CALIPER

The **vernier caliper** is a useful tool for measuring various dimensions on the automobile engine. The vernier caliper measures length to 0.001 of an inch. A vernier caliper can measure inside or outside dimensions and in some cases make depth measurements (***Figure 9–19***). Vernier calipers can also be used to measure the inside diameter with the telescoping gauges.

READING A VERNIER CALIPER

There are several steps to reading a vernier caliper. Referring to ***Figure 9–20***, the steps are as follows:

1. Read the number of whole inches on the main scale, left of the index zero on the vernier scale. In this case the reading is 1.000.
2. Read the number of major divisions on the main scale that also lie to the left of the index zero on the **vernier scale**. Each major division on the main scale is 0.100 of an inch. In this case there are five major divisions, so the reading is 0.500 of an inch. This reading is now added to the first reading for a total of 1.500 inches.
3. Read the number of minor divisions on the main scale between the number in step 2 (5, or 0.500 inch) and the index zero on the vernier scale. In this case there are three minor divisions. Each minor division equals 0.025 of an inch. Add this number (three minor divisions, or 0.075 of an inch) to the reading in step 2. The total so far is 1.575 inches.
4. Look at the vernier scale. Identify the lines or divisions that most perfectly coincide, or line up, with any graduation on the main scale. Each of these divisions on the vernier scale represents 0.001 inch. In this example, there are seven divisions from the index zero, so the reading would be 0.007 inch. This reading is then added to the result in step 3. The total reading is 1.582 inches. Once a vernier caliper scale is mastered, readings can be taken with speed and accuracy.

FEELER GAUGES

Feeler gauges are used to help check or adjust small clearances to a specific measurement. They are made of thin metal blades or wires, each of which is designed to be a different thickness in thousandths of an inch. The set of feeler gauges shown in ***Figure 9–21*** ranges in thickness from 0.0015 to 0.035 of an inch. When an adjustment is needed, such as the clearances between valves and rocker arms, the correct gauge is placed between the two objects. The clearance is adjusted so that a small amount of drag can be felt when pulling the feeler gauge out of the area of adjustment.

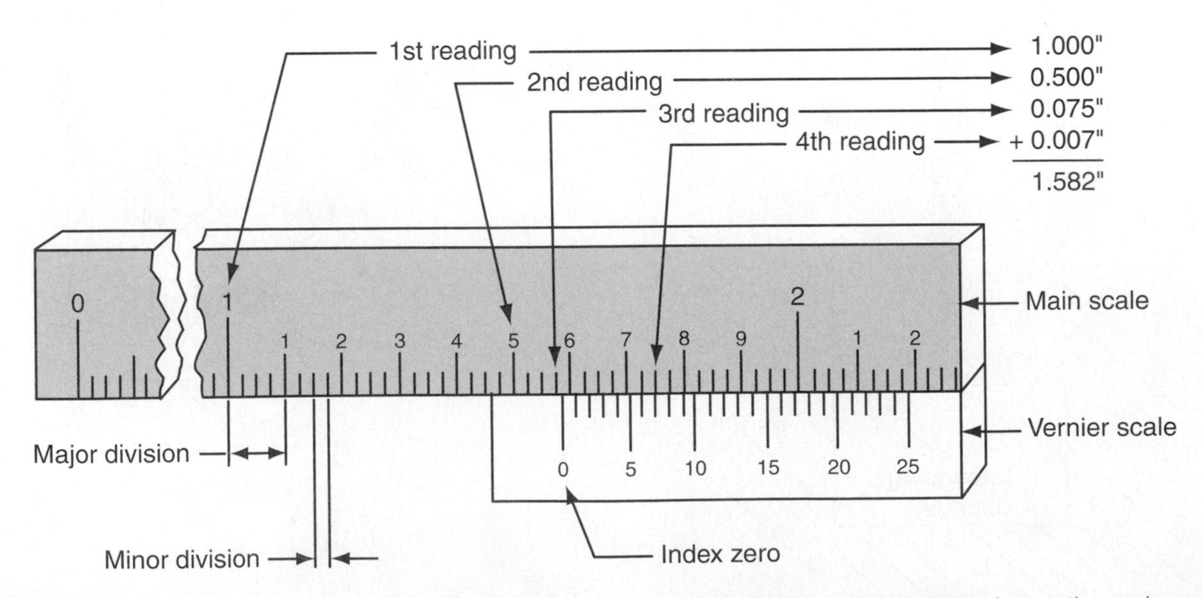

FIGURE 9-20 Vernier scales are much the same as those on a micrometer, except the readings are spread out on the vernier scale.

FIGURE 9-21 Feeler gauges are used to check or adjust clearances to a particular specification.

Some feeler gauges have a stepped clearance. For example, the feeler gauge may have 0.005-inch thickness at the end of the blade. Then, about 1/2-inch from the tip of the blade, the thickness changes to a thicker measurement. This type of feeler gauge is called a "go–no go" gauge. The clearance being measured should fit the 0.005-inch thickness but then be stopped by the thicker part of the blade.

Figure 9–22 shows various types and sizes of feeler gauges used on the automobile engine. Note that some of the feeler gauges are long, whereas others are short. The length of the gauge that should be used depends on the measurement being taken. Certain feeler gauges are made of nonmagnetic metal such as brass. These feeler gauges are used to measure clearances on magnetic pickup coils on many electronic systems. In addition, several wire feeler gauges (called wire gap gauges) are shown. Wire gap gauges are used mostly for setting spark plug clearances. Note that some feeler gauge sets include both feeler gauges and wire gap gauges.

FIGURE 9-22 Various feeler and wire gap gauges are used to check clearances on an automobile.

Car Clinic:
FEELER GAUGES

CUSTOMER CONCERN:
Correct Use of Feeler Gauge

A service technician seems to be getting different readings when using a feeler gauge on a set of contact points. What is a good method to make sure a feeler gauge is used correctly?

SOLUTION: A feeler gauge uses the technician's "feel" to get the correct reading. Often it is difficult to get an accurate reading using a feeler gauge. To increase accuracy when using a feeler gauge, slip the gauge into the clearance being checked. Now twist or tilt the feeler gauge from one side to the other. If there is movement, the clearance is still too high. Each time you use the feeler gauge, remember to twist or tilt the gauge from side to side to get an accurate reading.

WIRE GAUGES

Wire gap gauges, shown in ***Figure 9–23***, have different size wires to check specific dimensions. For example, common wire sizes include 0.020, 0.025, 0.028, 0.030, 0.032, 0.035, 0.040, and so on. If a spark plug gap dimension is stated as 0.055, then the service technician will try to insert the 0.055 gauge. If it does not fit, the clearance is too small and must be enlarged. If the wire slips through the gap too easily, the clearance must be reduced. It may take several adjustments to get the exact clearance needed.

DIAL INDICATORS

Figure 9–24 shows a **dial indicator** set. A dial indicator is a measurement tool used to determine clearances between two moving objects. For example, gear backlash (the clearances between the teeth on two gears in mesh) may have to be checked. Dial indicators are also used to check ball

WIRE GAUGES

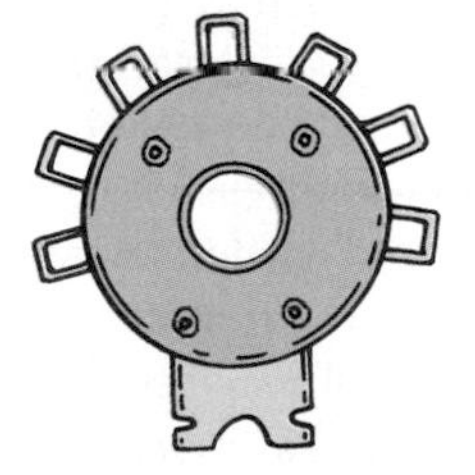

FIGURE 9-23 These wire gauges are used to measure small clearances such as spark plug gaps.

FIGURE 9-24 Dial indicators are used to measure small clearances such as crankshaft end play and flywheel runout.

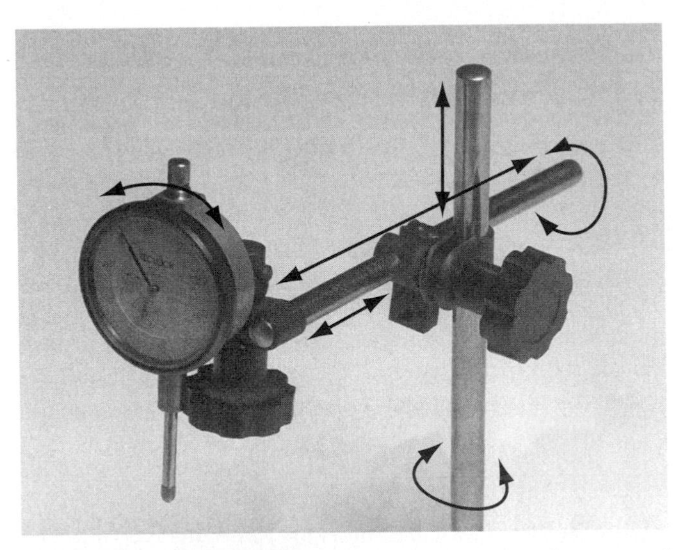

FIGURE 9-25 The dial indicator has attachments that are used to position the dial indicator in any position.

joints, tie-rods, cam and valve guide wear, crankshaft end play, and disc brake runout. Dial indicators measure clearances in increments of 0.001 of an inch.

There are several types of dial indicator attachments as part of the dial indicator stand. ***Figure 9–25*** shows an example of a dial indicator and the type of movements available from a particular stand. It is important to have these attachments so that the dial indicator can be positioned exactly in the right spot to allow accurate measurements to be taken. Some dial indicators also have magnets on the bottom of the attachments so that they can be easily held to a metal part.

FIGURE 9-26 This service technician is checking the axial movement (in-and-out) of the axle using a dial indicator.

A common adjustment that is taken with a dial indicator is in-and-out or axial movement. ***Figure 9–26*** shows an example of how a dial indicator is positioned to take in-and-out clearance on the rear axle. This type of dial indicator has a flexible attachment so that it can be placed in almost any position. To accurately read the axle movement, the dial indicator tip is first gently positioned on the axle. Then the dial indicator numbers are rotated on the face so that numbers read zero at the point of the needle. The axle is then moved gently in and out. As it is moved in and out, the dial indicator reads the clearance.

Figure 9–27 shows how runout on an axle is measured with a dial indicator. After the dial indicator is placed on the axle and adjusted to zero, the axle is gently rotated. As it is rotated, the service technician reads the amount of variation in 0.001-inch increments on the dial indicator face.

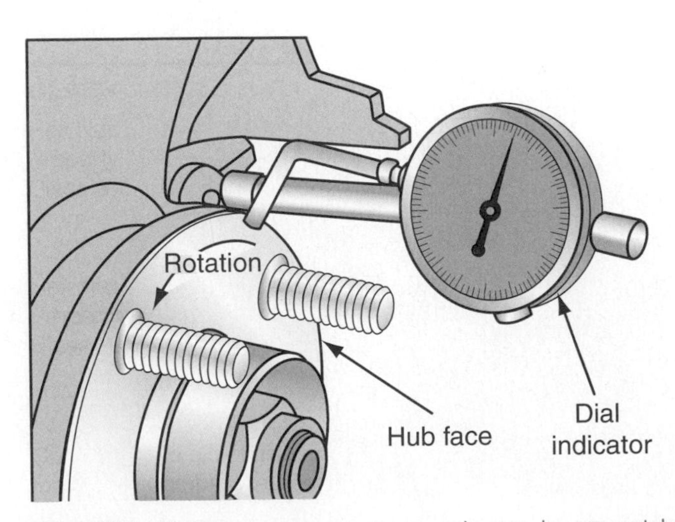

FIGURE 9-27 The runout on an axle can be accurately checked with a dial indicator.

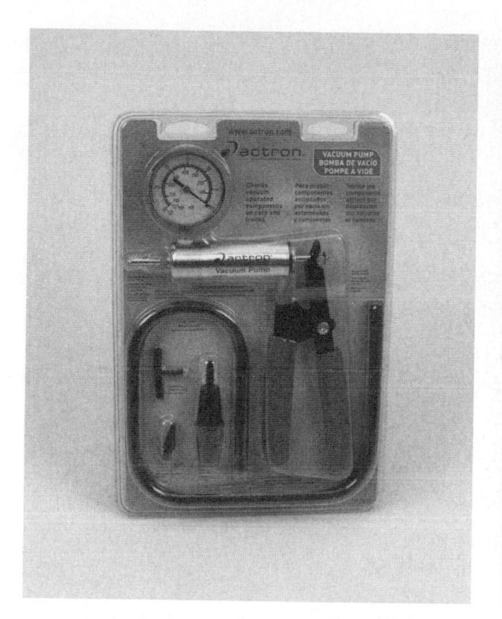

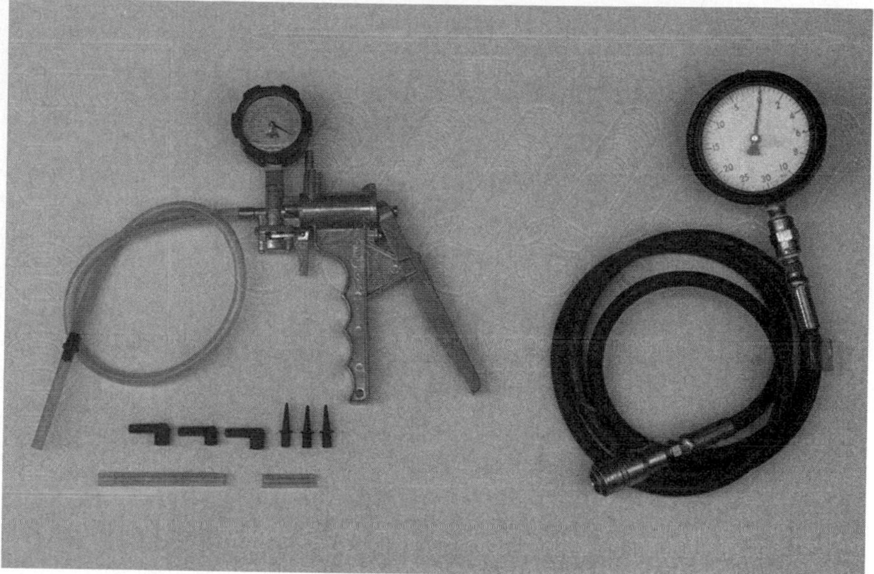

FIGURE 9-28 Pressure and vacuum gauges are used to take different readings on automobiles.

VACUUM AND PRESSURE GAUGES

Vacuum and **pressure gauges** are used to help troubleshoot, or, when necessary, to check various pressures and vacuum on the engine. ***Figure 9–28*** shows a complete set of pressure and vacuum gauges. Common gauges used by the service technician include the engine oil pressure gauge, transmission oil pressure gauge, engine vacuum gauge, fuel pump pressure gauge, fuel injection pressure gauge, the vacuum pump gauge, and the engine compression gauge. Service and diagnosis using vacuum and pressure gauges are discussed in many of the following chapters. One such example is shown in ***Figure 9–29***, in which the service technician is using a compression gauge to check the compression on an engine. Note the gauge and connections that are necessary for the compression check.

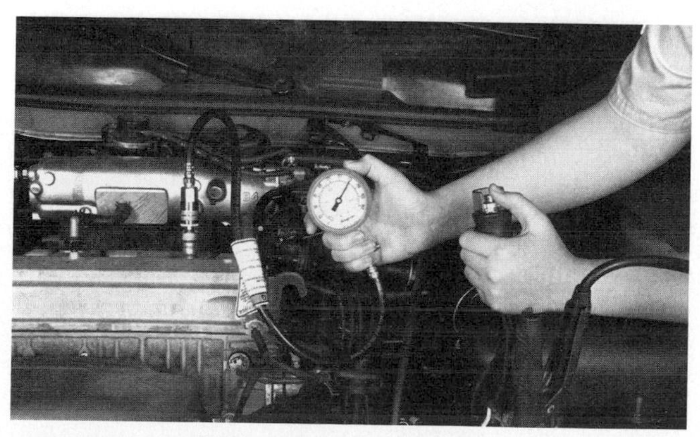

FIGURE 9-29 This service technician is using a pressure gauge to check the cylinder compression.

Safety Precautions

When working with measuring instruments in the automotive shop keep in mind these important safety suggestions.

1. Return all measuring tools to their designated toolbox or location on the wall after each use. Tools left on a vehicle may fall into moving parts of the engine. They could then fly out and cause injury.
2. When working with any measuring instruments in the shop, always wear approved safety glasses.
3. If you are not familiar with the correct use of any measuring instrument, always check with the instructor. Never use a tool that you don't know how to use correctly.
4. When using torque wrenches, always pull the wrench toward your body. With this technique, you get the maximum leverage and greatest safety margin.
5. While you are using pressure and vacuum gauges, the engine may need to be running. Always be careful where you place the gauges, and make sure the gauge connections, piping, and tubing are away from moving parts.

SUMMARY

The following statements will help to summarize this chapter:

- Service technicians use a number of measuring tools.
- There are two types of measuring systems: the USC, or English, system and the metric system.
- USC measurements include inches, feet, miles, quarts, pounds, ounces, and so on.
- All metric measurements are based on the meter and use a base of 10.
- Metric measurements include meters, decimeters, centimeters, millimeters, liters, and so on.
- Measuring tools are used to calculate various lengths, torque bolts, adjust clearances, and check pressures.
- Common measuring tools include micrometers, torque wrenches, vernier calipers, feeler gauges, dial indicators, and pressure and vacuum gauges.
- Micrometers measure lengths in 0.001 of an inch, and use a barrel and a spindle to read the measurement.
- Torque wrenches are used to tighten bolts and nuts, and are read in inch or foot pounds.
- Vernier calipers measure clearances in 0.001 of an inch, and use a vernier scale placed on the instrument.
- Feeler gauges are used to measure small clearances from 0.0015 to 0.080 of an inch, and measure spark plugs, valves, magnetic sensor clearances, etc.
- Dial indicators are used to measure clearances between two moving objects.
- Dial indicators can be used to measure runout on a shaft.
- Vacuum and pressure gauges are used to help the service technician troubleshoot specific pressures and vacuums that are produced in the engine.

TERMS TO KNOW

Can you explain each of the following terms? Review the chapter until you can use each term correctly.

Base of 10
Conversion factors
Cubic centimeter
Dial indicator
Feeler gauges
Gram
Liter
Meter
Metric system
Micrometers
Prefixes
Pressure gauges
Torque wrenches
USC measurements
Vacuum gauges
Vernier caliper
Vernier scale
Wire gap gauges

REVIEW QUESTIONS

Multiple Choice

1. Which of the following systems of measurement uses the base of 10?
a. USC
b. metric system
c. English system
d. SSC system
e. OSC system

2. What is the basic unit of length in the metric system?
a. Foot
b. Meter
c. Cubic centimeter
d. Liter
e. Inch

3. Which measurement tool is used to tighten bolts to correct specifications?
a. Dial indicator
b. Adjustable wrench
c. Torque wrench
d. Pressure gauge
e. Vacuum gauge

4. Which measurement tool can accurately measure 0.001 of an inch?
a. Vernier caliper
b. Vacuum gauge
c. Pressure gauge
d. Micrometer
e. A and D

5. Which of the following tools should be used to adjust or measure small clearances of 0.001 of an inch?
a. 3/8-inch socket
b. Torque wrench
c. Micrometer
d. Pressure gauge
e. Vacuum gauge

6. Which measuring tool has a scale in which two graduations or lines are lined up with each other to obtain the reading?
a. Micrometer
b. Vernier caliper
c. Feeler gauge
d. Torque wrench
e. None of the above

7. Which of the following wrenches uses a long pointer that points to a scale as the bolt is tightened?
a. Deflecting beam torque wrench
b. Dial torque wrench
c. Dial indicator
d. Vernier caliper
e. Micrometer

8. Which prefix represents the number 1,000?
a. Milli
b. Centi
c. Deci
d. Kilo
e. None of the above

9. Which of the following prefixes represents the number 0.001?
a. Milli
b. Centi
c. Deci
d. Kilo
e. Mega

10. When 1 foot is divided into 12 inches, a base of ________ is used.
a. 12
b. 10
c. 1/12
d. 120
e. 15

The following questions are similar in format to ASE (Automotive Service Excellence) test questions.

11. *Technician A* says that a feeler gauge can be used to measure small clearances of 0.001 of an inch. *Technician B* says that a micrometer can be used to measure clearances with an accuracy of 0.001 of an inch. Who is correct?
a. A only
b. B only
c. Both A and B
d. Neither A nor B

12. *Technician A* says the metric system uses a base of 12. *Technician B* says the USC system uses a base of 10. Who is correct?
a. A only
b. B only
c. Both A and B
d. Neither A nor B

13. *Technician A* says that the vernier scale on the vernier caliper can read degrees. *Technician B* says that the vernier scale on the vernier caliper can read in 0.001 inches. Who is correct?
a. A only
b. B only
c. Both A and B
d. Neither A nor B

14. *Technician A* says that a nonmagnetic feeler gauge should be used to measure clearances between the pickup coil and reluctor on electronic ignition engines. *Technician B* says that a nonmagnetic feeler gauge cannot be purchased. Who is correct?
a. A only
b. B only
c. Both A and B
d. Neither A nor B

Essay

15. Describe the relationship between a liter and a cubic centimeter.

16. What is the difference between a vernier caliper and a micrometer?

17. Which type of measuring tool would be used to measure small clearances such as valve clearances?

18. Describe the purpose and operation of a telescoping gauge.

19. Explain the procedure for using a wire gauge to measure a small clearance.

20. Explain the procedure for using a depth gauge.

Short Answer

21. When checking clearances on magnetic pickup coils on electronic ignition systems, always use a ________ type of feeler gauge.

22. The gauge used to measure the distance between two parallel surfaces is the ________ gauge.

23. The gauge used to measure small clearances such as the spark plug gap is called a ________ gauge.

24. The ________ gauge is used to measure small hole diameters such as the valve guide.

25. After a telescoping gauge is locked in place, a (an) ________ is used to measure the distance between the two spring-loaded plungers.

Applied Academics

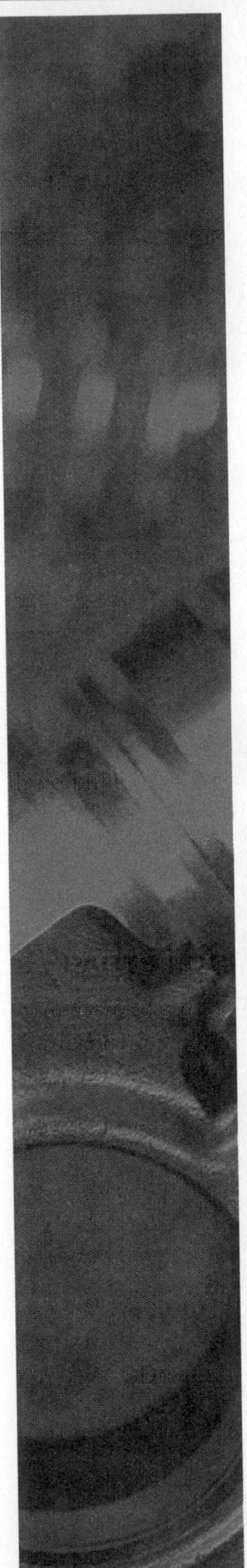

TITLE: Length Conversions

ACADEMIC SKILLS AREA: Science

NATEF Automobile Technician Tasks:

Convert measurements taken using either the USC or metric system to specifications stated in terms of the other system.

CONCEPT:

Automotive service technicians often work with both USC (English) and metric units of length. To help convert length measurements from one system to the other, conversion tables like the one shown in the accompanying figure are used. There are three simple steps to using this table. Say that you are interested in converting from inches to centimeters. First, under number 1 on the table, find the unit you wish to convert from—in this case, inches. Second, under the number 2 column on the right, find the row that has the unit you wish to convert to—in this case, centimeters. Follow the inch column down and the centimeter row across. Where they meet is the number to use to make the conversion—in this case, multiply the inches by 2.54 to get centimeters.

1. To Convert from — 3. Multiply by

INCH	FEET	YARD	ROD	MILE (statute)	MILE (naut.)	MILLI-METER	CENTI-METER	METER	KILO-METER	LIGHT YEAR	2. Into
1	12	36	198	63360	72962.4	.03937	.3937	39.37	39370	3.73×10^{17}	= Inch
.8333	1	3	16.5	5280	6060.2	.00328	.03281	3.2808	3280.8	3.10×10^{16}	= Feet
.02778	.3333	1	5.5	1760	2026.7	.00109	.01094	1.0936	1093.6	1.03×10^{16}	= Yard
.00505	.06061	.18182	1	320	368.49	.000199	.00199	.19884	198.84	1.88×10^{15}	= Rod
1.58×10^{-5}	.000189	.000568	.003125	1	1.1516	6.21×10^{-7}	6.21×10^{-6}	.00062	.62137	5.88×10^{12}	= Mile (statute)
1.37×10^{-5}	.000164	.000493	.00271	.86839	1	5.40×10^{-7}	5.40×10^{-6}	.00054	.53959	5.10×10^{12}	= Mile (naut.)
25.400	304.80	914.402	5029.2	1.61×10^{4}	1.85×10^{4}	1	10	1000	1×10^{4}	9.46×10^{18}	= Millimeter
2.540	30.480	91.4402	502.92	160935	185235	0.1	1	100	1×10^{5}	9.46×10^{17}	= Centimeter
.02540	.30480	.914402	5.0292	1609.35	1853.25	0.001	.01	1	1000	9.46×10^{15}	= Meter
2.54×10^{-5}	.000305	.000914	.00503	1.60935	1.85325	1×10^{-4}	1×10^{-5}	.001	1	9.46×10^{12}	= Kilometer
2.69×10^{-18}	3.23×10^{-17}	9.69×10^{-17}	5.33×10^{-16}	1.70×10^{-13}	1.96×10^{-13}	1.06×10^{-19}	1.06×10^{-18}	1.06×10^{-14}	1.06×10^{-13}	1	= Light Year

APPLICATION:

Can you determine how many centimeters are in 15 inches? Can you determine how many feet are in 5 meters? If you have a 13-centimeter measurement, how many inches are there in the same length?

Electrical Tools, Power and Pressing Tools, and Cleaning Equipment

CHAPTER 10

OBJECTIVES

After studying this chapter, you should be able to:

- Identify common electrical tools such as the volt-ohm-ammeter, testlight, dwell/tach meter, tachometer, timing light, and electronic analyzers and scanners.
- Describe various power and pressing tools including pullers; bushings, bearings, and seal installers; hydraulic presses; vehicle jacks; impact wrenches; shop cranes; grinders and wire wheels; and power drills.
- Analyze the uses of cleaning equipment in the automotive shop including parts cleaners, solvent tanks, spray washers, and other types of cleaners.

INTRODUCTION

In the automotive shop the service technician needs various types of electrical tools, pressing tools, power tools, and cleaning equipment to work effectively. These tools help the service technician complete work on a vehicle correctly, on time, and safely, while maintaining a clean and organized shop. This chapter introduces common electrical tools needed in the shop, along with power and pressing tools. In addition, cleaning equipment is also presented.

10.1 ELECTRICAL TESTING TOOLS

Many electrical tools are used to service electrical and electronic systems on today's automobiles. Electrical tools are used mostly for diagnosing various problems in many of the electrical systems. These tools make diagnostics easier and faster. Some of the more common electrical tools include voltmeters, ohmmeters, ammeters, testlights, tachometers, dwell/tach meters, timing lights, electronic engine analyzers and computers, and scanners.

VOLT-OHM-AMMETER

One of the most useful tools today is the **volt-ohm-ammeter (VOM)**. This testing instrument, also called the multimeter, shown in ***Figure 10–1***, can be used to test a variety of components, especially on electronically controlled engines. Depending on the manufacturer, year, and make of the multimeter, various styles can be purchased. Today, the most common type of multimeter has a digital readout. Generally, this tool is used to measure values such as the following:

- Voltage drop across an electrical component or circuit
- Internal resistance of any electrical component or circuit
- Current, or amperage, flowing through small electrical components
- Grounded circuits
- Open circuits
- High-resistance circuits

Most service technicians prefer using a digital volt-ohm-ammeter. Often called a **DVOM** (digital volt-ohm-meter),

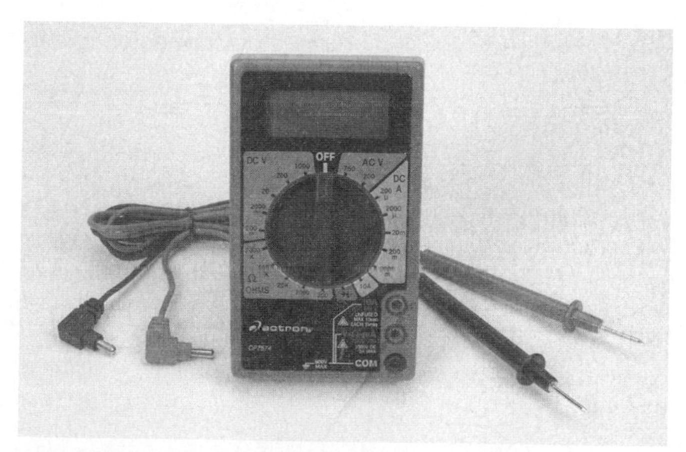

FIGURE 10–1 Multimeters are used in the automotive shop for testing electrical circuits for voltage, amperage, and/or resistance.

this meter reads out in a digital display. There are various ohm scales including 200, 2,000 (2K), 20,000 (20K), 200,000 (200K), 2,000,000 (2M), and 20,000,000 (20M). In addition there are various volt scales including 2 volts, 20 volts, and 200 volts.

An important characteristic of any volt-ohm-ammeter is internal resistance, referred to as internal impedance. The internal impedance of older analog VOMs was relatively low. This meant that when voltage and resistance readings were taken on a particular component, there was always high current flowing through the component. However, newer DVOMs have high impedance (10 mega-ohms or more). This means that much less current flows through the component that is being checked. This is important because of many of the computerized components of today's vehicles are sensitive and cannot handle the high currents associated with analog VOMs.

When using a digital volt-ohm-ammeter, it is important to connect the meter in the correct fashion. ***Figure 10–2*** shows how a DVOM is properly connected. When testing for voltage or voltage drop across a part of the circuit, the DVOM is always hooked in parallel as shown in Fig-

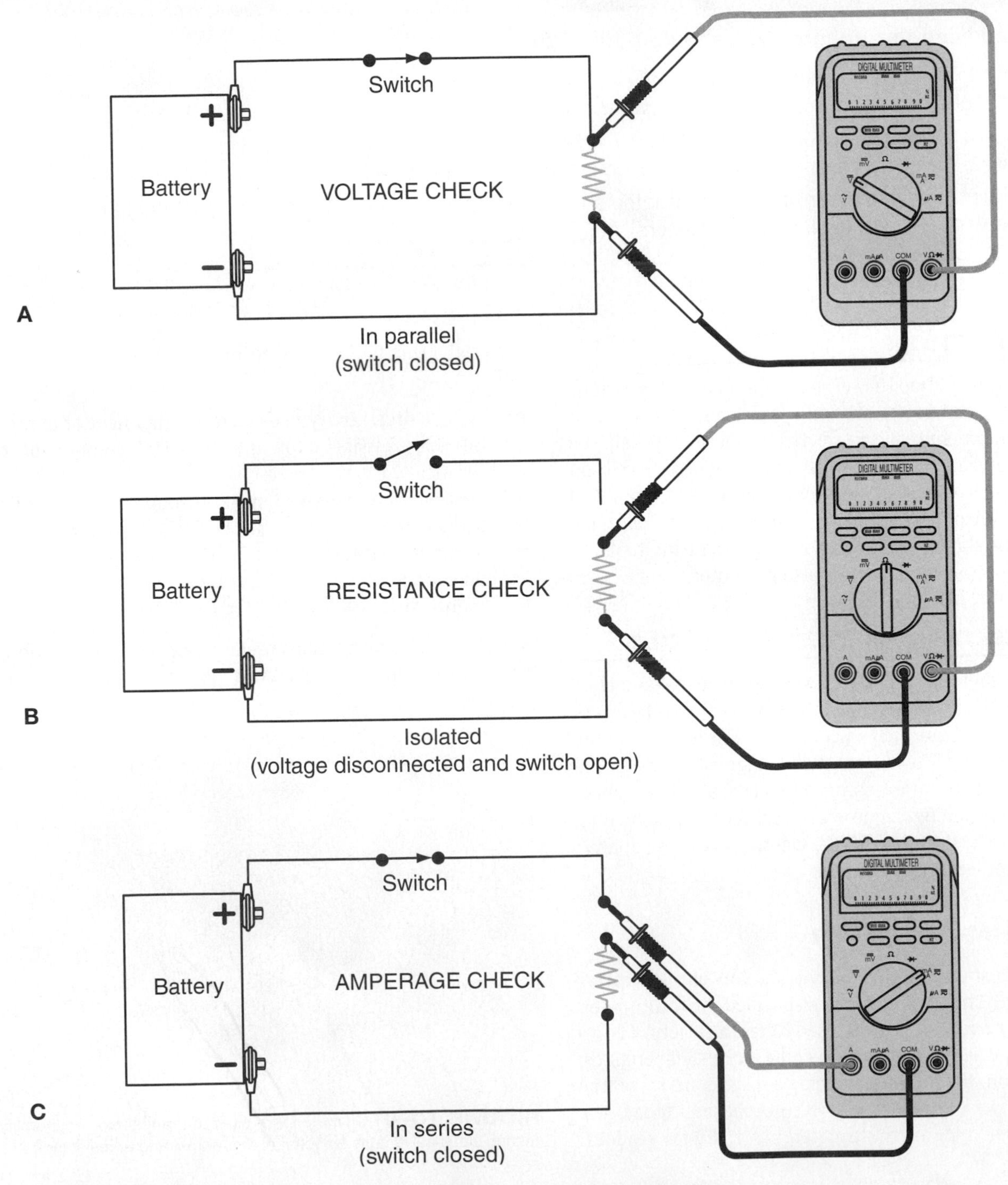

FIGURE 10–2 Meter hookups for checking voltage, resistance, and amperage are shown.

ure 10–2A. This means that the DVOM is connected across the resistor or load to be checked. Note that the switch is closed, which means that there is power to the resistor. When checking for resistance, the power must be removed from the circuit and the resistor must be isolated from the rest of the circuit. This is shown in Figure 10–2B. If there is any voltage or power to the resistor, the resistance readings will be incorrect. Figure 10–2C shows how to check for amperage. First, the circuit must be disconnected and the DVOM placed in series with the load. The power is then turned on (switch closed) and the reading is taken. Remember that when checking any voltage, resistance, or amperage, always set the meter to the correct scale.

TESTLIGHT

A **testlight**, shown in ***Figure 10–3***, is a simple testing instrument used to test if there is electricity at certain points in electrical circuits. Typically, one end of the testlight is grounded. The other end of the testlight has a sharp, pointed terminal. This pointed terminal can be inserted through wire insulation to touch the electrical wire. The two terminals are connected to a filament or bulb inside the testlight. The testlight will light up if there is voltage available where the pointed terminal is touching. When troubleshooting electrical circuits, the testlight can be helpful in finding the location of an electrical problem.

For example, say one of the taillights on a vehicle is not working. A testlight can be used to find out if there is any electricity flowing to the taillights. Referring to ***Figure 10–4***, note that the hot wire is coming from the battery. The electricity normally flows to the taillight and then to ground or back to the battery. If the taillight is not working, the test light can be inserted into the wire before the taillight. If the testlight turns on, it means that electricity is available before the taillight. If the testlight does not turn on, it means that electricity is not available in the wire before the taillight. This gives the service technician information to determine what area of the circuit to check further.

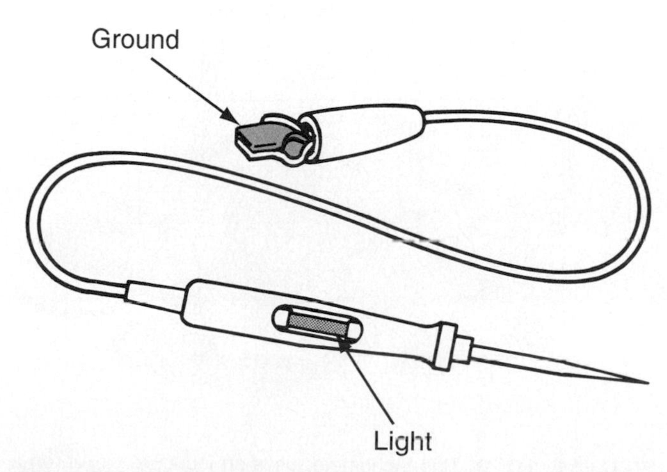

FIGURE 10–3 A testlight can be used to check for voltages in various parts of an electrical circuit.

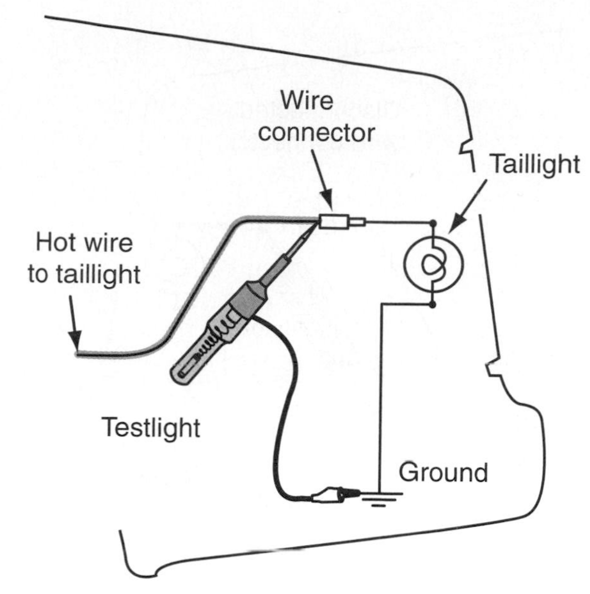

FIGURE 10–4 A testlight can be used to locate faulty components in a circuit like this simple taillight electrical circuit.

Note that the testlight is not recommended for troubleshooting electronic computerized circuits and controls.

CONTINUITY TESTER

Another type of circuit tester is the continuity tester. This tester is used to test for open circuits when there is no electricity flowing in the circuit. It looks similar to a regular testlight but has a battery inside the handle. The battery is used to check for **continuity** (open or completed circuit). When the probe is touched to the positive side, the lamp will light up if there is continuity within the circuit. When an open circuit exists, the light will not go on. ***Figure 10–5*** shows a typical continuity tester. Always remember to make sure no electricity is flowing in the circuit before using this tester.

Consider the previous example of the taillight that is not operating properly. The technician can then use a continuity tester to check for an open circuit between the wire

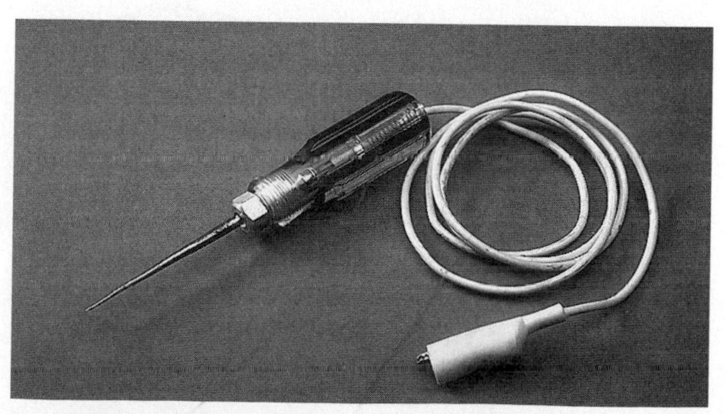

FIGURE 10–5 This continuity tester will light up when the circuit is complete. It has a battery inside to produce a small voltage.

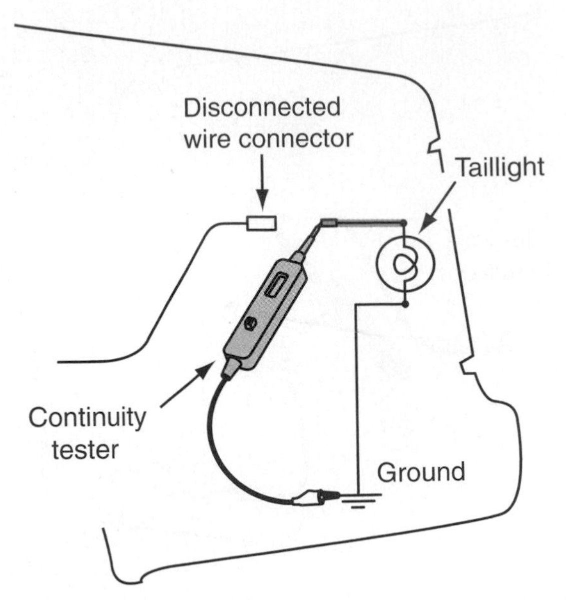

FIGURE 10–6 The continuity tester can be used to determine if a wire is broken as shown in this simple taillight electrical circuit.

connector and ground. The procedure for doing this is as follows. First, disconnect the wire connector to eliminate battery voltage. After being connected, the continuity tester will cause electricity to flow through the circuit as shown in ***Figure 10–6***. Depending on where the continuity tester is placed in the circuit, various circuit points of high resistance can be checked. If the light in the continuity tester lights up, there is a complete circuit. If the continuity tester does not light up, it means that there is a high resistance or open circuit such as broken wire or filament in the bulb.

LOGIC PROBE

In certain circuits, there is pulse or digital signal. These on-and-off signals transmit electronic information to different components on computer-controlled circuits. A logic probe is used to check the continuity of wires that carry digital signals. ***Figure 10–7*** shows a logic probe. When checking for continuity, three colored light emitting diodes (LEDs) are used. A red LED goes on to show there is a high voltage present in the wire. A green LED goes on to show that there is a low voltage present in the wire. A yellow LED goes on to show there is voltage pulse present. Note that both a yellow LED and a red or green LED can go on at the same time. This would indicate that there is a voltage pulse at that point in the circuit and either a high or low voltage.

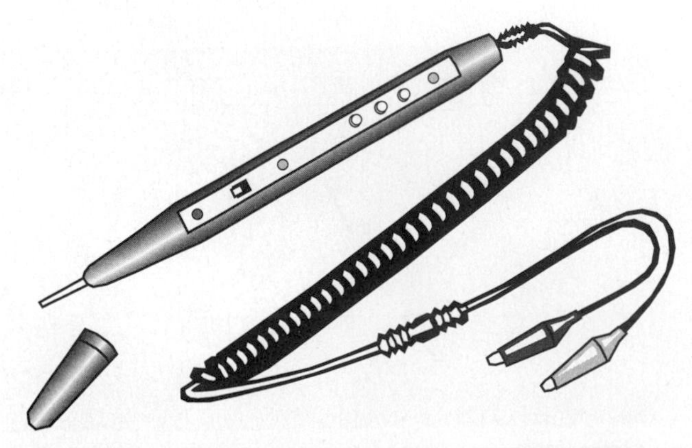

FIGURE 10–7 This logic probe can test for both high and low voltages, as well as pulses of voltage.

TACHOMETER

Tachometers are used to determine the speed, in revolutions per minute (rpm), at which an engine is operating. Most tachometers are part of other electrical analysis equipment used on automotive engines. Electrical tachometers generally measure the firing pulses sent to the spark plugs by the distributor. Other tachometers use a photoelectric or magnetic pickup.

Several types of tachometers are used in today's automotive shop. One common type is the inductive pickup tachometer (***Figure 10–8***). In this case, an inductive (electromagnetic) terminal clamps over the number 1 spark plug wire. As electricity flows through the spark plug wire, the inductive terminal picks up the magnetic pulse from the electrical spark signal. This is then transformed into an rpm reading for the engine.

Another type of tachometer, called a **photoelectric** tachometer, uses light pulses to determine the rpm of the engine (***Figure 10–9***). First, a piece of reflective tape is placed on the crankshaft, (or other rotating parts) most often on the front pulley or harmonic balancer. With the engine running, a light from the photoelectric tachometer is then pointed toward the reflective tape. The reflective tape reflects a series of light pulses back into the photoelectric tachometer. These light pulses are used to electronically calculate the exact rpm of any rotating part.

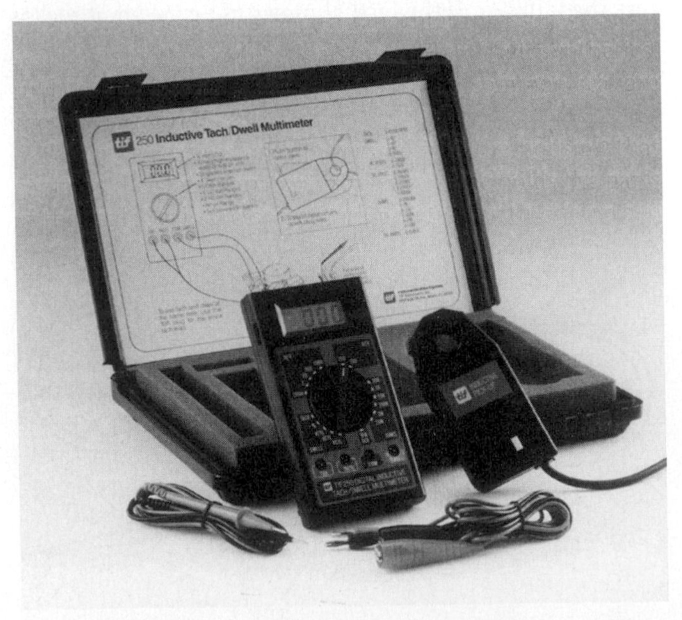

FIGURE 10–8 This tachometer uses an inductive pickup on a spark plug wire to determine the rpm of an engine. (Courtesy of TIF Instruments, Inc.)

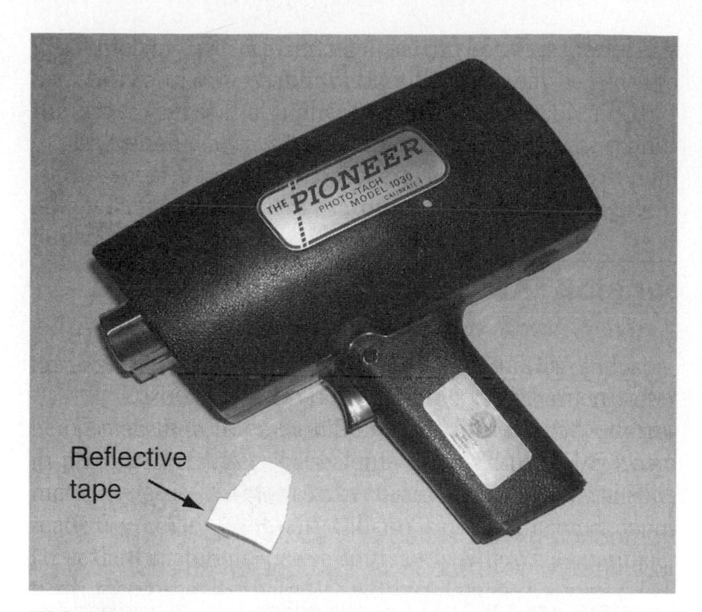

FIGURE 10–9 This photoelectric tachometer uses light reflective tape and light pulses to determine engine rpm.

DWELL/TACH METER

Older engines often use a *dwell/tach meter*. These meters are able to measure the dwell setting on engines that have conventional ignition systems. In addition, the meters are able to measure the rpm of the engine and the resistance of the ignition points.

TIMING LIGHT

The **timing light** shown in ***Figure 10–10*** is used to determine the exact timing in degrees rotation of the crankshaft. Timing lights are similar to strobe lights that are able to flash off and on rapidly. They sense exactly when cylinder 1 is firing. The firing of the cylinder creates a signal that is amplified enough to operate a light. The pulsing light is then used to monitor the exact timing of the engine ignition system.

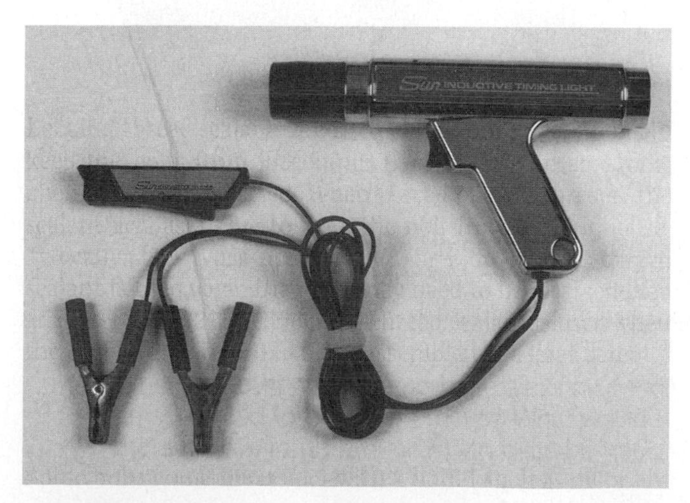

FIGURE 10–10 A timing light is used to determine the exact timing of the spark on the ignition system.

Timing lights such as these are used only on engines that have a mechanical distributor. Some newer engines do not have a distributor and use a distributorless ignition system. On these engines, the timing of the engine is all done through a computer.

Timing lights are an important tool in helping to tune up engines. A typical timing light has three leads. Two of the leads are attached to the battery to produce the voltage needed for the strobe light to work. The third inductive lead is attached to the number 1 spark plug wire. When electricity is sent through the number 1 spark plug wire, the inductive lead attached around the wire picks up an electromagnetic pulse. This pulse is then magnified by the timing light, so the timing light produces a pulse of high intensity light whenever the number 1 spark plug fires.

These high-intensity pulses of light are then pointed toward the crankshaft harmonic balancer. On the harmonic balancer there is a series of numbers and lines (***Figure 10–11***). The numbers and lines indicate the number of degrees of crankshaft rotation before the piston reaches top dead center. Note that there are numbers from 0 to 14 BTDC (before top dead center) and from 0 to 6 ATDC (after top dead center). Also note that there is a small timing mark on the spinning harmonic balancer. When the light flashes on these lines and on the small timing mark, it shows when the number 1 spark plug is firing in degrees rotation before top dead center. For example, say the strobe light flashes when the small timing mark is at the 10. This means that the number 1 spark plug is firing 10 degrees BTDC. If the strobe light flashes when the small timing mark is at 0, this means that the number 1 spark plug is firing at exactly top dead center, or TDC.

The service technician then checks the timing with manufacturer's specifications. If the timing is off, the service technician readjusts the timing to meet specifications. Timing is adjusted by rotating the distributor housing to the right or to the left while continuing to shine the timing

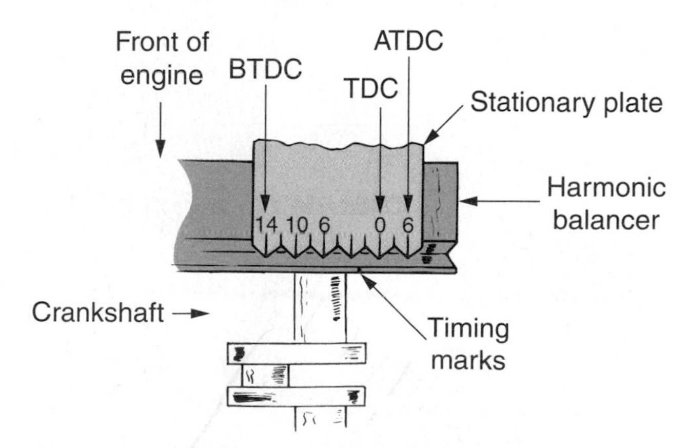

FIGURE 10–11 Timing of an engine with a distributor is done by shining a strobe light on the timing marks located on the harmonic balancer on the front of the engine.

light on the timing marks on the harmonic balancer. More information about engine timing can be found in other chapters in this textbook.

ADVANCE TIMING LIGHT

Some service technicians use an advance timing light. This timing light looks similar to a standard timing light. Digital advance timing lights can also be purchased. Advance timing lights have a dial and a scale or meter on the end. This scale or meter indicates the degrees of crankshaft timing advance. The knob on the end of the light can be turned to adjust the timing of the flash of light. The service technician points the flashing light at the timing marks on the engine. The dial is then turned until the timing marks appear to line up at the base timing position. At this point the dial or scale on the timing light will indicate the amount of timing advance.

MAGNETIC TIMING PROBE TESTER

Today, most engines use the distributorless ignition system. In order to monitor the timing of such systems, a magnetic timing probe tester is used. ***Figure 10–12*** shows a typical magnetic timing probe tester. This electronic tester is used to monitor the ignition timing of the engine. One lead on the magnetic timing probe tester is placed in a receptacle located on the front of the engine near the harmonic balancer (***Figure 10–13***). There is a small magnet placed in the harmonic balancer. As the harmonic balancer spins during engine operation, the magnetic lines from the magnet produce a small electrical signal in the inserted timing probe. The magnetic pulses are then used to read the crankshaft position, the engine rpm, and ignition timing on the engine.

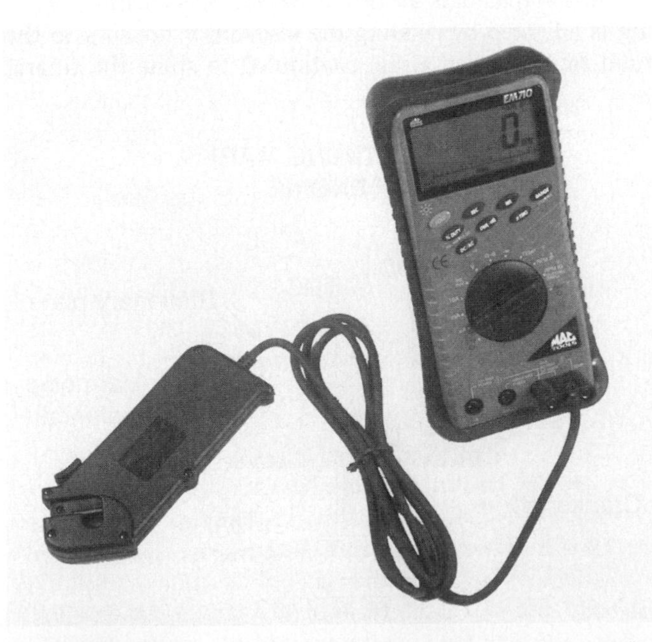

FIGURE 10–12 This magnetic timing probe tester has a probe that is inserted into a receptacle on the front of the engine near the harmonic balancer.

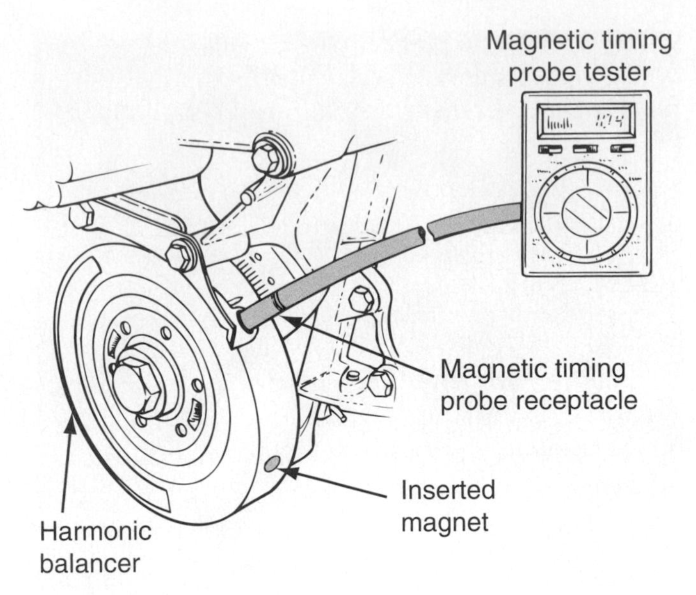

FIGURE 10–13 The receptacle located near the harmonic balancer holds the timing probe for monitoring crankshaft position, timing, and rpm.

HANDHELD DIAGNOSTIC COMPUTER

Since today's vehicles and engine systems use several computers, it is important to give the service technician effective tools to diagnose these electronic systems. One such diagnostic tester is called the Tech 2. It is a handheld diagnostic computer designed to diagnose most electronic systems on a vehicle. ***Figure 10–14*** shows an example of this tester. This diagnostic tester is attached to the vehicle through a data link connector known as a DLC. In addition, different program cartridges that have diagnostic data for specific vehicles can be inserted.

FIGURE 10–14 This handheld diagnostic computer helps to diagnose electronic systems on vehicles.

This type of diagnostic computer can interface with many of the computers used on an automobile today. The Tech 2 can interface with the following computers on the automobile:

ABS —Antilock Brake Systems Control Module

BCM —Body Control Module

DERM—Diagnostic Energy Reserve Module for SDI. R.

TCM —Transmission Control Module

VCM —Vehicle Control Module

A handheld diagnostic computer can help troubleshoot almost any of the electrical components on a vehicle. Although many of these computer modules may seem new, they are further explained in other chapters in this book. Note that this is just one type of electronic diagnostic tester. There are others that are also used. For example, there are analyzers that help to troubleshoot and diagnose only ignition systems.

FIGURE 10–15 Today's engines require sophisticated electronic engine analyzers to help solve many engine problems.

ELECTRONIC ENGINE ANALYZER

As automotive engines become more electronically controlled, testing tools must also be improved. ***Figure 10–15*** shows an **electronic engine analyzer**. This testing instrument is capable of taking many readings on an engine. Depending on the type, most electronic engine analyzers are capable of reading such items as:

- Engine rpm
- Firing sequence
- Voltage (kV) of each spark plug, under no load and load conditions
- Spark duration
- Oscilloscope patterns on the electronic ignition system
- Cranking voltages-current-resistances
- Timing
- Charging system output
- Comparative cylinder compression
- Dwell (if conventional ignition)
- Intake manifold vacuum
- Exhaust characteristics (carbon monoxide, hydrocarbons, and oxygen)

In addition, many newer electronic engine analyzers are now connected directly to computers to help the automotive technician troubleshoot an engine. Computer programs direct the technician through a series of tests to determine the exact failure on the engine.

SCANNERS

Various tool manufacturers produce and sell **scanners** to help automotive technicians troubleshoot electronic systems. Scanners are small handheld diagnostic instruments that guide the automotive technician through a series of diagnostic steps and checks to find faulty components or sensors. The technician will typically enter data into the scanner, such as the vehicle's VIN, make, model, year, and type of engine. The scanner then will read out troublecodes that have been stored in the engine's on-board computer when some electronic component has failed. The scanner is also able to provide a reading of the information that the vehicle's computer is receiving from its various input devices. The scanner can also provide information concerning the outputs that the computer is controlling. After the scanner gives an indication of a faulty part, the digital multimeter is typically used to perform the pinpoint diagnosis before the faulty part is replaced. ***Figure 10–16*** shows a typical handheld diagnostic scanner.

It should be noted that the companies that manufacture scanners are constantly working with automotive manufacturers to improve and expand scanners' capabilities. Depending on the year and type of car, various scanners are available to service technicians. In addition, automotive manufacturers are continually updating their vehicles' computer control systems on the automobile, so scanners must also be updated and expanded. ***Figure 10–17*** shows a scanner being used during a road test. This type of scanner is able

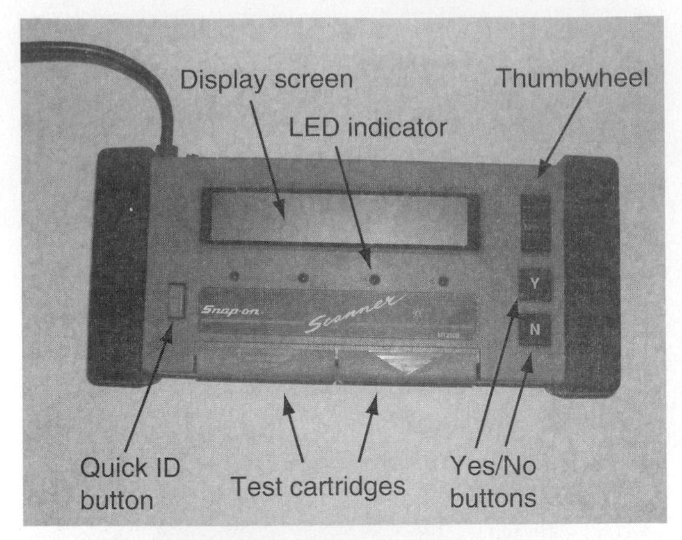

FIGURE 10-16 The electronic scanner helps to find problems in many electronic circuits.

FIGURE 10-17 This is a scan tool collecting engine data during a road test. (Courtesy of SPX Corporation)

to test the vehicle computer system during actual operation of the vehicle. The scanner has the ability to store or "freeze" data for analysis when the vehicle is returned to the shop.

10.2 POWER AND PRESSING TOOLS

The automotive service technician uses many power and pressing tools in the shop. These tools make the heavy jobs that must be done on the automobile much easier, safer, and quicker. Power and pressing tools include pullers, presses, impact wrenches, and bushing, bearing, and seal installers.

PULLERS

Pullers are used to accomplish three types of action: pulling an object off a shaft, pulling an object out of a hole, and pulling a shaft out of an object. These are illustrated in ***Figure 10–18***.

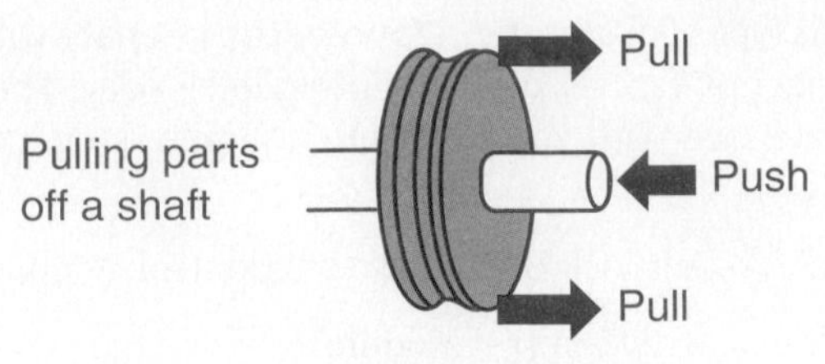

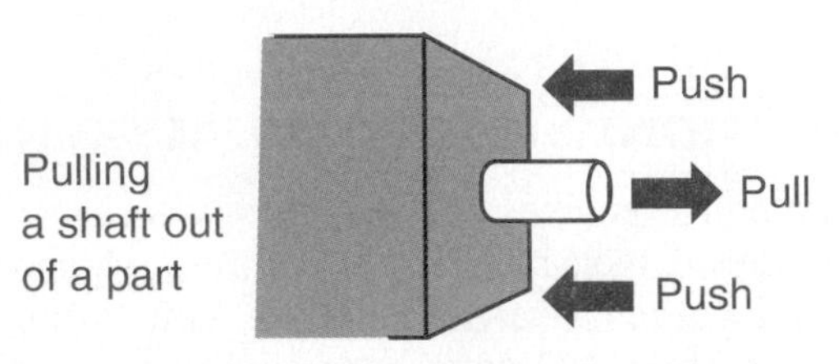

FIGURE 10-18 Pullers are used for many applications, but are designed to accomplish three specific actions.

The first example represents pulling a gear, wheel, or bearing off a shaft. The second example represents removing bearing cups, retainers, or seals from holes. The third action represents gripping a shaft and bracing against the housing to remove the shaft.

The examples shown in ***Figure 10–19*** illustrate how several puller applications work. In ***Figure 10–19A***, the puller is used to pull a gear off a shaft using internal puller jaws. ***Figure 10–19B*** shows how an outside puller jaw can remove a gear or other object from a shaft. ***Figure 10–19C*** shows how a slide hammer is used to pull a bearing from inside a shaft. ***Figure 10–19D*** shows how external jaws are used to remove a gear from the crankshaft. Obviously, there are many other applications for which pullers can be used.

Because there are so many types, styles, and sizes of seals, pulleys, and shafts, pullers must be designed to fit many applications. ***Figure 10–20*** illustrates a set of gear and bearing pullers that are commonly used on the automobile.

BUSHING, BEARING, AND SEAL INSTALLERS

Installing bushings, bearings, and seals can be a difficult job. During installation, these components must be aligned

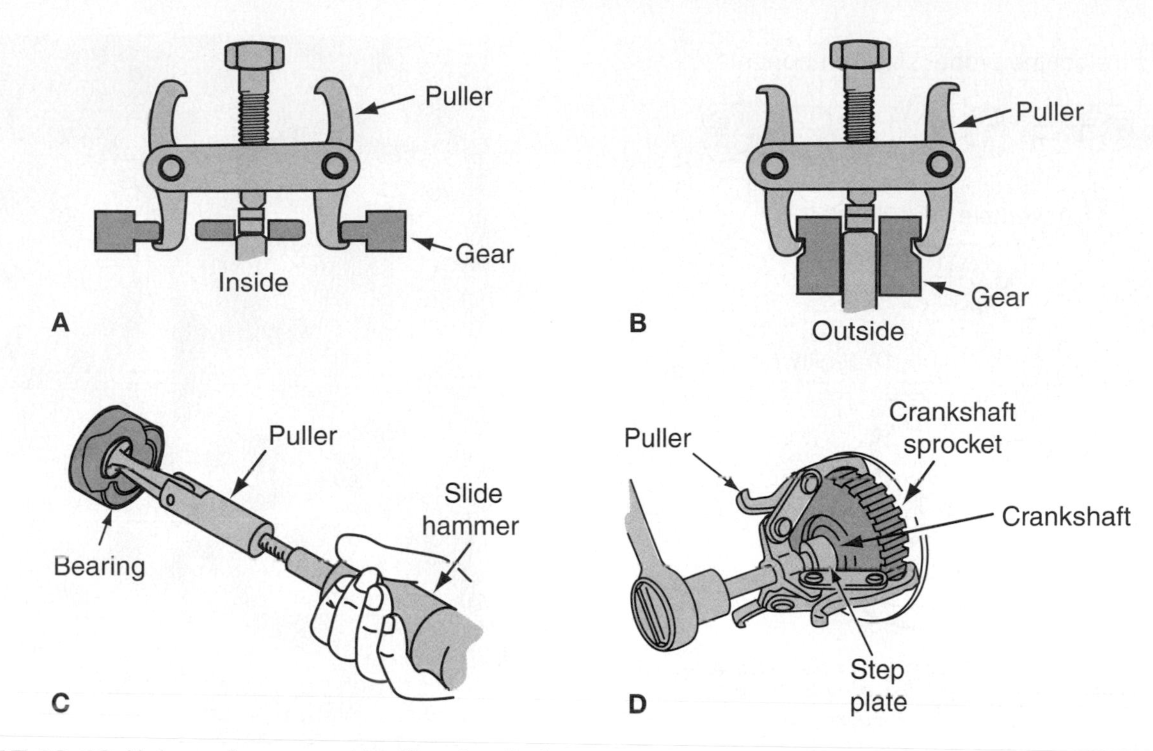

FIGURE 10-19 Various pullers are used for different applications.

correctly. Even pressure must be applied as the component is installed. **Bushing installers** are used to perform this job. ***Figure 10–21*** shows a bushing, bearing, and seal driver set. The circles that are shown are examples of the size of some of the **bushing pilots** used. Note that a side view of some of the bushing pilots is provided to show how they are shaped. Different pilots are used for different inside diameter bushings. These sets include discs and handles to provide a pilot. There are several spacers and drivers to help apply an even force on the part being installed. ***Figure 10–22*** shows a three-step process for installation. Discs range in size from 1/2 inch through 4 1/2 in diameter.

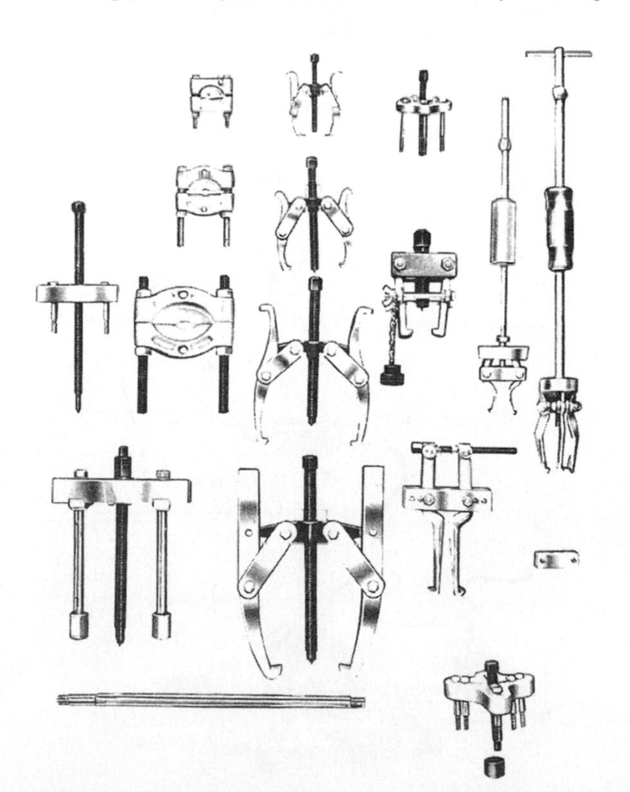

FIGURE 10-20 Many types of pullers are needed on the automobile. (Courtesy of SPX Corporation)

HYDRAULIC PRESSES

There are many times when working on automotive parts that pressing is required. For example, presses are used to work on rear axle bearings and piston pins, to press out

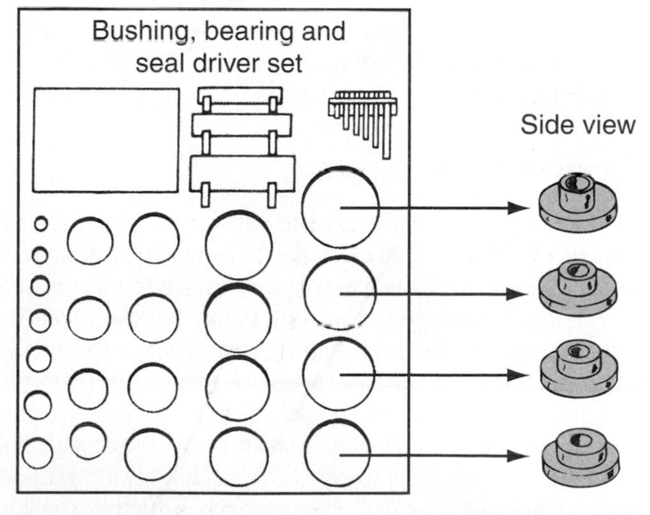

FIGURE 10-21 A bushing, bearing, and seal driver set is used to correctly install these components. Correct alignment and even distribution of forces are provided by these tools. (Courtesy of SPX Corporation)

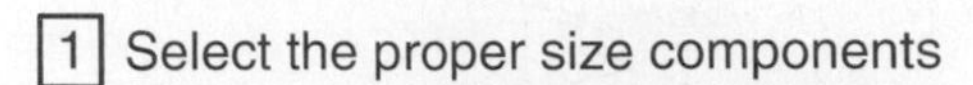

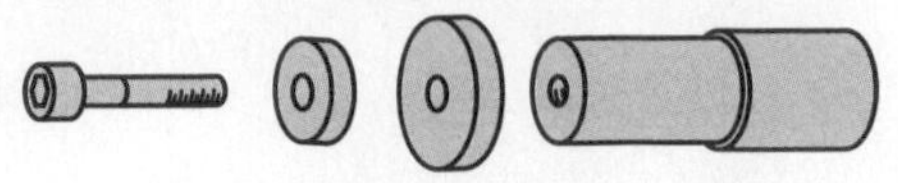

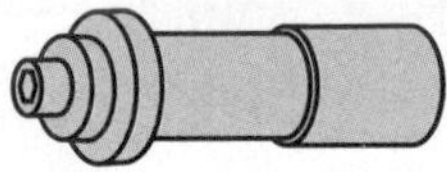

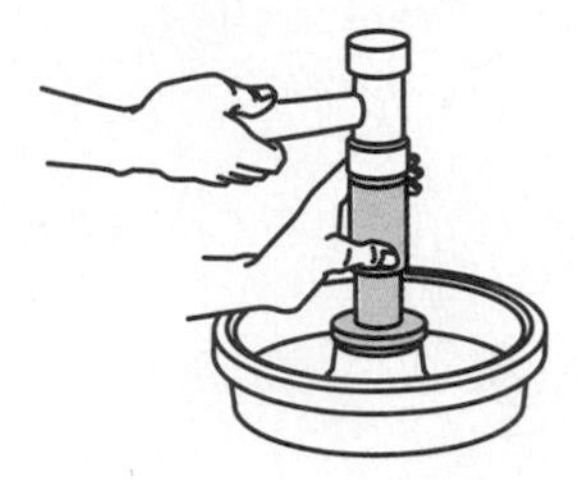

FIGURE 10–22 Installing bearings and/or seals with a driver set involves a three-step process.

studs, to straighten parts, and to press in bearings. These jobs are typically done with a hydraulic press. These presses are capable of producing 50 tons or more of pressure on a part. ***Figure 10–23*** is an example of a typical hydraulic press.

PNEUMATIC JACKS

There are several types of **pneumatic jacks** used to lift vehicles in the air for better service. Pneumatic jacks use compressed air to lift a vehicle. ***Figure 10–24*** shows an example of a typical pneumatic jack. Air from a compressor is attached to the jack. When air pressure is allowed into the central cylinder, the saddles lift the car up for service. The saddles are adjusted by hand to fit evenly under the rear of the car frame or suspension. Always remember to make sure the pneumatic jack has sufficient capacity for the vehicle being lifted.

HYDRAULIC JACKS

Hydraulic jacks are also used to lift a vehicle for service. ***Figure 10–25*** shows an example of a typical hydraulic jack. Generally, the saddle is placed either under the differential or under a supporting beam under the vehicle. Once the jack is in place, the service technician moves the handle back and forth to produce pressure inside the hydraulic cylinder. As the pressure increases, it lifts the saddle upward, lifting the vehicle. Note that some hydraulic jacks use a foot pedal to produce the pressure. Again, always remember to make sure the hydraulic jack has sufficient capacity for the vehicle being lifted. It is also a good idea to use jack stands for extra safety, when the vehicle is being held by a hydraulic jack.

FIGURE 10–23 Hydraulic presses are used for a variety of pressing jobs, including working on piston pins, axle bearings, and other heavy pressing applications.

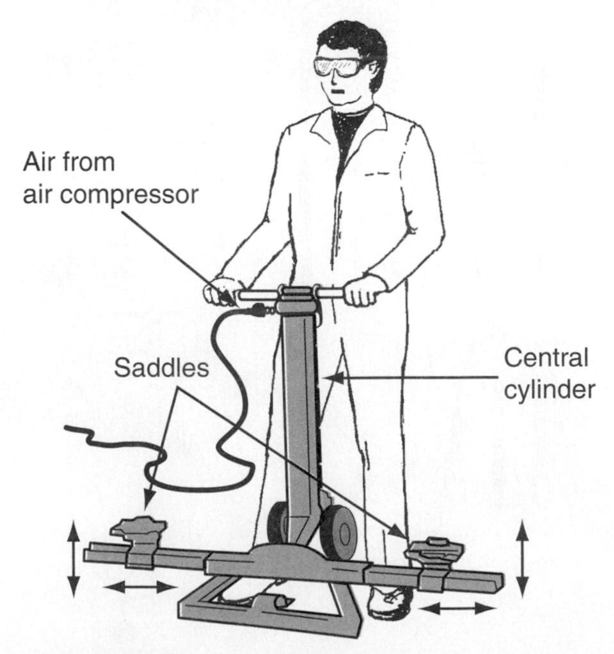

FIGURE 10–24 A pneumatic jack can be used to lift a vehicle into the air for easier and safer service.

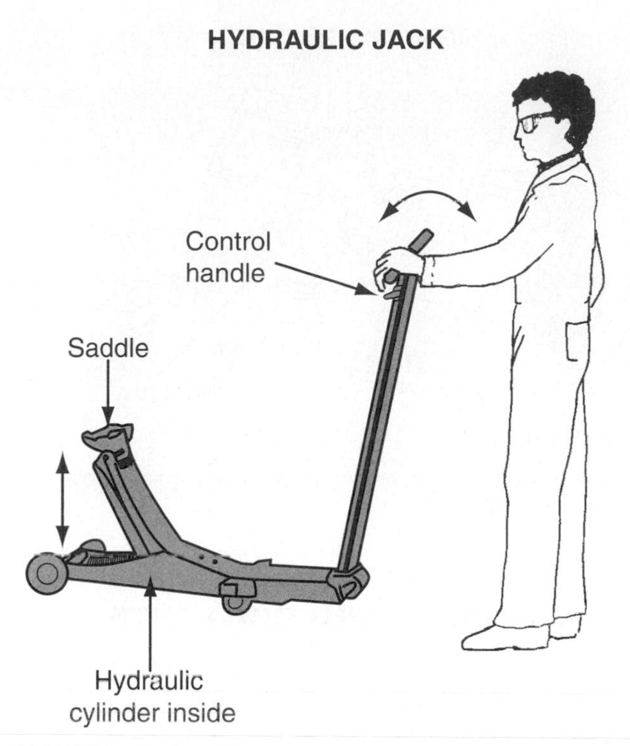

FIGURE 10–25 This hydraulic jack can be pumped up by using the control handle to lift a vehicle.

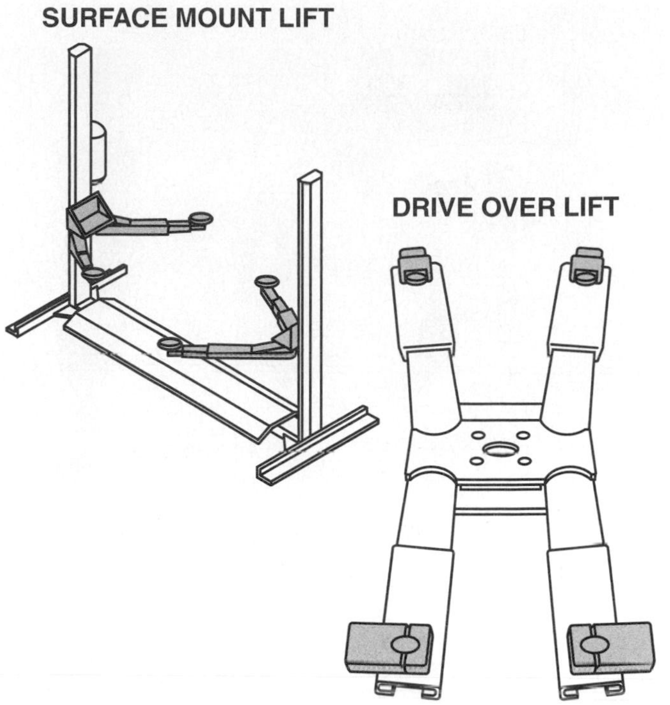

SURFACE MOUNT LIFT

DRIVE OVER LIFT

FIGURE 10–27 There are many types of vehicle lifts installed in automotive shops.

SHOP CRANES

Shop cranes are used to lift very heavy loads in the automotive shop. A typical shop crane is shown in ***Figure 10–26***. It is placed on a set of wheels and wheeled under the vehicle or part that must be lifted. The service technician then pumps a handle to produce hydraulic pressure. The hydraulic pressure is used to pump up a hydraulic cylinder. As the cylinder extends due to the high pressure, it lifts the upper telescoping boom. One common use of shop cranes is to lift engines out of the vehicle for service and complete engine overhaul. Always remember to make sure the crane has sufficient capacity for the part that is being lifted.

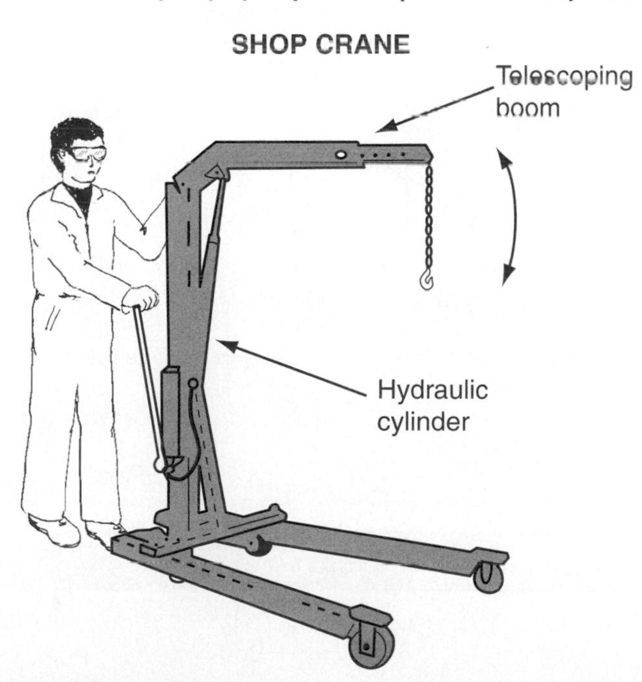

FIGURE 10–26 A shop crane uses hydraulic pressure to lift very heavy loads in the automotive shop.

VEHICLE HOISTS

There are a variety of vehicle hoists used in automotive shops today. Some are operated by air pressure, while others are operated by hydraulic oil pressure. ***Figure 10–27*** shows two examples of hoists used in the automotive shop. One type of lift is called a surface mount lift; the other is called a drive over lift. Although such lifts and hoists operate similarly, each has a specific operational procedure. Always check with your instructor to make sure you know the exact operation of the hoists in your shop.

IMPACT WRENCHES

An **impact wrench** is a powered wrench. It can be operated by using either compressed air (called pneumatics) or electrical power. The wrench works using the principle of impact rotation. Impact rotation is a pounding or impact force created to aid in loosening or tightening nuts or bolts. Impact wrenches speed up the process of tightening or loosening bolts and nuts. ***Figure 10–28*** shows an air-type impact wrench and several accessories. The drive can have 3/8-, 1/2-, 3/4-, or 1-inch ends. An internal valve regulates the power output. Also, both forward and reverse directions are selected easily.

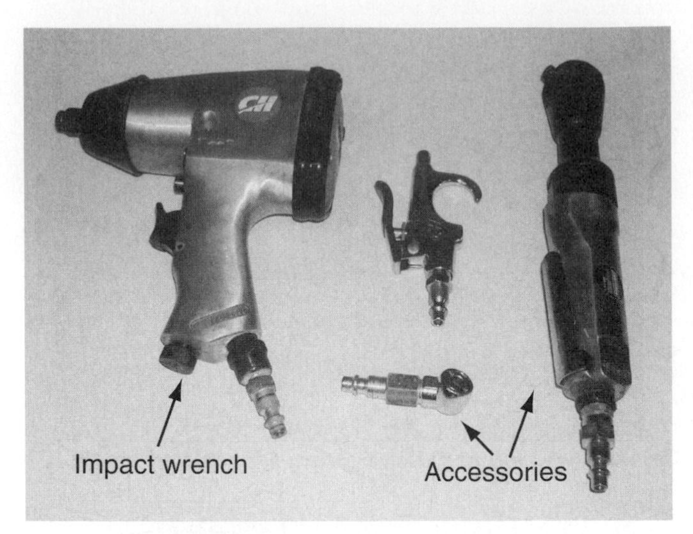

FIGURE 10–28 Impact wrenches are used to produce impact forces during rotation.

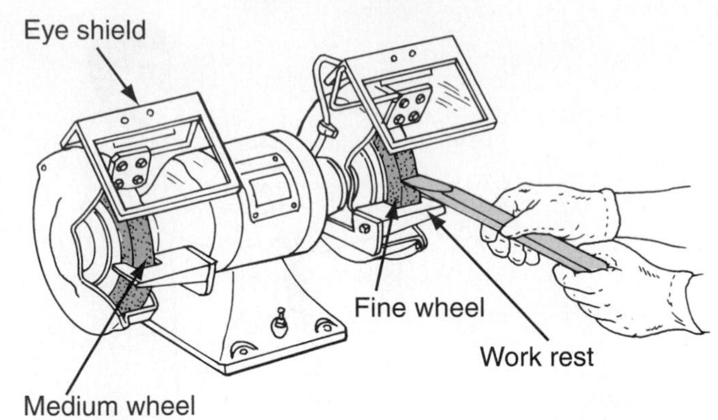

FIGURE 10–29 This bench grinder is used to keep metal tools sharp and in good condition.

GRINDERS

Grinders are used extensively in an automotive shop for sharpening tools (chisels, etc.), for grinding down sharp metal edges, and for a variety of cleaning operations. There are various grades of grinding wheels. Generally, coarse stones are used for rough work, whereas fine stones are used for finishing work. Grinding stones can be cleaned up with a dressing wheel. ***Figure 10–29*** shows a grinder being used to sharpen a chisel. The grinding stones can also be removed and replaced with wire wheels. Wire wheels are often used to clean valves, remove certain types of gasket materials, and for other cleaning needs. There are certain safety rules to follow when using a grinding wheel.

1. Always wear safety glasses when using a grinding wheel.
2. Never stand directly in line with the spinning grinding wheel.
3. Never use aluminum, wood, or other soft material on a grinding wheel. The soft material has a tendency to fill the wheel pores, damaging the grinding wheel.
4. Never grind on the side of the wheel unless the wheel has been specifically designed for such a purpose.
5. Always check the speed of the grinding wheel. Make sure never to run the grinder faster than the recommended wheel speed.
6. Never use a grinding wheel that has been dropped. If a grinding wheel has been dropped, there may be undetectable cracks. During operation, a cracked grinding wheel could fly apart.
7. Always make sure your safety eye shields are down and in place.
8. When grinding any component or part, always use gloves to protect your hands from cuts.
9. Make sure the part being ground sits solidly on the work rest.

POWER DRILLS AND DRILL BITS

Another type of power tool is the power drill (***Figure 10–30***). There are many styles and types of power drills used in the automotive shop. Since most drills today are reversible, both left- and right-handed drills have become popular. Remember these safety tips when using a power drill.

1. Always wear safety glasses when using a power drill.
2. Make sure all drill bits are sharpened correctly. ***Figure 10–31*** shows a typical drill bit and the associated terminology.

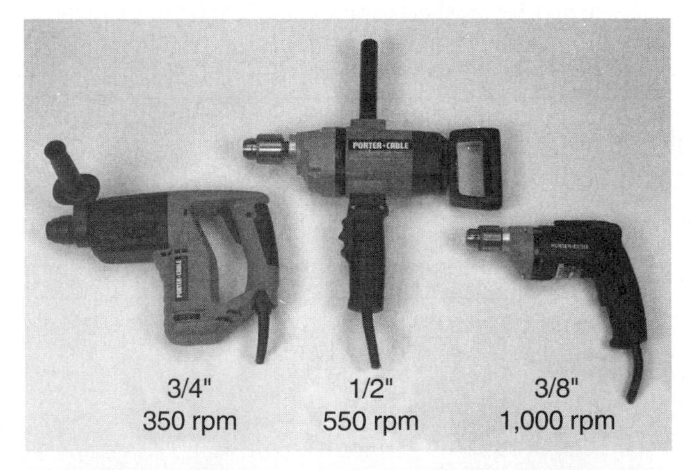

FIGURE 10–30 The automotive service technician uses a variety of power drills.

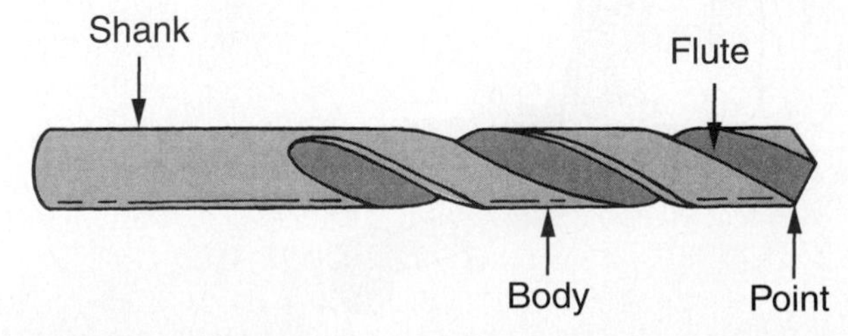

FIGURE 10–31 Always keep drill bits sharp and in good condition.

3. Never force a drill bit. It may break and cause serious injury to your hands.
4. Never wear loose-fitting clothes near the spinning drill.
5. Always unplug the drill before changing drill bits.
6. Make sure you are standing on a dry floor before using a drill. Never drill while standing in a pool of water.
7. Always make sure the piece to be drilled is held in a secure position. If not, the drill could catch the material causing it to spin around and seriously cut your hand.
8. Make sure the power drill is properly grounded.

10.3 CLEANING EQUIPMENT

PARTS CLEANERS AND SOLVENT TANKS

Often it is necessary to clean the oil and grease from engine parts. In addition, brake parts and other vehicle parts frequently need cleaning before being replaced. Parts cleaners and solvent tanks (also called cold solution cleaning) help the service technician with this task. ***Figure 10–32*** shows two types of parts washers. The larger one (***Figure 10–32A***) is for larger parts. The smaller one (***Figure 10–32B***) is used on a workbench for smaller parts. In most cases, the parts washers have a fluid pump to move solvent from a lower tank to the part being cleaned. In addition, each parts washer has a safety latch on the lid. This latch is a fusible link that will melt if a fire occurs in the tank. When the safety link melts, the tank lid is closed for safety purposes. Remember to follow these rules when washing parts in a parts clearer or solvent tank:

1. Always wear safety glasses when washing parts.
2. Avoid prolonged skin contact with any solvent.
3. Never smoke around a solvent tank.
4. Always keep the solvent tank in a well-ventilated area.
5. Use a nylon or copper brush for cleaning parts to prevent sparks.
6. Use only approved solvent in the solvent tank; never use gasoline to clean parts.
7. Always wash your hands after using cleaning solvent.
8. Solvents are considered a hazardous waste product and thus, must be disposed of properly.

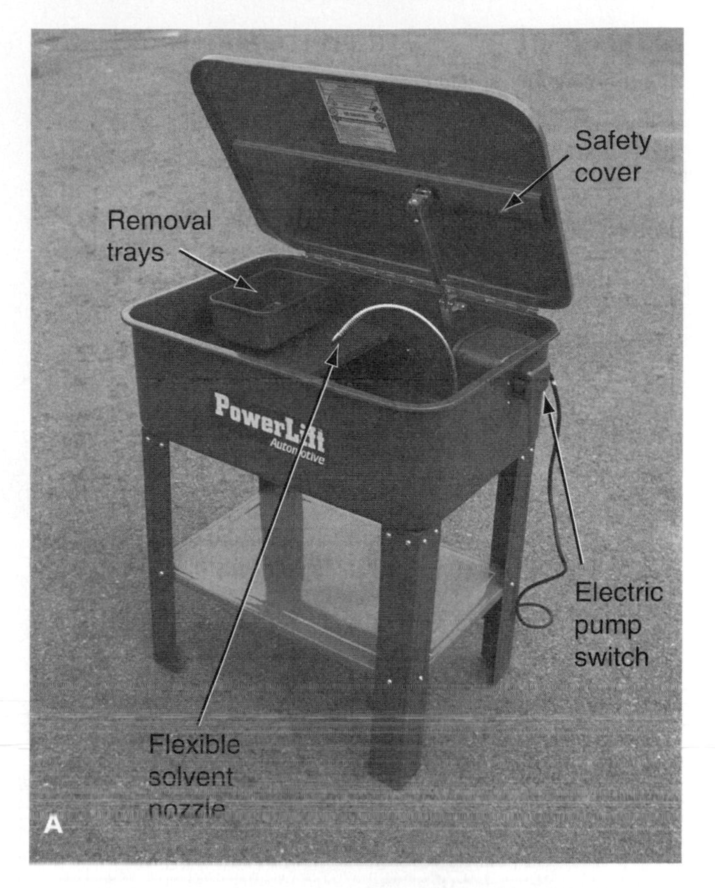

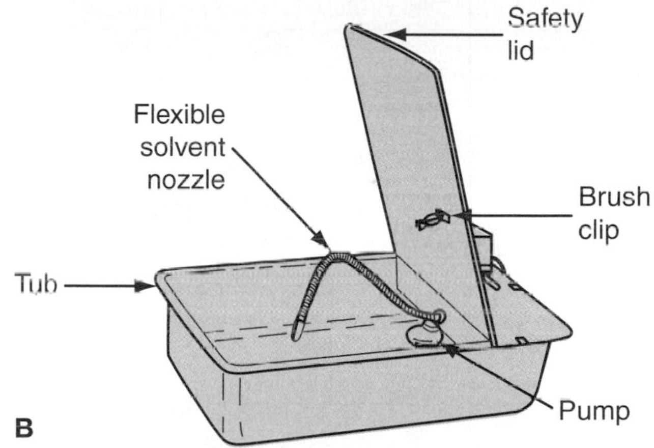

FIGURE 10–32 These parts washers are used to clean vehicle parts.

SPRAY AND PRESSURE WASHERS

Often it is necessary to clean large spots of grease and dirt from the vehicle before disassembly. A spray washer is used for this purpose. It uses high-pressure water to clean vehicle parts. ***Figure 10–33*** shows an example of a typical spray cleaner. In operation a water pump forces water or a solution of water and soap through a high-pressure hose. In some cases, the solution is heated as it goes through the system by a set of heating coils. This is called steam cleaning. The water and soap solution coming out of the spray nozzle is similar to that of a typical car wash.

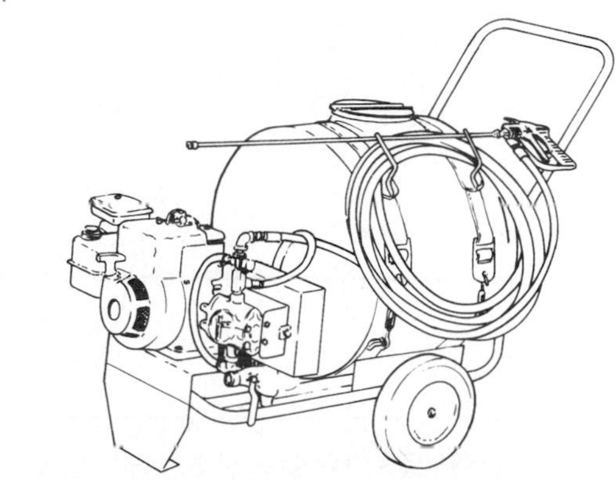

FIGURE 10–33 This high-pressure washer will remove heavy greases and dirt buildup before disassembly.

OTHER TYPES OF CLEANERS IN THE AUTOMOTIVE SHOP

There are many other types of cleaning tools used in the automotive shop. Depending on the exact nature of the shop, certain cleaning tools may or may not be available. Some of the more common cleaning tools include:

- Vacuum cleaner—used to clean vehicle interiors.
- Sandblaster—used to clean metal and other parts by blasting a spray of sand against the part in a closed container or room.
- Vapor cleaner—used to clean parts by heating a special chemical solution. The resulting vapors remove deposits from the part and suspend them in a bucket.
- **Cold soaking**—used to clean parts by soaking them in a special cleaning solution for a period of 20 to 30 minutes. For example, cold soaking is used to clean internal parts of the fuel system.

The most important thing to remember is to always follow the manufacturer's suggested procedure for any of these cleaning methods.

Service Manual Connection:

Because of the complexity of today's vehicles and engines, there are many special tools required in the automotive shop. Some of these additional tools are called "specialty tools." These special tools are often listed at the end of each section in manufacturers' service manuals. The tools are then mentioned in the individual procedures for performing certain service tasks on the automobile. At times, these special tools can be used only on one make or year of vehicle. Each special tool is usually referenced by number. An example of a set of special tools for basic engine service is shown in ***Figure 10-34***.

Illustration	Tool description
	Impact slide hammer
	Locknut pin remover
	Valve spring compressor set
	Seal replacer
	Crankshaft front seal replacer

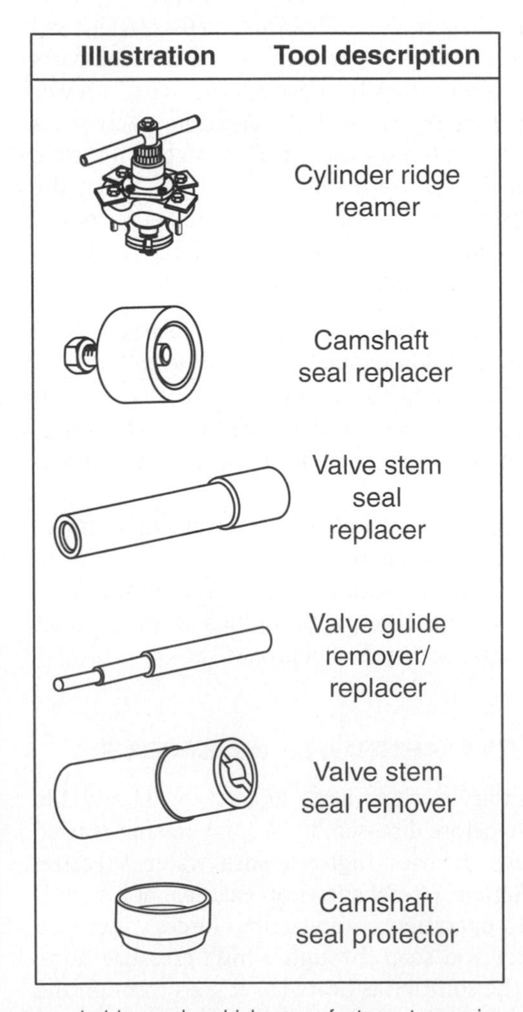

Illustration	Tool description
	Cylinder ridge reamer
	Camshaft seal replacer
	Valve stem seal replacer
	Valve guide remover/ replacer
	Valve stem seal remover
	Camshaft seal protector

FIGURE 10-34 Many special tools are recommended by each vehicle manufacturer to service vehicles correctly and according to specifications.

Safety Precautions

When working with electrical tools, power and pressing equipment, and cleaning tools in the automotive shop, keep in mind these important safety suggestions:

1. Make sure all tools used in the automotive shop are clean and free of grease and oil.
2. Remember to return all electrical tools, as well as power and pressing tools, to their designated toolbox or location on the wall after each use. Electrical tools left on the vehicle may fall into moving parts of the engine. They could fly out and cause serious injury.
3. When working with any tools in the automotive shop, always wear approved safety glasses.
4. If you are not familiar with the correct use of any tool, especially pressing and power tools, always check with the instructor. Never use a pressing tool that you don't know how to use correctly.
5. Pullers create extremely high pressures. Always make sure pullers are correctly installed for maximum safety.
6. Always use the correct puller to match the component or part being removed.
7. When installing bearings using a bearing and seal driver set, be careful not to pinch your fingers or the edges of your hand.
8. Hydraulic presses produce extremely high pressures. Always have the parts supported securely on the hydraulic press before removing bearings, gears, and so on.
9. When using any electrical test equipment, make sure the wires are not touching any moving parts such as fan blades, alternators, air compressors, water pumps, or timing belts.
10. When using any power tool, make sure the electrical cord is correctly grounded and there are no torn or frayed cords or exposed electrical wires.

SUMMARY

The following statements will help to summarize this chapter:

- Many electrical, power and pressing, and cleaning equipment tools are used in the modern automotive shop.
- Various electrical tools are used by service technicians to aid in diagnosis and troubleshooting.
- A multimeter, often called the DVOM (digital volt-ohm-ammeter) is used to check voltages, resistance values, and, in some cases, amperage.
- A continuity tester helps the service technician check for complete circuits.
- A testlight is used by the service technician to check for the location of an electrical problem.
- The dwell/tach meter, the timing light, and the magnetic timing probe tester help the service technician to tune up engines.
- Handheld scanners help the service technician identify faulty codes in computer systems.
- Pullers are used to remove gears, pull out bearings, and remove shafts from parts.
- Since there are so many bushings and bearings, special sets of bushing and bearing installers are used by the service technician.
- Various types of air and hydraulic presses are used to install bearings and other automotive components.
- In order to lift heavy objects in the automotive shop, hydraulic jacks and air/hydraulic hoists are used by the service technician.
- Other power tools often used by the service technician include impact wrenches, grinders, and power drills.
- A variety of cleaning tools help service technicians keep automotive parts clean.
- Spray and pressure washers are used to help clean automotive parts.
- There are many variations of tools used in the automotive shop. Successful service technicians make sure they know how to use each tool, as well as all safety procedures involved in using the tool.

TERMS TO KNOW

Can you explain each of the following terms? Review the chapter until you can use each term correctly.

Bushing installers
Bushing pilots
Cold soaking
Continuity
DVOM
Electronic engine analyzer
Impact wrench
Photoelectric
Pneumatic jacks
Pullers
Scanners
Tachometers
Testlight
Timing light
Volt-ohm-ammeter (VOM)

REVIEW QUESTIONS

Multiple Choice

1. Which of the following tools would be used to install a bearing on a shaft with a press fit?
a. Bearing puller
b. Hydraulic press
c. Impact wrench
d. Torque wrench
e. None of the above

2. Which of the following tools would be used to remove a gear from a shaft?
a. Puller
b. Hydraulic press
c. Impact wrench
d. Torque wrench
e. All of the above

3. Which of the following are controls that can be adjusted on the impact wrench?
a. Power or force
b. Forward or reverse
c. Angle of operation
d. A and B
e. All of the above

4. When using a bearing or bushing installer, it is important to:
a. Apply the correct distribution of forces
b. Make sure the bushing is not aligned
c. Never use a driver or spacer
d. Apply oil to the surface
e. None of the above

5. A DVOM is which type of meter?
a. A digital ohm meter
b. A digital volt meter
c. A meter that reads out in digital information
d. All of the above
e. None of the above

6. A circuit can be checked or tested for open circuits with a:
a. Continuity tester
b. Testlight
c. Tachometer
d. A and B
e. All of the above

7. An inductive-type tachometer has an inductive lead that is attached to:
a. The battery cable
b. The number 1 spark plug wire
c. The number 2 and 3 spark plug wires
d. All of the spark plug wires
e. The horn relay

8. A tachometer that reads light reflected from tape is called a (an):
a. Ohm meter
b. Ammeter
c. Photoelectric tachometer
d. Timing light meter
e. All of the above

9. The magnetic timing probe tester has a wire lead that is attached to a receptacle located:
a. Near the harmonic balancer
b. On the number 1 spark plug wire
c. On the battery
d. Near the front of the engine
e. A and D

10. The Tech 2 can be used to check and diagnose problems in which of the following modules?
a. ABS
b. TCM
c. PCM
d. All of the above
e. None of the above

The following questions are similar in format to ASE (Automotive Service Excellence) test questions.

11. *Technician A* says that to check the timing on distributor ignition systems, use a standard timing light. *Technician B* says to check the timing, use an advance timing light. Who is correct?
a. A only
b. B only
c. Both A and B
d. Neither A nor B

12. *Technician A* says that a vehicle can be lifted for service using a pneumatic jack. *Technician B* say that a vehicle can be lifted for service using a hydraulic jack. Who is correct?
    a. A only c. Both A and B
    b. B only d. Neither A nor B

13. *Technician A* says that to remove an engine from a vehicle, use a hydraulic shop crane. *Technician B* says that the shop crane is only to be used for very light applications. Who is correct?
    a. A only c. Both A and B
    b. B only d. Neither A nor B

14. *Technician A* says that when using a parts cleaner and solvent tank, always wear safety glasses. *Technician B* says that a parts cleaner has a heating element to produce super-heated solvent. Who is correct?
    a. A only c. Both A and B
    b. B only d. Neither A nor B

15. *Technician A* says that a multimeter can check only voltages. *Technician B* says that a multimeter can check amperage. Who is correct?
    a. A only c. Both A and B
    b. B only d. Neither A nor B

16. *Technician A* says that most tachometers measure the firing pulses sent to the spark plugs. *Technician B* says that most tachometers measure battery voltages. Who is correct?
    a. A only c. Both A and B
    b. B only d. Neither A nor B

17. *Technician A* says that scanners are handheld diagnostic instruments. *Technician B* says that scanners are able to identify trouble codes stored in the engine computer. Who is correct?
    a. A only c. Both A and B
    b. B only d. Neither A nor B

Essay

18. Describe the purpose of a shop crane.

19. Identify at least five safety rules to follow when using a parts cleaner and solvent tank.

20. Identify at least two types of pressure sprayers used in the automotive shop.

21. Describe the purpose for using a continuity tester.

22. Describe the procedure used for checking the engine timing using a timing light.

Short Answer

23. A ________ is used to check if electricity is available at different parts of an electrical circuit.

24. A magnetic timing probe is used to check and monitor which engine characteristics?

    ________ ________ ________

25. Electrical and computer circuits can easily be checked using a handheld ________.

26. To check the timing of an engine that has a distributor, use a ________.

27. Use a ________ to sharpen a chisel in the automotive shop.

Applied Academics

TITLE: Electrical Measurements

ACADEMIC SKILLS AREA: Science

NATEF Automobile Technician Tasks:

The technician uses precision electrical test equipment to measure current, voltage, resistance, continuity, and/or power.

CONCEPT:

When using electronic testing tools such as the DVOM, it is always important to make sure the meter is hooked up in the correct way. As mentioned in the chapter, when checking for voltage or voltage drop across a particular resistor, the voltmeter should be hooked in parallel. When checking for resistance, the resistor should be isolated or removed from the circuit and the meter hooked across the resistor. When checking for amperage in a circuit, the circuit should be opened and the ammeter should be hooked in series with the circuit, and then the power should be on.

APPLICATION:

Using the accompanying diagram, determine the position of the switch, the A, B, C, and D connectors, and the hook-up locations of the meter for the voltage check, the resistance, and the amperage check.

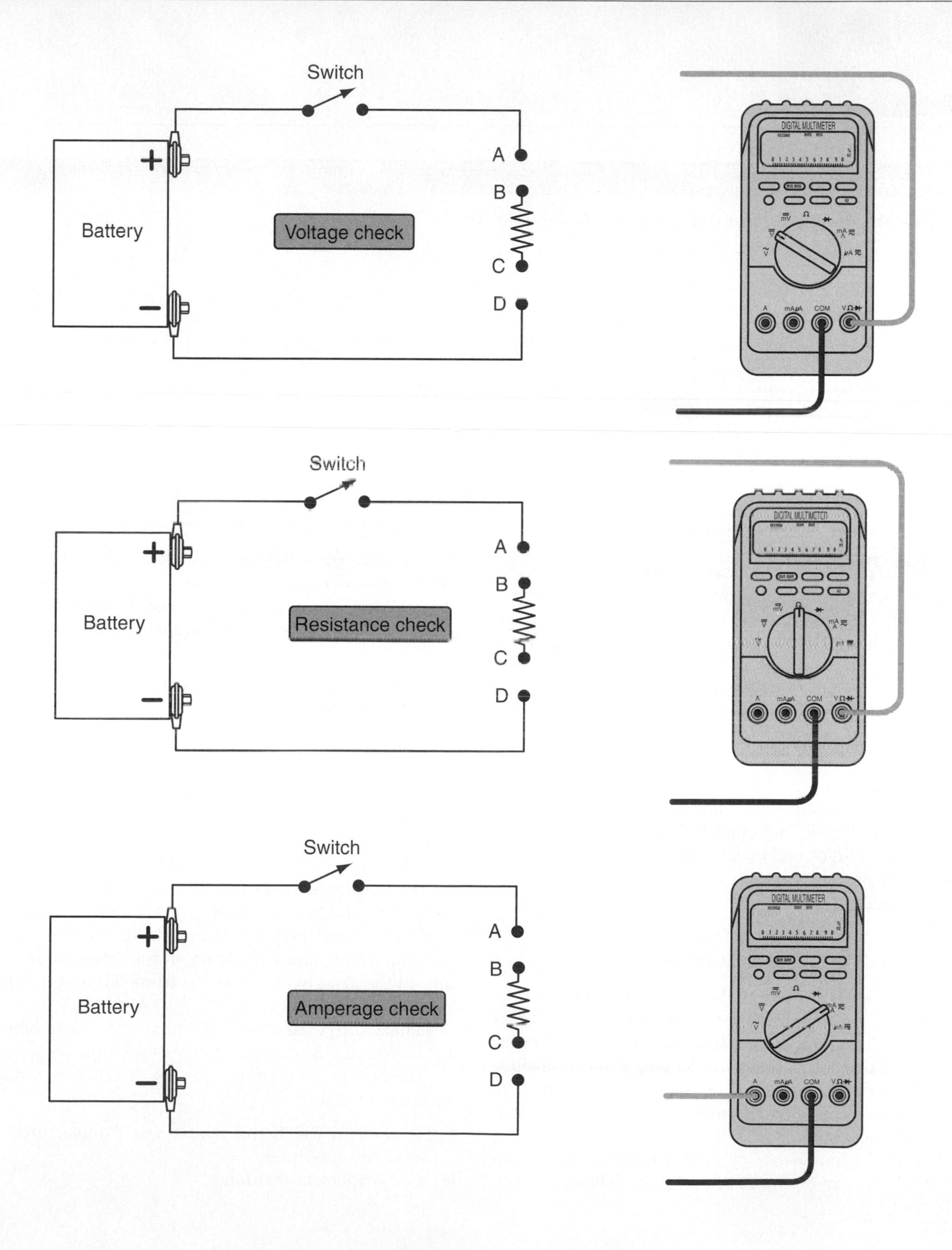
Switch
A
B
C
D
Battery
+
−
Voltage check
DIGITAL MULTIMETER
A
mA μA
COM
Switch
A
B
C
D
Battery
+
−
Resistance check
A
mA μA
COM
Switch
A
B
C
D
Battery
+
−
Amperage check
DIGITAL MULTIMETER
A
mA μA
COM

CHAPTER 11

Manuals and Specifications

OBJECTIVES

After studying this chapter, you should be able to:

- List the different types of information in service manuals.
- List different types of specifications.
- Know where to obtain technical data.
- Locate and identify a vehicle's VIN.
- State the common publishers of service data.
- Define the use of service bulletins.
- Identify computer information systems used by automotive service technicians.

INTRODUCTION

When an automobile is brought to the service shop for repair, it is the service technician's responsibility to repair the vehicle. Technicians must have many specifications and service procedures available. The technical data are different for each automobile. This information pertains to new and old vehicles, many manufacturers, both foreign and American, and many assorted problems. It is impossible for any person to remember all of this technical information and specifications.

Thus, service manuals become one of the most important tools used. It is important to know what is available, how to locate information, what this information means, and how to use specifications correctly. This chapter is designed to address these areas.

11.1 SERVICE MANUALS

Service manuals for automobiles include technical data, procedures, and service descriptions. This information is needed to troubleshoot and diagnose, service, and repair components on an automobile. Service manuals include everything from descriptions of how to remove a part, to finding a problem, or to determine the exact measurement for a clearance.

Typically, maintenance or service manuals do not include information on theory or operational characteristics. They are, however, designed and written to help the service technician service a vehicle. The best procedure and method of service are shown. In addition, special factory tools and their part numbers are provided in some service manuals.

VEHICLE IDENTIFICATION NUMBER (VIN)

The official vehicle identification number—or **VIN**—is needed to identify the exact type of vehicle being worked on. The VIN is also used for title and registration purposes. It is normally stamped on a metal tab that is fastened to the instrument panel close to the windshield. The VIN can be seen from the driver's side of the car and is visible from outside the vehicle (***Figure 11–1***). Besides the top of the dashboard, the VIN can also be found in other places. For example, depending on the vehicle and manufacturer, the VIN may be found on a label under the hood, in the driver's side door jam, on the glove box door, or the trunk lid to name a few other locations.

Manufacturers assign a VIN number to each vehicle manufactured. Depending on the manufacturer, the following information can (in most cases) be determined by reading the VIN number:

- Country in which the vehicle was manufactured or world manufacturer
- Corporation and division

FIGURE 11-1 The VIN can usually be found on the driver's side of the dashboard, and can be read through the window from the outside.

- Model or series of the vehicle
- Model year
- Body style
- Check digit
- Engine type
- Factory or plant code
- Vehicle serial number

Each vehicle service manual has a section in the front identifying the location of the VIN and describing exactly what information it contains. For example, ***Figure 11–2*** and ***Figure 11–3*** show examples of what each digit and number position means.

OTHER VEHICLE NUMBERS

Other numbers are also stamped on the vehicle to help the automotive technician identify other information. For example, various numbers are located on the vehicle to identify:

- Engine type
- Transmission type
- Axle/differential ratio
- Tire information
- Emission control information
- Final drive assembly
- Paint color codes

Each manufacturer uses a different system to identify this information. In addition, these numbers may be found in different locations on the vehicle. The exact locations of these numbers can be found in the front of service manuals for each vehicle. ***Figure 11–4*** shows one manufacturer's method of showing the vehicle numbers and their locations.

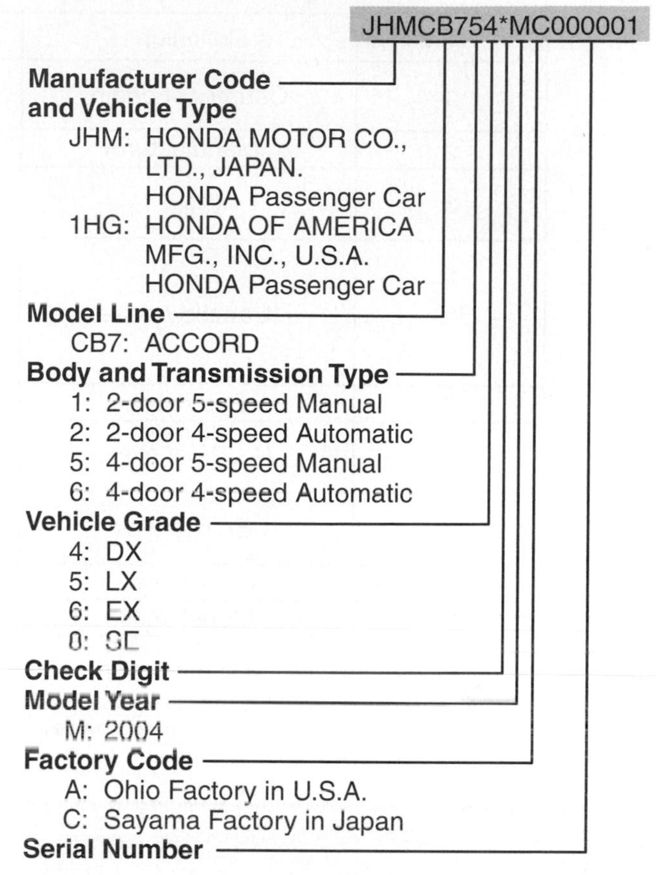

FIGURE 11-2 The VIN provides different types of information about a vehicle.

Car Clinic: VIN INFORMATION

CUSTOMER CONCERN:
Using the VIN
When looking up specifications in many service and maintenance manuals, you need to know the VIN. What is the VIN, and where is it located?

SOLUTION: The VIN is the vehicle identification number. In the specifications for a particular vehicle, publishers of service manuals use the VIN to identify the exact type of vehicle. In most cases, the VIN is located on the driver's side of the vehicle on the top of the dashboard. It can be read from outside the vehicle looking at where the windshield and dashboard meet. This number is used to identify various components on the vehicle such as model, body and transmission type, year manufactured, serial number, and so on. When looking up any specification for a vehicle, always have the VIN available and ready for reference.

VEHICLE IDENTIFICATION			
Position	Definition	Character	Description
1	Country of origin	1 2	U.S.A Canada
2	Manufacturer	G	General Motors
3	Make	1 4	Chevrolet Buick
4–5	Car line / series	B/N W/L C/W H/R	Caprice Classic LS Lumina Park Avenue Lesabre Limited
6	Body style	5	Four-door sedan
7	Restraint system	2	Active (manual) belts w/driver and passenger inflatable restraint system
8	Engine type	T M	LD9 (2.4L) L82 (3.1L)
9	Check digit	6	Check digit
10	Model year	W	2001
11	Plant location	4 6	Wilmington, DE Oklahoma City, OK
12–17	Plant sequence number	200001	Plant sequence number

FIGURE 11-3 The position of each number in a VIN indicates the type of information for a specific manufacturer. For example, a 4 in the third position of the VIN means the vehicle is a Buick.

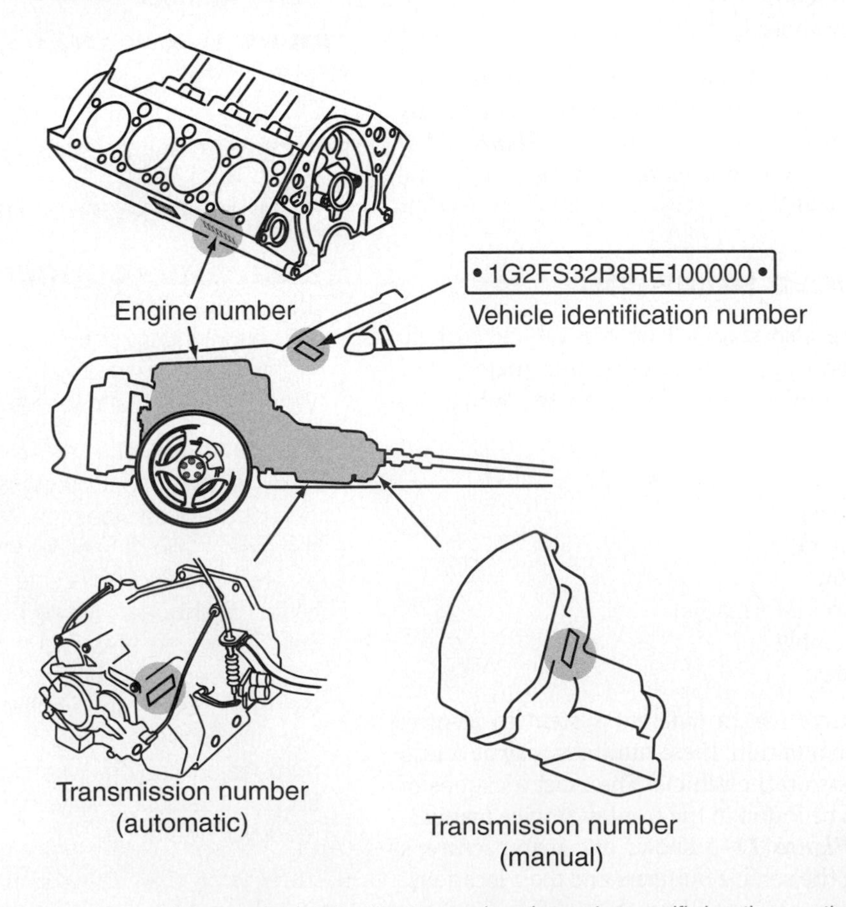

FIGURE 11-4 Other components on the vehicle can also be identified by numbers located at specific locations on the vehicle.

FULL SYSTEM DIAGNOSIS

The following diagnostic procedures are for fuel system problems and their effects on vehicle performance. Other systems of the vehicle can also cause similar problems and should be checked when listed on the chart. The problem areas described are:

1. Engine cranks normally. Will not start.
2. Engine starts and stalls.
3. Engine starts hard.
4. Engine idles abnormally and/or stalls.
5. Inconsistent engine idle speeds.
6. Engine diesels (after-run) when shut off.
7. Engine hesitates on acceleration.
8. Engine has less than normal power at low speeds.
9. Engine has less than normal power on heavy acceleration or at high speed.
10. Engine surges.
11. Poor gas mileage.

CONDITION (Problem)	POSSIBLE CAUSE (Diagnosis)	CORRECTION (Service)
Engine cranks normally. Will not start.	Improper starting procedure used.	Check with the customer to determine if proper starting procedure is used, as outlined in the owner's manual.
	No fuel in carburetor.	• Perform fuel pump flow test. • Inspect fuel inlet filter. If plugged, replace. • If fuel filter is okay, remove air horn and check for bent fuel lines.

FIGURE 11–5 Service manuals provide information about troubleshooting and diagnosing technical problems. The procedure to follow is: identify the condition (problem); select a possible cause (diagnosis); and then correct the problem (service).

DIAGNOSING INFORMATION

Because automobiles are so complex, service manuals include information to aid technicians in diagnosing and **troubleshooting** vehicles. Diagnosing or troubleshooting is defined as identifying and locating a problem, finding a cause, and correcting the problem. **Diagnosis** is one of the most important skills that the technician must have.

Figure 11–5 shows an example of a fuel system diagnostic sheet. Three things are given. The technician must first identify the condition (problem), and then select a possible cause (diagnosis). When the cause has been identified, the correction (service) is then provided. There are usually several causes for a particular condition. This gives the service technician the option of using experience when diagnosing a problem. This diagnostic guide is very easy to read. This guide also is especially helpful for identifying major problem areas on a vehicle.

A second method used to help diagnose a problem in a vehicle is shown in ***Figure 11–6***. Here a flowchart is shown. The service technician follows through each level, taking various readings and measurements. On the basis of the results, the technician follows the flowchart until an incorrect reading is obtained. Finally, a suggested repair helps solve the problem. This format has more detailed explanations than the diagnostic sheet, especially for more complex technical systems. It is very important not to skip any procedures when using this method of diagnosis.

PROCEDURE INFORMATION

The correct **service procedure** should always be followed when servicing an automobile. Suggested procedures are shown in the service manual to aid technicians. These procedures are used to speed up the technician's work and to make the job easier and safer. An example of a set of procedures is shown in ***Figure 11–7***. This procedure shows how to remove a starter on a particular vehicle. Notice the degree of detail. This service procedure will only work for the specific vehicle it was intended for. The service technician needs to have the correct service manual for the specific vehicle that is being serviced.

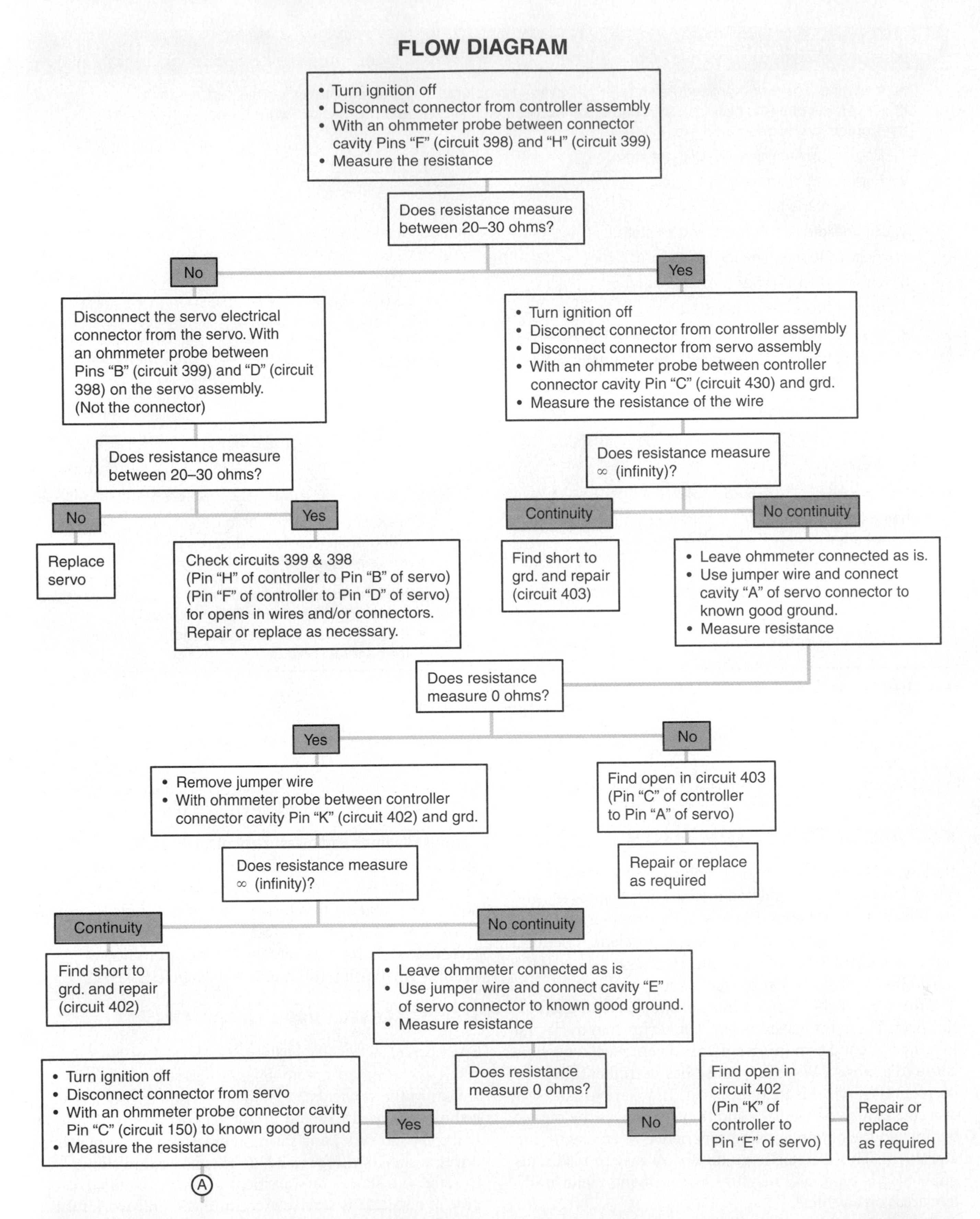

FLOW DIAGRAM
• Turn ignition off
• Disconnect connector from controller assembly
• With an ohmmeter probe between connector cavity Pins "F" (circuit 398) and "H" (circuit 399)
• Measure the resistance
Does resistance measure between 20–30 ohms?
No
Yes
Disconnect the servo electrical connector from the servo. With an ohmmeter probe between Pins "B" (circuit 399) and "D" (circuit 398) on the servo assembly. (Not the connector)
• Turn ignition off
• Disconnect connector from controller assembly
• Disconnect connector from servo assembly
• With an ohmmeter probe between controller connector cavity Pin "C" (circuit 430) and grd.
• Measure the resistance of the wire
Does resistance measure between 20–30 ohms?
Does resistance measure ∞ (infinity)?
No
Yes
Continuity
No continuity
Replace servo
Check circuits 399 & 398 (Pin "H" of controller to Pin "B" of servo) (Pin "F" of controller to Pin "D" of servo) for opens in wires and/or connectors. Repair or replace as necessary.
Find short to grd. and repair (circuit 403)
• Leave ohmmeter connected as is.
• Use jumper wire and connect cavity "A" of servo connector to known good ground.
• Measure resistance
Does resistance measure 0 ohms?
Yes
No
• Remove jumper wire
• With ohmmeter probe between controller connector cavity Pin "K" (circuit 402) and grd.
Find open in circuit 403 (Pin "C" of controller to Pin "A" of servo)
Does resistance measure ∞ (infinity)?
Repair or replace as required
Continuity
No continuity
Find short to grd. and repair (circuit 402)
• Leave ohmmeter connected as is
• Use jumper wire and connect cavity "E" of servo connector to known good ground.
• Measure resistance
• Turn ignition off
• Disconnect connector from servo
• With an ohmmeter probe connector cavity Pin "C" (circuit 150) to known good ground
• Measure the resistance
Does resistance measure 0 ohms?
Yes
No
Find open in circuit 402 (Pin "K" of controller to Pin "E" of servo)
Repair or replace as required
A

FIGURE 11-6 This flow diagram presents a detailed process for troubleshooting complex technical systems. After certain measurements are taken, the service technician makes a decision about what to do next. (Courtesy of MOTOR Publications, Auto Repair Manual, © The Hearst Corporation)

STARTER REMOVAL

This procedure requires the use of an engine hoist to lift the engine slightly.

1. Obtain the radio anti-theft code if applicable.
2. Disconnect the negative, then the positive battery cable.
3. Lift the coolant reservoir out of the way, then remove the battery and battery base.
4. Remove the alternator and belt.
5. Remove the left exhaust manifold cover from the engine assembly.
6. Remove the damper fork.
7. Disconnect the left lower ball joint from the suspension.
8. Remove the left drive shaft from the hub and transaxle assemblies.
9. Remove the transmission stop collar.
10. Remove the exhaust system Y pipe.
11. Remove the front motor mounting bolts.
12. Attach a suitable engine hoist and slightly lift the engine.
13. Remove the motor mount.
14. Disconnect the starter cable from the attaching stud and black/white wire.
15. Remove the starter.

FIGURE 11-7 The correct procedures for servicing most components on a vehicle are included in service manuals. Such information will help technicians perform service in a timely and safe manner.

SPECIFICATIONS

Specifications are an important part of service manuals. Specifications include the technical data, numbers, clearances, and measurements used to diagnose and adjust automobile components. Specifications can be referred to as specs. They are usually precise measurements under standard conditions, which are supplied by the automotive manufacturer. Examples of specifications include valve clearance, spark plug gaps, tire pressure, number of quarts of oil, ignition timing, and engine size. There are many other specifications as well.

Specifications are not necessarily the law. They should be used as guides to show the service technician how the automobile was set up when it was manufactured. A service technician working on a new car should follow the factory specifications as closely as possible. However, when an automobile gets older, some specifications may not be exactly right for best operation. In this case, the manufacturer's specifications must be considered along with the technician's experience. While this may require a departure from the factory's specifications, it is important to stay as close to the original specifications as possible. In addition, many service technicians subscribe to magazines that feature technical updates on older vehicles.

TYPES OF SPECIFICATIONS

Many types of specifications are needed in the automotive shop. Often, the specifications require the automotive technician to obtain certain information from the vehicle identification number (VIN). This information can then be used to help identify other types of specifications needed. The following is a list of common specifications found in many service manuals. Depending on the service manual, the publisher may or may not list the specifications using the six categories shown.

1. **General Engine Specifications**: These specifications identify the size and style of the engine for a group of vehicle years, say from 1999 to 2004. They include the vehicle model, VIN codes, engine displacement, number of cylinders, engine type, fuel system type, net horsepower, net torque, bore and stroke, compression ratio, and engine oil pressure as shown in ***Figure 11–8***. Each service manual publisher may include additional

GENERAL ENGINE SPECIFICATIONS

Year	Model	Engine ID/VIN	Engine Displacement Liters (cc)	No. of Cyl.	Engine Type	Fuel System Type	Net Horsepower @ rpm	Net Torque @ rpm (ft. lbs.)	Bore x Stroke (in.)	Compression Ratio	Oil Pressure @ rpm
20__	Integra	B18B1	1.8 (1834)	4	DOHC	PGM-FI	142@6300	127@5000	3.19x3.50	9.2:1	50@3000
	Integra GSR	B18C1	1.8 (1797)	4	DOHC	PGM-FI	170@7600	128@6200	3.19x3.43	10.0:1	50@3000
	Legend	C32A1	3.2 (3206)	6	SOHC	PGM-FI	200@5500	210@4500	3.54x3.31	9.6:1	50@3000
	Vigor	G25A1	2.5 (2451)	5	SOHC	PGM-FI	176@6300	170@3900	3.35x3.40	9.0:1	50@3000
20__	Integra	B18B1/①	1.8 (1834)	4	DOHC	PGM-FI	140@6300	127@5200	3.19x3.50	9.2:1	50@3000
	Integra GSR	B18C1/②	1.8 (1797)	4	DOHC	PGM-FI	170@7600	128@6200	3.19x3.43	10.0:1	50@3000
	Integra Type R	B18C5/②	1.8 (1797)	4	DOHC	PGM-FI	195@8000	130@7500	3.19x3.43	10.6:1	50@3000
	2.3CL	F23A1/YA3	2.3 (2254)	4	SOHC	PGM-FI	150@5700	152@4800	3.39x3.82	9.3:1	50@3000
	3.0CL	J30A1/YA2	3.0 (2997)	6	SOHC	PGM-FI	200@5600	195@4800	3.39x3.39	9.4:1	71@3000
	3.2TL	J32A1/UA5	3.2 (3210)	6	SOHC	PGM-FI	225@5500	216@5000	3.50x3.39	9.8:1	71@3000
	3.5RL	C35A1/KA9	3.5 (3474)	6	SOHC	PGM-FI	210@5200	224@2800	3.54x3.58	9.6:1	50@3000

FIGURE 11-8 One type of specification is called "General Engine Specifications." It gives engine specifications to help technicians identify the exact type of engine being serviced and its operational characteristics.

data in this section, such as firing order and engine numbers.

2. **Tune-Up Specifications**: These specifications help identify adjustments necessary for a correct tune-up on specific types of vehicles. Examples of engine tune-up specifications include engine spark plug gaps, ignition timing, fuel pump pressure, idle speed, and valve clearance as shown in ***Figure 11–9***. Individual service manual publishers may include additional specifications in this section, such as vehicle emission control information.
3. **Capacity Specifications**: These specifications include measurements needed to identify the capacity of different fluids on the vehicle. These include such specifications as cooling capacity, number of quarts of oil, fuel tank size, and transmission transaxle capacity. ***Figure 11–10*** provides a standard list of capacity specifications.
4. **Overhaul and Maintenance Specifications**: Often called "Engine Mechanical Specifications," such specifications are used to aid service technicians in the overhaul and maintenance of engines. Examples include specifications for compression, camshafts, valves, valve seats, valve springs, valve guides, rocker arms, cylinder blocks, pistons, piston rings, crankshafts, and other components of a specific engine. ***Figure 11–11*** is an example of a typical specification chart for a specific type of engine. These specifications help the technician determine how much wear has occurred. The service technician is then able to decide whether to

ENGINE TUNE-UP SPECIFICATIONS

Year	Engine Displacement Liters (cc)	Engine ID/VIN	Spark Plugs Gap (in.)	Ignition Timing (deg.) MT	Ignition Timing (deg.) AT	Fuel Pump (psi)	Idle Speed (rpm) MT	Idle Speed (rpm) AT	Valve Clearance In.	Valve Clearance Ex.
20__	1.6 (1595)	B16A2	0.047-0.051	16B	–	31-38	650-750	–	0.006-0.007	0.007-0.008
	1.6 (1590)	D16Y5	0.039-0.043	12B	12B	28-36	620-720	650-750	0.007-0.009	0.009-0.011
	1.6 (1590)	D16Y7	0.039-0.043	12B	12B	28-36	620-720	650-750	0.007-0.009	0.009-0.011
	1.6 (1590)	D16Y8	0.039-0.043	12B	12B	28-36	620-720	650-750	0.007-0.009	0.009-0.011
20__	1.6 (1595)	B16A2	0.047-0.051	16B	–	31-38	650-750	–	0.006-0.007	0.007-0.008
	1.6 (1590)	D16Y5	0.039-0.043	12B	12B	28-36	620-720	650-750	0.007-0.009	0.009-0.011
	1.6 (1590)	D16Y7	0.039-0.043	12B	12B	28-36	620-720	650-750	0.007-0.009	0.009-0.011
	1.6 (1590)	D16Y8	0.039-0.043	12B	12B	28-36	620-720	650 750	0.007-0.009	0.009-0.011

FIGURE 11-9 Engine tune-up specifications help the service technician tune up a vehicle. Such specifications help the service technician make proper adjustments during a typical tune-up.

CAPACITIES

Year	Model	Engine Displacement Liters (cc)	Engine ID/VIN	Engine Oil with Filter (qts.)	Transmission (pts.) 5-Spd	Transmission (pts.) Auto.	Transmission (pts.) CVT❶	Fuel Tank (gal.)	Cooling System (qts.)
20__	Civic HX	1.6 (1590)	D16Y5	3.5	3.8	5.8	3.9	11.9	4.5
	Civic DX, LX, CX	1.6 (1590)	D16Y7	3.5	3.8	5.8	—	11.9	4.4
	Civic EX	1.6 (1590)	D16Y8	3.5	3.8	5.8	—	11.9	4.3
	Civic del Sol S	1.6 (1590)	D16Y7	3.5	4.0	5.6	—	11.9	❶
	Civic del Sol Si	1.6 (1590)	D16Y8	3.5	4.0	5.6	—	11.9	8.2
	Civic del Sol VTEC	1.6 (1595)	B16A2	4.2	4.8	—	—	11.9	4.1
20__	Civic HX	1.6 (1595)	B16A2	4.2	4.8	—	3.9	11.9	4.1
	Civic DX, LX, CX	1.6 (1590)	D16Y5	3.5	3.8	5.8	—	11.9	4.5
	Civic EX	1.6 (1590)	D16Y7	3.5	3.8	5.8	—	11.9	4.4
	Civic Si	1.6 (1590)	D16Y8	3.5	3.8	5.8	—	11.9	4.3

❶ Continuously variable transmission

FIGURE 11-10 Capacity specifications tell the technician how much fluid is required in such components as the fuel tank, cooling system, engine, and transmission.

ENGINE MECHANICAL SPECIFICATIONS		
Description	**English Specifications**	**Metric Specifications**
Compression		
Pressure check @ 200 rpm wide open throttle		
Nominal	178psi	1230kPa
Minimum	135psi	930kPa
Maximum variation	28psi	200kPa
Cylinder head		
Warpage	0.02 in.	0.05mm
Height	5.589-5.593 in.	141.95-142.05mm
Camshaft		
Endplay	0.02 in.	0.5mm
Camshaft to holder oil clearance	0.006 in.	0.15mm
Total run-out	0.002 in.	0.04mm
Camshaft lobe height		
Intake		
Primary	1.3027 in.	33.088mm
Middle	1.4278 in.	36.267mm
Secondary	1.3771 in.	34.978mm
Exhaust	1.4964 in.	38.008mm
Primary	1.2907 in.	32.785mm
Middle	1.4063 in.	35.720mm
Secondary	1.3658 in.	34.691mm
Valves		
Clearance (cold)		
Intake	0.006-0.007 in.	0.15-0.19mm
Exhaust	0.007-0.008 in.	0.23-0.27mm
Stem O.D.		
Intake	0.2144 in.	5.445mm
Exhaust	0.2134 in.	5.420mm
Stem to guide clearance		
Intake	0.003 in.	0.08mm
Exhaust	0.004 in.	0.011mm
Valve seats		
Width		
Intake	0.079 in.	2.00mm
Exhaust	0.079 in.	2.00mm
Stem installed height		
Intake	1.5033 in.	38.185mm
Exhaust	1.4915 in.	38.885mm
Valve springs		
Free length		
Intake		
Outer	1.611 in.	40.92mm
Inner	1.445 in.	36.71mm
Exhaust	1.652 in.	41.96mm
Valve guides		
I.D.		
Intake	0.219 in.	5.55mm
Exhaust	0.219 in.	5.55mm
Installed height		
Intake	0.494-0.514 in.	12.55-13.05mm
Exhaust	0.494-0.514 in.	12.55-13.05mm
Rocker arms		
Arm-to-shaft clearance		
Intake	0.003 in.	0.08mm
Exhaust	0.003 in.	0.08mm

ENGINE MECHANICAL SPECIFICATIONS		
Description	**English Specifications**	**Metric Specifications**
Block		
Deck warpage	0.03 in.	0.08mm
Bore diameter	3.192 in.	81.07mm
Bore taper	0.002 in.	0.05mm
Reboring limit	0.01 in.	0.25mm
Piston		
Skirt (O.D. measured 0.8 in (21 mm) from bottom of skirt)	3.1878 in.	80.970mm
Clearance in cylinder	0.002 in.	0.05mm
Ring groove width		
Top	0.041 in.	1.05mm
Second	0.049 in.	1.25mm
Oil	0.112 in.	2.85mm
Piston rings		
Ring-to-groove clearance		
Top	0.005 in.	0.13mm
Second	0.005 in.	0.13mm
Ring end gap		
Top	0.024 in.	0.60mm
Second	0.028 in.	0.70mm
Oil	0.031 in.	0.80mm
Piston pin		
O.D.	0.8265-0.8268 in.	20.994-21.000mm
Pin-to-piston clearance	0.0004-0.0009 in.	0.010-0.022mm
Connecting rod		
Pin-to-rod clearance	0.0005-0.0013 in.	0.013-0.032mm
Small end bore diameter	0.8255-0.8260 in.	20.968-20.981mm
Large end bore diameter	1.89 in.	48.0mm
End-play installed on crankshaft	0.016 in.	0.40mm
Crankshaft		
Main journal diameter	2.1644-2.1654 in.	54.976-55.000mm
Rod journal diameter	1.7707-1.7717 in.	44.976-45.000mm
Rod/main journal taper	0.0001 in.	0.0025mm
Rod/main journal out or round	0.0001 in.	0.0025mm
Endplay	0.018 in.	0.45mm
Run-out	0.002 in.	0.04mm
Crankshaft bearing		
Main bearing-to-journal oil clearance	0.0020 in.	0.050mm
Rod bearing clearance	0.0020 in.	0.050mm
Oil pump		
Inner-to-outer rotor clearance	0.008 in.	0.20mm
Pump housing-to-outer rotor clearance	0.008 in.	0.21mm
Pump housing-to-outer rotor axial clearance	0.005 in.	0.12mm
Relief valve		
Check with the oil temp. at 176 deg. F or 80 deg. C		
Idle	10psi min.	70kPa min.
3000 rpm	50psi min.	340kPa min.

FIGURE 11–11 Engine mechanical specifications tell the service technician the exact measurement of the internal parts of an engine.

BREAK-IN SPEED LIMIT MPH (KM/H)						
		1st	2nd	3rd	4th	5th
Manual Transaxle	4-speed	0 to 22 (0 to 35)	12 to 37 (20 to 60)	20 to 55 (30 to 90)	25 to 75 (40 to 120)	
	5-speed	0 to 22 (0 to 35)	10 to 37 (15 to 60)	15 to 53 (25 to 85)	22 to 68 (35 to 110)	28 to 80 (45 to 130)
Automatic Transaxle		"1" Low 0 to 30 (0 to 50)		"2" Second 0 to 53 (0 to 85)		"D" Drive 0 to 75 (0 to 120)

FIGURE 11–12 Operational specifications show how the vehicle operates. In this case, the specifications show the break-in speed limit for operating a new vehicle.

PERFORMANCE SPECIFICATIONS						
	MPG (City Driving)	Acceleration 0–60 MPH (Sec.)	Brakes 60–0 MPH (Hot) (Ft.)	Handling (MPH)	Maneuverability (MPH)	Noise @ 60 MPH (dBA)
Chevrolet Camaro	16	10.5	192	61.9	27.7	71
Toyota in-line 6	21	11	172	66.5	28.5	71
Chevrolet Corvette V8	14	8.4	184	59.5	27.6+	74
Datsun turbo in-line 6	21	9.8	148	65+	27.6+	73
Mazda RX7 rotary	21	11.8	209	65+	27.6+	75
Porsche 924 turbo in-line 4	20	9.5	150	65+	27.6	74

FIGURE 11–13 Certain performance specifications compare vehicles on acceleration, miles per gallon, and so on.

replace the component in question. Usually maximum or minimum clearances are given for this purpose. In some cases, specifications are also given for a new vehicle or a used vehicle.

5. **Operational Specifications**: These specifications tell how the vehicle is to operate, what type of oil to use, and so on. Some of them are found in the owner's manual. For example, ***Figure 11–12*** shows the break-in speed limit taken from an owner's manual. Some specifications are also found in magazines and other technical literature. For example, ***Figure 11–13*** shows a performance comparison of several vehicles. Other specifications include tire inflation, type of gasoline to use, tire size, and general information for the operator of the vehicle.
6. **Torque Specifications**: It was mentioned in an earlier chapter that it is important to torque each bolt or nut correctly when replacing or installing a component on an automobile. Torque specifications are used for this purpose. ***Figure 11–14*** shows an example of several torque specifications on a particular manufacturer's engine. These torque specifications should be used in place of any standard bolt and nut torque specifications.
7. **System Specifications**: Often the service manual publisher will include additional specifications on specific systems within the automobile. For example, there may be additional specifications on the fuel system, cooling system, or suspension system that the service technician needs to know. ***Figure 11–15*** shows an example of a specific set of specifications for the brake system. Such specifications may include the master cylinder bore, the brake disc thickness, the diameter of the brake drum, the minimum brake lining thickness, and the brake caliper torque.

PROCEDURE FOR FINDING SPECIFICATIONS

There are many procedures for identifying and locating specifications in service manuals. Use the following general procedure for locating such information:

1. Identify the type, model, and year of the vehicle. Say, for example, engine rebuilding specifications are needed for a specific type of vehicle.
2. Locate the VIN on the vehicle for reference.
3. Select the correct service manual for the particular vehicle being serviced.
4. Refer to the first table of contents in the service manual, usually located on the first or second page.
5. Identify the system (section) that needs to be serviced such as fuel system, brakes, drive train, engine and engine overhaul, and so on, and turn to that section.
6. After turning to the correct section, read the section index for the required service operation, such as "CV

TORQUE SPECIFICATIONS		
Components	Ft. Lbs.	Nm
Transaxle Housing Mounting Bolts		
Manual	42-47	57-64
Automatic	40-43	54-59
Torque Converter to Drive Plate	100-104 inch lbs.	11.0-12.0
Rear Engine Stiffner	15-17	20-24
Stiffner to transaxle		
Manual	40-42	54-57
Automatic	29-32	39-43
Engine Mounting Brackets		
2.2, 2.3L	37-40	50-54
3.0L	24-28	32-38
Valve/Rocker Arm Cover		
Cover retainers	6.9-7.2	9.3-9.8
Cylinder Head		
Camshaft caps	6.7-7.2	9.0-9.8
Camshaft pulley	23-27	31-37
Rocker arm locknut	14-18	19-25
Cylinder head bolts		
Step 1	19-22	25-29
Step 2	59-63	80-85
Ignition System		
Distributor attaching bolt	13-17	17-24
Thermostat		
Mounting bolts	72-96 inch lbs.	7-11
Intake Manifold		
Throttle body	13-16	17-22
Manifold bolts	15-17	20-23
Idle air control valve	13-16	17-22
Idle air temperature sensor	3.9-4.0	5.2-6.0
Evap. purge control solenoid	7.3-8.0	9.9-11.0
Throttle body	14.6-17.0	19.7-22.0
Exhaust Manifold		
Manifold-to-block		
Self-locking nut	21-23	28-31
Manifold heat shield	16-17	21-24
Manifold down-pipe to manifold	40.0	54.0
Oxygen sensor	33.0	44.0
Catalytic converter		
Mounting bolts	11.0-12.0	15-16
Heat-shield/cover	7.2	9.8
Radiator		
Radiator mounting nuts	60-84 inch lbs.	7-9
Electrical Cooling Fan		
Fan motor-to-fan shroud nuts	44-66 inch lbs.	5-8
Fan assembly		
Nuts	35-41 inch lbs.	4-5
Screws	23-33 inch lbs.	3-4
Water Pump		
Water pump bolts	15-22	20-30
Oil Pan		
Oil pan bolts	8.5-8.7	12.0

FIGURE 11-14 Torque specifications help service technicians determine the proper torque for most bolts and nuts.

BRAKE SPECIFICATIONS

All measurements in inches unless noted

Year	Model		Master Cylinder Bore	Brake Disc: Original Thickness	Brake Disc: Minimum Thickness	Brake Disc: Maximum Runout	Brake Drum Diameter: Original Inside Diameter	Brake Drum Diameter: Max. Wear Limit	Brake Drum Diameter: Maximum Machine Diameter	Minimum Lining Thickness: Front	Minimum Lining Thickness: Rear	Brake Caliper: Bracket Bolts (ft. lbs.)	Brake Caliper: Mounting Bolts (ft. lbs.)
20__	Integra	F	NA	0.830	0.750	0.004	—	—	—	0.06	—	—	24
		R	NA	0.350	0.310	0.006	—	—	—	—	0.06	—	24
	Legend	F	NA	1.100	1.020	0.004	—	—	—	0.06	—	—	36
		R	NA	0.350	0.300	0.006	—	—	—	—	0.06	—	17
	Vigor	F	NA	0.910	0.830	0.004	—	—	—	0.06	—	—	36
		R	NA	0.390	0.310	0.004	—	—	—	—	0.06	—	17
20__	2.3CL	F	NA	0.910	0.830	0.004	—	—	—	0.06	—	80	54
		R	NA	0.390	0.310	0.004	—	—	—	—	0.06	28	18
	3.0CL	F	NA	0.980	0.910	0.004	—	—	—	0.08	—	80	54
		R	NA	0.390	0.310	0.004	—	—	—	—	0.08	28	18

FIGURE 11–15 Specific system specifications give the service technician detailed information, for example, about the brakes in the brake system.

Joint Overhaul"; turn to the particular page for the necessary service.

7. After reading, follow the service procedure as outlined in the service manual.
8. Pay close attention to any specifications and the wear limits provided. Also pay close attention to any footnotes given along with certain specifications. These footnotes give the service technician valuable information about specific vehicle characteristics that need to be followed carefully.

11.2 ADDITIONAL SOURCES OF INFORMATION

There are many sources from which specifications, service procedures, and troubleshooting information can be obtained. Both **independent publishers** and automotive manufacturers write service manuals. Parts manufacturers also distribute booklets with specifications and procedures so their parts will be installed correctly.

INDEPENDENT PUBLISHERS

One popular service and repair manual is a series published by the Hearst Corporation, referred to as Motor Manuals (***Figure 11–16***). In these manuals, all specifications, service procedures, and troubleshooting guides are printed. They are easy to read, and illustrate the service procedures clearly. These manuals usually include information about cars from the present back to about 7 years.

There are several sections in the basic auto repair manual. A troubleshooting section gives various tips on how to diagnose the major systems of the vehicle. Then there is an extensive section (vehicle information) on each of the major cars being produced. In each car section, there is also general service information that is common to a particular vehicle. Common service topics include alternator service, disc brakes, power brakes, transmissions, steering columns, and universal joints. These sections are very helpful in finding specific service information, tests to make in each area,

FIGURE 11–16 The Motor Manuals are commonly used manuals in the automotive shop. (Courtesy of MOTOR Publications, Auto Repair Manual, © The Hearst Corporation)

FIGURE 11–17 Chilton automotive service manuals offer a variety of technical data about automobiles. Each vehicle section is subdivided into technical systems such as brakes, engine, suspension, and so on.

FIGURE 11–18 Haynes Publishing provides specialized manuals on certain types of older vehicles. They can be purchased in most automotive parts stores.

and general disassembly and reassembly procedures. In addition, there are also several other sections on specifications and special service tools.

Motor also publishes a series of manuals that cover specific systems of the vehicle. Manuals are available on emission controls, automatic transmissions, air conditioners, heaters, and so on. These manuals are extremely helpful for service and diagnosis of special systems. Motor has a complete line of foreign car manuals as well.

Motor also publishes a monthly magazine entitled *Motor*. This magazine includes information about new automotive products, various special service and diagnosis information, special tools, and various interesting articles about performance and testing of vehicles.

Delmar Learning, the publisher of this textbook, also publishes Chilton auto repair manuals (***Figure 11–17***) which are another common source of technical information for automobile service shops. Auto manufacturers now produce automobiles in major lines, or "families." Internally, different models from a single manufacturer are quite similar. They share common systems and components. When dealing with large numbers of automobiles and components, it is easier sometimes to locate information about a technical system. Chilton manuals, for example, have sections on fuel systems, drivetrains, brakes, chassis, electrical, and so on. This information would be the same for one family of vehicles. Chilton manuals carry service information over a 5- to 7-year period.

Another company that prints service manuals is Haynes Publishing. Haynes Publishing publishes a series of service manuals for specific vehicles. Manuals are available that cover specifications, service procedures, and troubleshooting for a variety of specific vehicles. Haynes also publishes manuals for various imported vehicles. These manuals include both service procedures and principles of operation and theory. ***Figure 11–18*** shows some typical Haynes manuals.

MANUFACTURER'S SERVICE MANUALS

Automotive manufacturers also provide service manuals for their dealerships. **Manufacturer's service manuals** include principles of operation and some theory. These service manuals are written for one family or type of vehicle. ***Figure 11–19*** shows a typical service manual from one manufacturer. Common manuals are:

- Chassis service manual
- Body manual
- Electrical troubleshooting manual
- Owner's manual
- Wiring diagram manual
- Do-it-yourself service guide

FIGURE 11–19 Service manuals can also be obtained directly from manufacturers. These manuals usually cover a specific family of vehicles or a particular vehicle.

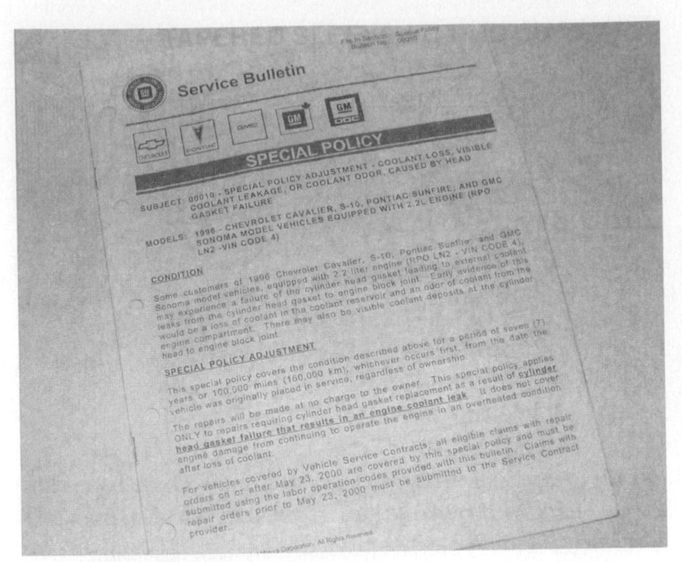

Service Bulletin

SPECIAL POLICY

SUBJECT: 00010 - SPECIAL POLICY ADJUSTMENT - COOLANT LOSS, VISIBLE COOLANT LEAKAGE, OR COOLANT ODOR, CAUSED BY HEAD GASKET FAILURE

MODELS: 1996 - CHEVROLET CAVALIER, S-10, PONTIAC SUNFIRE, AND GMC SONOMA MODEL VEHICLES EQUIPPED WITH 2.2L ENGINE (RPO LN2 -VIN CODE 4)

CONDITION

Some customers of 1996 Chevrolet Cavalier, S-10, Pontiac Sunfire, and GMC Sonoma model vehicles, equipped with 2.2 liter engine (RPO LN2 - VIN CODE 4), may experience a failure of the cylinder head gasket leading to external coolant leaks from the cylinder head gasket to engine block joint. Early evidence of this would be a loss of coolant in the coolant reservoir and an odor of coolant from the engine compartment. There may also be visible coolant deposits at the cylinder head to engine block joint.

SPECIAL POLICY ADJUSTMENT

This special policy covers the condition described above for a period of seven (7) years or 100,000 miles (160,000 km), whichever occurs first, from the date the vehicle was originally placed in service, regardless of ownership.

The repairs will be made at no charge to the owner. This special policy applies ONLY to repairs requiring cylinder head gasket replacement as a result of **cylinder head gasket failure that results in an engine coolant leak**. It does not cover engine damage from continuing to operate the engine in an overheated condition after loss of coolant.

For vehicles covered by Vehicle Service Contracts, all eligible claims with repair orders on or after May 23, 2000 are covered by this special policy and must be submitted using the labor operation codes provided with this bulletin. Claims with repair orders prior to May 23, 2000 must be submitted to the Service Contract provider.

FIGURE 11-20 Service bulletins update information and changes not included in service manuals.

These manuals can be obtained by writing to the manufacturer of the vehicle. Also, the owner's manual, which is usually kept in the vehicle, may have an address to contact for specific manuals.

FACTORY BULLETINS

There are numerous technical changes and recalls that occur on specific vehicles during a given year. Therefore, vehicle manufacturers often communicate to their dealerships and service technicians with factory bulletins. Typically, there are several types of bulletins including service bulletins, campaign bulletins, and recall bulletins. The **service bulletins**, similar to the one shown in ***Figure 11–20***, provide service technicians with important information about repair procedures and changes in specifications. Campaign bulletins are used to tell dealerships that a specific vehicle part is being sent to dealers and should be offered to the customer at no cost. Recall bulletins alert dealerships to recalls by vehicle manufacturers.

OWNER'S MANUALS

Each new vehicle includes an owner's manual or owner's guide for the car buyer (***Figure 11–21***). This manual contains only general information about the vehicle. For example, the following areas are generally found in most owner's manuals:

- Vehicle instrumentation
- Getting to know the vehicle
- Starting and operating the vehicle
- Specifications and capacities (basic)
- Servicing the vehicle (basic)
- Customer assistance
- Trailer towing

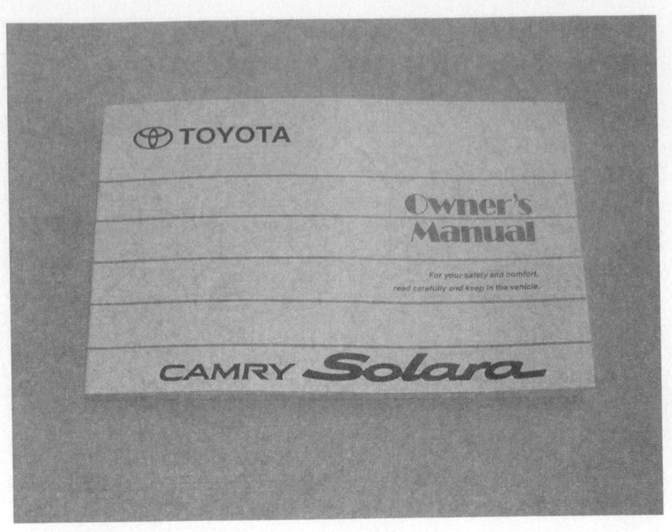

FIGURE 11-21 A vehicle's owner's manual provides valuable information about common service specifications such as vehicle instrumentation, specifications and capacities, trailer towing, and so on.

COMPUTERIZED INFORMATION SYSTEMS

The need for automotive service and diagnostic information is steadily increasing. Computers are available to provide this information. Over the past few years, several computerized networks have been developed to provide additional service and diagnostic information to automotive dealers, technicians, and service shops.

One such network is called the **Service Bay Diagnosis System**. This is an Internet system connected to the manufacturer's location or headquarters. When a service or diagnostic question needs to be answered, the vehicle serial number can be entered into the computer network. Through a series of computer prompts, various system codes can be entered. On the basis of this information, the computer tells the service technician what the problem is, if it is correctable, and helps to do a service bulletin search to find out if a past service bulletin can help solve the problem.

Another system that is available is **OASIS**, which stands for On-line Automotive Service Information System. This system is also capable of communicating with the vehicle manufacturer to help answer most diagnostic and service questions. Such information can be extremely useful, especially when working on the newer computerized/electronic engines. Again, the system works by entering the serial number of the vehicle and various codes to identify the problem. OASIS is more sophisticated than the Service Bay Diagnosis System; in some cases, the computer tells the service technician exactly what to replace and the repair procedures, on the basis of inputs given to the computer.

A third system often used in smaller automotive service facilities is the **AllData** Electronic Retrieval system. AllData is a system that incorporates all of the selected manufacturers' service manuals and all service bulletins published by each of those manufacturers. This information is contained in a single computer information terminal available to the

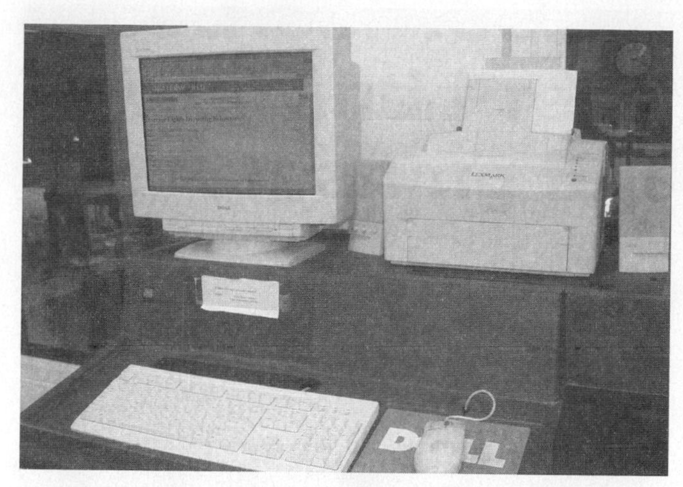

FIGURE 11–22 Dealer World provides a variety of automotive service data for specific dealerships of major vehicle manufacturers.

service technician. AllData has several databases in its CD-ROM system. A CD-ROM system uses discs like those in an audio CD player to store the data. The data on these discs include service and repair procedures, parts numbers, wiring diagrams, vacuum hose routings, flat rate information, and the latest TSB, or technical service bulletins.

Most automotive dealerships use the Internet to obtain service information. Each of the major vehicle manufacturers provide a direct Internet link from the service area in a dealership to the vehicle manufacturer. There are many names for such systems. One common system, called Dealer World, is shown in ***Figure 11–22***. This system provides the service technician with information such as:

- Recent service bulletins
- Technical data and specifications about specific vehicle systems
- Full-service manuals for specific vehicles
- Information on best procedures to use during service
- Recall information
- Parts numbers and associated information
- Wiring diagrams

These systems are provided to the dealerships by major vehicle manufacturers, and are not typically available to smaller service shops.

Car Clinic: NEW VEHICLE BREAK-IN PROCEDURE

CUSTOMER CONCERN:
Driving a New Car

A customer has been told several methods used to break in a new engine. The driver is keeping the engine at a constant 55 mph as much as possible. Also, it was suggested that the oil be changed after the first 500 miles. What is the best procedure for breaking in an engine?

SOLUTION: The best method is to follow the manufacturer's recommendation. This usually includes driving moderately between 50 and 60 miles per hour for about 2,000 miles under normal loads. Use the manufacturer's break-in oil. The oil is designed to aid the engine parts in wearing in to the mating surfaces. Also the engine needs about 50 to 60 cycles from cold to hot to help wear in the parts. Change the break-in oil according to the owner's manual for new cars.

SUMMARY

The following statements will help to summarize this chapter:

- Service manuals and specifications are important tools for service technicians.
- Manuals provide a variety of information for the service technician including diagnosis information, troubleshooting charts, vehicle specifications, and operator information.
- Common specifications found in a service manual include general engine specifications, tune-up specifications, and capacity and overhaul specifications.
- Torque specifications for bolts and nuts are also found in the service manual.
- Some common service manuals include Motor Manuals, Chilton automotive service manuals, and Mitchell manuals.
- Haynes Publishing also provides specific manuals that cover specifications, procedures, and troubleshooting for a specific type and year of vehicle.
- Automotive manufacturers publish their own service manuals for each family of vehicles they produce.
- Service bulletins give service technicians the most up-to-date information on vehicle operation.

TERMS TO KNOW

Can you explain each of the following terms? Review the chapter until you can use each term correctly.

AllData
Capacity specifications
Diagnosis
General engine specifications
Independent publishers
Manufacturer's service manuals
OASIS
Operational specifications
Overhaul and maintenance specifications
Service Bay Diagnosis System
Service bulletins
Service manuals
Service procedure
Specifications
Torque specifications
Troubleshooting
Tune-up specifications
VIN

REVIEW QUESTIONS

Multiple Choice

1. Which of the following information is not included in a service manual?
a. Specifications
b. Sales and promotion information
c. Disassembly procedures
d. Reassembly procedures
e. Fluid capacities

2. Which of the following information can be shown in a service manual using a flow diagram form?
a. Specifications
b. Troubleshooting and diagnostic information
c. Overhaul information
d. Special tools
e. Types of vehicles

3. Which of the following information is usually shown as measurements, clearances, and numbers?
a. Procedure information
b. Troubleshooting information
c. Specification information
d. Overhaul procedures
e. Special tools

4. Specifications that include spark plug gap, ignition timing, and other adjustments are called:
a. Tune-up specifications
b. Capacity specifications
c. Overhaul and maintenance specifications
d. Procedure specifications
e. None of the above

5. Specifications that show bearing clearances, shaft end play, and ring gaps are called:
a. Tune-up specifications
b. General engine specifications
c. Overhaul and maintenance specifications
d. Capacity specifications
e. Troubleshooting specifications

6. Torque specifications are identified as:
a. Torque on nuts and bolts
b. Engine torque at certain speeds
c. Torque applied to the crankshaft during operation
d. Torque produced from the generator
e. Torque applied on turns

7. Motor Manuals are published by:
a. The manufacturer of the automobile
b. An independent publisher
c. Dealerships
d. Mechanics
e. Universities

8. Which is not a common title of a service manual provided by a publisher?
a. Emission Control Manual
b. Chassis Service Manual
c. Wiring Diagram Manual
d. Water Pump Manual
e. All of the above

9. Approximately how many years do the Chilton and Motor manuals cover?
a. 1–3 years
b. 4–7 years
c. 15 years
d. 18 years
e. None of the above

10. Which of the following types of technical information is sent to the service dealers to update the service manuals?
a. Service bulletins
b. Update bulletins
c. New data bulletins
d. Technique bulletins
e. Sales bulletins

The following questions are similar in format to ASE (Automotive Service Excellence) test questions.

11. When repairing an engine, *Technician A* says overhaul procedures can be found in service manuals. *Technician B* says fluid capacities can be found in service manuals. Who is correct?

a. A only c. Both A and B
b. B only d. Neither A nor B

12. *Technician A* says that plug gaps, clearances, and torque on bolts cannot be found in a service manual. *Technician B* says that engine specifications for overhaul can be found in a service manual. Who is correct?

a. A only c. Both A and B
b. B only d. Neither A nor B

13. *Technician A* says that diagnosis information can be found in service manuals. *Technician B* says that sales information can be found in service manuals. Who is correct?

a. A only c. Both A and B
b. B only d. Neither A nor B

14. *Technician A* says that the VIN does not include any information about the engine in the vehicle. *Technician B* says that the VIN does not include the vehicle serial number. Who is correct?

a. A only c. Both A and B
b. B only d. Neither A nor B

15. *Technician A* says the VIN can be found on the dashboard on the driver's side. *Technician B* says the VIN can be found in the owner's manual. Who is correct?

a. A only c. Both A and B
b. B only d. Neither A nor B

16. *Technician A* says that there are numbers placed in the VIN to identify the type of transmission. *Technician B* says there are numbers placed in the VIN to identify axle and differential ratios. Who is correct?

a. A only c. Both A and B
b. B only d. Neither A nor B

Essay

17. What are examples of capacity specifications in the service manuals?

18. What is the purpose of service bulletins?

19. What are the major service manual publishers?

20. What is the purpose for having diagnosis sheets?

21. What type of specifications might be found under "General Engine Specifications"?

Short Answer

22. The number that gives the service technician information about model, year, body style, and so on, of the vehicle is called the ________.

23. The ________ provides general information about the vehicle such as vehicle instrumentation, servicing times, and trailer towing.

24. In order to make the service technician's job easier, several new ________ network systems are being used to provide service, diagnostic, and parts information.

25. The VIN is generally located on the ________ part of the vehicle.

26. To identify the exact information represented by the VIN, the service technician can look in the ________.

Applied Academics

TITLE: Automotive Service Training Centers

ACADEMIC SKILLS AREA: Language Arts

NATEF Automobile Technician Tasks:

The technician uses study habits and methods when consulting manufacturers' information on engine repair, for example, shop manuals, references, and computer databases.

CONCEPT:

Because there are so many new products and designs used on today's automobiles, there is a need for constant study and upgrading. Automotive service technicians must have a great deal of information available and must be able to learn new information quickly. Service technicians must have excellent study habits and be able to learn about new products, their operation, and servicing them quickly. Vehicle manufacturers offer a variety of service training centers for this type of learning to take place. These service training centers provide service technicians with specific, up-to-date, technical information about automobiles. They have experienced trainers to provide quality instruction. Some service training centers also provide technical advice and help customers with warranty problems. Service training centers offer various classes on new products. These classes typically are from 1 to 5 days or more in length, depending on the product.

APPLICATION:

Are your study habits refined enough to be a quality service technician? How well do you learn when classes last all day long for 5 days in a row? Do you have the ability to retain the information that is presented in a service training center program? Analyze your study habits and list five ways you can improve them.

Fundamentals of Automotive Engines

SECTION 2

Tara Kebabjian, 24

Customer Service Representative (CSR)

Valvoline Instant Oil Change

Latham, NY

Who inspired you to go into this field?

My best friend. She worked as a technician, and when a job opened up in the CSR area, I thought it sounded different from your average job. I thought to myself that if she was able to be a technician, then I would have no problem being a CSR.

What kind of education is necessary for this position?

A high school diploma. I took an automotive class in high school, and I've always been around it, but I'd say that more than half of the knowledge I have about cars I got from being here.

What's a typical day like for you?

When I arrive in the morning, I go over paperwork and deposits from the night before. Throughout the day, I'm dealing with the customers, as they come in requesting services and as they're billing out. I'm telling the technicians what the customers need. I'm also on the phone with suppliers. If we're really busy, in between, I'll go out and help in the garage, doing things like a basic oil change.

What's the most important skill needed for your job?

Communication. I have to be able to communicate to customers what the technicians are saying in a way that they'll be able to understand. It helps that I have a background in the automotive area because if a customer comes in with a question, I don't have to ask one of the techs. And then the customers see that I'm someone who knows what I'm talking about.

What do you wish you had learned more about?

Computers and math. I need to know these things in dealing with the suppliers. In this job, I don't go anywhere without a calculator!

What do you like about your job?

It's not typical 9-to-5 job behind a desk; I'm running in and out of the garage all day long. And I work with a lot of different personalities. Besides the eight technicians we have working at any time—who range in age from high school students to guys in their 50s—I put through more than 100 cars everyday, which means I see a lot of different people.

What's the worst part?

The weather! We're working in a garage straight through winter!

Converting Energy to Power

CHAPTER 12

OBJECTIVES

After studying this chapter, you should be able to:

- Define how energy is converted in the automobile.
- Classify engines according to their design.
- Analyze the major components needed to make an engine run.

INTRODUCTION

One way to analyze an automobile is to look at the energy in the vehicle. A car engine basically takes one form of energy and converts it to another form of energy. This chapter introduces you to how energy is converted by the engine. It also describes how engines are classified. In addition, this chapter introduces the basic parts and nomenclature of an automobile engine.

12.1 ENERGY CONVERSION

ENERGY DEFINED

The engine in an automobile is designed to accomplish one thing: to convert **energy** from one form to another. Energy is defined as the ability to do work. There is energy within the fuel that is put into the engine. The engine takes the energy from the fuel and converts it into a form of power. The power is used to propel the vehicle and to provide power for other uses on the automobile.

POWER DEFINED

Power is defined as a measure of the work being done. Power is the final output of the engine after it has converted the energy in the fuel into work. A more common term used today is *horsepower*. Horsepower is a measure of the work being done by the engine over a certain period of time.

FORMS OF ENERGY

Energy can take on one of six forms. Referring to ***Figure 12–1***, they include chemical, electrical, mechanical, thermal, radiant, and nuclear forms of energy. The automobile uses all of the preceding energy forms except nuclear energy.

1. *Chemical energy* is defined as energy contained in molecules of different atoms. Examples of chemical energy are gasoline, diesel fuel, coal, wood, chemicals inside a battery, and food.
2. *Electrical energy* is defined as the ability to move electrons within a wire. Electrical energy uses voltage, wattage, resistance, and so on for operation. Many of the components on a car, including the radio, horn, lights, and starter, utilize electrical energy.
3. *Mechanical energy* is defined as the ability to physically move objects. Examples include water falling over a dam, the ability to move a vehicle forward, and

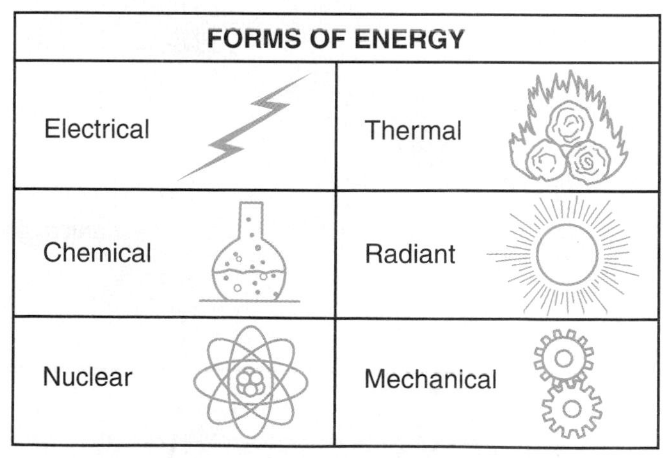

FIGURE 12–1 There are six forms of energy, including chemical, electrical, mechanical, thermal, radiant, and nuclear. All of these forms of energy are used on the automobile except nuclear energy.

gravity. The starter motor on a car takes electrical energy and converts it into mechanical energy to start the engine.

4. *Thermal energy* is defined as heat. This form of energy is released when fuel burns. The combustion of fuel produces thermal energy. The radiator removes excess thermal energy from the engine.
5. *Radiant energy* is defined as light energy. It is measured by frequencies. Examples of radiant energy include the energy coming to the Earth from the sun, the energy from a light bulb, and the energy from anything that glows.
6. *Nuclear energy* is defined as the energy that is released when atoms are split apart or combined. Nuclear energy is not used in the automobile.

ENERGY CONVERSION

Energy conversion is defined as changing one form of energy to another. Energy usually does not come in the right form. Therefore, it must be converted to a form that can be used. For example, a vehicle uses mechanical energy to go forward, electrical energy for the radio, and radiant energy from the light bulbs. Gasoline or diesel fuel is the main source of energy on a vehicle. It is in the form of chemical energy. The engine is designed to convert the chemical energy into the correct forms of energy needed.

Chemical to Thermal Conversion When any fuel is burned, it changes the energy from chemical (fuel) to thermal (heat). This process happens when the fuel burns in an engine (***Figure 12–2***). However, thermal energy is not really needed. Mechanical energy is what is needed from the engine to power the vehicle.

Thermal to Mechanical Conversion Once thermal energy is produced by burning fuel, the thermal energy causes rapid expansion of the gases within the engine. This rapid expansion is called mechanical energy. The combustion process on any engine converts chemical to thermal, and thermal to mechanical energy. This mechanical energy then pushes down the piston and connecting rod. This motion causes the crankshaft to turn, which is called torque. This is how mechanical energy is used to propel the vehicle.

Mechanical to Electrical Conversion The alternator (***Figure 12–3***) is designed to convert some of the mechanical energy into electrical energy. The electrical en-

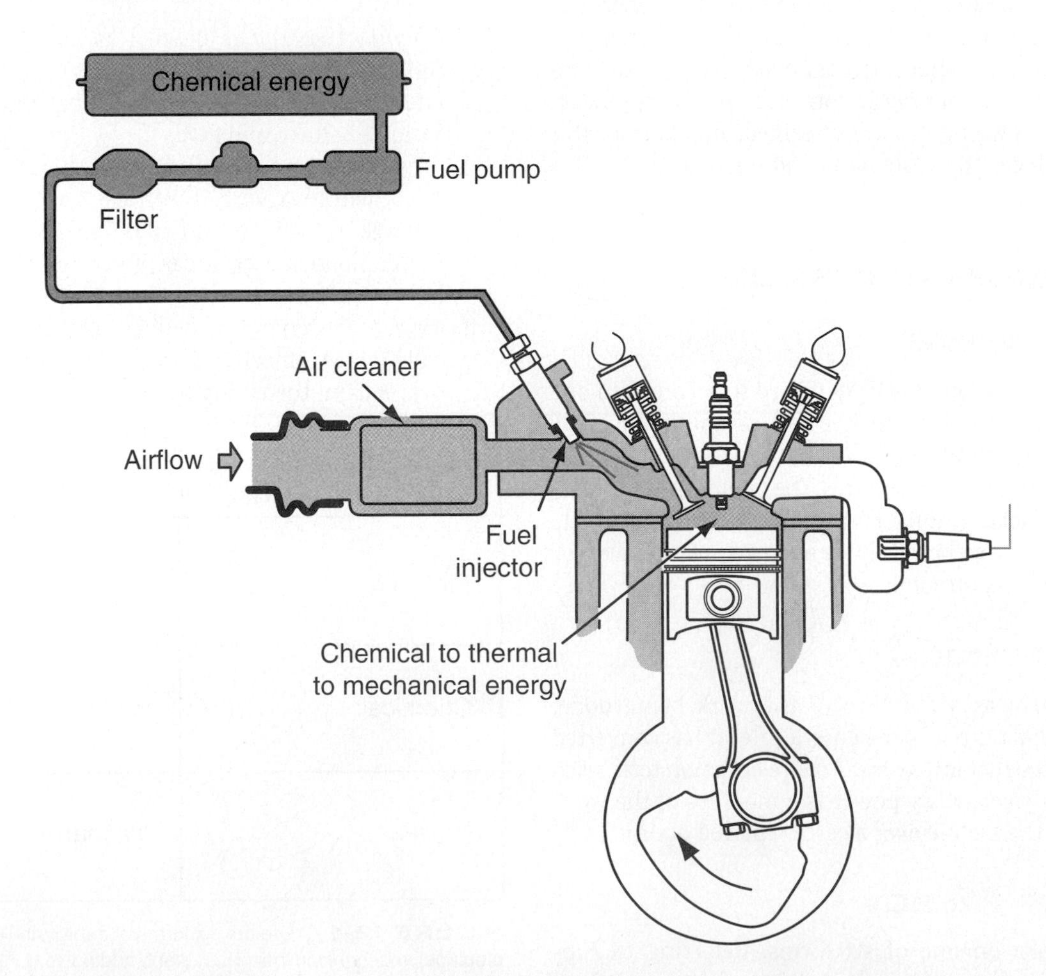

FIGURE 12–2 The internal combustion engine converts chemical energy in the fuel into thermal energy in the combustion area. The thermal energy is then converted into mechanical energy by the piston and crankshaft.

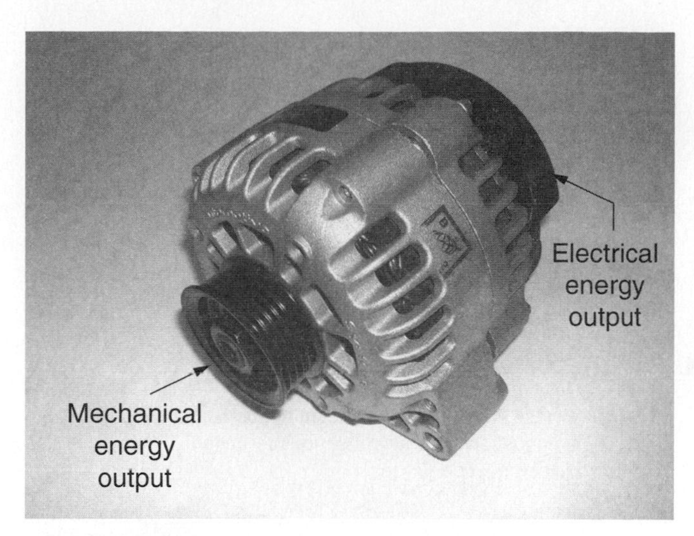

FIGURE 12-3 The alternator on an automobile engine is used to convert mechanical energy into electrical energy.

ergy is used to start the car, provide ignition, and operate the radio and other electrical appliances on the vehicle. The storage battery stores any excess electrical energy, for use if needed, when the car is not running, such as to power the starter motor.

Electrical to Mechanical Conversion The starter motor, ***Figure 12–4***, is designed to convert electrical energy into mechanical energy to crank the engine. This device is called a motor. All motors, including windshield wiper motors, heater fans, and starters, convert electrical energy to mechanical energy.

Electrical to Radiant Conversion Electrical to radiant conversion occurs when light bulbs are used. The energy coming out of a light bulb is radiant energy.

FIGURE 12-4 The starter motor on an automobile engine is used to convert electrical energy into the mechanical energy required to crank the engine during starting.

Electrical energy is used to operate a light bulb. A light bulb, then, converts electrical energy to radiant energy.

12.2 ENGINE CLASSIFICATIONS

Engines can be classified several ways. These include (1) by the location of the combustion, (2) by the type of combustion, and (3) by the type of internal motion.

INTERNAL COMBUSTION ENGINES

Engines can be classified by defining the location of combustion. **Combustion** is defined as ignition of the exact amount of fuel mixed with the correct amount of air inside an engine.

In an **internal combustion** engine (ICE), combustion occurs within the engine. The combustion process occurs directly on the parts that must be moved to produce mechanical energy. The fuel is burned within the engine (***Figure 12–5***). A gasoline engine is an internal combustion engine. Small lawn mower engines, snowmobile engines, and motorcycle engines are also internal combustion engines.

EXTERNAL COMBUSTION ENGINES

In an **external combustion** engine, the combustion is removed from the parts that must be moved (***Figure 12–6***). For example, the boiler in a steam engine is external. It is not touching the piston. Actually, the thermal energy in an

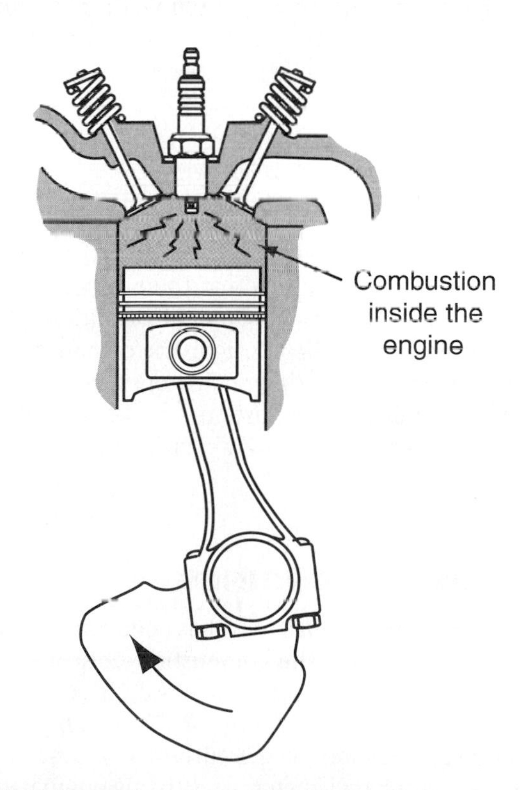

FIGURE 12-5 In an internal combustion engine, the combustion of gases occurs inside the engine, actually coming into contact with the moving parts.

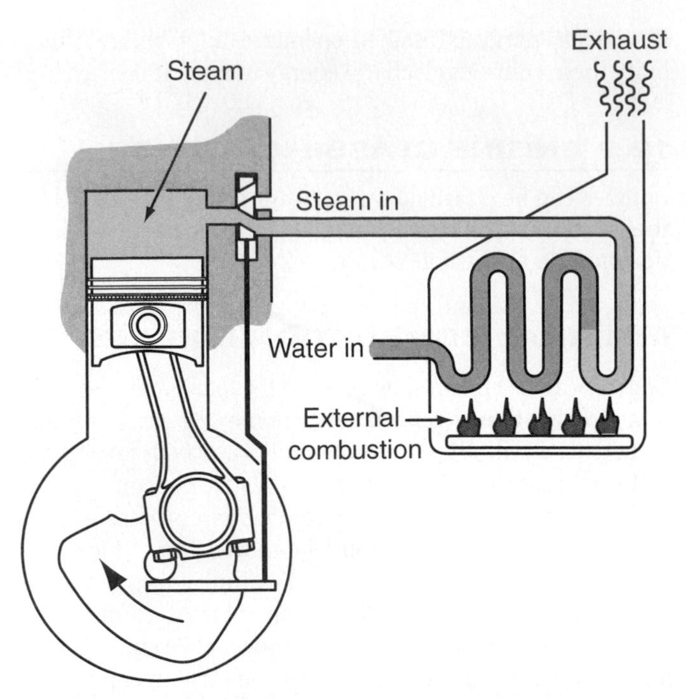

FIGURE 12–6 In an external combustion engine the combustion area is removed from the pistons and engine.

external combustion engine heats another fluid. In this case, it is water. Water, converted to steam, pushes against the piston.

There has been some research conducted to determine if an external combustion engine could work in an automobile application. So far, this type of engine has not proven successful in the automobile market but is only mentioned for comparison purposes.

INTERMITTENT COMBUSTION ENGINES

The second classification is by the type of combustion. **Intermittent combustion** means that the combustion within an engine starts and stops. A standard gasoline engine has an intermittent combustion design. The combustion starts and stops many times during engine operation. Diesel engines are intermittent combustion engines as well. Diesel engines have been used by several automobile manufacturers in the past years.

CONTINUOUS COMBUSTION ENGINES

A **continuous combustion** engine has combustion that continues all of the time. The combustion does not stop. It keeps burning continuously. A blowtorch is an example of continuous combustion (***Figure 12–7***). Engines that use continuous combustion include turbine engines (such as a helicopter engine), rocket engines, Stirling engines, and jet (or reaction) engines. Research has shown that turbines could be used in the automobile, but they are very costly for this purpose.

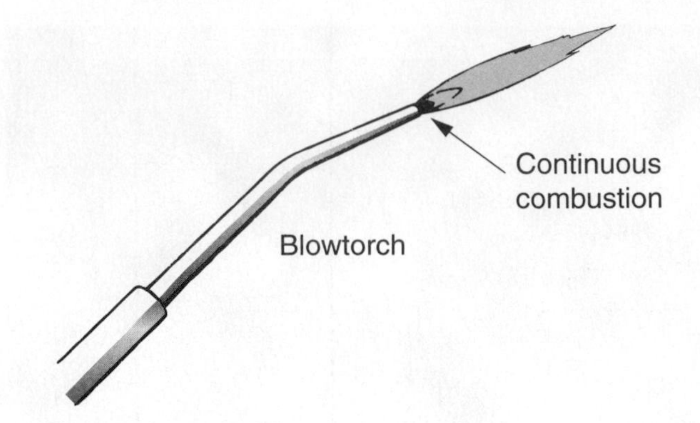

FIGURE 12–7 An example of continuous combustion is a blowtorch. The combustion is continuous, not intermittent. Jet engines use a continuous combustion process.

RECIPROCATING ENGINES

The third engine classification is by the type of internal motion. In a **reciprocating engine**, the motion produced from the energy within the fuel moves parts up and down. The motion reciprocates—moves back and forth or up and down. Gasoline and diesel engines are reciprocating engines. In this case, the power resulting from the burning of an air and fuel mixture starts the piston moving. The piston starts, then stops, then starts, then stops, and so on. In this engine, the reciprocating motion must then be changed to rotary motion. A crankshaft is designed to change this motion. ***Figure 12–8*** shows the up-and-down motion of the piston in the cylinder.

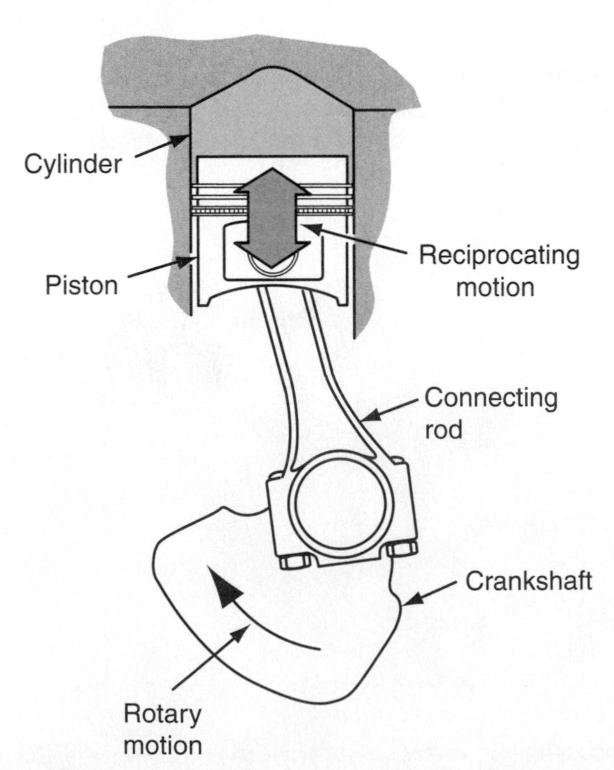

FIGURE 12–8 Gasoline and diesel engines are considered reciprocating engines. Reciprocating motion in these engines involves the parts moving up and down.

ROTARY ENGINES

In a **rotary engine**, the parts that are moving rotate continuously. For example, a turbine and a Wankel engine are considered rotary engines. The mechanical movement of the parts take the shape of a circle. In ***Figure 12–8***, the crankshaft is an example of rotary motion.

OTHER CLASSIFICATION METHODS

Engines can also be classified by the following four methods:

1. *By stroke*—There are two- and four-stroke engines.
2. *By cooling systems*—There are liquid-cooled and air-cooled engines.
3. *By fuel systems*—There are gasoline-fueled and diesel-fueled engines.
4. *By ignition systems*—There are spark-ignition and compression-ignition engines.

All of these methods of classifying engines can be combined with those previously mentioned.

CLASSIFICATION OF ENGINES USED IN AUTOMOBILES

Automobiles use engines with several of the classifications just listed. The gasoline and diesel engines used in cars are considered internal combustion, intermittent combustion, and reciprocating engine designs. If the rotary (Wankel) engine is used, it is considered an internal combustion, intermittent combustion, rotary design. Many alternative designs have been tested for use in automobiles. Gasoline, diesel, and rotary (Wankel) engines are the main ones used today. The gasoline engine is still the most popular form used. The gasoline engine is a four-cycle, spark-ignition engine. It can be either liquid-cooled or air-cooled.

12.3 BASIC ENGINE TERMINOLOGY

To understand the principles of automobile engines, certain parts must be defined. These parts are considered the major components of the engine. They include the cylinder block, cylinders, pistons, connecting rods and crankshaft, cylinder head, combustion chamber, valves, camshaft, flywheel, intake and exhaust manifolds, and the fuel system, including the carburetor or fuel injectors.

CYLINDER BLOCK

The **cylinder block** is considered the foundation of the engine (***Figure 12–9***). The cylinder block is most often made of cast iron or aluminum. All other components of the engine are attached to the cylinder block. The cylinder block has several internal passageways to let cooling fluid circulate around the block. It also has several large holes machined into the block where the combustion occurs.

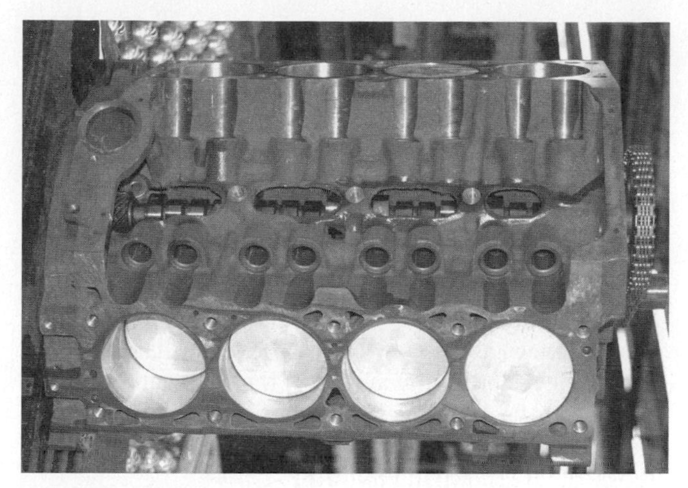

FIGURE 12-9 The cylinder block is the foundation of the engine. All other parts are attached to the cylinder block. (Courtesy of Jasper Engine and Transmissions)

CYLINDERS

The **cylinders** are internal holes machined into the cylinder block (***Figure 12–9***). These holes are used for combustion. The holes tell the number of cylinders used on an engine. For example, on small gasoline engines, such as lawn mowers, there is one cylinder. Automobiles usually use four, six, or eight cylinders. Some engines used on heavy equipment have as many as 24 cylinders.

PISTONS

The **piston** is the round object that slides up and down in a cylinder (***Figure 12–10***). There is one piston for each cylinder. Pistons are made of light material, such as high-quality aluminum, that can withstand high temperatures. When fuel and air ignite to cause expansion above the piston, this expansion forces the piston downward. The motion

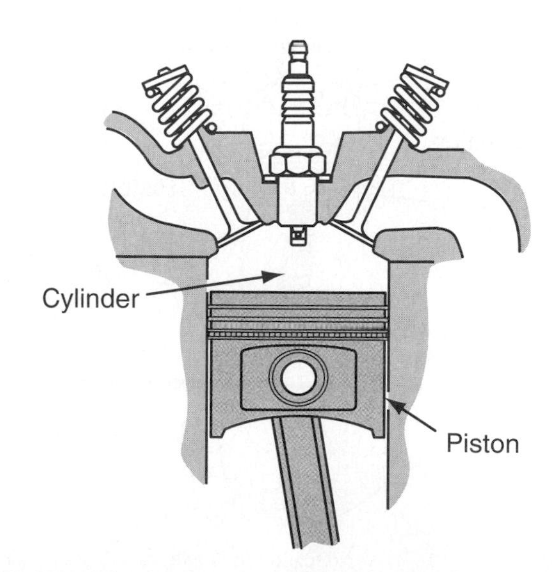

FIGURE 12-10 The piston slides up and down inside of the cylinder.

converts the energy in the fuel into mechanical energy (piston moving downward). The piston must also have seals or rings near the top to stop any combustion gases from passing by the piston on its sides.

CONNECTING ROD AND CRANKSHAFT

Attached to the bottom of the piston is the **connecting rod** (***Figure 12–11***). Its main purpose is to attach the piston to a device known as the **crankshaft**. The crankshaft is used to change the reciprocating motion of the piston and connecting rod to rotary motion. Rotary motion is used as the output power of the engine. The piston, connecting rod, and crankshaft are shown in ***Figure 12–12***.

CYLINDER HEAD

The **cylinder head** is the part that fits over the top of the cylinder block (***Figure 12–13***). It usually houses the ports and valves that allow fuel and air to enter into the cylinder. The spark plug is also attached to the cylinder head. The cylinder head is made of cast iron or aluminum. When it is bolted to the cylinder block, it seals the cylinders so that air and fuel in and out of the cylinder can be controlled.

COMBUSTION CHAMBER

The **combustion chamber** is where the combustion takes place inside the cylinder. When the cylinder head has been attached, the area inside of the cylinder head and block is called the combustion chamber. On some engines, the combustion chamber is located inside the head. Other engines

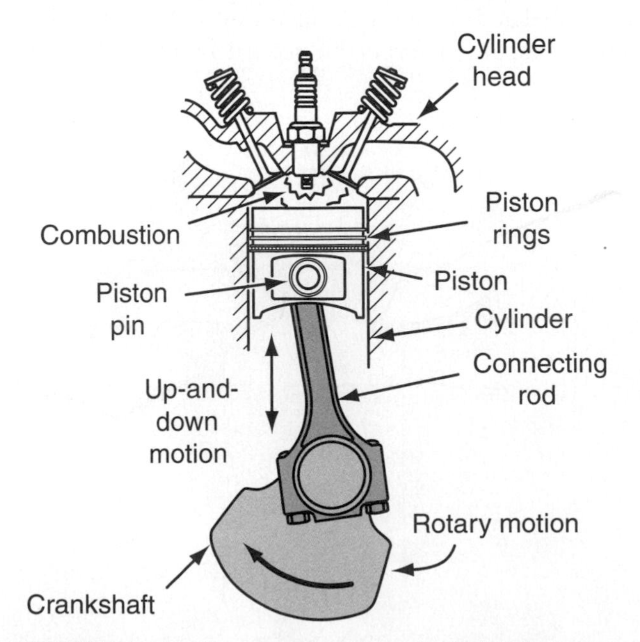

FIGURE 12–11 The connecting rod is attached to the bottom of the piston. The crankshaft changes the reciprocating motion to rotary motion.

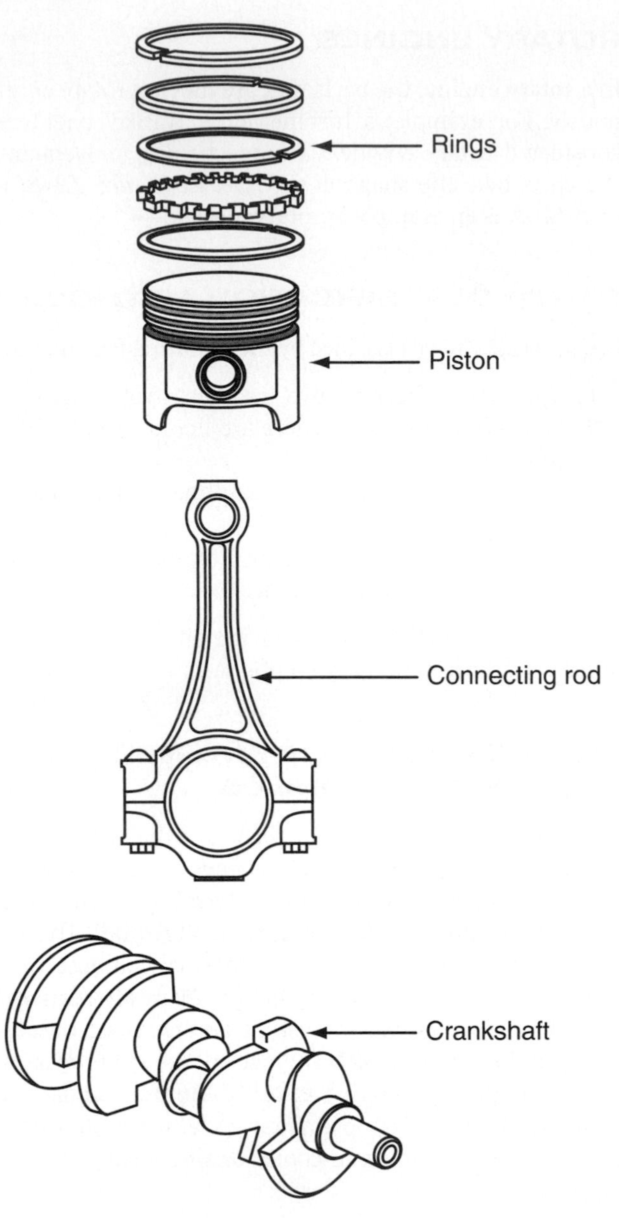

FIGURE 12–12 The piston, connecting rod, and crankshaft, shown here, for a four-cylinder engine are the parts that are used to change reciprocating motion to rotary motion.

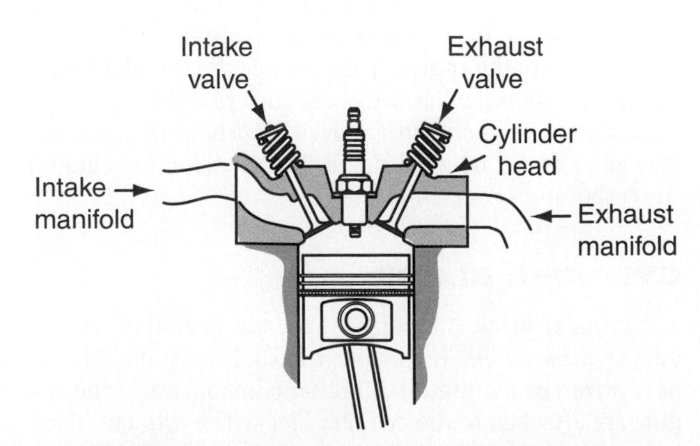

FIGURE 12–13 A cylinder head seals the top of the engine. It houses the intake and exhaust valves and ports, the spark plug, and the combustion chamber.

have the combustion chamber located inside the top of the piston. This is true on most diesel engines.

VALVES

Valves are placed inside the cylinder head to allow air and fuel to enter and burned gases to leave the combustion area. The valves are designed as shown in ***Figure 12–14***. Valves must be designed so that when they are closed, the port is sealed perfectly. They must also be designed so that they can be opened at exactly the right time. These valves are opened by using a camshaft and are closed by using springs. There is an intake valve to allow fuel and air to enter the cylinder. There is an exhaust valve to allow the burned gases to escape the cylinder.

VALVE AND CYLINDER HEAD ARRANGEMENTS

In the past, engines had two valves in each cylinder, one for the exhaust and one for the intake. Today, some engines use three or four valves per cylinder. If there are three valves per cylinder, and four cylinders, the engine is said to be a 12-valve engine. If there are four valves per cylinder and four cylinders, the engine is said to be a 16-valve engine. There are many arrangements. Three arrangements that are currently being used are:

1. Twelve-valve engines usually have three valves in each cylinder and four cylinders. They are called 3-valve engines. Generally, each cylinder has two intake valves and one exhaust valve. Two valves are used on the intake so that more air and fuel can be brought into the cylinder. Only one exhaust valve is needed because some of the energy in the air and fuel mixture has already been removed by the piston. Other three-valve arrangements are also possible, depending on the manufacturer. An in-line 6-cylinder engine with two valves for each cylinder would also be a 12-valve engine.
2. Sixteen-valve engines have four valves for each cylinder and four cylinders. They are called 4-valve engines. Generally, each cylinder has two intake and two exhaust valves. The purpose of having so many valves is to allow more air and fuel to enter smaller-sized engines, thus producing more power. A V8 engine with two valves for each cylinder would also be a 16-valve engine.
3. Twenty-four-valve engines have four valves in each cylinder and six cylinders. Each cylinder has two intake and two exhaust valves.

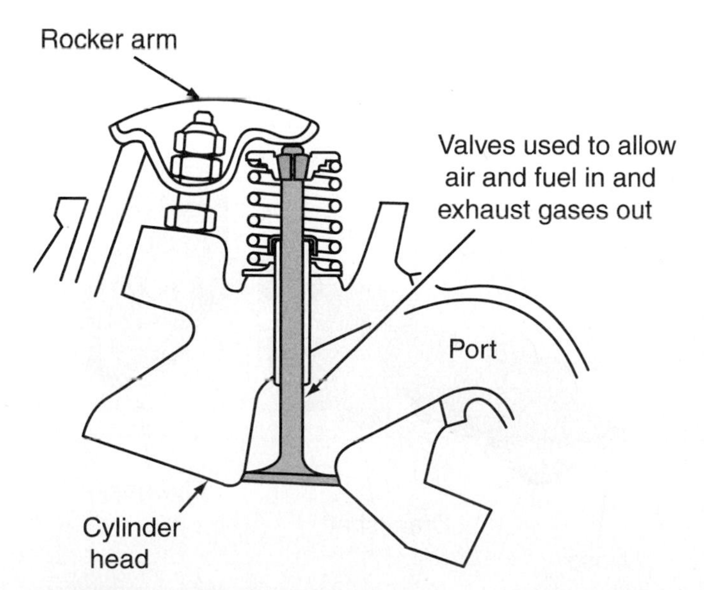

FIGURE 12–14 Valves open and close the ports in the combustion chamber to allow air and fuel intake into the engine and exhaust gases out of the engine.

CAMSHAFT

The **camshaft** is used to open the valves at the correct time. Cam lobes, or slightly raised areas, are machined on the camshaft to open the valves so that air and fuel can enter the cylinder. The valves are then closed by springs on each valve. The camshaft is driven by the crankshaft (***Figure 12–15***). The camshaft must be timed to the crankshaft so that the valves will open and close in correct time with the position of the piston. There is one lobe placed on the camshaft for each valve that must be opened and closed. The camshaft can also be placed or mounted directly on top of the cylinder head. This design is called an overhead camshaft (OHC).

Car Clinic:
VALVES

CUSTOMER CONCERN:
Sixteen- and 24-Valve Engines
Many manufacturers are designing engines using sixteen valves or twenty-four valves. What is the importance of using more valves in an engine?

SOLUTION: A 4-cylinder engine may use four valves for each cylinder, making up the sixteen valves. A 6-cylinder engine may use four valves for each cylinder, making up the twenty-four valves. More valves are used in engines for two reasons. One is to get more air and fuel into smaller engines. Another method would be to make the valves bigger. But if the valves were simply made bigger, they would be heavier and not able to move as fast. In that case, the engine could not run at as high an rpm. Thus efficiency would drop. The use of several smaller valves allows the same amount of air and fuel to enter and exhaust, and the smaller valves do not limit the rpm as do larger valves.

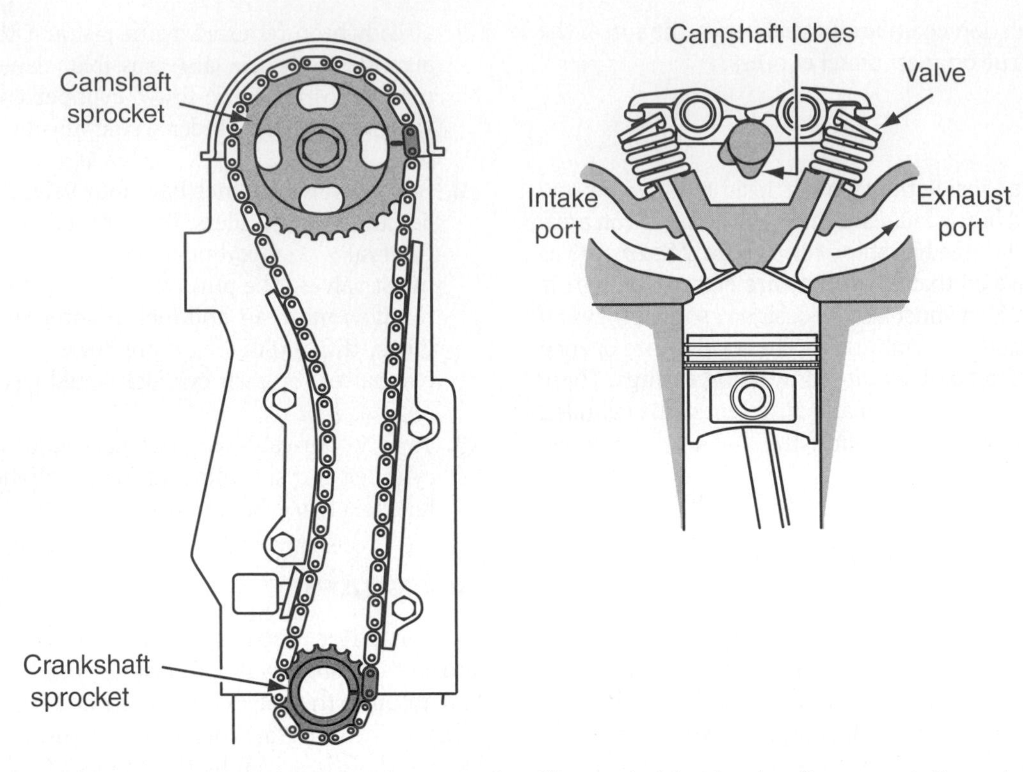

FIGURE 12–15 The camshaft opens the valves at the correct time. The camshaft is driven by either gears, belts, or chain drives from the crankshaft.

The overhead camshaft has several advantages. These include (1) fewer moving parts, (2) more precise valve operation, (3) higher rpm range, and (4) less frictional horsepower. Often only a single overhead camshaft is used for each bank of cylinders. This arrangement is referred to as an SOHC (single overhead camshaft) engine. When more valves (three or four) are used on each cylinder, then two camshafts may be needed. This arrangement is referred to as a DOHC (dual overhead camshaft) engine.

PARTS LOCATOR

*Refer to photo #2 in the front of the textbook to see what an **overhead camshaft** looks like.*

FLYWHEEL

The **flywheel** is located on the drive end of the crankshaft. It is designed to act as a weight to keep the crankshaft rotating once power has been applied to the piston. The flywheel is usually heavy. It smoothes out any intermittent motion from the power pulse (***Figure 12–16***).

INTAKE AND EXHAUST MANIFOLDS

The intake and exhaust manifolds are made up of ducts or piping that carry the air and fuel into the engine and the exhaust out of the engine. The design of the intake system is such that ducting allows the flow of air and fuel into the combustion chamber. On today's engines, there are also sensors that measure the amount of air going into the engine. The exhaust manifold is designed to allow hot combustion gases to pass safely out of the engine. In addition, there is an oxygen sensor that measures the amount of oxygen in the exhaust stream.

CARBURETOR

The **carburetor** (used on older vehicles) is placed on the engine to mix air and fuel in the correct proportion. This is called the fuel induction system. Air and fuel must be mixed

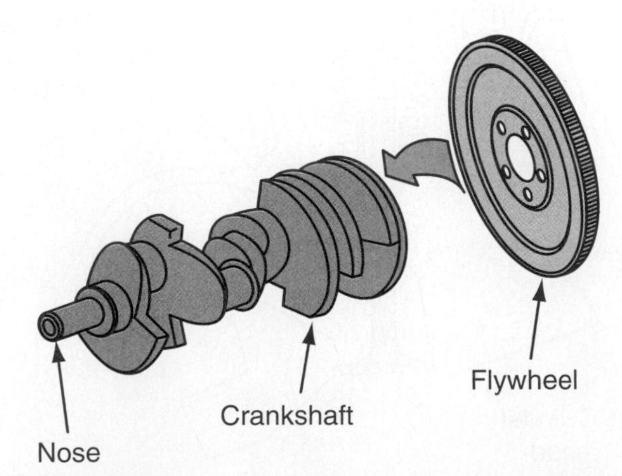

FIGURE 12–16 The flywheel is a heavy weight connected to the end of the crankshaft. It is used to smooth out any power pulses and helps return the piston to the top of the cylinder.

correctly for the engine to operate efficiently. The carburetor's job is to mix the air and fuel during cold weather, warm weather, high altitudes, high humidity, low-speed and high-speed conditions, and acceleration.

FUEL INJECTORS

Most of today's engines use fuel injectors rather than a carburetor. The fuel injectors do much the same job as a carburetor—mix the correct amount of air and fuel. However, the process of mixing air and fuel with injectors is much more precise than mixing with a carburetor. The fuel injectors are like small squirt guns, injecting gasoline into the intake of the engine so it can be mixed with the air coming into the engine.

SUMMARY

The following statements will help to summarize this chapter:

- An automobile engine is designed to convert the chemical energy in fuel into mechanical energy to move a vehicle forward or backward.
- Energy in an automobile engine can be in several forms including chemical, electrical, radiant, mechanical, and thermal.
- Gasoline engines can be classified as either internal or external combustion.
- In intermittent combustion engines, the combustion starts and stops.
- A continuous combustion engines means that the combustion continues all of the time.
- Reciprocating engines have motion that moves the parts up and down.
- Rotary engines have continuous rotation of parts.
- The cylinder block is the foundation of the engine.
- The connecting rod connects the piston to the crankshaft.
- The cylinder head fits over the top of the engine.
- The combustion chamber is where combustion of the fuel takes place.
- The valves open and close ports to let air and fuel in and exhaust gases out.
- The camshaft is used to open and close the valves.
- The flywheel is a weight that keeps the crankshaft rotating.
- The carburetor and fuel injectors are used to mix the air and fuel to the correct ratio.

TERMS TO KNOW

Can you explain each of the following terms? Review the chapter until you can use each term correctly.

Camshaft
Carburetor
Combustion
Combustion chamber
Connecting rod
Continuous combustion
Crankshaft
Cylinders
Cylinder block
Cylinder head
Energy
External combustion
Flywheel
Intermittent combustion
Internal combustion
Piston
Power
Reciprocating engine
Rotary engine
Valves

REVIEW QUESTIONS

Multiple Choice

1. The ability to do work is defined as ________.
   a. Power
   b. Energy
   c. Pressure
   d. Force
   e. Work done

2. A measure of the work being done is ________.
   a. Energy
   b. Power
   c. Thermal energy
   d. Force
   e. Work started

3. The energy used to start a vehicle is:
   a. Radiant energy
   b. Mechanical energy
   c. Thermal energy
   d. Electrical energy
   e. Forced energy
4. Which form of energy is the ability to move electrons in a wire?
   a. Thermal
   b. Electrical
   c. Mechanical
   d. Radiant
   e. None of the above
5. Into which type of energy does a gasoline engine convert chemical energy?
   a. Thermal
   b. Mechanical
   c. Nuclear
   d. A and B
   e. All of the above
6. The automotive gasoline engine is considered which type of engine?
   a. External combustion
   b. Continuous combustion
   c. Internal combustion
   d. All of the above
   e. None of the above
7. Other methods are used to classify engines. Which of the following method(s) is used to classify engines?
   a. By the cycles
   b. By the ignition system
   c. By the fuel system
   d. All of the above
   e. None of the above
8. The object that is the foundation of the engine and the point to which all other parts are attached is called the:
   a. Piston
   b. Crankshaft
   c. Fuel injector
   d. Cylinder block
   e. Cylinders
9. The part in the engine that moves up and down in a cylinder is called the:
   a. Piston
   b. Crankshaft
   c. Carburetor
   d. Flywheel
   e. Camshaft
10. Which part of the gasoline engine houses the valves?
    a. Carburetor
    b. Camshaft
    c. Cylinders
    d. Cylinder head
    e. Piston
11. Which object is used to keep the engine turning when the power is not being applied?
    a. Flywheel
    b. Crankshaft
    c. Camshaft
    d. Piston
    e. Connecting rod

The following questions are similar in format to ASE (Automotive Service Excellence) test questions.

12. *Technician A* says that an automobile engine converts chemical energy to thermal energy. *Technician B* says that an automobile engine converts thermal energy to mechanical energy. Who is correct?
    a. A only c. Both A and B
    b. B only d. Neither A nor B
13. *Technician A* says the gas engine used in a car is called an external combustion engine. *Technician B* says the gas engine used in a car is called a reciprocating engine. Who is correct?
    a. A only c. Both A and B
    b. B only d. Neither A nor B
14. *Technician A* says that engines today can have one valve per cylinder. *Technician B* says that engines today can have three or even four valves per cylinder. Who is correct?
    a. A only c. Both A and B
    b. B only d. Neither A nor B
15. *Technician A* says that an alternator converts mechanical energy to radiant energy. *Technician B* says that a starter motor converts electrical energy to mechanical energy. Who is correct?
    a. A only c. Both A and B
    b. B only d. Neither A nor B
16. *Technician A* says that the cylinder block is attached to the valves. *Technician B* says that the cylinder head is attached to the crankshaft. Who is correct?
    a. A only c. Both A and B
    b. B only d. Neither A nor B
17. *Technician A* says that the combustion chamber is where the air and fuel are ignited. *Technician B* says that the valves keep the crankshaft moving after the combustion process. Who is correct?
    a. A only c. Both A and B
    b. B only d. Neither A nor B
18. *Technician A* says that the piston is connected to the crankshaft by the connecting rod. *Technician B* says that the crankshaft continues turning between combustion pulses because of the flywheel. Who is correct?
    a. A only c. Both A and B
    b. B only d. Neither A nor B

Essay

19. Define the word *energy* and state how energy is used in a gasoline engine.

20. What are the six basic forms of energy?

21. Identify which types of energy are converted in the combustion process of a gasoline engine.

22. Describe the process of continuous combustion.

23. Describe why a gasoline engine is classified as an internal and intermittent combustion engine.

Short Answer

24. The term *power* is defined as a measure of __________.

25. Engines today can have __________, __________, or __________ valves per cylinder.

26. Energy is defined as the ability to do __________.

27. The type of energy that has the ability to move electrons is called ________ energy.

28. Heat is defined as ________ energy.

29. A gasoline engine has three classifications. They are ________, ________ combustion, and ________ design.

30. Both the carburetor and the ________ can be used to mix the air and fuel for combustion in a gasoline engine.

Applied Academics

TITLE: Power Conversion Units

ACADEMIC SKILLS AREA: Science

NATEF Automobile Technician Tasks:

Convert measurements taken using the English or metric system to specifications stated in terms of either system.

CONCEPT:

Service technicians often need to compare energy and power units to each other. For example, to understand a specific combustion chamber it may be necessary to know how many British thermal units (Btus) are in a gallon of gasoline or diesel fuel. It may also be necessary to convert power to watts or Btus. To help do these calculations, there are many conversion charts such as the one provided here. Knowing these conversions, the service technician can calculate from one unit to another.

POWER CONVERSION UNITS

1 BTU per Day =	1 Kilowatt-Hour per Year (kWh/yr) =	1 Watt (W) =	1 Kilowatt (kW) =	1 Megawatt (Mw) =
1 Btu/day	9.348 Btu/day	81.95 Btu/day	8.195×10^4 Btu/day	8.195×10^7 Btu/day
0.107 kWh/yr	1 kWh/yr	8.766 kWh/yr	8766 kWh/yr	8.766×10^6 kWh/yr
0.0122 W	0.1141 W	1 W	1000 W	10^6 W
1.22×10^{-5} kW	1.141×10^{-4} kW	0.001 kW	1 kW	1000 kW
1.22×10^{-8} Mw	1.141×10^{-7} Mw	10^{-6} Mw	0.001 Mw	1 Mw

Other units of power:
1 horsepower (hp) = 746 W
1 horsepower (hp) = 550 ft.-lb/second
1 Btu = 778 ft.-lb
1 gallon of gasoline = 100,000 Btu
1 gallon of diesel fuel = 130,000 Btu

APPLICATION:

Using the information in the chart, can you determine how many watts of power can be produced from an engine that has 230 horsepower? Can you determine which type of fuel—gasoline or diesel—has more energy in 1 gallon of fuel?

Gasoline Engine Principles

CHAPTER 13

OBJECTIVES

After studying this chapter, you should be able to:

- Specify the requirements for combustion, including such topics as air-fuel ratios, timing, bore and stroke, compression ratio, and engine efficiency.
- Compare the strokes of the four-stroke cycle engine.
- Describe the four-stroke timing diagram.

INTRODUCTION

The part in an automobile that powers the vehicle is called the engine. The engine is a device that is constructed to do one major thing. It converts energy in fuel into power so the vehicle can move. There are certain combustion requirements necessary for this to happen. This chapter will introduce you to the principles of combustion and the four-stroke engine design.

13.1 COMBUSTION REQUIREMENTS

AIR, FUEL, AND IGNITION REQUIREMENTS

The internal combustion engine has certain requirements for efficient operation. Any engine requires three things for its operation. There must be sufficient air for combustion, correct amounts of fuel mixed with the air, and some type of ignition to start combustion. When these three ingredients are present (***Figure 13–1***), combustion will take place. This combustion changes chemical energy in the fuel to thermal energy. The thermal energy created during combustion then causes rapid expansion of gases. This expansion (mechanical energy) pushes the piston downward. The downward force on the piston makes the crankshaft rotate. This rotary power can be used for pushing the vehicle forward. If any one of these three ingredients—air, fuel, or ignition—are missing, the engine will not run.

TIMING

Timing is the process of identifying when the air, fuel, and ignition combine to make combustion occur. This is done in relationship to the position of the piston and crankshaft. For the engine to operate efficiently, the air and fuel mixture must enter the cylinder at the correct time. This means that the intake valve must be opened and closed at the correct time. The exhaust valve must also be opened and closed at the correct time.

Ignition of the air and fuel must also occur at a precise time. The timing of the ignition changes with speed and load. When the intake and exhaust valves are correctly timed and the ignition occurs at the correct time, the conversion of chemical energy into mechanical energy will produce maximum power.

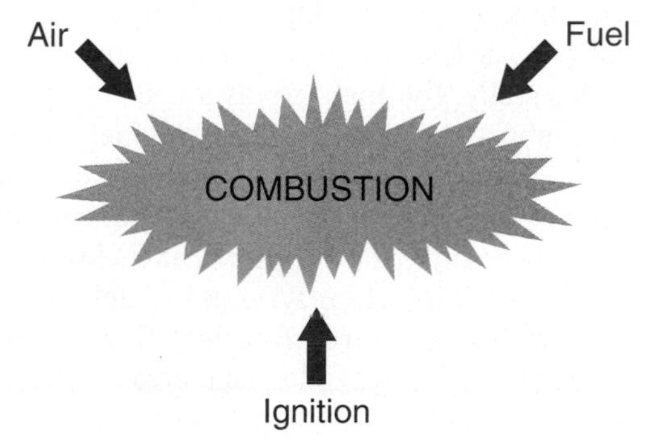

FIGURE 13–1 Three things are needed for combustion: air, fuel, and ignition. If one ingredient is missing, combustion cannot happen.

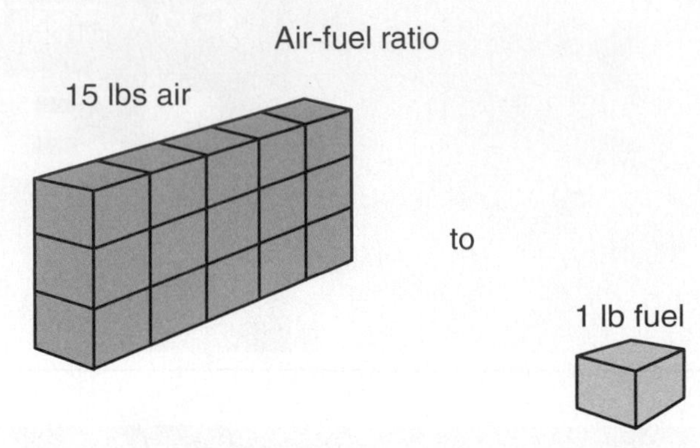

FIGURE 13-2 The air-fuel ratio is most efficient at 14.7 parts of air to 1 part of fuel. This ratio is measured by the weight of the air and fuel.

AIR-FUEL RATIO

Air-fuel ratio is the ratio of air to fuel mixed by the carburetor or fuel injectors. The most efficient air-fuel ratio is often called the **stoichiometric ratio**. The air and fuel must be thoroughly mixed. Each molecule of fuel must have enough air surrounding it to completely burn. If the two are not mixed in the correct ratio, engine efficiency will drop, and exhaust emission levels will increase.

The standard air-fuel ratio should be near 15 parts of air to 1 part of fuel. This measurement is calculated by weight. Actually, the most efficient ratio is stated as 14.7 to 1. For every pound of fuel used, 14.7 pounds of air is needed (***Figure 13–2***). In terms of volume, this is equal to burning 1 gallon of fuel in 9,000 gallons of air. Although 14.7 to 1 is the most efficient air-fuel (stoichiometric) ratio, there are times when this ratio is not effective. For example, more fuel is needed during starting and acceleration of the vehicle.

RICH AND LEAN MIXTURES

A low ratio of around 12 to 1 indicates a **rich mixture** of fuel. A mixture of 17 to 1 indicates a **lean mixture** (***Figure 13–3***). Generally, rich mixtures are less efficient during combustion. A rich mixture is used during cold weather and starting conditions. A lean mixture burns hotter than a rich mixture. Normally, the fuel acts as a coolant as it comes into the combustion process. With less fuel to cool, the combustion process gets hotter. This condition can cause severe damage to the pistons and valves if not corrected.

Much has been done to control the air-fuel ratio to exact requirements. Fuel injection systems are able to keep the mixture under better control. By controlling air-fuel mixtures accurately, fuel mileage can be increased well into the 25 to 35 miles per gallon range for modern engines.

TDC AND BDC

TDC stands for top dead center. **BDC** stands for bottom dead center. These two terms are used to help identify the position of the piston during some of the timing processes (***Figure 13–4***). TDC indicates the position of the piston when it is located at the top of its motion. When the piston is at the bottom of its travel, it is at BDC.

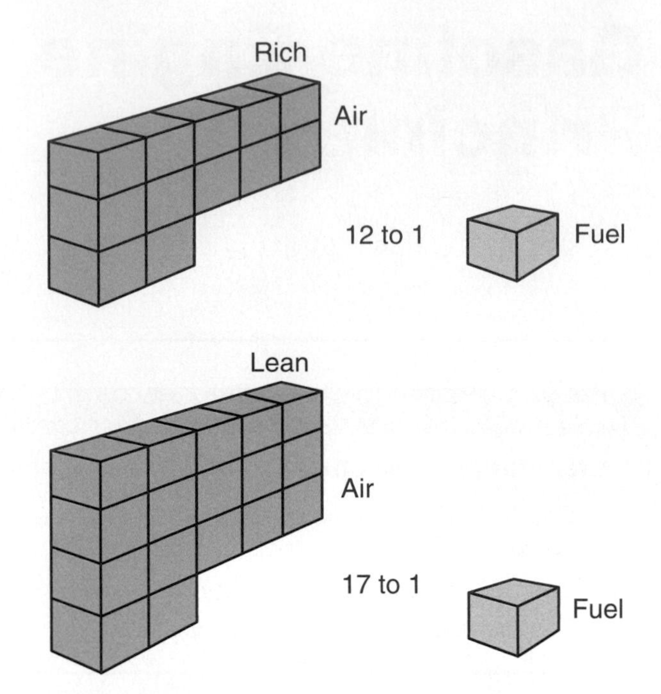

FIGURE 13-3 A rich mixture has a ratio of less air to 1 part of fuel (12 to 1). A lean mixture has more air per 1 part of fuel (17 to 1).

Car Clinic: RICH AIR-FUEL RATIO

CUSTOMER CONCERN:
Black Exhaust Smoke

A customer complains that there is black smoke during acceleration of the vehicle and that the gasoline mileage seems to be getting worse. What could be the problem?

SOLUTION: Black smoke coming from the exhaust usually means that there is a very rich mixture of gasoline and air. That is, the air-fuel ratio is very rich. This is often caused by an incorrectly adjusted air-fuel ratio. On fuel injection engines, this might indicate that the computer is not mixing the air and fuel to the correct ratio of 14.7 to 1. If this is the case, the computer and its associated components and sensors will need to be checked to determine how the air-fuel ratio is being adjusted to such a rich mixture.

BORE AND STROKE

The **bore** and **stroke** of an engine help identify its size. The bore of the engine is defined as the diameter of the cylinder (***Figure 13–5***). The stroke of the engine is a measurement

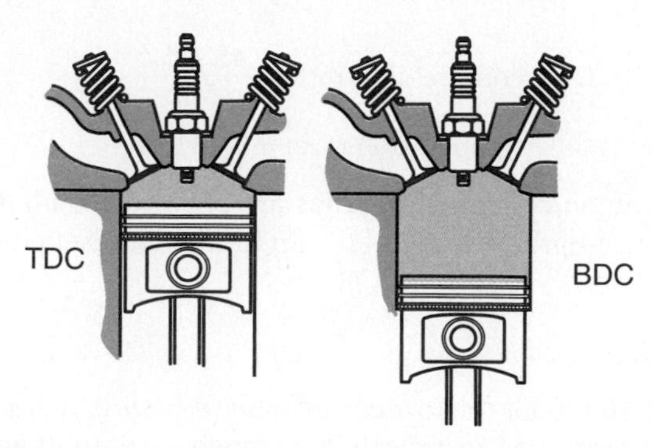

FIGURE 13–4 TDC means top dead center and is the highest point in the piston's travel. BDC means bottom dead center and is the lowest point in the piston's travel.

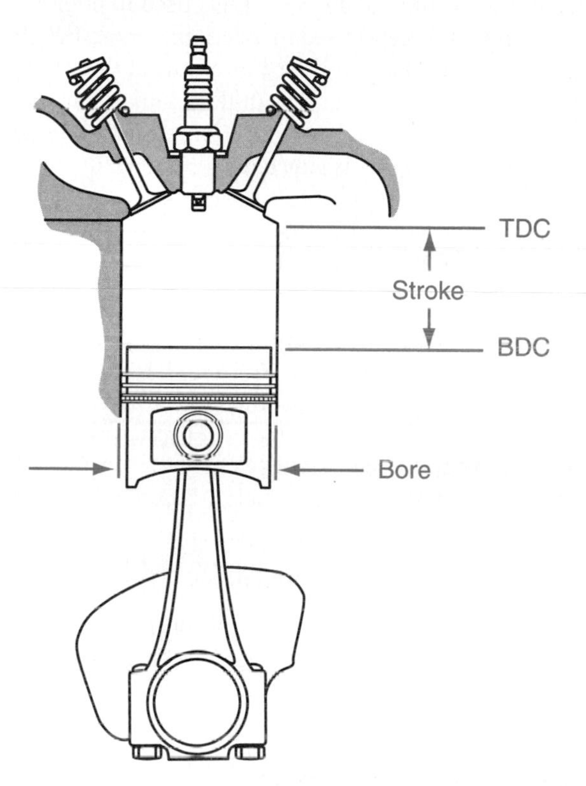

FIGURE 13–5 The bore and stroke of the engine help determine the size of the engine. Bore is the diameter of the cylinder, and stroke is the distance from TDC to BDC.

of the distance the piston travels from the top to the bottom of its movement. It is the distance from TDC to BDC.

The stroke is determined by the design of the crankshaft. The distance from the center of the crankshaft to the center of the crankpin is called the **throw** (***Figure 13–6***). If multiplied by 2, this dimension will be the same distance as the stroke. If the stroke is changed on the engine, the crankshaft will have a different length throw.

ENGINE DISPLACEMENT

Engine **displacement** is the volume of air in all of the cylinders of an engine. Each cylinder has a certain displacement. It can be determined by using the following formula:

$$\text{cylinder displacement} = 0.785 \times \text{bore}^2 \times \text{stroke}$$

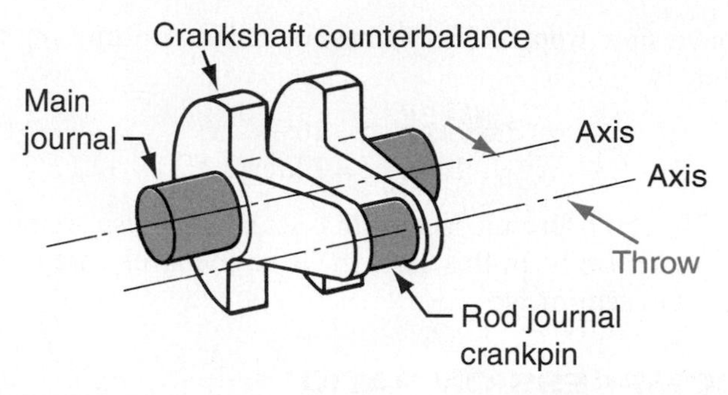

FIGURE 13–6 The distance from the center of the crankshaft to the center of the crankpin is called the throw. When this distance is multiplied by 2, the result is the stroke.

This formula will tell the exact cylinder displacement of one piston from top dead center to bottom dead center (***Figure 13–7***). If there is more than one cylinder, the total displacement would be multiplied by the number of cylinders.

From this information, and using this formula, what is the displacement of an engine that has six cylinders, a bore of 3.5 inches, and a stroke of 3.7 inches?

$$\text{Solution: total displacement} = 0.785 \times 3.5^2 \times 3.7 \times 6$$
$$\text{total displacement} = 213.5 \text{ cubic inches}$$

This formula calculates total displacement in cubic inches. Today, however, many engines are sized by cubic centimeters (cc or cm^3) and by liters. For example, today's engines are identified as 2.5 liters, 850 cc, and so on. The

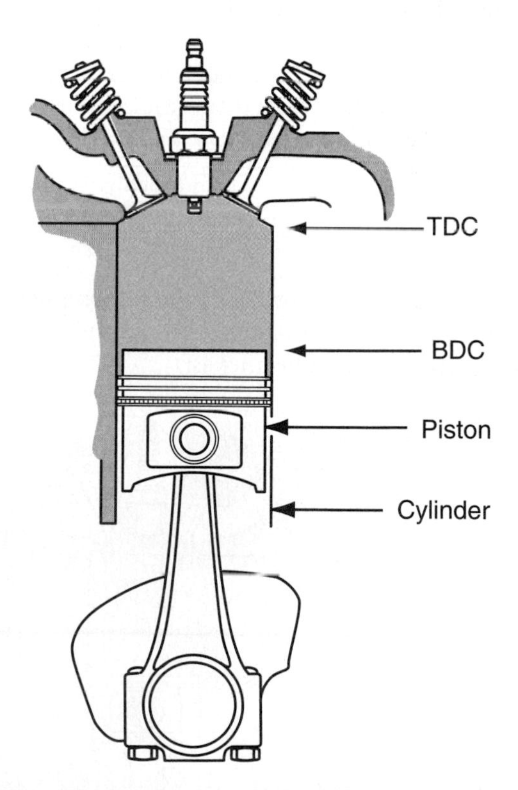

FIGURE 13–7 Cubic-inch displacement is the volume of the cylinder from BDC to TDC. It is also stated in cubic centimeters or liters.

conversion from cubic inches to cubic centimeters to liters is:

$$1 \text{ cubic inch} = 16.387 \text{ cc}$$
$$1{,}000 \text{ cc} = 1 \text{ liter}$$

The same formula is used to calculate the displacement in metric units. In this case, the bore and stroke are measured in centimeters.

COMPRESSION RATIO

During engine operation, the air and fuel mixture must be compressed. This will be covered in the discussion of the four-cycle principle later in this chapter. This compression helps squeeze and mix the air and fuel molecules for better combustion. Actually, the more the air and fuel are compressed, the better will be the efficiency of the engine.

Compression ratio is a measure of how much the air and fuel have been compressed. Compression ratio is defined as the ratio of the volume in the cylinder above the piston when the piston is at BDC to the volume in the cylinder above the piston when the piston is at TDC. The compression ratio is shown in ***Figure 13–8***. The formula for calculating compression ratio is:

$$\text{compression ratio} = \frac{\text{volume above the piston at BDC}}{\text{volume above the piston at TDC}}$$

In many engines, at TDC, the top of the piston is even or level with the top of the cylinder block. The combustion chamber volume is in the cavity in the cylinder head above the piston. This volume is modified slightly by the shape of the top of the piston. The combustion chamber volume must be added to each volume stated in the formula to give accurate results. In this case, the formula for compression ratio can be stated as follows.

$$\text{Compression ratio} = \frac{\text{total cylinder volume at BDC}}{\text{total combustion chamber volume at TDC}}$$

With this information, calculate the compression ratio if piston displacement is 45 cubic inches, and combustion chamber volume is 5.5 cubic inches.

$$\text{compression ratio} = \frac{45 + 5.5}{5.5}$$
$$\text{compression ratio} = 9.18 \text{ to } 1$$

Common compression ratios are anywhere from 8 to 1 on low-compression engines to 25 to 1 on diesel engines.

BMEP

BMEP stands for *brake mean effective pressure.* It is a theoretical term used to indicate how much pressure is applied to the top of the piston from TDC to BDC. It is measured in pounds per square inch. This term becomes very useful when analyzing the results of different fuels used in engines. For example, if diesel fuel is used in an engine, more BMEP will be produced and there will be more output power than if gasoline fuel were used. Also, as different injection systems, combustion designs, and new ignition systems are added, the BMEP of the engine is affected.

ENGINE EFFICIENCY

The term **efficiency** can be used to indicate the quality of different machines. Efficiency can also pertain to engines. Engine efficiency is a measure of the relationship between the amount of energy put into the engine and the amount of energy available out of the engine. The many types of efficiency will be discussed in a later chapter. For understanding basic engine principles, efficiency is defined as:

$$\text{efficiency} = \frac{\text{output energy}}{\text{input energy}} \times 100$$

For example, if there were 100 units of energy put into the engine, and 28 units came back out, the efficiency would be equal to 28%.

ENGINE IDENTIFICATION

Depending on the engine and the manufacturer, various styles and types of engines are found in automobiles. They are identified by stating the number of valves, the camshaft

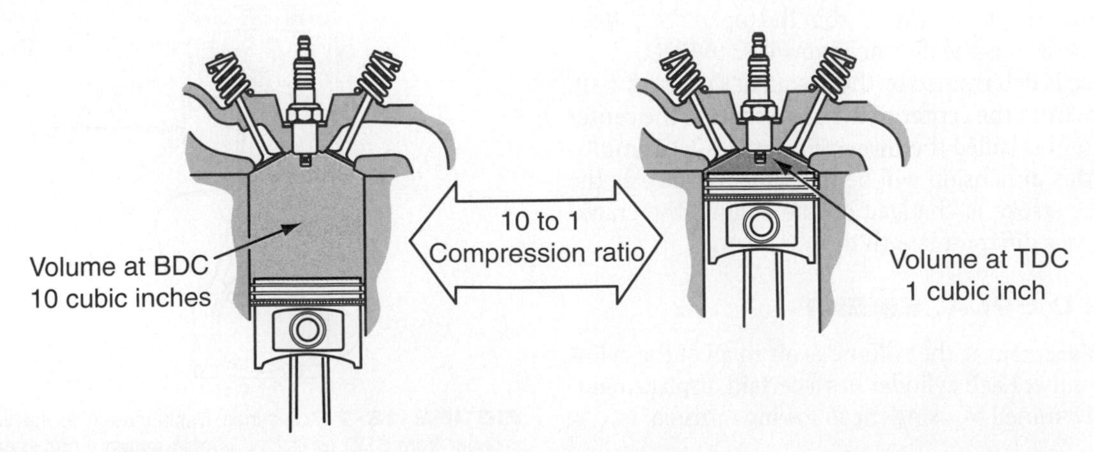

FIGURE 13–8 Compression ratio is the ratio of the volume above the piston at BDC to the volume above the piston at TDC.

arrangements, and the displacement of the engine. For example, the following engine identification styles are used:

- Twin-cams (two camshafts)
- 3.0 L V6 24-valve engine (3.0 liter, V6, 24 valves)
- DOHC 16-valve, 660 cc, L4 engine (dual overhead camshaft, with four valves per cylinder, 660 cc displacement, and four in-line cylinders)
- 2.2-L DOHC 16-valve turbo engine (2.2 liter, dual overhead camshafts, four valves per cylinder, turbocharged engine)
- 4.6-L SOHC V8 engine (4.6 liter, single overhead camshaft for each side of the V8 configuration)

13.2 FOUR-STROKE ENGINE DESIGN

Automotive vehicles use **four-stroke engines**. A four-stroke engine is sometimes referred to as a four-cycle engine. The terms *stroke* and *cycle* are often interchanged. Interchanging these terms causes some confusion. A stroke is defined as the movement of the piston from TDC to BDC. A cycle is defined as the events that occur in a certain number of degrees of crankshaft rotation, usually 360 degrees. A four-stroke engine has a very distinct operation. These four strokes are called the intake, compression, power, and exhaust strokes. In this section, the four-stroke gasoline engine will be explained.

INTAKE STROKE

Refer to ***Figure 13–9***. To start, the location of the piston is near TDC. Note that the intake valve is open. As the piston is cranked downward (called the intake stroke), air and fuel are drawn into the cylinder. This occurs because as the piston moves down, a vacuum is created. When any object is removed from an area, a vacuum is created. This vacuum (lower than atmospheric pressure) draws fresh air and fuel into the cylinder. It can also be said that the atmospheric pressure pushes the air and fuel into the cylinder.

The air is first drawn through the intake manifold. Here the air is mixed with the fuel by the fuel injector at the correct air-fuel ratio (14.7 to 1). When the piston gets to BDC, the intake valve starts to close. With the valve closed, the air and fuel mixture is trapped in the cylinder area.

COMPRESSION STROKE

The piston now travels from BDC to TDC with air and fuel in the cylinder. This action is called the compression stroke (***Figure 13–10***). The compression stroke takes the air-fuel mixture and compresses it according to the compression ratio of the engine. This compression causes the air and fuel to be mixed very effectively. Actually, the higher the compression ratio, the greater the mixing of air and fuel. This leads to improved engine efficiency.

It is very important that there be no leaks for the compression gases to escape. Leaks may occur in the valves, the gasket between the head and cylinder block, and past the rings on the piston. Note that at the end of the compression stroke, the crankshaft has revolved 360 degrees or one revolution of the crankshaft.

During the compression stroke, the air and fuel mixture is actually heated from the action of compression. It is like

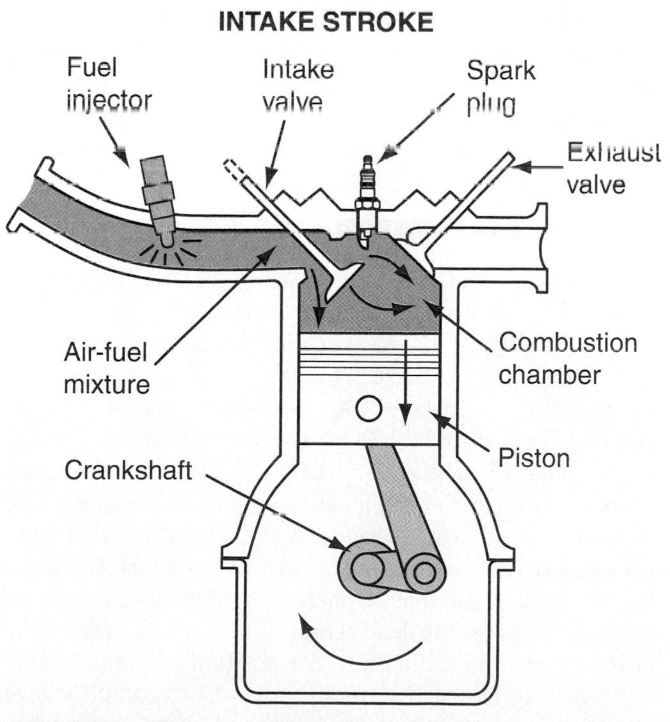

FIGURE 13–9 During the intake stroke, the piston moves down, bringing in fresh air and fuel. The intake valve is open until near BDC.

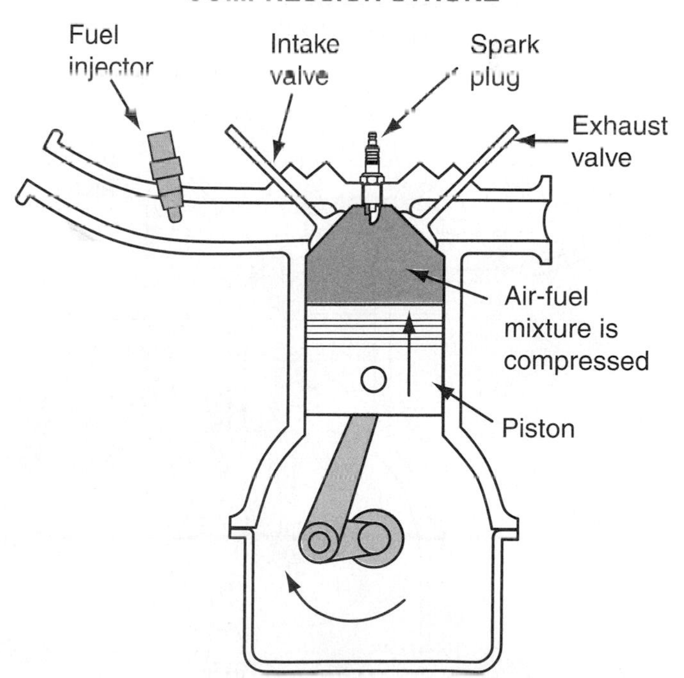

FIGURE 13–10 During the compression stroke, the piston moves from BDC to TDC. This action compresses the air and fuel in the cylinder.

using an air pump to pump up a tire. As the air at the bottom of the pump is compressed, the air gets hotter. If the compression ratio is too high, temperatures within the combustion chamber may ignite the fuel. This process is referred to as preignition and can cause pinging. This means that the explosion in the combustion chamber starts before the piston gets to TDC.

It would be very helpful if compression ratios were increased. However, as long as air and fuel are being compressed, the compression ratios must be low so the air and fuel will not preignite. The higher compression ratios of diesel engines will be discussed in Chapter 14.

POWER STROKE

During the power stroke (***Figure 13–11***), both the intake and exhaust valves remain closed. When the piston is coming up on the compression stroke, spark will occur very near TDC. At this point, air, fuel, and ignition are present. This combination causes the air and fuel mixture to burn rapidly. As it burns, the expanding gases push down on the top of the piston. This pressure pushes the piston downward through the power stroke. This is also when BMEP is created.

Again, it is very important for the entire combustion chamber to be sealed without any leaks. Leaks may allow some of the energy in the fuel to escape. This reduces the amount of power pushing down on the piston.

EXHAUST STROKE

The last stroke in the four-stroke design is called the exhaust stroke. The exhaust stroke (***Figure 13–12***) starts when the piston starts moving upward again. The crankshaft will continue to rotate because of the flywheel weight. At the beginning of the exhaust stroke, the exhaust valve opens. As the piston travels upward, it pushes the burned or spent gases out of the cylinder into the atmosphere.

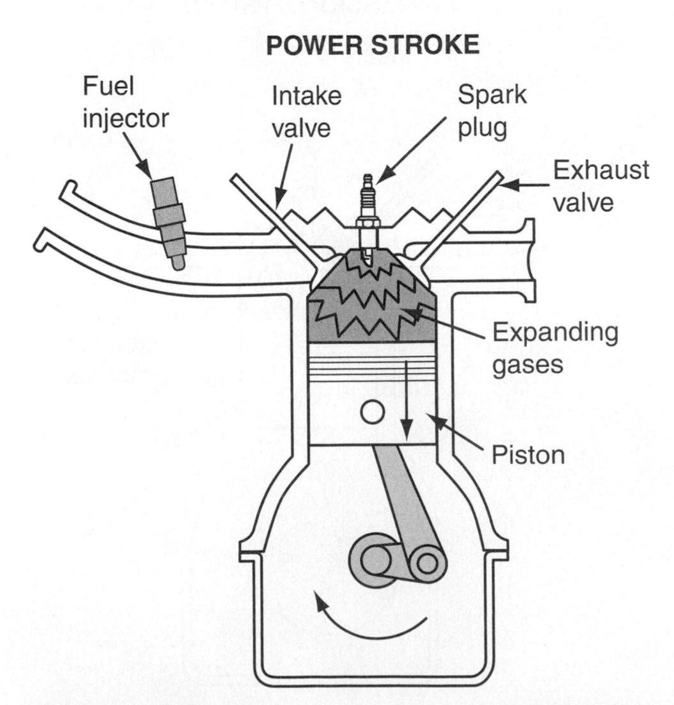

FIGURE 13–11 During the power stroke, ignition occurs slightly before TDC. The combustion and expansion of gases pushes the piston downward through the power stroke.

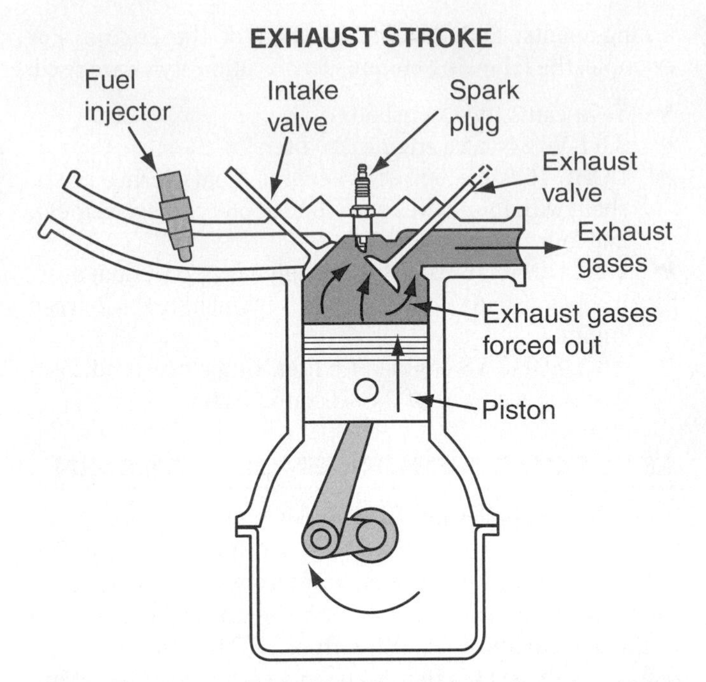

FIGURE 13–12 During the exhaust stroke, the exhaust valve opens. The upward motion of the piston pushes the exhaust gases out of the engine.

Near the top of the exhaust stroke, the exhaust valve starts to close. At this point, the intake valve is already starting to open for the next intake stroke. It is important to note that the crankshaft has revolved two revolutions at this point. Only one power stroke has occurred. If the engine is running at 4,000 rpm (revolutions per minute), then there are 2,000 power pulses for each cylinder per minute.

TIMING DIAGRAMS

A **timing diagram** is a method used to identify the time at which all of the four stroke events occur. A timing diagram is shown in ***Figure 13–13***. The diagram is set on a vertical and horizontal axis. There are 360 degrees around the axis. On the circle, events of the four-stroke engine can be graphed. One way to look at the diagram is to think of these events in terms of the position of the crankshaft and 360 degrees rotation. For example, at the top of the diagram, the piston would be located exactly at TDC. Any event that happens before TDC is referred to as BTDC (before top dead center). Any event that happens after top dead center is called ATDC (after top dead center). The mark at the bottom of the graph would illustrate the position of the piston at BDC. Two circles are shown to represent two complete revolutions of the crankshaft. During the four strokes of operation, the crankshaft revolves two complete revolutions, or 720 degrees of rotation.

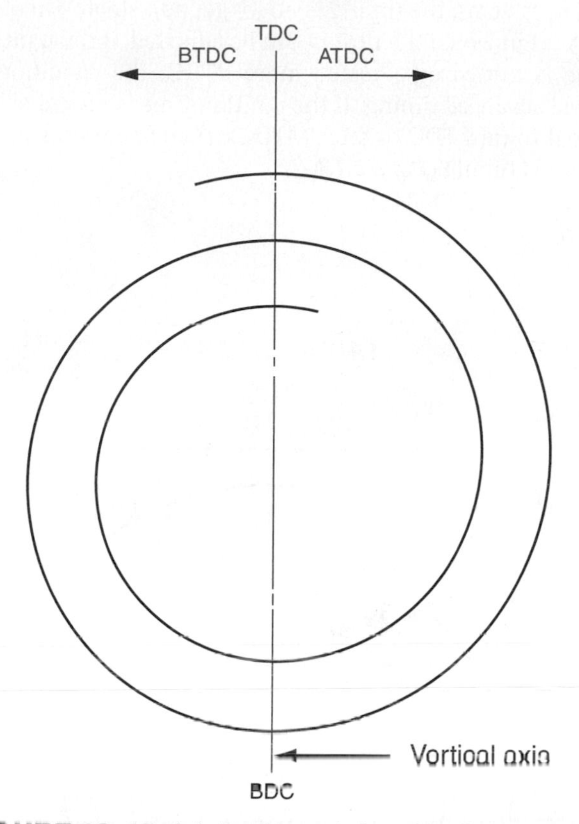

FIGURE 13–13 A timing diagram helps to identify when each of the four stroke events occur. The two revolutions of the 360-degrees rotation represents the revolutions of the crankshaft. BTDC stands for before top dead center, and ATDC stands for after top dead center.

CUSTOMER CONCERN:
Advance Timing an Engine

A customer says that when the timing was advanced on his engine, the engine ran better. If this is correct, should all engines be advance timed so they run better?

SOLUTION: Today's engines have a set timing depending on the type of ignition system that is used. Computerized ignition systems use a computer to advance the timing according to the load, speed, and other inputs. However, several years ago, most engines had a distributor that could be adjusted to time the engine. In fact, on some late-1970 and early-1980 model engines, the timing was actually retarded at the factory to produce less pollution when the car was idling. Some of these cars ran significantly better when the timing was advanced. Whether a car should be advanced depends on its year and make. The best advice is to always follow the manufacturer's recommendations when timing an engine.

FOUR-STROKE TIMING DIAGRAM

Referring to ***Figure 13–14***, follow through the four-stroke design on the timing diagram. Note that these events and degrees may vary with each engine and manufacturer. In addition, the degrees shown on the diagram may change with the engine rpm, the load on the engine, the type of fuel injection, variable valve timing, and other factors that may change the operation of the engine. The cycle starts with the intake valve opening slightly before TDC. It should be fully open at TDC. It takes this many degrees of crankshaft rotation to open the intake valve completely.

As the piston travels downward on the intake stroke, the intake valve starts to close shortly before BDC. It is fully closed slightly after BDC. At this point, the intake stroke is completed.

The compression stroke starts when the intake valve is fully closed. The piston travels upward, compressing the air and fuel mixture. As the piston is traveling upward, the air-fuel mixture is being mixed by the compression of gases. Also, the temperature is rising inside the combustion chamber. About 10 to 12 degrees before TDC, ignition from a spark plug occurs. The point of ignition is several degrees before TDC. It takes about 10 to 12 degrees for the explosion or expansion to actually build up to a maximum. At TDC,

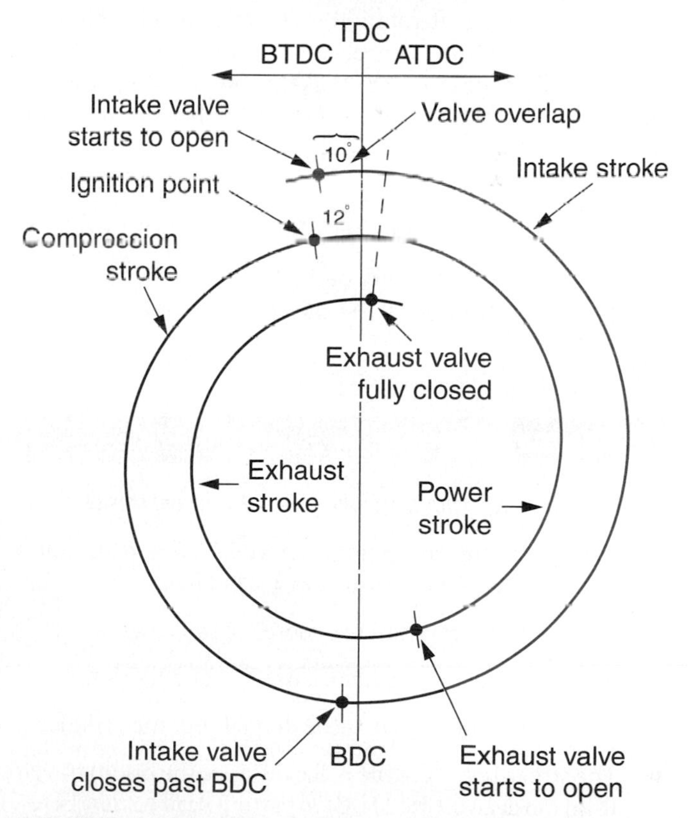

FIGURE 13–14 The timing diagram shows the events of the four-stroke engine. Intake, compression, power, and exhaust are plotted on the diagram.

the expansion is at a maximum point. Now the piston is ready to be pushed downward.

If the timing of the ignition were sooner, or more degrees before TDC, then the explosion would occur too soon. This would then reduce the BMEP during the power stroke. If the timing of the ignition were too late, or after TDC, then the BMEP would also be less. It is important for maximum power from the explosion of gases to occur just when the piston is at TDC.

The power stroke starts when the piston starts downward. In this case, the power stroke is shown on the inside circle of the timing diagram. As the explosion occurs, the gases expand very rapidly. This expansion causes the piston to be forced down. This action produces the power for the engine.

Near BDC, at the end of the power stroke, the exhaust valve starts to open. By the time the piston gets to BDC, the exhaust valve is fully open. As the crankshaft continues to turn, the piston travels upward. This action forces the burned gases out of the exhaust valve into the atmosphere. The exhaust valve is fully closed a few degrees after TDC. The time during which both the intake valve and the exhaust valve are open (near TDC) is called valve overlap.

Valve overlap is important for several reasons. First, it is important to keep the exhaust valve open as long as possible to make sure all of the exhaust gases have completely escaped from the cylinder. On the other hand, the intake valve needs to begin opening as soon as possible to get as much air and fuel as possible into the cylinder and combustion chamber. Also, it is important to note that it takes a certain amount of crankshaft rotation in degrees to get the intake valve from a "starting to open" position to a "fully opened" position. All of these factors make it important for the engine to have a certain amount of valve overlap for maximum efficiency.

ADVANCED AND RETARDED TIMING

On older engines, the only part of the timing that is adjustable is the timing of the ignition. However, on most computer-controlled ignition systems or distributorless ignition systems, the timing is no longer adjustable. On older engines in which the timing can be adjusted, if the ignition time is moved or adjusted more BTDC, the condition is called advanced timing. If the ignition time is moved or adjusted toward TDC or after (ATDC), the condition is called retarded timing (***Figure 13–15***).

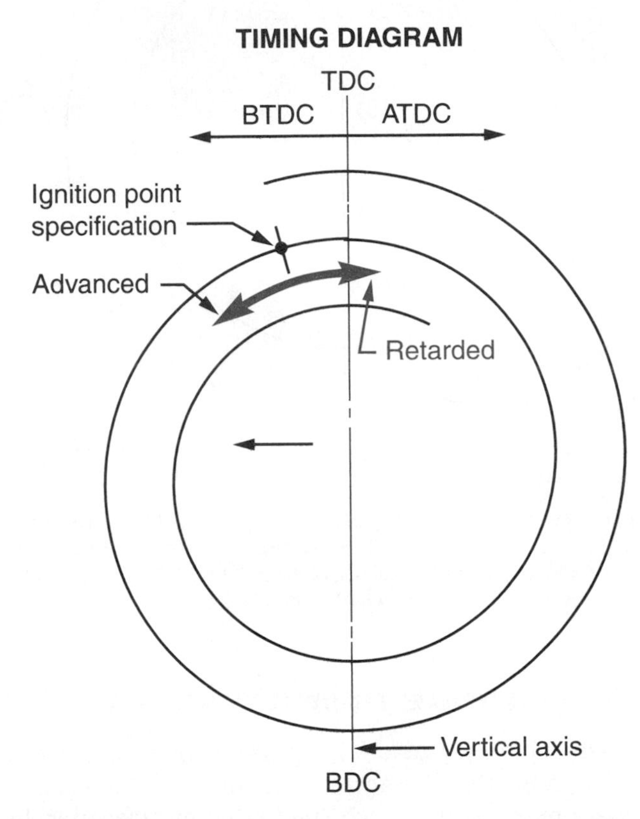

FIGURE 13–15 If the ignition timing moves farther before TDC, the engine is advanced. If the ignition timing moves toward TDC or after, the ignition is retarded.

SUMMARY

The following statements will help to summarize this chapter:

- For an engine to operate correctly the air-fuel ratio should be 14.7 parts of air to 1 part of fuel.
- A lean air-fuel mixture has more air and less fuel.
- A rich air-fuel mixture has more fuel and less air.
- The bore of the engine is the diameter of the cylinder.
- The stroke of the engine is the distance the piston travels from top dead center (TDC) to bottom dead center (BDC).
- Compression ratio is a measure of how much the air and fuel mixture has been compressed in the cylinder.
- Automobiles use a four-stroke engine that has intake, compression, power, and exhaust strokes.
- The intake stroke brings in fresh air and fuel.
- The compression stroke compresses the air-fuel mixture.
- The power stroke occurs when the air-fuel mixture is ignited.
- The exhaust stroke pushes the burned gases out of the engine.
- Both advanced and retarded timing can be observed on a timing diagram.

TERMS TO KNOW

Can you explain each of the following terms? Review the chapter until you can use each term correctly.

Air-fuel ratio
BDC
BMEP
Bore
Compression ratio
Displacement
Efficiency
Four-stroke engines
Lean mixture
Rich mixture
Stoichiometric ratio
Stroke
TDC
Throw
Timing
Timing diagram

REVIEW QUESTIONS

Multiple Choice

1. What is the most efficient air-fuel ratio for a gasoline engine?
a. 14.7 to 1
b. 18.3 to 1
c. 13.2 to 1
d. 12.1 to 3
e. 20.0 to 1

2. Which of the following is considered a lean mixture?
a. 13 to 1
b. 10 to 1
c. 17 to 1
d. 11 to 1
e. 12 to 1

3. The diameter of the cylinder is called the __________.
a. Stroke
b. Throw
c. Bore
d. Torque
e. Force

4. Engine displacement can be measured in:
a. Liters
b. Cubic inches
c. Cubic centimeters
d. All of the above
e. None of the above

5. When the volume above the piston at BDC is divided by the volume above the piston at TDC, the result is called:
a. Air-fuel ratio
b. Compression ratio
c. Fuel injection
d. Engine displacement
e. Rotary displacement

6. The pressure on top of the piston during the power stroke is called:
a. BMEP
b. Air pressure
c. Combustion
d. Force
e. All of the above

7. Which of the following is not called one of the strokes on the four-stroke engine?
a. Intake
b. Spark
c. Power
d. Exhaust
e. Compression

The following questions are similar in format to ASE (Automotive Service Excellence) test questions.

8. *Technician A* says the order of strokes on a four-stroke engine is intake, power, compression, and exhaust. *Technician B* says the order of strokes on a four-stroke engine is intake, compression, power, and exhaust. Who is correct?
a. A only
b. B only
c. Both A and B
d. Neither A nor B

9. *Technician A* says that the air-fuel ratio varies from 6 to 1 up to 185 to 1 on a standard gasoline engine. *Technician B* says that the air-fuel ratio is controlled by the crankshaft. Who is correct?
a. A only
b. B only
c. Both A and B
d. Neither A nor B

10. *Technician A* says that an engine needs air, fuel, and ignition for combustion to occur. *Technician B* says that an engine only needs ignition and air for combustion to occur. Who is correct?
a. A only
b. B only
c. Both A and B
d. Neither A nor B

11. *Technician A* says that a 12 to 1 air-fuel ratio is a lean mixture. *Technician B* says that a rich air-fuel mixture is 17 to 1. Who is correct?
a. A only
b. B only
c. Both A and B
d. Neither A nor B

12. *Technician A* says that to calculate engine displacement use the formula: .785 × bore^2 × stroke × the number of cylinders. *Technician B* says that engine displacement is calculated when the piston is at top dead center or TDC. Who is correct?
    a. A only
    b. B only
    c. Both A and B
    d. Neither A nor B

13. *Technician A* says that an engine identified as a 3.0 V6 24-valve engine means that the engine has 3.0 cubic inches per cylinder. *Technician B* says that this engine is a V6 cylinder arrangement. Who is correct?
    a. A only
    b. B only
    c. Both A and B
    d. Neither A nor B

14. *Technician A* says that the exhaust stroke occurs immediately before the power stroke. *Technician B* says that the intake stroke occurs immediately before the exhaust stroke. Who is correct?
    a. A only
    b. B only
    c. Both A and B
    d. Neither A nor B

Essay

15. Describe how the four cycles work on a standard four-stroke engine.

16. What is the definition of BMEP and why is it a useful term?

17. Identify the difference between TDC and BDC.

18. Define the term *engine efficiency*.

19. What is the total displacement of an engine that has a bore of 3 inches, a stroke of 4 inches, and is a 4-cylinder engine?

Short Answer

20. During the starting of an engine, the air-fuel ratio must be __________.

21. If the total volume above the piston at BDC is 60 cubic inches, and at TDC it is 5.2 cubic inches, the compression ratio is __________.

22. The distance from the center of the crankshaft to the center of the crankpin is called the __________.

23. On engines that have adjustable timing, if the timing changes to further before TDC, it is called __________ timing.

24. When the intake valve opens before the exhaust valve closes, as shown on a timing diagram, this is called valve __________.

Applied Academics

TITLE: Engine Displacement and Compression Ratio

ACADEMIC SKILLS AREA: Mathematics

NATEF Automobile Technician Tasks:

The service technician can determine the volume of a cylinder when the specification is in liters.

CONCEPT:

As mentioned in this chapter, the formula for engine displacement is:

$$\text{cubic inch displacement} = 0.785 \times \text{bore}^2 \times \text{stroke} \times \text{number of cylinders}$$

Also, 1 cubic inch = 16.387 cubic centimeter and 1,000 cc = 1 liter. Compression ratio is defined as the difference in volume from BDC to TDC including the volume in the cylinder head.

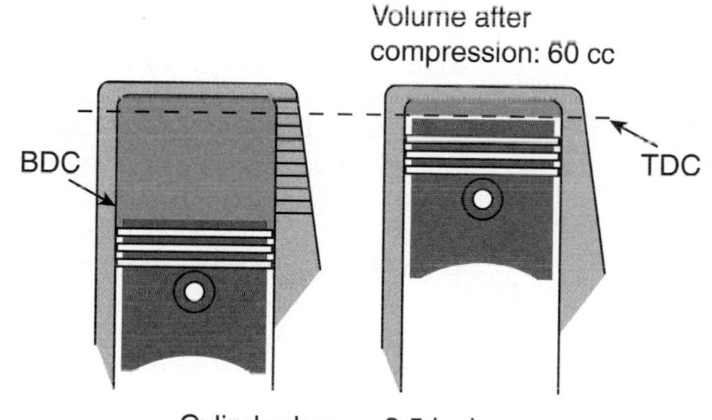

Cylinder bore = 3.5 inches
Piston stroke = 3.5 inches
Number of cylinders = 4

APPLICATION:

Using this information above and in the figure, determine the following:

1. What is the total cubic-inch displacement of the engine?
2. What is the total cubic-centimeters displacement of this engine?
3. What is the total liters displacement of this engine?
4. What is the compression ratio?

CHAPTER 14 Other Power Sources

OBJECTIVES

After studying this chapter, you should be able to:

- Identify the design and operating principles of the diesel engine.
- Compare the diesel engine to the gasoline engine.
- Describe the parts and operation of the rotary engine.
- Describe the principles and design of two-stroke engines.
- Describe the operation of hybrid vehicles.

INTRODUCTION

Several other types of power sources are also being used and tested in the automotive industry. Some, such as the diesel engine, have established and well-developed service procedures and are used in a variety of vehicles. There are other engine designs that are being considered for use in automobiles including the rotary engine and the two-stroke engine. Such engines continue to be tested, improved, and considered for use in the automotive market. Hybrid power sources are also becoming a reality. This chapter introduces the basic principles and designs of these engines and power sources.

14.1 DIESEL ENGINE PRINCIPLES

The **diesel engine** is much the same as the gasoline engine in many of its principles. Most diesel engines are four-stroke engines. However, there are also two-stroke diesel engine designs, which are discussed later in this chapter. Because of its characteristics, the diesel engine is usually used when heavier loads and torque requirements are anticipated. Most diesel engines are used in trucks, buses, and heavy-duty engine applications. ***Figure 14–1*** shows an example of a truck diesel engine. However, for several years (in the late 1970s and early 1980s), some automotive manufacturers developed a smaller V8 diesel engine for larger cars. During that time, diesel fuel cost significantly less per gallon than gasoline. But as gasoline prices remained stable and the cost of diesel fuel increased, the need for diesel engines in automobiles declined. In addition, the diesel engine had slower acceleration than a gas engine, was somewhat noisier, and produced unpleasant odors. Today however, some foreign car manufacturers, particularly Mercedes Benz, still use diesel engines in their cars. Diesel engines are also gaining popularity in many pickup trucks, especially those that are used to pull heavier campers and travel trailers across country.

The diesel engine is considered an internal combustion engine. It is also considered a compression-ignition rather than a spark-ignition engine.

The diesel engine has many of the same parts as a typical gasoline engine. Diesel engines have pistons, valves, crank-

FIGURE 14–1 Diesel engines are used mostly for heavy-duty applications. (Courtesy of Cummins, Inc.)

shafts, fuel systems, cooling systems, starting and charging systems, and pollution control systems. Since the diesel engine is used for heavier loads, these systems are more rugged and stronger than those in a gasoline engine. One big difference between diesel and gasoline engines is that the compression ratios are higher in diesel engines. In addition, the diesel fuel injection system uses high-pressure injection, whereas the gasoline engine injection system uses low-pressure injection.

DIESEL COMPRESSION RATIO

One major difference between a diesel engine and a gasoline engine is that the diesel engine has a very high compression ratio. Compression ratios from 25 to 1 up to 30 to 1 or higher are very common. This high compression ratio means that any fuel that is in the cylinder during compression will be ignited. Therefore, only air is brought into the cylinder during the intake stroke. No low-pressure port fuel injector is needed to mix the air and fuel. Fuel is injected directly into the combustion chamber in a diesel engine. With high compression ratios, temperatures inside the combustion chamber may be as high as 1,000 degrees Fahrenheit. This temperature would be high enough to ignite most diesel fuels. Because the fuel is ignited by the high temperatures produced by compression, the diesel engine is called a compression-ignition engine.

FUEL INJECTION

At TDC, or slightly before, a high-pressure fuel injector injects fuel directly into the combustion chamber on diesel engines. This design is called high-pressure **fuel injection.** A fuel injector is a device that pressurizes fuel to near 20,000 psi at the tip of the injector. This fuel is injected into the combustion chamber (***Figure 14–2***). At this point, all three ingredients are there to produce combustion. The power and exhaust strokes are the same as in the gasoline engine.

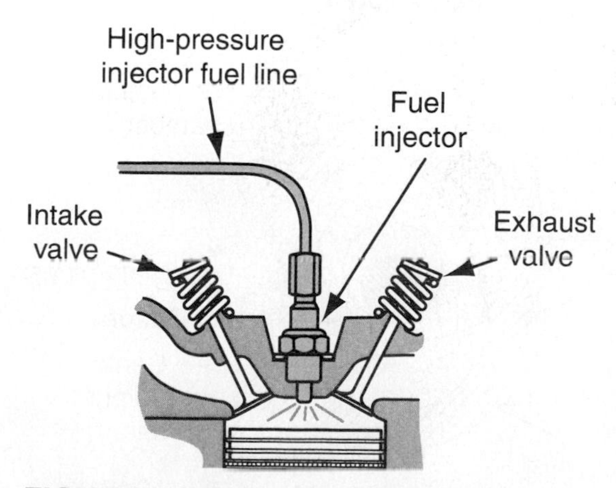

FIGURE 14–2 In a diesel engine, air and fuel are mixed near TDC. Fuel is injected into the combustion chamber under high pressure, and the heat of compression is used to ignite the fuel.

FOUR-STROKE DIESEL ENGINE

Most diesel engines are designed using the four-stroke sequence of events. ***Figure 14–3*** shows the four events on this type of engine. To begin, as the piston moves downward on the intake stroke, the intake valve is opened. Only air enters the combustion chamber at this time. The exhaust valve is closed on this stroke. When the piston begins to move upward on the compression stroke, the air is compressed. The temperature at the end of the compression stroke increases to above 1,000 degrees Fahrenheit because of the high compression ratio. At or shortly before TDC, high-pressure fuel is injected into the hot air. At this point, all three ingredients necessary for combustion are there. These are air, fuel, and ignition (the hot air). When combustion takes place, it pushes the piston down on the power

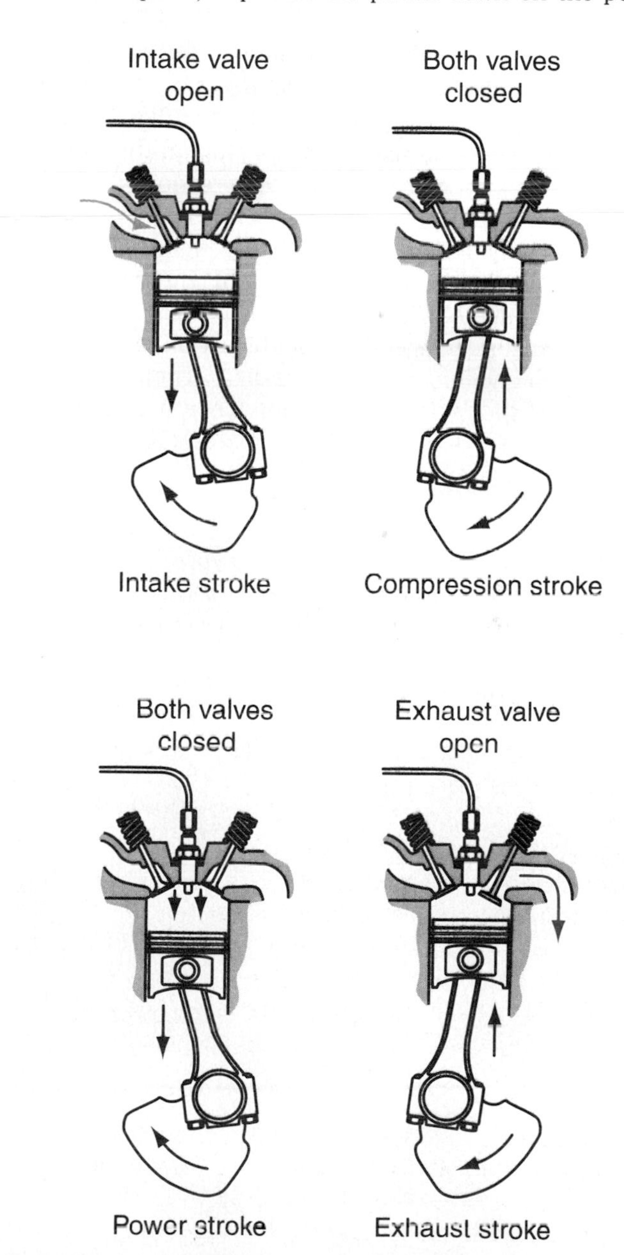

FIGURE 14–3 Most diesel engines use a four-stroke design with the intake, compression, power, and exhaust occurring through two crankshaft rotations.

stroke. As the piston continues to move upward on the exhaust stroke, the exhaust valve opens. The upward motion of the piston pushes out the spent or exhaust gases. The cycle then repeats itself.

COMPARISON OF DIESEL AND GASOLINE ENGINES

Figure 14–4 shows common comparisons between the diesel and gasoline four-stroke engines.

- The intake on the gasoline engine is an air-fuel mixture. The diesel engine has only air during the intake stroke.
- The compression pressures on the gasoline engine are lower because the compression ratios are lower. The compression temperatures on the gasoline engine are also lower.
- The air and fuel mixing point on the gasoline engine is at the carburetor or by the fuel injectors. The mixing point on the diesel engine is near TDC or slightly BTDC.
- Combustion is caused by a spark plug on the gasoline engine. The diesel engine uses compression ignition.
- The power stroke on the gasoline engine produces approximately 460 psi. On the diesel engine, the power stroke produces nearly 1,200 psi. There is more energy in diesel fuel than in gasoline.
- The exhaust temperature of the gasoline engine is much higher than that of the diesel engine, because some of the fuel is still burning when it is being exhausted. There is also a higher percentage of carbon monoxide in gasoline exhaust.

COMPARISON OF GASOLINE AND DIESEL ENGINES

	GASOLINE	DIESEL
Intake	Air-fuel	Air
Compression	8–10 to 1 130 psi 545°F	25–30 to 1 400–600 psi 1,000°F
Air-fuel mixing point	Carburetor or before intake valve with fuel injection	Near TDC by injection
Combustion	Spark ignition	Compression ignition
Power	464 psi	1,200 psi
Exhaust	1,300°–1,800°F CO = 3% No_x = Low	700°–900°F CO = 0.5% No_x = High
Efficiency	22–28%	32–38%

FIGURE 14–4 There are several differences between a gasoline engine and a diesel engine.

- One major drawback to the diesel engine is the amount of nitrogen oxides (NO_x) that are produced. Because of the high internal temperatures, more nitrogen oxides are produced by diesel engines compared to gasoline engines. Nitrogen oxides are considered greenhouse gases (gases that tend to make our environment similar to a greenhouse by causing the atmospheric temperature to increase slightly). Nitrogen oxides are also a major contributor to photochemical smog, though diesel engine manufacturers continue to make improvements to reduce these emissions.
- The efficiency of the diesel engine is about 10% higher than that of the gasoline engine. This is because compression ratios are higher in the diesel engine, and there is more energy in a gallon of diesel fuel than in a gallon of gasoline.

14.2 ROTARY DESIGN (WANKEL)

In the late 1960s, several new engine designs were introduced into the automotive market. One such engine was called the **rotary engine**. Although the rotary design has been in existence for some time, its popularity continues to vary. Improved designs continue to be developed even today. In this engine, a rotor, instead of a piston, is used to convert chemical energy into mechanical energy. The engine is an intermittent-combustion, spark-ignition, rotary design (not reciprocating).

ROTARY CYCLE OPERATION

Figure 14–5 shows position 1, which corresponds to the intake stroke of a standard reciprocating engine. The upper port is called the intake port. The lower port is called the exhaust port. There are no valves. The position of the center rotor opens and closes the ports much as a valve would. The rotor moves inside an elongated circle. Because of the shape of the circle or housing, certain areas are enlarged or com-

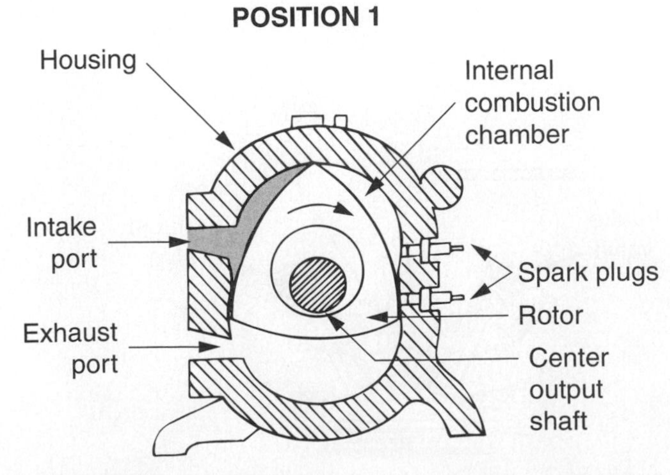

FIGURE 14–5 Intake on the rotary engine is produced when, in the shaded area, the leading edge of the rotor passes the intake port and leaves an opening that allows fuel and air to enter.

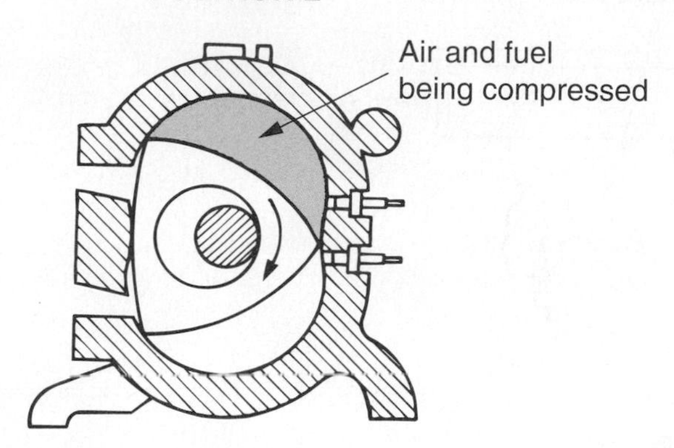

FIGURE 14-6 As the rotor continues to turn, the air and fuel are compressed to a standard compression ratio.

pressed during rotation. As the rotor is turned, an internal gear causes the center shaft to rotate. This rotation is the output power.

When the leading edge of the rotor face sweeps past the inlet port, the intake cycle begins. Gasoline and air (in a 14.7 to 1 air-fuel ratio) are drawn into the enlarging area. They continue to be drawn in until the trailing edge passes the intake port.

As the rotor continues to rotate, the enlarged area is now being compressed. This is position 2 (***Figure 14–6***), which corresponds to the compression stroke. The compression ratio is very close to that of a standard gasoline engine, because air and fuel are being mixed.

When the rotor travels to position 3 (the power stroke), shown in ***Figure 14–7***, the air and fuel are completely compressed. At this point, ignition occurs from two spark plugs. There are two spark plugs for better ignition. On some newly engineered rotary engines, three spark plugs are being used to improve combustion. All three ingredients are now available for correct combustion. The air and fuel ignite rapidly. Combustion causes expansion of gases. This expansion pushes the rotor face downward, causing the rotor to receive a power pulse.

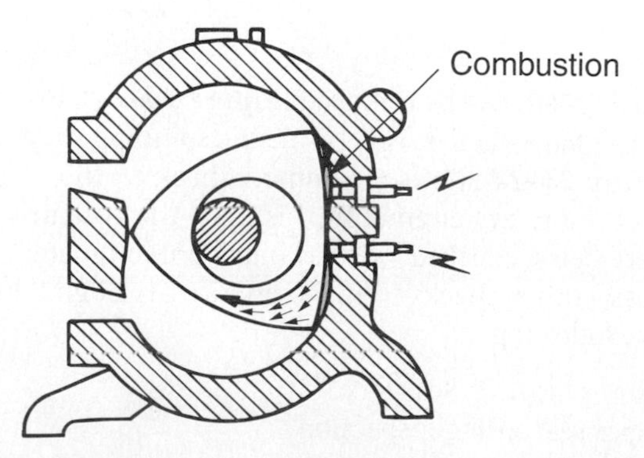

FIGURE 14-7 During the power stroke on a rotary engine, the expansion of gases causes pressure against the rotor face. This pressure causes the rotor to continue turning.

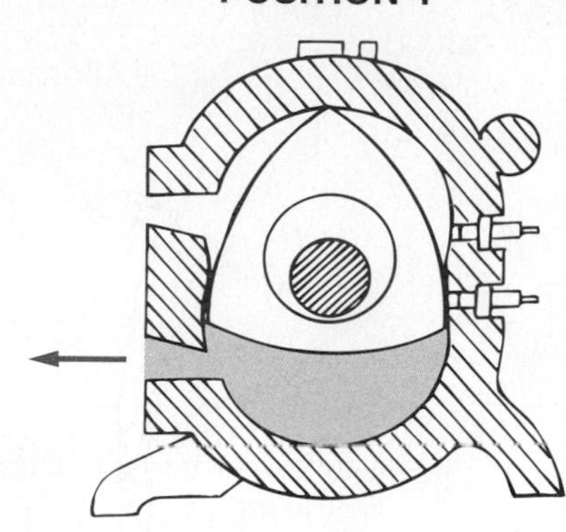

FIGURE 14-8 As the rotor continues to rotate, the leading edge uncovers an exhaust port. Exhaust gases are then forced out of the chamber.

Figure 14–8 illustrates position 4 (the exhaust stroke). As the rotor continues to travel or rotate, the leading edge uncovers the exhaust port. The rotor's movement within the housing causes the exhaust gases to be forced out of the engine.

So far, only one face of the rotor has been analyzed. Note that while intake is occurring in position 1, compression is occurring on another face of the rotor, and exhaust is occurring on the third face of the rotor. In this engine, there are three power pulses for each rotation of the rotor.

The rest of the rotary engine uses many of the same components as the standard gasoline engine. The fuel injection system is similar. The starter, alternator, and external components are the same. A complete rotary engine is shown in ***Figure 14–9***.

ROTARY ENGINE PARTS

The **rotor** is the center of the rotary engine. ***Figure 14–10*** shows a typical rotor. Most engines have two rotors that are identical. There is a recessed area on each of the three faces of the rotor. This recessed area is where the combustion takes place. Notice the internal gear located in the center of the rotor. This gear meshes with the center output shaft.

Since the rotor's motion produces a suction and pressure effect on the air and fuel, the rotor must be sealed on all sides. These seals function similar to piston rings, only the seals have a different shape. In ***Figure 14–10***, the service technician is checking the clearance between the side seal and the groove. ***Figure 14–11*** shows a drawing of the apex seals needed to seal one combustion chamber from another. If these seals are damaged, the compression from one combustion chamber will mix with another, causing the engine to be much less efficient.

The rotary engine is basically a series of housings and rotors stacked together. ***Figure 14–12*** shows an example of how one housing is removed from the rotor. After the bolts have been removed, the housing is lifted upward. When the housing has been removed, the center shaft, also called the

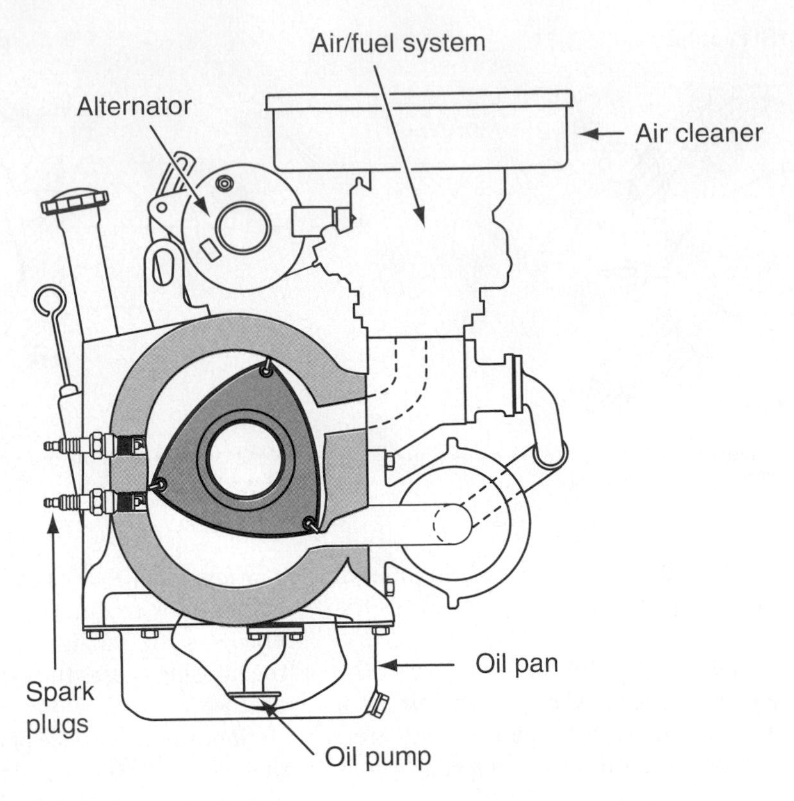

FIGURE 14–9 The rotary engine uses many of the same components as the gasoline reciprocating engine. This is a cross-cut section showing many internal parts.

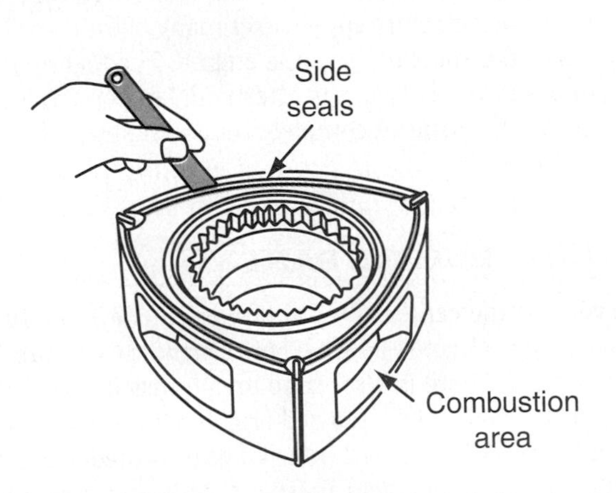

FIGURE 14–10 This rotary engine rotor is being checked for side seal clearance.

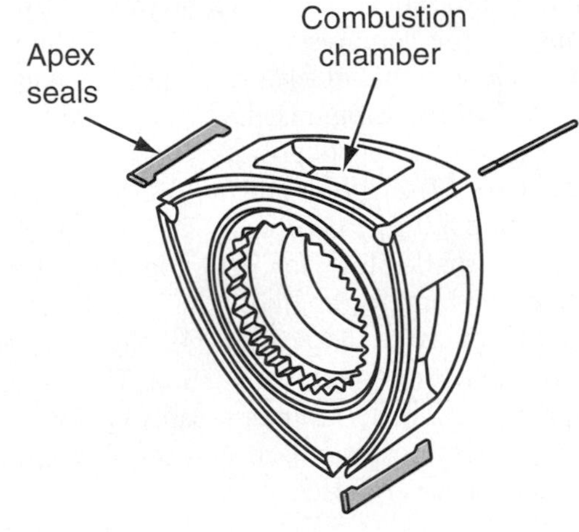

FIGURE 14–11 The apex seals help to keep the combustion gases from mixing with each combustion chamber.

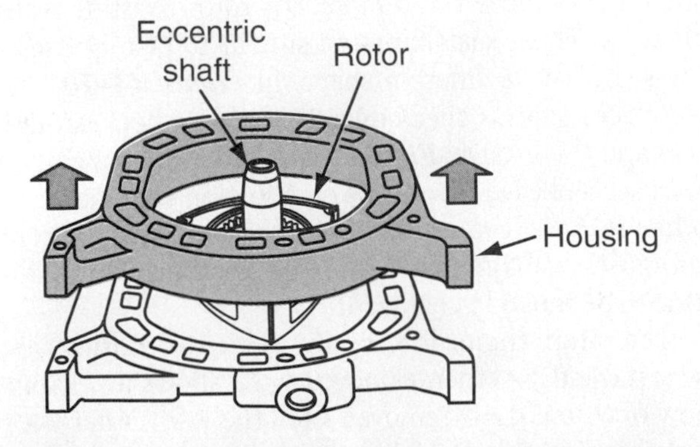

FIGURE 14–12 To disassemble a rotary engine, the housing lifts off the rotor.

eccentric shaft, can be removed (***Figure 14–13***). Notice the **eccentric journals** that ride inside the spinning rotor.

Figure 14–14 shows a common check on the eccentric shaft of the rotary engine. In this case, the eccentric journals are being checked with an outside micrometer. There are many other checks to make on a rotary engine including the following:

- Rotor bearing clearance
- Apex seal spring inspection
- Side seal spring inspection
- Rotor housing inspection
- Internal gear inspection

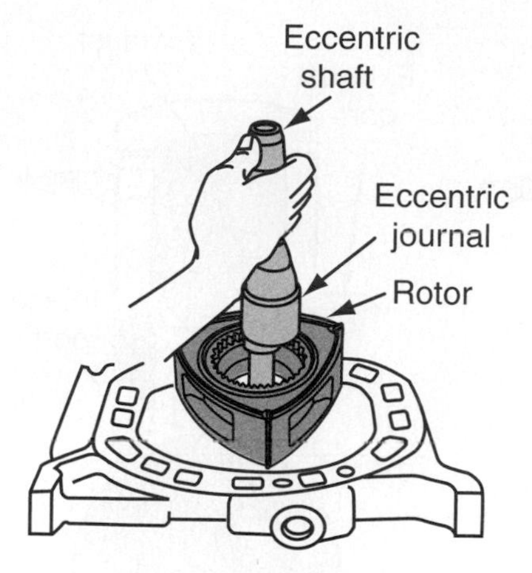

FIGURE 14–13 The eccentric center shaft sits inside the rotor.

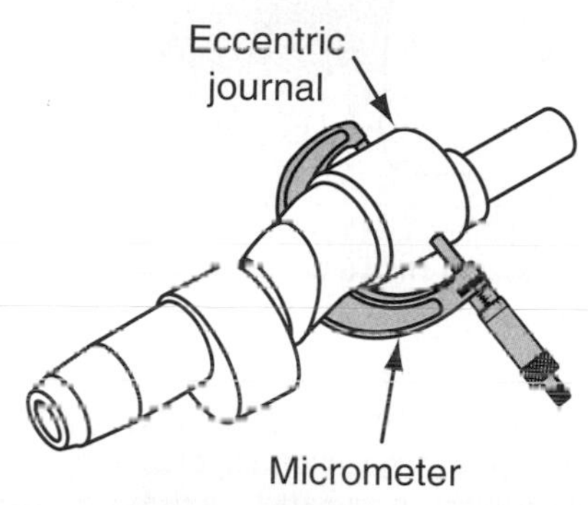

FIGURE 14–14 Just as with piston engines, the rotary engine must be checked according to manufacturer's specifications.

- Rotor inspection
- Front housing inspection

14.3 TWO-STROKE ENGINE DESIGN

In recent years most automotive manufacturers have been considering the use of another type of engine design for automobiles. This design is called the **two-stroke engine**. Two-stroke engines are commonly used on smaller power applications such as outboard motors, snowmobiles, and chain saws. Because it is simple and light, this design is being considered for automotive use.

TWO-STROKE ENGINE OPERATION

Figure 14–15 shows an example of a two-stroke engine operation. Note that many of the engine parts are the same as for the four-stroke engine. One difference is that the cylinder does not use the standard type of valves to allow air and fuel to enter the engine. In order to follow the operation, consider the events both above and below the piston. As the piston in ***Figure 14–15A*** moves upward, compression is produced above it. A vacuum is created in the crankcase area below the piston. The vacuum brings in a fresh charge of air and fuel past a reed valve. Note that oil must be added to the air-fuel mixture at this point because there is no oil in the crankcase as with four-stroke engines. The oil in the fuel acts as the lubricant. Oil can also be injected into the crankcase to obtain lubrication. Normally, an oil-gas ratio of from 20 to 1 to 50 to 1 or higher is used.

As the piston continues upward on the compression stroke, eventually a spark, and thus combustion occurs, as shown in ***Figure 14–15B***. The combustion pushes the piston downward. As the piston moves farther down, high pressure is created in the crankcase area. In the illustration, the reed valve is forced closed by the pressure, sealing the crankcase area. The reed valve is used for illustration only. The two-cycle engine being considered for the automobile will not use reed valves. It will use the position of the piston as a valve. When the piston gets low enough in its stroke, it eventually opens both the intake and exhaust ports, ***Figure 14–15C***. Ports are simply holes cut into the cylinder to allow air and fuel to enter and exhaust to escape.

When the ports are open, the crankcase pressure forces a mass of an air-fuel-oil mixture into the combustion chamber. This mass also helps to remove any exhaust gases by way of the exhaust port. As the piston starts upward, the ports are closed. Compression and power continue above the piston, while suction and a small pressure continue below the piston.

Car Clinic:
TWO-STROKE ENGINE

CUSTOMER CONCERN:
What are the advantages to the two-stroke engine?

SOLUTION: Most automotive manufacturers have research programs to evaluate and test different types of engines. One type being considered is the two-stroke engine. It is generally a very responsive engine because it has a power pulse on each piston, every revolution. It is able to go from idle to maximum rpm faster than the four-stroke engine. It is also much lighter because it has fewer parts. With a lighter engine, vehicle fuel mileage generally improves.

On the other hand, research must continue to improve the combustion efficiency, pollution characteristics, and air movement through the cylinder, and ways must be found to increase the two-stroke engine's BMEP.

TWO-STROKE TIMING DIAGRAM

A timing diagram can be used to illustrate the two-stroke engine. Remember, the vertical axis represents the TDC and BDC point on the piston and crankshaft. The events of the two-stroke process can be graphed on the timing diagram, as shown in ***Figure 14–16***. Although the two-stroke and four-stroke engine diagrams differ slightly, they show several common points.

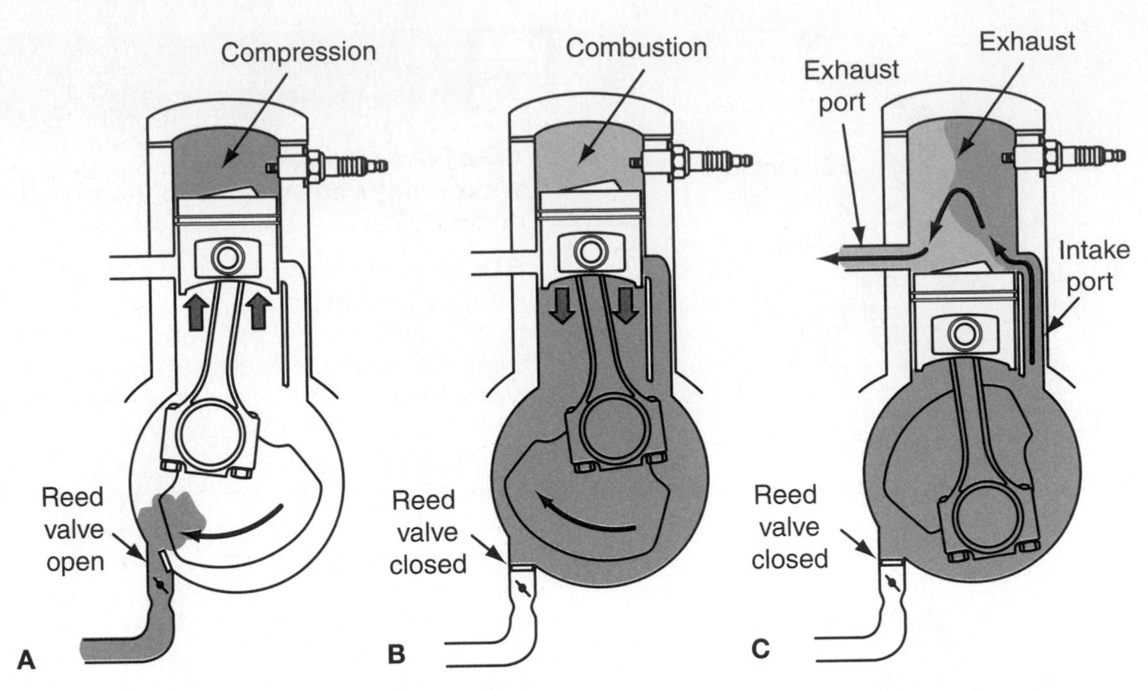

FIGURE 14–15 The two-stroke engine uses the pressure and vacuum created below the piston to draw a fresh charge of air, fuel, and oil into the combustion chamber.

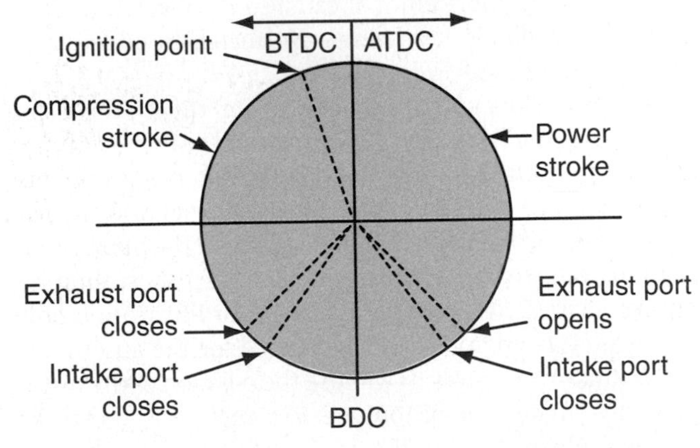

FIGURE 14–16 This timing diagram shows the events that occur during two-stroke engine operation.

1. Timing for ignition occurs slightly before TDC.
2. During the power stroke, the exhaust valve or port opens slightly before the intake valve or port.
3. During the compression stroke, the exhaust valve or port closes slightly after the intake valve or port.
4. When the piston is at the bottom of its travel, both intake and exhaust are occurring.

ADVANTAGES AND DISADVANTAGES OF TWO-STROKE ENGINES

There are several advantages and disadvantages of using the two-stroke engine in automobiles. Two advantages are:

1. Two-stroke engines are generally very responsive because there is a power pulse every revolution. Therefore, it takes less time to get from 500 rpm at idle to, say, 4,000 rpm at maximum speed.
2. Two-stroke engines usually weigh less than four-stroke engines, because they usually have fewer parts. With a lighter engine, vehicle fuel mileage generally improves.

The disadvantages of two-stroke engines are related to efficiency. They are less efficient than four-stroke engines for three reasons:

1. Poor movement of air and fuel. Air and fuel can enter the cylinder for only a very short period of time. Because less air and fuel can enter the engine, efficiency is reduced.
2. Poor combustion efficiency. The oil in the air-fuel mixture reduces combustion efficiency because of its burning characteristics.
3. Less BMEP (brake mean effective pressure). Total force during the power stroke is less because the power stroke is shorter. The power stroke ends when the exhaust port is opened.

Although many problems still exist, engineering efforts are being directed at improving efficiency, combustion, fuel injection, and scavenging (ease of air and exhaust movement).

LOOP-SCAVENGING TWO-STROKE ENGINE

The basic design of the two-stroke engine is continually being improved. **Loop scavenging** is a two-stroke design used to improve the ease with which air and fuel can enter and leave the cylinder. The main difference between loop scavenging and the basic design shown in ***Figure 14–15*** is that the reed valve system is replaced with an intake port lo-

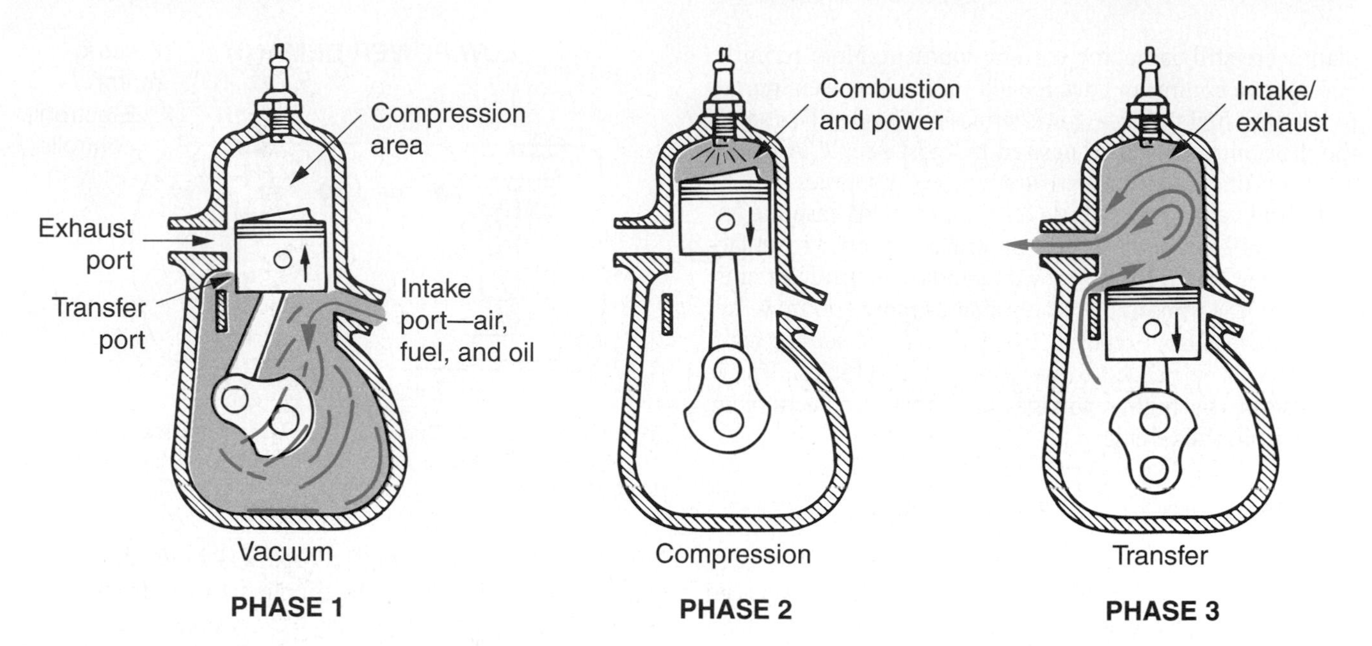

FIGURE 14–17 The loop-scavenging two-stroke engine eliminates the reed valve and uses an intake port opened and closed by the position of the piston.

cated on the bottom of the cylinder. The intake port is opened and closed by the position of the piston skirt (bottom of the piston). Events again occur both above and below the piston. *Figure 14–17* shows the operation of a loop-scavenging two-stroke engine.

The events occur in three phases. In phase 1, the piston moves upward, and the piston skirt opens the intake port in the lower right of the cylinder. The vacuum below the piston draws in fresh fuel, air, and oil for lubrication. During phase 2, the power stroke above the piston is occurring. Below the piston, low-pressure compression occurs on the air-fuel-oil mixture. During phase 3, the exhaust port and the intake port to the cylinder are opened. This opening causes the air-fuel-oil mixture below the piston to transfer to the combustion chamber. The spent exhaust gases in the cylinder are also exhausted to the atmosphere.

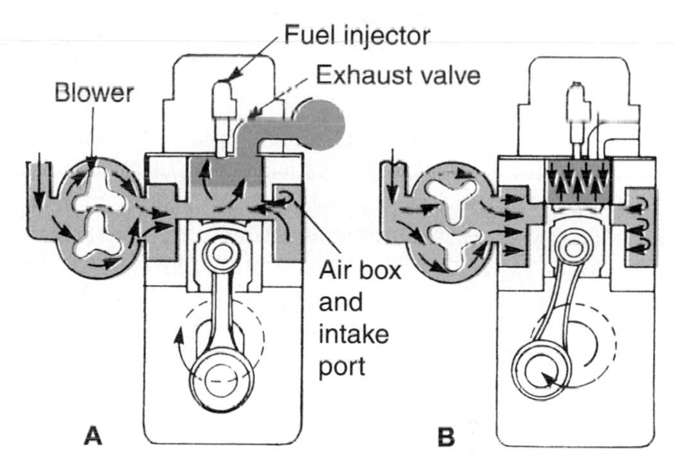

FIGURE 14–18 This two-cycle diesel engine uses a blower to force air into the combustion chamber.

TWO-STROKE DIESEL ENGINE

Another type of two-stroke cycle engine is manufactured by General Motors. Although this type of two-stroke engine is not used in automobiles, it is used in certain over-the-road trucks. This two-stroke cycle engine is shown in *Figure 14–18*. The major difference is that a blower is used to force air through an air box, intake ports, and into the cylinder. This can only occur when the piston is at BDC. This is shown in *Figure 14–18A*. The piston then comes up on the compression stroke and the intake ports are closed off by the position of the piston. When the piston moves upward, near TDC, high-pressure diesel fuel is injected into the very hot air. The fuel is then ignited by the heat of compression. The power stroke forces the piston downward. These events are shown in *Figure 14–18B*. At the end of the power stroke, the exhaust valve on top of the cylinder opens. At about the same time, the intake ports are uncovered by the position of the piston. The fresh air pressure from the blower helps to push the exhaust gases out through the exhaust valve.

The advantage of this engine design is that no oil is mixed with the fuel, as is the case with smaller two-stroke cycle engines. However, a significant amount of power is needed to turn the blower used to pump the air.

14.4 HYBRID POWER SOURCE

In the past several years, automotive engineers have been designing power sources that produce extremely low emissions but extremely high fuel mileage. For years, the battery-powered vehicle was considered to be the most energy-efficient, low-emission vehicle. However, most of the energy needed to charge the batteries came from coal-fired or nuclear-powered electrical generating plants. The net effect was that pollution and emissions from the power plant were still damaging the environment. More recently, automotive engineers have looked to hybrid power sources to increase fuel mileage, lower emissions, and still produce the drivability that is so desired by most people. A **hybrid** vehicle is one in which there are two power sources.

Hybrid vehicles today use a small, efficient gasoline engine, as well as a battery pack. Several automotive manufacturers now offer hybrid power sources in vehicles they produce. It is important to note that as more and more hybrid vehicles are purchased, specially trained service technicians will be needed to work on these vehicles. This is because of the higher voltages and improved technology used within the vehicle.

HYBRID VEHICLE OPERATION

In order to understand how a hybrid vehicle works, refer to ***Figure 14–19***. In this illustration, a battery pack is placed near the rear of the vehicle. In the front of the vehicle, there is a small 4-cylinder gasoline engine, an electric motor, a generator, an efficient transaxle system, and an electronic controller. The gasoline engine is typically a very small, efficient, and low-emission engine. A typical hybrid gasoline engine would be about 1.5 liters in size and be able to produce about 60 to 80 horsepower. The battery pack is used to store electrical energy when it is not needed for driving. The electric motor is also very sophisticated. The electric motor is able to act as a motor or as a generator, depending on the driving conditions. The transaxle is an electronic transaxle that can be controlled by the electronic controller. Finally, the electronic controller is designed to connect or disconnect these components together to get the highest fuel mileage and least emissions.

There are several modes of operation on the hybrid vehicle. Each mode of operation is described in more detail to help understand hybrid power sources.

Operational Mode 1—Low Power Demand In this mode of operation, there is a low power demand by the driver. This usually occurs during starting and low speeds. Referring to ***Figure 14–20***, during this mode of operation power is supplied to the front wheels from electricity provided by the battery pack. The electrical power is sent from the battery pack directly to the electric motor. The electric motor is then used to drive the transaxle and finally the front wheels. During this mode of operation, the gasoline engine is not running, so no pollution or emissions are produced. The electronic controller is used to turn the engine off and on, make the electrical connections from the battery pack to the motor, and to engage the motor to the transaxle.

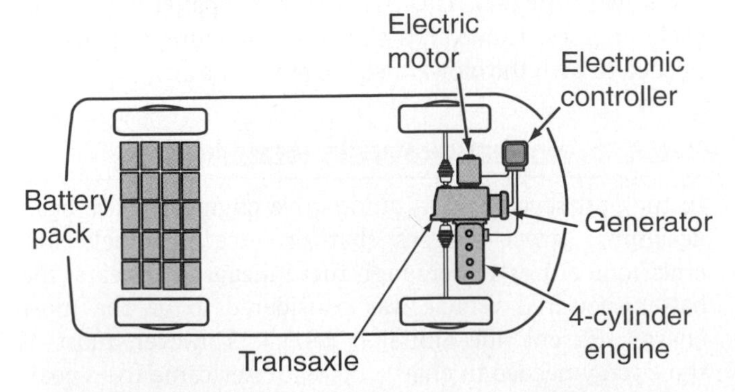

FIGURE 14-19 The major components of a hybrid vehicle are shown in this illustration.

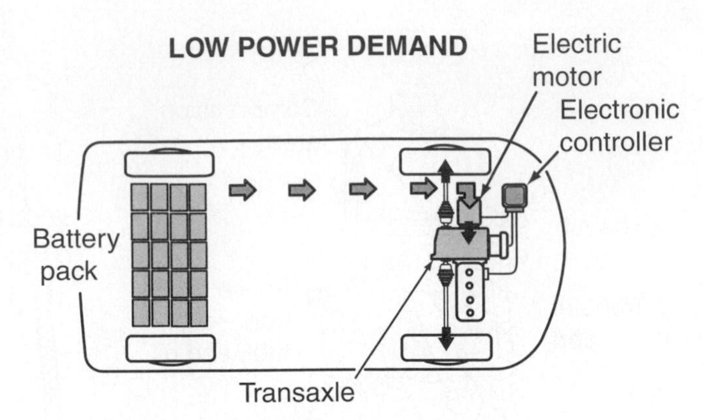

FIGURE 14-20 During low power demand, the battery pack supplies electrical energy to the electric motor, which uses its output to turn the wheels.

Operational Mode 2—Normal Driving During this mode of operation, the gasoline engine is started by the electronic controller and serves a dual purpose. The engine is used to run the generator that powers the electric motor that drives the front wheels. The output from the engine also is used to add power to the wheels through the transaxle. The exact ratio of power between the electric motor and the engine is precisely controlled by the electronic controller. The electronic controller also makes the necessary electrical connections to allow both the electric motor and engine to power the vehicle. ***Figure 14–21*** shows the power flow during this mode of operation.

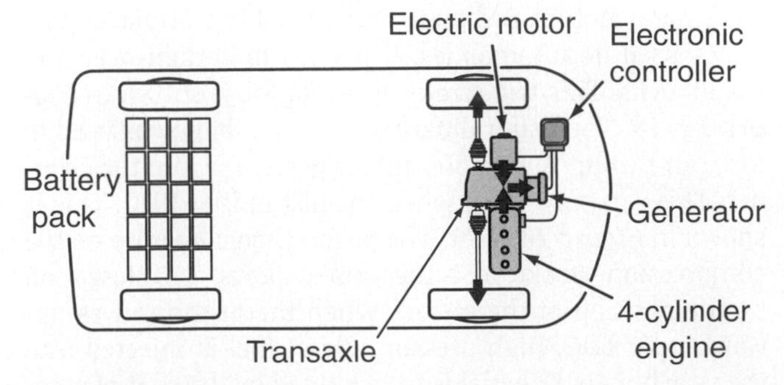

FIGURE 14-21 During normal driving conditions, the small 4-cylinder gas engine is used to provide the necessary power for moving the vehicle. Some of the mechanical energy from the gasoline engine is also converted to electrical energy through the generator and electric motor.

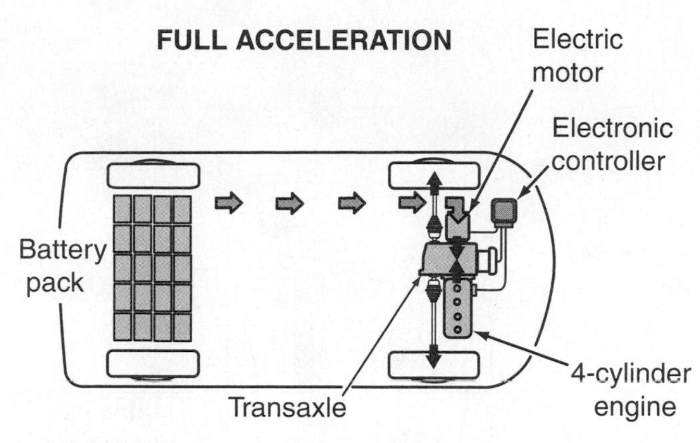

FIGURE 14-22 During acceleration, both the gasoline engine's mechanical energy and the battery pack's electricity are used to power the hybrid vehicle.

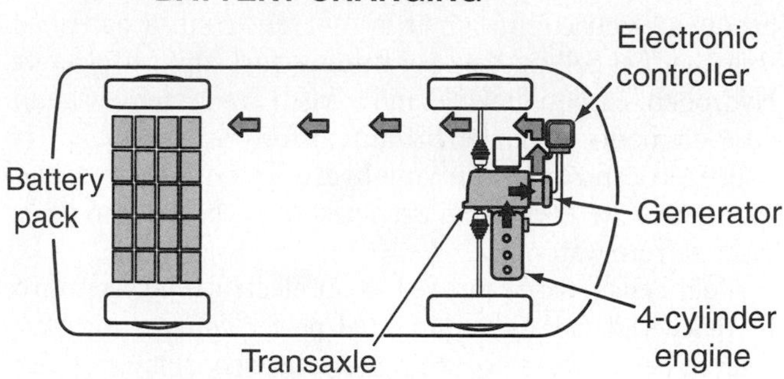

FIGURE 14-24 If the batteries need charging when the vehicle is stopped, the gasoline engine will operate the generator and charge the batteries.

Operational Mode 3—Full Acceleration During this mode of operation, power is needed to accelerate to higher speeds, pass vehicles on the road, and pull heavier loads. For this type of operation, it is necessary to have as much power available as possible. In order to do this, the electronic controller operates circuits to allow both electrical energy from the battery pack and mechanical energy from the engine to power the wheels. With the additional electrical power from the battery pack added to the engine power, acceleration occurs easily. Power comes from the engine and transaxle as well as from the battery pack, through the electric motor, and through the transaxle to the front wheels (***Figure 14–22***).

Operational Mode 4—Deceleration Deceleration occurs when the vehicle slows down to a lower speed or stops. During this time, the energy of the vehicle moving forward is typically lost to friction in the braking system. However, in a hybrid vehicle this energy is used to charge the batteries. Referring to ***Figure 14–23***, as the vehicle slows down, it creates mechanical energy. A good example of this energy is typically noticed when a person "downshifts" a manual transmission to slow down a vehicle. In the hybrid vehicle, this mechanical energy is used to turn the electric motor. However, when driven in this way, the electric motor acts as a generator. The generator produces electrical energy that is then used to charge the batteries.

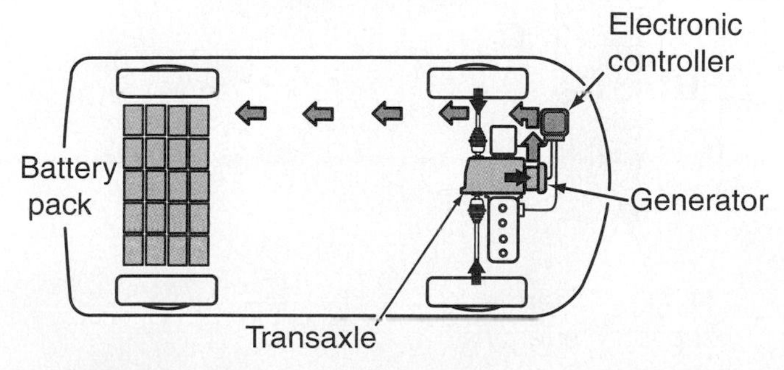

FIGURE 14-23 During deceleration, energy from the car's slowing down is converted to electrical energy to charge the batteries.

Operational Mode 5—Battery Charging When the vehicle is stopped, say at a stop sign in the city, there is no need to have mechanical power sent to the front wheel. However, if the batteries are in need of more electricity, the engine will automatically be turned on and charge the batteries. ***Figure 14–24*** shows that mechanical energy from the engine is used to turn the generator. The generator then sends electricity to the battery pack to charge the batteries.

FUEL MILEAGE

Fuel mileage on a hybrid is usually much higher than that of conventional engines. Typical fuel mileage for a hybrid is between 50 and 60 miles per gallon (mpg) for city driving and 40 to 50 mpg for highway driving conditions. Usually these statistics are reversed for conventional engines. But since the engine is usually off or only used for charging batteries in stop and go traffic, fuel mileage under city driving conditions is higher than the fuel mileage on the highway.

Fuel mileage on hybrid vehicles is also higher as compared to conventional engines under highway conditions because the electronic transaxle is very efficiently controlled. Some hybrid vehicles use an electronic **continuously variable transmission** (CVT). This means that there is no first, second, or third gear and so on. The transaxle has an infinite number of gear ratios. Depending on the driving conditions, the load, and so on, the electronic controller can tell the transaxle exactly what gear ratio to use to achieve maximum efficiency.

14.5 FUEL CELLS

Many technologies have been designed to save the earth's existing fuels. One such design is to create electrical energy by using fuel cells. A fuel cell is an electrical-type power source that uses hydrogen, carbon dioxide, and oxygen to produce electrical power. It is different from a battery in that the fuel cell needs a continuous fuel input to operate. Hydrogen, carbon dioxide, and oxygen are extremely abundant elements in our environment for creating fuels. The hydrogen can come from any hydrocarbon fuel, from natural gas to methanol, and even gasoline. Oxygen can be extracted from water.

Fuel cells have been used as an electrical power source in spacecraft and as commercial power sources for many years. Fuel cells are now being seriously considered as electrical power sources for vehicles and research continues to develop them for automotive applications. In fact, every major vehicle manufacturer has a research program to advance fuel cell technology for transportation.

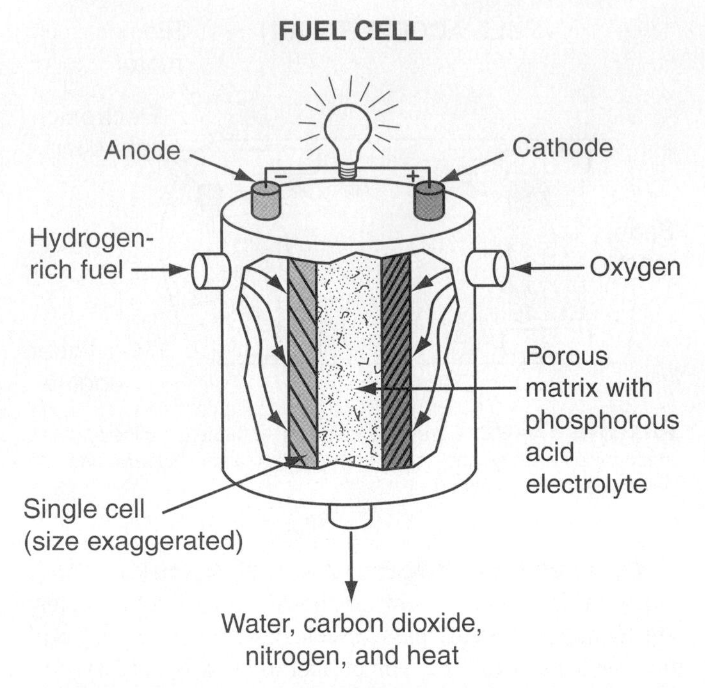

FIGURE 14-25 The major components of a simple fuel cell are shown in this illustration.

ADVANTAGES OF FUEL CELLS

Since the introduction of the hybrid vehicle and the use of battery-powered vehicles, fuel cell research has increased. The use of fuel cells in vehicles goes hand in hand with the development of hybrid vehicles. Some of the advantages of using fuel cells include:

- Higher efficiency compared to batteries
- Low maintenance requirements
- Easy sizing for different power ratings
- Individual fuel cells can be easily added together to increase power output
- Environmental safety
- No recharging required

FUEL CELL OPERATION

Fuel cells use an electrochemical process. The chemical energy that bonds atoms of hydrogen (in a hydrocarbon source) and oxygen (in air) is converted directly into electrical energy. The electrical energy can then be used to power a vehicle. ***Figure 14–25*** shows a simplified diagram of a fuel cell and its major components. Hydrogen, carbon dioxide, and oxygen are the inputs. A single fuel cell consists of an anode and a cathode plate with a chemical mixture between the two. Individual cells such as these generate about 0.5 to 1 volt and must be stacked and connected to boost voltage and amperage availability.

In operation, hydrogen, and oxygen from the fuel tank enter the cell. Referring to ***Figure 14–26***, the hydrogen enters on the left side of the fuel cell. The oxygen enters on the right side. The hydrogen flows through the anode plate, losing its electrons. This gives the plate a negative charge.

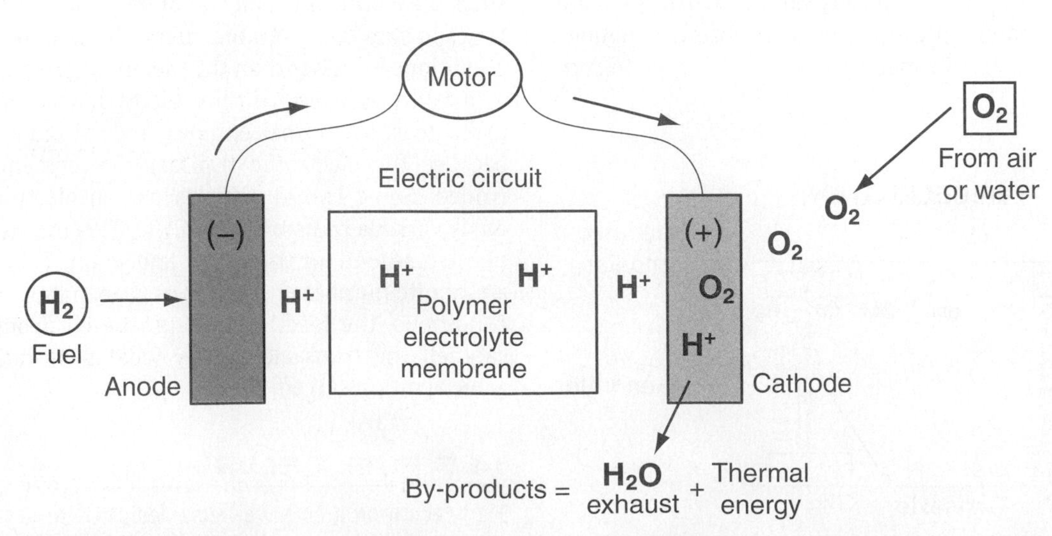

FIGURE 14-26 As hydrogen flows through the anode and oxygen flows through the cathode, a negative and positive charge is produced, which operates the motor.

When oxygen as a fuel flows through the cathode, oxygen molecules pick up electrons on the cathode plate. This action gives the cathode a positive charge. When an external load (such as the motor used to turn wheels) is connected to the two electrodes, excess electrons on the anode discharge through the load to the cathode. This action is similar to a typical battery.

During the discharge process, positive hydrogen ions are produced at the anode. An ion is an atom that is electrically charged as a result of having gained or lost electrons. The ions are transferred through an electrolyte solution and combine with oxygen ions from the cathode. This action produces water. The electrolyte solution is made of chemicals that are able to conduct electricity. The electrode plates are actually paper-thin carbon sheets that contain platinum and other chemicals to aid in ionization.

Since water and thermal energy (heat) are by-products, the water can be separated and the oxygen used again in the system. However, the fuel cell will continue to need additional hydrogen for efficient operation.

Research continues on the use of fuel cells in the transportation industry. It is anticipated that fuel cell technology will be used commercially along with hybrid vehicles within the next 10 years. New electrolytes, anode and cathode design, and hydrogen and oxygen delivery systems will continue to be developed.

Service Manual Connection

There are many important specifications to keep in mind when working on different types of engines. To identify various specifications, you will need to know the VIN number of the vehicle. Then, using the appropriate service manual, locate the type of vehicle that you need specifications for. Althoughthey may be titled differently, some of the more common braking system specifications are shown below for a rotary-type engine.

Rotary Engine Specifications

Common Specification	**Typical Example**
Displacement	1.3 Liters/ 80 cubic inches
Timing leading coil	5 degrees ATDC
Timing training coil	20 degrees ATDC
Compression ratio	9.0 to 1
Rotor bearing clearance	0.0010–0.0031 inch
Apex seal height	0.295 inch
Clearance between apex seal and groove	0.002–0.004 inch
Side seal protrusion from rotor	0.02 inch
Clearance between side seal and corner seal	0.002–0.006 inch

SUMMARY

The following statements will help to summarize this chapter:

- Various types of power sources are used on automobiles today and new sources are likely in the future.
- Today's diesel engines use higher compression ratios, and use the heat of compression to start combustion rather than a spark plug to ignite the fuel.
- The diesel engine does not use a low-pressure fuel injector, but instead uses high-pressure fuel injection.
- The rotary engine uses a triangular rotor to cause the intake, compression, power, and exhaust to occur within the engine.
- Air and fuel are mixed in a carburetor or by low-pressure fuel injection in a rotary engine.
- The two-stroke engine is being considered by several manufacturers for use as an automobile engine.
- The two-stroke engine is able to produce the intake, compression, power, and exhaust strokes in one revolution of the crankshaft, rather that two revolutions.
- The two-stroke engine uses a set of ports in the cylinder to allow air and fuel to enter the combustion chamber.
- General Motors manufactures a two-stroke diesel engine for use in heavier truck and industrial applications.
- The hybrid power source has become very popular in recent years as a low-emission, high-mileage vehicle.
- The hybrid vehicle uses batteries as well as a small gasoline engine to power the vehicle.
- The hybrid vehicle has several components that operate together including a small 4-cylinder gasoline engine, an electronic transaxle, a generator to produce electricity, an electric motor to power the wheels, a

battery pack to store electricity, and an electronic controller to control the system correctly and efficiently.

- Fuel cells are being researched for use in transportation vehicles.
- Fuel cells use hydrogen and oxygen as fuel to create electricity.
- Fuel cells produce water and thermal energy as by-products of electrical energy production.
- Fuel cell development parallels the continued development of hybrid vehicles.

TERMS TO KNOW

Can you explain each of the following terms? Review the chapter until you can use each term correctly.

Continuously variable transmission
Diesel engine
Eccentric journals
Fuel injection
Hybrid
Loop scavenging
Rotary engine
Rotor
Two-stroke engine

REVIEW QUESTIONS

Multiple Choice

1. What is the compression ratio on some diesel engines?
a. 25 to 1
b. 8 to 1
c. 10 to 1
d. 6 to 1
e. 5 to 1

2. The diesel engines uses ________ to get the fuel into the engine.
a. Fuel injection
b. Carburetors
c. Spark plugs
d. Camshafts
e. Electric motors

3. A diesel engine ignites the fuel from the:
a. Heat of compression
b. Spark plug
c. Fuel injection
d. Igniter
e. Exhaust from the first cylinder

4. How many power pulses are there on one rotor per rotor revolution on the rotary engine?
a. 1
b. 2
c. 3
d. 4
e. 5

5. What object opens and closes the ports on the rotary (Wankel) engine?
a. The valves
b. The position of the rotor
c. The carburetor
d. The spark plug
e. The camshaft

6. A hybrid vehicle is powered by:
a. One power source
b. Two power sources
c. Batteries and a turbine engine
d. A generator and a diesel engine
e. A two-stroke engine and a generator

7. A hybrid power source uses which of the following components?
a. Electronic controller
b. Small 4-cylinder gasoline engine
c. Battery pack for storing electricity
d. All of the above
e. None of the above

8. When a hybrid vehicle is accelerating, power comes to the front wheel from the:
a. Battery pack
b. Engine
c. Generator only
d. A and B only
e. All of the above

9. Which type of fuel is used for the negative side of a fuel cell?
a. Oxygen
b. Water
c. Plastic
d. Hydrogen
e. Ions

10. Which type of fuel is used for the positive side of a fuel cell?
a. Oxygen
b. Water
c. Plastic
d. Hydrogen
e. Ions

11. What are the by-products of a fuel cell's chemical action?
   a. Water
   b. Thermal energy
   c. Pure hydrogen
   d. Pure oxygen
   e. A and B of the above

12. A fuel cell produces which type of power for vehicle propulsion?
   a. Electrical power
   b. Mechanical power
   c. Chemical power
   d. Radiant power
   e. None of the above

The following questions are similar in format to ASE (Automotive Service Excellence) test questions.

13. *Technician A* says the compression ratio on a diesel engine is higher than on a gasoline engine. *Technician B* says the air-fuel ratio on a diesel engine is lower than on a gasoline engine. Who is correct?
   a. A only c. Both A and B
   b. B only d. Neither A nor B

14. *Technician A* says the gas engine used in a car is called a continuous combustion engine. *Technician B* says the gas engine used in a car is called an internal combustion engine. Who is correct?
   a. A only c. Both A and B
   b. B only d. Neither A nor B

15. *Technician A* says that the two-stroke engine uses a reed valve to control the pressure in the combustion chamber. *Technician B* says the two-stroke engine uses a reed valve to control the vacuum in the crankcase area. Who is correct?
   a. A only c. Both A and B
   b. B only d. Neither A nor B

16. *Technician A* says that the two-stroke engine is lighter than a four-stroke engine. *Technician B* says that the two-stroke engine is more responsive than a four-stroke engine. Who is correct?
   a. A only c. Both A and B
   b. B only d. Neither A nor B

17. When defining the points on a two-stroke diagram, *Technician A* says the intake ports close about 5 degrees BTDC. *Technician B* says the exhaust ports close about 5 degrees ATDC. Who is correct?
   a. A only c. Both A and B
   b. B only d. Neither A nor B

18. *Technician A* says that hybrid vehicles get better fuel mileage on the highway. *Technician B* says that hybrid vehicles get better fuel mileage in the city. Who is correct?
   a. A only c. Both A and B
   b. B only d. Neither A nor B

19. *Technician A* says the fuel cells produce hydrogen as a by-product. *Technician B* says that fuel cells do not need hydrogen for their operation. Who is correct?
   a. A only c. Both A and B
   b. B only d. Neither A nor B

Essay

20. List at least five differences between a diesel and a gasoline engine.

21. Describe the purpose of seals used on the rotor of the rotary engine.

22. What is the purpose of a blower on the two-stroke cycle diesel engine?

23. Describe how a hybrid vehicle gets power when it is accelerating

24. Identify the basic operation of how a fuel cell creates electricity.

Short Answer

25. The part on a gas turbine engine that produces the air pressure is called the ________.

26. The center shaft in a rotary engine is called the ________.

27. A two-stroke engine has ________ power pulse per cylinder, per revolution.

28. The part on a gas turbine engine that extracts exhaust heat and places this heat in the intake is called a ________.

29. As the rotor on a rotary engine turns, the ________ is turned, which produces the power.

30. On a hybrid vehicle power system, the ____________ controls the connections between the engine, battery pack, electric motor, and generator.

31. When using fuel cells, the anode is considered the ____________ charge.

32. A fuel cell needs both ______________ and ____________ as the input fuel for operation.

Applied Academics

TITLE: Hybrid Specifications

ACADEMIC SKILLS AREA: Language Arts

NATEF Automobile Technician Tasks:

The technician supplies clarifying information to customers, associates, the parts supplier, and the supervisor.

CONCEPT:

When looking over a set of specifications for a hybrid vehicle, the service technician comes across the mechanical and performance specifications. The following information was found in the sales literature for the hybrid vehicle:

HYBRID VEHICLES	
MECHANICAL/ PERFORMANCE	**SPECIFICATIONS**
Gasoline Engine	
Type:	Aluminum double overhead cam (DOHC) 16-valve VVT-i 4-cylinder
Displacement:	1.5 liters (1,497 cc)
Bore x stroke:	75.0 mm x 84.7 mm
Compression ratio:	13.0:1
Valvetrain:	Double overhead cam (DOHC) 16-valve with Variable Valve Timing with intelligence (VVT-i)
Induction system:	Multi-point EFI with Electronic Throttle Control System with intelligence (ETCS-i)
Ignition system:	Electronic, with Toyota Direct Ignition system (TDI)
Horsepower:	70 hp @ 4,500 rpm
Torque:	82 lb-ft. @ 4,200 rpm
Emissions rating:	Super Ultra Low Emission Vehicle (SULEV)
Electric Motor/Generator/Power Storage	
Motor type:	Permanent magnet
Power output:	33 kW/44 hp @ 1,040–5,600 rpm
Torque:	350 N-m/258 lb-ft. @ 0–400 rpm
Battery type:	Sealed Nickel-Metal Hydride (Ni-MH)[5]
Output:	273.6 V (228 1.2-V cells)
Brakes	
Type:	Power-assisted ventilated front disc/rear drum with standard Antilock Brake System (ABS) and regenerative braking

APPLICATION:

Using the information in this chapter about hybrid vehicles and the mechanical and performance specifications provided in the chart, can you successfully explain to a customer how the hybrid vehicle operates? Can you explain to the customer the difference between a hybrid vehicle and a standard gasoline-powered vehicle? Can you give the customer the definition of regenerative braking?

Engine Performance

CHAPTER 15

OBJECTIVES

After studying this chapter, you should be able to:

- Define the term *horsepower.*
- Compare the different types of horsepower.
- Relate torque to horsepower.
- Identify the effects of frictional losses on vehicle performance.
- Examine the use of dynamometers.
- Analyze performance charts.
- Compare different types of engine efficiency.

INTRODUCTION

Over the past years, automotive manufacturers have built many sizes and types of engines. These engines differ in the amount of power they can produce. Horsepower, torque, fuel consumption, and efficiency continue to improve. The purpose of this chapter is to familiarize you with the terms and methods used to measure engine performance and operation.

15.1 TYPES OF HORSEPOWER

Many types of horsepower are used for automobile engines. When comparing engines, brake horsepower is used. When discussing efficiency, frictional and indicated horsepower are used. When analyzing gasoline mileage, road horsepower is used. These and other horsepower definitions should be analyzed. To do this, the term *work* must first be discussed.

WORK

Work is the result of applying a force. This force is created by a source of energy. When the force moves a certain mass a certain distance, work is produced. Work is defined as shown in ***Figure 15–1***.

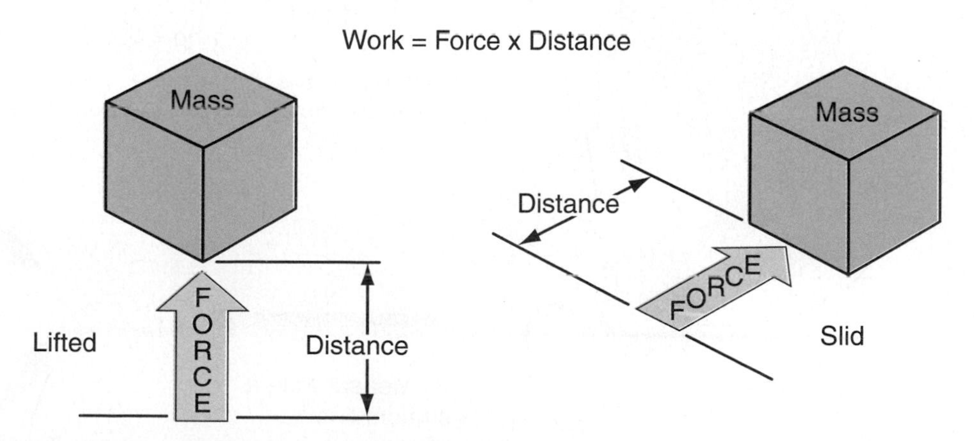

FIGURE 15–1 Work is defined as the result of moving a certain mass a certain distance. It is measured in foot-pounds, or ft.-lb. The movement can be lifting or sliding motion.

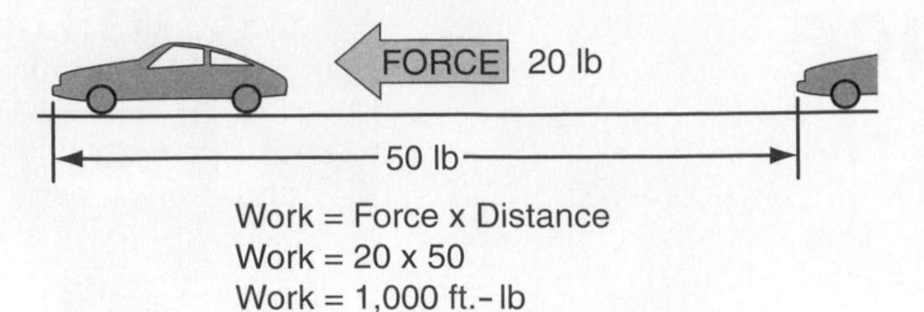

FIGURE 15–2 When a vehicle is pushed by 20 pounds of force a distance of 50 feet, 1,000 ft.-lb of work has been created.

Force is measured in pounds. Distance is measured in feet. When the two units are put together, foot-pounds are measured. Work, then, is measured in foot-pounds (ft.-lb). For example, as shown in ***Figure 15–2***, if a vehicle were moved 50 feet with a force of 20 pounds, then 1,000 ft.-lb of work would be produced.

TORQUE

Torque is one way to measure work. Torque is defined as twisting force (***Figure 15–3***). This force is produced in an engine because of the combustion of fuel. Combustion pushes the piston down. The piston causes the crankshaft to rotate, producing torque. This force causes the wheels to rotate.

Torque is actually available at the rear of the engine. Torque is expressed in foot-pounds (energy needed to move a certain number of pounds 1 foot). An engine is said to have 500 ft.-lb of torque at a certain speed. Speed on a gasoline or diesel engine is measured in **revolutions per minute (rpm)**. Torque can be measured directly from a rotating shaft by using a dynamometer. Dynamometers will be discussed later in this chapter.

HORSEPOWER DEFINED

Horsepower (hp) is also a measure of work. It is a unit of work or a measure of work done within a certain time. Horsepower also involves the rate at which work is being done. When anything is measured by a rate, time is considered. Therefore, horsepower is concerned with how long it takes to do certain work.

One definition of horsepower is the amount of work needed to lift 550 pounds 1 foot in 1 second (***Figure 15–4***). If this work is measured per minute (rather than per second), the other definition of horsepower is the amount of work needed to lift 33,000 pounds 1 foot in 1 minute. These two definitions are the standard way of defining horsepower. It is important to remember that the direction of motion when horsepower is applied is in a straight line. However, torque is always related to rotation (***Figure 15–5***).

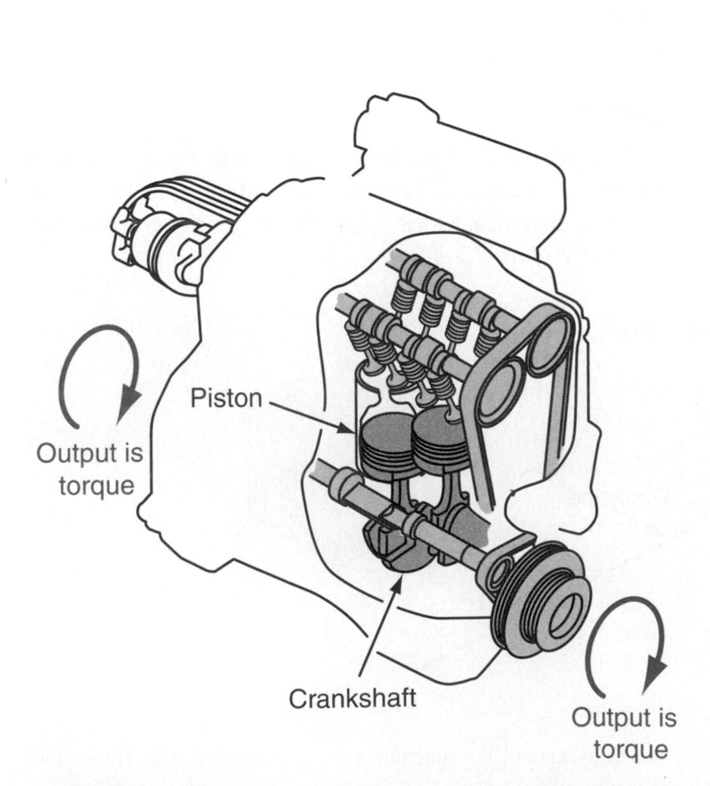

FIGURE 15–3 Torque is defined as twisting force. The work (output) that engines produce is measured as torque on the crankshaft.

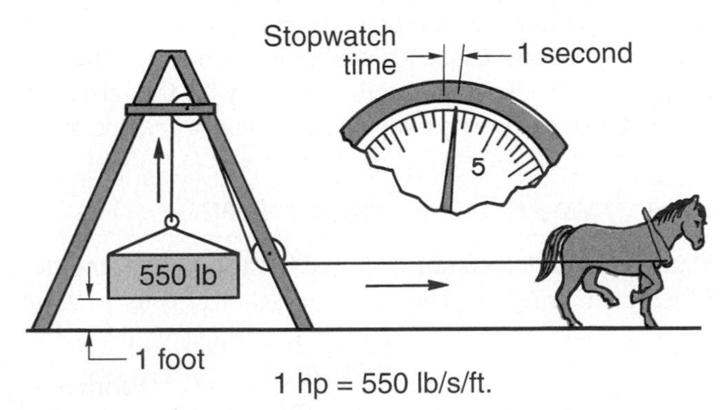

FIGURE 15–4 One definition of horsepower is the amount of work required to raise 550 pounds 1 foot in 1 second.

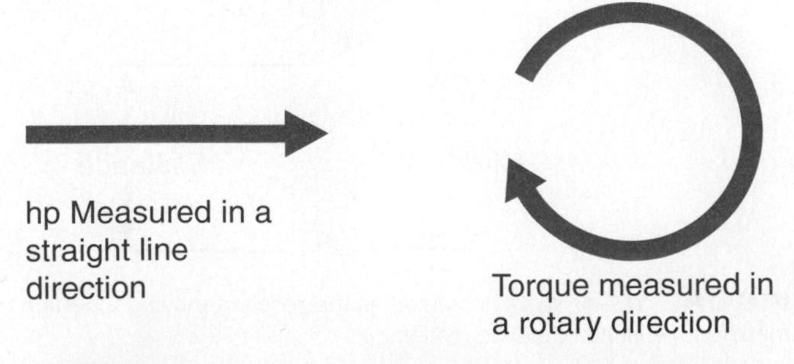

FIGURE 15–5 Horsepower is always a measure of work applied in a straight line. Torque is a measure of force in rotary motion.

SPECIFICATIONS
GENERAL ENGINE SPECIFICATIONS

Engine Liter/CID	Net hp @ rpm	Maximum Torque Ft.-Lbs @ rpm
3.8L/V6-232	140 @ 3800	215 @ 2400
3.8L/V6-232 SC❶	210 @ 4000	315 @ 2600
4.6L/V8-281❷	190 @ 4200	260 @ 3200
4.6L/V8-281❸	210 @ 4600	270 @ 3400
5.0L/V88-302 HO	200 @ 4000	275 @ 3000

❶ = Supercharged
❷ = Single exhaust
❸ = Dual exhaust

FIGURE 15–6 Horsepower and torque are often stated for each type of engine in many of the service manuals.

Figure 15–6 shows the horsepower and torque specifications from a typical service manual. Note that both the torque and horsepower are stated at a specific rpm.

BRAKE HORSEPOWER

Brake horsepower (bhp) is defined as the actual horsepower measured at the rear of the engine under normal conditions. It is called brake horsepower because a brake is used to slow down the engine crankshaft inside a dynamometer. Brake horsepower is often used to compare engines and their characteristics. Automotive manufacturers use brake horsepower to show differences between engines. For example, a 235-cubic inch engine will produce less bhp than a 350-cubic inch engine. Other factors that may change bhp include type of fuel system, quality of combustion, compression ratio, type of fuel, and air-fuel ratio.

INDICATED HORSEPOWER

Indicated horsepower (ihp) is defined as theoretical horsepower. Indicated horsepower has been calculated by the automotive manufacturers. Ihp represents the maximum horsepower available from the engine under ideal or perfect conditions. Ihp is calculated on the basis of engine size, displacement, operational speed, and the pressure developed theoretically in the cylinder. Indicated horsepower will always be more than bhp.

FRICTIONAL HORSEPOWER

Frictional horsepower (fhp) is defined as the horsepower used to overcome internal friction. Any time two objects touch each other while moving, friction is produced. Friction must be overcome with more energy. This happens within any engine. Sources of frictional horsepower include bearings, pistons sliding inside the cylinder, the piston moving upward on the compression stroke, the generator, fan, water pump, belts, air conditioner, and so on. ***Figure 15–7*** shows a dual overhead camshaft engine with many of the frictional horsepower components highlighted.

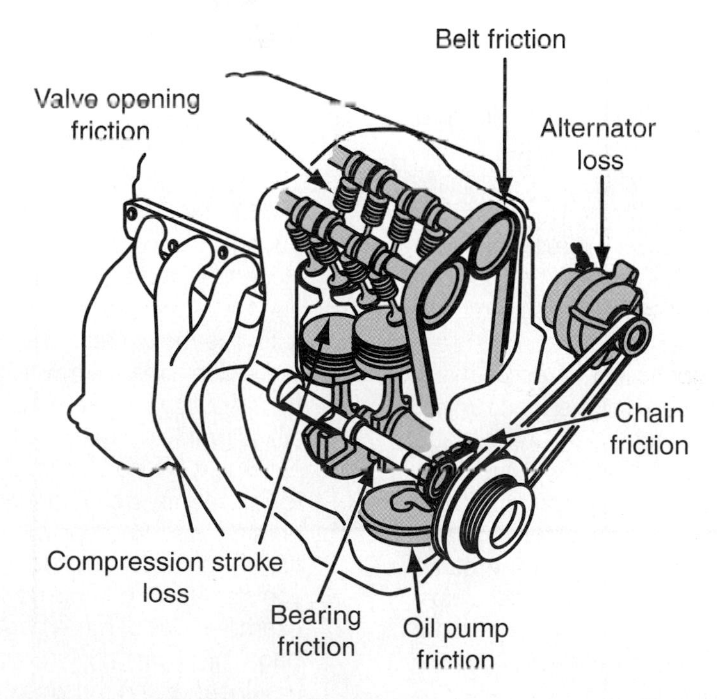

FIGURE 15–7 Frictional horsepower is created at various spots within an engine. Sources of frictional horsepower include bearing friction, the compression stroke, belts, the alternator, valves opening and closing, and so on.

Other sources of frictional horsepower losses on a vehicle outside the engine include the wind, tire rolling resistance, road conditions, and so on. All of these have a tendency to consume horsepower from the engine. They make up the frictional horsepower.

It is advantageous to reduce frictional horsepower as much as possible. The more frictional horsepower that the engine must overcome, the more energy will be needed to operate the vehicle. This means poorer fuel mileage for the vehicle.

Reducing Frictional Horsepower Frictional horsepower has been analyzed very carefully over the years. Research efforts found that poor gasoline mileage occurred because of large amounts of frictional horsepower. Considerable changes have been made in the automobile to reduce frictional losses. Some of these changes include the following:

- Reducing the rolling resistance on tires. Tires are designed by using computers to reduce the rolling resistance. Radial tires also reduce rolling resistance. Higher tire pressure also reduces rolling resistance.
- Reducing the air drag on a vehicle. Manufacturers have been making vehicles that have less wind resistance. Today, all vehicles have a specified **coefficient of drag.** The coefficient of drag is a measure of how much air is moved as the vehicle moves from one point to another. Generally, the coefficient of drag ranges from 1.00 to 0.00. The lower the number, the less wind resistance the vehicle has to overcome.
- Running the cooling fan on an electric motor, rather than from the engine. The fan now operates only when needed. Also, some fans turn off and on by using a clutch system.
- Making vehicles lighter. On the average, 1 mile per gallon (mpg) of fuel is lost for every 400 pounds on the vehicle. To help reduce vehicle weight, lighter materials are being used and smaller vehicles are being designed.
- Changing the undercarriage of the vehicle to reduce air drag on the bottom.
- Operating the fuel pump electrically rather than mechanically. Electric fuel pumps operate only when necessary, saving frictional engine losses that are typically produced by mechanical fuel pumps.
- Operating the power steering pump electrically. If the power steering pump is on only when there is a need for hydraulic pressure for power steering, then frictional horsepower is reduced.
- Operating the air-conditioning compressor pump only when needed. Now that there are electrical controls on the air-conditioning compressor, it can be operated only when needed.

Car Clinic: DRAG

CUSTOMER CONCERN:
Coefficient of Drag

In many automotive publications, you will see references to the coefficient of drag. What is the coefficient of drag, and why is it so important?

SOLUTION: The coefficient of drag is a measure of how much air is moved as a vehicle travels from one point to another at a specific speed. Generally, the coefficient of drag ranges from 1.00 to 0.00. The smaller the number, the less air is moved as the vehicle travels down the road. Today, it is not uncommon to have low coefficient of drag ratings such as 0.33, 0.38, 0.32, and 0.28.

In the calculation for coefficient of drag, many factors are included, such as vehicle speed, shape of the vehicle, and smoothness of the outer surface of the vehicle. Generally, as the coefficient of drag decreases, the vehicle is able to cut through the air more efficiently. This means less air is moved, and the engine has to work less to achieve the same speed. The result is that the vehicle will get better fuel mileage than a vehicle with a higher coefficient of drag, assuming, of course, all other factors are equal.

Car Clinic: HORSEPOWER

CUSTOMER CONCERN:
Reducing Frictional Horsepower

As frictional horsepower increases, fuel consumption will increase. What are some good methods used to reduce frictional horsepower in a typical vehicle?

SOLUTION: Frictional horsepower can be reduced in several ways. First, keep the windows closed. As the air rushes around the window opening, frictional horsepower increases. A second way is to use radial tires. There is less rolling resistance with radial tires than with the older four-ply tires. A third way is to make sure the front and rear wheels are aligned to specifications. If the front and rear alignment are out of line, the wheels may have more rolling resistance. The end result is more frictional horsepower. Another method used to reduce frictional horsepower is to reduce the weight in the vehicle. Never carry anything you do not need. For example, wood in the back of a pickup truck, sand bags for winter driving, and tools all make the vehicle heavier and thus increase frictional horsepower.

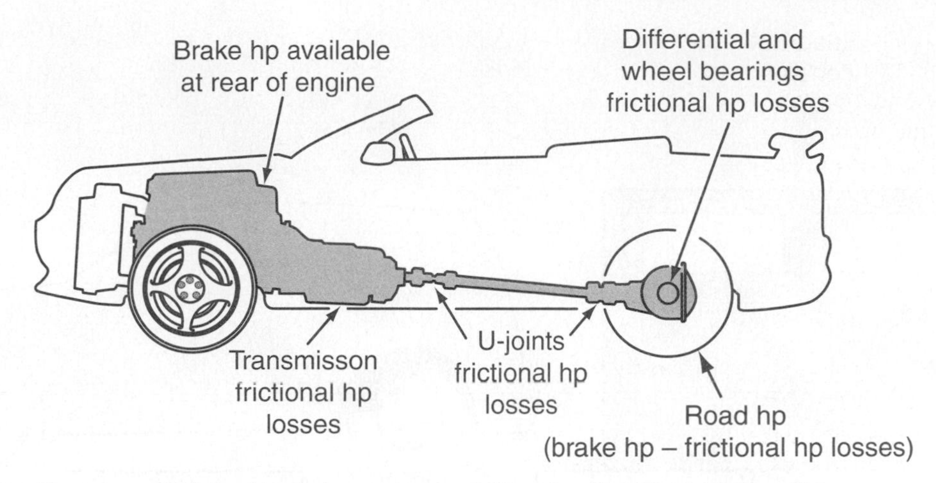

FIGURE 15–8 Road horsepower is defined as the horsepower available at the drive wheels of the vehicle.

Automotive design engineers are constantly looking for ways to reduce frictional horsepower. These and other new designs have allowed automotive manufacturers to improve gasoline mileage from 12–15 mpg to 35–45 mpg on some vehicles.

ROAD HORSEPOWER

Road horsepower is defined as the horsepower available at the drive wheels of the vehicle. Road horsepower will always be less than bhp. The difference between road horsepower and brake horsepower is the result of frictional horsepower. Frictional horsepower losses are produced from the friction in the transmission, drive shaft, and rear differential assemblies (***Figure 15–8***). Road horsepower can be shown as:

road hp = bhp – fhp through the drivetrain

15.2 DYNAMOMETERS

DYNAMOMETER DEFINED

At times it may be necessary to **load** the engine while troubleshooting an engine. Loading the engine is like pulling a trailer up a steep hill. One important reason to load the engine is to check emission of the engine during diagnostic procedures. Another reason to load an engine is that often problems occur in an engine only when it is loaded down or pulling a load such as going down the highway. For example, an engine may operate correctly at idle or at city speeds; however, a problem may be evident when the vehicle is traveling at 55 mph.

To load an engine, a dynamometer is used in the automotive service area. A **dynamometer** is a device attached to the back of the engine to absorb the power being created by the engine. When the engine is at idle, it is impossible to determine how much horsepower or torque can be produced. If an engine is run on a dynamometer, it can be loaded down to simulate actual driving conditions. The **engine dynamometer** measures brake horsepower and torque at the output of the engine.

Another type of dynamometer can be used to measure road horsepower. It is called a **chassis dynamometer** as shown in ***Figure 15–9***. A chassis dynamometer measures the horsepower and torque available at the drive wheels of the vehicle. This dynamometer measures road horsepower. ***Figure 15–10*** shows the components of a chassis dynamometer. In this case, the tires roll on two rollers. These are called the idle roller and drive roller. The power absorption unit absorbs the energy. This unit acts as the load on the vehicle. In order to load the vehicle, hydraulic fluid is allowed to enter the power absorption unit. As more fluid enters, it becomes more difficult to turn the drive roller. At this point, the engine begins to load down, acting like it is driving down a road. The amount of fluid in the power absorption unit determines the exact load on the vehicle. Both vehicle speed (rpm) and torque are measured on the scales shown.

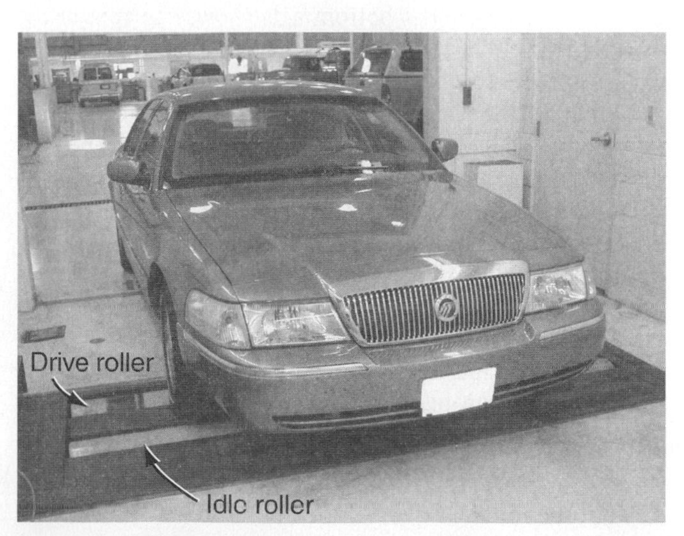

FIGURE 15–9 A chassis dynamometer is used to measure road horsepower directly from the wheels of the vehicle.

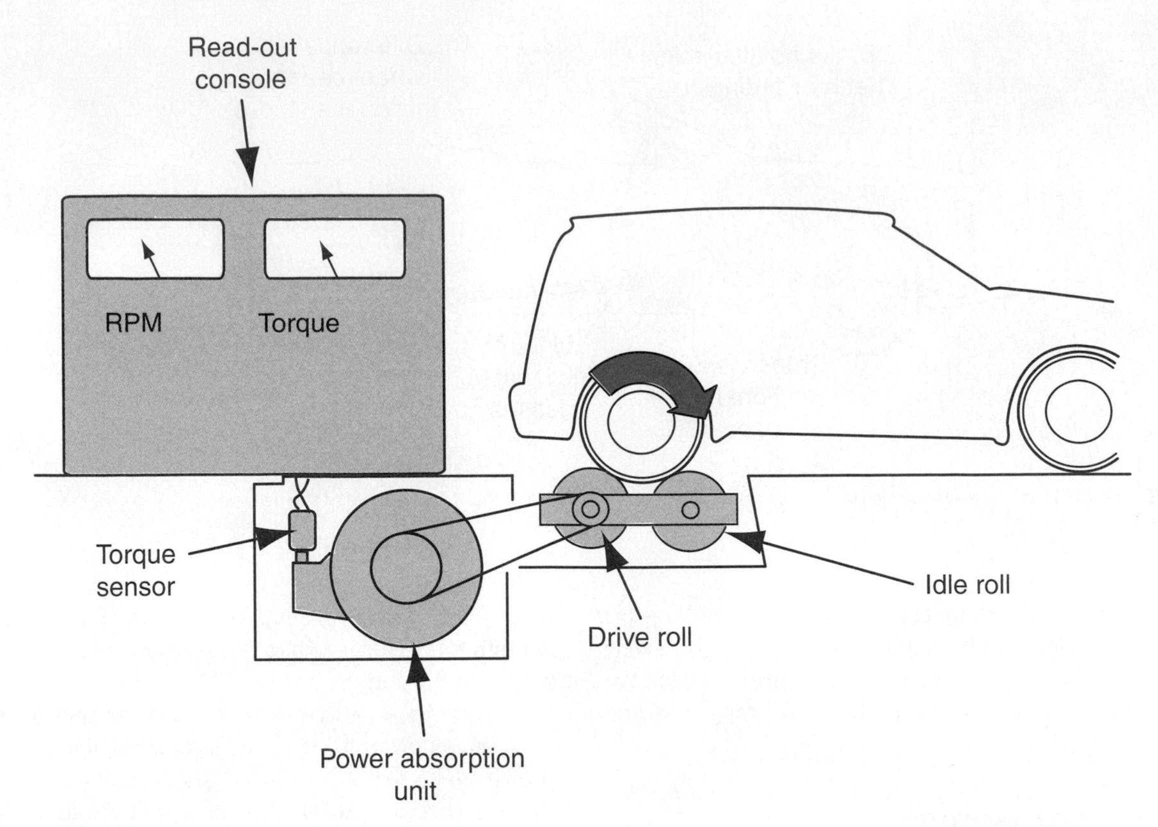

FIGURE 15–10 On a chassis dynamometer, the tires of the vehicle roll directly on the idle and drive roll. The absorption unit loads the system by increasing the fluid inside.

PERFORMANCE CHARTS

Gasoline and diesel engines have certain operating characteristics. This means that they have different torque, horsepower, and fuel consumption at different rpm. Using a dynamometer, a **performance chart** (also called a characteristic curve) can be developed. ***Figure 15–11*** shows a typical performance chart. The bottom axis shows the rpm of the engine. The left axis shows the bhp on the engine. The right axis shows the torque being produced on the engine. Also, note that on the bottom right, a fuel consumption scale is included. This indicates the amount of fuel used in pounds per bhp per hour. This unit is sometimes referred to as BSFC, or brake specific fuel consumption.

When the engine is loaded with a dynamometer, a certain maximum torque and horsepower can be produced at a specific rpm. For example, refer to ***Figure 15–11***. Four specific characteristics can be read from the performance chart. Take the specific reading of 2,500 rpm. At this speed, the engine is capable of producing the following output characteristics:

- rpm, 2,500
- Torque, 138 ft.-lb
- Horsepower, 58
- BSFC, 0.23 (lb/bhp/h)

It should be noted that a dynamometer can measure only the torque being produced at the rear of the engine or at the drive wheels. The dynamometer does not measure horsepower. The following formula determines horsepower readings so the curve can be made:

$$\text{horsepower} = \frac{\text{torque} \times \text{rpm}}{5{,}252}$$

The number 5,252 is called a constant and is related to the definition of one horsepower (33,000 1/ft.-lb per minute). Today, dynamometers are designed to read out both torque and horsepower digitally. Although the dynamometer can't really read horsepower, the machine automatically calculates the horsepower, which can in turn be read on a digital meter.

PROCEDURE FOR DYNAMOMETER TESTING

The procedure for dynamometer testing and producing a performance chart varies with the dynamometer manufacturer. However, most engine dynamometers have similar procedures. The general procedure used is as follows:

1. Start the engine and run at idle rpm.
2. Increase the load controls to produce maximum load.

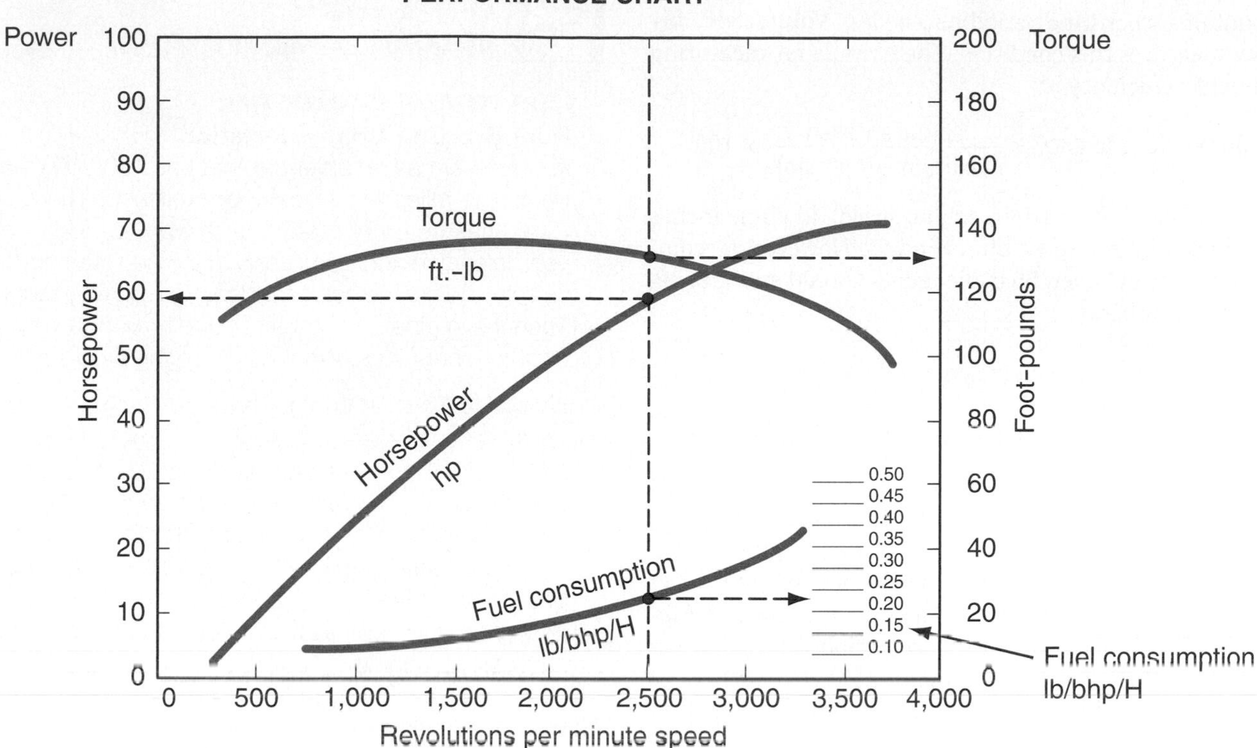

FIGURE 15-11 Performance charts show the amount of horsepower and torque that an engine can produce for a range of rpm. Fuel consumption is also shown and is measured in pounds per brake horsepower per hour.

3. Using the engine throttle, increase the speed by 100-rpm increments. At each point read the torque in foot-pounds.
4. Continue to increase the rpm until the maximum throttle has been reached.
5. Plot the torque curve using the torque data from each 100-rpm increment.
6. For each 100-rpm increment, calculate the horsepower using the formula previously stated.
7. Plot the horsepower curve with the calculated data for each 100-rpm increment.

15.3 ENGINE EFFICIENCY

The term **efficiency** means many things in the automotive field. Efficiency generally refers to how well a particular job can be done. It is usually expressed as a ratio of input to output. There are, however, other types of efficiency, including mechanical efficiency, volumetric efficiency, and thermal efficiency.

Efficiencies are expressed as percentages. They are always less than 100%. The difference between the efficiency and 100% is the percentage lost during the process.

MECHANICAL EFFICIENCY

One way to show efficiency is by measuring the mechanical systems of a machine. This method measures how efficient the mechanical systems are in a machine. The machine we are concerned with is the gasoline engine. **Mechanical efficiency** is a relationship between the theoretical (ihp) amount of work required to do a certain job and the actual (bhp) amount of work required to do the job. For example, if a certain car requires 185 actual horsepower and the theoretical horsepower required to do the same amount of work is 205, the mechanical efficiency can be calculated. The formula to calculate mechanical efficiency is:

$$\text{mechanical efficiency} = \frac{\text{actual horsepower}}{\text{theoretical horsepower}} \times 100$$

$$\text{mechanical efficiency} = \frac{185}{205}$$

$$\text{mechanical efficiency} = 90\%$$

The losses on any mechanical system are caused primarily by friction. If frictional horsepower can be reduced on an engine, the mechanical efficiency will increase. If frictional horsepower increases on an engine, the mechanical efficiency will decrease.

VOLUMETRIC EFFICIENCY

Another way to measure the efficiency of an engine is related to how easily air flows in and out of the engine. As the piston starts down on the intake stroke, air (and fuel) flow into the engine. As the engine increases in rpm, the intake valves are not open for as long a time. This means that the amount of air per time period may be less. **Volumetric efficiency** measures this condition. The formula for measuring volumetric efficiency is:

$$\text{volumetric efficiency} = \frac{\text{actual air used}}{\text{maximum air possible}} \times 100$$

For example, at a certain engine speed, 40 cubic inches of air-fuel mixture enter the cylinders. However, to completely fill the cylinder, 55 cubic inches should enter. Using these two numbers:

$$\text{volumetric efficiency} = \frac{40}{55} \times 100$$

$$\text{volumetric efficiency} = 90\%$$

One way to improve the volumetric efficiency is to improve the scavenging of the cylinder. This means improving the ease at which air and fuel can enter and exhaust is removed from the engine. Over a period of time, the valves may have a buildup of carbon deposits, as shown in ***Figure 15–12***. It is obvious that this condition will reduce the volumetric efficiency of the engine.

Other factors that affect volumetric efficiency include the following:

- Exhaust restriction
- Air cleaner restrictions
- Carbon deposits on cylinder and valves
- Shape and design of valves
- Amount of restriction in the intake and exhaust ports by curves (ports can be polished to reduce friction)
- Shape of the intake and exhaust manifolds
- Temperature of the air

CUSTOMER CONCERN:

Poor Starting Characteristics

A customer has an older car that has 100,300 miles on it. The mileage is mostly city mileage. The customer has never had any major work done on the vehicle engine and complains that when starting the vehicle, the engine has a tendency to run very rough. Then the engine smoothes out and runs okay during all other operations. What could be the problem?

SOLUTION: An engine that has this many miles on it, especially from city mileage, most likely will have a significant amount of carbon buildup inside the intake manifold and near the valves. When the engine is first started up, the carbon has a tendency to absorb some of the fuel before it gets to the combustion chamber. Since some of the fuel is being absorbed, it is not available for combustion. This causes the engine to run rough immediately after starting. The only way to fix the problem is to remove the carbon on the intake valve. This can be done by removing the head and cleaning the carbon off the valve.

THERMAL EFFICIENCY

A more specific form of efficiency is **thermal efficiency**. Thermal efficiency tells how effectively an engine converts the heat energy in its fuel into actual power at the output shaft. It takes into account all of the losses on the engine, including thermal losses, mechanical losses, and volumetric losses. For this reason, thermal efficiency is sometimes called overall efficiency. It is the most common form of efficiency used to compare engines.

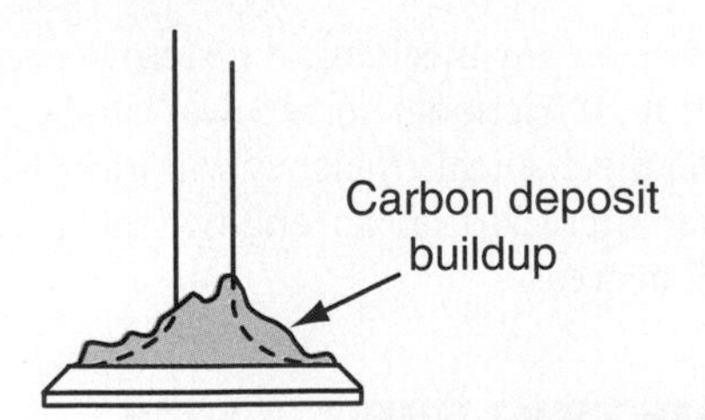

FIGURE 15–12 The volumetric efficiency of an engine can be reduced by restricting the airflow in and out of the engine. Carbon deposits can build up on the valve causing a decrease in volumetric efficiency.

Thermal efficiency is found by using the following formula:

$$\text{thermal efficiency} = \frac{\text{actual output}}{\text{heat input}}$$

When using this formula, always make sure the units of input and output are the same. The heat input is expressed in British thermal units (Btu). A Btu is an amount of heat needed to raise 1 pound of water 1 degree Fahrenheit. A gallon of gasoline has approximately 110,000 Btu.

To get the actual output in the same unit, note that 1 horsepower is equal to 42.5 Btu/min. Therefore, the formula can be shown as:

$$\text{thermal efficiency} = \frac{\text{bhp} \times 42.5 \text{ Btu/min}}{110{,}000 \text{ Btu/gal} \times \text{gal per min (gpm)}}$$

Relating this efficiency to a gasoline engine, approximately 30% of the input energy is available at the output. Referring to ***Figure 15–13***, the remaining part of the input energy is lost through various ways. The cooling system absorbs a certain percentage of the input energy. The exhaust system carries away a certain amount of energy. Eight percent of the input energy is lost through radiation. When all of these losses are added together, the output energy drops to about 30%.

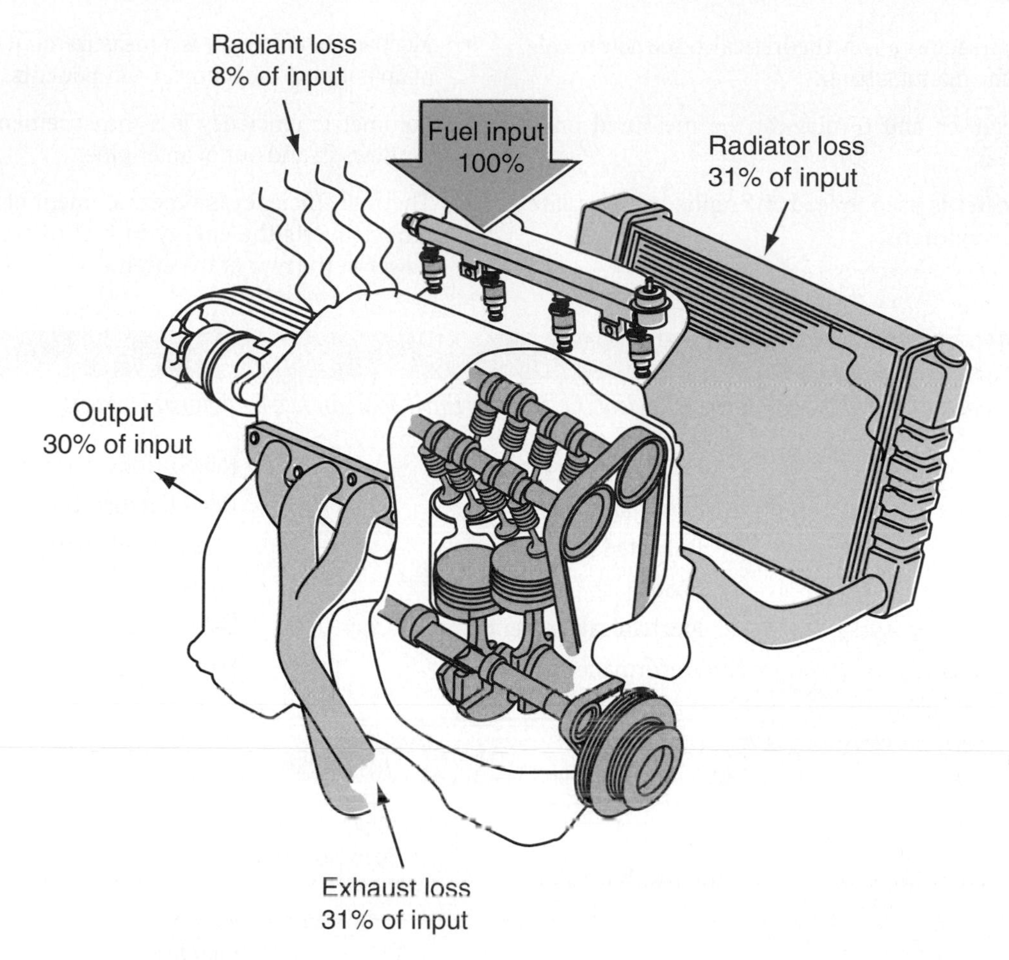

FIGURE 15–13 A gasoline engine loses much of its energy from combustion to other areas. About 31% is lost through the radiator. Another 31% is lost through the exhaust, while about 8% is lost through radiation. This leaves about 30% of the energy in the fuel for power.

Different machines have different efficiencies. ***Figure 15–14*** shows some of the more common machines and their thermal efficiencies. Diesel engines are about 10% more efficient than gasoline engines.

Many things affect efficiency. Some of these factors include the amount of energy in the fuel, the quality of combustion, the amount of frictional loss, and the mechanical quality of the machine. It is important to improve the efficiency of the automobile so that fuel mileage is maximized.

EFFICIENCIES OF DIFFERENT ENGINES	
Gasoline engine	28–32%
Diesel engine	35–40%
Aircraft gas turbine	33–35%
Liquid fuel rocket	46–47%
Rotary engine	22–25%
Steam locomotive	10–12%

FIGURE 15–14 Each of the engines shown has a different overall efficiency.

SUMMARY

The following statements will help to summarize this chapter:

- Engine performance terms include horsepower, torque, dynamometer, and engine efficiency.
- Work is defined as the result of applying a force. It is part of the definition of horsepower.
- Torque is defined as twisting force.
- When torque and rpm are known, the horsepower of an engine can be calculated.
- Horsepower is a measure of the amount of work being done.
- Brake horsepower is the horsepower at the rear of the engine.
- Road horsepower is the horsepower available at the drive wheels.
- Frictional horsepower is the horsepower that is consumed by friction inside the engine as well as in the transmission, differential, and so on.

- Indicated horsepower is a theoretical horsepower calculated by the manufacturer.
- Both horsepower and torque can be measured on a dynamometer.
- A dynamometer is used to load an engine and simulate actual road conditions.
- Mechanical efficiency is a measurement of the efficiency of an engine's mechanical components.
- Volumetric efficiency is a measurement of how easily air flows in and out of an engine.
- Thermal efficiency is a measurement of how well an engine converts the energy in fuel to actual mechanical power at the rear of the engine.

TERMS TO KNOW

Can you explain each of the following terms? Review the chapter until you can use each term correctly.

Brake horsepower
Chassis dynamometer
Coefficient of drag
Dynamometer
Efficiency
Engine dynamometer
Frictional horsepower
Horsepower
Indicated horsepower
Load
Mechanical efficiency
Performance chart
Revolutions per minute (rpm)
Road horsepower
Thermal efficiency
Torque
Volumetric efficiency
Work

REVIEW QUESTIONS

Multiple Choice

1. When force is multiplied by distance the result is called ________.
a. Work
b. rpm
c. Torque
d. Distance
e. Pressure

2. Which of the following work units is measured in a straight line?
a. Torque
b. Horsepower
c. rpm
d. All of the above
e. None of the above

3. Which of the following is true about horsepower and torque?
a. Torque is a measure of rotation; horsepower is measured in a straight line.
b. Horsepower is measured over a period of time.
c. Torque is measured in foot-pounds.
d. All of the above
e. None of the above

4. When 550 pounds are lifted in 1 ________ a distance of 1 foot, 1 horsepower is created.
a. Second
b. Minute
c. Hour
d. Day
e. Month

5. Horsepower measured at the rear of the engine is called:
a. Road horsepower
b. Brake horsepower
c. Indicated horsepower
d. Rear engine horsepower
e. Frictional horsepower

6. Horsepower measured at the drive wheels of a vehicle is called:
a. Road horsepower
b. Brake horsepower
c. Indicated horsepower
d. Frictional horsepower
e. Theoretical horsepower

7. Theoretical horsepower is referred to as:
a. Road horsepower
b. Frictional horsepower
c. Indicated horsepower
d. Brake horsepower
e. Torque

8. Horsepower lost because of friction in the engine is called ________ horsepower.
a. Frictional
b. Chassis
c. Indicated
d. Heat
e. All of the above

9. The greater the frictional horsepower, the
a. Better the gasoline mileage
b. Better the efficiency
c. Lower the gasoline mileage
d. More power available
e. Better the performance of the vehicle

10. Which chart is *not* shown on a performance curve?
a. Brake horsepower
b. Torque
c. Fuel pressure
d. Fuel consumption
e. Lb/bhp/h

11. How is a load applied to a vehicle to measure road horsepower?
a. By a chassis dynamometer
b. By an engine dynamometer
c. By an rpm gauge
d. By a pressure gauge
e. Road horsepower cannot be measured

12. Fuel consumption is measured by what unit?
a. Lb/ihp/h
b. Lb/bhp/h
c. Lb/fhp/h
d. Gallons per day
e. Quarts per hour

13. Which efficiency measures the airflow in an engine?
a. Volumetric
b. Thermal
c. Mechanical
d. Airflow
e. Electrical

14. Which efficiency measures the ratio of actual horsepower to theoretical horsepower?
a. Thermal
b. Mechanical
c. Volumetric
d. Indicated
e. Electrical

15. Which of the following could have a negative effect on volumetric efficiency?
a. An exhaust restriction
b. A dirty air cleaner
c. Heavy carbon deposits on the valve
d. All of the above
e. None of the above

The following questions are similar in format to ASE (Automotive Service Excellence) test questions.

16. *Technician A* says that as frictional horsepower increases, fuel consumption decreases. *Technician B* says as frictional horsepower increases, engine efficiency decreases. Who is correct?
a. A only c. Both A and B
b. B only d. Neither A nor B

17. *Technician A* says that if a vehicle has a dirty air cleaner, the volumetric efficiency will decrease. *Technician B* says that if a vehicle has a dirty air cleaner, the mechanical efficiency decreases. Who is correct?
a. A only c. Both A and B
b. B only d. Neither A nor B

18. *Technician A* says that torque and horsepower are different terms and have different meanings. *Technician B* says horsepower is measured per degrees rotation. Who is correct?
a. A only c. Both A and B
b. B only d. Neither A nor B

19. The actual horsepower is 80, and the theoretical horsepower is 100. *Technician A* says the mechanical efficiency of the engine is 80%. *Technician B* says the mechanical efficiency of the engine is 100%. Who is correct?
a. A only c. Both A and B
b. B only d. Neither A nor B

20. *Technician A* says the BSFC is a measure of piston size. *Technician B* says BSFC means brake specific flow condition. Who is correct?
a. A only c. Both A and B
b. B only d. Neither A nor B

21. *Technician A* says that 1 horsepower is equal to the amount of work necessary to raise 33,000 pounds 1 foot in 1 minute. *Technician B* says that 1 horsepower is equal to the amount of work necessary to raise 550 pounds 1 foot in 1 second. Who is correct?
a. A only c. Both A and B
b. B only d. Neither A nor B

22. *Technician A* says that as more horsepower is created, torque automatically drops. *Technician B* says that as more torque is created, horsepower automatically drops. Who is correct?
a. A only c. Both A and B
b. B only d. Neither A nor B

23. *Technician A* says that volumetric efficiency can be calculated by multiplying the actual horsepower by the actual air used. *Technician B* says that mechanical efficiency can be calculated by dividing the actual horsepower by the theoretical horsepower and then multiplying the result by 100. Who is correct?
a. A only c. Both A and B
b. B only d. Neither A nor B

Essay

24. Identify several ways in which frictional horsepower can be reduced.

25. What is the formula for calculating horsepower when both torque and rpm are known?

26. What is the difference between mechanical and volumetric efficiency?

27. Define thermal efficiency.

28. Define frictional horsepower.

Short Answer

29. If a vehicle is pushed by 50 pounds of force a total distance of 30 feet, the total foot-pounds of work done equals ________.

30. A measure of how much air is moved as the vehicle goes from one point to another is called the ________ of ________.

31. A dynamometer is producing 80 foot-pounds of torque at 3,600 rpm. The horsepower being produced under these conditions is ________.

32. When an engine is producing a BSFC of 24 lb/bhp/h at 42 hp, the engine is using ________ pounds of fuel each hour.

33. In a dynamometer test, the three readings taken include the ________, ________, and ________.

Applied Academics

TITLE: Performance Curves

ACADEMIC SKILLS AREA: Science

NATEF Automobile Technician Tasks:

The technician uses information from charts, tables, and graphs in service manuals to determine manufacturers' specifications for engine performance, systems operation, and appropriate repair/replacement parts and procedures.

CONCEPT:

Engine and vehicle performance can be tested while the vehicle is in the service area. A dynamometer is used to load the vehicle. The front drive wheels are placed on two rollers that are built into the floor. A hydraulic turbine is attached to the rollers under the floor. Increasing the amount of hydraulic fluid that the turbine must rotate through slows down the rollers. This causes a load to be applied to the drive wheels. Performance testing can provide the technician with important information when troubleshooting an engine. Dynamometers are used to produce performance curves such as the one illustrated here. In this case, after a dynamometer test was done, the actual horsepower and torque did not match up to the manufacturer's specifications.

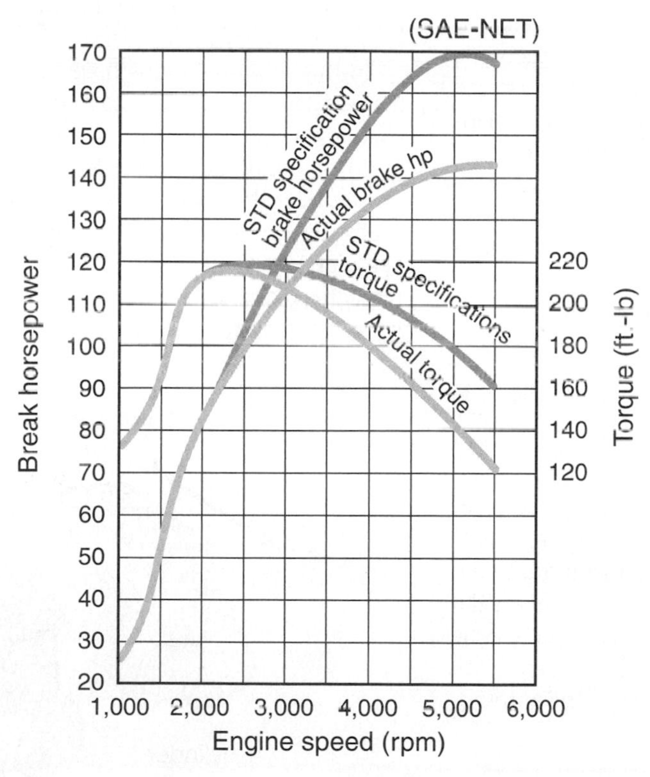

APPLICATION:

After studying the performance curve, determine what percentage of drop there was in engine horsepower and torque as compared to standard specification at 5,500 rpm. Thinking about volumetric efficiency, why might the horsepower and torque start to drop after 2,000 rpm? What would be the best rpm to operate this engine to get maximum torque? What would be the best rpm to operate this engine to get maximum horsepower?

CHAPTER 16 Engine Types

OBJECTIVES

After studying this chapter, you should be able to:

- Identify the difference between in-line, V, slant, and opposed cylinder arrangements.
- Define the variable displacement engine.
- Define the differences between the 4-, 6-, 8-, and 12-cylinder engines.
- Compare the differences between the L- and I-head designs.
- Analyze the difference between overhead, in-block, and dual camshaft design engines.
- Compare different valve arrangements for engines.
- Examine the operation of the stratified charged engine.

INTRODUCTION

Automobile engines can be designed in different styles, types, and configurations. Some engines may have an overhead cam, while others may have valves in the block. The principles of the engine design remain the same, but the location and configuration change. Some of the shapes the engine can take include in-line or V, slant or opposed, number of cylinders, types of head design, location of valves, and shape of the block. The purpose of this chapter is to investigate different types and styles of engines used in the automobile.

16.1 CYLINDERS AND ARRANGEMENT

Automobile engines use many styles, types, and configurations. **Configuration** means the figure, shape, or form of the engine. One style used is identified by its shape and number of cylinders. Depending on the vehicle, either an in-line, V, slant, or opposed cylinder arrangement can be used. In addition, engines come with a variety of cylinder numbers depending on the vehicle and the horsepower and torque requirements. The service technician will be exposed to 3-, 4-, 5-, 6-, 8-, 10-, and 12-cylinder engines in the automotive shop.

IN-LINE ENGINES

Engines can be designed with an **in-line** style. This means the cylinders are all placed in a single row (***Figure 16–1***). There is one crankshaft and one cylinder head for all of the cylinders. The block is a single cast piece with all cylinders located in an upright position.

In-line engine designs have certain advantages and disadvantages. They are easy to manufacture, which brings the cost down somewhat. They are very easy to work on and to perform maintenance on. In-line engines have adequate room under the hood, allowing the service technician more room to work on other vehicle parts. However, because the cylinders are positioned vertically, the front of the vehicle must be higher. This need affected the **aerodynamic** design of many older vehicles. The front of the vehicle could

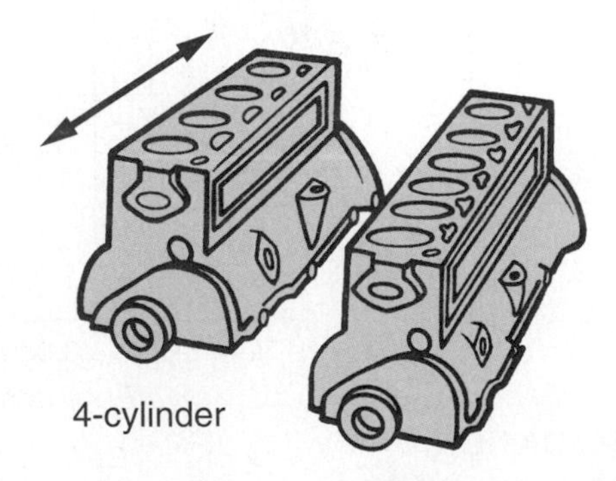

FIGURE 16–1 Engines can be designed using an in-line configuration. This means the cylinders are all in a line and in a vertical position.

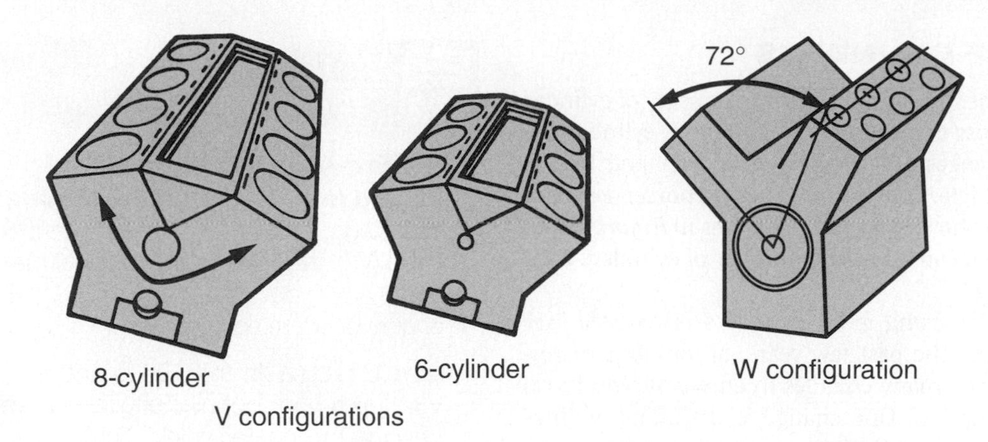

FIGURE 16–2 Engines can be designed in a V configuration. This means the cylinders are positioned in two rows in a V shape. Similar to a V configuration, the W configuration has double the pistons on each bank.

not be made lower as on other, newer vehicles. This meant that the aerodynamic design of the car could not be improved easily. Of course, this is not true today.

V CONFIGURATION ENGINES

The **V configuration** has two rows of cylinders (***Figure 16–2***). These cylinder rows are approximately 90 degrees from each other. This is the angle in most V configurations. There are other V configuration angles used on some engines. The most common angles are from 60 to 90 degrees. However, certain engine manufacturers may have used other angles as well.

This design utilizes one crankshaft that operates the cylinders on both sides. Two connecting rods are attached to each connecting rod journal on the crankshaft. However, there must be two cylinder heads for this type of engine.

One advantage of using a V configuration is that the engine is not as vertically high as with the in-line configuration. The front of a vehicle can now be made lower. This design improves the outside aerodynamics of the vehicle.

If eight cylinders are needed for power, a V configuration makes the engine much shorter and more compact. Manufacturers used to make in-line 8-cylinder engines. This made the engine rather long; it was hard to design the vehicle around this long engine. The long crankshaft also caused more torsional vibrations in the engine.

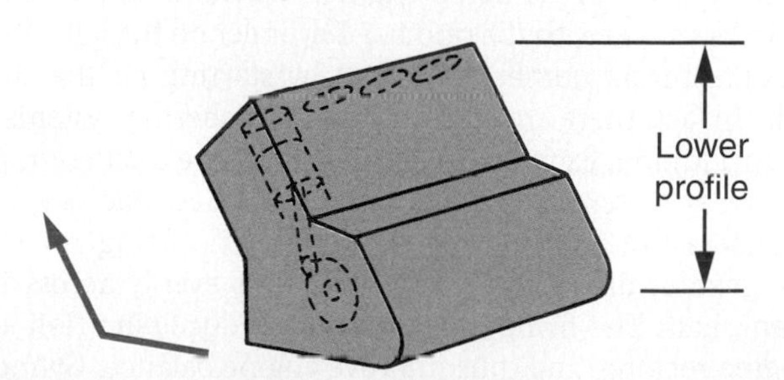

FIGURE 16–3 A slant cylinder configuration reduces the distance from the top to the bottom of the engine.

Some manufacturers have also designed a special V configuration called the W engine. This is also shown in ***Figure 16–2***. It looks like a V configuration, but there are twice as many pistons on each bank as compared to a V engine. This engine configuration is very compact, and has more power for a small-sized engine. The W configuration is used for heavy-duty applications in which a 10- or 12-cylinder power source is needed but compact size is also a requirement.

SLANT CYLINDER ENGINES

Another way of arranging the cylinders is in the **slant** configuration. This is shown in ***Figure 16–3***. It is much like an in-line engine except the entire block has been placed at a slant. The slant engine was designed to reduce the distance from the top to the bottom of the engine. Vehicles using the slant engine can be designed more aerodynamically.

OPPOSED CYLINDER ENGINES

Several manufacturers have designed **opposed cylinder** engines. An example is shown in ***Figure 16–4***. Opposed cylinder engines are used for applications in which there is very little vertical room for the engine. For this reason, opposed cylinder designs are commonly used on vehicles that have the engine in the rear. The angle between the two cylinders is typically 180 degrees. One crankshaft is used with two cylinder heads. There are two connecting rods attached to each journal on the crankshaft. Several car manufacturers, both foreign and American, have used this type of engine, mostly in smaller vehicles.

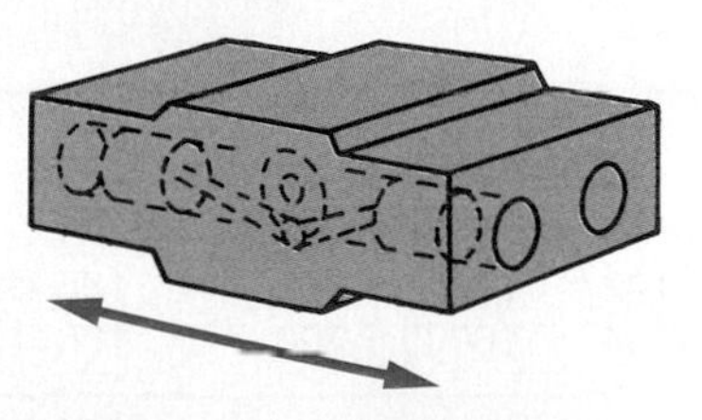

FIGURE 16–4 Opposed cylinder engines are used when there is very little vertical room for the engine.

NUMBER OF CYLINDERS

Automotive engines are designed using a variety of cylinder numbers. The most common are 4-, 6-, and 8-cylinder engines. The differences are in the horsepower and torque needed for the vehicle. For example, average horsepower and torque figures are shown for several engines in ***Figure 16–5***. The differences are caused by the number of cylinders used on the engine.

An advantage to using fewer cylinders is reduced **fuel consumption**. Over the past few years, automotive manufacturers have made many changes in engine design to reduce fuel consumption. One change is designing engines that have fewer cylinders. With current concerns for a clean environment and improved gasoline mileage, it is difficult to justify designing a vehicle with eight or more cylinders. Larger cylinder engines are especially hard to justify if they are only used for short distances and intercity travel. However, they are easier to justify if the larger engines are used for hauling trailers, construction work, or other heavy-duty applications. The horsepower and torque developed by larger engines are usually well above that needed for most driving applications. Applications that require heavy hauling, such as trailers and boats, may require the additional horsepower and torque. In fact, DaimlerChrysler has developed a V10 engine. This engine is used in larger pickup trucks for hauling heavy loads, such as trailers, motor homes, campers, and so on.

Some 5-cylinder and 3-cylinder engines have also been built for automobiles by both American and foreign manufacturers. Although these engines are not commonly found in the automotive field, they are used.

Combining the number and type of cylinders, manufacturers design the following common types of engines:

- In-line (4- and 6-cylinder engines, most common, as well as 3- and 5-cylinder engines)
- V configuration (6-, 8-, and 10-cylinder engines)
- Slant (4- and 6-cylinder engines)
- Opposed (4- and 6-cylinder engines)

There are also other less common engine configurations, which include:

- In-line (6- and 8-cylinder engines)
- V configuration (10- and 12-cylinder engines)

SPECIFICATIONS
GENERAL ENGINE SPECIFICATIONS

Liter/CID	Net hp @ rpm	Maximum Torque ft.-lbs @ rpm
2.3L/4-138	120 @ 5,200	140 @ 3,200
2.4L/4-146	150 @ 6,000	155 @ 4,400
3.1L/V6-192	150 @ 5,200	185 @ 4,000
3.3L/V6-204	160 @ 5,200	185 @ 2,000

FIGURE 16–5 The difference in engines of different cylinder numbers is the horsepower and the torque specifications for each engine.

Car Clinic: CYLINDER NUMBERS

CUSTOMER CONCERN:
V8 and In-Line 4-Cylinder Engines
Years ago, most manufacturers built 8-cylinder engines (V8). Why is the V8 engine not manufactured as much any more, and why has the in-line 4-cylinder engine become so popular?

SOLUTION: In the 1960s and early 1970s, the V8 was a very popular engine. However, it is not designed to be very fuel-efficient. During and after the energy crisis in 1973, the cost of gasoline increased rapidly, and the amount of gasoline available seemed to decline. We could no longer afford to drive cars that did not get good fuel mileage. So manufacturers started building smaller vehicles and smaller engines. Today, engines are very fuel-efficient, especially the in-line 4-cylinder engines. Cars with smaller engines cost less and make better use of the world's limited oil reserves.

VARIABLE DISPLACEMENT ENGINES

There have been several attempts to build engines that have a **variable displacement**. One method that has been tried is to start with an 8-cylinder engine then deactivate cylinders to reduce the displacement of the engine. Thus, if high power is needed, all cylinders (maximum displacement) would be in operation. If lighter loads are required, then several cylinders could be deactivated, thus running the engine on the remaining cylinders (minimum displacement). Several manufacturers have produced variable displacement engines. Additional research and testing are required to improve efficiency and please the customer with this approach.

CYLINDER NUMBERING AND FIRING ORDER

Engines can also be identified by the cylinder numbering and firing order. On all engines, the cylinders are numbered a certain way. On some V8 engines, the numbering on the cylinders starts with the number 1 cylinder on the left. Other manufacturers number cylinders by starting on the right side. In fact, there are many cylinder numbering systems.

In addition, each manufacturer may have a different firing order system. The firing order identifies exactly which cylinders fire and in what sequence. The firing order is designed to distribute the power pulses evenly across the crankshaft. The firing order helps to reduce vibration and engine rocking, and thus improve engine balance. Cylinder numbering and the firing order are always shown in service manuals. ***Figure 16–6*** shows some of the common cylinder

COMMON CYLINDER NUMBERING AND FIRING ORDER

IN-LINE

4-Cylinder		6-Cylinder	
①②③④		①②③④⑤⑥	
Firing Order	1-3-4-2 1-2-4-3	Firing Order	1-5-3-6-2-4

V CONFIGURATION

V6		V8	
⑤③①	Right Bank	①②③④	Right Bank
⑥④②	Left Bank	⑤⑥⑦⑧	Left Bank
Firing Order	1-4-5-2-3-6	Firing Order	1-5-4-8-6-3-7-2
②④⑥	Right Bank	①②③④	Right Bank
①③⑤	Left Bank	⑤⑥⑦⑧	Left Bank
Firing Order	1-6-5-4-3-2	Firing Order	1-5-4-2-6-3-7-8
①②③	Right Bank	②④⑥⑧	Right Bank
④⑤⑥	Left Bank	①③⑤⑦	Left Bank
Firing Order	1-2-3-4-5-6	Firing Order	1-8-4-3-6-5-7-2
①②③	Right Bank	②④⑥⑧	Right Bank
④⑤⑥	Left Bank	①③⑤⑦	Left Bank
Firing Order	1-4-2-3-5-6	Firing Order	1-8-7-2-6-5-4-3

FIGURE 16-6 Common cylinder arrangements and firing orders are shown in this chart.

arrangements and firing orders used on 4- and 6-cylinder in-line engines and V6 and V8 engines. Although there are others, the principle and purpose of having a firing order remains the same.

ENGINE IDENTIFICATION NUMBERS

As mentioned in an earlier chapter, part of the VIN number is used to identify exactly which engine is being used in a vehicle. However, an engine serial number and code number are often stamped on the engine block. The location of these numbers may vary, however; ***Figure 16–7*** shows some of the more common locations. The service manual will tell the service technician exactly where the serial and code number is located. The numbers are stamped in a machined location on the block.

Once the numbers are located, the service technician can use them to help identify the block being used in a particular vehicle. Note that most engine blocks also have a raised number on the block. This number is not the engine serial or code numbers. It is only a casting number and should not be mistaken for the engine serial and code numbers.

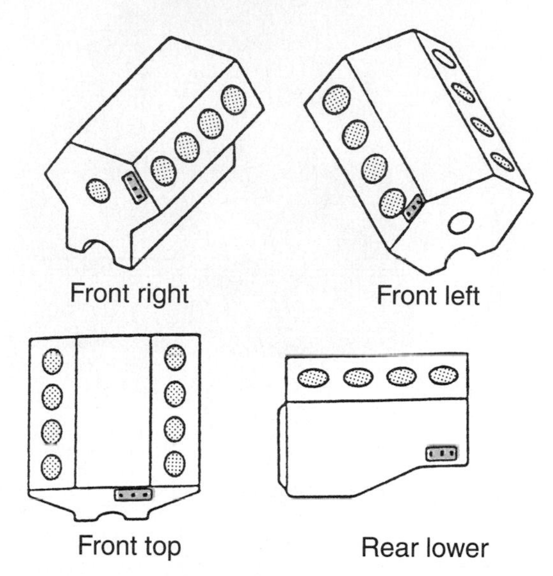

FIGURE 16-7 Cylinder block identification numbers can be located in various places on the cylinder block.

16.2 VALVE AND HEAD ARRANGEMENT

Engines used in the automobile can be designed with different valve and head designs. I- and L-head designs are used. Also, in-block, overhead, and dual camshafts are used. In addition, stratified charged engines are also used.

I-HEAD DESIGN

The **I-head** valve design is the most common arrangement of valves. Referring to ***Figure 16–8***, I-head means the valves are directly above the piston (overhead valves). The valves are located in the cylinder head. This design allows easy breathing of the engine. Air and fuel can move into and

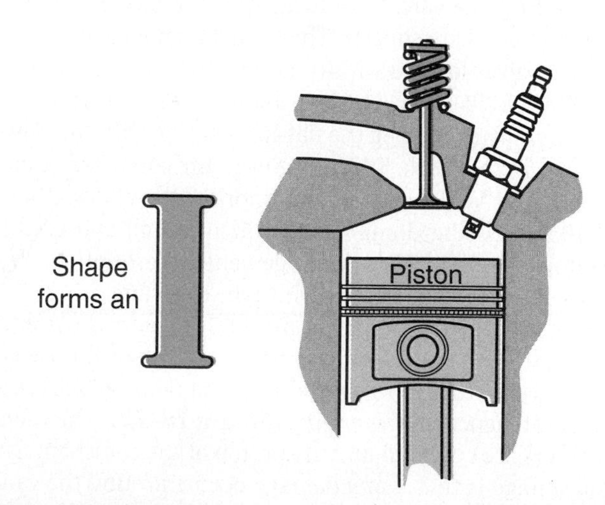

FIGURE 16-8 An I-head engine has the valves located above the piston in the head.

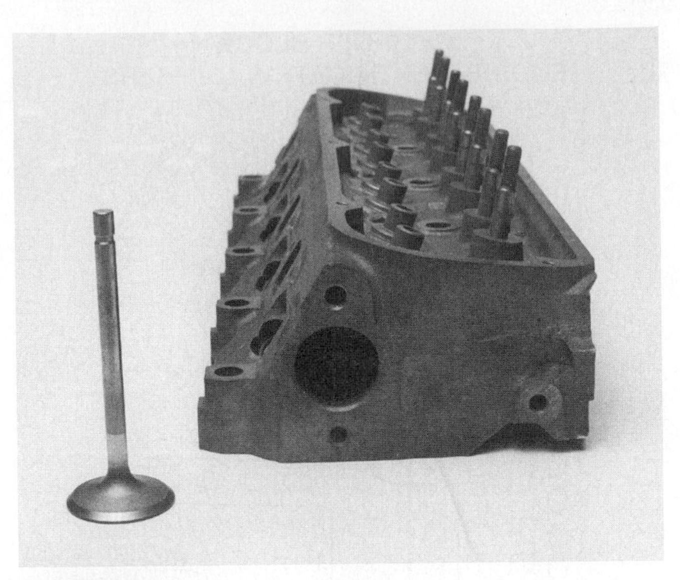

FIGURE 16–9 On an I-head design, the valves are located in the head, vertically.

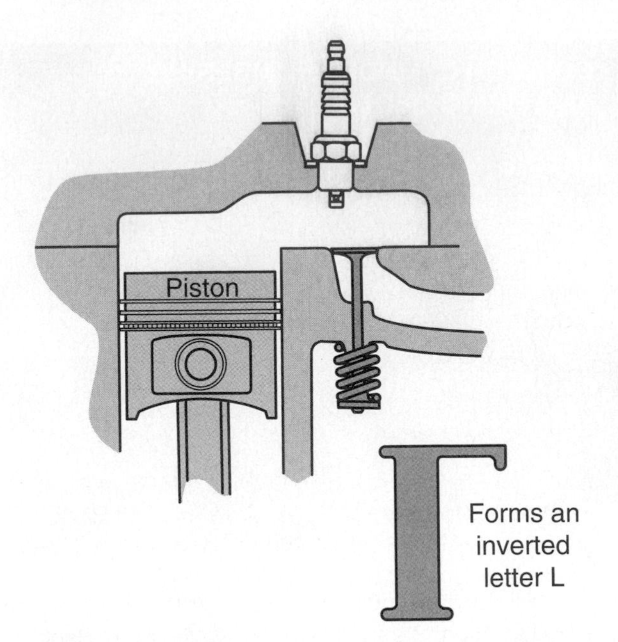

FIGURE 16–10 L-head engines have the valves located within the block. This engine is commonly called a "flat-head" engine.

out of the cylinder with little restriction. This process improves the volumetric efficiency of the engine. The I-head is also easy to maintain. For example, if a valve is damaged, it can be replaced easily. Adjusting the valves is also easier.

The valves can be on both sides of the piston, on top, or on one side only. ***Figure 16–9*** shows an example of a cylinder head with an I design. This type of valve arrangement is also called overhead valves, sometimes referred to as OHV. A rather complex mechanical system must be used to open the valves from the camshaft. It includes the **lifters**, **pushrods**, **rocker arms**, and valves. All of these parts cause certain known maintenance problems with such a system. These parts wear regularly and so the maintenance of these systems increases.

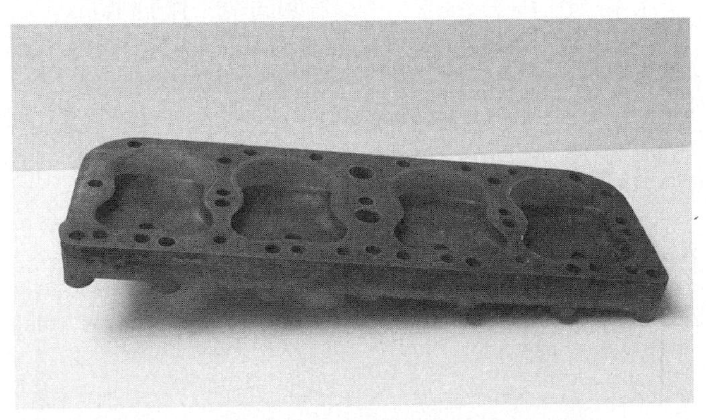

FIGURE 16–11 The L-head engine uses a flat head.

L-HEAD DESIGN

Another type of valve arrangement is called the **L-head** (***Figure 16–10***). The valves are located in the block. The inlet and outlet ports are shorter. The head does not have any mechanical valves located within its structure. This head is referred to as a "flat head." Older vehicles, especially from Ford Motor Company, utilized the flat-head design with the valves built within the block. These engines were commonly called the flat-head V8. They were common in the 1930s, 1940s, and 1950s. Flat-head engines are often rebuilt and used in customized street rods or vintage vehicle restoration. ***Figure 16–11*** shows the head on this type of engine.

In the L-head design, fewer mechanical parts are needed to operate the valves from the camshaft. Here the rocker arms and pushrods have been eliminated. This reduces wear and maintenance on the engine (***Figure 16–12***). The valves and lifters are operated directly on top of the camshaft. One disadvantage is that if any damage occurs around the valve port, the block may have to be replaced. It is also more difficult to adjust the valves when they are located in the block.

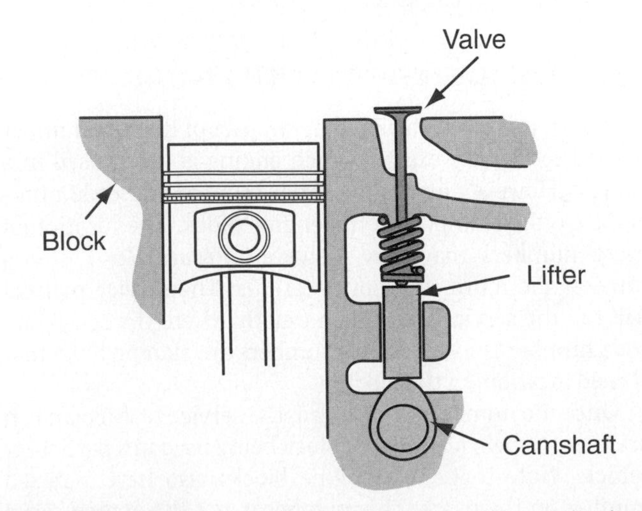

FIGURE 16–12 The L-head engine does not use pushrods or rocker arms. The valves are located in the block and ride directly above the camshaft.

FIGURE 16–13 Many engines have the camshaft located directly within the block.

IN-BLOCK CAMSHAFT

Some engines have the camshaft located directly within the block. This type of camshaft can be used on L- and I-head engines. ***Figure 16–13*** shows where the camshaft is located in this design. An advantage is that the camshaft can be driven directly off the crankshaft. A standard gear set or a gear and chain arrangement can be used (***Figure 16–14***).

A disadvantage of in-block camshafts is the linkage needed to open and close the valves. This includes the lifters, pushrods, and rocker arms. All of these parts can become worn, causing more chance of failure of the parts. Another disadvantage of this system is that the maximum rpm range of the engine is limited by the valve mechanism. The valves have a tendency to float (not close completely) at higher speeds.

This system must also have some way of accounting for the clearance between all of the valve system's moving parts. There is an adjustment called **valve clearance**. As the engine and parts heat up, the parts expand. If there were no valve clearances, the heated parts would expand and keep the valves open during the compression and power strokes. If this happened, the engine would not run correctly. The valve clearance can be accounted for by using hydraulic lifters or by adjusting the valve clearance. Valve clearance will be discussed in a later chapter.

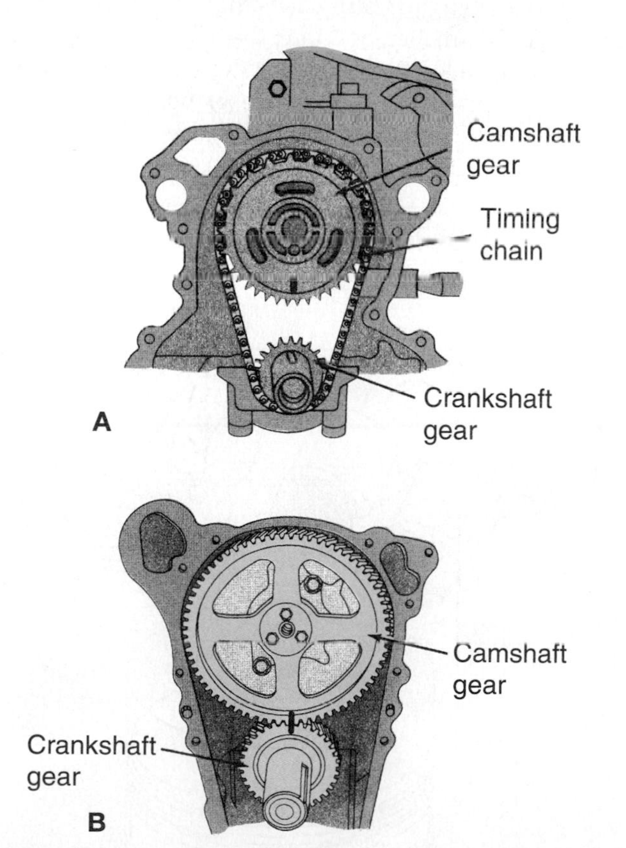

FIGURE 16–14 The camshaft is driven directly from the crankshaft when located within the block. Either gears or a chain drive can be used.

OVERHEAD CAMSHAFT

Nearly all manufacturers now use engines that have an **overhead camshaft**. The camshaft is placed directly above the valves in the head. This design is used on I-head valve designs (***Figure 16–15***). One advantage of this design is that the cam operates directly on the valves. There are fewer parts that can wear and cause failure. Also, having fewer valve train parts gives the engine a higher rpm range. In addition, there is less valve clearance needed for expansion as the engine heats up. There is also a more positive opening and closing of the valves. The valve movement, therefore, responds more quickly to the cam shape. This may not be true for in-block camshafts because of the extra valve components. The camshaft is now driven by gears, a chain, or a belt as shown in ***Figure 16–16***.

PARTS LOCATOR

Refer to photo #3 in the front of the textbook to see what a ***timing belt*** *looks like.*

FIGURE 16–15 Most engine manufacturers now use an overhead camshaft (OHC) with the camshaft mounted on top of the cylinder head.

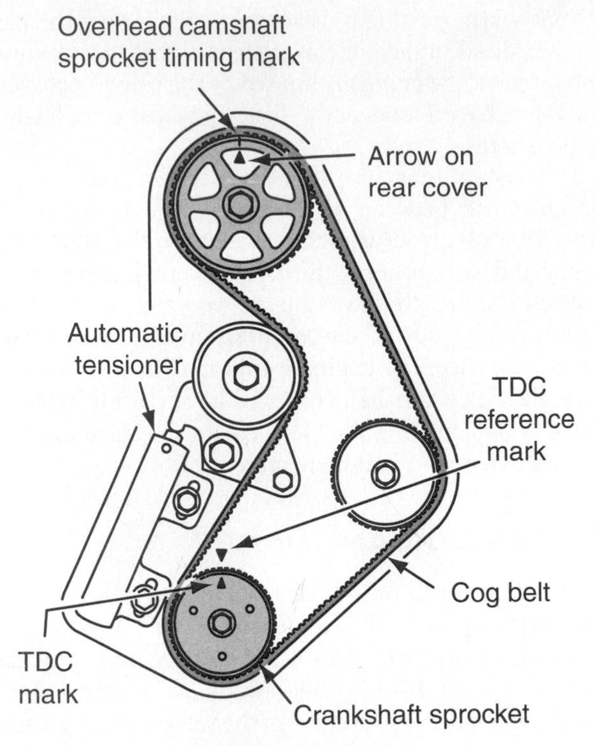

FIGURE 16–16 Cog belts can turn overhead camshafts. The cogs on the belt keep the cam and crankshaft in time with each other.

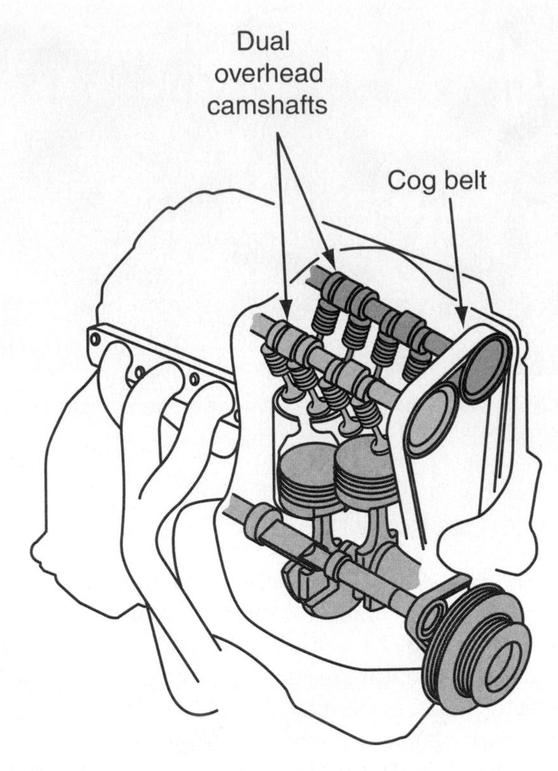

FIGURE 16–17 Many engines utilize twin, or dual, overhead camshafts.

DUAL CAMSHAFTS

Some engines use more than one camshaft. This type of engine is called a **dual camshaft** engine. Dual camshaft engines are designed so that one camshaft operates the intake valves, while the second camshaft operates the exhaust valves. Many foreign and domestic automotive companies manufacture engines using the dual overhead camshaft (DOHC) design. ***Figure 16–17*** shows an example of an engine with dual overhead camshafts.

Today, engine manufacturers are also looking for ways to improve combustion efficiency. To do this, some engine manufacturers have designed V-type engines that use two overhead camshafts on each side of the engine.

FOUR-VALVE ENGINES

Many engines used today have more than two valves per cylinder. When **turbochargers** (also known as **superchargers**) were first introduced into the automotive engine design, 4-valve engines became popular. Supercharging an engine means to force more air and fuel into the cylinders for more power. Supercharging will be discussed in a later chapter. **Four-valve heads** allow more air and fuel to get into the cylinder, thus increasing engine power. Because there are two intake valves and two exhaust valves within each cylinder, more intake air and fuel, as well as exhaust gases, can be moved through the engine. This allows 4-valve engines to operate with higher volumetric efficiency.

Four-valve engines also have other characteristics. The valves can be smaller, and thus the engine can operate at higher rpm ranges. Because of the success of the 4-valve engine, most engines today, whether or not they are supercharged, use the 4-valve design. ***Figure 16–18*** shows a typical 4-valve cylinder head.

4-VALVE HEAD
Exhaust
Intake

FIGURE 16–18 Four-valve heads are used to allow additional air and fuel to enter the engine.

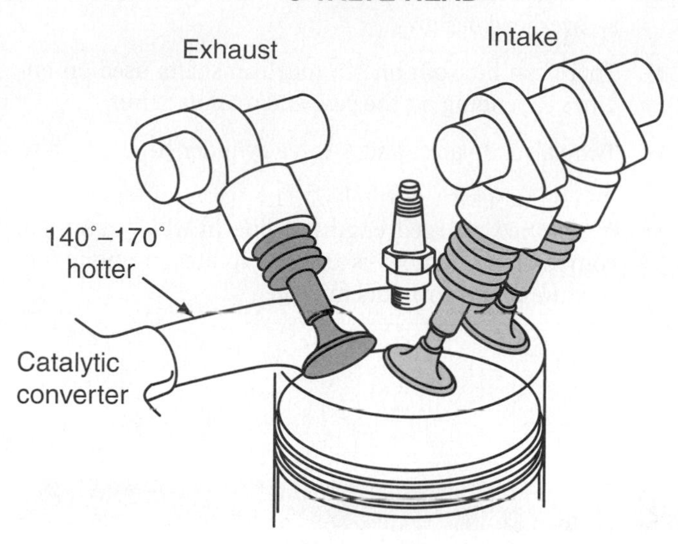

FIGURE 16–19 Three-valve engines use two intake valves and one exhaust valve.

THREE-VALVE ENGINES

Automotive engineers are continually trying to improve the efficiency of combustion on engines while still lowering the emissions. One way to increase efficiency and still reduce emissions is to use a 3-valve engine. ***Figure 16–19*** shows an example of a 3-valve engine. In this case, each cylinder has two intake valves and only one exhaust valve. Tests have shown that a catalytic converter needs very high exhaust temperatures to operate correctly. Generally, there is no problem getting high exhaust temperatures when the engine is at its operating temperature. However, during warmup, there is a period of time in which the exhaust gases are still too cool to allow the catalytic converter to work correctly. Thus, there is catalytic converter delay in removing emissions. Researchers found that this delay was caused by the use of two exhaust valves. When two exhaust valves are used, there is more exhaust heat lost between the engine and the catalytic converter. If only one exhaust valve is used, the exhaust temperature losses between the engine and the catalytic converter are reduced. In fact, the exhaust temperatures when a single exhaust valve is used are 140–170 degrees Fahrenheit hotter. This causes the catalytic converter to increase in temperature sooner, thus reducing emissions during starting and warmup times.

A second reason for using three valves per cylinder is that more room is made available for positioning the spark plug and injectors in the design of the cylinder head. By placing the fuel injectors and spark plugs in the best position, increased combustion efficiency can be realized.

STRATIFIED CHARGED ENGINES

Certain engines used in the automobile are called stratified charged engines. When the mixture within the combustion chamber is thoroughly mixed, the charge is called homogeneous. When the mixture is not evenly mixed, the charge is said to be **stratified**, or in layers. A **stratified charged engine** has a second, small area for combustion in the cylinder head (***Figure 16–20***). There is a spark plug in the chamber. A rich mixture of air and fuel enters the stratified chamber. A lean mixture of air and fuel enters the major combustion chamber. The overall air-fuel mixture is leaner. This is done primarily to reduce pollution from the engine. When the spark plug ignites the rich air-fuel mixture, this burning mixture is used to ignite the lean mixture.

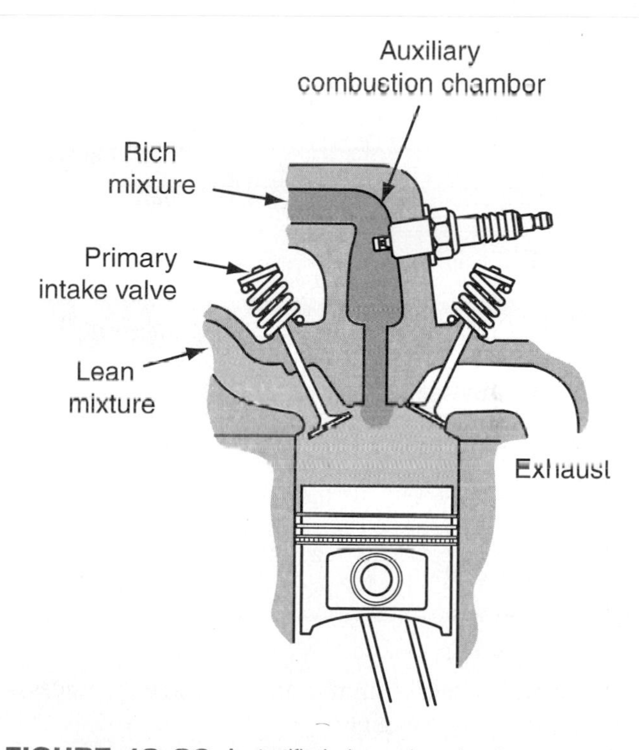

FIGURE 16–20 A stratified charged engine has an auxiliary combustion chamber used to ignite the air-fuel mixture. This arrangement allows a leaner mixture to be burned in the regular combustion chamber.

SUMMARY

The following statements will help to summarize this chapter:

- Today's automobiles use various engine and valve arrangements.
- Engines can be designed in an in-line or V configuration.
- Common in-line configurations are the 4- and 6-cylinder engines.
- Common V-type configurations include the V6 and V8 engines.

- Engine cylinders can be arranged in a slant configuration or an opposed cylinder arrangement.
- Some manufacturers also design engines to operate certain cylinders depending on the load needed. This is called a variable displacement engine.
- Valves and cylinder heads can also be designed differently.
- I-head engines have the valves located in the cylinder head.
- L-head engines have the valves located to the side of the pistons in the block.
- Camshafts can be located either below the valves or in an overhead design.
- There can be from one to four camshafts used on engines depending on the year and manufacturer.
- Two-valve, 3-valve, and 4-valve cylinder heads are also used.
- A stratified charged engine is one in which a special combustion chamber is used to ignite a rich-air fuel mixture to improve fuel mileage.

TERMS TO KNOW

Can you explain each of the following terms? Review the chapter until you can use each term correctly.

Aerodynamic
Configuration
Dual camshaft
Four-valve heads
Fuel consumption
I-head
In-line
L-head
Lifters
Opposed cylinder
Overhead camshaft
Pushrods
Rocker arms
Slant
Stratified
Stratified charged engine
Superchargers
Turbochargers
Valve clearance
Variable displacement
V configuration

REVIEW QUESTIONS

Multiple Choice

1. The ________ engine design has the cylinders in a vertical row.
a. Valve
b. In-line
c. V configuration
d. Rotary
e. V6

2. What are the most common angles between the sides of the V on a V-type engine?
a. 60–90 degrees
b. 150–160 degrees
c. 180–190 degrees
d. 195–205 degrees
e. 350–360 degrees

3. The slant engine is used because the engine:
a. Has a lower profile
b. Has more power
c. Is easier to manufacture
d. Is lighter
e. All of the above

4. Which type of engine would be best suited for placement in the rear of the vehicle?
a. Slant
b. Opposed
c. V configuration
d. In-line
e. Slant, V configuration

5. One of the biggest differences between 4-, 6-, and 8-cylinder engines is:
a. Each has different horsepower and torque
b. Only 6-cylinder engines can be used on small vehicles
c. Only four cylinders can be used in automobiles
d. Eight cylinders are lighter and more efficient
e. Four cylinders have more power

6. Which of the following is the most common valve arrangement today?
a. L-head
b. A-head
c. I-head
d. P-head
e. V-head

7. Which type of valve arrangement uses a rocker arm, pushrods, and lifters?
  a. L-head
  b. A-head
  c. I-head
  d. V-head
  e. T-head

8. The reason for having a 4-valve engine is:
  a. So more fuel and air can enter the engine
  b. To get better lubrication
  c. To reduce the power available on an engine
  d. To heat up the engine more
  e. To cool down the engine more

9. What type of engine uses a small additional combustion chamber with a rich air-fuel mixture?
  a. L-head
  b. Stratified charged engine
  c. F-head
  d. All of the above
  e. None of the above

The following questions are similar in format to ASE (Automotive Service Excellence) test questions.

10. *Technician A* says the type of engine most commonly used in vehicles is the in-line type. *Technician B* says the type of engine most commonly used in vehicles is the opposed type. Who is correct?
  a. A only c. Both A and B
  b. B only d. Neither A nor B

11. *Technician A* says that a stratified charged engine means the engine also has a turbocharger. *Technician B* says that a stratified charged engine means the engine has a precombustion chamber for burning more efficiently. Who is correct?
  a. A only c. Both A and B
  b. B only d. Neither A nor B

12. *Technician A* says that one reason for having a valve clearance is to help to lubricate the parts. *Technician B* says one reason for a valve clearance is to improve the efficiency of the engine. Who is correct?
  a. A only c. Both A and B
  b. B only d. Neither A nor B

13. *Technician A* says that a stratified charged engine is the same as a 4-valve engine. *Technician B* says that a stratified charged engine means the spark plugs are firing sooner before TDC. Who is correct?
  a. A only c. Both A and B
  b. B only d. Neither A nor B

14. *Technician A* says that in-line engines have pistons that are horizontal. *Technician B* says that an in-line engine can have the cylinders at a slant. Who is correct?
  a. A only c. Both A and B
  b. B only d. Neither A nor B

15. *Technician A* says that a variable displacement engine deactivates cylinders to reduce the total displacement. *Technician B* says that a variable displacement engine reduces the total displacement by changing the piston size. Who is correct?
  a. A only c. Both A and B
  b. B only d. Neither A nor B

16. *Technician A* says that the flat-head engine is called the L-head design. *Technician B* says that the flat-head engine has the valves above the pistons. Who is correct?
  a. A only c. Both A and B
  b. B only d. Neither A nor B

17. *Technician A* says that a 4-valve engine is designed to improve the volumetric efficiency. *Technician B* says that a 4-valve engine is designed to produce more horsepower. Who is correct?
  a. A only c. Both A and B
  b. B only d. Neither A nor B

18. *Technician A* says that a 3-valve engine has the two intake valves and one exhaust valve. *Technician B* says that the 3-valve engine heats the catalytic converter quicker when starting a cold engine. Who is correct?
  a. A only c. Both A and B
  b. B only d. Neither A nor B

Essay

19. What is the advantage of using an overhead camshaft type engine?

20. Compare the in-line and V configuration types of engines.

21. Compare the differences between the I-head and the L-head design of valve arrangements.

22. What is the purpose of having a slant configuration engine?

23. Describe the arrangement of valves and camshaft(s) on an overhead camshaft type engine.

Short Answer

24. An engine that has variable displacement is designed so that the ________ can be activated and deactivated.

25. An engine that has an auxiliary combustion chamber to premix the air and fuel is called a ________.

26. Dual camshafts are used on engines that have ________.

27. An advantage of having an engine with fewer cylinders is that there is________.

Applied Academics

TITLE: Variable Valve Lift

ACADEMIC SKILLS AREA: Science

NATEF Automobile Technician Tasks:

The technician can explain how levers are used to increase an applied force or distance.

CONCEPT:

Many manufacturers are now designing ways in which intake and exhaust valves can have a variable opening and closing amount depending on the throttle setting. This is called variable valve lift. Instead of using the throttle plate, the valves are used to control the amount of air going into the engine. As a throttle closes, it makes incoming air snake around it, thus wasting energy. Eliminating the throttle makes the engine more efficient because it doesn't have to pump as hard to get the needed amount of air. By varying the valve opening, the engine breathes much easier. In the figure, as the throttle is pressed, it causes the motor to rotate. As it rotates, the lever mechanism changes to vary the valve lift.

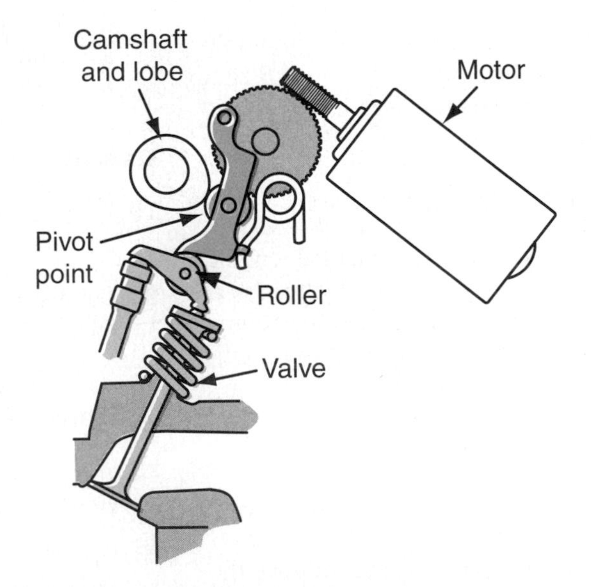

APPLICATION:

After reviewing the lever mechanism in the illustration, can you determine how the valve opens and closes by changing the lever position? How does the motor rotation change the position of the lever? Can you locate the contact point between the camshaft and the lever? Is this considered a first-, second-, or third-class lever?

Basic Engine Construction

CHAPTER 17

OBJECTIVES

After studying this chapter, you should be able to:

- Identify the major parts of the cylinder block.
- Recognize the purpose of core casting plugs.
- Identify cylinder block differences, such as aluminum versus cast iron, and types of sleeves and water jackets.
- List the parts of the crankshaft assembly and the purpose of crankshaft grinding.
- Describe the purpose of bearings and caps, oil passageways, vibration dampers, flywheels, and thrust surfaces on the crankshaft.
- Identify crankshaft seals and their purpose.
- List the parts of the piston and rod assembly
- Identify the purpose of oversize pistons and rings.
- State the purpose and operation of the rings, pistons, pins, and bearings.
- Describe the effect of pressures and temperatures on the design of the piston.
- Identify various problems, their diagnosis, and service tips and procedures to use when working on the cylinder block, pistons, crankshaft, and bearings.

INTRODUCTION

Basic engine construction is a prerequisite when studying the automotive power plant. This section will introduce the basic parts of the engine. The cylinder block, crankshaft, pistons, rods, and camshafts will be discussed.

17.1 CYLINDER BLOCK

BLOCK DESIGN

The engine block is the main supporting structure to which all other engine parts are attached. A cylinder block is a

Certification Connection

ASE Connection: The information in this chapter can help you prepare for the National Institute for Automotive Service Excellence (ASE) certification tests. The tests and content areas most closely related to this chapter are:

Test A1—Engine Repair

- **Content Area**—General Engine Diagnosis
- **Content Area**—Engine Block Diagnosis and Repair

NATEF Connection: Much of the information in this chapter is related to the NATEF tasks. The NATEF tasks and priority numbers most closely related to this chapter are:

1. Identify and interpret engine concern; determine necessary action. P-1
2. Diagnose engine noises and vibrations; determine necessary action. P-2
3. Perform cylinder power balance tests; determine necessary action. P-1

continued

Certification Connection *continued*

4. Perform cylinder compression tests; determine necessary action. P-1
5. Perform cylinder leakage tests; determine necessary action. P-1
6. Disassemble engine block; clean and prepare components for inspection and reassembly. P-2
7. Inspect engine block for visible cracks, passage condition, core and gallery plug condition, and surface warpage; determine necessary action. P-2
8. Inspect and measure cylinder walls for damage, wear, and ridges; determine necessary action. P-2
9. Deglaze and clean cylinder walls. P-2
10. Inspect crankshaft for end play, straightness, journal damage, keyway damage, thrust flange and sealing surface condition, and visual surface cracks; check oil passage condition; measure journal wear; check crankshaft sensor reluctor ring (where applicable); determine necessary action. P-2
11. Inspect and measure main and connecting rod bearings for damage, clearance, and end play; determine necessary action (including the proper selection of bearings). P-2
12. Identify piston and bearing wear patterns that indicate connecting rod alignment and main bearing bore problems; inspect rod alignment and bearing bore condition. P-3
13. Inspect and measure piston; determine necessary action. P-2
14. Remove and replace piston pin. P-1
15. Inspect, measure, and install piston rings. P-1
16. Inspect or replace crankshaft vibration damper (harmonic balancer). P-3
17. Assemble an engine using gaskets, seals, and formed-in-place (tube-applied) sealants, thread sealers, and so on, according to manufacturer's specifications. P-2

large cast iron or aluminum casting. It has two main sections: the cylinder section and the crankcase section. ***Figure 17–1*** shows a cylinder block for an 8-cylinder engine. The cylinder section is designed for the pistons to move up and down during operation. The surfaces are machined to allow the pistons to move with minimum wear and friction.

The **crankcase** section is used to house the crankshaft, oil pump, oil pan, and the oil during operation. Cooling passageways are built within the block. These passageways, also known as water jackets, surround the cylinders. They allow coolant to circulate throughout the cylinder area to keep the engine cool. There is also a drilled passageway within some blocks for the camshaft. Many oil holes are drilled internally so that engine parts can be adequately lubricated. Other holes are drilled to allow additional parts to be attached to the cylinder block.

Car Clinic: MOTOR MOUNTS

CUSTOMER CONCERN:
Engine Vibration

A customer has noticed that the engine seems to have excessive vibration at idle. At higher speeds solid vibration is also felt throughout the vehicle. What could be the problem?

SOLUTION: The most likely cause of engine vibration as described in this problem is a bad or broken motor or transmission mount. This assumes, of course, that the engine itself is operating correctly and that there is no engine miss due to ignition or fuel problems. Make a visual observation of all motor and transmission mounts and replace as necessary.

BLOCK MANUFACTURING

The first step in building a block is to design a pattern. Sand is then formed around the pattern. When the pattern is removed, sand cores are placed within the cavity. These sand cores will eventually be the cooling passageways and cylin-

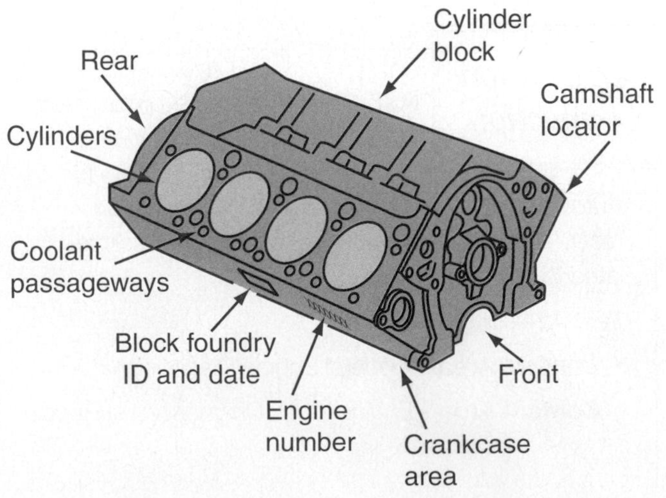

FIGURE 17–1 The parts of a V8 cylinder block are shown in this illustration.

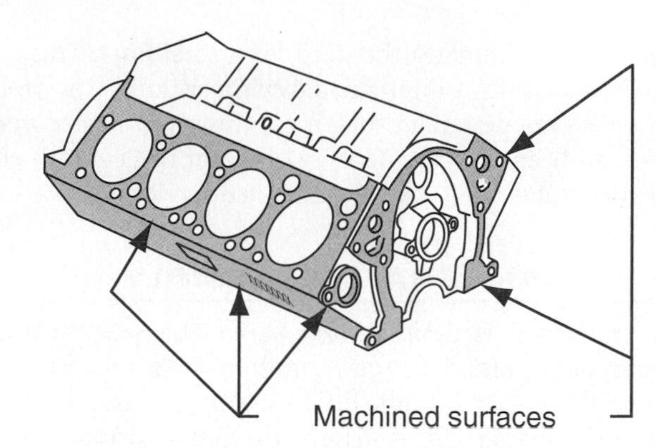

FIGURE 17–2 The cylinder block has many machined surfaces to which other parts are bolted.

ders. Molten metal is poured into the cavity made by the sand. After the metal has cooled, the sand is removed and the cores are broken so they can be removed. This design is called a **cast** block. The metal is usually a gray cast iron with several special metals added to it. The added metals increase the strength and wear characteristics of the block. The extra metals also help to reduce shrinkage and warpage from the heat produced by combustion.

Once the block is cast, and after it has been cooled and cured, its surfaces are machined so other parts can be attached to the block. These surfaces include the cylinders, top of the block (deck), camshaft bore, crankshaft bore, and oil pan surfaces. The front and rear of the block and engine mounts are also machined so that parts can be attached and sealed correctly (***Figure 17–2***).

Certain smaller engines can also be **die cast**. This means that the liquid metal is forced into a metal, rather than sand, mold. This kind of casting allows for smoother surfaces and more precise shapes to be made. Less machining is needed on this type of block.

CORE PLUGS

All cast iron cylinder blocks use **core plugs** (aluminum blocks do not have core plugs). These are also called freeze or expansion plugs. During the manufacturing process, sand cores are used. These cores are partly broken and dissolved when the hot metal is poured into the mold. However, holes have to be placed in the block to get the sand out of the internal passageways. These are called core holes. The holes are machined, and core plugs are placed into these holes (***Figure 17–3***).

Core plugs are made of soft metal. Under certain conditions, the core plugs can also protect the block from cracking. For example, if there is not enough antifreeze in the coolant during the winter, the coolant may freeze. As the liquid freezes, it expands. This expansion could cause the block to crack. However, if the expansion is near a core plug, the plug may pop out and possibly save the block from cracking. ***Figure 17–4*** shows a core plug pressed into a block.

Car Clinic: CORE PLUGS

CUSTOMER CONCERN:
Core Plugs Leak Coolant
A customer complains that coolant is leaking from the side of the cylinder block. The engine was recently overhauled, and the coolant has been leaking since the engine was rebuilt. What could be the problem?

SOLUTION: Cylinder block core plugs are usually replaced during an overhaul. Sometimes when new core plugs are installed, they are not inserted correctly. If they are inserted at a slight angle, they may leak engine coolant. Also, a sealant should be used during installation. Not using a sealant may also cause the core plugs to leak. To solve the problem, the leaky core plug must be removed and replaced correctly, using the recommended sealant.

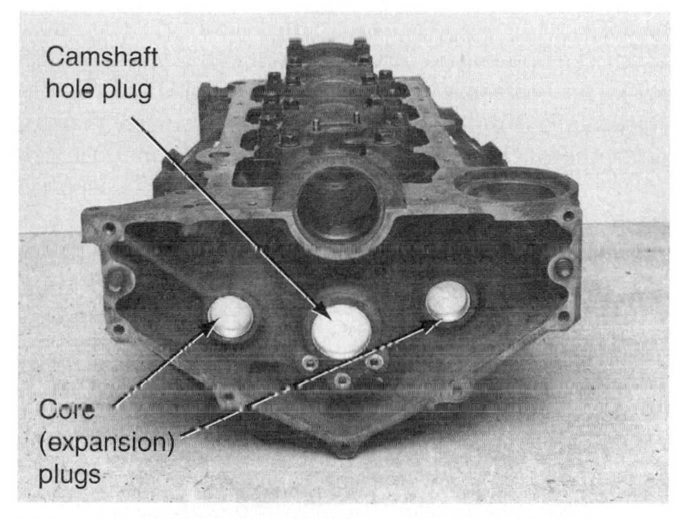

FIGURE 17–3 Core plugs, also called freeze or expansion plugs, are used to protect the block if the coolant freezes.

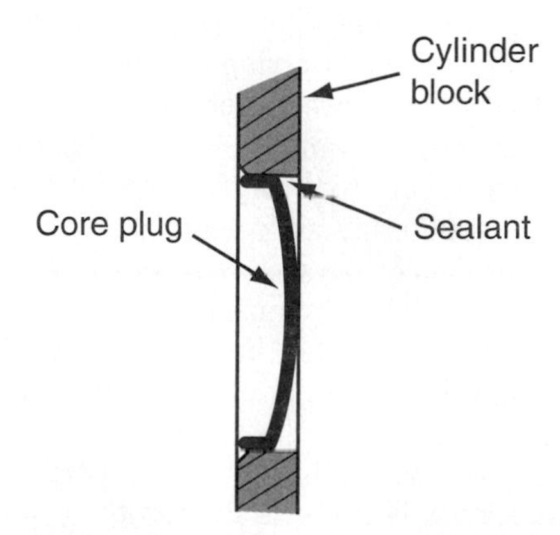

FIGURE 17–4 A core plug is pressed into the block with sealant.

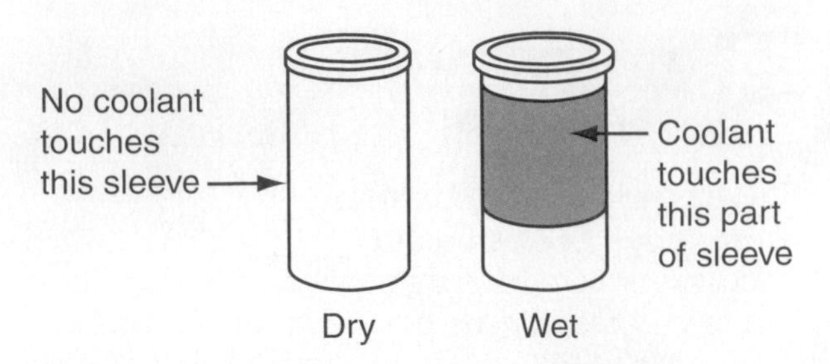

FIGURE 17-5 Cylinder sleeves are used on some engines. The sleeve is inserted into the block after it has been machined. There are both dry and wet sleeves.

CYLINDER SLEEVES

Some manufacturers use **cylinder sleeves**. Rather than casting the cylinder bores directly into the block, a machined sleeve is inserted. ***Figure 17–5*** shows a sleeve for a cylinder block. Sleeves are inserted after the block has been machined. The purpose of using a sleeve is that, if the cylinder is damaged, the sleeve can be removed and replaced rather easily. Blocks that don't have sleeves have to be bored out to remove any damage. After boring, larger pistons will be needed.

There are two types of sleeves: wet and dry sleeves. The dry sleeve is pressed into a hole in the block. It can be machined quite thin because the sleeve is supported from the top to the bottom by the cast iron block.

The wet sleeve is also pressed into the block. The cooling water touches the center part of the sleeve. This is why it is called a wet sleeve. It must be machined thicker than the dry sleeve because it is supported only on the top and the bottom. Seals must be used on the top and bottom of the wet sleeve to keep the water from leaking out of the cooling system. Wet sleeves are used on some larger diesel engines.

ALUMINUM VERSUS CAST IRON BLOCKS

Blocks can be made from either cast iron or aluminum. In the past, most blocks were made of cast iron. Cast iron improved strength and controlled warpage from heat. With the increased concern for improved gasoline mileage, however, car manufacturers are trying to make lighter vehicles. One way is to reduce the weight of the block. Aluminum is used for this purpose. Aluminum is a very light metal. Certain materials are added to the metal before it is poured into the mold. These materials are used to make the aluminum stronger and less likely to warp when heat from combustion is applied. Aluminum blocks must also have a sleeve or steel liner placed in the block. Steel liners are placed in the mold before the metal is poured. After the metal is poured, the steel liner cannot be removed.

Silicon is also added to the aluminum. Through a special process, the silicon is concentrated on the cylinder walls. This process eliminates the need for a steel liner. This design is called silicon-impregnated cylinder walls. One problem with this design is that it requires the use of very high-quality engine oils. Because of owner neglect, this engine does not usually reach its intended service life.

17.2 CRANKSHAFT ASSEMBLY

The crankshaft is designed to change the reciprocating motion of the piston to rotary motion. It is bolted to the bottom of the cylinder block. The crankshaft assembly includes the crankshaft, bearings, flywheel, harmonic balancer, timing gear, and front and rear seals. ***Figure 17–6*** shows a typical V8 crankshaft. There are four connecting rod journals with two connecting rods on each journal. In addition, five main bearing journals support the crankshaft in this block.

CRANKSHAFT DESIGN

The crankshaft is manufactured by either forging or casting. **Forged** steel crankshafts are stronger than cast iron crankshafts, but they cost more. Forging is a process during which metal is heated to a certain temperature, then stamped or forged into a particular shape. Casting involves heating the metal to its melting point and then pouring the liquid metal into a form made from sand. Because of the improvements in casting, more crankshafts are now being cast. Cast and forged crankshafts can be identified by separation or the parting lines (***Figure 17–7***). Forged crankshafts have a ground-off separation line. Cast crankshafts have a small parting line where the molds came together.

After the crankshaft is cast or forged, it must be **heat treated**. This means the outer surfaces must be made harder so that the crankshaft will not wear on the bearing surfaces. Heat treating is done by heating the outer part of the crankshaft to 1,600 to 1,800 degrees Fahrenheit. Then the metal is cooled rapidly in oil, water, or brine (salt water). The rapid cooling causes the outer part of the crankshaft (0.060 inch) to be hardened.

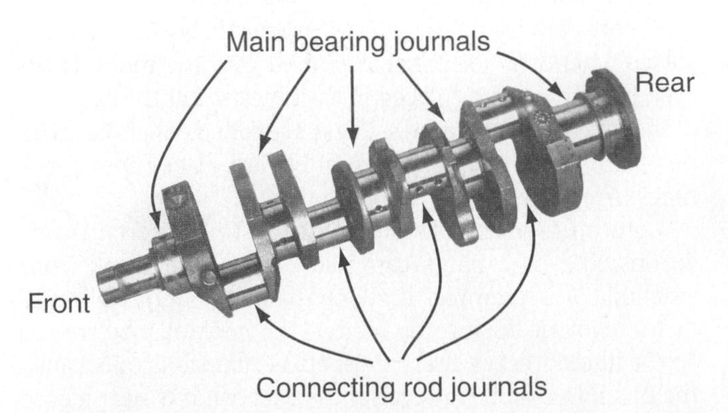

FIGURE 17-6 The crankshaft has connecting rod and main bearing journals.

FIGURE 17-7 Crankshafts can be either forged or cast. Cast crankshafts have a parting line as shown. Forged crankshafts have a ground-off separation line.

The parts of the crankshaft include:

- Throw—The distance from the centerline of the main journal on the crankshaft to the centerline of the connecting rod journal.
- The main journal—The part on the crankshaft that connects the crankshaft to the block.
- The connecting rod journal—The part on the crankshaft where the connecting rod is attached.
- The **counterweights**—Weights that are cast or forged into the crankshaft for balance. For each throw, there is a counterweight to balance the motion. Depending on the engine, the counterweight can be a weight or another connecting rod journal.
- Thrust surfaces—Surfaces machined on the crankshaft to absorb axial motion or thrust (***Figure 17–8***). Axial motion (back-and-forth motion on the crankshaft axis) is produced from the general motion of the crankshaft. It is also produced from using timing gears on the crankshaft that have angled teeth, as shown in ***Figure 17–9***. This type of gear is called a helical gear. As the engine increases and decreases in speed, the load on the helical gear moves the crankshaft back and forth on its axis. This motion is absorbed by machining thrust surfaces on the crankshaft and using a thrust bearing.

FIGURE 17-8 Thrust surfaces are machined into the crankshaft to absorb axial motion or thrust motion on the crankshaft.

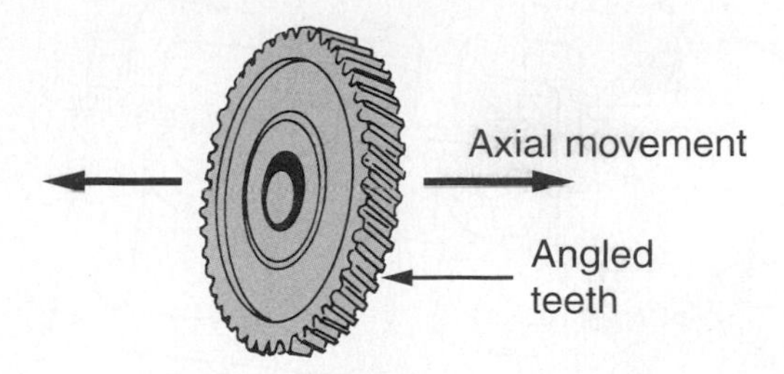

FIGURE 17-9 This helical gear has angled teeth that cause the crankshaft to move back and forth during operation.

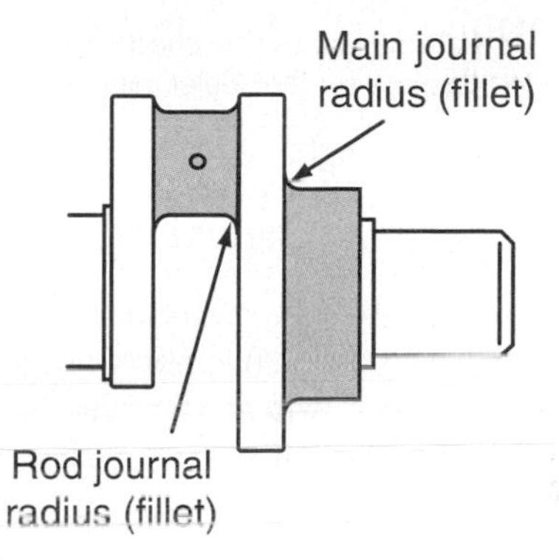

FIGURE 17-10 The crankshaft has a small radius (fillet) machined into the rod and main journals for strength.

- Drive flange—The end of the crankshaft that drives the transmission. A flywheel or flexplate is bolted to the drive flange. The transmission is attached to the flexplate or driven by the flywheel.
- **Fillets**—Small rounded areas (a radius) that help strengthen the crankshaft near inside corners. Stress tends to concentrate at sharp corners and drilled passageways. Fillets help reduce this stress (***Figure 17–10***).
- Nose—The end of the crankshaft that is used to drive the accessories such as the harmonic balancer, pulleys, alternator, air conditioning, and so on.

CRANKSHAFT ALIGNMENT

When the crankshaft fits into the block, it must be exactly in-line with the holes in the block. This is called **alignment**. When the block is bored for the crankshaft main bearing bore, a line boring machine is used. To do this, the main bearing caps are bolted and torqued to the block. The line borer then machines each main bearing area in-line with the others. Over a period of time, the block may warp. Warping could cause the alignment to be incorrect. This condition could cause excessive wear on certain parts of the crankshaft. Alignment can be checked by using a feeler gauge and a straightedge as shown in ***Figure 17–11***.

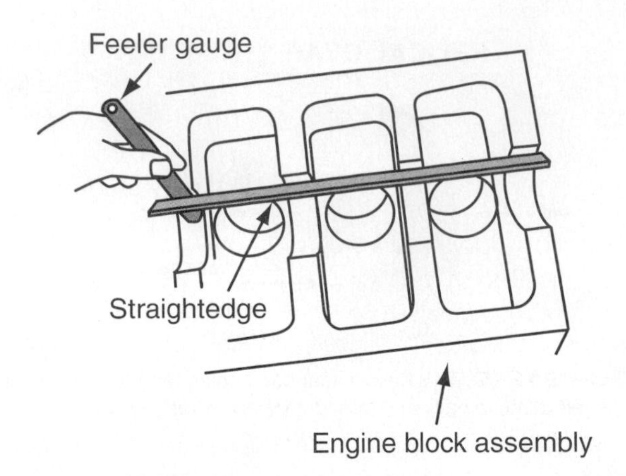

FIGURE 17-11 The block can be checked for correct alignment by using a straightedge and a set of feeler gauges.

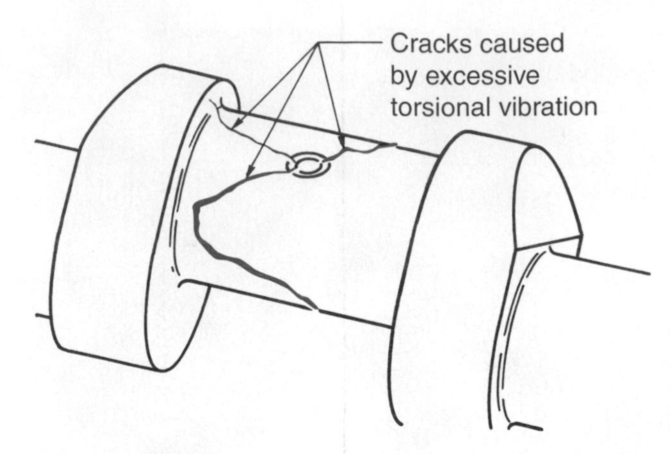

FIGURE 17-12 Torsional vibration within the crankshaft can cause severe cracking.

CRANKSHAFT VIBRATION

During normal operation, the crankshaft is twisted when it is being turned, producing constant vibration within the crankshaft. For example, when one cylinder is on a compression stroke, that part of the crankshaft tries to slow down. At the same time, other cylinders may have full pressure from the power stroke. This causes the crankshaft to partially twist and snap back during each revolution. This effect is called **torsional vibration.** Additional torsional vibration can be caused by using the wrong flywheel, converter drive plate, or torque converter in the transmission. ***Figure 17–12*** illustrates the results of too much torsional vibration. Crankshafts typically crack near the connecting rod journal of the number 1 cylinder.

VIBRATION DAMPERS

Vibration dampers are used to compensate for torsional vibration. The vibration damper is also called a harmonic balancer. It is constructed by using an inertia ring and a rubber ring. The inertia ring is used to help dampen the internal vibrations. The two are bonded together and attached to the front of the crankshaft. As the crankshaft twists back and forth, the inertia ring has a dragging or slowing down effect. As torsional vibrations occur, the rubber and inertia rings absorb the vibration. ***Figure 17–13*** shows two different types of vibration dampers. All have a rubber ring between the center and the outside pulley. Many vibration dampers also have steel interrupter blades that are used in computer systems to produce a signal to monitor crankshaft speed.

The weight of the inertia ring is sized to a particular engine. If the wrong vibration damper is used, it may cause vibrations in the crankshaft, which may in turn cause it to be damaged. Incorrect vibration dampers can be identified by observing the timing marks. The timing marks on the vibration damper may not line up correctly with the timing tag on the front of the block.

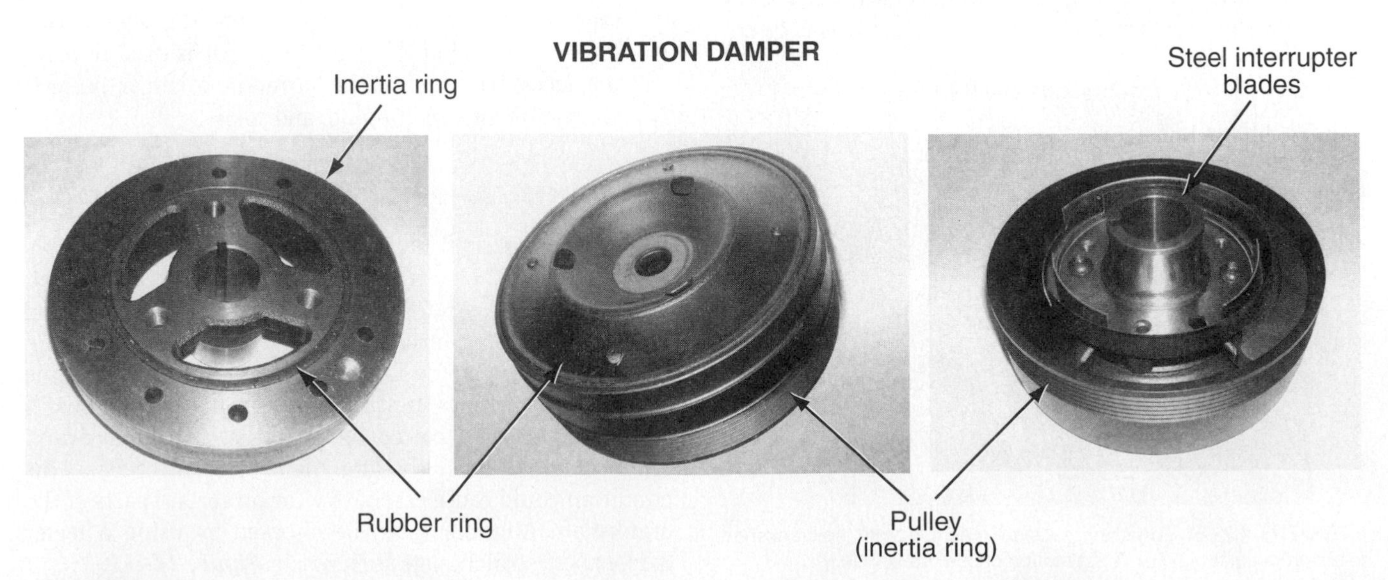

FIGURE 17-13 The vibration damper is used to absorb torsional vibration. It is placed on the front of the crankshaft.

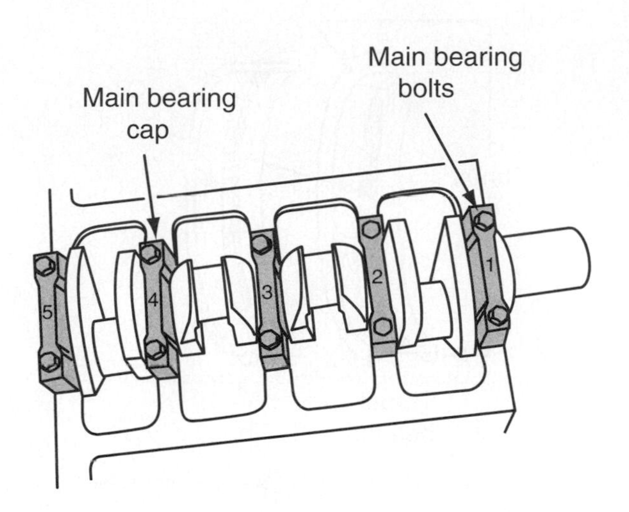

FIGURE 17–14 Main bearing caps and bolts are used to hold the crankshaft in place in the cylinder block.

BEARINGS AND BEARING CAPS

The crankshaft is held in place by the main bearings and caps. There is usually one more main bearing cap than the number of cylinders. Depending on the engine, however, there may be fewer main bearing caps than cylinders. The main bearing caps are bolted to the block. ***Figure 17–14*** shows the main bearing cap and bolts.

When in place, the main bearings hold the crankshaft securely in place to allow for rotation. On some high-performance gas engines and diesel engines, there may be four bolts, rather than two, holding the bearing cap to the block. The extra support is needed because these engines produce more torque and higher loads on the crankshaft.

Car Clinic: MAIN BEARINGS

CUSTOMER CONCERN:
Main Bearing Alignment

An engine was recently rebuilt. However, after about 200 miles, the engine is making loud knocking noises at all speeds. The knock sounds like bad bearings.

SOLUTION: When an engine is rebuilt, the crankshaft main bearings are replaced. Each main bearing cap must be replaced in its original position. During the manufacturing process, all main bearings are placed on the engine, and the main bearing bore is line bored. This means that the main bearing caps *must* be reinstalled to their original position. They are often marked with a number. If the main bearing caps are not installed to their original position, there is a good possibility that the main bearing will wear quickly. The solution to the problem is to replace the main bearings and make sure that the main bearing caps are installed to their correct position.

BEARING DESIGN

Insert bearings are placed between the main bearing caps and the crankshaft. As discussed in an earlier chapter, these bearings are designed as two-piece friction-type bearings. Main bearings have what is called spread. They are slightly larger than the housing into which they fit (***Figure 17–15***). The spread allows them to snap into place. Half the bearing is placed into the block. The other half is placed in the bearing cap.

The main bearings have several parts. All bearings have a steel backing. This provides the strength and support for the bearing. There are also several soft metals used on the bearing surface. Soft metals such as copper-lead, **babbitt**, aluminum, and tin allow a certain amount of dirt to be embedded into the soft metal. They also help the bearing to form to the shape of the crankshaft journals. ***Figure 17–16*** shows a common bearing with several metals in its design.

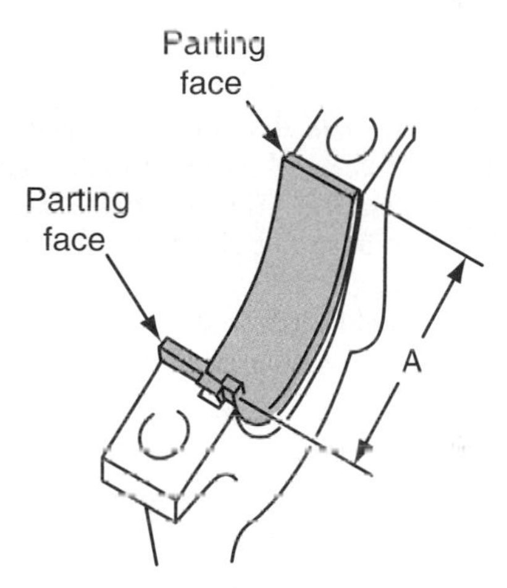

FIGURE 17–15 Spread (dimension A) helps keep the bearing in place in the block and bearing cap.

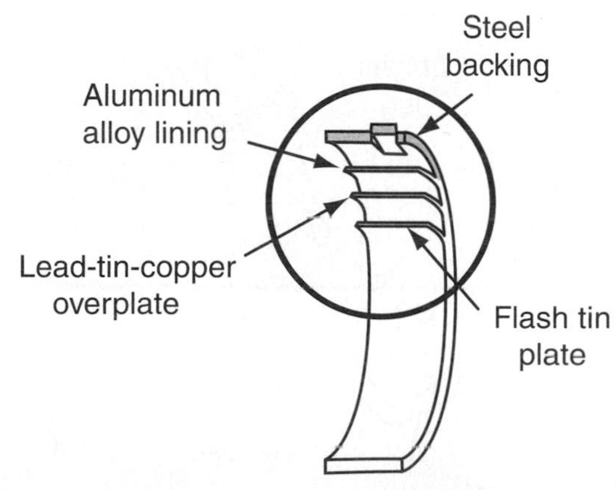

FIGURE 17–16 Main bearings have a steel backing and several soft metals such as copper, lead, tin, and aluminum.

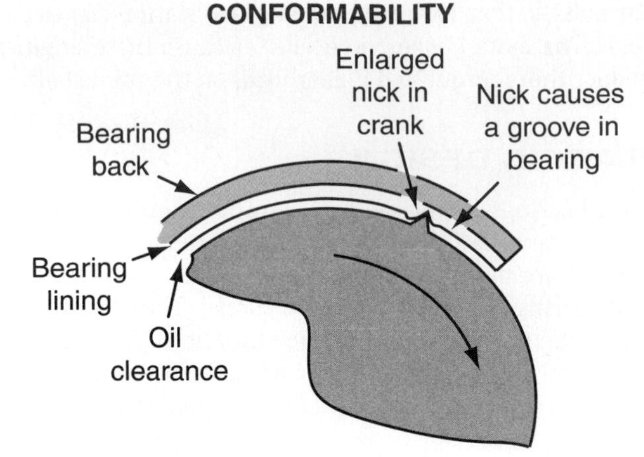

FIGURE 17-17 Bearings must be able to conform to the shape of the crankshaft. This capability is called conformability.

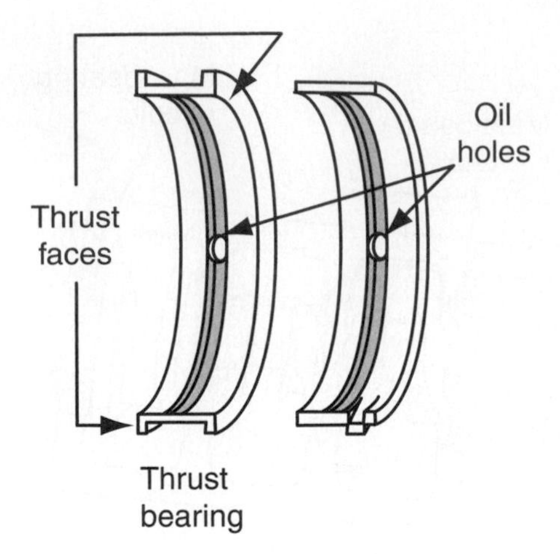

FIGURE 17-19 Oil holes and grooves are used to help lubricate the bearings.

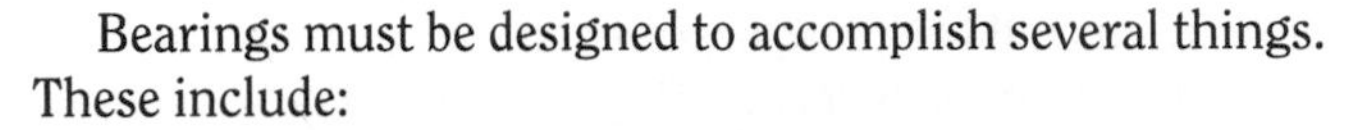

Bearings must be designed to accomplish several things. These include:

- Load-carrying capacity—The ability to withstand pressure loads from combustion
- Fatigue resistance—The ability to withstand constant bending
- Embedability—The ability to permit foreign particles to embed or be absorbed into the metal
- Conformability—The ability to be shaped to the small variations in the shape of the crankshaft (***Figure 17–17***)
- Corrosion resistance—The ability to resist the by-products of combustion that are carried to the bearing by the oil
- Wear rate—The ability to be strong enough to eliminate excessive wear, but soft enough not to wear the crankshaft journals

The bearing insert also has a locating lug (***Figure 17–18***), which holds the bearings in place within the block and cap so that the bearing cannot spin. Oil grooves and holes are also machined into the bearing insert. Oil passes from the center of the crankshaft through the hole in the bearing and circles around the crankshaft in the oil groove. This design provides complete lubrication of the crankshaft journals (***Figure 17–19***).

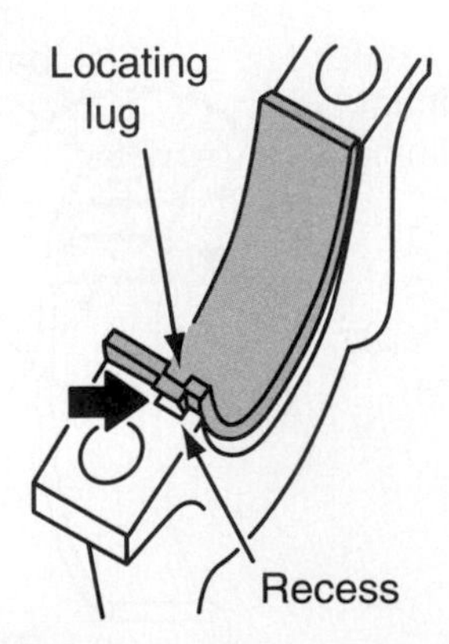

FIGURE 17-18 Bearings have locating lugs to keep them in place and keep them from spinning.

There must also be a clearance between the crankshaft journals and the insert bearings. This clearance is called the **main bearing clearance**. Main bearing clearance can be checked using **Plastigage**. Plastigage is a small, thin string of plastic with a predetermined diameter. A short piece of the Plastigage is placed between the crankshaft and the bearing. The bearing and cap are then torqued to their normal specifications. This torquing causes the thin Plastigage to flatten. The resulting width of the Plastigage is a measure of the bearing clearance. The procedure for using Plastigage is shown in the Problems, Diagnosis, and Service section of this chapter.

THRUST BEARINGS

The crankshaft moves back and forth axially (movement parallel to the axis of the crankshaft) during operation. This movement is often created from angled gears on the front of the crankshaft. This **axial motion**, called thrust, could cause the crankshaft to wear heavily on the block. Thrust bearings are used to compensate for this motion. One of the main bearings is designed for thrust absorption. The left bearing in ***Figure 17–19*** is a thrust bearing. The thrust bearing has a thrust face where a machined surface on the crankshaft rubs against it. On most engines, the thrust bearing works against the center main bearing. On some engines, however, the thrust bearing works against the flange on the rear main bearing. Other engines use separate thrust washers instead of the flanged type (***Figure 17–20***).

OIL PASSAGEWAYS

For the crankshaft to receive proper lubrication, oil must pass through the crankshaft to the bearings. ***Figure 17–21*** illustrates the internal passageways drilled for oil. Oil pressure from the lubrication system is fed through the main oil

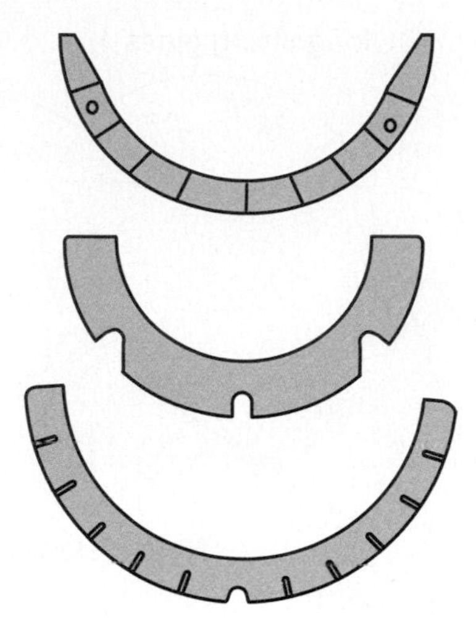

FIGURE 17–20 Separate thrust bearings are used on some older engines.

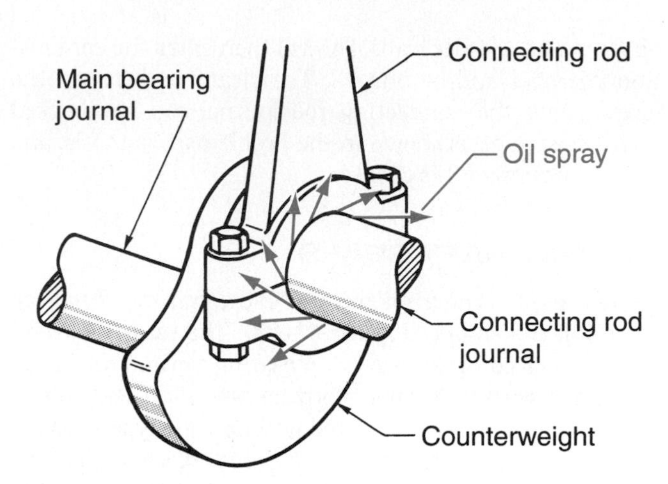

FIGURE 17–22 The bearing clearance determines how much oil flows past and out of the bearing area. Too much clearance will reduce oil pressure and increase the amount of oil on the cylinder walls.

galley to the block and then to each main bearing. The oil goes through the bearing insert and into a groove in the bearing. There is a drilled passageway from the main bearing through the crankshaft to the connecting rod bearing. Oil is then fed into the connecting rod bearings, where it eventually leaks out and sprays against the cylinder walls. The oil then drips back to the oil pan.

PARTS LOCATOR

Refer to photo #4 in the front of the textbook to see what the ***main oil gallery*** *looks like.*

BEARING CLEARANCE

The clearance between the bearing and the journal is called bearing clearance. This clearance is designed so that just the right amount of oil is allowed to flow through it so that there is no metal-to-metal contact (***Figure 17–23***). If the bearings wear, the clearance will increase. This condition may reduce oil pressure. If the connecting rod clearance is larger, more oil may pass through this clearance. With the correct bearing clearance, the amount of oil thrown off from the rotating shaft is minimal. When the clearance is doubled, oil throw-off is five times greater. As the clearance increases, oil throw-off is increased even more. Under these conditions, piston rings are unable to scrape this excessive

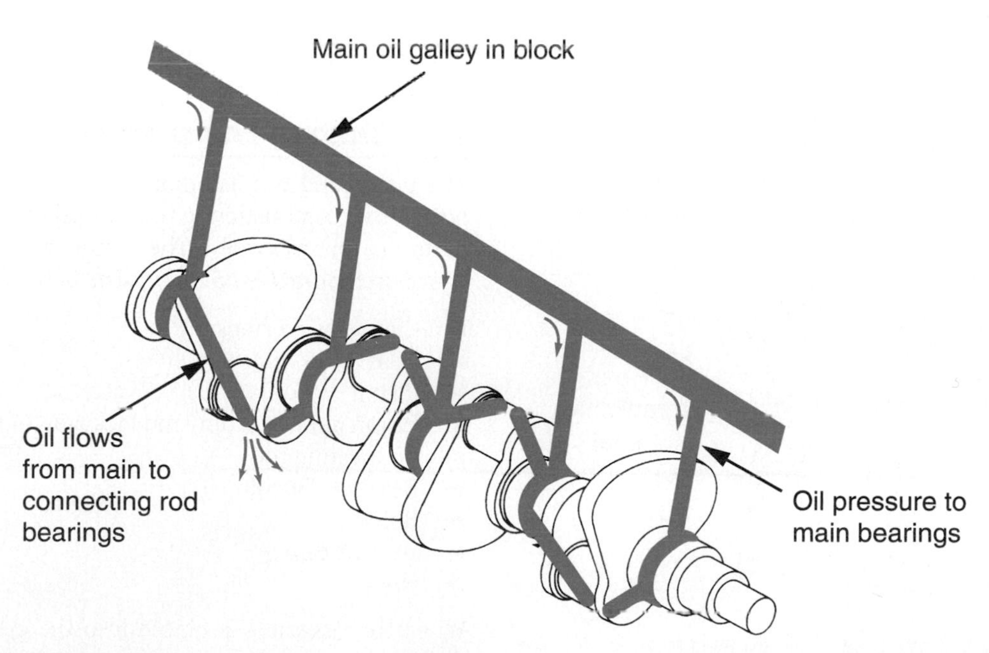

FIGURE 17–21 Oil flows from the main oil galley in the block to the main bearings, and then to the connecting rod bearings, through holes in the crankshaft.

oil from the cylinder walls. Oil will then enter the combustion chamber and be burned. The clearance between the bearing and the connecting rod journal can be checked using Plastigage as shown in the Problems, Diagnosis, and Service section of this chapter.

CRANKSHAFT SEALS

A seal is used at the front and rear of the crankshaft to keep the oil in the engine (***Figure 17–23***). The rear crankshaft seals are placed in the rear main bearing cap and the block. There are several designs. Both lip-type synthetic rubber seals and graphite-impregnated wick or rope-type seals are used on the rear of the crankshaft. Some are a two-piece design, others are a one-piece insert design. One-piece lip-type seals, or 360-degree seals, that ride on the vibration damper are used on the front of the engine.

When lip-type seals are used, the lip of the seal may wear a groove in the crankshaft over time. This may cause oil to leak out. Even if a new seal is installed, oil may continue to leak because of the wear on the crankshaft. This is often the case on the harmonic balancer. The front oil seal often rides on the harmonic balancer. To reduce the oil leak, the harmonic balancer may need to be replaced. However, repair sleeves may be used. A repair sleeve, as shown in ***Figure 17–24***, is inserted over the damaged surface area on the harmonic balancer. Now the new seal rides on the repair sleeve surface.

PARTS LOCATOR

Refer to photo #5 in the front of the textbook to see the location of the ***crankshaft seals****.*

CRANKSHAFT GRINDING

At times it may be necessary to grind the crankshaft or main bearing journals, or both, to a smaller size. Grinding would be necessary if there were excessive wear on the journal surfaces. Crankshafts are ground in special machine shops. When a crankshaft is ground, new bearings will be needed. Main and rod bearings are available in standard sizes and various undersizes, depending on the manufacturer. These bearings are called undersize because the diameter of the crankshaft journal is smaller after being ground down. Bearings can typically be purchased in 0.001-, 0.002-, 0.010-, 0.020-, 0.030-, and 0.040-inch undersizes. However, the exact size is determined by the manufacturer. The amount of undersize is stamped on the back of the bearing.

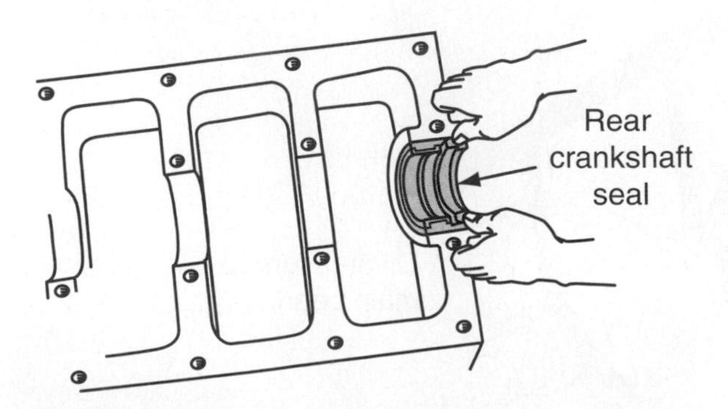

FIGURE 17-23 Crankshaft seals are used to keep oil inside the engine rather than leaking out around the front and back of the block.

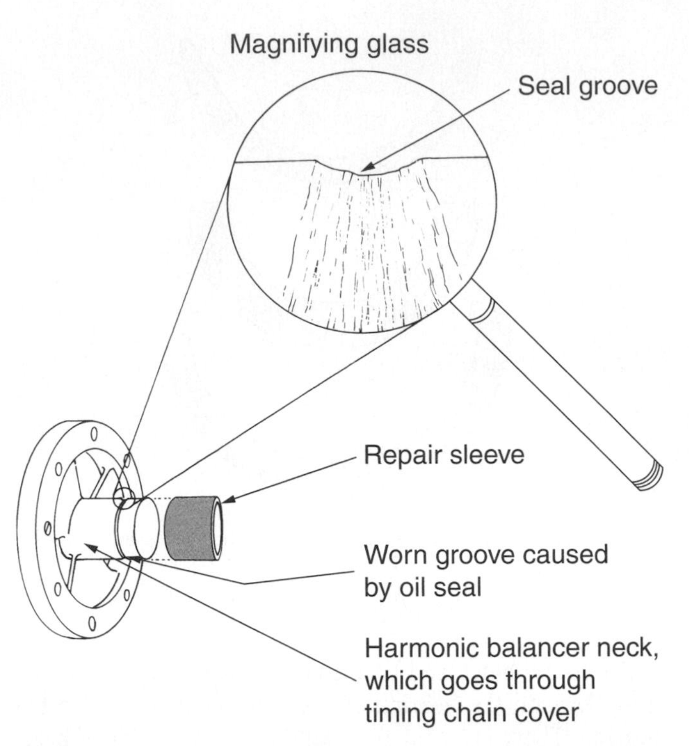

FIGURE 17-24 Repair sleeves can be installed on the harmonic balancer to cover a groove caused by the front crankshaft seal.

17.3 PISTON AND ROD ASSEMBLY

The piston and rod assembly is designed to transmit the power from combustion to the crankshaft. There are several parts on this assembly. The piston and rod assembly is shown in ***Figure 17–25***. It consists of the following parts:

- Compression rings
- Oil control rings
- Piston
- Piston pin (wrist pin) and lock ring (if used)
- Connecting rod
- Bearings (bushings)
- Bolts
- Bearing cap
- Nuts

When this assembly is placed into the cylinder block, the downward motion from combustion is transmitted to the crankshaft.

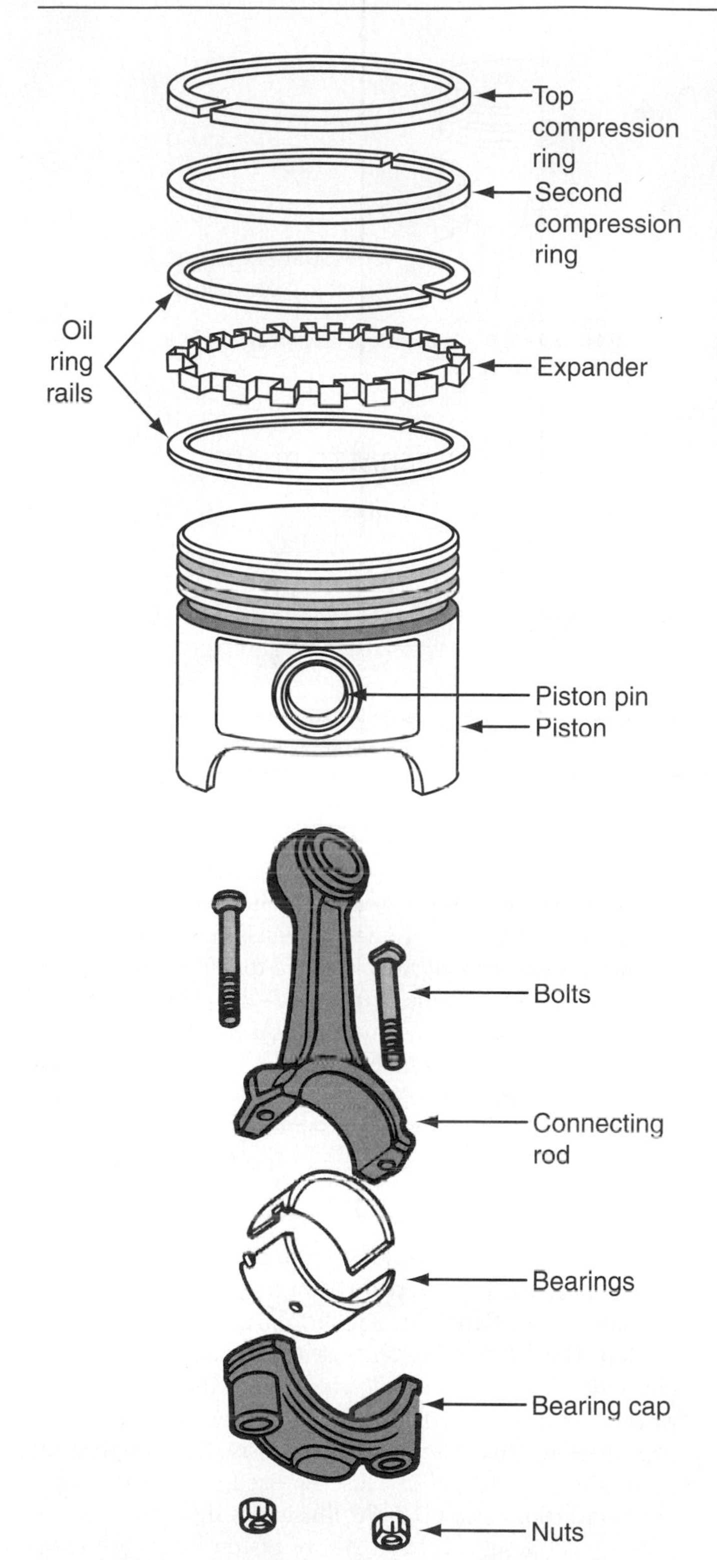

FIGURE 17–25 The parts of a complete piston and rod assembly are shown in this illustration.

PISTON PARTS

The piston is a hollow aluminum cylinder (***Figure 17–26***). It is closed on the top and open on the bottom. It fits closely within the engine cylinder or sleeve and is able to move alternately up and down in the cylinder bore. The piston serves as a carrier for the piston rings.

FIGURE 17–26 The piston, rings, and piston pins are shown in this photograph. (Courtesy of Arias Pistons)

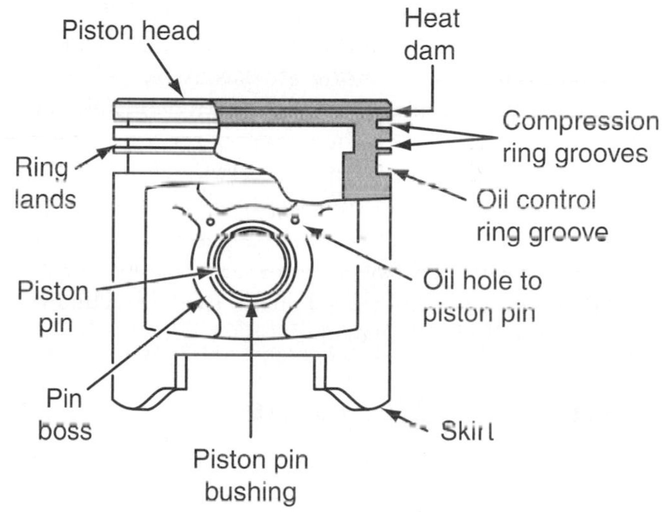

FIGURE 17–27 The piston has many parts. Some of the more important parts are shown in this illustration.

The common parts of the piston are shown in ***Figure 17–27***, and include the following:

- Land—That part of the piston above the top ring or between ring grooves. The lands confine and support the piston rings in their grooves.
- **Heat dam**—A narrow groove cut in the top land of some pistons to reduce heat flow to the top ring groove. This groove fills with carbon during engine operation and reduces flow to the top ring. Heat dams are also designed as cast slots in the piston.
- Piston head—The top piston surface against which the combustion gases exert pressure. The piston head may be flat, concave, convex, or of irregular shape.
- Piston pins (wrist pins or gudgeon pins)—Connections between the upper end of the connecting rod and the piston. They can be (1) anchored to the piston and floating in the connecting rod, (2) anchored to the connecting rod and floating in the piston, or (3) full floating in both connecting rod and piston. Number 3

requires a lock ring to hold the pin in place. Some piston pins and connecting rods are connected together using an interference fit. An **interference fit**, also called a press fit, is one in which the internal diameter of the connecting rod and the external diameter of the piston pin interfere with each other. This means the external diameter is larger than the internal diameter. Thus, the two parts must be pressed together during assembly.

- Skirt—That part of the piston located between the first ring groove above the wrist pin hole and the bottom of the piston. The skirt forms a bearing area in contact with the cylinder wall and is 90 degrees opposite the piston pin.
- Thrust face—The portion of the piston skirt that carries the thrust load of the piston against the cylinder wall.
- Compression ring groove—A groove cut into the piston around its circumference to hold the compression rings.
- Oil ring groove—A groove cut into the piston around its circumference. Oil ring grooves are usually wider than compression ring grooves. They generally have holes or slots through the bottom of the groove for oil drainage back to the crankcase area.
- Piston pin bushing—A bushing fitted between the piston pin and the piston. It acts as a bearing material and is used mostly on cast iron pistons. This bushing can also be located in the small end of the connecting rod assembly. It is usually made of bronze.

PISTON REQUIREMENTS

The piston assembly must be designed to operate under severe conditions. The temperatures produced on top of the piston are very high. This heat causes stress and expansion problems. The piston is moved up and down many times per minute, which produces high pressures and stress. To handle these conditions, most pistons are made of aluminum. Aluminum makes the piston lighter. However, some larger engines, especially certain diesels, may use a cast iron piston. In this case, the rpm will be lower. Lighter pistons can operate much more effectively in today's gasoline engines, which run in excess of 5,000 rpm.

PISTON EXPANSION

When combustion occurs on the top of the piston, some of the heat is transmitted down through the piston body. This causes the piston to expand. If the expansion were too great, the piston might wear the cylinder to a point of damage. To compensate for expansion, older pistons have a split skirt (***Figure 17–28***). When the piston skirt expands, the slot closes rather than increasing in size.

The T slot, which is also used on older engines, is another method of controlling expansion. In this case, the T slot tends to hold back the transfer of heat from the head to the skirt. It also allows for expansion within the slot.

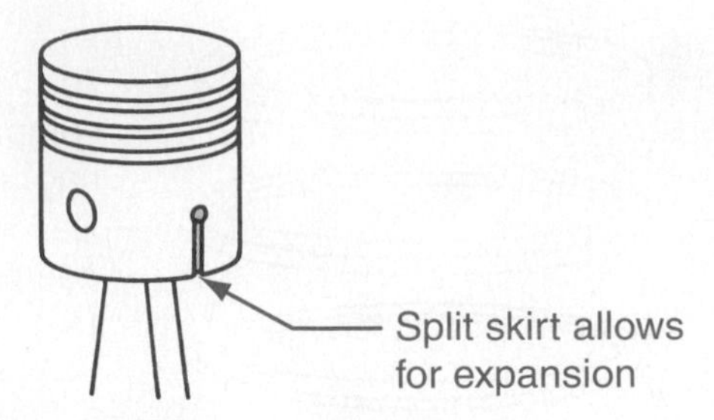

FIGURE 17–28 Older pistons have a split skirt to allow for expansion when heated.

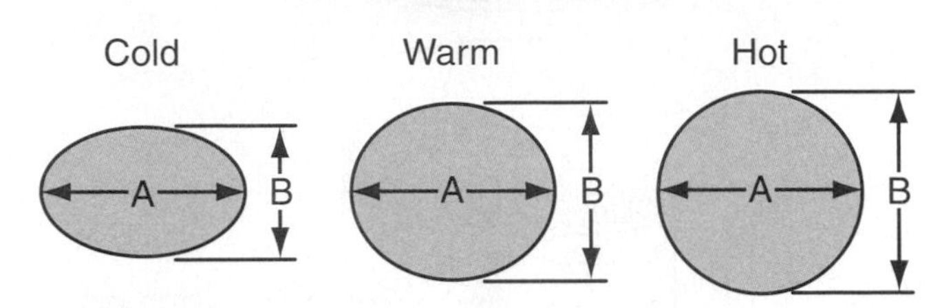

FIGURE 17–29 Cam ground pistons help to control expansion. As the piston heats up, it becomes round, which creates an accurate fit in the cylinder bore.

Some pistons use steel rings. These steel rings will not expand as much as the aluminum. The steel rings have a tendency to control or minimize expansion.

Cam ground pistons are also used to control the expansion in a gasoline engine (***Figure 17–29***). The pistons are ground in the shape of a cam or egg. As the piston heats up during operation, it becomes round. The piston is designed so that maximum expansion takes place on dimension B. Dimension A remains about the same.

PISTON HEAD SHAPES

The shape of the piston head varies according to the engine. Head shapes are used to create turbulence and change compression ratios. Generally, small, low-cost engines use the flat top. This head comes so close to the valve on some engines that there must be a recessed area in the piston for the valves. Another type of head is called the raised dome or pop-up head. This type is used to increase the compression ratio. The dished head can also be used to alter the compression ratio. ***Figure 17–30*** illustrates different types of piston head design. Other types of piston heads are used, but only for special applications.

PISTON SKIRT

Since the 1970s, it has become important to make engines as small as possible, yet still powerful. One way to do this is to keep the height of the piston and connecting rod to a minimum. This is done by shortening the connecting rod. A **slipper skirt** is used. Part of the piston skirt is removed so the counterweights on the crankshaft will not hit the

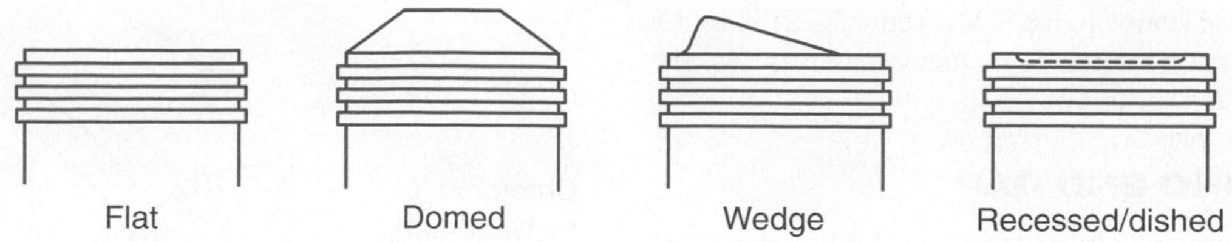

Flat Domed Wedge Recessed/dished

FIGURE 17-30 Pistons are designed with different head shapes. These are the most common.

piston. This design means there can be a smaller distance between the center of the crankshaft and the top of the piston. The output power of the engine is not affected because the bore and stroke still remain the same (***Figure 17–31***).

SKIRT FINISH

The surface of the skirt is somewhat rough. Small grooves are machined on the skirt so that lubricating oil will be carried in the grooves (***Figure 17–32***). This helps lubricate the piston skirt as it moves up and down in the cylinder. If the engine overheats, however, the oil will thin out and excessive piston wear may occur. Some pistons have an impregnated silicon surface on the skirt of the piston. Impregnated silicon (silicon particles placed into the external finish of the piston) helps to reduce friction between the skirt and the cylinder wall.

PISTON RINGS

There are two types of sealing problems on the piston. Combustion pressures and gases must not be allowed to escape past the rings. Escape of combustion pressures and gases is defined as **blow-by** (***Figure 17–33***). If blow-by occurs, there is a loss of power. If there is excessive blow-by, too much oil might be forced off the cylinder walls, causing excessive **scuffing** and wear on the cylinder walls and piston rings.

The rings must also keep the oil below the combustion chamber. If they do not, there may be excessive oil consumption. The moving piston is sealed with compression and oil control rings. The rings are slightly larger than the piston. When installed, they push out against the cylinder walls. Since they contact the cylinder wall, they seal against pressure losses and oil loss. Most engines use two compression

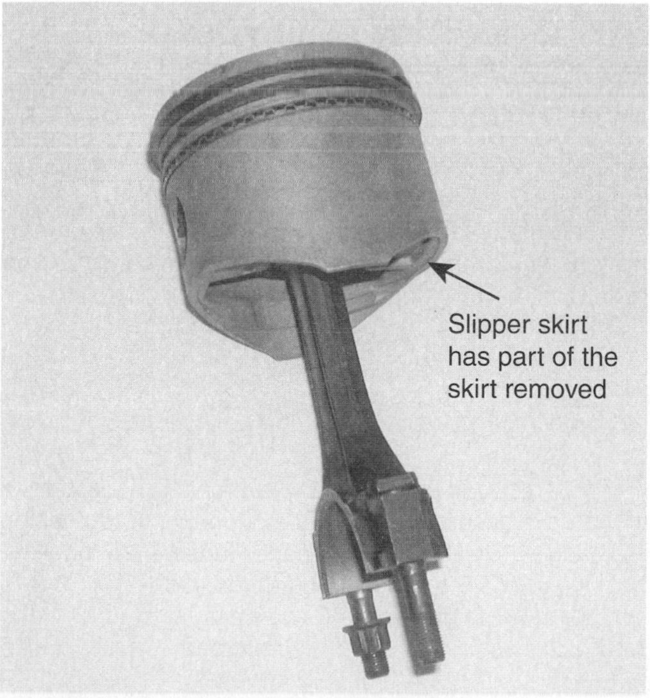

Slipper skirt has part of the skirt removed

FIGURE 17-31 A slipper skirt has part of the piston skirt removed so that the connecting rod can be made shorter, which in turn allows the engine to be smaller and lighter.

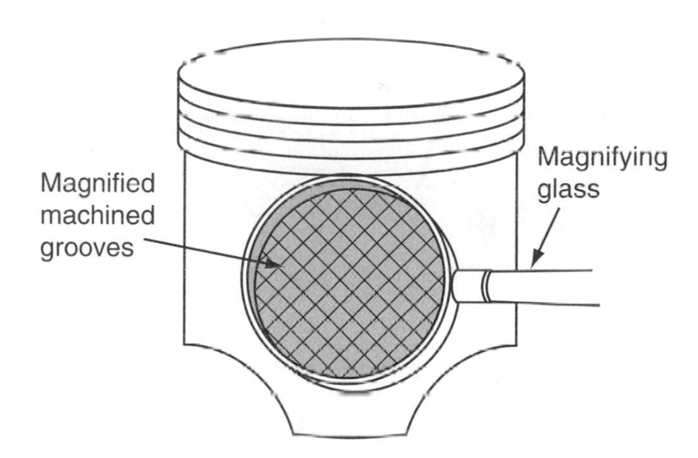

FIGURE 17-32 Machined grooves are placed on the side of a piston to hold a small amount of oil on its surface.

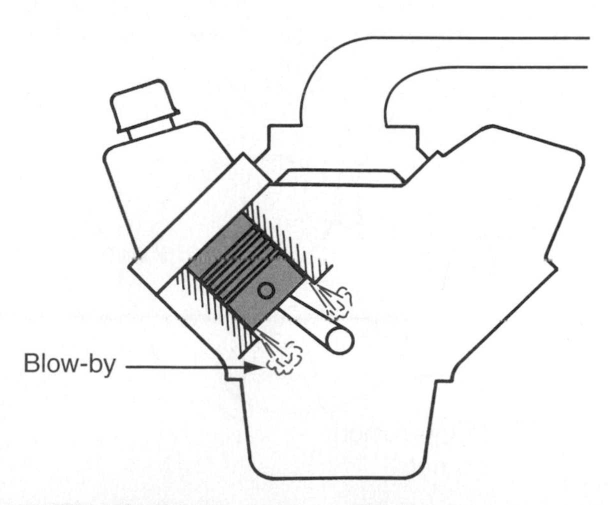

FIGURE 17-33 Blow-by is a term for the combustion pressure gases that escape past rings and into the crankcase area.

rings and one oil control ring. In certain diesel engines, however, more compression rings may be used to seal the higher pressures.

PISTON RING END GAP

When the compression ring is placed in the cylinder, there is a gap at the ends of the ring (***Figure 17–34***). This gap is referred to as piston ring end gap. The piston ring end gap gets smaller as the ring increases in temperature. If it is too small, the ring will bind in the cylinder and break, causing excessive scoring of the cylinder walls. If the piston ring end gap is too large, excessive blow-by will result. Piston ring end gap should be checked when rebuilding engines. ***Figure 17–35*** illustrates the measuring of piston ring end gap.

COMPRESSION RING MATERIAL

Compression rings are made of cast iron. This material is very brittle and can break easily if bent. However, the brittle material wears very well. Certain heavy-duty engines and some diesel engines use ductile iron as piston ring material. This material is stronger and resists breaking, but the cost of these rings is higher. Some high-quality piston rings have a fused outside layer of chromium or molybdenum. Chromium or molybdenum reduces wear on the rings and cylinder walls (***Figure 17–36***). Since chromium or molybdenum rings reduce wear, the break-in time for such rings is increased.

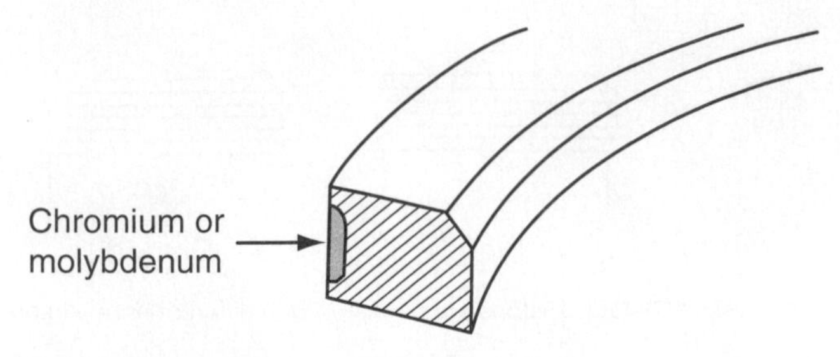

FIGURE 17-36 A chromium or molybdenum layer is placed on some compression rings to improve their wearing characteristics.

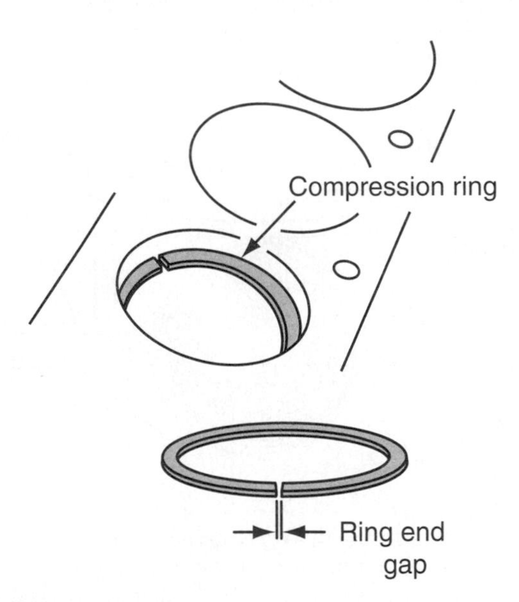

FIGURE 17-34 Piston ring end gap is the distance between the ends of the compression ring when it is placed in the cylinder.

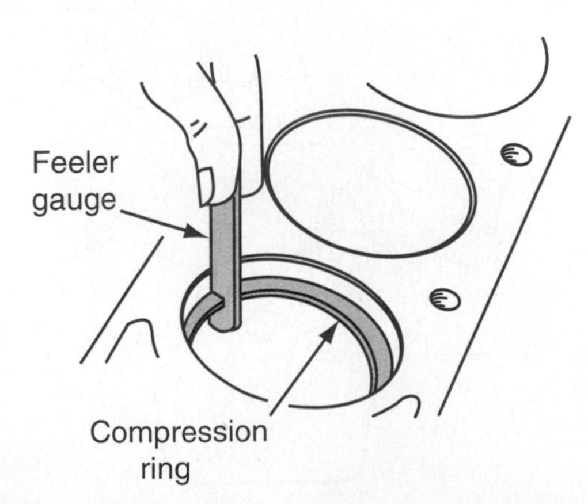

FIGURE 17-35 Piston ring end gap is checked using a feeler gauge.

RING DESIGN

There are several types of compression rings. ***Figure 17–37*** shows some of the more common rings. The plain or rectangular ring fits flat against the cylinder walls. The taper-faced ring improves scraping ability on the down stroke. Other rings such as the corner-grooved and reverse-beveled rings are designed as **torsional rings**. They have either a chamfer or counterbore machined into the rings.

Torsional rings are also shown in ***Figure 17–37***. Any chamfer or counterbore causes internal stresses in the ring.

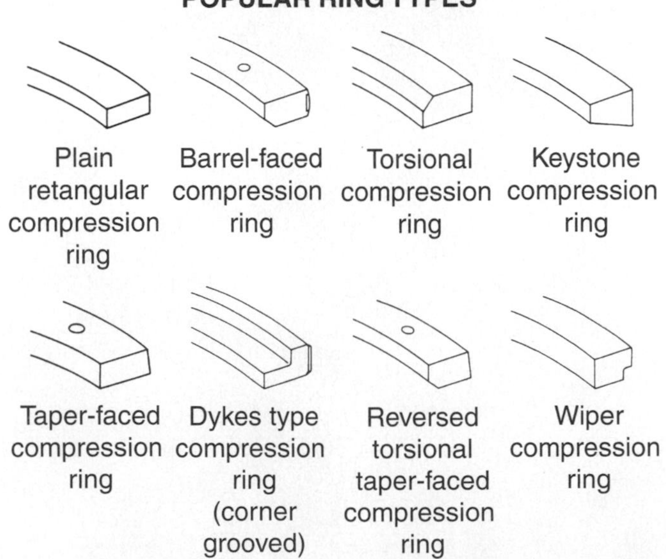

FIGURE 17-37 Compression rings can be designed in several ways. Note the different cross-sectional shapes. (Courtesy of Hastings Manufacturing Company)

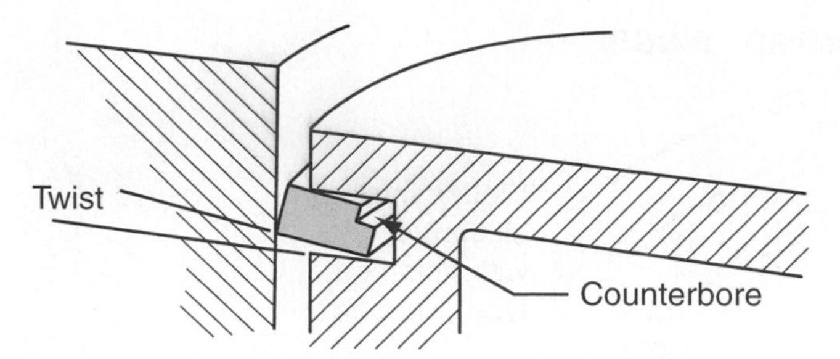

FIGURE 17-38 When a chamfer or counterbore is put on a compression ring, it causes the ring to twist in the groove. Twisting helps to seal the ring to the cylinder bore during operation.

These stresses cause the ring to twist slightly as shown in ***Figure 17–38***. This only happens when the ring is compressed inside the cylinder. Twist is used to form **line contact** sealing on the cylinder wall and the piston ring groove. Line contact improves the sealing and scraping characteristics of the rings. This is also called static tension.

When the ring is in this position, there is no downward pressure on the ring. This happens during the intake, compression, and exhaust strokes of the engine. High pressure is applied to the ring only on the power stroke. On the intake stroke, the twist forces the bottom corner of the ring to act as a scraper against the cylinder walls. This aids in removing any excess oil on the cylinder walls. On the compression stroke, the ring still retains the twist. This allows the ring to glide over any oil still on the cylinder walls rather than carrying it to the combustion chamber.

As the piston rises, compression pressures help flatten the ring for better sealing. On the power stroke, hot gases from the combustion chamber enter the ring groove. The ring is forced out and flat against the cylinder wall (***Figure 17–39***). Now there is a good seal during the power stroke. This process is also called dynamic sealing. During the exhaust stroke, static conditions are present again. The ring has a twist again. The twist causes the ring to glide again over any oil on the cylinder walls. Both the multiple-groove and radius or barrel-faced ring are designed to produce line contact as well. Better sealing characteristics are the result.

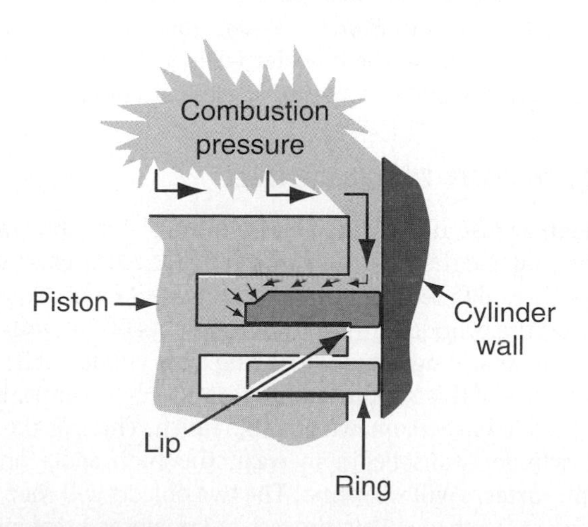

FIGURE 17-39 Pressure from combustion causes the rings to seal against the cylinder wall.

OIL CONTROL RINGS

When the engine is operating normally, a great deal of oil is thrown onto the cylinder walls. The connecting rods also splash oil on the walls. Some engines have a hole in the connecting rod to help spray oil directly on the cylinder walls. Oil on the cylinder walls aids lubrication and reduces wear. This oil, however, must be kept out of the combustion chamber.

Oil rings are made to scrape oil from the cylinder walls. They are also used to stop any oil from entering the combustion chamber and to lubricate the walls to prevent excessive wear.

All oil control rings are designed to scrape the oil off the walls on the downstroke (***Figure 17–40***). After being scraped off the cylinder walls, the oil passes through the center of the ring. It then flows through holes on the piston and back to the crankcase. This scraping process helps remove carbon particles that are in the ring area. The oil flow also helps cool and seal the piston.

Oil control rings are made of two, three, or four parts. These usually include an expander, a top rail, a spacer, and a bottom rail. On some rings, several of these parts may be built together in one piece. The expander is used to push the ring out against the cylinder walls. The top and bottom rails are used to scrape the oil off of the cylinder walls. These are sometimes called scraper rings. The spacer is used to keep the two scrapers apart. ***Figure 17–41*** shows a common type of oil control ring. On certain scraper rings, a

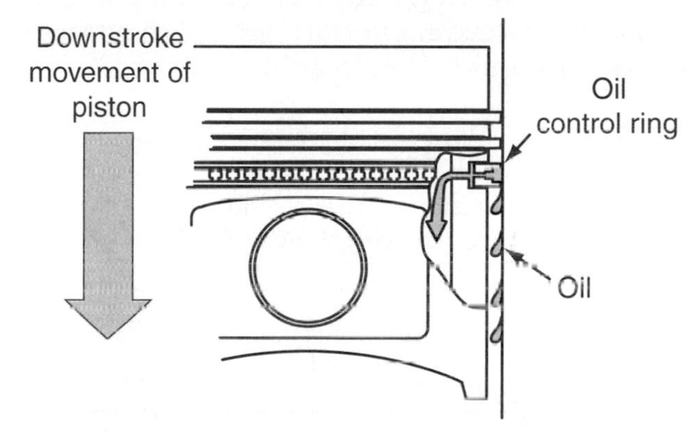

FIGURE 17-40 The oil being scraped off the cylinder wall passes through the ring, through holes in the piston, and back to the crankcase.

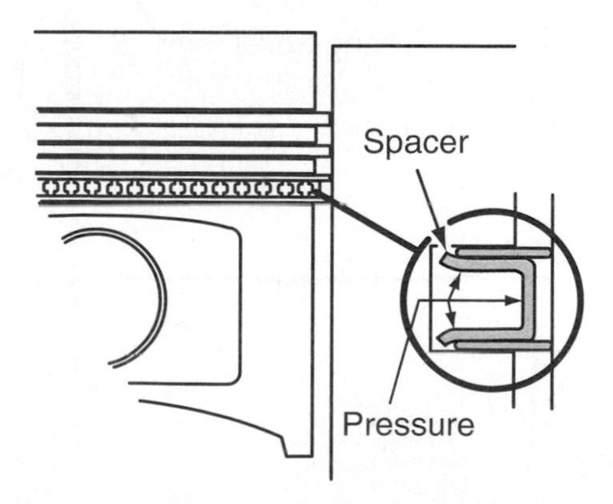

FIGURE 17-41 Oil control rings use a spacer to keep the two scrapers apart.

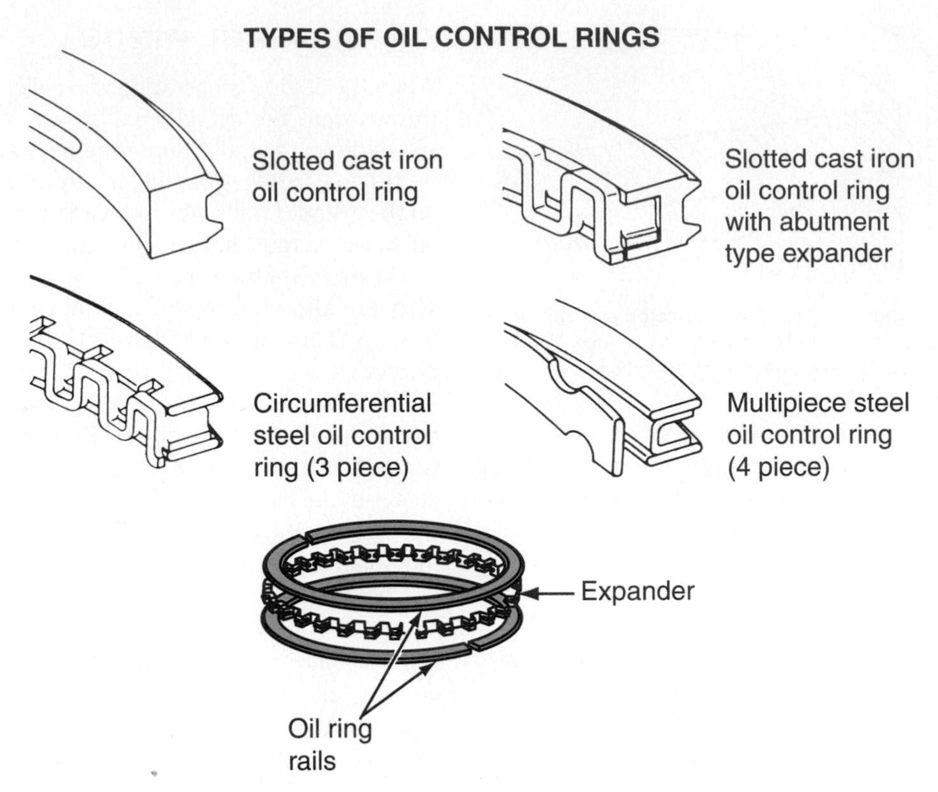

FIGURE 17-42 These are examples of different types of oil control rings.

chrome-plated section is used to improve wear characteristics. ***Figure 17–42*** shows several rings used on automobile engines and how they are put together.

CYLINDER WEAR

The motion of the piston and the position of the rings cause the cylinder to wear unevenly. ***Figure 17–43*** shows how a typical cylinder wears. The cylinder develops a taper. **Cylinder taper** is produced only where the rings touch the cylinder walls. The greatest amount of wear occurs near the top of the cylinder. The least amount of wear occurs near the bottom of the cylinder. This wear produces a ridge in the upper part of the cylinder bore. The ridge must be removed before the pistons are removed during an overhaul. If the ridge is not removed, the pistons will be damaged upon removal. New rings may also be damaged by hitting the bottom of the ridge after installation.

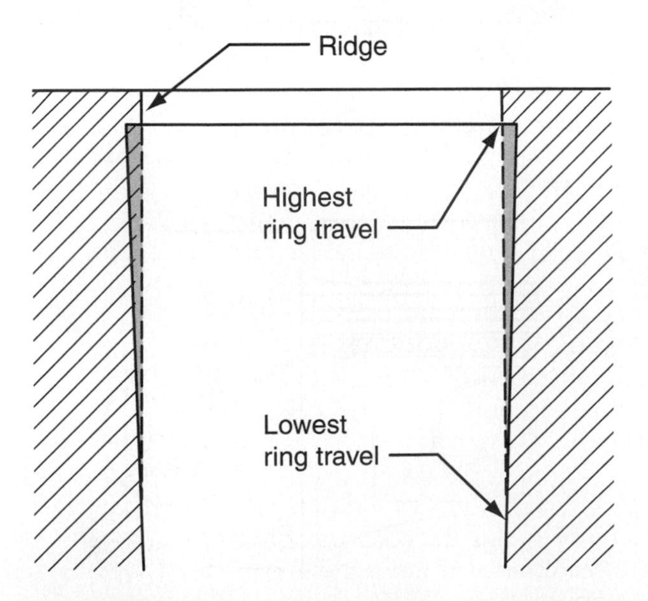

FIGURE 17-43 Over a period of time, the rings on the piston wear the cylinder bore to a tapered shape.

Manufacturer's specifications as listed in a vehicle's maintenance manual allow only a certain amount of taper on the cylinder walls. Too much taper affects the piston ring end gap as shown in ***Figure 17–44***. Too much end gap, as shown on the top of the cylinder walls, will produce excessive piston ring end gap and thus, excessive blow-by.

CYLINDER DEGLAZING

When new piston rings are placed in the piston, the outside of the ring and the cylinder wall will not have the exact same shape. This difference is shown in ***Figure 17–45***. The ring touches the cylinder only on the high spots of the ring. This causes poor sealing between the ring and cylinder walls.

Because of this condition, the ring and cylinder walls are designed to have a somewhat rough finish. Then, as the ring and cylinder walls begin to wear, the high spots on the rough surfaces will wear first. The two objects will then take on the same shape. This process is known as breaking in, seating in, or wearing in an engine. Some manufacturers

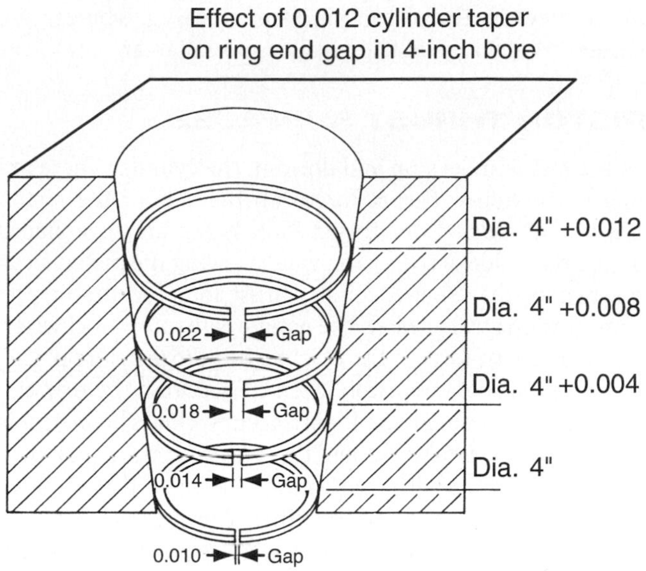

Effect of 0.012 cylinder taper on ring end gap in 4-inch bore

Dia. 4" +0.012
Dia. 4" +0.008
Dia. 4" +0.004
Dia. 4"
0.022 Gap
0.018 Gap
0.014 Gap
0.010 Gap

FIGURE 17-44 A tapered cylinder bore can change the ring end gap, producing excessive blow-by. (Courtesy of Hastings Manufacturing Company)

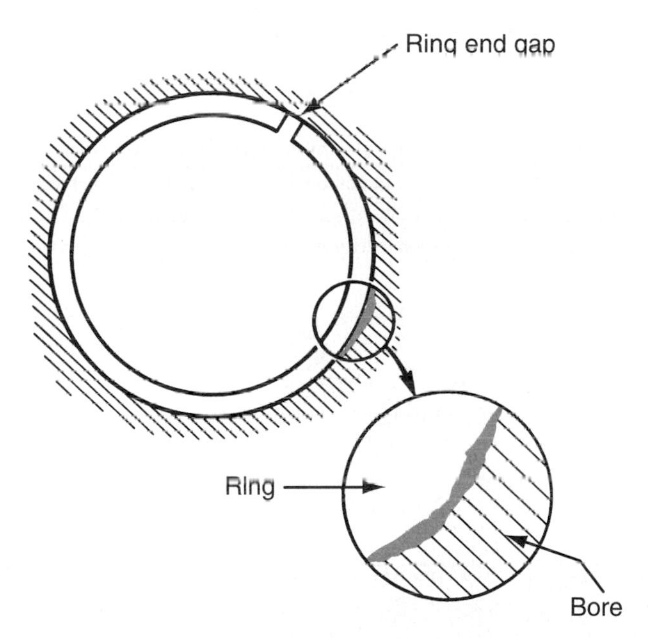

FIGURE 17-45 When new rings are placed in an engine, they are not seated. The shape of the ring is not exactly the same as the shape of the cylinder bore.

recommend that a certain distance in miles be driven for the "break-in period." For example, when a new car is purchased, the owner's manual may say to drive no faster than 50 miles per hour for the first 500 miles during the break-in period. This break-in period allows the piston rings and cylinder to wear correctly so that the rings seal effectively.

Deglazing is the operation used to make the cylinder rough (***Figure 17–46***). Deglazing also tends to help to retain some oil on the cylinder walls. The extra oil helps to lubricate the new piston rings and aids in the breaking-in process.

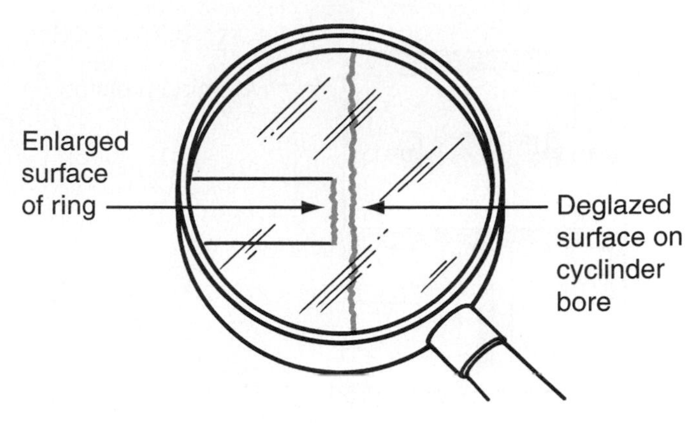

FIGURE 17-46 Deglazing a cylinder will aid in seating the rings.

Honing is another process used to improve the cylinder walls. Honing is somewhat different from deglazing. Deglazing is used to roughen up the cylinder walls to remove the glaze, whereas honing is used to make the cylinder rounder. Honing removes more material than deglazing, although the same size piston can be used. A hone is a very rigid tool, usually with four stones. There are only three stones on a deglazer. Note that cylinders can also be bored using a boring tool. In this case, boring removes more material so that larger pistons and rings must be used.

OVERSIZE PISTONS AND RINGS

If the ring end gap is too large, or if there is excessive wear on the cylinder bore, it is necessary to bore the cylinders to a larger size. If this must be done, both standard and oversize pistons and rings are available. The exact oversize ranges vary according to vehicle make and manufacturer. Common specifications include pistons that have 0.005-, 0.010-, 0.020-, and 0.030-inch oversizes. Common specifications for rings include 0.003-, 0.005-, 0.010-, 0.020-, and 0.030-inch oversizes. Each manufacturer may have different oversize ranges for both pistons and rings.

PISTON PINS

Piston pins, also known as wrist pins, are used to attach the piston to the connecting rod. ***Figure 17–47*** shows how the piston pin is positioned inside the piston. The piston pin can either be pressed in, as shown on the top view, or full floating, as shown on the bottom. When a press fit piston pin is used, it is pressed into the connecting rod and floats freely in the piston. When a full floating piston pin is used, two pin lock rings hold the piston in place.

Piston pins are made of high-quality steel in the shape of a tube. Piston pins are both strong and lightweight. They are **case-hardened** to provide long-wearing operation. Case-hardened means the outer surface of the piston pin has been hardened. The inside metal remains soft by comparison. Hardening a metal makes it very brittle and could cause the piston pin to break more easily. The softer metal on the

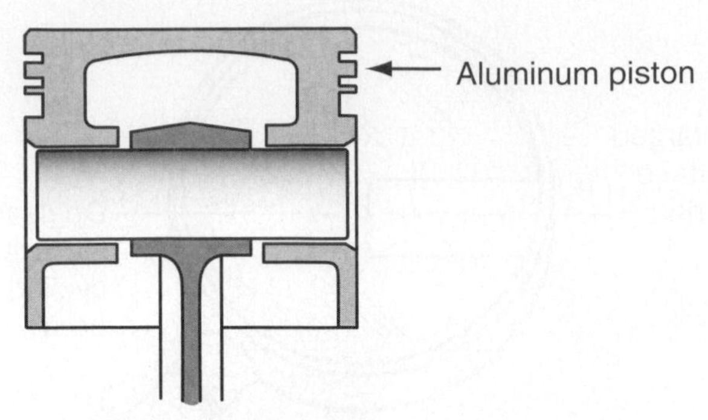

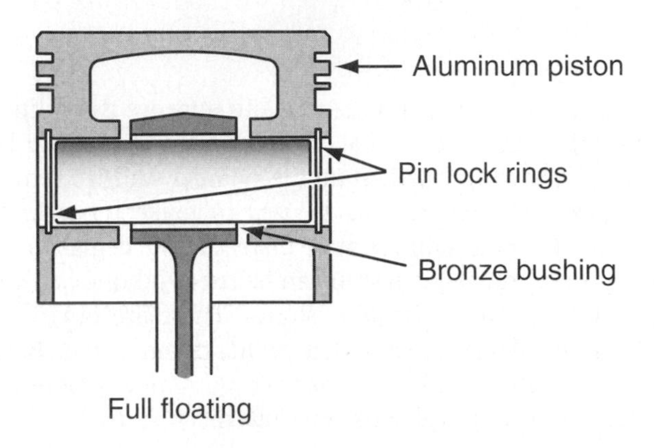

FIGURE 17-47 Piston pins are designed as either press fit or full floating.

inside prevents the piston pin from cracking, whereas the harder metal on the outside reduces overall wear.

PISTON THRUST SURFACES

As the piston moves up and down in the cylinder, there are **major** and **minor thrust** forces on the side of the piston (***Figure 17–48***). Minor thrust force is the pressure placed on the right side of the piston when viewing the piston from the rear of the engine. Minor thrust force occurs on the compression stroke, when the engine is turning clockwise.

Major thrust force is the reverse. It is the pressure placed on the left side of the piston when viewed from the rear and turning clockwise. Major thrust occurs when the piston is on the power stroke. When the crankshaft crosses over TDC, the piston shifts from minor to major thrust force. **Piston slap** is produced at this time. This means the piston slaps against the cylinder walls. Excessive piston slap occurs when there is too much clearance between the piston and the cylinder walls. This can cause noise and wear on the piston and the cylinder walls. Excessive piston slap can usually be eliminated by replacing the rings or by boring out the cylinder and installing an oversize piston and rings.

PISTON PIN OFFSET

To eliminate piston slap, the piston pin is located slightly off center. The piston pin is located closer to the minor thrust surface. Because the mechanics of movement have been changed, piston slap is reduced.

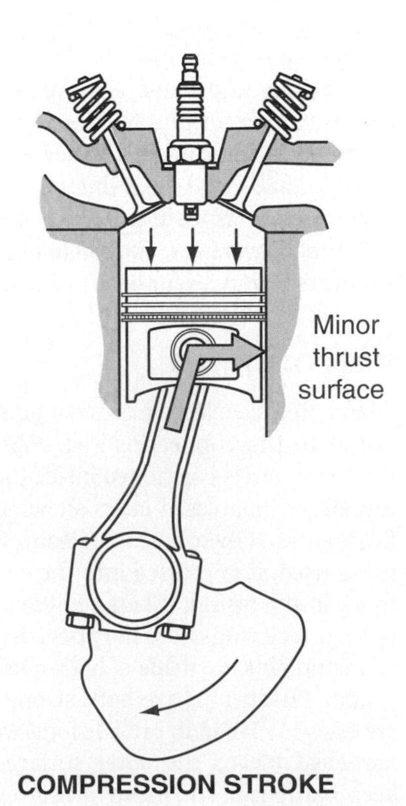

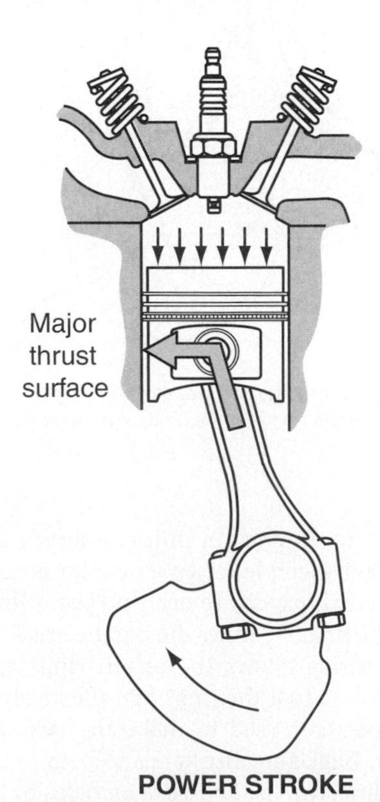

FIGURE 17-48 Major and minor thrust surfaces are built into the piston to absorb piston thrusts.

CONNECTING ROD

The connecting rods connect the piston to the crankshaft. One end is attached to the piston pin. The other end is attached to the crankshaft rod journal. ***Figure 17–49*** shows the parts of the connecting rod. These include the connecting rod, the rod cap, two bearing inserts, the connecting rod bolts, and the nuts. Surfaces called **bosses** are forged into the connecting rod to balance it. They are machined until a perfect balance is obtained.

The connecting rod and cap are line bored. The cap is attached to the connecting rod when the inside bore is machined. It is important to keep the connecting rod cap matched to the connecting rod during service. If they are ever mismatched, the bore may be incorrect. This will cause the connecting rod and bearing cap to be misaligned, causing damage or excessive wear on the bearings.

The caps and connecting rods are marked with numbers to keep the caps matched to the connecting rods (***Figure 17–50***). Always match these numbers when rebuilding an engine. On certain engines, the numbers have been omitted, whereas on others the same number may appear on the connecting rod. Always keep the cap matched to the correct connecting rod. If the caps are not matched to the original connecting rod, when tightened down, the crankshaft will lock up and not be able to turn. Some technicians also use a punch to mark each cap and cylinder rod accordingly. This ensures that the cap and connecting rod will be assembled correctly.

Some connecting rods are powder forged. This means that the iron powder, graphite, and copper are precast in the molds. The mixture is then heated to help mix the materials together. The end result is that each connecting rod is much closer in weight to the others. This makes the engine easier to balance. In addition, when powder-forged rods are made, the rod caps are not sawed off as in the production of conventional rods. To cut the connecting rod cap from the rod, two lines are scribed on the sides of the rods. The scribe line creates a stress point. When the rod caps are pulled apart from the connecting rod, the rod separates at the scribed lines. This causes each break to be different. However, each connecting rod now only fits with a particular connecting rod cap. This helps the service technician more easily match the cap and the rod. ***Figure 17–51*** shows an exaggerated break line between the connecting rod and the connecting rod cap.

Some connecting rods have an oil squirt hole in their lower section. This hole directs oil to the cylinder walls for improved lubrication. Other designs have a squirt hole in the cap mating surface. Some designs have no squirt hole.

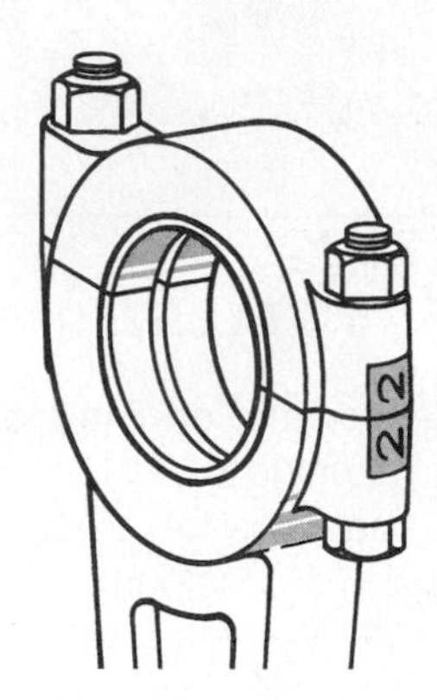

FIGURE 17–50 Connecting rod caps and connecting rods should always be kept together. Numbers are stamped on them to keep them in order. (Courtesy of Federal-Mogul Corporation)

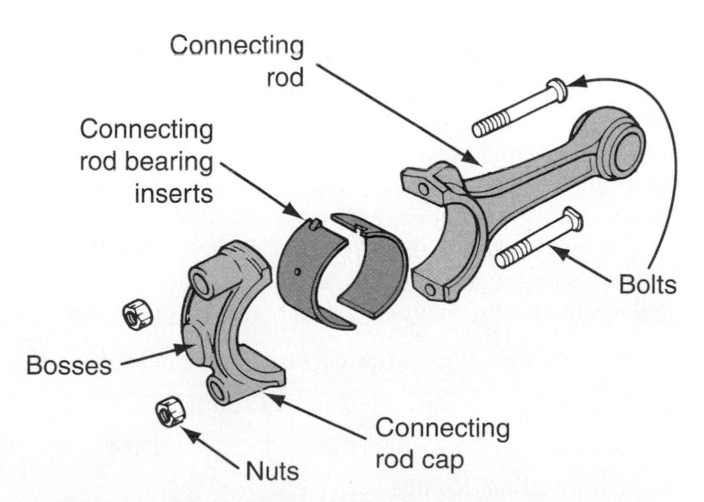

FIGURE 17–49 The connecting rod assembly includes the connecting rod, the connecting rod bearing cap, two bearing inserts, and the bolts and nuts. (Courtesy of Federal-Mogul Corporation)

CONNECTING ROD BEARINGS

The connecting rod bearings are designed the same as the main bearings. Each bearing is made of two pieces. Small locating lugs are used to locate the bearing in the proper position and to prevent the bearing from spinning in the bore. These lugs fit into a slot machined into the connecting rod cap and connecting rod. As in the main bearings, the connecting rods are made of steel back, copper and lead inside, and a thin coating of pure tin.

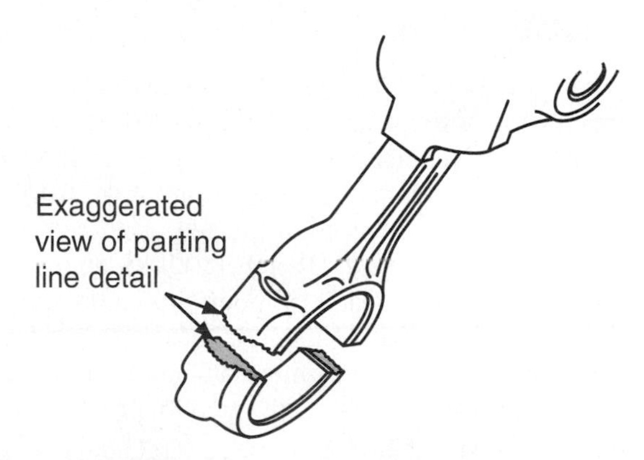

FIGURE 17–51 This connecting rod and connecting rod cap have been separated at a parting line that makes a jagged edge so that the cap cannot be mixed up with another cap.

Problems, Diagnosis, and Service

Safety Precautions

1. Many corners on the engine block may be sharp from being machined. Be careful not to cut your hands when handling the block. Use gloves when lifting or moving engine parts.
2. The clearances between the crankshaft and the block are very small. When turning the crankshaft during bearing installation, be careful not to get your fingers caught, as injury may result.
3. The engine crankshaft, pistons, and other parts are heavy. Be careful not to drop them on your toes or feet.
4. Always wear OSHA-approved safety glasses when working on the internal parts of the engine, especially during deglazing. Parts of the deglazer may break off and fly into your eyes.
5. When taking a compression test, make sure the engine will not start during cranking. High-pressure gases may escape from the cylinder being tested, and you may get a shock from the spark plug wires. Always disconnect the ignition system before cranking.
6. Make sure that all tools are clean and free of grease. Under high torque applications, a tool can become very slippery and slip off the bolt or nut.
7. Many bolts and nuts require precise torque readings. Always use a torque wrench to tighten bolts and nuts correctly.
8. When removing the vibration damper bolt, make sure the crankshaft can be held tight. Use a small piece of wood or other soft object to hold the crankshaft in place while loosening or tightening the vibration damper bolt.
9. When rebuilding an engine, always clean all parts before inspection. Parts that are dirty may easily slip and be dropped when being moved around.
10. When working with heavy parts, always lift the parts with your legs, not your back. In addition, if a part is too heavy, have a partner help. Always be careful not to set the heavy parts on your fingers.
11. When lifting a block, be sure to use the correct engine hoist.
12. Always use the correct tools for the job being done.

PROBLEM: Lack of Power, Excessive Oil Consumption, and/or Steady Engine Miss on One Cylinder

Engines, especially those with many miles on them, often lack power. In addition, these engines may also burn oil excessively. One area often identified as the problem is that of bad or worn rings.

DIAGNOSIS

Often an engine will show these signs of wear and poor performance. Some of the first steps to take are to check for a broken or bent valve. Also, there may be a blown head gasket. If there is no evidence of such damaged components, the problem may be worn rings.

Worn rings can be identified easily by checking the compression pressure within each cylinder and comparing the readings. Each manufacturer suggests maximum variations in cylinder pressure. For example, an engine that has a cylinder variation of more than 20% to 30% probably needs the piston rings replaced.

SERVICE

A compression test is made to determine the condition of the engine parts that affect cylinder pressure. This test should always be performed as part of a tune-up or whenever there is a complaint about poor engine performance. This check should also be made when there is excessive oil or fuel consumption. A compression test is a performance comparison between the cylinders, which is then compared with manufacturer specifications. The specifications are listed in many automotive maintenance and service manuals. All conditions during the test should be the same. These include the cranking speed, position of throttle and choke, and the temperature of the engine during the entire test. There are several types of compression tests: the dry and wet compression tests, and a running compression test. Preparation for the first two tests includes the following:

1. The engine must be at normal operating temperature.
2. Check to be sure the battery is fully charged.
3. Check the starter for operation at the same speed.
4. Remove all spark plugs.

CAUTION: *When removing the spark plug wires, twist and then pull only on the boot. Pulling on the wires may damage certain types of carbon-impregnated plug wires, which can cause the engine to misfire severely.*

5. Disable the ignition system to prevent high-voltage sparking.
6. Connect a remote starter switch so the engine can be cranked from the engine compartment.
7. Make sure the throttle plate is fully open.

The Dry Compression Test The dry compression test is performed to get a basic comparison between the cylinders and to determine the overall condition of the compression chamber. The dry compression test is performed as follows:

1. Hold the throttle linkage wide open.
2. Screw the compression tester into the spark plug hole (***Figure 17–52***). Do not use a wrench to tighten the compression tester.

CAUTION: *Be careful not to touch the hot exhaust manifold if the engine is at operating temperature.*

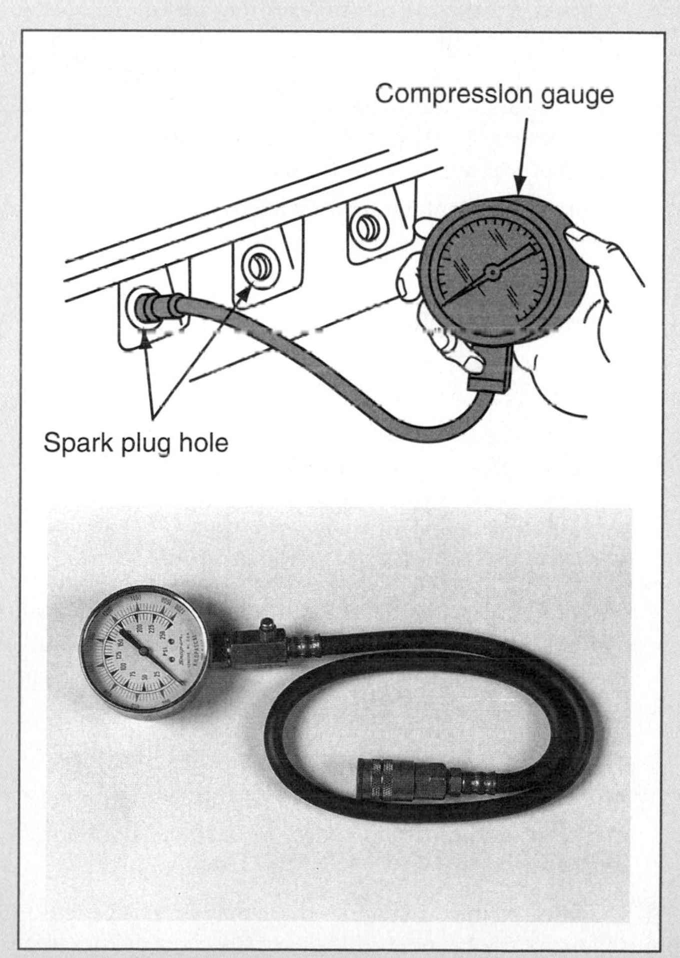

FIGURE 17-52 Both wet and dry compression tests can be made using the compression tester.

3. Crank the engine until the compression gauge reaches the highest reading. This should be reached with three or four revolutions of the engine. Write down the highest reading.

CAUTION: *Be careful not to get the remote starter wires or other tools in the fan area.*

4. Repeat this procedure with the same number of revolutions on all cylinders. Be sure to release the compression gauge after each test.
5. Compare the readings of all cylinders.
6. A 20–25% variation between the cylinder readings is usually satisfactory. If the readings vary more than that, the cause should be determined. Refer to the manufacturer's specifications for the exact amount of variation on each cylinder.

The Wet Compression Test The wet compression test is taken to determine if low readings are caused by compression leakage past the valves, piston rings, head gasket, or other leaks such as a crack or hole in the piston. The procedure for the wet compression test is as follows:

1. Squirt a small amount of oil into each cylinder through the spark plug hole.
2. Now perform the compression test as outlined in the dry test. The oil should temporarily form an improved seal between the piston and cylinder wall.
3. If the readings are 10 pounds or more higher than during a dry test, compression is probably leaking by the rings.
4. If there is no difference between the wet and dry test, or if the wet test produces readings of less than 10 pounds, the compression loss is probably caused by compression leaking past the valves or head gasket.
5. When taking a compression test on new rings, you may not be able to obtain the full compression until the rings are seated.

The Running Compression Test Many service technicians also perform the running compression test. A running compression test is used to check how well the engine breathes. Use the following general procedure to do a running compression test.

1. Remove one spark plug and ground the plug wire to prevent any powertrain control module damage. Also disconnect the fuel injector on the cylinder that is being checked.
2. Insert the compression tester into the spark plug hole.
3. After starting the engine, take a reading and write the result down.
4. Goose the throttle for a "snap" acceleration reading and write it down. To produce the snap acceleration manually, move the throttle as fast as possible without speeding up the engine. This forces the engine to take a "big gulp" of air.

5. Running compression at idle should be 50–75 pounds per square inch. The compression reading during the "snap" acceleration should be about 80% of the running compression.
6. If there is restricted air intake because the intake valves have excess carbon on them, or due to worn cam lobes or rocker arms, it will show during the snap acceleration reading. The reading will be must lower than the 80% of the running compression.

PROBLEM: Engine Needs Rebuilding

As an engine accumulates miles, many parts eventually wear out or break. Often it is necessary to rebuild an engine and to replace those parts that have excessive wear. In addition, many of the parts that are worn are part of the basic engine block, including the crankshaft, block, pistons, and seals.

DIAGNOSIS

Various signs and conditions help the automotive technician determine if an engine needs a complete overhaul. Six of the more common signs include the following:

1. Excessive oil consumption
2. Low or uneven compression readings
3. Lack of power and performance
4. Excessive mileage on the vehicle
5. Dirty oil from combustion blow-by
6. Engine missing on one or more cylinders

SERVICE

There are many important service procedures and suggestions for rebuilding an engine.

Cylinder Block

1. Always deglaze the cylinder bore when using new piston rings. A deglazing tool is shown in ***Figure 17–53***.

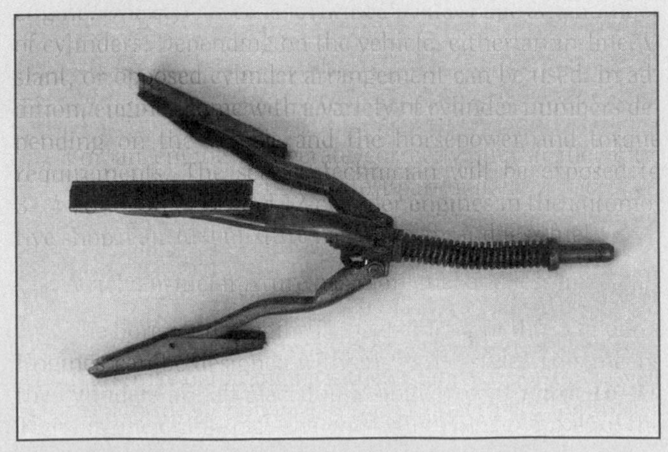

FIGURE 17-53 This deglazer is used to roughen up the cylinder walls to aid in break-in when rebuilding an engine.

   a. With the piston removed from the cylinder, place the deglazer in the cylinder bore.
   b. Place a small amount of cutting oil in the cylinder.
   c. Using a drill attached to the deglazer, rotate slowly, making sure the deglazer does not hit against the crankshaft. If it does hit the crankshaft, the deglazer stones may break.

CAUTION: *Keep your fingers away from the spinning deglazer. Also, never spin the deglazer outside the cylinder; it may separate and be thrown into your body.*

   d. While rotating the deglazer, move it up and down (about 45 degrees) to obtain a cross-hatch pattern in the cylinder bore.
   e. Remove the deglazer and clean the bore.
   f. Check the bore for any spots on the cylinder bore that were not deglazed evenly.
   g. If there are spots that have not been deglazed, deglaze again for a short time. If the spots are too large, however, the cylinder bore may have to be honed. Honing takes more material off than deglazing. Again, make sure the cylinder bore has been completely cleaned after deglazing, using clean rags.

2. Always check the top of the block for flatness. Checking the flatness of the block determines if there has been warpage of the block. Warpage can be created from excessive heat. If the block is slightly warped, the head gasket may not seal correctly and will leak. Use the following procedure to check for flatness on the block:
   a. Use a long, steel, straight-edged ruler.
   b. Place the straightedge on the top of the block.
   c. Using a feeler gauge, check for clearance under the straightedge. If the space is greater than the specifications listed in the repair manual, the block may have to be machined to produce a flat surface again.
3. Always check the block for small cracks between the cylinder walls and the coolant passageways. Excessive heat can cause the block to crack at this point, and coolant water can get into the cylinder.
4. When installing core plugs, make sure each plug is inserted evenly. If it is not inserted evenly, there is a good possibility that the core plug will leak engine coolant. Use a small amount of sealant between the block and the plug.
5. Remember to check the oil passageways in the block for foreign matter blocking the holes (***Figure 17–54***).

CAUTION: *Be careful not to cut or scrape your hands on the sharp corners of the block.*

6. As indicated in this chapter, there may be a ridge left on the top of the cylinder wall where the rings did not wear the cylinder material away. The ridge should be removed before the pistons are removed. If the ridge is not re-

moved first, when the pistons are removed, the rings may be broken or the piston surfaces scratched or damaged. This ridge can be removed using a tool called a ridge reamer. It is a device that has a cutting tool that is inserted, tightened down, and turned inside of the cylinder. As the tool is turned, the tool bit cuts the ridge away from the top of the cylinder. ***Figure 17–55*** shows an example of ridge reaming.

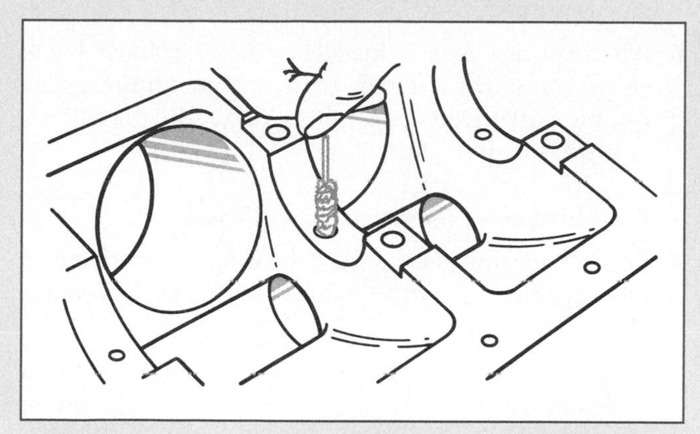

FIGURE 17-54 All oil passages on the crankshaft should be cleaned before reassembling the engine. (Courtesy of Federal-Mogul Corporation)

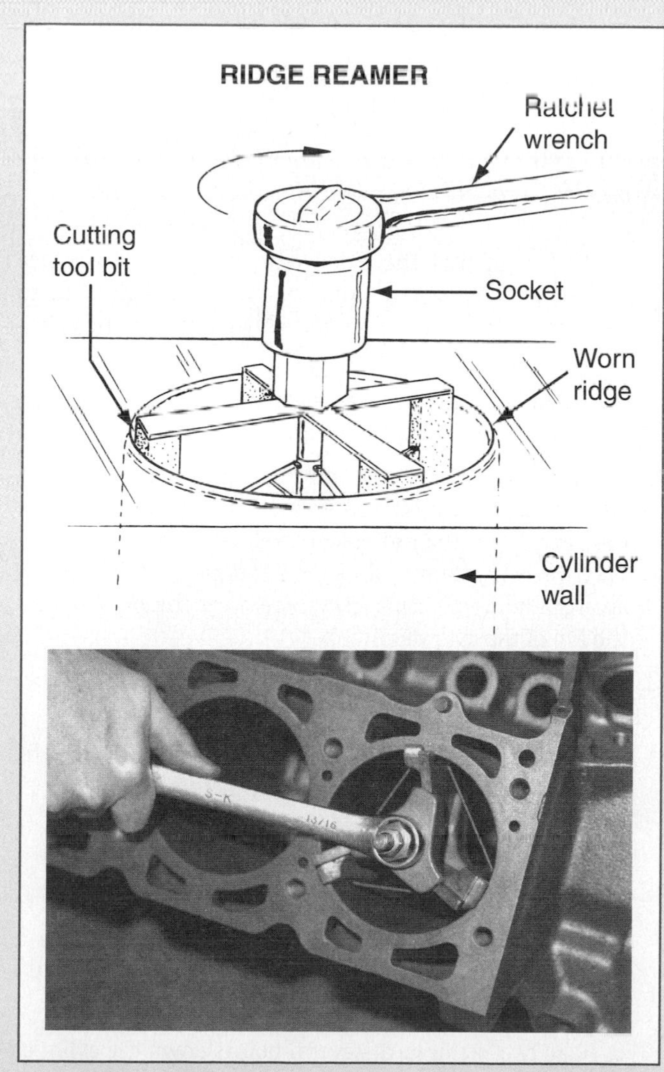

FIGURE 17-55 A ridge reamer can be used to remove the ridge worn into the top of a cylinder block.

The ridge reamer is a tool that has several points of contact with the cylinder walls. One of the points has a cutting tool at its end. The ridge reamer is inserted into the cylinder bore and slightly tightened. Then, a standard box wrench or socket and ratchet wrench is used to turn the tool. As the tool is turned, it slowly cuts away the ridge on the top of the cylinder. After the ridges have been removed and the piston removed, the cylinder wall should be checked for wear, taper, and other damage. Note that if the cylinder walls are too badly damaged and out of specifications, the cylinders may have to be bored to a larger size.

Crankshaft and Main Bearings

1. When installing main bearing caps, make sure the caps are placed in the correct order. The caps should be inserted exactly in the same cylinder and position as when disassembled.
2. Always check bearing clearance with Plastigage (***Figure 17–56***). Use the following procedure when using Plastigage:

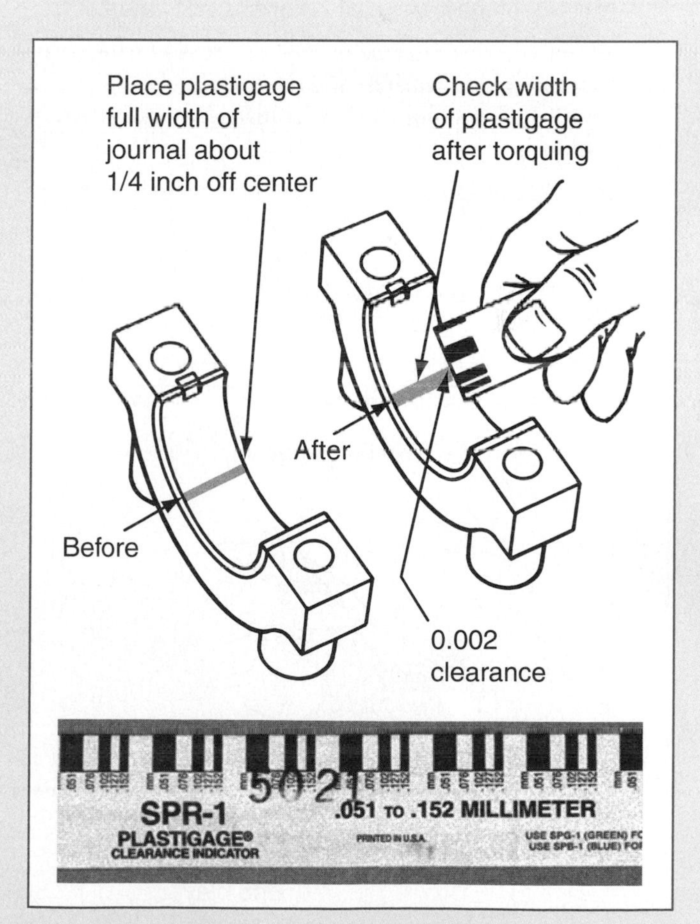

FIGURE 17-56 Plastigage is used to check the clearance of the main and connecting rod bearings. As the cap is torqued to specifications, the small plastic strip flattens out. The width of the flattened strip determines the clearance.

a. Place a small amount of Plastigage on the bearing journal.

b. Place the bearing and bearing cap on the journal and torque to service manual specifications.

c. Remove the bearing cap and bearing.

d. Measure the width that the Plastigage has been flattened. This measurement can be made using a vernier caliper or the gauge furnished on the Plastigage container.

e. The width of the flattened Plastigage is a measure of the bearing clearance.

3. When installing the crankshaft, use a feeler gauge or dial indicator to check for crankshaft end play (***Figure 17–57***).

a. First, put the main bearings in place and install the crankshaft. Then install the main bearing caps and tighten to the correct torque specification.

CAUTION: *The crankshaft is very heavy. Be careful not to pinch your fingers as the crankshaft is being set into the block.*

b. Move the crankshaft fully to one end of its end play.

c. Select the correct size feeler gauge to measure the amount of movement axially on the crankshaft.

d. Check the clearance between the thrust bearing (usually the center bearing) and the thrust surfaces on the crankshaft and compare with manufacturer's specifications.

4. Care must always be taken to install the main bearings in the correct piston. All oil holes must be correctly lined up with the oil passages in the block assembly for correct installation. Also, the small notches on both bearing segments (used to stop the bearings from spinning) must be installed on the same side. It's always a good idea to use a bottoming tap to clean the threads in the bolt holes before installing the main bearing caps.

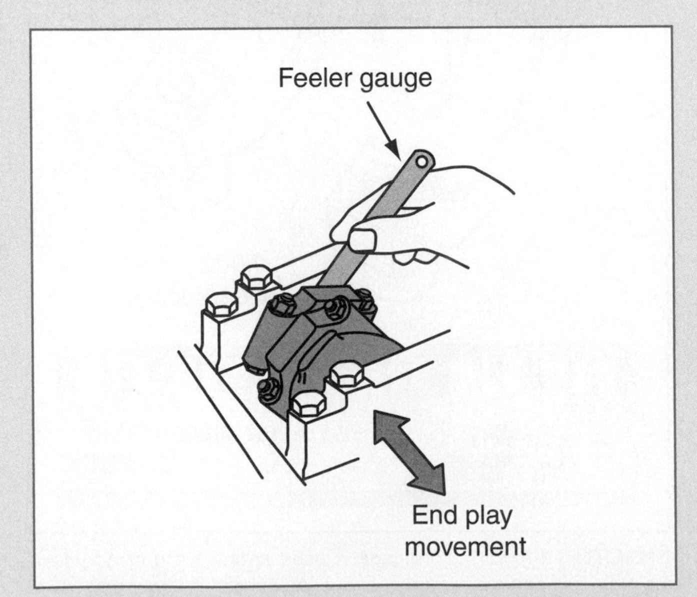

FIGURE 17-57 A feeler gauge can be used to check for crankshaft end play.

5. When rebuilding the engine, make sure to replace both the front and the rear crankshaft seals. Replacing these seals will ensure that there are no oil leaks after the engine is running.

6. Whenever any part is installed on the cylinder block, always torque the fasteners to the correct torque specifications. To get accurate readings, make sure the bolt and hole threads are clean.

Pistons and Connecting Rods

1. Before removing pistons from the cylinder bore, remove the ridge that has been produced on top of the cylinder with a ride reamer.

a. A ridge reamer such as the one shown at the top of ***Figure 17–55*** can be used to remove the ridge on top of the cylinder bore.

b. Bolt the ridge reamer on the block according to the manufacturer's recommendations.

c. Adjust the ridge reamer to take a very small cut.

d. After getting the feel of rotating the ridge reamer, adjust it to take a larger cut.

CAUTION: *If the ridge reamer is too hard to turn, too much of a cut is being made. Readjust as necessary.*

e. Be careful that the ridge reamer does not produce chatter. This chatter may cause vertical grooves to be cut into the cylinder, possibly damaging the cylinder bore.

f. Continue removing the ridge until there is no ridge left on the cylinder wall.

2. When pistons are being removed from the cylinder bore, cover the connecting rod bolts with a piece of rubber hose to protect the rod bolt threads and crankshaft journal from being damaged. Using a soft piece of wood, push against the connecting rod to remove the piston from the top of the cylinder block.

3. When the pistons are removed, check for cracking between the piston pin and the top of the piston.

4. After the pistons have been removed, always check the piston pin for excessive wear in the piston.

5. When pistons have been removed from the cylinder bore, check the amount of cylinder taper on the bore.

a. Cylinder bore taper can be checked by using a bore gauge.

b. Taper can also be checked using an inside diameter micrometer.

c. Take two readings at several points down the cylinder (take one reading, then rotate the gauge 90 degrees and take a second reading).

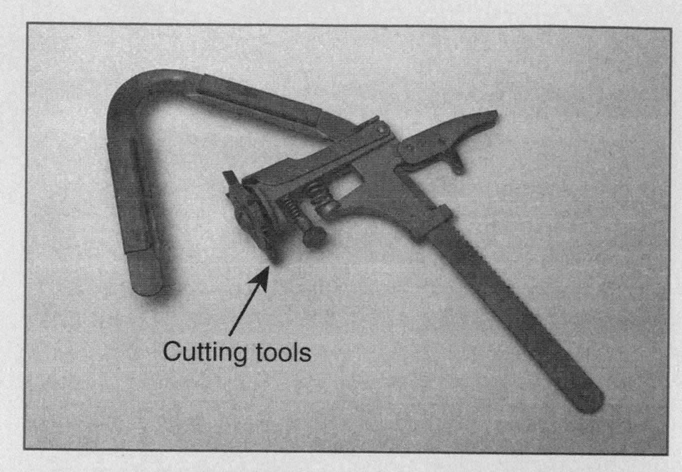

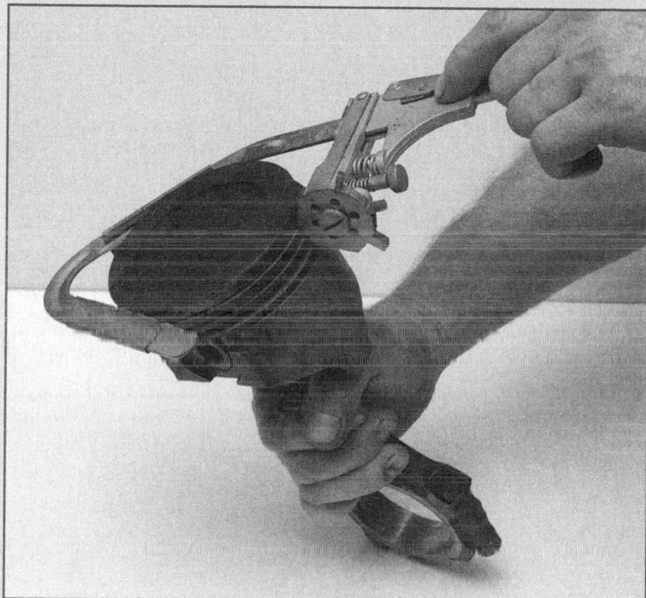

FIGURE 17–58 A ring groove cleaner is used to remove carbon from the ring grooves.

6. Use a ring groove cleaner (***Figure 17–58***) to clean the carbon built up in the ring grooves before installing new piston rings. Cleaning the carbon from the ring grooves ensures that the piston rings will be able to move adequately for proper sealing.
   a. Select the correct size cutting tool for the ring groove.
   b. Place the piston connecting rod in a vise so the piston is stable.
   c. Place the ring groove cleaner in place with the cutting tool in the ring groove.
   d. Rotate the ring groove cleaner until all carbon is removed from the groove.

CAUTION: *Always use protective gloves when cleaning the ring grooves since the tool may stick during rotation and injure your hands.*

   e. Continue cleaning each groove until the carbon has been cleaned from all ring grooves.
   f. Clean the piston completely before installing new rings.
7. Always check piston ring end gap on new piston compression rings. End gap measurements are typically about 0.004 inch for each inch diameter of the piston.
   a. Place the compression ring near the top of the cylinder bore without the piston.
   b. Make sure the compression ring is placed evenly in the bore.
   c. Using a feeler gauge, measure the gap produced by the ring in the cylinder and compare it with the manufacturer's specifications. If the end gap is too large, the cylinder may have to be bored to a larger diameter, thus requiring a different set of pistons and rings.
8. When installing the rings on the pistons, make sure the rings are placed in the correct groove and installed with the correct side up. In addition, the ring end gaps should be correctly positioned. Generally the ring end gaps should be placed at between 90 and 120 degrees from the other gaps.

CAUTION *Always use a ring expander to expand the rings over the piston, being careful not to break the brittle compression rings.*

9. When installing the pistons after they have been removed and cleaned, replace them in the same cylinder and in the correct position. The service manual gives the correct position for each piston during installation.
10. Always check the connecting rod bearing clearance with Plastigage. Use the same procedure as with the main bearings.
11. When installing pistons, use a piston ring compressor (***Figure 17–59***)
    a. Make sure there is sufficient lubrication on the new rings and cylinder walls before installing the piston into the cylinder.
    b. Place the bearings on the connecting rods.
    c. Put short rubber hoses over the ends of the connecting rod bolts. These small rubber hoses will protect the bolts and crankshaft from marks and scratches during installation.
    d. Place the connecting rod gently in a vise.
    e. Make sure the end gaps of the rings are not aligned together. The end gaps are typically staggered around the piston (90 to 120 degrees apart).
    f. Place a sufficient amount of oil on the piston and rings. Then place the piston ring compressor around the top of the piston and rings.
    g. Tighten the ring compressor to compress the rings.
    h. Place the assembly in the cylinder bore, making sure the piston and connecting rod are in the correct position.

FIGURE 17-59 The ring compressor is used to compress the rings so the piston can be inserted into the cylinder bore.

i. Push the piston down into the bore with a soft rubber hammer until all the rings are inside the cylinder. The ring compressor will relax somewhat when the rings are inside the cylinder.

CAUTION: *Do not force the piston into the cylinder. There should be little resistance as the piston goes into the cylinder. If the piston is forced into the cylinder bore, the rings may break. If there is resistance, the ring compressor may not be aligned correctly, or it may not have compressed the rings enough. Remove the assembly and start over again.*

12. Care must always be taken to install the connecting rod bearings in the correct position. All oil holes must be correctly lined up with the oil passages in the crankshaft for correct installation. Also, the small notches on both bearing segments (used to stop the bearings from spinning) must be installed on the same side.

Service Manual Connection

There are many important specifications to keep in mind when rebuilding an engine. These specifications can be found in most service manuals. To identify the specifications for an engine, you will need to know the VIN (vehicle identification number) of the vehicle, the type and year of the vehicle, and the type of engine. Although they may be titled differently, some of the more common specifications found in service manuals are listed here. Note that these specifications are typical examples. Each vehicle and engine will have different specifications.

Common Specification	Typical Example
Compression test	The lower cylinder must be within 20–25% of the highest
Connecting rod cap bolts (torque)	45 ft.-lb
Crankshaft end play	0.0035–0.0135 in.
Flywheel to crankshaft (torque)	60 ft.-lb
Main bearing cap bolts (torque)	100 ft.-lb
Main bearing clearance	0.0005–0.0015 in.
Piston clearance	0.0010–0.0020 in.
Piston ring end gap (compression rings)	0.010 in.
Piston ring end gap (oil)	0.015 in.
Rod bearing clearance	0.0005–0.0026 in.
Vibration damper (torque)	300–310 ft.-lb

SUMMARY

The following statements will help to summarize this chapter:

- The basic engine includes the cylinder block, crankshaft, bearings, seals, pistons, rods, and the camshaft.
- The cylinder block provides the basic structure of the engine and is the foundation to which all other parts are attached.
- Many passageways are drilled in the cylinder block to allow for oil to be added in order to lubricate parts as well as for different parts to be attached.
- Cylinder blocks are made by pouring liquid metal into a specially shaped mold.
- Special metals are added to the molten metal to make it stronger and able to withstand more wear.
- Core, or expansion, plugs are installed into the cylinder block to protect it from cracking if the coolant freezes and expands.
- Cylinders are machined into the block so that the pistons can fit correctly.
- Some engines have wet sleeves or dry sleeves for the pistons to move in.
- The crankshaft assembly is used to convert the downward motion of the piston to rotary motion for power.
- The crankshaft assembly includes the bearings, flywheel, vibration damper, timing gears, and sprockets.
- The crankshaft can be either forged or cast.
- Vibration dampers are used to reduce internal vibration in the crankshaft.
- Bearings on the crankshaft must be designed to conform to the shape of the crankshaft, to carry heavy loads, and to be able to resist fatigue and corrosion.
- Thrust bearings are used on the crankshaft to absorb axial thrust forces.
- The crankshaft has holes drilled from each main bearing journal to the connecting rod journal for lubrication.
- Correct bearing clearances are critical for proper engine operation.
- Seals on both ends of the crankshaft are needed to keep oil from leaking out.
- The piston and rod assembly includes the compression and oil control rings, the piston, the piston pin, the connecting rod, the bolts and nuts, and the bearing inserts.
- The piston, which is made of aluminum and certain alloys, is designed to withstand many forces including top pressure during combustion and side thrust forces.
- Pistons are also designed to handle expansion for heat.
- Piston rings are used on the piston to help seal the pressures of compression and combustion.
- Generally there are two compression rings and one oil control ring on each piston.
- During normal operation, the cylinder bore wears in the shape of a taper.
- The taper of the bore should be checked to make sure it is within specifications.
- A ring groove cleaner is used to remove carbon from the ring grooves.
- When installing a piston into the cylinder bore, always compress the rings using a ring compressor.
- The ridge reamer is used to remove the upper ridge, which is formed on the top, inside area of the cylinder walls.
- Plastigage is used to check the clearances of the main and connecting rod bearings.
- Cylinder compression checks can help determine if the rings are bad on an engine.

TERMS TO KNOW

Can you explain each of the following terms? Review the chapter until you can use each term correctly.

Alignment
Axial motion
Babbitt
Blow-by
Bosses
Cam ground pistons
Case-hardened
Cast
Core plugs
Counterweights
Crankcase
Cylinder sleeves
Cylinder taper
Die cast
Fillets
Forged
Heat dam
Heat treated
Interference fit
Line contact
Main bearing clearance
Major thrust
Minor thrust
Piston slap
Plastigage
Scuffing
Slipper skirt
Torsional rings
Torsional vibration

REVIEW QUESTIONS

Multiple Choice

1. The part of an engine that is the basic structure to which other parts are bolted is called the:
   a. Crankcase
   b. Crankshaft
   c. Cylinder block
   d. Camshafts
   e. Cylinder head

2. When a cylinder block is cast, the internal passageways have a ________ core.
   a. Metal
   b. Sand
   c. Paper
   d. Plastic
   e. Cast

3. ________ are used to seal up holes (used to remove sand cores during casting) in the block and can protect the block at times from cracking if the coolant freezes.
   a. Core plugs
   b. Metal seals
   c. Rubber seals
   d. Sand plugs
   e. All of the above

4. Which of the following is a type of sleeve or liner used on a cylinder block?
   a. Wet sleeve
   b. Dry sleeve
   c. Cast in liner
   d. All of the above
   e. None of the above

5. Cylinder blocks can be made of:
   a. Cast iron
   b. Brass
   c. Aluminum
   d. Plastic
   e. A and C

6. The device used to change reciprocating motion to rotary motion on the engine is called the:
   a. Crankshaft
   b. Piston
   c. Piston rings
   d. Connecting rod bearings
   e. Cylinder head

7. Thrust surfaces are machined on the crankshaft to:
   a. Absorb rotary thrust
   b. Absorb axial thrust
   c. Help balance the crankshaft
   d. Improve friction
   e. Reduce oil flow

8. Which device is used to reduce torsional vibration?
   a. Crankshaft
   b. Piston rings
   c. Vibration damper
   d. All of the above
   e. None of the above

9. Main bearings are designed for:
   a. Fatigue resistance
   b. Embedability
   c. Conformability
   d. All of the above
   e. None of the above

10. The small lug on the bearing insert is used to:
    a. Keep the bearing from spinning in the bore
    b. Increase lubrication to the bearings
    c. Keep the bearing in balance
    d. Help in manufacturing
    e. None of the above

11. ________ are used to absorb the axial thrust produced on the crankshaft.
    a. Thrust weights
    b. Thrust bearings
    c. Expansion rings
    d. Core plugs
    e. Radial bearings

12. As the amount of main bearing clearance increases, which of the following happens?
    a. Oil pressure increases
    b. Oil pressure decreases
    c. Blow-by decreases
    d. Blow-by increases
    e. Compression increases

13. A/An ________ is used to reduce the heat flow to the top piston ring.
    a. Oil control ring
    b. Heat dam
    c. Piston pin
    d. Compression ring
    e. Cylinder heat sink

14. The part that connects the piston to the connecting rod is called a/an ________.
    a. Compression ring
    b. Heat dam
    c. Wrist pin
    d. Expansion pin
    e. Connecting pin

15. Pistons are designed to have a major and minor ________.
   a. Pressure area
   b. Heat sink
   c. Wear surface
   d. Thrust surface
   e. Speed surface

16. Which of the following are used as a means to control the expansion of the piston (both new and older pistons)?
   a. Cam-shaped pistons
   b. Split or slotted skirt
   c. Oil control ring gap
   d. T slot in piston
   e. A, B, and D of the above

17. As the end gap of the compression rings increases,
   a. Blow-by increases
   b. Power from combustion increases
   c. Oil consumption decreases
   d. Compression ratio increases
   e. All of the above

18. Piston rings that have a chamfer or counterbore cut into them are called:
   a. Oil control rings
   b. Torsional rings
   c. High-pressure rings
   d. Low-pressure rings
   e. Lubrication rings

19. The oil scraped off of the cylinder walls by the oil control rings is:
   a. Sent back to the crankcase through small tubes
   b. Sent back to the crankcase through small holes in the ring groove
   c. Held in the ring groove for further lubrication
   d. Not used again
   e. Sent to the cylinder head

20. The greatest amount of cylinder wear is found:
   a. Near the top of the cylinder bore
   b. In the middle of the cylinder bore
   c. On the bottom of the cylinder bore
   d. On the extreme bottom of the cylinder bore
   e. None of the above

21. To get new piston rings to fit the cylinder bore correctly, it is necessary to:
   a. Bore all cylinders
   b. Deglaze each cylinder
   c. Polish each cylinder
   d. Replace the cylinder each time
   e. Insert a new piston each time

22. Which of the following tools is used to check the clearance between the main bearing and the main bearing journals?
   a. Plastigage
   b. Micrometer
   c. Ruler
   d. Dial indicator
   e. Vernier caliper

The following questions are similar in format to ASE (Automotive Service Excellence) test questions.

23. A piston is removed from the vehicle. After careful inspection, carbon is observed under the rings and in the ring grooves. *Technician A* says the carbon in the ring grooves should be removed and new rings should be installed. *Technician B* says the piston can and should be replaced in the cylinder bore without removing the carbon buildup. Who is correct?
   a. A only c. Both A and B
   b. B only d. Neither A nor B

24. On a compression test, the psi readings on all four cylinders are within 30% of each other. *Technician A* says all cylinders are OK. *Technician B* says all are bad. Who is correct?
   a. A only c. Both A and B
   b. B only d. Neither A nor B

25. After removing the cylinder heads, a small ridge is noticed on the top of the cylinder. *Technician A* says to leave the ridge there and remove it with a cylinder deglazer after the pistons have been removed. *Technician B* says to remove the ridge with a ridge reamer before removing the pistons. Who is correct?
   a. A only c. Both A and B
   b. B only d. Neither A nor B

26. *Technician A* says it is not OK to exchange vibration dampers from different engines. *Technician B* says it is OK to exchange vibration dampers from different engines. Who is correct?
   a. A only c. Both A and B
   b. B only d. Neither A nor B

27. *Technician A* says that core plugs protect the block from cracking. *Technician B* says they may pop out if the coolant freezes. Who is correct?
   a. A only c. Both A and B
   b. B only d. Neither A nor B

28. *Technician A* says that axial motion on the crankshaft is caused from using straight-toothed gears. *Technician B* says that thrust surfaces are not needed to absorb axial motion on the crankshaft. Who is correct?
   a. A only c. Both A and B
   b. B only d. Neither A nor B

29. *Technician A* says that Plastigage is used to check valve clearance. *Technician B* says that Plastigage can be used to check main bearing clearance. Who is correct?

a. A only
b. B only
c. Both A and B
d. Neither A nor B

30. *Technician A* says that to determine the clearance with Plastigage, measure its length. *Technician B* says that Plastigage cannot be used to measure connecting rod clearance. Who is correct?

a. A only
b. B only
c. Both A and B
d. Neither A nor B

Essay

31. Explain how a forged crankshaft can be identified.

32. List the parts of the crankshaft.

33. What is torsional vibration?

34. What are some of the characteristics in the design of bearings?

35. What is the purpose of a thrust face on a piston?

36. What would be the result if the piston ring end gap were too small during installation?

Short Answer

37. The crankshaft can be ________ to make the journals a smaller size.

38. When the external diameter of the piston pin interferes with the internal diameter of the connecting rod, it is called a(an) ________ fit.

39. To help reduce friction between the piston skirt and the cylinder wall, some pistons have ________.

40. Three common sizes for oversize pistons include ________, ________, and ________.

41. When completing a cylinder compression, if the cylinder variation is more than ________, the piston rings need to be replaced.

Applied Academics

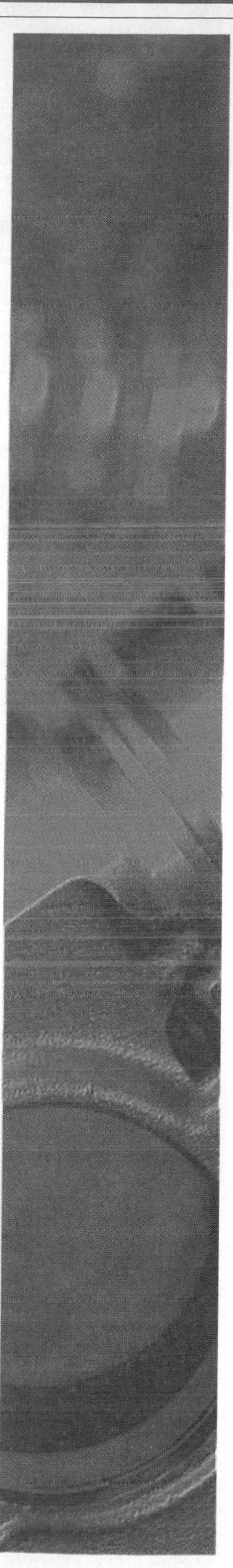

TITLE: Crankshaft Throw

ACADEMIC SKILLS AREA: Science

NATEF Automobile Technician Tasks:

The automotive technician can explain how engine rotational motion is changed to linear motion and the need for balance in rotating systems.

CONCEPT:

The crankshaft throw, which determines the length of the piston stroke, is designed into the crankshaft. As shown in the accompanying illustration, the position of the connecting rod throw on a crankshaft will vary with different engines. The in-line 4-cylinder engine has the throws positioned 180 degrees from each other. The V8 engine has the throws positioned 90 degrees from each other. In the V6 engine the throws are 120 degrees from each other. The number of cylinders, the firing order, and the degrees between the cylinder banks determine these positions. Most V8 engines have the cylinder banks set at 90 degrees. Most V6 engines have the cylinder banks set at 60 degrees. On some engines, the connecting rod journals are splayed, that is, spread out or set apart. This is done with the connecting rod journals as shown in the V6 splayed crankshaft, because on some V6 engines the cylinder banks are set at 90 degrees apart to make the engine lower. The firing order is also different. Thus, the connecting rod journals must be splayed to accommodate the different firing order.

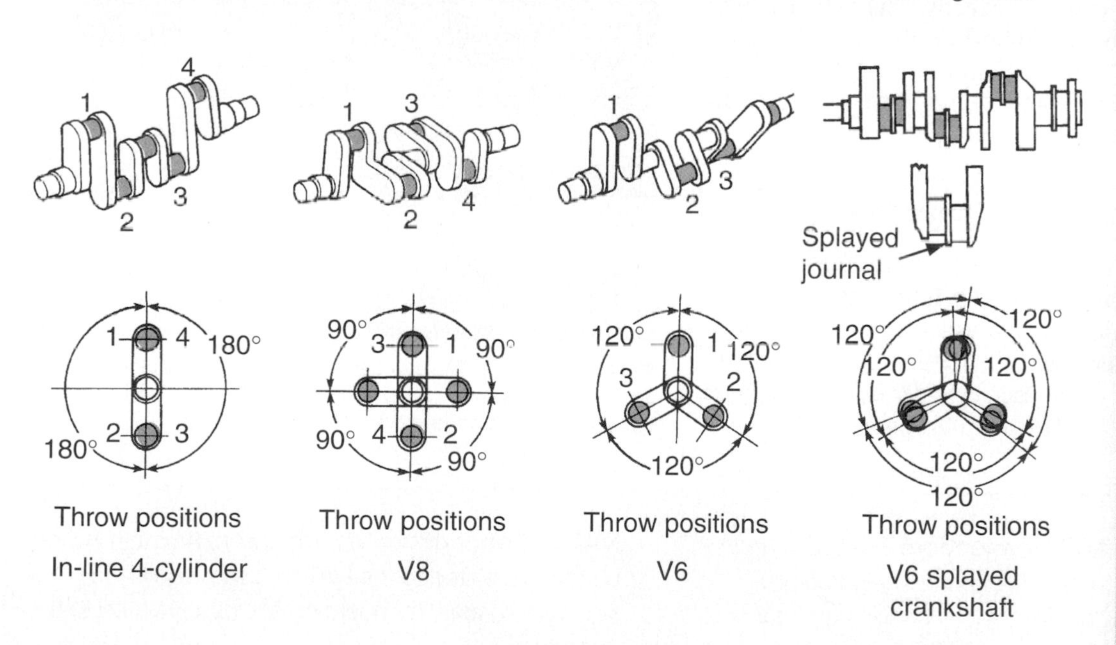

APPLICATION:

Why are crankshaft journals positioned differently on different engines? How does the throw determine the length of the stroke of the engine? What is the relationship between the design of the crankshaft and vibration?

CHAPTER 18 Cylinder Heads and Valves

OBJECTIVES

After studying this chapter, you should be able to:

- Describe the purpose of the cylinder head.
- List all parts on the cylinder head.
- Identify the designs used for combustion chambers.
- List the parts of the valve assembly.
- Analyze the purposes of the valve guides, seals, seats, springs, keepers, retainers, and rotators.
- Analyze several problems, their diagnosis, and service procedures used on the cylinder head and valve assemblies.

INTRODUCTION

All automotive engines use a cylinder head and a set of valves to operate the engine correctly. The cylinder head acts as a cover or top to the combustion chamber. The valves allow the air, fuel, and exhaust to move through the engine at the correct time. This chapter will acquaint you with the parts used on the cylinder head and valves.

18.1 CYLINDER HEAD DESIGN

The cylinder head has several purposes. It acts as a cap or seal for the top of the combusion chamber. It holds the valves, and it has ports to allow air, fuel, and exhaust to move through the engine. The cylinder head on many engines also contains the combustion chamber for each cylinder or piston. ***Figure 18–1*** shows a cylinder head for a multicylinder engine.

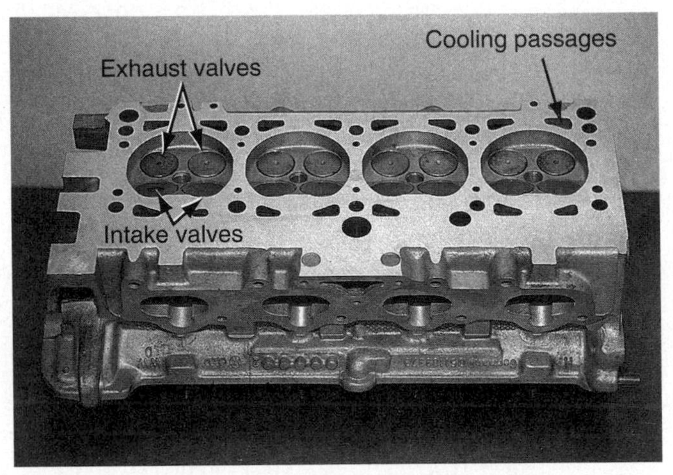

FIGURE 18–1 The cylinder head is used to seal off the top of the combustion chamber and to hold valves, allow air and fuel to flow through, and to hold the injector and/or spark plugs.

Certification Connection

ASE Connection: The information in this chapter can help you prepare for the National Institute for Automotive Service Excellence (ASE) certification tests. The tests and content areas most closely related to this chapter are:

Test A1—Engine Repair

- **Content Area**—Cylinder Head and Valve Train Diagnosis and Repair

NATEF Connection: Much of the information in this chapter is related to the NATEF tasks. The NATEF tasks and priority numbers most closely related to this chapter are:

1. Remove cylinder head(s), visually inspect cylinder head(s) for cracks; check gasket surface areas for warpage and leakage; check passage condition. P-2

2. Install cylinder heads and gaskets; tighten according to manufacturer's specifications and procedures. P-1
3. Inspect valve springs for squareness and free height comparison; determine necessary action. P-2
4. Replace valve stem seals on an assembled engine; inspect valve spring retainers, locks, and valve grooves; determine necessary action. P-2
5. Inspect valve guides for wear; check valve stem-to-guide clearance; determine necessary action. P-3
6. Inspect valves and valve seats; determine necessary action. P-3
7. Check valve face-to-seat contact and valve concentricity (runout); determine necessary action. P-3
8. Check valve spring assembly height and valve stem height; determine necessary action. P-3
9. Inspect hydraulic or mechanical lifters; determine necessary action. P-2
10. Adjust valves (mechanical or hydraulic lifters). P-1

CYLINDER HEAD MANUFACTURING

Cylinder heads can be made from cast iron or aluminum. Aluminum is used to make the engine lighter, but it transfers heat more rapidly and expands more than cast iron with the addition of heat. This may cause warpage. Both aluminum and cast iron objects are made by pouring hot liquid metal into a sand mold.

The cylinder head must have coolant passages. This means that sand cores also have to be used in the casting process. In addition, passages must be cast for intake and exhaust ports. ***Figure 18–1*** shows an example of some of the ports that are cast into the cylinder head.

After the cylinder head has been cast, it must be machined. Areas are machined so that intake and exhaust manifolds can be attached, valves can be seated, spark plugs and injectors can be installed, and a good seal can be provided to the block.

Car Clinic: CYLINDER HEADS

CUSTOMER CONCERN:
Engine Misses When Cold

A vehicle misses very severely when it is cold. The car has about 80,000 miles on it. After the engine warms to operating temperature, the miss disappears. The problem started at about 45,000 miles.

SOLUTION: Misses such as this are usually caused by a cracked cylinder head. When the engine warms up, the metal expands, the crack is sealed, and the miss disappears. Give the engine a cylinder leakdown test when it is cold to see if the compression is leaking. If the cylinder head is cracked, it will need to be replaced.

INTAKE AND EXHAUST PORTS

Intake and exhaust ports must be cast into the cylinder head. These ports are made so the air and fuel can pass through the cylinder head into the combustion chamber. It would be ideal if one port could be used for each valve. Because of space, however, ports are sometimes combined. These ports are called **siamese ports** (***Figure 18–2***). Siamese ports can be used because each cylinder uses the port at a different time.

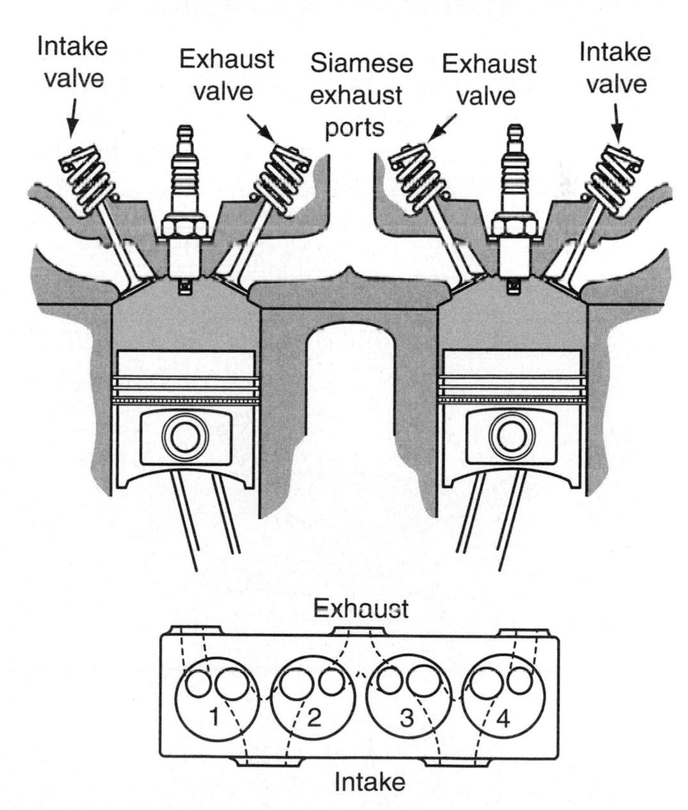

FIGURE 18–2 Siamese ports are used on the cylinder head. This means that two cylinders will feed the same exhaust or intake ports.

Cross-flow ports are used on many engines. Cross-flow heads have the intake and exhaust ports on opposite sides.

COOLANT PASSAGES

Large openings that allow coolant to pass through the head are cast into the cylinder head. Coolant must circulate throughout the cylinder head so excess heat can be removed. The coolant flows from passages in the cylinder block through the head gasket and into the cylinder head. Depending on the engine configuration, the coolant then passes to other parts of the cooling system.

COMBUSTION CHAMBER

The shape of the combustion chamber affects the operating efficiency of the engine. Two types of combustion chamber designs are commonly used. They are the **wedge-shaped combustion chamber** and the **hemispherical combustion chamber**.

Several terms are used to describe combustion chambers. **Turbulence** is a very rapid movement of gases. When gases move, they make contact with the combustion chamber walls and pistons. Turbulence causes better combustion because the air and fuel are mixed better. **Quenching** is the cooling of gases by pressing them into a thin area. The area in which gases are thinned is called the quench area.

WEDGE COMBUSTION CHAMBER

The wedge-shaped combustion chamber was used on cars until about 1968 (***Figure 18–3***). As the piston comes up on the compression stroke, the air and fuel mixture is squashed in the quench area. The quench area causes the air and fuel to be mixed thoroughly before combustion. This helps to improve the combustion efficiency of the engine. Spark plugs are positioned to get the greatest advantage for combustion. When the spark occurs, smooth and rapid burning moves from the spark plug outward. The wedge-shaped combustion chamber is also called a turbulence-type combustion chamber. On newer model cars, the quench area has been reduced, which helps reduce exhaust emissions.

HEMISPHERICAL COMBUSTION CHAMBER

The hemispherical combustion chamber gets its name from its shape. *Hemi* is defined as half, and *spherical* means circle; therefore, the combustion chamber is shaped like a half-circle. This type of chamber is also called the hemi-head. The valves are located as shown in ***Figure 18–4***. One distinct advantage is that larger valves can be used. This increases the amount of air and fuel that can enter the engine, thus improving volumetric efficiency.

The hemispherical combustion chamber is considered to be a nonturbulence-type combustion chamber. Little or no turbulence is produced in this chamber as the air and fuel enter. The air and fuel mixture is compressed evenly on the compression stroke. When flat-top pistons are used, little turbulence can be created. The spark plug is located directly in the center of the valves. Combustion radiates evenly from the spark plug, completely burning the air-fuel mixture.

One of the more important advantages of the hemispherical combustion chamber is that air and fuel can enter the chamber very easily. The wedge combustion chamber restricts the flow of air and fuel to a certain extent. This restriction is called **shrouding**. ***Figure 18–5*** shows the valve very close to the side of the combustion chamber, which causes the air and fuel to be restricted. Volumetric efficiency is reduced. Hemispherical combustion chambers do not have this restriction. Hemispherical combustion chambers are used on many high-performance applications. This is especially true when large quantities of air and fuel are needed in the cylinder.

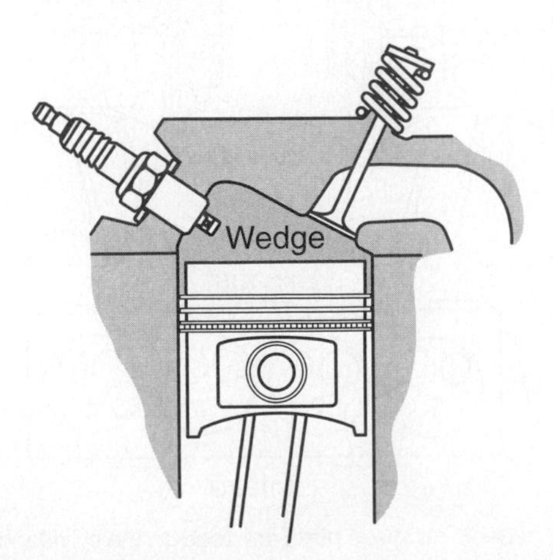

FIGURE 18–3 The wedge combustion chamber is shaped like a wedge to improve the turbulence within the chamber.

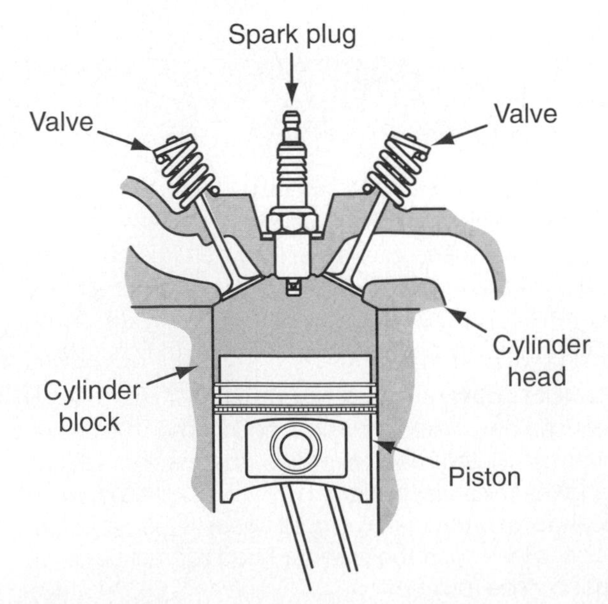

FIGURE 18–4 The hemispherical combustion chamber is shaped like a half-circle. The valves are placed on both sides of the spark plug.

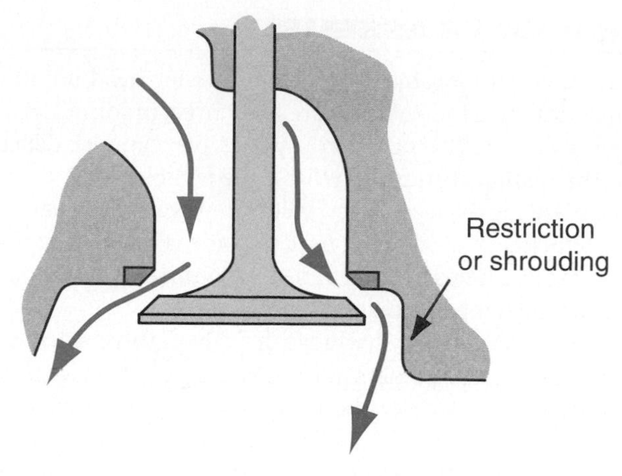

FIGURE 18–5 Shrouding is defined as a restriction in the flow of intake gases caused by the shape of the combustion chamber.

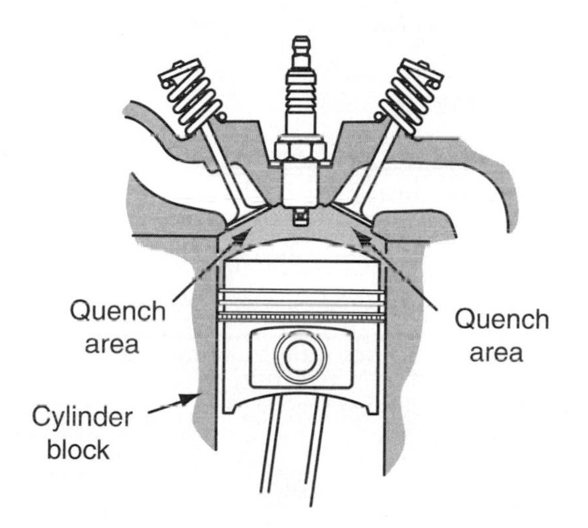

FIGURE 18–6 The hemispherical design of the combustion chamber improves efficiency by producing a quench area.

Many high-performance engines use a domed piston. This type of piston has a quench area to improve turbulence (***Figure 18–6***). Several variations of this design are used by different engine manufacturers.

PENTROOF COMBUSTION CHAMBER

Today, many engines use four valves per cylinder. Two valves are used for intake and two valves are used for exhaust. To aid in this design, a pentroof combustion chamber is often used. As shown in ***Figure 18–7***, it looks like an inverted V configuration. With this design, there is room for the intake and exhaust valve. In addition, the design produces good turbulence and lower emissions.

DIESEL COMBUSTION CHAMBER

Diesel combustion chambers are different from gasoline combustion chambers. Diesel fuel burns differently, so the combustion chamber must be different. Three types of combustion chambers are used in diesel engines: the open combustion chamber, the precombustion chamber, and the turbulence combustion chamber.

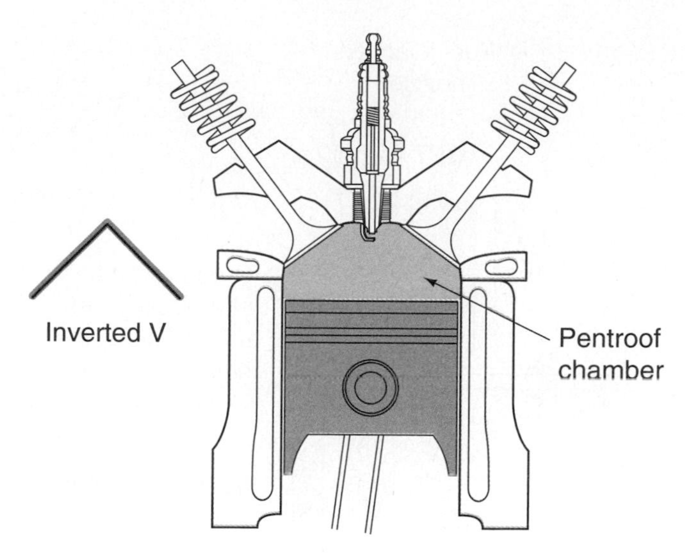

FIGURE 18–7 The Pentroof combustion chamber is shaped like an inverted V and is used when there are four valves on each combustion chamber.

The open combustion chamber has the combustion chamber located directly inside the piston. ***Figure 18–8*** shows the open combustion chamber with diesel fuel being injected directly into the center of the chamber. The shape of the chamber and the quench area produces turbulence.

A **precombustion chamber**, shown in ***Figure 18–9***, was used on both gas and diesel engines in the past. A smaller, second chamber is connected to the main combustion chamber. On the power stroke, fuel is injected into the small chamber. Combustion is started and then spreads to the main chamber. The precombustion chamber has a

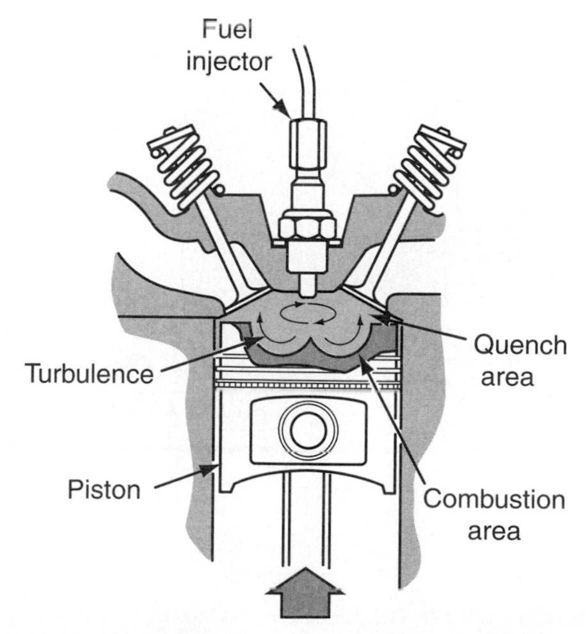

FIGURE 18–8 Diesel engines have an open combustion chamber that is located directly within the top of the piston.

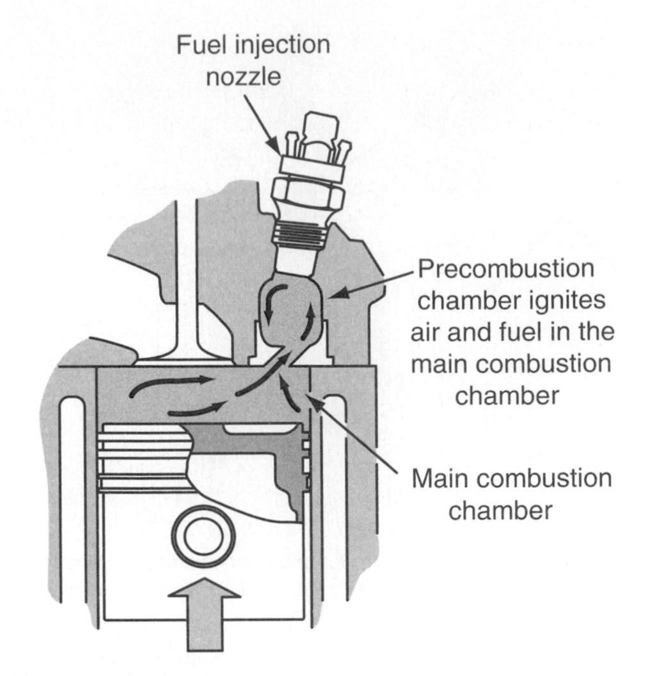

FIGURE 18–9 Precombustion chambers are used to ignite air and fuel in a small prechamber. The combustion in this chamber then ignites the air and fuel in the main combustion chamber.

very rich mixture, but the main chamber can be very lean. The overall effect is a leaner engine, producing better fuel economy.

The turbulence combustion chamber is shown in ***Figure 18–10***. The chamber is designed to create an increase in air velocity or turbulence in the combustion chamber. The fuel is injected into the turbulent air and burns more completely.

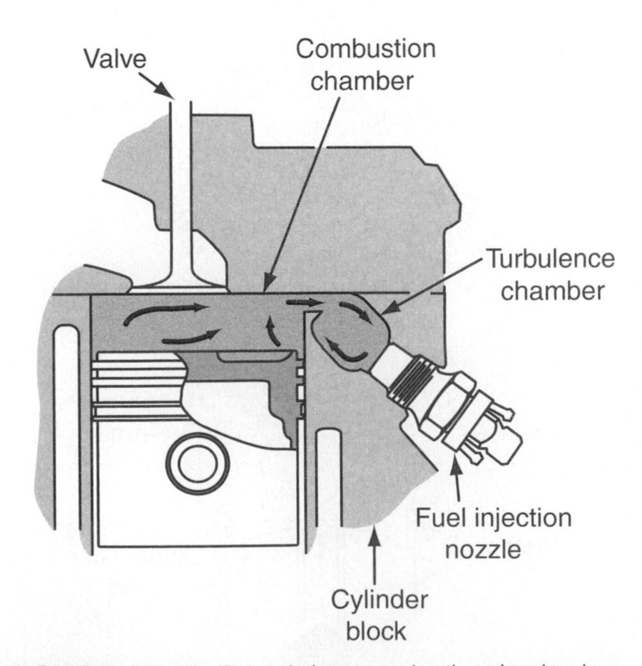

FIGURE 18–10 The turbulence combustion chamber is used on certain diesel applications to increase turbulence of air and fuel.

18.2 VALVE ASSEMBLY

The valves are located within the cylinder head on all engines designed today. There are two, three, or four valves for each cylinder. The number of valves per cylinder depends on the manufacturer and year of the vehicle. Most engines today use four valves per cylinder. However, three-valve arrangements are also being used in some vehicles. The number of valves helps to determine the volumetric efficiency as well as the temperature of the exhaust.

The valve assembly includes the valves, valve seats, valve guides, springs, retainers, and seals. The intake valve is usually larger in diameter than the exhaust valve because the intake valve and port handle a slow-moving air-fuel mixture. On the exhaust stroke, the gases move more easily from the pressure of the piston forcing them out. These valves are called **poppet-type valves**. The exhaust valve is usually smaller in diameter than an intake valve.

VALVE PARTS

The valve has several parts. ***Figure 18–11*** shows the parts of the valve. The head of the valve is the part that is inside the combustion chamber. It must withstand extremely high temperatures, in the range of 1,300 to 1,500 degrees Fahrenheit.

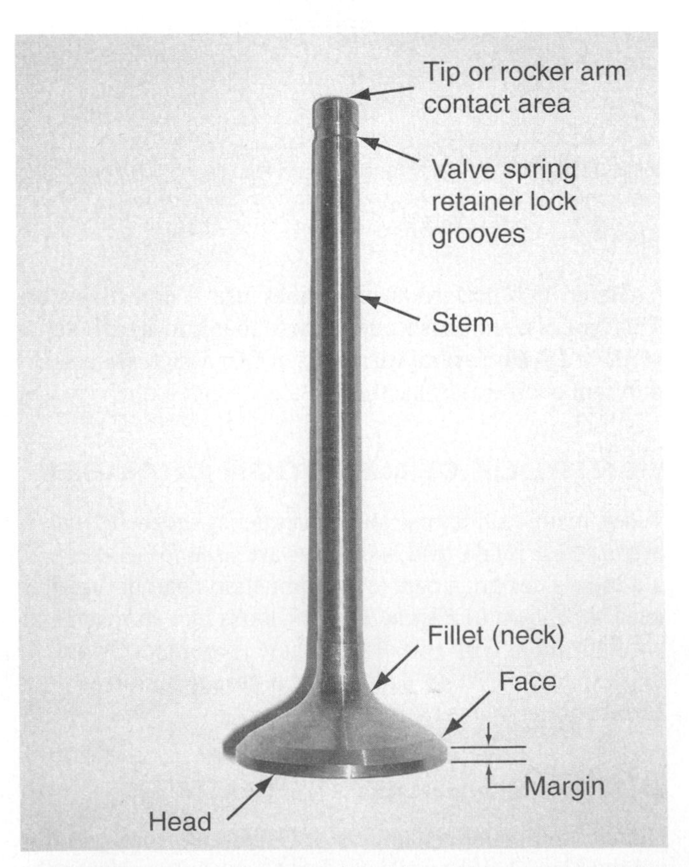

FIGURE 18–11 The parts of a typical valve are shown in this illustration.

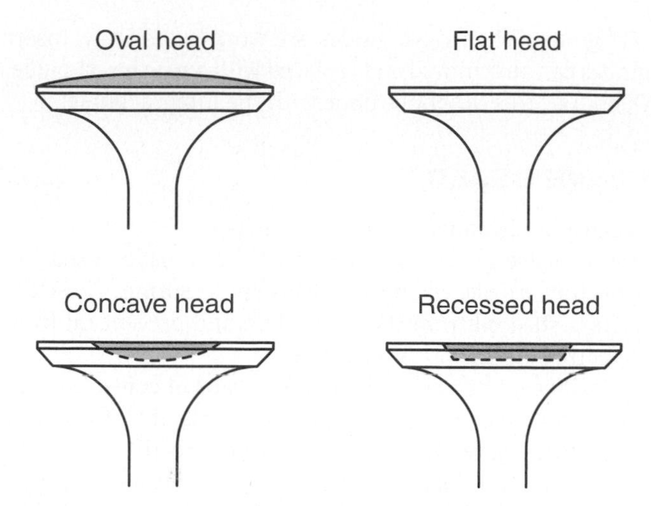

FIGURE 18-12 Valves are designed with different head shapes for different purposes.

The valves can be designed with different head shapes. ***Figure 18–12*** shows some of the shapes used on engines today. The more metal on the head of the valve, the more rigid the valve. Less metal means the valve will be able to conform to the seat more effectively. These valves are said to be elastic. Rigid valves last longer, but they don't seal as well. Elastic valves seat better, but they may not last as long.

The **valve face**, shown in ***Figure 18–11***, is the area that touches and seals the valve to the cylinder head. This is the area that must be machined if the valve is damaged. The fillet is the curved area between the stem and the inner edge of the face. The fillet provides extra strength for the valve. The valve stem is used to support the valve in the cylinder head. The valve spring retainer lock grooves are used to keep the spring attached to the valve during operation. The valve margin is the distance between the face of the valve and the head. The margin is reduced whenever the valve is machined or ground down. If this margin is too small (see manufacturer's specification), the valve may burn easily.

Figure 18–13 shows an example of two valves. The valve margin is shown before grinding and after grinding. Note how the thickness of the valve margin is reduced. If this margin is reduced too much, there will not be enough metal to transfer the heat away. This will cause the valve to heat up and burn. When a valve burns, it allows the combustion gases to leak during the compression and combustion processes.

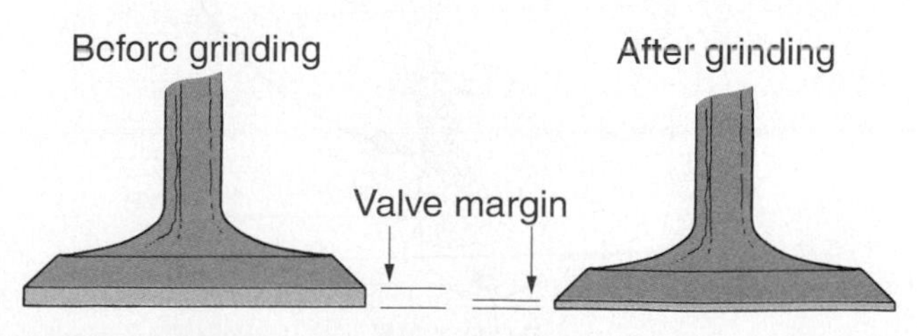

FIGURE 18-13 This illustration shows a valve margin before and after grinding.

VALVE MATERIAL

The valve is made of very strong metal with nickel, chromium, and small amounts of manganese and other materials. The metal must be able to transfer heat very rapidly. If heat is not transferred rapidly, the valve will burn and become damaged. Some exhaust valves also use a metallic sodium inside the stem. The sodium becomes a liquid at operating temperature. The liquid sodium then helps to transfer the heat from the stem to the valve guide more rapidly. ***Figure 18–14*** shows a sodium-filled valve.

VALVE GUIDES

The **valve guide** is the hole that supports the valve in the cylinder head. It acts as a bushing for the valve stem to slide in. The valve guide is part of the cylinder head. The valve guide helps to support and center the valve so that correct **seating** can be obtained. It also helps to dissipate the heat produced within the combustion chamber through the valve. ***Figure 18–15*** shows a valve guide and how heat is transferred to the guide, to the cylinder head, and finally to the cooling system.

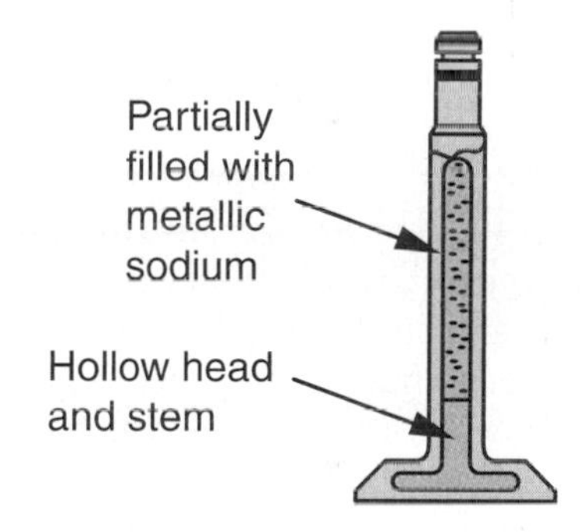

FIGURE 18-14 Sodium-filled valves are used to help cool the valve.

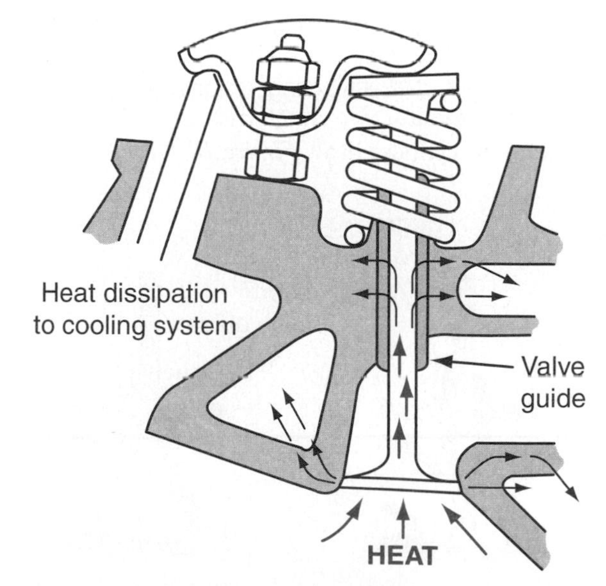

FIGURE 18-15 The valve guide helps to dissipate heat from the valve.

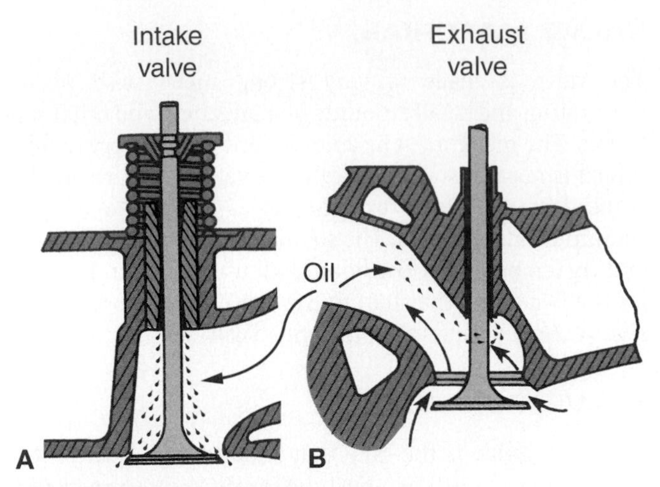

FIGURE 18-16 When valve guides are worn, oil can pass by the valve and into the intake or exhaust stream.

The clearance between the valve stem and the guide is very important. Generally, the clearance is between 0.001 and 0.004 inch. If the valve guides are worn and there is greater clearance, several things may happen. The valve may leak air, causing the air-fuel ratio to be altered. Oil may also leak past the valve guide, causing high oil consumption. The valve may also not seat evenly, causing the valve seat to wear rapidly.

Figure 18–16 shows an example of how oil might get past the valve guides. In ***Figure 18–16A***, an intake valve guide has too large a clearance. Since there is a vacuum on the intake valve, any oil that passes by the valve guide will be drawn into the combustion chamber and burned. In ***Figure 18–16B***, an exhaust valve guide has too large a clearance. In this case, as the oil passes by the valve guide, the oil is drawn into the exhaust stream, burned, and sent out with the exhaust. In either case, there would be increased oil consumption, as well as burned oil in the exhaust.

There are two types of valve guides: integral and insert. **Integral guides** are machined directly into the cylinder head. ***Figure 18–17A*** shows an example of this type of guide. **Insert guides** are small cast cylinders that are pressed into the cylinder head. ***Figure 18–17B*** shows an insert guide. When valve guides are worn excessively, insert guides can be removed and replaced with a new insert guide. Of course, this cannot be done with the integral guide.

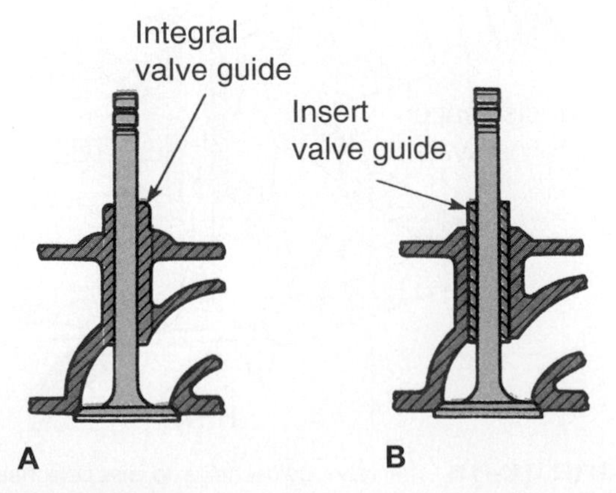

FIGURE 18-17 Both integral and insert valve guides are used on engines.

VALVE SEALS

Because a clearance is necessary between the valve stem and the valve guide, oil control methods must be used. Oil deflectors are placed on the valve stem or spring. These deflectors shed oil from the valve stem and prevent oil from collecting on the top of the guide.

Positive guide seals are another means of controlling oil flow. Positive guide seals are small seals that fit snugly around the valve stem. The seal is held to the stem with small springs or clamps. These seals restrict most oil that would normally pass through the valve guide.

Some car manufacturers use **passive seals**, including O-rings and umbrella-type valve stem seals. These seals are used mostly on new engines. They do not work as well on older engines that have more wear on the valve stem. ***Figure 18–18*** shows both types of valve seals.

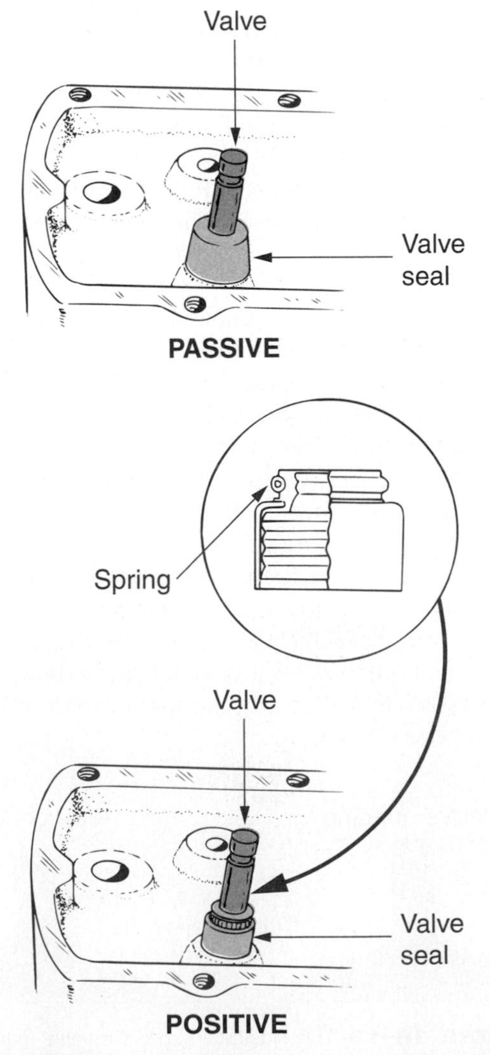

FIGURE 18-18 Valve seals are used to stop oil from going down the valve guide.

Car Clinic: VALVE STEM SEALS

CUSTOMER CONCERN:
Car Smokes When Started

A vehicle produces blue smoke when the engine is first started. Once the engine is running and at operating temperature, the smoking stops. What could be the problem?

SOLUTION: One common cause of this problem is oil seeping past the exhaust valves when the engine is shut down. The oil then sits in the exhaust manifold until the engine is started. When the engine is started, the oil is burned, producing the blue smoke. Replacing the valve stem seals should solve the problem.

Car Clinic: VALVE SEALS

CUSTOMER CONCERN:
Engine Misses on One Cylinder

A customer complains that the engine on her vehicle seems to be missing and lacking power. The engine seems to be missing only on one cylinder. In addition, there is a slight rhythmic puffing sound coming from the exhaust.

SOLUTION: When an engine has a burned exhaust valve, it can often be heard at the exhaust. Each time the piston in the cylinder that has a burned valve comes up on the compression stroke, the piston has a tendency to push the gases out the exhaust through the burned valve. This is heard easily by placing your hand about an inch from the end of the exhaust pipe in the stream of the exhaust. If there is a burned valve, you will be able to hear a slight puffing sound coming from the exhaust. The puffing sound is the piston pushing the air and fuel out through the burned valve. If this is the case, the cylinder head will need to be disassembled and the valve will need to be ground or replaced according to correct service procedures and specifications.

VALVE SEATS

Valve seats are circular surfaces that are machined into the cylinder block or head. The valve face seals or seats against the valve seat (***Figure 18–19***). These seats provide a surface for the intake and exhaust valves to seal for gas leakage. The seats also help to dissipate the heat built up in the valve.

There are two types of valve seats: integral and insert. As with valve guides, the integral type is cast directly as part of the cylinder head. The insert type uses a metal ring as the seat. It is pressed into the cylinder block and ground to the correct angle. The insert type of valve seat is used most often on engines with aluminum cylinder heads. Insert valve seats can be made from cast iron, hardened cast iron, high-chrome steel, and **stellite** (very hard steel).

Valve seats are ground to a specific angle for correct operation. They are either 30 or 45 degrees. An **interference angle** is very common when grinding the valves. An interference angle is obtained by grinding the valve face about 1 degree less than the valve seat. This is shown in ***Figure 18–20***. In this example, the valve is ground to 44 degrees, whereas the seat is ground to 45 degrees. The interference angle tends to cut through any deposits that have been formed on the seat as the valve closes. It also produces a more positive seal. As the engine is run and the valve seats wear, the interference angle is gradually eliminated. The result is good **line contact**, which helps transmit excess heat away from the valve.

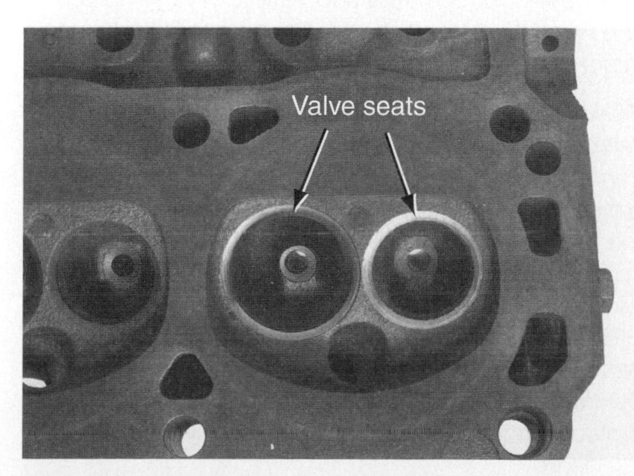

FIGURE 18-19 Valve seats help the valves seal to the cylinder head. They are machined into the cylinder head.

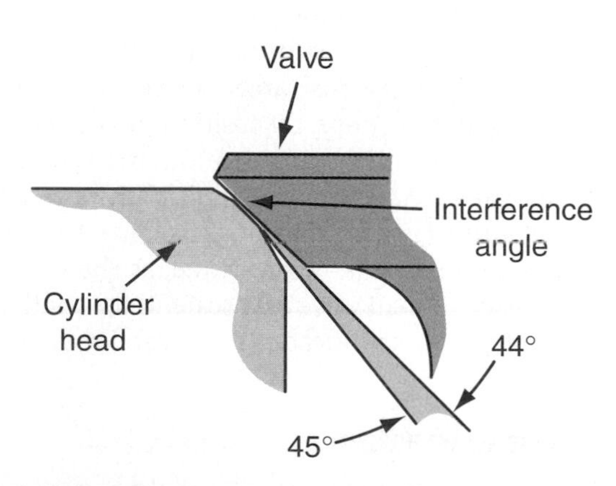

FIGURE 18-20 An interference angle is cut between the valve face angle and the seat angle. This helps the valves to seat to the cylinder head faster.

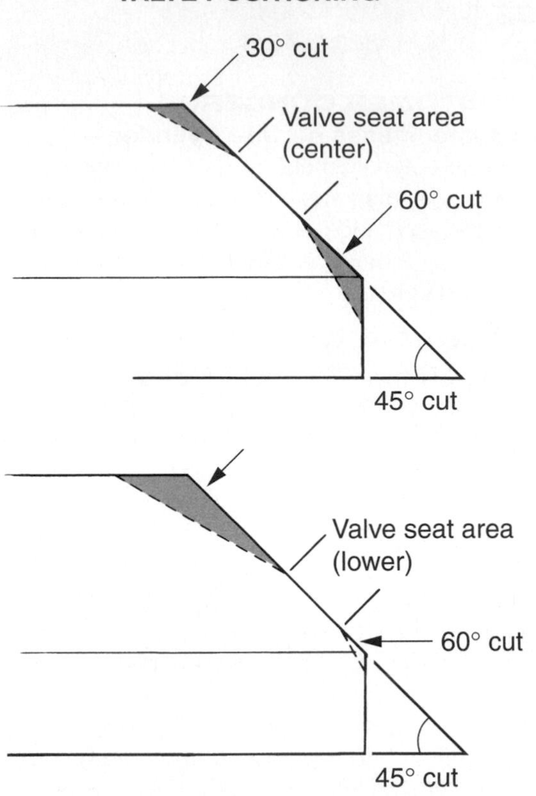

FIGURE 18–21 The position of the seat can be raised or lowered by grinding a 30-degree and 60-degree cut.

VALVE SEAT POSITIONING

The valve seat area can be raised or lowered. This is done by removing or grinding above and below the seat. ***Figure 18–21*** shows an example of positioning the valve seat. First the seat is ground to the standard specification, say 45 degrees. Then a 30-degree grinding stone is used to cut off the top of the valve seat. Then, a 60-degree stone is used to cut off the bottom of the valve seat. By cutting the right amount off the top and bottom, the position of the valve seat can be adjusted. Also, by using these two stones, the exact length of the valve seat area can be ground to the manufacturer's specifications. By positioning the valve seat, the valve position can also be raised or lowered as necessary.

Once the valve seat has been positioned correctly, it is important to make sure the valve seat has the correct width. Valve seat width can be measured, as shown in ***Figure 18–22***, using a small valve seat scale. If the width is incorrect or not within specifications, the service technician

PARTS LOCATOR

*Refer to photo #6 in the front of the textbook to see what **valve seats** look like when positioned correctly.*

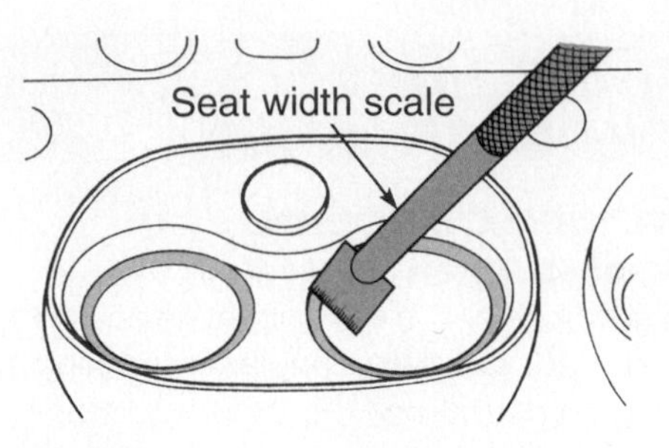

FIGURE 18–22 The valve seat width can be measured with a small seat width scale.

will then need to regrind the seat to its proper width, still making sure the seat is positioned correctly.

VALVE SPRINGS

The valve springs are designed to keep the valves closed when the camshaft is not lifting the valves. The valve springs are held to the valve stem with various types of **keepers**. The spring must be designed to close the valve correctly. If the valve spring is weak, there may be **valve float**. Valve float means the valve stays open slightly longer than it is designed to do. This usually happens when the valve springs are weak and the engine is operating at high speeds. **Valve bounce** can also occur if the spring is weak and operated at high speeds. Valve bounce occurs when the valve slams against the seat, causing it to bounce slightly.

Valve springs are made from several types of wire materials. These include carbon wire, chrome vanadium, and

CUSTOMER CONCERN:

Engine Misses on One Cylinder

An engine has developed a steady miss on one cylinder. The engine has 45,000 miles on it. The engine developed the miss suddenly. The spark plugs and wires have been checked and are all OK. What might be a good thing to check?

SOLUTION: The key is that the engine has developed a steady miss and that it happened suddenly. A common cause of this problem is a broken valve spring. To check which cylinder is missing, run the engine and carefully remove and replace each spark plug wire with the appropriate spark plug wire pliers. Note which cylinder shows no change in the RPM. This is the cylinder that probably has a broken valve spring. Remove the valve cover and check the condition of the valve spring. Use tools designed to replace the valve spring without removing the cylinder head.

chrome silicon. The stress, load temperature, and aging qualities determine exactly what type of material is used.

VALVE SPRING VIBRATION

During normal operation, valve springs may develop a vibration known as **harmonics**. At times, harmonics cause the spring to function incorrectly. Several designs are used to reduce harmonics. These include using stronger springs, variable **pitch** springs, dual springs, varying the outside diameter of the spring, and placing small vibration dampers inside the springs. ***Figure 18–23*** shows the variable pitch spring. The end with the closer spacing should always be installed toward the cylinder head. ***Figure 18–24*** shows a valve spring with a damper spring that can be inserted inside the valve spring.

VALVE KEEPERS AND RETAINERS

Keepers and retainers are used to keep the valve spring secured to the valve stem. The retainer acts as a washer and seat for the top of the spring (***Figure 18–25***). The retainer sits on top of the spring. The valve then passes through the retainer.

Valve keepers are used to hold the retainers to the valve stem. Valve keepers are designed as tapered keys or locks. As the spring pressure pushes up on the retainer, the keepers are pinched or wedged into the retainer. This action causes the spring to be firmly attached to the valve during all operation. Several types of keepers are used, but the most common is the split type. The split type is easy to remove yet maintains a positive lock. ***Figure 18–26*** shows a more complete illustration of the valve keepers. Note that on some valves an O-ring is placed directly below the two valve keepers. This is done to stop oil from going down to the valve guide.

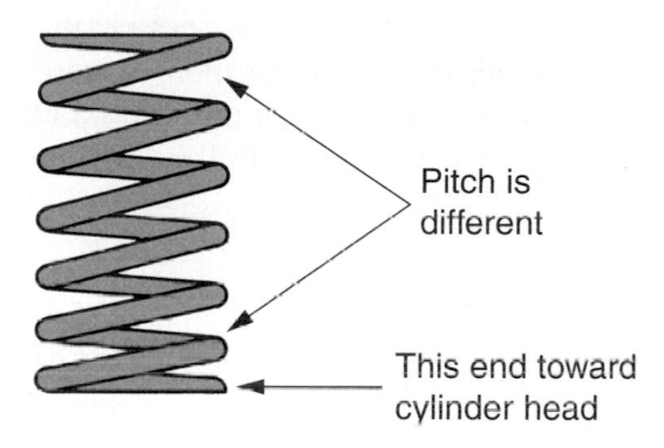

FIGURE 18–23 Variable pitch springs reduce vibration within the spring during operation.

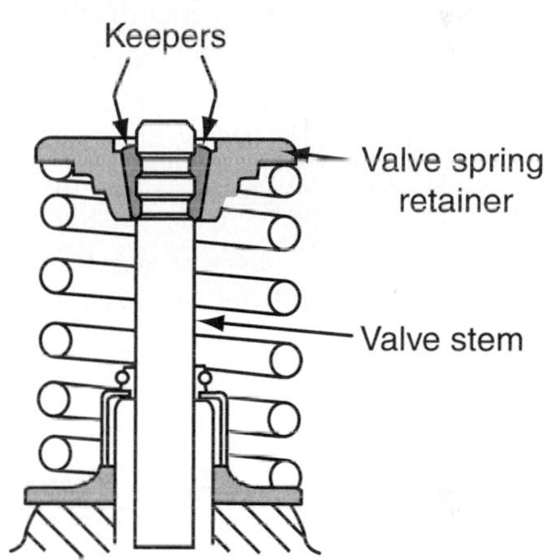

FIGURE 18–25 The valve is held in place with a valve spring retainer.

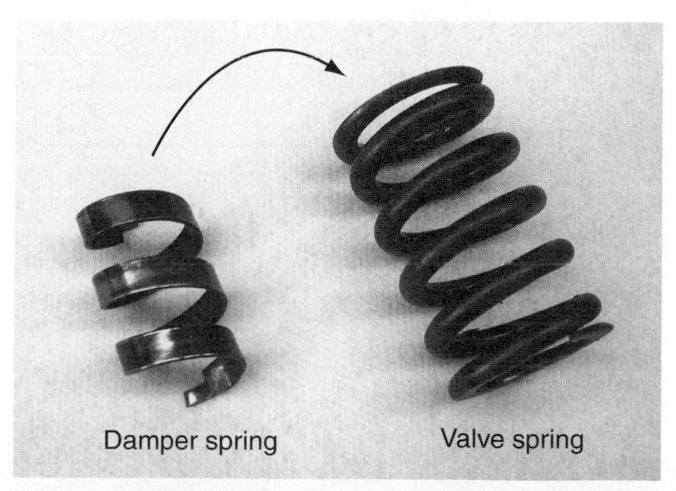

FIGURE 18–24 A damper spring placed inside the valve spring reduces harmonics.

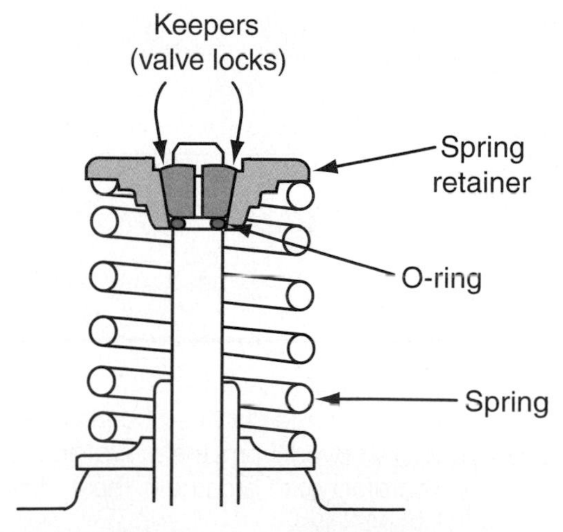

FIGURE 18–26 The O-ring located just below the valve keepers helps to keep oil off the valve stem.

VALVE ROTATORS

Valve rotators are used on certain engines. If valves are rotated a small amount each time they are opened, valve life will be extended. Rotating the valves:

- Minimizes deposits of carbon on the stem of the valve
- Keeps the valve face and seat cleaner
- Prevents valve burning caused by localized hot spots
- Prevents valve edge distortion
- Helps to maintain uniform valve head temperatures
- Helps to maintain even valve stem tip wear
- Helps to improve lubrication on the valve stem

There are several types of valve rotators. Most operate on the principle that as the valve spring is compressed, small balls inside the rotator roll up an **inclined surface**. This action causes the valve to rotate slightly as it is being compressed. ***Figure 18–27*** shows a ball bearing–type valve rotator. In this case, the assembly is located in place of the valve spring retainer.

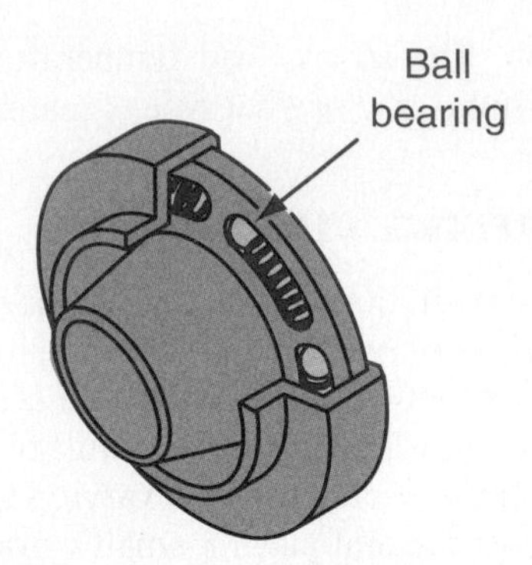

FIGURE 18–27 Valve rotators have ball bearings and a spring that cause the valve to rotate slightly each time it is compressed.

NUMBERS OF VALVES PER CYLINDER

In the past few years, automotive manufacturers have been experimenting with different numbers of valves per cylinder. In the past, most engines had one intake and one exhaust valve for each cylinder. Today, however, many smaller engines have two intake valves and two exhaust valves per cylinder. This design helps to make smaller engines more powerful. Also, three-valve engines are used today to increase the exhaust temperatures of the catalytic converter, see Chapter 16.

Recently, other designs have also been tested. For example, one manufacturer is testing a 5-valve cylinder head. This design uses three intake valves and two exhaust valves per cylinder. Five valves are better than two, three, or four valves per cylinder because they provide an optimal flow path for air and fuel into the cylinders. This flow path, in engineering terms, is called the **curtain area**. The curtain area gives engineers an indication of how evenly the air and fuel flow into the cylinder. The greater the curtain area, the more efficient the combustion. Some engineering reports suggest that the use of five valves, rather than two, three, or four valves, increases the curtain area efficiency by about 30%.

Problems, Diagnosis, and Service

Safety Precautions

1. The edges of the cylinder head may be very sharp because of the machined surfaces. Be careful not to cut your fingers when handling the cylinder head.
2. During cleaning, large chunks of carbon and soot are usually removed from the valve and cylinder head. Be careful not to breathe in the carbon dust, and always wear safety glasses to protect your eyes.
3. When using a toxic cleaning solution, always wear gloves to protect your hands.
4. When valves and valve springs are being removed or installed, the valve springs are depressed and under high pressure. Make sure the valve spring compressor is secured correctly on the valve so

the spring will not slip under pressure. Your eyes and face may be injured if you are hit by a spring that has slipped out of the grasp of the tool.

5. Always wear safety glasses around the shop. These glasses should be OSHA-approved safety glasses with side protectors. Wear the glasses at all times while in the shop, even while looking up specifications, reviewing manuals, or working on the cylinder head components.
6. Make sure that all tools used are clean and free of grease. Under high torque specifications, such as for cylinder head bolts, tools can become slippery and slip off the bolts. This is especially true when using a spring compressor. Always make sure the spring compressor is used correctly.
7. Many of the bolts and nuts require high torque specifications. Always use the correct torque wrench and specifications when working with cylinder heads.
8. When rebuilding the cylinder head, always clean the parts before inspection. Parts that are dirty may slip easily and be dropped.
9. When grinding valves or using other electrical equipment, never string an extension cord across the floor.
10. Make sure all electrical tools are grounded.
11. Never use electrical tools while standing in wet areas.
12. Always wear protective gloves when lifting or moving cylinder heads.

PROBLEM: Bad Cylinder Head

Loss of power in one cylinder, white exhaust smoke, loss of coolant and sometimes a blown head gasket.

DIAGNOSIS

This type of problem is often associated with a crack in the cylinder head or possible warpage of the cylinder head. A crack in the block may also cause coolant to enter the combustion chamber and cause white exhaust smoke. In addition, a warped cylinder head may cause the head gasket to blow. Both of these problems could cause the lack of power in a specific cylinder. Such problems can be diagnosed by performing wet and dry compression tests.

SERVICE

1. With the cylinder head removed, check between each combustion chamber and the valve guides for possible cracks. In most cases, the cracks cannot be observed unless the area has been cleaned with a wire brush. Cracks can be very small and at times may be very difficult to notice.
2. Check the cylinder head for warpage by using a straightedge and feeler gauge.
   a. First clean the cylinder head and remove all dirt and grease.
   b. Using a steel straightedge as shown in ***Figure 18–28***, lay the straightedge across the length of the head.
   c. Now try to slip a feeler gauge between the head and the straightedge. These spots show that the head is warped.
   d. Start with a 0.003-inch feeler gauge, then go to a larger feeler gauge, if possible.
   e. Check several spots on the length as well as the width of the head. Also check the warp diagonally.
   f. A typical specification for maximum warp is 0.003 inch for any 6-inch length or 0.007 inch overall. Note that not all manufacturers have the same specifications. Generally, maximum warp is between 0.003 and 0.007 inch.
   g. If the cylinder head is warped beyond the manufacturer's specifications, the cylinder head can be machined. This process is called milling the cylinder head. Generally, a specialized machine shop has the correct tools to mill a cylinder head. If the cylinder head is milled (approximately 0.010 inch), the compression ratio will increase slightly. On overhead camshaft engines, milling the head may change the belt or chain tension on the camshaft.

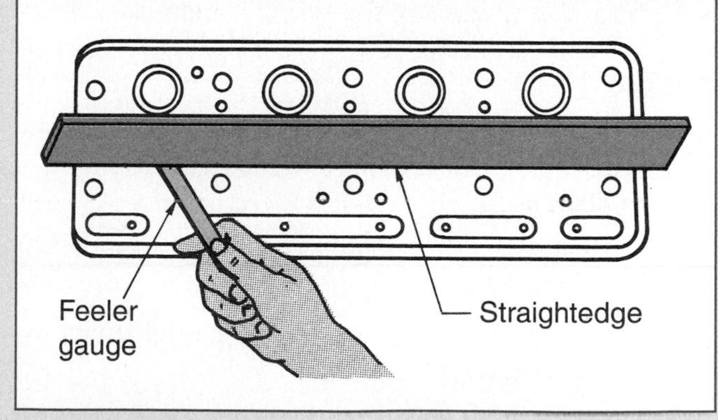

FIGURE 18–28 Check cylinder head flatness using a straightedge and a feeler gauge.

PROBLEM: Loss of Power

Loss of engine power, engine misses on one or more cylinders, and loss of compression on one or more cylinders.

DIAGNOSIS

After an engine has had many miles on it, the intake and exhaust valves often begin to lose compression. In addition, the valve lifters may be worn, causing the valves not to open as far as they should. The valve springs may also be broken or show signs of weakening. There are parts and components on the cylinder head and valve mechanism that can be worn or damaged. Some of the more important components include the intake and exhaust valves, the valve seats, camshaft, valve springs, rocker arms, lifters, and pushrods.

SERVICE

Disassembly of Cylinder Head

1. After the cylinder head has been removed from the engine, thoroughly clean the lifters and pushrods.
2. Remove the valve rocker arm mechanisms and clean each thoroughly.
3. Check the rocker arms for wear on the valve and the pushrod end. If the signs of wear are evident on the valve end, some rocker arms can be ground to eliminate the wear. Maximum metal removal is approximately 0.010 inch.
4. Check each pushrod for straightness. This can be done by rolling the pushrod along a flat surface.
5. Check the valve guides carefully. Note if the valve stem-to-guide clearance is out of manufacturer's specification. If it is, on certain engines the guide can be reamed to the next oversize and the appropriate oversize valve installed. Depending on the year of the vehicle and the manufacturer, valves are available in standard sizes and oversizes of, for example, 0.00295, 0.0059, and 0.00984 inch. Always check the manufacturer's service manual to determine if this procedure is possible.
6. Remove the valve using the spring compressors. There are several valve removing tools available as shown in ***Figure 18–29***.
   a. Place the cylinder head in a position so that the compressor tools can be applied to each valve as shown in ***Figure 18–30***.
   b. Make sure the spring compressor is matched to the size of the valve.
   c. Slowly compress the valve spring by tightening the valve spring compressor.
   d. When the valve spring is fully compressed, carefully remove the valve keepers and valve retainer.
   e. Slowly release the valve spring compressor.

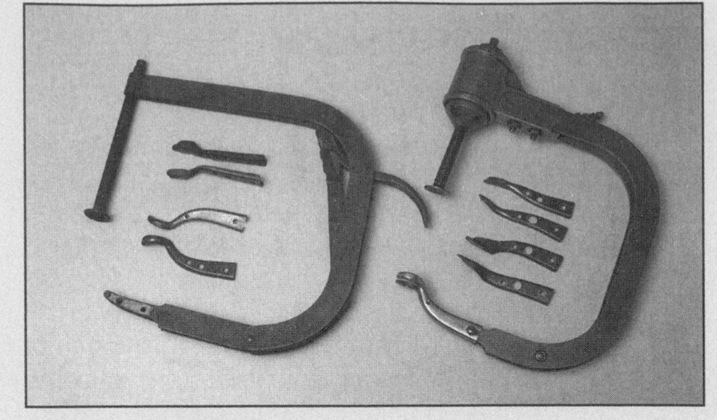

FIGURE 18–29 These valve-removing tools are used in valve removal and installation. (Courtesy of Snap-on Tools, www.snapon.com)

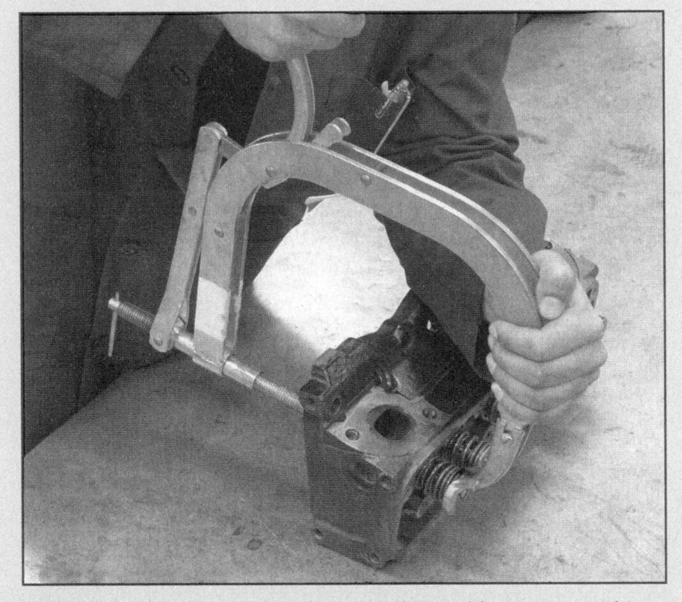

FIGURE 18–30 This valve-removing tool is compressed over the spring to remove the valve keepers and valve spring retainers.

CAUTION: *Valve springs are under high pressure. Be careful not to release the compressor too fast as the spring may be forced out rapidly and cause injury.*

   f. Check for any metal burrs on the end of the valve stem caused by the rocker arm hitting the valve. Carefully remove these burrs with a file before removing the valve. If the burrs are not removed, they may scratch the inside of the valve guide and cause oil leakage after the engine has been rebuilt.
   g. Now carefully remove each valve by pushing it out through the valve guide.
7. Completely remove all carbon deposits from the cylinder head and combustion chamber. On aluminum heads, be careful not to damage or scratch the softer aluminum.
   a. Using a small punch, chip away the larger particles of carbon.

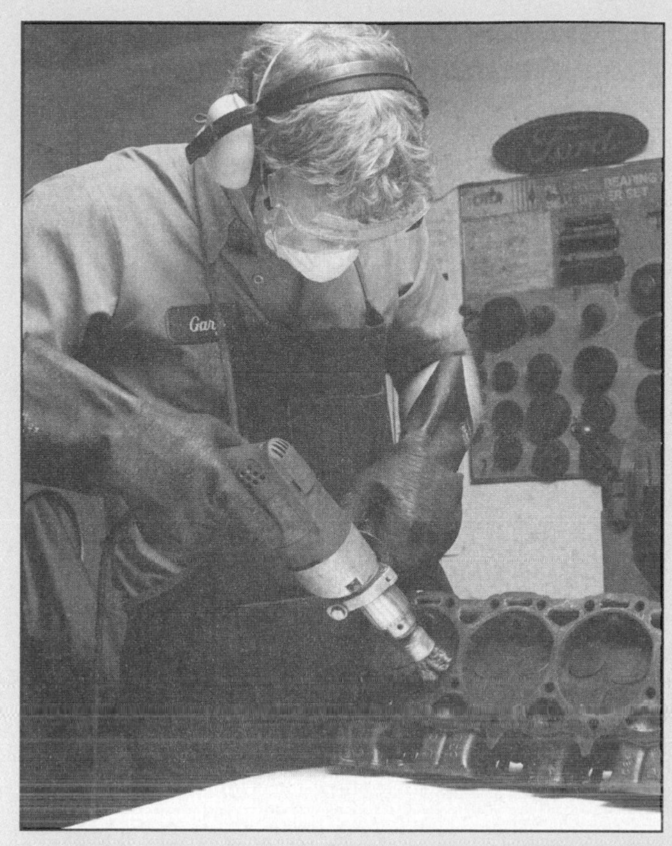

FIGURE 18–31 Clean the combustion chamber by using a drill and wire brush.

b. Using an electrical drill with a coarse wire brush, clean the carbon deposits from the combustion chamber (***Figure 18–31***).

CAUTION: *Be sure to wear a face mask and safety glasses during this process.*

c. Be careful not to damage the seat area of the valves with the wire wheel.

8. Completely remove all carbon deposits from the valves.
   a. Chip off the large particles of carbon with a steel punch. Be careful not to chip the valve surface, especially at the fillet.
   b. Using a wire brush on a power grinder, clean off the remaining valve deposits.

CAUTION: *Be sure to wear a face mask and safety glasses during this process.*

   c. Do not clean the stem of the valve with the wire brush.
   d. Soak the valve in cleaning solvent to soften the remaining carbon and varnish. Use a polishing grade of emery cloth (not coarser than #300 grit) to clean the remaining varnish.

Checking and Grinding Valves and Seats

1. Once the valves have been removed, check the valve face and valve seats for indications of burned surfaces.

FIGURE 18–32 A valve-grinding machine is used to grind the valves correctly. (Courtesy of Sioux Tools, Inc.)

2. Using a valve grinder as shown in ***Figure 18–32***, grind each valve according to the manufacturer's specifications. Most valves are ground at either a 44-, 45-, or 46-degree angle. Other engines may have a 30-degree angle. Refer to the manufacturer's specifications to determine the valve face angle for your vehicle. Generally, the procedure is as follows:
   a. Use the valve-grinding machine by first dressing the stone correctly.
   b. Place the valve in the holding chuck, and tighten securely.
   c. Turn on the machine and move the valve closer to the wheel. Make sure the coolant valve is turned on to allow coolant to flow over the stem.
   d. Advance and grind the valve slowly, making sure the valve is not turning blue. If it does, the valve is too hot and too much metal is being ground off at one time.
   e. Continue grinding the valve until all pits and signs of wear have been removed.
3. After the valve has been ground, check the valve margin and compare it with the manufacturer's recommended specifications.
   a. Use a vernier caliper to check the margin.
   b. The minimum width of valve margins is 1/32 inch. If the width is less than 1/32 inch, the valve should be discarded and replaced with a new valve.
4. Grind the valve seats using a valve seat grinder. Follow the tool manufacturer's procedure to grind the seats correctly. Generally, a 45-degree angle stone is placed on the sleeve of the valve seat grinder. Then a pilot rod is placed down into the valve guide. The valve seat grinder and stone are then placed over the guide and carefully brought down to touch the valve seat.

If the manufacturer calls for an interference fit, different degree angle stones must be used to grind off the upper and lower part of the seat. The valve seat must be a specified width. To adjust the width, use a 30-degree or 60-degree stone as described earlier in this chapter.

Checking and Grinding Valve Stems

1. Check the tip of the valve stem for wear from the rocker arms.
   a. If the stem is not square or has nicks and is rounded off, reconditioning is needed.
   b. Use the valve-grinding machine by first dressing the stone correctly.
   c. Place the valve in the holding clamp and tighten the clamp.
   d. Turn on the machine and move the valve closer to the wheel. Make sure the coolant valve is turned on to allow coolant to flow over the stem.
   e. Advance and grind the stem slowly, making sure the valve is not turning blue. If it is, the valve is too hot and too much metal is being ground off at one time.
   f. Grind the stem until all pits and signs of wear have been removed.
   g. If the chamfer has been removed on the valve stem, it may have to be reground. Make sure the chamfer is no wider than 1/32 inch.
2. If the valve face and stem have been ground, the valve will sit at a different height when replaced in the cylinder head. This difference may cause the clearance to change slightly, especially on overhead camshafts. Follow the manufacturer's procedure to check the installed spring height if necessary.

Checking Valve Guides

1. Check the valve guides for wear according to the manufacturer's specifications. These specifications are listed in many of the automotive maintenance manuals.
   a. Check the valve guide by determining the maximum dimension at the port end or top end of the guide.
   b. Place a telescoping or small-hole gauge inside the guide.
   c. After removing the gauge, measure it with a micrometer.
   d. Measure the smallest valve stem diameter on the valve. This is usually found where the valve stem touches the top of the valve guide.
   e. Subtract the two readings. This value is called the valve clearance.
   f. Average specifications are: intake, 0.001–0.003 inch; exhaust, 0.0015–0.0035 inch. Note that these readings may vary with different manufacturers and engine makes.

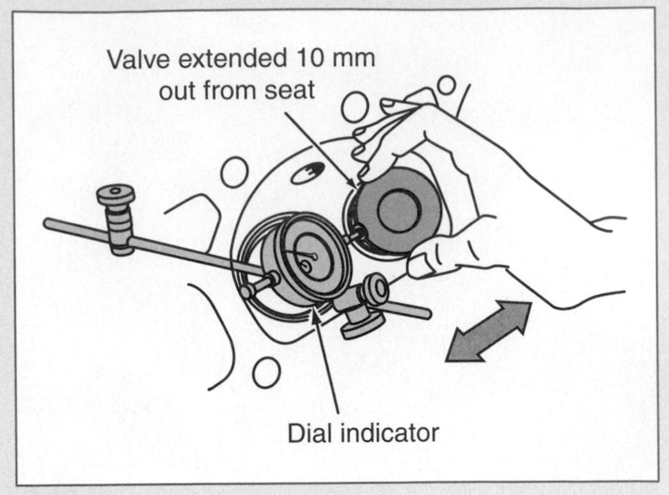

FIGURE 18–33 To check valve stem clearance, use a dial indicator against the valve.

   g. A dial indicator can also be used to check the valve guide clearance. With the dial indicator touching the valve, move the valve back and forth to read the amount of clearance, as shown in ***Figure 18–33***.
   h. If they are out of specifications, the valve guides may need to be reamed as shown in ***Figure 18–34***.

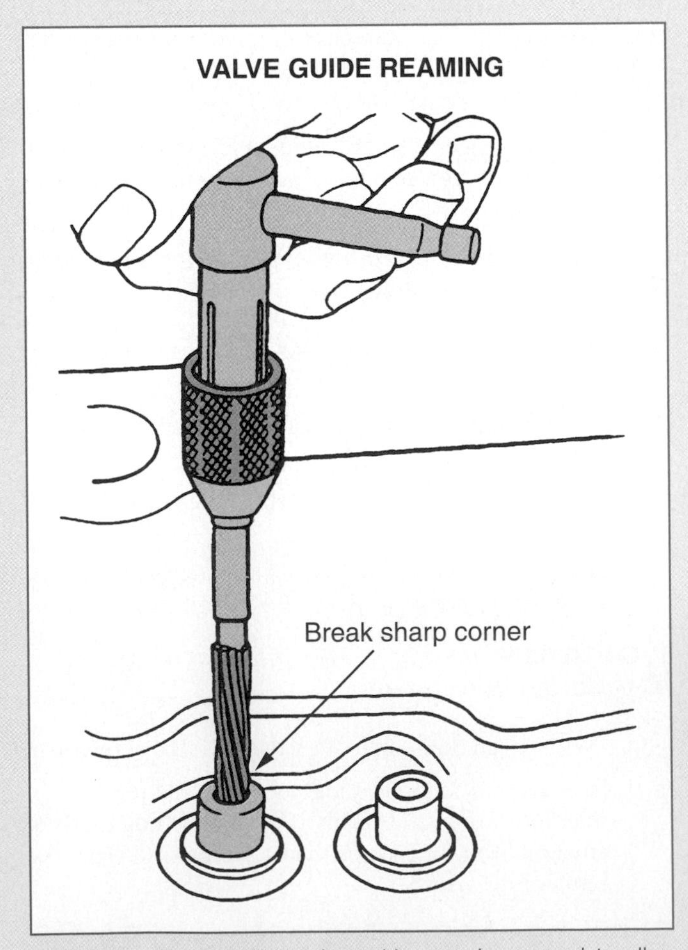

FIGURE 18–34 The valve guides can be reamed to allow an oversize valve to be used. (Courtesy of American Honda Motor Co., Inc.)

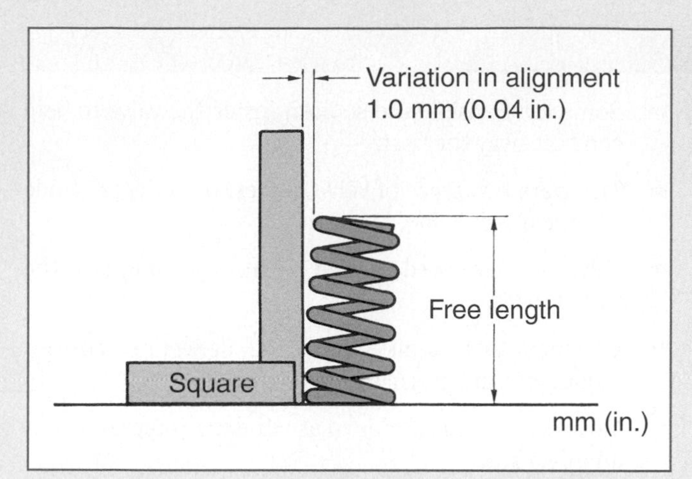

FIGURE 18-35 Use a square to check spring alignment.

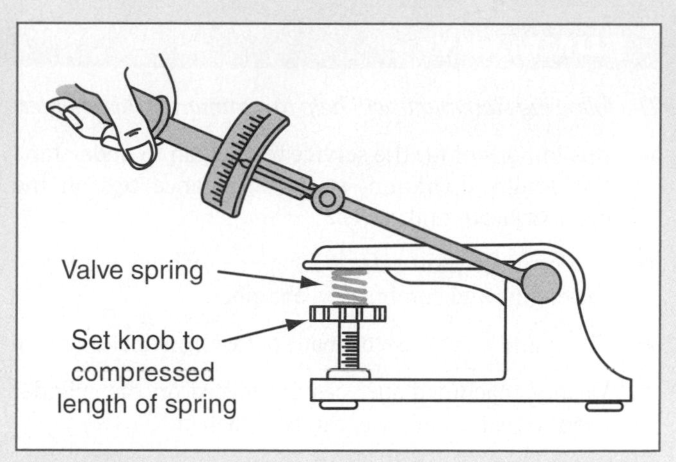

FIGURE 18-36 Valve spring tension (compression) can be checked on a valve spring tension checker.

Checking Valve Springs

1. During disassembly of the cylinder head, check each valve spring for possible breakage. Replace as necessary.
2. Check the valve springs for alignment and squareness.
   a. Stand the valve spring on a flat surface.
   b. Hold the bottom of the spring against a square as shown in ***Figure 18–35***.
   c. Using a feeler gauge or vernier caliper, measure the distance away from the square, the gap, and from the spring top to the square.
   d. Compare the distance measured with the manufacturer's specification. The maximum distance allowed varies, but generally it is in the range of 1/16 to 5/64 inch.
3. Check the valve springs for the correct valve spring tension. A valve spring tension tester is normally used as shown in ***Figure 18–36***.
   a. After placing the spring in the tester, compress each spring the same amount. There is usually a stop on the handle of the spring tension tester.
   b. Read the tension on the dial and compare the reading with the manufacturer's specifications. The valve spring tension should usually be within 10% of the specifications listed in the manual. A typical compression pressure reading would be 180 pounds of compression at 1.09 inches of movement.

Service Manual Connection

There are many important specifications to keep in mind when rebuilding the cylinder head and valve mechanism. To identify the specifications for your engine, you will need to know the VIN (vehicle identification number) of the vehicle, the type and year of the vehicle, and the type of engine. Although they may be titled differently, some of the more common cylinder head and valve specifications (not all) found in service manuals are listed below. Note that these specifcations are typical examples. Each vehicle and engine will have different specifications.

Common Specification	Typical Example
Cylinder head warpage limit	0.006 inch
Seat angle	45 degrees
Seat width (intake)	0.069–0.091 inch
Valve margin (intake)	1/32 inch
Valve spring out-of-square limit	0.060 inch
Valve stem diameter (intake)	0.3159–0.3167 inch
Valve stem to guide clearance	0.0008–0.0027 inch

In addition, the service manual will give specific directions for many service procedures. Some of the more common procedures include:

- Adjusting valves
- Cylinder head replacement
- Pushrod service
- Replacing the camshaft
- Rocker arm service
- Valve guide service

SUMMARY

The following statements will help to summarize this chapter:

- It is important for the service technician to understand the design, operation, and maintenance tips on the cylinder heads and valves.
- The cylinder head acts as a cap, or seal, for the top of the engine and combustion chamber.
- The cylinder head can be made of cast iron or aluminum.
- Various machined surfaces are placed on the cylinder head so that other parts can be attached.
- The cylinder head has both intake and exhaust valves and ports to allow fuel and air to enter and leave the engine.
- Exhaust, intake, and cooling passageways are cast into the cylinder head.
- There are several types of combustion chambers, including the wedge-shaped, the hemispherical, and the pentroof combustion chamber.
- Combustion chambers are designed to reduce pollution, increase turbulence, and quench the air-fuel ratio.
- Diesel engines have the combustion chamber built into the piston head.
- The valve assembly consist of the valves, valve seats, seals, guides, springs, and retainers.
- Valves can be made of several materials including nickel, chromium, and manganese.
- Some valves use liquid sodium inside the valve to help conduct away the heat.
- There are two types of valve guides: the integral guide and the insert guide.
- Valve seals are used to stop oil from seeping past the valve guides.
- Common valve seals include oil deflectors, positive guide seals, and passive guide seals.
- Valve seats can be designed as either an integral seat or an insert seat.
- When checking valve seats, check for cracks and excessive wear.
- Valve seats can be ground so that there is an interference fit and the width of the valve seat is within manufacturer's specifications.
- Valve springs are used to keep the valves closed and are held in place by retainers and keepers.
- Valve rotators are used to slightly rotate the valve each time it is opened.
- There are many checks to make on the cylinder head and valves, including valve spring tension, valve spring alignment, valve guide wear, valve stem wear, cylinder head flatness, valve margin, and valve seat width, among other checks.

TERMS TO KNOW

Can you explain each of the following terms? Review the chapter until you can use each term correctly.

Curtain area
Harmonics
Hemispherical combustion chamber
Inclined surface
Insert guides
Integral guides
Interference angle
Keepers
Line contact
Passive seals
Pitch
Poppet-type valve
Precombustion chamber
Quenching
Seating
Shrouding
Siamese ports
Stellite
Turbulence
Valve bounce
Valve face
Valve float
Valve guide
Wedge-shaped combustion chamber

REVIEW QUESTIONS

Multiple Choice

1. Which of the following is used to seal the top of the cylinder block and hold the valves?
a. Valve guides
b. Cylinder head
c. Piston
d. Piston rings
e. Crankshaft

2. Two cylinders drawing air and fuel from the same port are called:
a. Double porting
b. Multiple porting
c. Siamese ports
d. Single ports
e. All of the above

3. Which type of combustion chamber is designed as a half circle?
a. Hemispherical combustion chamber
b. Wedge-shaped combustion chamber
c. Open combustion chamber
d. Precombustion chamber
e. Closed combustion chamber

4. When gases are cooled by pressing them in the cylinder area, ________ has occurred.
a. Turbulence
b. Quenching
c. Combustion
d. Cool down
e. Convection

5. Which type of combustion chamber is called a turbulence-type combustion chamber?
a. Hemispherical
b. Wedge
c. Precombustion
d. Open
e. Closed

6. The________ combustion chamber is a second small chamber used to ignite a rich mixture of fuel.
a. Hemispherical
b. Wedge
c. Precombustion
d. Closed
e. Open

7. The valve assembly consists of:
a. Valve guides
b. Valve seats and valves
c. Springs, retainers, and seals
d. All of the above
e. None of the above

8. The ________ on the valve makes direct contact with the valve seat.
a. Valve stem
b. Valve face
c. Valve margin
d. Valve bottom
e. Valve side

9. The distance between the valve face and the valve head is called the:
a. Stem
b. Fillet
c. Margin
d. Seat
e. Guide

10. Which of the following valve parts are worn when oil leaks into the exhaust stream?
a. Valve guides
b. Valve margin
c. Valve fillet
d. Valve seat
e. Valve head

11. Which type of valve seat can be removed and replaced if it is defective?
a. Insert type
b. Integral type
c. Ground seats
d. Hemi-seat
e. Wedge seat

12. When the valve is ground at a different angle than the seat by 1 degree, a/an ____________ is obtained.
a. Wrong angle
b. Interference angle
c. Poor contact
d. Side angle
e. Offset angle

13. Which of the following are used to reduce harmonics in valve springs?
a. Variable spring pitch
b. More than one spring
c. Small spring vibration damper
d. All of the above
e. None of the above

The following questions are similar in format to ASE (Automotive Service Excellence) test questions.

14. *Technician A* says valve springs should be checked for weight. *Technician B* says valve springs should be checked for color. Who is correct?
a. A only
b. B only
c. Both A and B
d. Neither A nor B

15. *Technician A* says that to position the valve seat correctly for maximum sealing, grind the bottom of the seat using a 30-degree stone. *Technician B* says to use a 60-degree stone. Who is correct?
    a. A only
    b. B only
    c. Both A and B
    d. Neither A nor B

16. *Technician A* says that engines can have four valves per cylinder. *Technician B* says that engines can have up to eight valves per cylinder. Who is correct?
    a. A only
    b. B only
    c. Both A and B
    d. Neither A nor B

17. *Technician A* says that the top of the cylinder head should be checked for roughness. *Technician B* says that the top of the cylinder head should be checked for flatness. Who is correct?
    a. A only
    b. B only
    c. Both A and B
    d. Neither A nor B

18. *Technician A* says that if the valve guides are worn and out of specifications, oil may leak into the exhaust. *Technician B* says that if the exhaust valves are burned, compression gases will leak into the exhaust. Who is correct?
    a. A only
    b. B only
    c. Both A and B
    d. Neither A nor B

19. *Technician A* says that on some engines, valve guides can be bored to different oversize specifications. *Technician B* says that valves cannot be serviced on all engines. Who is correct?
    a. A only
    b. B only
    c. Both A and B
    d. Neither A nor B

20. *Technician A* says that when removing valves, use a valve spring compressor. *Technician B* says to always remove burrs on the end of the valve stem before removing it from the valve guide. Who is correct?
    a. A only
    b. B only
    c. Both A and B
    d. Neither A nor B

Essay

21. What is quenching in reference to the combustion chamber?

22. What is the purpose of a precombustion chamber?

23. What should be done to the valve if the valve face is damaged?

24. What will be the result if the valve stem-to-guide clearance is too large?

25. What is the purpose of the interference angle on valves?

26. What is the purpose of valve rotators?

Short Answer

27. A crack in the cylinder head can cause the coolant to ________.

28. Two common problems that valves have after extended operational time are ________ and ________.

29. Always check the valves for indications of ________.

30. Valve guides can be checked by using a/an ________.

Applied Academics

TITLE: Valve Positioning

ACADEMIC SKILLS AREA: Mathematics

NATEF Automobile Technician Tasks:

The service technician can visually perceive the geometric figure relationship of systems and subsystems that require geometric alignment.

CONCEPT:

Valve positioning is a critical component of reconditioning the cylinder head. The valve cannot be too high, too low, and/or too narrow or wide. It is critical that the valve be positioned according to the manufacturer's specifications. The figure shows examples of how the valve should be positioned within the cylinder head and four examples of incorrect positioning. The information in this chapter indicated a procedure to adjust the valve positioning correctly by cutting a 30-degree and/or 60-degree angle on the valve seat.

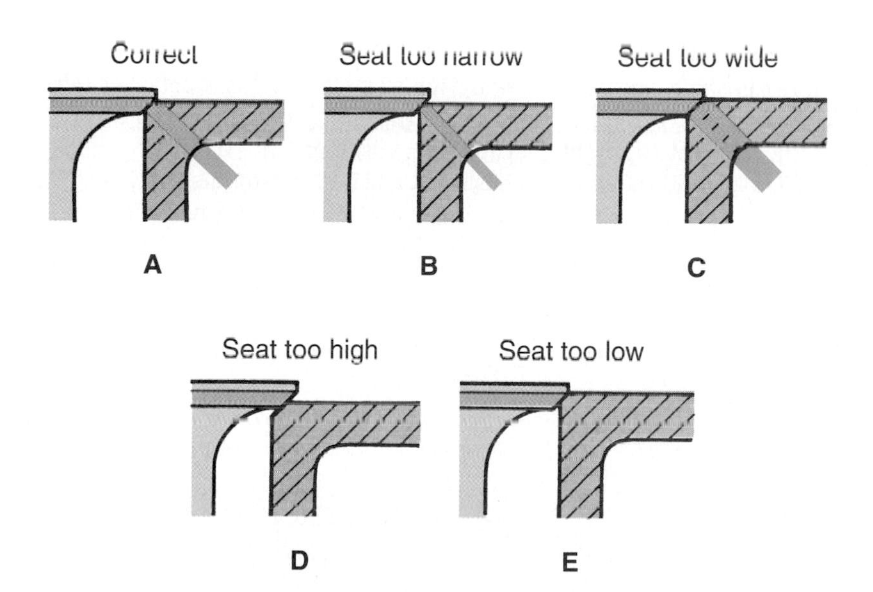

APPLICATION:

Review the information in this chapter about positioning a valve correctly in the seat. What type of grinding procedure and angles would be needed to correct the problems identified in illustrations B and C? What type of grinding procedure and angles would be needed to correct the problems identified in illustrations D and E? Can you identify the problems that may occur with each of the incorrect valve positioning examples?

CHAPTER

19

Camshaft and Valve Drives

OBJECTIVES

After studying this chapter, you should be able to:

- Define the purpose and parts of the camshaft, including the thrust plate and bushings.
- Analyze the design of cam lobes.
- Determine how to time the camshaft to the crankshaft.
- Identify the parts in the valve operating mechanism, including the lifter, pushrods, and rocker arms.
- Describe the operation of variable valve timing.
- Analyze the diagnosis and service procedures for several problems of the camshaft and valve drive mechanisms.

INTRODUCTION

All automotive engines use one or more camshafts and various types of valve drive mechanisms to work correctly. The camshaft is used to open and close the valves at the right time. The valve drive mechanism is used to connect the camshaft to the valves so the valves can operate correctly. Some of the topics studied in this chapter include camshaft design, lobe design, timing of the camshaft to the crankshaft, and camshaft bushings. In addition, lifters, cam followers, pushrods, and rocker arms are presented. Variable valve timing is also explained as a means to improve engine efficiency.

Certification Connection

ASE Connection: The information in this chapter can help you prepare for the National Institute for Automotive Service Excellence (ASE) certification tests. The tests and content areas most closely related to this chapter are:

Test A1—Engine Repair

- **Content Area**—General Engine Diagnosis: Removal and Reinstallation (R & R)
- **Content Area**—Cylinder Head and Valve Train Diagnosis and Repair

NATEF Connection: Much of the information in this chapter is related to the NATEF tasks. The NATEF tasks and priority numbers most closely related to this chapter are:

1. Inspect and replace timing belts (chains), overhead cam drive sprockets, and tensioners: check belt/chain tension; adjust as necessary. P-1
2. Inspect the camshaft for runout, journal wear, and lobe wear. P-2
3. Inspect camshaft bearing surface for wear, damage, out-of-round, and alignment; determine necessary action. P-3
4. Establish camshaft timing and cam sensor indexing according to manufacturer's specifications and procedures. P-1
6. Inspect camshaft drives (including gear wear and backlash, sprocket and chain wear); determine necessary action. P-2
7. Inspect hydraulic or mechanical lifters; determine necessary action. P-2
8. Adjust valves (mechanical or hydraulic lifters). P-1
9. Inspect pushrods, rocker arms, rocker arm pivots, and shafts for wear, bending, cracks, looseness, and blocked oil passages (orifices); determine the necessary action. P-2

19.1 CAMSHAFT ASSEMBLY

The camshaft is used to open and close the intake and exhaust valves throughout the four strokes of the engine. To do this, the camshaft is driven by the crankshaft. The camshaft assembly includes the camshaft, camshaft timing gear, camshaft bearings, and timing chain, belt, or gears, if used.

VALVE TIMING

The exact time at which both the intake and exhaust valves open and close is critical to maximum engine efficiency. For example, it is very important for the intake valves to open slightly before the exhaust valves close. ***Figure 19–1*** shows a timing diagram of when the valves open and close throughout the four strokes of engine operation. In this figure, there are two complete revolutions of the crankshaft. As shown in Chapter 13, TDC is shown directly at the top of the diagram. BDC is shown at the bottom of the diagram and represents 180 degrees. The intake, compression, power, and exhaust strokes are shown in terms of degrees of rotation. Note that as the piston comes up on the exhaust stroke, the intake valve starts to open before the exhaust valve is closed. This valve overlap is important for correct engine operation. The exact amount of overlap on engines will vary. Overlap provides more efficient cylinder filling at high rpm. However, overlap also causes lower engine vacuum and poorer low-end performance, idle quality, and low-speed fuel economy. Also note that the exhaust valve opens 57 degrees before bottom center. Of course, when the exhaust valve opens determines when the power stroke effectively ends.

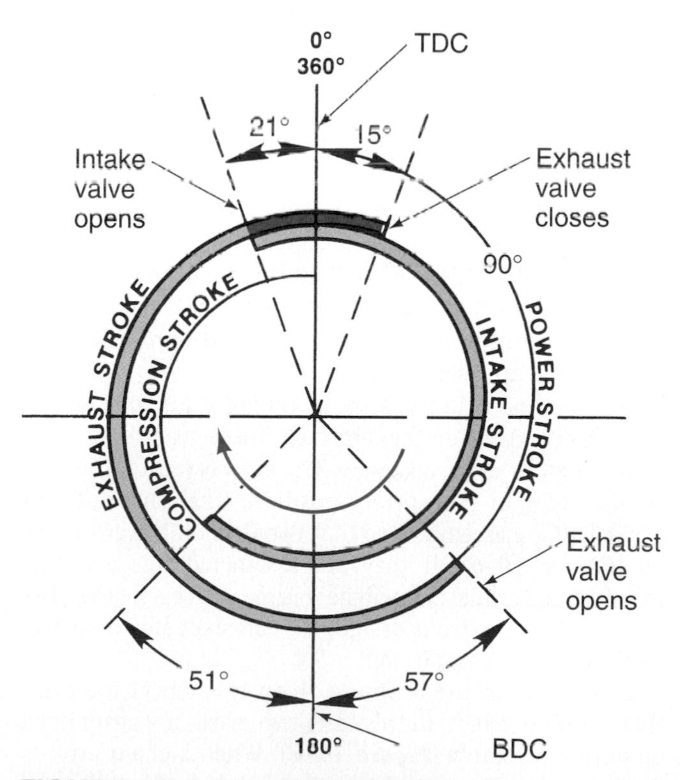

FIGURE 19–1 A timing diagram helps to show the amount of valve overlap between the intake and exhaust valves.

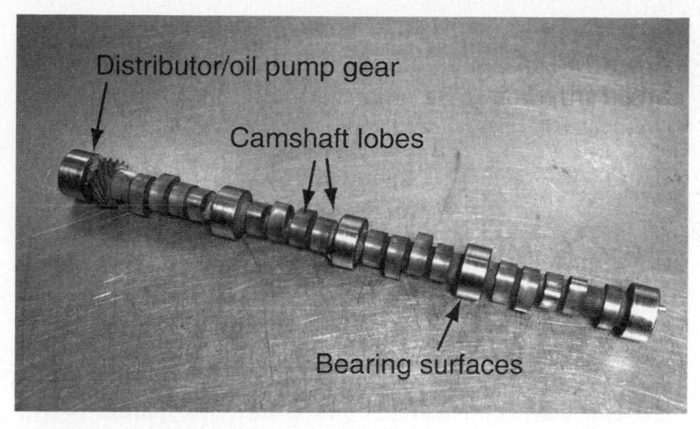

FIGURE 19–2 Cam lobes are machined on the camshaft to lift the valves open at the right time.

CAMSHAFT DESIGN

The camshaft is a long shaft that fits into the block or head on overhead cam engines. The cams, or **lobes**, are machined on the camshaft for each cylinder. As the camshaft turns, the lobes open and close the valves at the right time (***Figure 19–2***). Several bearing surfaces are also machined on the camshaft. These surfaces are used to support the camshaft at several places along its length. The typical camshaft is made of cast or forged steel. The cam and bearing surfaces are hardened to provide protection from excessive wear. In addition, a gear is placed on the camshaft. This gear is used to drive the distributor (if used) and the oil pump.

CAMSHAFT THRUST

Camshafts need some means to control the shaft end thrust. One method is to use a **thrust plate** between the camshaft gear and a flange machined into the camshaft. The thrust plate is bolted to the block to contain any thrust movement of the camshaft. On certain overhead camshaft designs, the thrust plate is bolted directly to the cylinder head. ***Figure 19–3*** shows how the clearance between the camshaft and the thrust plate is checked.

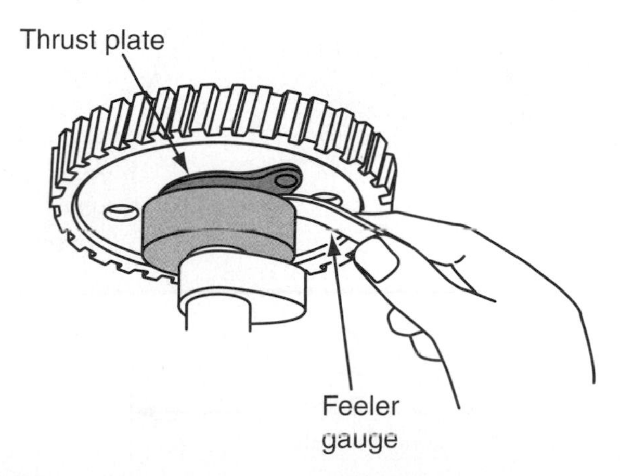

FIGURE 19–3 The clearance between the thrust plate and the end of the camshaft is shown.

CAMSHAFT LOBE DESIGN

There are many designs used on camshaft lobes. The contour of the camshaft lobe determines how and when the valves open and close. Camshaft lobes play an important part in volumetric efficiency. The speed and amount of opening, and the speed of closing are controlled by the shape of the cam lobe.

The camshaft lobe has several design components (***Figure 19–4***). The amount of lift on the valves is determined by the shape of the camshaft lobe. The lift is important because it indicates how far the valves will open. Camshaft lift is how high the cam lobe reaches above its base diameter. The camshaft duration is a measure of how long the valve stays open. The shape of the cam lobe, where it starts on the base diameter and where it ends, determines the camshaft duration. Camshaft duration is measured in degrees of rotation. The open and close ramps on the camshaft lobe lift and allow the valve to close. The lobe can have several shapes, depending on the speed needed to open or close the valve. For example, on certain camshafts it may be desirable to open the valve gradually and close it slowly without bouncing.

Figure 19–5 shows several exaggerated variations of ramps. The nose is defined as the top of the cam lobe. Its shape determines how long the valve will remain fully open. It can have different shapes, depending on how long the valve needs to be fully open. The heel of the camshaft is the bottom part of the camshaft shape. When the lifter or valve is moving on the heel, the valve is fully closed. The shape of these parts determines the specific characteristics—time and speed—of the valve opening process.

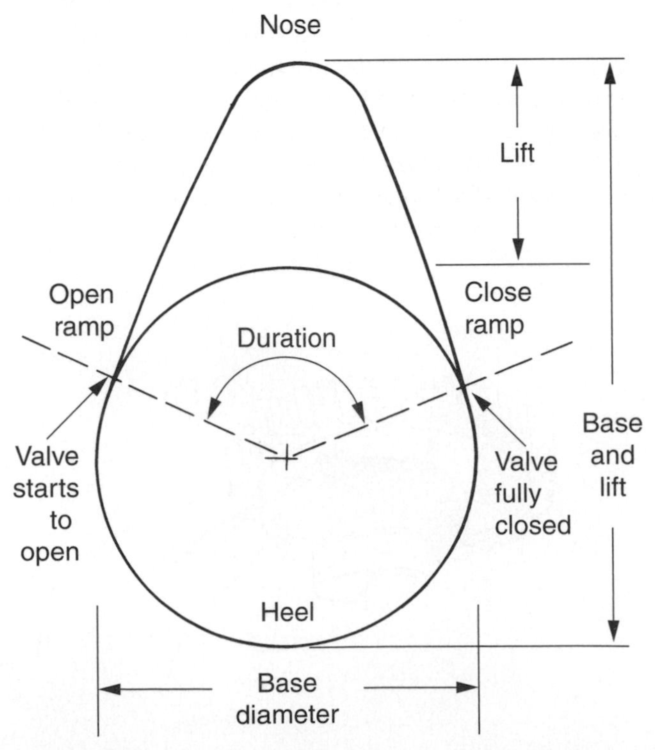

FIGURE 19–4 The design components of a typical camshaft lobe are shown.

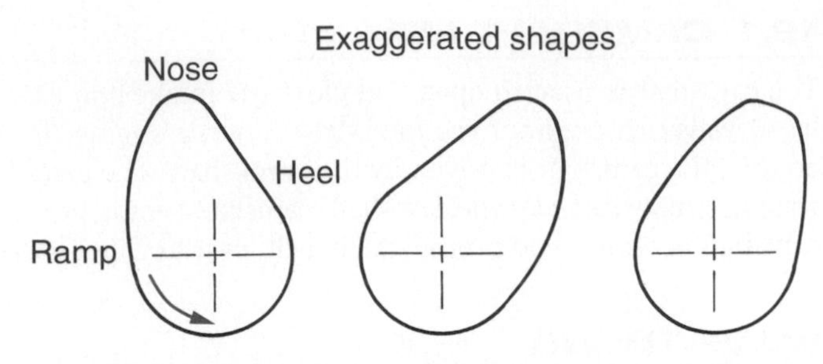

FIGURE 19–5 The shape of the camshaft lobe is designed to raise and lower the valve at a specific time.

Car Clinic: CAMSHAFT WEAR

CUSTOMER CONCERN:
Camshaft Damage
A car has about 60,000 miles on it. During this time the camshaft has been replaced two times. Why would a camshaft go bad in so few miles?

SOLUTION: Several conditions cause excessive camshaft wear. The friction and bearing loads on a camshaft are very great. Under these conditions motor oil may be good for only about 5,000 miles. Beyond this point the oil breaks down and cannot lubricate under high stresses. Frequent oil changes will improve this situation. Several manufacturers have also reduced the valve spring tension so as not to wear the camshaft as much. Remember also that high rpm on the engine can also increase the stresses on the camshaft.

CAMSHAFT TIMING TO CRANKSHAFT

For the valves to open and close in correct relation to the position of the crankshaft, the camshaft must be timed to the crankshaft. This means that the shafts must be assembled so that the lobes open the valves at a precise time in relation to the position of the piston and crankshaft. Several methods are used to do this.

One method is to use a set of timing gears, one on the crankshaft and one on the camshaft. These gears are located on the shaft by using a **keyway**. The keyway locates the gear on the shaft in the correct position. The camshaft and crankshaft are assembled so that two dots on the gears line up (***Figure 19–6***). If they are assembled this way, the camshaft and crankshaft will be in time with each other. Because of the four-stroke design, the camshaft always rotates half as fast as the crankshaft.

Some engines use a timing chain to connect the camshaft and crankshaft. In this case, two marks are again lined up during assembly (***Figure 19–7***). When a chain drive is used on some engines, a spring-loaded damper pad is used to keep the chain tight.

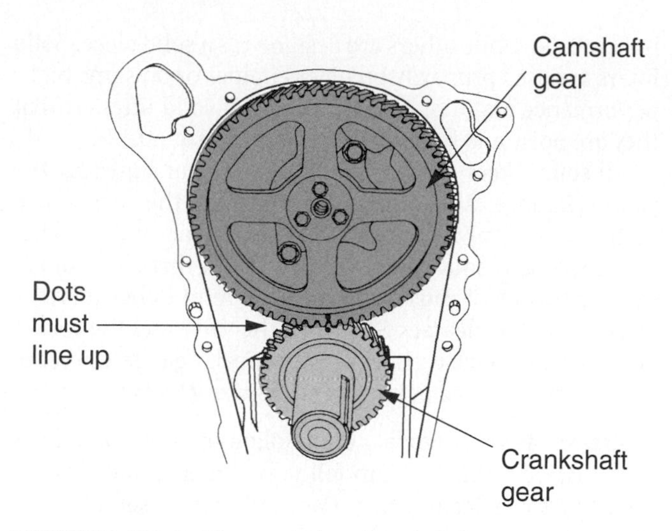

FIGURE 19–6 The camshaft and crankshaft gears must be timed to each other. Timing marks are lined up to ensure correct installation.

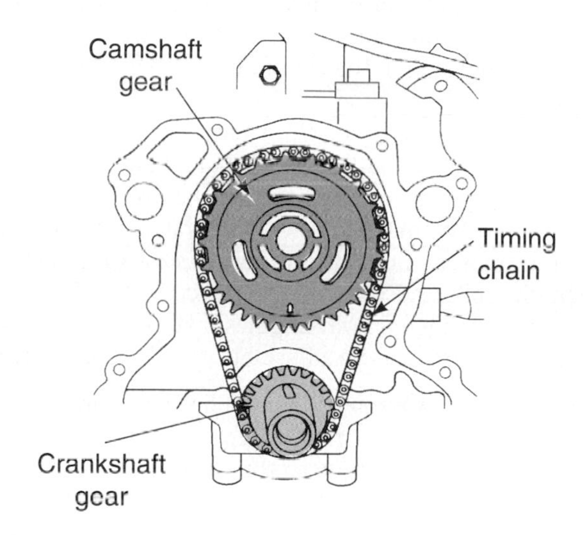

FIGURE 19–7 When a chain connects the camshaft and crankshaft, the timing marks must be lined up during installation.

Overhead camshafts are also timed by lining up marks on the shafts, as shown in ***Figure 19–8***. It is important to always review the procedures listed in the maintenance manual when timing the camshaft and crankshaft. A chain tension device is used to keep the chain tight.

If the marks are not lined up correctly after reassembly, the camshaft and crankshaft will not be in time with each other. If this happens, when the engine is reassembled and cranked over, the valves will not open and close at the proper time. As a result of this, the piston may come up and bend a valve. In some cases, when a chain is worn or loose, it may jump a tooth. This would also disrupt the engine timing and possibly result in valves being bent.

CAMSHAFT BUSHINGS

The camshaft is supported in the cylinder block by several bushings. These bushings are friction-type bearings and are called camshaft bearings by some service technicians. They are designed as one piece and are typically pressed into the camshaft bore in the block (***Figure 19–9***).

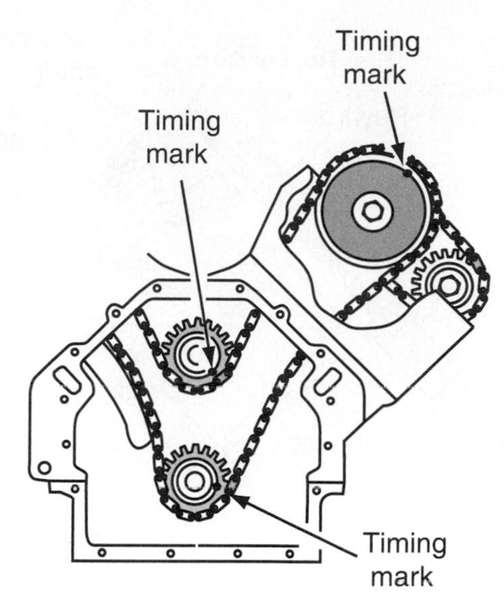

FIGURE 19–8 The timing marks are shown for an overhead camshaft in a V configuration engine.

FIGURE 19–9 A camshaft bearing is a full round design pressed into the block.

PARTS LOCATOR

Refer to photo #7 in the front of the textbook to see what ***camshaft bushings*** *look like.*

19.2 VALVE DRIVE MECHANISM

The valve operating mechanism consists of the lifters that ride on the camshaft, the pushrods, and the rocker arms. These components make up the mechanism used to transfer the camshaft lift to the valve assembly (***Figure 19–10***).

As the engine heats up during operation, the parts of the valve mechanism expand. Because of this expansion, there must be a clearance in the valve mechanism. This clearance

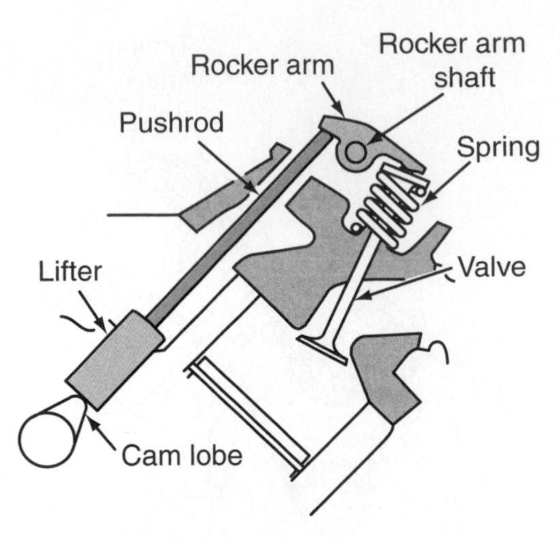

FIGURE 19-10 The lifters, pushrods, and rocker arms make up the valve operating mechanism.

is called **valve train clearance**. All engine valve mechanisms must account for this clearance.

LIFTERS

The valve lifters, which are also called **tappets**, are designed to follow the shape of the camshaft and lobes. The valve lifters ride directly on the camshaft. Lifters are used to change the rotary motion of the camshaft to reciprocating motion in order to open and close the valve.

Several types of lifters are used. They include solid lifters, cam followers (or roller lifters), and hydraulic lifters.

Solid Lifters Solid lifters transfer motion as a solid piece from the camshaft to the pushrod. Solid lifters are designed as lightweight cylinders. Some are designed as a hollow tube, while others are designed as a solid piece. Solid lifters are used primarily on older engines or on some high-performance engines. A disadvantage of solid lifters is that they are noisy and have a distinct tapping sound.

All solid lifters must have some means of adjusting the valve clearance. Adjustments are made on the rocker arm section of the valve mechanism. On one end of the rocker arm there is a small adjustment. As the threaded adjustment is turned, it causes the rocker arm to either increase or decrease the clearance between the valve and the rocker arm. The clearance is checked with a feeler gauge and set to manufacturer's specifications (see ***Figure 19–11***).

Cam Followers Cam followers are sometimes used instead of lifters. Cam followers are also called roller lifters by some technicians. Cam followers resemble solid lifters. The only difference is that followers have rollers that roll on the camshaft, rather than metal that slides on the camshaft. Cam followers reduce friction and more evenly distribute the load. They are used primarily on high-compression engines such as diesel and racing engines. The roller is on one end of the cam followers. The other end is machined so the pushrod can be inserted. Cam followers also have a small groove machined into the side. Generally, there is a small pin in the lifter bore that matches the groove. This design helps eliminate the possibility of the cam followers spinning in its bore. Several designs are used. One of the most common is shown in ***Figure 19–12***. In this design, the roller rides directly on the camshaft. The pushrod sits in the cam follower. Engines that use cam followers of this design also have an adjustable valve train clearance.

Hydraulic Lifters Hydraulic lifters are used to reduce noise and to control the valve clearance. Hydraulic lifters use oil pressure to keep the valve clearance at zero. The main parts of the hydraulic lifter are shown in ***Figure 19–13***. The lifter body houses the internal parts of the hydraulic lifter. The plunger moves up and down inside

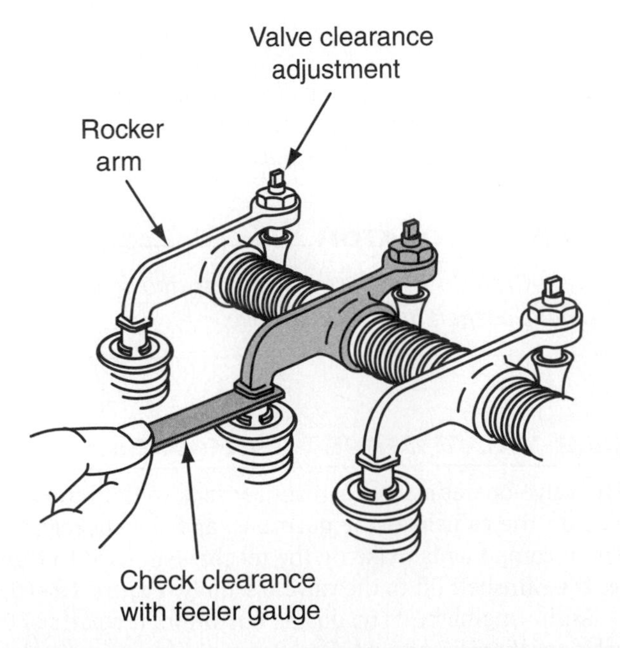

FIGURE 19-11 When solid lifters are used, the valves must be adjusted using a feeler gauge.

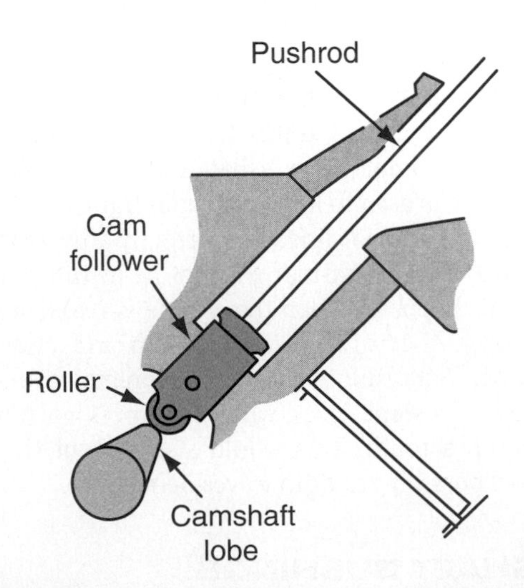

FIGURE 19-12 Cam followers or roller lifters are sometimes used in place of solid lifters. They have a roller that rides on the camshaft.

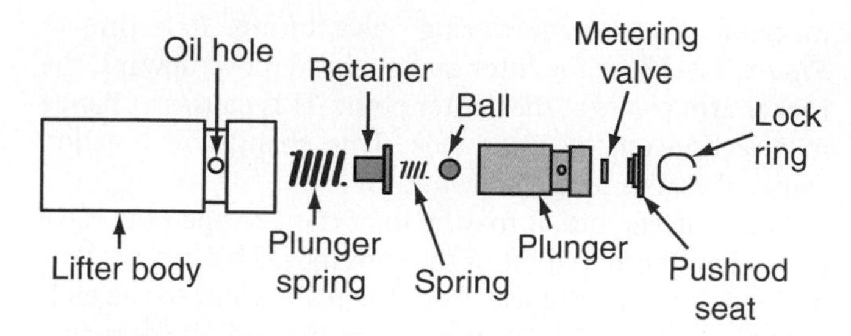

FIGURE 19–13 The parts of a typical hydraulic lifter are shown in this illustration.

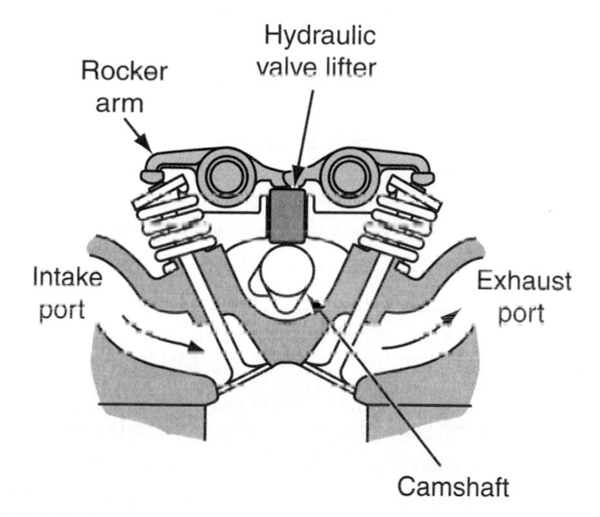

FIGURE 19–14 On an overhead cam engine, the lifters are set directly above the camshaft. No pushrod is needed.

the body. ***Figure 19–14*** shows the hydraulic lifter operation for an overhead cam engine. The rocker arms push directly on the lifter.

The hydraulic lifter operates as shown in ***Figure 19–15***. Referring to ***Figure 19–15A***, before the cam lobe starts to lift, the plunger is pushed upward by the internal return spring. Oil is fed from the lifter bore into the recessed area called the **oil relief**. The oil pressure comes from the oil system in the engine. During this time, the high-pressure chamber is filled with oil. Any clearance within the valve mechanism is now taken up by the oil pressure. Note that the oil passes by the check ball. The check ball allows oil to pass in one direction only. If oil tries to flow from the high-pressure chamber into the plunger, it will be stopped. Some lifters use a check valve rather than a check ball, but the principle is still the same.

When the cam lobe starts to raise the lifter (***Figure 19–15B***), the valve spring pressure, which is felt through the rocker arm, attempts to keep the plunger from moving upward. The body of the lifter, however, is raised by the cam lobe. This causes the pressure in the high-pressure chamber inside the lifter to increase. As the cam lobe continues to lift the body of the lifter, the high pressure locks the system, and the valve is forced to open. A small amount of oil then leaks from the clearance between the plunger and the body, causing the plunger to move down slightly. This is called leak-down. Leak-down is very important. It is controlled precisely by the clearance between the plunger and the body. Its purpose is to allow the valve spring pressure to push the plunger down adequately into the body.

As the cam lobe continues to turn, the pressure on the body is reduced. This is because the cam lobe has passed the

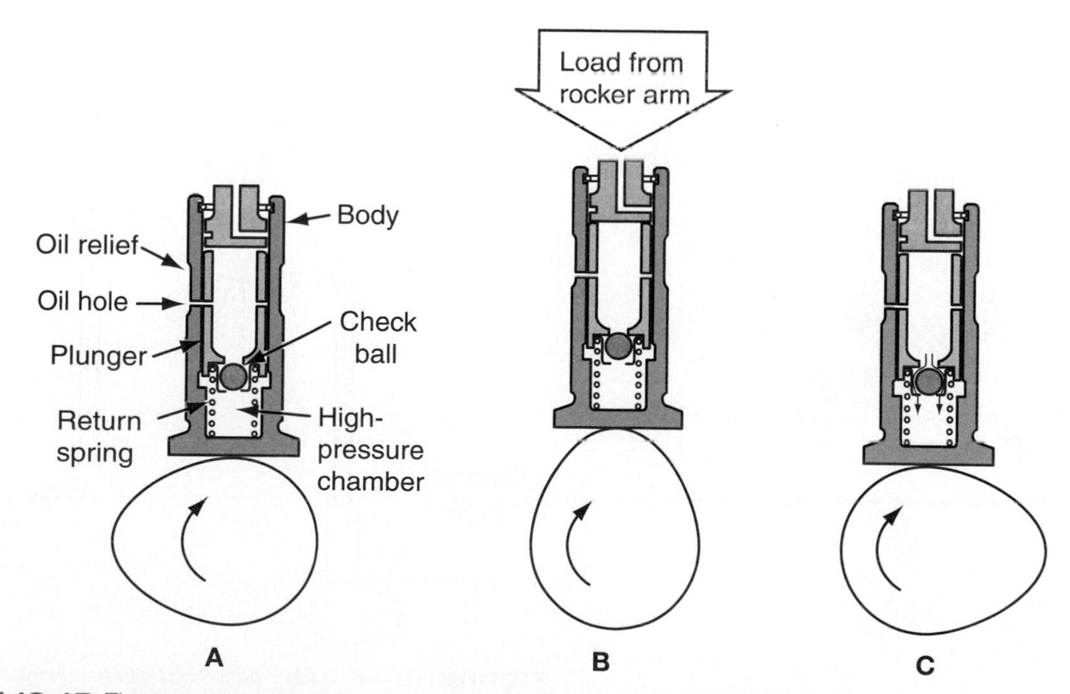

FIGURE 19–15 The operation of a hydraulic lifter is shown. View A shows the operation before the cam lobe lifts. View B shows the operation during cam lobe lifting. View C shows the operation after the cam lobe has been lifted and is going down.

lifter. As the pressure decreases, the pressure from the valve spring forces the valve to close. The plunger and the body of the lifter now return to the original position. Again, any clearance in the valve mechanism is taken up by the oil pressure. When the lifter is at the position shown in ***Figure 19–15C***, it is ready to repeat the cycle.

PUSHRODS

Pushrods are designed to be the connecting link between the rocker arm and the valve lifter. They are made to be as light as possible. Some are designed to have small convex balls on the end. These small balls ride inside the lifter and the rocker arm. Pushrods are used only on engines that have the camshaft placed within the block. Overhead camshafts do not need pushrods.

Pushrods are either solid or hollow. On certain engines, the pushrods have a hole in the center to allow oil to pass from the hydraulic lifter to the upper portion of the cylinder head. Rather than having a convex ball, the end of the pushrod on solid lifters is concave. This end then fits into a ball on the rocker arm. ***Figure 19–16*** shows examples of different types of ends on pushrods.

ROCKER ARMS

Rocker arms are designed to do two things: (1) change the direction of the cam lifting force, and (2) provide a certain **mechanical advantage** during valve lifting. Referring to ***Figure 19–17***, as the lifter and pushrod move upward, the rocker arm pivots at the center point. This causes a change in direction on the valve side. This change in direction causes the valve to open downward.

On some engines, it may be important to open the valve more than the actual lift of the cam lobe. This can be done by changing the distances from the **pivot point** to the ends of the rocker arm. The distance from the ends of the rocker arm to the pivot point is not the same (***Figure 19–18***). Note that the distance from point A to the pivot point is less than the distance from point B to the pivot point. The ratio between these two measurements is called rocker arm ratio. In this example, the ratio is 1.5 to 1. This means that the valve will open 1.5 times more than the actual lift on the cam lobe.

Rocker arms are designed and mounted in several ways. Some are designed to fit on a rocker arm shaft. Springs, washers, individual rocker arms, and bolts are used in this type of assembly. Other rocker arms are placed on studs that are mounted directly in the cylinder head.

Some overhead camshaft engines use rocker arms in such a way that the camshaft rides directly on top of the rocker arm (***Figure 19–19A***). With this type of mechanical camshaft arrangement, the valve will open slightly more than the camshaft lobe height. In ***Figure 19–19B***, the camshaft lobe rides on the rocker arm, but there is an additional hydraulic lash adjuster on the right side of the rocker arm. In this case, the hydraulic lash adjuster works the same as a typical hydraulic lifter, taking up any clearance that exists between the valve and rocker arm. Some overhead camshaft engines have done away with the rocker arm completely. In this type of camshaft arrangement, the camshaft lobes ride directly on top of the valves.

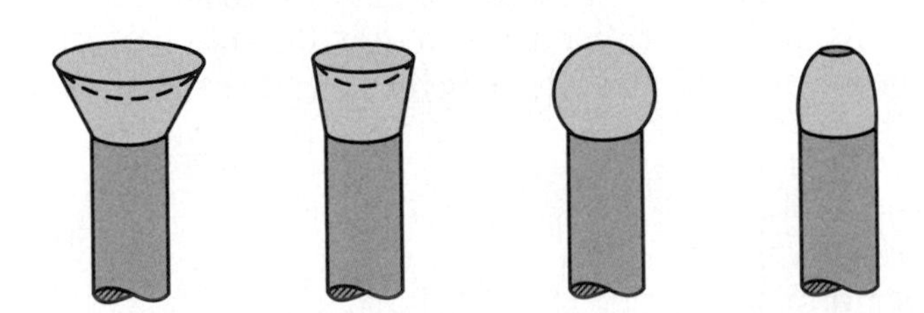

FIGURE 19-16 Several types of ends are used on pushrods.

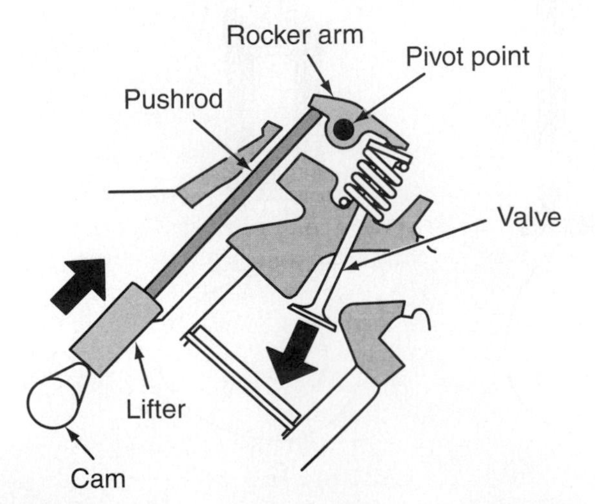

FIGURE 19-17 Rocker arms are used to change the direction of motion on the valve operating mechanism.

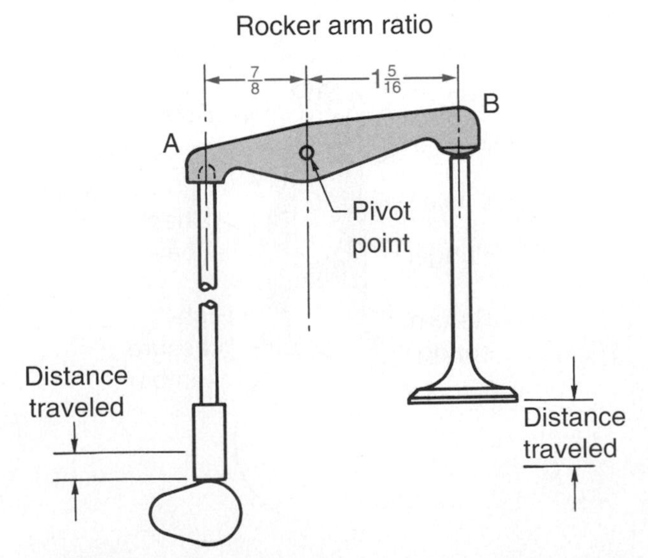

FIGURE 19-18 Rocker arms have different lengths from their center to their ends. This difference produces a rocker arm ratio, which is used to open the valve more than the cam lift.

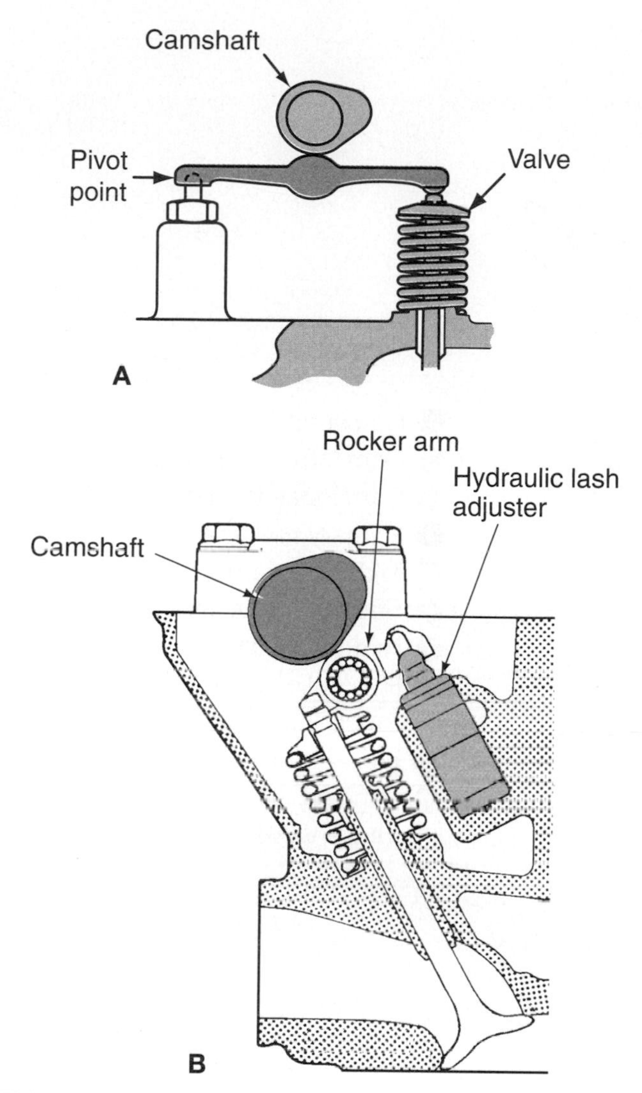

FIGURE 19–19 Certain engines have rocker arms that work directly under the camshaft. In some cases, there is also a hydraulic lash adjuster.

Figure 19–20 shows two examples of how the valves and camshafts work together. In ***Figure 19–20A***, there are two camshafts located above the valves. Also note the intake and exhaust rocker arms that transfer the motion from the cam lobes to the valves. In ***Figure 19–20B***, note that the two camshafts ride directly on the top of the valves. The only thing between the valves and the camshaft lobes are the hydraulic lash adjusters used to adjust the valve clearance.

VARIABLE VALVE TIMING

Recent engine designs, such as hybrid vehicle engines, have included a valve mechanism called **variable valve timing**. Variable valve timing is incorporated so that the total lift of the valve from its seat can be varied. In addition, the duration of the valve opening can also be changed. By changing the lift and duration of the valve, the engine power characteristics also change. Variable valve timing allows for the engine to change combustion characteristics (air-fuel amounts, combustion efficiency, and so on) throughout the rpm range of the engine. The result is a more powerful, more economical, and cleaner-running engine.

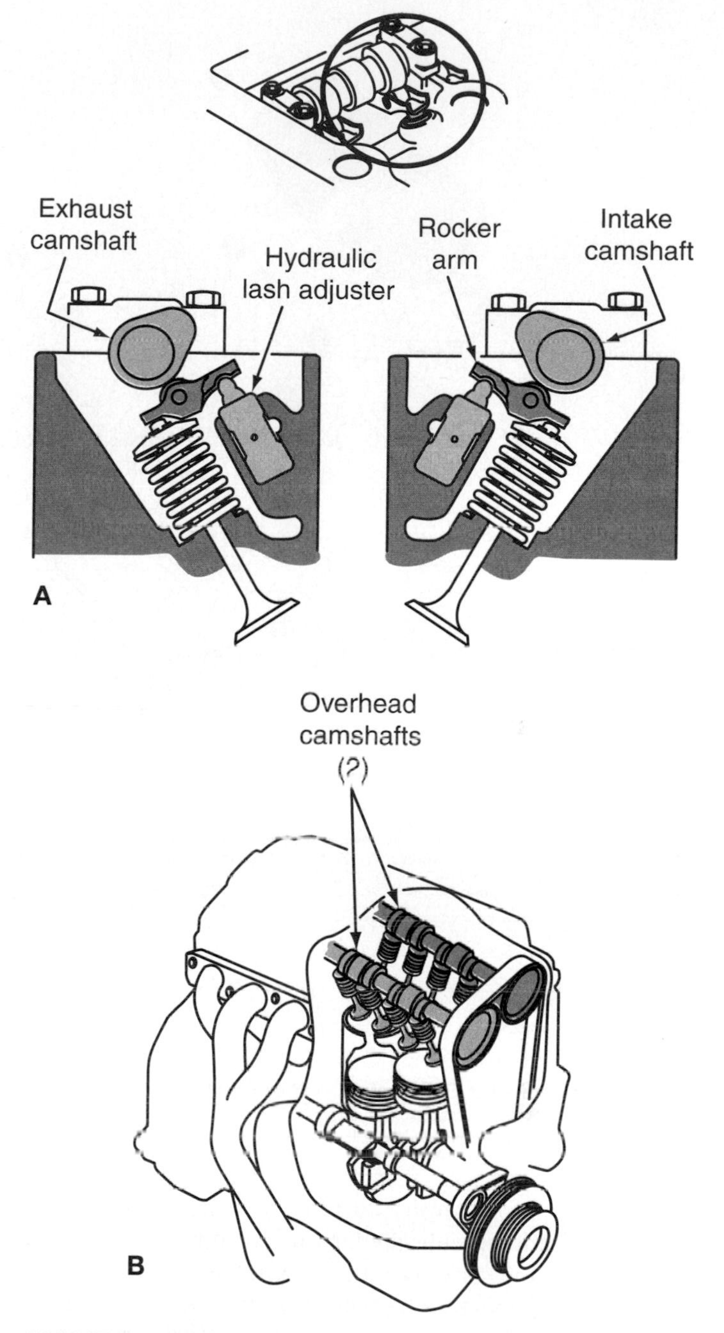

FIGURE 19–20 Valves can be connected to the camshaft using several methods.

In one system, the valves are adjusted according to the load sensed by the engine computer. For example, when the engine computer says that more power is called for, it adjusts the variable valve mechanism to get the maximum valve lift for the longest duration.

Mechanically, several systems are being used. One system uses the engine computer in conjunction with a control solenoid (a solenoid is used to convert electrical energy to mechanical energy). The control solenoid controls the amount

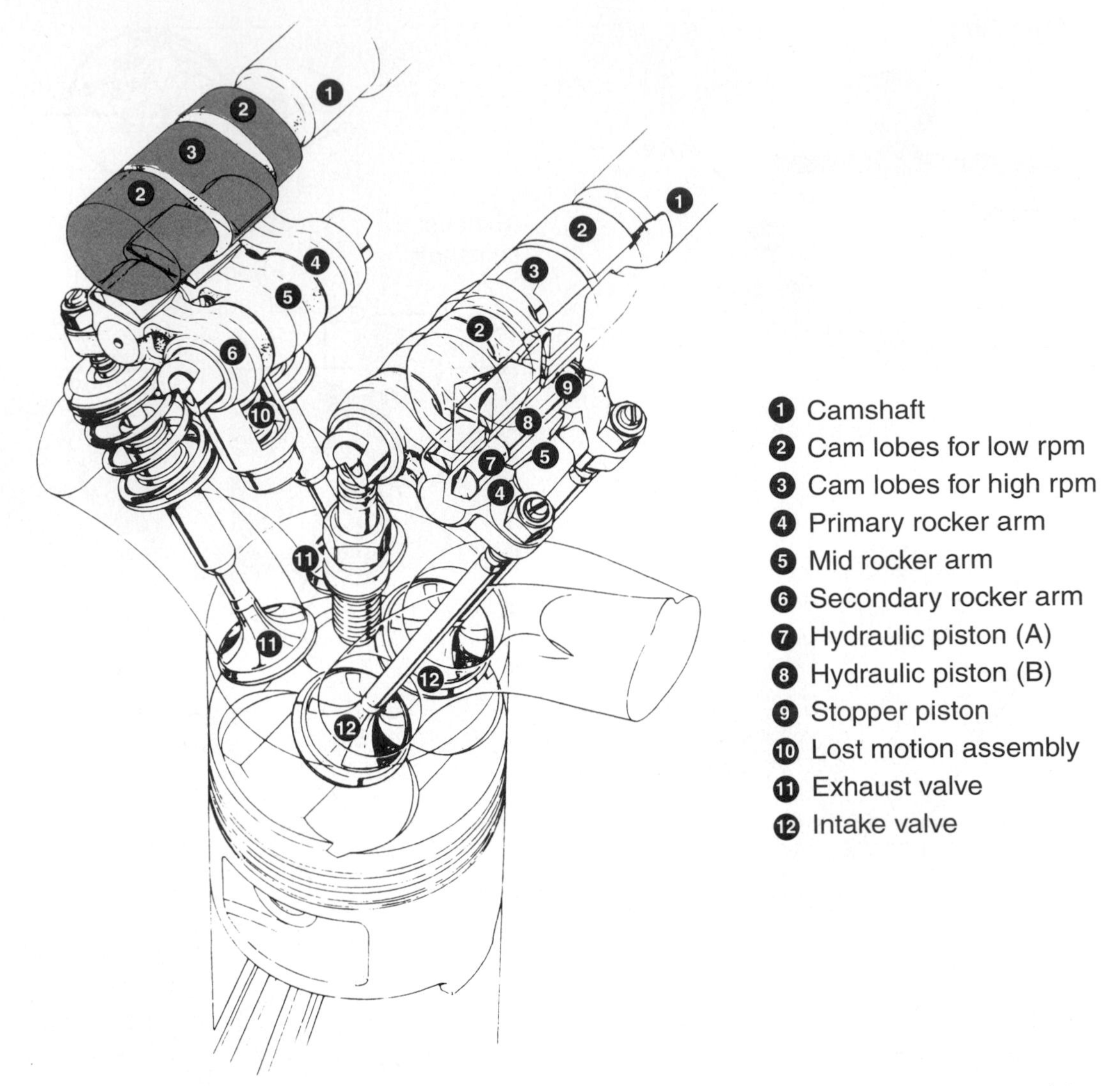

FIGURE 19-21 This illustration shows a variable valve timing engine with three cam lobes per cylinder. (Courtesy of American Honda Motor Co., Inc.)

of oil going into the hydraulic lifter. Depending on this flow, the lifter will hydraulically lock up to control the lift and duration of the valve opening. Another system has several cams located above each valve. A locking mechanism locks up the one that is needed for optimum efficiency. A third system has a hydraulic mechanism placed on the camshaft to allow it to be rotated relative to the drive sprocket. This produces a 20-degree shift in intake timing. As more efficiency and cleaner-burning engines are demanded, more variable valve systems will be designed in the future.

Figure 19–21 shows an example of an engine that has variable timing and uses several cam lobes for each valve. In general, it would be ideal if the engine could be designed with the high rpm performance of a racing engine and the low rpm performance of a standard engine. This would result in a maximum performance engine with a wide power band as shown in ***Figure 19–22***. On the bottom axis, rpm is shown. On the left axis, engine output torque is shown. Note that from low to high rpm, the torque band is much wider with variable valve timing. One big difference between a racing engine and a standard engine is the timing of the intake and exhaust valve and the degree of valve lift. Racing engines have longer intake and exhaust valve timing

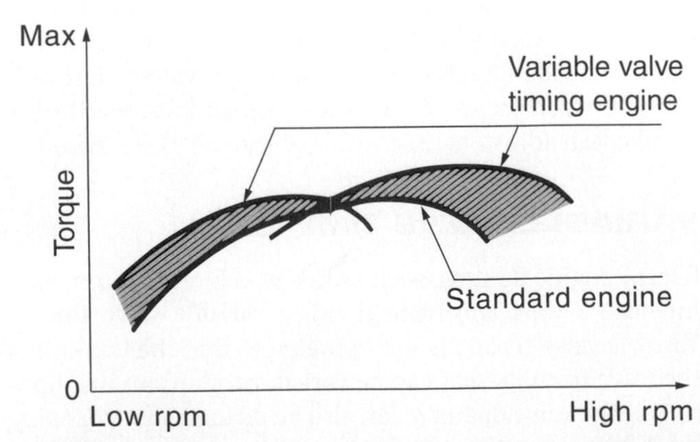

FIGURE 19-22 When variable valve timing is used, the torque or power band is much wider. (Courtesy of American Honda Motor Co., Inc.)

	Racing engine	Variable valve timing (VTEC)	Standard engine
Valve timing (exhaust / intake valve lift)	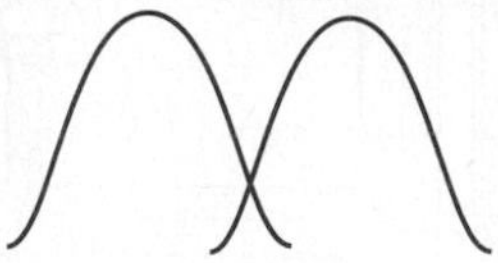 High profile	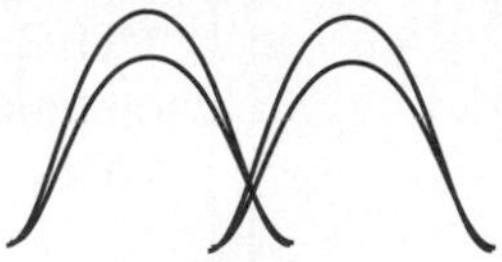 High and low profile	 Low profile
Max. power	Excellent	Excellent	Poor
Low rpm torque	Poor	Excellent	Excellent
Idle stability	Poor	Excellent	Excellent

FIGURE 19–23 The differences in valve lift for racing, VTEC, and standard engines are shown.

and higher valve lift compared to standard engines. With this engine, both conditions are met. This engine is called the VTEC or variable timing and electronic control engine.

Figure 19–23 shows the valve lift of the engine under three conditions. They are of a racing engine, variable valve timing and electronic control (VTEC) engine, and the standard engine. Note that the VTEC engine basically has the valve lift of both engines. The racing engine cam profile is designed to have maximum power. However, the low rpm torque and the idling stability are not maximized. With the standard engine, the cam profile is designed to maximize the low rpm torque and the idling stability; however, it is very difficult to get maximum power from this profile. The VTEC engine combines the best of both cam profiles. It combines maximum power with low rpm torque and idling stability.

In operation (see ***Figure 19–24***), there are three cam lobes for each set of intake and exhaust valves. During standard engine conditions, cam lobes A and B operate the two

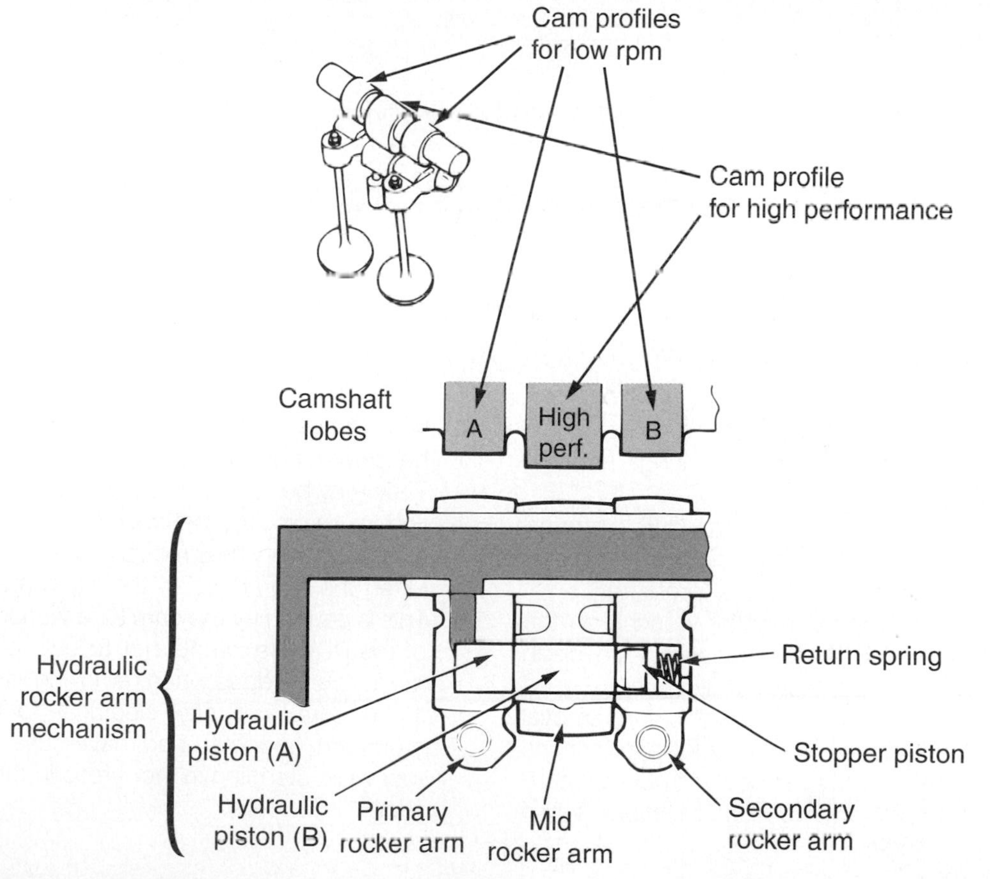

FIGURE 19–24 Hydraulic pressure is used to connect the mid rocker arm to the primary and secondary rocker arms during high-performance operation.

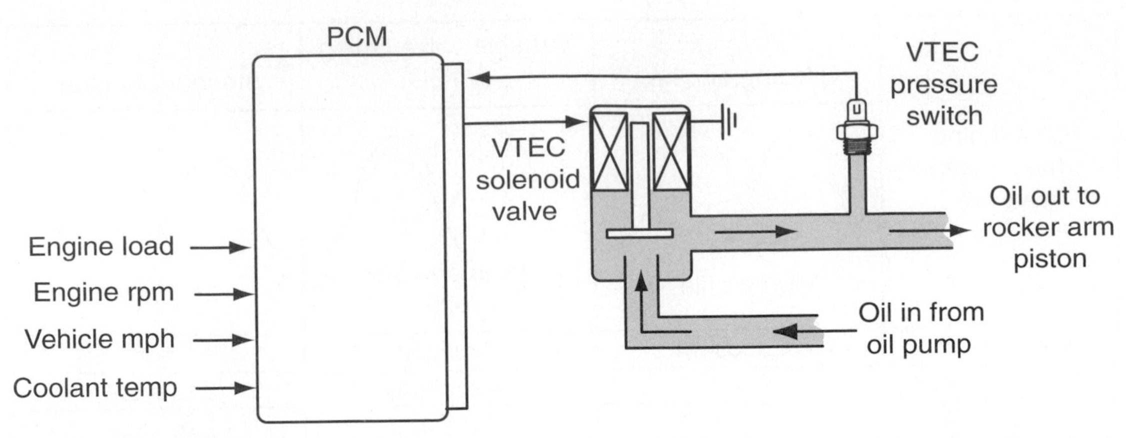

FIGURE 19-25 The engine control module (ECM) controls a solenoid valve to apply oil pressure to the rocker arm mechanism.

valves individually. The center cam lobe and the mid rocker arm are not being used. The center cam lobe is spinning and moving the mid rocker arm, but is not connected to the two outside (primary and secondary) rocker arms. When high performance is needed, the center cam, which is the racing or high-performance cam lobe, is used. During this time, the mid rocker arm is hydraulically connected to the outside rocker arms. This is done by applying hydraulic pressure to a piston inside the three cam followers. Now all three rocker arms are connected together. Thus, the center cam lobe is operating the valves for high performance. During this time, the other two outside camshaft lobes are spinning, but not touching the outside rocker arms.

Hydraulic pressure to the mechanism is controlled by the powertrain control module (PCM) shown in ***Figure 19–25***. Depending on engine speed, engine load, vehicle speed, and engine coolant temperature, an electronic signal is sent to a solenoid valve. The solenoid valve opens or closes an oil passageway from the oil pump to the camshaft rocker arm piston to either lock up or run separately.

Problems, Diagnosis, and Service

Safety Precautions

1. When installing a camshaft, be careful not to pinch your fingers between the cam bearings and the cylinder block holes.
2. Always wear safety glasses when working around the automotive shop. These should be OSHA-approved safety glasses with side protectors.
3. Make sure that all tools used for working on the camshaft and valve drive mechanism are clean and free of dirt and grease. All of the bolts used during the disassembly of the engine and removal of the camshaft require exact torque specifications during reassembly. Always follow the service manual's recommended torque specifications when rebuilding the engine.
4. When timing an engine camshaft to the crankshaft, be careful not to cut your fingers on the gears. They may be sharp. Also be careful not to let your fingers be pinched between the belt and the gear, or serious injury may result.
5. Often the top edges of the lifters are very sharp. This is caused by the cam lobe wearing on the top of the lifter. Be careful not to cut your fingers on these sharp edges when removing lifters.
6. Make sure that any electrical tools used are grounded correctly, and never use an electrical tool while standing in wet areas in the automotive shop.

PROBLEM: Camshaft Worn

Loss of power in two cylinders.

DIAGNOSIS

This type of problem is often associated with one or two worn camshaft lobes. If one or two camshaft lobes are worn, these valves will not be opening as far as necessary for correct engine operation. This will cause less air and fuel to enter these cylinders. Thus, volumetric efficiency is reduced. In this case, the camshaft needs to be removed and checked for wear on the lobes. While the engine is disassembled, the lifters should also be checked against the manufacturer's specifications.

SERVICE

Camshaft

1. With the camshaft removed from the engine, check each cam lobe for excessive wear by checking the lift of each lobe with a dial indicator or micrometer.
    a. Remove the camshaft from the engine.
    b. Place the camshaft in a set of V blocks so it can be rotated.
    c. Using a dial indicator, rotate the cam to determine the height of the cam lobe.
    d. If a dial indicator is not available, a micrometer can be used. Check the smallest diameter of each cam, and subtract this dimension from the largest diameter at the cam height (***Figure 19–26***). Compare each cam lobe and the lift. Compare the wear with the manufacturer's specifications. The maximum lobe wear allowed is usually about 0.005 inch.

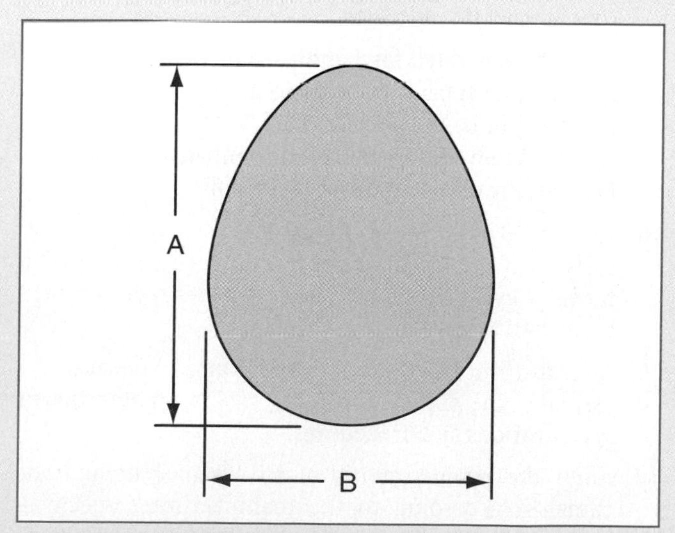

FIGURE 19-26 The height of the cam (cam lobe lift) can be measured subtracting dimension B from dimension A.

2. With the camshaft removed from the engine, check for excessive wear or scraping on the thrust plate.
3. The end play of the camshaft should also be checked. To complete this check, the camshaft must be placed in the block.
    a. Place a dial indicator on the end of the camshaft.
    b. Move the camshaft back and forth along its axis, and measure the amount of end play. Typical readings will vary; however, camshaft end play will be from 0.001 inch to as high as 0.010 inch on some engines.
4. When installing the camshaft, make sure the timing marks are aligned according to the service manual specifications. Automotive maintenance manuals have drawings showing how the timing marks should look. Although each manufacturer's procedure may be different, the camshaft must be timed correctly to the crankshaft.

Lifters

1. If the lifters are to be reused, they should be marked so that they can be installed so they operate on their original cam lobes.
2. Check each lifter for wear at the bottom or the part that rides directly on the camshaft. The lifter bottom should show no signs of wearing, scuffing, scoring, or pitting. Lifters that are concave or flat as shown in ***Figure 19–27*** must be replaced. Lifters that are slightly convex are not worn. Replace lifters if wear is evident.
3. The outer surface of the lifter should have no signs of wearing. If there is wear, scuffing, or scoring on the outside of the lifter, also check the lifter bores for wear. Replace the lifter if wear is noticeable.

PARTS LOCATOR

*Refer to photo #8 in the front of the textbook to see what **worn lifters** look like.*

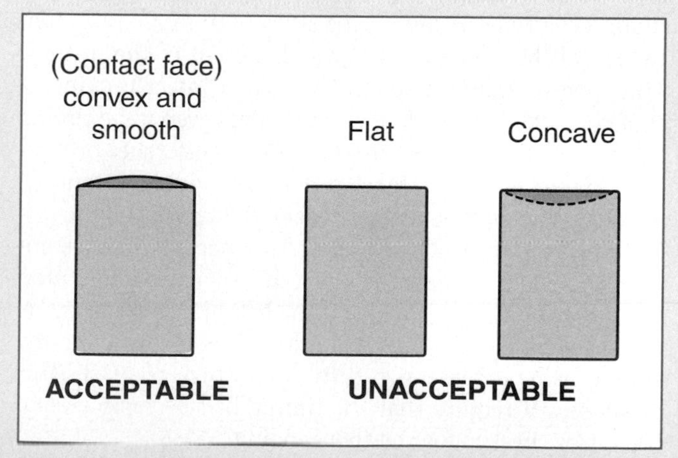

FIGURE 19-27 Acceptable and unacceptable lifter wear are shown in this illustration.

4. If in doubt about lifter quality, leak-down tests can be performed.
   a. Two systems can be used to complete a leak-down test. One system uses a special testing fixture, the other can be done on a workbench. If a fixture is available, follow the recommended procedure in the operator's manual.
   b. If a testing fixture is not available, disassemble the lifter, removing the plunger spring and oil.
   c. Reassemble the lifter without the plunger spring and snap ring.
   d. Depress the plunger until a light resistance is felt. Note the position of the plunger. Now rapidly depress and release the plunger.
   e. Observe the position the plunger returns to. If it stays down or does not return to the previously noted position (in step d above), the leak-down rate is excessive. The lifter should then be replaced.

CAUTION: *It is acceptable to replace old lifters and still use the old camshaft. However, never use old, worn lifters on a new camshaft.*

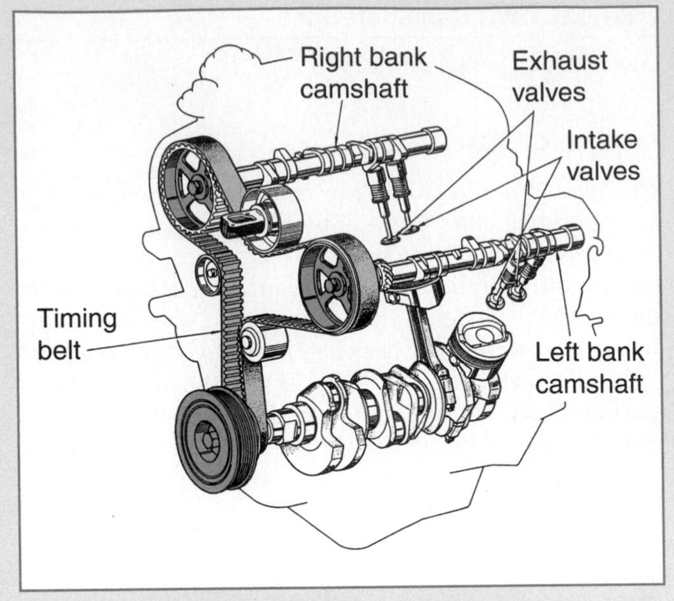

FIGURE 19–28 The timing belt needs to be replaced at specific mileage intervals as suggested by the manufacturer.

PROBLEM: Broken Timing Belt

An in-line 4-cylinder overhead cam engine has stopped running abruptly, and now doesn't even crank over completely.

DIAGNOSIS

When a timing belt is not replaced at the mileage specified by the manufacturer or is worn out, the belt will often break. If a timing belt has become weakened due to excessive wear, the belt may break under rapid acceleration conditions. If this happens, the camshaft and crankshaft get completely out of time. In most engines, when the camshaft stops rotating, the crankshaft will still turn a few more revolutions as the engine stops. If this happens, often valves will be bent by the upward motion of the pistons. This type of engine is often called a "contact" engine. A "noncontact" engine will not bend valves even if the timing belt breaks. Engines can also be called interference or noninterference engines. An interference engine is one in which contact will be made between the valves and the piston when the timing belt breaks. In a noninterference engine, there is no contact between the valves and the piston when the timing belt breaks. If the belt breaks on an interference engine, the valves will most likely be bent, and the engine will probably not turn over when cranked. This problem will require that the timing belt be replaced and the valves checked for possible bending.

Figure 19–28 shows an example of a timing belt that could break if it is not replaced according to the manufacturer's specified mileage period.

SERVICE

1. Following the manufacturer's recommended procedure, remove the outer covers of the timing belt.
2. Loosen all belt tension devices.
3. Remove the old belt and check to see if it is broken. The belt may be broken in half or may only have the ridges worn away.
4. Remove the valve cover to expose the overhead camshaft.
5. Remove the camshaft from the cylinder head.
6. Remove the cylinder head according to the manufacturer's recommendations.
7. Inspect each valve for bending and replace all valves that have been bent. In order to do this procedure, you will need to bend the valve back to its original condition as closely as possible. First, remove the valve keepers and springs using the correct tools.
8. Now remove the valve. Be careful that the bent valve doesn't scrape the inside of the valve guide during removal. Never reuse a bent valve. Always replace a bent valve with a new valve.
9. After the bent valve has been removed and replaced, reassemble the engine according to the manufacturer's specifications and procedures.
10. Align the timing marks of the crankshaft and the camshaft according to the manufacturer's specifications. Remember that each engine may be different.
11. Timing is done by positioning the crankshaft at TDC for the number 1 cylinder. On some engines there is an

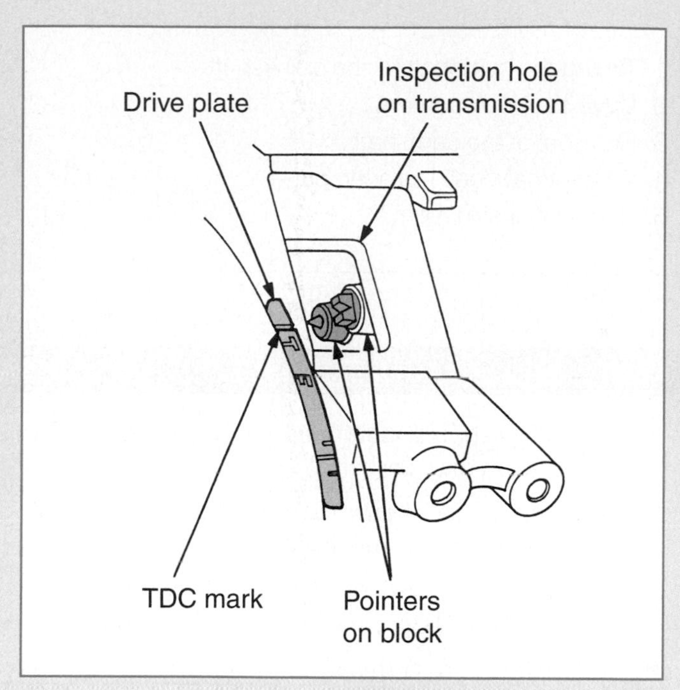

FIGURE 19–29 When replacing a timing belt, put the engine crankshaft at TDC by looking through the inspection hole on the transmission housing. (Courtesy of American Honda Motor Co., Inc.)

inspection hole on the transmission that shows when the crankshaft is at TDC for the number 1 cylinder. ***Figure 19–29*** shows an example of a typical timing alignment of the crankshaft at TDC.

12. After the crankshaft is correctly aligned, rotate the camshaft until the marks line up correctly, as shown in ***Figure 19–30***.
13. Now install the belt.
14. After the belt is correctly installed, reassemble the engine according to the manufacturer's recommended procedure.

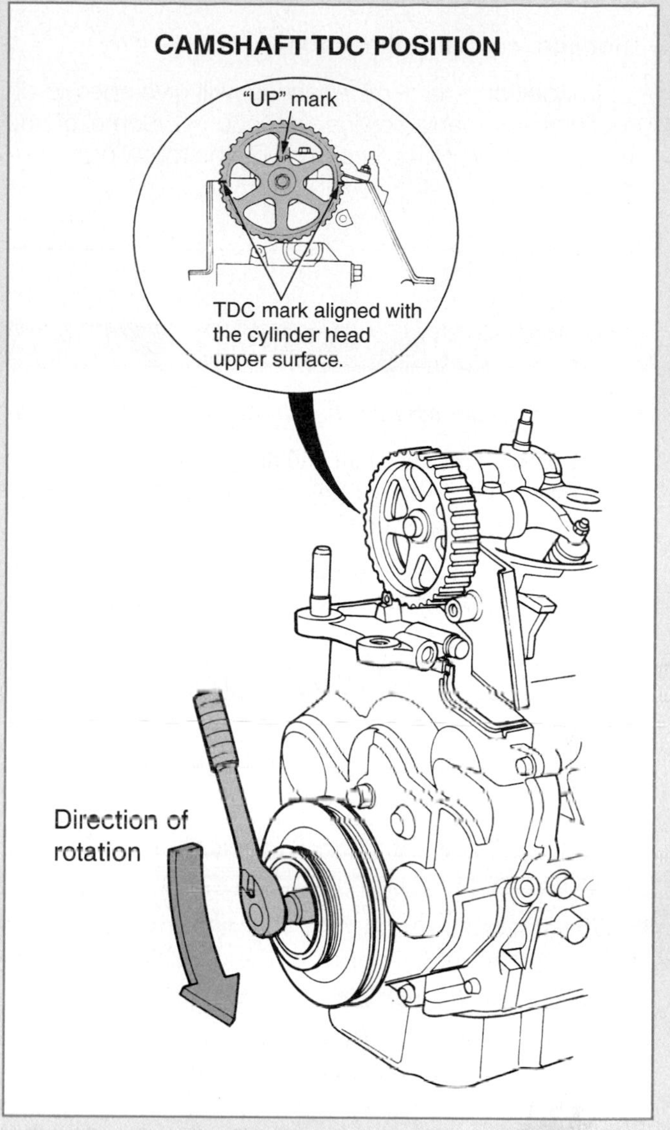

FIGURE 19–30 When replacing a timing belt, the camshaft must be positioned correctly. (Courtesy of American Honda Motor Co., Inc.)

Service Manual Connection

There are many important specifications to keep in mind when working with the camshaft or valve drive mechanisms. To identify the specifications for your engine, you will need to know the VIN (vehicle identification number) of the vehicle, the type and year of the vehicle, and the type of engine. Although they may be titled differently, some of the more common camshaft and valve drive mechanisms are listed here. Note that these specifications are typical examples. Each engine and vehicle may have a different specification.

Common Specification	Typical Example
1. Camshaft bearing clearance	0.0008–0.0028 inch
2. Camshaft journal diameter	1.2582–1.2589 inches
3. Lifter-to-bore clearance	0.009–0.0026 inch
4. Camshaft lobe total height (intake)	1.5168–1.5267 inches
5. Camshaft end play	0.002–0.006 inch
6. Rocker arm-to-shaft clearance (exhaust)	0.0007–0.0021 inch

continued

Service Manual Connection *continued*

In addition, the service manual will give specific directions for many service procedures. Some of the more common procedures include the following:

1. Timing the camshaft to the crankshaft
2. Valve lifter replacement
3. Removal of the camshaft
4. Timing chain/belt replacement
5. Camshaft installation

SUMMARY

The following statements will help to summarize this chapter:

- The parts of the camshaft and drive mechanism include the camshaft, rocker arms, camshaft bearings, pushrods, lifters, and timing belts.
- The cams, or lobes, on the camshaft are used to open and close the intake and exhaust valves at the correct time.
- The shape of the cam on the camshaft determines how quickly the valves open and close.
- The camshaft is driven by the crankshaft through gears, belts, or chains.
- The camshaft and crankshaft should always be timed to each other.
- The lifters, pushrods, and rocker arms are used to change the rotary motion of the camshaft to reciprocating motion to open and close the valves.
- Because of the heat of expansion, there must be a valve train clearance that can be either manually or automatically accounted for.
- Cam followers resemble solid lifters but use a roller that rides on the camshaft.
- Hydraulic lifters absorb the valve train clearance by using engine oil pressure inside of the lifter body.
- Pushrods often have small convex balls inserted on each of their ends to allow for proper alignment.
- Rocker arms are designed to change the direction of the cam lifting force and provide the mechanical advantage for lifting the valve.
- Variable valve timing and electronic control (VTEC) engines use three cam lobes that allow for both low rpm and load performance as well as high rpm and load performance.
- A computer controls oil that is sent to the VTEC system to use either low- or high-rpm cam profiles.
- Service information is available to check for camshaft lobe wear, camshaft bearing clearance, camshaft end play, and timing belt replacement, as well as other checks recommended by a vehicle's manufacturer.

TERMS TO KNOW

Can you explain each of the following terms? Review the chapter until you can use each term correctly.

Keyway
Lobes
Mechanical advantage
Oil relief
Pivot point
Tappets
Thrust plate
Valve train clearance
Variable valve timing

REVIEW QUESTIONS

Multiple Choice

1. Camshaft thrust is absorbed by the use of a:
a. Thrust plate
b. Thrust gear
c. Center cam thrust bearing
d. Special washer
e. Thrust bearing

2. Camshafts are driven from the crankshaft by:
a. Timing belts
b. Timing chains
c. Timing gears
d. Any of the above
e. None of the above

3. Which of the following lifters has "leak-down"?
   a. Hydraulic lifters
   b. Cam followers
   c. Solid lifters
   d. Rollers
   e. Solid rollers

4. Which of the following lifters will produce the least wear on the camshaft?
   a. Hydraulic lifters
   b. Cam followers
   c. Solid lifters
   d. Hollow lifters
   e. Slide lifters

5. Rocker arms are designed in an I-head engine to do which of the following?
   a. Change the rotary motion of the camshaft to reciprocating motion for the valves
   b. Reduce wear on the valve face
   c. Hold the valve in place in the cylinder head
   d. Control the speed of the camshaft
   e. All of the above

6. Rocker arm ratio is used on valve mechanisms to:
   a. Open the valve a certain distance
   b. Reduce wear on the valve stem
   c. Increase oil sealing on the valve stem
   d. Increase the valve size
   e. Decrease the valve size

7. When installing the timing gear and camshaft:
   a. Always time the gears according to the manufacturer's specification
   b. Never worry about correct timing of the two gears
   c. Always force the gears together with a rubber hammer
   d. Put the gears 90 degrees apart from each other
   e. Bolt the gears together

8. Valve mechanisms can wear on:
   a. The valve stem
   b. The valve face and seat
   c. The valve guide
   d. All of the above
   e. None of the above

The following questions are similar in format to ASE (Automotive Service Excellence) test questions.

9. *Technician A* says a leak-down test can be performed on pushrods. *Technician B* says a leak-down test can be performed on lifters. Who is correct?
   a. A only c. Both A and B
   b. B only d. Neither A nor B

10. After checking the lift on the lobes of a camshaft, one lobe is found to be 0.002 inch less than the other lobes. *Technician A* says the camshaft need not be replaced. *Technician B* says the camshaft should be replaced. Who is correct?
   a. A only c. Both A and B
   b. B only d. Neither A nor B

11. *Technician A* says the camshaft cannot wear and thus need not be checked. *Technician B* says the camshaft lobes will wear and thus should be checked for lift. Who is correct?
   a. A only c. Both A and B
   b. B only d. Neither A nor B

12. *Technician A* says that cam followers should be able to spin in their bore. *Technician B* says that cam followers do not touch the cam. Who is correct?
   a. A only c. Both A and B
   b. B only d. Neither A nor B

13. *Technician A* says that today's engines now have variable valve timing. *Technician B* says that mechanical and hydraulic mechanisms are used to control variable valve timing. Who is correct?
   a. A only c. Both A and B
   b. B only d. Neither A nor B

14. *Technician A* says that variable valve timing means that the engine has a broader power band. *Technician B* says that variable valve timing has the timing and cam lobes of a racing engine and a standard engine. Who is correct?
   a. A only c. Both A and B
   b. B only d. Neither A nor B

Essay

15. What is the definition of rocker arm ratio?

16. What is the purpose of variable valve timing?

17. Identify the operation of a variable valve timing engine that uses three cam lobes to operate each set of valves.

18. Why do some engines not need to use pushrods?

19. What is the difference between a cam follower and a solid lifter?

Short Answer

20. A common problem with camshafts is that the camshaft lobes will often ________.

21. The ________ of the lifter should always be checked for wear.

22. When a timing belt breaks, it is a good idea to also check for bent ________.

23. When you install a new timing belt, the crankshaft must be positioned at ________ for the number 1 cylinder.

24. A loss of power in two cylinders may indicate that the lobes on the ________ may be worn seriously.

Applied Academics

TITLE: Solenoid-Controlled Valves

ACADEMIC SKILLS AREA: Science

NATEF Automobile Technician Tasks:

The technician can explain the relationship between electrical current in a conductor and a magnetic field produced in a coil such as the starter solenoid.

CONCEPT:

The best possible solution to variable valve timing electronic control (VTEC) is to have an infinitely variable time in which the valve can open. The VTEC systems that are now being used have camshafts with both high-rpm and low-rpm ability, switching between two camshaft profiles. One system that is being tested uses a solenoid to control each valve separately. The illustration below shows an example of a cutaway solenoid placed directly on top of the valve. This solenoid is designed to open and close each valve at precisely the right time, the right amount, and the right speed. An on-board computer controls its action. One advantage of this design is that there is no need for a throttle plate to control the amount of air and fuel coming into the engine. The amount of air and fuel entering the engine is controlled by the amount of opening on the intake valves. Another advantage is the elimination of the frictional losses on the valve train because there is no need for lifters, rocker arms, pushrods, and so on. On the downside, the biggest problem is that using a solenoid to open and close valves does not work as effectively at higher rpm. In order for such a system to work, extremely powerful solenoids must be used on each valve to control its opening and closing at higher speeds.

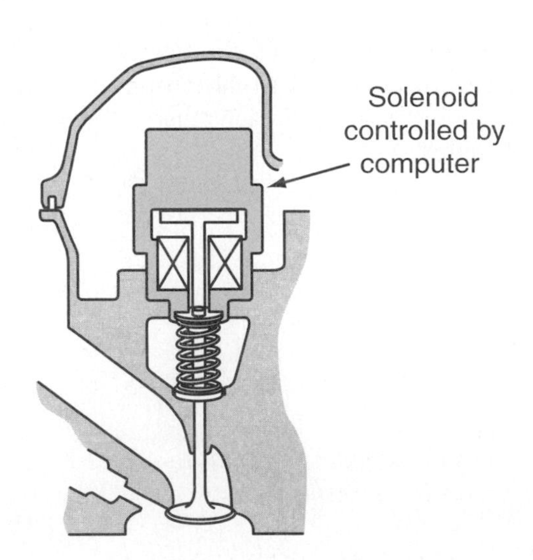

APPLICATION:

Can you explain how a solenoid operates to control the opening and closing of a valve? What component of an electrical circuit would need to vary if the valve were to open faster or slower? What effect would there be on engine performance if the engine did not use a throttle plate or camshaft to open and close the valves?

Mechanical/Fluid Engine Systems

SECTION 3

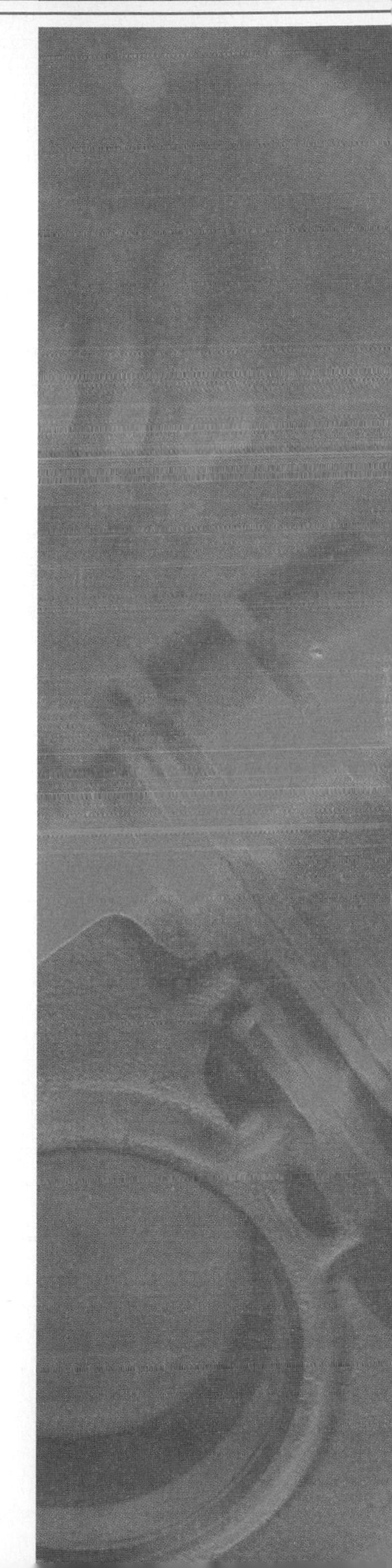

Greg Gillum, 28

Cardinal Chevrolet Cadillac

Technician (ASE Master)

Hazard, KY

What path did you take to your current position?

When I was 16, I had to take an elective, so I chose auto tech, and then took it again in my senior year. I decided after high school to go to Nashville Auto Diesel College to get my associate's degree. My first job was with a dealer in Savannah, Georgia—it was a huge operation, with thirty technicians. I then came to Cardinal where there are only seven techs; a few years later I left to become a technical trainer with DaimlerChrysler for a few years. I was sent to Belgium for one of them. Then I came back to work here.

Which did you like better—the big shop or the small one?

I can see the benefits of both ends. In Savannah, that was more of a specialty shop, and since you'd see the same problems every day, it became a lot easier to identify the problem. But at someplace like this, you get to learn a little bit of everything, and become more well rounded.

What's your typical day like?

We're dispatched out on a job-per-job basis, and each day varies. It can be everything from electronics to driveability problems to engine overhaul to oil change.

What's the advantage of being at a dealer shop?

First of all, you have the undying support of the manufacturer. And they provide you with the latest and greatest information and constant training. We have instructors at training centers in Atlanta, and we also have distance learning via satellite. They'll educate us up to ten times a year on new product information. And you're working on just one line all the time, which makes the job easier.

What's the disadvantage of being at a dealer shop?

At the dealership, you have to fix the problem. You can't say to the customer, "You'll have to take it somewhere else," like they do at some aftermarket shop.

What's the most frustrating part of your job?

It used to be that you were sent to vocational school and auto classes if you couldn't hack it in high school. These days, it takes a higher caliber person to be able to comprehend automobiles, but there's still a lack of respect for what we do.

Where do you see yourself in a few years?

I'd like to go back to technical training—it was a great experience and the traveling was great.

Principles of Lubrication

CHAPTER 20

OBJECTIVES

After studying this chapter, you should be able to:

- Define the purposes of the lubricating system.
- Identify the contaminants within the engine that must be removed by the lubricating system.
- Analyze the characteristics of lubricating oil.
- Compare the different ways oil can be classified.
- Compare the advantages and disadvantages of using synthetic oils.
- Identify various fluid recommendations.

INTRODUCTION

The automobile engine and other mechanical systems have many moving parts. These parts must be lubricated adequately in order for the engine and other parts to operate correctly. To keep all automobile parts lubricated, high-quality lubrication oil and grease are used. Improved lubricating oils are the main reason that engines (if serviced properly) in the modern automobile can operate for over 100,000 miles without internal engine repair in most cases. The goal of this chapter is to define the purpose of the lubrication system and identify contaminants, oil characteristics, classifications of oil, and fluid recommendations.

20.1 THE PURPOSES OF THE LUBRICATION SYSTEM

Lubricating oil is used in gasoline and diesel engines for many purposes. Lubricating oil is designed to do four major things within the engine. These are to lubricate, cool, clean, and seal various parts within the engine (***Figure 20–1***).

OIL LUBRICATES

The most important function of the oil is to lubricate the parts that are moving close together. As engine parts move close together, they produce friction. **Friction** is defined as

Certification Connection

ASE Connection: The information in this chapter can help you prepare for the National Institute for Automotive Service Excellence (ASE) certification tests. The tests and content areas most closely related to this chapter are:

Test A1—Engine Repair

- **Content Area**—Lubrication and Cooling Systems Diagnosis and Repair

NATEF Connection: Much of the information in this chapter is related to the NATEF tasks. The NATEF tasks and priority numbers most closely related to this chapter are:

1. Perform oil pressure tests; determine necessary action. P-1
2. Perform oil and filter change. P-2

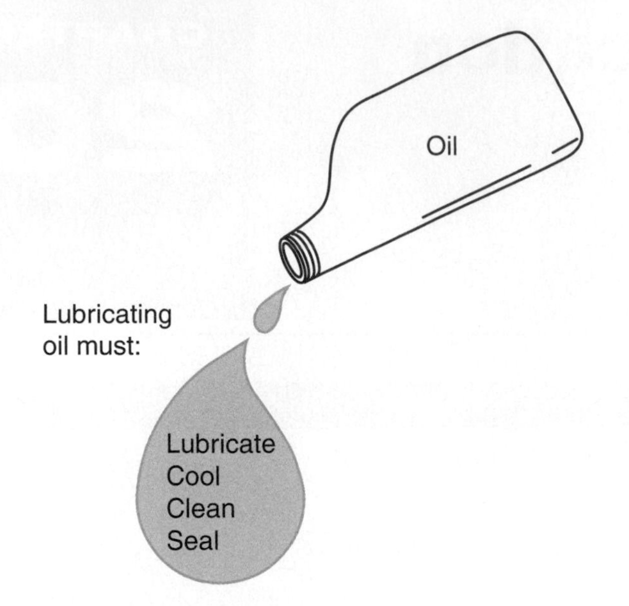

FIGURE 20-1 The purpose of any lubrication system is to lubricate, cool, clean, and seal within the engine.

resistance to motion between two bodies in contact with each other. In an earlier chapter, frictional horsepower—(the horsepower used to overcome friction)—was discussed. Lubricating oil helps to reduce frictional horsepower. If used correctly, lubricating oil can substantially reduce wear between two parts.

OIL COOLS

Lubricating oil comes directly in contact with vital internal moving engine parts. Lubricating oil is designed to effectively carry away the excess heat that is produced. As the lubricating oil passes over a hot engine part, it removes a great amount of heat. The heat (now in the oil) is transferred to the cooler crankcase oil located in the oil pan. Here heat is dissipated through the oil pan to the outside air. If an engine lacks lubricating oil, the engine may overheat. This is because there is not enough oil to remove the excess heat.

OIL CLEANS

It is also very important to keep all engine parts clean and free of carbon and dirt. Clean engine parts ensure proper circulation and cooling of the engine. As particles of dirt get into the engine, the lubricating oil helps to remove these particles and send them to an oil filter. Most of the physical dirt that gets into the engine comes from blow-by. Dust and dirt also come in through the air filter. This dirt must be removed from the engine and sent to the oil filter.

OIL SEALS

Lubricating oil inside the engine produces a film on the moving parts. This film acts as a protective sealing agent in the vital ring zone area. Oil located on the cylinder walls and rings helps to produce a seal between the two. The oil actually takes up space between the two moving parts. Without this oil, compression gases would escape into the crankcase area. As was discussed in an earlier chapter, cylinders and rings are deglazed or honed. Honing helps keep the oil on the walls so that sealing can be effective. It is necessary to keep all foreign materials out of this area so that proper sealing can be obtained.

FLUID FRICTION

Oil, like other matter, is composed of molecules. These oil molecules exhibit two important characteristics for engine lubrication.

1. Oil molecules will stick to metal surfaces more readily than to other oil molecules.
2. Oil molecules will slide against each other freely.

If a film of oil is between a crankshaft and a bearing, as shown (exaggerated) in ***Figure 20–2***, the top layer of oil molecules will stick to the crankshaft. The bottom layer of oil molecules will stick to the bearing. As the crankshaft moves, the internal layers of oil molecules slide against each other. This action produces less friction than two metal pieces that slide against each other without the oil film.

20.2 CONTAMINANTS IN THE ENGINE

Many **contaminants** get into the engine oil. These contaminants must be controlled. This means that either the oil must be designed to reduce the amount of contaminants in the engine or the oil must be filtered to reduce the amount of contaminants. To understand the lubricating system better, several contaminants should be defined. These include dust and dirt particles, carbon, water, fuel, oil oxidation, and acid buildup.

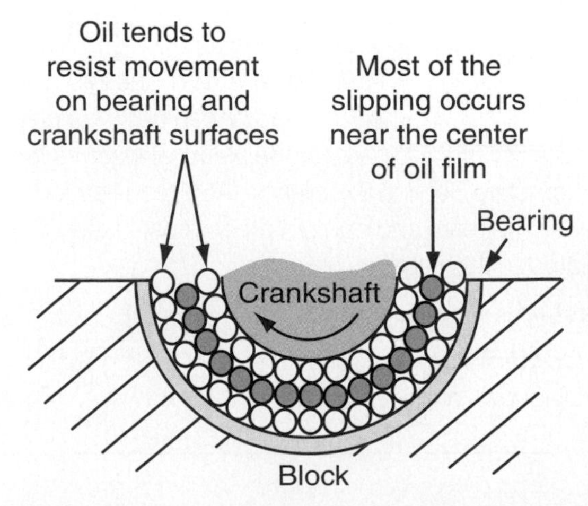

FIGURE 20-2 Oil acts like many small ball bearings. When movement occurs between the crankshaft and the bearing, the internal layers of oil slide against each other, reducing friction.

CUSTOMER CONCERN:
Removing Engine Sludge
An engine has a heavy amount of sludge inside. What is the best way to remove sludge from an engine?

SOLUTION: The best way to remove sludge from an engine is to disassemble the engine, clean all the parts, replace worn or damaged parts, and rebuild the engine. Chemicals on the market that remove sludge may cause oil contamination. In addition, there may be severe wear problems in the engine. Removing the sludge will not fix these problems. To avoid sludge buildup, change the oil frequently and be sure that the thermostat is the correct temperature and is operating properly.

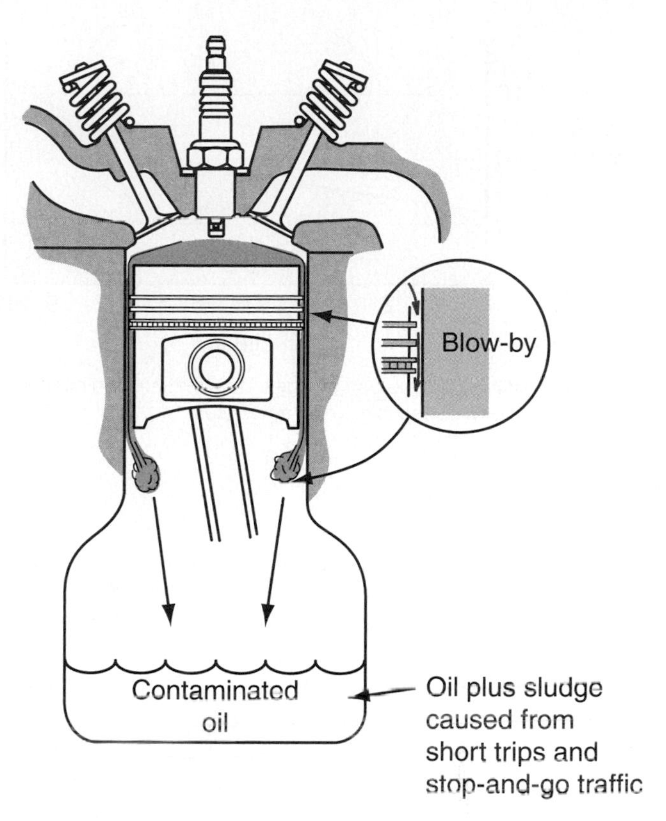

FIGURE 20-3 Oil is contaminated by the blow-by from the engine.

ROAD DUST AND DIRT

Road dust and dirt can get into the engine from the outside. Road dust and dirt enter the engine through the air cleaner and travel into the combustion chamber. They then pass by the rings during blow-by and into the crankcase area. After the dirt gets into the crankcase, it settles in the oil. The typical gasoline engine takes in an excess of 10,000 gallons of dust-carrying air for each gallon of gasoline it consumes. This dust and dirt can be very harmful to the engine. It is the job of the oil system to remove these contaminants.

CARBON AND FUEL SOOT

Many particles of carbon and soot are not burned completely in the combustion process. Pressures produced by combustion force some of this unburned fuel and carbon past the rings and into the crankcase area (***Figure 20–3***). This process is known as blow-by. Again, if carbon and soot particles were to remain in the oil, the life of the engine could be reduced drastically. The oil must be filtered to remove these particles.

WATER CONTAMINATION

Water can contaminate the engine lubrication system. Water can contaminate the lubricating oil because of leaks in the cooling system or from water vapor. Water vapor is a combustion by-product. The blow-by process described earlier, forces the water vapor into the crankcase area. As the water vapor enters the crankcase, it cools and condenses into a liquid. Water vapor can also enter the crankcase through several pollution control devices. Air usually has moisture in it. As outside air flows through the crankcase, the moisture from the air is condensed and forms a liquid. Water within the crankcase can produce problems with **sludges** and acids. Sludges are produced when the oil becomes thickened with water and other contaminants.

FUEL CONTAMINATION

The performance of motor oil is seriously affected when unburned or partially burned fuel enters the crankcase. Faulty operation of the fuel system (rich mixture), including the fuel injectors and the pressure regulator, as well as poor combustion, bad timing, or worn pistons and rings, can all cause fuel to enter the crankcase. Faulty injection and worn engine parts produce the same condition in diesel engines.

Fuel **dilution** of crankcase oil may occur in any engine. The amount may vary from a mere trace to as much as 50%. The presence of fuel in excess of 5% gasoline or 7% diesel fuel may lead to rapid engine wear and deterioration of oil. Fuel contamination of 5% changes the qualities of motor oil. It makes it thinner and less able to stick to moving parts. This change reduces the ability of the oil to lubricate.

As shown in ***Figure 20–4***, when there is as little as 1% to 3% fuel dilution of the oil, the SAE viscosity of the oil is around 30. However, as more fuel is added to the oil and the percent of dilution is around 10%, the SAE viscosity drops to 8–10. This drop in viscosity makes the oil much thinner and less able to lubricate, cool, clean, and seal.

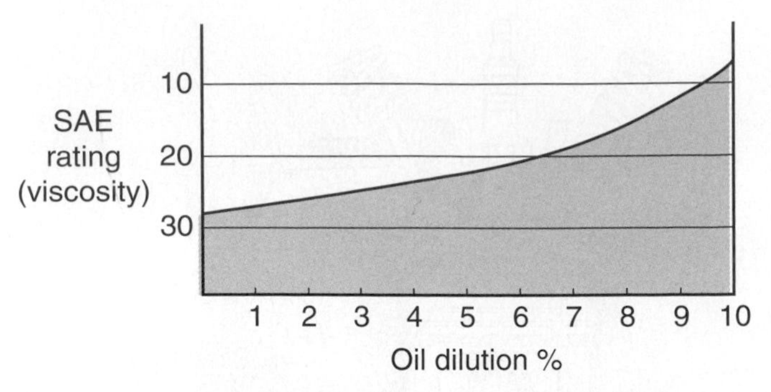

FIGURE 20-4 As the percentage of fuel that contaminates the oil increases, the SAE viscosity rating decreases.

OIL OXIDATION

Oil **oxidation** occurs within the engine during normal operation. Oil oxidation is defined as the combining of hydrocarbons and other combustion products with oxygen. This is a normal process or by-product of combustion. The result is the actual production of acids within the crankcase. This process produces oil with high **acidity**. Oil oxidation is enhanced or increased in the presence of certain types of metals used in the engine. In fact, certain metals act as a **catalyst**, an agent that speeds up the process, to promote further oxidation and add more corrosive acids to the oil. Increases in oil temperature and in water in the crankcase also enhance oil oxidation. Once the process of oil oxidation starts, the situation becomes progressively worse because it is self-accelerating.

ACIDS IN THE ENGINE

Certain **organic** acids are highly corrosive. Other organic acids tend to form gums and lacquers within the engine crankcase. If allowed to become concentrated from not changing oil regularly, the organic acids will attack certain bearing metals. The acids in the oil cause pitting on the metal surfaces and, eventually, failure of the parts. As shown in ***Figure 20–5***, the pitting often occurs on the crankshaft journals. If this happens, friction increases between the crankshaft and the connecting rod bearings. The additional friction often will cause the bearings to wear prematurely.

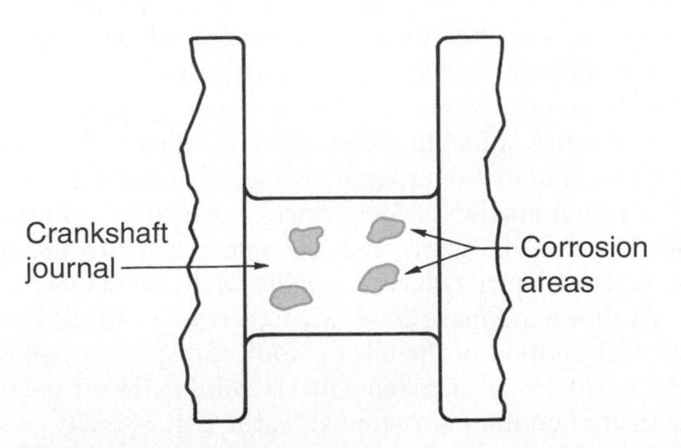

FIGURE 20-5 Acids that get into the engine oil tend to cause corrosion and pitting.

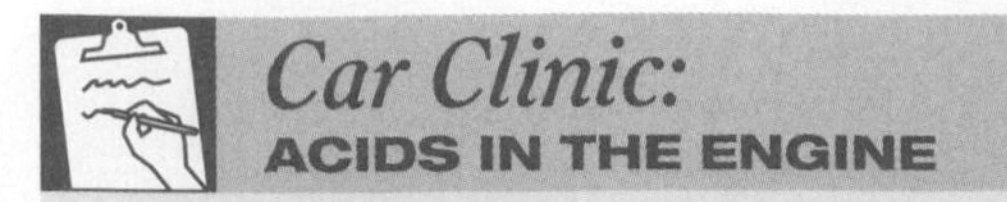

Car Clinic: ACIDS IN THE ENGINE

CUSTOMER CONCERN:
Oil Leaks When Car Sits Overnight
Oil leaks from the vehicle when it sits overnight in a driveway. The valve cover gaskets have been replaced several times, but the leak always comes back. The air cleaner has also been changed.

SOLUTION: A chronic oil leak like this can be caused by too much crankcase pressure. Check the crankcase pressure and compare it with the manufacturer's specifications. If it is above specifications, perform a cylinder leak-down test. This test will check for leaky piston rings. If the piston rings are in good condition, check the condition of the PCV system for correct venting. Crankcase pressure can also be increased when the oil return holes are plugged due to acid and sludge buildup in the oil system. Fixing these problems should solve leaking valve cover gaskets.

When there are acids in the oil, the pitting often occurs when the engine is being stored. Consequently, it is a good idea to always change the oil in the engine of any engine that is going to be stored for any length of time. When the oil is changed, the acids are removed and pitting will not occur.

The acids also react with the remainder of the oil to form soft masses. These are referred to as sludges. Sludges cause engine trouble by settling in the oil passageways, in the oil pan, in the filters, and in the oil coolers. The heavier oxidation products form hard **varnish** deposits on pistons, valve stems, and other engine parts. The sludges, acids, and varnish obviously reduce engine life. Therefore, it is very important that oxidation be reduced as much as possible.

20.3 OIL CHARACTERISTICS

As has been discussed, oil has several major jobs to do within the engine. Oil must reduce contaminants within the engine and cool, lubricate, and seal under various conditions.

VISCOSITY

The **viscosity** of lubricating oil is defined as the oil's **fluidity** or thickness at a specific temperature. Viscosity is also defined as resistance to flow. Viscosity is measured by a device known as the **Saybolt Universal Viscosimeter** (***Figure 20–6***). Viscosity is determined at a specific oil temperature. The most common temperatures are 0, 150, and 210

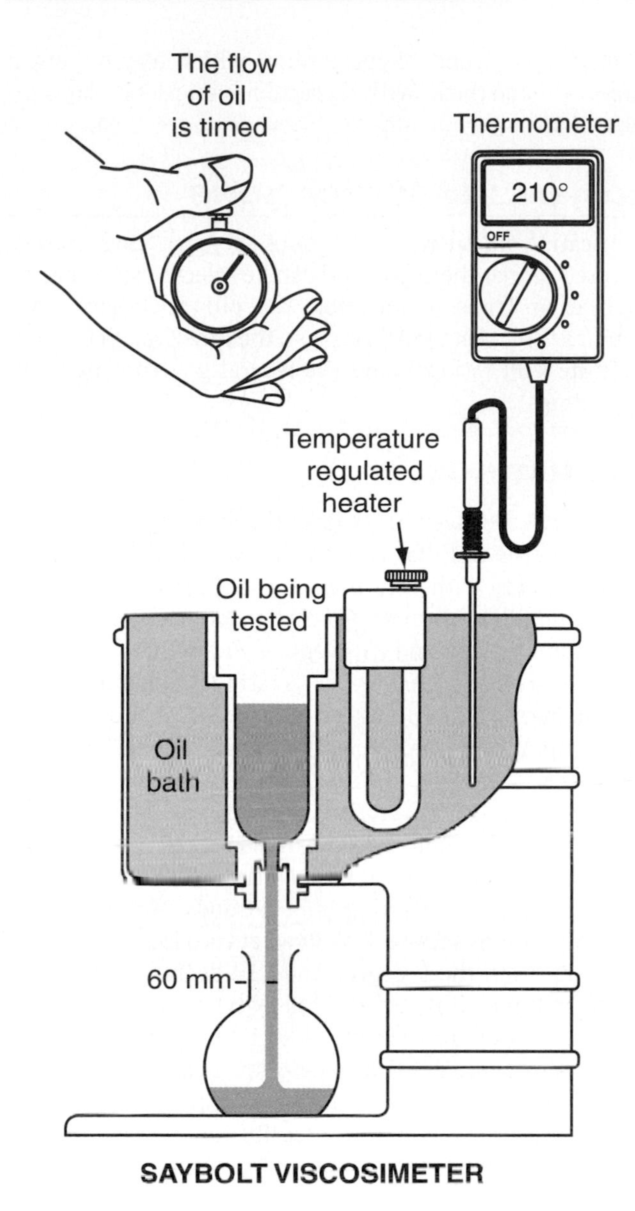

FIGURE 20–6 A Saybolt Universal Viscosimeter measures the thickness of the oil.

degrees Fahrenheit. A sample of oil is drained from the viscosimeter into a receiving flask. This is done at a specific temperature. The time required for the sample to completely drain is recorded in seconds. The viscosity is given in Saybolt Universal Seconds (SUS) at a particular temperature.

VISCOSITY INDEX

As an oil becomes cooler, it becomes less fluid. As an oil becomes hotter, it becomes thinner. The **viscosity index** (VI) is used to control oil thickness at different temperatures. The viscosity index is a measure of how much the viscosity of an oil changes with a given temperature. Chemicals are added to oil so that as the temperature of the oil changes, the viscosity will not change as much. Usually the higher the viscosity index, the smaller the relative change in viscosity with temperature.

POUR POINT

Oil must flow as a liquid in order to be used. Extremely cold conditions may increase oil viscosity until the oil cannot flow. The temperature at which oil ceases to flow is defined as the **pour point**. As the oil approaches its pour point, it becomes thicker and more difficult to pump. **Additives** are put into the oil to control the pour point.

OXIDATION INHIBITORS

Oxidation inhibitors reduce oxygen. Reduction of oxygen reduces the formation of sludges and varnish produced within the engine. As indicated earlier, any oxygen in the crankcase area may increase oil oxidation in the engine and thus produce acids. Chemicals that help eliminate or absorb oxygen are added to oil during the refining process. If the oxygen is absorbed, less is available to produce oil oxidation. In turn, fewer acids are produced. Note that if oil is not changed regularly, oxidation may increase.

DETERGENTS AND DISPERSANTS

Detergents are added to engine oil to help clean particles, dust, dirt, and other foreign materials from the engine. Detergents in the oil loosen and detach deposits of carbon, gum, and dirt. The oil suspends these particles. This characteristic is called **suspendability**. The oil then carries the loosened materials to the oil filter. Particles that are smaller than the filter size stay in the oil until the oil is changed. Note, however, that detergents also cause a certain amount of increased wear on the engine.

Dispersants are also added to oil. A chemical is added to the oil to keep the small dirt particles dispersed, or separated from each other. Without dispersants, particles tend to collect and form larger particles. These particles, which are heavier and harder to suspend in oil, can block oil passageways and filters. As shown in ***Figure 20–7***, the dirt par-

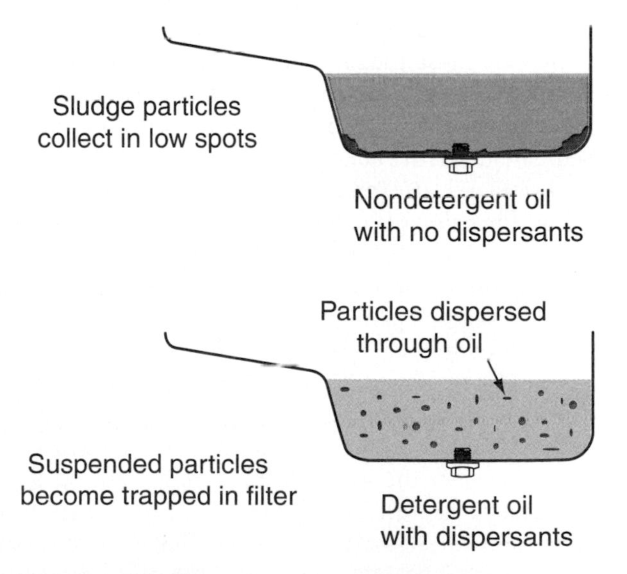

FIGURE 20–7 Detergents and dispersants added to the oil help to keep dirt and soot particles separated from each other so they can be filtered out more easily.

ticles are suspended and remain suspended until they eventually go through the oil filter. At that time, the dirt and soot particles are removed by the filter. Dispersants greatly increase the amount of contaminants the oil can carry and continue to function correctly.

Petroleum manufacturers are now placing more emphasis on dispersants than on detergents. It is felt that if contaminants can be kept suspended in the oil, and eventually removed by the oil filter, they will not deposit on engine parts. This means there is less need for detergent chemicals in the oil.

ENERGY CONSERVING OILS

Today, there is a need to improve fuel consumption. Fuel consumption can be reduced by using oils identified as energy-conserving oils. These oils have various types of additives that are designed to reduce friction. If friction is reduced, there is less frictional horsepower being consumed in the engine. The end result is that fuel mileage will increase. There are two types of energy-conserving (EC) oils. They are ECI and ECII. ECI oils provide about a 1.5% increase in fuel economy. ECII oils provide about 2.7% increase in fuel economy. Energy-conserving oils are identified on the containers as either ECI or ECII.

PARTS LOCATOR

*Refer to photo #9 in the front of the textbook to see what an **ECII oil container** looks like.*

OTHER ADDITIVES

Antifoaming additives are added to the lubricating oil to reduce foaming of the oil. During normal operation, the oil is pressurized and pumped through the engine. This constant pumping may cause the oil to foam. Any foam produced in the engine reduces lubricating, sealing, cooling, and cleaning. Hydraulic lifters also operate poorly with oil that has foamed. Chemicals are added to the oil to reduce foaming.

Corrosion and *rust* **inhibitors** are added to oil to reduce, or inhibit, corrosion of engine parts. Chemicals are added to help neutralize any acids produced. Rust inhibitors are also used to help remove water vapors from engine parts. Oil can then coat the engine parts correctly.

Antiscuff additives are put in the oil to help polish moving parts. This is especially important during new engine break-in periods.

Extreme-pressure resistance additives are added to lubricating oil to keep the oil molecules from splitting apart under heavy pressures. During operation, oil is constantly being pressurized, squeezed, and so on. This may cause the oil molecules to separate and reduce the oil's lubricating properties. The added chemicals react with metal surfaces to form very strong, slippery films, which may be only a molecule or so thick. With these additives, protection is increased during moments of **extreme pressure (EP)**.

20.4 OIL CLASSIFICATIONS

Lubricating oils are rated by various agencies. These ratings are used so that the correct oil can be selected for a specific application. Three organizations rate oil: the Society of Automotive Engineers (SAE ratings), the American Petroleum Institute (API ratings), and the federal government (military ratings).

SAE RATINGS

SAE ratings are given in terms of the viscosity or thickness of the oil. The SAE has applied the Saybolt Universal Seconds (SUS) test to their ratings. SAE ratings are determined for different temperatures. Oil is tested for viscosity at 0 degrees, 150 degrees, and 210 degrees Fahrenheit. The most common are 0 and 210 degrees. A rating that has a letter W after it means that the viscosity is tested at 0 degrees. If there is no W after the rating, the oil has been tested at 210 degrees. For example, in ***Figure 20–8***, a 10W oil has a rating of 6,000 to 12,000 seconds at 0 degrees. This means it took this many seconds for the sample of oil to flow through the SUS test. When the oil temperature is increased to 210 degrees, the oil has a rating of 39 seconds. Multigrade oils that have ratings such as 10W30 meet the viscosity requirements for both the 0 degree and 210 degree ranges. This means that the oil is less affected by temperature changes. Viscosity index improvers are added to make this possible.

It is sometimes difficult to determine exactly which type of oil viscosity to use. Manufacturers include various charts to help determine the best viscosity for the best service. ***Figure 20–9*** shows such a chart. For example, SAE 10W30 oils should be used in temperatures from about –10 degrees

VISCOSITY RANGE, SAYBOLT UNIVERSAL SECONDS

SAE Number	Seconds at 0°F		Seconds at 210°F	
	Min.	Max.	Min.	Max.
5W		4,000	39	
10W	6,000	12,000	39	
20W	12,000	48,000	39	
20			45	58
30			58	70
40			70	85
50			85	110

FIGURE 20–8 Viscosity is measured in seconds. Oil must have a certain flow rate in seconds for each SAE rating.

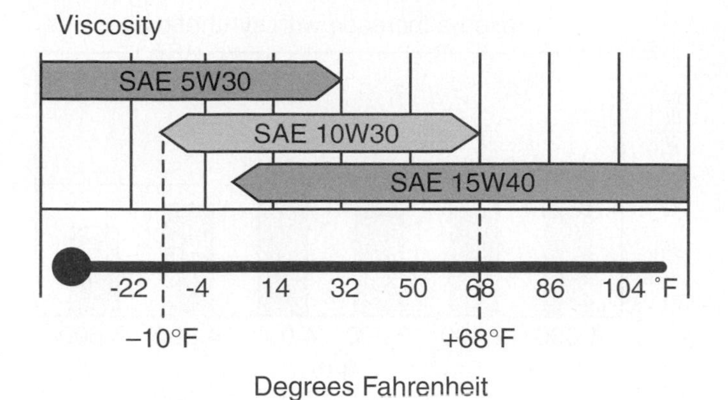

FIGURE 20-9 Car manufacturers recommend certain viscosity ratings for different temperatures.

to about 68 degrees Fahrenheit. Other oils are shown for different temperature conditions. The manufacturer's recommendations should always be followed to determine which oil is correct for a vehicle's engine. Failure to do so will void the new car warranty. If there is any question as to the viscosity to use in an engine, refer to the vehicle's owner's manual. Most owner's manuals contain a section on the oil viscosity to use at different temperature conditions.

API RATINGS

The American Petroleum Institute (API) rates oil on the basis of the type of engine service the oil should be used in. Many API ratings for gasoline engines have been used in the past. The following API ratings that have either been used or are still being used today include SA, SB, SC, SD, SE, SF, SG, SH, SJ, and SL. Over the years, API continually improves the oil service qualities in terms of reducing engine deposits, withstanding higher temperature and pressure requirements, and providing better rust and corrosion protection. Thus, the oil service ratings continue to change and be updated. The only current ratings today include the SJ and SL ratings. All other gasoline ratings are now obsolete. Note that the most recent service ratings include the performance properties of each earlier category. The SJ and SL ratings can be used as follows:

- SJ rating-to be used for 2001 and older automotive engines.
- SL rating-to be used for all automotive engines presently in use. Introduced in July 1, 2001, the SL oils are designed to provide better high-temperature deposit control and lower oil consumption. Some of these oils may also meet the latest ILSAC specification and/or qualify as Energy Conserving oils.

In terms of API categories for diesel engines, the ratings are different. The following ratings that have either been used or are still in use today include CA, CB, CC, CD, CE, CF, CF-2, CF-4, CG-4, CH-4, CI-4. These ratings have also continually been updated. The ratings identified as CA, CB, CC, CD and CE are now obsolete. However, the remaining ratings are still current and can be used as follows:

- CF rating-introduced in 1994, is for off-road, indirect injected and other diesel engines. Can be used in place of CD oils.
- CF-2 rating-introduced in 1994, is for severe duty, two stroke engines.
- CF-4 rating-introduced in 1990, is for high-speed, four-stroke naturally aspirated and turbocharged engines.
- CG-4 rating-introduced in 1995, is for severe duty, high speed, four-stroke engines meeting the 1994 emission standards.
- CH-4 rating-introduced in 1998, is for high speed, four-stroke engines designed to meet the 1998 exhaust emission standards.
- CI-4 rating-introduced in 2002, is for high-speed, four-stroke engines designed to meet 2004 exhaust emission standards that were implemented in 2002.

To keep updated on information about the American Petroleum Institute service ratings contact http://www.api.org/eolcs.

In the early 1990s, American and Japanese automotive manufacturers formed the International Lubricant Standardization and Approval Committee (ILSAC). The committee sets manufacturers' minimum lubrication standards for gasoline-fueled engines (GF). The American Petroleum Industry (API) is the organization that administers this system of standards. Oil containers that list the letters ILSAC GF contain oil that meets the tough standards set by this organization.

As automotive manufacturers design engines with higher temperatures and closer clearances, the service ratings of oil used will have to be changed. Other ratings will eventually be introduced for vehicles that require even more protection against sludge, varnish, wear, and engine deposits.

MILITARY RATINGS

The third rating is associated with military testing and performance. The federal government contracts with an engine manufacturer to build engines for a particular application. When this is done, a military specification is issued for the type of oil recommended in that specific application. Some of the tested ratings are:

- MIL-L2104 B, MIL-L46152, or MIL-L-6082
- Supplement 1
- Series 3

From these tests, an engine manufacturer then recommends a specific type of oil for its engines in the military application. SAE and API ratings are usually included with this recommendation. If there is a military rating stamped on an oil container, it can be assumed that this oil has been tested for a specific application with military equipment.

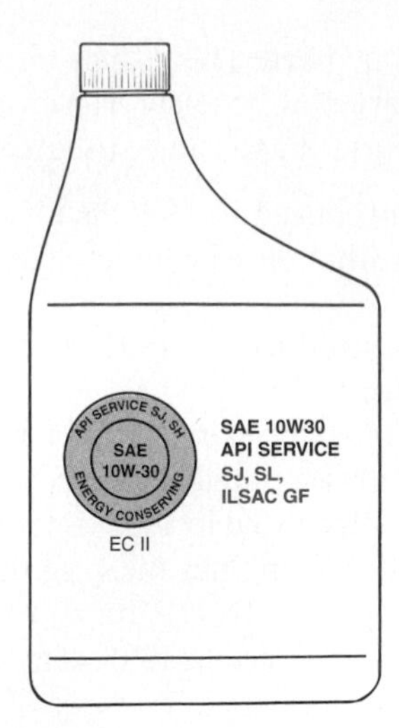

FIGURE 20–10 Each oil container has different ratings printed on the side.

Most of these ratings are normally printed on the oil container. ***Figure 20–10*** shows a typical container with the ratings printed on the side. If the oil container does not have an API rating, the oil does not meet the API specifications.

20.5 SYNTHETIC LUBRICANTS

In recent years, the use of **synthetic** lubricants has increased. Synthesize means to combine parts into a whole or to make complex compounds from a series of individual molecules. Synthetic lubricants are made chemically by mixing alcohol, various acids, other chemicals, and hydrocarbons together. The hydrocarbons can be taken from coal, oil, natural gas, wood, or any agricultural resource. The result is a synthesized product that is capable of meeting and exceeding the lubrication needs of various engines. Synthetic oils are manufactured by combining specific molecules into an end product tailored to do a specific job. They are designed to meet or exceed the SAE and API recommendations.

ADVANTAGES OF SYNTHETIC OILS

Synthetic oils were initially designed for jet engine use because of the higher temperatures and pressures. They are now used in automobile applications because of their ability to withstand high temperatures with little change in the oil viscosity. There are several advantages and disadvantages of using synthetic oil in automotive engines. Some advantages are:

- Increased thermal and oxidation stability. For example, synthetic oils can operate effectively from –60 degrees Fahrenheit to +400 degrees Fahrenheit. Oxidation occurs at about 50 degrees higher than in conventional oils.
- Less evaporation. Only about 1% of the oil evaporates over a standard period of time as compared with about 25% for standard lubrication oils.

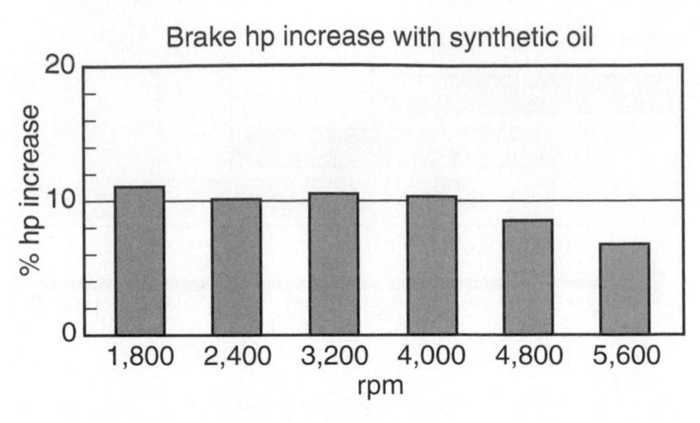

FIGURE 20–11 Use of synthetic oils reduces frictional losses and thus increases available brake horsepower.

- Less viscosity change with temperature. This improves cold starting ability and fuel mileage during cold weather.
- Improved fuel mileage. Synthetic oils have increased lubricating properties. The results are less frictional resistance and more horsepower. ***Figure 20–11*** shows horsepower increases at different rpm. For example, if the engine is running at 1,800 rpm, there is about an 11% increase in brake horsepower. If the engine is running at 4,800 rpm, there is about an 8% increase in brake horsepower.
- Reduced oil consumption. Synthetic oils increase sealing characteristics.
- Cleaner engine parts. Less maintenance is required when engine parts remain cleaner.
- Not affected by fuel contamination.

Car Clinic: SYNTHETIC OIL

CUSTOMER CONCERN:
Mixing Synthetic Oil with Regular Oil
A customer has been using synthetic oil for about 15,000 miles in the vehicle. Now it is noticed that the engine is using a small amount of oil. Is it acceptable to mix regular oil with synthetic oil to keep the oil at the correct level?

SOLUTION: Some manufacturers of synthetic oil recommend that synthetic oil not be mixed with regular oil, while others say it is all right to mix the two. Always check with the oil manufacturer if you are unsure. More important, if the engine is using oil, it may have to be rebuilt. New rings may have to be put on the pistons to eliminate oil consumption. Also, check for valve guide seals leaking into the intake or exhaust. A compression leak-down test will help determine which cause is producing the oil burning.

DISADVANTAGES OF SYNTHETIC OILS

Although synthetic oils seem to be better than conventional oils, there are several disadvantages that should be mentioned. These include:

- Poor break-in characteristics. Because of the increased lubrication property, break-in will not occur as with regular oils. Synthetic oils should not be used until the vehicle has reached about 6,000 miles.
- Ineffectiveness in older engines. Synthetic oils should not be used in engines with bad rings, valve seals, or seals. If used on such engines, the oil may be consumed more rapidly.
- Higher cost. The cost per quart is still higher for synthetic oil than for conventional motor oil. Over the life of the vehicle, however, the total cost for lubrication will be about the same. The savings come from cleaner engines and better fuel mileage. Oil change periods range in the area of 25,000 miles.
- Inability to be mixed. It is not advisable to mix different brands of synthetic oils. This may cause some inconvenience when trying to buy a particular brand.

Continued research is being done on synthetic oils. As the temperatures of engines increase in the future, synthetic lubricants will have to be reevaluated to see if they can be used more effectively than conventional oils. Each automobile owner will have to decide about the use of synthetic lubricants. It is important to check the vehicle warranty as well. In some cases the use of a synthetic oil may void the warranty.

20.6 GREASES

The term *grease* has long been employed to describe lubricants that are in a semifluid state and are used to reduce friction. Grease is made by adding various thickening agents such as soap to a liquid lubricant such as oil. Greases are commonly used on automotive chassis parts such as the universal joints, wheel bearings, and front-end steering parts.

Although greases are suitable for many lubrication jobs, several modifying agents are often added to greases to improve certain properties. Examples of properties that may be improved include: tackiness, oxidation stability, consistency stability, load-carrying capacity, and rust prevention.

Greases can be identified by several methods. One method is by measuring the consistency of grease (***Figure 20–12***). Consistency is a measure of relative hardness. This property is expressed in terms of the American Society for Testing and Materials (ASTM) penetration or **National Lubricating Grease Institute (NLGI)** grade or consistency number.

The consistency of a grease is an important factor in its ability to lubricate, seal, and remain in place. It is also an important factor in how easily the grease can be applied and removed. Most automotive greases are in the NLGI number 2 grade or range. This means the grease ranges from soft to medium consistency. The NLGI number is sometimes referred to as the **EP (extreme pressure)** number or grade.

The ASTM penetration is a numerical statement of the actual penetration of a grease sample at a specific temperature. It is measured in tenths of a millimeter using a standard testing device under standard conditions. The higher the penetration value, the softer the grease, and the greater the penetration.

CLASSIFICATION OF GREASES BY NLGI CONSISTENCY NUMBERS

NLGI[a] Number	ASTM[b] Worked Penetration @ 25°C (77°F) (tenths of a millimeter)	
000	445–475 (Semifluid)	**More**
00	400–430 (Semifluid)	**penetration**
0	355–385 (Soft)	
1	310–340	
2	265–295	
3	220–250	
4	175–205	
5	130–160	**Less**
6	85–115 (Hard)	**penetration**

[a]*National Lubricating Grease Institute (NLGI)*
[b]*American Society for Testing and Materials (ASTM)*

FIGURE 20–12 Greases are classified by their penetration number.

20.7 RECOMMENDATIONS FOR FLUID CHANGE INTERVALS

OIL RECOMMENDATIONS

Car manufacturers recommend that oil and other lubricants be changed on a regular basis. As contaminants begin to build up in an engine, the oil has less and less lubricating ability. Eventually, the oil will no longer be able to do its job of cooling, cleaning, lubricating, and sealing. At this point, if the engine continues to operate with the dirty lubricant, the engine components may begin to wear. Eventually parts will be damaged and will need replacement. To avoid this condition, automobile manufacturers list various recommendations for the type of oils and lubricants to be used and the appropriate change periods. ***Figure 20–13*** shows a typical example of one car manufacturer's recommendations for the type of oil to use in the vehicle. The information includes the engine, the capacity, and the classification of oil or lubricant recommended.

LUBRICANT		
Item	**Capacity**	**Classification**
Engine oil Drain and refill		API grade SJ, SL, Energy Conserving II or ILSAC multigrade and recommended viscosity oil with SAE 5W30 being the preferred engine oil
w/oil filter change	4.7 liters (5.0 US qts. 4.1 Imp. qts)	
w/o oil filter change	4.5 liters (4.8 US qts. 4.0 Imp. qts)	

FIGURE 20–13 This chart shows an example of the lubricants recommended by a particular car manufacturer.

OIL CHANGE INTERVALS

Each car manufacturer also recommends a certain time to change oil or other lubricants. The exact time will depend on the type of service the vehicle performs (severe, moderate, regular, etc.), the type of vehicle, and the year of the vehicle. For example, some automotive manufacturers recommend that the oil be changed every 6 months and every 3,750 miles for severe duty. Other manufacturers recommend higher mileage intervals, say at 7,500 miles. The exact mileage interval will depend on the manufacturer and the type of duty or severity of driving conditions. ***Figure 20–14*** shows a typical example of a maintenance schedule for both severe-duty and regular-duty conditions. Severe duty includes conditions such as the following:

- Towing a trailer or a camper or using a cartop carrier
- Operating on dusty, rough, muddy, or salt-spread roads
- Repeated short trips of less than 5 miles and outside temperatures remaining below freezing
- Extensive idling and/or low-speed driving for a long distance such as police, taxi, or door-to-door delivery use

Regular duty is considered conditions other than those listed above.

Looking at the schedule in ***Figure 20–14***, it can be seen that the recommended engine oil change period for severe duty of this vehicle is every 6 months or every 3,750 miles, whichever comes first. For regular duty, the recommendation is every 12 months or 7,500 miles. This car manufacturer also is recommending that the oil filter be changed at the same interval as the lubricant. It is extremely important

TYPICAL MAINTENANCE SCHEDULE

Severe Duty

Service Interval*	Odometer Reading																Months
× 1,000 km × 1,000 miles	6 3.75	12 7.5	18 11.25	24 15	30 18.75	36 22.5	42 26.25	48 30	54 33.75	60 37.5	66 41.25	72 45	78 48.75	84 52.5	90 56.25	96 60	
Engine Components and Mission Control Systems																	
Engine oil	R	R	R	R	R	R	R	R	R	R	R	R	R	R	R	R	R: Every 6 months
Engine oil filter	R	R	R	R	R	R	R	R	R	R	R	R	R	R	R	R	R: Every 6 months

Regular Duty

Service Interval*	Odometer Reading								Months
× 1,000 km × 1,000 miles	12 7.5	24 15	36 22.5	48 30	60 37.5	72 45	84 52.5	96 60	
Engine Components and Emission Control Systems									
Engine oil	R	R	R	R	R	R	R	R	R: Every 12 months
Engine oil filter	R	R	R	R	R	R	R	R	R: Every 12 months

*Use odometer reading or months, whichever comes first.
R = Replace, change or lubricate.

FIGURE 20–14 Maintenance schedules show the intervals for oil changes.

to follow the recommended oil change period for each type of vehicle. This is especially true for new cars. If the recommended oil changes are not performed, certain engine repairs may not be covered by the manufacturer's warranty. Always check the vehicle's owner's manual to determine the proper oil change intervals. In most cases, there will be specific recommendations for oil change intervals for each vehicle.

SUMMARY

The following statements will help to summarize this chapter:

- The lubrication system is designed to lubricate, cool, clean, and seal internal parts of the engine.
- As oil passes through the oil filter, dirt and soot particles are removed.
- Some of the contaminants in an engine include dust and dirt from the road, carbon and fuel soot from combustion, fuel, and acids from oil oxidation.
- All contaminants in an engine reduce the oil's ability to lubricate, cool, clean, and seal.
- The oil oxidation process in the crankcase of the engine produces acids in the oil.
- Acids in oil can produce corrosion and damage to the engine parts.
- SAE rates oil by its viscosity at 0 degrees, 150 degrees, and 210 degrees Fahrenheit.
- Viscosity is defined as the thickness of oil at different temperatures.
- Pour point is a measure of the temperature at which oil is too thick to pour.
- Detergents and dispersants are added to oil to help with the cleaning process.
- Other chemicals added to oil include antifoaming additives, corrosion and rust inhibitors, antiscuff additives, and extreme-pressure resistance additives.
- The American Petroleum Institute rates oil according to the type of service in which it is used.
- The military also rates certain oils for use in specific applications.
- Synthetic oils are very stable and have many advantages as well as disadvantages.
- Grease is identified by its consistency, which is a measure of its relative hardness.

TERMS TO KNOW

Can you explain each of the following terms? Review the chapter until you can use each term correctly.

Acidity
Additives
Catalyst
Contaminants
Dilution
Dispersants
Extreme pressure (EP)
Fluidity
Friction
Inhibitors
National Lubricating Grease Institute (NLGI)
Organic
Oxidation
Oxidation inhibitors
Pour point
Saybolt Universal Viscosimeter
Sludges
Suspendability
Synthetic
Varnish
Viscosity
Viscosity index

REVIEW QUESTIONS

Multiple Choice

1. An engine lubricating system is designed to:
a. Cool the engine
b. Clean and seal the engine
c. Lubricate the engine
d. All of the above
e. None of the above

2. Which of the following is not an oil contaminant in an engine?
a. Road dust
b. Carbon from combustion
c. Wheel bearing grease
d. Oil oxidation
e. Road dirt

3. The viscosity of oil will ________ as fuel enters the oil.
a. Increase
b. Decrease
c. Remain the same
d. Heat up
e. Thicken

4. Acids ________ in an engine's lubricating system.
a. Cannot be produced
b. Can be produced
c. Form from the bearing noise
d. Form from vibration
e. None of the above

5. Viscosity is defined as the ________ of oil.
a. Acidity
b. Pour point
c. Oxidation
d. Thickness
e. Temperature

6. At what temperatures are oils normally tested for viscosity?
a. 0 and 100 degrees Fahrenheit
b. 0, 150, and 210 degrees Fahrenheit
c. 210 degrees Fahrenheit only
d. 80 and 100 degrees Fahrenheit
e. 1,000 to 1,005 degrees Fahrenheit

7. Chemicals added to oil to keep small particles of dirt separated from each other are called:
a. Detergents
b. Dispersants
c. Viscosity index improvers
d. Filters
e. Acids

8. Which organization classifies oil by its viscosity?
a. American Petroleum Institute
b. Military
c. Society of Automotive Engineers
d. All of the above
e. None of the above

9. An oil with a 10W30 label means the oil has a/an ________ at 0 degrees.
a. API 30 rating
b. SAE rating of 10
c. SAE rating of 30
d. API CD rating
e. API SD rating

10. When an oil is manufactured by combining specific molecules into an end product, the oil is said to be:
a. Synthetic
b. Oxidized
c. High in viscosity
d. Low in viscosity
e. High in acidity

The following questions are similar in format to ASE (Automotive Service Excellence) test questions.

11. A car must be put away for 4 months in storage. *Technician A* says the oil should be changed before storage because of acid buildup in the oil. *Technician B* says keep the old oil in the engine and then change it after the car is taken out of storage. Who is correct?
a. A only
b. B only
c. Both A and B
d. Neither A nor B

12. *Technician A* says that greases that have a higher grade such as 4, 5, or 6, are thinner. *Technician B* says that 4, 5, and 6 grade greases have poorer lubrication quality. Who is correct?
a. A only
b. B only
c. Both A and B
d. Neither A nor B

13. *Technician A* says that synthetic oil should never be used as a break-in oil. *Technician B* says that synthetic oil should never be used in an old engine. Who is correct?
a. A only
b. B only
c. Both A and B
d. Neither A nor B

14. *Technician A* says that an engine should have its oil changed every 6 months for severe duty. *Technician B* says that the engine oil never needs to be changed on new engines. Who is correct?
a. A only
b. B only
c. Both A and B
d. Neither A nor B

15. *Technician A* says that to improve fuel mileage, use an ECII oil. *Technician B* says that certain additives can be added to oil by the vehicle owner for extreme pressure conditions. Who is correct?
a. A only
b. B only
c. Both A and B
d. Neither A nor B

Essay

16. What four things must an oil lubricant do in an engine?

17. List at least three contaminants that can make the oil dirty.

18. How can fuel contaminate oil?

19. What is oil oxidation, and what are the results of oil oxidation?

20. Define two methods used to classify oil.

21. What is an oil dispersant?

22. List three advantages and three disadvantages of using a synthetic oil.

Short Answer

23. When hydrocarbons combine with oxygen in the oil, the process is called ________.

24. An engine oil is designed to cool, clean, lubricate, and ________.

25. Water can contaminate the oil because of leaks in the ________ system.

26. Pitting and corrosion on the crankshaft can be caused from ________ building up in the lubrication system.

27. The additive put into oil to help polish moving parts is called a/an ________ additive.

28. For the newest vehicles on the road, an API rating of ________ should be used.

29. Oil change intervals for cars depend on the year of the vehicle, the type of vehicle, and the type of ________.

30. A synthetic oil should never be used for older engines or for ________.

Applied Academics

TITLE: Used Motor Oil

ACADEMIC SKILLS AREA: Science

NATEF Automobile Technician Tasks:

The technician evaluates the waste products resulting from an engine repair task and handles the disposal of materials in accordance with applicable federal, state, and local rules and regulations.

CONCEPT:

Used motor oil contains pollutants including organic chemicals and metals. When disposed of improperly—in the trash, on the ground, or in a sewer system—the pollutants may reach lakes, rivers, or the groundwater. For that reason, disposing of motor oil in this way is illegal. Between 5 and 10 million gallons of used motor oil are disposed of improperly each year in any one state. On a national level, the estimate is about 400 million gallons of oil each year.

Recycled oil can be a valuable source of energy. It can be burned as an industrial fuel or cleaned and reused. Dealers that sell motor oil should have a sign posted indicating the nearest place that accepts used oil. Many communities provide collection tanks for used oil. Some full-service stations accept used motor oil as well. To recycle your oil, pour the used motor oil into an oil container. Seal the container, label it, and take it to a used motor oil collection center. Transmission and brake fluids can be mixed with the oil for recycling. However, never mix solvents, gasoline, or antifreeze with used oil. Once the oil has been contaminated with these products, it is difficult or impossible to recycle.

APPLICATION:

Why is it important to recycle used motor oil? How can you convince your coworkers that recycling is important? Can you think of a way to encourage the general public to recycle their used oil?

Lubrication System Operation

CHAPTER 21

OBJECTIVES

After studying this chapter, you should be able to:

- Identify the parts of the lubrication system used on gasoline and diesel engines.
- Follow the flow of oil through an engine.
- Describe the use and purpose of oil filters.
- Identify the purpose, design, and operation of oil gauges.
- Identify common problems, their diagnosis, and service procedures for the lubrication system.

INTRODUCTION

An automobile's lubrication system is critical to the vehicle's operation. Without a properly operating lubrication system, the engine would cease to run. This chapter discusses how the oil flows through an engine, the basic parts of a lubrication system, the electrical circuits used to control oil sensors and gauges, and basic lubrication system service.

21.1 LUBRICATION SYSTEM PARTS AND OPERATION

The lubrication system is composed of several parts, including the oil pan, oil pump, main oil galleries, oil filters, oil pressure regulators, oil coolers, and oil sensors. These parts are needed to make the lubrication system operate correctly.

OIL PAN

The purpose of the oil pan, or **sump**, is to hold the excess oil during operating and nonrunning conditions. This reservoir for the oil is located at the bottom of the engine. After the parts of the engine are lubricated, gravity causes the oil to flow back to the oil pan. A plug in the bottom of the oil pan is used to drain the oil. The cooling of the oil takes place within the oil pan.

Certification Connection

ASE Connection: The information in this chapter can help you prepare for the National Institute for Automotive Service Excellence (ASE) certification tests. The tests and content areas most closely related to this chapter are:

Test A1—Engine Repair

- **Content Area**—Lubrication and Cooling System Diagnosis and Repair

NATEF Connection: Much of the information in this chapter is related to the NATEF tasks. The NATEF tasks and priority numbers most closely related to this chapter are:

1. Inspect oil pump gears or rotors, housing, pressure relief devices, and pump drive; perform necessary action. P-2
2. Perform oil pressure tests; determine necessary action. P-1
3. Inspect, test, and replace oil temperature and pressure switches and sensors. P-2
4. Perform oil and filter change. P-2

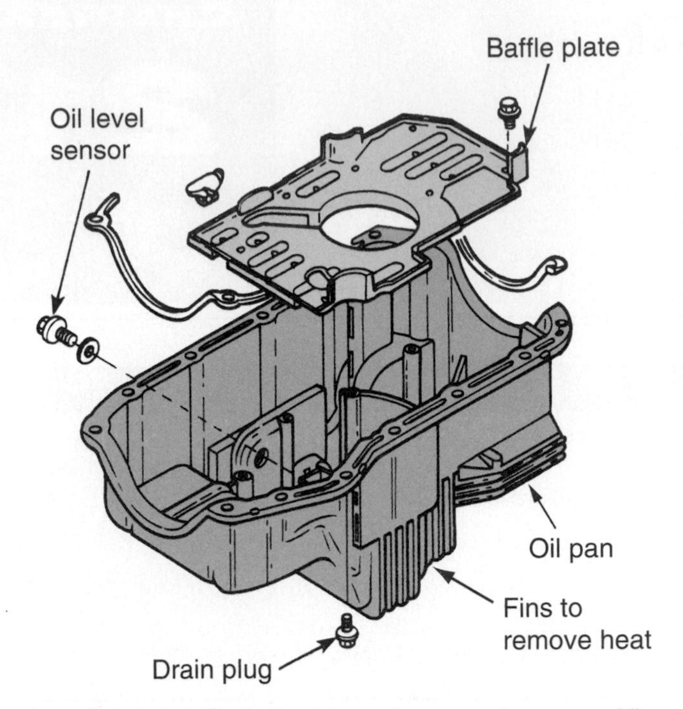

FIGURE 21–1 The baffle plate used on some oil pan assemblies helps to keep the oil from splashing, which can cause air bubbles to get into the oil and produce foam.

In some oil pans a baffle is placed above the oil pan. Shown in ***Figure 21–1***, this baffle is used to keep the oil from splashing around when the vehicle is either starting or stopping fast or if the vehicle is on a rough road. If the oil is allowed to splash around too much, air bubbles may get into the oil and cause foaming. Foam does not lubricate, cool, clean, or seal an engine the way oil is designed to do. Also notice that the oil pan has a series of fins located on the outside. These fins help to dissipate heat so that the oil does not get too hot.

OIL FLOW

Each engine manufactured has a certain flow pattern to get the oil from the oil pan to the various parts that need lubrication. Flow patterns are different for each type of engine and for each year manufactured. One type of flow pattern is diagramed in ***Figure 21–2***. The flow of oil starts at the oil pan where the oil is stored and cooled. From there the oil is drawn into the oil pump through the oil screen. In this case, the oil pump is located in the front of the engine on the crankshaft. Other oil pumps are driven from the camshaft or through the distributor shaft. After the oil is pressurized, it is sent to the oil filter to be cleaned. From the oil filter, the oil is

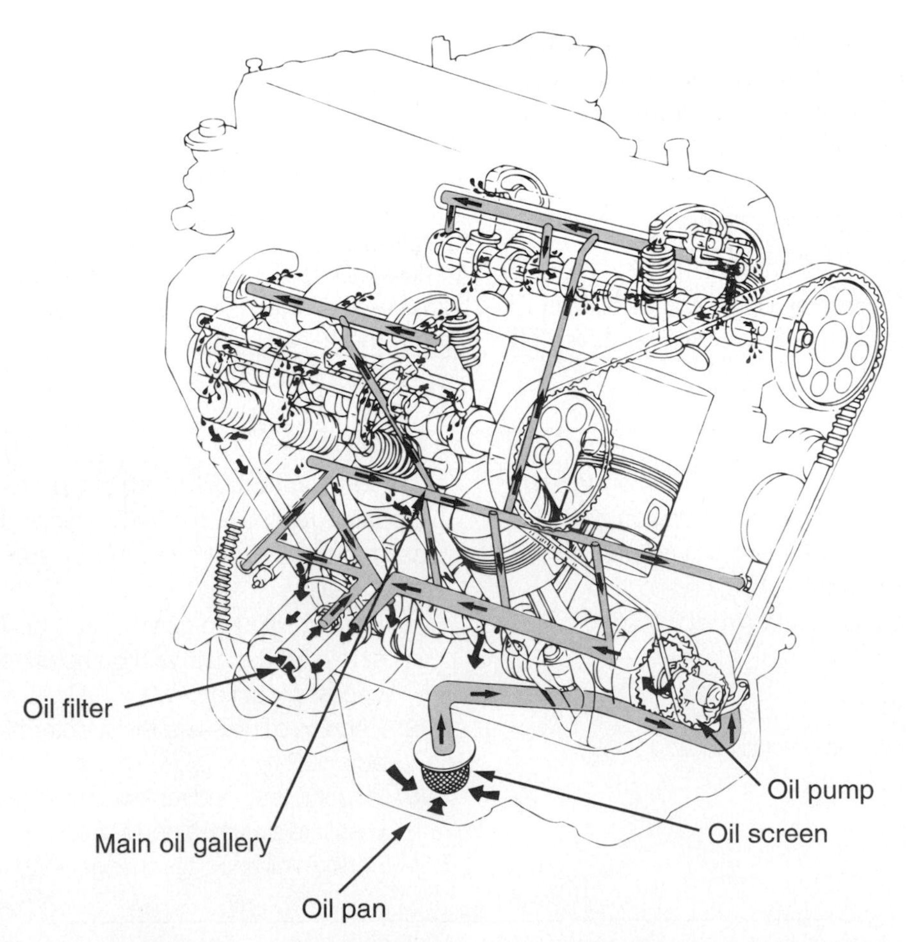

FIGURE 21–2 Oil flows from the oil pan, through the pump, through the filters, and into the main oil gallery. From there the oil is sent to all the parts of the engine. (Courtesy of Nissan North America, Inc.)

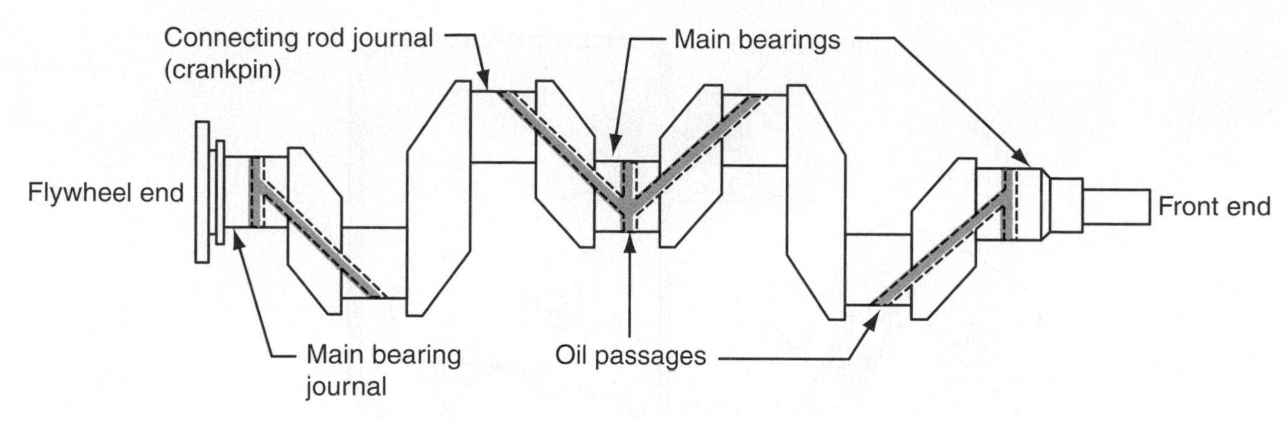

FIGURE 21-3 Oil from the main oil gallery flows through passages into the crankshaft. The oil at the main bearings then flows to the connecting rod bearings.

sent into passages in the block of the engine. These are referred to as the main **oil gallery**. From the main oil gallery, oil is sent to all other parts of the engine. There is an oil pressure sensor on the main oil gallery. The oil pressure sensor is used to sense the oil pressure. It is electronically or mechanically connected to the oil light or gauge on the dashboard.

Oil is sent through drilled passages that come into contact with the main oil gallery to the crankshaft. Here the oil is used to lubricate the main and connecting rod bearings. ***Figure 21–3*** shows oil holes drilled in a crankshaft. Oil from the main bearings feeds to the connecting rod bearings and then drops back to the oil pan, or sump.

From the main oil gallery, the oil is also sent through drilled passages to the head. On some engines, the oil flows from the main oil gallery, through the lifters, through the pushrods, and up to the rocker arms. Here the oil is used to lubricate the camshafts, rocker arms, and valves. The oil then drains back to the oil pan through holes in the block and cylinder heads.

A second oil flow diagram is shown in ***Figure 21–4***. The oil flow progresses as follows:

1. Oil is drawn from the oil pan, through a wire mesh filter (oil pump screen) into the oil pump.
2. Oil is pressurized by the oil pump and sent into the oil filter to be cleaned. Drilled internal passageways are used.
3. Oil is sent to a main oil gallery to lubricate both the crankshaft and the camshaft. Drilled internal passageways are used.
4. From the main oil gallery, oil is sent to the camshafts. Drilled internal passageways are used.

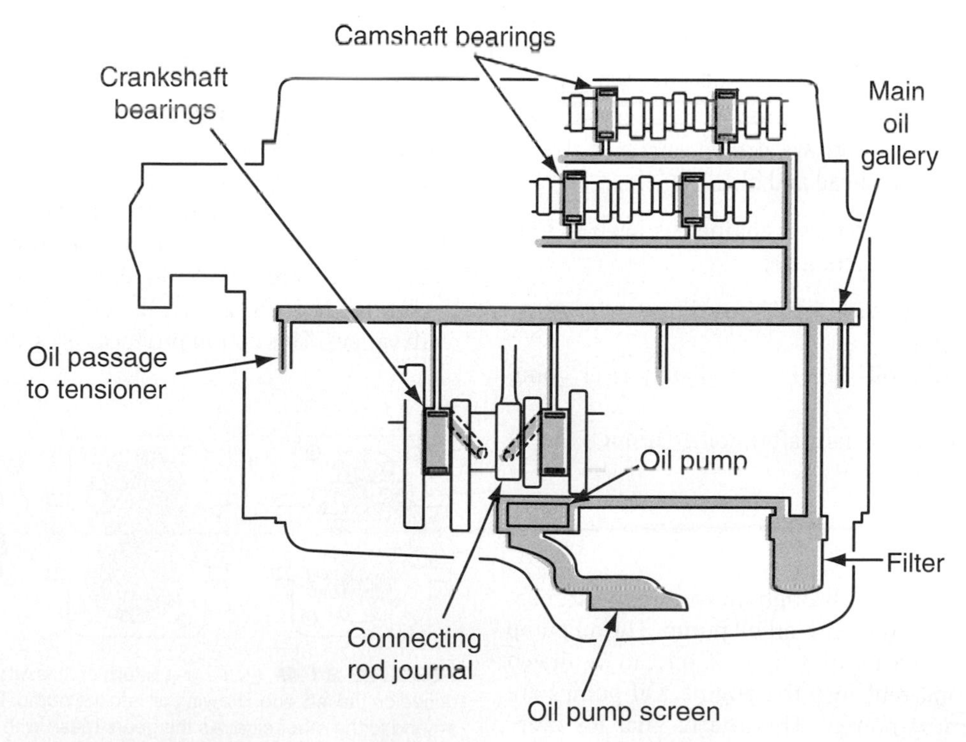

FIGURE 21-4 A typical oil flow diagram for a V6 or a V8 engine.

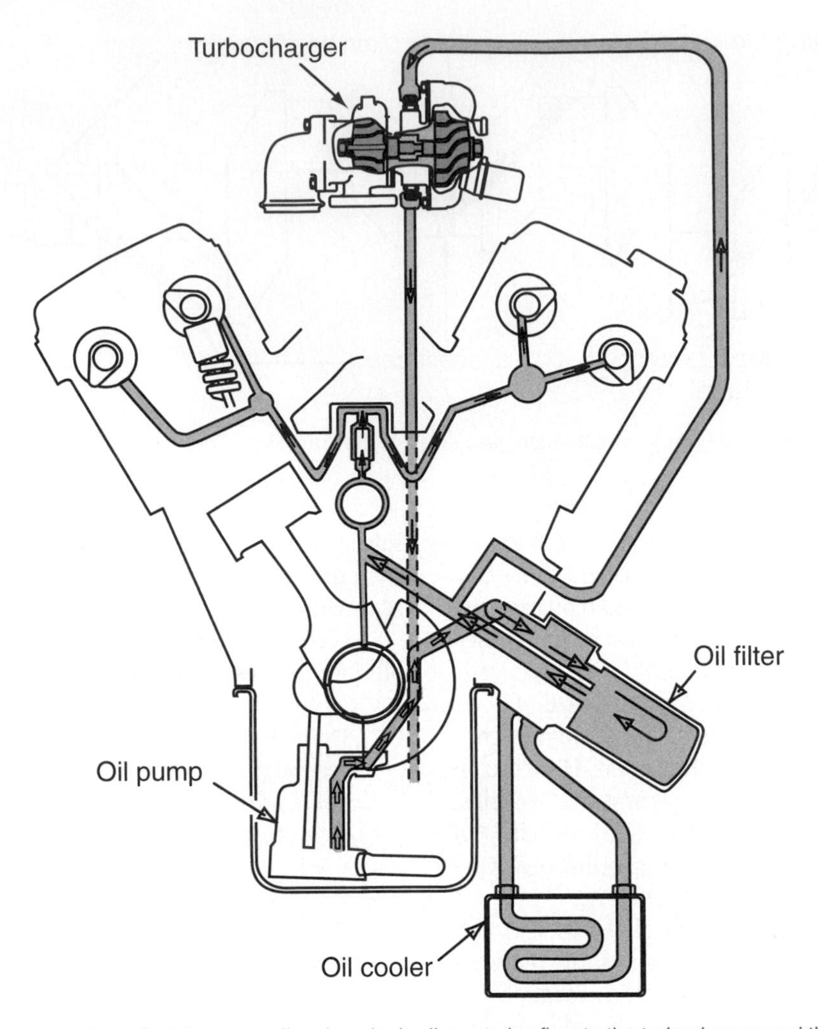

FIGURE 21–5 On certain engines (in this case a diesel engine), oil must also flow to the turbocharger and the oil cooler.

5. From there oil is sent to lifters.
6. At this point the oil is no longer under pressure, and it drains back to the oil pan through internal passageways. Some passageways are drilled, while some are cast into the cylinder head and block.

An oil flow diagram for a diesel engine is shown in ***Figure 21–5***. Some differences include:

1. Oil is sent to the turbocharger immediately after going through the filter.
2. Oil is sent to the oil cooler immediately after going through the filter.
3. Oil is sent to the camshafts after going through the oil filter.

OIL PUMPS

For the oil to be distributed throughout the engine, it must be pressurized. This is done with an oil pump. The oil pump is located in the crankcase area so that oil can be drawn from the oil pan and sent into the engine. Oil pumps are **positive displacement** pumps. This means that for every revolution on the pump a certain volume of oil is pumped. Therefore, as rpm increases, oil pressure and volume also increase.

There are several types of pumps. ***Figure 21–6*** shows how a standard gear-type pump operates. In this case, the oil is drawn in on the left side due to the vacuum that is created when each tooth is removed from a valley on the gears. The oil is then carried around to the other side of the pump as the gears continue to rotate. As the gear teeth mesh on the right side of the pump, the oil is forced out of the gear teeth valleys. This action produces pressure.

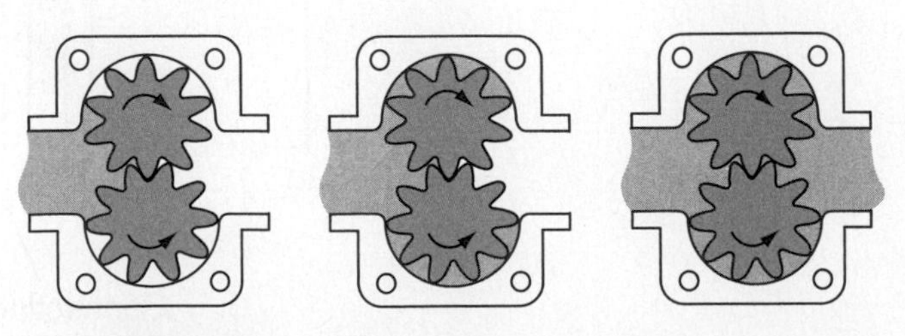

FIGURE 21–6 As the gears turn on this oil pump, suction is produced on the left side, drawing oil into the pump. The oil is then carried around to the other side. As the gears mesh with each other, the oil is squeezed out and pressurized.

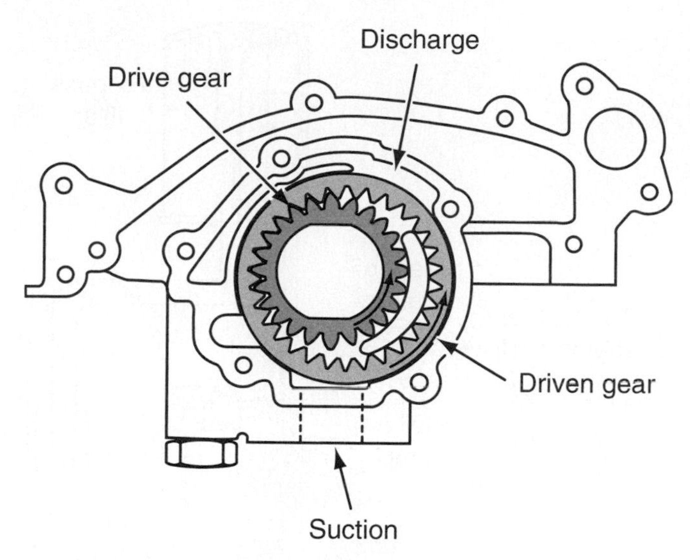

FIGURE 21–7 The gears of an eccentric oil pump are shown in this illustration.

An **eccentric** gear-type pump operates on a similar principle. (Eccentric means not having the same center.) ***Figure 21–7*** shows two gears. The centers of the gears are not at the same position. As the inside drive gear is turned by the crankshaft, a suction is produced between the teeth on the bottom, and pressure is produced between the teeth on the top. Once again, the suction is created when a tooth on the inside drive gear comes out of the valley on the outside driven gear. Pressure is created when a tooth on the inside drive gear fills a valley on the outside driven gear, forcing oil out and creating pressure.

Figure 21–8 shows a typical oil pump driven by the crankshaft. The pump is an eccentric type. As the inside gear

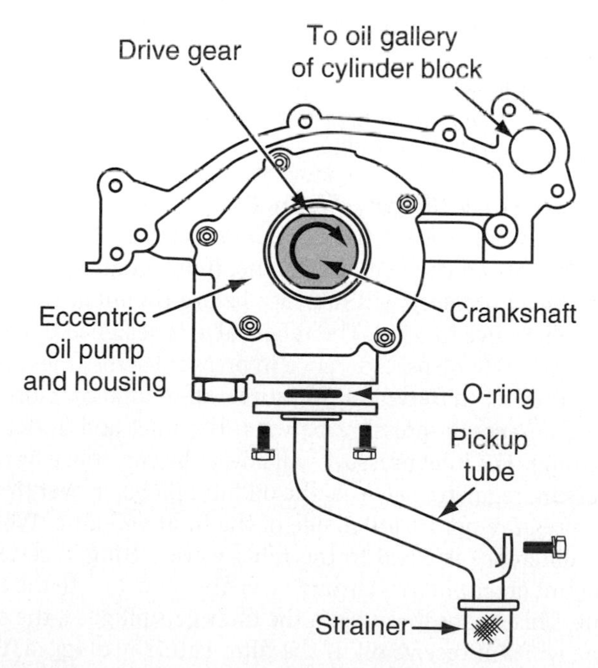

FIGURE 21–8 The eccentric pump is driven directly by the crankshaft.

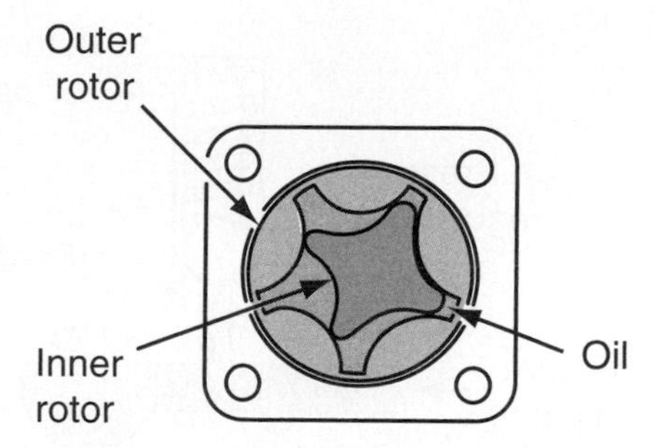

FIGURE 21–9 The rotor oil pump uses an inner rotor with four lobes. As they rotate, the outer rotor is also turned, causing suction and pressure.

is turned, the vacuum produced draws oil from the oil pan through the oil screen, through the tube, and into the oil pump. The oil is then pressurized and sent into the engine.

PARTS LOCATOR

*Refer to photo #10 in the front of the textbook to see what an **oil pump** looks like.*

Figure 21–9 shows another type of positive displacement oil pump called a rotor oil pump. This type of oil pump uses an inner rotor with four lobes and an outer rotor with five valleys. A shaft connected to the crankshaft drives the inner rotor. The inner rotor drives the outer rotor. As the inner rotor rotates and the teeth come out of mesh with the outer rotor, suction and a vacuum are created in a manner similar to the eccentric gear oil pump.

OIL PRESSURE REGULATING VALVE

Because all oil pumps are positive displacement types, a pressure regulator valve must be used. A pressure regulator valve is used to keep the pressure within the oil system at a constant maximum value. As the rpm of the engine changes, the amount of pressure produced by the oil pump also changes. In addition, as the oil gets thicker because of cold weather, oil pressure may increase. The pressure regulator valve maintains a constant specified pressure. Whenever the pressure exceeds this specified value, the regulator valve opens to reduce the pressure. Normal oil pressures vary according to the engine manufacturer. Check the engine manufacturer's specifications to determine the oil pressure for a particular engine.

Figure 21–10 shows how an oil regulator valve works. As the oil pressure increases in the pump, the pressure pushes against a ball or valve held in place by a spring. When the oil pressure is greater than the spring tension, the ball lifts off its seat. At this point, some of the oil is returned to the suction side of the oil system or to the oil pan. This reduces the pressure, which seats the ball again. The spring tension is designed to set the oil pressure at the manufacturer's specifications. If a stronger spring is used,

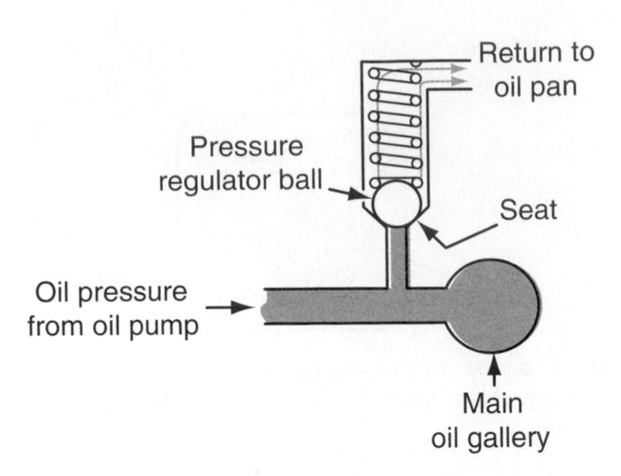

FIGURE 21-10 During normal operation, the oil is sent into the main oil gallery. As oil pressure increases, the ball is lifted off its seat, against the spring pressure. Some of the oil is then returned to the oil pan, reducing and controlling the oil pressure.

the oil pressure will increase. If the spring pressure is less or the spring is broken, the oil pressure will be less. Pressures are normally about 40–60 psi for passenger cars and light trucks. Larger engines may have pressures in the range of 40–70 psi.

PARTS LOCATOR

*Refer to photo #11 in the front of the textbook to see what an **oil pressure regulator valve** looks like.*

OIL FILTERS

Oil must be filtered and cleaned constantly within the engine. As was mentioned in Chapter 20, several contaminants get into the oil. Oil filters are used to clean the dirt particles out of the oil. ***Figure 21–11*** shows a typical oil filter on an engine.

FIGURE 21-11 Oil filters are used to remove dust, dirt, and sludge particles from engine oil.

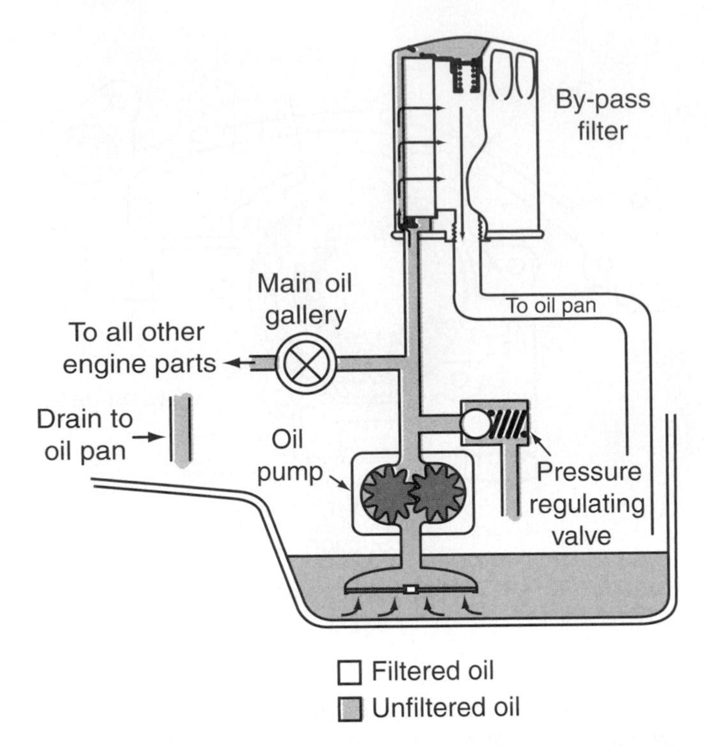

FIGURE 21-12 A by-pass oil filter is used on some engines. In this system, only a small amount of oil is filtered. The filter is a by-pass line around the oil system.

There are two types of oil filtering systems: the full-flow system and the by-pass system. Years ago, only the by-pass system was used. However, because of a greater need to filter engine oil, full-flow systems are now being used. Full-flow systems filter all of the oil before it enters the engine. By-pass systems filter only part of the oil during operation.

By-Pass Filter ***Figure 21–12*** shows the by-pass system. Approximately 90% of the oil is pumped directly to the engine. Only about 10% is sent into the oil filter to be cleaned. If this filter becomes plugged, no oil can be filtered. Oil will still be pressurized, however, and sent into the engine. Certain diesel applications use this system along with a full-flow filter.

Full-Flow Filter ***Figure 21–13*** shows the full-flow system. In this system, all of the oil must pass through the oil filter before entering the engine. If dirt plugs the oil filter, the oil pressure will increase before the filter, causing the relief valve to open. The relief valve is designed to open at about a 5 to 40 psi difference in pressure across the filter. This is called **differential pressure**. Differential pressure is the difference in pressure between the inlet and outlet of the filter. The inlet pressure will always be controlled by the pressure regulator valve. As the oil filter plugs up over time, the pressure on the other side of the filter will drop. When the difference is equal to the relief valve setting, a certain amount of oil will pass through the filter and go into the engine. This means that even if the filter gets plugged, the engine will still receive oil. If the filter is totally plugged, the oil pressure will be slightly less than normal pressure. Remember that the two types of filters are not interchangeable.

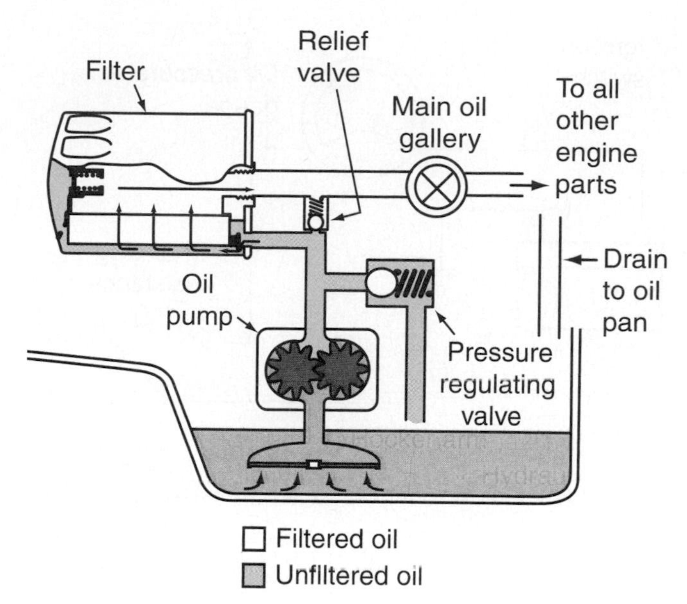

FIGURE 21–13 When a full-flow oil filter is used, all of the engine oil must go through the filter before entering the engine passageways.

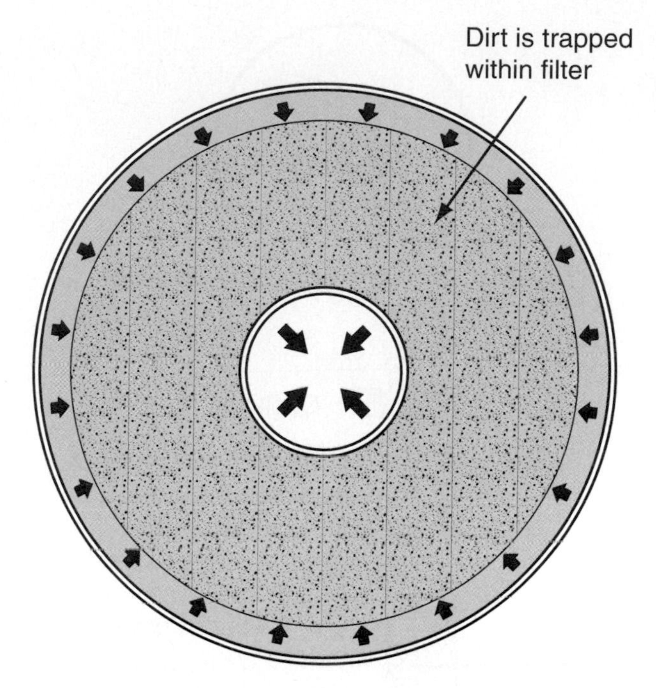

FIGURE 21–15 A depth filter is made of cotton threads and other fibers. The dirt particles are trapped within the filter.

Filter Design Oil filters come in a variety of shapes and sizes. Two common types are surface filters and depth filters. The surface filter is shown in ***Figure 21–14***. The oil flows over the surface of the paper material. The contaminants are trapped on the surface of the paper, but the oil flows through microscopic pores in the paper.

The depth filter material is a blend of cotton thread and various fibers. ***Figure 21–15*** shows a depth filter. As the oil and dirt flow through the filter, contaminants are trapped inside the filter material. Depth filters are used less today because they produce more restriction to the oil flow. Today's engines need more oil flow to aid in cooling the engine.

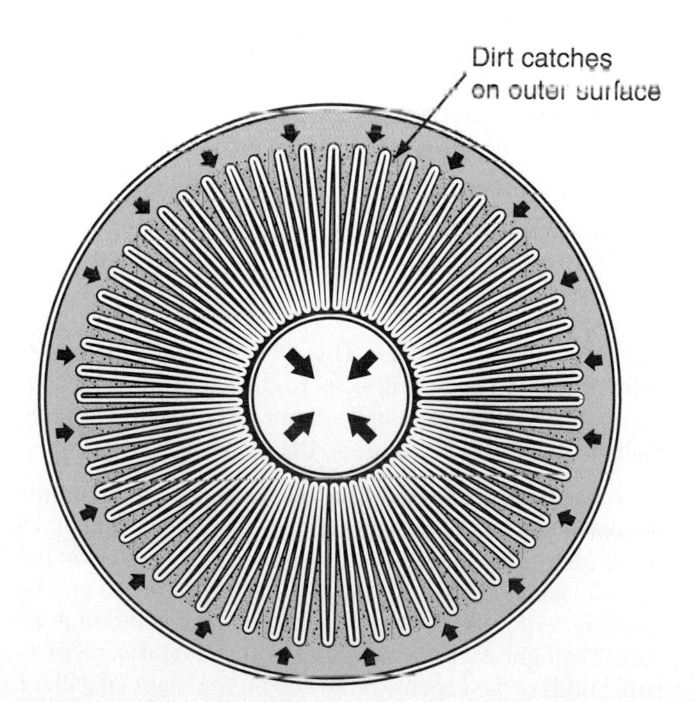

FIGURE 21–14 A surface filter catches the dirt particles on the surface of the filter material.

OIL COOLERS

Oil coolers are used on certain heavy-duty gasoline engines and many diesel engines. An oil cooler is a device that helps keep the oil cool. Oil temperature should be in the range of 180 to 250 degrees Fahrenheit. Under normal conditions, oil is cooled by having the right amount of oil in the oil pan. When excess temperatures occur, however, this cooling may not be enough. Oil coolers are used then.

Common oil coolers are designed with many copper tubes that are sealed together. The assembly is then sealed in a shell-type housing. This type of device is called a liquid-to-liquid heat exchanger. This means that the hotter liquid transfers some of its heat to a cooler liquid. Coolant from the engine cooling system passes from the top to the bottom of the oil cooler through the shell. As the coolant passes over the hotter oil-filled copper tubes, the heat in the oil is transferred to the coolant. Although the two fluids do not mix, the result is that the engine coolant, in effect, cools the oil. The coolant then expels the excess heat out through the radiator. ***Figure 21–16*** shows how a typical oil cooler works.

OIL SENSORS AND GAUGES

Oil pressure sensors are used to indicate if the oil system has the right amount of pressure. Oil pressure is usually sensed or measured directly from the main oil gallery. This information is sent to the dashboard on the vehicle so the

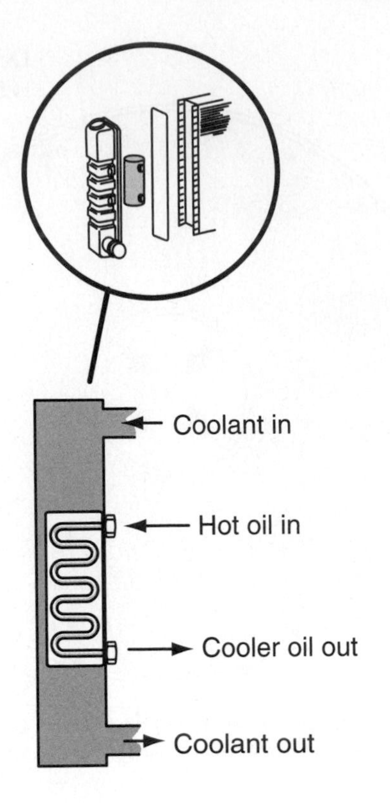

FIGURE 21-16 This oil cooler uses the principle of a liquid-to-liquid heat exchanger to remove heat from the oil.

operator can read the pressure. Two types of systems are commonly used. These are the pressure gauge and the oil indicator light. The pressure gauge reads the pressure of the oil within the system. The oil indicator light system has a light that goes on when oil pressure is low, usually at 7 psi or below.

PARTS LOCATOR

*Refer to photo #12 in the front of the textbook to see what **oil sensors** look like.*

Electric Oil Gauge *Figure 21–17* shows how an electric oil pressure gauge works. As oil pressure is sensed from the main oil gallery, the pressure moves an arm in such a way that it changes the resistance of the circuit. This change in resistance causes the oil pressure gauge to read differently for different pressures. The oil sensor is connected to the main oil gallery. Wires are then connected from the sensor to the electric oil gauge on the dashboard.

Bourdon Gauge The Bourdon pressure gauge is also used to measure and read oil pressure. ***Figure 21–18*** shows a typical Bourdon gauge. The tube is made of thin brass. The free end is connected to an indicating needle on the gauge dial. As pressure in the oil system increases, the gauge tends to become straighter. This causes the needle to

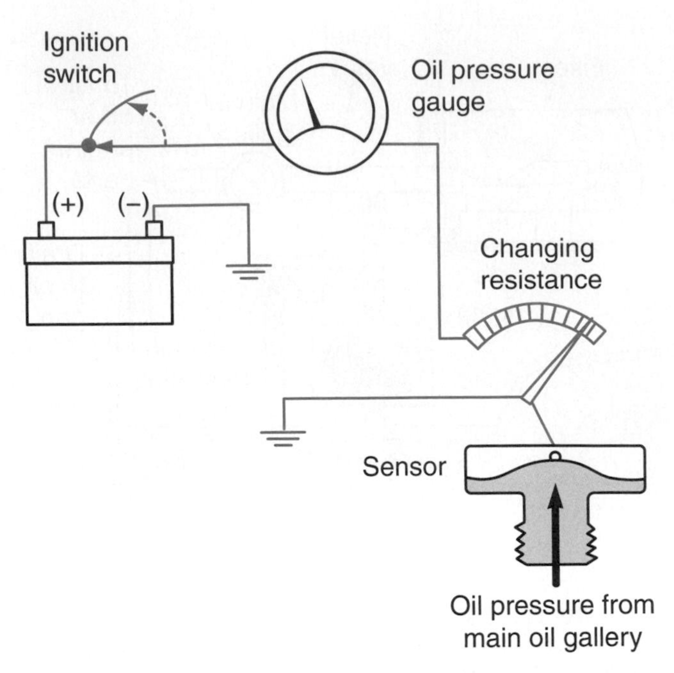

FIGURE 21-17 An electric oil pressure gauge senses oil pressure. As the pressure changes, resistance within the circuit also changes, causing the needle to give different readings.

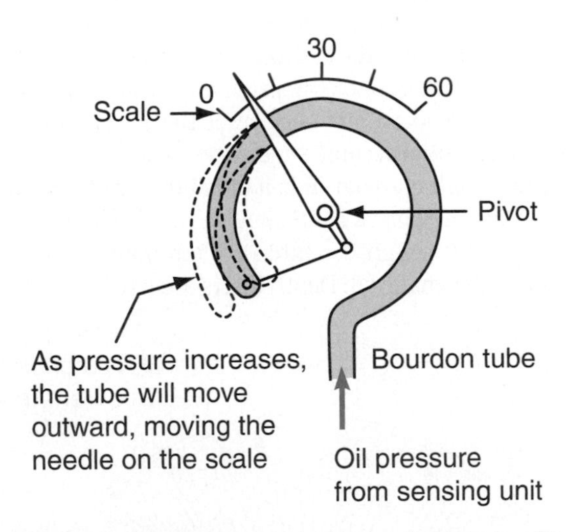

FIGURE 21-18 As pressure increases, the tube in a Bourdon pressure gauge becomes straighter, giving a different reading.

read differently on the dial. This type of gauge reading requires the oil to be sent directly to the gauge. A small copper or plastic tube is usually connected from the engine main oil gallery to the gauge on the dashboard.

Indicator Light The oil indicator light simply goes on when the oil pressure is low. The light is connected to the ignition switch. There is also a pressure switch on the main oil gallery. When the ignition switch is turned on, the pressure switch is still closed because there is no oil pressure. The light is on then. When the engine starts, oil pressure builds up and opens the switch on the main oil gallery. Opening this switch turns off the oil light. ***Figure 21–19*** shows two types of oil sensors.

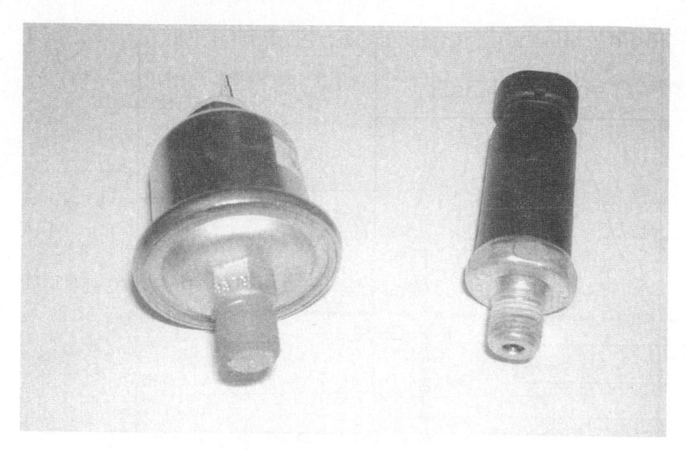

FIGURE 21-19 Oil sensors are used to sense the pressure in the main oil gallery. As pressure increases, a switch in the sensor opens. When pressure drops, the switch closes and turns on the oil indicator light.

Oil Pressure Indicator System *Figure 21–20* shows a common oil pressure indicator system. These electrical diagrams are used to help troubleshoot the oil pressure indicator system. Notice that the wires are all identified by **color codes.** In addition, the wires are numbered. There are also several reference points in the circuit that lead the service technician to other circuits, photos, and specific pages within a manufacturer's service manual for a particular vehicle.

The circuit works in the following way. There is a low oil pressure indicator light (located in the center on the right side of the diagram). This is the oil light on the dashboard and instrument panel of the vehicle. It works in two ways. First, it flashes continuously following a momentary loss of oil pressure. Second, it goes on and stays on with a complete loss of engine oil pressure.

When the engine first starts, before oil pressure rises above 4.3 psi, voltage is applied to the closed and grounded engine oil pressure switch (located on the bottom right of the circuit). To do this, electricity comes from fuse 1, over to the YEL wire, through the low oil pressure indicator light, to the YEL/RED wire, and finally through the closed engine oil pressure switch. This turns on the low oil pressure indicator light. One switch is for a V6 engine while one is for the in-line 4-cylinder engine.

When the engine starts and oil pressure increases to above 4.3 psi, the engine oil pressure switch moves to the left and opens. This action stops the electricity flowing through the low oil pressure indicator light. The light now turns off.

If the oil level is low or the oil pump and regulator are not working correctly in the engine, the oil pressure may drop momentarily (more than 0.5 second) and then come back. The engine oil pressure switch will then close and open momentarily. In this case, the integrated control unit (left center of diagram) senses this momentary loss of oil pressure, through the YEL/RED wire on its right side. When this happens, even though the oil pressure has been restored, the low oil pressure indicator flasher circuit turns the low oil pressure indicator light on and off, in a blinking fashion. This indicates that the oil may be low and needs attention. During this condition, electricity is flowing from the fuse, over to the YEL wire, through the low oil pressure indicator light, down and to the left, into the low oil pressure indicator flasher circuit and through the BLK wire to ground. The light will flash on and off until the ignition is turned off. Note that the flashing feature will not work for the first 30 seconds of engine operation.

If the engine oil pressure falls below 4.3 psi and does not increase, the engine oil pressure switch will stay closed. The low oil pressure indicator light will then come on and stay on.

Electrical Oil Temperature Gauge Some vehicles have a gauge on the dashboard that shows the temperature of the oil in the engine. Oil temperature is very important when driving under severe conditions. If the oil temperature increases too much, the oil may break down and lose its lubricating properties.

The oil temperature sensor is attached to the main oil gallery. Oil from the crankcase is sent to this sensor. Inside the sensor, there is a **variable resistor**. As the temperature of the oil increases or decreases, the resistance of the sensor changes. Thus, the current in the electrical circuit is being controlled by the variable resistor. When the oil temperature is low, the sending unit resistance is high. So the current flow to the gauge is restricted and the pointer on the gauge moves very little. As oil temperature increases, the sending unit resistance decreases. Now the current flow through the gauge increases. The increased current causes the pointer on the gauge to move more, indicating a higher oil temperature.

Oil Change Indicator Some vehicles, especially those with an electronic instrument panel, may have an oil change indicator gauge. This gauge is controlled by the body control module (BCM). This is a computer that monitors various electrical inputs. Then, based on these inputs, the computer sends a signal to a gauge so it can be read. In this case, the coolant temperature, engine speed, and vehicle speed are continuously being monitored. These three inputs give the computer an indication of the type of driving conditions of the vehicle. Based on this information, the computer makes a decision about the condition of the oil. When an oil change is necessary, the computer turns on a light on the dashboard that says "Change Oil."

Some engines also have an oil level indicator system. As shown in ***Figure 21–21***, an oil level sensor is located on the side of the oil pan assembly. The oil level sensor is used to sense if there is oil at the level that the sensor has been placed. If there is oil at this level, the sensor acts as an open circuit keeping the oil level indicator light on the dashboard off. If the oil level drops below the sensor, it acts as a closed circuit and turns on the oil level indicator light on the dashboard.

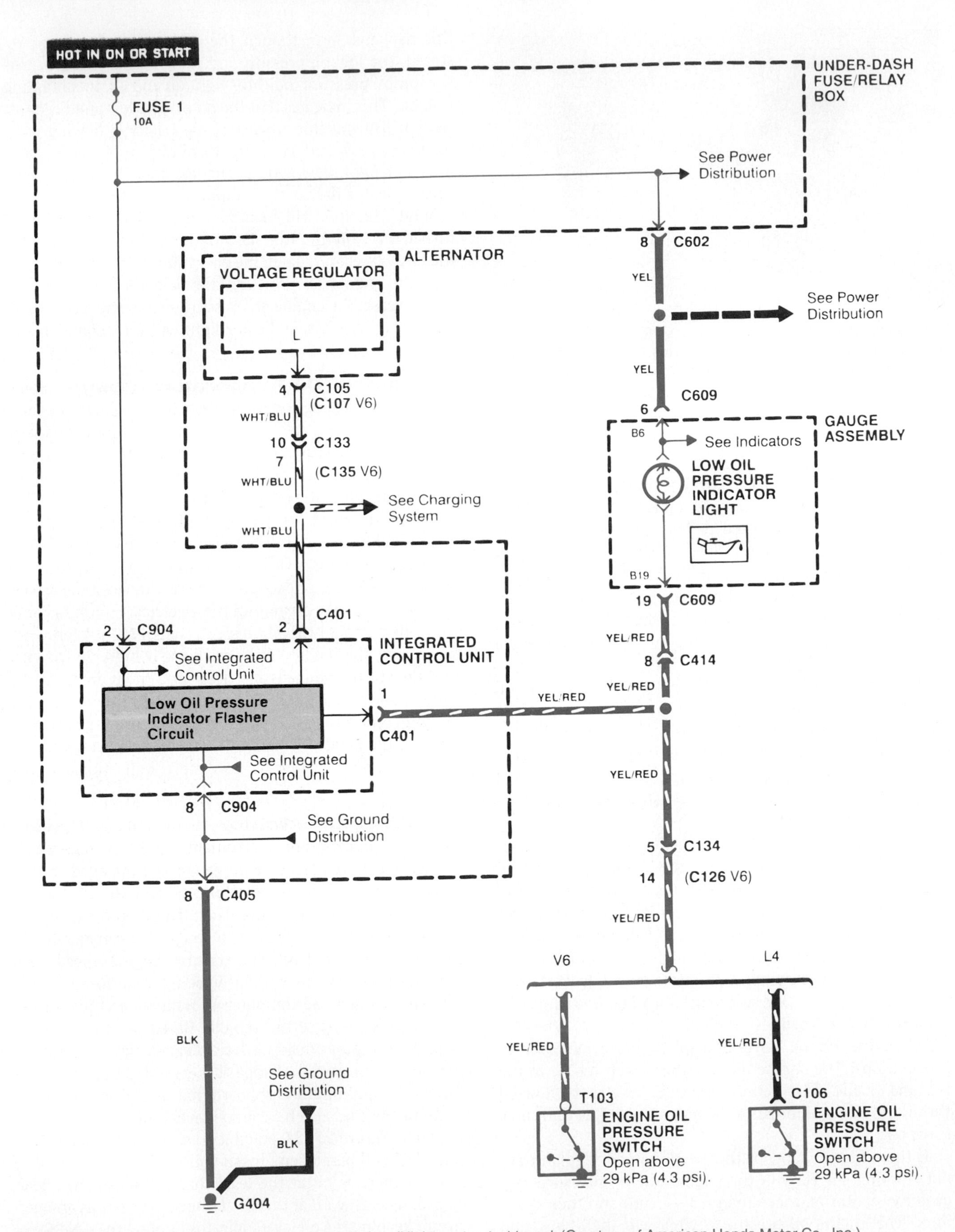

FIGURE 21-20 This electrical circuit operates the oil light on the dashboard. (Courtesy of American Honda Motor Co., Inc.)

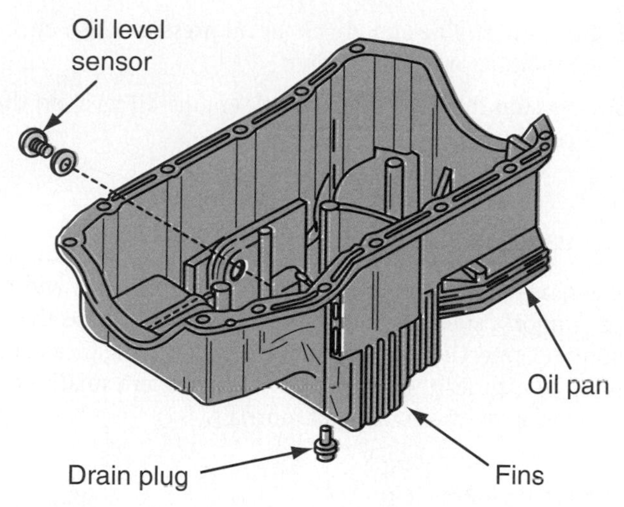

FIGURE 21-21 On some oil pans, there is an oil level sensor to tell the operator that the oil level is low and that oil needs to be added or changed.

TURBOCHARGED ENGINES AND OIL PRESSURE

On turbocharged engines, oil pressure from the main oil gallery is used to lubricate the spinning compressor and turbine blades. If this oil pressure is not available, the turbocharger bearings will lose lubrication and possibly be damaged. To avoid this, keep in mind the following two suggestions:

1. Never accelerate a turbocharged engine and then immediately shut the engine off. The oil pressure will be shut down, but the turbocharger will slowly spin to a stop. During this time, the turbocharger will not have adequate oil for its operation.
2. When changing the oil or filter, remember that the oil also has been drained from the turbocharger. To get oil back to the turbocharger after an oil change, it is good practice to disconnect the ignition system and crank the engine for a few (5 to 10) seconds. Then reconnect the ignition and start the engine. The additional cranking will ensure oil pressure to the turbocharger.

Problems, Diagnosis, and Service

Safety Precautions

1. A vehicle must be lifted up whenever the oil is changed. When jacking up a vehicle with a hydraulic lift, always use jack stands to support the vehicle in case the hydraulic lift should fail.
2. When you change oil or oil filters, oil often spills on the floor. Clean up oil spills immediately so that you won't slip and fall.
3. Always use proper clothing (shop coats and so on) when working on the lubrication system. Oil spilled on clothing, especially diesel oil, may damage the clothing.
4. When running the engine to check oil pressure, make sure the exhaust fumes are adequately removed by the exhaust system in the building.
5. Used oil from the engine is very toxic and should not be disposed of on driveways or in fields. All used oil from engines should be recycled. Many states have specific recycling requirements for oil. In many states, retail outlets that sell oil must either recycle the used oil or provide information noting where it can be recycled in the area. Check with your local or state pollution control agency if you are in doubt about how to recycle used oil.
6. Always keep all tools clean and free from oil.
7. Make sure you wear safety glasses in the service area.
8. Always use the correct tools when working on the lubrication system.
9. Never remove an oil pressure sensor or gauge from an engine while the engine is running.

PROBLEM: Changing Oil

The oil has not been changed in a vehicle for a long period of time. If not changed at regular intervals, sludge, varnish deposits, and engine wear could result.

DIAGNOSIS

Refer to the manufacturer's recommended change periods. Depending on the vehicle year and make, the change periods in miles will vary. Most newer vehicles should have the engine oil changed every 3,000 to 7,500 miles, depending on the severity of service.

SERVICE

When changing the oil, make sure you know the correct quantity of oil. When the oil filter is changed, an additional quart of oil typically must be added.

1. Bring the engine up to operating temperature.
2. Jack up the vehicle and place the jack stands securely under the car.
3. Remove the oil plug.
4. Let the old oil drain into an appropriate oil pan.
5. Remove the oil filter and drain it, as shown in ***Figure 21–22***.
6. Make sure all the oil has been drained before tightening the oil plug.

CAUTION: *Be careful not to strip the threads on the oil plug during installation.*

7. Select the correct oil to use. Make sure the oil viscosity is correct for the application and weather conditions. Again, always refer to the manufacturer's specifications, as using the incorrect oil viscosity may void the manufacturer's warranty. Make sure the API ratings are correct for the application and year of the vehicle.
8. Before installing a new filter, put oil on the oil filter seal. Then fill the new filter with the correct type of oil. This procedure will eliminate any air in the lubrication system during initial starting. Now replace the oil-filled filter on the engine block. Make sure the oil filter is tight. Be careful not to strip the oil filter threads when installing a new filter.
9. Remove the vehicle from the jack stands.
10. Add the correct quantity of oil to the engine.

FIGURE 21–22 When changing the oil and filter, always remove the filter and drain it into the appropriate container.

11. Start the engine and check for oil pressure. Also check for oil leaks around the filter.
12. Now stop the engine and check engine oil level on the dipstick.

PROBLEM: Acid Buildup in the Oil

An engine has been stored for over a year. When the engine was run for a short period of time, it was suspected that a main or connecting rod bearing was bad. During disassembly of the engine, it was noticed that there were small spots of corrosion on the crankshaft journals.

DIAGNOSIS

Used oil that sits in an unused engine may have a buildup of acids over a period of time. The acids come from the combustion process and enter the oil reservoir because of blowby past the rings.

SERVICE

To prevent acids from developing in the oil, always change the used oil in an engine before the engine is stored. Do this if the engine is to be stored more than 3 months at a time.

PROBLEM: Gasoline in the Oil

Oil pressure seems to have been dropping continuously over a period of several weeks. In addition, the oil seems to be getting thinner.

DIAGNOSIS

These two problems generally suggest that there may be fuel in the oil. Fuel can be detected in the oil by a gaseous smell or reduced viscosity. Fuel can get into the oil in two ways:

1. If the fuel system is continually too rich, fuel may seep into the oil and make it thinner. This could be caused by a faulty fuel regulator or faulty injectors.
2. If the engine is flooded continuously when starting, fuel may also seep into the crankcase.

SERVICE

If fuel is found in the oil, determine where it is coming from.

1. Check the fuel pressure regulator for correct pressure.
2. Check the fuel injectors for correct operation, making sure they shut off completely and that they are injecting the correct amount of fuel.
3. Check for fault codes to determine if all fuel management sensors are working correctly.

4. Make sure that the operator does not pump the throttle when starting since this will cause flooding of the fuel system and excess fuel in the lubrication system.
5. Once the problem has been found and repaired, drain the contaminated oil. Also remove and replace the filter.
6. Now replace with the correct type and quantity of oil.

PROBLEM: Water in the Oil

The engine oil appears to be light brown or whitish in color.

DIAGNOSIS

When the oil becomes light brown or whitish, water may have entered the lubrication system. Water can get into the oil from the cooling system. The water usually comes from a cracked cylinder block or head, or by a blown head gasket.

SERVICE

If coolant is getting into the lubricating system, the cylinder head can be checked by performing a compression test on the cylinder. A low compression reading may indicate a cracked head or a broken head gasket. If this is the case, the engine will need to be disassembled and the cylinder head or gasket replaced. If the cylinder compression test is OK, then the crack most likely is in the cylinder block. In this case, the engine will need to be disassembled and a new block installed.

PROBLEM: Low Oil Pressure

The operator of the vehicle has noticed that the oil pressure is low.

DIAGNOSIS

Various components may be damaged and cause low oil pressure.

- A broken spring on the regulator valve could be a cause of low oil pressure.
- If the oil pressure dropped suddenly, check the oil level first, then check the oil pressure sensor and the gauge for correct operation.
- Low oil pressure can also be caused by worn main and connecting rod bearings. Check the age and number of miles on the engine.
- Low oil pressure can be caused by a faulty sending unit or gauge.

SERVICE

1. If a broken spring on the regulator valve is suspected, remove the filter, valve, and spring from the engine. Replace the spring with a new one.
2. If the oil sensor is bad, replace it with a new one. This can be done by unscrewing the oil sensor from its location on the main oil gallery. Make sure the electrical or mechanical connections have been removed before removing the oil pressure sensor. The oil sending unit can be checked by comparing the resistance of the sensor with and without oil pressure. For example, with oil pressure, there should be 10,000 ohms of resistance. Without oil pressure, there should be about 5 ohms of resistance. Check the manufacturer's specifications for the exact readings.
3. If the main and connecting rod bearings are suspected, the engine must be disassembled. To check the clearance on the main and connecting rod bearings, use Plastigage.

PROBLEM: Oil Leaks

An engine has developed small oil leaks.

DIAGNOSIS

Oil leaks can occur throughout the engine. When working on the lubrication system, visually check for leaks at the following possible trouble spots:

1. The oil plug, which may be stripped or loose (***Figure 21–23***)
2. The gaskets around the valve covers
3. All of the bolts that come into contact with the crankcase. Oil can leak around the threads of the bolts. These bolts should have copper, felt, or other sealing washers to stop leaks. These are more common on older engines.
4. The front or rear main seals on the crankshaft
5. The filter, which may be loose or sealed incorrectly (***Figure 21–24***)

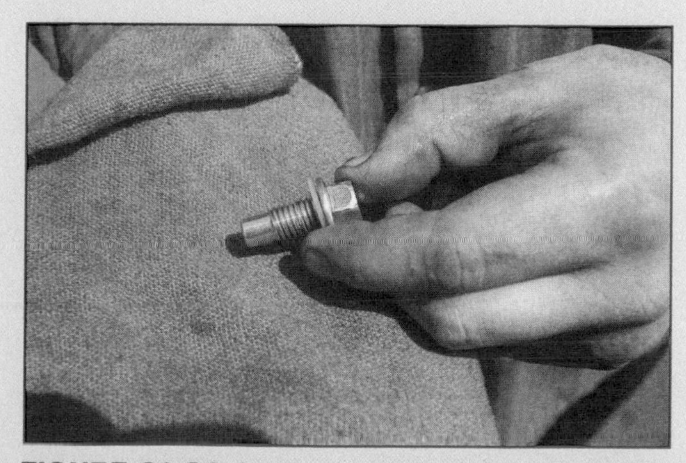

FIGURE 21-23 Always check the oil plug threads for damage before installation. If damaged, replace the plug to eliminate oil leaks. Also, always replace the O-ring or seal to prevent any leakage.

FIGURE 21-24 When looking for leaks, check the oil filter gasket for proper sealing. If the filter is leaking, the gasket should be somewhat irregular or should show where it was misaligned to cause the leak.

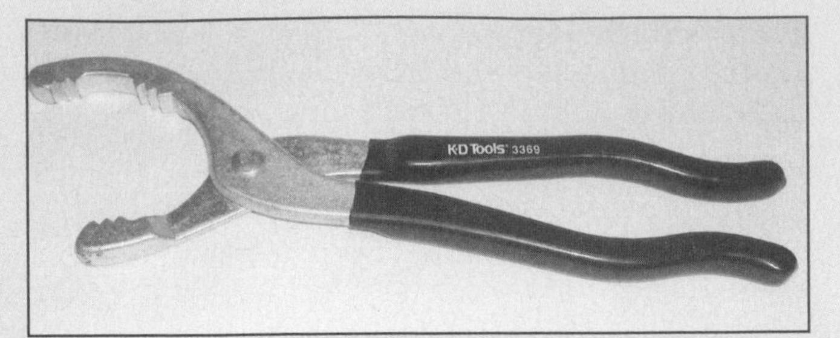

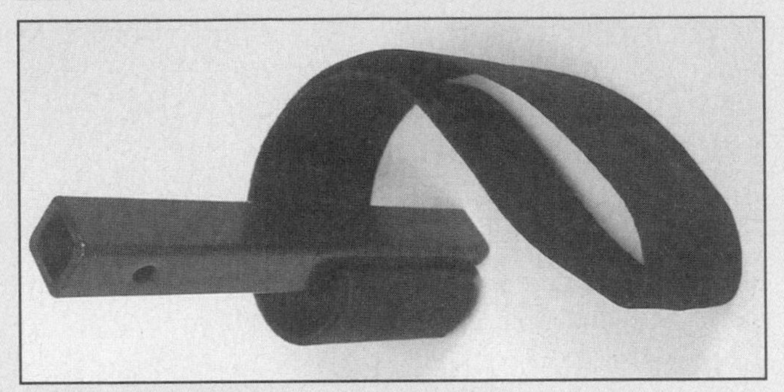

FIGURE 21-25 Oil filter wrenches such as these are used to remove oil filters from the engine.

SERVICE

The type of service will depend on the exact location of the leak.

1. If the oil plug is stripped, a self-tapping oil plug can be installed to help seal this area.
2. If the gaskets on the valve covers are leaking, these covers will need to be removed and replaced with new gaskets.
3. If bolts are contacting the crankcase area and are leaking, thread sealers can be used to help seal the bolts.
4. If the rear or front main seals are leaking, they must be replaced. Follow the manufacturer's recommended procedure to replace the crankshaft seals.
5. Check the oil filter for correct installation. Make sure the oil filter was not cross-threaded during installation. If it was, remove the oil filter and replace it with a new one. Oil filters can be removed easily using an oil filter wrench. There are many types of oil filter wrenches. Some service technicians prefer to use a chain wrench while others prefer to use a strap to loosen or tighten the oil filter. ***Figure 21–25*** shows two types of oil filter wrenches used in the automotive shop.

There also may be other conditions that cause oil leaks from the engine. Use the chart shown in ***Figure 21–26*** to help diagnose and service the oil system. This is a typical diagnosis and service chart provided in many service manuals. The left column states a problem, the middle column provides the diagnosis, and the right column presents the service suggestions for this type of problem.

PROBLEM: Excessive Oil Consumption

The engine uses an excess amount of oil.

DIAGNOSIS

Oil consumption in the engine can be caused by several internal conditions:

1. Oil can be consumed because the valve guides are worn. Oil then seeps down into the intake valve and is burned in combustion. Bluish exhaust smoke indicates that oil is being burned.
2. Oil can be consumed because the oil control rings are worn. When the rings are worn, the oil is not scraped completely from the cylinder walls. This oil is then burned in the combustion chamber and exhausted. Again, a bluish exhaust is evidence of oil burning in the combustion chamber.
3. To determine if the oil consumption is from the exhaust valves or from the rings, perform a compression leakdown test as described in an earlier chapter.

SERVICE

1. If the valve guides are leaking oil, the cylinder head must be rebuilt. Follow the manufacturer's recommended procedure to rebuild the cylinder head.
2. If the oil control rings are bad, the rings must be replaced. Follow the manufacturer's recommended procedure to replace the piston rings.
3. There may be many other reasons for having excessive oil consumption. ***Figure 21–27*** shows a typical service manual troubleshooting chart for identifying the problem, diagnosis, and service suggestions when there is excessive oil consumption.

PROBLEM	DIAGNOSIS	SERVICE
External oil leaks	(1) Fuel pump gasket broken or improperly seated.	(1) Replace gasket.
	(2) Cylinder head cover RTV sealant broken or improperly seated.	(2) Replace sealant; inspect cylinder head cover sealant flange and cylinder head sealant surface for distortion and cracks.
	(3) Oil filler cap leaking or missing.	(3) Replace cap.
	(4) Oil filter gasket broken or improperly seated.	(4) Replace oil filter.
	(5) Oil pan side gasket broken, improperly seated or opening in RTV sealant.	(5) Replace gasket or repair opening in sealant; inspect oil pan gasket flange for distortion.
	(6) Oil pan front oil seal broken or improperly seated.	(6) Replace seal; inspect timing case cover and oil pan seal flange for distortion.
	(7) Oil pan rear oil seal broken or improperly seated.	(7) Replace seal; inspect oil pan rear oil seal flange; inspect rear main bearing cap for cracks, plugged oil return channels, or distortion in seal groove.
	(8) Timing case cover oil seal broken or improperly seated.	(8) Replace seal.
	(9) Excess crankcase pressure because of restricted PCV valve.	(9) Replace PCV valve.
	(10) Oil pan drain plug loose or has stripped threads.	(10) Repair as necessary and tighten.
	(11) Rear oil gallery plug loose.	(11) Use appropriate sealant on gallery plug and tighten
	(12) Rear camshaft plug loose or improperly seated.	(12) Seat camshaft plug or replace and seal, as necessary.
	(13) Distributor base gasket damaged.	(13) Replace gasket.

FIGURE 21-26 When checking for external oil leaks, use these diagnosis and service suggestions.

PROBLEM	DIAGNOSIS	SERVICE
Excessive oil consumption	(1) Oil level too high. Check for contamination (fuel, coolant).	(1) Drain oil to a specified level.
	(2) Oil with wrong viscosity being used.	(2) Replace with specified oil.
	(3) PCV valve stuck closed.	(3) Replace PCV valve.
	(4) Valve stem oil deflectors (or seals) are damaged, missing, or incorrect type.	(4) Replace valve stem oil deflectors.
	(5) Valve stems or valve guides worn.	(5) Measure stem-to-guide clearance; repair as necessary.
	(6) Poorly fitted or missing valve cover baffles.	(6) Replace valve cover.
	(7) Piston rings broken or missing.	(7) Replace broken or missing rings.
	(8) Scuffed piston.	(8) Replace piston.
	(9) Incorrect piston ring gap.	(9) Measure ring gap; repair as necessary.
	(10) Piston rings sticking or excessively loose in grooves.	(10) Measure ring side clearance; repair as necessary.
	(11) Compression rings installed upside down.	(11) Repair as necessary.
	(12) Cylinder walls worn, scored, or glazed.	(12) Repair as necessary.
	(13) Piston ring gaps not properly staggered.	(13) Repair as necessary.
	(14) Excessive main or connecting rod bearing clearance.	(14) Measure bearing clearance; repair as necessary.

FIGURE 21-27 When there is excessive oil consumption, use these diagnosis and service suggestions.

PROBLEM: Using Synthetic Oil

An engine has been using synthetic oil, and oil consumption has increased.

DIAGNOSIS

Keep the following three rules in mind when using synthetic oils:

1. Never use a synthetic oil in an old engine that has an excess amount of wear on the rings.
2. Never mix synthetic oils.
3. Never use synthetic oils for break-in periods.

SERVICE

If there is excessive consumption when synthetic oils are used, remove the synthetic oil and replace with conventional oil with the correct SAE and API ratings.

PROBLEM: Checking for Low Oil Pressure

An engine has developed low oil pressure.

DIAGNOSIS

Low oil pressure could mean several things are wrong. Although the oil pump or oil regulator may not be working correctly, the problem could also be a damaged or broken oil sending unit. It is always a good idea to double-check to see if the oil pressure at the main oil gallery is actually low. If it is not low after checking the oil pressure, then the oil sending unit may be broken or damaged.

SERVICE

Oil pressure can be checked at the main oil gallery by using the following procedure:

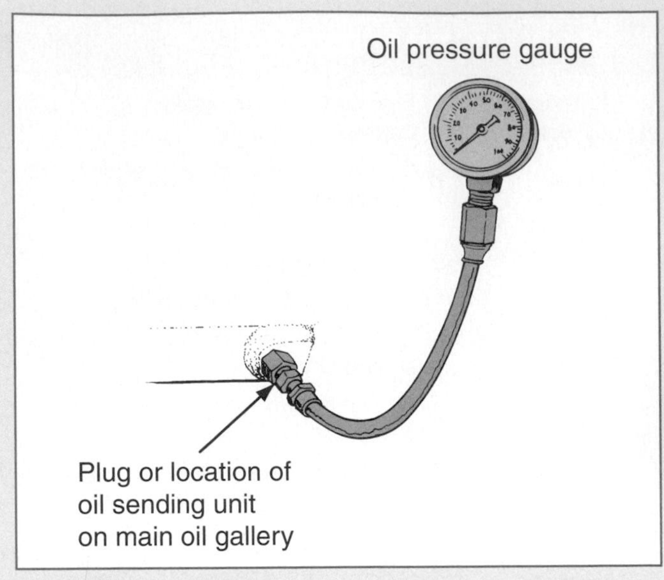

FIGURE 21–28 An oil pressure gauge can be attached to the main oil gallery to check and verify engine oil pressure.

1. First bring the engine up to operating temperature.
2. Now shut the engine off.
3. Remove an accessible oil plug or remove the oil pressure sending unit from the main oil gallery.
4. Insert an oil pressure gauge as shown in ***Figure 21–28***.
5. Now turn the engine on and check the oil pressure on the gauge. If the oil pressure is within specifications, this shows that the oil pressure sending unit may not be working correctly. In many cases, replacing the oil sending unit will correct the problem.
6. There may be many other reasons for low oil pressure. ***Figure 21–29*** shows a typical troubleshooting chart that would be found in a service manual identifying the problem, diagnosis, and service suggestions when there is low oil pressure.

PROBLEM	DIAGNOSIS	SERVICE
Low oil pressure	(1) Low oil level.	(1) Add oil to correct level.
	(2) Inaccurate gauge, warning lamp or sending unit.	(2) Inspect and replace as necessary.
	(3) Oil excessively thin because of dilution, poor quality, or improper grade.	(3) Drain and refill crankcase with recommended oil.
	(4) Excessive oil temperature.	(4) Correct cause of overheating engine.
	(5) Oil pressure relief spring weak or sticking open.	(5) Remove and inspect oil pressure relief valve assembly.
	(6) Oil inlet tube and screen assembly has restriction or air leak.	(6) Remove and inspect oil inlet tube and screen assembly. (Fill inlet tube with lacquer thinner to locate leaks.)
	(7) Excessive oil pump clearance.	(7) Inspect and replace as necessary.
	(8) Excessive main, rod, or camshaft bearing clearance.	(8) Measure bearing clearances; repair as necessary.

FIGURE 21–29 If an engine has low oil pressure, use these diagnosis and service suggestions.

PROBLEM: High Oil Pressure

An engine has high oil pressure.

DIAGNOSIS

High oil pressure can be caused by several engine components failing. The most common failure is a sticky or broken pressure regulator valve or pressure relief valve in the oil system. This is especially true if the valves stick in the closed position. When there is high oil pressure, oil can be washed off the bearings and other parts before it has time to lubricate, cool, clean, and seal the parts. High oil pressure also may cause leaks and could also damage the oil filter.

SERVICE

As mentioned earlier in this chapter, the pressure regulator and pressure relief valves are used to control the oil pressure. If the pressure relief valve sticks in the closed position and the oil filter is extremely dirty, high oil pressure may result. The relief valve is often located in the housing near the oil filter. Remove the filter and locate the relief valve. Often a small nut can be removed, and the spring and plunger (ball) can be removed. Inspect the spring and plunger (ball) and replace if there are damaged or broken parts on the valve.

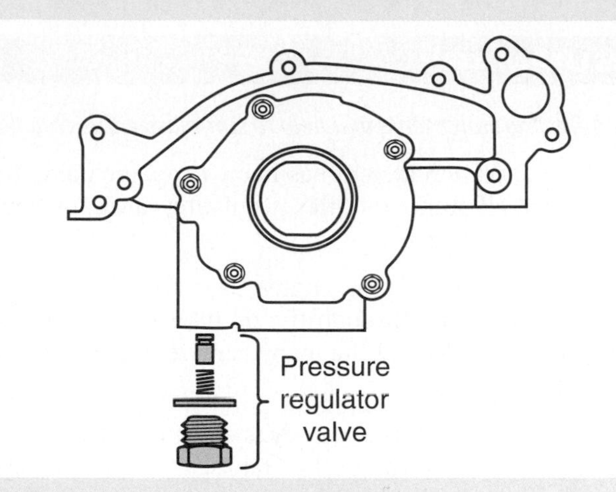

FIGURE 21-30 The pressure regulator valve can be found on the housing for the oil pump.

The pressure regulator valve can be checked in the same way. The pressure regulator valve is often located near the oil pump. In ***Figure 21–30***, the pressure regulator valve is located on the bottom of the oil pump housing. It can be checked by loosening the nut and removing the spring and plunger. Check for broken parts and replace where necessary.

There are also other causes of high engine oil pressure. The chart in ***Figure 21–31*** shows the diagnosis and service necessary for this problem.

PROBLEM	DIAGNOSIS	SERVICE
High oil pressure	(1) Improper oil viscosity.	(1) Drain and refill crankcase with correct viscosity oil.
	(2) Oil pressure gauge or sending unit inaccurate.	(2) Inspect and replace as necessary.
	(3) Oil pressure relief valve sticking closed.	(3) Remove and inspect oil pressure relief valve assembly.

FIGURE 21-31 If an engine has high oil pressure, use these diagnosis and service suggestions.

Service Manual Connection

There are many important specifications to keep in mind when working on a lubrication system. To identify the specifications for your engine, you will need to know the VIN (vehicle identification number) of the vehicle, the type and year of the vehicle, and the type of engine.

Although they may be titled differently, some of the more common lubricating specifications (not all) found in service manuals are listed below. Note that these specifications are typical examples. Each vehicle and engine will have different specifications.

Common Specification	Typical Example
Engine oil refill	4 quarts (add 1 quart with filter change)
Normal oil pressure	30–80 lb
Oil pump gear backlash	0.005–0.009 inch
Oil pump gear end play	0.001–0.004 inch
Oil pump cover screws torque	105 in. lb
Ring end gap (oil control ring)	0.15 inch

In addition, the service manual will give specific directions for various service procedures. Some of the more common procedures include the following:

- Crankshaft rear oil seal, replace
- Oil pan, replace
- Oil pump, replace

SUMMARY

The following statements will help to summarize this chapter:

- The lubrication system has many parts, including the oil pan, oil pump, oil filter, regulating valves, oil coolers, and oil sensors.
- Oil flows through the engine from the oil screen to the pump, then through the oil filter to the main oil gallery, and then to the many parts that need lubrication.
- Oil from the main oil gallery is sent to the main bearings and then through the crankshaft to the connecting rod journals.
- The oil pump is a positive displacement pump, which means it pumps a specific volume and pressure for each rpm of the engine.
- There are several types of oil pumps, including the gear-type, rotor-type, and eccentric-type oil pumps.
- The two types of filters used on the oil system are the by-pass and the full-flow filter. Today's engines use the full-flow filter.
- The oil coolers on some heavy-duty gasoline engines and on many diesel engines use a liquid-to-liquid heat exchanger.
- Sensors and gauges are used to monitor the oil pressure, oil temperature, and oil level in the oil pump.
- Mechanical as well as electrical gauges are used to monitor the oil system.
- A Bourdon gauge straightens as the pressure increases, causing a needle to move, which indicates oil pressure.

TERMS TO KNOW

Can you explain each of the following terms? Review the chapter until you can use each term correctly.

Color codes
Differential pressure
Eccentric
Oil gallery
Positive displacement
Sump
Variable resistor

REVIEW QUESTIONS

Multiple Choice

1. Which type of oil pump has two gears that have different center points?
a. All oil pumps
b. Eccentric oil pumps
c. Differential pressure pumps
d. Centrifugal pumps
e. None of the above

2. Which type of filter forces only a small portion of the oil through the filter?
a. Full-flow filter
b. By-pass filter
c. Differential pressure filter
d. Easy-flow filter
e. None of the above

3. Which device controls the pressure of the oil directly from the oil pump?
a. Oil relief valve
b. Oil pressure regulator valve
c. Oil screen
d. Oil pump
e. Oil pan

4. The difference between the input pressure of the oil filter and the output pressure of the filter is called:
a. Pressure equalness
b. Filter difference
c. Pressure differential
d. Input versus output pressure
e. None of the above

5. Which type of filter catches the dirt and dust particles on the outside of the filter surface?
a. Depth filter
b. Round filter
c. By-pass filter
d. Surface filter
e. None of the above

6. The oil cooler takes heat from the oil and transfers it to the ________.
a. Cooling system
b. Engine block
c. Oil sump
d. All of the above
e. None of the above

7. Which type of oil gauge straightens a curved tube as the oil pressure increases?
   a. Electric oil gauge
   b. Bourdon gauge
   c. By-pass filter gauge
   d. Curved-tube gauge
   e. Low-pressure differential gauge

8. If the oil pressure drops suddenly, the problem may be the:
   a. Oil crankshaft
   b. Oil regulator valve
   c. Oil pan
   d. Oil passages
   e. Oil seals

9. There is no problem when using a full-flow filter for a by-pass filter, or a by-pass filter for a full-flow filter.
   a. False, the two should never be mixed.
   b. True, it is OK to interchange the filters.
   c. True, but always use a clean filter if they are changed.
   d. False, the two can be mixed only after 5,000 miles.
   e. False, the two can be mixed only after 10,000 miles.

The following questions are similar in format to ASE (Automotive Service Excellence) test questions.

10. Fuel has been found in the oil. *Technician A* says it could be because of too rich a fuel mixture. *Technician B* says it could be caused by a bad diaphragm in the mechanical fuel pump. Who is correct?
   a. A only c. Both A and B
   b. B only d. Neither A nor B

11. *Technician A* says excessive oil consumption can be caused by bad rings. *Technician B* says excessive oil consumption is caused by bad valve guides. Who is correct?
   a. A only c. Both A and B
   b. B only d. Neither A nor B

12. Oil pressure in a car with enough oil in the engine has dropped. *Technician A* says the first thing to check is the oil sensor. *Technician B* says the oil pump is bad and should be replaced immediately. Who is correct?
   a. A only c. Both A and B
   b. B only d. Neither A nor B

13. *Technician A* says that when the oil gets to be light brown or whitish, fuel is in the oil. *Technician B* says that whitish oil causes increased oil leakage. Who is correct?
   a. A only c. Both A and B
   b. B only d. Neither A nor B

14. *Technician A* says that high oil consumption can be caused by a poorly adjusted carburetor. *Technician B* says that high oil consumption can be caused by having bad main bearings. Who is correct?
   a. A only c. Both A and B
   b. B only d. Neither A nor B

Essay

15. What is the difference between a by-pass filter and a full-flow filter?

16. Describe how a mechanical/electrical oil pressure indicator system operates.

17. What is the purpose of an oil pressure regulator valve?

18. Describe how an electrical oil temperature gauge operates.

19. What causes oil to become light brown or whitish in color?

Short Answer

20. When oil has gasoline mixed with it, the oil pressure has a tendency to ________.

21. Under normal driving conditions, newer vehicles should have the oil changed approximately every ________ miles.

22. When oil sits in the engine for a long period of time, the oil has a tendency to build up ________.

23. Two common problems that cause the oil pressure to drop are a damaged ________ and ________.

24. An oil pressure sending unit can be checked by checking the oil pressure in the main oil gallery with a ________.

25. On turbocharged engines, always make sure to idle the engine before shutting it off so that the turbocharger ________ are not damaged.

Applied Academics

TITLE: Synthetic Oil and High Temperature

ACADEMIC SKILLS AREA: Science

NATEF Automobile Technician Tasks:

The technician can explain the dynamic control properties of a hydraulic system.

CONCEPT:

One problem associated with regular oils used in automobiles is oxidation. As the temperature of the engine increases, generally above 250 degrees Fahrenheit, oils have a tendency to oxidize. As the oil oxidizes, its volume is reduced and there is less oil available for lubrication. A synthetic oil has less of a tendency to oxidize when exposed to high temperatures. This is shown in the accompanying charts. The chart on the left shows that a synthetic oil gets almost twice as many miles per quart as compared to standard or mineral oils. The graph on the right shows why this happens. With the oil held at 300 degrees Fahrenheit, regular oils lose more than 30% of their volume over a 200-hour time period. On the other hand, synthetic oils lose less than 5% over the same time period. As engines operate at higher and higher temperatures, the use of synthetic oils becomes more justified.

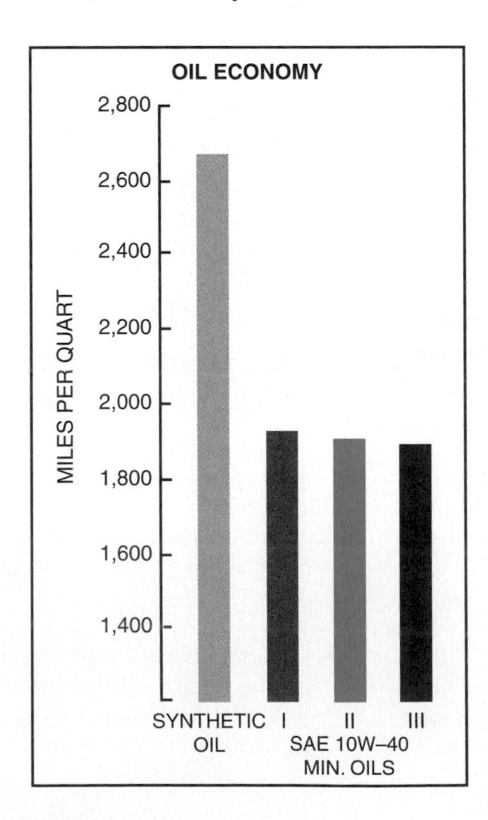

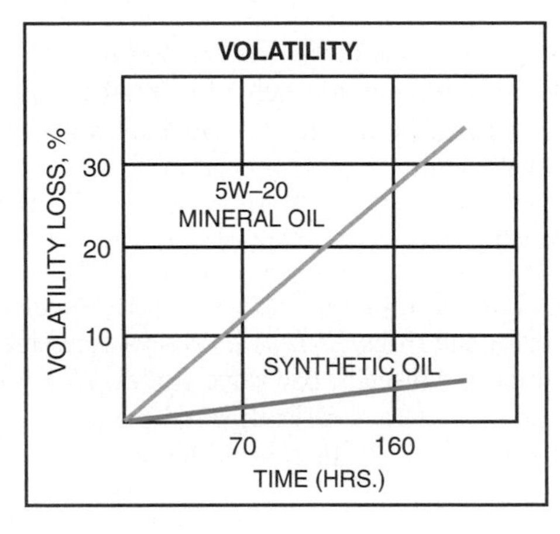

APPLICATION:

Do you think the cost of synthetic oils justifies its use in today's automotive engines? If regular oils lose 30% of their volume over time, how does this affect lubricating quality?

Cooling System Principles and Operation

CHAPTER 22

OBJECTIVES

After studying this chapter, you should be able to:

- Identify the purposes of the cooling system.
- Compare the ways in which heat can be transferred.
- Compare the different types of cooling systems.
- Define the characteristics of coolant and antifreeze.
- Define the proper procedure to dispose of antifreeze.
- Describe the operation of water pumps.
- State the purpose and operation of thermostats and pressure caps.
- Describe the operation of expansion tanks.
- State the purpose and operation of radiators.
- Compare the operation and design of fans, shrouds, and belts.
- Describe the operation of electrical circuits used to control the cooling system.
- Identify various problems, their diagnosis, and service procedures for the cooling system.

INTRODUCTION

The cooling system is one of the more important systems in the automotive engine. If the cooling system is not operating correctly, the engine may be severely damaged. To understand the cooling system, it is important to study coolant characteristics. The various parts of the cooling system, including the water pump, thermostat, radiators, hoses, pressure caps, fans, and temperature indicators, must also be studied.

22.1 PRINCIPLES OF THE COOLING SYSTEM

PURPOSE OF THE COOLING SYSTEM

The purpose of the cooling system is to do three things. The first is to maintain the highest and most efficient operating temperature within the engine. The second is to remove excess heat from the engine. The third is to bring the engine up to operating temperature as quickly as possible. If the

Certification Connection

ASE Connection: The information in this chapter can help you prepare for the National Institute for Automotive Service Excellence (ASE) certification tests. The tests and content areas most closely related to this chapter are:

Test A1—Engine Repair

- **Content Area**—Lubrication and Cooling System Diagnosis and Repair

NATEF Connection: Much of the information in this chapter is related to the NATEF tasks. The NATEF tasks and priority numbers most closely related to this chapter are:

1. Perform cooling system, cap, and recovery system tests (pressure, combustion leakage, and temperature); determine necessary action. P-1

continued

Certification Connection *continued*

2. Inspect, replace, and adjust drive belts, tensioners, and pulleys; check pulley and belt alignment. P-1
3. Inspect and replace engine cooling and heater system hoses. P-1
4. Inspect, test, and replace thermostat and housing. P-2
5. Test coolant; drain and recover coolant; flush and refill cooling system with recommended coolant; bleed air as required. P-1
6. Inspect, test, remove, and replace water pump. P-1
7. Remove and replace radiator. P-2
8. Inspect and test fans(s) (electrical or mechanical), fan clutch, fan shroud, and air dams. P-2
9. Inspect auxiliary oil coolers; determine necessary action. P-3

engine is not at the highest operating temperature, it will not run efficiently. Fuel mileage will decrease, and wear on the engine components will increase.

In heavy-duty driving, an engine could theoretically produce enough heat to melt an average 200-pound engine block in 20 minutes. Even in normal driving conditions, combustion gas temperatures may be as high as 4,500 degrees Fahrenheit. Lubricated parts such as pistons may even run 200 degrees Fahrenheit or more above the boiling point of water (212 degrees Fahrenheit).

HEAT REMOVAL

Within the gasoline or diesel engine, energy from the fuel is converted to power for moving the vehicle. Not all of the energy, however, is converted to power. Referring to

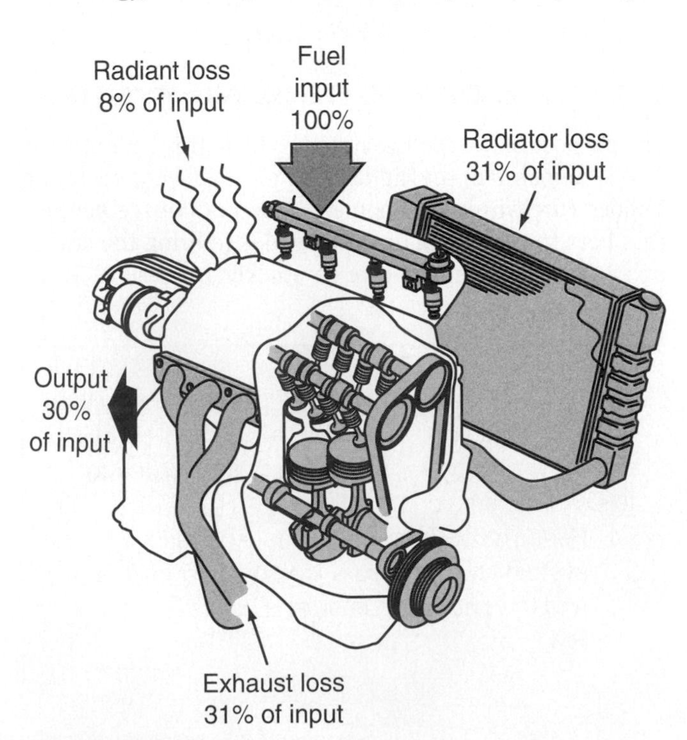

FIGURE 22–1 Approximately 31% of the heat generated by the engine must be removed by the cooling system.

Figure 22–1, of the energy going into the engine, about 30% is used to push the vehicle. About 8% of the heat generated by the fuel is lost through radiation, and 31% is sent out through the exhaust system. The remaining 31% must be removed by the cooling system. If this is done correctly, the temperature of the engine will be at its highest efficiency.

If the engine temperature is too high, various problems will occur. These include the following:

- Overheating of lubricating oil, which will result in the lubricating oil breaking down
- Overheating of the parts, which may cause loss of strength of the metal
- Excessive stresses between engine parts, which may cause increases in friction that cause excessive wear.

If the engine temperature is too low, various problems will occur. These include:

- Poorer fuel mileage. The combustion process will be less efficient.
- Increases in carbon buildup. As the fuel enters the engine, it will condense and cause excessive buildup on the intake valves.
- Increases in varnish and sludges within the lubrication system. Cooler engines enhance the buildup of sludges and varnishes.
- Loss of power. If the combustion process is less efficient, the power output will be reduced.
- Incomplete burning of fuel. This will cause fuel to dilute the oil and cause excess engine wear.

HEAT TRANSFER

The cooling system works on the principles of **heat transfer**. Heat will always travel from a hotter to a cooler object. Heat transfers in three ways: conduction, convection, and radiation. **Conduction** is defined as transfer of heat between two solid objects. For example, in ***Figure 22–2***, heat is transferred from the valve stem to the valve guide. Since both objects are solid, heat is transferred from the hotter valve stem to the cooler valve guide by conduction. Heat is also

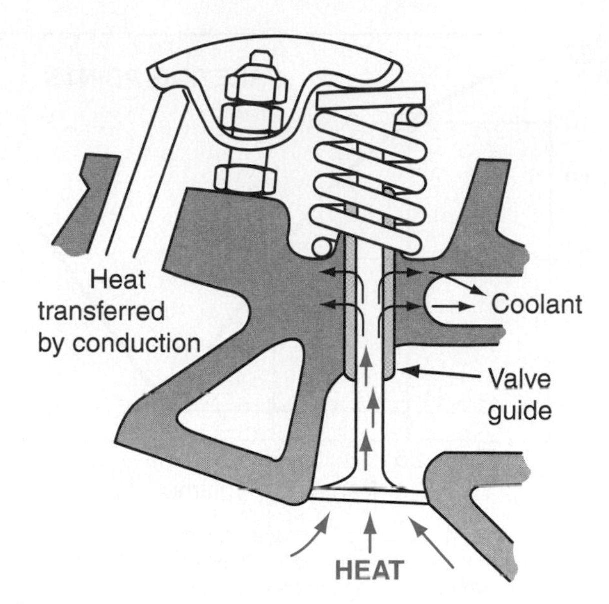

FIGURE 22–2 Heat is transferred by conduction from the valve stem to the valve guide. Both objects are solid.

transferred from the valve guide to the cylinder head by conduction.

Heat can be transferred by **convection**. Convection is defined as the transfer of heat by the circulation of the heated parts of a liquid or gas. The hot cylinder block transfers heat to the coolant by convection. Convection also occurs when the hot radiator parts transfer heat to the cooler air surrounding the radiator.

Radiation is another way that heat is transferred. Radiation is defined as the transfer of heat by converting heat energy to radiant energy. Any hot object will give off radiation. The hotter the object, the greater the amount of radiant energy. When the engine is hot, some of the heat is converted to radiation (about 8% in ***Figure 22–1***). The cooling system relies on these principles to remove the excess heat within an engine.

22.2 TYPES OF COOLING SYSTEMS

Engine manufacturers commonly use two types of cooling systems: the air-cooled system and the liquid-cooled system.

AIR-COOLED ENGINES

Several manufacturers have designed engines that are **air-cooled**. Certain foreign manufacturers still use air-cooled engines. Air-cooled engines have fins or ribs on the outer surfaces of the cylinders and cylinder heads. These fins are cast directly into the cylinders and heads. The fins increase the surface area of the object which, in turn, increases the amount of convection and radiation available for heat transfer. The heat produced by combustion transfers from the internal parts of the engine by conduction to the outer fins. Here the heat is dissipated to the passing air. In some cases, cylinders are separated and are designed to increase air circulation around the cylinders.

Air-cooled engines require air circulation around the cylinder block and heads. Some sort of fan is usually used to move the air across the engine. A shroud is also used in some cases to direct or control the flow of air across the engine. Air-cooled engines usually do not have exact control over engine temperature; however, they do not use a radiator and water pump. This may reduce maintenance on the engine over a long period of time.

LIQUID-COOLED ENGINES

In a **liquid-cooled** engine, the heat from the cylinders is transferred to a liquid flowing through jackets surrounding the cylinders. The liquid then passes through a radiator. Air passing through the radiator removes the heat from the liquid to the air. Liquid-cooling systems usually have better temperature control than air-cooled engines. They are designed to maintain a coolant temperature of 180 to 250 degrees Fahrenheit.

LIQUID COOLANT FLOW

Figure 22–3 shows the parts of a liquid-cooling system. When the vehicle is started, the coolant pump begins circulating the coolant. The coolant goes through the cylinder block from the front to the rear. The coolant circulates around the cylinders as it passes through the cylinder block.

The coolant then passes up into the cylinder head through the holes in the head gasket. From there, it moves forward to the front of the cylinder head through internal passages. These passages permit cooling of high-heat areas such as the spark plug and exhaust valve areas.

As the coolant leaves the cylinder head, it passes to the thermostat. As long as the coolant temperature remains low, the thermostat stays closed. Under these conditions,

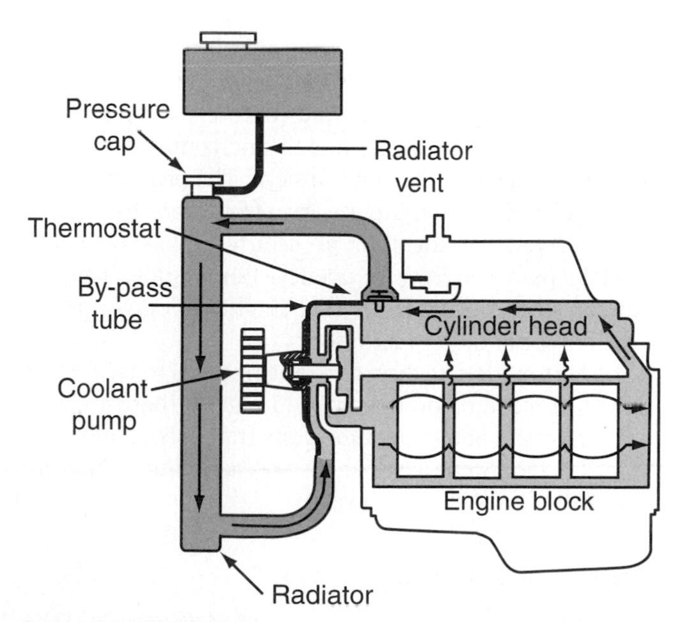

FIGURE 22–3 Coolant is pumped from the water pump, through the cylinder block and head, through the thermostat into the radiator, and back to the water (coolant) pump.

the coolant flows through the **by-pass tube** and returns to the pump for recirculation through the engine. As the coolant heats up, the thermostat gradually opens to allow enough hot coolant to pass through the radiator. This will maintain the engine's highest operating temperature.

From the thermostat, the coolant flows to the internal passages in the radiator. These are tubes in the core with small fins on them. The coolant is cooled by the air passing through the radiator. From there it returns to the outlet of the radiator and back to the pump. It then continues its circulation through the engine.

22.3 COOLANT CHARACTERISTICS

ANTIFREEZE

Water has been the most commonly used engine coolant, because it has good ability to transfer heat and can be readily obtained. Water alone, however, is not suitable for today's engines for a number of reasons. Water has a freezing point of 32 degrees Fahrenheit. Engines must operate in colder climates. Also, water has a boiling point of 212 degrees Fahrenheit. Engine coolant temperature often exceeds this point. In addition, water can be very **corrosive** and produce rust within a coolant system.

To overcome these problems, **antifreeze** is added to the coolant. Ethylene glycol–type antifreeze coolant is the most common type used. When purchased on the market, this antifreeze includes suitable corrosion inhibitors. The best percentage of antifreeze to water to use is about 50% antifreeze mixed with 50% water.

ANTIFREEZE AS A HAZARDOUS WASTE PRODUCT

Antifreeze is considered a **hazardous waste** product. That is, it is harmful to people and to the environment. Therefore, it must be disposed of according to certain procedures. Antifreeze should not be allowed to enter sewer systems. Antifreeze has various heavy metals and contains contaminants that will harm the environment. The procedures and laws for disposing of antifreeze vary from state to state. In order to determine the exact procedure in your state, call the state pollution control agency. Some states have an office of waste management that might also give specific directions for antifreeze disposal.

The best method to dispose of antifreeze is to recycle the product. Depending on the city and region, there are various private and public organizations that collect and recycle antifreeze. Some companies are recycling significant quantities of antifreeze each year.

FREEZING POINTS

Figure 22–4 shows what happens to the freezing point of a coolant when different percentages of antifreeze are used. For example, when 100% water is used, the freezing point is 32 degrees Fahrenheit. When 25% antifreeze and 75% water are used, the freezing point of the coolant is about 10 degrees Fahrenheit. When the ratio is 50–50, the freezing point is about –33 degrees Fahrenheit. At 68% antifreeze, the freezing point of the coolant is about –92 degrees Fahrenheit. As the amount of antifreeze percentage increases from this point, the freezing point goes back toward 0 degrees Fahrenheit.

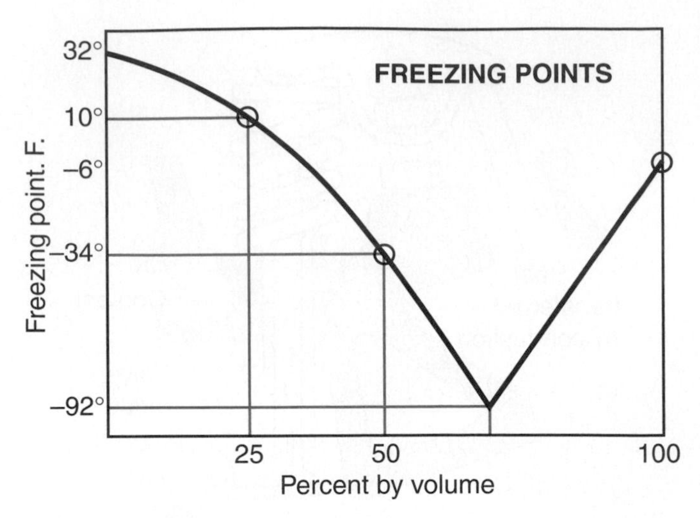

FIGURE 22–4 As the percentage of antifreeze increases, the freezing point of the solution is lowered. This is true up to the point at which there is 68% antifreeze and 32% water. Beyond this point, the freezing temperature starts to increase.

BOILING POINTS

The addition of antifreeze in the cooling system increases the boiling point. The **boiling point** of a fluid is the temperature at which a liquid becomes a vapor. Any coolant that becomes a vapor has very poor conduction and convection properties; therefore, it is necessary to protect the coolant from boiling. This protection provides a greater margin of safety against engine cooling system overheating failure.

Figure 22–5 shows how boiling points increase with increases in percentages of antifreeze. For example, when

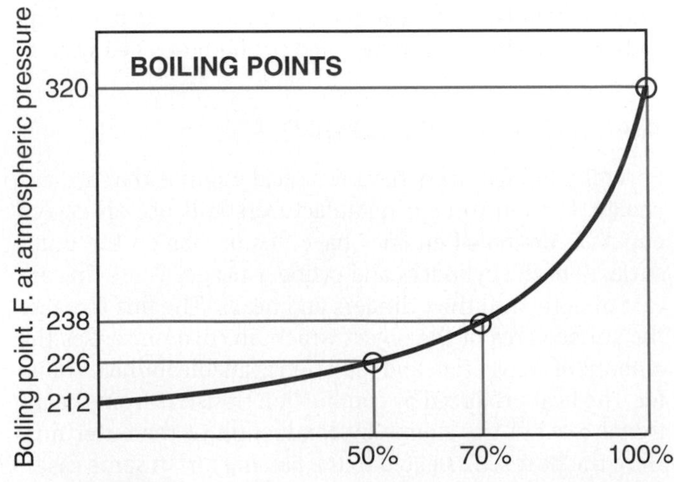

FIGURE 22–5 As antifreeze is added to the coolant, its boiling point rises.

there is no antifreeze, the boiling point is at 212 degrees Fahrenheit. When there is 50% antifreeze in the coolant, the boiling point increases to 226 degrees Fahrenheit. If there is 70% antifreeze, the boiling point increases to about 238 degrees Fahrenheit. At 100% antifreeze, the boiling point is 320 degrees Fahrenheit. It can be seen that antifreeze protects the coolant during both summer and winter operation.

CORROSION

Corrosion in the cooling system can be very damaging to the engine. Corrosion can be produced in several ways. Direct attack occurs when water in the coolant mixes with oxygen from the air. This process can produce rust particles, which damage water pump seals and cause increased leakage. Electromechanical attack is a result of using different metals in an engine. In the presence of the coolant, different metals may set up an electrical current in the coolant. If this occurs, one metal may deteriorate and deposit itself on the other metal. For example, a core plug may deteriorate to a point of causing leakage. **Cavitation** is a high shock pressure developed by collapsing vapor bubbles in the coolant. These bubbles are produced by the rapid spinning of the water pump impeller. The shock waves erode nearby metal surfaces such as the pump impeller.

Mineral deposits such as calcium and silicate deposits are produced when hard water is used in the cooling system. Both deposits restrict the conduction of heat out of the cooling system (***Figure 22–6***). For example, if there were a 1/16-inch mineral deposit on a 1-inch thick piece of cast iron, the heat transfer ability would be the same as 4 1/4 inches of cast iron. This means that when mineral deposits build up inside the engine cooling system, it becomes very difficult to transfer the excess heat out. The deposits cover the internal passages of the cooling system, also causing uneven heat transfer out of the engine (***Figure 22–7***). When an ethylene glycol antifreeze solution is added to the coolant, many of these corrosion problems are eliminated. Chemicals are added to the antifreeze to reduce corrosion. It is usually not necessary to add any corrosion inhibitor to the cooling system. In some cases, mixing different corrosion inhibitors produces unwanted sludges within the cooling system.

Car Clinic: ANTIFREEZE

CUSTOMER CONCERN:
Rust in the Cooling System

An engine has rust in the antifreeze. The antifreeze looks very brown, and water pump bearings keep going out. Since this is an older vehicle, there is no overflow tank. What could be the problem?

SOLUTION: Rust in the cooling system and antifreeze is normally caused by air getting into the radiator when the engine cools. As the coolant cools, it shrinks. An overflow tank normally has sufficient coolant in it to keep the radiator full even at cold temperatures. Without an overflow tank, air will enter the cooling system through the radiator cap. The rust in the coolant also causes wear on the water pump seal and the bearings. The best solution is to purchase an overflow tank and install it on the cooling system. Keep the overflow tank filled to the mark shown on the tank.

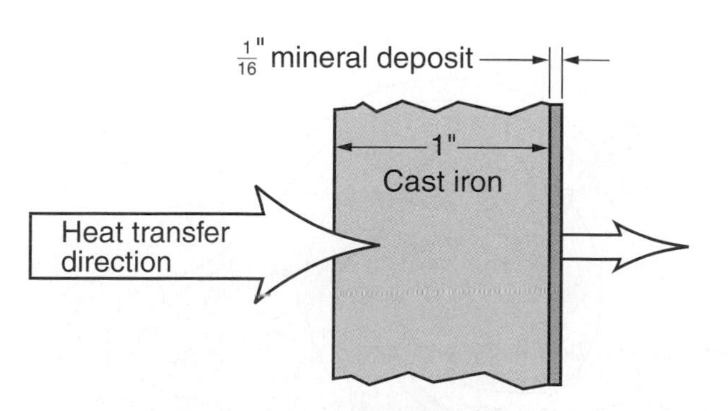

FIGURE 22-6 Mineral deposits inside the cooling system restrict the conduction of heat.

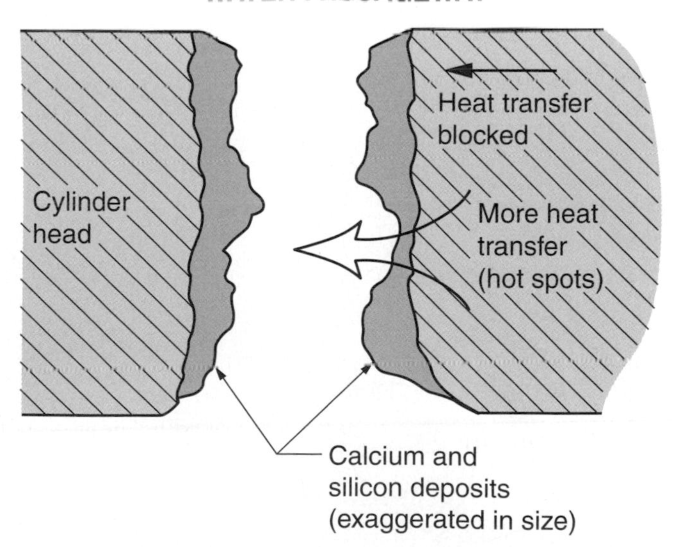

FIGURE 22-7 When minerals such as calcium and silicon build up inside the cooling system, certain areas may not transfer heat while other areas may transfer heat more rapidly. This can lead to hot spots in the engine.

COOLING SYSTEM LEAKS

Because a full cooling system is essential, leakage ranks very high on the list of cooling system problems. There are numerous sources of leaks. Some leaks can be corrected by tightening hose clamps, but a large percentage of leaks are from small pinholes. These are commonly found in the radiator, freeze plugs, heater core, or around gaskets. To eliminate leaks, certain antifreezes have stop-leak protection. These products are designed to seal the common pinhole leaks. This will prevent inconvenient breakdown and costly repair bills.

22.4 COOLING SYSTEM PARTS AND OPERATION

Various parts are used to operate the cooling system correctly. These include the water pump, water jackets, thermostats, radiators, transmission coolers, pressure caps, expansion tanks, fans, shrouds, belts, and temperature indicators.

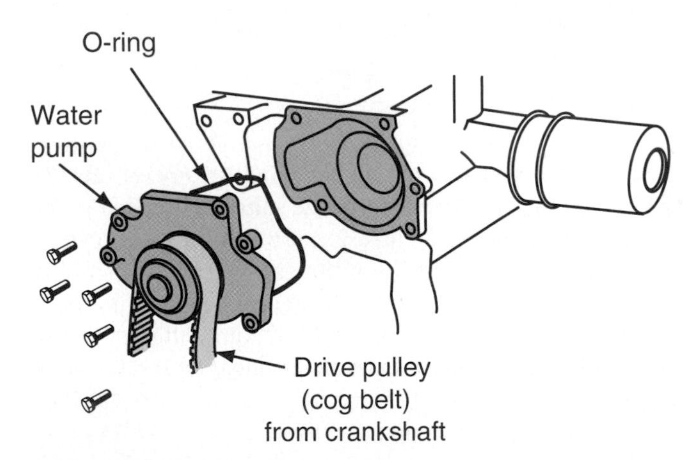

FIGURE 22-8 A cog belt driven off the crankshaft drives this water pump.

WATER PUMP

The purpose of the water pump is to circulate the water through the cooling system. The pump is located on the front of the engine. In most vehicles, it is driven by a serpentine or cog belt that is attached to the crankshaft. As the crankshaft turns, the belt turns the pump, causing coolant to be circulated (***Figure 22–8***).

The coolant pump is called a **centrifugal pump**. This means that as coolant is drawn into the center of the pump, centrifugal forces pressurize water and send it out into the cooling system (***Figure 22–9***). Centrifugal forces throw the coolant outward from the impeller tips. This type of pump will pump only the coolant that is required by the system. Therefore, it does not require a pressure regulator or relief valve.

The pump consists of a housing, a bearing on a shaft, the impeller, and seal. The housing has a coolant inlet and outlet. The seal is used to keep the water inside the pump. The bearing is used to support the shaft on which the impeller rides (***Figure 22–10***). This assembly is bolted to the cylinder block. A gasket is used to keep the coolant from leaking. On some engines, there is also a hub where a fan and pulley can be connected to the shaft. Other engines have the fan mounted directly on the radiator. In this case, the fan is operated electrically. Only the drive belt pulley is connected to the water pump shaft.

There is a small vent hole on the bottom of the water pump housing. This vent hole allows coolant to escape if the seal leaks. Today it is easier to simply replace a damaged water pump with a rebuilt or new pump than to replace the seals inside the water pump. The old pump can then be returned to get a "core charge" so it can be rebuilt by a manufacturer.

PARTS LOCATOR

Refer to photo #13 in the front of the textbook to see what a ***water pump*** *looks like.*

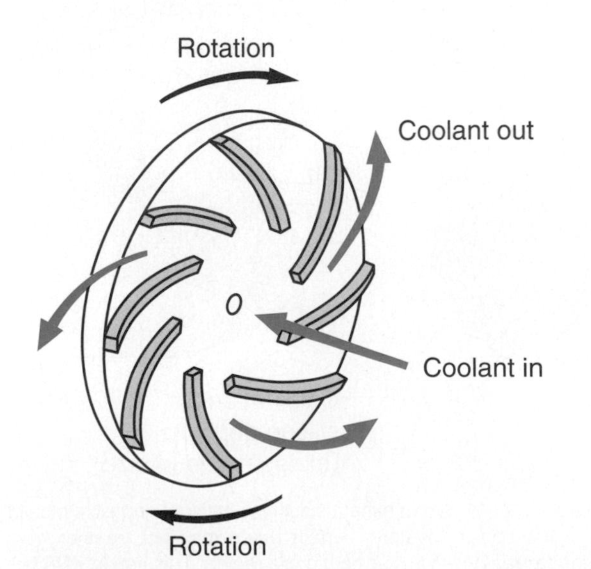

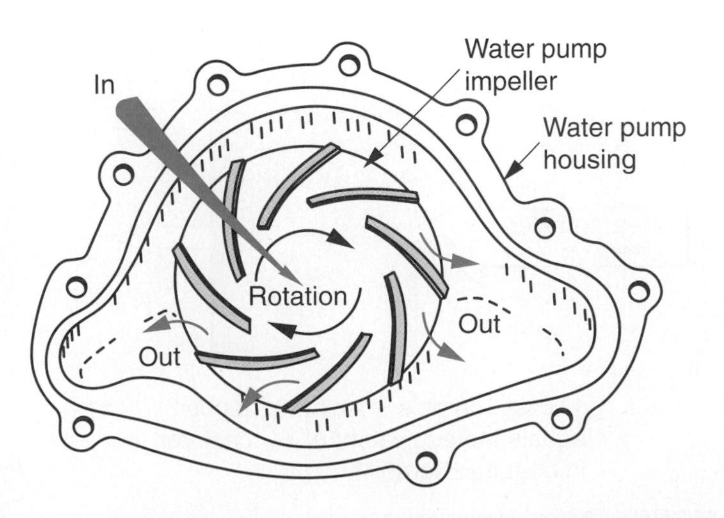

FIGURE 22-9 A centrifugal pump draws the fluid into the center of the vanes, and the centrifugal forces throw the fluid outward.

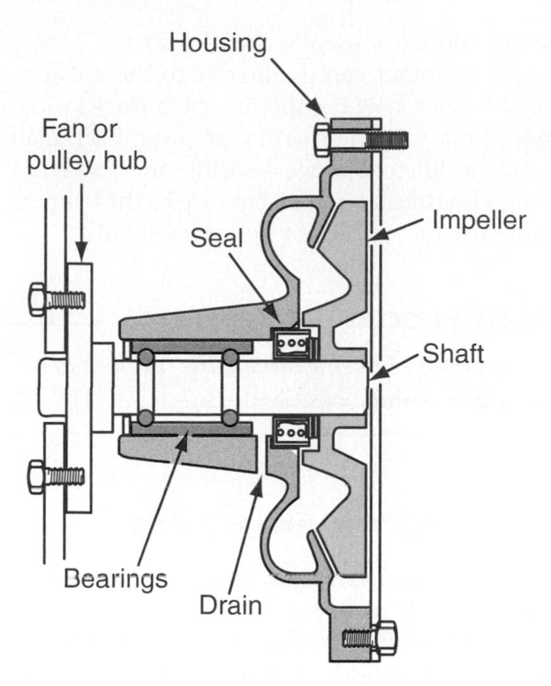

FIGURE 22–10 The water pump consists of a housing, bearings, shaft, and seals.

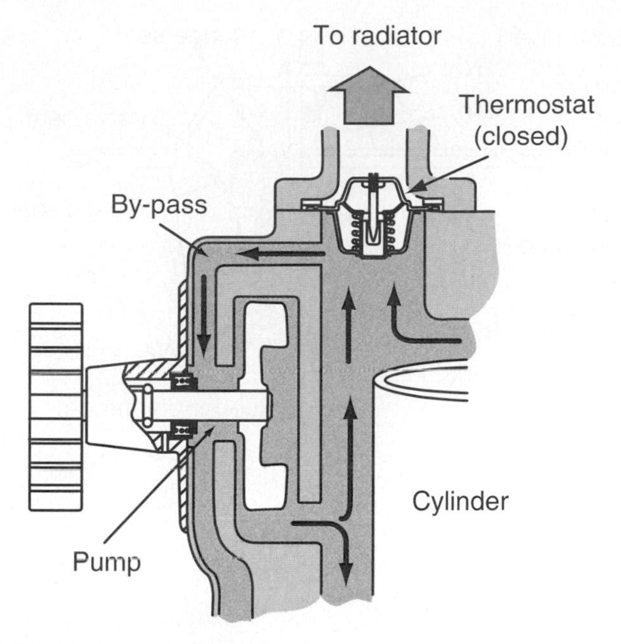

FIGURE 22–12 The thermostat is used to open a passageway to the radiator or to send the coolant through a by-pass tube.

WATER JACKETS

Water jackets are defined as the open spaces within the cylinder block and cylinder head where coolant flows. These water jackets are designed to allow coolant to flow to the right spots so that maximum cooling can be obtained around the cylinders. ***Figure 22–11*** shows internal water jackets used on a cut-a-way view of an engine.

THERMOSTATS

The thermostat is one of the most important parts of the cooling system. The purpose of the **thermostat** is to keep the engine coolant at the most efficient temperature. The thermostat is used to bring the coolant temperature up to operating temperature as quickly as possible. It is designed to sense the temperature of the coolant. If the coolant remains cold, the thermostat will be closed (***Figure 22–12***). The coolant then goes to the by-pass tube. When the thermostat is closed, a small amount of coolant flows into the radiator to be cooled. The remaining coolant flows through the by-pass tube. This coolant is recirculated without being cooled. If the engine is under a heavier load, more cooling is necessary. When the temperature of the coolant increases to the thermostat opening temperature, the thermostat opens slightly. As the temperature of the coolant increases further, the thermostat opens more. This allows more coolant to reduce its temperature through the radiator. When the engine is under full load, the thermostat is fully open. The maximum amount of coolant is sent to the radiator for cooling, and a small amount of coolant continues to flow through the by-pass tube.

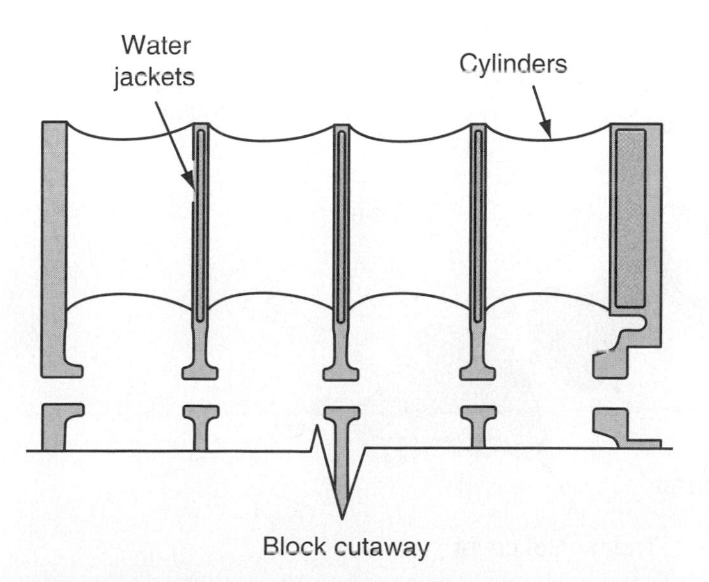

FIGURE 22–11 Water jackets are used to allow coolant to flow around the cylinders within the engine.

Thermostats operate on a very simple principle. A wax pellet material within the center of the thermostat expands and causes the mechanical motion that opens the thermostat (***Figure 22–13***). This allows coolant to pass through to the radiator. It should be noted that the thermostat is opened only partially when the temperature reaches its opening point. As the coolant temperature increases, the thermostat opens further. Eventually, the coolant is hot enough to cause the thermostat to open fully to get maximum cooling.

Thermostats are designed to open at different temperatures. Common thermostat temperatures are 180, 195, and 210 degrees Fahrenheit. Always follow the manufacturer's recommendations for determining the correct thermostat temperature.

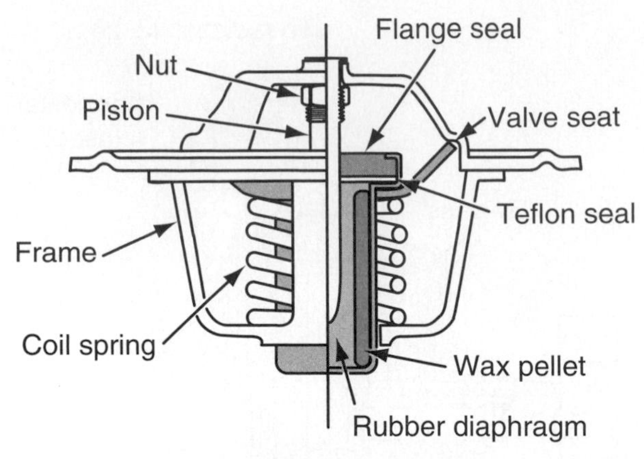

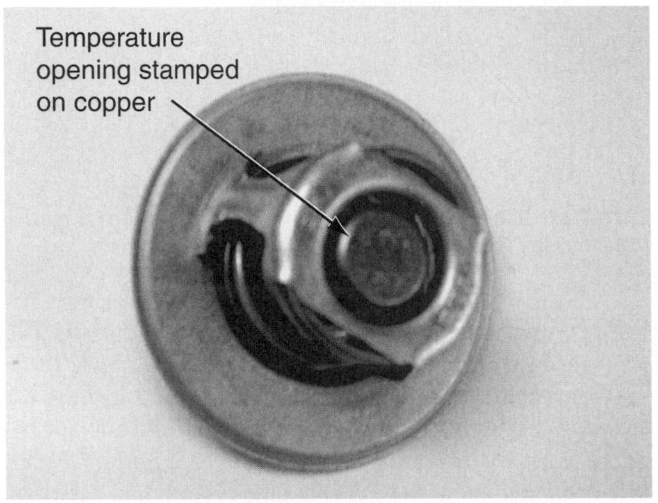

FIGURE 22-13 As the temperature of the wax pellets increases, the coil spring and valve lift so that coolant can pass through to the radiator.

The thermostat is usually placed near the front of the engine so its output can go directly to the radiator. ***Figure 22–14*** shows how the thermostat is placed within the thermostat housing and thermostat cover. Each manufacturer uses a different shape housing and gaskets. When replacing a thermostat, remember to keep the temperature-sensitive valve toward the hot side of the engine.

PARTS LOCATOR

*Refer to photo #14 in the front of the textbook to see what a **thermostat** looks like.*

TWO-STAGE THERMOSTATS

Some thermostats have two stages of operation. ***Figure 22–15*** shows an example of a two-stage thermostat. Notice the small subvalve located on the outer ring of the thermostat. This subvalve begins to open at 175 degrees Fahrenheit. This action allows a small amount of coolant to pass through the thermostat and to the radiator to be cooled. Then, when the coolant reaches approximately 195 degrees Fahrenheit, the main valve opens to allow more coolant to be sent to the radiator. At 212 degrees Fahrenheit both the subvalve and the main valve are fully open. This type of thermostat allows for more precise and even control of engine coolant temperature. For example, under certain engine conditions, coolant temperatures often have peak hot periods. During this time, the coolant increases in temperature rapidly. This thermostat has a tendency to reduce

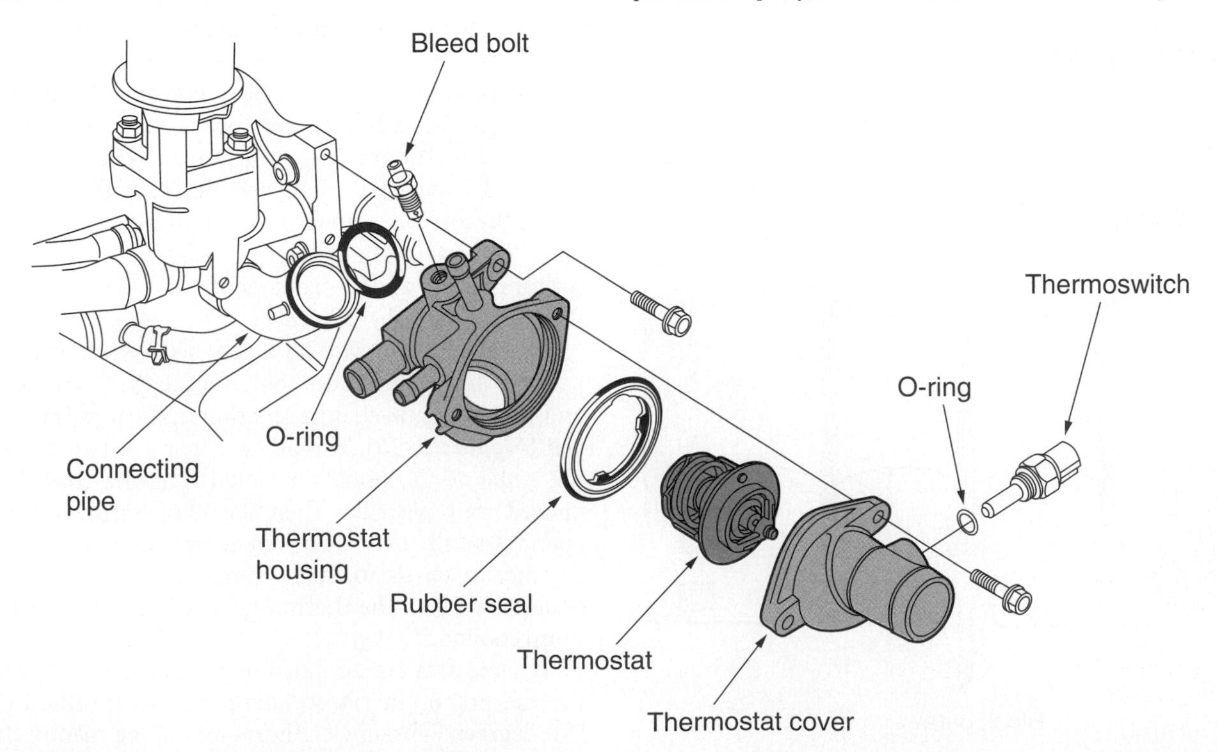

FIGURE 22-14 The thermostat is held in place between the thermostat housing and the thermostat cover. (Courtesy of American Honda Motor, Co., Inc.)

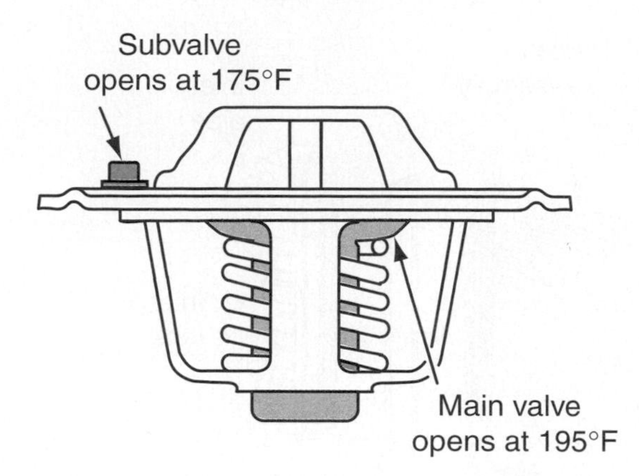

FIGURE 22-15 This two-stage thermostat controls coolant temperature more precisely than other thermostats.

these peaking temperatures and even out the coolant temperatures.

RADIATORS

The purpose of the radiator is to allow fresh air to reduce the temperature of the coolant. This is done by flowing the coolant through tubes. As the coolant passes through the tubes, air is forced around the tubes. This causes a transfer of heat from the hot coolant to the cooler air. This process is called heat exchange. In this case, heat is exchanged from a liquid coolant to the air. This is called a liquid-to-air **heat exchanger**.

Figure 22–16 shows an internal diagram of a typical radiator. Note that the coolant flows through the tubes, and air flows through the air fins. In ***Figure 22–16A*** notice that there are two sets of coolant passages. Each set of coolant passages is called a core. This is a two-core radiator. In ***Figure 22–16B*** there are five sets of coolant passages. This is a five-core radiator. The number of cores on a radiator depends on the size and load of the engine. In larger engines that are used for heavy loads, a four- or five-core radiator may be used to make sure the engine is properly cooled. In smaller engines intended for smaller loads, one- or two-core radiators may be adequate. Always be careful not to overload an engine with a one- or two-core radiator. The amount of heat that can be removed from this type of radiator might not be enough to cool the engine under heavy loads such as pulling a boat trailer.

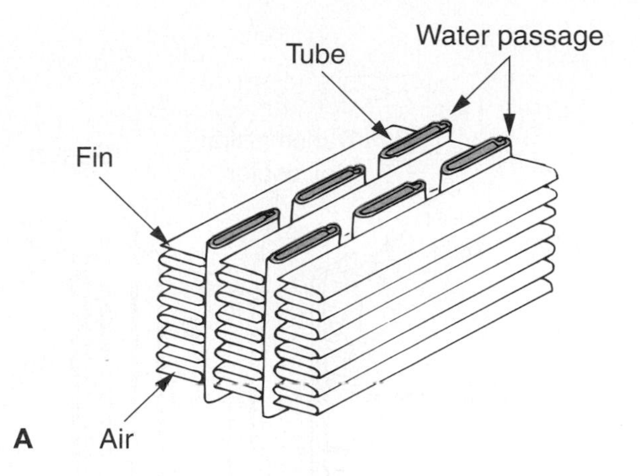

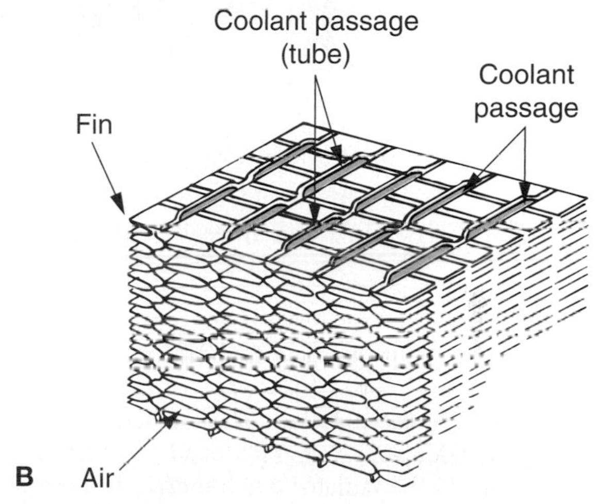

FIGURE 22-16 Inside the radiator, there are small tubes through which the coolant flows. The air fins help to remove the heat to the surrounding air. (Courtesy of SAE, reprinted with permission from 1983 SAE Handbook, © 1983 SAE International)

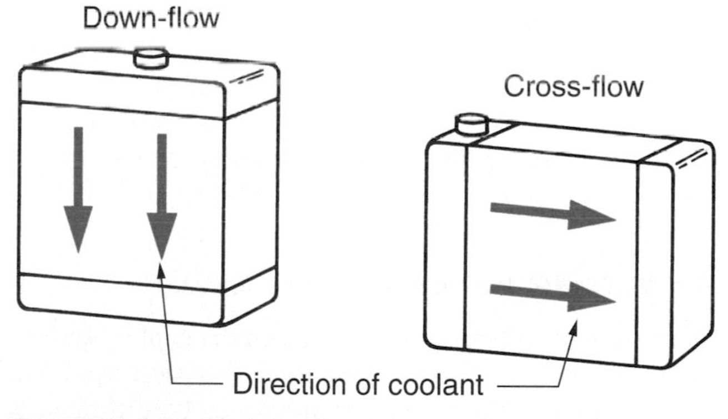

FIGURE 22-17 These illustrations show the direction of flow in down-flow and cross-flow radiators.

DOWN-FLOW AND CROSS-FLOW RADIATORS

Two types of radiators are commonly used in the automobile: the down-flow radiator and the cross-flow radiator. In the down-flow radiator, coolant flows from the top of the radiator to the bottom. In the cross-flow radiator, the coolant flows from one side of the radiator to the other side. ***Figure 22–17*** shows both a down-flow radiator and a cross-flow radiator. The down-flow radiator is found on most older vehicles. Because newer vehicles are lower in front, the cross-flow radiator has been used on most vehicles manufactured since 1970. Some American and foreign manufacturers, however, still use the down-flow radiator.

RADIATOR PARTS

Radiators are made of several parts. The tubes mentioned earlier are called the **core**. The core is made of the center

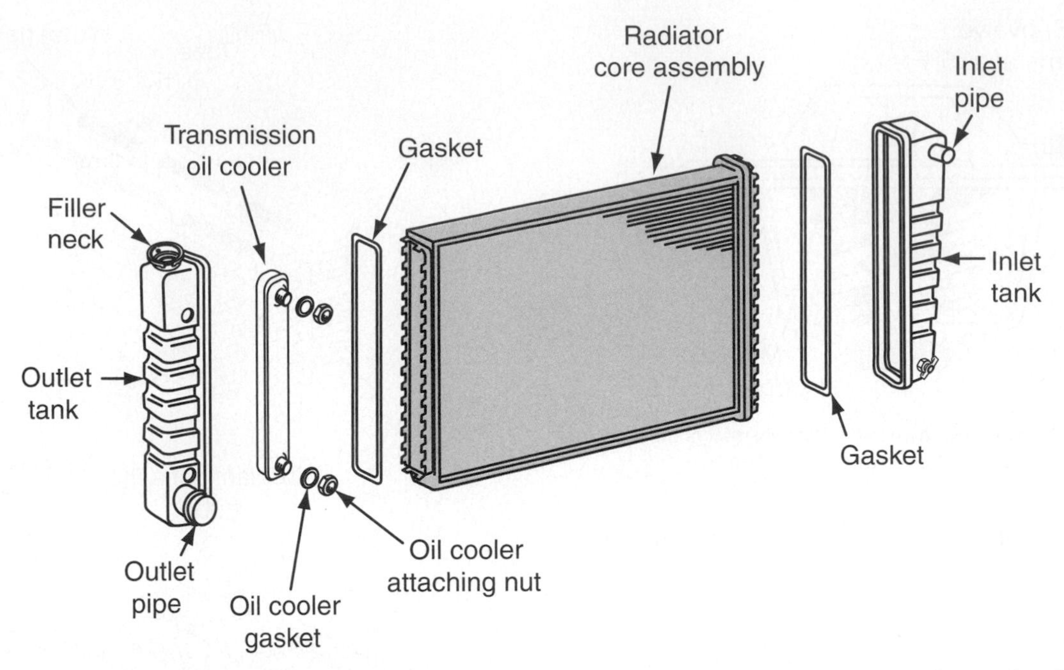

FIGURE 22-18 The parts of a standard radiator are shown in this illustration. (Courtesy of General Motors Corporation, Service Technology Group)

coolant tubes and the fins that surround the tubes. The core on the radiator is made of brass, copper, or aluminum or some combination of those metals. These metals are used because they transfer heat very rapidly to the fins. This characteristic makes the radiator much more efficient. Depending on the size of the engine and the cooling requirements, one, two, three, four, or five cores can be used. There are also inlet and outlet tanks. They can be made of metal or plastic. These tanks hold the coolant before it goes into the radiator or into the block. The inlet tank has a hose connection to allow coolant to flow from the engine into the radiator. The outlet tank has a hose connection to allow coolant to pass back to the engine. In addition, there is a filler neck attached to one of the tanks. The radiator pressure cap is placed here. ***Figure 22–18*** shows the common parts of the radiator.

RADIATOR HOSES AND CLAMPS

There are several types of hoses used on the cooling system. ***Figure 22–19*** shows some of the more common types. The most common is the molded or preformed type of radiator hose. There are many shapes and sizes of this type of hose, which is used for the upper and lower radiator connections. Each manufacturer has a specific shape of hose for each style of engine and vehicle. The hoses are made of impregnated butyl or neoprene rubber and fiber. Some hoses have metal reinforcement. The flexible hose type allows the hose to be bent slightly to fit a particular application. When using this type of hose, make sure there are no kinks or severe bends that could block the coolant flow. Some hoses also have a wire spring placed inside the hose. This prevents the top hose from collapsing when the engine cools down and there is a defective radiator cap. Some lower radiator hoses also have a spring inside. This is to stop them from collapsing during operation from the pull or suction of the water pump during high speeds or loads. ***Figure 22–20*** shows how the radiator hose is connected using different types of clamps, depending on the automobile manufacturer.

Cooling system hoses have a tendency to wear out. With extended use, hoses often become very soft. This is an indication that the hose is deteriorating. At this point, the hose

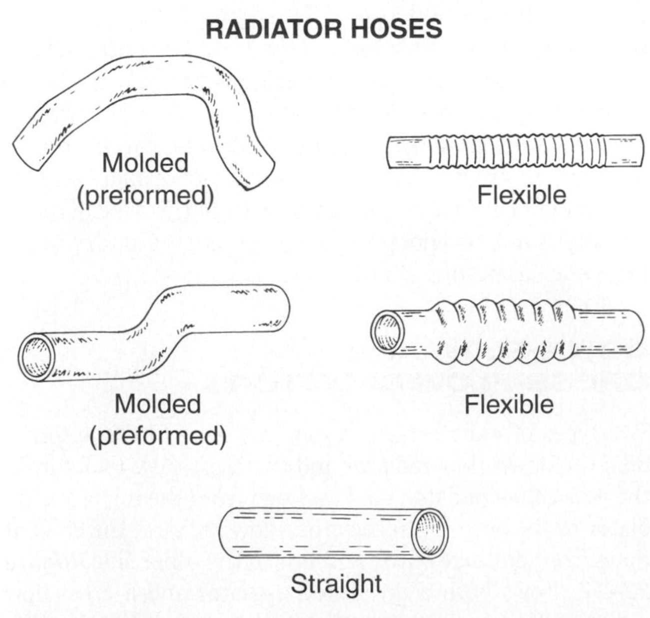

FIGURE 22-19 Many types of radiator hoses are used on vehicles today.

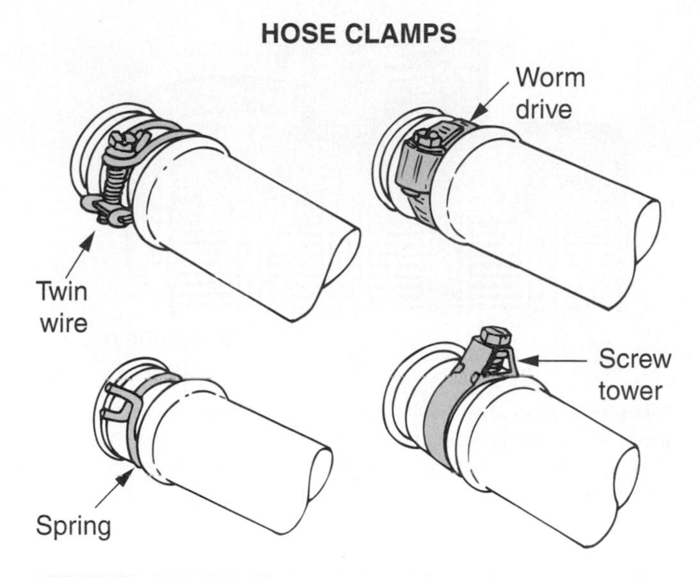

FIGURE 22-20 Various types of hose clamps are used on cooling system hoses.

should be replaced and all other hoses should be checked for soft spots, small leaks, and other signs of wear and replaced if necessary.

TRANSMISSION COOLERS

Vehicles that have automatic transmissions must have some means of cooling the transmission fluid. If the transmission fluid gets too hot, the transmission may be severely damaged. Transmission fluid is cooled by passing the fluid out of the transmission, into a tube inside the radiator outlet tank, then back to the transmission. The liquid in the radiator is cool enough to lower the temperature of the transmission fluid. This heat transfer is done by using a liquid- (transmission fluid) to-liquid (engine coolant) heat exchanger (***Figure 22–21***).

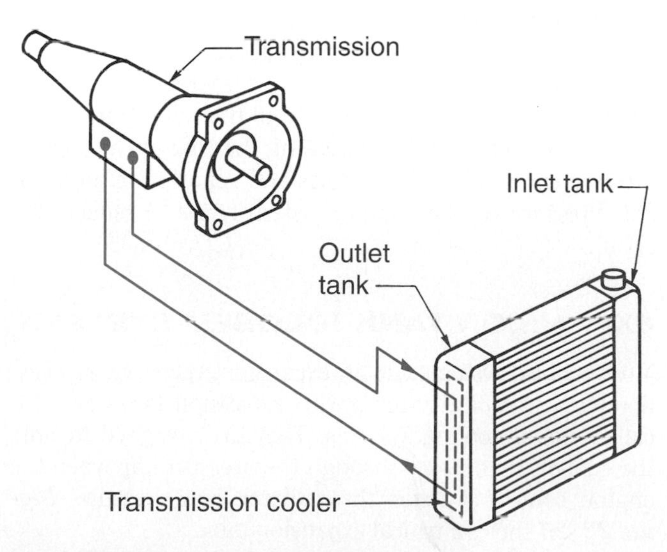

FIGURE 22-21 Fluid from the automatic transmission is sent through metal tubes to the radiator to be cooled.

PRESSURE CAPS

Pressure caps are placed on the radiator to do several things. They are designed to:

- Increase the pressure on the cooling system
- Reduce cavitation
- Protect the radiator hoses
- Prevent or reduce surging

It is very important to maintain constant **pressure** in the cooling system. The pressure should be approximately 14 to 17 pounds per square inch (psi). Pressure caps are placed on the radiator to maintain the correct pressure on the cooling system.

Pressure on the cooling system changes the boiling point. As pressure is increased, the boiling point of the coolant also increases. This is shown in ***Figure 22–22***. The bottom axis shows pressure. The left vertical axis shows the boiling point in degrees Fahrenheit. Different solutions of antifreeze are also shown. For example, using pure water, the boiling point at 0 **psig** (pounds per square inch, on a gauge) is 212 degrees Fahrenheit. If the pressure is increased to 15 psig, the boiling point increases to about 250 degrees Fahrenheit. If a pressure cap keeps the coolant at 15 psig with 50% antifreeze and 50% water, the boiling point increases to 266 degrees Fahrenheit.

BOILING POINTS OF AQUEOUS
"PRESTONE II" WINTER-SUMMER CONCENTRATE SOLUTIONS
AT VARIOUS PRESSURES

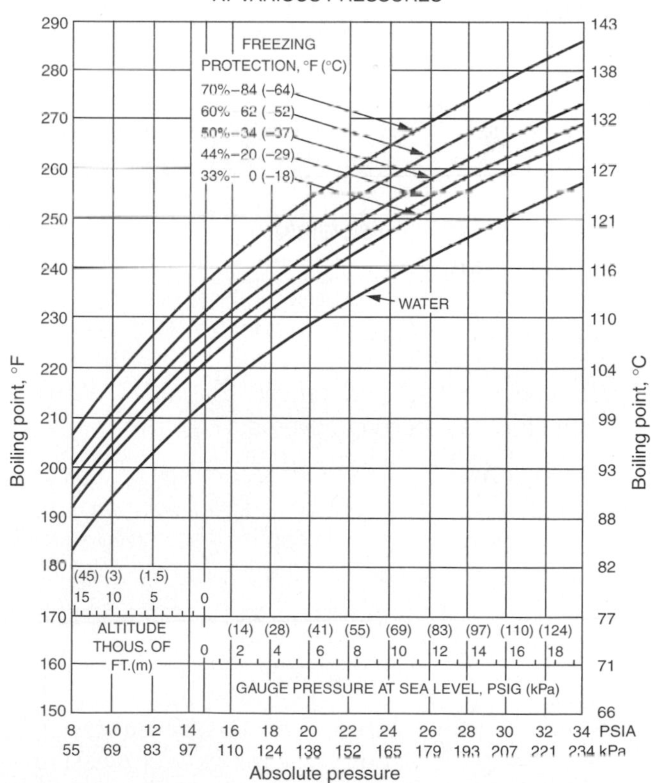

FIGURE 22-22 As pressure and the amount of antifreeze are increased, the boiling point of the coolant increases.

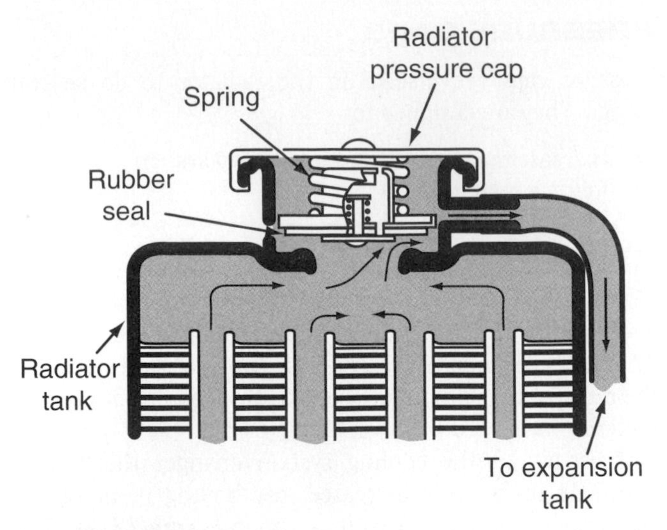

FIGURE 22–23 As the pressure increases in the radiator, the spring compresses and allows coolant to flow into the overflow tank.

Figure 22–23 shows how a pressure cap maintains the constant pressure. As the coolant increases in temperature, it begins to expand. As it expands, the coolant cannot escape. The spring holds a rubber seal against the filler neck. This keeps the fluid in the cooling system and increases the pressure. When the pressure reaches 17 psig, the rubber seal is lifted off the filler neck against spring pressure. The coolant then passes through the pressure cap to a tube that is connected to a recovery or overflow tank. This type of system is called a closed system. An open system allows the coolant to pass through the pressure cap directly to the road surface.

Car Clinic: RADIATOR PRESSURE CAP

CUSTOMER CONCERN:
Engine Overheats

A car has recently been overheating on the highway. The problem started suddenly. One person says the problem is the thermostat. But the thermostat was checked and proved to be OK. What might cause an engine to overheat?

SOLUTION: One of the more common causes of an engine overheating is a bad radiator pressure cap. The radiator pressure cap is designed to keep 15 pounds of pressure on the cooling system. As the pressure of the cooling system increases, the boiling point also increases. If the pressure on the cooling system has been reduced (bad radiator pressure cap), the boiling point is also reduced. Using a pressure cap and cooling system tester, check the cap. It sounds as if the radiator pressure cap cannot hold pressure on the cooling system.

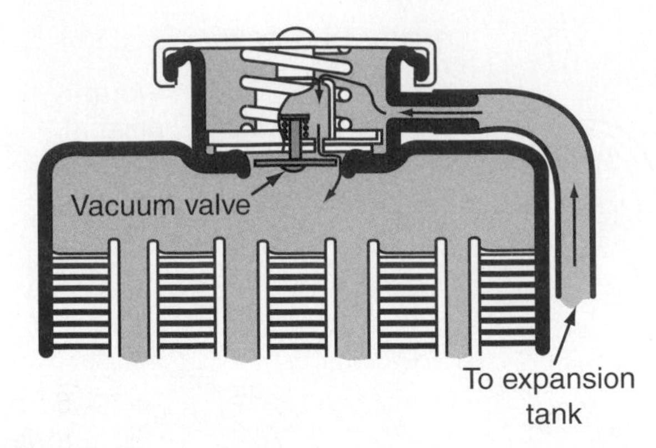

FIGURE 22–24 As the coolant cools down, its volume is reduced, which creates a vacuum in the radiator. The vacuum opens the small vacuum valve and lets coolant from the overflow tank back into the radiator.

The pressure cap also protects the hoses from expanding and collapsing. When the engine is shut down, the coolant starts to cool. As it cools, the coolant volume reduces. Eventually, a **vacuum** is created in the cooling system. This means that the pressure outside the radiator is greater than the pressure inside the radiator. This difference in pressure causes the hoses to collapse. Continued expanding and collapsing of the hoses cause them to crack and eventually leak. The pressure cap has a vacuum valve that allows atmospheric pressure to seep into the cooling system when there is a slight vacuum.

During operation, a small spring holds the vacuum valve closed. When there is a vacuum inside the cooling system, the vacuum valve is pulled down and opened as shown in *Figure 22–24*. The vacuum is then reduced within the cooling system. When the vacuum valve opens, the vacuum also draws in coolant from the overflow tank. So, as the coolant in the engine continues to cool down and its volume continues to reduce, additional coolant is brought in rather than air.

Increasing the pressure also reduces cavitation. Cavitation is the production of small vacuum bubbles by the water pump action. Increased pressure reduces this action.

Pressure on the cooling system also reduces **surging**, which is a sudden rush of water from the water pump. This can be caused by rapidly increasing the rpm of the engine. Surging can produce air bubbles and agitation of the coolant. Pressure on the cooling system tends to reduce this action.

EXPANSION TANK (CLOSED SYSTEM)

Most cooling systems use an expansion, recovery, or overflow tank. Cooling systems with expansion tanks are also called closed cooling systems. They are designed to hold the coolant that passes through the pressure cap when the engine coolant reaches the radiator cup pressure. *Figure 22–25* shows a typical expansion tank.

As the engine warms up, the coolant expands. This expansion eventually causes the pressure cap to release. The

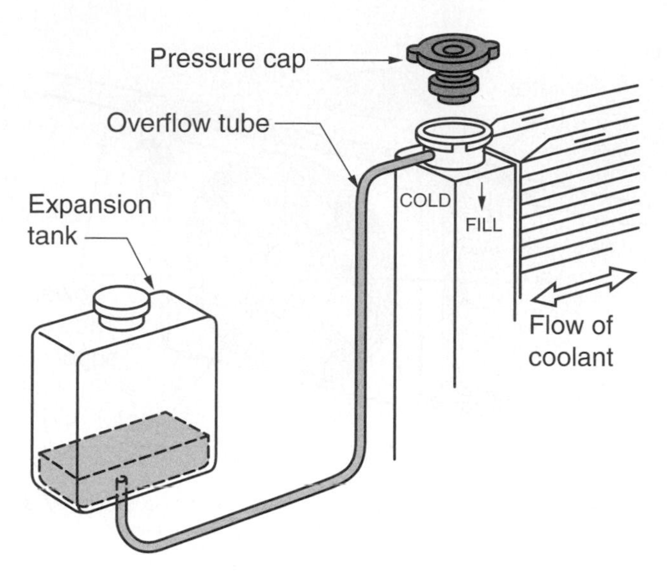

FIGURE 22-25 The expansion tank holds coolant that comes out of the radiator.

coolant that might be passed to the atmosphere is now sent to an expansion tank. When the engine is shut down, the coolant begins to reduce in volume. Eventually, the vacuum spring inside the pressure cap opens. When this happens, the coolant from the expansion tank is drawn back into the cooling system. (On open systems, those without an expansion tank, air is drawn into the cooling system.) The major advantage of using the closed system is that air never gets into the cooling system. Air in the cooling system can cause rust and corrosion, which can damage the cooling system components.

When additional coolant is needed, it is added to the expansion tank. There is a small plastic cap on the expansion tank that can be removed (***Figure 22–26***). Coolant can be added when the engine is hot because the pressure cap does not have to be removed to add the coolant.

FIGURE 22-26 Coolant can be added directly to the expansion tank.

> **PARTS LOCATOR**
>
> *Refer to photo #15 in the front of the textbook to see what an **expansion tank** looks like.*

FAN DESIGNS

The purpose of the fan is to draw air through the radiator for cooling during low-speed and idle operation. A fan is not needed at faster speeds because air is pushed through the radiator by the vehicle speed. On older vehicles, the fan was driven on the same shaft as the water pump. A belt was used to turn the water pump and fan. Because the fan produces frictional horsepower losses, it has been designed to operate only at certain times on newer vehicles, using an electric motor. It is estimated that the fan on the cooling system can absorb up to 6% of the engine horsepower.

FAN BLADES

There are usually four to six blades on the fan. The fan blades are usually spaced unevenly (***Figure 22–27***). This is done to reduce vibration and fan noise. If the blades are evenly spaced, they may produce fan noise and vibration that can annoy the vehicle operator. The fan blades slapping the air actually make the noise. If the blades are spaced irregularly, the noise is broken up and reduced.

VARIABLE-SPEED FANS

The fan does not need to be turned when the engine is cold or running at high speeds. Variable-speed fans are used to control the time when the fan should turn. ***Figure 22–28*** shows a typical variable-speed fan clutch. As air passes through the radiator, its temperature is sensed on the front of the fan clutch. During cold engine operation, the fan is

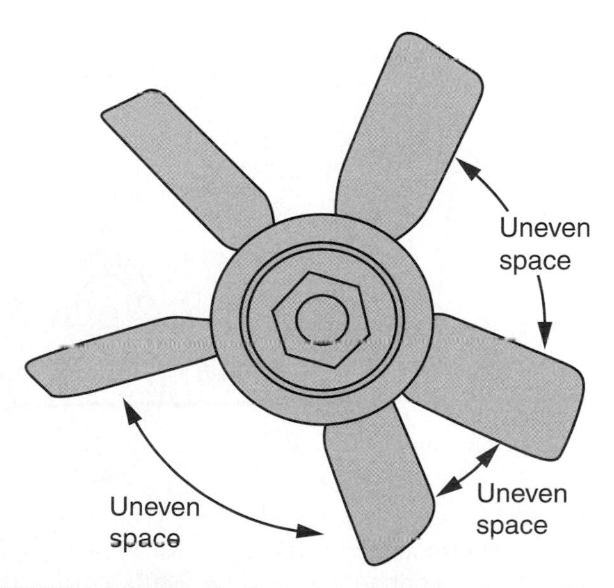

FIGURE 22-27 Fan blades are spaced unevenly to reduce vibration and fan noise.

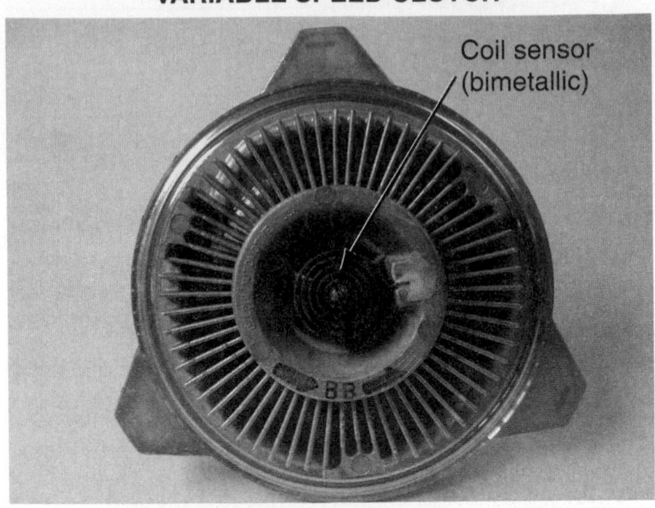

FIGURE 22-28 A variable-speed fan senses the temperature of the air coming through the radiator. When hot air comes through, the inside clutch engages and causes the fan to turn.

allowed to slip. When the temperature increases to the correct point, a **bimetallic strip** (the thermostatic coil) on the front of the clutch expands. This expansion causes a shaft or valve inside the clutch to turn slightly. As the shaft turns, it opens a hole to allow silicon oil to enter a fluid coupling. This oil causes the fan to lock up against a clutch and start turning. There are several variations of this design; however, the result is the same. ***Figure 22–29*** shows the internal parts of a typical variable-speed fan.

ELECTRIC FANS

Another method used to turn the fan involves a small motor (***Figure 22–30***). In this system, the fan is turned on and off according to the engine coolant temperature. When the coolant reaches a certain temperature, usually around 200 degrees Fahrenheit, a sensor tells the electric fan to turn on. If the temperature is less than 200 degrees Fahrenheit, the fan remains off. The advantage of this system is that

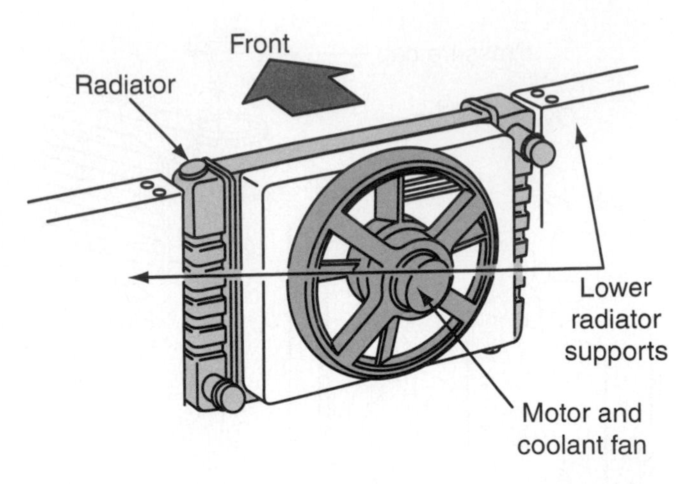

FIGURE 22-30 On many vehicles, the fan is driven by an electric motor, which reduces frictional horsepower losses.

there are no fan belts. There is no frictional horsepower loss, as with belt-driven fans. There is, however, a small power loss because the alternator must charge the battery more often. This is because the power used to turn the electric motor comes from the electrical system of the vehicle.

Figure 22–31 shows a typical electrical circuit used to turn the electric fan off and on at the correct time. This is the type of circuit diagram found in many service manuals that service technicians can use to troubleshoot electrical circuits. Although there are many parts to the circuit, such as wire color, pin locations, and numbers to locate the parts, the circuit operates as follows. The cooling fan motor is located on the right side of the circuit. In order for the fan motor to start, the engine cooling fan relay (center left) must operate to close the switch. When this switch is closed, the fan motor turns on. The switch is closed by oper-

PARTS LOCATOR

*Refer to photo #16 in the front of the textbook to see what an **engine coolant fan relay** looks like.*

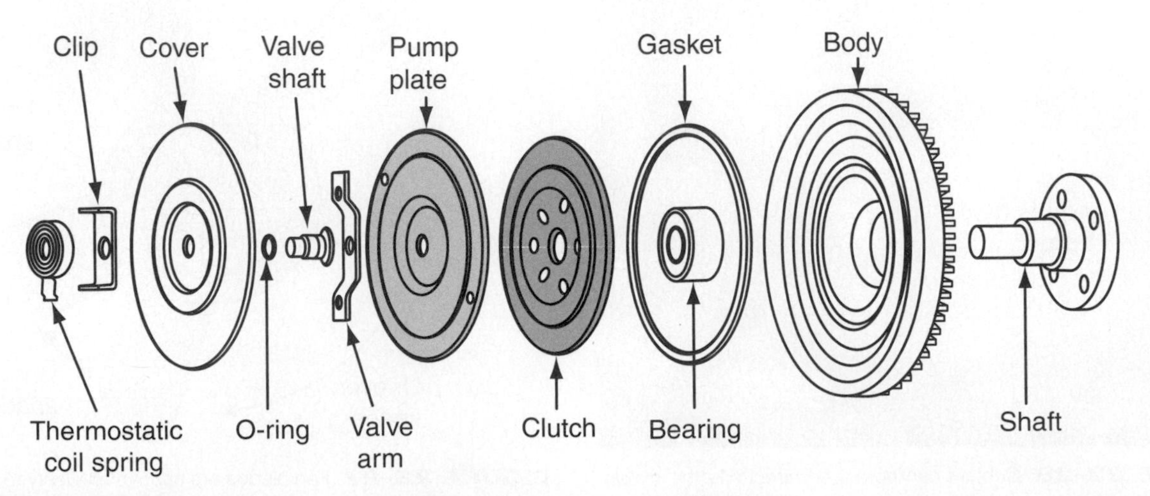

FIGURE 22-29 The internal parts of the variable-speed fan are shown in this illustration.

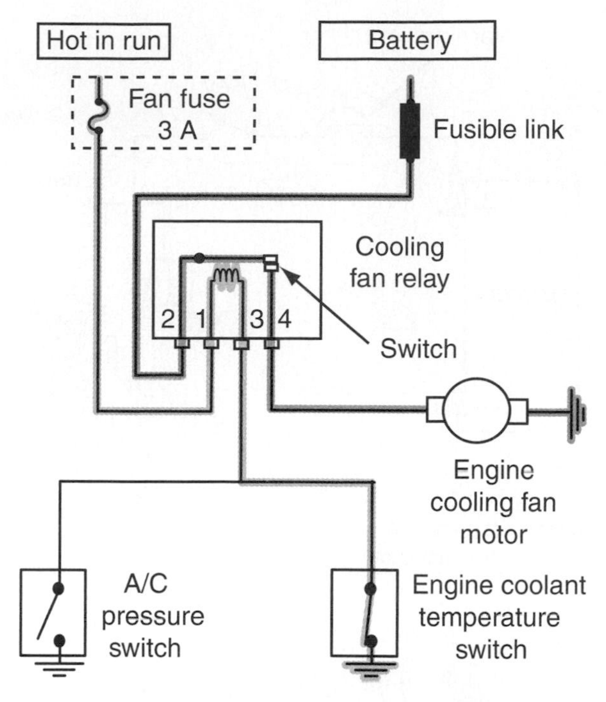

FIGURE 22-31 To start the engine cooling fan motor, the engine coolant temperature switch must close. This energizes the engine cooling fan relay, completing the circuit for the fan motor.

ating a second circuit. The circuit starts at the fan fuse, then goes through the engine cooling fan relay (coil) and then to the engine coolant temperature switch. This switch is located on the bottom right of the circuit. The engine coolant temperature switch closes because the engine temperature has increased to a certain point. When this switch closes, the coil in the engine cooling fan relay is energized and starts the engine cooling fan motor. Each manufacturer has different circuits; however, the basic principle is the same.

PARTS LOCATOR

*Refer to photo #17 in the front of the textbook to see what an **engine cooling fan motor** looks like.*

FLEXIBLE-BLADE FANS

Some engine manufacturers use a flexible-blade fan. Flexible blades are used to reduce frictional horsepower loss from the fan. The flexible blades are made of fiberglass or metal. As the speed of the engine and vehicle increases, the blades flatten. The blades then move less air, causing less frictional horsepower to be lost. ***Figure 22–32*** shows an example of how the blade flattens out. The engine in ***Figure 22–32A*** is at a lower speed, so the pitch is greater. As the engine and vehicle increase in speed, there is less need for the fan to pull air through the radiator. At this point, because the fan pitch is less (***Figure 22–32B***) there is less frictional horsepower loss.

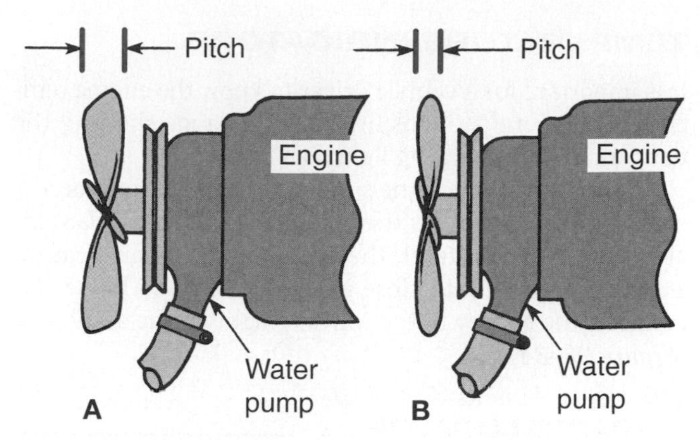

FIGURE 22-32 As the flexible-blade fan turns faster, the pitch is reduced so that frictional horsepower losses are reduced.

FAN SHROUDS

Certain vehicles use a fan **shroud.** A fan shroud is used to make sure the fan pulls air through the entire radiator evenly. If a fan shroud is not used, there may be hot spots in the radiator (***Figure 22–33***). For example, the fan normally pulls air through the radiator directly in front of the blades. Very little air moves through the corners of the radiator. Using a fan shroud causes air to be pulled evenly through the entire surface area of the radiator.

FAN BELTS

The fan belt is used to turn the water pump and the fan. Most fan belts are V belts. Friction is produced between the sides of the belt and the pulley. Using the V belt produces a larger area of contact. This means that a more positive connection can be made between the belt and the pulley.

Other types of belts are also being used today. One popular type is the flat serpentine belt, also called a poly-V belt. It is about 1 1/2 inches wide and has several V grooves on one side. It is flat on the other side. On some engines, this belt is used to reverse the direction of a water pump. This is done because objects can be driven from either side of the belt.

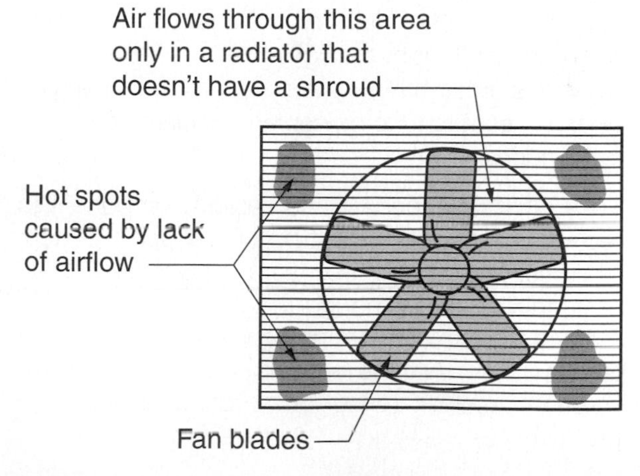

FIGURE 22-33 When a fan shroud is not used, hot spots can be created in the corners of the radiator.

TEMPERATURE INDICATORS

It is important for a vehicle driver to know the engine temperature. Several systems, including a gauge showing the temperature or a warning light, can be used.

The warning light is the simplest system. An indicator is placed in the coolant. As the temperature of the coolant increases to a certain level, the heat from the engine causes an electrical circuit to close inside the sensor. This sensor connects the battery to the warning light on the dashboard (***Figure 22–34***).

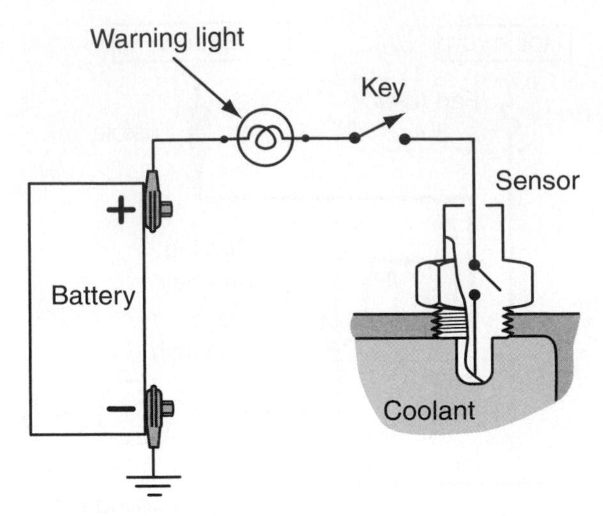

FIGURE 22–34 A warning light operates when the coolant reaches a certain temperature.

PARTS LOCATOR

*Refer to photo #18 in the front of the textbook to see what a **coolant temperature sensor** looks like.*

A second type of temperature indicator reads the actual coolant temperature and displays this on the dashboard of the vehicle (***Figure 22–35***). The **Bourdon tube** gauge operates in this manner. As pressure changes within the Bourdon tube, the needle reads differently on the dial. Pressure is produced in the tube inside a sensor, which is immersed in the coolant. As the temperature of the coolant increases, a liquid inside the sensor vaporizes. The vapor produces a pressure that is transmitted to the gauge through a tube so a dial can be moved.

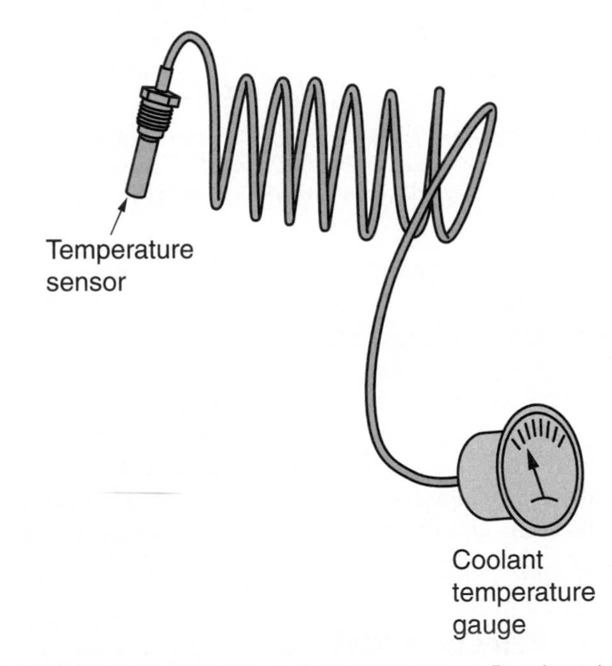

FIGURE 22–35 This coolant gauge uses a Bourdon tube to move the needle on the gauge.

When an electrical unit is used, it has a sensor placed in the coolant. As the temperature of the engine coolant increases, the electrical resistance of the sensor decreases. As the temperature of the coolant decreases, the electrical resistance of the sensor increases. This increase or decrease in resistance causes the needle in the temperature gauge to read the different coolant temperatures.

Car Clinic: COOLANT TEMPERATURE SENSOR

CUSTOMER CONCERN:
Hot Engine

A car with 45,000 miles on it seems to be overheating. There is a "hot" light on the dashboard. On the highway at 60 mph, the light goes on and then off. Does this mean the engine is very hot? The engine doesn't seem to lose power, and coolant never has to be added.

SOLUTION: The first and easiest component to replace is the coolant temperature sensor or switch. At times, these coolant sensors or switches will operate incorrectly. Although the sensor or switch is reading "too hot," the engine may be at the correct temperature. Test the sensor, and if defective replace the coolant sensor or switch to eliminate the problem.

TWO-SPEED FAN MOTOR

Certain engines use a two-speed fan motor. Two speeds are needed most often for engines that have air conditioning (A/C) or for engines that are under excessive loads on hot days. In order to control the fan, an electrical circuit is used. ***Figure 22–36*** shows a two-speed fan motor circuit.

In operation, the engine cooling fan motor (shown in the center right of the circuit) has two separate windings

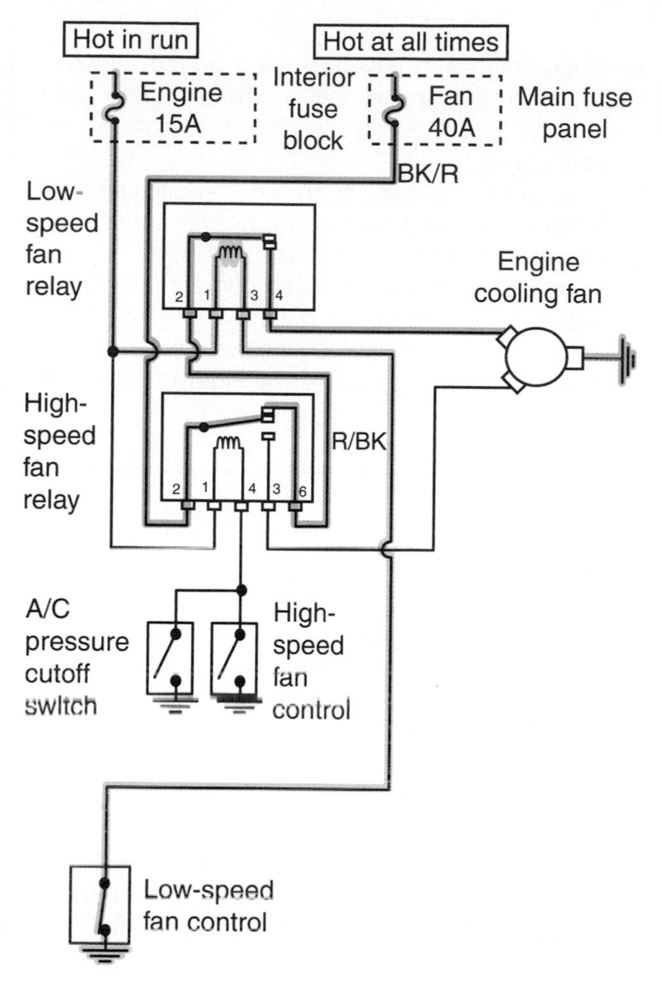

FIGURE 22-36 Some cooling systems use a two-speed fan to help control the cooling fan speed.

inside the motor. Electricity going through the low-speed fan relay is for the low-speed operation. Electricity going through the high-speed fan relay is for the high-speed operation. To get the fan to operate at low speed, first the low-speed fan control switch (on the left bottom of the circuit) must close. This will close when the coolant temperature is 207 degrees Fahrenheit. When this closes, electricity flows from the Hot (in the upper left-hand corner of the circuit) through the low-speed fan speed relay coil. When the coil is energized, it pulls the switch above and closes it. Electricity then flows from the Hot (on the upper right side of the circuit) through the BK/R wire, through the high-speed fan control relay switch, to the R/BK wire (right side) through the low-speed fan control relay switch, to the upper connection of the engine cooling fan motor and to ground. Now the motor is running at the low speed.

When the A/C is turned on or if the high-speed fan control switch closes, the coil inside the high fan control relay is energized. This closes its switch and allows electricity to flow to the lower connection of the engine cooling fan motor. Now the fan is operating at the higher speed.

COOLANT LEVEL INDICATORS

All vehicles today have an overflow tank on the cooling system. Often there is a coolant level sensor on this tank. The sensor has a small float that senses the coolant level inside the overflow tank. When the coolant drops below a certain level, the float also drops. When it drops to a certain point, an electrical contact is made. When this happens, the electrical circuit turns on a "low coolant" light on the dashboard.

Problems, Diagnosis, and Service

Safety Precautions

1. When working on the cooling system, remember that at operating temperature the coolant is extremely hot, approximately 220 degrees Fahrenheit or hotter. Touching the coolant or spilling the coolant on the body can cause serious injury.
2. When working on the coolant system, for example when replacing the water pump or thermostat, you may spill some coolant on the floor. The antifreeze in the coolant causes it to be very slippery. Immediately wipe up any coolant that spills to reduce or eliminate the chance of injury.
3. Always wear proper clothing and eye protection when using coolant additives to remove silicate and calcium deposits; they may be very corrosive.
4. When working on the fan, water pump, or belts, makes sure the engine ignition system is off or the battery is disconnected, or both.

continued

Safety Precautions *continued*

5. Whenever a vehicle is running, always keep fingers, tools, and clothing away from the moving fan.
6. When working on the water pump, fan, or belts, be careful not to scrape your knuckles against the radiator, as injury may result.
7. Never remove the pressure cap when the engine is hot. Removing the pressure cap releases the pressure on the cooling system. This reduces the boiling point of the coolant. The coolant will then boil violently and burn or injure a person's hands or face.
8. When using a chemical cooling system cleaner, be careful not to get the chemical in your eyes or on your skin. Always follow the manufacturer's instructions.
9. Keep antifreeze away from children and animals. Antifreeze has a sweet taste, and, if consumed, can lead to death in as little as 24 hours. If antifreeze is ingested, immediately contact a doctor or poison control center.
10. Always dispose of used antifreeze in an approved container. Never pour used antifreeze down the water drain.

PROBLEM: Correct Amount of Antifreeze

A customer has asked a service technician to check a vehicle radiator for the correct amount of antifreeze.

DIAGNOSIS

It is important to have the correct amount of antifreeze in the cooling system for both summer and winter operation. Antifreeze protects the cooling system from corrosion in the summer and from freezing in the winter. The correct amount of antifreeze can be checked by using a cooling system hydrometer. A typical hydrometer is shown in ***Figure 22–37***. A refractometer can also be used to check the amount of antifreeze. This meter measures deflection of light through the antifreeze. Also a litmus test (using litmus test strips) can be used for checking temperature and corrosion conditions.

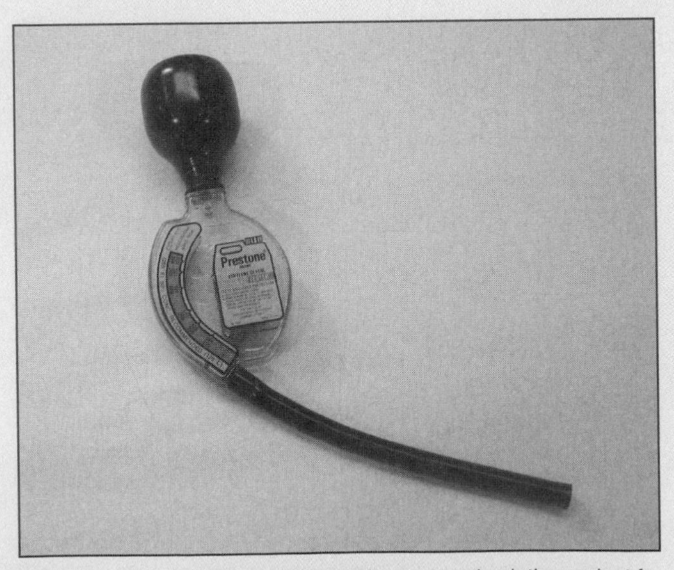

FIGURE 22–37 A hydrometer is used to check the coolant for protection against freezing.

SERVICE

CAUTION: *Never remove the radiator cap when the engine is at operating temperature.*

1. Remove the radiator cap and draw coolant up into the hydrometer.
2. Read the hydrometer to determine the exact freezing point. Because there are several types of hydrometers, read the directions to find out what the reading means. Usually, the hydrometer indicates the lowest temperature at which the coolant is protected before it will freeze.

PROBLEM: Corrosion in the Cooling System

The coolant in the radiator has turned brownish, and the water pump seal and bearings need to be replaced.

DIAGNOSIS

If the coolant has a rusty color, corrosion may be building up in the cooling system. Corrosion can easily cause the water pump seal and bearings to be damaged. The rust in the system acts as an abrasive on the water pump seal. Corrosion is usually caused by the presence of oxygen in the cooling system. Oxygen can get into the system if the expansion tank is plugged.

SERVICE

1. Drain the coolant from the system.
2. Check the expansion tank for any plugged hoses or leaks.
3. Completely clean the expansion tank.
4. Flush the cooling system completely. Remember that antifreeze must be disposed of according to the state pollution control regulations. Never drain the coolant into the sewer system.

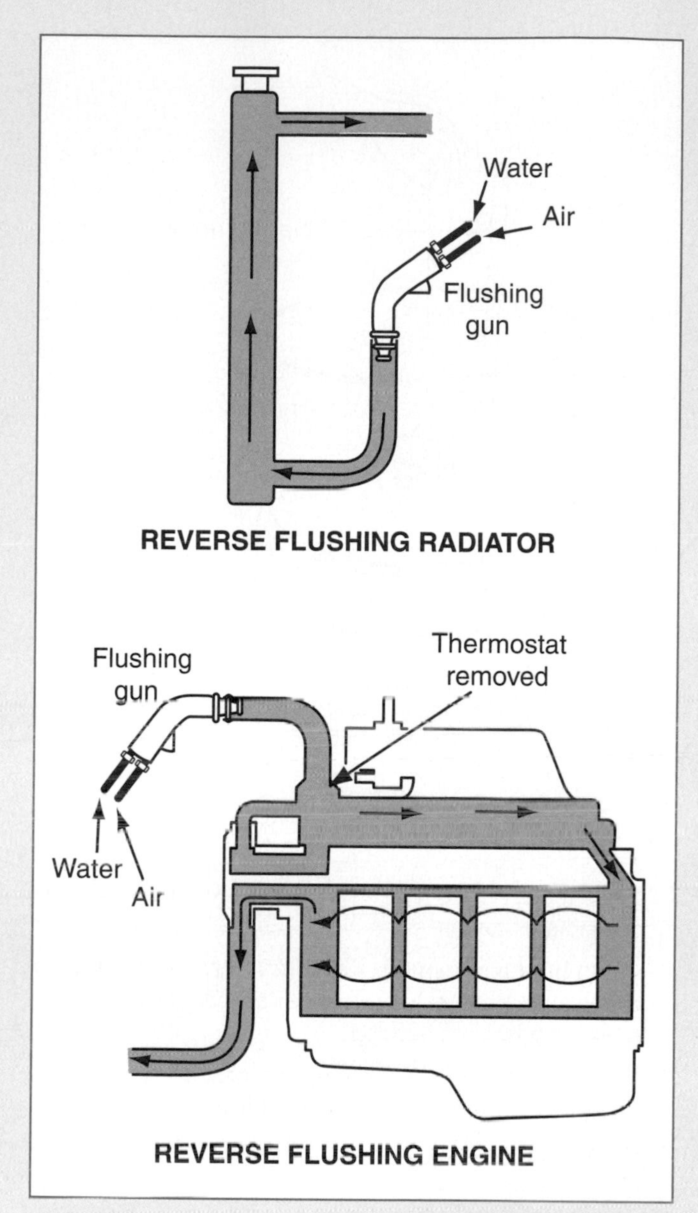

FIGURE 22–38 When flushing a cooling system, always reverse flush the radiator and engine.

5. The most effective way to flush a cooling system is to reverse flush the system. This is done by flushing with regular water and air in a reverse direction as shown in ***Figure 22–38***. In some automotive shops there is a coolant flushing machine. It provides the water pressure and air necessary to remove most corrosion in the cooling system while flushing. When flushing the radiator, force the water and air into the lower hose of the radiator and let the water and corrosion come out the top hose of the radiator. When flushing the engine block, first remove the thermostat. Then force the water and air into the thermostat outlet and let the water and corrosion come out the water pump inlet. Always be careful not to put too much pressure on the system as it might damage the radiator. Note that reverse flushing is not recommended on plastic or aluminum radiators.
6. Remove the water pump if the seal has been leaking, and replace it with a new or rebuilt water pump.
7. Add the correct amount of antifreeze and water, and refill the cooling system.
8. After the cooling system has been filled, it is necessary to bleed the air out of the system to make sure there are no air pockets. One simple method is to leave the radiator cap off the radiator and run the engine until the thermostat opens. The coolant will then circulate through the radiator, and any trapped air bubbles will escape through the radiator cap opening. Some manufacturers suggest other methods to bleed the cooling system, so always follow the manufacturer's suggested procedure for the particular vehicle you are working on.

PROBLEM: Engine Overheats—Hard Water

A customer has complained that the engine overheats easily in hot weather. The problem has been getting worse as time goes on.

DIAGNOSIS

Many problems can cause an engine to overheat. One could be the type of water used in the cooling system. If hard water is used, a significant amount of silicate and calcium can build up on the system. Silicate and calcium buildup have a tendency to restrict heat transfer out of the radiator. This restriction could cause the engine to overheat easily. Also, when there is corrosion buildup, the temperature sensors may read the wrong temperature.

SERVICE

Check for silicate and calcium buildup by looking down into the radiator and observing the tubes (***Figure 22–39***). Deposits can be seen building up around each tube. If there appears to be silicate and calcium buildup, the radiator and engine will have to be cleaned. Radiator shops are often

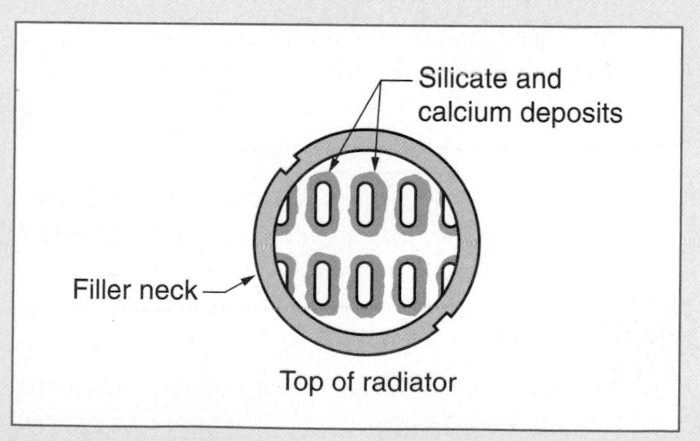

FIGURE 22–39 Deposits of calcium and silicate can easily be seen on down-flow radiators by looking directly into the filler neck. The calcium and silicate deposits will be built up around the core tubes.

fully equipped to clean the internal passageways of the radiator. If there is equipment in the shop, follow the manufacturer's suggested procedure for cleaning radiators and flushing the engine block. If you drain the coolant from the cooling system, make sure to dispose of the coolant correctly. Antifreeze is considered a hazardous waste product.

PROBLEM: Engine Overheats—Faulty Thermostat

A customer has complained that the engine has overheated. The problem came on very rapidly.

DIAGNOSIS

If the engine has overheated and the problem came on rapidly, the thermostat may be faulty. If the thermostat remains closed, no coolant can pass through the thermostat to the radiator.

SERVICE

To remove and check the thermostat use the following procedure:

CAUTION: *Make sure the cooling system is at room temperature before starting work.*

1. Drain the cooling system and save all coolant.
2. Remove the upper radiator hose from the thermostat.
3. Remove the thermostat housing so the thermostat can be removed.
4. Check the thermostat as shown in ***Figure 22–40***.
   a. Suspend the thermostat in a pan of hot water with a thermometer in the water.
   b. Increase the temperature of the water in the pan until the thermostat starts to open. The opening can be observed visually. A thermostat should still be closed at about 10 degrees Fahrenheit below the number stamped on the thermostat.
   c. The thermostat should be fully open about 25 degrees Fahrenheit above the number stamped on it.

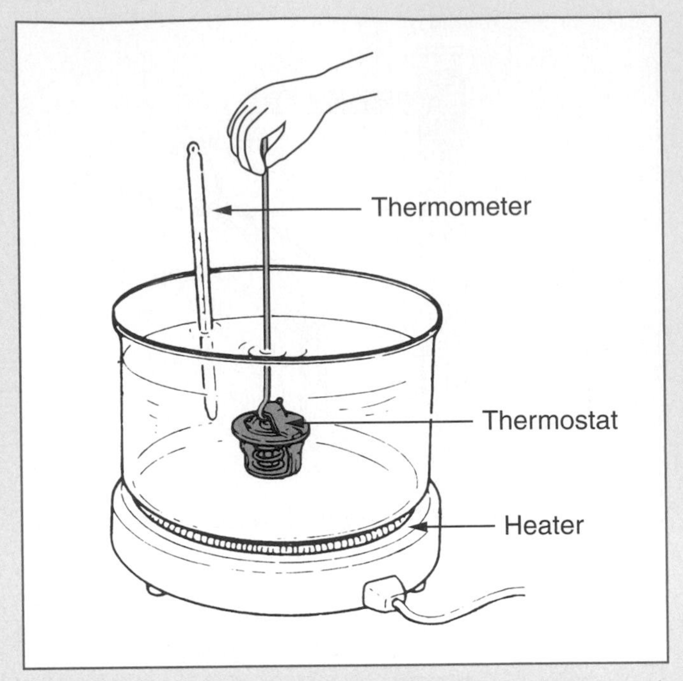

FIGURE 22-40 The thermostat can be checked by suspending it in a pot of water. Increase the temperature of the water and observe when the thermostat opens. (Courtesy of American Honda Motor, Co., Inc.)

PROBLEM: Leaking Antifreeze

A customer has complained that the cooling system has a leak, and antifreeze is always found under the car after the vehicle sits overnight.

DIAGNOSIS

Cooling systems can be diagnosed for leaks by using a pressure tester (***Figure 22–41***).

1. Remove the radiator cap.

CAUTION: *Be sure the engine is cool, not hot or at operating temperature.*

2. Place the pressure tester on the radiator and seal it like the radiator cap.
3. Pump up the pressure tester to the pressure reading on the cap. Do not overpressurize the cooling system. It may cause leaks to develop.
4. With pressure on the system, observe the hoses, water pump, radiator, and so on for small leaks.
5. If no external leaks are evident but the gauge refuses to hold pressure for at least 2 minutes, there may be an in-

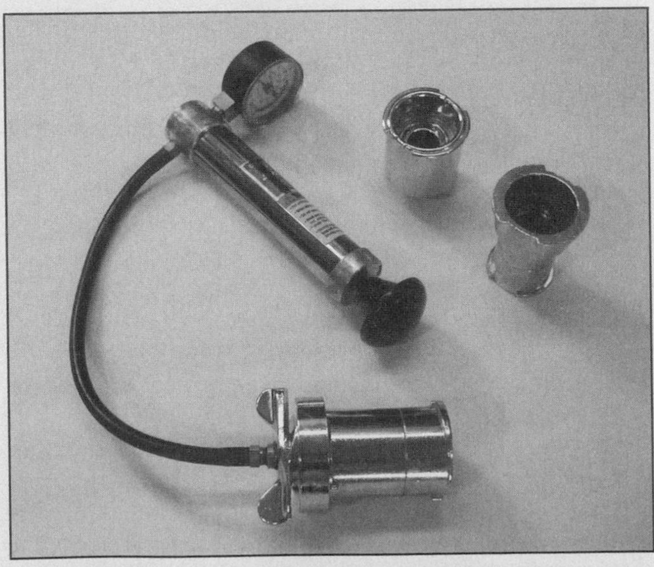

FIGURE 22-41 The cooling system can be checked for small leaks by putting the system under pressure.

ternal coolant leak. If this is the case, the engine must be disassembled, and the block and head must be checked for cracks.

6. Another method used to check for leaks in the cooling system is by using a black light. A black light is light that emits a certain frequency or intensity. When a dye is added to the coolant, the black light is able to easily detect small leaks that may occur in the cooling system.

SERVICE

If a leak is visually spotted, the component will need to be replaced. If coolant is leaking from the water pump, the water pump seal is damaged. If leakage occurs, replace the water pump using the following general procedure:

1. Remove the radiator cap.

CAUTION: *Be sure the engine is cool, not hot or at operating temperature.*

2. Drain the coolant from the radiator and cylinder block. Save the coolant if it is to be used again. If not, remember that antifreeze is considered a hazardous waste product and must be disposed of correctly.
3. Remove the belts by loosening the appropriate component on the belts. This may include the alternator, air-conditioning compressor, and so on.
4. On some engines, the radiator may have to be removed to get at the water pump.
5. Remove the fan if it is driven by the water pump.
6. Remove the bolts holding the water pump to the engine cylinder block.
7. Scrape the old gasket completely off the block.
8. Replace the old gasket with a new gasket.
9. Replace the water pump with a new or rebuilt pump.
10. Reverse the previous procedure to complete the assembly.
11. Before installing the fan belts, check the inside surface for wear as shown in ***Figure 22–42***. Twist the belt to see if there are any cracks on the inside surface.
12. Always check the reinforcement springs inside the lower radiator hose. The spring is there on some vehicles to prevent the rubber hose from collapsing because of water pump suction. If the lower hose is replaced, make sure the spring is not deformed, missing, or out of position. Without the spring, the vehicle is likely to overheat at cruising speed (***Figure 22–43***).
13. Check the front of the radiator. Keep the front of the radiator clean and free of dirt, bugs, and other debris.
14. After all the parts have been inspected and replaced, add the coolant back into the cooling system. Extra antifreeze may have to be added.

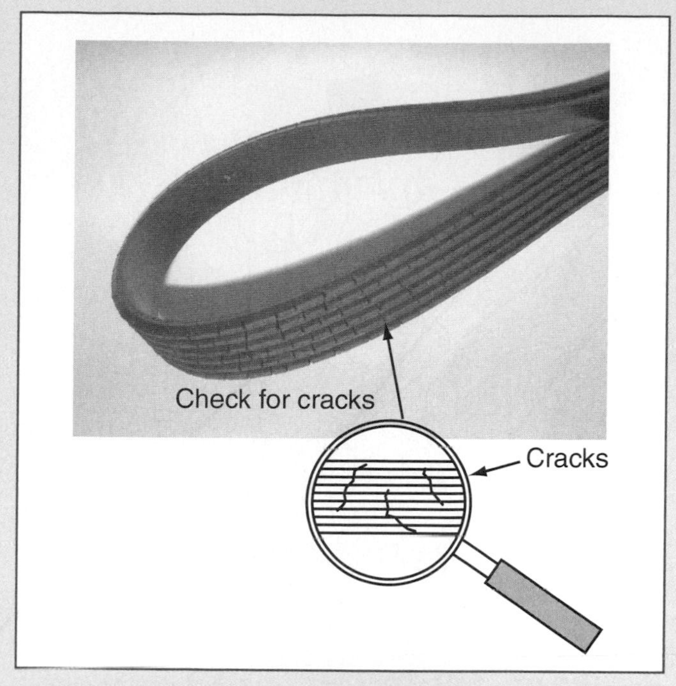

FIGURE 22–42 Fan belts should be checked for cracks and excessive wear. Bend a belt backward to check for cracks.

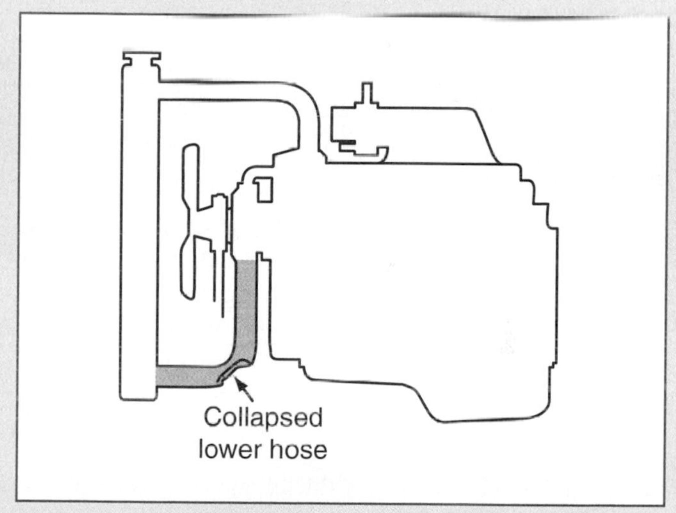

FIGURE 22–43 The lower radiator hose may collapse at cruising speed because of the suction of the water pump. This often occurs when there is a restriction in the radiator. A spring is placed inside the hose to prevent the hose from collapsing.

15. Check the fan belts.
    a. Always check the fan belts for wear and correct tension.
    b. Tension on the fan belts can be checked by using the gauge shown in ***Figure 22–44***. Position the gauge between the two pulleys as shown. The tighter the belt, the less it will bend from pressure created by the gauge. Read the gauge to determine the tightness of the belt.
16. After all parts have been inspected and reassembled, run the engine and pressure test for leaks again.
17. When finished, remove the pressure tester and replace the pressure cap.

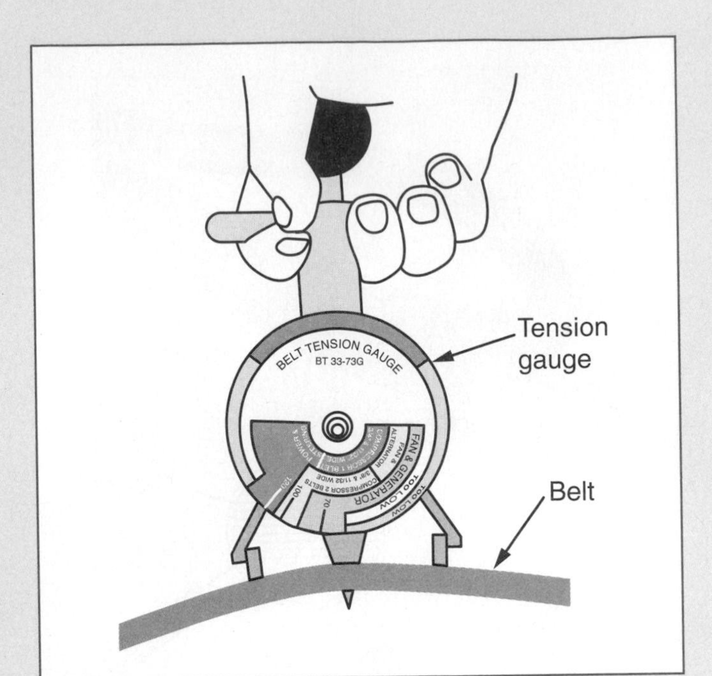

FIGURE 22-44 Fan belts are checked for tension by using this belt tension gauge.

PROBLEM: Faulty Radiator Cap

A customer has complained that the engine has overheated. The problem seemed to have happened rapidly.

DIAGNOSIS

If the engine has overheated and the problem came on rapidly, the radiator cap may be faulty. If the radiator cap doesn't hold the pressure on the cooling system, the boiling point will be reduced, possibly causing the engine to overheat.

SERVICE

To check the radiator cap use the following procedure:

CAUTION: *Make sure the cooling system is at room temperature before starting work.*

1. Remove the radiator cap.
2. Place the pressure cap on the pressure tester as shown in ***Figure 22–45***. Seal the cap on the pressure tester by turning, as if on a radiator.
3. Increase the pressure on the cap, using the pump on the pressure tester.
4. Note the pressure at which the pressure cap releases. As the pressure is increased by the pumping, the pressure will eventually not be able to go higher.
5. Make sure the cap releases pressure at the pressure setting shown on the cap.
6. If the pressure is released at a lower point, the cap is defective. Replace the cap if this occurs.

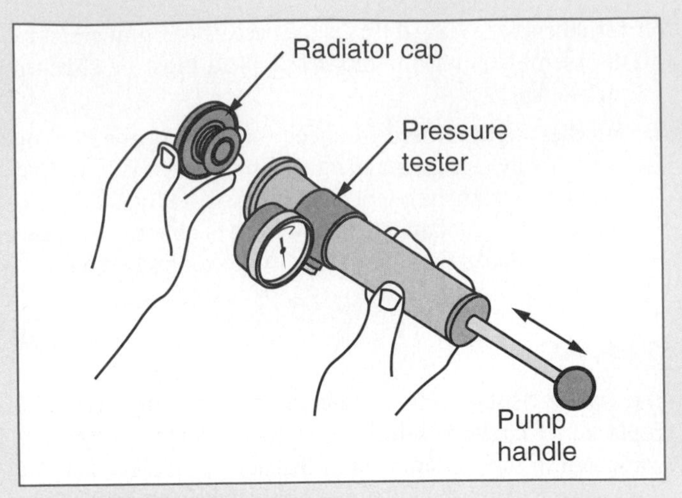

FIGURE 22-45 The radiator pressure cap tester is used to determine exactly how much pressure will open the pressure cap.

7. Check the pressure cap vacuum valve also. One method used to check this valve is to observe the upper radiator hose. If the upper radiator hose collapses after the engine has cooled down, the vacuum valve may be defective or clogged. When clogged or damaged, the valve will not allow pressure in the cooling system to equalize when being cooled down. If this is the case, replace or clean the vacuum valve to remedy the problem (***Figure 22–46***).

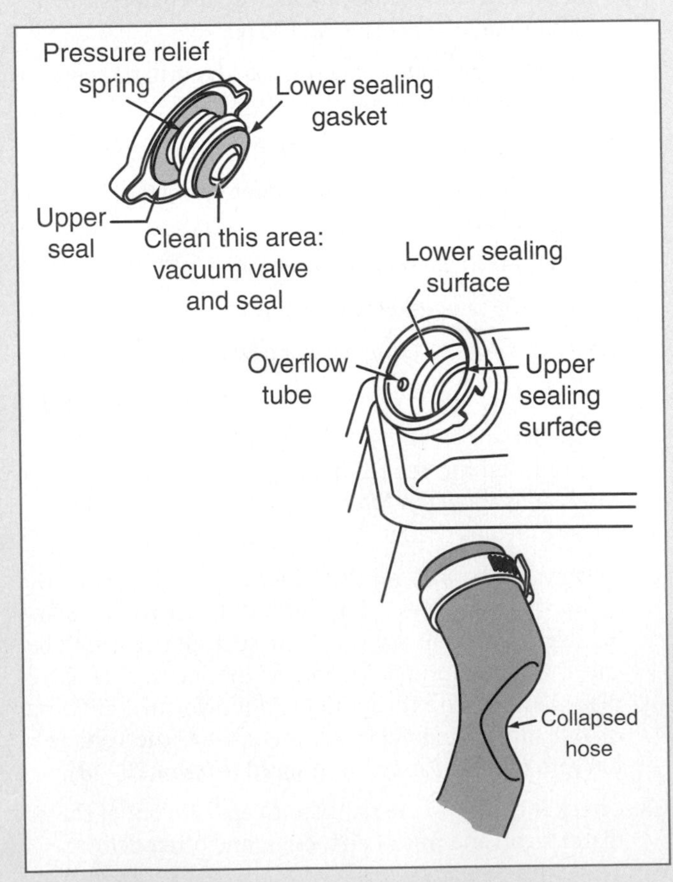

FIGURE 22-46 If the pressure cap vacuum valve is defective or clogged, the upper radiator hose will collapse when the engine cools down.

Service Manual Connection

There are many important specifications to keep in mind when working with the cooling system. To identify the specifications for your engine, you will need to know the VIN (vehicle identification number) of the vehicle, the type and year of the vehicle, and the type of engine. Although they may be titled differently, some of the more common cooling system specifications (not all) found in service manuals are listed below. Note that these specifications are typical examples. Each vehicle and engine will have different specifications.

Common Specification	Typical Example
Belt tension	
New	165 lb
Used	100 lb
Cooling system capacity	
Without air conditioner (A/C)	15.3 quarts
With air conditioner (A/C)	15.9 quarts
Radiator cap relief pressure	16 lb
Thermostat opening temperature	196°

In addition, the service manual will give specific directions for various service procedures. Some of the more common procedures include:

- Checking the variable-speed fan operation
- Setting belt tension
- Thermostat removal and replacement
- Water pump replacement

SUMMARY

The following statements will help to summarize this chapter:

- The cooling system is designed to maintain the most efficient operating temperature within the engine.
- The cooling system uses three means of heat transfer to remove heat from the engine: conduction, convection, and radiation.
- Automobile engines use either an air-cooled or liquid-cooled cooling system to remove the excess heat.
- Liquid-cooled engines use a mixture of antifreeze and water to cool the engine.
- Coolant is pumped by the water pump through internal passageways inside of the engine.
- A thermostat is used to bring the coolant up to operating temperature as quickly as possible.
- Antifreeze is designed to keep the coolant from freezing in cold weather operation and to increase the coolant's boiling point in hot weather conditions.
- Pressure provided by the radiator cap helps to increase the boiling point of antifreeze.
- The water pump is used to circulate the coolant throughout the system.
- When the pressure gets higher than the radiator cap setting, coolant flows into an expansion tank.
- Radiators are designed as either cross-flow or down-flow configurations.
- Several types of fans are used on the cooling system including the plastic, variable-speed, and electric fans. All are designed to reduce frictional losses from fan operation.
- Temperature indicators and electrical circuits are used to determine the engine coolant temperature.
- Always check the belt tension on the water pump.
- The pressure cap and the radiator can be checked for leaks by using a coolant pressure tester.
- The thermostat can be checked by placing it in a container of water, increasing the water temperature, and observing at what temperature the thermostat opens.
- Always check hoses and belts on the cooling system for cracks, leaks, and weak spots.

TERMS TO KNOW

Can you explain each of the following terms? Review the chapter until you can use each term correctly.

Air-cooled
Antifreeze
Bimetallic strip
Boiling point
Bourdon tube
By pass tube
Cavitation
Centrifugal pump
Conduction
Convection
Core
Corrosive

continued

Hazardous waste
Heat exchanger
Heat transfer
Liquid-cooled
Pressure
Psig
Radiation
Shroud
Surging
Thermostat
Vacuum

REVIEW QUESTIONS

Multiple Choice

1. Which of the following is not a purpose of the cooling system?
   a. Keeping the engine temperature as low as possible
   b. Removing excess heat from the engine
   c. Bringing the temperature to operating range as quickly as possible
   d. Operating the engine at the best operating temperature for highest efficiency
   e. Protecting the internal parts from overheating

2. Which of the following is transfer of heat when both objects are solid?
   a. Conduction
   b. Convection
   c. Radiation
   d. Surging
   e. Vacuum

3. Heat transfer by ________ means to move heat by circulation of air or liquid.
   a. Conduction
   b. Convection
   c. Radiation
   d. Surging
   e. Vacuum

4. The ________ engine uses air passing over fins to cool the engine.
   a. Liquid-cooled
   b. Air-cooled
   c. Conduction
   d. Radiant
   e. Closed-loop

5. If the coolant is not hot enough to be cooled, it is sent to the ________.
   a. Radiator
   b. Pressure cap
   c. By-pass tube
   d. Differential regulator
   e. Coolant sensor

6. Antifreeze is used to protect the cooling system against:
   a. Freezing
   b. Boiling
   c. Corrosion
   d. All of the above
   e. None of the above

7. Which two deposits build up inside the cooling system?
   a. Calcium and water
   b. Silicate and rust
   c. Calcium and silicate
   d. Water and rust
   e. Nitrogen and water

8. What is the most common mixture for antifreeze and water?
   a. 20% water, 80% antifreeze
   b. 50% water, 50% antifreeze
   c. 90% water, 10% antifreeze
   d. 30% water, 70% antifreeze
   e. None of the above

9. The water pump can be considered a/an ________ type pump.
   a. Positive displacement
   b. Centrifugal
   c. Eccentric
   d. All of the above
   e. None of the above

10. The water pump is often driven by a:
    a. Belt driven from the starter
    b. Belt driven from the oil pump
    c. Belt driven from the crankshaft
    d. Belt driven from the transmission
    e. None of the above

11. When the engine gets hot enough, the automotive thermostat:
    a. Closes
    b. Opens
    c. Blocks off the radiator
    d. Tells the radiator to shut down
    e. None of the above

12. The thermostat is ________.
    a. Open all of the time
    b. Closed all of the time
    c. Opened more and more as the coolant temperature increases
    d. Opened to allow coolant to flow to the transmission
    e. None of the above

13. The radiator is considered a/an ________ type of heat exchanger.
    a. Air-to-liquid
    b. Liquid-to-air

c. Air-to-air
d. All of the above
e. None of the above

14. Which type of radiator is used in vehicles that are lower in front?
a. Down-flow
b. Cross-flow
c. Centrifugal-flow
d. Up-flow
e. Side-flow

15. Which of the following is not a part of the radiator?
a. Inlet tank
b. Outlet tank
c. Core
d. Water pump
e. Cooling baffles, fins, and tubes

16. Pressure on the cooling system will protect the cooling system from:
a. Cavitation
b. Increased operating temperatures
c. Cold weather operation
d. All of the above
e. A and B

17. What is the standard pressure on a cooling system?
a. 60–70 pounds
b. 10 pounds
c. 14–15 pounds
d. 2–4 pounds
e. None of the above

18. When an expansion tank is used on the cooling system:
a. Overflow coolant from the pressure cap enters the expansion tank
b. Coolant is drawn back into the radiator when the engine cools down
c. Coolant can be added to the expansion tank
d. All of the above
e. None of the above

19. Which system will produce more rust and corrosion in a cooling system?
a. Cooling system without an expansion tank
b. Cooling system with an expansion tank
c. Cooling system without a cross-flow radiator
d. Cooling system with a radiator cap
e. Cooling system with a thermostat

20. What is a big disadvantage of using a fan to draw air through the radiator?
a. Fans make oil cool faster
b. Fans reduce transmission heat
c. Fans consume frictional horsepower
d. All of the above
e. None of the above

21. The electric-driven fan is turned off and on by a sensor that reads:
a. Temperature of the engine coolant
b. Pressure of the coolant
c. Speed of the engine
d. Pressure of the lubrication
e. Amount of coolant flow

22. Which type of temperature indicator uses a curved tube that straightens when pressure increases?
a. Electrical sensor
b. Bourdon tube gauge
c. Silicon oil fan clutch
d. All of the above
e. None of the above

23. A fan shroud is used to:
a. Reduce any hot spots in the corners of the radiator
b. Protect the fan from damage
c. Protect the radiator from damage
d. Contain leaks in the cooling system
e. Protect the hoses

The following questions are similar in format to ASE (Automotive Service Excellence) test questions.

24. *Technician A* says that brownish coolant is a sign of rust in the cooling system. *Technician B* says that brownish coolant is caused by the addition of antifreeze in the coolant. Who is correct?
a. A only
b. B only
c. Both A and B
d. Neither A nor B

25. *Technician A* says the thermostat can be checked only by keeping it in antifreeze. *Technician B* says to check the thermostat, remove it, and suspend it in water with a thermometer while heating up the water. Who is correct?
a. A only
b. B only
c. Both A and B
d. Neither A nor B

26. After checking the engine, it was found that the coolant was boiling at 212 degrees Fahrenheit. *Technician A* says this is normal. *Technician B* says the thermostat should be checked with a pressure tester. Who is correct?
a. A only
b. B only
c. Both A and B
d. Neither A nor B

27. An engine has a large amount of silicon and calcium buildup in the cooling system. *Technician A* says the silicon is there because of the high quality of the antifreeze. *Technician B* says the calcium is there to lubricate the aluminum cylinder block. Who is correct?
a. A only
b. B only
c. Both A and B
d. Neither A nor B

28. *Technician A* says that antifreeze should be treated as a hazardous waste product. *Technician B* says the antifreeze should be dumped into the water drain. Who is correct?
    a. A only
    b. B only
    c. Both A and B
    d. Neither A nor B

29. *Technician A* says that rust in the cooling system can easily damage the water pump bearings. *Technician B* says that rust in the cooling system can damage the water pump seal. Who is correct?
    a. A only
    b. B only
    c. Both A and B
    d. Neither A nor B

30. An engine has a leak in the cooling system. *Technician A* says to check the cooling system with a pressure tester. *Technician B* says to check the pressure relief valve in the radiator cap with a torque wrench. Who is correct?
    a. A only
    b. B only
    c. Both A and B
    d. Neither A nor B

31. An engine is overheating. *Technician A* says that the problem could be a bad thermostat. *Technician B* says the problem could be silicate and calcium deposits inside the cooling system. Who is correct?
    a. A only
    b. B only
    c. Both A and B
    d. Neither A nor B

Essay

32. What problems can occur when the engine temperature is too high?

33. What problems can occur when the engine temperature is too low?

34. What is the difference between convection, conduction, and radiation forms of heat transfer?

35. What is the purpose of the by-pass tube on the cooling system?

36. What happens to the freezing point of coolant if the percentage of antifreeze is too high?

37. Describe the purpose and operation of a thermostat.

38. Describe the purpose and operation of a pressure cap.

Short Answer

39. Because antifreeze is considered a hazardous waste product, it should always be ________.

40. The temperature-sensitive valve on a thermostat should always be placed toward the ________ side of the engine.

41. Generally, ________ or ________ type materials are used as the fins on a radiator.

42. On most newer vehicles, the radiator fan is controlled or turned off and on by ________.

43. To check the freezing point of the antifreeze, use a ________.

Applied Academics

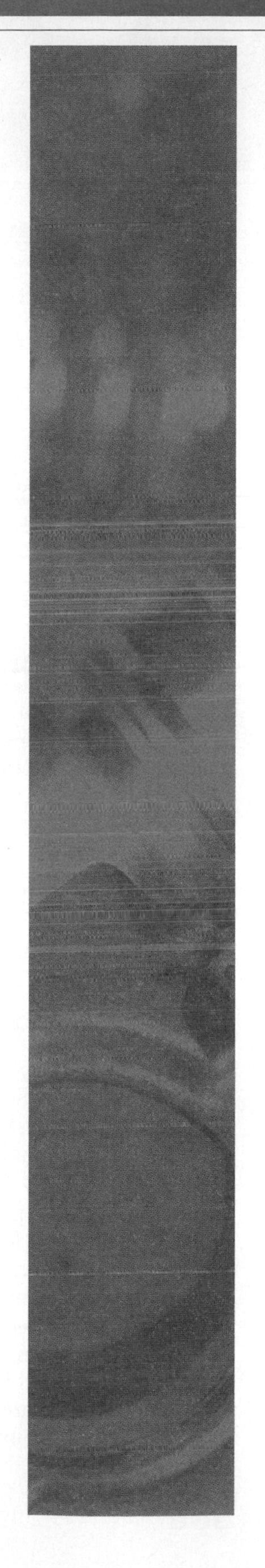

TITLE: Radiator Cores

ACADEMIC SKILLS AREA: Science

NATEF Automobile Technician Tasks:

The service technician is able to explain the concept of heat transfer in terms of conduction, radiation, and convection in automotive engine systems.

CONCEPT:

The thermal energy or heat from an engine's cooling system is transferred to the radiator during normal operation. The purpose of the radiator is to transfer the thermal energy or heat from the coolant to the outside air. To do this, heat must transfer from the coolant to the internal core tubes through conduction. Heat transfer then occurs from the core tubes to the radiator fins through conduction. Heat is then transferred to the air by radiation.

The number of cores inside a radiator will vary based on how much heat needs to be transferred out of the coolant. Depending on the size of the engine and its use, the radiator can have from one to five or more radiator cores. For example, a typical small engine may have only one core. However, a vehicle that is used for towing may have two to three or more cores, depending on the size of the engine. The cross-flow radiator shown in the illustration has a three-core radiator. It is always a good idea to know how many cores a radiator has. For example, if a vehicle has a small engine and a single-core radiator, it should not be used for towing a heavy trailer, even if the engine has the necessary power. Both the engine and the transmission could overheat, damaging both very seriously.

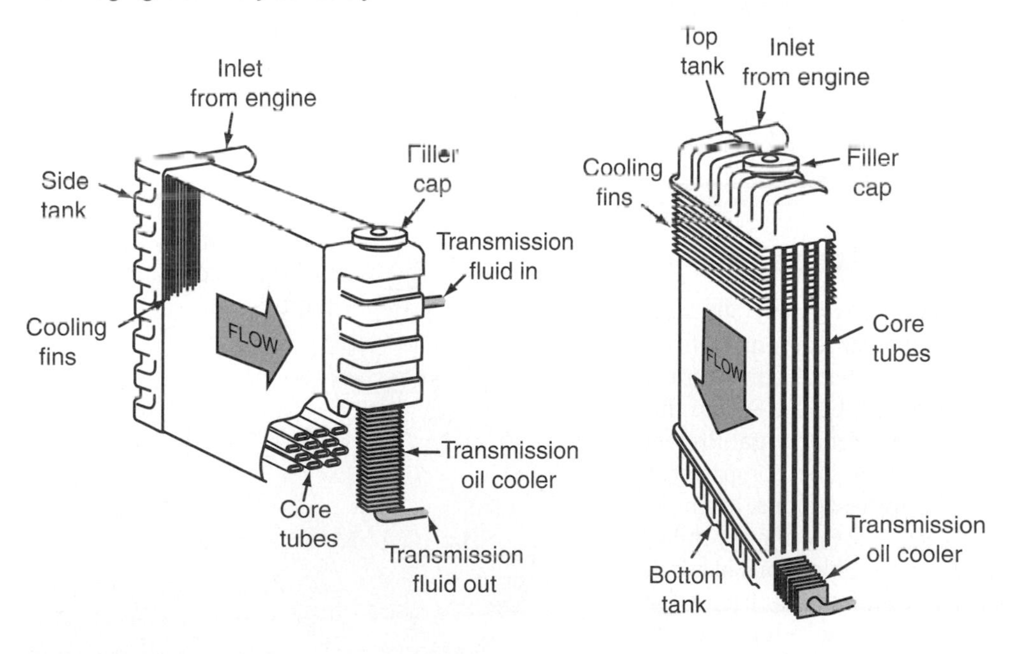

APPLICATION:

Why would the transmission overheat if the engine overheats due to overtaxing a single-core radiator? Can you look at a radiator in a vehicle and see how many cores it has? Can you look through the radiator fill tube to see the number of cores? Why does a three-core radiator transfer more heat than a single-core radiator?

CHAPTER 23 Fuel Characteristics

OBJECTIVES

After studying this chapter, you should be able to:

- Define the refining process and determine the heating values of different fuels.
- State the characteristics and properties of gasoline.
- Describe the characteristics and properties of diesel fuels.
- Recognize the characteristics of gasohol as a fuel.
- Compare liquid petroleum gas as a fuel with diesel fuel and gasohol fuel.
- Describe the characteristics of various alternative fuels.

INTRODUCTION

The correct operation and driving characteristics of automotive engines are directly related to the type of fuel that is used. Today's automotive engines use gasoline as the primary fuel. However, diesel fuel is available for diesel engines. In addition, gasohol and other fuels are increasingly being used to see if they are efficient and clean-burning. This chapter is related to the study of these fuels.

23.1 REFINING PROCESSES

HYDROCARBONS AND CRUDE OIL

Gasoline is a complex mixture of approximately 300 various chemical ingredients, mainly hydrocarbons, refined from crude oil for use as fuel in engines. Refiners must meet gasoline standards set by the American Society for Testing and Materials (ASTM), the Environmental Protection Agency (EPA), some states, and their own companies.

Fuels used in vehicles are called **hydrocarbons**. In simple terms, hydrogen and carbon molecules are combined chemically to make different fuels. These combinations of molecules are called hydrocarbons. The oil industry begins the manufacture of automotive fuels by first exploring for crude oil. Crude oil is the thick brown and black, slippery liquid from which all fuels are made. It contains thousands of different chemical combinations of hydrogen and carbon (***Figure 23–1***).

Crude oil is usually located from 5,000 to 25,000 feet below ground. It is the oil company's job to find the oil and then get it out of the ground. Large offshore and land oil rigs are used to pump the crude oil to the surface (***Figure 23–2***).

After crude oil has been pumped from the ground, it is transported to a refinery to be processed into different fuels. After the fuels have been refined, they are shipped to the distributors for sale to the public.

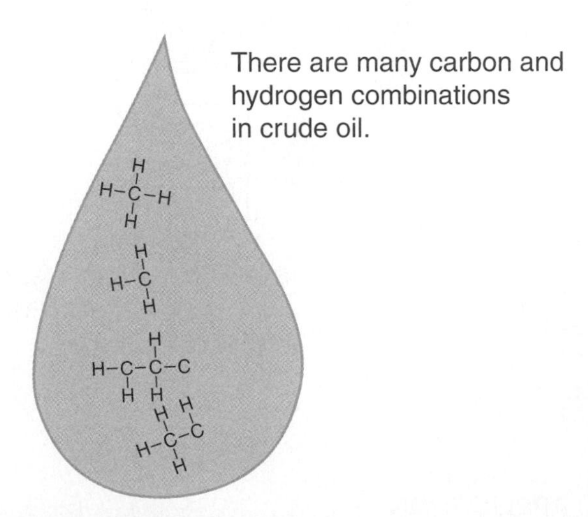

There are many carbon and hydrogen combinations in crude oil.

FIGURE 23–1 Crude oil contains many combinations of hydrogen and carbon. Different combinations make different fuels and other petroleum products.

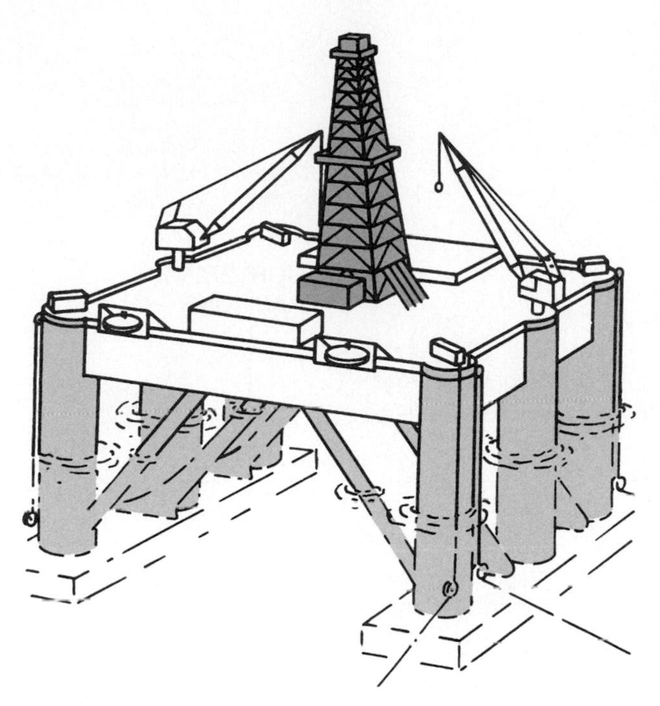

FIGURE 23-2 Large oil rigs such as this one remove crude oil from the earth.

FIGURE 23-3 At a refinery, crude oil is separated into many products used in the automotive industry. (Courtesy of API)

BOILING POINTS

The **boiling point** of a hydrocarbon is defined as the temperature at which the hydrocarbon starts to become a vapor. If the vapor is cooled below its boiling point, the vapor will **condense**, or become a liquid again. Hydrocarbon boiling points range from –250 degrees Fahrenheit to more than 1,300 degrees Fahrenheit. Most hydrocarbons have a range of boiling points. For example, gasoline molecules start to boil from about 195 degrees Fahrenheit to 390 degrees Fahrenheit. Other fuels have different boiling points.

DISTILLATION

Crude oil is converted to usable fuels in a refinery. ***Figure 23–3*** shows a typical refinery. One of the major processes done in a refinery is that of distillation. **Distillation** is the process that separates the different hydrocarbons in the crude oil. Certain hydrocarbons from crude oil are used to make gasoline. Other hydrocarbons are used to make diesel fuels. Others are used to make many other products. ***Figure 23–4*** shows some of the products that can be made from crude oil.

Distillation is done by heating all of the crude oil and the hydrocarbons to the highest boiling point. At the bottom of the column in ***Figure 23–5*** all hydrocarbons are vaporized. As the vapors rise to the top of the column, they start to cool. The hot vapors are cooled near their boiling points. At this point, they are condensed back to a liquid. For example, at different temperatures on the column, vapors cool to

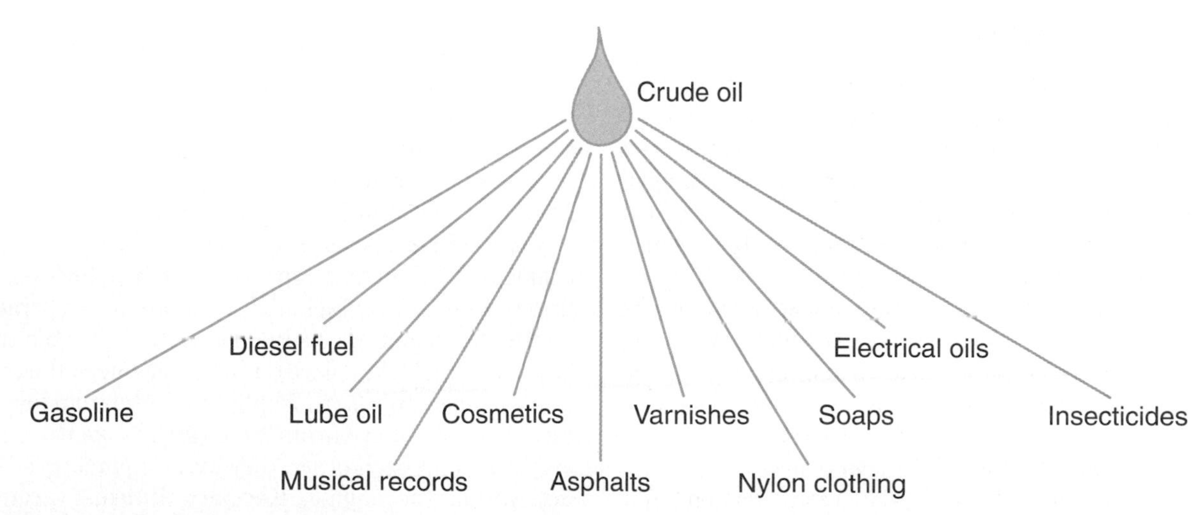

FIGURE 23-4 Many products, including automotive fuels, can be extracted from crude oil.

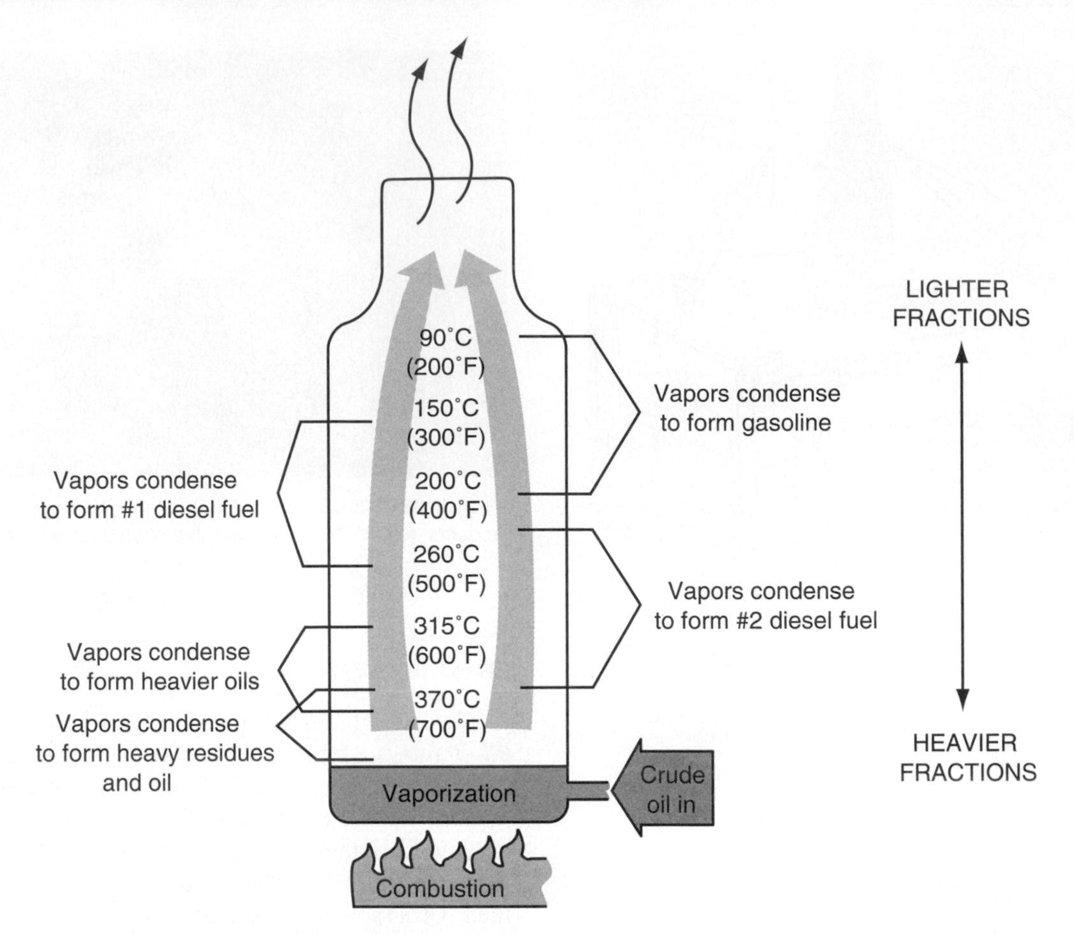

FIGURE 23-5 Distillation vaporizes all hydrocarbons in crude oil. The vapors rise, cool, and are condensed into many products used in the automotive industry.

form gasolines, diesel fuels, heavier oils, and different residues. This column is also called a fractionating column.

From this point, the fuels are cleaned, and chemicals are added to improve the fuels. They are now ready to be transported to distributors and service stations.

HEATING VALUES

Each of the fuels that are refined has a different heating value. Heating values are measured in British thermal units, or **Btu**. It is defined as the amount of heat needed to raise 1 pound of water 1 degree Fahrenheit. For example, 1 Btu is about equal to the heat given off when one wooden match burns completely. A home on a cold winter day may require 30,000 Btu per hour to be heated.

Gasoline and diesel fuels are compared by the number of Btus they contain if converted to heat. The higher the heating value of a fuel, the better the combustion and thus the better the fuel mileage. For example, a gallon of gasoline contains about 110,000–120,000 Btu. Diesel fuels typically contain about 130,000–140,000 Btu per gallon.

The heating value of a fuel is related to its boiling point during distillation. Normally, the higher the boiling point of a hydrocarbon, the higher the heating value of the fuel. Diesel fuels have a higher boiling point than gasoline. This means that diesel fuels have more Btu per gallon than gasoline.

23.2 PROPERTIES OF GASOLINE

KNOCKING

This term is used when studying fuel and different gasoline qualities. Knocking is a process that happens within the combustion chamber. It sounds like a small ticking or rattling noise within the engine. It can be very damaging to the pistons and rings as well as to the spark plug and valves. Another name for knocking is detonation.

Figure 23–6 shows what happens inside a combustion chamber when detonation occurs. When the spark plug fires to ignite the air-fuel mixture, it produces a flame front. Shortly after ignition, a second explosion (flame front) and ignition of fuel takes place on the other side of the combustion chamber. This flame front is produced from low-octane fuel combusting, or burning, too early. When these two energy fronts hit each other, they cause a **pinging**, or knocking, within the engine. Knocking is usually caused by poor-quality fuel. The temperatures within the combustion chamber are high enough to cause the air-fuel mixture to ignite without the spark plug. Fuels are made to have

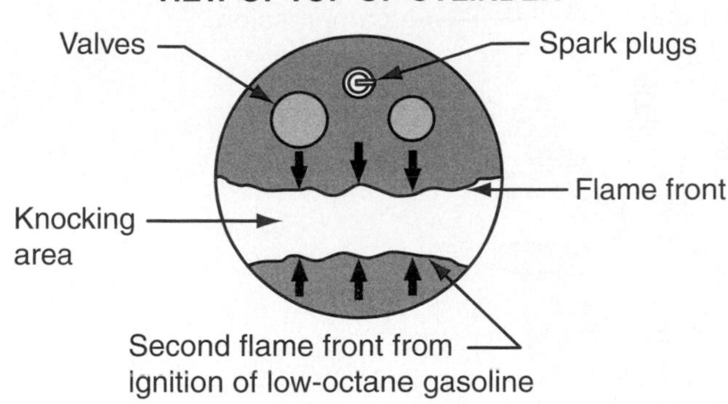

FIGURE 23-6 Detonation results when two flame fronts hit each other. One is from the spark plug combustion. The other is from early fuel ignition caused by poor fuel quality.

antiknocking characteristics. Actually, antiknocking fuels need higher temperatures to start burning.

OCTANE NUMBER

Different types of gasoline are identified by the **octane number**. The octane number reflects a fuel's resistance to burning. For example, the higher the octane number of a gasoline, the higher the temperature needed to ignite the fuel. The lower the octane number, the easier the fuel can start burning. Standard octane ratings range from 85 to 90 for regular gasoline. Premium gasoline or high-test gasoline has an octane range from 90 to 95. Gasolines that have higher octane ratings tend to reduce knocking and detonation.

TYPES OF OCTANE RATINGS

Three types of octane ratings are used in the petroleum industry. They include the research octane, motor octane, and road octane. Research octane is a laboratory measure of gasoline and its antiknock characteristics. The octane is tested under mild engine operating conditions, low speed, and low temperatures. Motor octane is a laboratory measure of the antiknock characteristics of the fuel under severe engine operation. This includes high speed and high temperatures within the engine. Road octane represents actual road driving conditions. Road octane is the one posted on gasoline pumps. It is calculated by taking the average of the research and motor octanes. On gasoline pumps, road octane is stated as research octane plus motor octane divided by 2. This is illustrated on the pumps as:

$$\text{Octane} = \frac{R + M}{2}$$

where:

R = Research octane
M = Motor octane

Until about the 1970s, the octane numbers of gasoline were increasing. This was because the automotive manufacturers were constantly increasing compression ratios. Higher compression ratios usually meant more efficient combustion. As the compression ratios increased, so did the compression temperatures. If a lower-octane fuel was used in these higher-compression ratio engines, knocking always occurred. **Regular gasoline** was used for low-compression engines. **High-test** was used for higher-compression engines.

OCTANE REQUIREMENTS

The octane requirements of automobiles vary a great deal. In fact, as an engine gets older, the octane requirement may increase somewhat. This is because when the engine is new, there is a set compression ratio. As the engine gets older, however, carbon buildup may increase the compression ratio. This means that the engine will need a higher-octane gasoline. ***Figure 23–7*** illustrates why this happens. A typical compression ratio is found by comparing the cubic inches at BDC to the number of cubic inches at TDC. This is shown in ***Figure 23–7A***. In ***Figure 23–7B*** carbon buildup has been calculated into the compression ratio. If only .3 cubic inch buildup is formed on the piston head, the compression ratio increases from 8 to 1 up to 11 to 1.

Many other factors also influence the octane needed in an engine. These include:

- *Air temperature*—The higher the air or engine temperature, the greater the octane requirements.
- *Altitude*—The higher the altitude, the lower the octane.
- *Humidity*—The lower the humidity, the greater the octane needed in the engine.
- *Spark timing*—The more advanced the spark timing, the greater the octane needed.
- *Fuel settings*—The leaner the air-fuel ratio, the higher the octane needed.
- *Method of driving*—If the vehicle is accelerated rapidly, higher octane is needed. Start-and-stop driving also increases combustion deposits and buildup. Higher octane will then be needed.

ADDING LEAD TO GASOLINE

In the past, one method used to increase the octane of gasoline was by adding a chemical called TEL, which stands for **tetraethyl lead**. Tetramethyl lead, called TML, is also used to increase the octane rating. These chemicals were added during the refining process. They also acted as a lubricant on the valves and valve guides. It was found, however, that TEL and TML are dangerous pollutants. The exhaust gases from cars were producing lead pollution. Today many pollution control devices, particularly catalytic converters, should not be used with leaded fuels.

REMOVING LEAD FROM GASOLINE

In the early 1970s, TEL and TML were removed from gasoline because of pollution. The result was called low-lead or unleaded gasoline. Other chemicals were added to the gasoline to aid in lubricating the valves.

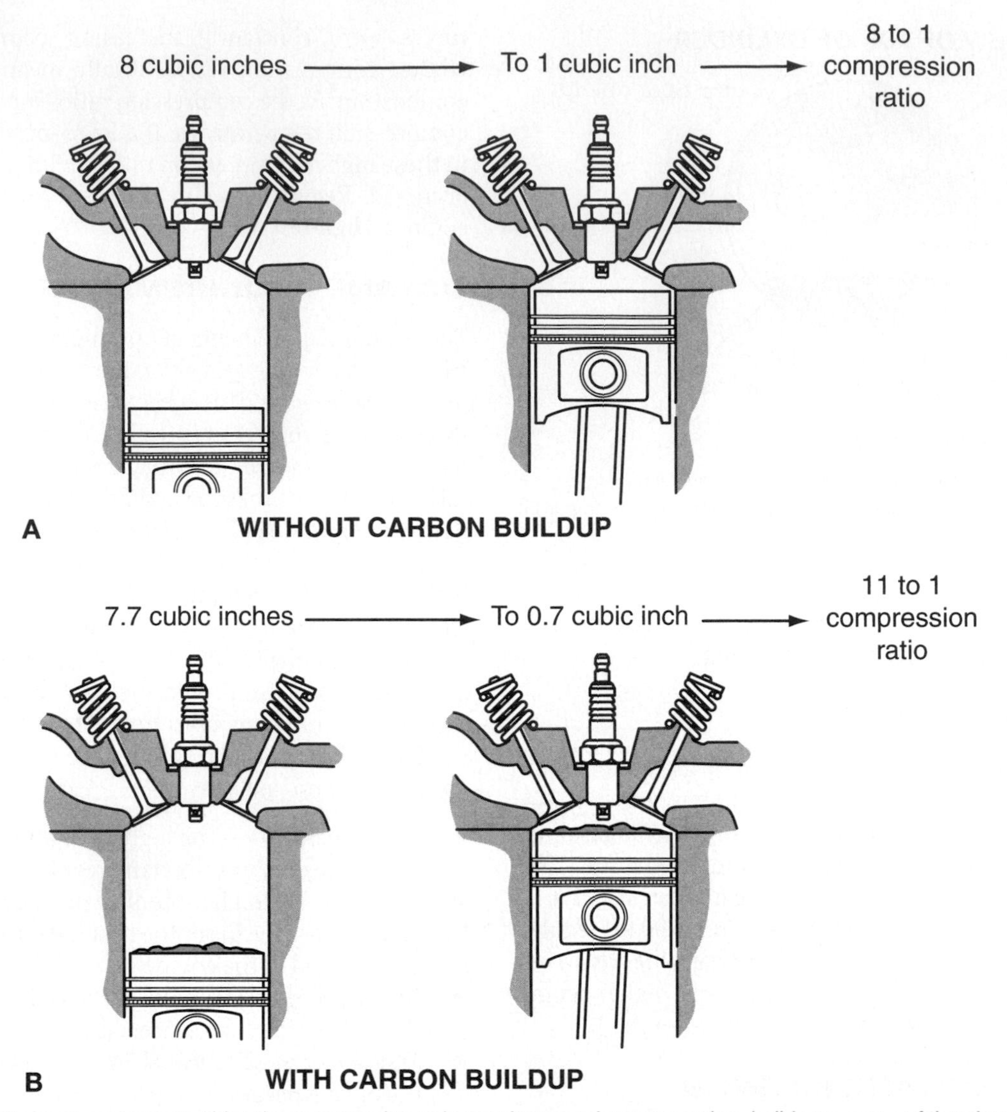

FIGURE 23-7 As an engine gets older, the compression ratio may increase because carbon builds up on top of the piston.

When the lead was removed from gasoline, the octane ratings fell. Therefore, the automobile manufacturing companies had to reduce their compression ratios. Compression ratios dropped from an average of 12 to 1, down to 8.5 to 1. This made engines less efficient. Recent engines still have compression ratios from 8 to 1 up to about 9 to 1. One advantage to lowering compression ratios is that the internal combustion temperatures are lower. This means fewer **nitrogen oxides** are produced within the exhaust. Nitrogen oxides are also a source of pollution. Nitrogen oxides are expressed as NO_x.

The problem with compression ratios and octane ratings exists only in engines that compress an air and fuel mixture. Diesel engines do not have this problem. Diesel engines have increased their compression ratios to as high as 25 to 1. This can be done because the compression stroke on a diesel engine compresses only air. Fuel is then injected near TDC. It is nearly impossible for a diesel engine to detonate, because fuel does not enter the cylinder until the piston is near TDC (***Figure 23–8***).

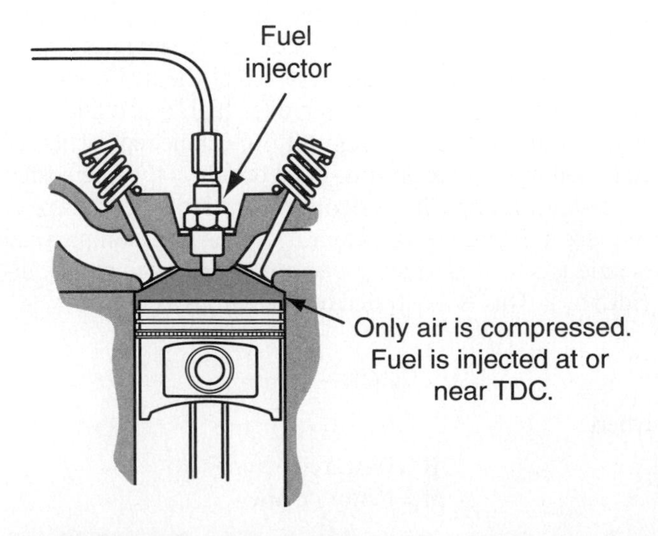

FIGURE 23-8 In a diesel engine, only air is compressed. There is no chance that the fuel will ignite too soon.

CUSTOMER CONCERN:
Using Unleaded Fuel in Older Cars
An older (classic) vehicle has always used a leaded fuel. Is it OK to run unleaded fuel in such an engine? Could unleaded fuels damage the engine?

SOLUTION: The biggest problem in using an unleaded fuel in older engines is that it shortens the life of the exhaust valves. The lead in leaded fuels lubricates the valves. In engines designed to use unleaded fuel, the valves are harder, so they don't need as much lubrication. If the engine is going to be used for a long time, try changing the valves to a harder type if they are available. Also, an oiler kit that oils the valves from the top could improve the situation. A lead substitute can also be added to help correct the problem.

VOLATILITY

Gasoline must vaporize in order to burn effectively. If a fuel does not vaporize completely before going into the combustion chamber, combustion efficiency and fuel efficiency, in turn, will decrease. **Volatility** is defined as the ease with which a gasoline vaporizes. Highly volatile fuels **vaporize** very easily. Fuels that are used in cold weather must be highly volatile. In warm weather, however, gasoline should be less volatile to prevent formation of excess vapor. Excess vapor causes a loss of power or stalling due to vapor lock. Gasolines vary in volatility seasonally.

VAPOR LOCK

Fuels that have high volatility also have a tendency to cause a condition known as **vapor lock**. Vapor lock is a vapor buildup that restricts the flow of gasoline through the fuel system. Vapor lock occurs from the heating of fuel, causing it to turn to a vapor. During very warm operating conditions, vapor lock can occur in fuel lines that are near exhaust systems or near other heat sources. When vapor lock occurs, the fuel vapors cannot be pumped by the fuel pump. This action blocks the fuel to the fuel injection system. To eliminate the potential of vapor lock, additional vapor lines are added to the fuel system to relieve and remove the vapor from the system.

GASOLINE ADDITIVES

Various additives are put into fuel to change its characteristics. These additives have different properties and uses. Many are added during the refining processes; others can be added by the consumer.

CUSTOMER CONCERN:
Eliminate Vapor Lock
A customer has frequently complained about vapor lock. How can vapor lock be reduced?

SOLUTION: Vapor lock is normally caused when fuel is heated to its vapor point in the lines of the fuel system. Methods used to reduce vapor lock include:

1. Use neoprene and impregnated fabric lines because they are less of a heat sink than metal lines.
2. Try to direct the hot air away from the area of vapor lock.
3. Devise shields to eliminate hot exhaust heat from radiating to the fuel lines.
4. Check the vapor point on the specific fuel that is being used.

Each vehicle will be somewhat different, so the best solution will depend on the car and the exact cause of the problem.

Ethyl alcohol or methanol can be put into gasoline as an anti-icing additive. It prevents gasoline from freezing. Winter additives, including **anti-icers**, stop fuel lines and other fuel system components from icing up in cold weather. For example, some anti-icers help keep ice from building up on the throttle plates. If ice does build up, missing and stalling will result. The ice is formed when moist air hits a throttle plate that has been cooled by the vaporization of gasoline. The ice then restricts the flow of air into the engine.

Oxidation inhibitors are added to gasoline that is being stored for long periods of time. During storage, harmful gum deposits form because of the reaction of certain gasoline chemicals with each other and with oxygen. Oxidation inhibitors help the gasoline to remain more stable. This helps to reduce gum, varnishes, and other deposits in the gasoline.

Gasoline is in constant contact with various metals found in the fuel lines, fuel tank, fuel injectors, and so on. **Rust inhibitors** are added to gasoline to inhibit any reaction between the fuel and the metal. Without rust inhibitors, small abrasive particles will form, plugging the fuel system and filters.

Detergents are added to gasoline to clean certain components inside the fuel system and engine. For example, a fuel that has detergents helps to keep the fuel injectors clean during normal operation.

Ignition control and combustion modifiers are also added to gasoline. These additives help prevent spark plug fouling and **preignition**. Preignition is caused by glowing carbon

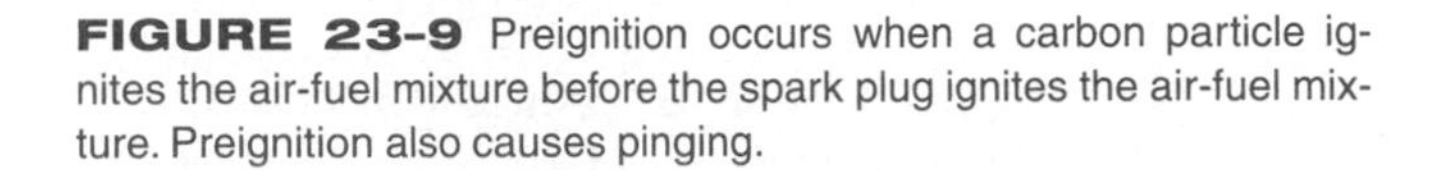

FIGURE 23-9 Preignition occurs when a carbon particle ignites the air-fuel mixture before the spark plug ignites the air-fuel mixture. Preignition also causes pinging.

deposits in the combustion chamber. These deposits ignite the air-fuel mixture before the spark plug fires (***Figure 23–9***). This causes a pinging or knocking sound as does detonation. Other additives include antirust, antigum, and antiwear chemicals.

ENGINE RUN-ON

Another characteristic of combustion caused by poor-quality gasoline is called **postignition**. Postignition is when the engine runs after the ignition has been shut off (***Figure 23–10***). For example, a poor-quality gasoline may have many glowing deposits inside the combustion chamber during operation. When the ignition is shut off, these glowing deposits act as the ignition source for the air and fuel still in the combustion chamber. Postignition is also called dieseling or run-on.

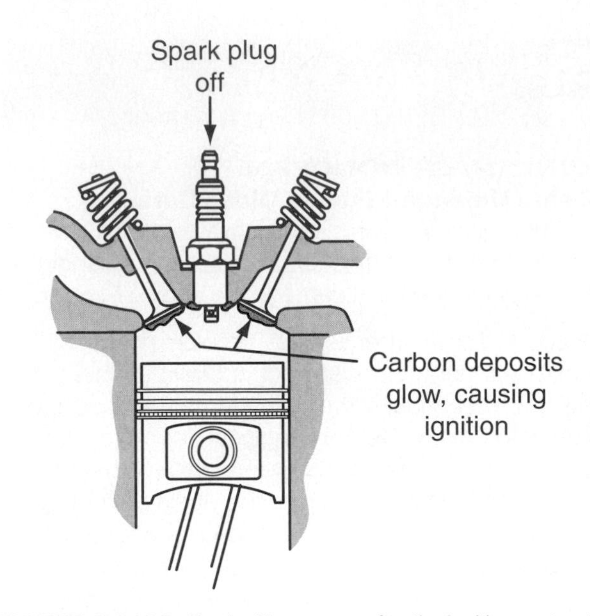

FIGURE 23-10 Postignition occurs after the ignition system has been shut off. A glowing particle of carbon ignites the air-fuel mixture.

23.3 PROPERTIES OF DIESEL FUEL

Diesel fuels are also a mixture of hydrogen and carbon molecules. However, diesel fuels have more energy per gallon than gasoline. In addition, diesel fuels burn with higher efficiency during combustion.

Diesel fuel is part of a group of fuel oils called distillates. This group of fuel oils makes up such products as jet fuels, kerosene, home fuel oil, and diesel fuels.

Three properties are important in the selection of diesel fuel. These include cetane number, distillation end point, and sulfur content. Other characteristics, including pour point, flash point, viscosity, and ash content, are also considered.

CETANE NUMBER

There is a delay between the time that fuel is injected into the cylinder and the time that the hot gases ignite. This time period or delay is expressed as a **cetane number**. Cetane numbers range from 30 to 60 on diesel fuels. The cetane number is an indication of the ignition quality of diesel fuel. The higher the cetane number, the better the ignition quality of the fuel. High cetane numbers should be used for starting in cold weather. **Ether**, with a cetane number of 85 to 96, is often used for starting diesel engines in cold weather. If a low cetane number is used in a diesel engine, some of the fuel may not ignite. The fuel will then accumulate within the cylinder. When combustion finally does occur, this excess fuel will explode suddenly. This may result in a knocking sound in a gasoline engine.

DISTILLATION END POINT

Fuels can be burned in an engine only when they are in vaporized form. The temperature at which a fuel is completely vaporized is called the **distillation end point**. The distillation end point should be low enough to permit complete vaporization at the temperature encountered in the engine. This means that for engines operating at reduced speeds and loads or in cold weather, lower distillation end points will give better performance.

SULFUR CONTENT

The **sulfur** content in a diesel fuel should be as low as possible. Sulfur is part of crude oil when it is taken out of the ground. Refining is designed to remove as much sulfur as is practical. There is a direct link between the amount of sulfur present in the fuel and the amount of corrosion and deposit formation within the engine. Sulfur is also contained in the exhaust gases. These gases can pollute the air and cause a significant amount of **acid rain**. Certain tests have shown that increasing sulfur content from 0.25% to 1.25% increases engine wear by 135%. Engine wear is most noticeable on the cylinder walls and piston rings.

POUR POINT

Diesel fuels are also rated as to their pour point. Pour point is defined as the temperature at which fuel stops flowing. For cold weather operation, the pour point should be about 10 degrees Fahrenheit below the **ambient** (surrounding air) temperature at which the engine is run.

CLOUD POINT

The **cloud point** of a diesel fuel is defined as the temperature at which wax crystals start to form in the fuel. Diesel fuels tend to produce wax crystals in very cold weather. The wax crystals plug up fuel filters and injectors.

VISCOSITY

As we saw in the lubricating system, viscosity is a measure of the resistance to flow. The fuel should have a viscosity that allows it to flow freely in the coldest operation. The viscosity of fuel oil also affects the size of the fuel spray droplets from the fuel injection nozzles. The higher the viscosity, the larger the droplets. The size of the droplets, in turn, affects the **atomization** qualities of the fuel spray.

Fuel oil viscosity is normally checked at 100 degrees Fahrenheit. It is measured in centistokes. This is another unit used to measure viscosity. It is used instead of the Saybolt Universal Second. Additives are incorporated into the fuel during refining to keep the viscosity at the correct level.

FLASH POINT

The **flash point** of a diesel fuel is defined as the fuel's ignition point when exposed to an open flame. It is determined by heating the fuel in a small enclosed chamber. The temperature at which vapors ignite from a small flame passed over the surface of the liquid is the flash point temperature. The flash point should be high enough for the fuel to be handled safely and stored without danger of explosion.

ASH CONTENT

The ash content of diesel fuels is a measure of the impurities, which include metallic oxides and sand. These impurities cause an abrasive action on the moving parts of the engine. The amount of ash content should be kept to a minimum.

ASTM DIESEL FUEL CLASSIFICATION

Diesel fuels are classified as either No. 1-D or No. 2-D. The characteristics of each are shown in ***Figure 23–11***. These are the American Society for Testing and Materials (ASTM) classifications.

ASTM CLASSIFICATION OF DIESEL FUEL OILS

	NO. 1-D	NO. 2-D
Flash Pt.; °F Min.	100	125
Carbon Residue; %	0.15	0.35
Water and Sediment; (% by volume) Max.	Trace	0.10
Ash; % by Wt.; Max.	0.01	0.02
Distillation, °F 90% Pt.; Max.	550	640
Min.	–	540
Viscosity at 100°F; centistrokes Min.	1.4	2.0
Max.	2.5	4.3
Sulfur; % Max.	0.5	0.7
Cetane No; Min.	40	40

FIGURE 23–11 The ASTM classifications for the two types of diesel fuel. (Copyright ASTM. Reprinted with permission)

Grade No. 1-D fuels have the lowest boiling ranges (are the most volatile). These fuels also have the lowest cloud and pour points. Fuels within this classification are suitable for use in high-speed diesel engines in services involving frequent and relatively wide variations in loads and speeds. This may include stop-and-go bus and door-to-door operations. They are also used where abnormally low fuel temperatures are encountered.

Grade No. 2-D fuels have higher and wider boiling ranges than 1-D fuels. These fuels normally have higher cloud and pour points than 1-D fuels. They are used in diesel engines in services with high loads and uniform speeds. They are also used in climates where cold starting and cold fuel handling are not severe problems. These fuels satisfy the majority of automotive diesel applications.

23.4 GASOHOL (E10) AS A FUEL

Gasohol is a term used to describe a motor fuel that blends 90% gasoline and 10% alcohol (***Figure 23–12***). It is also

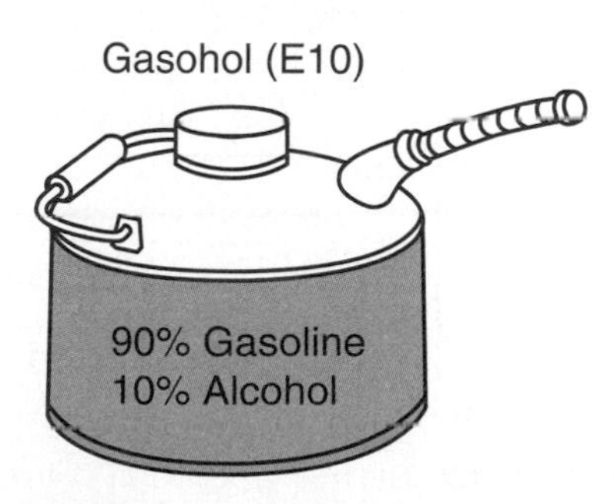

FIGURE 23–12 Gasohol is made from 90% gasoline and 10% alcohol.

called E10. The alcohol in gasohol is called **ethanol**. Ethanol is produced by distilling agricultural crops such as corn, wheat, timber, and sugar cane. The purpose of using alcohol is that for every gallon of gasohol sold, 10% of the fuel is renewable, whereas gasoline is a nonrenewable source of fuel. The study and use of E10 grew out of the energy shortage of 1973.

GASOHOL CHARACTERISTICS

Overall, the Btus in a gallon of E10 are slightly less than in gasoline. Gasoline has about 115,000 Btu per gallon. Ethanol contains only 75,000 Btu per gallon. By adding ethanol to gasoline, the total Btu content is slightly less than that of gasoline. Alcohol, however, has a slightly higher octane than gasoline. The octane of E10 is normally about 3 to 4 points higher than that of gasoline. The increase in octane reduces or eliminates engine knock and ping. E10's higher octane also reduces engine postignition and run-on. Testing has shown that when E10 is used, cleaner burning occurs in the combustion chamber. This tends to reduce harmful carbon deposits and soot buildup, which, in turn, promotes longer engine life. The alcohol in E10 is also an anti-icer (it prevents freezing in cold weather). E10 has been used in many parts of the country. Car manufacturers vary in their recommendations concerning the use of gasohol.

Car Clinic: ALCOHOL FUEL

CUSTOMER CONCERN:
Alcohol in the Fuel
Many car owners are concerned about the use of alcohol in fuel such as gasohol. Is there any problem with running high levels of alcohol in the fuel?

SOLUTION: Alcohol may cause engine damage. The amount of damage depends on the percentage of alcohol in the gasoline. Anything above 10% could damage the engine. At 15%, alcohol may damage aluminum parts and cause steel to begin rusting. Alcohol may also dilute the engine oil considerably. Always be aware of how much alcohol is being added to the fuel. If a vehicle has been using gasoline for a long period of time and the owner switches to gasohol (10% alcohol), the alcohol may loosen some gum deposits in the fuel tank and lines, and it has a tendency to plug up fuel filters.

23.5 LIQUID PETROLEUM GAS

LPG CHARACTERISTICS

Another type of fuel that can be used in the automobile is called liquid petroleum gas, or LPG. LPG is also a hydrocarbon fuel and a by-product of the distillation process. It is made primarily of **propane** and butane. One major difference of LPG is its boiling point (***Figure 23–13***). LPG has a very low boiling point. In normal ambient pressures, LPG has boiled and is in the vapor form. If the fuel is put under a pressure, the boiling point can be raised. The fuel can then be stored in a pressurized container as a liquid (***Figure 23–14***). This makes storage much easier but possibly more dangerous in accidents. When the fuel is allowed to come out of the container, it turns back to a vapor.

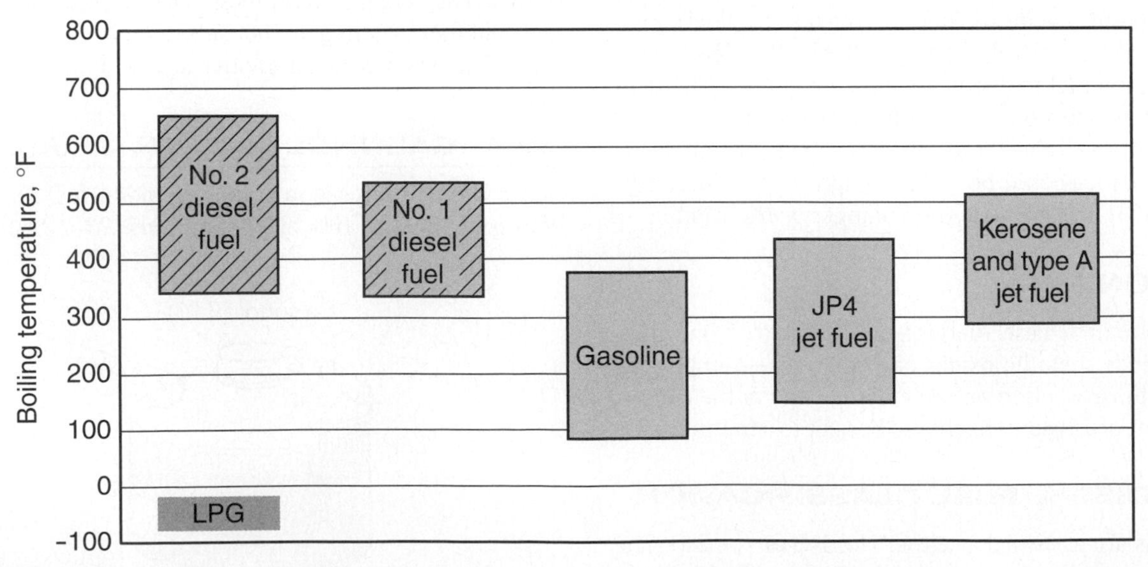

FIGURE 23-13 The boiling point of LPG is considerably lower than that of other fuels. At ambient pressure and temperature, LPG fuel has already boiled and is in a vaporized state.

FIGURE 23–14 LPG is stored in a pressurized container to keep it in a liquid form.

LPG also has a higher octane rating (about 100 octane) than gasoline. This means that compression ratios could be increased slightly to improve efficiency. However, because the Btu content of LPG per gallon is less, there is slightly less power than with gasoline. Many vehicle manufacturers are now marketing multifuel and flexible-fuel vehicles today to allow for LPG use. Always follow the manufacturer's fuel use recommendations.

ADVANTAGES AND DISADVANTAGES OF LPG

One of the big advantages of using LPG is that it is extremely clean and pollution-free. This is the reason many lift trucks use LPG fuels inside buildings. The exhaust is not as dangerous as the carbon monoxide that is produced by gasoline. Large fleet operations are probably the biggest users of LPG fuels in the automotive industry. In these operations, the fuel can be containerized and sold in larger quantities. However, the fuel systems must be altered slightly to allow LPG to be used. Regulators, converters, and several other components must be installed to operate the system correctly.

23.6 ALTERNATIVE FUELS

OXYGENATED FUELS

Gasoline producers have continued to research and design better fuels based on the need for pollution control. Out of this research, oxygenated gasoline has developed. Oxygenated gasoline is considered a finished motor gasoline that has an increase in the oxygen content. By definition, oxygenated fuels must have an oxygen content of greater than 2.7% by weight. Generally, about 7% to 8% of the volume is oxygen. Its main purpose of development was to help reduce the carbon monoxide (CO) emissions for gasoline vehicles. In fact, gasohol that is added to gasoline is considered an oxygenated fuel.

In order to add the necessary amount of oxygen to gasoline, several chemicals and other fuels have been added to oxygenated gasoline. The most common additives include ethanol, **methanol**, MTBE (methyl tertiary butyl ether), and ETBE (ethyl tertiary butyl ether). All these additives increase the percentage of oxygen in gasoline. The end result is that carbon monoxide (CO) emissions are reduced. Ethanol and methanol are very similar in chemical makeup and are both alcohol based. The only difference is the number of carbon and hydrogen molecules in each. The best additive to use depends on the time of the year (winter or summer), the location, and the cost of each additive.

Oxygenated fuels can be used in most modern vehicles. However, when MTBE is used, there is some danger of increased corrosion with certain metals. In addition, since the chemicals that are being added to the gasoline have fewer Btus per gallon, there will be a slight decrease in fuel economy. Studies suggest about a 2–3% drop in fuel mileage. However, there is about a 2-to-3-point increase in the octane rating of these fuels (***Figure 23–15***).

REFORMULATED FUELS

The Clean Air Act Amendments of 1990 required the use of reformulated gasoline (RFG) in nine major metropolitan

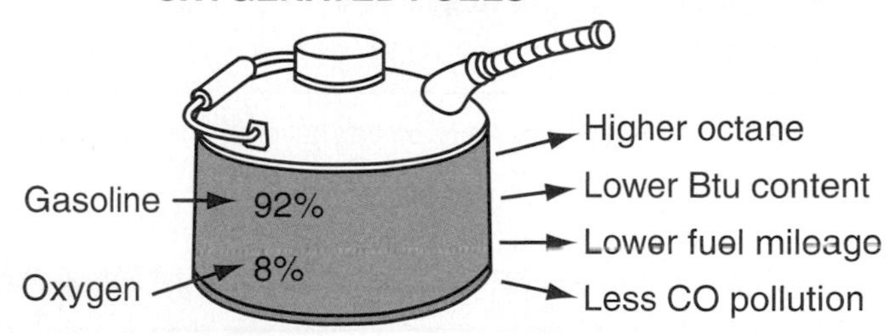

FIGURE 23–15 Oxygenated fuels use a mixture of gasoline and methanol or another additive to increase the percentage of oxygen in the fuel.

areas of the United States. These areas were chosen because they had serious air pollution that contributed to reduction in the ozone layer of the atmosphere. When RFG is used in automobiles, fewer air toxins and ozone-forming pollutants are produced.

The two most common RFG fuels are called E85 and M85. E85 is a blend of 85% ethanol and 15% gasoline; M85 is a blend of 85% methanol and 15% gasoline. With the use of ethanol and methanol, there is enough oxygen in the fuel to reduce carbon monoxide and still reduce ozone pollutants. However, because of the corrosive nature of ethanol and methanol, these reformulated fuels can only be used in vehicles that are specifically designed for their use.

Many vehicle manufacturers have designed alternative fuel vehicles. RFG fuels are considered alternative fuels because the alcohol additive can be produced from nonpetroleum products such as wood, corn, agricultural waste, grain, and so on. In order to identify which vehicles can use alternative fuel, refer to the vehicle manufacturer. They should be able to provide a list of alternative fueled vehicles that they produce. This information can also be obtained by going to the Internet and searching under methanol, ethanol, RFG, and/or Alternative Fuels Data Center.

One of the complaints about RFG is that of fuel mileage. Since there is a significant amount of methanol and ethanol in the fuel, the overall Btu content of the fuel is less. Referring to ***Figure 23–16***, note that a gallon of gasoline has about 110,000 Btu available. Both the E85 and the M85 have less Btu content. The other comparisons help to better understand the differences between alternative fuels and gasoline, including cost and pollution control.

USING ETHANOL-BASED FUELS

Ethanol fuel blends are approved under the warranties of all domestic and foreign automobile manufacturers that market vehicles in the United States. In fact, all U.S. vehicle manufacturers recommend the use of oxygenated fuels, such as ethanol, because of their clean air benefits.

Ethanol is also considered a good cleaning agent. It can loosen contaminants and residues that have been deposited in a vehicle's fuel system and that have collected in fuel filters. Occasionally, this can be a problem when ethanol is used in older vehicles. It can easily be corrected by changing fuel filters. The problem seldom occurs in today's computerized vehicles.

All alcohols have the ability to absorb water. Since ethanol blends contain at least 10% ethanol, condensation of water in the fuel system is absorbed and does not have the opportunity to collect and freeze.

Ethanol blends can eliminate the need for adding a gas line antifreeze.

Ethanol can be used as a fuel in both older and new engines. However, for engines older than 1969, with nonhardened valve seats, a lead substitute should be added to the fuel to prevent premature valve seat wear.

Typically, valve burning on today's vehicles is decreased when ethanol is used. This is because ethanol burns cooler that gasoline. Many high-performance racing engines use pure alcohol for this same reason. When operating correctly, modern computerized vehicles will perform better than noncomputer-equipped vehicles. The improved performance is due to the computerized fuel management system, which is able to make more adjustments with changes in operating conditions and fuel type.

	GASOLINE	E85	M85
Btus/gallon	110,000	80,500	65,000
Costs/gallon	1.50–2.00	More	Less
Main fuel source	Petroleum	Corn, grains, or agricultural waste	Natural gas, coal, or biomass
Efficiency compared to gasoline	100%	70%	57%
Pollution	High	Low	Low

FIGURE 23–16 Comparisons among gasoline and two alternative fuels, also known as reformulated gasoline.

SUMMARY

The following statements will help to summarize this chapter:

- Gasoline is considered a hydrocarbon, or a series of hydrogen and carbon chemical combinations.
- Refineries separate crude oil hydrocarbons into different products including gasoline, oil, diesel fuel, and so on.
- Knocking, a characteristic of burning gasoline, occurs when poor quality fuel ignites on one side of the combustion chamber and the spark plug ignites the other side.
- Octane is a measure of the fuel's ability to resist burning.

- Lead was used years ago as an additive to increase octane.
- Gasoline has many additives including anti-icers, detergents, lubricants for valves, and combustion modifiers.
- Poor quality gasoline can cause postignition, or continued engine running after the ignition system has been shut off.
- Diesel fuels are rated by the cetane number.
- Some characteristics of diesel fuel include the distillation end point, sulfur content, pour point, cloud point, flash point, and ash content.
- Gasohol, a mixture of 90% gasoline and 10% alcohol, is used in some vehicles.
- Liquid petroleum gas, or LPG, is another hydrocarbon used as a fuel in some vehicles.
- Oxygenated fuel has at least 2.7% oxygen in the fuel and is used to reduce carbon monoxide emissions.
- Oxygenated fuels use methanol, ethanol, MTBE, or ETBE additives to increase the percentage of oxygen in the fuel.
- Reformulated gasoline, known as RFG, combines 85% methanol or ethanol with 15% gasoline and can be used only in certain vehicles specifically designed for these fuels.
- Ethanol fuels act as a good cleaner and can also absorb moisture in the fuel system.

TERMS TO KNOW

Can you explain each of the following terms? Review the chapter until you can use each term correctly.

Acid rain
Ambient
Anti-icers
Atomization
Boiling point
Btu
Cetane number
Cloud point
Condense
Distillation
Distillation end point
Ethanol
Ether
Flash point
High-test
Hydrocarbons
Methanol
Nitrogen oxides
Octane number
Oxidation inhibitors
Pinging
Postignition
Preignition
Propane
Regular gasoline
Rust inhibitors
Sulfur
Tetraethyl lead
Vaporize
Vapor lock
Volatility

REVIEW QUESTIONS

Multiple Choice

1. Crude oil is primarily made of what molecules?
a. Nitrogen and oxygen
b. Carbon and hydrogen
c. Sulfur and nitrogen
d. Nitrogen and sulfur
e. Sulfur and carbon

2. What is the major difference between different fuels?
a. Boiling points
b. Acid levels
c. Degree of lubrication
d. Density
e. Thickness

3. The process of separating crude oil into different fuel is called ________.
a. Combustion
b. Condensing
c. Distillation
d. Departing
e. None of the above

4. When two flame fronts hit each other, a ________ sound is heard.
a. Knocking
b. Soft
c. Cracking
d. Hissing
e. None of the above

5. The resistance to burning of gasoline is defined as:
a. Cetane
b. Octane
c. Butane
d. Methane
e. Ethane

6. Which octane is a result of calculating (R + M)/2?
a. Road octane
b. Motor octane
c. Research octane
d. All of the above
e. None of the above

7. Which of the following will influence the octane number needed in an engine?
a. Fuel settings
b. Air temperature
c. Altitude
d. All of the above
e. None of the above

8. Which chemical was added to gasoline to increase the octane number in leaded gasoline?
a. Sulfur lead
b. Distillation chemicals
c. Tetraethyl lead
d. Nitrogen
e. Oxygen

9. When lead was removed from gasoline, what had to be done with the compression ratios?
a. They were increased
b. They were kept the same
c. They were reduced or lowered
d. They were doubled
e. None of the above

10. The condition that occurs when the engine ignition is shut off and the engine continues to run is called ________.
a. Pinging
b. Knocking
c. Early firing
d. Postignition
e. Run-on

11. The measure of a diesel fuel's quality is called:
a. Ash point
b. Cetane number
c. Sulfur content
d. Methane number
e. Octane number

12. The temperature at which a diesel fuel is completely vaporized is called the:
a. Distillation end point
b. Flash point
c. Pour point
d. Fire point
e. Boiling point

13. The term *ambient* is defined as:
a. Surrounding
b. Closed in
c. Open
d. End point
e. Cetane point

14. A diesel fuel's ignition point when exposed to open flame is called its:
a. Pour point
b. Flash point
c. Distillation point
d. Burning point
e. Firing point

15. What are the two types of diesel fuel classifications?
a. 1-D and 2-D
b. 2-D and 4-D
c. 3-D and 4-D
d. 4-D and 5-D
e. 4-D and 1-D

16. Gasohol is a mixture of what percentages of gasoline and alcohol?
a. 10% gasoline and 90% alcohol
b. 20% gasoline and 80% alcohol
c. 30% gasoline and 70% alcohol
d. 40% gasoline and 60% alcohol
e. None of the above

17. Gasohol has ________ ratings compared to gasoline.
a. Lower octane
b. Higher octane
c. The same octane
d. Higher cetane
e. Lower cetane

18. If put under a pressure, LPG will change to a ________.
a. Liquid
b. Gas
c. Solid
d. Semisolid
e. Semiliquid

The following questions are similar in format to ASE (Automotive Service Excellence) test questions.

19. *Technician A* says gasoline has an octane rating. *Technician B* says gasoline has a cetane rating. Who is correct?
a. A only
b. B only
c. Both A and B
d. Neither A nor B

20. *Technician A* says that resistance to burning is measured with a cetane rating on gasoline. *Technician B* says that resistance to burning has no type of measurement for gasoline. Who is correct?
a. A only
b. B only
c. Both A and B
d. Neither A nor B

21. *Technician A* says that preignition can be caused by glowing deposits inside the combustion chamber. *Technician B* says that preignition can be caused by bad lubrication. Who is correct?
a. A only c. Both A and B
b. B only d. Neither A nor B

22. *Technician A* says that lead has recently been added to fuel to improve nitrogen oxides. *Technician B* says that lead has recently been added to fuel to lower the compression ratio. Who is correct?
a. A only c. Both A and B
b. B only d. Neither A nor B

23. *Technician A* says that vapor lock occurs when fuel is too hot. *Technician B* says that vapor lock is caused by a lack of oxidation. Who is correct?
a. A only c. Both A and B
b. B only d. Neither A nor B

24. *Technician A* says that anti-icer could be used in the winter. *Technician B* says that anti-icer is a cleaning agent in the fuel. Who is correct?
a. A only c. Both A and B
b. B only d. Neither A nor B

25. *Technician A* says that the lack of oxidation inhibitors will cause the fuel filter to plug up. *Technician B* says that oxidation inhibitors are added to gasoline that is being stored. Who is correct?
a. A only c. Both A and B
b. B only d. Neither A nor B

26. *Technician A* says that as carbon builds up on top of the piston, the compression ratio decreases. *Technician B* says that as carbon builds up on top of the piston, detonation decreases. Who is correct?
a. A only c. Both A and B
b. B only d. Neither A nor B

27. *Technician A* says that oxygenated fuels add oxygen to the fuel to reduce the amount of carbon monoxide. *Technician B* says that oxygenated fuels can only be used on certain vehicles. Who is correct?
a. A only c. Both A and B
b. B only d. Neither A nor B

28. *Technician A* says that reformulated fuels, known as RFGs, can only be used in certain vehicles. *Technician B* says that RFG fuels help to reduce ozone-forming pollutants. Who is correct?
a. A only c. Both A and B
b. B only d. Neither A nor B

Essay

29. What does the term *distillation* mean?

30. What is knocking?

31. How does octane of a fuel relate to the burning of fuel?

32. Why are anti-icers added to fuel?

33. State several advantages and disadvantages of using gasohol.

34. What are two advantages and two disadvantages of using reformulated gasoline as a fuel?

Short Answer

35. When storing gasoline, it is a good idea to add a/an ________ to the fuel.

36. To prevent gasoline from freezing in cold climates, add ________ to the fuel tank.

37. To help keep the fuel injector system clean, always use a fuel to which ________ has been added.

38. Vapor lock can be developed in the fuel system when fuel lines are ________ to the heat source.

39. One of the major processes done in a refinery is that of ________.

40. Reformulated gasoline has __________ Btu per gallon than gasoline.

Applied Academics

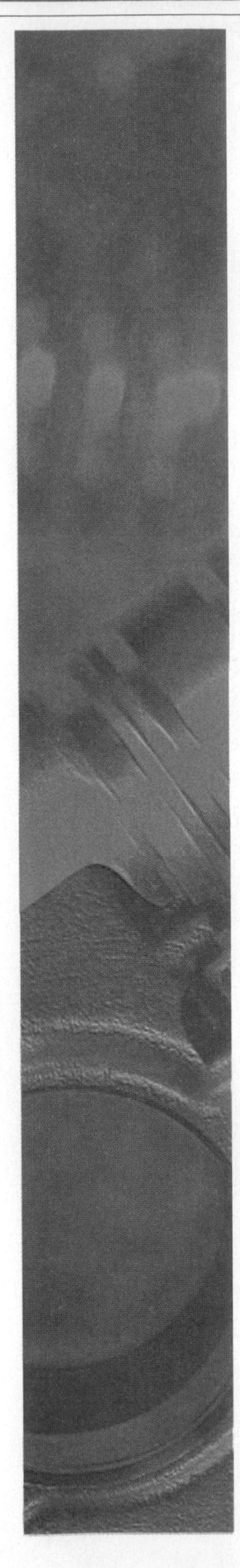

TITLE: Alternative Fuels

ACADEMIC SKILLS AREA: Language Arts

NATEF Automobile Technician Tasks:

The technician interprets charts, tables, and graphs to determine manufacturers' specification for engine operation to identify out-of-tolerance systems and subsystems.

CONCEPT:

The use of alternative fuels is gaining popularity among the automotive industry. Several types are considered alternative fuels as shown in the accompanying table. A comparison of each liquid or gas fuel is shown with its chemical structure, primary components, main fuel source, energy content per gallon, and energy ratio compared to gasoline.

ALTERNATIVE FUELS

	Compressed Natural Gas (CNG)	Ethanol (E85)	Liquefied Natural Gas (LNG)	Liquefied Petroleum Gas (LPG)	Methanol (M85)
Chemical structure	CH_4	CH_3CH_2OH	CH_4	C_3H_8	CH_3OH
Primary components	Methane	Denatured ethanol and gasoline	Methane that is cooled cryogenically	Propane	Methanol and gasoline
Main fuel source	Underground reserves	Corn, grains, or agricultural waste	Underground reserves	A by-product of petroleum refining or natural gas processing	Natural gas, coal, or woody biomass
Energy content per gallon	29,000 Btu	80,460 Btu	73,500 Btu	84,000 Btu	65,350 Btu
Energy ratio compared to gasoline	3.94 to 1 or 25% at 3,000 psi	1.42 to 1 or 70%	1.55 to 1 or 66%	1.36 to 1 or 74%	1.75 to 1 or 57%
Liquid or gas	Gas	Liquid	Liquid	Liquid	Liquid

APPLICATION:

Can you determine which type of alternative fuel has the most energy per gallon? How does its energy content compare to gasoline? Which type of alternative fuel would produce the best fuel mileage? Can you describe in simple words to a customer what "energy ratio compared to gasoline" means? Which fuel do you think would be the best for our environment?

Fuel Delivery Systems

CHAPTER 24

OBJECTIVES

After studying this chapter, you should be able to:

- Identify the total fuel flow on an automobile engine.
- Analyze the parts and operation of the fuel tank and fuel metering parts.
- Recognize the parts and operation of mechanical and electrical fuel pumps.
- State the purpose and operation of fuel filters.
- Identify common problems, their diagnosis, and service procedures for the basic fuel delivery system components.

INTRODUCTION

The fuel system is critical to engine operation. This chapter is about the fuel delivery system and its operation. The fuel delivery system consists of the fuel tank, lines, pumps, filters, electronic carburetors, and injectors. However, because electronic carburetors and injection systems have become rather complex, a separate chapter is devoted to each. This chapter deals specifically with the basic components of the fuel delivery system needed to get the fuel to the electronic carburetor and the fuel injectors.

24.1 TOTAL FUEL FLOW

The fuel system is designed so that fuel can be stored in a fuel tank and be ready for delivery to the engine. ***Figure 24–1*** shows a total fuel system flow of an engine with an electronic carburetor. The fuel in the fuel tank is ready to be used when the engine needs it. The fuel pump, which is electrically or mechanically driven, draws fuel from the tank and sends it to the electronic carburetor. The purpose of the electronic carburetor is to keep the fuel at the proper air-fuel ratio. It is very important to maintain a ratio of 14.7

Certification Connection

ASE Connection: The information in this chapter can help you prepare for the National Institute for Automotive Service Excellence (ASE) certification tests. The tests and content areas most closely related to this chapter are:

Test A1—Engine Repair

- **Content Area**—Fuel, Electrical, Ignition, and Exhaust System Inspection and Service

Test A8—Engine Performance

- **Content Area**—Fuel, Air Induction, and Exhaust System Diagnosis and Repair

NATEF Connection: Much of the information in this chapter is related to the NATEF tasks. The NATEF tasks and priority numbers most closely related to this chapter are:

1. Inspect and test mechanical and electrical fuel pumps and pump control systems for pressure, regulation, and volume; perform necessary action. P-1
2. Replace fuel filters. P-1
3. Check fuel for contaminants and quality; determine necessary action. P-3

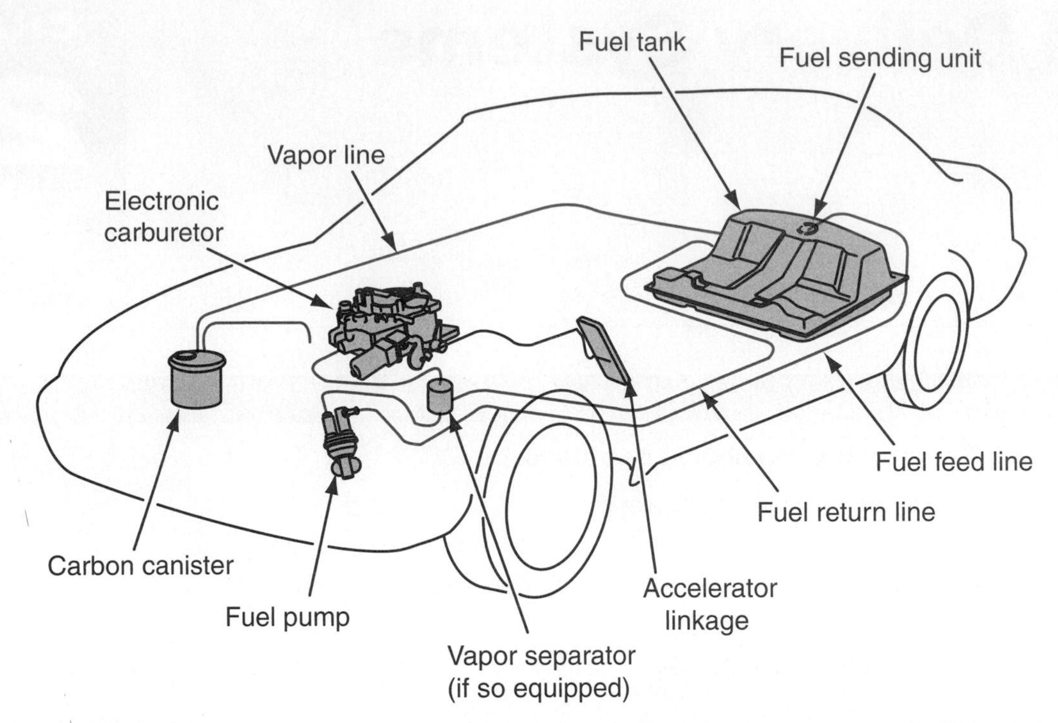

FIGURE 24–1 The fuel system consists of the fuel tank, fuel lines, fuel pump, carburetor or fuel injectors, fuel filter, charcoal canister, and vapor separator.

parts air to 1 part fuel. The air-fuel ratio must be accurate throughout engine operation.

Figure 24–2 shows the total fuel system flow of a fuel-injected engine. The fuel is stored in the fuel tank. The fuel pump is located in the fuel tank. Fuel is pressurized and sent to the fuel feed pipe, through a fuel filter, and into the fuel injectors. A **pressure regulator valve** keeps the pressure constant on the system.

Other parts of the gasoline fuel system include the fuel vapor line, the carbon canister, and the fuel filter. The fuel vapor line is used to send any **fuel vapors** back to the fuel tank. The carbon canister is a pollution control device that holds excess vapors from the tank and electronic carburetor. The carbon canister is studied in detail in the chapter on emission control systems. The fuel filter, which is located directly before the electronic carburetor or fuel

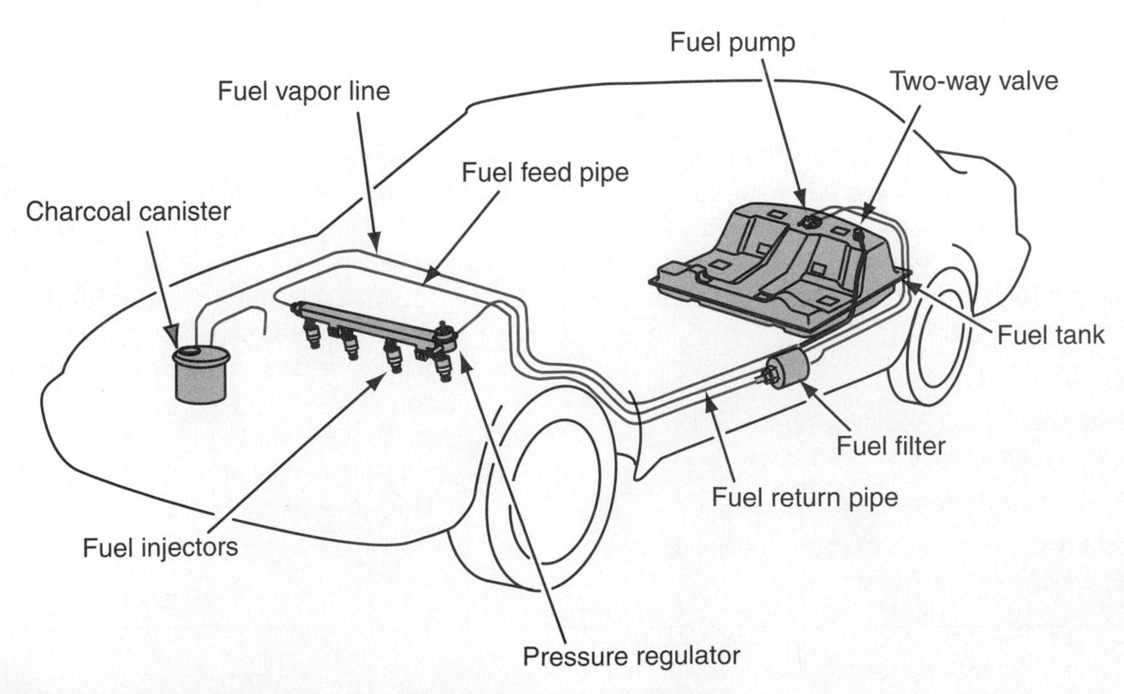

FIGURE 24–2 Total fuel flow for a fuel-injected engine is shown in this illustration.

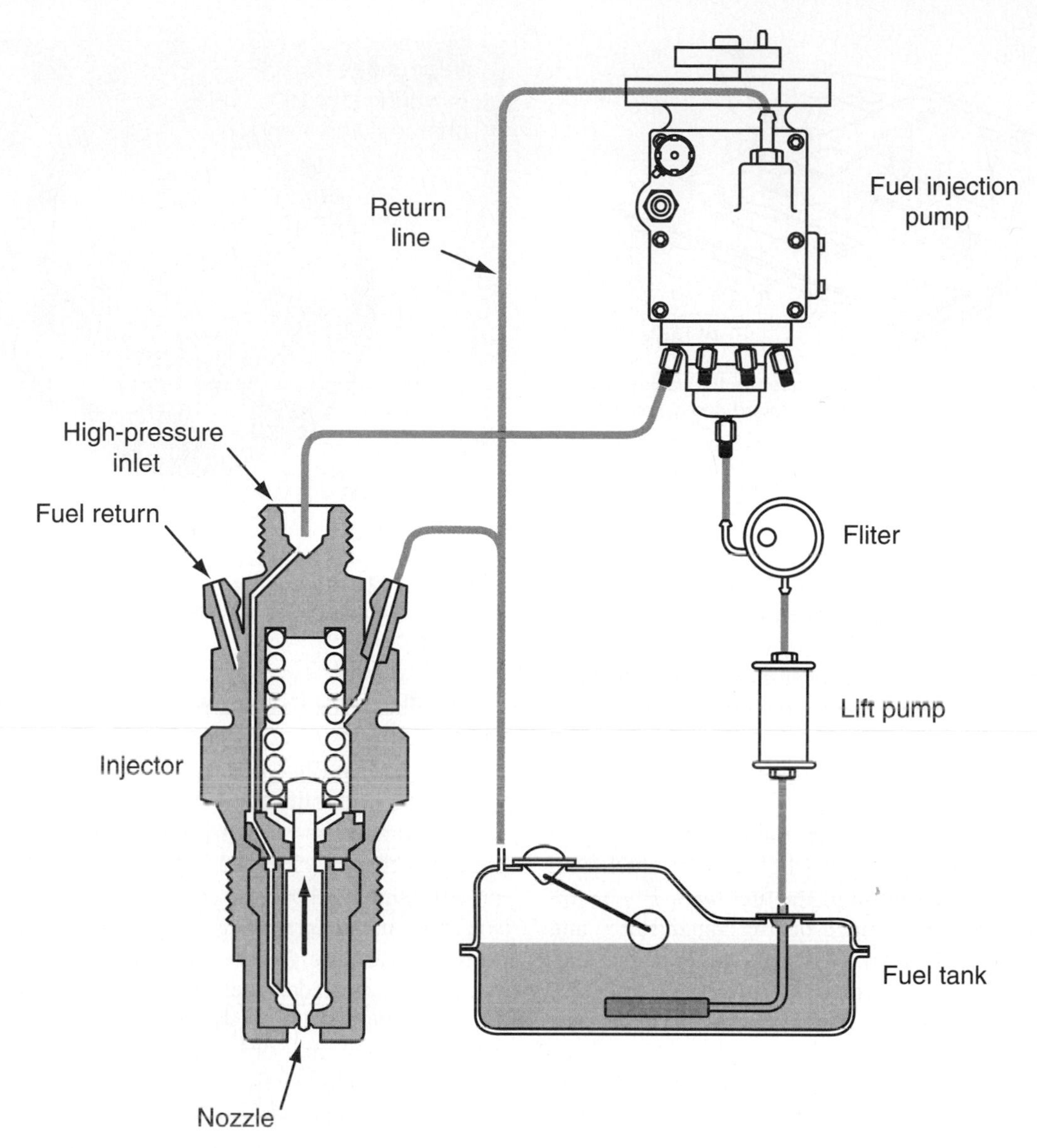

FIGURE 24–3 A diesel fuel system has a tank, mechanical pump, fuel filter, and fuel injectors.

injectors, is used to keep the fuel clean and free of contaminants.

Figure 24–3 shows a diesel fuel system. Here the fuel goes to the **fuel injector** pump, rather than to the carburetor then to the fuel nozzles. All fuel injector systems on diesel engines must have a fuel return line to the fuel tank. This is also shown. Any fuel not used by the engine is returned to the fuel tank or to the suction side of the fuel pump to be used again.

24.2 FUEL TANK

The fuel tank is made to hold or store the excess fuel needed by the automobile. It is usually located in the rear of the vehicle, but in cars that have rear-mounted engines, it is located in the front part of the vehicle. ***Figure 24–4*** shows a typical fuel tank. The shape of the fuel tank depends on the physical design of the vehicle. The fuel tank must fit around the frame and still be protected from impacts.

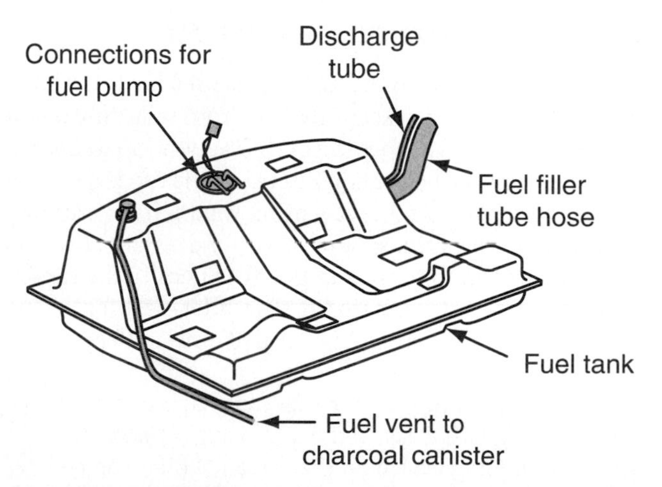

FIGURE 24–4 The fuel tank consists of many parts that are designed to fit around the frame of the vehicle.

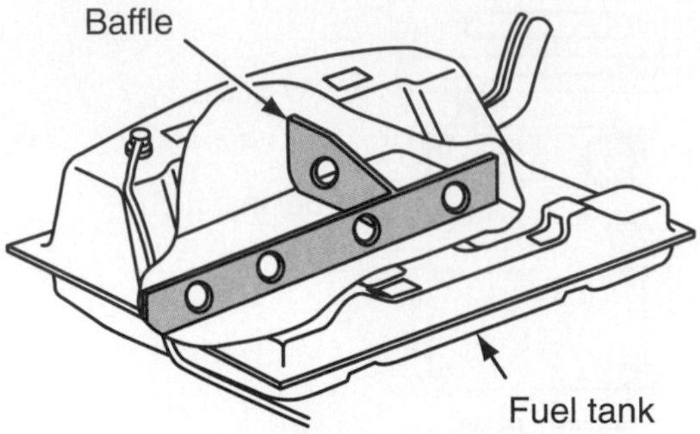

FIGURE 24-5 Baffles are used inside the fuel tank to reduce sloshing and moving of the fuel during cornering and acceleration or deceleration.

The size of the fuel tank depends on the driving range of the vehicle. Because today's cars get better fuel economy than older cars, fuel tanks are smaller. Generally, tank sizes range from 10 to 25 gallons.

The fuel tank is made of thin sheet metal or plastic. Various parts of the fuel tank are welded or soldered together to make the completed assembly. A lead-tin alloy is placed on the sheet metal to keep gasoline from rusting and corroding the tank.

Inside the fuel tank there are several baffles as shown in ***Figure 24–5***. These baffles are used to keep the fuel from sloshing and splashing around in the fuel tank. The baffles help to restrict fuel movement during rapid starts and stops, cornering, and so on. Without the baffles, small bubbles caused by the constant agitation or movement might form in the fuel. If these bubbles were to get into the fuel system, the engine would perform poorly.

Many fuel tanks also have an expansion tank built into the existing tank. The expansion tank is not filled during a normal fill-up at a gasoline station. When the tank is full and the fuel expands because it becomes warmer (during hot summer days) fuel enters the expansion tank rather than overflowing.

FUEL CAP AND FILLER NECK

The fuel, or filler, cap on the fuel tank is used for several reasons. First, the fuel cap keeps the fuel from splashing out of the tank. Second, the cap releases the vacuum created when the fuel is removed by the engine. Third, the cap is designed to release pressure while preventing vapors from escaping directly into the atmosphere. When necessary, the cap passes vapors through tubes in the filler neck to the carbon canister, where they are stored. The filler neck on newer vehicles has a restrictor door so that only unleaded fuel nozzles can be used.

Many designs are used for fuel caps. One common type seals the tank with a cap that has threads (***Figure 24–6***). A ratchet tightening device on the threaded filler cap reduces the chances of incorrect installation. If the wrong cap is used, vapor control may not operate correctly.

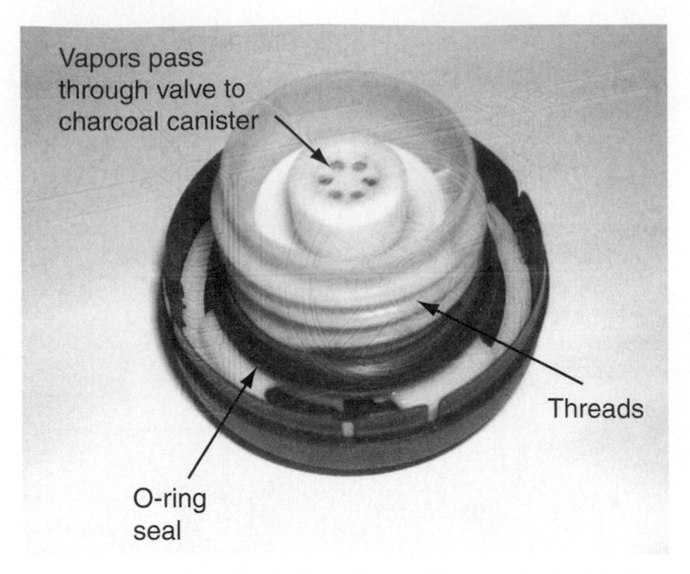

FIGURE 24-6 The fuel cap is used to seal the fuel tank so pressure can be released without allowing vapors to escape directly into the atmosphere.

The filler neck has internal threads. The cap contains a plastic center extension. This acts as a guide when inserted into the filler neck. As the cap is turned, a large O-ring is seated upon the filler neck flange. After the seal and flange make contact, the ratchet produces a clicking noise. This indicates that the seal has been set. Now the vacuum and pressure lines will operate correctly.

A **filler neck restrictor** is used on all vehicles that require unleaded gasoline. ***Figure 24–7*** shows the neck restrictor. Today, only unleaded fuels are available. However, several years ago there was still the possibility of pumping regular leaded gasoline into vehicles. At that time, the neck restrictor was used to allow the consumer to pump only unleaded gas in the vehicle. The smaller hole (produced by the re-

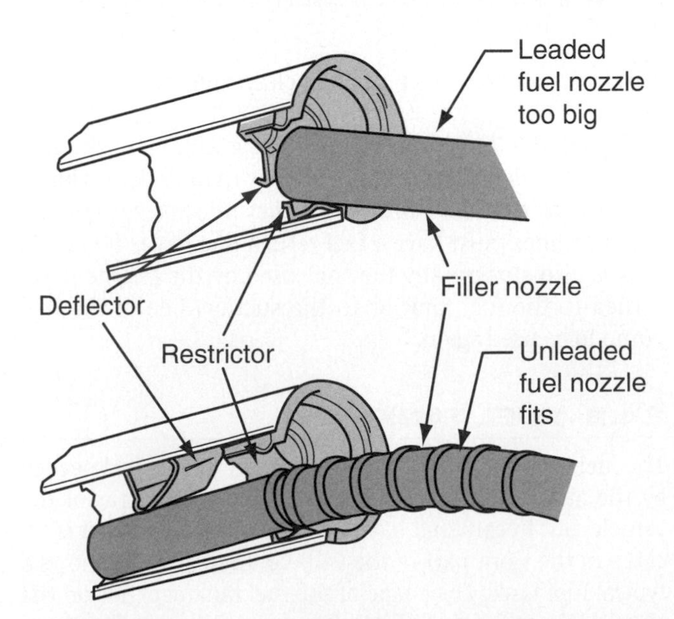

FIGURE 24-7 This restrictor allows only an unleaded fuel nozzle to fit into the filler neck.

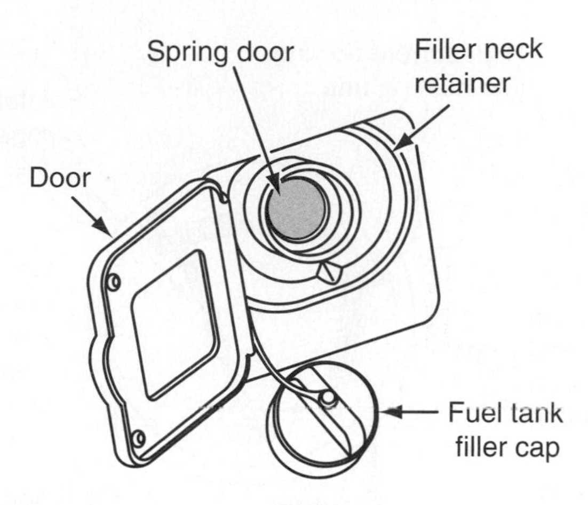

FIGURE 24-8 The small door on the filler neck has a spring to keep it closed when the fuel nozzle has been removed.

strictor) prevented the customer from using the larger nozzle at the fuel pump. The larger nozzle was used only to deliver the leaded gasoline.

Figure 24–8 shows an example of the filler pipe from the outside of the vehicle. The small door on the inside tube has a spring that keeps it closed when the fuel nozzle is removed. The fuel filler cap is a threaded screw-on design. It is often secured to the vehicle by means of a tether or strap. The fuel tank filler cap allows venting of the fuel tank for the first three-quarter turn and for pressure/vacuum relief functions.

CAUTION: *Remember that when the fuel filler cap is removed, there may be a pressure inside the fuel tank. Remove the fuel tank filler cap slowly, allowing a hissing sound to be heard. Wait until the hissing stops before completely removing the fuel tank filler cap. If not, fuel may spray out and cause personal injury.*

Car Clinic:
FUEL TANK CAP

CUSTOMER CONCERN:
Engine Misses at High Speed

A customer complains that when running at full throttle up a slight grade, the vehicle begins to slow down, sputters, and misses. Several items have been repaired or replaced. These include the electronic carburetor, timing, fuel pump, filters, and the exhaust gas recirculation (EGR) valve. What other items should be checked?

SOLUTION: It sounds as if the fuel tank is developing a vacuum, and the fuel is being restricted to the electronic carburetor. Check the fuel tank cap. The caps are designed to vent the tank of fumes and to allow air to come into the tank as the fuel is removed. Also check the carbon canister for damage or lines that are crimped.

PARTS LOCATOR

Refer to photos #19, #20, and #21 in the front of the textbook to see what the ***carbon canister, fuel injectors,*** *and* ***fuel tank inlet*** *look like.*

FUEL METERING UNIT

The operator must know how much fuel is in the tank. This is measured with a fuel metering unit. Many styles are used. One type has a hinged float inside the tank. As the float changes position with different levels of fuel, the sensing unit changes its resistance. The resistance then changes the current in the circuit and the position of the needle on the dashboard gauge. ***Figure 24–9*** shows a typical in-tank fuel metering unit.

Inside the fuel level sensor there is a variable resistor. As the fuel level changes, more or less resistance is put into the fuel sensing circuit. ***Figure 24–10*** shows how the electrical circuit operates. The design uses an electrical bimetallic strip (thermostatic strip) that can be bent or

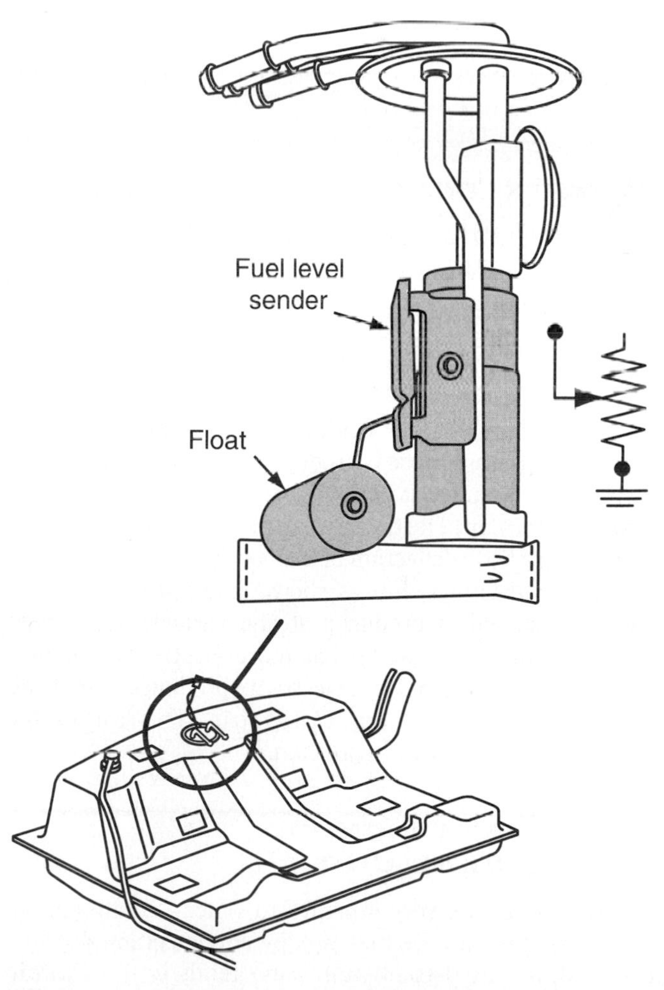

FIGURE 24-9 The fuel metering unit has a float that changes resistance as the fuel level changes.

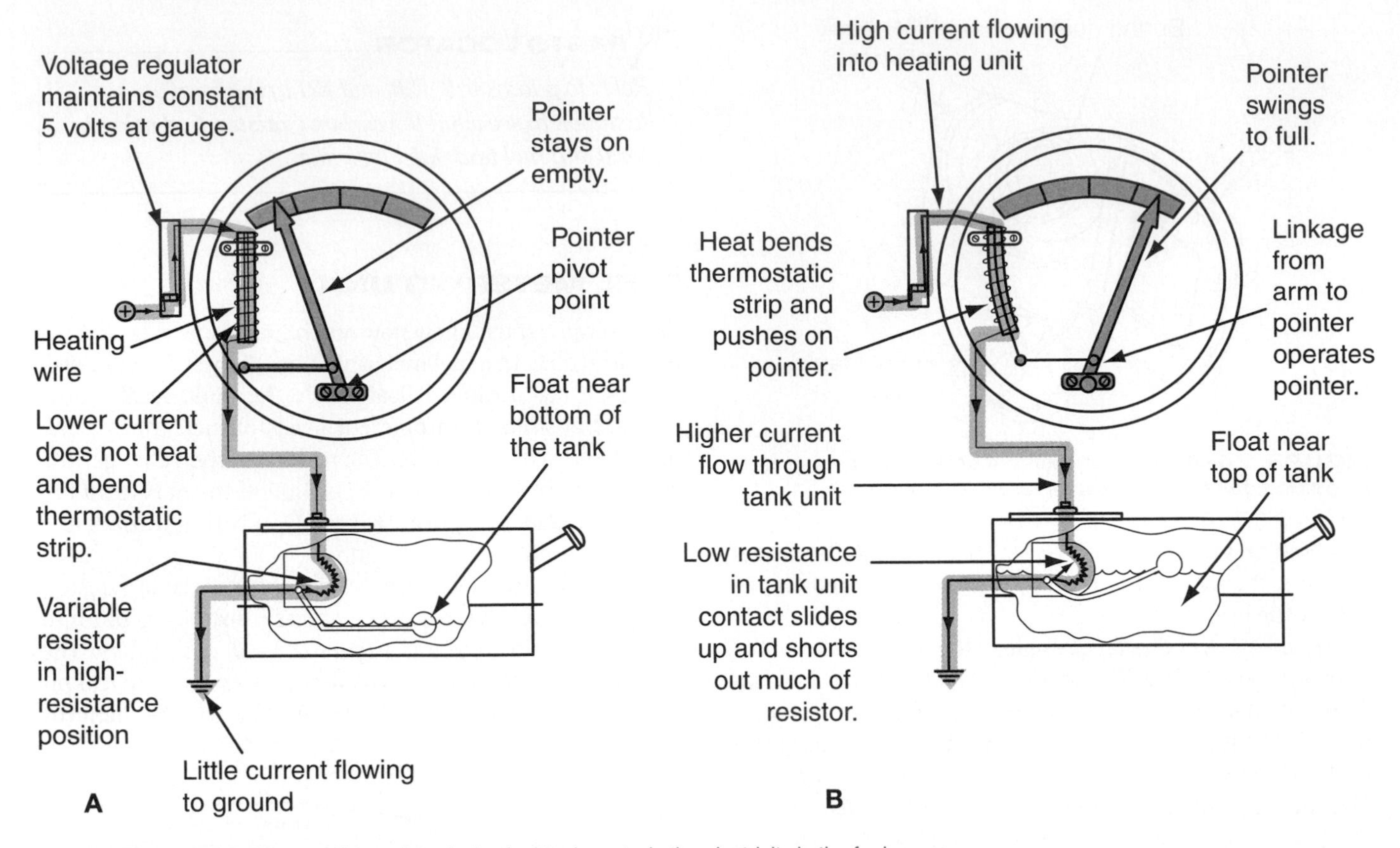

FIGURE 24–10 The variable resistor in the fuel tank controls the electricity in the fuel gauge.

PARTS LOCATOR

*Refer to photo #22 in the front of the textbook to see what a **fuel metering unit** looks like.*

warped by heat. Heat is produced by the electricity flowing in the circuit.

When the tank is empty or low in fuel as shown in ***Figure 24–10A***, high resistance is produced at the variable resistor in the fuel tank. Consequently, very little electricity can flow in the circuit. The thermostatic strip remains cool, and there is very little deflection of the pointer.

When the tank is full as shown in ***Figure 24–10B*** a lower resistance is produced at the variable resistor in the fuel tank. This means that more electricity can flow in the bimetallic strip. More electricity produces more heat. The heat bends the thermostatic strip. This causes the pointer to move to the full position.

DIESEL FUEL PICKUP AND SENDING UNIT

Diesel engines are very sensitive to water contamination. Water can get into the fuel tank by **condensation**. When a fuel tank is only partially full, water tends to form within the tank. The moisture in the air within the tank condenses on the inside tank wall. This occurs in the fuel tank or in any storage tank between the refinery and the vehicle.

By law in many states, water in diesel fuel should be no more than one-half of 1%. That quantity of water will be absorbed by the fuel. Higher levels may have a damaging effect on the injectors.

Hydrocarbon fuels are lighter than water. This means that water sinks to the bottom of the tank. Pickup systems are made to keep water out of the fuel. Some use a "sock" to absorb the water. Others use a detector to determine if water should be drained from the system. ***Figure 24–11*** shows a common type of diesel fuel pickup with a sock to absorb water.

PORTABLE FUEL STORAGE TANKS

At times, there may be a need to remove the gasoline or diesel fuel from the vehicle tank. Often this is necessary when the fuel tank needs to be removed and serviced. Before removing any fuel tank it is necessary to pump the fuel out of the tank and into a portable fuel storage container. ***Figure 24–12*** shows a typical portable fuel storage tank. In operation, a fuel hose is inserted into the fuel tank. A pump on top of the tank is used to transfer the fuel from the fuel tank to the portable storage tank. A ground wire is also used to eliminate sparks. Always follow the manufacturer's recommendations when using a portable fuel storage tank.

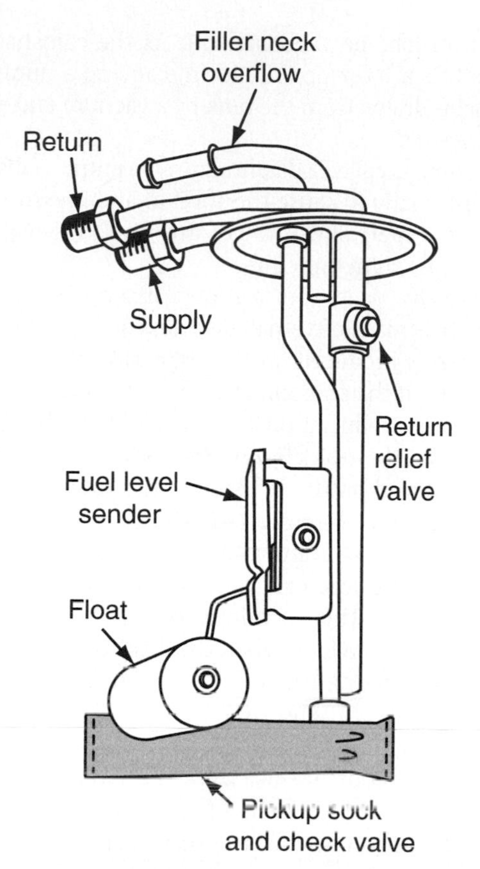

FIGURE 24–11 This fuel pickup/sending unit uses a sock to absorb water.

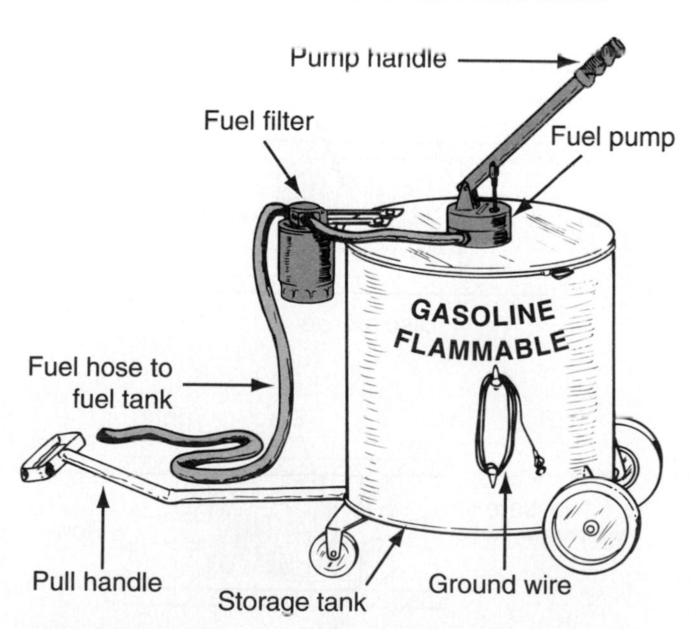

FIGURE 24–12 A portable storage tank is used to remove and store fuel from a vehicle.

FUEL LINE COUPLINGS

There are many types of fuel line couplings. One of the most common is the double flare fitting. The double flare fitting is made in several steps. First, the line is cut using a pipe cutting tool such as a tool cutter. After the line has been cut and cleaned, the end is inserted into a flaring tool bar, as shown in ***Figure 24–13***. An adapter is placed inside the tube and pressed down on its top. The top of the tube is turned or bent inward slightly from the pressing down of the adapter. The adapter is then removed and a flaring cone is pressed into the top of the tube, causing it to be flared outward. This double flare type of fitting helps to produce strong, leakproof fuel line connections.

24.3 FUEL PUMPS

The purpose of the fuel pump is to transfer the fuel from the tank to the electronic carburetor or fuel injectors. Gasoline engine fuel pumps are either mechanical or electrical. Diesel engines have a complex high-pressure fuel pump.

MECHANICAL FUEL PUMPS

For many years, automotive manufacturers used mechanical fuel pumps on all vehicles. As more electronic components

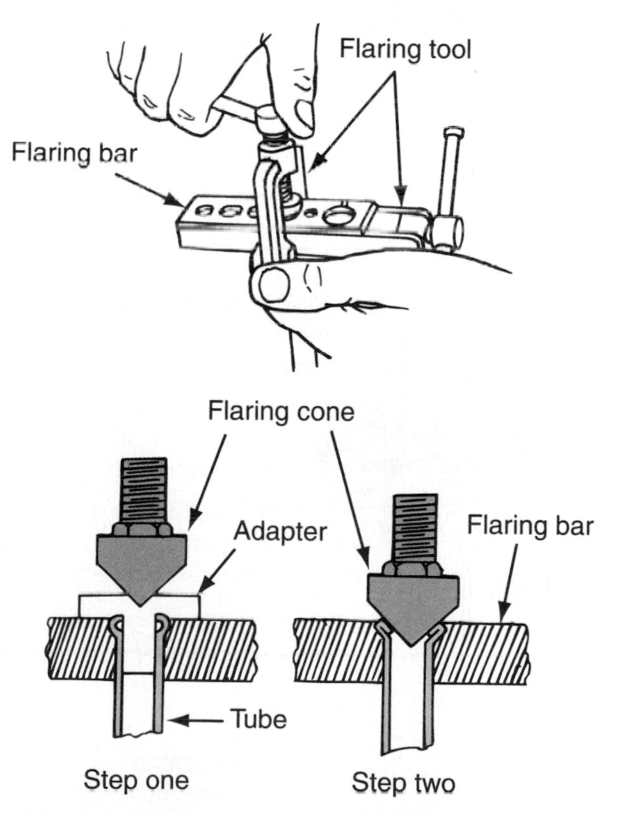

FIGURE 24–13 A double flare connection is often used on metal fuel line connections.

were designed for vehicles, more electrical fuel pumps were used. Electric fuel pumps are used on vehicles today. However, there is still a need to understand the basic operation of mechanical fuel pumps.

A typical mechanical fuel pump and an internal diagram is shown in ***Figure 24–14***. The unit is driven by a specially shaped cam lobe on the camshaft. As the camshaft turns, the lobe lifts a lever up and down, causing a pumping action. Fuel is drawn from the tank by a vacuum and is sent to the fuel system.

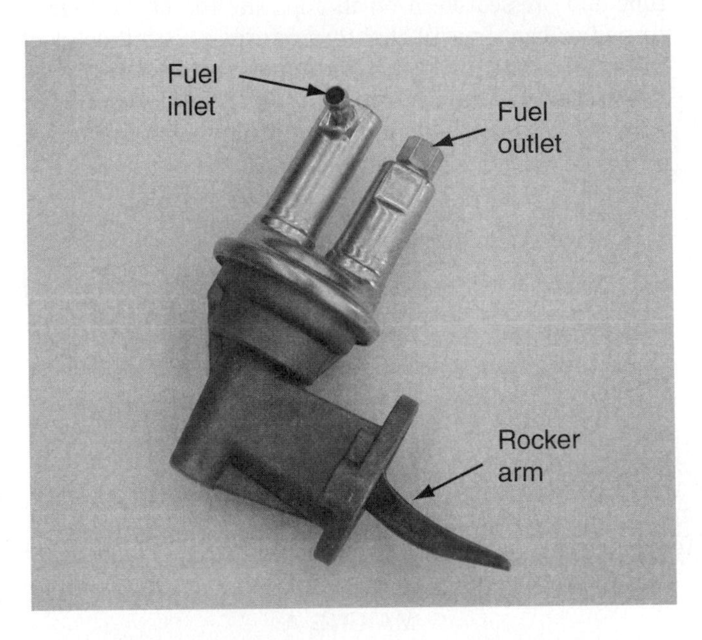

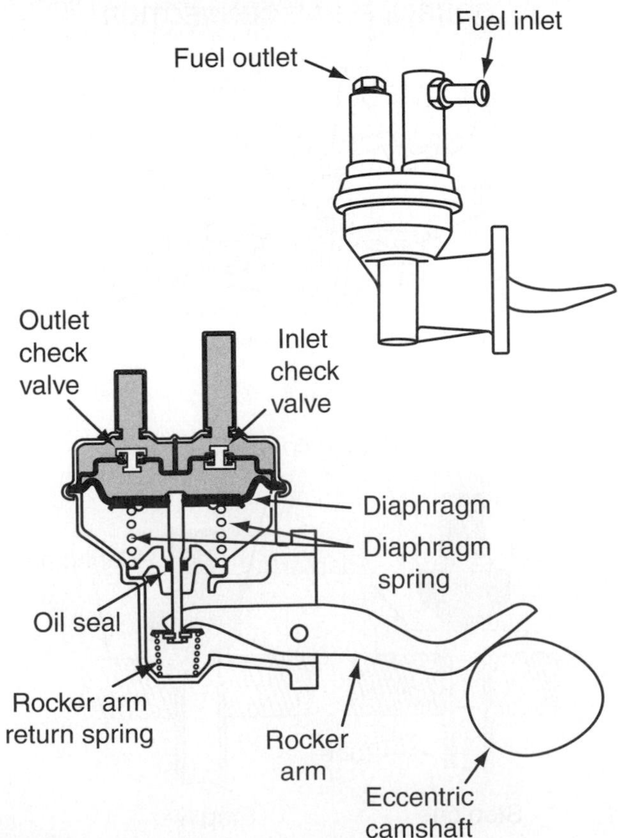

FIGURE 24–14 A diaphragm is used to cause suction and a pressure. The fuel is drawn in and pressurized, and then it is sent through the check valve to the electronic carburetor or fuel injectors.

This pump is called a diaphragm-type pump. A **diaphragm** is used internally to cause the suction and pressure needed to move the fuel. Check valves are used to keep the fuel moving in the right direction.

Check valve operation is shown in ***Figure 24–15***. The check valve is made of a small disc held on a seat by a spring. If a suction is produced on the right side of the valve, the disc will lift off the seat and draw fuel into the pump. When a pressure is produced on the back side of the disk, it is forced against the seat (***Figure 24–15A***). Also, when a pressure is produced on the left side of the valve, the disc will open to allow flow (***Figure 24–15B***).

The same check valve action occurs inside the fuel pump. When the eccentric cam lifts the diaphragm, a suction is produced. This suction causes the inlet check valve to open and draw in fuel. When the diaphragm moves in the opposite direction, a pressure is produced inside the fuel pump. This pressure closes the inlet check valve and forces the fuel through the outlet check valve. This action causes a suction, which draws fuel from the tank. This creates a pressure, which forces the fuel to the carburetor.

Fuel pumps can be checked for two specifications, pressure and capacity. The pressure created by a mechanical fuel pump is generally low. Each vehicle is different, but average fuel pump pressures range from 3 to 10 pounds per square inch. The capacity of a fuel pump measures how much volume the fuel pump can deliver in a certain period of time, generally 30 seconds. Average capacity ratings are from 0.5 pint to 1 pint per 30 seconds.

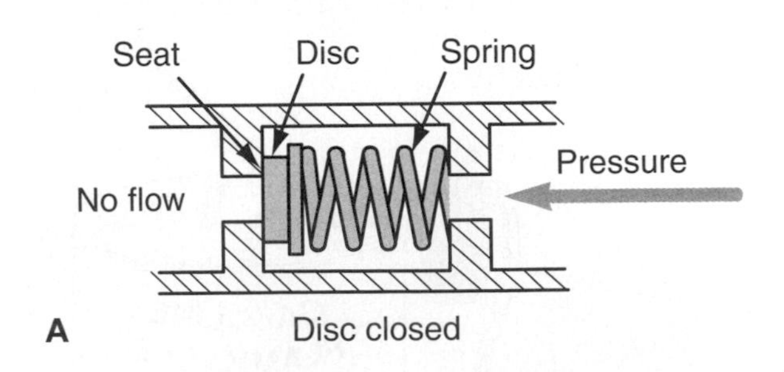

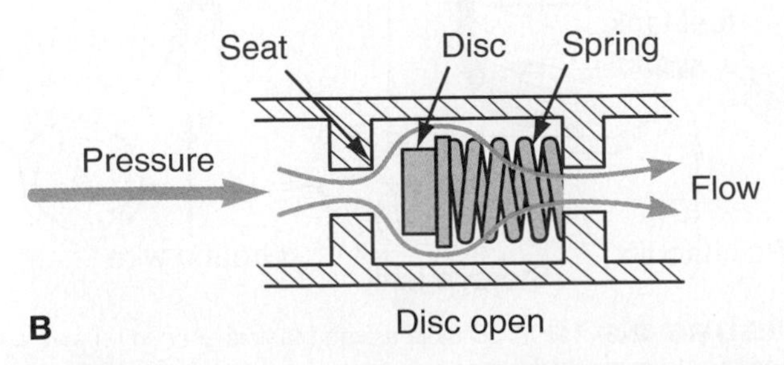

FIGURE 24–15 Check valves are used inside a mechanical fuel pump to control the suction and pressure of the fuel.

Car Clinic: MECHANICAL FUEL PUMP

CUSTOMER CONCERN:
Oil Level Increases
When checking the oil, a customer has noticed that the oil level has increased. How can the level of oil be higher without adding oil to the crankcase?

SOLUTION: The level cannot increase unless some type of fluid is being added. There are two types of fluid that can enter the oil: engine coolant and fuel. Engine coolant can enter the oil through a cracked block. A more likely cause, however, is that fuel (gasoline) is being added to the oil. One way to check for this is to smell the oil. If fuel is leaking into the oil, there will be an odor of gasoline. As fuel is added, the oil level appears to increase. Fuel can be added to the oil in two major ways. The fuel pump (if mechanical and located on the side of the engine) can have a leak in the diaphragm. Fuel then passes through the diaphragm into the crankcase area. A second way in which fuel can be added to oil is by having the electronic carburator or fuel injectors extremely rich during starting. A heavily choked engine will cause raw fuel to slip past the rings and into the crankcase area. To eliminate the problem, either replace the mechanical fuel pump or readjust the fuel system according to the manufacturer's specifications. If the vehicle uses injectors, the computer controls will have to be checked for a rich mixture.

VAPOR RETURN LINE

Some mechanical fuel pumps have a small connection on the fuel pump called a vapor return. The purpose of this line is to return fuel vapor that has built up in the fuel pump to the fuel tank. As the engine and underhood temperatures increase, vapors may develop in the fuel lines. Vapors are removed at this point.

ELECTRICAL FUEL PUMPS

Electrical fuel pumps are now used instead of mechanical fuel pumps. Electrical fuel pumps have certain advantages over the mechanical type. Electrical fuel pumps work independently on electrical current. They provide fuel when the switch is turned on. Mechanical fuel pumps must have the engine cranking or turning before they deliver fuel. Another advantage is that electrical fuel pumps can be located farther away from the engine. This means that heat from the engine will not produce vapors in the fuel pump. Therefore, there is less risk of vapor lock. Electrical fuel pumps also work well in conjunction with computer-controlled vehicles. In addition, electrical fuel pumps consume less frictional horsepower. Electrical fuel pumps also produce more pressure than the typical mechanical pump. Depending on the engine and manufacturer, average pump pressures range from 15 to 60 psi.

Several styles of the electrical fuel pump are used, including the bellows type and the impeller or roller vane type. The year of manufacture and the type of vehicle will determine which type is used.

The electrical circuit used to operate the electrical fuel pump is shown in ***Figure 24–16***. Electricity starts at the

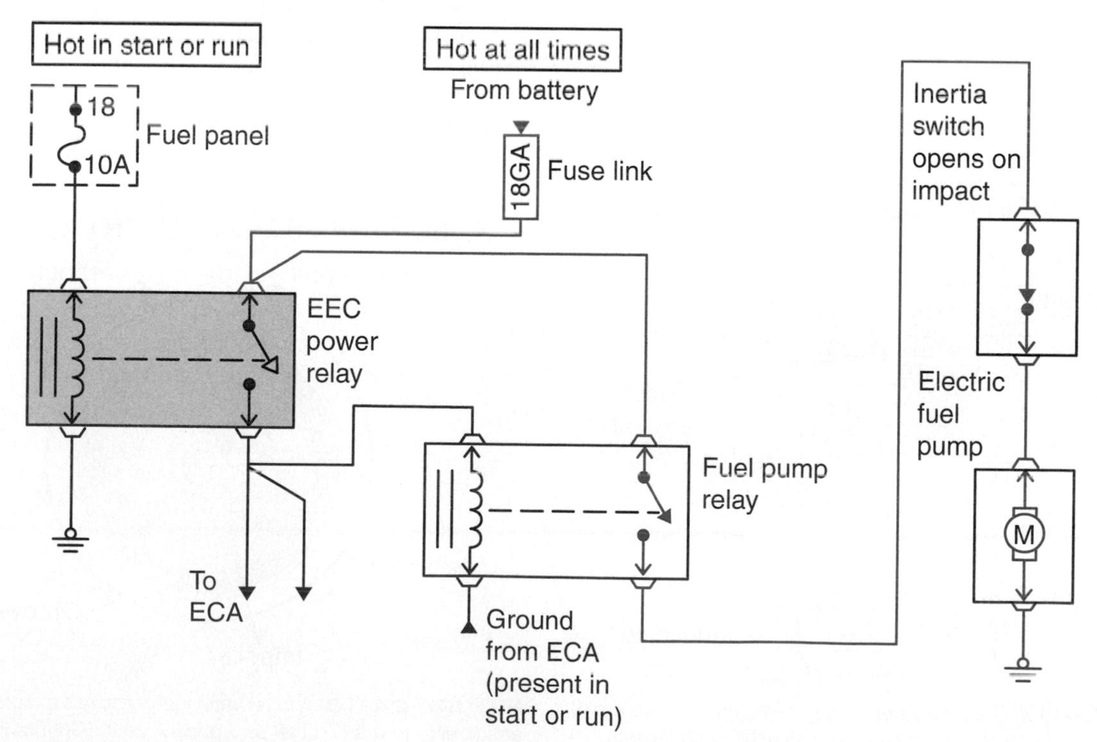

FIGURE 24–16 The circuit to operate the electrical fuel pump has an inertia switch to break the circuit during sudden impacts (accidents).

battery and flows to the fuse link (center). The fuse link is connected to the fuel pump relay. When the ignition switch is turned on, the coil in the ECC power relay is energized. This causes the ECC power relay switch to close. When the power relay switch closes, the fuel pump relay coil is then energized. When energized, the fuel pump relay switch closes. Electricity then flows through an **inertia switch**, which is normally closed. From there the electricity flows into the electrical fuel pump to turn the fuel pump motor.

PARTS LOCATOR

Refer to photo #23 in the front of the textbook to see what a ***fuel system inertia switch*** *looks like.*

The inertia switch is a safety switch that opens the circuit whenever there is sudden impact from a vehicle crash or accident. This keeps the fuel pump from running and spilling gasoline if the car is in an accident.

BELLOWS-TYPE ELECTRICAL FUEL PUMP

In the bellows-type electrical fuel pump, a metal **bellows** is used instead of a diaphragm. A bellows can cause a vacuum or pressure when stretched or compressed. The bellows is moved back and forth by the action of a solenoid. See ***Figure 24–17***. When the electrical current is sent to the magnetic coil, the **armature** is drawn downward. This action causes the metal bellows to expand and stretch. Fuel is drawn in by a vacuum. When the armature reaches its lowest point, the electrical current on the magnetic coil is removed by being grounded. The return spring then pushes the bellows upward. This causes a pressure that forces the fuel out of the fuel pump and into the carburetor.

ROLLER VANE FUEL PUMP

A second type of fuel pump used on many vehicles is the roller vane electrical fuel pump. This pump is located inside the fuel tank. There are several variations of this type of pump. ***Figure 24–18*** shows an example of such a pump. The electric motor assembly is operated by electrical current. As the motor turns an impeller, fuel is drawn in at the inlet port of the pump. It is pressurized and sent out the discharge port for delivery to the engine. The impeller at the inlet end serves as a vapor separator. The unit operates at

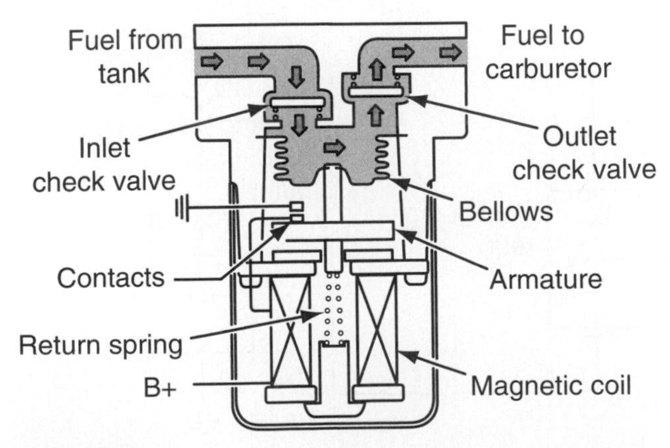

FIGURE 24–17 A bellows electric fuel pump uses a solenoid to raise and lower a bellows. This action causes suction and pressure to be created.

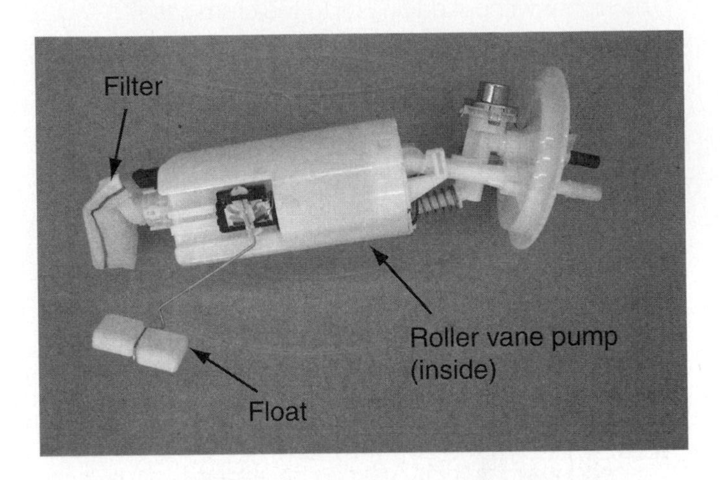

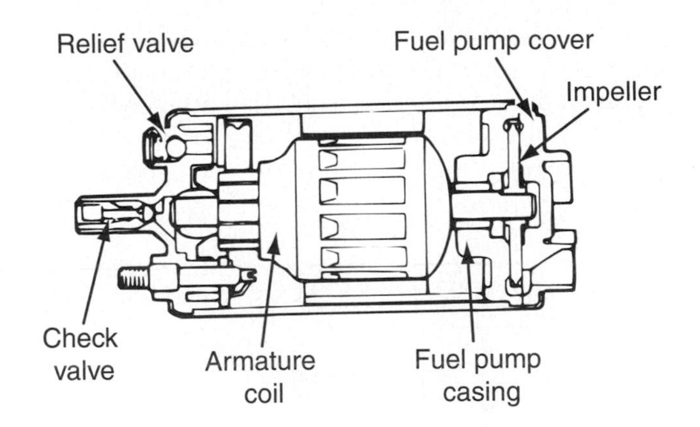

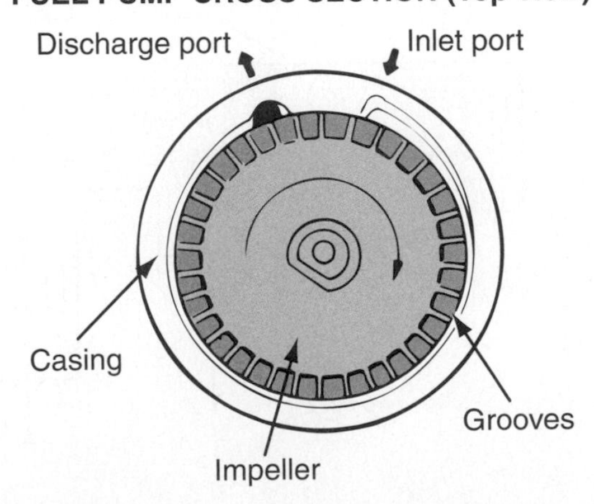

FIGURE 24–18 A roller vane fuel pump has a motor that rotates an impeller to draw in fuel and pressurize it. (Courtesy of American Honda Motor, Co., Inc.)

approximately 3,500 rpm. A pressure relief valve keeps fuel pump pressure at a constant pressure such as 60 to 90 psi. The fuel pump delivers more fuel than the engine can consume even under the most extreme conditions. The impeller is designed so that it produces suction on one side and a pressure on the other side. The inlet from the fuel tank is connected to the suction side. The outlet from the pump is sent to the fuel injectors.

FUEL PRESSURE REGULATORS

Fuel systems that have electrical fuel pumps and fuel injectors use a fuel pressure regulator. ***Figure 24–19*** shows such a regulator. It is most often located in the front or center of the engine attached to the injector fuel rails. ***Figure 24–20*** shows a pressure regulator attached directly to the fuel rail. The fuel from the electric fuel pump flows into the fuel rail and to the injectors. The fuel pressure regulator controls the pressure inside the fuel rail. It is always located near the injectors in the fuel circuit. The fuel pressure regulator regulates the fuel that comes out of the fuel pump, through the fuel filter, and into the injector. It controls the fuel pressure at the fuel injectors. The purpose of the regulator is to keep the pressure of the fuel at the injectors between 15 and 60 psi. Some manufacturers have lower fuel injector pressure.

PARTS LOCATOR

*Refer to photo #24 in the front of the textbook to see what a **fuel pressure regulator** looks like.*

The fuel pressure regulator operates internally as shown in ***Figure 24–21***. The fuel pressure regulator contains a

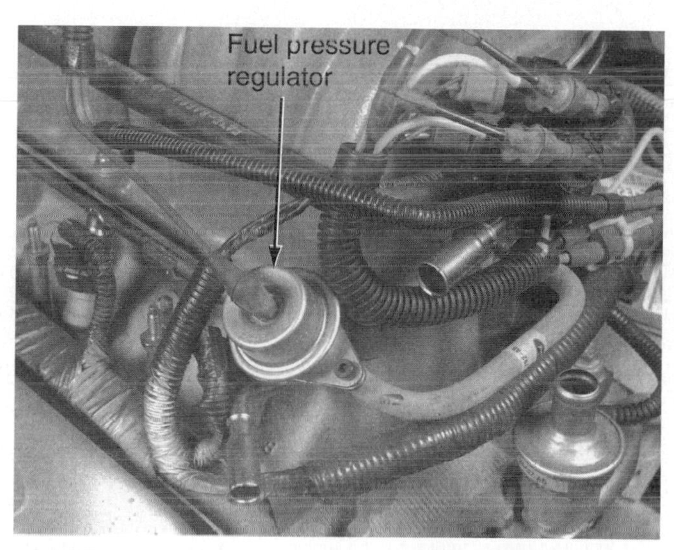

FIGURE 24–19 A fuel pressure regulator controls the pressure of fuel going to the injectors.

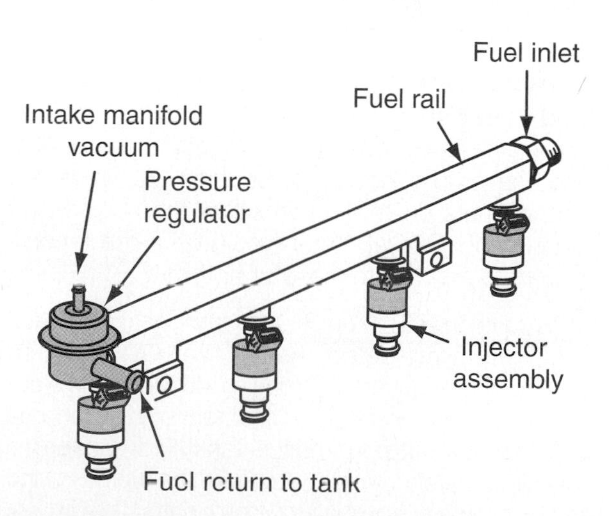

FIGURE 24–20 The fuel pressure regulator is located in a position where it can regulate the fuel going to the injectors.

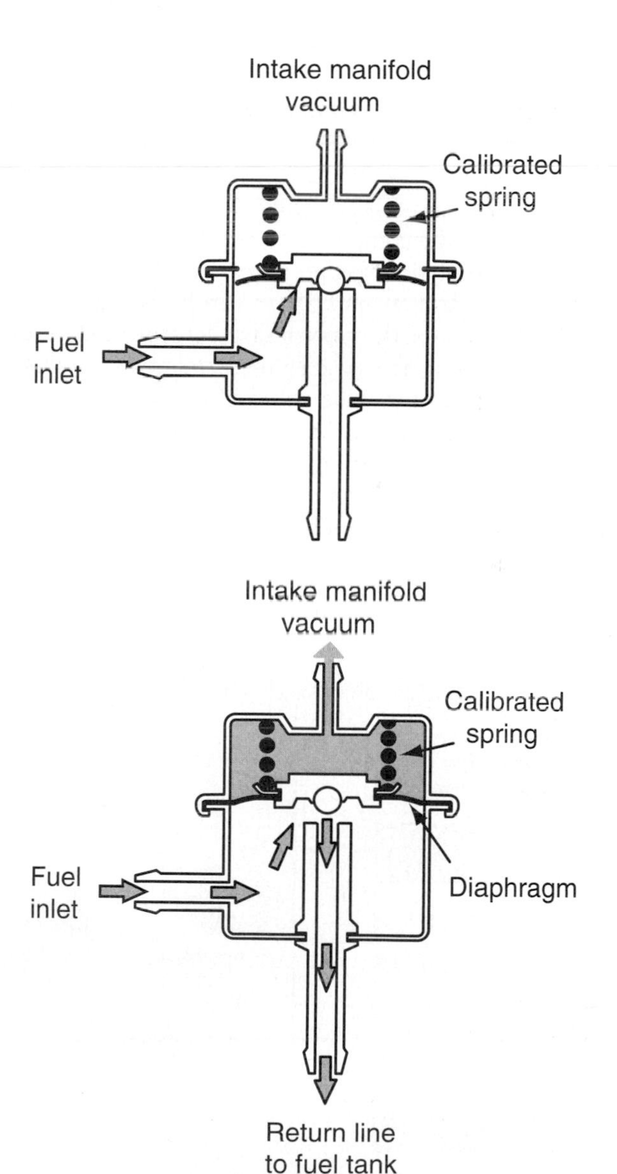

FIGURE 24–21 Fuel pressure is regulated by using a diaphragm and spring. As fuel pressure increases, the spring lifts and allows fuel to pass through the fuel return line, back to the fuel tank.

pressure chamber that is separated by a diaphragm. A **calibrated** spring is placed in the vacuum chamber side. Fuel pressure is regulated when the fuel pump pressure overcomes the spring pressure. At this point, a small relief valve in the center opens. This action passes excess fuel back to the fuel tank. Vacuum action on the top side of the diaphragm, along with spring pressure, controls fuel pressure. A decrease in vacuum creates an increase in fuel pressure. A small increase in the intake manifold vacuum creates a decrease in fuel pressure. As an example, when the engine is under heavy load, vacuum is decreased. This condition requires more fuel. A decrease in vacuum allows more pressure to the top side of the pressure relief valve. This increases fuel pressure, and thus more fuel is delivered.

COMPUTERS AND FUEL CONTROL

Manufacturers are using computers to control various functions of the engine. Fuel systems are also being controlled by computers. One such system is diagrammed in ***Figure 24–22***, which shows how a computer is used to control several systems. The left side of the diagram shows the signals being fed into the computer. The right side of the diagram shows what systems are being controlled by the computer. In this case, the computer is operating the fuel pump. It could also be controlling other systems as well.

Operation of the fuel pump is controlled by the computer, also known as the powertrain control module or PCM. Several signals are fed into the computer to control the fuel pump. These include the engine speed, temperature, a start signal, the throttle valve idle position, and the battery voltage. The computer then takes each of these signals and operates the fuel pump accordingly. More information will be given about computer-controlled systems in later chapters.

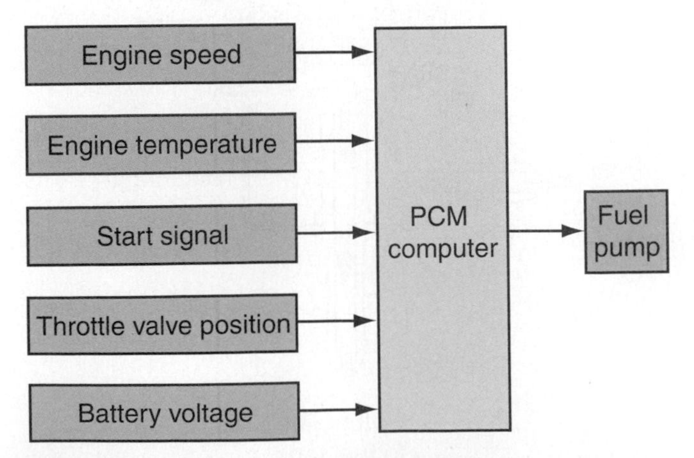

FIGURE 24-22 On computer-controlled engines, the computer controls the fuel pump. Various inputs help determine the correct operation of the fuel pump.

24.4 FUEL FILTERS

PURPOSE OF FUEL FILTERS

Most internal combustion engines consume a mixture of fuel and air to produce power. The key word in the operation of any such engine is *cleanliness*. Any fuel system using an electronic carburetor or fuel injectors has many small passages and delicate parts that can be damaged by dirt particles. A dirty electronic carburetor or fuel injector can cause erratic performance or complete engine shutdown.

Diesel engines use a fuel injector under very high pressure. Many small openings in the tip of the injectors will be damaged if dirt particles get into the system.

SOURCES OF GASOLINE CONTAMINANTS

Contaminants may enter the fuel system from various sources. These include:

- Unfiltered fuel that is pumped into the vehicle tank
- Loose tank caps or faulty sealing gaskets
- Rust, a powerful abrasive, that flakes off from the fuel tank and lines
- Contaminants or dirt particles left in the tanks or lines during manufacturing and assembly

GASOLINE FUEL FILTERS

Depending on the manufacturer and the year of the vehicle, some engines have one or two filters in the fuel system. The first filter, ***Figure 24–23***, located in the gasoline tank, is made of fine woven fabric. This filter prevents large pieces of contaminant from damaging the fuel pump. The tank filter also prevents most water from going to the electronic carburetor or fuel injectors.

Car Clinic:
FUEL FILTERS

CUSTOMER CONCERN:
Bad Fuel Filter
An engine is becoming consistently more erratic, and the engine power is reducing. At times the engine shuts down. It seems that the engine isn't getting enough fuel. What would be a quick component to check with these symptoms?

SOLUTION: The quickest component to check with these symptoms is the fuel filter. Over a period of time, the fuel filter in the main fuel line to the electronic carburetor or fuel injectors may become plugged with dirt from the fuel tank. First, perform a fuel pressure check. If low, replace the fuel filter, and the problems will probably be eliminated.

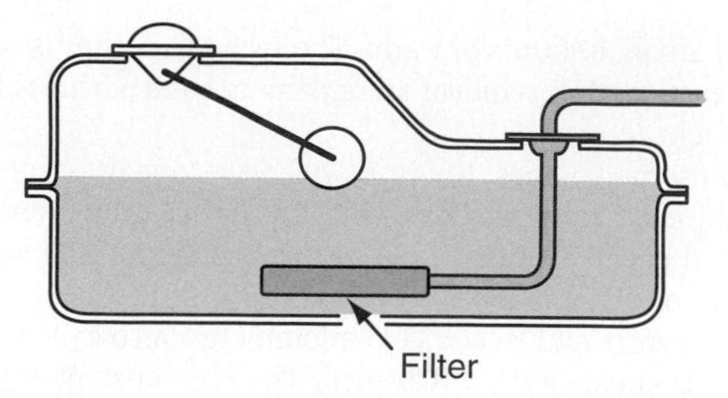

FIGURE 24-23 The first filter in a gasoline engine is located inside the fuel tank.

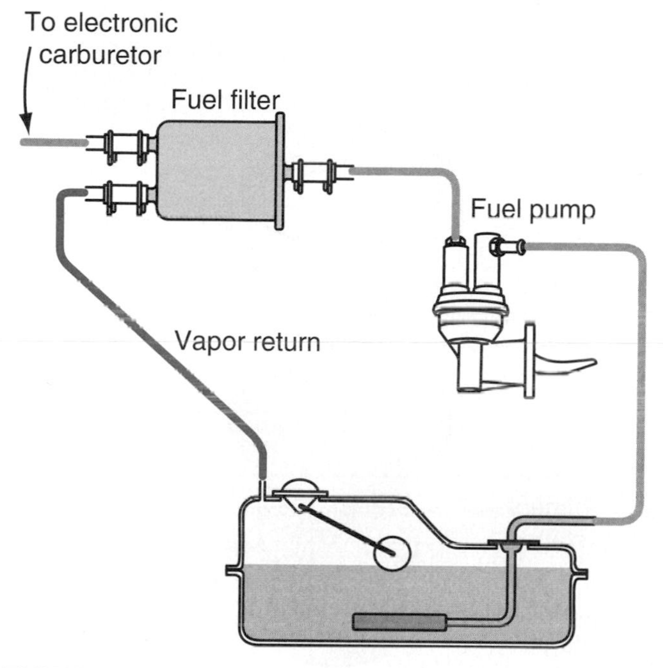

FIGURE 24-24 An in-line filter is used to filter out small particles of dirt and rust. A vapor return line helps to reduce vapor buildup in the system.

The second filter is found in one of several locations. An in-line filter is shown in ***Figure 24–24***. Fuel is drawn from the tank into the mechanical fuel pump. The fuel pump pressurizes the fuel and sends it to the filter. A vapor return line is used on some vehicles to return vapor to the fuel tank.

Gasoline from the pump flows to the inlet fitting of the filter. Gasoline used by the electronic carburetor passes through the filter and out the center fitting. A small amount of gasoline will exit through the second outlet and return to the tank. The recirculation of gasoline through the vapor line cools the gas and prevents vapor lock.

Figure 24–25 shows the in-carburetor type of fuel filter. This filter has small pleated paper filters with a gasket to provide positive sealing.

PARTS LOCATOR

*Refer to photo #25 in the front of the textbook to see what **fuel filters** look like.*

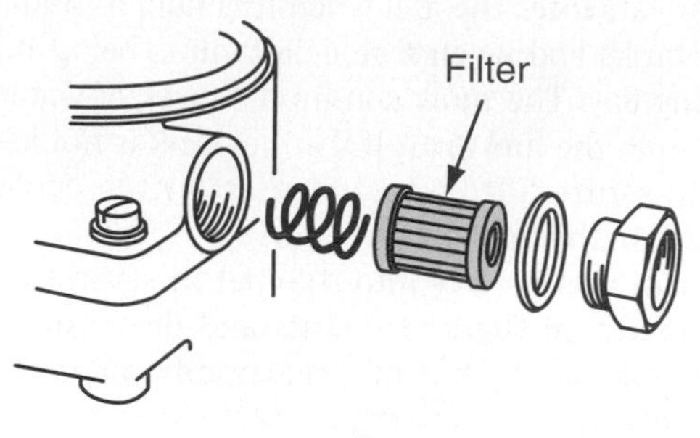

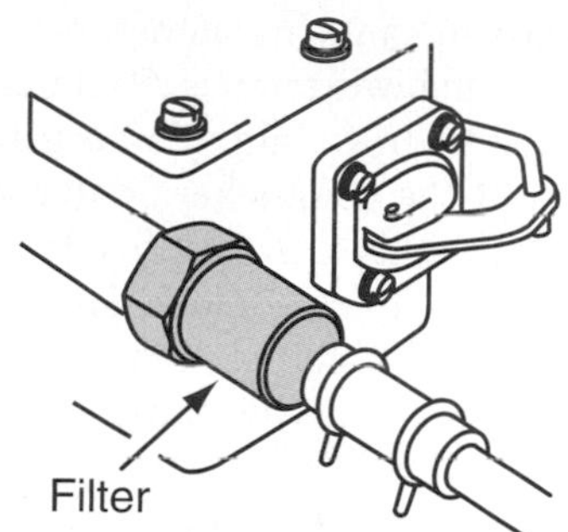

FIGURE 24-25 The filter is attached directly to the electronic carburetor on some engines.

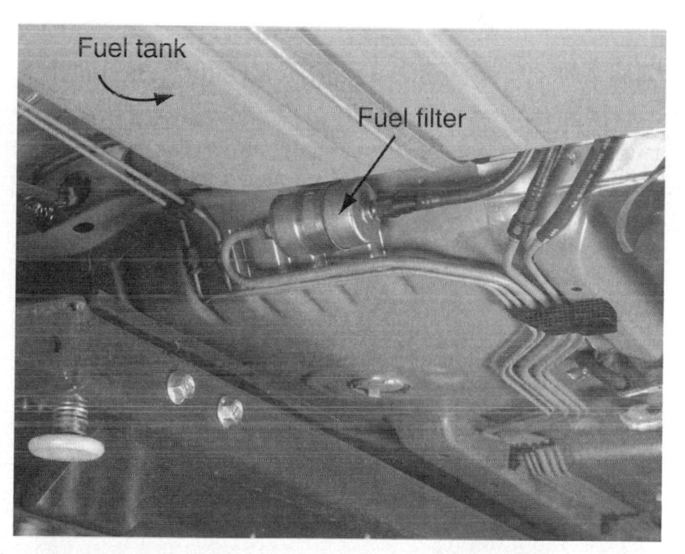

FIGURE 24-26 This in-line fuel filter is located under the vehicle near the fuel tank.

Many cars today have an in-line filter between the fuel tank pump assembly and the fuel injectors. ***Figure 24–26*** shows such a filter located under the vehicle, usually near the fuel tank.

DIESEL FUEL CONTAMINANTS

When diesel fuel is being shipped from the refinery, it can pick up certain fuel contaminants. These include rust, dirt, and water.

Rust usually comes from large storage tanks or vehicle tanks. Rust occurs when there are low fuel levels in the tank over a long period of time. Rust is an abrasive and can damage the injection system.

Water can enter the fuel when it is held in underground storage tanks and when a vehicle tank is being filled on a wet, rainy day. The most common source of water is condensation in the fuel tank. If the fuel tank is not kept filled, warm, moisture-filled air condenses on the cooler inside metal wall of the fuel tank.

Dirt can find its way into the fuel in several ways. Dirt can be found on the tank spouts and dispensing nozzles. The vent system must be in good condition and checked as a source of dirt.

Two other contaminants are sometimes found in fuel. These are bacteria and wax crystals. Bacteria growth takes place in diesel fuel when certain microorganisms begin to grow. They grow at the point where water and diesel fuel meet in the tank. The tank is dark inside, and if the fuel moves very little, conditions are ideal for bacteria to grow. As the bacteria grow, they form a "slime" that will eventually be carried to the filter. This slime causes the filter to plug prematurely.

Wax crystals form when the diesel fuel reaches its cloud point temperature. As the wax crystals begin to form, they plug the filters. Less fuel will be sent to the engine, which could cause stalling or complete shutdown.

DIESEL FUEL FILTERS

The most desirable arrangement for diesel engines is the two-filter system. However, many diesel engines used in automotive applications have only one filter, which is called the primary filter, shown in ***Figure 24–27***. The primary filter will catch most of the solid contaminants and remove small amounts of water. The secondary filter is much more efficient. It removes all remaining solid particles from the fuel.

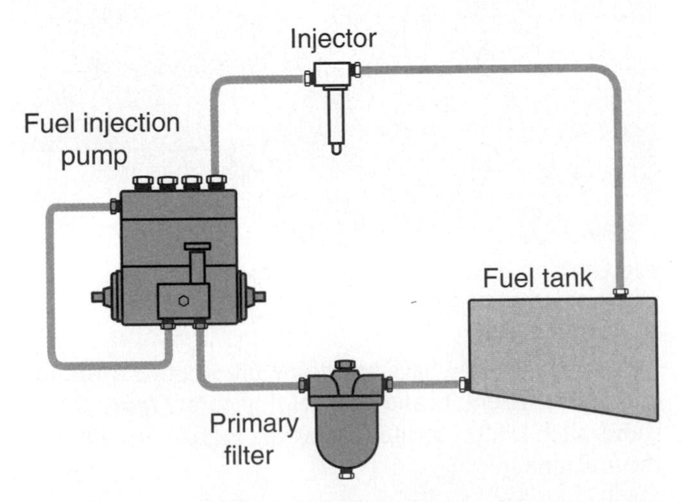

FIGURE 24–27 It is generally desirable to use a primary filter in the diesel fuel system.

On some engines, the primary and secondary stages are combined (***Figure 24–28***). This is a surface-type filter with pleated paper. The first stage consists of about 400 square inches of filtering area. It removes 94% of the particles 10 **micrometers** and larger. One micrometer (also called a micron) is equal to 0.000039 inch. The second-stage filter is made of the same paper material. It consists of about 200 square inches of filtering surface. This stage is 98% effective in filtering the fuel already filtered by the first stage.

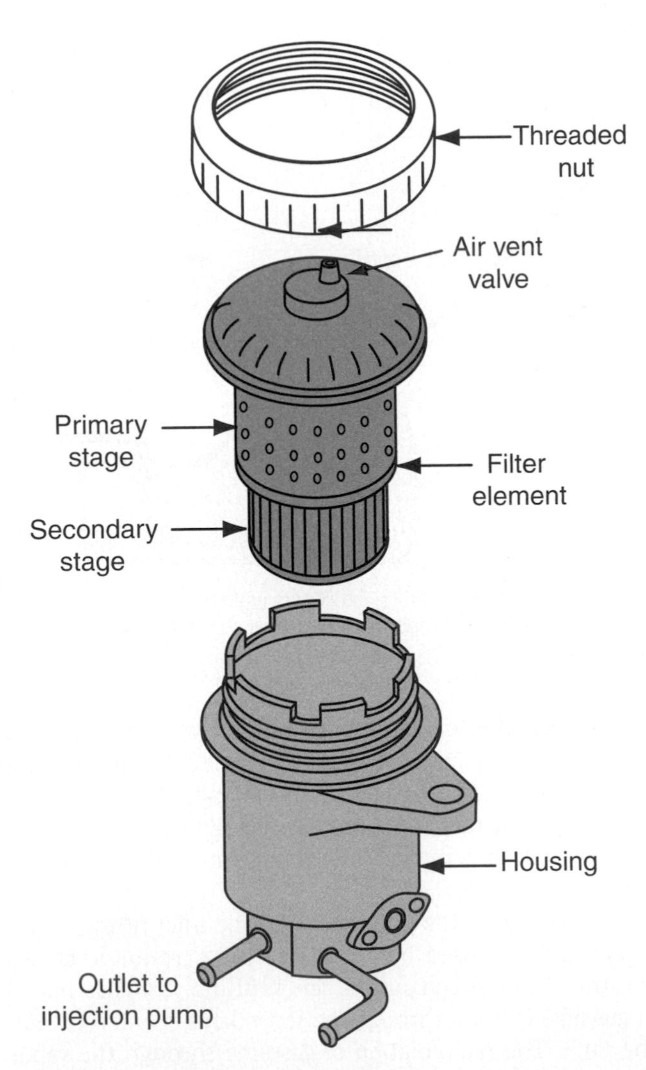

FIGURE 24–28 In some automotive diesel fuel systems, the primary and secondary filters are combined into a single filter.

Problems, Diagnosis, and Service

Safety Precautions

1. Gasoline is very toxic and dangerous to the skin and eyes. Always wear OSHA-approved safety glasses and correct clothing to protect your skin from gasoline. If gasoline does get into your eyes, wash them with warm water immediately or use an eyewash fountain.
2. Before working with the fuel system, disconnect the battery so there is no possibility of producing a spark that would ignite gasoline.
3. Never remove a fuel filter when the engine is at operating temperature. When you remove the filter, gasoline could spill onto the hot manifold and cause a fire.
4. When checking the fuel pump pressure, make sure there is no electrical spark that could ignite the fumes of the open container of gasoline.
5. When working with gasoline, always use approved containers and maintain safety rules.
6. Avoid welding on or near the fuel tank. Send the fuel tank to a repair shop equipped to handle this service.
7. When checking the electrical components in the fuel system, always use a battery-operated tester to eliminate any sparks. Sparks could cause a fire if near gasoline fumes.
8. If even a small quantity of gasoline must be stored for a short period of time, store the fuel in an OSHA-approved container. Never store gasoline in an open container. The possibility of a fire exists because the gasoline fumes can travel several feet from their source. They could be ignited by a spark several feet from the source.
9. When working on fuel injection systems, be sure to relieve the fuel pressure while the engine is off.
10. Always keep an approved fire extinguisher nearby when working on the fuel system.
11. Whenever gasoline is spilled, immediately wipe it up and dispose of the rags in a sealed container.
12. Never clean any automotive parts with gasoline.
13. Never smoke in or around an automobile or shop area.
14. Always be aware of the possibility of spontaneous combustion. When oil- or gasoline-soaked rags are left in a pile, they may ignite due to spontaneous combustion. Always make sure that such rags are stored in an approved, sealed, fireproof container.
15. Never use a trouble light with an exposed bulb when working on the fuel system. If the bulb breaks, it could ignite the fuel.

PROBLEM: Faulty Fuel Pump

The engine has been losing power in a vehicle with a mechanical fuel pump.

DIAGNOSIS

Often, mechanical fuel pumps will fail. In some cases, the fuel pump cannot transfer enough fuel to operate the engine. If the fuel pump is suspected, perform a fuel pump pressure test as shown in ***Figure 24–29*** as follows:

1. Disconnect the fuel pipe at the carburetor inlet.
2. Attach the correct pressure gauge and hose between the electronic carburetor inlet and the disconnected fuel pipe.
3. Start and run the engine.
4. Check the automotive service manual to see if the pressure is within the limits stated in the specifications for that type and year of vehicle.
5. The pressure should remain constant or return very slowly to zero when the engine is shut off.

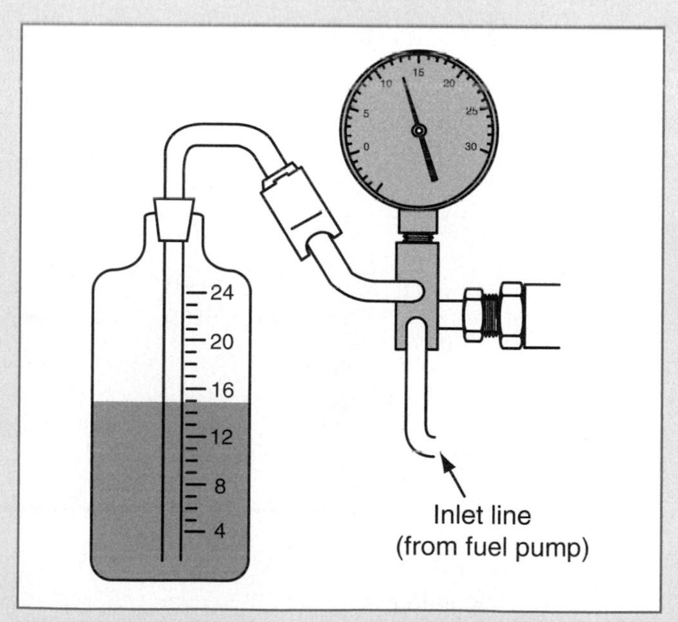

FIGURE 24–29 Use a gauge and pint size container to test fuel pump pressures. Run the engine, measure the amount of fuel in the container, and compare it to manufacturer's specifications.

6. Make sure the hose is placed inside the pint container to check the capacity.
7. Run the engine at idle speed and note how long it takes to fill the container.
8. Depending on the pump being tested, it should take about 20 to 30 seconds to fill the container.

SERVICE

If the fuel pump does not meet the specifications stated in the service manual, replace it using the following general procedure:

1. Disconnect the negative side of the battery to eliminate the potential for sparks.
2. Disconnect the input and output fuel lines from the fuel pump. Use an open end (flare) wrench to loosen the fuel line connections, as shown in ***Figure 24–30***. On some flexible lines, the clamp must be removed before the line can be removed.
3. Remove any emission control or vacuum lines that are attached and label them for reassembly.
4. Using the correct size socket and ratchet wrench, remove the two bolts holding the fuel pump to the block.
5. After the fuel pump has been removed, make sure the gasket is also removed. In some cases, the gasket must be scraped from the block to eliminate the possibility of leakage later.
6. Replace with a new fuel pump. Make sure new gaskets have been installed.
7. In some cases, it may be difficult to push the fuel pump into place against the block. If this is the case, the eccentric cam is in a position to lift the fuel pump lever. Rotate the engine slightly to eliminate this problem.

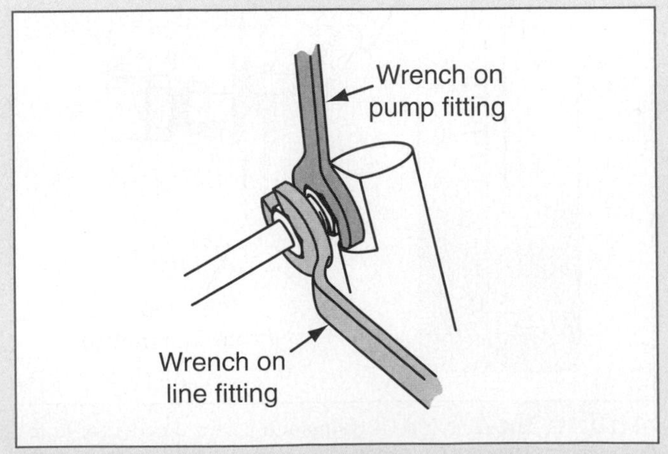

FIGURE 24-30 When removing an in-line filter, always use two wrenches.

8. After the fuel pump has been installed correctly, tighten the bolts to the correct torque specification.
9. Reassemble the fuel lines and vacuum hoses as necessary.

PROBLEM: Faulty Fuel Pump Diaphragm

The level of oil has been increasing recently in a vehicle with a mechanical fuel pump.

DIAGNOSIS

An increased level of oil in the lubrication system may indicate a bad mechanical fuel pump. If the diaphragm is cracked, fuel will leak through the diaphragm and into the crankcase. This condition can be detected by smelling the oil. If it has a gasoline smell, the fuel pump may be damaged and should be replaced. A fuel pressure and capacity test would also show reduced pressures and capacities.

SERVICE

Follow the service procedure just explained to remove and replace the mechanical fuel pump.

PROBLEM: Plugged Fuel Filter

An engine has stopped completely. Before it stopped, it hesitated and operated sluggishly.

DIAGNOSIS

This type of problem is most likely caused by a plugged fuel filter. The fuel filter increasingly blocks the fuel to the electronic carburetor or fuel injectors. Eventually, the filter totally blocks the fuel to the engine.

SERVICE

Remove and replace the fuel filter, using the following general procedure:

1. Disconnect the negative terminal of the battery to avoid any sparks.
2. Disconnect the fuel lines at the fuel inlet nut.
3. Remove the fuel inlet nut from the electronic carburetor.
4. Remove the filter and spring.
5. Make sure there are no dirt particles around the filter element (see ***Figure 24–31***).
6. Obtain a new filter, and soak it in gasoline for a short period of time.
7. Some filters have a filter check valve. Make sure the check valve is replaced correctly inside the filter.
8. Install the spring, filter, gasket, and fuel inlet nut.

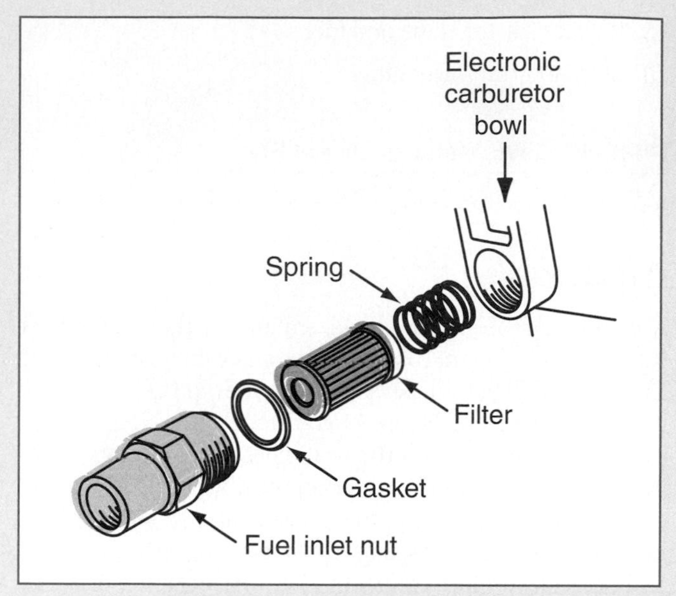

FIGURE 24–31 This fuel filter can be replaced by removing the assembly from the side of the carburetor.

9. Be careful not to strip the threads when inserting the fuel inlet nut. It is very easy to misalign the threads, causing them to be damaged. Always start the fuel inlet nut with your fingers. Use a flare wrench to tighten the nut to the correct torque specifications.

PROBLEM: Low Fuel Pressure

An engine is sluggish and seems to lack power. The vehicle has an electrical fuel pump and injectors.

DIAGNOSIS

Problems such as this generally suggest that the fuel pressure feeding into the injectors is not high enough. Fuel pressure can be checked on the fuel manifold. Each manufacturer has a different method for checking the fuel pressure. Sometimes, a service bolt is removed and a fuel pressure gauge is installed. The engine is run, and fuel pressure is measured and compared with the manufacturer's specifications. If the pressure is higher than specified, check for the following:

- A pinched or clogged fuel return hose or piping
- A faulty pressure regulator

If the pressure is lower than specified, check for the following:

- A clogged fuel filter
- Faulty pressure regulator
- Leakage in the line
- The fuel pump

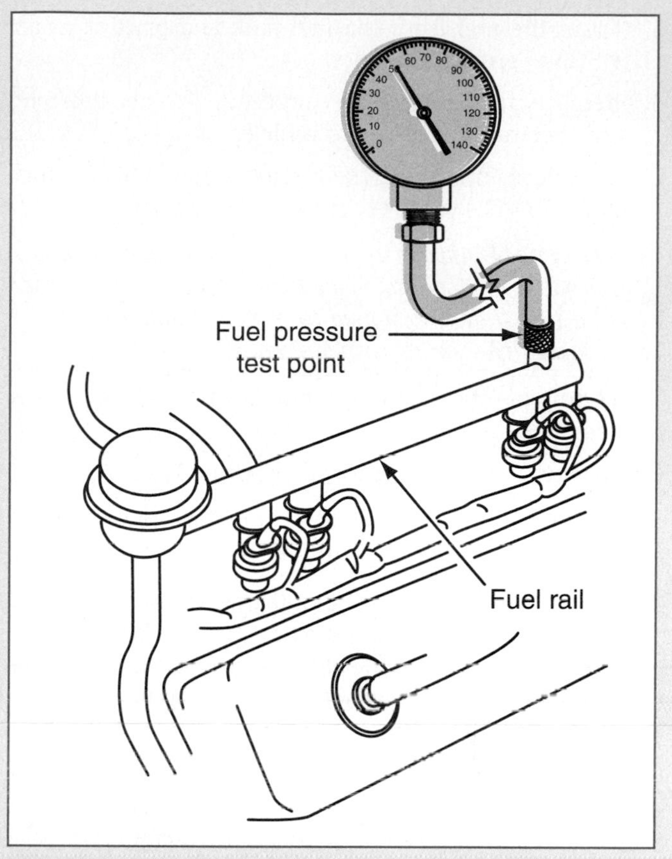

FIGURE 24–32 Fuel pressure can be checked on fuel injection engines by attaching a fuel pressure gauge to the fuel manifold.

SERVICE

To check the fuel pressure regulator use the following general procedure:

1. Attach the pressure gauge as shown in ***Figure 24–32***.
2. Start and run the engine at idle.
3. Check that the fuel pressure rises when the vacuum hose from the regulator is disconnected.
4. If the fuel pressure does not rise, check to see if it rises with the fuel return hose slightly pinched.
5. If the fuel pressure still does not rise, replace the pressure regulator.
6. If the fuel pressure regulator is operating correctly, the problem is a poor electrical fuel pump.

Replace the electrical fuel pump using the following general procedure:

1. The entire electrical fuel pump must be replaced.
2. Remove the negative battery cable to eliminate the possibility of sparks.
3. Relieve any fuel pressure in the system.
4. Lift the vehicle on an appropriate hoist.

5. Drain the fuel from the fuel tank and place it in an OSHA-approved container.
6. Remove the fuel tank. This procedure may be different and will vary with different vehicles.
7. Loosen, then remove, the assembly that holds the fuel pump in the fuel tank as shown in ***Figure 24–33***.

CAUTION: *Always use a brass punch or drift so as not to produce any sparks. Also, several tool manufacturers sell a tool that is designed to remove sending units by using a ratchet or a spanner wrench.*

8. Lightly tap the lock ring tab in a counterclockwise direction.
9. Remove the fuel pump and tank sending unit.

CAUTION: *Be careful not to produce any electrical sparks, as there are still gasoline vapors in the fuel tank.*

10. Inspect hoses, sender, and pump assembly for signs of deterioration. The fuel pump can be checked again by applying 12 volts to the electrical wires. Never operate an electrical fuel pump out of the gasoline container for more than 30 seconds. Listen for bearing noise caused by damaged parts.
11. Reverse the procedure to install a new fuel pump and sending unit assembly. Make sure the O-rings and locking assembly are correctly in place.
12. Replace the fuel tank and fuel.
13. Start and run the engine.

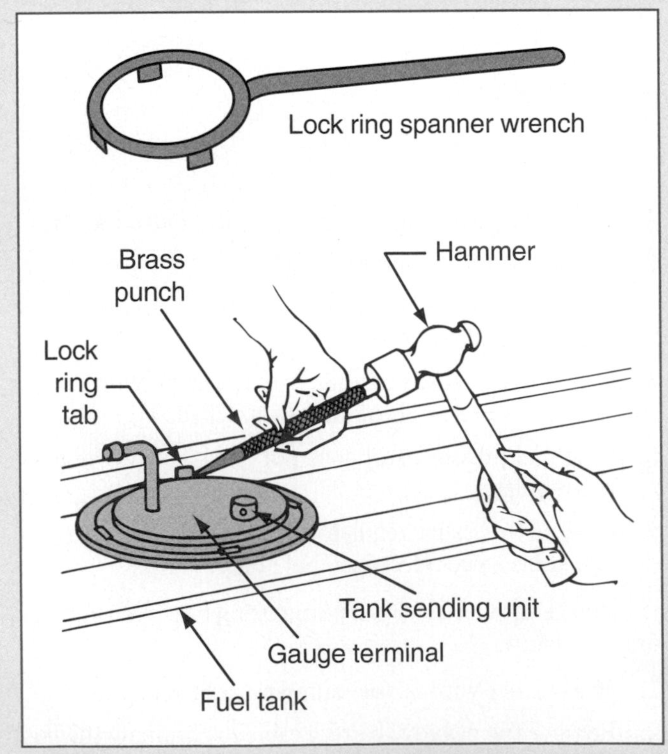

FIGURE 24–33 Remove the electrical fuel pump from the tank by using a lock ring spanner wrench or by lightly tapping the lock ring tab on top of the fuel tank with a brass punch. Always use a brass punch to eliminate the possibility of producing sparks.

PROBLEM: Water in Diesel Fuel

A diesel engine is operating sluggishly and irregularly.

DIAGNOSIS

A common problem on diesel engines is that water gets into the fuel system. Also the filters get dirty and restrict the fuel flow. If you suspect there is water in the diesel fuel, drain the water out before operation. Fuel is lighter than water so the water will settle to the bottom. A bottom drain can be used to drain the water out of the system. On some diesel fuel systems there are valves called petcocks at the bottom of the fuel tank and at the bottom of the primary and secondary filter canisters. The petcocks are used to remove the water from the fuel system. They should be drained regularly.

SERVICE

To replace the fuel filters, use the following general procedure:

1. Disconnect the negative side of the battery.
2. Locate the primary and/or secondary filters.
3. Using the correct wrench size, carefully remove the filters. Be careful not to spill the fuel that is in the filter canisters.
4. Clean the fuel filter canisters using an approved cleaning solvent as shown in ***Figure 24–34***.
5. Replace with new filters. It is best to fill the filter canisters with fuel before reinstalling them on the engine.

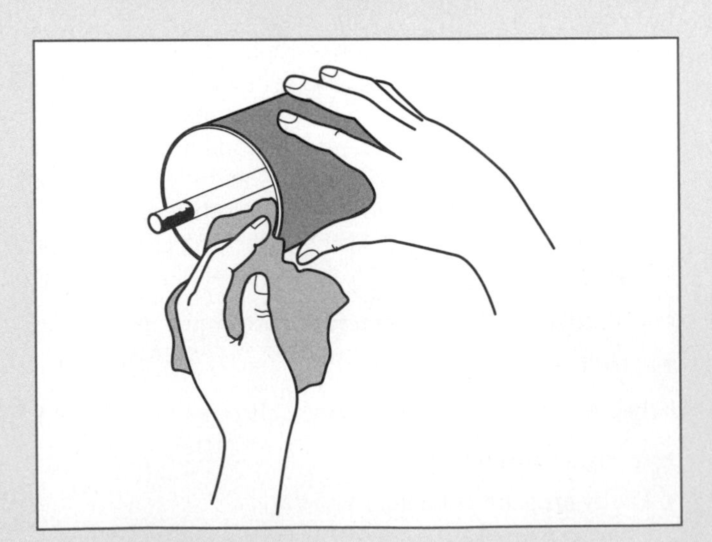

FIGURE 24–34 Always clean the inside of the container before replacing diesel fuel filters.

Filling them could save having to prime the engine later. Be careful not to spill any of the fuel during assembly.

PROBLEM: Faulty Electrical Fuel Pump

A customer has complained that the engine does not run. It is suspected that there is no fuel getting from the tank to the fuel injectors.

DIAGNOSIS

If you suspect that a problem exists with the fuel pump, check that the fuel pump is actually able to run. When the fuel pump is on, the service technician can hear a humming noise from the fuel tank. Place your ear near the fuel fill port with the fuel fill cap removed. The fuel pump should run for about 2 seconds or slightly more when the ignition switch is first turned on. If there is no noise, check to see if there is voltage getting to the fuel pump. This can be done by removing the access panel in the trunk of the car. Make sure the ignition switch is off, then remove the connector to the fuel pump. Check to see if there is voltage at the fuel pump electrical connectors. If there is voltage at the connector, this indicates that the fuel pump is not operating correctly and should be replaced.

SERVICE

On some vehicles, the fuel supply system can also be diagnosed by checking a series of electrical pin connectors. To begin, identify the type of self-diagnostic trouble code that is flashing on the dashboard. ***Figure 24–35*** shows a typical trouble code sheet for one vehicle manufacturer. Notice that the trouble code 43 means that there is a problem with the fuel supply system.

At this point a test harness can be connected to the powertrain control module (PCM). ***Figure 24–36*** shows an example of the test harness connected to the PCM. Often the PCM is located on the right kick panel in the front seat. Once the test harness is connected to the computer, various circuits can be tested for voltage readings. At this point, the service technician will need to have a service manual to determine exactly which pin holes need to be checked for each specific electrical circuit. In this particular example, the manufacturer recommended that a jumper wire be connected between point A6 and A26. Then, using a digital voltmeter, check the voltage between points A26 and D14. This should indicate a certain battery voltage as recommended by the manufacturer (***Figure 24–37***). As electrical circuits become more and more complex, this type of troubleshooting and diagnostic check will become common on all electrical circuits.

DIAGNOSTIC TROUBLE CODE (DTC)	SYSTEM INDICATED
0	Powertrain control module (PCM)
1	Heated oxygen sensor (HO_2S)
3, 5	Manifold absolute pressure (MAP) sensor
4	Crankshaft position (CKP) sensor
6	Engine coolant temperature (ECT) sensor
7	Throttle position (TP) sensor
8	Top dead center (TDC) position sensor
9	No. 1 cylinder position (CYP) sensor
10	Intake air temperature (IAT) sensor
12	Exhaust gas recirculation (EGR) valve lift sensor
13	Barometric pressure (BARO) sensor
14	Idle air control (IAC) valve
15	Ignition output signal
16	Fuel injector
17	Vehicle speed sensor (VSS)
20	Electrical load detector (ELD)
21	Variable valve timing & valve lift electronic control (VTEC) solenoid valve
22	Variable valve timing & valve lift electronic control (VTEC) pressure switch
23	Knock sensor (KS)
30	A/T FI Signal A
31	A/T FI Signal B
41	Heated oxygen sensor (HO_2S) heater
43	Fuel supply system

FIGURE 24–35 When troubleshooting the fuel delivery system, first check the engine trouble code. (Courtesy of American Honda Motor Co., Inc.)

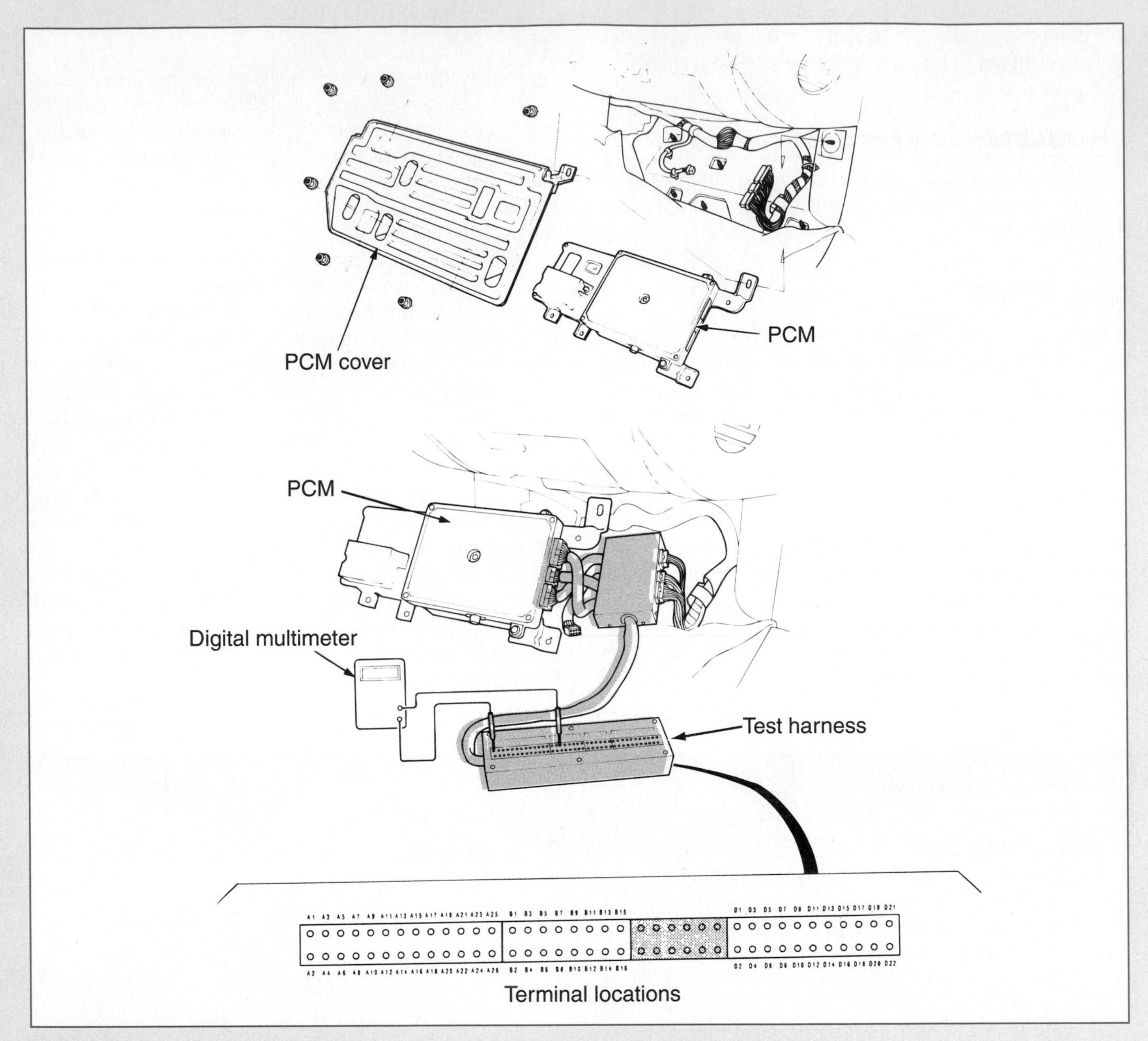

FIGURE 24-36 A test harness can be connected to the computer (PCM) to measure numerous voltages in many circuits. (Courtesy of American Honda Motor Co., Inc.)

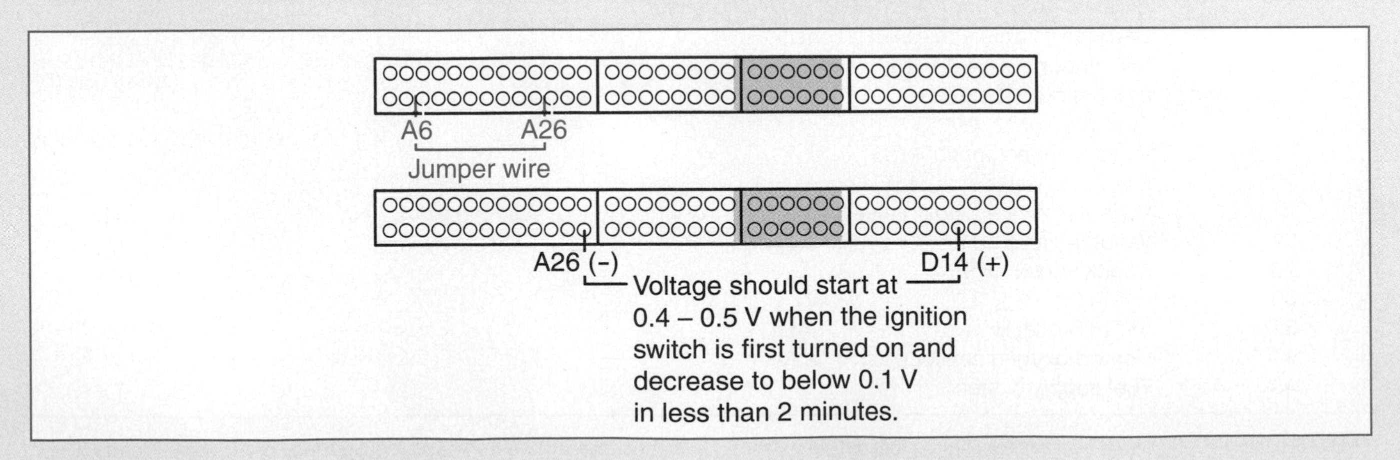

FIGURE 24-37 A digital voltmeter is used to test various electrical circuits. Always follow the service manual for correct procedures. (Courtesy of American Honda Motor Co., Inc.)

Service Manual Connection

There are many important specifications to keep in mind when working with the fuel delivery system. To identify the specifications for your engine, you will need to know the VIN (vehicle identification number) of the vehicle, the type and year of the vehicle, and the type of engine. Although they may be titled differently, some of the more common fuel system specifications (not all) found in service manuals are listed below. Note that these specifications are typical examples. Each vehicle and engine may have different specifications.

Common Specification	Typical Example
Fuel injector resistor	5–7 ohms
Fuel pressure regulator	34–41 psi
Fuel pump pressure	6–8 lb

In addition, the service manual will give specific directions for various service procedures. Some of the more common procedures include:

- Checking fuel pressure regulator
- Fuel filter replacement
- Fuel pump replacement
- Fuel pump testing
- Relieving fuel pressure
- Replacing fuel injectors
- Troubleshooting the fuel system

SUMMARY

The following statements will help to summarize this chapter:

- The main purpose of the fuel system is to supply fuel to the electronic carburetor or fuel injectors so that 14.7 parts of air mix with 1 part of fuel.
- The fuel delivery system includes the fuel tank, the fuel pump, and the fuel filters.
- The fuel tank is made to hold excess fuel for use when needed.
- The fuel tank must operate both under a vacuum and a pressure.
- A fuel-metering unit is located inside the tank to determine the amount of fuel available.
- Diesel fuel tanks have a water sensor to determine if there is water in the fuel tank.
- Fuel pumps are used to transfer the gasoline from the tank to the electronic carburetor or fuel injectors.
- The mechanical fuel pump uses a diaphragm that moves up and down to create suction on one side and pressure on the other side.
- An electric fuel pump can be either a bellows type or a roller vane type.
- The bellows-type electric fuel pump uses a bellows that moves up and down by an electric solenoid, causing the suction and pressure needed to transfer the fuel.
- The roller vane type of electric fuel pump acts as a motor that produces a suction on one side and a pressure on the other side to transfer the fuel.
- Vapor lines are used to return vapor to the fuel tank or to the charcoal canister.
- A pressure regulator is used on many fuel injection systems to regulate the fuel in the fuel rail.
- Computers are used to control when and how long electric fuel pumps stay on.
- Fuel filters are used to stop any contamination from getting into the fuel system.
- Fuel filters are used in fuel tanks, electronic carburetor input lines, and fuel delivery lines from the fuel tank to the injectors.
- Fuel pressure can be checked by using a fuel pressure gauge attached to the fuel rails.
- Fuel can leak into the oil through a defective mechanical fuel pump diaphragm.
- The service technician can check for correct fuel supply by using the troubleshooting codes and electrical checks as outlined in the service manual.
- Fuel pump pressure can be checked by using a gauge and a pint size container to measure how much fuel is being pumped. That measurement is then compared to the manufacturer's specifications.

TERMS TO KNOW

Can you explain each of the following terms? Review the chapter until you can use each term correctly.

Armature
Bellows
Calibrated
Condensation
Diaphragm
Filler neck restrictor
Fuel injector
Fuel vapors
Inertia switch
Micrometers
Pressure regulator valve

REVIEW QUESTIONS

Multiple Choice

1. Which of the following is not part of the basic fuel delivery system flow?
   a. Fuel tank
   b. Catalytic converter
   c. Fuel pump
   d. Fuel filter
   e. Vapor line

2. Which system always has a fuel return line back to the fuel pump or tank?
   a. Diesel fuel system
   b. Gasoline fuel system
   c. Contaminated fuel system
   d. All of the above
   e. None of the above

3. The fuel cap is designed to:
   a. Help produce high vacuum in the fuel tank
   b. Help produce high pressure in the fuel tank
   c. Relieve the pressure and vacuum in the fuel tank
   d. Clean the fuel
   e. Filter the fuel

4. The fuel metering unit changes the ________ in an electrical circuit when the float changes position.
   a. Resistance
   b. Voltage
   c. Calibration
   d. Filters
   e. Wattage

5. Water in diesel fuel will ________.
   a. Increase the horsepower
   b. Damage the fuel system
   c. Cause vibration
   d. All of the above
   e. None of the above

6. A mechanical fuel pump is most often driven from the:
   a. Flywheel
   b. Distributor shaft
   c. Camshaft
   d. Alternator
   e. Valve system

7. Vapor in the gasoline fuel system is caused by:
   a. Rapidly cooling down the temperature of gasoline
   b. Heating gasoline to its boiling temperature
   c. Pumping gasoline through small holes
   d. Mixing it with nitrogen
   e. None of the above

8. Which of the following is a type of electrical fuel pump?
   a. Bellows-type
   b. Roller vane type
   c. Mechanical diaphragm
   d. All of the above
   e. A and B

9. What is the purpose of using a fuel pressure regulator?
   a. It regulates the oil pressure
   b. It regulates the fuel pressure at the injectors
   c. It controls the thickness of the fuel in cold weather
   d. It controls the temperature of the fuel
   e. None of the above

10. Which of the following is a contaminant that is of concern in diesel fuel?
   a. Wax crystals
   b. Microorganisms
   c. Rust
   d. All of the above
   e. None of the above

11. Fuel filters are ________ inside the fuel pump housing.
   a. Never placed
   b. Always placed
   c. Sometimes placed
   d. Dissolved
   e. Expanded

12. What is the most desirable arrangement of fuel filters on a diesel engine?
   a. Only one filter should be used
   b. Two filters should be used
   c. Three filters should be used
   d. No filters should be used
   e. Five or more filters should be used

13. One micrometer (micron) is equal to ________ inch(es).
a. 0.0045
b. 0.000039
c. 1.3
d. 1.35
e. 0.0039

14. Which filter has the smallest micrometer (micron) size holes?
a. The primary filter
b. The secondary filter
c. Both filters have the same micrometer size
d. The combining filter
e. Filters are not rated by micrometer size

The following questions are similar in format to ASE (Automotive Service Excellence) test questions.

15. *Technician A* says to check fuel pump pressure, use a compression gauge. *Technician B* says to check fuel pump pressure, disconnect the fuel line to the electronic carburetor, and use a pint jar and pressure gauge. Who is correct?
a. A only c. Both A and B
b. B only d. Neither A nor B

16. An engine seems to be running irregularly and stalls often. *Technician A* says the fuel filter needs to be replaced. *Technician B* says the main fuel line is broken. Who is correct?
a. A only c. Both A and B
b. B only d. Neither A nor B

17. A gasoline engine hesitates and runs sluggishly. *Technician A* says the problem could be the use of high-octane fuel. *Technician B* says the problem could be a dirty fuel filter. Who is correct?
a. A only c. Both A and B
b. B only d. Neither A nor B

18. There is an increase in the oil level in a gasoline engine. *Technician A* says fuel could be leaking into the oil through the fuel pump. *Technician B* says fuel cannot get into the oil, and probably too much oil was put into the crankcase. Who is correct?
a. A only c. Both A and B
b. B only d. Neither A nor B

19. *Technician A* says the fuel pressure for a mechanical fuel pump should be about 5 to 6 pounds. *Technician B* says the fuel pressure for a mechanical fuel pump should be checked with a vacuum meter. Who is correct?
a. A only c. Both A and B
b. B only d. Neither A nor B

20. *Technician A* says the fuel pressure for an electrical fuel pump should be about 5 to 6 pounds. *Technician B* says the fuel pressure for an electrical fuel pump is measured with a voltmeter. Who is correct?
a. A only c. Both A and B
b. B only d. Neither A nor B

21. *Technician A* says that the restrictor on the filler neck is to keep unleaded fuel out of the gasoline tank. *Technician B* says the filler neck has hoses to vent the fuel tank. Who is correct?
a. A only c. Both A and B
b. B only d. Neither A nor B

22. *Technician A* says the inertia switch is used to shut the fuel off to the injection system during impacts or accidents. *Technician B* says the inertia switch is used to regulate pressure in the fuel pump during slow deceleration. Who is correct?
a. A only c. Both A and B
b. B only d. Neither A nor B

Essay

23. Why is there a vacuum release on the fuel cap?

24. Describe the operation of the diaphragm-type fuel pump.

25. What are some advantages of using an electrical fuel pump?

26. Why do some cars have a fuel pressure regulator?

27. List the types of contaminants that can enter the fuel system.

28. What is a micrometer as related to fuel filters?

29. Describe the procedure used to check fuel pump pressure.

30. Describe how to test a fuel system using a digital voltmeter and a test harness.

Short Answer

31. A(n) ________ valve is used to keep the pressure constant on fuel injection systems.

32. Fuel pumps can be checked for two specifications including ________ and ________.

33. When inserting the fuel inlet nut on the carburetor, always be careful not to ________ the threads on the nut.

34. If the fuel pressure from an electrical fuel pump is lower than specifications, check for ________ and ________.

35. To keep the fuel from splashing around in the fuel tank ________ are placed inside the fuel tank.

Applied Academics

TITLE: Alternative Fuel and the Environment

ACADEMIC SKILLS AREA: Science

NATEF Automobile Technician Tasks:

The service technician develops and maintains an understanding of all federal, state, and local rules and regulations regarding environmental issues related to the work of the automotive technician. The technician uses government impact statements, media information, and general knowledge of pollution and waste management to correctly use and dispose of products that result from the performance of a repair task.

CONCEPT:

Alternative fuels offer a significant emission benefit. All the alternative fuels reduce ozone-forming tailpipe emissions. The accompanying chart shows the percentage of combined carbon monoxide (CO) and nitrogen oxide NO_X emission for each alternative fuel when compared to gasoline also referred to as RFG or reformulated gasoline. For example, the emissions from compressed natural gas (CNG) vehicles are estimated to be 20% compared to 100% emissions from vehicles using RFG. This means that compressed natural gas vehicles demonstrate an 80% reduction in ozone-forming emissions.

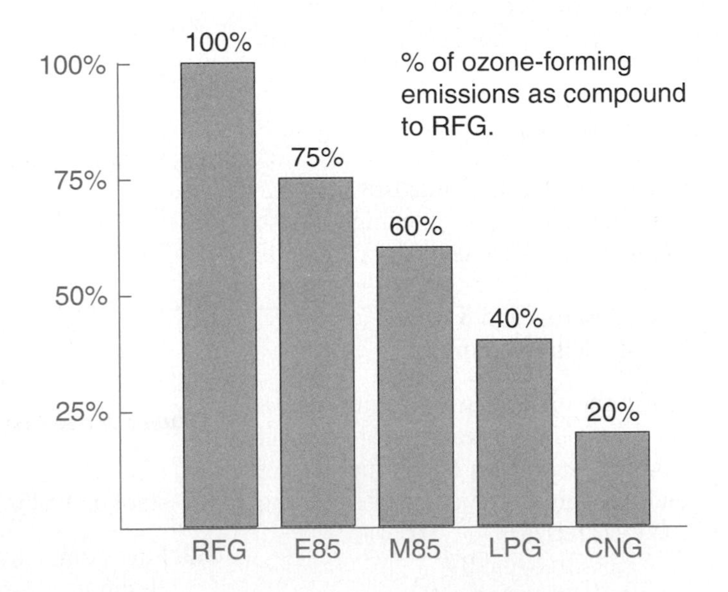

APPLICATION:

If reformulated gasoline is 100%, how much less ozone-forming emissions are being produced when using E85 fuel? What effect would using alternative fuels have on the overall environment in the future? Can you describe why alternative fuels are so much better for the environment as compared to RFG? Why don't more consumers use alternative fuels? Search the Web site http://www.afdc.doe.gov to identify which vehicles can use alternative fuels.

Electronic Feedback Carburetors

CHAPTER 25

OBJECTIVES

After studying this chapter, you should be able to:

- State the basic principles of fuel systems.
- Identify the advantages of electronic feedback carburetors.
- Define both closed and open loop computer operation.
- Describe the term *pulse width* in relationship to computers.
- Analyze the operation of various computer inputs and outputs in relation to the electronic computer.
- Identify various types of electronic carburetor circuits.
- State various problems, their diagnosis, and service procedures regarding electronic feedback computers.

INTRODUCTION

All vehicles manufactured today use fuel injection. However, there are still some vehicles on the road that use electronic carburetors that need to be serviced. It is important to note that the technology developed for mechanical carburetors led to electronic carburetors, and that electronic carburetors led to the development of fuel injection. This chapter is about electronic carburetors.

Electronic carburetors are controlled by a computer. Carburetors that are controlled by computers are called electronic feedback carburetors. This means that a computer is sensing various inputs. Then, based on these inputs, the computer tells the carburetor how much fuel to put into the engine. This chapter describes the basic operating principles of electronic feedback carburetors. In addition, many of the computer components introduced in

Certification Connection

ASE Connection: The information in this chapter can help you prepare for the National Institute for Automotive Service Excellence (ASE) certification tests. The tests and content areas most closely related to this chapter are:

Test A1—Engine Repair

- **Content Area**—Fuel, Electrical, Ignition, and Exhaust Systems Inspection and Service

Test A8—Engine Performance

- **Content Area**—Fuel, Air Induction, and Exhaust System Diagnosis and Repair

NATEF Connection: Much of the information in this chapter is related to the NATEF tasks. The NATEF tasks and priority numbers most closely related to this chapter are:

1. Diagnose hot or cold no-starting, hard starting, poor drivability, incorrect idle speed, poor idle, flooding, hesitation, surging, engine misfire, power loss, stalling, poor mileage, dieseling (post ignition), and emission problems on vehicles with carburetor-type fuel systems; determine necessary action. P-3
2. Check idle speed and fuel mixture. P-2
3. Adjust idle speed and fuel mixture. P-3

this chapter will help you understand the basic fuel injection system in the chapter that follows.

25.1 CARBURETION SYSTEM PRINCIPLES

AIR-FUEL RATIO

Research has shown that if a gasoline engine operates on an accurate and precise air-fuel ratio, engine efficiency will be improved. However, many things cause the air-fuel ratio to be upset or changed from the optimum 14.7 to 1. The following factors may change the air-fuel ratio:

- Air density and altitude
- Acceleration of the vehicle
- Deceleration of the vehicle
- Temperature of the air
- Moisture content of the air
- Speed of the engine or vehicle
- Load on the engine
- Overall condition and efficiency of the engine

The fuel system must be designed to operate under different conditions. ***Figure 25–1*** shows how the air-fuel ratio changes with different speeds. Air-fuel ratios range from 9 to 1 during idle to 18 or 19 to 1 during deceleration. When the vehicle is started and at idle, the air-fuel ratio is about 12 to 1. This is a very rich mixture and produces poor fuel economy. As the engine speed is increased to move the vehicle to 40 mph, the air-fuel ratio settles at about 15 to 1. As the speed of the vehicle increases to 60 mph or above, the air-fuel ratio drops off again to about 12 to 1. Also, when the vehicle is accelerated or decelerated, the air-fuel ratio changes accordingly. Deceleration produces a leaner mixture, while acceleration produces a richer mixture.

ATOMIZATION

Fuel systems operate on the principle that as more air flows into the engine, more fuel is added. However, the fuel must be in an **atomized** state. Fuel that is atomized is in very small mistlike droplets. The fuel is then **vaporized**. When a liquid is changed to a vapor, it is vaporized. Vaporization occurs after the fuel is atomized. This can be done by increasing the temperature of the fuel. Fuel in a vaporized state mixes very well with the air that is going into the engine. ***Figure 25–2*** shows how atomized particles of fuel need to be surrounded by 14.7 parts of air for the most efficient combustion. Fuel that is in a liquid state, not vaporized or atomized, mixes poorly with the air passing into the engine.

VENTURI

On electronic carbureted engines, air is drawn into the engine by the action of the piston moving downward on the intake stroke. When the piston moves down and the intake valve is opened, a **vacuum** is produced. This vacuum causes air to be drawn or pulled into the engine. The air, however, passes through the electronic carburetor and **venturi** as it

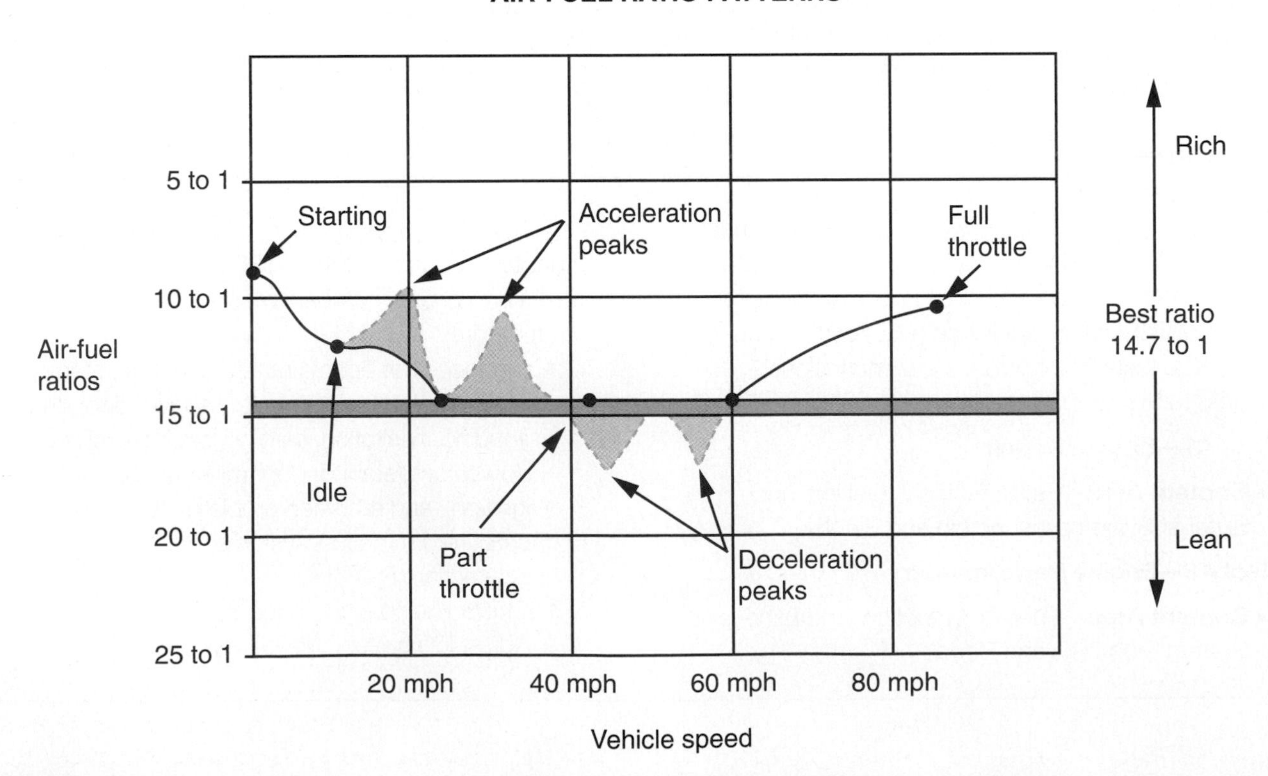

FIGURE 25–1 Air-fuel ratios will change with different driving speeds. Acceleration and deceleration peaks are also shown.

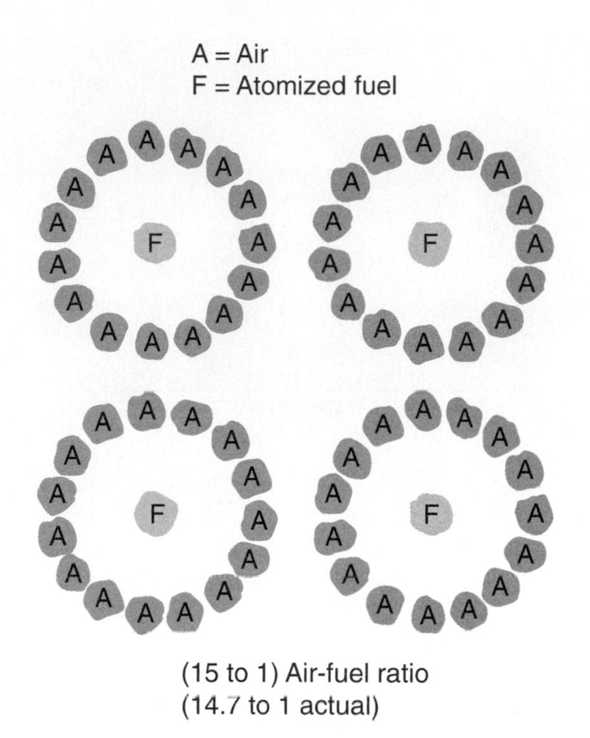

FIGURE 25-2 When air and fuel are mixed correctly, 14.7 parts of air will surround 1 part of fuel. The fuel mixes best with the air when in the atomized state.

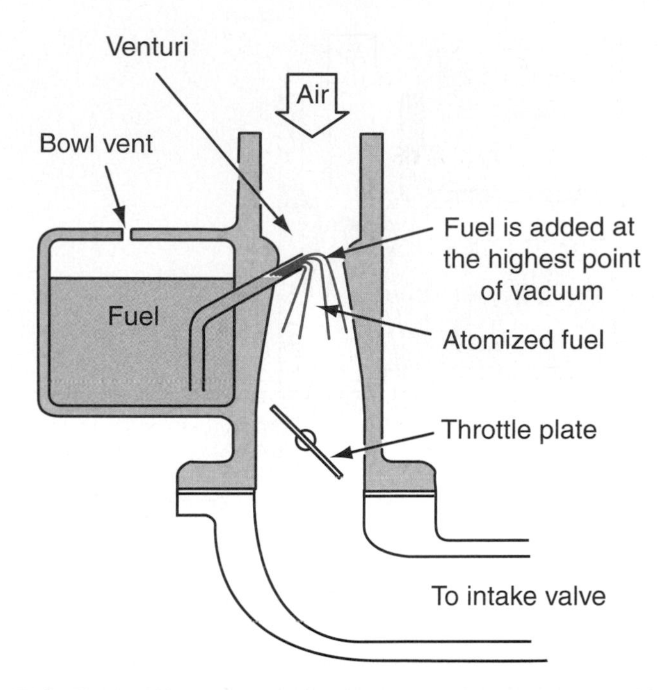

FIGURE 25-4 The vacuum that is produced at the venturi is used to draw in the fuel from the carburetor.

goes into the engine. (The vacuum can also be called a pressure differential. At the venturi, there is actually a lower pressure than above the venturi. For simplicity, the word *vacuum,* rather than pressure differential, will be used throughout the chapter.)

A venturi is a streamlined restriction that partly restricts airflow into the engine (***Figure 25–3***). Air is drawn into the engine by the intake manifold vacuum. As the air enters the venturi, it is forced to speed up, or increase in velocity, in order to pass through the restriction. This restriction causes an increase in vacuum by the venturi. The vacuum is also felt slightly below the major restricted area, and it continues to be reduced farther down the bore.

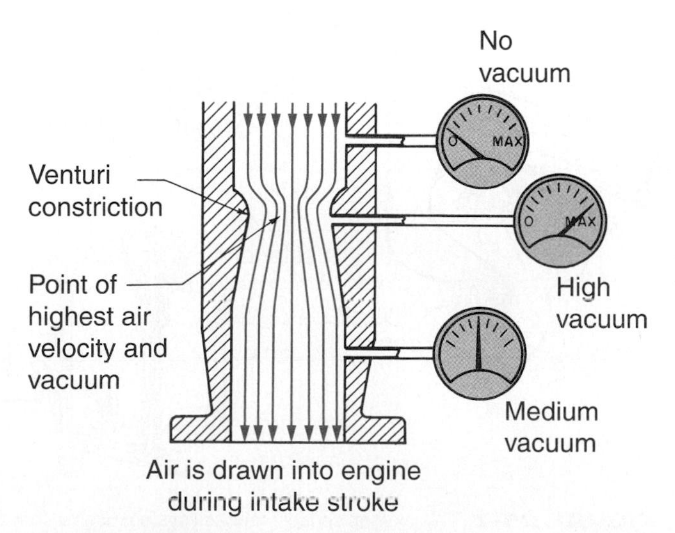

FIGURE 25-3 A venturi is a restriction in the path of airflow. A vacuum is produced at the point of greatest restriction.

As the engine speed increases during acceleration, more air goes into the engine. This causes the venturi vacuum to increase because the greater the velocity of air passed through the venturi, the greater the vacuum. The vacuum produced at the venturi is used to draw in the correct amount of fuel. As the vacuum increases, more fuel is drawn in. As it decreases, less fuel is drawn in.

The venturi also aids fuel atomization and vaporization by exposing the fuel to air. The fuel is added in the center of the strongest vacuum point of the venturi. ***Figure 25–4*** shows fuel being added to the airflow through the electronic carburetor. A discharge tube is located near the venturi. As the air flows through the venturi, the vacuum draws the fuel from a bowl into the stream of air going into the engine.

THROTTLE PLATE

The flow of air and fuel through the electronic carburetor is controlled by the throttle plate (***Figure 25–5***). The throttle plate, which is made of a circular disc, is placed directly in the flow of air and fuel below the venturi. Its purpose is to control the amount of air and fuel that enters the engine. The throttle plate is connected to the driver's throttle. As the driver's foot is depressed, the throttle plate opens to a vertical position. During this condition, there is very little restriction of air and fuel. This is a maximum load and speed condition. As the driver's foot is removed, a spring closes the throttle plate and restricts the amount of air and fuel going into the engine. This is a low-speed and load condition.

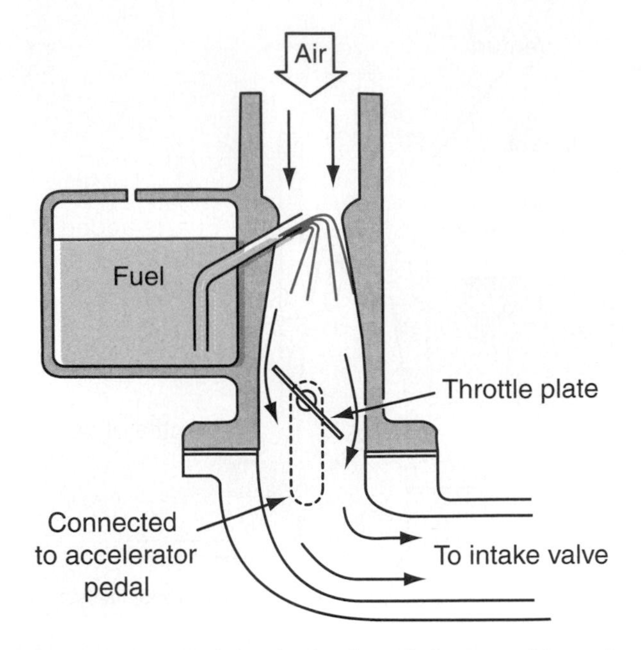

FIGURE 25-5 A throttle plate is put in the base of the carburetor to control the amount of air and fuel flowing into the engine.

FIGURE 25-6 A computer is used to control and manage fuel. This computer is part of a system called computer command control, manufactured by General Motors.

25.2 ELECTRONIC CARBURETION PRINCIPLES OF OPERATION

INTRODUCTION TO ELECTRONIC FEEDBACK CARBURETORS

An electronic feedback carburetor is one in which the amount of fuel being sent through the carburetor is controlled by an on-board computer. In recent years, computers have been used to control various components in the engine. This is also true in the fuel system. Control of fuel by computers is referred to as fuel management. A computer is placed on board the vehicle. During normal operation, many sensors electronically feed electrical signals into the computer, which acts as a sort of brain. On the basis of these input signals, the computer sends out a signal to operate a certain component.

PARTS LOCATOR

Refer to photo #26 in the front of the textbook to see what a ***computer*** *looks like.*

In the past there were many names used for the fuel management computer. However, over the past few years, all manufacturers have begun to use the same name. Today, the most common name for such computers is the PCM, or powertrain control module. ***Figure 25–6*** shows an example of the PCM. Basically, the PCM is used to control the amount of fuel that is sent through the electronic carburetor. The big advantage of this type of system is that the air-fuel ratio mixture is held much closer to the 14.7 to 1 ratio. This ensures that exhaust emissions and fuel consumption are reduced.

Figure 25–7 shows a schematic of a simple fuel management system. The PCM is designed to manage the fuel so that higher gasoline mileage can be obtained while producing fewer emissions. A more complex schematic is shown in ***Figure 25–8***. In this system, there are several inputs from the engine that feed signals into the PCM. Based on these inputs, the PCM operates a solenoid to control the fuel entering the electronic carburetor. However, before studying these systems, two terms must be defined. These are *closed* and *open loop operation*.

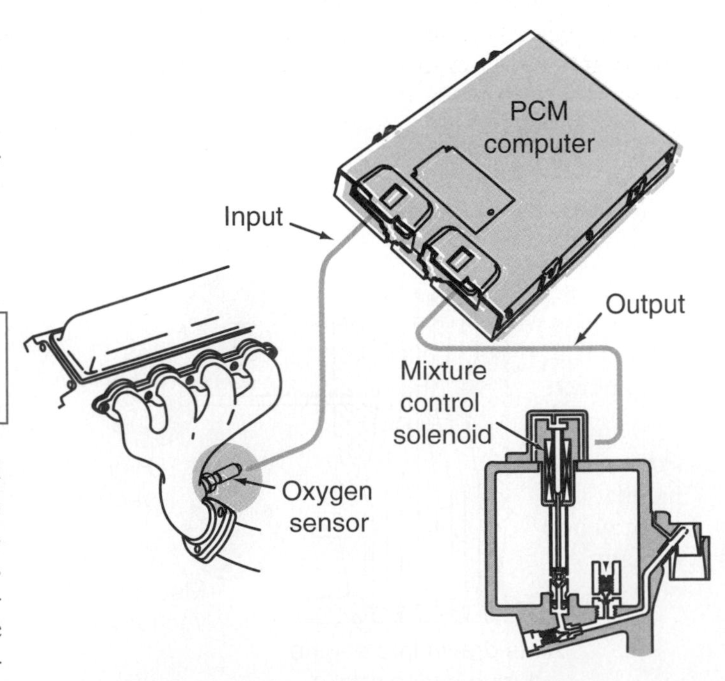

FIGURE 25-7 This oxygen sensor feedback system uses a computer and an oxygen sensor to control the fuel flow in the carburetor.

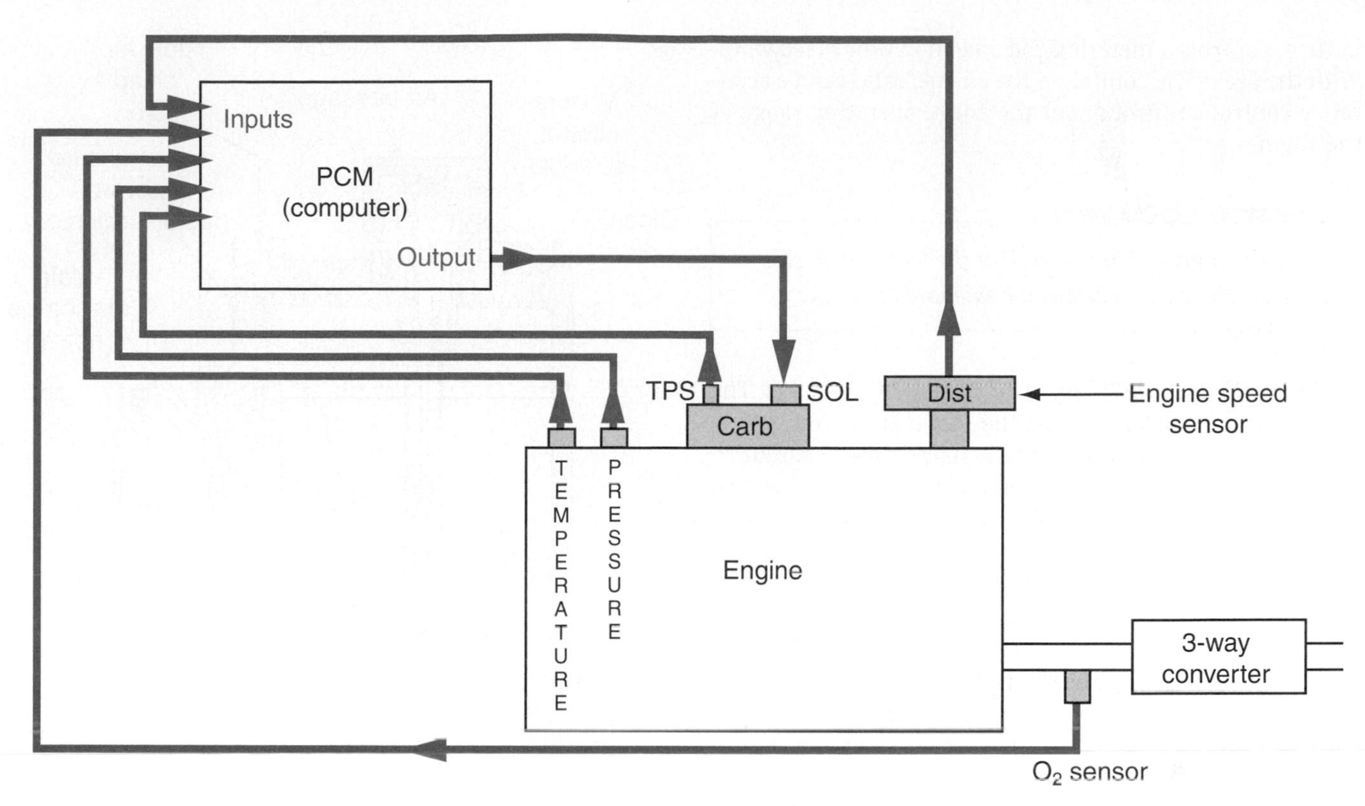

FIGURE 25–8 This schematic shows the inputs and outputs on a fuel management system.

CLOSED AND OPEN LOOP

Computer systems used on automobiles can operate in either a closed or open loop mode. In the **closed loop** mode, an oxygen sensor is placed in the exhaust system (see ***Figure 25–8***) to get an indication of the air-fuel ratio of the engine. The term *closed loop* describes the relationship between three components: the oxygen sensor, the computer, and the fuel-metering control device. The oxygen sensor sends electrical signals telling the computer what the air-fuel mixture is. The computer sends a command to the fuel-metering control device to adjust the air-fuel ratio as close to 14.7 to 1 as possible. This adjustment usually causes the air-fuel ratio to go too far in the opposite direction. The oxygen sensor senses this overadjustment and sends a signal to the computer to adjust back toward the 14.7 to 1 ratio. This cycle occurs so quickly that the air-fuel ratio never gets very far from the optimum of 14.7 to 1 in either direction. This cycle repeats continuously.

The closed loop mode is the most efficient operating mode. The computer is designed and programmed to keep the system in the closed loop as much as possible. The computer reads three criteria to keep the system in closed loop:

1. The oxygen sensor must reach a temperature of at least 600 degrees Fahrenheit.
2. The engine coolant temperature must reach a predetermined temperature of 150 degrees Fahrenheit.
3. The engine must have been running for a certain period of time. The time varies from system to system. The range of time is from several seconds to several minutes.

Open loop operation is used whenever the optimum air-fuel ratio will not work. For example, during starting and warm-up conditions a richer mixture is needed. Also, a richer mixture is needed at wide-open throttle. During these conditions, the computer uses coolant temperature, engine load, barometric pressure, and engine speed to determine exactly what the air-fuel ratio should be. On the basis of these inputs, the computer sends a command to the fuel mixture control inside the carburetor to set a predetermined air-fuel ratio (not 14.7 to 1). The air-fuel ratio will not change until one of the inputs changes. In the open loop mode, the computer does not use the oxygen sensor for input. Thus, open loop operation is generally not as efficient as closed loop operation. In addition, more emissions are produced. Thus, the computer is designed to put the system into the closed loop mode as soon as possible. Certain failures in the system may cause the computer to go out of closed loop operation.

25.3 INPUT AND OUTPUT CONTROLS

CARBURETOR SOLENOID

An electrical solenoid in the carburetor controls the air-fuel mixture. The solenoid is connected to and controlled by the PCM. The PCM sends a signal to the solenoid. The solenoid, in turn, controls a metering rod and an idle air bleed valve. With the use of the computer, the air-fuel ratio can be accurately controlled throughout the entire operating range of the engine.

PARTS LOCATOR

*Refer to photo #27 in the front of the textbook to see what an **electronic carburetor solenoid** looks like.*

Figure 25–9 shows a mixture control (M/C) solenoid. The mixture control solenoid uses the solenoid to control fuel flow from the bowl to the main discharge tube or nozzle in the electronic carburetor. The idle circuit air bleed is also controlled by the solenoid. The solenoid coil and plunger are mounted vertically in the carburetor. The stem and valve end reach to the bowl floor where the solenoid controls a passage between the bowl and the discharge tube. It acts as a metering valve that opens and closes. This happens at a rapid rate of 10 times per second.

The upper end of the solenoid rod plunger opens and closes the idle air bleed. When the solenoid is energized, the plunger moves down. This opens the idle air bleed and closes the discharge tube passage. Both the idle and main metering systems become lean or rich together.

Figure 25–10 shows how the mixture control valve is placed within the electronic carburetor. In most cases, the M/C solenoid is placed in the center of the float bowl area. The throttle plate and venturis are located on the right side of the carburetor. Note that when the solenoid plunger moves downward, the idle air circuit opens. When the plunger moves upward, the idle air circuit closes.

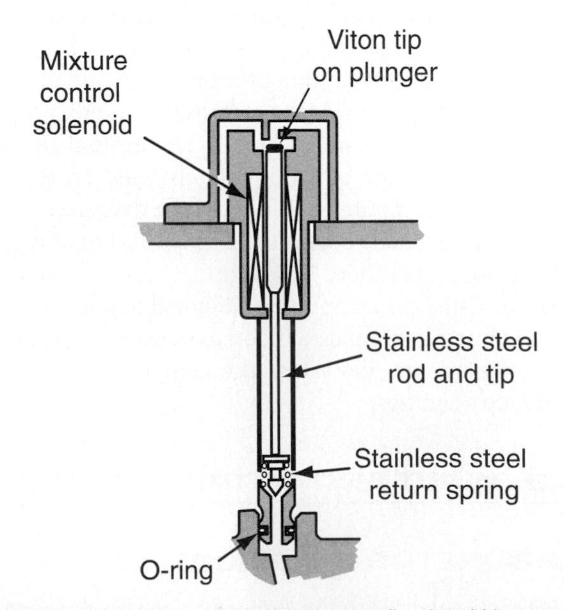

FIGURE 25–9 The M/C (mixture control) solenoid is controlled by the computer. It keeps adjusting the size of the main jet for different air-fuel conditions.

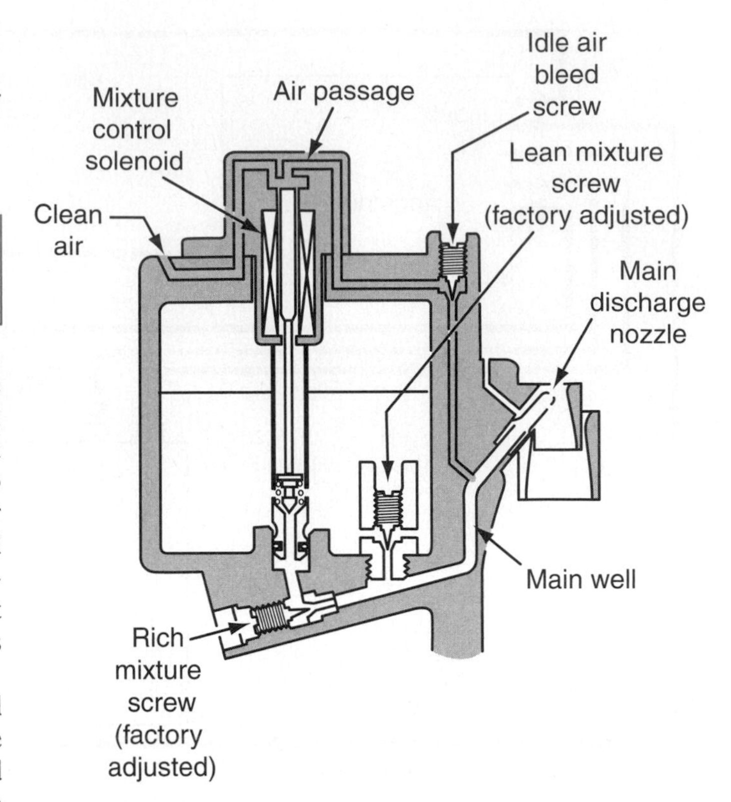

FIGURE 25–10 When the M/C solenoid is energized, the plunger moves down, opening the air circuit.

M/C SOLENOID PULSE WIDTH

As previously mentioned, the M/C solenoid may open and close many times per second. The length of time that the solenoid valve stays energized is referred to as the **pulse width**. In addition, these systems also have a duty cycle. The **duty cycle** is the percentage of on-time compared to the total cycle time. To complete a cycle, the solenoid must go from off, to on, and then to off again. On one manufacturer's electronic feedback carburetor system, when the solenoid is not being energized, there is a maximum amount of fuel going through the valve. Then, as the M/C solenoid is energized and is closed, there is no fuel flowing through the valve. ***Figure 25–11*** shows the basic definitions of pulse width, cycle, and duty cycle. In addition, three different pulse widths are shown. The first pulse width is very short. This means that the M/C solenoid is open, then closes for a short period of time, then opens again. Under these conditions, there is a very rich mixture. On the second pulse width, the M/C solenoid is open for a longer period of time. It opens, then closes for a little longer period of time, then opens again. Under these conditions, the air-fuel mixture is about 14.7 to 1. On the third pulse width shown, the M/C solenoid is open for only a short period of time. The M/C solenoid opens, then closes for a long period of time, then opens again. Under these conditions, the air-fuel ratio is very lean.

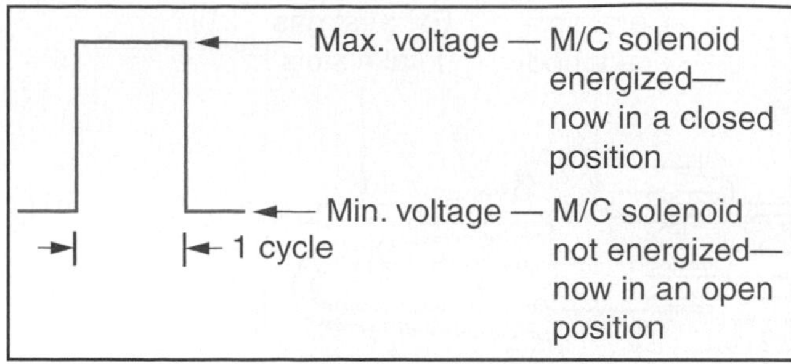

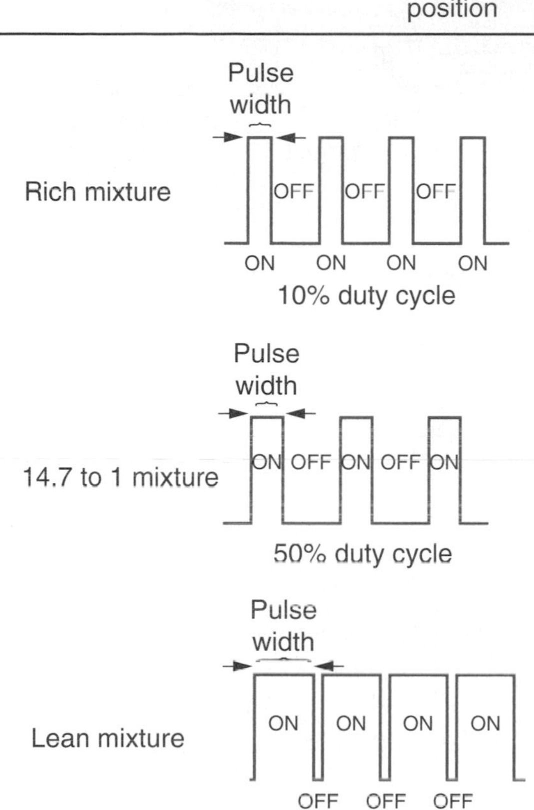

FIGURE 25–11 The pulse width determines the amount of fuel going through the carburetor.

OXYGEN SENSOR

A sensor is located in the exhaust stream close to the engine. It is known as an oxygen sensor. It measures the amount of oxygen in the exhaust gas. ***Figure 25–12*** shows an oxygen sensor. There is a direct relationship between the air-fuel mixture and the amount of oxygen in the exhaust gas. The oxygen sensor determines whether the exhaust is too rich or too lean. It sends a low-voltage (below 450 millivolts) signal to the PCM when the mixture is lean. A high-voltage (above 450 millivolts) signal is sent to the PCM when the mixture is rich.

The PCM then signals the mixture control (M/C) solenoid to deliver a richer or leaner mixture. As the electronic carburetor makes an air-fuel change, the oxygen sensor immediately senses that change and again signals the PCM. This goes on continually during the engine operation.

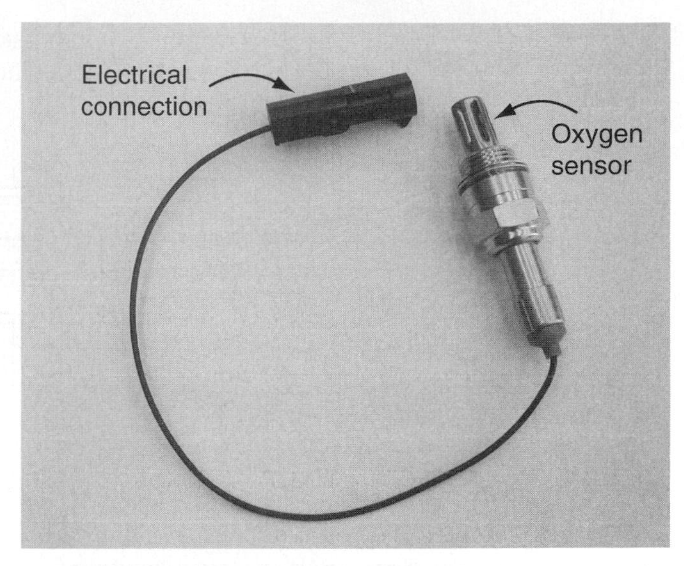

FIGURE 25–12 This oxygen sensor, which is inserted into the exhaust stream, is used to determine the amount of oxygen in the exhaust. This information is then sent to the computer as a signal to help control the air-fuel mixture.

This process is called closed loop operation. Closed loop operations deliver an accurate 14.7 to 1 air-fuel ratio to the engine.

OXYGEN SENSOR OPERATION

Most automotive oxygen sensors are made of zirconia. ***Figure 25–13*** shows all of the parts of a typical oxygen sensor. The ceramic material (zirconia) on the tip of the sensor produces a voltage based upon the amount of oxygen in the exhaust stream. It does this by comparing the amount of oxygen in the exhaust to the amount of oxygen in the surrounding air. Note that the hollow center terminal is open to the atmosphere. When the exhaust is lean (excess air), the sensor produces a low voltage, near zero. When the exhaust is rich (excess fuel), it produces high voltage (up to 1 volt). ***Figure 25–14*** shows a chart with the voltage (millivolts) of the oxygen sensor varying over time. The high points on the line represent times in which the exhaust is rich. The bottom points of the line represent times in which the exhaust is lean.

For the sensor to work correctly, it needs a good source of outside air and a temperature of at least 500 degrees Fahrenheit. The outside air is used as a reference. The oxygen sensor gets its heat from the exhaust gas temperature. Newer sensors contain an electric heater that helps them heat up faster, even above the exhaust gas temperature. Many heated sensors are also waterproof, and so must receive outside air through an electrical lead fed into the sensor. Although only a small amount of air is needed, special care must be taken to not damage the air path.

In operation, the PCM compares the voltage from the oxygen sensor to the values programmed into the computer.

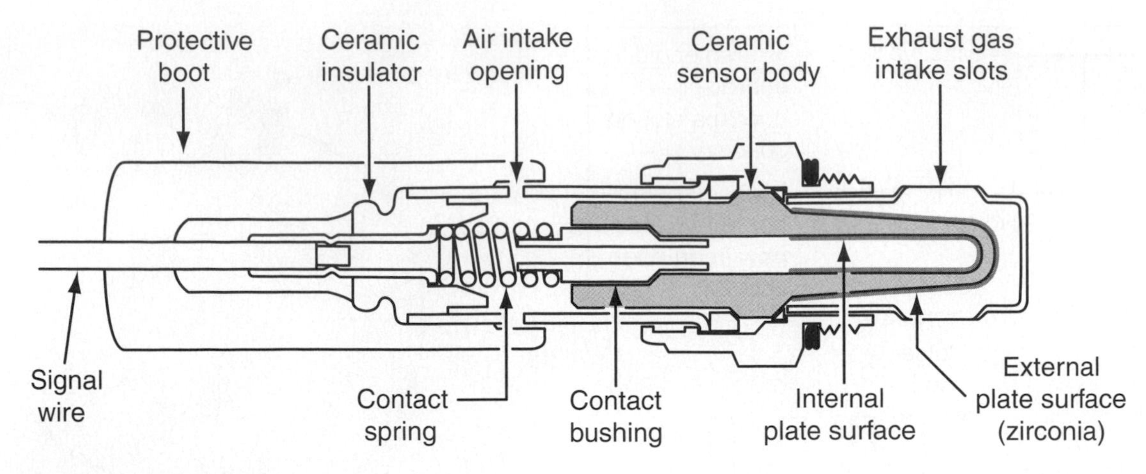

FIGURE 25-13 This illustration shows the parts of an oxygen sensor.

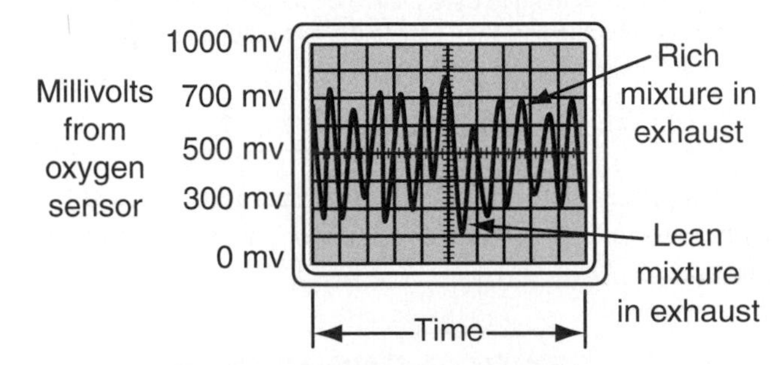

FIGURE 25-14 The oxygen sensor produces voltage peaks and valleys, depending on the amount of oxygen in the exhaust.

If the air-fuel ratio is lean, the computer adds fuel in the carburetor. If the air-fuel ratio is rich, the computer subtracts fuel in the carburetor.

TYPES OF OXYGEN SENSORS

There are several types of oxygen sensors being used today. Sensors are identified based on the way they get heat for the element and the way the outside air is protected. ***Figure 25-15*** shows an example of several types of oxygen sensors. Most early oxygen sensors were of the unheated type. In this case, the exhaust gases provided the heat to bring the sensor element up to its operating conditions. Many newer sensors are of the heated type. An electric heater inside the sensor provides added heat to the zirconia element. This helps the element come up to temperature faster, operate better in colder climates, give improved fuel control, and make it less sensitive to contaminants. The heater power comes directly from the vehicle electrical system. It is usually turned on with the ignition switch. On some vehicles, relays are used to turn the heater off and on under certain conditions.

Zirconia oxygen sensors need only a small amount of air. However, there needs to be enough clean outside air for them to work correctly. Most unheated sensors have an open path for air to flow over them. Also, they are placed in a position where water will not splash on them. On some vehicles, the sensor could be splashed with water. In this case, a water-resistant oxygen sensor is used. Added shielding helps keep water out of the sensor, while still making it easy for air to pass through.

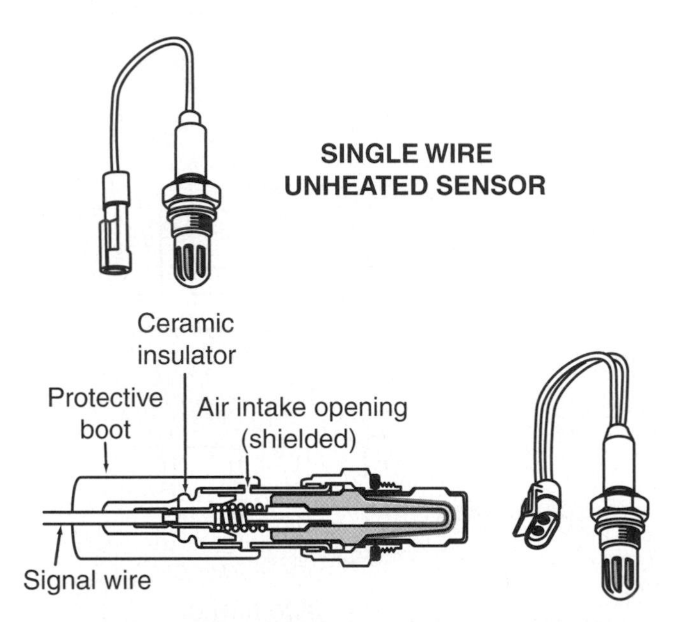

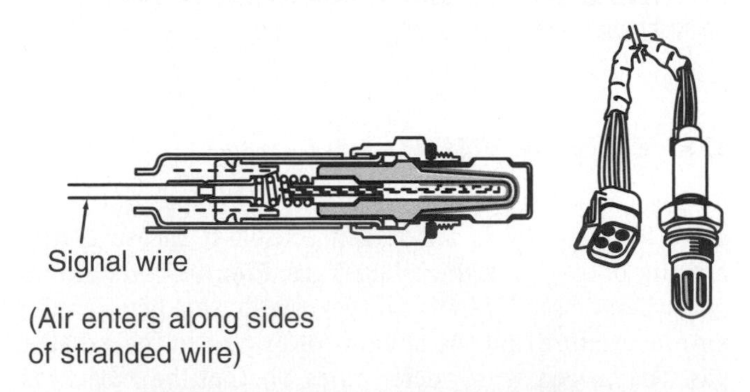

FIGURE 25-15 There are several types of oxygen sensors.

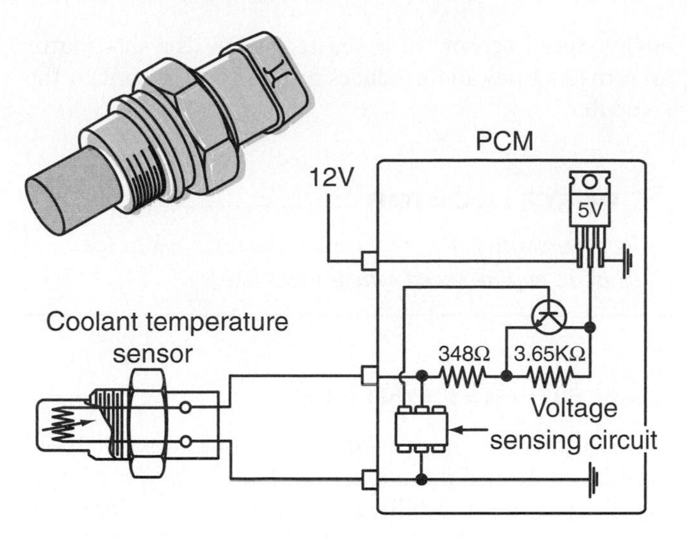

FIGURE 25–16 A coolant temperature sensor is used to sense the temperature of the engine. This information is sent to the computer to help control the air-fuel ratio.

TEMPERATURE SENSOR

A temperature sensor is also used with the PCM (***Figure 25–16***). The sensor is located in the cooling system and is connected to the PCM. Whenever the engine is cold, there is no need for the oxygen sensor to control the air-fuel ratio. Under these conditions, the PCM tells the electronic carburetor to deliver a richer mixture. The mixture is based on what has been programmed into the PCM and what other sensors are telling the computer. This is an open loop operation. The sensor's resistance is lowered as coolant temperature increases. The resistance is raised as the temperature decreases.

After the engine reaches operating temperature, the temperature sensor signals the PCM to read what the oxygen sensor is providing. If other requirements are met, closed loop operation begins.

PARTS LOCATOR

*Refer to photo #28 in the front of the textbook to see what a **temperature sensor** looks like.*

PRESSURE SENSOR

The load on the engine also affects the air-fuel mixture needed in the engine. As a load is placed on the engine, a richer air-fuel mixture is needed. This condition can be measured by using a sensor connected to the intake manifold vacuum, as shown in ***Figure 25–17***. This sensor detects changes in the manifold pressure. As the pressure changes, a flexible resistor attached to a diaphragm also changes its resistance. This pressure change causes a voltage change that the PCM can read. This voltage change signals the PCM that there is an increase in load. The PCM takes this signal into account to determine the exact air-fuel ratio required.

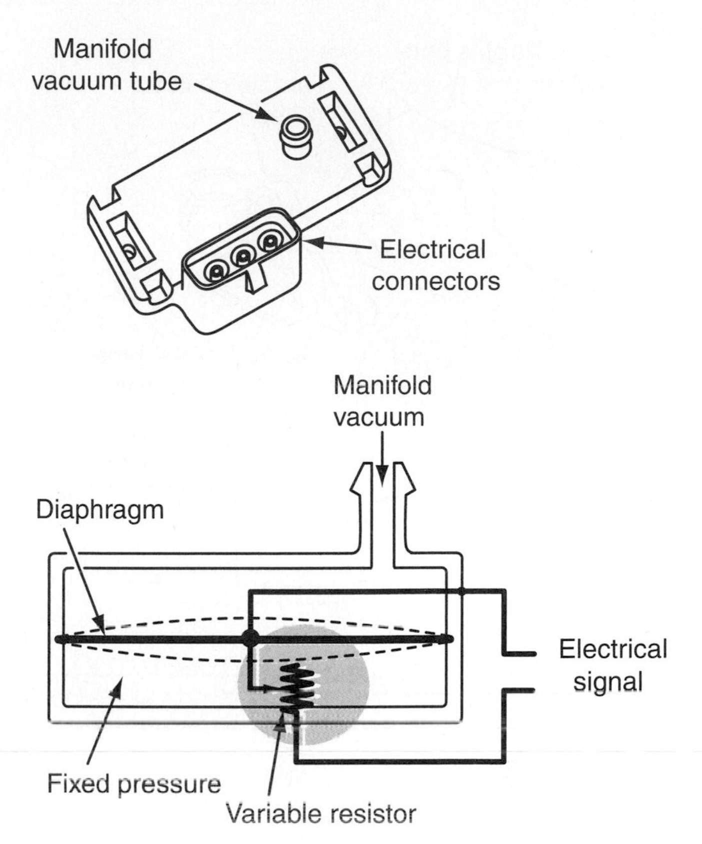

FIGURE 25–17 A manifold vacuum sensor located in the intake manifold detects changes in manifold vacuum.

THROTTLE POSITION SENSOR

The position of the throttle opening is another factor in determining what air-fuel ratio is needed. The more the throttle is open, the richer the mixture required by the engine. ***Figure 25–18*** shows the throttle position sensor (TPS). It is a variable resistor that sends a signal to the PCM. Depending on the position of the throttle, the PCM will signal the electronic carburetor solenoid to increase or decrease the air-fuel mixture.

The throttle position sensor is a variable resistor mounted on the electronic carburetor. As the position of the throttle changes, the voltage also changes. At closed throttle, the voltage is about 1 volt or less. As the throttle opening increases, the voltage increases to about 5 volts at wide-open throttle.

PARTS LOCATOR

*Refer to photo #29 in the front of the textbook to see what a **throttle position sensor** looks like.*

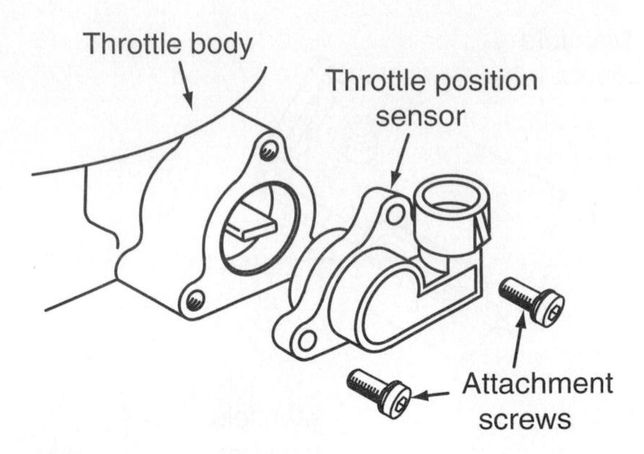

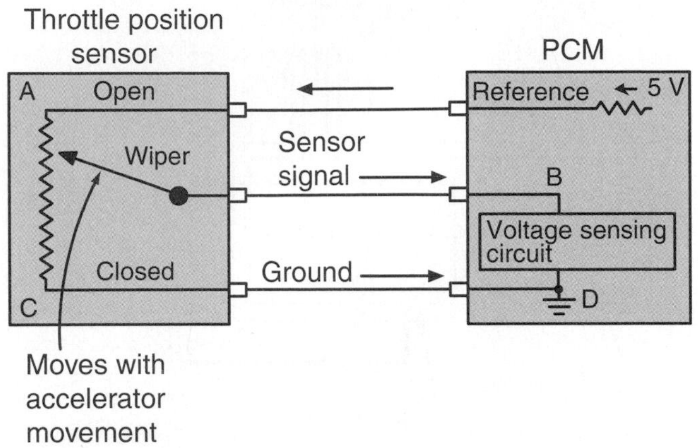

FIGURE 25-18 This throttle position sensor (TPS) is placed on the carburetor to sense the amount of throttle opening.

ENGINE SPEED SENSOR

Engine speed also has a direct bearing on the air-fuel mixture. When the engine is operating at a low speed, less fuel is needed. When the engine operates at a higher rpm, more fuel is needed. This adjustment is made by using an engine speed sensor. A tachometer signal from the distributor is sent to the PCM. This signal tells the PCM the rpm of the engine. The computer considers this signal when setting the exact air-fuel ratio needed. ***Figure 25–19*** shows the engine speed sensor. It is located inside the distributor on some engines and produces a signal that is sent to the computer.

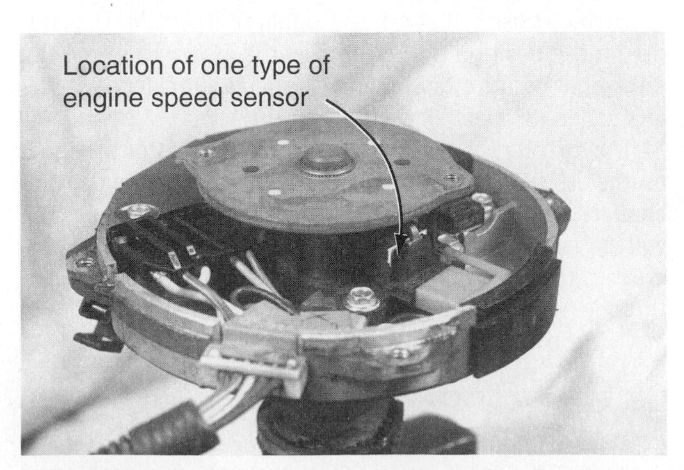

FIGURE 25-19 The speed of the engine can be sensed by using a signal generated in the distributor.

PARTS LOCATOR

Refer to photo #30 in the front of the textbook to see what an ***engine speed sensor*** *looks like.*

IDLE SPEED CONTROL

Since the increase in emission control standards, idle speed must be controlled more precisely. The computer controls the idle speed for drivability as well as for fuel economy and emission control. This is done by using the PCM and a reversible electric motor. See ***Figure 25–20***. The PCM maintains a selected idle speed regardless of the load imposed on the engine. A plunger that acts as a movable idle stop changes the idle speed. The plunger is positioned by a small electric motor. A throttle contact switch tells the PCM to operate only when the throttle lever is closed. When the throttle lever moves away from the throttle contact switch, the PCM is instructed not to operate. The driver now has control of the engine speed.

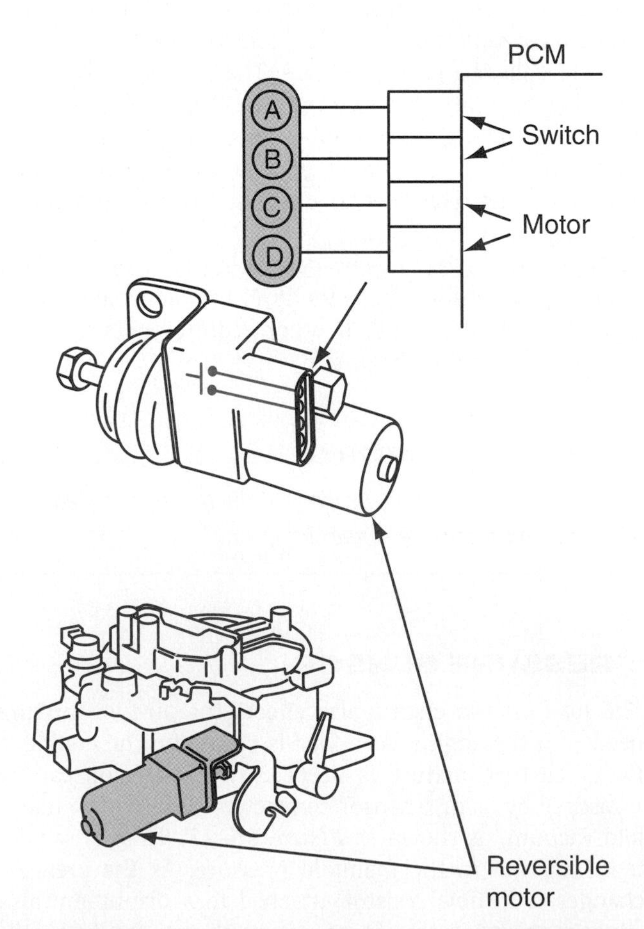

FIGURE 25-20 The computer operates the idle speed control motor. The motor positions an inside plunger to set the proper idle.

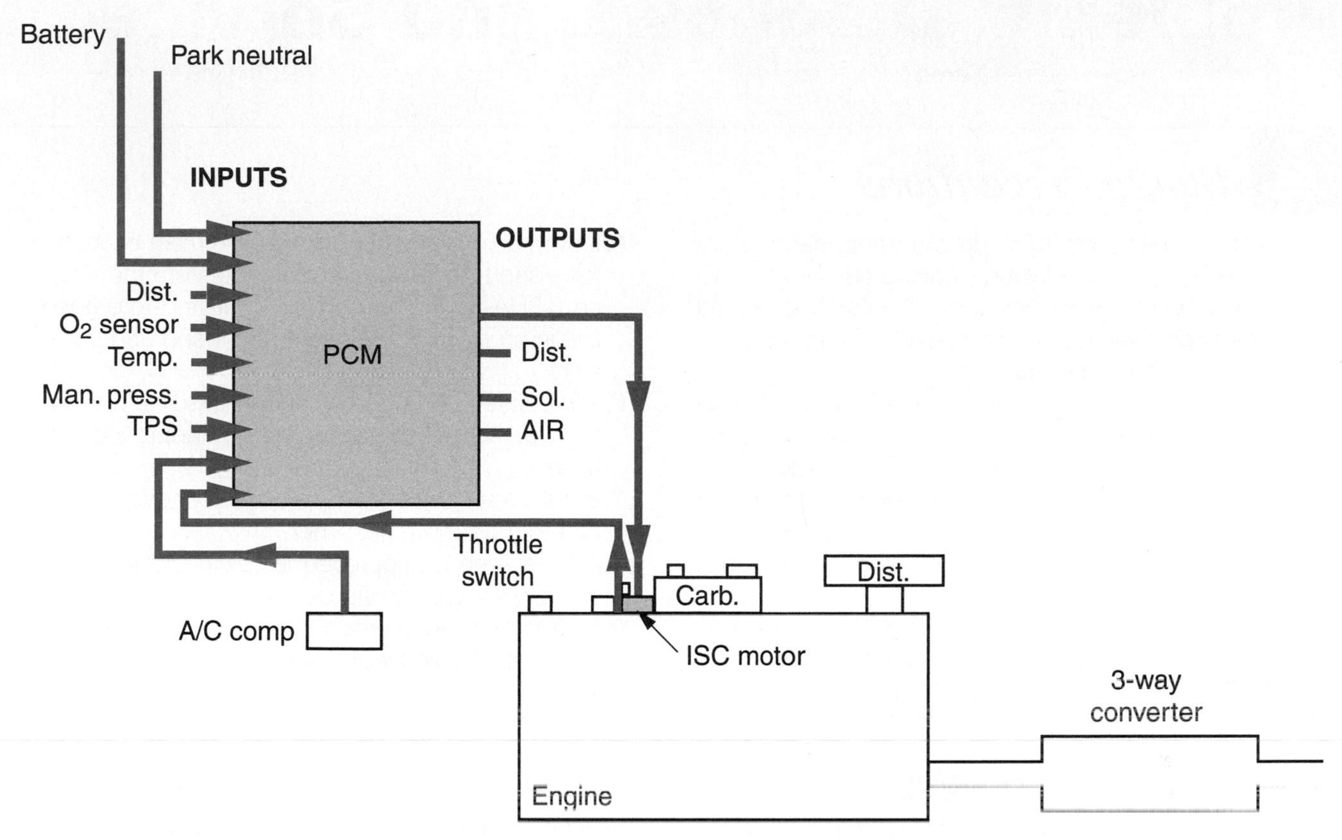

FIGURE 25-21 This schematic shows the inputs and outputs used to control the idle speed control (ISC) motor.

Figure 25–21 shows a schematic of the PCM and identifies the inputs and the outputs to the ISC (idle speed control) electric motor. Note that the throttle position sensor (TPS) is considered an input to the PCM. Also, many of the inputs are the same as those used with the electronic carburetor solenoid. The following controls are considered input signals to the PCM:

- Distributor—sensing rpm
- Oxygen sensor—sensing oxygen in exhaust
- Temperature sensor—sensing coolant temperature
- Pressure sensor—sensing manifold pressure and load
- Throttle position sensor—sensing throttle position

The **battery signal** is used to sense the system's operating voltage. If the voltage signal falls below a predetermined level, the PCM will instruct the idle speed control plunger to extend. This will increase the engine speed, which will, in turn, increase the alternator speed so that alternator output will increase.

The **park/neutral switch** tells the PCM when the transmission has shifted. When the transmission shifts, the load on the engine changes. The idle speed must then be changed. This prevents different idle speeds at neutral, drive, and reverse conditions.

PARTS LOCATOR

*Refer to photo #31 in the front of the textbook to see what an **idle speed control** looks like.*

Problems, Diagnosis, and Service

Safety Precautions

1. The cleaner used to clean electronic carburetors is extremely toxic. When cleaning carburetors, always wear protective gloves, OSHA-approved safety glasses, and protective clothing to eliminate any possibility of injury.
2. When setting any adjustment on an electronic carburetor, always put the vehicle in park, with the emergency brake on and the wheels blocked.
3. When checking the electrical components on or near a carburetor, always use a battery-operated tester to eliminate any sparks. Sparks can ignite fumes and cause a fire.
4. Make sure there are no unshielded flames, such as a cutting torch, around or near a vehicle when the electronic carburetor is being serviced.
5. If even a small quantity of gasoline must be stored for a short period of time, always store the fuel in an OSHA-approved container. Never store gasoline in an open container. It might spill and cause a fire.
6. Immediately wipe up any spilled gasoline and dispose of the rags in a sealed OSHA-approved container.
7. Never clean any external electronic feedback carburetor parts with gasoline.
8. Always keep an approved fire extinguisher nearby when working on an electronic feedback carburetor.
9. Never smoke in or around a vehicle or when working on electronic feedback carburetor systems.

PROBLEM: Carburetor-Computer Controls

On the dashboard of a vehicle equipped with a computer to manage fuel, a Check Engine light comes on.

DIAGNOSIS

Today, there are many electrical components used on computerized systems. Often these components fail. The type of failure is usually an open, a short, or a voltage value that stays too high or too low for too long. The computer is able to monitor these faults.

There are two methods used to diagnose these faults in computerized control systems.

1. *Demand diagnostics*—This method performs various diagnostic procedures only on demand from an outside control. In this case, a service technician initiates the computer to go through certain diagnostic procedures. These procedures are stored in the computer memory. When variations from standards are found, service codes are signaled to the service technician. Various scanners and electronic testers are also used to perform similar diagnostics.
2. *Ongoing diagnostics*—Most manufacturers have ongoing diagnostic systems. Problems are signaled to the driver by a flashing Check Engine light on the dash. A trouble code associated with that problem is also stored in the computer's RAM (random access memory). As long as power is supplied to the computer, the codes remain stored in the memory.

If the problem is intermittent, the Check Engine light will go out. However, the trouble code will remain in the memory as long as power is applied.

SERVICE

Assessing stored **trouble codes**and other diagnostic information can be done easily. The system used for electronic carburetor troubleshooting on many vehicles is called On Board Diagnostics I (OBDI). An updated version of that system, OBDII, is discussed later in the textbook. The technician grounds a test lead under the dash. The connector is often called the assembly line communication link (ALCL). It is also called the assembly line diagnostic link (ALDL). The location of the ALCL will vary with each manufacturer. Some are under the dash, whereas others are under the hood. ***Figure 25–22*** shows one example on the passenger

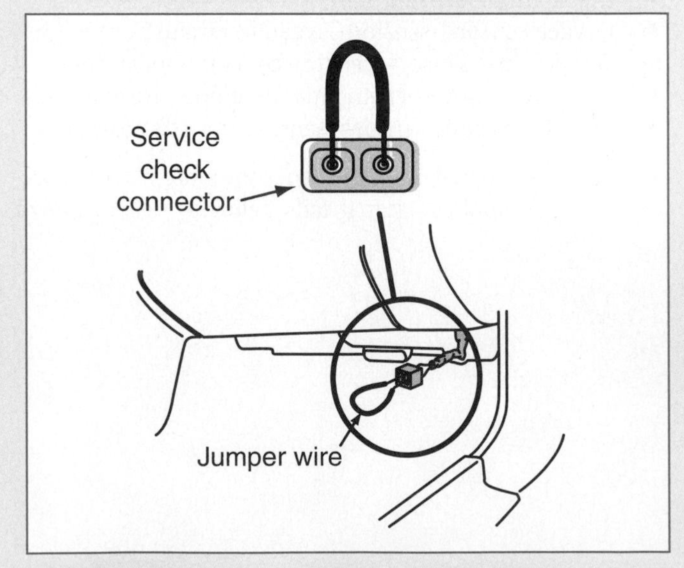

FIGURE 25–22 To determine the trouble code on some vehicles, use a jumper wire on the service check connector.

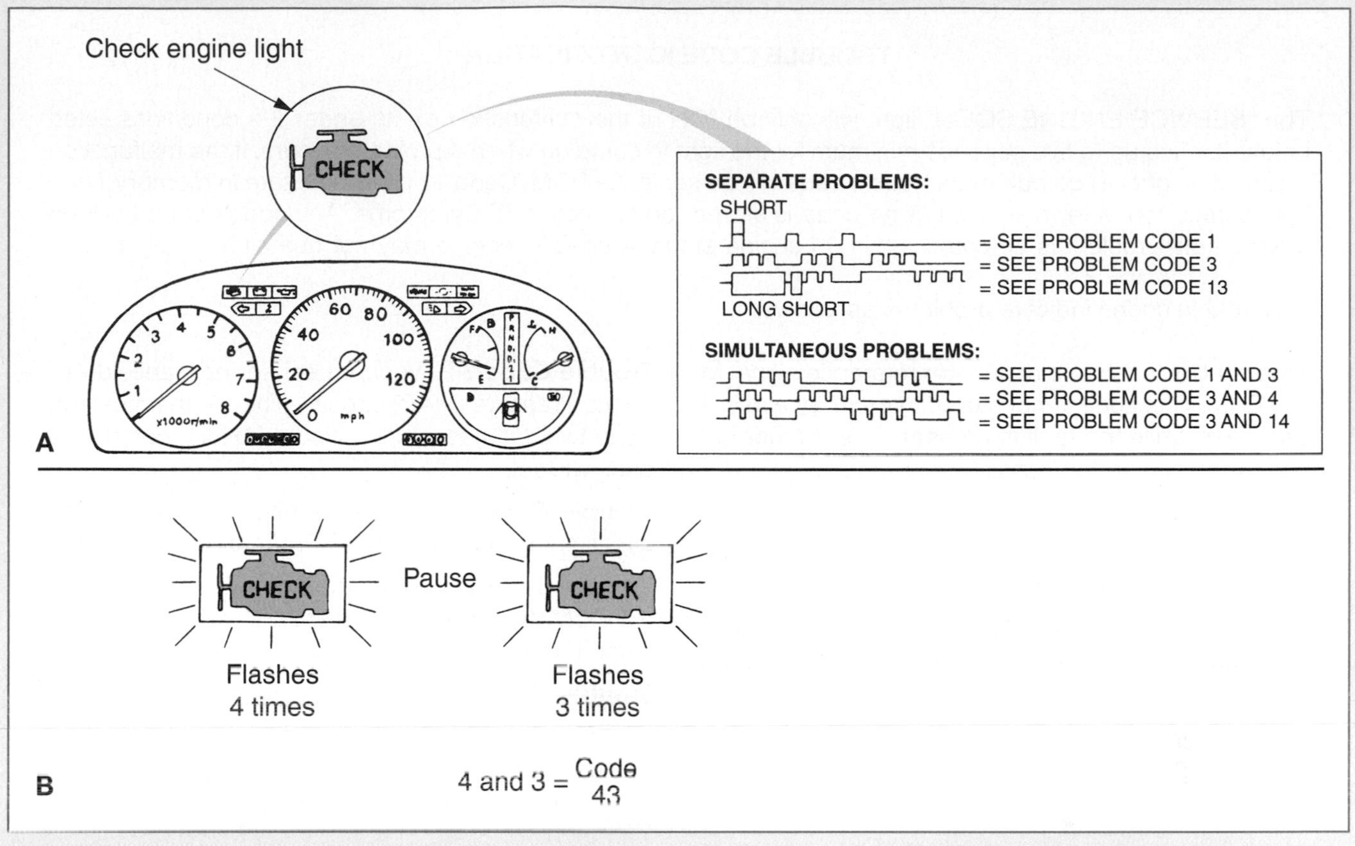

FIGURE 25-23 These Check Engine lights tell the driver that a trouble code has been placed in the computer. The trouble code is identified by observing the on-and-off sequence of the Check Engine light. ((A) Courtesy of American Honda Motor Co., Inc.)

side of the vehicle. The Check Engine light shown in ***Figure 25–23*** then flashes on and off in a certain sequence. The exact sequence depends on the engine. Two systems are shown for comparison. The system shown in ***Figure 25–23A*** is able to show both separate problems and simultaneous problems. For example, this manufacturer uses long blinks for the first digit in the trouble code and short blinks for the second digit. ***Figure 25–23B*** also shows a graphical representation of the blinking light. In ***Figure 25–23B*** only one code is shown using a series of blinks and pauses.

There are certain codes to help diagnose the fuel system. For example, if the light in ***Figure 25–23A*** flashes four times, pauses for a moment, then flashes three more times, it means the trouble code 43 has been stored. ***Figure 25–24*** identifies the problem for code 43 as the fuel supply system. Each manufacturer also has different codes.

Figure 25–25 shows another manufacturer's trouble code numbers. In this case, a trouble code of 43 would mean there is a problem with the electronic spark control system. These trouble code charts are located in the appropriate service manuals for each vehicle that the service technician is troubleshooting.

SELF-DIAGNOSIS INDICATOR BUNKS	SYSTEM INDICATED
0	ECU
1	Oxygen content
3, 5	Manifold absolute pressure
4	Crank angle
6	Coolant temperature
7	Throttle angle
8	TDC position
9	No. 1 cylinder position
10	Intake air temperature
12	Exhaust gas recirculation system
13	Atmospheric pressure
14	Electronic air control
15	Ignition output signal
16	Fuel injector
17	Vehicle speed sensor
20	Electric load detector
30	A/T FI signal A
31	A/T FI signal B
41	Oxygen sensor heater
43	Fuel supply system

FIGURE 25-24 After determining the code from the Check Engine light, refer to this chart to determine the system fault. (Courtesy of American Honda Motor Co., Inc.)

TROUBLE CODE IDENTIFICATION

The "SERVICE ENGINE SOON" light will only be "ON" if the malfunction exists under the conditions listed below. It takes up to five seconds minimum for the light to come on when a problem occurs. If the malfunction clears, the light will go out and a trouble code will be set in the PCM. Code 12 does not store in memory. If the light comes "on" intermittently, but no code is stored, go to section B–Symptoms. Any codes stored will be erased if no problem is detected within 50 engine starts. A specific engine may not use all available codes.

The trouble codes indicate problems as follows:

Trouble Code 12: No distributor reference signal to the ECM. This code is not stored in memory and will only flash while the fault is present. This is a normal code with ignition "on," engine not running.

Trouble Code 13: Oxygen Sensor Circuit—The engine must run up to four minutes at part throttle, under road load, before this code will set.

Trouble Code 14: Shorted coolant sensor circuit—The engine must run two minutes before this code will set.

Trouble Code 15: Open coolant sensor circuit—The engine must run five minutes before this code will set.

Trouble Code 21: Throttle Position Sensor (TPS) circuit voltage high (open circuit or misadjusted TPS). The engine must run 10 seconds, at specified curb idle speed, before this code will set.

Trouble Code 22: Throttle Position Sensor (TPS) circuit voltage low (grounded circuit or misadjusted TPS). Engine must run 20 seconds, at specified curb idle speed, to set code.

Trouble Code 23: M/C solenoid circuit open or grounded.

Trouble Code 24: Vehicle speed sensor (VSS) circuit—The vehicle must operate up to two minutes, at road speed, before this code will set.

Trouble Code 32: Barometric pressure sensor (BARO) circuit low.

Trouble Code 34: Vacuum sensor or Manifold Absolute Pressure (MAP) circuit—The engine must run up to two minutes, at specified curb idle, before this code will set.

Trouble Code 35: Idle speed control (ISC) switch circuit shorted. (Up to 70% TPS for over 5 seconds.)

Trouble Code 41: No distributor reference signal to the ECM at specified engine vacuum. This code will store in memory.

Trouble Code 42: Electronic spark timing (EST) bypass circuit or EST circuit grounded or open.

Trouble Code 43: Electronic Spark Control (ESC) retard signal for too long a time; causes retard in EST signal.

Trouble Code 44: Lean exhaust indication—The engine must run two minutes, in closed loop and at part throttle, before this code will set.

Trouble Code 45: Rich exhaust indication—The engine must run two minutes, in closed loop and at part throttle, before this code will set.

Trouble Code 51: Faulty or improperly installed calibration unit (PROM). It takes up to 30 seconds before this code will set.

Trouble Code 53: Exhaust Gas Recirculation (EGR) valve vacuum sensor has seen improper EGR control vacuum.

Trouble Code 54: M/C solenoid voltage high at ECM as a result of a shorted M/C solenoid circuit and/or faulty ECM.

FIGURE 25-25 General Motors uses this chart to identify the trouble code flashed on the dashboard.

PROBLEM: Faulty Oxygen Sensor

A vehicle with an electronic feedback carburetor is not performing correctly.

DIAGNOSIS

When the oxygen sensor is not operating correctly, it may be sending the computer incorrect signals, thus altering the air-fuel ratio.

SERVICE

The first step is to remove the oxygen sensor. Often they are difficult to remove after being in the engine exhaust manifold for a long period of time. If it is difficult to remove, apply some type of penetrating oil to the sensor thread area. Be careful not to get any on the upper part of the sensor where the reference air enters. Sensors are easier to remove from a warm exhaust than from a cold one. Work the sensor back and forth to help remove it without damage to the thread.

Once removed, there are several checks that can be performed on an oxygen sensor. One simple check is to determine if there is a short to ground in the oxygen sensor. To complete this check, use a digital ohmmeter. Place one ohmmeter lead on one of the heater terminals and the other on the case of the sensor. This should normally be an open circuit. If there is continuity, there is a short to ground.

Another simple service check is to determine the heater resistance. Heater resistance can be measured with an ohmmeter according to the specifications given in the manufacturer's manual. Use the following procedure to check the oxygen sensor:

1. Be sure the sensor is at room temperature.
2. The heater on the oxygen sensor should be shut off.
3. Using a digital ohmmeter, check the resistance between the signal (+) and ground (–) terminals of the sensor case. The resistance should be greater than 20 megohms (***Figure 25–26***).

The best way to diagnose oxygen sensor performance is to follow the recommendations in the vehicle service manual. Each manufacturer has a different procedure to use and a special electrical tool to measure performance. In addition, some manufacturers also recommend checking the output voltage of the oxygen sensor. Always follow the exact procedure recommended for checking oxygen sensor voltage. ***Figure 25–27*** shows a typical oxygen sensor diagnosis chart.

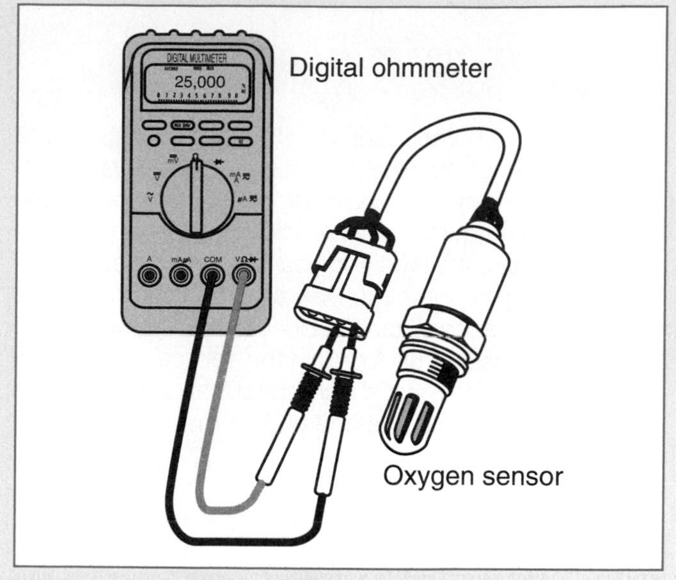

FIGURE 25-26 The oxygen sensor can be checked for heater resistance.

CAUTION: *Oxygen sensors can be damaged if exposed to silicon gasket material.*

OXYGEN (O_2) SENSOR

PROBLEM	DIAGNOSIS	SERVICE
Low voltage	• Shorted wire • Internal metal chips	• Replace sensor
Low rich or high lean voltage	• Silicone contamination	• Correct source of silicone in engine, replace sensor
Slow response time	• Lead, oil, or fuel contamination	• Use low lead fuel only
Low or negative voltage	• Oil contamination/water splash/ vehicle rust proofing on sensor	• Eliminate oil source, use caution when rust proofing
Slow response time or limited voltage range	• Engine oil problem or fuel control problem	• Correct engine problem
Low voltage, symptom of open or short circuit	• Water intrusion, sensor in splash prone area	• Replace sensor
Open circuit or negative voltage—boot may be burned	• Broken element continuous bias voltage (450 mv)	• Replace sensor
Voltage may go negative during normal lean operation	• Exhaust leak • Sensor internal failure	• Repair exhaust leak • Replace sensor
Loss of output due to overheating, high resistance, loss of lower shield, too hot	• Sensor location or engine problem	• Move sensor and correct problems • Use high temp sensor
System lean, voltage stays near bias (450 mv)	• Too cold. Sensor location or engine application	• Move sensor closer to engine • Use heated sensor
System rich (sensor indicates lean). Engine misses	• Engine problem	• Correct cause of misfire

FIGURE 25-27 Various oxygen sensor problems are shown along with possible causes and suggested corrections.

Service Manual Connection

There are many important specifications to keep in mind when working with electronic feedback carburetor systems. To identify the specifications for your engine, you will need to know the VIN (vehicle identification number) of the vehicle, the type and year of the vehicle, and the type of engine. Although they may be titled differently, some of the common feedback carburetor system specifications (not all) found in the service manuals are listed here. Note that these specifications are typical examples. Each vehicle and engine may have different specifications.

Common Specification	Typical Example
Operation temperature of oxygen sensor	500 degrees F
Resistance of oxygen sensor element	20 megohms
Oxygen sensor voltage (rich)	650 millivolts
Fuel pressure regulator	39 psi

In addition, the service manuals give specific directions for various service procedures. Some of the more common procedures include:

- Adjusting speed idle
- Electronic carburetor overhaul
- M/C solenoid removal/replacement

SUMMARY

The following statements will help to summarize this chapter:

- Computers control electronic feedback carburetors.
- The powertrain control module (PCM) uses various input data to control several outputs on the electronic feedback carburetors.
- Input signals can include a battery signal, park/neutral signal, distributor signal, (engine speed), oxygen sensor signal, engine temperature signal, manifold pressure signal, and throttle position sensor signal.
- Outputs on this system can include the idle speed control (ISC) motor and the mixture control (M/C) solenoid.
- Electronic feedback carburetors operate in either a closed loop or open loop mode.
- Engines generally operate in the closed loop mode.
- Engines operate in the open loop mode primarily during starting of the vehicle.
- The M/C solenoid opens and closes a jet in the electronic carburetor as fast as ten times per second.
- *Pulse width* is a term used to indicate the length of time the M/C solenoid is on.
- The oxygen sensor senses the amount of oxygen in the exhaust, which in turn tells the computer to increase or decrease the fuel by changing the pulse width of the M/C solenoid.
- The oxygen sensor can be checked for resistance and voltage drop.
- Computerized fuel management systems such as electronic feedback carburetion rely on a Check Engine message to a dashboard light, which tells the driver a trouble code has been placed in the computer.
- Once a trouble code is accessed, the service technician uses a trouble code chart to determine the exact problem in the system.

TERMS TO KNOW

Can you explain each of the following terms? Review the chapter until you can use each term correctly.

Atomized
Battery signal
Closed loop
Duty cycle
Open loop
Park/neutral switch
Pulse width
Trouble codes
Vacuum
Vaporized
Venturi

REVIEW QUESTIONS

Multiple Choice

1. The main purpose of the electronic carburetor is to create an air-fuel ratio of:
   a. 12.3 to 1
   b. 14.7 to 1
   c. 20 to 1
   d. 25.3 to 1
   e. 30.2 to 1

2. Atomization is used to help ________ the fuel.
   a. Cool
   b. Vaporize
   c. Weight
   d. Pressurize
   e. Reduce the weight of

3. The restriction in the flow of air in an electronic carburetor is called a ________.
   a. Needle valve
   b. Pressure
   c. Venturi
   d. Dashpot
   e. None of the above

4. When the computer sends a set of programmed instructions to the electronic carburetor to tell it what to do, it is said to be in:
   a. Open loop operation
   b. Closed loop operation
   c. Intermediate loop operation
   d. High-speed loop
   e. Low-speed loop

5. The computer on the PCM system controls:
   a. The metering control solenoid
   b. The idle speed control motor
   c. The choke plate
   d. All of the above
   e. A and B

6. Which of the following is not used as an input on the PCM when it is used for fuel management?
   a. Coolant temperature
   b. Rpm
   c. Throttle position
   d. Coolant fan speed
   e. Exhaust oxygen

7. When an electronic feedback carburetor is used, the most common mode of operation is the:
   a. Open loop
   b. Closed loop
   c. Continuous loop
   d. Varied loop
   e. All of the above

8. On an electronic feedback carburetor, the M/C solenoid can be pulsed as often as ________ times per second.
   a. 1
   b. 10
   c. 100
   d. 1,000
   e. All of the above

9. Oxygen sensors need both an outside airflow and ________ to operate correctly.
   a. Heat
   b. Oil
   c. Engine coolant
   d. All of the above
   e. None of the above

10. A wide pulse width on an electronic feedback carburetor means that the air-fuel mixture is:
    a. Rich
    b. Normal
    c. Lean
    d. Hot
    e. Cooled with engine coolant

11. The duty cycle is defined as the percentage of ________ time compared to the total cycle time of the M/C solenoid.
    a. On
    b. Off
    c. Grounded
    d. Nonenergized
    e. None of the above

12. At the tip of some oxygen sensors, there is a ceramic material called:
    a. Plutonium
    b. Zirconia
    c. Alpha particles
    d. All of the above
    e. None of the above

13. Which of the following is not a type of oxygen sensor?
    a. Water-resistant sensor
    b. Heated sensor
    c. Single wire and unheated sensor
    d. Engine coolant sensor
    e. Heated and waterproof sensor

The following questions are similar in format to ASE (Automotive Service Excellence) test questions.

14. *Technician A* says that the heated oxygen sensor needs outside air for correct operation. *Technician B* says that the unheated oxygen sensor has three wires going into its top. Who is correct?
    a. A only
    b. B only
    c. Both A and B
    d. Neither A nor B

15. *Technician A* says that an ohmmeter can be used to check an oxygen sensor. *Technician B* says that a voltmeter can be used to check an oxygen sensor. Who is correct?
    a. A only
    b. B only
    c. Both A and B
    d. Neither A nor B

16. *Technician A* says that voltage being produced inside an oxygen sensor can go as high as 650 millivolts. *Technician B* says that voltage being produced inside an oxygen sensor can be as low as 200 millivolts. Who is correct?
    a. A only
    b. B only
    c. Both A and B
    d. Neither A nor B

17. *Technician A* says that trouble codes on an electronic carburetor systems are determined by a flashing light on the dashboard. *Technician B* says that trouble codes can be identified by grounding a service check connector under the dashboard. Who is correct?
    a. A only
    b. B only
    c. Both A and B
    d. Neither A nor B

18. *Technician A* says that closed loop means that the computer uses the oxygen sensor for an input. *Technician B* says that open loop means that the engine temperature is below 150 degrees Fahreneheit. Who is correct?
    a. A only
    b. B only
    c. Both A and B
    d. Neither A nor B

Essay

19. What are three input signals to the PCM on electronic feedback carburetors?

20. Explain the term *pulse width* and state how it is used on electronic feedback carburetors.

21. Define the term *duty cycle* in relationship to electronic feedback carburetors.

22. Describe a simple check that can be done with a digital ohmmeter on an oxygen sensor.

23. Describe the difference between closed and open loop operation on an electronic feedback carburetor.

Short Answer

24. There are two methods used to diagnose faults in electronic feedback carburetor systems. These are ________ diagnostics and ________ diagnostics.

25. As a load is placed on the engine, the air-fuel ratio must become ________.

26. To operate correctly, oxygen sensors need both ________ and ________.

27. The part of the oxygen sensor that senses the amount of oxygen in the exhaust is made of a ceramic material called ________.

28. Oxygen sensors are easiest to remove from an engine when the engine is ________.

Applied Academics

TITLE: Duty Cycles

ACADEMIC SKILLS AREA: Language Arts

NATEF Automobile Technician Tasks:

The technician interprets charts/tables/graphs to determine manufacturer's specifications for engine operation to identify out-of-balance systems and subsystems.

CONCEPT:

The duty cycle of an electronic component is defined as the percentage of time that the component is operational. For example, on an electronic carburetor and the mixture control (M/C) jet, the percentage of time the solenoid is on is called the "duty cycle." Duty refers to how much work time, or duty, the solenoid accomplishes. When control voltage from the computer is applied only one-half of the time in each second, the duty cycle is 50%. If the solenoid voltage is applied and the M/C jet is closed for 90% of the time, or a 90% duty cycle, fuel flow will be small in volume. This is shown in the accompanying diagram on the left. If the duty cycle is say 10%, the solenoid voltage is open for only 10% of the time causing larger fuel volume, as shown in the diagram on the right.

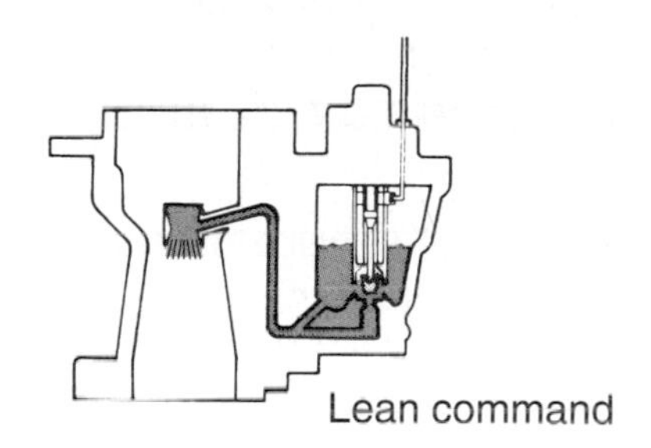

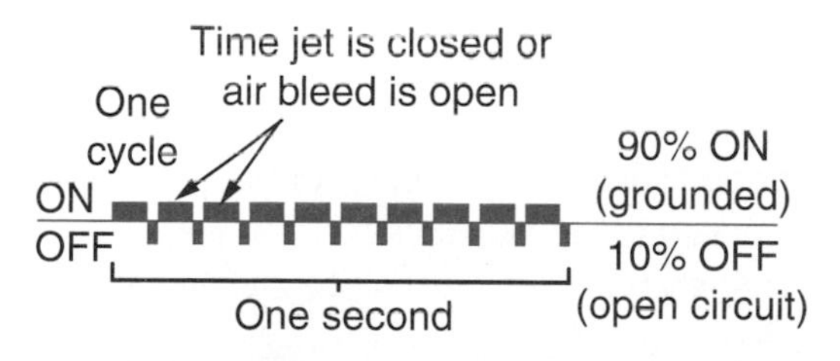

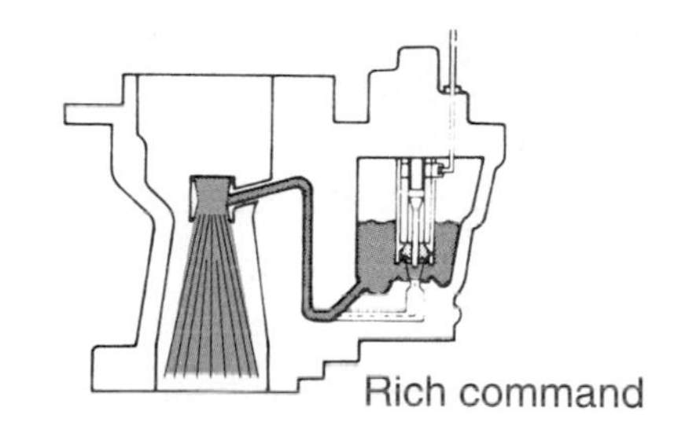

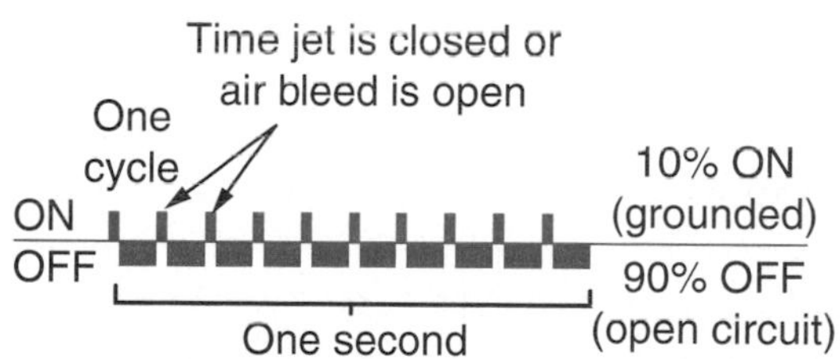

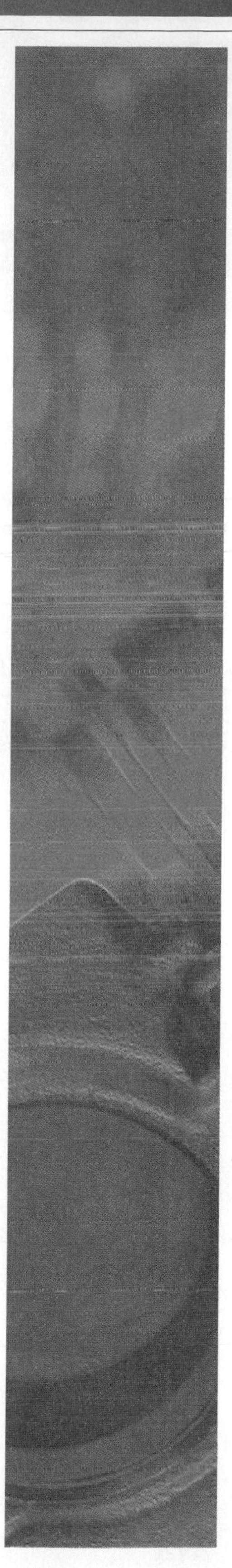

APPLICATION:

What is the position of the M/C jet (closed or open) if the duty cycle is 100%? What is the position of the M/C jet (closed or open) if the duty cycle is 0%? When there is no voltage applied to the M/C solenoid, is the jet opened or closed?

CHAPTER 26

Gasoline Fuel Injection Systems

OBJECTIVES

After studying this chapter, you should be able to:

- Define the purposes of using gasoline fuel injection systems on engines.
- State the different types of gasoline fuel injection systems used.
- Analyze the throttle body fuel injection system.
- Analyze the types of sensors used on computerized gasoline fuel injection systems.
- Analyze the port injection system.
- Describe the different modes of operation on fuel injection systems.
- State the various problems, diagnosis, and service procedures used on gasoline injection systems.

INTRODUCTION

In the past few years, many changes have occurred in the design of fuel systems. One change is to use fuel injection rather than carburetion to mix the fuel. Fuel injection has always been used in diesel engines. It has been used on gasoline engines since the mid- to late 1980s. The reason for using fuel injection is to control the air-fuel ratio of the engine more precisely. The purpose of this chapter is to study different gasoline fuel injection systems used on automobile engines.

26.1 CLASSIFICATIONS OF FUEL INJECTION SYSTEMS

DIRECT AND INDIRECT FUEL INJECTION

Fuel injection systems can be divided into two major types: (1) the high-pressure, or direct, injection system used on diesel engines and (2) the low-pressure, or indirect, system used on gasoline engines (***Figure 26–1***). Direct fuel injection means the fuel is injected directly into the combustion

Certification Connection

ASE Connection: The information in this chapter can help you prepare for the National Institute for Automotive Service Excellence (ASE) certification tests. The tests and content areas most closely related to this chapter are:

Test A1—Engine Repair

- **Content Area**—Fuel, Electrical, Ignition, and Exhaust Systems Inspection and Service

Test A8—Engine Performance

- **Content Area**—Fuel, Air Induction, and Exhaust System Diagnosis and Repair

NATEF Connection: Much of the information in this chapter is related to the NATEF tasks. The NATEF tasks and priority numbers most closely related to this chapter are:

1. Diagnose hot or cold no-starting, hard starting, poor drivability, incorrect idle speed, poor idle, flooding, hesitation, surging, engine misfire, power loss, stalling, poor mileage, dieseling, and emission problems on vehicles with injection-type fuel systems; determine necessary action. P-1
2. Inspect and test fuel injectors. P-2
3. Check idle speed and fuel mixture. P-2
4. Adjust idle speed and fuel mixture. P-3

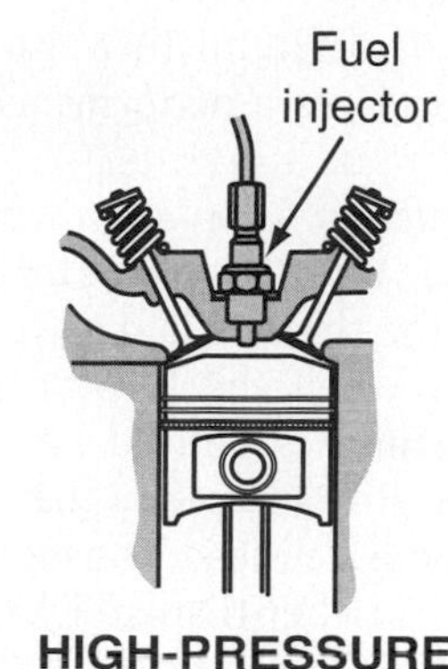

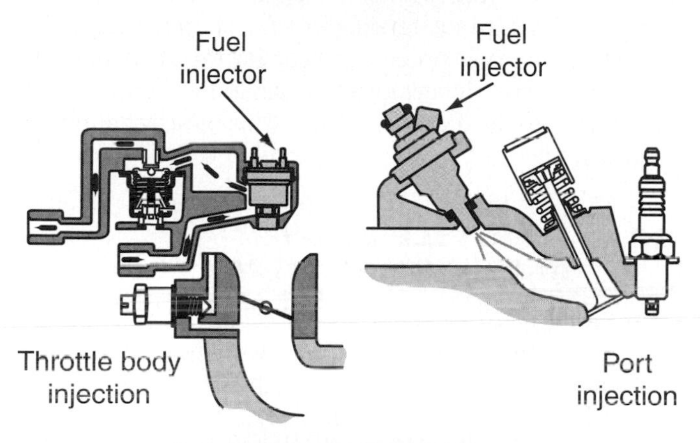

FIGURE 26-1 There are several types of fuel injection systems. Fuel can be injected into the combustion chamber, into the port, or into the throttle body.

chamber. Indirect fuel injection means the fuel is injected either into the port before the intake valve or into the intake manifold by a **throttle body** injector. When fuel is injected into the combustion chamber, the pressure of the injection must be increased to high values. High-pressure injection is necessary because the injection occurs during the compression stroke on the diesel engine. The injection pressure must be much higher than the compression pressure for correct atomization. Indirect pressure injection systems inject fuel either into the port before the intake valve or into the throttle body. The area of injection is low in pressure, and thus low-pressure injection can be used.

This chapter introduces the basic principles, parts, and nomenclature of indirect fuel injection systems used on gasoline engines.

PORT AND THROTTLE BODY FUEL INJECTION

Two types of indirect fuel injection are used. When the fuel is injected into the port, it is called port fuel injection (PFI). In this case, there is one fuel injector for each cylinder and set of valves. When the fuel is injected into the center of the throttle body where the carburetor used to be, it is called throttle body injection (TBI). In this case, only one fuel injector is used on the system. The fuel injector feeds all of the cylinders of the engine. These two systems are also called multiple-point (port) and single-point (throttle body) fuel injection.

TIMED AND CONTINUOUS FUEL INJECTION

Fuel injection systems are also classified by the type of injection action. Some injection systems are defined as timed fuel injection. This means that fuel injection occurs at a precise time. Both gasoline and diesel engines use the timed injection system. Diesel engines have high-pressure timed injection systems. Gasoline engines have low-pressure timed injection systems.

Continuous injection is another type of injection. Certain gasoline engines have used continuous fuel injection in the past. This injection system is designed to spray a continuous flow of fuel into the engine. Diesel engines do not have a continuous injection system.

In addition to these classifications, fuel injection can be either mechanical or electronic. Diesel engines utilize mechanical fuel injection. Initially, gasoline engines used mechanical fuel injection. However, because of stricter **emission standards** and the increased use of computers on gasoline engines, today's fuel injection systems are totally electronic.

26.2 FUEL INJECTION REQUIREMENTS

FUEL INJECTION DEFINED

Fuel injection is defined as the process of injecting fuel before the valves so that a 14.7 to 1 air-fuel ratio can be maintained in the combustion chamber. Carburetors have been used in the past to mix the correct amount of fuel with the air. With the increased emphasis on pollution and emissions, however, more precise methods are needed. Fuel injection can precisely measure the amount of fuel to maintain the exact air-fuel ratio.

Fuel injection systems operate in conjunction with electronic computers. As discussed in the previous chapter, the powertrain control module (PCM) takes in data from many sensors. On the basis of this information, the PCM tells the fuel injector to inject the proper amount of fuel at a specific time.

FUEL INJECTION AND AIR-FUEL RATIO

When the air-fuel ratio is at 14.7 to 1, conditions are ideal for complete combustion. Complete combustion helps to ignite the mixture, ensuring release of all the heat

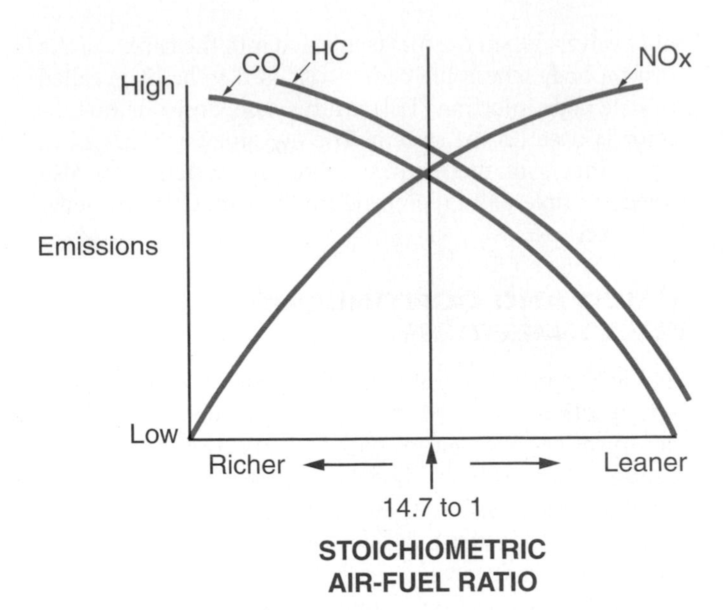

FIGURE 26–2 When the air-fuel mixture is lean, more nitrogen oxides (NO_x) are produced. When the air-fuel mixture is rich, there are more carbon monoxide (CO) and hydrocarbon (HC) emissions.

energy in the fuel. If combustion is complete, very little unburned fuel is left. The 14.7 to 1 air-fuel ratio is known as the **stoichiometric ratio**. This is the best ratio for achieving both optimum fuel efficiency and optimum emission control under ideal conditions.

There are three primary pollutants caused by poor combustion: carbon monoxide (CO), hydrocarbons (HC), and nitrogen oxides (NO_x). These pollutants will be discussed in detail later in this textbook. As air-fuel ratios are changed, these pollutants increase or decrease. ***Figure 26–2*** shows the relationship between air-fuel ratios and pollution characteristics. Referring to this chart, it is easy to understand why it is very important to maintain a precise 14.7 to 1 air-fuel ratio. The stoichiometric ratio is the optimum ratio to minimize undesirable emissions.

NONSTOICHIOMETRIC AIR-FUEL RATIO CONDITIONS

There are certain operating conditions when special air-fuel ratio requirements have priority over those of emission control. These conditions include times when the engine is cold and times when there is low manifold vacuum. Low manifold vacuum is a result of an increased load put on the engine.

Cold engine operation occurs whenever the engine is below normal operating temperature and is started or operated. During this condition, the cold inner surfaces of the intake manifold cause some of the fuel in the air-fuel mixture to condense. If the intake air is also below the desirable temperature of 70 degrees Fahrenheit, its ability to vaporize the fuel and mix with it is reduced. During cold engine operation, a richer mixture is needed. The rich mixture replaces the fuel that is lost through condensation or poor vaporization. The rich mixture also aids prompt starting and smooth responsive performance during warm-up periods.

Low manifold vacuum is produced when the load on the engine is increased. When the operator needs more power from the engine, the throttle is pushed down. This opens the throttle plate. When the throttle plate is open, the **intake manifold vacuum** is reduced. Any reduction in manifold vacuum is a positive indication that the engine is being asked to take on an added load. During this time, a richer mixture is needed. A richer-than-stoichiometric ratio helps provide excess fuel for the increased load. This condition is similar to acceleration. Once the engine and load stabilize, the air-fuel mixture is returned to the optimum ratio. Although carburetion systems provide for this increased air-fuel ratio, electronic fuel injection can more precisely limit enrichment to the degree needed and for the exact duration required.

26.3 THROTTLE BODY INJECTION SYSTEMS

After the introduction of electronic feedback carburetors, the next type of fuel injection that was developed was throttle body injection. For several years during the 1980s, throttle body injection was the preferred type. Since that time, port fuel injection is now used in all engines. However, throttle body injection is presented here because it is possible you will encounter these systems in older vehicles.

The TBI system uses a computer to control the amount of fuel injected into the manifold. It is considered an indirect type of injection. Air is drawn into the engine and passes by the injector nozzle. The exact amount of fuel necessary for the conditions of operation is added.

Figure 26–3 shows a throttle body injector. It is typically located on top of the throttle body housing. The spray from the injector is being injected into the venturi and throttle plate area.

FIGURE 26–3 The throttle body injector is positioned above the throttle body and injects fuel directly into the venturi.

FUEL INJECTOR

A typical TBI system uses a solenoid-operated fuel injector controlled by a computer. The throttle body injector is centrally located on the intake manifold, usually near the top center of the fuel charging assembly. Here air and fuel are mixed correctly. Fuel is supplied to the injector from the electric fuel pump located in the fuel tank. The incoming fuel is directed to the low end of the injector assembly. An electrically operated solenoid valve in the injector is used to control fuel delivery from the injector nozzle. As shown in ***Figure 26–4***, when there is no electrical current from the PCM to the solenoid, a spring keeps a parabolic-type metering valve closed inside the injector nozzle. This prevents fuel from flowing through the nozzle.

When the solenoid valve is energized by the computer, the spring-loaded metering valve moves up to its full open position. Fuel under pressure from the fuel pump is injected in a **conical** spray pattern into the throttle body bore. The throttle body bore is located directly above the **throttle plate**. The volume of fuel flow is changed by varying the length of time the injector is held open by the PCM.

PARTS LOCATOR

*Refer to photo #32 in the front of the textbook to see what a **throttle body fuel injector** looks like.*

INJECTOR PULSE WIDTH OR DURATION

The length of time the injector is open (turned on) and emitting fuel is called the **pulse width**. The pulse width is measured in milliseconds (ms). The injector is pulsed electronically by the distributor for each piston stroke. The correct amount of fuel is **metered** into the engine by controlling when and how long to pulse or turn on the injector. This is controlled by the computer.

FUEL PRESSURE REGULATOR

The fuel pressure must be regulated in order for the TBI to work correctly. This is done by a fuel pressure regulator.

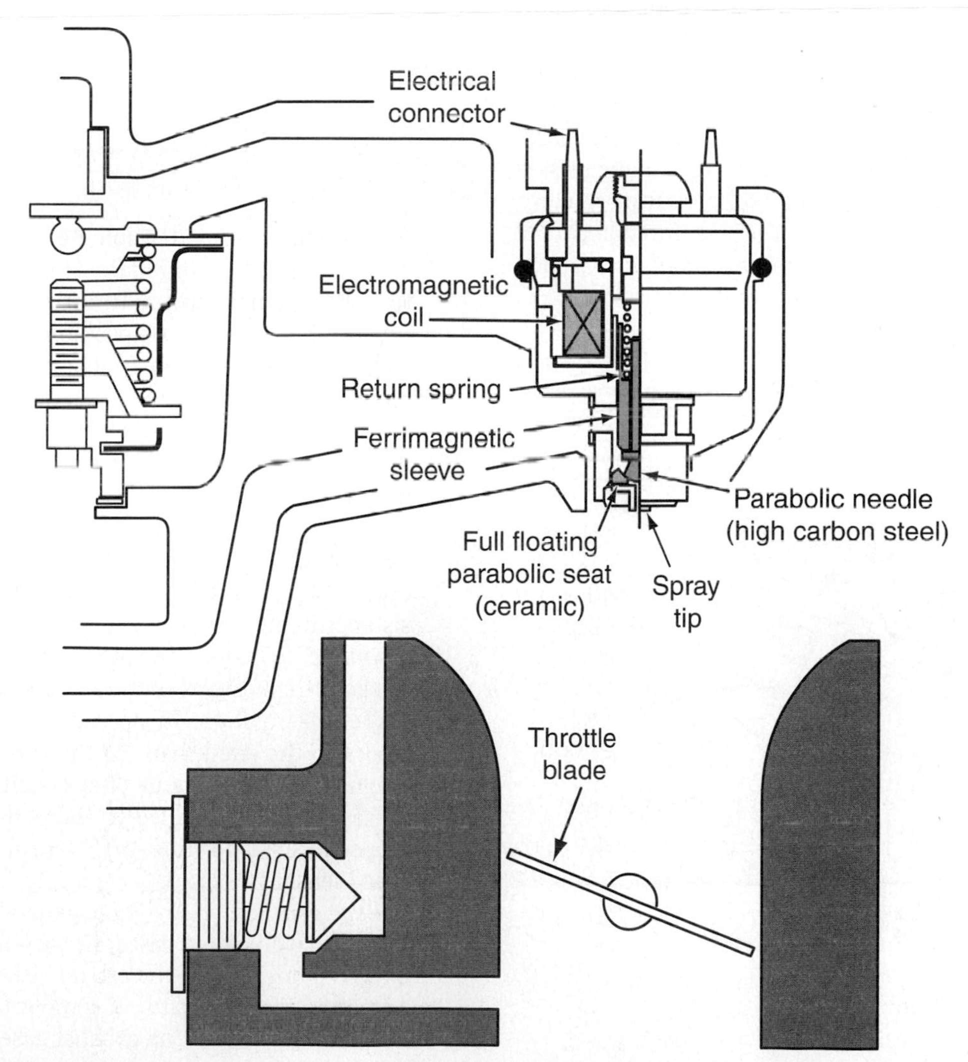

FIGURE 26–4 The throttle body injector is operated by a solenoid that, when energized, causes a ball valve to come off its seat and inject fuel.

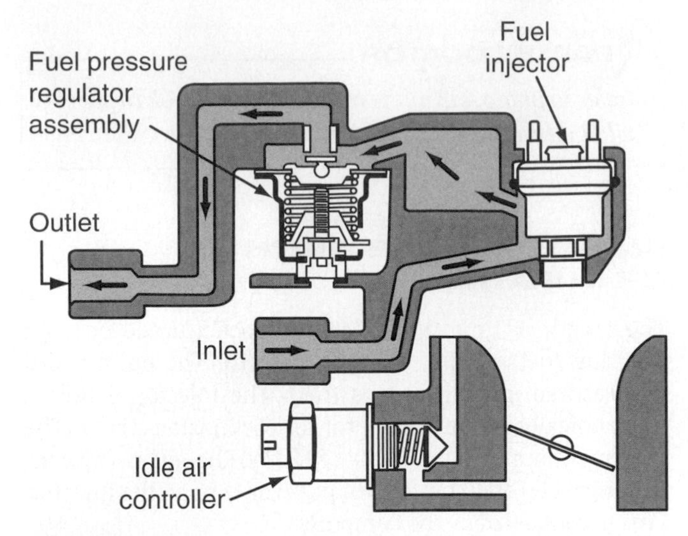

FIGURE 26–5 The fuel pressure regulator is used to control the pressure of the fuel at the throttle body injector. If fuel pressure increases too high, the fuel will pass by the regulator and back to the fuel tank.

Figure 26–5 shows how the fuel pressure regulator is built into the fuel circuit. The throttle body injector is located directly above the throttle body bore. Fuel from the fuel tank is pressurized and sent into the fuel inlet. From the fuel inlet the fuel is sent to the TBI. There is usually an excessive amount of fuel available at the injector. Excess fuel passes through the injector and into the fuel pressure regulator assembly. The fuel pressure regulator is an integral part of the throttle body injection unit. ***Figure 26–6*** shows such a unit.

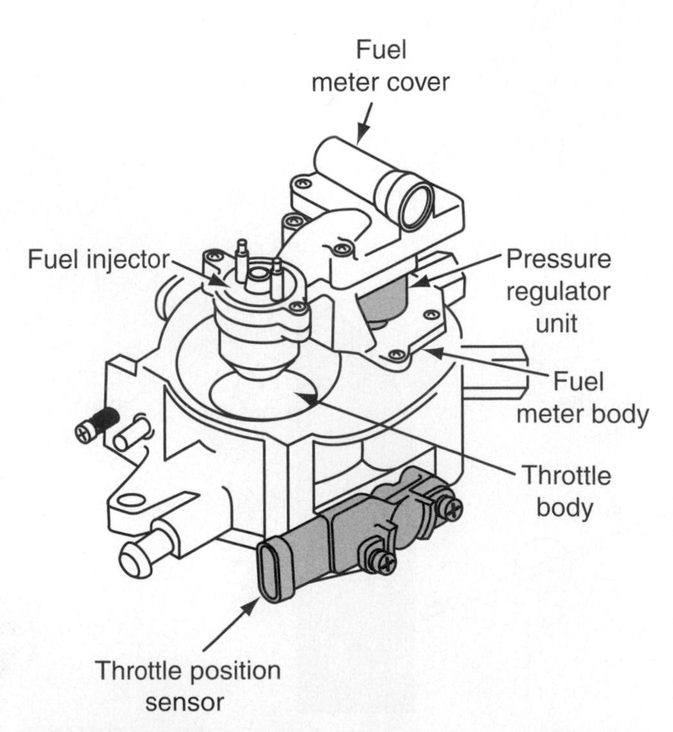

FIGURE 26–6 The fuel pressure regulator can be designed as an integral part of the throttle body injection unit.

The fuel pressure regulator is a mechanical device that maintains approximately 11 psi across the tip of the injector. The exact pressure depends on the manufacturer. Some are as high as 36 psi. When the fuel pressure exceeds the regulator setting, it pushes the spring-loaded diaphragm down. This uncovers the fuel return port. As the fuel pressure drops below the regulator setting, the spring tension pushes the diaphragm up. This closes off the fuel return port to maintain the pressure of the fuel.

VACUUM ASSIST

A vacuum assist is used on some, but not all, fuel pressure regulators. The vacuum assist is used to change the fuel pressure during different operating conditions. For example, when the throttle plate is open, a greater vacuum is felt on the injector tip. Therefore, less fuel pressure is needed to make the fuel flow at the same rate. A vacuum line is connected from above the throttle plate to the bottom of the fuel pressure regulator. This vacuum line reduces the spring pressure, which causes the fuel pressure regulator to open sooner or at a lower pressure.

SENSOR DESIGNS

Various sensors are used to feed information into the computer on the computerized injection system. The following is a list of five types of sensors used.

1. *Potentiometer*—A potentiometer is a sensor that can change the voltage in relationship to mechanical motion. For example, as the throttle is moved to change the fuel setting, a throttle position sensor changes voltage drop. The computer is able to read the change in voltage drop, thus affecting the fuel setting.
2. *Thermistor*—A **thermistor** is a sensor that can change its electrical resistance on the basis of a change in temperature. For example, the computer needs to know the air temperature outside. The air charge temperature (ACT) sensor is able to tell the computer the outside temperature on the basis of the electrical resistance in the circuit.
3. *Magnetic pickup*—A **magnetic pickup** sensor is used to sense the position of the crankshaft within its 360 degrees of rotation. In operation, a small magnet is placed in the crankshaft. As the crankshaft spins, the magnet induces a charge of electricity. This charge or pulse of electricity is fed to the computer to operate the crankshaft position (CP) sensor and the vehicle speed sensor (VSS).
4. *Voltage generator*—A voltage generator sensor is able to produce a voltage based on various inputs such as oxygen. For example, the exhaust gas oxygen sensor is a small voltage generator. It is mounted in the exhaust flow. It is able to pick up exhaust gases that are lean or rich. It is very sensitive to the presence of oxygen. Lean mixtures produce smaller output voltages (0 to

0.4 volt), while richer mixtures produce larger output voltages (0.6 to 1.0 volt).

5. *Frequency generator*—A frequency generator is a sensor that can read various pressures and convert these pressures to an electrical voltage, read by the computer. The conversion from pressure to voltage is done through an electrical generating device using capacitors, oscillators, and amplifiers. The computer needs to know the manifold absolute pressure (MAP) and the barometric pressure (BP). Frequency generators are used to read these pressures for the computer.

THROTTLE POSITION SENSOR (TPS)

The TBI system also uses a throttle position sensor (TPS). The throttle position sensor was introduced in the chapter on electronic feedback carburetors. In this system, the throttle position sensor works in the same way. ***Figure 26–6*** shows the throttle position sensor located on the bottom of the throttle body assembly.

The throttle position sensor contains a variable resistor that is used to regulate an input voltage on the basis of the angle of the throttle valve. The PCM uses this signal as a reference to determine idle speeds and air-fuel ratios.

IDLE AIR CONTROL (IAC)

On certain throttle body injection systems, an idle air control (IAC) is used. Some manufacturers call this system the automatic idle speed (AIS) motor. The purpose of the idle air control system is to control engine rpm at idle, while preventing stalls due to changes in engine load. Changes in engine load may be caused by accessory loads, such as air conditioning, during idle.

An IAC assembly motor is mounted on the throttle body unit. It provides control of by-pass air around the throttle valve. By extending or retracting a **pintle**, a controlled amount of air is routed around the throttle valve (***Figure 26–7***).

If the rpm of the engine is lower than desired during operation, more air is diverted around the throttle valve. This increases the rpm. If the rpm is higher than desired, less air is diverted around the throttle valve. This decreases rpm. The PCM monitors the manifold vacuum and adjusts the fuel delivery as idle requirements change.

During idle, the PCM uses the information from several input signals to calculate the desired pintle position. If the rpm drops below a value stored in the PCM's memory, and the throttle position sensor indicates that the throttle is closed, the PCM calculates the desired pintle position. The PCM will increase or decrease rpm to prevent stalling.

THROTTLE BODY INJECTOR (TBI) CONTROLS

Throttle body injection systems utilize several sensors that send information to the PCM. These sensors include manifold absolute pressure (MAP), oxygen in the exhaust (oxygen sensor), coolant temperature (coolant sensor), engine speed (rpm), and throttle position (***Figure 26–8***). Information sent to the PCM from these sensors tells the TBI unit exactly how much fuel should be metered at a specific time. Other inputs are also used to provide information to the PCM. These include a park/neutral safety switch, brake switch, and air-conditioning switch.

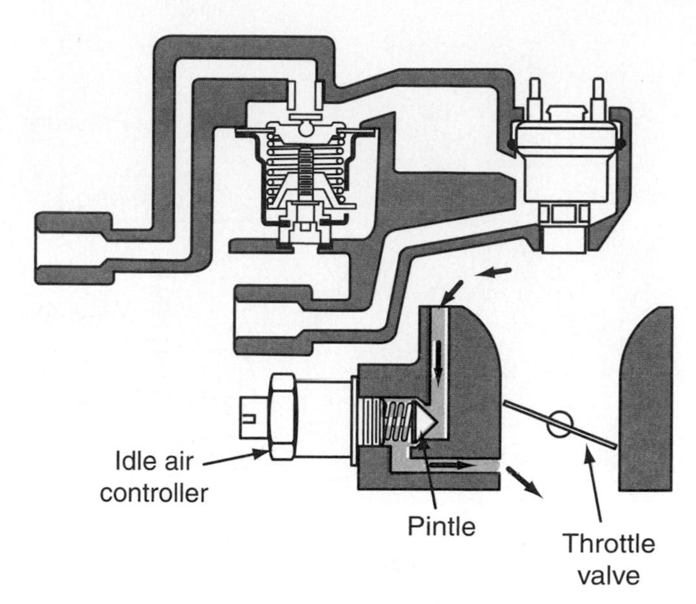

FIGURE 26–7 Idle air control (IAC) is achieved by controlling the air passing around the throttle valve.

MANIFOLD ABSOLUTE PRESSURE (MAP)

Manifold absolute pressure (MAP) is a sensor used to measure the absolute pressure (vacuum) inside the intake

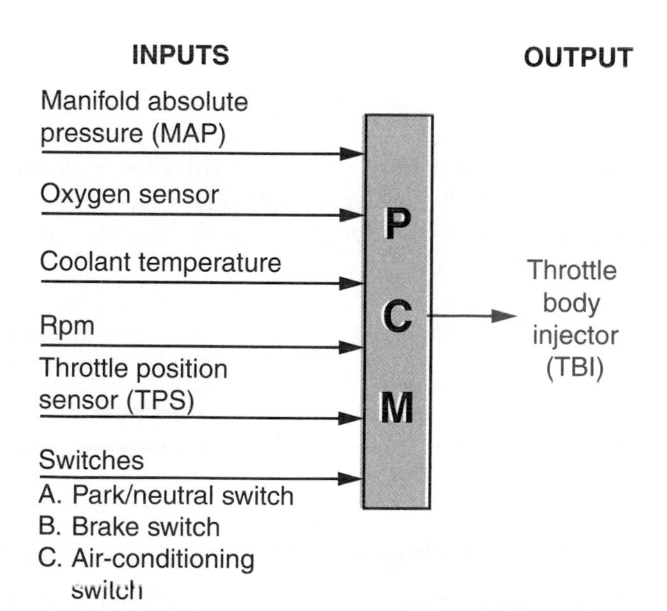

FIGURE 26–8 The PCM controls the throttle body injector. Several inputs are needed to obtain the correct control of the TBI.

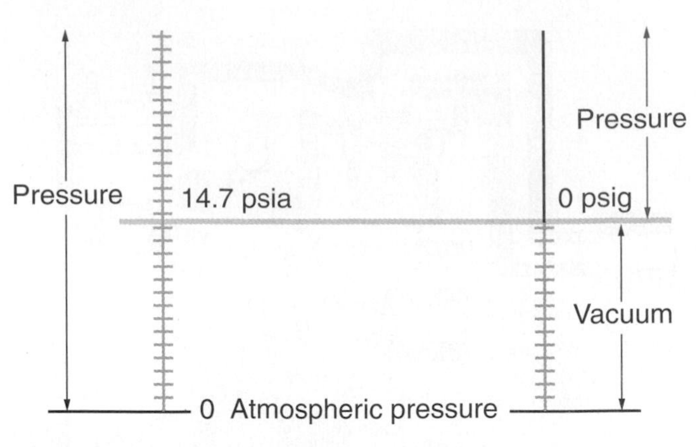

FIGURE 26-9 Manifold absolute pressure can be understood by comparing different types of pressure scales; psig and psia scales are used. A vacuum on the psig scale is a pressure on the psia scale.

FIGURE 26-10 The manifold absolute pressure sensor is used to determine the amount of vacuum in the intake system.

manifold. In order to understand MAP, absolute pressure and gauge pressure must be studied.

Gauge pressure is defined as pressure on a scale (psig) starting with zero at atmospheric pressure or 14.7 atmospheres. Absolute pressure is defined as pressure on a scale (psia) starting with zero at zero atmospheric pressure. ***Figure 26–9*** shows the differences between absolute pressure and gauge pressure. Zero psig is the same as 14.7 psia. One advantage of using the absolute scale is that there are no vacuum readings. A vacuum on the psig scale is a pressure on the psia scale.

Intake manifold vacuum can now be stated as a pressure. MAP is a pressure reading on the psia scale. It would be considered a vacuum on the psig scale. Note that as a manifold pressure increases, vacuum decreases. Manifold absolute pressures are used in the study of computer-controlled combustion systems.

The MAP sensor uses a flexible-type resistor. When the resistor is flexed because of an increase in manifold pressure, its resistance changes. This change causes a voltage drop that the PCM uses to control the TBI. ***Figure 26–10*** shows a typical MAP sensor. ***Figure 26–10A*** shows the typical location of a MAP sensor while ***Figure 26–10B*** shows the actual MAP sensor.

PARTS LOCATOR

*Refer to photo #33 in the front of the textbook to see what a **manifold absolute pressure (MAP) sensor** looks like.*

CROSSFIRE INJECTION

Certain engines utilize a pair of throttle body injection units. These are mounted on the front and rear of a single manifold cover. This arrangement allows each TBI unit to supply the correct air-fuel mixture through a crossover port. The port is located inside the intake manifold and feeds the cylinders on the opposite side of the engine, thus the term **crossfire injection**.

MODES OF OPERATION

All electronic fuel injection systems operate in either a closed or open loop mode. During closed loop operation, the computer uses various inputs, including the oxygen sensor, to produce an air-fuel ratio close to 14.7 to 1. During the open loop mode, the computer is programmed to provide an air-fuel ratio best suited for conditions such as starting and wide-open throttle. In addition, many other modes of operation are designed into the system for starting, initial running, flooding, and other conditions. The exact titles vary with each manufacturer and the year of the vehicle. Some common terms for modes of operation include the following:

- *Synchronized mode*—In the synchronized mode, the throttle body injector is pulsed once for each reference pulse from the distributor. The pulses of fuel injection are synchronized with the distributor pulses. The injector sprays once for each firing of the cylinder. All closed loop operation is in the synchronized mode.

- *Nonsynchronized mode*—In this mode, the throttle body injector is pulsed every 12.5 milliseconds. The pulses are independent of the distributor reference pulses. This mode is generally used when the engine is under special conditions such as wide-open throttle (open loop). During this condition, the injector must open and close extremely rapidly. The injector is not mechanically able to open and close this fast, so the computer takes over and pulses the injector every 12.5 milliseconds.
- *Cranking mode*—During the cranking mode, an enriched air-fuel ratio is required. In order to provide the enriched mixture, the computer senses the outside air temperature and sets the pulse width of the injector accordingly. For example, when the outside air temperature is very cold, the pulse width is longer, making an enriched air-fuel ratio. If the outside air temperature is warmer, the pulse width is calibrated to a short period of time, reducing the enriched mixture.
- *Clear flood*—At times the engine may be flooded and not able to start. At this point, if the throttle is pushed down, say to 80% of wide-open throttle, the computer goes into the clear flood mode. In the clear flood mode, the computer calibrates the injector pulse to produce a lean air-fuel ratio, say 20 to 1. The computer keeps the system in this mode until the engine starts or the throttle is moved to less than 80% of wide-open throttle.
- *Run*—Once the engine is started and running at curb idle, say 600 rpm, the computer immediately puts the system into open loop operation. (Other chapters discuss open and closed loop operation.) During curb idle, the computer is monitoring three conditions to get the system into closed loop operation as soon as possible.

  1. The oxygen sensor must reach 570 degrees Fahrenheit.
  2. The engine coolant temperature must reach 150 degrees Fahrenheit.
  3. A specified amount of time must have passed since the engine was started.

When these three conditions are met, the computer puts the fuel injection system into its closed loop mode of operation. At this point, the computer is using the oxygen sensor to help make decisions about the air-fuel ratio.

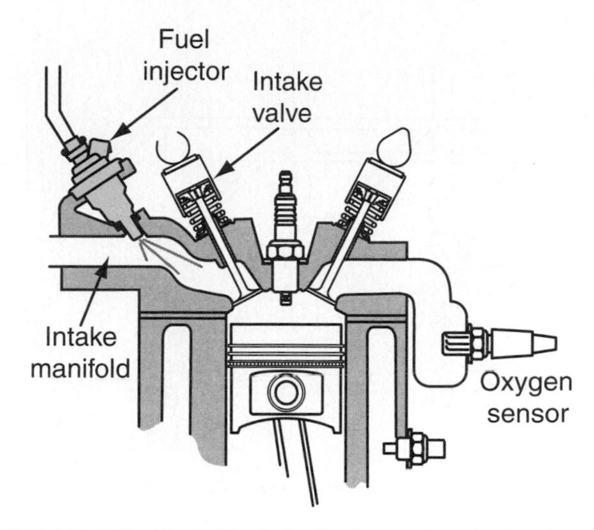

FIGURE 26–11 Port fuel injection has an injector located near the intake port. The fuel is sprayed directly into the port before the intake valve.

26.4 PORT FUEL INJECTION SYSTEMS

Port fuel injection is another way of electronically injecting fuel into a gasoline engine. Port fuel injection is designed to have a small fuel injector placed near the intake port of each cylinder (***Figure 26–11***). Each injector is controlled by the PCM, and the metering is based on several inputs, as with throttle body injection.

One of the biggest advantages of using port injection over TBI is the ability to get the same amount of fuel to each cylinder. When a throttle body injector or carburetor is used, the intake manifold acts as a sorting device to send fuel to each cylinder. As air flows through the curved ports in the intake manifold, air, being lighter than fuel, is sorted evenly. However, fuel, which is heavier than air, tends to collect at certain spots inside the intake manifold. This causes some cylinders to be richer than others during operation (***Figure 26–12***).

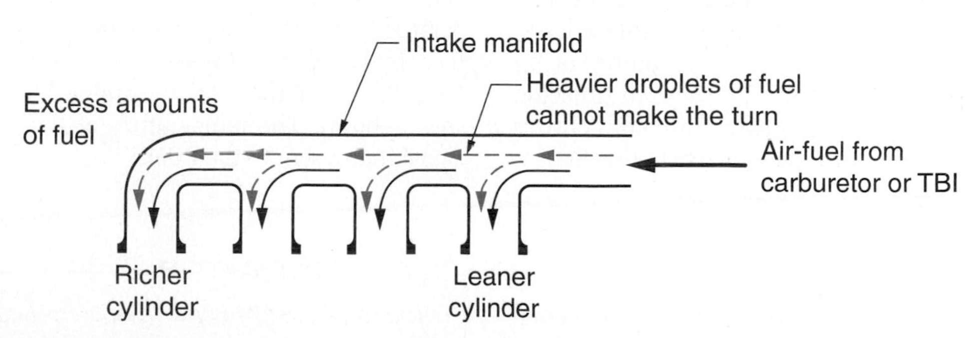

FIGURE 26–12 When throttle body injection or carburetion is used, fuel delivered to the cylinders may vary because of the design of the manifold. This problem can be overcome by using port fuel injection.

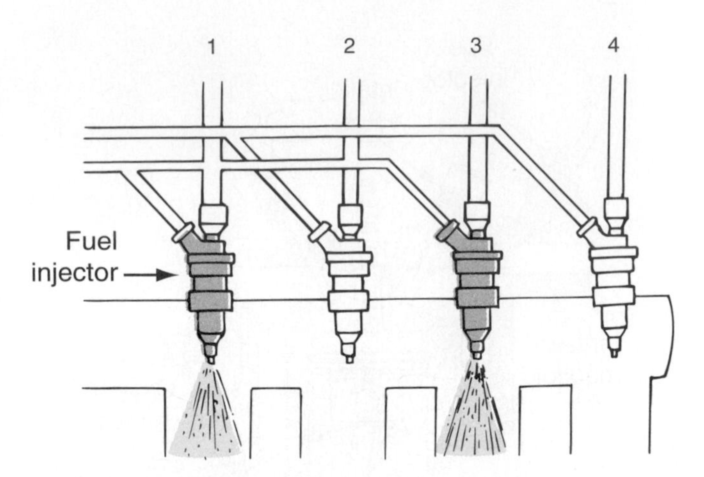

FIGURE 26-13 On some engines, groups of fuel injectors are fired at the same time.

TYPES OF PORT FUEL INJECTION

Several types of port fuel injection (PFI) are used. One type of PFI uses double-fire fuel injection. This means that all injectors are pulsed one time for each engine revolution. Two injections of fuel are mixed with incoming air to produce the charge for each combustion cycle. On some engines, groups of injectors are fired at the same time. In ***Figure 26–13***, injectors 1 and 3 are being fired or injected together. Then injectors 2 and 4 will be fired or injected together.

A second type of PFI is referred to as sequential fuel injection (SFI). On some vehicles, the injectors are pulsed sequentially (one-by-one) in sparkplug firing order.

A third type of PFI is called the tuned port injection (TPI). Tuned ports are used to regulate the airflow through the intake manifold to the intake port. **Tuned ports** are designed with equal and minimum restriction. This design ensures that the same amount of air will be delivered to each cylinder.

PORT FUEL INJECTION FLOW DIAGRAM

Although PFI systems may be designed differently by other manufacturers, the basic principles remain the same. The port fuel injector is located just in front of the intake valve. It is controlled by the PCM. Fuel from the fuel pump and filter in the tank is sent through a fuel filter into the injector. As with the TBI system, a fuel pressure regulator controls the pressure of the fuel sent to the injector. Other components that have already been studied as part of the electronic carburetion or TBI section include the oxygen sensor, coolant temperature sensor, throttle position sensor, and the idle air control motor.

As with throttle body injection, port fuel injection vehicles send various inputs to the PCM. ***Figure 26–14*** shows a schematic of the type of inputs used. The exact type of inputs depend on the type of vehicle, engine, and the

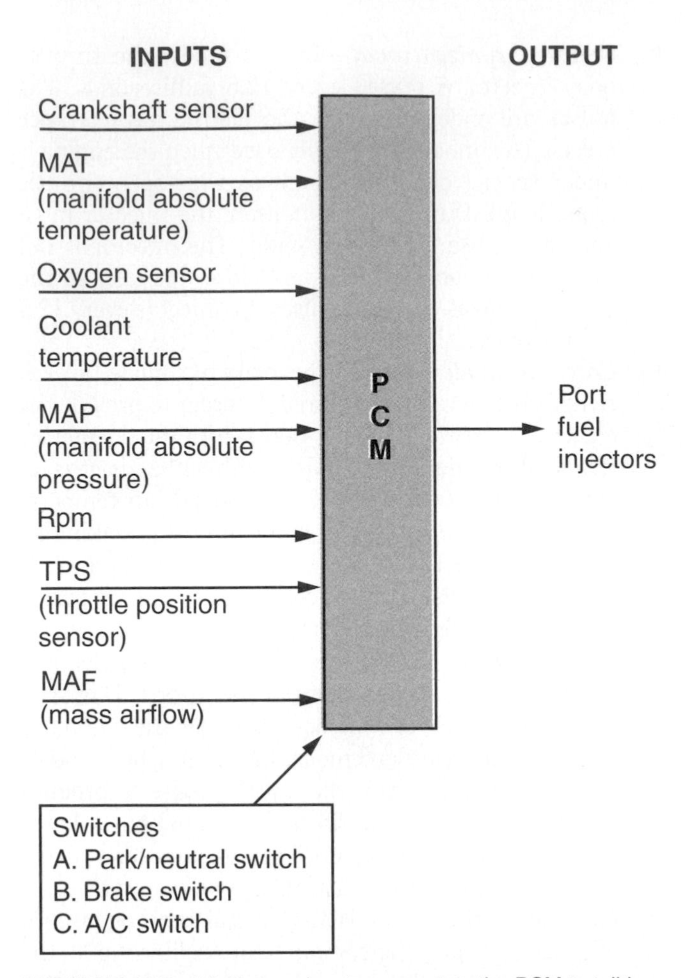

FIGURE 26-14 Various inputs are sent to the PCM to tell how much fuel should be injected by the port fuel injectors.

year. The overall effect of these inputs is to tell the computer the operating conditions of the engine. In turn, the PCM tells the fuel injectors how much fuel to inject. This is done in most fuel injection engines by varying the length of the pulse width.

INTAKE AIR TEMPERATURE (IAT)

In order for the PCM to meter the correct amount of fuel during all driving conditions, the temperature of the air coming into the intake manifold is monitored. As the temperature of the air changes, the amount of oxygen per cubic foot also changes. In ***Figure 26–15***, the intake air temperature (also called the MAT or manifold air temperature) sensor is shown. The temperature of the incoming air is a signal that is sent to the PCM.

PARTS LOCATOR

*Refer to photo #34 in the front of the textbook to see what a **manifold absolute temperature (MAT) sensor** looks like.*

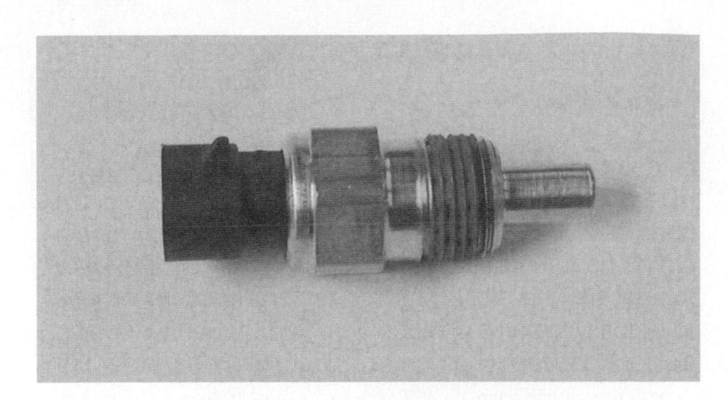

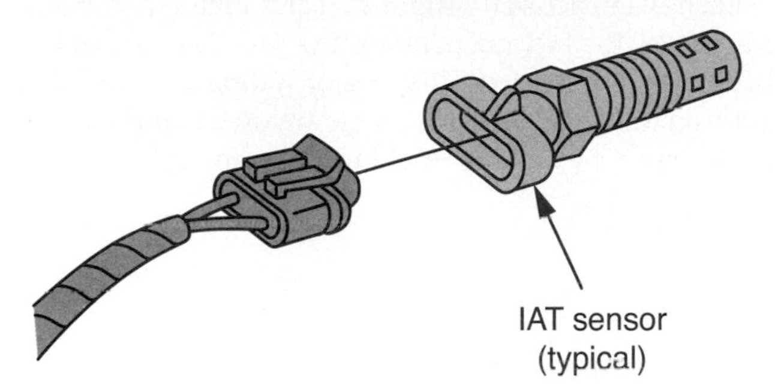

FIGURE 26-15 This is an example of a typical MAT, also called the intake air temperature sensor.

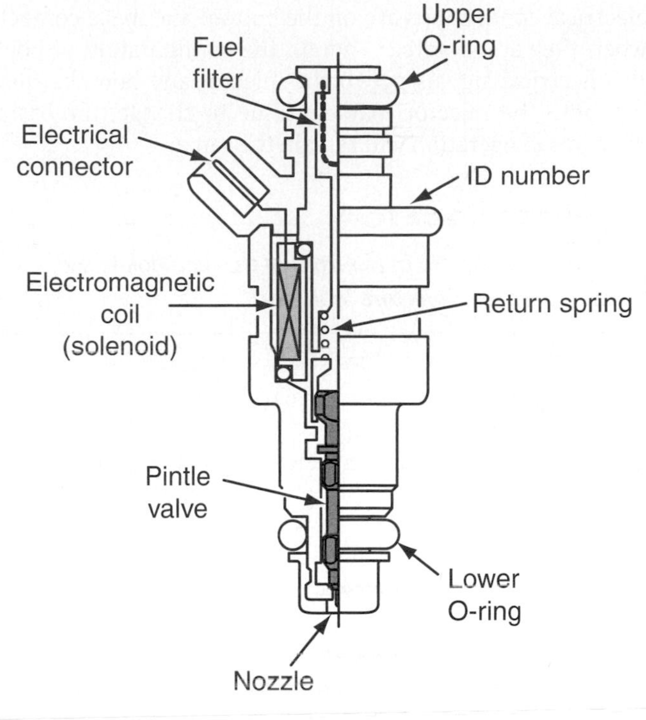

FIGURE 26-17 The operation of a typical port fuel injector uses a solenoid that raises a plunger and needle valve to inject the fuel.

PORT FUEL INJECTORS

The fuel injectors used with PFI inject fuel directly before the intake valve (***Figure 26–16***). The nozzle spray angle is 25 degrees. Two O-rings are used for installation. One O-ring is used to seal the injector nozzle to the intake manifold. The second O-ring is used to seal the injector with the fuel inlet connection. Both O-rings also prevent excessive injector vibration.

The injector operates as shown in ***Figure 26–17***. It is a solenoid-operated injector, consisting of a valve body and a **needle valve** that has a specially ground pintle. A solenoid winding electromagnetic coil is operated electrically by the PCM. When an electric pulse is sent to the injector, the magnetic field inside the solenoid lifts the needle valve from its seat. The PCM controls the length of the pulse, which establishes the pulse width or the amount of injection. A small spring closes the pintle valve when the electrical signal is removed.

Figure 26–18 shows two types of port fuel injectors. The most common port fuel injector is shown in ***Figure 26–18A***. ***Figure 26–18B*** shows that injectors are being designed to be smaller and more efficient. In this case, the

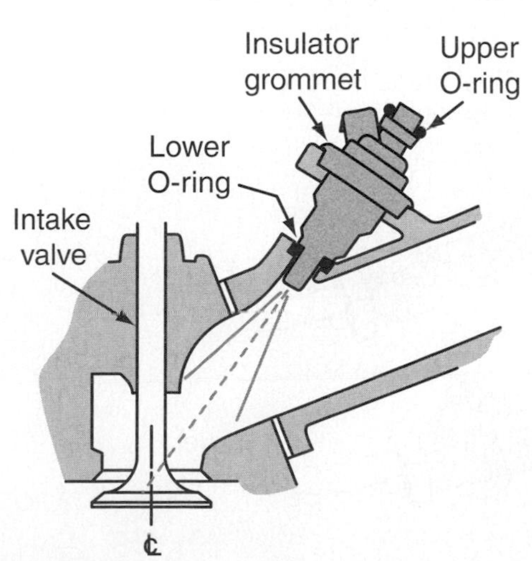

FIGURE 26-16 Fuel is injected at the correct angle in the intake port.

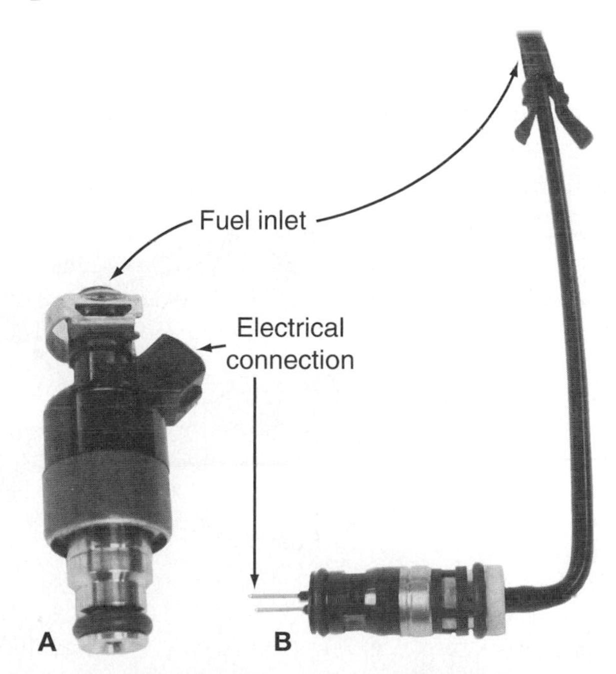

FIGURE 26-18 Different types of port fuel injectors.

electrical connections are on the bottom and make contact when they are inserted. Though the configuration of how the electrical signals get to the injector and how the fuel gets into the injector may continue to change, the basic principle of operation will remain the same.

PARTS LOCATOR

Refer to photo #35 in the front of the textbook to see what ***port fuel injectors*** *look like.*

FUEL RAIL

Fuel must be sent to each injector from the fuel pump. The fuel is sent into a fuel rail. From there, the fuel is distributed to each injector. The pressure regulator shown in ***Figure 26–19*** keeps the pressure inside the fuel rail at approximately 35 psi on this particular system. The injectors are locked in place using retainer clips.

PARTS LOCATOR

Refer to photo #36 in the front of the textbook to see what ***fuel rails*** *look like.*

Each manufacturer designs its parts differently. In ***Figure 26–20***, the fuel rail contains the fuel injectors and the fuel pressure regulator. This type of system is mounted directly on the cylinder head. O-rings are used to prevent leakage between the fuel rail, fuel injectors, and cylinder head.

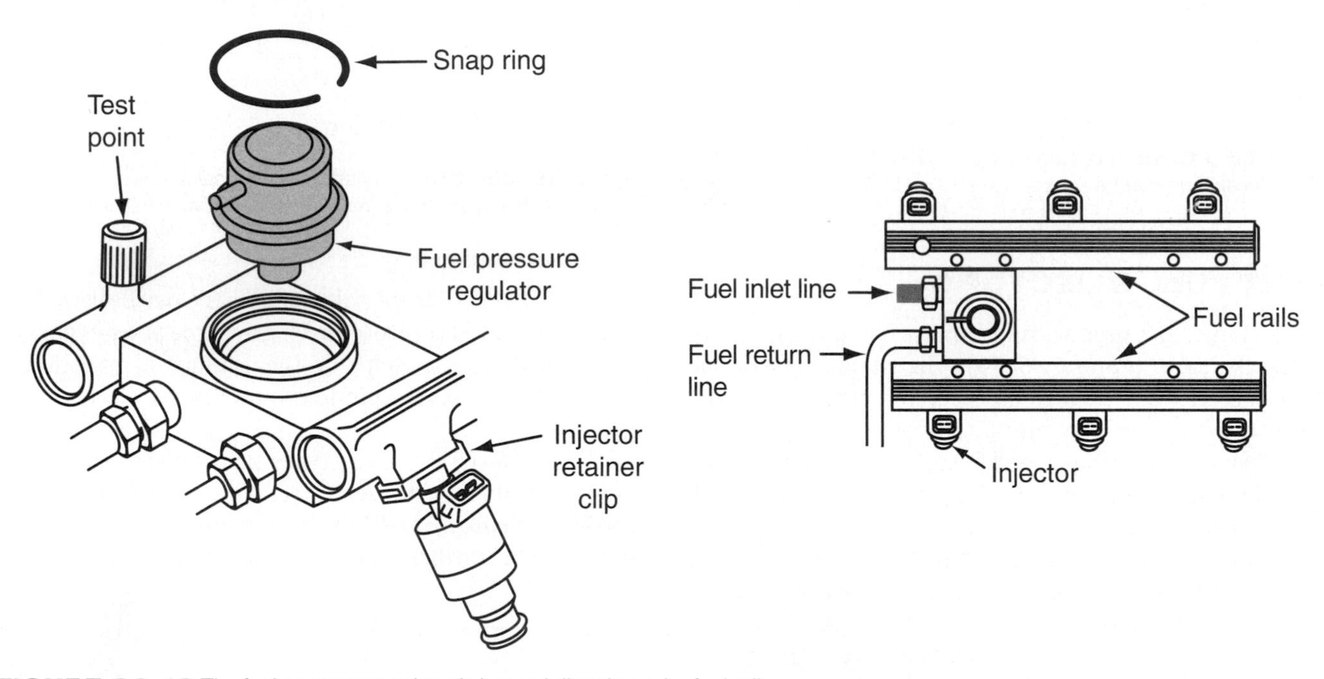

FIGURE 26–19 The fuel pressure regulator is located directly on the fuel rail.

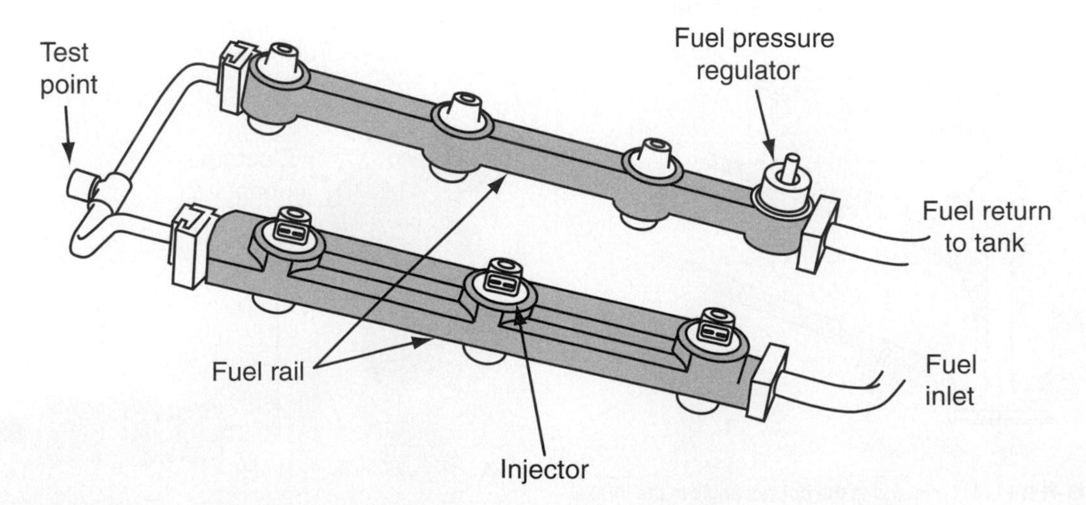

FIGURE 26–20 This fuel rail is located on top of the cylinder head.

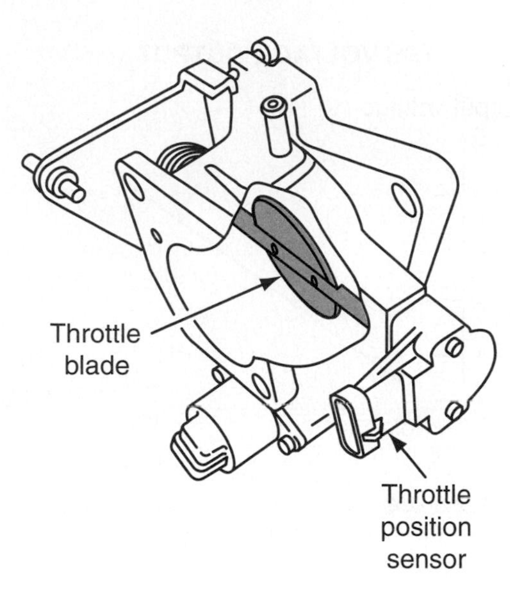

FIGURE 26-21 The throttle body consists of the throttle blade (valve), which controls the airflow, and the throttle position sensor.

THROTTLE BLADE

The throttle blade controls the volume of air that enters the engine (***Figure 26–21***). The throttle blade is controlled by the position of the operator's foot. As the foot is depressed, the blade opens and allows more air into the engine. As the foot is released, the blade closes and reduces the amount of air allowed into the engine. Some engines also have a coolant passage that allows engine coolant to flow through the throttle body unit. The purpose of this coolant flow is to increase the temperature of the incoming air to assist in preventing throttle blade icing during cold weather operation.

The throttle body also supports and controls the movement of the throttle position sensor (TPS). As stated earlier, the TPS is used to send a signal to the PCM. This signal gives the position of the throttle under all operating conditions. ***Figure 26–22*** shows two views of a typical throttle body assembly. Note the location of the throttle position sensor and the idle air control attached to the assembly.

PARTS LOCATOR

*Refer to photo #37 in the front of the textbook to see what a **throttle blade** looks like.*

IDLE AIR CONTROL (IAC)

Attached to the throttle body is an idle air control (IAC). The IAC is used to control the amount of air during idle conditions. The PCM controls idle speed by moving the IAC valve in and out as shown in ***Figure 26–23***. The PCM does this by sending the IAC voltage pulses called "counts" to the proper motor winding. The motor shaft and valve move a given distance for each count received.

For example, to increase the idle speed, the PCM sends enough counts to retract the IAC valve and allow more air

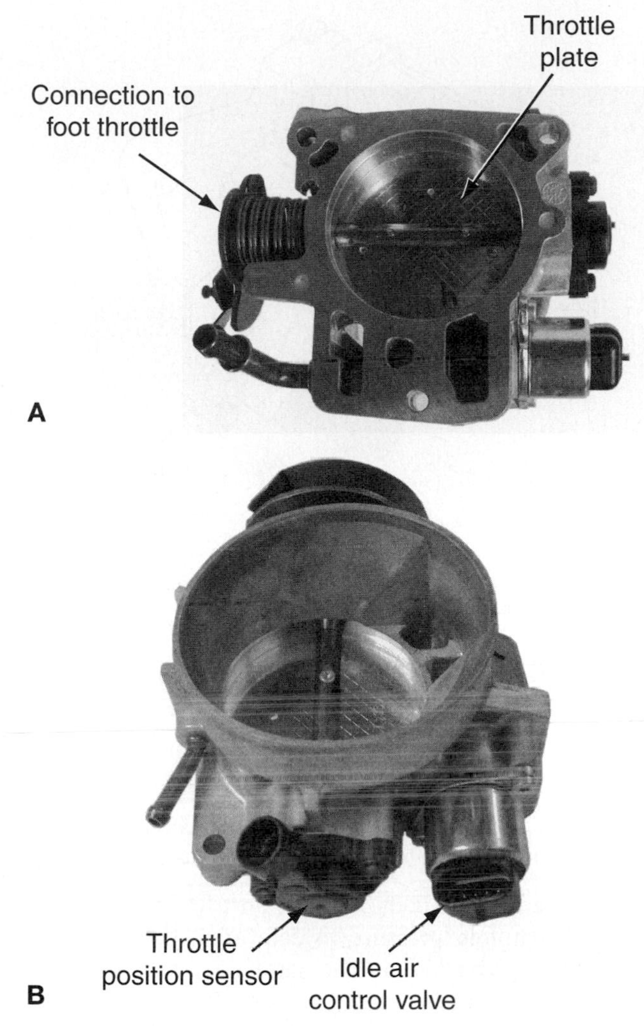

FIGURE 26-22 Here are two views of a typical throttle body assembly.

to flow around the throttle blade. The increase in airflow into the engine causes the idle to increase. (A corresponding increase in fuel metered by the PCM must also occur.) To decrease the idle speed, the PCM sends the correct number of

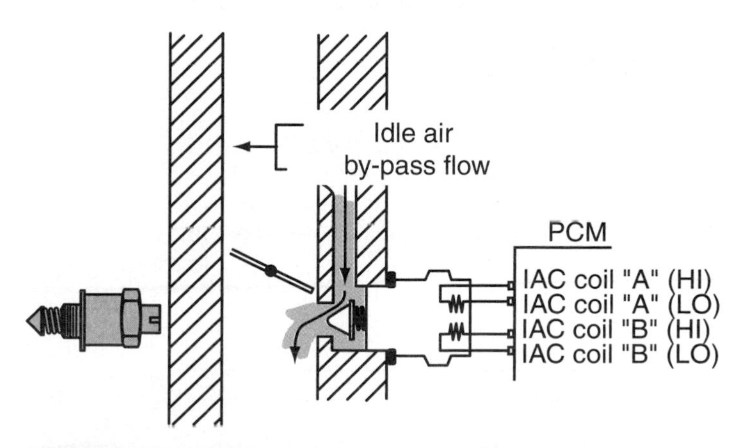

FIGURE 26-23 The idle air control is attached to the throttle body. The PCM controls the motor windings to open or close the IAC valve.

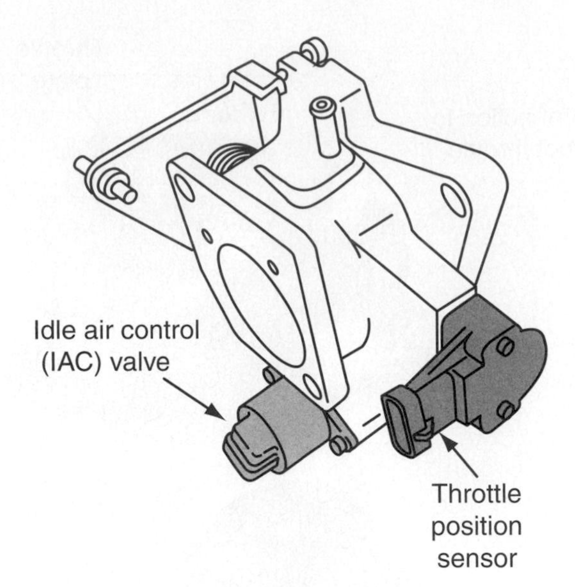

FIGURE 26–24 On this throttle body, the throttle position sensor and the IAC valve are attached.

counts to the IAC to extend the valve and reduce airflow. ***Figure 26–24*** shows the throttle body unit with the IAC valve and the throttle position sensor.

MANIFOLD ABSOLUTE PRESSURE (MAP)

As with the throttle body injector, port fuel injection also uses the manifold pressure to determine the amount of fuel needed in the engine. The manifold absolute pressure (MAP) sensor is used to tell the PCM the amount of load on the engine. As load is applied, the manifold absolute pressure increases. The MAP sensor is located on the intake manifold. It senses the pressure/vacuum inside the intake manifold. The sensor is a resistive device that changes resistance with different absolute pressures. Because the resistance is changed, the voltage signal also changes. In ***Figure 26–25***, voltage is shown in the vertical column.

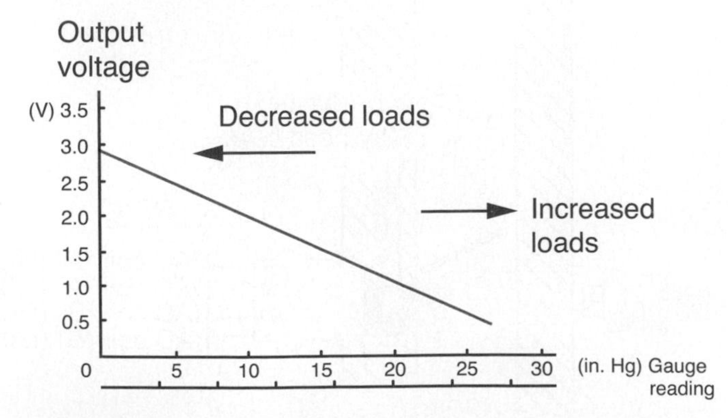

FIGURE 26–25 As the manifold absolute vacuum increases, the voltage signal sent to the PCM becomes lower. (Courtesy of American Honda Motor Co., Inc.)

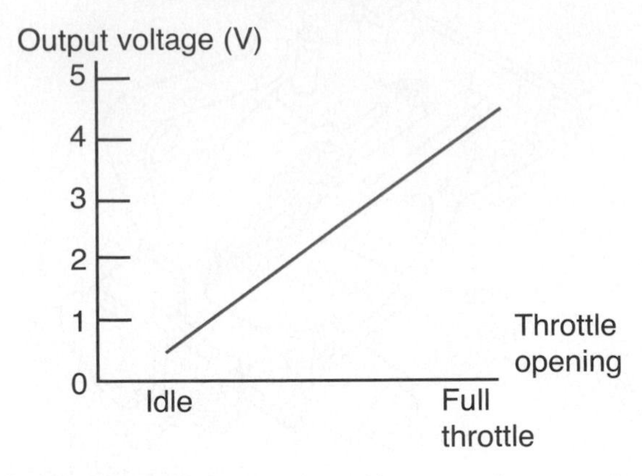

FIGURE 26–26 The throttle position sensor increases the output voltage as the throttle is opened wider. (Courtesy of American Honda Motor Co., Inc.)

Absolute pressure is shown on the horizontal column. Note that as the absolute pressure increases under increased loads, the voltage signal sent to the PCM is lower. Although the voltage may vary with each manufacturer, the principle is the same. Checking the MAP voltages helps to determine its condition.

THROTTLE POSITION SENSOR (TPS)

The throttle position sensor (TPS) is a device that measures the opening of the throttle. The wider the throttle is open, the greater the amount of fuel needed. This sensor is a potentiometer which is operated by the throttle blade shaft. The throttle position sensor is not adjustable, but it does send a voltage signal to the PCM. ***Figure 26–26*** is a graph that indicates the relationship between the position of the throttle and the voltage signal. Note that when the throttle plate is fully open, the output voltage is at a maximum, about 4 to 5 volts. As the throttle is returned to idle, the output voltage is reduced to below 1 volt.

MASS AIRFLOW (MAF)

Mass airflow (MAF) is defined as the total amount (mass) of air going into the engine. This amount depends on the altitude of the vehicle, temperature of the air, density of the air, and moisture in the air. All of these variables cause the air-fuel ratio to change. In order to maintain a stoichiometric air-fuel ratio of 14.7 to 1, these variables must be known. They are measured by a mass airflow sensor.

Figure 26–27 shows an MAF sensor. It consists of the following six parts:

1. A flow tube that houses the parts
2. A sample tube that directs air to a sensor
3. A screen that breaks up the airflow
4. A ceramic resistor (thermistor) that measures the temperature of the incoming air (air temperature sending resistor)

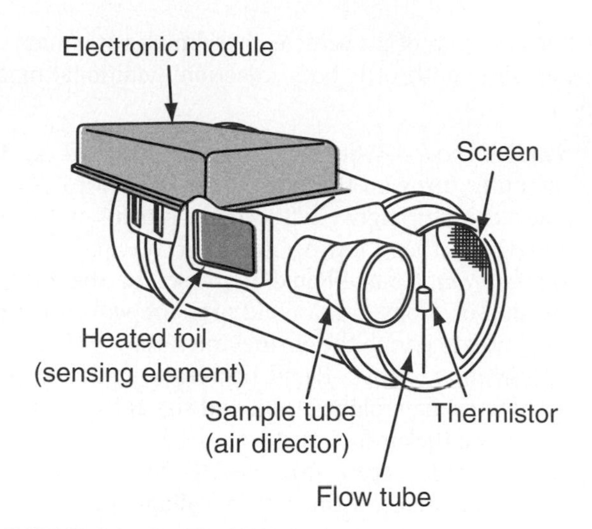

FIGURE 26–27 The MAF sensor determines the exact amount (mass) of air flowing into the engine.

5. A heated foil sensing element that senses air mass
6. An electronic module that determines the mass airflow from the sensing element

This type of sensor can compensate for altitude and humidity. In operation, air mass is determined by measuring the amount of electrical power needed to keep the temperature of the sensing element 75 degrees Celsius (167 degrees Fahrenheit) above the incoming air. As air enters the unit, it passes over and cools the sensing element. When the element is cooler, it requires more electrical power to keep the element at 75 degrees Celsius above the incoming air temperature. The electrical power requirement is a measure or indication of the mass airflow. The power is converted to a digital signal as a frequency. This signal is sent to the PCM and used to calculate engine load. Using mass airflow, engine temperature, and rpm, the PCM can calculate the exact amount of fuel to be metered to provide a stoichiometric ratio of 14.7 to 1. ***Figure 26–28*** shows the location of a typical MAF sensor on the air inlet tube between the air cleaner and the throttle body assembly.

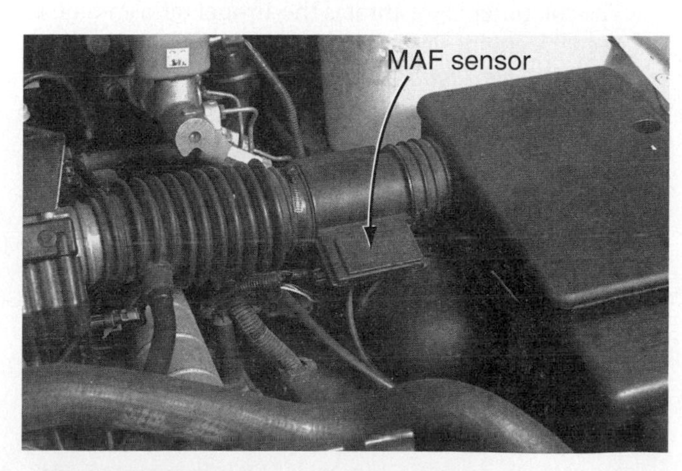

FIGURE 26–28 The MAF sensor is usually located on the intake tubes between the air cleaner and the throttle body assembly.

PARTS LOCATOR

*Refer to photo #38 in the front of the textbook to see what a **mass airflow (MAF) sensor** looks like.*

CRANKSHAFT POSITION (CKP) SENSOR

Today's computerized fuel injection systems often need a reference to the position of the crankshaft. The position of the crankshaft helps to determine such things as firing order, degrees before top dead center (BTDC), and when the number 1 piston is at top dead center (TDC). On certain types of port fuel injection systems (such as sequential firing fuel injection) this information is needed by the computer. Today, there are several techniques used to determine crankshaft position. One of the more common is shown in ***Figure 26–29***. In this particular case, the sensor is located on the transaxle housing. The sensor detects the passing of slots on the flywheel. This flywheel contains 12 slots arranged in three groups of 4 (***Figure 26–30***). Each group is 120 degrees apart. When the flywheel metal passes by the

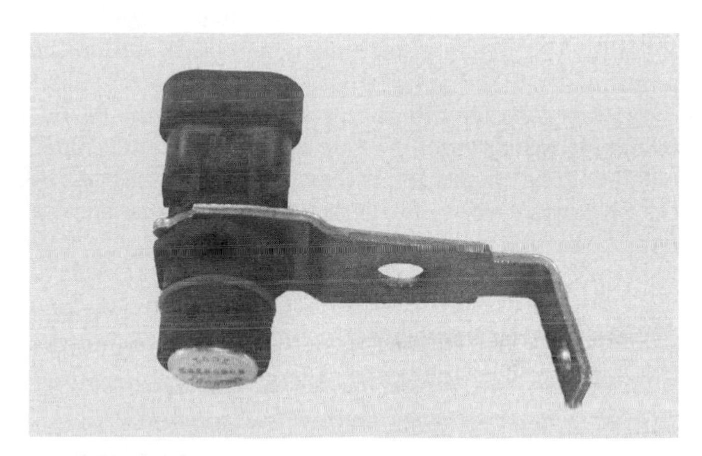

FIGURE 26–29 The CKP sensor generates a small voltage as the crankshaft is turned.

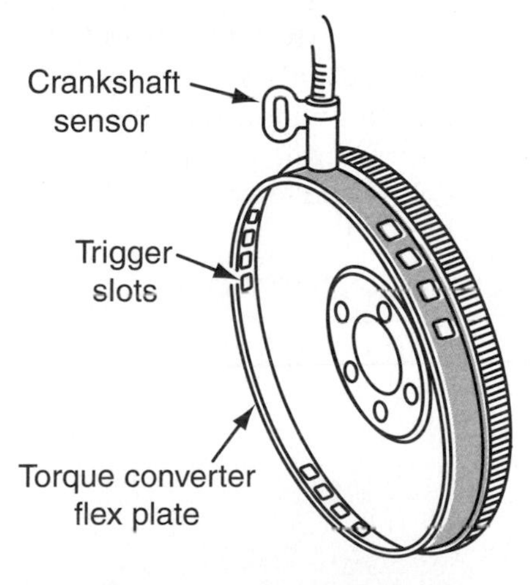

FIGURE 26–30 A CKP sensor senses the holes or slots in the torque converter flex plate.

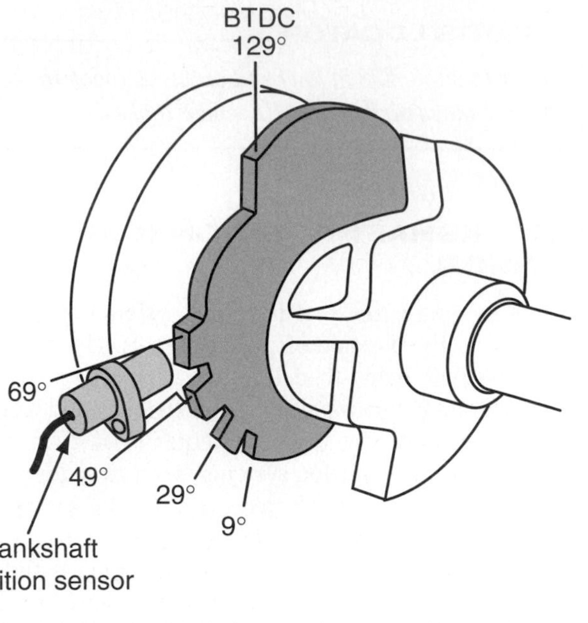

FIGURE 26-31 This CKP sensor detects machined notches on the crankshaft.

CKP sensor, the sensor is on and the signal voltage is low, about 0.5 volt. When a slot passes by the CKP sensor, the sensor is off and the voltage goes to a high of about 5 volts.

Another common CKP sensor is shown in ***Figure 26–31.*** In this particular case, the CKP sensor detects machined notches in the crankshaft. As the machined notches pass by the CKP sensor, the sensor is being turned on and off. The on and off voltages are then sent to the PCM.

PARTS LOCATOR

*Refer to photo #39 in the front of the textbook to see what a **crankshaft position (CKP) sensor** looks like.*

The first notch in the group of four is at 69 degrees BTDC. The second notch is at 49 degrees BTDC. The third notch is at 29 degrees BTDC and the fourth notch is at 9 degrees BTDC. From this information, the engine computer can determine when a pair of cylinders will reach TDC. Each group of slots relates to a pair of cylinders.

MODES OF OPERATION

Like throttle body injection systems, port fuel injection systems operate in various modes. Depending on the manufacturer and the year of the vehicle, these modes are much the same as those of throttle body injection. Additional modes include:

- *Starting mode*—When the ignition is turned on, the computer turns on the fuel pump relay. If it does not read a cranking signal in 2 seconds, it turns off the fuel pump relay, thus shutting off the fuel pump.
- *Acceleration mode*—Rapid increases in the throttle opening or drops in manifold pressure will signal the computer to enrich the air-fuel mixture.
- *Deceleration mode*—Rapid decreases in the throttle opening or manifold pressure will signal the computer to lean out the air-fuel mixture.
- *Battery voltage correction mode*—On some port fuel injection systems, if the battery voltage drops below a specific value, the computer will compensate for a weak ignition spark in three ways:
  1. Enrich the air-fuel mixture
  2. Increase the throttle opening to increase the idle slightly
  3. Increase the ignition dwell
- *Fuel cutoff mode*—When the engine is shut off, the injectors could continue to pulse as the engine rpm slows to a stop. This continued pulsing may cause dieseling. To avoid the problem, the computer is designed to immediately stop pulsing the injectors the second the ignition is shut off.
- *MPG lean cruise*—When the engine is at cruising conditions, the computer is able to improve the fuel economy. This is done by taking the system out of closed loop operation and leaning out the air-fuel mixture to improve fuel economy.
- *Modular strategy or mode*—At times it may be necessary to operate the engine during special conditions. Three of these conditions include:
  1. Cold engine
  2. Overheated engine
  3. High altitude

  During these conditions, it may be necessary to compensate air-fuel ratios just to keep the vehicle drivable. The computer compensates the air-fuel ratio enough to keep the engine operating.
- *Limited operational strategy (LOS) mode*—At times, a system component failure may prevent the engine from operating in a normal closed loop mode. When this happens, the computer enters an alternate mode designed to protect other system components, keeping the vehicle drivable until service can be completed. This mode is variously referred to as "limp home mode," "backup mode," or "fail safe mode."

Problems, Diagnosis, and Service

Safety Precautions

1. Always disconnect the battery and place the vehicle in park when working on any part of the fuel system.
2. When running an engine during a check of a specific part of the fuel injection system, always make sure that the exhaust fumes are being properly vented, the brake is on, and the vehicle is in park.
3. Never work with the electrical components when gasoline has been spilled or there are gasoline fumes nearby.
4. When checking the electrical components on or near the fuel injection system, always use a battery-operated tester to eliminate any sparks. Sparks near gasoline fumes can cause a fire.
5. Make sure there are no unshielded flames, such as from a cutting torch, around or near a vehicle when the fuel injection system is being serviced.
6. When working on the fuel system, never use a drop light to inspect fuel system components; use a flashlight to reduce the possibility of sparks.
7. Immediately wipe up any spilled gasoline and dispose of the rags in an OSHA-approved, sealed container.
8. Never smoke in or around an automobile service shop.
9. Always store the fuel in an OSHA-approved container. Never store gasoline in an open container.
10. Always keep an approved fire extinguisher nearby when working on the fuel injection system.
11. Always wear OSHA-approved safety glasses when working on the fuel injection system.
12. When working on vehicles that have computers, always follow the manufacturer's suggested service procedure. Standard testlights and unfused jumper wires may damage the computer. Never use these tools unless recommended by the manufacturer.

Many computer controls are used on fuel injection systems. Each year the manufacturers improve these systems. In addition, each car manufacturer uses a different set of procedures for troubleshooting and for diagnosing problems in its computerized fuel injection system. The following diagnoses and service hints are general in nature, but they give an idea of how to diagnose and troubleshoot fuel injection systems.

PROBLEM: Faulty PROM

A vehicle's Check Engine light on the dashboard has come on. After grounding the test terminal under the dashboard or using a scan tool to determine the code, it was found that the PROM needs to be checked.

DIAGNOSIS

The computer used on each vehicle has a PROM, which stands for "programmable read only memory." The PROM is a manufacturer-programmed set of data to put into the computer. It includes programmed data related to the vehicle's weight, engine size, transmission, axle ratio, and so on. If any diagnostic procedure calls for the computer to be replaced, the calibration unit (PROM) should be checked to see if it is the proper one and if it is installed correctly. If it is the correct one, the PROM should be removed and placed in the new computer.

Car Clinic: PROM

CUSTOMER CONCERN:
Idle Too High

A car that has a 2.5-liter engine with fuel injection has just been purchased. The vehicle should normally get about 35 mpg, but it is only getting about 25 mpg. The idle is very high, and the car is very hard to slow down. What can be done?

SOLUTION: GM cars that use a computer and fuel injection typically have few controls for idle. On some vehicles, the programmable read only memory (PROM) may be programmed for different altitudes. The engine is getting too much fuel and is running very rich. In this case, the PROM should be replaced. Go to the manufacturer to get the PROM; the manufacturer is the only source for the correct PROM.

SERVICE

1. Locate the computer in the vehicle and remove it.
2. Make sure the ignition switch is in the off position and the battery is disconnected.

3. Remove the cover over the PROM.
4. Using a special PROM removal tool, remove the PROM carefully.

CAUTION: *Avoid touching the PROM with your hands, as the oils on your hand can affect the PROM operation.*

5. Make sure all of the electrical pins and terminals are straight.
6. Install the PROM in the new computer. Be sure that the reference marks match between the computer and the PROM.
7. Make sure the PROM is securely fastened and fully inserted into the socket.
8. Reinstall the computer into the mounts, and connect the electrical connectors to the computer.

PROBLEM: Open Coolant Sensor Circuit

A vehicle's Check Engine light on the dashboard has come on.

DIAGNOSIS

1. Using a test wire, ground out the test terminal under the dashboard of the vehicle.
2. Read the code flashing on the testlight.
3. The code flashes once, then pauses, then flashes five times.
4. Using the list codes in ***Figure 26–32***, you determine the problem to be an open coolant sensor circuit (code 15).

SERVICE

Use the diagnostic information in ***Figure 26–33*** to complete the service procedure on the coolant sensor circuit. To check the engine coolant temperature (ECT) sensor, resistance checks can be made. The resistance of the ECT changes with the temperature. As the temperature increases, the resistance of the sensor decreases. The amount of resistance causes the voltage drop across the ECT to change. Use the chart in ***Figure 26–33*** to identify the resistances at different temperatures of coolant. When checking the voltage drop, use the following general procedure:

1. Run the engine for 5 minutes in the closed loop or until the Check Engine light comes on.
2. Now, with the engine off but with the ignition switch on, disconnect the coolant sensor from the sensor connector terminals. Check the voltage between the sensor connector terminals.
3. If it is under 4 volts (low resistance) the problem is not the ECT sensor. If it is over 4 volts, check the resistance of the ECT sensor. It should be under 1,000 ohms on a warm engine.
4. Replace the coolant sensor as necessary.

CODE	CIRCUIT AFFECTED
12	NO DISTRIBUTOR (TACH) SIGNAL
13	O_2 SENSOR NOT READY
14	SHORTED COOLANT SENSOR CIRCUIT
15	OPEN COOLANT SENSOR CIRCUIT
16	GENERATOR VOLTAGE OUT OF RANGE
18	OPEN CRANK SIGNAL CIRCUIT
19	SHORTED FUEL PUMP CIRCUIT
20	OPEN FUEL PUMP CIRCUIT
21	SHORTED THROTTLE POSITION SENSOR CIRCUIT
22	OPEN THROTTLE POSITION SENSOR CIRCUIT
23	EST/BY-PASS CIRCUIT PROBLEM
24	SPEED SENSOR CIRCUIT PROBLEM
26	SHORTED THROTTLE SWITCH CIRCUIT
27	OPEN THROTTLE SWITCH CIRCUIT
28	OPEN FOURTH GEAR CIRCUIT
29	SHORTED FOURTH GEAR CIRCUIT
30	ISC CIRCUIT PROBLEM
31	SHORTED MAP SENSOR CIRCUIT
32	OPEN MAP SENSOR CIRCUIT
33	MAP/BARO SENSOR CORRELATION
34	MAP SIGNAL TOO HIGH
35	SHORTED BARO SENSOR CIRCUIT
36	OPEN BARO SENSOR CIRCUIT
37	SHORTED MAT SENSOR CIRCUIT
38	OPEN MAT SENSOR CIRCUIT
39	TCC ENGAGEMENT PROBLEM
44	LEAN EXHAUST SIGNAL
45	RICH EXHAUST SIGNAL
51	PROM ERROR INDICATOR
52	PCM MEMORY RESET INDICATOR
53	DISTRIBUTOR SIGNAL INTERRUPT
60	TRANSMISSION NOT IN DRIVE
63	CAR AND SET SPEED TOLERANCE EXCEEDED
64	CAR ACCELERATION EXCEEDS MAX. LIMIT
65	COOLANT TEMPERATURE EXCEEDS MAX. LIMIT
66	ENGINE RPM EXCEEDS MAXIMUM LIMIT
67	SHORTED SET OR RESUME CIRCUIT

FIGURE 26-32 Use these codes to determine which circuit is affected after noting the code number. (Courtesy of MOTOR Publications, Auto Engine Tune-Up Electronics Manual © Hearst Corporation)

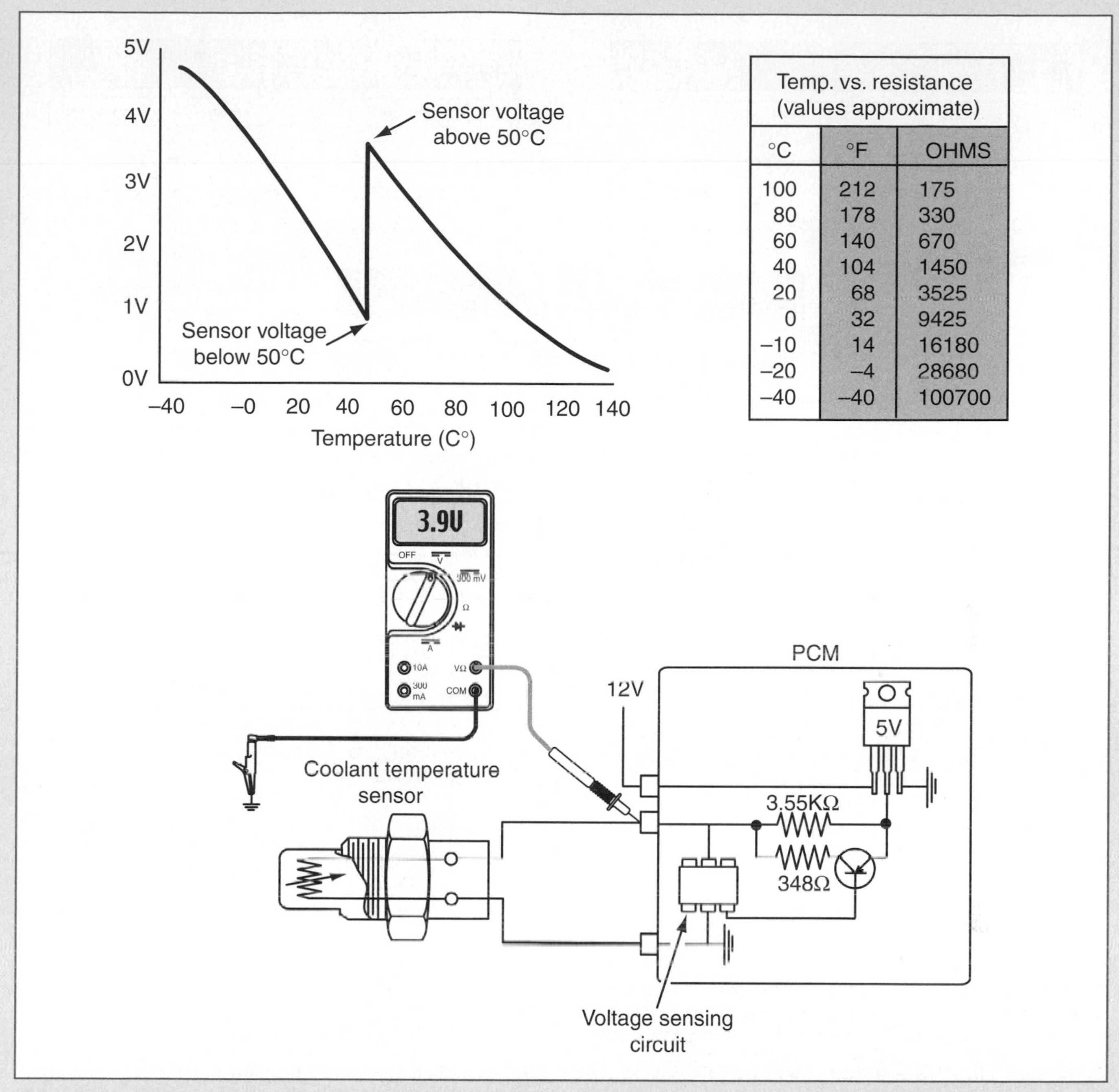

Temp. vs. resistance (values approximate)

°C	°F	OHMS
100	212	175
80	178	330
60	140	670
40	104	1450
20	68	3525
0	32	9425
–10	14	16180
–20	–4	28680
–40	–40	100700

FIGURE 26-33 The ECT sensor can be checked for resistance by checking the voltage drop of the sensor.

PROBLEM: Idle Speed Control

A vehicle's Check Engine light on the dashboard has come on. After the lead in the test terminal under the dashboard was grounded, (or using a scan tool), the problem was identified as being the idle speed control.

DIAGNOSIS

Certain manufacturers use a pinpoint test along with a set of codes. Pinpoint tests are used to get more specific readings on individual components in the system. Voltmeter readings are taken at different points and compared with the manufacturer's specifications. The procedures and readings can be found in many of the service manuals used today. Since there are so many variations, it is important to follow the exact procedure in the manual for the particular year and make of the vehicle.

SERVICE

1. Locate the year and model of the vehicle.
2. After determining the problem component (idle speed control) find the pinpoint procedure for this component.

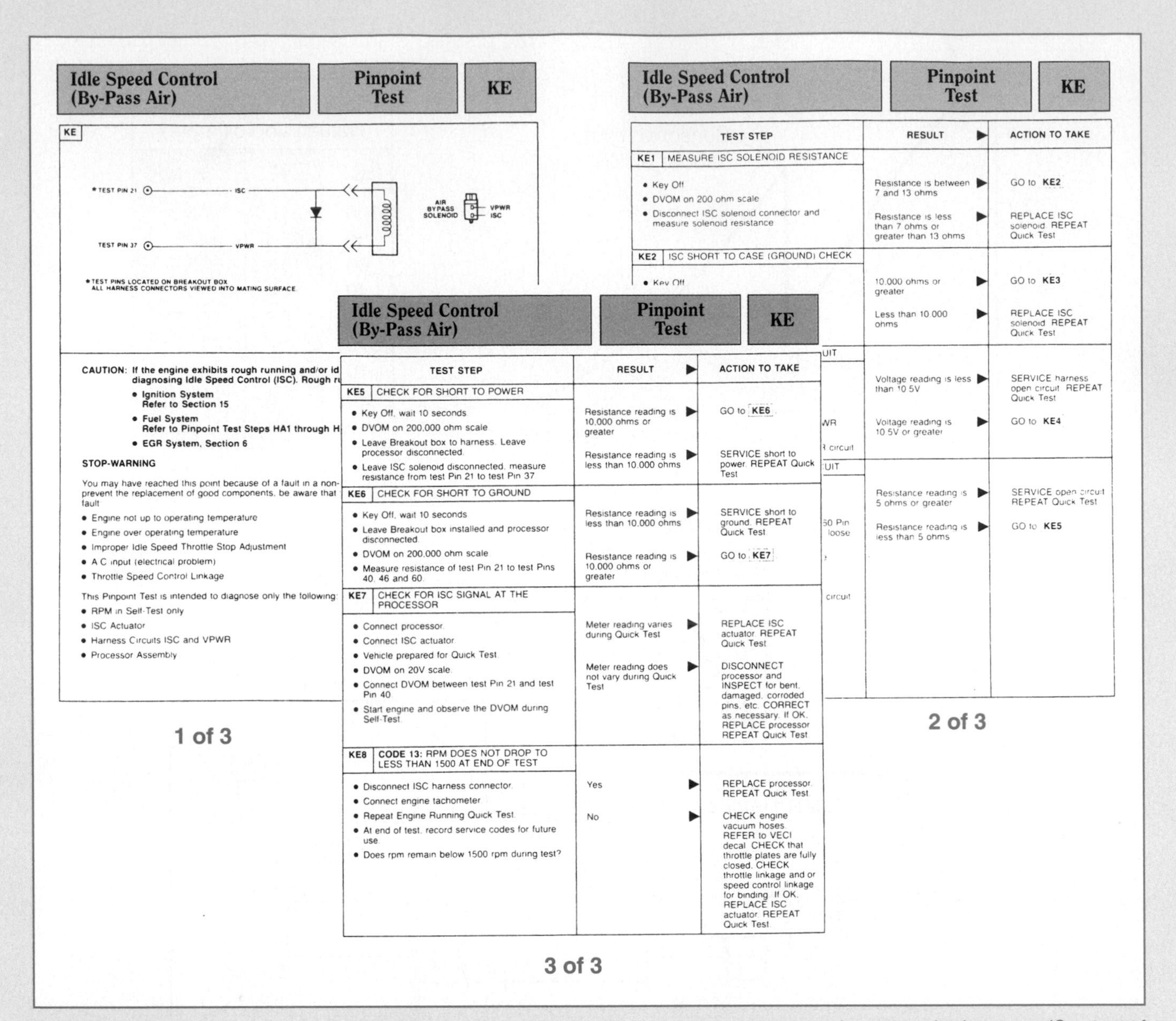

Idle Speed Control (By-Pass Air) | **Pinpoint Test** | **KE**

KE

CAUTION: If the engine exhibits rough running and/or id... diagnosing Idle Speed Control (ISC). Rough r...

- **Ignition System** Refer to Section 15
- **Fuel System** Refer to Pinpoint Test Steps HA1 through H...
- **EGR System, Section 6**

STOP-WARNING

You may have reached this point because of a fault in a non-... prevent the replacement of good components, be aware that ... fault

- Engine not up to operating temperature
- Engine over operating temperature
- Improper Idle Speed Throttle Stop Adjustment
- A C input (electrical problem)
- Throttle Speed Control Linkage

This Pinpoint Test is intended to diagnose only the following:

- RPM in Self-Test only
- ISC Actuator
- Harness Circuits ISC and VPWR
- Processor Assembly

1 of 3

Idle Speed Control (By-Pass Air) | **Pinpoint Test** | **KE**

TEST STEP	RESULT ▶	ACTION TO TAKE
KE1 MEASURE ISC SOLENOID RESISTANCE		
• Key Off • DVOM on 200 ohm scale • Disconnect ISC solenoid connector and measure solenoid resistance	Resistance is between 7 and 13 ohms ▶ Resistance is less than 7 ohms or greater than 13 ohms ▶	GO to **KE2** REPLACE ISC solenoid. REPEAT Quick Test
KE2 ISC SHORT TO CASE (GROUND) CHECK		
• Key Off	10,000 ohms or greater ▶ Less than 10,000 ohms ▶	GO to **KE3** REPLACE ISC solenoid. REPEAT Quick Test
...UIT		
...WR ...R circuit	Voltage reading is less than 10.5V ▶ Voltage reading is 10.5V or greater ▶	SERVICE harness open circuit. REPEAT Quick Test GO to **KE4**
...UIT		
...60 Pin ... loose ...circuit	Resistance reading is 5 ohms or greater ▶ Resistance reading is less than 5 ohms ▶	SERVICE open circuit. REPEAT Quick Test GO to **KE5**

2 of 3

Idle Speed Control (By-Pass Air) | **Pinpoint Test** | **KE**

TEST STEP	RESULT ▶	ACTION TO TAKE
KE5 CHECK FOR SHORT TO POWER		
• Key Off, wait 10 seconds. • DVOM on 200,000 ohm scale. • Leave Breakout box to harness. Leave processor disconnected. • Leave ISC solenoid disconnected, measure resistance from test Pin 21 to test Pin 37	Resistance reading is 10,000 ohms or greater ▶ Resistance reading is less than 10,000 ohms ▶	GO to **KE6** SERVICE short to power. REPEAT Quick Test
KE6 CHECK FOR SHORT TO GROUND		
• Key Off, wait 10 seconds • Leave Breakout box installed and processor disconnected. • DVOM on 200,000 ohm scale. • Measure resistance of test Pin 21 to test Pins 40, 46 and 60.	Resistance reading is less than 10,000 ohms ▶ Resistance reading is 10,000 ohms or greater ▶	SERVICE short to ground. REPEAT Quick Test. GO to **KE7**
KE7 CHECK FOR ISC SIGNAL AT THE PROCESSOR		
• Connect processor. • Connect ISC actuator. • Vehicle prepared for Quick Test. • DVOM on 20V scale. • Connect DVOM between test Pin 21 and test Pin 40. • Start engine and observe the DVOM during Self-Test.	Meter reading varies during Quick Test ▶ Meter reading does not vary during Quick Test ▶	REPLACE ISC actuator. REPEAT Quick Test DISCONNECT processor and INSPECT for bent, damaged, corroded pins, etc. CORRECT as necessary. If OK, REPLACE processor. REPEAT Quick Test.
KE8 **CODE 13:** RPM DOES NOT DROP TO LESS THAN 1500 AT END OF TEST		
• Disconnect ISC harness connector. • Connect engine tachometer. • Repeat Engine Running Quick Test. • At end of test, record service codes for future use. • Does rpm remain below 1500 rpm during test?	Yes ▶ No ▶	REPLACE processor. REPEAT Quick Test. CHECK engine vacuum hoses. REFER to VECI decal. CHECK that throttle plates are fully closed. CHECK throttle linkage and or speed control linkage for binding. If OK, REPLACE ISC actuator. REPEAT Quick Test.

3 of 3

FIGURE 26–34 Use the manufacturer's pinpoint tests to help troubleshoot various components in computerized systems. (Courtesy of MOTOR Publications, Auto Engine Tune-up & Electronics Manual, © The Hearst Corporation)

3. As shown in ***Figure 26–34***, there may be several pinpoint procedure tests (1 of 3, 2 of 3, 3 of 3).
4. Follow the procedure as stated on the pinpoint procedure. Note that the pinpoint procedure shows the appropriate tests, the possible results, and the actions to perform based on the result. Always follow the manufacturer's suggested test to determine the problem. Then replace the faulty components as necessary.

PROBLEM: Faulty ISC or TPS

A vehicle with port fuel injection continually stalls.

DIAGNOSIS

For diagnosing various problems in port fuel injection systems such as "continually stalling," troubleshooting guides are available for reference in many of the service manuals. ***Figure 26–35*** is an example of a typical Troubleshooting Guide for Port Fuel Injection Systems. In this particular case, the problem may be in the idle speed control (ISC), the throttle position sensor (TPS), or the MAP sensor.

SERVICE

The exact service will depend on the components that need to be tested, serviced, or replaced. Follow the manufacturer's recommended procedure for service and replacement of the faulty component. If the procedure suggests that vacuum and pressure hoses be inspected for leaks, various vacuum hose diagrams are also available. Service and repair manuals have drawings for specific vehicles, models, years, and hose connections. ***Figure 26–36*** shows an example of the vacuum hose routing for a specific type of vehicle.

TROUBLESHOOTING GUIDE FOR PORT FUEL INJECTION SYSTEMS		
PROBLEM	**DIAGNOSIS**	**SERVICE**
Hard start, hot	Bleeding injector	Inspect injector for dripping; service, or replace as required.
	Leaking intake manifold gasket or base gasket	Replace defective gasket.
	MAP sensor	Check MAP sensor and vacuum hose; service or replace as required.
	Pressure regulator	Check pressure regulator for setting and bleed down; service or replace as required.
Rough idle, hot	MAP sensor	Check MAP sensor and vacuum hose; service or replace as required.
	CTS	Check coolant level; service or replace sensor.
	TPS	Check TPS; adjust or replace as required.
	Injector	Check injector for variation in spray pattern; clean or replace injector as required.
	Oxygen sensor	Replace oxygen sensor.
	Defective computer	Replace computer.
	ISC/IAC	Check idle speed control device; service or replace as required.
Stalling	ISC/IAC	Check idle speed control device; service or replace as required.
	TPS	Check TPS; adjust or replace as required.
	MAP sensor	Check MAP sensor and vacuum hose; service or replace as required.

FIGURE 26–35 This is typical troubleshooting guide to help the service technician locate and solve problems in computerized port fuel injection systems.

Several companies also manufacture special diagnostic instruments for checking computerized vehicle systems. ***Figure 26–37*** shows such an instrument. These test instruments can be connected directly into the computer system (ALCL) test terminal (under the dashboard or in the engine compartment) to make rapid and accurate diagnostic checks. These handheld testers are also known as scanners.

At times all of the components may check out OK, but a problem still exists in the vehicle. When this happens, the fault may be in the wiring harness used to hold all of the

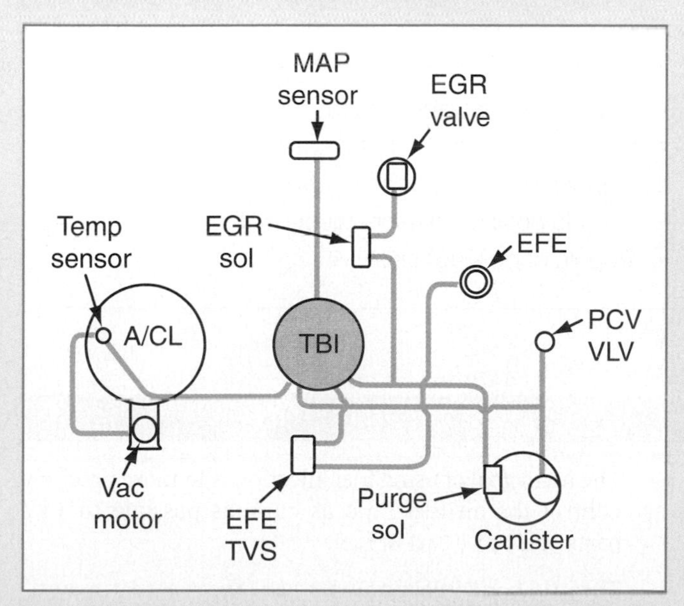

FIGURE 26–36 Always refer to the manufacturer's specifications manual for routing of vacuum hoses. Check for leaks and for damaged or broken parts.

FIGURE 26–37 Various manufacturers sell test instruments to diagnose computer-controlled systems. These instruments can save the service technician valuable time and money.

wires together. ***Figure 26–38*** shows a typical wiring harness used on a computerized fuel management system. Often, this wiring harness will vibrate against the vehicle frame and cause one of its circuits to stop operating. The exact point of contact is very difficult to locate. Visually inspect the wiring harness to make sure all the electrical connections are complete. You may have to refer to recent manufacturer's service bulletins to determine exactly where the wiring harness is rubbing against the frame.

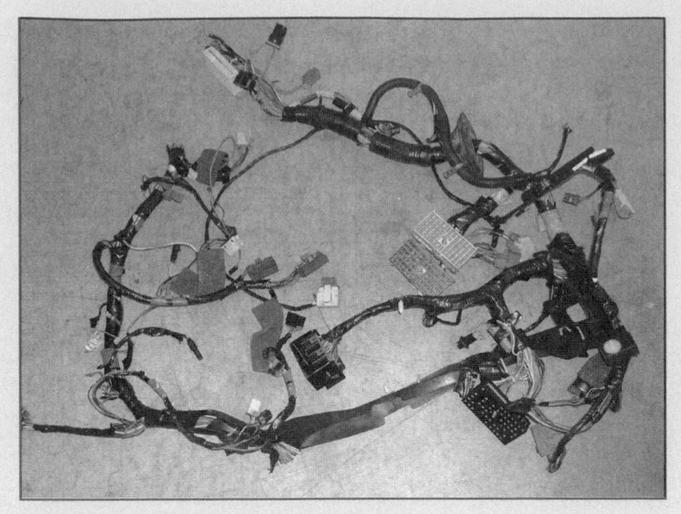

FIGURE 26–38 The wiring harness may often short out against the frame of the vehicle, causing irregular vehicle operation.

Service Manual Connection

There are many important specifications to keep in mind when working with injection systems. To identify the specifications for your engine, you will need to know the vehicle identification number (VIN) of the vehicle, the type and year of the vehicle, and the type of engine. Although they may be titled differently, some of the more common injection system specifications (not all) found in service manuals are listed here. Note that these specifications are typical examples. Each vehicle and engine may have different specifications.

Common Specification	Typical Example
Curb idle speed	
Manual transmission	900 rpm
Automatic transmission	850 rpm in neutral
Fuel control solenoid	28–66 ohms
Fuel pressure regulator	36 psi

Common Specification	Typical Example
Idle speed control motor adjustment	1.06 volts
Injector resistance	1.5–2.5 ohms
Oxygen sensor voltage	0–0.99 volts
Throttle body injector torque	200 in.-lb
Throttle position sensor pinpoint test	More than 1,900 ohms

In addition, the service manual will give specific directions for various service and testing procedures. Some of the more common procedures include:

- Coolant sensor replacement
- Pinpoint test, barometric pressure
- Pinpoint test, idle speed control
- Throttle position sensor adjustment
- Trouble code diagnosis

SUMMARY

The following statements will help to summarize this chapter:

- Fuel injection is the process of injecting fuel into the flow of air going into the engine.
- The most common types of fuel injection for gasoline engines are the throttle body fuel injection and the port fuel injection.
- The main goal of using fuel injection is to more precisely control the air-fuel ratio as close as possible to 14.7 parts of air to 1 part of fuel.
- Throttle body injection uses one injector that is positioned on top of the intake manifold.

- Throttle body injection uses a computer to control when and how long the injection takes place.
- Many sensors are used on a throttle body injection system, including a throttle position sensor, an idle air control valve, a manifold absolute pressure sensor, and an oxygen sensor, among others.
- The port fuel injector uses an injector for each cylinder.
- Port fuel injectors are placed just before the intake valve in the intake manifold.
- The port fuel injector uses a solenoid-operated valve with a pintle to inject the fuel.
- Many sensors are used on port fuel injection systems, including the manifold air temperature sensor, the idle air control valve, the cold start injector, the mass airflow indicator, the crankshaft position sensor, the intake air temperatures sensor, the camshaft sensor, the knock sensor, and the oxygen sensor, among others.

TERMS TO KNOW

Can you explain each of the following terms? Review the chapter until you can use each term correctly.

Conical
Crossfire injection
Emission standards
Gauge pressure
Intake manifold vacuum
Magnetic pickup
Metered
Needle valve
Pintle
Pulse width
Stoichiometric ratio
Thermistor
Throttle body
Throttle plate
Tuned ports

REVIEW QUESTIONS

Multiple Choice

1. Gasoline fuel injection is considered:
a. Indirect
b. High-pressure
c. Direct
d. Nonmetered
e. Governed

2. Which type of fuel injection has one fuel injector feeding all pistons?
a. Port fuel injection
b. Direct fuel injection
c. Throttle body injection
d. All of the above
e. None of the above

3. The correct stoichiometric ratio is:
a. 12.2 to 1
b. 14.7 to 1
c. 16.7 to 1
d. 18.7 to 1
e. 19.2 to 1

4. Metering of throttle body fuel injection is controlled by the:
a. Starting system only
b. Fuel pump
c. Computer
d. Cooling system
e. Lubrication system

5. The fuel injector in the throttle body fuel injection system is opened and closed by using:
a. Battery voltage
b. A solenoid
c. The operator's foot
d. Mechanical linkage from the carburetor
e. Lubricating oil pressure

6. The amount of fuel injection from a throttle body injector is measured by the:
a. Pulse width
b. Cooling system
c. Size of the battery
d. Temperature of the fuel
e. Speed of the tires

7. The fuel pressure regulator in throttle body fuel injection is located on the:
a. Cooling thermostat
b. Fuel pump
c. Fuel rails
d. Exhaust manifold
e. Throttle body

8. The throttle position sensor is used as a/an ________ the PCM.
a. Input to
b. Output from
c. Frequency sensor from
d. Voltage from
e. Speed sensor from

9. Which of the following units bypasses air around the throttle plate on the TBI system?
a. Throttle position sensor
b. Gauge pressure unit
c. Manifold absolute pressure
d. Idle air control
e. Absolute pressure unit

10. Which of the following measures absolute pressure inside the manifold?
a. MAF
b. IAC
c. MAP
d. TPS
e. None of the above

11. Which pressure scale starts at zero, when the atmospheric pressure is at 14.7 psi?
a. Psig
b. Psia
c. CCC
d. Port pressure
e. None of the above

12. Which fuel system would eliminate fuel droplets forming inside the intake manifold because the fuel is heavier than air?
a. Throttle body
b. Port
c. Carburetors
d. Central fuel system
e. Controlled-combustion fuel system

13. Manifold air temperature is used as an ________.
a. Input to the PCM
b. Output from the PCM
c. Input to the MAP
d. Output from the MAP
e. None of the above

14. The port fuel injector opens and closes because of the operation of a ________.
a. Mechanical stop
b. Set of rollers
c. Regulator
d. Solenoid
e. High-pressure diaphragm

15. Which system has a sensing element that must be kept 75 degrees Celsius (167 degrees Fahrenheit) above the temperature of the incoming air?
a. Mass air temperature
b. Manifold absolute pressure
c. Idle air control
d. Mass airflow
e. None of the above

16. When diagnostic work is done on a computerized fuel-injection system, a series of ________ are used.
a. Codes
b. Subsystems
c. Voltage readings that are displayed on the dashboard
d. Special instructions that are displaced on the dashboard
e. None of the above

The following questions are similar in format to ASE (Automotive Service Excellence) test questions.

17. *Technician A* says when the air-fuel ratio is constant during starting, the computer system is in a closed loop mode. *Technician B* says when the air-fuel ratio is constant during starting, the computer system is in an open loop mode. Who is correct?
a. A only c. Both A and B
b. B only d. Neither A nor B

18. *Technician A* says the PROM is a set of data that tells the fuel injector when to shut off. *Technician B* says the PROM is a speed sensor. Who is correct?
a. A only c. Both A and B
b. B only d. Neither A nor B

19. The air-fuel ratio is being held constant. *Technician A* says the system is in a closed loop mode. *Technician B* says the system is in an open loop mode. Who is correct?
a. A only c. Both A and B
b. B only d. Neither A nor B

20. *Technician A* says that the PROM is a set of programmable data for each specific car. *Technician B* says that the PROM is the same for all cars being manufactured. Who is correct?
a. A only c. Both A and B
b. B only d. Neither A nor B

21. *Technician A* says that handheld scanners can be connected to the ALCL. *Technician B* says that handheld scanners are used to check fuel pressures. Who is correct?
a. A only c. Both A and B
b. B only d. Neither A nor B

22. *Technician A* says that during closed loop operation, the oxygen sensor is not used as an input. *Technician B* says that open loop operation doesn't start until the engine reaches 150 degrees Fahrenheit. Who is correct?
a. A only c. Both A and B
b. B only d. Neither A nor B

23. *Technician A* says that the computer enriches the fuel during cold starting. *Technician B* says that the computer has no effect on the air-fuel ratio during closed loop operation. Who is correct?
a. A only c. Both A and B
b. B only d. Neither A nor B

24. *Technician A* says that a coolant sensor is considered a thermistor type of sensor. *Technician B* says that the magnetic pickup sensor is used to sense barometric pressure. Who is correct?
    a. A only
    b. B only
    c. Both A and B
    d. Neither A nor B

Essay

25. What is the difference between direct and indirect fuel injection?
26. What is the difference between port and throttle body injection?
27. Describe the difference between continuous and timed injection.
28. Describe the purpose and operation of the idle air control.
29. What is the purpose of the fuel rails?
30. Describe the purpose and operation of the mass airflow sensor.

Short Answer

31. The crankshaft position sensor and the vehicle speed sensor use a ________ pickup to produce the electronic signal.
32. When the fuel injectors are being pulsed independently from the distributor reference pulses, the system is in the ________ mode.
33. When the throttle is pushed to 80% of wide open during starting, the computer system is in the ________ mode.
34. The abbreviation PROM stands for ________ ________ ________ ________.
35. Often, when a problem occurs in a computerized fuel injection system, the computer goes into ________ ________.

Applied Academics

TITLE: Direct Fuel Injection

ACADEMIC SKILLS AREA: Language Arts

NATEF Automobile Technician Tasks:

The service technician uses computerized and print databases to obtain system information.

CONCEPT:

During the past several years, many automotive manufacturers have tested direct fuel injection systems. As mentioned in this chapter, direct fuel injection systems are used on diesel engines and inject the fuel slightly BTDC under extremely high pressures. However, the gasoline direct fuel injection systems that are being tested inject fuel directly into the cylinder at low pressures similar to port fuel injection. This can be done because of the time in which the direct fuel injection occurs.

There are times in the four-cycle events in which fuel can be directly injected into the cylinder under low pressure. Fuel can be injected into the cylinder directly only on the intake stroke. There is a vacuum inside the cylinder during the intake stroke. If injection takes place at this time, low-pressure direct fuel injection can be accomplished. The intake valve is only used to allow air into the cylinder. A computer precisely controls the fuel injection system. Direct fuel injection systems promise to improve fuel economy and to reduce emissions significantly in the future.

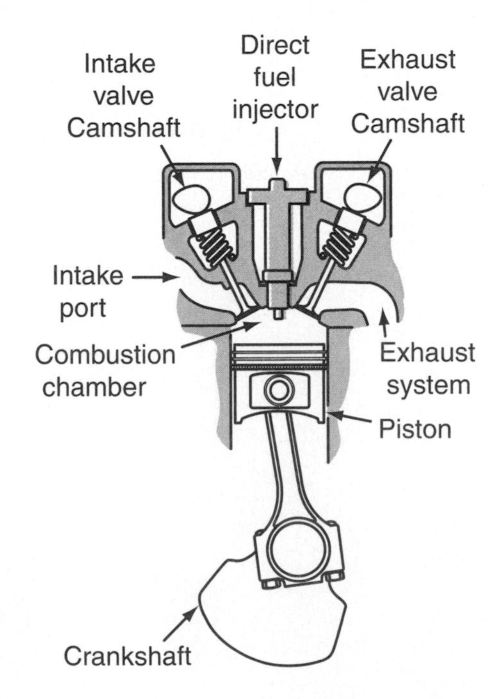

APPLICATION:

Search the Internet to find out more information about low-pressure direct fuel injection systems for gasoline engines. Go to Orbital Engine Corporation Limited at http://www.orbeng.com to find out more information about low-pressure direct fuel injection. What links can you find that will give more information on low-pressure direct fuel injection?

Air Intake and Exhaust Systems

CHAPTER 27

OBJECTIVES

After studying this chapter, you should be able to:

- Analyze the use and operation of air filter systems.
- Define the use and operation of intake manifolds.
- Define the operation of exhaust systems.
- Describe various problems, their diagnosis, and service procedures on the intake and exhaust systems.

INTRODUCTION

For an engine to operate correctly, air must flow into and out of the engine without restriction. Air intake and exhaust systems are designed to clean the air coming in and reduce the noise coming out. This chapter deals with the components and operation of the air intake and exhaust systems.

27.1 AIR INTAKE SYSTEMS

PURPOSE OF AIR FILTERS

The average gasoline engine brings in and exhausts approximately 10,000 gallons of air for every gallon of fuel consumed. Intake air must be clean. **Airborne** contaminants can shorten engine life or even cause premature failure. For

Certification Connection

ASE Connection: The information in this chapter can help you prepare for the National Institute for Automotive Service Excellence (ASE) certification tests. The tests and content areas most closely related to this chapter are:

Test A1—Engine Repair

- **Content Area**—Fuel, Electrical, Ignition, and Exhaust Systems Inspection and Service

Test A8—Engine Performance

- **Content Area**—Fuel, Air Induction, and Exhaust System Diagnosis and Repair

NATEF Connection: Much of the information in this chapter is related to the NATEF tasks. The NATEF tasks and priority numbers most closely related to this chapter are:

1. Inspect the integrity of the exhaust manifold, exhaust pipes muffler(s), catalytic converter(s), tailpipe(s), and heat shield(s); perform necessary action. P-2
2. Perform exhaust system backpressure test; determine necessary action. P-1
3. Diagnose emission and drivability problems resulting from malfunctions in the intake air temperature control system; determine necessary action. P-3
4. Inspect and test components of the intake air temperature control system; perform necessary action. P-3

example, dirt that gets into the engine causes excessive wear on the rings, pistons, bearings, and valves. Depending on the amount of dirt that gets in, engine life can be shortened by one-third to one-half.

As was indicated when studying fuel systems, the correct amount of air is very important for proper engine operation. Engines must breathe freely to provide maximum power. Air filters are used on engines to trap contaminants, yet provide a free flow of air into the engine. Air filters that are dirty and not replaced can cause large restrictions to the air. This condition will cause the engine to run excessively rich. Fuel mileage can be substantially reduced by a dirty air cleaner. Exhaust emissions will also be increased.

The type of dirt and contaminants that enter the engine are determined by how and where the engine is used. The most common contaminants are leaves, insects, exhaust soot, dust, and road dirt. The geographic location of the vehicle also affects the amount of airborne contaminants. In open country or areas with unpaved roads, conditions tend to be dusty; there may be more dirt than there is in a city or an area where the roads are paved. All of these variables determine the type of filter used and the change periods recommended.

DRY-TYPE AIR FILTERS

The dry-type air filter is made of a paper element. ***Figure 27–1*** shows an example of two shapes of dry-type air filters. These filters permit air to flow into the engine with little resistance. However, they trap and hold contaminants inside the paper. When the dry-type air filter becomes plugged with dirt, it is replaced with another filter.

There are several types of dry-type air filters. The light-duty paper-type air cleaner is shown in ***Figure 27–2***. These filters are generally used on passenger vehicles and small pickup trucks. The filter element is made of paper pleats. These filters are usually small because of space restrictions under the hood. The efficiency of this type of filter is nearly 98% for most driving conditions. Because of the high efficiency, the dry air filter has replaced the old oil bath air cleaner. The filter can be designed in many shapes to fit different types of air cleaners. The design depends on the type

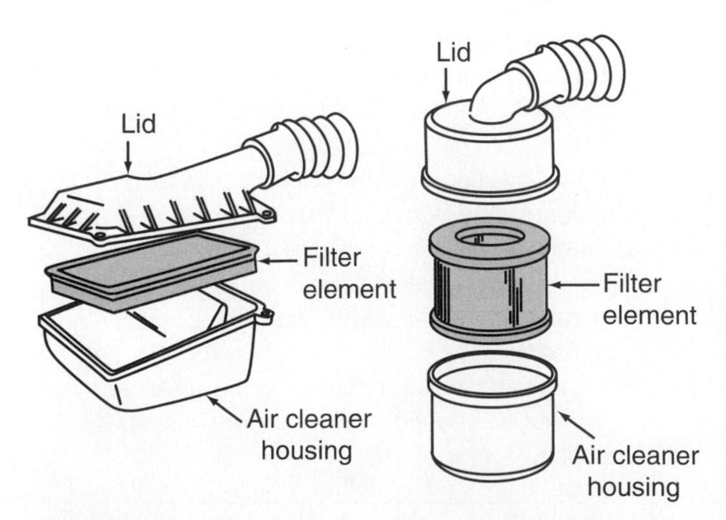

FIGURE 27–1 There are many styles and shapes of dry-type air filter elements.

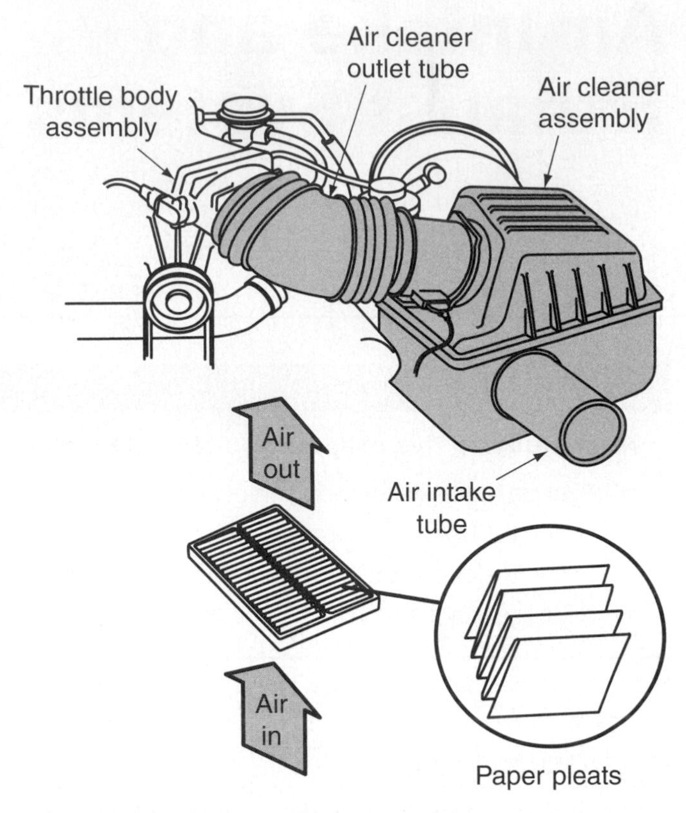

FIGURE 27–2 This air filter is made of pleated paper housed in a frame made of plastic or metal. It seals on the top and bottom.

Car Clinic: AIR FILTER

CUSTOMER CONCERN:
Air Intake Restriction

Customers often complain that their vehicles' engines are not running correctly. Technicians usually say that the problem may be in the air intake. What are the symptoms of an air intake restriction?

SOLUTION: One of the most common problems today in all automobiles is air intake restrictions. If the owner does not change the air filter at regular intervals, the dirty air filter restricts the air as it tries to enter the engine. Common symptoms of an air intake restriction are:

- The engine is hard to start.
- The engine has a loss of power.
- The coolant temperature decreases.
- The exhaust smoke increases in density and is black.
- The oil consumption and fuel usage increase significantly.

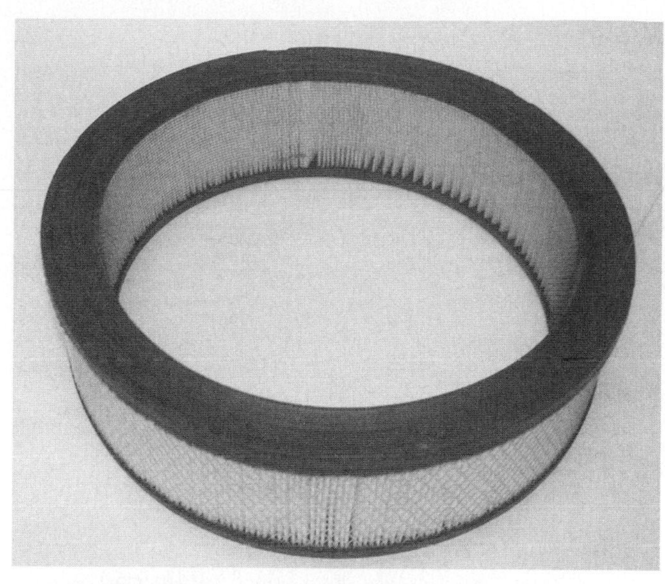

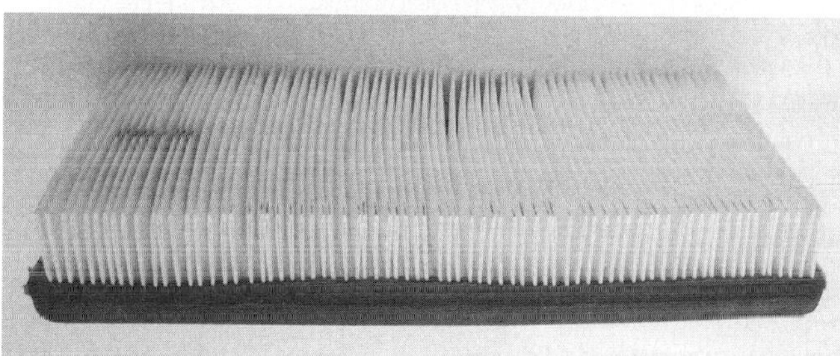

FIGURE 27-3 Here are two examples of dry-type, pleated air cleaner elements.

of ducting, housing used on the engine, and whether the engine has a carburetor or uses fuel injection. ***Figure 27–3*** shows two typical paper air filters.

A second type of light-duty air cleaner is called the polyurethane filter. This filter is sometimes called the foam filter (***Figure 27–4***). It is normally placed on the outer cover over a dry-type paper element. It consists of a polyurethane wrapper stretched over a metal support. The material has thousands of pores and interconnecting strands that create a mazelike contaminant trap. It may be used dry or with a thin coat of oil. In both cases, this filter has about the same efficiency as a paper-type element. The advantage to this filter is that it can be removed, cleaned, and reused. Many polyurethane filters are used as **aftermarket** equipment. Aftermarket equipment parts are sold to consumers by local parts dealers.

A series of special heavy-duty filters are used on certain equipment. Heavy-duty air cleaners are used in very dirty and contaminated areas. However, because automobiles are normally driven on paved roads, there is less call for heavy-duty air cleaners in cars.

PARTS LOCATOR

*Refer to photos #40 and #41 in the front of the textbook to see what **air cleaners** look like.*

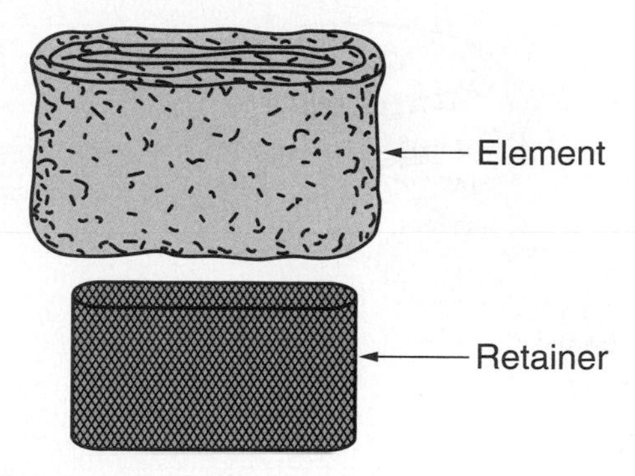

FIGURE 27-4 The polyurethane-wrapped filter stretches over a metal or platic support frame.

INTAKE DUCTING

On vehicles that have larger V8 engines, air cleaners are typically placed directly on top of the intake manifold. ***Figure 27–5*** shows such an example. The air is drawn into the fresh-air intake, through a temperature-controlled valve, and into the air cleaner. From the air cleaner, the air is drawn directly into the intake manifold.

When manufacturers started using fuel injection and other configurations on the automobile, the intake ducting was also changed. Intake ducting is also different on engines that are **turbocharged**. ***Figure 27–6*** shows two examples of different ducting. The air cleaners are light-duty paper elements. They are located for easy maintenance and service. The intake ducting has several clamps and rubber ducting to get the air into the engine.

PARTS LOCATOR

*Refer to photo #42 in the front of the textbook to see what **intake ducting** looks like.*

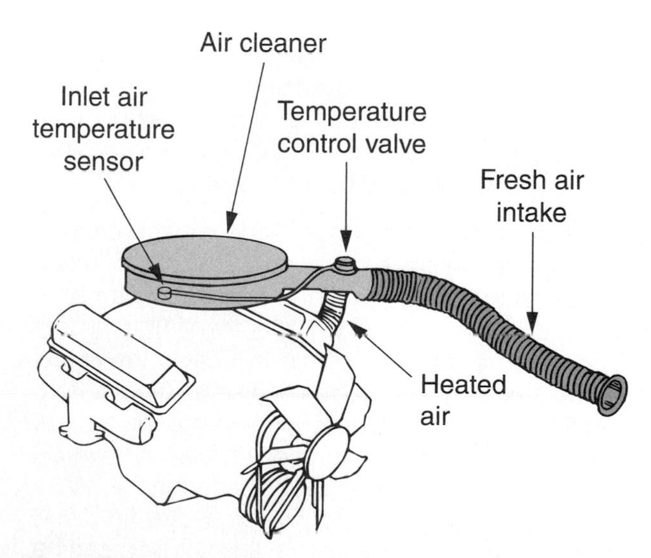

FIGURE 27-5 Intake ducting can take many forms. A simple type with the air cleaner mounted directly on top of the engine is shown here. (Courtesy of DaimlerChrysler Corporation)

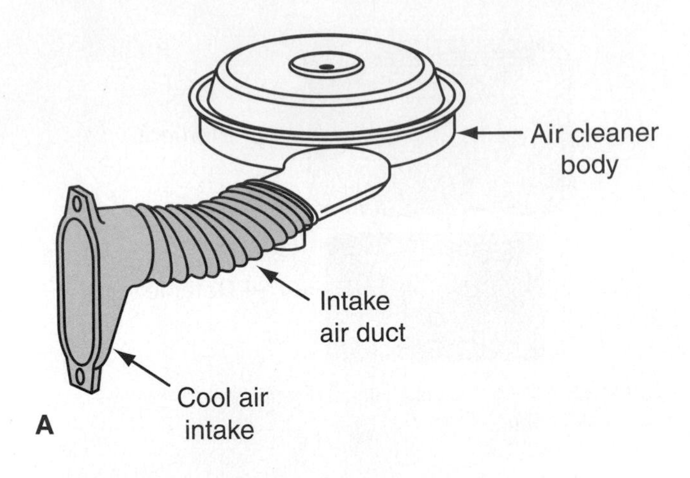

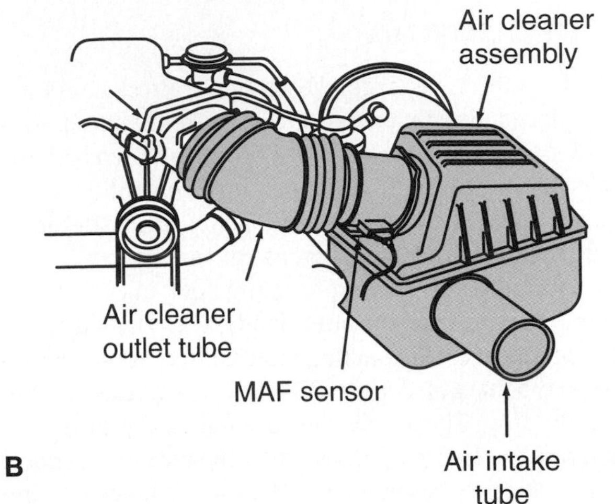

FIGURE 27–6 Intake ducting will change as different types of engines are designed. (A) The ducting for a large V configuration engine. (B) The ducting for a fuel-injected engine.

Car Clinic: INTAKE DUCTING

CUSTOMER CONCERN:
Loose Air Intake Connections

A customer complains that an engine has poor performance and that the Check Engine light often comes on. What could be the problem?

SOLUTION: Poor performance is often created by loose connections (clamps, etc.) on the air intake hoses and ducting. When this happens, extra air is being drawn into the engine after the MAP (manifold absolute pressure) sensor. In this case, the actual amount of air going into the engine is different from what the computer senses is going into the engine. The result is that an improper air-fuel ratio will be produced, causing poor engine performance. Make sure that all air ducting connections are tight and that there are no places where additional air can be drawn into the engine.

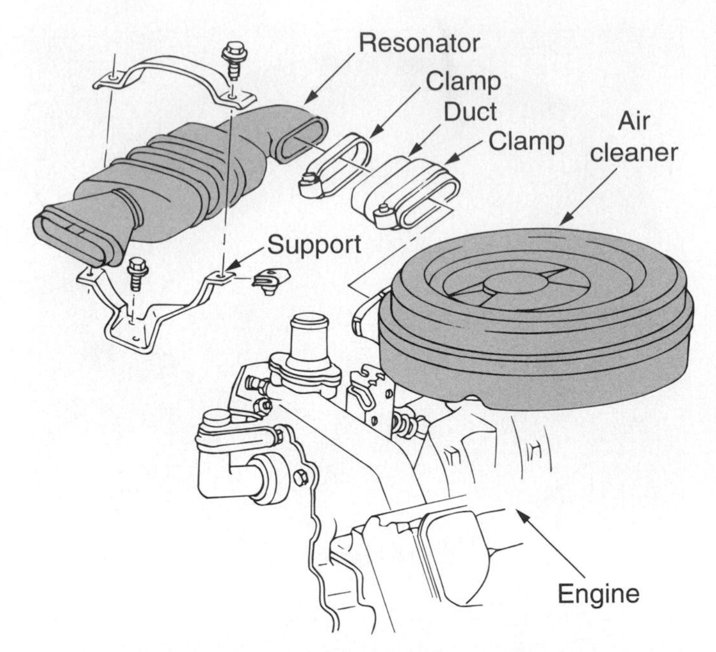

FIGURE 27–7 A resonator is used to reduce engine intake noise.

RESONATOR

Certain engine applications use a **resonator** on the intake ducting. ***Figure 27–7*** shows a resonator used in a diesel engine application. The purpose of the resonator is to reduce induction noise produced on the intake system.

INTAKE MANIFOLDS

An intake manifold is used to transfer or carry the air or fuel or both from the air cleaner to the intake valve. ***Figure 27–8*** shows a typical intake manifold for a V6 gasoline engine. On 4- and 6-cylinder engines, the intake and exhaust manifolds form an assembly.

The intake manifold is designed to deliver the right amount of air and fuel to each cylinder under all driving conditions. It would be best if all intake ports were the same

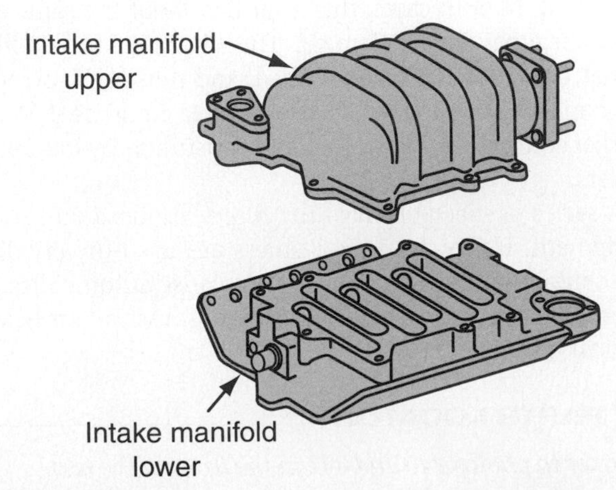

FIGURE 27–8 A typical V6 intake manifold for a gasoline engine is shown in this illustration.

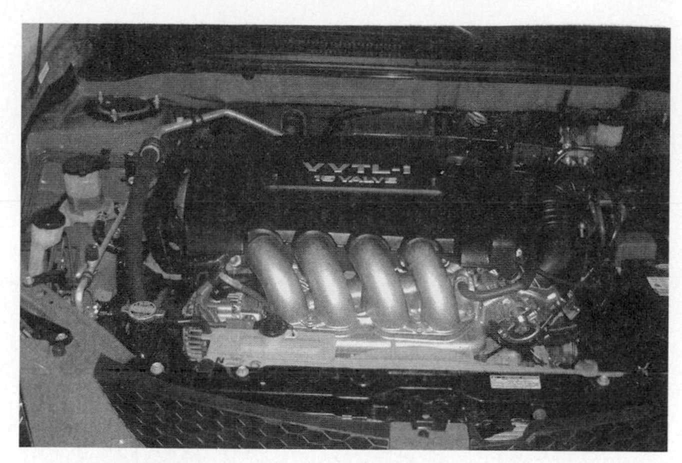

FIGURE 27–9 A 4-cylinder engine intake manifold with tuned ports is shown here.

length (tuned ports). However, on many engines this design is compromised to reduce the cost. The most common manifolds are made of a one-piece casting of either cast iron or aluminum. ***Figure 27–9*** shows an example of a 4-cylinder engine with a one-piece aluminum cast intake manifold. Note that the ports are all the same length (tuned ports) so that each cylinder gets the same amount of air and fuel.

On 4-cylinder engines, the intake manifold has either four **runners** or two runners that break into four near the intake manifold (***Figure 27–10***). On in-line 6-cylinder engines there are either six runners or three that branch off into six near the intake manifold. On V-configuration engines (V6 and V8), both open and closed intake manifolds are made. Open intake manifolds have an open space between the bottom of the manifold and the valve lifter valley. Closed intake manifolds act as the cover to the intake lifter valley.

Some intake manifolds are made of several pieces. The manifold shown in ***Figure 27–11*** is a two-piece manifold made of aluminum. This manifold was designed to increase the horsepower rating over older manifolds. The horsepower increase was achieved by designing the shape and size of the intake manifold to maximize airflow into the engine. Note also the position of various emission controls and computer sensors.

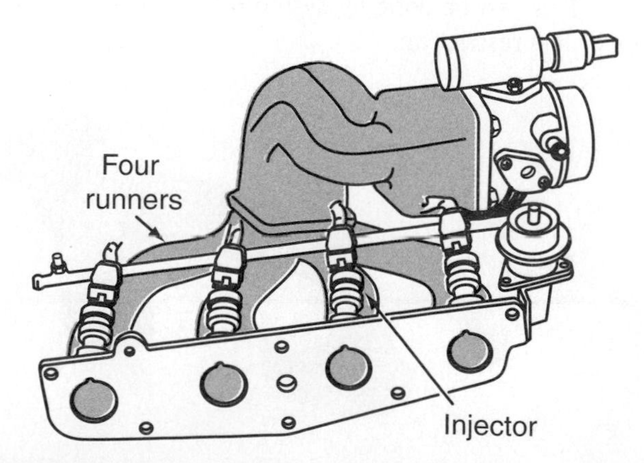

FIGURE 27–10 The intake manifold for an in-line, 4-cylinder engine can have two or four runners. Here two runners form into four runners.

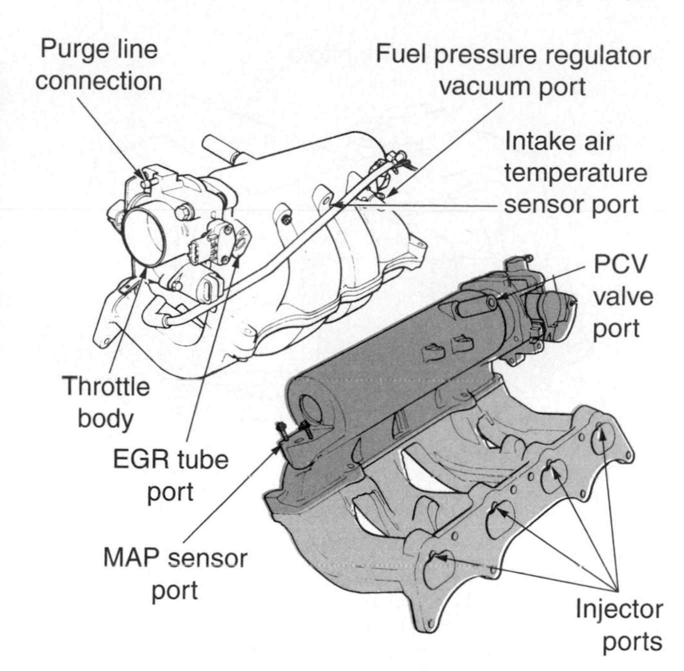

FIGURE 27–11 This two-piece intake manifold is designed to increase horsepower by maximizing the airflow through the manifold. (Courtesy of DaimlerChrysler Corporation)

PARTS LOCATOR

*Refer to photo #43 in the front of the textbook to see what an **intake manifold** looks like.*

WET AND DRY MANIFOLDS

Manifolds can also be either wet or dry. Wet manifolds have coolant passages cast directly into the manifold. Dry manifolds do not have coolant passages.

EXHAUST CROSSOVER INTAKE MANIFOLD

On some V6 and V8 type manifolds, there is an exhaust crossover passage. This passage allows the exhaust from one side of the engine to cross over through the intake manifold to the other side to be exhausted. The exhaust crossover provides heat to the base of the throttle body to improve the vaporization of the fuel while the engine is warming up (***Figure 27–12***). The crossover also reduces fuel icing.

INTAKE AIR TEMPERATURE (IAT) SENSOR

Most engines today use an intake air temperature (IAT) sensor. Its purpose is to sense, or measure, the temperature of the air entering the engine. The sensor is threaded into the

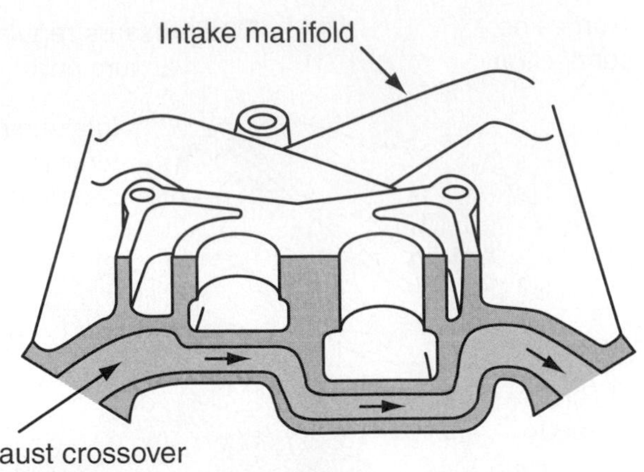

FIGURE 27–12 A crossover manifold is used to transfer the exhaust from one side of the engine to the other to be exhausted. The exhaust heat causes better vaporization of the fuel in the carburetor.

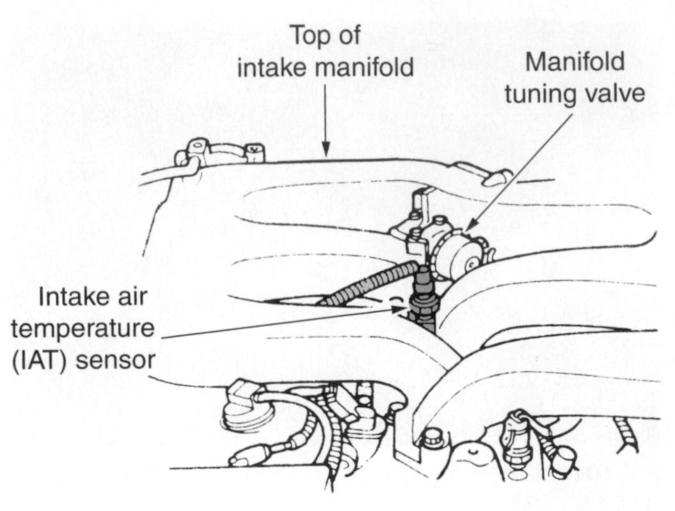

FIGURE 27–14 The IAT senses the air temperature before it enters the combustion process. (Courtesy of DaimlerChrysler Corporation)

top or side of the intake manifold. In some cases, the IAT is mounted on the air cleaner housing as shown in ***Figure 27–13***. The IAT acts as an input to the computer. The computer, such as the PCM, needs this data for calculations to adjust injector pulse width and spark advance. ***Figure 27–14*** shows a typical installation of the IAT. In this case, it is located on the top of the intake manifold.

The IAT sensor provides a voltage signal to the computer. The voltage varies according to the resistance of the sensor. In this case, the sensor resistance is inversely proportional to the air temperature. This means that as the temperature of the air input increases, the resistance of the sensor drops. Generally, the voltage produced by the IAT is between 0.5 volt and 5 volts.

PARTS LOCATOR

*Refer to photo #44 in the front of the textbook to see what an **intake air temperature (IAT) sensor** looks like.*

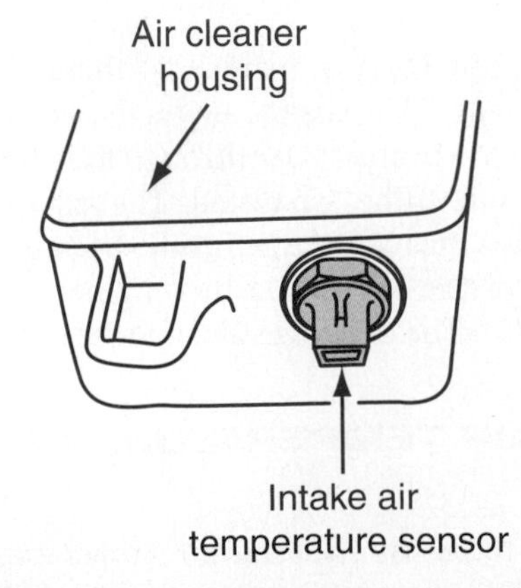

FIGURE 27–13 This intake air temperature (IAT) sensor is located on the air cleaner housing.

27.2 EXHAUST SYSTEMS

The exhaust system collects the high-temperature gases from each combustion chamber and sends them to the rear of the vehicle to be **dispersed**. An exhaust manifold, heat riser, mufflers, and pipes are used to accomplish this. A catalytic converter is also used.

EXHAUST MANIFOLD

The exhaust manifold is connected to the cylinder head of the engine. The exhaust gases from the exhaust valves pass directly into the exhaust manifold. The exhaust manifold is made of cast iron or steel piping that can withstand rapid increases in temperature and expansion (***Figure 27–15***). Under full-load conditions the exhaust manifold may be red hot, yet cold water from the road can be splashed on the manifold while driving.

Volumetric efficiency was defined in an earlier chapter as the efficiency of air moving in and out of the engine. Exhaust manifolds can be designed to improve volumetric efficiency. This can be done by designing the manifolds with more or less restriction.

FIGURE 27–15 This is one of two exhaust manifolds used on a V8 engine.

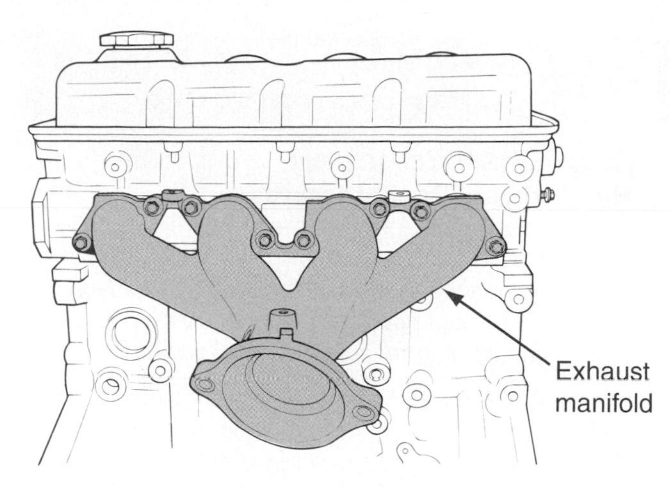

FIGURE 27–16 This is an example of a four-runner exhaust manifold. (Courtesy of DaimlerChrysler Corporation)

Several types of exhaust manifolds are used. Four-cylinder engines use either three- or four-runner manifolds. On three-runner manifolds, the two center cylinders feed one runner. Four-runner manifolds have a runner for each cylinder, and provide better volumetric efficiency. The four-runner exhaust manifold shown in ***Figure 27–16*** is made of nodular cast iron (cast iron that has had certain chemicals or minerals added to the metal to increase its strength). This exhaust manifold is encased by a double-layered heat deflection shield. On 6-cylinder engines, the exhaust manifold is either a four or six runner. Again, the four-runner manifold has two of the center cylinders feeding one runner. On V6 and V8 engines, there is an exhaust manifold on each side.

PARTS LOCATOR

*Refer to photo #45 in the front of the textbook to see what **exhaust manifolds** look like.*

Headers may be used on some original equipment high-performance engines. Headers are welded steel tubing used for exhaust. They are designed to allow a smooth, even flow of exhaust gases out of the engine. This smooth flow ensures that each cylinder has equal exhaust backpressure. It also ensures that each cylinder is completely cleaned of exhaust (scavenged). It has been found that headers improve high-speed and load performance. They have a lesser effect on normal driving performance. ***Figure 27–17*** shows a set of tuned headers used on some high-performance engines. Note that because of the design and layout of the exhaust tubes, the exhaust from one cylinder will not mix with another until it meets at the connection to the muffler.

HEAT RISER

On many older engines, a type of heat riser is attached to the exhaust manifold. The heat riser is a valve. Its purpose is to restrict the exhaust gases during starting and warm-up periods. This restriction tends to raise the engine to operating temperature more quickly, which aids in the vaporization of fuel. On in-line engines, the heat riser also helps to vaporize the fuel during cold starting.

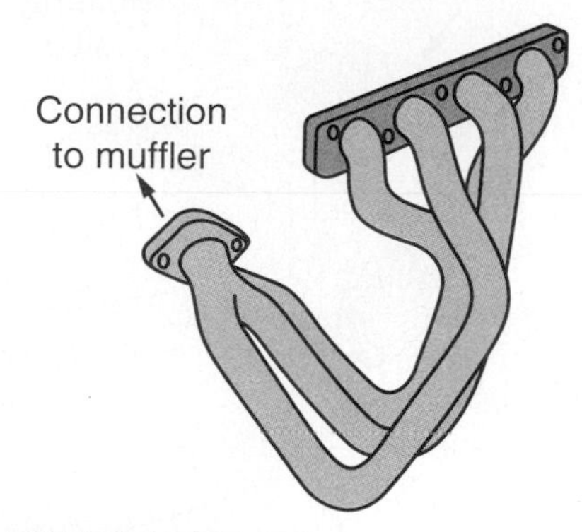

FIGURE 27–17 This exhaust arrangement has tuned headers. In this type of exhaust system, the exhaust from each cylinder is not mixed with the others until they reach the connection to the muffler.

The heat riser is controlled by a bimetal spring. When the engine is cold, the spring and a counterweight cause a valve in the exhaust manifold to close. As the spring heats up, it relaxes. This causes the counterweight to open the valve and allow normal exhaust. ***Figure 27–18*** shows how the exhaust gases are routed to improve vaporization.

EXHAUST PIPING

The exhaust pipe is the connecting pipe between the exhaust manifold and the muffler or catalytic converter. Many types of exhaust piping are used on vehicles. The shape depends on the configuration of the engine, size of the engine, and undercarriage of the car. Exhaust piping can also be single or dual design.

EXHAUST MUFFLER

The muffler is used to dampen the exhaust sound of the engine. Two types are generally used. One uses a series of baffled chambers to reduce the sound. The other uses a

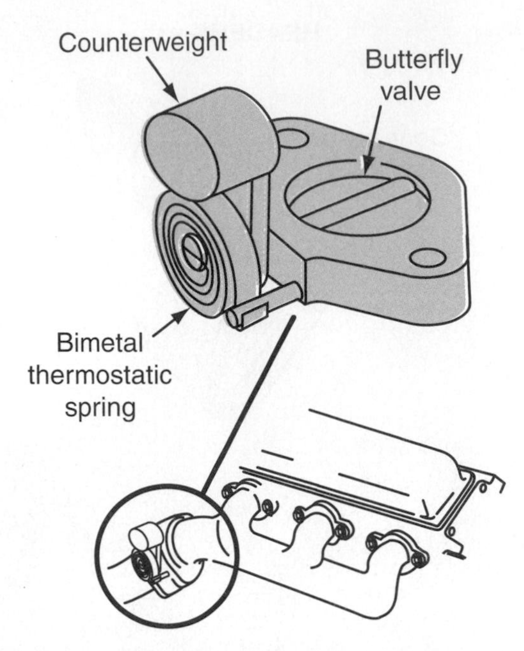

FIGURE 27-18 A heat riser is used to increase vaporization of the fuel during starting. During cold starting, exhaust gases are routed internally to increase the temperature at the base of the carburetor.

perforated straight pipe enclosed in fiberglass and a shell. The straight pipe, which is also called a "glass pack," reduces exhaust backpressure, but it does not reduce the sound as much as the baffle type. Glass packs are illegal in most states because they alter emissions considerably.

RESONATOR

A resonator is another type of muffler. Most of the noise from an exhaust system is sound vibration. These vibrations cause louder noise. Resonators provide additional sound protection at critical points in the exhaust flow. They are used to absorb excessive sound vibration.

PARTS LOCATOR

*Refer to photo #46 in the front of the textbook to see what an **exhaust resonator** looks like.*

TAILPIPE

The tailpipe is a tube that is used to carry the exhaust gases from the muffler or resonator to the rear of the automobile. Many shapes and sizes are used, depending on the vehicle. The tailpipe is supported by a series of hangers that allow the exhaust system to flex and move during driving. Rubber connectors help isolate vibration from the rest of the vehicle.

TOTAL EXHAUST SYSTEM

Figure 27–19 shows a total exhaust system. Many of these parts, such as the catalytic converter and oxygen sensors,

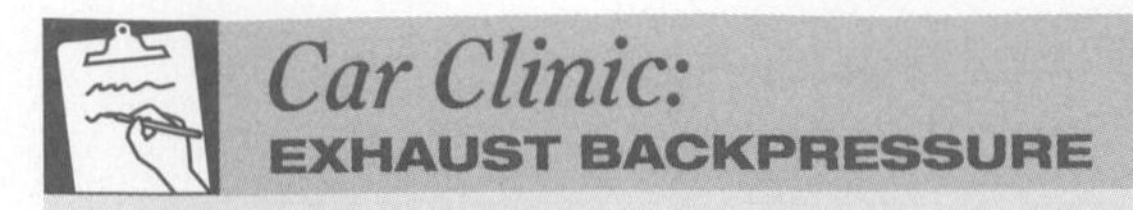

Car Clinic: EXHAUST BACKPRESSURE

CUSTOMER CONCERN:
High Exhaust Backpressure
A customer owns a car with a diesel engine. Recently, an automotive technician suggested that the rough idle noticed by the driver could be caused by high exhaust backpressure. What are the effects of high exhaust backpressure on diesel engines?

SOLUTION: High exhaust backpressure is developed by some type of restriction in the exhaust system. A rusted muffler, plugged tailpipe, or bent exhaust system may have caused the high backpressure. The symptoms of high backpressure include:

1. Higher engine temperature
2. A decrease in power
3. Denser exhaust smoke
4. Rough idle
5. Excessive carbon buildup on the valves, injectors, and pistons
6. Contaminated oil

If these symptoms appear, the exhaust system should be inspected and repaired or replaced.

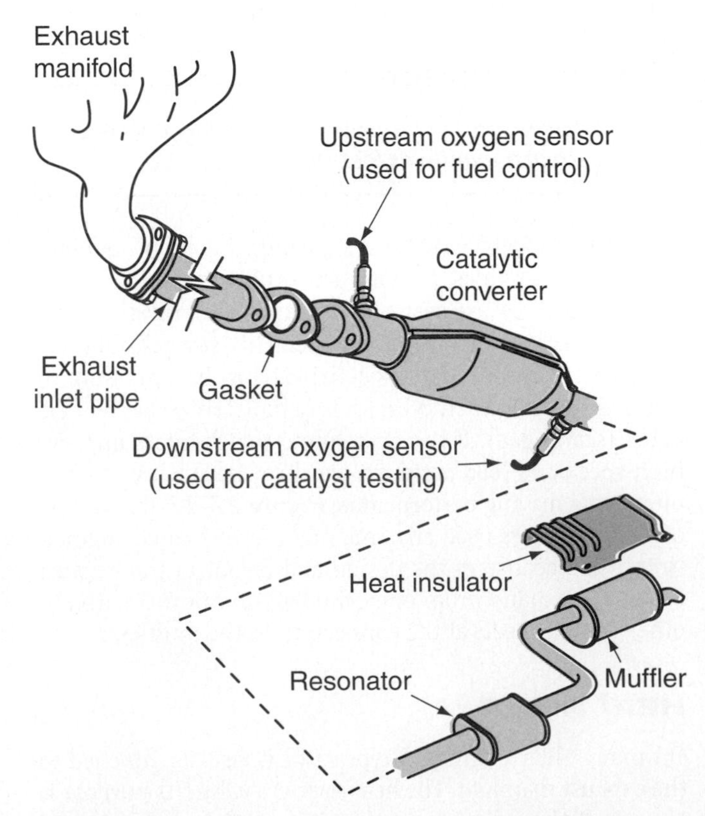

FIGURE 27-19 A complete exhaust system is shown in this illustration.

are introduced in other chapters. The basic parts of the exhaust system include:

- Exhaust inlet pipe
- Upstream oxygen sensor
- Catalytic converter
- Downstream oxygen sensor
- Muffler inlet pipe and resonator
- Muffler
- Heat insulators

These major parts are connected together using gaskets, exhaust system hangers and insulators, bolts and nuts, and attached pipes. Although each manufacturer will have a different exhaust system configuration, the basic parts will remain the same.

Problems, Diagnosis, and Service

Safety Precautions

1. Never replace an air cleaner when the engine is running.
2. Always wear OSHA-approved safety glasses when working on the exhaust or air intake system.
3. When checking air intake ducting, especially near the exhaust manifold (heated air for intake), be careful not to touch the hot manifold and burn your fingers.
4. When working on the exhaust system, parts may be very hot. Touching them can cause serious injury. Be especially careful with the exhaust systems used on turbocharged engines.
5. Never smoke in or around an automobile service shop.
6. Always keep an approved fire extinguisher nearby when working on the air cleaner or fuel system.
7. Be careful not to cut your hands on the sharp edges of the air intake housing and ductwork.

PROBLEM: Dirty Air Cleaner

An engine is losing power slightly, and excessive black smoke is coming from the exhaust.

DIAGNOSIS

Excessive black smoke may indicate a dirty air cleaner. A dirty air cleaner restricts the amount of air that can enter the engine. Less air creates a very rich air-fuel ratio. The rich mixture produces excessive black smoke.

SERVICE

Use the following service procedures and suggestions when working with the air cleaner and intake ducting.

1. Depending on the driving conditions, replace dirty air cleaners at least at every tune-up.
2. Remove the air cleaner cover by removing the wing nuts or other housing.
3. Remove the air cleaner element.
4. Clean the air cleaner housing when replacing the filter. All dirt and grease should be removed from the housing using a solvent that is not flammable.
5. If you have a polyurethane filter around the outside of the air cleaner, use soap and water to clean the filter. Be sure to dry the polyurethane filter before installation.
6. Use a light coat of oil on the polyurethane filter. The light coat of oil will help to capture more of the dirt and dust particles that flow through the air cleaner element.
7. While the air cleaner is removed, inspect all intake ducting for cracks or leaks and bad seals and gaskets, and replace where necessary. Any leak that occurs in the air inlet may also bring in dirty, unfiltered air. The dirty air can damage the engine.
8. Always use the correct filter suggested by the manufacturer.
9. Replace the filter in the same position as the old filter.
10. Tighten all wing nuts and other fasteners securely when replacing the air cleaner. Be careful not to tighten the wing nut too much, as the ductwork could be damaged.

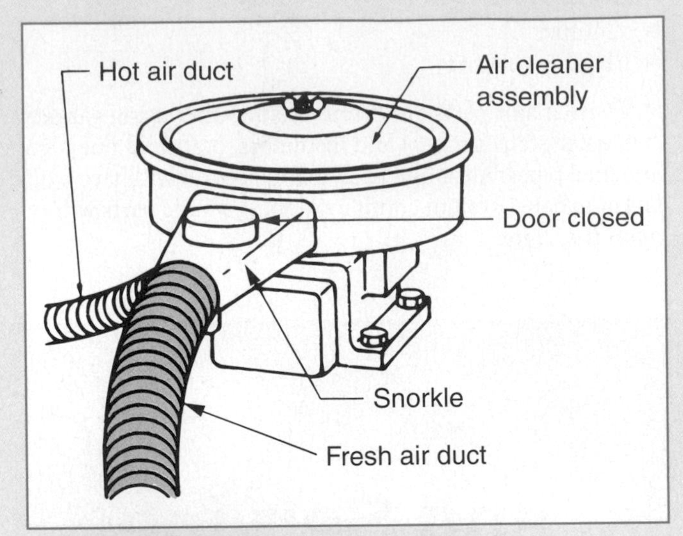

FIGURE 27-20 The fresh air duct should be checked for cracks. This is done by removing the duct and slightly bending it to observe cracks and damaged sections.

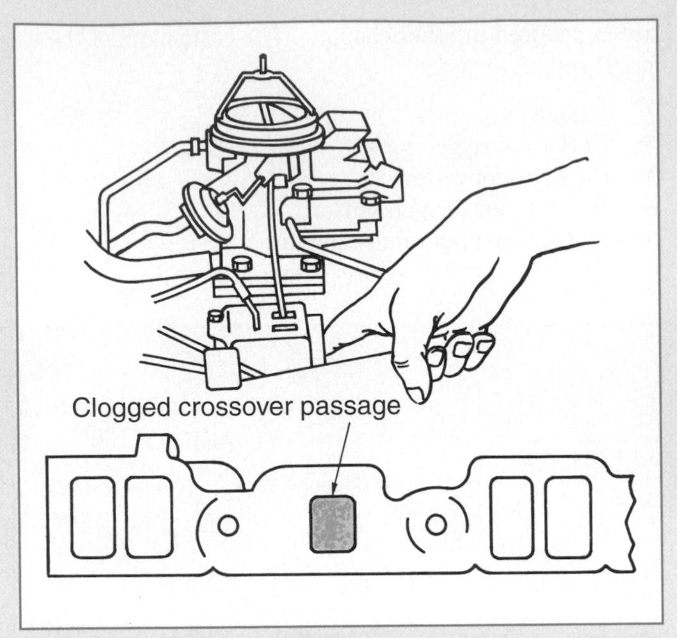

FIGURE 27-21 The exhaust crossover passage may become plugged. To check it, touch your finger to the intake manifold where the crossover passage is located. If it is cold when the engine is hot, the passage is plugged.

PROBLEM: Damaged Fresh Air Duct

An engine is hard to start. In addition, there is hesitation and poor drivability.

DIAGNOSIS

The fresh air duct shown in ***Figure 27–20*** is designed to deliver cool air to the engine when the air under the hood is hot. Bringing air that is too hot into the engine may cause starting or drivability problems such as vapor lock or hesitation.

SERVICE

1. Check the fresh air duct by unclamping and removing it. Bend the duct slightly to inspect between the bellows for tears or rips.
2. Make sure clamps are not damaged and that they seal the fresh air duct securely.
3. If the duct appears to be torn or ripped, replace as necessary.

PROBLEM: Plugged Crossover Passage

An engine hesitates and has poor fuel economy.

DIAGNOSIS

These two problems suggest that the intake manifold crossover passage may be plugged. During normal operation, this crossover passage has air and fuel passing around it. The exhaust heat nearby aids in fuel vaporization. However, the passage can become plugged because of being too cool. This causes carbon to build up significantly in the passageway. To check for blockage:

1. Let the engine idle for several minutes.
2. Carefully touch the manifold crossover area with your finger.
3. If the manifold is relatively cool, the passage is probably plugged. If it is not plugged, the crossover area should be hot.

CAUTION: *Be careful not to touch the manifold too long because it may burn your fingers.*

Figure 27–21 shows an example of a service technician touching the crossover passage area.

SERVICE

If the passage is plugged, use the following general procedure to remove the intake manifold and clean the passageway. The exact procedure will depend on the engine and manufacturer.

1. Remove all linkages and electrical connections to the carburetor or throttle body injection system. Mark and label accordingly.
2. Remove the bolts holding the intake manifold to the block.
3. Once removed, check for carbon buildup in the crossover passageway. Clean the carbon out of the passageway using a wire brush or other suitable tool.
4. Remove and clean all gasket material on both the block and the intake manifold.
5. Replace all intake manifold gaskets with new ones.
6. Carefully place the intake manifold gaskets on the block.
7. Secure all bolts and torque to specifications.
8. Replace all linkages and wires.

PROBLEM: Exhaust Restriction

An engine has a loss of power, and an exhaust sound is heard near the engine.

DIAGNOSIS

Often a loss of power will occur when the exhaust is restricted. The restriction can be caused when a vehicle hits a curb (or other object) bending the exhaust pipe inward. The effect is that the exhaust is restricted and cannot be removed from the engine adequately. In addition, various cracks may form around and near the exhaust manifold. Depending on the type of exhaust manifold, either a weld may have broken or the exhaust manifold may have cracked. Check the following when checking for cracks and for restrictions in the exhaust system:

CAUTION: *Never run an engine in an unventilated area. When in a building, always connect the exhaust pipe to the building exhaust system to remove all gases.*

1. Check the exhaust manifold for cracks. Cracks generally occur around each runner on the exhaust manifold.
2. Check the heat riser for movement. The heat valve should be able to move freely. On some engines, the heat riser has rusted closed and may cause a restriction, longer warm-ups, and hesitation when cold.
3. Check the exhaust manifolds and exhaust pipes for any leaks or dents in the system that may restrict the flow of gases.
4. Exhaust restrictions can be found by checking the exhaust backpressure. Each manufacturer has a different procedure for checking exhaust backpressure. Generally, a gauge is placed on the exhaust system near the exhaust manifold. Exhaust backpressure is then measured. Typically, if the exhaust backpressure is above 2 psi, an exhaust restriction is indicated. Of course, the backpressure specifications will be different for each type of vehicle and manufacturer. ***Figure 27–22*** shows one manufacturer's suggested diagnosis and test procedure for an exhaust restriction.

PINPOINT TEST A: RESTRICTED EXHAUST SYSTEM TEST

TEST STEP	RESULT	▶ ACTION TO TAKE
A1 PERFORM VACUUM TEST		
• Attach Vacuum/Pressure Tester 164–RO253 or equivalent to the intake manifold vacuum source. • Connect Rotunda 88 Digital Multimeter 105–R0053 or equivalent to read engine speed (rpm). • Set the parking brake.	Yes	▶ No restriction in the exhaust system. GO to the Diagnostic Routines, Section 4A (1.9L) in the Powertrain Control/Emissions Diagnosis Manual.
• Start the engine and gradually increase the engine speed to 2,000 rpm with the transaxle in NEUTRAL (N) M/T, PARK (P) on A/T. • **Is the manifold vacuum above 406.4 mm-Hg (16 in-Hg)?**	No	▶ GO to **A2**.
A2 PERFORM VACUUM TEST—EXHAUST DISCONNECTED		
• Turn the engine OFF.	Yes	▶ GO to **A3**.
• Disconnect the exhaust system at the exhaust manifold. • Repeat test Step A1. • **Is the manifold vacuum above 406.4 mm-Hg (16 in-Hg)?**	No	▶ GO to **A4**.
A3 PERFORM VACUUM TEST—THREE-WAY CATALYTIC CONVERTER ON/MUFFLER OFF		
• Turn the engine OFF.	Yes	▶ REPLACE the muffler.
• Reconnect the exhaust system at the exhaust manifold. • Disconnect the muffler. • Repeat test Step A1. • **Is the manifold vacuum above 406.4 mm-Hg (16 in-Hg)?**	No	▶ REPLACE the Three Way Catalytic Converter and INSPECT the muffler to be sure TWC debris has not entered the muffler.
A4 CHECK EXHAUST MANIFOLD RESTRICTION		
• Remove the exhaust manifold. Inspect the ports for casting flash by dropping a length of chain into each port. NOTE: Do not use a wire or light to check ports. The restriction may be large enough for them to pass through but small enough	Yes	▶ GO to the Diagnostic Routines, Section 4A (1.9L) in the Powertrain Control/Emissions Diagnosis Manual.
to cause excessive backpressure at high engine rpm. • **Is the exhaust manifold free of casting flash?**	No	▶ REPLACE the exhaust manifold.

FIGURE 27-22 This series of tests will help the service technician diagnose and test an exhaust system.

SERVICE

When servicing the exhaust system, use the following general procedures. Specific procedures will vary depending on the vehicle and the manufacturer.

CAUTION: *Never work on an exhaust system that is hot. Always wait until the engine has cooled to surrounding temperature.*

1. To remove the exhaust manifold, remove the wire attached to the oxygen sensor, if used.
2. Remove the bolts holding the exhaust manifold to the cylinder head.
3. Remove the exhaust manifold.
4. Check the entire exhaust manifold for cracks.
5. Check the exhaust manifold gaskets for breaks.
6. If there is a crack in the manifold, repair and replace as necessary.
7. Check the exhaust pipe for cracks, leaks, and dents.
8. If the exhaust pipe is damaged, replace as necessary.
9. Check and replace any tail pipe or exhaust pipes that are broken or that leak.
10. Many exhaust systems are supported by free-hanging rubber mountings. This arrangement permits some movement of the exhaust system, but it does not permit transfer of noise into the passenger compartment. Annoying rattles and noise vibrations in the exhaust system are usually caused by misalignment of parts. Loosen all bolts and nuts. Then realign the exhaust system parts. Working from the front to the rear of the engine, tighten each part.

Service Manual Connection

There are many important specifications to keep in mind when working with air induction and exhaust systems. To identify the specifications for your engine, you will need to know the vehicle identification number (VIN) of the vehicle, the type and year of the vehicle, and the type of engine. Although they may be titled differently, some of the more common air intake and exhaust system specifications (not all) found in service manuals are listed below. Note that these specifications are typical examples. Each vehicle and engine may have different specifications.

Common Specification	Typical Example
Exhaust backpressure	2 psi
Intake air temperature voltage	0.5–5 volts
Exhaust clamp nuts	25–35 ft.-lb
Catalytic converter to muffler nuts	30–41 ft.-lb
Intake manifold vacuum	406.4 mm mercury

In addition, the service manual will give specific directions for various service and testing procedures. Some of the more common procedures include:

- Muffler removal and replacement
- Catalytic converter replacement
- Exhaust cleaning and inspection
- Replacement of exhaust hanger insulators
- Exhaust system alignment
- Exhaust shield inspection

SUMMARY

The following statements will help to summarize this chapter:

- The air intake and exhaust system includes the air filters, intake ducting, exhaust manifolds, and heat risers, mufflers, and heat insulators.
- Air filters are designed to keep dirt and other contaminants out of the engine.
- Most air cleaners are about 98–99% efficient.
- Both dry-type and polyurethane-type filters are used today.
- The intake ducting is used to transfer the air from outside the engine to the filter and finally to the engine.
- Some intake ducting systems include a resonator to reduce the intake noise.
- The intake manifold is used to transfer the air and fuel to the intake valves.
- Both wet and dry intake manifolds are used on engines today.
- The exhaust system includes the exhaust pipes, the exhaust resonator, the catalytic converter, and the muffler.

TERMS TO KNOW

Can you explain each of the following terms? Review the chapter until you can use each term correctly.

Aftermarket
Airborne
Dispersed
Headers
Resonator
Runners
Turbocharged

REVIEW QUESTIONS

Multiple Choice

1. If dirt gets into the engine, the engine life may be:
a. Increased by one-half
b. Decreased by one-third
c. Affected very little—nothing to worry about
d. Increased by one-third
e. None of the above

2. A dirty air cleaner will produce:
a. White exhaust smoke
b. More power in the engine
c. Black smoke
d. A rich mixture
e. Both c and d

3. One of the disadvantages of a paper-type air cleaner is that:
a. It must be cleaned with oil
b. It must be replaced and cannot be cleaned
c. It usually doesn't fit correctly
d. It must be replaced every 2,000 miles
e. It cannot be removed

4. Which of the following is a type of dry air cleaner?
a. Paper
b. Polyurethane
c. Runner
d. Timed
e. Both a and b

5. Which type of air cleaner can be cleaned?
a. Paper
b. Polyurethane
c. Runner
d. Timed
e. Both a and b

6. Which of the following is used primarily to reduce noise on the intake system?
a. Intake manifold
b. Intake ducting
c. Intake resonator
d. Intake filter
e. Catalytic converter

7. How many runners may an intake manifold for a 4-cylinder engine have?
a. One
b. Two
c. Three
d. Four
e. Both b and d

8. Steel tubes of the same length that are welded into an exhaust manifold are called:
a. Turbochargers
b. Headers
c. Exhaust resonators
d. All of the above
e. None of the above

9. Which device is used to block off the flow of exhaust gases to improve vaporization during cold starting?
a. Wastegate
b. Blower
c. Turbocharger
d. Intake actuator
e. Heat riser

10. After approximately how many miles should the air cleaner be replaced (under normal driving conditions)?
a. 4,000
b. 10,000
c. 50,000
d. 80,000
e. At every tune-up

The following questions are similar in format to ASE (Automotive Service Excellence) test questions.

11. *Technician A* says that an exhaust restriction can be checked by testing for exhaust backpressure. *Technician B* says that an exhaust restriction can be checked by observing the timing. Who is correct?
a. A only
b. B only
c. Both A and B
d. Neither A nor B

12. *Technician A* says that all air cleaners can be cleaned with gas and replaced. *Technician B* says that there are no air cleaners on older or newer vehicles that can be cleaned. Who is correct?

a. A only
b. B only
c. Both A and B
d. Neither A nor B

13. An engine is lacking power. *Technician A* says it could be caused by a dirty air cleaner. *Technician B* says it could be caused by a restriction in the air intake system. Who is correct?

a. A only
b. B only
c. Both A and B
d. Neither A nor B

Essay

14. What is the difference between a wet- and dry-type manifold?

15. Explain the operation of the intake air temperature (IAT) sensor.

16. Describe the purpose of the resonator used on an exhaust system.

Short Answer

17. When the exhaust is producing black smoke, this may be an indication of a dirty ________.

18. If the intake manifold crossover passage is plugged, it will cause the engine to ________.

19. An exhaust restriction can be found by checking and measuring the exhaust ________.

20. As the air temperature increases on the IAT, the voltage signal to the PCM ________.

21. Exhaust manifolds can be made of cast iron or ________ piping.

Applied Academics

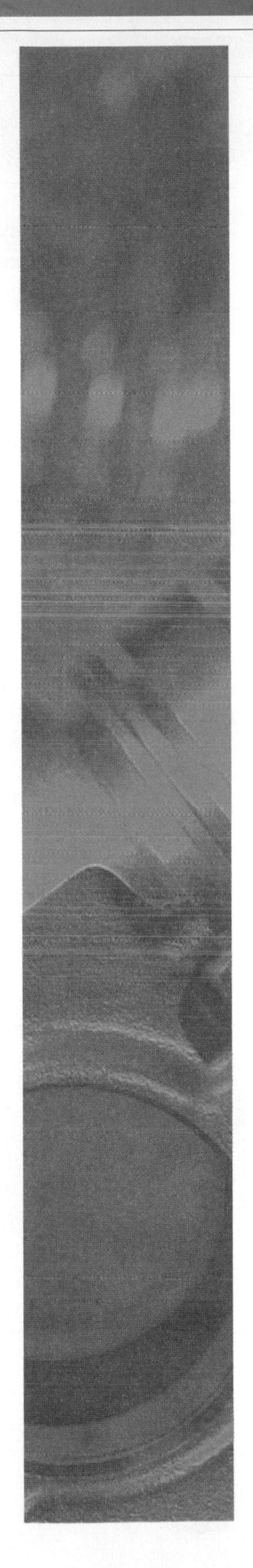

TITLE: Intake Air Temperature Sensor

ACADEMIC SKILLS AREA: Science

NATEF Automobile Technician Tasks:

The technician can describe the role of thermistors in controlling engine performance and how ECT and IAT sensors modify powertrain control module (PCM) outputs.

CONCEPT:

The intake air temperature (IAT) sensor is used on engines today to provide an input to the PCM (powertrain control module) for fuel management. The IAT is called a thermistor. A thermistor is an electrical sensor that changes its resistance with any corresponding changes in the air temperature surrounding it. The IAT is designed so that as the temperature of the intake air increases, its resistance value in ohms decreases. This is shown in the accompanying table. Note that as the air temperature increases, the resistance of the IAT decreases. As the resistance value changes, the PCM senses a corresponding voltage drop. The voltage drop is one of many inputs that tells the PCM how to control its output.

MANIFOLD/INTAKE AIR TEMPERATURE SENSOR TEMPERATURE VERSUS RESISTANCE AND VOLTAGE DROP (APPROXIMATE)

°C	°F	OHMS	VOLTAGE DROP ACROSS SENSOR
-40	-40	100,000	4.95
-8	+18	15,000	4.68
0	32	9,400	4.52
10	50	5,700	4.25
20	68	3,500	3.89
30	86	2,200	3.46
40	104	1,500	2.97
50	122	1,000	2.47
60	140	700	2.00
70	158	500	1.59
80	176	300	1.25
90	194	250	0.97
100	212	200	0.75

APPLICATION:

Can you identify why the voltage drops with each corresponding drop in resistance in the IAT? If the resistance drops with an increase in temperature, does the current flowing through the IAT increase or decrease? How does the current that is flowing through the IAT change the voltage drop?

CHAPTER 28

Turbochargers and Supercharging Systems

OBJECTIVES

After studying this chapter, you should be able to:

- State the definition of supercharging.
- Identify the parts and operation of the mechanical Roots-type blower.
- Identify the parts and operation of a turbocharger.
- Identify common problems associated with turbochargers.

INTRODUCTION

Many automotive engines today use superchargers to increase the amount of air going into the engine. Supercharging helps to increase the horsepower characteristics of the engine. The most common type of supercharger is called a turbocharger. In addition, on some racing engines, mechanical Roots-type blowers are used to increase the amount of air going into the engine. This chapter introduces the principles of supercharging and presents the basic operation of turbochargers and mechanical blowers.

28.1 SUPERCHARGING

With the increased use of fuel injection and computer-controlled combustion systems, turbochargers have become common components of gasoline and diesel engines. In the past, only large engines had turbochargers. Today, with the precise control afforded by computers, turbochargers are making smaller engines more efficient and capable of producing more power. In order to study turbochargers, supercharging must first be defined.

SUPERCHARGING DEFINED

When the piston moves downward on the intake stroke, a vacuum is created. This vacuum causes air and fuel to be drawn into the engine. This design is called a **naturally aspirated**, or normally aspirated, engine. The amount of air entering the engine is based on atmospheric pressure. Most engines are considered to be naturally aspirated. However, in smaller engines that are naturally aspirated, there may be a lack of power for certain driving conditions.

To overcome this lack of power, an engine can be supercharged. **Supercharging** an engine means delivering a greater volume of air to the cylinders than is delivered from

Certification Connection

ASE Connection: The information in this chapter can help you prepare for the National Institute for Automotive Service Excellence (ASE) certification tests. The tests and content areas most closely related to this chapter are:

Test A1—Engine Repair

- **Content Area**—Fuel, Electrical, Ignition, and Exhaust Systems Inspection and Service

Test A8—Engine Performance

- **Content Area**—Fuel, Air Induction, and Exhaust System Diagnosis and Repair

NATEF Connection: Much of the information in this chapter is related to the NATEF tasks. The NATEF tasks and priority numbers most closely related to this chapter are:

1. Test the operation of turbocharger/supercharger systems; determine necessary action. P-3

the suction of the pistons alone. Then the engine is not naturally aspirated; it is supercharged. When more air is forced into the cylinders, there must be a corresponding increase in fuel to maintain a 14.7 to 1 air-fuel ratio. If both these conditions occur, then an increase in power will result. In some cases, up to 50% more power can be obtained by supercharging an engine.

BLOWER

Either a blower or turbocharger can be used to supercharge an engine. A blower is a mechanical air pump that forces air into the engine. It is driven by a set of gears or belts from the crankshaft. It produces a substantial frictional loss on the engine because it requires horsepower from the engine to operate.

ROOTS-TYPE BLOWER

There are several types of mechanical blowers used on engines. One of the most common, named after its inventor, is called the Roots-type blower. This blower is found on some heavy-duty diesel engines and many high-performance and racing engines. Its purpose is to force large amounts of air into the engine. Then, with a corresponding increase in fuel, the engine has increased its horsepower. ***Figure 28–1*** shows an example of a Roots-type blower on a diesel V6 engine. On V6 and V8 engines, it is usually located between the two cylinder banks. It is driven either by gears or often by a cog-type belt off the crankshaft. Inside the blower, there are two long rotors that spin but do not touch each other. These rotors are shown in ***Figure 28–2***. These rotors act like a positive displacement air pump, drawing in huge amounts of air, and forcing this air into the intake manifold.

One of the biggest disadvantages of the Roots-type blower is that it consumes a large amount of frictional horsepower from the crankshaft. This means that engines that use Roots-type blowers usually do not have good fuel economy, but they do have large increases in horsepower and torque. Thus, they are used mainly on sports vehicles and for heavy-duty and racing applications.

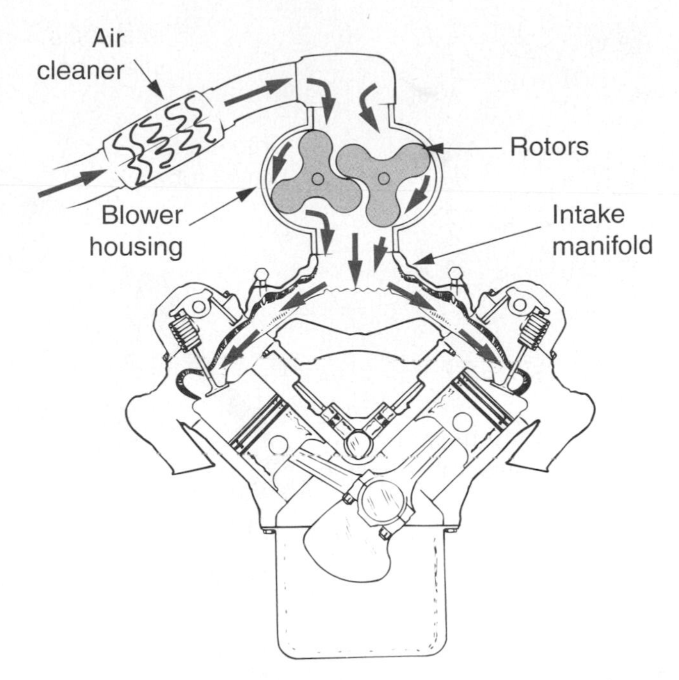

FIGURE 28–2 The Roots-type blower uses two 3-lobed rotors spinning rapidly to force air into the cylinders.

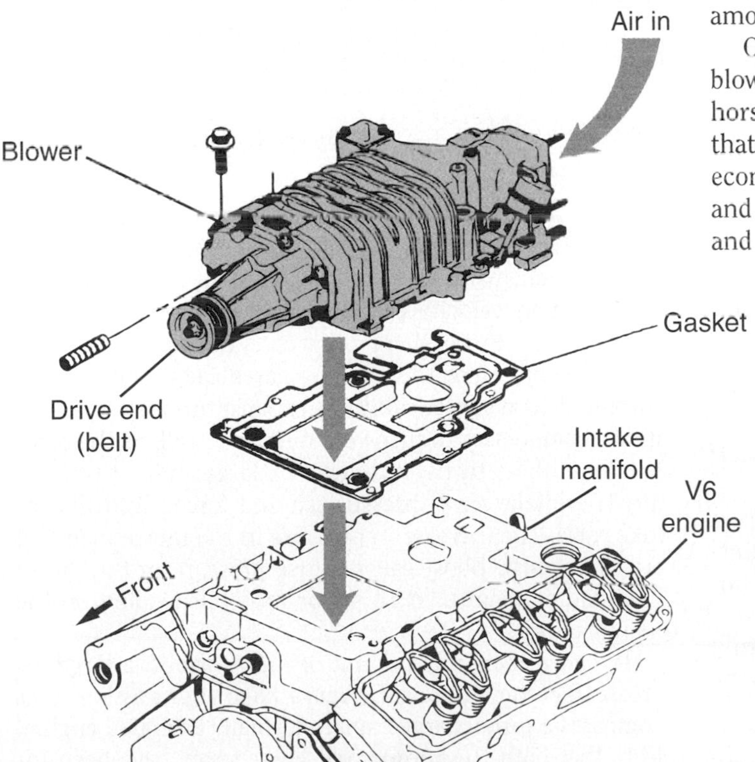

FIGURE 28–1 A Roots-type blower forces large amounts of air into an engine to increase horsepower.

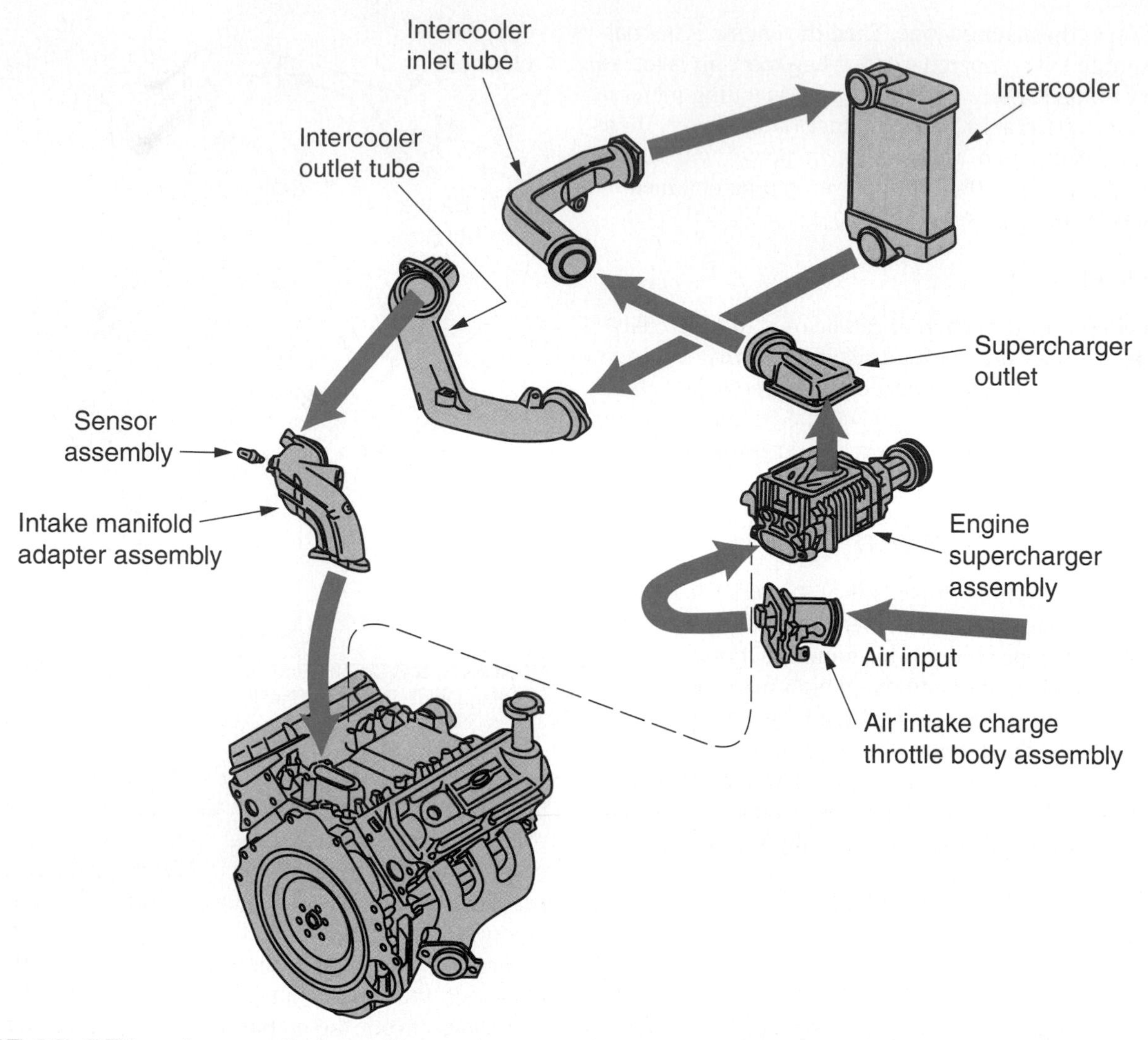

FIGURE 28–3 This engine uses a Roots-type blower to increase performance of the engine. The major components are shown.

Figure 28–3 shows the major components that are used on an engine that has a Roots-type blower. This particular system is used on high-performance diesel engines. The blower in this case is called a supercharger. The airflow through the system is as follows:

1. First, the air comes into the air intake charge throttle body assembly.
2. The air then flows to the lower section of the blower/supercharger.
3. After the air goes through the blower/supercharger, it goes through tubing to the intercooler. This intercooler helps to reduce the air temperature for increased performance.
4. From the intercooler, the air flows through ducting to the intake manifold adapter assembly, and into the engine.

28.2 TURBOCHARGING PRINCIPLES

A turbocharger is a device that uses the exhaust gases, rather than engine power, to turn an air pump or compressor. The air pump then forces an increased amount of air into the cylinders. Both diesel and gasoline engines in the automotive market use turbochargers. ***Figure 28–4*** shows a typical schematic of air and exhaust in a turbocharged engine. High velocity exhaust gases pass out of the exhaust valves. From there they pass through a **turbine**-driven pump. Here the exhaust gases cause the exhaust turbine to turn very rapidly. The exhaust turbine causes the intake **compressor** to turn very rapidly as well. As the compressor turbine turns, it draws in a large amount of fresh air. The intake air is pressurized and forced into the intake valve. The increase in pressure in the intake manifold is called **boost**. Boost can produce pressure in the intake manifold of about 6 to 10 psi or more, depending on the manufacturer.

If a corresponding amount of fuel is added, a large increase in power will result. ***Figure 28–5*** shows a chart that compares a turbocharged and a normally aspirated engine. Note that both the torque and horsepower have been increased at all rpm. For example, at 5,000 rpm the normally aspirated engine produces about 80 hp. At this rpm, the turbocharged engine can produce about 140 hp. ***Figure 28–6*** shows a turbocharger used on an automobile.

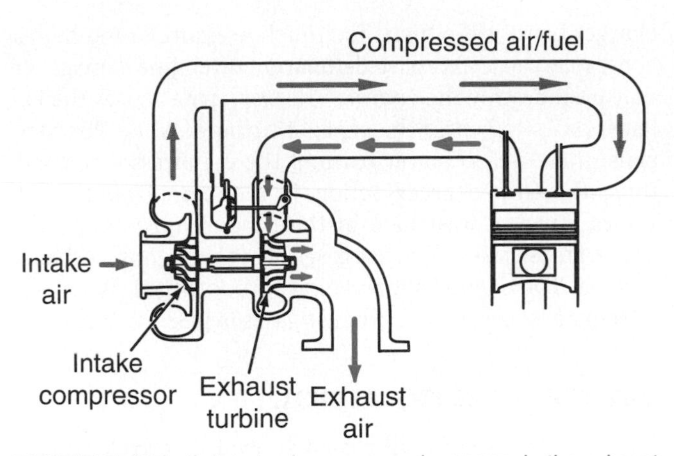

FIGURE 28-4 A turbocharger uses the energy in the exhaust gases to turn an air pump or compressor. The compressor forces extra air into the engine for increased performance.

FIGURE 28-6 A complete turbocharger for an automobile engine is shown in this illustration.

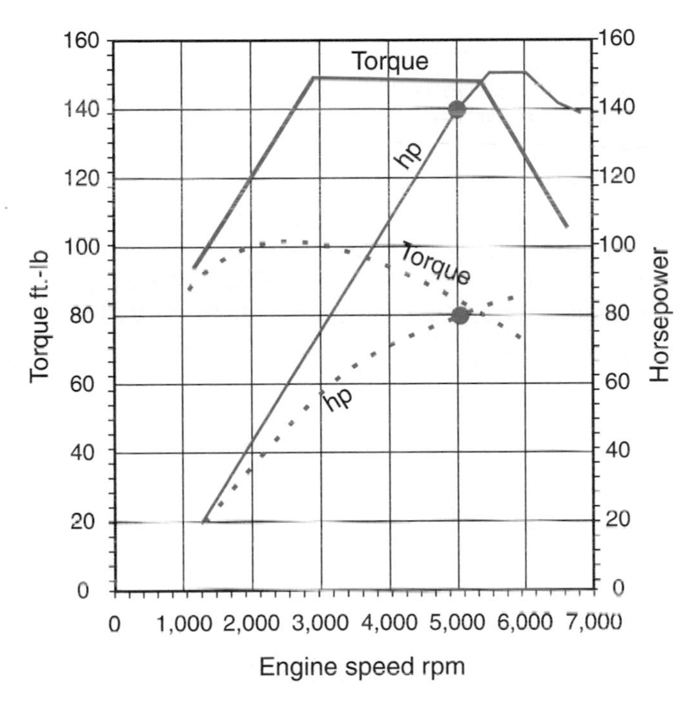

FIGURE 28-5 This chart shows the increase in power and torque when an engine is turbocharged.

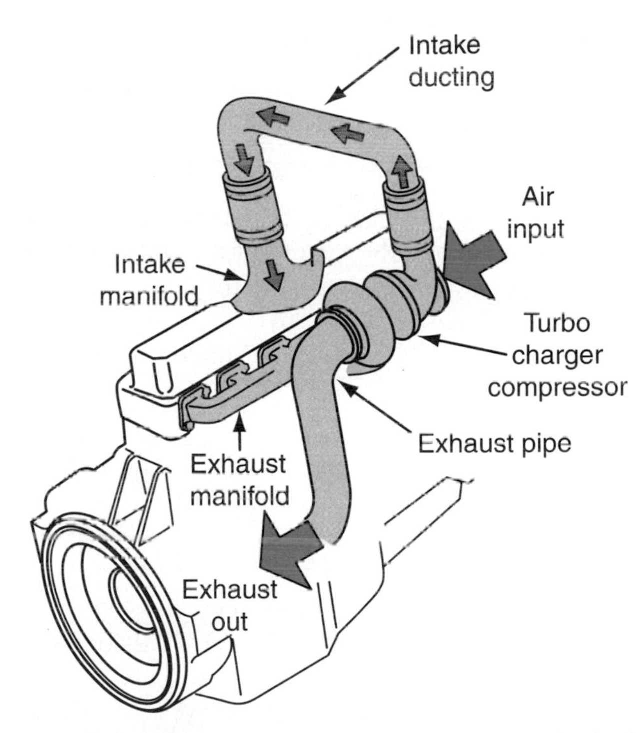

FIGURE 28-7 When a turbocharger is used, the intake ducting will be different from a typical nonturbocharged engine. The intake and exhaust ducting on this diesel engine are designed so that the turbocharger can be located on the side of the engine. (Courtesy of General Motors Corporation, Service Technology Group)

TURBOCHARGED ENGINE CHANGES

A few internal changes are necessary on turbocharged engines. These include strengthening the pistons, using different piston rings, and making sure the bearings can withstand the extra load. However, turbocharged and nonturbocharged engines share basically the same compression ratios and emission and electronic control devices.

TURBOCHARGER DUCTING

The inlet ducting is changed when a turbocharger is used (***Figure 28–7***). The exhaust gases pass through the exhaust manifold and into the exhaust turbine on the turbocharger. From the exhaust turbine, the exhaust gases are sent through the exhaust system into the environment. As the compressor turbine turns, air is drawn through the air cleaner into the intake manifold. The pressurized air is sent through ducting to the throttle body unit. Here the air and fuel are mixed in the correct proportions.

PARTS LOCATOR

*Refer to photo #47 in the front of the textbook to see what **turbocharger ducting** looks like.*

TURBOCHARGER LAG

One problem associated with a turbocharged engine is called **lag**. Lag is defined as the time it takes for the turbocharger to increase the power. It is the delay between a rapid throttle opening and the delivery of increased boost. There is a lag between the time the operator calls for the extra power and the actual power produced. This is because it takes time for the turbine speed to increase and produce the necessary power. Turbochargers on automobiles operate at about 10,000 rpm at idle. They run most efficiently at about 100,000 to 150,000 rpm under maximum boost. It takes time for the turbocharger to increase to this speed for best efficiency.

TURBOCHARGER WASTEGATE

A wastegate is connected to the turbocharger. The wastegate is used to bypass the exhaust gases when the turbocharger boost is too high. Too much pressure or too high a boost may cause excessive detonation or engine damage, or may even destroy the engine. The wastegate causes the exhaust gases to bypass the exhaust turbine. When this happens, there is less power turning the compressor turbine; thus, the turbocharger action is reduced. ***Figure 28–8*** shows a typical wastegate in the exhaust flow. When the wastegate is closed, exhaust gases pass through the exhaust turbine. The engine is now being turbocharged. When the wastegate is opened, the exhaust gases bypass the turbine.

WASTEGATE CONTROL

The wastegate is normally closed. It opens to bypass exhaust gases and prevent an overboost condition. The wastegate opens when vacuum is applied to the **actuator**. The actuator is controlled by a wastegate control valve that is pulsed on and off by the PCM (***Figure 28–9***). Under normal driving conditions, the control solenoid is energized 100% of the time. This means the exhaust gases pass through the exhaust turbine. During rapid acceleration, there may be an increase in boost pressure. As the boost increases, it is sensed by the manifold absolute pressure (MAP) sensor. The PCM now pulses the wastegate control valve above a boost of 15 psi. With the wastegate pulsing on and off, the manifold pressure decreases. If an overboost condition does occur, the PCM will also reduce the fuel delivery. ***Figure 28–10*** shows the wastegate actuator attached to the turbocharger.

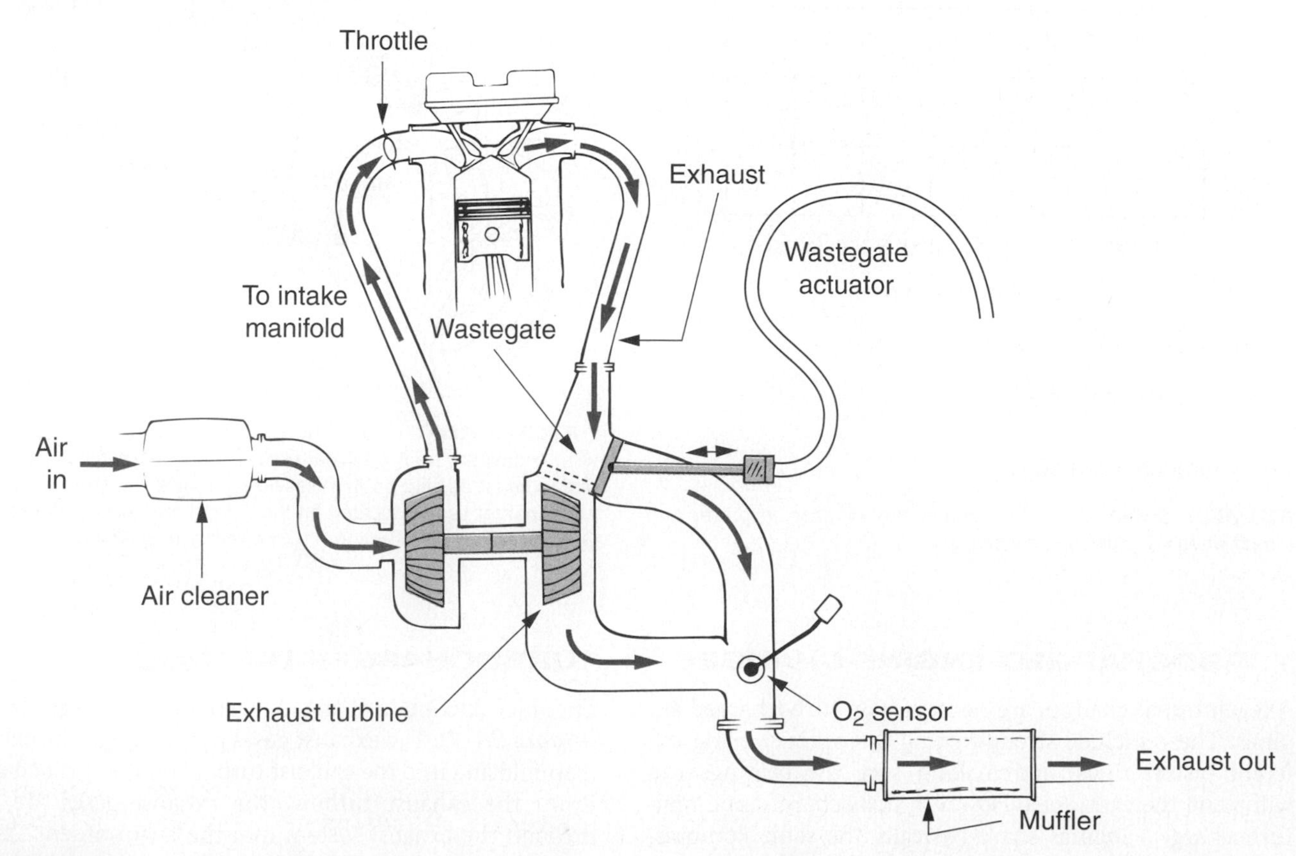

FIGURE 28-8 A wastegate is used to bypass part of the exhaust gases during times of high boost.

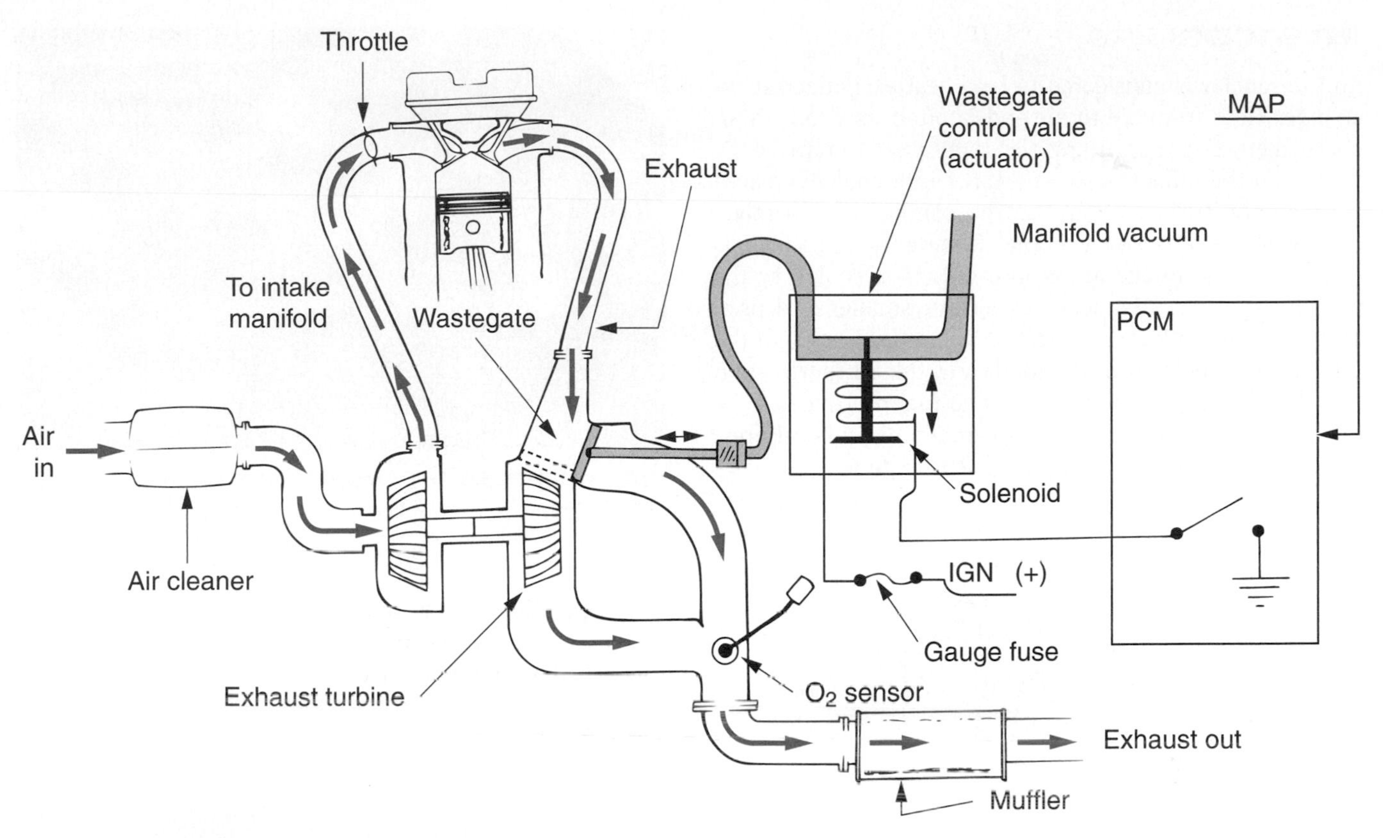

FIGURE 28-9 The wastegate is controlled by a wastegate control valve, solenoid, and manifold vacuum along with the powertrain control module (PCM) signals.

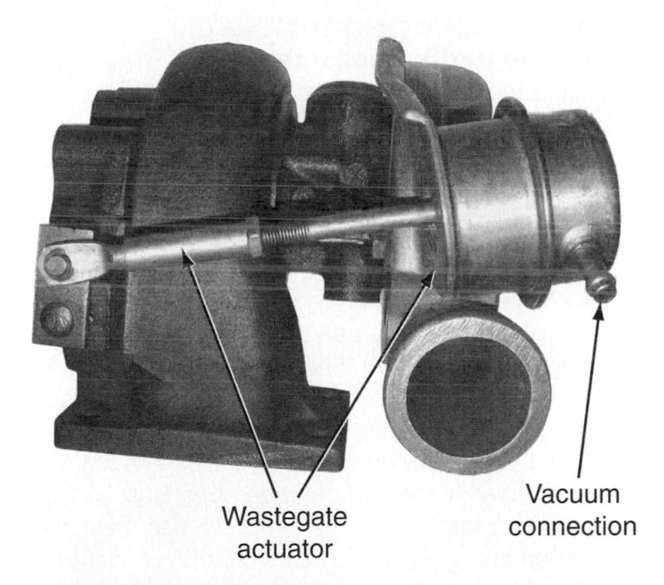

FIGURE 28-10 The wastegate actuator located on the turbocharger controls the wastegate.

TURBOCHARGER CONSTRUCTION

Figure 28-11 shows a complete turbocharger. The shafts on the intake and exhaust turbines are connected together. The intake turbine is designed to act as a centrifugal compressor. The exhaust turbine acts as a fan, causing the shaft to turn from the exhaust gases. A housing surrounds both turbines. Flanges are attached to the housing for mounting.

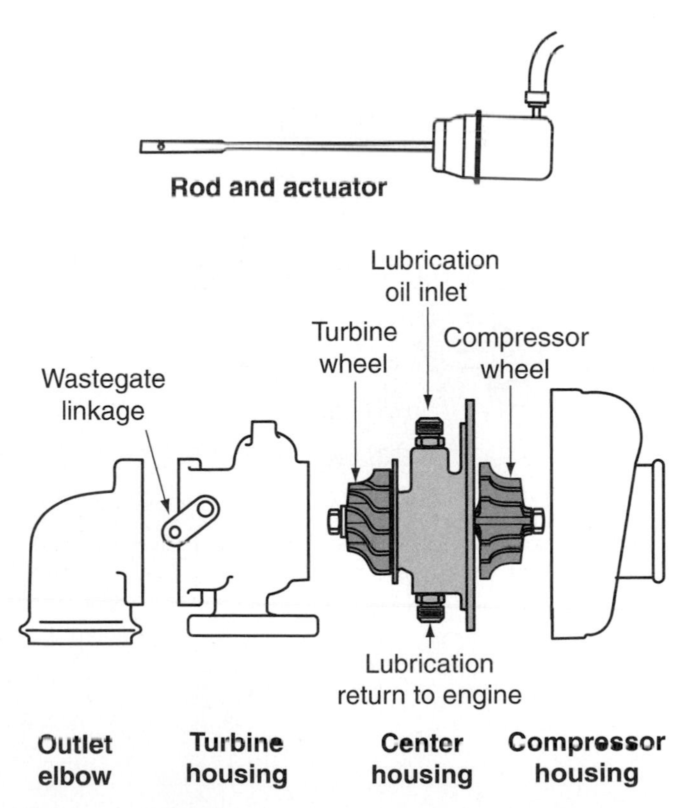

FIGURE 28-11 A disassembled view of a turbocharger.

INTERCOOLERS

An **intercooler** is considered to be a heat exchanger. It exchanges heat from air to air and is often used on turbocharged engines. The intercooler is designed to remove the heat from the intake air. In effect, the intercooler is placed between the turbocharger and the combustion chamber. ***Figure 28–12*** shows an example of where the intercooler is located. As the intake or compressed air is cooled by the outside air, it has a tendency to become smaller or denser. This condition allows more air to be compressed into the intake. As it cools down the intake air, the air increases in density. This allows for more air and fuel molecules to be forced into the combustion chamber at a given boost pressure. The result is more power from the engine.

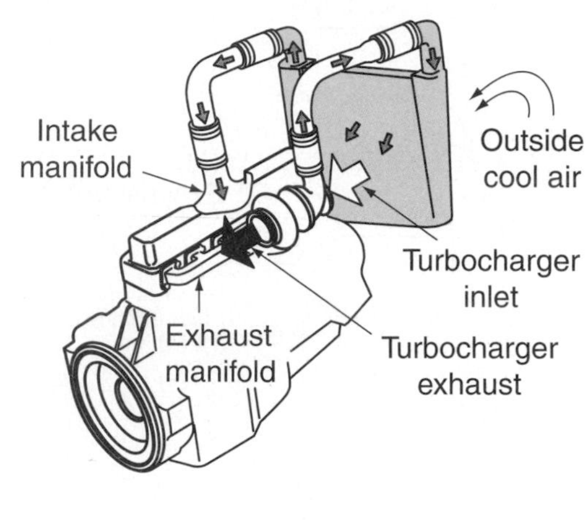

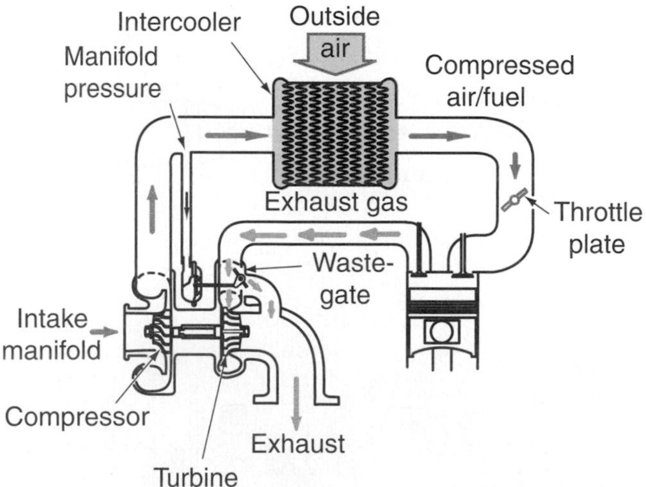

FIGURE 28–12 The intercooler, located on the compression side (air intake) of the airflow, uses outside air to cool the compressed air.

FIGURE 28–13 An intercooler can be used on turbocharged engines to reduce the temperature of the intake air going into the combustion chamber.

In operation, as the turbocharged air is pressurized, it increases in temperature. However, when this air is fed through the intercooler, fresh air from outside can somewhat cool down the intake air. ***Figure 28–13*** shows a typical example of an intercooler on an engine.

Car Clinic: TURBO LUBRICATION

CUSTOMER CONCERN:
Stopping the Turbocharger

What is the best way to shut off an engine with a turbocharger? Some say to just shut the engine down after any rpm, whereas others say to let the engine idle for a short period of time. Who is right?

SOLUTION: Turbocharged engines should be shut down by first idling the engine for a short period of time. This procedure lets the turbocharger in the turbine slow down after being at high speed. During this slow-down time, the engine is still running and is thus sending oil to the turbocharger bearings. If the engine is shut down immediately after high-speed operation, the oil pressure will not be available at the turbocharger bearings, which may cause damage. This problem has become common enough for the marketing of "add-on" oilers that continue to circulate oil after the engine has been turned off.

Problems, Diagnosis, and Service

Safety Precautions

1. Never work on a turbocharger when the engine is running.
2. Always wear OSHA-approved safety glasses when working on a turbocharger.
3. When working on a turbocharger, remember that parts may be very hot. Touching them may cause serious injury.
4. Be careful not to cut your hands on the sharp edges of the turbocharger's parts.

PROBLEM: Damaged Turbocharger

Various problems can occur with a damaged turbocharger. Some of the more common problems include:

- Engine lacks power
- Black exhaust smoke
- Excessive oil consumption
- Blue exhaust smoke
- Noisy turbocharger
- Oil leaks at compressor or turbine seal or at both

DIAGNOSIS

Be aware of different sounds made by the turbocharger. Different noise levels during operation may signal air restriction or dirt built up in the compressor housing. Some of the common diagnostic checks to determine turbocharger problems include:

1. Restricted air intake duct to turbocharger.
2. Air leak in duct from compressor to intake manifold.
3. Restricted exhaust system.
4. Air leak at intake manifold to engine mating surface.
5. Restricted turbocharger center housing.
6. Dirt on compressor wheel and/or diffuser vanes.
7. Damaged turbocharger.
8. Exhaust gas leak in turbine inlet exhaust manifold or exhaust manifold.
9. Restricted turbocharger oil drain line.
10. Restricted turbocharger center housing.
11. Restricted PCV system.
12. Incorrect wastegate boost pressure. Overboost can be caused by:
    a. A sticking wastegate or wastegate actuator
    b. A control valve stuck in the closed position
    c. A cut or pinched vacuum hose
    d. A fault within the computer

Underboost can be caused by:

a. The wastegate sticking open
b. The control valve sticking open
c. A faulty computer (PCM) unit

SERVICE

CAUTION: *A turbocharged engine has exhaust pipes located high in the engine compartment. Care must be taken to avoid accidental contact with hot exhaust pipes resulting in injury.*

Various service procedures are used to correct problems dealing with turbochargers. Because of the detail and complexity of these procedures, only general service procedures are listed here.

1. The wastegate-boost pressure test is used to determine the amount of boost pressure created under acceleration conditions. Follow the manufacturer's recommended procedure for testing wastegate-boost pressure. Generally, the wastegate control valve can be checked by using a vacuum tester. For example, the actuator should begin to move at 4 psi and obtain full travel at 15 psi. Readings will be determined by the manufacturer's recommendations.
2. Inspecting the turbocharger's internal condition.
    a. Check for mechanical movement of the wastegate.
    b. Spin the compressor wheel, checking for binding, drags, and other poor conditions.
    c. Inspect internal housing for sludge and dirt and clean accordingly.
    d. Inspect compressor oil seal for damage or leakage. Replace as necessary.
    e. Check turbocharger main shaft for radial and axial clearances as described in the service manuals.

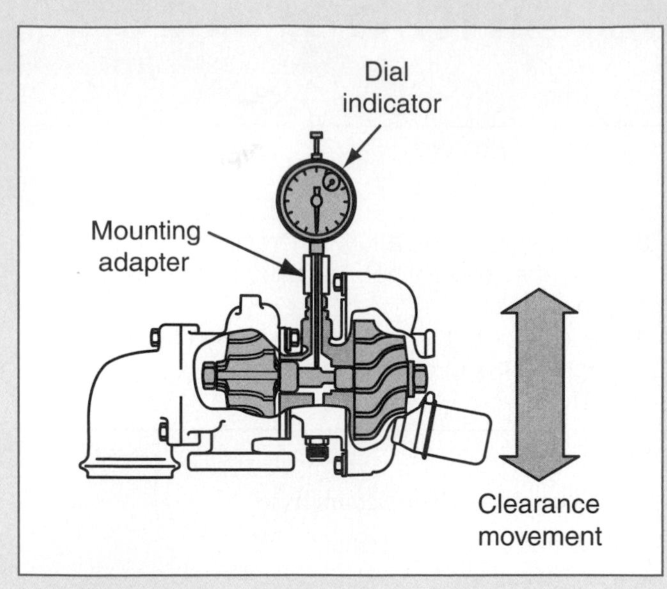

FIGURE 28–14 The journal bearings can be checked by using a dial indicator.

f. Check the journal bearing clearances. As shown in ***Figure 28–14***, a dial indicator is used to check the journal bearings. The dial indicator is positioned so that it can touch the main turbocharger shaft between the two bearings. Then the service technician moves the shaft up and down from either the compressor or turbine end. As the shaft is moved, the dial indicator will read the clearance or wear on the journal bearings. Compare the readings to the manufacturer's recommended clearances. Replace the journal bearings if the clearances are out of specifications.

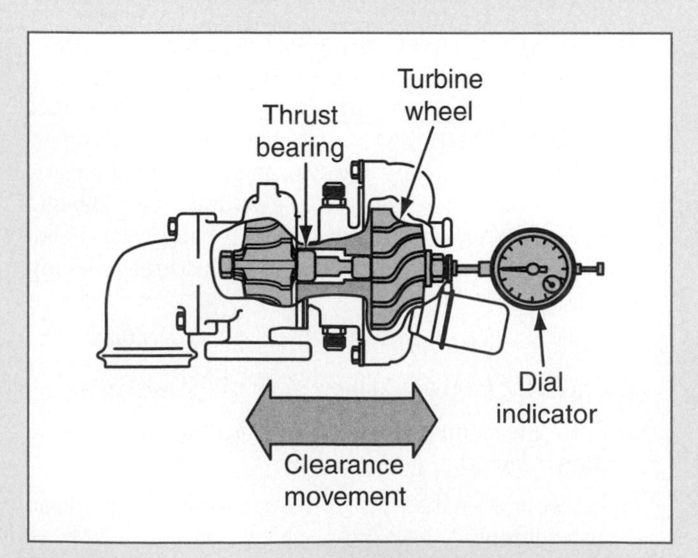

FIGURE 28–15 The thrust bearings can be checked by using a dial indicator.

g. Check the thrust bearing clearances. As shown in ***Figure 28–15***, a dial indicator is used to check the thrust bearing clearance. The dial indicator is positioned on the end of the main turbocharger shaft. The shaft is the moved back and forth along its axis to determine the thrust bearing clearances. Compare the readings to the manufacturer's recommended specifications. Replace the thrust bearings if the clearances are out of specifications.

h. Check the external condition of the compressor and turbine blades as shown in ***Figure 28–16***. Checks should be made for cracked, broken, or bent blades. Replace turbine or compressor blades as necessary.

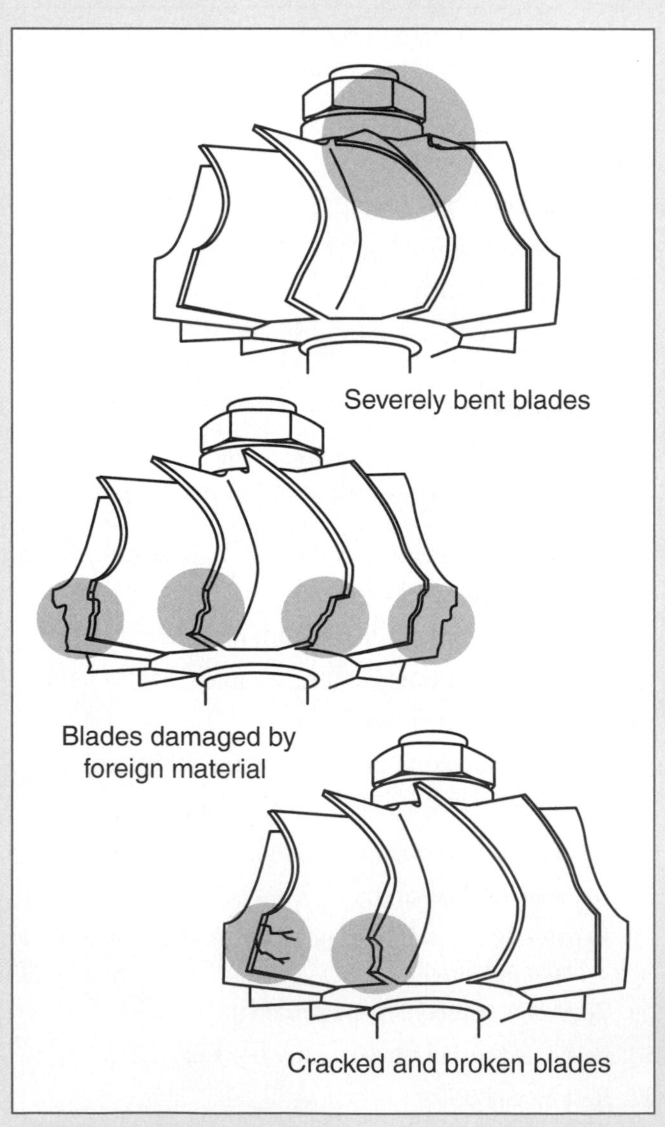

FIGURE 28–16 Always check for cracked, broken, or bent turbine and/or compressor blades when inspecting turbochargers.

Service Manual Connection

There are many important specifications to keep in mind when working with turbocharger systems. To identify the specifications for your engine, you will need to know the vehicle identification number (VIN) of the vehicle, the type and year of the vehicle, and the type of engine. Although they may be titled differently, some of the more common air intake and exhaust system specifications (not all) found in service manuals are listed below. Note that these specifications are typical examples. Each vehicle and engine may have different specifications.

Common Specification	Typical Example
Turbocharger bearing axial clearance	0.001–0.003 in.
Turbocharger bearing radial clearance	0.003–0.006 in.
Turbocharger to exhaust manifold torque	16–19 ft.-lb
Turbocharger rpm	120,000
Wastegate boost pressure	9 psi

In addition, the service manual will give specific directions for various service and testing procedures. Some of the more common procedures include:

- Turbocharger assembly/replace
- Turbocharger internal inspection
- Turbocharger wastegate-boost pressure test
- Wastegate actuator replace

SUMMARY

The following statements will help to summarize this chapter:

- Supercharging is a process in which air is forced into an engine to gain more horsepower and torque.
- Supercharging an engine can be done by using a turbocharger or blower.
- A turbocharger uses the energy in the hot exhaust gases to turn a turbine, which in turn rotates a compressor that forces more air into the engine.
- A blower uses a set of three-lobed rollers driven from the crankshaft to force additional air into an engine.
- A supercharged engine needs stronger pistons, stronger rings, and higher quality bearings that can withstand the extra loads.
- When the boost pressure gets too high on a turbocharged engine, the exhaust gases are allowed to pass by the turbine by means of a wastegate.
- The wastegate is controlled by vacuum and signals from the PCM.
- Common turbocharger service checks include checking the clearances of the journal bearings and the thrust bearings.
- Always check the turbocharger compressor and turbine blades for cracks, bends, or breaks.

TERMS TO KNOW

Can you identify each of the following terms? Review the chapter until you can use each term correctly.

Actuator
Boost
Compressor
Intercooler
Lag
Naturally aspirated
Supercharging
Turbine

REVIEW QUESTIONS

Multiple Choice

1. An engine that uses atmospheric pressure to force the air into the engine is called a ________ engine.
a. Supercharged
b. High vacuum
c. Naturally aspirated
d. Low pressure
e. None of the above

2. Which type of component uses exhaust gases to turn a turbine that forces air into the cylinder?
a. Blower
b. Wastegate
c. Turbocharger
d. Intake manifold
e. Runner

3. The wastegate bypasses exhaust gases when:
a. Turbine speed is low
b. Turbine speed is high
c. Too much fuel is added
d. Too much boost pressure is sensed by the MAP
e. The load is removed

4. On a turbocharged engine:
a. The pistons must be strengthened
b. The rings must be strengthened
c. The bearings must be strengthened
d. All of the above
e. None of the above

5. On a turbocharged engine, the delay between a rapid throttle opening and the delivery of boost is called:
a. Turbocharger efficiency
b. Volumetric efficiency
c. Turbocharger lag
d. Wastegate control
e. Speed control

6. The Roots-type blower is considered a ________.
a. Turbocharger
b. Supercharger
c. Cooler
d. Naturally aspirated engine
e. All of the above

7. The Roots-type blower uses two long ________ to produce the additional air pressure.
a. Rotors
b. Slides
c. O-rings
d. Pistons
e. Gears

8. Turbochargers on automobiles normally run most efficiently between ________ rpm.
a. 10 and 15
b. 100 and 150
c. 1,000 and 1,500
d. 100,000 and 150,000
e. 1,000,000 and 1,500,000

The following questions are similar in format to ASE (Automotive Service Excellence) test questions.

9. A turbocharged engine is exhausting black smoke. *Technician A* says it could be caused by air restriction. *Technician B* says it could be caused by a bad PCM (powertrain control module). Who is correct?
a. A only
b. B only
c. Both A and B
d. Neither A nor B

10. An engine with a turbocharger is lacking power. *Technician A* says the problem may be a broken compressor blade. *Technician B* says the problem may be dirt in the compressor. Who is correct?
a. A only
b. B only
c. Both A and B
d. Neither A nor B

11. An engine with a turbocharger needs to be inspected. *Technician A* says that the radial clearances should be checked on the main shaft. *Technician B* says that axial clearances should be checked on the main shaft. Who is correct?
a. A only
b. B only
c. Both A and B
d. Neither A nor B

12. *Technician A* says the wastegate operation can be checked with a dial indicator. *Technician B* says it can be checked with a micrometer. Who is correct?
a. A only
b. B only
c. Both A and B
d. Neither A nor B

13. *Technician A* says that a turbocharger with a broken compressor blade will not affect engine performance. *Technician B* says that a turbocharger with a bent blade need not be replaced. Who is correct?
a. A only
b. B only
c. Both A and B
d. Neither A nor B

Essay

14. Describe the purpose and operation of a turbocharger.

15. What is the definition of supercharging?

16. What is the purpose and operation of a wastegate on a turbocharger?

17. Describe the operation of a Roots-type blower.

18. Define the term *turbocharger lag* and identify why it is a concern.

Short Answer

19. An engine with a turbocharger can produce increased ________ and ________.

20. An intercooler is considered a/an ________ to ________ heat exchanger.

21. The wastegate on a turbocharger is controlled by the ________.

22. When inspecting a turbocharger, always check for cracked, broken, or bent ________.

23. The Roots-type blower uses ________ long spinning rotors to produce additional air pressure.

Applied Academics

TITLE: Turbochargers and Exhaust Temperature

ACADEMIC SKILLS AREA: Science

NATEF Automobile Technician Tasks:

The technician interprets charts, tables, and/or graphs to determine manufacturer's specifications for system operation to identify out-of-tolerance systems and/or subsystems.

CONCEPT:

Heat in the exhaust system has a tendency to degrade both the engine and the exhaust system. As more horsepower is required from engines, the exhaust temperatures often are well above 1,200 degrees Fahrenheit. When an engine has a turbocharger, generally the exhaust temperatures drop slightly as shown in the accompanying chart. The engine without a turbocharger has exhaust temperatures ranging from 1,125 to 1,250 degrees Fahrenheit. On the other hand, a typical turbocharged engine has exhaust temperatures as low as 980 to 1,110 degrees Fahrenheit. The end result of these lower exhaust temperatures is that the exhaust valves and the exhaust system will have a longer life.

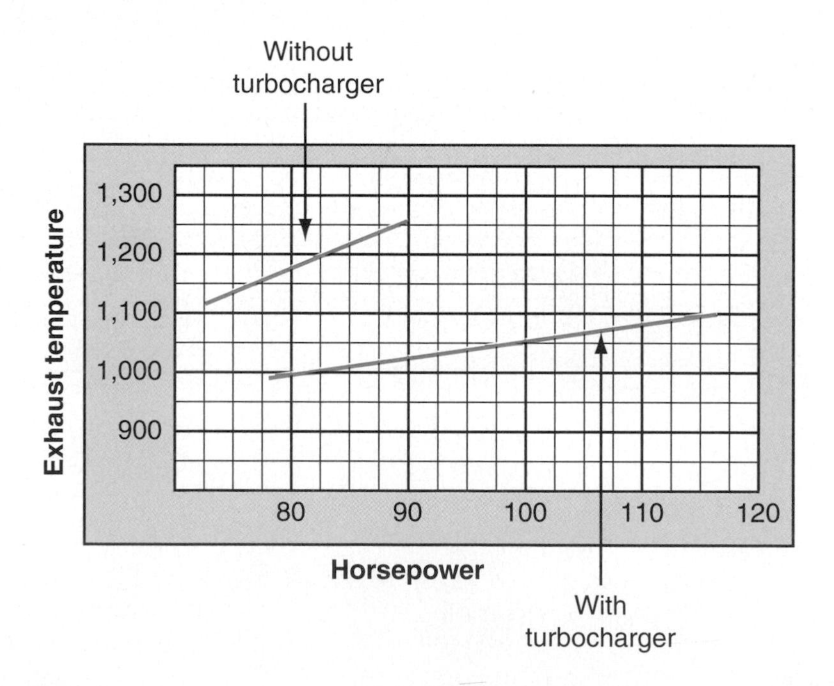

APPLICATION:

What effect do lower exhaust temperatures have on the emissions from the engine? Why will the exhaust components last longer if the exhaust temperatures are lower? Why are the exhaust temperatures lower when a turbocharger is used on an engine?

Electrical Engine Systems

SECTION

4

Kelly Pietras, 28

Design Release Engineer

General Motors

Detroit, MI

How did you decide to go into automotive?

I've always had a passion for cars—but I didn't decide on this career until college. It was a great job market for engineers.

What's your educational background?

I have a bachelor's degree in mechanical engineering from Lawrence Technological University.

What do you do?

I'm responsible for certain parts in the transmission, including shafts and hubs. I review changes to proposed design improvements, and track them from paper to development. I work with someone trained in CAD (computer aided design) to create the new design. I supervise the testing of the parts and also go to suppliers to see what kinds of new technology they're offering. We're currently working on a completely new transmission for the 2006 model year.

What kind of improvements have you made on it?

People are looking for improved fuel economy, improved performance, and the ability to accelerate more quickly.

Do you feel the transmission when you ride in a car ordinarily?

Yes! When you become highly familiar with the parts, you become attuned to everything that's going on inside. I'm always analyzing and thinking how I can improve it.

What skills are necessary for your job?

Definitely knowledge in math and physics, because I have to calculate things like fatigue damage. I spend a lot of time calculating part interfacings. Communication, since I spend about 70% of my time in meetings. Oh, and the ability to multitask—I've always got a hundred projects going at once!

What's the best part of your job?

It's very nice as someone who creates something on paper to have it end up in vehicles. When I see someone driving a car I had something to do with, that feels good.

CHAPTER 29

Electrical Principles

OBJECTIVES

After studying this chapter, you should be able to:

- Define electricity in terms of voltage, amperage, and resistance.
- Calculate both Ohm's and Watt's Laws.
- Define voltage drop, simple circuits, and symbols.
- Analyze series, parallel, and series-parallel circuits.
- Analyze electrical schematics.
- Apply magnetism principles to electromagnetic induction.
- Explain the principles of a simple generator.

INTRODUCTION

Electricity, electronics, and computers continue to control more and more of the automobile. The powertrain control module (PCM) and the oxygen feedback system discussed in other chapters are examples of electronics and computers used in the automobile. In addition, storage batteries, ignition systems, charging systems, and starting systems use electricity to operate. Therefore, the study of automotive technology must include a discussion of basic electrical principles. This chapter will help you understand electricity and how it is applied to automotive systems.

29.1 INTRODUCTION TO ELECTRICITY

ATOMIC STRUCTURE

The heart of all information concerning electricity is in the study of **atoms** and atomic structure. Everything—water, trees, buildings—is made up of atoms. They are very small, about a millionth of an inch across. There are millions of atoms in a single breath of air.

The structure of atoms can be illustrated as shown in ***Figure 29–1***. Each atom has at its center a nucleus that contains both **protons** and neutrons. The nucleus is the

Certification Connection

ASE Connection: The information in this chapter can help you prepare for the National Institute for Automotive Service Excellence (ASE) certification tests. The tests and content areas most closely related to this chapter are:

Test A6—Electrical/Electronic Systems

- **Content Area**—General Electrical/Electronic System Diagnosis

NATEF Connection: Much of the information in this chapter is related to the NATEF tasks. The NATEF tasks and priority numbers most closely related to this chapter are:

1. Identify and interpret electrical/electronic system concern; determine necessary action. P-1
2. Diagnose electrical/electronic integrity for series, parallel, and series-parallel circuits using principles of electricity (Ohm's Law). P-1
3. Use wiring diagrams during diagnosis of electrical circuit problems. P-1

4. Check electrical circuits with a testlight, determine necessary action. P-2
5. Measure source voltage and perform a voltage drop test in electrical/electronic circuits using a voltmeter; determine necessary action. P-1
6. Measure current flow in electrical/electronic circuits and components using an ammeter; determine necessary action. P-1
7. Check continuity and measure resistance in electrical/electronic circuits and components using an ohmmeter; determine necessary action. P-1
8. Locate shorts, grounds, opens, and resistance problems in electrical/electronic circuits; determine necessary action. P-1

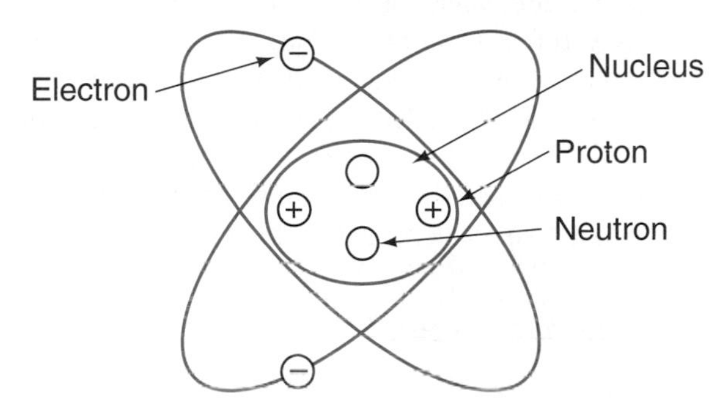

FIGURE 29–1 An atom has protons, neutrons, and electrons.

major part of the atom. Protons are said to carry a positive charge (+). Neutrons carry no charge and are not considered in the study of electricity. Also present in the atom are **electrons**, which orbit the nucleus. Electrons are very light in comparison to the nucleus. They carry a negative charge (–).

There is an attraction between the negative electrons and the positive protons. The attractive force and the centrifugal forces cause the electron to orbit the nucleus or protons in the center (***Figure 29–2***).

The number of electrons in all orbits and the number of protons in the nucleus will try to remain equal. If they are equal, the atom is said to be balanced or neutral. ***Figure 29–3*** shows several atoms for comparison.

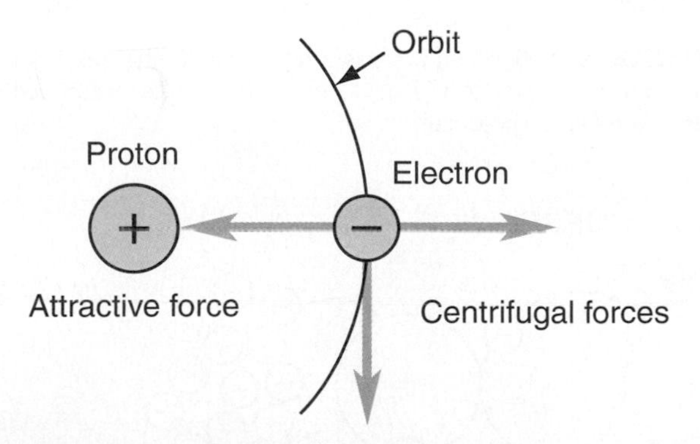

FIGURE 29–2 When an attractive force equals the centrifugal force, the electron will orbit the proton.

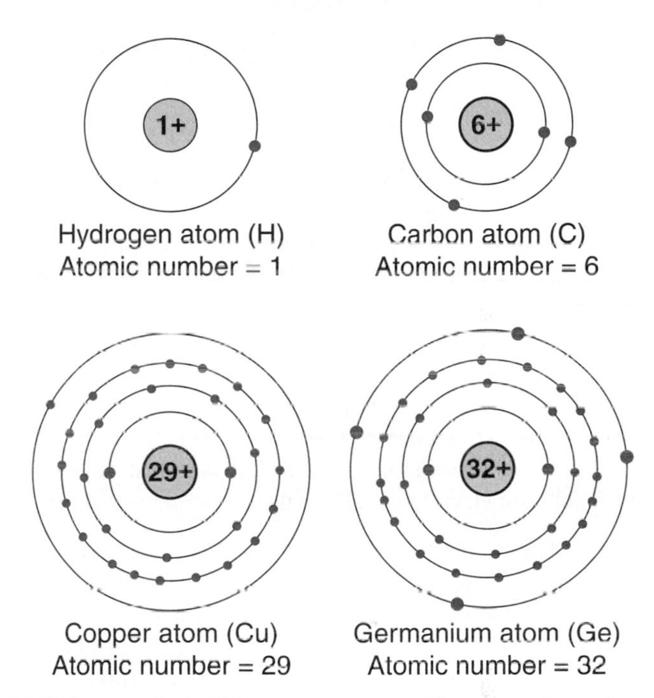

FIGURE 29–3 Different atoms have different numbers of protons and electrons. Hydrogen is the simplest atom; germanium is much more complex

VALENCE RING

In the study of electricity, we are concerned only with the electrons in the outer orbit. The outer orbit of the atom is called the **valence ring**. It holds the outermost electrons. Actually, electrons in the valence ring can easily be added or removed. Generally, an atom with several electrons missing will try to gain or capture other electrons in order to balance itself. Also, if an atom has an excess amount of electrons in the valence ring, it may try to get rid of these electrons. This will also help balance the atom.

Certain materials can lose or gain electrons rather easily. This depends on the number of electrons needed in the valence ring to balance the atom. If an atom loses electrons easily, the material is called a good **conductor**. If the atom cannot lose electrons easily, the material is called a good **insulator**. Insulators and conductors are defined as follows:

- Three or fewer electrons—conductor
- Five or more electrons—insulator
- Four electrons—semiconductor

Materials that have four electrons in the outer orbit can be considered either a conductor or an insulator. These materials are called **semiconductors**. They are used in solid-state components, which will be discussed in the next chapter.

ELECTRICITY DEFINED

Electricity can be defined as the movement of electrons from atom to atom. This can happen only in a conductor. An example is shown in ***Figure 29–4***. Copper atoms are shown with only the valence ring. Copper is a good conductor. If an excess amount of positive charges, or protons, is placed on the left, one of the positive atoms will try to pull the outer electron away from the copper atom on the far left.

This action will make the far left copper atom slightly positively charged. There is one more proton than electron in the total atom. This positively charged atom will then pull an electron from the one on its right side. This atom also becomes positively charged. Overall, as this action continues to happen, electrons flow from the right to the left. Remember, there must be an abundant amount of protons on the left and electrons on the right.

Electron Theory In *Figure 29–4* the electrons were flowing from a negative point to a positive point. When electricity is defined this way, it is called the **electron theory**. This means that electricity flows from a negative point to a more positive point. This is one method of defining the direction of electrical flow.

Conventional Theory Electricity can be defined another way. Electricity can be defined as flow from a positive point to a more negative point. This is called the **conventional theory**. For example, referring to ***Figure 29–4***, while negative charges are flowing from right to left, positive charges are flowing from left to right. This means that electrical charges could also flow from positive to negative. In the automotive field, this method has been used to define the direction of electrical flow.

It is important to be consistent with the method you choose. If electron theory is used, stay with electron theory. If conventional theory is used, stay with conventional theory.

AMPERAGE DEFINED

The measurement of the amount of electrons flowing from a negative point to a positive point in a given time period is called **amperage**. Amperage, or current, is analogous to water flowing through a pipe (electrons flowing through a wire). Amperage is defined as the amount of electrons passing any given point in the circuit in one second. One ampere is the equivalent of 6.28 billion, billion electrons passing a given point in one second. The number 6.28 billion, billion electrons (6,280,000,000,000,000,000) can also be written as 6.28×10^{18}. The letter used to identify amperage, or current, is *I*, which stands for the *i*ntensity of current flow. It is measured in units called amperes. Note that amperage may also be identified by the letter *A*.

VOLTAGE DEFINED

Voltage is defined as the push or force used to move the electrons. Referring to ***Figure 29–4***, the difference between the positive and negative charges is called voltage. This difference in charges has the ability to move electrons through the wire. ***Figure 29–5*** illustrates the definition of voltage.

Other terms are used to describe voltage. They include potential difference, electromotive force (emf), and electrical pressure. Voltage can be compared to a water system: Water pressure is used to push water through a pipe; voltage is used to push electrons through a wire. Voltage is represented by the letter *E*, which stands for *e*lectromotive force, and is measured in units called volts. In some instances, voltage may also be represented by the letter *V*.

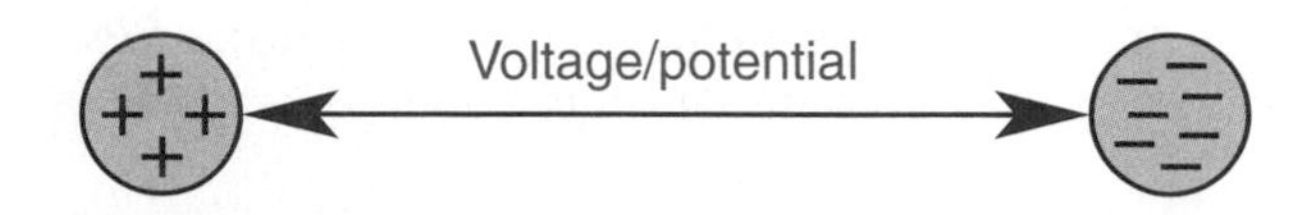

FIGURE 29–5 When there is a difference in charges from one end of the conductor to the other, the difference is called voltage, often referred to as potential.

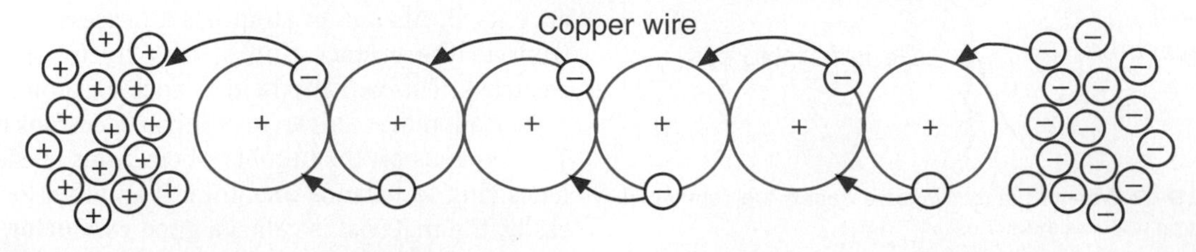

FIGURE 29–4 When positive charges are placed on one end of a copper wire and negative charges are placed on the other end, electrons will move through the wire. This is called the flow of electrons, or electricity.

RESISTANCE DEFINED

The third component in electricity is **resistance**. Resistance is defined as opposition to current flow. Because of their atomic structure, certain materials offer poor conductivity. This will slow down the electrons. Actually, as the electrons move through a wire, they bump into other atoms in the conductor. As this occurs, the material heats up and causes even more resistance.

Various types and values of resistors are designed to control the flow of electrons. This depends on how much current is needed to flow through a circuit. Resistance is identified by the letter *R,* which stands for *r*esistance to electron flow. It is measured in units called ohms. To compare resistance to a water circuit, flow valves and faucets control or restrict water flow; resistance in an electrical circuit controls or restricts electron flow.

OHM'S LAW

The three electrical components just described interact with each other. For example, if the resistance decreases and the voltage remains the same, the amperage will increase. If the resistance stays the same and voltage increases, the amperage will also increase. These relationships can be identified by a formula called **Ohm's Law**. Ohm's Law is a mathematical formula that shows how voltage, amperage, and resistance work together. The triangle shown in ***Figure 29–6*** is a graphical way of showing this formula. It shows that if two electrical components are known, the third can easily be found.

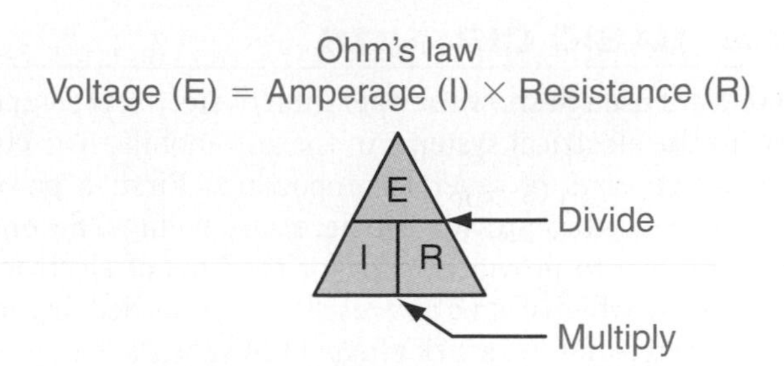

FIGURE 29–6 Ohm's Law states that Voltage (E) = Amperage (I) × Resistance (R).

For example, if resistance (R) and voltage (E) are known, cover the unknown (I) to see the formula. The amperage can be found by dividing the resistance into the voltage. If the amperage and resistance are known, voltage can be found by multiplying the amperage by the resistance. In actual practice, many electrical circuits are designed so that control of the current, or amperage, can be obtained by changing the voltage or resistance. In addition, when using the PCM (powertrain control module), a certain resistance will set up a voltage signal to be sent to the PCM.

Car Clinic: OHM'S LAW

CUSTOMER CONCERN:
Key Off, Battery Drain

On a computer-controlled vehicle there is a small amount of battery drain when the key is in the OFF position. What is causing this drain, and is it normal?

SOLUTION: With the increasing use of electronics and computer memories, there will normally be a small amount of battery drain when the key is in the OFF position. Use a 1-ohm, 10-watt resistor hooked in series with the negative battery terminal and cable, and a digital volt-ohmmeter (DVOM) to check the battery drain. Measure the voltage drop across the resistor. If there is a voltage drop, then there is a small amount of current drain with the key in the OFF position. The current can be calculated using Ohm's Law. Divide the voltage drop by 1 ohm (the resistance of your test resistor) to get the current drain. An acceptable battery drain when the key is OFF should be around 100 milliamps.

WATT'S LAW

Wattage is another term used to help analyze electrical circuits. It is calculated using **Watt's Law**. Wattage is a measure of the power (P) used in the circuit. Wattage is a measure of the total electrical work being done per unit of time. When voltage (E) is multiplied by amperage (I), the result is wattage (P). Wattage, which is a measure of electrical power, may also be referred to as kilowatts (kW). One thousand watts equals 1 kilowatt. ***Figure 29–7*** illustrates the relationship between voltage, amperage, and wattage. If the amperage and wattage are known, cover the voltage to see the formula. If the voltage and wattage are known, amperage can be calculated.

Wattage (or watts) is used to describe batteries and battery performance. For example, batteries are often rated by using watt-hours. A watt-hour is the amount of power or watts that can be produced from a battery for a period of 1 hour. It is typically used as an indicator of the amount of power that a particular battery is capable of producing. Wattage is also used to rate light bulbs such as 5 watts or a 60-watt light bulb.

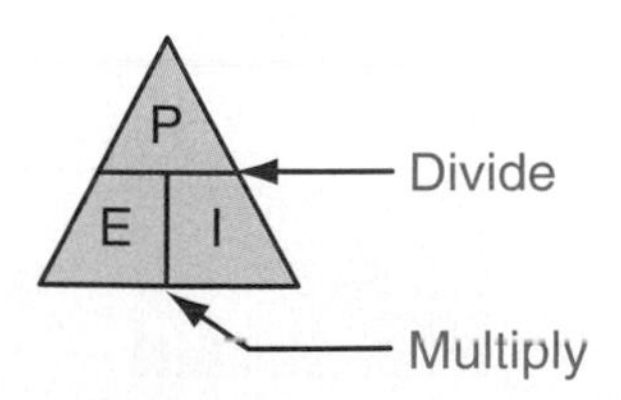

FIGURE 29–7 Watt's Law states that Wattage (P) = Voltage (E) × Amperage (I).

29.2 BASIC CIRCUITS

Circuits are used to show the operation of electrical components in the electrical systems in the automobile. A circuit normally consists of several components. First, a power source is needed to provide the necessary voltage. Second, wire is needed to provide a path for the flow of electrons. Third, a load, which can be any resistance, is needed. Lights, radios, starter motors, spark plugs, PCM sensors, batteries, and wiper motors are all examples of loads. The load provides the resistance in the circuit.

SIMPLE CIRCUIT

The simplest circuit consists of a power source (a battery), a single unit or load to be operated (a light), and the connecting wires. ***Figure 29–8*** shows a simple circuit. Note that the wires must be connected to complete the circuit. In this case, electricity flows from the positive terminal on the battery, through the wire to the light, and back to the negative terminal of the battery.

VOLTAGE DROP

When testing and troubleshooting an electrical circuit, the technician usually measures **voltage drop**. Voltage must be present for amperage to flow through a resistor. Voltage is dropped across each resistor that it pushes amperage through. To determine the voltage drop across any resistor, simply use Ohm's Law. In this case, however, use only the voltage, amperage, and resistance at that particular resistor. Voltage drop at any resistor is shown as:

$$\text{Voltage drop} = \text{Resistance} \times \text{Amperage } (I \times R)$$
(at any one resistor)

Voltage drop can be measured by using a voltmeter. Usually, the voltmeter leads are placed across the component to be checked or across the component to ground.

OPENS, SHORTS, AND GROUNDS

Electrical systems may develop an open, shorted, or grounded circuit. Each of these conditions will render the circuit ineffective.

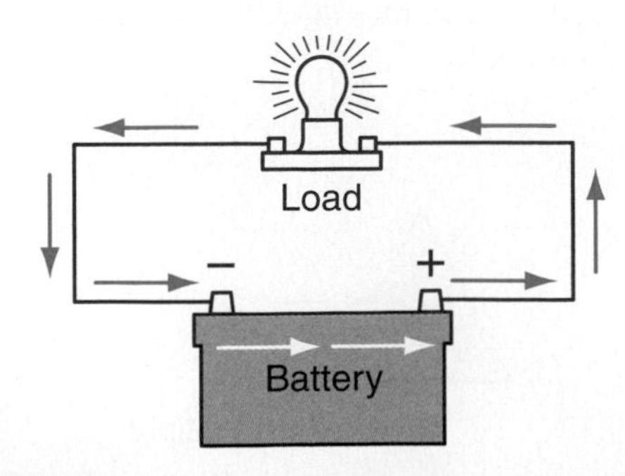

FIGURE 29–8 A simple circuit shows amperage flowing through the light to make it operate.

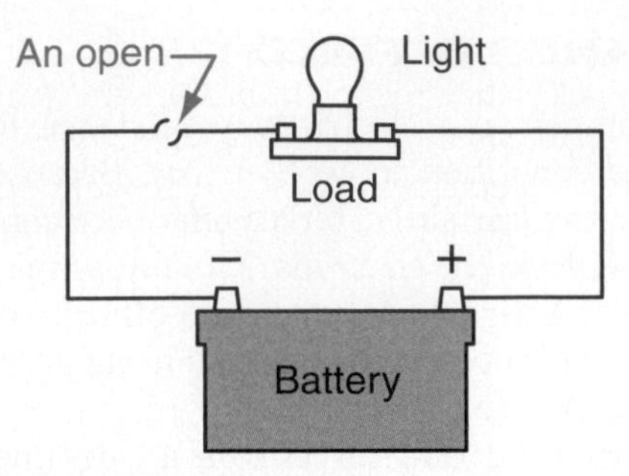

FIGURE 29–9 An open circuit contains a break that stops all current from flowing through the circuit.

An **open circuit** is one that has a break in the connection (***Figure 29–9***). This is called a break in continuity or an open. If the circuit is open, there is not a complete path for the current to flow through. An open circuit acts the same as if the circuit had a switch in the open position. Voltage drop across an open circuit is always the same as the source, or maximum, voltage.

A shorted circuit is one that allows electricity to flow past part of the normal load. An example of this is a shorted coil (***Figure 29–10***). The internal windings are usually insulated from each other. However, if the insulation breaks and allows the windings to touch each other, part of the coil will be bypassed. Any load can be partially or fully bypassed by having a shorted circuit. If a load is fully bypassed, the voltage dropped across the load will be zero.

A **grounded circuit** is a condition that allows current to return to the battery before it has reached its intended destination. An example is a grounded taillight (***Figure 29–11***).

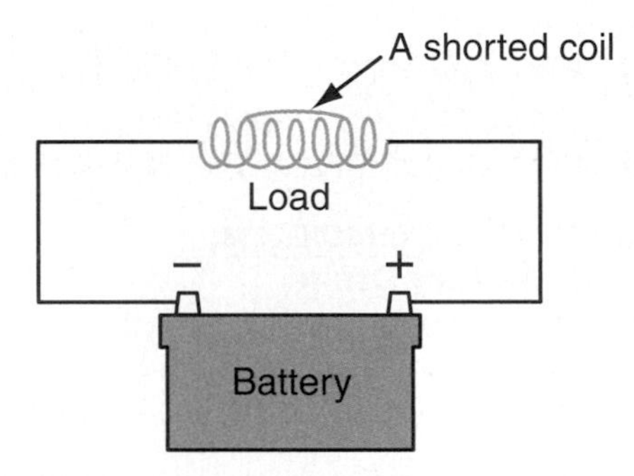

FIGURE 29–10 A shorted circuit will cause the amperage to bypass part of the load.

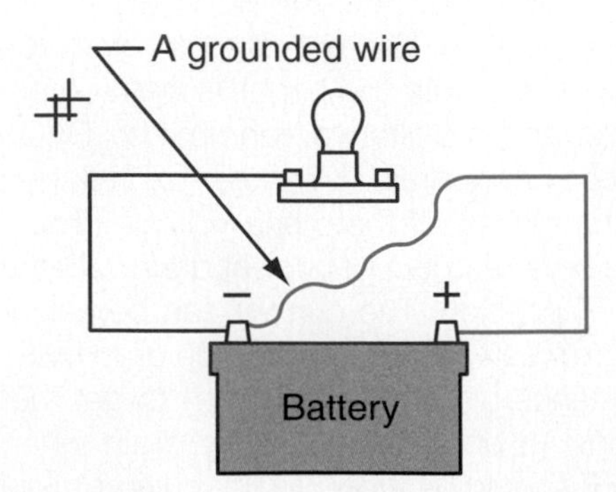

FIGURE 29–11 A grounded wire can cause excessive drain on the battery.

If a wire leading to the light were broken and touching the frame, the electricity would be grounded back to the battery. Grounded circuits can cause excessive current to be drained from the battery.

SERIES CIRCUIT

A **series circuit** consists of two or more resistors connected to a voltage source with only one path for the electrons to follow. An example is shown in ***Figure 29–12***. The series circuit has two resistors placed in the path of the electrons. The resistors are shown as R_1 and R_2. (The jagged line is a symbol used to represent a resistor.) All of the amperage that comes out of the positive side of the battery must go through each resistor, then back to the negative side of the battery. In a series circuit, the resistors are added together to get the total resistance (R total).

Series circuits are characterized by the following four facts:

1. The resistance is always additive. R total is equal to $R_1 + R_2 + R_3$, and so on.
2. The amperage through each resistor is the same. Amperage is the same throughout the circuit.
3. The voltage drop across each resistor will be different if the resistance values are different.
4. The sum of the voltage drops of all the resistors equals the source voltage.

PARALLEL CIRCUIT

Parallel circuits provide two or more paths for the current to flow through. Each path has separate resistors and operates independently from the other parallel paths. In a parallel circuit, amperage can flow through more than one resistor at a time. An example of a parallel circuit is shown in ***Figure 29–13***. Note that if one branch of the circuit breaks or has an open circuit, the remaining resistors can still operate. The resistance in a parallel circuit is calculated by the formula

$$R\ \text{total} = \frac{1}{1/R_1 + 1/R_2 + 1/R_3 + 1/R_4,\ \text{etc.}}$$

Total resistance in a parallel circuit will always be less than the resistance of the smallest resistor. Most circuits on the automobile are parallel circuits. If more resistors are added to the circuit, the total resistance will decrease. A parallel circuit has the following four characteristics:

1. The total resistance is less than the resistance of the lowest resistor.
2. The amperage flowing through the resistors is different for each if the resistance values are different.
3. The voltage drop across each resistor is the same. This is also the source voltage.
4. The sum of the separate amperages in each branch equals the total amperage in the circuit.

SERIES-PARALLEL CIRCUIT

A series-parallel circuit is designed so that both series and parallel combinations exist within the same circuit. ***Figure 29–14*** shows four resistors connected in a series-parallel circuit. To calculate total resistance in this circuit, first calculate the parallel portions, then add the result to the series portions of the circuit. Total current flow will be determined by the total resistance and the total source voltage.

GROUND SYMBOL

A ground symbol is sometimes used when analyzing a circuit. This means that the circuit is connected to the steel structure of the vehicle. The ground symbol is shown in ***Figure 29–15***. The symbol indicates that the electricity is

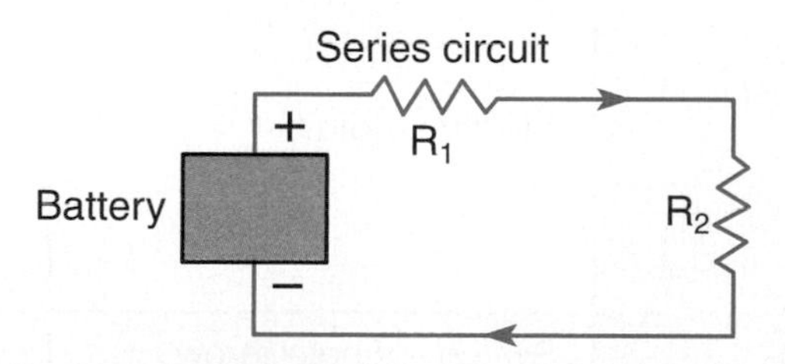

FIGURE 29–12 A series circuit allows only one path for the current to flow through.

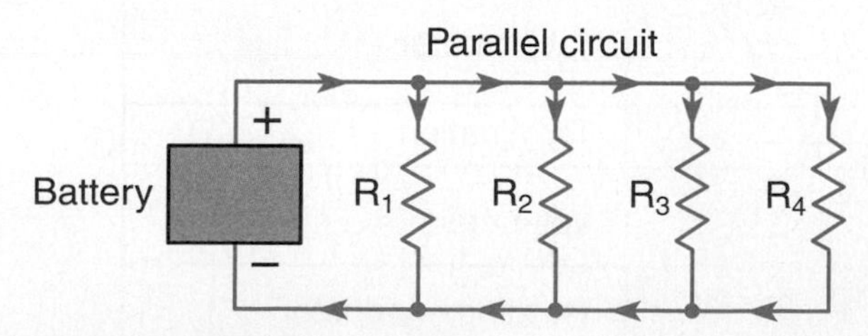

FIGURE 29–13 A parallel circuit has more than one path for the current to flow through.

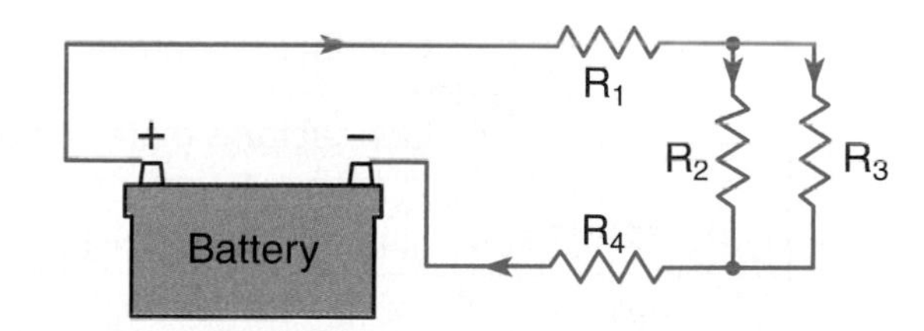

FIGURE 29–14 A series-parallel circuit, or a combination circuit, includes both series and parallel circuits.

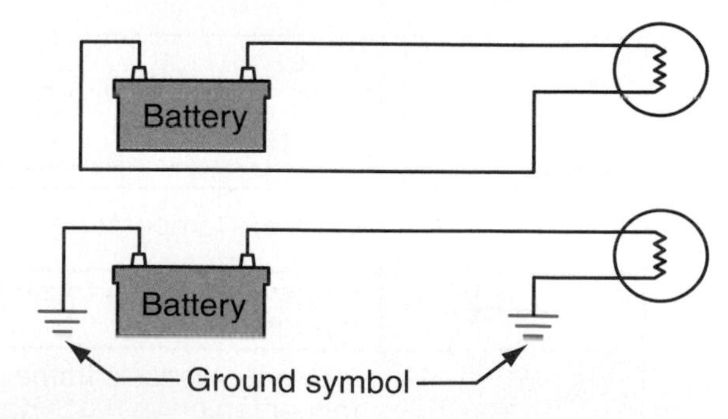

FIGURE 29–15 Ground symbols are used to make a circuit easier to read.

returning to the battery through the frame of the vehicle. Any steel structure of the vehicle can actually act as the wire that returns electricity to the battery. This could include the body sheet metal, frame, engine block, or transmission case.

ELECTRICAL SYMBOLS

A more complete listing of symbols used in automotive circuits is provided in ***Figure 29–16***. Many symbols are used in the wiring diagrams in this textbook and in the automotive repair manuals. It is important to become familiar with these symbols in order to analyze more involved circuits.

TESTING ELECTRICAL CIRCUITS

Once an electrical circuit is understood, it can be easily tested for faulty components, opens, grounds, shorts, and other problems. Electrical circuits are commonly tested using a voltmeter, an ohmmeter, or an ammeter.

AUTOMOTIVE ELECTRICAL SYMBOLS			
SYMBOL	REPRESENTS	SYMBOL	REPRESENTS
ALT	Alternator	HORN	Horn
A	Ammeter		Lamp or bulb (preferred)
	Battery – one cell		Lamp or bulb (acceptable)
	Battery – multicell	MOT	Motor – electric
12 V + −	The long line is always positive polarity	−	Negative
+ BAT −	Battery – voltage box	+	Positive
	Bimetal strip		Relay
	Cable – connected		Resistor
	Cable – not connected		Resistor – variable
	Capacitor	IDLE STOP	Solenoid – idle stop
	Circuit breaker	B SOL STARTING MOTOR	Starting motor
	Connector – female contact		Switch – single throw
	Connector – male contact		Switch – double throw
	Connectors – separable – engaged	TACH	Tachometer
	Diode		Termination
HEI	Distributor	V	Voltmeter
	Fuse	OR	Winding – inductor
FUEL	Gauge – fuel		
TEMP	Gauge – temperature		
	Ground – chassis frame (preferred)		
	Ground – chassis frame (acceptable)		

FIGURE 29–16 Various symbols are used in electrical circuits.

When using a voltmeter keep these rules in mind:

- Always ground the negative lead and probe the circuit with the positive lead.
- When connecting the test lead, the power should be off. After connecting the leads, turn the power on.
- Voltmeters are always connected in parallel with the circuit.
- Use a voltmeter when testing for an open circuit or failed component.
- Use a digital volt-ohmmeter with a 10-megohm or higher impedance rating when testing a circuit that has solid-state components.

When using an ohmmeter keep these rules in mind:

- Always select the highest range before making your ohmmeter check.
- When checking solid-state components, always use a digital ohmmeter with a 10-megohm impedance rating to prevent damage to the components.
- Either the component being checked should be removed from the circuit, or that part of the circuit should be disconnected.

When using an ammeter keep these rules in mind:

- Always start with the highest scale on the ammeter and work down until you are on the correct scale.
- Always connect the ammeter in series with the circuit.

CAPACITANCE

A capacitor is used on certain electrical circuits in the automobile. A capacitor is an electrical device used to store electrical energy. It is like a miniature battery. The electrical energy is stored in a capacitor as voltage. A capacitor uses the principle of capacitance to store electricity. Capacitance is the ability of an electrical circuit to collect a charge of electricity. When a capacitor is used in an electrical circuit, it is often used to absorb voltages as the voltage changes. For example, at times in an electrical circuit there may be damaging voltage spikes. In this case, a capacitor would be inserted into the circuit to absorb the voltage spikes and protect the circuit from damage. ***Figure 29–17*** shows several examples of capacitors that are used in automobile circuits.

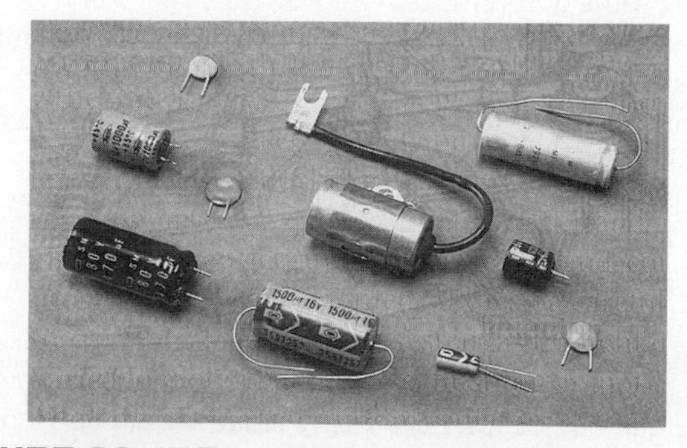

FIGURE 29-17 There are many styles and sizes of capacitors.

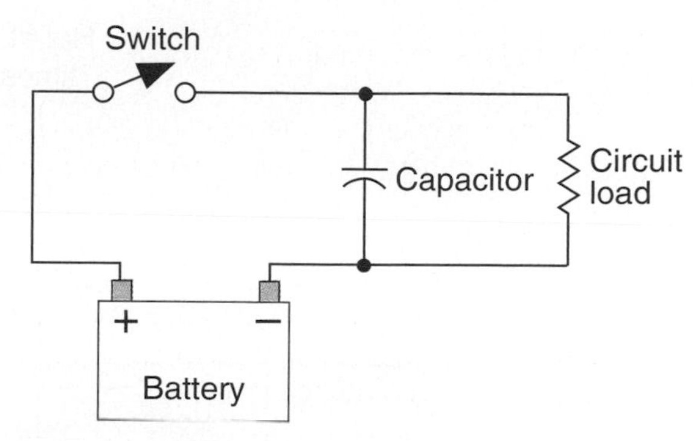

FIGURE 29-18 When the switch is closed, the electricity will flow to the circuit load and also will charge up the capacitor.

Figure 29–18 shows a circuit with a capacitor connected in parallel across a circuit load. The symbol for the capacitor is a straight line and a curved line as shown. In operation, when the switch is closed, the circuit load is operated normally. However, when the switch is closed, the capacitor also charges up and absorbs positive charges on the straight line and negative charges on the curved line, similar to a small battery. Depending on the type and size of the capacitor, it can be used to absorb voltage spikes or could be used for some other voltage need within the circuit. For example, say there was a need to operate a circuit for a short period of time after the switch is opened (a second or two). If the switch were opened, the capacitor could still discharge through the circuit load for a short period of time or until the capacitor is fully discharged.

ELECTRICAL SCHEMATIC

The electrical schematic is the service technician's key to solving many of the electrical problems in circuits. A schematic can be considered a road map used to follow the current through the circuit. The schematic shows the symbols used to indicate a junction, splice, a male or female connector, components, fusible links, fuses, and any other elements in the circuit. It tells the service technician how current gets from one point to another, through switches, fuses, and so on.

Remember that a schematic is not drawn to scale. For example, both a 6-foot and a 4-foot length of wire in the vehicle may be represented by a 2-inch line on the schematic. Also, the actual location and physical appearance of the components are not the same as in the vehicle. In most cases, schematics are read from the top down or from left to right. ***Figure 29–19*** shows a schematic used to analyze an antilock brake system (ABS). This schematic uses some of the following symbols.

- Wires are color-coded (for example, BK/GN means black with green strip).
- A number in front of the color shows the metric size of the wire.
- Relays such as the ABS relay and ABS control module are boxed in.

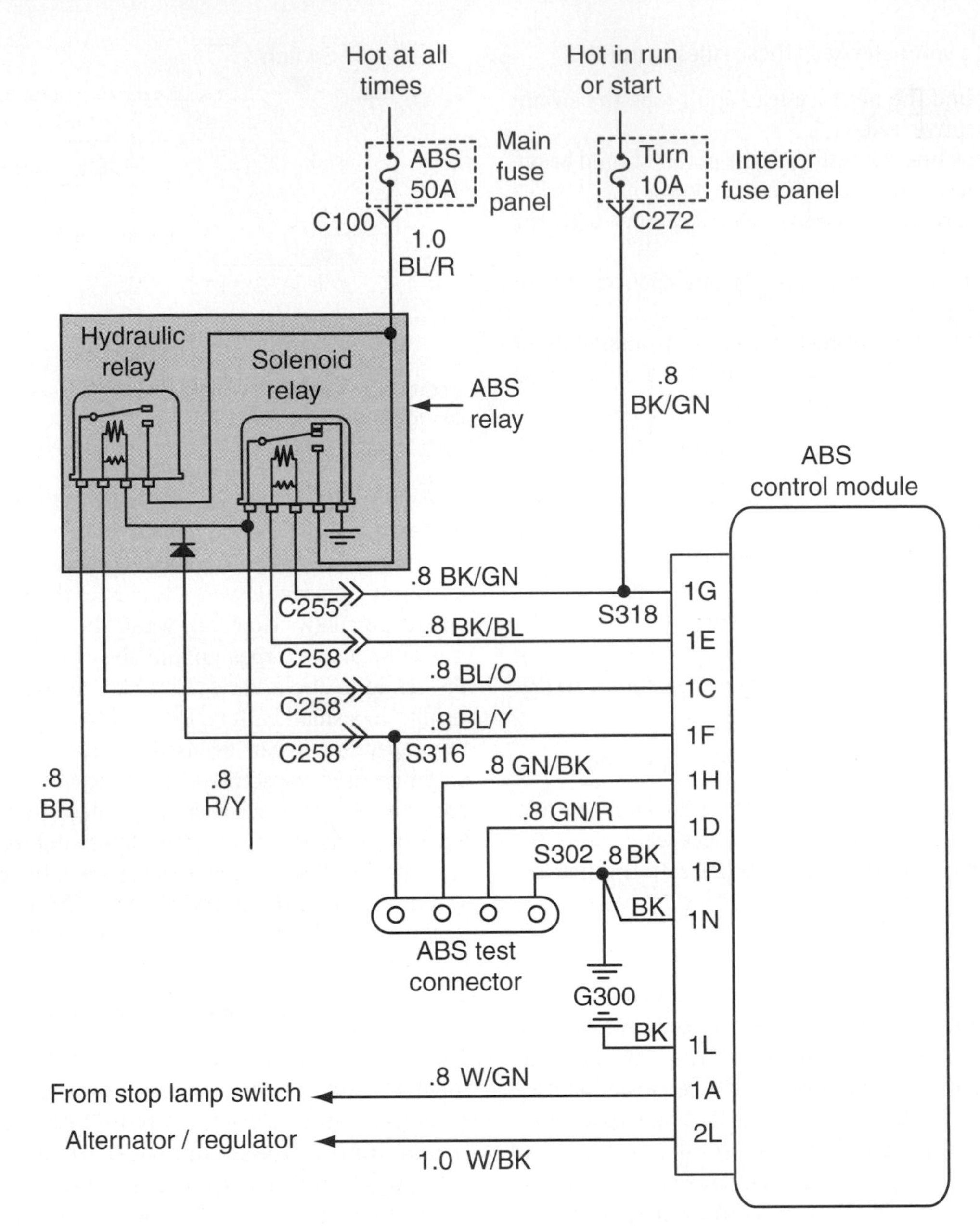

FIGURE 29-19 A schematic such as this is a tool the service technician can use to troubleshoot electrical circuits in automobiles.

- Test connector points are shown for reference.
- Fuses are enclosed in a dotted box.
- Certain wires have arrows to show direction of electricity.
- Ground points are shown by G300, G100, and so on.
- Connector points are shown by C255, C258, and so on.
- In-line splices are shown by S316, S318, and so on.

A more complete list of symbols used on schematics is shown in ***Figure 29–20***.

29.3 MAGNETISM

One area of study in automotive systems is that of magnetism. The principles of magnetism are integrated into motors, alternators, solenoids, and other electrical systems in the automobile. Magnetism can be best understood by observing some of its effects.

The effects of magnetism were first observed when fragments of iron ore, referred to as lodestones, were attracted to pieces of iron. It was further discovered that a long piece of iron would align itself so that one end always pointed toward the Earth's north pole. This end of the bar was called the north (N) pole, and the other end was called the south (S) pole. The bar was called a bar magnet.

DOMAINS

Inside the bar magnet are many small **domains.** Domains are minute sections in the bar where the atoms line up to

SYMBOLS USED IN WIRING DIAGRAMS			
+	Positive		Temperature switch
—	Negative		Diode
	Ground		Zenner diode
	Fuse		Motor
	Circuit breaker	C101	Connector 101
	Condenser	→	Male connector
Ω	OHMS	≻	Female connector
	Fixed value resistor		Splice
	Variable resistor	S101	Splice number
	Series resistors		Thermal element
	Coil		Multiple connectors
	Open contacts	88:88	Digital readout
	Closed contacts		Single filament bulb
	Closed switch		Dual filament bulb
	Open switch		Light emitting diode
	Ganged switch (N.O.)		Thermistor
	Single pole double throw switch		PNP bipolar transistor
	Momentary contact switch		NPN bipolar transistor
	Pressure switch		Gauge

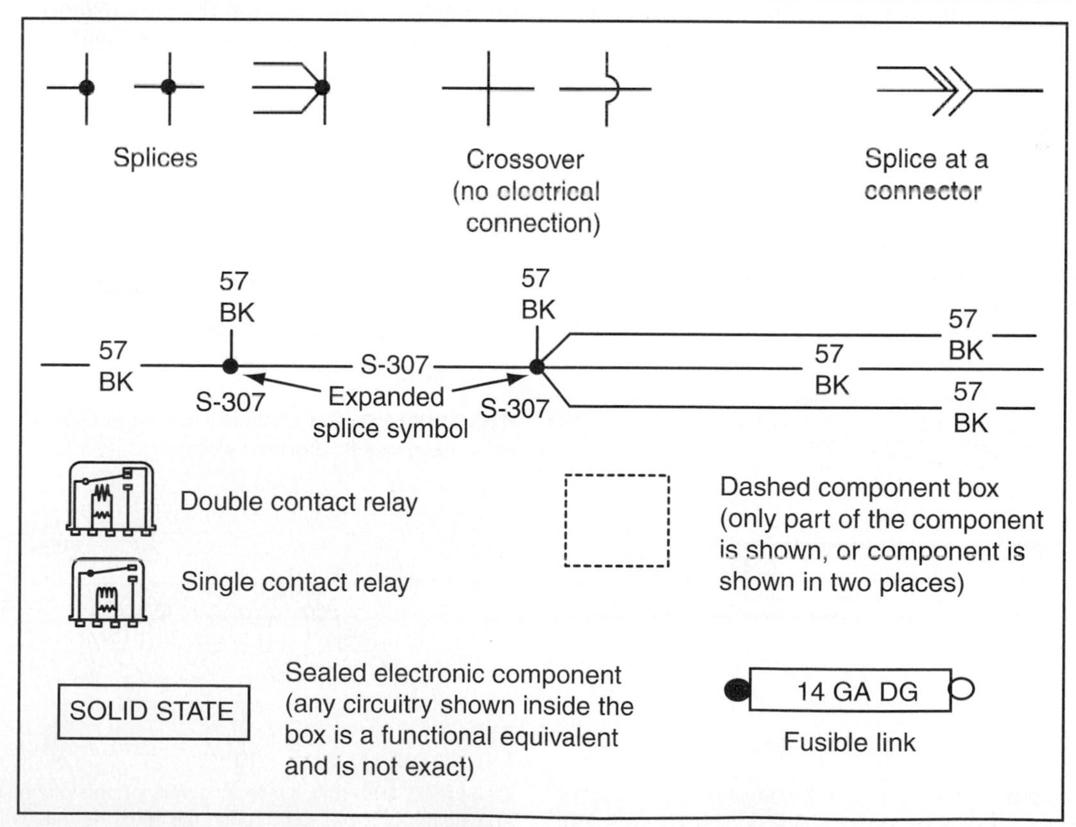

FIGURE 29-20 These are some common symbols used on schematics in service manuals.

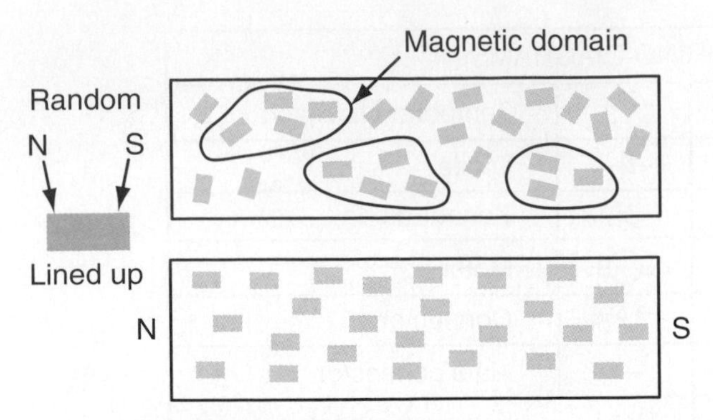

FIGURE 29-21 There is no magnetism in a metal bar when the domains are not lined up. When they are lined up, magnetism is produced.

produce a magnetic field. Most of the domains must be lined up in the same direction in the bar magnet to form a magnetic field. ***Figure 29–21*** shows a bar of metal with the domains located randomly and a bar with the domain lined up.

LINES OF FORCE

Magnets can be further defined by the **lines of force** being produced. The magnetic field is defined as invisible forces that come out of the north pole and enter the south pole (***Figure 29–22***). The shape of the magnetic lines of force can be illustrated by sprinkling iron filings on a piece of paper on top of the bar magnet. When the paper is tapped, the iron filings align to form a clear pattern around the bar magnet. Note that the lines of force never touch each other. Also note that the lines of force are more concentrated at the ends of the magnet.

REPULSION AND ATTRACTION

If two bar magnets are placed together at unlike poles, they will snap together. If the ends have the same poles, they will repel each other. This is shown in ***Figure 29–23***.

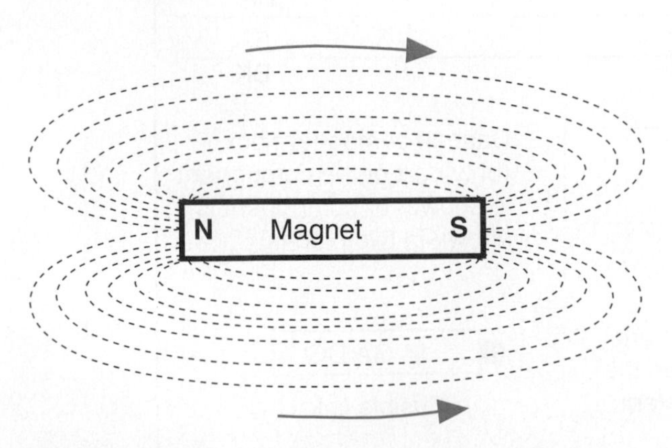

FIGURE 29-22 A bar magnet and iron filings show the invisible lines of force around a magnet. Magnetic lines of force always flow from the north pole to the south pole.

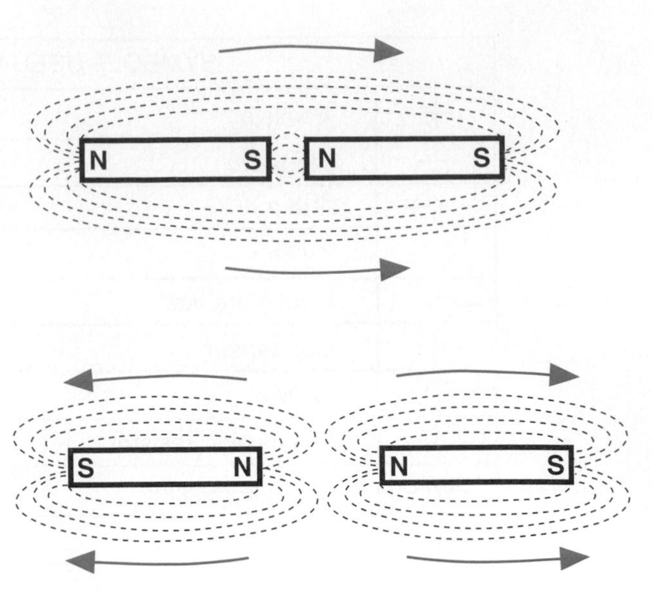

FIGURE 29-23 Unlike poles attract each other, whereas like poles repel each other.

ELECTROMAGNETISM

The bar magnet that was mentioned previously is called a permanent magnet. There are also temporary magnets that can be made from electricity. This can be done by wrapping an electrical wire around an unmagnetized bar to make an electromagnet. This is called **electromagnetism.**

When any wire has electricity flowing through it, a magnetic field develops around the wire (***Figure 29–24***). If the wire is then placed in the shape of a coil, as shown in ***Figure 29–25***, the magnetic field in the center of the coil will

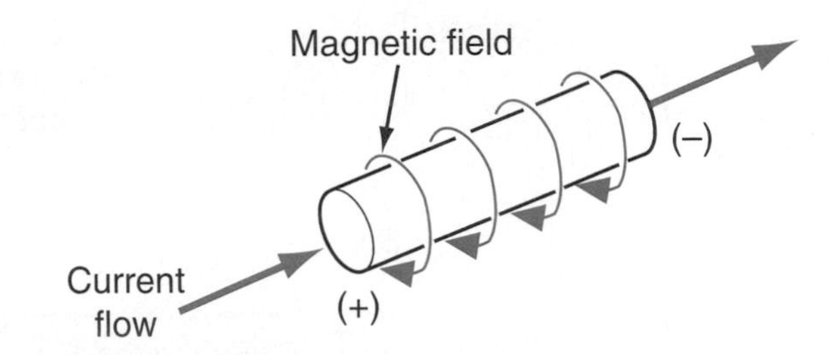

FIGURE 29-24 Electricity flowing through a wire produces a small magnetic field around the wire.

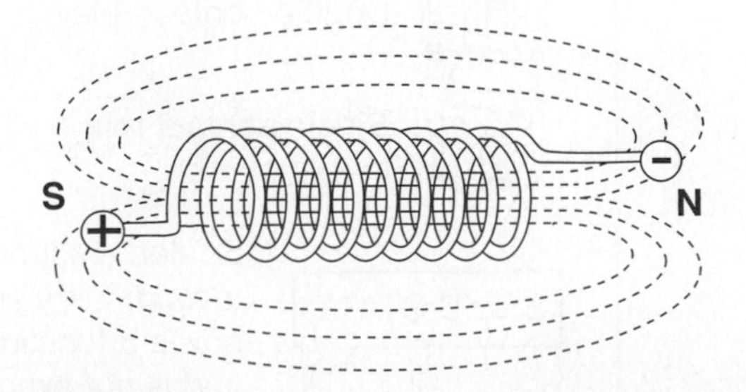

FIGURE 29-25 If a conductor is formed into the shape of a coil and electricity is passed through the wire, the magnetic field is additive in the center of the coil.

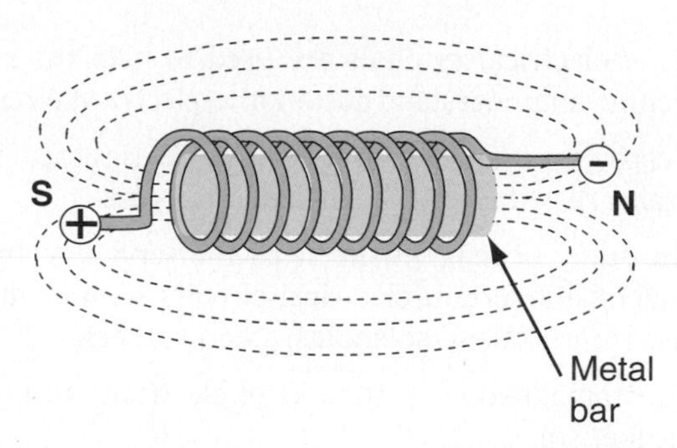

FIGURE 29-26 The strength of a magnetic field in a coil of wire can be increased by inserting an iron bar.

be additive. If a nonmagnetized bar is placed in the center of the coil, the bar will also be magnetized (***Figure 29–26***).

ELECTROMAGNETIC INDUCTION

Through experimentation, it was discovered that a conductor moving across or cutting a magnetic field would produce a voltage. This is called **electromagnetic induction**. Actually, an electromotive force (emf) will be induced or generated within the wire. Internally, a generator is converting mechanical energy to electrical energy. In the simplest form, ***Figure 29–27*** shows how voltage is produced with a magnet.

The direction of current flow produced by the voltage can be reversed if the movement is also reversed. In fact, if the wire were moved back and forth rapidly, the result would be alternating current.

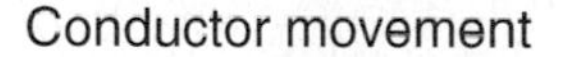

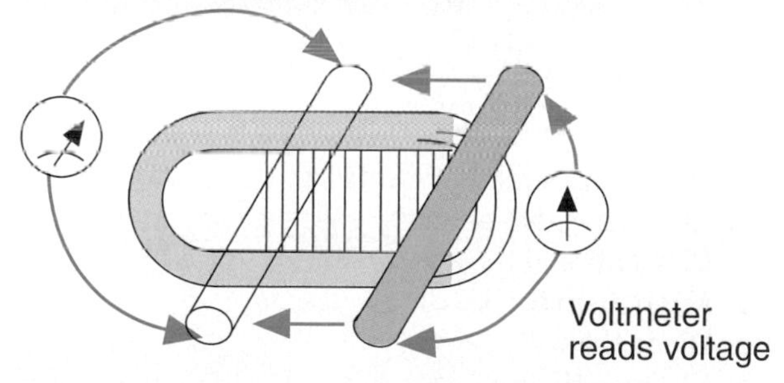

FIGURE 29-27 As a wire is moved through or cuts through a magnetic field, voltage is produced in the wire.

Figure 29–28 shows a simple generator used to produce voltage in a wire. Note that rather than using permanent magnets, electromagnets are used. Also, 12 volts have been applied to produce the electromagnets. This voltage is called the field voltage. If the center wire moves up, a voltage will be produced on the voltmeter. If the wire moves down, a reverse voltage will be produced. This means the positive and negative points will have been reversed.

Note that three things are necessary to induce a voltage in a generator. These include:

1. A magnetic field producing lines of force
2. Conductors that can be moved
3. Movement between the conductors and the magnetic field so that the lines of force are cut

If any one of the preceding factors increases or decreases, the induced voltage also increases or decreases. Note also that a conductor can be stationary while the lines of force are moved. This will still induce a voltage because the lines of force will be cut.

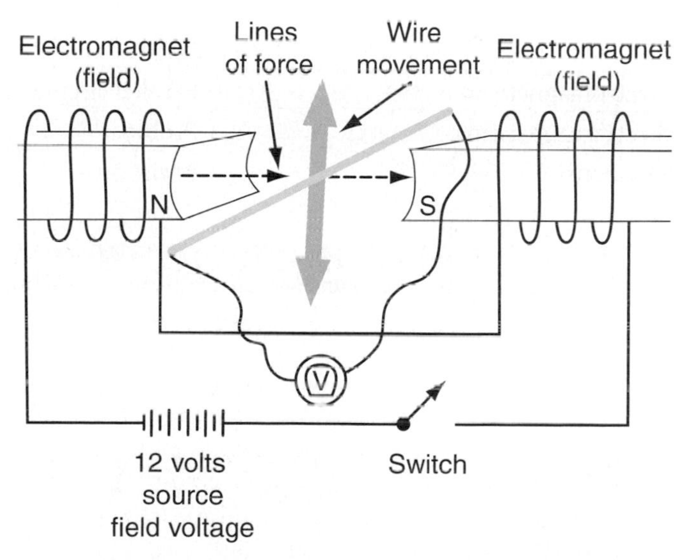

FIGURE 29-28 If a conductor is moved perpendicular to the lines of force from the magnetic, a voltage will be induced in the conductor.

SUMMARY

The following statements will help to summarize this chapter:

- The study of electricity and its associated theories is a basic part of understanding the complete automobile and its systems.
- The study of electricity involves an understanding of the atom, including protons, neutrons, and electrons.
- Electricity is the movement of electrons from atom to atom.
- Amperage is the number of electrons per second that flow from atom to atom.
- Voltage is the push or force needed to move the electrons.

- Resistance is the opposition to a current flow.
- Ohm's Law states that 1 volt will push 1 ampere through a resistance of 1 ohm.
- Voltage drop is the voltage lost through a resistor and can serve as an excellent troubleshooting tool for the service technician.
- When troubleshooting electrical circuits, always check for opened, shorted, or grounded circuits.
- The three circuits in an automobile electrical system are the series, parallel, and combination (series-parallel) circuit.
- Many electrical symbols are used to help the service technician understand automotive electrical circuits.
- A capacitor is an electrical device that uses capacitance to absorb voltages for a particular purpose.
- The study of magnetism helps the service technician understand certain electrical circuits such as motors, generators, relays, solenoids, among others.
- Electromagnetism is the use of electricity to produce magnetism.
- Electromagnetic induction is a process of producing electricity by moving an electrical conductor perpendicular to magnetic lines of force.

TERMS TO KNOW

Can you explain each of the following terms? Review the chapter until you can use each term correctly.

Amperage
Atoms
Conductor
Conventional theory
Domains
Electricity
Electromagnetic induction
Electromagnetism
Electrons
Electron theory
Grounded circuit
Insulator
Lines of force
Ohm's Law
Open circuit
Parallel circuits
Protons
Resistance
Semiconductors
Series circuit
Valence ring
Voltage
Voltage drop
Wattage
Watt's Law

REVIEW QUESTIONS

Multiple Choice

1. Which of the following have a negative charge in the atom?
a. Protons
b. Electrons
c. Neutrons
d. Watts
e. Volts

2. Which of the following materials has been identified as being a good conductor?
a. A material with four electrons in the valence ring
b. A material with three electrons in the valence ring
c. A material with five or more electrons in the valence ring
d. Glass
e. Wood

3. Which theory says that electricity flows from a positive point to a negative point?
a. Conventional theory
b. Magnetism theory
c. Diode theory
d. Electron theory
e. Valence ring theory

4. Which theory says that electricity flows from a negative point to a positive point?
a. Conventional theory
b. Magnetism theory
c. Diode theory
d. Electron theory
e. Valence ring theory

5. Pressure, or push, on the electrons is defined as:
a. Amperage
b. Voltage
c. Wattage
d. Resistance
e. Magnetism

6. Wattage is found by multiplying the voltage by the:
a. Resistance
b. Power
c. Amperage
d. Protons
e. Neutrons

7. Current is also called:
a. Amperage
b. Voltage
c. Wattage
d. Resistance
e. Magnetism

8. Which circuit has only one path for the electricity to flow in?
a. Series-parallel circuit
b. Parallel circuit
c. Series circuit
d. Open circuit
e. None of the above

9. A break in a wire is referred to as a/an:
a. Short
b. Open
c. Ground
d. Series circuit
e. Parallel

10. What is necessary to produce an induced voltage?
a. A magnetic field
b. A conductor
c. Movement between a conductor and magnetic field
d. All of the above
e. None of the above

11. A wire with electricity flowing through it:
a. Has no magnetic field around it
b. Has a magnetic field around it
c. Is usually considered an open circuit
d. Is usually considered grounded
e. Has maximum wattage

The following questions are similar in format to ASE (Automotive Service Excellence) test questions.

12. *Technician A* says that voltage drop is the same as resistance. *Technician B* says that voltage drop is the amount of voltage dropped across each resistor. Who is correct?
a. A only c. Both A and B
b. B only d. Neither A nor B

13. There is a break in an electrical wire. *Technician A* says this is called a short. *Technician B* says this is called grounded. Who is correct?
a. A only c. Both A and B
b. B only d. Neither A nor B

14. *Technician A* says that a voltmeter must be hooked up in series in a circuit. *Technician B* says that an ammeter must be hooked up in series. Who is correct?
a. A only c. Both A and B
b. B only d. Neither A nor B

15. *Technician A* says that on an electrical schematic, ground wires are shown. *Technician B* says that an electrical schematic does not show the color code on wires. Who is correct?
a. A only c. Both A and B
b. B only d. Neither A nor B

Essay

16. Define Ohm's Law.

17. Define and state the difference between an open, a short, and a ground.

18. Define voltage drop.

19. What are three characteristics of a series circuit?

20. What are three characteristics of a parallel circuit?

Short Answer

21. If an electrical circuit has 12 volts and 18 amperes, the circuit resistance is ________.

22. If the resistance of a component in an electrical circuit is 320 ohms and 0.014 ampere is flowing through that component, the voltage drop across the component is ________.

23. What is the total resistance of a parallel circuit that has five resistors, with each resistor having a resistance of 7 ohms? ________

24. What is the total resistance of a series circuit with the resistors of R_1 = 4 ohms, R_2 = 4 ohms, R_3 = 6 ohms, R_4 = 8 ohms, R_5 = 2 ohms, R_6 = 12 ohms? ________

25. The three things necessary to produce a voltage include a magnetic field, conductors that can be moved, and ________.

Applied Academics

TITLE: Ohm's Law

ACADEMIC SKILLS AREA: Science

NATEF Automobile Technician Tasks:

The technician can demonstrate an understanding of and explain the use of Ohm's Law in verifying circuit parameters (resistance, voltage, amperage).

CONCEPT:

The circuit shown here can be used to help understand Ohm's Law. Note that there are four resistors. Three 9-ohm resistors are in parallel with each other. The other 3-ohm resistor is in series with the three parallel resistors. The resistors are identified as R_1, R_2, R_3, and R_4. Also, there are 12 volts applied to the total circuit identified as E_t. At each resistor, there is a current, identified as I_1, I_2, I_3, or I_4. There is also a voltage drop that occurs at each resistor, identified as E_1, E_2, E_3, or E_4.

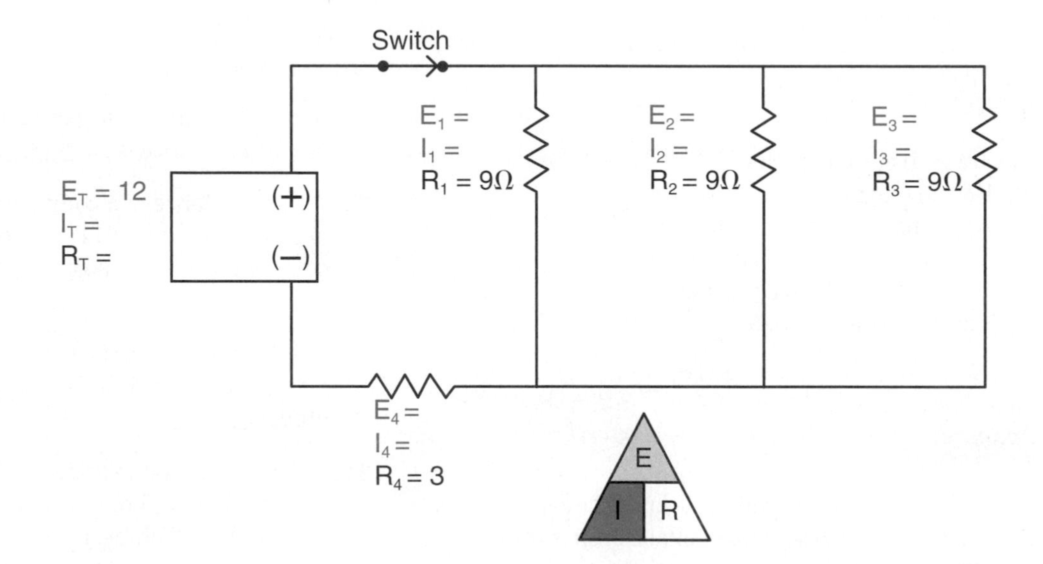

APPLICATION:

Based on the information given in the circuit and applying the principles of Ohm's Law, calculate the following: (1) total resistance or R_t, (2) total current or I_t, (3) voltage drop at each resistor (E_1, E_2, E_3, and E_4), and (4) current flowing through each resistor I_1, I_2, I_3, and I_4. Remember that Ohm's Law can be calculated for the totals and/or at the individual resistors.

Computer Principles

OBJECTIVES

After studying this chapter, you should be able to:

- Identify the fundamentals of basic electronics and solid-state circuitry.
- Describe the use of diodes, transistors, integrated circuits, and microprocessors in the automobile.
- Identify basic computer terminology.
- Describe how computers communicate.
- Define the different types of computer memory.
- Identify automotive computer locations and names.
- Examine the major inputs and outputs used on an automotive computer.
- Explain computer terminology such as binary codes, interfaces, and data links.
- Describe how to use fault codes.
- Define the purpose of using pinpoint tests.
- Identify how to clear the memory on a computer.
- Examine how to test and replace a computer.

Certification Connection

ASE Connection: The information in this chapter can help you prepare for the National Institute for Automotive Service Excellence (ASE) certification tests. The tests and content areas most closely related to this chapter are:

Test A1—Engine Repair

- **Content Area**—Fuel, Electrical, Ignition, and Exhaust System Inspection and Service

Test A6—Electrical/Electronic Systems

- **Content Area**—General Electrical/Electronic System Diagnosis

Test A8—Engine Performance

- **Content Area**—Computerized Engine Controls Diagnosis and Repair
- **Content Area**—Engine Electrical Systems Diagnosis and Repair

NATEF Connection: Much of the information in this chapter is related to the NATEF tasks. The NATEF tasks and priority numbers most closely related to this chapter are:

1. Use wiring diagrams during diagnosis of electrical circuit problems. P-1
2. Inspect and test switches, connectors, relays, and solid-state devices; determine necessary action. P-1
3. Retrieve and record stored OBDI diagnostic trouble codes; clear codes. P-2
4. Retrieve and record stored OBDII diagnostic trouble codes; clear codes. P-1

INTRODUCTION

Today's automobiles use many computers to control various engine and vehicle functions. Computers control the fuel, air, cooling, starting, charging, emission, and ignition systems. In addition, computers control transmissions, braking systems, all-wheel drive systems, suspension systems, and air bag systems. Furthermore, computers will play a significant role in controlling hybrid cars as they continue to be improved and incorporated into our society. Because of the extensive use of computers in vehicles, it is important to have a basic knowledge of computers, computer terms, and computer logic. This chapter will help you understand solid-state electrical components, computers, their basic terminology, and how they are applied to an automobile.

30.1 SOLID-STATE COMPONENTS

SEMICONDUCTORS

One area of electrical study that has continued to grow is that of electronics. Electronics is the study of solid-state devices such as diodes, transistors, and integrated circuits. Solid-state devices are those that have no moving parts except internal electrons. They are used in computer circuits.

To begin, let's review the study of conductors and insulators presented in the previous chapter. It was mentioned that any material that has four electrons in its outer orbit is called a semiconductor (***Figure 30–1***). This means that it could be either a good conductor or a good insulator.

Semiconductors are also called solid-state devices. Because of their characteristics, semiconductors are often used as switches. Circuits can be turned off and on by semiconductors with no moving parts.

DIODES

The **diode** is a semiconductor that permits current to flow through a circuit in one direction but not in the other. ***Figure 30–2*** shows a circuit with a diode. Say the alternator produces a current that flows back and forth 60 times per second. This is called ac (alternating current) voltage. In the battery, however, direct current (dc) voltage is used to charge the battery. The battery is considered the load. The diode can be used to convert the ac to dc. In ***Figure 30–2***, electricity is able to flow only in one direction. If electricity tries to flow in the opposite direction, it will be stopped. There will be no current flow in the reverse direction.

CUSTOMER CONCERN:
Checking the Condition of a Diode
A service technician has been working on an electrical circuit and thinks that a diode is bad. How can a typical diode be checked easily?

SOLUTION: A diode is an electrical device that allows electricity to flow in one direction but stops the current flow in the opposite direction. The condition of a diode can be checked by using an ohmmeter. Check the resistance of the diode in one direction. Then check the resistance of the diode in the other direction. One direction should have zero or very little resistance. The other direction should then have an infinite amount of resistance. If both directions have zero resistance, the diode is bad. If both directions have infinite resistance, the diode is also bad.

TRANSISTORS

The **transistor** is also a type of semiconductor. In this case, the transistor has some semiconductor material added to it. A circuit using a transistor is shown in ***Figure 30–3***. In operation, the circuit to be turned off and on is identified as circuit A. Circuit B is the controlling circuit. The transistor has three wires, called the base, the emitter, and the collector.

If a small amount of current flows (when the switch is closed) in the emitter-to-base circuit, the resistance be-

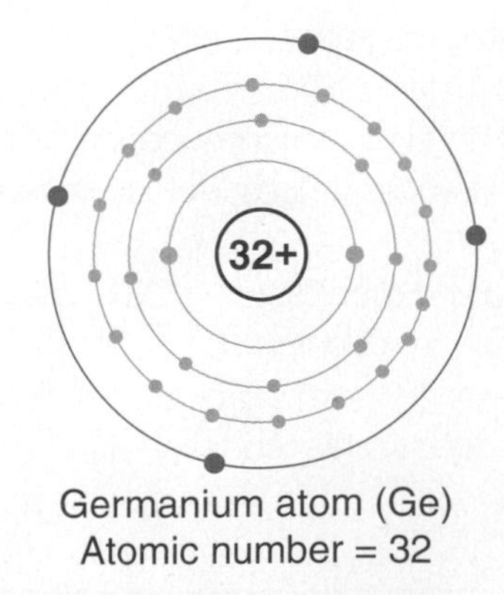

FIGURE 30–1 Semiconductors have only four electrons in the outer orbit.

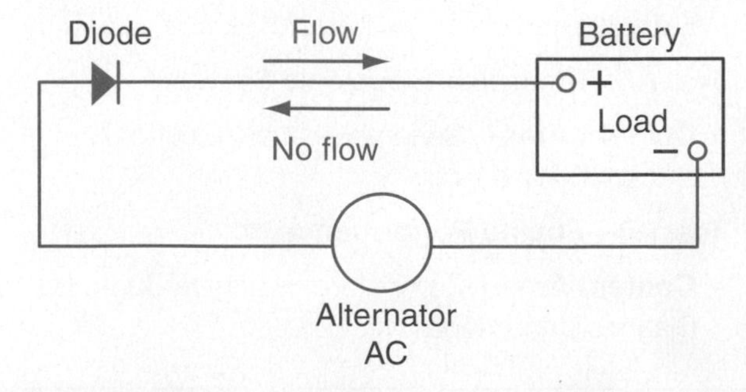

FIGURE 30–2 A diode in a circuit allows electricity to flow in only one direction.

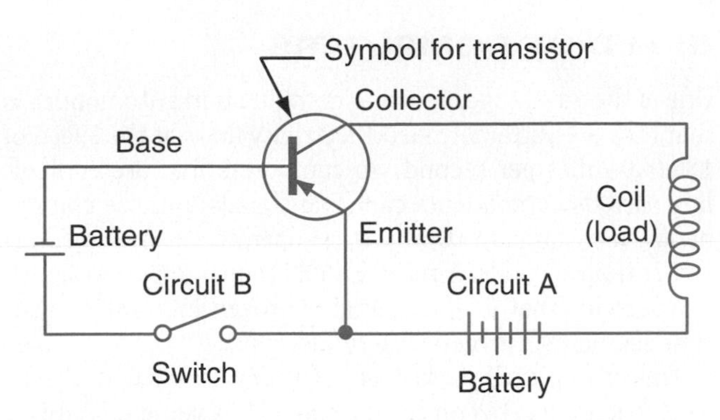

FIGURE 30–3 A transistor can control the on-off sequence of the coil. Circuit B controls circuit A.

tween the emitter and collector circuits will be zero. Circuit A is then turned on to operate the coil. (This could be a primary coil in an ignition system.) When the current stops flowing in circuit B (switch open), the resistance between the emitter and collector is very high. This resistance shuts off circuit B.

Transistors can also be used to amplify the on-off sequence of a signal. A small amount of current flowing on and off in circuit B can control a large amount of current flowing in circuit A. The on-off sequence will be amplified from circuit B to circuit A.

INTEGRATED CIRCUITS

Over the past several years, engineers have found ways to make diodes and transistors extremely small. Because they are smaller, many diodes, transistors, and other semiconductors can be placed on a board called an **integrated circuit** (IC). These circuits may contain many semiconductors. More recently, the chip has been designed to incorporate even smaller components on the integrated board.

The integrated circuit manufacturing process uses an etching and photographic process to produce thousands of components on a silicon slice. ***Figure 30–4*** shows an example of a cylindrical silicon crystal. As the crystal is cut into small chips, each slice is etched through a photographic process and made into thousands of transistors and other semiconductor components. In fact, a single chip may contain more than 1 million semiconductor components. The chips, about 1/4 inch or less in diameter, are then placed into a housing that has the necessary connections to make a complete circuit. The completed chip is then installed into the microprocessor or computer used in the automobile. This type of manufacturing allows computers to handle more than 1 million instructions per second.

Some applications that use integrated circuits and chips in the automobile include solid-state ignition, electronic fuel injection, electronic engine systems, computer-controlled combustion, and speed and cruise controls. Other circuits are constantly being developed to further control automobiles with electronic components. Many of these will be discussed in other chapters in this book.

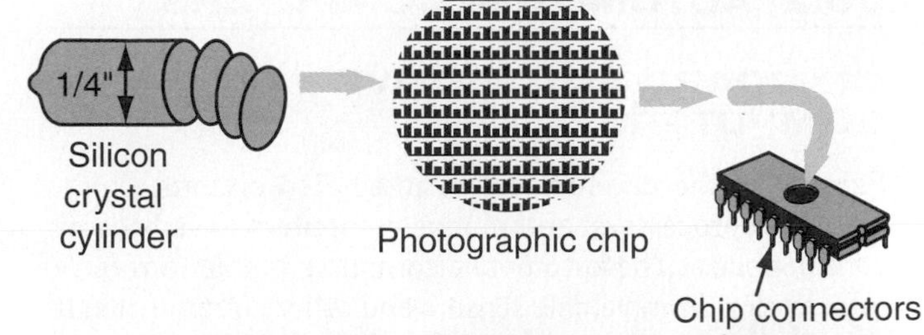

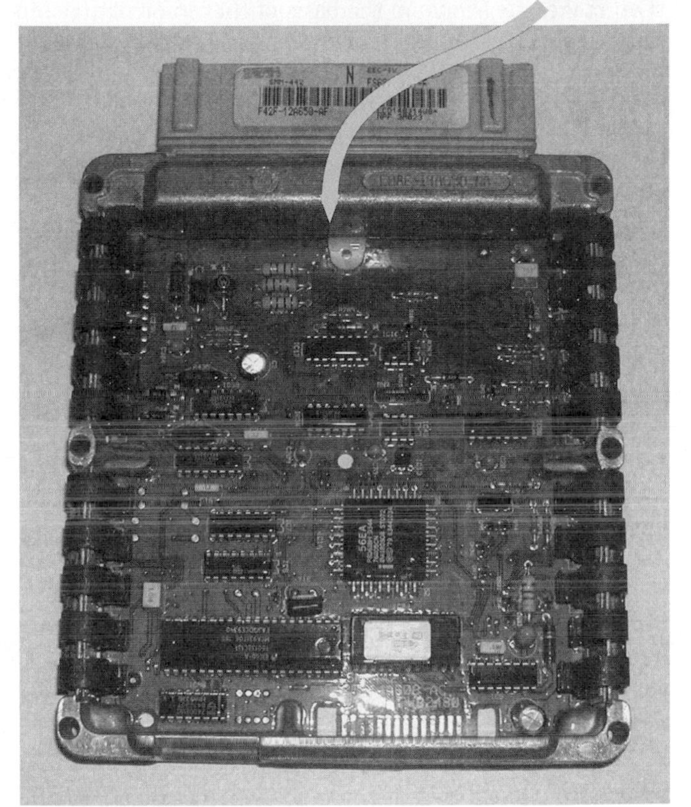

FIGURE 30–4 Silicon crystals are made into chips, which are etched with a photographic process. The chips are then placed in a housing with the proper connectors as part of the total computer circuitry.

MICROPROCESSORS

With the addition of integrated circuits, computers, and chips, the automobile is using more and more **microprocessors**. Microprocessors are small computers that can be used for a variety of tasks. They contain logic and control circuits. Various sensors relay input information in the form of electrical pulses to the microprocessor. Engine speed, temperature, outside weather conditions (barometric pressure), load and weight distribution of the vehicle, vehicle speed, throttle position, and so on can be fed into the microprocessor. The microprocessor uses this input information to control other electrical circuits in the vehicle to achieve optimum performance.

30.2 AUTOMOTIVE COMPUTERS

ON-BOARD AUTOMOTIVE COMPUTERS

Because of the development of integrated circuits, chips, and microprocessors, automotive computers have become commonplace. The automotive computer is able to receive information from vehicle sensors and other components. It then makes decisions on the basis of that information and takes actions as a result of those decisions. The microprocessor inside the computer does the calculations and makes decisions of exactly what to control.

When automobile computers do not work, they are generally replaced. The automotive service technician must first confirm whether the computer is working correctly or not. The troubleshooting processes discussed in earlier chapters help to do this. However, there are various terms, checks, and systems that will help the service technician understand and work with these computers.

COMMUNICATING WITH A COMPUTER

Computers communicate by voltage, or electrical, signals. Voltage signals transmit information in three ways:

1. By changing voltage levels or amounts
2. By changing the shape of the voltage pulses
3. By changing the speed at which the signals switch levels

In addition, voltage causes current to flow to working devices such as solenoids, relays, and lamps.

SPEED OF COMPUTERS

One of the advantages of using computers in automobiles is that they are extremely fast. Electricity flows at the speed of 186,000 miles per second. So computers that are controlling various components can send signals to these components very quickly. Computers operate in milliseconds rather than seconds. A **millisecond** is equal to a thousandth of a second. That is, if a second is broken into 1,000 equal time segments, each one is a millisecond.

One example of how fast a computer works is with the air bags that are located on the dashboard of a vehicle. Think of a vehicle that has a head-on collision going 35 miles per hour. When the front of the vehicle hits the other vehicle, the collision is instantly sensed by a sensor near the front bumper. The sensor then sends a signal to the computer. The computer in turn sends a signal to the air bags to inflate. As they inflate, the passengers are protected from hitting the dashboard. All of this must happen before the momentum of the head-on crash moves the passenger into the dashboard. This must happen in milliseconds. Computers are the only known method to achieve this speed.

COMPUTER LOCATIONS AND NAMES

Today numerous computers are used in automobiles. The number of computers and the type depend on the manufacturer, the year of the vehicle, and the model of the vehicle. ***Figure 30–5*** shows a view of a vehicle with many of the computers used today. Note that the location of each computer may be different for each manufacturer, year, and

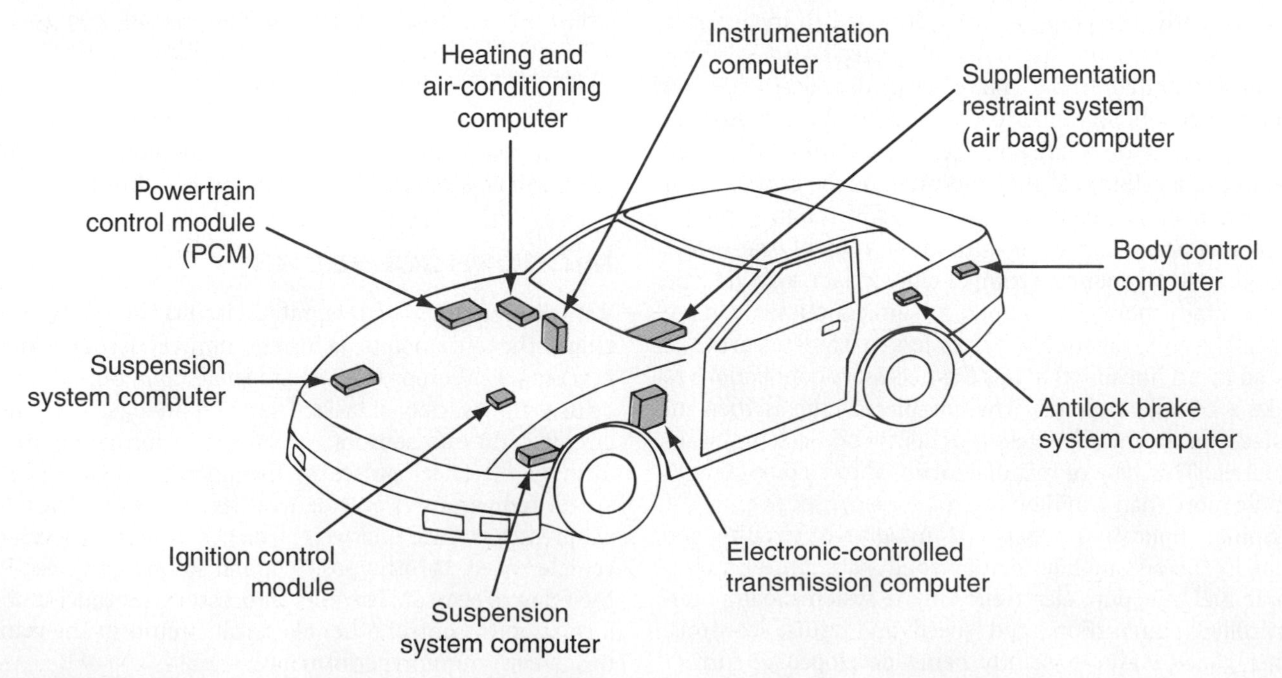

FIGURE 30–5 There are many computers used in today's automobiles.

model of vehicle. In addition, certain functions may be combined into one computer rather than by using several computers. For example, the body control computer, also called the **body control module**, may operate several functions, such as the suspension system controls and the supplemental restraint system.

In the past, there were many names used in the service manuals to identify computers used on automotive vehicles. Some of the more common ones included:

- ECU—electronic control unit
- ECM—engine control module
- ECM—electronic control module
- ECA—electronic control assembly
- ECCS—electronic computer control system
- ECECS—electronic constant engine control system
- BCM—body control module

Today, however, the most common computer name is the PCM. The PCM refers to the powertrain control module, sometimes called the power control module. The two terms are used interchangeably within the automotive industry. The PCM is the main computer used to control the engine and transmission functions.

PARTS LOCATOR

*Refer to photos #48, #49, #50, and #51 in the front of the textbook to see what a **vehicle computer** looks like.*

COMPUTER INPUTS AND OUTPUTS

Computers are much like a human brain. Think of the sensation in the tip of your finger when you touch a hot frying pan. The nerves in the tip of your finger act as an input to the brain. The brain then sends a signal back as a reflex to pull your finger away from the hot pan. This is called the output. Computers work in much the same way. Various inputs from an automobile are sent to the computer. These inputs come from various sensors and switches. Then, based on these inputs, the computer makes a decision and controls some type of output. The outputs are usually in the form of electrical signals to control a solenoid, a relay, or some display on the dashboard.

Figure 30–6 shows an example of the overall system. A simple example would be the use of fuel injection. Although there are more, this example shows three sensors sending input to the computer. They include the coolant temperature sensor, the oxygen sensor, and the vehicle speed sensor. The oxygen sensor measures the oxygen in the exhaust stream after combustion. The coolant sensor measures the temperature of the engine coolant. The vehicle speed sensor measures the speed of the vehicle. Based on these three inputs, the computer then makes a decision about how much fuel to put through the fuel injector.

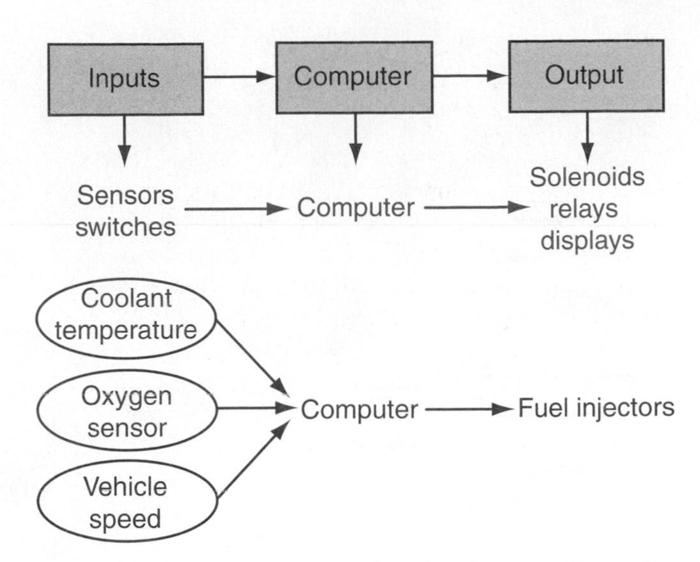

FIGURE 30–6 A computer reads various inputs and then makes a decision based on these inputs so that an output can be controlled.

TYPES OF INPUT DEVICES

Various types of sensors are used to feed information into the automotive computer. ***Figure 30–7*** shows the symbol for each. The following are the five most common types of sensors used throughout the automobile.

1. *Potentiometer*—This type of sensor can change voltage in relationship to mechanical motion. For example, as the throttle is moved to change the fuel setting, the throttle position sensor changes the voltage to the computer. The computer is able to read the change in voltage and so control some type of output.
2. *Variable resistor*—This sensor changes its electrical resistance on the basis of a change in temperature. For example, the computer often needs to know the engine coolant temperature. As the coolant increases in temperature, the resistance of the sensor changes. The computer is then able to sense this change in resistance. This type of variable resistor is also called a thermistor.
3. *Magnetic pickup*—This type of sensor is used to measure the speed of a rotating object. In operation, a small magnet is placed in the rotating object. As the shaft spins, the small magnet induces a voltage in a small coil. This charge or pulse of electricity in the coil can be sensed by the computer. A good example of a **magnetic pickup sensor** would be the vehicle speed sensor. As the wheels on the vehicle rotate, they produce a small charge or pulse of electricity. The computer can then take this pulse and determine the speed of the rotating wheel.
4. *Voltage generator*—This type of sensor produces a voltage based on certain inputs, such as oxygen. For example, the oxygen sensor is basically a small voltage generator. It is mounted in the flow of exhaust from the engine. Since this sensor is very sensitive to the presence of oxygen, it can develop a voltage. Lean

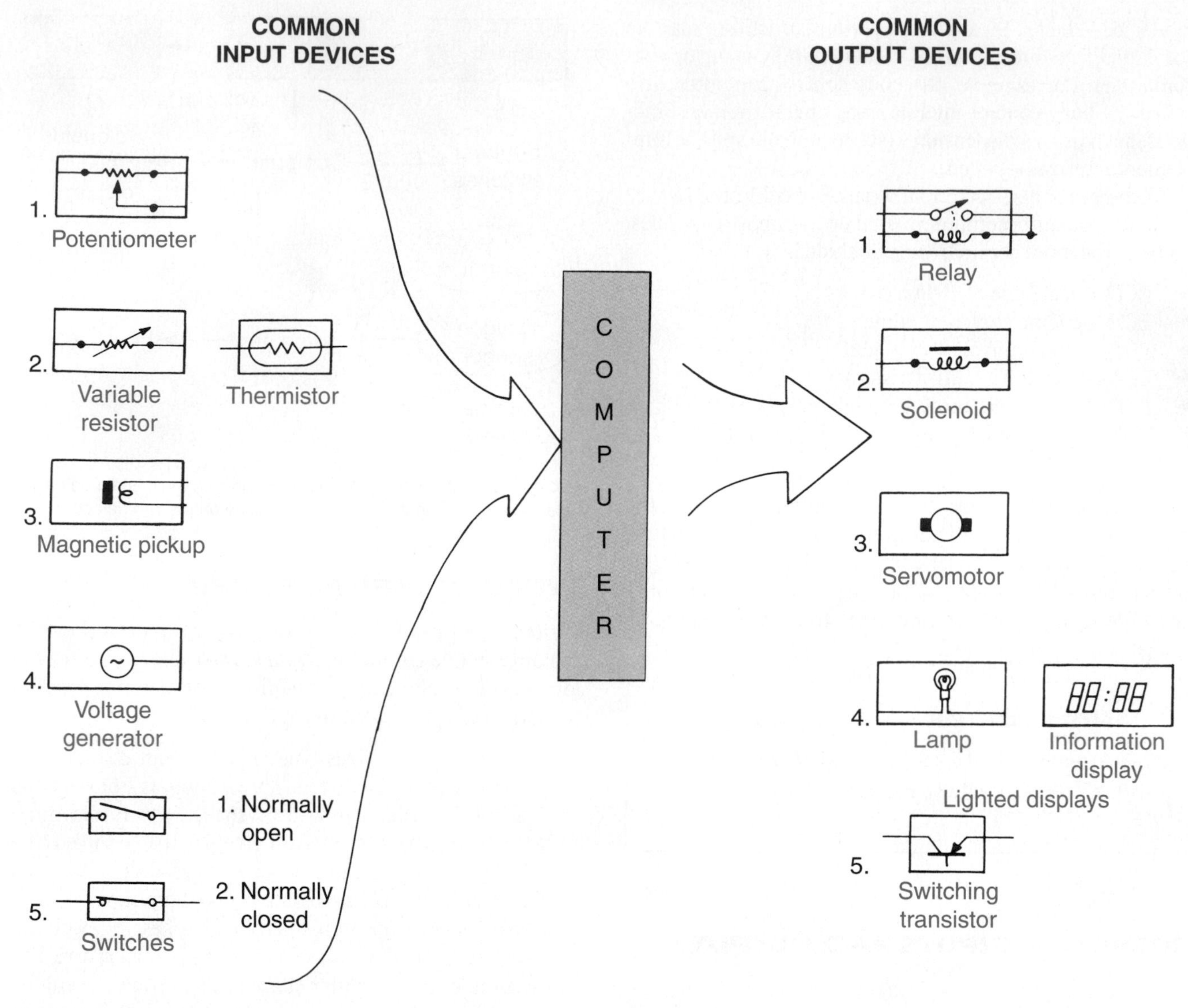

FIGURE 30-7 Symbols for common input and output devices are shown for reference.

mixtures produce a low voltage, while rich mixtures produce a high voltage.

5. *Switch*—This type of sensor is the most common and easiest to understand. It simply indicates an on-off position. For example, at times the computer needs to know if the air-conditioning system is off or on.

TYPES OF OUTPUT DEVICES

A computer is used to control various functions. ***Figure 30–7*** includes symbols for each output device. There are many types of output devices, but certain common ones include:

- *Relay*—This type of output device uses a small coil to control the opening and closing of another circuit. For example, the computer needs to control the off and on of a coolant fan. A **relay** is used for this purpose.
- *Solenoid*—This type of output device uses a coil of wire to move a mechanical rod or metal part inside the coil with the use of magnetism. One common type of **solenoid** is the action of a port fuel injector or the mixture control solenoid in an electronic carburetor.
- *Servomotor*—This type of output device is a motor used to turn something. The motor could be used in a variety of applications such as the fuel pump motor.
- *Lighted display*—This type of output device activates a light. For example, the liquid crystal displays or other light devices on a dashboard are such outputs.
- *Switching transistor*—This type of output device is a transistor that is energized by the computer. The computer energizes the base current so that a circuit with a greater current can be controlled. This type of output is similar to a relay, only the device is a solid-state component rather than a mechanical device.

ANALOG AND DIGITAL COMPUTER SIGNALS

Computer communications involve two kinds of signals: analog and digital signals. An **analog signal** is a continuously variable signal and can be represented by any voltage within a given range. For example, temperature- and pressure-sensing signals (coolant temperature, oil pressure) give off an analog signal. Analog signals are generally used when a certain condition is gradually or continuously changing.

The second type of signal, called **digital signal**, has specific values, usually two. There is no ability for the signal to represent any in-between values. Digital signals are useful when there is a choice between only two alternatives such as 1 or 0, yes or no, on or off, high or low, and so on. Digital signals are often shown graphically as a square wave signal. ***Figure 30–8*** shows a graphical comparison of analog and digital signals.

BINARY CODES

A computer is able to string a series of digital signals into useful combinations of binary numbers. **Binary codes** are made up of combinations of 1 and 0. Each digital signal (1 or 0) is called a *bit*. For example, voltage above a certain given value converts to a 1. Voltage below a certain value converts to a 0. A combination of eight bits is called a *byte*.

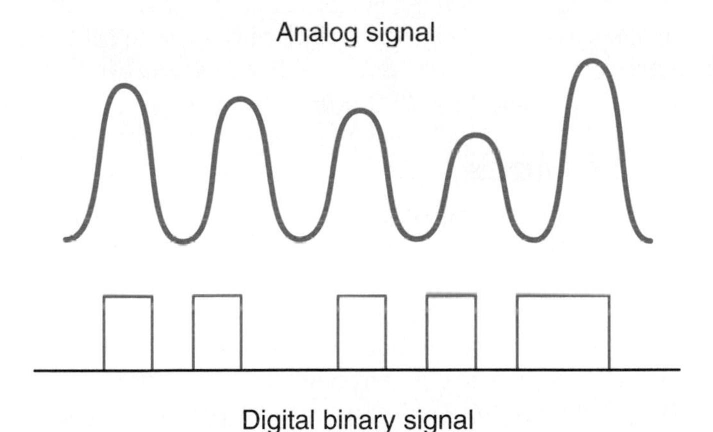

FIGURE 30–8 Analog signals are continuously changing, whereas digital signals are either off or on.

Figure 30–9 shows how a typical byte can be determined. For each bit, 1 or 0 is represented. Eight bits are shown making up the total binary number of 11001011. Under each bit a power is also shown. The power, a numerical value, is used to represent each bit that has a 1. The power starts from the right side and increases to the left. Starting with 1 on the right side, each number is doubled. This is called the power. To determine the numerical value of the byte, multiply each bit (1 or 0) by the power, then add the powers together. In this case the numerical value of the binary number 11001011 is 203.

Automotive computers started out using eight-bit microprocessors. However, computers that have sixteen bits or thirty-two bits are now common. These computers are able to communicate much more information than the old eight-bit computer.

Binary codes also enable computers to perform mathematical functions very rapidly. The ability of these computers to manipulate large series of digital signals almost instantaneously makes them ideal for automotive use. In order for a computer to be able to perform mathematical functions, there must be a relationship between decimal numbers and binary numbers. ***Figure 30–10*** shows the binary number made of zeros and ones for each number from 1 to 10.

DECIMAL	BINARY
0	0
1	1
2	10
3	11
4	100
5	101
6	110
7	111
8	1000
9	1001
10	1010

FIGURE 30–10 Each decimal number on the left has a corresponding binary number on the right.

Byte
Bit

Digital signal →								
Binary number →	1	1	0	0	1	0	1	1
Power →	128	64	32	16	8	4	2	1
Conversion →	128	64	0	0	8	0	2	1 = 203

FIGURE 30–9 Digital signals, or bytes, can be converted to numeric values.

COMPUTER MEMORY

The microprocessor in a computer is not able to store any information. However, memory chips are placed in the computer to allow it to store binary code information until it is needed. Basically, three types of memory are used in computers. The exact titles will vary; however, the following are very common:

- ROM—read-only memory
- RAM—random access memory
- PROM—programmable read-only memory

ROM (read-only memory) contains permanent information that cannot be changed. When the computer is manufactured, the programs that control the microprocessor are stored in the ROM. The ROM is soldered into the computer and generally cannot be removed by a service technician.

RAM (random access memory) can be both read from and written into. It is often used for temporary storage of information that changes. For example, if the computer receives information from the coolant temperature sensor, it may be stored in RAM and then read out each time it is needed. The RAM can also be called the "learning" portion of the computer or microprocessor.

There are two types of RAM—a nonvolatile RAM and a volatile RAM. The nonvolatile RAM holds its information even after the power has been removed. A good example of the type of information stored in a nonvolatile RAM is the information necessary for digital displays on the dashboard. Volatile RAM is erased when the power is shut off. In an automotive computer system, a volatile RAM can be connected to the battery through a fuse or fusible link after the ignition has been shut off. This arrangement is called keep-alive RAM. However, if the battery goes dead the keep-alive RAM will lose its information.

PROM (programmable read-only memory) is much like ROM only it can be removed from the computer and is easily accessed. The PROM chip shown in ***Figure 30–11*** can be removed from the computer, and a new chip with a different programming can be installed. Other types of memory may also be referenced in the automotive literature, including:

- KAM (keep-alive memory), similar to RAM.
- E-PROM, a PROM that reverts to an unprogrammable state when any light strikes it. A piece of Mylar tape is used to cover a small window. When the tape is removed, an ultraviolet light strikes the circuit and erases its memory.
- EE-PROM is an electrically erasable (EE) PROM. In this case, the memory is changed or erased one bit at a time. Some manufacturers use this type of memory to store information about the vehicle mileage, vehicle identification number, and other options.

CLOCK GENERATOR/PULSES

A great deal of information must be transmitted by the computer. To aid in this process, a precise time interval is needed to allow a computer to transmit each bit of information. A clock generator is installed in the computer that provides periodic pulses like a clock that ticks thousands of times each second. The microprocessor, its memories, and its interfaces all use these pulses as a reference to time the digital signals. The generated clock pulses are sometimes referred to as the baud rate. A computer that has a baud rate of 5000 can transmit 5,000 bits per second.

FIGURE 30–11 This is an example of a PROM chip.

INTERFACES

Computers in automobiles also need several support functions. One is called the interface. An interface circuit is used to allow the computer to read input signals to the computer. A second interface is also used to produce the necessary output from the computer to other components such as the carburetor and fuel injector. The major function of the interface is to translate any analog signals coming into the computer to digital signals. In addition, the interface also changes any digital outputs from the computer to analog signals to operate and control components on the engine.

DATA LINKS

Computers send and receive digital signals through what are called **data links**. Data links allow the computer to communicate with both input and output sensors. In addition, computers can communicate with one another using data links. Some data links transmit data in only one direction. Other data links can transmit in both directions.

To get an idea how the flow of information in a computer occurs, refer to ***Figure 30–12***. Data are brought in on the left side of the diagram. The data come from various input sensors such as the exhaust oxygen sensor and the throttle position sensor. Other sensors for things such as engine

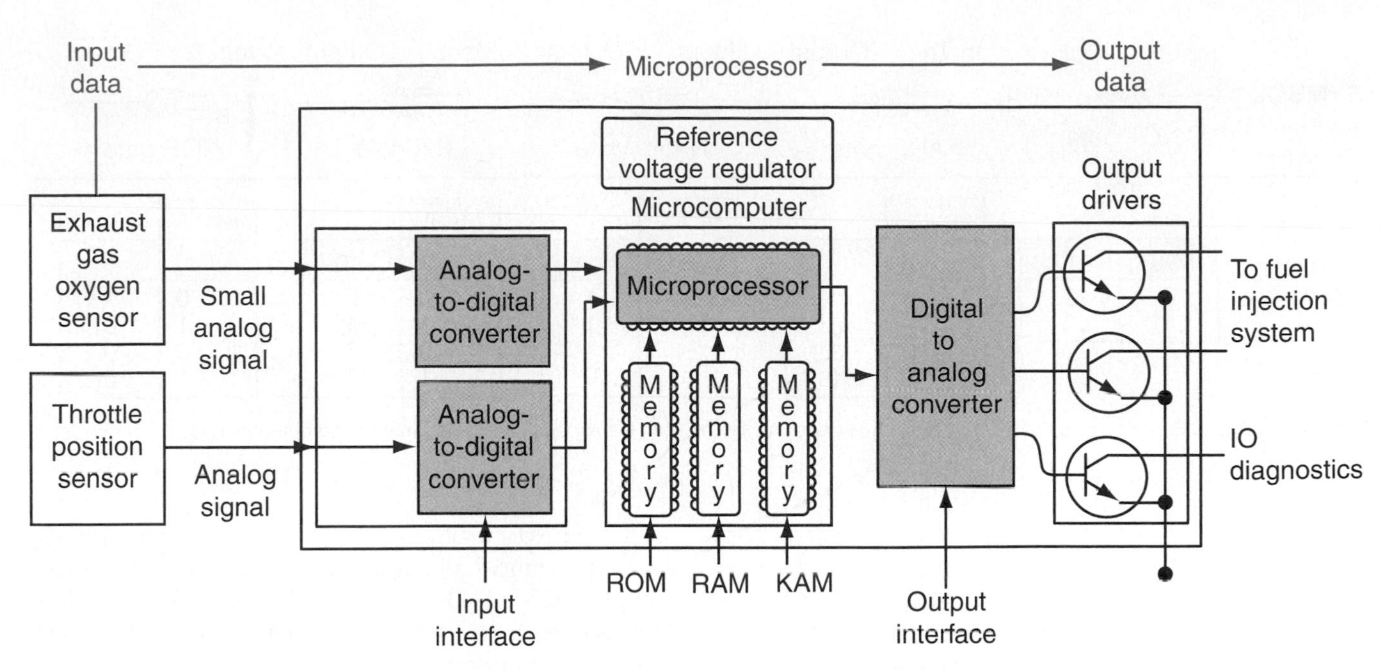

FIGURE 30–12 This diagram shows how all of the computer components operate together.

speed, intake air temperature, and coolant temperature can also be inputs. The input data, which are analog signals, are then converted to digital signals. The digital signals are then sent to the microprocessor. On the basis of the memory data such as the ROM, RAM, and KAM, a certain result is sent out of the microprocessor. This output signal goes through the output interface and is finally sent to the component that is being controlled. In this case, it is the fuel injection system. All of the systems shown are controlled by the clock pulses in the reference voltage regulator.

FUZZY LOGIC COMPUTERS

Today, automotive manufacturers are developing more advanced computers. Computers are now being designed that utilize "fuzzy logic." These computers are able to adjust to many of the varying driving and operating conditions of an automobile. Most computers can really only understand or distinguish between a 0 and a 1. The fuzzy logic computers are able to understand more than just 0 and 1. They can understand unclear conditions and make adjustments in the outputs of the computer as necessary.

GATING CIRCUITS

Many of today's electrical automotive circuits use logic gates or gating circuits. Gating circuits are those that produce a desired output voltage after a given input has been introduced. Gating circuits are made of many transistors put together to produce the desired output. Generally, there are one or more inputs and usually only one output. Gating circuits help the service technician understand the operation of complex computer electrical circuits. Although it is out of the scope of this textbook to troubleshoot such circuits, it is important to understand the basic gate circuits and their inputs and outputs.

The most common gate circuits are called the NOT, AND, NAND, OR, and NOR gates. (As electrical circuits become more and more complex, other gate circuits will be incorporated as well.) The electrical input and output of these gates are identified by truth tables. A truth table uses binary numbers (0 and 1) to help identify the input and outputs. Remember that in the binary system, a 0 usually means a voltage reading below a certain set value. A 1 means a voltage above a certain value. ***Figure 30–13*** shows the electrical symbol for each of these five gates and the truth table for each.

1. NOT Gate—A NOT gate is an electrical device in a circuit that reverses the binary input (say, 1) to a binary output (0). Sometimes this type of gate is called an inverter. So, if there is a 1 on the input, then there is always a 0 on the output. Also, when there is a 0 on the input, the output is always a 1.
2. AND Gate—The AND gate has two inputs and one output. In an AND gate, the only way there can be a 1 for an output is to have both of the inputs at 1.
3. NAND Gate—The NAND gate also has two inputs and one output. The NAND gate operates exactly the opposite of the AND gate. When either of the inputs (A or B) has a 0, the output is 1. Only when both inputs are 1 is the output a 0.
4. OR Gate—The OR gate only needs one of its inputs to produce a 1 for its output. However, if there are no 1s on the input, the output will be 0 as well.
5. NOR Gate—The NOR gate operates just the opposite of the OR gate. When both inputs are 0, the output is 1. If any input has a 1, the output goes to 0.

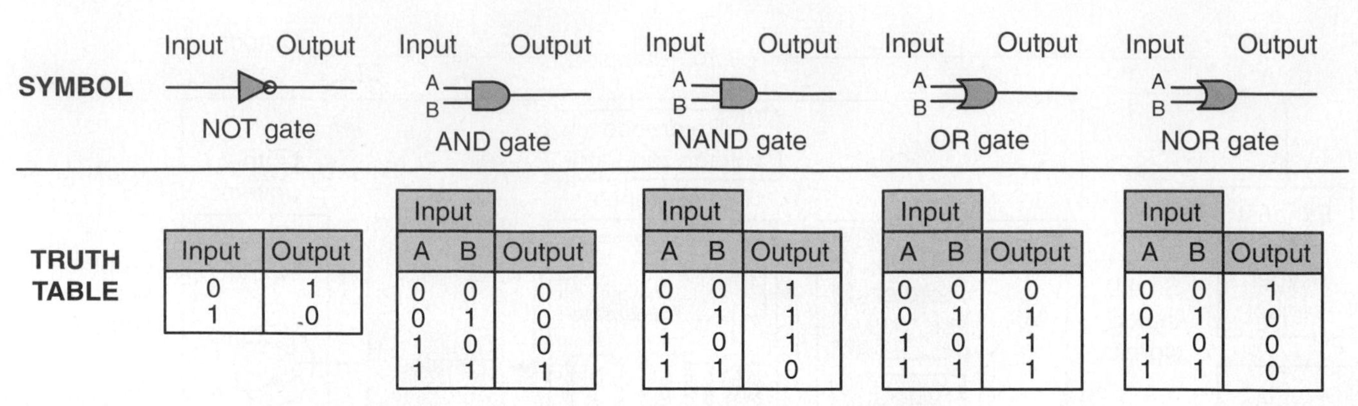

NOT gate

Input	Output
0	1
1	0

AND gate

Input A	Input B	Output
0	0	0
0	1	0
1	0	0
1	1	1

NAND gate

Input A	Input B	Output
0	0	1
0	1	1
1	0	1
1	1	0

OR gate

Input A	Input B	Output
0	0	0
0	1	1
1	0	1
1	1	1

NOR gate

Input A	Input B	Output
0	0	1
0	1	0
1	0	0
1	1	0

FIGURE 30–13 The most common gate circuits are shown with the truth tables representing the inputs and outputs of each.

These five gate circuits can be combined together to make almost any type of logic circuit. ***Figure 30–14*** shows a power window circuit in which several of these gate circuits are employed. The circuit is designed to operate the power windows on a vehicle. The gates used in this circuit are of the NOT, NOR, and AND type of gate circuits. By studying the input and output characteristics of these gate circuits, the service technician will be able to identify how various outputs control motors, door locks, seat belts, and other circuits work.

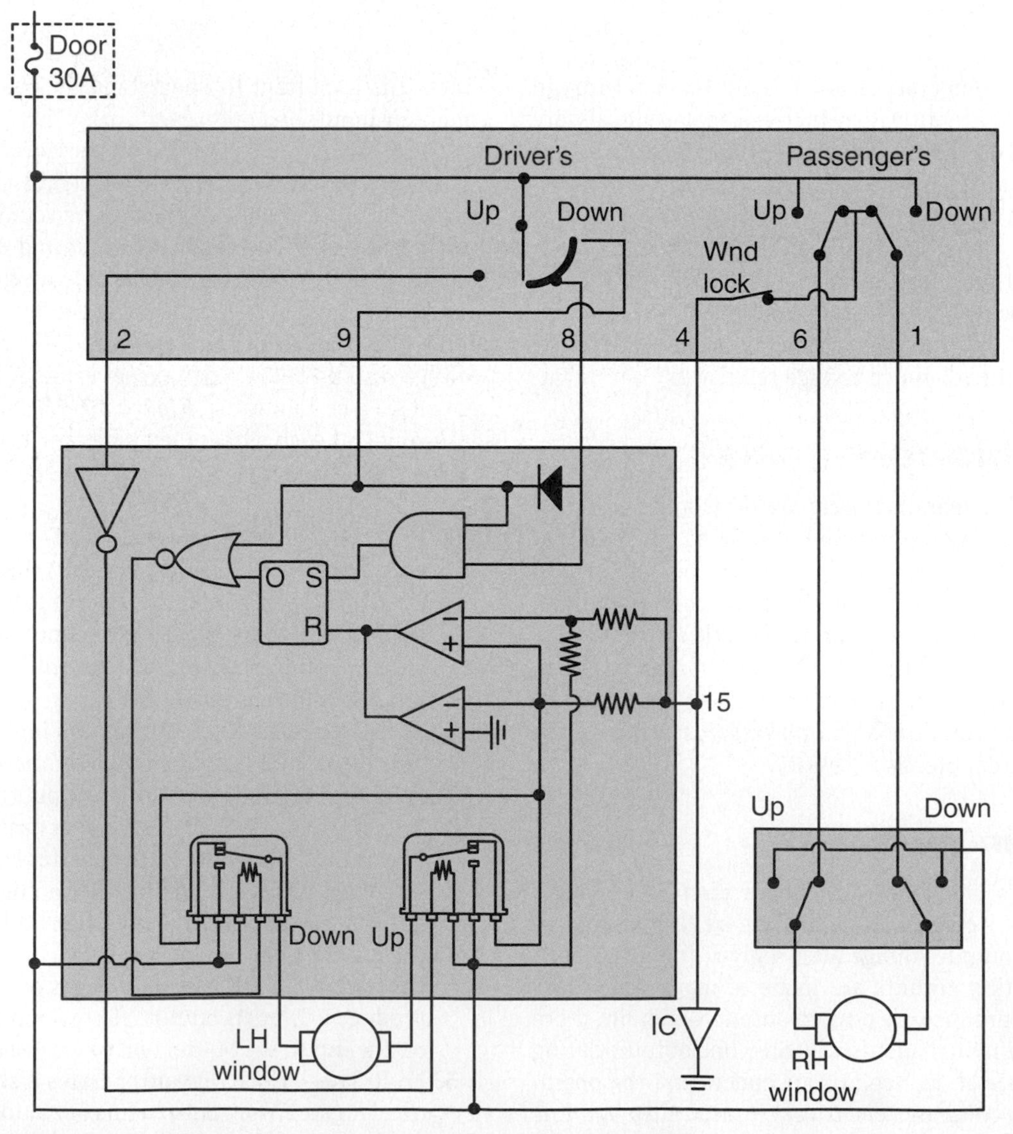

FIGURE 30–14 These NOT, NOR, and AND gates are used to help control power windows.

30.3 WORKING WITH COMPUTER SYSTEMS

FAULT CODES

As mentioned earlier, computers use many inputs and control many outputs. In order for the service technician to effectively troubleshoot these systems, **fault codes**, or trouble codes, are used. When a problem occurs a certain number of times in any of the inputs or outputs, a fault or trouble code is stored in the computer memory. At this point, a dashboard warning light will go on. For example, the warning light may read Check Engine or Service Engine Soon. This warning light tells the driver and the service technician that there is a problem with one or more of the inputs or outputs. This light on the dashboard is now also called a malfunction indicator lamp, known as an MIL.

PARTS LOCATOR

Refer to photo #52 in the front of the textbook to see what a ***check engine*** *light on a dashboard looks like.*

DIAGNOSTIC CIRCUIT CHECK/ ON BOARD DIAGNOSTICS (OBD)

At this point the service technician needs to complete a **diagnostic circuit check**. Sometimes this is called the self-diagnosis check or is referred to as On Board Diagnostics (OBD). The diagnostic circuit check is used to determine which input or output devices are operating incorrectly or are damaged. Each automotive manufacturer has a diagnostic connector that is attached to the electrical wiring harness. This connector is called the assembly line diagnostic link (ALDL). (Some manufacturers call this connector the **data link connector** or DLC.) The ALDL or DLC looks like an electrical connector that has several pins on it. The pins are identified by either a letter or a number. Generally there are twelve- or sixteen-pin connectors as shown in ***Figure 30–15***.

The ALDL or DLC can be located in several places. Depending on the vehicle, the ALDL may be located under the hood on the right or left side of the engine, behind the glove compartment, under the dashboard below the steering column, on the right side panel on the passenger seat, or in some other location.

PARTS LOCATOR

Refer to photos #53 and #54 in the front of the textbook to see what a ***DLC connector*** *looks like.*

The purpose of the ALDL or DLC is to access fault or trouble codes stored in the computer. Each code indicates a different problem. ***Figure 30–16*** shows an example of one manufacturer's typical fault or trouble code chart. For example, a fault code of 21 indicates a problem with the throttle position sensor (TPS). Note that there are many fault or trouble code charts. Trouble code charts are available for each of the computers that are used on the vehicle. For example, ***Figure 30–17*** shows an example of a diagnostic trouble code (DTC) for the antilock braking system, known as the ABS. In this case, the trouble codes have a main code and a subcode.

Once the ALDL has been located, the service technician needs to determine which fault or trouble codes have been stored in the computer memory. There are several ways to access the codes. For example, some manufacturers use a blinking light on the dashboard to tell the service technician

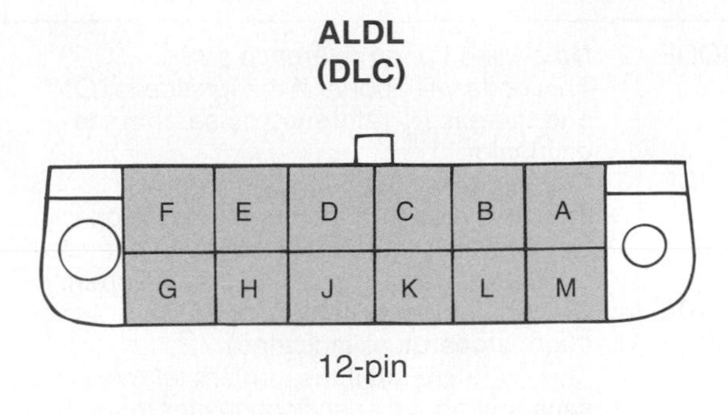

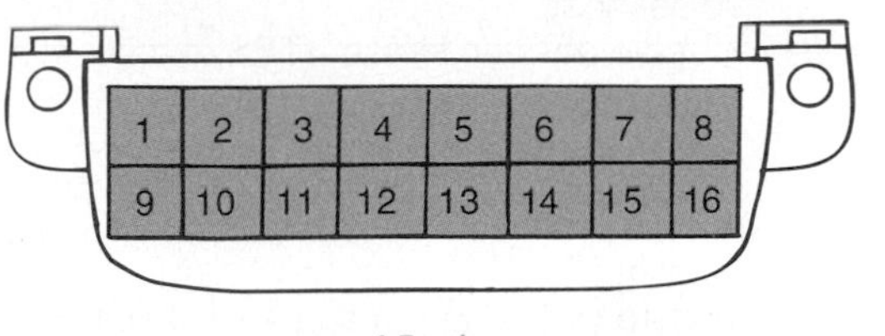

FIGURE 30–15 This ALDL (assembly line diagnostic link), or DLC (data link connector), is the connection for the scanner that is used to help determine the trouble codes held in a computer's memory.

CODE 12	No engine speed reference pulse This code will appear if the ignition is "ON" and there is no reference pulse from the distributor.
CODE 13	Oxygen sensor circuit open circuit This indicates that the engine has properly warmed up, other conditions have been met, but the oxygen sensor circuit is still open.
CODE 14	Coolant temperature sensor (CTS) (high temperature indicated) This code will set if the coolant temperature sensor is above specifications for more than 90 seconds.
CODE 21	Throttle position sensor (TPS) circuit (signal voltage high) This code will set if the TPS circuit voltage remains high while other signals (engine rpm and manifold vacuum) indicate closed throttle.
CODE 23	Mixture control (M/C) solenoid circuit (signal voltage low) This code will set if the M/C solenoid stays low and does not vary.
CODE 24A	Vehicle speed sensor (VSS) circuit This code will set if the vehicle is in gear, the engine is above idle, the throttle is open but the indicated speed is below 5 mph.
CODE 24B	Park/neutral (P/N) circuit This code will set if the P/N signal is compared to other inputs and it does not change to another range when the vehicle moves.
CODE 31	Canister purge solenoid circuit This code will set if the voltage in this circuit does not reflect the command given by the PCM.
CODE 33	Manifold absolute pressure (MAP) sensor circuit This code will set when voltage is high indicating that there is low vacuum.
CODE 35	Idle speed control (ISC) switch This code will set if the ISC switch is shorted. Conditions 50% throttle for 2 seconds.
CODE 41	No distributor reference pulse This code will set if there is no reference pulse but engine vacuum is present.
CODE 42	Electronic spark timing (EST) circuit This code will set when the ECM detects an open or grounded by-pass circuit or EST circuit.
CODE 44	Oxygen sensor circuit (lean exhaust indicated) This code will set whenever the system is in "Closed Loop." TPS voltage is within a specified range and the ECM detects a lower O voltage than specified for a designated time.
CODE 51	PROM error This could indicate a faulty, incorrectly installed, or incorrect application.

FIGURE 30–16 Fault codes help the service technician troubleshoot a vehicle's computer systems.

DIAGNOSTIC TROUBLE CODE (DTC)		PROBLEMATIC COMPONENT/ SYSTEM
MAIN CODE	SUBCODE	
1	—	ABS pump motor over-run
1	2	ABS pump motor circuit problem
1	3	High pressure leakage
1	4	Pressure switch
1	8	Accumulator gas leakage
2	1	Parking brake switch-related problem
3	1 2 4	Pulser(s)
4	1 2 4 8	Wheel sensor
5	— 4 8	Wheel sensor(s)
6	— 1 4	Fail-safe relay
7	1 2 4	Solenoid related problem

FIGURE 30–17 These trouble codes identify problems in the antilock braking system for the service technician. (Courtesy of American Honda Motor Co., Inc.)

which fault or trouble code has been stored. One manufacturer uses a jumper wire to connect two terminals (say terminals A and B) within the ALDL. With the key in the ON position, the Check Engine light will begin to flash off and on in a certain sequence.

The sequence and number of flashes determine the exact trouble code. For example, if the fault code is 4, the light will blink four consecutive times. Then the light will pause for about 2.5 seconds and blink four times again. This will continue. To read a two-digit fault code such as fault code 21,

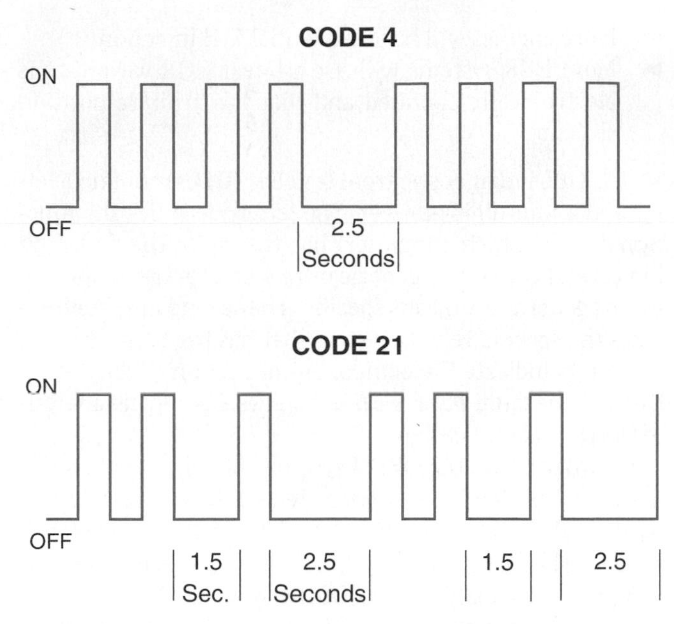

FIGURE 30–18 Trouble codes can be read from the computer by observing the number and sequence of flashes of the Check Engine light on the dashboard.

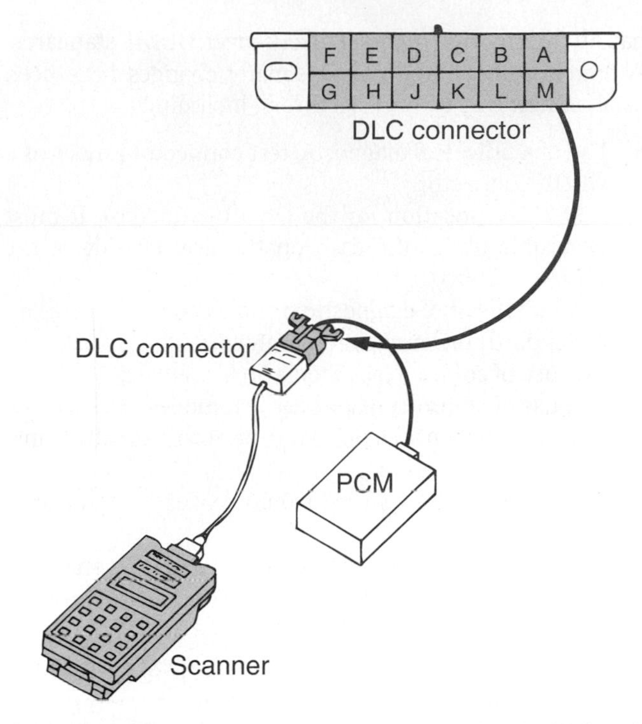

FIGURE 30–19 A scanner can be connected to the DLC to determine trouble codes.

the light will flash two times, pause for 1.5 seconds, then flash 1 time. Then there will be another pause of 2.5 seconds. ***Figure 30–18*** shows an example of the flashing lights graphically. If there are two fault codes stored in the computer, there will be a 2.5 second pause between each code.

Car Clinic:

ADLC

CUSTOMER CONCERN:

Check Engine Light Is On

A customer complains that the Check Engine light comes on and stays on each time the car is started. The car is a 1995 Buick. How can the customer tell if there are trouble codes in the computer?

SOLUTION: General Motors vehicles have the ADLC (assembly data link connector) located below the steering column on the bottom of the dashboard. To check the trouble codes, turn on the ignition and use a jumper wire to connect terminals A and B together. This will cause the Check Engine light to flash on and off in a certain sequence and number of times. When this is done, the light flashes ON three times, then pauses for 1.5 seconds, then flashes ON three more times. A scan tool can also be connected into the ADLC. In this case, the scan tool will read Code 33. The code stored in the computer is Code 33. Checking a service manual, you will find that this code means that the manifold absolute pressure (MAP) sensor needs to be checked for resistance and possible replacement.

Other methods used by manufacturers to access the fault codes include the following:

1. Use a jumper wire to connect a particular terminal to ground. Light flashes are then converted to a specific fault code.
2. Connect an analog (needle-type) voltmeter (not a digital voltmeter) between the positive terminal of the bat tery and a specific terminal on the DLC. Then read the needle fluctuations, which are converted to fault codes like a flashing light bulb.
3. Turn the vehicle ignition off and on a specified number of times and in a certain sequence. This will activate the flashing lights to determine the trouble code.
4. Use a test light connected to a specific terminal on the DLC. The light flashes are then converted to specific codes.
5. Connect a scanner to the DLC as shown in ***Figure 30–19***. Read the display, which gives the exact fault code numbers.

ON BOARD DIAGNOSTICS GENERATION TWO (OBDII)

In the early 1980s, General Motors pioneered the use of On Board Diagnostics (OBD). As mentioned earlier, OBD is used to help service technicians diagnose and repair computer-controlled systems on the automobile. In 1988, California required that all automotive manufacturers provide systems that could identify faults in the computer control system. This system was called On Board Diagnostics Generation One (OBDI). Since 1996, federal law has required that all automotive manufacturers meet OBDII standards. To match OBDII requirements, many changes have been made on new cars by manufacturers, including:

- Use of a universal diagnostic test connector known as a OBDII connector.
- A standard location for the OBDII connector. It must be visible under the dash on the driver's side of the vehicle
- A standard list of diagnostic trouble codes
- A standard communication protocol
- The use of common scan tools on all vehicles
- The use of common diagnostic test modes
- The vehicle identification must be automatically transmitted to the scan tool
- The ability to clear stored trouble codes from the scan tool
- The ability to record and store in memory a snapshot of the operating conditions that existed when a fault occurred
- The ability to store a code wherever the exhaust gas has poor quality
- The use of a standard glossary of terms, acronyms, and definitions in all electronic control systems

To meet the requirements of OBDII, many engines require additional hardware changes. Some of them include:

- More vehicles will have both a MAP and a MAF sensor.
- More vehicles will use heated oxygen sensors.
- All vehicles will have a preconverter oxygen sensor.
- All vehicles will have a postconverter oxygen sensor.
- More engines will have sequential fuel injection.
- More EGR systems will use a linear EGR valve that is electronically operated and that has a pintle position sensor.

The OBDII diagnostic trouble codes (DTCs) contain a letter and a four-digit number. The letter identifies the function device (which computer) has the fault. The first digit indicates if the DTC is generic or manufacturer specific. A 0 means generic, a 1 means specific. The second number indicates the specific vehicle system that has the fault. The last two digits indicate the component or section of the system that has the fault. ***Figure 30–20*** shows a graphical analysis of the DTC identification.

In order to use the OBDII system, scanning tools can be used. In the past, the popular Tech 1 scanning tool was used. If the Tech 1 scanning tool is used, an adapter is needed. However, the Tech 2 shown in ***Figure 30–21***, is completely compatible with OBDII systems. Although this is just an introduction to OBDII, there is a great deal of additional information available on OBDII and the Tech 2 scanner. Refer to appropriate service manuals and manufacturers for more information.

PINPOINT TESTING

After the service technician has performed the diagnostic circuit check or self-diagnosis check, the next step is to locate and identify the specific input or output device that might be defective. This process is called **pinpoint testing**. Pinpoint testing is the process used to check relays, sole-

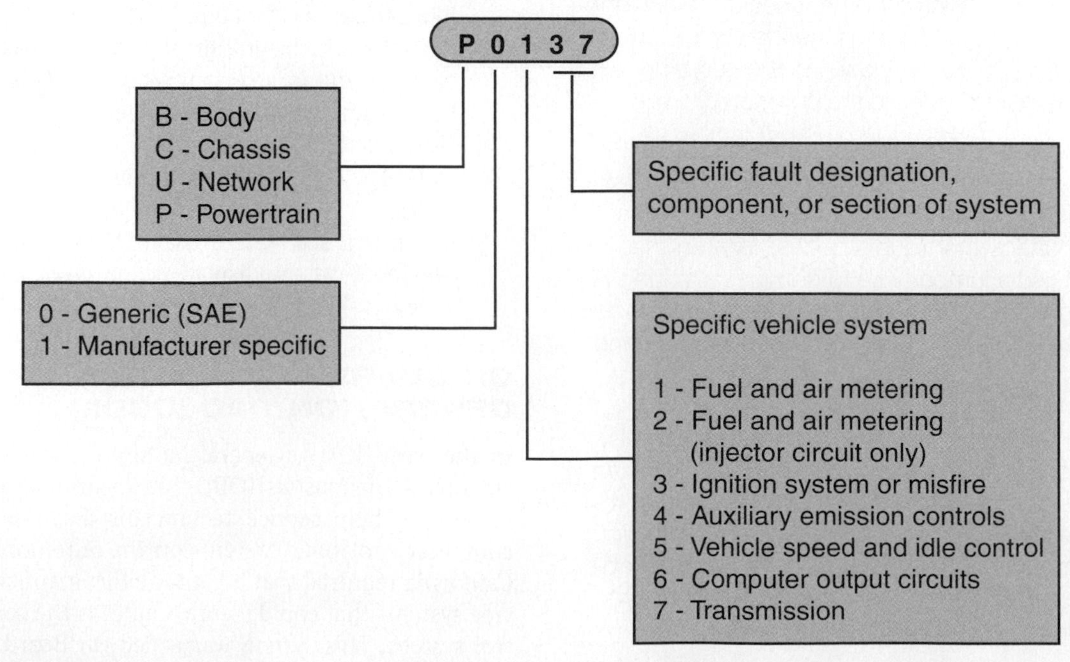

FIGURE 30–20 With the introduction of OBDII, trouble codes have now been standardized across the industry.

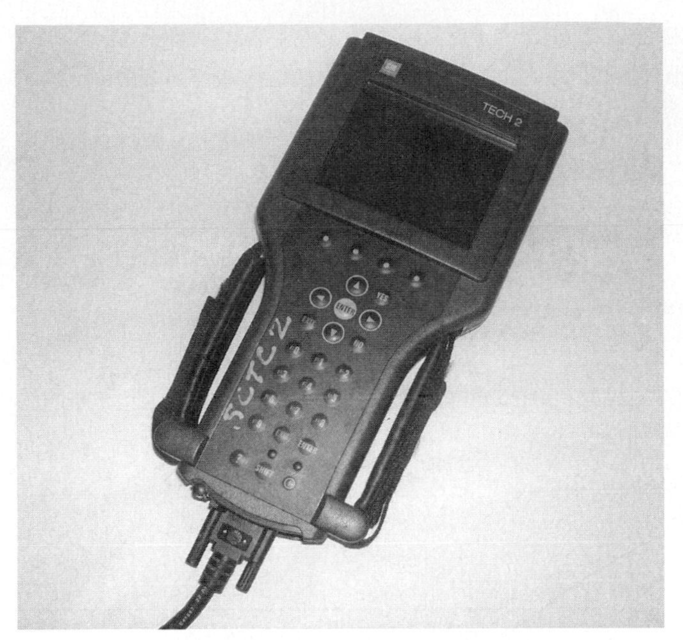

FIGURE 30-21 This diagnostic scanning tool is fully OBDII compliant.

noids, switches, actuators, and connectors for faults. At this point, the service manual must be followed exactly. Each pinpoint test is different and there are many pinpoint tests. ***Figure 30–22*** shows an example of two pinpoint tests recommended for the air-conditioning system for a particular manufacturer. Note that there are literally hundreds of pinpoint tests available. Pinpoint testing helps save money for the consumer because many times it saves buying unnecessary parts.

CLEARING MEMORY

Any time a code has been stored in the computer memory, it is stored in the keep alive memory (KAM). The KAM is similar to the RAM because it stores data on a temporary basis. When an electrical component has been replaced, the KAM must be cleared. There are several methods used to clear the KAM. One common method is to disconnect the negative battery cable for a certain length of time, say 5 minutes. Then reconnect the battery and drive the vehicle a certain distance. Other manufacturers suggest that the service technician disconnect the PCM power feed at the battery terminal pigtail. Others suggest removing the fuse on the positive connection at the battery. Some manufacturers suggest that the scanning tool be used to clear the KAM on OBDII-compliant systems.

CAUTION: *It is very important to follow the manufacturer's exact procedure to clear the KAM. Instructions can be found in the appropriate service manual or the scanner manual. Failure to follow the correct procedure may damage the PCM.*

PINPOINT TEST G: A/C BLOWER MOTOR DOES NOT OPERATE PROPERLY

TEST STEP	RESULT	▶ ACTION TO TAKE
G1 CHECK BLOWER MOTOR SWITCH		
• Perform the Blower Motor Switch component test in this section. • **Is the blower motor switch OK?**	Yes No	▶ GO to **G2**. ▶ REPLACE the blower motor switch.
G2 CHECK WIRES TO A/C BLOWER MOTOR RESISTOR		
• Key off. • Disconnect the A/C blower motor resistor connectors, A/C blower motor connector, and the blower motor switch connector. • Measure the resistance of the following wires between the following components:	Yes No	▶ REPLACE the A/C blower motor resistor. ▶ SERVICE the wire(s) in question.

Wire	Component	Component
BL/BK	A/C Blower Motor	A/C Blower Motor Resistor
BL	A/C Blower Motor Resistor	Blower Motor Switch
BL/R	A/C Blower Motor Resistor	Blower Motor Switch
BL/W	A/C Blower Motor Resistor	Blower Motor Switch
BL/Y	A/C Blower Motor Resistor	Blower Motor Switch

• Measure the resistance of the same wires between the A/C blower motor resistor and ground.
• **Is the resistance less than 5 ohms between components and greater than 10,000 ohms between the A/C blower motor resistor and ground?**

continued

FIGURE 30-22 Pinpoint tests are used to troubleshoot and diagnose specific parts of vehicle systems.

PINPOINT TEST H: NO OPERATION IN LOW BLOWER SETTING		
TEST STEP	**RESULT ▶**	**ACTION TO TAKE**
H1 CHECK BLOWER MOTOR SWITCH		
• Perform the Blower Motor Switch component tests in this section. • **Is the blower motor switch OK?**	Yes No	▶ GO to **H2**. ▶ REPLACE the blower motor switch.
H2 CHECK WIRE BETWEEN A/C BLOWER MOTOR RESISTOR AND BLOWER MOTOR SWITCH		
• Key off. • Locate and disconnect the A/C blower motor resistor connector C1. • Locate and disconnect the blower motor switch connector. • Measure the resistance of the "BL/Y" wire between the A/C blower motor resistor connector C1 and the blower motor switch connector. • **Is the resistance less than 5 ohms?**	Yes No	▶ REPLACE the A/C blower motor resistor. ▶ SERVICE the "BL/Y" wire.

FIGURE 30-22 *(continued)*

TESTING SENSORS AND ACTUATORS

Each input and output for the computer can be tested. If it is found that a particular sensor or actuator is not functioning properly, the service technician must go to the specific service manual to identify the procedure and exact pinpoint test. Generally, most input sensors and output components can be checked by measuring resistances, voltages, and continuity. In many of the other chapters in this textbook, more specific troubleshooting and diagnostic procedures are given for sensors and actuators.

TESTING THE COMPUTER

For the most part, computers used on vehicles today are not serviced. When it is suspected that the computer has a fault, the unit is generally replaced. However, some manufacturers have several simple voltage tests that can be performed. Before working with any automotive computer, remember that computers are very sensitive to static voltage changes. The sign shown in ***Figure 30–23*** suggests that any static electricity could damage internal components. To eliminate the possibility of static electricity, the service technician should always wear an antistatic wrist strap as shown.

Tests are made with a digital voltmeter. Often there is a small voltage signal (called a reference voltage, of about 5 volts) sent to certain types of sensors. This reference voltage is used to make them operate. Then, depending on the resistance of the sensor, a certain voltage drop is present at the sensor. This often helps the service technician in diagnosing problems. This voltage can be checked and compared to the manufacturer's specifications. Before checking the voltage, make sure to follow the manufacturer's procedure exactly.

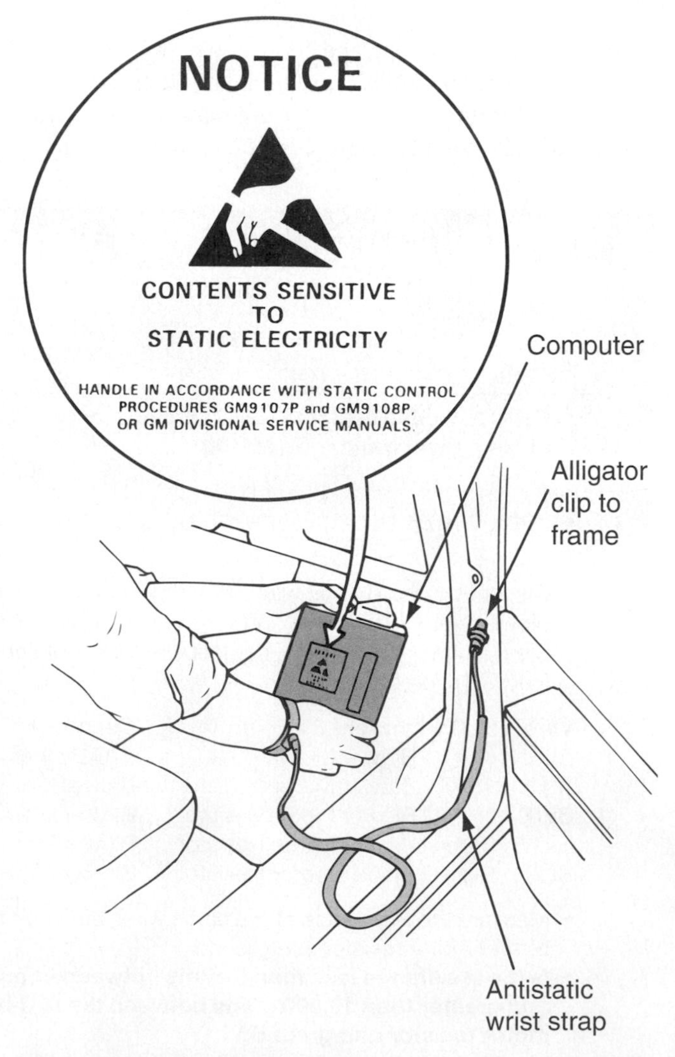
NOTICE
CONTENTS SENSITIVE TO STATIC ELECTRICITY
HANDLE IN ACCORDANCE WITH STATIC CONTROL PROCEDURES GM9107P and GM9108P, OR GM DIVISIONAL SERVICE MANUALS.
Computer
Alligator clip to frame
Antistatic wrist strap

FIGURE 30-23 When working with computers, always wear an antistatic wrist strap to avoid static electricity.

COMPUTER REPLACEMENT

Automotive computers can be replaced. If there is a faulty computer, be sure to replace it with the exact same computer. However, the PROM, which is considered the calibration unit of the computer, needs to be removed from the old computer and placed in the new computer. The PROM seldom will be faulty. To remove the PROM, always follow the manufacturer's suggested procedure. Basically, first disconnect the battery and then remove the computer, making sure to wear an antistatic wrist strap. After the computer has been removed, remove the input and output wiring harnesses. Remove the PROM cover and remove the PROM (with a special tool if necessary). Pull the PROM straight out of the old computer, disconnecting it from the pins on the computer. Place the PROM in the new computer and install the new computer in the vehicle. There may be reference marks to aid in installation of the PROM into the new computer.

SUMMARY

The following statements will help to summarize this chapter:

- Semiconductors are those electrical components that have four electrons in their outer orbit.
- Semiconductors such as diodes and transistors are used to make up an integrated circuit.
- An integrated circuit (IC) chip is made of a silicon wafer that is etched through a photographic process and placed in a housing that has connectors.
- Several integrated circuits are put together to make a microprocessor.
- Computers for automobiles have an input, a processor, and an output.
- Computers communicate with electrical signals.
- Both analog and digital signals can be used in automotive computers.
- Signals are strung together using binary codes.
- Binary codes are stored in memory components called RAM, ROM, KAM, PROM, and so on.
- To aid in proper functioning, computers use an interface to change analog signals to binary signals and binary signals to analog signals.
- The most common name for onboard computers in automobiles is the PCM, or powertrain control module.
- Computers are also used on the automobile for control of the transmission, body, suspension, air bag system, brake system, ignition system, heating and air conditioning, and instrumentation.
- Inputs to the computer are generally supplied by potentiometers, variable resistors, magnetic pickups, voltage generators, and switches.
- Outputs from the computer are generally used to control relays, solenoids, servomotors, lighted displays, and switch transistors.
- Fault codes are used to identify a problem with one of the inputs or outputs from the computer.
- On Board Diagnostics and OBDII systems help service technicians troubleshoot and diagnose computer inputs and outputs.
- After faulty input or output components have been identified, pinpoint tests are used to check specific components for voltage and resistance readings.

TERMS TO KNOW

Can you explain each of the following terms? Review the chapter until you can use each term correctly.

Analog signal
Binary codes
Body control module
Data links
Data link connector
Diagnostic circuit check
Digital signal
Diode
Fault codes
Integrated circuit
Magnetic pickup sensor
Microprocessors
Millisecond
Pinpoint testing
PROM
RAM
Relay
ROM
Solenoid
Transistor

REVIEW QUESTIONS

Multiple Choice

1. Which of the following devices can amplify an off-on signal?
   a. Diode
   b. Resistor
   c. Transistor
   d. Voltmeter
   e. Resistance meter

2. Which of the following devices causes electricity to flow in only one direction?
   a. Diode
   b. Resistor
   c. Transistor
   d. Voltmeter
   e. Resistance meter

3. Which part on a transistor is considered to be the control part?
   a. Base
   b. Emitter
   c. Collector
   d. Negative terminal
   e. Positive terminal

4. Which of the following is a method used to communicate with a computer?
   a. By changing voltage levels
   b. By changing the shape of the voltage pulse
   c. By changing the speed of the voltage signal
   d. All of the above
   e. None of the above

5. Computers can read which of the following numbers in the binary code?
   a. 0 and 1
   b. 1 and 2
   c. 2 and 3
   d. All of the above
   e. None of the above

6. A continuously variable signal to a computer is called a/an ________ signal.
   a. Digital
   b. Alpha
   c. Beta
   d. Analog
   e. Fifth

7. What is the most common name for the computer used in today's vehicles for engine and transmission control?
   a. ECU
   b. ECA
   c. ECM
   d. BCM
   e. PCM

8. A potentiometer provides a form of ________ to the computer.
   a. Output
   b. Analog
   c. Input
   d. Binary
   e. Relay

9. Solenoids, relays, servomotors, and switching resistors are controlled ________ by the computer.
   a. Outputs
   b. Analogs
   c. Inputs
   d. Binaries
   e. Relays

10. To determine a trouble or fault code, the service technician must use the ________ to determine the exact code.
    a. Output
    b. DLC
    c. OBDII connector
    d. All of the above
    e. b and c of the above

The following questions are similar in format to ASE (Automotive Service Excellence) test questions.

11. *Technician A* says that inputs such as vehicle speed and engine temperature are fed to the microprocessor. *Technician B* says that inputs such as engine speed and barometric pressure are fed to the microprocessor. Who is correct?
    a. A only   c. Both A and B
    b. B only   d. Neither A nor B

12. *Technician A* says that a transistor is the same as a diode. *Technician B* says that a transistor is the same as a resistor. Who is correct?
    a. A only   c. Both A and B
    b. B only   d. Neither A nor B

13. *Technician A* says that RAM memory in the computer is temporary storage. *Technician B* says that ROM memory is temporary storage. Who is correct?
    a. A only   c. Both A and B
    b. B only   d. Neither A nor B

14. *Technician A* says that RAM memory in the computer can read digital signals with an input interface. *Technician B* says that the output interface is needed to convert analog signals to digital signals. Who is correct?
    a. A only   c. Both A and B
    b. B only   d. Neither A nor B

15. *Technician A* says that to access the trouble codes, use the OBDII connector. *Technician B* says that to access the trouble codes, use a scanner attached to the DLC. Who is correct?

a. A only c. Both A and B
b. B only d. Neither A nor B

16. *Technician A* says that after a sensor has been replaced, the KAM needs to be cleared. *Technician B* says that testing a sensor can be done by following the pinpoint test. Who is correct?

a. A only c. Both A and B
b. B only d. Neither A nor B

17. *Technician A* says that the PROM should be replaced new every time the computer is removed from the vehicle. *Technician B* says that the PROM can be removed from an old computer and put into a new computer. Who is correct?

a. A only c. Both A and B
b. B only d. Neither A nor B

18. *Technician A* says that a common form of output of the computer is to control a relay. *Technician B* says that a common form of input to the computer is from a variable resistor. Who is correct?

a. A only c. Both A and B
b. B only d. Neither A nor B

Essay

19. Identify two reasons why computers are used in automobiles.

20. What are five types of input sensors used with the automotive computer?

21. What are five types of output devices used with the automotive computer?

22. What is the purpose of pinpoint testing?

23. Define the purpose and operation of a diode.

24. What is the difference between a diode and a transistor?

Short Answer

25. Computers communicate with two types of signals, which are called ________ and ________ signals.

26. When working with a computer in an automobile, always use a(n) ________ to avoid static electricity.

27. When performing on board diagnostic checks, the scanner should be connected to the ________.

28. When the Check Engine light flashes on three times, then pauses for 1.5 seconds, then flashes on four times, the trouble code in the computer is number ________.

29. Computers that read varying conditions rather than a yes/no or 0/1 are called ________ computers.

30. The type of sensor that measures the speed of a rotating shaft would be a ________ sensor.

Applied Academics

TITLE: Fuse Types and Styles

ACADEMIC SKILLS AREA: Science

NATEF Automobile Technician Tasks:

The service technician can describe and explain the role of a fuse or fusible link when used as a protective device in an electrical or electronic circuit of the engine.

CONCEPT:

Along with a growing use of computers on automobiles, there are also an increasing number of electrical circuits. As mentioned throughout this textbook, each type of electrical circuit has a specific purpose and operation. As protection, various types of fuses are used on the automobile electrical circuits. When working on electrical circuits, the technician may encounter any of the fuses shown in the accompanying table. The fuses are typically located on the fuse panel. The fusible link is often located directly in the wire from the positive side of the battery. It is located there because it would be impractical to run the wire through the fuse panel. Fusible links are often found in the engine compartment near the battery.

When using electrical circuits, the service technician may encounter different symbols for each type of fuse. A good practice in troubleshooting any electrical circuit is to always check the fuse, fusible link, or circuit breaker first before proceeding to other diagnostic procedures.

ILLUSTRATION	SYMBOL	PART NAME	ABBREVIATION
		Fuse	Fuse
		Medium current fuse	M-fuse
	or	Fusible link	FL
		Circuit breaker	CB

APPLICATION:

Can you define the purpose of a fuse in an electrical circuit? What would happen if fuses were not used in an electrical circuit? Why should only the correct size (amperage) be used when replacing a fuse or fusible link?

Automotive Batteries

CHAPTER

31

OBJECTIVES

After studying this chapter, you should be able to:

- Identify the purpose of the automotive battery.
- Analyze the internal parts, construction, and operation of the battery, including chemical action and specific gravity.
- Determine the methods used to rate batteries.
- Identify the methods used to test and maintain batteries.
- Analyze various battery problems, diagnosis, and service procedures.

INTRODUCTION

The battery is a very important part of the automobile. It is used to supply electricity for many systems and components within the vehicle. The purpose of this chapter is to study the theory, operation, and maintenance of the automotive storage battery.

31.1 BATTERY DESIGN AND OPERATION

PURPOSE OF THE BATTERY

The purpose of the battery is to act as a reservoir for storing electricity. The battery receives, stores, and makes electrical energy available to the automobile. It is called a storage battery because it stores, or holds, electricity. The battery is considered an **electrochemical** device. This means that it uses chemicals to produce and store electricity. Energy in the battery is stored in a chemical form. The chemical energy is released as electrical energy for use in the automobile. The purpose of the battery is to provide sufficient electrical energy to crank the starter and operate the ignition system, computers, solenoids, lights, and other electrical components. Batteries are also used in hybrid vehicles to provide power for propulsion. In this case, several batteries are needed to provide the necessary power.

TYPES OF BATTERIES

Batteries can be subdivided into two major groups: primary and secondary batteries. **Primary batteries** are those that

Certification Connection

ASE Connection: The information in this chapter can help you prepare for the National Institute for Automotive Service Excellence (ASE) certification tests. The tests and content areas most closely related to this chapter are:

Test A6—Electrical/Electronic Systems

- **Content Area**—Battery Diagnosis and Service

NATEF Connection: Much of the information in this chapter is related to the NATEF tasks. The NATEF tasks and priority numbers most closely related to this chapter are:

1. Perform battery state-of-charge test; determine necessary action. P-1
2. Perform battery capacity test; confirm proper battery capacity for vehicle application; determine necessary action. P-1

continued

Certification Connection *continued*

3. Inspect, clean, fill, and replace battery. P-2
4. Perform slow/fast battery charge. P-1
5. Inspect and clean battery cables, connectors, clamps, and holddowns; repair or replace as needed. P-1
6. Start a vehicle using jumper cables and a battery or auxiliary power supply. P-1

are nonrechargeable. Examples of primary batteries include those used in flashlights, calculators, smoke alarms, and radios. Various metals and chemicals are used to manufacture primary batteries and produce the necessary power.

Secondary batteries are those that can be discharged and recharged repeatedly. This can be done by reversing the normal current flow through the battery. The automobile battery is called a secondary battery because it can be charged and discharged many times.

BATTERY CELLS

Batteries are made by putting together a number of cells. A **battery cell** is that part of a battery that stores chemical energy for later use. In its simplest form, a battery cell consists of three components: a positive plate (one type of metal), a negative plate (another type of metal), and an **electrolyte**, or acidic solution. When these three components, two dissimilar metals and an electrolyte solution, are placed together, electricity can be produced.

All batteries, both primary and secondary, have cells. The difference between battery cells is the type of metals and electrolyte chemicals used in the cells. For example, a 2-volt primary battery cell uses different metals and chemicals than a 9-volt primary battery cell.

LEAD-ACID BATTERY

Figure 31–1 shows a simple battery cell. The positive plate is made of a metal called lead peroxide (PbO_2). The negative plate is made of a metal called sponge lead (Pb). The electrolyte solution is made of a mixture of sulfuric acid (H_2SO_4) and water. This is called a lead-acid battery cell. The metals are made of different types of lead, and the electrolyte is made of acid. When these three components are arranged as shown, approximately 2.1 to 2.5 volts are produced across the positive and negative plates. If an electrical circuit were placed across this voltage source, electrons would flow from the negative plate, through the circuit, back to the positive plate.

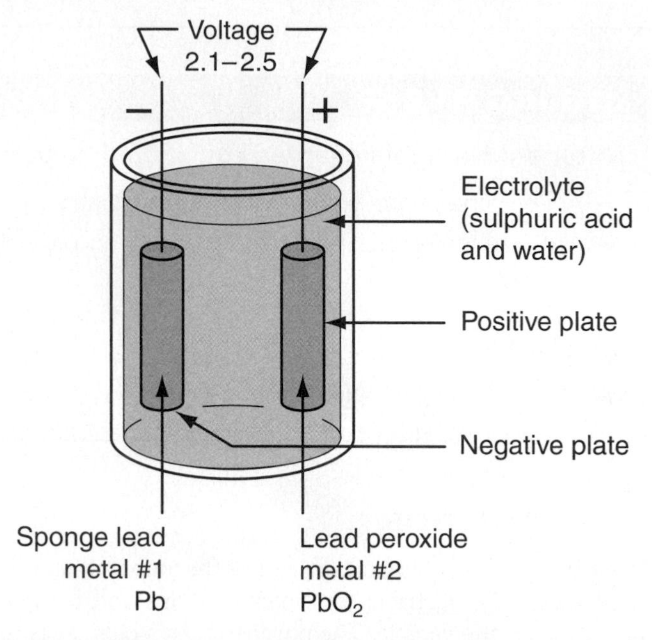

FIGURE 31-1 Battery cells consist of three components. These include two different metals and an electrolyte solution. When these two metals are placed together in the electrolyte solution, a voltage can be produced across the metals.

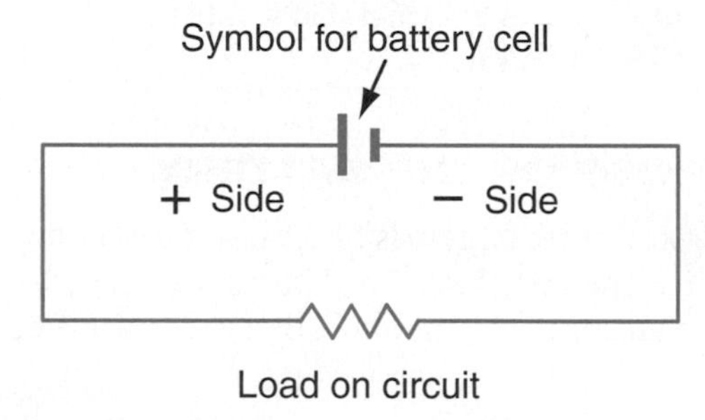

FIGURE 31-2 The symbol for a battery cell. The long line represents the positive (+) plate. The short line represents the negative (–) plate.

BATTERY CELL SYMBOL

The electrical symbol for a battery cell is shown in ***Figure 31–2***. The longer line represents the positive plate. The shorter line represents the negative plate.

COMBINING BATTERY CELLS

Automobiles today require 12–14 volts to operate. Battery cells can be connected to produce different voltages. When battery cells are connected in series, the voltage is additive. For example, in order to produce a 6-volt lead-acid battery, three 2-volt cells are connected in series. A 12-volt battery has six 2-volts cells connected in series (***Figure 31–3***).

Whenever battery cells are placed in parallel, the voltage remains the same. However, the amperage that the battery can produce is increased. This is called **amperage capacity**.

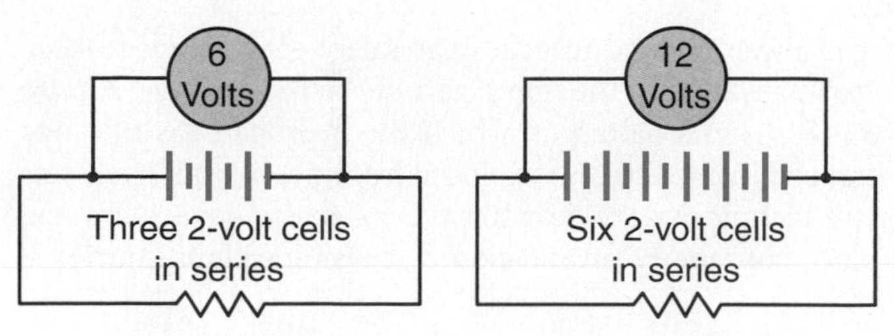

FIGURE 31–3 When three 2-volt cells are placed in series, the total voltage is 6 volts. When six 2-volt cells are placed in series, the total voltage is 12 volts.

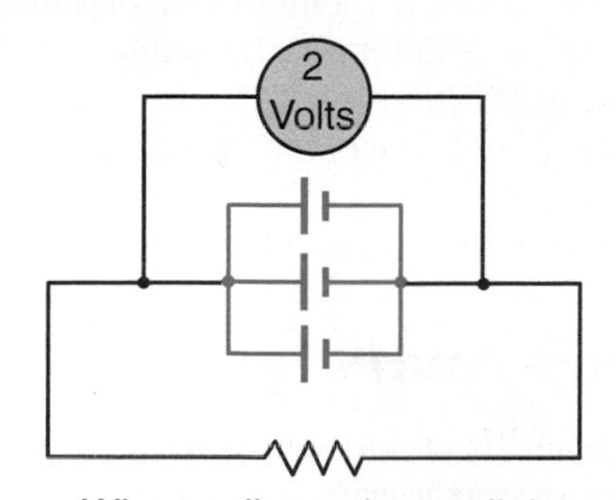

FIGURE 31–4 When 2-volt cells are connected in parallel, the voltage remains the same, but the amperage capacity increases.

This means that the battery can produce a certain amount of amperage over a longer period of time than a single cell could produce (***Figure 31–4***).

As stated earlier, to make a 12-volt battery there needs to be six cells connected in series with each other. In ***Figure 31–5***, six cells are placed in series with each other. Then, to increase the amperage capacity of this 12-volt battery, additional plates are added to each cell in parallel with each other.

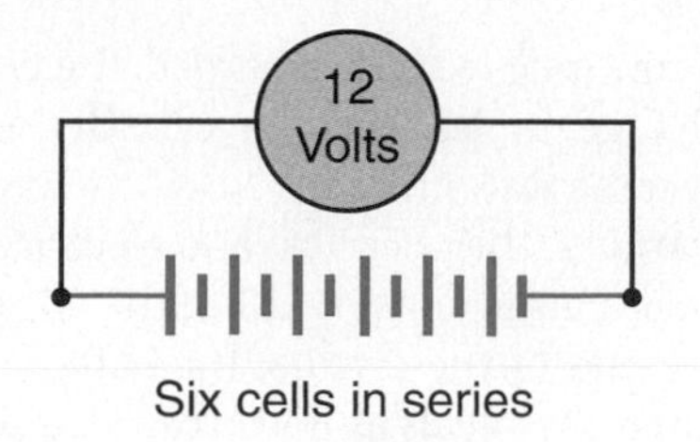

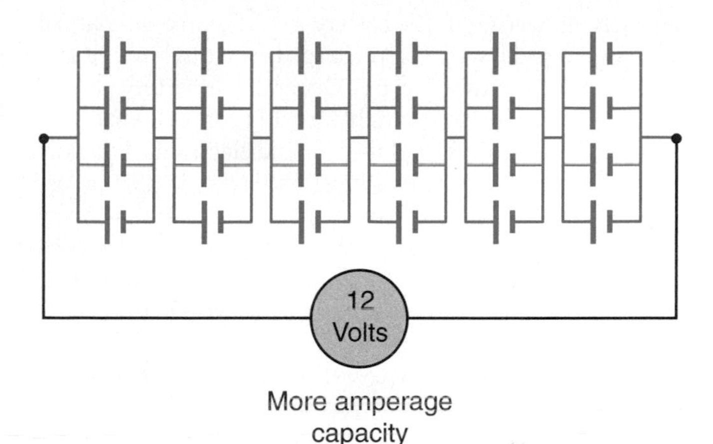

FIGURE 31–5 To make a 12-volt battery with high capacity, six cells are used with additional plates put in parallel with each cell.

CHEMICAL ACTION IN A BATTERY

Figure 31–6 shows how chemical reactions occur inside the lead acid battery cell. During **discharging**, or when a load is applied, the lead peroxide (PbO_2) is separated. The sulfuric acid (H_2SO_4) and water are also separated. Some of the sulfate (SO_4) combines with the negative-plate sponge lead (Pb). Some of the sulfate (SO_4) combines with the lead (Pb) part of the positive plate. The remaining oxygen (O_2) combines with the hydrogen (H_2) left from the electrolyte.

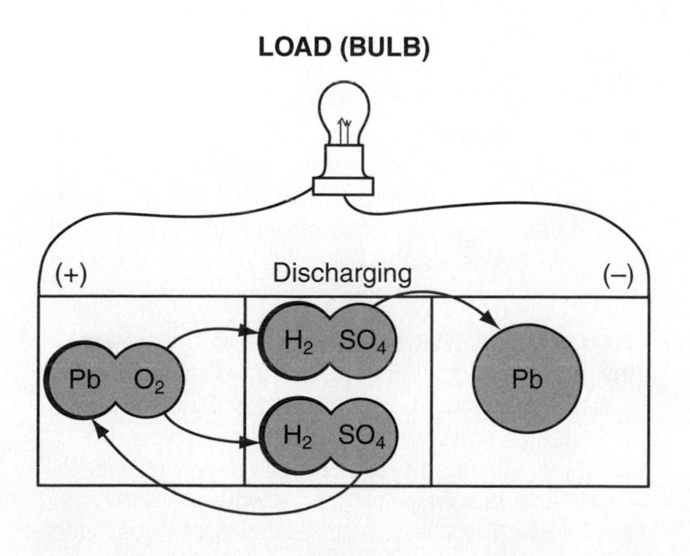

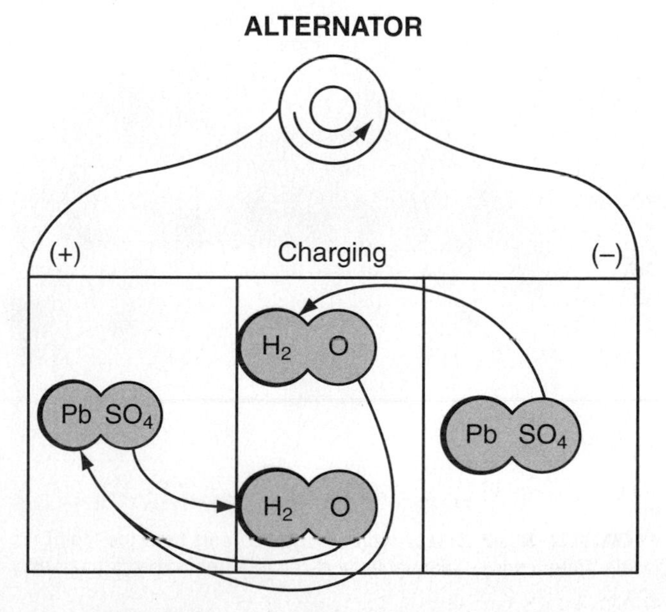

FIGURE 31–6 The chemical action that takes place in a battery during discharging and charging.

If the battery continues to be discharged, the two metals will eventually change to lead sulfate ($PbSO_4$) and the electrolyte will become water (H_2O).

During **charging**, the chemical and electrical actions are just the reverse. This is done by using the alternator on the vehicle. At the end of the charge, the battery cell changes back to the same chemicals as it was before discharge.

When a battery is discharged and recharged, it has gone through a **cycle**. For example, if the vehicle's lights had been left on overnight, the battery would be in a discharged state and would have to be recharged. The discharging and recharging constitute a cycle. Generally, a battery can go through only so many cycles before it becomes damaged. A damaged battery loses its ability to hold electricity. In a normal automotive application, most batteries are able to last between 3 and 7 years. Of course, it depends on how many and how severe the cycles were. A battery can be designed using different chemicals to allow more cycles in its lifetime. This type of battery is called a **deep cycle battery**.

EXPLOSIVE GASES

Hydrogen and oxygen gases are released during the process of charging and discharging a conventional battery. The gases escape the battery through the vent caps. The more rapidly a battery is charged or discharged, the greater the possibility of producing these gases. These gases can be very explosive.

CAUTION: *It is always important to keep any sparks or open flame away from a battery that is being charged or discharged. Always charge batteries in a well-vented room.*

Maintenance-free batteries are designed to reduce the amount of hydrogen gas that is produced.

There are times when a battery may produce excessive amounts of hydrogen gas during rapid charging or discharging. The vent plugs may be dirty and plugged. In either case, it is always a good idea to wear safety glasses and to keep sparks away from the upper part of the battery case. ***Figure 31–7*** shows a battery in which the hydrogen gas that was trapped inside the case exploded from nearby sparks. If service technicians are near this type of explosion, acid can explode onto their skin and clothes, causing serious injuries.

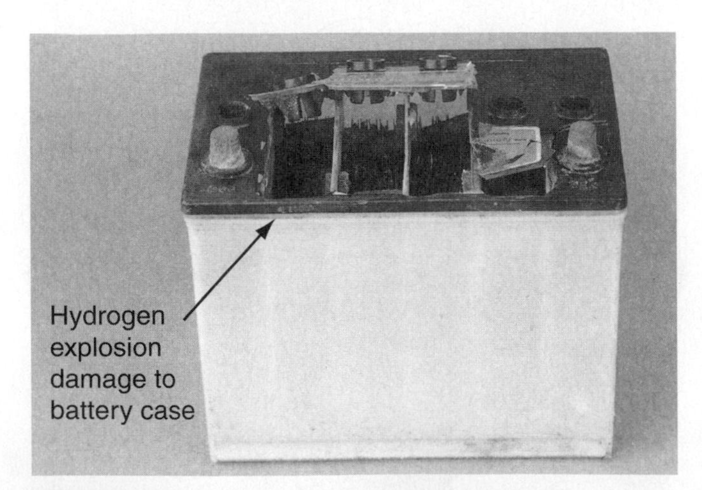

FIGURE 31-7 When hydrogen gas collects near the top of the inside battery case, it can ignite and cause an explosion from nearby sparks.

ELECTROLYTE

The electrolyte in a battery is a combination of water and sulfuric acid. Normally, the ratio of acid to water is 40% acid to 60% distilled water. This combination of water and acid is the best electrolyte solution for the battery. During normal operation, the water will evaporate. The acid does not usually evaporate. Distilled water is the only ingredient that should be added to a battery. The right amount of acid will always be in the battery unless it is tipped over and spilled.

SPECIFIC GRAVITY

The condition of the battery electrolyte can be checked by measuring its **specific gravity**. Specific gravity is defined as the weight of a solution compared with the weight of water. The specific gravity of water is rated as 1.000. Any solution heavier than water is expressed in terms of the ratio of its density to the density of water. Sulfuric acid is heavier than water. Adding sulfuric acid to water causes the density to exceed that of water, or 1.000.

Figure 31–8 shows how the specific gravity of water and acid change when they are mixed. The specific gravity of water (1.000) mixed with acid (1.835) is equal to an electrolyte solution of 1.270.

The condition or state of charge can be determined by measuring the specific gravity of the electrolyte solution.

Car Clinic: SPECIFIC GRAVITY

CUSTOMER CONCERN:
Frozen Battery

A car has been stored over the winter in an outside garage. During the winter an inspection was made, and it was found that the battery had cracked and the fluid inside had frozen. What could cause a frozen battery?

SOLUTION: Batteries are able to freeze only if the battery is discharged severely. Electrolyte has a lower freezing point than water. The electrolyte changes to water when the battery is discharged. Then the water is able to freeze. When water freezes, it expands, cracking the case. Always store the battery inside a building to prevent this problem, and keep the battery charged during storage.

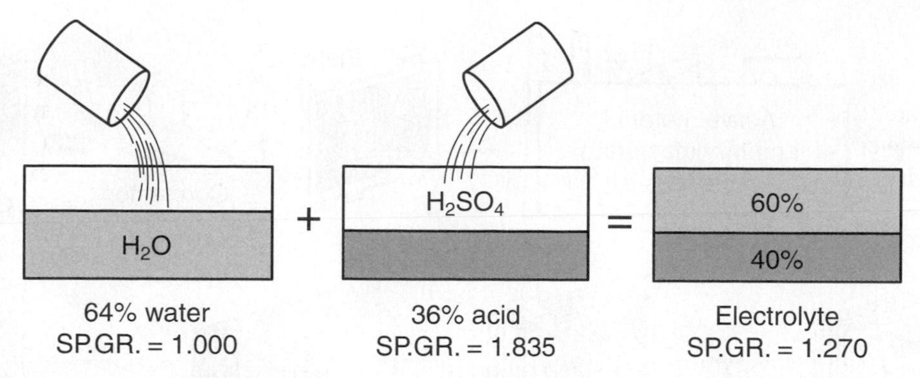

FIGURE 31–8 The ratio of water to acid is about 60% water to 40% acid. The specific gravity of the two solutions combine together to have a specific gravity of about 1.270.

SPECIFIC GRAVITY READINGS AT 80°F	
1.260–1.280	Full charged
1.235–1.260	3/4 charged
1.205–1.235	1/2 charged
1.170–1.205	1/4 charged
1.140–1.170	Poorly charged
1.110–1.140	Dead

FIGURE 31–9 The specific gravity reading can determine the condition of the charge on a battery.

Figure 31–9 shows specific gravity readings for different battery conditions. Specific gravity is measured with a hydrometer, which is discussed later in this chapter.

LOW-MAINTENANCE BATTERIES

In a standard battery, antimony (a metallic substance) is added to the lead plates to strengthen them. Antimony also has a lower resistance to charging. Because of these characteristics, the battery produces more hydrogen gas during charging and discharging. Calcium and other materials such as cadmium or strontium are added to low-maintenance batteries' plates (formerly called maintenance-free batteries) to reduce gassing. A trace of tin is also included. These low-maintenance lead-calcium batteries normally discharge less gas than lead-acid batteries unless they are grossly overcharged. They do not have filler caps and never need water.

The purity of the lead-calcium plates in a low-maintenance battery reduces self-discharging and makes the battery harder to overcharge. (Overcharging is a major cause of water loss in a battery.) Because of these characteristics, the low-maintenance battery has no filler holes and smaller vent holes than a standard battery. The vent holes are located below a small cover on the top or side of the battery. Most of the battery breathing takes place through a series of **baffles** inside the battery. These baffles collect acid vapor and condense it back into the solution. Corrosion outside the battery is also reduced. Acid vapors that help cause outside corrosion are kept inside the battery.

BATTERY CONSTRUCTION

Figure 31–10 shows the internal parts and construction of a typical battery. Following the diagram, first a grid is manufactured as the base of the positive and negative plates. Lead peroxide, sponge lead, calcium, antimony, and other materials (called the active materials) are manufactured and spread on the grid. The result is a completed plate.

Connectors are used to assemble the plates into a group of cells. Note that the plates are placed in parallel to increase the amperage capacity. Both positive and negative groups are manufactured. Separators are then placed between the groups to keep the plates from touching each other. One cell, or element, is now complete. This cell is capable of producing about 2.1 to 2.5 volts. Amperage capacity can be increased by adding more plates to each cell.

Once the elements, or cells, are completed, they are connected in series and placed in a plastic container. A one-piece cover is placed on top of the container and sealed. Here the battery posts are on top. Many batteries have the battery posts on the side. The electrolyte is added, and the vent caps are installed. The battery is now complete. ***Figure 31–11*** shows a completed and assembled battery. Note the number of plates and cell partitions that are used to increase the amperage capacity.

Several types of battery terminals are used on automotive batteries. Older batteries use a post or top terminal. However, newer vehicles use batteries with a side terminal, as shown in ***Figure 31–12***. Some specialty vehicles and certain imported batteries use a L terminal. There was a serious safety problem when using the top terminal battery connection. Service technicians often placed tools on top of the battery when working on the vehicle. At times, tools accidentally shorted between the two terminals, causing excessive drain on the battery, possible explosions, and damage to the tools.

There has been significant research on the grid designs. For example, the way in which the grids are configured can

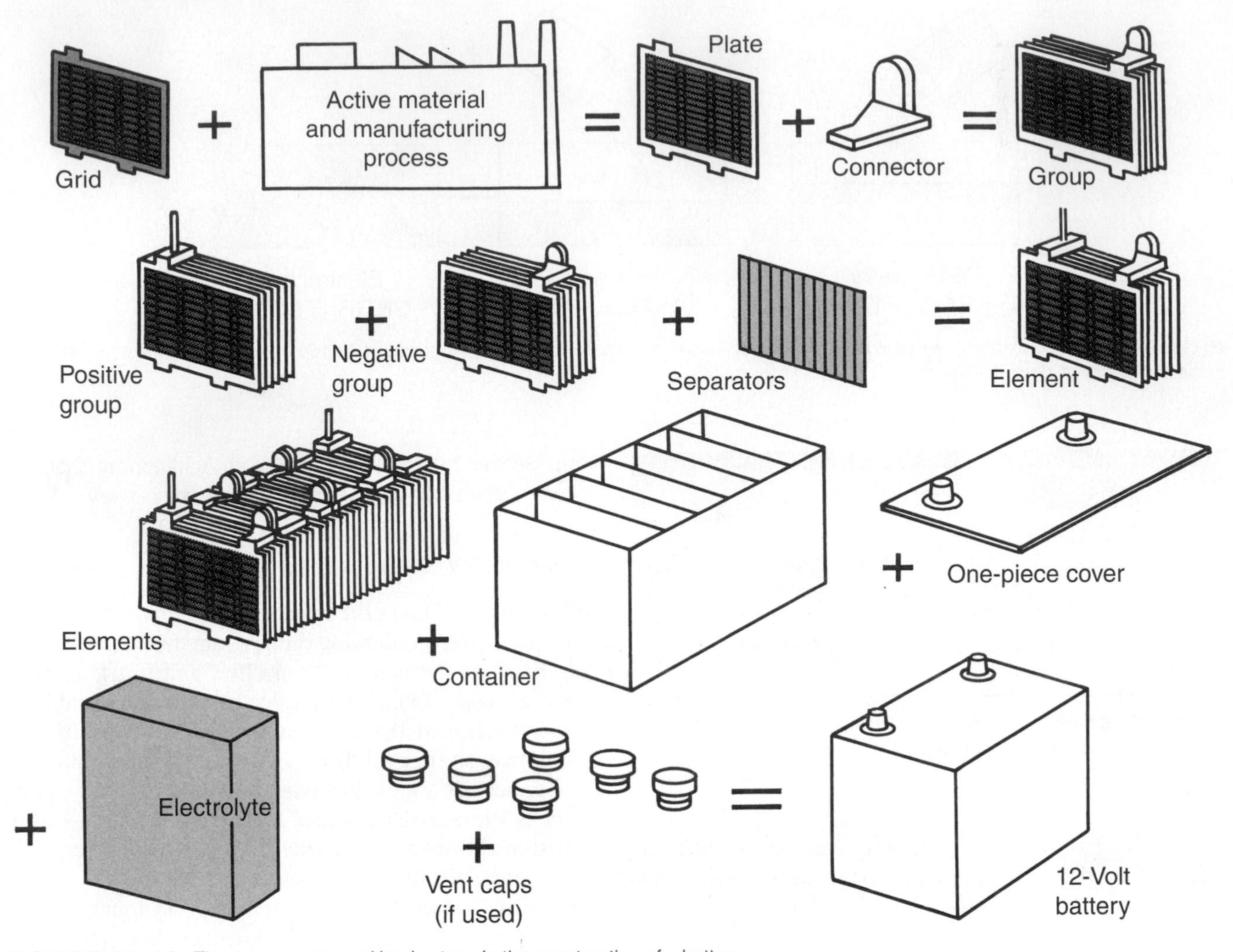

FIGURE 31-10 The components and basic steps in the construction of a battery.

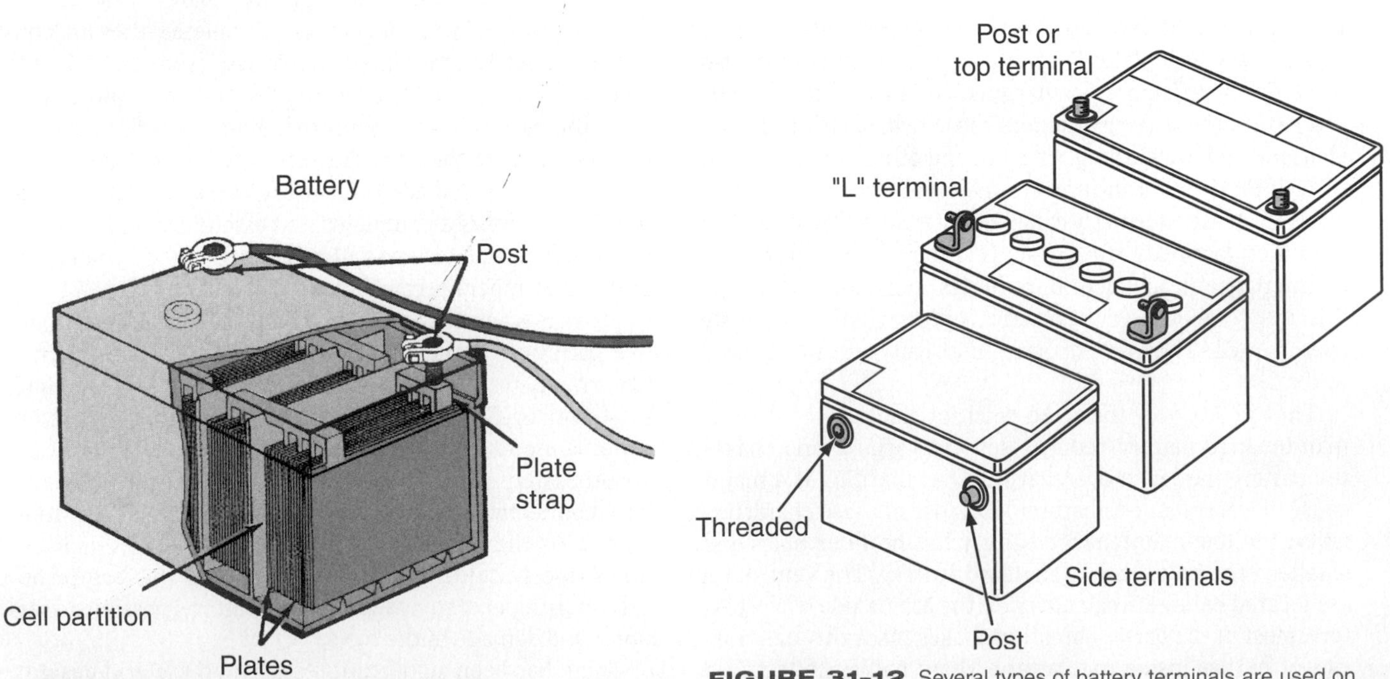

FIGURE 31-11 A cutaway of an assembled battery.

FIGURE 31-12 Several types of battery terminals are used on automotive batteries today.

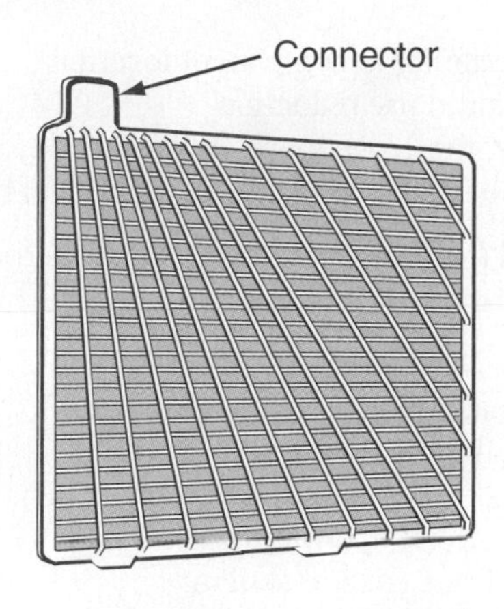

FIGURE 31–13 A low-maintenance battery has a grid designed to reduce electrical resistance.

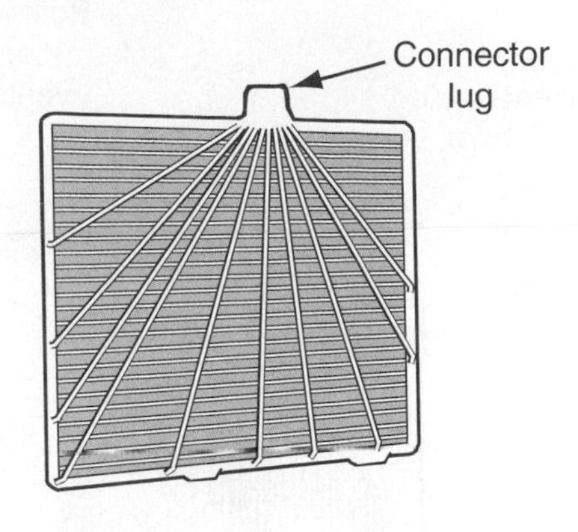

FIGURE 31–14 The hybrid grid uses a special design to help reduce resistance and so reduce gassing.

improve the charging and discharging by reducing resistance. ***Figure 31–13*** shows a grid designed for a low-maintenance battery. Note that the vertical grid bars are at an angle and point toward the connector, which helps to reduce the resistance of the grid.

Today, manufacturers are experimenting with other component materials and battery construction to increase the **cell density** of the battery. Cell density is a measure of how many watts can be discharged per hour, per pound of battery. For example, a lead-antimony battery has a cell density of 3–4 watt-hours per pound, whereas a lead-calcium battery has a cell density of 6–8, and a zinc-calcium battery has a cell density of 10–15. Future batteries may have cell densities of 45–50 watt-hours per pound.

HYBRID BATTERIES

Hybrid batteries combine the advantages of low-maintenance batteries with deep cycle batteries. The grid design of a hybrid battery consists of small amounts of antimony alloy on the positive plates and calcium alloy on the negative plates. This allows the battery to withstand deep cycling while retaining amperage capacity and cranking performance. Grid construction is different also. The grid has the output connector or lug located near the top center of the grid (***Figure 31–14***). The grid bars are then designed to maximize the battery's efficiency by reducing the resistance of the grid, as with low-maintenance batteries.

RECOMBINATION BATTERIES

Newer chemicals are being tested and used in batteries. Some manufacturers use an electrolyte in the form of a gel to further reduce gassing. This type of battery is called a recombination battery. The separators are placed between the grids and have very low electrical resistance. Because of this design, output voltage and current are higher than in a conventional or low-maintenance battery. The following are some advantages of the recombination battery:

- If cracked, will not spill electrolyte because of the gel form
- Can be installed in any position, even upside down
- Is corrosion free
- Has no electrolyte loss
- Can last four times longer that a conventional battery
- Can withstand deep cycling without damage
- Can be rated over 800 cold cranking amperes

GAS VENTS AND LIQUID/GAS SEPARATORS

Gassing occurs when a battery is constantly charged and discharged. The gases are mostly hydrogen and oxygen. This is normal, but it continuously reduces the level of the electrolyte. In a conventional battery, water is added to restore the proper electrolyte level.

Low-maintenance batteries are more resistant to gassing. This is because of the use of lead calcium and other materials rather than lead antimony. The manufacturing process for making the battery grids also helps to reduce gassing. So, the low-maintenance battery uses much less water and the holes and filler caps can be eliminated.

However, even with a low charging rate, a small amount of gassing still occurs in a low-maintenance battery. This small amount of gas escapes through a series of passages and vents in the end of the cover of the battery. ***Figure 31–15*** shows an example of the cover of a low-maintenance battery. There are built-in flame arresters to reduce the possibility of explosion.

To escape, the gasses rise and pass through the liquid/gas separator area. Any liquid electrolyte carried with the gases is trapped in the liquid/gas separator. It then flows back down into the cells. This process almost totally eliminates electrolyte loss.

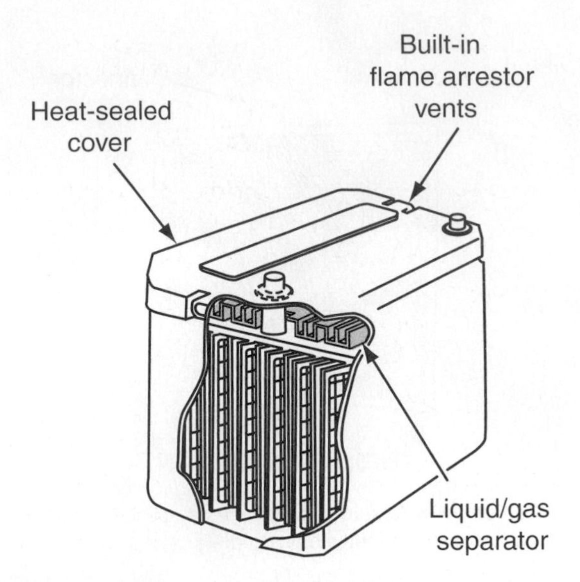

FIGURE 31–15 Gases produced from charging a battery flow through the liquid/gas separator to separate liquid electrolyte from the gases.

42-VOLT BATTERY SYSTEM

During the next few years, the automotive field will conduct a great deal of research and design on the use of a 42-volt battery system for cars. Today, automotive electrical systems operate between 12 and 14 volts; thus, a 12-volt battery is used. However, all automotive manufacturers plan to go to a new battery system over the next 5 to 10 years. The new voltage of the system will be between 36 and 42 volts, which represents major changes for all vehicle manufacturers.

As automotive vehicles become more and more advanced, additional electrical components are added to make them more efficient. For example, years ago a mechanical cooling fan was used. Now all vehicles use an electrical cooling fan. Using electrical rather than mechanical drives has three advantages:

1. More flexibility in component location
2. Significant noise reduction
3. Running the component only at needed times, which helps the engine consume less frictional horsepower

Additional loads continue to be placed on the automotive electrical system, including:

- Electrically heated catalytic converters
- Electromagnetic valves
- Electrically operated water pumps
- Electrically operated distributors
- Electrical braking (brake by wire)
- Electrical ride control systems
- Electrically assisted power steering
- Electrical navigation equipment
- Electrically heated windshields and seats
- Electrically operated air conditioning

In addition, more computers will be added to the vehicle to make the total automobile safer, more user-friendly, and more convenient for the passengers. All of these electrical loads have a tendency to increase the total amperage of the vehicle's electrical system. The electrical loads will increase even more with the demands of hybrid vehicles, which use electricity for propulsion.

The total electrical load on an automobile is measured in watts at peak power. For example, vehicles today use between 1,000 to 1,500 watts of peak power or more, depending on the year of the vehicle and the number of electrical loads placed on its system. This means that in today's 12-volt system, there are between 80 to 125 amperes flowing through the wires to the battery during peak loads. To handle this much amperage, wires must be very large. The wires going to and from the battery terminals are presently able to handle this amount of amperage. However, the peak power in watts is expected to increase significantly in the future. ***Figure 31–16*** illustrates manufacturers' projections for increased electrical demand. Note that in 2010, it is

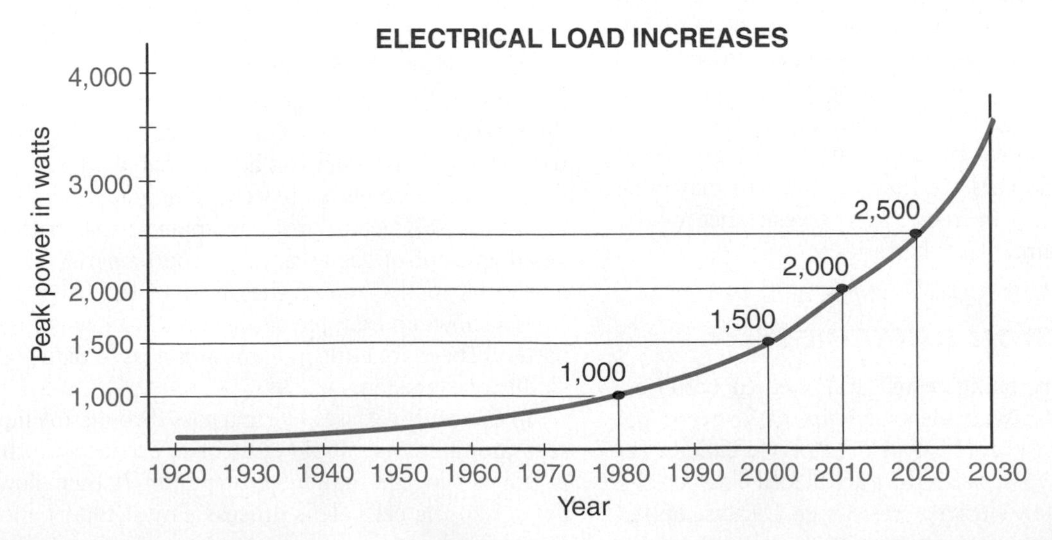

FIGURE 31–16 As more electrical demands are placed on the automotive battery system, the demand for peak power in watts will continue to increase as well.

anticipated that the average electrical demand will be near 2,000 watts of peak power. By 2030, the demand may be over 4,000 watts needed for peak power. In fact, some experts are suggesting much higher peak wattage ratings in the future. Using a 12-volt system would mean that wire size would need to double or triple to handle such increased amperage.

To help prevent this problem, vehicle manufacturers have made a commitment to change the electrical system to use a 42-volt battery. Of course, this means that the automotive battery will be much larger and heavier. However, this will be offset by additional engine efficiency improvements, increased fuel mileage, reduced pollution, more computer control, increased safety, and more passenger conveniences. In addition, smaller and more manageable wiring systems are planned, which will make installation of components much easier. On the other hand, every electrical motor, solenoid, light, computer, and so on will need to be redesigned to handle the increased voltage.

31.2 BATTERY RATINGS

Several ratings are used to identify the size and amperage capacity of the battery. Certain battery applications require cranking ability, while others may require long periods of high-amperage output. Because of these application differences, batteries are rated by (1) cold-cranking performance, (2) reserve capacity, (3) ampere-hours, and (4) watt-hours.

COLD-CRANKING PERFORMANCE RATING

The cold-cranking performance rating is a measure of a battery's ability to crank an engine under cold weather conditions. During cold weather operation, the engine is more difficult to crank. It takes much more amperage to crank a cold engine than to crank a warm engine. This rating indicates the number of cranking amperes the battery can deliver at 0 degrees Fahrenheit for a period of 30 seconds. During this time, the cell voltage cannot drop below 1.2 volts, or the total battery voltage below 7.2 volts. This rating is listed, for example, as 500 CCA (cold-cranking amperes). When buying a battery, you should compare this rating with other batteries in the same price range.

RESERVE CAPACITY RATING

The reserve capacity rating is expressed in minutes. It measures a battery's ability to provide emergency power for ignition, lights, and so on if the charging system is not working. This rating involves a constant discharge at normal temperatures. The reserve capacity rating is defined as the number of minutes a fully charged battery at 80 degrees Fahrenheit can be discharged at 25 amperes and still maintain a minimum voltage of 1.75 volts per cell or 10.5 volts for the total battery. The higher the reserve capacity rating, the longer the battery can provide the emergency power for lights and accessories. A typical reserve capacity rating might be 125 minutes.

AMPERE-HOUR RATING

The ampere-hour rating is another measure of a battery's capacity. It is obtained by multiplying a certain flow of amperes by the time in hours during which current will flow. This is usually expressed as a "20-hour rating." Again, the test is run at 80 degrees Fahrenheit until the cell voltage drops below 1.75 volts or the total battery voltage drops below 10.5 volts.

The amount of amperage used in this test is 1/20 of the published 20-hour capacity in ampere-hours. For example, if a battery is rated at 105 ampere-hours, its discharge amperage is 5.25 (1/20, or 0.05, of 105 equals 5.25). To pass the test, this battery would be discharged at 5.25 amperes for 20 hours and still have a total voltage above 10.5.

WATT-HOUR RATING

The watt-hour rating is measured by multiplying the ampere-hour rating by the voltage of the battery. It is much the same as the ampere-hour rating, except watts are measured instead of amperes.

31.3 BATTERY MAINTENANCE AND TESTING

HYDROMETER TESTING

A **hydrometer** is used to measure the specific gravity of batteries that have filler caps (although such batteries are used less and less). ***Figure 31–17*** shows a typical hydrometer. The hydrometer has a weighted float inside a glass tube.

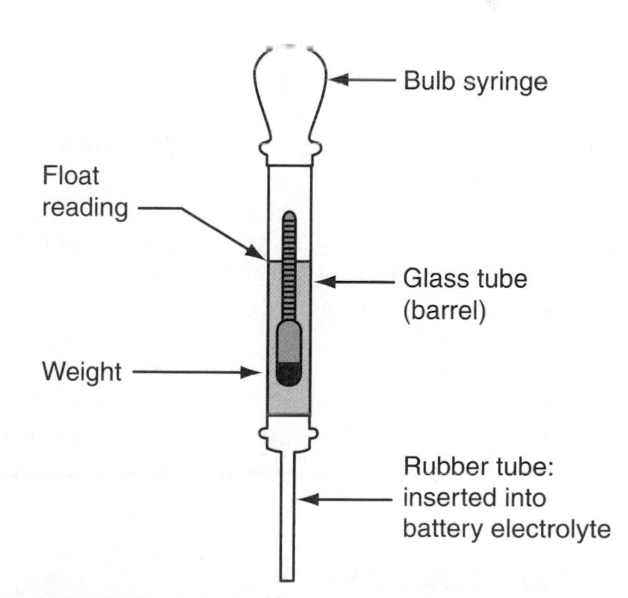

FIGURE 31–17 A hydrometer is used to measure the specific gravity of the battery electrolyte. Fluid is drawn into the glass tube by the bulb syringe, and the float reading tells the specific gravity of the electrolyte.

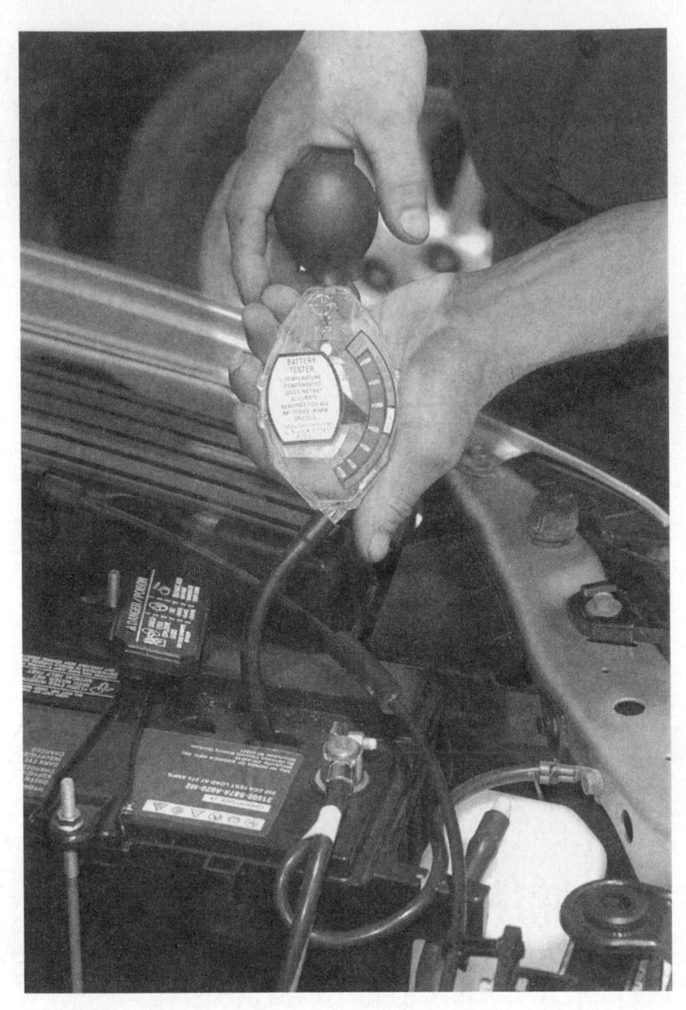

FIGURE 31-18 The specific gravity reading is measured on a hydrometer by the floating needle.

A bulb syringe is used to draw the electrolyte into the glass tube. Some hydrometers also have a temperature indicator or thermometer inside the float.

The object of the hydrometer is to determine the density of the electrolyte. The heavier, or denser, the electrolyte, the greater the percentage of acid. The lighter, or less dense, the electrolyte, the less the percentage of acid. If the electrolyte is denser, the float inside the hydrometer will not sink as deep into the solution. If the electrolyte is less dense, the float will sink deeper into the solution. The amount the float sinks into the solution is a measure of the specific gravity.

Figure 31–18 shows how a hydrometer is used. The electrolyte solution is drawn into the plastic tube. The inside of the float has a specific gravity scale to show the exact reading. Never take specific gravity readings immediately after the battery has been filled with water. The water and acid will not be completely mixed, and the reading will be incorrect.

TEMPERATURE AND SPECIFIC GRAVITY

The temperature of the electrolyte has an effect on the exact specific gravity. Specific gravity readings must be compensated for temperature. For each 10 degrees above 80 degrees Fahrenheit, 0.004 points must be added to the specific gravity reading. For every 10 degrees below 80 degrees Fahrenheit, 0.004 points must be subtracted from the specific gravity reading. For example, ***Figure 31–19*** shows a hydrometer correction scale. If the electrolyte solution is at 100 degrees Fahrenheit and the specific gravity is measured at 1.218, the correct specific gravity will be 1.226. (For each 10 degrees above 80 degrees Fahrenheit, 0.004 are added; 1.218 + 0.008 points = 1.226 corrected specific gravity reading.)

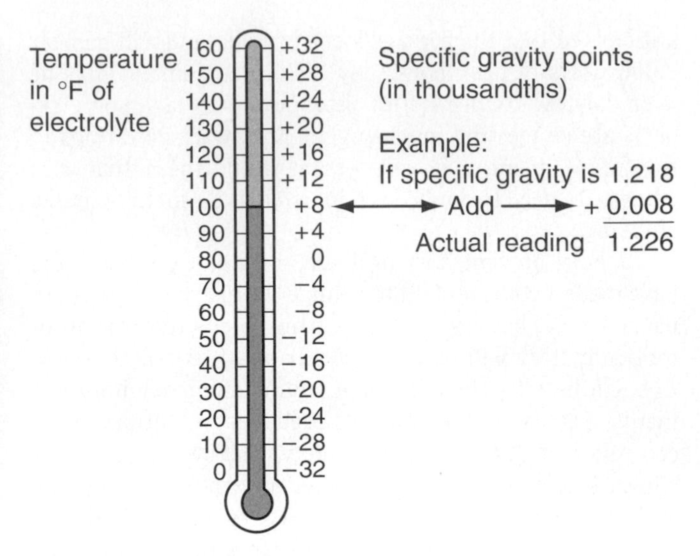

FIGURE 31-19 Specific gravity readings must be corrected for temperature according to this table.

BATTERY FREEZING

The freezing point of the electrolyte depends on its specific gravity. A fully charged battery should never freeze. As the battery discharges and the specific gravity gets close to 1.000 or pure water, the chance of the electrolyte freezing increases. If the electrolyte freezes, the battery case may crack and damage the battery. The table shown in ***Figure 31–20*** indicates the freezing temperatures of electrolytes at various specific gravities.

VALUE OF SPECIFIC GRAVITY	FREEZING TEMP. DEG. F.	VALUE OF SPECIFIC GRAVITY	FREEZING TEMP. DEG. F.
1.100	18	1.220	–31
1.120	13	1.240	–50
1.140	8	1.260	–75
1.160	1	1.280	–92
1.180	–6	1.300	–95
1.200	–17		

FIGURE 31-20 The freezing temperature of the electrolyte changes with different specific gravity readings.

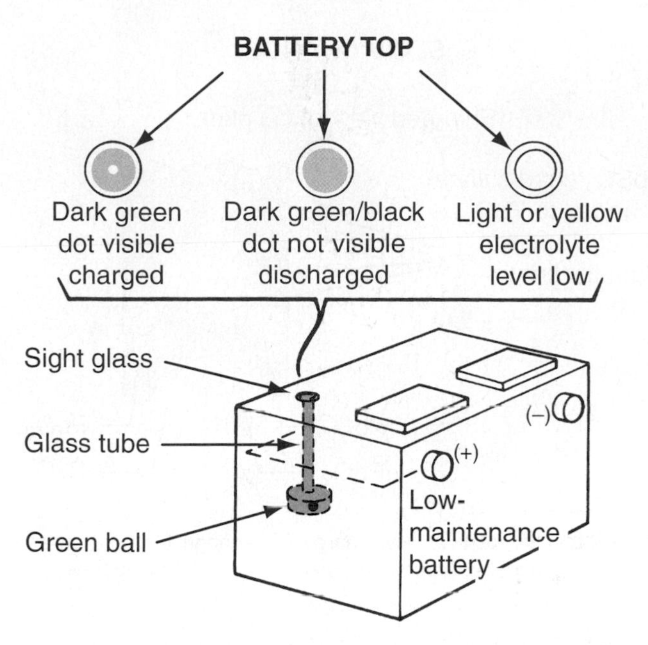

FIGURE 31–21 Specific gravity can be measured with a built-in hydrometer. (Courtesy of DaimlerChrysler Corporation)

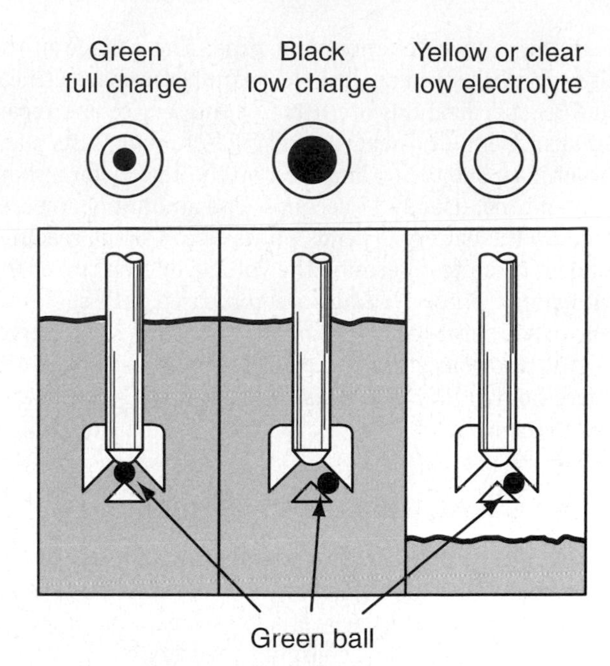

FIGURE 31–22 This built-in hydrometer uses a small ball. A green ball indicates the battery is fully charged.

SPECIFIC GRAVITY ON LOW-MAINTENANCE BATTERIES

Because there are no filler vents on a low-maintenance battery, another method is used to measure specific gravity, (***Figure 31–21***). On certain batteries, specific gravity is determined by a float device built into the battery. A small colored ball floats when the specific gravity is correct or in the right range. If the specific gravity is below the required level, the ball sinks. If a green dot is seen, the specific gravity is correct. If the indicator is black, the specific gravity is low. The battery should be charged. If the indicator shows light yellow, the battery electrolyte is too low. The battery may be damaged and should be replaced.

Other low-maintenance batteries have a test indicator that is also located on the top of the battery. When the test indicator shows "OK," the battery is satisfactory. This is shown by a blue or other color on the test indicator. If the test indicator is colorless, charging is necessary.

PARTS LOCATOR

Refer to photo #55 in the front of the textbook to see what a ***sight glass on a battery*** *looks like.*

Figure 31–22 shows how a built-in hydrometer works. When the battery is fully charged or at least at 65% of its charge, the ball rises to its highest point in the hydrometer. If the battery is in need of a charge, the small green ball will drop away from the center of the plastic tube. When this happens, the indicator will display black. If there is a lack of electrolyte, the indicator will display yellow or clear. When yellow or black are displayed, always tap the top of the built-in hydrometer to make sure there are no gas bubbles causing the yellow or black color.

On older maintenance-free batteries, the sight glass only indicated the specific gravity reading of one cell. Today's low-maintenance batteries are designed to give an indication of the specific gravity reading in all cells. When reading the hydrometer, make sure the top of the battery is clean. Remember to always look straight down when viewing the hydrometer. Use a flashlight in low-light conditions.

BATTERY OPEN CIRCUIT TEST

An open circuit battery voltage test can be done by measuring total battery voltage across the positive and negative terminals. This is done after the battery has been stabilized or charged fully. This test will quickly determine the general state of charge and battery condition. This test uses a voltmeter placed across a battery that is disconnected. The voltmeter should read more than 12 volts. If there is a bad cell, the overall voltage will be less than that. ***Figure 31–23*** shows the relationship between open circuit voltage and percent of charge.

LOAD TESTS

Several types of load tests can be performed on a battery. A light load test places a small 10- to 15-ampere load on the

OPEN CIRCUIT VOLTAGE

Voltage	11.7	12.0	12.2	12.4	12.6 or more
% charge	0%	25%	50%	75%	100%

FIGURE 31–23 When testing for open circuit voltage, use these voltage readings to determine the percent of charge.

battery for a specific amount of time. The voltage of the cells is then checked against the manufacturer's specification. The specifications are listed in the service and repair manuals. The high-discharge and cold-cranking tests place a larger load, 300 to 500 amperes, on the battery for a short period of time, usually 15 seconds. The amount of amperes placed on the battery depends on its size. Voltage readings are again taken to determine the voltage of each cell or the total battery. These readings are compared with specifications to determine the condition of the battery. The reserve-capacity test places the same load on the battery as the ampere-hour rating test. This rating is a time or minutes measurement.

SLOW CHARGING BATTERIES

Batteries can be brought back to fully charged state by two methods: the slow and fast charges. The slow charge is defined as charging the battery at a low ampere rating over a long period of time. Slow charging a battery is much more effective than fast charging. Slow charging causes the lead sulfate of the discharged battery plates to convert to lead peroxide and sponge lead throughout the thickness of the plate (***Figure 31–24***).

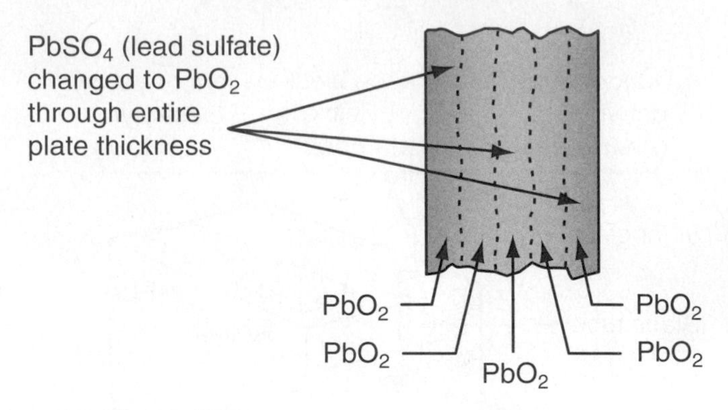

FIGURE 31-24 A slow charge will change the lead sulfate back to lead peroxide throughout the entire thickness of the plate.

The slow charge ampere rate is 1 ampere per positive plate in one cell. For example, if the cell has eleven plates, there are six negative and five positive plates. The slow charge ampere rate would then be 5 amperes until the battery is fully charged. This may be in excess of 24 hours.

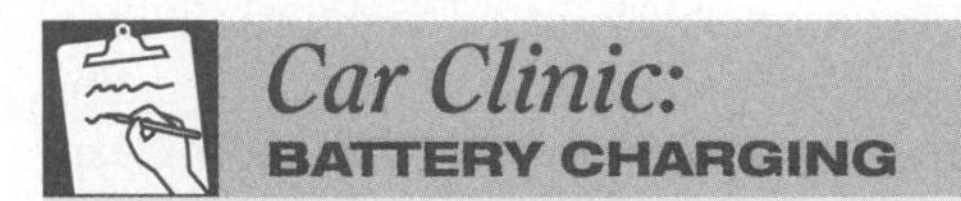

Car Clinic: BATTERY CHARGING

CUSTOMER CONCERN:
Electronics Damaged by Use of Jumper Cables

A vehicle had a dead battery and needed to be jump-started by another vehicle. The jump start damaged several electronic components. What caused the electrical components to be damaged?

SOLUTION: While they were being connected to the battery cables, the jumper cables were momentarily reversed, and a spark was produced at the battery terminal. Reversing the polarity caused excessive reverse voltage to be applied to many electronic components and circuits. This reverse voltage may have caused some of the electronic components to be damaged. When using jumper cables, it is very important never to reverse the polarity, even for a brief moment. Vehicles today have numerous electronic circuits that can be easily damaged. Always follow the *exact* recommended procedure when using jumper cables to jump-start another vehicle. Also note that some manufacturers recommend that you do not jump-start other vehicles because of possible damage to computers.

FAST CHARGING BATTERIES

Batteries can also be charged using a higher rate of amperes. Fast charging a battery changes the lead sulfate only on the outside of the battery plates. The internal parts of the plate are still lead sulfate (***Figure 31–25***). Fast charges are used to "boost" the battery for immediate cranking power. ***Figure 31–26*** shows a table of time and charge rates for different specific gravity readings. Use this table only as a guideline.

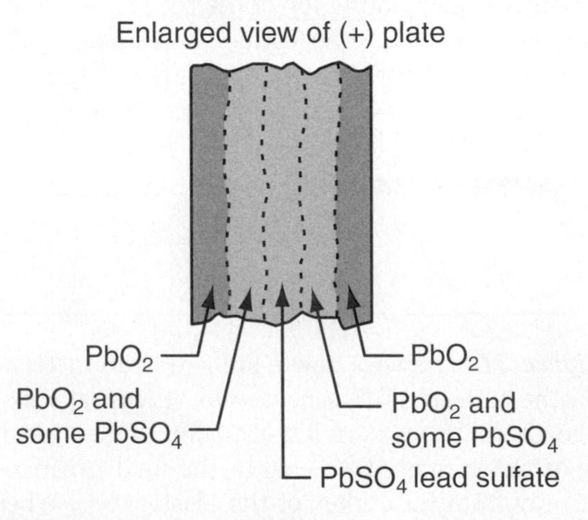

FIGURE 31-25 When a battery is fast charged, only the outer surface of the plate changes to lead peroxide. The inner part of the plate is still lead sulfate.

BATTERY HIGH RATE CHARGE TIME SCHEDULE					
SPECIFIC GRAVITY READING	CHARGE RATE AMPERES	BATTERY CAPACITY—AMPERE-HOURS			
		45	55	70	85
Above 1.225	5	★	★	★	★
1.200–1.225	35	30 min.	35 min.	45 min.	55 min.
1.175–1.200	35	40 min.	50 min.	60 min.	75 min.
1.150–1.175	35	50 min.	65 min.	80 min.	105 min.
1.125–1.150	35	65 min.	80 min.	100 min.	125 min.

★ Charge at 5-ampere rate until specific gravity reaches 1.250 @ 80°F.

FIGURE 31-26 This table shows the charging rates for different specific gravity ratings and battery capacity ratings. Use this chart as a guideline when fast charging a battery.

Problems, Diagnosis, and Service

Safety Precautions

1. The battery acid used in batteries is very toxic and corrosive. Never let the battery acid get on your clothes or skin. Battery acid can be neutralized by mixing baking soda with the acid. Always have a box of baking soda nearby when working on batteries. If battery acid gets into the eyes, flush immediately with tap water for at least 15 to 20 minutes. Seek medical attention immediately.
2. Never place tools on the top of the battery, especially on older batteries that have the terminals on top. If the terminals are shorted by a tool, sparks will be produced and excessive current will heat the tool to its melting point, possibly causing injury to a person.
3. When batteries are discharged and charged, hydrogen gas, which can be highly explosive, is produced. Keep sparks from the electrical system away from the battery. Also, if a battery cable has a loose connection, when moved, it too can cause sparks that can ignite the hydrogen gas.
4. Do not use an excessive charge rate or charge batteries with cells that are low on electrolyte.
5. When working around batteries, always wear protective clothing.
6. Batteries are very heavy. Always lift the battery using the correct lifting tool, as back injury may occur during lifting.
7. When cleaning battery posts, be careful that the corrosion built up on the posts does not get onto your skin or clothes. It is very acidic and may cause acid burns as well as ruin your clothes.
8. Never smoke in or around an automobile or when working on its battery.
9. Because batteries are explosive, always keep an approved fire extinguisher nearby when working on batteries.
10. Always wear safety glasses when working on batteries.
11. Battery acid is considered a hazardous waste product. Batteries should not be thrown away, as damage to the environment may result. Stores that sell batteries normally take old batteries and properly dispose of them. When buying a new battery, bring your old battery to the seller for proper disposal.

PROBLEM: Corroded Battery Terminals

The engine in a vehicle is not able to start. The battery seems to be in good condition and is still operating the accessories.

DIAGNOSIS

Often, dirty battery terminals restrict the heavy flow of electricity, during starting, from the terminal to the battery post. Refer to ***Figure 31–27***. Corrosion can build up on the top- and side-type battery terminals. To check the condition of the battery post and connector, a voltmeter can be used. Check the voltage drop across or between the battery post and the connector. Any voltage drop indicates that a resistance is present between the two, and corrosion exists.

SERVICE

1. Always use the correct battery tools when working on automotive batteries (***Figure 31–28***). To clean the battery post, use the following general procedure:

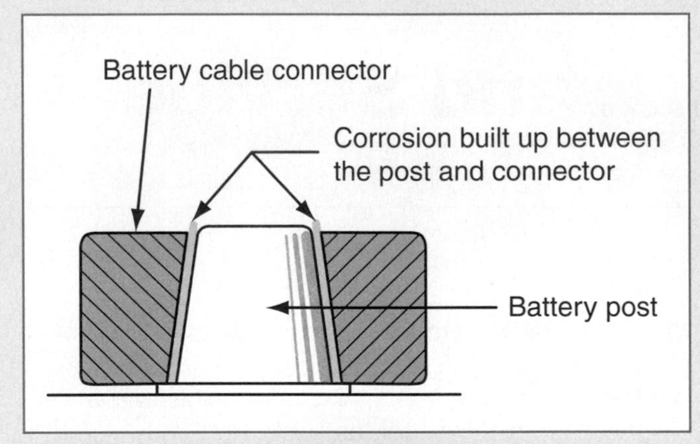

FIGURE 31–27 Corrosion can develop between the battery post and the cable connector. Use a wire brush to clean off the corrosion on the post and the cable connector.

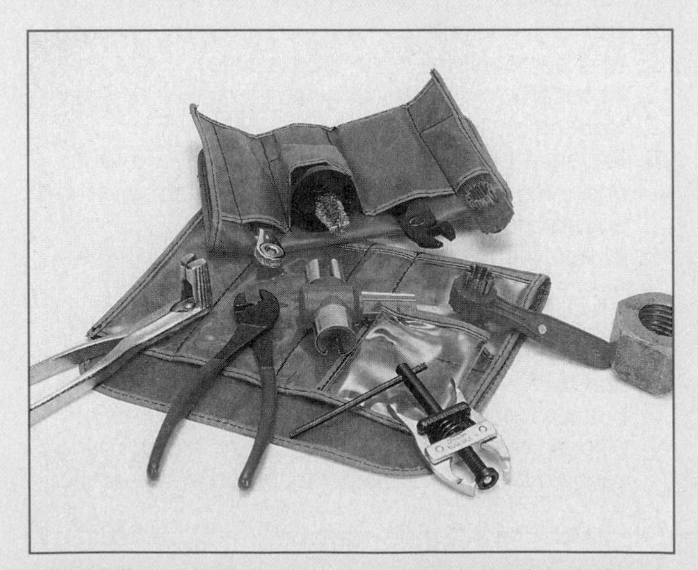

FIGURE 31–28 Always use the proper tools when working on a battery. (Courtesy of Snap-on Tools Corporation, www.snapon.com)

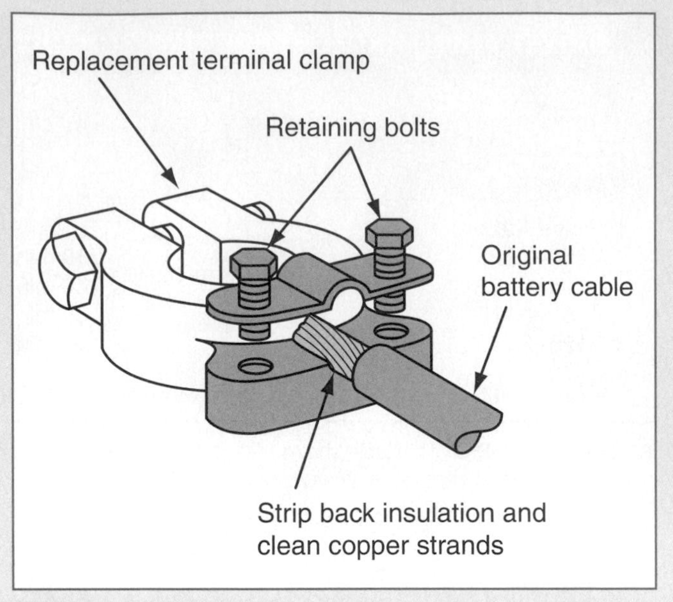

FIGURE 31–29 The cable connectors on the ends of battery cables can be replaced. Remove the old connector, cut back the wire insulation, and replace the connector with a new one. After replacement, put a light coat of grease on the connector to cut down on corrosion.

   a. Remove the negative terminal, then the positive terminal of the battery.
   b. Using a wire brush, clean both the post and the connectors, removing all corrosion.
   c. Clean the remaining corrosion using a small amount of a solution of baking soda and water.
2. Battery terminal clamps that are corroded beyond repair can be replaced. Remove the clamps from the battery post. Cut off the old clamp and strip about 3/4 inch of insulation from the cable. Clean the exposed copper wire strands and install a top- or side-mounted replacement clamp or terminal (***Figure 31–29***).

PROBLEM: Battery Overcharging

A battery continually needs water to maintain the correct electrolyte level.

DIAGNOSIS

A battery that continually needs water is being overcharged by the alternator or is overcharging because it will not accept a charge.

SERVICE

Various checks need to be completed:

1. The regulator on the charging system needs to be checked for correct voltage and current output (see Chapter 34).
2. Perform a load test on the battery. ***Figure 31–30*** shows one manufacturer's procedure for testing a battery. In

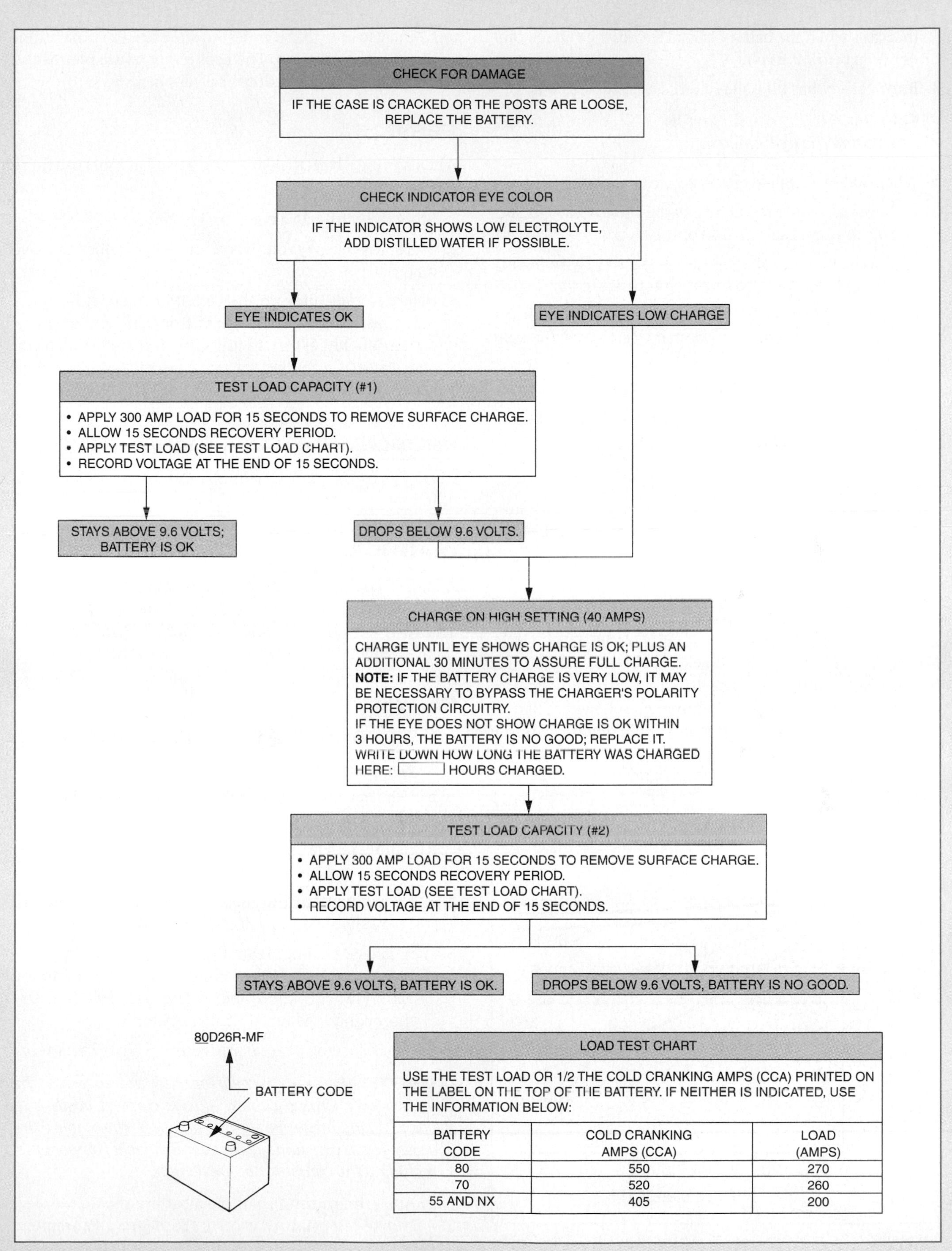

BATTERY CODE	COLD CRANKING AMPS (CCA)	LOAD (AMPS)
80	550	270
70	520	260
55 AND NX	405	200

FIGURE 31-30 This procedure can be used to check a battery. Note that after each load test, the battery voltage must be at least 9.6 volts. (Courtesy of American Honda Motor Co., Inc.)

this case, when the battery is load tested, the voltage cannot drop below 9.6 volts.

3. If the battery has fill holes, check the specific gravity.

CAUTION: *Be careful not to spill any electrolyte when using the hydrometer.*

4. When adding water, keep these rules in mind:
   a. Never add sulfuric acid to a battery to increase its specific gravity readings. Add only water.
   b. Check the level of water in the battery by removing the vent caps (if not a low-maintenance battery).
   c. Using distilled water, add water to each cell until the water is level with the lower inside ring of the vent hole.
   d. Make sure the level of water is the same in each cell.
   e. Replace the vent caps, and dry off any water spilled on the top of the battery.

PROBLEM: Storing a Battery

A battery needs to be stored over a long period of time. What procedure should be followed?

DIAGNOSIS

A battery will self-discharge while in storage. As shown in ***Figure 31–31***, the specific gravity of the electrolyte solution falls as the time in storage increases. If the temperature during storage is 80 degrees Fahrenheit, follow that particular line to determine the drop in specific gravity. Say, for example, the specific gravity is 1.270 when a battery is put in storage. After 50 days, the specific gravity has dropped to about 1.234. After 90 days, the specific gravity has dropped to 1.215. Also note that it is best to store a battery that is not in use in a cool location. The cool temperature retards the chemical reaction and reduces self-discharging.

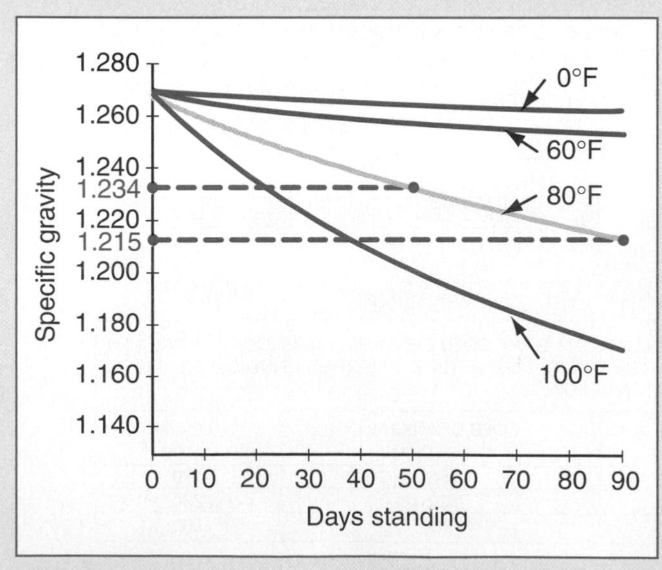

FIGURE 31-31 Batteries will self-discharge depending on the temperature around the battery. Colder conditions slow down the self-discharge rate.

SERVICE

When storing a battery over a long period of time, use three general rules:

1. Store the battery in a cool, dry location.
2. Store the battery on wood or other nonconductive material.
3. Before storage, prepare the battery as follows: Keep the outer case of the battery clean from dirt and grime. A certain amount of battery leakage can occur through dirt and grime. Clean the battery case with a solution of water and baking soda.

PROBLEM: Dirty Battery Casing and Top

A battery seems to be losing its charge when it sits for a long time.

DIAGNOSIS

Acid, dirt, corrosion, or cracks on the top of the battery can cause a slow discharge or leakage of the energy stored in the battery. If enough of this leakage occurs, there is a good chance the battery will be too weak to start the engine.

SERVICE

1. Use the following general procedures to check for leakage:
   a. Make sure the engine is off.
   b. Using a voltmeter, connect the black clip to the negative side of the battery.
   c. Lightly touch the red clip to various parts of the battery top and sides (not the battery posts).
   d. Carefully watch the meter for any voltage readings. There should be no voltage readings.
   e. A voltage reading, regardless of how small, indicates leakage. To correct the leakage, thoroughly clean the battery with a water and baking soda solution. Dry thoroughly, and retest the battery for leakage.
2. Use the following general procedure to charge a battery.

CAUTION: *Be extremely careful not to cross the jumper or charging cables. Always connect positive to positive and negative to negative. If the cables are crossed, you can damage various electrical components, including the computer in the vehicle.*

   a. At no time during charging a battery should the electrolyte temperature exceed 125 degrees Fahrenheit. (The battery will feel very hot to the touch.)
   b. Make sure the top of the battery is clean.

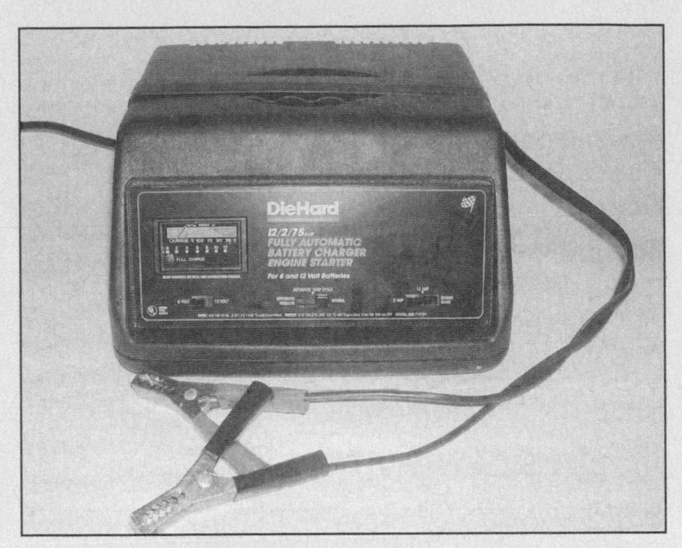

FIGURE 31–32 Each battery charger has a slightly different procedure to charge batteries. Always follow the battery charger manufacturer's recommended procedure.

c. Remove the vent caps, if applicable, to allow gases to escape easily.

d. Determine if the battery will be slow or fast charged. This will determine the amperage rating needed.

e. Each battery charger manufacturer may have a slightly different procedure. ***Figure 31–32*** shows a typical battery charger.

f. With the battery charger disconnected from the power (110 volts), attach the positive lead of the charger to the positive terminal of the battery. If the battery terminal posts on the top cannot be identified as positive or negative, the larger-diameter top terminal post is usually the positive side. It may also have a plus sign (+) near it or the wires may be painted red.

g. Attach the negative lead of the battery charger to the negative terminal on the battery.

h. At this point, plug in the charger to 110 volts and set the amperage readings (according to the specific charger instructions).

i. Some battery chargers may also have a timer that should now be set.

j. The time required to charge a battery will vary depending on several factors:

- Size of battery—The larger the battery, the more time for charging.
- Temperature—It takes longer to charge a colder battery than a warmer battery.
- Charger capacity—Battery chargers vary as to how much current they can put into the battery. Small chargers can only put around 10 to 12 amperes into the battery. Other chargers can put up to 30 amperes in.
- State-of-charge—The more discharged the battery, the greater the time needed for charging.

k. After the battery is charged, shut the charger off.

l. Remove the power cord (110 volts).

m. Now remove the positive and negative terminal leads between the battery and the battery charger.

CAUTION: *The purpose of having the charger unplugged from the terminal leads is to prevent an explosion. Any small spark could ignite explosive gases around the battery.*

PROBLEM: Jump-Starting a Battery

A vehicle has been left with the lights on while parked. The battery is now dead and will not start the vehicle.

DIAGNOSIS

Batteries that have been temporarily discharged may be jump-started by another vehicle's battery if recommended by the manufacturer. The battery voltage of the second car should be enough to jump-start the discharged battery.

CAUTION: *When jump-starting a battery that has a computer in the vehicle, make sure the battery cables do not get crossed. If they do, they could produce a spark and damage the computer beyond repair.*

SERVICE

Use the following general procedure in an emergency for jump-staring a battery:

1. ***Figure 31–33*** shows an example of the connections for jump-starting a battery.
2. Make sure the engine is off.
3. Connect the positive terminal of the discharged battery to the positive terminal of the good battery. It is very important that the positive terminals are connected together.

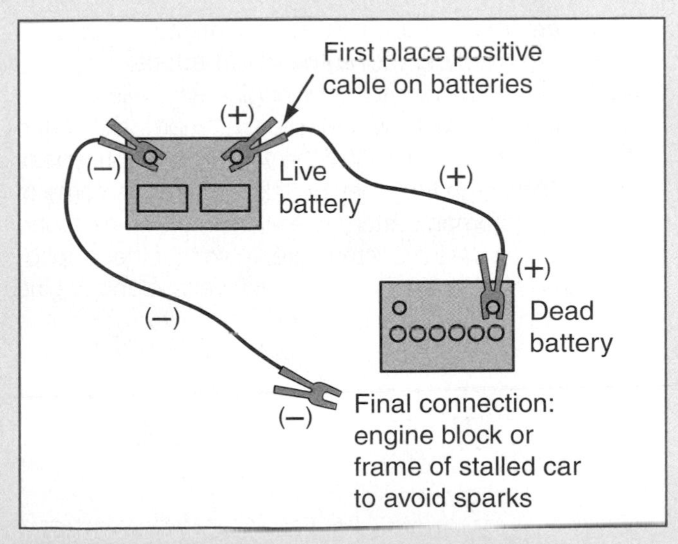

FIGURE 31–33 When boosting or jump-starting a car, always connect the jumper cables as shown. Make a final connection between the engine block or frame to avoid sparks around or near the battery.

4. The negative cable should be connected to the engine block, ground, or frame to avoid sparks around or near the battery.

CAUTION: *Do not connect directly to the negative terminal of the dead or discharged battery.*

5. Now start the engine of the live battery. The live battery's vehicle should be kept running during starting to avoid draining its battery.
6. In some cases, if the battery is extremely discharged, it may have to accept a charge for several minutes before starting.
7. Now reverse the directions exactly when removing the jumper cables.

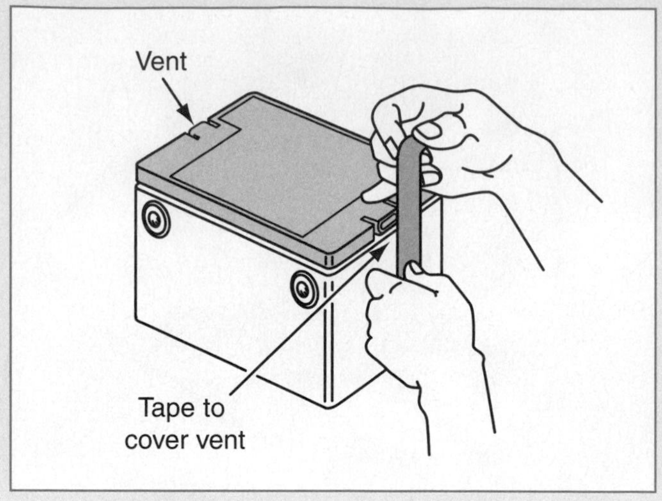

FIGURE 31-34 Always tape the vent holes before cleaning a battery with a mixture of baking soda and water.

PROBLEM: Cleaning the Top of a Low-Maintenance Battery

A vehicle has a low-maintenance battery that has a very dirty top cover. The owner suspects that the dirt on top of the battery may cause premature discharging.

DIAGNOSIS

Batteries that have dirt and grease on the top of the cover may discharge prematurely. The top of the battery needs to be washed with a mixture of baking soda and clean water.

SERVICE

The first step before cleaning is to cover the vent holes with a small piece of tape as shown in ***Figure 31–34***. Then clean the top of the battery with a solution of baking soda and water. After the top has been cleaned and dried, remove the small pieces of tape to allow the battery to breath correctly.

Service Manual Connection

There are many important specifications to keep in mind when working with automotive batteries. To identify the specifications for your engine, you will need to know the VIN (vehicle identification number) of the vehicle, the type and year of the vehicle, and the type of engine. Although they may be titled differently, some of the more common battery specifications (not all) found in service manuals are listed here. Note that these specifications are typical examples. Each vehicle and engine may have different specifications.

Common Specification	Typical Example
Ampere-hour rating	105 ampere-hours
Cold-cranking amperes (CCA)	550 amperes
Hydrometer reading	1.280
Test load	270 amperes
Voltage after test load applied	9.6 volts minimum

In addition, the service manual will give specific directions for various service and testing procedures. Some of the more common procedures include:

- Battery charging (procedure/time/amperage)
- Battery diagnosis
- Battery replacement
- Cold-cranking tests
- Light load tests
- Measuring specific gravity

SUMMARY

The following statements will help to summarize this chapter:

- The purpose of a storage battery is to store electrical energy in a chemical form.
- A primary battery cannot be recharged.
- Secondary batteries can be recharged.
- Batteries are manufactured with different voltages and amperage capacities by connecting cells in series or parallel connections.
- The most common type of automobile battery is the lead-acid battery.
- When acid is mixed with water it is called an electrolyte.
- During charging and discharging, a conventional battery will give off hydrogen gases.
- During discharging, the two dissimilar metals turn into lead sulfate.
- During charging, the lead sulfate turns the metals back into their original form.
- As a battery discharges, the electrolyte turns to water.
- Specific gravity is a measure of the amount of acid in the electrolyte.
- A low-maintenance battery reduces the amount of hydrogen gassing.
- Other batteries, such as the hybrid battery and the recombination battery, are being used to further reduce hydrogen gassing.
- Automotive manufacturers have decided to change the 12-volt battery system to a 42-volt system so that more electrical loads can be used on automotive vehicles, while still managing to keep the wiring relatively small.
- There are several types of battery ratings including cold-cranking performance rating, reserve capacity rating, watt-hour rating, and ampere-hour rating.
- A battery can be tested for specific gravity using a battery hydrometer.
- Load tests are used to measure a battery's ability to handle an electrical load.
- The reserve capacity test places a continuous load on a battery over a certain period of time.
- A battery can be either slow charged or fast charged, depending on its state of charge and the time available.
- Always use the proper tools when working on and around a battery.

TERMS TO KNOW

Can you explain each of the following terms? Review the chapter until you can use each term correctly.

Amperage capacity
Baffles
Battery cell
Cell density
Charging
Cycle
Deep cycle battery
Discharging
Electrochemical
Electrolyte
Hydrometer
Primary batteries
Specific gravity

REVIEW QUESTIONS

Multiple Choice

1. The positive plate of a standard lead-acid battery is made of:
   a. Lead oxide
   b. Lead peroxide
   c. Sponge lead
   d. Sulfuric acid
   e. None of the above

2. The negative plate of a standard lead-acid battery is made of:
   a. Lead oxide
   b. Lead peroxide
   c. Sponge lead
   d. Sulfuric acid
   e. None of the above

3. A battery is able to store ______ energy for use later as electrical energy.
   a. Radiant
   b. Chemical
   c. Thermal
   d. Nuclear
   e. Mechanical

4. Which type of battery can be recharged?
   a. Primary battery
   b. Antimony battery
   c. Secondary battery
   d. Sulfur battery
   e. None of the above

5. As a lead-acid battery is discharged, the negative plate is converted to:
   a. Lead sulfate
   b. Sponge lead
   c. Water
   d. Sulfuric acid
   e. None of the above

6. As a lead-acid battery is discharged, the positive plate is converted to:
   a. Lead sulfate
   b. Sponge lead
   c. Water
   d. Sulfuric acid
   e. None of the above

7. When battery cells are placed in series, the voltage of the battery will:
   a. Be reduced to zero
   b. Be additive
   c. Remain the same
   d. Increase by one half of each cell voltage
   e. None of the above

8. When battery cells are placed in parallel, the amperage capacity of the battery will:
   a. Be reduced to zero
   b. Be increased
   c. Remain the same
   d. Decrease
   e. None of the above

9. What is the ratio of water to sulfuric acid in a lead-acid battery?
   a. 10% water, 90% acid
   b. 20% water, 80% acid
   c. 30% water, 70% acid
   d. 50% water, 50% acid
   e. None of the above

10. What is the specific gravity of a fully charged battery?
   a. 1.000–1.002
   b. 1.128–1.130
   c. 1.270–1.295
   d. 1.940–2.300
   e. 1.750–2.100

11. Cell density of a battery is a measure of the:
   a. Electrolyte amount
   b. Watt-hours/pound from the battery
   c. Amperage capacity
   d. Active materials in the battery
   e. Specific gravity

12. Which battery rating is used to measure the battery's ability to provide electricity when the engine is shut off?
   a. Cold-cranking performance
   b. Reserve capacity rating
   c. Ampere-hour rating
   d. Watt-hour rating
   e. None of the above

13. The object of a hydrometer test is to determine the ________ of the electrolyte in the battery.
   a. Voltage
   b. Amperage
   c. Wattage
   d. Specific gravity
   e. Amount of lead sulfate

14. What is the purpose of changing to a 42-volt battery system?
   a. Make the engine more powerful
   b. Handle additional electrical loads in the future
   c. Allow the engine to run faster
   d. Keep the battery lighter
   e. Keep the electrical wires larger

15. In order to reduce gassing in a hybrid battery, its:
   a. Grids are smaller
   b. Grids are designed to reduce voltage
   c. Grids are designed to reduce resistance
   d. Grids are larger
   e. Grids have no negative plates

16. What correction is needed to compensate for the temperature of the electrolyte when taking specific gravity readings?
   a. Increase 0.004 for each 10 degrees above 80 degrees Fahrenheit
   b. Increase 0.008 for each 10 degrees above 80 degrees Fahrenheit
   c. Decrease 0.004 for each 10 degrees above 80 degrees Fahrenheit
   d. Decrease 0.004 for each 1 degree above 80 degrees Fahrenheit
   e. None of the above

17. Which type of charge will give a battery the deepest and most effective charge?
   a. Fast charge
   b. Boost charge
   c. Intermediate charge
   d. Slow charge
   e. No charge

18. What should be used when battery acid gets into the eyes?
   a. Water only
   b. Lead sulfate
   c. Sulfuric acid
   d. Baking soda and water
   e. Sponge lead

19. When a battery is stored or not used, it is better to store the battery in a ________ location.
    a. Warm
    b. 80 degrees or above
    c. Cool
    d. Damp, moist
    e. None of the above

The following questions are similar in format to ASE (Automotive Service Excellence) test questions.

20. *Technician A* says that low-maintenance batteries cannot be charged. *Technician B* says that low-maintenance batteries can be checked for specific gravity using a handheld hydrometer. Who is correct?
    a. A only
    b. B only
    c. Both A and B
    d. Neither A nor B

21. Battery acid has just been spilled on a person's clothing. *Technician A* says to wash the acid off with baking soda and water. *Technician B* says to wash the acid off with water. Who is correct?
    a. A only
    b. B only
    c. Both A and B
    d. Neither A nor B

22. While being charged, a battery becomes very hot. *Technician A* says the heat is normal, and there is no need for concern. *Technician B* says that the battery could be overcharging, and the amount of charge should be reduced. Who is correct?
    a. A only
    b. B only
    c. Both A and B
    d. Neither A nor B

23. *Technician A* says the larger battery terminal is positive and the smaller terminal is negative. *Technician B* says the smaller battery terminal is positive and the larger terminal is negative. Who is correct?
    a. A only
    b. B only
    c. Both A and B
    d. Neither A nor B

24. *Technician A* says that a battery can be checked for leakage around a dirty battery case. *Technician B* says that a battery case can be cleaned with baking soda. Who is correct?
    a. A only
    b. B only
    c. Both A and B
    d. Neither A nor B

25. *Technician A* says the amount of a charge is determined by the size of the jumper cables. *Technician B* says the time of charge is determined by the size of the battery. Who is correct?
    a. A only
    b. B only
    c. Both A and B
    d. Neither A nor B

26. A battery continually needs additional water. *Technician A* says it's OK to add acid to the battery. *Technician B* says the battery may not be able to accept a charge any longer. Who is correct?
    a. A only
    b. B only
    c. Both A and B
    d. Neither A nor B

27. A battery will not start the engine, but the lights and horn still work. *Technician A* says the problem is a corroded battery cable connector. *Technician B* says the problem is that the electrolyte has a specific gravity of 1.280. Who is correct?
    a. A only
    b. B only
    c. Both A and B
    d. Neither A nor B

Essay

28. What is the difference between primary batteries and secondary batteries?

29. What is a definition of the specific gravity of a battery?

30. Define cell density of a battery.

31. What is the definition of the cold-cranking performance rating of a battery?

32. What is the purpose of a hydrometer when used with a battery?

33. What happens to the specific gravity as temperature increases?

34. What are some of the reasons for changing to a 42-volt battery system?

Short Answer

35. At what amperage could a 120-ampere-hour-rated battery be discharged for a period of 20 hours? ________

36. A battery has a specific gravity of 1.165 and has an electrolyte temperature of 100 degrees Fahrenheit. The actual specific gravity of the battery is ________.

37. On low-maintenance batteries, the vent holes are located ________.

38. What is the watt-hour rating of a 12-volt battery that has a 100 ampere-hour rating? ________

39. At the completion of a battery open circuit test, the voltmeter should read ________ volts across the battery terminals.

40. The low-maintenance battery uses ________ added to the grids to reduce gassing.

Applied Academics

TITLE: Cell Density of Batteries

ACADEMIC SKILLS AREA: Science

NATEF Automobile Technician Tasks:

The service technician demonstrates an understanding of the electrochemical reactions that occur in wet and dry cell batteries.

CONCEPT:

Researchers are continually looking for methods to make new and more efficient batteries. The need for better batteries has increased along with interest in battery-powered and hybrid-powered vehicles. It is estimated that such vehicles will account for 20% of all new automotive sales in the early part of the twenty-first century. Several major battery characteristics are being researched. These include the chemicals in the cell, the voltage being produced by the cell, and the cell density. The accompanying table compares six types of batteries. The cell density is measured in watt-hours per kilogram of weight, or Wh/kg. The batteries are compared to the lead-acid batteries that are commonly used in vehicles today. Much of the new research about batteries comes from the lessons learned with computer and cell phone batteries. Battery research has resulted in smaller batteries and increased cell density. This will become extremely important in cars of the future.

BATTERY CELL COMPARISON

CELL TYPE	NOMINAL VOLTAGE	STORAGE DENSITY
Lead Acid	2.1 volts	30 Wh/kg
Nickel Cadmium	1.2 volts	40 to 60 Wh/kg
Nickel Metal Hydride	1.2 volts	60 to 80 Wh/kg
Circular Lithium Ion	3.6 volts	90 to 100 Wh/kg
Prismatic Lithium Ion	3.6 volts	100 to 110 Wh/kg
Polymer Lithium Ion	3.6 volts	130 to 150 Wh/kg

APPLICATION:

Why are watt-hours per kilogram of weight so important for future batteries? Why is the lead-acid battery still being used so extensively in the automotive market? What would be some advantages to a vehicle if the watt-hours per kilogram increased four times as much?

Conventional Ignition System Principles and Spark Plugs

CHAPTER 32

OBJECTIVES

After studying this chapter, you should be able to:

- Identify the parts and operation of the conventional ignition system that use contact points.
- Define the operation of the primary and secondary circuits.
- Examine the operation and purpose of advance mechanisms.
- Identify spark plug design and operation.
- State common problems, their diagnoses, and service procedures used on the ignition system.

INTRODUCTION

In order for proper combustion to take place, the air and fuel mixture in the engine must be ignited. The ignition system is designed to produce a spark in the combustion chamber at a precise moment. This chapter discusses the principles, requirements, and components that make up the conventional ignition system. These include the coil, points, advance mechanisms, and spark plugs.

32.1 CONTACT POINT (CONVENTIONAL) IGNITION SYSTEMS

PURPOSE OF THE IGNITION SYSTEM

Today, all engines use electronic and computer-controlled ignition systems. The information on the conventional ignition systems in this chapter (Sections 32.1 and 32.2) is

Certification Connection

ASE Connection: The information in this chapter can help you prepare for the National Institute for Automotive Service Excellence (ASE) certification tests. The tests and content areas most closely related to this chapter are:

Test A1—Engine Repair

- **Content Area**—Fuel, Electrical, Ignition, and Exhaust System Inspection and Service

Test A8—Engine Performance

- **Content Area**—Ignition System Diagnosis and Repair

NATEF Connection: Much of the information in this chapter is related to the NATEF tasks. The NATEF tasks and priority numbers most closely related to this chapter are:

1. Diagnose ignition system related problems such as no-starting, hard starting, engine misfire, poor drivability, spark knock, power loss, poor mileage, and emission concerns on vehicles with distributor ignition (DI) systems; determine necessary action. P-1
2. Inspect and test ignition primary circuit wiring and solid-state components; perform necessary action. P-2
3. Inspect, test, and service distributor. P-3
4. Inspect and test ignition system secondary circuit wiring and components; perform necessary action. P-2
5. Inspect and test ignition coil(s); perform necessary action. P-1
6. Check and adjust ignition system timing and timing advance/retard (where applicable). P-3

presented to help the service technician understand the basic requirements of the ignition system and electronic ignition systems, as well as to service older engines that use conventional ignition systems.

The air and fuel in the combustion chamber must be ignited at a precise point in time during the four-stroke cycle. This is done by causing an electrical spark to jump across a gap on a spark plug. About 5,000 to 50,000 volts are needed to force the electrical current to jump across the spark plug gap. However, there is only a 12-volt battery within the automobile. The ignition system is used to increase the voltage to the necessary amount at the right time for the spark to occur. In addition, the time of spark must also be altered as speed and load increase or decrease. Advance and retard mechanisms are used to accomplish this.

The ignition system has developed and adapted as the automobile has changed. The conventional ignition system using a mechanical set of points and condensers has been updated to an electronic ignition system using semiconductors and transistors. From this development, electronic ignition systems now work directly with computer-controlled systems. In order to analyze these ignition systems, the conventional ignition system will be briefly analyzed to help the service technician understand the demands on newer computer-controlled systems.

CONVENTIONAL IGNITION SYSTEM COMPONENTS AND OPERATION

There are two separate circuits in the ignition system: the primary circuit and the secondary circuit. The **primary circuit** is considered the low-voltage circuit. Low voltage is battery voltage, or about 12 volts. The **secondary circuit** is called the high-voltage circuit. Components in this circuit operate at voltages between 5,000 and 50,000 volts, depending on the type of system being used.

The primary circuit operates in the sequence shown in ***Figure 32–1***.

1. The battery is used to provide a source of electrical energy needed to operate the system. The negative side is grounded to the frame, and the positive side is fed directly to the ignition switch.
2. The ignition switch connects or disconnects the flow of electricity to the ignition system. The ignition switch directs the current through a by-pass route during cranking and through the ballast resistor during normal operation. The ballast resistor can also be a resistive-type wire.
3. The ballast resistor controls the current flow to the coil during normal operation. The resistor reduces the voltage available to the coil at low engine speeds. It in-

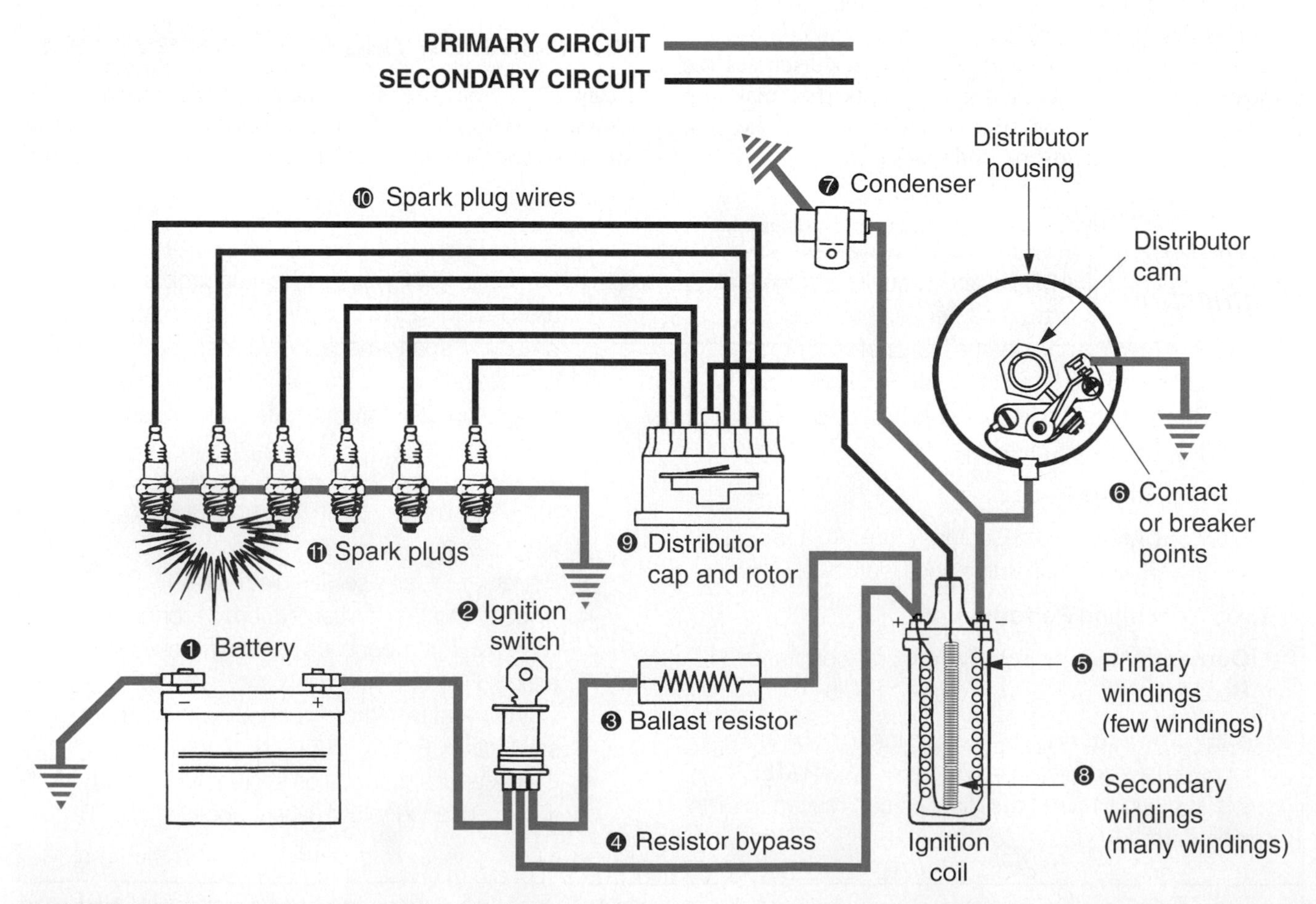

FIGURE 32–1 Both primary and secondary circuits are used in an automotive ignition system. The primary circuit operates on 12 volts, and the secondary circuit operates on 5,000 to 50,000 volts.

creases the voltage at higher rpm when the voltage requirement increases.

4. The resistor by-pass circuit is used only when the engine is being cranked. During this time, more voltage is needed at the coil to produce spark. When the operator stops cranking the engine, the electrical current flows through the ballast resistor.
5. The primary coil windings are used to convert the electrical energy into a magnetic field. When electricity passes through the primary windings, a strong magnetic field is produced. This magnetic field also surrounds the secondary windings.
6. **Breaker points** are used to close and open the primary circuit. As the distributor shaft rotates, it also causes a distributor cam to rotate. The **distributor cam** causes a small set of points to open and close at each cam lobe position. When the points are closed, current flow in the primary circuit causes the magnetic field to build up in the coil. When the points are opened, the current flow stops. This causes the magnetic field in the coil to collapse. The sudden collapse of the primary magnetic field produces a strong induced voltage in the secondary windings.
7. The condenser is used to reduce the amount of arcing when the points open the primary circuit. Whenever an electrical circuit is broken, it produces arcing. This arcing can cause the points to be pitted or corroded. The condenser helps to reduce corrosion on the points.

   The secondary circuit is called the high-voltage circuit. Depending on the system, the voltage in the secondary circuit may be as high as 50,000 volts. The components in the secondary circuit operate as shown in ***Figure 32–1***.
8. The secondary coil windings are used to capture the voltage produced by the collapsing primary magnetic field. There is approximately a 12 to 20,000 volt (1 to 1,666) ratio of windings between the primary and the secondary circuits. This means that for each volt in the primary windings, there are about 1,666 volts in the secondary windings. This will produce 20,000 volts on the secondary circuit. This high voltage available at the secondary windings is fed to the coil tower when the primary magnetic field collapses.
9. The distributor cap and rotor are used to distribute the surges of high voltage available to the coil tower. The high-voltage surge is sent from the coil to the center of the distributor by the coil wire. The surges are then directed, one at a time, to each outer terminal of the distributor cap by the rotor. The rotor is turned by the distributor shaft.
10. The spark plug wires are used to connect the high-voltage surge to each spark plug. They are arranged in the firing order of the engine.
11. The spark plugs provide a predetermined gap within the combustion chamber so that each time a high-voltage surge is delivered, a quality spark will occur in the combustion chamber.

DWELL

Dwell is defined as the length of time the points remain closed. ***Figure 32–2*** shows the dwell on a conventional ignition engine. As the distributor cam rotates, the points are opened and closed. Dwell is important because during dwell the magnetic field is building up in the primary windings. As the dwell time increases, the magnetic field buildup increases, producing a greater secondary voltage. ***Figure 32–3*** shows the different dwells for different cylinder engines. Four-cylinder engines have a greater amount of time that the points remain closed (greater dwell). This causes a greater magnetic field buildup around the primary coil, which produces a greater secondary voltage. With the conventional ignition system, there is only a certain amount of dwell that can be achieved because of the 360 degrees rotation of the cam. However, with electronic ignition systems, the dwell can be increased much more (electronically) to maximize efficiency and increase the spark voltage.

DISTRIBUTOR

The distributor is used to hold many of the ignition components. It has a center shaft that is driven by the main camshaft of the engine. This is done by using a small helical gear that meshes with a similar gear on the camshaft. On

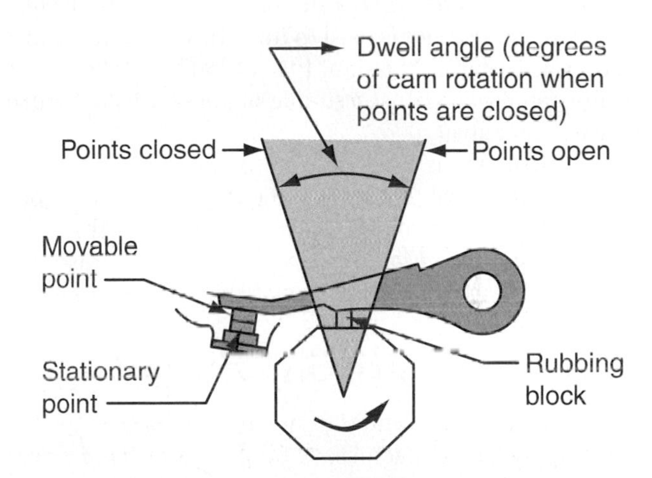

FIGURE 32-2 Dwell is the length of time the points are closed in degrees of rotation.

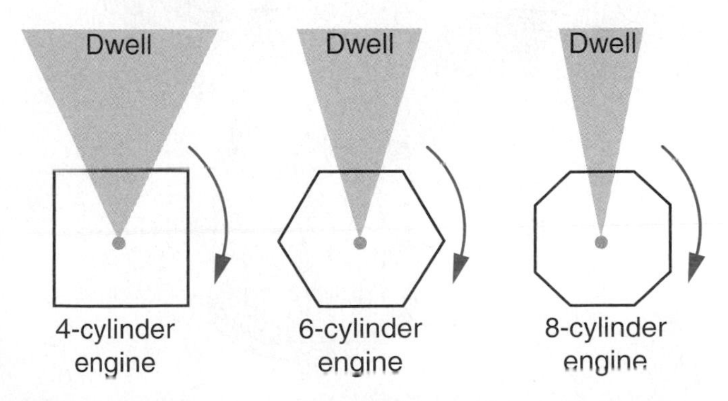

FIGURE 32-3 Four-cylinder engines have a greater dwell time than 8-cylinder engines.

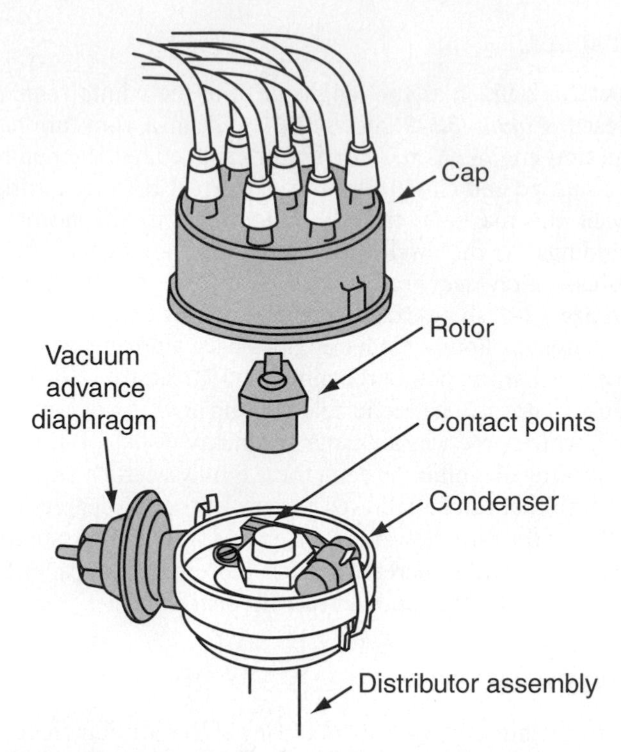

FIGURE 32-4 The distributor holds the primary ignition parts, including the rotor and the distributor cap.

many engines, there is also a slot on the bottom of the distributor shaft. This slot is used to turn the oil pump. It must be aligned during installation. In addition to a helical gear, certain manufacturers use a simple slot or offset slot to turn the distributor shaft.

The distributor is used to turn the distributor cam. In addition, the distributor holds the contact points, condenser, and advance mechanisms, and supports the rotor and distributor cap. ***Figure 32–4*** shows a typical distributor and related parts for a conventional ignition system.

DISTRIBUTOR ROTOR AND CAP

The distributor rotor and cap are used to distribute the secondary voltage to each spark plug. ***Figure 32–5*** shows two rotors and a cap used on an 8-cylinder engine. They are made from insulating materials that are easily shaped, such as Bakelite, plastic, or epoxy. Conductors are placed inside the material to allow high-voltage electricity to pass through.

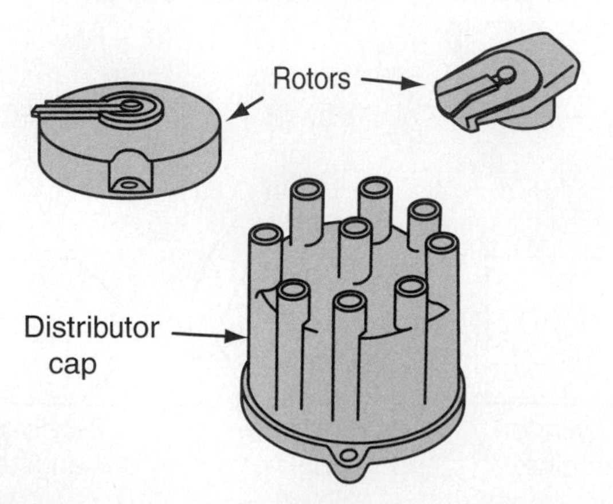

FIGURE 32-5 These are examples of rotors and a distributor cap used on older engines.

The high voltage produced each time the points open is sent from the high-voltage terminal on the coil to the center of the distributor cap, which is called the coil tower. The high voltage is then sent inside the cap to the rotor, which is being turned by the distributor shaft. As the rotor revolves inside the cap, it distributes the high voltage to each cylinder according to the firing order of the engine. There are many shapes and sizes of rotors used on older engines.

32.2 MECHANICAL ADVANCE MECHANISMS

PURPOSE OF ADVANCING THE SPARK

Ignition timing is defined as the time in degrees of crankshaft rotation that the spark occurs during idle. This is called the base or initial timing. The initial timing is adjusted when the engine is at idle. It is usually adjusted several degrees before top dead center (BTDC). As the engine speed and load increase, the timing must also increase.

As the engine speed increases, the piston moves faster and the time of spark must also be advanced. This is because the crankshaft will move farther during the time the combustion occurs. There must be enough time for the combustion to be complete. In ***Figure 32–6***, when the engine is at 1,200 rpm, the spark occurs about 6 degrees BTDC. At 23 degrees after top dead center (ATDC), the combustion ends. This produces an even power pulse to the piston on the power stroke. As the engine is run faster, say at 3,600 rpm, the timing must be increased to as far as 30 degrees BTDC so the combustion can end at 23 degrees ATDC. Of course, these figures will vary with different engines. The principle, however, is the same.

Load also affects when the timing should occur. If the load is increased, more air and fuel are needed in the combustion chamber. So it takes longer for the combustion to occur, and the timing must be advanced.

INITIAL TIMING

The initial timing is set by adjusting the distributor. When the distributor is placed in the engine, the distributor shaft gear meshes with the engine camshaft. The distributor housing can, however, be rotated. When the housing is rotated, the timing of the engine will be either advanced or retarded. Timing is set by using a timing light attached to the number 1 cylinder. When the number 1 cylinder fires, the timing light flashes. If the light is pointed toward the timing marks on the crankshaft of the engine, the exact timing of the engine can be determined (***Figure 32–7***).

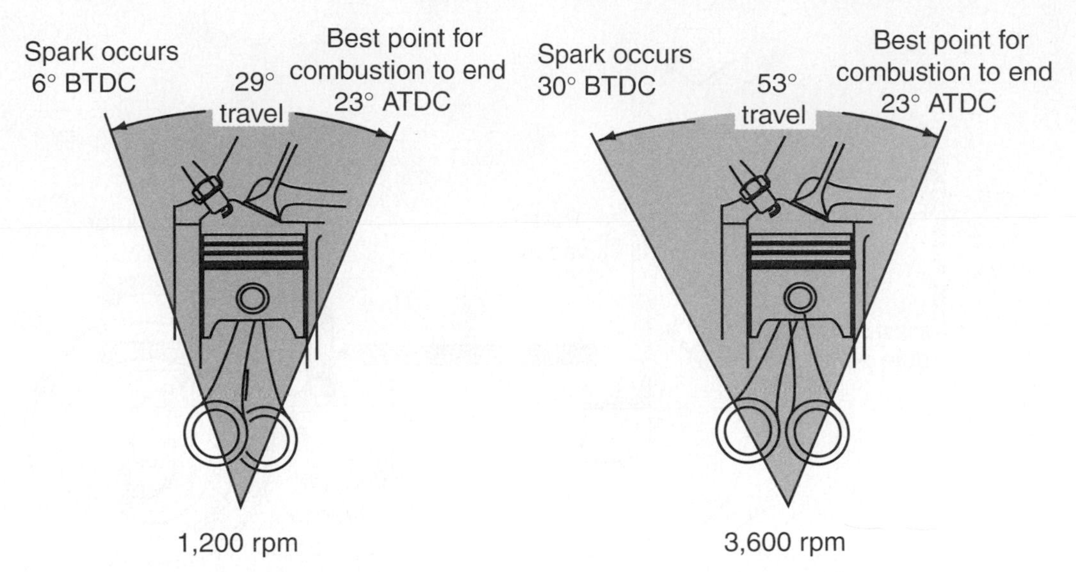

FIGURE 32–6 The speed of the engine determines how much spark advance is needed. As the engine speeds up, more advance is required to complete the combustion on time.

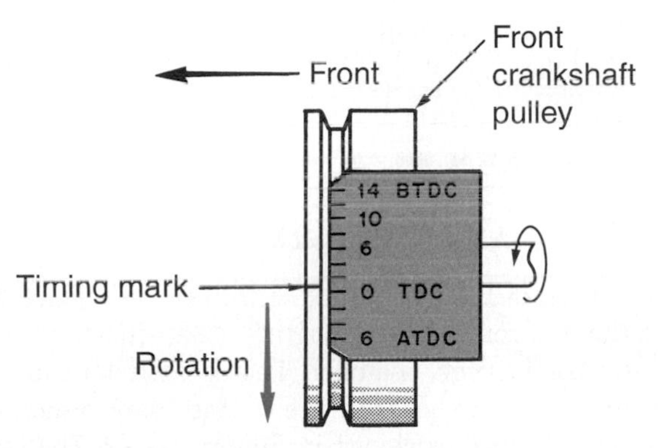

FIGURE 32–7 Timing marks are used to help determine the spark advance of the engine.

CENTRIFUGAL ADVANCE

The centrifugal advance increases or decreases the timing based on the engine speed. There are two weights on the distributor shaft (***Figure 32–8***). Springs are used to hold them inward. As these weights turn, their centrifugal force causes them to spin outward. The outward movement of the weights causes the time in which the points open and close to be different. On some engines, weight movement causes the breaker plate (the plate the points are attached to) to move so the points open earlier. On other engines, the movement of the weights causes the position of the distributor cam to change. This causes the cam to open the points at a different time. ***Figure 32–9*** shows the effect of engine rpm and the increase in timing advance.

VACUUM ADVANCE

The vacuum advance is used to increase the timing of the engine as the load increases. As load is applied to the engine, the intake manifold vacuum is reduced. The change in the intake manifold vacuum is used to help advance the timing.

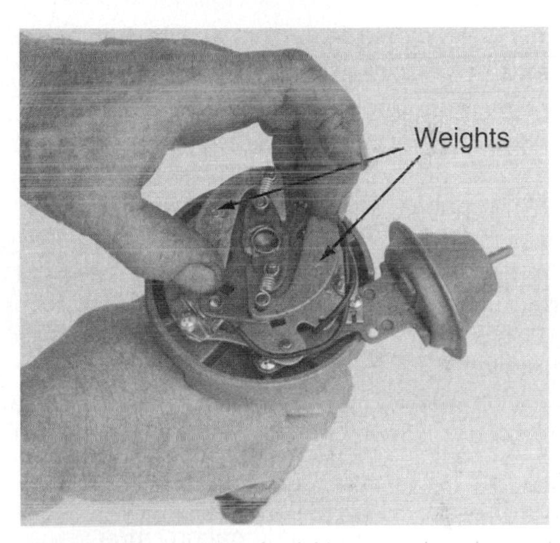

FIGURE 32–8 Centrifugal weights are used to advance the timing of the spark on the basis of rpm.

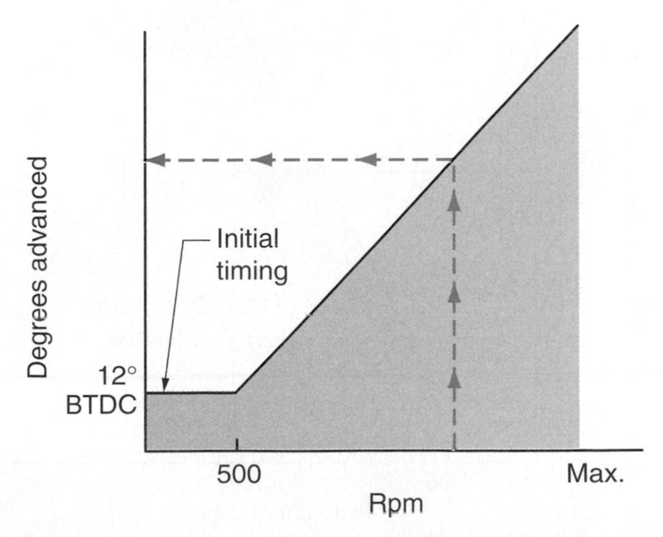

FIGURE 32–9 This chart shows the effect of engine rpm and the increased timing of the spark. The initial timing is set at idle.

FIGURE 32-10 A vacuum is used to rotate the breaker plate so the timing can be retarded or advanced on the basis of load.

A diaphragm is attached to the side of the distributor. One side of the diaphragm is mechanically attached to the breaker plate by a small rod. The other end of the diaphragm is attached to the intake manifold vacuum. The intake manifold vacuum used is considered a **ported vacuum**. A ported vacuum is present only when the engine is above idle. It is taken from slightly above the throttle plate when the throttle plate is in the closed position. A spring is used to force the diaphragm and breaker plate to a retarded position. As the engine throttle is opened for increased load, the vacuum pulls back on the diaphragm against spring pressure. ***Figure 32–10*** shows how the breaker plate advances the engine spark. ***Figure 32–11*** shows how the vacuum advance is added to the centrifugal advance to obtain the total advance of the engine.

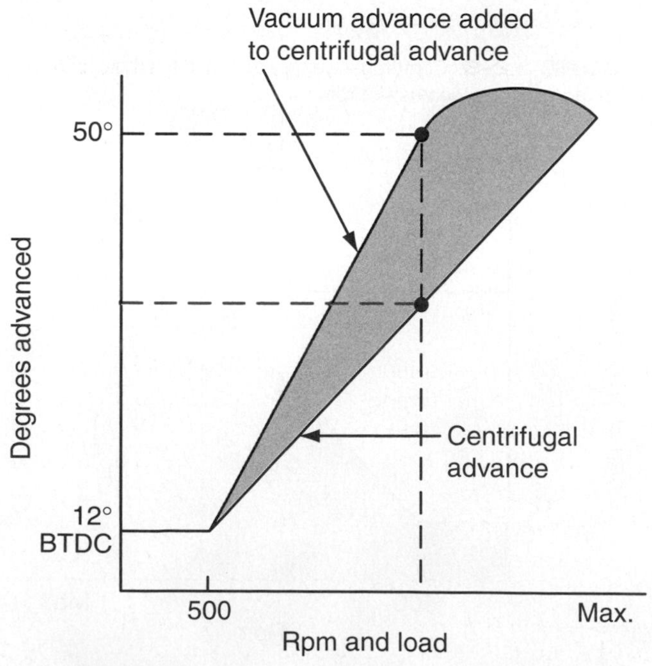

FIGURE 32-11 Vacuum advance is added to centrifugal advance during engine operation.

32.3 SPARK PLUGS

SPARK PLUG DESIGN

The purpose of the spark plug is to provide a place for a spark that is strong enough to ignite the air-fuel mixture inside the combustion chamber. This is done by causing a high voltage to arc across a gap on the spark plug. Spark plugs are designed as shown in ***Figure 32–12***. The center electrode, often made of copper or platinum or a copper-

Car Clinic:
ADVANCE

CUSTOMER CONCERN:
Spark Knock

A GM engine constantly knocks when the vehicle goes up hill. This started at about 50,000 miles. Using a higher-octane fuel helps a little, but the engine manufacturer says a lower-octane fuel is OK. Many ignition parts have already been replaced with no improvement.

SOLUTION: On GM engines, the ignition is controlled by the PROM in the vehicle computer. The original PROM may have allowed too much advance on the ignition under engine load and octane conditions. The new PROM advances timing less at low intake manifold pressures, when knock is hard to control. Check with the manufacturer and, if recommended, replace the PROM.

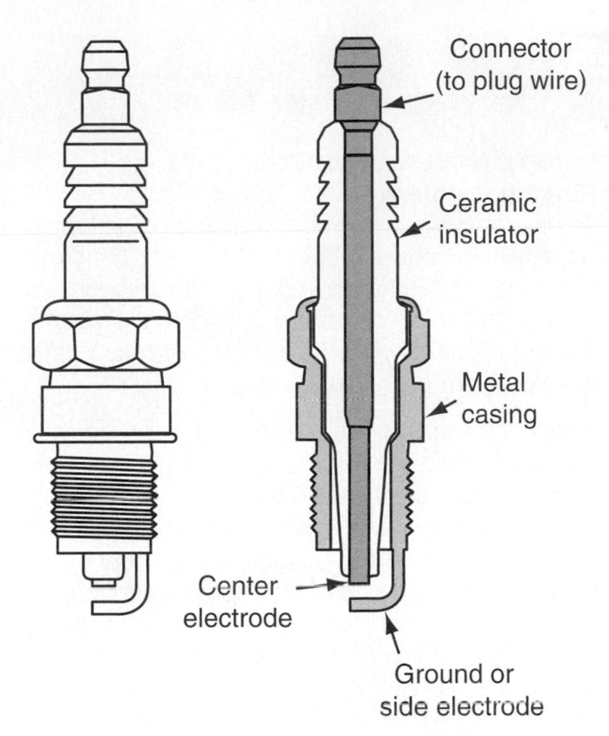

FIGURE 32–12 A spark plug is made of a center electrode, a ceramic insulator, a metal casing, and a side electrode.

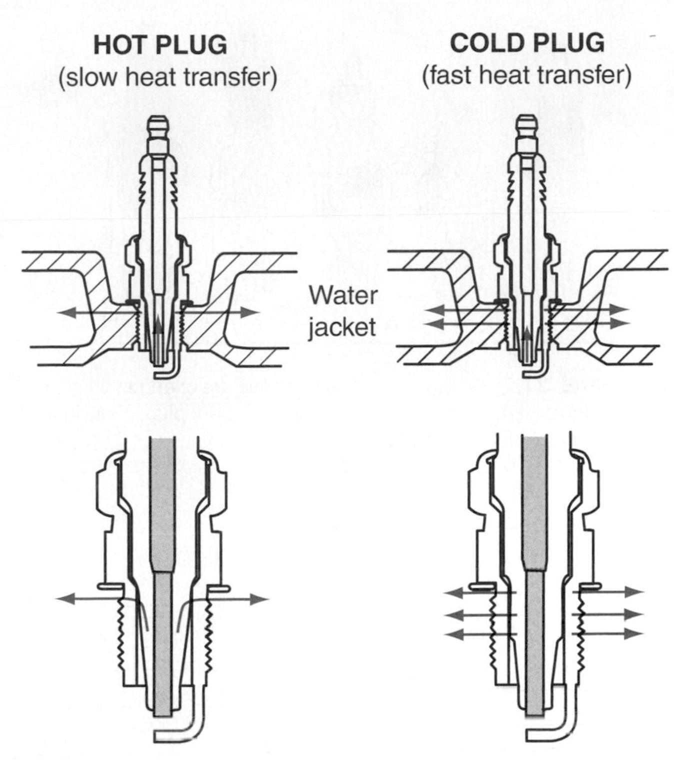

FIGURE 32–13 Spark plug heat ranges are based on the length of time it takes to remove the heat from the tip of the spark plug.

platinum alloy, is a thick metal wire that runs through the center of the plug. Its purpose is to conduct electricity from the high-voltage wire to the combustion chamber area. This insulator is a porcelain-like casing that surrounds the center electrode. The upper and lower portions of the center electrode are exposed. The metal casing is a threaded casing used for installing the spark plug into the cylinder head. It has threads and is hex-shaped to fit into a spark plug socket. The side electrode is a short, thick wire made of nickel alloy. It extends about 0.020–0.080 inch away from the center electrode. Its position creates the gap for the spark to jump across.

PARTS LOCATOR

Refer to photo #56 in the front of the textbook to see what a ***spark plug tip*** *looks like.*

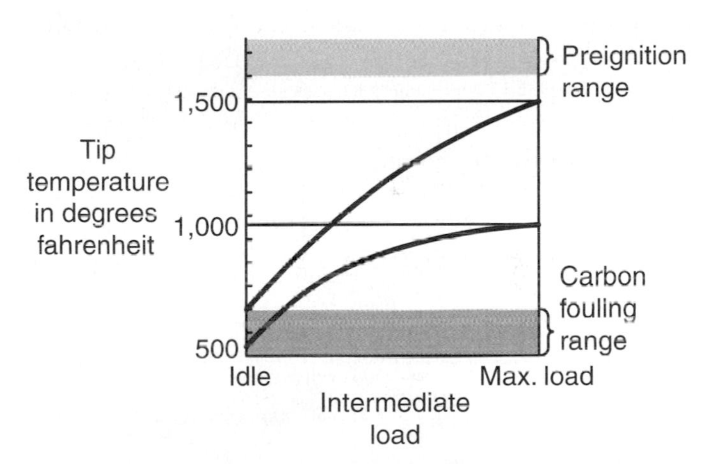

FIGURE 32–14 When plugs are too cold, they may foul out. If plugs are too hot, they may cause preignition.

SPARK PLUG HEAT RANGE

The heat range of a spark plug refers to its thermal characteristics. The thermal characteristics of a plug are a measure of how fast the plug can transfer combustion heat away from its firing end to the cylinder head of the engine (***Figure 32–13***). Plugs are considered to be cold or hot. There is a certain amount of thermal temperature at the time of the spark. If the plug tip temperature is too cold, the plug may foul out with carbon, oil, and other combustion deposits. If the plug tip temperature is too hot, preignition occurs and the plug and piston may be damaged. ***Figure 32–14*** shows the effect on the engine when the plugs are too hot or cold. The tip temperature is shown as the vertical axis. The load on the engine is shown as the horizontal axis. When too cold a plug is used, carbon and oil fouling occur at idle conditions. When the plug is too hot, electrode burning and preignition occur at full load.

The spark plug heat range is changed by changing the length of the insulator nose. Hot plugs have relatively long insulator noses with a long heat flow path to the cylinder head. Cold plugs have a short insulator nose with a short heat flow path to the cylinder head. ***Figure 32–15*** shows the difference in design of hot and cold plugs.

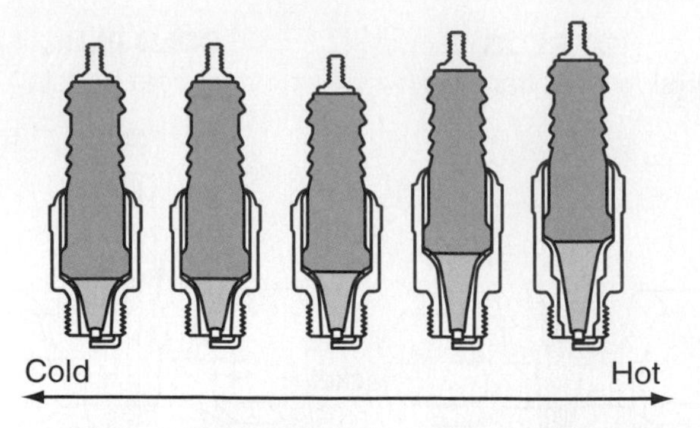

FIGURE 32-15 Spark plug heat ranges are changed by changing the length of the insulator nose inside the spark plug. (Courtesy of Federal-Mogul Ignition Company)

FACTORS THAT AFFECT SPARK PLUG TEMPERATURES

Many operational factors also affect the temperature of spark plugs. ***Figure 32–16*** shows some of the factors. Insulator tip temperature is shown on the left axis of each graph. Different factors are listed on the bottom axis of each graph. Each graph shows the type of relationship. Ignition timing causes spark plug temperature to change. ***Figure 32–16A*** shows that as the engine is overadvanced, tip temperature increases. ***Figure 32–16B*** shows that as coolant temperature increases, the tip temperature also increases. ***Figure 32–16C*** show that as more detonation occurs from lower-octane

Car Clinic: SPARK PLUG TEMPERATURE RANGE

CUSTOMER CONCERN:
Pinging Problem

An older 4-cylinder engine has a problem with pinging at high speeds and hot temperatures. The timing has been checked and is right on specifications. All of the advance mechanisms, including the weights and the vacuum systems, have been checked. What other problem might cause pinging?

SOLUTION: The only items not checked were the spark plugs. Check the spark plugs and make sure they are the correct ones for the engine. If a spark plug range is too hot, the tip of the plug might not be cooling enough. This could cause pinging at high speeds and temperatures. The only other problem might be a large amount of carbon buildup on the top of the piston. This could cause the compression ratio to increase, which would cause compression temperatures to increase and cause preignition.

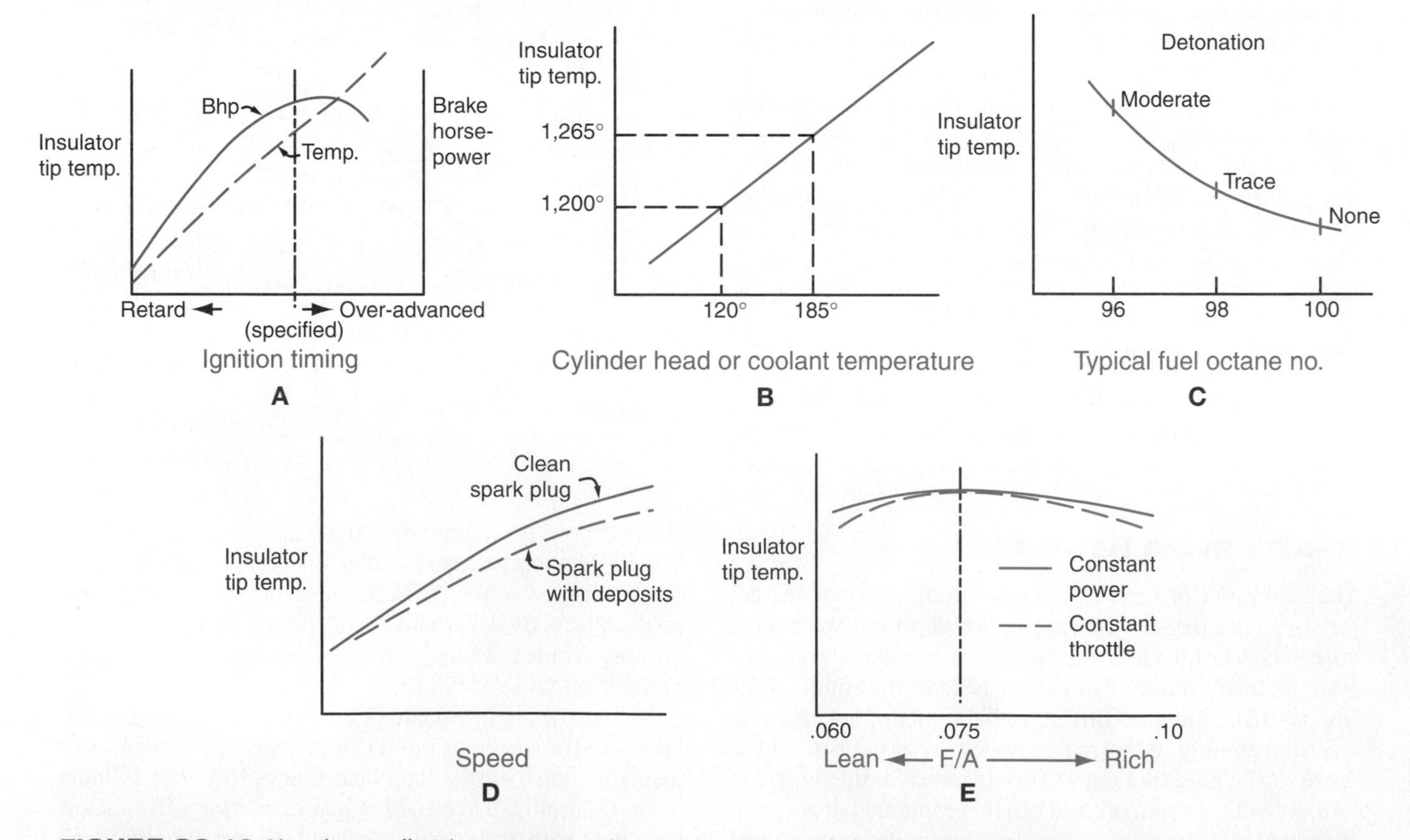

FIGURE 32-16 Many factors affect the temperature of a spark plug. These include (A) advance, (B) coolant temperature, (C) detonation, (D) condition of spark plugs, and (E) air-fuel ratio. (Courtesy of Federal-Mogul Ignition Company)

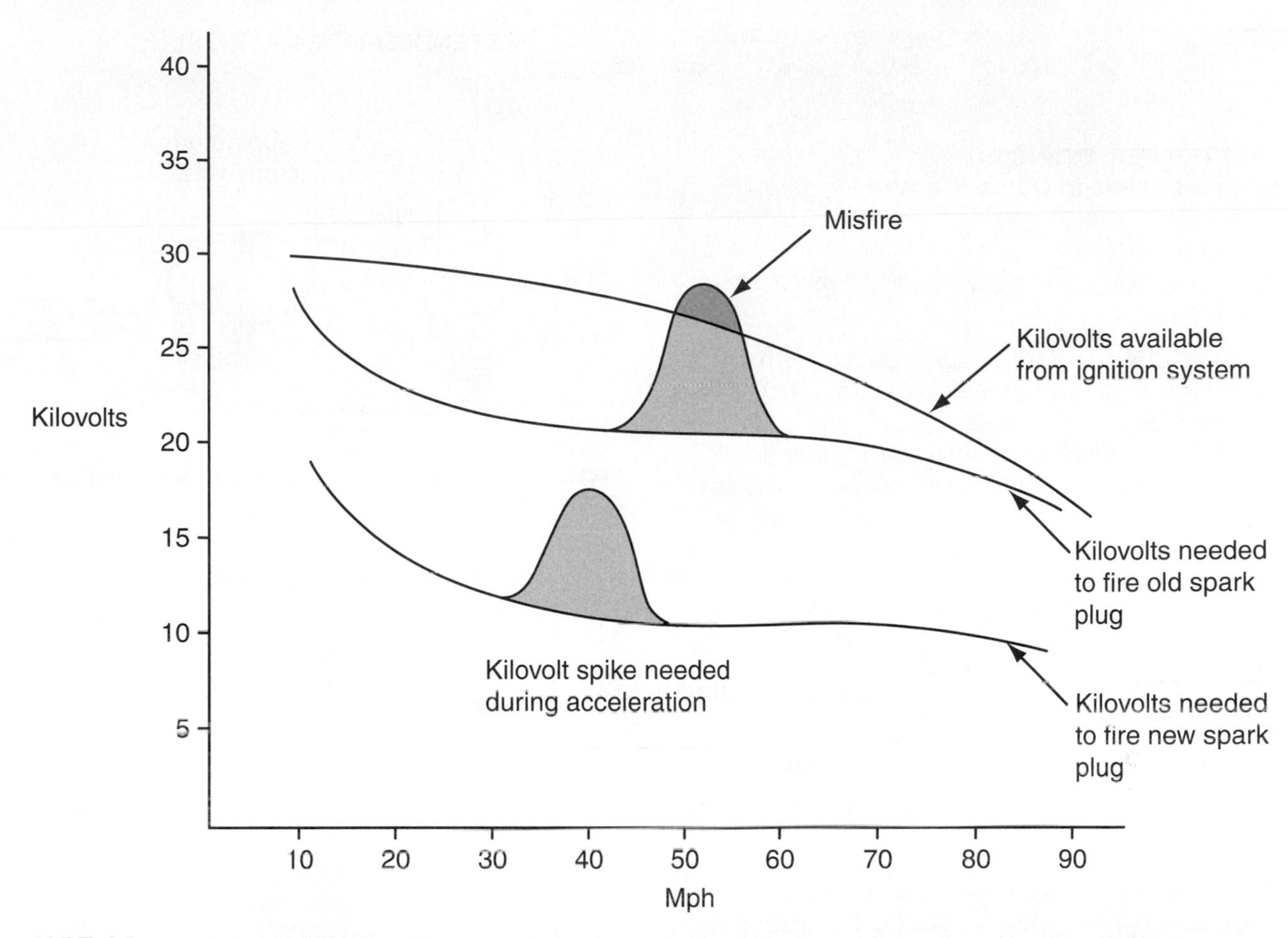

FIGURE 32–17 When old plugs are used, voltage requirements are higher. There may be times when older plugs misfire.

fuel, tip temperature also increases. ***Figure 32–16D*** shows that clean spark plugs have higher tip temperatures than spark plugs with deposits. ***Figure 32–16E*** shows that as the air-fuel ratio changes from rich to lean, insulator tip temperatures also change.

FACTORS THAT AFFECT SPARK PLUG VOLTAGES

The voltage required by spark plugs also changes with various factors. ***Figure 32–17*** shows how new and old plugs require a different voltage provided by the ignition system. On the chart, the top line is the kilovolts available from the ignition system. The bottom line is the kilovolts required to fire a new spark plug. When old plugs are used, the voltage required to produce the spark is greater. Older plugs may also misfire at certain times as shown on the middle line.

Other factors also affect the voltage requirements. ***Figure 32–18*** shows some of these factors. ***Figure 32–18A*** shows that as compression pressures increase, voltage requirements also increase. ***Figure 32–18B*** shows that as plug gap is widened, voltage requirements increase. ***Figure 32–18C*** shows that as the throttle or load is increased, the voltage requirements also increase. ***Figure 32–18D*** shows that as the engine timing is advanced, voltage requirements decrease.

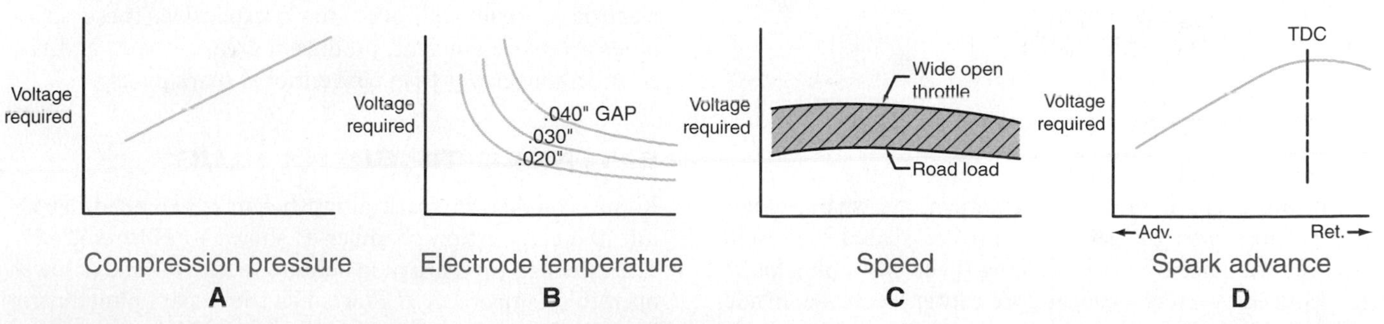

FIGURE 32–18 Many factors affect the voltage required of the spark plug. These include (A) compression pressures, (B) plug gap, (C) speed/load, and (D) timing advance. (Courtesy of Federal-Mogul Ignition Company)

Car Clinic: PLUG WIRES

CUSTOMER CONCERN:
Engine Misses In Damp Weather
An engine operates fine in dry weather. When it rains or the weather is very damp, the engine starts to run very rough. The customer also notices a loss of power.

SOLUTION: The most common problem that causes an engine to misfire in damp weather is poor plug wires. If the wires are bad (have poor insulation) they will short between each other and to ground when the surrounding air has high moisture content. The best way to check this is to run the engine at night (so you can see the shorted sparks) in damp weather. However, to verify that the spark plug wires are bad, hook up an electronic engine analyzer to the engine.

VARIATIONS IN SPARK PLUG DESIGN

Automotive manufacturers require many variations in the spark plug. These include:

- The number of threads on the spark plug
- The design of the gasket for sealing the spark plug to the cylinder, or the use of a tapered seat rather than a gasket
- The type of resistor element inside the plug to eliminate interference in radios and TVs
- The type of electrode used to establish the spark plug gap

Manufacturers base their spark plug numbering system on these and other design features. Each spark plug manufacturer uses a different system. Consider for example, the spark plug manufacturer's number R V 18 Y C 4. These numbers indicate the following:

- R = a resistor-type plug
- V = the type of shell design
- 18 = the heat range
- Y = firing end design
- C = firing end design
- 4 = wide gap designation

EXTENDED-LIFE SPARK PLUGS

Many manufacturers today recommend the use of extended-life spark plugs. An extended-life spark plug is shown in ***Figure 32–19***. It has a nickel-plated shell to improve resistance to corrosion. Low-speed spark plug fouling is reduced by using a copper-core center electrode. In addition, center and side electrodes are **platinum** tipped. This reduces spark erosion and corrosiveness in the combustion chamber. In some cases, manufacturers recommend that these spark plugs be changed every 60,000 to 80,000 miles. However, platinum-tip spark plugs are usually priced higher than standard spark plugs. ***Figure 32–20*** shows another type of platinum-tip spark plug. In this case, the center electrode is positioned between two side or ground electrodes. Again, although more expensive, these spark plugs are more efficient, produce a greater spark, and last considerably longer than conventional spark plugs.

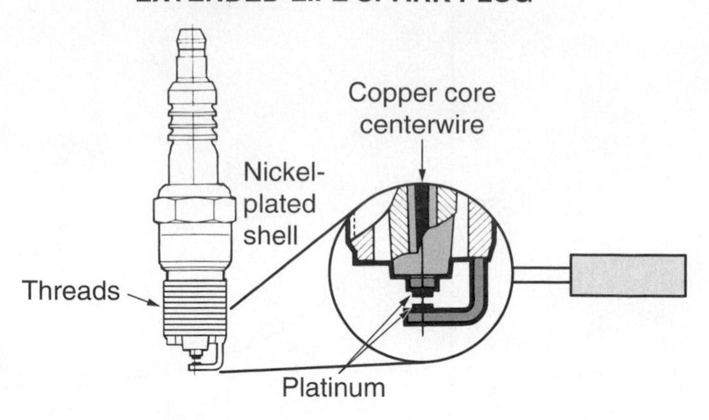

FIGURE 32–19 The extended-life spark plug uses platinum-tip electrodes.

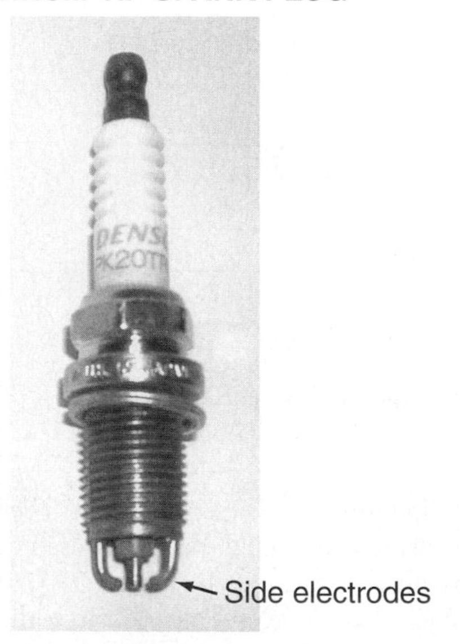

FIGURE 32–20 This platinum-tip spark plug has two side electrodes to improve spark as well as extend its life.

EXTENDED-TIP SPARK PLUGS

In an extended-tip spark plug, the tip is extended deeper into the combustion chamber as shown in ***Figure 32–21***. The extended-tip spark plug helps to reduce fouling at lower operating temperatures. It acts like a hot spark plug during low speed and operates normally under higher operating temperatures and speeds.

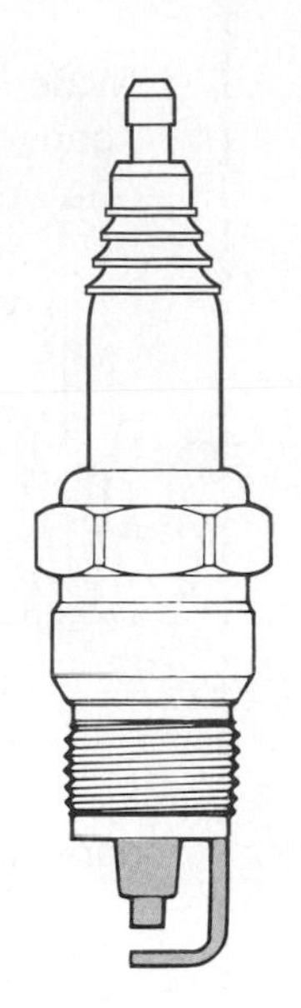

FIGURE 32-21 The extended-tip spark plug helps to reduce fouling at low operating temperatures.

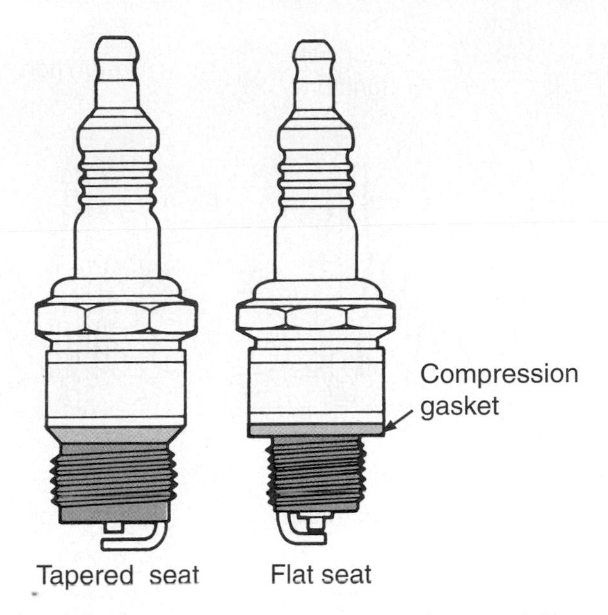

FIGURE 32-23 Both tapered seats and flat seats (with compression gaskets) are used on spark plugs today.

WIDE-GAP SPARK PLUGS

Lean air-fuel mixtures are being used to help lower vehicle emission levels. A conventional spark plug with a gap of 0.035 inch may not provide the necessary spark for these leaner mixtures. Thus, manufacturers have developed high-energy ignition systems and computerized ignition systems. These systems use a redesigned plug with a much wider spark plug gap as shown in ***Figure 32–22***. The wider gap (0.060 or 0.080 inch) helps to deliver better engine performance with fewer emissions. Note that conventional spark plugs should not be regapped to the wider gaps. The spark plug's life will be shortened considerably.

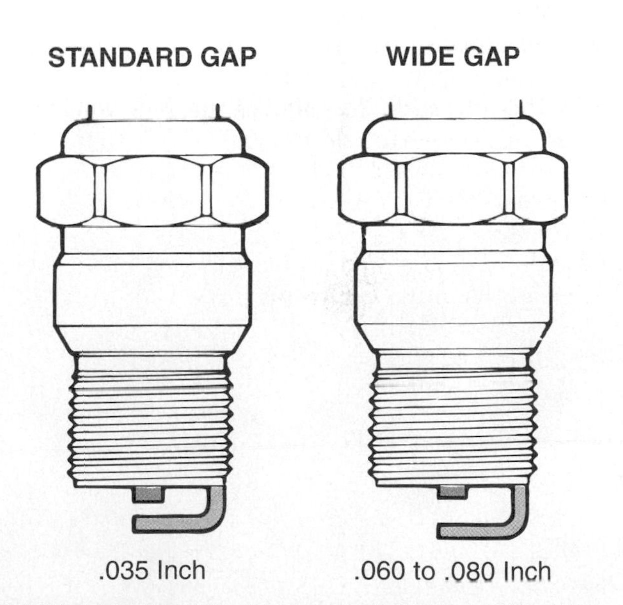

FIGURE 32-22 The wide-gap spark plug works effectively in a lean air-fuel mixture.

SPARK PLUG SEATS

Two types of seats are used on spark plugs. The seat is designed to seal the compression in the combustion chamber rather than leaking past the spark plug. Some spark plugs use a flat seat and a compression washer. This is a very common type of seat and seal on spark plugs. When tightened to the correct torque specifications, the gasket compresses and helps to seal the compression inside the cylinder and combustion chamber. The tapered seat, shown in ***Figure 32–23***, seals the spark plug to the cylinder head. This type of tapered seat does not require the use of a compression washer. As this spark plug is tightened to its correct torque, the tapered seat is forced into the cylinder head, causing it to seal tightly. It should be noted that newer spark plug designs are being incorporated into vehicles to ensure that the electrode angle is always placed in the same position in the combustion chamber, as the plug is being tightened down.

HOW SPARK PLUGS MISFIRE

Spark plugs misfire for a variety of reasons. When a spark plug misfires, it causes poor fuel economy, poor performance, and increased emissions. The eight examples shown in ***Figure 32–24*** show the difference between varying spark plug conditions and misfire.

1. *Normal ignition*—A high-voltage pulse travels down the center electrode, arcs across the gap to the side electrode, and properly ignites the air-fuel mixture.
2. *Preignition*—The spark plug overheats to the point where the air-fuel mixture ignites before the high-voltage pulse arcs across the electrodes. This condition causes engine damage and loss of power and performance.

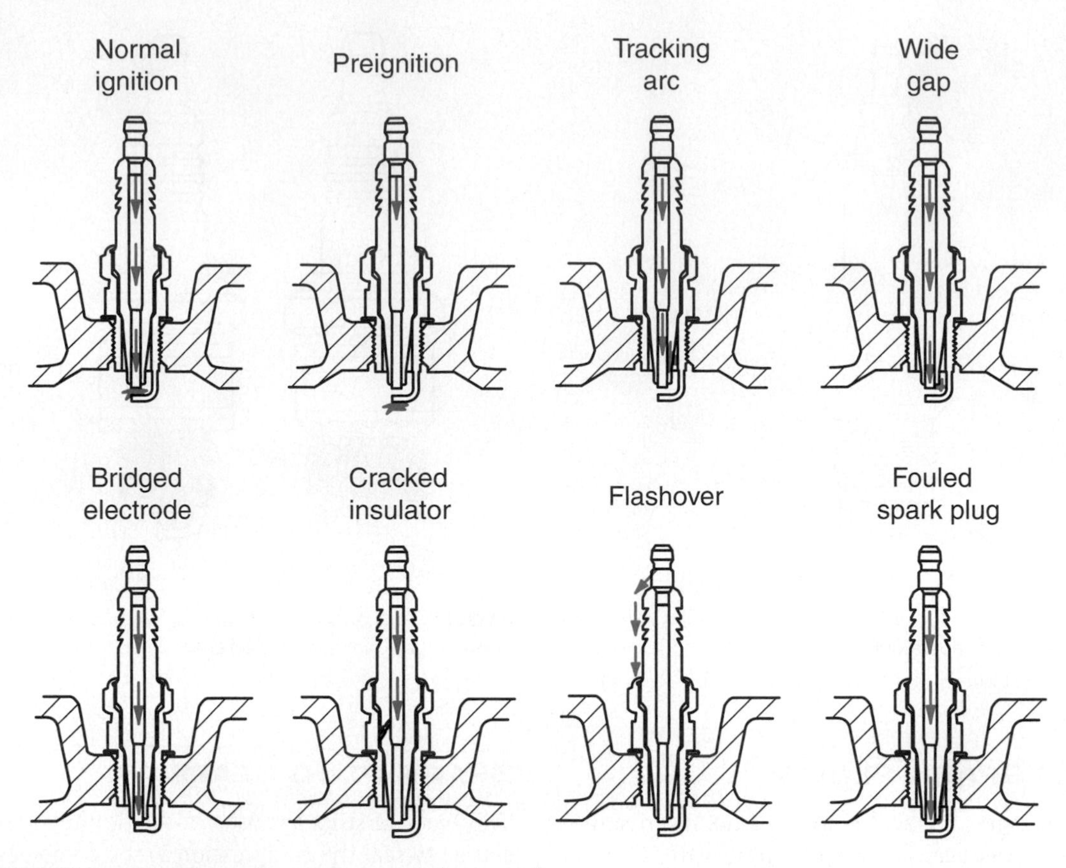

FIGURE 32-24 These conditions can cause a spark plug to misfire.

3. *Tracking arc*—High-voltage arcs between a fouling deposit on the insulator tip and the spark plug shell. This causes a loss of power and excessive exhaust HC levels.
4. *Wide gap*—Spark plug electrodes are worn so that the high-voltage charge cannot arc across the electrodes under high load. Fuel remains unburned and a power loss results.
5. *Bridged electrodes*—Fouling deposits between the electrodes "ground out" the high voltage needed to fire the spark plug. The arc does not occur and the air-fuel mixture is not ignited.
6. *Cracked insulator*—A crack in the upper or lower portion of the spark plug insulator can cause the high-voltage charge to "ground out." The spark plug does not ignite and the air-fuel mixture is not ignited.
7. *Flashover*—A damaged spark plug hood, along with dirt and moisture, permits the high-voltage charge to short over the insulator to the spark plug shell or the engine.
8. *Fouled spark plug*—Deposits that have formed on the insulator tip become conductive and provide a "shunt" path to the shell. This prevents the high voltage from arcing between the electrodes.

Problems, Diagnosis, and Service

Safety Precautions

1. When replacing parts or doing other work on the ignition system, always make sure the ignition system is off. If it is left on, a spark may be produced, causing gasoline fumes to ignite, causing a safety hazard.
2. If it is necessary to crank the engine when working on the distributor, disconnect and ground the coil wire to make sure you will not get a shock from the high-voltage wire.
3. Never use only your hands to pull high-tension wires off the spark plug when the engine is running. Use the correct pliers to remove the plug wires.
4. When removing spark plug wires and spark plugs, always remove them after the engine has cooled down. The spark plugs and wires are often located near the exhaust manifold, which can burn your hands if they are hot.
5. When working on an ignition system when the engine is running, always keep your hands away from the spark plug wires. In addition, be careful not to drop any tools into the fan or pulleys on the front of the vehicle.
6. Always use the correct tools when working on an ignition system.
7. Always wear OSHA-approved safety glasses when working on an ignition system.

PROBLEM: Open Circuit in Primary Coil

An engine was running fine, then suddenly stopped running. Although the engine can crank, it will not fire the spark plugs. The vehicle has a conventional ignition system.

DIAGNOSIS

A common problem is to have the ignition coil primary develop an open circuit because of high current or getting too hot. This causes the total ignition system to shut down. The service technician can check for this problem by using an ohmmeter. An open circuit will show infinite resistance on the primary. Check the resistance between the positive side of the primary circuit and the negative side of the primary circuit.

SERVICE

If it is determined that the ignition coil is bad, it must be replaced. Use the following general procedure to replace the ignition coil.

1. Disconnect the battery.
2. Disconnect both the negative and positive connections to the primary coil.
3. Carefully remove the secondary wire on the top of the coil.
4. Remove the coil and replace with a new one.
5. Reverse the procedure and test accordingly.
6. When reconnecting the wires, make sure the positive side of the coil primary is connected to the ignition switch. The coil will then operate with the correct polarity. If it is connected backward, the coil output will be reduced by approximately one-half. Whatever the vehicle ground polarity, that side of the coil goes toward the ignition points.

PROBLEM: Engine Misfires

A vehicle has a conventional ignition system. The engine is missing badly.

DIAGNOSIS

Various parts may be damaged on a conventional ignition system. Use the following diagnostic checks to determine the condition of the ignition system:

1. As contact points become pitted, the engine may misfire erratically. This is because the points are not breaking the primary circuit fast enough.
2. If a ballast resistor or resistive wire is shorted out, the points may burn prematurely because of the increased amount of current in the primary circuit.
3. If the condenser is bad, the points may corrode faster.
4. Check the distributor cam lobes for excessive wear. If these cam lobes are worn, the points will open and close at the incorrect time.
5. Check the distributor bearings for wear. Bad bearings will affect the mechanical opening and closing of the points.
6. Check for moisture or moisture spots on the inside top of the distributor cap. Moisture condensing inside the

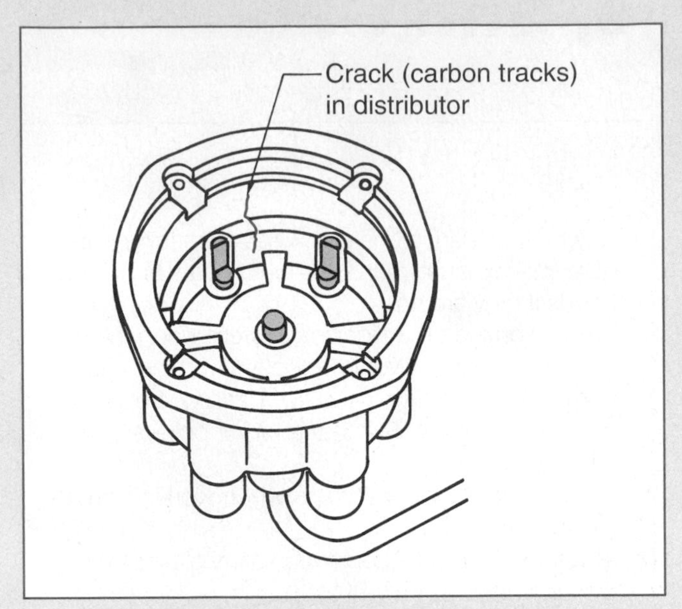

FIGURE 32–25 Always check the inside top of the distributor cap for small cracks or carbon tracks, which can cause the engine to misfire.

distributor cap may also cause the engine to misfire or not start at all.

7. A cracked distributor cap can cause the engine to misfire or not start at all. A cracked distributor cap can be found by removing the distributor cap and carefully observing the inside of the cap for small cracks (***Figure 32–25***).
8. Check the springs on the centrifugal advance. If these springs break, the engine will advance at the wrong time. Always check the spring weights for the correct tension and lubricate all moving parts.
9. Check the vacuum diaphragm for leakage. A leaking vacuum diaphragm on the distributor prevents the vacuum advance from operating. This causes a lack of power, poor fuel economy, and increased exhaust emissions. Use a vacuum tester to check the diaphragm as shown in ***Figure 32–26***.

SERVICE

The procedure will depend on the exact nature of the fault. Refer to the service manual for the manufacturer's suggested procedure for the defective part.

PROBLEM: Damaged Spark Plug Wires

An engine has a very irregular misfire. The misfire seems to increase in wet or moist conditions. Often when the hood of the vehicle is lifted at night, one can see sparks coming from the spark plug wires.

DIAGNOSIS

If sparks are continuously jumping from one spark plug wire to another, the insulation could be bad. The wires will have to be replaced.

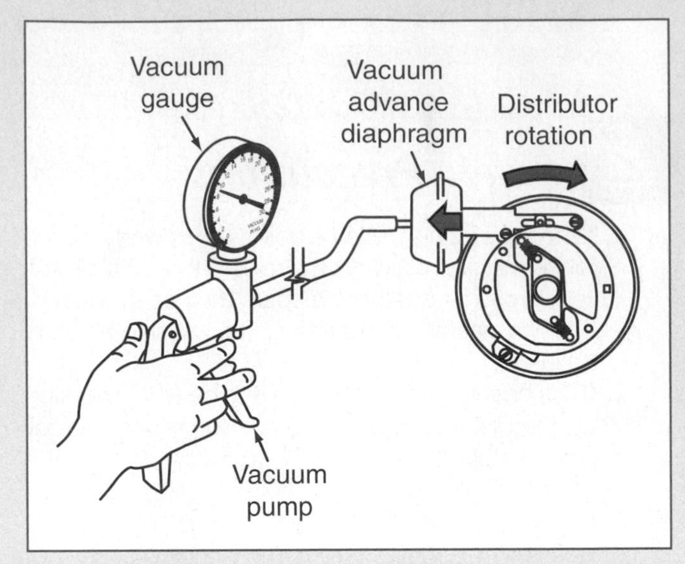

FIGURE 32–26 Use a vacuum gauge to test the condition of the vacuum advance.

SERVICE

Spark plug wires must often be cut to the correct length after purchasing the new set. Replace only one spark plug wire at a time so that the wires don't get mixed up. Use the following procedure if the plug wires have been mixed up and cannot be replaced in the correct order.

CAUTION: *Carelessly pulling spark plug wires off the spark plugs, bending them sharply, or stretching them may damage the new wires and cause the engine to misfire.*

1. With all plug wires removed, remove the number 1 spark plug. The number 1 spark plug can be determined by looking in the maintenance manual.
2. Slowly crank the engine until the number 1 spark plug is at top dead center (TDC) on the compression stroke. Top dead center can be determined by observing the timing marks and lining them up correctly. When turning the engine to get to TDC on the compression stroke, a slight pressure should be felt if a finger is placed over the spark plug hole.
3. Remove the distributor cap and notice the location of the rotor. At this point (TDC for the number 1 spark plug), the rotor should be pointing at the number 1 spark plug wire on the distributor cap.
4. Place a spark plug wire on the correct hole on the distributor and on the number 1 spark plug.
5. With the distributor cap removed, observe the direction of rotation of the rotor as the engine is cranked.
6. Place the distributor cap back on the distributor.
7. Using the firing order listed on the intake manifold or in the maintenance manual, continue placing the spark plug wires on the distributor and the corresponding

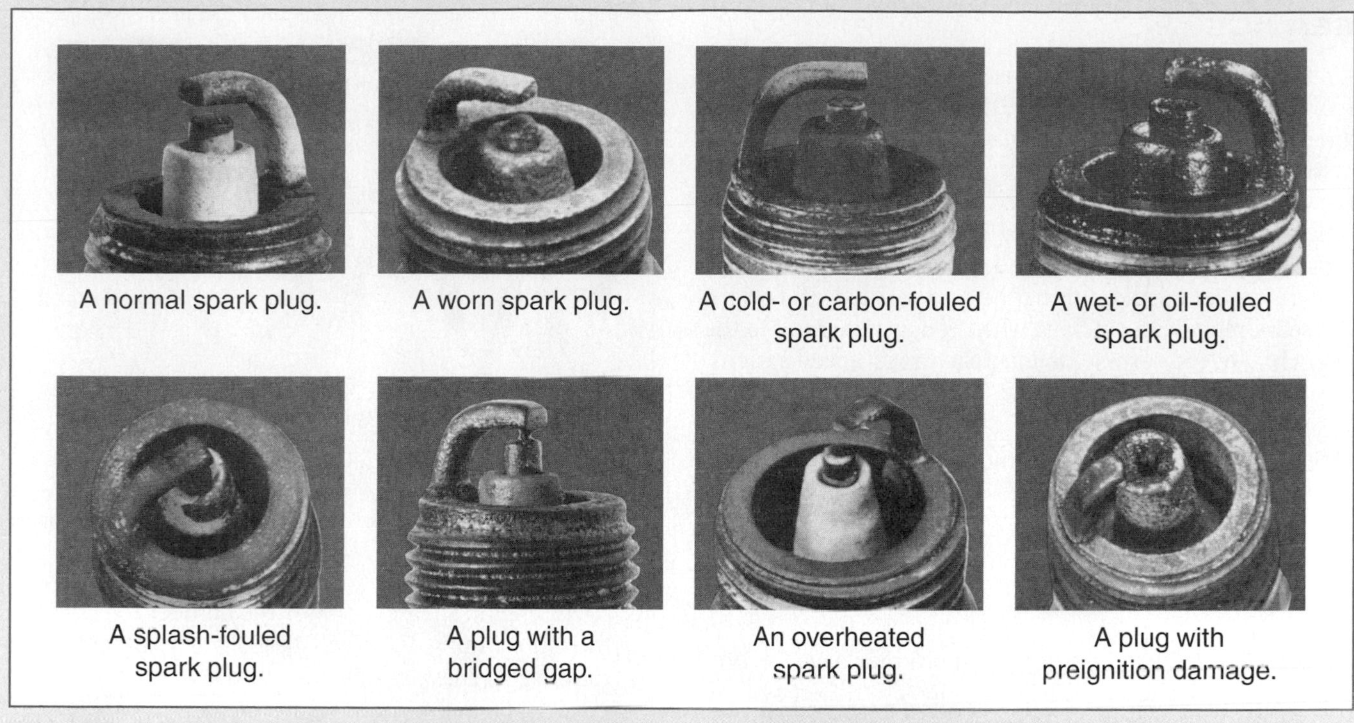

FIGURE 32-27 The condition of the spark plugs can help the service technician to identify many problems with an engine. (Courtesy of Champion Spark Plug Company)

spark plug. For example, if the firing order is 1, 5, 3, 6, 2, 4, then after number 1 fires, the next cylinder to fire is number 5. Place a spark plug wire from the next hole in the distributor cap (remember the distributor rotation) to spark plug number 5. Continue this procedure until all spark plug wires have been replaced.

PROBLEM: Damaged Spark Plugs

An engine has a steady miss in one or more cylinders.

DIAGNOSIS

Use the following diagnostic checks to determine the condition of the spark plugs:

1. Often the spark plugs may be damaged, or they may need to be replaced. ***Figure 32–27*** shows different spark plug conditions. When diagnosing the exact problem with spark plugs, match the spark plugs with the correct characteristics. Replace the spark plugs if necessary.
2. Check the spark plug heat range against the manufacturer's specifications. When purchasing new spark plugs, make sure the number on the spark plug is the same as the number recommended by the manufacturer.
3. Always make certain the plug gaps are adjusted to the manufacturer's specifications. These can be found in the service and repair manuals.
4. Always check to make sure that the correct spark plug has been installed in the engine. Make sure that the threads on the spark plug are the correct length. Both short-reach and long-reach spark plug threads are used. As shown in ***Figure 32–28***, if short-reach threads are used when long-reach threads are recommended, the spark will be too far away from the combustion chamber. This can cause a misfire. If long-reach threads are used when short-reach threads are recommended, the piston can damage the spark plug by hitting it. The piston may also be damaged.
5. Note that electronic ignition systems use wide spark plug gaps.
6. Always follow the recommended diagnosis and service procedures listed in the service and repair manuals to troubleshoot and diagnose electronic and computer-controlled spark systems.

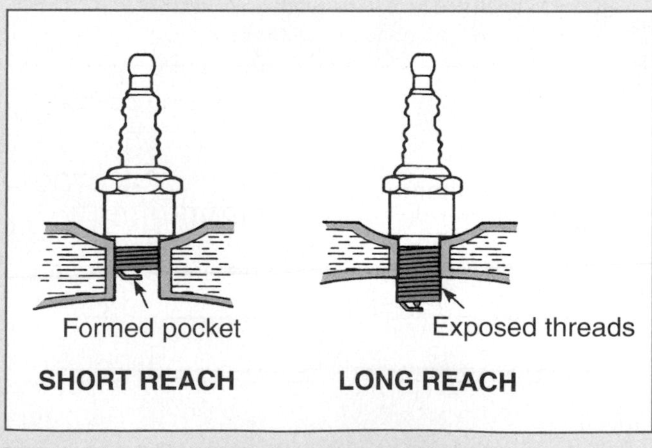

FIGURE 32-28 Always make sure the correct spark plug is used with the correct number and length of threads.

SERVICE

After determining the exact cause of the spark plug failure, replace the plugs as necessary. Depending on the exact engine and manufacturer, the replacement of spark plugs can be very easy or extremely difficult. Make sure to use the correct tools to remove and replace necessary components in order to get at the spark plugs. Remember that when removing spark plugs, always use a spark plug socket that has a small rubber cushion inside the socket. This will prevent the spark plug from cracking when you apply torque to the wrench. Always torque to manufacturer's specifications, normally around 12 to 25 pound-feet.

PROBLEM: Incorrect Ignition Timing

An engine is running but produces a pinging or knock especially during acceleration.

DIAGNOSIS

Often an engine that has a ping or knock may have the timing slightly advanced. On a conventional ignition system, timing can be changed easily by readjusting the position of the distributor in the block.

SERVICE

Use the following general procedure to set the timing of an engine:

1. Use a timing light to check the timing of an engine (***Figure 32–29***).
2. Attach the clip-on leads to the battery, making sure to attach positive to positive and negative to negative.
3. Place the inductive lead over the number 1 cylinder spark plug wire.
4. Slightly loosen the bolt that holds down the distributor. Do not remove the bolt, as the distributor will be too loose.
5. Start the engine and allow it to idle. On some engines you may have to remove and plug the vacuum advance lines to disable the vacuum advance system.

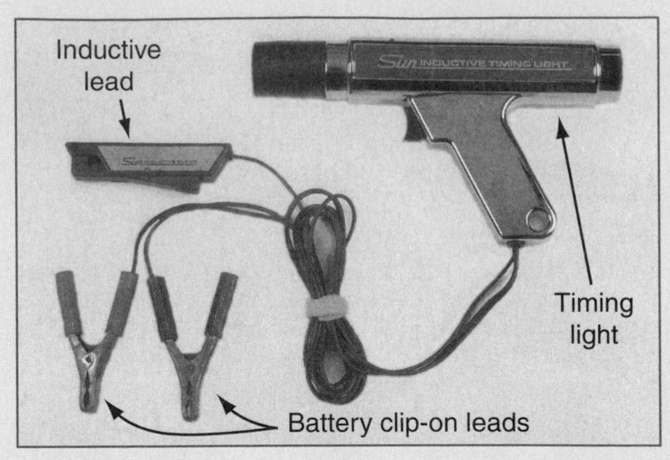

FIGURE 32–29 A timing light is used to adjust the time of the spark. The light is attached to the battery and to the number 1 spark plug wire. When the plug fires, the light flashes to show the timing marks.

6. As the engine idles, every time the number 1 spark plug fires, the timing light will flash. When the engine is running, the flashes look like a strobe light.
7. Now carefully shine the timing light on the timing marks on the front of the engine during idle.

CAUTION: *Be careful not to get the wires near the spinning fan or near the electric fan.*

Figure 32–30 shows several examples of the timing marks on different engines.

8. When the light shines on the timing marks on the front of the engine during idle, the amount of advance or retarded condition can be observed.
9. With the distributor base loosened, as the distributor is turned, advancing or retarding the ignition can be easily observed by the timing light flashes.
10. Set the timing marks (by turning the distributor) to the exact amount of advance BTDC, as suggested by the manufacturer.
11. Shut the engine off and tighten down the distributor base.
12. Remove the timing light. The ignition advance should now be set correctly.

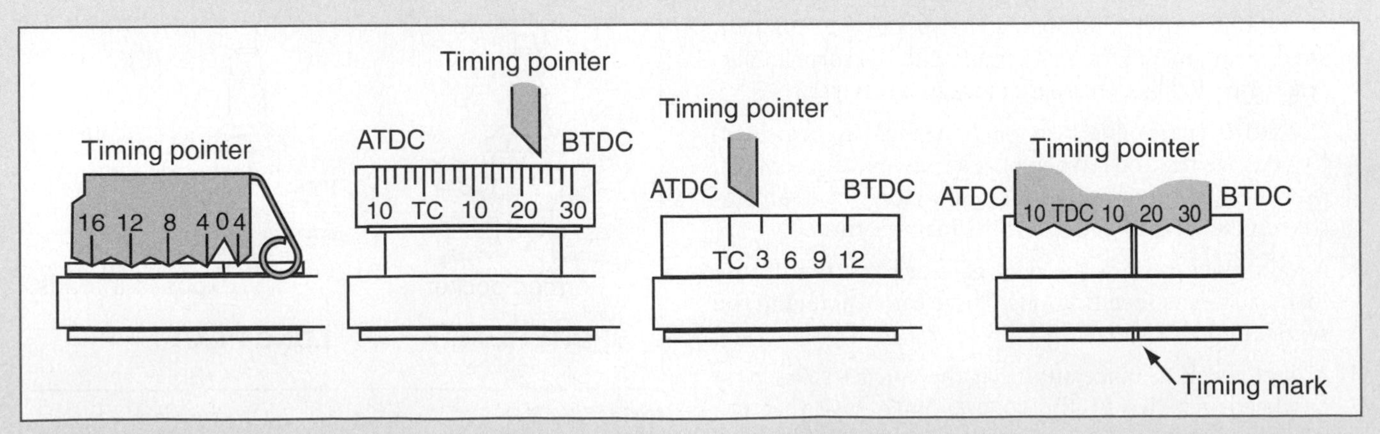

FIGURE 32–30 These are examples of how the timing marks might look on different vehicles.

Service Manual Connection

There are many important specifications to keep in mind when working with ignition systems. To identify the specifications for your engine, you will need to know the VIN (vehicle identification number) of the vehicle, the type and year of the vehicle, and the type of engine. Although they may be titled differently, some of the more common ignition system specifications (not all) found in service manuals are listed below. Note that these specifications are typical examples. Each vehicle and engine may have different specifications.

Common Specification	Typical Example
Full advance	10 degrees at 3,000 rpm
Firing order	1-3-4-2
Ignition timing BTDC (before top dead center)	
Manual transmission	10 degrees
Automatic transmission	12 degrees
Spark plug gap	0.060 inch
Spark plug torque	10–15 ft.-lb
Starting of centrifugal advance	0 degrees at 600 rpm

In addition, the service manual will give specific directions for various services and testing procedures. Some of the more common procedures include:

- Ignition lock release
- Ignition switch replace
- Ignition timing

SUMMARY

The following statements will help to summarize this chapter:

- The main purpose of an ignition system is to produce between 20,000 and 50,000 volts for the spark plug to ignite the fuel in the combustion chamber.
- The primary circuit operates on 12 volts.
- The secondary circuit operates on high voltage in the range of 20,000 to 60,000.
- On a conventional ignition system, the primary circuit components include the ignition switch, primary coil windings, resistive wire, condenser, and contact points.
- On a conventional ignition system, the secondary circuit components include the high-voltage windings, the coil, the rotor, distributor cap, spark plug wires, and spark plugs.
- The ignition switch is used to turn the ignition system on and off.
- The coil is used to convert 12 volts to 20,000–60,000 volts.
- The breaker points on a conventional ignition system switch the primary circuit off and on.
- On the conventional ignition system, the length of time the points are closed is defined as dwell.
- The rotor and distributor cap are used to distribute the high-voltage spark to the cylinders.
- The spark plug, located in the combustion chamber, provides the spark for the combustion process.
- Spark timing must be changed when the speed or load is changed.
- Both vacuum and centrifugal advance systems are used to advance the spark timing.
- Spark plugs have heat ranges, which is the length of time it takes to transfer the heat from the tip of the spark plug.
- Spark plug temperature can be affected by many conditions, including the amount of advance, the coolant temperature, gasoline octane, rpm of the engine, and the lean/rich condition of the air-fuel mixture.
- Spark plug voltage can be affected by many conditions, including the compression temperature, electrode temperature, engine rpm, and spark advance.
- There are many types of spark plug variations, including the thread design, the design of the gasket, the resistor in the spark plug, and the electrode used to establish the spark.
- Extended-life spark plugs use a nickel-plated shell to improve their resistance to corrosion.
- Spark plugs can misfire from various conditions, including preignition, tracking arc, wide gap, bridged electrode, cracked insulator, flashover, and fouled spark plugs.
- There are various service procedures used on conventional ignition systems, including replacement of the wires, vacuum test on the advance systems, open primary windings, engine misfires, and timing problems.

TERMS TO KNOW

Can you explain each of the following terms? Review the chapter until you can use each term correctly.

Breaker points
Distributor cam
Dwell
Platinum
Ported vacuum
Primary circuit
Secondary circuit

REVIEW QUESTIONS

Multiple Choice

1. The primary circuit on a conventional ignition system operates on ________ volts.
a. 5
b. 7
c. 12
d. 20,000
e. 50,000

2. Common voltages on the secondary circuit on a conventional ignition system operates at ________ volts.
a. 5
b. 7
c. 12
d. 20,000–60,000
e. 70,000

3. Which of the following components is not part of the primary circuit on a conventional ignition system?
a. Condenser
b. Rotor
c. Points
d. Ballast resistor
e. Large-size windings in the coil

4. On a conventional ignition system, which of the following components is not part of the secondary circuit?
a. Points
b. Condenser
c. Ballast resistor
d. All of the above
e. None of the above

5. Dwell is defined as:
a. The length of time in degrees the points are closed
b. The length of time in degrees the points are open
c. Point gap in thousandths of an inch
d. The voltage at the secondary circuit
e. The voltage at the primary circuit

6. Which type of advance is used to increase the timing as speed increases?
a. Contact point advance
b. Centrifugal advance
c. Vacuum advance
d. Ported advance
e. Ignition advance

7. Which type of advance is used to increase the timing as load increases?
a. Contact point advance
b. Centrifugal advance
c. Vacuum advance
d. Ported advance
e. Piston advance

8. A plug that is fouling out with carbon and soot may be too:
a. Hot a plug
b. High a voltage on the plug
c. Cool a plug
d. Long a plug
e. Large a plug

9. The tip temperature of the plug will increase with:
a. An overadvanced engine
b. A decrease in coolant temperature
c. Clean spark plugs
d. All of the above
e. None of the above

10. The spark plug voltage required will increase with:
a. Decreases in compression
b. Decreases in spark plug gap
c. Decreases in load on the engine
d. All of the above
e. None of the above

The following questions are similar in format to ASE (Automotive Service Excellence) test questions.

11. *Technician A* says that as the contact points become pitted, the gasoline mileage goes down. *Technician B* says that as the contact points become pitted, the voltage to the spark plug is decreased. Who is correct?
a. A only
b. B only
c. Both A and B
d. Neither A nor B

12. *Technician A* says the moisture inside a distributor cap will have no effect on the ignition system operation. *Technician B* says that moisture inside a distributor cap will cause the spark plug voltage to increase and improve ignition system operation. Who is correct?
a. A only
b. B only
c. Both A and B
d. Neither A nor B

13. *Technician A* says that the resistance of the spark plug wire can be checked with an ohmmeter. *Technician B* says that the resistance of the spark plug wire does not affect the spark. Who is correct?
    a. A only c. Both A and B
    b. B only d. Neither A nor B

14. An engine starts but seems to miss on number 3 cylinder. *Technician A* says the problem may be number 3 spark plug. *Technician B* says the problem is a bad distributor bearing. Who is correct?
    a. A only c. Both A and B
    b. B only d. Neither A nor B

15. Sparks are noticed jumping from spark plug wire to wire. *Technician A* says the problem is too high a voltage to the spark plugs. *Technician B* says the problem is the ignition switch. Who is correct?
    a. A only c. Both A and B
    b. B only d. Neither A nor B

Essay

16. What is the difference between the primary and secondary ignition circuits?

17. Define the term *dwell.*

18. State two types of advance, and identify how they operate.

19. Describe spark plug heat ranges.

20. Identify two things that affect spark plug temperature.

Short Answer

21. An extended-life spark plug uses a material called ________ on the tips of the electrodes.

22. The average gap for an extended-gap spark plug is ________ inches.

23. A ________ type of spark plug misfire causes the voltage arc to pass from the top of the center electrode, over the outside of the spark plug, to the engine.

24. Two types of spark plug misfire in which the air-fuel mixture will not be ignited are the ________ and ________.

25. An open circuit on the ignition coil primary will show ________ ohms when tested with an ohmmeter.

26. If there is a slight knock or pinging heard in the engine, the ignition timing may be too far ________.

Applied Academics

TITLE: Lab Scope

ACADEMIC SKILLS AREA: Science

NATEF Automobile Technician Tasks:

The service technician uses precision electrical test equipment to measure current, voltage, resistance, continuity, and/or power.

CONCEPT:

Computerized technologies on engines often require engine electronic ignition analyzers to help the service technician diagnose and repair problems. The most common choice for the service technician today is the lab scope (A) shown in the accompanying figure. This analyzer is able to check emissions, timing, ignition components, and computer systems. Some tool manufacturers also have handheld lab scopes.

The scope, also called an oscilloscope, is a visual voltmeter. An oscilloscope converts electrical signals into a visual image representing voltage changes over a specific period of time. The voltage is displayed on a line called the waveform (B) also shown in the figure. The waveform is displayed on a voltage (vertical) and time (horizontal) axis. A typical waveform is shown for the secondary ignition circuit.

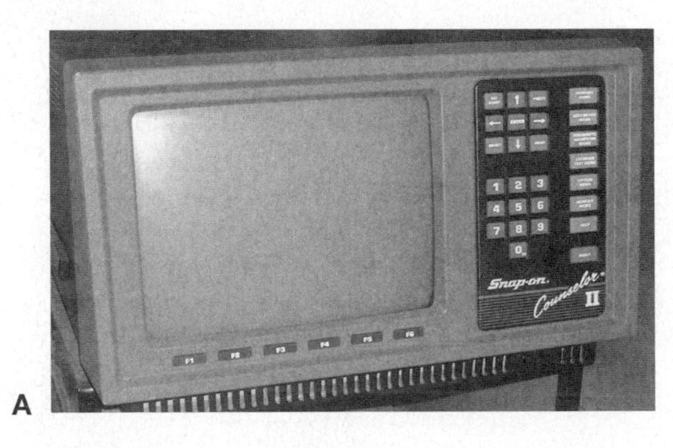

A

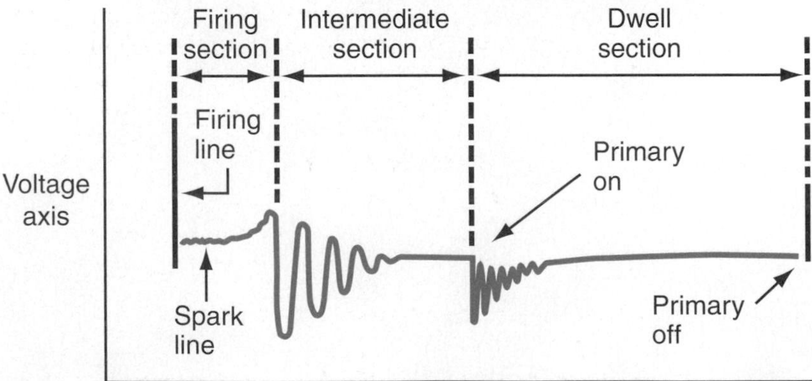

B

APPLICATION:

Why would it be important to use a lab scope on the ignition system? Why is the time important when defining the waveform?

Electronic and Computerized Ignition Systems

CHAPTER 33

OBJECTIVES

After studying this chapter, you should be able to:

- Analyze the parts and operation of the electronic spark control ignition system.
- Define the purpose and operation of computerized ignition systems.
- Identify the input sensors used on the PCM for computer-controlled ignition systems.
- State common problems, their diagnosis, and service procedures used on the electronic and computerized ignition systems.

INTRODUCTION

The ignition system on automotive engines has continued to become more electronically sophisticated and complex. Today's vehicles exclusively use electronic and computer-controlled ignition systems to energize the spark plugs. These systems are much more efficient and precise than the old conventional ignition system. This chapter introduces the electronic ignition system and describes the parts and operation of computerized ignition systems. In addition, various problems, their diagnoses, and service procedures are presented.

Certification Connection

ASE Connection: The information in this chapter can help you prepare for the National Institute for Automotive Service Excellence (ASE) certification tests. The tests and content areas most closely related to this chapter are:

Test A1—Engine Repair

- **Content Area**—Fuel, Electrical, Ignition, and Exhaust System Inspection and Service

Test A8—Engine Performance

- **Content Area**—Ignition System Diagnosis and Repair

NATEF Connection: Much of the information in this chapter is related to the NATEF tasks. The NATEF tasks and priority numbers most closely related to this chapter are:

1. Diagnose ignition system–related problems such as no-starting, hard starting, engine misfire, poor drivability, spark knock, power loss, poor mileage, and emission concerns on vehicles with electronic ignition (distributorless) systems; determine necessary action. P-1
2. Inspect and test ignition system secondary circuit wiring and components; perform necessary action. P-2
3. Inspect and test ignition coil(s); perform necessary action. P-1
4. Check and adjust ignition system timing and timing advance/retard (where applicable). P-3

33.1 ELECTRONIC IGNITION SYSTEMS

PURPOSE OF ELECTRONIC IGNITION SYSTEMS

Two of the biggest problems with the conventional ignition system are the wear on the points and the speed at which the primary current is stopped. Conventional ignition systems use a set of contact points. With the introduction of solid-state components and transistors in the 1970s, however, all vehicle manufacturers converted the conventional ignition system to solid-state systems. These ignition systems were developed and improved over a period of time. Systems were called capacitive discharge ignition systems, breakerless ignition systems, solid-state ignition systems, electronic ignition systems, and high-energy ignition—**HEI**—systems. But they all had one thing in common. They used semiconductors (transistors) in circuits to open the primary circuit faster. The secondary circuit remained much the same, except that higher voltages were produced.

TRANSISTORS USED IN IGNITION SYSTEMS

Transistors are used to open and close the primary side of the ignition system. It was found that a transistor can open and close a circuit much faster than a set of mechanical breaker points. The magnetic field will then collapse much faster, producing a higher voltage in the secondary circuit. The secondary voltage can be as high as 80,000 volts with a transistor ignition system. This improves the combustion, performance, and emission characteristics of the engine.

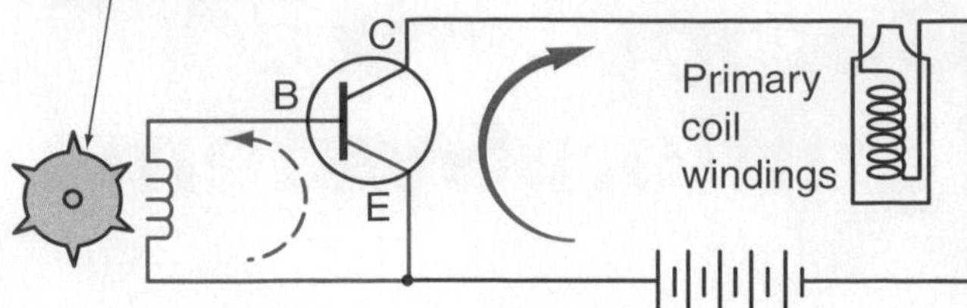

FIGURE 33–1 The transistor in a solid-state ignition system is used to turn off and on the primary side of the ignition coil. A small signal voltage triggers the transistor.

TRIGGERING THE PRIMARY CIRCUIT

Figure 33–1 shows how a simple transistor is used to trigger the primary side of the ignition coil. A small signal voltage is used to trigger the emitter-base circuit. This voltage signal is produced by a pickup coil and **reluctor** that act much like a small generator. The pickup voltage is a precisely timed signal. It triggers the electronic circuitry and transistors in the control unit. This interrupts the current flowing in the primary circuit, causing the ignition coil's magnetic field to collapse. The difference is that the emitter-collector circuit can be stopped very rapidly. This causes the magnetic field to collapse rapidly, producing a higher secondary voltage. The electronic circuit is much more complex than just one transistor. Dwell and timing advance can also be designed electronically into the system. ***Figure 33–2*** compares some different types of reluctors used over the years to trigger the transistor.

Figure 33–3 shows an example of a typical HEI module. A complex electronic circuit inside turns the primary igni-

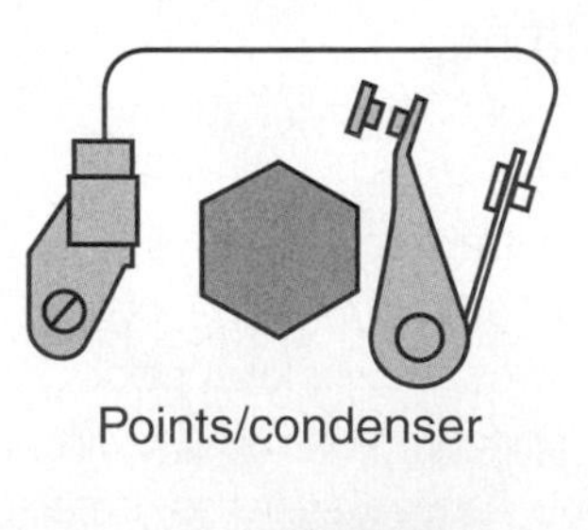

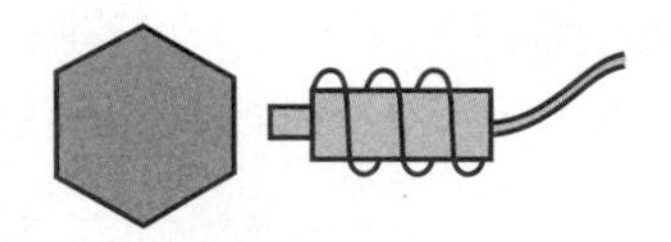

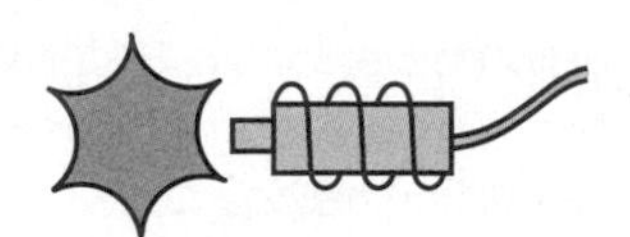

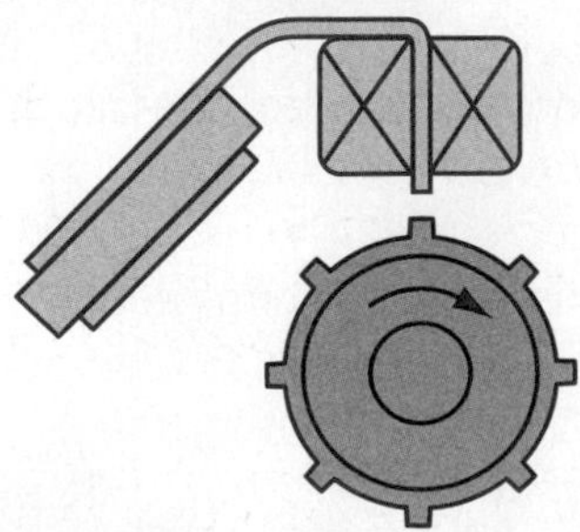

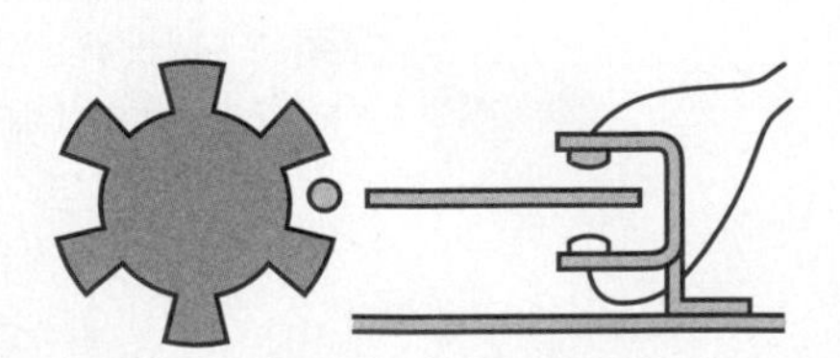

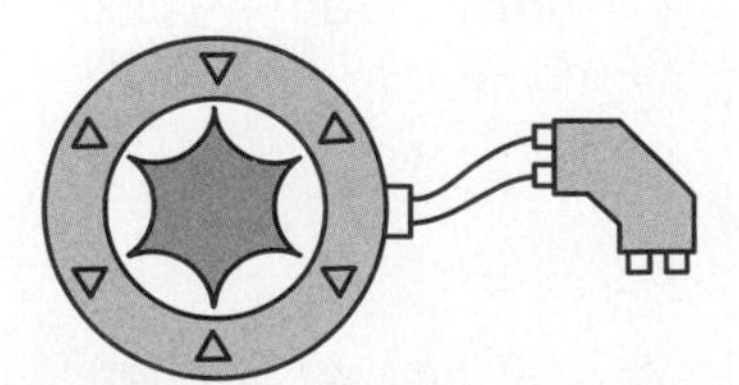

FIGURE 33–2 Many types of reluctors are used to trigger the transistor on engine ignition systems.

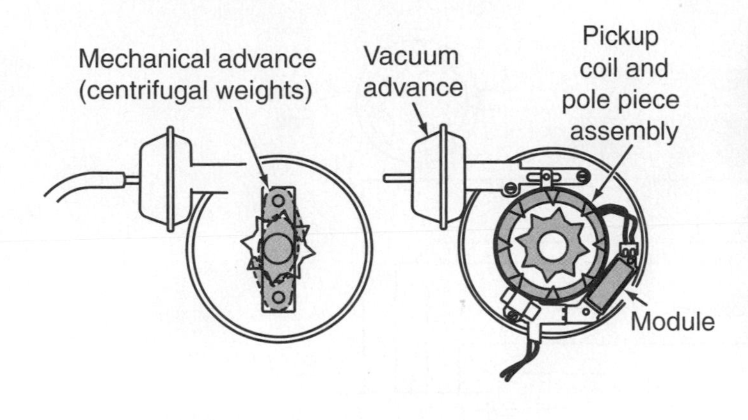

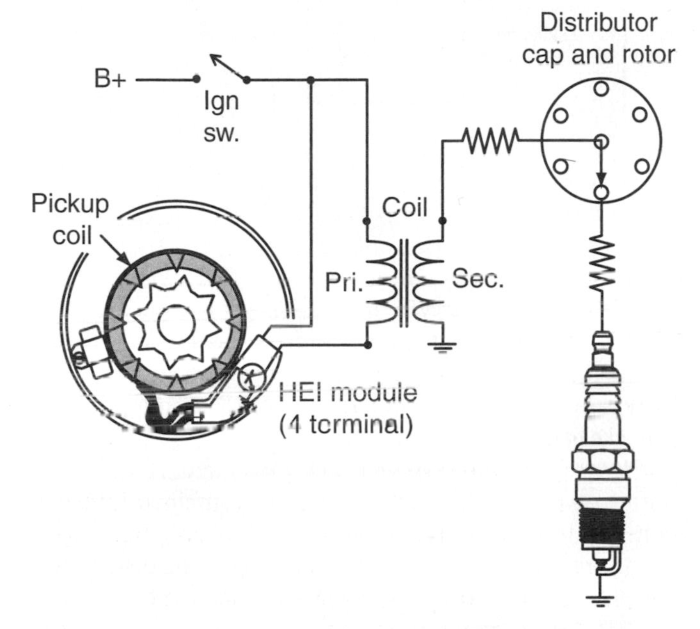

FIGURE 33-3 The control unit in a high-energy ignition (HEI Module) system uses the signal produced from the reluctor to turn a transistor on and off. The transistor controls the primary coil current.

tion coil on and off. The input includes the pickup coil signal, and the output is the primary side of the ignition coil. A mechanical vacuum advance and a centrifugal advance are still used on this system.

PARTS LOCATOR

*Refer to photo #57 in the front of the textbook to see what an **electronic control unit (transistorized)** looks like.*

33.2 COMPUTERIZED IGNITION SYSTEMS

There have been various names applied to electronic and computerized ignition systems. Over the past few years, each vehicle manufacturer designed and incorporated many new designs for these systems. Some were called distributorless ignition systems while others were called computer-

OLD NAME	NEW NAME
Distributorless Ignition System (DIS)	Electronic Ignition System (EI)
Direct Ignition System (DIS)	Electronic Ignition System (EI)
High Energy Ignition (HEI)	Distributor Ignition System (DI)
Computer-Controller Coil Ignition (C_3I)	Electronic Ignition System (EI)
Electronic Spark Timing (EST)	Ignition Control (IC)

FIGURE 33-4 The names listed under "New Name" have become the industry standard, replacing the various names that were used for electronic and computer ignition systems.

controlled coil ignition systems. The ignition systems presented in this section show the progression of design from ignition systems that still had mechanical parts to ignition systems that are totally controlled by the computer.

In 1993, the Society of Automotive Engineers established a standard that said certain electrical and electronic components and systems should share the same name among the car manufacturers if they have the same function. ***Figure 33–4*** shows some of the old names and the new name used as an industry standard. In order to eliminate confusion and maintain consistency, car manufacturers now use the new terminology. In the following text, both the new and old names will be referenced for ease of learning. Also, various types of computerized ignition systems and their associated sensors and operation will be addressed.

ELECTRONIC SPARK TIMING (EST)/ IGNITION CONTROL (IC)

Electronic spark timing (EST), now called ignition control (IC), uses many signals sent into the computer to electronically control or advance the spark. The electronic advance is much more exact and reliable than the mechanical advance mechanisms. ***Figure 33–5*** shows how the timing of the spark occurs in the HEI module system mentioned at the beginning of this chapter. To help understand the operation, a relay with a double set of contact points is shown in the HEI module. Solid-state circuitry is used in the module, but adding the relay makes it easier to visualize how EST or IC functions.

During cranking, the relay is in the de-energized position. The pickup coil is connected to the base of the transistor. When the pickup coil applies a positive voltage to the transistor, it turns on. When the voltage is removed, the transistor turns off. It then accomplishes what the contact points did in the old ignition systems. When the transistor

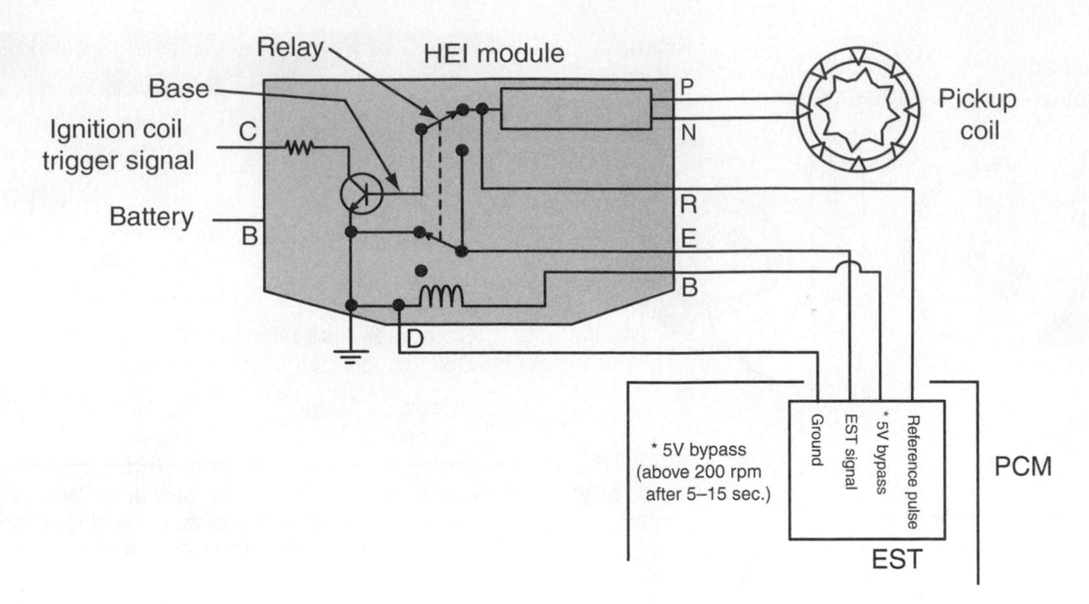

FIGURE 33–5 This circuit shows how timing of the spark is produced during cranking of the engine.

turns on, current flows through the ignition coil trigger wire to the primary winding of the ignition coil. When the transistor turns off, the primary current stops and a spark is developed. The EST circuit is located inside the powertrain control module (PCM). Several inputs are also shown for reference. The condition shown is for starting the engine. Timing is not being electronically controlled at this point.

Figure 33–6 shows how the timing is controlled when the engine is running. At about 200 rpm, the PCM applies about 5 volts to the bypass line. This voltage enters the HEI module at pin B and energizes the relay, causing it to shift. This is actually done electronically. The EST signal from the PCM is connected directly to the transistor base. The HEI system is controlled by the signal from the PCM. The time at which the spark occurs is determined by a circuit in the PCM based on the many inputs to the PCM. Timing is controlled electronically.

Figure 33–7 shows a typical HEI system used with EST. Note that the distributor and coil are combined into one unit. The unit is referred to as an HEI-EST distributor with an internal coil. The coil is placed on top of the distributor. The electronic control unit is placed inside the distributor. Electrical wires from the control module are attached to the connectors.

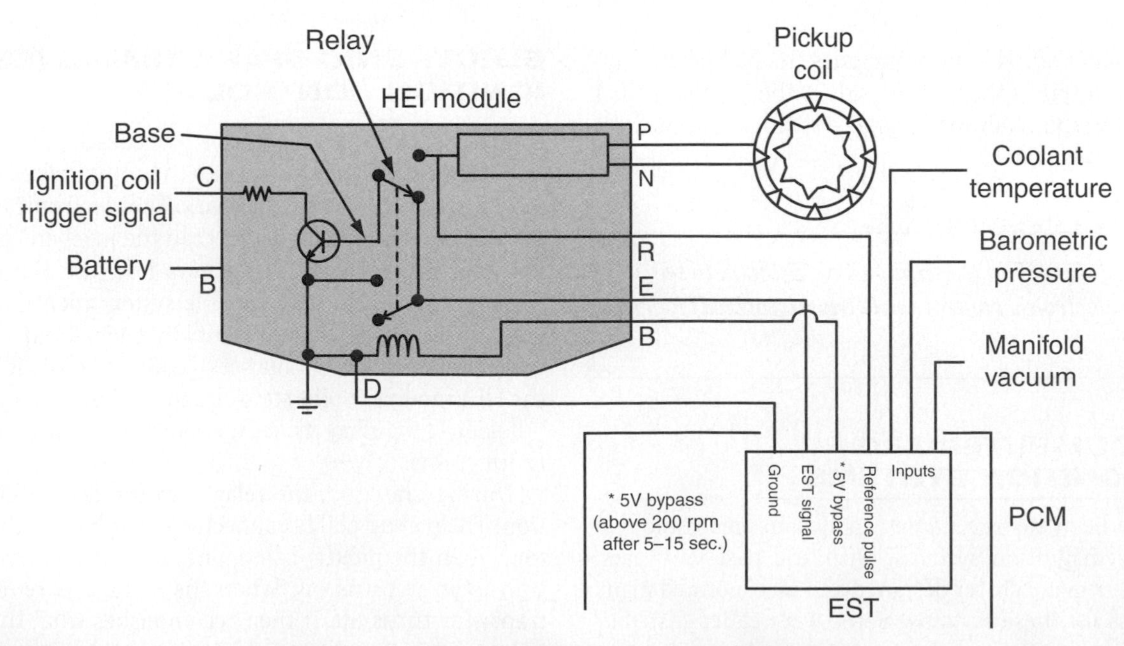

FIGURE 33–6 This circuit shows how timing of the spark is produced while the engine is running.

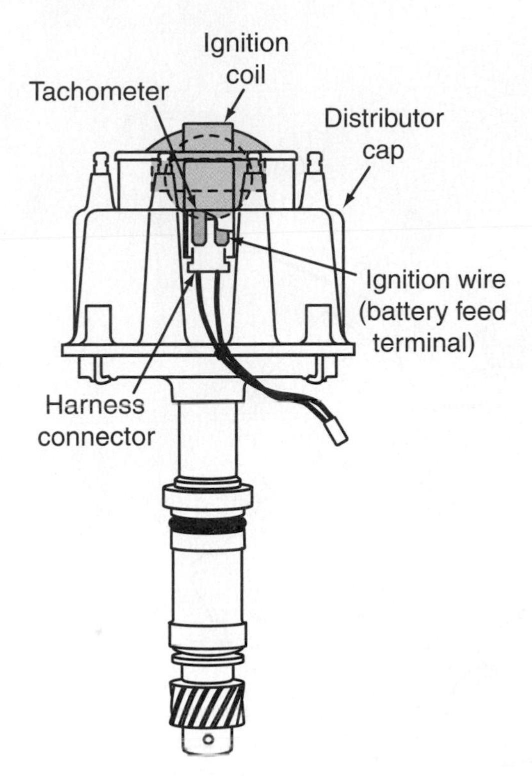

FIGURE 33-7 In this HEI—high-energy ignition—system, the coil and the distributor are combined into one unit.

COMPUTER-CONTROLLED SPARK

The next logical step in the development of electronic ignition systems is to increase the number of sensors sending input to the computer to control the spark. One example of such a system is the computer command control and the electronic engine control system. These systems use several inputs from different sensors. These inputs are combined, and electronic decisions are made on the amount of fuel to be added to the engine and, in this case, the amount of spark advance to be used. The centrifugal weights and the vacuum advance are controlled by the computer. ***Figure 33–8*** shows a similar system in block diagram. It is called the electronic constant engine control system. There are several inputs to the computer. These include engine speed, amount of air, temperature of the engine, throttle position, vehicle speed, start signal, engine knocking, and battery voltage. On the basis of these inputs, the power transistor in the electronic ignition system is operated to give the correct spark timing.

COMPUTER-CONTROLLED COIL IGNITION (C^3I) SYSTEM/ ELECTRONIC IGNITION (EI)

Another advance in ignition systems is the computer-controlled coil ignition C^3I system, now called electronic ignition (EI). It is considered to be a **distributorless ignition system**. This system consists of a PCM, ignition (coil) module, and electromagnetic camshaft and crankshaft position sensors. This system has eliminated the distributor and conventional ignition coil. It uses a microprocessor that receives and alters information from the crankshaft and camshaft position sensors. This information is processed to determine the proper firing sequence. The system then triggers each of three interconnected coils on 6-cylinder engines (two on 4-cylinder engines) to fire the spark plugs. Ignition timing is again determined by the PCM, which monitors crankshaft position, engine rpm, engine temperature,

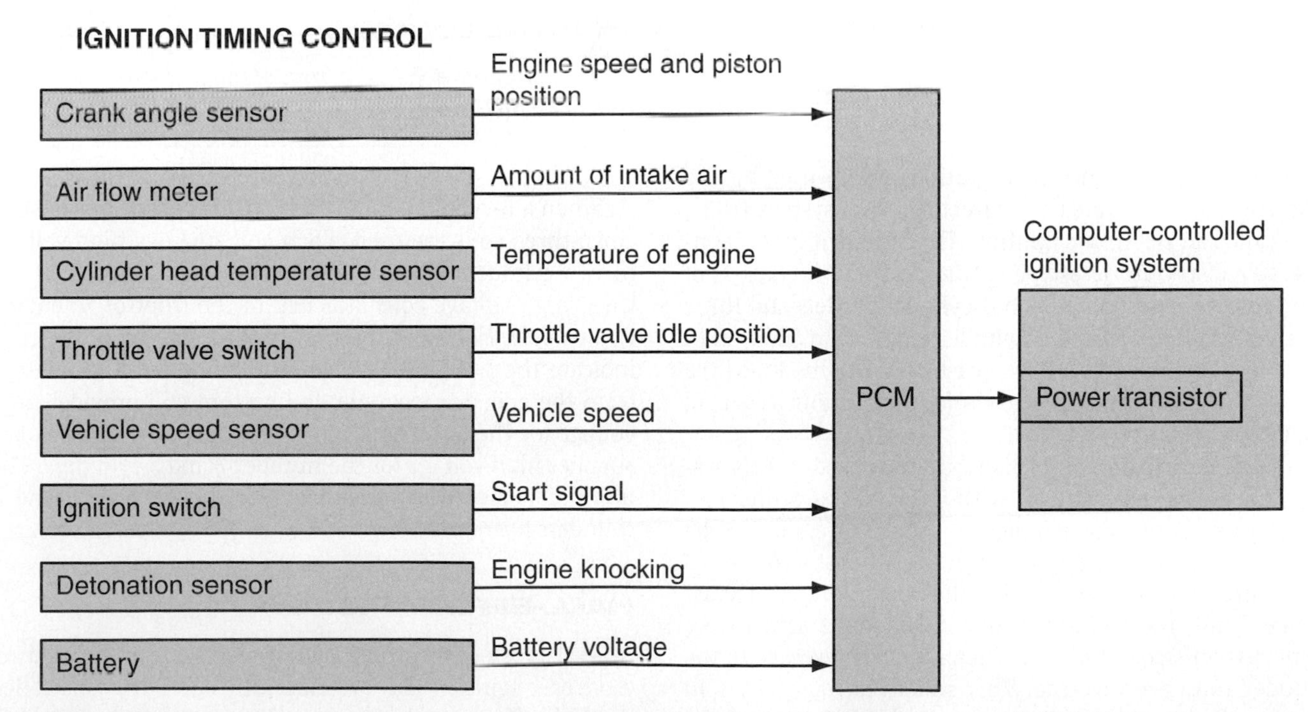

FIGURE 33-8 A computer-controlled spark system uses many inputs to electronically control the exact timing of the spark.

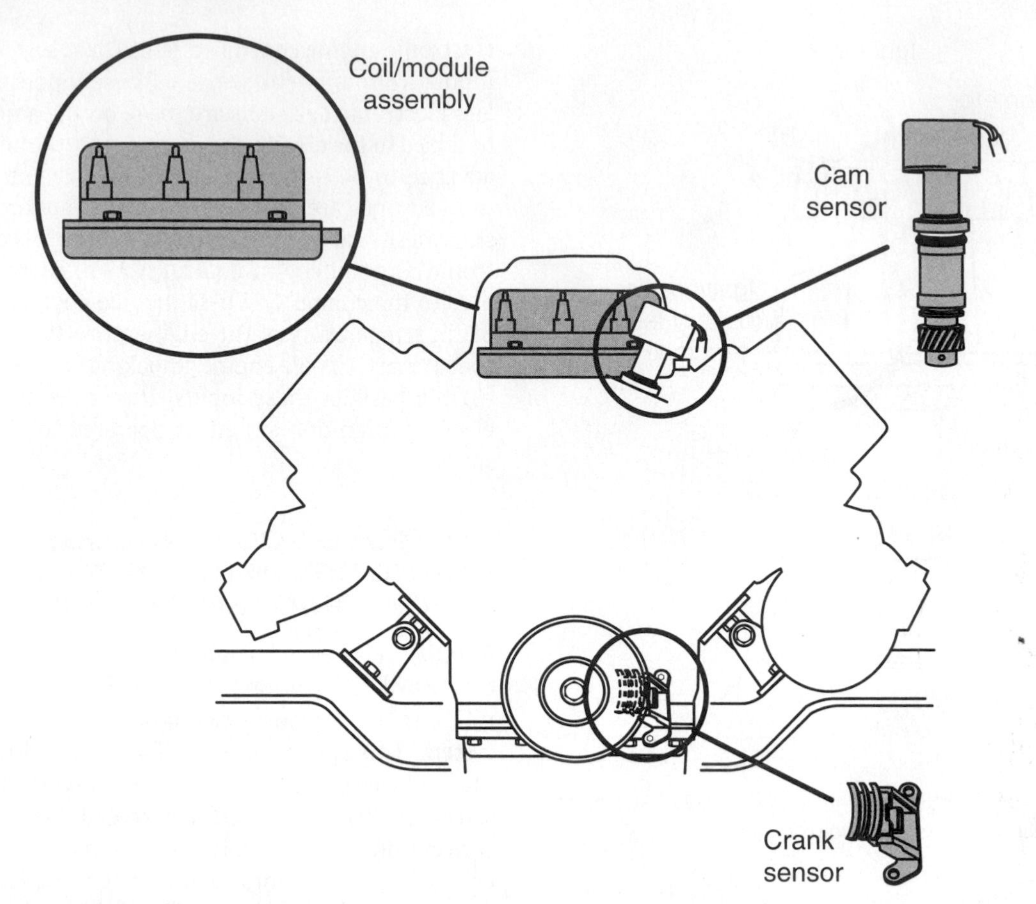

FIGURE 33-9 This computer-controlled coil ignition (C^3I) system has eliminated the distributor and conventional ignition coil. It uses several sensors that feed information to the ignition module, which in turn triggers the coils to produce the high voltage.

and the amount of air the engine is consuming. It then signals the ignition module to produce the necessary spark at the right time. ***Figure 33–9*** shows a typical C^3I ignition module installation.

DIRECT IGNITION SYSTEM (DIS), GENERAL MOTORS

Another computer-controlled ignition system used by one manufacturer was called the direct ignition system (DIS). Now called an electronic ignition (EI) system, it is also considered a distributorless-type ignition system. It consists of two separate ignition coils on 4-cylinder models and three separate ignition coils on 6-cylinder engines. In addition to the coils, the following are also used: (1) a DIS module, (2) a crankshaft sensor, and (3) electronic spark timing as part of the PCM.

A **waste spark** method of distribution is used on this system. Each cylinder is paired with its opposing cylinder in firing order. Thus, one cylinder on the compression stroke fires, while its opposing cylinder on the exhaust stroke fires. It requires less voltage to fire the plug on the exhaust stroke. Thus, most of the available voltage is sent to the compression-stroke cylinder. The process is reversed as the cylinder roles are reversed. This system is represented in ***Figure 33–10***, which shows a 4-cylinder DIS view of the components. Note that there are two coils, one for cylinders 2 and 3, one for cylinders 1 and 4.

PARTS LOCATOR

Refer to photo #58 in the front of the textbook to see what a ***distributorless ignition system coil*** *looks like.*

When a 6-cylinder engine uses the DIS system with DIS coils, three coils are used. Each coil supplies high voltage to two cylinders. ***Figure 33–11*** shows an example of the three high-voltage coils attached to the front of an engine in a vehicle. Note that the numbers on the top of the coils indicate the two cylinders that are being fed high voltage from the coil. For example, the bottom coil provides high voltage for the number 1 and 4 cylinders. The middle coils supplies high voltage for the number 2 and 5 cylinders. The upper coil provides the voltage for the number 6 and 3 cylinders.

HALL-EFFECT PICKUP

Several types of electrical pickup devices can be used on electronic ignition (EI) systems. One such device is called the **Hall-effect** switch or sensor. Its purpose is to provide a

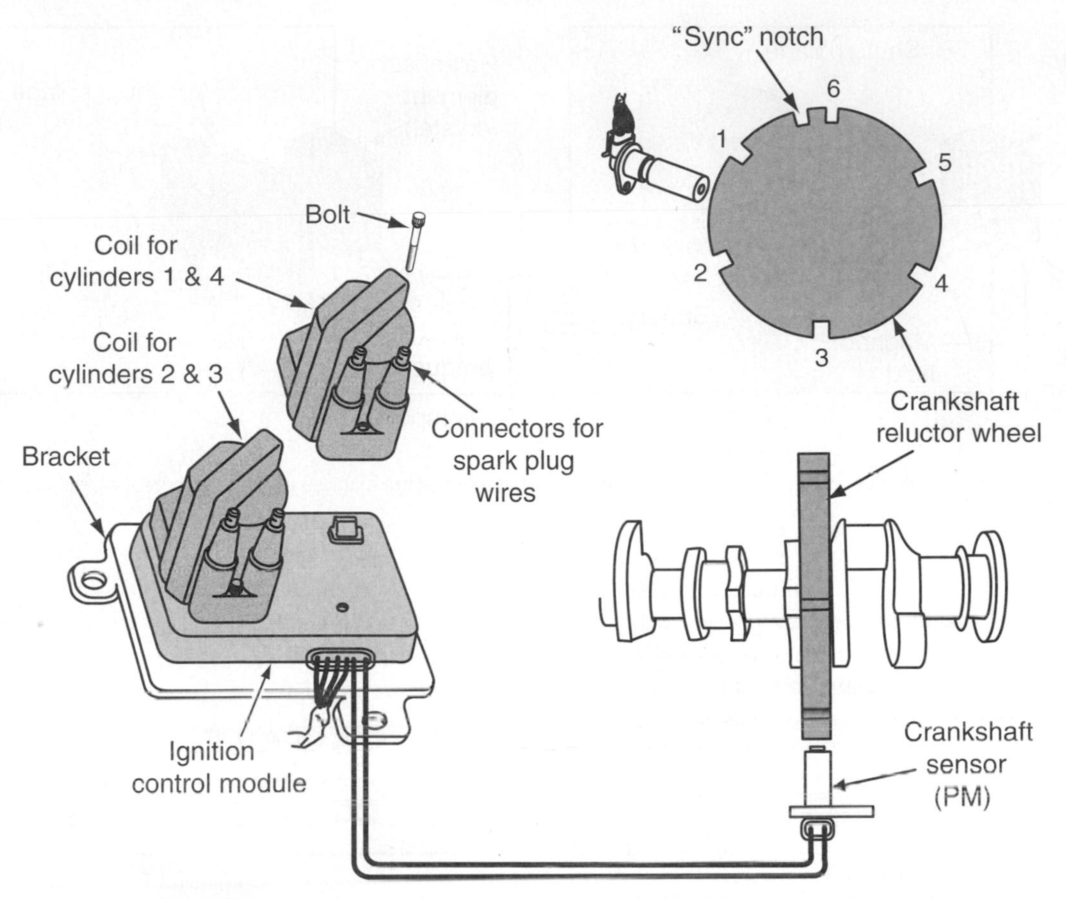

FIGURE 33–10 This direct ignition system (DIS) uses two coils, one for firing cylinders 2 and 3, and one for firing cylinders 1 and 4.

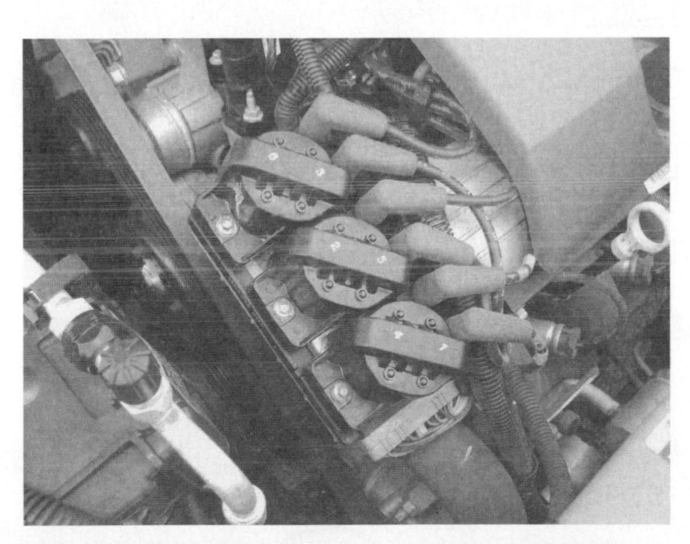

FIGURE 33–11 On a 6-cylinder engine, three DIS coils are used to provide the spark. Each coil supplies high voltages to two cylinders and spark plugs.

signal each time a piston reaches TDC. Hall-effect sensors are also able to read various positions within the 360 degrees rotation of the crankshaft. The signal produced by the switch is then used for ignition timing. One advantage of the Hall-effect sensor is that it provides a digital signal for the computer. ***Figure 33–12*** shows a typical Hall-effect pickup located inside a distributor. There are several arrangements, but Hall-effect sensors consist of three parts. There is a permanent magnet, a shutter wheel, and a Hall-effect element. The permanent magnet is stationary. The Hall-effect element, or pickup, also called a crystal, is also stationary. The shutter wheel is rotated by the distributor

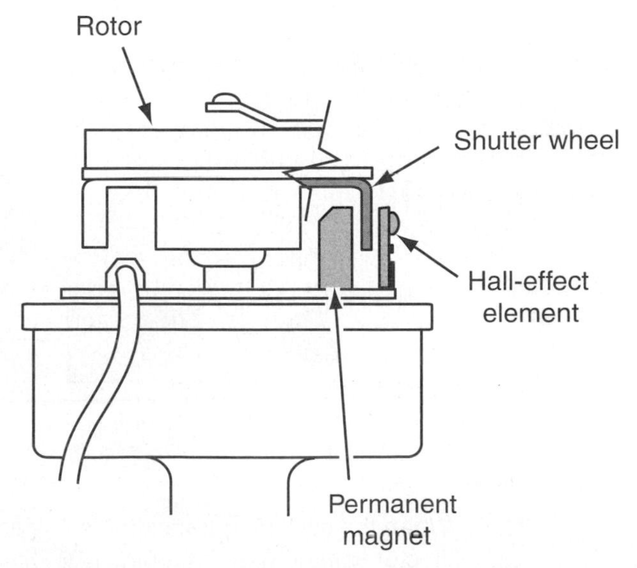

FIGURE 33–12 This Hall-effect sensor is used to sense and determine the position of the crankshaft for the computer.

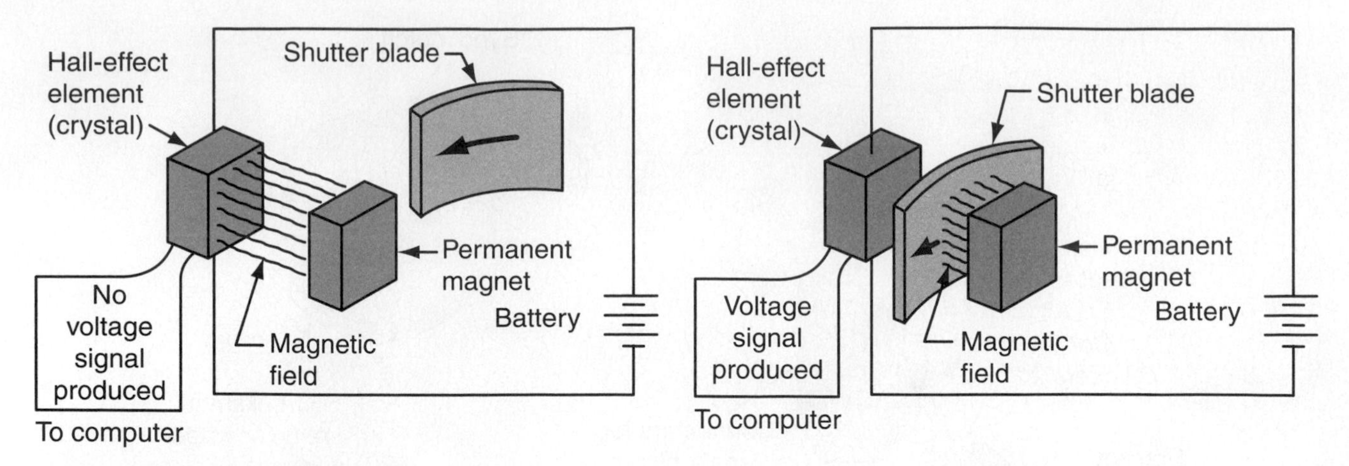

FIGURE 33–13 As the shutter blade passes in front of the magnet, it reduces the magnetic field, thus increasing the voltage and signal of the Hall-effect element.

shaft. As shown in ***Figure 33–13***, a voltage is applied to the Hall-effect element producing zero or no output voltage. This is because when the shutter blade is not between the magnet and the crystal, the magnetic field has the effect of reducing the signal. This produces a weak voltage in the crystal (about 0.4 volt). However, when the shutter blade (vane) is between the magnet and the Hall-effect crystal, the magnetism is reduced, producing a higher output voltage (12 volts) in the Hall-effect element. Through internal circuitry, the signal is then amplified and sent to the computer. ***Figure 33–14*** shows the Hall-effect pickup and the shutter blades.

PARTS LOCATOR

*Refer to photo #59 in the front of the textbook to see what a **Hall-effect sensor** (in distributor) looks like.*

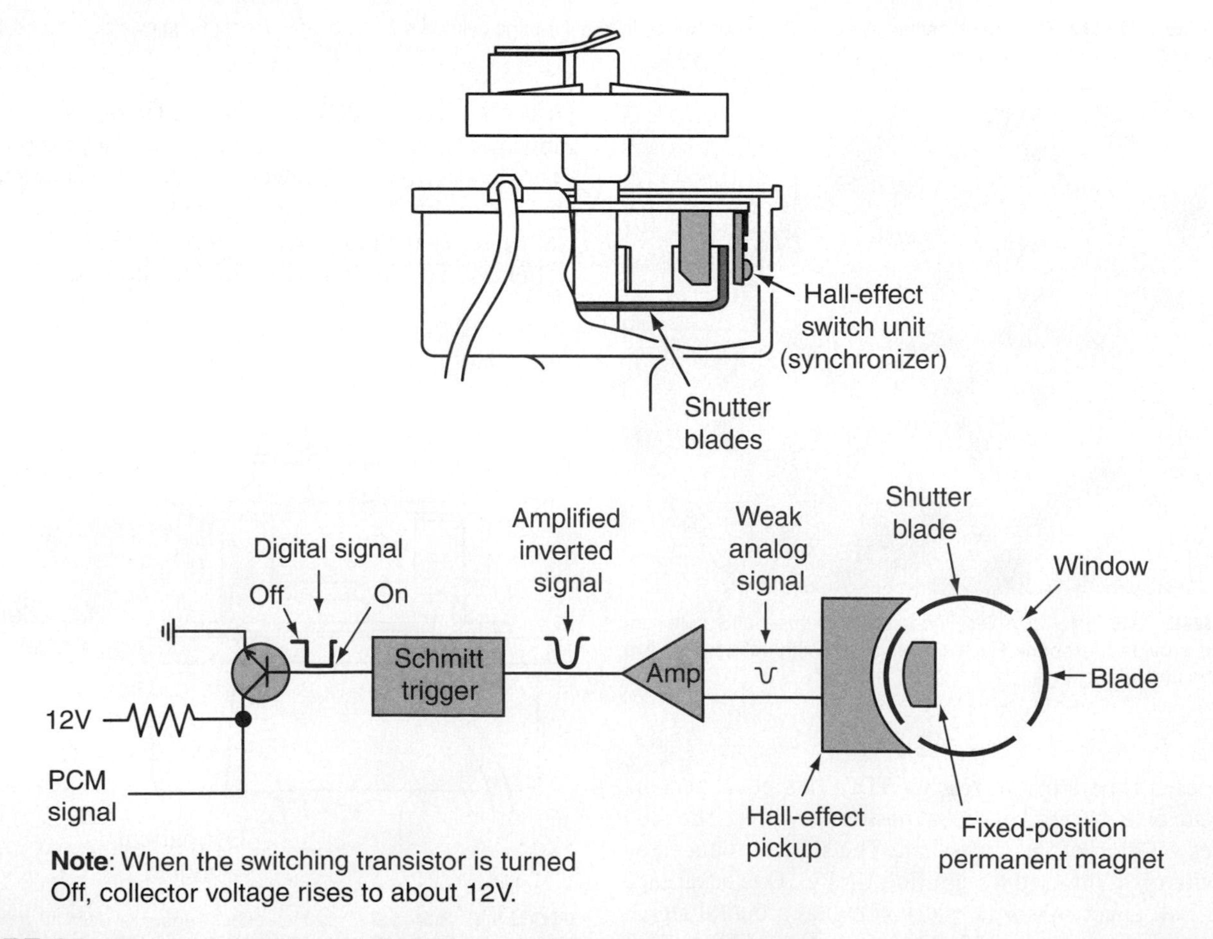

FIGURE 33–14 The location of the Hall-effect pickup and related components are shown.

CRANKSHAFT POSITION SENSOR

On some electronic ignition (EI) systems, a crankshaft position (CKP) sensor is used to trigger the ignition system or the computer. It is also used on many computer systems to provide a signal for ignition timing and engine speed. For example, Ford Motor Company uses a profile ignition pickup (PIP) for this purpose. ***Figure 33–15*** shows a typical crankshaft position sensor. It has three parts: the pulse ring, the crankshaft position sensor, and the related parts assembly. The pulse ring is mounted and rotates on the crankshaft. Its purpose is to act as a trigger wheel to provide engine speed information. The crankshaft position sensor is mounted near the pulse ring. Its purpose is to pick up a magnetic signal from the pulse ring and send the signal to the electronic or computer ignition system. It does much the same as the magnetic pickup on older EI systems. A clamp assembly holds the sensor securely in the front of the cylinder block. ***Figure 33–16*** shows an actual crankshaft sensor. The magnetic sensor is embedded in the plastic housing. The electrical connections to the computer are located on the top. The holddown bracket keeps the crankshaft sensor secure and attached to the engine.

Other types of crankshaft position sensors are also used. They can be designed to read various positions on the 360 degrees rotation of the crankshaft. One is shown in ***Figure 33–17***. It is called a permanent magnet pulse generator. Depending on the manufacturer and the year of the computer, the PCM is able to read and separate each of the pulses. Note that other types of crankshaft position sensors were presented in earlier chapters as well.

PARTS LOCATOR

Refer to photo #60 in the front of the textbook to see what a ***crankshaft position sensor*** *looks like.*

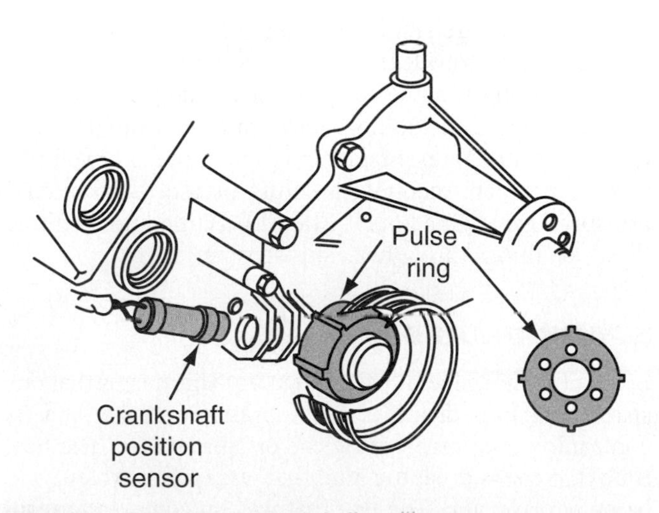

FIGURE 33–15 The crankshaft position sensor is located near the front of the crankshaft and is able to determine the crankshaft position for the computer.

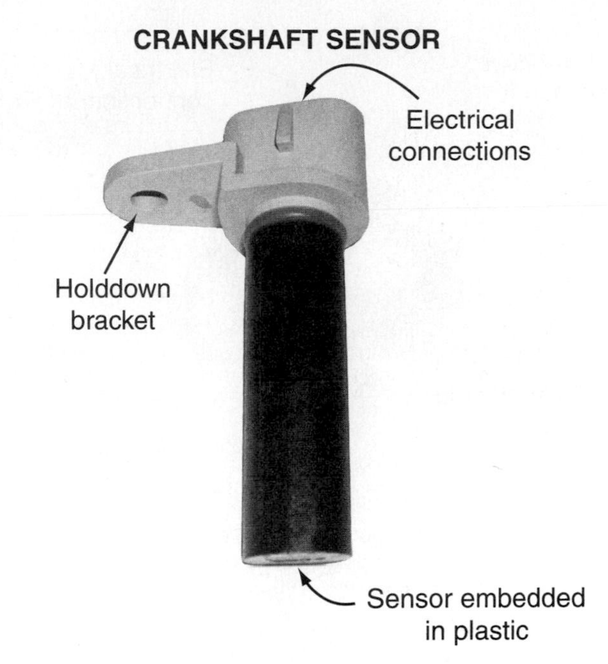

FIGURE 33–16 This is a typical crankshaft sensor.

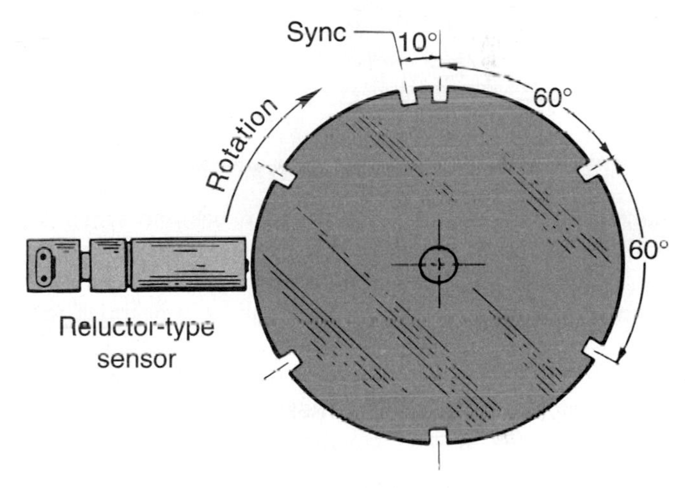

FIGURE 33–17 The permanent magnet pulse generator is able to read various positions of the crankshaft for the computer.

DETONATION SENSORS

The use of computers to control spark timing can be improved if the computer can sense detonation during the combustion process. If detonation or "knock" can be sensed, the optimum ignition timing can be controlled by the computer. A knock sensor, or **detonation sensor**, is a device that is able to pick up small, high-frequency vibrations in the engine. When knock does occur (from poor fuel, cylinder temperature, and so on), a high-frequency vibration is caused by the spark knock. This high frequency (5,000–6,000 hertz, or cycles per second) can be converted to a voltage signal for the computer. The knock sensor uses a piezoelectric crystal to sense knock. This crystal is able to produce an electrical signal (about 0.3 volt) when a vibration occurs. The knock sensor is generally threaded into a hole near the combustion chamber (***Figure 33–18***).

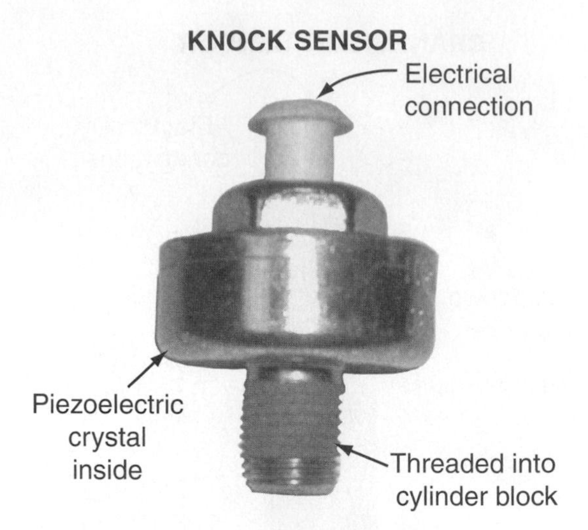

FIGURE 33-18 This knock sensor, or detonation sensor, is used to sense if the engine is pinging or has detonation. It is threaded into the cylinder block near the top of the piston.

*Refer to photo #61 in the front of the textbook to see what a **denotation sensor** looks like.*

CAMSHAFT POSITION SENSOR

Many of the EI systems use a **camshaft position sensor** in addition to the crankshaft sensor. The camshaft position sensor, such as these shown in ***Figure 33–19***, is used most often to identify the position of the number 1 piston. These cam sensors are similar in design to the Hall-effect crankshaft sensors. However, they look somewhat different. On most engines the camshaft sensor is mounted near or on the timing cover. As the camshaft turns, the signal generated is sent to the PCM where the signal is used to turn an internal transistor off and on. This, in turn, fires the coil(s) at the proper time. The signal is also used to sequence fuel injection. ***Figure 33–20*** shows one example of typical camshaft position sensors.

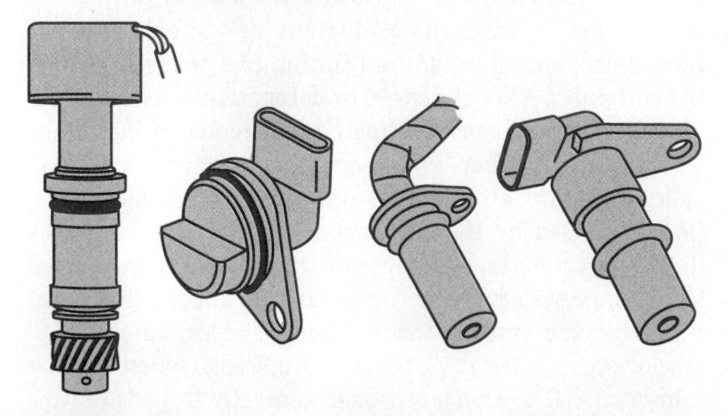

FIGURE 33-19 Examples of camshaft position sensors.

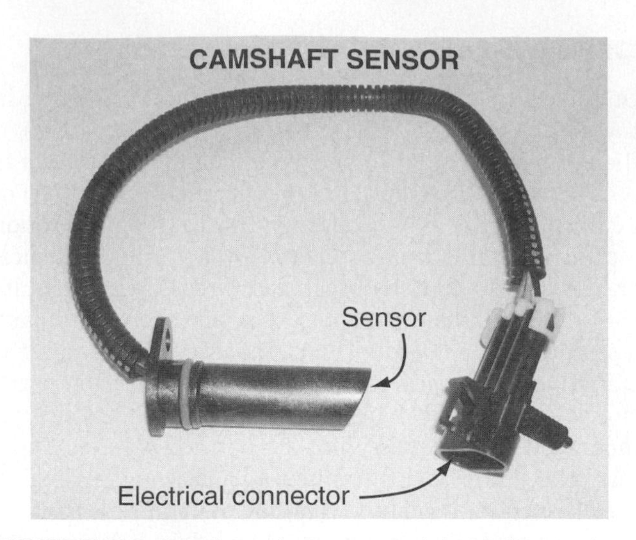

FIGURE 33-20 A camshaft sensor and electrical connectors.

*Refer to photo #62 in the front of the textbook to see what a **camshaft position sensor** looks like.*

33.3 DISTRIBUTORLESS IGNITION SYSTEM (DIS), (FORD)

It is beyond the scope of this book to discuss all the variations on electronic ignition (EI) systems in detail. However, to aid understanding of the basic concepts, the distributorless ignition system (DIS) by Ford Motor Company is further discussed.

IGNITION SYSTEM PURPOSE

As discussed earlier, the purpose of an ignition system is to ignite the air-fuel mixture. The ignition spark must be present at the correct time and in the correct sequence. The computer in Ford's system, called the EEC IV and now the PCM, is used to provide this information. In this system, the ignition timing is adjusted constantly. Many factors affect ignition timing, including engine rpm, coolant temperature, EGR flow rate, intake air temperature and volume, throttle position, manifold absolute pressure, barometric pressure, and engine knock. The EEC IV module monitors these conditions through various sensors.

SPARK PLUG FIRING

Spark plugs are generally fired in pairs. This means that one spark plug fires during the compression stroke, and its companion plug fires during the exhaust stroke. The next time that coil is fired, the plug that was on the exhaust will be on compression. The one that was on compression will be on exhaust. The spark in the exhaust cylinder is wasted, but little of the coil energy is lost.

IGNITION SYSTEM COMPONENTS

Several components are used on Ford's DIS system. These include the:

- Battery
- Ignition switch
- DIS module
- EEC IV module (PCM)
- Spark plugs
- Spark plug wires
- PIP sensor (profile ignition pickup, or crankshaft sensor)
- CID sensor (cylinder identification sensor)
- Ignition coils (contains both primary and secondary coils)

Many of these components have been discussed previously, thus only certain components will be further defined. ***Figure 33–21*** shows several of these components and their location on the engine.

PIP SENSOR

The function of the profile ignition pickup (PIP), or crankshaft sensor, is to detect the position and speed of the crankshaft. The PIP sensor uses a Hall-effect type of pickup, shown in ***Figure 33–22***. It is mounted to the engine block near the crankshaft pulley and hub assembly. It consists of three vanes that interrupt the magnetism. The digital signal produced (also shown) is on for 60 degrees, then off for 60 degrees, on 60 degrees, off 60 degrees, on 60 degrees, and finally off for 60 degrees. This makes up one complete crankshaft revolution of 360 degrees. Note also that the leading edge of the digital signal (the first On signal) always occurs at 10 degrees BTDC. This is the base timing of the engine.

PARTS LOCATOR

Refer to photo #63 in the front of the textbook to see what a ***profile ignition pickup (PIP)*** *looks like.*

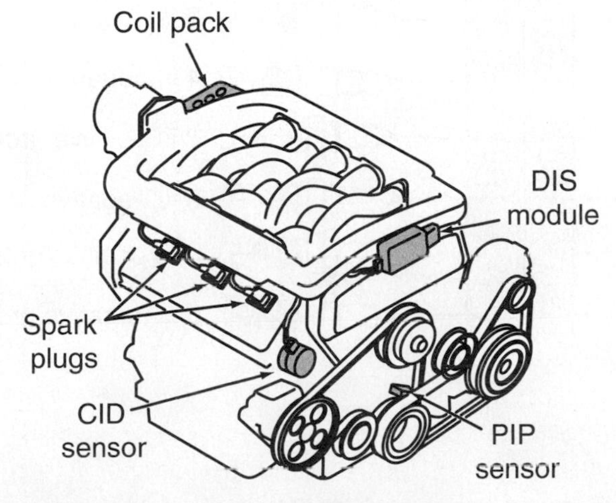

FIGURE 33–21 Ford's distributorless ignition system components are shown for general reference.

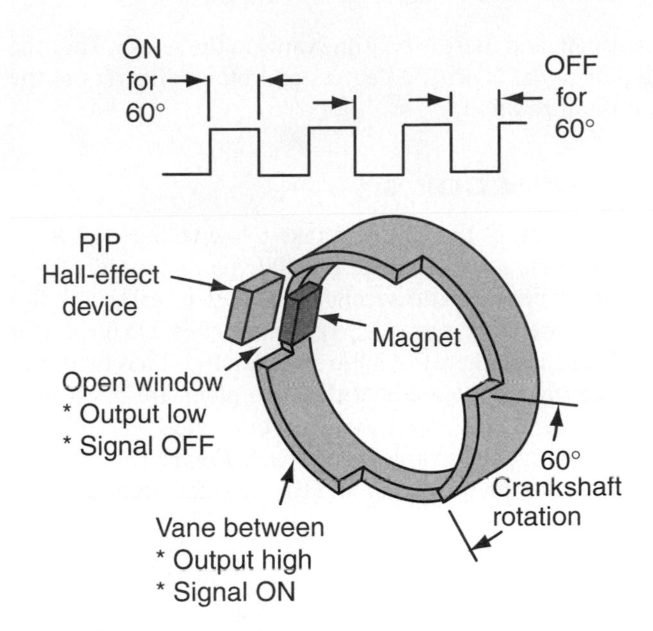

FIGURE 33–22 The PIP uses a Hall-effect pickup to generate a digital signal.

CID SENSOR

The function of the cylinder identification (CID) sensor is to detect the position of the engine camshaft. Camshaft position is used to identify when piston 1 is 26 degrees ATDC of its compression stroke. The DIS module uses the CID signal to select the proper coil to fire. The CID sensor is much like the PIP sensor. It uses a Hall-effect pickup as shown in ***Figure 33–23***. The differences are that the CID is driven by the

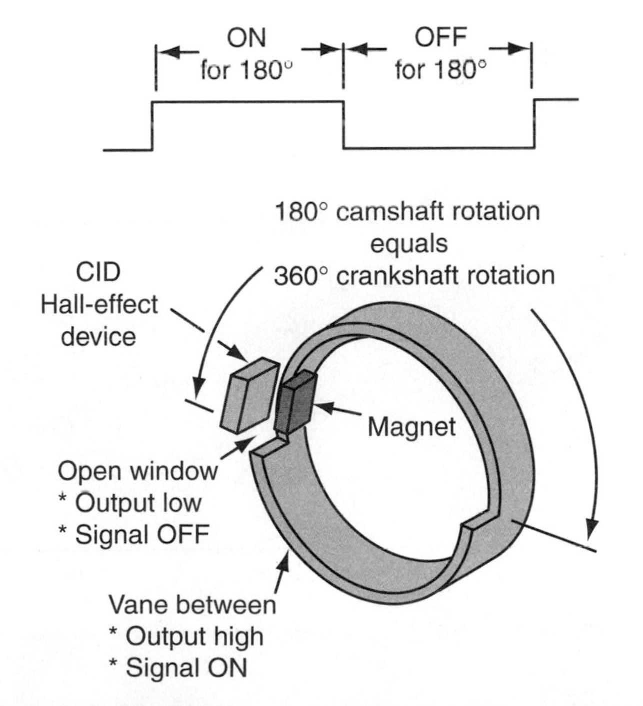

FIGURE 33–23 The cylinder identification (CID) sensor uses a Hall-effect pickup driven from the camshaft to determine the location of the number 1 piston.

camshaft, and there is only one vane on the sensor. Thus the digital signal is on 180 degrees and off 180 degrees of the camshaft rotation.

IGNITION COILS

The function of the coil is to take a low voltage (12 volts) and produce a high voltage (60,000 volts or more). There is both a primary and secondary circuit in each coil. If a 4-cylinder engine is used, there are two coils. If a 6-cylinder engine is used, there are three coils. Each coil has two high-voltage towers. (Remember that two plugs are being fired each time.) Each tower supplies one plug. All coils are mounted together in a single coil pack. ***Figure 33–24*** shows a coil pack for a 6-cylinder and the internal circuits of that coil pack.

DIS MODULE

The four functions of the DIS module are to:

1. Select which coil to fire
2. Control current in the primary coils
3. Generate a diagnostic signal for the EEC IV
4. Provide an ignition ground circuit

The DIS module has several inputs and outputs. The inputs include:

- Battery voltage
- CID sensor input
- PIP sensor input
- Spark timing signal from the EEC IV

The outputs include:

- An ignition ground
- Ground path for the primary circuit for coil 1
- Ground path for the primary circuit for coil 2
- Path for the diagnostic circuit and tachometer

In its basic operation, a spark timing signal is sent from the EEC IV to the DIS. On the basis of the information from the various sensors, the DIS opens the ground path to the primary of one of the coils. Note that in each coil, there is a primary and secondary circuit. When the spark timing signal is shut off in the DIS, the primary circuit is broken rapidly. This causes the primary circuit to collapse rapidly, causing the secondary voltage to be produced. The secondary voltage is then sent out to two spark plugs for ignition. The system is actually much more complex electronically. However, more detail is beyond the scope of this book.

PARTS LOCATOR

*Refer to photo #64 in the front of the textbook to see what an **ignition coil pack for a DIS ignition system** looks like.*

CABLELESS EI SYSTEM

To eliminate problems with the secondary wires that go to the spark plugs, some manufacturers use cableless EI systems. ***Figure 33–25*** illustrates several parts of the cableless system. Two coils, located in top of the housing assembly, feed four spark plugs that are located directly below. The secondary voltage output of the two coils is distributed within the housing assembly to the spark plug connectors. The spark plug boot retainer and inside spark plug rubber boot is then attached to the housing assembly by two plastic clips. When the assembly is put in place, it is pushed down onto the spark plugs. The spark plugs then fit inside the rubber boot and make electrical contact with the spark plug connectors.

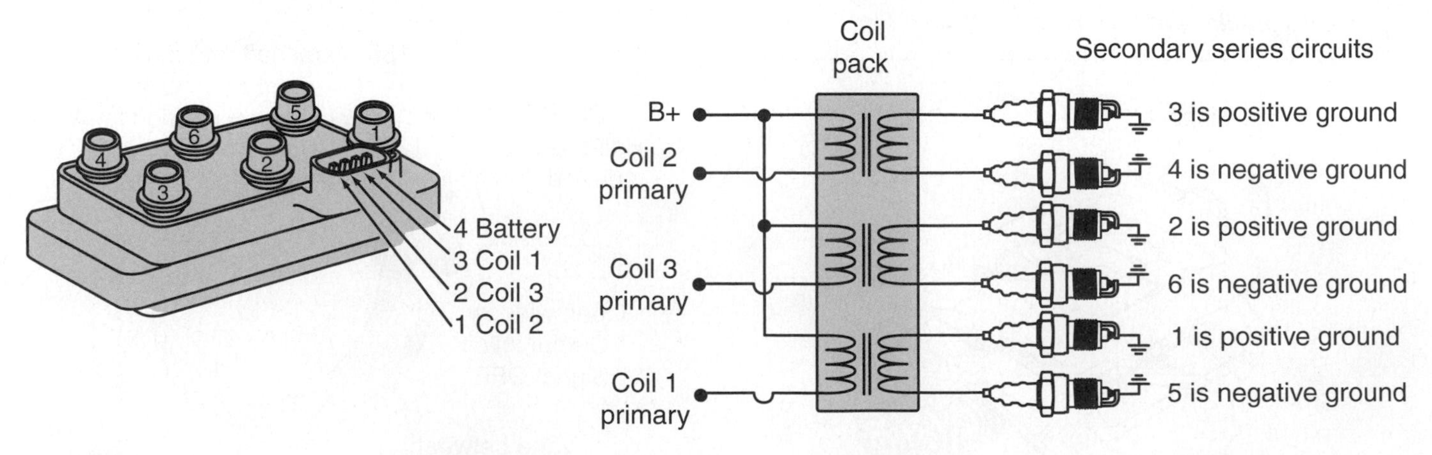

FIGURE 33-24 An ignition coil pack, along with its internal circuitry.

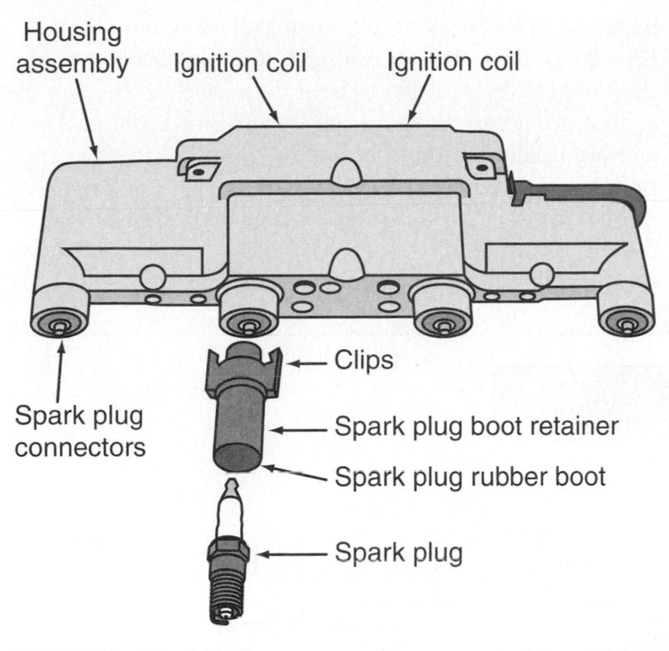

FIGURE 33-25 These components are part of the cableless ignition system used by some manufacturers.

Car Clinic: SPARK PLUG WIRES

CUSTOMER CONCERN:
Fouled Plug
A customer has an engine that consistently fouls out on the number 5 cylinder. The vehicle has 76,000 miles on it. The number 5 plug is very wet and carbon-filled. Even a new plug fouls out in a short time. Why would only one plug foul out and not the others?

SOLUTION: The most common cause of having just one plug foul out is a bad spark plug or spark plug wire. Since the plug has been changed, it would be wise to check the spark plug wires. This can be done with an electronic engine diagnosis scope or analyzer. When was the last time the spark plug wires were changed? The life of spark plug wires is about 60,000 miles, depending on the vehicle. They do eventually go bad. Replace the spark plug wires with a new set to eliminate the problem.

Problems, Diagnosis, and Service

Safety Precautions

1. When working on electronic ignition (EI) systems, always make sure the ignition switch is off. If it is left on, a spark may occur, causing gasoline fumes to ignite.
2. If it is necessary to crank the engine when working on an electronic ignition, always be careful of any moving parts (fan blades, etc.) that might be turning. They can cause serious injury.
3. Always use a spark plug wire pliers to remove spark plugs wires. If one is not available, always pull the spark plug wires from the spark plug boot. If just the wire is pulled, the wire may break and cause the engine to misfire.
4. When removing spark plug wires from an EI system, be careful of the hot parts. To avoid severe burns, wait until the engine cools down before removing the spark plugs and wires.
5. When working on EI systems when the engine is running, keep your fingers and hands away from the spark plug wires. There may be a break in the wire and this may cause you to get a serious shock.
6. To further eliminate the possibility of electrical shock, never crank an engine with one of the spark plug wires off.
7. Always use the correct tools when working with electronic and computerized ignition systems.
8. Always wear OSHA-approved safety glasses when working on electronic and computerized ignition systems.
9. Always use the correct tools when working on electronic and computerized ignition systems.

PROBLEM: Electronic Ignition Fault Parts

An engine with an electronic ignition (EI) system is misfiring badly.

DIAGNOSIS

Many engines today use EI systems. Diagnosis procedures are different for each of these systems. In addition, there is a constant flow of service bulletins that help the service technician troubleshoot and diagnose the ignition system. ***Figure 33–26*** shows an example of one of many troubleshooting charts available to help the service technician diagnose problems. As each step is performed, the problem will eventually be found by taking voltage and resistance readings and testing for spark.

Many of the ignition system components are checked on the electronic oscilloscope. Several procedures can, however, be followed when diagnosing an EI system.

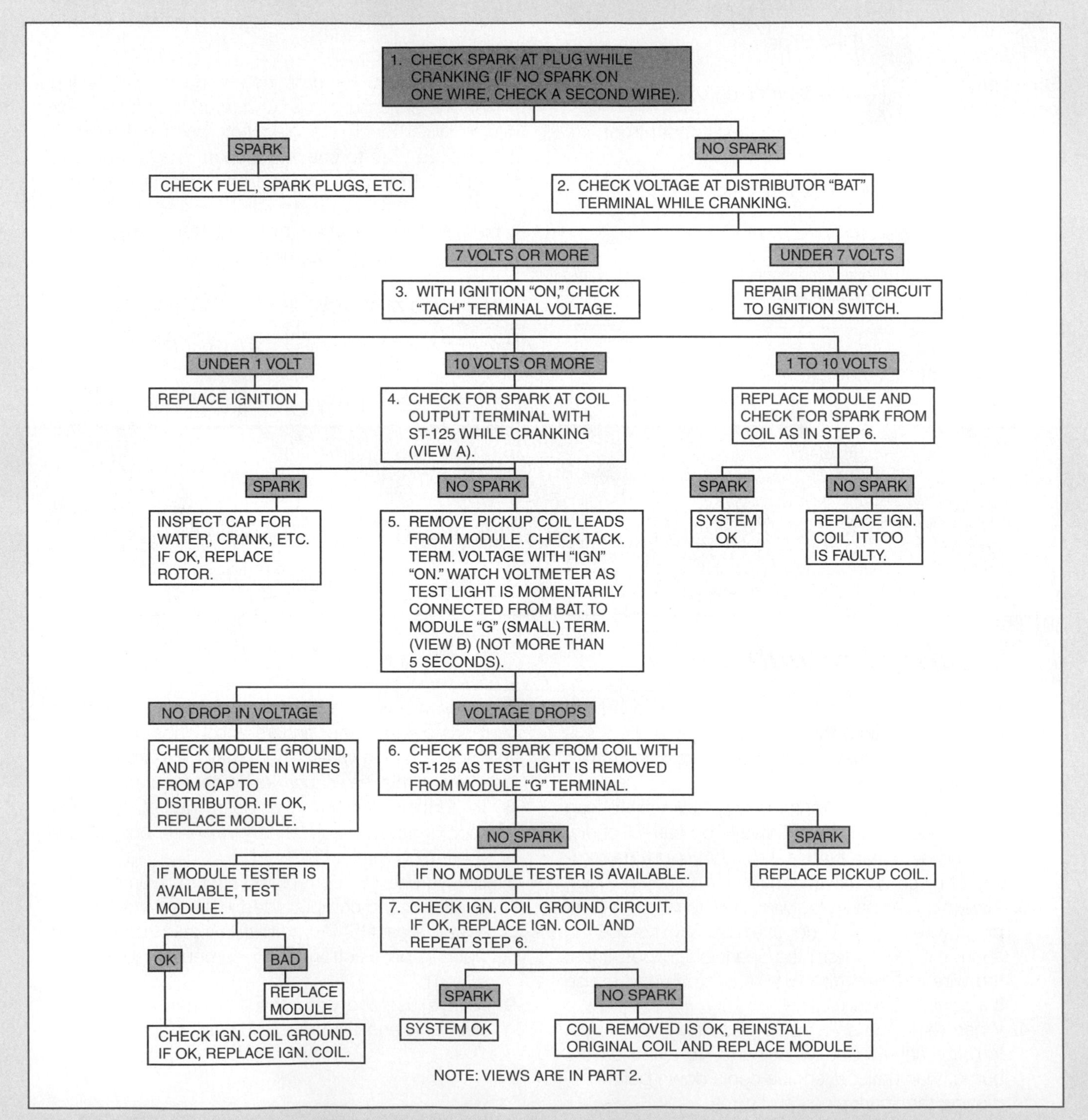

FIGURE 33–26 Because of the complexity of many of the ignition systems, various diagnosis charts are available in service and repair manuals. Follow the correct diagnosis chart to help troubleshoot an electronic ignition system. (Courtesy of MOTOR Publications, Auto Engine Tune-up and Electronics Manual, © The Hearst Corporation)

1. Using an ohmmeter, check the spark plug wire resistance. The resistance should be 5,000 ohms per inch or less. If the resistance of the wire is greater, replace the wire.
2. Spark plug wires should be checked for road salt, deposits, dirt, damaged boots, and cuts and punctures.
3. Various tests can be made to check voltage to the spark plug. Because several methods can be used, follow the specific manufacturer's recommendation to determine if spark exists.
4. In many maintenance and service manuals, specific connections are shown for ohmmeter tests. ***Figure 33–27A*** shows how to test an HEI system for the primary and secondary coils using the ohmmeter. ***Figure 33–27B*** shows how to check the electronic pickup using an ohmmeter. Compare the results with those suggested in the service manuals.

SERVICE

The service for each problem will vary. Follow the service procedures in the service manual for the exact type of ignition system.

PROBLEM: No Distributor Reference Pulse

A vehicle that has a C^{3}I distributorless ignition (EI) system has a Check Engine light on.

DIAGNOSIS

After grounding the "test" terminal under the dashboard or after using a scanner, it is determined that Code 12 is the problem. Code 12, according to this manufacturer, is "No distributor reference pulses to the PCM."

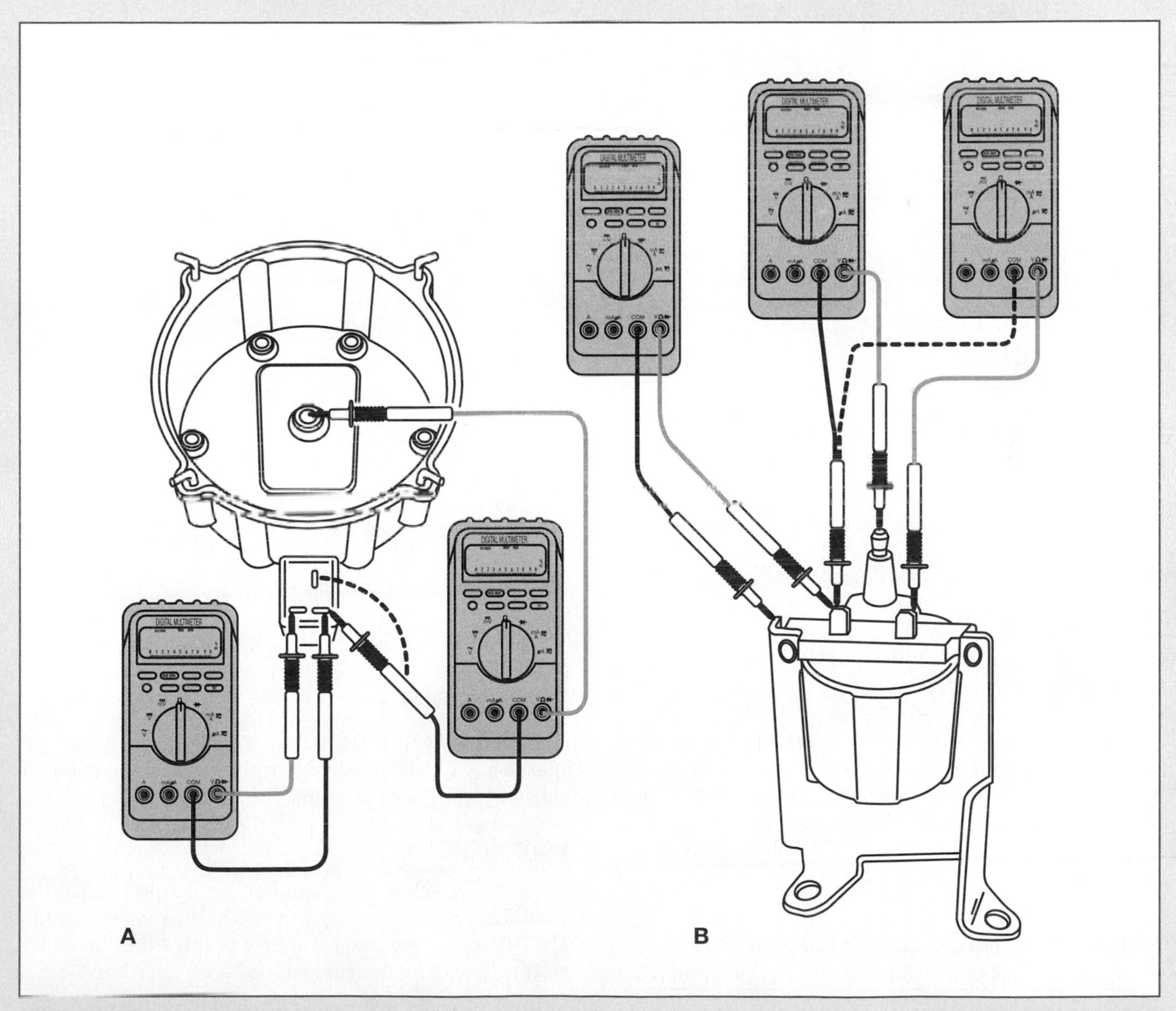

FIGURE 33–27 (A) This ohmmeter check will help to determine the condition of the primary and secondary coil. (B) This check will help determine the condition of the electronic distributor pickup.

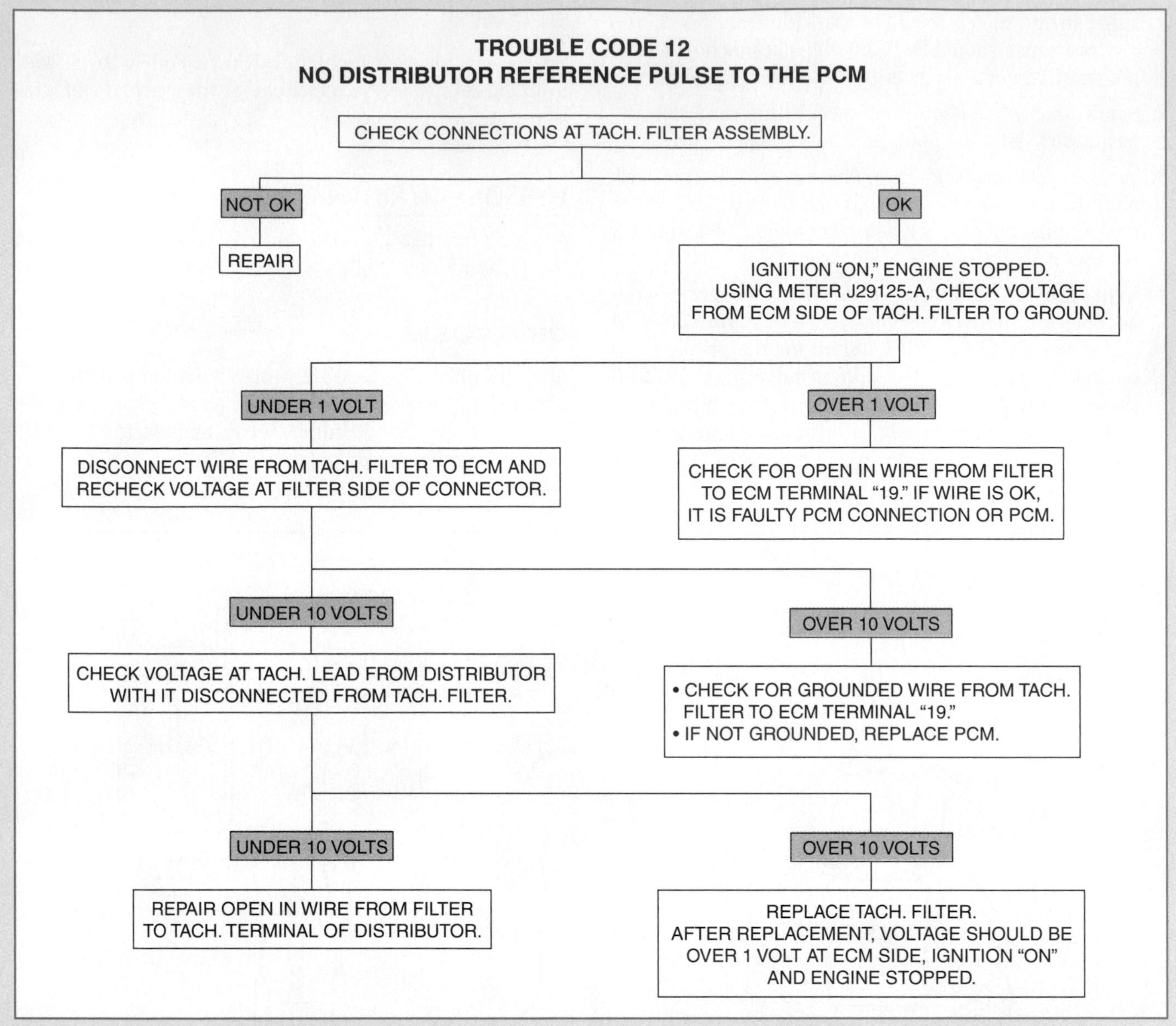
TROUBLE CODE 12
NO DISTRIBUTOR REFERENCE PULSE TO THE PCM
CHECK CONNECTIONS AT TACH. FILTER ASSEMBLY.
NOT OK
REPAIR
OK
IGNITION "ON," ENGINE STOPPED.
USING METER J29125-A, CHECK VOLTAGE
FROM ECM SIDE OF TACH. FILTER TO GROUND.
UNDER 1 VOLT
DISCONNECT WIRE FROM TACH. FILTER TO ECM AND
RECHECK VOLTAGE AT FILTER SIDE OF CONNECTOR.
OVER 1 VOLT
CHECK FOR OPEN IN WIRE FROM FILTER
TO ECM TERMINAL "19." IF WIRE IS OK,
IT IS FAULTY PCM CONNECTION OR PCM.
UNDER 10 VOLTS
CHECK VOLTAGE AT TACH. LEAD FROM DISTRIBUTOR
WITH IT DISCONNECTED FROM TACH. FILTER.
OVER 10 VOLTS
• CHECK FOR GROUNDED WIRE FROM TACH.
FILTER TO ECM TERMINAL "19."
• IF NOT GROUNDED, REPLACE PCM.
UNDER 10 VOLTS
REPAIR OPEN IN WIRE FROM FILTER
TO TACH. TERMINAL OF DISTRIBUTOR.
OVER 10 VOLTS
REPLACE TACH. FILTER.
AFTER REPLACEMENT, VOLTAGE SHOULD BE
OVER 1 VOLT AT ECM SIDE, IGNITION "ON"
AND ENGINE STOPPED.

FIGURE 33–28 This diagnosis chart can be used to help the service technician diagnose and repair Code 12, "No distributor reference pulses to the PCM," on a C³I ignition system. (Courtesy of MOTOR Publications, Auto Engine Tune-up and Electronics Manual, © The Hearst Corporation)

SERVICE

At this point it will be necessary to refer to the manufacturer's diagnosis chart shown in ***Figure 33–28***. Follow the exact diagnosis chart as illustrated until the problem is found and corrected. (For each code listed by the manufacturer there is a diagnosis chart to follow in order to aid the service technician in repairing the problem.)

PROBLEM: No Crankshaft Position Sensor Signal

A vehicle with an electronic ignition (EI) system has a Check Engine light on.

DIAGNOSIS

After grounding the "test" terminal under the dashboard, or after using a scanner, it is determined that the crankshaft position sensor has a problem.

SERVICE

On some vehicles, the manufacturer suggests that the crankshaft position sensor can be checked for resistance. The following is an example of how to check the crankshaft position sensor for one car manufacturer.

1. Disconnect the negative terminal of the battery.
2. Disconnect the wires from the crankshaft position sensor.

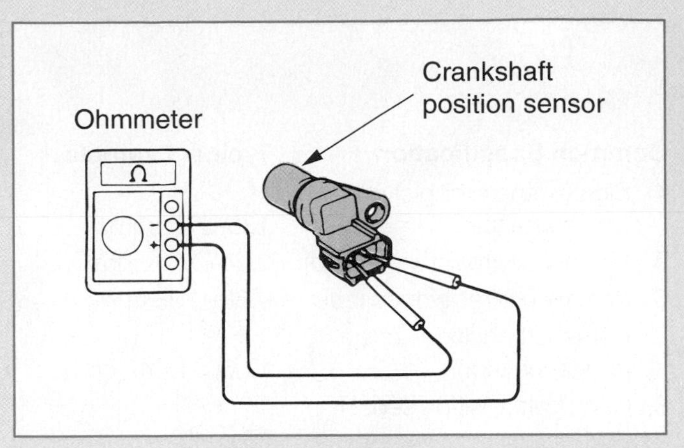

FIGURE 33–29 The crankshaft position sensor can be checked for resistance.

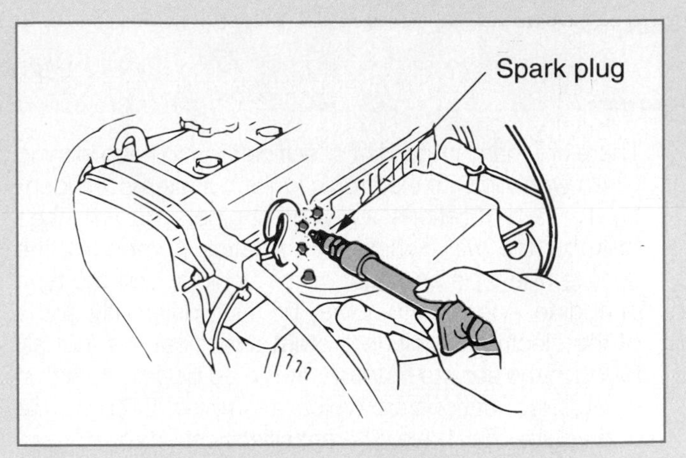

FIGURE 33–30 Check the spark by holding the spark plug tip about 1/4 inch from the engine and observing the spark while cranking the engine.

3. Remove the crankshaft position sensor by removing the bolt holding it to the engine.
4. Check the crankshaft position sensor for correct resistance.
   a. Using an ohmmeter, measure the resistance between the two terminals as shown in ***Figure 33–29***. The resistance should be the following:

      Resistance (Cold—14–122 degrees Fahrenheit) 1,630 to 2,740 ohms.

      Resistance (Hot—122–212 degrees Fahrenheit) 2,060 to 3,225 ohms.
   b. If the resistance is not as specified, replace the crankshaft position sensor with a new one.
5. Replace the new crankshaft position sensor by reversing the preceding procedure.

PROBLEM: High-Tension Ignition Wires Damaged

A vehicle with an electronic ignition (EI) system has a steady miss on one cylinder.

DIAGNOSIS

To determine which cylinder is misfiring, check that spark occurs on each cylinder. Disconnect all high-tension spark plug wires from the spark plugs. Hold the end approximately 1/4 inch from the body ground as shown in ***Figure 33–30***. Now crank the engine to see if spark occurs. To prevent gasoline from being injected by the injectors during this test, disconnect the wires to the injectors first. Also, crank no more than 10 seconds per cylinder. Check the resistance of the wires in which there was no spark created.

SERVICE

The following is a typical procedure used to test the resistance of the high-tension spark plug wires.

1. When removing high-tension wires, disconnect by pulling on the rubber boot. Do not pull on the high-tension wires in the middle of the wire. This may damage the conductors inside.
2. Using an ohmmeter, measure the resistance of the wire(s) as shown in ***Figure 33–31***. The maximum resistance on each wire should be 25,000 ohms per foot of wire.
3. If the resistance is greater than the maximum, check the terminals for corrosion. If necessary, replace the high-tension wire(s) with new ones.
4. After replacing the high-tension wires, start the engine and again check for misfire.

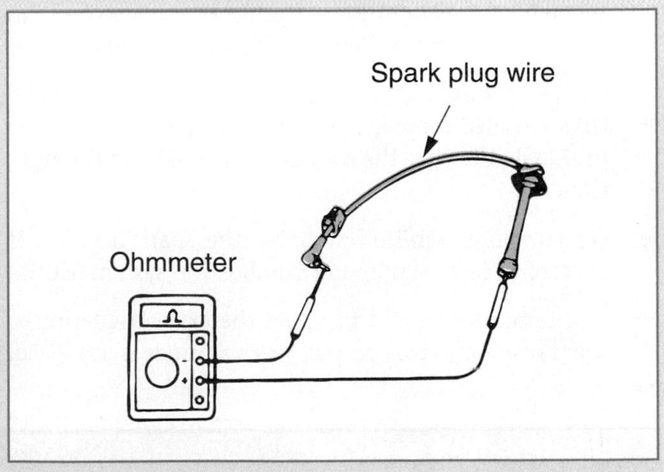

FIGURE 33–31 High-tension cords can be checked for resistance using an ohmmeter.

Service Manual Connection

There are many important specifications to keep in mind when working with electronic ignition systems. To identify the specifications for your engine, you will need to know the VIN (vehicle identification number) of the vehicle, the type and year of the vehicle, and the type of engine. Although they may be titled differently, some of the electronic ignition system specifications (not all) found in the service manuals are listed below. Note that these specifications are typical examples. Each vehicle and engine may have different specifications.

Common Specification	Typical Example
1. High-tension wire resistance	25,000 ohms
2. Crankshaft position sensor resistance (cold)	1,630–2,740 ohms
3. Gap between signal rotor and pickup coil	0.008–0.016 inches
4. Signal generator pickup coil resistance	185–275 ohms
5. Primary coil resistance (cold)	0.54–0.84 ohm
6. Primary coil resistance (hot)	0.68–0.98 ohm
7. Camshaft position sensor resistance (hot)	1,060–1,645 ohms
8. Camshaft position sensor bolt torque	69 in.-lb

In addition, the service manual will give specific directions for various service and testing procedures. Some of the common procedures include:

- Checking spark plugs
- Pinpoint tests for various input sensors
- Ignition coil inspection
- Distributor inspection
- Inspect signal generator

SUMMARY

The following statements will help to summarize this chapter:

- Vehicle manufacturers have changed the ignition system from the conventional to electronic- and computer-controlled ignition systems.
- Electronic ignition systems use transistors to quickly open and close the primary ignition system.
- When electronic ignition systems are used, voltages in the secondary can be between 20,000 to 60,000 volts or more.
- On some electronic ignition systems, a reluctor is used to signal or trigger the transistor on and off at the right time.
- On computer ignition systems, the spark advance is controlled by the computer combustion system (CCS).
- The computer is used to signal the exact triggering of the transistor so that correct spark advance is produced.
- The amount of advance is determined by several inputs to the computer, including engine speed, the amount of air going into the engine, the throttle position, an engine knocking signal, a start signal, the battery voltage, the engine coolant temperature, camshaft position sensor signals, and the crankshaft position sensor signal.
- The distributorless ignition system (DIS) uses a crankshaft sensor, cylinder identification sensors, the DIS module, and a Hall-effect system to operate the system.
- Some manufacturers use a cableless second voltage ignition system, which has a direct connection from the coils to the spark plugs.
- Various tests can be done on electronic and computer ignition systems, including resistances in the primary and secondary windings, voltage readings on the primary circuit, resistance of the high-tension wires, and the resistances of the crankshaft and camshaft sensors.

TERMS TO KNOW

Can you explain each of the following terms? Review the chapter until you can use each term correctly.

Camshaft position sensor
Detonation sensor
Distributorless ignition
Hall-effect
HEI
Reluctor
Waste spark

REVIEW QUESTIONS

Multiple Choice

1. What component is used to open and close the ignition primary circuit on an electronic ignition system?
a. A condenser
b. A diode
c. A computer
d. A transistor
e. A set of contact points

2. What is an average voltage on the secondary circuit for an electronic ignition system?
a. 10,000 volts
b. 20,000 volts
c. 12 volts
d. 60,000 volts
e. 6 volts

3. Which of the following is used to trigger the electronic circuit to open and close the primary windings in the coil?
a. Conductor
b. Reluctor
c. Condenser
d. Transistor
e. Secondary coil windings

4. The spark on a computer-controlled ignition system is timed by the:
a. Centrifugal weights
b. Vacuum advance
c. Computer or electronic control module
d. Springs on the weights
e. None of the above

5. When diagnosing the secondary circuit, which of the following should be checked?
a. Moisture in the distributor cap
b. Cracks in the distributor cap
c. Condition of the spark plug wires
d. All of the above
e. None of the above

6. The electronic ignition system that has the distributor and coil combined into one unit is called the:
a. DCI system
b. HEI system
c. Reluctor system
d. All of the above
e. None of the above

7. Which of the following is not used as a sensor on electronic ignition (EI) systems?
a. Airflow sensor
b. Throttle position sensor
c. Camshaft position sensor
d. Detonation sensor
e. Tire rotation sensor

8. On some electronic ignition systems, two spark plugs are firing at the same time, one on compression and one on exhaust. This type of system is referred to as a:
a. Throw away system
b Waste spark system
c. Double up system
d. All of the above
e. None of the above

9. The purpose of the Hall-effect switch or sensor is to signal each time the ________ reaches TDC.
a. Cylinder
b. Crankshaft
c. Camshaft
d. Piston
e. Spark plug

10. The sensor that has a shuttle blade passing in front of a magnet is called the:
a. Hall-effect sensor
b. Cooling sensor
c. Charging sensor
d. Throttle speed sensor
e. Airflow sensor

11. The sensor that is able to pick up small, high-frequency vibrations in the engine is called a/an:
a. Hall-effect sensor
b. Cooling sensor
c. Detonation sensor
d. Throttle speed sensor
e. Airflow sensor

The following questions are similar in format to ASE (Automotive Service Excellence) test questions.

12. *Technician A* says that when the coil fires on a DIS system, it actually fires two cylinders at the same time. *Technician B* says that when the coil fires on a DIS system, both cylinders are on the compression stroke at the same time. Who is correct?
a. A only c. Both A and B
b. B only d. Neither A nor B

13. *Technician A* says that the Hall-effect pickup is not used on electronic or computerized ignition systems. *Technician B* says the Hall-effect pickup is used as a sensing device on the PIP. Who is correct?
a. A only c. Both A and B
b. B only d. Neither A nor B

14. *Technician A* says that cracks that occur in a distributor cap may cause a misfire. *Technician B* says that only voltage tests can be made on the distributor and coil. Who is correct?
a. A only c. Both A and B
b. B only d. Neither A nor B

15. *Technician A* says that the Hall-effect pickup is a device that uses voltage to produce a digital signal for the EEC IV. *Technician B* says that the Hall-effect pickup uses a magnet. Who is correct?
a. A only
b. B only
c. Both A and B
d. Neither A nor B

16. When checking an engine, it is suspected that the spark plug wires may be damaged. *Technician A* says that the spark plug wires can be checked by measuring the voltage going through them. *Technician B* says that spark plug wires can be checked for resistance. Who is correct?
a. A only
b. B only
c. Both A and B
d. Neither A nor B

17. The crankshaft position sensor is suspected of being damaged or broken. *Technician A* says to check the voltage output of the sensor. *Technician B* says to check the resistance of the sensor. Who is correct?
a. A only
b. B only
c. Both A and B
d. Neither A nor B

Essay

18. What is the purpose of a camshaft position sensor?

19. Explain how a detonation sensor works.

20. What is the waste spark system of ignition firing?

21. On electronic ignition (EI) systems, why is there no need to time the engine?

22. What is a reluctor and why is it used?

23. State the purpose of electronic spark timing.

Short Answer

24. The ________ sensor is able to provide a digital signal within the 360 degrees of crankshaft rotation.

25. The sensor used to provide a digital signal for timing and engine speed is called the ________ position sensor.

26. The knock sensor can also be called a(n) ________ sensor.

27. An open circuit on the ignition coil primary will show ________ ohms when tested with an ohmmeter.

Applied Academics

TITLE: Diagnostic Thought Process

ACADEMIC SKILLS AREA: Language Arts

NATEF Automobile Technician Tasks:

The service technician adapts a reading strategy for all written materials, for example, customer notes, service manuals, shop manuals, technical bulletins, and computer/data feed readouts, and so on, that will help identify solutions for engine performance problems.

CONCEPT:

For the contemporary automotive service technician, the thought process for diagnosing any problem is very complex. To help understand this process look at the accompanying diagram. For any valid diagnosis to occur, the service technician must first think about and define the type of problem that has occurred in the automobile. After the problem has been clearly defined, the service technician thinks about the information available in various service manuals, computer databases, and so on. Such information can provide various schematics and troubleshooting trees to help solve the problem. If all else fails, the service technician can also call the vehicle manufacturer and ask for technical assistance. When diagnosing an automobile problem, the contemporary service technician must be able to call on all of these resources.

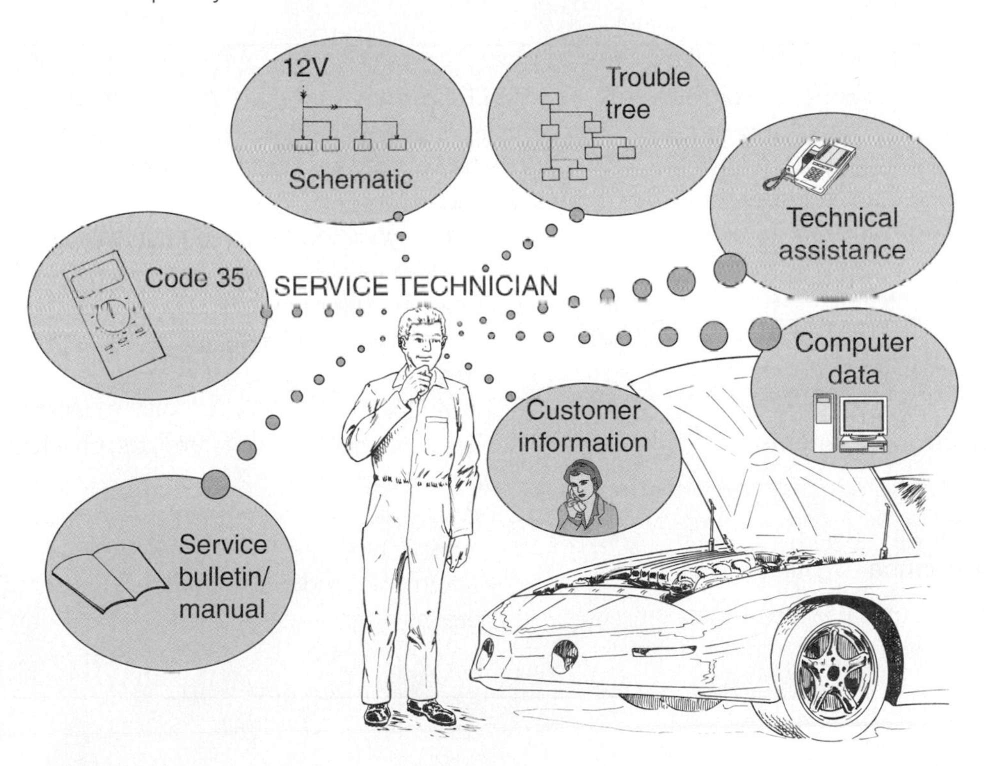

APPLICATION:

Why is it important to first clearly define the problem in an automobile? Can you identify why customer information is important in the process of diagnosing a problem? At what point do you think a part or component needs to be replaced when diagnosing an automobile problem?

CHAPTER 34

Charging Systems

OBJECTIVES

After studying this chapter, you should be able to:

- Identify the purpose of the charging system.
- Analyze the principles of converting the mechanical energy of the engine to electrical energy for charging.
- Define the parts and operation of dc, or direct current, alternators.
- State the operation of dc regulation on an alternator.
- Identify how three-phase voltages from the alternator are rectified.
- State the operation of solid-state electronic and computerized regulation systems.
- Identify the purpose of high-output alternators.
- Identify basic problems, their diagnosis, and service procedures on the charging system.

Certification Connection

ASE Connection: The information in this chapter can help you prepare for the National Institute for Automotive Service Excellence (ASE) certification tests. The tests and content areas most closely related to this chapter are:

Test A1—Engine Repair

- **Content Area**—Fuel, Electrical, Ignition, and Exhaust System Inspection and Service

Test A6—Electrical/Electronic Systems

- **Content Area**—Charging System Diagnosis and Repair

Test A8—Engine Performance

- **Content Area**—Engine Electrical Systems Diagnosis and Repair (Charging System)

NATEF Connection: Much of the information in this chapter is related to the NATEF tasks. The NATEF tasks and priority numbers most closely related to this chapter are:

1. Perform charging system output test; determine necessary action. P-1
2. Diagnose charging system for the cause of undercharge, no-charge, and overcharge conditions. P-1
3. Inspect, adjust, or replace generator (alternator) drive belts, pulleys, and tensioners; check pulley and belt alignment. P-2
4. Remove, inspect, and install generator (alternator). P-1
5. Perform charging circuit voltage drop tests; determine necessary action. P-1

INTRODUCTION

The purpose of this chapter is to study the charging system. This includes the study of basic generators, alternators, regulation, and troubleshooting of the charging system.

34.1 GENERATOR PRINCIPLES

PURPOSE OF THE CHARGING SYSTEM

All automobiles use a charging system to convert the mechanical energy of the engine to electrical energy. The electrical energy is used to operate the vehicle during normal driving conditions and to charge the battery. Each time electrical energy is removed from the battery, the battery must be recharged. If it is not recharged, the battery will eventually become completely discharged. The charging system is used to prevent this.

In addition to charging, it is important to be able to regulate the amount of charge. If too little charge is produced, the battery will have a low charge. If the charging system puts too much back into the battery, the battery may be overcharged and damaged. The regulation of the charge ensures that the exact amount of electricity needed is being generated.

PRODUCING ELECTRICITY

A generator or alternator uses three things to change mechanical energy to electrical energy. These are a magnetic field, a conductor or wire, and movement between the two. This principle is called electromagnetic induction. ***Figure 34–1*** shows a simple generator used to produce electricity. When wire is wound around a metal core, a magnetic field is produced. The magnetic lines of force are moving from the north pole to the south pole. When a wire or conductor is moved to cut the magnetic field, a voltage is produced within the wire. This voltage is used to push electrons back to the battery for charging. It is also used to supply current to the rest of the vehicle during normal operation.

Generated voltage (and, therefore, available current) can be increased by increasing any one of the three components in the generator. If the magnetic field is increased, the voltage will increase. If the amount of copper wire is increased, the voltage will increase. If the speed of motion is increased, the voltage will increase.

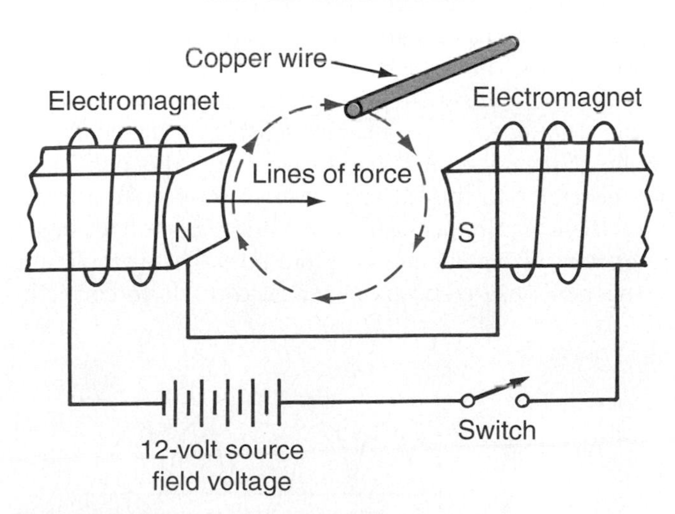

FIGURE 34–1 A simple generator is made of a set of poles, copper wire, and movement between the poles and the wire, in which the wire cuts the magnetic lines of force. All three are needed to produce a voltage.

DIFFERENCE BETWEEN GENERATORS AND ALTERNATORS

For years, the automobile charging system used a generator to charge the battery. Today's automobiles use an alternator. The difference between a generator and alternator is the method of physical construction. A generator has stationary poles, and the wire moves across the field. An alternator has moving poles (magnetic field) and a stationary wire. Although both produce electricity, the alternator is much more efficient.

There are also several minor differences that will be discussed in this chapter. Generators use brushes and a **commutator** to remove the produced voltage from the rotating windings. The alternator uses diodes to **rectify** the voltage to dc.

DC GENERATORS

All vehicles manufactured today use an alternator to produce the necessary voltage. However, by studying the basic parts of a generator, the purpose of many of the alternator's parts can be more easily understood. ***Figure 34–2*** shows a typical dc generator. It is driven by a pulley and belt from the crankshaft.

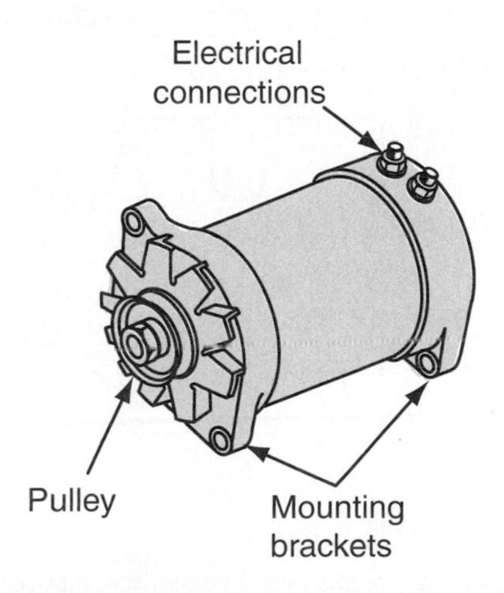

FIGURE 34–2 A dc generator is used on older vehicles to produce a voltage for charging the battery.

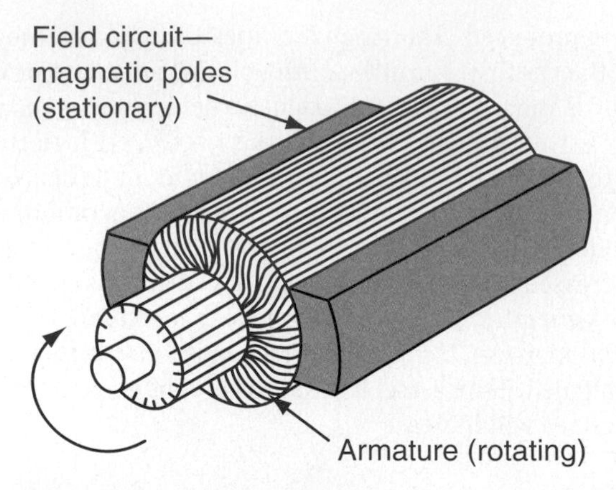

FIGURE 34-3 A dc generator has several parts, including the poles (which are stationary) and the armature (copper wire), which rotates inside the poles.

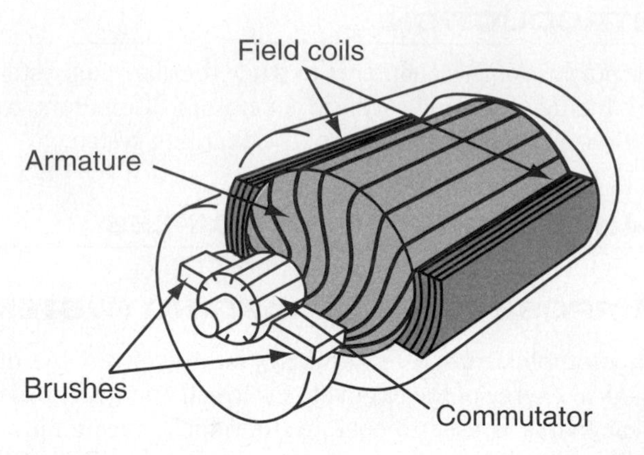

FIGURE 34-5 The generator is made of an armature, field coils, the commutator, and the brushes.

Dc Generator Parts The dc generator is made of several parts. These are shown in ***Figure 34–3***. The poles are stationary and produce the magnetic field needed to generate electricity. This circuit is called the **field circuit**. Rotating inside the field poles is an **armature**. Voltage is produced in the armature as it rotates within the magnetic field.

Ac and Dc Voltages The voltage produced within the armature is ac, or alternating current. Ac voltage is continually changing back and forth as shown in ***Figure 34–4***. The current flows first in a positive direction, then in a negative direction. Dc voltage is continuous, or flows in one direction. Dc voltage always flows in a positive or negative direction.

Rectifying Ac Voltage The ac voltage within the armature must be converted to dc voltage for the battery and other automotive circuits. A commutator and a set of brushes are used to accomplish this. The commutator is also called a split-ring commutator. Ac voltage is changed to dc voltage when it goes from the commutator to the brushes. The brushes are made of carbon and rub against the commutator. The voltage at the brushes is dc voltage that can be used to charge the battery or operate the vehicle circuits. ***Figure 34–5*** shows the armature, the commutator, and the brushes.

A graph of the voltage rectified by the split-ring commutator is shown in ***Figure 34–6***. Note that although the voltage is pulsating from zero to a maximum point, it continually goes in the same direction. It does not reverse direction as ac voltages do.

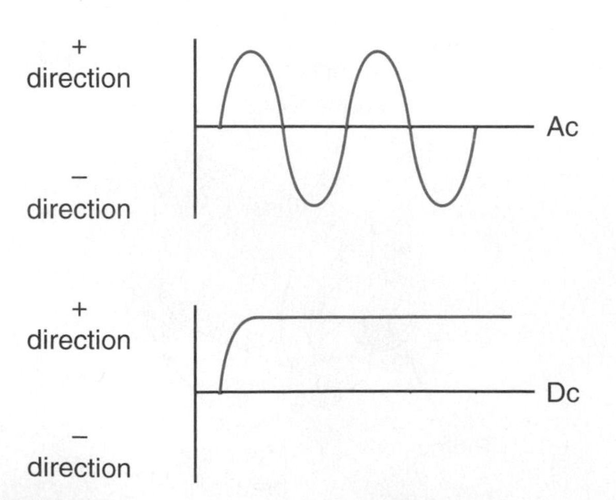

FIGURE 34-4 Ac voltages alternate from a positive to a negative direction and back. Dc voltages always have current flow in only one direction. The illustrations show the differences between ac and dc voltages.

Purpose of Regulating the Generator The output of the dc generator must be regulated and controlled. This is done to protect the generator, the battery, and the electrical circuits in the automobile from too much voltage or current. It is very important to keep the voltage output of the generator slightly above 12 (about 13.5) volts. If the voltage is too high, the light bulbs can burn out and the battery can overheat. Too low a voltage causes poor charging of the battery and incorrect operation of the electrical circuits. The voltage regulator is used for this control.

A dc generator system uses several components for regulation. The voltage and amperage must be regulated. In addition, it is important that the electricity flows only from the generator to the battery, never in reverse. To accomplish these requirements, the generator uses a voltage regulator, current regulator, and cutout relay. These regulating components use coils and electrical contacts to correctly

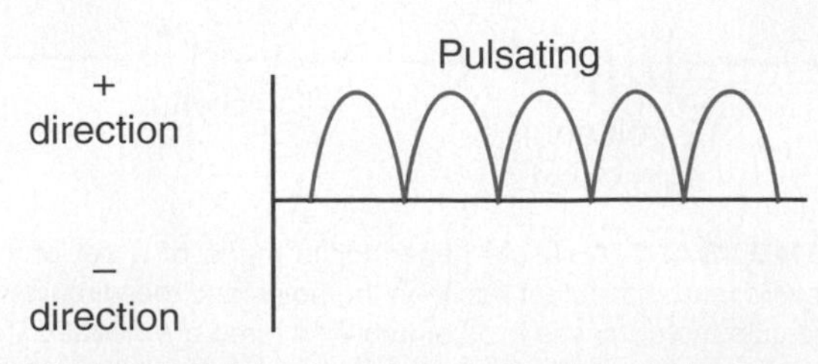

FIGURE 34-6 The dc voltage that is changed by the split-ring commutator is pulsating dc.

control the generator system. Since these components are no longer used on today's vehicles, they are not covered in this textbook.

34.2 ALTERNATORS

Today's vehicles use an alternator to charge the battery and operate the electrical circuits. The alternator is much more efficient than a generator. Alternators are much smaller, lighter in weight, and produce more current than generators. The alternator has a set of rotating poles and a stationary set of windings. In addition, there is no split-ring commutator. Solid-state diodes are used to convert ac to dc voltages. The alternator is made of a stator, rotor, and slip-ring and brush assembly. Many modern (late-model) alternators have the regulator built into the housings as a complete unit.

STATOR

The **stator** (***Figure 34–7***) is made of a circular, **laminated** iron core. There are three separate windings wound on the core. The windings are arranged so that a separate ac voltage **waveform** is induced in each winding as the rotating magnetic field cuts across the wires.

ROTOR

The **rotor** (***Figure 34–8***) is made of a coil of wire wound around an iron core on a shaft. When current is passed through the windings, the assembly becomes an electromagnet. One side is a north pole, and the other side is a south pole. Iron claws are placed on both ends. Each projection has the same polarity as the ends of the coil. When the claws are meshed together from each side, pairs of north and south poles are formed around the circumference of the rotor (***Figure 34–9***). The number of north and south poles is determined by the manufacturer. Four, six, and seven sets of poles are common in alternators today. The poles are designed to rotate inside the stator, producing the voltage needed to charge the battery.

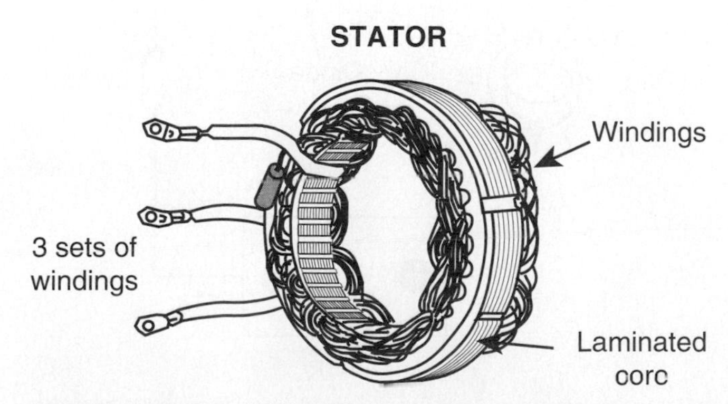

FIGURE 34–7 Alternators use a stator, or stationary set of windings, to generate voltage.

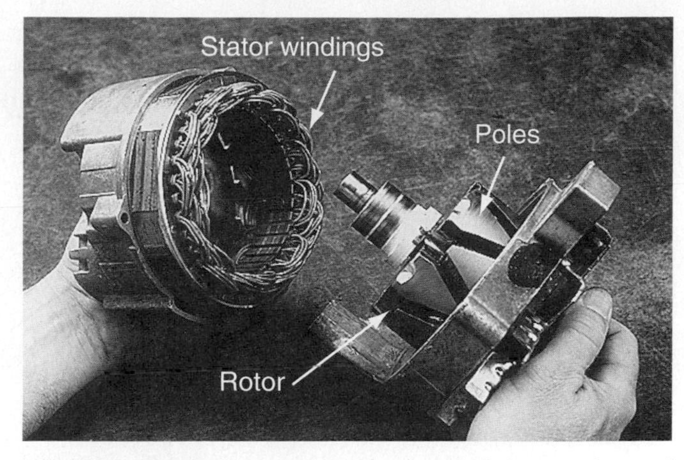

FIGURE 34–8 The rotor is made of a coil of wire wound around a rotating iron core.

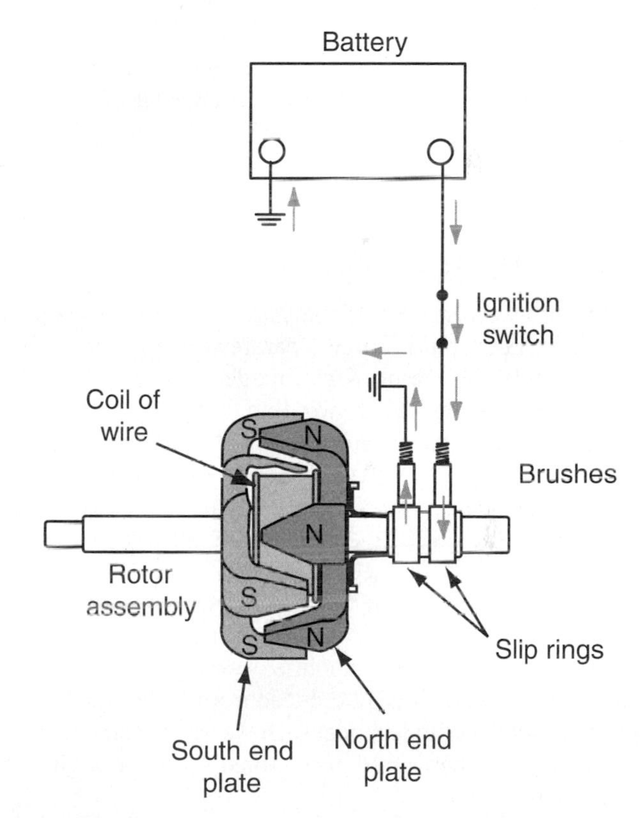

FIGURE 34–9 The coil of wire on the rotor has two end plates meshed together. Seven sets of north and south poles produce the magnetic fields.

SLIP-RING AND BRUSH ASSEMBLY

The ends of the rotor coil are connected to **slip rings** that are mounted on the shaft. Current is supplied from the battery through the brushes and slip rings to energize the rotor field windings. This produces the magnetic field needed for making the north and south poles in the alternator.

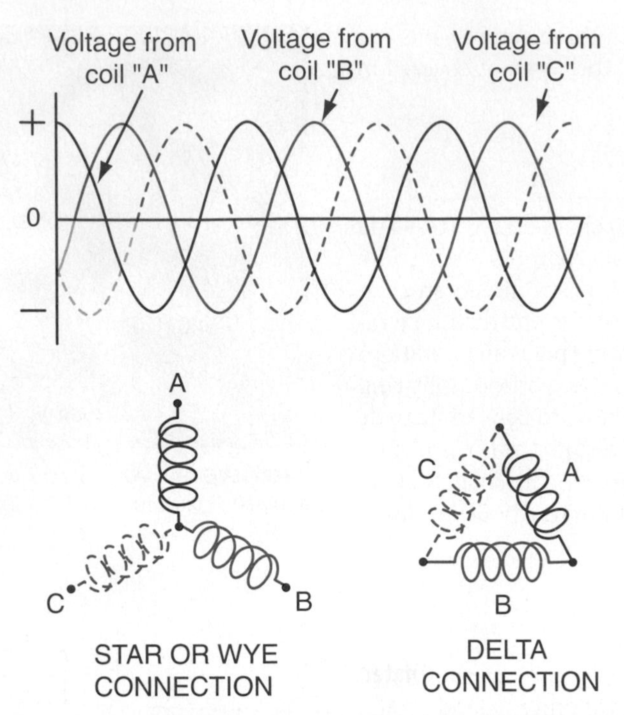

FIGURE 34–10 The alternator produces a three-phase voltage, as shown in this graph. Each phase is produced by a separate set of stator windings. Also, the stator can be connected as a star, or wye, connection or as a delta connection.

THREE-PHASE VOLTAGE

As the rotor revolves inside the stator windings, voltage is produced in the stator. Three separate windings are spaced evenly inside the stator. The voltage produced is called **three-phase** voltage. Each winding produces a separate voltage. This is shown in ***Figure 34–10***. The voltages, just like the windings, are 120 degrees apart from each other.

The three-phase stator windings can be connected in one of two ways. The star connection is also called the wye connection. The delta connection is used for higher-output alternators. Each winding produces a separate voltage. In ***Figure 34–10***, the solid black voltage line is produced from coil A. The solid color voltage line is produced from coil B. The dashed voltage line is produced from coil C. Each voltage is separate and independent. However, it is still ac voltage and must be converted, or rectified, to dc for use in the automobile.

RECTIFYING THREE-PHASE AC VOLTAGE

Six diodes are used to convert, or rectify, the three-phase voltage in the alternator stator windings to dc. Three diodes are positive, and three diodes are negative. Diodes allow current to flow in only one direction. ***Figure 34–11*** shows a diode in a circuit. Note that there is a battery, a switch, and a light as the load. The diode only allows electricity to flow from the positive (P) to the negative (N) materials inside the diode. If the voltage on the battery were reversed through the diode, current would not flow from N to P, because a diode only allows current flow in one direction.

Figure 34–12 shows the electrical symbol for a diode and the direction that current is allowed to flow. A cutaway of a diode is also shown for reference. This type of diode is used in many alternators to rectify the voltage from ac to dc.

If a single-phase ac voltage were applied to a circuit with a diode, the ac voltage would be rectified to dc voltage. Re-

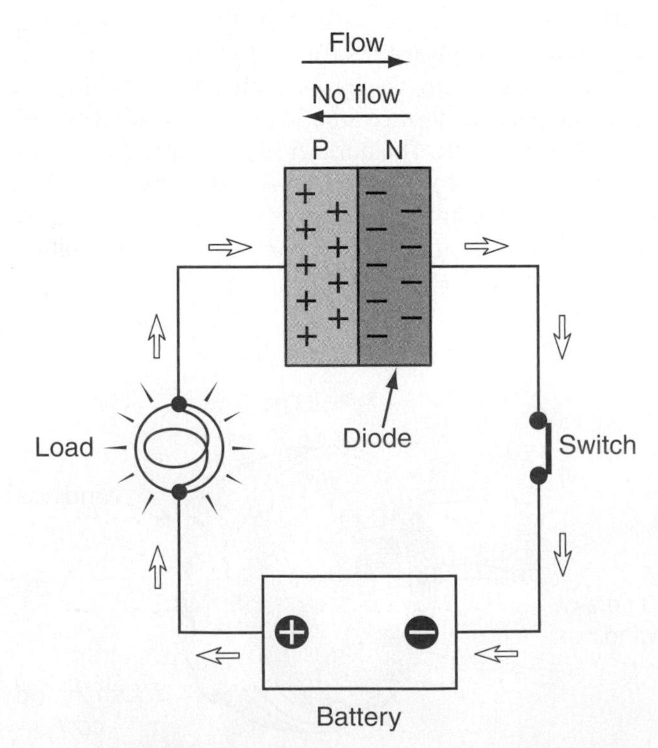

FIGURE 34–11 Diodes are used to convert ac voltage to dc for use in an automobile. Diodes allow current to flow in only one direction.

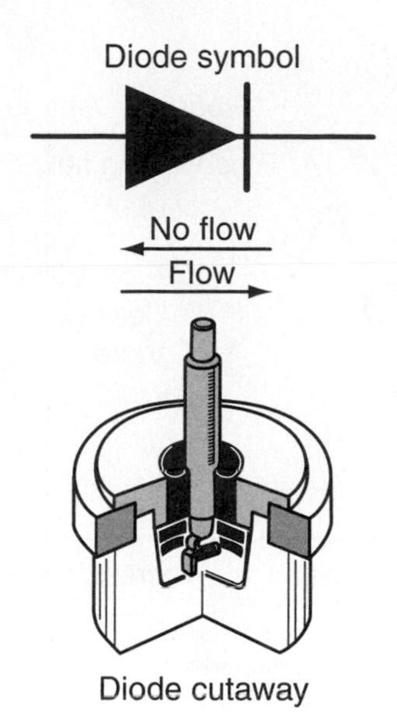

FIGURE 34-12 A diode symbol and a cutaway view of a diode.

ferring to ***Figure 34–13***, there is an ac voltage being produced from the alternator windings on the left side. A diode has been placed in the circuit as well as a load (battery). Because the diode allows electricity to flow in one direction, only the top half of the ac voltage goes into the battery for charging. The bottom half is stopped by the diode. The result is called pulsating dc voltage. One problem with this type of rectifier is that the bottom half of the voltage has been lost. Rectifiers are designed to capture the bottom half of the voltage. When three phases need to be rectified, six diodes are used.

Figure 34–14 shows a circuit that uses six diodes to fully rectify three-phase ac voltage to dc. Coils A, B, and C produce an ac voltage. Each voltage is sent to the rectifier. No matter where the current enters the rectifier, it is sent out on top to the positive side of the battery. The current at the battery has been fully rectified to dc.

In ***Figure 34–14***, the first half-cycle produced from coil A passes current from point 1 to diode C to the positive side of the battery. The current returns to coil A through diode D and coil B. On the negative half-cycle, coil A passes current from point 2, through coil B, through diode A, to the positive side of the battery. The current returns from the battery through diode F to point 1 on coil A.The voltage produced by coil A is now rectified to dc. It goes through the battery in the same direction.

Coil B works the same way. The first half-cycle produced by coil B passes current from point 4 to diode A to the positive side of the battery. The current returns to coil B through diode E and coil C. On the negative half-cycle, coil B passes current from point 3, through coil C, to diode B, to the positive side of the battery. The current returns from

PARTS LOCATOR

*Refer to photo #65 in the front of the textbook to see what a **charging system rectifier** looks like.*

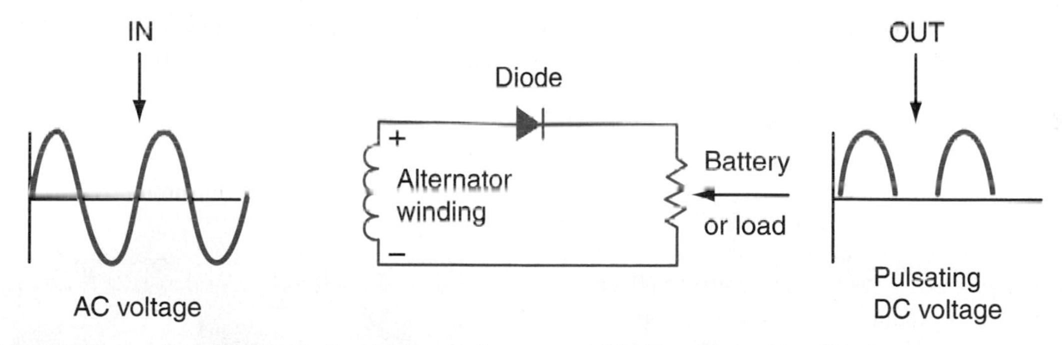

FIGURE 34-13 When a diode is placed in the circuit, ac voltages are rectified to pulsating dc voltages.

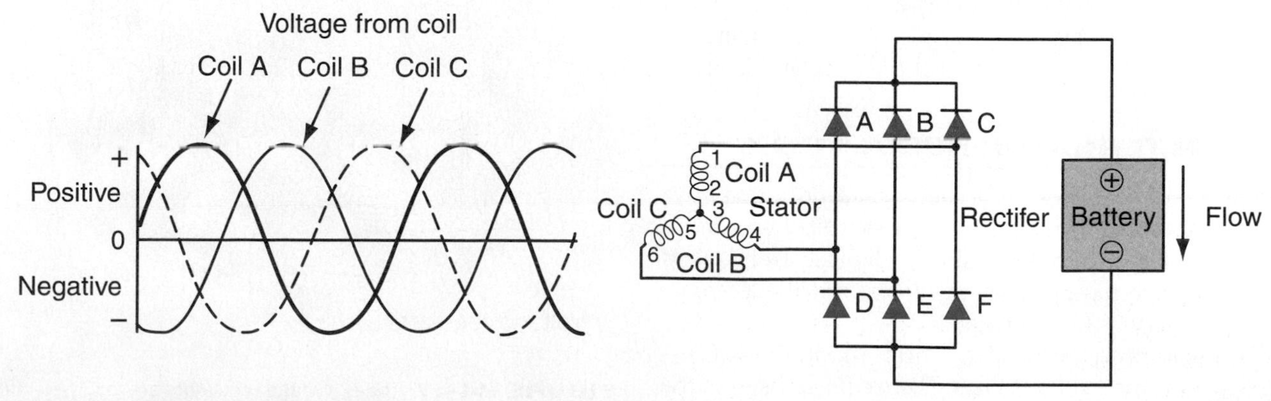

FIGURE 34-14 Six diodes are used to convert three-phase ac to pulsating dc.

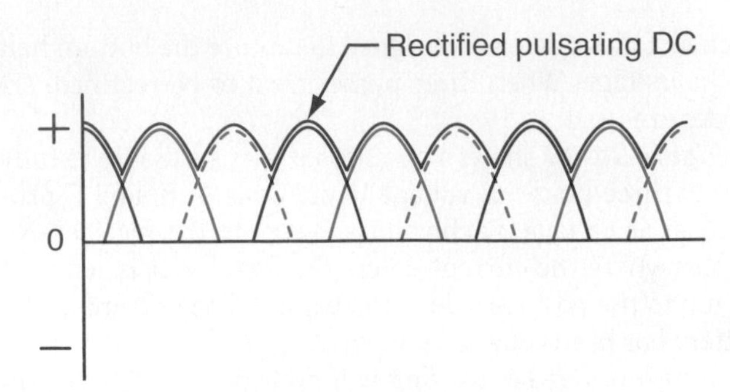

FIGURE 34–15 The voltage that is rectified on an alternator is called pulsating dc. It is made by rectifying all three phases of ac to dc in the rectifier.

the battery through diode D to point 4 on coil B. The voltage produced by coil B is now rectified to dc.

The flow through the rectifier can be followed the same way for coil C. It is also rectified to dc. ***Figure 34–15*** shows the resultant dc voltage wave that is produced by rectifying three-phase ac voltage. It is called pulsating dc and is considered **full-wave rectification**.

HIGHER-OUTPUT ALTERNATORS

Alternators are generally rated by the amount of amperage that they can produce. Voltage generally remains at 12 to 14 volts. As more and more electrical components are placed on vehicles, amperage demand continues to increase. Thus, higher-output alternators are being used on many vehicles today. For example, average alternator output ratings ranged between 40 and 60 amperes several years ago. Today some vehicles require alternators that can produce 80 to 130 amps. In order to produce the higher amperage, the alternator rotor is made larger. This allows the magnetic field to be stronger. In addition, more stator windings are put into the alternator, thus increasing its amperage output.

There are certain charging systems that allow the voltage to increase as well. For example, on some vehicles that use heated windshields, 50 to 70 volts are needed to heat the windshield. In order to produce this voltage, all other circuits are operated directly from the battery. Then, the voltage of the alternator can be increased by the voltage regulator or computer to the necessary voltage. Each system is different, so it is necessary to refer to the manufacturer's service manual to diagnose and service problems.

34.3 ALTERNATOR REGULATION

Many types of regulators have been used on automotive charging systems. As more solid-state circuitry was developed, the regulation systems slowly changed from relays and coils to electronics. Today regulators are all electronic and require very little maintenance.

All regulators are designed to control the amount of current sent into the field windings. ***Figure 34–16*** shows the relationship between rotor speed, field current, and the regulated voltage. The solid line represents the regulated voltage from the alternator. The dotted line represents the field current. As the rotor speed increases, the field current is reduced to keep the regulated voltage controlled. In electronic regulators, the switching of the field current is controlled by transistors turning off and on.

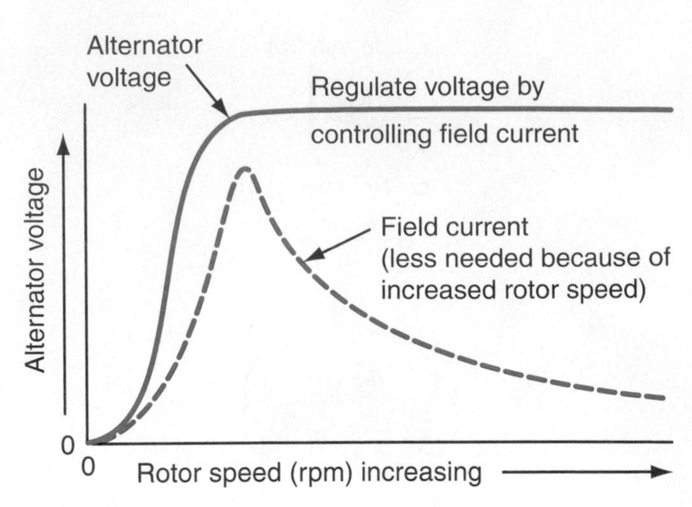

FIGURE 34–16 A graph showing the relationship between rotor speed, current in the field windings, and output voltage. As the rotor speed increases, field current is reduced to keep the regulated voltage controlled.

ELECTRONIC REGULATORS

As solid-state circuitry was developed, voltage regulators eventually became totally electronic. ***Figure 34–17*** shows an alternator with the voltage regulator built directly into

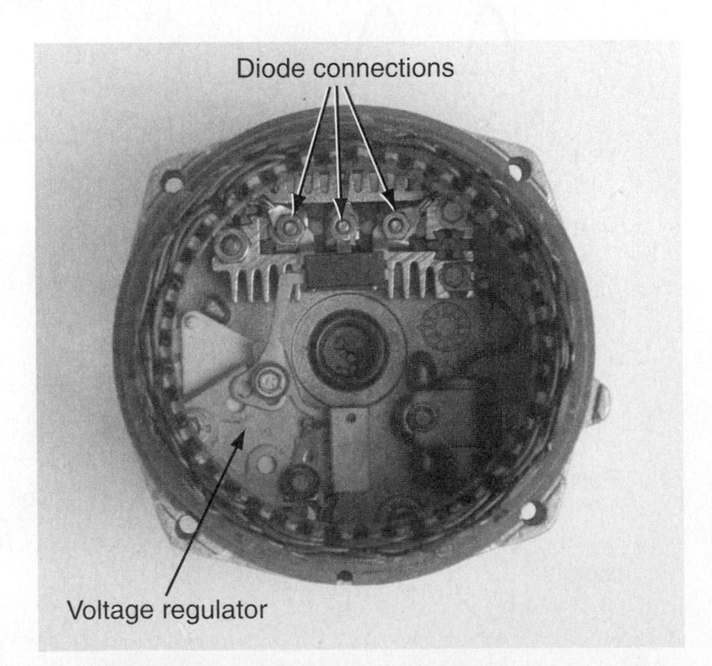

FIGURE 34–17 Today, voltage regulators are totally electronic. Transistors turn the field windings off and on for control.

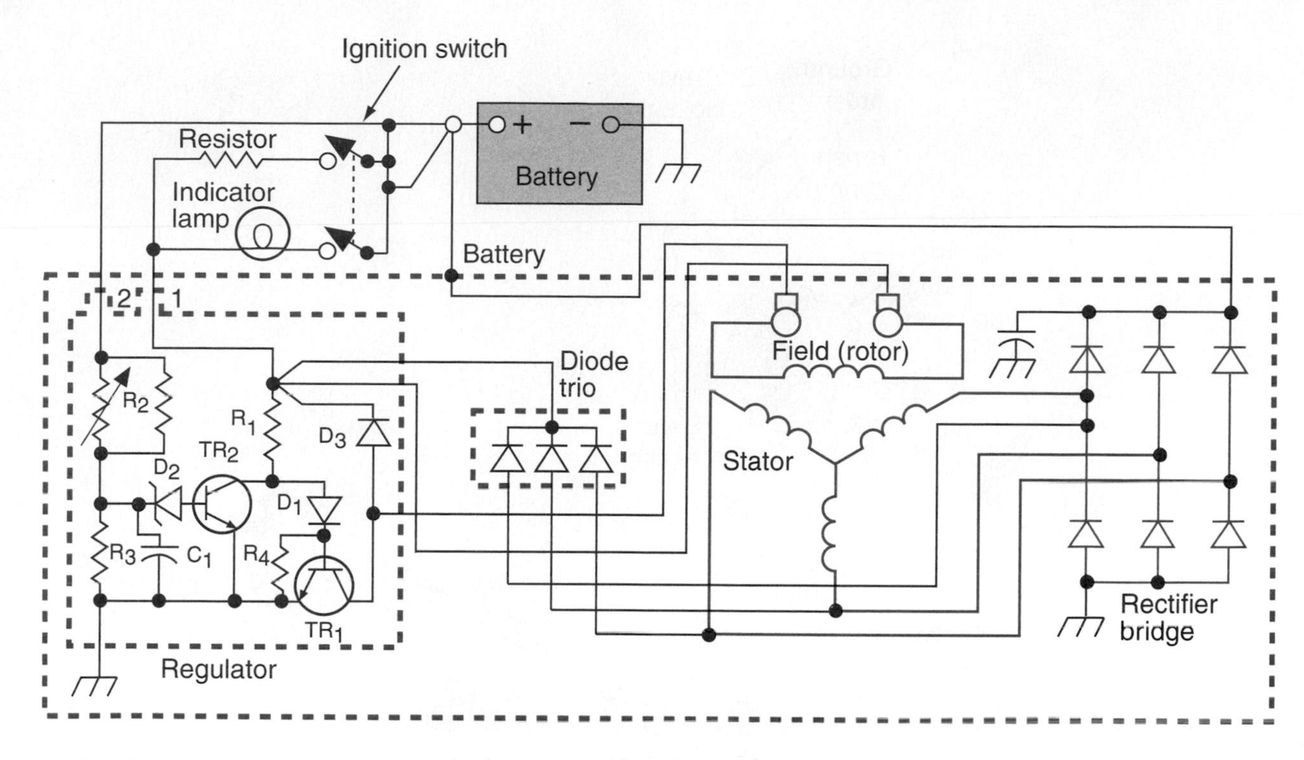

FIGURE 34–18 A complete circuit of an alternator, including the field windings, electronic voltage regulator, and diodes for rectification.

the alternator. ***Figure 34–18*** shows a typical charging circuit for the alternator. The action of the regulator is similar to relay types of years ago, except transistors are used to turn the field windings on and off.

There are several components in this circuit. On the right side of the circuit, six diodes are used to convert the ac voltages from the stator to dc. The dc voltage is then sent out of the regulator (dotted line) and to the positive terminal on the battery. This voltage and current is then used to charge the battery with the correct amount of charge. As mentioned, the amount of charge or regulation of the voltage is controlled by the amperage being fed into the field, or rotor. The amperage to the field comes out of the regulator from the TR_1 transistor. This transistor turns off and on to control the field amperage. In fact, depending on the conditions, this transistor can turn off and on from 10 to 7,000 times per second. This quick cycling provides very accurate control of the field current through the rotor.

A **zener diode** is used in the circuit to tell the regulator exactly when the alternator is producing too much voltage. The zener diode is identified as D_2. When the zener diode turns on, transistor TR_1 turns off, shutting off the current to the field windings. The reverse happens when the voltage drops to an acceptable level. Although the action is more complicated, the concept of controlling the field current is the same.

There are an additional three diodes, called the diode trio, also included in this circuit. These diodes rectify ac current from the stator to dc current that is applied to the field windings.

In operation, when the ignition switch is in the Run position and the engine is not running, battery voltage and current lights the indicator lamp, flows to the common point above R_1, and eventually energizes the field coils. When the engine starts to run, the stators produce voltage. Depending on the load on the battery, the zener diode will tell the regulator circuit how fast to turn off and on. The end result is regulation of the voltage and current by controlling the current going to the field (rotor) windings. There are variations of this regulation circuit, however, the basic concepts remain the same for each.

CUSTOMER CONCERN:
Overcharged Battery

The battery on a customer's car seems to always overcharge. The battery seems very hot after operation, and the customer has replaced the battery once. What could be the problem?

SOLUTION: The problem is most likely tied to the voltage regulator. The voltage regulator is used to monitor and control the amount of current and voltage going back into the battery during charging. If the voltage regulator is overcharging, it should be replaced. This should solve the problem.

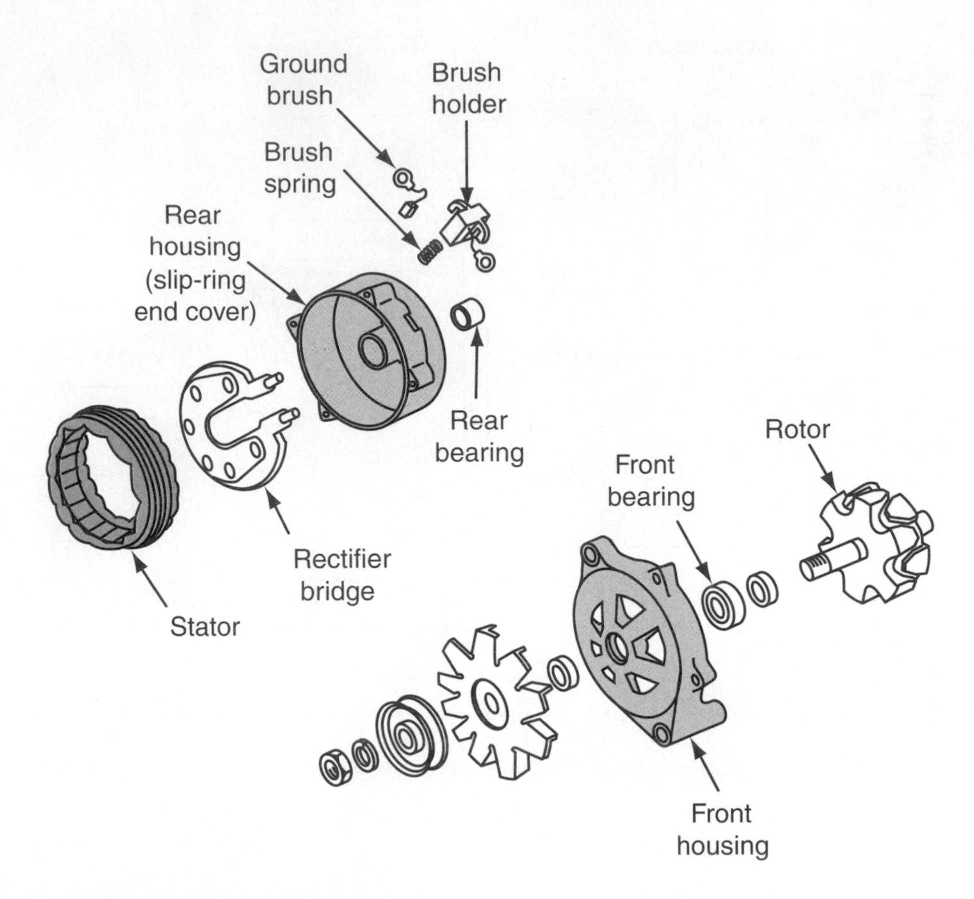

FIGURE 34-19 An exploded diagram of an alternator.

ALTERNATOR CONSTRUCTION

Alternators today are designed as complete integral units. ***Figure 34–19*** is an exploded diagram of an alternator. Note the position of each part previously discussed in this chapter. Housings are used on both ends for support. They also include the necessary bearings for supporting the rotor. The rotor has six fingers on the North Pole and six fingers on the South Pole. The rectifier bridge and stator fit inside the slip-ring end cover. Note the location of the brush spring and brush holder attached to the rear housing.

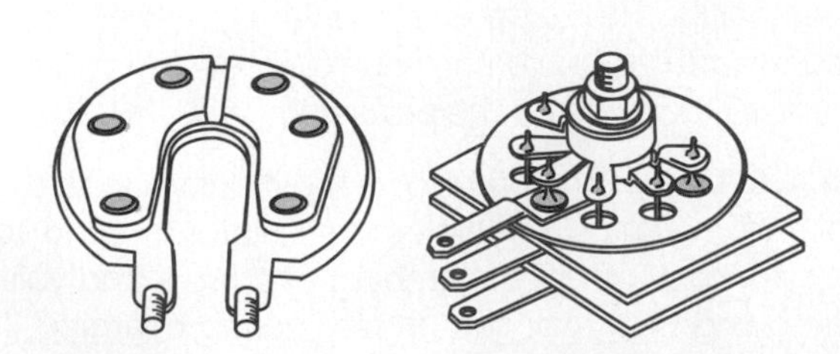

FIGURE 34-20 Rectifier packs are designed in many styles. Usually the diodes are grouped together on a plate for correct operation and polarity.

RECTIFIER ASSEMBLY

Figure 34–20 shows just two of many types of rectifier assemblies. Rectifiers with diodes are constructed on plates of metal. One plate holds one polarity of diodes, and another plate holds the opposite polarity of diodes. Depending on the manufacturer and the circuit, different diode combinations are used.

34.4 COMPUTER-CONTROLLED CHARGING CIRCUITS

COMPUTERIZED REGULATION

Beginning in 1985, some manufacturers began using a computer to regulate the charging circuit. The computer is now able to control the alternator's output. The computer essentially replaced the voltage regulator. In order to regulate the alternator output, the internal circuitry for regulation is much the same. ***Figure 34–21*** shows one vehicle manufacturer's system used to control the alternator. In this system, the PCM (powertrain control module) monitors various sensors. Various components on the engine will require or demand more output from the alternator. Many of these components have sensors that tell the computer that more output is needed by the alternator. In addi-

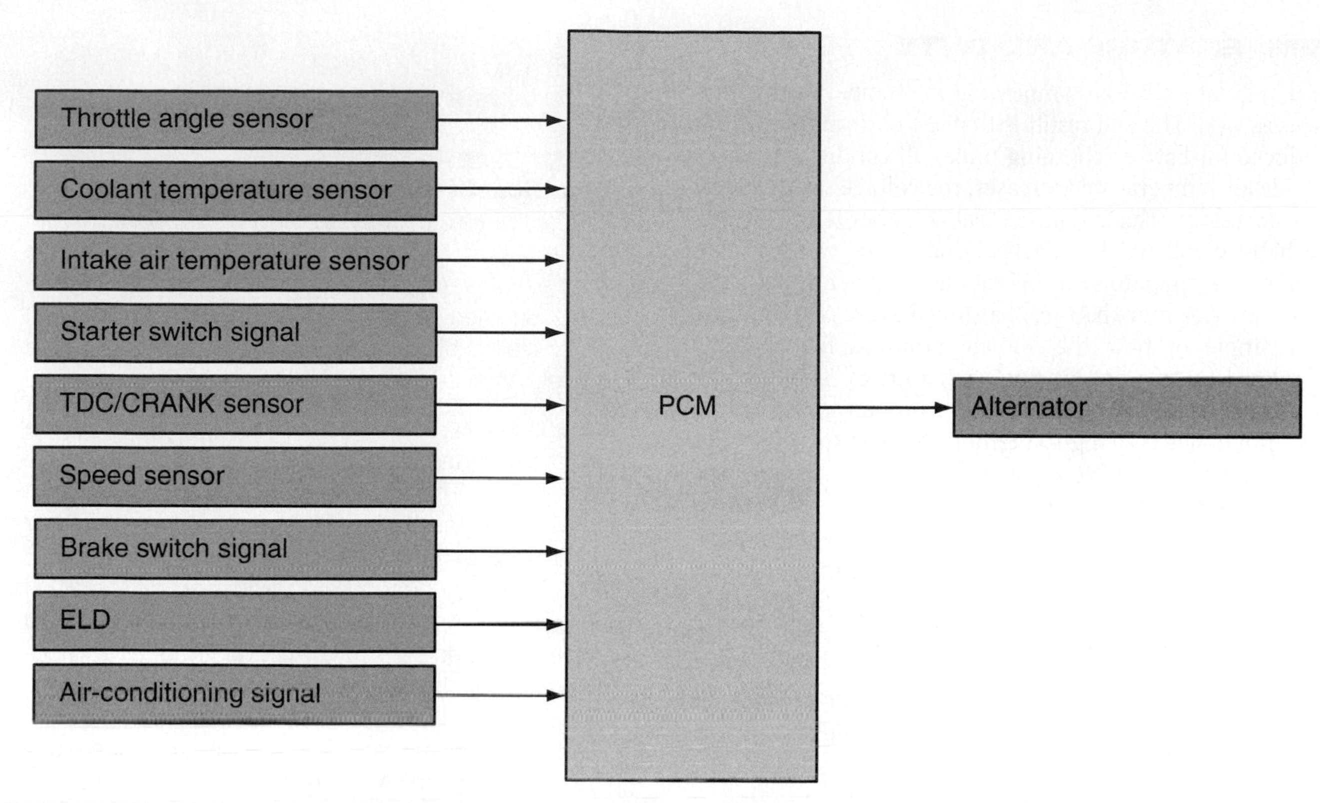

FIGURE 34–21 Many vehicles today use the on-board computer to regulate the output of the alternator. Various sensors are fed into the PCM to control the output of the alternator. (Courtesy of American Honda Motor Co., Inc.)

tion, there is also an **electric load detector**, or ELD). This sensor is used to detect the exact amount of electrical load required by the various components on the engine. The result is improved fuel economy and more efficient operation of the charging system.

VOLTAGE AND DUTY CYCLES

On computerized charging systems, the regulator and control module switch the field voltage on and off at a fixed frequency of about 400 cycles per second. Voltage control is obtained by varying the on-off time of the field current. Thus, at low speeds, the field may be turned on 90% of the time and off 10% of the time. This is called the duty cycle. This results in a relatively high average field current, which, when combined with the alternator speed, produces the desired system voltage.

As alternator speed increases, less field current may be needed to generate the desired system voltage. The duty cycle changes to reduce the average field current. For example, at high engine speeds, the regulator may be on only 10% of the time and off 90% of the time. This duty cycle will change as operating factors change. The result is just the right amount of field current to generate the required system voltage.

Figure 34–22 illustrates how the duty cycle can change. Basically the field current can be controlled anywhere in between the two conditions. When the field current is on for 90% of the time, the average field current is very high. When the field current is on for only 10% of the time, the average field current is low.

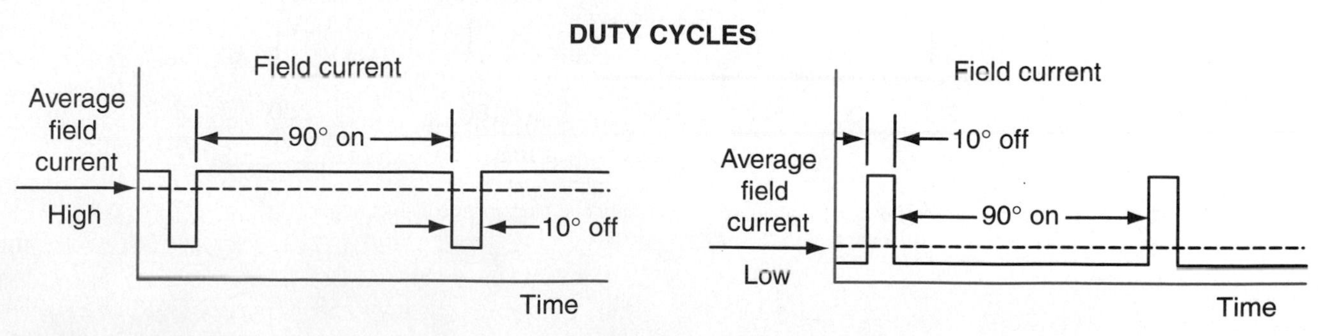

FIGURE 34–22 The on-off duty cycle of the alternator can be changed to meet the exact charging need.

TEMPERATURE AND DUTY CYCLES

Most regulators today compensate for temperature differences as well. The end result is that the optimum voltage is produced for battery charging under all conditions. As the outside air temperature increases, the voltage setting needs to be decreased. This is done so that at higher temperatures, the battery will not overcharge. Under cold weather conditions, the regulator will operate at a higher voltage setting to provide more battery charging. ***Figure 34–23*** shows an example of how the voltage required for charging changes with different outside temperatures. The horizontal axis represents different operating temperatures in Fahrenheit, with the highest temperature on the right side. Colder temperatures are shown on the left side. The vertical axis represents the battery charging voltage. The curved line shows average voltages required at different temperatures. For example, if the outside air temperature is 77 degrees Fahrenheit, about 14.4 volts are needed for charging. If the outside temperature drops to –31 degrees Fahrenheit, then 14.85 volts are needed. The graph also shows much higher temperatures because on a high 90-degree day, the temperatures on black tar roads may be as high as 160 degrees or more.

Car Clinic:
LOOSE WIRE CONNECTION

CUSTOMER CONCERN:
Electrical System Totally Fails

The entire electrical system on a new vehicle with 500 miles on it began to lose voltage. While driving on the highway at night, the owner noticed that the lights seemed to become dimmer. The radio and other electrical components also seemed to lose power. What could be the problem?

SOLUTION: Since the vehicle is new, one of the electrical connections on the alternator may have come loose. The first and most obvious connection to check is the main wire between the alternator and the battery (red). If this connection is broken or faulty, any electrical current produced by the alternator will not get to the battery or the other electrical systems. When the battery is drained of all its electrical energy, the electrical circuits also lose power. Check the BAT wire to make sure that it has a solid electrical connection. Also check the alternator drive belt tension. A loose belt will slip and reduce alternator output.

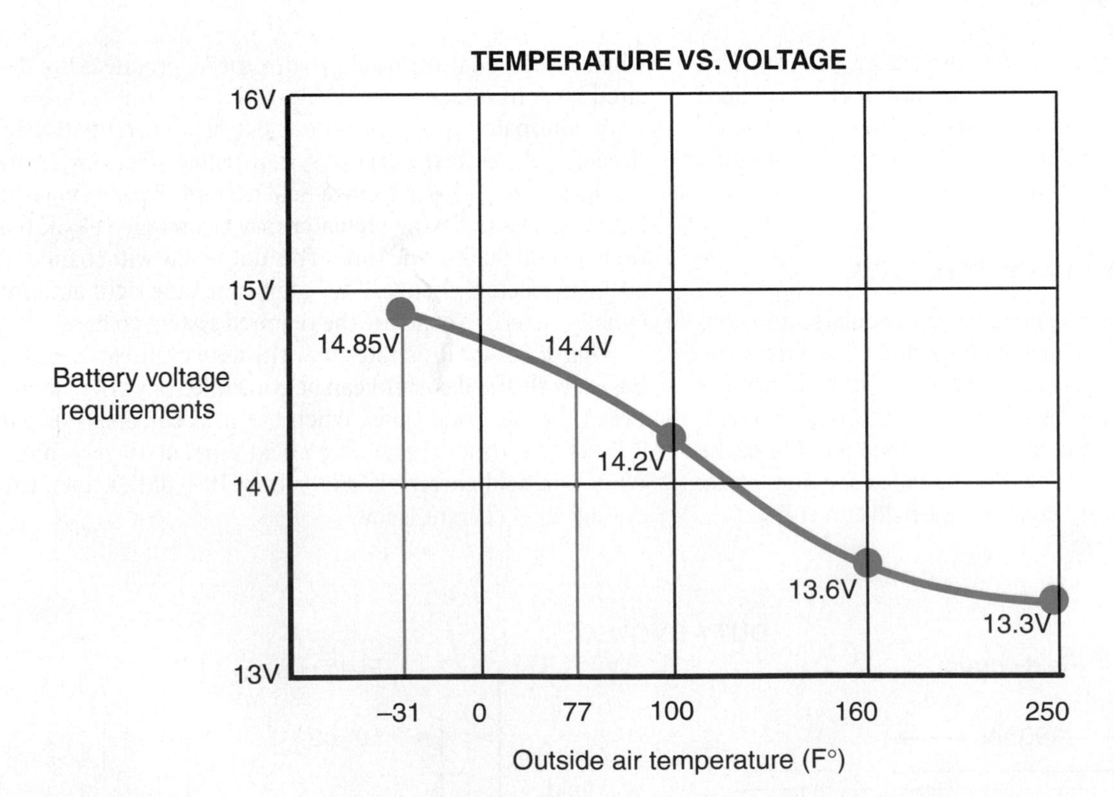

FIGURE 34–23 As the outside air temperature changes, the necessary charging voltage also changes.

CHARGING SYSTEM ELECTRICAL CIRCUIT

Today's charging systems are integrated into the powertrain control module (PCM) and other circuits as well. ***Figure 34–24*** shows a complete charging system. The following notes about the circuit will help to better explain computerized charging systems:

- Note the shaded boxes. The bottom one represents the charging unit and voltage regulator and the windings and diodes.
- The upper-left shaded box represents the underhood fuse relay box. Throughout the circuit, references are made to photos. This particular vehicle manufacturer includes photos to help the service technician find and locate the specific parts of the circuit.
- Note that the PCM not only uses all of the standard sensors stated previously as inputs, but also uses the ELD unit input and the alternator output signal. As mentioned earlier, the ELD unit detects the required electrical load. The alternator output is a voltage signal needed to determine what the alternator output is at any one particular time.
- Depending on all of the stated inputs, including those from the electrical load detector unit and all of the PCM inputs (including inlet air temperature), the voltage regulator controls the field windings and the duty cycle to produce the exact voltage needed at a particular time.
- The charging system indicator light is controlled by the output of the voltage regulator.

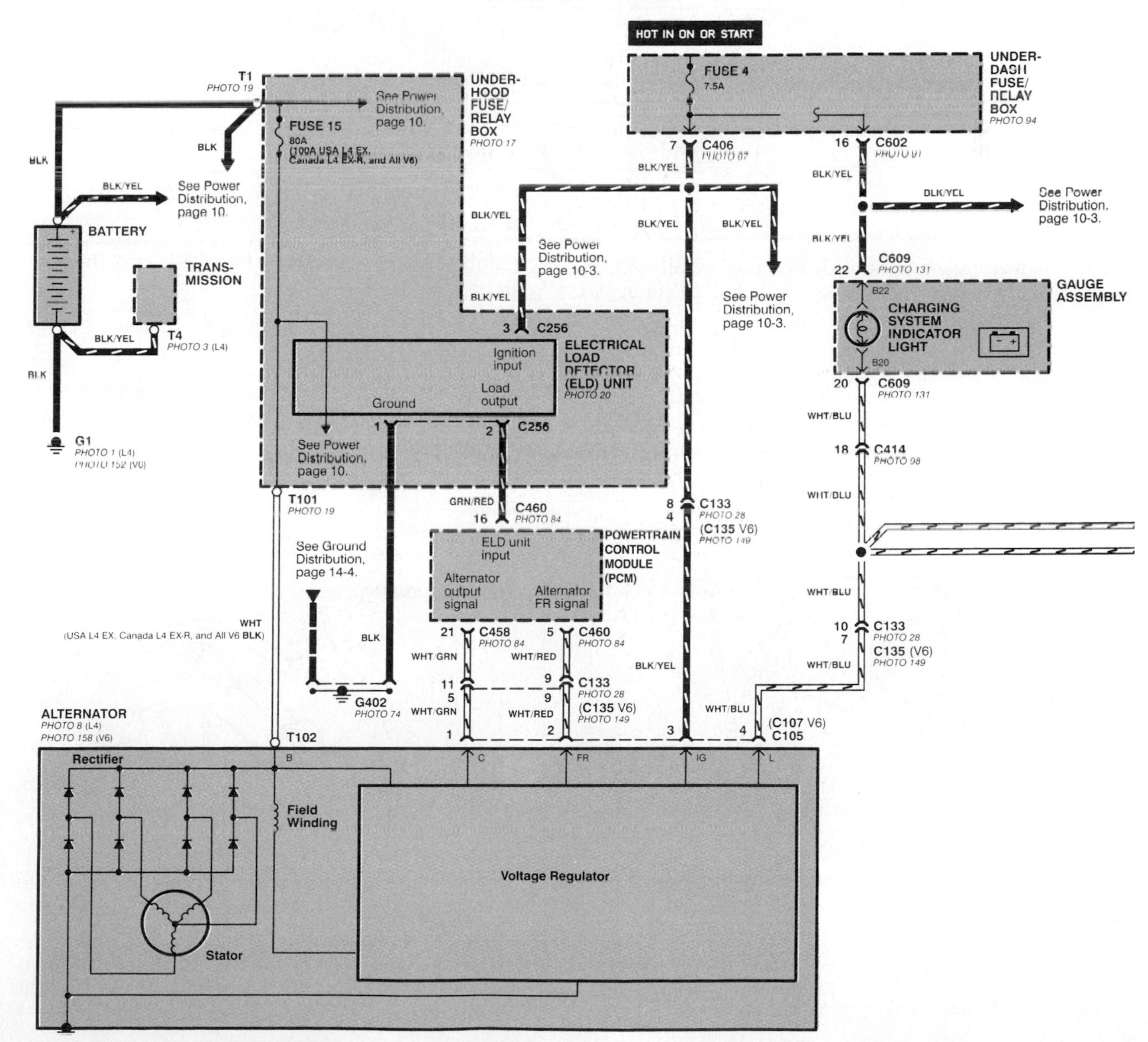

FIGURE 34–24 The charging system can be controlled by the PCM. (Courtesy of American Honda Motor Co., Inc.)

Problems, Diagnosis, and Service

Safety Precautions

1. When working on the charging system, remove rings, bracelets, and other jewelry to eliminate the possibility of getting an electrical shock.
2. When the engine is running, always be careful that tools, wires, and other loose equipment do not get caught in the alternator fan.
3. When handling an alternator, be careful not to cut your hands on the sharp edges of the parts.
4. When tightening the belt on the alternator, always watch your hands so they do not get pinched or caught near the belt.
5. Never work on the alternator while the engine is running.
6. When working on the alternator, always make sure the battery is disconnected.
7. Always wear OSHA-approved safety glasses when working on the alternator.

PROBLEM: Incorrect Alternator Belt Tension

A squealing sound comes from the front of the engine during acceleration.

DIAGNOSIS

Often the squealing sound is made because the alternator belts are not tightened to the correct tension, or the belts have been tightened too much. If they are too tight, there will be excess pressure on the alternator bearings. The bearings may then have to be replaced.

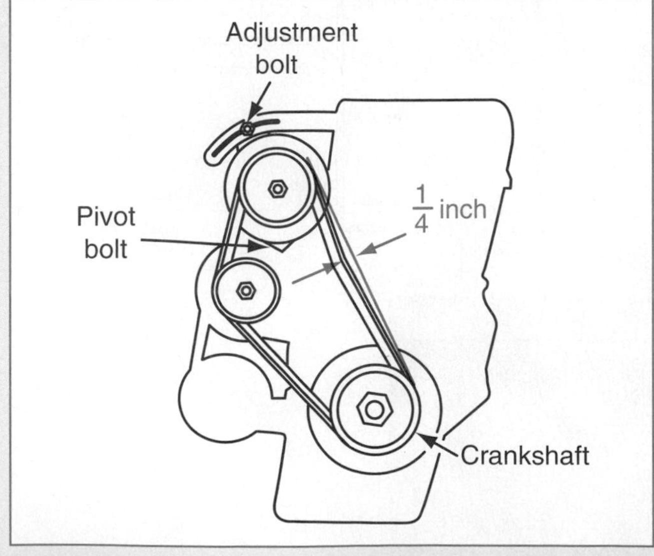

FIGURE 34-25 When checking the charging circuit, always make sure the belt tension on the alternator is correct.

SERVICE

Figure 34–25 shows one manufacturer's procedure for checking belt tension.

1. Loosen the alternator adjustment bolt and the pivot bolt so the alternator can move.
2. Lightly pry back the alternator, tightening the belt tension.
3. Make sure the belt is loose enough to produce the specific deflection between the alternator and the crankshaft.
4. On certain vehicles, there is a 3/8-inch square opening in the belt tension bracket. This square opening, shown in ***Figure 34–26***, allows the use of either a torque wrench or drive socket. This approach eliminates stress on the components.)

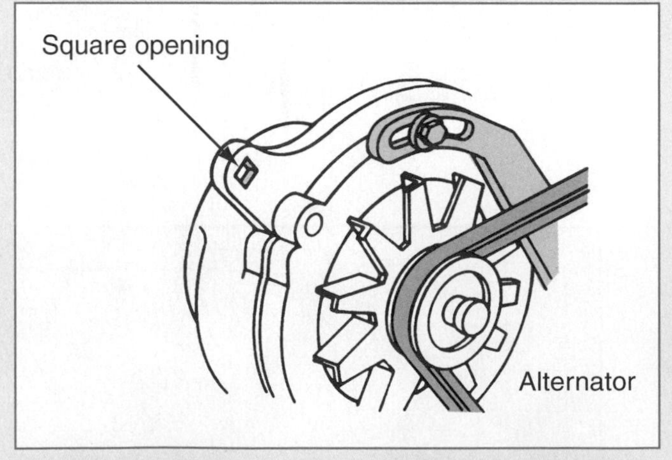

FIGURE 34-26 The alternator can be moved for tightening the belt tension by placing a ratchet in the square opening on the alternator housing.

5. Holding the alternator at this position, tighten the adjustment bolt and the pivot bolt to the manufacturer's torque specifications.

PROBLEM: Defective Diodes

An alternator on an engine seems to rattle during operation.

DIAGNOSIS

A rattling noise from the alternator can be caused by several things. Generally, the problem is caused by worn or dirty bearings, loose mounting bolts, a loose drive pulley, a defective stator, or defective diodes. If any of these is the cause, it will be necessary to remove the alternator and perform various procedures.

SERVICE

1. Disconnect the battery ground cable.
2. Remove the alternator wires and external connections.
3. Remove the pivot and adjusting bolts from the alternator.
4. Carefully remove the alternator from the engine.
5. Check the alternator for bad bearings. Bad bearings can be determined by rotating the armature and feeling for rough turning. Bearings should be replaced if there is any evidence of wear.
6. Refer to the manufacturer's procedure to disassemble the alternator. ***Figure 34–27*** shows a completely disassembled alternator. Use the following general procedure to disassemble the alternator:

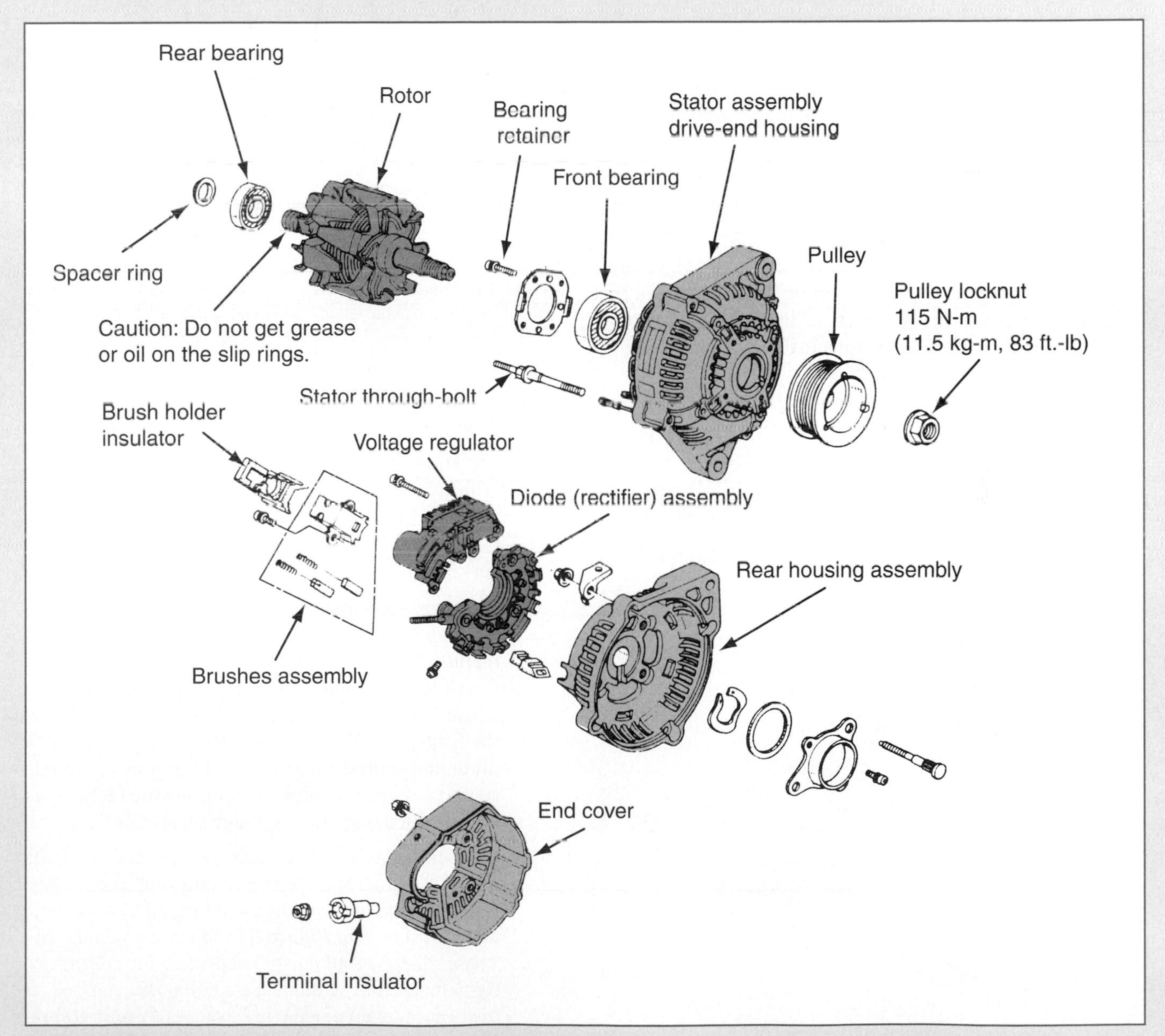

FIGURE 34–27 When disassembling an alternator, follow the manufacturer's recommended procedure. (Courtesy of American Honda Motor Co., Inc.)

a. Carefully remove the bolts that hold the alternator to the brackets on the engine.

b. Scribe marks on the alternator front and rear housings to retain the proper position during reassembly. On some alternators, the regulator can be removed from the rear of the assembly by removing several small screws.

c. Remove the through-screws that hold the front and rear housing together.

d. Pry the assembly apart, making sure to keep the stator assembly with the rear half.

e. Remove the center rotor assembly from the end housing.

f. Remove the regulator and other parts of the assembly as necessary.

g. Reverse this procedure for reassembly.

7. The rotor can be checked for shorts or opens with an ohmmeter as shown in ***Figure 34–28***. The ohmmeter is used to check for shorts between the rotor shaft, frame poles, and windings.

8. A stator can be checked for shorts or opens as shown in ***Figure 34–29***. The ohmmeter is used to check each winding.

9. Inspect the brushes for excessive wear, damage, or corrosion. If there is any doubt about their condition, replace the brushes. Also measure the brushes for wear and compare with the manufacturer's specifications. If the brushes have worn down below the service limit, replace them.

10. Bad diodes can also cause a rattling noise during operation. This is because if one diode has shorted out, one-third of the alternator stator will not be charging or under load. Thus, in one revolution of the alternator, two-thirds of the revolution is under load while one-third is not. This causes a vibration to occur in the rotor shaft, which sounds like bad bearings or a rattling noise.

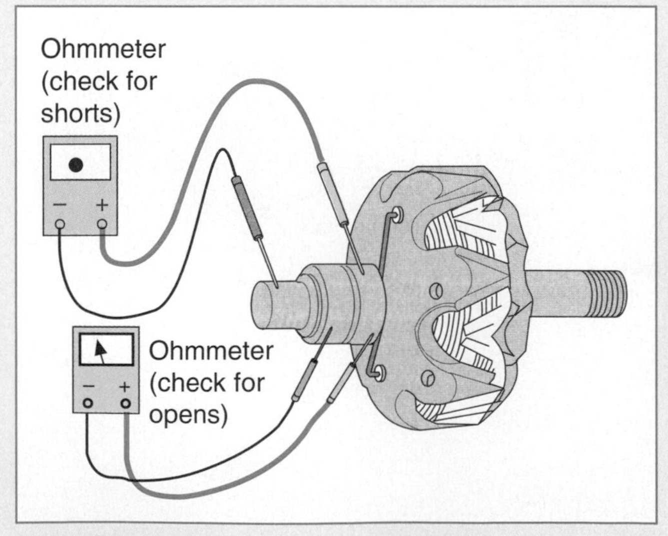

FIGURE 34–28 The rotor can be checked by using an ohmmeter. Grounds, shorts, and opens can be checked.

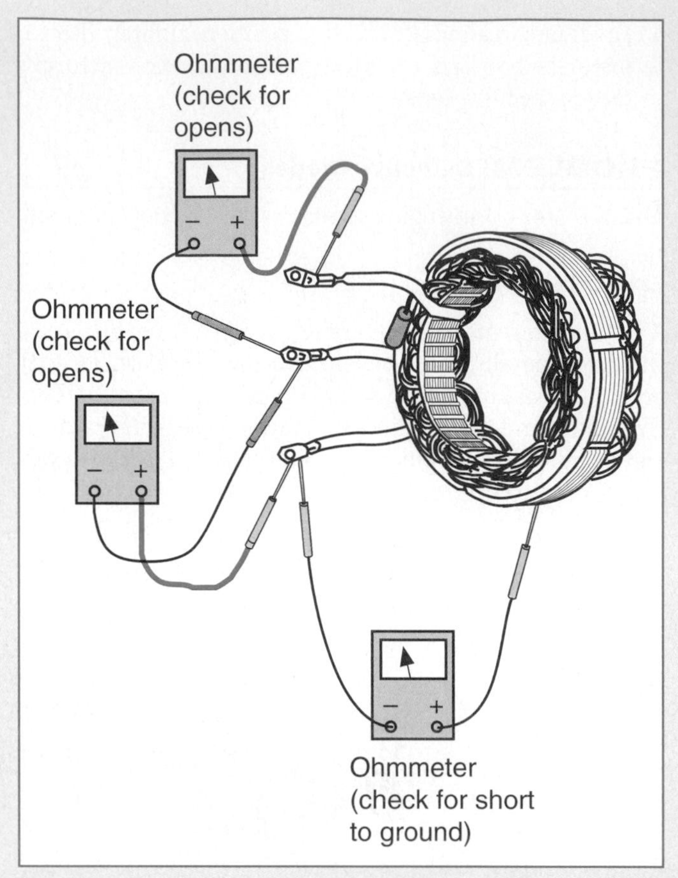

FIGURE 34–29 The stator can be checked using an ohmmeter. All three windings should be checked for grounds and opens.

a. An ohmmeter is used to check for defective diodes. When placed across the diode, high resistances should be noted in one direction and low resistances should be noted in the reverse direction. If both readings are very low or very high, the diode is defective.

b. On some vehicles, six diodes are used. On others, eight may be used. ***Figure 34–30*** shows how one manufacturer recommends checking these diodes. Check for continuity in each direction, between the B and P terminals (of each diode pair), and between the E (ground) and P terminals (of each diode pair). All diodes should have continuity in only one direction. If any of the diodes fails, replace the rectifier assembly. Diodes are not available separately.

c. On some regulators, a diode trio is also used. Its purpose is to rectify current entering the voltage regulator and field or rotor windings. The schematic diagram shown in ***Figure 34–18*** shows a diode trio. These diodes should also be checked for continuity. Remember that a diode can either be shorted or opened. An open diode will have a high (infinite) resistance in both directions. A shorted diode will have a low (zero) resistance in both directions.

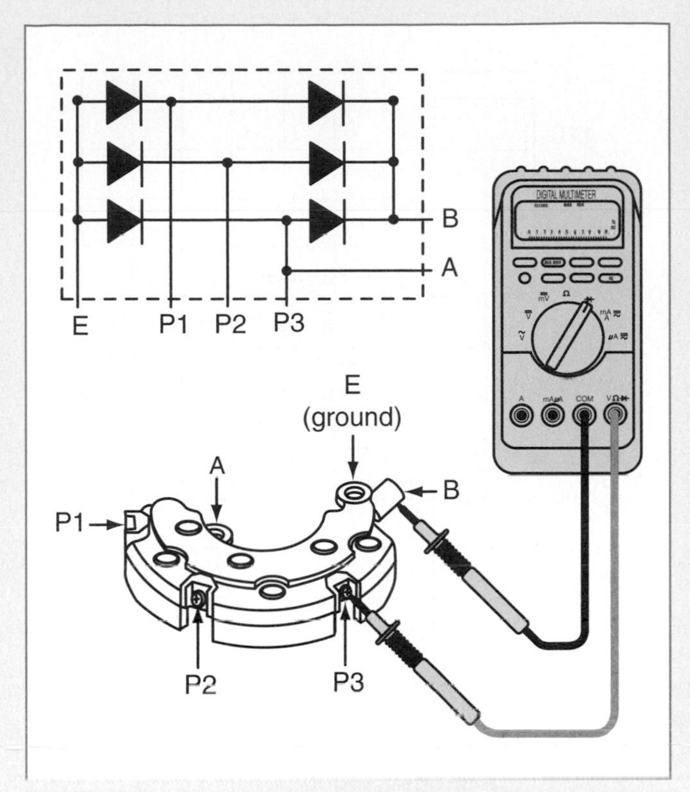

FIGURE 34-30 When checking diodes, always check each diode for continuity. If any one diode is either shorted or opened, replace the entire rectifier assembly.

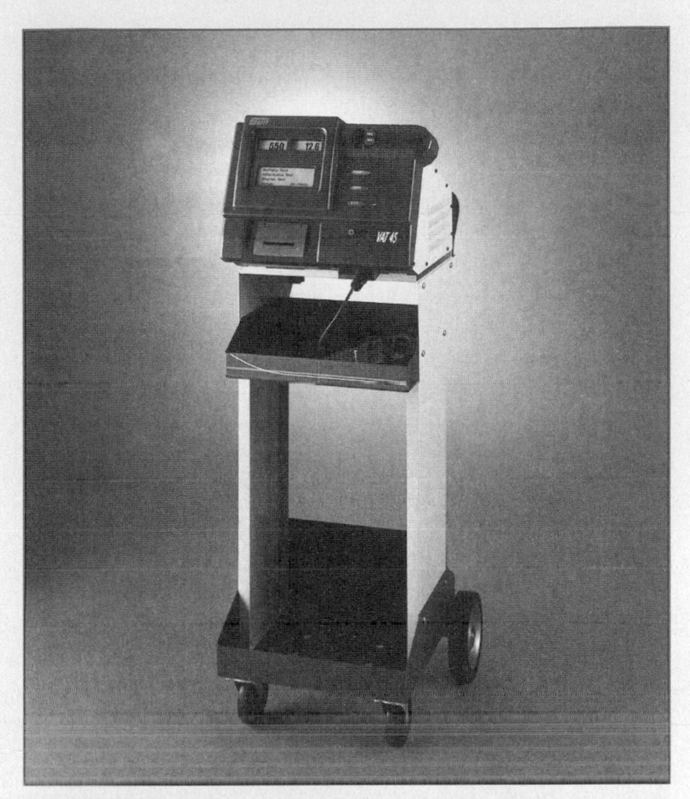

FIGURE 34-31 Charging systems can be checked using this voltage-amperage tester (VAT). Both voltage and amperage can be checked to determine the charging system's condition. (Courtesy of Snap-on Tools Corporation, www.snapon.com)

PROBLEM: Faulty Alternator or Regulator

The battery on a vehicle does not seem to be getting a full charge from the alternator. At times during operation, the headlights seem to dim.

DIAGNOSIS

Various tests can be run on the alternator when it is on the vehicle. Four of them are:

1. Voltmeter test of charging system
2. Alternator amperage output under maximum load using the VAT (voltage-amperage tester)
3. Regulator voltage test, used to check the regulator, using the VAT
4. Regulator by-pass test, used to short out the regulator to find the fault, using the VAT

The first test can be done using a voltmeter. Use the following general procedure to test the voltage output using a voltmeter only.

1. Run the engine at idle for about 15 minutes to get the engine warm.
2. Stop the engine, and place a voltmeter on the alternator. Clip the black lead to the negative side of the battery. Attach the positive side of the meter to the positive post on the battery.

CAUTION: *Keep wires and body parts away from any part of the engine while it is running.*

3. Start the engine and gradually increase the engine speed to about 1,800 rpm. Note the voltage reading on the meter scale.
4. If the meter pointer climbs steadily and comes to an abrupt halt within a range of 13.5 to 15.5 volts for a 12-volt system, the entire charging system is most likely operating correctly.
5. If the meter climbs but stops below the correct range of 13.5 to 15.5 volts or above it, or if the meter fluctuates and will not stop, the regulator is most likely defective.
6. If the meter does not climb to the correct range, the alternator is most likely defective.

The other tests are done with a VAT, shown in ***Figure 34-31***. Each manufacturer of these testing instruments has specific procedures for testing the charging system.

SERVICE

Use the following general procedure to test the charging system with a VAT.

1. First make sure the battery is in good shape.
2. ***Figure 34-32*** shows a typical hookup between the VAT, the battery, and the alternator. Connect as shown. Put selector switch on the "starting" position.

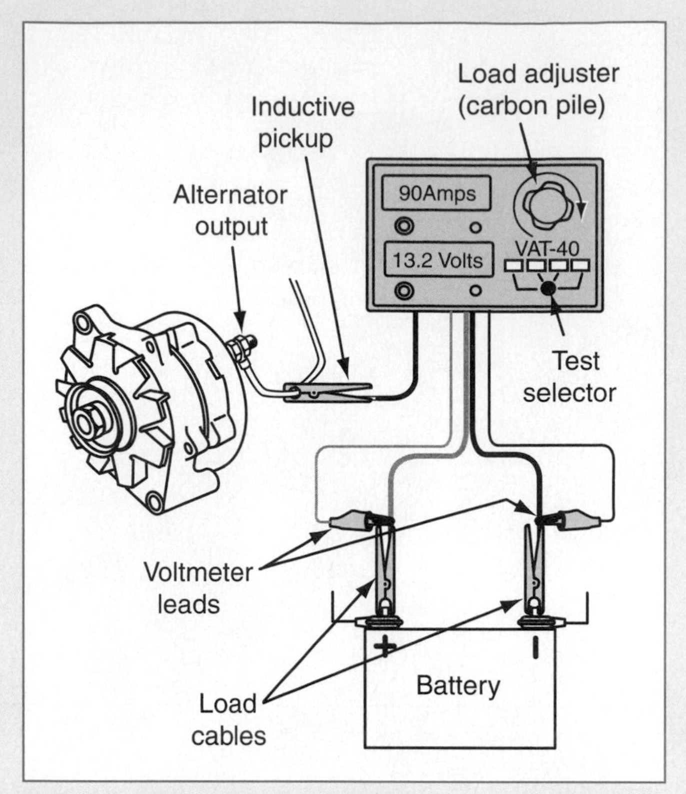

FIGURE 34–32 The VAT should be connected as shown to check the output of the charging system.

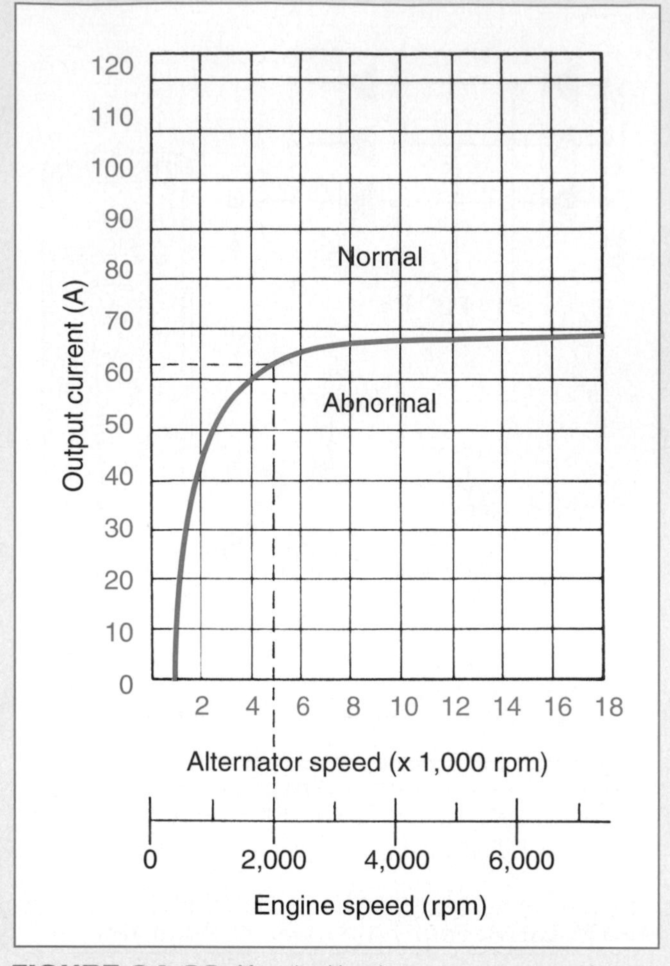

FIGURE 34–33 After checking the amperage output of the alternator, compare the readings with this graph to determine if the alternator is faulty. (Courtesy of American Honda Motor Co., Inc.)

3. Start the engine, and turn off all the accessories.
4. Move the selector switch to the "charging" position. Remove the inductive pickup, and zero the ammeter on the VAT. Now reconnect the inductive pickup to the alternator output wire.
5. Increase the engine speed to 2,000 rpm, and hold it at this rpm. Make sure the cooling fan is not on.
6. Now apply a "load" with the carbon pile, so the voltage drops back to not less than 12 volts.
7. At this point, check the maximum amperage reading and compare it with the chart in ***Figure 34–33***. You should subtract from 5 to 10 amperes from the maximum reading because of engine operation. For example, if the engine is running at 2,000 rpm, the alternator speed is about 5,000 rpm. At this alternator speed the alternator should produce about 63 amperes minimum (after subtracting 5 to 10 amperes). If the reading is below this number, the alternator needs to be replaced or rebuilt.
8. Check the regulator voltage test by turning the selector knob to the correct setting.
9. Run the engine again at 2,000 rpm, and check the voltage on the VAT. Generally, the voltage should be between 13.5 and 14.5 volts. This is the normal voltage output for a fully charged battery. If the voltage is steady and within this range, the regulator is OK. If the voltage is higher or lower than this, the voltage regulator should be replaced.
10. A regulator by-pass test can also be performed. This test is done by shorting out the regulator and operating the alternator. Direct battery voltage is used to excite the rotor field. Each manufacturer suggests different ways of shorting out the regulator. Some involve shorting a test tab to ground on the rear of the alternator. Others require the use of a jumper wire to connect the battery voltage to the field. Check the service manual for the correct procedure. Now perform the voltage and current test previously mentioned. If the charging voltage and current increase to normal levels with the regulator bypassed, the problem is most likely the regulator. If the voltage and current remain the same when the regulator is bypassed, then the problem is most likely the alternator.

Service Manual Connection

There are many important specifications to keep in mind when working with charging systems. To identify the specifications for your engine, you will need to know the VIN (vehicle identification number) on the vehicle, the type and year of the vehicle, and the type of engine. Although they may be titled differently, some of the more common charging system specifications (not all) found in service manuals are listed below. Note that these specifications are typical examples. Each vehicle and engine may have different specifications.

Common Specification	Typical Example
Alternator adjusting bolt torque	18 ft.-lb
Alternator belt tension (deflection)	0.33–0.47 in.
Alternator brush length	0.22 inch, lower limit
Alternator output amperage at 12 volts	65 amperes hot
Alternator voltage	12.5–14.5 volts

In addition, the service manual will give specific directions for various service and testing procedures. Some of the more common procedures include:

- Alternator belt adjustment
- Alternator overhaul
- Alternator regulator testing
- Alternator replacement
- Brush inspection
- Charge warning light test
- Checking output voltage of alternator
- Rectifier test
- Rotor slip-ring test
- Stator test

SUMMARY

The following statements will help to summarize this chapter:

- Before alternators, vehicles used dc generators to produce voltage to charge the battery.
- A dc generator system uses a series of relays to control the voltage and the amperage output.
- Three things are needed to produce electricity, including a magnetic field, electrical wires, and having the electrical wires move or cut across the magnetic field.
- If the magnetic field, number of wires, or speed of movement increase, the voltage output of the generator or alternator will increase.
- An alternator is much like a generator, but it is much more efficient.
- An alternator has stationary wires and the magnetic field moving on an armature.
- There are several sets of magnetic fields on an armature, with a common configuration of seven north and seven south poles.
- An alternator uses a slip-ring commutator to connect electricity to the rotating fields.
- An alternator produces a three-phase voltage as opposed to the single-phase voltage produced by a dc generator.
- Three-phase voltages are rectified using sets of six diodes that produce full-wave three-phase rectification.
- Alternator output is regulated by controlling the amount of amperage to the field (rotating) windings.
- Voltage regulation on vehicles today is done mostly by solid-state electronics and uses diodes, resistors, transistors, and zener diodes.
- Computerized charging systems use an on-board computer called the PCM to regulate the voltage and amperage out of the alternator.
- Various inputs are used on computerized charging and regulation systems, including the throttle position sensor, coolant temperature sensor, intake air temperature sensor, start signal, air-conditioning signal, brake switch signal, speed sensor, and crankshaft sensor.
- During servicing of alternators, four major symptoms can be observed, including (1) the battery is getting a low charge, (2) the battery getting too high a charge, (3) the indicator lamp may be faulty, and (4) the alternator may be noisy and have damaged parts.
- Various faults can occur in a charging system, including bad diodes; grounded, open, or shorted field windings or stationary windings; bad bearings; or a faulty voltage regulator.

TERMS TO KNOW

Can you explain each of the following terms? Review the chapter until you can use each term correctly.

Armature
Commutator
Electric load detector
Field circuit
Full-wave rectification
Laminated
Rectify
Rotor
Slip rings
Stator
Three-phase
VAT
Waveform
Zener diode

REVIEW QUESTIONS

Multiple Choice

1. The charging system is used to convert ________ energy to electrical energy.
a. Radiant
b. Mechanical
c. Chemical
d. Nuclear
e. Thermal

2. What is necessary to produce a voltage in an alternator or generator?
a. Conductors
b. Magnetic field
c. The conductors cutting the magnetic field
d. All of the above
e. None of the above

3. In order to increase the output of a generator or alternator:
a. Increase the magnetic field strength
b. Increase the amount of wire or conductors
c. Increase the speed of movement between the magnetic field and the wire
d. All of the above
e. None of the above

4. The field circuit in a generator is also called:
a. The magnetic core
b. Field poles
c. Slip-ring commutator
d. Diode
e. Voltage circuit

5. Which component is used to extract the voltage produced in the armature on a dc generator?
a. Slip-ring commutator
b Diode
c. Split-ring commutator
d. Laminations
e. Stator

6. Voltage produced inside the dc generator or alternator is always:
a. Dc, and must be converted to ac
b. Ac, and must not be converted to dc
c. Dc, and must not be converted to ac
d. Ac, and must be converted to dc
e. Greater than the battery voltage

7. A voltage regulator on a charging system is always controlling the:
a. Diodes
b. Ac voltage
c. Strength of the magnetic field
d. Speed of operation
e. Zener diode

8. Which of the following components are rotating on an alternator?
a. Magnetic north and south poles
b. Slip rings
c. Magnetic pole fingers
d. All of the above
e. None of the above

9. The alternator produces:
a. Single-phase voltages
b. Double-phase voltages
c. Three-phase voltages
d. Dc without diodes
e. Ac without rectification being needed

10. How many diodes are used on an alternator?
a. 1
b. 3
c. 4
d. 5
e. 6

11. Regulators used in current charging systems today:
a. Use two relays
b. Are all solid state
c. Use one relay
d. Have no voltage regulator
e. Use three relays

12. A zener diode is used to:
a. Close off the field current
b. Tell the transistors when there is too much voltage
c. Generate voltage
d. Rectify three-phase voltages
e. Rectify single-phase voltages

13. Which of the following is not a problem in a charging system?
a. Diodes shorting out
b. Field windings grounding
c. Bad bearings
d. Too high rpm
e. Brushes wearing down

The following questions are similar in format to ASE (Automotive Service Excellence) test questions.

14. *Technician A* says that if a constant 11 volts are being produced by the alternator, the system is all right. *Technician B* says that if a constant 14 volts are being produced by the alternator, the system is damaged and not operating correctly. Who is correct?
a. A only c. Both A and B
b. B only d. Neither A nor B

15. A noise seems to be coming from an alternator, as if a bearing is damaged. *Technician A* says that the bearings should be checked and possibly replaced. *Technician B* says that the diodes should be checked. Who is correct?
a. A only c. Both A and B
b. B only d. Neither A nor B

16. An alternator does not have enough current output. *Technician A* says that the problem is bad brushes. *Technician B* says the stator is mechanically reversed. Who is correct?
a. A only c. Both A and B
b. B only d. Neither A nor B

17. An alternator is not charging correctly. *Technician A* says the problem is in the brushes and they may be worn. *Technician B* says the problem is that the alternator frame has too much magnetism. Who is correct?
a. A only c. Both A and B
b. B only d. Neither A nor B

18. *Technician A* says that the six diodes can be checked with an ohmmeter. *Technician B* says that the diodes can also be checked with a voltmeter. Who is correct?
a. A only c. Both A and B
b. B only d. Neither A nor B

19. *Technician A* says the charging system voltage can be checked with a voltmeter. *Technician B* says the charging system output can be checked with a VAT. Who is correct?
a. A only c. Both A and B
b. B only d. Neither A nor B

20. An alternator is producing a rattling noise when in operation. *Technician A* says the problem is faulty bearings. *Technician B* says the problem is a faulty diode. Who is correct?
a. A only c. Both A and B
b. B only d. Neither A nor B

21. *Technician A* says that charging voltage can be checked using VAT. *Technician B* says that a VAT checks for continuity. Who is correct?
a. A only c. Both A and B
b. B only d. Neither A nor B

Essay

22. What three things are necessary to produce a voltage?

23. What is the difference between a dc generator and an alternator?

24. Give the definition of *rectifying*.

25. What is the purpose of a stator?

26. What is a zener diode, and how is it used in an electronic regulator?

27. Describe how to check the output of an alternator.

Short Answer

28. Higher-output alternators generally have higher ________ ratings.

29. On computerized regulation systems, a(n) ________ module is used to control the exact amount of electrical load.

30. A squealing sound from the alternator system means that the ________ needs to be adjusted.

31. A bad diode in the regulator may cause the alternator ________ to be damaged.

32. Two tests that can be performed on the alternator when it is in the vehicle are the ________ and the ________.

Applied Academics

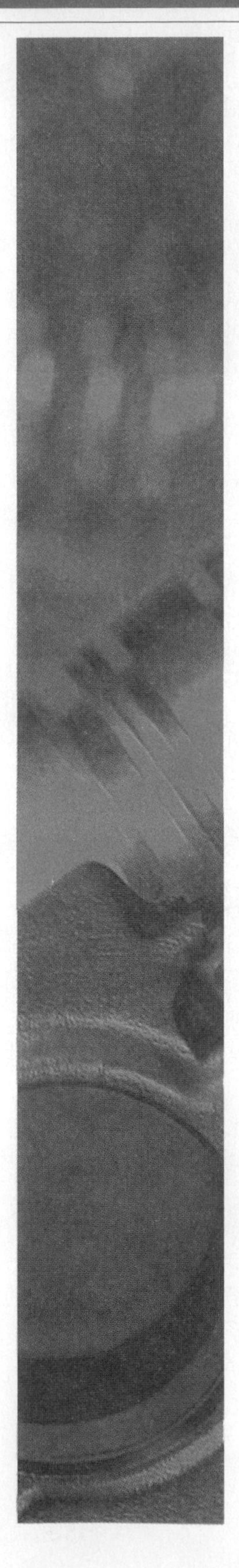

TITLE: Alternator Output

ACADEMIC SKILLS AREA: Science

NATEF Automobile Technician Tasks:

The service technician can explain how the use of rotating electrical fields can generate electricity.

CONCEPT:

One of the most important characteristics of an alternator is the amount of current or amperage that it produces. In past years, alternators were able to produce maximum current ratings of 30–50 amperes. Today, however, with the increased amount of electrical devices being placed on the automobile and the engine, amperage ratings from the alternator have been designed to be much higher. Electrical loads, such as heated windshields, more computers, electric seats and windows, heated seats, and so on, all demand more current from the alternator. It is not uncommon for standard vehicles today to have alternator amperage ratings of 100 to 150 amperes. In fact, on some vehicles amperage ratings are higher than 200 amperes. The amperage rating of any alternator is either stamped on the alternator housing or written on a tab attached to the alternator.

Amperage is also a function of the alternator's speed and the engine rpm. As shown in the graph, as the engine rpm increases, the amperage output capacity of the alternator also increases. For example, if an engine is running at 1,300 rpm, the amperage available from this particular alternator would be about 105 amperes. Of course, the amperage output depends on the resistance of the external loads (heated windshields, heated seats, computer demand, etc.)

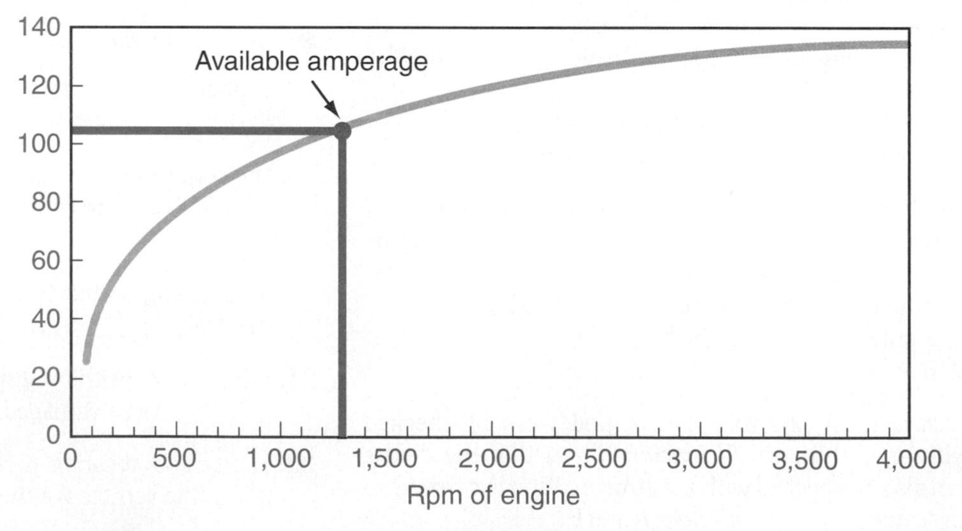

APPLICATION:

Can you think of new electrical components that have been added to vehicles that could demand more current or amperage? What would happen if there were more current or amperage demand than an alternator could provide at a certain rpm? What would be the effect on the alternator belt, if the alternator was constantly being overloaded?

Starting Systems

CHAPTER

35

OBJECTIVES

After studying this chapter, you should be able to:

- Identify the principles of starter motors.
- List the parts of the starter motor and state their purpose.
- Compare the operation and parts of different starter drive and clutch mechanisms.
- Describe the purpose and operation of the solenoid and related circuits.
- Describe the purpose of a crank signal on computerized vehicles.
- Describe various problems, their diagnosis, and service and repair procedures.

INTRODUCTION

The starting system is a type of electrical circuit that converts electrical energy into mechanical energy. The electrical energy contained in the battery is used to turn a starter motor. As this motor turns, the engine is cranked for starting. This chapter is concerned with the starting system, its components, operation, and service.

35.1 STARTER SYSTEM OPERATING PRINCIPLES

PURPOSE OF THE STARTER SYSTEM

Although energy cannot be created, it can be changed from one form to another. In the starter system, the starter motor is used to convert electrical energy from the battery

Certification Connection

ASE Connection: The information in this chapter can help you prepare for the National Institute for Automotive Service Excellence (ASE) certification tests. The tests and content areas most closely related to this chapter are:

Test A6—Electrical/Electronic Systems

- **Content Area**—Starting System Diagnosis and Repair

Test A8—Engine Performance

- **Content Area**—Engine Electrical Systems Diagnosis and Repair (Starting System)

NATEF Connection: Much of the information in this chapter is related to the NATEF tasks. The NATEF tasks and priority numbers most closely related to this chapter are:

1. Perform starter current draw tests; determine necessary action. P-1
2. Perform starter circuit voltage drop tests; determine necessary action. P-1
3. Inspect and test starter relays and solenoids; determine necessary action. P-2
4. Remove and install a starter in a vehicle. P-1
5. Inspect and test switches, connectors, and wires of starter control circuits; perform necessary action. P-2
6. Differentiate between electrical and engine mechanical problems that cause a slow-crank or no-crank condition. P-2

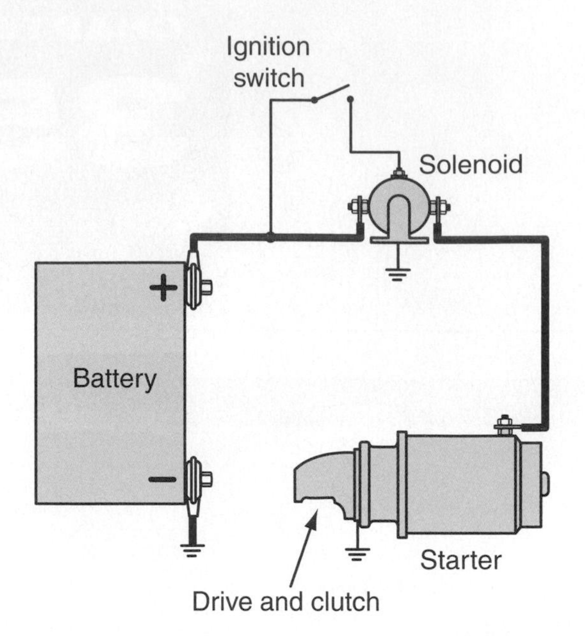

FIGURE 35–1 The starter system is made of the starter, the battery, the solenoid, the ignition switch, and the starter drive mechanism.

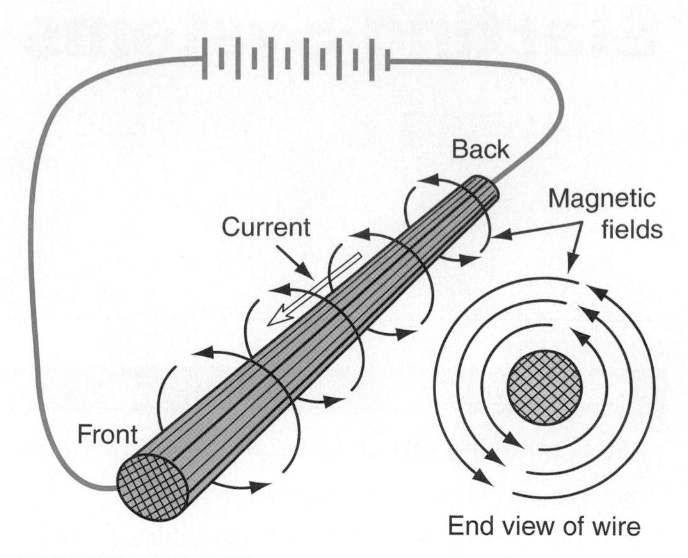

FIGURE 35–2 When current passes through a conductor, a magnetic field is built up around the conductor.

to mechanical energy to crank the engine. Several parts are needed to do this. These include the starter motor, a drive and clutch mechanism, and a solenoid that is used to switch on the heavy current in the circuit.

A considerable amount of mechanical power is necessary to crank and start a car engine. About 2 horsepower, or approximately 250 to 500 amperes of electricity, is normally needed. The amperage is higher on a diesel engine, but the basic principles are the same. Because of the high amounts of current, heavy electrical cables must be used to carry this current. The ignition switch is used to energize the heavy-current circuits. ***Figure 35–1*** shows a complete electrical circuit of the starting system. It includes the starter, the battery, the solenoid, the drive and clutch mechanism, electrical wire, and the ignition switch. Each component will be studied in more detail in this chapter.

MAGNETIC FIELD AROUND A CONDUCTOR

When current is passed through a conductor, a magnetic field is built up around the conductor. This is shown in ***Figure 35–2***. When current passes from the back of the wire to the front, a magnetic field is created. It circles around the conductor in a counterclockwise direction. The lines of magnetic force are formed in a definite pattern. If the current is reversed, the magnetic lines of force will flow in a reverse or clockwise direction.

PRODUCING MOTION FROM ELECTRICITY

A starter motor has a north pole and a south pole. This is shown at the top of ***Figure 35–3***. The north and south poles produce a magnetic field. The lines of force between the two poles flow from the north pole to the south pole. A wire carrying electricity is placed within the magnetic field. When this is done, the two magnetic fields below the wire travel in the same direction. Above the wire, the lines of force travel in opposite directions. The result is a stronger magnetic field on the bottom of the wire and a weaker magnetic field on the top. This difference in magnetic fields pushes the wire upward.

There is another way to define this movement. Magnetic lines of force traveling in the same direction tend to repel each other. Magnetic lines of force are often compared to a stretched elastic band. If there is any distortion, the bands (magnetic lines of force) will try to realign themselves. This causes the wire to move away from the area where the lines of force are concentrated.

In a starter, the wire is wound into loops. ***Figure 35–4*** shows how a loop of wire would be turned if it were placed within a magnetic field. The electricity flows into point B

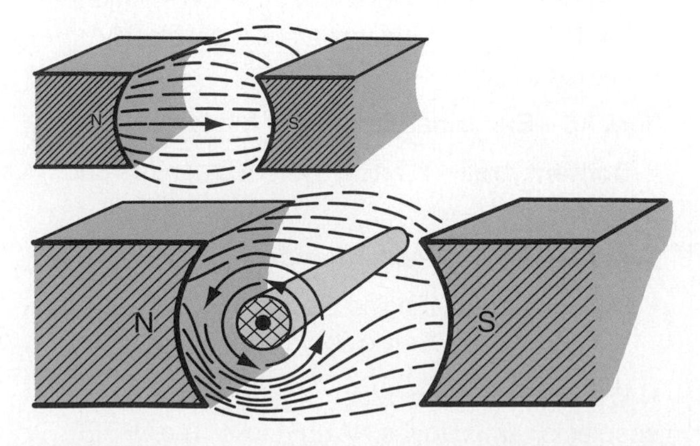

FIGURE 35–3 A north pole and a south pole are needed to create the action in a motor. If the current-carrying conductor is placed within the magnetic field, the wire tends to move. This movement is caused by differences in the magnetism on top of and below the wire.

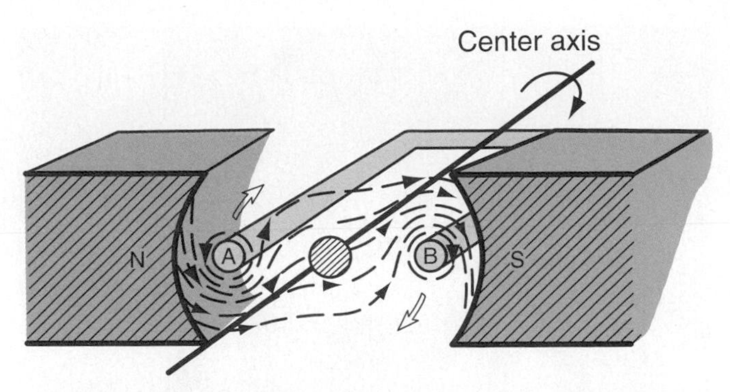

FIGURE 35–4 When a loop of wire with current flowing in the wire is placed within a magnetic field, the loop will turn because of the direction of the magnetic fields.

and loops around to come out at point A. The coil of wire rotates on the center axis. The direction of the magnetic field around the left and right conductors is opposite. When this condition occurs, the magnetic lines of force strengthen below the left side and above the right side. The concentration of magnetic lines of force causes the wire loop to turn in a clockwise direction on the center axis. When electricity is passed through the wire, electrical energy is converted to mechanical energy.

THE PRINCIPLE OF COMMUTATION

In ***Figure 35–4***, when the loop has rotated one-half revolution, point A is positioned at the south pole and point B is positioned at the north pole. Because the magnetic field around point B is clockwise, magnetic distortion occurs above B and below A. This means that the wire loop will reverse direction to return to the original position.

To keep the loop turning in the same direction, the electrical current in the loop must be reversed at precisely the right time. This is done by using a **commutator**. Commutation, or current reversal, is achieved by joining the ends of the loop to two metal segments. The contact surface between the battery and the segments is formed by the brushes.

Figure 35–5 shows a simple starter motor with a commutator and brushes. Note that both ends of the loop of wire are connected to different sections of the commutator. In ***Figure 35–5A***, the right side of the loop is connected, through the upper commutator 2, to the positive side of the battery. The left side of the loop is connected to the lower commutator ring 1. The brushes are riding on the circular commutator rings. Under this condition, electricity flows from the positive side of the battery, through the top commutator ring 2, through the loop of wire, through the lower commutator ring 1, and finally, back to the negative side of the battery. The direction of the magnetic field around the wire causes the loop to rotate 180 degrees clockwise.

Referring to ***Figure 35–5B***, after 180 degrees of rotation, commutator ring 1 is now positioned on top and commutator ring 2 is now positioned on the bottom. Once again, current from the battery flows into the top commutator ring, through the wire loop, back out the lower commutator ring, and back to the negative side of the battery. Once again, the direction of the magnetic field around the wire causes the loop to rotate 180 degrees clockwise.

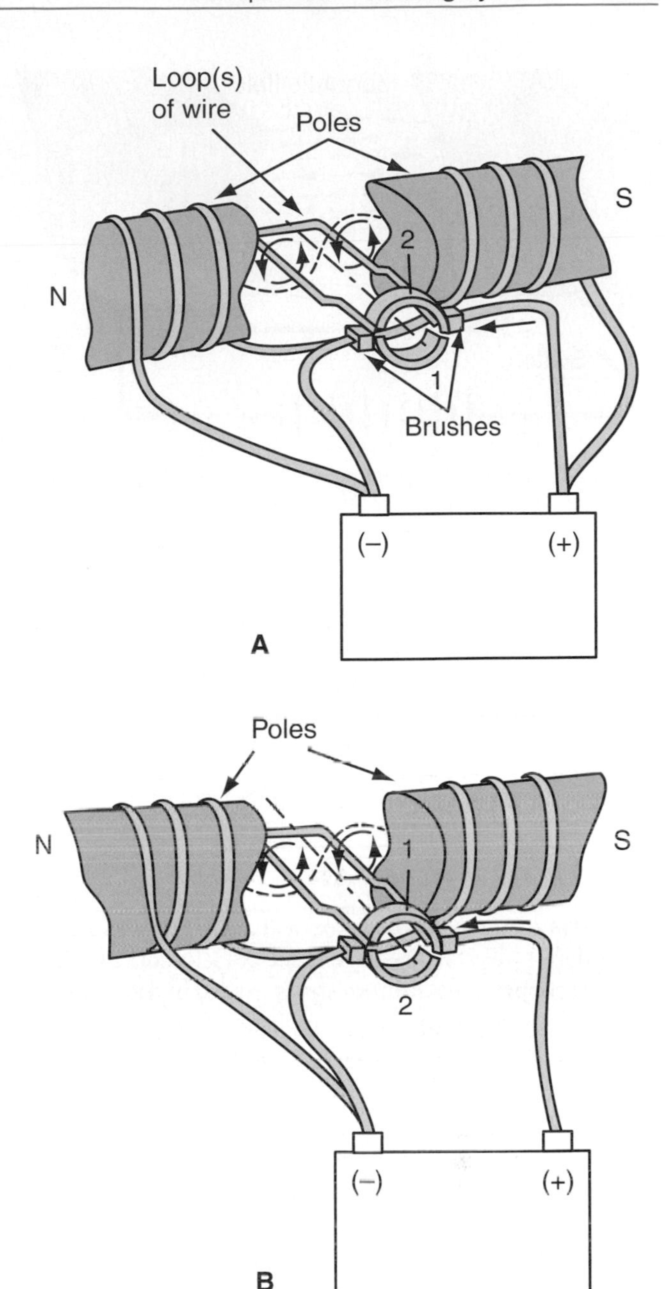

FIGURE 35–5 A simple starter motor consists of a set of poles, a loop of wire, and a commutator.

PRODUCING THE NORTH AND SOUTH POLES

The north and south poles are created by using soft iron that can be easily magnetized. This is done by wrapping a coil of wire around the soft iron core and passing current through the wire. These coils are called field coils or poles. The soft iron core is specially shaped to concentrate the

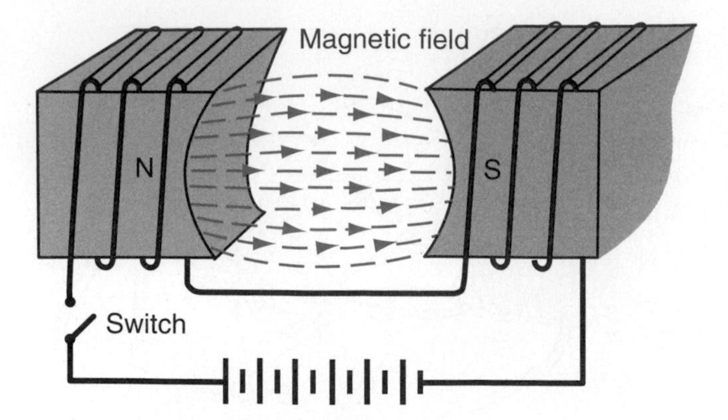

FIGURE 35-6 All starter motors have a set of north and south poles that are produced by winding wire around a soft iron core.

magnetic lines of force in the space provided. ***Figure 35–6*** shows the main magnetic fields. In this case, the north pole is formed on the left, and the south pole is formed on the right. The lines of magnetic force always flow from the north pole to the south pole.

35.2 STARTER MOTORS

PARTS OF THE STARTER

The parts of the starter motor will now be considered in more detail. All starters have a set of field coils. Coils are made of copper or aluminum strips wound in the correct direction to produce the poles. Typically, four field coils are used on a starter. ***Figure 35–7*** shows an example of the field coils. These coils are assembled over the soft iron core. The iron core, or shoe, is attached to the inside of a heavy iron frame on the motor (***Figure 35–8***). The iron frame and **pole shoes** provide a place for the field coils. They also provide a low-resistance path for the magnetic lines of force.

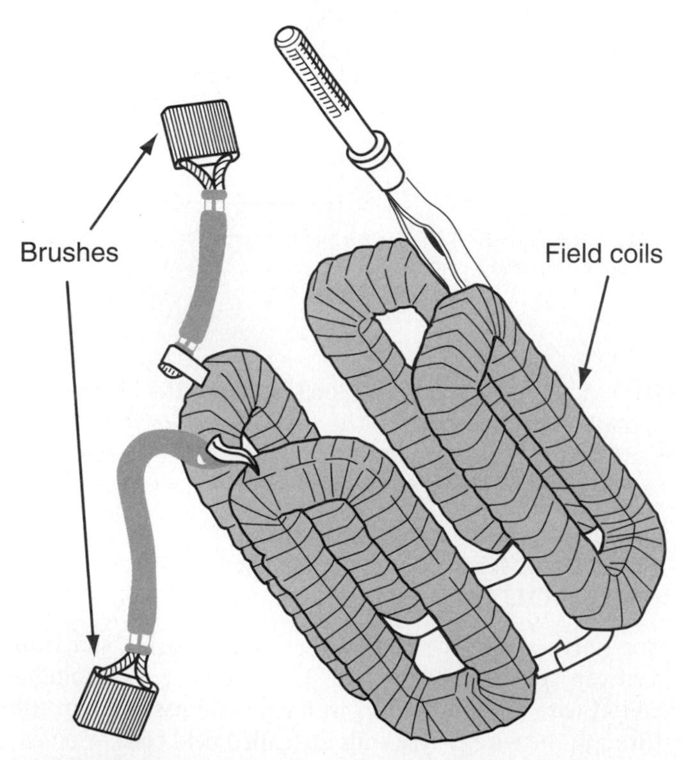

FIGURE 35-7 Four field coils are used on a starter motor.

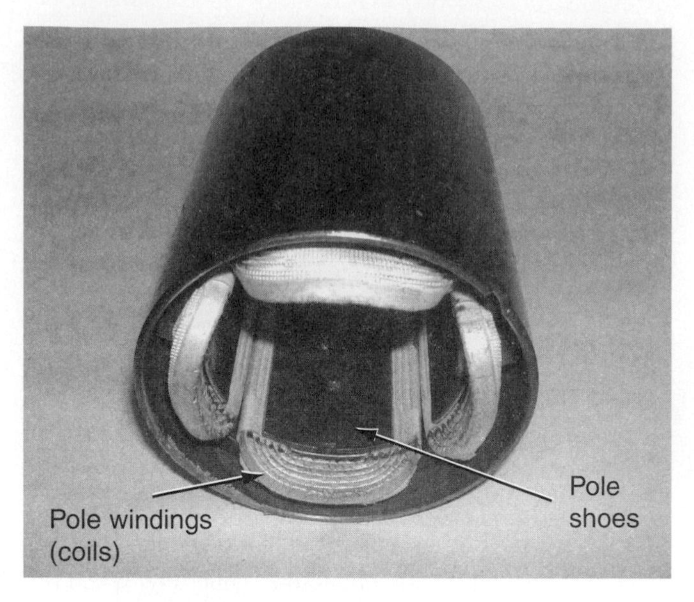

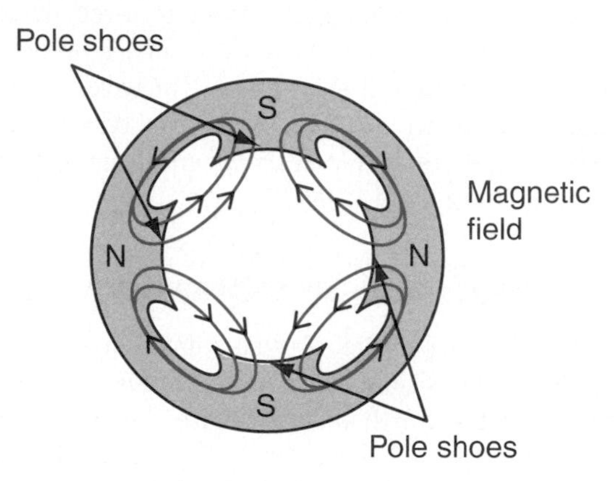

FIGURE 35-8 The four field coils are attached to the starter frame. The shoes provide a place for the field coils to be wound around.

SERIES WOUND FIELD CIRCUITS

The four field coils in the starter can be connected in several ways. Two common connections are the series and the series-parallel wound field. The one selected is determined by the application, engine speed, torque requirements on the starter, cable size, battery capacity, and current-carrying capacity of the brushes.

Figure 35–9 shows the series wound circuit. Here the field coils are connected in series or one after the other. The four field coils are wound in different directions to produce alternate north and south poles. The electrical current flows from the battery into each coil, then to the **armature** windings. All four field coils are in series with the armature windings.

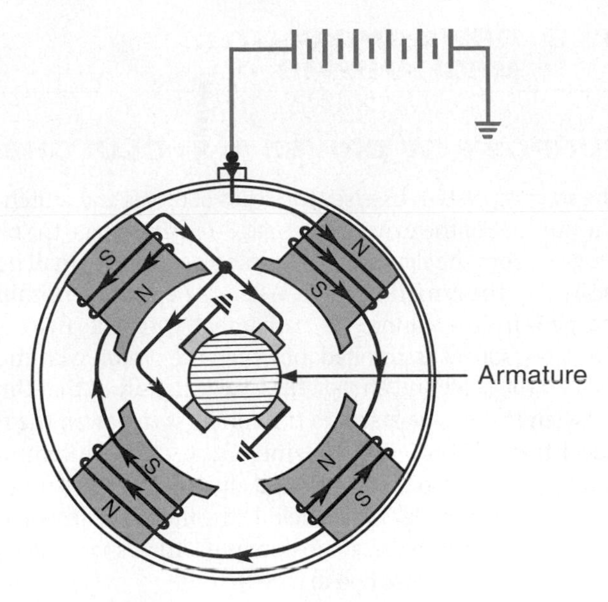

FIGURE 35–9 A series wound circuit in the starter has all four coils in series with the armature.

SERIES-PARALLEL WOUND FIELD CIRCUITS

Another type of electrical circuit used in starters is called the series parallel field connection. This arrangement is shown in ***Figure 35–10***. The electrical current comes into the starter and splits. The field circuit consists of two paths for the current. The field windings, however, are still in series with the armature, while one field coil is in parallel with the armature windings. This is sometimes called a **shunt** coil connection. It is used to control the maximum speed of the starter during operation. In the automobile, shunt type motors have not been used for starter motors in several years. However, the shunt motor may be found in wiper motors, power window motors, power seat motors, and so on.

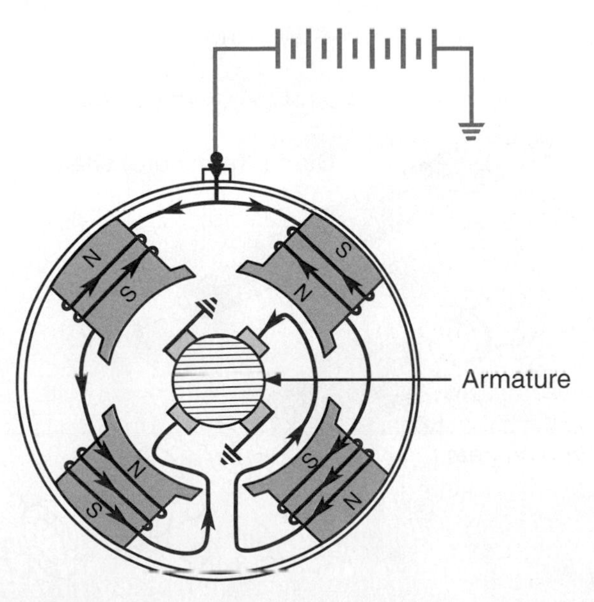

FIGURE 35–10 A series-parallel wound circuit in the starter has two paths for the current to flow through the poles.

PERMANENT MAGNET STARTING MOTORS

One of the more recent changes in starter motors is the permanent magnet starting motor. In this starter motor, permanent magnets are used as field coils. In contrast to electromagnets that have been used on starter field coils in the past, permanent magnets do not require the use of electricity to be energized. Permanent magnet starter motors are typically lighter and stronger. Most of the testing and service procedures are the same as other designs. Permanent magnet motors also have a planetary gear system to reduce the speed of the motor and gain torque for cranking the engine.

ARMATURES

The armature is the rotational part of the starter. Earlier in this chapter, the armature was called the loop of wire. Actually, there are many loops of wire in the armature. Each wire loop is connected to the copper segments arranged in a barrel shape (***Figure 35–11***). The segments are insulated from each other and from the armature shaft. These segments of copper form the commutator and provide the running surface for the brushes.

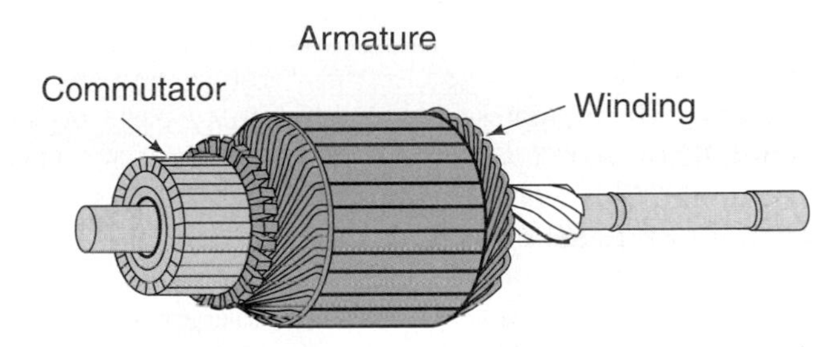

FIGURE 35–11 The armature in a starter is the rotational part that has many loops of wire with a commutator.

CUSTOMER CONCERN:

Bad Starter Bushings

A vehicle has 60,000 miles on it. No work has been done on the starter system. There have been times recently when the engine would not even crank over. It sounds as if the solenoid is trying to engage the starter pinion, but the starter just won't turn over.

SOLUTION: Quite often the starter bushings will wear. When the bushings wear, the armature sits lower in the starter field windings, often scraping on them. Remove the starter and check the no-load speed test. This should show that the starter bushings are bad and should be replaced.

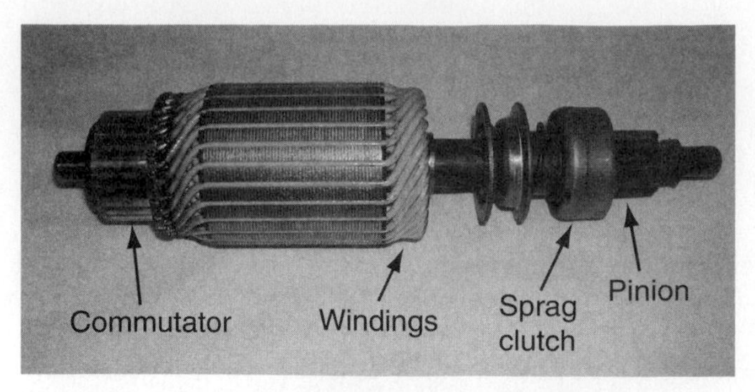

FIGURE 35-12 An armature and associated parts used in a starter motor.

The armature assembly consists of a stack of iron **laminations**, or layers, located on a steel shaft. The steel shaft is supported by bearings located on the end plates of the starter. The iron laminations are used to concentrate the magnetic field. Laminations are also used to reduce the heating effect of **eddy currents**. Eddy currents are small electrical currents that are produced inside the iron core. The windings are made of heavy copper ribbons that are inserted into the slots in the iron laminations. ***Figure 35–12*** shows a typical armature used in a starter motor.

BRUSHES

Brushes are used to make electrical contact between the armature, which rotates, and the stationary parts of the circuit. Brushes are made from a high percentage of copper and carbon. This material minimizes electrical losses due to overheating. ***Figure 35–13*** shows an example of the brushes held in place on the starter end plate. Each brush uses a small metal spring to force the brush against the commutator during operation. There are typically four brushes. Two brushes are used to feed electricity into the armature. Two brushes are grounded and used to return the electricity to ground, which is the negative side of the battery.

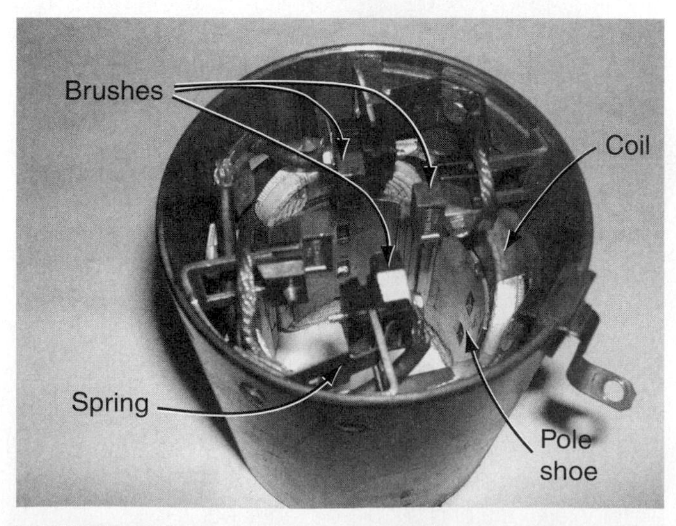

FIGURE 35-13 Brushes are held in place on the end plate of the starter frame.

35.3 DRIVE AND CLUTCH MECHANISMS

PURPOSE OF DRIVES AND CLUTCHES

The starting system uses several types of drives and clutches. The purpose of the drive is to engage and disengage the **pinion gear** from the flywheel. The pinion gear is the small gear located on the armature shaft. When the engine is running, the pinion gear cannot be in contact with the flywheel. When the starter is cranked, however, the pinion gear must slide on the shaft and engage the flywheel, only at that time.

When the engine starts to turn on its own power, the flywheel turns faster. Now the pinion gear must be removed rapidly. Several types of drives and clutches are used for this purpose. Two popular types include the inertia drive and the overrunning clutch drive. Most manufacturers use a design similar to one of these two drives.

INERTIA DRIVES

Not long ago, **inertia drives** were commonplace. Inertia drives are also called Bendix drives. Although they may differ considerably in appearance, each drive operates on the principle of **inertia**. Inertia is a physical property of an object. An object at rest tends to stay at rest. An object in motion tends to remain in motion. Inertia is used to move the pinion gear to engage the engine ring gear when the starter motor is energized.

Figure 35–14 shows an inertia drive system. There is a pinion gear, a screw sleeve, the armature shaft, and springs. All the parts of the drive mechanism are assembled onto the armature drive shaft. The drive shaft has a spline that meshes with the internal spline on the screw sleeve. Thus, the screw sleeve can move along the axis of the armature shaft.

When the armature turns during starting, the armature shaft and the screw sleeve rotate. The inertia of the pinion (remaining at rest) causes the screw sleeve to turn inside

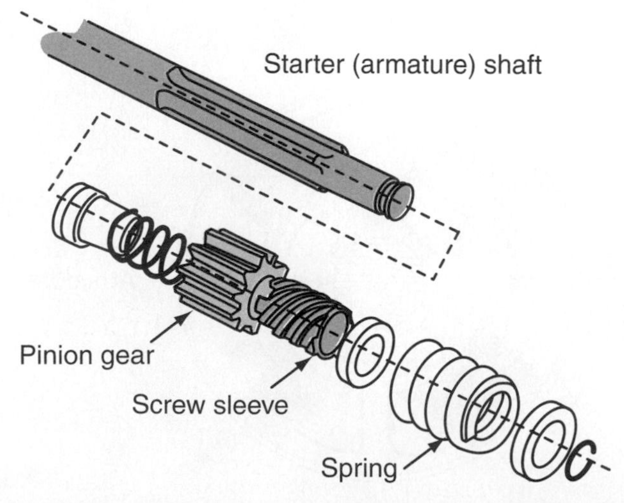

FIGURE 35-14 An inertia drive system is used on some older starter systems to engage the pinion gear with the ring gear.

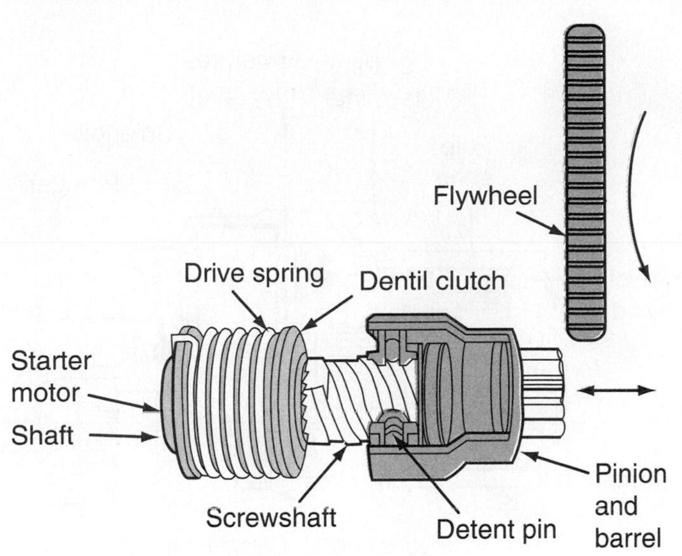

FIGURE 35–15 This inertia drive system has a pinion and barrel assembly that moves to engage or disengage the pinion with the flywheel ring gear.

the pinion. This causes the pinion to move out and engage with the flywheel ring gear. As soon as the engine fires and runs under its own power, the flywheel is driven faster by the engine. As the pinion gear is forced faster, it is moved back along the screw sleeve and out of engagement with the flywheel. At this point, the starter switch is released. The spring is used to absorb excessive twisting of the shaft. On certain drives, a compression spring is used to reduce the shock at the moment of engagement.

A second inertia drive is shown in ***Figure 35–15***. In this figure, note that the pinion and barrel assembly will move to the right when the starter motor begins to turn. After the flywheel begins to spin faster, the pinion and barrel assembly moves back to the left on the screwshaft, taking the pinion gear out of mesh with the flywheel.

OVERRUNNING CLUTCH DRIVES

The most common type of starter drive is the **overrunning clutch drive**. This type of drive is also referred to as a roll-

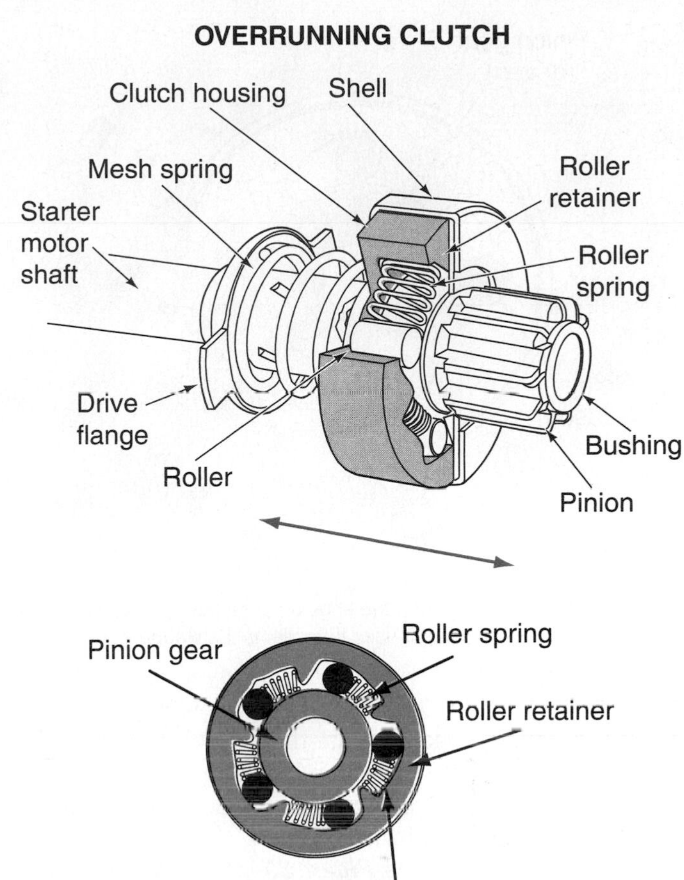

FIGURE 35–16 A roller clutch, also called the overrunning clutch, is used on some starter systems to engage the pinion gear with the ring gear.

type drive, or **sprag** (small angular wedges) drive. The sprag drive uses a series of thirty or so sprags, rather than rollers. It is commonly used on diesel engines and on some gasoline engines. The roller clutch drive is shown in ***Figure 35–16***. It consists of a drive and driven member and a series of cylindrical rollers. These rollers are placed in wedge-shaped tracks in the clutch housing. As shown in ***Figure 35–17***, when the armature shaft rotates during cranking, the small

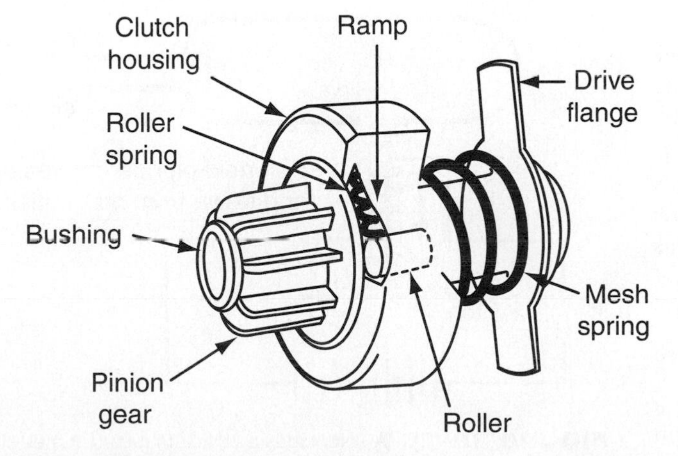

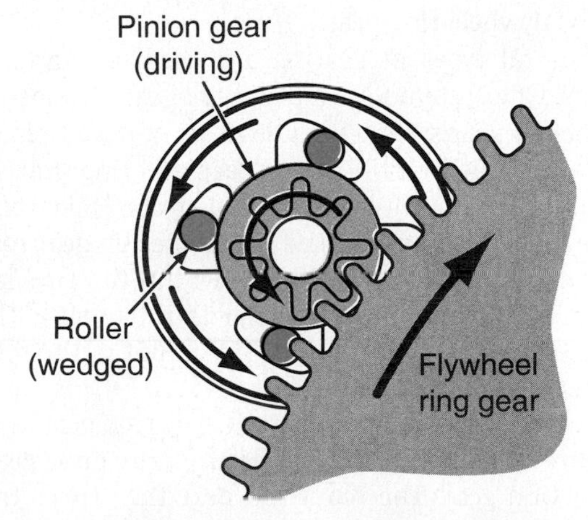

FIGURE 35–17 When the armature shaft rotates during cranking, small rollers become wedged, causing the pinion gear to lock up and rotate the ring gear.

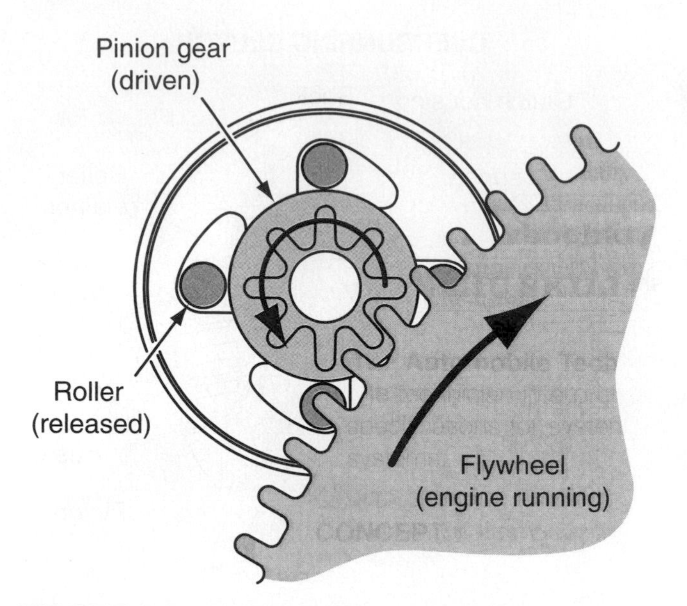

FIGURE 35-18 When the engine starts, the flywheel spins the pinion gear faster, which releases the roller in the wedge.

rollers become wedged against the collar attached to the pinion gear. This wedging action locks the pinion gear with the armature shaft. The pinion gear now rotates with the shaft, cranking the engine.

When the engine starts, ***Figure 35–18***, the flywheel spins the pinion faster than the armature. This action releases the rollers, unlocking the pinion gear from the armature shaft. The pinion then "overruns" safely and freely until it is pulled out of mesh when the ignition switch is released. The overrunning clutch is moved in and out of mesh by linkage operated by the solenoid.

GEAR REDUCTION STARTER DRIVES

On many vehicles, a gear reduction system is used between the starter and the pinion gear. Its purpose is to increase the amount of torque during starting of the engine. The gears are designed to reduce the speed of the pinion gear (keeping the starter at the same speed), thus producing more torque at the flywheel ring gear.

Several types of gearing are used on starter systems today. One system uses a planetary gear arrangement. The planetary gear system includes a set of planet pinion gears, a planet pinion carrier, a sun gear, and ring gear. In operation, one of the three gears needs to be locked up. In this case, the ring gear is locked up. As the sun gear rotates, the planet pinion gears also rotate within the ring gear. This causes the planet pinion carrier to rotate also. The planet pinion carrier is attached to the output of the system and turns the overrunning clutch.

Another system uses a single reduction gear as shown in ***Figure 35–19***. In this case, the armature drive gear turns a reduction gear. The reduction gear then turns the pinion shaft, overrunning clutch, and drive pinion.

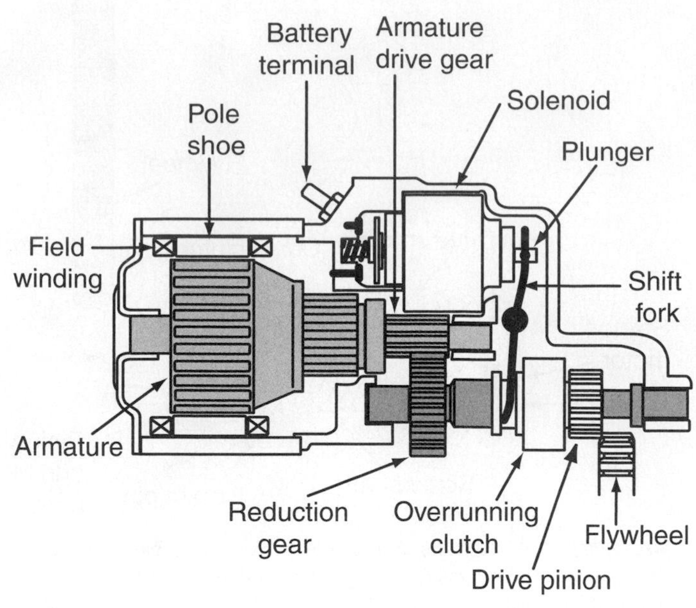

FIGURE 35-19 A reduction gear is used to reduce the speed of the pinion, thus increasing the torque to the flywheel ring gear.

35.4 SOLENOIDS

DEFINITION OF A SOLENOID

A **solenoid** (also called a relay) is an electromechanical device that switches electrical circuits on and off. ***Figure 35–20*** shows the operation of a solenoid. When current flows through the electrical coil, a magnetic field is created inside the coil. If a soft iron core of metal is placed near the center of the coil, the metal core tends to center itself inside the coil. If a spring is used to hold the metal core outside the center, it again tries to center itself. Thus, when current passes through the coil, the metal core will move. This process, then, converts electrical energy into mechanical energy. The movement of the metal core is used to open and close the electrical circuits. Therefore, a solenoid is typically considered an electromechanical switch.

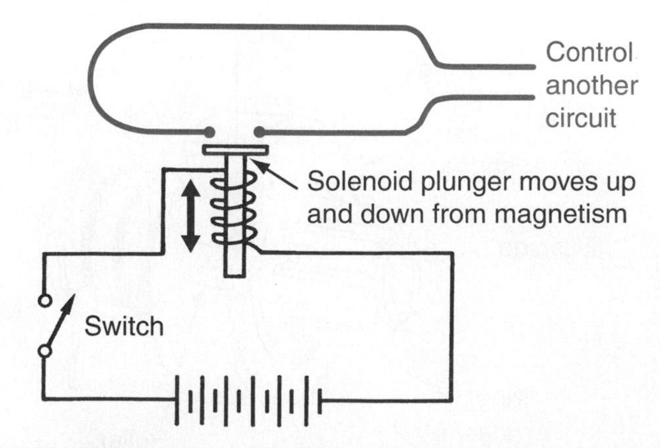

FIGURE 35-20 A solenoid is a device that converts electrical energy to mechanical energy by the use of magnetism.

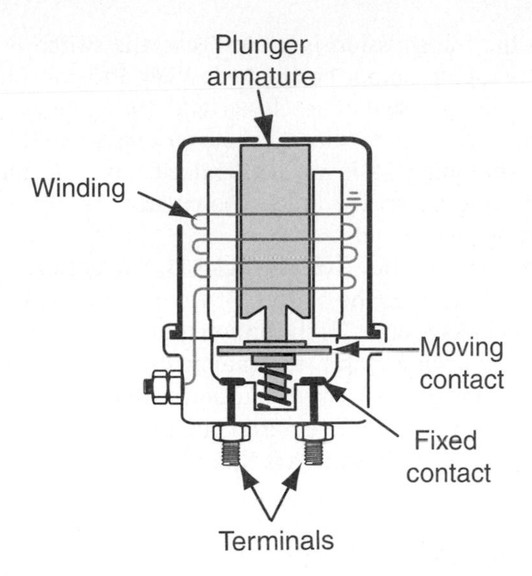

FIGURE 35–21 When the solenoid is operated, a plunger or armature moves down and makes contact with two fixed contacts used to turn the starter motor on.

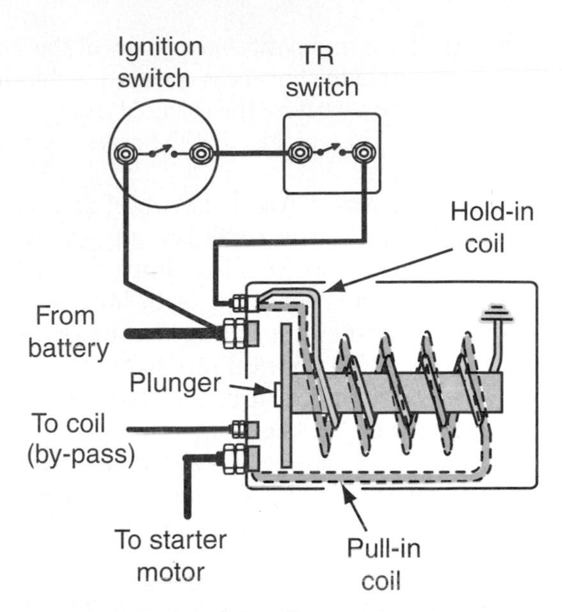

FIGURE 35–22 The two windings used in a solenoid close the plunger and hold it in

SOLENOID OPERATION

Figure 35–21 shows the internal operation of a solenoid used on some vehicles. The solenoid is used to start and stop the heavy current that flows to the starter motor through the terminals during cranking. An electrical winding is placed around the iron core. This core is called the armature or plunger. On the bottom of the armature is a set of moving contacts. On the bottom of the solenoid is a set of fixed contacts connected to the heavy-current terminals. When the solenoid windings are energized, the center armature is forced downward magnetically. This causes an electrical connection between the two fixed contact points. When the current is stopped in the windings, the armature returns to its original position. Some older armatures can also be pushed manually to accomplish the same result.

PARTS LOCATOR

*Refer to photo #66 in the front of the textbook to see what a **starting system solenoid** looks like.*

CLOSING AND HOLD-IN COILS

Many solenoids contain two coils. One is called the closing, or pull-in, coil. One is called the hold-in coil. This is shown in the electrical circuit in ***Figure 35–22***. The heavy-gauge winding is called the closing coil, or pull-in winding. It has low resistance. The finer or thinner coil in the center is the hold-in winding. It has higher resistance.

The closing coil has less resistance. This means that there will be more current and more force in the coil. This force is needed along with the hold-in coil to move the center plunger. Once the plunger movement has been completed, much less magnetism is needed to hold the plunger in. When the contact disc touches the terminals, the pull-in coil is shorted out, thus no current flows through the coil. This reduces current draw on the battery during cranking.

SOLENOID DRIVE LINKAGE

On many vehicles, there is a linkage attached to the opposite end of the plunger on the solenoid. This linkage is used to engage the pinion with the flywheel gear. ***Figure 35–23*** shows an example of this linkage. Movement of the plunger

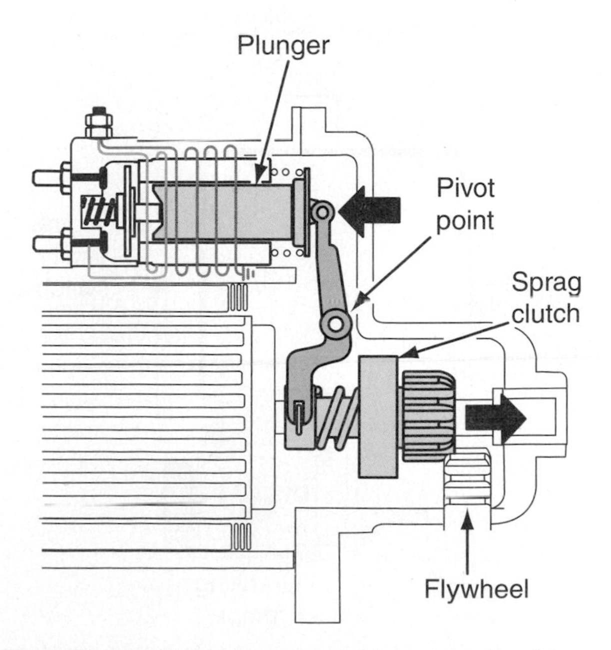

FIGURE 35–23 Linkage is used to engage the pinion gear with the flywheel gear.

results in a similar but opposite movement of the pinion gear. The movement of the linkage is designed so the pinion gear meshes completely before the solenoid makes contact with the terminals. This ensures that the gears are meshed before the starter turns.

On occasion, the pinion gear and flywheel gear may not instantly mesh, because the teeth butt up against each other. A spring on the overrunning clutch assembly will overcome this problem. When the teeth butt up against each other, the plunger and operating lever move their full distance. However, the pinion gear may still not be meshed. When the starter begins to turn, the teeth immediately mesh and cause the engine to crank.

The gear teeth on the pinion gear also have a slight edge, or **chamfer**, cut on the end of each tooth. This also helps the pinion and ring gear to mesh more smoothly.

35.5 STARTER CIRCUIT OPERATION

SIMPLE STARTER CIRCUIT

Now that all of the components have been studied on the starting system, the complete circuit can be analyzed. ***Figure 35–24*** shows the complete operation of the starting circuit. Electrical energy is made available by the battery. A large, heavy electrical wire is connected from the battery to the starter solenoid. The ignition switch is also connected to the positive side of the battery.

A neutral safety switch is also included in the circuit. This switch remains open except when the transmission is in park neutral, or when the clutch pedal is depressed. When the transmission is in any gear, the switch is open and the engine cannot be cranked. When the transmission is in park, the switch is closed and the engine can be cranked. At this time, electrical current passes to the solenoid. The solenoid is now energized, and the plunger moves. This causes the linkage to engage the pinion gear and crank the engine.

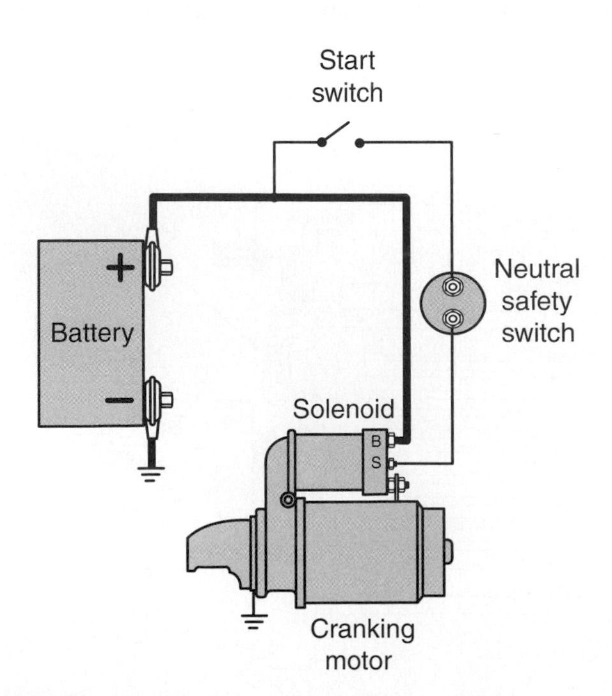

FIGURE 35–24 The complete electrical circuit of the starter system.

When the ignition switch is closed, current causes the solenoid to engage the starter. At the same time, the mechanical linkage engages the pinion gear with the ring gear. When the engine starts, the pinion gear is pushed away from the ring gear. When the ignition switch is released, the solenoid also releases. The starter circuit is now ready to operate again when the switch is turned to the crank position.

COMPLETE STARTER CIRCUIT OPERATION

To gain a complete understanding of how a starter works on today's automobile, refer to ***Figure 35–25***. This circuit shows all of the components of a typical starting circuit. The following notes should be helpful in understanding the complete circuit:

- Note that the positive terminal of the battery is connected to the starter solenoid at connection T3. The starter motor will not turn until the pull-in and hold-in coils are energized. When the pull-in and hold-in coils are energized, the switch will pull to the right and close. This, in effect, will cause the starter motor to turn.
- To get the starter solenoid energized, several conditions must be present. Note that first the electricity must go through the electrical load detector (ELD) on the upper left of the circuit. When the starter motor is in the cranking mode, this unit detects the additional amount of current passing through the ELD. The ELD is also connected to the charging circuit so that the field current in the charging circuit is energized to a maximum point to produce the maximum charging voltage.
- Following the wires out of the ELD, note that the ignition switch must also be in the start position. If the ignition switch is in any other position, there will be no electricity going to the starter relay, directly below.
- If the ignition switch is in the start position, electricity then flows two ways. First, electricity flows to the starter relay contacts. To energize the starter relay, electricity also flows through the underdash fuse/relay box.
- The electricity is also sent to the automatic transmission (A/T) gear position switch. If the automatic transmission is placed in the P (park) or N (neutral) position, electricity will flow through the switch down to the coil inside the starter relay. If the automatic transmission is placed in any gear, including reverse, the starter relay will not be energized and the starter will not crank.

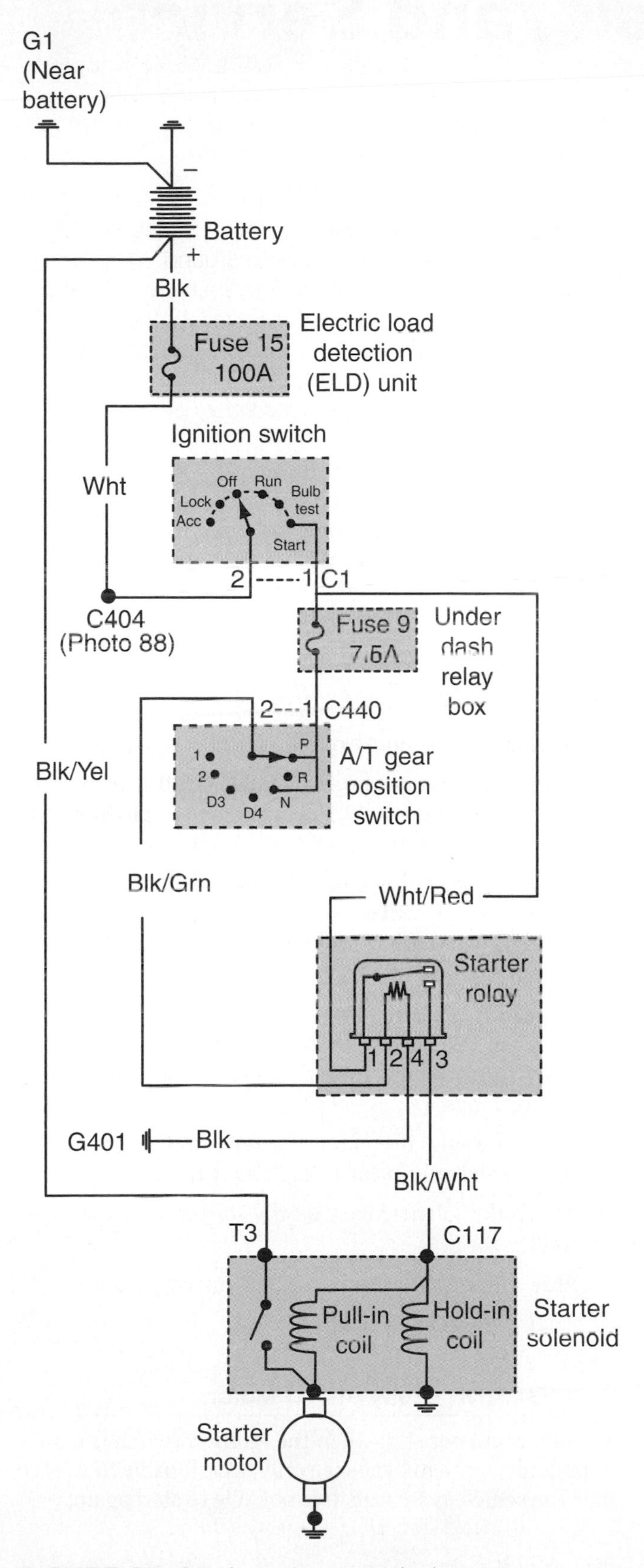

FIGURE 35-25 A complete starter circuit.

> **PARTS LOCATOR**
>
> *Refer to photo #67 in the front of the textbook to see what a **neutral position switch** looks like.*

- In summary, in order to start the starter motor on an automatic transmission, the transmission must be in either N or P, and the ignition switch must be on start. When this happens, the starter solenoid hold-in and pull-in coils will be energized, causing high current to pass from the battery through the starter motor.
- Note that there are photo numbers located in different parts of the circuit. These photo numbers are there so the service technician can reference a photo of this part in the service manual. Then the service technician can easily go to the vehicle and locate the part correctly.

STARTING SYSTEM COMPUTER SIGNAL

On engines that use computerized control, a start or **crank signal** is sent to the computer. Its purpose is to tell the computer that the engine is starting to crank. The computer will then enrich the air-fuel mixture accordingly, to the injectors. On some engines, a cold start injector is used during starting. In order to get the crank signal to the computer, a wire is fed from the starter solenoid circuit to the computer.

Problems, Diagnosis, and Service

Safety Precautions

1. Some starters are very heavy. When installing a starter, be careful not to pinch your fingers between the block and the starter frame.
2. Often the pinion gear or the flexplate ring gear is worn and has very sharp edges. Be careful not to cut your fingers on the sharp edges of a worn gear.
3. In order to remove a starter, the vehicle must be put on jack stands or a hoist. Follow the manufacturer's procedure when lifting a car on the hoist.
4. When lifting a car on a small hydraulic or air jack, always support the vehicle with additional jack stands for extra safety.
5. When removing a starter, first disconnect the battery terminals (negative first) to eliminate the possibility of the engine cranking over or accidentally arcing from the battery cable to ground.
6. When testing a starter on a "free speed test," make sure the starter is fastened securely to the bench. A great deal of torque is usually produced when the starter first begins to spin.
7. Always wear OSHA-approved safety glasses when working on a starter.
8. Always use the correct tools when working on a starter.
9. Always remove rings, bracelets, and other jewelry to eliminate the possibility of getting an electrical shock.

PROBLEM: Faulty Solenoid

The engine in a vehicle will not start. When the ignition switch is engaged, the solenoid seems to click, but the starter will not turn.

DIAGNOSIS

If the solenoid engages but the starter does not turn, the solenoid moving contact may be corroded. On certain starters, this moving contact can be cleaned and replaced in the solenoid. To check, use a battery jumper cable to bypass the solenoid. Connect the jumper cable across the heavy-current lines, between the input and output of the solenoid. If the starter turns, then the solenoid is defective. Often the problem occurs because each time the solenoid is turned on, a copper washer makes contact with two electrical connections. When the solenoid is turned off, there is always a small amount of electrical arcing on the washer and contacts. This causes corrosion and eventually increases the resistance enough so that electrical contact cannot be made.

SERVICE

Use the following general procedure to remove the solenoid and replace or repair the moving contact.

1. Disconnect the battery.
2. Generally the solenoid is located either directly on top of the starter or on the side of the engine compartment on the wheel well. Remove all wires from the solenoid and mark them for reassembly later.
3. Remove the solenoid from the vehicle or starter.
4. Some solenoids can be disassembled. If this is the case, disassemble the solenoid, being careful not to break any electrical connections.
5. Using sandpaper or a file or both, clean both internal heavy-current contacts in the solenoid. Also clean the small washer that makes contact between the two heavy-current contacts.
6. While the solenoid is apart, both the hold-in and pull-in coils can be checked for resistance and voltage drops. Refer to the manufacturer's specifications for the correct readings.
7. After cleaning the contact points and checking the hold-in and pull-in coils, reassemble the solenoid.
8. Attach the solenoid back on the starter or to the wheel well.
9. Attach all electrical wires to their correct location.
10. Engine should now start.

PROBLEM: Faulty Starter Motor

An engine will not start when the ignition switch is closed. At first, the problem happened only occasionally. Now, each time the vehicle is stopped, it is not able to start again.

DIAGNOSIS

Often the starter has a bad set of brushes, bad bearings, a bad electrical contact, or windings in the starter have either shorted out or grounded. This means the starter must either be replaced or rebuilt. Several checks can be made on the starter to determine if it needs to be rebuilt.

1. A voltmeter can be used to check various points in the starter circuit. Refer to ***Figure 35–26***.
2. The voltage drop in the starter can be checked with a voltmeter during cranking. Excessive voltage drop could mean the resistance in the starter circuit is too high. Several checks can be made. ***Figure 35–27*** shows three checks. The voltage drop can be checked between the vehicle frame and the grounded battery terminal post (A), between the vehicle frame and the starter field frame (B), or between the ungrounded battery terminal post and the battery terminal on the solenoid (C). If any of these readings shows more than a 0.2-volt drop when the starting motor is cranking, the electrical connections

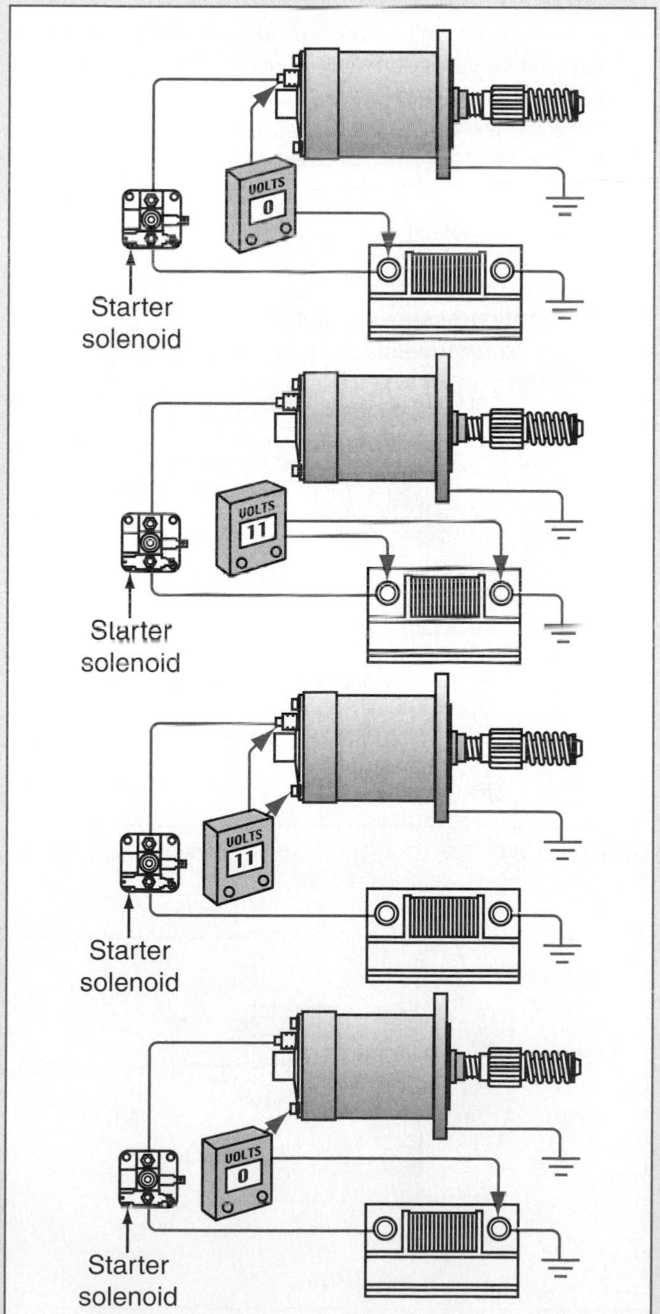

FIGURE 35–26 Several voltage checks can be made to test the voltage drop on the starter circuits.

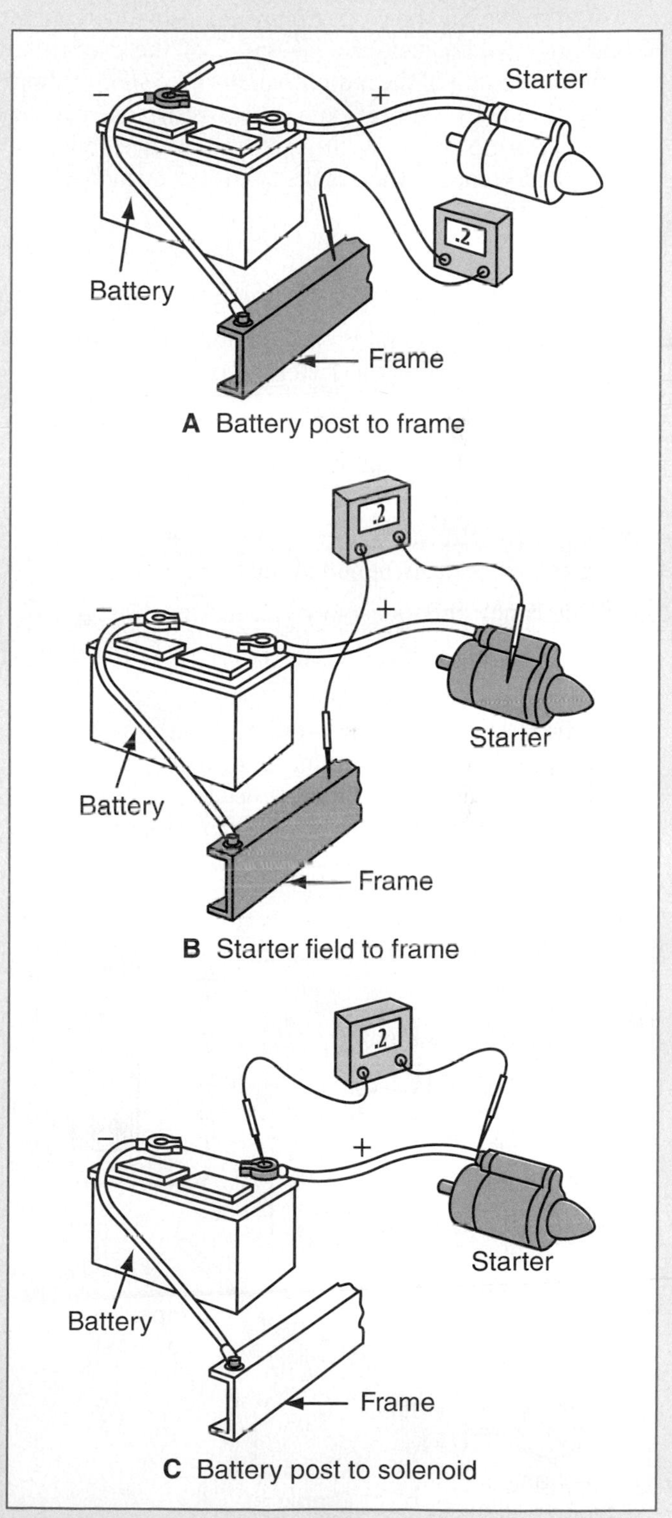

FIGURE 35–27 Several voltage drops can be checked to determine the condition of the engine. Voltage can be checked (A) between the frame and the negative battery terminal, (B) between the frame and the starter field frame, and (C) between the positive side of the battery and the solenoid terminal.

will need to be cleaned. A light coating of grease on the battery cables and terminal clamps will retard further corrosion. If the starter still does not crank the engine, the starter will need to be rebuilt.

3. The free speed test can also be performed on the starter. Using the setup shown in ***Figure 35–28***, a tachometer can be used to measure the speed of the starter per minute. Failure of the motor to perform to specifications may be due to bad bearings or high-resistance connections. During this test, check the amperage, volts, and rpm and compare the results with the manufacturer's specifications.

SERVICE

Use the following general procedure and checks when disassembling and repairing a starter motor.

1. Disconnect the battery.
2. Using the correct tools, remove the starter from the engine. Often this procedure is listed in the service manuals.
3. If the solenoid is attached to the starter, make sure all wires have been disconnected and marked for reassembly.
4. Disassemble the starter by removing the through bolts.
5. Remove both ends of the starter to expose the internal parts.
6. If the starter appears inoperative, the field coils can be checked for shorts, opens, or grounds. Use an appropriate ohmmeter to measure the resistance of the field coils. Compare the results with manufacturer's specifications.

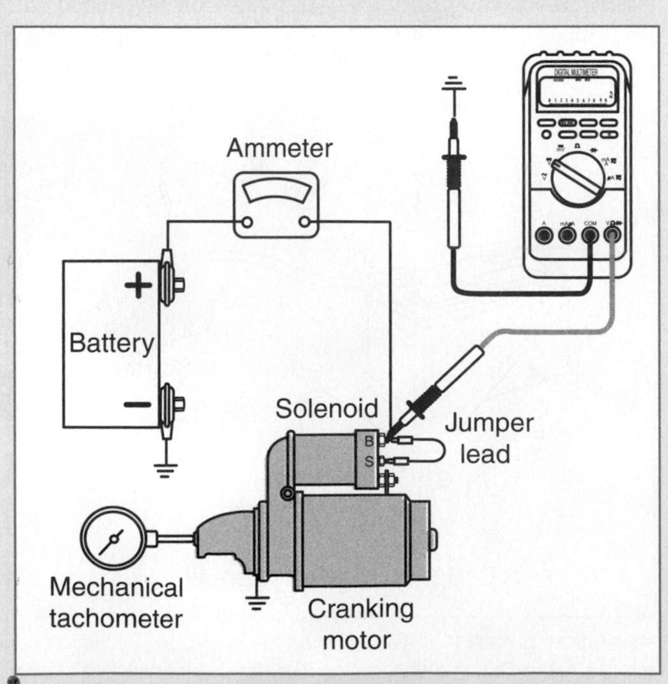

FIGURE 35–28 The free speed test can be performed on a starter when it is out of the vehicle. Voltage, amperage, and rpm are compared with specifications given by the manufacturer.

7. Check the armature end play and compare the readings with the manufacturer's specifications. Shims may be needed to correct excessive end play. Depending on the manufacturer, the clearance should be between 0.010 and 0.140 inch (***Figure 35–29***). Also, make sure that during assembly of the starter, the small friction disc on the end of the pinion gear is replaced in the same position.
8. Check the starter brushes for excessive wear. Check the manufacturer's specifications for the amount of wear allowed on the brushes. Use a vernier caliper as shown in ***Figure 35–30*** to check the brush length. Replace the brushes if necessary.
9. Check the brushes to be sure the brush springs are providing the correct tension. Depending on the manufacturer, the tension may be from 30 to 100 or more ounces (2 to 6 lbs). Check the manufacturer's specifications for the exact amount required for each starter. If the tension is not within the specifications, replace the springs. ***Figure 35–31*** shows an example of how to check spring tension.

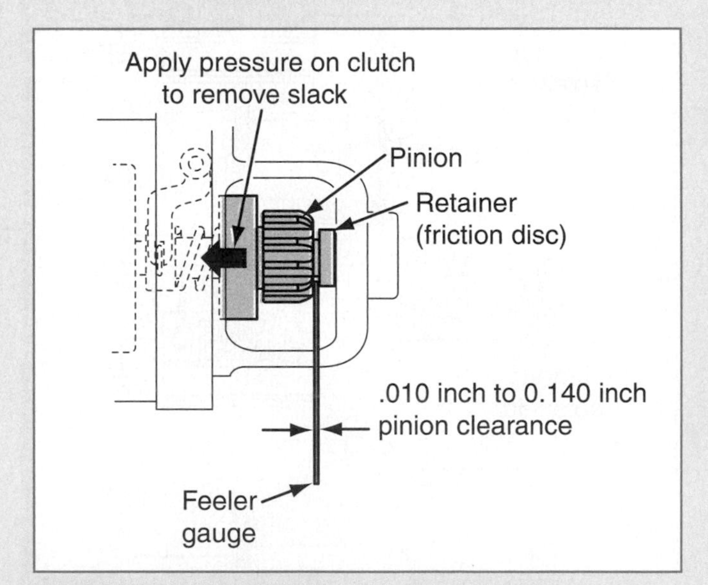

FIGURE 35–29 The starter can be checked for end play between the pinion gear and the retainer.

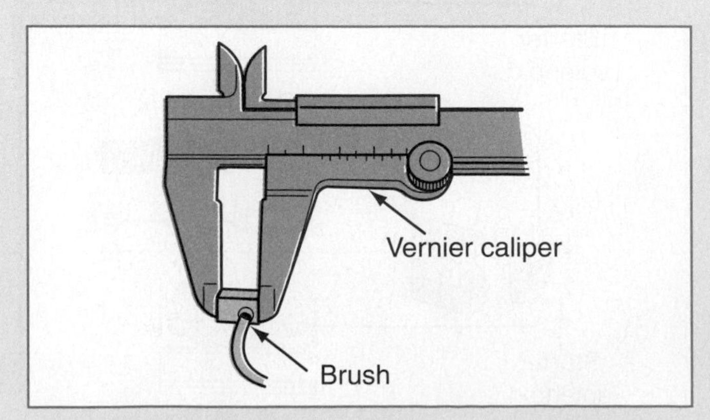

FIGURE 35–30 Use a vernier caliper to check the starter brushes for excessive wear. (Courtesy of American Honda Motor Co., Inc.)

10. With the starter disassembled, check the roller clutch drive for smoothness, instantaneous lockup, and free movement along the shaft. Check for broken or distorted springs on the starter linkage mechanism. Also check the condition of the pinion gear and the ring gear on the flywheel as shown in ***Figure 35–32***. If any of these conditions exist, replace the roller clutch or spring mechanism, or both.
11. Bushings in the end plates wear and need to be replaced in the starter. If the bushings wear excessively, the starter armature may rub on the pole shoes, causing damage to the starter. If the bushings are bad, the starter will require very high current draw during cranking. A growling or scraping sound may also be heard during cranking. Replace the bushings if wear is suspected.
12. The starter armature can be checked for short circuits by using a growler (***Figure 35–33***). A growler is a tool that creates a magnetic field around the armature. A hacksaw blade is placed on top of the armature. When it vibrates, it indicates a short in the armature.
13. Check the armature commutator segments for evidence of burning. This is done by looking for burned copper on the commentator. If there is evidence of burning, the commutator should be checked for out-of-roundness using a dial indicator as shown in ***Figure 35–34***. If so, the commutator can be cut down on a lathe or sanded lightly with #500 or #600 grit sandpaper. The normal procedure when using a lathe is to take light cuts until the worn or bad spots are removed. After cutting down the commutator, remove the burrs with sandpaper.
14. There may be other checks that can be done on the starter circuits and components. These depend on the car manufacturer and the style of the starter. After you have made all the checks on the starter, make sure to clean all parts before reassembly.
15. Lubricate all bushings in the starter with high-temperature grease before assembly.
16. Reassemble the starter as recommended by the manufacturer.
17. Start and run the engine.

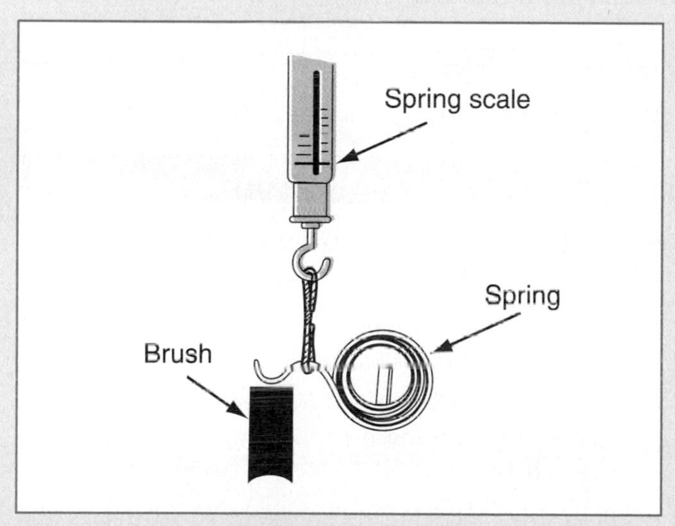

FIGURE 35–31 Use a small spring scale to check the spring tension for the starter motor brushes. (Courtesy of American Honda Motor Co., Inc.)

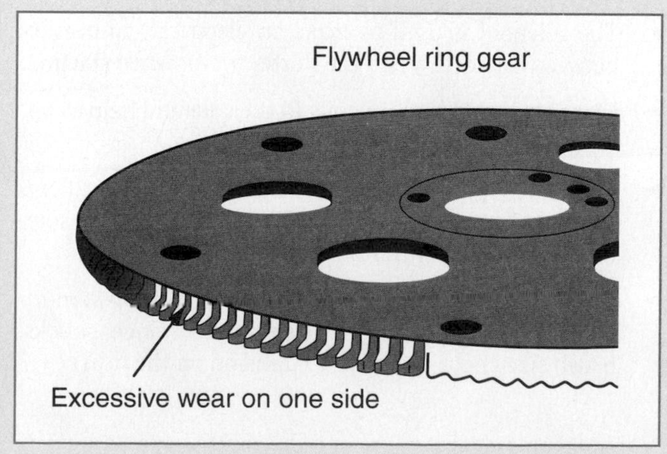

FIGURE 35–32 When rebuilding the starter, always check the condition of the pinion gear and the ring gear on the flywheel.

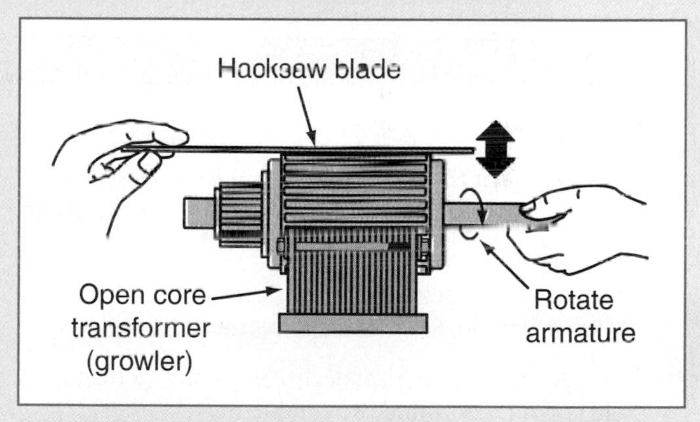

FIGURE 35–33 A growler is used to check for shorts in the armature.

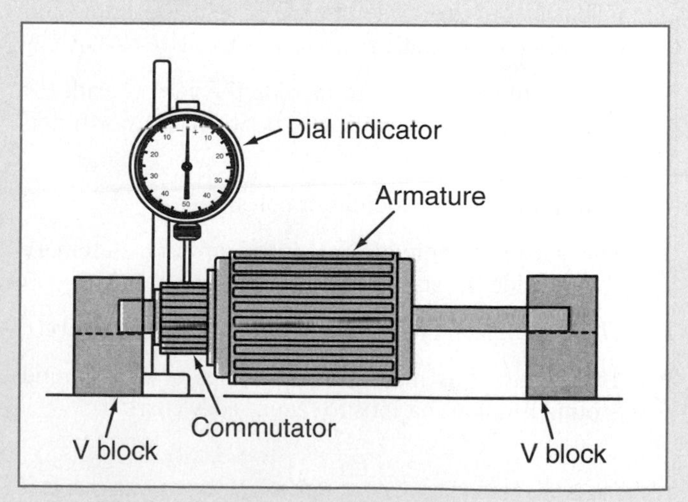

FIGURE 35–34 If the commutator is burned or worn, it can be checked for out-of-roundness using a dial indicator.

Service Manual Connection

There are many important specifications to keep in mind when working with starting systems. To identify the specifications for your engine, you will need to know the VIN (vehicle identification number) of the vehicle, the type and year of the vehicle, and the type of engine. Although they may be titled differently, some of the more common starting system specifications (not all) found in service manuals are listed below. Note that these specifications are typical examples. Each vehicle and engine may have different specifications.

Common Specification	Typical Example
Cranking current	350 amperes
Cranking voltage	8.5 volts
Free speed test	77 amperes
Starter brush length	
New	0.59–0.61 inch
Service limit	0.39 inch
Starter brush spring tension	40–80 ounces
Starter commutator runout	
New	0–0.001 inch
Service limit	0.002 inch

In addition, the service manual will give specific directions for various service and testing procedures. Some of the more common procedures include:

- Overrunning clutch inspection
- Solenoid replacement
- Starter brush replacement
- Starter overhaul
- Starter replacement
- Testing starter current

SUMMARY

The following statements will help to summarize this chapter:

- The purpose of the starting system is to convert electrical energy in the battery to mechanical energy to crank the engine.
- The starter system uses a starter motor, ignition switch, solenoid, and starter drive mechanism.
- Whenever a wire carries electrical current, a magnetic field is built up around the conductor.
- When a conductor that has a surrounding magnetic field is placed in another magnetic field, the aiding and canceling of magnetic fields causes the wire to move.
- A commutator is used to provide the correct path for electricity into a loop of wire between the north and south poles.
- A starter has four windings or poles.
- The starter windings are wound around stationary shoes inside the starter.
- The rotating part of the starter is called the armature.
- The armature is made of many loops of wire wound around iron laminations to reduce eddy currents.
- Brushes are used on a starter motor to connect the electricity to the rotating armature windings.
- The starter uses several drive mechanisms, including the Bendix and the sprag, or overrunning clutch.
- The overrunning clutch uses a set of rollers that become wedged in the drive housing when the starter begins to turn.
- Both drive mechanisms are designed to engage the pinion gear with the flywheel ring gear during starting and to release the pinion gear when the engine begins to run.
- The solenoid is used to make an electrical connection between the battery and the starter motor when starting.
- The pull-in and hold-in coils in the solenoid help to operate the plunger correctly.
- When the ignition switch is closed, electricity flows from the battery, through the switch, through the solenoid, to the starter motor.
- Several checks can be made on the starting system including voltage drops, free speed, clearance checks, brush size checks, and spring tension on the brushes.

TERMS TO KNOW

Can you explain each of the following terms? Review the chapter until you can use each term correctly.

Armature
Chamfer
Commutator
Crank signal
Eddy currents
Inertia
Inertia drives
Laminations
Overrunning clutch drive
Pinion gear
Pole shoes
Shunt
Solenoid
Sprag

REVIEW QUESTIONS

Multiple Choice

1. The starter system converts:
a. Mechanical energy to chemical energy
b. Thermal energy to electrical energy
c. Mechanical energy to electrical energy
d. Electrical energy to mechanical energy
e. Thermal energy to mechanical energy

2. Any wire that has electrical current passing through it:
a. Moves toward a north pole
b. Moves toward a south pole
c. Has a magnetic field around it
d. Has no magnetic field around it
e. Reduces its internal resistance

3. What causes the movement to occur on the armature of a motor?
a. Two magnetic fields opposing and aiding each other
b. Thermal energy forces
c. A north pole pushing against a south pole
d. Gravity
e. Eddy currents

4. The ________ keep(s) the current flowing in the correct direction in the armature.
a. Eddy currents
b. Stator
c. Commutator
d. Brushes
e. Solenoids

5. The north and south poles in a starter are:
a. Always touching each other
b. Made from electromagnets
c. Repelling each other
d. Turned off during starting
e. Operated in pulses

6. The ________ hold(s) the field windings.
a. Solenoid
b. Brushes
c. Hold-in coil
d. Pole shoes
e. Sprags

7. Eddy currents can be reduced in the armature by using a core made of:
a. Plastic
b. Silicon
c. Copper
d. Carbon
e. Metal laminations

8. Which type of starter drive mechanism uses the principle that an object tends to remain at rest or stay in motion?
a. Overrunning clutch
b. Roller clutch
c. Inertia drive
d. Sprag drive
e. All of the above

9. Which type of starter drive mechanism uses a series of sprags?
a. Overrunning clutch
b. Inertia drive
c. Bendix drive
d. Roller clutch
e. None of the above

10. On the overrunning clutch, when the engine starts and turns faster than the starter:
a. The rollers are not being wedged in
b. The pinion gear moves out of mesh with the ring gear
c. The starter will not be forced to turn faster
d. All of the above
e. None of the above

11. A solenoid is used to convert:
a. Mechanical energy to thermal energy
b. Electrical energy to chemical energy
c. Electrical energy to mechanical energy
d. All of the above
e. None of the above

12. When the plunger is energized and the contact disc makes electrical contact inside the solenoid:
a. The hold-in winding releases
b. The starter motor begins to turn
c. The pinion gear is disengaged
d. The ring gear stops turning
e. Brushes release contact on the armature

13. Which coil in the solenoid is shorted out when the starter motor begins to turn?
a. Field coils
b. Hold-in coil
c. Pull-in coil
d. All of the above
e. None of the above

14. Which of the following is not considered a service check on the starter system?
a. Checking the voltage drop of the starter
b. Checking for opens on the field poles
c. Checking the amount of wear on the brushes
d. Checking the operation of the diodes in the circuit
e. Checking the condition of the commutator

15. Which of the following is not considered a service check on the starter system? Checking the:
a. Bushings for wear
b. Circuit for grounds and opens
c. Drive mechanism for voltage drop
d. Commutator for signs of wear
e. Drive mechanism for damaged brushes

The following questions are similar in format to ASE (Automotive Service Excellence) test questions.

16. *Technician A* says that if the commutator is slightly burned, run the starter until it becomes inoperative. *Technician B* says that if the commutator is slightly burned, it may be machined on a lathe to remove the burn spots. Who is correct?
a. A only c. Both A and B
b. B only d. Neither A nor B

17. When tested on a growler, the hacksaw blade vibrates at a specific spot. *Technician A* says that the armature has a short in it. *Technician B* says that the field poles have a short in them. Who is correct?
a. A only c. Both A and B
b. B only d. Neither A nor B

18. The spring tension on the brushes on the starter is weak and below specifications. *Technician A* says that the springs should be bent to obtain more tension. *Technician B* says that the weaker spring tension will not cause any damage and that they need not be replaced. Who is correct?
a. A only c. Both A and B
b. B only d. Neither A nor B

19. The armature of the starter has been rubbing on the field coils. *Technician A* says the problem could be bad bearings. *Technician B* says the problem is misalignment of the solenoid. Who is correct?
a. A only c. Both A and B
b. B only d. Neither A nor B

20. A set of starter brushes has worn down to the service limit. *Technician A* says to replace the starter brushes with new ones. *Technician B* says the starter brush springs should be checked for correct tension. Who is correct?
a. A only c. Both A and B
b. B only d. Neither A nor B

21. In a cranking test on the starter, the voltage reading is about 9 volts. *Technician A* says to replace the starter. *Technician B* says the amperage should be about 15 amperes. Who is correct?
a. A only c. Both A and B
b. B only d. Neither A nor B

22. Various electrical checks can be made on a starter. *Technician A* says that one test is a voltage drop on the ground line. *Technician B* says that one test is battery voltage on the load. Who is correct?
a. A only c. Both A and B
b. B only d. Neither A nor B

23. *Technician A* says that armature end play always needs to be checked during starter overhaul. *Technician B* says that resistance of the field coils should be checked. Who is correct?
a. A only c. Both A and B
b. B only d. Neither A nor B

Essay

24. Describe how magnetism is produced in a coil of wire.

25. What is the definition of commutation?

26. What is the purpose of the pole shoes on a starter?

27. What is the purpose of using laminations of steel on the armature?

28. Describe the operation of the inertia drive clutch.

29. What is the purpose of the solenoid?

30. What is the purpose of the hold-in and pull-in coil on the solenoid?

Short Answer

31. A(n) ________ is cut on the end of the pinion gear to allow the teeth to mesh more easily.

32. On certain vehicles a(n) ________ system is used between the starter and the pinion gear to increase torque.

33. On computerized engines, the crank signal causes the air-fuel ratio to become ________ during cranking.

34. Voltage drop in the starter can be checked using a(n) ________.

35. To measure the speed of the starter in rpm, a(n) ________ test can be performed.

Applied Academics

TITLE: Starter Motor Counter Electromotive Force

ACADEMIC SKILLS AREA: Science

NATEF Automobile Technician Tasks:

The service technician can explain the relationship between electrical current in a conductor and a magnetic field produced in a coil such as the starter solenoid.

CONCEPT:

Starter motors have what is called **c**ounter **e**lectro**m**otive **f**orce (cemf). Cemf is produced because as the motor operates, the armature windings are forced to rotate around a magnetic field. When this happens, all three conditions are there to produce an induced voltage in the armature. These include a conductor, a magnetic field, and motion between the two. Thus, as a starter motor turns, it also produces a voltage and current (cemf) in the armature windings that is opposite in direction to the battery current when operating the starter. The cemf opposes the current necessary to produce the torque when cranking the starter motor. The higher the cemf, the less torque produced inside the starter motor. This concept is graphically illustrated in the charts that follow.

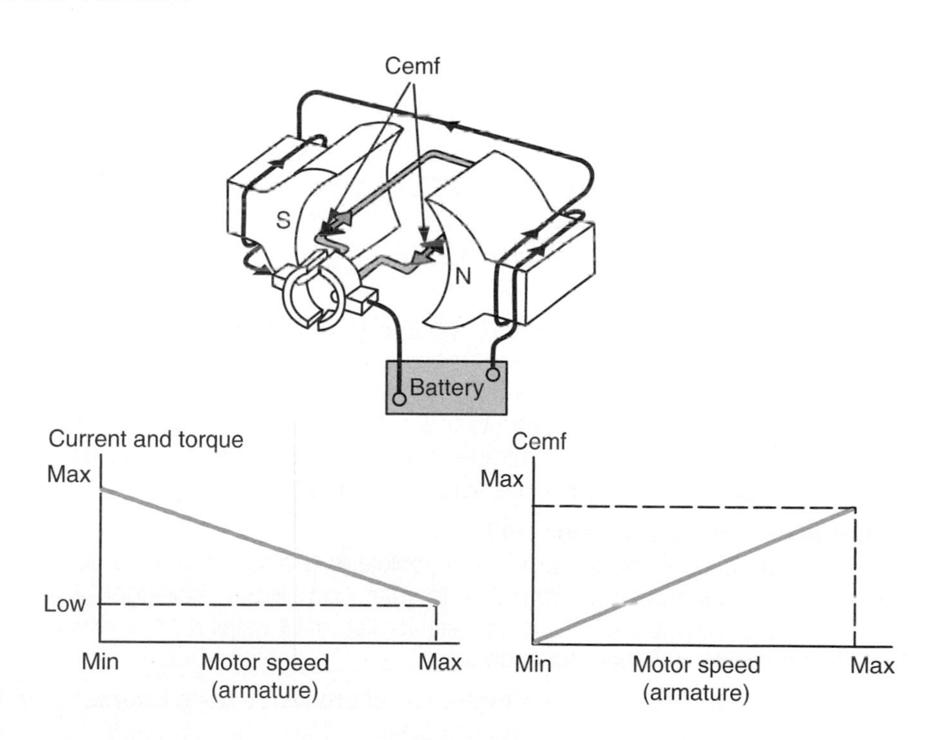

APPLICATION:

What three things are necessary to produce a voltage in current in an operating starter? What causes the cemf to increase or decrease in an operating starter motor? What effect does the cemf have on the torque of a starter motor during operation?

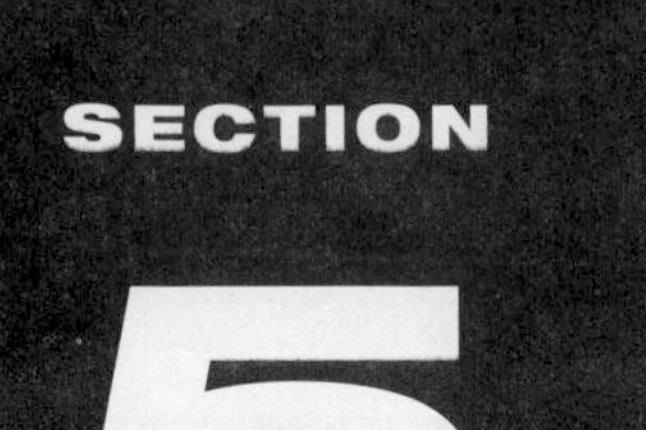

SECTION 5 Emission and Control Systems

Michele Winn, 31

Technician/Shop Manager

Linder Technical Services (Fuel Injector Shop)

Indianapolis, IN

How did you first become interested in the automotive industry?

I was a senior in high school and buying a car was a very big deal to me! My father helped me pick one out, helped pay for it, and offered to pay the insurance with one stipulation: I had to maintain the car and pay for any repairs that were needed. That started my quest for knowledge.

What's your educational background?

After high school, I went on to college to study business administration. After 2 years of college, I still hadn't found what I was looking for, so I quit. A few years later, I attended Lincoln Technical Institute and graduated with an associate's degree in Automotive Service. I currently have six ASE certifications: Brakes, Steering & Suspension, Heating and A/C, Electrical, Engine Repair and Engine Performance. I recently tested for the ASE L1 (Advanced Driveability) but haven't received the results yet.

Do you continue to do training?

Absolutely! As quickly as things are changing, ongoing training is a must! Part of my job here is to conduct training courses on a monthly basis. As a teacher, you're forced to know the subject matter in and out and you end up learning new things every time you teach. In addition to teaching classes, I try to attend around 60 hours of training each year, mostly at trade shows across the country sponsored by groups like the ASA (Automotive Service Association) of Ohio.

What do you like about your job?

Every day is different because each vehicle is different. Even the same make and model vehicles may differ from year to year. And I like a challenge! Most of the time when cars come to me, they have been looked at by several other shops, so I like trying to fix a problem that others couldn't.

What advice do you have for a high school student who is interested in this profession?

Get familiar with trade publications and read articles on how different vehicle systems work and how to properly diagnose and repair them. Get a job in a local shop, even if all they will allow you to do is sweep the floor. Work hard at what they will allow you to do and it will pay off. Also, find the smartest technician in the shop and become his friend. During slow times, ask questions. This is a great way to learn! Decide what you like and specialize. As complex as vehicles are becoming, I believe having a specialty such as fuel injection or electrical will put you in demand.

Characteristics of Air Pollution

CHAPTER 36

OBJECTIVES

After studying this chapter, you should be able to:

- Compare the definitions and effects of the major types of air pollution produced by the automobile.
- Identify how pollution is being controlled through certification of new cars, regulatory agencies, emissions requirements, exhaust gas analyzers, and fuel economy standards.

INTRODUCTION

Before the late 1960s, society placed very little emphasis on pollution and environmental damage. All industries, including the automotive manufacturers, were producing goods with little concern for how the waste products affected the environment. Any product that is manufactured produces waste. By the mid-1960s, there was serious concern about how these pollutants could be reduced. In the early 1960s, government emission standards were set. Because of these standards, automobile emission controls were developed to reduce pollution. This chapter is about the types of pollutants being produced by the automobile and how our society is improving the air quality in the environment.

36.1 TYPES OF POLLUTANTS

CARBON MONOXIDE

Carbon monoxide is a potentially deadly poison gas that is colorless and odorless. When people inhale carbon monoxide in small quantities, it causes headaches and vision difficulties. In larger quantities, it causes sleepiness and, in some cases, death.

Carbon monoxide (CO) is a by-product of incomplete combustion. Carbon monoxide emissions increase as the combustion process becomes less efficient. During perfect combustion the by-products are water (H_2O) and carbon dioxide (CO_2). Carbon monoxide forms in the engine exhaust when there is insufficient oxygen to form carbon dioxide. Thus, whenever the engine operates in a rich air-fuel mixture, increased CO is the result. The best way to reduce CO is to increase the amount of oxygen in the combustion chamber. Carbon monoxide is not one of the chemicals that produces photochemical smog.

In general, carbon monoxide concentration is much lower in diesel exhaust than in gasoline exhaust. ***Figure 36–1*** shows the CO percentages in gasoline and diesel exhaust for different air-fuel ratios. Note that CO is considerably lower in diesel than in gasoline engines. This is because diesel engines are typically operated with a leaner air-fuel ratio than gasoline engines.

HYDROCARBONS

Hydrocarbons (HC) are another type of pollutant produced by the automobile combustion process. Hydrocarbons are also called **organic materials**. Fossil fuels are made of various hydrogen and carbon molecules. Unburned hydrocarbons emitted by automobiles are largely unburned portions of fuel. Any fuel that is partially burned contains

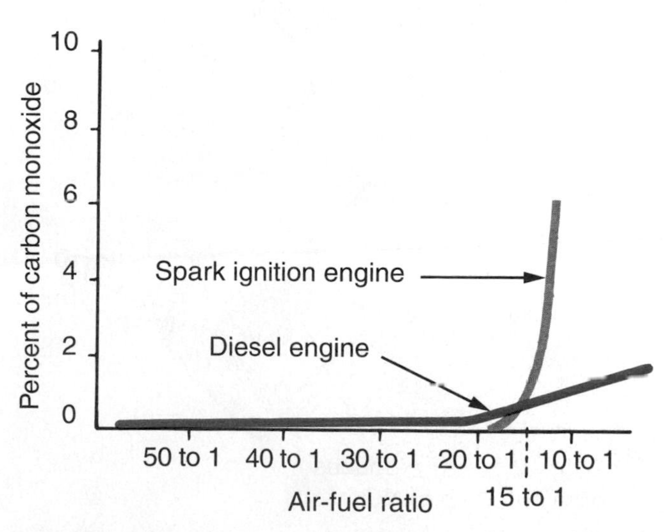

FIGURE 36–1 As the air-fuel ratio changes, the percentage of carbon monoxide in engine exhaust also changes.

hydrocarbons. For example, gasoline in the combustion chamber burns very rapidly. Some of the hydrogen and carbon molecules near the sides of the combustion chamber may get partially burned. This action produces HC in the exhaust. All petroleum products produce small traces of HC during the combustion process. For example, gasoline in the fuel tank also gives off traces of HC.

Most hydrocarbons are poisonous at concentrations above several hundred parts per million. Although they are not as dangerous by themselves as CO, hydrocarbons are the main ingredient in the production of photochemical smog.

NITROGEN OXIDES

Nitrogen oxides (NO_x) are formed by a chemical union of nitrogen molecules with one or more oxygen molecules. Nitrogen oxides form freely under extreme heat conditions. As the combustion process becomes leaner, combustion temperatures typically increase. Higher temperatures cause nitrogen oxides to be produced. When the combustion temperatures reach 2,200 to 2,500 degrees Fahrenheit, the nitrogen and oxygen in the air-fuel mixture combine to form large quantities of nitrogen oxide (***Figure 36–2***).

Nitrogen oxide by itself does not appear to have any important harmful air pollution effects. However, nitrogen oxide reacts with hydrocarbons to form harmful irritating **oxides** and gives photochemical smog its characteristic light brown color. Nitrogen oxide is also considered a major greenhouse gas.

Gasoline engines that are running very lean with higher temperatures produce nitrogen oxides. Diesel engines generally produce higher concentrations of nitrogen oxides because their internal temperatures are usually higher than gasoline engines. The best method of reducing the emission of nitrogen oxides is to reduce the temperature of the combustion process. Doing this, however, will result in less-efficient burning and increases in HC and CO emissions.

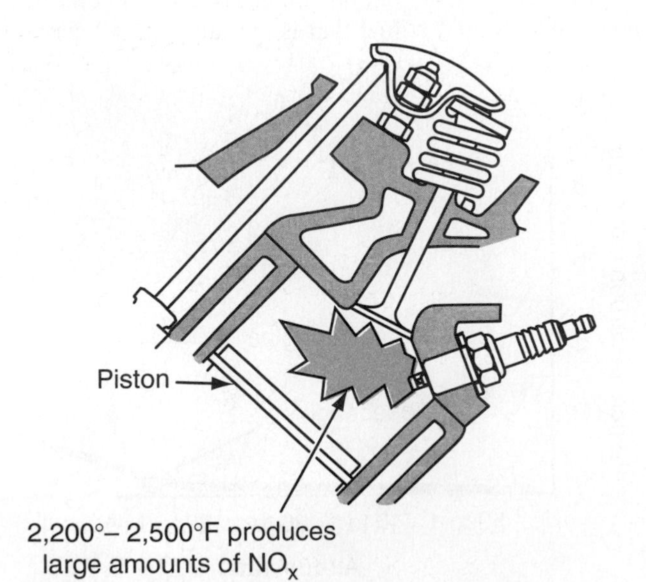

FIGURE 36–2 Whenever combustion temperatures increase to the 2,200–2,500 degree Fahrenheit range, larger quantities of NO_x are produced.

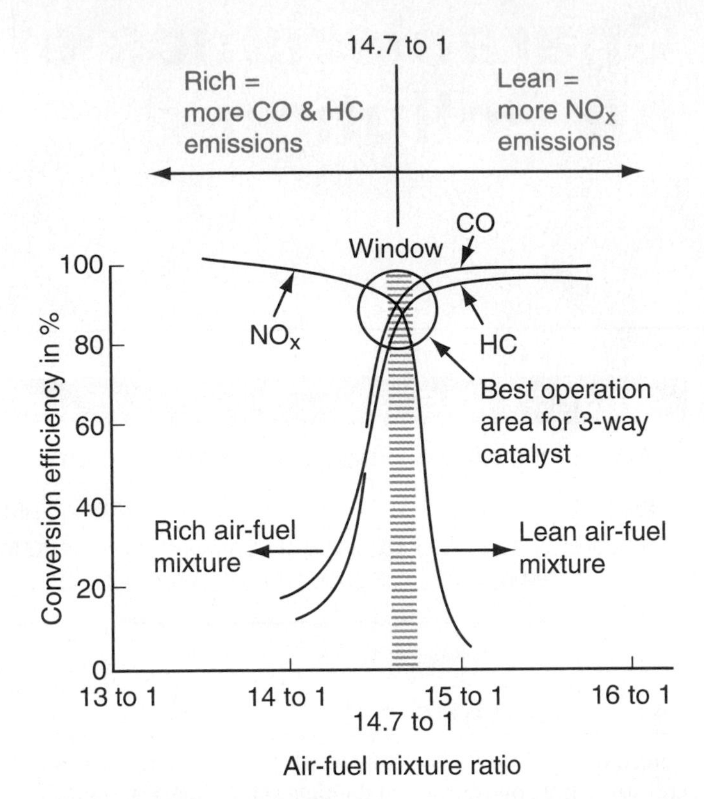

FIGURE 36–3 Whenever steps are taken to reduce HC and CO emissions, NO_x emissions increase. It is very important to keep the air-fuel ratio close to 14.7 to 1 for more effective emission control.

To illustrate this, ***Figure 36–3*** shows how the three emissions discussed are affected by the air-fuel ratio and conversion efficiency. For example, as the air-fuel ratio becomes leaner, nitrogen oxide conversion efficiency decreases, thus producing more of this particular pollution form. The CO and HC emissions increase (conversion efficiency decreases on the chart) as the air-fuel mixture becomes richer. By observing this chart, it can be seen that the best air-fuel ratio for any engine is in the "window" area, or close to 14.7 to 1. Today's engines use computer controls to obtain an accurate air-fuel ratio.

SULFUR DIOXIDE

Sulfur dioxide (SO_2) is another form of air pollution. Sulfur is present in many of the hydrocarbon fuels that are used in engines and power plants. Diesel fuel has slightly more sulfur than gasoline; therefore, sulfur emissions are greater with diesel engines than with gasoline engines. When sulfur gets into the atmosphere, it breaks down and combines with water in the air to produce sulfuric acid. This acid is very corrosive and produces the commonly known acid rain form of pollution. NO_x also contributes significantly to the acid rain problem.

CUSTOMER CONCERN:
Sunlight and Photochemical Smog

In discussions about pollution and the environment, often the term *photochemical smog* is used. Exactly what is photochemical smog?

SOLUTION: Photochemical smog is a by-product of combining three ingredients. When nitrogen oxides (NO_x) and hydrocarbons (HC) are mixed in the presence of sunlight, the result is photochemical smog. Thus, the sunlight helps to produce the photochemical smog. Nitrogen oxides and hydrocarbons are both by-products of gasoline and diesel engine combustion. Photochemical smog is dangerous because it irritates the eyes and corrodes metals and other materials. It is also dangerous to one's health.

PARTICULATE MATTER

Particulates are defined as a form of solid air pollution, such as microscopic solid particles of dust, soot, and ash. They can be solid or liquid matter that floats in the atmosphere. Examples of particulate substances are lead and carbon produced by burning leaded gasoline. These particulates are absorbed directly into the body and can cause severe health hazards.

Engines that use unleaded fuel produce very little particulate matter. Fuel systems that are set very rich, however, may also produce carbon particulates that could be a health hazard.

Particulate matter is damaging because it reduces air visibility and allows less sunlight to reach the ground. Particulates carry damaging materials such as sulfuric acids to the surfaces they strike. There is also some evidence that particulate matter is having an effect on weather patterns.

ADDITIONAL DIESEL POLLUTANTS

The diesel engine uses different fuels, has different compression ratios, and uses a different ignition system than the gasoline engine. Because of these characteristics, diesel engines present two other emission concerns. Diesel engines in general have an exhaust odor that can be considered a nuisance. These exhaust odors have recently been measured so that proper technological steps can be taken to reduce them. In addition, smoke is evident on many diesel engines. The blue-white smoke is caused by vaporized droplets of lubricating or fuel oil. This indicates that maintenance is required. The black-gray smoke often seen on diesel vehicles is caused by unburned carbon particles. This is caused largely by inaccurate air-fuel ratios. Black smoke will increase as the load is increased. In some cases, smoke charts have been developed to measure the amount of smoke being emitted by diesel engines.

36.2 CONTROLLING POLLUTION

NEW CAR CERTIFICATION

One method of reducing pollution from automobiles is to set certain emission standards for new vehicles. Standards are established by the **Environmental Protection Agency (EPA)**. Families of cars are given emissions tests. They are tested for HC, CO, and NO_x emissions. Tests are performed under cold start, normal, and hot start conditions. If the vehicle passes the emission standards set by the EPA, it can be offered for sale to the public.

PROGRESS IN REDUCING EMISSIONS

Much has been done recently to reduce automobile emissions. From 1970 to 1980, a number of **emission control** devices were placed on all vehicles. These devices drastically reduced the amount of CO, particulate matter, HC, and NO_x in the atmosphere. In fact, there was approximately a 37% drop in CO, a 50% drop in particulate matter, and an 11% drop in HC emissions for this 10-year period.

Although these emissions were reduced, NO_x increased during this period. When pollution standards were started in 1968, the automotive manufacturers worked to reduce HC and CO. As these were reduced, however, the engine combustion process became leaner, increasing the production of NO_x. Thus, efforts continue to be made to reduce NO_x in today's vehicles.

As more and more computer controls were put on engines, emissions were further reduced. Today's engines are very efficient, controlling all emissions (including nitrogen oxides) well within the standards set by the federal government.

Another factor that helps to reduce exhaust emissions is keeping engines properly maintained. During a recent EPA test, cars were tested for carbon monoxide and hydrocarbon emissions before and after maintenance. ***Figure 36–4*** shows the results. After maintenance, CO emissions dropped an average of 66%.

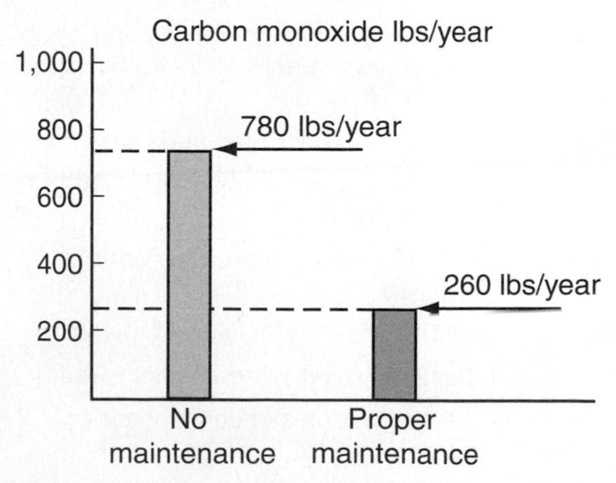

FIGURE 36–4 In one EPA test, cars checked at random produced 779 pounds of carbon monoxide per year. After proper maintenance, their CO emissions dropped an average of 66%.

REGULATING AGENCIES

Several regulating agencies have been established to help identify standards and regulations for automobiles. California became the first state to use air pollution standards. California also established an Air Resources Board (ARB), which helps to enforce the emission standards. California standards are typically much stricter than those of other states.

In 1963, the Clean Air Act was passed by the U.S. Congress. This act set allowable levels of HC, CO, and NO_x. Since that time, many federal automotive emission milestones have occurred. The original Clean Air Act of 1963 has been amended several times. ***Figure 36–5*** shows some of the more important federal automotive emission legislation that has been enacted over the years. Note that in 1970 and in 1990 additional amendments were added to the original Clean Air Act. Some of the features of the amendments of 1990 include:

- Stricter tailpipe emission standards for cars, trucks, and buses
- Expansion of inspection and maintenance (I/M) programs with more stringent testing
- Attention to fuel technology (the development of alternate fuels as a potential area of emission reduction)
- Greater focus on efforts to reduce total mileage growth in vehicle travel
- Study of nonroad engines (e.g., boats, farm equipment, home equipment, construction equipment, lawn mowers, etc.) to determine necessary regulations
- Mandatory alternative transportation programs (for example, car pooling) in heavily polluted cities

Since that time, the Clean Air Act has proposed many changes and continued to reduce emission levels. These changes and the time proposed for implementation are listed in ***Figure 36–6***.

YEAR	EMISSION STANDARDS
1963	• Original Clean Air Act (CAA) brought into law • PCV systems introduced on vehicles
1968	• HC and CO exhaust controls added to vehicles
1970	• Clean Air Act amendments to current policies • Environmental Protection Agency (EPA) formed
1971	• Evaporative emission standards enacted • EVAP systems added to vehicles
1972	• First inspection and maintenance (I/M) program introduced
1973	• NO_x exhaust standards enacted • EGR systems added to vehicles
1975	• First catalytic converter (oxidizer)
1979	• Close loop computer control added to vehicles (California)
1980 1/2	• Close loop computer control and 3-way catalytic converters added to certain vehicles (federal)
1989	• Gasoline volatility standards enacted
1990	• New Clean Air Act amendments to current policies
1995	• I/M 240 testing for gasoline vehicles required
1996	• OBDII vehicle compliance required
1999	• Tier 2 emission standards approved
2004	• Tier 2 emission standards phase-in begins
2009	• Tier 2 emission standards full compliance

FIGURE 36–5 Various emission standards and programs.

CLEAN AIR ACT	
1992	Nationwide limits on maximum gasoline vapor pressure set. Oxygen content regulation for 39 high-pollution areas.
1993	Production of leaded gasoline vehicles halted. Reduction of sulfur by 80% in diesel fuels required.
1994	Phase-in of tighter tailpipe standards for light-duty vehicles. First I/M 240 (Inspection and Maintenance program) in some areas. Phase-in of cold-temperature CO standard for light-duty vehicles. Trucks and buses required to meet tighter diesel particulate standard.
1995	Reformulated gasoline required in nation's worst (9) polluted cities. Warranty covering emission control systems required. I/M 240 testing begins in most nonattainment areas.
1996	I/M 240 testing operating in nonattainment areas. Phase-in of California's "Clean Fuel" pilot programs. Lead banned in motor vehicle fuel. New cars and light trucks meet tighter tailpipe and cold CO standard.
1998	Clean fuel fleet programs begin in 19 states.
2001	Second phase of California's "Clean Fuel" program begins.

FIGURE 36–6 Changes and implementation times are shown for the Clean Air Act.

The EPA is the agency responsible for enforcing the Clean Air Act. Many states have passed their own laws. These laws must be approved by the EPA. These laws were based on different altitude operation, such as in mountainous regions. As a result, there were several emission standards. These included (1) California vehicles, (2) federal 49 state vehicles, and (3) high-altitude vehicles.

The manufacturers must be able to meet the standards set by the EPA or other agency. It is up to the manufacturer, however, to determine the type of technology used to reduce emissions.

EMISSION REQUIREMENTS

The Clean Air Act of 1963 set a national goal of clean and healthy air. Specific responsibilities were set for government and private industry to reduce emissions. Today's vehicles average between 70% and 90% less pollution over their lifetimes than the cars produced and manufactured in the 1970s. The average per-vehicle emission of hydrocarbons in grams per mile is shown by year, from 1960 to 2010, in ***Figure 36–7***. Note the significant drop and leveling off of average emissions through this time period.

Unfortunately, other factors continued to increase the pollution from cars. While each vehicle now produces fewer emissions per mile, the number of miles driven every year has increased. ***Figure 36–8*** shows how the vehicle miles driven increases over the same time period. From 1960 to today, mileage driven has more than doubled. This increased mileage has resulted in many parts of the country not meeting clean air standards. The EPA has identified these as "Nonattainment Areas." They are classified according to the level of air pollution. ***Figure 36–9*** shows several maps of the United States. Each map shows the nonattainment areas for a particular type of air pollution. Particulate matter, carbon monoxide, sulfur dioxide, and ozone maps are shown. Although helpful in identifying areas and cities within the United States that are not meeting clean air standards, note that these maps change from month to month. The ones shown here serve as examples of how such data is represented.

Because of such data, various standards have been placed on car manufacturers. ***Figure 36–10*** shows how the car manufacturers must change to meet air quality standards. Over the years, Congress has approved several sets of standards. These standards are referred to as Tier 0, Tier 1, and Tier 2 standards. Tier 0 and Tier 1 standard data is shown for reference. Tier 2 standards only show NO_x, sulfur, and particulate matter. They took effect in 2004, and the car manufacturers must be in complete compliance by 2009. There are four important components of such standards:

1. The readings are measured in g/mi or grams per mile.
2. The Tier 2 standards are for all passenger and light-duty trucks.
3. Both federal and California standards are shown for Tier 0 and Tier 1 standards, but for Tier 2, all standards are the same.

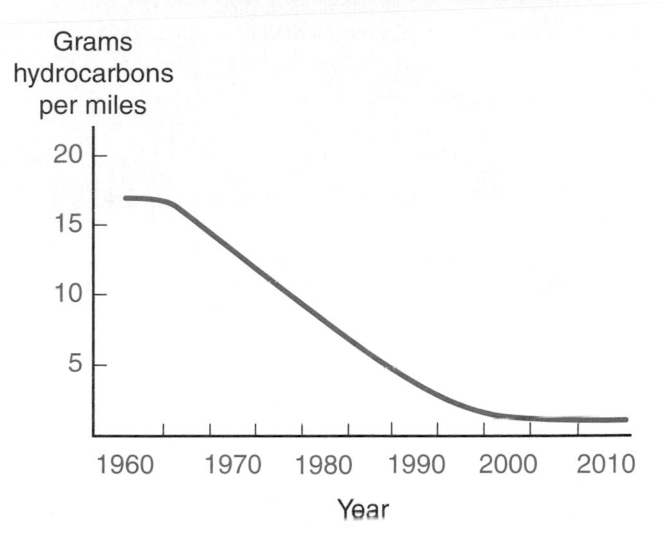

FIGURE 36–7 Average pollution per vehicle emissions in hydrocarbons.

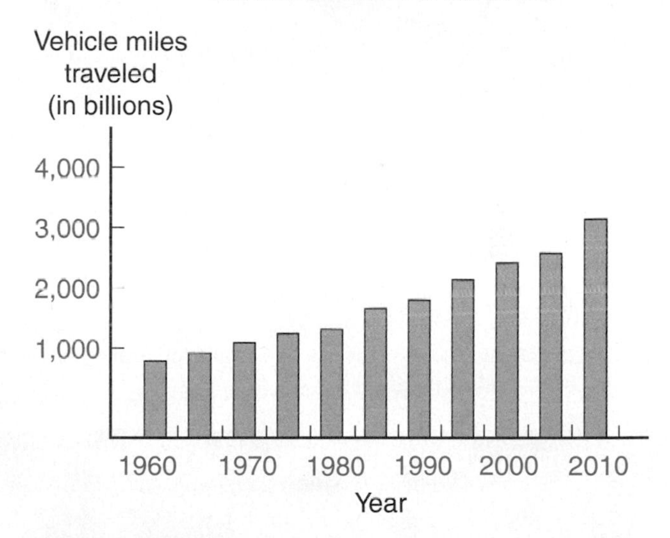

FIGURE 36–8 Because vehicle miles traveled each year continue to increase, some areas in the United States are not meeting clean air standards.

4. Reading this chart, if a passenger vehicle has 30,000 miles on it, outside of California, the Tier 1 standards for NO_x are 0.4 g/mi. If a person has a light-duty truck built after 2004, the vehicle will have to have a NO_x reading of 0.07 g/mi.

Emission control standards and other emission control information are often placed under the vehicle's hood (***Figure 36–11***). These are used to identify the type of vehicle and to help service technicians identify the amount of emissions and vacuum routing diagrams for each vehicle. These labels are often referred to as **vehicle emission control information (VECI)**.

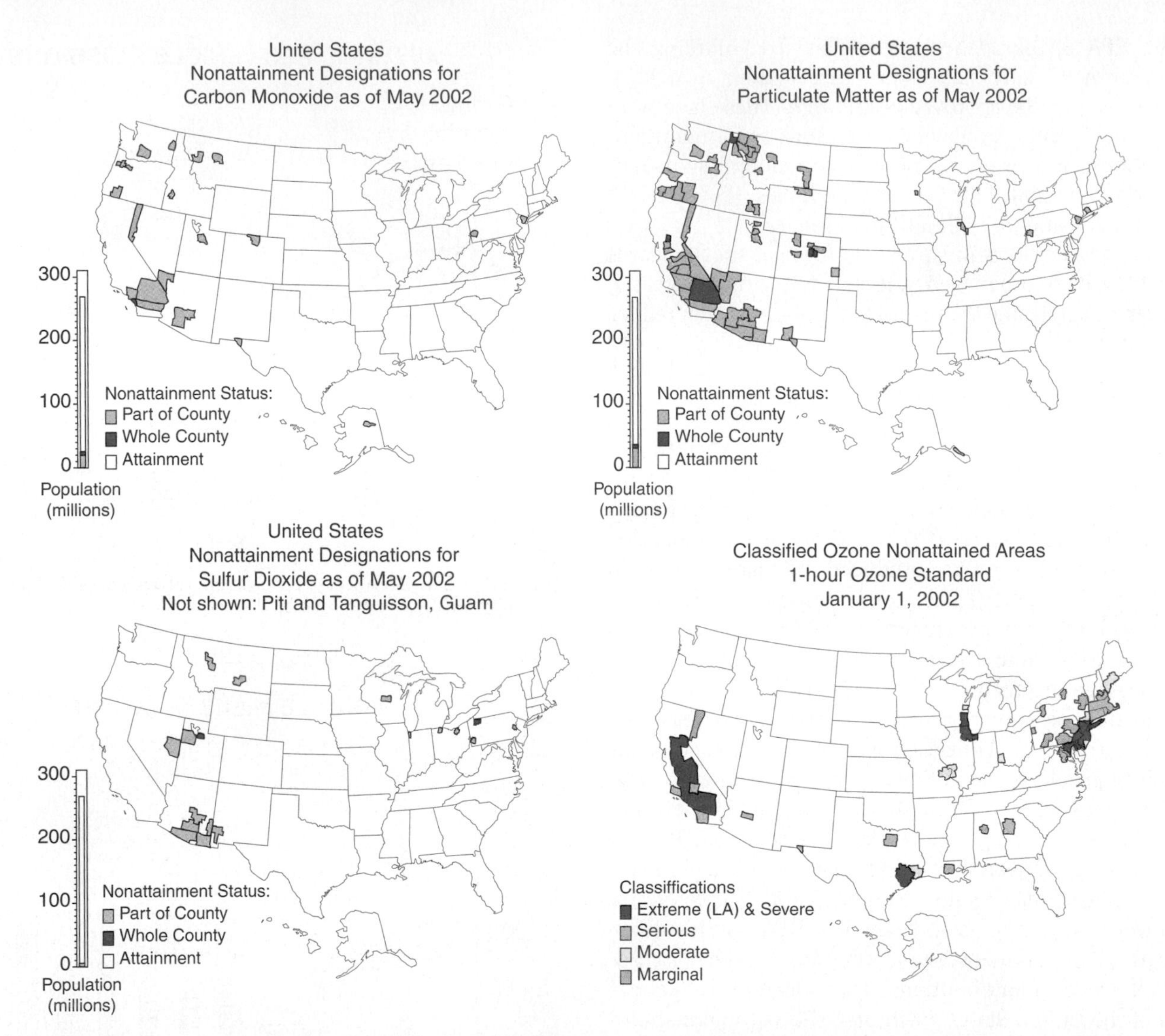

FIGURE 36-9 The nonattainment areas are not meeting the clean air standards due to the increased mileage of each vehicle being driven.

EPA TIER STANDARDS EXHAUST EMISSION STANDARDS

Vehicle Type	Emission Category		Certification Standards—g/mi					
			5 Yrs/50,000			10 Yrs/100,000		
			HC	CO	NO_x	HC	CO	NO_x
Passenger cars	Tier 0	Fed.	0.41	3.4	1.0	—	—	—
		CA	0.41	7.0	0.4	—	—	—
	Tier 1	Fed.	0.41	3.4	0.4	—	4.2	0.6
		CA	—	3.4	0.4	—	4.2	0.6
Light-duty trucks	Tier 0	Fed.	—	—	—	—	—	—
		CA	0.41	9.0	0.4	—	—	—
	Tier 1	Fed.	—	3.4	0.4	—	4.2	0.6
		CA	—	3.4	0.4	—	4.2	0.6

Vehicle Type		NO_x	Sulfur Standard	Particulate Matter
New passenger cars and light-duty trucks	Tier 2 fed.	.07 g/mi	120 ppm 30 ppm by 2006	.08 g/mi

FIGURE 36-10 Emission standards cover the maximum grams per mile for HC, CO, and NO_x emissions.

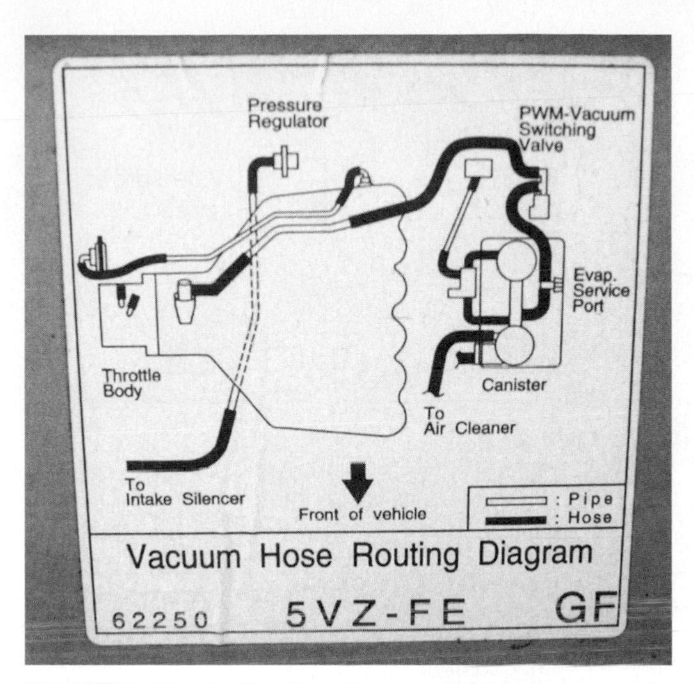

FIGURE 36–11 Certain pollution control information can be found under the hood of a vehicle.

INFRARED EXHAUST ANALYZER

It is very important to keep the engine emission levels within federal and state standards. These emission levels can be tested by using an **infrared exhaust analyzer**. There are three types of infrared exhaust analyzers. There is the two-gas analyzer, the **four-gas analyzer**, and the five-gas analyzer. The two-gas analyzer is able to read the amount of HC and CO emissions. The four-gas analyzer gives much more information. This analyzer is able to read the levels of four chemicals:

1. HC (hydrocarbons)
2. CO (carbon monoxide)
3. CO_2 (carbon dioxide)
4. O (oxygen)

The five-gas analyzer can read nitrogen oxides as well.

Although O is not considered a dangerous emission, it provides useful data about combustion efficiency. Also, CO_2 is considered a major greenhouse gas. The five-gas analyzer is used on all newer vehicles. It provides accurate information for the service technician to evaluate the condition of and adjustments needed on the engine. The following are examples of the type of problems and conditions that can be diagnosed using an exhaust gas analyzer:

- Ignition system condition and problems
- Vacuum hose condition and leaks
- Air filter condition
- Catalytic converter condition
- Computer control system problems
- Fuel injection or carburetor condition
- Valve and piston conditions
- Condition of pollution devices (e.g., PCV, EGR)

I/M 240 TESTING

As listed in the information about the Clean Air Act, there is a mandated testing procedure called **I/M 240 testing**. This test is used to measure the emissions of the vehicle under specific driving conditions on a dynamometer. Refer to ***Figure 36–12***. The I/M 240 test begins by driving the vehicle onto the dynamometer, securing the vehicle with restraints, properly placing the exhaust collection device,

FIGURE 36–12 The I/M 240 test puts a vehicle on a dynamometer to test the engine against various emission standards.

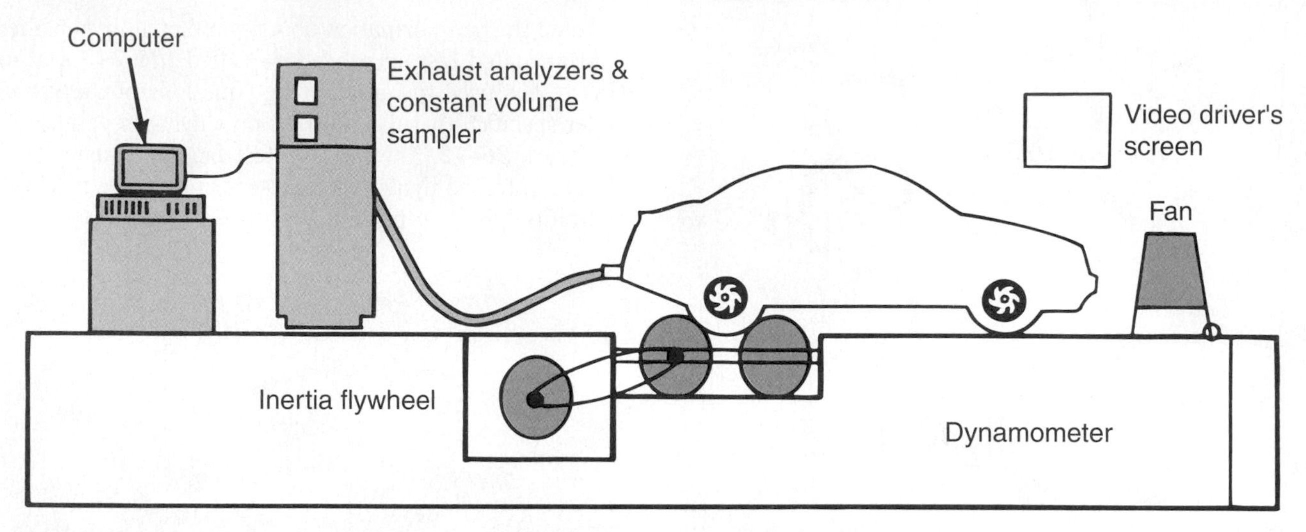

FIGURE 36–13 The dynamometer and associated testing equipment are shown for the I/M 240 test.

and positioning the auxiliary engine cooling fan. An inspector then "drives" the vehicle according to a prescribed cycle that is shown on the video screen in front of the vehicle. ***Figure 36–13*** shows the complete I/M 240 testing equipment. The inspector follows a specified driving cycle by accelerating and braking just as if the vehicle were being driven on a city street. A cursor on the video screen indicates vehicle speed. The inspector is required to keep the cursor within a tolerance band on the video trace. Referring to ***Figure 36–14***, first the inspector accelerates to 20 miles per hour, then slows down to about 14 miles per hour, then increases back to 20 miles per hour, and so on. During this I/M 240 test, the exhaust analyzer and computers monitor the exhaust gas output of the vehicle. At this point, the exhaust emissions can be compared to the EPA standards.

To determine the emission levels, second-by-second emission measurements are monitored by the computer. The computer continually assesses the emission levels and applies a sophisticated pass-fail equation. This equation determines if the exhaust emissions are clean or very dirty (high emissions). If the computer determines that the engine is very clean, the test ends after phase one (about 97 seconds). If the computer determines that the engine is not meeting EPA standards, then the test is continued up to 240 seconds (thus, the test name I/M 240). At this point, the test has been run long enough to accurately determine if the vehicle passes or fails the EPA standards.

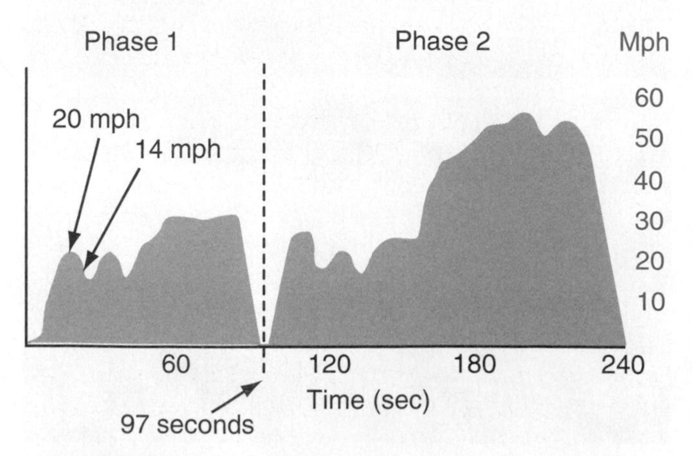

FIGURE 36–14 This is an example of an I/M 240 driving test cycle, shown on a video screen, over a certain time in seconds.

Older vehicles that had fewer emission controls do not need to meet standards as strict as the late-model vehicles. Always check the correct service manual to determine the exact amounts of CO, HC, and NO_x acceptable for each year of vehicle.

STOICHIOMETRIC AIR-FUEL RATIO

Technology and research have continued to control pollutants and reduce exhaust emissions. This research discovered that to effectively reduce gasoline engine emissions, the air-fuel ratio must be as close as possible to 14.7 to 1. The term most often used to describe this ideal ratio is *stoichiometric*. An air-fuel ratio of 14.7 to 1 provides the best control of the three major pollutants in the exhaust (before the catalytic converter). To help understand how various pollutants are affected by the air-fuel ratio, refer to ***Figure 36–15***. This graph was developed under specific engine operation. On the bottom horizontal axis the air-fuel ratio is shown. Lean ratios are shown to the right and richer ratios are shown to the left. Five gases are shown for reference. These include carbon monoxide (CO), nitrogen oxide (NO_x), hydrocarbons (HC), carbon dioxide (CO_2), and oxygen (O). All five gases exist in the exhaust of the gasoline engine. The dangers of CO, NO_x, and HC were discussed earlier in this chapter. In addition, CO_2, which is normally not considered dangerous, is considered a greenhouse gas and, thus, is also

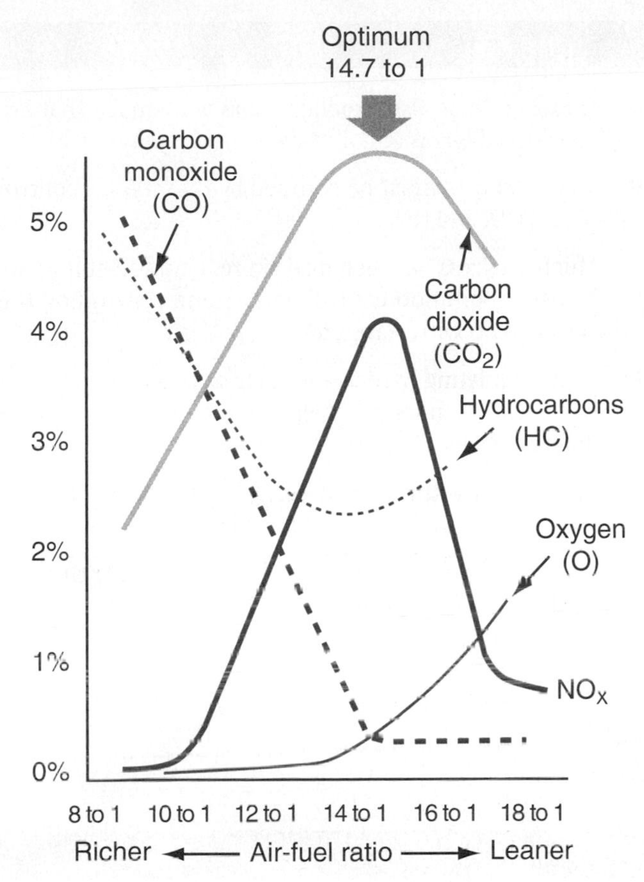

FIGURE 36–15 The most ideal air-fuel ratio is 14.7 to 1. The exact air-fuel ratio determines the level of O, HC, CO, NO_x, and CO_2.

thought of as a contributor to environmental damage. Oxygen, on the other hand, is not necessarily dangerous. However, it is used to help determine air-fuel ratios with the PCM. Various scales for these gases are also included on the vertical axis.

Several conclusions can be made by reading this chart. Some include:

- NO_x and CO_2 are the highest at or near a 14.7 to 1 ratio.
- The leaner the engine air-fuel ratio, the lower the CO.
- HC are lowest at or near the 14.7 to 1 ratio.
- O continues to increase as the air-fuel ratio gets leaner.

Based on these conclusions, engine control systems can be designed to maximize the air-fuel ratio and thus minimize vehicle emissions.

FUEL ECONOMY STANDARDS

As the emission requirements became stricter, the fuel economy of vehicles was reduced. In 1973, the United States faced an "energy crisis." There were severe supply and demand problems, especially in obtaining petroleum for gasoline. The supply of gasoline was much less than the demand. This resulted in federally mandated economy standards.

FUEL ECONOMY STANDARDS

MODEL YEAR	MPG	TOTAL IMPROVEMENT OVER THE 1974 MODEL YEAR
1978	18.0	50%
1979	19.0	58%
1980	20.0	67%
1981	22.0	83%
1982	24.0	100%
1983	26.0	116%
1984	27.0	125%
1985	27.5	129%
1986	26.0	116%
1987	26.0	116%
1988	26.0	116%
1989	26.5	120%
1990	27.5	129%
1992	27.5	129%
1994	27.5	129%
1996	27.5	129%
1997	27.5	129%
1998	27.5	129%
1999	27.5	129%
2000	27.5	129%
2001	27.5	129%
2002	27.5	129%
2003	27.5	129%
2004	27.5	129%
2005	27.5	129%
2006	27.5	129%
2007	27.5	129%

FIGURE 36–16 Automotive manufacturers must meet corporate average fuel economy (CAFE) standards.

Congress enacted the Energy Policy Conservation Act. This act set standards for the manufacturers to follow. These are referred to as **CAFE (corporate average fuel economy)** standards. ***Figure 36–16*** shows the corporate average fuel economy for cars through several years. Since 1990, the CAFE standards have changed very little. While certain vehicle lines may obtain less than the standard, others in the fleet will obtain higher than the standard. However, while the CAFE standards have remained constant in the late 1990s and the early 2000s, there is much debate about changing these standards. Many organizations would like to see these standards made more stringent; others are opposed. Thus, continued debate will occur over the limits and when these standards will change.

In order for the manufacturers to meet both the fuel economy and emission level standards, it is necessary to control HC, CO, and NO_x emissions. This must be done while maintaining or increasing fuel economy standards and performance. This is accomplished with the use of electronic fuel-air control, the catalytic converter, and the addition of port fuel injection on engines.

SUMMARY

The following statements will help to summarize this chapter:

- Pollution from automotive exhaust emissions is considered a harmful contamination of our environment.
- The combustion of hydrocarbon (HC) fuels is considered a major source of air pollution.
- Carbon monoxide (CO), an air pollutant, is a colorless, odorless gas that is produced by a lack of oxygen in the combustion process.
- Nitrogen oxides (NO_X) are produced when there are high temperatures within the combustion process, usually above 2,200 degrees Fahrenheit.
- HC are produced when there is incomplete combustion of fuel.
- Sulfur dioxides (CO_2) are produced from the acid in crude oil before refining.
- Sulfur dioxides and nitrogen oxides contribute to the acid rain problem in the environment.
- Diesel engines also produce odors and smoke that are considered forms of pollution.
- Today, all cars must be certified by the EPA for control of HC, CO, and NO_X.
- Much progress has been made in reducing all pollutants by using pollution control devices and controlling the combustion with computers.
- Many regulating agencies and acts have been established by the U.S. Congress to help enforce the regulations in different states.
- Each manufacturer must design and manufacture vehicles that pass the standards set by these agencies.
- Federal corporate average fuel economy (CAFE) standards have been implemented to improve the fuel mileage of vehicles.

TERMS TO KNOW

Can you explain each of the following terms? Review the chapter until you can use each term correctly.

CAFE (corporate average fuel economy)
Carbon monoxide
Emission control
Environmental Protection Agency (EPA)
Four-gas analyzer
Hydrocarbons
I/M 240 testing
Infrared exhaust analyzer
Nitrogen oxides
Organic materials
Oxides
Particulates
Vehicle emission control information (VECI)

REVIEW QUESTIONS

Multiple Choice

1. Which of the following emissions from the automobile is an odorless, tasteless gas that can cause death?
a. CO
b. NO_x
c. HC
d. SO_x
e. None of the above

2. Which of the following emissions from the automobile is a contributor to acid rain?
a. CO
b. NO_x
c. HC
d. All of the above
e. None of the above

3. Which of the following emissions from the automobile increases with combustion temperatures above 2,200 degrees Fahrenheit?
a. CO
b. NO_x
c. HC
d. SO_x
e. Particulates

4. As combustion efficiency increases to reduce HC and CO emissions, what happens to NO_x?
a. NO_x increases
b. NO_x decreases
c. There is no effect on NO_x
d. NO_x is completely eliminated
e. None of the above

5. Another term for HC is ________.
   a. CO
   b. Organic materials
   c. NO_x
   d. Particulates
   e. SO_x

6. Lead in gasoline from years ago is considered what type of pollution?
   a. NO_x
   b. CO
   c. HC
   d. Particulate
   e. Photochemical smog

7. What is the CAFE standard for a vehicle manufactured in 2003?
   a. 22.5 mpg
   b. 27.5 mpg
   c. 31.5 mpg
   d. 38.5 mpg
   e. CAFE has nothing to do with miles per gallon

8. What agency will help enforce emission standards on automobiles?
   a. U.S. Congress
   b. U.S. CAFE
   c. EPA
   d. ARB
   e. None of the above

9. The original Clean Air Act was instituted in which year?
   a. 1945
   b. 1963
   c. 1977
   d. 1980
   e. 1995

10. Since 1960, average per vehicle HC emissions have dropped steadily, but the ________ have increased.
    a. Vehicle miles traveled
    b. CO emissions
    c. NO_x
    d. All of the above
    e. None of the above

The following questions are similar in format to ASE (Automotive Service Excellence) test questions.

11. *Technician A* says that as the engine air-fuel ratio becomes leaner, NO_x increase. *Technician B* says that as the engine air-fuel ratio becomes richer, HC increase. Who is correct?
    a. A only   c. Both A and B
    b. B only   d. Neither A nor B

12. *Technician A* says that NO_x, CO, and SO_2 emissions are all checked before the manufacturer sells the vehicle. *Technician B* says that NO_x, CO, and HC emissions are all checked before the manufacturer sells the vehicle. Who is correct?
    a. A only   c. Both A and B
    b. B only   d. Neither A nor B

13. A four-gas exhaust analyzer is being used on an engine. *Technician A* says that CO and HC can be tested. *Technician B* says that CO_2 and O can be tested. Who is correct?
    a. A only   c. Both A and B
    b. B only   d. Neither A nor B

14. *Technician A* says that there are federally mandated standards for emissions amounts. *Technician B* says that there are state standards for emission amounts. Who is correct?
    a. A only   c. Both A and B
    b. B only   d. Neither A nor B

15. *Technician A* says that the exhaust emission levels can be tested with a voltmeter. *Technician B* says that exhaust emission levels can be tested with an infrared exhaust analyzer. Who is correct?
    a. A only   c. Both A and B
    b. B only   d. Neither A nor B

16. *Technician A* says that to check exhaust emissions on older vehicles, the four-gas analyzer should be used. *Technician B* says that to check the exhaust emissions on newer vehicles, the two-gas analyzer should be used. Who is correct?
    a. A only   c. Both A and B
    b. B only   d. Neither A nor B

17. *Technician A* says that the I/M 240 test is a measure of engine fuel consumption. *Technician B* says that the I/M 240 test has a phase 1 and phase 2. Who is correct?
    a. A only   c. Both A and B
    b. B only   d. Neither A nor B

18. A vehicle is being given the I/M 240 test. *Technician A* says the test measures emissions over different driving conditions. *Technician B* says the test measures emissions when the engine is at idle only. Who is correct?
    a. A only   c. Both A and B
    b. B only   d. Neither A nor B

Essay

19. What is carbon monoxide?

20. What three forms of air pollution are commonly checked on a car?

21. What does CAFE stand for?

22. What is the most efficient air-fuel ratio?

Short Answer

23. The gas from an engine that is considered odorless, colorless, and deadly is called ________.

24. When combustion temperatures reach 2,200 degrees Fahrenheit or more, ________ emissions are produced.

25. The black-gray smoke often seen on diesel vehicles is caused by unburned ________ particles.

26. The federal emission standards for passenger cars in 2006 is ________ grams per mile of CO.

27. The maximum length of time for the I/M 240 test is ________.

Applied Academics

TITLE: The Greenhouse Effect

ACADEMIC SKILLS AREA: Science

NATEF Automobile Technician Tasks:

The service technician develops and maintains an understanding of all federal, state, and local rules and regulations regarding environmental issues related to automobiles. The technician uses such things as government impact statements, media information, and general knowledge of pollution and waste management to correctly use materials and dispose of waste products that result from the performance of a repair task.

CONCEPT:

One major environmental issue facing our society today is called the greenhouse effect. The greenhouse effect is defined as a process in which the temperature of the Earth's atmosphere continually increases. It is also called global warming. Research has shown that global warming can cause major weather patterns to change, growing seasons to be altered, and so on. Global warming is caused by increasing certain gases in the atmosphere. These gases are shown in the table that follows. Burning any fossil fuel, such as gasoline, diesel fuels, natural gas, and so on, produces various greenhouse gases. Unfortunately, because automobiles use refrigerants and burn oil (gasoline), they have become a major contributor to global warming. Automobiles produce greenhouse gases such as carbon dioxide, methane, and nitrogen oxide. In addition, the refrigerants used in air-conditioning systems produce chlorofluorocarbons, known as CFCs. That is why it is important for the service technician to be able to check and service the automobile engine and other systems, thereby reducing pollution to our environment.

Gas	Atmospheric Concentration (ppm)	Annual Increase (percent)	Life Span (years)	Relative Efficiency (CO_2 = 1)	Current Greenhouse (percent)	Principal Sources of Gas
Carbon dioxide	351.3	0.4	x^1	1	57	
Fossil fuels					44	Coal, oil, natural gas
Biological					13	Deforestation
Chlorofluorocarbons	0.000225	5	75–111	15,000	25	Foams, aerosols, refrigerants, solvents
Methane	1.675	1	11	25	12	Wetlands, rice, fossil fuels, livestock
Nitrous oxide	0.31	0.2	150	230	6	Fossil fuels, fertilizers, deforestation

Carbon dioxide is a stable molecule with a 2–4 year average residence time in the atmosphere.

APPLICATION:

Which greenhouse gas do you think is the most detrimental to our environment? When gasoline is burned, which type of greenhouse gases are produced? Which greenhouse gas do you think is the least detrimental to our environment? Which type of greenhouse gas has the longest life span in years?

CHAPTER 37

Emission Control Systems

OBJECTIVES

After studying this chapter, you should be able to:

- Explain how the PCV system works.
- Describe how evaporative emission controls operate.
- Identify the purpose of the carbon canister.
- Examine how intake and exhaust emissions are controlled.
- Describe several devices that are used to control combustion efficiency.
- Examine electronic and computer control of emissions.
- Identify common problems, their diagnosis, and service procedures used on pollution control devices.

INTRODUCTION

During the mid-1960s and 1970s, our society became very concerned about the amount of pollution produced by both industries and automobiles. Because of this interest, it became mandatory that emission controls be installed on all automobiles. The automobile has been analyzed very carefully to determine the types of pollution that it produces. Usually the fuel tank, exhaust gases, crankcase area, and fuel system produce the majority of pollution in the automobile. Automobile emissions can be controlled with various devices. This chapter examines various emission control systems that have been used in the past few years.

Certification Connection

ASE Connection: The information in this chapter can help you prepare for the National Institute for Automotive Service Excellence (ASE) certification tests. The test and content area most closely related to this chapter are:

Test A8—Engine Performance

- **Content Area**—Emission Control Systems Diagnosis and Repair

NATEF Connection: Much of the information in this chapter is related to the NATEF tasks. The NATEF tasks and priority numbers most closely related to this chapter are:

1. Diagnose oil leaks, emissions, and drivability problems resulting from malfunctions in the positive crankcase ventilation (PCV) system; determine necessary action. P-2
2. Inspect, test, and service PCV filter/breather cap, valve, tubes, orifices, and hoses; perform necessary action. P-2
3. Diagnose emissions and drivability problems caused by malfunctions in the exhaust gas recirculation (EGR) system; determine necessary action. P-1
4. Inspect, test, service, and replace components of the EGR system, including EGR tubing, exhaust passages, vacuum/pressure controls, filters and hoses; perform necessary action. P-2
5. Inspect and test electrical/electronic sensors, controls, and wiring of EGR system; perform necessary action. P-2
6. Diagnose emission and drivability problems resulting from malfunctions in the secondary air injection and catalytic converter systems; determine necessary action. P-2

7. Inspect and test mechanical components of secondary air injection systems; perform necessary action. P-3
8. Inspect and test electrical/electrically operated components and circuits of air injection system; perform necessary action. P-3
9. Inspect and test catalytic converter performance. P-1
10. Diagnose emission and drivability problems resulting from malfunctions in the intake air temperature control system; determine necessary action. P-3
11. Inspect and test components of intake air temperature control systems; perform necessary action. P-3
12. Diagnose emission and drivability problems resulting from malfunctions in the early fuel evaporation control system; determine necessary action. P-3
13. Inspect and test components of early fuel evaporation control systems; perform necessary action. P-3
14. Diagnose emission and drivability problems resulting from malfunctions in the evaporative emission control system; determine necessary action. P-1
15. Inspect and test components and hoses of evaporative emission control systems; perform necessary action. P-2
16. Interpret evaporative emission-related diagnostic trouble codes (DTCs); determine necessary action. P-1

37.1 POSITIVE CRANKCASE VENTILATION (PCV) SYSTEM

PURPOSE OF CRANKCASE VENTILATION

Figure 37–1 shows where most of the pollution comes from in an automobile. During normal engine operation, a considerable amount of dirty air also passes through the engine crankcase. This air is a result of a process called **blow-by**. Blow-by is a product of the combustion process (***Figure 37–2***).

Every time combustion occurs, a certain amount of blow-by (from combustion) escapes past the piston rings. This blow-by produces a small **crankcase pressure**. The gases from blow-by are very acidic. If they are allowed to stay in the crankcase area, the acids attack the oil and metal within the engine. To help prevent this on older engines, air was drawn into the engine through the oil filler cap. The air flowed through the crankcase area, picking up the acidic gases. It was then directed out through a tube into the atmosphere. This tube was referred to as the road draft tube, or breather tube. Obviously, these gases are a large source of air pollution from the automobile.

PARTS LOCATOR

*Refer to photo #68 in the front of the textbook to see what a **PCV valve** looks like.*

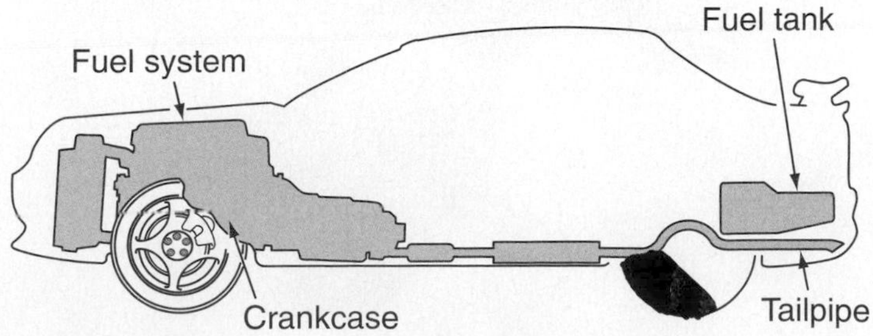

FIGURE 37–1 Sources of pollution on the automobile include the fuel system, crankcase, tailpipe, and fuel tank.

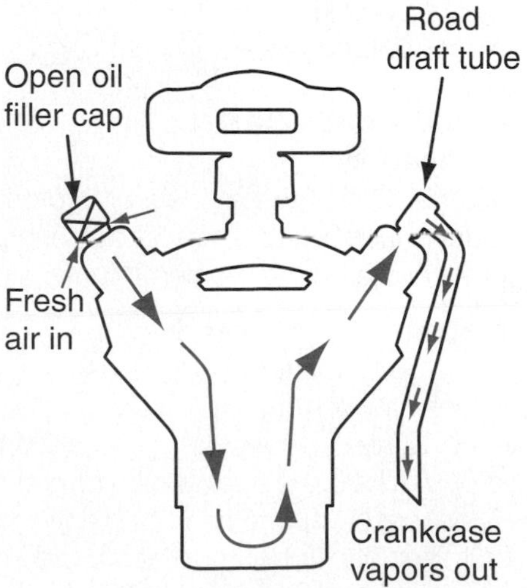

FIGURE 37–2 Crankcase ventilation in older vehicles was accomplished by passing the vapors in the crankcase out of the engine through a draft tube.

POSITIVE CRANKCASE VENTILATION SYSTEM OPERATION

Crankcase emissions are easily controlled by the positive crankcase ventilation (PCV) system (***Figure 37–3***). In this system, any crankcase vapors produced are directed back into the air intake system to be reburned.

With this system, which is called a closed system, air is drawn through the air filter, into the engine valve compartment and crankcase. These vapors are then drawn up through a vacuum-and-spring-controlled ventilating valve (PCV valve) and into the intake manifold. The vapors are then mixed with the intake air-fuel mixture and burned in the combustion process.

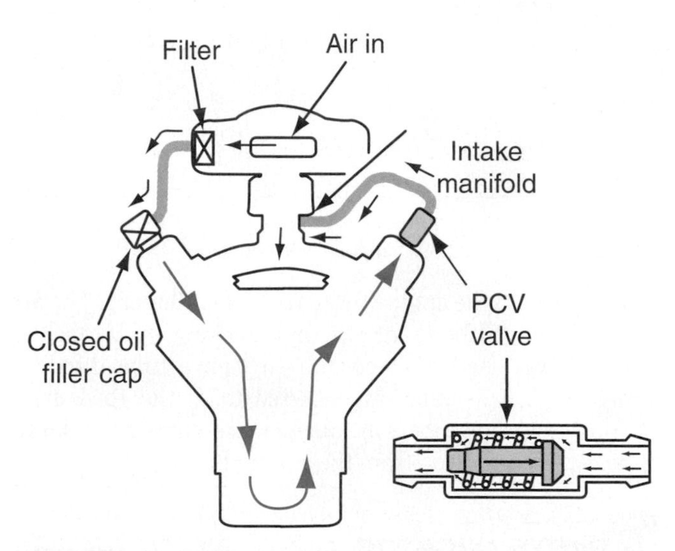

FIGURE 37–3 A closed PCV system brings air through the closed oil filler cap and into the crankcase. The vapors are then sent through the PCV valve and back into the intake manifold to be reburned.

Car Clinic: PCV SYSTEM

CUSTOMER CONCERN

Oily Air Cleaner

A customer says that the air cleaner is getting oil in it. The filter must be replaced about every 1,500 miles. What could be the problem?

SOLUTION: The oil gets into the air cleaner through the PCV system. Crankcase vapors are normally sent though the PCV valve to the carburetor to be burned. However, if the PCV valve is plugged or blocked, the engine cannot breathe correctly. Crankcase vapors now come through the closed oil filter cap tube into the air cleaner. This causes the filter to be soaked with oil. Check the PCV valve by shaking it. It should rattle if it is not plugged. Replace if necessary.

Not all PCV systems were designed like this. Earlier systems, which were called open systems, did not bring the fresh air through the air cleaner assembly. The air came through the open oil filler cap. Any air that entered the open system contained dirt and other materials. The open system was replaced by the closed system.

PCV VALVE OPERATION

If crankcase vapors are allowed to flow into the intake during all loads and engine rpm, the crankcase gases will upset the air-fuel ratios. Therefore, a PCV valve is placed in the flow just before the intake valves to control the flow. A typical PCV valve is shown in ***Figure 37–4***. ***Figure 37–5*** shows the internal operation of the PCV valve. Two forces are working against each other. Vacuum from the intake manifold is working against spring pressure inside the valve. When the engine is stopped, no intake manifold vacuum exists (***Figure 37–5A***). During this time, the PCV valve is held closed by the force of the internal spring. The tapered valve is moved fully to the bottom, closing off the entrance to the valve.

When the engine is at idle or is decelerating, vacuum in the intake manifold is very high (***Figure 37–5B***). The tapered valve plunger is drawn up. This action closes off the metered opening. Little crankcase vapor is allowed to enter the intake manifold.

When the engine operates at normal loads and speeds, the vacuum in the intake manifold drops (***Figure 37–5C***). This drop allows the spring to push the plunger down. This action causes the metered opening to increase in size. The amount of crankcase vapor sent back to the intake manifold is now increased.

During acceleration or heavy loading, the intake manifold vacuum is very low (***Figure 37–5D***). The spring moves the tapered valve open even more. Now the maximum amount of crankcase vapor enters the intake. Therefore, when the engine is at low speeds, little crankcase vapor is

FIGURE 37–4 A PCV valve is located in the positive crankcase ventilation system to control the flow of vapors back to the carburetor.

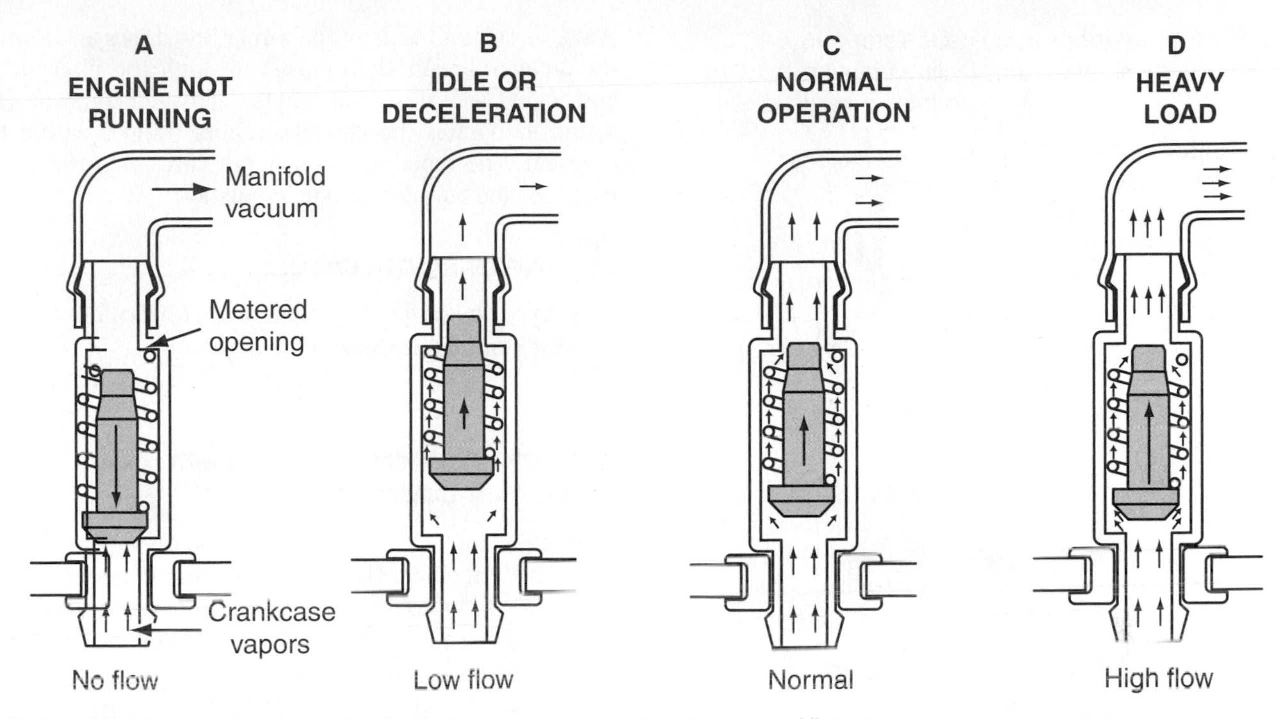

FIGURE 37–5 This is a typical PCV valve. Spring pressure pushes the valve down, while intake manifold vacuum pulls the valve up.

sent to the intake manifold. As the engine speed and load increase, more and more of these vapors are allowed to enter the intake manifold.

37.2 EVAPORATIVE EMISSION CONTROLS

Any gasoline released from the fuel tank in a liquid or vapor form is also pollution. These liquids and vapors are made of hydrocarbons. A variety of devices are being used to help reduce these emissions. These systems are called evaporative emission controls. ***Figure 37–6*** shows the typical parts of the evaporative emission controls.

CARBON CANISTER

The **carbon canister** is used to store any vapors from the fuel tank. Refer to ***Figure 37–7***. The carbon canister is made of a bed of activated charcoal (carbon). Warm weather can cause the fuel to expand in the fuel tank. This expansion causes the vapors to be forced out of the tank and into the carbon canister. The carbon absorbs or stores gasoline

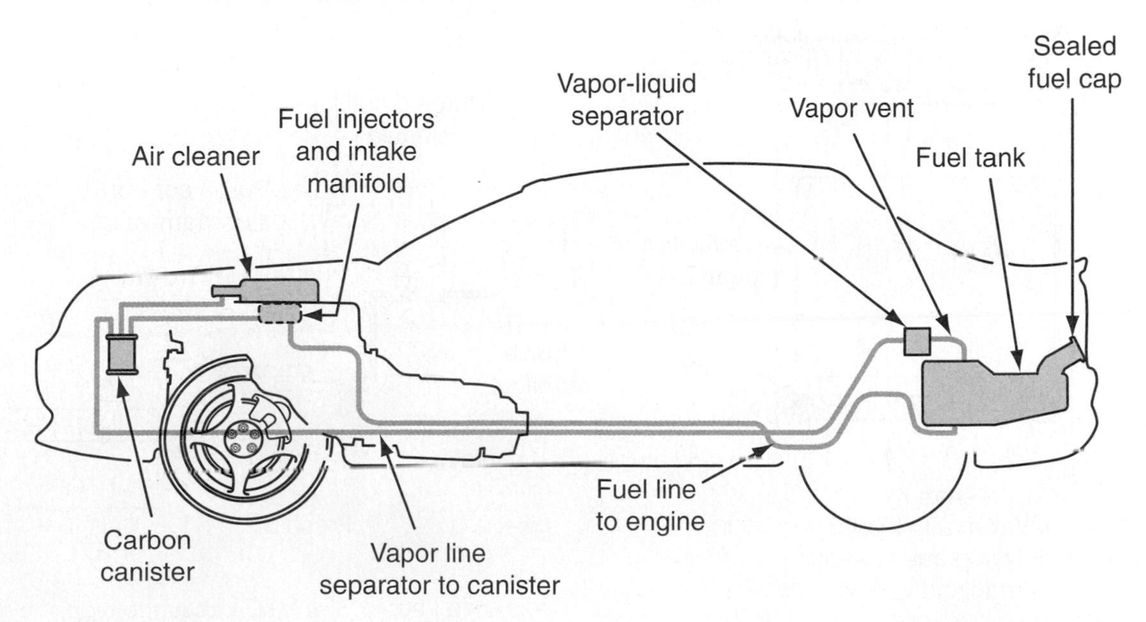

FIGURE 37–6 In a typical evaporative emission control system, vapors and liquids from the fuel tank and carburetor (if used) are controlled.

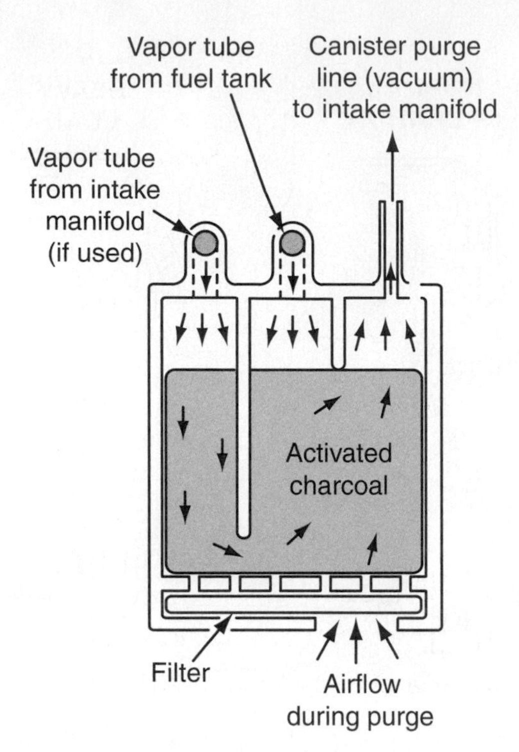

FIGURE 37-7 A schematic of a charcoal (carbon) canister used to store fuel vapors.

vapors from the fuel tank and intake manifold when the engine is not running.

When the engine is running, the clean vapors in the carbon canister are sent back to the intake manifold to be burned. As shown in ***Figure 37–8***, when the engine is operating, a vacuum inside of the **purge** line draws air through the air cleaner. Air then passes through the filter at the bottom of the canister and into the activated charcoal. This action evaporates the gasoline vapors trapped within the charcoal. The remaining fumes are carried to the intake manifold and burned during combustion.

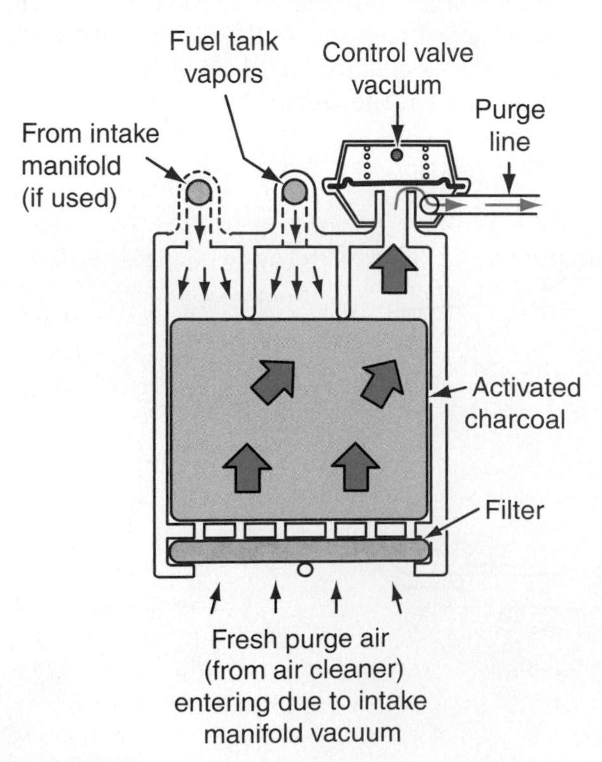

FIGURE 37-8 Arrows show the flow of air and vapors during canister purging or when the engine is running.

PARTS LOCATOR

Refer to photo #69 in the front of the textbook to see what a ***carbon canister*** *looks like.*

COMPUTER-CONTROLLED CARBON CANISTER

More and more emission control systems are being controlled by computers. The process of purging the carbon canister can also be controlled by the computer. ***Figure 37–9*** illustrates one such system. The fuel vapors from the fuel tank are first sent to the carbon canister for storage. When pressure in the fuel tank is higher than the set value, the two-way valve opens and allows the vapors to enter the canister. Canister purging is done by drawing fresh air from the bottom of the canister, through the canister, through the purge control diaphragm valve, then into a port near the throttle (butterfly valve). From this point, the vapors are sent into the combustion chamber for burning.

The amount of flow is controlled by the purge control diaphragm valve and the purge cutoff solenoid valve. Depending on various input signals to the powertrain control unit (PCM), the computer controls a signal sent to the purge cutoff solenoid valve. This signal, in turn, controls the vacuum sent to the purge control diaphragm valve. The position of this valve (controlled by the vacuum) then controls the amount of vapors being sent back to the throttle body.

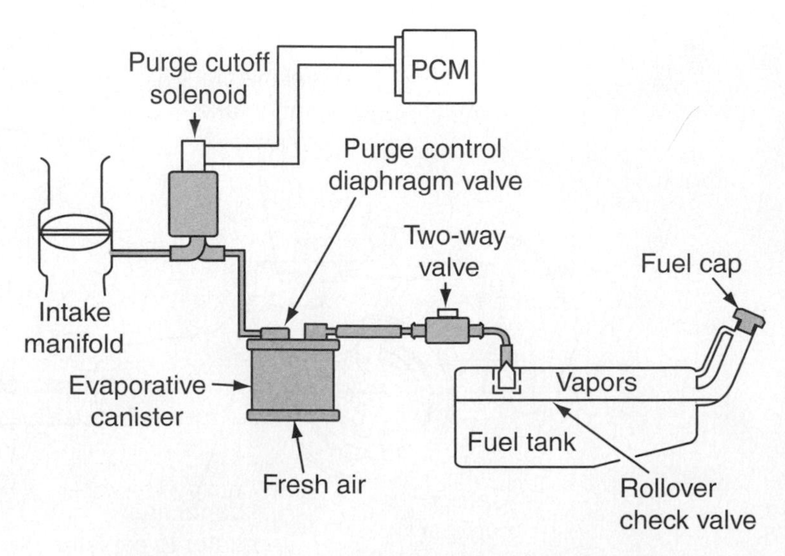

FIGURE 37-9 This system uses a computer to control the purging of the fuel vapors from the carbon canister to the intake of the engine.

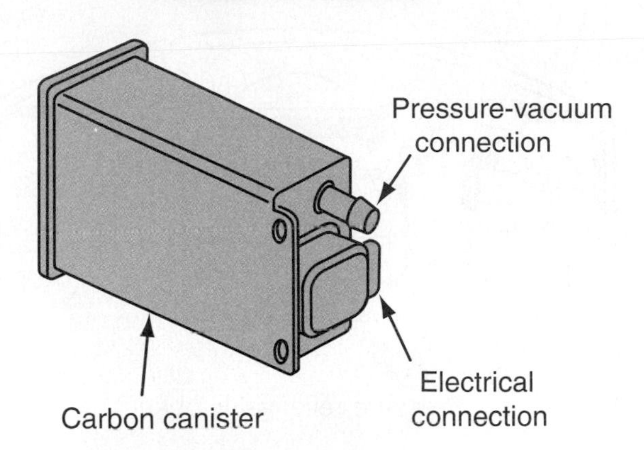

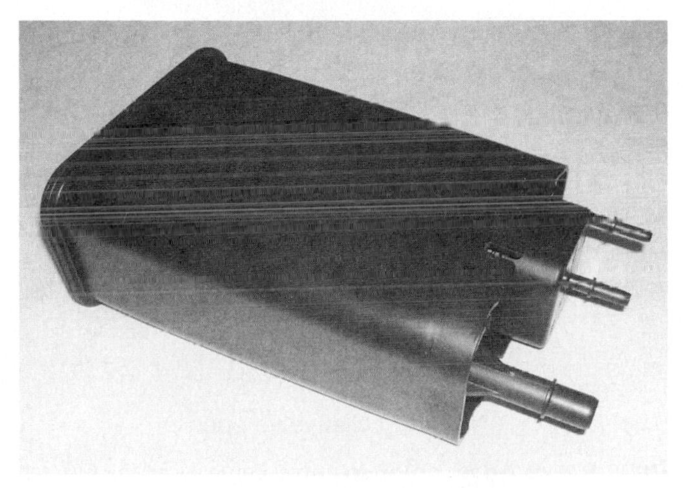

FIGURE 37–10 On some newer vehicles, the carbon canister is attached to the fuel tank.

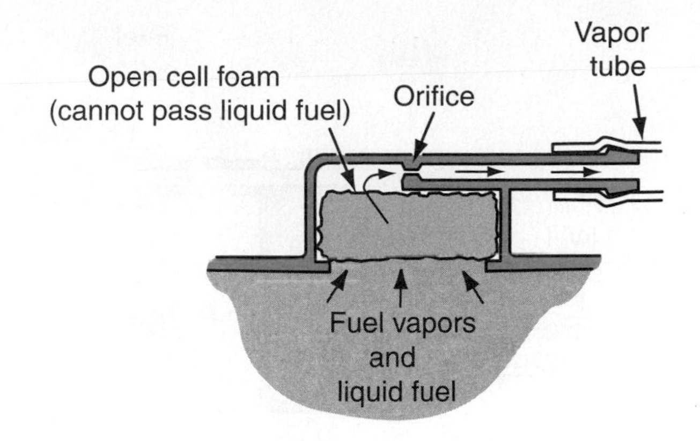

FIGURE 37–11 This vapor-liquid separator system uses a foam to separate the liquid and vapors.

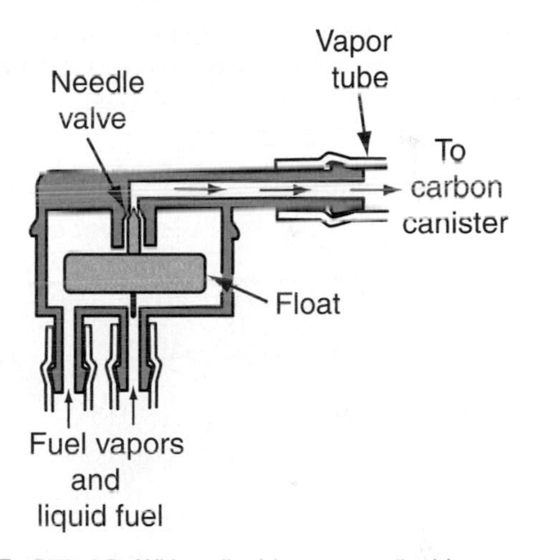

FIGURE 37–12 With a float-type vapor-liquid separator, fuel entering the valve raises the float. The raised float forces the needle valve to seal off the vent line to the carbon canister.

On many newer vehicles, the carbon canister is located directly on the fuel tank rather than in the front of the vehicle. ***Figure 37–10*** shows what this type of carbon canister looks like. Both electrical and vacuum connections are still used and the principle of operation remains the same.

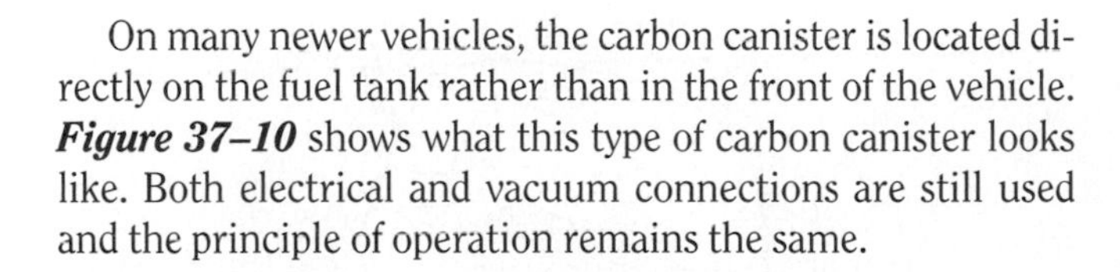

VAPOR-LIQUID SEPARATOR

It is important that only vapors, not liquid fuel, enter the carbon canister. Liquid fuel may leak out when the engine is stopped, or excess fuel may enter if a carburetor is used. Other conditions may also cause liquid gasoline to enter the carbon canister. For example, a full fuel tank that expands from heat or a vehicle that accidentally rolls over may cause fuel to enter the carbon canister. To prevent this, a vapor-liquid separator is used.

One type of vapor-liquid separator passes the vapors through open-cell foam. Liquid gasoline cannot pass through the foam. As shown in ***Figure 37–11***, the vapors will pass through the open-cell filter material and be sent to the carbon canister. The separator is usually located directly on or near the fuel tank. ***Figure 37–6***, shown earlier in this chapter, depicts the location of the vapor-liquid separator near the fuel tank.

A second type of vapor-liquid separator is the float type. ***Figure 37–12*** shows an example of this type of separator. As vapors come into the lower two inlets, they are easily passed through the center and out to the carbon canister. The vapors do not affect the position of the center float. However, when fuel enters from the fuel tank, the float rises and closes off the needle valve assembly at the top. This stops all fuel flow to the carbon canister.

ROLLOVER CHECK VALVE

Another method of controlling the liquid gasoline on some vehicles is by using a rollover check valve. This valve is located in series with the main fuel or vapor line from the fuel tank. If the vehicle rolls over in an accident, a small stainless steel ball forces the plunger to close off the line (***Figure 37–13***). On many of today's vehicles, this valve is located directly on top of the fuel tank to also control fuel vapor. ***Figure 37–14*** shows an example of this type of rollover valve and the connecting vapor line.

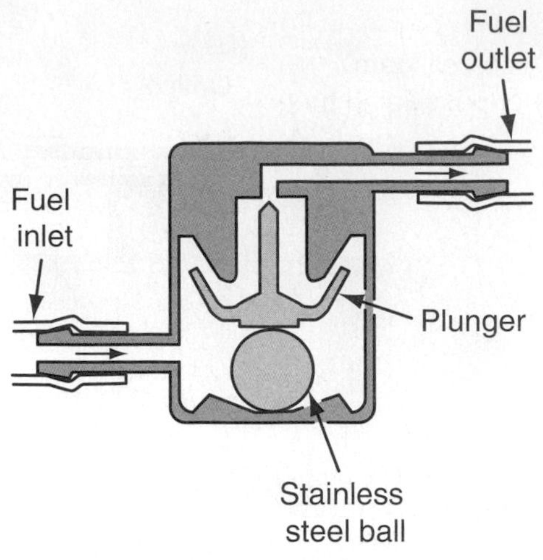

FIGURE 37–13 A rollover check valve (stainless steel ball) will stop any liquid fuel from entering the carbon canister if a vehicle rolls over in an accident.

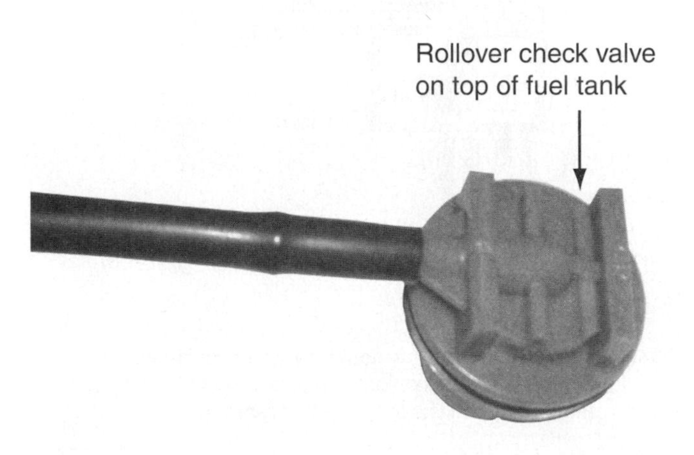

FIGURE 37–14 This rollover check valve is located directly on top of the fuel tank.

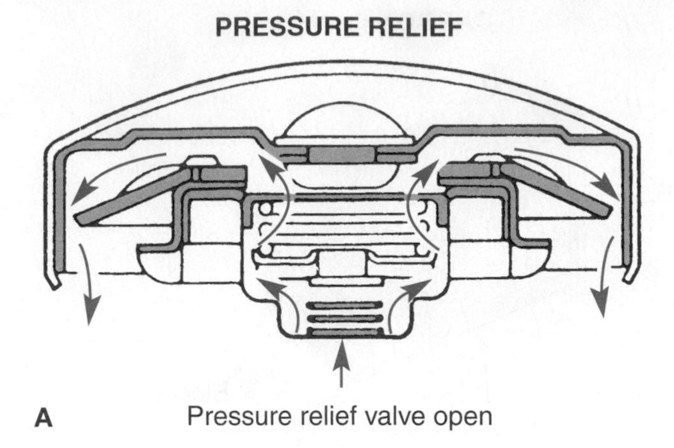

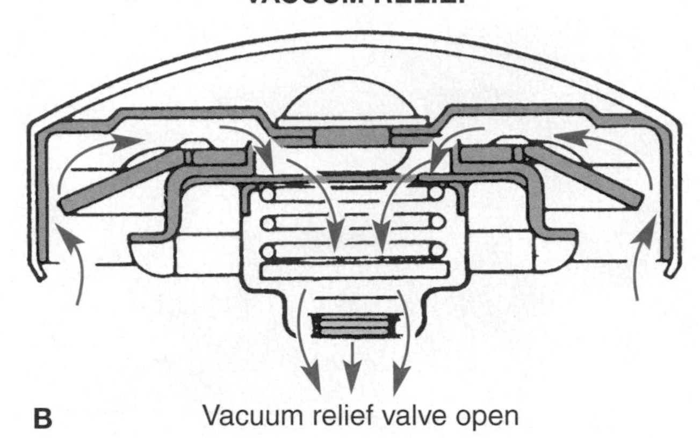

FIGURE 37–15 The pressure-vacuum relief on the fuel cap opens with either high pressure or vacuum in the fuel tank. This action prevents the tank or other components in the fuel system from being damaged.

FUEL TANK

Most fuel tanks are sealed with a special pressure-vacuum relief filler cap. This is shown in ***Figure 37–15***. The relief valves in the cap are used to regulate pressure or vacuum in the fuel tank. Excess pressure or vacuum could be caused by some malfunction in the evaporative emission control system or by excess heating or cooling. The fuel tank is allowed to "breathe" through the filler cap when gasoline expands or contracts from heating or cooling.

When pressure in the fuel tank rises above 0.8 psi, the pressure cap opens to let the pressure in the fuel tank escape (***Figure 37–15A***). If the vacuum in the fuel tank is greater than 0.1 Hg (inches of mercury), the valve allows the vacuum to be released (***Figure 37–15B***).

Some fuel tanks are designed to allow for expansion when the tank has been filled. The tank shown in ***Figure 37–16*** has a built-in expansion tank. The orifices leading to the expansion tank are so small that it takes 10 to 15 minutes for gasoline from the main tank to enter the expansion tank. The expansion tank area will accept some fuel from the main fuel tank after it is filled at the fuel pump. This also provides room for fuel expansion and vapors in the main tank.

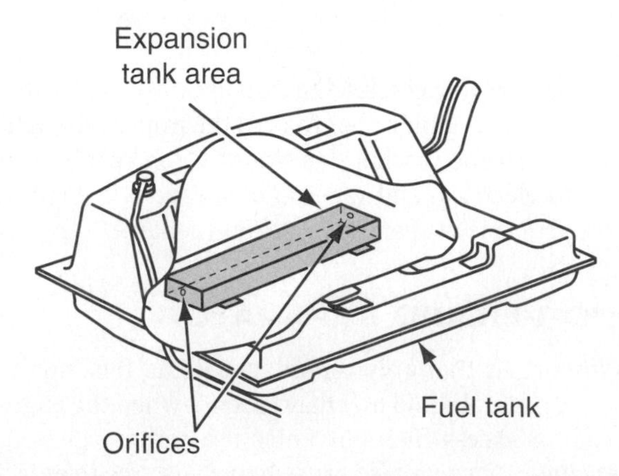

FIGURE 37–16 An expansion tank is used on certain vehicles to allow for fuel expansion when the tank is filled with gasoline.

VAPOR CONTROL FROM THE CARBURETOR

On older vehicles that use a carburetor, the carburetor float bowl is also vented into the carbon canister. Fuel vapors are produced in the float bowl, especially when the engine is shut off. During this time, heat is transferred from the engine to the carburetor. This heat causes excess vapor to be produced inside the carburetor bowl. ***Figure 37–17*** shows the carburetor bowl area connected with a line to the carbon canister.

When the engine is shut off, vapors from the float bowl are sent to the bowl vent valve. This valve is open when the engine is off. The carburetor vapors now go directly to the carbon canister to be stored. When the engine starts, a vacuum line from the carburetor closes off the bowl vent valve. The valve stops the vapors coming from the carburetor to the carbon canister.

When the engine is running, the idle purge line draws stored vapors out of the carbon canister to the carburetor. These vapors are burned during combustion.

Some carburetors use an insulator (***Figure 37–18***) to reduce heat transfer from the engine intake manifold to the carburetor and float bowl. The insulator is normally placed directly below the base of the carburetor on the intake manifold.

PARTS LOCATOR

*Refer to photo #70 in the front of the textbook to see what a **carburetor heat gasket** looks like.*

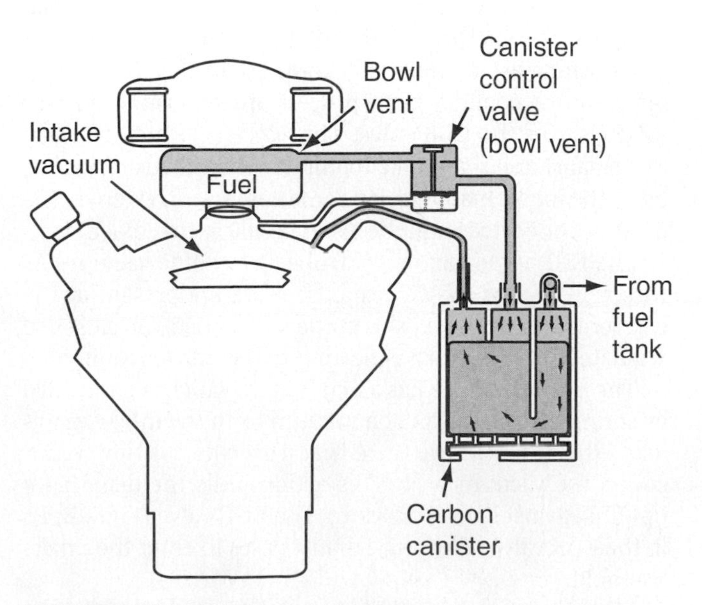

FIGURE 37–17 This complete evaporative control system uses a bowl vent valve.

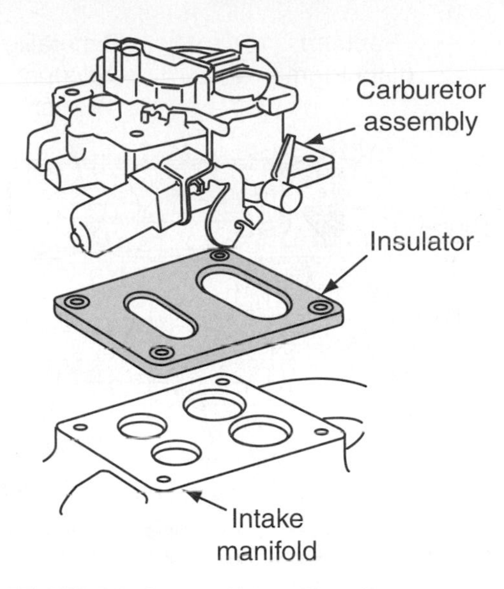

FIGURE 37–18 Some vehicles with carburetors use an insulator to reduce heat transfer from the intake manifold to the carburetor. This reduces the formation of vapors in the carburetor float bowl.

37.3 INTAKE AND EXHAUST EMISSION CONTROLS

In addition to the pollution controls already mentioned, certain changes can be made to the intake and exhaust systems. These particular systems alter or change the intake temperatures, the exhaust hydrocarbons (HC), the nitrogen oxides (NO_x), and the carbon monoxide (CO).

HEATED AIR INTAKE

Automotive manufacturers have made the air-fuel ratio mixture leaner over the past few years. Leaner air-fuel mixtures have lower HC and CO emissions. This is true especially during idle conditions. Faster warm-ups have also helped reduce emissions. To aid in faster warm-up on engines, a heated air intake system is used. During starting, a richer mixture is used compared to what is actually needed. If the amount of time the rich mixture is used can be reduced, less emissions will result. It is also known that a lean air-fuel mixture will ignite easier when the air is warm.

To accomplish this, air going into the intake manifold can be preheated. ***Figure 37–19*** shows how a heated air intake system operates. This system is also referred to as a thermostatically controlled air cleaner. When heated air is required, the air control motor attached to the air cleaner assembly closes off cold air coming into the engine. All air coming into the engine passes over the exhaust manifold and through the hot air pipe. The area where the air is heated is called the **heat stove**.

The motor is operated by the vacuum of the engine and a bimetallic strip. When the engine is off, no vacuum is produced in the vacuum chamber. This is shown in ***Figure 37–20C***. In this position, the control damper assembly is closing off the hot air pipe. When the engine starts, vacuum is produced. The vacuum lifts the diaphragm and closes

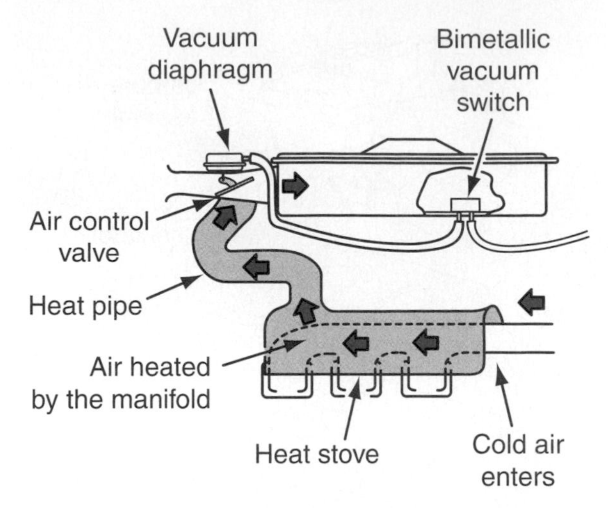

FIGURE 37–19 Heated air is obtained on a thermostatically controlled air cleaner by drawing the air going into the air cleaner across the exhaust manifold.

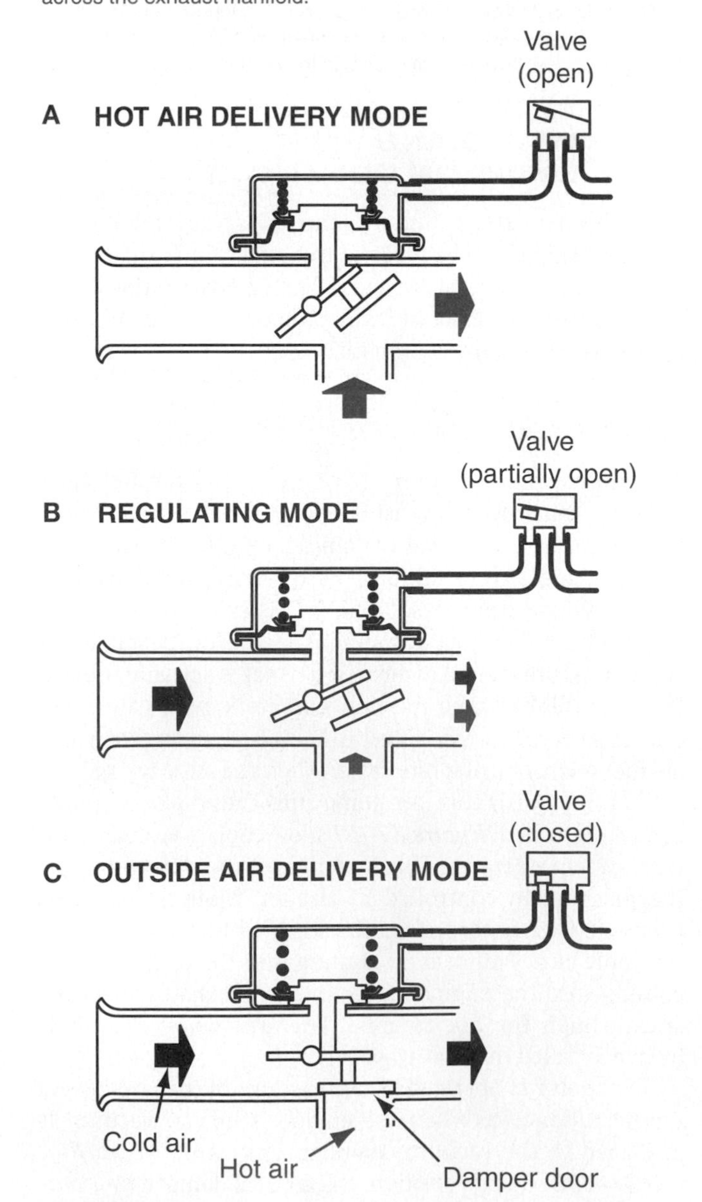

FIGURE 37–20 The vacuum chamber and the temperature sensor control when and how much warm air enters the engine.

off the air inlet. Now only warm air drawn over the exhaust manifold can enter the intake manifold (***Figure 37–20A***).

Temperature can also be used to control the direction of airflow. This is done by a temperature sensor. When the engine is operating, and the underhood temperature is below 85 degrees Fahrenheit, the temperature-sensing spring closes the air bleed valve. This admits full engine vacuum to the vacuum chamber. Now all outside air below 85 degrees Fahrenheit (cold air) is shut off. Only warm air can enter the carburetor (***Figure 37–20A***).

When the underhood temperature reaches about 135 degrees Fahrenheit, the bimetallic strip spring in the temperature sensor opens the air bleed valve. This reduces any vacuum in the vacuum chamber. The diaphragm spring now pushes the control damper assembly in a position to close off the hot air pipe (***Figure 37–20C***).

When the air temperature under the hood is between 85 and 135 degrees Fahrenheit, the control damper assembly is between a fully open and fully closed position (***Figure 37–20B***). The actual opening and closing temperatures can vary somewhat, depending on the manufacturer. Also, when the engine is accelerated heavily, vacuum to the chamber will drop off instantly. This causes the spring to snap the damper downward, closing off the hot air tube. This action will allow maximum airflow through the **snorkel tube**. Under heavy acceleration, engines need all the intake air they can get for good performance.

EXHAUST GAS RECIRCULATION (EGR)

When combustion temperatures are in the range of 2,200 to 2,500 degrees Fahrenheit, nitrogen mixes with oxygen and produces oxides of nitrogen (NO_x). This type of emission has a detrimental effect on the environment. The method used to reduce NO_x is to cool down the combustion process. This is done by using an **exhaust gas recirculation (EGR)** valve.

In operation, part of the exhaust gas (usually less than 10%) is sent back through the intake manifold. The exhaust gases, which are considerably cooler than the combustion temperature, cool down the process of combustion. ***Figure 37–21*** shows the EGR valve. This valve is located between the exhaust and the intake manifold. The exhaust is picked up by the intake manifold. On some engines, an external line must be connected from the exhaust side to the intake side.

The EGR valve can be controlled by engine vacuum. As shown in ***Figure 37–22***, exhaust gases are present at the base of the EGR valve. The **pintle** valve opens or closes to regulate the exhaust gases flowing to the intake manifold.

The EGR valve also has a diaphragm, which is controlled by spring pressure against a vacuum from the intake manifold. When the throttle valve is at a specific position, it uncovers the vacuum port. This action pulls the diaphragm upward against spring pressure. The EGR valve is now open in the EGR valve, allowing exhaust gases to enter the intake manifold.

When the engine is at idle, the throttle valve has not yet uncovered the port. During this condition, no vacuum is

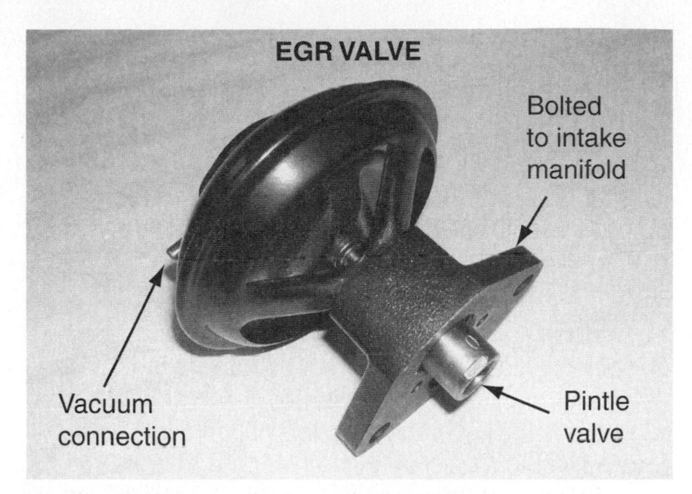

FIGURE 37–21 The EGR valve located on the intake manifold recirculates exhaust gases back into the intake for cooler combustion and fewer NO_x emissions.

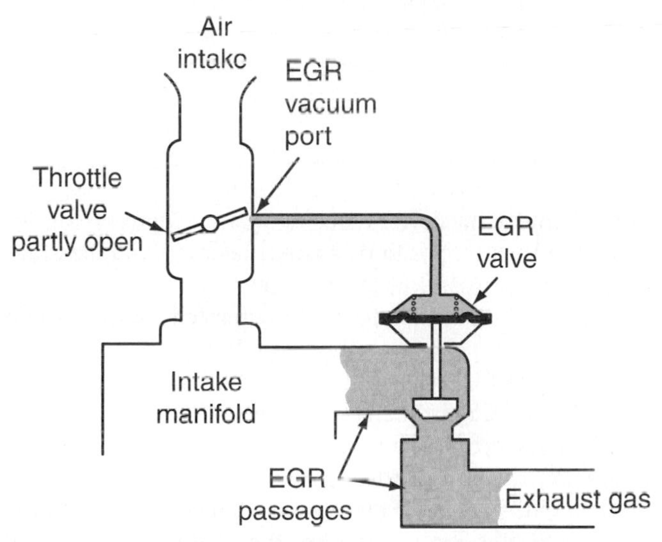

FIGURE 37–22 The EGR valve controls the amount of exhaust flowing back into the intake manifold.

available to operate the EGR valve. The EGR valve remains closed during idle or heavy deceleration.

A temperature-sensing control is sometimes placed between the EGR valve and the intake manifold. Exhaust gas recirculation is not needed below certain temperatures. If it is allowed to occur, it will cause poor engine performance. ***Figure 37–23*** shows the location of the temperature control. The EGR temperature control senses the temperature of the engine coolant. When the engine temperature reaches a certain point, the EGR temperature control opens and allows the vacuum to operate the EGR valve.

Some EGR valves also have a back-pressure **transducer** that senses pressure in the exhaust manifold to change the amount of opening.

PARTS LOCATOR

Refer to photo #71 in the front of the textbook to see what an ***EGR valve*** *looks like.*

Car Clinic:
EGR VALVE

CUSTOMER CONCERN
Erratic Engine Operation
An engine with 22,000 miles on it has developed several problems. The system uses computer-controlled EFI (electronic fuel injection). The complaints are that the engine stops after cold starting, stops at idle after deceleration, surges during cruise, and has a rough idle. What would be a good starting point for troubleshooting?

SOLUTION: Erratic engine operation such as this indicates a bad or inoperative EGR valve. The symptoms suggest the EGR valve is providing too much flow of exhaust back into the intake. Too little EGR flow would produce problems with spark knock, engine overheating, and emission test failure. Make sure the vacuum system on the EGR valve operates. Use a vacuum pump to check the diaphragm. Also make sure the EGR valve passages are clean of carbon buildup. Repair or replace the EGR valve to eliminate the erratic engine operation.

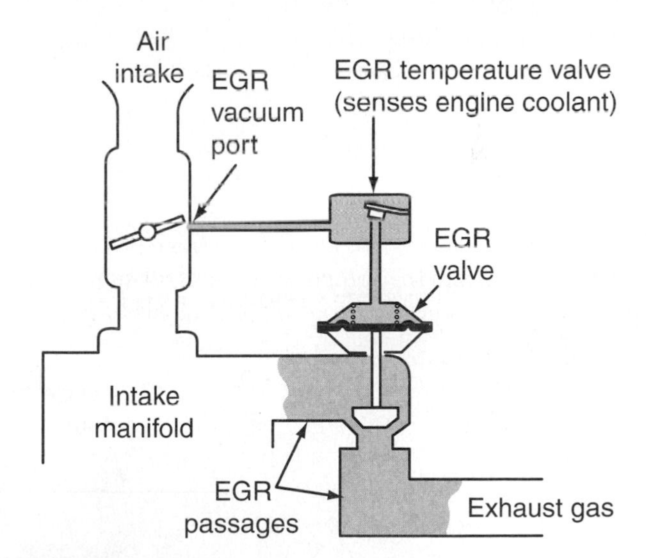

FIGURE 37–23 A coolant temperature valve can also control the EGR valve. The EGR valve is designed to be closed below a certain temperature in the engine.

COMPUTER-CONTROLLED EGR SYSTEM

With the use of computers controlling engine functions, the EGR valve can also be controlled. Normally, the EGR valve is either open or closed. With computer control, the EGR valve can be either open, closed, or anywhere in between. The result is more precise control of the EGR valve. This results in improved engine efficiency.

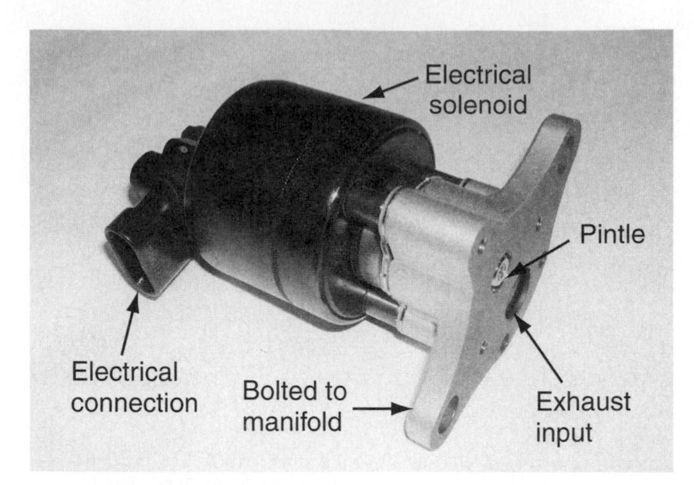

FIGURE 37-24 Vehicles with computer control use an integrated electronic gas-recirculating valve.

Car Clinic: EGR VALVE

CUSTOMER CONCERN
Deceleration Problem
A vehicle makes a jerking motion while decelerating. The problem happens only when the engine is warmed up.

SOLUTION: Normally, any jerking during deceleration is caused by incorrect air-fuel mixtures. The jerking could also be caused by too much ignition timing advance. A third possible cause might be the EGR valve. A quick way to check the EGR valve is to disconnect the valve and run the engine. If it still has a jerking motion on deceleration, reconnect the EGR valve (if it's OK) and check the advance. Try retarding the ignition timing slightly and see if the jerking stops. If the jerking is still there, the air-fuel ratio will have to be checked with an exhaust gas analyzer.

Several designs are used to control the EGR valve position. However, a PCM is needed to control the EGR valve. The computer can control the EGR valve by several methods. Three popular methods are:

1. Bleeding off the vacuum from the intake manifold using a solenoid
2. Using an EGR control solenoid to control the vacuum
3. Using solenoids directly attached to the EGR valve. One manufacturer calls this an IEEGR (integrated electronic exhaust gas recirculating) valve

Figure 37–24 shows an example of an electrically controlled EGR valve. Notice that the exhaust enters the EGR valve and is allowed to pass through the pintle into the intake manifold.

No matter which system is used, the solenoids are operated by signals from the computer. The output signal to the solenoids is controlled by several input signals to the computer. Some of the devices used to produce input signals include the throttle position sensor, the engine temperature sensor, and the manifold pressure sensor.

Figure 37–25 illustrates a computerized control system for the EGR valve. In operation, the PCM contains memories for ideal EGR valve lifts for various operating conditions. The EGR valve lift sensor detects the amount of EGR valve lift and sends this information to the PCM. The PCM then compares it with the ideal EGR valve lift (determined by the input signals). If there is any difference between the two, the PCM reduces current to the EGR control solenoid valve. This reduced current in turn reduces the vacuum ap-

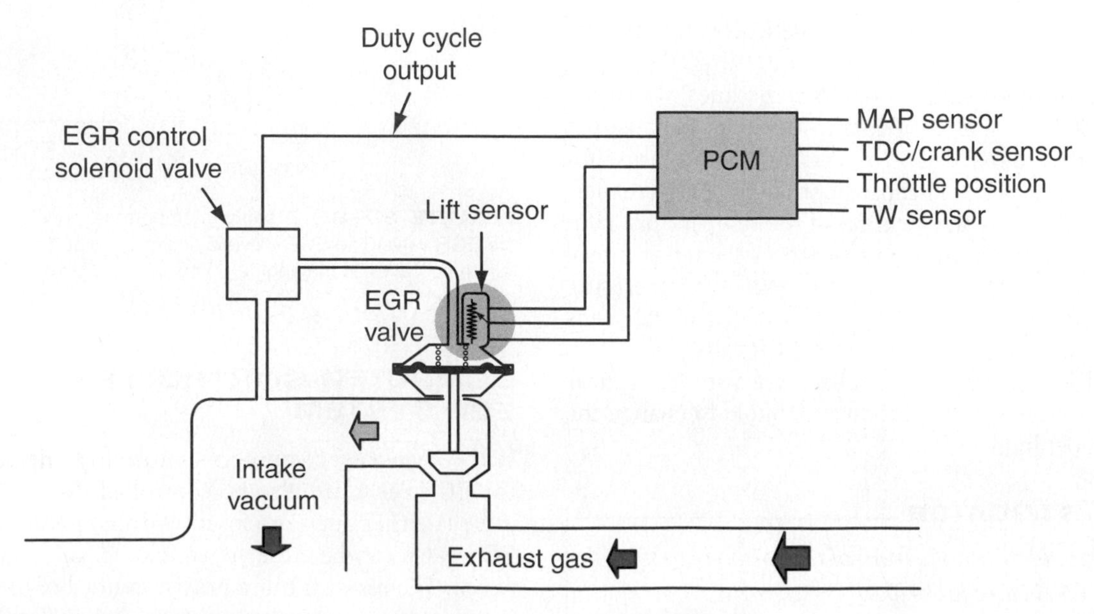

FIGURE 37-25 This system uses a computer to control the movement of the EGR valve.

plied to the EGR valve. The result is that the EGR changes the amount of exhaust gases passing back into the intake airflow.

AIR INJECTION SYSTEM

On some vehicles, another method is used to reduce the amount of HC and CO in the exhaust. This is done by forcing fresh air into the exhaust system after the combustion process. The additional fresh air causes further oxidation and burning of the unburned HC and CO. The process is much like blowing on a dwindling fire. When oxygen in the air combines with the HC and CO, the burning is enhanced and the HC and CO concentrations are reduced. This allows them to oxidize and produce harmless water vapor and CO_2. The following examples show some of the parts and components of the air injection systems.

AIR INJECTION SYSTEM OPERATION

This method of cleaning the exhaust does not affect the efficiency of combustion. However, in the past a small amount of frictional horsepower (up to 3 hp) was needed to operate the air injection pump. This acts as a form of horsepower loss to the engine. The systems that are used on newer vehicles use electrical energy to turn the air pump.

These systems, which are referred to as **air injection reactors (AIR)**, thermactors, air guards, and so on, use an air pump, air manifolds, and valves to operate. A typical mechanical system used several years ago is shown in ***Figure 37–26***.

The air pump is driven by a belt from the crankshaft. The air pump produces pressurized air that is sent through the exhaust manifold to the injection tubes located at each cylinder. The air is then injected directly into the exhaust flow. ***Figure 37–27*** shows where the air is injected into the exhaust flow. This is only one setup. The air can be injected at the base of the exhaust manifold as shown, or it can be injected directly through the head at the exhaust port. One other way to use this air is to inject it into the exhaust catalytic converter in the exhaust system.

AIR PUMP

The air pump is a rotating vane (also called an eccentric-type) **positive displacement pump** (***Figure 37–28***). A relief valve is placed on the pump to control the amount of pressure the pump can develop. When pressure inside the pump exceeds a predetermined level, the relief valve opens and allows the excess pressure to escape.

As each rotating vane passes the intake chamber, a vacuum is produced. This vacuum draws in a fresh charge of

PARTS LOCATOR

Refer to photo #72 in the front of the textbook to see what an ***air injection pump*** *looks like.*

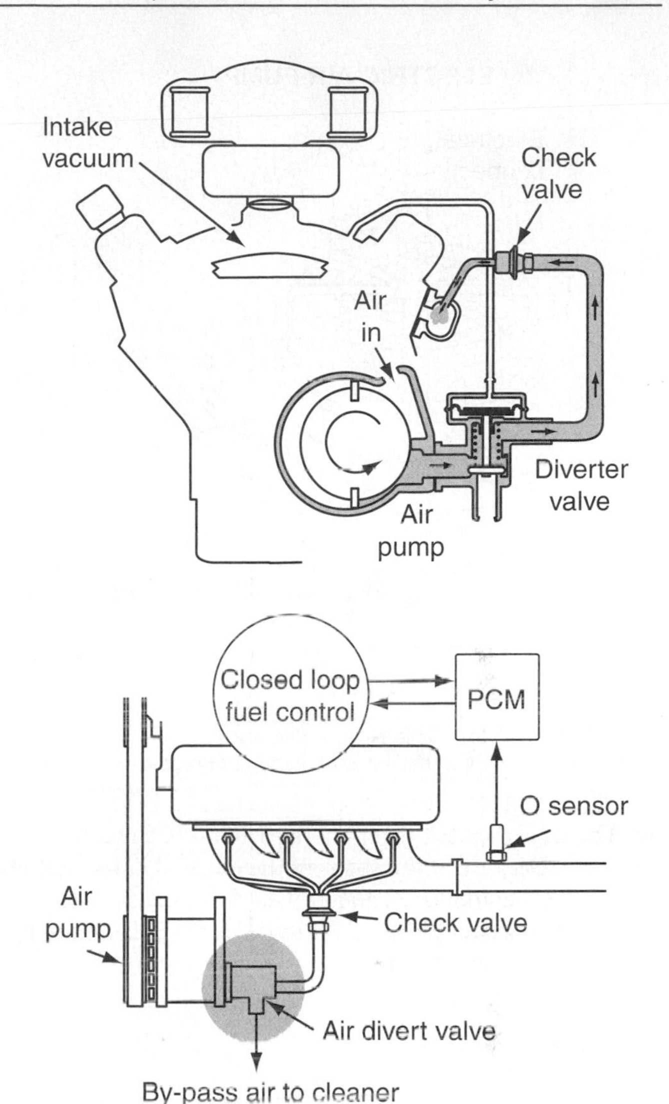

FIGURE 37–26 A typical mechanical air injection system injects air (oxygen) directly into the exhaust flow. This air assists the burning of any unburned hydrocarbons and carbon monoxide in the exhaust system.

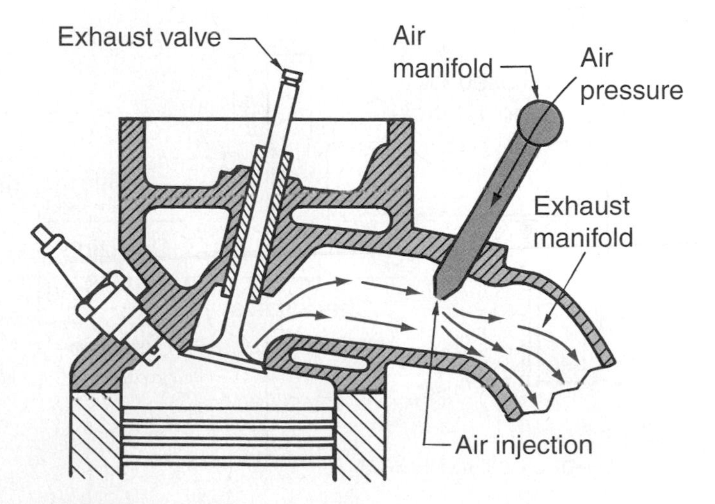

FIGURE 37–27 Air is injected directly into the exhaust gases in an air injection system. This extra fresh air causes the unburned hydrocarbons and carbon monoxide to burn more completely, reducing emissions.

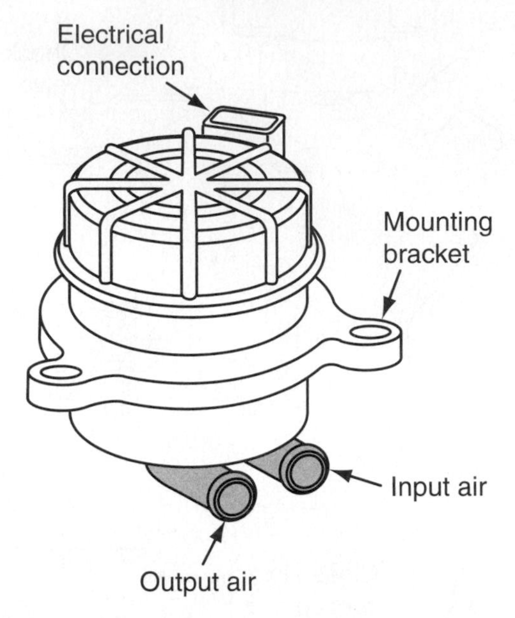

FIGURE 37–28 This positive displacement, electrically operated air pump uses internal vanes to produce air pressure.

air. The vane carries the air charge around to the compression area. Because of the shape of the internal housing, the air charge is compressed and sent to the exhaust chamber. The pressurized air is then sent to the manifold and finally to the exhaust manifolds.

CHECK VALVE

A check valve is also placed in the air injection system. This check valve is used to prevent backflow of exhaust gases. Backfire may occur whenever the exhaust pressure is greater than the air pump pressure. A check valve is shown in ***Figure 37–29***.

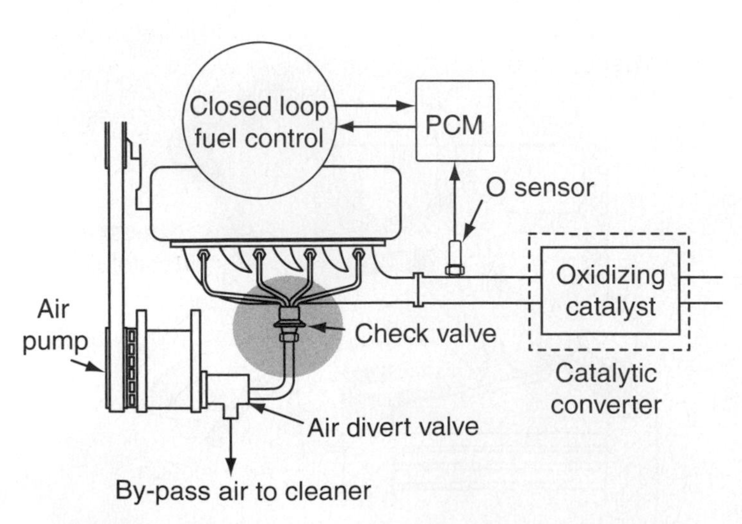

FIGURE 37–29 Check valves are used on some air injection systems to control any reverse flow of exhaust gases. The check valve allows airflow in one direction but not in the other.

DIVERTER VALVE

The diverter valve is used on some older vehicles to bypass the air from the air injection pump when the throttle is closed quickly. It is also called a deceleration or gulp valve. During this condition, there is a high intake manifold vacuum. The high vacuum produces a rich air-fuel mixture from the carburetor. The mixture is too rich to be ignited in the combustion chamber. It will, therefore, go out the exhaust valve and into the exhaust manifold. If excess air were pumped into this rich mixture, violent burning action might be produced. This could cause a backfire and possibly damage the exhaust system.

The diverter valve operates as shown in ***Figure 37–30***. During normal operation, pressurized air enters the inlet and goes directly to the exhaust manifold through the outlet. When deceleration occurs, the high vacuum is felt at the vacuum signal line. This vacuum causes the metering

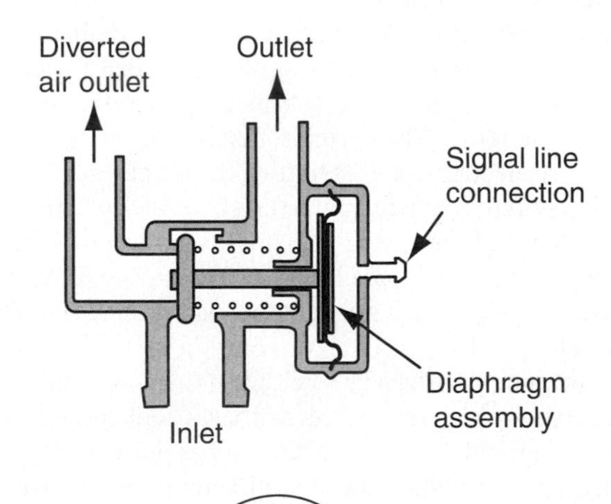

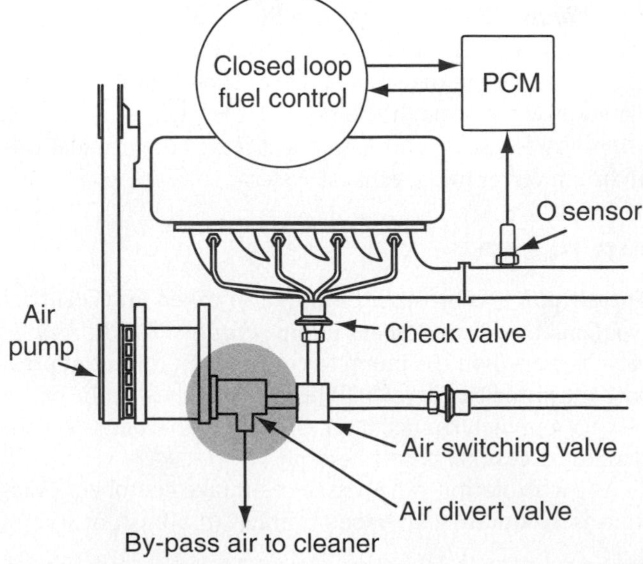

FIGURE 37–30 A diverter valve, or deceleration valve, is used on some vehicles to stop the flow of pumped air to the exhaust during deceleration periods. This action eliminates the possibility of backfire damage.

valve to rise, closing off the line to the outlet and the exhaust manifold. Pressurized air is then directed internally to the diverted air outlet.

ASPIRATOR VALVE

The **aspirator valve** is a system that permits air to be injected into the exhaust without the use of an air pump. This unit consists of a steel tube containing a one-way aspirator valve that is attached to the exhaust manifold. This unit (also called a pulse air injector system) uses exhaust pressure pulsations to draw fresh air into the exhaust system. The additional air reduces CO and, to a lesser degree, HC emissions.

The aspirator valve is shown in ***Figure 37–31***. During the exhaust pulses, when the pressure in the exhaust system is positive (view A), the aspirator valve is closed. Air is not sent into the exhaust to reduce emissions. When the exhaust has a negative pressure during the pulses (view B), the aspirator valve opens and allows a small amount of air to enter the exhaust. This small amount of fresh air is first brought through the carburetor. It has an effect similar to that of injecting air into the exhaust. This action reduces emissions.

CATALYTIC CONVERTERS

Catalytic converters provide another method of treating exhaust gases. Catalytic converters are located in the exhaust system between the engine and the muffler. They are used to convert harmful pollutants such as HC, CO, and NO_x into harmless gases. In operation, the exhaust gases pass over a large surface area that is coated with some form of **catalyst**. A catalyst is a material that causes a chemical reaction without becoming part of the reaction process. The catalyst is not chemically changed in the process.

The catalyst used on a catalytic converter depends on the exact type of pollutant being removed. When exhaust gases are passed through a bed of platinum- or palladium-coated pellets or through a coated honeycomb core, the HC and CO react with the oxygen in the air. The result is the formation of water and CO_2. When the metal rhodium is used as the catalyst, the NO_x in the exhaust gases is reduced to harmless nitrogen and oxygen. In this case, rhodium acts as a reducing catalyst.

The reaction within the catalyst produces additional heat in the exhaust system. Temperatures up to 1,600 degrees Fahrenheit are normal. This additional heat is necessary for the catalyst to operate correctly. Because of these high temperatures, catalytic converters are made of stainless steel. Special heat shields are used to protect the underbody from excessive heat. Each car manufacturer has its own unique heat shielding.

It is important that only unleaded fuel be used with a catalytic converter. Leaded gasoline will destroy the effectiveness of the catalyst as an emission control device. Under normal conditions, the catalytic converter will not require maintenance. It is important, however, to keep the engine properly tuned. If it is not properly tuned, engine misfiring can cause overheating of the catalyst. This heat can cause damage to the converter. This situation can also occur during engine testing. If any spark plug wires have been removed and the engine is allowed to idle for a prolonged period of time, damage may occur.

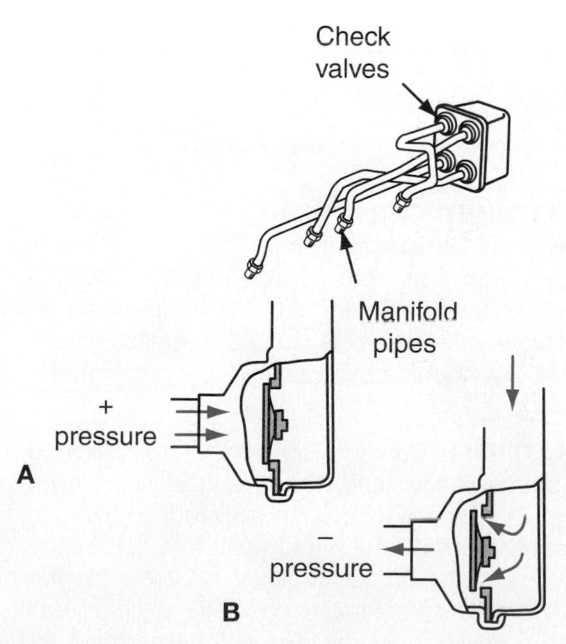

FIGURE 37–31 The aspirator air injection system uses positive and negative pressure pulses to operate.

TYPES OF CATALYTIC CONVERTERS

There are two types of catalytic converters: the two-way converter and the three-way converter. Both types can employ either a **monolith** or a pellet design. The pellet converter used on older vehicles consists of two louvered sheet-metal retainers. These retainers house 100,000 to 200,000 ceramic pellets, sometimes called beads. The monolith converter used mostly today can have a catalyst made of either ceramic or metal. With both, the catalyst is held in place by metal shells. The monolith and pellet designs are shown in ***Figures 37–32*** and ***Figure 37–33***, respectively.

Two-Way Catalytic Converter The two-way catalytic converter reduces CO and HC particles. It does not reduce any NO_x emissions. ***Figures 37–32*** and ***Figure 37–33*** are examples of the two-way catalytic converter. Here, only platinum and palladium are used as catalysts to reduce HC and CO.

Three-Way Catalytic Converter The three-way catalytic converter is designed to reduce NO_x emissions as well. An additional catalyst bed coated with platinum and

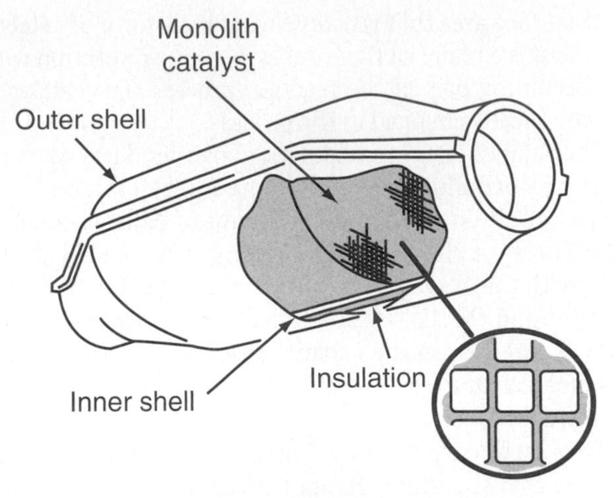

FIGURE 37-32 The two-way monolith catalytic converter uses a catalyst made like a honeycomb to allow gases to pass through.

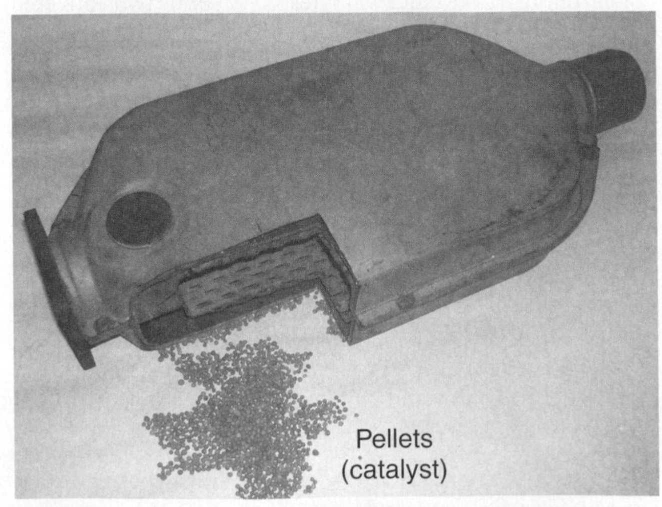

FIGURE 37-33 The two-way pellet-type catalytic converter on some older vehicles uses a catalyst shaped into pellets. Exhaust gases are forced through the catalyst pellets.

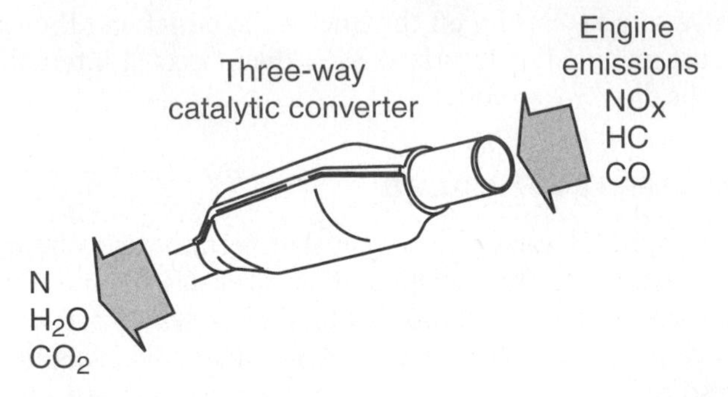

FIGURE 37-34 The inputs and outputs of a three-way catalytic converter.

rhodium is used. The bed not only helps reduce HC and CO but also lowers the levels of NO_x emissions. As shown in ***Figure 37–34***, HC, CO, and NO_x are the inputs to the three-way catalytic converter. The outputs are generally water (H_2O), carbon dioxide (CO_2), and nitrogen (N). The catalyst promotes chemical reactions that oxidize hydrocarbons (HC) and carbon monoxide (CO). The converter changes them into water vapor (H_2O) and carbon dioxide (CO_2). In addition, the NO_x is reduced to nitrogen (N).

Figure 37–35 shows a three-way monolith catalytic converter. The front bed, or inlet, is treated with platinum and rhodium and is termed a reducing catalyst. The rear bed is coated with palladium and platinum and is referred to as the oxidizing catalyst.

In ***Figure 37–35***, exhaust gases first pass through the reducing catalyst. This causes the levels of NO_x to be reduced. Pressurized air from the air injection system is forced into the space between the catalyst beds. Extra air supplies additional oxygen. The extra air causes more oxidation of the gases.

Car Clinic: CATALYTIC CONVERTER

CUSTOMER CONCERN:
Converter Temperatures

A customer has complained that the catalytic converter is running very hot. The temperature on the converter is about 800 to 900 degrees Fahrenheit. Is this normal or is something wrong with the engine?

SOLUTION: Catalytic converters are designed to operate at this temperature and at times even hotter. The heat is needed to help reduce the NO_x emissions. However, if the temperature gets hotter and the catalytic converter glows red, there may be a problem. The first thing to check for is a blockage in the catalytic converter. This will cause the temperature to increase significantly.

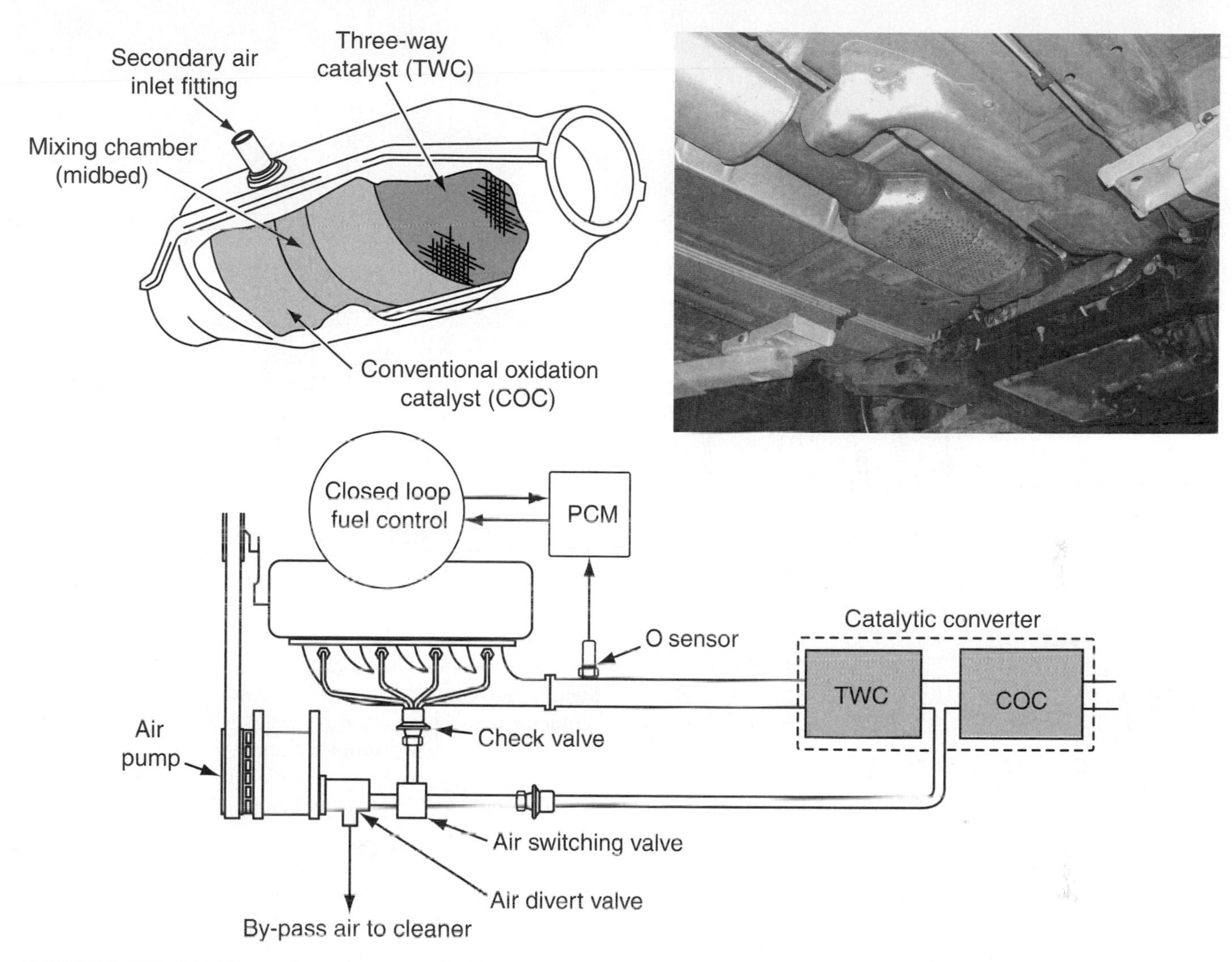

FIGURE 37–35 When using a three-way, dual-bed, monolith catalytic converter, secondary air from the air injection system is used to cause greater oxidation of the emissions.

As the treated exhaust gases from the first bed continue, they eventually pass through the conventional oxidation catalyst made of palladium and platinum. Here, HC and CO emissions are reduced.

37.4 CONTROLLING COMBUSTION AND AIR-FUEL MIXTURES

Many other types of emission control systems are used on today's automobiles. These systems are designed to change or control the air-fuel ratio or the combustion process, or both. Knowledge of the fuel and ignition system is necessary to understand this information.

IDLE LIMITERS

Most engines with carburetors have idle limiter screws. Their purpose is to ensure that the carburetor will deliver a leaner air-fuel mixture, especially during idle. The idle mixture adjustment screw is adjusted leaner by the manufacturer. Then, the idle limiter cap is installed. The cap permits only a small idle mixture adjustment. The cap should be removed only during carburetor repair. ***Figure 37–36*** shows the idle limiter screws in one model of carburetor. On later-model carburetors, the idle mixture screw is completely

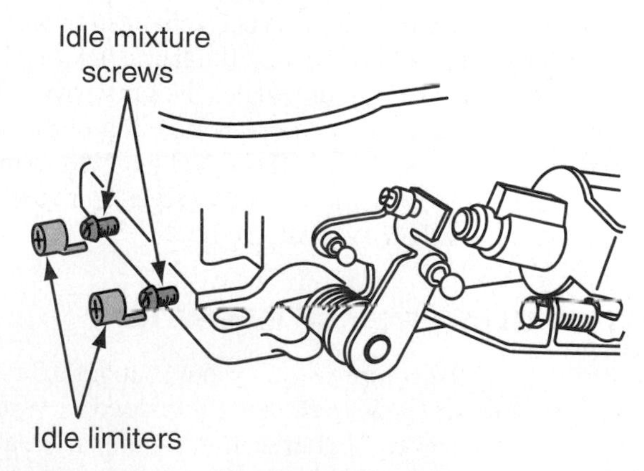

FIGURE 37–36 The idle limiter is located at the base of the carburetor. It ensures a leaner air-fuel mixture during idle conditions. This idle limiter mixture screw is sealed, so adjustments cannot be made.

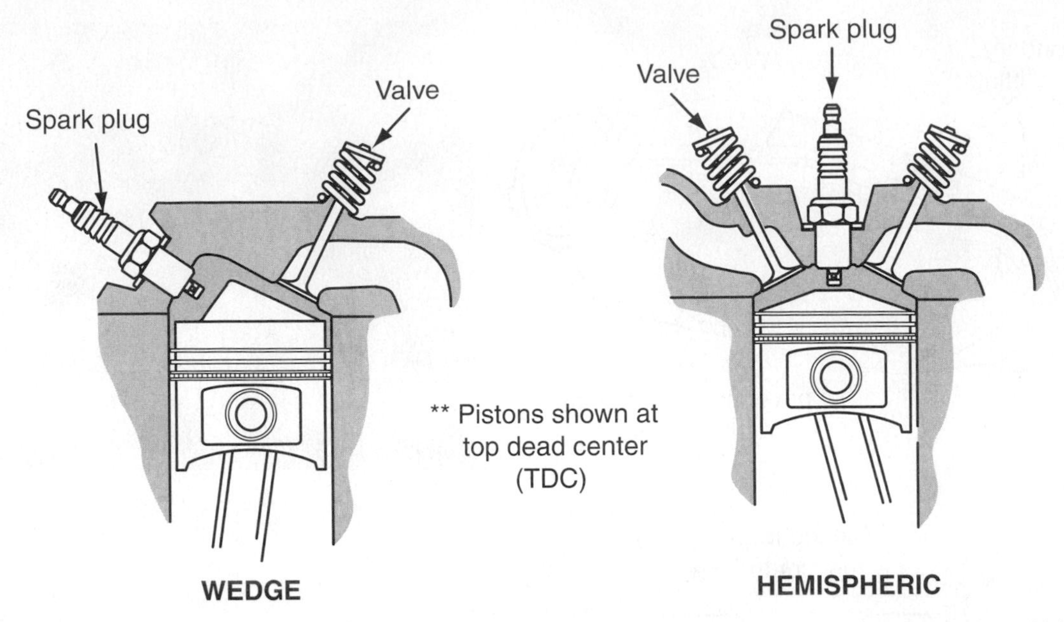

FIGURE 37–37 The hemispheric combustion chamber has a lower S-V (surface-to-volume) ratio than the wedge, which produces less pollution in the cylinder.

sealed. No adjustment is possible on such carburetors except by authorized technicians.

COMBUSTION CHAMBER

Combustion chambers have also been changed to produce fewer emissions. During combustion, the layer of air-fuel mixture next to the cooler cylinder head and piston head typically does not burn completely. The metal surfaces chill this area below the combustion temperature. A certain amount of unburned fuel is swept out of the cylinder on the exhaust stroke.

These emissions can be controlled by reducing the S-V ratio of the combustion chamber. The S-V ratio is the ratio between the surface (S) area and the volume (V) of the combustion chamber. A wedge-design combustion chamber has the highest S-V ratio. The hemispheric combustion chamber has a lower S-V ratio. ***Figure 37–37*** shows the wedge and the hemispheric combustion chambers.

Combustion chambers have also been changed to reduce close clearances that tend to quench the flame before all of the air-fuel mixture is burned. Quench heights have been increased. This permits more complete burning of the air-fuel mixture in these areas. On late-model engines, manufacturers have gone to a more "open" style combustion chamber to reduce exhaust emissions.

LOWER COMPRESSION RATIOS

On older engines, the compression ratio was about 9.5 to 1. In fact, compression ratios were being increased each year to produce more power. Higher compression ratios also produced higher combustion temperatures. These higher temperatures then increased the amount of NO_x being produced. Because of this problem, automobile manufacturers have lowered the compression ratios to about 8.5 to 1. Reducing the compression ratio reduces peak combustion temperatures, which, in turn, reduces the amount of NO_x produced. However, reducing compression ratios also reduces engine performance and efficiency to some extent.

INTAKE MANIFOLDS

Intake manifolds have been modified to ensure more rapid vaporization of fuel during warm-up periods. The exhaust crossover flow area of the intake manifold between the inlet and exhaust gases has been made thinner. The time required to get the heat from the exhaust gases into the inlet gases has been reduced.

IGNITION TIMING

Ignition timing of the automobile engine is very important to the reduction of emissions. For example, during part throttle operation on older engines, the distributor vacuum advance normally advances the ignition timing. This provides more time for the leaner air-fuel mixture to burn. The added time, however, also allows more NO_x to develop. Systems have been incorporated to control the advance under certain driving conditions such as reverse, neutral, or low forward speeds.

In the past, several systems have been designed to control the spark advance more precisely. For example, a **transmission-regulated spark (TRS)** was used to regulate the vacuum advance during low speeds. Also, electronic control of spark and timing was used. This latter system used a magnetic pickup coil sensor in the distributor to help control the spark advance. Currently, on-board computers, called the PCM, or powertrain control module, control spark advance.

IGNITION TIMING CONTROL

The use of computers to control spark advance has led to the development of computer command control systems, EFI (electronic fuel injection) systems, ECCS (electronic concentrated engine control) systems, and EI (electronic ignition) systems used on automobiles today. These systems provide very sensitive and precise control of air-fuel ratios and ignition system timing advance.

A computer system diagram is shown in ***Figure 37–38***. This system is capable of sensing up to eleven variables and controlling six functions with the information. Systems include injection timing, injection of air into the exhaust manifold, injection of air into the catalytic converter, idle speed, and others. Precise control of these systems makes possible a sizable reduction in emissions.

The computer used on automobiles is also tied into some of the existing emissions control equipment. These include carbon canister purge control, EGR control, and EFE control.

1. The carbon canister purge is controlled by the computer. A solenoid is placed in the manifold vacuum purge line. The computer controls when the purging takes place. For a more detailed description, refer to Section 37.2, Evaporative Emission Controls.
2. The EGR valve can also be controlled by the computer. On the basis of the coolant temperature, mass airflow, engine rpm, and throttle position, the computer will cycle the EGR valve open and closed. For a more detailed description, refer to Section 37.3, Intake and Exhaust Emission Controls.
3. The **EFE (early fuel evaporation)** control is used to cause exhaust gases on one side of a V engine to be directed over to the other side. The excess heat causes the fuel to evaporate faster, reducing emissions. The EFE system is also controlled by the computer.

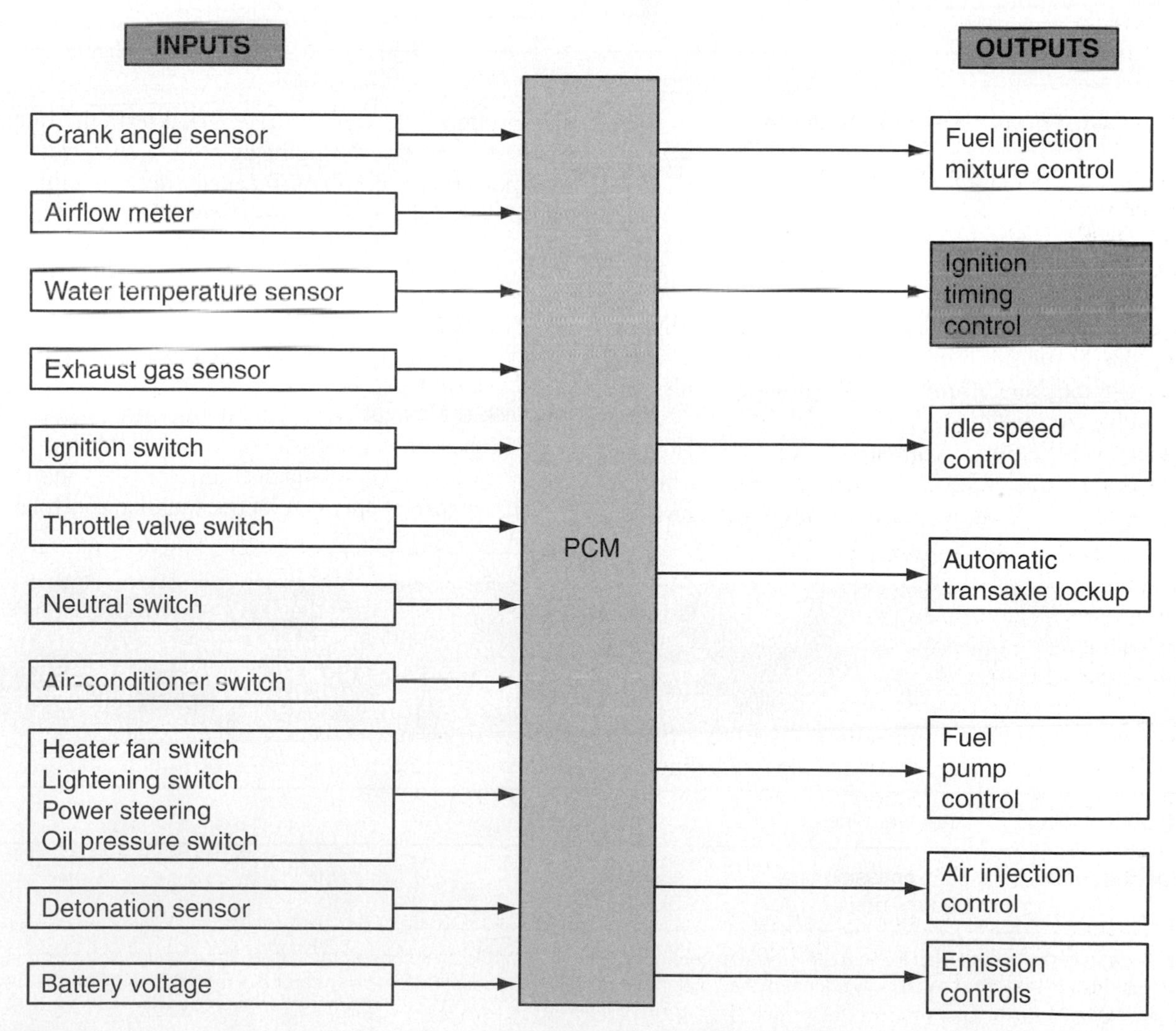

FIGURE 37-38 This diagram shows a closed loop electronic control system that monitors and controls a wide range of units and functions to help reduce emissions.

Problems, Diagnosis, and Service

Safety Precautions

1. After the engine has been running, the EGR valve may be very hot. Let the engine cool down to avoid burning your hands.
2. Never weld anything on or near the gasoline tank, even if the fuel has been removed. There may still be gasoline fumes inside the tank that could cause an explosion.
3. When checking a heated intake air system, be careful not to burn your hands. The tube is connected to the exhaust manifold and can be very hot.
4. The catalytic converter normally operates at a higher temperature than the exhaust. Always be extra careful when working around the catalytic converter to eliminate the possibility of severe burns.
5. Always use compressed air and an air hose to blow out valves, tubes, or other clogged passageways on emission control devices. Never blow these out with your mouth as you may be exposed to toxic liquids.
6. At times when diagnosing the emission control system, the engine must be running. Be careful not to drop anything into the spinning parts or in the fan blades. Remember also to stay clear of the electric fan. It may go on unexpectedly, causing injury.
7. Always wear OSHA-approved safety glasses when working on emission control systems.
8. Always use the correct tools when working on emission control systems.
9. When running the engine, be sure to properly vent exhaust fumes.

PROBLEM: Faulty EGR or PCV Valve

An engine is running rough, has a very irregular and rough idle, and stalls.

DIAGNOSIS

Often the EGR or the PCV valve may be defective. If either is defective, the air-fuel ratio may be off and will produce an irregular or rough idle. ***Figure 37–39*** shows an emission system troubleshooting guide.

On most vehicles with computer-controlled systems, trouble codes are shown on the instrument panel when a fault is detected. If a scanner is used or if the service technician grounds a trouble code test lead terminal under the dashboard, the Check Engine light will flash a trouble code that indicates the problem area. Refer to the manufacturer's service manual for the correct identification of codes for each vehicle.

SERVICE

1. When servicing the emission control systems, always check the condition of emission control vacuum hoses and tubes for leaks, small cracks, or bad sealing. Any leak could cause a pressure or vacuum leak. This, in turn, may cause incorrect operation of the emission control devices.

SYMPTOM	SUB SYSTEM	CATALYTIC CONVERTER	EGR SYSTEM	POSITIVE CRANKCASE VENTILATION SYSTEM	EVAPORATIVE EMISSION CONTROLS
Rough idle			①	②	
Frequent stalling	After warming up		①		
Poor performance	Fails emission test Loss of power	① ①			②

FIGURE 37–39 To help troubleshoot the emission control system, look for the symptom on the left. Then read across to the most likely problem. Number 1 is the first or most likely source of the problem. Number 2 is the second most likely source of the problem. (Courtesy of American Honda Motor Co., Inc.)

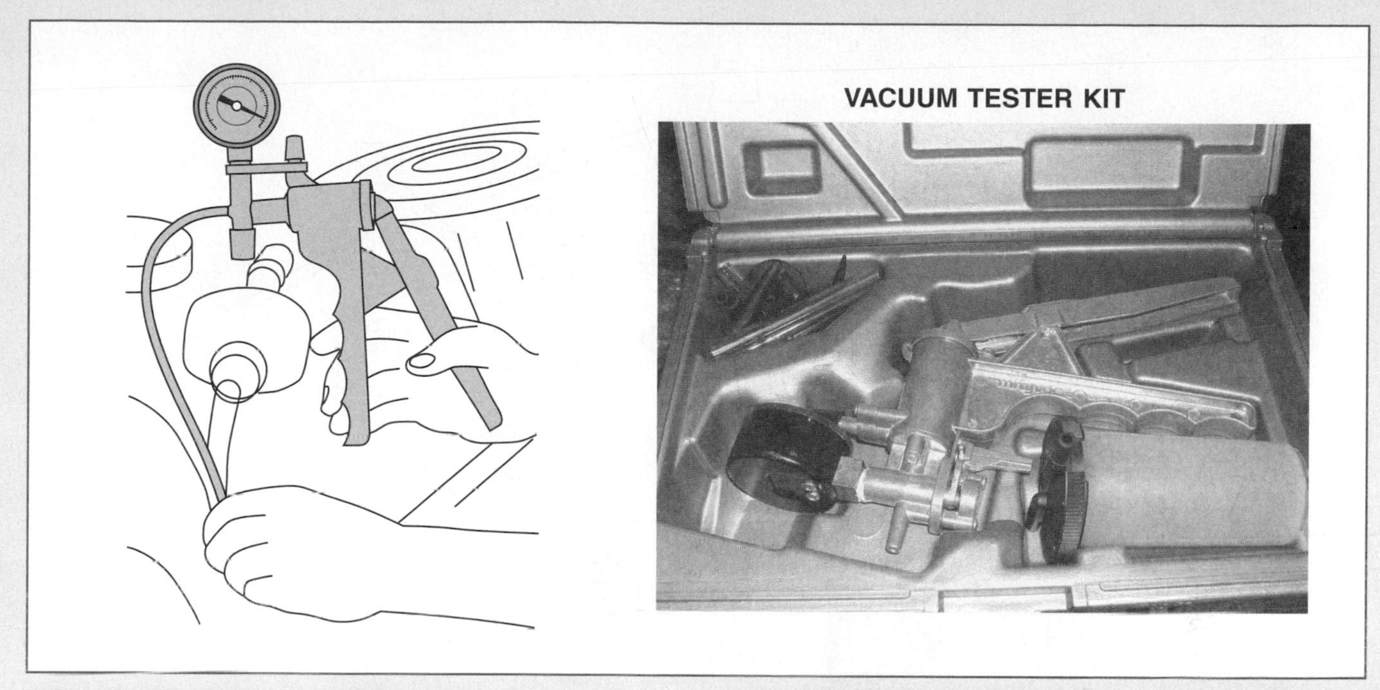

FIGURE 37–40 A vacuum tester can be used to check for vacuum leaks and the condition of diaphragms and other pollution control equipment.

2. Hoses, vacuum diaphragms, and tubes can be checked with a vacuum tester as shown in ***Figure 37–40***. To check a component, remove the hose attached to the device to be checked. Using the vacuum tester, apply vacuum to the component by pumping the vacuum tester. The device should be able to hold a vacuum without leaking. Leakage can be observed easily on the vacuum scale.
3. Always check the specification manual for the correct hookup of vacuum hoses. Most service manuals have sections that show the vacuum hose routing. ***Figure 37–41*** shows two typical examples of vacuum hose routing diagrams like those in service manuals.
4. Since there are several types of EGR valve systems, refer to the manufacturer's specifications and procedure for correct testing and service of each. On some engines, the Check Engine light (malfunction indicators lamp, or MIL) will come on to indicate a certain trouble code. For example, on vehicles made by one manufacturer, Code 12 indicates "Exhaust Gas Recirculating System." Check the EGR valve using the following general procedure:
   a. Place a finger under the EGR valve and lift up on the diaphragm plate.

   CAUTION: *Be careful not to burn your fingers or hands.*

   The plate should move up and down freely without binding or sticking. If it doesn't move freely, replace the valve.

   b. Check the vacuum by connecting a vacuum gauge to the EGR valve signal line using a T fitting. Start the engine and run it at part throttle. The vacuum gauge should read at least 5 inches of Hg (check the manufacturer's specifications for each vehicle) at part throttle and at normal engine operating temperature. If the gauge reads less, check the hoses and vacuum source, and repair as needed.
   c. With the engine running as in step b, disconnect the vacuum hose to the EGR valve. The diaphragm plate should move, and the engine speed should increase. If the plate does not move, replace the EGR valve.
   d. Reconnect the vacuum hose. The EGR valve diaphragm plate should move, and the engine speed should decrease.
   e. If the diaphragm moves but there is no change in the engine speed, check the manifold passages for blockage. If the passages are clean, the EGR valve is defective.
   f. The intake manifold can be cleaned with the EGR valve removed. Use a suitable screwdriver, scraper, and wire brush to remove deposits.
   g. On certain vehicles, there is a replaceable filter on the EGR valve. Replace this filter at the manufacturer's recommended intervals.
   h. Replace either the EGR valve assembly or the filter before starting the engine.
5. The PCV valve may also be defective, thus producing a rough idle. Use the following general procedures when servicing the PCV valve:
   a. Check the PCV valve for correct operation by removing the valve, running the engine, and observing a

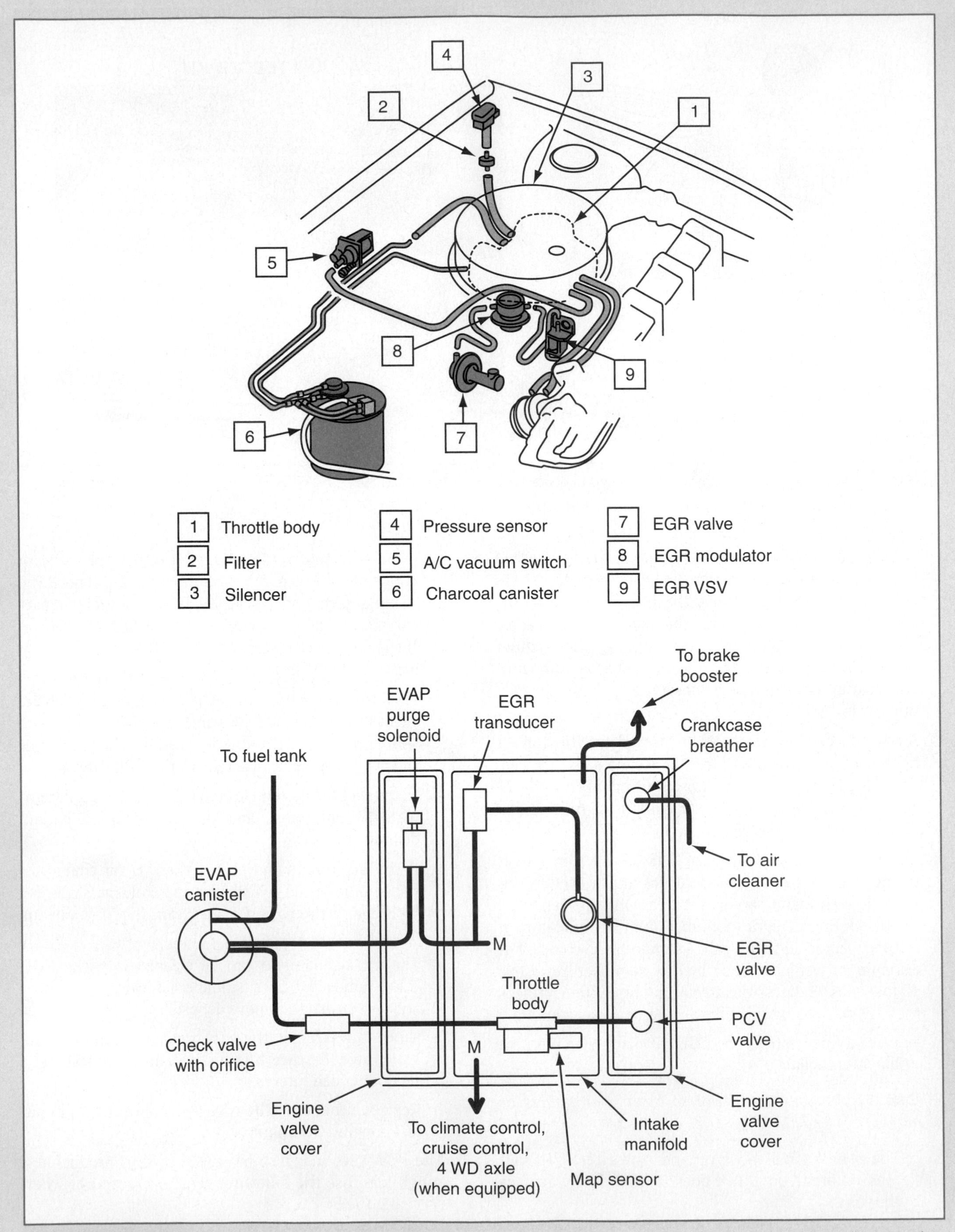

FIGURE 37-41 Vacuum hose routing diagrams are available in many maintenance and service manuals. Refer to these diagrams when checking for correct vacuum connections.

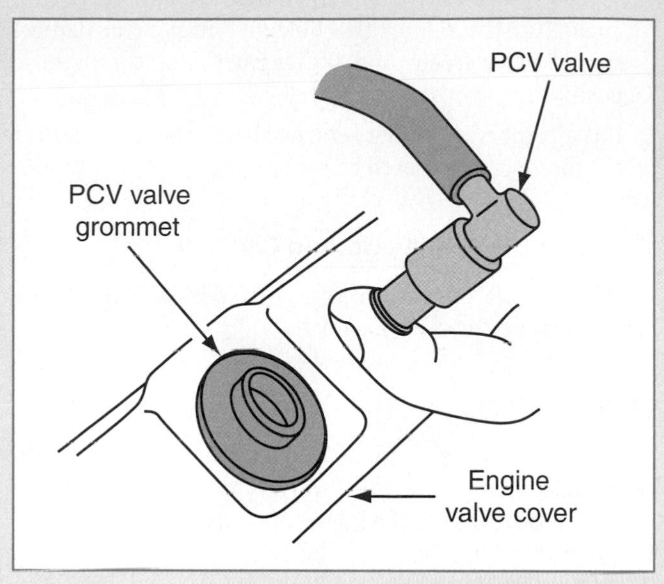

FIGURE 37–42 When checking the PCV valve, remove the valve and run the engine at idle. A small vacuum should be felt at the end of the PCV valve.

small vacuum (***Figure 37–42***). A hissing sound should also be heard. If there is no hissing sound, no vacuum, or if there is a slight pressure, the PCV valve may be damaged or plugged. If this is the case, remove the valve and replace it with a new one.

b. A plugged PCV valve can cause rough idle, stalling at low speeds, oil leaks, oil in the air cleaner, or sludge in the engine. The PCV valve can easily be checked by shaking it lightly. When it is shaken, a PCV valve should make a slight clicking noise, indicating that the valve is free to move. If there is no clicking, the valve may be plugged or inoperative. Replace the valve if necessary.

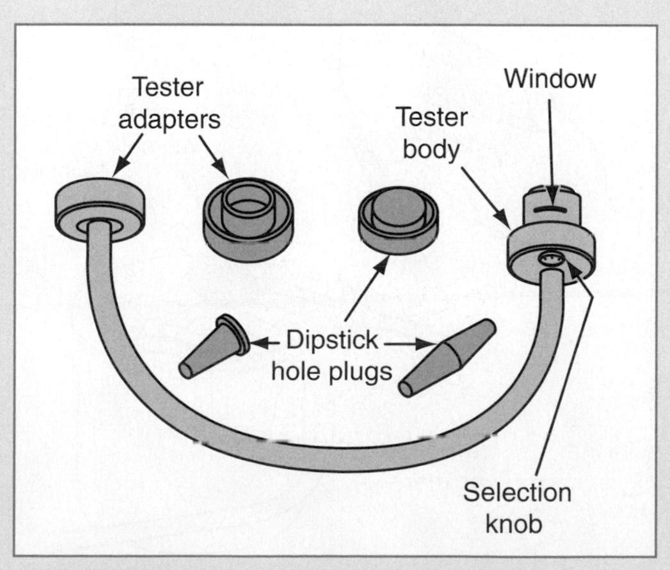

FIGURE 37–43 This PCV system tester measures vacuum in the system. It reads out the condition using several colors to indicate whether the system is operating correctly or not.

c. The PCV system can also be checked using the PCV system tester, shown in ***Figure 37–43***). Other testers can also be used. This tester is connected into the PCV vacuum system while it is operating. On the basis of the amount of vacuum, various colors will read out on the tester body. Green indicates the system is operating correctly. Yellow indicates the system is partially plugged. Red indicates the system is fully plugged. The system is then serviced accordingly.

PROBLEM: Faulty Fuel Tank Cap

The fuel tank on a vehicle is beginning to collapse.

DIAGNOSIS

A collapsing fuel tank usually indicates that the fuel tank cap is defective. A defective cap causes the tank to collapse because as the fuel is removed during normal driving, no air can get into the tank to replace the volume of fuel.

SERVICE

To remedy the problem, remove the filler cap and replace it with a new one.

PROBLEM: Faulty Diverter Valve

The engine produces a backfire.

DIAGNOSIS

Often an engine with a diverter valve may backfire if the valve is operating incorrectly. The diverter valve may be putting excess air into the combustion chamber. The result would be backfiring.

SERVICE

Use the following general procedure to test the diverter valve:

1. Start the engine and run it until it reaches normal operating temperature.
2. Disconnect the vacuum hose from the diverter valve to ensure the presence of a vacuum at the hose.
3. Reconnect the hose to the valve. Air should be vented through the diverter valve for at least 1 second.
4. Momentarily accelerate the engine to full throttle while observing the diverter valve. Each time the engine is accelerated, air should be discharged from the diverter valve muffler for about 1 second. If the diverter valve fails to perform as outlined, the valve is defective and should be replaced with a new one.

PROBLEM: Defective Air Injection Reactor (AIR) Pump

The air injection reactor pump is making loud noises.

DIAGNOSIS

The air injection reactor (AIR) pump often will become defective. The result is that there may be excessive belt, bearing, or knocking noises from the pump.

SERVICE

Use the following general procedures to service the AIR pump:

1. Check for the correct tension on the AIR pump belt. This can be done using the belt tension checker.
2. There should be no air leaks in the AIR system components. Look for air leaks at the hoses and the air injection tubes on the exhaust manifold. If there are any leaks, replace the components as necessary.
3. Check the mounting bolts for looseness and to see if the pump is aligned correctly.
4. After removing the belt, check for a seized air pump. Replace if necessary.
5. Check for loose air pump hoses.
6. Check for a defective pressure relief valve.
7. Inspect the check valves for airflow in only one direction. Replace as necessary.

PROBLEM: Damaged Heated Air Intake

An engine usually runs rough, hesitates, or stalls during cold starting conditions.

DIAGNOSIS

The heated intake air system may be damaged and operating incorrectly. If it is damaged, the intake manifold will not get warm air (drawn across the exhaust manifold) during cold starting. The result is that the air-fuel ratios may be slightly off during cold starting.

SERVICE

When this system is defective, several checks can be made:

1. Check the hoses that feed the vacuum to the vacuum motor for cracks and leaks.
2. Check the hose that feeds the warm air from the exhaust manifold to the air cleaner housing. This tube often rips, breaks, or cracks, depending on its type.
3. The vacuum motor can be checked with the vacuum tester.
4. First disconnect the vacuum hoses that lead to the vacuum motor. When a vacuum from the vacuum tester is applied to the motor, the linkage and control damper should move freely, and the damper should remain in position.
5. If the damper assembly returns slowly, there is a leak in the motor and it should be replaced.

PROBLEM: Faulty Carbon Canister

Fuel is leaking from the carbon canister. What is the necessary service on the carbon canister?

DIAGNOSIS

Often, the carbon canister is overlooked when a vehicle is serviced. If it is not operating correctly or if fuel is leaking, there may be cracks in the hoses, the purge valve may be defective, or the filters may need to be replaced.

SERVICE

Use the following suggestions when inspecting the carbon canister:

1. Replace the filter on the carbon canister according to the manufacturer's recommended time intervals.
2. Check the carbon canister for cracks or fuel leaking from the fittings. Repair and replace where necessary.
3. Check the carbon canister using a vacuum pump as shown in ***Figure 37–44***. Refer to the exact manufacturer's procedure to perform this test. Replace as necessary.

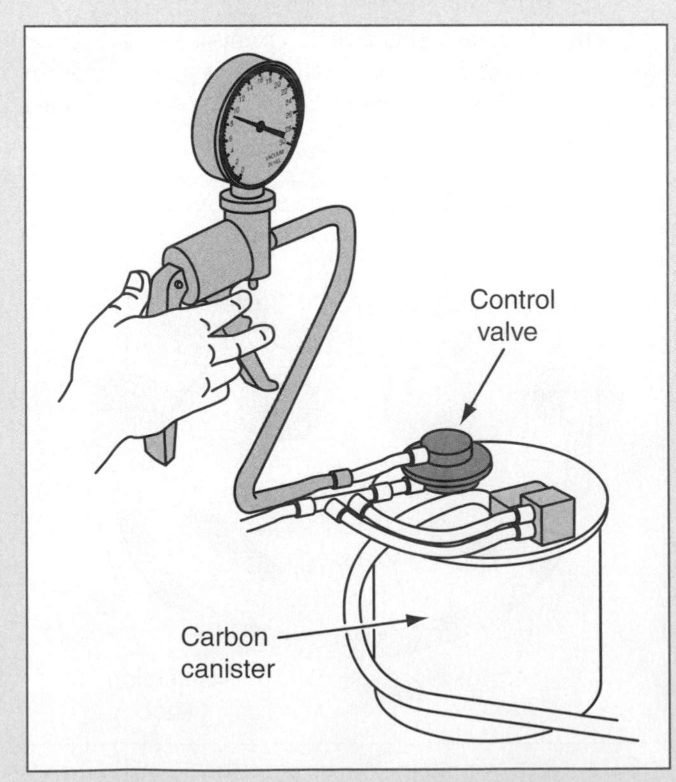

FIGURE 37–44 Use a vacuum tester to check the condition of the carbon canister.

PROBLEM: Faulty Catalytic Converter

An engine has a significant power loss at lower speeds.

DIAGNOSIS

At times, the catalytic converter can become plugged because of the use of leaded fuel in the engine. The honey comb core may also disintegrate as shown in ***Figure 37–45***.

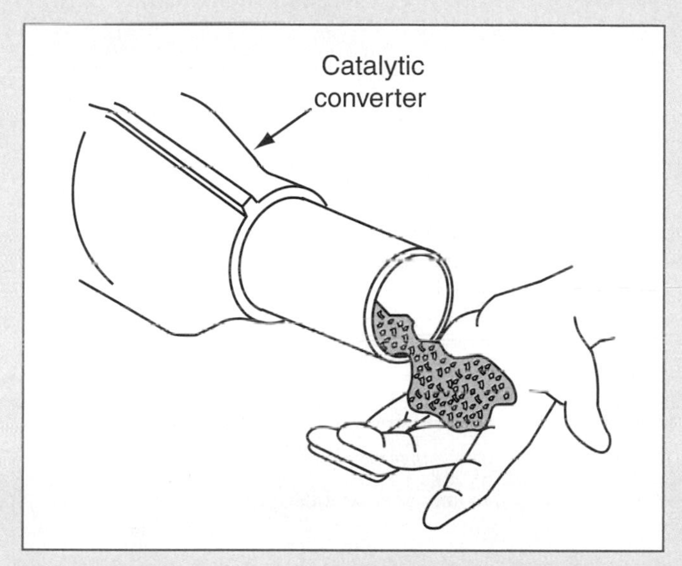

FIGURE 37–45 This catalytic converter's honeycomb core has disintegrated due to overheating and is blocking the exhaust flow from the engine.

SERVICE

Use the following procedures to service the catalytic converter.

1. Check to see if the customer has been using leaded fuel in the engine. Also check to see if the restrictor ring has been removed from the fuel inlet. If leaded fuel has been used or if the restrictor has been removed, the catalytic converter will need to be replaced.
2. A rotten egg smell is often produced by vehicles that have a catalytic converter. This odor is generally caused from too rich a mixture in the fuel system. If this smell persists, check the fuel injection system or try using another type of fuel to eliminate the odor.
3. Pellets coming out of the exhaust indicate that the catalytic converter is damaged internally. Replace as necessary.

PROBLEM: Faulty Exhaust Gas Recirculation (EGR) System

The malfunction indicator lamp (MIL) indicates the diagnostic trouble code (DTC) 12. This code means there is a problem in the exhaust gas recirculation (EGR) system.

DIAGNOSIS

Certain manufacturers use the PCM to control the opening and closing of the EGR valve. The system, as shown in ***Figure 37–46***, is composed of the EGR valve, EGR vacuum

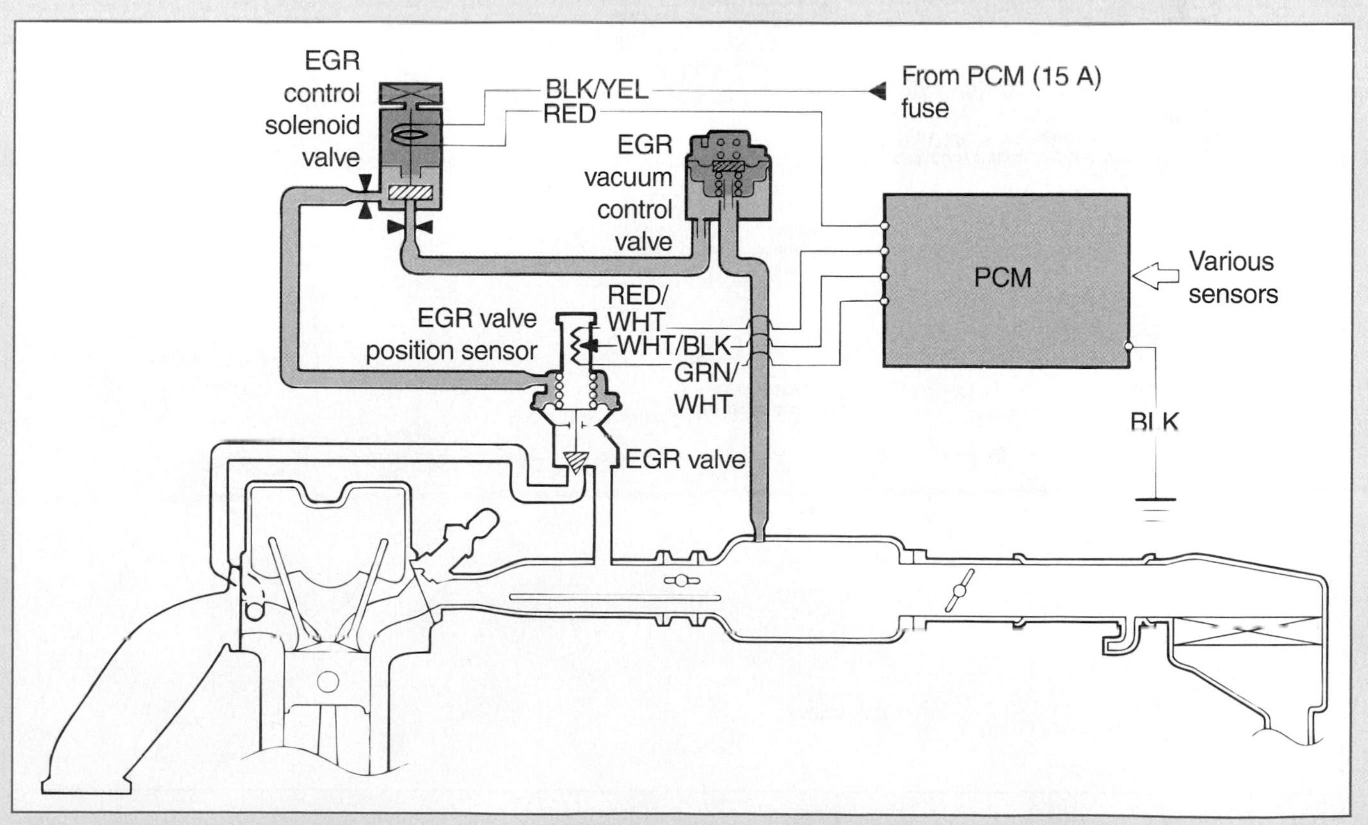

FIGURE 37–46 This EGR valve is controlled by the PCM. (Courtesy of American Honda Motor Co., Inc.)

control valve, EGR control solenoid valve, PCM, and various sensors. In this system, the PCM memory contains ideal EGR valve lifts for varying operating conditions. The EGR position sensor detects the amount of EGR valve lift and sends this information to the PCM. The PCM then compares it with the ideal EGR valve lift. The ideal lift is determined by signals sent from the other sensors. If there is any difference between the ideal and actual valve lift, the PCM varies current to the EGR control solenoid valve to further regulate vacuum applied to the EGR valve.

SERVICE

Each manufacturer will present a detailed procedure to check the various components of such system. In these procedures, each of the system components is checked for vacuum, electrical resistance, and so on. It is beyond the scope of this textbook to include detailed service procedures for each part. The exact procedure may be three to four pages in length. However, ***Figure 37–47*** shows a portion (the beginning) of such a service procedure as an example only.

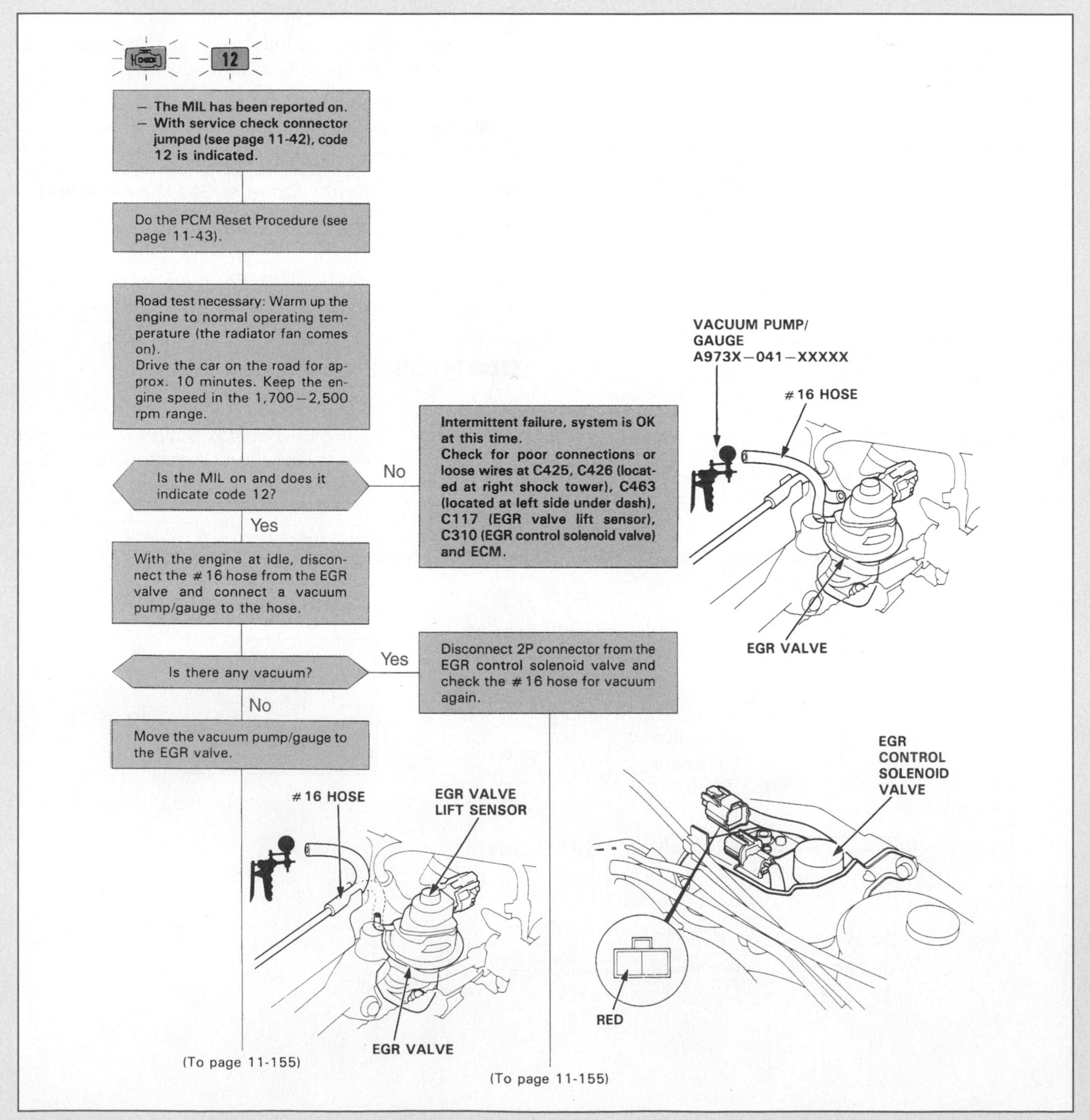

FIGURE 37–47 Detailed service procedures like those shown here help the service technician to work on complicated emission control systems. (Courtesy of American Honda Motor Co., Inc.)

Service Manual Connection

There are many important specifications to keep in mind when working with emission control systems. To identify the specifications for your engine, you will need to know the VIN (vehicle identification number) of the vehicle, the type and year of the vehicle, and the type of engine. Although they may be titled differently, some of the more common emission control system specifications (not all) found in service manuals are listed below. Note that these specifications are typical examples. Each vehicle and engine may have different specifications.

Common Specification	Typical Example
Air pump drive belt (on belt tension gauge)	3/8 inch
Carbon monoxide (CO)	3.4 grams per mile
Filler cap relief valve opening	0.8 psi
Hydrocarbons (HC)	0.41 gram per mile
Nitrogen oxides (NO_x)	1.0 grams per mile
Thermostatic air cleaner check valve closing temperature	80 degrees F

In addition, the service manual will give specific directions for various service and testing procedures. Some of the more common procedures include:

- Carbon canister servicing
- Catalytic converter replacement
- Exhaust gas recirculating valve testing
- Oxygen sensor replacement
- Ported vacuum switches testing
- Positive crankcase ventilation system servicing
- Vacuum control valves testing
- Vacuum hose routings

SUMMARY

The following statements will help to summarize this chapter:

- Emission control systems include the positive crankcase ventilation system, evaporative systems, intake and exhaust controls, and combustion control.
- The positive crankcase ventilation system is used to prevent crankcase vapors from entering the atmosphere.
- The positive crankcase ventilation system uses a PCV valve to control the blow-by gases that are sent back to the intake manifold for further burning.
- Automobiles also produce evaporative emission vapors from gasoline.
- A carbon canister system is used to control gasoline vapors to help reduce HC emissions into the air.
- The charcoal in the carbon canister absorbs fuel vapors.
- Other evaporative emission control systems include the vapor-liquid separator, rollover check valve, and fuel expansion tanks.
- A heated air intake system is used to heat the air going into the engine to improve engine efficiency during cold weather starting.
- The exhaust gas recirculation (EGR) valve is used to control nitrogen oxides by sending some of the exhaust back into the intake for more complete burning.
- The air injection reactor system uses an air pump to force air into the exhaust system to reduce emissions.
- A catalytic converter is used on today's vehicles to change the chemistry of the exhaust, thus reducing HCs, CO, and NO_x.
- Two types of catalytic converters are the two-way and three-way converters.
- Both pellet and monolith designs are used for catalytic converters.
- Combustion can also be controlled as part of emission control systems.
- Combustion control is accomplished by systems such as the idle limiters on carburetors, reshaping the combustion chamber for improved emission control, changing the compression ratios, and controlling the timing by computers.

TERMS TO KNOW

Can you explain each of the following terms? Review the chapter until you can use each term correctly.

Air injection reactors (AIR)
Aspirator valve
Blow-by
Carbon canister
Catalyst
Catalytic converters
Crankcase pressure
EFE (early fuel evaporation)
Exhaust gas recirculation (EGR)
Heat stove
Monolith
Pintle
Positive displacement pump
Purge
Snorkel tube
Transducer
Transmission-regulated spark (TRS)

REVIEW QUESTIONS

Multiple Choice

1. Crankcase pressures are produced by:
a. The cooling system
b. The ignition system
c. Power output
d. Blow-by
e The starter

2. The PCV valve ________ the amount of vapor being burned in the combustion chamber during low speeds.
a. Increases
b. Decreases
c. Doubles
d. Has no effect on
e. None of the above

3. The carbon canister stores:
a. Fuel vapor from the fuel tank
b. Fuel vapor from the carburetor
c. Blow-by vapors
d. All of the above
e. A and B

4. The carbon canister:
a. Stores fuel vapors
b. Has a filter
c. Has a purge line back to the carburetor
d. All of the above
e. None of the above

5. Which of the following stops raw fuel from entering the carbon canister?
a. Check valve
b. Vapor-liquid separator
c. Air pump
d. PCV valve
e. Idle limiter screws

6. Which device is used to allow a place for expanding fuel in a full tank of gasoline in warm weather?
a. Check valve
b. Rollover valve
c. Expansion tank
d. PCV valve
e. AIR system

7. Warming up the intake air during cold starting will:
a. Reduce emissions
b. Increase emissions
c. Reduce fuel consumption
d. Cause rough idling
e. Increase power

8. A heat stove is used on what pollution control device?
a. AIR
b. PCV
c. Thermostatically controlled air cleaner
d. Transmission-controlled spark
e. CCC

9. Which of the following pollution control devices is designed to reduce the NO_x emissions?
a. AIR
b. PCV
c. EGR
d. TCS
e. None of the above

10. On the EGR system, ________ is/are sent back to the intake.
a. Exhaust gases
b. Blow-by
c. Extra spark
d. Liquid coolant
e. Catalyst

11. Which pollution control system forces air into the exhaust stream to help burn emissions?
a. AIR
b. PCV
c. EGR
d. TCS
e. CCC

12. Which of the following will have little or no effect on the combustion process?
a. PCV
b. Carbon canister
c. AIR
d. Evaporative emission systems
e. Heated air intake systems

13. What type of system allows air to be injected into the exhaust without the use of a positive displacement pump?
a. CCC
b. Aspirator system
c. Carbon canister
d. All of the above
e. None of the above

14. Which of the following systems change(s) the exhaust system to reduce emissions?
a. Two-way catalytic converter
b. Three-way catalytic converter
c. Diverter valve
d. AIR system
e. A, B, and D

15. Which of the following can be found in a catalytic converter?
a. Monolith design
b. Pellet design
c. Air connection from the AIR system
d. Platinum catalyst
e. All of the above

16. Which catalytic converter system uses air from the AIR system?
a. Two-way converter
b. Three-way converter
c. Four-way converter
d. All of the above
e. None of the above

17. Which of the following has/have been changed to enhance vaporization of the fuel?
a. Combustion chambers
b. Intake manifolds
c. Exhaust flow
d. Catalytic converters
e. PCV systems

18. Which of the following is/are part of the computer-controlled systems to reduce pollution?
a. EGR valves
b. Carbon canister
c. EFE (early fuel evaporation)
d. All of the above
e. None of the above

The following questions are similar in format to ASE (Automotive Service Excellence) test questions.

19. *Technician A* says that the PCV valve should not rattle when it is shaken. *Technician B* says that if the PCV valve rattles when shaken, the PCV valve is not plugged. Who is correct?
a. A only
b. B only
c. Both A and B
d. Neither A nor B

20. *Technician A* says that the EGR valve can be tested by pushing up inside the valve and seeing if the diaphragm moves freely. *Technician B* says that the EGR valve can be checked using the vacuum tester. Who is correct?
a. A only
b. B only
c. Both A and B
d. Neither A nor B

21. *Technician A* says that the charcoal canister has no filter to be changed. *Technician B* says that the charcoal canister has a filter that should be changed at regular recommended intervals. Who is correct?
a. A only
b. B only
c. Both A and B
d. Neither A nor B

22. *Technician A* says that a collapsed fuel tank is caused by too much vacuum in the tank. *Technician B* says a collapsed fuel tank is caused by a bad fuel tank cap. Who is correct?
a. A only
b. B only
c. Both A and B
d. Neither A nor B

23. *Technician A* says computer controls can be used on the EGR system. *Technician B* says computer controls can be used on the PCV system. Who is correct?
a. A only
b. B only
c. Both A and B
d. Neither A nor B

24. Fuel is leaking near the front of the engine. *Technician A* says the problem could be the carbon canister. *Technician B* says the problem could be the EGR valve system. Who is correct?
a. A only
b. B only
c. Both A and B
d. Neither A nor B

25. A vehicle has a plugged catalytic converter. *Technician A* says the problem is caused by using leaded fuel. *Technician B* says the problem could be a plugged PCV system. Who is correct?
a. A only
b. B only
c. Both A and B
d. Neither A nor B

26. *Technician A* says a common problem with the AIR system is oil leakage from the bearings. *Technician B* says a common problem with the AIR system is loose drive belts. Who is correct?
    a. A only
    b. B only
    c. Both A and B
    d. Neither A nor B

Essay

27. Describe the purpose and operation of the PCV system.

28. Describe the purpose and operation of the EGR system.

29. What is the carbon canister used for?

30. What is a heat stove used for on a pollution control system?

31. Describe the operation of a catalytic converter.

32. Define the S-V ratio when dealing with combustion chambers.

33. List several ways in which pollution has been reduced by changing the combustion chamber or air-fuel mixtures, or both.

Short Answer

34. The on-board computer (PCM) on a vehicle is used to ________ the carbon canister.

35. With the use of the on-board computer (PCM), the EGR valve can be either open, closed, or ________.

36. If the PCV valve is bad, it most likely will produce a ________ idle.

37. If the heated intake air system is damaged, the engine will run rough during ________ conditions.

38. If the catalytic converter is plugged, the engine will have a ________.

Applied Academics

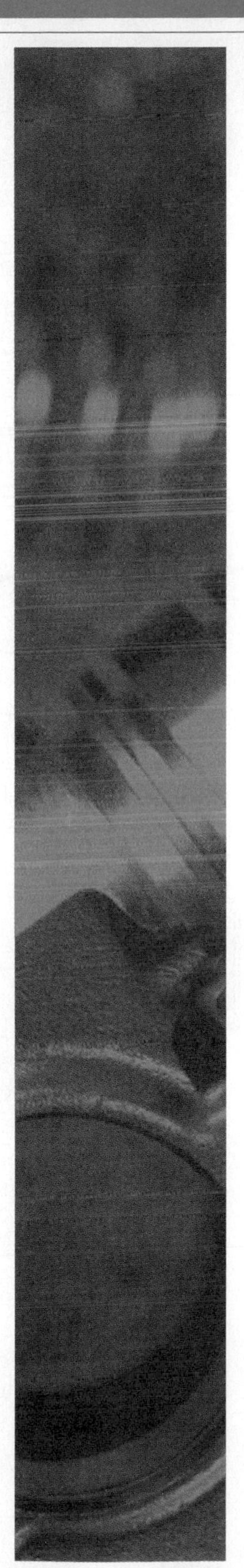

TITLE: Carbon Monoxide and Safety

ACADEMIC SKILLS AREA: Science

NATEF Automobile Technician Tasks:

The service technician follows all safety regulations and related procedures while performing an engine performance task.

CONCEPT:

It is very important when running a vehicle in the shop to take necessary safety precautions regarding carbon monoxide. Carbon monoxide is found in automobile engine exhaust. It displaces oxygen in the bloodstream, which results in carbon monoxide poisoning. The concentration of carbon monoxide in the air is measured in parts per million (ppm). This means that in one million parts of air, a certain number of the parts are carbon monoxide. The accompanying figure shows how a person is affected when exposed to different levels of carbon monoxide. The horizontal axis shows the duration of exposure in hours. The vertical axis shows the type of health effects that exposure to carbon monoxide can have on a person's health. Each curved line represents a different concentration of carbon monoxide. For example, if a person is exposed to a concentration of 100 ppm carbon monoxide for less than 1 hour, no symptoms are noticeable. If the time of exposure at this concentration is increased to 10 hours, the person will experience headaches and reduced mental ability.

HEALTH EFFECTS OF EXPOSURE TO CARBON MONOXIDE

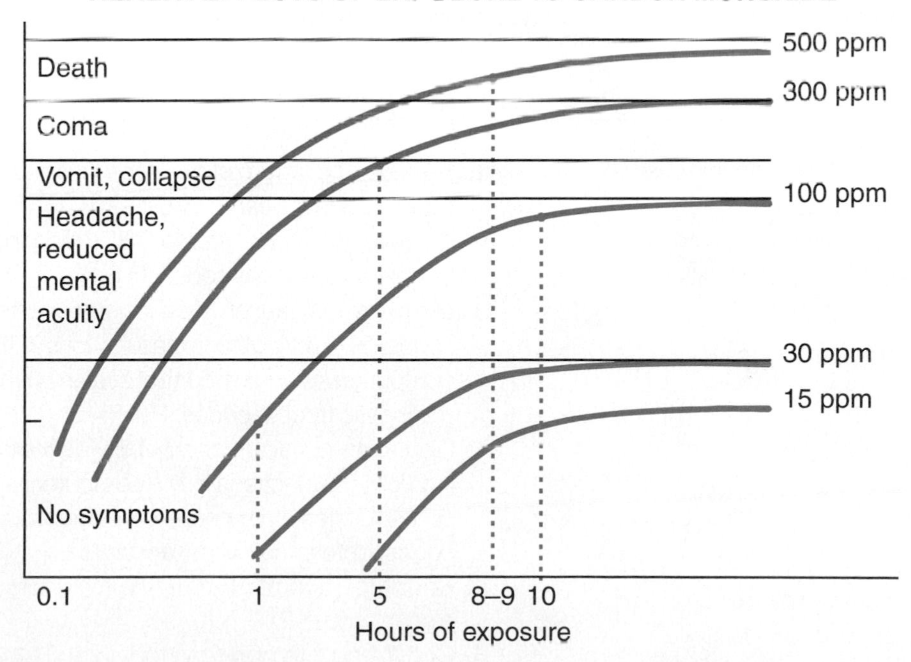

APPLICATION:

If a person is exposed to 500 ppm of carbon monoxide, how long will it take for that person to go into a coma? How long would a person have to breathe carbon monoxide at 300 ppm to come close to death? What can you do to guard against carbon monoxide poisoning in the automotive shop?

CHAPTER 38

Computerized Engine Control Systems

OBJECTIVES

After studying this chapter, you should be able to:

- Identify the design and operation of engine computer input devices.
- Describe the operation of the computer output devices.
- Analyze common procedures used by the service technician to identify and diagnose problems and service the input and output devices used on the engine computer.

INTRODUCTION

Engines used on today's automobiles use computers to control many of their systems and components. The computer is used to control fuel, ignition, cooling, and many other functions. In order for the computer to operate correctly, there are many inputs to the computer and many outputs controlled by the computer. This chapter details many of the input and output devices used by the computer. In addition, this chapter discusses some common problems, their diagnosis, and service procedures that are used on these input and output components.

38.1 NEED FOR COMPUTERS ON THE ENGINE

This chapter discusses the various types of input and output components used on computerized engine controls. Many of these components have been mentioned in other chapters. However, this chapter will give more detail about such

Certification Connection

ASE Connection: The information in this chapter can help you prepare for the National Institute for Automotive Service Excellence (ASE) certification tests. The tests and content areas most closely related to this chapter are:

Test A8—Engine Performance

- **Content Area**—Computerized Engine Controls Diagnosis and Repair
- **Content Area**—Engine Electrical Systems Diagnosis and Repair

NATEF Connection: Much of the information in this chapter is related to the NATEF tasks. The NATEF tasks and priority numbers most closely related to this chapter are:

1. Retrieve and record stored OBDI diagnostic trouble codes; clear codes. P-2
2. Retrieve and record stored OBDII diagnostic trouble codes; clear codes. P-3
3. Diagnose the causes of emissions or drivability concerns resulting from malfunctions in the computerized engine control system with stored diagnostic trouble codes. P-1
4. Diagnose emissions or drivability concerns resulting from malfunctions in the computerized engine control system with no stored diagnostic trouble codes; determine necessary action. P-1
5. Check for module communication errors using a scan tool. P-2
6. Obtain and interpret scan tool data. P-1
7. Access and use service information to perform step-by-step diagnosis. P-1

input and output devices. Computer terminology and operation will not be discussed. This information has already been presented in a previous chapter.

As mentioned earlier, the computer used to control engine functions such as fuel and ignition is called the PCM, or powertrain control module. Although manufacturers have called the engine computer by different names, all references to the engine computer in this chapter will be to the PCM.

Over the past several years, many terms and abbreviations have been standardized. In fact, to help service technicians, the Society of Automotive Engineers issued directive J1930, which identified various terms used in the past by all manufacturers and provided standard acronyms or abbreviations. The J1930 terminology list is included as Appendix C in this textbook as a reference tool for finding correct terms, acronyms, and abbreviations.

ENGINE COMPUTER SYSTEMS

Over the past few years, the computer has been used on automobiles to control many functions. One function that the computer can easily control is the operation of the engine. In order to control the operation of the engine, the computer control system must be able to monitor many conditions. Then, based on these conditions, the computer makes decisions about how to control various components.

There are three major stages involved in engine computer control. These include inputs, processing, and outputs. The inputs come from various types of sensors; the processing is done by the computer; and the outputs convert electrical signals from the computer into mechanical actions. Refer to ***Figure 38–1***. The input stage is when data are supplied by the sensors. Data are collected by many sensors placed throughout the automobile. During the processing stage, the computer interprets various types of data and determines the type of output needed. During the output stage, the computer sends an electrical signal to some actuator to move or change something.

For a simple example of the input, processing, and output stages, think of your body. The five senses of the body supply the input. Your brain is the processor. Your muscles are the actuators. If you are driving down the road and another car pulls out in front of you, all three stages begin to operate. First, your eyes sense that there is a vehicle pulling out in front of you. Your eyes are a sensor providing input. Then your brain receives the input from your eyes and makes a decision about how to avoid an accident. Your brain is like the computer (processor). To avoid an accident, your brain sends electrical signals (output) to your leg and foot to push the brake pedal. The muscles in your leg and foot are the output actuators that respond to the brain's instructions.

INPUTS AND OUTPUTS FOR ENGINE CONTROL SYSTEMS

To understand the total engine computer control system, look at ***Figure 38–2***. This drawing shows the many types of sensors that provide inputs to the computer, as well as the various types of actuators that can be controlled. Not every engine or vehicle has all types of sensors. Nor does every vehicle have all of the suggested output devices. Generally, the newer the vehicle, the more input sensors and the more output actuators used. The exact types of inputs and outputs depend on four things including:

1. The year of the vehicle
2. The manufacturer of the vehicle
3. The type of ignition system (electronic ignition, transistorized ignition, etc.)
4. The type of fuel system (electronic carburetor, throttle body, port injection)

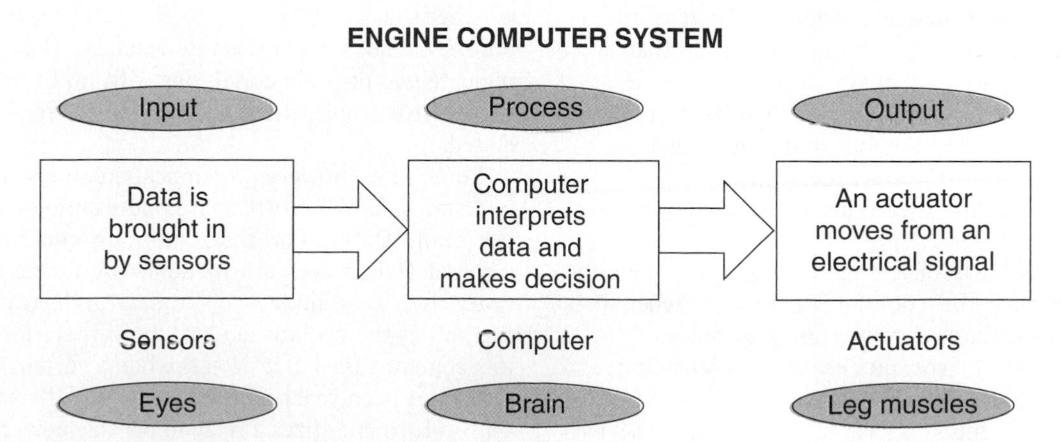

FIGURE 38–1 The computer system that controls engine functions has three stages: input, processing, and output.

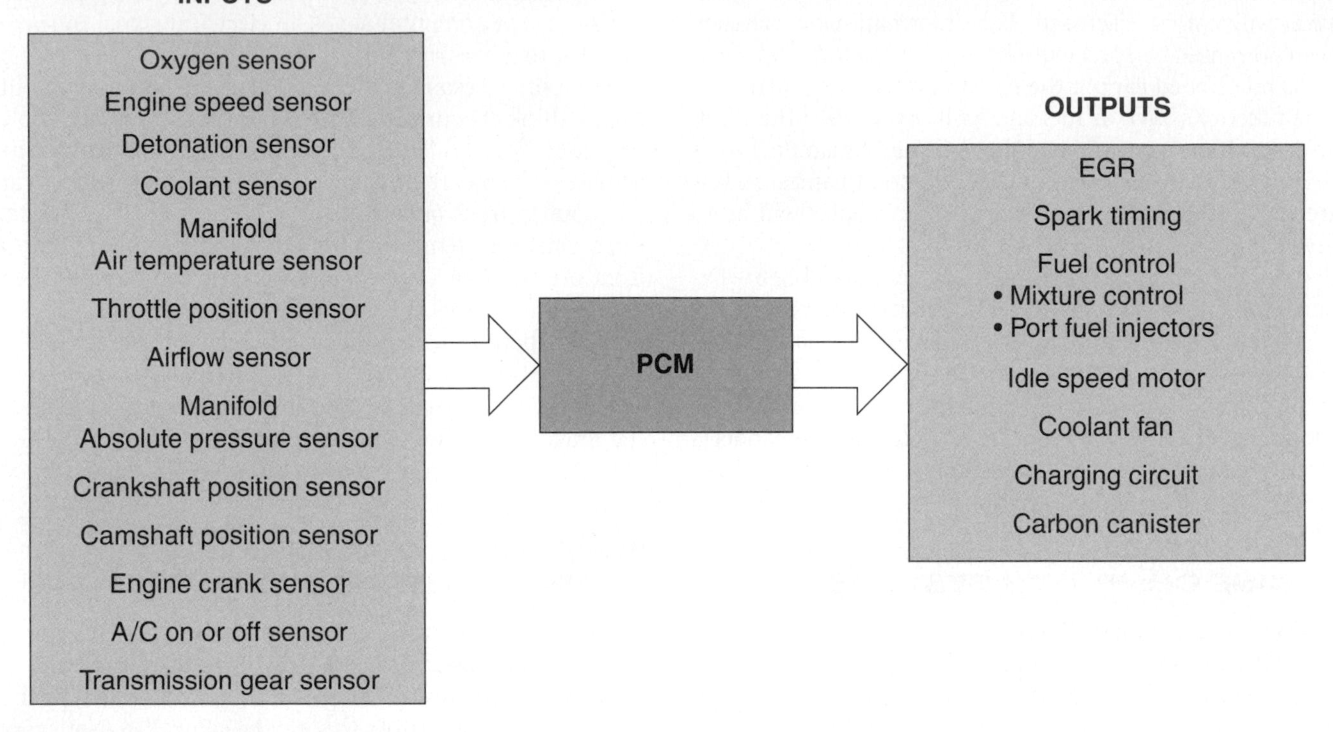

FIGURE 38-2 Although there are others, these are the common input and output devices that are used on a computerized engine control system.

ADVANTAGES OF USING ENGINE COMPUTERS

There are many advantages to using a computer to control engine functions. Some of them include the following:

- The air-fuel ratio is controlled as closely as possible to 14.7 to 1 under all types of operating conditions, such as cold starting, warm operation, high altitude, moist conditions, and so on.
- The engine is operated much more efficiently at cold temperatures, for example during starting.
- All emission control devices are operated more efficiently.
- Timing is controlled much more precisely than with conventional timing.
- The engine is able to make rapid changes in timing in response to changing driving conditions.
- The engine gets much better fuel mileage.
- The emissions are much lower compared to older engines.
- There is less waste of energy. For example, the computer can turn the cooling fan on only when it is needed, rather than running a fan at all times.
- Computers operate very quickly and can sense and send signals several hundred times or more per second.
- The engine is much more responsive and the performance of the engine increases across all ranges of operation.

38.2 INPUT COMPONENTS (SENSORS)

INTRODUCTION TO SENSORS

A sensor is defined as a **transducer**. A transducer is an electrical device that converts a physical quantity into an electrical equivalent. A door jam switch, for example, could be thought of as a sensor or a transducer. It converts the status of the door, either open or closed, to an electrical signal. Another example is a trunk lid switch on a car. It indicates whether the deck lid is open or closed. These types of transducers are considered **discrete sensors**. That is, they only indicate two possible conditions with no in-between. With the above example, the two conditions were either open or closed.

Sometimes, however, a physical quantity has more than two states. Temperature is a good example. Temperature has many states. For these differing conditions, several types of transducers are manufactured. One design indicates when a certain level of temperature is reached. A good example is the coolant fan switch. This sensor is mounted in a coolant passage. It detects when a certain temperature range has been reached. When reached, the switch closes. This in turn energizes a relay to provide power to the cooling fan motor. This is an example of an engine operating condition sensor.

Often it is necessary to measure a quantity that not only constantly changes but also has many values between its two extremes. There are certain electrical/electronic sensors that provide a variable quantity. Variable resistors are constructed so some kind of mechanical connection is made between the **variable resistor** and the item that is being monitored. As the item changes, the mechanical connection changes the electrical resistance of the variable resistor. This in turn changes the voltage drop and current in the circuit.

TYPES OF SENSORS

In addition to the switch and variable resistor, there are other types of sensors. The following is a list of the most common types of sensors used to provide inputs for the PCM and engine. ***Figure 38–3*** shows the electrical symbol for each including the switch and variable resistor.

1. *Switch*—This type of sensor is the most common and easiest to understand. It simply indicates an on-off position. For example, at times the computer needs to know if the air-conditioning system is off or on.
2. *Variable resistor*—This sensor changes its electrical resistance on the basis of a change in temperature. For example, the computer often needs to know the outside air temperature. As the temperature increases, the resistance of the sensor changes. The computer is then able to sense this change in resistance. This type of variable resistor is also called a thermistor.
3. *Potentiometer*—This type of sensor can change voltage in relationship to mechanical motion. For example, as the throttle is moved to change the fuel setting, the throttle position sensor changes the voltage to the computer. The computer is able to read the change in voltage and so control some type of output.

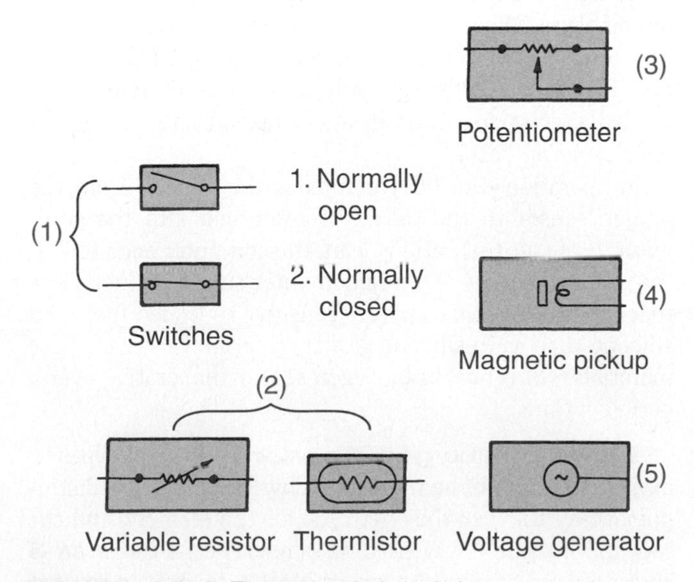

FIGURE 38-3 These electrical symbols represent the most common types of input sensors used on the automobile.

Car Clinic: VOLTAGE DROP

CUSTOMER CONCERN
What Is the Difference between Resistance and Voltage Drops?
Many input sensors need to have their resistance checked. Yet many manufacturers suggest checking the voltage drop. What is the difference between resistance and voltage drop?

SOLUTION: Resistance is measured in ohms. Resistance is considered a load on an electrical circuit. The greater the load, the less amperage that can flow in the circuit. Voltage drop is defined as the result of the amperage flowing through a resistor multiplied by the resistance in ohms. For example, if a resistor of 5 ohms has 2 amperes flowing through it, the voltage drop is (5 × 2 = 10 volts). Generally, as resistance increases, the voltage drop also increases. On many input sensors, voltage drop across the resistor (input sensor) is between 1 and 4.5 volts.

4. *Magnetic pickup*—This type of sensor is used to measure the speed of a rotating object. In operation, a small magnet is placed in the rotating object. As the shaft spins, the small magnet induces a voltage in a small coil. This charge or pulse of electricity in the coil can be sensed by the computer. A good example of a **magnetic pickup sensor** would be the vehicle speed sensor. As the wheels on the vehicle rotate, they produce a small charge or pulse of electricity. The computer can then take this pulse and determine the speed of the rotating wheels.
5. *Voltage generator*—This type of sensor produces a voltage based on certain inputs. For example, the oxygen sensor is a small voltage generator. It is mounted in the flow of exhaust from the engine. Since this sensor is very sensitive to the presence of oxygen, it can develop a voltage. Lean mixtures produce a low voltage, whereas rich mixtures produce a high voltage.

OXYGEN SENSOR

The oxygen sensor is one of the most important sensors feeding information to the PCM. It is a voltage generator sensor. Its purpose is to determine the amount of oxygen in the exhaust. This sensor continuously gives feedback to the PCM on how well the fuel system is delivering the best air-fuel ratio mixture to the engine. Most automotive oxygen sensors are made of **zirconia.** ***Figure 38–4*** shows all of the parts of a typical oxygen sensor.

The ceramic material (zirconia) on the tip of the sensor produces a voltage based upon the amount of oxygen in the

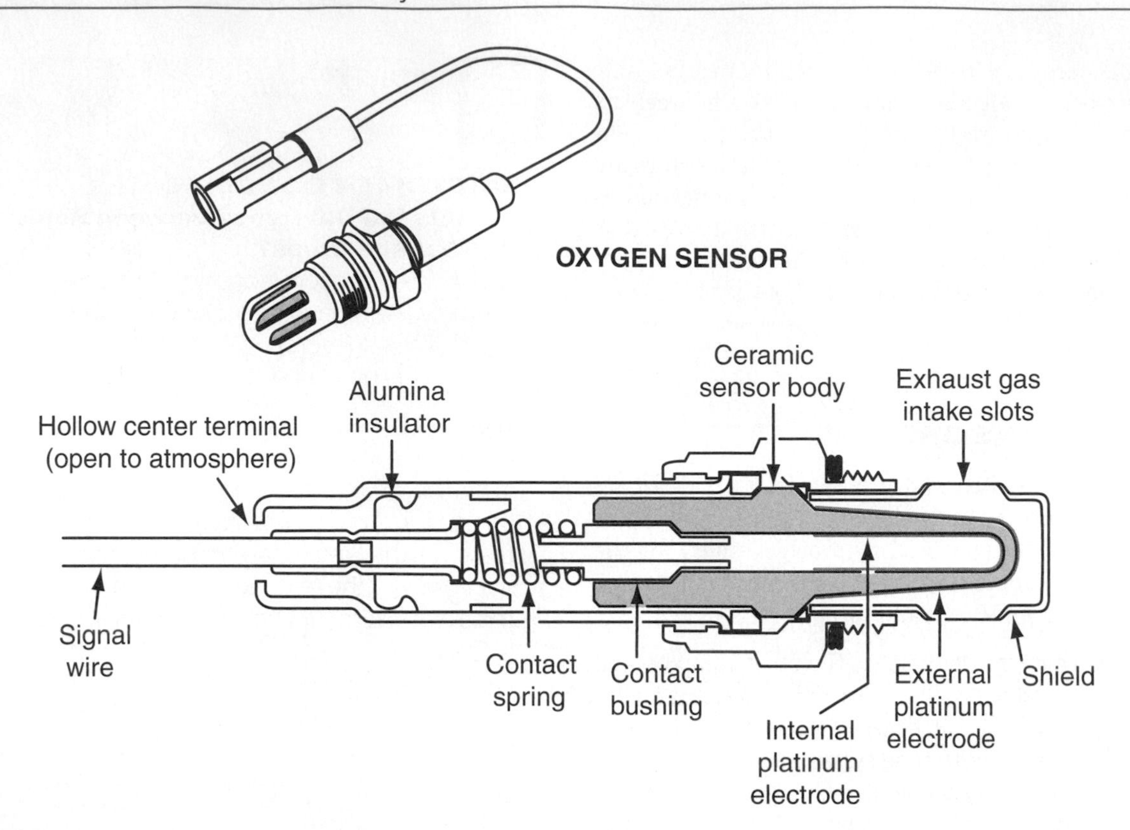

FIGURE 38-4 The parts of an oxygen sensor.

exhaust stream. The zirconia is sandwiched between two layers of **platinum**. One of the platinum plates is exposed to the exhaust gases in the exhaust manifold. The other platinum plate is exposed to outside air. This causes oxygen ions to build up on the platinum plates. An ion is an electrically charged atom. An oxygen ion has two excess electrons. This gives it a negative charge. Naturally, the platinum plate that is exposed to the outside air comes in contact with more oxygen. Therefore, it will build up more oxygen ions than the side exposed to the exhaust gases.

The difference in the number of ions on the two plates causes an electrical potential or voltage. Note that the hollow center terminal is open to the atmosphere. When the exhaust is lean (excess air) the sensor produces a low voltage, near zero. When the exhaust is rich (excess fuel), it produces high voltage (up to one volt). In summary:

Lean air-fuel mixture = High oxygen content in exhaust and low oxygen sensor voltage

Rich air-fuel mixture = Low oxygen content in exhaust and high oxygen sensor voltage

When speaking of high and low voltage in this case, it should be noted that the oxygen sensor is capable of producing only very small voltages, from about 0 to 1 volt. Generally, high voltage means anything over 450 millivolts (mv). Low voltage means anything under 450 millivolts (mv).

For the sensor to work correctly, it needs a good source of outside air and a temperature of between 500 and 600 degrees Fahrenheit. When this point is reached, the oxygen sensor will send a signal to the PCM. During the time the sensor is cold, the PCM is in the **open loop** mode. This means that there is no feedback going to the PCM telling it that the air-fuel ratio is too lean or rich. After the oxygen sensor reaches its operating temperature, the system goes into the **closed loop** mode. At this point the O sensor sends a signal to the PCM to adjust fuel delivery.

Outside air is used as a reference. Some oxygen sensors get their heat from the exhaust gas temperature. Newer sensors contain an electric heater that helps them heat up faster and reach temperatures even above the exhaust gas temperature. Many heated sensors are also waterproof, so they must receive outside air through an electrical lead fed into the sensor. Although only a small amount of air is needed, special care must always be taken to not damage or restrict the air path.

In operation, the PCM compares the voltage from the oxygen sensor to the values programmed into the computer. If the air-fuel ratio is lean, the computer adds fuel. If the air-fuel ratio is rich, the computer subtracts fuel. Note that the oxygen sensor is very sensitive to leaded fuels and silicone such as in silicone gasket material. Either of these materials will render the oxygen sensor inoperative over a period of time.

Types of Oxygen Sensors Several types of oxygen sensors are being used today. The types are distinguished by the way they get heat for the element and the way the outside air is used as a reference. ***Figure 38–5*** shows several types of O sensors. Most early oxygen sensors were of the unheated type. In this case, the exhaust gases provided the heat to bring the sensor element up to its oper-

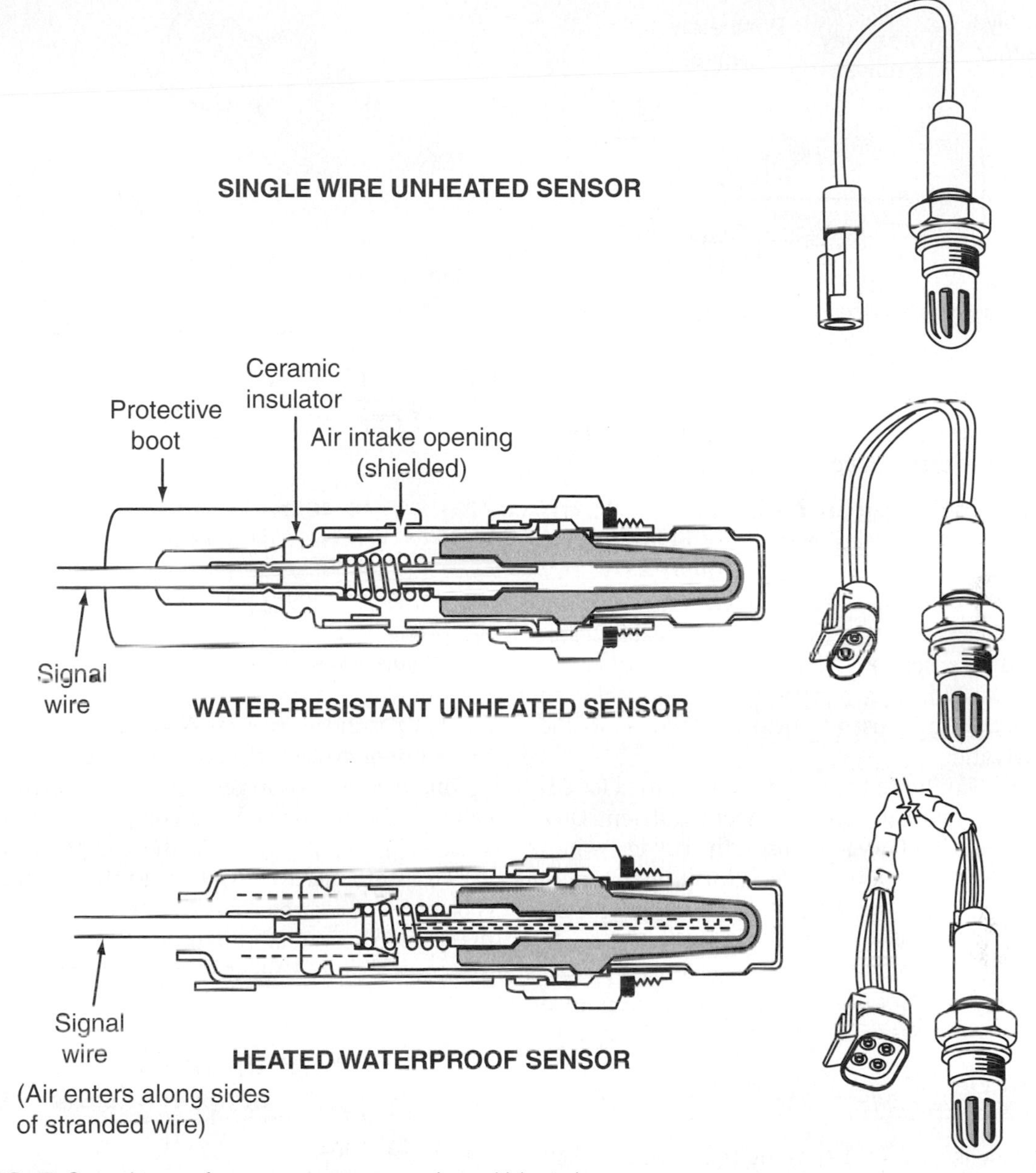

FIGURE 38–5 Several types of oxygen sensors are used on vehicles today.

ating conditions. In many newer sensors, an electric heater provides added heat to the zirconia element. This helps the element come up to temperature faster, operate better in colder climates, and give improved fuel control. It is also less sensitive to contaminants. The heater power comes directly from the vehicle electrical system. It is usually turned on with the ignition switch. On some vehicles, relays are used to turn the heater off and on under certain conditions. ***Figure 38–6*** shows a typical heated oxygen sensor.

Zirconia oxygen sensors need only a small amount of air to work correctly, however, the air needs to be fresh outside air. Most unheated sensors have an open path for air to flow over the sensor. Also, they are placed in a position where water will not splash on them. On some vehicles, the sensor is in a position where water could splash on it. In this case, a water-resistant oxygen sensor is used. Added

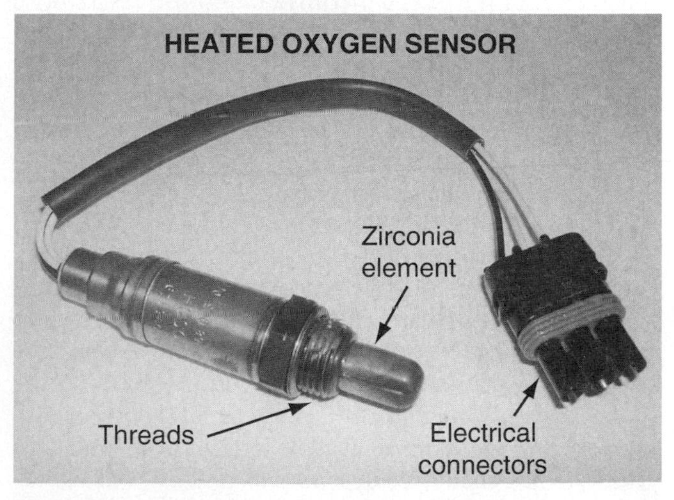

FIGURE 38–6 A typical heated oxygen sensor.

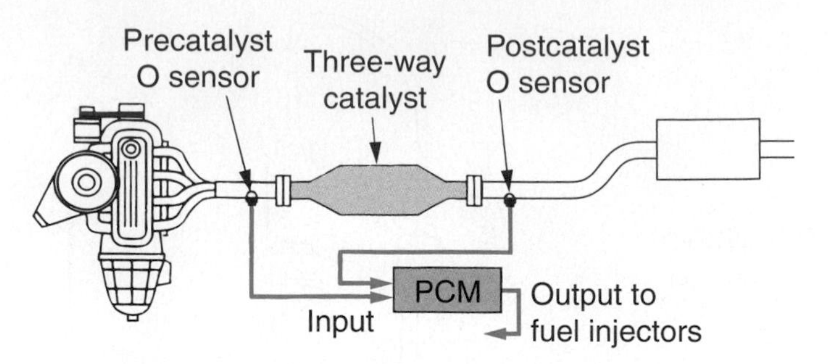

FIGURE 38–7 Some vehicles use an oxygen sensor after the catalytic converter.

shielding helps keep water out of the sensor while still making it easy for air to pass through.

Postcatalyst Oxygen Sensor In 1994, certain vehicles began to use an oxygen sensor in the exhaust, after the catalytic converter as shown in ***Figure 38–7***. This sensor is used to check the performance of the catalytic converter and the exhaust system. Leaks in the exhaust system ahead of this sensor can cause improper emission system performance. This oxygen sensor can help detect such problems. It may also be used to help the PCM adjust the engine air-fuel ratio.

This sensor is similar to the oxygen sensors used for engine control, however, their voltage output is different. On a typical exhaust manifold oxygen sensor, the voltage swings up and down between about 0.8 and 0.2 volt one or more times per second. The signal from the postcatalyst oxygen sensor moves much slower. It is not unusual for the signal voltage to stay at either high or low voltage for several seconds or even minutes. This type of voltage output is normal and should not be a concern.

COOLANT SENSOR

A coolant sensor is used to determine the engine coolant temperature. A signal is then sent to the PCM to let the computer know the engine temperature. The design of a coolant sensor is that of a thermistor. In theory, as any wire changes in temperature, it also has a corresponding change in electrical resistance. Unfortunately, a piece of copper wire's resistance does not change in a predictable manner. The wire resistance changes radically as compared to an actual sensor. To level out the change in resistance, the thermistor was designed. It has a much more consistent relationship between temperature and resistance.

There are two type of thermistors. One increases resistance as the temperature increases. Another decreases resistance as the temperature increases. This type is shown in ***Figure 38–8***. The thermistor is generally threaded into the intake manifold so that one end of the sensor is touching the engine coolant. This type, which is used most often, is called a negative temperature coefficient (NTC) sensor. This type of sensor has about 20,000 ohms of resistance at –4 degrees Fahrenheit, and 1,200 ohms at 248 degrees Fahrenheit. ***Figure 38–9*** shows this relationship on a chart. On the horizontal axis, temperature is shown. On the vertical axis, resistance is shown in ohms. Note that as the temperature of the sensor or coolant increases, the resistance decreases.

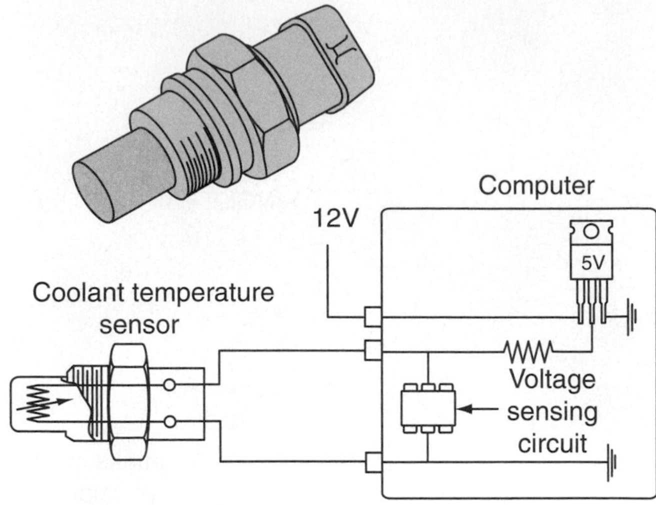

FIGURE 38–8 This engine coolant temperature (ECT) sensor changes resistance as the coolant temperature increases.

This type of sensor generally has two wires connected between the sensor and the computer. One wire is used to carry a voltage signal to the computer. The other is used as a ground. In operation, as the temperature of the coolant increases, the resistance of the sensor decreases. As the resistance decreases, the voltage drop across the sensor decreases as well. Voltage drop is high when the sensor is

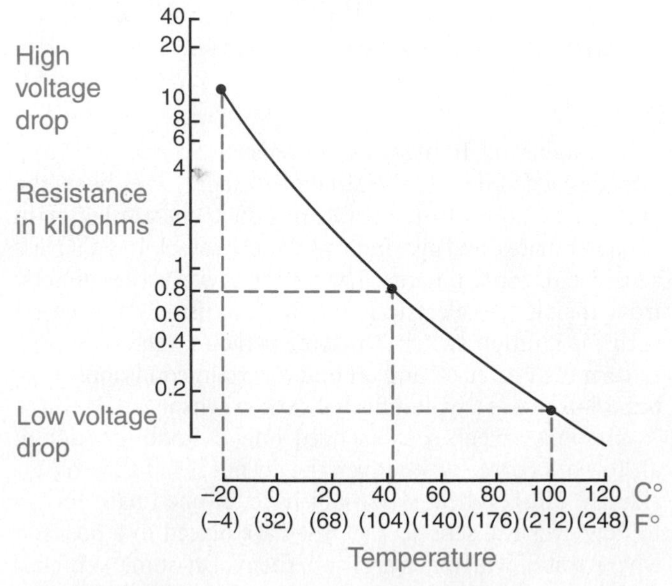

FIGURE 38–9 This graph shows that as the coolant temperature decreases, the sensor's resistance increases.

FIGURE 38–10 This is a typical coolant temperature sensor.

CUSTOMER CONCERN:
Rough Idle

A Ford Escort seems to have a rough idle. The customer complains that the idle is fast, then slows down rapidly. After the engine heats up, the idle increases in speed again and drops off to a slow idle. What could be the problem?

SOLUTION: At times, the throttle position sensor (TPS) goes bad. A TPS is a variable resistor that is called a potentiometer. Its purpose is to sense the opening and closing position of the throttle (which is controlled by the driver). If the wires in the potentiometer become damaged, the voltage drop read by the PCM is incorrect. Therefore, the computer will try to continually adjust the idle because the voltage drop readings are constantly changing. Replacing the TPS should solve the problem.

cold, or about 4.5 volts. When the engine coolant temperature is hot, the resistance of the sensor is low, so the voltage drop is about 0.3 volt.

Figure 38–10 shows a typical coolant temperature sensor used on an engine. The electrical connection is used to send an electrical signal to the computer. The sensor is threaded into the block so that the sensor element touches the coolant.

AIR CHARGE TEMPERATURE (ACT) SENSOR

Many engine control systems use an air charge temperature (ACT) sensor, now called an intake air temperature (IAT) sensor. This sensor measures the temperature of the incoming air. Cold air intake is much denser than warm air, therefore, a richer air-fuel ratio is needed. When the IAT sensor indicates colder air temperature, the computer then provides a richer air-fuel mixture.

The IAT is much the same as the engine coolant temperature sensor just discussed. It is a thermistor that has resistance and voltage drop readings the same as the coolant sensor. As the air temperature increases, the resistance of the IAT decreases. The result of this resistance decrease is that the voltage drop also decreases. The computer senses the drop in voltage and changes the air-fuel ratio accordingly.

The IAT sensor is threaded into the intake manifold and, in some cases, mounted in the air cleaner. The lower portion of the sensor needs to stick down into the airflow passages as the air comes into the engine.

PARTS LOCATOR

Refer to photo #73 in the front of the textbook to see what an ***intake air change temperature (IAT) sensor*** *looks like.*

THROTTLE POSITION SENSOR (TPS)

Another sensor used on computer-controlled engines is the throttle position sensor (TPS). This sensor monitors and senses the position of the throttle on either carburetor or fuel-injected engines. The TPS is considered a **potentiometer**. This means that as the position of the throttle changes, the resistance inside the potentiometer is also changing. The changing resistance means that the voltage drop across the sensor will also change.

On fuel injection systems, the sensor is positioned on the end of the throttle shaft in the throttle body, which controls the amount of air going into the engine. There are both three- and four-wire throttle position sensors.

On the three-wire sensor, one wire is a ground, one wire comes from the computer to supply a constant 5-volt reference voltage, and one wire is the TPS signal back to the computer. A typical TPS has a resistance of about 1,000 ohms at the idle position. Resistance is about 4,000 ohms at wide-open throttle. In terms of voltage drop, there is about a 0.5 volt drop at idle and about a 4.5 volt drop at wide-open throttle. The four-wire TPS has an additional wire used as an idle switch.

When the engine is accelerated, the computer must be able to adjust the air-fuel ratio accordingly. The computer is also sensitive to how rapidly the throttle is opened. ***Figure 38–11*** shows a typical throttle position sensor internal circuit.

THROTTLE POSITION SENSOR (TPS)

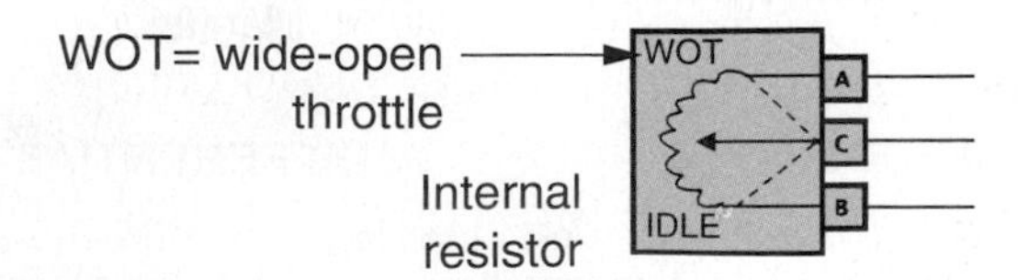

FIGURE 38–11 The throttle position sensor (TPS) is a potentiometer used to indicate the position of the throttle.

MANIFOLD ABSOLUTE PRESSURE (MAP) SENSOR

The manifold absolute pressure (MAP) sensor is used on most computer-controlled engines to determine how much load is being placed on the engine. This is necessary because if the load is heavy, there must be more fuel put into the engine. If the load is light, less fuel is needed. There are two types of sensors used for reading manifold absolute pressure. One is called an absolute sensor and one is called a differential sensor. ***Figure 38–12*** compares the two types. The **absolute sensor** compares the intake manifold pressure to a reference or fixed pressure sealed inside the sensor. The **differential sensor** compares the intake manifold pressure to atmospheric or barometric pressure. As the two pressures work against each other, the center diaphragm moves slightly because of the difference in pressures.

On some vehicles, the diaphragm is a pressure-sensitive disk capacitor. When the center diaphragm moves, a digital voltage signal of varying frequency is produced. So when the engine is idling and at high intake manifold vacuum (about 17 to 18 inches of mercury), a signal of about 95 hertz is produced. When the engine is under heavy load and at a low intake manifold (about 2 inches of mercury), a signal of about 160 hertz is produced. Based on this changing frequency signal, a richer or leaner air-fuel ratio is produced.

On other vehicles, the diaphragm is made of silicon. As the pressure changes, the signal voltage also changes from about 1 volt at idle (low load) to about 4.5 volts at heavy-load conditions. ***Figure 38–13*** shows a typical MAP sensor, noting the vacuum port to read intake manifold vacuum.

The MAP sensor can be checked for output voltages in relationship to a vacuum applied to the sensor diaphragm.

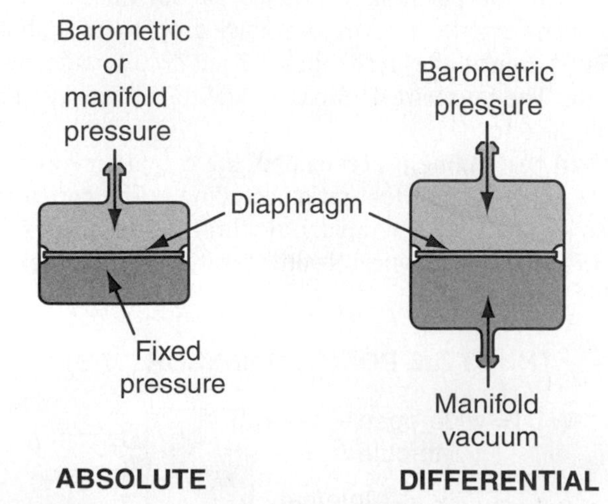

FIGURE 38–12 A manifold absolute pressure (MAP) sensor works by comparing two pressures. If there is a difference in pressure, the center diaphragm moves, causing a voltage signal to be generated for the computer.

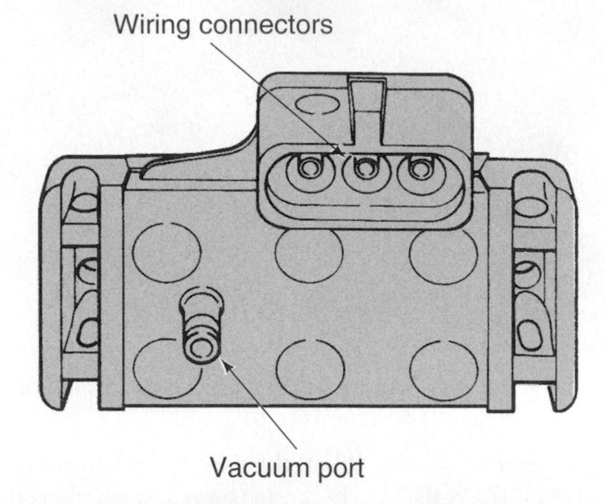

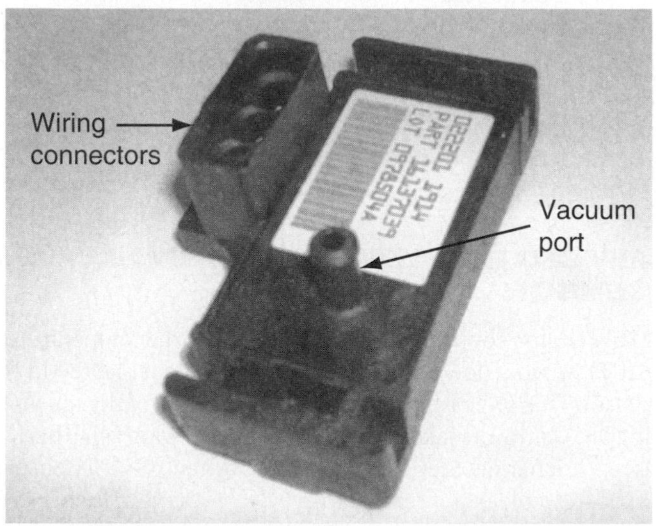

FIGURE 38–13 This manifold absolute pressure (MAP) sensor is used to determine the load on an engine. (Top: Courtesy of DaimlerChrysler Corporation)

Always follow the manufacturer's suggested procedure and specifications when checking the MAP sensor.

MASS AIRFLOW (MAF) INDICATOR

On most fuel injected engines, a mass airflow (MAF) indicator is used. This input sensor tells the computer how much total air or mass air is entering the engine for combustion.

There are several types of MAF sensors used on engines today. One is the vane-type MAF sensor. This type uses a plate moving in the intake airstream tube. As greater amounts of air come into the engine (under heavier loads), the vane moves more. The moving vane is attached to a variable resistor. The resistance of this resistor is an indication of the amount of air going into the engine.

A second, more popular, type is the heated resistor MAF. In this instance, a heated resistor is placed in the center of the airstream. Some MAFs use an electric grid in place of

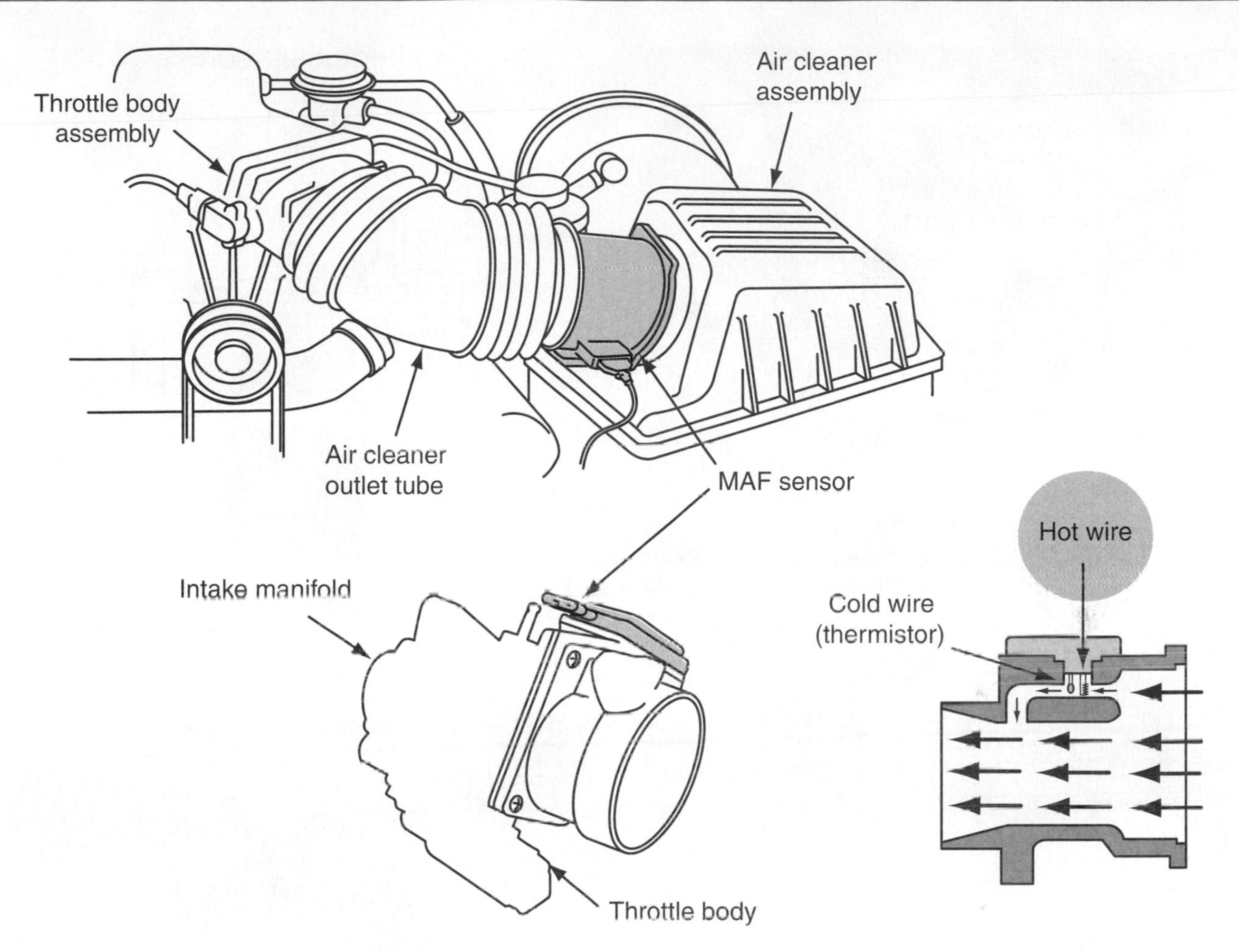

FIGURE 38–14 The computer keeps the temperature of the hot wire in the MAF sensor about 392 degrees Fahrenheit above the temperature of the cold wire.

the resistor. With the ignition switch on, a voltage is applied to the resistor or electric grid. The resistor or electric grid is heated by the electric current. The computer maintains a specific temperature on the resistor (grid). When more air comes into the engine, say during heavy acceleration, the cool air coming into the engine has a tendency to cool the resistor (grid), dropping its resistance. At this point the computer increases the current to the resistor (grid) to keep it at a specified temperature. So the amount of current going into the resistor (grid) is directly proportional to the amount of air flowing into the engine. These adjustments all happen in a few milliseconds.

A third type of MAF is shown in ***Figure 38–14***. This system is the hot wire-type MAF sensor. It is similar to the heated resistor type, only it has an additional cold wire. Both the hot and cold wires are placed in the intake airflow. The cold wire senses the incoming air temperature, so it is called the cold wire. Electric current is sent from the computer to the hot wire to keep it at a specified number of degrees above the temperature of the cold wire. The temperature difference is about 392 degrees Fahrenheit hotter than the cold wire. So the amount of electric current sent to the hot wire varies depending on the temperature of the cold wire. The advantage of this type of sensor over the heated resistor is that it is now sensitive to the temperature of the air coming into the engine as well. The signal or amount of electricity being sent into the hot wire is directly proportional to both the temperature and the amount of air going into the engine. ***Figure 38–15*** shows an example of a typical mass airflow indicator. The top view shows the wires that are heated by electrical current. The bottom view shows the screen used on the input side of the MAF sensor.

On the vane-type MAF, drops in both the signal voltage and the output voltage can be checked and compared to manufacturer's specifications. In addition, the variable resistance can be checked as well. On some heated resistor and hot wire MAFs, the manufacturer recommends that a voltage frequency test be performed. This test determines if the voltage frequencies are erratic or smooth. A special voltage frequency meter is needed. Always follow the manufacturer's exact procedure and specifications when performing this test.

PARTS LOCATOR

Refer to photo #74 in the front of the textbook to see what a ***mass airflow indicator*** *looks like.*

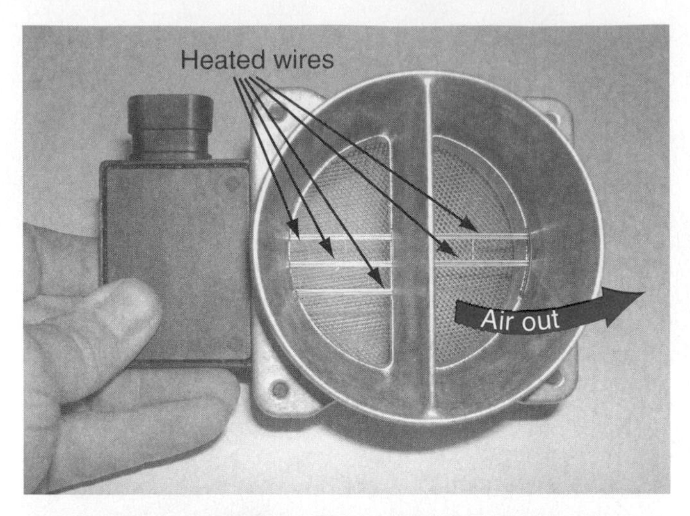

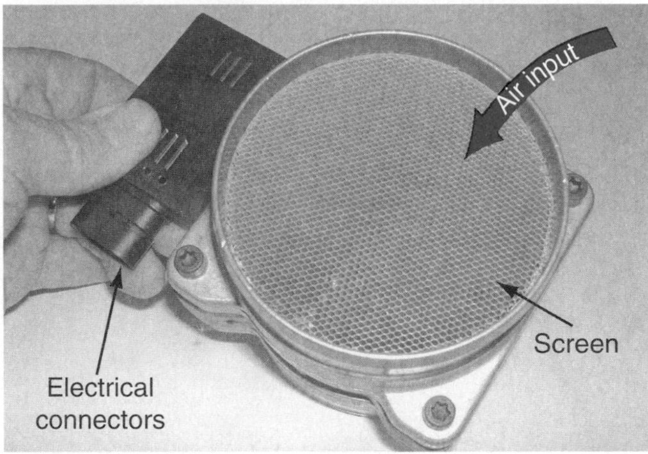

FIGURE 38–15 The internal parts of a mass airflow sensor.

KNOCK SENSOR (DETONATION SENSOR)

The computer used to control the ignition system needs to know if the engine timing is too advanced and, consequently, is causing a knocking or detonation. If there is a slight knocking during combustion, the computer needs to reduce the spark advance to eliminate the knocking or detonation.

When the engine knocks or detonates, a small vibration is produced in the engine block. The knock sensor (some engines have two knock sensors) changes this vibration to a voltage signal. The signal is then sent back to the computer so timing adjustments can be made on the ignition system. The knock sensor contains a **piezoelectric** sensing element. A piezoelectric element is a sensitive crystal that produces electricity when there is a mechanical stress on it. ***Figure 33–16*** shows an example of the internal parts of the knock sensor showing the piezocrystal location. In operation, the vibration produced from knocking and detonation causes a mechanical stress on the piezocrystal. The mechanical stress generates a voltage signal. A typical knock sensor produces a voltage signal of about 300 millivolts to 500 millivolts, depending upon the degree of knock or detonation. ***Figure 38–17*** shows a typical knock sensor.

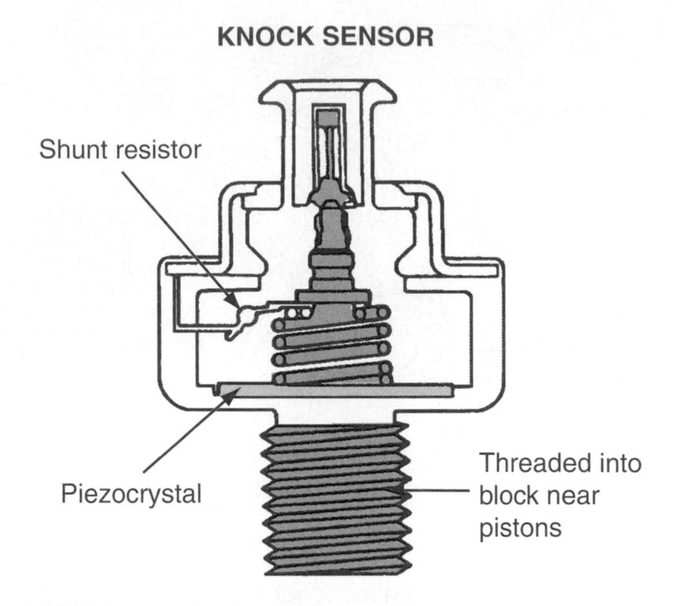

FIGURE 38–16 A knock sensor senses knock, or detonation, so the PCM can readjust the timing advance.

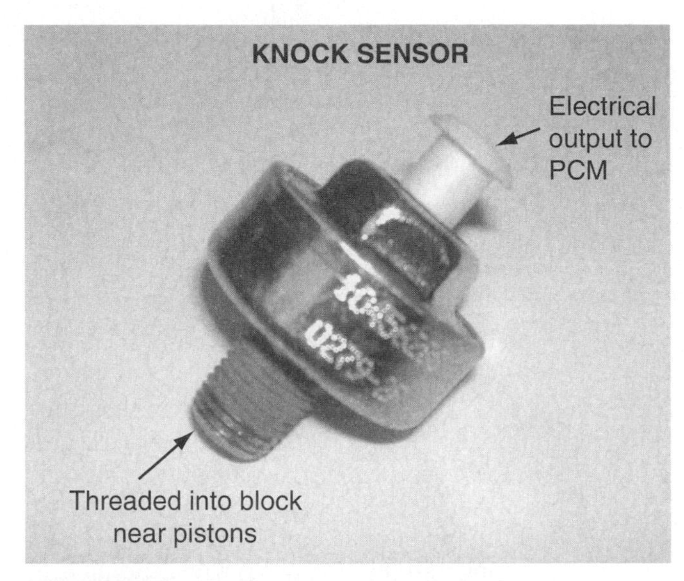

FIGURE 38–17 The parts of a typical knock sensor.

VEHICLE SPEED SENSOR (VSS)

The vehicle speed sensor (VSS) tells the computer how fast the vehicle is moving down the road. The VSS is often connected to the speedometer cable or on the output speed of the transmission. On some vehicles, it is mounted on the transaxle. One type of VSS operates from a magnet that is driven from the speedometer drive. As the magnet turns, it acts like a small generator that produces alternating current (ac) voltages. The ac signal is sent to the computer to

PARTS LOCATOR

*Refer to photo #75 in the front of the textbook to see what a **vehicle speed sensor** looks like.*

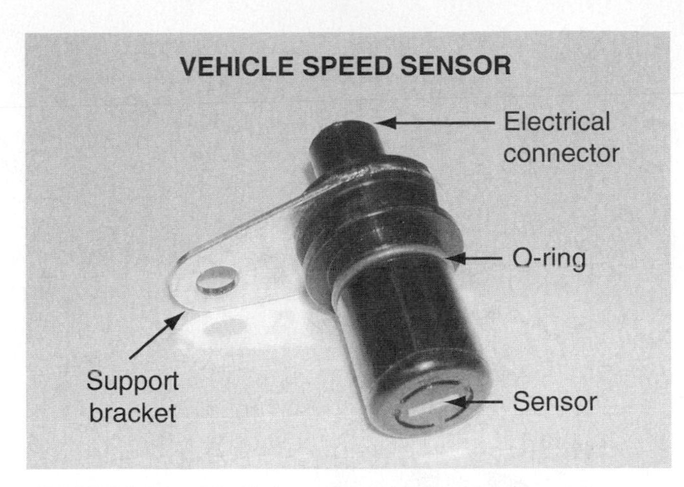

FIGURE 38–18 The parts of a typical vehicle speed sensor.

determine the exact speed of the vehicle. As the speed of the vehicle increases, the ac signal also increases in frequency and voltage. The computer can read the changes in frequency and voltage and make adjustments as necessary.

Figure 38–18 shows a typical vehicle speed sensor. This particular sensor is attached to the back of the transmission housing, near the speedometer location. It is designed to sense the speed of the output of the transmission. The bottom of the sensor is located close to a magnet placed on the output shaft of the transmission. An O-ring is used to keep any transmission fluid from leaking out. The signal that is produced from the sensor is sent to the computer through the electrical connector on top.

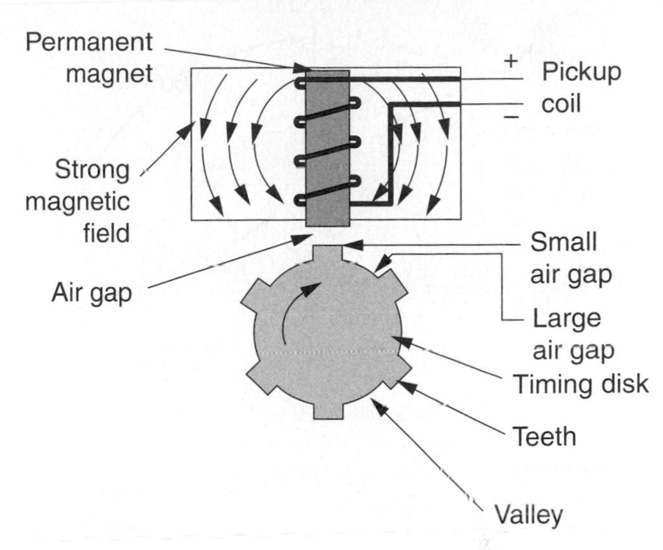

FIGURE 38–19 The knobs on the timing disk cause the air gap to change, thus changing the magnetic field.

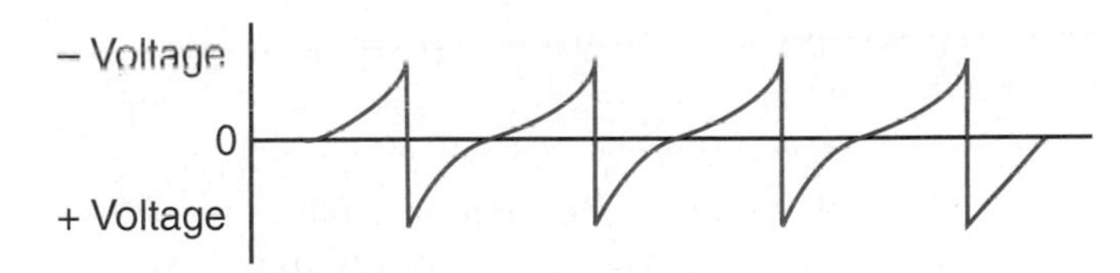

FIGURE 38–20 This is a sine wave that is produced from a magnetic pulse generator or a variable reluctance sensor.

CRANKSHAFT POSITION (CKP) SENSOR

Fuel injection and ignition systems today often need a reference to the position of the crankshaft in relationship to rotational degrees. The position of the crankshaft helps to determine such things as firing order, degrees before TDC, and when number 1 piston is at TDC. On most fuel injection systems (such as sequential firing fuel injection), this information is also needed by the computer. Such sensors can broadly be defined as variable reluctance or Hall-effect sensors. Most often, the crankshaft position sensor uses the variable reluctance type system. However, some electronic ignition systems such as the C^{3}I use the Hall-effect type.

Variable Reluctance CPK Sensor The variable reluctance sensor is categorized as a magnetic pickup sensor. This is because it has a permanent magnet surrounded by a winding of wire. The sensor is mounted in a fixed position on the engine block. The tip of the sensor protrudes into the crankcase at a distance of 0.05 plus or minus 0.02 inch from the crankshaft. ***Figure 38–19*** shows its theory of operation. As the timing disk rotates, it is constantly changing the air gap. For example, when one of the small teeth moves past the permanent magnet, the air gap is very small. When the valleys on the timing disk pass the permanent magnet, the air gap is large. As the air gap changes, the magnetic field continues to change as well. In the position shown, the magnetic field is the strongest. The end result is that a voltage signal is produced, as shown in ***Figure 38–20***. In this particular case, notches are used rather than teeth, however, the result is the same. As the notch passes by the magnetic sensor, a positive and a negative voltage is produced, which is sent to the PCM.

Figure 38–21 shows an example of variable reluctance sensor. Since this type of sensor is mounted in a fixed position, it is not adjustable. The magnetic field of the sensor is affected by a machined ring located on the crankshaft. This ring is called a reluctor ring (similar to the timing disk). The machined ring has notches that cause the magnetic field to vary. The varying magnetic field causes fluctuations in the voltage signal.

In operation, as the crankshaft rotates, the notches in the reluctor ring change the size of the gap between the sensor and the metal surface. When the notch widens the gap, the magnetic field in the sensor changes. This changing magnetic field produces a voltage signal that is induced in the sensor winding. ***Figure 38–22*** shows an example of the sensor output as it passes a notch in the reluctor ring. Many manufacturers use similar approaches to develop a crankshaft position signal. One of the major differences is

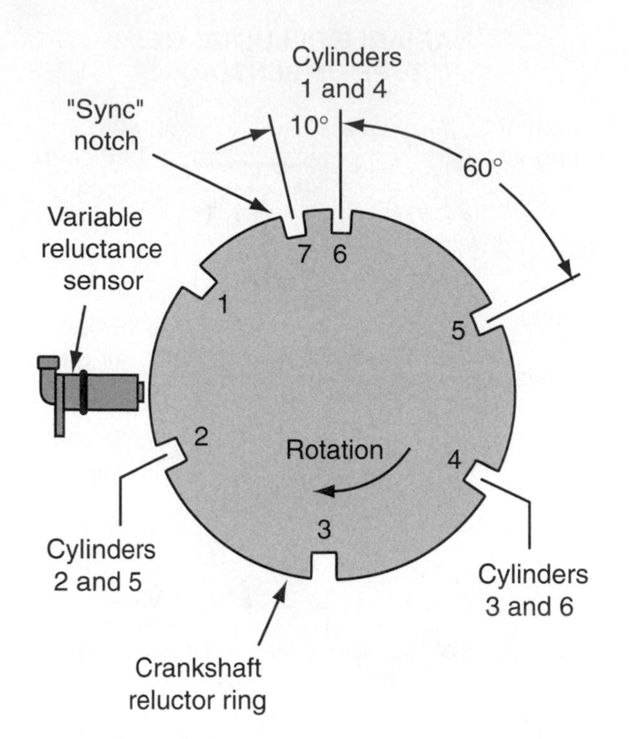

FIGURE 38–21 This type of computer sensor is called a variable reluctance sensor.

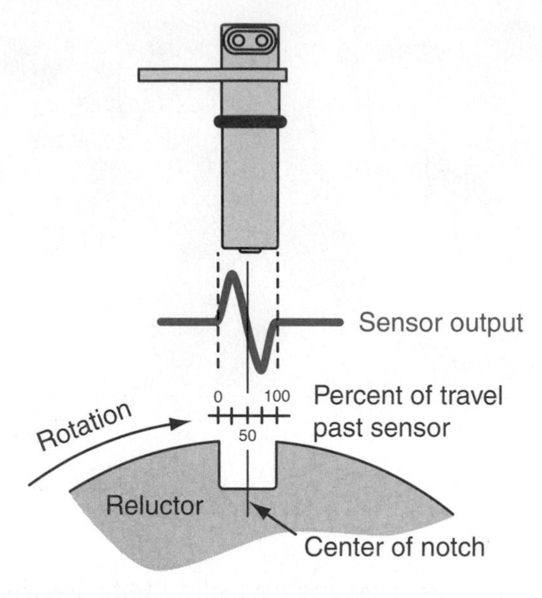

FIGURE 38–22 The voltage signal is produced when a notch passes the magnetic sensor.

the number and type of notches used to produce the signal. Some have only seven as shown. Others may have 254 notches, while others have 68 notches. It all depends on the manufacturer and the type of system used. However, the principle of variable reluctance sensors remains the same.

Another type of common CKP sensor is shown in ***Figure 38–23***. In this particular case, the sensor is located on the transaxle housing. The sensor detects the passing of slots on the flywheel. This flywheel contains twelve slots arranged in three groups of four. Each group is 120 degrees apart. As the flywheel metal is passing by the CKP sensor, the sensor is on and the signal voltage is low, about 0.5 volt. When a slot passes by the CKP sensor, the sensor is off and the voltage goes to a high of about 5 volts.

Another common CKP sensor is shown in ***Figure 38–24***. In this particular case, the crankshaft position sensor is sensing machined notches in the crankshaft. As the machined notches pass by, the crankshaft position sensor is being turned on and off. The on and off voltages are then sent to the PCM.

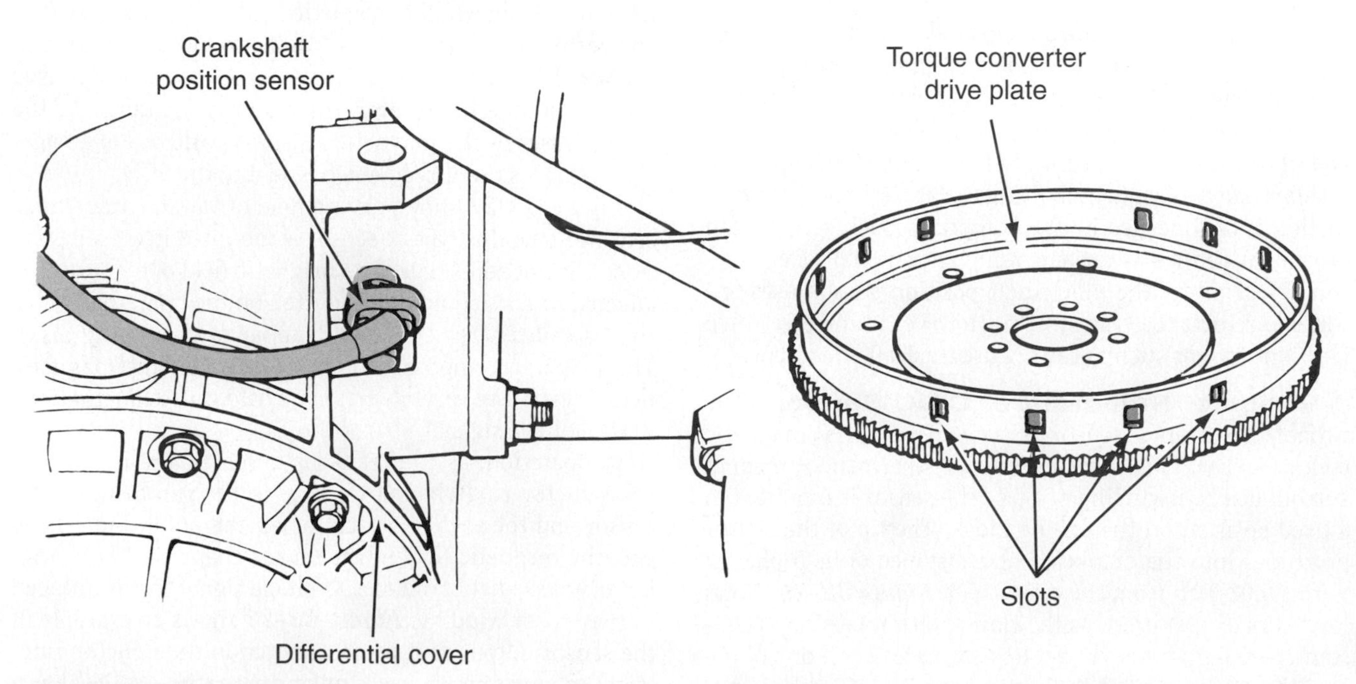

FIGURE 38–23 Some crankshaft positions (CKP) sensors sense the holes (slots) in the torque converter drive plate. (Courtesy of DaimlerChrysler Corporation)

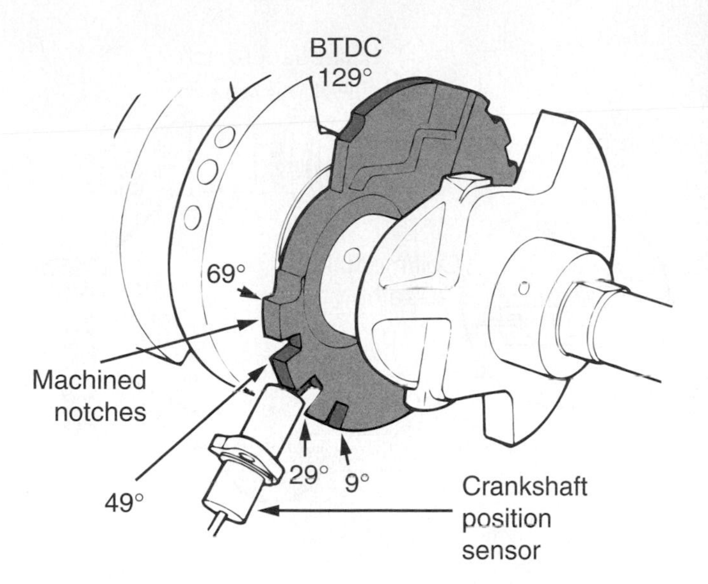

FIGURE 38-24 This crankshaft position (CKP) sensor senses machined notches on the crankshaft. (Courtesy of DaimlerChrysler Corporation)

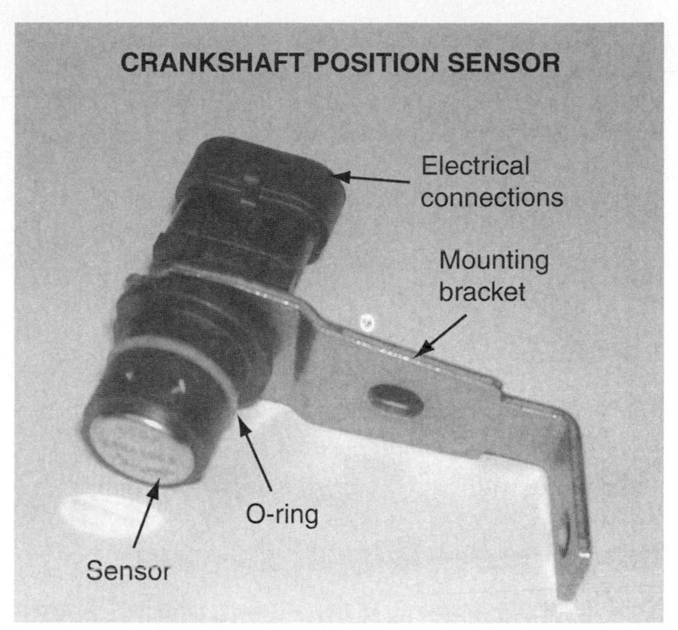

FIGURE 38-25 The parts of a typical crankshaft position sensor.

The first notch in the group of four is at 69 degrees BTDC. The second notch is at 49 degrees BTDC. The third notch is at 29 degrees BTDC and the fourth notch is at 9 degrees BTDC. From this information, the engine computer can determine when a pair of cylinders will reach TDC. Each group of slots relates to a pair of cylinders. ***Figure 38–25*** shows a typical crankshaft sensor and associated parts.

Crankshaft Position (CKP) Sensor Hall-Effect The Hall effect, also called the Hall voltage principle, was discovered by a physicist named Dr. Edward Hall. He discovered that when a magnetic field is introduced perpendicular to a current flowing through a solid conductor, a measurable voltage is induced at the sides of the conductor at right angles to the main current flow.

Figure 38–26 shows how the Hall-effect sensor works in principle both in the on and off position. There are three major parts to a Hall-type sensor. First, there is a permanent (fixed) magnet located on the left side of the sensor. It has a north and south pole. In addition, there is a semiconductor wafer located on the right side. The third part is a steel interrupter blade (sometimes called a vane) that spins between the wafer and permanent magnet. When the vane is not between the wafer and the permanent magnet, the

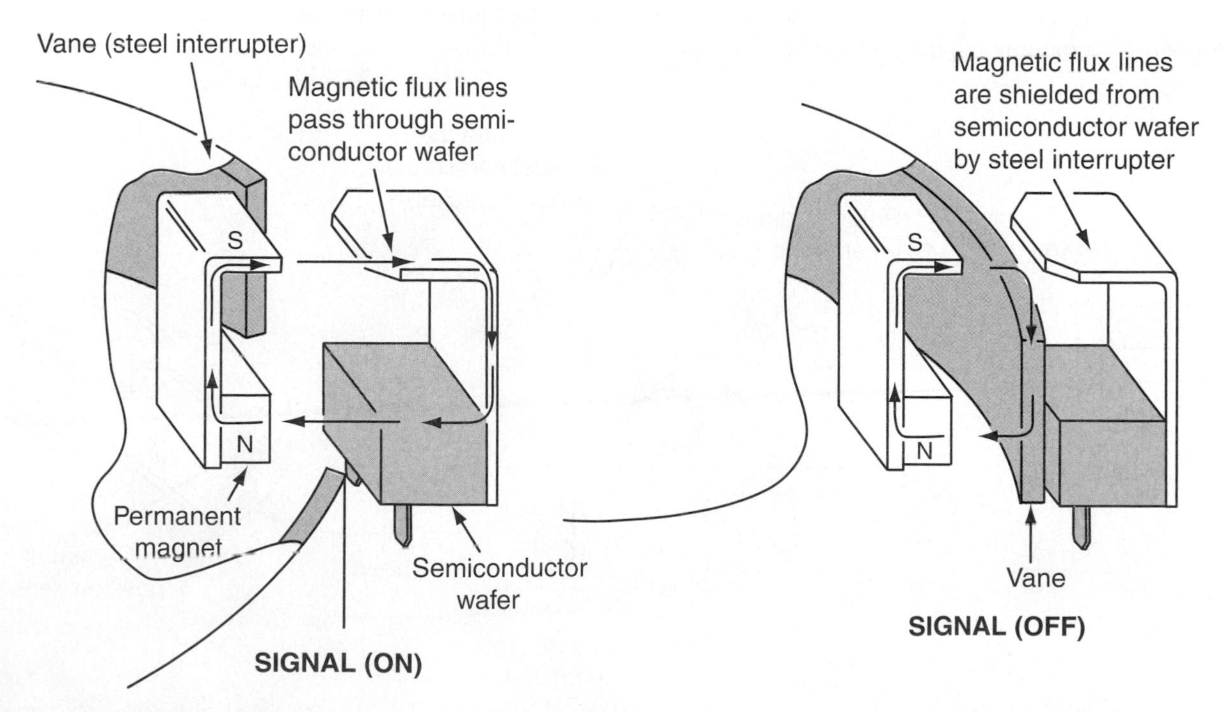

FIGURE 38-26 As the vane passes between the magnet and wafer, the Hall-effect sensor is turned off.

permanent magnet's lines of force induce a voltage in the semiconductor wafer. The Hall switch is now on. When the vane is positioned between the wafer and the permanent magnet, it interrupts the magnetic field. The Hall switch is now off and there is no voltage signal sent to the PCM. The on and off voltage signal is sent to a transistor and the signal is amplified in the PCM. Note that there are many physical designs used for the three parts. However, all Hall-effect sensors operate on basically the same principle.

PARTS LOCATOR

*Refer to photo #76 in the front of the textbook to see what a **Hall-effect crankshaft position sensor** looks like.*

CAMSHAFT POSITION SENSOR

Most electronic ignition (EI) systems use a **camshaft position sensor** in addition to the crankshaft sensor. The camshaft position sensor is used most often to identify the position of the number 1 piston. Many of these cam sensors are similar in design to the Hall-effect crankshaft sensors, however, they look somewhat different. On most engines, the camshaft sensor is mounted near or on the timing cover. As the camshaft turns, the signal generated is sent to the PCM where the signal is used to turn the internal transistors off and on. This, in turn, fires the coil(s) at the proper time. The signal is also used to sequence fuel injection. ***Figure 38–27*** shows an example of a typical camshaft position sensor and where it is located on the engine.

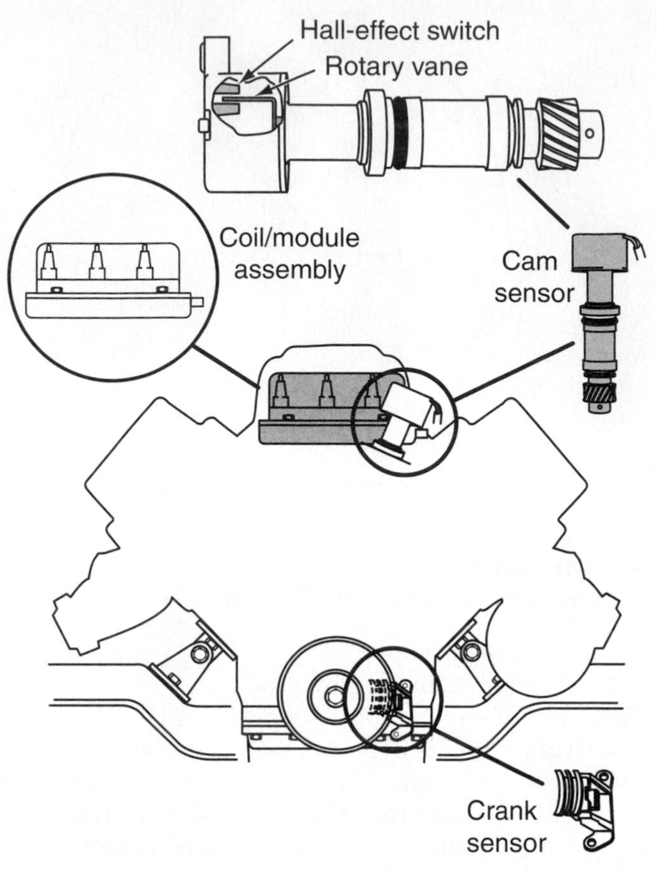

FIGURE 38–27 The camshaft position sensor is located on the front of this engine.

On some V8 engines, the camshaft position sensor works as a variable reluctance sensor. As shown in ***Figure 38–28***, there is a reluctor pin attached to one of the camshafts, in this case, the rear exhaust camshaft. The sensor, which is a magnetic sensor, is mounted in the cylinder head. In operation, as the reluctor pin rotates past the magnetic sensor, it creates an ac voltage signal. This voltage signal, as shown in ***Figure 38–28***, is then sent to the ignition coil control module. The signal is created every camshaft revolution or every two crankshaft revolutions.

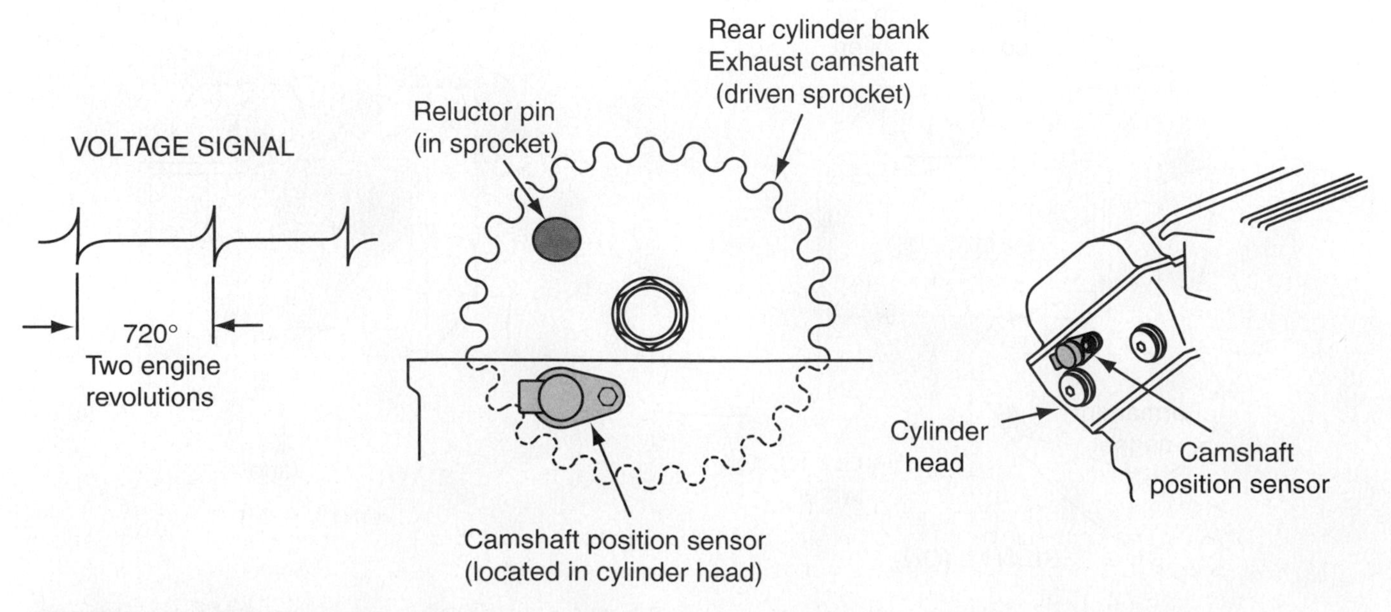

FIGURE 38–28 The camshaft sensor uses a reluctor pin located on one of the exhaust camshafts.

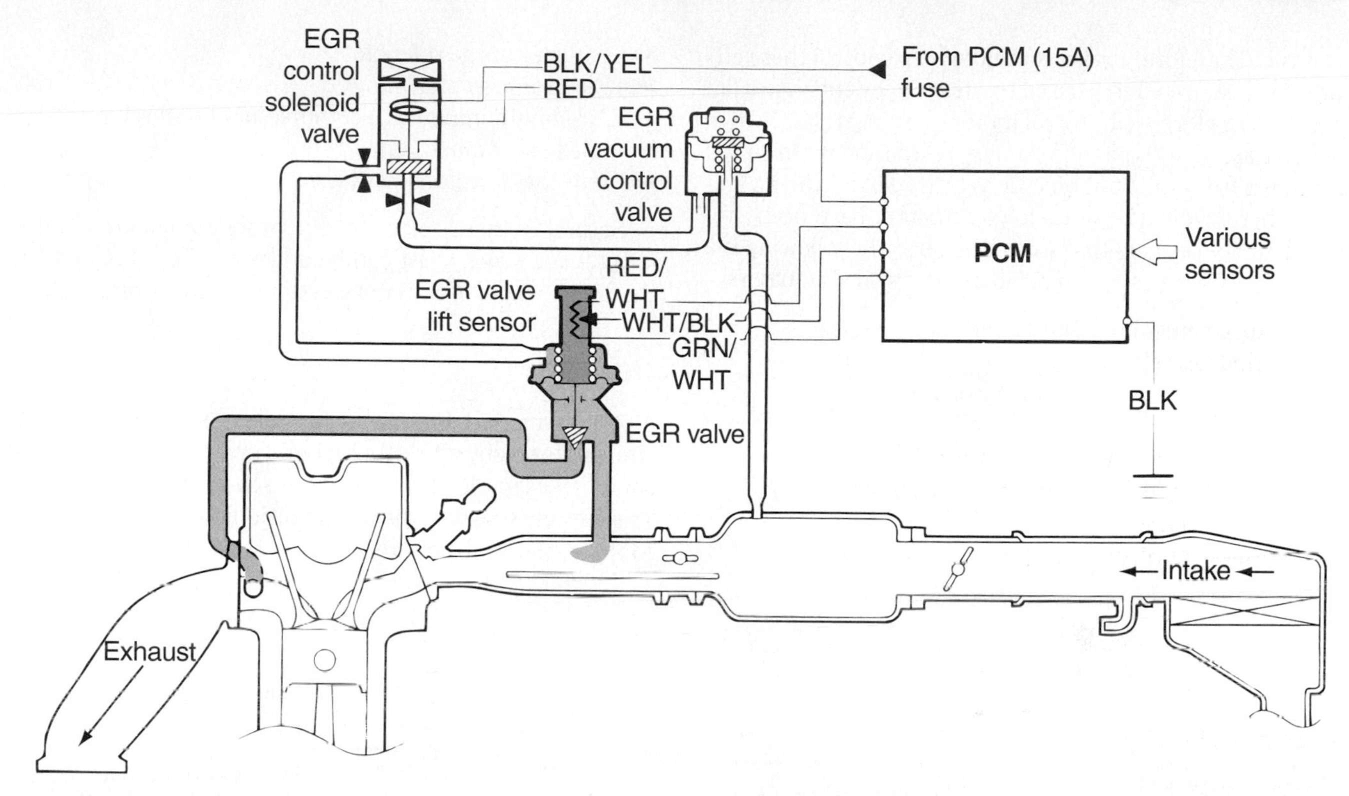

FIGURE 38–29 The EGR position sensor, also called an EVP sensor, is located on the top of the EGR valve. (Courtesy of American Honda Motor Co., Inc.)

EGR VALVE POSITION (EVP) SENSOR

To operate correctly, the PCM needs to know the position of the EGR valve. The exhaust gas recirculation (EGR) valve position (EVP) sensor acts as a feedback to the PCM to tell the computer the position of the EGR valve. The sensor is located generally on top of the EGR valve as shown in ***Figure 38–29***. The sensor is a potentiometer. When the EGR valve is open, the potentiometer center needle moves, which changes the resistance of the sensor. This causes a higher voltage drop signal that is sent to the computer. When the EGR valve is closed, the potentiometer center needle moves the opposite way, and a lower voltage drop signal is sent to the computer. In summary:

EGR valve open = High resistance, high voltage drop
EGR valve closed = Low resistance, low voltage drop

The resistance and voltage signals of the potentiometer readings changes from about 5,000 ohms and 4.5 volts when in the open position to about 3,000 ohms and about 0.8 volts when in the closed position.

PARK/NEUTRAL SWITCH

On computerized engine controlled vehicles, there is also a switch that signals the computer about the position of the gear shift. The computer needs to know what gear the transmission is in so it can adjust the idle speed. The switch can be in the off or on position. It is controlled by the position of the transmission linkage. When the transmission is in park or neutral, the switch is closed. When the transmission is in any gear such as drive, reverse, and so on, the switch is open. A closed switch sends a voltage signal generally below 1 volt to the computer. An open park/neutral switch sends a voltage of about 5 volts to the computer.

A/C SWITCH

The computer also needs to know whether the air conditioning is off or on. To indicate this condition, an A/C switch is used. When the A/C switch is on, the computer needs to know this condition, so that the idle speed can be increased. When the A/C is off, the computer changes or reduces the idle speed accordingly. The switch is often located under the dash of the vehicle and can be checked for continuity in both the off and on position. Again, depending on the position of the switch, a corresponding voltage drop signal will be sent to the PCM.

38.3 OUTPUT COMPONENTS (ACTUATORS)

The majority of the actuators controlled by the PCM are either relays, solenoids, or motors. Solenoids are used where the amount of current in the circuit is low, about 0.75 ampere or less. Relays handle greater currents. For example, relays handle such outputs as the cooling fan, air-conditioning compressor clutch, the early fuel evaporator grid, and others. The current flowing in these actuators is much higher. Generally, the PCM is not capable of handling such high currents. Therefore, a relay is used to help control these circuits. Motors are often used to rotate components like a fan motor or an idle speed control motor.

There are many types of solenoids, relays, and motors used as outputs from the PCM on each vehicle. Not all can be presented here. Only selected output actuators have been presented, to help explain the basic principles. The following is a partial list of some common computer-controlled actuators:

- Mixture control (M/C) solenoid
- Cooling fan relay
- Variable valve timing control solenoid
- Air injection reactor (AIR) solenoid
- Early fuel evaporation (EFE) relay
- Exhaust gas recirculation (EGR) solenoid
- Torque converter clutch (TCC) solenoid
- Idle speed control (ISC) motor
- Canister purge solenoid
- Electronic spark timing
- Air-conditioning (A/C) compressor clutch relay
- Idle load compensator
- Port fuel injector solenoids

Several of these computer actuators are detailed in this chapter. However, most can be understood by studying the three basic types of actuators: relays, solenoids, and motors.

RELAYS

A relay is an electrical device that uses low current to control another circuit with higher current. ***Figure 38–30*** shows an example of the electrical schematic using a relay. The circuit is part of the air-conditioning system controls. In the center of the circuit, there is a shaded area called the

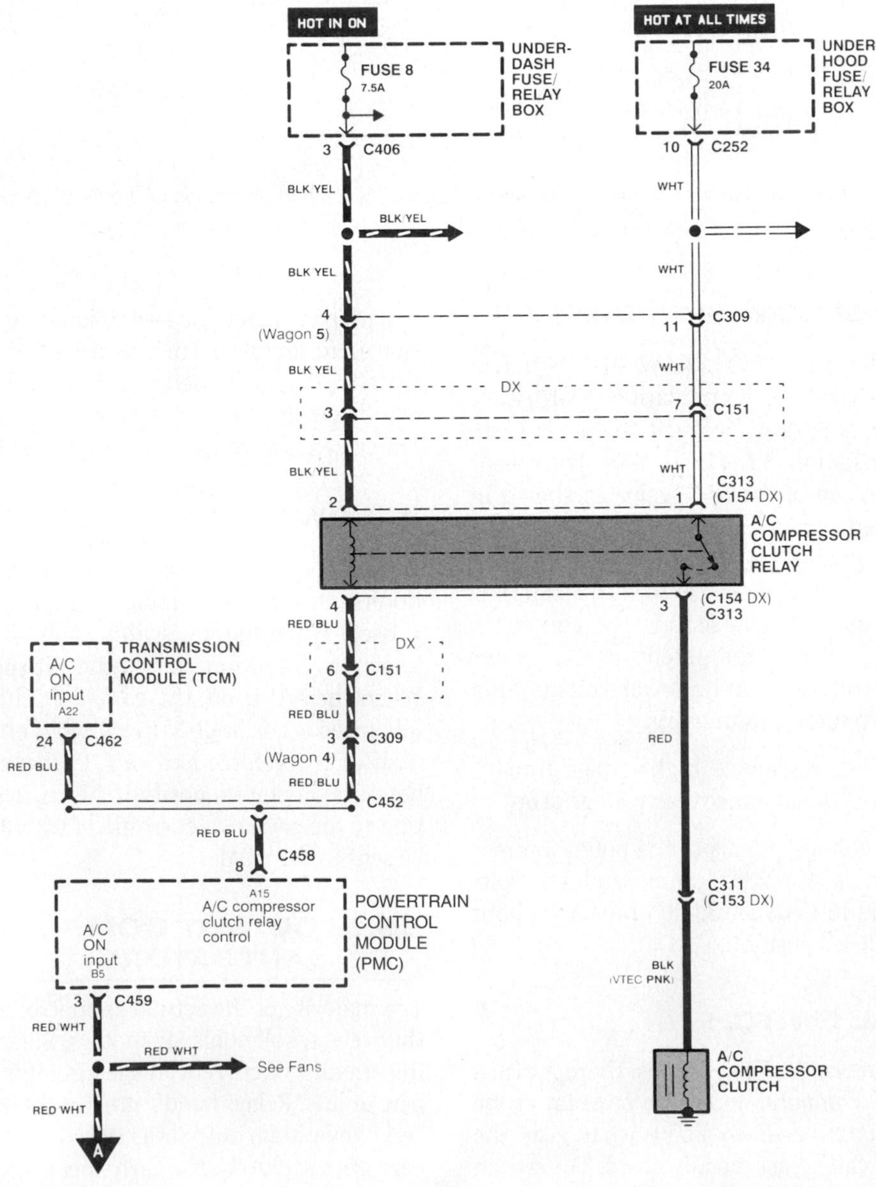

FIGURE 38–30 The A/C compressor clutch relay is energized by a low current circuit, but it controls a high-current circuit, the A/C compressor clutch. (Courtesy of American Honda Motor Co., Inc.)

A/C compressor clutch relay. The relay works in the following manner: The powertrain control module (PCM) on the bottom of the circuit turns on the low current electricity that flows through the left side of the relay (the relay coil). When this occurs, the relay is energized and the magnetism from the coil closes the switch on the right side. This switch and circuit are able to carry the higher current that is needed to turn on the A/C compressor.

Figure 38–31 shows two common relays. The top view is an air compressor relay used to turn the air compressor on and off. Note the electrical connections at the base of the relay. The bottom view shows a cooling fan relay used to turn the cooling fan off and on at the proper time. These relays are often located under the vehicle hood with easy access. ***Figure 38–32*** shows another cooling fan relay. Note that on the side of the relay there is a schematic to help the service technician identify the type of circuits located inside the relay.

SOLENOIDS

A solenoid can also be thought of as an output actuator from the computer. A solenoid has a coil of wire and a metal core in the center of the coil. When electricity flows through the coil of wire, the magnetic field in the center of the coil causes the metal in the center to move within the center of the coil. For example, refer to ***Figure 38–33***. In this illustration there is a coil of wire and a metal core that is pulled to the right by a spring. When the switch is closed, the coil produces a strong magnetic field in its center. When this happens, the metal core in the center of the coil moves to center itself in the magnetic field. So the metal core moves to the left, against spring pressure. When the switch

FIGURE 38–31 The air compressor relay and the cooling fan relay are examples of relays that are used on computerized engine control systems.

FIGURE 38–32 This is a coolant fan relay that has small circuits printed on the housing.

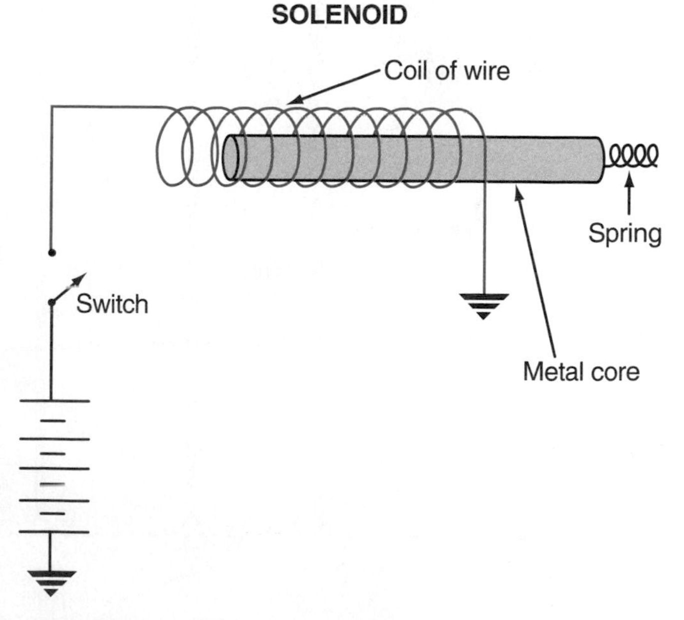

FIGURE 38–33 A solenoid is used to move a metal core in the center of a coil of wire, changing electrical energy to mechanical energy.

is opened, the magnetic field is stopped and the spring pulls the metal core back to the right.

Figure 38–34 shows a typical fuel injector circuit, which uses four solenoids to operate the fuel injector. On the bottom right side of the circuit, there are four solenoids. The symbol for a solenoid is a coil of wire and two lines. The lines next to the coil represent the metal core. Each solenoid controls a fuel injector. When electricity (controlled by the center relay in the diagram below) flows through each coil of wire, the metal core moves to open the fuel injectors.

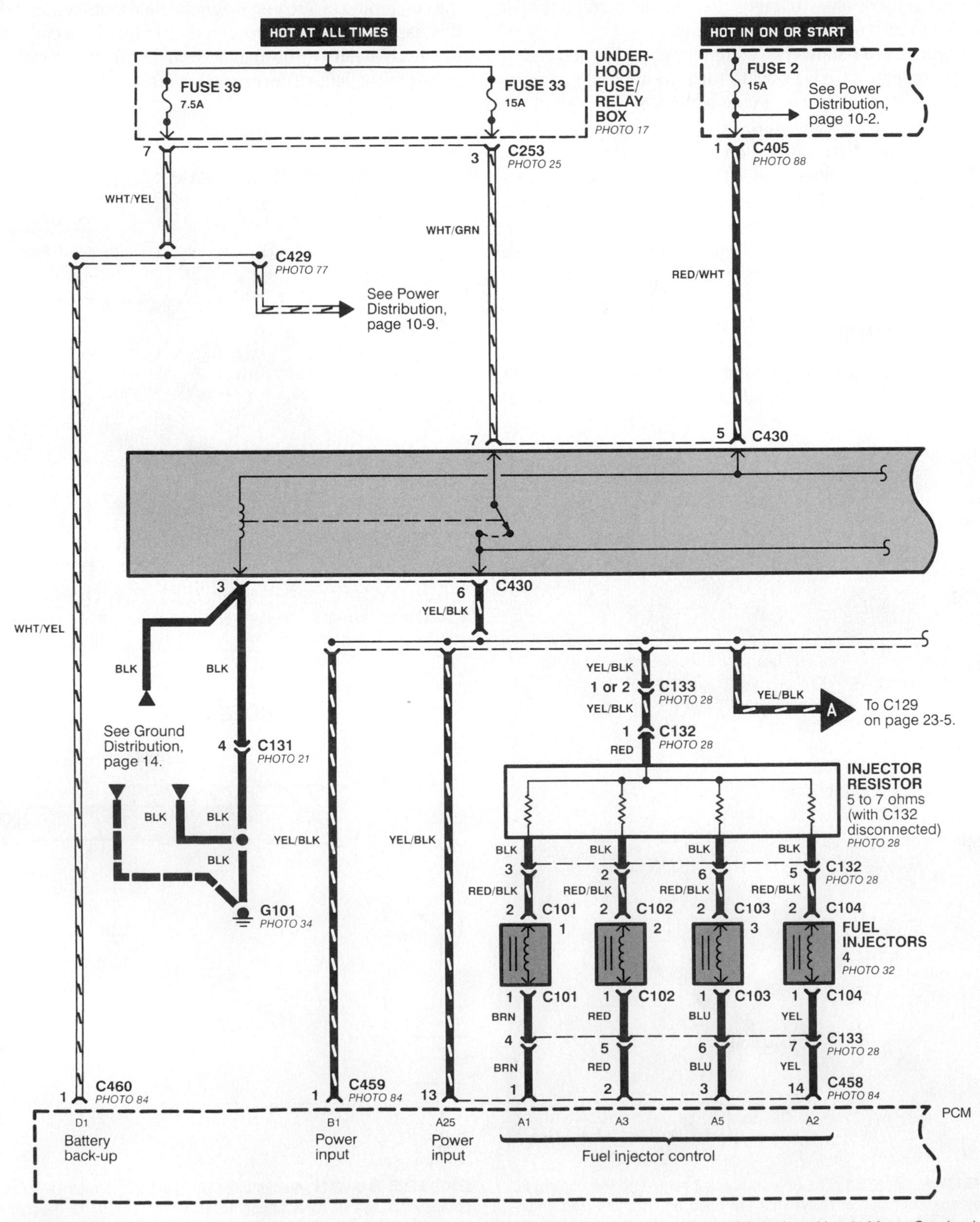

FIGURE 38–34 The fuel injector uses solenoids to monitor the fuel input to the engine. (Courtesy of American Honda Motor Co., Inc.)

MOTORS

Motors are yet another type of actuator controlled by the computer. Motors can be used to turn fans, to move a component a certain number of degrees, or to turn on an air-conditioning compressor. There are many uses for motors. For example, the circuit in ***Figure 38–35*** is for the fuel pump motor. When the switch in the relay above the pump is closed, the motor turns so that fuel is pumped to the fuel injectors.

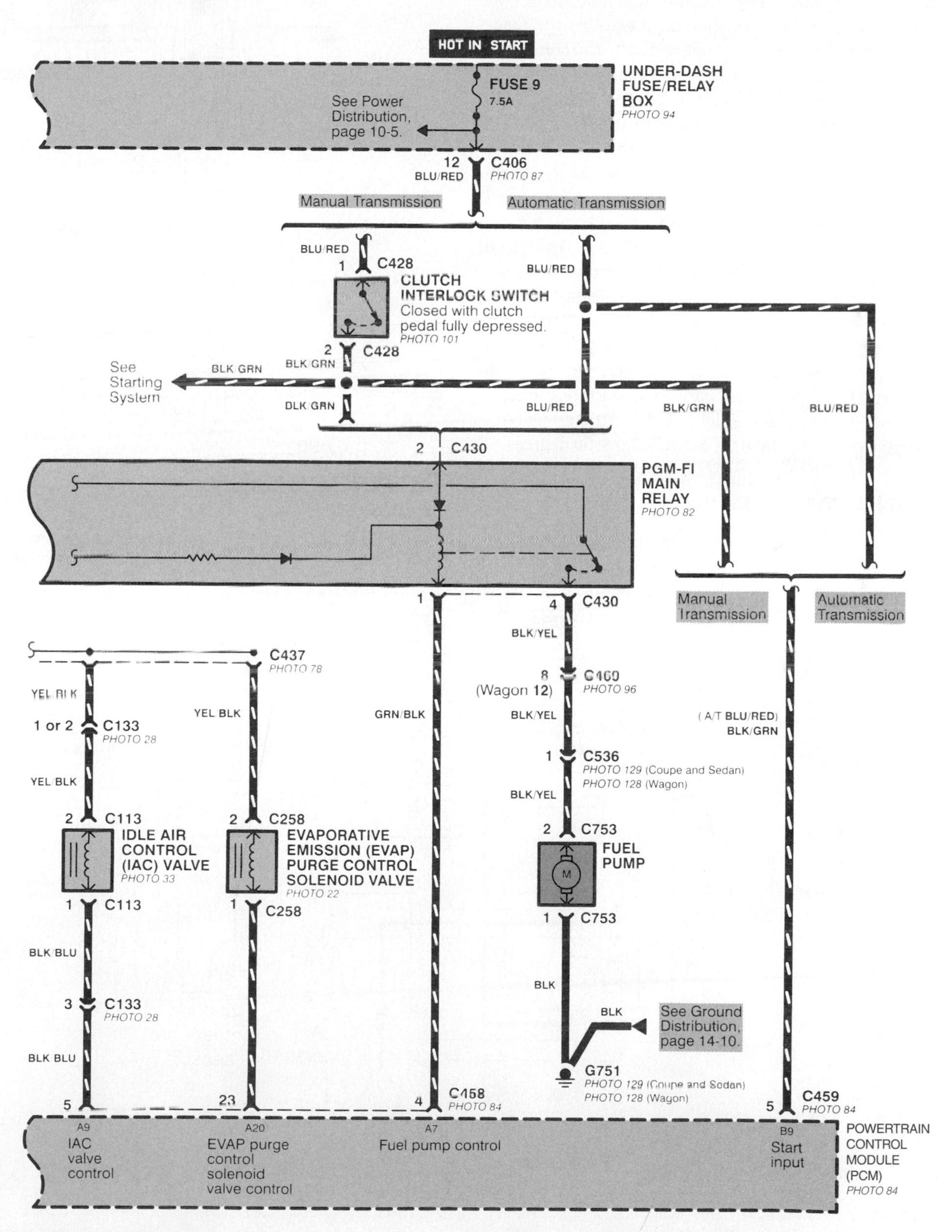

FIGURE 38-35 The fuel pump is a motor actuator controlled by the computer. (Courtesy of American Honda Motor Co., Inc.)

M/C SOLENOIDS

A mixture control (M/C) solenoid is used on electronically controlled carburetors. The M/C solenoid is located inside the carburetor bowl and regulates the amount of fuel entering the main metering circuit in the carburetor.

Figure 38–36 shows a typical M/C solenoid. In operation, the center plunger is pushed downward by the magnetic force when the solenoid is energized by the computer. In the down position, it restricts the amount of fuel flowing into the main metering circuit. When the solenoid is de-energized, the plunger rises because of spring tension.

The PCM switches the M/C solenoid on and off ten times per second. The amount of on and off time can be determined by using a dwell meter. A reading of 60 degrees dwell means the solenoid is energized 100% of the time. A reading of 0 degrees dwell means the solenoid is de-energized 100% of the time. So, a reading of 30 degrees means the solenoid is energized 50% of the time. If the carburetor is operating correctly, the dwell readings of the M/C solenoid should be between 10 and 50 degrees. An increase in dwell means there is a lean air-fuel ratio. A decrease in the dwell means there is a rich air-fuel ratio. On some vehicles, a trouble code will show for a dwell below 10 degrees. Another trouble code will show for a dwell above 50 degrees.

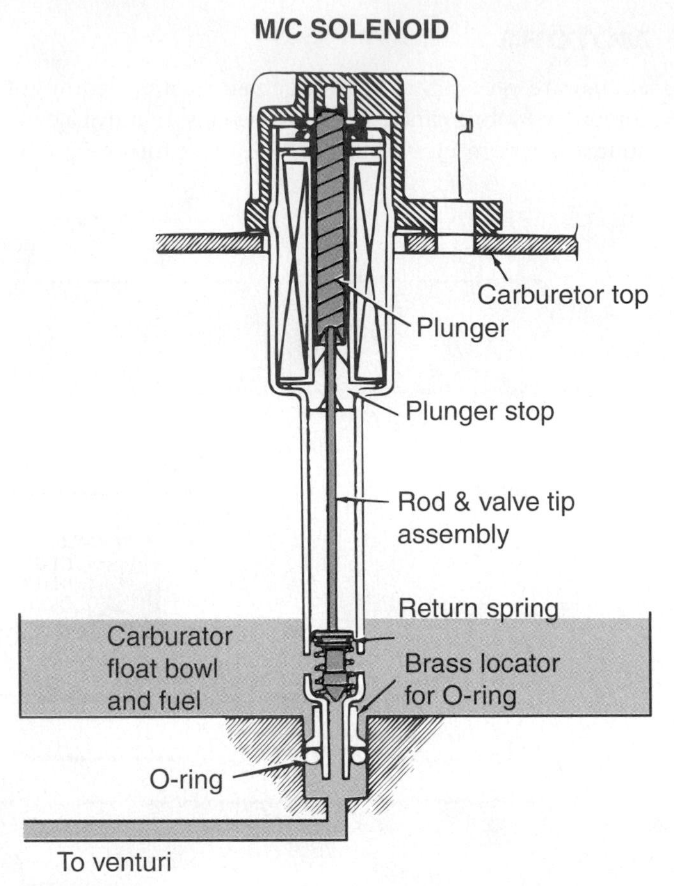

FIGURE 38–36 The M/C solenoid controls the amount of fuel going into the main metering circuit on an electronic carburetor.

COOLING FAN RELAY

The cooling fan relay is another type of output from the PCM. In this case, the cooling fan relay, which is controlled by the PCM, operates the fan motor. To understand how the relay is used to control the cooling fan, refer to ***Figure 38–37***. This

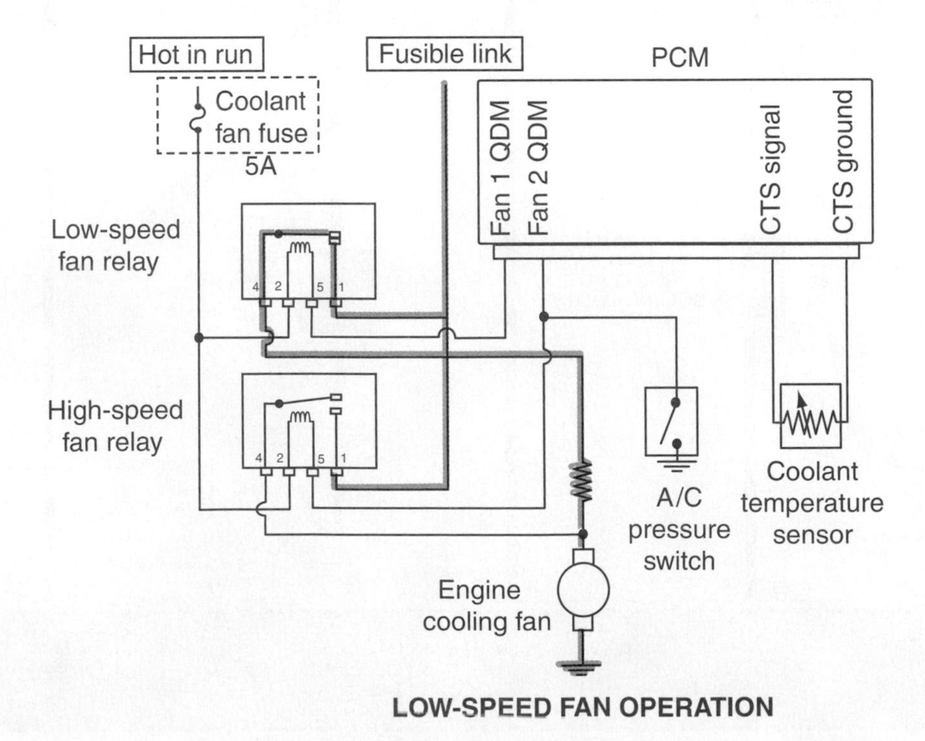

FIGURE 38–37 This two-speed fan circuit uses two relays to control fan motor speed.

electrical schematic shows the use of a two-speed cooling fan. This electrical circuit has two relays, one for high-speed operation and one for low-speed operation. If the PCM sends an electrical charge to the low-speed fan relay, the low-speed fan relay is energized. This action causes the switch inside the relay to close. The closed switch allows electricity to flow through a resistor and then into the fan motor, located just before the cooling fan motor. The resistor reduces the current and causes the cooling fan motor to operate at a slower speed. When the PCM sends an electrical charge to the high-speed fan relay, its switch will close. When closed, electricity flows through the relay switch and directly into the fan motor by passing the resistor.

PARTS LOCATOR

Refer to photo #77 in the front of the textbook to see what a ***cooling fan relay*** *looks like.*

VARIABLE VALVE TIMING CONTROL SOLENOID

More vehicle manufacturers are now using variable valve timing systems. These systems control the opening and closing of the valves using a different timing profile. These systems use two types of camshafts, one shaped for low rpm and one shaped for high rpm. In order to change cam lobes, oil pressure from the main oil galley is sent to the camshafts. A variable timing and electronic control (VTEC) solenoid is used to turn the oil pressure off or on, depending on which cam is needed.

Figure 38–38 shows such a circuit. In operation, the computer is sensing various inputs such as engine speed, engine load, vehicle speed, and engine coolant temperature. Depending on the conditions of these inputs, the PCM controls the VTEC solenoid valve. Normally, the VTEC solenoid is not energized and the solenoid valve is closed. When the computer senses that a different cam is needed, the PCM energizes the VTEC solenoid valve. The oil pressure is then used to move a piston inside the camshaft so as to change to a different cam lobe.

IDLE SPEED CONTROL (ISC) MOTOR

Another output solenoid controlled by the PCM is called the idle speed control (ISC) motor or the idle air control (IAC) valve. This valve controls the amount of air that is sent into the engine during idle conditions. Although there are several variations, ***Figure 38–39*** shows how this valve changes the amount of air bypassing into the intake manifold. The PCM controls the amount of electric current sent to the IAC. When the IAC valve is activated, the valve opens to maintain the proper idle speed. For example, if the air-conditioning compressor is turned on, the idle speed needs to be increased. The PCM sends additional current to the valve so it opens further, letting more air in, and thus increasing the idle speed. On some valves, a rotational solenoid is used, which allows the opening to be varied slightly.

Figure 38–40 shows a typical idle speed control motor. Note that the ISC valve, located at the left, is moved by the motor to increase or decrease the amount of air that is bypassing the throttle plate and being sent into the intake manifold. An O-ring is used to seal the motor so that no outside air comes into the intake manifold when the ISC is mounted.

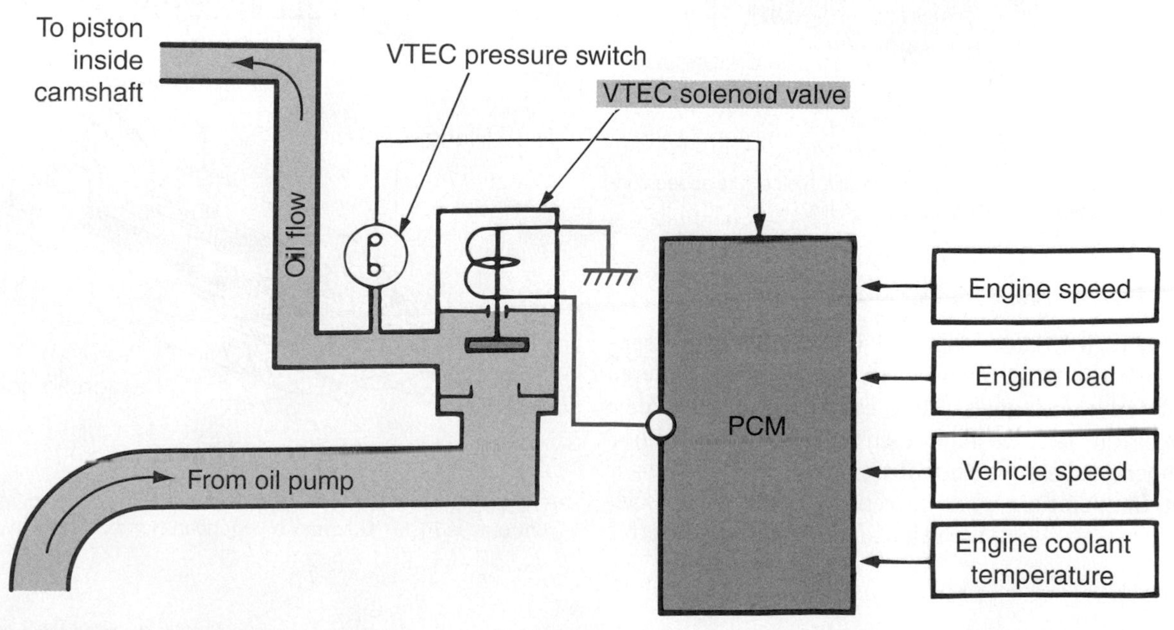

FIGURE 38–38 The solenoid in a variable valve timing control system is another output controlled by the PCM.

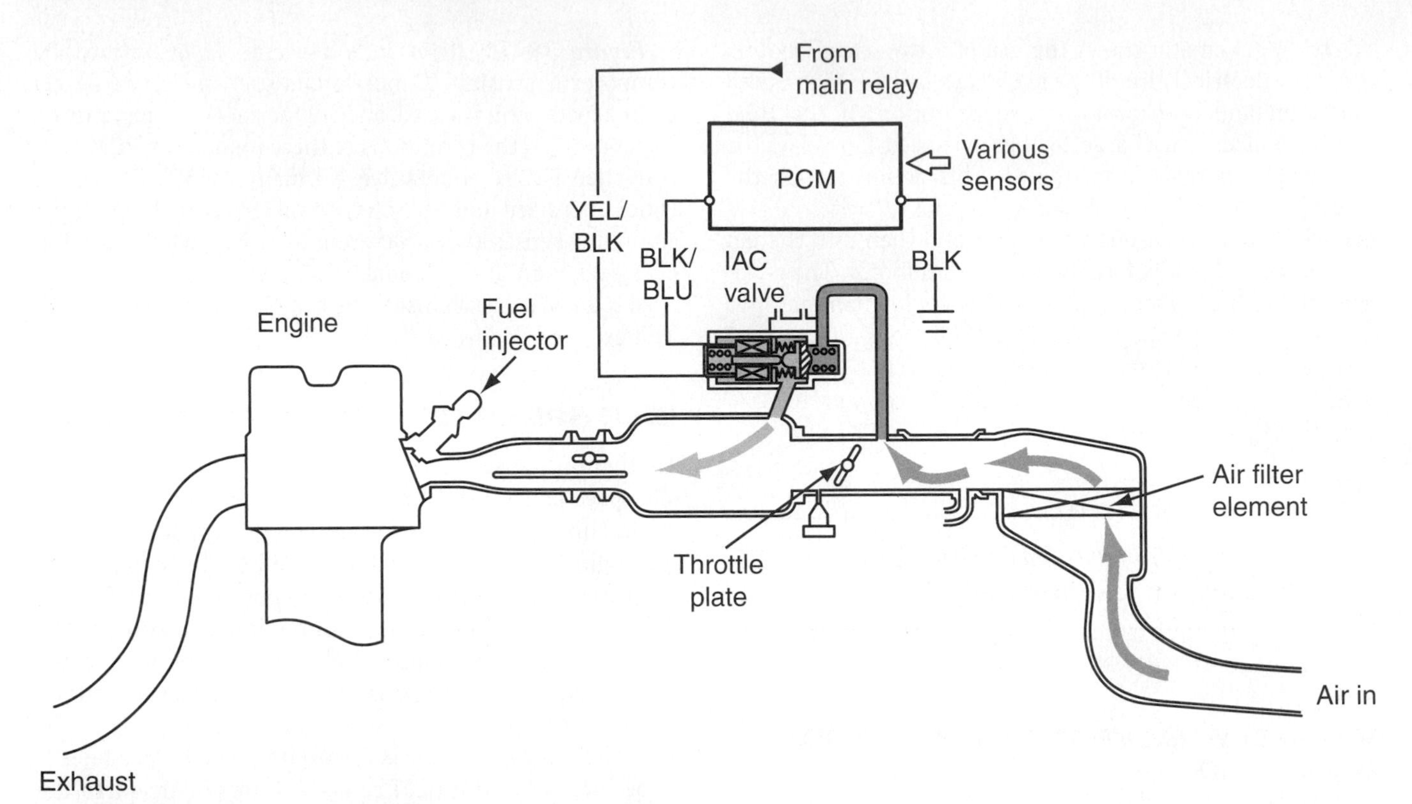

FIGURE 38–39 The idle air control (IAC) valve, controlled by the PCM, lets the right amount of air pass through the throttle plate to control idle speed.

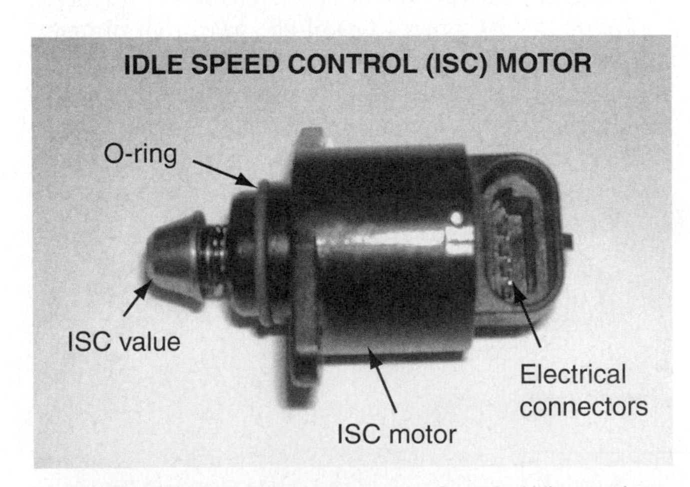

FIGURE 38–40 This is an example of a typical idle speed control (ISC) motor, which is controlled by the PCM.

PORT FUEL INJECTOR SOLENOIDS

Most engines today use port fuel injectors that are controlled from the PCM. Many fuel injectors are considered constant stroke solenoids. They consist of a solenoid, a plunger needle valve, and a housing. ***Figure 38–41*** shows the plunger needle valve, the solenoid coil, and the plunger housing. In operation, when current is applied to the solenoid coil from the PCM, the valve lifts up off its seat. At this point, pressurized fuel is injected into the intake manifold. The needle valve lift and fuel pressure are constant. Therefore, the amount of fuel being injected is determined by the length of time that the valve is open. The length of time the valve is open is determined by the duration of current supplied by the PCM.

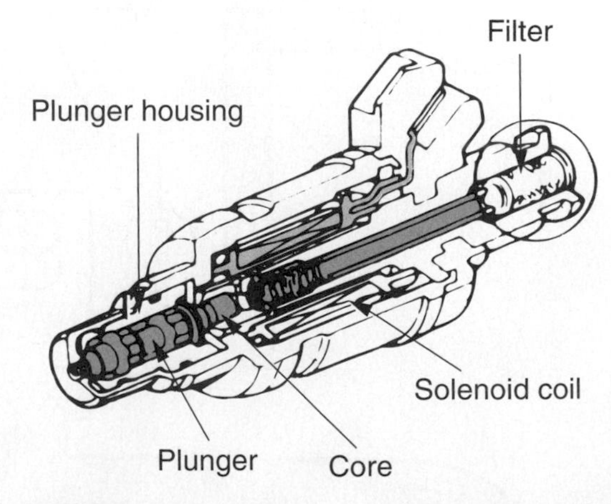

FIGURE 38–41 Control of fuel injectors is considered an output from the PCM. (Courtesy of American Honda Motor Co., Inc.)

Problems, Diagnosis, and Service

Safety Precautions

1. Whenever working with sensors and actuators, always be sure to wear OSHA-approved safety glasses.
2. When removing or installing any input sensor or output actuator, make sure to use the correct tools.
3. Always follow the manufacturer's recommended removal and replacement procedures when working with input sensors and output actuators.
4. Whenever disconnecting wires from input sensors or output actuators, always make sure to mark the wires so that you know how they should be reattached.
5. Many of the input sensors and output actuators are made of plastic. Always handle each component with care.
6. Always use the recommended parts when replacing input sensors and output actuators.
7. Often a digital meter is needed to check resistances and/or voltage drops on input sensors and output actuators. Always be careful not to get the meter wires tangled in moving parts.
8. When necessary, use a grounding strap to make sure your body is grounded to the frame of the vehicle.
9. Be careful not to cut your hands on any of the sharp parts of the sensors and actuators.

PROBLEM: Faulty Oxygen Sensor

It is suspected that an oxygen sensor is not operating correctly. After doing a diagnostic check with the Check Engine light, the oxygen sensor code keeps coming on.

DIAGNOSIS

Special laboratory testing equipment is required to determine if the oxygen sensor is performing correctly. However, certain tests that can be made in the vehicle will give a good indication of its operation. The heater resistance can be measured with an ohmmeter. An ohmmeter check can determine if the sensor's signal terminal is shorted to ground. First, be sure the sensor is at room temperature and that the heater has been turned off for at least 30 minutes. Under these conditions, the resistance between the signal (+) and ground (–) terminal of the sensor case should be greater than 20 megohms.

The output voltage can be measured on the sensor. Be sure to use a 10 megohm digital volt-ohmmeter. The digital voltmeter should be connected from the oxygen sensor signal wire to ground. The voltage should be cycling from low to high voltage somewhere between 0 and 1 volt.

SERVICE

Generally, there are three possibilities for oxygen sensor voltages.

1. Low oxygen sensor voltage output (lean air-fuel mixture)—This usually indicates that the sensor voltage signal is low. The sensor voltage is below 0.3 volt (300 millivolts) for a long period of time as shown in ***Figure 38–42A***. Possible causes might include a leaking vacuum hose, low fuel pressure, or a leak in the exhaust system above the oxygen sensor.
2. High oxygen sensor voltage output (rich air-fuel mixture)—This usually indicates that the sensor circuit is functioning properly. However, the rich mixture causes the sensor voltage to stay above 0.06 volt (600 millivolts) for a long period of time, as shown in ***Figure 38–42B***. Possible causes might include leaky fuel injectors, a continuously purging vapor canister, the EGR valve not seating, or a silicone-contaminated sensor.

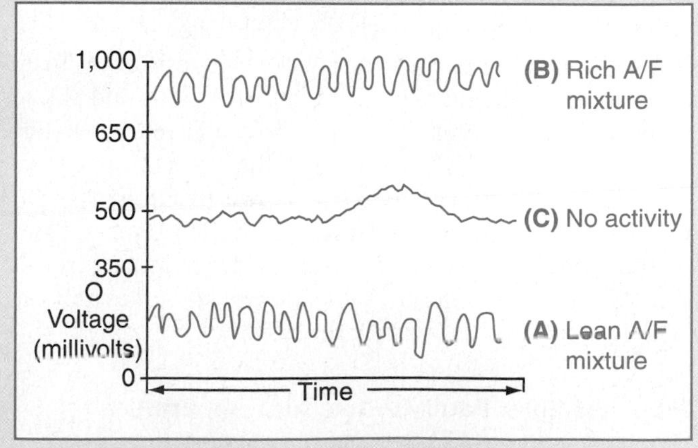

FIGURE 38-42 This voltage graph shows O sensor voltages under different conditions.

3. Oxygen sensor voltage output (no activity detected)—This indicates that the sensor has been between 0.3 volt and 0.6 volt (300–600 millivolts) for a long period of time as shown in ***Figure 38–42C***. Possible causes might include high resistance at the oxygen sensor connector or an opening in the sensor wiring harness.

PROBLEM: Faulty Engine Coolant Temperature Sensor

It is suspected that the engine coolant temperature (ECT) sensor is not operating correctly. After doing a diagnostic check with the Check Engine light, the ECT sensor code keeps coming on.

DIAGNOSIS

When the engine coolant temperature sensor is faulty, it can cause several problems. These problems occur because both fuel and air decisions are made by the computer based on the temperature of the engine coolant. The following problems may be noticed:

- Hard engine starting
- Rich or lean air-fuel ratios
- Improper operation of emission devices
- Increased fuel consumption
- Engine stalling, hesitation, or poor acceleration
- Improper converter clutch lockup

The ECT can be checked for voltage drop and resistance readings.

SERVICE

To check voltage drop readings, use a digital voltmeter. With the ECT installed on the engine, check the voltage drop across the two ECT wires when the engine is hot. Compare the voltage readings at specific temperatures to the manufacturer's specified voltages. Readings will generally be between 0.3 volt (hot) to 4.5 volt (cold).

The ECT sensor can also be checked for resistance. To do this, the ECT must be removed from the engine and placed in a container of water. Attach an ohmmeter to the two terminals from the ECT. Then increase the water temperature, checking with a thermometer as shown in ***Figure 38–43***. Compare the resistance readings at different temperatures to the specified resistances stated by the manufacturer. If the resistances at different temperatures do not match the vehicle specifications, replace the ECT.

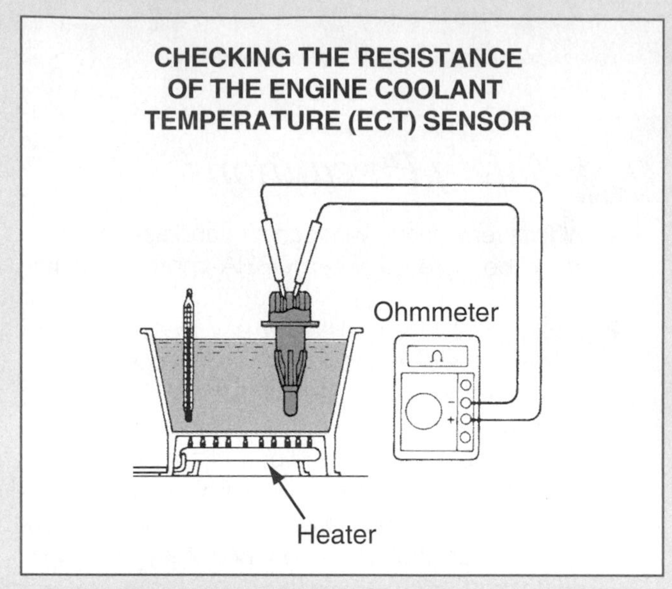

FIGURE 38–43 An engine coolant temperature (ECT) sensor can be checked by putting it in a container of water and measuring its resistance at different temperatures.

PROBLEM: Faulty Intake Air Temperature (IAT) Sensor

It is suspected that the IAT sensor is not operating correctly. After doing a diagnostic check with the Check Engine light, the IAT sensor code keeps coming on.

DIAGNOSIS

When the intake air temperature sensor is faulty, it may cause several problems. These problems occur because both fuel and air decisions are made by the computer based upon the temperature of the air coming into the engine. The following problems may be noticed:

- Hard engine starting
- Rich or lean air-fuel ratios
- Engine stalling or surging
- Increased fuel consumption
- Acceleration stumbling

The IAT sensor can be checked for voltage and resistance readings.

SERVICE

To check voltage drop readings, use a digital voltmeter. With the IAT installed on the engine, check the voltage drop across the two IAT wires when the engine is hot. Compare the voltage readings at specific temperatures to the manufacturer's specified voltages. Readings will generally be between 0.3 volt (hot) to 4.5 volt (cold). ***Figure 38–44*** shows an example of the type of voltage readings that should be read. For example, if the coolant temperature is 60 degrees Fahrenheit, the voltage reading should be about 3.67 volts. As the temperature gets hotter, the voltage readings should decrease. For example, at 220 degrees Fahrenheit, the voltage reading should be about 0.48 volts.

The IAT sensor can also be checked for resistance. To do this, the IAT must be removed from the engine and placed in a container of water. Attach an ohmmeter to the two ter-

CHARGED TEMPERATURE SENSOR TEMPERATURE VS. VOLTAGE CURVE	
TEMPERATURE	**VOLTAGE**
–20°F	4.81 V
0°F	4.70 V
20°F	4.47 V
40°F	4.11 V
60°F	3.67 V
80°F	3.08 V
100°F	2.51 V
120°F	1.97 V
140°F	1.52 V
160°F	1.15 V
180°F	0.86 V
200°F	0.65 V
220°F	0.48 V
240°F	0.35 V
260°F	0.28 V

FIGURE 38–44 The intake air temperature (IAT) sensor can be checked for voltage drop at different temperatures. (Courtesy of DaimlerChrysler Corporation)

minals from the IAT. Then increase the water temperature, checking with a thermometer as shown in ***Figure 38–45A***. Compare the resistance readings at different temperatures to the specified resistances stated by the manufacturer as shown in ***Figure 38–45B***. As an example, if the temperature of the water is heated to 140 degrees Fahrenheit, the resistance should be about 0.5 ohms. If the resistances at different temperatures do not match the vehicle specifications, replace the IAT.

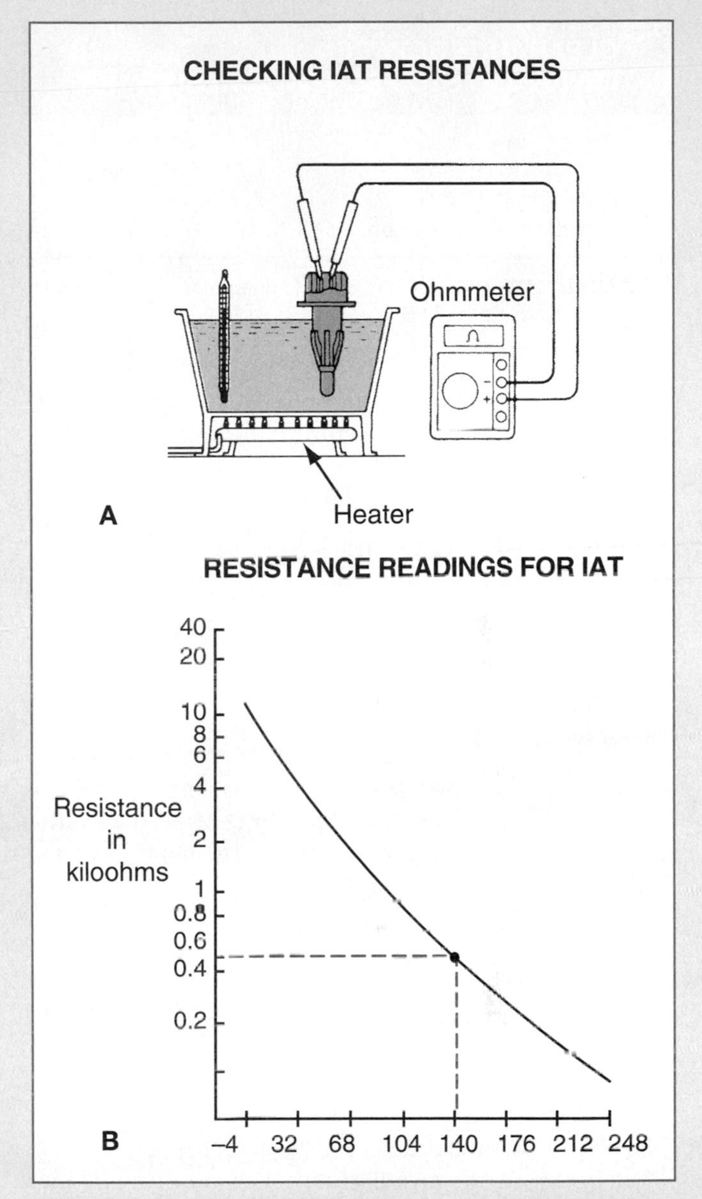

FIGURE 38–45 The IAT sensor can be checked for resistance.

PROBLEM: Faulty Throttle Position Sensor (TPS)

It is suspected that the throttle position sensor (TPS) is not operating correctly. After doing a diagnostic check with the Check Engine light, the TPS sensor code keeps coming on.

DIAGNOSIS

A faulty TPS can cause several problems. These problems occur because both fuel and air decisions are made by the computer based upon the position of the throttle. The following problems may be noticed:

- Acceleration stumbling
- Engine stalling
- Improper idle speed, or hunting

The TPS can be checked for voltage readings.

SERVICE

To check voltage from the computer, turn on the ignition. Now check for voltage from the 5-volt reference wire to ground. Always refer to the vehicle manufacturer's specifications. If these readings are near 5 volts, check the condition of the TPS voltage drop signal back to the computer. With the ignition switch on, connect a voltmeter from the sensor signal wire to ground. Slowly open the throttle and read the voltmeter. The readings should increase gradually. Typical readings would be about 0.5 volts at idle to about 4.5 volts at wide-open throttle. If the TPS does not have these specified readings or if the readings are erratic, replace the TPS.

The TPS can also be checked for resistance at different throttle position settings. For example, ***Figure 38–46*** shows suggested resistance readings at different throttle positions. At idle (0 inches), the resistance should be between 0.28 and 6.4 kilohms. At wide-open throttle, the resistance should be between 2.0 and 11.6 kilohms.

Note that some TPSs have an adjustment that can be made. This adjustment is done by turning the TPS in two

CLEARANCE BETWEEN LEVER AND STOP SCREW	RESISTANCE
0 mm (0 in.)	0.28–6.4 kΩ
0.35 mm (0.014 in.)	0.5 kΩ or less
0.70 mm (0.28 in.)	Infinity
Throttle valve fully open	2.0–11.6 kΩ

FIGURE 38–46 Use this chart to determine resistance readings when checking the TPS.

slotted holes that hold it on the engine (***Figure 38–47***). Follow the specific manufacturer's recommended procedure to adjust this type of TPS.

PROBLEM: Faulty Knock Sensor

It is suspected that the knock sensor is not operating correctly. After doing a diagnostic check with the Check Engine light, the knock sensor code keeps coming on.

DIAGNOSIS

A faulty knock sensor can cause several problems. These problems occur because the computer controls the ignition system and the amount of advance on the ignition system. The following problems may be noticed:

- Engine detonation
- Reduced spark advance
- Reduced fuel economy

SERVICE

The procedure for checking the knock sensor may be different for each engine manufacturer. Also, on many engines, the coolant may need to be drained before the knock sensor can be removed. Use the following general procedure to check the condition of the knock sensor.

1. Disconnect the knock sensor wiring connector.
2. Now turn on the engine ignition switch.
3. Check the voltage drop from the sensor wire to ground. The voltage should be about 4 to 6 volts. If the voltage drop is higher, there may be a problem with the computer wires or the computer.
4. Connect an ohmmeter from the knock sensor terminal to ground.
5. Typical knock sensors should have between 3,300 ohms to 4,500 ohms. Always check the manufacturer's recommended resistance first. If the knock sensor has too high or too low a resistance compared to the manufacturer's specifications, replace the knock sensor.

When replacing the knock sensor, always torque the sensor to the exact torque specifications. If it is overtorqued, the knock sensor may become too sensitive and cause reduced spark advance and, thus, poor fuel economy. If the knock sensor is undertorqued, the knock sensor may have too little sensitivity and detonation may occur. This may reduce engine performance.

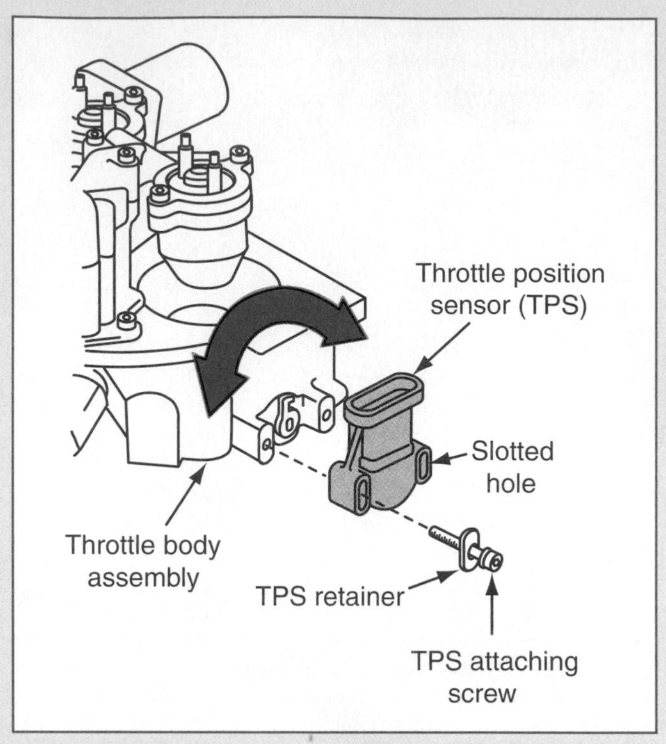

FIGURE 38–47 This type of throttle position sensor is adjustable by turning the TPS in the slotted holes.

Service Manual Connection

There are many important specifications to keep in mind when working with computerized engine controls. To identify the specifications for your engine, you will need to know the VIN (vehicle identification number) of the vehicle, the type and year of the vehicle, and the type of engine. Although they may be titled differently, some of the common computerized engine control specifications (not all) found in the service manuals are listed below. Note that these specifications are typical examples. Each vehicle and engine may have different specifications.

Common Specification	Typical Example
Engine coolant temperature sensor resistance	269,000 ohms cold
1,200 ohms hot	
MAP signal voltage	1–1.5 volts at idle to 4.5 volts at wide-open throttle
Knock sensor voltage	300 to 500 millivolts
Heated oxygen sensor resistance	10–40 ohms
Crankshaft position sensor resistance	350–700 ohms

In addition, the service manuals give specific directions for various service procedures. Some of the more common procedures include:

- M/C solenoid removal/replacement
- MAP removal/replacement
- Cooling fan relay removal/replacement
- Fuel pump removal/replacement
- Park/neutral switch troubleshooting and diagnosis

SUMMARY

The following statements will help to summarize this chapter:

- Sensors are used to monitor many of the conditions that affect the operation of the engine.
- Inputs from the sensors are fed into the PCM, which in turn operates or controls the outputs.
- The outputs or actuators help to control fuel, ignition, cooling system fans, air compressors, and so on.
- There are five types of input sensors, including the switch, variable resistor, potentiometer, magnetic pickups, and voltage generators.
- Common input sensors include the oxygen sensor, coolant temperature sensor, intake air temperature sensor, throttle position sensor, manifold absolute pressure sensor, mass airflow sensor, knock sensor, vehicle speed sensor, crankshaft position sensor, camshaft position sensor, and the EGR valve position sensor.
- Computerized engine output controls are categorized as either solenoids, relays, or motors.
- Common output actuators include the mixture control solenoid, cooling fan relay, variable valve timing control solenoid, air injector reactor solenoid, early fuel evaporative relay, exhaust gas recirculation solenoid, torque converter clutch solenoid, idle speed control motor, canister purge solenoid, electronic spark timing, air-conditioning compressor clutch relay, and the port fuel injector solenoid.
- The most common checks that a service technician makes on the input sensors and output actuators are voltage drop readings and resistance readings of the sensors and actuators.

TERMS TO KNOW

Can you explain each of the following terms? Review the chapter until you can use each term correctly.

Absolute sensor
Camshaft position sensor
Closed loop
Differential sensor
Discrete sensors
Magnetic pickup sensor
Open loop
Piezoelectric
Platinum
Potentiometer
Transducer
Variable resistor
Zirconia

REVIEW QUESTIONS

Multiple Choice

1. Computers use ________ sensors to determine various conditions for the engine to run correctly.
   a. Output
   b. Actuator
   c. Input
   d. Mechanical
   e. None of the above

2. Which of the following is not an advantage of using a computer to control engine functions?
   a. Air-fuel ratio closely controlled
   b. More efficient engine operation
   c. Reduce emissions
   d. Reduced fuel mileage
   e. Engine is more responsive

3. Another name for an engine sensor is a/an ________.
   a. Transducer
   b. Fan
   c. Electronic actuator
   d. Alpha
   e. Motor

4. Which of the following types of sensors is designed to give an off-on signal to the computer?
   a. Variable resistor
   b. Switch
   c. Potentiometer
   d. Voltage generator
   e. None of the above

5. Which of the following types of sensors is designed to produce a voltage signal to the computer?
   a. Variable resistor
   b. Switch
   c. Potentiometer
   d. Voltage generator
   e. None of the above

6. Which of the following types of sensors is used as a crankshaft sensor?
   a. Variable resistor
   b. Switch
   c. Potentiometer
   d. Magnetic pickup
   e. None of the above

7. When there is no feedback going to the PCM telling it that the air-fuel ratio is too lean, the computerized engine control system is in the ________ mode.
   a. Open loop
   b. Closed loop
   c. Intermediate loop
   d. All of the above
   e. None of the above

8. The tip of an oxygen sensor is made of which type of material?
   a. Aluminum
   b. Zirconia
   c. Plastic
   d. Rubber
   e. Glass

9. Which of the following sensors tells the computer the position of the driver's foot on the throttle?
   a. Coolant temperature sensor
   b. Throttle position sensor
   c. Air charge temperature sensor
   d. Knock sensor
   e. Crankshaft position sensor

10. Which of the following sensors compares barometric pressure to manifold vacuum?
    a. Coolant temperature sensor
    b. Throttle position sensor
    c. Air charge temperature sensor
    d. Manifold absolute pressure (MAP) sensor
    e. Crankshaft position sensor

11. Which of the following sensors use a piezocrystal to measure small vibrations?
    a. Coolant temperature sensor
    b. Throttle position sensor
    c. Air charge temperature sensor
    d. Knock sensor
    e. Crankshaft position sensor

12. Which of the following is not considered an output actuator on computerized engine control systems?
    a. Mixture control solenoid
    b. Port fuel injector solenoid
    c. Cooling fan relay
    d. Piston shape relay
    e. Variable valve timing control solenoid

13. Which of the following output actuators are used on a computerized engine control system?
    a. Relays
    b. Solenoids
    c. Motors
    d. All of the above
    e. None of the above

14. The fuel pump is considered which type of output actuator?
    a. Relay
    b. Solenoid
    c. Motor
    d. All of the above
    e. None of the above

15. Which of the following output actuators allows a certain amount of air to bypass the throttle plate, and is controlled by the PCM?
    a. M/C control solenoid
    b. Idle air control valve
    c. Fuel injector solenoid
    d. Air injector reactor solenoid
    e. Canister purge solenoid

The following questions are similar in format to ASE (Automotive Service Excellence) test questions.

16. *Technician A* says that the crankshaft position sensor can be a magnetic type sensor. *Technician B* says that the crankshaft position sensor can be a variable reluctance type sensor. Who is correct?
    a. A only c. Both A and B
    b. B only d. Neither A nor B

17. *Technician A* says that when a vane is used to interrupt a magnetic field, this type of sensor is called the Hall-effect sensor. *Technician B* says that the Hall-effect sensor uses a fixed magnet and a semiconductor wafer to operate. Who is correct?
    a. A only c. Both A and B
    b. B only d. Neither A nor B

18. *Technician A* says the camshaft position sensor is most often used to identify the position of number 1 piston. *Technician B* says the camshaft position sensor works in conjunction with the EGR valve position sensor to determine crankshaft position. Who is correct?
    a. A only c. Both A and B
    b. B only d. Neither A nor B

19. The engine coolant temperature sensor is faulty. *Technician A* says that this will cause the engine to increase fuel consumption. *Technician B* says this problem will cause the engine to be hard starting. Who is correct?
    a. A only c. Both A and B
    b. B only d. Neither A nor B

20. The intake air temperature (IAT) sensor is faulty. *Technician A* says that this will cause the engine to stall or surge. *Technician B* says this will cause the engine rpm to increase by 500 rpm at idle. Who is correct?
    a. A only c. Both A and B
    b. B only d. Neither A nor B

Essay

21. Explain how the input, processing, and output systems on a computer relate to a person's senses, brain, and muscles.

22. There are many inputs and outputs used on computerized engine control systems. Describe what type of variables will determine the exact type of input and output used.

23. Describe the five major types of sensors used to gather inputs on a computerized engine control system.

24. Describe how an oxygen sensor works.

25. Explain the operation and use of the Hall-effect sensor.

Short Answer

26. The Hall-effect sensor is able to send a/an ________ signal to the PCM.

27. The park/neutral switch is used as an input to the computer so it can adjust the ________ on the engine, when the transmission is in gear.

28. The on and off sequence of the cooling motor fan is controlled by the ________ relay.

29. The port fuel injectors use a(n) ________ type of actuator to operate correctly.

30. The ________ sensor can be checked by placing it in hot water and checking its resistance at different temperatures.

Applied Academics

TITLE: Electrical Connectors

ACADEMIC SKILLS AREA: Science

NATEF Automobile Technician Tasks:

The service technician can explain the conductivity problems in a circuit when connectors corrode due to electrochemical reactions.

CONCEPT:

Many types of electrical connectors and connections are used on computerized engine control systems. In some cases, electrical connectors may have one or two wires. In other cases, an electrical connector may have as many as six or seven wires. One problem with electrical connections is that they often corrode because of electrochemical reactions. Over time, as electricity passes through a connector, the electricity eventually causes a change in the chemistry of the metal inside of the connector. This electrochemical reaction can cause the connector to corrode and the resistance to change at the point of the electrical connection. Also, connectors can corrode if not protected from atmospheric conditions. The end result is a high resistance that is produced within the connector. To help reduce this possibility, various types of electrical connectors are used, as shown in the figure that follows. Each type of electrical connector may be shaped differently and will need to be disconnected and reconnected in different ways. They are designed to keep the electrical connection inside a plastic container to reduce corrosion. When servicing any electrical circuit, always use the intended connector for the circuit being repaired. Also, the plastic connectors become brittle over time, so make sure to disconnect and connect each type of connector the correct way. In many cases, there is a small tab that, when pulled up or pushed down, will release the connector.

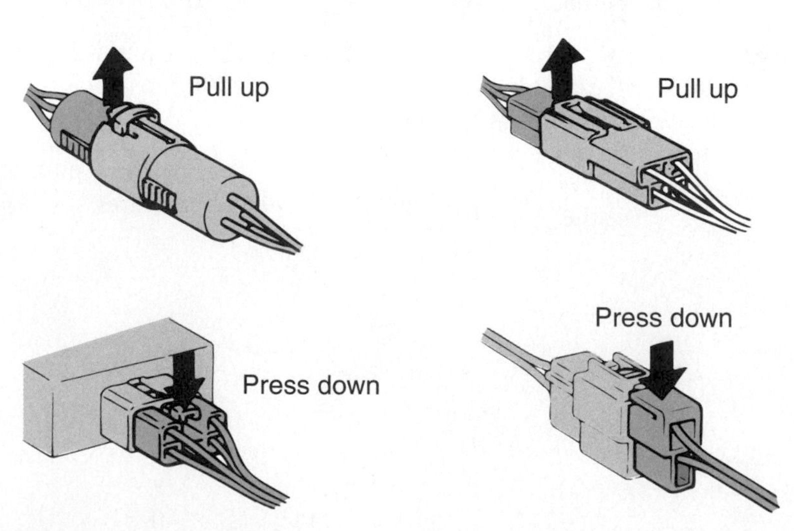

APPLICATION:

Can you determine how corrosion can cause the resistance to increase on a connector? What happens if the resistance of a connector increases too much? Why would it be a good idea to keep all electrical connectors clean?

Power Transmission Systems

SECTION

6

Mike Gibson

Aamco Transmissions

Lead Technician/Primary Transmission Rebuilder

Winston Salem, NC

How did you first get interested in this field?

I didn't have any auto experience when I got out of the Army, but when I was going through the help wanted ads, this was the only job that appealed to me. So I started as an apprentice.

What's the biggest misperception of your work?

People think this is a dirty job—but if it's done right, you can eat off the floor here.

What kind of education do you need for this position?

You don't necessarily need pretraining, though it does give you basic proficiency. But you really need ongoing learning because this field is increasingly technical, and if you don't attend classes, you're likely to fall behind.

What's your most important tool?

I keep my laptop on my toolbox, so I can refer back to databases from the manufacturer for things like wiring diagrams. It helps me do my job faster, and the more I can produce for the company, the more they'll pay me.

What's your typical day like?

It starts with a team meeting where the boss decides what projects we'll be assigned. Then I spend the rest of the day tearing down and rebuilding transmissions: cleaning the parts, finding the problems, then putting them back together. (I can do about 2½ in a day.) Then I move it on to a technician to be installed.

What's frustrating about your job?

When a customer brings in a car that doesn't work because it's been repaired by a bargain shop—then you end up fixing the last guy's job and dealing with a customer who's angry with our whole business.

What advice do you have for a high school student interested in this career?

Apply at any name shop—you'll learn more and get trained to do the job right. When you go into a place that's dumpy and filthy, ask yourself, do you really want to work like that? I think their top level is our bottom level.

Automotive Clutches

CHAPTER 39

OBJECTIVES

After studying this chapter, you should be able to:

- Identify the purpose of using a mechanical clutch.
- Describe the parts of a standard clutch system.
- Examine the parts of the clutch linkages.
- Define the purpose and function of the flywheel.

INTRODUCTION

Torque that is produced at the end of the crankshaft by the engine must be transmitted to the driving wheels. To accomplish this, torque must first pass through the clutch and then into the transmission. The clutch is used to mechanically engage and disengage the transmission. This chapter is about clutch systems and how they operate.

39.1 CLUTCH SYSTEMS

PURPOSE OF CLUTCHES

All standard or manual transmissions have a **clutch** to engage or disengage the transmission. While the engine is running, there are times when the driving wheels must not turn. The clutch is used as a mechanism to engage or disengage the transmission and driving wheels. If clutches were not used, every time the vehicle came to a stop, the engine would have to stop. Since this is not practical, the operator can engage and disengage the clutch when needed.

The clutch is designed to engage the transmission gradually. This will eliminate jumping abruptly from no connection at all to a direct solid connection to the engine. This is done by allowing a certain amount of slippage between the input and the output shafts on the clutch. Several components are needed to do this. These include the pressure plate, driven plate or friction disc, flywheel, clutch release bearing, clutch fork, and clutch housing. These are shown in ***Figure 39–1***.

Certification Connection

ASE Connection: The information in this chapter can help you prepare for the National Institute for Automotive Service Excellence (ASE) certification tests. The test and content area most closely related to this chapter are:

Test A3—Manual Drive Train and Axles

- **Content Area**—Clutch Diagnosis and Repair

NATEF Connection: Much of the information in this chapter is related to the NATEF tasks. The NATEF tasks and priority numbers most closely related to this chapter are:

1. Diagnose clutch noise, binding, slippage, pulsation, and chatter; determine necessary action. P-1
2. Inspect clutch pedal linkage, cables, automatic adjuster mechanism brackets, bushings, pivots, and springs; perform necessary action. P-1

continued

Certification Connection *continued*

3. Inspect hydraulic clutch slave and master cylinders, lines, and hoses; determine necessary action. P-1
4. Inspect release (throw-out) bearing, lever, and pivot; determine necessary action. P-1
5. Inspect and replace clutch pressure plate assembly and clutch disc. P-1
6. Bleed clutch hydraulic system. P-1
7. Inspect, remove, or replace pilot bearing or bushing (as applicable). P-1
8. Inspect flywheel and ring gear for wear and cracks; determine necessary action. P-1
9. Inspect engine block, clutch (bell) housing, transmission/transaxle case mating surfaces, and alignment dowels; determine necessary action. P-3
10. Measure flywheel runout and crankshaft endplay; determine necessary action. P-3

Figure 39–2 shows the basic principle of engaging a clutch. The left side is considered the drive member, or input, to the clutch. It is composed of the flywheel and **pressure plate**. The output is the center driven member, or **friction disc**. The output of this shaft drives the manual transmission. When the pressure plate is withdrawn, the engine input can revolve freely and is disconnected from the driven member and the transmission. When the pressure plate moves in the direction of the arrows, however, the friction disc is squeezed between the two and forced to turn at the same speed as the input, or drive member. ***Figure 39–3*** shows the actual components including the flywheel, pressure plate, and fiction disc.

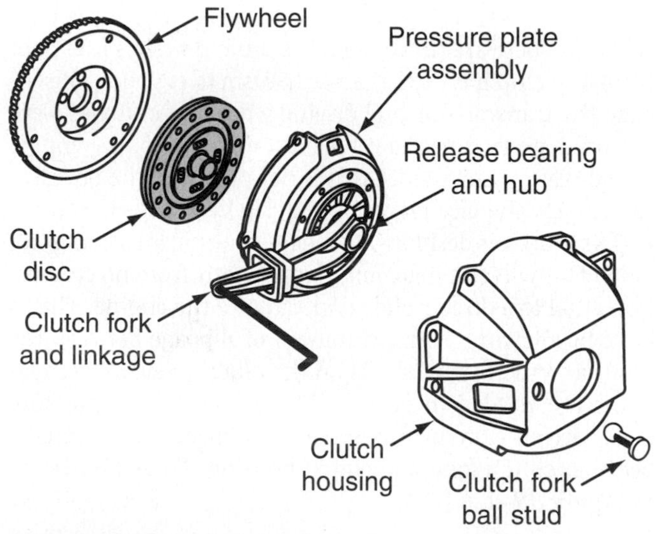

FIGURE 39–1 The main parts of a clutch.

PRESSURE PLATE ASSEMBLY

The pressure plate is designed to squeeze or clamp the friction disc between itself and the flywheel. The pressure plate has several components (***Figure 39–4***). They include the cover, a series of springs, release fingers, and the pressure plate. The pressure plate and the flywheel have smooth machined surfaces to clamp the clutch disc. The pressure plate cover is bolted to the flywheel and turns at exactly the same speed as the flywheel. The springs are used to squeeze or clamp the friction disc between the two plates. Several types

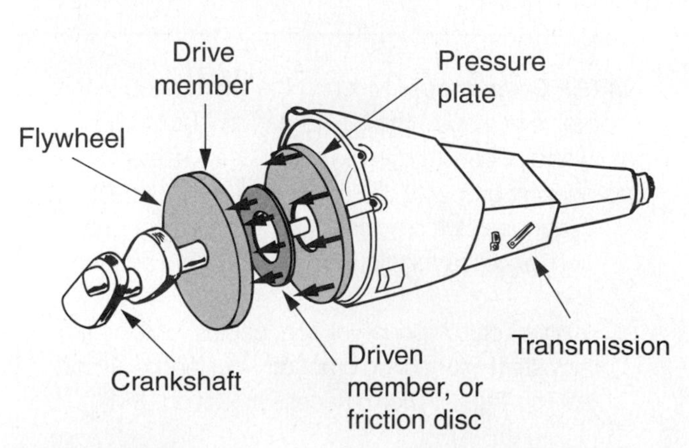

FIGURE 39–2 When the clutch is engaged, the driven member is squeezed between the two drive members. The transmission is now connected to the engine.

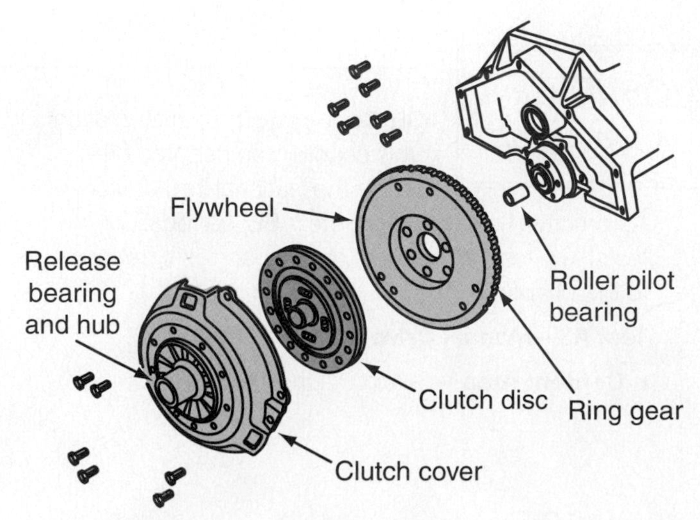

FIGURE 39–3 The components of a typical clutch system.

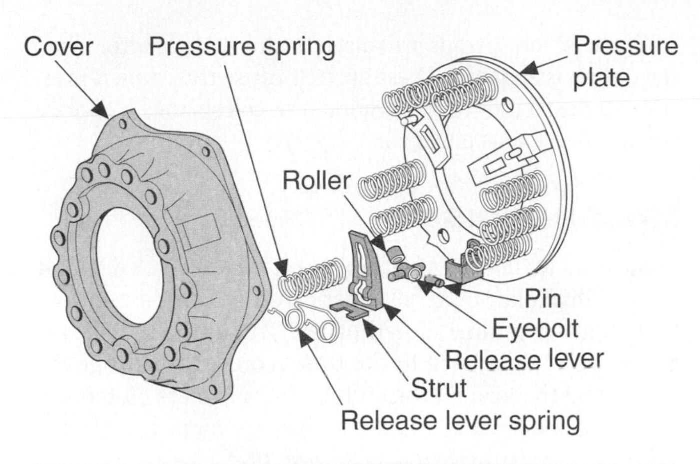

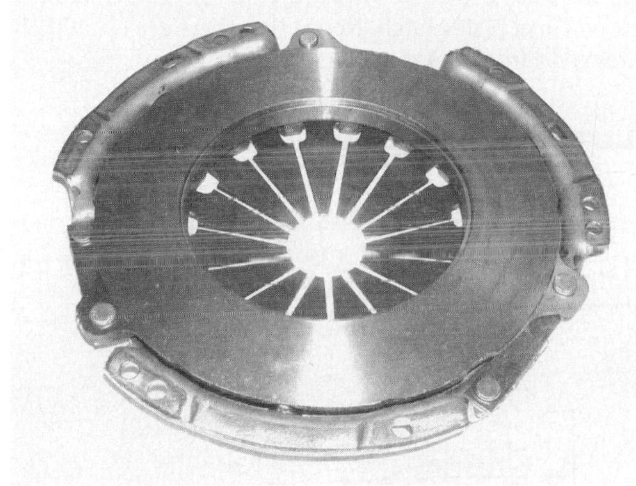

FIGURE 39-4 The pressure plate is constructed of several parts.

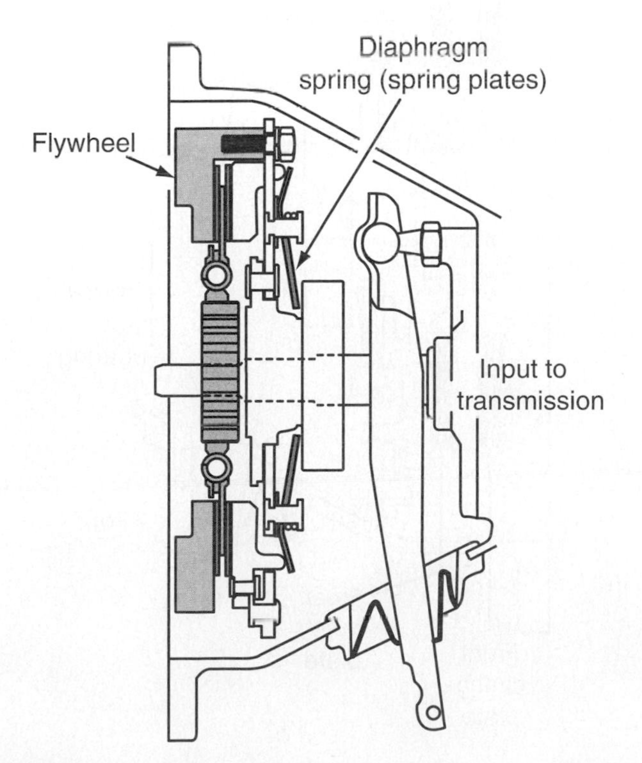

FIGURE 39-5 The diaphragm spring is also used to help force the pressure plate against the clutch driven disc.

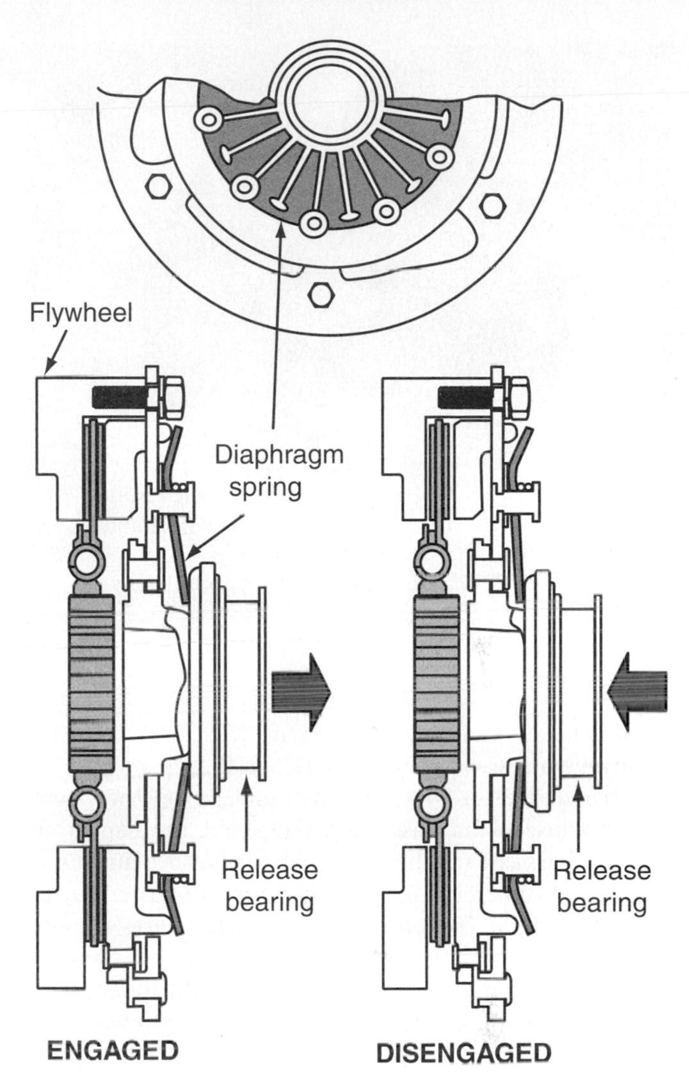

FIGURE 39-6 All clutches must have some type of linkage or retracting (diaphragm) springs to engage and disengage the clutch.

of springs are used. The **diaphragm spring** shown in ***Figure 39–5*** is the type often used. It consists of a series of spring plates used to force or clamp the friction disc.

Pressure plates also use a linkage to engage or disengage the friction disc (***Figure 39–6***). When the release bearing and diaphragm springs are pushed in to the left, the pressure plate surface is moved to the right. This action disengages the friction disc. When there is no force on the release levers, the diaphragm springs force the pressure plate to tighten down and squeeze the friction disc to make it rotate.

FRICTION DISC

The friction disc is the output of the clutch system. When the clutch assembly spins, it drives the manual transmission. The clutch is made of several parts. These are shown in ***Figure 39–7***. The center of the clutch has a spline hole for the transmission input shaft connection. The grooves on both sides of the clutch disc lining prevent sticking of the plate to the flywheel and pressure plate. Frictional facings

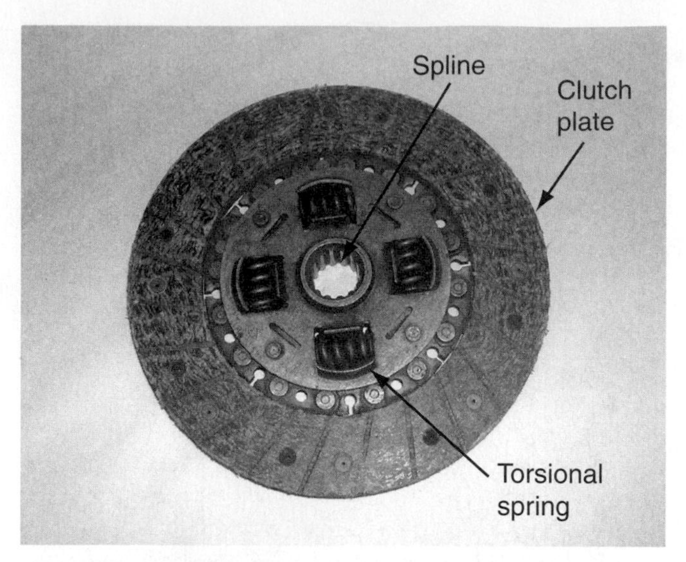

FIGURE 39–7 The clutch disc has torsional springs to eliminate jerky starting during engagement.

are attached to each side of the clutch disc. These are made of various materials. Several years ago they were made of cotton and asbestos fibers, woven or molded together. However, because of increased concern about the dangerous effects of **asbestos**, other materials are also being used. Fiberglass is now gaining in popularity as one material used on friction discs. On some clutches, copper wires are also woven into the material for additional strength.

The clutch has a flexible center to absorb the **torsional vibration** of the crankshaft. Steel compression springs permit the disc to rotate slightly in relation to the pressure plate.

Car Clinic: CLUTCH SURFACE

CUSTOMER CONCERN:
Clutch Scratching Sound

A vehicle has a standard transmission with a clutch. The driver says that a harsh scratching sound is heard when the clutch pedal is released. What could be the problem?

SOLUTION: When a clutch is used excessively, often it will wear down. If it wears down far enough, rivets that hold the clutch material to the clutch frame will begin to scratch the clutch surface area on the flywheel. When this happens, the rivets will damage the clutch surface area and make it much rougher. When this happens, the clutch abrasive material wears even farther, causing more wear on the flywheel. In this case the flywheel will need to be replaced or possibly machined to a smooth surface again and the clutch disc replaced.

The cushion springs are raised to eliminate chatter when the clutch is engaged. These springs cause the contact pressure on the facings to rise gradually as the springs compress when the clutch is engaged.

MULTIPLE DISC CLUTCHES

Some vehicles use two clutch plates. These are called double plate or dual plate clutches. There are also some clutch systems that have three clutch plates. The number of clutch plates is largely related to the time required to engage the clutch and the load on the clutch. A single plate clutch will engage sooner. A dual plate clutch takes more time to engage and carry heavier loads. ***Figure 39–8*** shows a cutaway view of a dual plate clutch. Notice that there are two clutch plates and a front and rear pressure plate.

CLUTCH RELEASE BEARING (THROW-OUT BEARING)

The clutch release bearing, also called a throw-out bearing, is a ball-thrust bearing held within the clutch housing. The

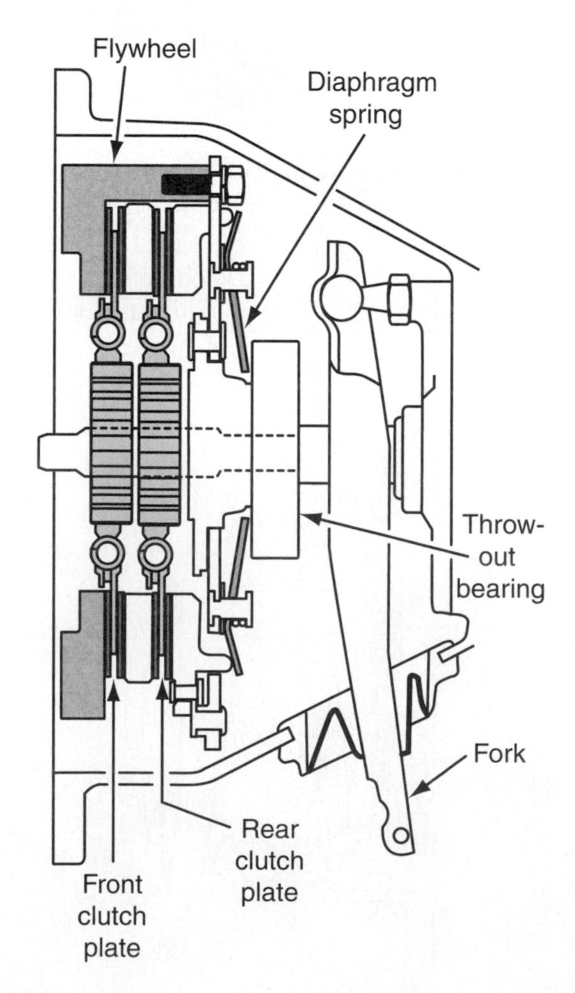

FIGURE 39–8 A dual plate clutch has two clutch discs plus a front and rear pressure plate.

CUSTOMER CONCERN:

Clutch Chatter

An older vehicle with about 70,000 miles on it has a standard or manual transmission. When the clutch pedal is released, a noticeable chatter is felt in the vehicle. What could be the problem?

SOLUTION: Clutch chatter is most often caused by oil getting onto the clutch and pressure plate surface. The friction and heat of the clutch against the flywheel and pressure plate causes the oil to become very sticky. This causes the chatter when the clutch pedal is released and the clutch tries to engage. Most often the oil comes from a leaky rear main seal on the engine. Replace the rear main seal on the engine to eliminate this problem.

bearing is moved by the clutch pedal and linkage to engage or disengage the pressure plate. When the clutch pedal is pressed down, the clutch release bearing pushes against the revolving pressure plate release levers. This action disengages the clutch. When the clutch pedal is released, the clutch release bearing moves back. The springs in the pressure plate now engage the clutch disc (***Figure 39–9***).

CLUTCH LINKAGE

The purpose of the clutch linkage is to engage and disengage the clutch. As the operator moves the clutch pedal down, the clutch release bearing pushes against the release levers to disengage the clutch. Depending on the design of the vehicle, several clutch linkages are used. These include the rod and lever, the cable type, and the hydraulic linkage. A simplified rod and lever system is shown in ***Figure 39–10***. As the clutch is pushed in, the bottom of the clutch fork moves to the left. Note the pivot points. When the bottom of the clutch fork moves to the left, the

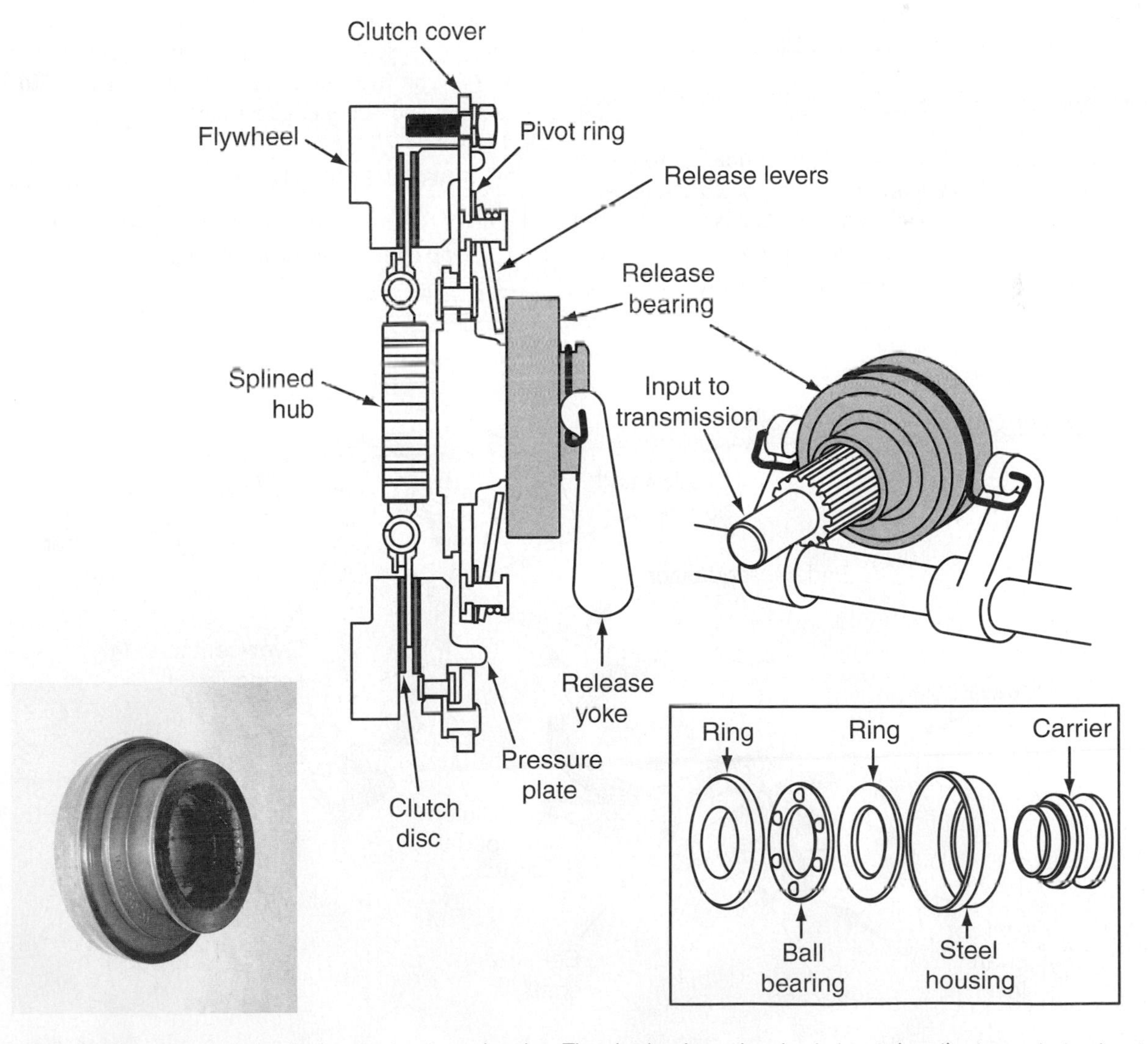

FIGURE 39-9 The main elements in the clutch release bearing. The clutch release bearing is located on the transmission input shaft sleeve.

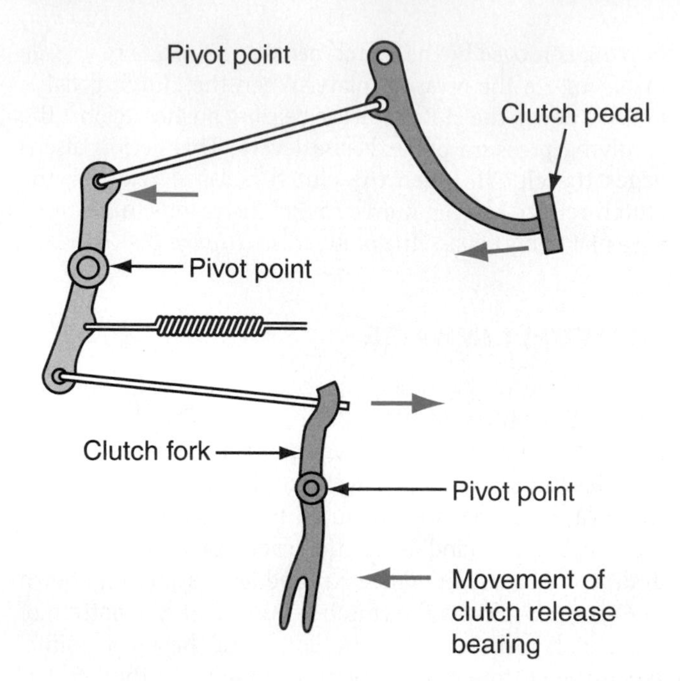

FIGURE 39-10 This rod and lever system is used to engage and disengage the clutch.

clutch release bearing disengages the clutch from the pressure plate.

In an actual clutch linkage system, similar movements occur. Referring to ***Figure 39–11***, when the clutch pedal is pushed in to disengage the clutch, the pedal is moved in the direction shown at point A. This movement causes the pedal-to-equalizer rod to move in the direction shown at arrow B. Again, this movement causes the equalizer shaft lever to move in the direction shown at arrow C. The end result is that the release lever moves as shown at the arrow at point D, disengaging the clutch.

The cable type of linkage uses a flexible cable connected from the clutch pedal to the clutch fork. As the clutch pedal is pressed down, the flexible cable forces the clutch fork to move. ***Figure 39–12*** shows how the cable is attached to the clutch fork.

Hydraulic Clutch Linkage The hydraulic system consists of master cylinder and an actuator called the **slave cylinder**. When pressure is applied to the clutch pedal, hydraulic pressure is built up in the master cylinder. The pressure is sent through hydraulic tubing to the slave cylinder. Here the pressure is used to move the clutch fork to engage or disengage the clutch disc. ***Figure 39–13*** shows the components of a hydraulic clutch linkage system. The system is similar to a braking system in that it has a master cylinder, a pedal to operate the hydraulic system, a slave cylinder, and a reservoir to hold extra hydraulic fluid. In operation, when the clutch pedal is depressed, it produces a hydraulic pressure in the hydraulic lines. The hydraulic pressure is sent to the slave cylinder shown in ***Figure 39–14***. The hydraulic pressure is then used to move the clutch in and out of engagement.

PARTS LOCATOR

Refer to photo #78 in the front of the textbook to see what a ***clutch linkage*** *looks like.*

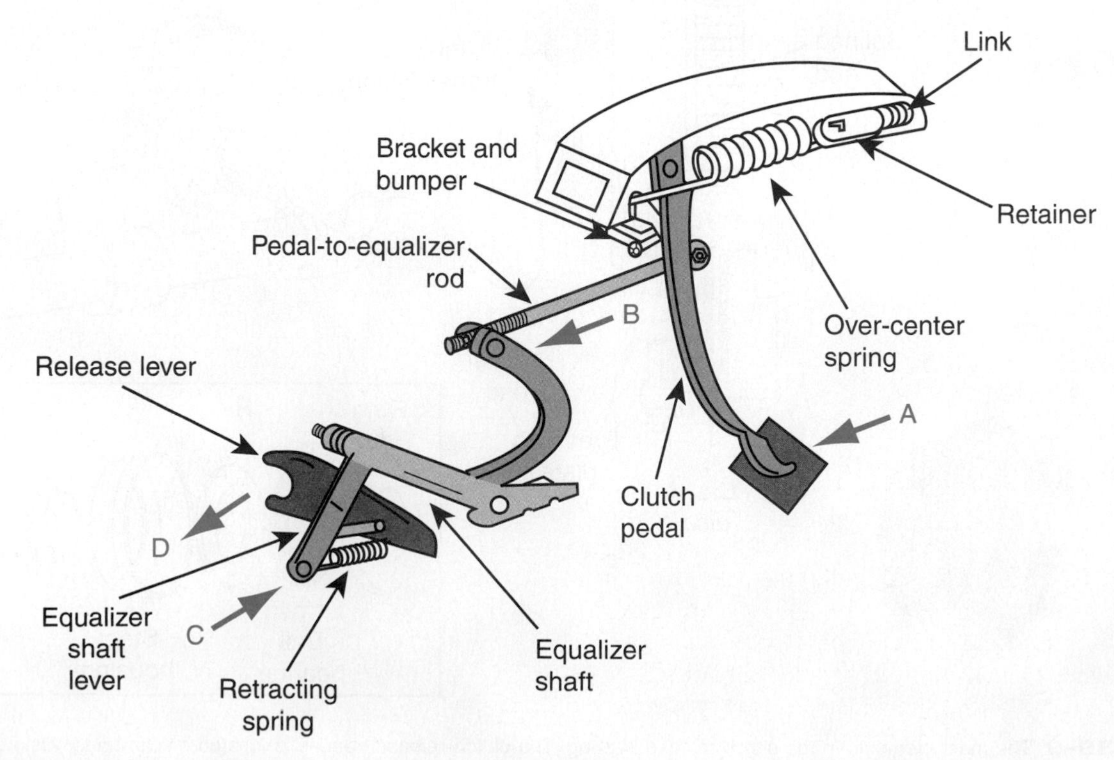

FIGURE 39-11 The clutch mechanism, along with the direction of movement when disengaging the clutch.

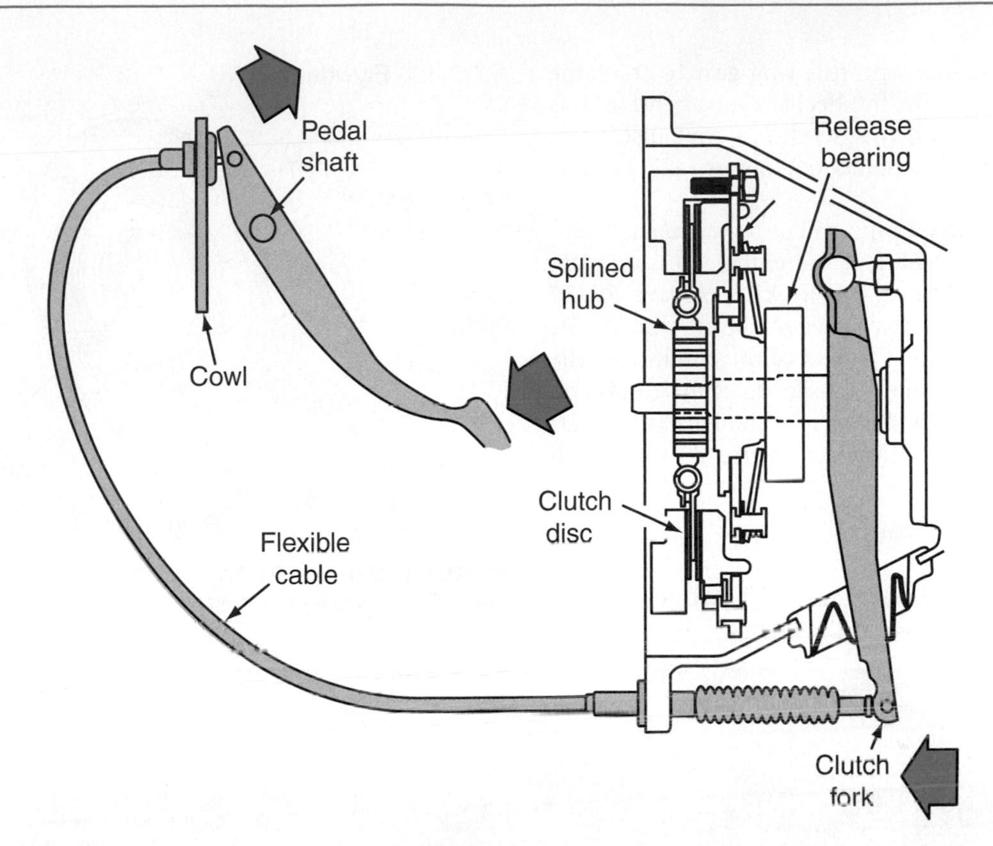

FIGURE 39-12 A flexible cable can be used to disengage the clutch.

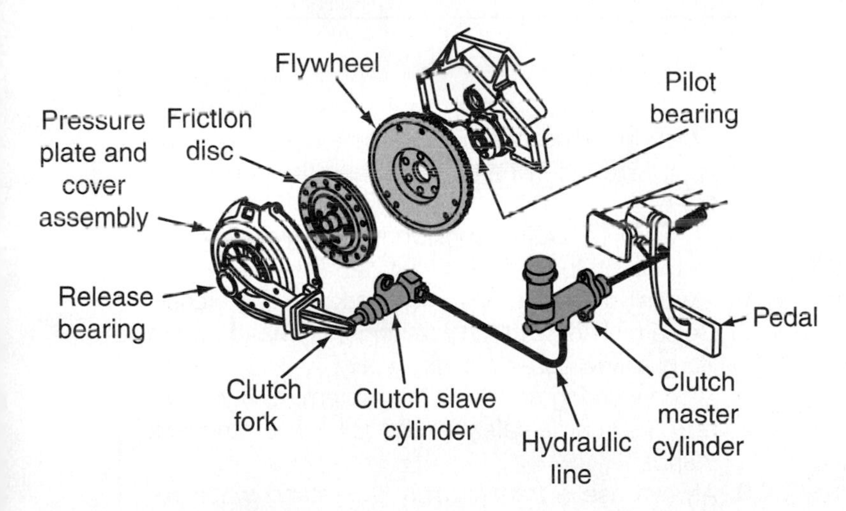

FIGURE 39-13 Clutch linkage can also be designed as a hydraulic system.

39.2 FLYWHEEL DESIGN

The flywheel is attached to the engine and spins at the same speed as the engine crankshaft. It is normally made of cast iron. There is a percentage of graphite mixed in with the metal. The graphite is used to help lubricate the clutch against the flywheel. A smooth surface is machined into the flywheel where the clutch surface touches the flywheel.

The flywheel has a ring gear attached to its circumference. This ring gear, which can be replaced, is used in the starting system. It can also have timing marks or notches

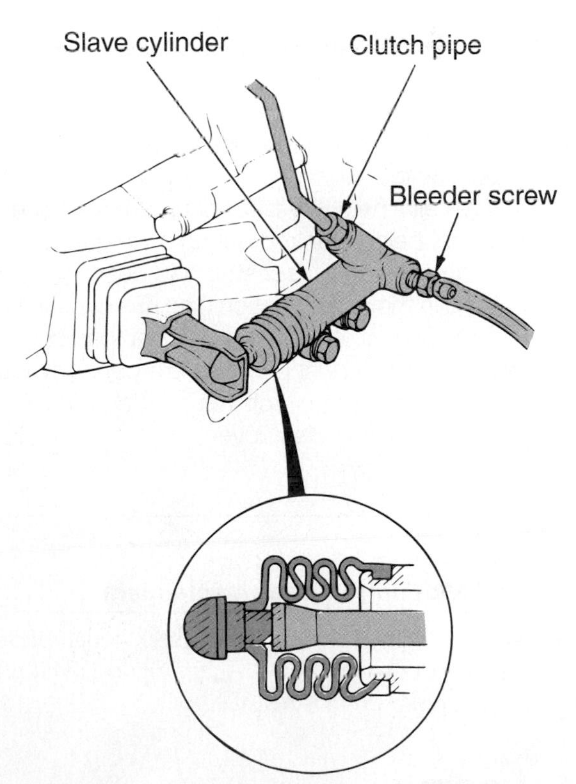

FIGURE 39-14 This slave cylinder is used in the clutch hydraulic system to engage and disengage the clutch. (Courtesy of American Honda Motor Co., Inc.)

on it. The starter meshes with this ring gear to crank the engine during starting. The flywheel is also very heavy. It is designed to be heavy to help smooth out any pulses from the four-cycle intake, compression, power, and exhaust strokes.

The flywheel is bolted to the end of the crankshaft with several large bolts. In addition, the clutch pressure plate and cover are also bolted to the flywheel. ***Figure 39–15*** shows how the flywheel is positioned on the crankshaft. The eight center bolts hold the flywheel to the crankshaft. The bolt holes and three alignment pins for the clutch and pressure plate are also shown. A ring gear holder has been attached to help hold the flywheel secure during clutch installation.

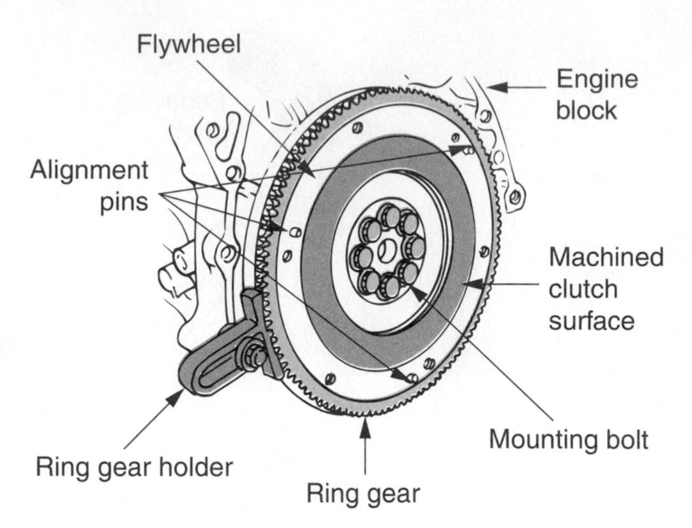

FIGURE 39–15 The flywheel is bolted to the end of the crankshaft. (Courtesy of American Honda Motor Co., Inc.)

Problems, Diagnosis, and Service

Safety Precautions

1. When working under the vehicle on the clutch, make sure that no oil or antifreeze has been spilled. You may slip and seriously injure yourself.
2. Always disconnect the battery cables from the battery before working on any part of the clutch. This will eliminate any possibility of the engine accidentally being cranked during maintenance.
3. On many vehicles, asbestos has been used as a material in the clutch friction disc. Be careful not to breathe any of the asbestos dust particles into your lungs. The dust particles are generally found in and around the clutch housing and inside the clutch and friction disc cover.
4. When installing clutches, be careful not to pinch your fingers between the part and the housing or flywheel.
5. Always wear OSHA-approved safety glasses when working on the clutch.
6. Always use proper tools when working with clutches.
7. Be careful not to crush your fingers or hands when lifting a flywheel as it is very heavy.
8. When working on the clutch system, always support the vehicle safely on the hoist or on the jack stands.
9. Always use a transmission jack stand when removing a transmission. The transmission is very heavy and can easily be dropped.

PROBLEM: Faulty Clutch Mechanism

As the vehicle is accelerated and the clutch pedal released, the clutch seems to be slipping. In addition, as the clutch is engaged, a scraping sound is heard.

DIAGNOSIS

This problem is usually caused by several defective components. The clutch disc may slip because:

1. It is worn down and cannot be adjusted any further.
2. The clutch linkage is not adjusted correctly.

The scraping sound may come from two sources:

1. The clutch release (throw-out) bearing may be damaged. This is normally damaged by constantly applying a slight pressure to the clutch pedal. This causes the clutch release bearing to spin constantly.
2. The clutch disc may be worn so badly that the rivets are scraping on the clutch pressure plate or on the fly-

CLUTCH TROUBLE DIAGNOSIS		
PROBLEM	DIAGNOSIS	SERVICE
Fails to release (pedal pressed to floor)	• Improper linkage adjustment • Improper pedal travel • Loose linkage or worn cable • Faulty pilot bearing • Faulty drive disc • Fork off ball stud • Clutch hub binding on spline • Clutch disc warped • Pivot rings loose, or damaged	• Adjust linkage • Trim bumper and adjust • Replace as necessary • Replace bearing • Replace disc • Install properly and lube • Repair or replace gear or disc • Replace disc • Replace cover and pressure plate
Slipping	• Improper adjustment • Oil soaked driven disc • Worn or damaged facing • Warped flywheel or pressure plate • Weak diaphragm spring • Drive plate not seated in • Driven plate overheated	• Adjust linkage • Install new disc and correct leak • Replace disc • Replace flywheel or pressure plate • Replace pressure plate • Make 30 to 40 normal starts • Allow to cool
Grabbing (chattering)	• Oil on facing • Worn spine on clutch • Loose engine mounts • Warped flywheel or pressure plate • Resin on flywheel or pressure plate	• Install new disc and correct leak • Replace clutch • Tighten or replace mounts • Replace flywheel or pressure plate • Sand off or replace component
Rattling or transmission click	• Weak retracting springs • Release fork loose • Oil in drive plate damper • Driven plate damper spring bad	• Replace pressure plate • Check ball stud and retainer • Replace driven disc • Replace driven disc
Release bearing noise (clutch fully engaged)	• Improper adjustment • Bearing binding on retainer • Insufficient fork spring tension • Fork improperly installed • Weak return spring	• Adjust linkage • Clean and lubricate • Replace fork • Install properly • Replace spring
Noisy	• Worn release bearing • Fork off ball stud • Pilot bearing loose	• Replace bearing • Install properly and lube • See bearing fits section
Pedal stays on floor when disengaged	• Bind in cable • Weak pressure plate springs • Springs over traveled	• Replace cable • Replace pressure plate • Adjust linkage
Hard pedal effort	• Bind in linkage • Worn driven plate • Friction in cable	• Lube and free linkage • Replace driven plate • Replace cable

FIGURE 39–16 Use this chart to determine the cause of various clutch problems.

wheel. ***Figure 39–16*** shows a clutch trouble diagnostic chart.

If there are other problems with the clutch system, refer to this chart for the condition, probable cause, and correction.

SERVICE

If the clutch linkage is correctly adjusted, the clutch and related components will need to be removed, serviced, and replaced if necessary. Each clutch system to be disassembled uses a different procedure. Refer to the service manual for the correct procedure. Use the following general procedure to remove and disassemble the clutch and related components.

1. Disconnect the battery ground cable.
2. Using the proper safety precautions, position the car so that the transmission can be removed.
3. Remove the drive shaft(s).

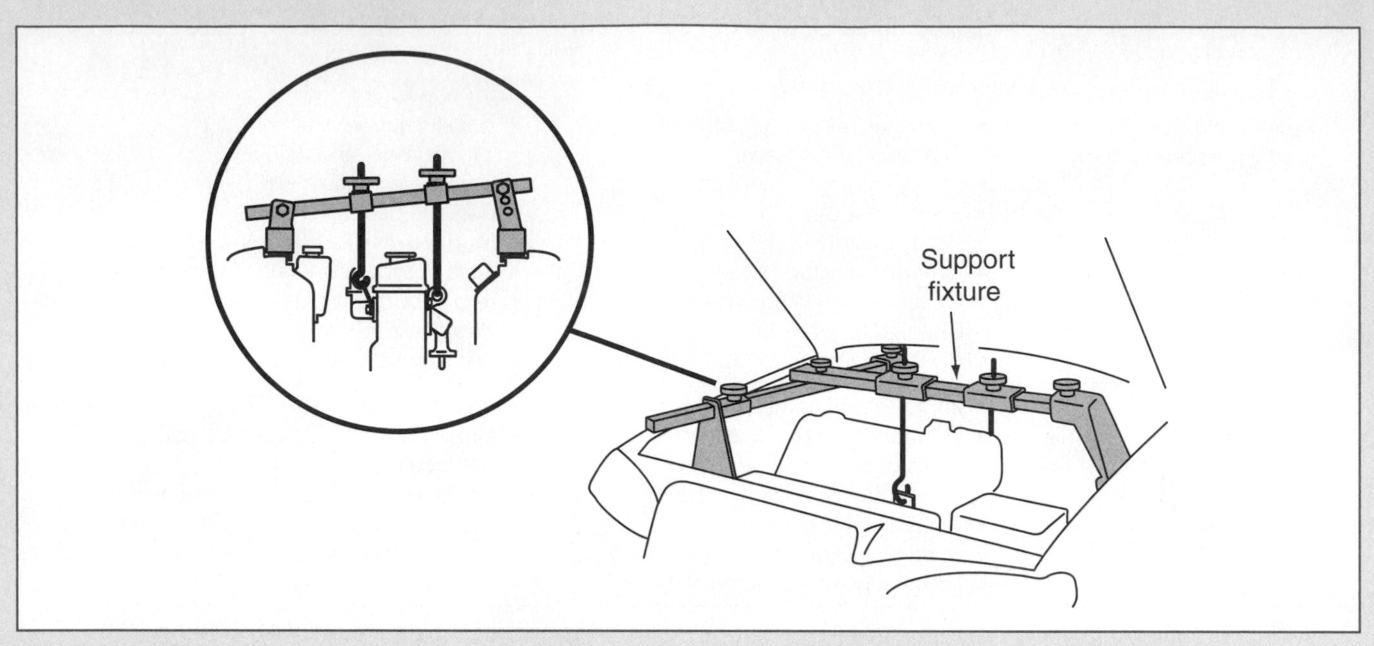

FIGURE 39–17 When a manual transmission is removed, the engine may need to be supported by special engine support fixtures.

4. Check to see if the rear of the engine is supported by the transmission. If it is not, the engine must be supported while the transmission is being worked on. ***Figure 39–17*** shows an engine support used to support the rear of the engine.
5. Drain the gear oil from the transmission.
6. Disconnect the linkage used to shift the transmission.
7. Remove the speedometer cable from the transmission.
8. Remove any electrical wires attached to the transmission and mark for reassembly.
9. Support the transmission with a transmission jack stand.
10. Remove the bolts holding the transmission to the clutch housing. The transmission should now be easily removed and set on the workbench.
11. Disconnect the clutch linkage from the clutch fork.
12. At this point, the clutch release bearing should be in a position to be removed easily.
13. Remove the bolts holding the clutch housing to the engine block.
14. At this point, the clutch pressure plate assembly should be observable.
15. Insert the clutch plate alignment tool into the clutch plate and crankshaft to support its weight when the bolts are removed. Using the correct wrench size, remove the clutch pressure plate.

CAUTION: *The pressure plate is very heavy. Be careful not to drop the pressure plate on your feet. Also, use gloves to protect your fingers.*

The clutch disc should now be loose and easily removed.

CAUTION: *Be careful not to breathe any of the asbestos particles from the clutch disc.*

16. Check the condition of the clutch release (throw-out) bearing. If there is any restriction while turning the bearing, replacement will be necessary. The bearing should turn smoothly.
17. Check the following conditions of the clutch disc.
    a. First check to see if the clutch is worn beyond specifications. Using a vernier caliper, check the thickness of the clutch disc as shown in ***Figure 39–18***. If the thickness is less than the service limit, replace the clutch disc.

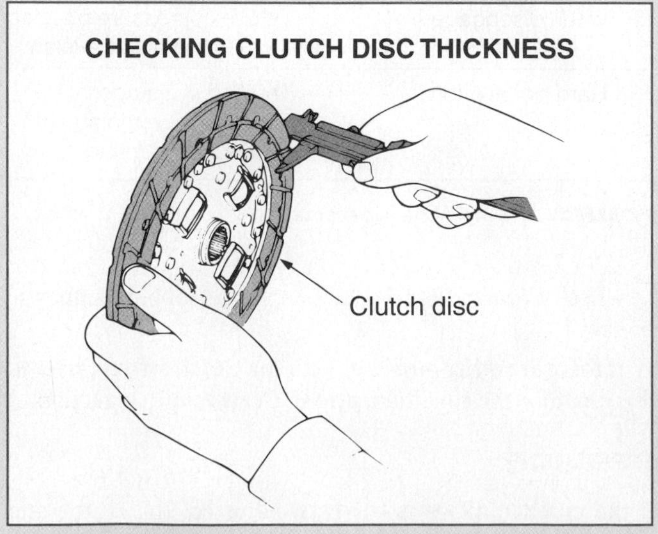

FIGURE 39–18 The clutch disc can be checked for thickness using a vernier caliper. (Courtesy of American Honda Motor Co., Inc.)

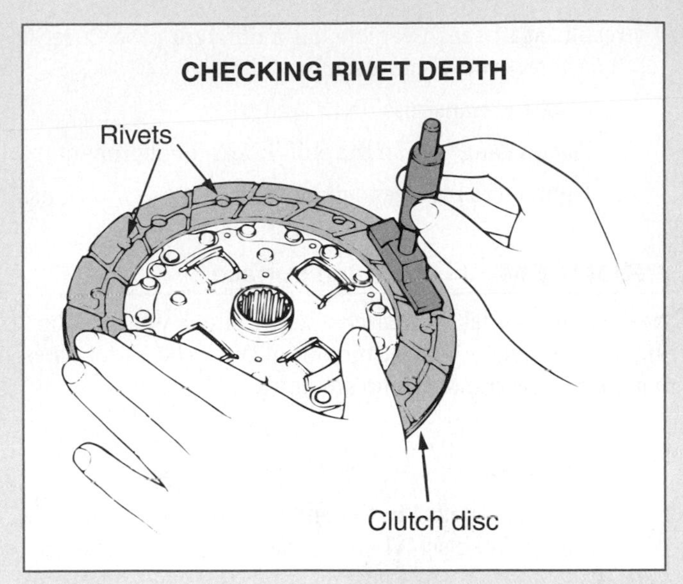

FIGURE 39-19 The depth of the rivets can be checked for thickness using a depth micrometer. (Courtesy of American Honda Motor Co., Inc.)

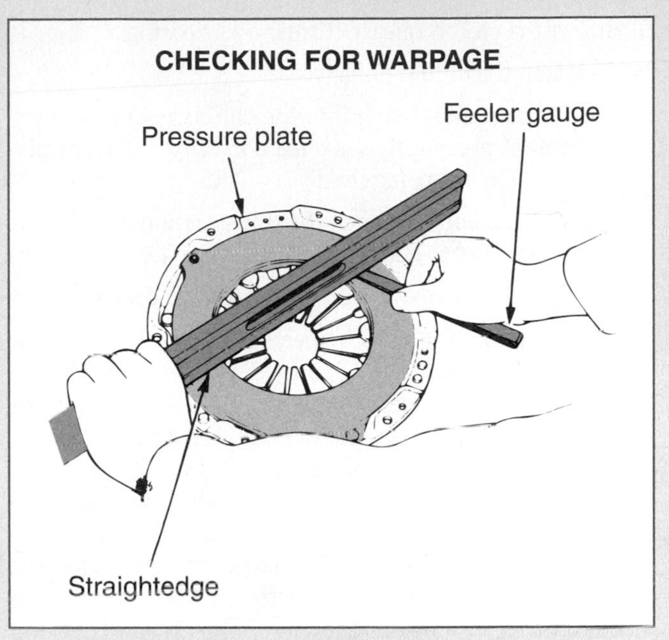

FIGURE 39-20 The pressure plate can be checked for warpage by using a straightedge. (Courtesy of American Honda Motor Co., Inc.)

b. Use a depth gauge to check the depth of the rivets as shown in ***Figure 39–19***. The measurement should be from the lining surface to the rivets, on both sides.

18. Check the condition of the pressure plate and flywheel surface. If the friction disc is worn too thin, it may score and scratch the pressure plate and flywheel surfaces. This will cause grooves to be cut into the pressure plate. If they are scored and scratched, the pressure plate and flywheel will have to be machined or replaced.

    Also check the pressure plate for warpage as shown in ***Figure 39–20***. Use a straightedge and measure clearances under the straightedge with a feeler gauge. If the warpage is more than the service limit, replace the pressure plate.

19. Check for excessive wear or bent diaphragm spring fingers when the clutch is disassembled. If the spring fingers are worn, the pressure plate must be replaced.

20. Clean all parts and begin reassembly.

21. Attach the friction disc and pressure plate to the flywheel. The friction disc and pressure plate must be correctly aligned before installing the transmission. Use the following procedure to align the disc correctly.

    a. Place the friction disc in the correct position on the flywheel.

    b. Using the correct bolts, attach the pressure plate. However, do not tighten the bolts at this point. The clutch disc must still be loose enough to be moved slightly for alignment.

    c. Using the correct alignment tool (it should look the same as the transmission mainshaft), align the pressure plate assembly and clutch disc to the flywheel. The alignment tool positions the disc with the flywheel so that the transmission mainshaft can be inserted easily into the clutch disc and flywheel. ***Figure 39–21*** shows how the alignment tool will position the friction disc with the pressure plate. Make sure the end of the alignment tool is completely inserted into the end of the crankshaft before tightening the pressure plate bolts.

    d. Using the correct torque specifications, tighten the pressure plate bolts.

    e. Remove the alignment tool.

22. Replace the flywheel housing and torque to specifications.

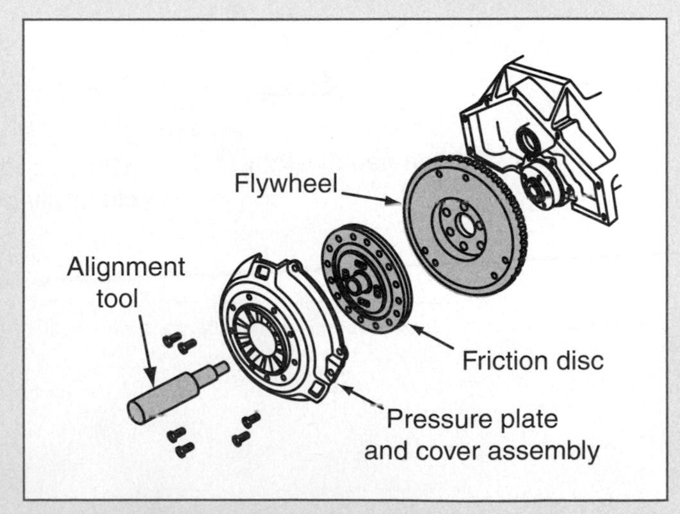

FIGURE 39-21 Before tightening the pressure plate, insert the correct alignment tool. This will help align the pressure plate, friction disc, and flywheel correctly so that the transmission can be inserted easily.

23. Insert the clutch release (throw-out) bearing.
24. Reattach the clutch linkage.
25. Insert the transmission into the clutch assembly. If the alignment procedure was done correctly, the transmission will slip into place easily.
26. Tighten the bolts between the transmission and the flywheel housing to stated torque specifications.
27. Replace all linkages, wires, driveshafts, and so on.
28. At this point the clutch mechanism needs to be adjusted. If the clutch linkage is adjusted incorrectly, the clutch may not engage completely, and it will slip or be hard to shift. Each vehicle has a different procedure for adjusting the clutch mechanism. ***Figure 39–22*** illustrates several common measurements taken. Refer to the manufacturer's specifications and the suggested procedures to adjust clutch linkage. In most cases, the clutch linkage is adjustable. Often the adjustment can be located just before the clutch fork while checking the clutch pedal movement.
29. Adjust the shifting linkage according to manufacturer's specifications. Also make sure the linkage is not damaged or bent. If the linkage is not adjusted properly, the transmission may slip out of gear or hard shifting may result. Again, each vehicle has a different procedure. A typical procedure, however, includes:
    a. Place the transmission in neutral.
    b. Place a gauge pin in the shift levers for alignment.
    c. Tighten the linkage to manufacturer's specifications.

PROBLEM: Oil on Clutch Surface

As the clutch pedal is released and the clutch engaged during acceleration, a jerking motion occurs. The clutch does not engage the transmission smoothly.

DIAGNOSIS

If the rear main oil seal in the engine leaks oil, the oil may get on the friction disc. When this happens, the oil becomes very sticky and causes a jerking motion during release of the clutch and engagement of the transmission.

SERVICE

To repair, the rear main seal on the engine must be replaced. Refer to the manufacturer's specifications and procedures for details.

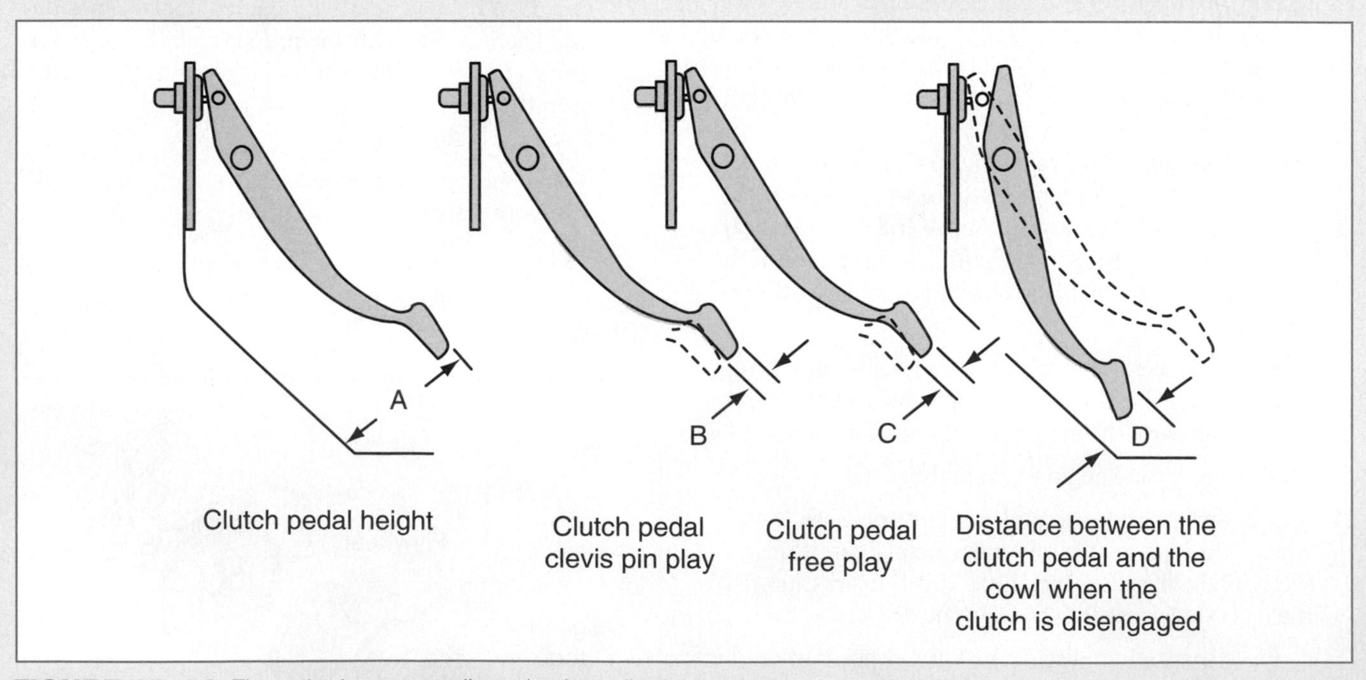

FIGURE 39–22 These checks are normally made when adjusting the clutch.

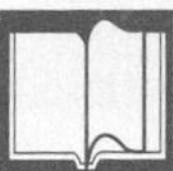

Service Manual Connection

There are many important specifications to keep in mind when working with clutch systems. To identify the specifications for your engine, you will need to know the VIN (vehicle identification number) of the vehicle, the type and year of the vehicle, and the type of engine. Although they may be titled differently, some of the common clutch system specifications (not all) found in the service manuals are listed below. Note that these specifications are typical examples. Each vehicle and engine may have different specifications.

Common Specification	Typical Example
Clutch pedal adjustment free play	7/8–1 1/8 inches
Clutch pedal height	9.5–8.7 inches
Clutch disc thickness service limit	.24 inch
Rivet depth	0.05–0.07 inch
Diaphragm spring fingers height service limit	0.03 inch
Pressure plate runout service limit	0.006 inch

In addition, the service manuals give specific directions for various service procedures. Some of the more common procedures include:

- Clutch alignment procedure
- Checking release bearing
- Inspecting release fork
- Clutch pedal adjustment

SUMMARY

The following statements will help to summarize this chapter:

- The purpose of the clutch is to engage and disengage the manual transmission.
- The clutch system is composed of the friction disc, the pressure plate, the clutch linkage, and the clutch release bearing.
- When the clutch is depressed, the linkage moves the clutch release bearing and releases the pressure plate so that the transmission becomes disengaged with the engine.
- When the operator releases the clutch pedal, the springs on the pressure plate squeeze the friction disc between the flywheel and the pressure plate, engaging the transmission again.
- When the friction disc is turning at the same speed as the engine, the transmission is engaged.
- Some vehicles use two friction discs to handle more load.
- On some vehicles, rather than using mechanical linkages, a flexible cable can be used to connect the clutch pedal to the clutch release bearing.
- Some vehicles use a hydraulic system with a slave cylinder to transfer the clutch pedal motion to the clutch release bearing.
- There are many service procedures that can be done on the clutch system, including linkage adjustment, clutch replacement, pressure plate wear or damage, damaged clutch release bearing, and damaged flywheel.
- Always follow the manufacturer's suggested service procedure when working on a clutch system.

TERMS TO KNOW

Can you explain each of the following terms? Review the chapter until you can use each term correctly.

Asbestos
Clutch
Diaphragm spring
Friction disc
Pressure plate
Slave cylinder
Torsional vibration

REVIEW QUESTIONS

Multiple Choice

1. The purpose of the clutch system is to:
   a. Start the engine
   b. Engage the crankshaft to the oil pump
   c. Engage the engine to the transmission
   d. Reduce the speed of the overdrive
   e. Increase the speed of the synchronizers

2. What device squeezes the friction disc against the flywheel?
   a. Torsional springs
   b. Clutch linkage
   c. Clutch release bearing
   d. Pressure plate
   e. Synchronizers

3. The clutch release bearing pushes against the ________ to disengage the friction disc.
   a. Pressure plate release levers
   b. Pressure plate springs and linkage
   c. Flywheel
   d. Transmission case
   e. Torsional springs

4. ________ are placed on the friction disc to eliminate a jerky or chatter motion when the clutch is engaged.
   a. Asbestos springs
   b. Torsional springs
   c. Splines
   d. Gears
   e. Pressure plates

5. Which device is used to push the clutch release bearing in and out of engagement?
   a. Clutch disc
   b. Torsional springs
   c. Clutch fork
   d. Synchronizers
   e. Pressure plate

6. If the clutch chatters or causes a jerky motion when engaged, the problem could be caused by:
   a. A leaky rear transmission seal
   b. Bad synchronizers
   c. Oil leaking from the engine onto the friction disc
   d. A broken extension housing
   e. A bad clutch release bearing

7. A ________ is used on a hydraulic clutch linkage, to push the clutch fork.
   a. Hydraulic bolt
   b. Slave cylinder
   c. Guide pin
   d. All of the above
   e. None of the above

The following questions are similar in format to ASE (Automotive Service Excellence) test questions.

8. *Technician A* says that if oil leaks on the clutch disc, the clutch will slip during engagement. *Technician B* says that if oil leaks on the clutch disc, the clutch will chatter during engagement. Who is correct?
   a. A only c. Both A and B
   b. B only d. Neither A nor B

9. When adjusting the clutch mechanism, *Technician A* says to adjust the throw-out bearing. *Technician B* says to adjust the flywheel. Who is correct?
   a. A only c. Both A and B
   b. B only d. Neither A nor B

10. A vehicle has a grinding noise during clutch engagement. *Technician A* says the problem is most likely the synchronizer. *Technician B* says the problem is a badly worn clutch scraping against the pressure plate. Who is correct?
    a. A only c. Both A and B
    b. B only d. Neither A nor B

11. *Technician A* says that an important check on a clutch disc is its thickness. *Technician B* says that a clutch can be checked for circular dimension. Who is correct?
    a. A only c. Both A and B
    b. B only d. Neither A nor B

12. A clutch seems to be scraping when being engaged. *Technician A* says that the rivets may be touching the pressure plate. *Technician B* says that the rivets may be touching the flywheel surface. Who is correct?
    a. A only c. Both A and B
    b. B only d. Neither A nor B

13. *Technician A* says that the pressure plate can be checked for runout. *Technician B* says that the pressure plate should be checked for the correct amount of lubricant. Who is correct?
    a. A only c. Both A and B
    b. B only d. Neither A nor B

Essay

14. Explain how a hydraulic clutch linkage system operates.

15. What is the purpose of a pressure plate?

16. Define torsional vibration.

17. What is the purpose of the clutch release bearing?

Short Answer

18. Asbestos is being replaced with ________ on clutch discs.

19. A scraping sound during clutch engagement may be caused by ________.

20. Oil on the clutch disc and pressure plate will cause the vehicle to ________ when the clutch is engaged.

Applied Academics

TITLE: Coefficient of Friction

ACADEMIC SKILLS AREA: Science

NATEF Automobile Technician Tasks:

The service technician can explain the function of friction on the clutch to transmit power to the transmission.

CONCEPT:

The term *coefficient of friction* is important to the operation of any clutch system. The coefficient of friction is a measure of how much friction is produced by a particular material. For example, say that a flywheel is placed on top of a metal workbench as shown in the figure that follows. It takes 5 pounds of pressure to push it across the metal countertop. Now put a rubber mat on the workbench and place the flywheel on top of the mat. To push the flywheel now takes 20 pounds of pressure. The rubber mat has more grip and has a higher coefficient of friction than the metal workbench. When related to clutch materials, a higher coefficient of friction means that it will take more load on the clutch to cause it to slip.

There are several types of friction materials used on clutches to maximize the coefficient of friction. The materials are also designed to dissipate heat rapidly. Common materials are categorized as organic, Kevlar composites, bronze metallic, and sintered iron. The exact type will depend on the application, the load, and the need to dissipate heat at a certain rate.

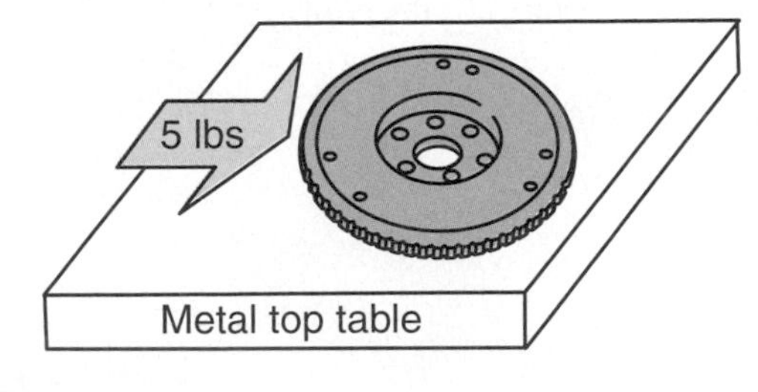

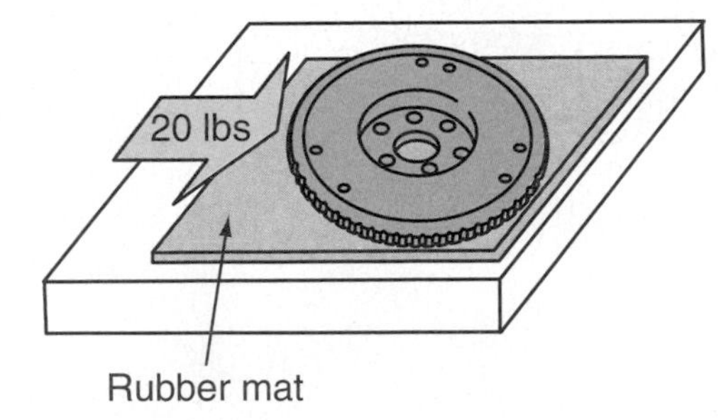

APPLICATION:

Why is the coefficient of friction important with clutches? As the heat increases in clutch materials, how do you think the coefficient of friction will change? How would a clutch material that has a greater coefficient of friction affect the smoothness when the clutch pedal is released?

Manual Transmissions

CHAPTER 40

OBJECTIVES

After studying this chapter, you should be able to:

- Define the purpose of the standard, or manual, transmission.
- Analyze the purpose of different gear ratios.
- Describe the operation and gear selection of the manual transmission.
- State the purpose and operation of synchronizers.
- Identify transmission lubricants.
- Compare different types of transmissions.
- Identify the parts and operation of transaxle systems.
- Describe the operation of linkages and accessories used on manual transmissions.
- State common problems, their diagnosis, and service suggestions pertaining to manual transmissions.

INTRODUCTION

Torque that is produced at the end of the crankshaft by the engine must be transmitted to the driving wheels. To accomplish this, torque must first pass through the clutch and transmission. This chapter describes manual transmissions, transaxles, their parts, and their service.

40.1 DESIGN OF TRANSMISSIONS

PURPOSE OF THE TRANSMISSION

The purpose of a transmission is to apply different torque forces to the driving wheels. Vehicles are required to perform under many types of loads. Stopping and starting,

Certification Connection

ASE Connection: The information in this chapter can help you prepare for the National Institute for Automotive Service Excellence (ASE) certification tests. The tests and content areas most closely related to this chapter are:

Test A3—Manual Drivetrain and Axles

- **Content Area**—Transmission Diagnosis and Repair
- **Content Area**—Transaxle Diagnosis and Repair

NATEF Connection: Much of the information in this chapter is related to the NATEF tasks. The NATEF tasks and priority numbers most closely related to this chapter are:

1. Remove and reinstall transmission/transaxle. P-1
2. Disassemble, clean, and reassemble transmission/transaxle components. P-2
3. Inspect transmission/transaxle case, extension housing, case mating surfaces, bores, bushings, and vents; perform necessary action. P-3
4. Diagnose noise, hard shifting, jumping out of gear, and fluid leakage concerns; determine necessary action. P-2
5. Inspect, adjust, and reinstall shift linkages, brackets, bushings, cables pivots, and levers. P-2
6. Inspect and reinstall powertrain mounts. P-3
7. Inspect and replace gaskets, seals, and sealants; inspect sealing surfaces. P-2

continued

Certification Connection *continued*

8. Remove and replace transaxle final drive. P-3
9. Inspect, adjust, and reinstall shift cover, forks, levers, grommets, shaft sleeves, detent mechanism, interlocks, and springs. P-2
10. Measure end play or preload (shim or spacer selection procedure) on transmission/transaxle shafts; perform necessary action. P-2
11. Inspect and reinstall synchronizer hub, sleeve, keys (inserts), springs, and blocking rings. P-1
12. Inspect and reinstall speedometer drive gear, driven gear, vehicle speed sensor (VSS), and retainers. P-2
13. Diagnose transaxle final drive assembly noise and vibration concerns; determine necessary action. P-3
14. Remove, inspect, measure, adjust, and reinstall transaxle final drive pinion gears (spiders), shaft, side gears, side bearings, thrust washers, and case assembly. P-2
15. Inspect lubrication devices (oil pump or slingers); perform necessary action. P-3
16. Inspect, test, and replace transmission/transaxle sensors and switches. P-1

heavy loads, high speeds, and small loads are examples of the different demands placed on the vehicle. The transmission is designed to change the torque applied to the driving wheels for different applications. In addition, the transmission is used to reverse the vehicle direction and to provide neutral (no power) to the wheels. Normally, less torque is needed with higher speeds and smaller loads. When slower speeds and higher loads are used, more torque is needed.

LEVERS AND FORCES

In order to understand how the gears in a transmission work, levers and forces must first be defined. A **lever** is a device that changes forces and distances. For example, in ***Figure 40–1A***, a lever has an input, a **fulcrum**, and an output. When a force is applied to one side over a certain distance, a resulting force is produced with a certain distance. Force times distance on the input will always equal force times distance on the output. Therefore, work input (which is equal to force × distance) always equals work output.

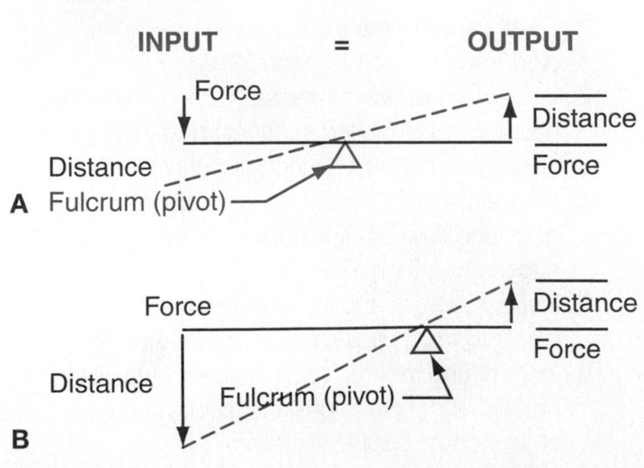

FIGURE 40–1 When the pivot point is moved to the right as in B, increased forces can be obtained. However, the distance traveled is less. Force times distance input will always equal force times distance output.

If the fulcrum is positioned farther to the right as shown in ***Figure 40–1B*** several things occur. First, the work input will again equal the work output. However, the output force will be much greater, while the output distance will be much smaller. Also, the input distance will be much greater. By changing the position of the fulcrum, different forces can be obtained while giving up distance.

TORQUE AND TORQUE MULTIPLICATION

The principle of levers can also be applied to gears. A gear can be considered a set of spinning levers. A set of gears can increase or decrease torque the same way that levers increase or decrease force. This is shown in ***Figure 40–2A***. One lever is pushing a second lever. The length of each lever determines how much force will be created on them. For example, if the short lever pushes against the longer lever, the longer lever will move less distance but will have more

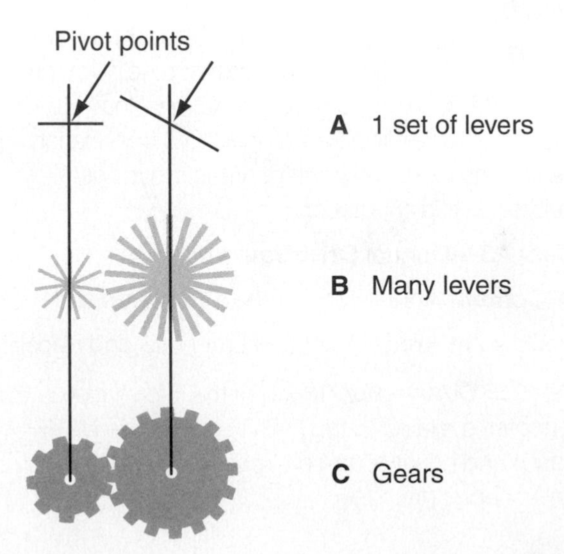

FIGURE 40–2 Gears can be used to increase torque just as levers are used to increase force.

force. If a series of levers is attached as shown in ***Figure 40–2B***, then continuous torque is developed. When applied to gears as shown in ***Figure 40–2C***, a small gear can increase torque in a larger gear. However, as with levers, the larger gear moves less distance than the input gear.

GEAR RATIOS

The amount of torque increased from the input to the output depends on the relative size of the gears. The difference in size between the input and output gears is called the *gear ratio*. The best way to show a gear ratio is by counting the number of teeth on each gear. ***Figure 40–3*** shows a set of gears. The input gear has twelve teeth and the output gear has twenty-four teeth. It will take two revolutions of the input gear to get the output gear to turn once. The speed of the output gear will be half the speed of the input gear; however, the torque will be doubled on the output. The gear ratio is stated as 24 to 12, or 2 to 1. This means that the input gear will turn two times while the output gear will turn one time.

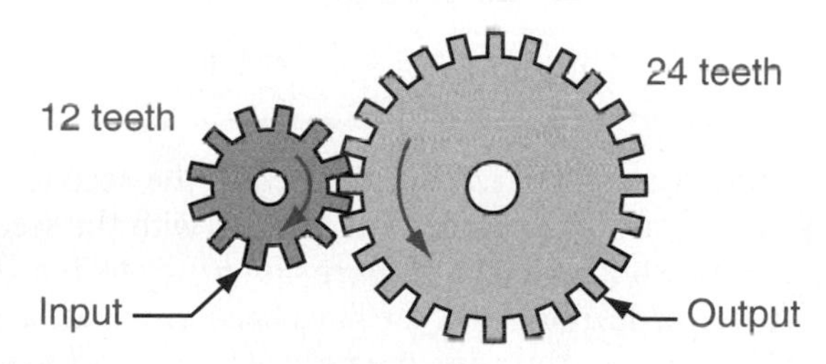

FIGURE 40–3 Gear ratios are determined by comparing the number of teeth on the output with the number of teeth on the input. This gear ratio is 24 to 12, or 2 to 1.

MANUAL TRANSMISSION PARTS

The transmission is a case of gears located behind the clutch. The output of the clutch drives the set of gears. The case that houses the gears is attached to the clutch housing. There are several shafts with different-sized gears inside the case. As the gears are shifted to different ratios, different torques can be selected for different operational conditions. ***Figure 40–4*** shows a simplified diagram of the internal parts of a three-speed manual transmission. A three-speed manual transmission is used to help simplify the operation. Although there are many parts in the transmission, the main ones include the drive, or input gear, the counter shaft gear (also called the cluster gear) with four gears on it, the main shaft with two gears on it, and the reverse idler gear. The counter shaft gears all turn at the same speed. The low and reverse gear and the second and high-speed gear are able to slide on the main shaft spline.

GEAR SELECTION

First gear is selected by connecting the gears so as to produce the greatest torque and the lowest output speed. This condition is used to start the vehicle moving, to go up a very steep

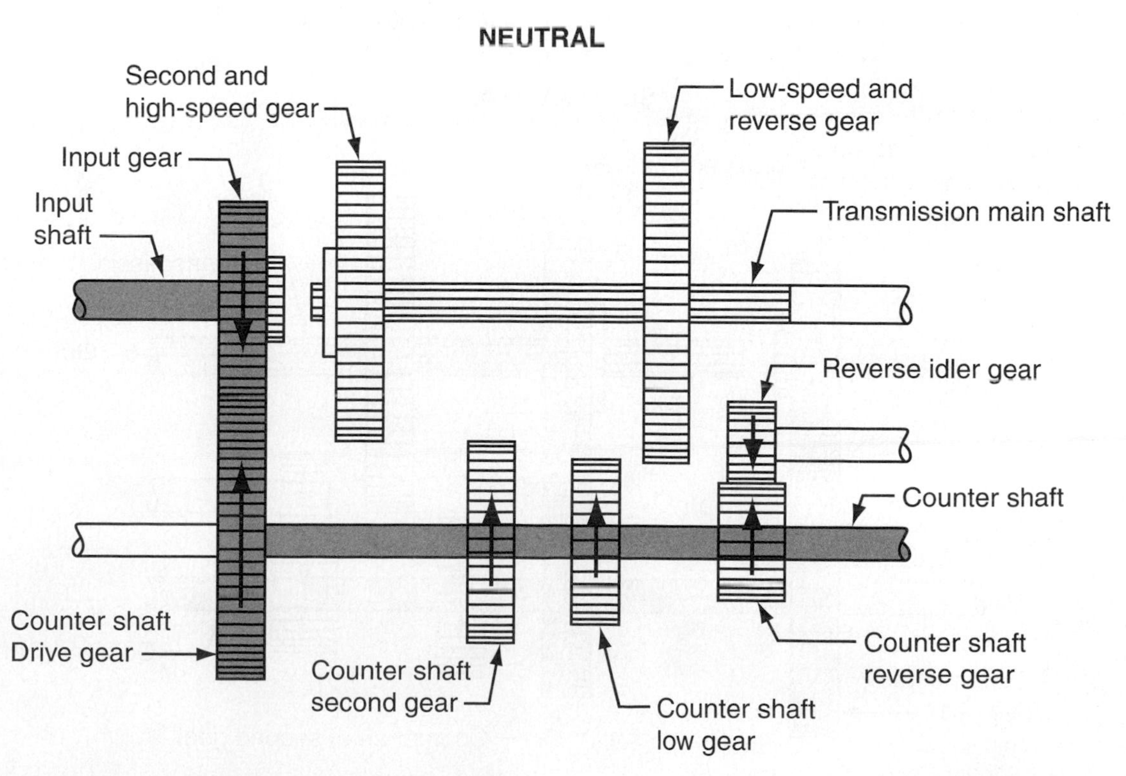

FIGURE 40–4 The main gears and shafts of a transmission.

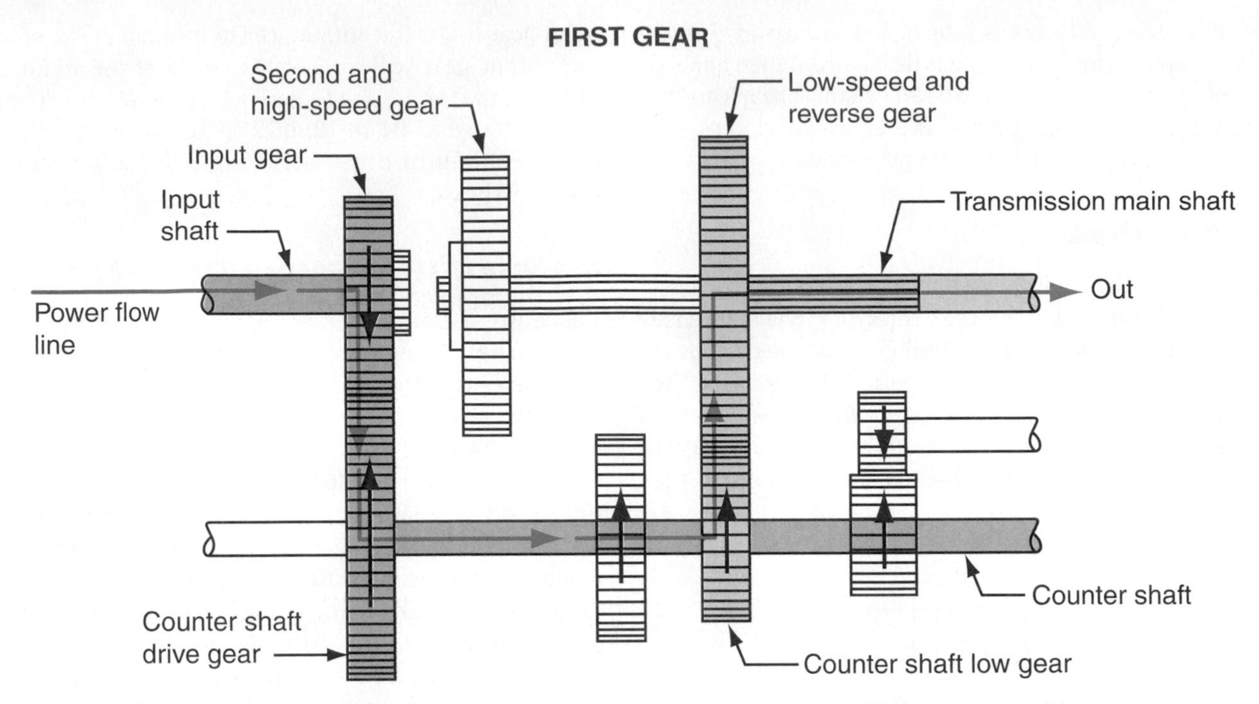

FIGURE 40–5 Low, or first, gear is selected by moving the low and reverse gear in mesh with the counter shaft low gear.

hill, or to pull a heavy load. This is done as shown in ***Figure 40–5***. When the shift lever is moved to low or first gear, the low and reverse gear on the main shaft meshes with the low gear on the counter shaft. Power is now transmitted from the input shaft to the counter shaft. From the counter shaft, the power is transmitted to the main shaft. Because of the gear ratios, the input shaft speed is about three times as fast as the speed of the output main shaft. However, the output shaft has about three times as much torque as the input shaft.

Second gear is selected by connecting the second and high-speed gear on the main shaft in mesh with the second gear on the counter shaft (***Figure 40–6***). Note that the gears used for low or first gear have been disengaged and are not in mesh. In this case, the ratio of input shaft to output shaft is about 2 to 1. The speed has been increased slightly, while torque has been decreased.

Third gear is selected by connecting the second and high-speed gear directly to the input shaft. This causes the

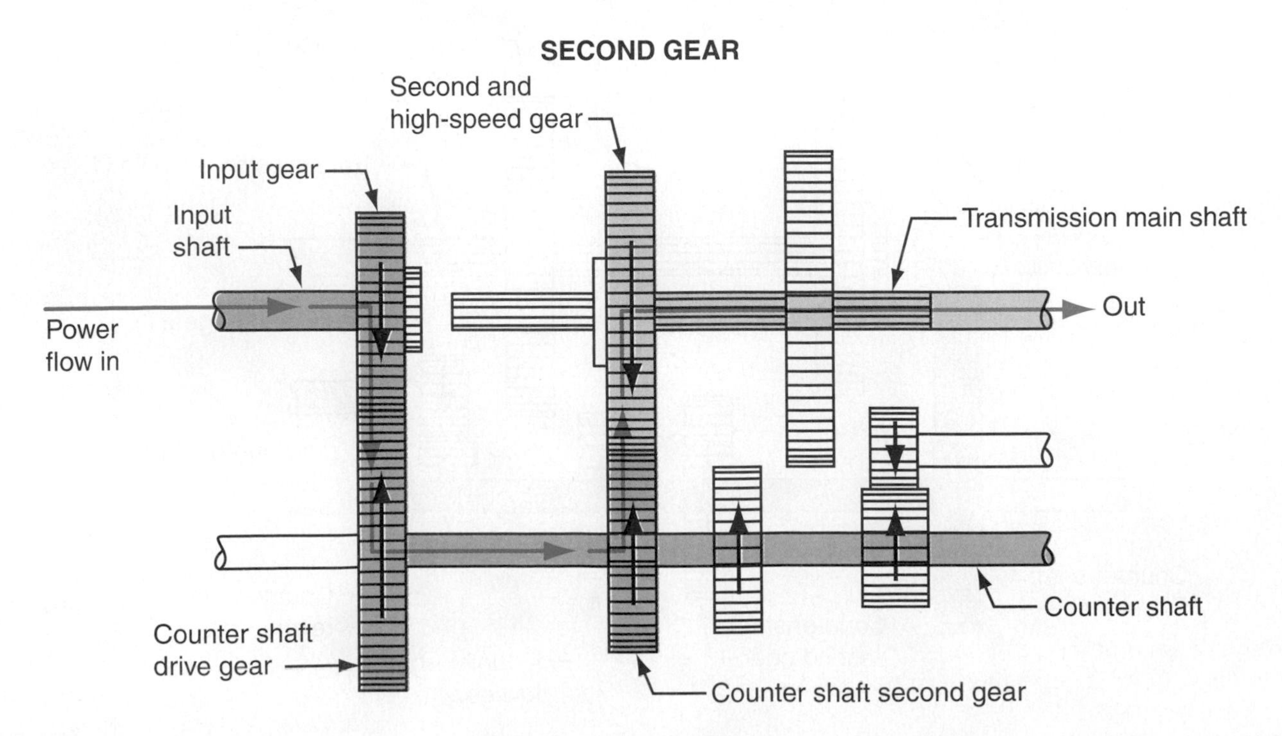

FIGURE 40–6 Second gear is selected by meshing the second and high-speed gear with the input gear.

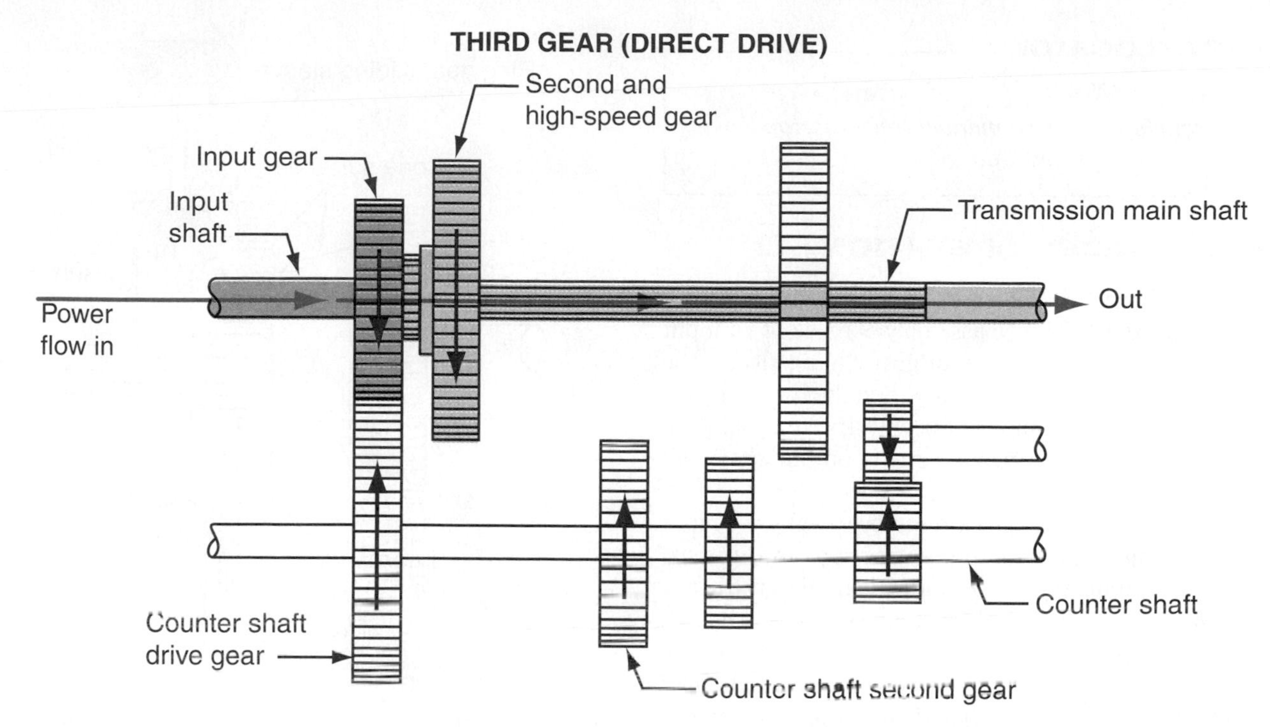

FIGURE 40-7 Third gear is selected by connecting the second and high-speed gear with the input gear.

input and output to rotate at the same speed (***Figure 40–7***). The ratio is now 1 to 1. In this condition, maximum speed with minimum torque is produced.

Reverse gear is selected by connecting the low-speed and reverse gear with the reverse **idler gear**. When the extra idler gear is added to the gear train, the direction of the output speed reverses. Reverse is considered a very low-speed, high-torque gear arrangement.

TYPES OF GEARS

Several types of gears are used in manual transmissions. Spur gears are found in older transmissions. These gears are typically very noisy. In order to reduce noise, helical gears are now used. Helical gears have the teeth at an angle. This produces a smoother meshing of teeth between the two gears. It also causes more surface contact between the teeth, which makes them much stronger. Most manual transmissions use helical gears. A third type of gear is called the internal gear. The gear teeth are machined inside a ring. A fourth type of gear is called the compound gear. A compound gear consists of several gears of different sizes placed on a shaft. All the gears spin at the same speed. The counter shaft gear in a manual transmission is a compound gear. ***Figure 40–8*** shows examples of different types of gears.

PURPOSE OF SYNCHRONIZERS

During normal shifting patterns, gears must be engaged and disengaged. Most gears, however, are not turning at the same speed before being engaged. To eliminate gears clashing with each other during normal downshifting or upshifting, **synchronizers** are used. Synchronizers are used in all forward speeds in today's manual transmissions. There are several types of synchronizers, but they all work on a similar principle. If the gear speeds can be synchronized or put at the same speed, they will mesh easily and there will be little clash or clatter.

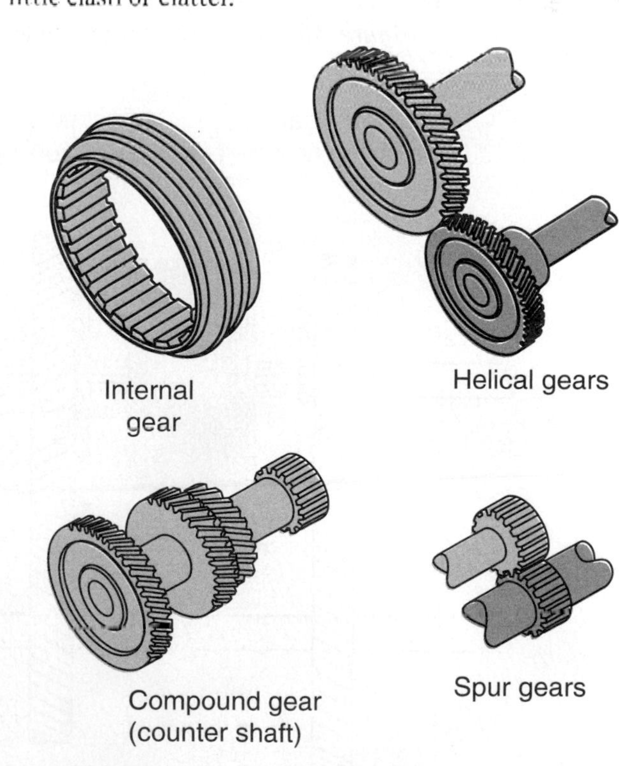

FIGURE 40-8 Several types of gears are used in manual transmissions depending on the style and year of the transmission. These may include the spur, helical, internal gears, and compound gear.

PARTS LOCATOR

*Refer to photo #79 in the front of the textbook to see what a **synchronizer in a manual transmission** looks like.*

SYNCHRONIZER OPERATION

Figure 40–9 shows how the principle of a synchronizer operates. In ***Figure 40–9A***, the assembly is made of an input gear, several cone surfaces, a ring gear sliding sleeve, the hub, several spring-loaded steel balls, and the internal gear. The goal is to mesh the input gear with the internal gear. If the internal gear is turned, then the output shaft will also turn.

For ease of understanding, start out by viewing the input gear as spinning and the output shaft, hub, and internal gear as not spinning. As the shift fork is moved to the left, it moves the ring gear sliding sleeve and the hub to the left. The hub and the ring gear are splined together. As the movement to the left continues, the two cone surfaces eventually begin to rub together. This action brings both gears to the same speed. The internal gear is still not in mesh with the input gear at this point. As the shift fork is moved farther to the left, the ring gear sliding sleeve continues to move while the hub keeps the cone surfaces touching each other. As the ring gear continues to move left, it eventually meshes easily with the input gear because both are spinning at the same speed (***Figure 40–9B***).

In most transmissions, the synchronizers are moved rather than the gears. This type of transmission is called a constant mesh type. ***Figure 40–10*** shows this principle.

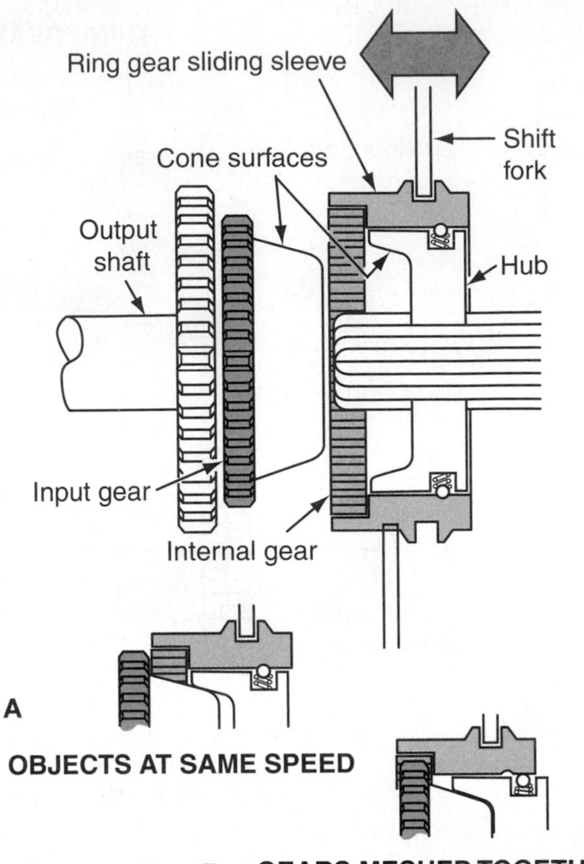

FIGURE 40–9 On a synchronizer, as the hub and internal gear assembly are moved to the left, the cone surfaces begin to touch. This brings both gears to the same speed. As the shift fork continues to move, it eventually slides the ring gear sliding sleeve in mesh with the input gear.

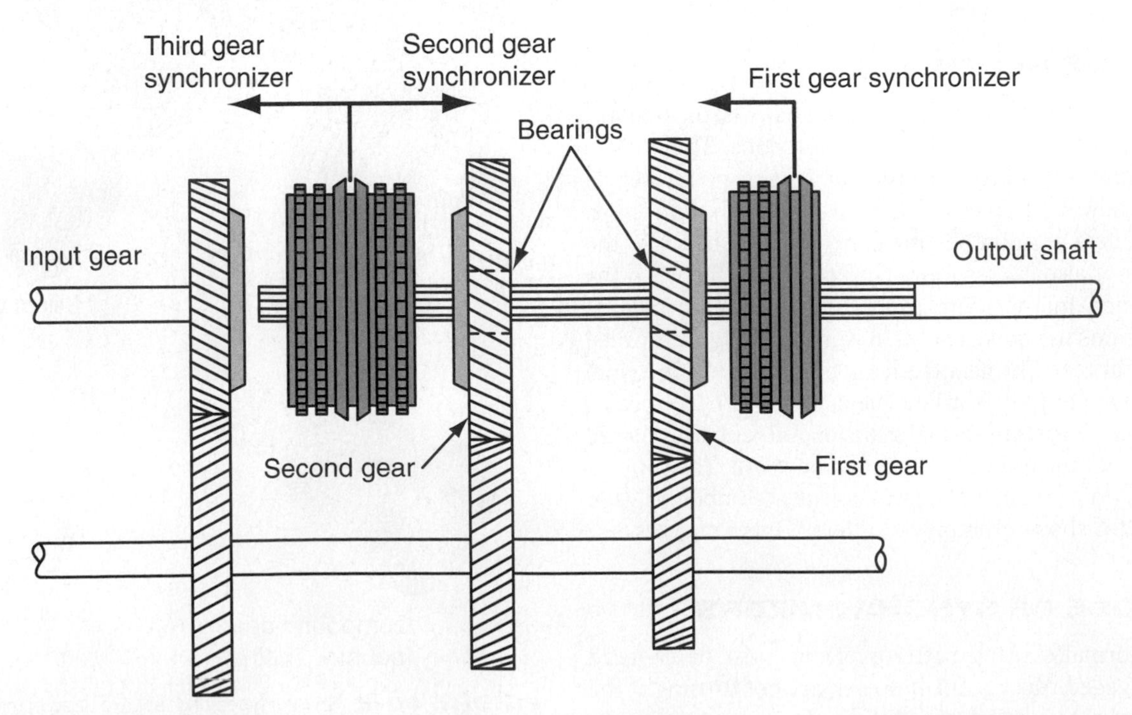

FIGURE 40–10 In most manual transmissions, the gears are always meshed and spinning on a bearing. However, the synchronizers actually engage and disengage the gears to the main output shaft for correct operation.

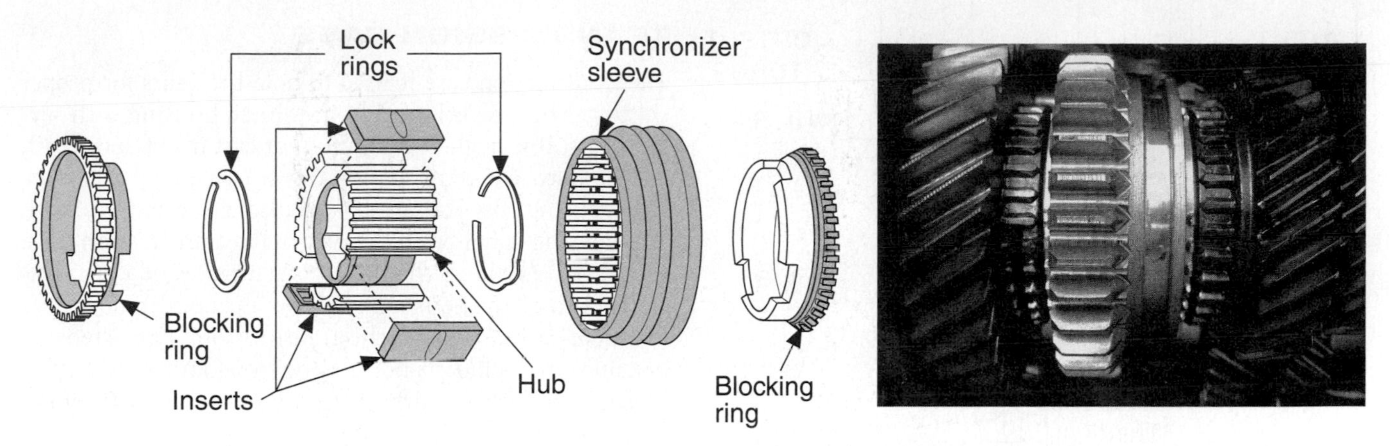

FIGURE 40–11 This is a disassembled synchronizer with its associated parts.

The gears are meshed at the start, but they are spinning on bearings on the output shaft. When the first gear synchronizer is moved to the left, it brings the output shaft to the same speed as the first gear. When the second gear synchronizer is moved to the right, it brings the second output shaft to the same speed as second gear. When the third gear synchronizer is moved to the left, the output shaft is brought to the same speed as the input gear. For simplicity, the reverse gear is not shown here.

Although many synchronizers differ in appearance, the basic principle is the same. Most synchronizers use a synchronizer ring (called a blocking ring) to replace the internal cone. Keys, rather than the small spring-loaded balls, are used on some synchronizers. ***Figure 40–11*** shows a disassembled synchronizer.

When two synchronizers are placed together, some of the parts are combined as shown in ***Figure 40–12***. In this arrangement, note that the synchronizer sleeve and hub are combined together to reduce its size. As the shift levers move the synchronizer sleeve to the left, gear B is turned.

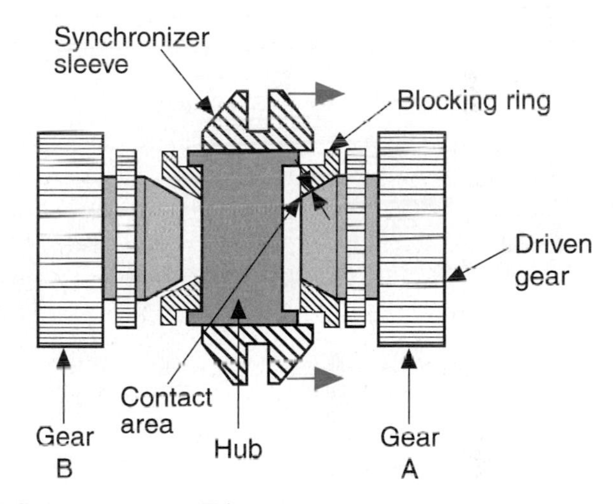

FIGURE 40–12 This cutaway shows a complete synchronizer for first and second gears.

As the shift lever moves the synchronizer sleeve to the right, gear A is turned.

Car Clinic: SYNCHRONIZERS

CUSTOMER CONCERN:
Poor Shifting

A customer indicates that the five-speed manual transmission has been difficult to downshift into first gear and is getting worse. The vehicle is used mostly in the city. Downshifting into first gear makes driving much easier in traffic. It is becoming more difficult to make the downshift, and gears are clashing.

SOLUTION: Manual transmissions have synchronizers to enable ease of downshifting. The synchronizers are used to get both gears at the same speed before engaging them. Replace the first gear synchronizer, and the problem should be eliminated.

OVERDRIVE SYSTEMS

The purpose of an **overdrive** system in a transmission is to decrease the speed of the engine during high vehicle speed operation. If this is done, the engine fuel mileage will improve; however, the torque at high speeds will be reduced. On transmissions without overdrive, the highest gear ratio is usually 1 to 1. On overdrive transmissions, the ratio may be 0.8 to 1 as shown in ***Figure 40–13***. This means the engine crankshaft has slowed down slightly. This is typically done by adding another gear in the gear train. The size of the gear and the direction of power through the transmission cause the engine to slow slightly at the higher vehicle speeds. On older transmissions, the overdrive gear was engaged by mechanical means. On modern transmissions, a synchronizer clutch or electrical solenoid is used to engage the overdrive gear. If the transmission has an overdrive gear, the gear shift knob often has the letters OD, or a similar representation, stamped on it.

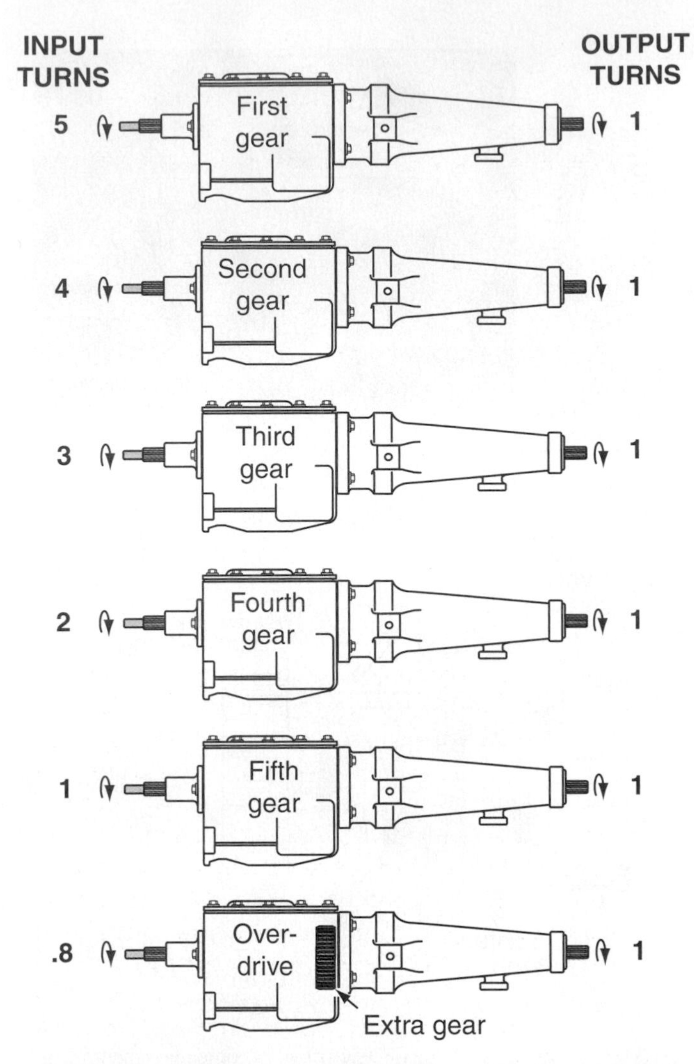

FIGURE 40-13 An overdrive gear has a ratio of approximately 0.8 to 1. This means that the input shaft is going slower than the output shaft. This ratio reduces engine speed and improves fuel mileage; however, torque at high speeds is reduced.

TRANSMISSION CASE

The transmission case is used to hold the gears for proper shifting. The case is bolted to the clutch housing with several bolts. It is made of aluminum or cast iron for strength and support. The transmission case also includes an extension housing that contains the transmission output shaft. The housing is bolted to the rear of the transmission case. ***Figure 40–14*** shows the transmission case, and the extension housing. The housing is used as an engine mount base and holds a rear seal to the transmission. The extension housing also includes a hole for the speedometer cable. The speedometer cable is driven from the output shaft by the speedometer gear.

TRANSMISSION LUBRICATION

Many transmissions use gear oil for lubrication. Gear oil is a type of lubrication used to keep gears well lubricated. Each manufacturer typically recommends its own type of gear oil. In addition, gear oils have different SAE viscosities. Typical viscosity ranges recommended by manufacturers are from 75 to 90 or more, depending on the outside temperature. The viscosity recommendation increases as the outside temperature increases.

Today, manufacturers are also recommending motor oils for use in manual transmissions. For example, it is not uncommon for a manufacturer to recommend engine oil 5W-30 for the transmission fluid. The exact viscosity is determined by the outside temperature, as with engine oil. In addition, some manufacturers also recommend automatic fluid such as Dexron III ATF. Always refer to the vehicle owner's manual for the correct type of transmission lubrication to use.

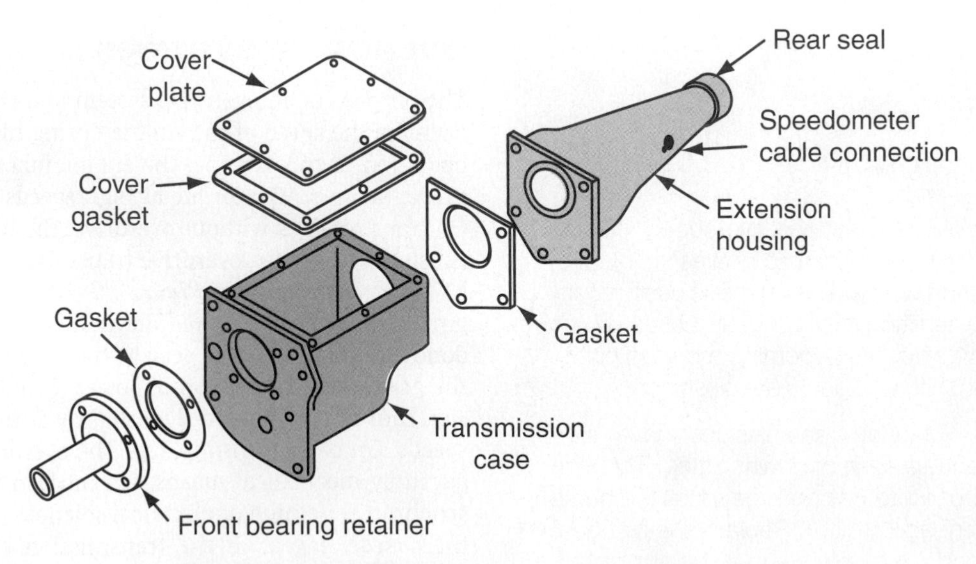

FIGURE 40-14 A complete transmission case and extension housing.

Car Clinic: SEALS AND LUBRICATION

CUSTOMER CONCERN:

Leaky Transmission Fluid

A customer has a vehicle with a standard transmission and complains that whenever the car sits in the driveway overnight, there is always a small spot of oil on the driveway. What could be the problem?

SOLUTION: At the end of the manual transmission, there is a seal that holds oil inside the transmission. This seal is pressed into the extension housing of the manual transmission. When the drive shaft is inserted into the transmission, this seal slides against the drive shaft slip yoke to help seal the transmission oil inside the transmission. If this seal is damaged or does not seal correctly, the oil may leak out of the transmission. Check to see if transmission oil is leaking out of this seal. If it is, there should be a noticeable amount of oil around the seal and on the end of the drive shaft. If the seal is bad, replace the seal to stop the leaking of transmission oil.

40.2 TYPES OF TRANSMISSIONS

PURPOSES OF DIFFERENT TRANSMISSIONS

Now that basic transmission components have been studied, other styles and types can be discussed. Vehicle designs have been changing over the past several years. Front-wheel drive vehicles are now very popular. In addition, many manufacturers are using smaller engines. Because of these changes, several types of transmissions have been developed. These include four-speed, five-speed, and transaxle transmissions.

FOUR-SPEED TRANSMISSIONS

The four-speed manual transmission has an additional gear set to produce four forward gears rather than three. With four gears, a wider range of torque characteristics is available. Increased torque range is usually needed for smaller engines. ***Figure 40–15*** shows an example of a four-speed transmission. The major difference is that the counter shaft has four forward gears rather than three. Synchronizers are also used to aid in shifting.

FIVE-SPEED TRANSMISSIONS

Five-speed transmissions are used on many vehicles today. With five gears, the range of torque is increased so that a smaller engine can be used. ***Figure 40–16*** shows an example of a five-speed transmission. In order to get five speeds, an additional set of gears is added to the transmission. Synchronizers are again used to engage and disengage the gears. In each gear shown, the transmission of mechanical energy is from the input shaft, to the counter shaft, to the specific gear on the main or upper shaft, through the synchronizer, and to the output shaft. In this particular case, the fifth gear is considered an overdrive gear. The fourth gear is the direct drive gear. The arrows help to show which gears are used for each shift condition.

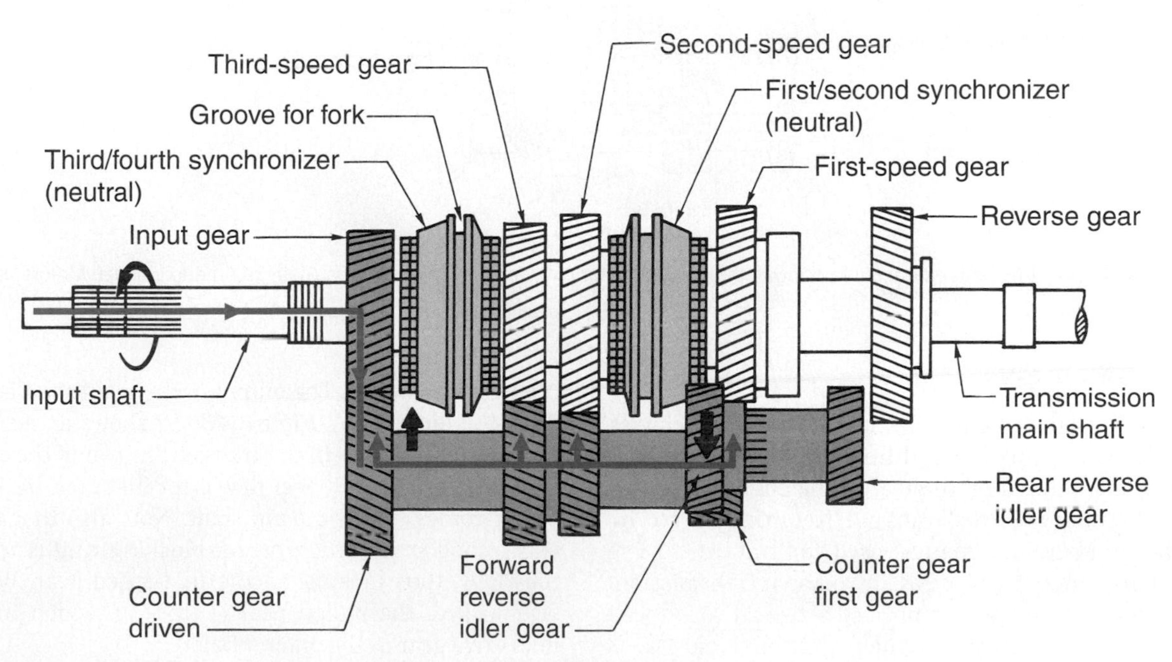

FIGURE 40–15 A four-speed transmission has an extra set of gears on the main shaft and counter shaft.

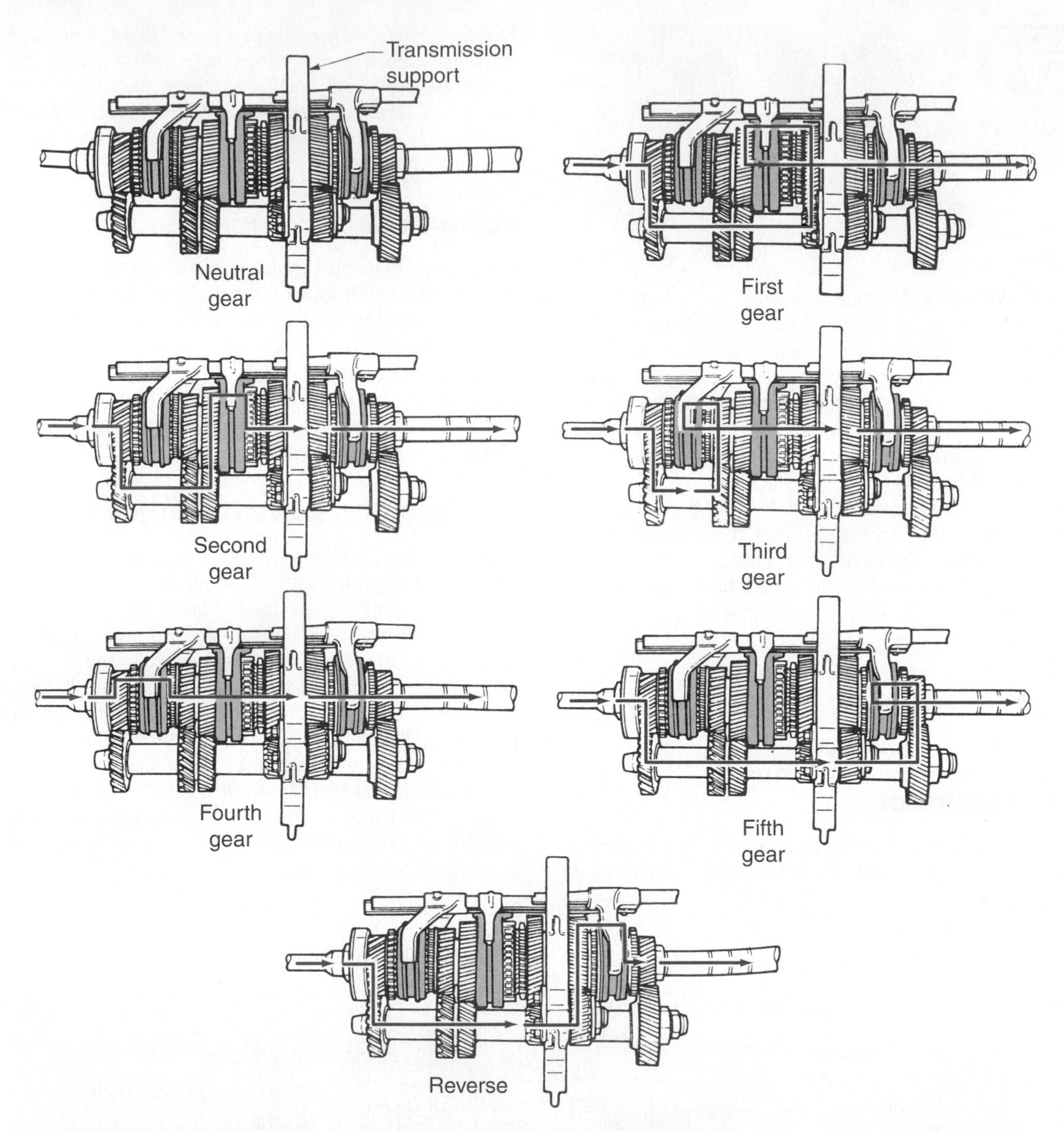

FIGURE 40–16 A five-speed transmission, with the power flow for each for the five gears. (Courtesy of American Isuzu Motors, Inc.)

TRANSAXLE SYSTEM

The **transaxle** system is used on front-wheel drive vehicles that have the transmission and final drive gearing placed together. It is mounted on the rear of the engine in such a way that the output of the transaxle feeds directly to the front wheels. Many designs are used for transaxles. The principles are much the same as any manual transmission. ***Figure 40–17*** shows an example of a typical four-speed transaxle. The counter shaft, called the input cluster, is attached directly to the clutch system. The gears are in constant mesh with synchronizers that are used to engage and disengage each gear. The output of the main shaft is used to drive the differential. ***Figure 40–18*** shows an example of how the power flows in the transaxle in two of the gears. In ***Figure 40–18A***, the power flow is for first gear. In this case, power comes into the main shaft. Note also that the first and second speed synchronizer blocking ring is moved to the right, thus locking up the first-speed gear. With this gear locked, the pinion gear is turning, which turns the final drive gear in the differential.

Referring to ***Figure 40–18B***, the transaxle has now been shifted to fourth gear. In this gear, the main shaft is driving

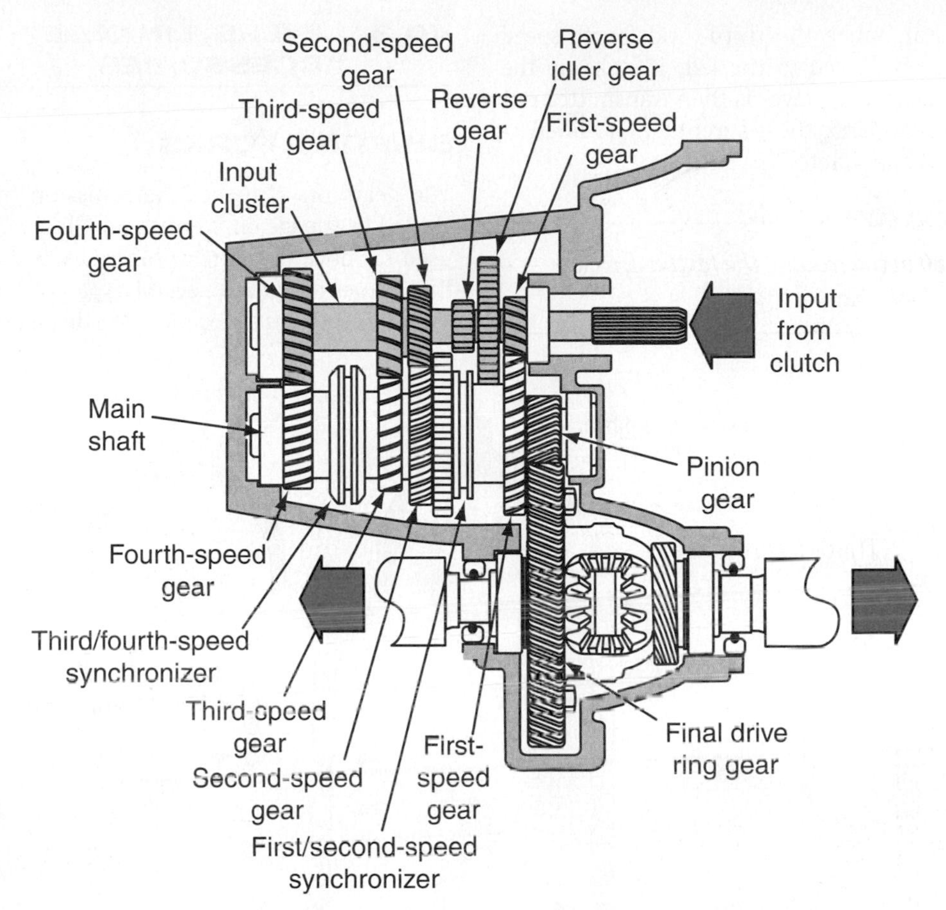

FIGURE 40-17 A four-speed transaxle has the differential attached to the manual transmission. Most front-wheel drive vehicles have similar designs.

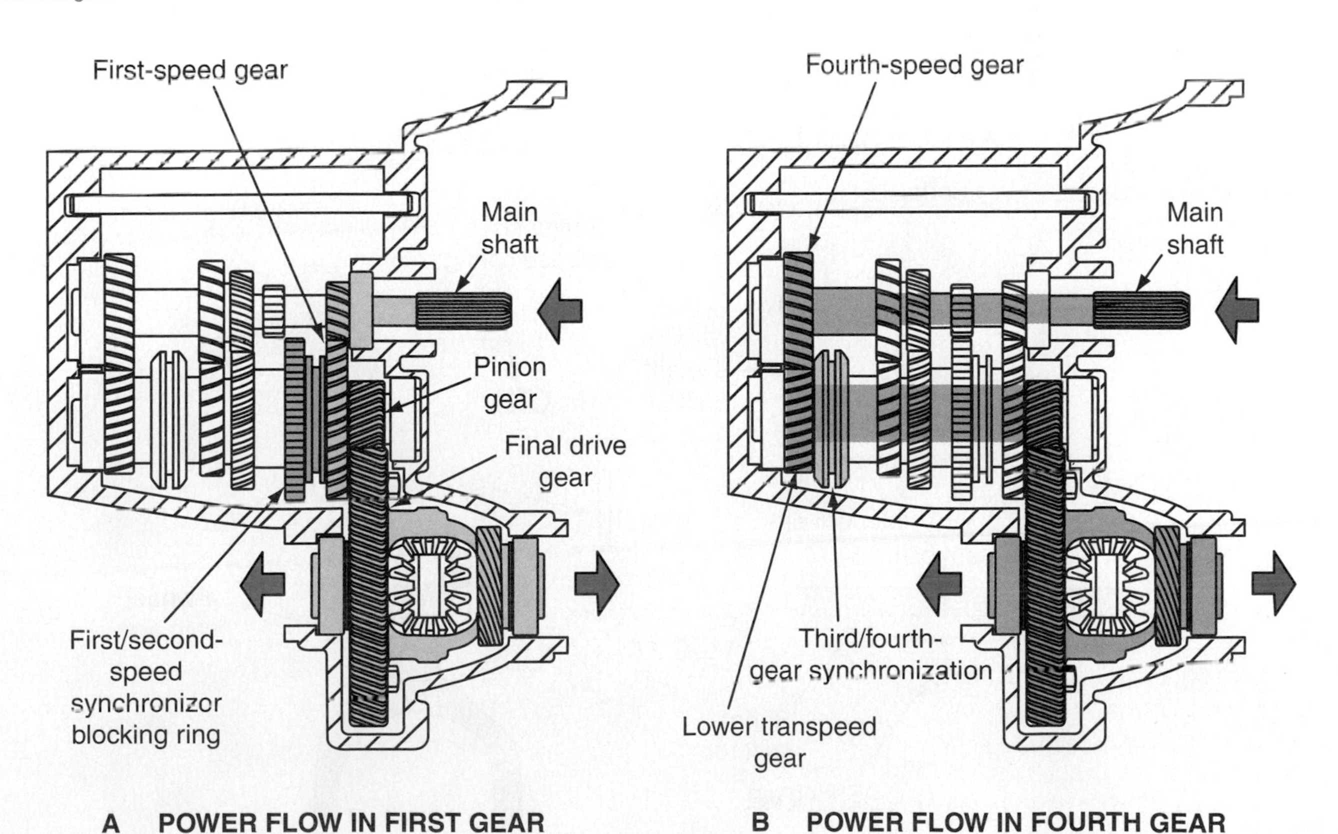

FIGURE 40-18 This view of a four-speed transaxle system in both first gear and fourth gear shows the power flow through the transaxle.

the fourth-speed gear. Since the third- and fourth-speed synchronizer has been moved to the left, it locks up the lower fourth-speed gear. The power is then transmitted out to the final drive gear and into the differential. The differential is discussed in a later chapter.

PARTS LOCATOR

Refer to photo #80 in the front of the textbook to see what a ***transaxle*** *looks like.*

40.3 LEVERS, LINKAGES, AND ACCESSORIES

SHIFTING FORKS

The gears on a manual transmission must be moved or shifted to engage different gears. Shifting forks or arms are used to move the synchronizers back and forth to engage the gears. There are several types of shifting forks. ***Figure 40–19*** shows an example of a disassembled shifting fork

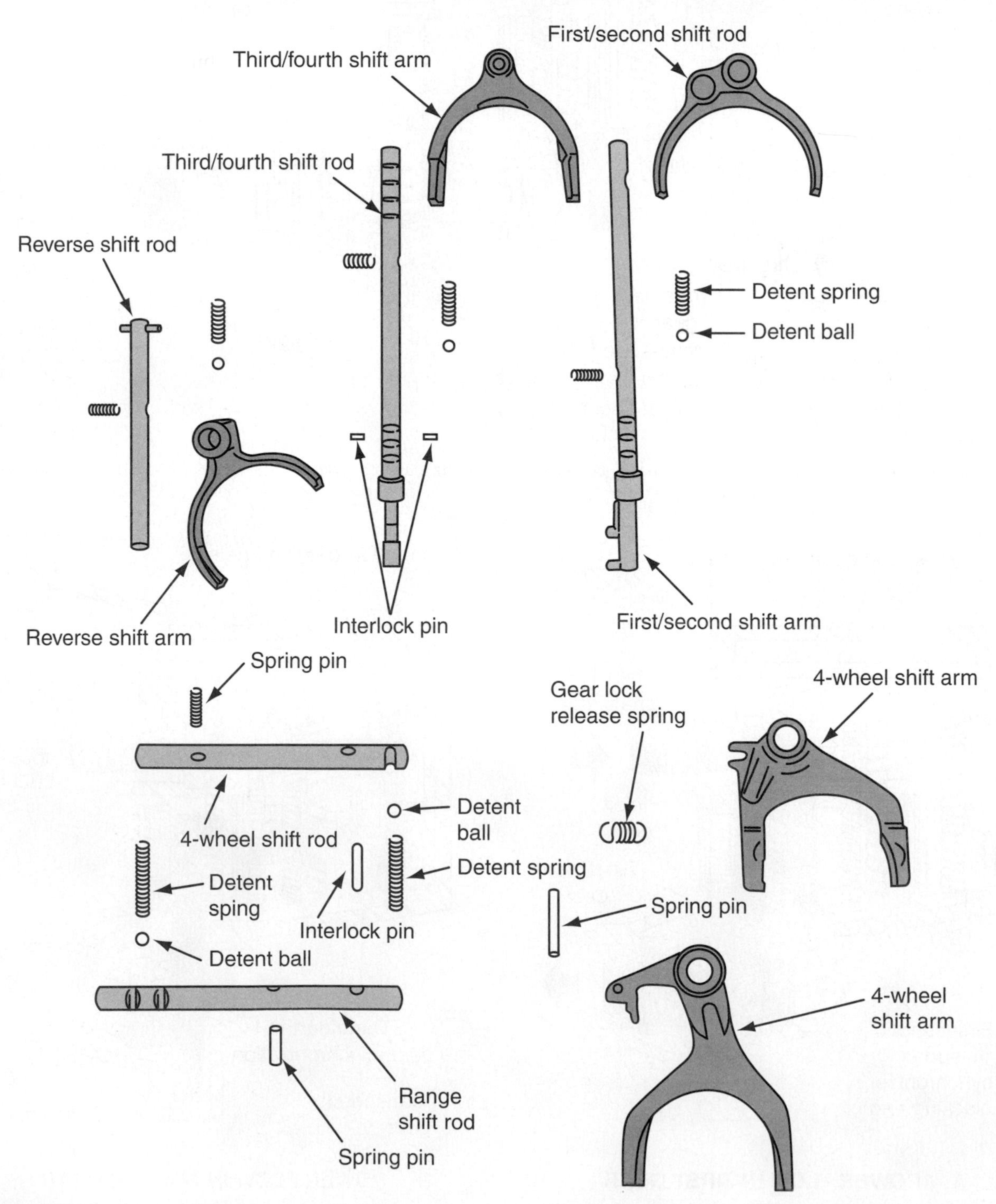

FIGURE 40–19 Clutch forks are used to move the synchronizers.

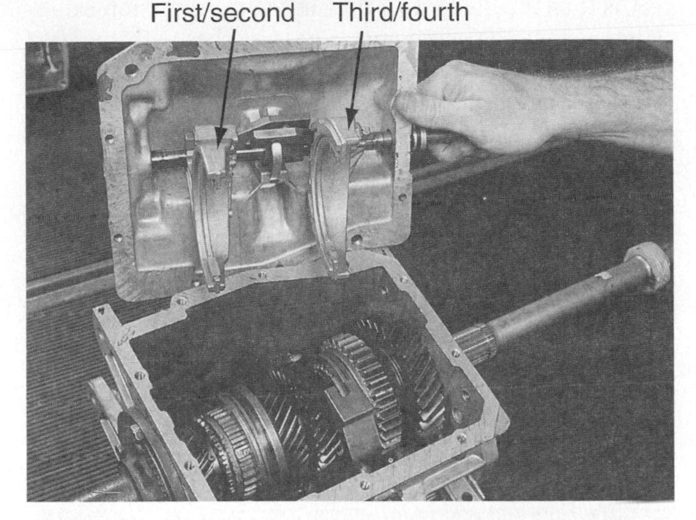

FIGURE 40-20 First/second gear and third/fourth gear shift fork assemblies.

mechanism. It is positioned in such a manner that it fits over the synchronizer assembly.

Figure 40–20 shows the first/second and third/fourth gears shifting fork assembly as a whole. As the operator moves the shift linkage, the fork moves the synchronizer for a specific gear.

Detent springs, balls, and rollers are used to hold the shifting fork in the correct gear position to ensure full engagement. ***Figure 40–21*** shows examples of detent springs and balls and rollers used on a manual transmission. A small groove is cut in the shift fork shaft. As the shift fork moves to its fully engaged position, the detent spring forces the detent roller down into the groove. The pressure from the spring holds the shift fork in place so it does not become disengaged from the gear.

SHIFT MECHANISMS

There are many designs of shifting mechanisms. The style and design depend on the position of the shift lever, the vehicle style, the year of the vehicle, the manufacturer, the number of gears, and the type of transmission. Certain vehicles have the shift lever on the steering wheel column, while others have a floor shift system. Both mechanical linkages and cables can be used to transfer the shift lever movement to the shifting forks. ***Figure 40–22*** shows a typical floor shift mechanism. As the shift (change) lever is moved to select the correct gear, one of the two shift or select cables is moved. To select between a group of forward gears or reverse gear the select cable is moved. To shift from one forward gear to another, the shift cable is moved. Each cable is fed through a tube attached to the frame. This allows it to slide in and out of the tube. The moving cable is then

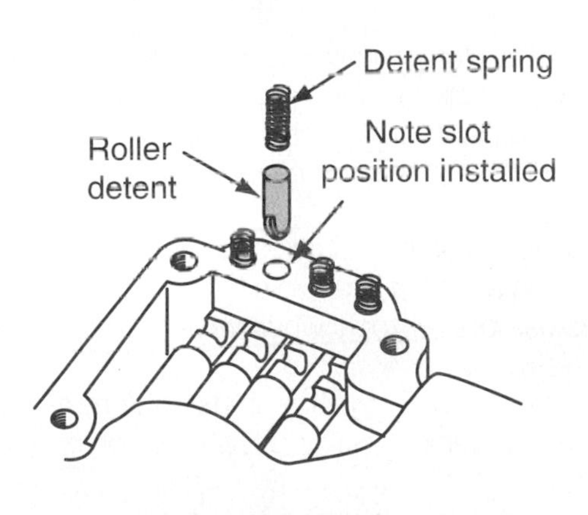

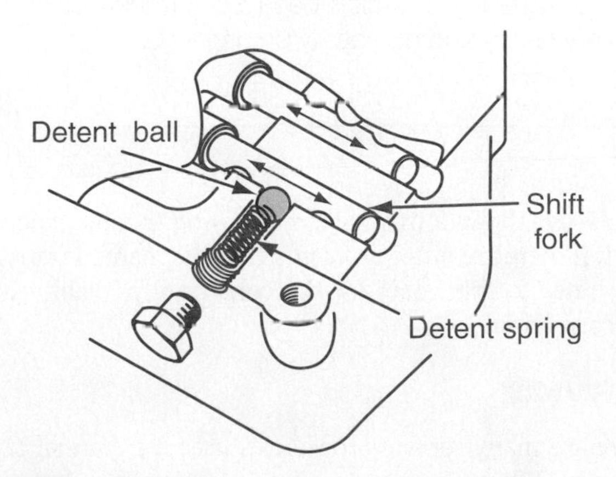

FIGURE 40-21 These detent springs, balls, and rollers are used to hold the shift fork in gear.

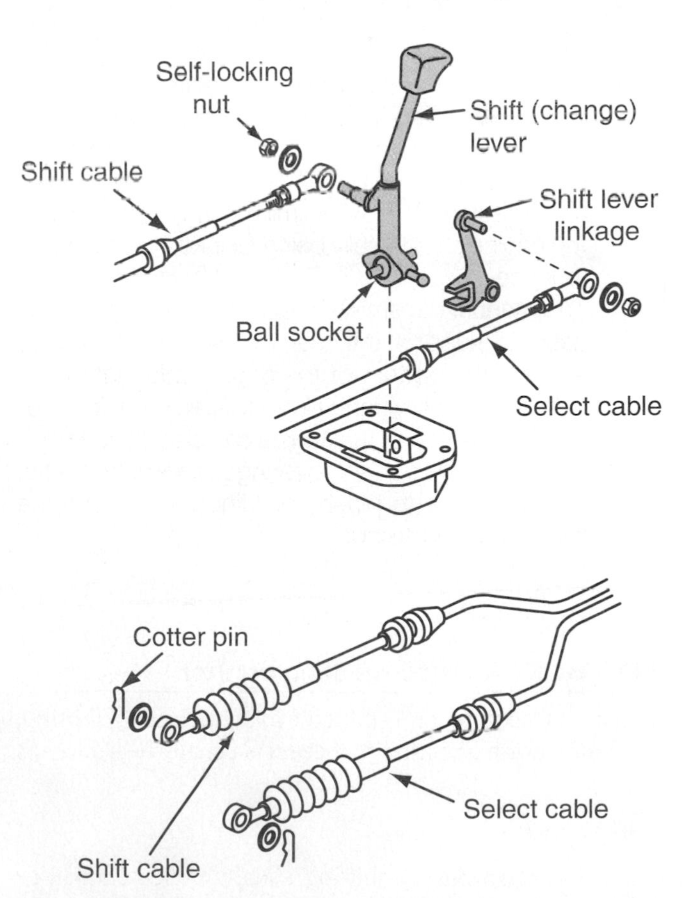

FIGURE 40-22 A complete shift mechanism for a five-speed transaxle with a floor shift.

attached to the transmission shift forks to shift the gears. Refer to the specific maintenance manual specifications for the type and correct adjustment of the shift mechanisms.

PARTS LOCATOR

*Refer to photo #81 in the front of the textbook to see what a **shifting mechanism on a manual transmission** looks like.*

OTHER ACCESSORIES

The speedometer in older vehicles is mechanically driven from the output shaft on the manual transmission. A small drive gear is attached to the output shaft. The speedometer gear is then inserted into a hole in the transmission extension housing. The speedometer gear meshes with the gear attached to the transmission output shaft. On most vehicles, there is also an electrical hookup for the backup light switch and overdrive solenoid.

Today, most vehicles use an electrical speed sensor rather than a mechanical cable to sense the output speed of the transmission. With the increased use of computers on today's vehicles, this type of speed sensing is more reliable and has fewer service problems compared to the mechanical gear and cable. In addition, this speed signal is used as an input by on-board computers for various purposes, including such systems as fuel management, electronic transmissions, ignition, and so on.

Problems, Diagnosis, and Service

Safety Precautions

1. When working on the transmission from under the vehicle, make sure that no oil or antifreeze has been spilled. You may slip and seriously injure yourself.
2. Always disconnect the battery cables from the battery before working on any part of the manual transmission. This will eliminate any possibility of the engine accidentally being cranked during maintenance.
3. The manual transmission external housing is often used to support the rear of the engine. Always support the bottom of the engine adequately before removing the transmission. Special engine supports on the top of the engine can also be used.
4. When installing gears, bearings, and gear shafts, be careful not to pinch your fingers between the part and the housing.
5. When removing or installing snap rings, always use snap ring pliers. Snap rings have a high degree of spring tension. If not installed with the correct tool, a snap ring may fly off and injure your eye, face, or body.
6. Always wear OSHA-approved safety glasses when working on a manual transmission.
7. Always use proper tools when working on manual transmissions.
8. When working on a manual transmission, always support the vehicle safely on a hoist or on jack stands.
9. Always use a transmission jack stand when removing a transmission because a transmission is very heavy and can easily be dropped.

PROBLEM: Defective Synchronizer

A manual transmission is difficult to shift. During shifting, a grinding noise or clashing of gears is usually heard.

DIAGNOSIS

Gears that clash during shifting usually indicate a bad or defective synchronizer. The transmission will need to be removed, disassembled, and overhauled accordingly. ***Figure 40–23*** shows a manual transmission trouble diagnosis chart. If there are other problems with the manual transmission, refer to this chart for the condition, probable cause, and correction.

SERVICE

There are many service procedures used on manual transmissions. When working on the transmission, always use the correct tools. Common tools, other than the standard

TROUBLE DIAGNOSIS		
PROBLEM	**DIAGNOSIS**	**SERVICE**
Hard shifting	1. Clutch. 2. Synchronizers worn or broken. 3. Shift shafts or forks worn. 4. Incorrect lubrication fluid.	1. Adjust. 2. Replace. 3. Replace. 4. Replace.
Slips out of gear	1. Shift shafts worn. 2. Bearings worn. 3. Drive gear retainer broken or loose. 4. Excessive play in synchronizers.	1. Replace. 2. Replace as necessary. 3. Tighten or replace retainer. 4. Replace.
Noisy in all gears	1. Insufficient lubricant. 2. Worn counter gear bearings. 3. Worn or damaged drive gear and counter gear. 4. Damaged drive gear or main shaft. 5. Worn or damaged counter gear.	1. Fill to correct level. 2. Replace counter gear bearings and shaft. 3. Replace worn or damaged gears. 4. Replace damaged bearings or drive gear. 5. Replace counter gear.
Noisy in neutral	1. Damaged drive gear bearing. 2. Damaged or loose pilot bearing. 3. Worn or damaged counter gear. 4. Worn counter gear bearings.	1. Replace damaged bearing. 2. Replace pilot bearing. 3. Replace counter gear. 4. Replace counter gear bearings and shaft.
Noisy in reverse	1. Worn or damaged reverse idler gear or idler bushing. 2. Worn or damaged reverse gear. 3. Damaged or worn counter gear.	1. Replace reverse idle gear assembly. 2. Replace reverse gear. 3. Replace counter gear assembly.
Leaks lubricant	1. Excessive amount of lubricant in transmission. 2. Loose or broken drive gear bearing retainer. 3. Drive gear bearing retainer gasket damaged. 4. Center support gaskets either side. 5. Rear extension seal. 6. Speedometer driven gear.	1. Drain to correct level. 2. Tighten or replace retainer. 3. Replace gasket. 4. Replace gaskets. 5. Replace. 6. Replace O-ring seal.

FIGURE 40–23 When troubleshooting a manual transmission, always refer to the manufacturer's problems, diagnosis, and service information. (Courtesy of American Isuzu Motors Inc.)

wrenches, include a snap ring pliers, gear pullers, drift punches, and a hammer. Use the following general service procedures when replacing synchronizers or rebuilding manual transmissions.

1. Remove the transmission, following the manufacturer's suggested procedure.
2. With the transmission on a workbench, remove the bolts holding the side and top covers on the transmission.
3. Remove the main drive gear (***Figure 40–24***). Snap rings may have to be removed first. ***Figure 40–25*** shows the main shaft removed from the transmission case.
4. Remove the counter shaft. This may require removing the woodruff key from the rear of the transmission case.

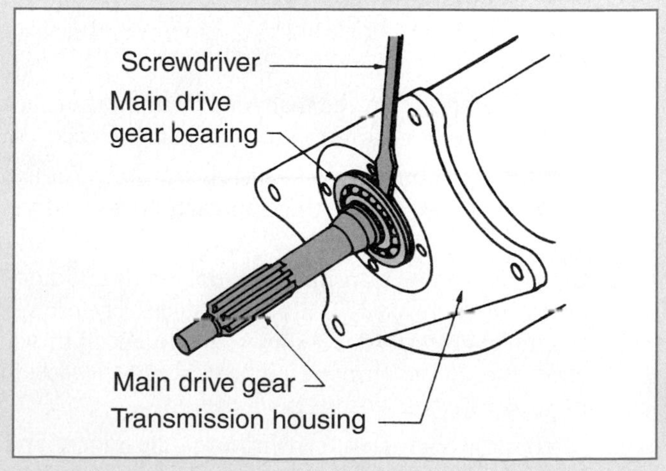

FIGURE 40–24 To disassemble the transmission, the main drive gear needs to be removed first.

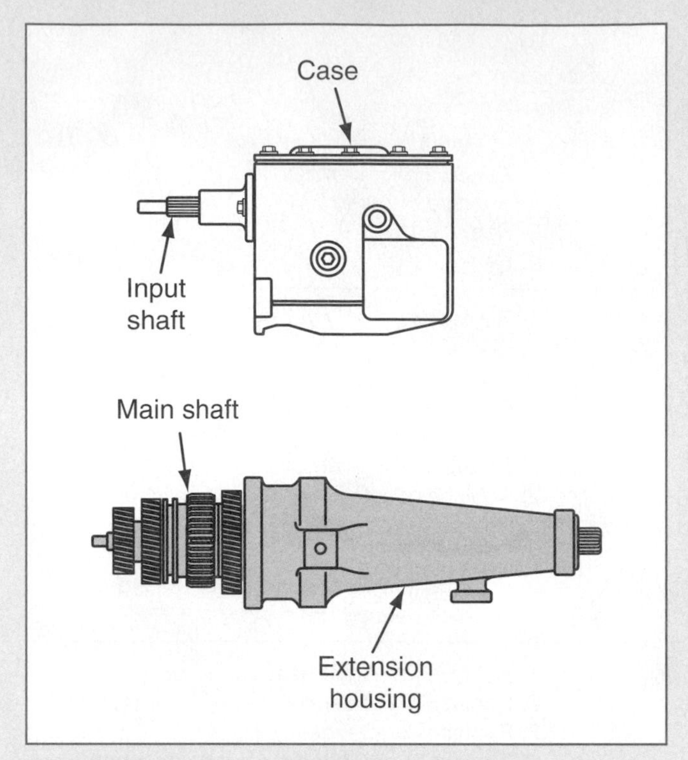

FIGURE 40–25 The main shaft can be removed from the transmission case as shown.

5. Remove all other gears in the transmission. ***Figure 40–26*** shows various gears being removed from a transaxle system.
6. Follow the manufacturer's recommended procedure to remove and replace the synchronizer assembly.
7. With the transmission disassembled, various checks can be performed. Although there are many procedures, the following are common checks on most transmissions:
   a. Check for wear and missing or damaged teeth on all gears and synchronizers. Check all internal bearings. Check the synchronizer sleeves to see that they slide freely on their hubs. A small presence of metal in the transmission fluid may indicate gear or synchronizer wear. Replace where necessary.
   b. Inspect the reverse gear bushing for wear and replace if necessary.
   c. Remove all gears on the main and counter shafts according to manufacturer's recommended procedures.
   d. Check the runout of the shafts, as shown in ***Figure 40–27***, and compare to the manufacturer's specifications.
   e. On many manual transmissions, the thrust clearance needs to be checked with a dial indicator before reassembly. ***Figure 40–28*** shows how to check thrust clearance. If the thrust clearance is not correct, it must be adjusted during reassembly.
   f. Often gear end play, synchronizer clearance, and bearing position need to be adjusted during reassembly. The adjustments are normally controlled by the thickness of various shims, snap rings, and thrust washers. For example, as shown in ***Figure 40–29***, various snap rings are available with different markings. Depending on the clearance needed, a certain snap ring is used to maintain the clearance.

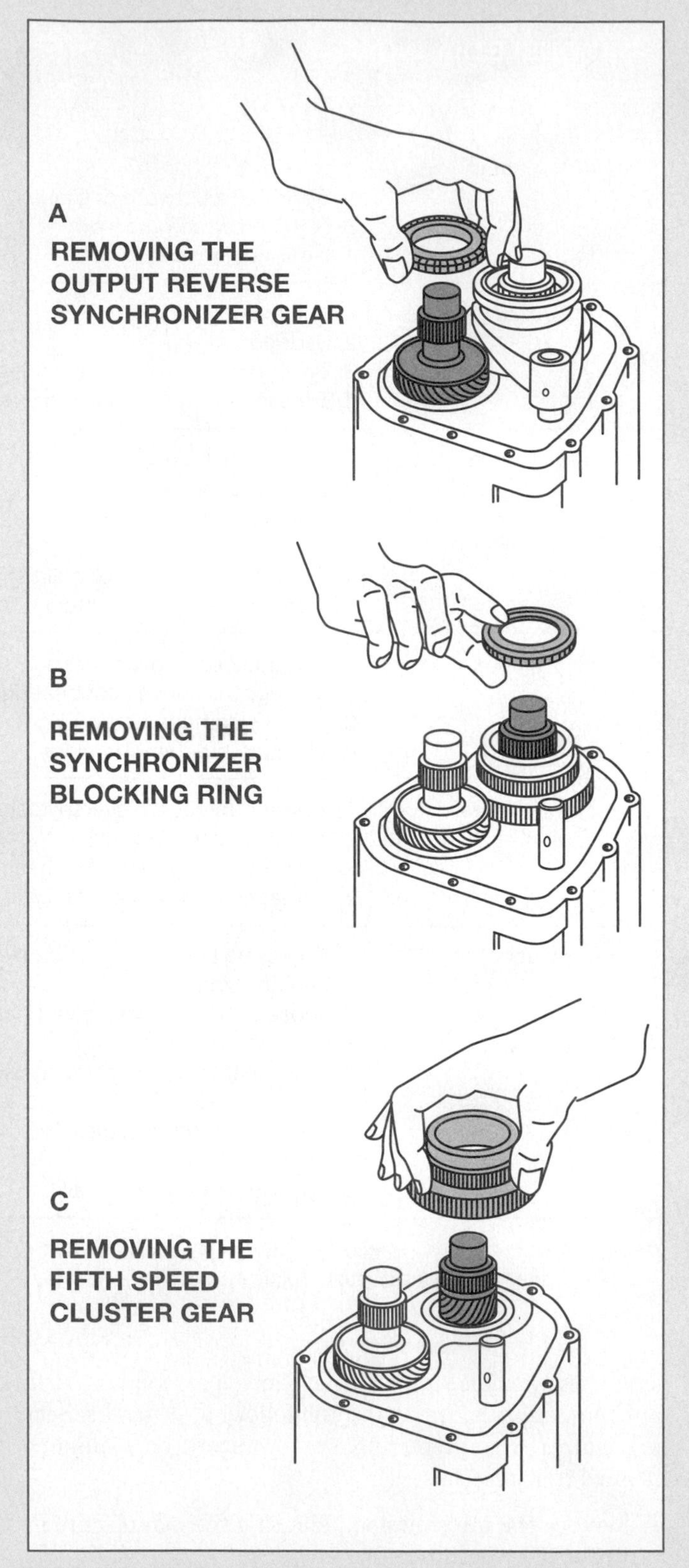

FIGURE 40–26 Views A, B, and C show various gears being removed from a front-wheel drive manual transaxle.

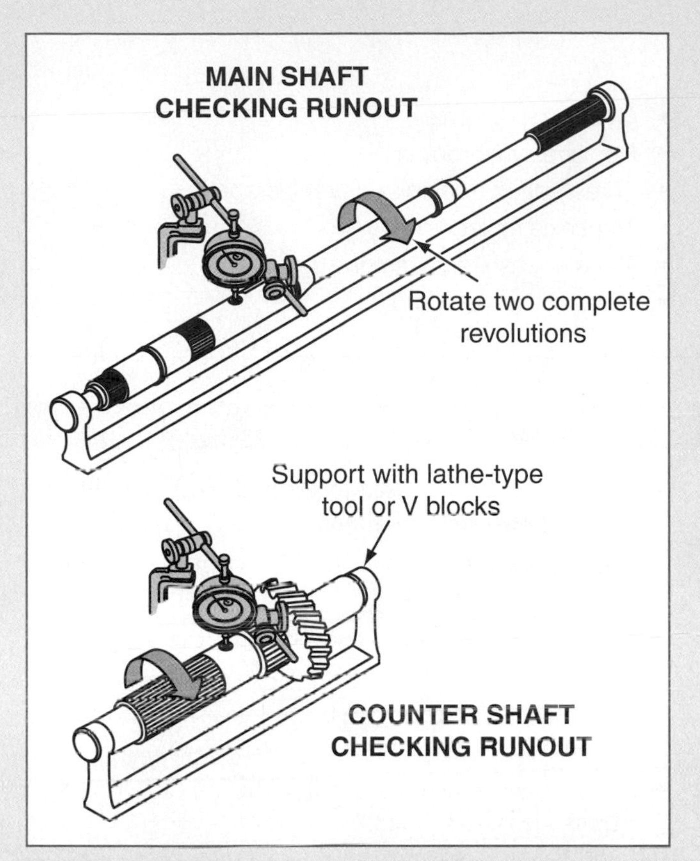

FIGURE 40-27 The main shaft and counter shaft runout can be checked with a dial indicator when the transmission is disassembled.

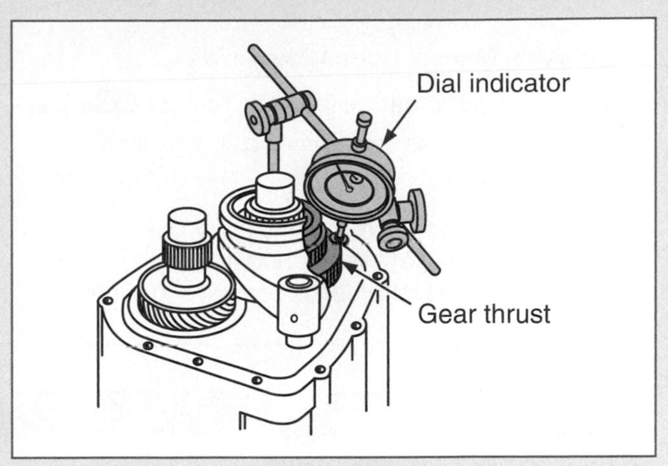

FIGURE 40-28 Gear thrust clearance can be checked with a dial indicator.

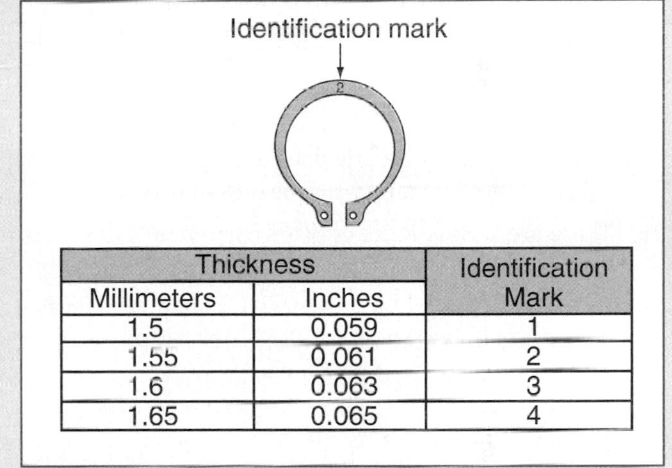

Thickness		Identification Mark
Millimeters	Inches	
1.5	0.059	1
1.55	0.061	2
1.6	0.063	3
1.65	0.065	4

FIGURE 40-29 End play on gears and synchronizers can be adjusted using different width snap rings.

Service Manual Connection

There are many important specifications to keep in mind when working with manual transmissions. To identify the specifications for your engine, you will need to know the VIN (vehicle identification number) of the vehicle, the type and year of the vehicle, and the type of engine. Although they may be titled differently, some of the more common manual transmission specifications (not all) found in service manuals are listed here. Note that these specifications are typical examples. Each vehicle and engine may have different specifications.

Common Specification	**Typical Example**
Bearing preload/control end play	0.001–0.005 inch
Gear ratios	
First	4.14 to 1
Second	2.50 to 1
Third	1.48 to 1
Fourth	1.00 to 1
Fifth	0.86 to 1
Reverse	3.76 to 1
Output shaft end play	0.002 inch
Shifter shaft thrust washer	0.1181 inch or 3.0 millimeters

continued

Service Manual Connection *continued*

Common Specification	Typical Example
Snap ring thickness	0.0689 inch
Transmission fluid capacity	3 1/4 pints

In addition, the service manual will give specific directions for various service and testing procedures. Some of the more common procedures include:

- Oil seal replacement
- Replace synchronizer
- Three-speed transmission replacement
- Transaxle replacement
- Transfer case replacement
- Transmission assembly

SUMMARY

The following statements will help to summarize this chapter:

- Transmissions are used to produce different torque capabilities at the driving wheels of a vehicle.
- High torque is needed during starting, going up a hill, and under heavy loads.
- Lower torque is needed at higher speeds.
- A transmission is designed so that, depending on the load and speed, different torque output can be used.
- There are several types of gears in a manual transmission, including spur, helical, and internal gears.
- Gears can change torques because they act like two levers working together.
- Manual transmissions typically have three, four, or five speeds.
- Depending on the gear that the vehicle is in, there is a certain gear ratio established, such as a 5:1, 4:1, 3:1, and so on, of input to output.
- Synchronizers are used to help eliminate the clashing and grinding of gears during meshing.
- Synchronizers are used to get both gears turning the same speed before meshing.
- Synchronizers use cones to help get the gears rotating at the same speed.
- An overdrive system is used to get better fuel mileage on many manual transmissions.
- An overdrive system has a 0.8 to 1 gear ratio from input to output.
- When the engine is mounted in the front of a vehicle, a transaxle is used.
- A transaxle has a manual transmission and a differential in the same casing.
- All transmissions have some form of linkage between the shift lever and the transmission synchronizer forks.
- Both mechanical and cable linkages are used on manual transmissions.
- Possible areas of diagnosis and service on a transmission include the synchronizers, bearings, linkages, and gears.

TERMS TO KNOW

Can you explain each of the following terms? Review the chapter until you can use each term correctly.

Detent springs
Fulcrum
Idler gear
Lever
Overdrive
Synchronizers
Transaxle

REVIEW QUESTIONS

Multiple Choice

1. The purpose of the standard or manual transmission is to:
a. Increase torque output at the right time
b. Change the speed of the output shaft
c. Decrease the input shaft at high speeds in overdrive
d. All of the above
e. None of the above

2. On a set of gears, if the input gear has twelve teeth and the output gear has twenty-four teeth, what is the gear ratio?
a. 3 to 1
b. 2 to 1
c. 1 to 1
d. 1 to 3
e. 1 to 0.2

3. If a smaller gear is used to drive a larger gear, the:
a. Torque output will be decreased
b. Torque output will be increased
c. Speed output will be increased
d. Input speed will be decreased
e. None of the above

4. When reverse gear is selected in a transmission:
a. Two extra gears are added to the gear train
b. Three extra gears are added to the gear train
c. One extra gear is added to the gear train
d. The overdrive gear is always used
e. The clutch is always disengaged

5. What is the gear ratio for a three-speed transmission in first gear?
a. About 1 to 4
b. About 1 to 5
c. About 1 to 3
d. About 3 to 1
e. About 2 to 1

6. In high, or third gear, on a three-speed transmission, the counter shaft:
a. Is driving all gears
b. Is freewheeling and not driving any gears
c. Is driving the second and first gears
d. All of the above
e. None of the above

7. Which type of gear has the teeth at an angle to the gear axis, rather than parallel to the gear axis?
a. Spur gear
b. Helical gear
c. Spur gear that is internal
d. Synchronizer internal gear
e. Overdrive gear

8. In current transmissions that use synchronizers, the counter shaft gears in the transmission:
a. Are not in mesh during shifting
b. Are always in mesh during shifting
c. Are not spinning
d. Must first engage the reverse gear
e. Mesh only after the synchronizer engages

9. The center of a synchronizer part that slides on the main shaft is called the:
a. Cone
b. Synchronizer ring
c. Synchronizer sleeve
d. Hub
e. Key

10. Which of the following is not part of a synchronizer?
a. Cone surface
b. Shift fork
c. Blocking ring
d. Hub
e. Roller bearing

11. The shift forks on a synchronized transmission control or move the:
a. Hub
b. Synchronizer sleeve
c. Synchronizer keys
d. Synchronizer cones
e. Splines

12. What would be a common gear ratio in highest gear with a transmission using an overdrive system?
a. 1 to 1
b. 1 to 1.8
c. 1 to 2
d. 2 to 1
e. 0.8 to 1

13. Which type of transmission would be used on a front-wheel drive vehicle?
a. Two-speed
b. Six-speed
c. Transaxle
d. All of the above
e. None of the above

14. The shift lever for the operator of the vehicle:
a. Moves the shifting forks
b. Has several types of linkages used to shift the transmission
c. Can be on the column of the steering wheel or on the floor
d. All of the above
e. None of the above

15. Which component is not attached to the transmission?
a. Speedometer cable drive
b. Backup lights
c. Oil seals
d. Oil pan
e. Extension housing

16. Gear clashing during shifting may be caused by:
a. A bad speedometer cable
b. A bad motor mount
c. Damaged or worn synchronizers
d. A worn clutch release bearing
e. A leaky rear seal

The following questions are similar in format to ASE (Automotive Service Excellence) test questions.

17. *Technician A* says that gear clashing is caused by misalignment of the shift linkage. *Technician B* says that gear clashing is caused by a bad synchronizer. Who is correct?
a. A only
b. B only
c. Both A and B
d. Neither A nor B

18. During assembly of the transmission to the clutch assembly, the transmission cannot be inserted into the clutch assembly. *Technician A* says that the transmission main shaft is damaged. *Technician B* says that the clutch disc is not properly aligned. Who is correct?
a. A only
b. B only
c. Both A and B
d. Neither A nor B

19. A small amount of ground metal filings are found in a transmission. *Technician A* says that they are clutch particles, not filings. *Technician B* says that these filings are metal and indicate possible wear. Who is correct?
a. A only
b. B only
c. Both A and B
d. Neither A nor B

20. *Technician A* says the gear end play on a manual transmission is set by different thickness snap rings. *Technician B* says that bearings must be adjusted when reassembling a manual transmission. Who is correct?
a. A only
b. B only
c. Both A and B
d. Neither A nor B

21. *Technician A* says that runout can be checked on the main shaft. *Technician B* says that runout can be checked on bearings. Who is correct?
a. A only
b. B only
c. Both A and B
d. Neither A nor B

Essay

22. Describe the gear ratios for first, second, third, and fourth gears.

23. Describe the purpose and basic operation of a synchronizer.

24. Why are overdrive systems used on transmissions?

25. What is a transaxle?

Short Answer

26. A popular type of transmission lubricant is identified as ________.

27. The ________ springs are used to hold shifting forks in the correct position.

28. Gear thrust clearance can be checked with a ________.

29. Transaxles are used on ________ -wheel drive vehicles.

30. The main shaft and counter shaft runout on a manual transmission can be checked using a ________.

Applied Academics

TITLE: Gear Tooth Design

ACADEMIC SKILLS AREA: Science

NATEF Automobile Technician Tasks:

The service technician is able to identify and define terms that specifically relate to manual drivetrain and axle systems, diagnosis, service, and repair.

CONCEPT:

The design of gear teeth used on a manual transmission is very scientific and precise. A great deal of engineering goes into the exact shape of the tooth on a gear. The figure that follows shows some of the more common measurements and terminology of standard gear teeth. The gear teeth are designed so that minimum contact and wear will occur as the gears mesh with one another. In the figure, a mating tooth is also shown. Note where the gear teeth touch each other during meshing. By studying gear design, it becomes apparent that proper lubrication and a clean environment are necessary to reduce gear wear.

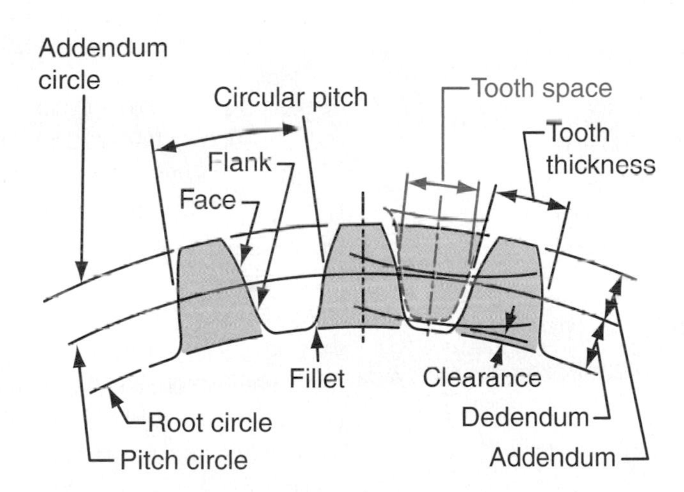

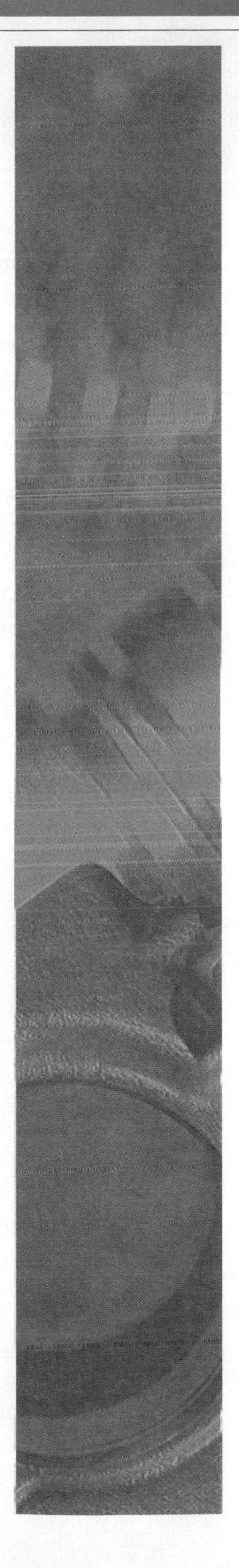

APPLICATION:

How would small particles of dirt in the lubricant affect the wear on two mating gear teeth? What affect would a bad bearing have on the clearances between two mating teeth during gear meshing? What affect would the lack of lubrication have on the meshing of two gears?

CHAPTER 41

Automatic Transmissions

OBJECTIVES

After studying this chapter, you should be able to:

- Identify the purpose and operation of the torque converter and lockup system.
- Explain the purpose and operation of the planetary gear system used on automatic transmissions.
- Analyze the different types of clutches and bands used on automatic transmissions.
- State the purpose and basic operation of the hydraulic systems used on automatic transmissions.
- Define the purpose and operation of various standard control devices used on automatic transmissions.
- Identify common problems, their diagnosis, and service suggestions for automatic transmissions.

Certification Connection

ASE Connection: The information in this chapter can help you prepare for the National Institute for Automotive Service Excellence (ASE) certification tests. The tests and content areas most closely related to this chapter are:

Test A2—Automatic Transmission/Transaxle

- **Content Area**—General Transmission/Transaxle Diagnosis
- **Content Area**—Transmission/Transaxle Maintenance and Adjustment
- **Content Area**—In-Vehicle Transmission/Transaxle Repair
- **Content Area**—Off-Vehicle Transmission/Transaxle Repair

NATEF Connection: Much of the information in this chapter is related to the NATEF tasks. The NATEF tasks and priority numbers most closely related to this chapter are:

1. Identify and interpret transmission/transaxle concerns; ensure proper engine operation; determine necessary action. P-1
2. Research applicable vehicle and service information, such as transmission/transaxle system operation, vehicle service history, service precautions, and technical service bulletins. P-1
3. Diagnose fluid usage, level, and condition concerns; determine necessary action. P-1
4. Diagnose electronic, mechanical, hydraulic, and vacuum control system concerns; determine necessary action. P-1
5. Diagnose noise and vibration concerns; determine necessary action. P-2
6. Diagnose transmission/transaxle gear reduction/multiplication concerns using driving, driven, and held member (power flow) principles. P-1
7. Service transmission; perform visual inspection; replace fluids and filters. P-1
8. Inspect, adjust, or replace (as applicable) vacuum modulator; inspect and repair or replace lines and hoses. P-1
9. Inspect, repair, and replace governor assembly. P-3
10. Inspect and replace external seals and gaskets. P-2
11. Inspect extension housing, bushings, and seals; perform necessary action. P-3

12. Inspect, leak test, flush, and replace cooler, lines, and fittings. P-2
13. Inspect and replace speedometer drive gear, driven gear, vehicle speed sensor (VSS), and retainers. P-2
14. Diagnose noise and vibration concerns; determine necessary action. P-2
15. Remove and reinstall transmission and torque converter (rear-wheel drive). P-2
16. Remove and reinstall transaxle and torque converter assembly. P-1
17. Disassemble, clean, and inspect transmission/transaxle. P-1
18. Inspect, measure, clean, and replace valve body (includes surfaces and bores, springs, valves, sleeves, retainers, brackets, check-balls, screens, spacers, and gaskets. P-2
19. Inspect servo bore, pistons, seals, pin, spring, and retainers; determine necessary action. P-3
20. Inspect accumulator bore, piston, seals, spring, and retainer; determine necessary action. P-3
21. Assemble transmission/transaxle. P-1
22. Inspect converter flexplate, attaching parts, pilot, pump drive, and seal areas. P-2
23. Inspect, measure, and reseal oil pump assembly and components. P-1
24. Measure end play or preload; determine necessary action. P-1
25. Inspect, measure, and replace thrust washers and bearings. P-2
26. Inspect oil delivery seal rings, ring grooves, and sealing surface areas. P-2
27. Inspect bushings; determine necessary action. P-2
28. Inspect and measure planetary gear assembly (includes sun, ring gear, thrust washers, planetary gears, and carrier assembly); determine necessary action. P-2
29. Inspect case bores, passages, bushings, vents, and mating surfaces; determine necessary action. P-2
30. Inspect transaxle drive, link chains, sprockets, gears, bearings, and bushings; perform necessary action. P-2
31. Inspect, measure, repair, adjust, or replace transaxle final drive components. P-2
32. Measure clutch pack clearance; determine necessary action. P-1
33. Inspect roller and sprag clutch, races, rollers, sprags, springs, cages, and retainers; replace as needed. P-1
34. Inspect bands and drums; determine necessary action. P-2

INTRODUCTION

The automatic transmission is designed to shift gears automatically from low to overdrive without the driver's control. Several major components are used in an automatic transmission. These include the torque converter, planetary gear system, hydraulic components, different types of clutches and bands, and various controlling parts. This chapter explains how these components are used in the automatic transmission.

41.1 BASIC DESIGN AND REQUIREMENTS

PURPOSE OF THE AUTOMATIC TRANSMISSION

The purpose of the automatic transmission is to connect the rotational forces of the engine to the drive wheels and to provide correct torque multiplication. This must be done at varying loads, speeds, and driving conditions. For example, to get a vehicle started, low speed and high torque are required. As the speed requirements increase, the torque requirements decrease. The automatic transmission is designed to produce the correct torque needed for varying driving conditions.

MAJOR PARTS OF THE AUTOMATIC TRANSMISSION

A torque converter is used for the same reason a clutch is used in a standard transmission. The torque converter connects the engine to the transmission gearing. At times, the engine and gears must be directly connected. At other times (e.g., at a stoplight), the two must be disconnected. In addition, the torque converter multiplies the torque sent to the transmission for varying loads.

The planetary gear system is used to produce the correct gear ratio for different torque and speed conditions. Low speeds, drive, and reverse gear ratios can be achieved by using the planetary gear system.

Various hydraulic controls are used to lock up parts of the gear system in an automatic transmission. Clutches, bands, and other components help to control which gears are used in the planetary gear system.

An automatic transmission uses several hydraulic control valves to operate the clutches and bands for correct operation. Complex hydraulic circuits have been incorporated into the automatic transmission to accomplish this goal.

41.2 TORQUE CONVERTER

In a standard transmission, a clutch is used to engage and disengage the engine from the transmission. This mechanism is called a friction drive. In the automatic transmission, a **fluid coupling** is used to engage and disengage the engine from the transmission. This coupling is called the **torque converter**. It uses a hydraulic, or fluid, coupling.

FLUID COUPLING

A common way of describing a fluid coupling is by using two electric fans. In ***Figure 41–1***, one fan produces a pressure and blows the air against the other fan. The air pressure produces enough energy to rotate the second fan. This action couples the input and output. The first fan is called the pump, and the second fan is called the **turbine**. The faster the pump turns, the better the fluid lockup between the input and output. In the actual torque converter, a pump and turbine are used, and transmission oil, rather than air, is used.

PARTS OF THE TORQUE CONVERTER

Several parts make up a torque converter to produce the fluid or, in this case, hydraulic coupling. ***Figure 41–2*** shows the major parts of a typical torque converter. These include the pump (also called an impeller), the turbine, and the guide wheel or **stator**. Note that each part is made of a series of vanes to direct oil through the torque converter. These parts are contained in a sealed housing that is completely filled with transmission fluid. Motion and power are transferred by the pressure of the mass of the flowing oil. There is no direct mechanical contact between the input and the output drives, only a fluid connection. Because of this type of connection, torque converters operate essentially wear-free.

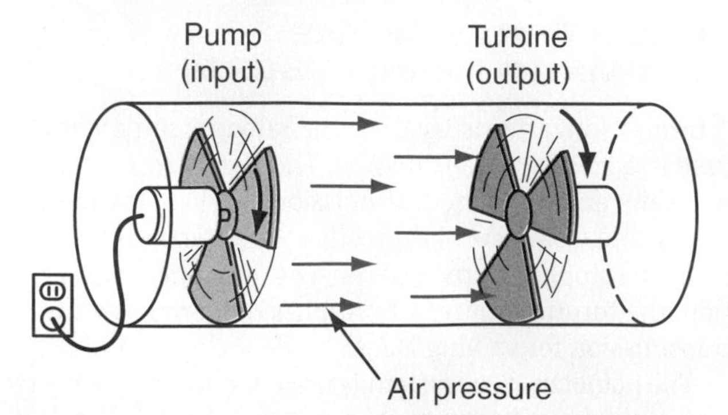

FIGURE 41–1 A torque converter produces a fluid coupling in a manner similar to the operation of two fans. The pressure from the pump fan causes the blades on the second fan, or turbine, to turn. The input and output are now connected as a fluid coupling.

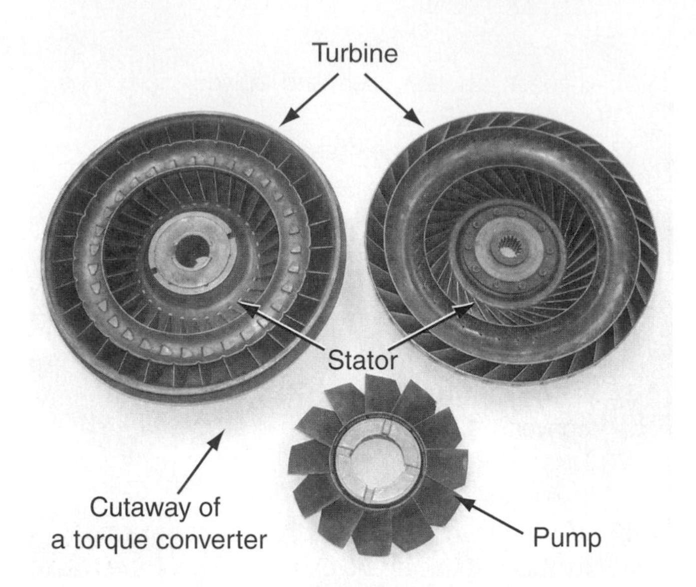

FIGURE 41–2 The major parts of the torque converter are the pump, turbine, and guide wheel or stator.

OPERATION OF THE TORQUE CONVERTER

Figure 41–3 shows the operation of the torque converter. The entire torque converter is bolted to the engine crankshaft through a flexplate (***Figure 41–4***). As the engine turns, the entire outside housing of the torque converter turns. The pump vanes are attached directly to the inside of the housing of the torque converter. This connection causes the pump vanes to turn. As the pump rotates, the oil in the pump vanes is forced to the **periphery**, the outer edges of the pump, by centrifugal force. The oil now reaches the turbine vanes with high velocity (just like the air hitting the fan as shown in ***Figure 41–1***). The high-velocity oil causes the turbine vanes to rotate and to lock up hydraulically with the pump.

Within the turbine vanes, the oil flow is changed to mechanical rotation by the sharply curved vanes of the turbine. The turbine is connected to the output drive of the torque converter. The stator or guide wheel is used to direct the oil flow back to the pump in the most efficient manner. In effect, the oil flows in a circular fashion throughout the torque converter as shown in ***Figure 41–5*** (much like a spring bent into a circle).

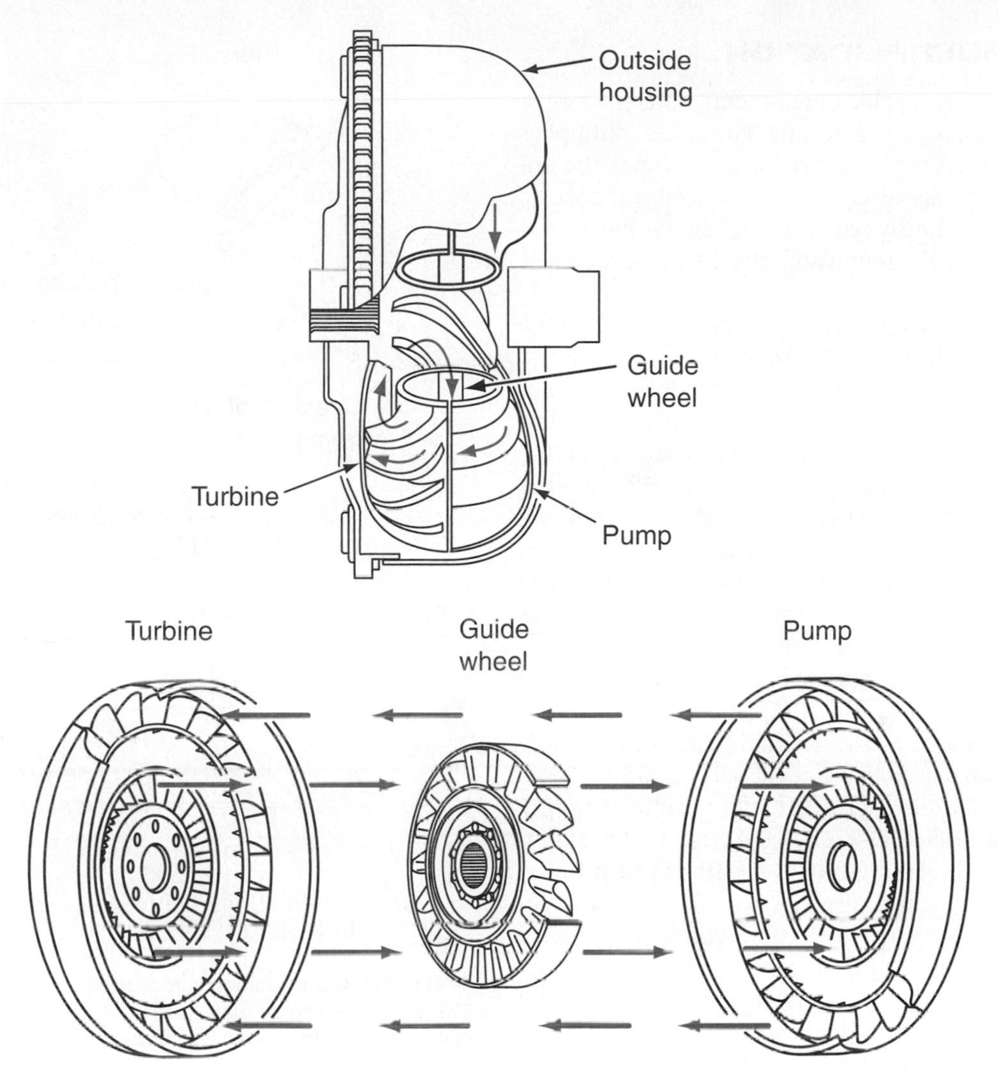

FIGURE 41–3 As the pump rotates, oil is sent to the turbine, through the stator or guide wheel, and back to the pump. This action produces a hydraulic coupling.

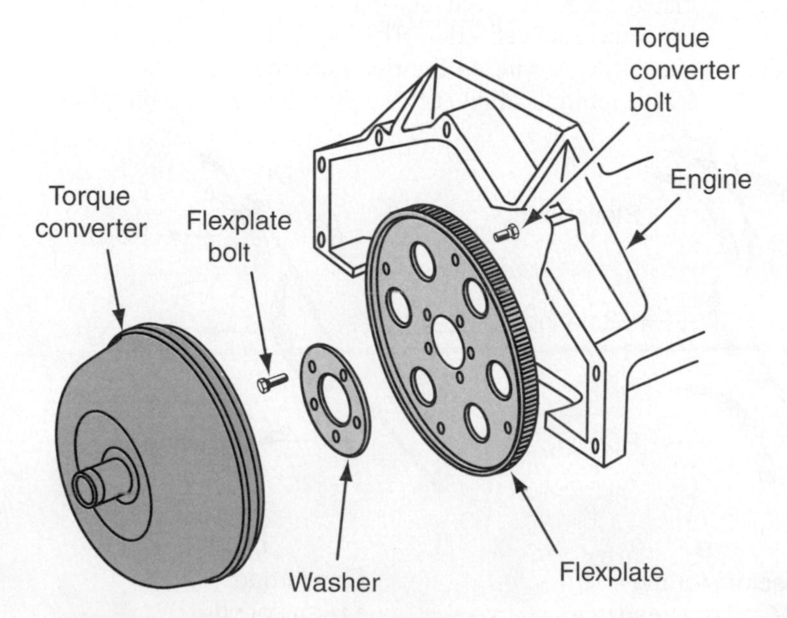

FIGURE 41–4 The flexplate connects the engine crankshaft to the torque converter.

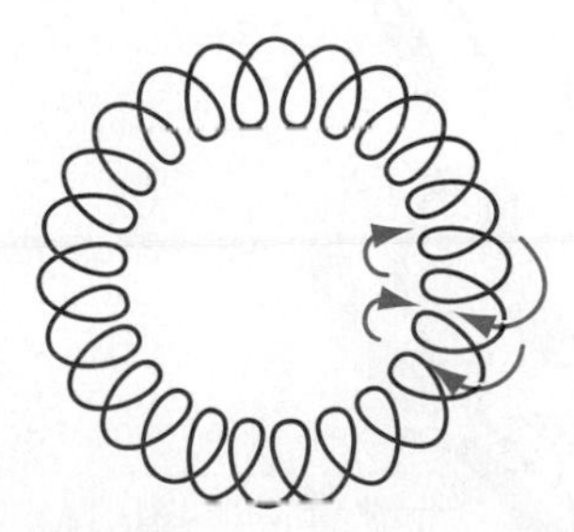

Oil flows inside the torque converter from the pump, to the turbine, to the stator, and back to the pump.

FIGURE 41–5 Oil flows in a circular fashion throughout the torque converter.

TORQUE MULTIPLICATION

A torque converter is capable of producing different torque ratios. When starting, the torque converter multiplies torque. The torque may be increased 2–3.5 times the engine torque. As the speed of the turbine blades increases, the torque is continuously reduced until the torque ratio is 1 to 1. At this point, the pump and turbine are spinning at the same speed.

Torque on the turbine wheel is a direct result of how the oil is deflected from the blades. In ***Figure 41–6***, when oil hits the turbine vane, the vane is forced to deflect. It has a greater **deflection angle** when the turbine vane is stationary or at low speeds. The greater the deflection angle, the greater the torque on the turbine wheel. When the turbine wheel is stopped, there is a maximum deflection angle. As the turbine blade increases in speed, the oil is not deflected as much. This is because the turbine blade is now moving to the right. Since the turbine blade is also moving, there is less deflection. Less deflection means less torque multiplication.

Figure 41–7 shows a graphic of the oil flow through all three components inside the torque converter. ***Figure 41–7A*** shows the pump rotating with the turbine and guide wheel (stator) stopped. Notice that there is maximum angle of deflection of oil, causing greater torque multiplication.

The oil flows from the torque converter to the stator. Here the oil flow direction wants the stator to turn to the left, but the stator cannot turn that way. The stator is designed so that it can rotate only to the right. It is supported by a one-way, freewheeling sprag or roller clutch drive system (***Figure 41–8***). The sprag clutch drive system allows rotation in one direction but locks up in the opposite direction. The oil then is redirected to the correct angle to reenter the pump.

As the speed of the turbine increases, the angle of deflection decreases, producing less torque multiplication. This is shown in ***Figure 41–7B***. The stator is still not moving because there is still pressure to move it to the left. Notice that the pattern of oil flow is becoming straighter.

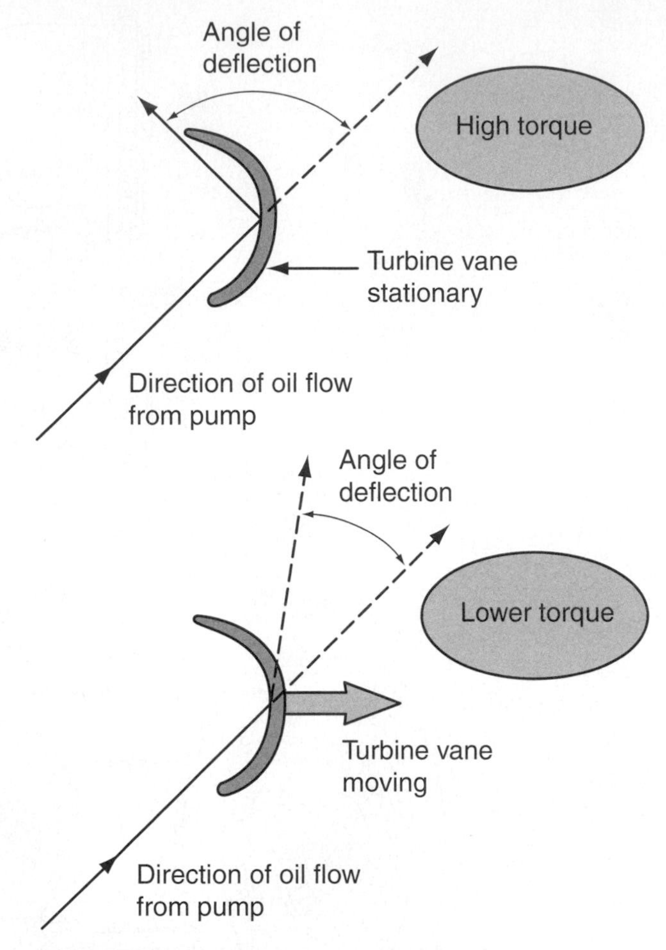

FIGURE 41–6 Torque is multiplied inside the torque converter. The greater the angle of deflection of oil, the greater the torque. As the speed of the turbine increases, torque decreases.

Figure 41–7C shows the pump and turbine at approximately the same speed. During this condition, the oil flow has little or no deflection. The speed of the oil and the turbine blade, moving to the right, are the same. There is no torque multiplication at this point. However, the oil flow is

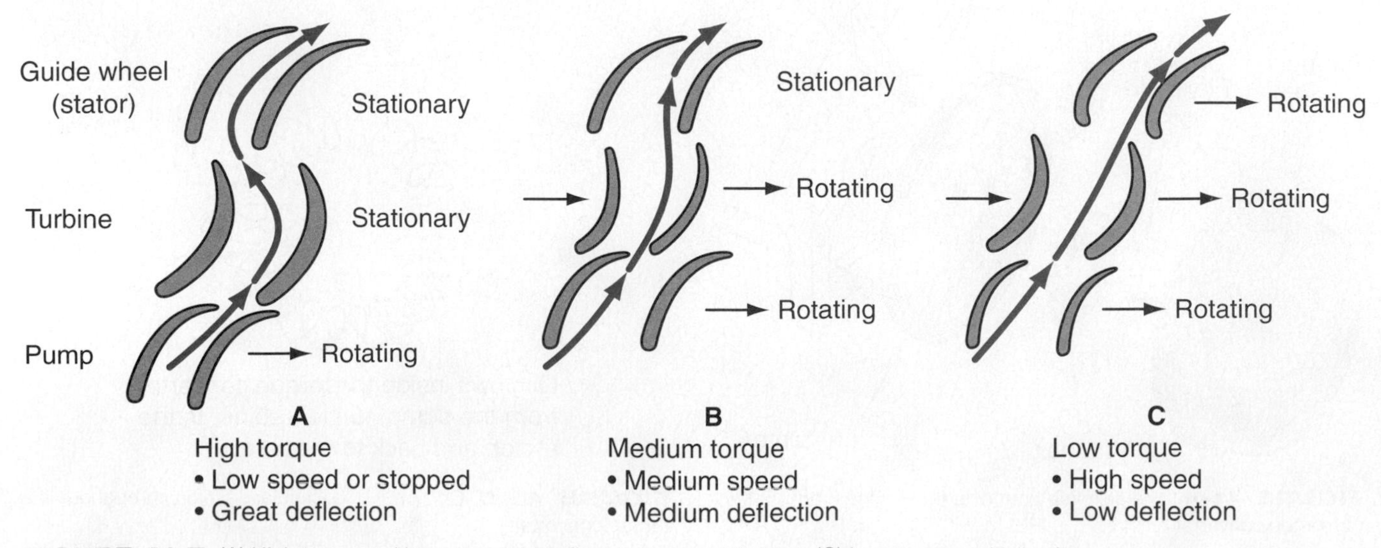

FIGURE 41–7 (A) High torque multiplication, (B) Medium torque multiplication, (C) Low torque multiplication

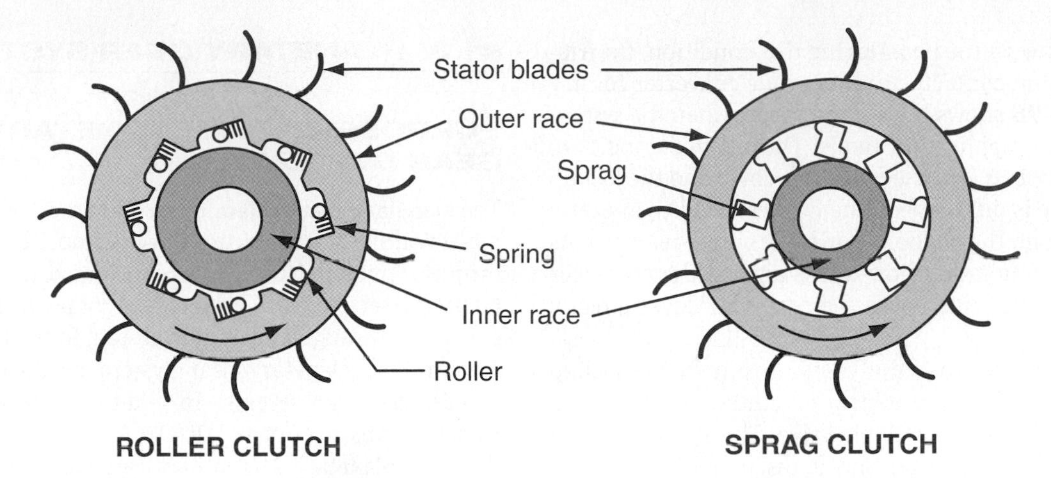

FIGURE 41–8 The stator must turn in one direction only. It is prevented from turning in the other direction by either a roller or sprag clutch drive system. When the inner hub is held stationary, the outer stator vanes can rotate only in a counterclockwise direction. When forced in the opposite direction, the stator vanes cannot turn.

now hitting the back of the stator or guide wheel blades. If left this way, the oil will not flow back to the pump correctly. To eliminate this problem, the stator starts to turn to the right. The freewheeling action of the stator is now in effect.

TORQUE CONVERTER LOCKUP

Torque converters always operate with some loss. There is usually a certain slip between the turbine speed and the pump speed. The pump speed may be 2–8% faster than the turbine speed. In older transmissions, this loss is not accounted for, but it does reduce gasoline mileage.

Today's automatic transmissions have a lockup system between the torque converter turbine and the engine. The effect of locking up these two components is similar to direct drive. Actually, the torque converter is no longer a part of the power flow. It is being bypassed. The overall effect is that during high gear or speed operation, there is no transmission slippage, and fuel efficiency improves.

Figure 41–9 shows a schematic of the parts and oil flow through the torque converter during clutch release and clutch application. Note that several parts are added to the torque converter. A lockup plate is placed between the torque converter turbine and the torque converter cover. This plate is made of a friction disc and a damper spring. On the right or outside of the lockup plate assembly is the friction disc. This friction disc matches up with a machined surface on the inside of the torque converter cover. A damper spring (not shown) is used to push the lockup plate to the turbine. It is used to help dampen engine power pulses coming from the drivetrain.

There are two modes of operation. The first mode operates during the first several gears (not high gear). During this time, the **lockup clutch** is released or in the clutch off position. Refer to ***Figure 41–9A***. Oil pressure from the transmission is sent through the clutch shuttle valve to the area between the lockup plate and the torque converter cover. The effect is that the oil allows for a certain amount

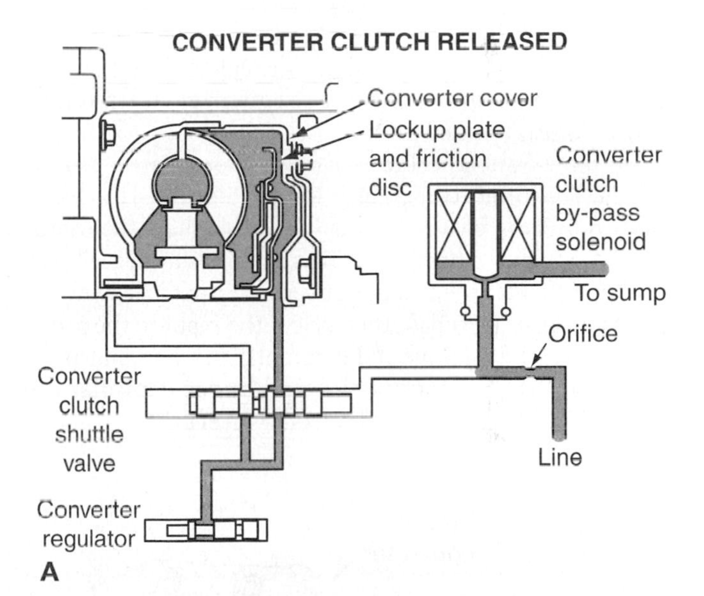

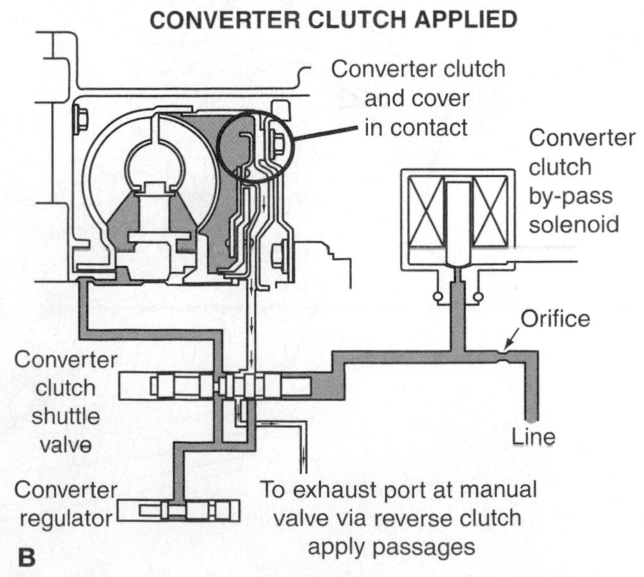

FIGURE 41–9 Most torque converters have a lockup mechanism. Its purpose is to lock up the torque converter in high gear to improve fuel economy.

of slippage between the two. During this condition, the friction disc is not in contact with the torque converter cover.

Figure 41–9B shows the second mode, when the system is in the clutch application mode. During this condition, the oil pressure between the converter cover and the lockup plate assembly is discharged. The converter oil then exerts pressure through the piston, against the converter cover. As a result, the friction disc now works against the converter cover. This action firmly locks the converter cover directly to the turbine.

The oil pressure from the transmission to the lockup clutch is controlled by a converter clutch shuttle valve. This valve is also called an apply valve. The shuttle valve is controlled by the computer and is discussed later in this chapter.

TORQUE CONVERTER LOCATION

On rear-wheel drive vehicles, the torque converter is located directly behind the engine between the flexplate and the transmission gears. The torque converter and transmission are one combined unit.

On some front-wheel drive cars, a different arrangement is used. The torque converter is located alongside the transmission. A large heavy chain is used to connect the torque converter to the transmission. This chain is shown in ***Figure 41–10***. Note that the torque converter output is driving the chain and gear. The chain then drives the input to the planetary gears and final drive and differential. This arrangement is used so that the transmission and torque converter can easily be positioned to allow for front-wheel drive.

41.3 PLANETARY GEAR SYSTEM

PURPOSE OF THE PLANETARY GEAR SYSTEM

The **planetary gear system** is used to produce different gear ratios within the automatic transmission. The output of the torque converter (turbine) is considered the input to the planetary set of gears. Power flow comes from the torque converter, through a hydraulic clutch, to the planetary gear system. The planetary gear system produces either low speeds, drive, or reverse. In addition, certain automatic transmissions may have D1, D2, D3, and overdrive. More than one planetary gear system can also be used on transmissions to produce more gear ratios.

PARTS LOCATOR

*Refer to photo #82 in the front of the textbook to see what a **planetary gear system** looks like.*

PARTS OF THE BASIC PLANETARY GEAR SYSTEM

All planetary gear systems have several common parts. They include the:

1. Sun gear
2. Ring gear
3. Planet gears (also called pinion gears)
4. Planet gear carrier

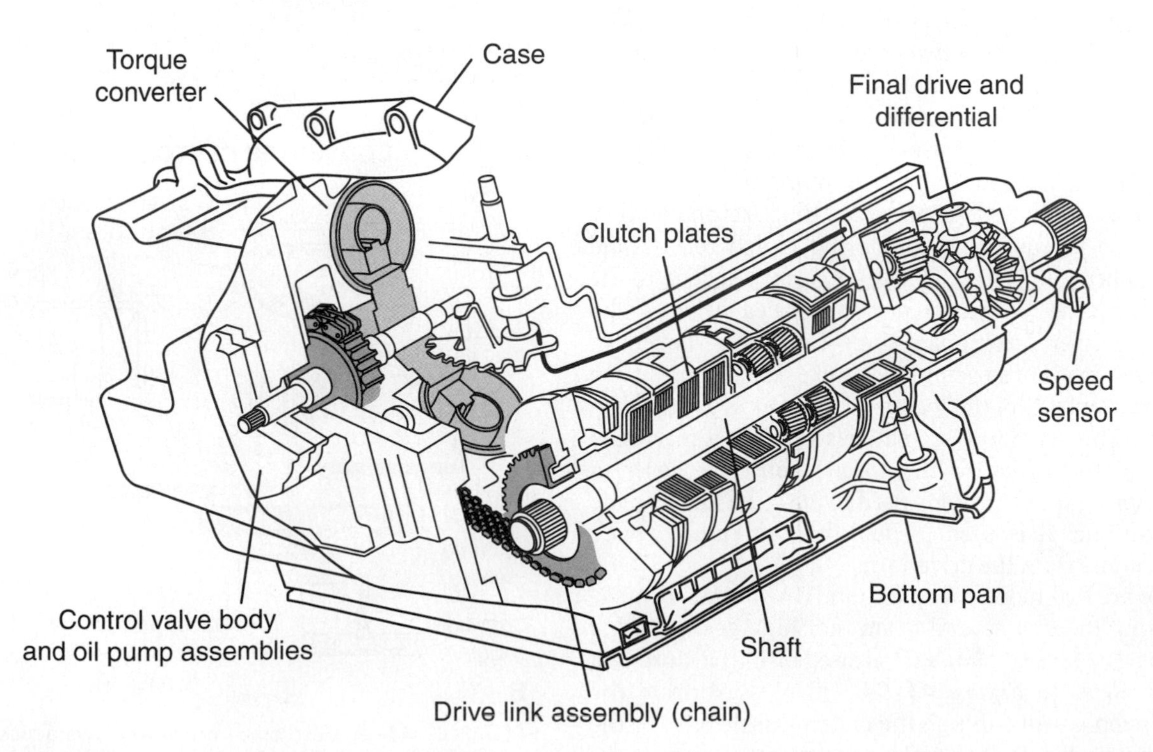

FIGURE 41–10 On front-wheel drive vehicles, a heavy chain (also called the drive link assembly) is used on some transmissions to connect the torque converter to the transmission case.

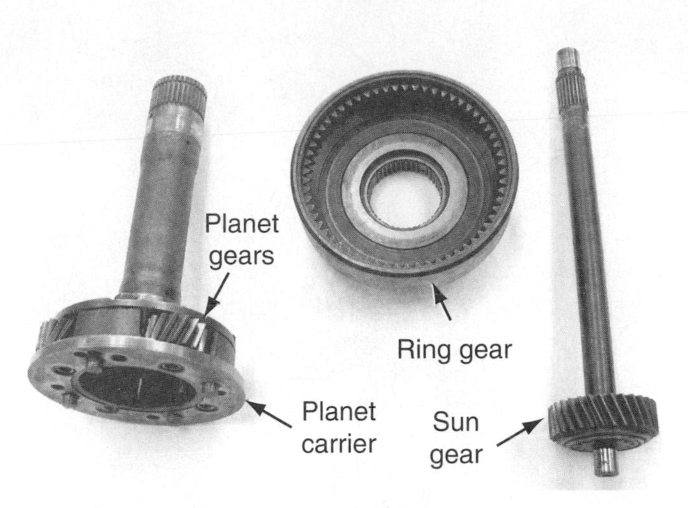

FIGURE 41–11 A planetary gear system is made of a sun gear, a ring gear, several planet gears, and a planet carrier.

Figure 41–11 shows a simple planetary gear system. The sun gear is the center gear. The planet (pinion) gears are held in position by the **planet carrier.** The ring gear is an internal gear that surrounds all of the planet gears. All gears are helical in design and are in constant mesh.

OPERATION OF THE PLANETARY GEAR SYSTEM

To select different gear ratios, one of three gears, the sun, ring, or planets, must be held stationary. The input and output then occur on the remaining gears. For example, if the ring gear is held stationary and the input is the sun gear, the carrier is the output with a lower speed. The most common arrangement for low gear is to have the sun gear held stationary, the input on the ring gear, and the output on the carrier.

Eight combinations can be produced in a planetary gear system. These are shown in ***Figure 41–12***. For example, look at gear combination number 3. In this case, the sun gear is held stationary. The ring gear is the input and the carrier is the output. Under these conditions, the output speed will be lower than the input speed. This means that the transmission is in a low gear for starting or low-speed acceleration. In addition, direct drive occurs when any two gears of the planetary gear system are turned at the same speed. The third gear must then rotate at the same speed. In this condition, the planet gears do not rotate on their shafts. The entire unit is locked together to form one rotating part. Neutral occurs when all members of the planetary gear system are free. When the transmission is placed in park, a small pawl is engaged with the teeth on the planet carrier.

Overdrive conditions can also be obtained by using planetary gears. Overdrive can be used when the sun gear or the ring gear is held stationary. Input would be as shown in ***Figure 41–12***, condition 1 or 2.

VARIATIONS IN THE PLANETARY GEAR SYSTEM

There are several variations to the basic planetary gear system. One type of planetary gear system is called the simple type. It has two sets of planet gears. Another type of planetary gear system uses two sets of planet gears, two sun gears, and a planet carrier. ***Figure 41–13*** shows this system, called the compound planetary gear system. On this

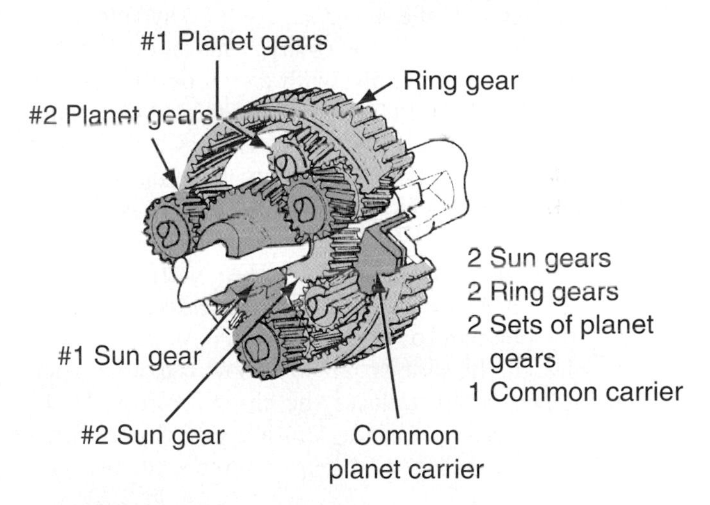

FIGURE 41–13 A compound planetary gear system has two sets of planet gears and two sun gears. (Courtesy of DaimlerChrysler Corporation)

GEAR COMBINATIONS

GEAR	1	2	3	4	5	6	7	8
RING	Output	Hold	Input	Hold	Input	Output	Hold any two gears	Free all gears
CARRIER	Input	Input	Output	Output	Hold	Hold		
SUN	Hold	Output	Hold	Input	Output	Input		
SPEED CHANGE (Input to Output)	Increase	Increase	Lower	Lower	Increase Reverse	Lower Reverse	Direct Drive	Neutral

FIGURE 41–12 Eight combinations of gear ratios can be produced in a planetary gear system. By holding one gear stationary and changing the input and output, various gear ratios can be achieved.

compound planetary gear system, note the location of the #1 and #2 sun gears. Also, note that there are two sets of planet gears. The ring gear for the second set of planet gears is also shown.

41.4 CLUTCHES, BANDS, AND SERVO PISTONS

PURPOSE OF CLUTCHES AND BANDS

The planetary gear system provides the gear ratios, but the clutches and bands control which gears are locked up or released. The clutches and bands are used to lock up the correct gear to get the right gear ratio. Depending on which clutch or band is activated, one member of the planetary gear system is held, while the other is driven.

MULTIPLE-DISC CLUTCH

One common clutch used in automatic transmissions is called the **multiple-disc clutch** (***Figure 41–14***). It is made of a series of friction discs placed between steel discs or plates. The exact number of discs depends on the vehicle manufacturer, the load, and so on. The friction discs are the drive component and the steel discs are the driven (output) component. ***Figure 41–15*** shows a typical friction disc. The disc is made of a steel center with a friction material on both sides. The friction discs or composition-faced plates have rough gripping surfaces. The steel discs have smooth metal surfaces. These two components make up the input and output of the clutch. The clutch pack also has a piston and return springs.

When fluid pressure is applied to the clutch, the piston moves and compresses the clutch pack together. The action locks up the input and output of the clutch. When the pressure is released, the springs help to remove the pressure on the discs. This action unlocks the clutch. ***Figure 41–16*** shows an exploded view of the multiple-disc clutch with a clear view of the springs that help to remove the pressure on the discs. The springs are located inside the clutch pack housing.

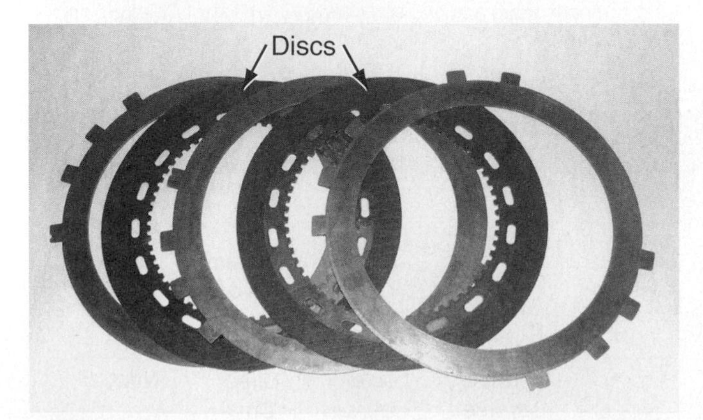

FIGURE 41–14 Multiple-disc clutches use both friction discs and steel discs. When pressure is applied, the two lock up and transmit the necessary torque.

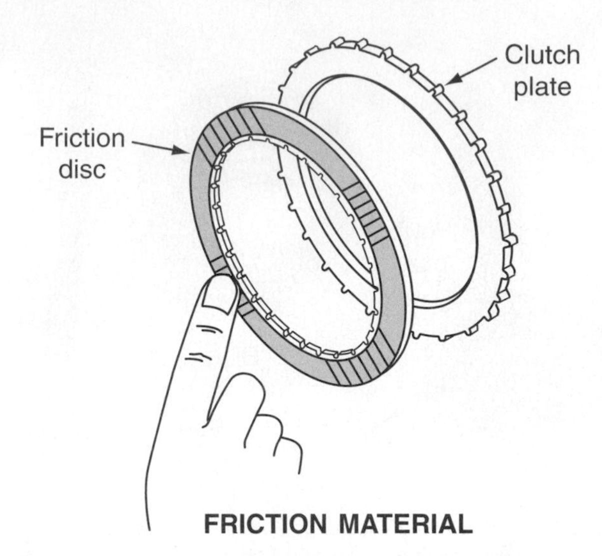

FIGURE 41–15 Friction material is placed on a friction disc to allow for friction.

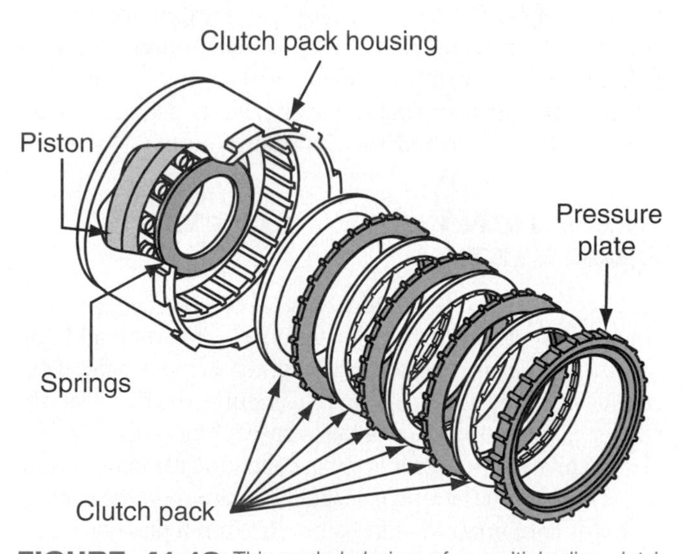

FIGURE 41–16 This exploded view of a multiple-disc clutch shows the piston, springs, and other operational parts.

This type of clutch could be used to connect two shafts together. For example, the torque converter output must be connected and disconnected to the planetary gear system. The mechanism that makes this connection is called the forward clutch, and it uses a multiple-disc clutch.

TRANSMISSION BAND

Another type of clutch is called the **transmission band** (***Figure 41–17***). This transmission band is located around the clutch housing or drum. As the band is tightened down, the clutch housing is held stationary. The band is tightened or loosened to hold or free the clutch housing or drum. If the clutch housing is attached to the ring gear, then control of the ring gear is accomplished by this type of transmission band.

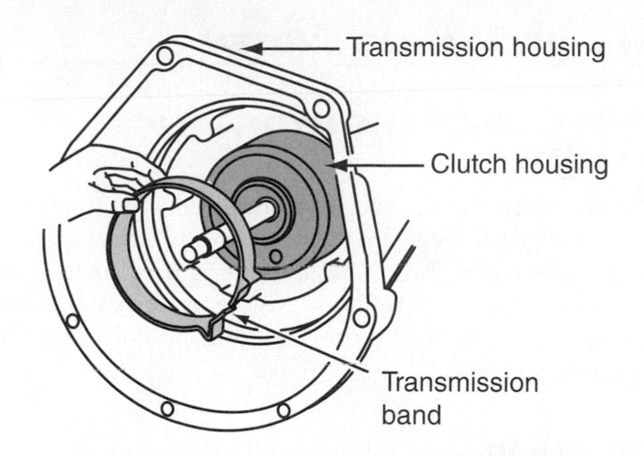

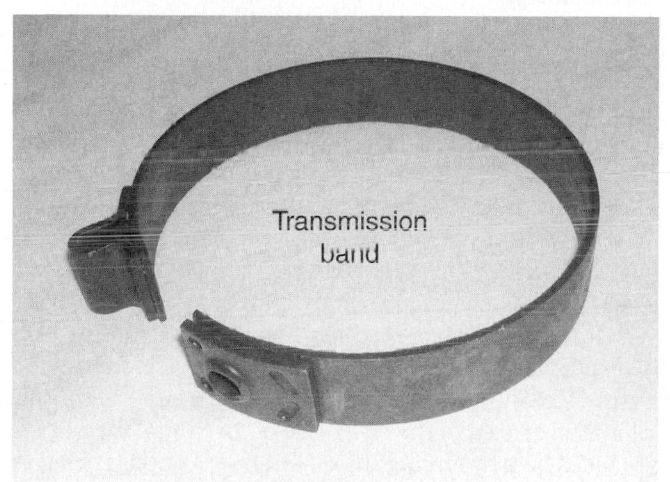

FIGURE 41–17 A transmission band is used to lock up a clutch housing.

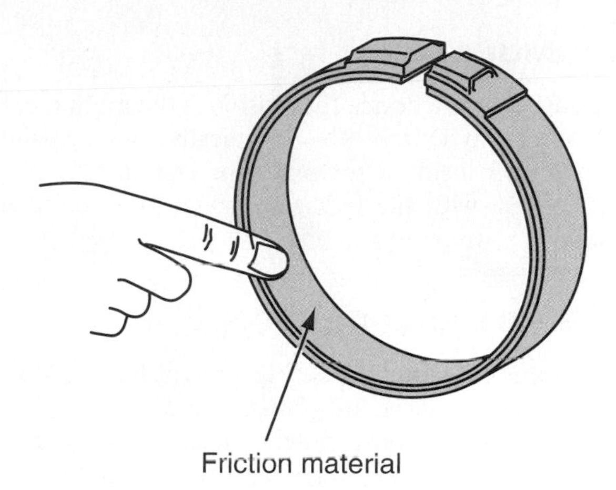

FIGURE 41–18 This transmission band has a friction material to allow for friction.

The inside of the band is coated with a friction material. The friction material is used to help grip the clutch housing of the drum. ***Figure 41–18*** shows the friction material located on the inside of the band.

SERVO PISTON

The **servo** piston is used to control the transmission band operation. In ***Figure 41–19A***, the band is controlled by a servo piston. The servo piston is made of the case, piston, piston stem, spring, and cover. When oil pressure from the hydraulic system inside of the transmission is applied, it pushes the servo piston to the right. This action causes the piston stem to move and tighten the band. When oil pressure is removed, the spring pushes the servo piston back and releases the band. The band clearance can be changed by turning the adjusting screw.

Figure 41–19B shows a second type of servo piston. In this case, as the oil is applied from the transmission hydraulic system, it pushes the servo piston down. This action causes a lever to move downward. The lever has a pivot point so that the right side of the lever moves upward. This causes the band to tighten around the clutch housing.

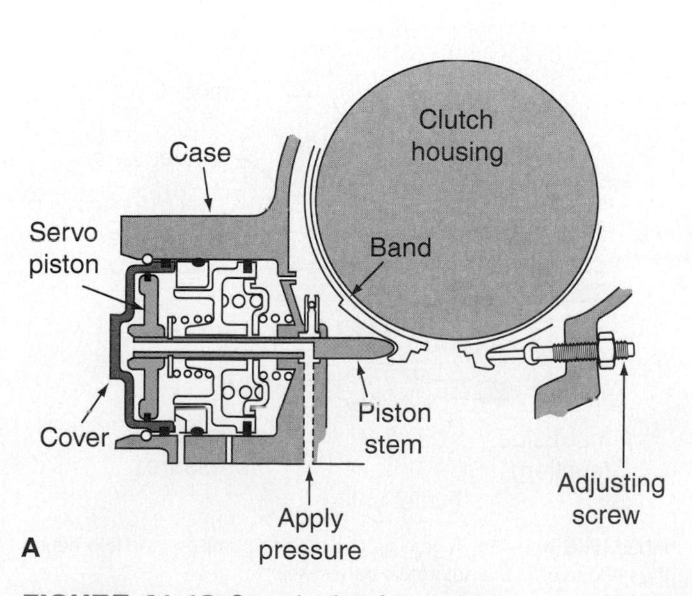

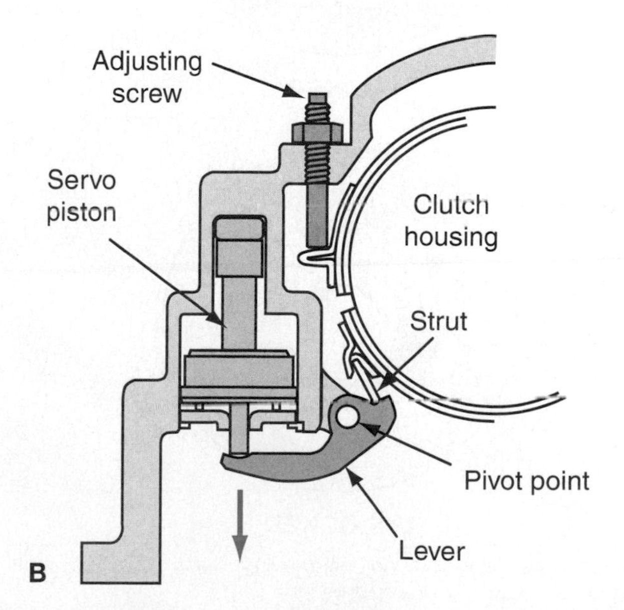

FIGURE 41–19 Several styles of servo pistons can be used to operate the transmission band.

ACCUMULATOR

An **accumulator** is a device that cushions the motion of the clutch and servo actions. This is typically done by using a smaller piston inside a servo piston. The smaller piston makes contact with the linkage quicker, thus acting as a cushion.

OVERRUNNING CLUTCH

The overrunning clutch is used to prevent backward rotation of certain parts of the planetary gear system during shifting. The overrunning clutch shown in ***Figure 41–20*** is made of the inner hub, outer cam, and a series of rollers and springs. It operates the same as the sprag drive mentioned earlier, allowing rotation in only one direction.

It is difficult to get a smooth upshift from low to drive during shifting. If one band releases slightly before the second band engages, the engine may have a rapid increase in rpm during shifting. The overrunning clutch provides smooth engagement and disengagement without rapid increase in rpm. This improves the shift quality and timing from low to drive gear.

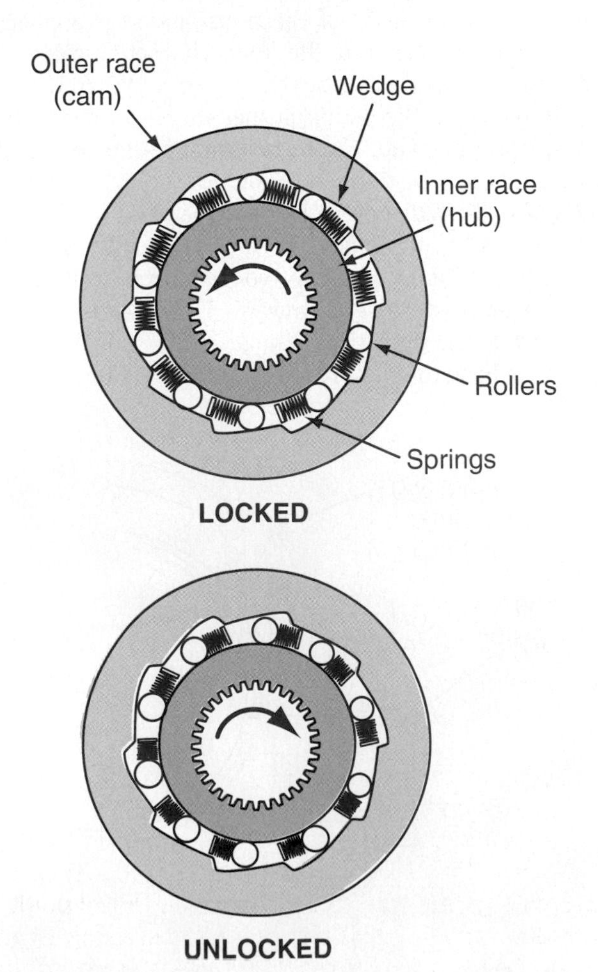

FIGURE 41–20 An overrunning clutch assembly provides smooth shifting from low to drive.

41.5 HYDRAULIC SYSTEM

PURPOSE OF THE HYDRAULIC SYSTEM

The automatic transmission is controlled by the hydraulic system. When the operator shifts into drive, low, reverse, or any other gear, hydraulic pressure is used to lock up different clutches and bands on the planetary gear system. The torque converter also uses transmission fluid from the hydraulic system.

OIL PUMP

All pressurized oil in the automatic transmission is produced by the transmission oil pump. It creates the pressure used to lock up the clutches or operate the servo to lock up the transmission bands. It also sends oil to the torque converter.

Several types of oil pumps are used on automatic transmissions. These may include the gear, vane type, or rotor type. These are all positive displacement pumps. ***Figure 41–21*** shows how a common transmission oil pump operates. It uses a gear-type, eccentric pump. As the inside gear is turned, a suction is created on the left side. Automatic transmission fluid is drawn in between the gear teeth. The oil is then carried to the other side of the pump. Pressure is produced here by forcing the oil out from between the gear teeth. The relief valve is used to bypass high-pressure oil back to the suction side. This happens only when higher rpm is producing oil pressure above the recommended pressure.

The oil pump most often is built into the front of the transmission case, directly behind the torque converter. The pump is driven by the torque converter. The torque converter is inserted into the transmission body. Two **tangs**, or feet, on the inner gear come into contact with the torque

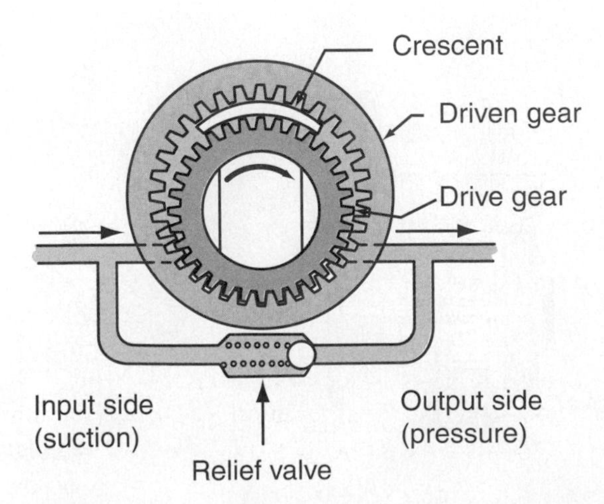

FIGURE 41–21 A gear-type, eccentric pump is used to produce the needed oil in the automatic transmission.

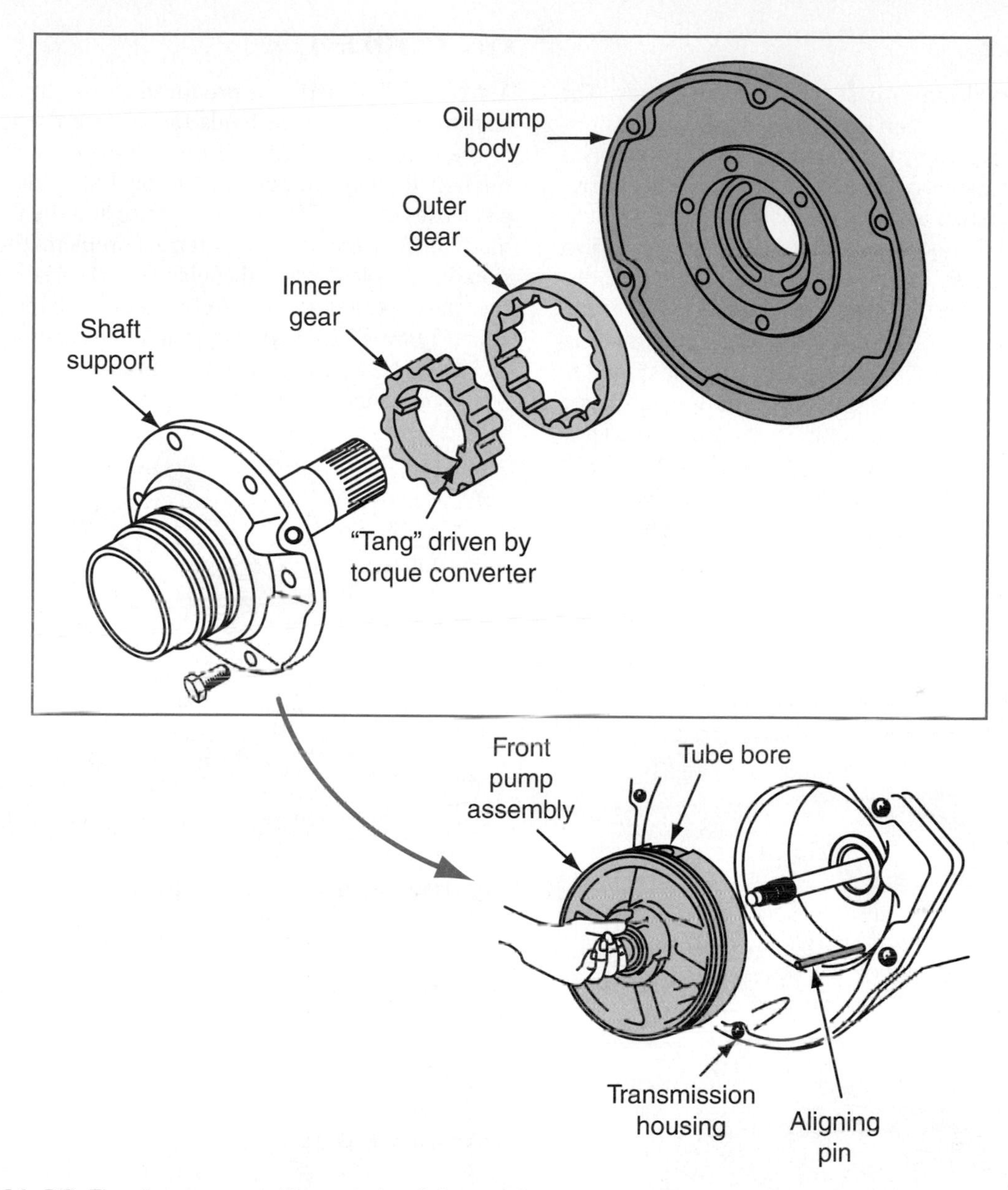

FIGURE 41–22 The oil pump on most transmissions is located directly in front of the transmission. It is driven by the torque converter.

converter. ***Figure 41–22*** shows an oil pump and body near the front of the transmission.

AUTOMATIC TRANSMISSION FLUID

In the transmission fluid there are high shearing and pressure conditions. The automatic transmission fluid must withstand extreme pressure, friction, and shearing stresses. In addition, both cold and hot conditions, which might change viscosity, may exist. Because of all these conditions, the automatic transmission fluid must have the following:

- Corrosion inhibitors
- Detergents
- Pour-point depressants
- Friction modifiers
- Antifoam agents
- Viscosity-index improvers

Because of these special characteristics and requirements, only the recommended fluid should be used. The fluid is generally reddish or brownish. It is often identified as **automatic transmission fluid (ATF)**. Two types of automatic transmission fluid are recommended by most vehicle manufacturers. One type is referred to as Dexron III AFT. Another type is called Mercon V ATF. Always follow the manufacturer's recommendation when selecting ATF.

VALVE BODY

The oil from the oil pump is sent to the **valve body**. The valve body is often located on the underside of the transmission case and is covered by the transmission oil pan. This is true when the transmission is used on a rear-wheel drive vehicle. However, when used on a front-wheel drive vehicle, the valve body can be located either under the gears as just mentioned or vertically on the side or back of the transmission. Inside of the valve body, there are a series of valves and worm tracks or small oil passageways. These worm tracks carry the transmission oil to the proper hydraulic circuit to engage or disengage various clutches. The separator is a plate with holes that feed the oil from the worm tracks to the specific holes in the transmission housing. A series of valves, controls, and springs inside the valve body control the shifting of the automatic transmission. ***Figure 41–23*** shows a typical valve body.

PARTS LOCATOR

Refer to photo #83 in the front of the textbook to see what an ***automatic transmission valve body*** *looks like.*

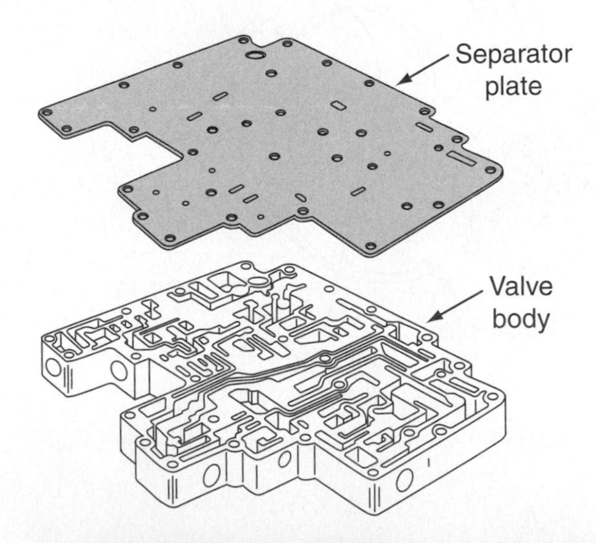

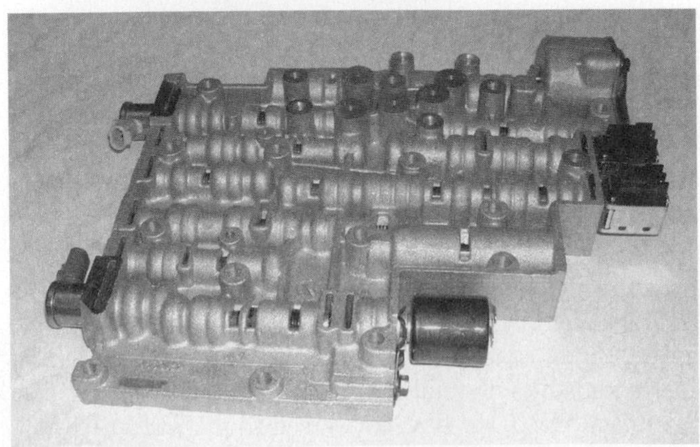

FIGURE 41–23 The valve body and separator plate are used to house the transmission controls. The separator plate is also used to direct the oil to the proper part for correct operation.

OIL COOLER

A great deal of friction is produced within the clutches on transmissions. Friction tends to increase the temperature of the transmission fluid. If the temperature of the transmission fluid is allowed to get too high, the lubricating properties of the ATF may deteriorate and the transmission may fail. To overcome this potential problem, the transmission oil is cooled by an oil cooler. Oil is typically sent out of the transmission to a heat exchanger placed inside the radiator (***Figure 41–24***). Transmission fluid operates about 40 to 50 degrees Fahrenheit hotter than engine coolant. Most heat is produced by the friction in the clutches and the

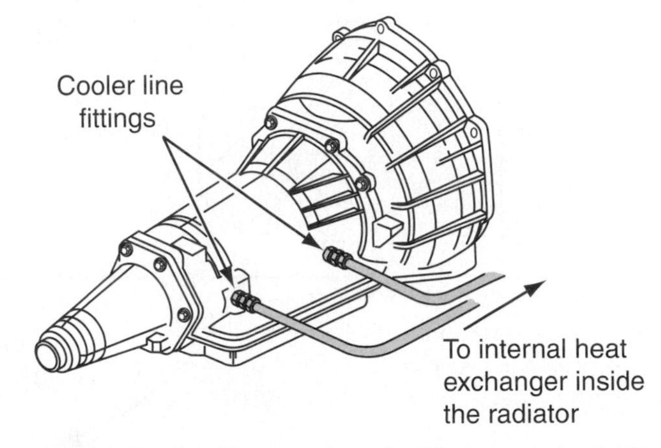

FIGURE 41–24 The cooler lines direct the transmission fluid to the heat exchanger in the radiator.

CUSTOMER CONCERN:

Towing Problem

A person would like to use a vehicle with a 6-cylinder engine to tow a trailer and boat. What problems might occur using the standard radiator and transmission?

SOLUTION: The most common problem when towing a boat or trailer is engine overheating. The engine must do more work, consequently, it will run hotter and the transmission will be under more stress. The transmission is cooled by the engine coolant. If the engine is running hotter, the transmission fluid is also hotter. Damage to the transmission could result.

The best protection is to have the manufacturer install a towing package. The most common package includes an extra transmission cooler placed on the front of the radiator. Some manufacturers also recommend changing the rear end differential to a different ratio. The cost is much higher, but the engine will not be loaded as much.

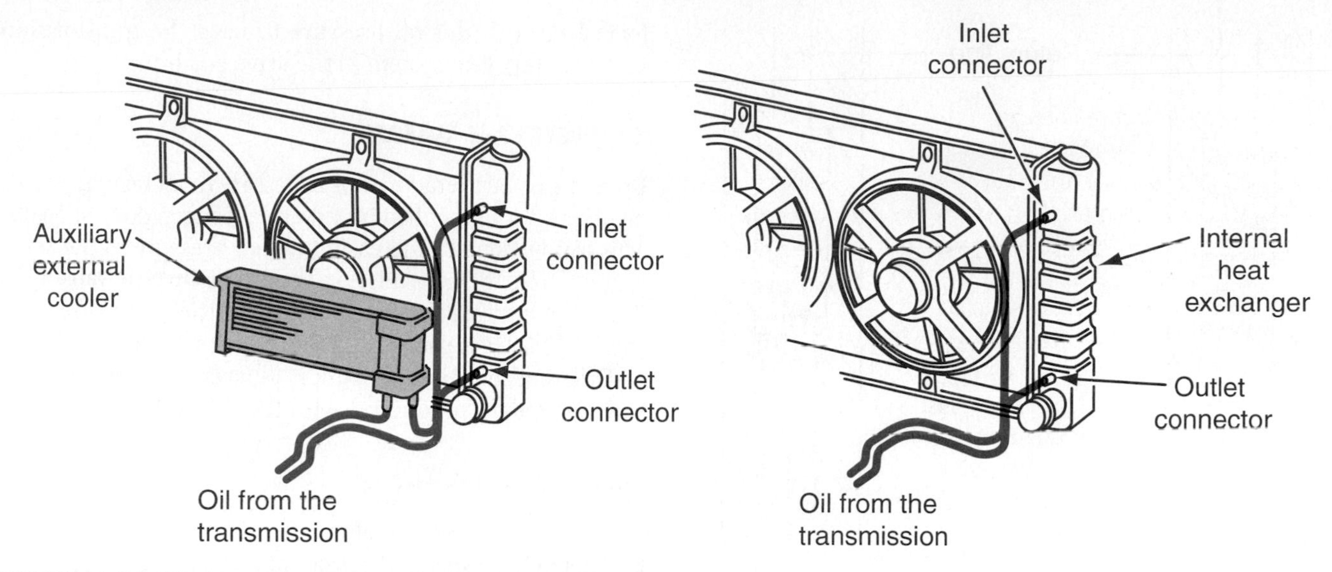

FIGURE 41-25 Two styles of external heat exchangers for transmission cooling.

torque converter. Engine coolant absorbs thermal energy from the transmission to cool the fluid. Vehicles that have excess loads placed on them may also have an additional external or auxiliary cooler on the front of the radiator. ***Figure 41–25*** shows both the internal heat exchanger in the radiator as well as the auxiliary or external cooler.

41.6 TRANSMISSION CONTROLS

PURPOSE OF TRANSMISSION CONTROLS

Several controls are used to lock up or release the clutches and transmission bands to control the shifting. In addition, the torque converter clutch must be controlled. The type of valves used are determined by the year and manufacturer of the transmission. Pressure regulator valves, kickdown valves, shift valves, governors, throttle valves, vacuum modulators, and manual valves are used. A working knowledge of these valves will help you to visualize the internal operation of different transmissions. Certain valves are also used to increase the smoothness and quality of shifting.

PRESSURE REGULATOR VALVE

A **pressure regulator valve** is used to regulate the amount of oil pressure. Various oil pressures are required within the transmission. Certain pressures are needed to control locking of clutches and other valves. The pressure regulator valve uses a **spool valve** to control the oil pressure as shown in ***Figure 41–26***. The design of this control valve uses a spool in the center of the valve. The spool is held in place by the balance between spring pressure one way and oil pressure the other. As the pressure increases from the transmission oil pump, oil pressure eventually pushes the spool valve to the right, which opens a port. When the port is opened, some of the oil is returned back to the oil pump. This action reduces or controls the oil pressure from the pump. There are several designs of spools valves but their basic operation is the same.

There are also other types of pressure controls valves used in automatic transmissions. The valves shown in ***Figure 41–27*** use either a ball check valve or a poppet valve operation. In both cases, as oil pressure is felt below the check ball or below the poppet valve, no oil flow is allowed. The check ball or poppet valve is held down by the spring tension above. However, if the oil pressure increases too much, the ball or poppet valve will lift off its seat, allowing oil to flow past the valve and back to the source.

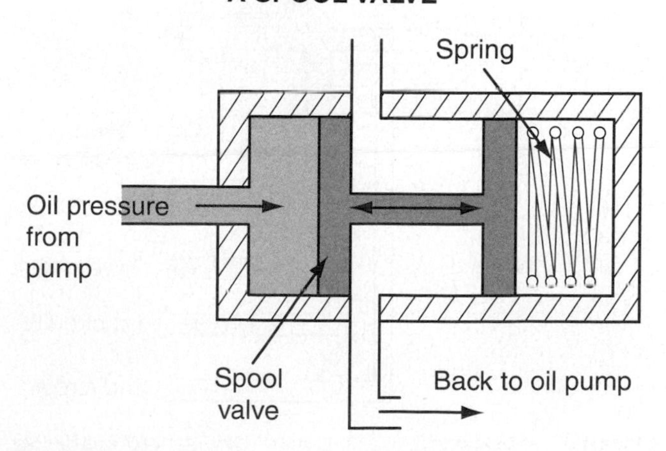

FIGURE 41-26 This simple spool valve controls oil pressure from the transmission oil pump.

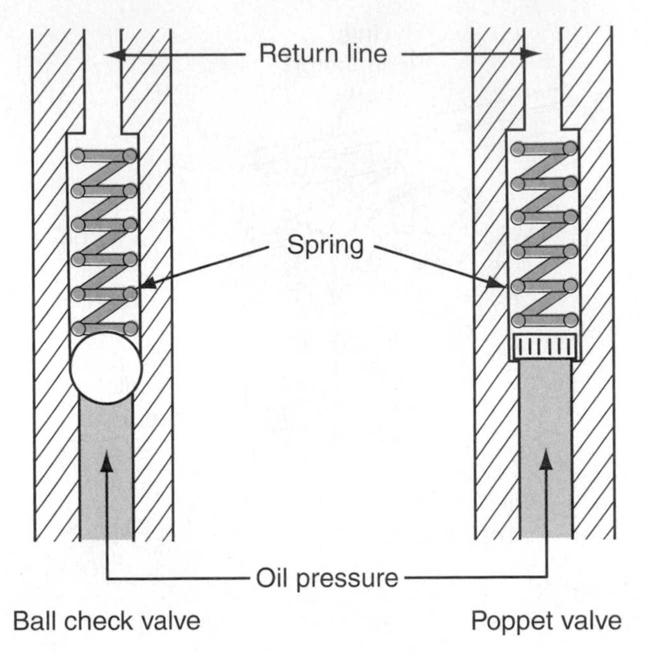

FIGURE 41-27 Both the ball check valve and the poppet valve are used to control transmission fluid pressure.

MANUAL SHIFT VALVE

Manual shifting is done by placing the vehicle shift lever in the proper position inside the car. Depending on the vehicle, park, reverse, neutral, drive, and low are common positions to place the shift lever. Many cars also have a D1 and D2. When the operator shifts the lever, a set of shift linkages causes a **manual shift valve** to move inside the valve body. ***Figure 41–28*** shows the valve that is moved. It is very similar to a spool valve.

Note that as the manual shift valve is moved right or left, various oil passageways are opened or closed. In its present position, oil comes in from the oil pump, through the spool valve, and into the drive circuit. The clutches are now locked up with this oil pressure to keep the transmission and planetary gear system in the drive position.

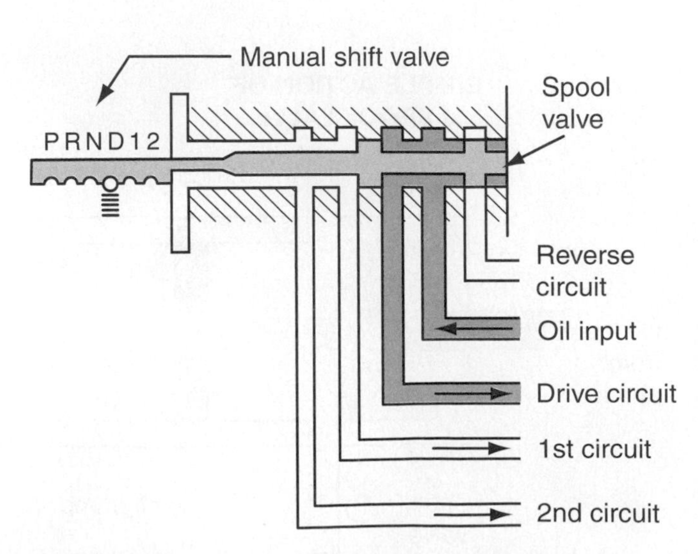

FIGURE 41-28 When the gearshift lever is moved, a manual shift valve is operated inside the transmission valve body. The movement of this valve sets different hydraulic circuits into operation.

THROTTLE VALVE

Under heavy acceleration, it is necessary to increase the force on the transmission bands and clutches to reduce slippage. This can be done by increasing the oil pressure to the servo that controls the transmission band. The **throttle valve (TV)** is used to increase pressure. As the throttle is increased, the throttle linkage is used to change the position of the valve.

Different positions produce different pressures. This valve is used in conjunction with the shift valve.

GOVERNOR

A governor is used to help shift the transmission on the basis of vehicle speed. For example, when the vehicle is in low or first gear, it must automatically shift to second or drive at a certain speed. Also, as the vehicle slows to a stop, the transmission must downshift. This is done by using a governor assembly as shown in ***Figure 41–29***. The speed of the vehicle is sensed by the governor from the output shaft on the transmission. A set of small weights moves out and in from centrifugal force. Centrifugal force works against a spring force. These two forces acting against each other cause the valve to open and close a set of hydraulic ports. The output oil pressure controlled by the governor is sent to the **shift valve** to activate the necessary shift oil circuits (***Figure 41–30***).

SHIFT VALVE

The throttle valve and the governor do not actually send oil to the clutches. The oil is first sent to the shift valve. Here, governor oil pressure works against throttle valve oil pressure to position the shift valve. ***Figure 41–31*** shows a simplified example of this valve. Governor oil pressure enters on the left side. Throttle valve pressure enters at the top. These two pressures balance the valve. Anytime this balance is disturbed, the shift valve moves upward or downward. Line pressure (pressure used to control the clutches) is also directed to the the shift valve on the bottom. This pressure

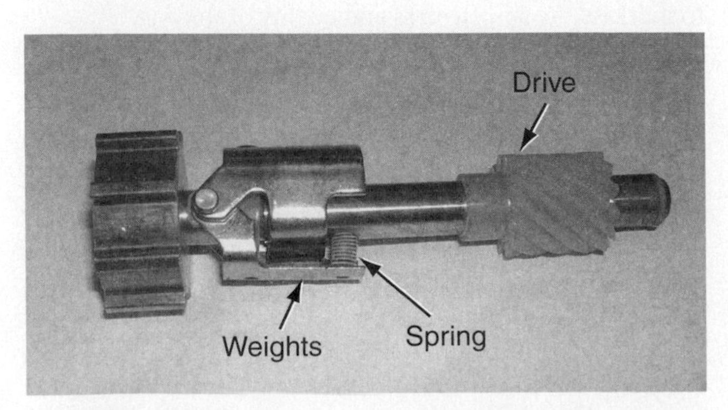

FIGURE 41-29 A governor is used to sense engine speed and to shift the transmission accordingly.

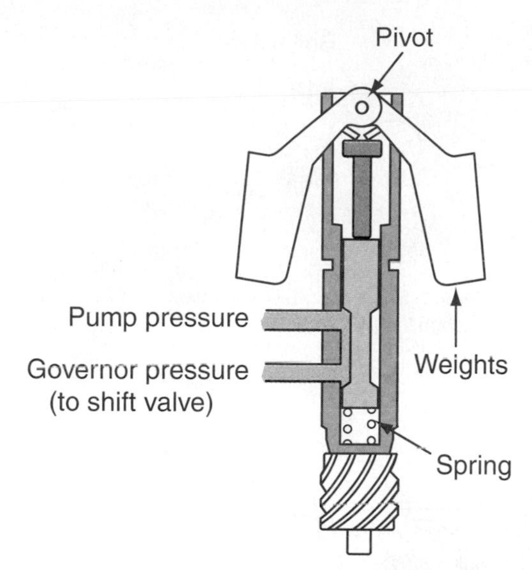

FIGURE 41–30 As the weights move outward due to increases in transmission speed, the spool valve in the center is positioned to allow hydraulic pressure to shift gears.

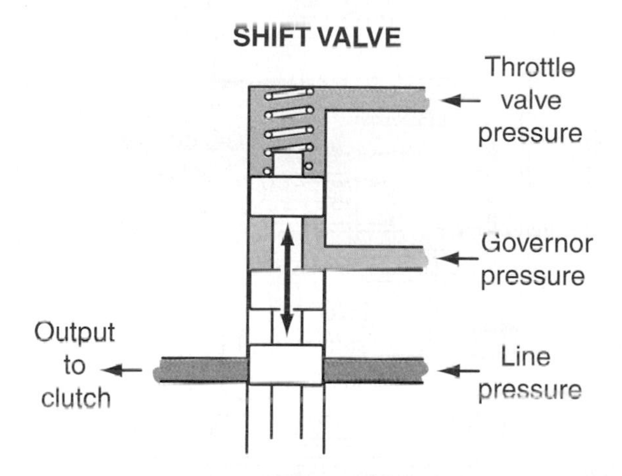

FIGURE 41–31 A shift valve is controlled by the throttle valve pressure working against the governor oil pressure. Its position tells exactly when and to which clutch the oil pressure should be sent.

does not enter into shift valve movement. This pressure is simply ready to be directed to the correct clutch in the planetary gear system to lock up or release the correct clutch. The shift valve is actually more complicated than this and has several outputs to different clutches.

VACUUM MODULATOR

Another way in which engine load can be determined to help shifting is by using a **vacuum modulator** valve instead of the throttle valve. Both valves are used in vehicles today.

As the throttle opening is increased, the intake manifold vacuum decreases. The intake manifold vacuum is sent to the transmission and is hooked to a vacuum diaphragm. Spring pressure acting against the vacuum causes a rod to be moved. The rod controls the position of the throttle valve. ***Figure 41–32*** shows a vacuum modulator.

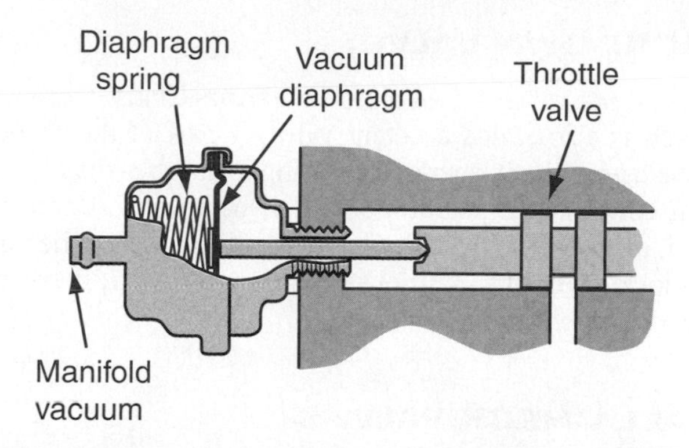

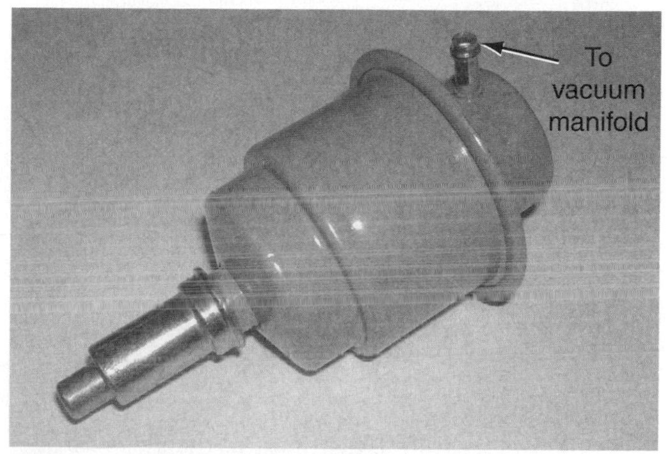

FIGURE 41–32 The vacuum modulator is used to sense engine load on the basis of the intake manifold vacuum.

PARTS LOCATOR

Refer to photo #84 in the front of the textbook to see what a ***vacuum modulator on an automatic transmission*** *looks like.*

Car Clinic: VACUUM MODULATOR

CUSTOMER CONCERN:
Hard Shifting and Loss of Transmission Fluid
A customer complains of continually adding transmission fluid to an automatic transmission that does not show signs of leaking. Also, the customer complains of very hard downshifting during closed throttle.

SOLUTION: The vacuum modulator on the automatic transmission cushions shifting and is connected to a vacuum source on the engine intake manifold. When the modulator diaphragm ruptures, the shifting will become hard and engine vacuum will draw transmission fluid into the engine where it is burned in combustion with the fuel. The solution is to replace the vacuum modulator.

KICKDOWN VALVE

A kickdown valve is used on some transmissions. This valve, which is also called a detent valve, is part of the throttle valve assembly. It operates by using a variable-shaped cam that controls the position of the throttle valve. Under full load, or when the throttle is pushed all the way to the floor, the kickdown valve positions the throttle valve in such a position as to shift to a lower gear.

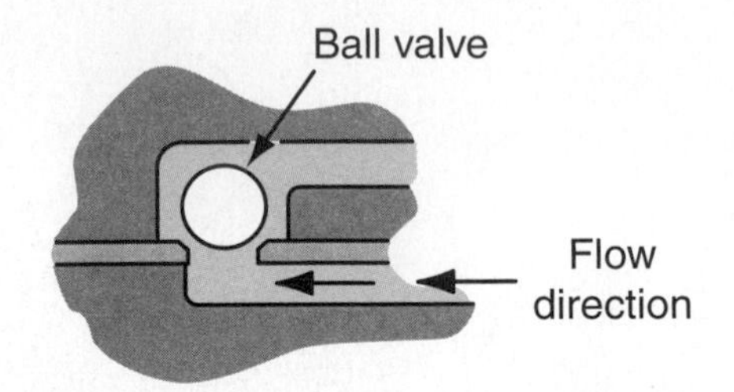

FIGURE 41–33 Ball check valves are used to control the direction of oil flow through the passageways.

BALL CHECK VALVES

Ball check valves are used to control the direction of oil flow in the transmission. Check valves prevent flow until a certain pressure is reached. They also close passages to prevent back flow of transmission fluid. ***Figure 41–33*** shows the action of a ball check valve.

OIL FLOW CIRCUITS

Many oil flow circuits are outlined in vehicle service manuals. ***Figure 41–34*** shows a typical oil flow circuit. It is not necessary to understand the complete flow and operation of this diagram. It is more important to observe how all the components work together to control the fluid in the automatic transmission. Note that most of the major components have already been discussed. These components include the oil pump, manual and shift valves, accumulator, regulator valve, servo, governor, oil cooler, clutches, torque converter, pressure relief valve, ball check valves, and converter clutch control valve.

TORQUE CONVERTER CLUTCH VALVE

As mentioned earlier, the torque converter is locked up at certain times. The oil pressure used to lock up the torque converter is controlled by:

1. A **torque converter clutch (TCC) valve**, and
2. The converter signal, controlled by the converter clutch shift valve.

Figure 41–35 illustrates the operation of these valves. To begin, when the TCC is not energized its upper spring pressure positions the valve so that oil pressure is sent to the lockup plate, through the torque converter, and back to the TCC valve. The arrows show the direction of flow of oil through the circuit. At this point, the converter clutch shift valve is not sending a hydraulic signal to the TCC.

If the position of the converter clutch shift valve is changed, oil pressure would be sent to the right side of the TCC. The pressure pushes the TCC to the left into a new position. In this position, the resulting oil pressure would be used to move the clutch lockup plate assembly to the right. This action would lock up the torque converter.

The TCC is also controlled by the on-board computer. The powertrain control module (PCM), for example, con-

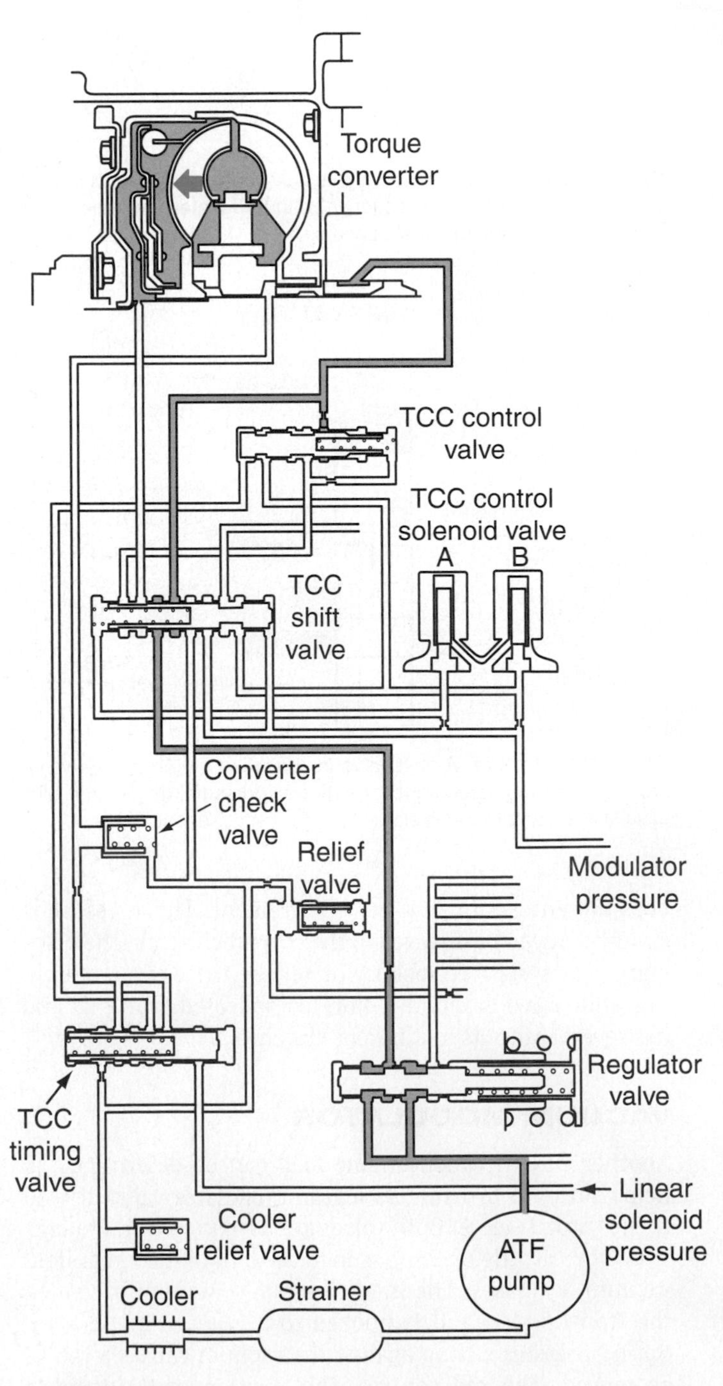

FIGURE 41–34 Oil flow circuits are used to help understand the complete oil flow in an automatic transmission.

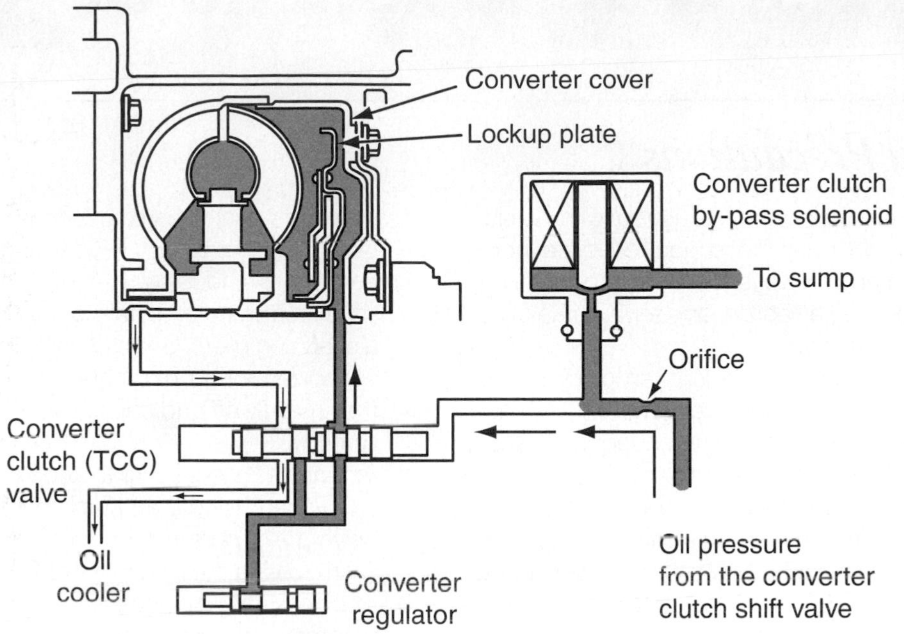

FIGURE 41–35 The converter clutch valve (or torque converter clutch, TCC) is used to control the oil flowing to the lockup torque converter during release and apply conditions.

trols the operation of the by-pass solenoid. Normally, the solenoid is positioned so the oil is bled off to the sump. When all conditions are met, the PCM energizes the solenoid. The result is that the converter signal pressure is no longer bled off. Now the oil pressure is able to build up and move the TCC. The result is that the TCC is positioned to apply oil so the clutch is locked up. Although the circuitry is more complicated, ***Figure 41–36*** shows the common inputs to the PCM used to control the TCC solenoid.

NEUTRAL START SWITCH

On most vehicles, there is an electrical neutral start switch. This switch protects the vehicle from starting when the shift linkage is in any position except neutral or park. If the vehicle were able to start in a forward gear, a possible safety hazard could exist.

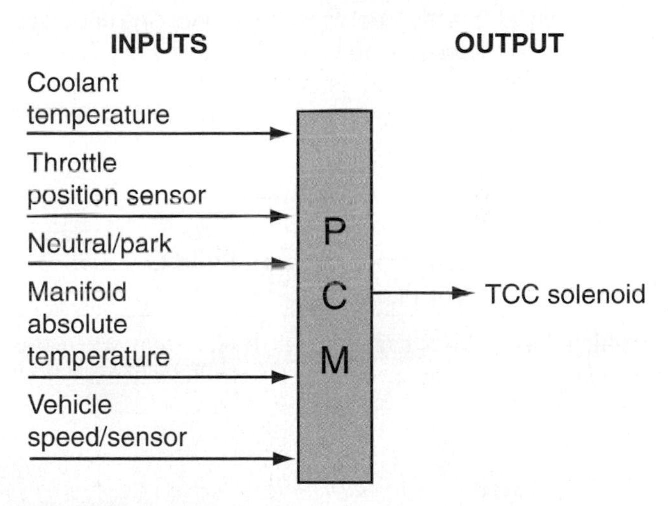

FIGURE 41–36 The torque converter clutch (TCC) solenoid is controlled by several inputs.

Problems, Diagnosis, and Service

Safety Precautions

1. When you are working on an automatic transmission, the vehicle must be supported on a hoist or hydraulic stand. Always follow the correct procedure when lifting a vehicle, and always use extra jack stands for safety.
2. The automatic transmission is often used as the rear support for the engine. Never remove the automatic transmission without first adequately supporting the engine in the vehicle.
3. When you remove the automatic transmission, you may spill ATF onto the floor. To eliminate any possibility of slipping, which could cause serious injury, wipe up any spilled fluid immediately.
4. Transmission fluid operates above the temperature of the radiator. Be very careful when checking the fluid when the vehicle has just been stopped. A serious burn could result.
5. When disconnecting the torque converter from the flywheel, the flywheel must be turned. The ring gear on the flywheel is very sharp and you could accidentally scrape your knuckles and hands. Be very careful, and use the correct tools when removing the torque converter.
6. When installing gears, bearings, and gear shafts, be careful not to pinch your fingers between the part and the housing.
7. When removing or installing snap rings, always use a snap ring pliers. Snap rings have a high degree of spring tension. If not installed with a correct tool, they may fly off and cause eye, face, or other bodily injury.
8. At times it is necessary to keep the vehicle running and in gear to test an automatic transmission. Always make sure the emergency brake is applied when testing the transmission.
9. At times it may be necessary to run the engine and automatic transmission on a lift, in gear. Be careful of the rotating front or rear wheels while servicing the vehicle in this mode of testing.
10. Always wear OSHA-approved safety glasses when working on an automatic transmission.
11. Always use the proper tools when working on an automatic transmission.
12. Always use the special tools that are recommended for service on an automatic transmission.

PROBLEM: Low Transmission Fluid Level from Leaks

A vehicle has a slight transmission leak. Lately, when the engine is cold, the transmission seems to slip when accelerating.

DIAGNOSIS

When the transmission fluid is low and the transmission is cold, there is the possibility of slippage in the transmission bands. If the fluid level gets too low, complete loss of power may occur, especially when the transmission is cold. This condition will further damage the transmission internally. The leak is most likely in the front or rear seal on the transmission.

Always check the transmission fluid for the correct level. The fluid level should be checked when the transmission is at operating temperature (about 200 degrees Fahrenheit) and in park or neutral while the engine is running (***Figure 41–37***).

CAUTION: *When adding fluid, do not overfill because foaming and loss of fluid may occur through the vent.*

SERVICE

If the transmission fluid is leaking from the front, the front seal is damaged or not sealing. The front seal is located behind the torque converter on the oil pump housing. Replace the seal according to the manufacturer's recommended procedure. Use the following general procedure to remove the transmission.

CAUTION: *Remove the ground cable on the battery before starting removal.*

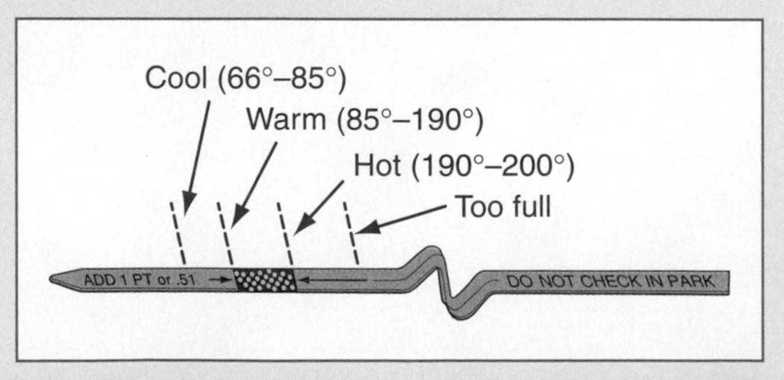

FIGURE 41–37 Follow the manufacturer's recommendations when checking the fluid level.

1. Disconnect the battery cable and remove the transmission dipstick.
2. Lift the car up on a suitable car hoist or jack. Remember to follow the safety guidelines when operating the car hoist.
3. Remove the drive shaft from between the transmission and the differential. On front-wheel drive vehicles, remove the front axles.
4. Remove the shift linkage connected to the transmission.
5. Remove and mark all electrical connections to the transmission.
6. Remove the transmission cooler lines.
7. Remove the speedometer cable.
8. Remove any vacuum connections to the transmission and mark for reassembly.
9. Drain the fluid from the transmission into a suitable container by removing the drain plug (if equipped) or by removing the transmission oil pan.
10. Remove the flywheel undercover.
11. Remove the bolts that hold the torque converter to the flywheel. There are usually three bolts. Mark the flywheel and converter for reference during installation.
12. Using a suitable jack, lift the rear of the engine to take the weight off the rear transmission mount.
13. Remove the transmission mount from the transmission.
14. Place a transmission jack stand under the transmission to support it once it has been disconnected from the engine block.
15. Attach an engine support fixture as shown in ***Figure 41–38***. This fixture is used to support the engine while the transmission is removed from the vehicle.
16. Slowly and carefully lower the engine until the bolts between the transmission torque converter housing and the block of the engine can be removed.

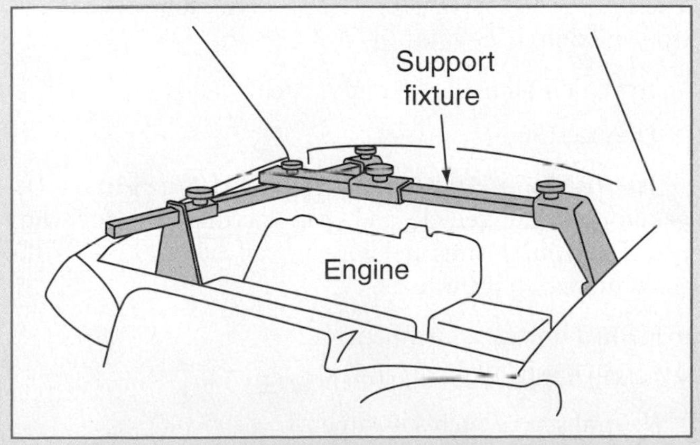

FIGURE 41–38 Always install an engine support fixture before removing the automatic transmission.

CAUTION: *Lowering the engine too far could press certain parts against the firewall and cause them to break or be damaged.*

17. After the bolts between the transmission housing and the block have been removed, slowly lower the transmission jack stand and transmission away from the engine.

CAUTION: *Be careful not to drop the torque converter because it is very heavy and filled with transmission fluid.*

18. Use a suitable converter-holding tool to secure the converter.
19. With the transmission on the bench, remove the torque converter and place it aside.
20. The front seal is used to seal the transmission fluid where the torque converter inserts into the transmission. Remove the front seal using a suitable tool or seal puller.
21. Install a new front seal. Make sure a suitable liquid sealant has been placed around the outside of the seal.
22. To install the transmission, reverse the removal procedure.

PROBLEM: Dirty Transmission Fluid

A vehicle's transmission fluid has been checked and seems to be dirty. There are no apparent driving problems. However, there is varnish or gum on the dipstick.

DIAGNOSIS

It is recommended that the automatic transmission fluid and filter be changed at regular intervals. Refer to the owner's manual in the vehicle to determine when to change the fluid and filter.

SERVICE

Use the following general procedure to change the fluid and filter on the transmission.

1. Using a suitable jack, support the vehicle so that work can be done easily on the bottom of the transmission. The vehicle is usually placed on a car hoist.

CAUTION: *Follow all safety procedures when lifting a vehicle up on a hoist.*

2. Drain the transmission fluid into a suitable container by loosening the oil pan and letting the transmission fluid drain out. There may also be a drain plug or filler tube connection that can be removed. Make sure to dispose of the automatic transmission fluid according to state or local regulations.

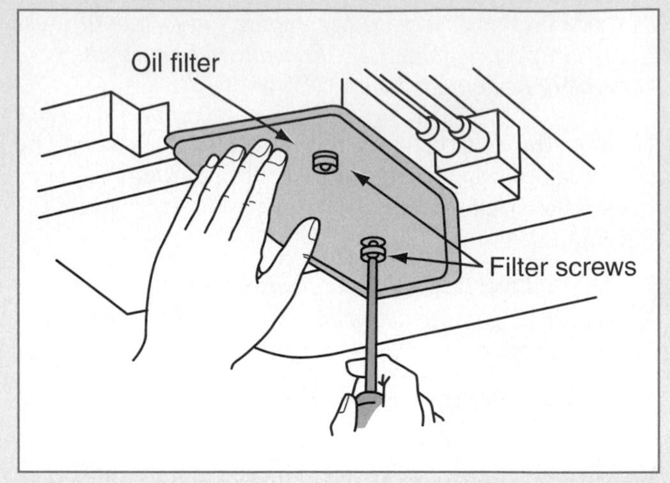

FIGURE 41–39 Replace the filter and oil at the recommended intervals. Always replace the filter and gasket in their proper positions.

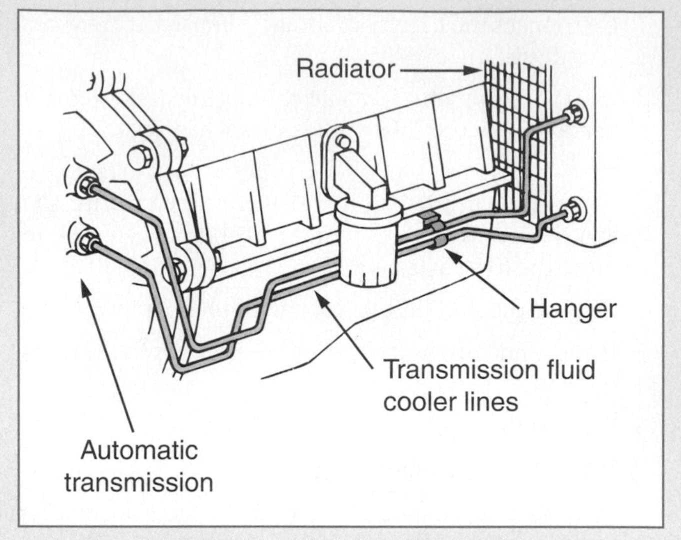

FIGURE 41–40 Inspect the transmission fluid cooler lines for damage or leakage. Support the lines with the correct type of hanger.

3. Using the correct size socket wrench, remove the transmission oil pan. At this point, the transmission filter can be observed.
4. Remove the oil filter and replace it with a new one. Replace the oil filter and gaskets in their proper positions. Failure to do so may damage the transmission (***Figure 41–39***).
5. To reassemble, reverse the preceding procedure.

PROBLEM: Damaged Clutches or Bands

A vehicle has hard shifting from first to second gear.

DIAGNOSIS

Generally, this type of problem (and other shift problems) indicate damaged clutches or bands. Several diagnostic checks can be done. When diagnosing transmission shift problems, always check the following:

1. Periodically inspect the automatic transmission cooler lines for signs of damage. Worn or crimped lines may cause ATF loss or restriction. Prevent future damage by supporting the lines with the proper clamp or hanger as shown in ***Figure 41–40***.
2. Check the ATF for a burned odor, darkened color, or metal particles. A burned odor may indicate overheating and, possibly, damaged clutches. Metal particles may indicate damaged gears or extremely worn clutches.
3. If the engine cooling system is not working properly (overheating), the transmission may operate at higher temperatures as well. Higher oil temperature may damage the clutches and transmission bands. Check the cooling system if overheating is suspected.
4. Bubbles in the transmission fluid indicate the presence of a high-pressure leak in the transmission. A leak such as this may be in the valve body, valve, or hydraulic passageways. If such a leak is suspected, the transmission hydraulic system will need to be checked and repaired. Normally, the valve body can be replaced as a unit.
5. A poorly tuned engine that has low or insufficient vacuum may cause shifting problems. This is because several components on the transmission operate off a vacuum. Before testing any part of the automatic transmission, make sure that the engine vacuum is correct.
6. Inspect all linkages to the transmission for binding and damage. Incorrect linkage adjustments may cause bad shifting.
7. Numerous diagnosis flowcharts are available for further transmission diagnosis. ***Figure 41–41*** is an example of one such chart. Charts are usually available for each make and year of transmission.

SERVICE

There are two general categories of service for an automatic transmission.

1. In-vehicle adjustments and service
2. Overhaul

In-Vehicle Adjustments and Service Depending on the vehicle and manufacturer, the following types of adjustments and service can be made with the transmission in the vehicle:

- Manual linkage adjustments
- Downshift linkage adjustments
- Neutral start switch adjustments
- Bands adjust
- Valve body replace
- Governor replace

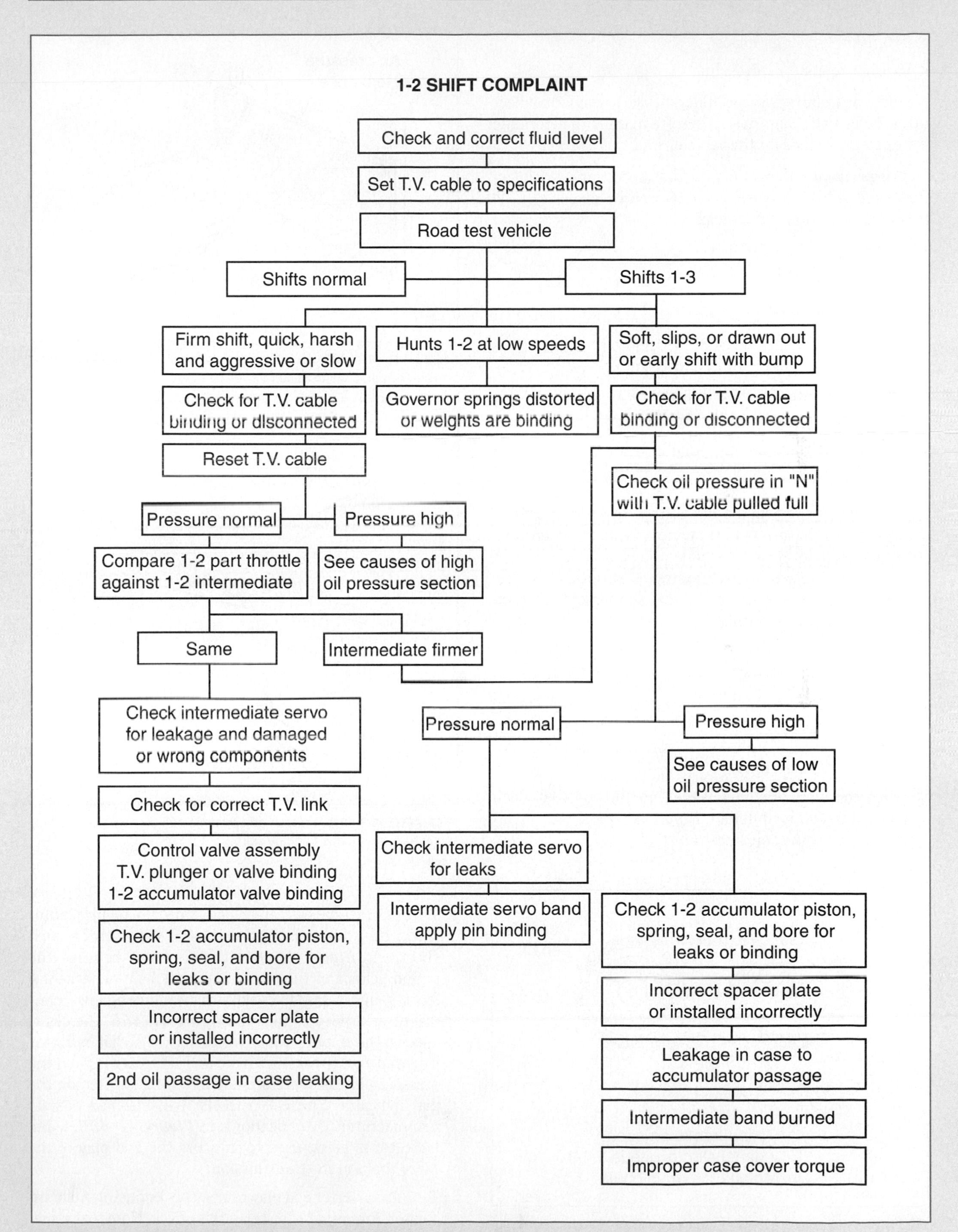
1-2 SHIFT COMPLAINT
Check and correct fluid level
Set T.V. cable to specifications
Road test vehicle
Shifts normal
Shifts 1-3
Firm shift, quick, harsh and aggressive or slow
Hunts 1-2 at low speeds
Soft, slips, or drawn out or early shift with bump
Check for T.V. cable binding or disconnected
Governor springs distorted or weights are binding
Check for T.V. cable binding or disconnected
Reset T.V. cable
Check oil pressure in "N" with T.V. cable pulled full
Pressure normal
Pressure high
Compare 1-2 part throttle against 1-2 intermediate
See causes of high oil pressure section
Same
Intermediate firmer
Check intermediate servo for leakage and damaged or wrong components
Pressure normal
Pressure high
See causes of low oil pressure section
Check for correct T.V. link
Control valve assembly T.V. plunger or valve binding 1-2 accumulator valve binding
Check intermediate servo for leaks
Intermediate servo band apply pin binding
Check 1-2 accumulator piston, spring, seal, and bore for leaks or binding
Check 1-2 accumulator piston, spring, seal, and bore for leaks or binding
Incorrect spacer plate or installed incorrectly
Incorrect spacer plate or installed incorrectly
Leakage in case to accumulator passage
2nd oil passage in case leaking
Intermediate band burned
Improper case cover torque

FIGURE 41–41 Flowcharts are used to help diagnose problems in automatic transmissions.

- Extension housing replace
- Vacuum regulator valve adjust

Each procedure will be different with each manufacturer. Refer to the appropriate service manual to determine the exact procedure for the vehicle.

Overhaul Depending on the vehicle and manufacturer, the following types of service can be made with the transmission out of the vehicle:

- Transmission disassembled
- Governor assembly
- Front end play check
- Oil pump disassembly and assembly
- Forward clutch disassembly and assembly
- Sun gear and sun drive shell service
- Valve body service
- Low and reverse roller clutch service
- Transmission assembly

Each procedure will be different with each manufacturer. Refer to the appropriate service manual to determine the exact procedure for the vehicle.

The following are selected examples of both in-vehicle and out-of-vehicle service checks and adjustments that are done on automatic transmissions:

1. On front-wheel drive vehicles that use a chain drive, check the chain for excessive wear (***Figure 41–42***).
2. Various air pressure checks are recommended during disassembly of automatic transmission. These checks determine the condition of various passageways and components (***Figure 41–43***).
3. Often the manual shift linkage must be adjusted. Each procedure will be different. ***Figure 41–44*** shows a typical linkage being adjusted.

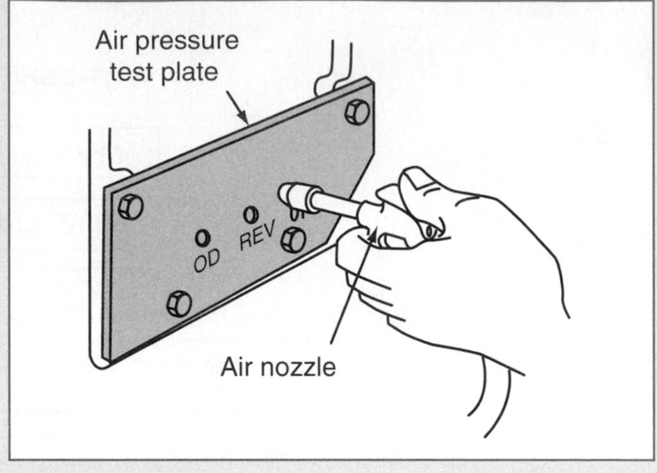

FIGURE 41-43 Air passageways in the transmission can be checked by using an air hose with 40 psi.

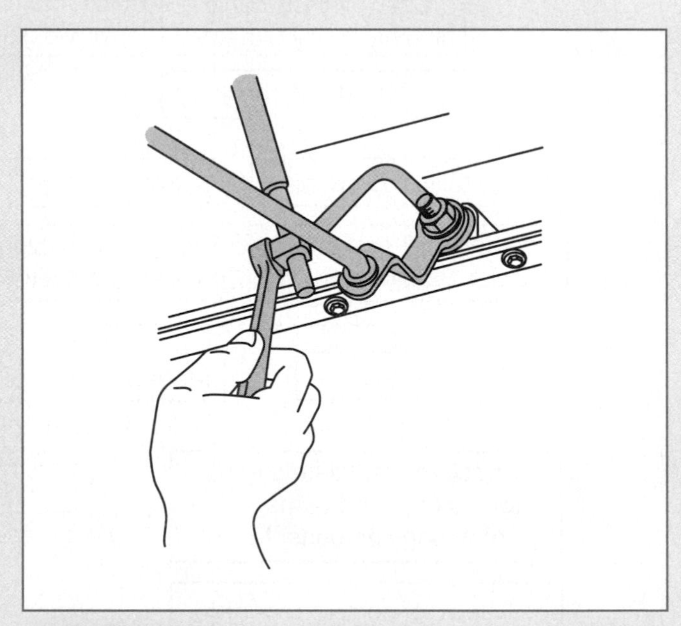

FIGURE 41-44 Various adjustments can be made on the linkage of each automatic transmission.

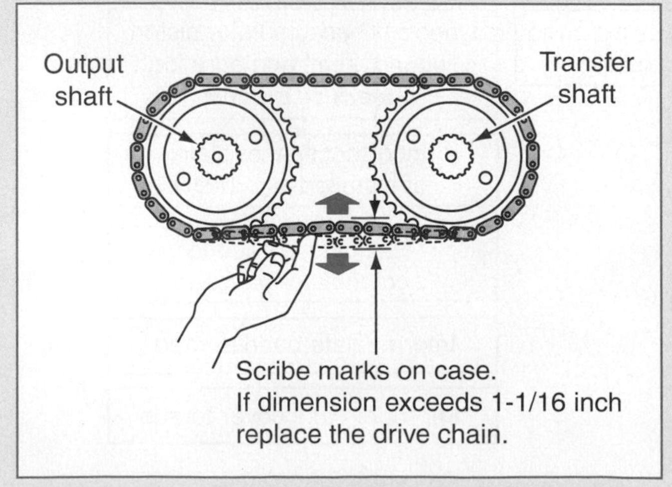

FIGURE 41-42 Check front-wheel drive transmissions for chain wear and replace the chain if necessary.

4. Various other checks can be performed on a typical automatic transmission. ***Figure 41–45*** shows three such checks. In ***Figure 41–45A*** the clearance between the pinion gears and the planet carrier is being checked. A feeler gauge is used to check this clearance against manufacturer's specifications. In ***Figure 41–45B***, the clearance on the clutch pack is being checked. A dial indicator is placed on the pressure plate and set to zero. Then the pressure plate is depressed and a reading is taken on the dial indicator. The clearance is then checked against manufacturer's specifications. In ***Figure 41–45C***, a dial indicator is being used to measure the end play of the input shaft on the transmission.
5. Torque converter end play can be checked using a dial indicator. ***Figure 41–46*** shows this check. Note that a special tool is used to perform this test.

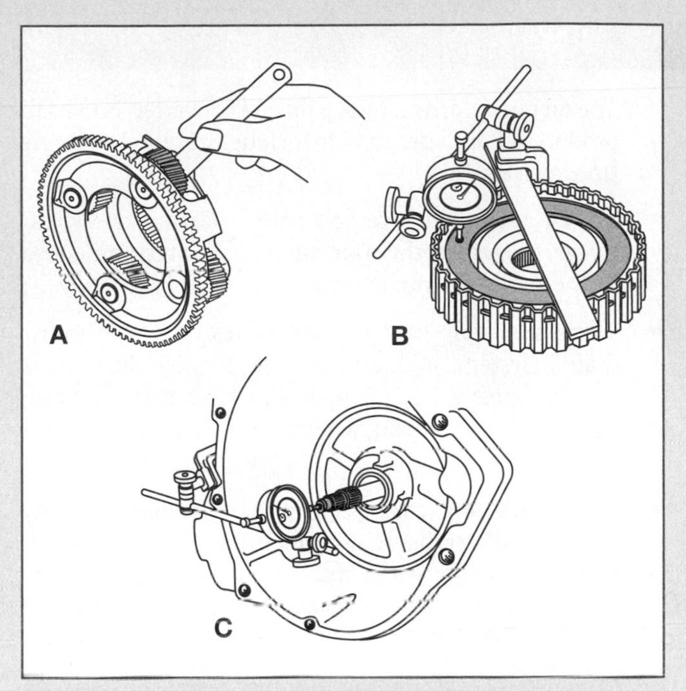

FIGURE 41-45 (A) Measuring planetary gear wear clearances, (B) Measuring clutch pack clearances, (C) Measuring end play on the transmission main shaft.

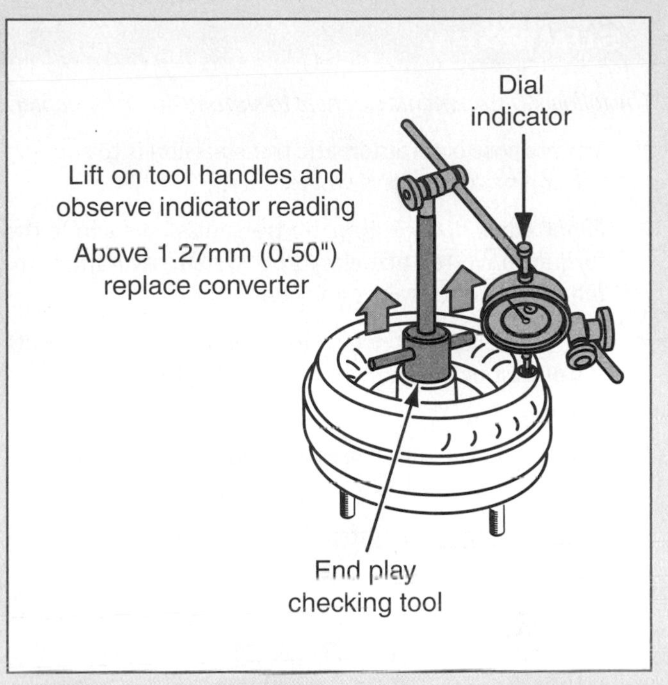

FIGURE 41-46 Use a dial indicator to check torque converter end play.

Service Manual Connection

There are many important specifications to keep in mind when working with automatic transmissions. To identify the specifications for your engine, you will need to know the VIN (vehicle identification number) of the vehicle, the type and year of the vehicle, and the type of transmission. Although they may be titled differently, some of the more common automatic transmission specifications (not all) found in service manuals are listed below. Note that these specifications are typical examples. Each vehicle and engine may have different specifications.

Common Specification	Typical example
Bands adjustment	4 1/4 turns
(front) after 10 ft.-lb of torque is applied (backoff)	4 1/4 turns
Downshift linkage adjustment	0.050–0.070 inch
Fluid change period	30,000 miles
Oil pan bolts, torque	12 ft.-lb
Oil pressure (minimum throttle valve)	66–74 psi
Oil pump clearance, gears to cover	0.0005–0.0035 inch
Transmission fluid capacity	
Pan only	4.0 quarts
Total	12.0 quarts

In addition, service manuals provide specific directions for various service and testing procedures. Some of the more common procedures include:

- Adding fluid to dry transmission
- Bands adjustment
- Governor replacement
- Oil pump replacement
- Speedometer gear replace
- Throttle downshift linkage adjustment
- Transmission replacement

SUMMARY

The following statements will help to summarize this chapter:

- The purpose of an automatic transmission is to connect the engine to the drive wheels.
- Major parts of an automatic transmission include the torque converter, planetary gear system, hydraulic system, and various valves and controls.
- The torque converter is a fluid coupling that connects the output of the engine to the gearing inside the transmission.
- The torque converter uses a turbine, pump, and stator to operate correctly.
- The planetary gear system is used to produce different gears for the transmission.
- The planetary gear system uses a sun gear, planet gears and carrier, and a ring gear to produce different gear ratios.
- There are several types of clutches used to lock up one of the gears in the planetary gear system, including the multiple-disc system and the transmission band.
- To aid in smooth shifting, an overrunning clutch is used on transmissions.
- The oil pump, driven by the torque converter, is used to produce oil pressure to help lock up the clutches in the transmission.
- Since there are high temperature, shearing, and stress forces, always use the transmission fluid recommended by the vehicle manufacturer.
- There are many controls and valves used on the hydraulic system, including the spool valve, ball check valves, poppet valves, manual shift valve, governor, vacuum modulator, kickdown valve, and torque converter valve.
- There are several major considerations when servicing an automatic transmission, including keeping the filters clean, keeping the fluid at the correct level, and checking many of the physical specifications of the automatic transmission.
- Common checks on the automatic transmission include checking the clutch pack for correct clearances, checking end play, adjusting linkages, checking clearances on the planetary gear system, and checking torque converter end play.

TERMS TO KNOW

Can you explain each of the following terms? Review the chapter until you can use each term correctly.

Accumulator
Automatic transmission fluid (ATF)
Deflection angle
Fluid coupling
Lockup clutch
Manual shift valve
Multiple-disc clutch
Overdrive
Periphery
Planet carrier
Planetary gear system
Pressure regulator valve
Servo
Shift valve
Spool valve
Stator
Tangs
Throttle valve (TV)
Torque converter
Torque converter clutch (TCC) valve
Transmission band
Turbine
Vacuum modulator
Valve body

REVIEW QUESTIONS

Multiple Choice

1. Which of the following is not considered a major part in an automatic transmission?
   a. Torque converter
   b. Planetary gear system
   c. Oil pump
   d. Power steering gear
   e. Hydraulic valves

2. Which of the following parts is/are used as the input inside the torque converter?
   a. Turbine
   b. Pump
   c. Stator
   d. Seal
   e. All of the above

3. The output of the torque converter is taken off the:
a. Turbine
b. Stator
c. Pump
d. Seal
e. All of the above

4. The coupling between the engine and transmission on the automatic transmission is considered a/an:
a. Air coupling
b. Fluid coupling
c. Electrical coupling
d. Mechanical coupling
e. All of the above

5. As the turbine speed decreases, the angle of deflection:
a. Increases
b. Decreases
c. Remains the same
d. Causes the stator speed to change
e. Causes the pump speed to increase

6. Which of the following is a part of a planetary gear system?
a. Sun gear
b. Planet carrier
c. Planet gears
d. Ring gear
e. All of the above

7. The planetary gear system is able to produce a/an ________ in output speed.
a. Increase
b. Decrease
c. Reverse direction
d. All of the above
e. None of the above

8. Which of the following uses a set of friction discs and steel discs pressed together to engage a clutch?
a. Transmission band
b. Overrunning clutch
c. Multiple-disc
d. Servo
e. Accumulator

9. What component aids in the operation of the transmission band?
a. Servo piston
b. Accumulator
c. Vacuum modulator
d. Multiple-disc
e. Torque converter

10. What component is used to help smooth the shifting from one gear to another?
a. Overrunning clutch
b. Servo
c. Sun gear
d. Ring gear
e. Torque converter

11. The oil pump on the automatic transmission uses the ________ type of design.
a. Eccentric
b. Vane
c. Gear
d. All of the above
e. None of the above

12. The oil pump on the transmission housing is located:
a. Inside the planetary gear system
b. On the hydraulic circuit body
c. On the front of the transmission case
d. On the drive shaft
e. None of the above

13. Which component is used to turn the oil pump on an automatic transmission?
a. Planetary gear system
b. Accumulator
c. Servo
d. Crankshaft
e. Torque converter

14. Which component in the automatic transmission houses the valves and contains the many oil passageways?
a. The accumulator
b. The modulator valve
c. The valve body
d. The planetary gear system
e. None of the above

15. Which of the following components helps to cool the transmission fluid?
a. Radiator
b. Torque converter
c. Accumulator
d. Planetary gear system
e. Valve body

16. Which of the following transmission control valves is used to measure vehicle speed and, in turn, direct oil flow to shift gears?
a. Manual shift valve
b. Shift valve
c. Throttle valve
d. Governor
e. Modulator valve

17. Which of the following transmission control valves uses intake manifold vacuum to measure vehicle load?
a. Shift valve
b. Modulator valve
c. Manual shift valve
d. Throttle valve
e. Governor

18. Which type of valve is used to control or direct oil flow and prevent backflow of oil?
a. Ball check valve
b. Governor
c. Throttle valve
d. Shift valve
e. Modulator valve

19. If the transmission fluid has a burned odor, the trouble may be in the ________ on the automatic transmission.
a. Torque converter
b. Clutches
c. Seals
d. Governor
e. Shift valve

20. What parts may need service on the automatic transmission?
a. Filter may be dirty
b. Clutches may be worn
c. Seals may leak
d. All of the above
e. None of the above

The following questions are similar in format to ASE (Automotive Service Excellence) test questions.

21. *Technician A* says that if the engine coolant is too hot, the transmission can overheat and be damaged. *Technician B* says that the cooling system has no effect on the operation or temperature of the transmission. Who is correct?
a. A only c. Both A and B
b. B only d. Neither A nor B

22. Bubbles are noticed in the transmission fluid. *Technician A* says that there is too much transmission fluid, and foam is being produced. *Technician B* says that there is an internal leak in the hydraulic system. Who is correct?
a. A only c. Both A and B
b. B only d. Neither A nor B

23. Transmission fluid is leaking from the front of the transmission housing. *Technician A* says to remove and replace the transmission filter. *Technician B* says to replace the oil pump. Who is correct?
a. A only c. Both A and B
b. B only d. Neither A nor B

24. *Technician A* says the transmission fluid level should be checked when the transmission is at operating temperature. *Technician B* says it should be checked when the transmission is cold. Who is correct?
a. A only c. Both A and B
b. B only d. Neither A nor B

25. *Technician A* says that adjusting the shift linkage is considered an in-vehicle adjustment. *Technician B* says that checking the snap ring end play is an in-vehicle adjustment. Who is correct?
a. A only c. Both A and B
b. B only d. Neither A nor B

26. *Technician A* says that when the apply circuit is energized on the TCC system, the torque converter is locked up. *Technician B* says that when the release circuit is energized on the TCC system, the torque converter is not locked up. Who is correct?
a. A only c. Both A and B
b. B only d. Neither A nor B

27. A vehicle's ATF has a burnt smell. *Technician A* says the clutches are bad. *Technician B* says the bands are bad. Who is correct?
a. A only c. Both A and B
b. B only d. Neither A nor B

Essay

28. What is the purpose of the torque converter?

29. What is the deflection angle on automatic transmissions?

30. Describe the purpose and operation of a planetary gear system.

31. What are the purposes of clutches and bands in an automatic transmission?

32. What is the purpose of the valve body in an automatic transmission?

33. What is the purpose of using a governor in an automatic transmission?

34. Describe the purpose and operation of a neutral start switch.

Short Answer

35. On automatic transmission lockup systems, the torque converter is locked to the ________.

36. Lockup on the torque converter is accomplished by ________ pressure.

37. The two most common types of ATF include ________ and ________.

Applied Academics

TITLE: Continuously Variable Transmission (CVT)

ACADEMIC SKILLS AREA: Science

NATEF Automobile Technician Tasks:

The service technician can explain how pulleys can be used to increase an applied force or distance in an automatic transmission and transaxle system.

CONCEPT:

The continuously variable transmission (CVT) is a new type of automatic transmission. This type of transmission has no gears for changing torque and speed requirements. Instead, a variable pulley system or cone system is used to change speed ratios as the engine rpm changes. The two variable pulleys are connected together by a heavy-duty belt. As shown in the figure that follows, as the drive pulley or cone increases in speed, the running diameter of the pulley increases. At the same time as the driven pulley increases in speed, its running diameter gets smaller. The combination of these variable diameter pulleys or cones produces an infinite number of input to output speed ratios. The principles of the CVT have been designed into hybrid vehicles and other smaller vehicles on the road today. For example, the Saturn will have exclusive rights in GM to use the CVT. With variable speed ratios, the engine will be able to operate at the most efficient rpm. This will also optimize fuel mileage. In a test comparing a CVT with a five-speed transmission, the CVT used 11+% less fuel and produced 33% fewer hydrocarbon emissions. In addition, the carbon monoxide emissions dropped by 20%. The transmission and the variable pulleys are controlled by a computer to get the best input-output speed ratio.

APPLICATION:

When the drive pulley has a small pulley diameter and the driven pulley has a large pulley diameter, what happens to the output speed and torque capabilities? When the drive pulley has a large pulley diameter and the driven pulley has a smaller pulley diameter, what happens to the output speed and torque capabilities? Why is a continuously variable transmission important in a hybrid vehicle?

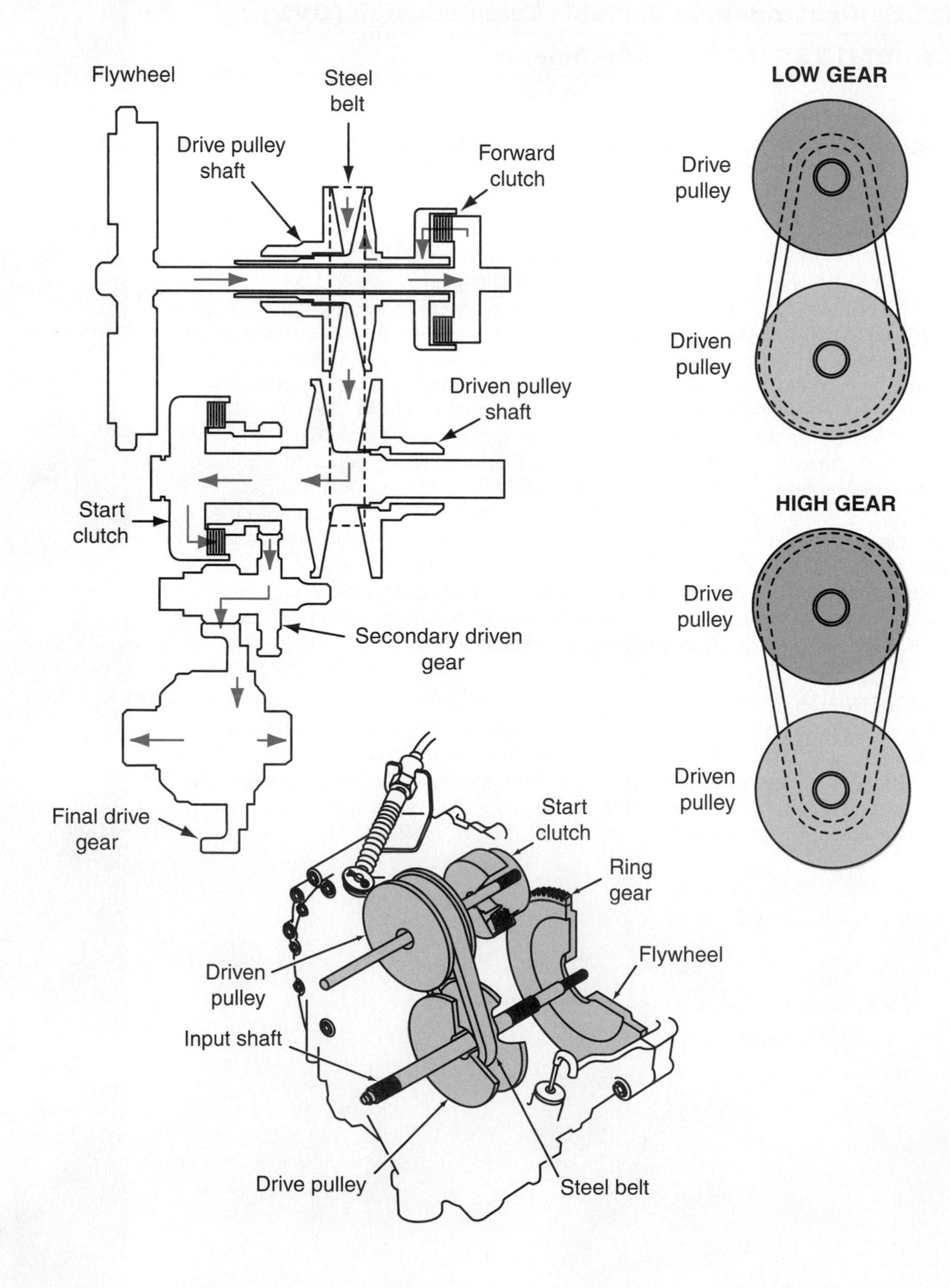
Flywheel
Steel belt
Drive pulley shaft
Forward clutch
Driven pulley shaft
Start clutch
Secondary driven gear
Final drive gear
LOW GEAR
Drive pulley
Driven pulley
HIGH GEAR
Drive pulley
Driven pulley
Start clutch
Ring gear
Flywheel
Driven pulley
Input shaft
Drive pulley
Steel belt

Electronic and Computer-Controlled Transmissions

CHAPTER 42

OBJECTIVES

After studying this chapter, you should be able to:

- Identify the purpose for using electronic computer-controlled transmissions.
- State the purpose of the powertrain control module (PCM).
- Identify the purpose and operation of the PCM inputs on electronic computer-controlled transmissions.
- Identify the purpose and operation of the computer output devices used on electronic computer-controlled transmissions.
- Define various problems, their diagnosis, and service procedures used on electronic computer controlled transmissions.

INTRODUCTION

Over the past few years, automotive manufacturers have developed electronic and computer-controlled automatic transmissions. These transmissions improve the operation and control of automatic transmissions. This chapter introduces the principles, parts identification, and operation of electronic computer-controlled automatic transmissions.

42.1 PURPOSE OF COMPUTER-CONTROLLED TRANSMISSIONS

Computers and electronics have become more integrated into automatic transmissions. In the past several years, most manufacturers have developed transmissions that use a computer to control the shifting and also to control the locking up of the torque converter. Automatic transmissions that

Certification Connection

ASE Connection: The information in this chapter can help you prepare for the National Institute for Automotive Service Excellence (ASE) certification tests. The test and content area most closely related to this chapter are:

Test A2—Automatic Transmission/Transaxle

- **Content Area**—General Transmission/Transaxle Diagnosis (Electronic Systems)

NATEF Connection: Much of the information in this chapter is related to the NATEF tasks. The NATEF tasks and priority numbers most closely related to this chapter are:

1. Inspect and replace speedometer drive gear, driven gear, vehicle speed sensor (VSS), and retainers. P-2
2. Diagnose electronic transmission control systems using a scan tool; determine necessary action. P-1
3. Diagnose electronic, mechanical, hydraulic, and vacuum control system concerns; determine necessary action. P-1

use hydraulics and mechanical valves to control the shift patterns are typically slower and less responsive. For example, every driver has had an automatic transmission that seems to jump from gear to gear. Because of the very nature of hydraulics, there will always be some rough shifting, additional wear, and a loss of reliability for mechanical transmissions. Electronically controlled transmissions use computers to:

- Reduce rough shifting
- Reduce wasted power in the transmission
- Reduce operation costs
- Reduce manufacturing costs
- Reduce the complexity of hydraulic circuitry
- Increase transmission reliability

42.2 OPERATION OF ELECTRONICALLY CONTROLLED TRANSMISSIONS

There are many types and styles of computer-controlled automatic transmissions. For consistency, they will all be referred to as electronically controlled transmissions. This chapter introduces the basic theory, operation, and initial service of the electronically controlled transmission.

PURPOSE OF THE PCM

The **powertrain control module (PCM)** is the name of the computer used on today's vehicles to work in conjunction with the electronically controlled transmission. Although some vehicle manufacturers may call this computer by a different name, they still operate in much the same manner. For example, some manufacturers call this computer the TCM or transmission control module. The TCM or PCM is often located under the dashboard, behind the right-side kick panel on the passenger's side. The PCM receives various electronic inputs from sensors located on the vehicle and transmission. The PCM then processes that information to determine the vehicle operating conditions. Depending on these conditions, the PCM controls the following:

- Transmission upshifts and downshifts by operating a pair of electronic shift solenoids in an on-off pattern
- Transmission shift feel by electronically controlling the pressure control solenoid (PCS), which adjusts to line pressure
- Torque converter clutch (TCC) "apply and release" timing and TCC apply "feel" in some applications through the TCC control solenoid(s)

The electronic control of these transmission operating characteristics provide for consistent and precise shift points and improved shift quality.

PARTS LOCATOR

Refer to photo #85 in the front of the textbook to see what a ***powertrain control module*** *looks like.*

TRANSMISSION MODES OF OPERATION

Electronic control allows the transmission to compensate for specific driving conditions. Depending on the manufacturer, electronically controlled transmissions have various dashboard controls that use one or more of the following modes of operation available on the vehicle:

1. **Performance mode**—In the performance mode, the PCM commands the power control solenoid to increase hydraulic line pressure. This causes the transmission to have a firmer shift pattern. The PCM also extends the shifts slightly to provide for greater acceleration before upshifting to the next higher gear.
2. **Winter mode**—Some vehicles have a winter mode for the transmission to operate in. In this mode, the PCM commands the transmission to begin moving the vehicle in second or third gear. This occurs especially if the tires begin to slip on snowy or icy pavement. This is useful because in first gear, there is much more torque, and high torque can cause the tires to break loose easier in slippery conditions. The winter mode makes it easier and safer to move away from a stoplight when the road is covered with ice or snow.
3. **Manual shifting mode**—When a vehicle is in the manual shifting mode, the driver is able to shift the automatic transmission similarly to a standard transmission. The driver can use the selector level as the "stick shift." This is done by placing the level in either manual first, manual second, manual third, or overdrive (fourth) gear. Note that if the operator attempts to stay in first or second gear with excessive rpm, the PCM overrides the manual mode, protecting the transmission.

Vehicles that do not have a winter mode may use the manual shifting mode in winter. When the lever is placed in the manual second mode, the transmission operates like the winter mode. First gear is prevented with the gear selector in the manual second mode gear range. This forces the transmission to begin moving the vehicle in second gear, rather than in first.

BASIC OPERATION

Figure 42–1 shows the types of inputs and outputs used on an electronically controlled transmission. This schematic shows a typical front-wheel drive vehicle, the engine, and the PCM connections. There are various input signals that are sent to the PCM. The exact number and type of inputs depend on the vehicle manufacturer, the year and type of vehicle, and the type of transmission being used. In this particular case, ten input sensors are shown. Many of these sensors have been discussed in other chapters in this text, including the MAP, A/C switch, ECT, TPS, VSS, engine speed (ignition module), and others. These and similar inputs provide signals to the PCM so it can make decisions about its output.

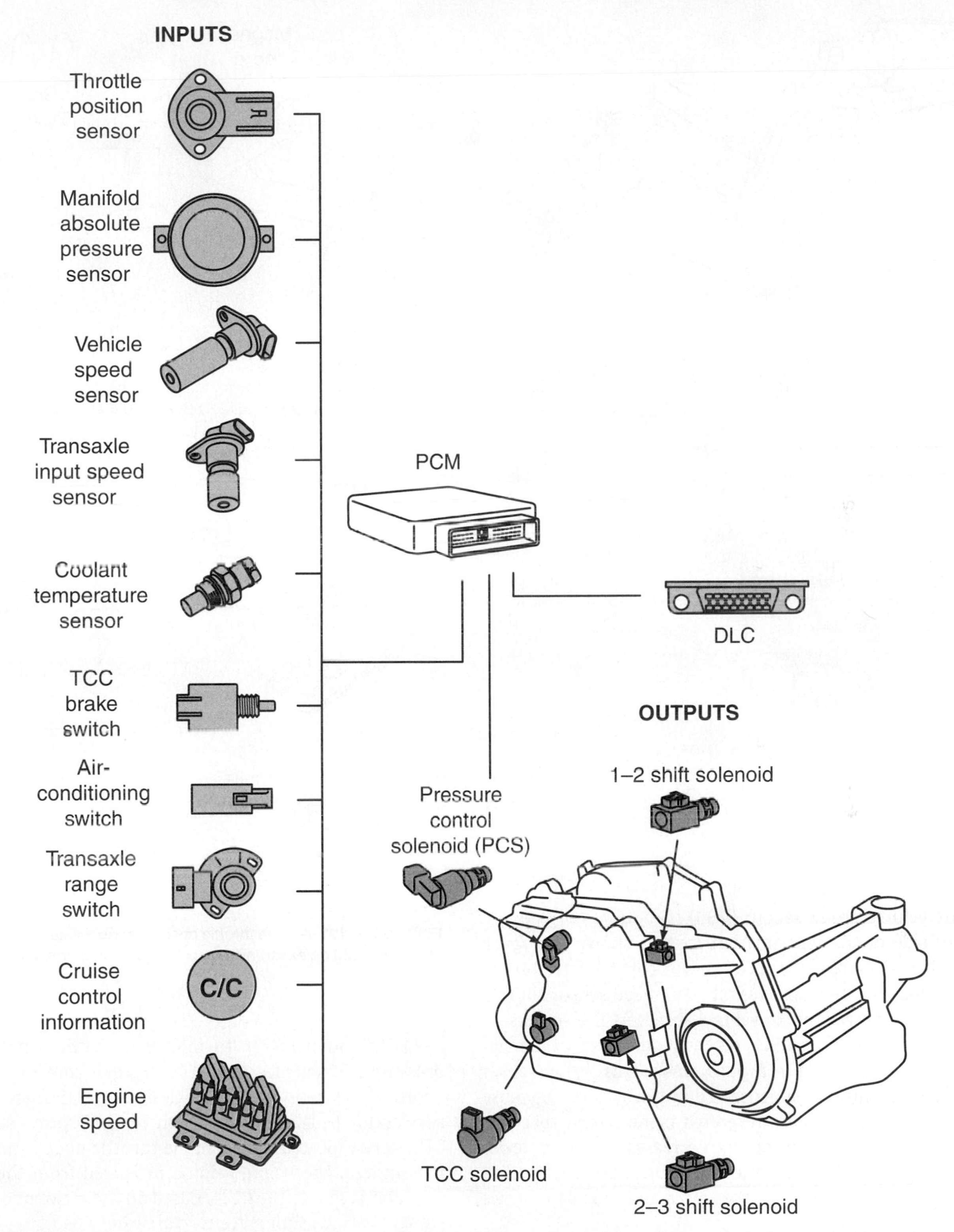

FIGURE 42-1 The electronic computer-controlled automatic transmission uses a variety of inputs to the PCM to control four outputs.

The most common types of outputs include the following:

- Pressure control solenoid (PCS)
- 1–2 shift solenoid ("A")
- 2–3 shift solenoid ("B")
- TCC control solenoid

SHIFT TIMING

Although there are many inputs to the PCM, in order to further understand the shift timing, only the vehicle speed sensor (VSS) and the throttle position sensor (TPS) are explained in this section. Other input sensors will be explained later in this chapter.

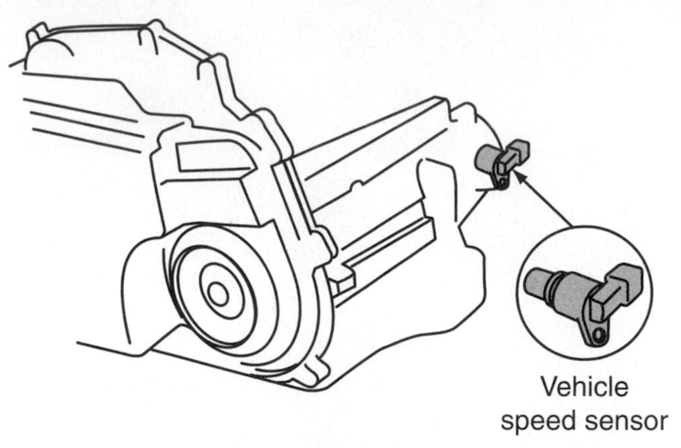

FIGURE 42–2 The vehicle speed sensor (VSS) is located at the output of the transmission.

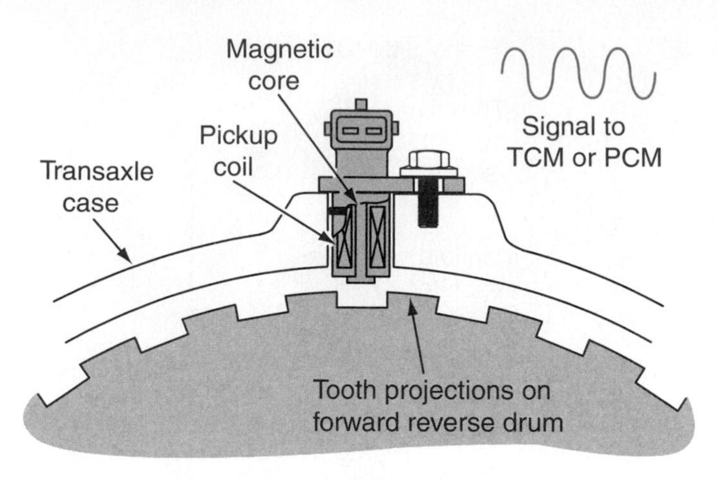

FIGURE 42–3 As the tooth projections pass the magnetic core and pickup coil, a speed signal is produced that, in turn, is sent to the PCM.

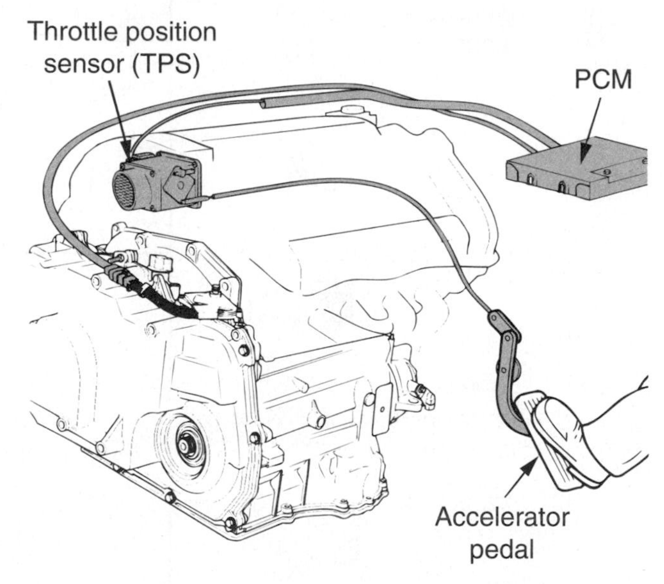

FIGURE 42–4 The throttle position sensor is a major input sensor to the PCM on electronic computer-controlled transmissions.

The **vehicle speed sensor** (VSS) is mounted directly on the outside of the transmission case as shown in ***Figure 42–2***. It is operated by a speed sensor rotor located on the output shaft of the transmission. The speed sensor rotor is always rotating relative to vehicle speed. As the speed sensor rotor rotates, the splines on the rotor pass by the magnetic pickup as shown in ***Figure 42–3***. This action causes an electrical pulse to be generated in the sensor. The pulse is sent to the PCM and is interpreted as the vehicle speed. As the vehicle speed increases or decreases, more or fewer rotor teeth pass the magnetic pickup in a given time. The PCM reads the frequency of the signal and thus determines vehicle speed.

The throttle position sensor (TPS) is a measure of how far the driver has depressed the accelerator pedal. The PCM needs this information to modify the shift patterns, line pressure and shift feel, and TCC (torque converter clutch) control. This sensor is considered a potentiometer. As the accelerator pedal is moved as shown in ***Figure 42–4***, the TPS sensor located at the engine throttle plate is moved. As it is moved, the voltage signal produced from the potentiometer is sent to the PCM. Based on these two major input signals, various shift patterns can be obtained.

PARTS LOCATOR

Refer to photo #86 in the front of the textbook to see what a ***vehicle speed sensor on an electronic transmission*** *looks like.*

UPSHIFTING

To help understand the principles of operation for an electronically controlled transmission, consider the process of **upshifting**. For example, think of two different drivers accelerating away from a stop sign. Driver #1 accelerates the vehicle slowly with a 10% throttle position. Driver #2 accelerates the vehicle rapidly with a 50% throttle position. In

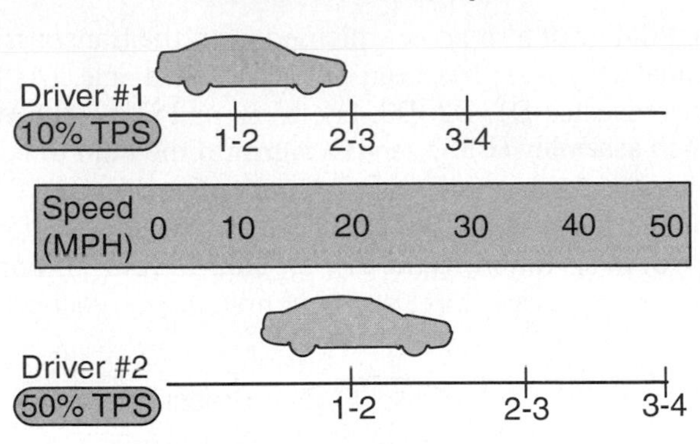

FIGURE 42–5 If Driver #2 has 50% throttle position, the shift pattern from first to fourth gear will be delayed.

both cases, the PCM can read the vehicle speed and the amount of throttle. ***Figure 42–5*** shows the shift patterns for each driver in relation to the speed of the vehicle. For example, Driver #1 will shift from first to second at about 10 miles per hour. Driver #2 will shift from first to second at about 20 miles per hour. Driver #2 will shift into third and fourth gear later as well. The reason for the difference is that the PCM extends the shift time for Driver #2 because the signal from the TPS is different (50% throttle).

Another example in ***Figure 42–6*** shows what happens when the throttle position sensor senses a decrease in the throttle position at higher speed. For example, let's say Driver #2 is traveling at 40 miles per hour and is still in third gear. When Driver #2 eases off the throttle to say 10%, the transmission will upshift immediately into fourth gear. Many drivers experience this shift pattern when they are accelerating and then decelerate to get the transmission to go into 4th gear. All of these shifting patterns are primarily controlled by the TPS and VSS.

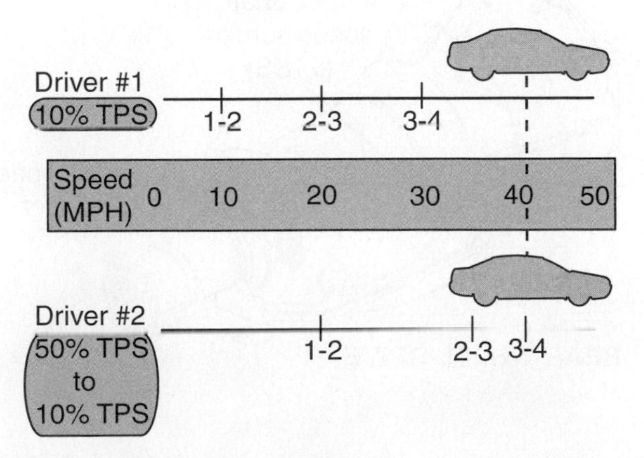

FIGURE 42–6 If Driver #2 is at 40 mph and in third gear when the vehicle is decelerated to 10% TPS, the transmission will immediately shift to fourth gear.

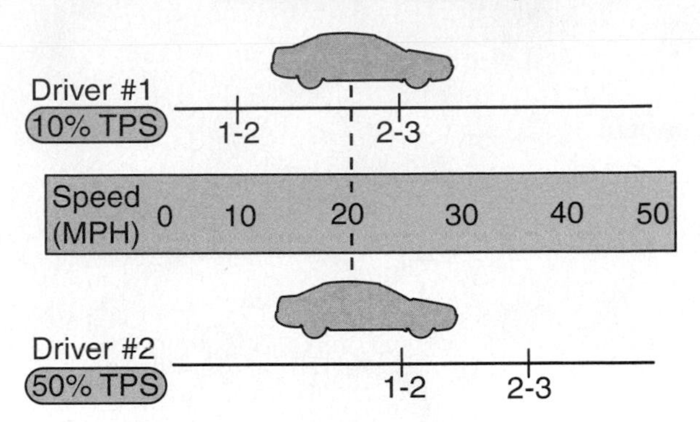

FIGURE 42–7 If Driver #1 accelerates from 10% to 50% TPS, the transmission will downshift to first gear.

DOWNSHIFTING

Downshifting is also controlled by the VSS and the TPS. To help understand the operation during downshifting, refer to ***Figure 42–7***. Let's say that again there are two drivers. Driver #1 is again at 10% throttle and going 20 miles per hour. Driver #2 is again at 50% throttle and going 20 miles per hour. Driver #1 is in second gear. Driver #2 is still in first gear. These gears have again been determined by the VSS and TPS signals into the PCM. At this point, if Driver #1 pushes hard on the throttle (say to pass another vehicle), the TPS is now at 50%. Under these conditions, the PCM immediately downshifts into first gear. Now Driver #1 and Driver #2 are operating at the same gear. This downshifting action produces high performance and acceleration on command (at the 50% TPS). Of course, the same type of transmission action will occur in higher gears and at higher speeds. Also note that if Driver #2 decreases the throttle position to 10%, the transmission will upshift to second gear.

42.3 ADDITIONAL PCM INPUTS

Many other inputs are used in an electronically controlled transmission. Some are considered internal sensors, whereas others are considered external sensors. Each of these sensors helps the PCM to determine the exact vehicle and transmission operating conditions. The result is precise control of the transmission for both shift timing and shift feel.

TRANSMISSION INPUT SPEED SENSOR (TISS)

Many electronic transmissions use a TISS or transmission input speed sensor (***Figure 42–8***). This sensor is typically located immediately after the torque converter and measures the input speed of the transmission. With this input, the PCM can determine both the input speed of the transmission (TISS) and the output speed of the transmission (VSS). Based on these two speeds, the PCM can determine

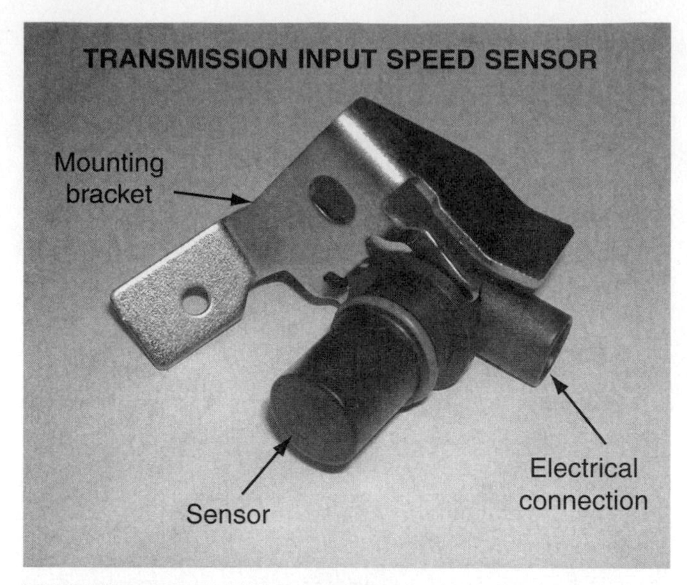

FIGURE 42-8 This transmission input speed sensor (TISS) measures the input speed of the transmission.

which gear the transmission is operating in at any specific time.

The TISS produces a speed signal similar to the VSS. It uses a toothed gear in front of the transmission. A magnetic pickup is then mounted near the toothed gear to produce the speed signal. ***Figure 42–9*** shows the location of the TISS on both front-wheel drive and rear-wheel drive transmissions.

PARTS LOCATOR

*Refer to photo #87 in the front of the textbook to see what a **transmission input speed sensor** looks like.*

PRESSURE SWITCH ASSEMBLY (PSA)

The PCM must also know which position the transmission manual shift lever has been put in. For example, has the driver selected D1, D2, D3, D4, N, R, or P? The **pressure switch assembly (PSA)** sends a signal to the PCM to tell it which gear has been selected. To accomplish this, the PSA is mounted directly on the valve body as shown in ***Figure 42–10***. In operation, as the manual shift valve is positioned in one of the gears, a certain fluid pressure is produced in

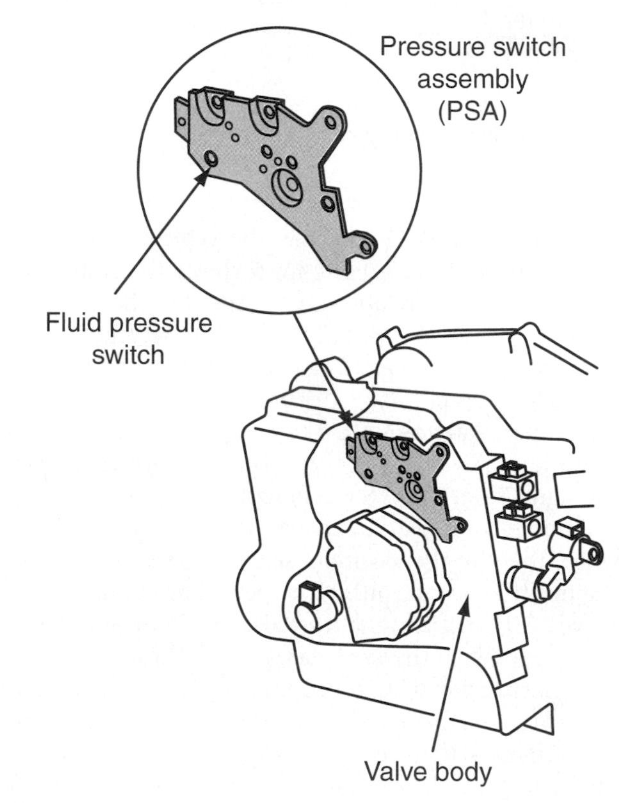

FIGURE 42-10 The pressure switch assembly (PSA) tells the PCM which gear has been selected by the driver.

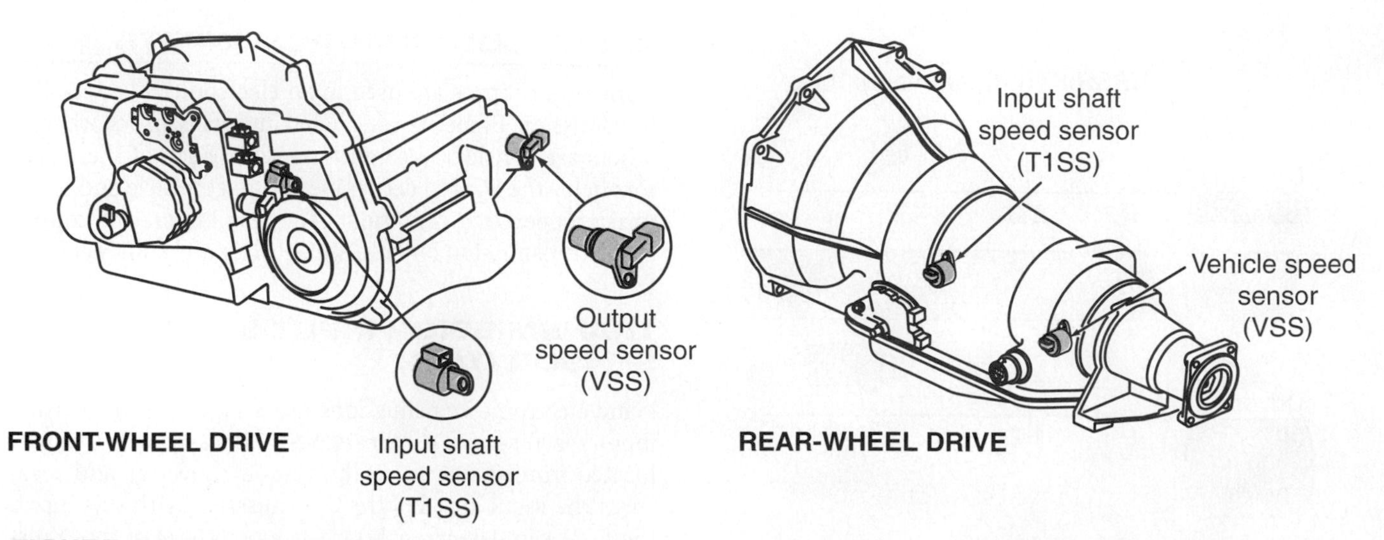

FIGURE 42-9 Transmission input speed sensors (T1SS) help tell the PCM which gear the transmission is in. By comparing transmission input speed data with transmission output speed data from a vehicle speed sensor, or VSS, the PCM can determine which gear the transmission is operating in.

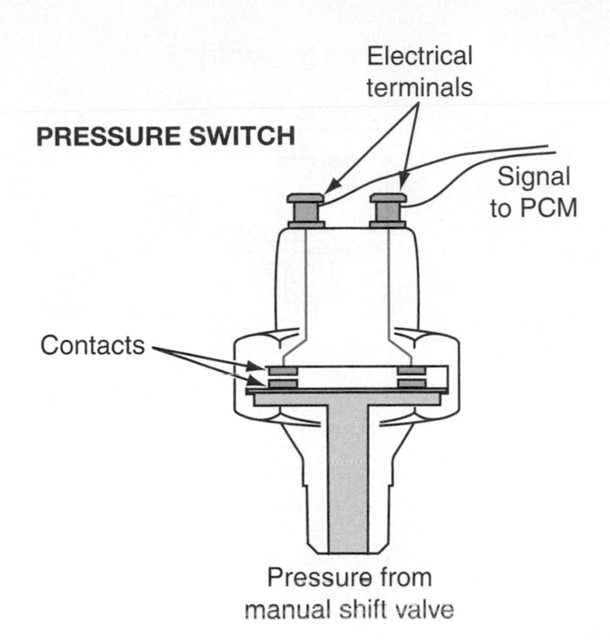

FIGURE 42–11 The internal components of the pressure switch.

the hydraulic system. The various fluid pressures from the manual shift valve act on fluid pressure switches. These switches open and close electrical contacts. ***Figure 42–11*** shows the internal components and operation of the pressure switch. As the pressure from the manual shift valve increases, the bottom set of contacts lifts upward and makes contact with the top contacts. This completes the upper circuit and the voltage drop across the pressure switch drops to zero. This voltage drop is then sensed by the PCM. The combination of open and closed switches determines the signal that is sent to the PCM. The PCM interprets this signal and determines which gear has been selected by the driver.

PARTS LOCATOR

*Refer to photo #88 in the front of the textbook to see what a **pressure switch assembly on an electronic transmission** looks like.*

TRANSMISSION FLUID TEMPERATURE (TFT) SENSOR

The shift feel is an important characteristic of an electronically controlled transmission. Shift feel is defined as the smoothness of the shift from one gear to another. Shift feel will change with increasing and decreasing transmission fluid temperature. Therefore, the PCM needs to know the temperature of the transmission fluid. The temperature of the fluid is determined by using the transmission fluid temperature (TFT) sensor. This sensor is a thermistor-type sensor. It is able to change its resistance based upon the increasing and decreasing temperature. It is mounted generally in a location where it can be touching the transmission fluid. This is generally on the valve body. Based on the fluid temperature, the PCM adjusts the shifting so that a gentle shift is made.

PARTS LOCATOR

*Refer to photo #89 in the front of the textbook to see what a **transmission fluid temperature sensor** looks like.*

ENGINE COOLANT TEMPERATURE (ECT) SENSOR

As with regular automatic transmissions, it is important that the torque converter clutch (TCC) engage (apply) and disengage (release) at the correct time. It is also important that the TCC applies only when the engine is warm. Therefore, the ECT sensor is used as an input to the PCM so it can apply the TCC at the right time. Engine coolant should reach a minimum temperature before the TCC is applied.

ENGINE SPEED SENSOR

The PCM monitors the engine speed to help determine the right shift patterns and the right time to apply and release the TCC.

Car Clinic:
SPEED SENSOR

CUSTOMER CONCERN:
Transmission Jerks Hard When Shifting
The driver who has an electronic transmission in a Honda Prelude complains that the transmission seems to shift very hard. The vehicle actually jerks when shifting from one gear to another.

SOLUTION: Although there may be other problems, one of the more common problems is a faulty speed sensor. If the speed sensor is faulty, the PCM or TCM does not know the speed of the vehicle. Therefore, it cannot be used as an input to the computer. Check the voltage drop and resistance of the speed sensor and compare these readings to the manufacturer's specifications. If the speed sensor is faulty, replace it.

BRAKE SWITCH SENSOR

When the brake is applied, it is important for the transmission to release the TCC. This is the reason that the PCM monitors the brakes. A brake switch sensor is used to tell the PCM when the brakes are applied. When the brakes are applied, the PCM signals the TCC to release.

MANIFOLD ABSOLUTE PRESSURE (MAP) SENSOR

The load on the engine is an important characteristic concerning transmission shifting. When a load is applied to the engine, the upshift patterns are slightly delayed. When there is no load on the engine, the upshift patterns are more rapid. To help determine when shift patterns should occur with different loads, the PCM senses the intake manifold pressure. This is done with the use of the MAP. Changes in the intake manifold absolute pressure are used by the PCM to adjust hydraulic line pressure and shift timing.

A/C SWITCH SIGNAL

When the air-conditioning system has been turned on, it increases the load on the engine. The PCM needs to know when the A/C has been turned on. The A/C switch signals the PCM that it has been turned on. The PCM uses this information to adjust the hydraulic line pressure and shift timing.

CRUISE CONTROL INFORMATION

The PCM also needs to know when the cruise control system has been engaged. For example, consider a vehicle traveling along a rolling terrain at 50 miles per hour with the cruise control engaged. Under these conditions, the transmission will tend to downshift when going up a hill. It will also tend to upshift when going down a hill. If the road has a series of ups and downs, the constant shifting may become annoying to the driver. The PCM prevents this type of needless shifting when the cruise control is engaged.

42.4 PCM OUTPUTS

As mentioned earlier, the PCM controls four primary outputs:

1. 1-2 **shift solenoid**, identified as solenoid A
2. 2-3 shift solenoid, identified as solenoid B
3. Pressure control solenoid (PCS)
4. TCC control solenoid

Each of these outputs will be further discussed to help explain the full operation of the electronically controlled transmission.

PURPOSE OF SHIFT SOLENOIDS

Each electronically controlled transmission uses a shift solenoid to control the hydraulic circuits for each of the gears in the transmission. The solenoid controls the flow of fluid through the shift valve and eventually to the clutches. The valve can be in either the off or on position. There is no other condition the valve can be in. ***Figure 42–12*** shows a typical shift solenoid in the off and on positions.

Referring to the top of ***Figure 42–12***, the solenoid is in the on position and the solenoid is not energized. During this condition the metering ball is positioned in such a manner as to close off the flow of transmission fluid through the metering ball. Under these conditions, the shift valves can use the hydraulic pressure created by the pump to operate clutches. When the solenoid is energized, the metering ball rises slightly. This action causes the transmission fluid to flow through the metering ball and be exhausted to the sump. Under these conditions, the hydraulic fluid cannot operate clutches because the fluid has been returned to the sump.

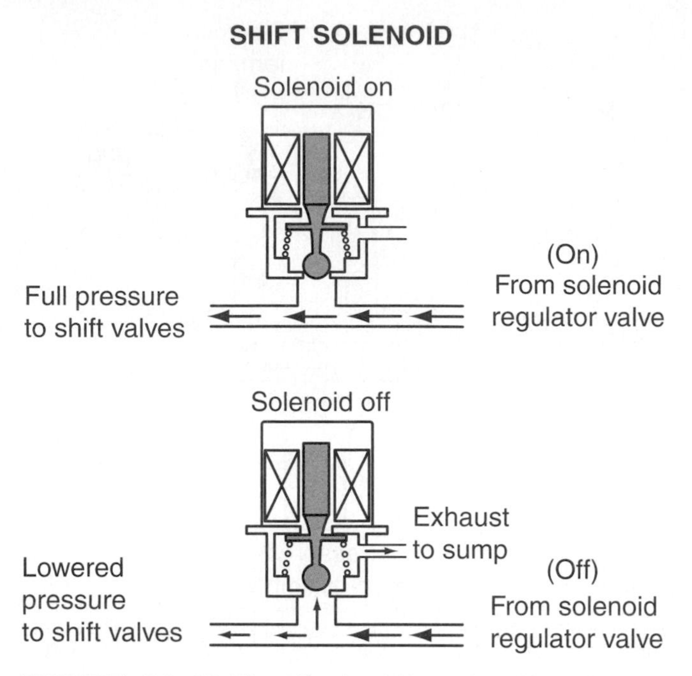

FIGURE 42–12 The shift solenoid is used to either exhaust or block the fluid flow, using a metering ball and plunger.

Figure 42–13 shows a typical example of a shift solenoid used in an electronic transmission. Notice the location of the coil as well as the O-ring used to seal in automatic transmission fluid.

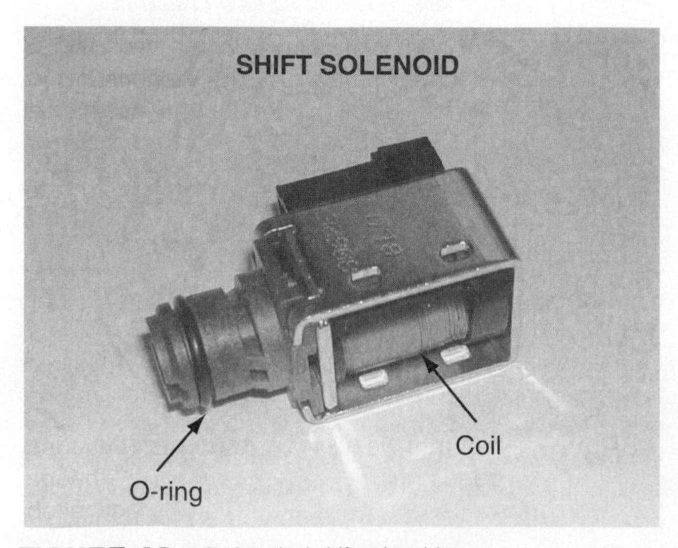

FIGURE 42–13 A typical shift solenoid.

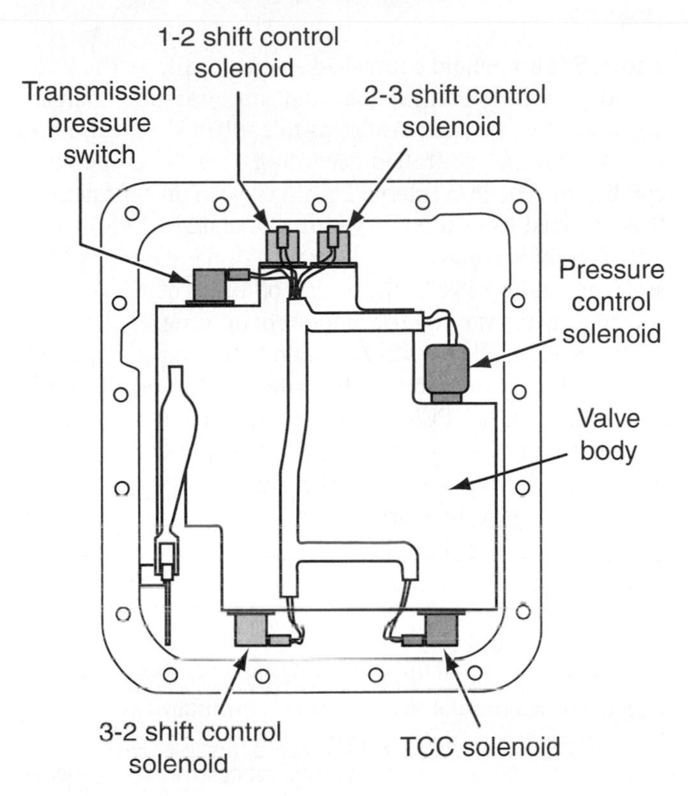

FIGURE 42-14 Shift solenoids are typically located on the transmission valve body.

Shift solenoids are typically located on the valve body as shown in ***Figure 42–14***. In most cases, when the transmission fluid pan is removed, the shift solenoids can be observed attached to the valve body.

GEAR	SOLENOID	
	A	B
1st	On	Off
2nd	Off	Off
3rd	Off	On
4th	On	On

FIGURE 42-15 When shift solenoids are turned off or on, various hydraulic circuits are operated to shift from one gear to another.

SHIFT SOLENOIDS A AND B

For the electronically controlled transmission, the shift valves operate in various on-off combinations to send pressurized transmission fluid to different clutches for first, second, third, and fourth gear. The chart in ***Figure 42–15*** shows the combination of on-off shift valves for each gear. For example, in order to set up the hydraulic circuits for second gear, both solenoids A and B must be in the off position.

The hydraulic circuit for second gear is shown in ***Figure 42–16***. Several points can be made about the circuit.

1. Note that the manual shift valve located on the top of the diagram is in D4.
2. Note that shift valve A (located in the center of the diagram) is in the off position.
3. Note that shift valve B (located on the center bottom of the diagram) is in the off position.
4. Note that there are actually two separate hydraulic circuits for each shift valve. One circuit is a control circuit called the signal fluid pressure. It is identified by a

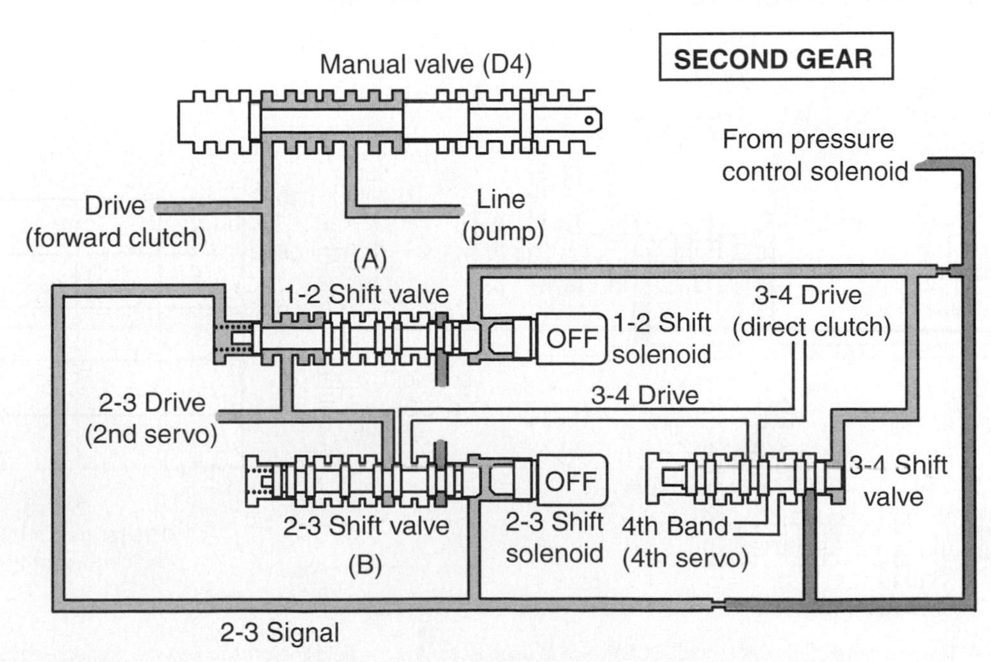

FIGURE 42-16 When both shift solenoids are in the off position, fluid is sent to the second gear clutch.

lighter shade of color. The other is the drive pressure used to lock up various clutches. It is identified by a darker shade of color.

5. With both shift valves in the off position, as the **signal fluid** comes into the 1-2 shift solenoid and the 2-3 shift solenoid, the signal fluid is allowed to be exhausted through the solenoid.
6. The lack of signal pressure on the right side of each shift valve causes them to be positioned in such a way that drive pressure is sent to the second gear clutch.
7. Under these conditions, drive pressure comes from the pump, through the manual shift valve, down and through the 1-2 shift valve, then over to the second gear clutch. The fluid tries to go through the 2-3 shift valve, but due to its position, the fluid cannot go through it.

PARTS LOCATOR

Refer to photo #90 in the front of the textbook to see what ***shift solenoids A and B*** *look like.*

TRANSMISSION SHIFT FEEL

The term *shift feel* is often used to describe how hard the shift feels from one gear to another. Basically, shift feel is defined as the rate or speed that the fluid pressure is applied to the clutches and band. Shift feel can be controlled by adjusting the pump pressure inside the transmission. The PCM can easily control the pump or line pressure by controlling the PCS or pressure control solenoid.

PRESSURE CONTROL SOLENOID (PCS)

The PCS is a solenoid controlled electronically by the PCM. Its purpose is to control the fluid line pressure (increase and decrease line pressure) so as to control shift feel. Fluid line pressure is controlled according to vehicle operating conditions. The PCS solenoid is not only an on-off solenoid. It is referred to as a pulse width modulated (PWM) valve. This type of valve gets an electronic signal from the PCM and continuously cycles the on and off times of a device (the solenoid) while varying the amount of on time.

As shown in ***Figure 42–17***, when higher clutch pressure is needed (say for faster clutch application and increased holding force), the PCM decreases the current to the PCS. The result is that the torque signal coming out of the PCS is increased. The increased torque signal moves the boost valve to increase line pressure at the pressure regulator valve. The relationship between control pressure after the pressure regulator and input current to the PCS is also shown. Note that as PCS current decreases, control pressure increases and vice versa.

The torque signal fluid pressure is also sent from the PCS to the accumulator valve. The accumulator valve helps to regulate accumulator pressure. Increasing accumulator pressure creates a more firm shift, especially during heavy acceleration.

PARTS LOCATOR

Refer to photo #91 in the front of the textbook to see what a ***pressure control solenoid*** *looks like.*

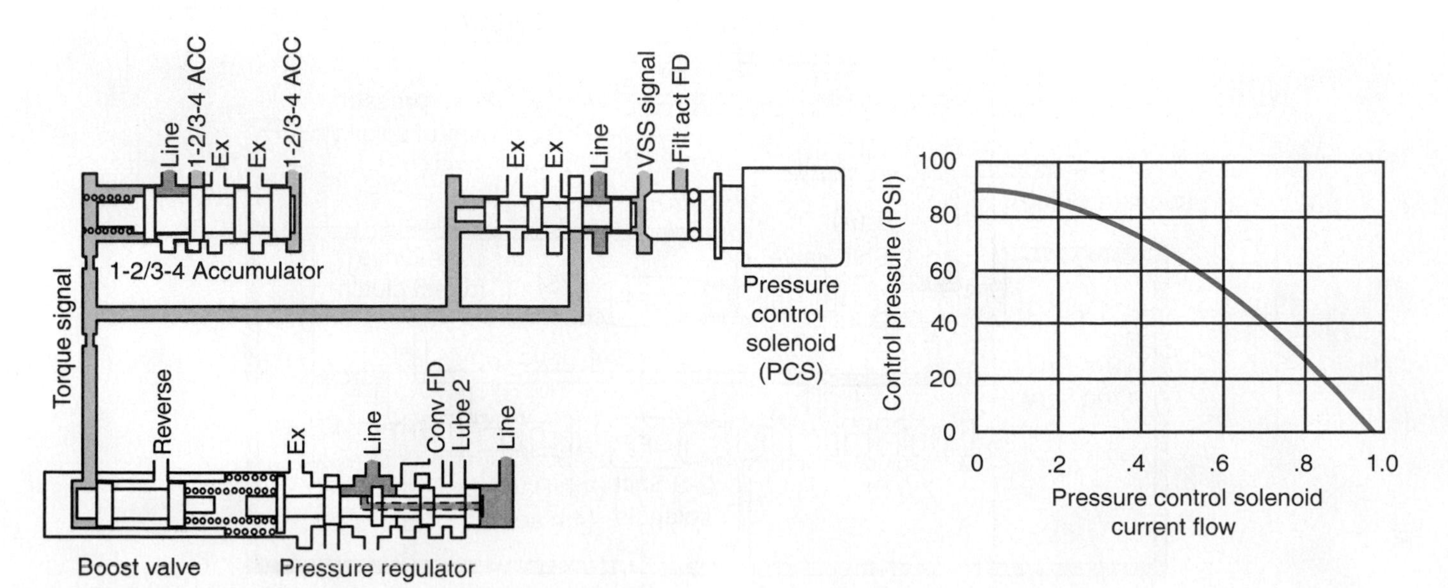

FIGURE 42–17 The pressure control solenoid (PCS) valve is used to control the fluid line pressure in the pressure regulator and to adjust shift feel of the transmission.

TORQUE CONVERTER CLUTCH (TCC) SOLENOID

As you recall, it is important to lock up the torque converter at specific times to reduce slippage and to improve fuel economy. There are several types of TCC lockup systems used on electronically controlled transmissions. This system uses an on-off solenoid similar to the one shown in ***Figure 42–18***. This system simply controls the timing for when the TCC is released and locked up. ***Figure 42–19*** shows an example of hydraulic circuits using the TCC solenoid. When the TCC solenoid is on, the signal fluid is blocked from exhausting through the valve. With the signal fluid pressure blocked, the converter clutch valve is positioned in such a way that the regulated line pressure passes through the valve, directly to the torque converter. This pressure causes the torque converter to lock up. Although not shown in the diagram, when the solenoid is off, signal fluid exhausts through the solenoid. Then spring pressure positions the TCC valve so that the fluid pressure is released and the torque converter is in the release position.

PARTS LOCATOR

*Refer to photo #92 in the front of the textbook to see what a **torque converter clutch solenoid** looks like.*

Car Clinic: TCC

CUSTOMER CONCERN:
TCC Does Not Engage or Apply

A vehicle has a transmission that has a lockup torque converter. The driver complains that the engine seems to be hunting or surging when the vehicle is at normal highway speed. The symptoms seem to be worse when the vehicle is pulling a trailer. What could be the problem?

SOLUTION: Occasionally, there will be a problem with the TCC system. If certain parts are faulty, the electronic control circuits may not be able to keep the torque converter locked up when it should. This causes the TCC to apply and release when the TCC should be locked or in the apply mode. The effect is that the car seems to surge. Actually the torque converter clutch is applying and releasing when it should be in the apply mode. This particular problem is solved by checking the connections between the PCM and the transmission. If even one connection, say the TCC solenoid wire, is not making good contact, there can be a problem.

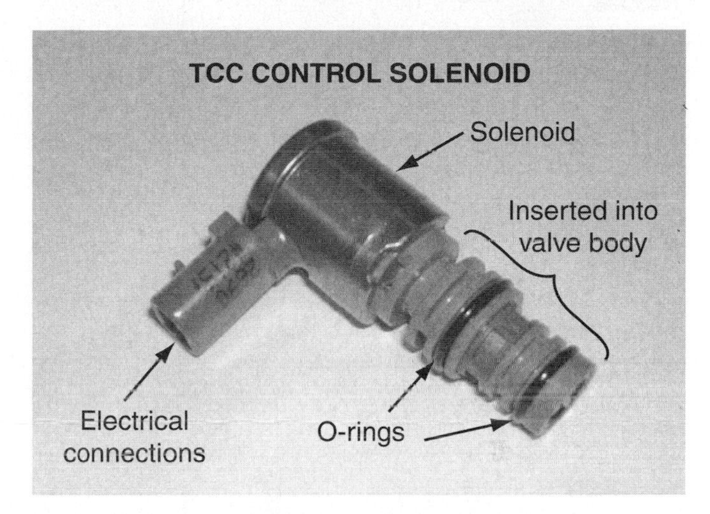

FIGURE 42–18 A typical torque converter clutch (TCC) control solenoid.

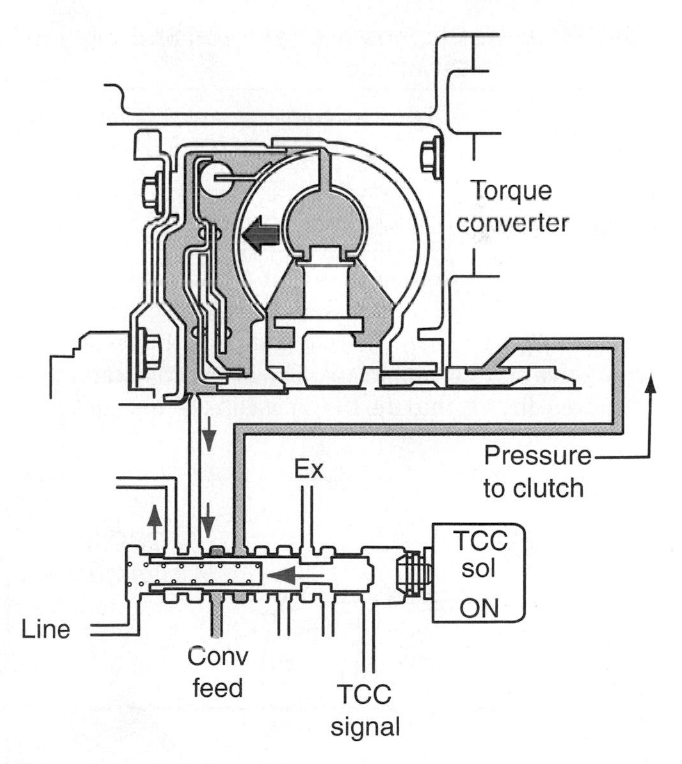

FIGURE 42–19 The TCC solenoid controls the fluid pressure to the torque converter so it can be applied or released.

Problems, Diagnosis, and Service

Safety Precautions

1. When working with electronic transmissions, the vehicle must be supported on a hoist or hydraulic stand. Always follow the correct procedure when lifting the vehicle, and always use an extra jack stand for safety.
2. The electronic transmission is often used as the rear support for the engine. Never remove the electronic transmission without first adequately supporting the engine in the vehicle.
3. When disconnecting the torque converter from the flywheel, the flywheel may need to be turned. The ring gear on the flywheel may be very sharp and could accidentally scrape your knuckles and hands. Be very careful and use the correct tools when removing the torque converter.
4. When working with electronic transmissions, take care not to spill transmission fluid on the floor. To eliminate any possibility of slipping, which could cause serious injury, wipe up any spilled ATF immediately.
5. The transmission operates above the operating temperature of the engine and radiator. Be careful when checking the transmission fluid or working on the transmission immediately after the vehicle has been run. A serious burn could result.
6. Always wear OSHA-approved safety glasses when working on electronically controlled transmissions.
7. Always use the proper tools when working on the electronically controlled transmission. Often special tools are recommended by the manufacturer.

PROBLEM: Diagnostic Codes and Shifting Problems

The Check Engine light has come on and it is found that code 22 has been stored in the PCM.

DIAGNOSIS

When a diagnostic trouble code (DTC) is stored in the PCM, it can easily be checked by using a diagnostic scan tool as shown in ***Figure 42–20***. The diagnostic scan tool is standard automotive service equipment. The diagnostic scan tool is connected directly into the DLC (data link connector).

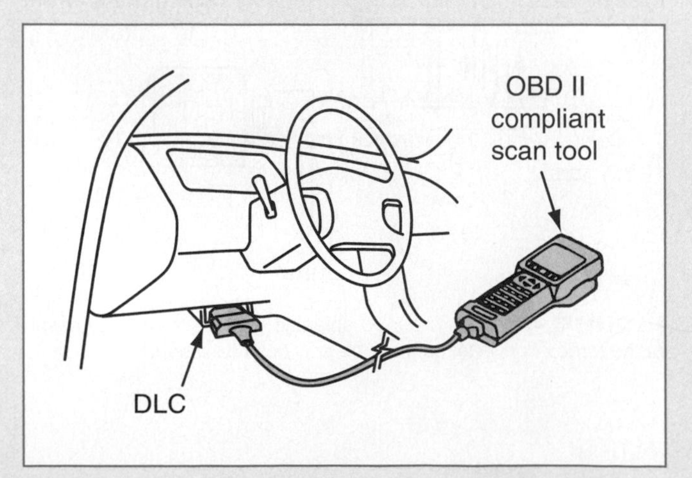

FIGURE 42–20 A scan tool can be directly connected to the data link connector (DLC) to determine transmission problems or trouble codes.

Depending on the year of the vehicle, it will either be OBDI- or OBDII-compliant. OBD, which stands for On Board Diagnostics, has become the industry standard. Vehicles up to 1995 are usually OBDI-compliant, however, the codes are not standardized between vehicle manufacturers. Vehicles from 1996 on are OBDII-compliant and have a set of standardized diagnostic codes. OBDI diagnostic codes usually are identified as a number only. OBDII codes are preceded by the letter P. For example, P0783 is identified as 3-4 Shift Malfunction. Always refer to the manufacturer's service manual to get the correct diagnostic code and to determine if the vehicle is OBDI- or OBDII-compliant.

SERVICE

Each vehicle manufacturer has a different set of diagnostic or fault code numbers for each problem identified in the electronic transmission. Always use the vehicle's service manual to determine the exact code numbers. ***Figure 42–21*** shows each of the OBDI fault codes for one manufacturer. As an example, if fault code 22 is identified on the DLC, the service technician knows that the 2/4 clutch pressure is too low.

Other manufacturers use a series of blinking lights to determine the transmission problem. For example, ***Figure 42–22*** shows a Symptom-to-Component Chart for the electrical system on an electronic transmission. After determining the number of indicator light blinks, possible causes can be determined. Then, follow the manufacturer's suggested troubleshooting chart and/or pinpoint tests to service and repair the problem.

FAULT CODE NUMBER	CONDITION	FAULT CODE NUMBER	CONDITION
21	OD Clutch—pressure too low	38	Partial torque converter clutch out of range
22	2/4 clutch—pressure too low	47	Solenoid switch valve stuck in the LR position
23	2/4 clutch and OD clutch—pressures too low	50	Speed ratio default in reverse
24	L/R clutch— pressure too low	51	Speed ratio default in 1st
25	L/R clutch and OD clutches—pressures too low	52	Speed ratio default in 2nd
26	L/R clutch and 2/4 clutches—pressures too low	53	Speed ratio default in 3rd
27	OD, 2/4, and L/R clutches—pressures too low	54	Speed ratio default in 4th
31	OD clutch pressure switch response failure	60	Inadequate LR element volume
32	2/4 pressure switch response failure	61	Inadequate 2/4 element volume
33	2/4 and OD clutch pressure response failures	62	Inadequate OD element volume
37	Solenoid switch valve stuck in the LO position		

FIGURE 42-21 Each manufacturer has a series of fault codes for its electronic transmission that can be identified by a scanner. (Courtesy of DaimlerChrysler Corporation)

SYMPTOM-TO-COMPONENT CHART				
Number of D_4 indicator light blinks while Service Check Connector is jumped.	**D_4 indicator light**	**Possible Cause**	**Symptom**	**Refer to page**
1	Blinks	• Disconnected lock-up control solenoid valve A connector • Short or open in lock-up control solenoid valve A wire • Faulty lock-up control solenoid valve A	• Lock-up clutch does not engage. • Lock-up clutch does not disengage. • Unstable idle speed.	14–38
2	Blinks	• Disconnected lock-up control solenoid valve B connector • Short or open in lock-up control solenoid valve B wire • Faulty lock-up control solenoid valve B	• Lock-up clutch does not engage.	14–39
3	Blinks or OFF	• Disconnected throttle position sensor (TPS) connector • Short or open in TP sensor wire • Faulty TP sensor	• Lock-up clutch does not engage.	14–40
4	Blinks	• Disconnected vehicle speed sensor (VSS) connector • Short or open in VSS wire • Faulty VSS	• Lock-up clutch does not engage.	14–41
5	Blinks	• Short in A/T gear position switch wire • Faulty A/T gear position switch	• Fails to shift other than 2nd↔4th gears. • Lock-up clutch does not engage.	14–42
6	OFF	• Disconnected A/T gear position switch connector • Open in A/T gear position switch wire • Faulty A/T gear position switch	• Fails to shift other than 2nd↔4th gears. • Lock-up clutch does not engage. • Lock-up clutch engages and disengages alternately.	14–44
7	Blinks	• Disconnected shift control solenoid valve A connector • Short or open in shift control solenoid valve A wire • Faulty shift control solenoid valve A	• Fails to shift (between 1st↔4th, 2nd↔4th or 2nd↔3rd gears only). • Fails to shift (stuck in 4th gear)	14–46
8	Blinks	• Disconnected shift control solenoid valve B connector • Short or open in shift control solenoid valve B wire • Faulty shift control solenoid valve B	• Fails to shift (stuck in 1st or 4th gears).	14–47
9	Blinks	• Disconnected countershaft speed sensor connector • Short or open in the countershaft speed sensor wire • Faulty counter shaft speed sensor	• Lock-up clutch does not engage.	14–48

FIGURE 42-22 On some manufacturer's electronic transmissions, a series of blinking lights assists the service technician in determining possible causes of problems. (Courtesy of American Honda Motor Co., Inc.)

There are many troubleshooting charts that can be followed in each vehicle's service manual. ***Figure 42–23*** shows an example of one such troubleshooting chart. Note that in the troubleshooting chart there are various times at which voltages must be checked. Note that this particular chart identifies terminals that are accessible to check different voltages. For example, note that as part of the troubleshooting chart the service technician needs to check the voltages between terminals D18 and A25/A26. Based on the voltage reading, the service technician is directed to either repair the open or short in the circuit or continue checking other circuits for problems.

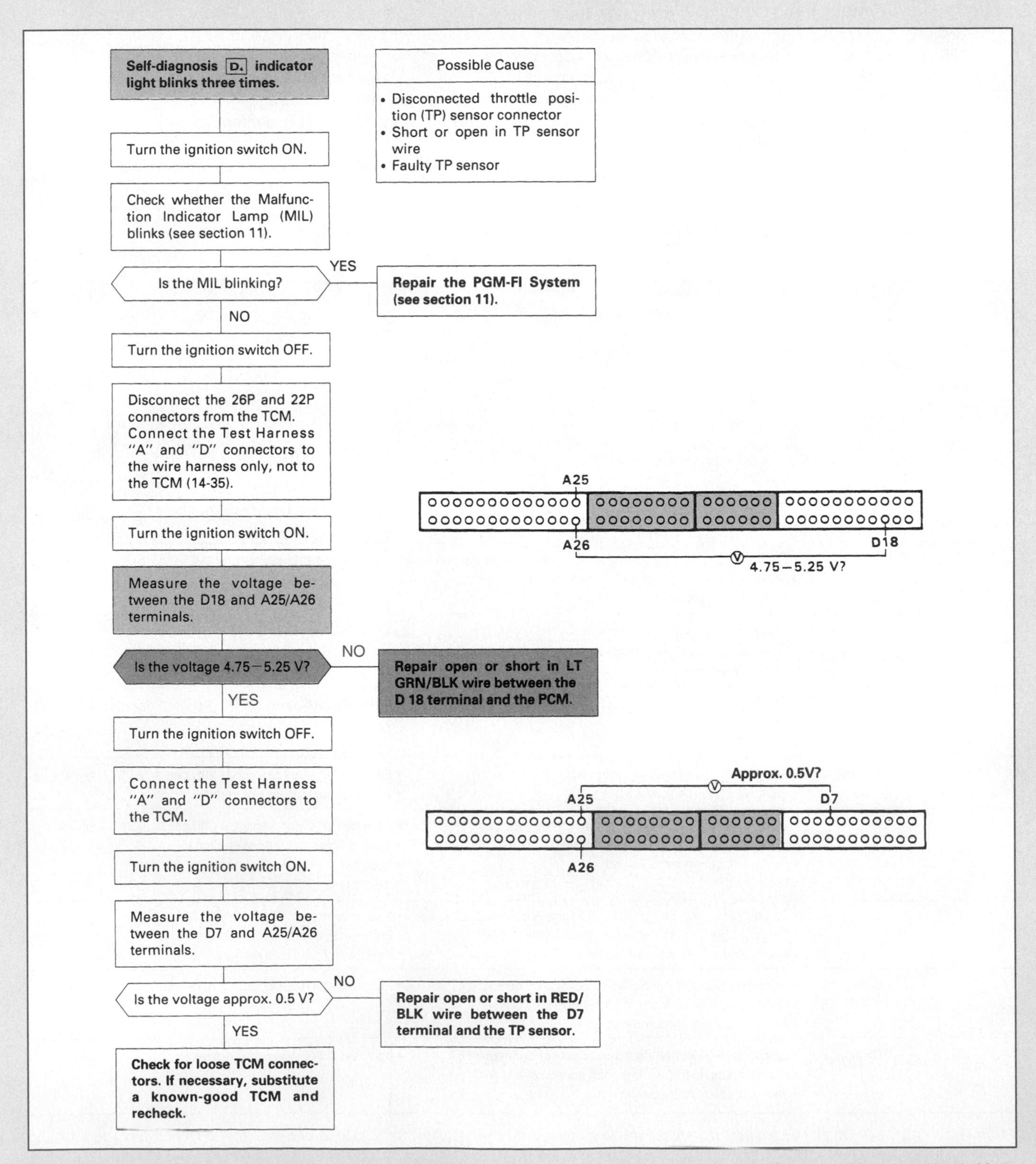

FIGURE 42-23 Follow the manufacturer's troubleshooting chart to service electronic transmissions. (Courtesy of American Honda Motor Co., Inc.)

Service Manual Connection

There are many important specifications to keep in mind when working with electronic transmissions. To identify the specifications for your engine and transmission, you will need to know the VIN (vehicle identification number) of the vehicle, the type and year of the vehicle, and the type of engine. Although they may be titled differently, some of the common computerized engine control specifications (not all) found in the service manuals are listed here. Note that these specifications are typical examples. Each vehicle and engine may have different specifications.

Common Specification	Typical Example
Throttle position sensor voltage	4 volts fully opened 1.6–2.2 volts half throttle
Shift solenoid A resistance	12–24 ohms
Shift solenoid B resistance	12–24 ohms
Voltage drop on shift solenoids	12 volts
Brake switch voltage drop	12 volts

In addition, service manuals provide specific directions for various service procedures. Some of the more common procedures include:

- Replace mainshaft speed sensor
- Test lockup control solenoid
- Road testing
- Pressure testing

SUMMARY

The following statements will help to summarize this chapter:

- The transmission computer is called the powertrain control module (PCM).
- Various types of inputs determine the operating characteristics of the engine, which enable the PCM to make decisions about when to shift and what gear to shift into.
- The two most important inputs to the PCM are the throttle position sensor (TPS) and the vehicle speed sensor (VSS).
- Other input sensors include the transmission input speed sensor, the pressure switch assembly, the transmission fluid temperature, the engine coolant temperature sensor, the engine speed sensor, the brake switch sensor, the manifold absolute pressure sensor, the A/C switch signal, and the cruise control switch.
- The PCM controls several types of outputs including the shift solenoids, the fluid pressure, and the torque converter clutch lockup.
- Many service procedures are listed in various vehicle maintenance manuals for electronically controlled transmissions. Always follow the manufacturer's suggested service to determine fault codes for the PCM and troubleshooting guides to repair faulty parts.

TERMS TO KNOW

Can you explain each of the following terms? Review the chapter until you can use each term correctly.

Downshifting
Manual shifting mode
Performance mode
Powertrain control module (PCM)
Pressure switch assembly (PSA)
Shift solenoid
Signal fluid
Upshifting
Vehicle speed sensor
Winter mode

REVIEW QUESTIONS

Multiple Choice

1. Electronically controlled transmissions are used to:
a. Reduce rough shifting
b. Reduce wasted power in the transmission
c. Reduce cost of operation
d. All of the above
e. None of the above

2. The computer most often used on electronically controlled transmissions is currently called the:
a. Powertrain control module (PCM)
b. Engine control module (ECM)
c. Engine coolant temperature (ECT)
d. Transmission control module (TCM)
e. Both a and d

3. When an electronically controlled transmission is in the winter mode of operation, the transmission:
a. Stays in first gear
b. Starts out in second gear
c. Cuts out the fourth gear
d. All of the above
e. None of the above

4. Which of the following is not an input sensor for electronically controlled transmissions?
a. Throttle position sensor
b. Vehicle speed sensor
c. Transmission input speed sensor
d. Knock sensor
e. Manifold absolute pressure sensor

5. Which of the following is considered an output of the PCM?
a. Cruise control
b. Air conditioning
c. Shift solenoid A
d. TCC solenoid
e. Both C and D

6. The ________ uses a speed sensor ring and is located on the output of the transmission.
a. Throttle position sensor
b. Engine coolant sensor
c. Transmission hydraulic speed sensor
d. Vehicle speed sensor
e. A/C switch

7. Increasing or decreasing the transmission fluid temperature has an effect on the:
a. Shift feel
b. Speed sensor
c. Engine coolant temperature
d. Speed of the engine
e. None of the above

8. The pressure switch assembly is used to tell the PCM:
a. Which gear it is in
b. How fast the vehicle is going
c. The temperature of the transmission fluid
d. The position of shift solenoid A
e. The position of shift solenoid B

9. The brake switch sensor is used so that the ________ is released when the brakes are put on.
a. TCC
b. PCM
c. TISS
d. VSS
e. TPS

10. To improve the shift feel of an electronically controlled transmission, the PCM controls the fluid pressure by controlling the:
a. TCC
b. PCM
c. PCS
d. VSS
e. TPS

The following questions are similar in format to ASE (Automotive Service Excellence) test questions.

11. Driver #1 is going 10 miles per hour and is at 10 percent TPS. Driver #2 is going 10 miles per hour and is at 50% TPS. Under these conditions, *Technician A* says that Driver #1 is shifting into second gear. *Technician B* says that Driver #2 will still be in first gear. Who is correct?
a. A only
b. B only
c. Both A and B
d. Neither A nor B

12. *Technician A* says that the PCM only controls one output. *Technician B* says that the PCM is used to control shift solenoids. Who is correct?
a. A only
b. B only
c. Both A and B
d. Neither A nor B

13. *Technician A* says that to check the diagnostic trouble codes, use a scan tool. *Technician B* says that some vehicles use a blinking light to determine the problem with the electronically controlled transmission. Who is correct?
a. A only
b. B only
c. Both A and B
d. Neither A nor B

14. *Technician A* says that the shift solenoids are controlled by the signal pressure. *Technician B* says that the shift valves have fluid drive pressure applied to them. Who is correct?
a. A only
b. B only
c. Both A and B
d. Neither A nor B

15. *Technician A* says that when a shift solenoid is off, it exhausts the signal pressure. *Technician B* says that when the shift solenoid is on, it closes off the signal fluid. Who is correct?
    a. A only
    b. B only
    c. Both A and B
    d. Neither A nor B

Essay

16. Describe the operation of a shift solenoid.

17. What is the purpose of using the TPS and VSS sensor on an electronically controlled transmission?

18. Explain how the PCM controls the TCC.

19. Why is it important to know the transmission fluid temperature on an electronically controlled transmission?

20. Why is there a speed sensor on both the input and output of an electronically controlled transmission?

Short Answer

21. If a driver has a 50% throttle position, the car must go much ________ to upshift.

22. There are typically ________ types of hydraulic pressure that are being applied to the shift solenoids and valve assembly.

23. The TCC solenoid is used to apply and ________ the fluid pressure to the torque converter clutch.

24. Another name for the PCM is the ________.

25. How many shift valves are in a typical electronically controlled transmission?

Applied Academics

TITLE: Torque Converter Clutch (TCC) and Pulse Width Modulation (PWM)

ACADEMIC SKILLS AREA: Science

NATEF Automobile Technician Tasks:

The service technician demonstrates an understanding of the role mechanical transducers play in sending an electrical control signal to modify the operating characteristics of the automatic transmission and transaxle.

CONCEPT:

As discussed in this chapter, the torque converter clutch (TCC) is engaged and disengaged by the powertrain control module (PCM). However, the rate and speed of engagement is often too fast. The result of rapid engagement is a transmission that feels too rough. To help offset any undesirable feel on the transmission, some PCMs use a more sophisticated TCC solenoid. This system uses pulse width modulation (PWM). For this application, the TCC solenoid controls the "apply" and "release" pressures to smooth out the TCC engagement. The accompanying graph shows how the TCC applies fluid pressure changes over a time period (within a few seconds). The bottom line represents time. The vertical line represents the percent of duty cycle. Rather than immediately engaging the TCC, the pressure increases over a longer period of time. This is shown in the graph from point A to point B. At point C the TCC is fully applied. During release, the TCC releases the pressure on the torque converter over a period of time (from E to F). The net effect is smoother apply and release of the torque converter clutch.

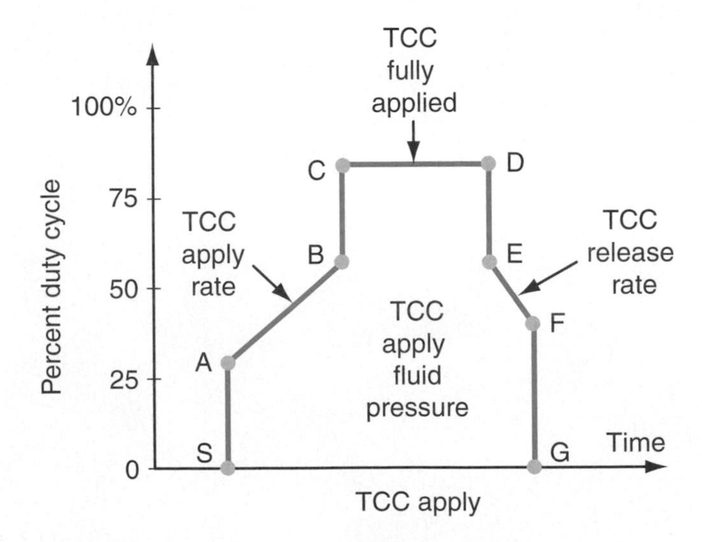

APPLICATION:

Why is time such an important consideration when engaging the torque converter clutch? What do you think a driver would feel if the TCC was engaged over a longer period of time?

Drive Lines, Differentials, and Axles

CHAPTER 43

OBJECTIVES

After studying this chapter, you should be able to:

- Identify the parts and operation of rear-wheel drive shafts and universal joints.
- Identify the parts and operation of front-wheel drive shafts and constant velocity universal joints.
- Identify the parts and operation of differentials, including the nonslip (limited slip) differentials.
- State the correct lubrication used in differentials.
- Describe common problems, their diagnosis, and service tips used with drive lines, differentials, and axles.

Certification Connection

ASE Connection: The information in this chapter can help you prepare for the National Institute for Automotive Service Excellence (ASE) certification tests. The tests and content areas most closely related to this chapter are:

Test A3—Manual Drive Train and Axles

- **Content Area**—Drive Shaft, Half Shaft, and Universal Joint/Constant Velocity (CV) Joint Diagnosis and Repair (Front- and Rear-Wheel Drive)
- **Content Area**—Rear-Axle Diagnosis and Repair

NATEF Connection: Much of the information in this chapter is related to the NATEF tasks. The NATEF tasks and priority numbers most closely related to this chapter are:

1. Diagnose constant velocity (CV) joint noise and vibration concerns; determine necessary action. P-1
2. Diagnose universal joint noise and vibration concerns; perform necessary action. P-1
3. Replace front-wheel drive (FWD) front wheel bearing. P-1
4. Inspect, service, and replace shafts, yokes, boots, and CV joints. P-1
5. Inspect, service, and replace shaft center support bearings. P-3
6. Check shaft balance, measure shaft runout, and measure and adjust driveline angles. P-2
7. Diagnose noise and vibration concerns (differential); determine necessary action. P-2
8. Diagnose fluid leakage concerns (differential); determine necessary action. P-2
9. Inspect ring gear and measure runout; determine necessary action. P-2
10. Measure and adjust drive pinion depth. P-2
11. Check ring and pinion tooth contact patterns; perform necessary action. P-1
12. Inspect and flush differential housing; refill with correct lubrication. P-2
13. Inspect and replace drive axle shaft wheel studs. P-2
14. Remove and replace drive axle shafts. P-1
15. Inspect and replace drive axle shaft seals, bearings, and retainers. P-2
16. Measure drive axle flange runout and shaft end play; determine necessary action. P-2

INTRODUCTION

To get a vehicle moving, power or torque must be transferred from the transmission to the drive wheels. This is done by using drive lines, differentials, and axles. The purpose of this chapter is to define the parts, operation, and diagnosis and service of these components.

43.1 REAR-WHEEL DRIVE SHAFTS AND OPERATION

PURPOSE OF THE DRIVE SHAFT

The torque that is produced from the engine and transmission must be transferred to the rear wheels to push the vehicle forward and backward. The **drive shaft** must provide a smooth, uninterrupted flow of power to the axles. The drive shaft and differential are used to transfer this torque. ***Figure 43–1*** shows the location of the drive shaft and differential.

The drive shaft has several functions. First, it must transmit torque from the transmission to the axle. During operation, it is necessary to transmit maximum low-gear torque developed by the engine. The drive shaft must also be capable of rotating at the very fast speeds required by the vehicle.

The drive shaft must also operate through constantly changing angles between the transmission and the differential and axles. The rear axle is not attached directly to the frame of the vehicle. It rides suspended by the springs and travels in an irregular floating motion. As the rear wheels roll over bumps in the road, the differential and axles move up and down. This movement changes the angle between the transmission and the differential.

The length of the drive shaft must also be capable of changing while transmitting torque. Length changes are caused by axle movement due to torque reaction, road deflections, braking loads, and so on. ***Figure 43–2*** shows the movements that the drive shaft undergoes during normal driving operation.

PURPOSE OF UNIVERSAL JOINTS

Universal joints (U-joints) are used to permit the drive shaft to operate at different angles. ***Figure 43–3*** shows a drawing and a photograph of a simple universal joint. Note that the rotation can be transmitted when two shafts are at different angles. This type of universal joint is called the cross and bearing type **Cardan universal joint**, or four-point joint. It consists of two **yokes** and a journal assembly with four **trunnions**. A yoke is a Y-shaped assembly that is used to connect the U-joint together. The trunnion is a protrusion on the journal assembly. The journal assembly is also called a **cross** and bearing assembly or spider.

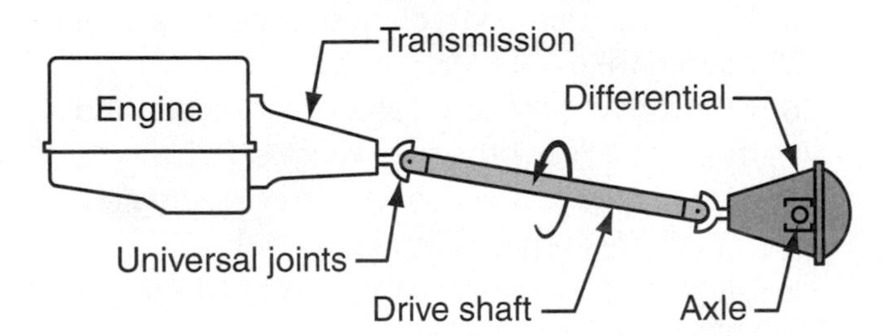

FIGURE 43–1 The drive shaft is used to transmit the power from the engine and transmission to the differential and axles on rear-wheel drive vehicles.

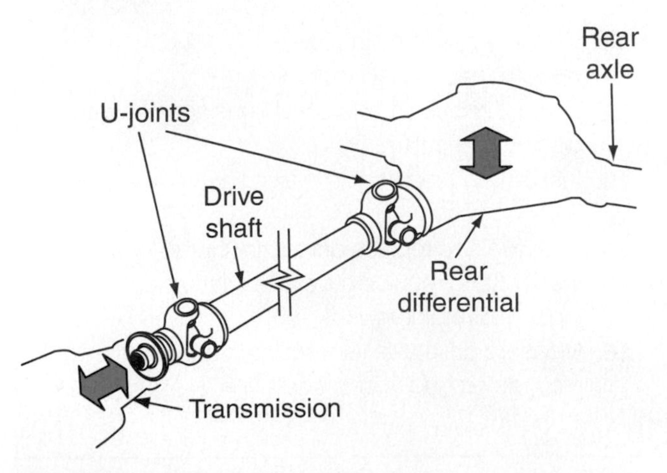

FIGURE 43–2 The drive shaft is designed to allow for both up and down motion on the differential and shortening and lengthening between the differential and transmission.

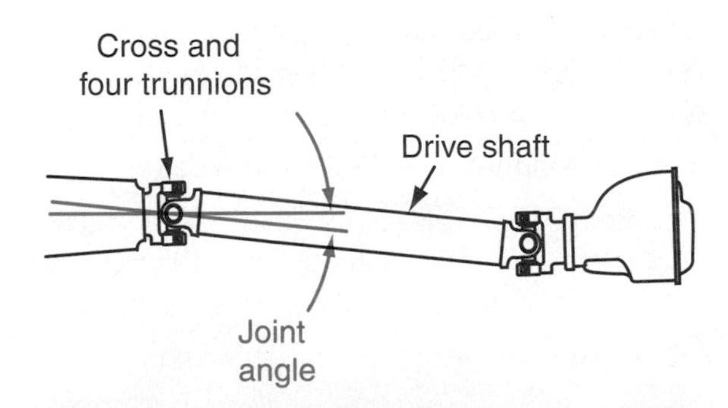

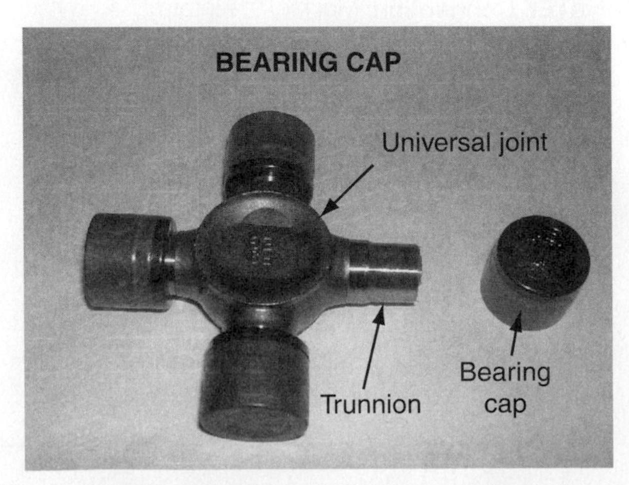

FIGURE 43–3 Universal joints allow the drive shaft to turn through different angles.

UNIVERSAL JOINT PARTS

The universal joint is made of several parts. The center of the universal joint is called the cross and bearing assembly. Its purpose is to connect the two yokes. The yokes are attached directly to the drive shafts. Four **bearing caps** are placed on the universal joint. Each cap is placed on a trunnion part of the universal joint. Each bearing cap is a needle-type bearing that allows free movement between the trunnion and yoke. The needle bearing caps are attached to the yokes by several methods. They can be pressed into the yokes, bolted to the yokes, or held in place with bolts, nuts, U-bolts, or metal straps. Snap rings are also used to hold bearing caps in place (***Figure 43–4***). On most replacement universal joints, there is a lubrication fitting to put grease into the needle bearings.

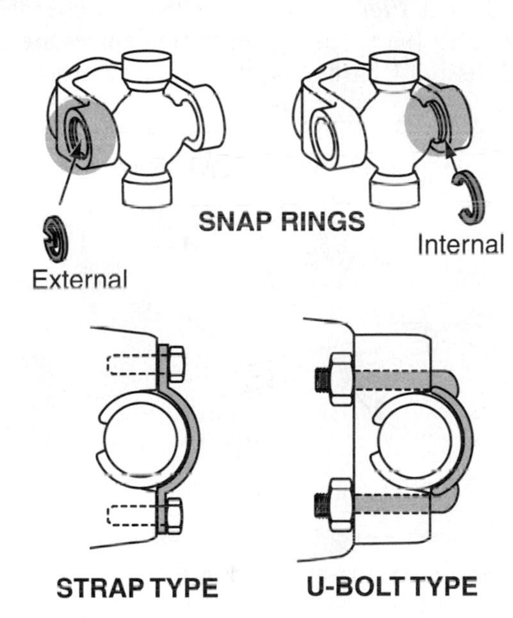

FIGURE 43–4 Universal joints are held in the yoke in several ways, including snap rigs, U-bolts, and straps.

Car Clinic:
4-JOINTS

CUSTOMER CONCERN:
Noise in Drive Shaft

A car has developed a squeaky noise in the drive shaft. The noise is not heard all the time. However, when the car is shifted into reverse and then into forward again, a slight clunk is heard and the sound returns.

SOLUTION: The most common cause of noise in the drive shaft is a bad universal joint. When it is worn, the U-joint squeaks. The squeak comes from the lack of lubrication in the U-joint needle bearings. Remove the drive shaft and check the U-joints. When moving the U-joints to check them, look for any irregular bumps or motion. Replace the U-joints, if necessary.

SLIP JOINTS

The drive shaft must also be able to lengthen and shorten during operation with irregular road conditions. A **slip joint** is used to compensate for this motion. The slip joint is usually made of an internal and external spline. It is located on the front end of the drive shaft and is connected to the transmission (***Figure 43–5***). The slip joint can also be placed in the center of the drive shaft.

DRIVE SHAFT VIBRATION

Vibration is the most common drive shaft problem. It can be either transverse or torsional. As shown in ***Figure 43–6***, transverse vibration is the result of an unbalanced condition acting on the shaft. This condition is usually caused by dirt or foreign material on the shaft, and it can cause a rather noticeable vibration in the vehicle.

Torsional vibration occurs from the power impulses of the engine or from improper universal joint angles. It produces a noticeable sound disturbance and can cause a mechanical shaking. In excess, both types of vibration can damage the universal joints and bearings.

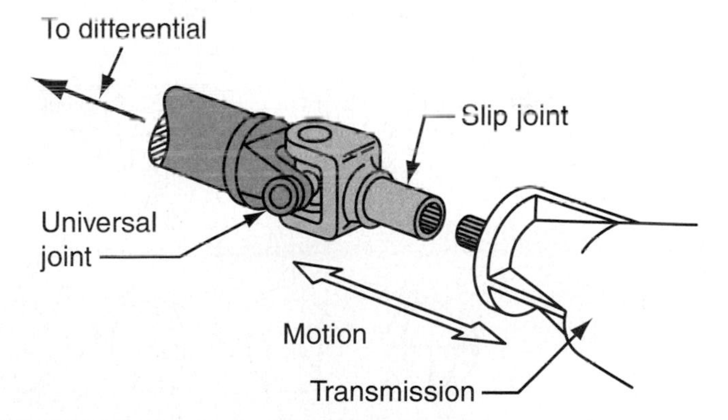

FIGURE 43–5 The slip joint is designed to allow the drive shaft to shorten and lengthen when the differential goes over irregular road surfaces.

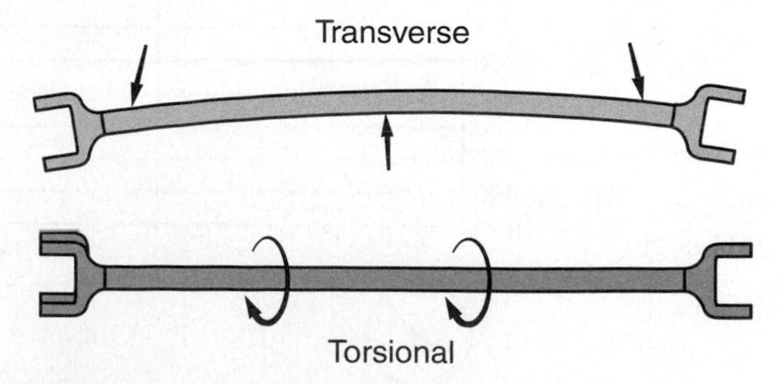

FIGURE 43–6 Vibration in the drive shaft can be either transverse or torsional. Both types of vibration can damage the universal joints.

DRIVE SHAFT VELOCITY AND ANGLES

The popular cross and bearing assembly universal joint has one disadvantage. When the universal joint transmits torque through an angle, the output shaft increases speed and slows down twice in each revolution of the shaft. The rate at which the speed changes depends on the steepness of the universal joint angle. The speed changes are not normally visible during rotation. However, they may be felt as torsional vibration due to improper installation, steep and/or unequal operating angles, and high speeds.

Figure 43–7 shows how the output shaft changes speed. The input path and bearings rotate in a circular motion when viewed from the end of the drive shaft. The output path and bearings rotate in an elliptical motion or path. The output path is viewed at an angle instead of straight on because it is at an angle from the input shaft. Therefore it looks like an ellipse.

In operation, the input shaft speed has a constant velocity. The output shaft speed accelerates and decelerates (catching up and falling behind in rotation) during one complete revolution. Refer to ***Figure 43–8***. From 0 to 90 degrees rotation, the output shaft accelerates. From 90 to 180 degrees rotation, the output shaft decelerates. From 180 to 270 degrees rotation, the output shaft accelerates again. From 270 to 360 degrees rotation, the output shaft decelerates. The input and output shafts complete one rotation at exactly the same time. The greater the output angle, the greater the change in velocity of the output shaft for each revolution.

The torsional vibrations mentioned earlier travel down the drive shaft to the next universal joint. At the second universal joint angle, similar acceleration and deceleration occur. However, these take place at equal and reverse angles to the first joint (***Figure 43–9***). Now the speed changes cancel each other when the two operating angles are equal. Drive shafts must have at least two universal joints, and

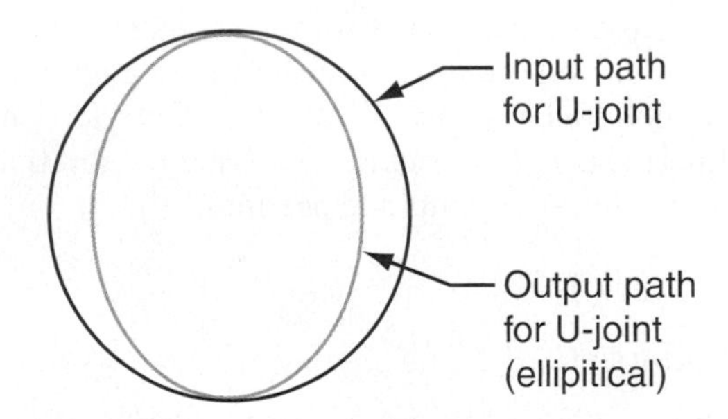

FIGURE 43-7 The input path of the universal joint is circular. The output path of the universal joint is elliptical. The elliptical path causes the output shaft to accelerate and decelerate within each shaft revolution.

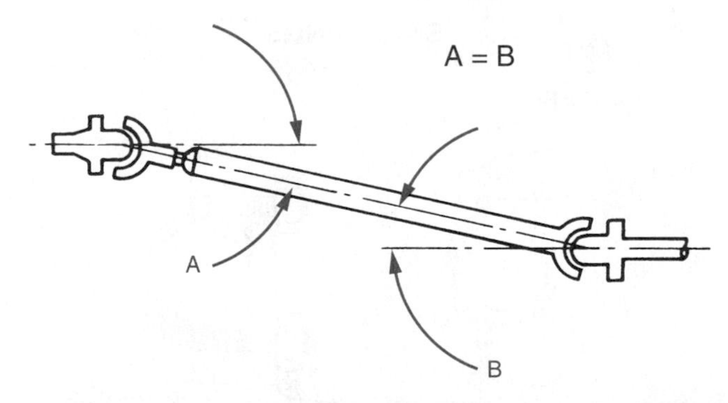

FIGURE 43-9 To offset the velocity changes in a drive shaft, a second universal joint is used at an equal and opposite angle. These operating angles are called companion angles.

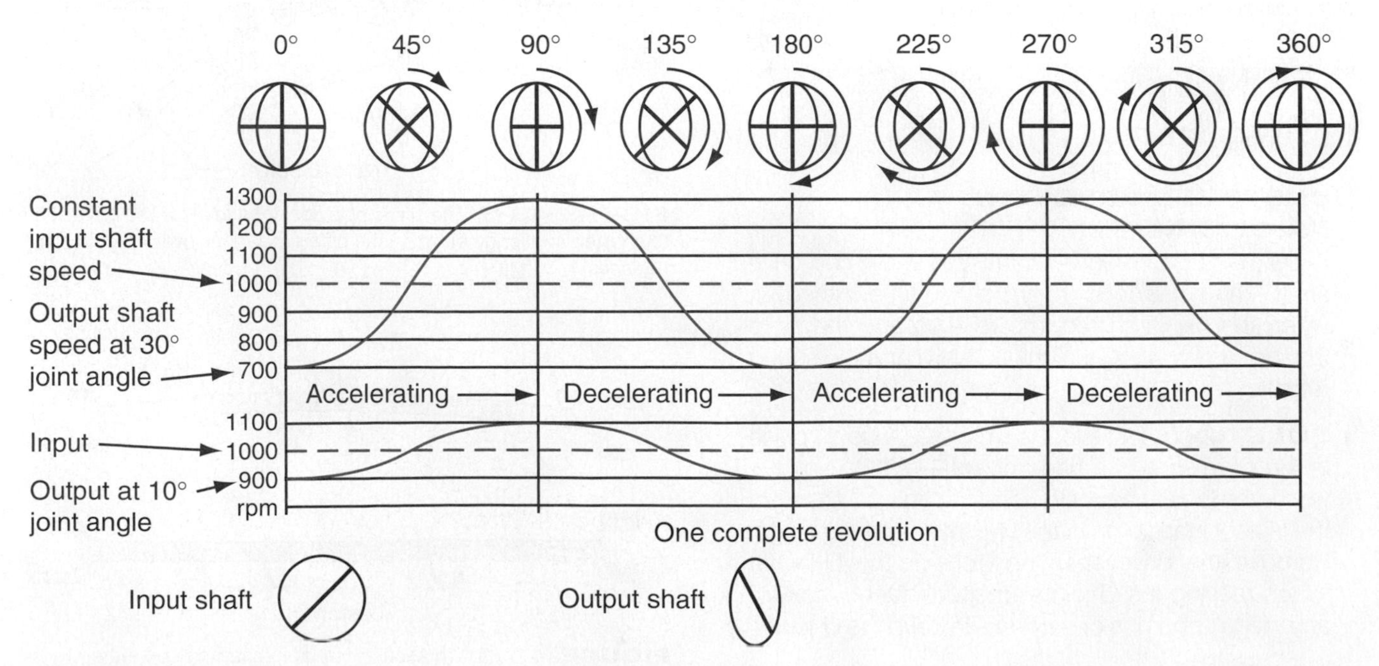

FIGURE 43-8 Within one revolution of a drive shaft with the universal joint at an angle, the speed accelerates and decelerates two times. The greater the angle of the universal joint, the greater the change in speed.

operating angles (also called companion angles) must be small and equal. Any variations from this can cause excessive vibration and needle bearing and trunnion wear.

DRIVE SHAFT CONSTRUCTION

There are several designs for drive shafts. The type depends on several factors. These include the length of the vehicle, the amount of power the engine must transmit, the rpm of the drive shaft, and the year of the vehicle. ***Figure 43–10*** shows a typical drive shaft. In most cases, a universal joint is placed on both sides. The most common location for the slip joint is on the output of the transmissions. On certain vehicles, the slip joint is located in the center of two drive shafts.

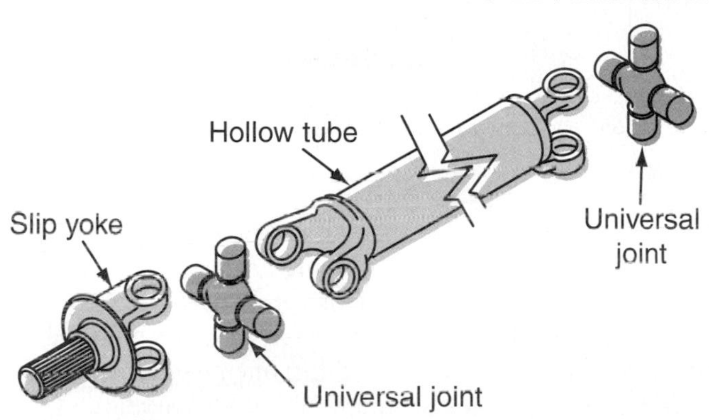

FIGURE 43–10 These parts comprise a standard drive shaft.

CENTER BEARING

On many large vehicles, it may be necessary to use a center bearing for support on the drive shaft. ***Figure 43–11*** shows a typical center bearing assembly. When a center bearing is used, there will be two drive shafts and usually more than two universal joints. The bearing assembly supports the end of the first drive shaft by being bolted directly to the frame of the vehicle.

*Refer to photo #93 in the front of the textbook to see what a **center bearing assembly** looks like.*

43.2 FRONT-WHEEL DRIVE SYSTEMS

REQUIREMENTS OF FRONT-WHEEL DRIVE SYSTEMS

A front-wheel drive vehicle presents several unique problems in the design of drive shafts and joints. The drive shaft must be able to do three things. First, it must allow the front wheels to turn for steering. Second, it must be able to telescope, an action similar to that of a slip joint on rear-drive vehicles. Third, it must transmit torque continuously without vibration. Because of these characteristics, a cross and bearing universal joint would not work well, especially

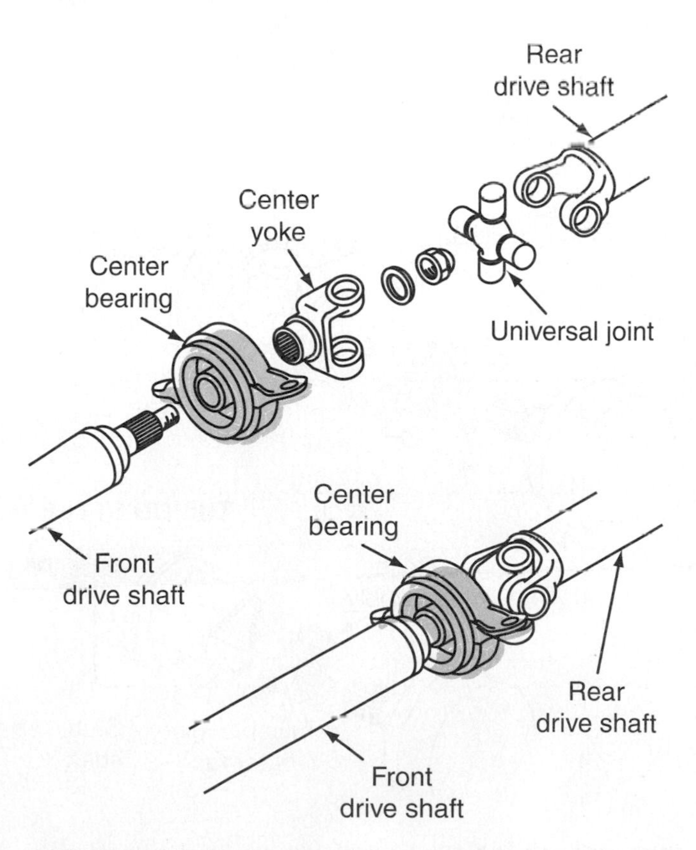

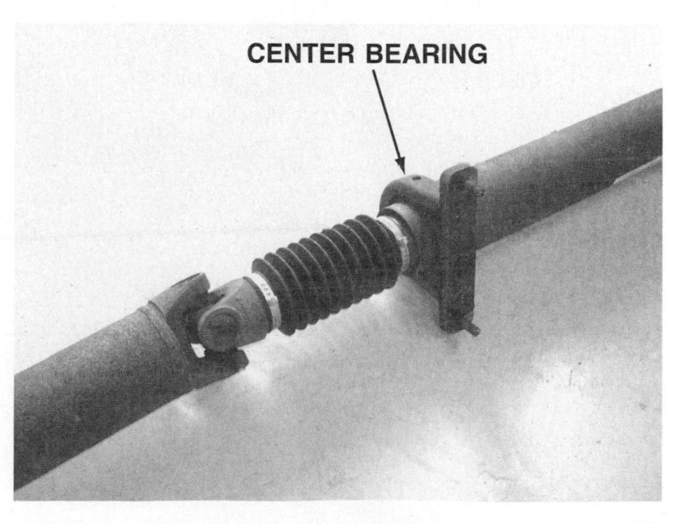

FIGURE 43–11 The center bearing is used to support longer drive shaft systems. The center bearing assembly is attached directly to the frame.

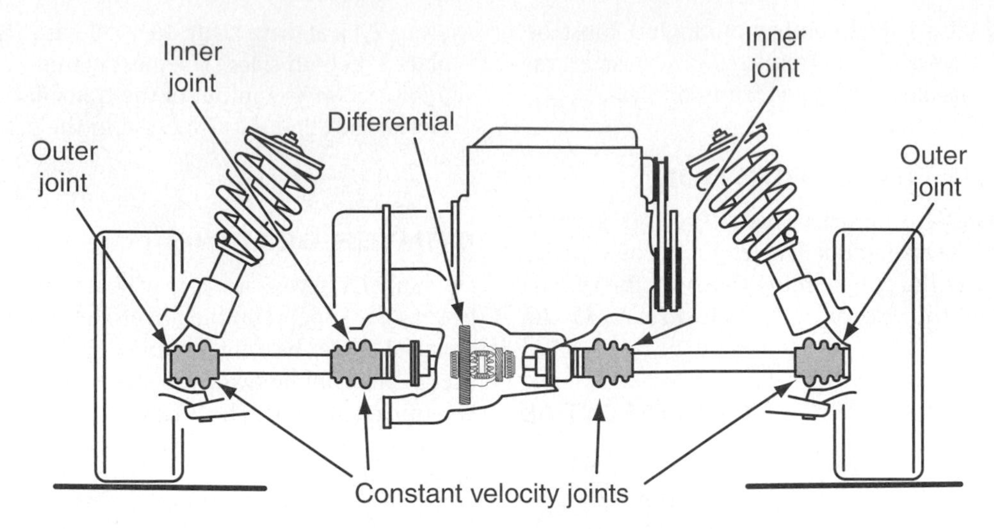

FIGURE 43–12 Front-wheel drive systems use constant velocity (CV) joints on both axles to transmit power to the wheels.

during sharp turning. A **constant velocity (CV) joint** is used instead (***Figure 43–12***).

CONSTANT VELOCITY JOINTS

The constant velocity joint is designed much the same as a set of bevel gears. Balls and grooves are used, rather than gears (***Figure 43–13***). Balls and grooves connect the input and output shafts. If the balls are placed in elongated grooves, the result is a CV joint. This type of joint does not produce the speed or velocity changes on the output shaft that a cross and bearing-type universal joint produces. As you recall, when a cross and bearing universal joint is used, there are several speed variations within each revolution of the drive shaft. With constant velocity joints, the speed of the drive shaft is continuous within one revolution. Therefore, there is much less wear and smoother operation.

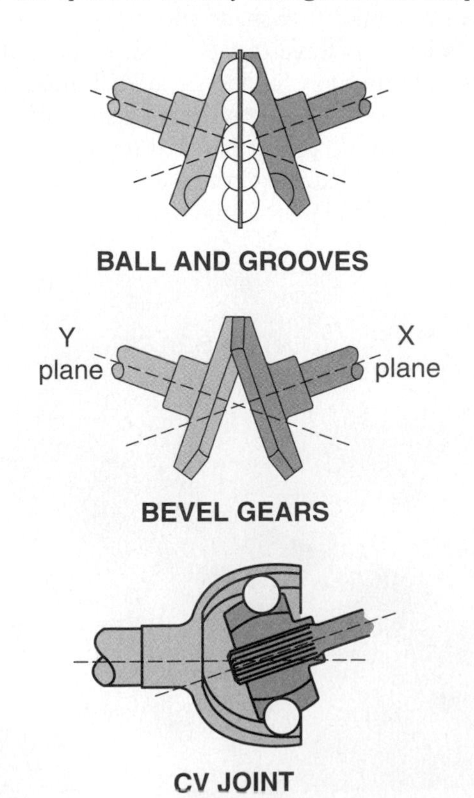

FIGURE 43–13 CV joints act much the same as bevel gears or ball and groove connections.

TYPES OF CV JOINTS

There are two types of CV joints. ***Figure 43–14*** shows the ball-style and the tripod-type CV joints. The ball style uses a series of balls, a cage, and inner and outer races. The **tripod** uses a tulip assembly and three rollers. Both perform well in front-wheel drive cars.

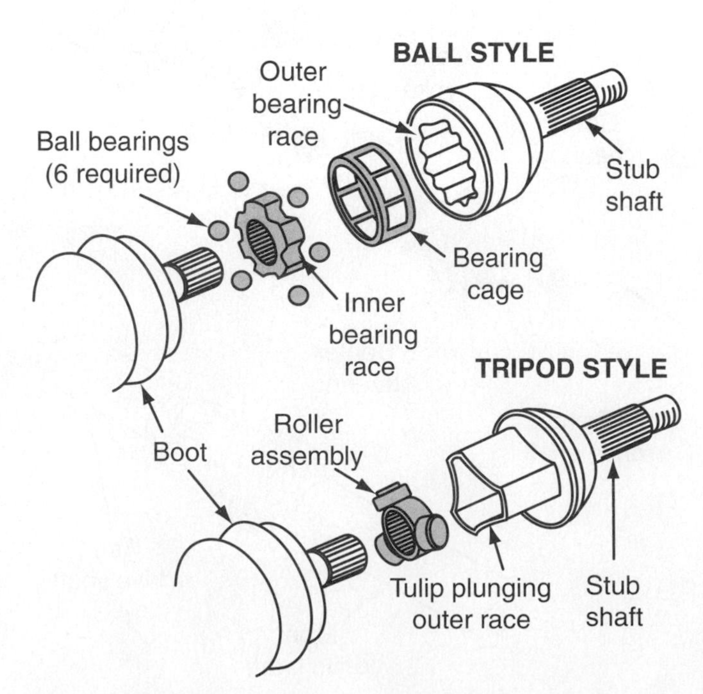

FIGURE 43–14 Two types of CV joints are commonly used—the ball-style and the tripod-type.

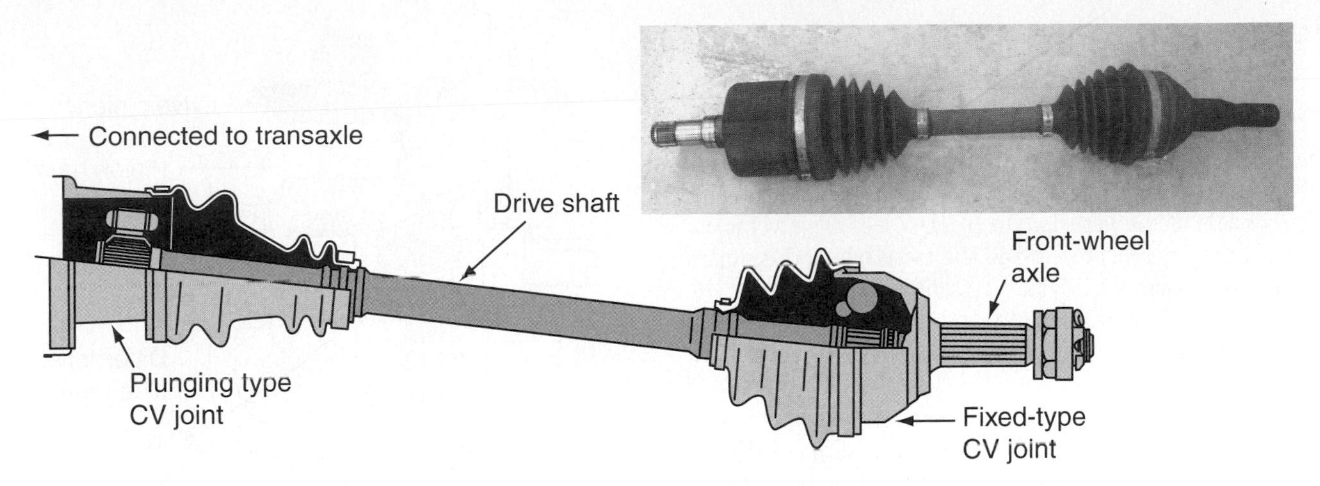

FIGURE 43–15 The plunging CV joint is used on the transaxle end. The fixed-type CV joint is used on the wheel end.

The typical front-wheel drive car uses two drive shaft assemblies. One assembly drives each wheel. Each assembly has a CV joint at the wheel end. This CV joint is called the fixed joint or outboard joint. A second joint on each shaft is located at the transaxle end. This CV joint is called the inboard or plunging joint. This may be either a ball or a tripod CV joint. It allows the slip motion that is required when the drive shaft shortens or lengthens because of irregular surfaces (***Figure 43–15***).

OTHER APPLICATIONS FOR CONSTANT VELOCITY

On certain makes of cars, (such as some foreign vehicles) the engine is located in the rear. On these cars, constant velocity joints may be used in the rear. ***Figure 43–16*** shows such a system. Two drive shafts are used on this system. In addition, two constant velocity joints are used on each drive shaft.

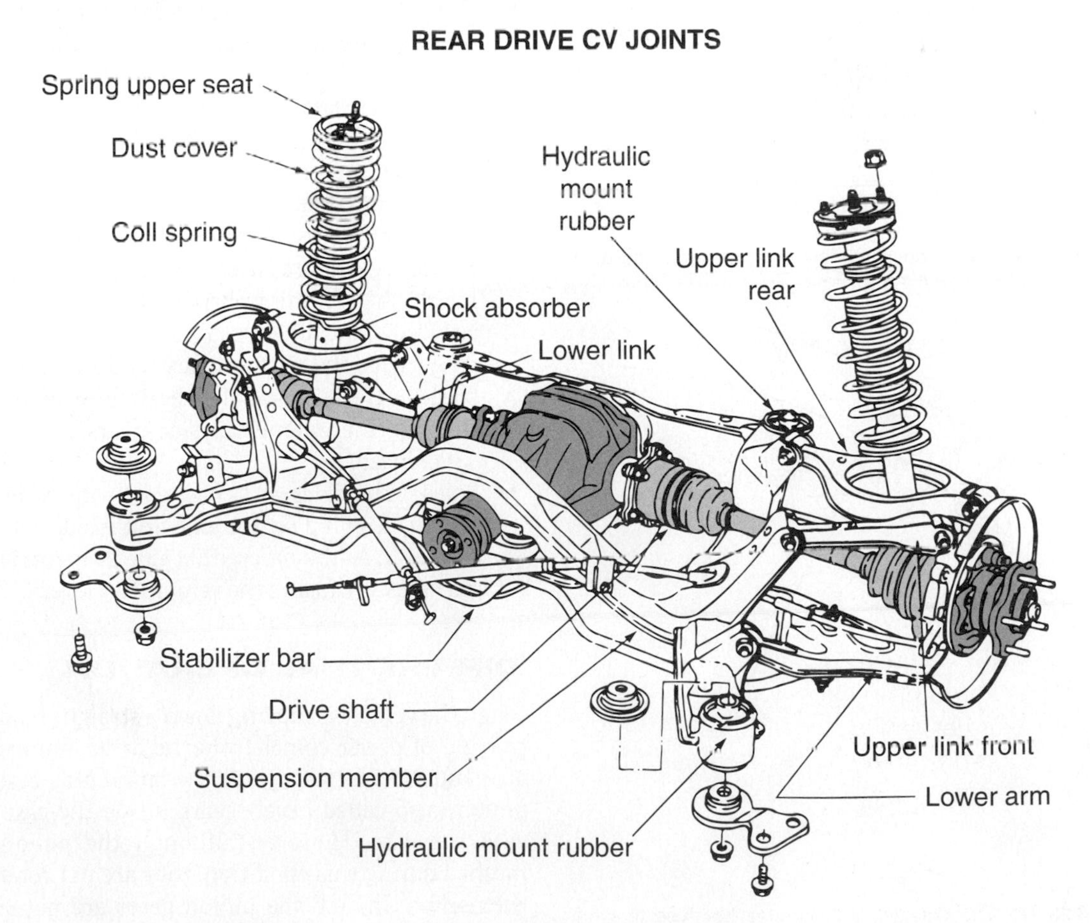

FIGURE 43–16 Constant velocity joints are also used on certain types of rear engine vehicles. (Courtesy of Nissan North America, Inc.)

43.3 DIFFERENTIALS AND AXLES

PURPOSE OF THE DIFFERENTIAL

The purpose of the **differential**, shown in ***Figure 43–17***, is to transmit the torque from the drive shaft to the axles and drive wheels of the vehicle. On front-wheel drive vehicles, the differential is located inside the transaxle and is a part of the total assembly. Torque is transmitted from the engine, through the transmission, and to the differential. The differential then splits the torque and sends it to the drive wheels.

In addition, the differential allows the rear wheels to turn at different speeds during cornering. ***Figure 43–18*** shows why the wheels turn at different speeds on corners. The inside rear wheel turns at a smaller radius than the outside rear wheel. The differential is designed to keep the power transmitting equally to both wheels while they are traveling at different speeds.

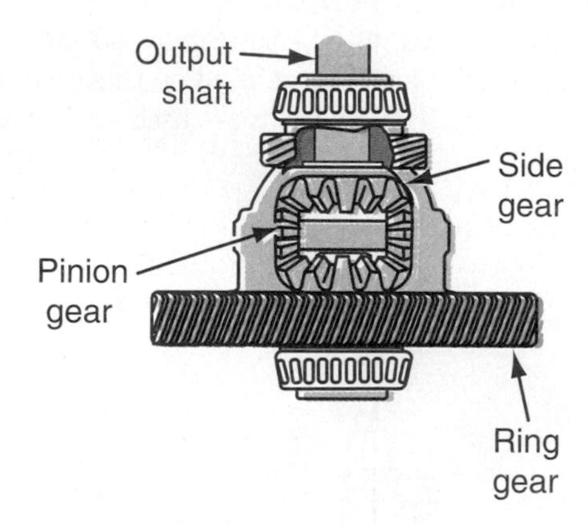

FIGURE 43–17 A differential transmits torque from the drive shaft to the axles and drive wheels.

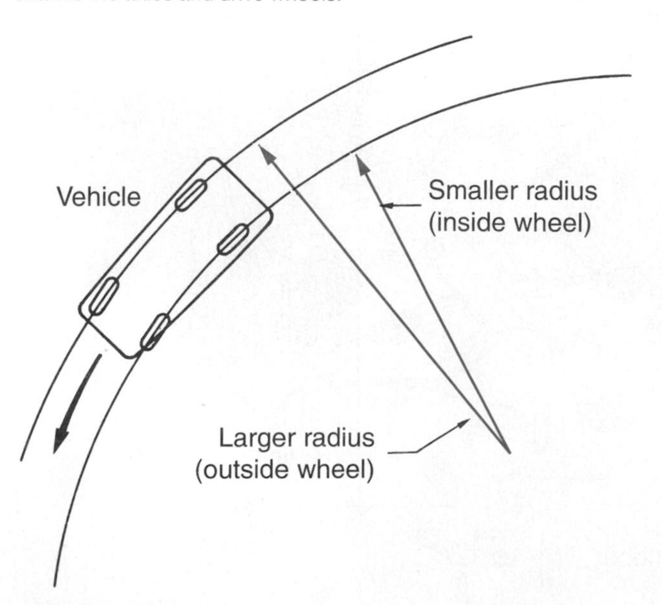

FIGURE 43–18 The differential is needed because the rear wheels turn at different speeds when going around a corner.

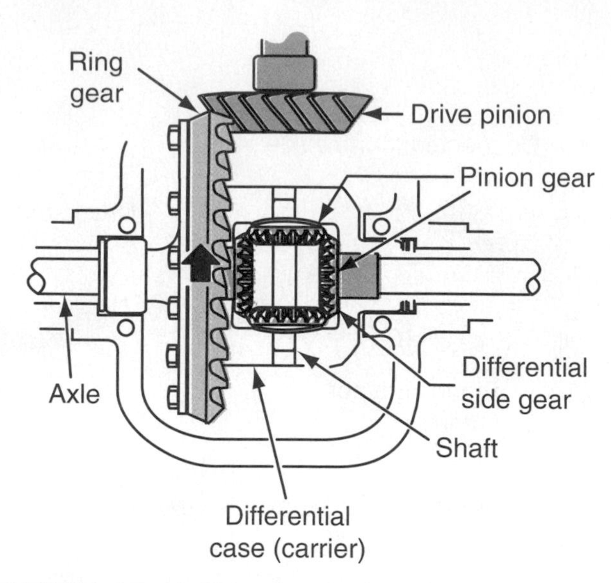

FIGURE 43–19 The main parts of the differential.

DIFFERENTIAL MAIN PARTS

Figure 43–19 shows the main parts of the differential:

- Drive pinion—The drive pinion is the main input shaft to the differential. It is driven from the vehicle drive shaft.
- Ring gear—The ring gear is driven from the drive pinion. Its purpose is to drive the remaining parts of the differential.
- Differential case—The differential case or carrier holds several bevel gears. The entire differential case is driven from the ring gear, which is bolted to it.
- Two differential side gears and two pinion gears—These four gears are placed inside the differential case. All four gears are meshed together. The pinion gears have a shaft running through their center. The shaft is secured to the differential case. Thus, as the differential case turns, the shaft rotates (end to end) at the same speed as the differential case. Because all four gears are meshed, the differential side gears also rotate at the same speed.
- Axles—The axles are attached to the differential side gear by a spline on the axle and inside the differential side gear. As the differential side gear rotates, the axles also rotate, causing the vehicle to move.

DIFFERENTIAL OPERATION

When the vehicle is moving down a straight road, the transmission of power comes from the drive pinion to the ring gear and differential case. As the differential case turns, the pinion (also called bevel) gears inside the case also move with the case. However, although the pinion gears are meshed during this condition, they are not rotating among themselves. In fact, the pinion gears are not spinning on their shafts at this point.

As the car goes around a left corner, the left axle slows down and the right axle speeds up. During this slowing down and speeding up, the pinion gears inside the differential case begin to rotate among themselves. The pinion gears are now rotating or turning on the shaft. The pinion gear is said to walk around the differential side gears. When the vehicle returns to a straight line, both axles are spinning at the same speed, but the gears are no longer turning among themselves. As the car goes around a right corner, the right axle slows down and the left axle speeds up. Again, the pinion gears inside the differential case begin turning among themselves. (The pinion gear walks around the differential side gears.) This is necessary so that the axles will spin at different speeds while still transmitting power.

Car Clinic: DIFFERENTIAL

CUSTOMER CONCERN:
Rear End Whine

A car has a noticeable whine in the rear end or differential at 30 to 45 mph. When the engine decelerates, the whine stops. What could be the problem?

SOLUTION: The whine is probably coming from the ring and pinion gear in the differential. Check the differential ring and pinion gear. Check for wear and backlash specifications. The ring and pinion gear will probably have to be replaced.

COMPLETE DIFFERENTIAL

Figure 43–20 shows a complete differential. In addition to the parts already mentioned, several other parts are included. These are:

- Pinion gears—This differential shows four pinion gears for more torque-carrying capabilities.
- Adjusters—As they are turned in and out, adjusters move the entire differential case (gear case) from side to side. This adjustment is used to change the gear backlash between the drive pinion and the ring gear, and to preload bearings.
- Bearings—Various bearings are used to support each shaft.
- Thrust washers—Thrust washers are used to absorb any side thrust produced by the gears in the differential.
- Carrier—The carrier housing is used to hold all the differential parts together.

LIMITED SLIP DIFFERENTIAL

On a standard differential, equal torque is transmitted to the rear wheels if the loads on the two wheels are the same. However, if one wheel hits a patch of ice or other slippery surface, the torque on the two wheels will be different. In this case, one wheel may spin freely while the other wheel produces torque.

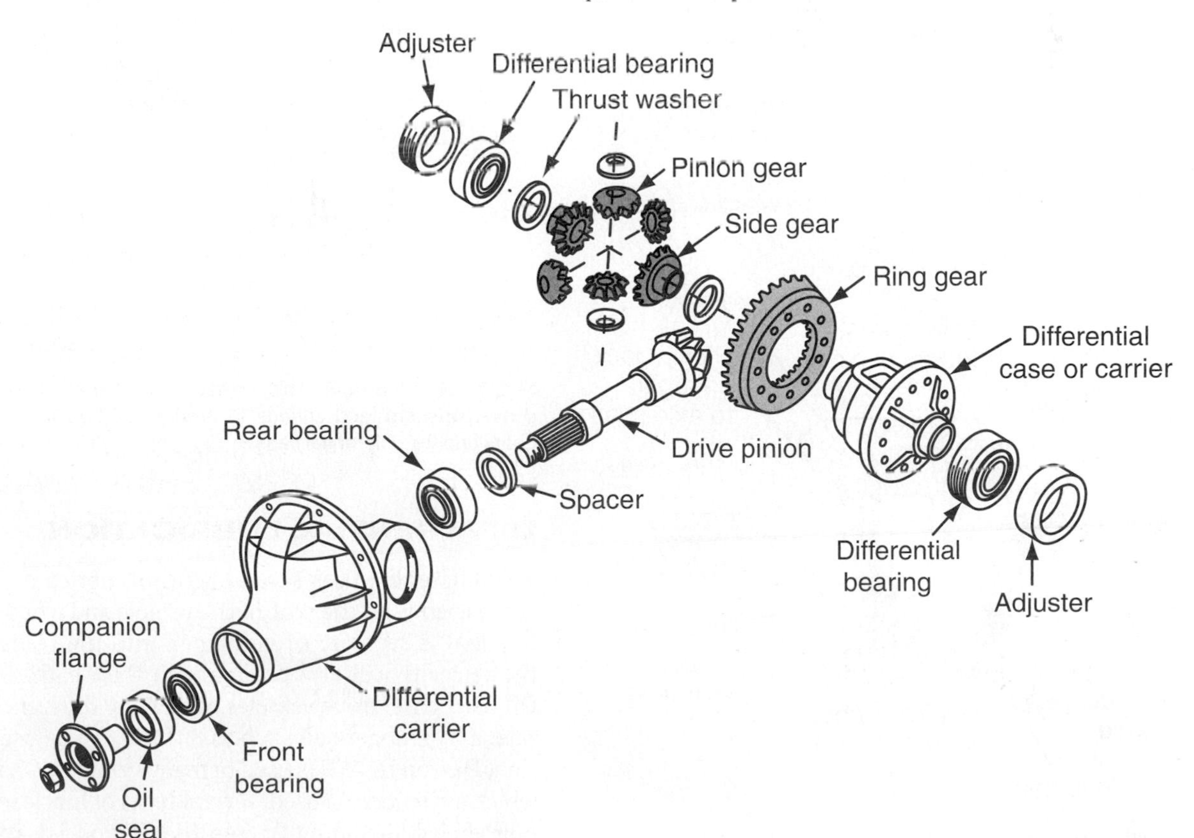

FIGURE 43–20 The differential assembly uses four pinion gears for more torque-carrying capabilities.

To eliminate this condition, **limited slip differentials** are used. Limited slip differentials use a set of clutches or cones to lock up both wheels. The clutches or cones apply a pressure to the side gears. This additional pressure prevents one wheel from spinning more rapidly than the other. The clutch assembly or cone assembly is located inside the differential case, usually between the side gears and the carrier housing or differential case.

Figure 43–21 shows a limited slip differential system. Note that there are additional clutches, forming a clutch pack, located between the side gears and the differential case. When pressurized by a spring, these clutches help to lock the side gears to the differential case so that they will not spin on their axes.

Although there are many designs, the principles of operation remain the same. Springs or some other force must push clutches or cones against a surface to lock the side gears to the differential case.

Another illustration of the clutches is shown in ***Figure 43–22***. The clutch discs are made of steel and have a friction lining. Between the clutch discs there are clutch plates. These are steel plates without the friction lining. The combination of these parts is called a clutch pack. In operation, the clutch discs are connected to the side gears in the differential. The clutch plates are connected to the differential case. Pressure is kept on the clutch pack by springs. As long as there is pressure on the clutch pack the side gears are locked up, which results in limited slip.

Today, many new cars use a cone-type limited slip differential as shown in ***Figure 43–23***. As with the clutch pack, the cones are located between the side gears and the differential case. The outside surface of the cones also has a friction material placed on it. Springs force the cones into the case cap, causing friction to be developed between the side gears and the differential case.

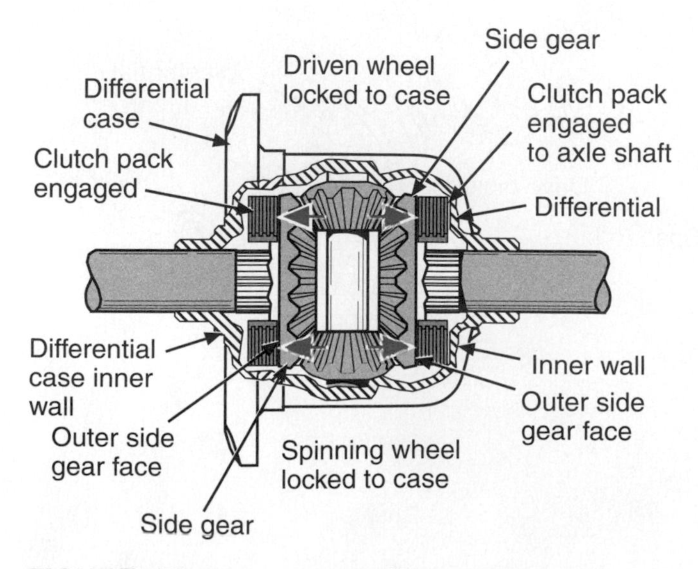

FIGURE 43–21 A limited slip differential uses a clutch plate, springs, and a pressure ring to lock up the two side gears.

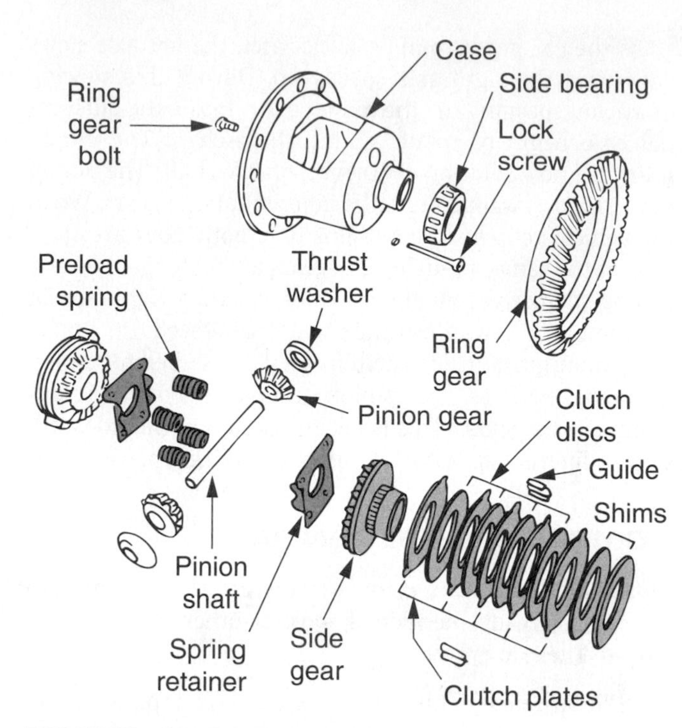

FIGURE 43–22 This limited slip differential uses a set of clutches to produce friction between the side gears and the differential case.

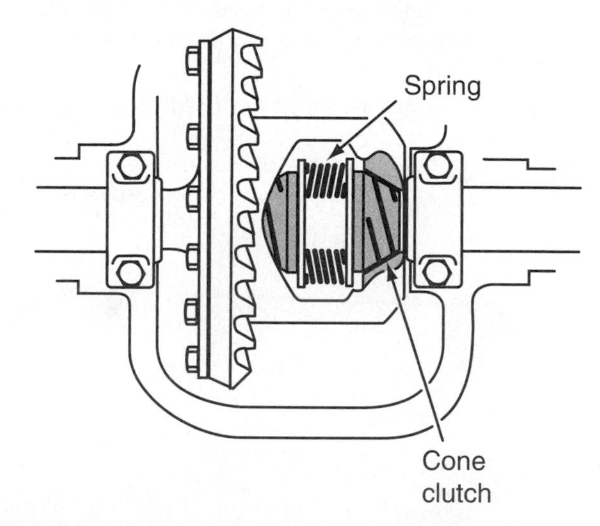

FIGURE 43–23 This limited slip differential uses a set of cones, washers, and springs to produce friction between the side gears and the differential case.

DIFFERENTIAL LUBRICATION

The differential uses several types of lubricants. The exact type depends on the year of the vehicle and whether the differential is separate or combined into the transaxle. Also, the viscosity will also change with the weather condition. On rear-wheel drive vehicles, where the differential is separate, a **hypoid gear oil** such as 80W-90 can be used. In addition, Dexron III ATF is used in many vehicles. On vehicles in which a transaxle is used, several types of lubricants are recommended, including Dexron III ATF and 5W-30 grade engine oil.

LIMITED SLIP LUBRICANT

On certain axles equipped with locking clutches or limited slip differentials (Dana Trac-Lok, Sure-Grip, Traction-Lok, and others) a **friction modifier** is added to the gear oil. Generally, 4 ounces of special friction lubricant are needed on these differentials. For example, one manufacturer recommends replacing 4 ounces of standard differential lubricant with 4 ounces of Mopar hypoid gear oil additive modifier with every refill. Without the correct amount of modifier, the slip between the axles will be incorrect. Too much modifier allows the clutch packs (***Figure 43–24***) to slip too much. Too little modifier makes the differential chatter because it becomes too sticky. To determine the exact type of differential lubricant, refer to the service manual or owner's manual.

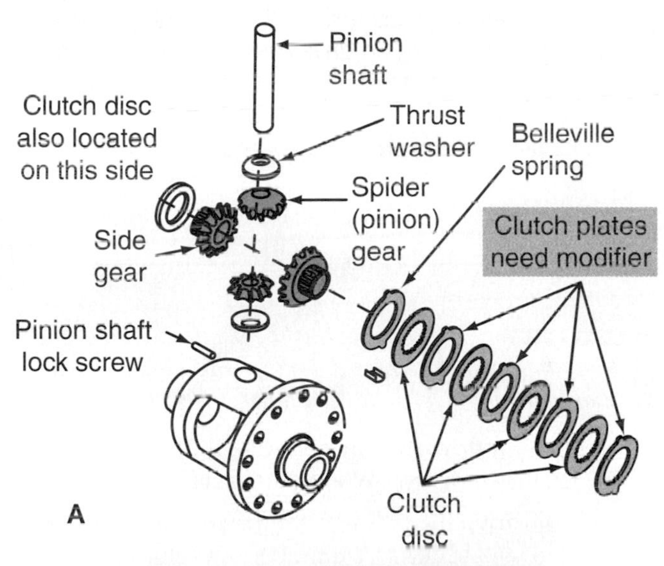

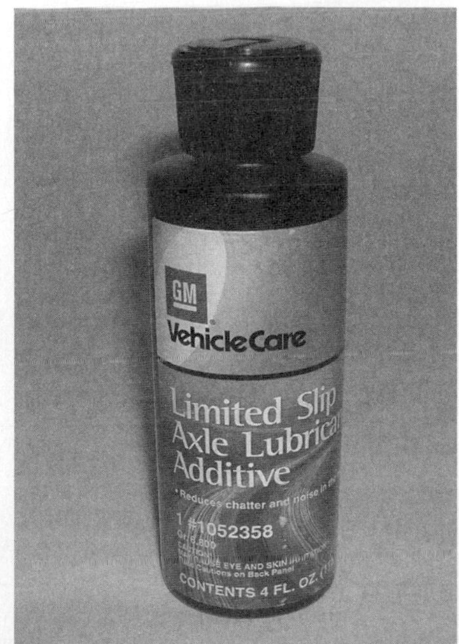

FIGURE 43–24 Limited slip differentials need a special friction modifier added to the lubricant to aid the set of clutch discs on both sides of the differential.

REAR AXLE

The rear axle is used for several purposes. On rear-wheel drive vehicles the rear axle is used to connect the differential to the rear wheels. The rear axle shown in ***Figure 43–25*** is attached to the side gears inside the differential case. The axle is connected to the side gear by internal and external splines. As the side gear turns inside the differential case, the axles also turn. The axle is supported in the axle tube as shown in ***Figure 43–26***. Bearings, gaskets, shims, oil seals, bearing retainers, bearing collars, and spacers are used to support the shaft and keep the differential lubricant from leaking out of the assembly.

On front-wheel drive vehicles, the rear axle is considered a dead axle. This means that the axle is used only to support the vehicle rather than transmit power. This is in contrast to a live axle, which is used to transmit power and torque to the wheels. In a dead axle, bearings are used to support the vehicle. However, since there is no differential, the axle is not connected to a differential and is not used to transmit power.

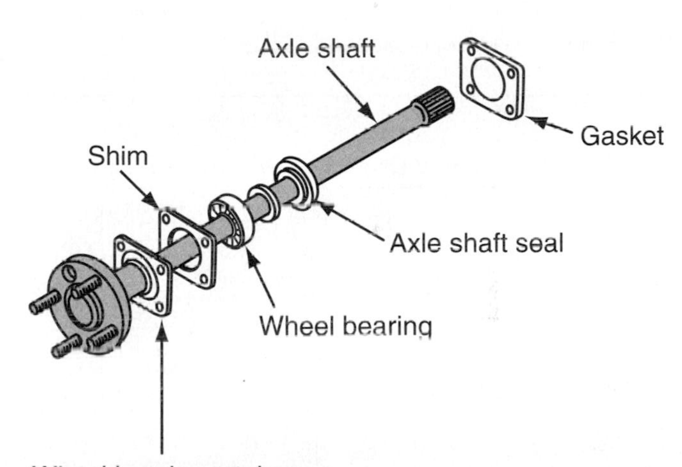

FIGURE 43–25 An axle is used to transmit the power from the differential side gears to the wheels.

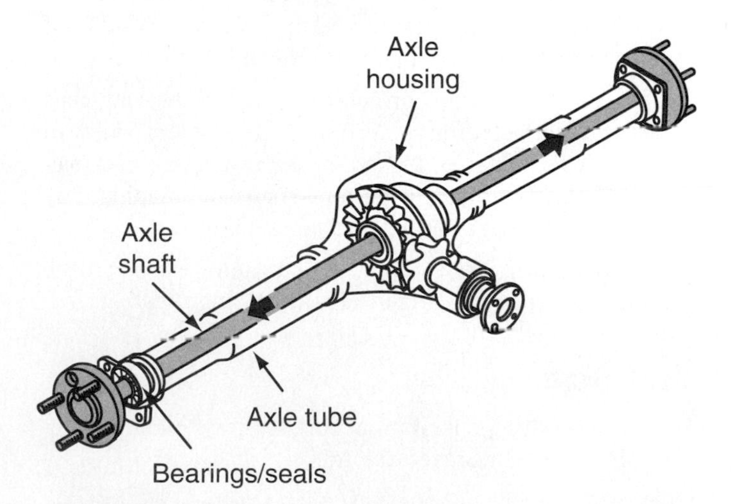

FIGURE 43–26 The axle fits inside the axle tube.

Problems, Diagnosis, and Service

Safety Precautions

1. To work on the drive line, differential, and axle, you must lift a vehicle. Use the correct procedure when lifting the vehicle on a hoist. Also, always use an extra jack when jacking up a vehicle with a hydraulic or air jack. Always block the front wheels to eliminate the possibility of the vehicle moving.
2. When universal joints are being replaced, high mechanical pressures are needed to remove and install the components. Make sure that all components are securely fastened to the tool to eliminate the possibility of injury to hands and face.
3. When installing gears, bearings, gear shafts, or seals, be careful not to pinch your fingers between the part and the housing.
4. Differentials are very heavy. Be careful not to crush your fingers or hand when the differential is being moved during disassembly.
5. When removing or installing snap rings, always use a snap ring pliers. Snap rings have a high degree of spring tension. If not installed with the correct tool, they may fly off and cause eye, face, or bodily injury.
6. Always wear OSHA-approved safety glasses when working on drive lines, differentials, and axle components.
7. Never try to lift heavy drive lines or differentials without proper lifting equipment.
8. Immediately wipe up any oil that is spilled on the floor when working on drive lines, differentials, and axle components to prevent slips and falls.
9. Always use the proper tools when working on drive lines, differentials, and axle components.

PROBLEM: Worn Universal Joints

A vehicle with a rear drive has a noticeable clunk when shifting from forward to reverse. In addition, there seems to be a small ticking sound when the vehicle is in forward gear and low speeds.

DIAGNOSIS

The most likely cause of this problem is bad or worn universal joints. There are several diagnostic checks that tell if the universal joints are bad or worn.

1. Check the universal joint trunnion for brinelling. Brinelling is the process of producing grooves in the trunnion from the needle bearings (***Figure 43–27***). If brinelling has occurred, replace the universal joints.
2. If there is no obvious brinelling, move the bearing caps around on the trunnion. Also move the universal joint in the drive shaft yoke. If there is any restriction of movement, small bumps during movement, or scraping during movement, replace the universal joints.
3. Check the universal joints for end **galling** (wear on the end of the trunnion and inside the bearing caps).

SERVICE

Use the following general procedure to disassemble and reassemble with new universal joints.

CAUTION: *The vehicle must be positioned so that the drive shaft can be removed. Follow all safety procedures when putting the vehicle on a hoist.*

1. Remove the drive shaft by removing the two U-bolts that attach the rear universal joint to the differential yoke.
2. Pull the drive shaft away from the yoke and pull backward to remove the assembly. Certain vehicles have two drive shafts that must be removed. In this case the center bearing will need to be removed along with the drive shafts.
3. Remove the snap rings and retainer plates that hold the bearings in the yoke and drive shaft.

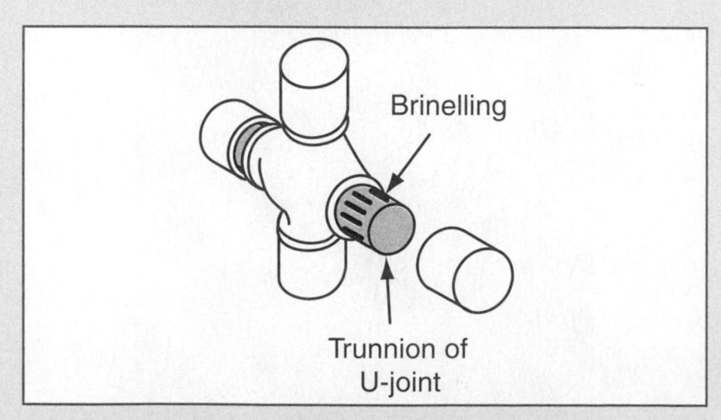

FIGURE 43–27 Brinelling (grooves from needles wearing into the trunnion) is caused by improper angles, lack of lubrication, or too much load.

4. Use special pressing tools to remove the universal joint bearings from the yoke and the drive shaft. Depending on the manufacturer of the tools, they allow you to press out one side of the bearings into a hollow tube. Then, using the correct pressing tools, the spider and remaining bearing are pressed in the opposite direction to remove them from the drive shaft. Note that the universal joint bearings are pressed very tightly into the drive shaft yoke, and so a great deal of pressure is needed to remove them. Always follow the manufacturer's suggested procedure when removing and replacing universal joints.
5. The spider or cross of the U-joint should now be easily removed.
6. Obtain the correct replacement kit from the parts center. Since there are many types and sizes of U-joints, it is possible to accidentally obtain the wrong replacement kit. Always double-check the numbers on the universal kit to make sure they are correct.
7. Begin reassembly by packing the bearing caps with the correct type of grease. Note that certain universal joints do not require grease. Refer to the manufacturer's recommendation. Note also that too much grease may damage the seals on the U-joint.
8. Using the correct pressing tool and a vise, press one bearing cap part way into the drive shaft yoke. Position the spider in the partially installed bearing cap.

CAUTION: *Make sure that no needle bearings have fallen into the center of the bearing cap. The grease should be used to hold them in place (**Figure 43–28**).*

9. Now position the second bearing cap in the drive shaft yoke. Place the assembly in the vise so that the bearing caps can be pressed in by the jaws of the vise.
10. As the vise is tightened, press the bearing caps into the drive shaft. Make sure the spider and trunnion are in such a position as to avoid binding or damage to the needle bearings.
11. Use the correct pressing tool to push the bearing caps far enough into the yoke so the snap rings can be installed.
12. When installing the snap rings, make sure they are completely touching the trunnion. Note that some late model cars use an injected nylon retainer on the universal joint bearing cap. When service is necessary, pressing the bearing cap out as described previously will shear the nylon retainer. Replacement U-joints must be the steel snap ring-type. See ***Figure 43–29***, which shows the retaining rings being installed. If the bearing caps cannot be pushed in far enough to allow the retainers to be inserted, a needle bearing has been caught in the base of the cap. The U-joint must be removed and replaced again.
13. Lubricate the spider or cross and bearing caps if required (***Figure 43–30***).

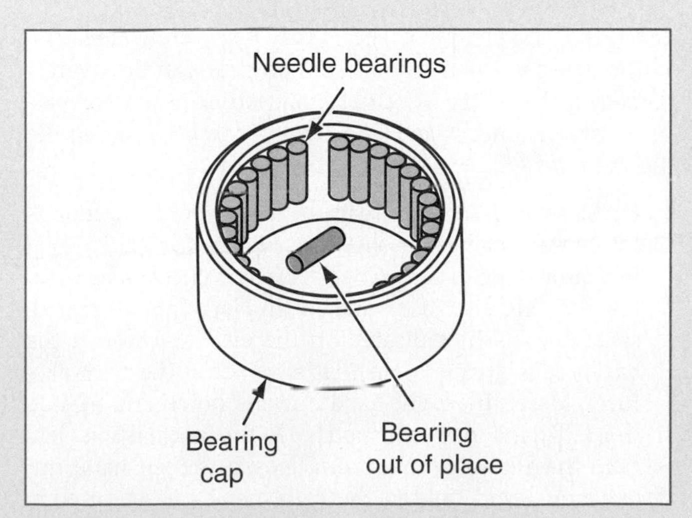

FIGURE 43–28 Make sure that all needle bearings are lined up inside the cap.

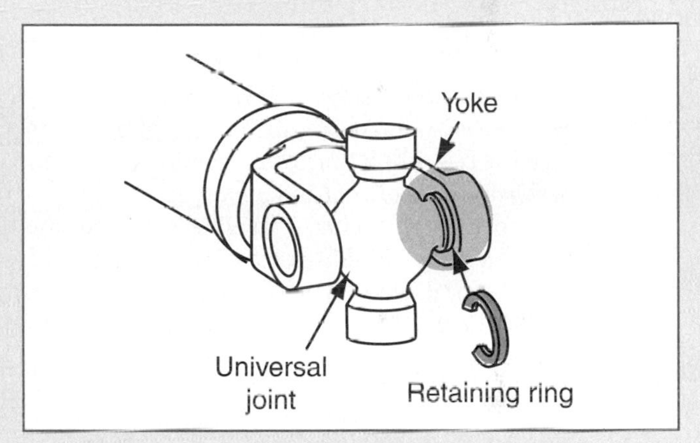

FIGURE 43–29 Retaining rings should be installed on the universal joints to hold the universal joint in place.

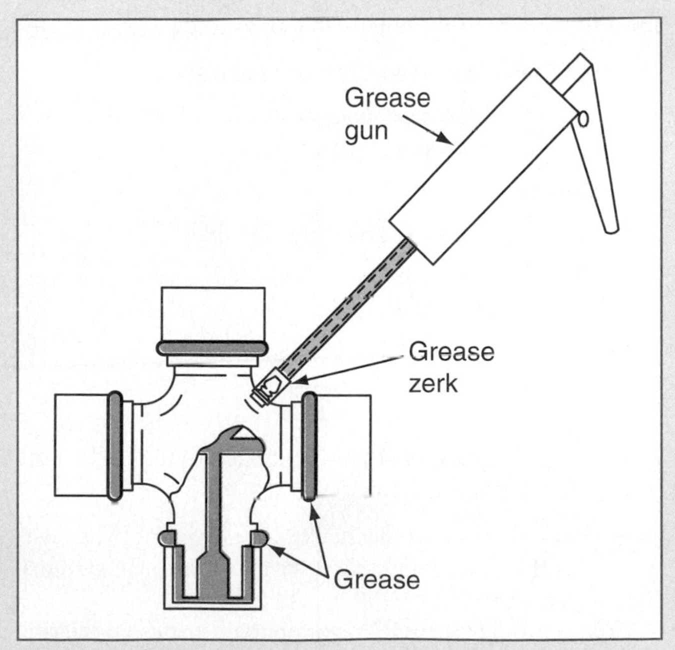

FIGURE 43–30 Universal joints should be lubricated before installation. Grease must flow from all four bearing seals.

14. Replace any other U-joints that are worn.
15. Install the drive shaft and test the vehicle.

PROBLEM: Faulty Differential

A vehicle has developed a humming noise in the differential.

DIAGNOSIS

A humming noise in the differential usually indicates that the differential needs to be checked and inspected for gear backlash, gear wear, or bad bearings. When the differential is noisy or is suspected of being damaged, first make sure the noise is not caused from any of the following:

- Road noise
- Tire noise
- Front bearing noise
- Transmission and engine noise
- Drive shaft and U-joint noise

If tests show that the differential or rear axle is noisy, the noise is most likely coming from one of several bad parts. These include the rear wheel bearings, differential side gear and pinion, side bearing, or ring and pinion gear.

Also check to see if the vehicle has different-sized tires on the rear wheels. This can cause excessive wear in the differential, because the differential acts as if the vehicle is cornering. Limited slip differentials can also develop excessive wear on the clutches or cones when different-sized tires are used.

SERVICE

Use the following general procedure to disassemble and inspect the differential components:

1. Lift the vehicle with an appropriate hoist.
2. Loosen the differential cover bolts and allow lubricant to drain into a suitable container.
3. Remove the housing cover.
4. Remove the rear axles and propeller shaft.
5. Scribe reference marks on differential bearing and bearing caps. These are to be used during reassembly. Remove bearing cap bolts.
6. Using a suitable tool, pry differential case, bearing races, and shims out of the housing.
7. Additional disassembly procedures can be found in the vehicle's service manual.

Perform the following checks, inspection, and adjustments on the differential. Note that on transaxle systems, the service is similar.

Figure 43–31 shows the wear characteristics on the ring gear and what adjustments to make to fix the problems. Also check the condition of the pinion gear. In extreme

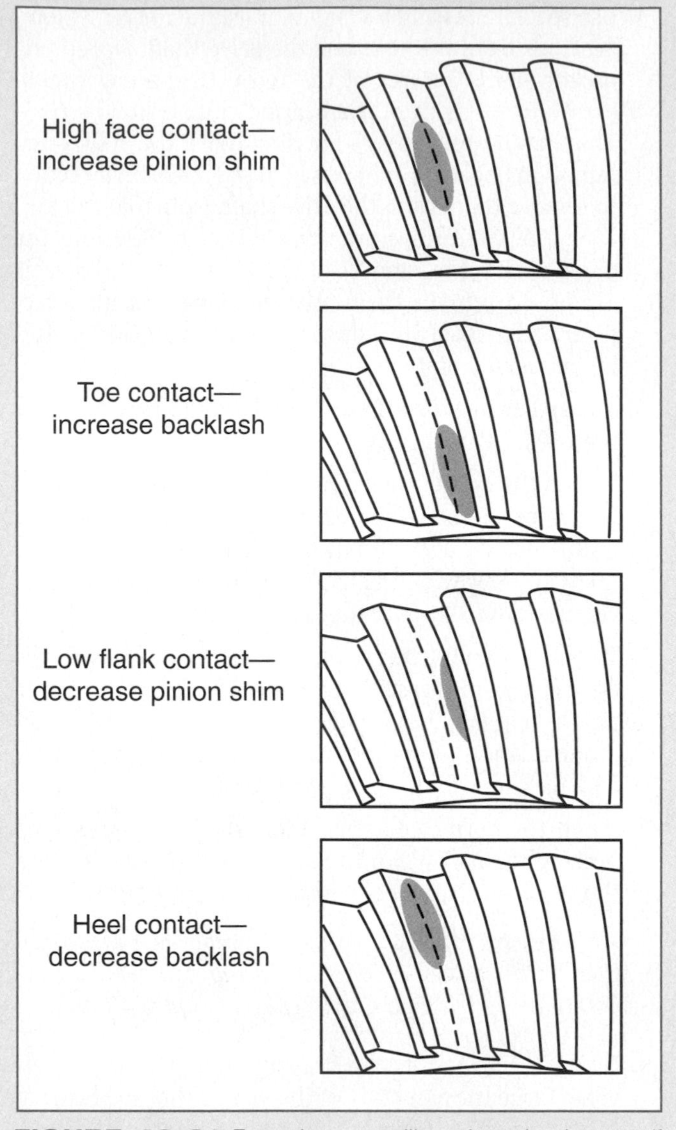

FIGURE 43–31 Excessive wear will result on the ring gear if the differential backlash is incorrect. Adjust the backlash by using shims and adjusters.

cases, the ring and pinion gears may be severely damaged. Without proper lubrication, a pinion gear can be severely damaged. On the basis of these and other gear wear patterns, several checks and adjustments can be made on the differential.

1. Backlash adjustment is made by mounting a dial indicator on the ring gear as shown in ***Figure 43–32***. The dial indicator is set to zero. The backlash is checked by moving the ring gear back and forth. Note the amount of space (backlash) indicated on the dial indicator. If the backlash is greater than that allowed by the manufacturer, loosen the right-hand nut one notch and tighten the left-hand nut one notch. If the backlash is less than the allowable minimum, loosen the left-hand nut one notch and tighten the right-hand nut one notch. The adjustment nuts are located on the sides of the bearing caps.

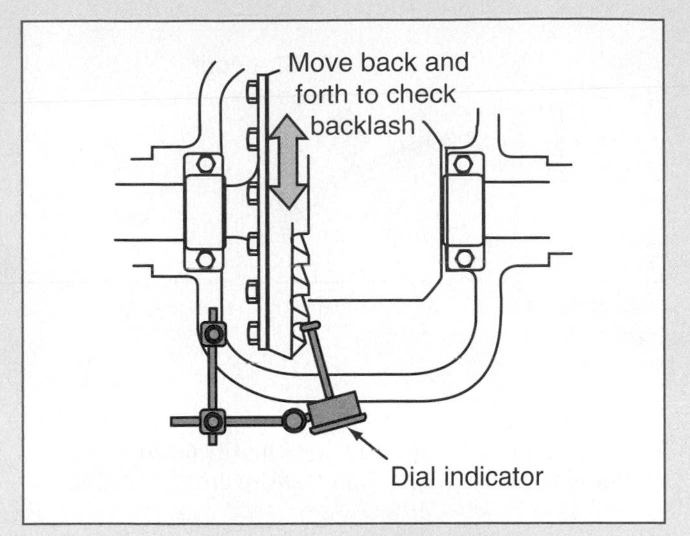

FIGURE 43–32 A dial indicator should be used to check the backlash between the ring gear and the pinion.

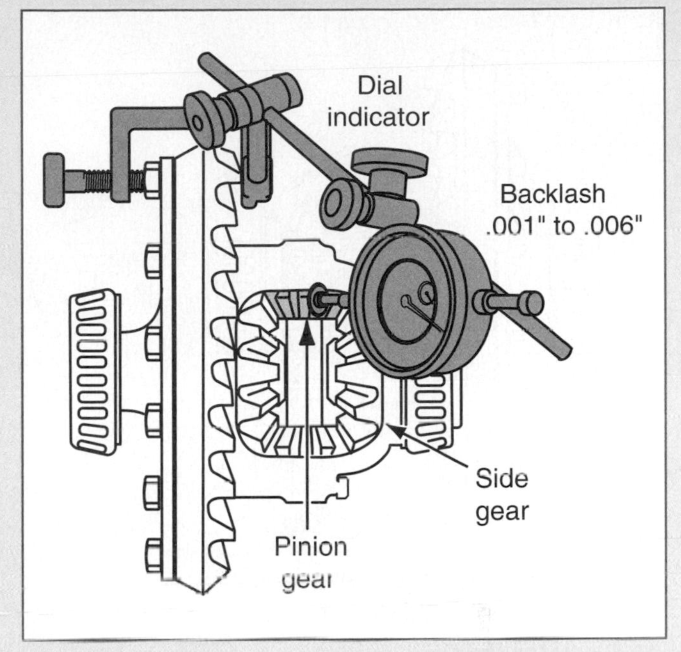

FIGURE 43–34 Gear backlash can be measured between the side and pinion gears in the differential case using a dial indicator.

2. The position of the drive pinion can also be checked and adjusted. A shim pack is placed between the pinion head and the inner race of the rear bearing. Increasing the shim pack will move the pinion close to the ring gear. Decreasing the shim pack will move the pinion away from the ring gear (***Figure 43–33***).
3. Check the clearance between the side gear and pinion gear in the differential case using a dial indicator (***Figure 43–34***). Clearance should generally be in the range of 0.001–0.006 inch. If clearance is more than the maximum, shims must be added. If the clearance is less than the minimum, shims need to be removed. Generally, a 0.002-inch shim will change the clearance about 0.001 inch.

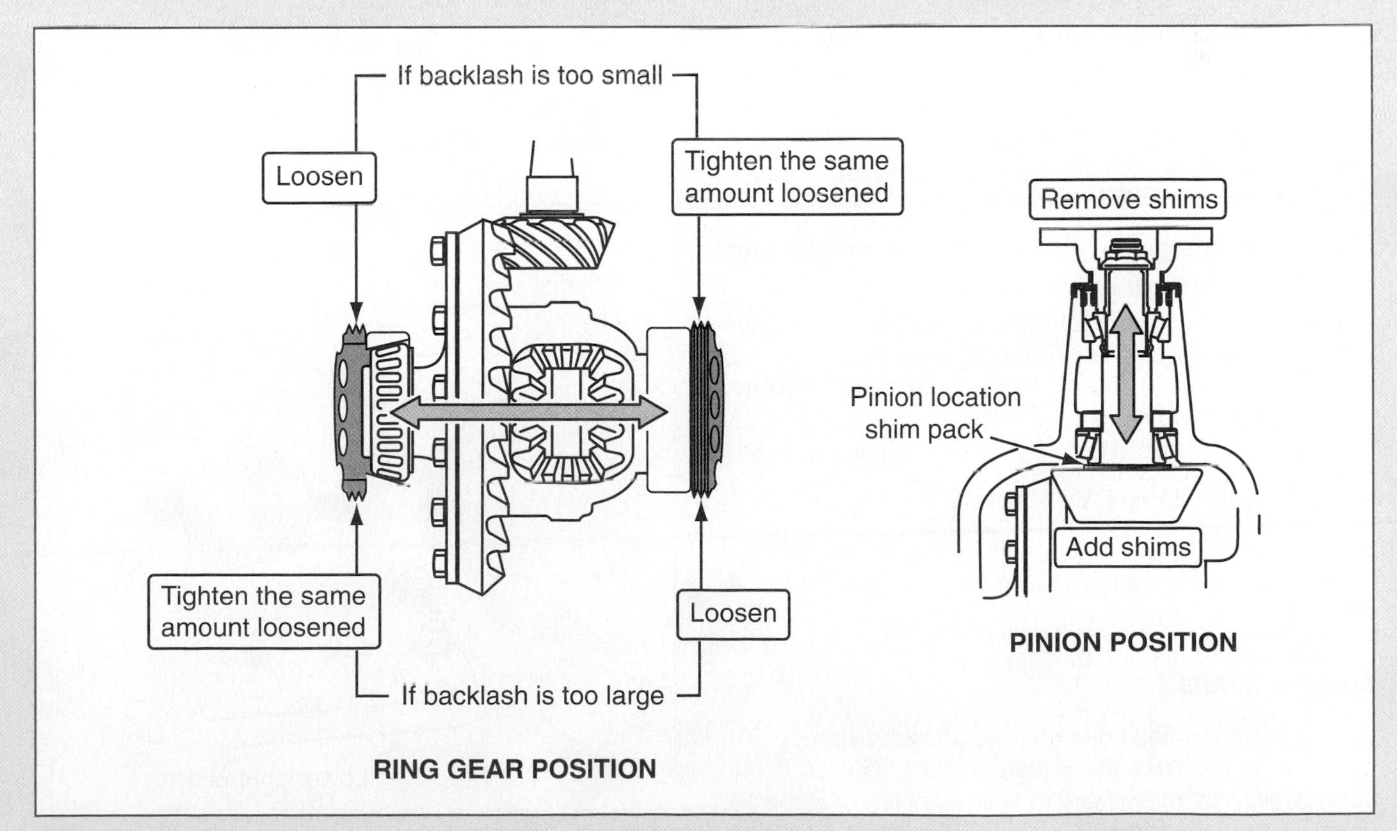

FIGURE 43–33 This drawing shows the direction of movement when adjusting the ring gear and pinion gear. Shims are used to position the pinion gear, and bearing adjusting nuts are used to position the ring gear.

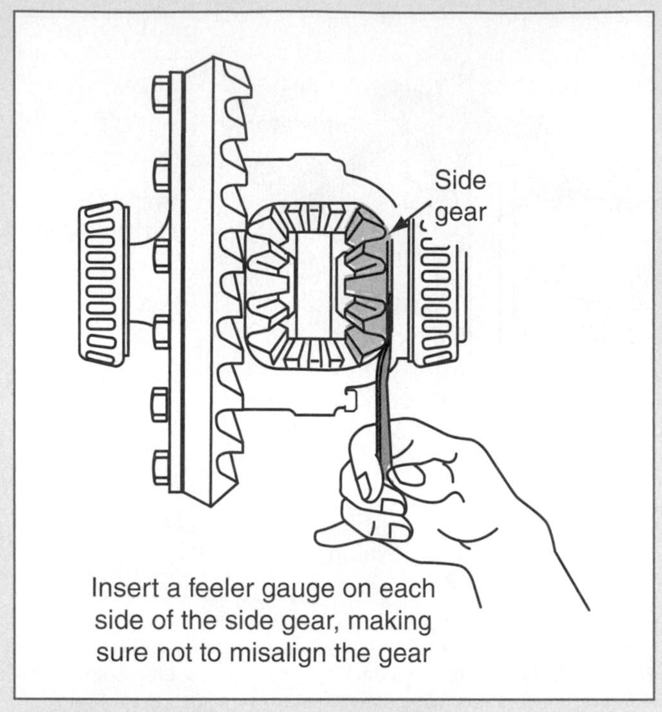

FIGURE 43-35 Side gear-to-case clearance is checked by using feeler gauges.

4. During assembly of the differential it may be necessary to check the pinion depth. Various special tools are available or a dial indicator can be used. Shims are used to control the position of the pinion gear. Follow the manufacturer's specific procedures to perform this task.
5. Check the side gear-to-case clearance by using a set of feeler gauges. Common measurements are between 0.000 and 0.006 inch. If the clearance exceeds specifications, the differential case will need to be replaced (***Figure 43–35***).
6. Follow the manufacturer's recommended procedure for reassembly.
7. Make sure the correct lubricant is used in the differential. Also, limited slip differentials require about 4 ounces of friction modifier. Refer to the manufacturer's recommendations.

PROBLEM: Worn CV Joints

A front-wheel drive vehicle is producing a humming noise during driving. Also, a popping or clicking sound is heard on sharp turns.

DIAGNOSIS

Generally, a humming noise indicates the early stages of insufficient or incorrect lubrication at the CV joints. If the noise is more of a vibration, then one or both of the drive shafts may be bent. ***Figure 43–36*** shows an exaggerated view of a drive shaft that is bent and has excessive axial runout.

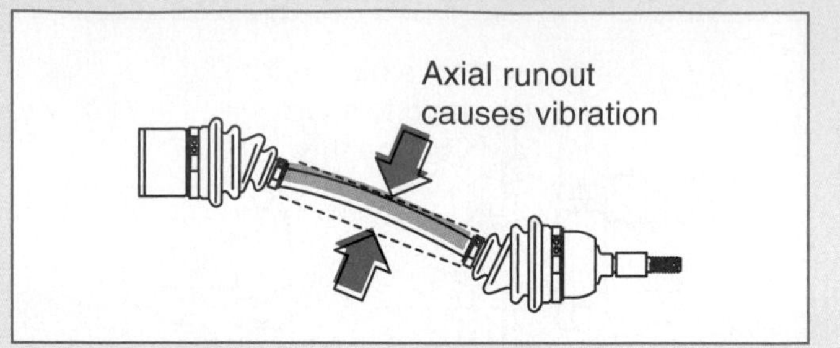

FIGURE 43-36 A bent front-wheel drive shaft can cause vibration during operation.

The popping or clicking noise indicates possible CV joint wear in the outer, or wheel-end, joint. The clicking noise is produced when the CV joint is extended during turning.

SERVICE

Various service checks, inspection, and measurements can be taken to service CV joints. Use the following general procedure when servicing CV joints.

1. Remove the wheel, damper fork, and steering knuckle from the front of the brake and strut assembly (***Figure 43–37***). Also drain the lubricant from the transaxle.
2. Using a suitable pry bar, remove the shaft from the transmission differential case as shown in ***Figure 43–38***. Do not pull on the drive shaft, as the CV joint may come apart. Mark the position of the CV joint to ensure correct reassembly later.
3. Inspect the condition and straightness of the drive shafts. A straightedge can be used to check for straightness. There should be no bend or axial runout. If there is, the axle must be replaced.

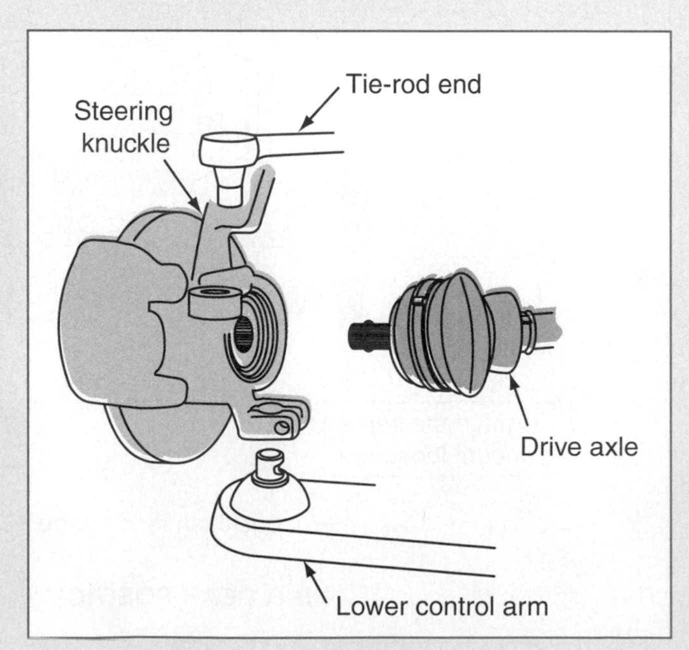

FIGURE 43-37 In order to remove the CV joints, the front wheel, brake assembly, and strut must be moved.

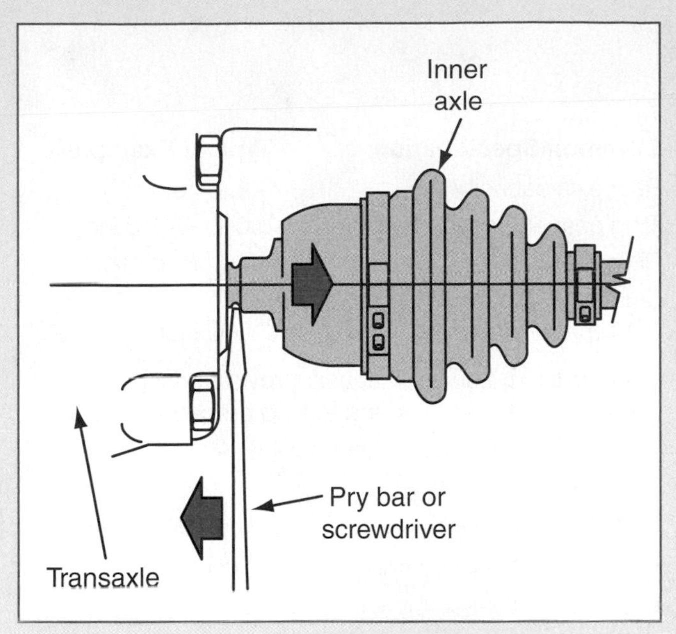

FIGURE 43-38 To service CV joints, the drive shaft must be pried from the transaxle.

4. Once the drive shaft is removed, carefully work the CV joint back and forth about 40 degrees while rotating the shaft. If there are any bumps, scrapes, or resistance while moving the CV joint throughout its range of motion, it must be replaced.
5. Remove the rubber boot by disconnecting the boot band or clamp. If welded, it may have to be cut. Take care not to cut or rip the boot, as it will leak grease.
6. Mark the spider and drive shaft so they can be reinstalled in their original position.
7. If the CV joint is OK, clean all parts thoroughly.
8. Thoroughly pack the CV joints and boots with high-quality grease (***Figure 43–39***). Use the manufacturer's recommended grease type.

FIGURE 43-39 When servicing CV joints, the boots and universal joints must be packed with the recommended grease.

9. Assemble the CV joint back onto the drive shaft according to the manufacturer's recommended procedure. ***Figure 43–40*** shows how the parts are assembled. Make sure there is adequate lubrication for the CV joint.
10. Install the boot and clamp accordingly.
11. Reinstall the drive shaft and CV joints into the transaxle, making sure the snap ring is pushed into the snap ring groove (***Figure 43–41***).
12. Refill the transaxle with the recommended lubricant and test it.

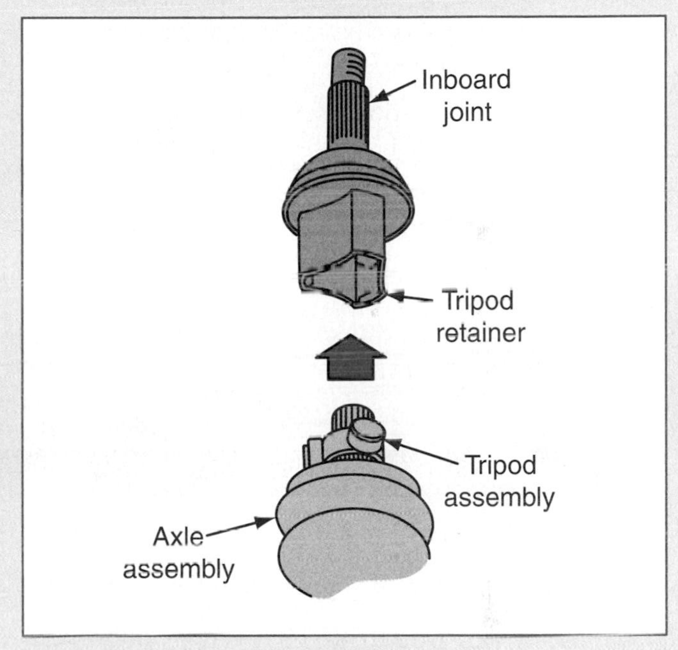

FIGURE 43-40 During assembly of the CV joint, the axle, boot, and joint are pushed up into the inboard joint. Make sure that the tripod assembly slides easily into the tripod retainer.

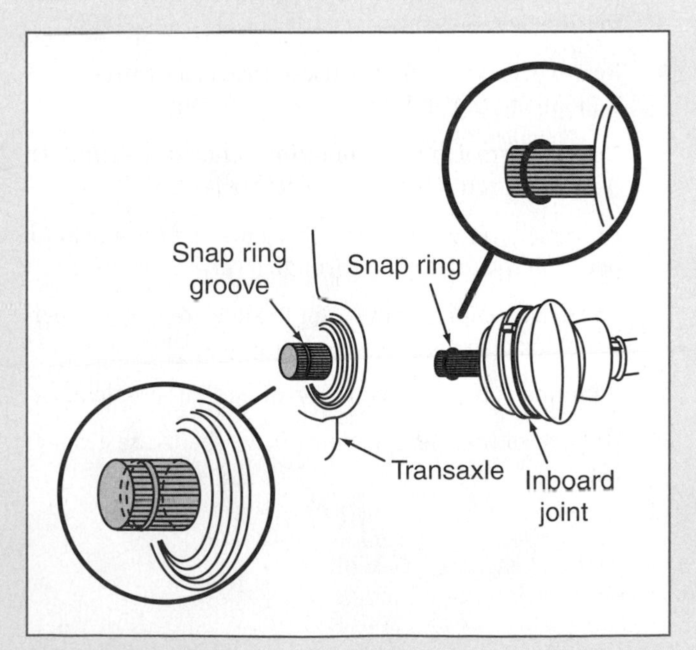

FIGURE 43-41 When inserting the drive shaft into the transaxle, make sure the snap ring engages with the snap ring groove.

Service Manual Connection

There are many important specifications to keep in mind when working with drive lines, differentials, and axles. To identify the specifications for your engine, you will need to know the VIN (vehicle identification number) of the vehicle, the type and year of the vehicle, the type of engine, and the type of transmission. Although they may be titled differently, some of the more common drive line, differential, and axle specifications (not all) found in service manuals are listed below. Note that these specifications are typical examples. Each vehicle may have different specifications.

Common Specification	Typical Example
Differential carrier backlash	0.003–0.006 inch
Nominal pinion locating shim	0.030 inch
Pinion bearing preload (new with seal)	16–29 in.-lb
Pinion nut torque	140 in.-lb
Rear axle capacity	4.4 pints
Ring gear and pinion backlash	0.008–0.015 inch
Ring gear runout	0.004 inch max.
Side bearing clearance shim thickness	0.002 inch

In addition, service manuals provide specific directions for various service and testing procedures. Some of the more common procedures include:

- Backlash inspection
- Bearing replacement
- Differential disassembly and reassembly
- Drive shaft assembly and greasing
- Drive shaft removal
- Oil seal removal
- Pinion gear removal

SUMMARY

The following statements will help to summarize this chapter:

- Torque is transmitted through the drivetrain to the wheels.
- To get torque to the wheels on front-wheel drive vehicles, power must be transmitted from the transmission, to the differential, to the drive shaft(s), and finally to the axles and wheels.
- Rear-wheel drive vehicles use a drive shaft with universal joints to transmit the power to the differential.
- During normal driving conditions, the drive shaft must transmit torque through a variety of angles.
- Universal joints are used to allow transmission of power in the drive shaft through a variety of angles.
- The drive shaft has a slip joint to allow for correct operation during irregular road conditions.
- Vibration is the most common drive shaft problem.
- Because of their length, some drive shafts use a center bearing.
- Front-wheel drive systems use constant velocity (CV) joints on their drive shafts.
- Constant velocity universal joints keep the speed of the input and output shaft exactly the same.
- The differential is used to transmit equal torque to both of the drive wheels.
- The differential includes a ring gear, pinion gear, differential case, bevel gears, and side gears.
- When a vehicle goes around a corner, the differential gears and the pinion gears work together to keep torque equal on both the inner and outer wheels.
- The limited slip differential is designed to provide equal torque to the drive wheels when one wheel is slipping on ice or snow.
- On a limited slip differential, clutches are used to lock up the two side gears that are connected to the axle.
- Axles are used to connect the differential output to the drive wheels.

TERMS TO KNOW

Can you explain each of the following terms? Review the chapter until you can use each term correctly.

Bearing caps
Cardan universal joint
Constant velocity (CV) joint
Cross
Differential
Drive shaft
Friction modifier
Galling
Hypoid gear oil
Limited slip differentials
Slip joint
Tripod
Trunnions
Yokes

REVIEW QUESTIONS

Multiple Choice

1. Which of the following is used to transmit the torque to the differential?
a. Axles
b. Universal joints
c. Drive shaft
d. Axle bearings
e. Both B and C

2. Which of the following is/are used to allow the drive shaft to rotate through changing angles?
a. Differential side gears
b. Differential ring gear
c. Universal joint
d. Slip joints
e. None of the above

3. Which of the following are used to allow the length of the drive shaft to change (lengthen and shorten) during operation?
a. Differential side gears
b. Bearings
c. Universal joints
d. Differential ring gears
e. Slip joints

4. Which of the following is/are part of the universal joint?
a. Yoke
b. Trunnion
c. Needle bearings
d. All of the above
e. None of the above

5. ________ vibration is caused by the power pulses of the engine.
a. Transverse
b. Elongated
c. Trunnion
d. Torsional
e. Velocity

6. One major disadvantage of Cardan universal joints is that:
a. The input shaft continually slows down during operation.
b. The output shaft continually slows down during operation.
c. There is always excessive wear on the tripods.
d. The output shaft changes velocity in each revolution because of the angle of operation.
e. The input shaft changes velocity in each revolution because of the angle of operation.

7. A ________ is used on vehicles that require a long drive shaft distance.
a. Center bearing
b. Single universal joint
c. Double slip joint
d. Constant velocity joint
e. All of the above

8. Front-wheel drive systems:
a. Use constant velocity joints
b. Use two drive shafts
c. Must have a slip joint on each shaft
d. All of the above
e. None of the above

9. Which type of joint on a drive system does not have a velocity change within one revolution?
a. Cross/bearing type
b. Constant velocity
c. Cardan type
d. All of the above
e. None of the above

10. The fixed-type constant velocity joint is used:
a. On the inboard side of the drive shaft
b. On the outboard side of the drive shaft
c. On rear-wheel drive systems
d. On the inside of the differential
e. On rear axles

11. Which of the following is/are not a part of the differential?
  a. Pinion gear
  b. Ring gear
  c. Side and bevel gears
  d. Universal joints
  e. Differential case

12. The differential is used on the drivetrain of a vehicle:
  a. To transmit equal power during cornering
  b. To allow for changing angles of the drivetrain
  c. To reduce the velocity changes during rotation
  d. To support the axles only
  e. None of the above

13. When the vehicle is moving in a straight line and there is equal resistance at each wheel:
  a. The differential case is turning
  b. The side gears and bevel gears are in mesh but are not rotating with each other
  c. The ring gear is turning
  d. All of the above
  e. None of the above

14. The adjusters on the differential are used to change the backlash between the pinion gear and the ________.
  a. Side gears
  b. Axles
  c. Differential case
  d. Bevel gears
  e. Ring gear

15. To eliminate one wheel spinning on ice or other slippery surface, the _______ is/are used.
  a. Double ring gear
  b. Universal joints
  c. Limited slip differential
  d. Spacer
  e. Bearing

16. The axle and differential side gears are attached by:
  a. Bolts
  b. A strong welded section
  c. An internal and external spline
  d. Grease seals
  e. Gaskets

17. The process of needle bearings producing grooves in the trunnion is called:
  a. Metal wear
  b. Brinelling
  c. End galling
  d. Yoke
  e. Torsional vibration

18. Which of the following is not a service concern on drive shafts and differentials?
  a. Correct lubrication in U-joints
  b. Adjustment between ring gear and pinion gear
  c. Backlash adjustment
  d. End galling
  e. Piston scraping

The following questions are similar in format to ASE (Automotive Service Excellence) test questions.

19. *Technician A* says that backlash can be checked on the differential by using a vernier caliper to measure the correct clearance. *Technician B* says that backlash can be checked by using a dial indicator. Who is correct?
  a. A only
  b. B only
  c. Both A and B
  d. Neither A nor B

20. *Technician A* says that too much lubrication can damage the seals in the U-joint. *Technician B* says that too much lubrication can damage the spider in the U-joint. Who is correct?
  a. A only
  b. B only
  c. Both A and B
  d. Neither A nor B

21. A humming noise is heard in the rear end of the vehicle. *Technician A* says the differential backlash may be out of adjustment. *Technician B* says the differential ring gear may be damaged or worn. Who is correct?
  a. A only
  b. B only
  c. Both A and B
  d. Neither A nor B

22. *Technician A* says that the U-joints can be replaced by using common tools in the automotive shop. *Technician B* says that the universal joints are replaced by using special presses and hydraulic jacks. Who is correct?
  a. A only
  b. B only
  c. Both A and B
  d. Neither A nor B

23. A vehicle with front-wheel drive has a vibration coming from the front end. *Technician A* says the problem is a shock absorber. *Technician B* says the problem is a bent drive shaft. Who is correct?
  a. A only
  b. B only
  c. Both A and B
  d. Neither A nor B

24. A vehicle has a clunk when shifting from reverse to forward. *Technician A* says the problem is the end play in the CV joint. *Technician B* says the problem is a bad universal joint. Who is correct?
  a. A only
  b. B only
  c. Both A and B
  d. Neither A nor B

25. A differential needs to be inspected and possibly rebuilt. *Technician A* says it is important to check the ring gear-to-pinion gear backlash. *Technician B* says it is important to check the side gear-to-pinion gear clearance. Who is correct?
  a. A only
  b. B only
  c. Both A and B
  d. Neither A nor B

26. A vehicle has a limited slip differential. *Technician A* says there is a set of clutches in the differential. *Technician B* says that a friction modifier is needed in the lubricant. Who is correct?

a. A only
b. B only
c. Both A and B
d. Neither A nor B

Essay

27. Define the purpose of universal joints.

28. What is the purpose of a slip joint?

29. What is the difference between constant velocity joints and Cardan universal joints?

30. State the purpose of a differential.

31. What is a limited slip differential?

32. What is galling on universal joints?

Short Answer

33. Each bearing cap is placed on the ________ of the universal joint.

34. On rear-wheel drive vehicles the differential lubrication used is ________.

35. Lubricants that are used for limited slip differentials typically have a(n) ________ modifier added to the lubricant.

36. The process of producing grooves in the trunnion from the needle bearings on universal joints is called ________.

37. A humming noise in the rear axle area can mean that the ________ is/are worn.

Applied Academics

TITLE: Hypoid Gear Ratios

ACADEMIC SKILLS AREA: Science

NATEF Automobile Technician Tasks:

The service technician can describe how torque transmitted to the manual drivetrain and axle relates to the force that moves the vehicle.

CONCEPT:

The differential on a rear drive vehicle uses a hypoid gear. A hypoid gear, shown in the figure that follows, has the centerline of the pinion and ring gears at different levels. The pinion gear is dropped down slightly so that the drive shaft of the vehicle can be lower. If the drive shaft is lower, there is a lower hump on the tunnel inside the passenger compartment. Hypoid gears run very quietly and allow several teeth to absorb the drive force at one time.

The teeth on both the pinion and ring gears are curved, causing a wiping action during meshing. The number of teeth on both the pinion and ring gears determines the ratio between the two. For example, if the pinion gear has eleven teeth and the ring gear has thirty-two teeth, the gear ratio is 2.91 to 1. This means that the pinion gear must turn 2.91 times to get the ring gear to turn 1 time. Common ratios for differential hypoid gears are 2.91 to 1, 3.00 to 1, 3.5 to 1, and 4.5 to 1. Changing hypoid gear ratios affects both torque and fuel economy. Generally, as the ratios increase, more torque output is available. As the ratios decrease, fuel economy improves.

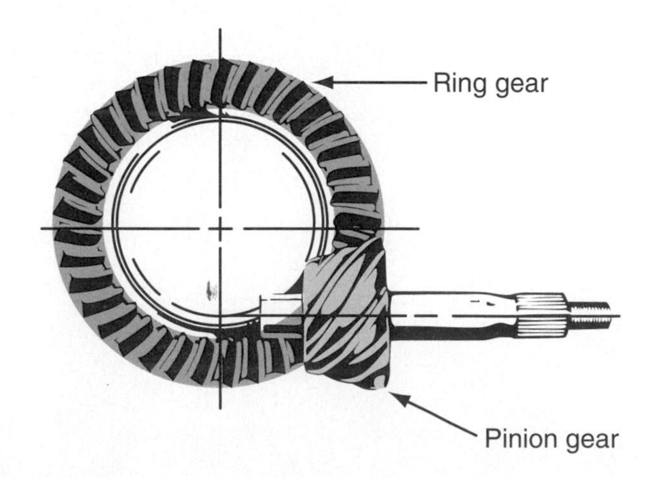

APPLICATION:

Why would the torque availability increase as the gear ratios of the differential increase? Why would fuel economy improve as torque ratios decrease? What is the differential ratio if the ring gear has thirty-five teeth and the pinion gear has eleven teeth?

Four-Wheel/All-Wheel Drive Systems

CHAPTER 44

OBJECTIVES

After studying this chapter, you should be able to:

- Define the principles of operation of four-wheel/all-wheel drive systems.
- Identify the different operation modes of four-wheel/all-wheel drive systems.
- Describe the main parts and operation of transfer cases.
- Examine the operation of various types of couplings and differentials used on four-wheel/all-wheel drive systems.
- Identify, diagnose, and service various problems dealing with four-wheel/all-wheel drive systems.

INTRODUCTION

Many vehicles today have either four-wheel drive or all-wheel drive systems. Vehicles with four-wheel drive are much easier to operate on slippery, rough terrain and in muddy or snowy conditions. This chapter introduces the principles of four-wheel drive. The basic parts and operation of transfer cases and couplings are also discussed. In addition, various problems, their diagnoses, and service procedures are included to help the service technician work on these systems.

44.1 PRINCIPLES OF FOUR-WHEEL DRIVE SYSTEMS

Most vehicles today have either front- or rear-wheel drive systems. However, an increasing number of vehicles have power sent to all four wheels. These vehicles are called either four-wheel drive (4WD) systems, or all-wheel drive (AWD) systems. In either case, when the power of the engine is transmitted to all four wheels, the vehicle has increased traction. Increased traction is needed in snowy, icy, muddy, or slippery conditions, or for **off-road** operation in

Certification Connection

ASE Connection: The information in this chapter can help you prepare for the National Institute for Automotive Service Excellence (ASE) certification tests. The test and content area most closely related to this chapter are:

Test A3—Manual Drivetrain and Axles

- **Content Area**—Four-Wheel/All-Wheel Drive Component Diagnosis and Repair

NATEF Connection: Much of the information in this chapter is related to the NATEF tasks. The NATEF tasks and priority numbers most closely related to this chapter are:

1. Diagnose noise, vibration, and unusual steering concerns; determine necessary action. P-3
2. Remove and reinstall transfer case. P-3
3. Disassemble, service, and reassemble transfer case and components. P-3
4. Inspect front-wheel bearings and locking hubs; perform necessary action. P-3
5. Check drive assembly seals and vents; check lube level. P-3
6. Diagnose test, adjust, and replace electrical/electronic components of four-wheel drive systems. P-3

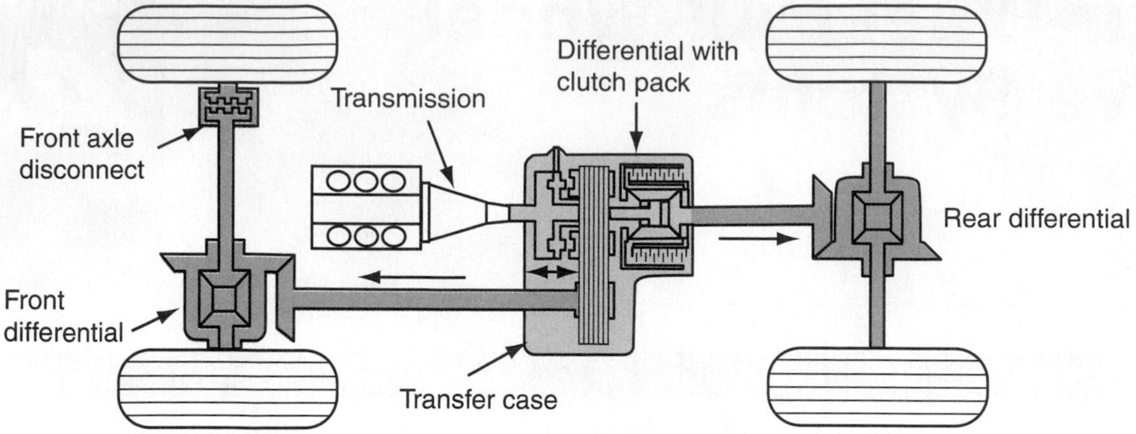

FIGURE 44–1 The major parts of a 4WD system. (Courtesy of DaimlerChrysler Corporation)

rough terrain. Such systems provide maximum traction between the tires and the road surface. The most common types of vehicles that use four-wheel drive systems are the sport utility vehicle (SUV) and the pickup truck. In addition, some regular vehicles, such as station wagons, vans, and passenger cars, are incorporating all-wheel drive systems.

4WD SYSTEMS DEFINED

The power transfer in a typical four-wheel drive system is shown in ***Figure 44–1***. The 4WD system uses a **transfer case** and additional drive shafts to transfer the power to all four wheels. The transfer case is located immediately behind the transmission. The purpose of the transfer case is to split the power and torque from the engine. Some of the torque is transferred to the rear differential, and some of the torque is transferred to the front differential. In this system, all wheels are driven basically with an equal amount of torque. The driver has the option of using a selector lever or a dashboard button or switch that can make the system operate in either two-wheel or four-wheel drive modes.

Newer vehicles are now able to shift from 2WD to 4WD **on the fly**. This means that while the vehicle is moving down the road, the operator can shift from 2WD to 4WD without stopping. To shift from 2WD to 4WD on older vehicles, the car had to be stopped and locked in. The key in defining 4WD systems is that the operator has a choice of either 2WD or 4WD and that a transfer case allows either mode of operation.

AWD SYSTEMS DEFINED

AWD systems are most often used on standard-sized vehicles and are only used for increasing traction on highways and roads. They are most often used on front-wheel drive vehicles that use a transaxle system. They are typically not used for off-road conditions. ***Figure 44–2*** shows a typical example of how the power is transferred to all four wheels. In this case, a center differential (rather than a transfer case) is used to split the power to all four wheels. The operator of an AWD vehicle does not have the choice of 2WD or 4WD operation. The system operates in 4WD all of the time.

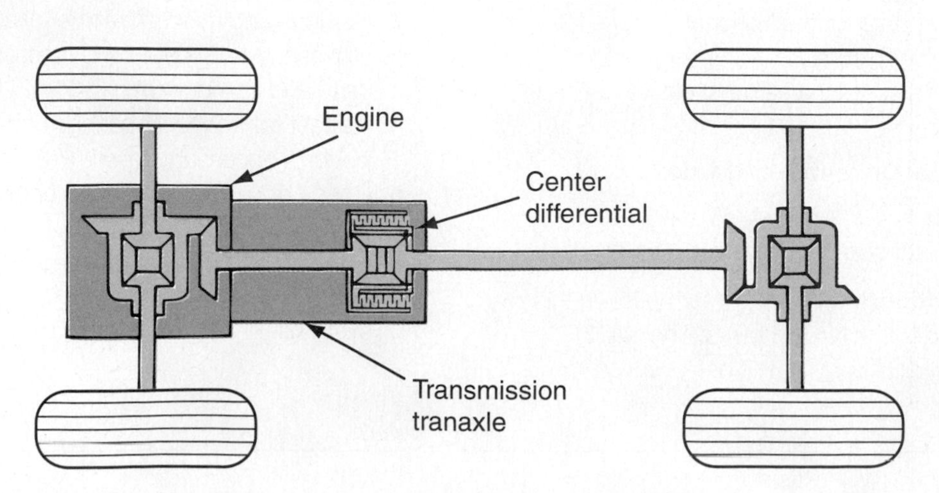

FIGURE 44–2 The parts of an AWD system. (Courtesy of DaimlerChrysler Corporation)

In operation, when one wheel slips the system automatically transfers torque to the other axles that have better traction. Although there are several variations of this design, the basic purpose and operation of each AWD system will be similar.

DISADVANTAGES OF 4WD AND AWD

There are some disadvantages to using 4WD and AWD systems. Generally, when in 4WD or AWD operation, fuel mileage is reduced. This is due to the additional friction produced from the tires and additional rotating parts. Also, the extra transfer cases, differentials, and so on, add additional weight to the vehicle, which also has a tendency to reduce fuel mileage.

TIRE SCUFFING

When a vehicle is driving in 4WD there also is a problem of **tire scuffing** when cornering. As any vehicle turns a corner, the outer wheels must turn in a larger circle than the inner wheels (***Figure 44–3***). This means that as the car turns a corner, the outer wheels will go faster than the inner wheels. This is true on all vehicles. On 2WD systems, this difference in speed on the drive wheels is accounted for by using a differential. However, in a 4WD system, imagine what would happen if all four wheels were connected or locked together on dry pavement. As the vehicle turns a corner, some of the tires would have to slip or scuff along the pavement. This would cause excessive wear on the tires.

Tire scuffing is only a problem when driving in 4WD on paved surfaces that are not slippery, muddy, icy, or snowy. Tire scuffing is not a problem on slippery, muddy, icy, or snowy conditions when in 4WD, because under these conditions the tires are allowed to slip on the road. Thus, all full-time 4WD systems need special types of equipment and components to compensate for or reduce the problem of tire scuffing. The addition of equipment to eliminate tire scuffing costs additional money, adds weight to the vehicle, and increases frictional horsepower from the engine. The end result is that fuel mileage is decreased.

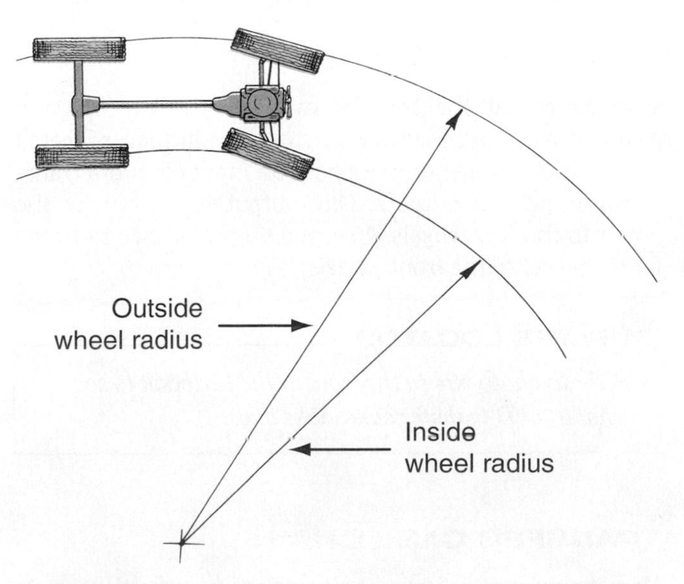

FIGURE 44-3 The outer wheels on a car that are turning a corner must go in a larger radius or arc than the inner wheels.

PART-TIME AND FULL-TIME 4WD SYSTEMS

Some vehicles have what is called "full-time" 4WD. This type of system keeps all four wheels connected together during all driving conditions. For example, some off-road jeeps have full-time 4WD systems. If a vehicle with **full-time 4WD** is operated on paved road surfaces, tire scuffing can be a serious problem. To help eliminate some of the tire scuffing, these full-time 4WD systems may have viscous couplings in the transfer case or specially designed differentials to allow the wheels to spin at different arcs and rates of speed.

On the other hand, a very popular type of 4WD system today is the "part-time" 4WD. Part-time 4WD systems are designed for driving on slippery surfaces, such as mud or snow, or for driving in off-road conditions, or for use under normal city and highway driving conditions. When the vehicle does not need 4WD, it can be put back into 2WD. When the part-time 4WD system is on, the front and rear wheels are locked together. The driver now has excellent traction. However, when the vehicle is driving down the road and there is no mud, snow, or ice, the driver can use a shift lever or button to change the operation to 2WD.

4WD MODES OF OPERATION

Each manufacturer may have different modes of operation for its 4WD systems. The exact type and mode of operation is determined by the vehicle manufacturer, the model of the vehicle, and the year of the vehicle. The following are the most common 4WD modes of operation:

- 2WD mode—In this mode, torque is transmitted to the rear wheels of the vehicle. There is no torque being transferred to the front wheels. This is the most fuel-efficient mode of operation for vehicles with part-time 4WD. In this mode, the transmission operates through its normal gears. Power is transmitted through the transmission gears, directly through the transfer case and to the rear wheels. Typical conditions for this mode of operation would be normal driving conditions on dry pavement, at normal highway or city speeds.
- 4WD H (HI)—In this mode, torque is transmitted to the front wheels as well as to the rear wheels. This mode is less efficient than the 2WD mode. However, it may be needed when there are normal driving speeds on slippery road surfaces. Under these conditions, 4WD may need to be engaged. So the 4WD HI mode is used in icy and snowy conditions, generally at normal or slightly reduced speeds due to poor driving conditions.
- 4WD L (LO)—This mode is often used for off-road conditions, at reduced speeds, in rough terrain and severe

Car Clinic: 4WD SHIFTING

CUSTOMER CONCERN:
Shifting from 2WD to 4WD and Back

Often it is necessary to shift from 2WD to 4WD while moving down the road. Can this be done, and will it hurt any components on the 4WD system?

SOLUTION: The exact procedure for shifting will depend on the model of the vehicle, the year of the vehicle, and the manufacturer. For example, on newer vehicles, most manufacturers say that shifting from 2WD to 4WD and back can be made at speeds of less than 50 miles per hour. However, it is always a good idea to check the operator's manual. The recommended procedure may also be different for automatic transmissions and manual transmissions. On old vehicles, this convenience is most likely not possible. On many older vehicles, the manufacturer says to stop the vehicle, then make the shift from 2WD to 4WD. In addition, some manufacturers are still using locking hubs that must also be engaged and disengaged. The best recommendation is to read the operator's manual for your particular vehicle. Then you can be assured that there will be no damage to your 4WD system.

traction conditions. The transfer case generally has two gears, a HI and LO. These two gear ratios are produced through either gearing (like a manual transmission) or planetary gear systems (like an automatic transmission). In **4WD L (LO)** mode, torque is also transmitted to the rear wheels and the front wheels but at a reduced speed and increased torque. This mode is even less fuel-efficient than the 4WD HI mode.

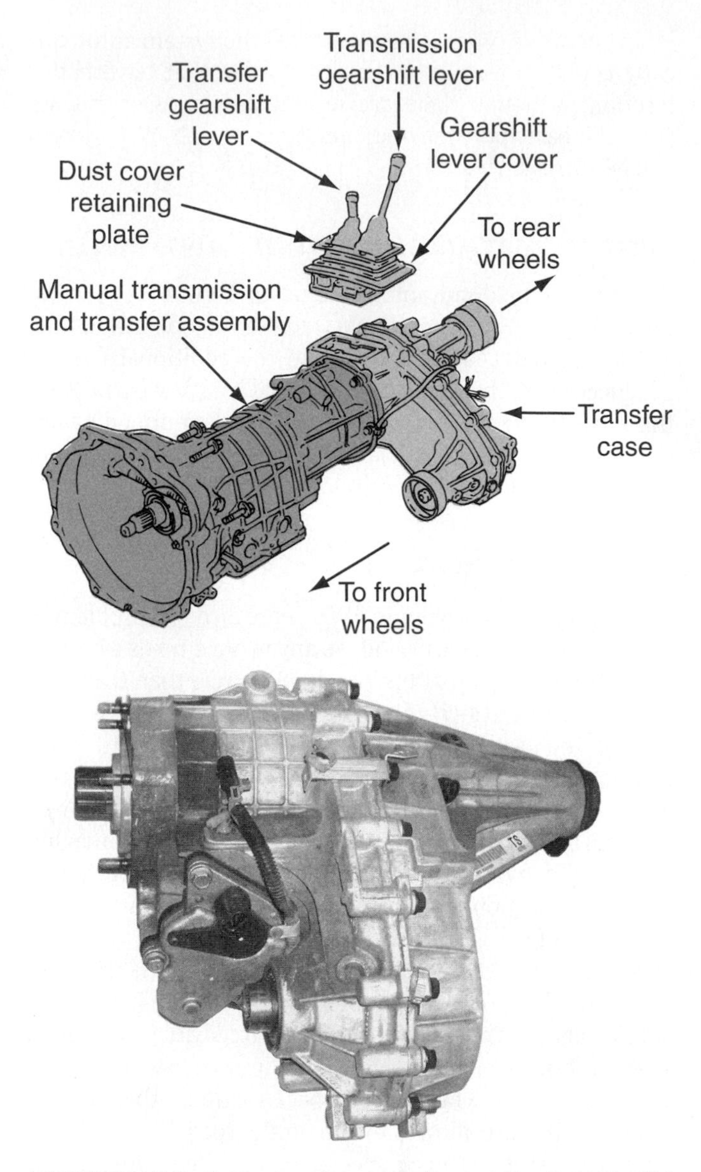

FIGURE 44–4 The transfer case is located directly behind the transmission. (Courtesy of Mitsubishi Motor North America, Inc.)

44.2 FOUR-WHEEL DRIVE COMPONENTS

There are many variations and designs used on 4WD systems. Some systems use viscous couplings, some use limited slip differentials, some use planetary gear systems. Because there are so many variations for 4WD systems, only the major components will be discussed. Some of the more common components include the transfer case, locking hubs, viscous couplings, shift controls, and limited slip differentials.

PURPOSE OF THE TRANSFER CASE

The purpose of the transfer case is to transfer the torque from the transmission to both the front wheels and the rear wheels during 4WD operation. ***Figure 44–4*** shows a typical transfer case attached to the rear of an automatic transmission. Note that the transfer case has the transmission gearshift lever and the transfer case gearshift lever located on the top. All transfer cases have an input from the transmission and two outputs. One output is to transfer the power to the rear wheels. A second output is used to transfer the power to the front wheels.

PARTS LOCATOR

*Refer to photo #94 in the front of the textbook to see what a **4WD transfer case** looks like.*

TRANSFER CASE CHAIN

Many transfer cases use a chain drive to transfer the torque from the rear wheels to the front wheels. ***Figure 44–5***

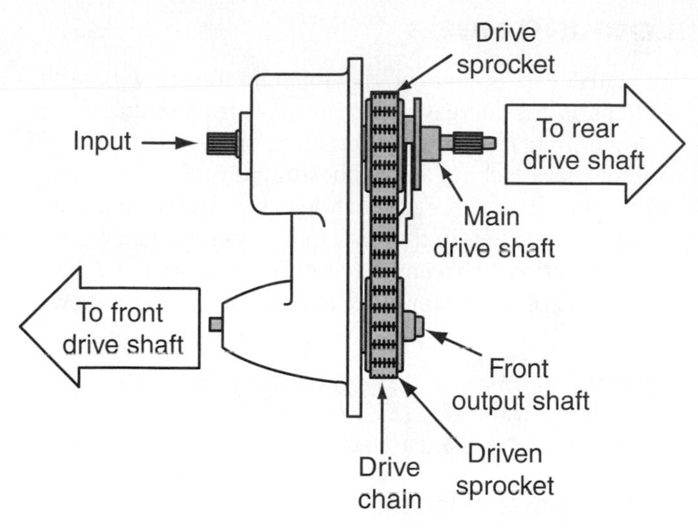

FIGURE 44–5 The drive chain is used to connect the rear-wheel output drive to the front output shaft.

shows the drive chain on a typical transfer case. The shaft on the upper left is the main drive shaft. The transmission drives this shaft on one end. The rear wheels are driven from the other end of the shaft. Also, attached to this drive shaft is a drive sprocket. The drive sprocket is used to also turn the chain. The other end of the chain is attached to the driven sprocket. This sprocket is attached to the front wheels, through the front output shaft.

TRANSFER CASE GEARS

Some transfer cases do not use a chain to split the torque or power. A set of gears is used as shown in ***Figure 44–6***. In this particular design, the input comes into the transfer case on the upper-left shaft. The torque or power is transmitted directly through this shaft to the rear axle drive shaft. In addition, there is a center idler gear that connects the main shaft to the front axle shaft to produce the 4WD operation. With these conditions, the 4WD system is in the HI mode. If the two sliding gears are moved to the right and engage the right side of the idler gear, the 4WD system will be in the LO mode of operation.

FIGURE 44–6 Gears can be used to engage and disengage 4WD. In addition, movement of sliding gears produces the ratio for HI and LO ranges.

VISCOUS COUPLINGS

As mentioned earlier, vehicles that use full-time 4WD operate well on rough terrain. However, if used on dry pavement, there is a tendency for the wheels to scuff as they go around a corner. To eliminate this problem, a **viscous coupling** can be used. A viscous coupling is often used between the front and the rear wheels of the vehicle. It is basically a drum filled with a thick (viscous) fluid. Inside the drum there is an input and an output shaft as shown in ***Figure 44–7***. A set of circular plates is attached to the input shaft, while another set of circular plates is attached to the output shaft. The two circular plates rotate within each other, while the viscous material has a tendency to lock up the two sets of circular plates.

In operation, as the demand for more torque on one axle increases, there is greater slippage between the two sets of circular plates. The slippage causes friction and the fluid begins to increase in temperature. As the fluid heats up, the

PARTS LOCATOR

*Refer to photo #95 in the front of the textbook to see what a **viscous coupling for a 4WD vehicle** looks like.*

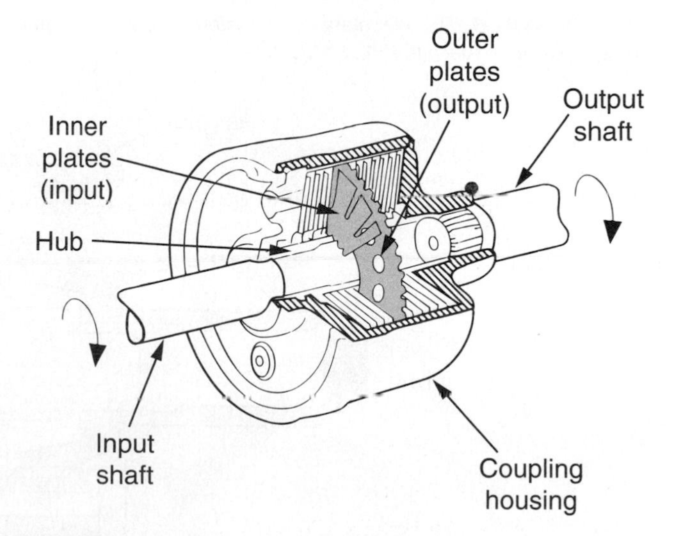

FIGURE 44–7 The viscous coupling locks up the input shaft to the output shaft with a viscous (thick) fluid. (Courtesy of Volkswagen of America, Inc.)

viscosity of the fluid increases. This causes the fluid to get thicker. The thicker fluid increases the friction between the two circular plates. The net result is that the torque between the front and rear axles splits according to the actual needs of the vehicle.

PLANETARY GEAR SYSTEMS

Some 4WD systems (in particular those that use a chain drive) use a planetary gear system to produce the HI and LO gearing needed for different modes of operation. As discussed in an earlier chapter, a planetary gear system is used on automatic transmissions to produce different gear ratios. The same is true inside the transfer case. The planetary gear system is used to produce both the HI and LO gearing for 4WD system operation.

In review (***Figure 44–8***), the planetary gear system uses a ring gear, a planet pinion carrier, planet gears, and a sun gear to produce the different gear ratios. In operation, the sun gear is always driven by the input from the transmission. ***Figure 44–9*** shows a cutaway view of the planetary drive system. To produce the various modes of operation, a sliding clutch mechanism releases or locks up the chain drive sprocket. To get the LO and HI speed 4WD modes, power is extracted from different gears of the planetary gear system.

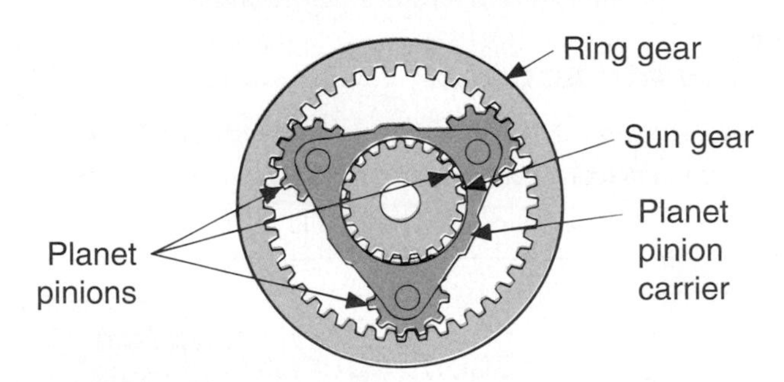

FIGURE 44–8 The planetary gear system is used on some 4WD systems to produce the HI and LO gearing.

LOCKING HUBS

As stated earlier, one of the main problems with 4WD systems is the increased friction and, thus, reduced fuel mileage. To help reduce this problem, some part-time 4WD systems have locking and unlocking hubs. For a moment, let's think about a vehicle that does not use locking hubs. When the vehicle is in the 2WD mode, there is no mechanical connection between the transfer case and the front wheels. All of the power from the engine is used to power

Car Clinic: LOCKING HUBS

CUSTOMER CONCERN:

Locking Hubs Do Not Work

A driver has complained that the fuel mileage on the 4WD system is getting worse and worse. The particular vehicle uses locking hubs to engage and disengage the front drive system from the 4WD system. The driver says that since the vehicle was purchased, there has been no service on the locking hubs. What could be the problem?

SOLUTION: One of the problems with using a 4WD system with locking hubs is that on some vehicles the hubs must always be engaged and disengaged manually. If the locking hubs are not working, this means that when the vehicle is in 2WD, the front axles, differential, and front drive shaft are still turning (being driven by the front wheels). If the locking hubs are dirty and need grease, they may not engage and disengage properly. This can cause the front axles, differential, and front drive shaft to continue turning while in 2WD, thus causing poor gasoline mileage. Cleaning out and repacking the front wheels and locking hubs will most likely solve the problem.

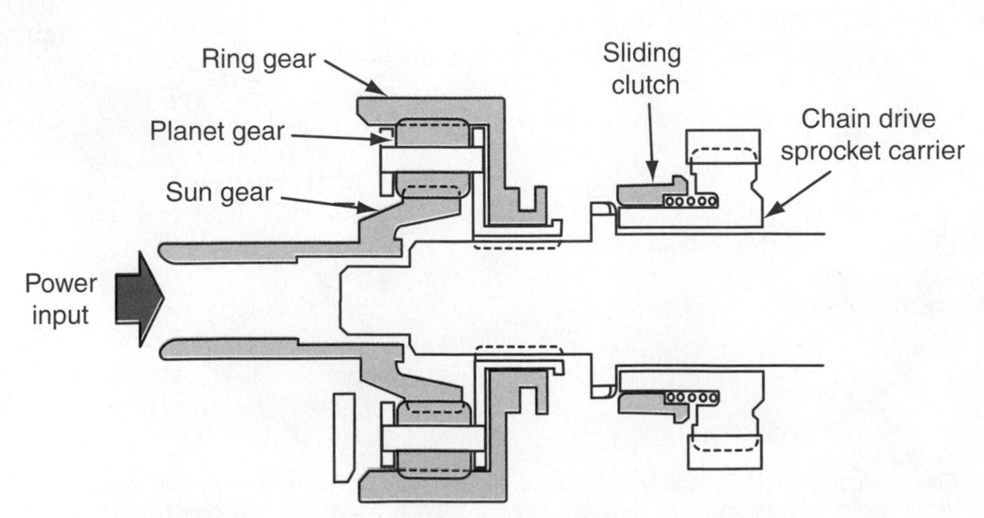

FIGURE 44–9 To obtain the HI and LO gearing, power is extracted off different gears in the planetary gear system.

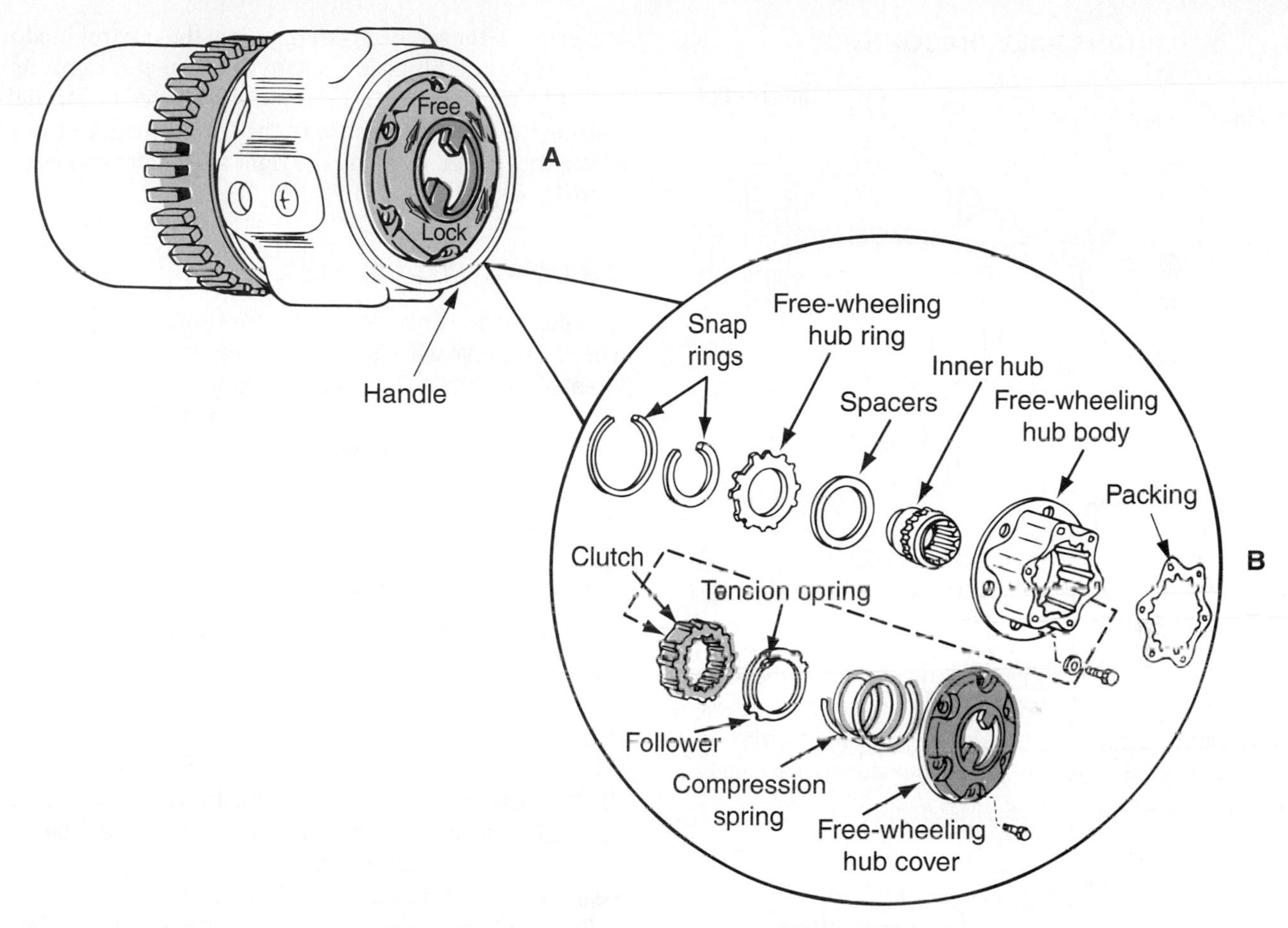

FIGURE 44–10 To lock or unlock the hubs, the outside handle is turned to engage or disengage a clutch mechanism. (Courtesy of Mitsubishi Motor Sales of America, Inc.)

the back wheels. However, the two front wheels are turning (without power). If the two front wheels are turning, then the front axles, the front differential, and the front drive shaft are all turning as well. In effect, these components are being driven by the turning motion of the front wheels. This action is often called **backdrive**. Backdrive is when the turning of the front wheels acts as the power source to drive the front axles, the front differential, and the front drive shaft. When these front components are turning, they produce friction, and such friction reduces fuel mileage.

To overcome this problem, such 4WD vehicles use some type of locking and unlocking hub design. If the hubs on the front wheels could be disconnected from the front axles, front differential, and front drive shaft, these three components would stop spinning. The net effect would be less friction and improved fuel economy.

There are two types of **locking hubs**: manual and automatic. Manual locking hubs are those in which the driver must manually stop the vehicle and turn a small shift knob on the end of each front wheel hub. ***Figure 44–10*** shows a typical example of a manual hub. The internal parts of the hub lock up the axle by using a spring and clutch mechanism that engages and disengages a sliding gear to the axle shaft.

The automatic locking hub is more convenient than the manual locking hub. In this case the driver does not have to stop the vehicle, or manually turn anything. When the driver selects 4WD operation, the hubs automatically lock up.

PARTS LOCATOR

*Refer to photo #96 in the front of the textbook to see what a **front axle disconnect** looks like.*

FRONT AXLE DISCONNECT

Many vehicles use a front axle disconnect system (***Figure 44–11***). This system has a small splined collar on one of the front axles. The front axle is made of two parts, the inner axle and the outer axle. When the sleeve is moved to connect the two axle parts together, the system locks up. When the sleeve is moved to the left, the front axle disconnects. A small vacuum motor is used to move the splined collar to the left or to the right. In the disconnect position, the ring gear, differential case, and pinion gears all stop turning. This action causes the front drive shaft to also stop spinning. The net effect is again reduced friction and increased fuel economy in 2WD.

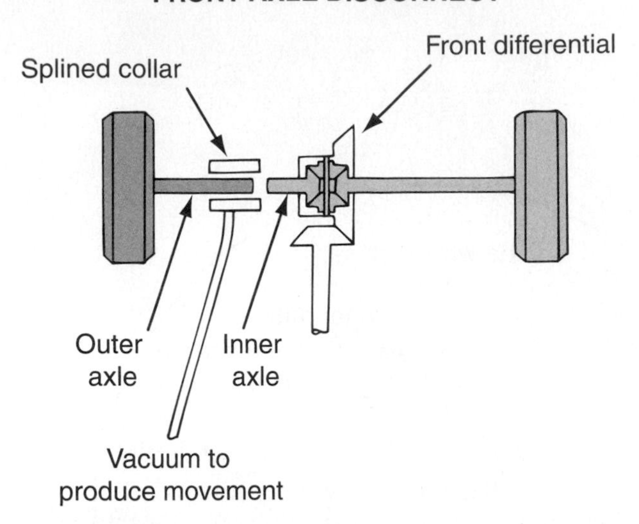

FIGURE 44–11 A splined collar is used to engage or disengage the front axle with the front differential.

In practice, the front axle locking system is located inside the front differential. ***Figure 44–12*** shows how the system operates. In the top drawing, the shift mechanism is in 2WD. In this mode, the left-side output shaft of the differential is not connected to the left axle shaft. In the lower drawing, an actuator has been energized by the control module. This causes the plunger shaft to move to the right. As it moves to the right, the shift fork also moves to the right, causing the sleeve to mesh with the output shaft. In this position, the output shaft and the right axle shaft are now connected together.

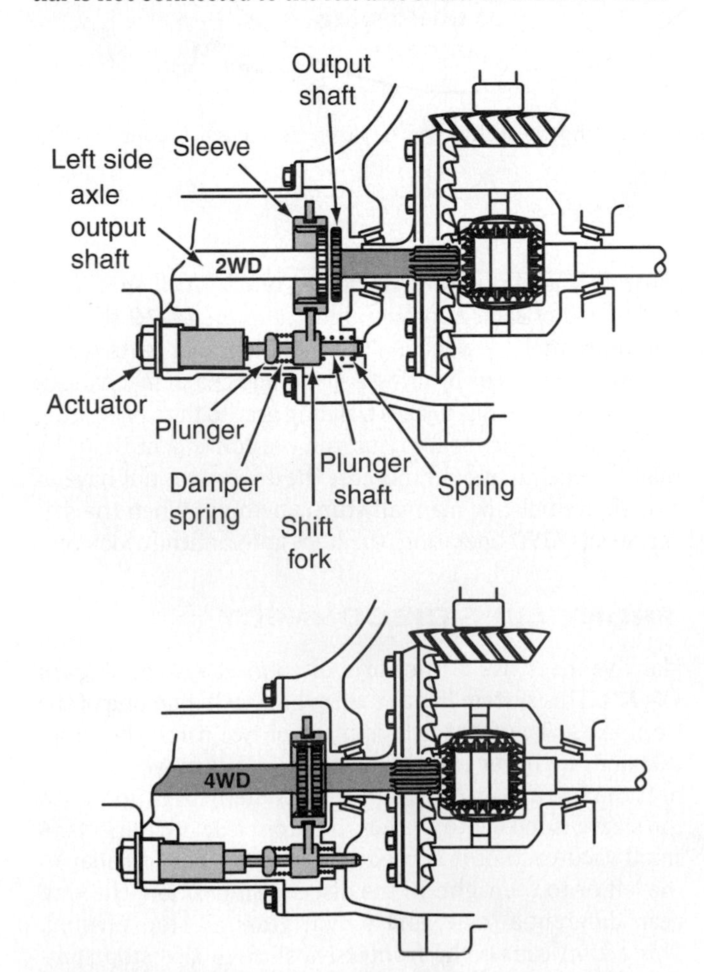

FIGURE 44–12 When the shift fork is moved to the right, both axles are connected together.

DIFFERENTIALS

In order for power to be equally transmitted to both front wheels and rear wheels when driving around corners, differentials are used. As discussed in an earlier chapter, the differential is used to help equalize the torque when turning. So, in most 4WD systems, there is a rear and a front differential. The differentials operate the same as discussed previously.

INTERAXLE DIFFERENTIAL

When driving a vehicle that is in the 4WD mode of operation, the rear wheels often try to spin faster than the front wheels, or vice versa. When this happens, and if the front wheels were locked (by the transfer case) to the rear wheels, **drive line windup** would occur. Drive line windup means that because of different loads on the front and back wheels, internal twisting forces are produced within the total drive train. These twisting or torsion forces can cause excessive wear and premature failure on many of the 4WD parts.

To eliminate the problem of drive line windup, a third differential is often used. This differential, called the interaxle differential, is placed inside the transfer case. The interaxle differential is located on the rear drive output. In this position, the interaxle differential is essentially placed between the front and rear torque or power forces. ***Figure 44–13*** shows its location. So, when there are torsional or twisting forces that occur between the rear and front differentials, the slippage may now occur within the interaxle differential. The net result is that there is much less wear on the internal parts of the 4WD system.

ALL-WHEEL DRIVE COMPONENTS

As mentioned earlier, AWD systems cannot select either 2WD or 4WD modes of operation. AWD systems are in 4WD all of the time. Vehicles with AWD are most often used exclusively on highways and under normal driving (not off-road) conditions. AWD systems are designed to increase vehicle traction on poor or slippery roads.

The biggest advantage of AWD systems is that they are designed to transfer the majority of the engine torque to the four wheels in equal amounts. This gives the vehicle more control in slippery and snowy conditions. Then, if one tire begins to slip more than the other three, the torque on that wheel is reduced and the torque on the other three wheels is increased. This is often referred to as on-demand 4WD. To accomplish this, AWD systems use a viscous clutch to help transfer the torque to all wheels.

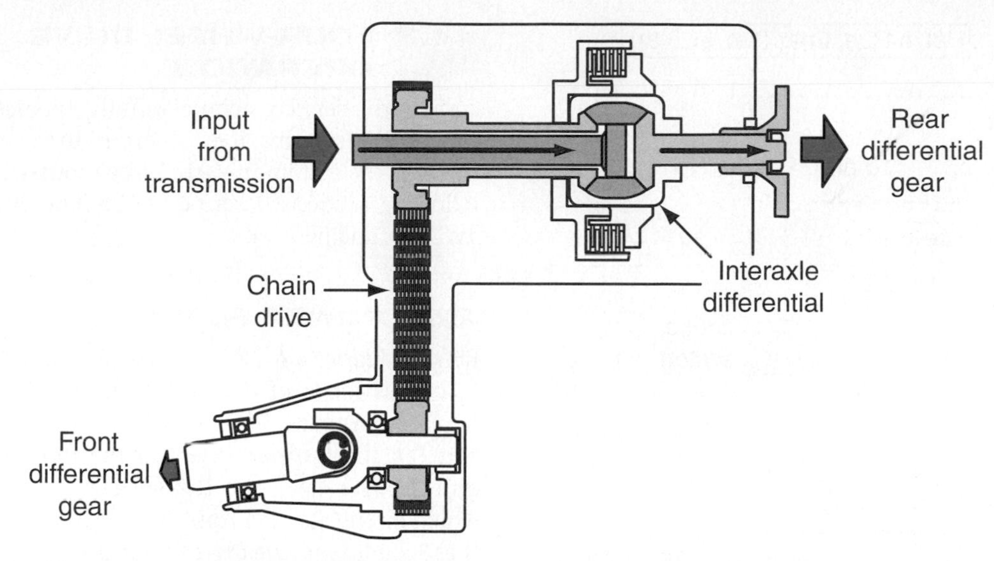

FIGURE 44–13 The interaxle differential is used to reduce drive line windup on 4WD systems.

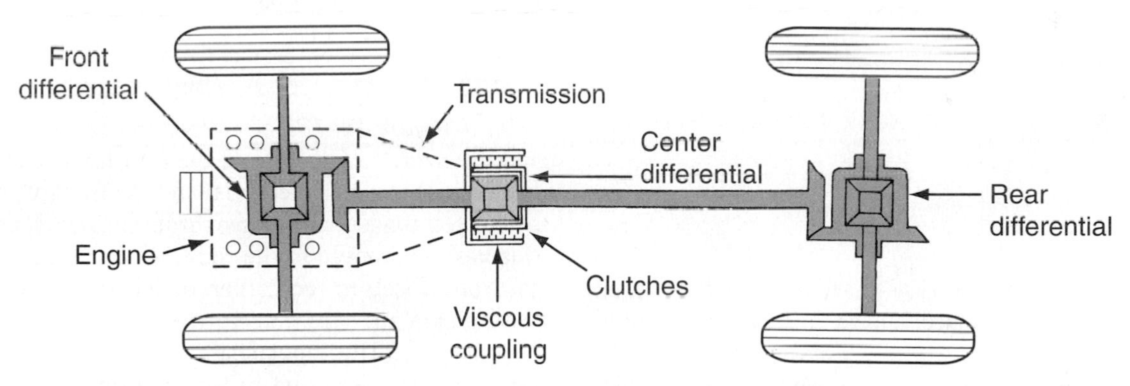

FIGURE 44–14 A center differential is used on some AWD vehicles.

Figure 44–14 shows the power flow components of the AWD system. In this case the vehicle uses a regular transmission. Then, from the transmission, the power is sent to a viscous coupling and differential. The output of the viscous coupling and differential goes to the front wheel differential and also the rear wheel differential.

ELECTRONICALLY CONTROLLED 4WD AND AWD

With the addition of electronic controls, 4WD and AWD systems are now using electronically controlled clutches and transfer cases. ***Figure 44–15*** shows how a clutch system operates. In this particular case, a series of input sensors is

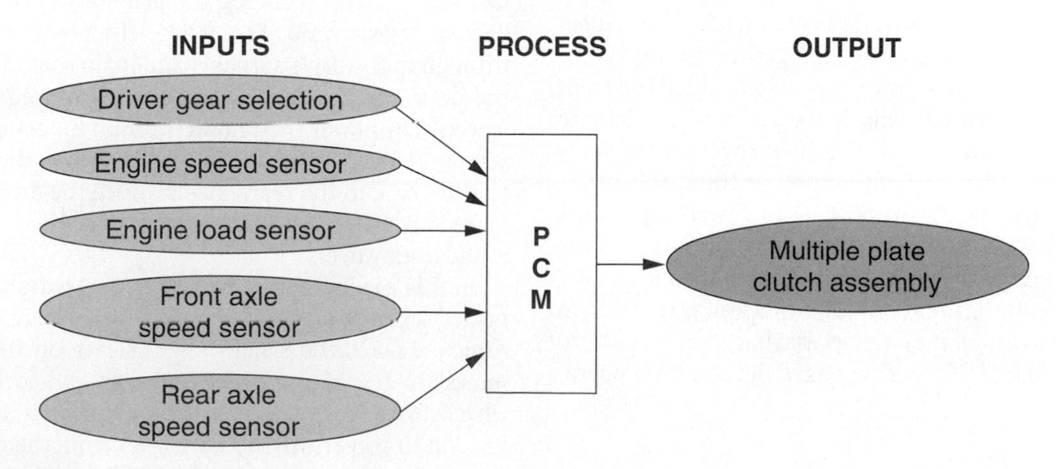

FIGURE 44–15 An electronically controlled AWD system has various input sensors used to control the hydraulic pressure in the multiple plate clutch assembly.

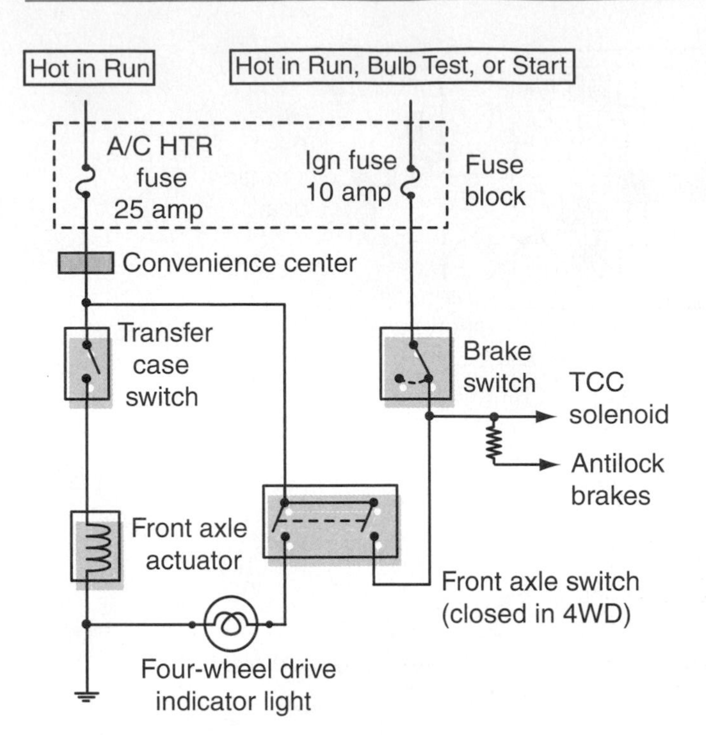

FIGURE 44–16 This 4WD electrical circuit shows the major components used to control 4WD systems.

used. Such input sensors as gear selection, engine speed, engine load, front axle speed, and rear axle speed are used. Based on these inputs, the powertrain control module (PCM) determines how much to lock up the multiple plate clutch assembly. This, in effect, determines how much torque is sent to the front or rear wheels.

Figure 44–16 shows an electrical circuit for a 4WD system. The transfer case electrical circuit operates as follows:

1. When the ignition switch is in the run position and the 4WD engaged, electricity flows from the fuse block, through the convenience center, and into the transfer case switch.
2. Since the transfer case switch is closed (in 4WD), the front axle actuator solenoid is energized. This causes the front axles to be connected together during 4WD.
3. With the front axles engaged, a switch called the front axle switch is closed. This action causes electricity to flow through the switch, and then through the four-wheel drive indicator light located on the dashboard.
4. With the front axle switch closed, electricity also is able to flow to the antilock braking circuit and the TCC solenoid. Electricity flows to the antilock braking system to disengage the rear antilock brakes while in 4WD. In addition, it is important that the torque converter clutch (TCC) is disengaged during 4WD operation as well.

44.3 FOUR-WHEEL DRIVE OPERATION

To help the service technician fully understand the operation of 4WD systems, it is important to see how power flows through the system in each of the modes of operation. The following sections describe power flow in 2WD, 4WD HI, 4WD LO, and neutral.

2WD POWER FLOW

Refer to ***Figure 44–17*** to follow the power flow.

In this mode of operation, the front axles are disconnected to reduce friction. Also, there is no connection between the transfer case and the front differential. The synchronizer sleeve on the left side is moved to the left, which disengages the front wheels from each other. Under these conditions, the drive pinion and ring gear are stationary. The left and right axles are not connected together but turn according to the speed of each individual front wheel.

4WD HI POWER FLOW

Refer to ***Figure 44–18***.

In this mode of operation, the transfer case is now connected to the front differential and the front axles are now connected together. The synchronizer sleeve on the left side has been moved to the right, which engages and locks the front wheels to each other. Under these conditions, the drive pinion and ring gear turn by the power through the transfer case. The input shaft going into the differential turns the ring gear, and in turn the two front axles turn. The left and right axles are now connected together through the differential and turn from the power produced by the engine.

4WD LO POWER FLOW

In this mode, power flows from the transmission, into the transfer case, and into the planetary gear system. Shifting into 4WD LO causes the power to flow through the planetary gear system, reducing the planetary output to about a 2.69 to 1 gear ratio. From here the power flows directly through to the drive sprocket and main shaft. In this mode, the drive sprocket is also engaged or locked to the front wheels. The power then flows through the synchronizer assembly (which is engaged), and then out the rear of the transfer case to the rear wheels. Since the drive sprocket is locked up, power also flows through the drive chain and out to the front wheels.

In this mode of operation, the front axles are now connected together just as in the 4WD HI mode of operation. In ***Figure 44–19***, the synchronizer sleeve on the left side is moved to the right, which engages and locks the front wheels to each other. Under these conditions, the drive pinion and ring gear turn by the power from the transfer case. The input shaft going into the differential turns the ring gear, and in turn the two front axles turn. The left and right

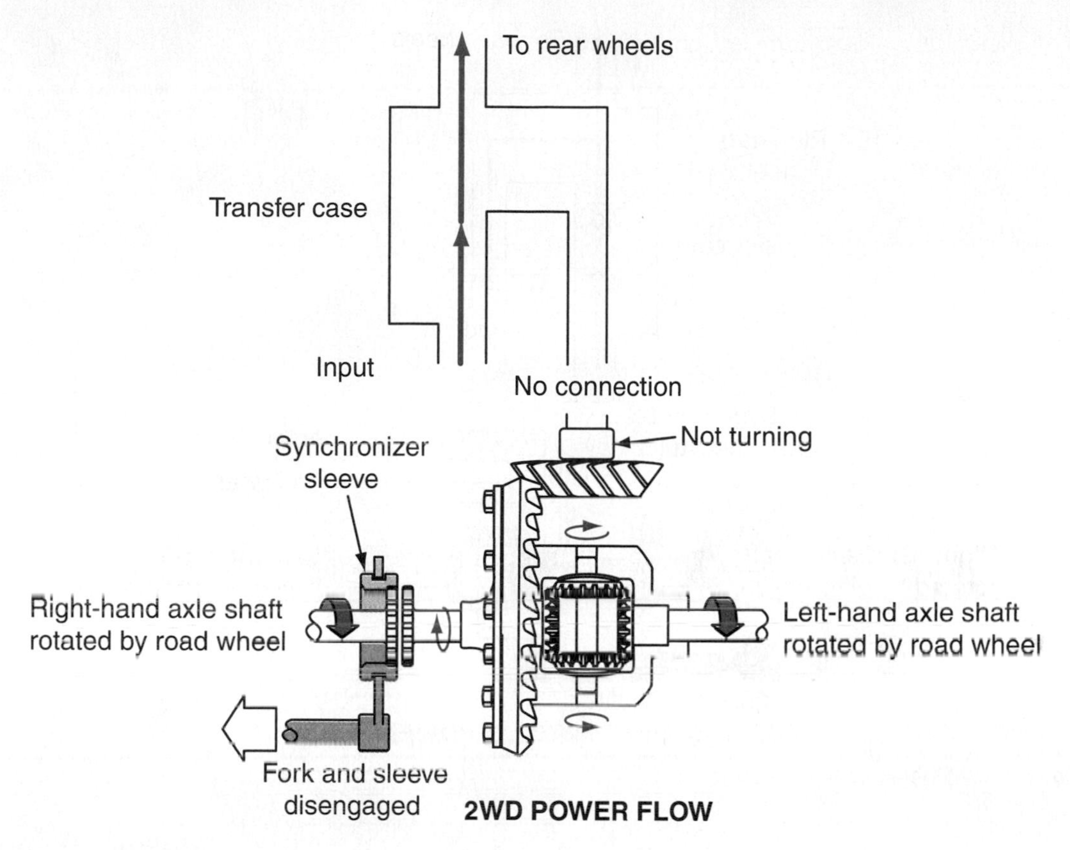

FIGURE 44-17 2WD power flow is shown.

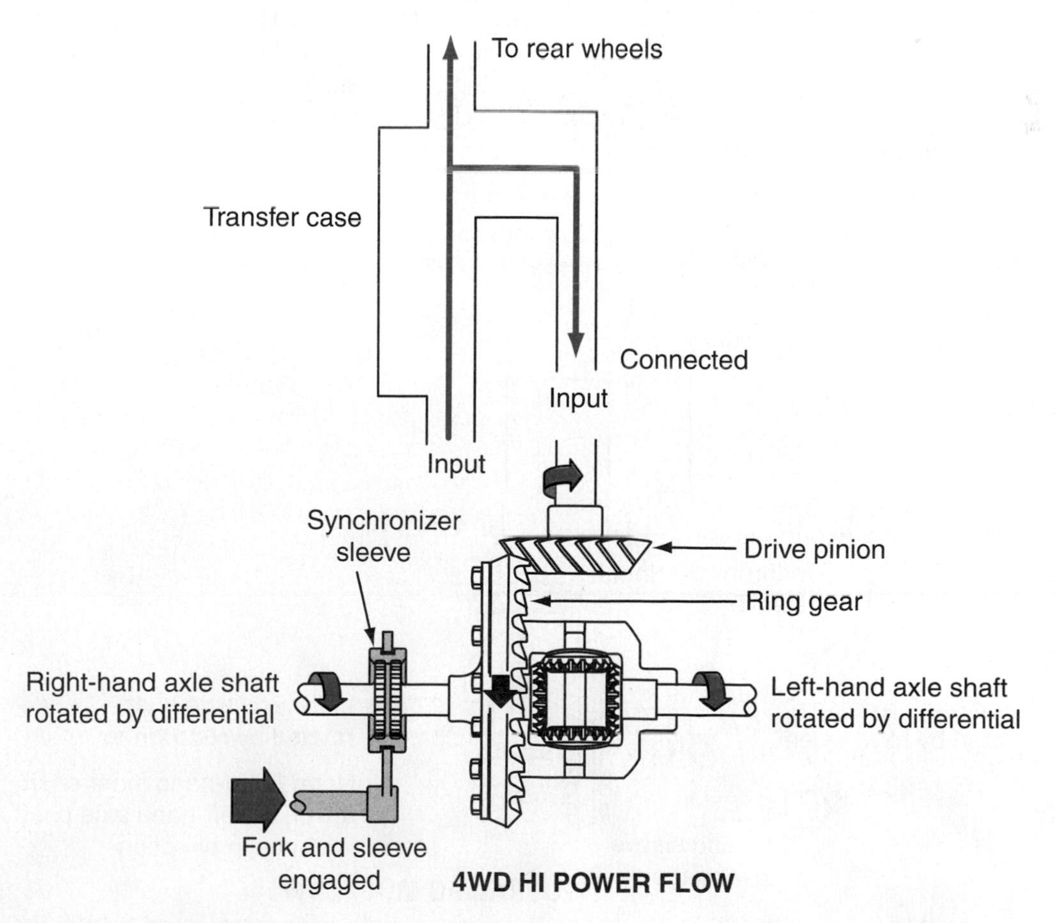

FIGURE 44-18 4WD HI power flow is shown.

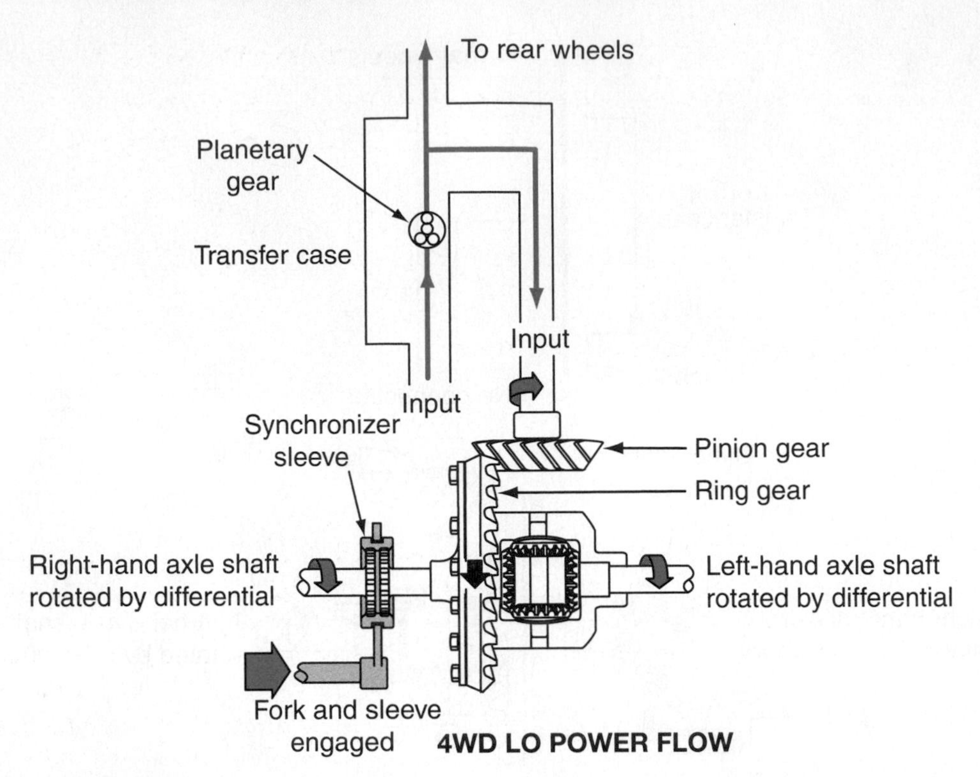

FIGURE 44–19 4WD LO power flow is shown.

axles are now connected together through the differential and turn from the power produced by the engine.

NEUTRAL POWER FLOW

In this mode of operation, the transfer case has been disconnected from the front differential. The front axles are also disconnected to reduce friction if the vehicle is moving. In ***Figure 44–20***, the synchronizer sleeve on the left side is moved to the left, which disengages the front wheels from each other. Under these conditions, the drive pinion and ring gear are stationary. The left and right axles are not connected together but turn according to the speed of each individual front wheel.

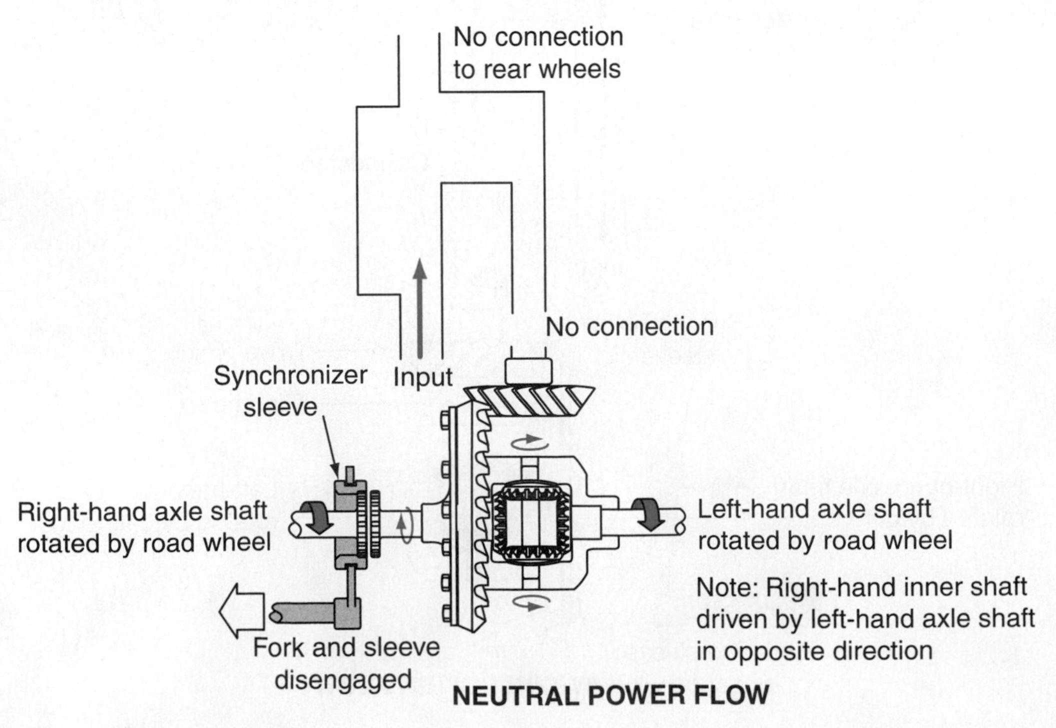

FIGURE 44–20 Neutral power flow is shown.

Problems, Diagnosis, and Service

Safety Precautions

1. Whenever working with 4WD and AWD drive systems, always be sure to wear OSHA-approved safety glasses.
2. When removing or installing any component on a 4WD/AWD drive system, make sure to use the correct tools.
3. Always follow the manufacturer's recommended removal and replacement procedures when working with 4WD/AWD drive systems.
4. Whenever disconnecting wires from input sensors or other electrical equipment on 4WD/AWD drive systems, always be sure to mark the wires so that you know how they should be reattached.
5. Many of the parts on a 4WD/AWD drive system, such as transfer cases and differentials, are very heavy. Always use the correct lifting devices and be careful not to strain your back when lifting these parts.
6. Always use the recommended parts when replacing any components on 4WD/AWD drive systems.
7. If a vehicle must be put on a jack, always use an approved-type jack and be sure to support the vehicle correctly and safely.

PROBLEM: AWD Noise

An AWD system on a vehicle seems to be making a noise while the wheels are rotating.

DIAGNOSIS

As with other drive systems, many problems are possible with AWD and 4WD systems. ***Figure 44–21*** shows a typical troubleshooting chart that will help the service technician determine the cause of the problem. In this case, the problem that produces the noise may be a brake dragging, a bent axle shaft, or a worn or scarred axle shaft bearing.

There are many other types of problems that can also be identified with troubleshooting charts. For example, ***Figure 44–22*** shows a typical troubleshooting chart for an all-wheel drive system when the rear wheels are not overrunning. Always follow the manufacturer's suggested procedure for troubleshooting each individual component problem.

SERVICE

After each component has been checked, the damaged component will need to be replaced. To replace the component, always follow the manufacturer's suggested procedure. Since there are many procedures listed in a typical service manual, make sure to identify the correct component, the correct model, etc., before beginning any service.

PROBLEM	DIAGNOSIS	SERVICE
AXLE SHAFT		
Noise while wheels are rotating	Brake drag Bent axle shaft Worn or scarred axle shaft bearing	Replace
Grease leakage	Worn or damaged oil seal Malfunction of bearing seal	Replace
DRIVE SHAFT		
Noise	Wear, play or seizure of ball joint Excessive drive shaft spline looseness	Replace
DIFFERENTIAL (CONVENTIONAL DIFFERENTIAL)		
Constant noise	Improper final drive gear tooth contact adjustment Loose, worn or damaged side bearing Loose, worn or damaged drive pinion bearing	Correct or replace
	Worn drive gear, drive pinion Worn side gear spacer or pinion shaft Deformed drive gear or differential case Damaged gear	Replace
	Foreign material	Eliminate the foreign material and check; replace the parts if necessary.
	Insufficient oil	Replenish
Gear noise while driving	Poor gear engagement Improper gear adjustment Improper drive pinion preload adjustment	Correct or replace
	Damaged gear	Replace
	Foreign material	Eliminate the foreign material and check; replace the parts if necessary.
	Insufficient oil	Replenish
Gear noise while coasting	Improper drive pinion preload adjustment Damaged gear	Correct or replace Replace
Bearing noise while driving or coasting	Cracked or damaged drive pinion rear bearing	Replace
Noise while turning	Loose side bearing Damaged side gear, pinion gear, or pinion shaft	Replace
Heat	Insufficient gear backlash Excessive preload	Adjust
	Insufficient oil	Replenish
Oil leakage	Clogged vent plug	Clean or replace
	Cover insufficiently tightened Seal malfunction	Retighten, apply sealant, or replace the gusset
	Worn or damaged oil seal	Replace
	Excessive oil	Adjust the oil level
DIFFERENTIAL (LIMITED SLIP DIFFERENTIAL)		
Abnormal noise during driving or gear changing	Excessive final drive gear backlash Insufficient drive pinion preload	Adjust
	Excessive differential gear backlash	Adjust or replace
	Worn spline of a side gear	Replace
	Loose companion flange self-locking nut	Retighten or replace

FIGURE 44-21 Use troubleshooting charts like this one to help diagnose various problems with an AWD system. (Reproduced from MOTOR Light Truck and Auto Repair Manual, © The Hearst Corporation)

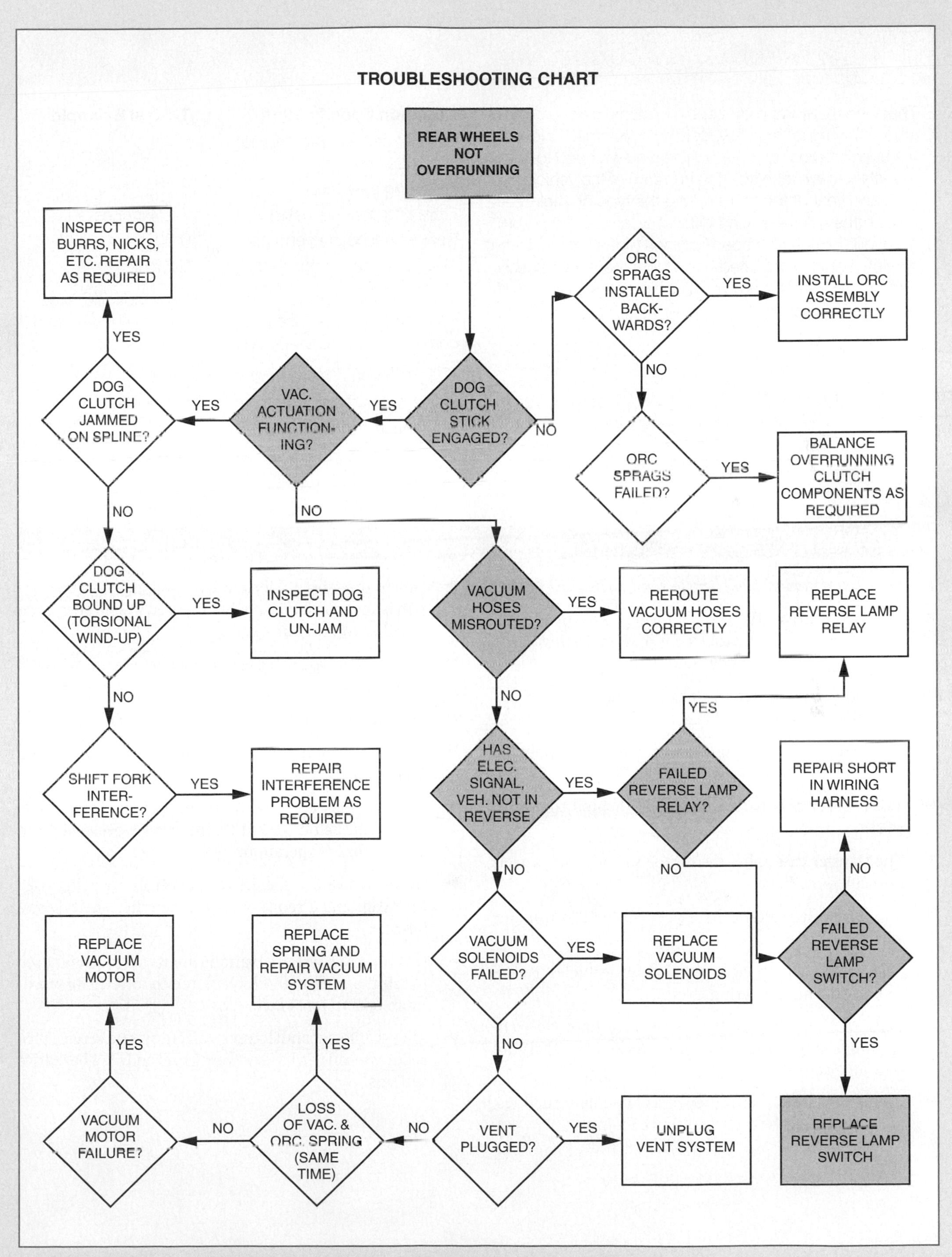

FIGURE 44-22 Troubleshooting charts help the service technician diagnose and solve AWD problems.

Service Manual Connection

There are many important specifications to keep in mind when working with four-wheel drive systems. To identify the specifications for your engine, you will need to know the VIN (vehicle identification number) of the vehicle, the type and year of the vehicle, and the type of engine. Although they may be titled differently, some of the common four-wheel drive specifications (not all) found in the service manuals are listed here. Note that these specifications are typical examples. Each vehicle may have different specifications.

Common Specification	Typical Example
Transfer case to transmission torque	25–35 ft.-lb
Front ring gear ratio	2.76 to 1
Front ring gear diameter	7 1/4 inches
Rear wheel bearing end play	0.002 inch
Rear axle gear backlash	0.2 inch

In addition, service manuals provide specific directions for various service procedures. Some of the more common procedures include:

- Transfer case disassembly
- Transfer case diagnostic code identification
- Planetary ring gear removal

SUMMARY

The following statements will help to summarize this chapter:

- Four-wheel drive (4WD) and all-wheel drive (AWD) systems are used to increase traction when driving in icy, snowy, slippery, or muddy conditions.
- Increased traction is produced by driving all four wheels from the engine.
- Four-wheel drive systems use a transfer case to split the torque or power.
- The transfer case is located directly behind the transmission.
- The transfer case splits the torque to both back wheels or to all four wheels.
- All-wheel drive systems have power directed to all four wheels.
- One problem with four-wheel drive vehicles is tire scuffing during cornering.
- Because the outer and inner wheels have different turning radii, and the wheels are locked together, tire scuffing occurs.
- Components in the transfer cases, viscous couplings, and differentials help to reduce tire scuffing.
- One of the most important components of a 4WD system is the transfer case, which is used to split the power to either two or four wheels.
- The transfer case can use either a chain drive or a set of gears.
- The transfer case often uses HI and LO gears.
- Locking hubs are another common component on 4WD systems.
- Locking hubs are used to disconnect the front drivetrain from being turned by the front wheels when in 2WD modes of operation.
- On today's vehicles, electronic controls are being used to engage and disengage various clutches on 4WD systems.
- On computer systems, various inputs are read and used to tell the computer which clutch to lock up for maximum operational efficiency.
- Always follow manufacturer's diagnosis and service procedures when working on four-wheel and all-wheel drive systems.

TERMS TO KNOW

Can you explain each of the following terms? Review the chapter until you can use each term correctly.

4WD L (LO)
Backdrive
Drive line windup
Full-time 4WD
Locking hubs
Off-road
On the fly
Tire scuffing
Transfer case
Viscous coupling

REVIEW QUESTIONS

Multiple Choice

1. The transfer case on a 4WD system is located:
a. In front of the transmission
b. Directly behind the transmission
c. Only on the front axle
d. Only on the rear axle
e. On the rear drive shaft

2. Which of the following is not a common mode of operation with a 4WD system?
a. 2WD
b. 4WD HI
c. 4WD LO
d. 2WD LO
e. None of the above

3. Which of the following is a major disadvantage of 4WD systems?
a. Tire scuffing on dry pavement
b. Increased frictional loads
c. Reduced fuel mileage
d. All of the above
e. None of the above

4. Which of the following is not true of 4WD and AWD systems?
a. All-wheel drive systems are in a four-wheel drive mode all of the time.
b. 4WD systems have only one gear to operate in.
c. AWD systems use a viscous coupling.
d. 4WD systems have a HI and LO gear mode of operation.
e. 4WD systems can now be shifted from 2WD to 4WD on the fly.

5. Tire scuffing occurs because the:
a. Vehicle always goes in a straight line
b. Vehicle needs to go around corners
c. Tires are always flatter on 4WD systems
d. Transfer case is never locked up
e. Driver is going too slowly

6. AWD systems are considered:
a. Part-time 4WD
b. Full-time four-wheel drive
c. Good for off-road use
d. Better for pickup trucks and sports utility vehicles
e. Useful for rough terrain operation

7. When driving down the highway with dry pavement conditions, a 4WD vehicle should always be in which gear?
a. 2WD
b. 4WD HI
c. 4WD LO
d. Any of the above
e. None of the above

8. When driving in off-road conditions over rough terrain, a 4WD vehicle should always be in which gear?
a. 2WD
b. 4WD HI
c. 4WD LO
d. Any of the above
e. None of the above

9. A planetary gear system can be used inside the ________ to develop different gearing ratios for 4WD systems.
a. Transfer case
b. Front axle
c. Rear axle
d. Differential
e. Viscous coupling

10. Locking hubs should always be disconnected or unlocked when the vehicle is in which driving condition?
a. Slippery road conditions
b. Muddy off-highway conditions
c. Dry highway pavement, without slippery conditions
d. Off-road muddy conditions
e. All of the above

The following questions are similar in format to ASE (Automotive Service Excellence) test questions.

11. *Technician A* says that fuel mileage will increase when using 4WD systems. *Technician B* says that 2WD systems have a HI or LO mode of operation. Who is correct?
a. A only
b. B only
c. Both A and B
d. Neither A nor B

12. *Technician A* says that AWD systems can be put into either part-time or full-time operation. *Technician B* says that AWD systems should not be used in off-road conditions. Who is correct?
a. A only
b. B only
c. Both A and B
d. Neither A nor B

13. *Technician A* says that the purpose of the transfer case is to transfer power to the rear wheels when in 2WD. *Technician B* says that the purpose of the transfer case is to transfer power to the rear and front wheels in 4WD modes of operation. Who is correct?
a. A only
b. B only
c. Both A and B
d. Neither A nor B

14. *Technician A* says that there is a differential on a 4WD system. *Technician B* says that up to three differentials can be used on certain 4WD systems. Who is correct?
a. A only
b. B only
c. Both A and B
d. Neither A nor B

15. *Technician A* says that some 4WD systems use a computer and various input sensors for its control. *Technician B* says that locking hubs are only used on AWD systems. Who is correct?
a. A only
b. B only
c. Both A and B
d. Neither A nor B

Essay

16. Describe why locking hubs are used on some 4WD vehicles.

17. What is the purpose of using two differentials on a 4WD system?

18. Explain the operation of the front axle disconnect system.

19. Define the major components in an AWD system.

20. Explain the purpose of using a planetary gear system in a 4WD system.

Short Answer

21. A 4WD system can use either gears or a(n) ________ inside the transfer case to transfer the power to the front wheels.

22. A(n) ________ coupling uses a thick fluid that increases its viscosity as its temperature increases.

23. Name one type of axle input sensor on an electronically controlled 4WD system. ________

24. One way to stop the front axles from spinning when in 2WD modes of operation is to have either manual or automatic ________.

25. One of the biggest disadvantages of a 4WD system is that the ________ will decrease.

Applied Academics

TITLE: Gear Wear and Alignment

ACADEMIC SKILLS AREA: Science

NATEF Automobile Technician Tasks:

The service technician can describe circular motion in a part of a vehicle as it relates to problems in the manual drivetrain and axle.

CONCEPT:

Four-wheel drive components have many gears that can wear due to the lack of lubrication and misalignment. Prevention of wear requires the correct type of lubrication, as well as the proper alignment of the gears. The wear patterns on gear teeth are a good indicator of incorrect alignment of gears. When checking wear on gears, look carefully at the type of wear pattern on the teeth. The contact area should be a shiny spot that shows the actual area of contact between the teeth as shown in the accompanying illustration. Depending on the shape and position of the contract area, the service technician can determine how the teeth of two gears are meshing. Often, if the contact pattern is not in the acceptable region, adjustments can be made to reposition the gears so correct contact can be made. Often adding or removing shims in the differential case makes such adjustments.

Description	Accept	Reject
Desired contact pattern		
End contact pattern		
Traveling contact pattern (moves from side to side)		
High contact pattern		
Low contact pattern		

APPLICATION:

Why is it important to have the contact area correct on gears? What type of gear pattern is affected when the gears become farther and farther apart? As gears become misaligned from one another, what type of gear pattern is affected?

SECTION

7 Vehicle Suspension and Control Systems

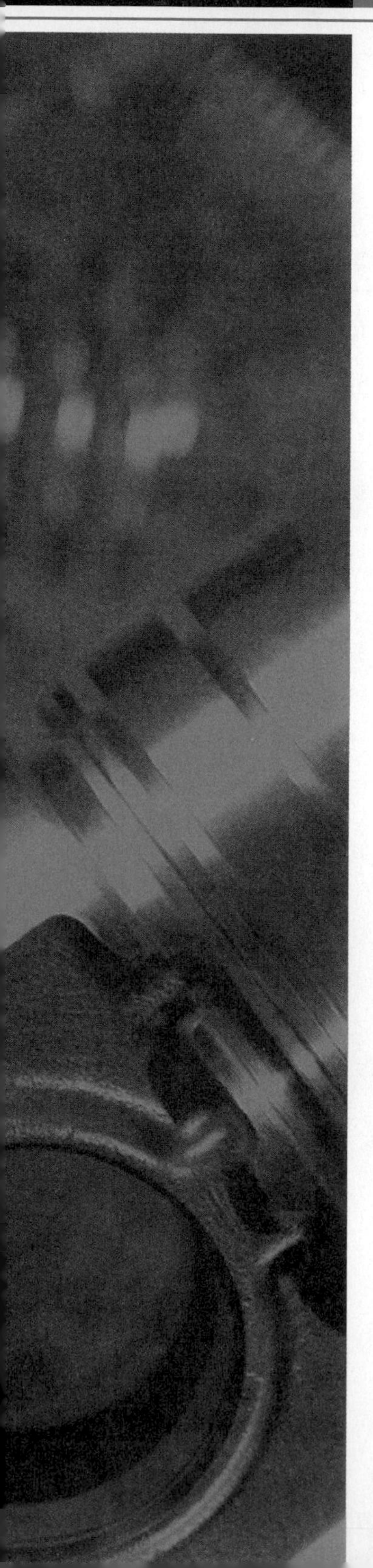

Deon Doyle, 29

Auto Technician

STS Tire and Auto Center

New Brunswick, NJ

How did you get interested in automotive service?

I enjoy helping people. The automotive repair business is a business of trust. I am a trustworthy person. Everyone owns a car, and most car owners really don't know that much about their cars. Car owners want to do business with someone they can trust.

What's your educational background?

I attended Lincoln Technical Institute. I'd never "turned a wrench" until I went there. I also continue learning through the STS career development center, which teaches me about the latest technology in tires and service and helps me keep on top of the business.

What's a typical day like?

Yesterday I was at work from 6:45 A.M. until 8:00 P.M. I worked on ten vehicles—three brake jobs, a window switch, a wiper blade change, and oil changes. My day usually ends at 6:00 P.M., but we had a customer who was stranded, so I stayed until her car was fixed—we do whatever it takes!

What skills are necessary for your job?

You must do the work correctly and at the agreed-upon price, so people will bring their cars back to you. And you have to hustle. Our customers' time is very important.

What's the best part of your job?

I like helping customers. Here's a recent example: I was at a basketball game the other night when a woman came up to me and gave me a big hug and told me what a good job I did on her car! That gave me a really good feeling.

Where do you see yourself in a few years?

I'm already working toward my ASE certifications. I am looking forward to becoming a master tech, and one day, a service manager.

Standard Braking Systems

CHAPTER 45

OBJECTIVES

After studying this chapter, you should be able to:

- Identify the principles of friction, hydraulic circuits, and basic braking system operation.
- State the name and operation of all braking system components.
- Analyze the purpose and operation of power brake systems.
- Analyze various problems, their diagnosis, and service tips and procedures used on braking systems.

INTRODUCTION

The automobile braking system is used to control the speed of the vehicle. The braking system must be designed to enable the vehicle to stop or slow down at the driver's command. The brake system is composed of many parts, including friction pads on each wheel, a master cylinder, wheel cylinders or calipers, and a hydraulic control system. This chapter discusses the standard braking system used on the automobile.

Certification Connection

ASE Connection: The information in this chapter can help you prepare for the National Institute for Automotive Service Excellence (ASE) certification tests. The tests and content areas most closely related to this chapter are:

Test A5—Brakes

- **Content Area**—Hydraulic System Diagnosis and Repair
- **Content Area**—Drum Brake Diagnosis and Repair
- **Content Area**—Disc Brake Diagnosis and Repair
- **Content Area**—Power Assist Units Diagnosis and Repair
- **Content Area**—Miscellaneous System Diagnosis and Repair

NATEF Connection: Much of the information in this chapter is related to the NATEF tasks. The NATEF tasks and priority numbers most closely related to this chapter are:

1. Identify and interpret the brake system concern; determine necessary action. P-1
2. Diagnose pressure concerns in the brake system using hydraulic principles (Pascal's Law). P-1
3. Measure the brake pedal height; determine necessary action. P-2
4. Check the master cylinder for internal and external leaks and proper operation; determine necessary action. P-2
5. Remove, bench bleed, and reinstall the master cylinder. P-1
6. Diagnose poor stopping, pulling, or dragging concerns caused by malfunctions in the hydraulic system; determine necessary action. P-1
7. Inspect the brake lines, flexible hoses, and fittings for leaks, dents, kinks, rust, cracks, bulging, or wear; tighten loose fittings and supports; determine necessary action. P-2

continued

Certification Connection *continued*

8. Select, handle, store, and fill the brake fluids to proper level. P-1
9. Inspect, test, and/or replace metering (hold-off), proportioning (balance), pressure differential, and combination valves. P-2
10. Inspect, test, and adjust the height (load) sensing proportioning valve. P-3
11. Inspect, test, and/or replace components of the brake warning light system. P-3
12. Bleed (manual, pressure, vacuum, or surge) the brake system. P-3
13. Flush the hydraulic system. P-3
14. Diagnose poor stopping, noise, pulling, grabbing, dragging, or pedal pulsation concerns; determine necessary action. P-1
15. Remove, clean (using proper safety procedures), inspect, and measure the brake drums; determine necessary action. P-1
16. Remove, clean and inspect the brake shoes, springs, pins, clips, levers, adjusters/self-adjusters, other related brake hardware, and backing support plates; lubricate and reassemble. P-1
17. Remove, inspect, and install the wheel cylinders. P-2
18. Install the wheel, torque the lug nuts, and make final checks and adjustments. P-1
19. Remove the caliper assembly from mountings; clean and inspect for leaks and damage to the caliper housing; determine necessary action. P-1
20. Clean and inspect the caliper mounting and slides for wear and damage; determine necessary action. P-1
21. Remove, clean, and inspect the pads and retaining hardware; determine necessary action. P-1
22. Disassemble and clean the caliper assembly; inspect parts for wear, rust, scoring, and damage; replace the seal, boot, and damaged or worn parts. P-2
23. Reassemble, lubricate, and reinstall the caliper, pads, and related hardware; seat the pads, and inspect for leaks. P-1
24. Inspect the vacuum-type power booster unit for vacuum leaks; inspect the check valve for proper operation; determine necessary action. P-2
25. Diagnose wheel bearing noises, wheel shimmy, and vibration concerns; determine necessary action. P-1
26. Remove, clean, inspect, repack, and install wheel bearings and replace seals; install hub and adjust wheel bearings. P-1
27. Check the parking brake cables and components for wear, rusting, binding, and corrosion; clean, lubricate, or replace as needed. P-2
28. Check the parking brake operation; determine necessary operation. P-1
29. Replace the wheel bearing and race. P-1

45.1 BRAKING SYSTEM PRINCIPLES

FRICTION

Friction is defined as a resistance to motion between two objects. When two surfaces rub against each other, there is friction (***Figure 45–1***). The amount of friction depends on two things: the roughness of the surfaces and the amount of pressure between the two surfaces.

HEAT ENERGY

When there is friction, **kinetic energy** (energy in motion) is converted to thermal (heat) energy. The greater the amount of kinetic energy that must be brought to rest, the greater the amount of heat produced. The energy of motion, or kinetic energy, depends on the weight of the vehicle and the speed of the vehicle. Brakes must also be able to remove the heat that is produced.

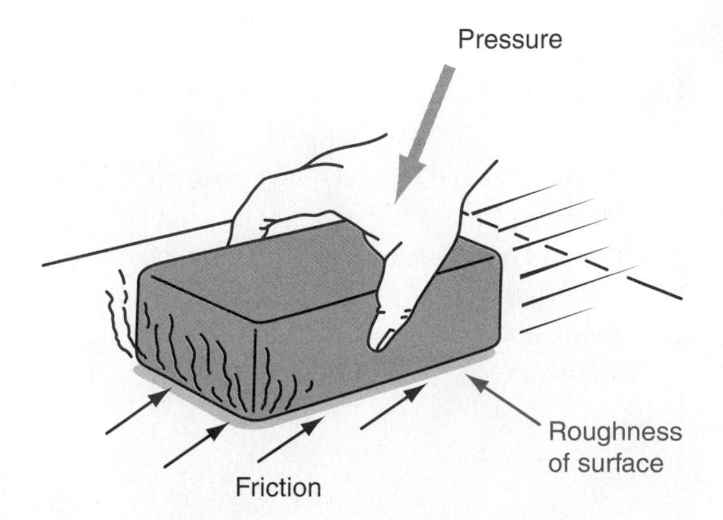

FIGURE 45–1 When two surfaces rub together, friction is produced. Brakes produce friction to stop or slow down a vehicle.

FRICTION AND BRAKING SYSTEMS

In any braking system, the amount of friction is controlled by the operator. By varying friction, the vehicle can be stopped, and its speed can be modified on curves, grades, and in different driving conditions. Control of friction is obtained by forcing a stationary brake shoe or pad against a rotating drum or disc. As the driver presses harder on the brake pedal, friction increases.

As the wheel is slowed down by the brake friction, the tire is also slowed down. However, friction is also produced between the tire and the road. The friction on the brakes must be matched by the friction of the tires and the road. If the tires on the road cannot produce the friction, the tires will lock up and skid. A car stops better if the wheels are not locked. Locked wheels can produce dangerous results, especially since there is no driver control of the friction between the tires and the road. **Computer-controlled brakes** are also being used to control the slowing down of each wheel without skidding.

BASIC OPERATION OF DRUM BRAKES

A drum brake assembly consists of a cast drum that is bolted to and rotates with the wheel. Inside the drum, there is a **backing plate** that has a set of **brake shoes** attached to it. Other components are also attached to the backing plate, including a hydraulic cylinder and several springs and linkages. The brake shoes are lined with a frictional material. The frictional material contacts the inside of the drum when the brakes are applied (***Figure 45–2***). When the brakes are applied, the brake shoes are forced out and produce friction against the inside of the drum.

SHOE ENERGIZATION

When the brakes are applied, it is important for the shoe to be **self-energizing**. When the brake shoe is engaged, the frictional drag acting around the shoe tends to rotate the shoe about its pivot point as shown in ***Figure 45–3***. When the drum rotates in the same direction, the frictional drag between the two causes the shoe to become tighter against the inside of the drum. This action is called self-energizing.

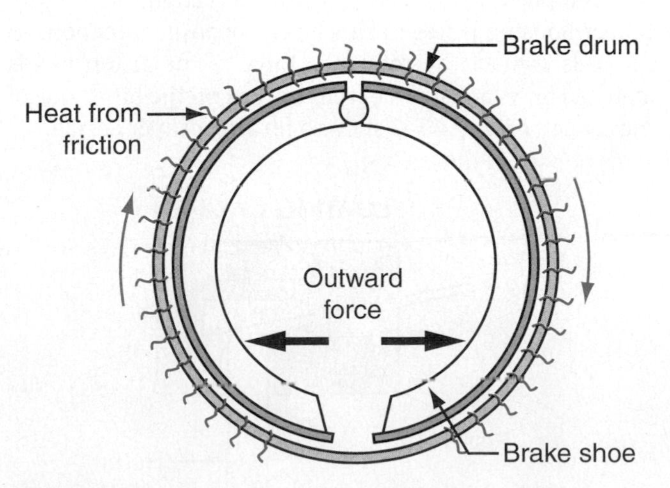

FIGURE 45–2 On a drum brake system, the shoes are forced outward against a brake drum to produce the necessary friction.

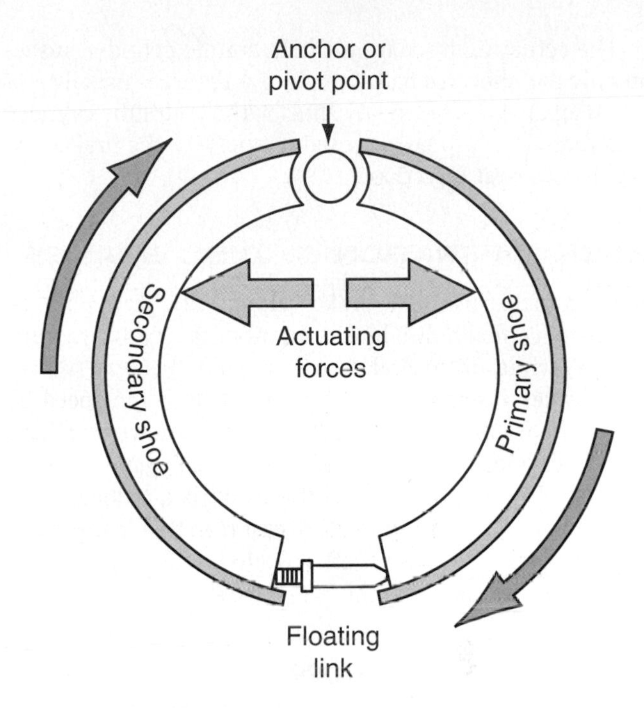

FIGURE 45–3 When the brakes are applied, the primary shoe reacts first. Then the primary shoe pushes against the secondary shoe to produce additional friction.

SERVO-TYPE BRAKES

A servo is a device that converts a relatively small force into a larger force. In most vehicles today, **servo brakes** are used to cause the brake shoes to move outward from a hydraulic pressure inside a cylinder. The pressure is produced by the operator's foot. The motion is the outward push of the brake shoes against the drum.

PRIMARY AND SECONDARY SHOE OPERATION

In a drum brake system, there are primary and secondary shoes as shown in ***Figure 45–3***. When the brakes are applied, the **primary shoe** reacts first. It has a weaker return spring. The shoe lifts off the anchor and contacts the drum surface. As the shoe begins to contact the drum, the shoe is energized, forcing it to rotate deeper into the drum, producing increased friction.

During this time, there is also action on the **secondary shoe**. As the primary shoe moves, it tends to push or move the secondary shoe at the bottom. This motion forces the secondary shoe in the same direction as the drum. Note that the secondary shoe cannot move upward because it is forced against the anchor. This causes the secondary shoe to also be energized. The servo brake system acts or behaves as if it were one continuous shoe.

The actuating force from the hydraulic cylinder pushes on only one shoe. A small amount of pressure is usually produced against the secondary shoe by the hydraulic cylinder. This force is not, however, used to energize the brakes, except in a reverse direction.

BASIC OPERATION OF DISC BRAKES

Many vehicles use **disc brakes** along with drum brakes. On many vehicles, disc brakes are used on the front of the vehicle, while drum brakes are used on the rear wheels. Disc brakes resemble the brakes used on a ten-speed bicycle. The friction is produced by pads, as shown in ***Figure 45–4***. These pads are squeezed or clamped against a rotating disc. The disc, also called the rotor, is attached to the rim and tire. The rotor is made of cast iron that is machined on both sides. The pads are attached to metal plates that are actuated by pistons from the hydraulic system.

CALIPER OPERATION

The pistons in a disc brake system are contained or held in place by the **caliper**. The caliper does not rotate because it is attached to the vehicle's frame. The caliper is a housing that contains hydraulic pistons and cylinders. It also contains seals, springs, and fluid passages that are used to produce the movement of the piston and pads.

The pads act perpendicular to the rotation of the rotor (***Figure 45–5***). This is different from the drum brake system. Disc brakes are said to be non-self-energized. This means that they require more force to achieve the same braking effort. For this reason, disc brakes are usually used with power brakes.

Fixed Caliper Design There are two types of caliper designs: the fixed caliper and the floating caliper. The **fixed caliper** design has the caliper assembly attached directly to the frame or steering components. Each pad is actuated by a piston. ***Figure 45–6*** shows the fixed caliper design.

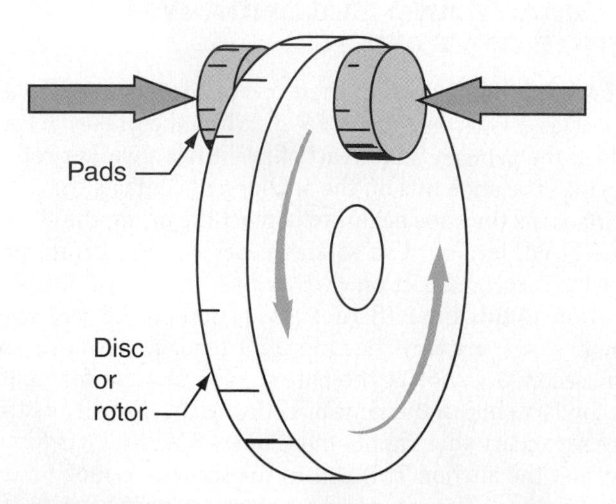

FIGURE 45-4 A disc brake system uses two pads reacting against a rotor to produce the friction necessary to stop the vehicle.

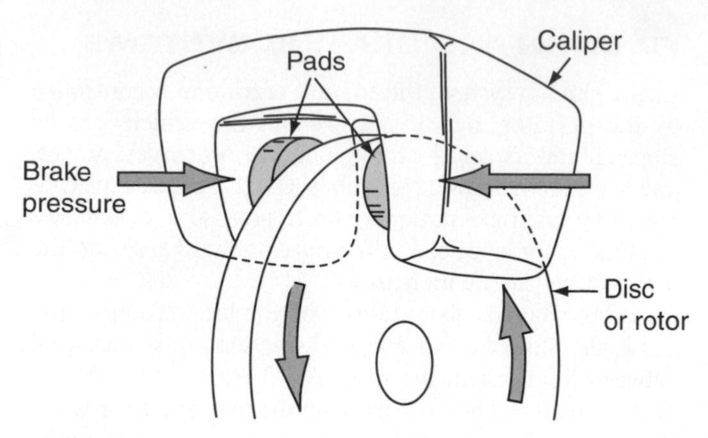

FIGURE 45-5 On a disc brake system, the pads work perpendicular to the rotor.

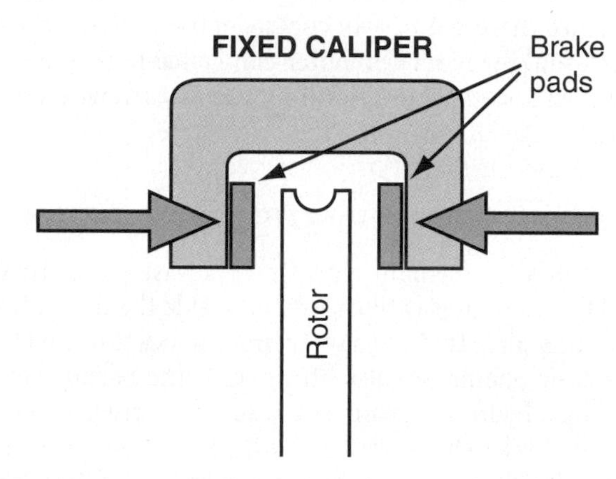

FIGURE 45-6 The fixed caliper remains stationary, and pads move in and out to produce friction. There are two pistons on this system.

Floating Caliper Design In the **floating caliper** design, the main housing of the caliper allows it to slide in and out a small amount on the mountings. There is a piston on only one side. The other has only a friction pad. When the brakes are applied, the hydraulic pressure within the cylinder pushes the piston in one direction. The entire caliper housing is free to slide in the opposite direction. As the pads contact the rotor, the force of the piston pad is matched by an equal force from the pad on the other side of the caliper. ***Figure 45–7*** shows a floating caliper design.

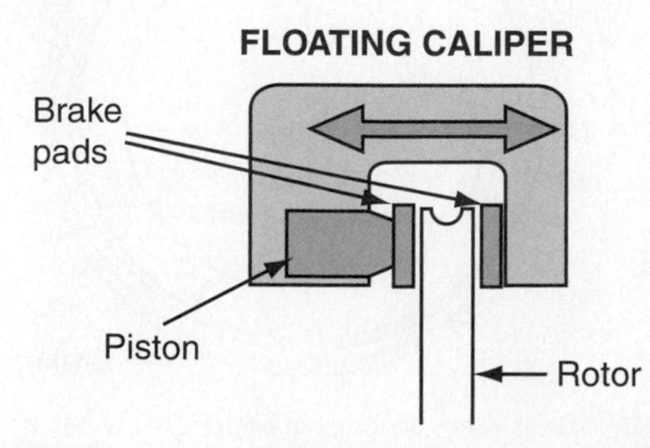

FIGURE 45-7 A floating caliper is able to slide in and out a small amount on its mountings. There is only one piston on this system.

FLUIDS

Fluids play an important part in braking systems. Brake fluid is used to transfer the motion of the operator's foot to the cylinders and pistons at each brake. Fluids cannot be compressed, while gases are compressible, as shown in ***Figure 45–8***. Any air in the brake hydraulic system will compress as the pressure increases, which reduces the amount of force that can be transmitted. This is why it is very important to keep all air out of the hydraulic system. To do this, air must be removed from the brakes. This process is called **bleeding** the brake system.

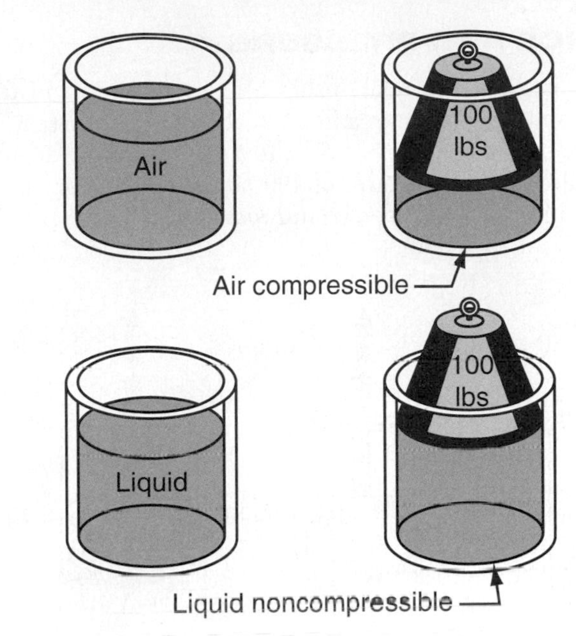

FIGURE 45–8 Hydraulic fluid is used in brake systems because it is noncompressible. Air, however, is compressible and must be removed from the hydraulic system.

HYDRAULIC PRINCIPLES

The automotive braking system uses hydraulic pressure to transfer the force of the operator's foot to press the friction surfaces together. In ***Figure 45–9***, when the foot pedal is pressed, a pressure is built up in the master cylinder. This pressure is then transferred throughout the hydraulic lines to each wheel cylinder. Note that the pressure at each point in the system is the same.

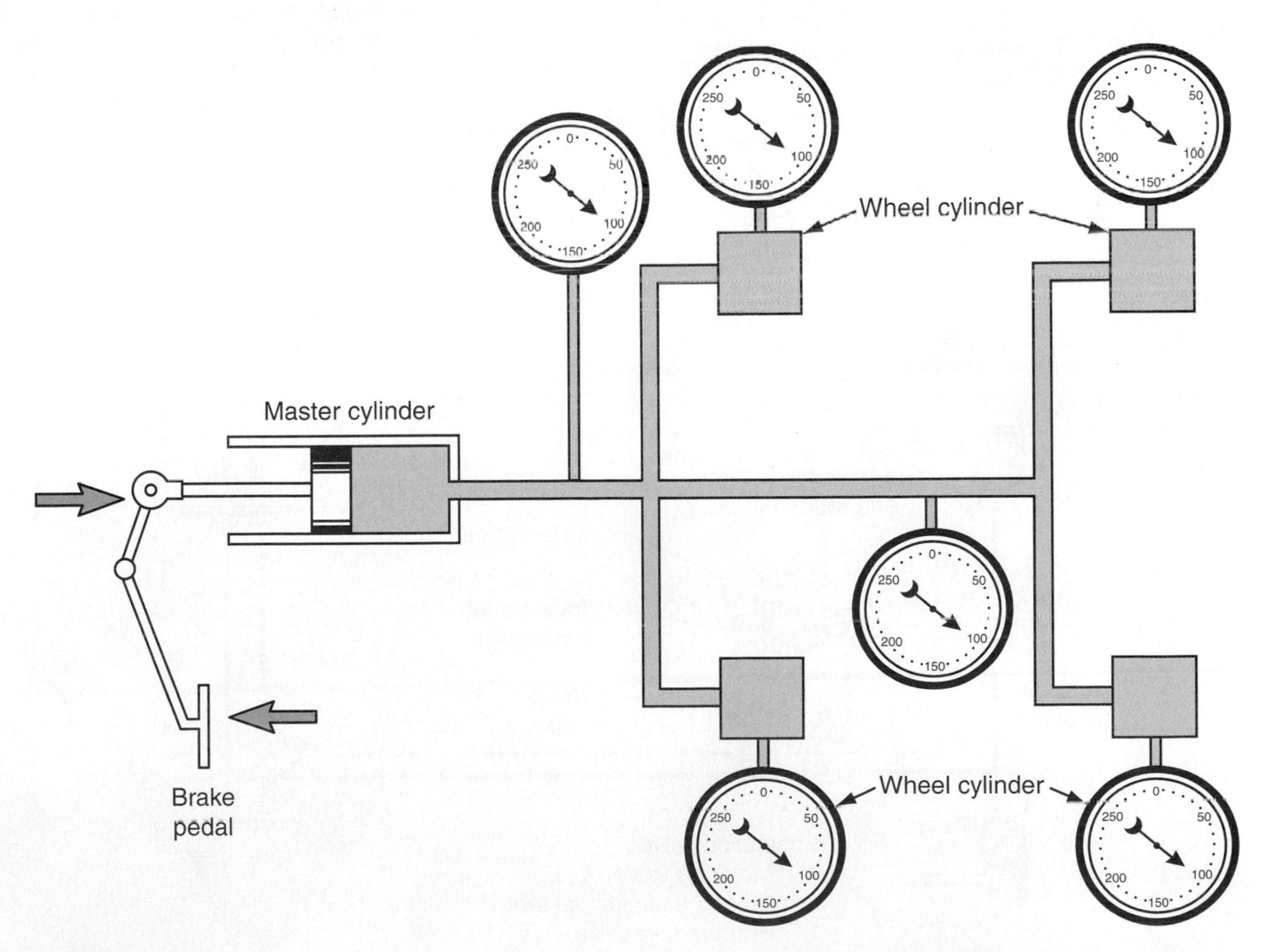

FIGURE 45–9 When a driver pushes the brake pedal, hydraulic pressure increases and is sent to each wheel cylinder to operate the brake mechanism.

FORCE AND PRESSURE

There is a specific relationship between the force of the pedal and the piston area in a closed **hydraulic system**. If a force of 100 pounds is applied to a piston with an area of 1 square inch, a pressure of 100 pounds per square inch is produced. Also, as shown in ***Figure 45–10***, if there are other pistons in the hydraulic system, they may produce different pressures because of their size. A 1/2-square-inch piston produces 50 pounds of force. The 1-square-inch piston produces 100 pounds of force. The 2-square-inch piston produces 200 pounds of force. This example shows that a certain force applied to a hydraulic system can produce different forces, depending on the piston size.

Very little movement of fluid occurs in the hydraulic system. It is the pressure and the forces that do the job needed in the braking system. In actual practice, the fluid is used only to transfer the force and pressure of the operator's foot to the piston and friction pads.

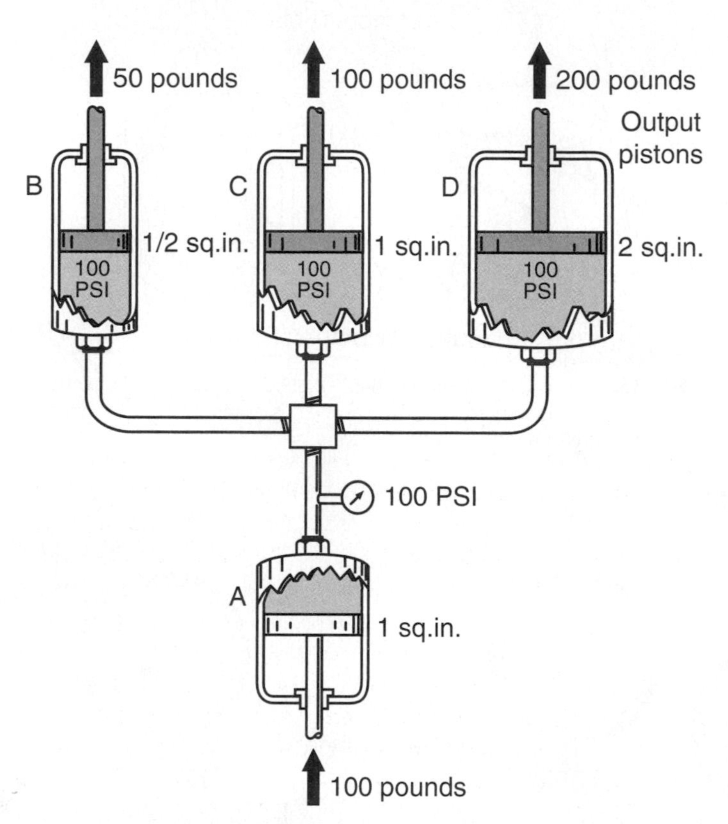

FIGURE 45–10 Pressure in the hydraulic system can be changed by varying the size of the piston.

45.2 BRAKING SYSTEM COMPONENTS AND OPERATION

TOTAL SYSTEM OPERATION

A common brake system is shown in ***Figure 45–11***. In this system, drum and disc brakes are used. The system starts at the brake pedal, which is attached to the **master cylinder**. The master cylinder is used to produce the necessary pressure in the hydraulic system. Hydraulic lines are connected from the master cylinder, through the combination (metering) valve to the individual wheels. Here the hydraulic pressure is sent to each **wheel cylinder**, which finally moves the drum or disc brake mechanism.

In addition to the hydraulic system, all vehicles have a parking brake system that uses mechanical linkage to the rear wheels. Either a pedal mechanism or a hand parking brake can be used.

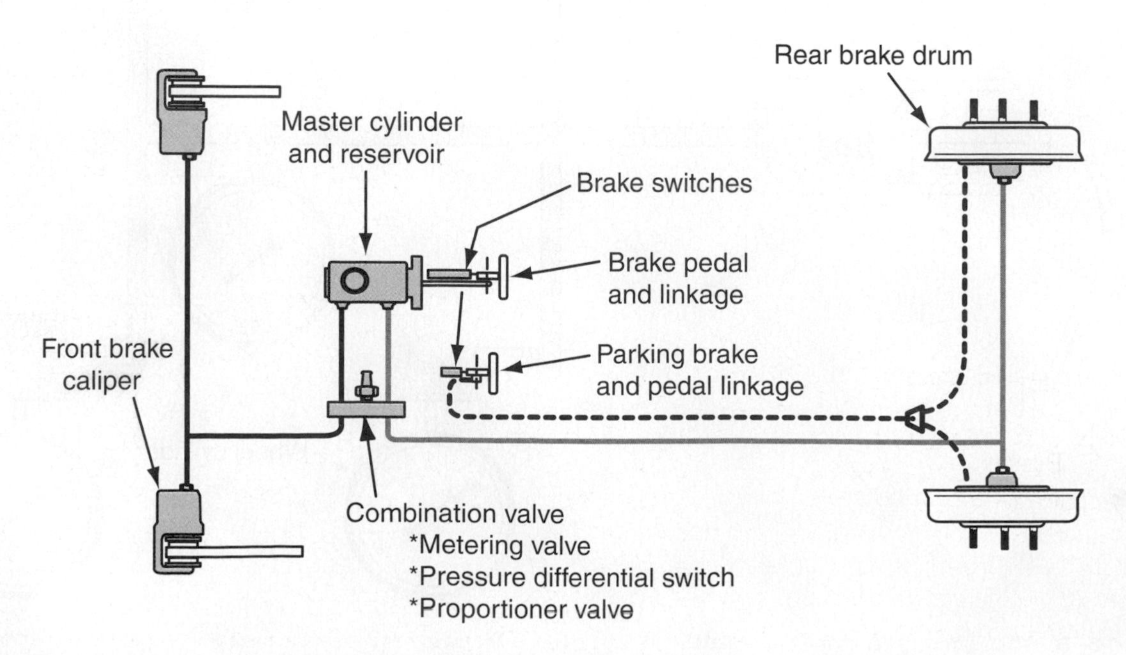

FIGURE 45–11 A common brake system and the major parts are shown on the vehicle.

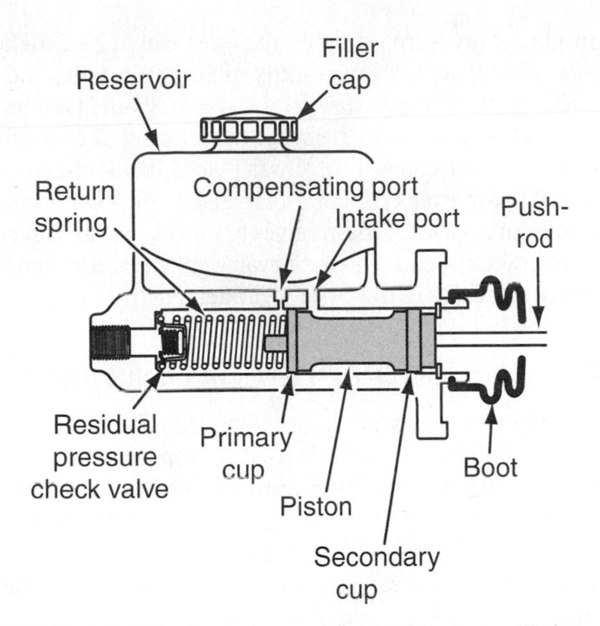

FIGURE 45–12 The main components of a master cylinder.

MASTER CYLINDER OPERATION

The purpose of the master cylinder is to convert the mechanical force of the operator's foot to hydraulic pressure. The main components of the master cylinder are shown in ***Figure 45–12***. Although many designs are used for master cylinders, the principles remain the same. The important parts include:

- Pushrod, which is moved by the movement of the operator's foot
- Piston, which produces the pressure
- Primary and secondary sections (cups), which help produce the pressure
- Return spring, which returns the pedal after braking
- Compensating and intake ports, which enhance the speed of the fluid flow

PARTS LOCATOR

Refer to photo #97 in the front of the textbook to see what a ***master cylinder*** *looks like.*

FORWARD STROKE

When the operator pushes the brake pedal, the pushrod and piston move to the left as shown in ***Figure 45–13***. This action causes the pressure on the left side of the piston to be increased. Pressure is only produced when the primary cup passes the compensating port. The vacuum produced on the backside is relieved because brake fluid is drawn in from the reservoir above the piston through the intake port. The reservoir is also needed because as brake pads and linings wear, there is more fluid displacement during braking. The reservoir holds this extra amount of fluid when needed.

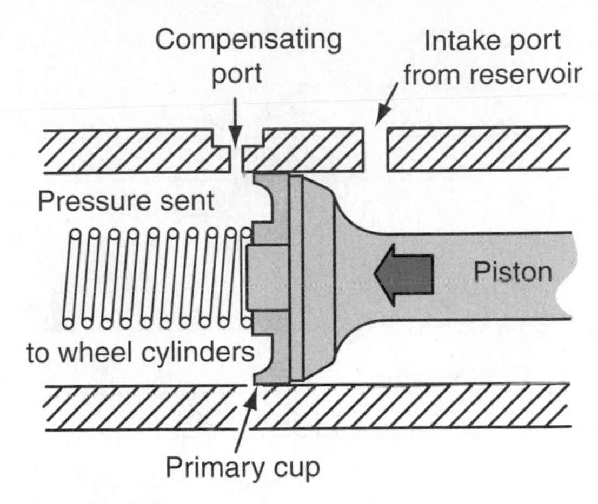

FIGURE 45–13 Once the pushrod moves past the compensating port, pressure starts to build up. This pressure is then sent to the wheel cylinders.

RETURN STROKE

During the return stroke, the brake pedal is pushed back to its original position. The return spring is used to move the piston back. During this action, a low pressure is created on the left side of the piston (***Figure 45–14***). The piston moves back faster than the fluid coming from the brake lines. If the operator immediately reapplied the brakes, there would not be enough fluid for them to operate correctly. In order to remedy this, fluid must be able to flow from the secondary to the primary port of the master cylinder.

The shape of the primary cup allows this to happen. ***Figure 45–15*** shows that as the piston moves to the right during the return stroke, a certain amount of fluid passes

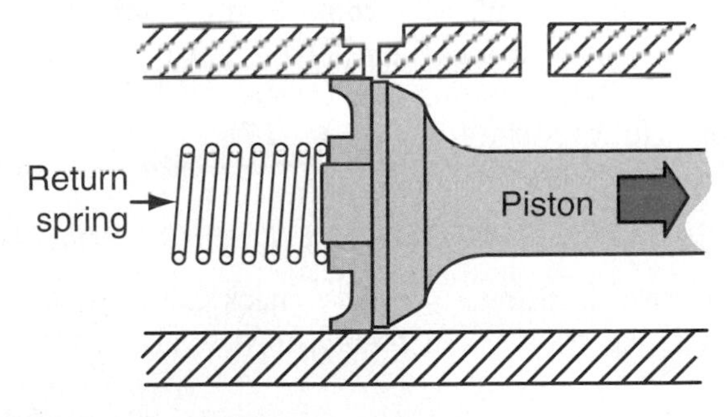

FIGURE 45–14 The return spring is used to push the master cylinder piston back to its original position.

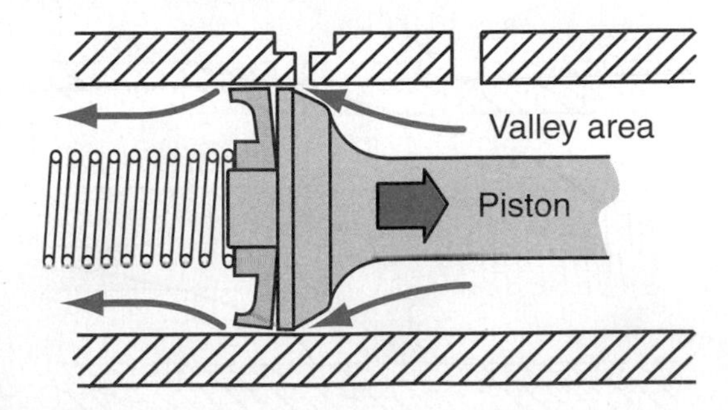

FIGURE 45–15 As the piston moves back during the return stroke, a small amount of brake fluid will pass around the primary cup to equalize the pressure.

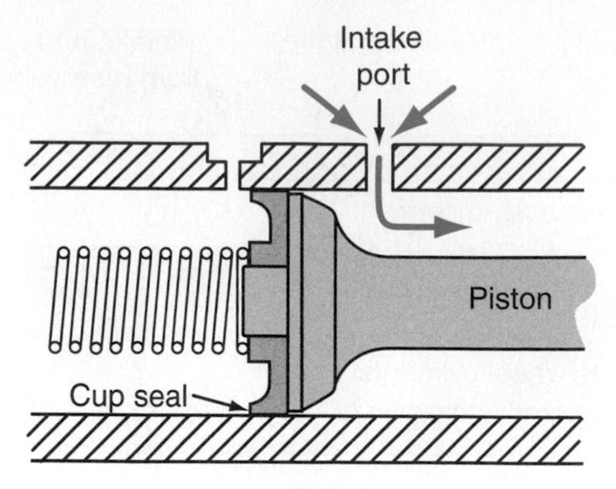

FIGURE 45-16 Any extra brake fluid needed on the right side of the primary cup will be admitted through the intake port.

around the primary cup. This can happen only when there is a lower pressure on the left side of the cup. The piston cup serves as a one-way valve. Under these conditions, the piston, pushrod, and brake pedal will return very quickly, allowing for successive rapid brake strokes.

To make up for the extra brake fluid passed to the left of the piston, a certain amount of fluid is drawn from the reservoir into the secondary area of the piston (***Figure 45–16***).

COMPENSATING PORT

Now that the piston has been fully returned, the fluid from the brake lines will continue to enter the area left of the piston. This area is now, however, full of brake fluid from the return stroke. If there were no place for the fluid from the lines to go, the brakes would not release as necessary. Another passage, called the **compensating port**, allows the excess fluid to return to the reservoir when the pedal is released. The compensating port is uncovered only when the piston is fully returned (***Figure 45–17***).

RESIDUAL PRESSURE CHECK VALVE

In the past, a **residual pressure** check valve was used on drum brake systems. In theory, it was felt that a small amount of pressure on the brake lines would be beneficial. This could reduce the possibility of air getting into the system. Also, a slight pressure on the system would take up any slack in the linkages in the brake mechanism at each wheel. A small pressure check valve was placed inside the primary area of the master cylinder. Since disc brakes don't employ shoe return springs, these valves are omitted. Also, in some drum brake systems, the check valve has been eliminated by improved design of the wheel cylinders.

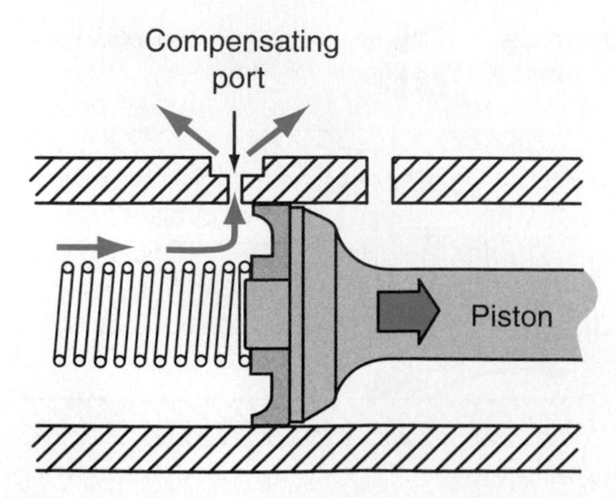

FIGURE 45-17 The compensating port allows excess fluid to return to the reservoir when the pedal is released.

RESERVOIR DIAPHRAGM GASKET

A flexible rubber gasket is used between the reservoir and the master cylinder cap. It is used to stop moisture and dirt from getting into the brake fluid reservoir. The reservoir must be vented to the atmosphere because of the rising and falling of the brake fluid level during brake operation. The diaphragm gasket separates the brake fluid from the air above it, while remaining free to move up and down with fluid level changes. ***Figure 45–18*** shows a typical reservoir diaphragm gasket.

DUAL MASTER CYLINDER

The dual master cylinder provides two separate and distinct pressure chambers in one bore. This design was required by federal law in the late 1960s. Should a failure occur in one master cylinder piston, the second piston will still work. ***Figure 45–19*** shows a typical dual master cylinder in the

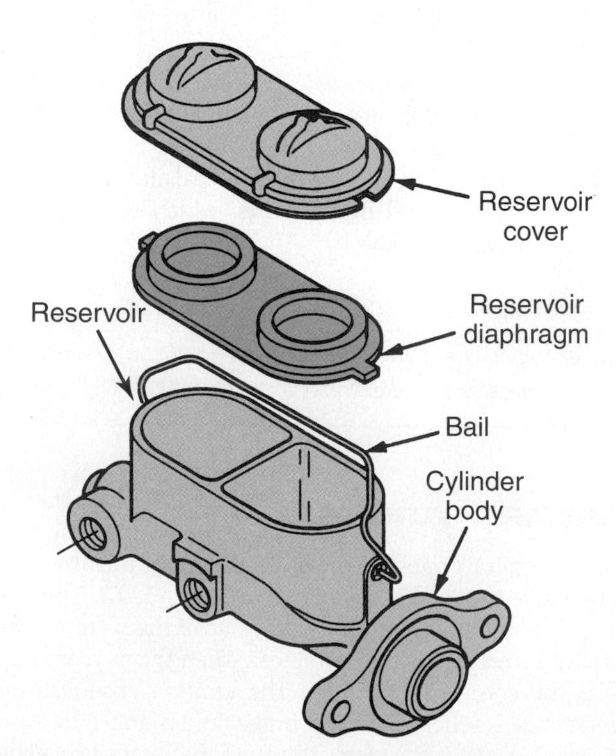

FIGURE 45-18 The reservoir diaphragm gasket keeps dirt out of the reservoir and allows the brake fluid to rise and fall during normal braking operation. It is also vented to the atmosphere. (Courtesy of Federal-Mogul Products, Inc.)

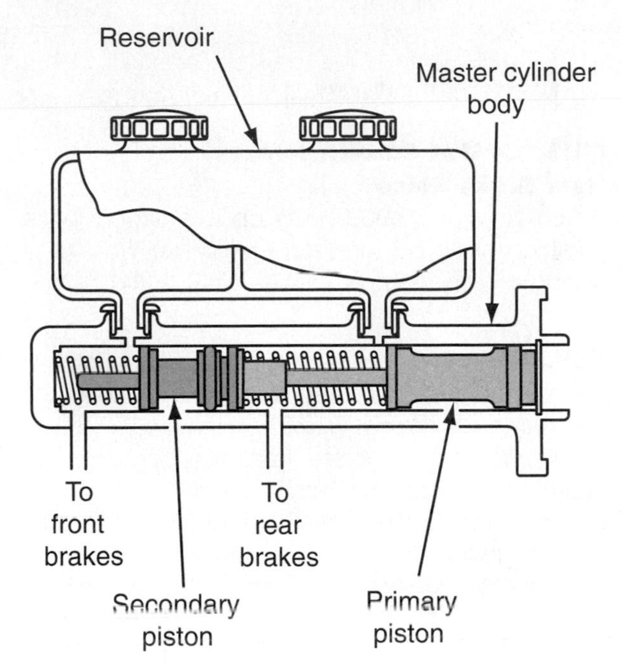

FIGURE 45–19 The dual master cylinder has both a primary and secondary piston to produce pressure. If one master cylinder section fails, the other will stop the vehicle.

applied position. One chamber is used for the front brakes, and the other chamber is used for the rear brakes. Note that on this system the by-pass holes act as the compensating port. The operation of the dual master cylinder is the same as that of the single cylinder just described, except that two cylinder pressures are being developed. One cylinder is actuated by the pushrod. The second cylinder is operated by a spring and the "plug" of fluid between the two. This system is sometimes called a **tandem master cylinder**.

DIAGONAL BRAKE SYSTEM

Another variation in brake systems is the diagonal brake system. In the diagonal system, the right-front and left-rear brakes are connected to one chamber of the master cylinder. The left-front and right-rear brakes are connected to the other master cylinder chamber. The purpose of this system is again to make sure that there is still braking on two wheels if the master cylinder fails to work.

ADDITIONAL BRAKE SYSTEM COMPONENTS

Four other components are used on brake fluid systems, depending on the type of vehicle and the year the vehicle was manufactured. They include the quick-take-up master cylinder, warning lights switches, and proportioning and metering valves.

1. Quick-take-up master cylinder—This system has a master cylinder with two different bore sizes. Its purpose is to displace a larger amount of brake fluid during the initial stages of brake pedal movement. This helps to take up shoe return spring linkage more quickly.
2. Warning light switches—A warning light switch is cast directly into the master cylinder. This switch senses the pressure in the master cylinder. If half of the system has failed, the pressure in the other half will be greatly increased. This pressure trips an electrical switch and lights up a warning light.
3. Proportioning valve—The **proportioning valve** is used to proportion the pressure to the rear brakes and the front brakes. It is located in the brake line after the master cylinder. The harder the brakes are applied, the more weight shifts to the front of the vehicle. If the pressure is equal to all wheels, the back wheels may lock up, causing loss of vehicle control. As the force on the brake pedal increases, the proportioning valve causes the pressure to the rear brakes to be less than to the front brakes. This action reduces the possibility of rear wheel skidding.
4. Metering valve—The metering valve is used on systems that have both disc and drum brakes. The metering valve keeps the front discs from operating until the rear drums have started to work. This is needed because the disc system operates faster than drum brakes. In operation, the fluid to the front disc brakes must go through the metering valve. The metering valve acts like a regular valve. It holds back the fluid to the front brakes until a certain amount of pressure has been developed. When this pressure is reached, the metering valve opens, and the system operates normally.

BRAKE LINES

Brake lines are used to carry brake fluid and pressure from the master cylinder to the individual cylinders. Brake lines are made of double-walled, rust-resisting steel except where they have to flex. Flexing usually occurs between the chassis and the front or rear wheels. Here flexible high-pressure hoses are used. All brake lines are designed for high pressure by using double-flared ends and connectors.

BRAKE LININGS

Brake linings provide the friction against the drum to stop the car. There are many kinds of linings. The lining is attached to the shoe either by riveting or by bonding (***Figure 45–20***). The primary shoe has the shorter length lining. The secondary shoe has a full-length lining because it carries a bigger load. In addition, most brake linings used today are ground so that they are slightly thicker at the center (often called brake shoe arching). This design improves the ease with which the lining comes in contact with the drum. When the shoe pressure is increased, the lining and shoe flex slightly to produce full contact. ***Figure 45–21*** shows a typical brake lining as well as the thickest part of the lining.

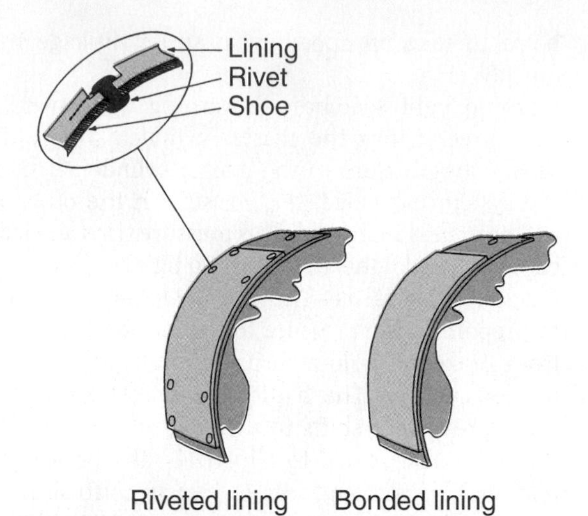

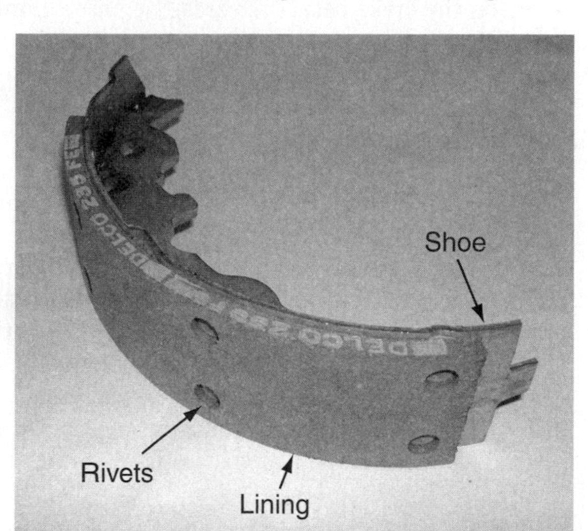

FIGURE 45-20 Both a riveted lining and a bonded lining.

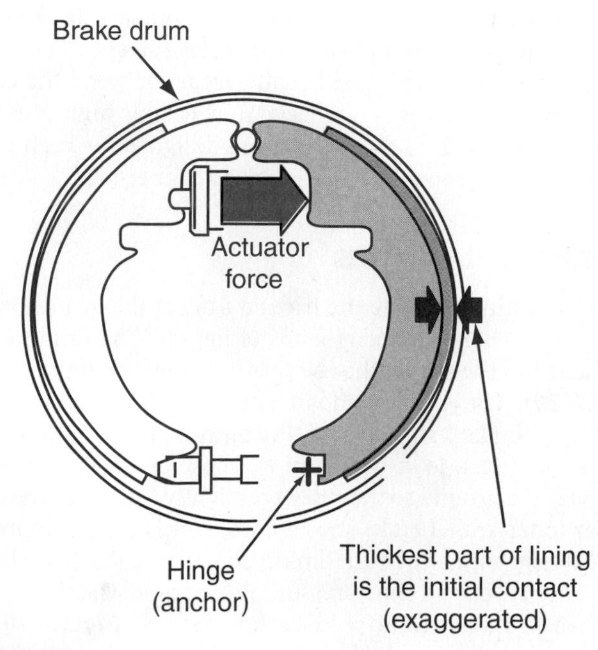

FIGURE 45-21 When a brake shoe is applied on a drum brake system, the initial point of contact is near the center.

Car Clinic: BRAKE LININGS

CUSTOMER CONCERN:

Rear Brake Noise

A vehicle with 80,000 miles on it is producing a scraping noise on the rear right brake when it is raining or very damp outside. What could be the problem?

SOLUTION: Brake linings can become glazed and covered with road dirt and grease under wet and damp conditions. Glazed linings are very hard, which can produce noises. In addition, some of the brake retainer parts and mechanisms may be worn or locked up. Remove the rear brake and check all parts, including the drums, for wear and correct operation. Also, consider replacing the brake linings.

DRUM BRAKE WHEEL CYLINDER

The purpose of the wheel cylinder is to convert hydraulic pressure to mechanical force. ***Figure 45–22*** shows a typical wheel cylinder kit for a drum brake. The wheel cylinder kit contains the parts that are replaced when the wheel cylinder is serviced. The assembly includes two pistons, two cups, two boots, a bleeder screw, and an internal spring. When two pistons are used, it is called a duo servo system. When the brakes are applied, hydraulic pressure inside the wheel cylinder forces both pistons outward, causing the brakes to be applied.

PARTS LOCATOR

*Refer to photo #98 in the front of the textbook to see what a **drum brake wheel cylinder** looks like.*

DUO SERVO DRUM BRAKE ASSEMBLY

There are many variations on the drum brake assembly. One common type, the duo servo drum brake, is shown in ***Figure 45–23***. The parts include:

- Primary shoe—produces friction for stopping (the forward shoe)
- Secondary shoe—produces friction for stopping (the rear shoe)
- Return springs—pull the shoes back away from the drum after the brakes have been released
- Wheel cylinder—produces the mechanical motion to move the brake shoes
- Holddown spring and cup—hold the brake shoes against the backing plate

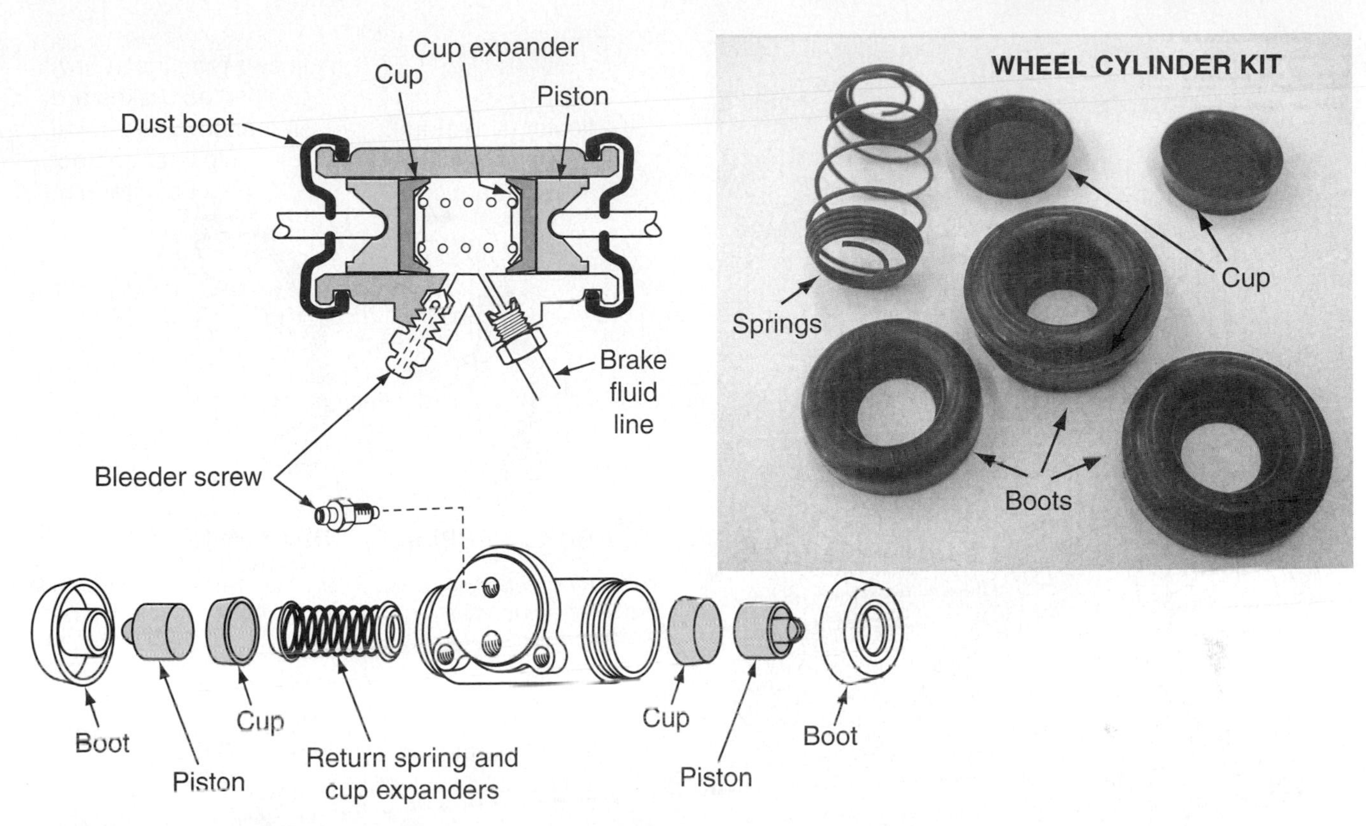

FIGURE 45-22 The standard wheel cylinder, with each part identified, is shown along with a wheel cylinder kit.

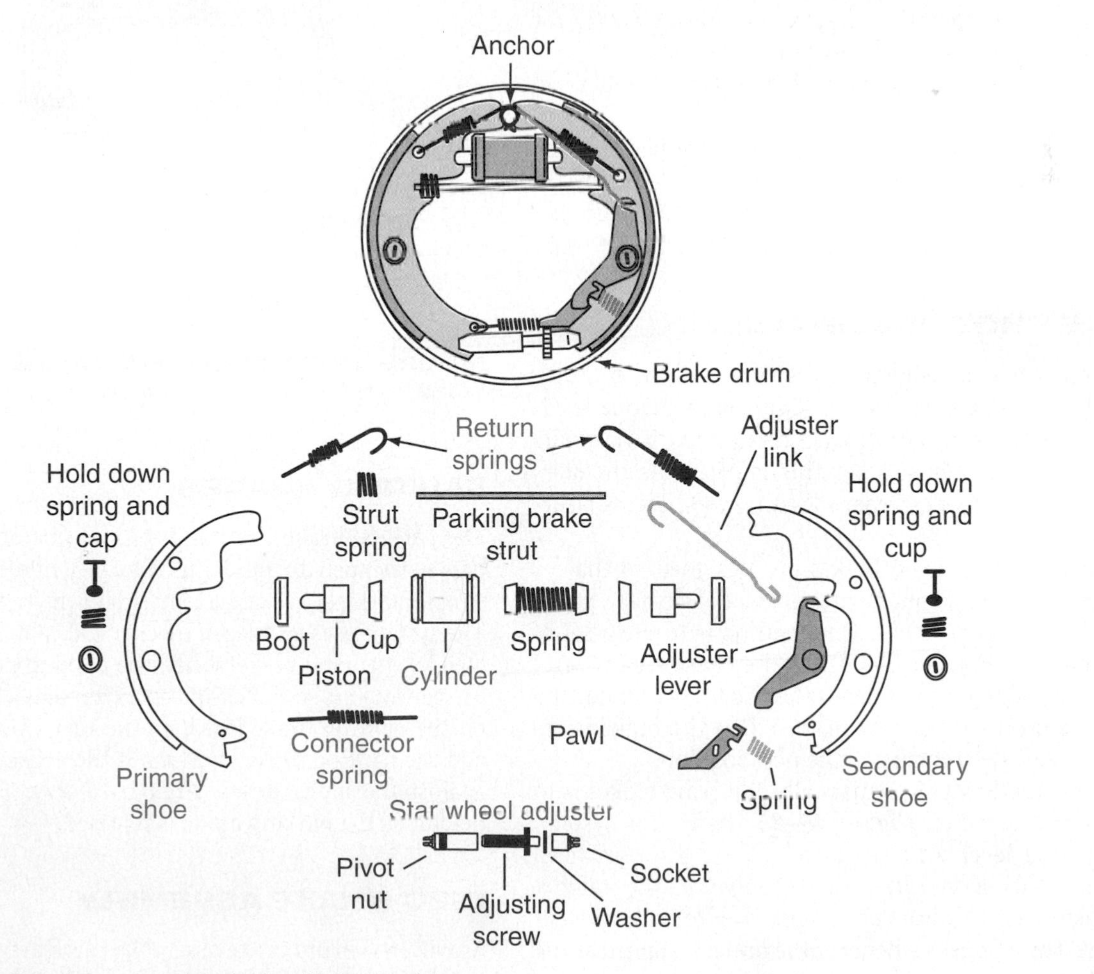

FIGURE 45-23 A duo servo drum brake with its associated parts.

CUSTOMER CONCERN:
Car Pulls to Left
Recently, a vehicle has been pulling to the left when the brakes are applied. The brake fluid seems to be down a little. The vehicle has drum brakes on both the front and rear and has 56,000 miles on it. What could be the problem?

SOLUTION: The most likely cause of this problem is that the front wheel cylinder on the left side is leaking. The brake fluid leaking out of the wheel cylinder is getting on the brake shoes and drum. The heat from the friction of the brakes causes the brake fluid to become very sticky. This causes the left front wheel to brake more than the right, causing the vehicle to pull to the left. Remove, service, and replace the front wheel cylinder with a new wheel cylinder kit to eliminate the problem.

- Anchor—used for self-energization and as a stop for the brake shoes
- Connector spring—holds the brake shoes together on the bottom
- Starwheel adjuster—adjusts the distance between the brake shoe linings and the drum
- Brake drum—absorbs the friction produced by the brake shoes and reduces the speed of the wheels

BRAKE SHOE ADJUSTMENT

Drum brakes may be adjusted either manually or by automatic adjusters to compensate for lining wear. Manually adjusted brakes have an adjusting screw, which is normally called a starwheel adjuster, for this purpose. As the starwheel is turned by an external adjusting tool, excess clearance is removed.

Automatically adjusted brakes are designed so that as the shoes move in and out during normal operation, excess lining clearance is removed. One method is to use a set of eccentric cams (***Figure 45–24***). As the brake shoes travel outward, the adjuster pin follows the shoe. This rotates the adjuster cam on the backing plate. When the brake is released, the adjuster remains in the new position.

A second method of automatically adjusting brakes is to use a ratchet adjuster. ***Figure 45–25*** shows this system. The adjusting lever acts like a ratchet on the starwheel. Each time the brake shoe moves outward, the ratchet mechanism tries to advance the starwheel to make the adjustment. This happens whenever the brake is applied as the vehicle is moving in reverse.

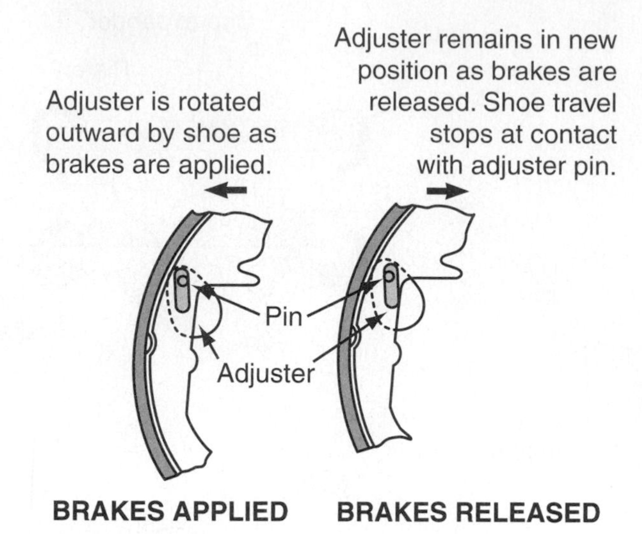

FIGURE 45–24 A set of eccentric cams is used on some brake systems to adjust the brakes automatically. (Courtesy of Honeywell International, Inc.)

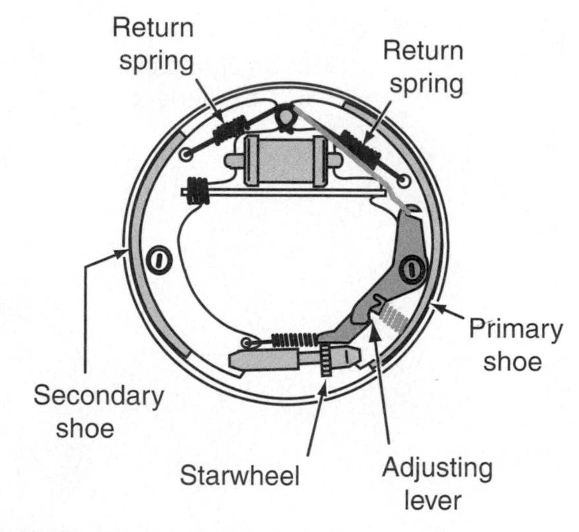

FIGURE 45–25 Certain types of brake systems use an adjusting lever that works against the starwheel to adjust the brakes.

PARKING BRAKES

The parking brake is a hand- or foot-operated mechanical brake designed to hold the vehicle while it is parked. A simple parking brake system is shown in ***Figure 45–26***. The system uses a series of mechanical cables that are operated by the hand brake. When the parking brake is applied, the parking brake cables and equalizer apply a balanced pull on the parking brake levers in the rear wheels. The levers and the parking brake strut move the brake shoes outward against the brake drum (***Figure 45–27***). This position is held until the parking brake is released.

DISC BRAKE ASSEMBLY

As with the drum brake assembly, there are many arrangements for disc brake assemblies. ***Figure 45–28*** shows

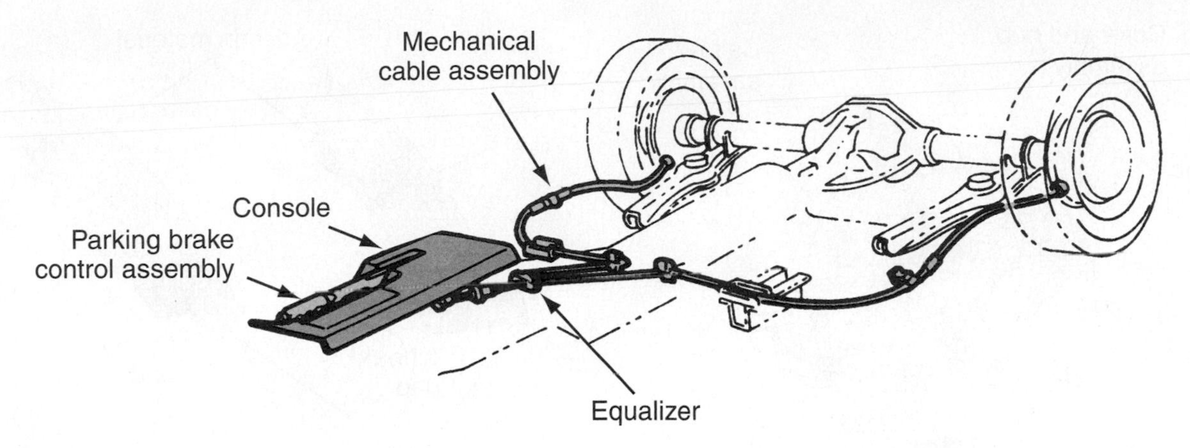

FIGURE 45-26 A typical integral parking brake system.

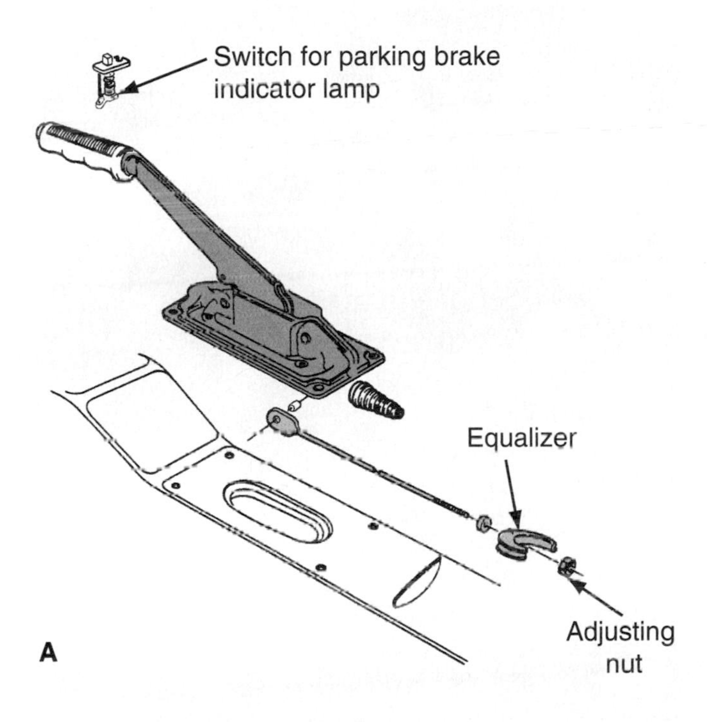

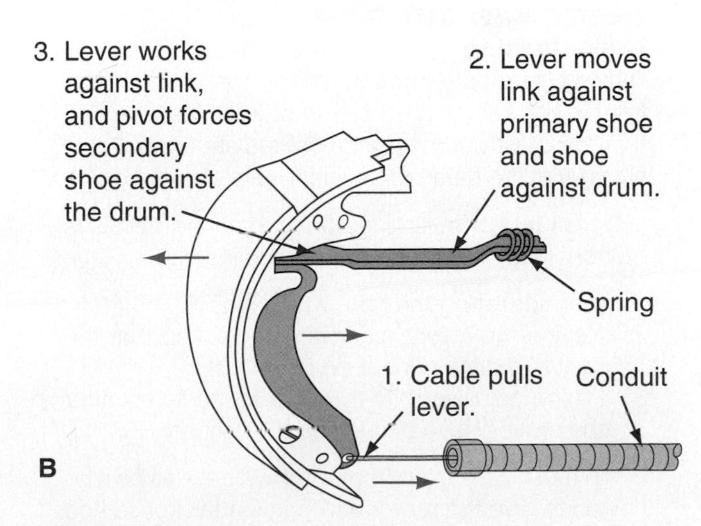

FIGURE 45-27 The parking brake lever works against the secondary shoe. (Courtesy of Volkswagen of America, Inc.)

a floating-caliper type of disc brake assembly. The parts include:

- Inboard and outboard pad—produce the friction against the rotor
- Rotor—attached to the wheel and used to absorb the friction to slow down the wheel
- Piston—produces the pressure from the hydraulic system to force the shoes against the rotor
- Piston seal—seals the brake fluid inside the piston bore
- Boot—keeps dust and dirt out of the piston bore
- Mounting bracket—holds the assembly on the vehicle
- Bleeder screw—used to remove air from the hydraulic fluid

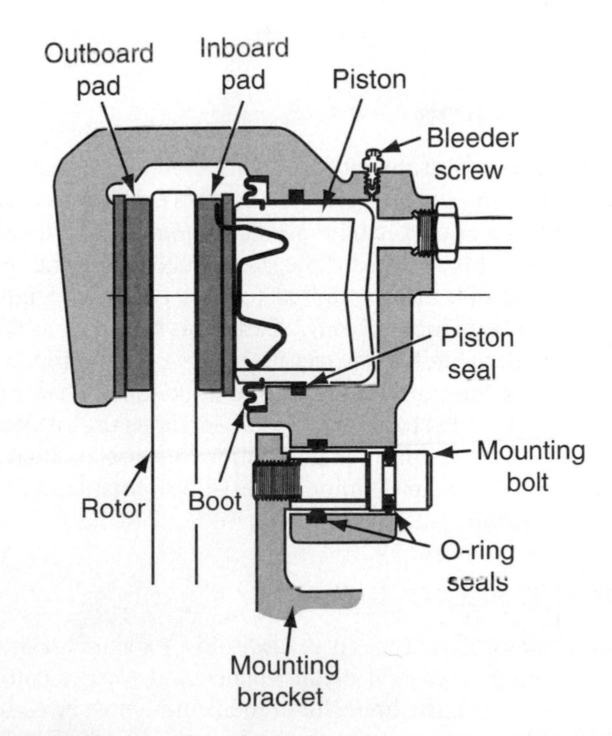

FIGURE 45-28 A floating-caliper disc brake assembly, with all parts identified.

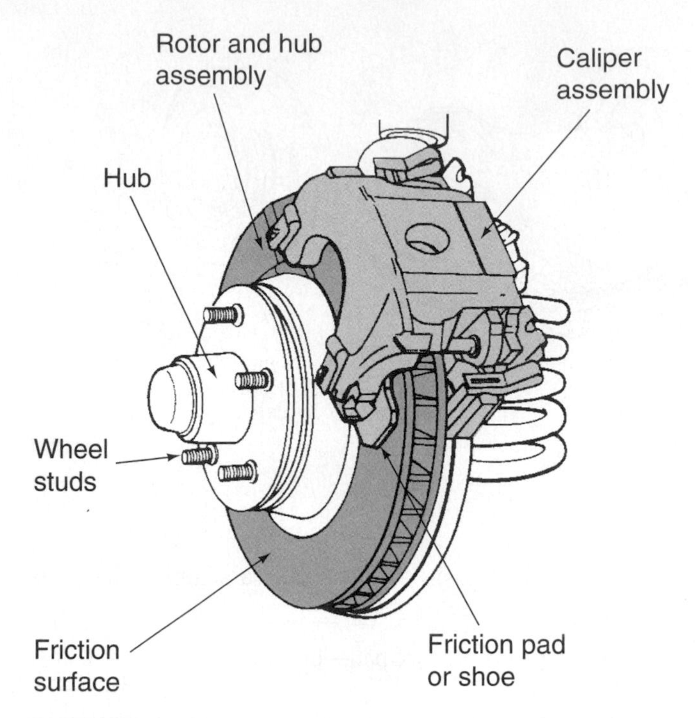

FIGURE 45–29 A complete assembly of a disc brake system and discs.

Although the parts on other vehicles may look different, the principles of operation are still much the same. ***Figure 45–29*** shows a complete assembly of a disc brake system.

PARTS LOCATOR

*Refer to photo #99 in the front of the textbook to see what a **disc brake assembly** looks like.*

BRAKE PADS

As with brake shoes, many designs are used for brake pads. Basically, a brake pad is a steel plate with a friction material lining bonded or riveted to its surface. ***Figure 45–30*** shows a typical set of brake pads. Note the metal backing and the friction material. Also, note that there is typically an outboard pad and an inboard pad. Asbestos has been used as the friction material in the past because of its excellent friction qualities and long service life. However, the health hazards of asbestos have led to a drastic reduction in its use, or even its removal from most brake friction materials. Today's basic types of disc pad lining material are organic, semi-metallic, metallic, and synthetic.

BRAKE FLUID

A wide variety of materials is used within a standard braking system. Several types of metal, rubber, and plastics come into contact with the brake fluid. Brake fluid must be compatible with all materials in the brake system and maintain stability under varying conditions, both in temperature and

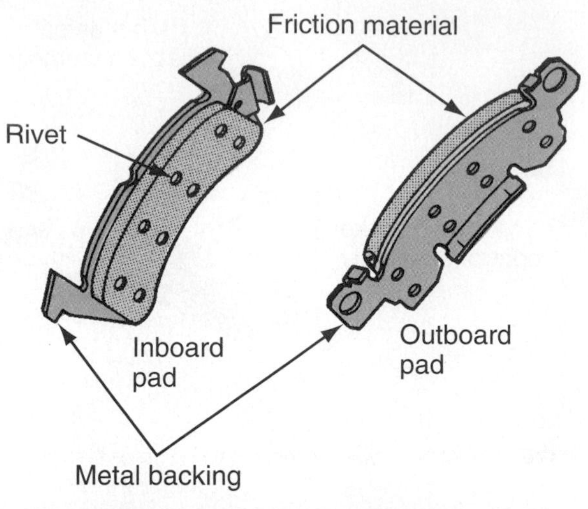

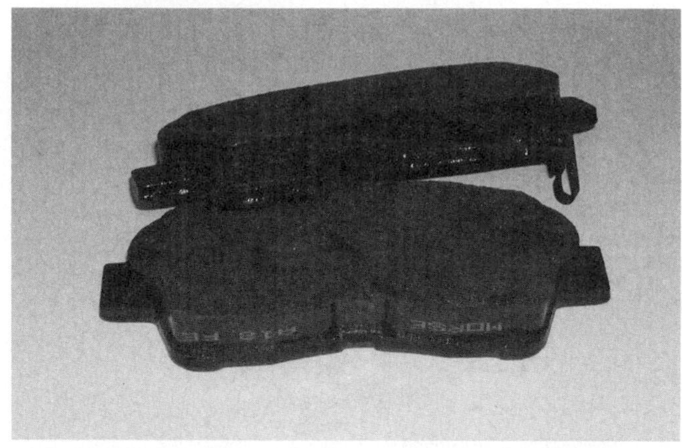

FIGURE 45–30 Both the inboard and outboard disc pads are shown here.

Car Clinic: WARPED ROTORS

CUSTOMER CONCERN:
Brake Pulsing

What is the most common reason the rotors on the front disc brakes warp? A customer has noticed that there is a pulsing as the brakes are applied or a shimmy of the steering on light brake application.

SOLUTION: Several factors can cause the discs on front disc brakes to warp. These include:

1. Overtightening the front wheel lugs
2. Having the rotors machined too far so that the heat of friction cannot be removed
3. Hitting cold water in the road immediately after the brakes have been used excessively

In most cases, the front rotor will have to be machined to give the motor a true, even thickness and eliminate runout.

pressures. Because of these conditions, brake fluid must possess the following six characteristics:

1. Viscosity—must be free-flowing at all temperatures.
2. High boiling point—must remain liquid at high operating temperatures without vaporization.
3. Noncorrosive—must not attack metal, plastic, or rubber parts.
4. Water tolerance—must be able to absorb and retain moisture that collects in the system. This characteristic is called hygroscopic. Water causes pitting in the brake system.
5. Lubricating ability—must lubricate pistons and cups to reduce wear and internal friction.
6. Low freezing point—must meet a certain freezing point as established by Federal Motor Vehicle Safety standards.

It is best to refer to the vehicle manufacturer's recommendations to determine the exact type of brake fluid to use. The Department of Transportation (DOT) specifies brake fluid for vehicles. Manufacturers recommend a specific DOT specification. For example, when a vehicle has a combination of disc and drum brakes, generally a DOT-3 is recommended. DOT-4 is recommended for vehicles with four-wheel disc brakes. Heavy-duty applications may require a silicon-type DOT-5 brake fluid (***Figure 45–31***).

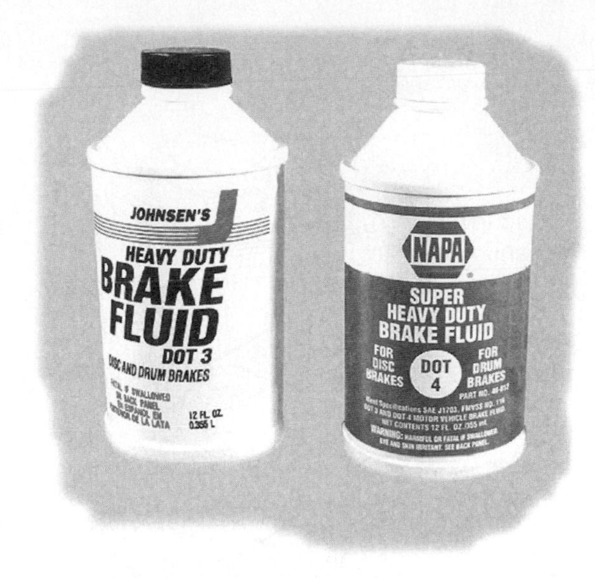

FIGURE 45–31 Always use the recommended brake fluids in the hydraulic system.

BRAKE LIGHT SWITCHES

The brake light switch is a spring-loaded electrical switch that comes on when the brake pedal is depressed. There are generally two types of switches.

1. Mechanically operated switches—used on most recent model vehicles
2. Hydraulically operated switches—used on older vehicles

The mechanically operated brake light switch is operated by contact with the brake pedal. It is usually attached to a bracket on the brake pedal. The hydraulic switch is operated by hydraulic pressure developed in the master cylinder. In both types, there is no electrical current through the switch when the brakes are not being applied. When the brakes are applied, the circuit through the switch closes and causes the brake light to come on. ***Figure 45–32*** shows examples of various mechanically and hydraulically operated switches. Note that these are special sockets for removal and installations.

PARTS LOCATOR

*Refer to photo #100 in the front of the textbook to see what a **brake light switch** looks like.*

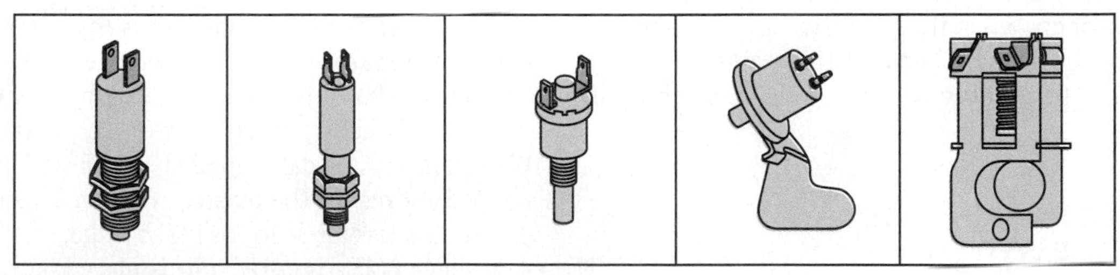

MECHANICALLY OPERATED SWITCHES

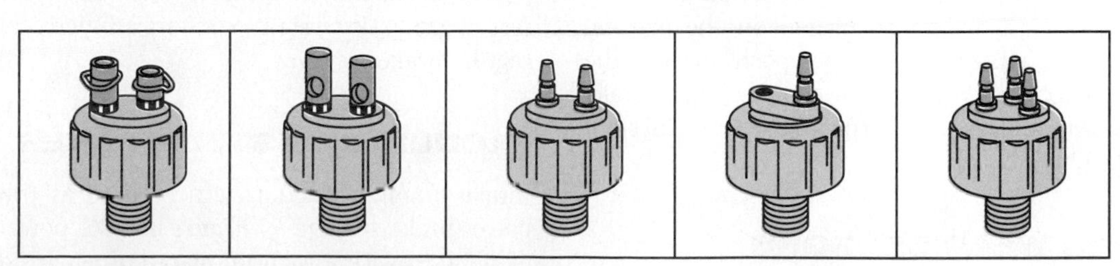

HYDRAULICALLY OPERATED SWITCHES

FIGURE 45–32 Many brake light switches are used. Both the mechanically and hydraulically operated switches are used, depending on the year that the vehicle was manufactured. (Courtesy of Honeywell International, Inc.)

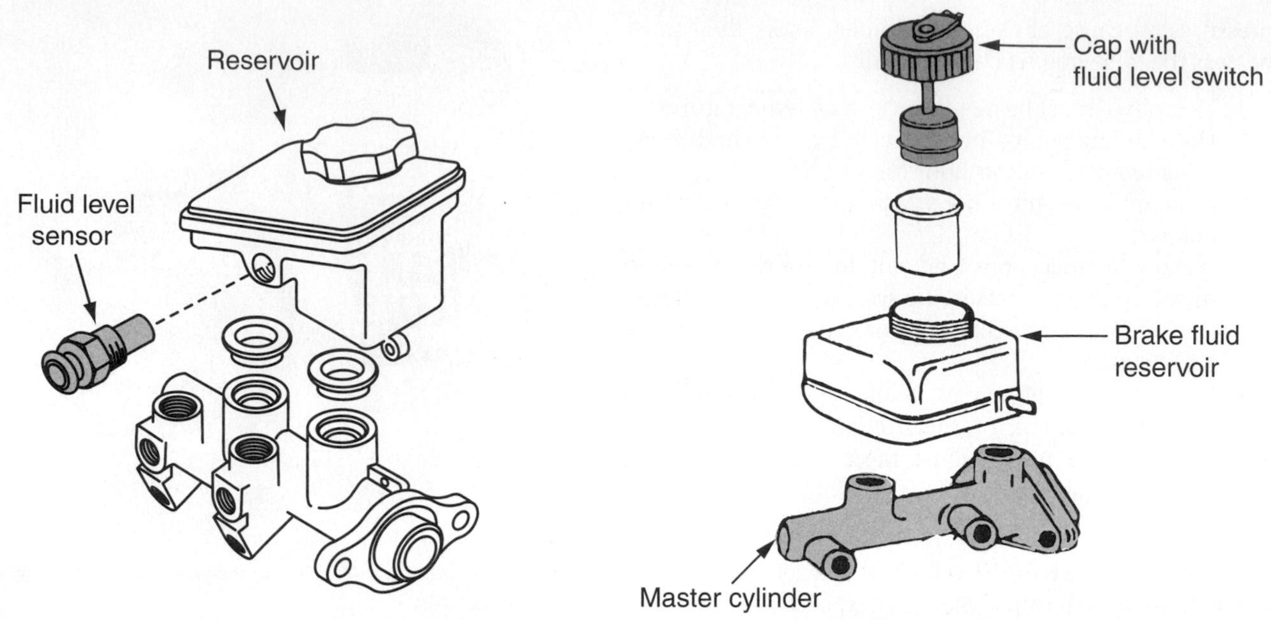

FIGURE 45–33 Fluid level sensors are most often located in the brake fluid reservoir assembly.

MASTER CYLINDER FLUID LEVEL SENSOR

On some vehicles there is a master cylinder fluid level sensor. The purpose of this sensor is to tell the operator of the vehicle that the master cylinder fluid level is getting low and should be serviced. One type of fluid level sensor is shown in ***Figure 45–33***. This sensor is placed inside the master cylinder reservoir assembly. The sensor can also be located on top of the fluid level cap. In each case, the sensor uses a float mechanism that closes a set of contacts when the fluid level is low. ***Figure 45–34*** shows an example of the brake indicator electrical circuit. In this case, the brake fluid level switch is located on the bottom of the circuit. When brake fluid is at the correct level in the reservoir, the switch remains open. When the fluid level drops to a predetermined level, the switch closes. At this point, the Brake System light on the dashboard in the gauge assembly will light up.

PARKING BRAKE INDICATOR LAMP AND SWITCH

On many vehicles there is also a parking brake indicator lamp and switch. This switch is located on the parking brake assembly. It is in the normally open position when the parking brake is not being applied. When the parking brake is applied, the switch closes and turns on the Brake System light on the dashboard.

BRAKE PAD WEAR SENSORS

Some vehicles have a brake pad wear sensor. This sensor is located in one of the brake pads. As the brake pads wear down, the brake pad material wears away and eventually an electrical wire is exposed. When the wire touches the metal rotor, an electrical circuit is completed. This circuit is used to alert the driver that the brake pads are worn and need service. ***Figure 45–35*** shows the brake pad and the brake pad sensor.

PARTS LOCATOR

*Refer to photo #101 in the front of the textbook to see what a **brake pad wear sensor** looks like.*

45.3 POWER BRAKES

Power brakes are used today on many passenger vehicles. Power brakes are designed to have an extra pressure called a booster. The boost is produced either by a vacuum or by hydraulic fluid acting as an extra force for the brake pedal. When the brake pedal is applied, the booster unit multiplies the pedal force for the master cylinder. This means the operator puts less force on the brake pedal, making it easier to stop the car. The booster unit is placed between the brake pedal and the master cylinder. The master cylinder and the rest of the brake system parts are all identical to those in a regular brake system.

VACUUM-ASSISTED BRAKES

Intake manifold vacuum can be used as the booster in a power brake system. In ***Figure 45–36***, power brakes use a diaphragm with a vacuum placed on each side. The center shaft of the diaphragm is connected to the master cylinder. If a vacuum is placed on both sides of the center diaphragm, the diaphragm will not move. However, if the vacuum is

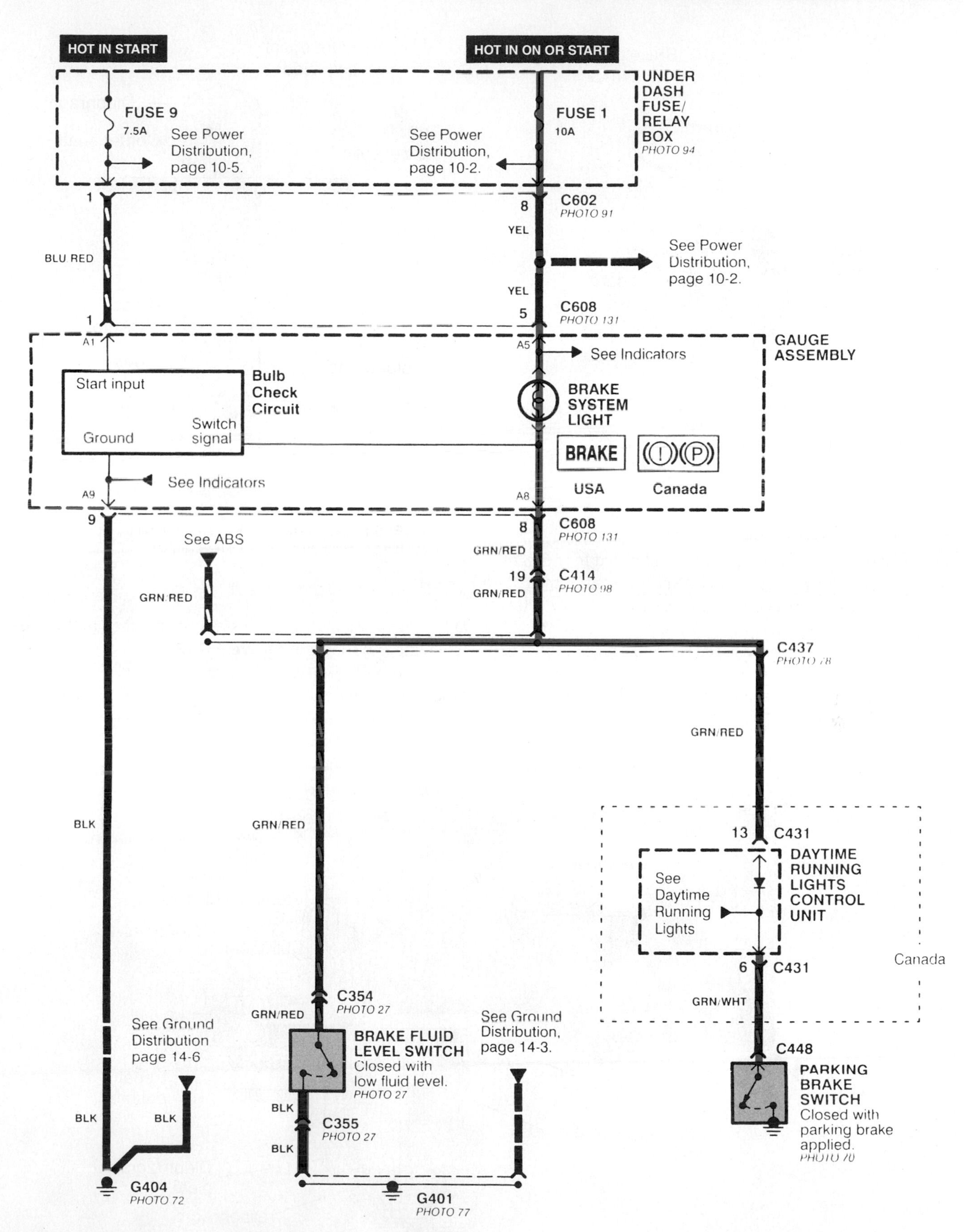

FIGURE 45–34 The Brake System light will turn on when the parking brake remains on if there is a lack of brake fluid and the brake fluid level switch is closed. (Courtesy of American Honda Motor Co., Inc.)

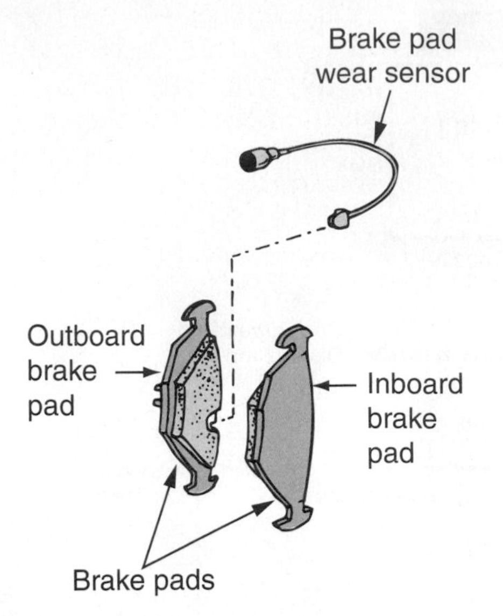

FIGURE 45-35 This brake pad sensor is used to alert the driver that the brake pads are wearing out. (Courtesy of BMW NA, LLC)

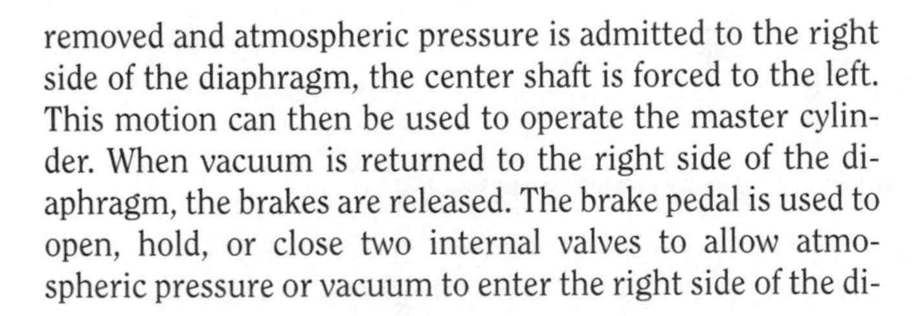

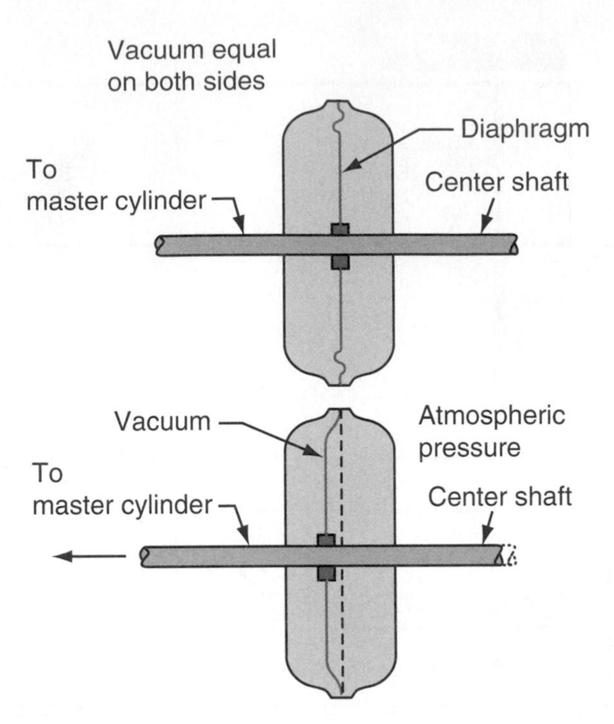

FIGURE 45-36 Power brakes use a vacuum to produce the push against the pushrod on the master cylinder.

removed and atmospheric pressure is admitted to the right side of the diaphragm, the center shaft is forced to the left. This motion can then be used to operate the master cylinder. When vacuum is returned to the right side of the diaphragm, the brakes are released. The brake pedal is used to open, hold, or close two internal valves to allow atmospheric pressure or vacuum to enter the right side of the diaphragm. An example of a booster unit attached to the master cylinder is shown in ***Figure 45–37***.

The power brake system operates in one of three modes: hold, apply, and release. Normally, the entire booster is

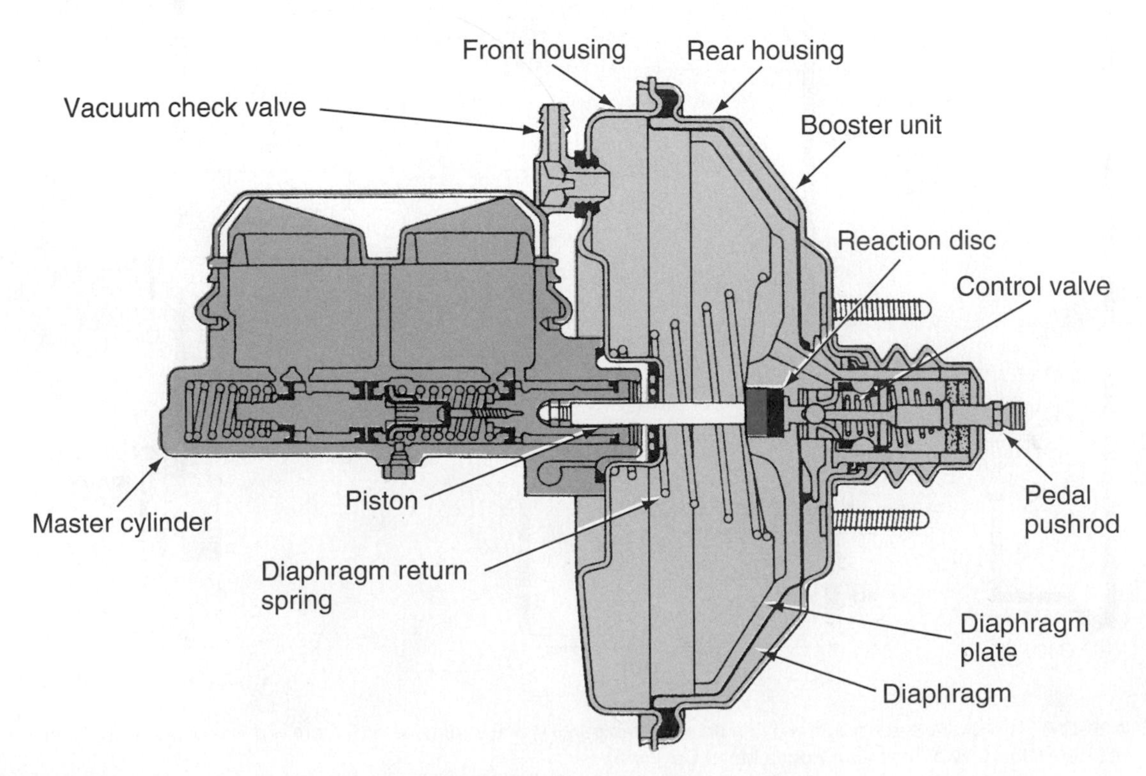

FIGURE 45-37 The booster unit for power brakes is attached directly to the master cylinder.

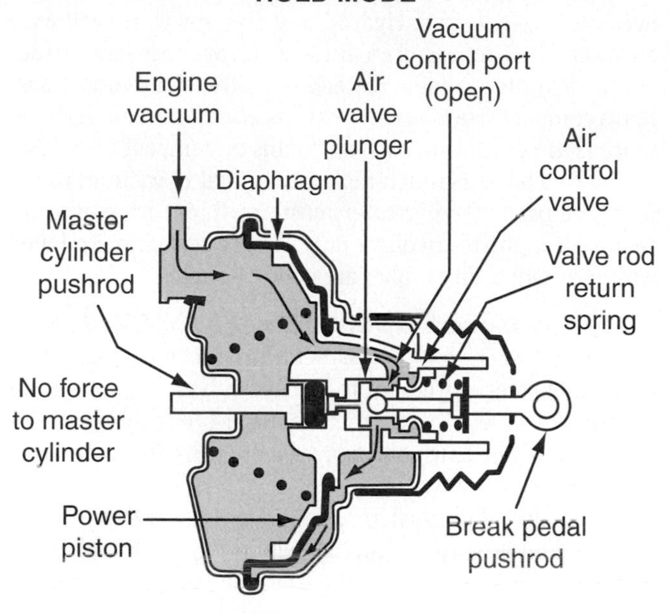

FIGURE 45-38 During the hold mode, a vacuum is applied to both sides of the diaphragm.

under a vacuum. ***Figure 45–38*** shows the internal parts and their arrangement during this time, which is called the "hold" condition. Note that engine vacuum is felt on both sides of the diaphragm. This is because the air valve plunger and the air control valve are in position to allow the engine vacuum to be routed to both sides of the diaphragm.

In the "apply" mode when the operator applies the pressure to the brake pedal, the brake pedal pushrod is moved to the left. This action causes the air valve plunger to move to the left as shown in ***Figure 45–39***. With the air valve plunger and the air control valve in this position, the vacuum on the right side of the diaphragm is replaced with atmospheric pressure. On the left side of the diaphragm, the vacuum remains. The vacuum on the left side now causes the diaphragm to move to the left, producing an additional force to the master cylinder pushrod and piston.

In the "release" mode when the operator releases the brake, the master cylinder pushrod spring inside of the master cylinder moves the master cylinder pushrod to the right. This action causes the air valve plunger and air control valve to move back to the position shown in ***Figure 45–40***. This position keeps the vacuum on both sides of the diaphragm equal so there is no additional pressure to the master cylinder pushrod.

Vacuum Check Valve The vacuum is obtained from the intake manifold on the engine. Since fluctuations occur in the intake manifold vacuum, there is a reservoir in the system to act as a storage for vacuum. The reservoir is the large canister surrounding the diaphragm (***Figure 45–41***). Also, there is a check valve between the manifold and the reservoir. This check valve prevents the vacuum from escaping the reservoir during conditions of wide-open throttle. The check valve is also a safety device in the event of a leaking supply line or other vacuum failure.

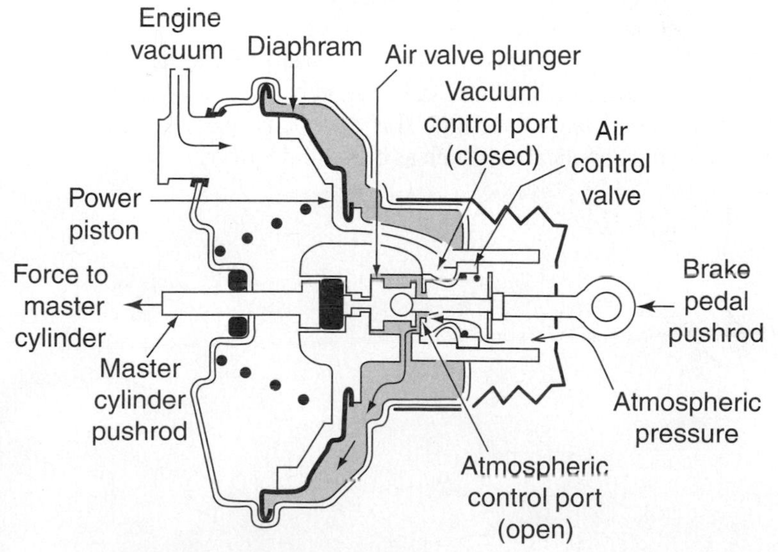

FIGURE 45-39 During the apply mode, the vacuum on the right side of the diaphragm is removed and replaced with atmospheric pressure.

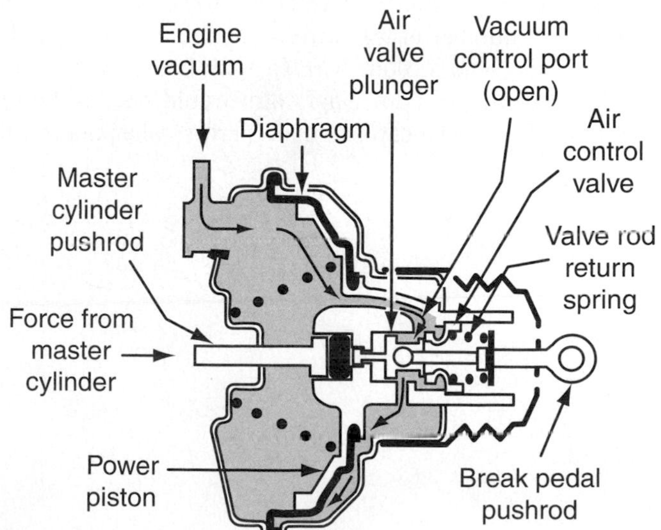

FIGURE 45-40 During the release mode, the vacuum is equal on both sides of the diaphragm.

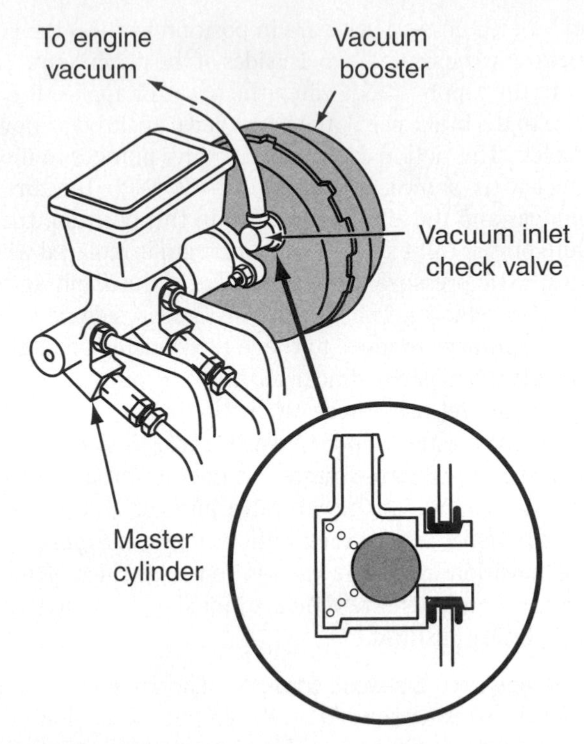

FIGURE 45–41 Vacuum for the power brakes is taken directly from the intake manifold.

HYDRO-BOOST BRAKES

Hydro-boost brakes are designed to use hydraulic pressure from the power steering pump to boost the pressure for the master cylinder. This system does the same job as a vacuum booster and is connected to the brake system in much the same way. One reason the hydro-boost unit is used is because of federal regulations which require that vehicles stop in a fewer number of feet with less pressure on the brake pedal. This could be done with vacuum boost systems, but the size of the vacuum diaphragm would have to be increased. The trend today is to make parts smaller, not larger.

In certain applications, hydro-boost has many advantages over vacuum boosters. Hydro-boost systems work well with diesel and turbocharged engines, which at times have inadequate vacuum available. In addition, the hydro-boost system's compact size allows it to be installed where underhood space is at a premium, such as in vans or compact cars. Also, because its boost is much higher than that of vacuum units, it can be used where greater master cylinder pressures are required. Light-to-medium duty trucks and cars equipped with four-wheel disc brakes are good examples.

Hydro-Boost Operation ***Figure 45–42*** shows a hydro-boost system. Pressure from the power steering pump and reservoir is sent to the hydro-boost system. Note its small size compared to a vacuum-boost system. Hydraulic fluid is used to multiply the pressure for the master cylinder.

The hydro-boost system uses a spool valve that is built into the unit. The spool valve shown in ***Figure 45–43*** is operated by the movement of the brake pedal. The position of the spool valve directs the high-pressure fluid either back to the steering system or to the power piston. This directs the high-pressure fluid to a cavity behind the power piston. The pressure forces the power piston forward, applying the pressure to the output pushrod. The output pushrod operates the master cylinder.

VARIATIONS IN POWER BRAKE SYSTEMS

Although the principles remain the same, there are two variations for power-assisted brakes. These variations are determined by the manufacturer, the type of application, and the year of production. These variations are:

1. Tandem power head—a power brake booster with two diaphragms in tandem or series. This provides additional boost to the master cylinder.
2. Dual power brake system—a power brake system that uses both a vacuum-assisted and hydraulically assisted (hydro-boost) design. This system is used on heavy-duty applications such as buses and trucks.

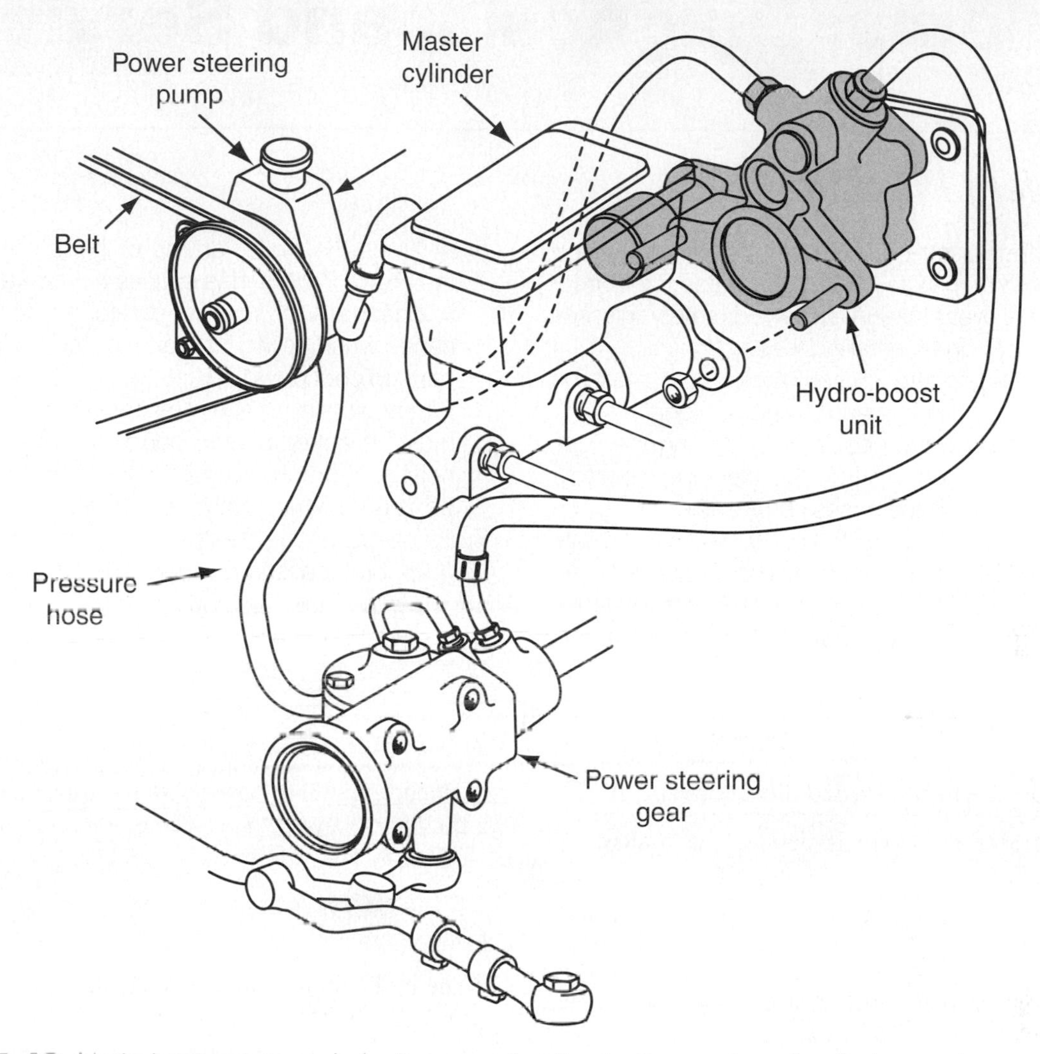

FIGURE 45-42 A hydro-boost system uses hydraulic pressure from the steering pump to produce the increased pressure working on the master cylinder.

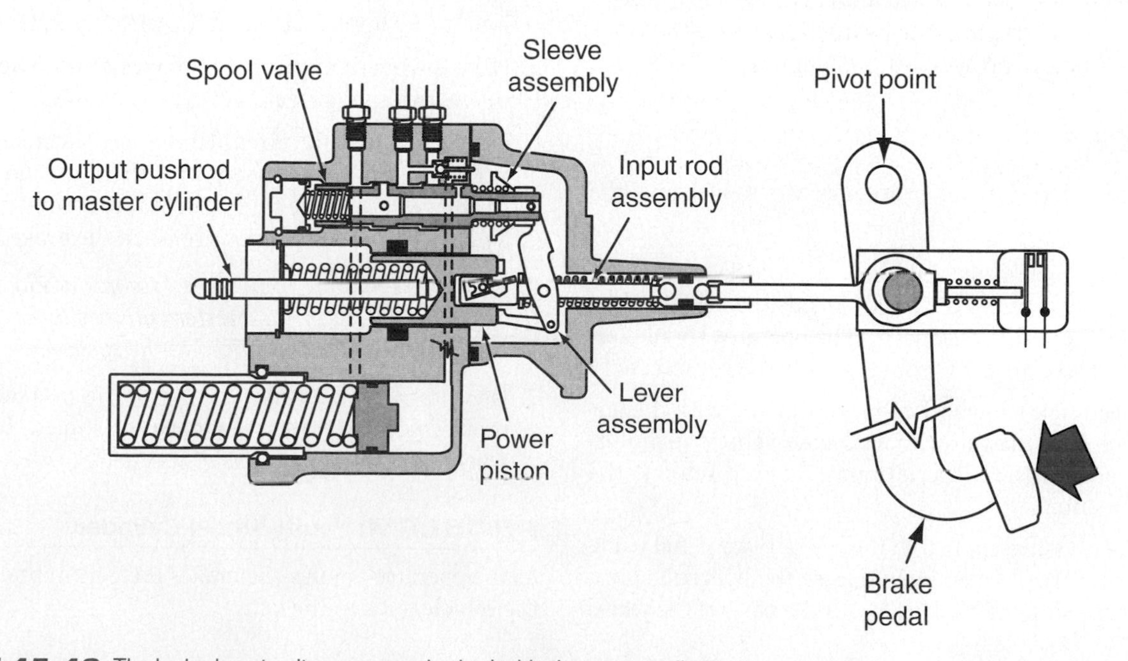

FIGURE 45-43 The hydro-boost unit uses a spool valve inside the master cylinder.

Problems, Diagnosis, and Service

Safety Precautions

1. The springs used on drum brakes are under high pressure during installation. Always use the correct tools and wear OSHA-approved safety glasses when removing or installing these springs.
2. When bleeding the brakes, brake fluid must be forced out of the hydraulic system, along with the air. To eliminate the possibility of spilling the fluid, use a hose connected from the bleeding valve to a canister to catch the excess brake fluid.
3. Always wipe up any spilled brake fluid to eliminate the possibility of injuries from slips and falls.
4. Be careful not to crush your fingers or hand when the disc brake assembly is being moved during disassembly.
5. At times the brake drums or pads may drag on the drum or rotor. This causes the brake assembly to become very hot. Be careful not to burn your hands when servicing brake systems. Wait for the parts to cool down.
6. When working on a brake system, the vehicle must be jacked up and correctly supported with a hoist or hydraulic or air jack. Use the correct safety procedure when lifting the vehicle.
7. Be careful not to breathe the dust particles left in a drum brake assembly when removing the brakes. The dust may contain asbestos, which can seriously injure your lungs.

PROBLEM: Air in the Hydraulic System

When the operator of a vehicle applies the brakes, the brakes seem to be very spongy and soft.

DIAGNOSIS

Spongy or soft brakes are an indication of air in the hydraulic system. When the brakes are applied the air compresses, producing a spongy feeling. Air can get into the brakes during service or if one of the wheel cylinders is leaking. Often older vehicles that are stored during the winter leak fluid from a wheel cylinder. Then, when the brakes are pumped, air enters into the hydraulic system. To get rid of air in the brakes, the hydraulic system must be bled.

SERVICE

Use the following general procedure when bleeding the brakes:

1. With the vehicle placed properly on a jack stand, have another person pump the brakes inside the vehicle, then press down on the brake pedal. The operator should feel a spongy brake pressure from the air being compressed.
2. Start bleeding each wheel cylinder, starting with the farthest from the master cylinder. Some vehicle manufacturers recommend a special procedure, so refer to the service manual.
3. With the pressure applied to the brake system and using the correct size flare wrench, release the hydraulic pressure by opening the bleed valve on the back of the wheel cylinder. A rubber hose should be attached to the end of the bleed valve so the hydraulic brake fluid can be directed into a container.

CAUTION: *Be careful not to get any brake fluid in your eyes.*

The brake fluid should have small bubbles of air mixed with it at this time.

4. When the valve is open, the operator's foot should go to the floor. When the pedal is completely to the floor, tighten the bleed valve before the operator lets the pedal spring back. If the valve is not closed first, air will be drawn back into the hydraulic system at the wheel cylinder.
5. Have the operator pump the brakes again. There should be less of a spongy feeling.
6. Continue bleeding air on the wheel cylinder farthest from the master cylinder until there are no more air bubbles in the brake fluid. Also, there should be a firmer and more solid brake pressure when the brake is applied.

CAUTION: *It is important to continually check the brake fluid level in the master cylinder to be sure there is always enough brake fluid in the system.*

7. Depending on the exact nature of the problem, one or more wheel cylinders may have to be bled. Bleed each wheel cylinder as previously described.

PROBLEM: Faulty Wheel Cylinder

As the operator applies the brakes on a drum brake system, the vehicle pulls to the left.

DIAGNOSIS

One of the most common problems that causes a vehicle to pull to the left or right is a leaky wheel cylinder. A leaky wheel cylinder will usually have brake fluid on the inside surface of the tire. This can be visually checked. The leaky wheel cylinder will also cause the wheel to grab first when the brakes are applied. As hydraulic fluid gets on the brake lining, it becomes sticky. This causes the vehicle to pull to the left. Any of the four wheel cylinders can leak.

SERVICE

Use the following general procedure to replace the wheel cylinder on drum brakes.

1. With the car's emergency brake on, jack up the vehicle following the safety guidelines.
2. Remove the hubcap and remove the wheel.
3. Check the brake hoses for leaks, cuts, cracks, twists, and loose supports. Replace where necessary.
4. Remove the brake drum. If it is the front brake drum, the front wheel bearings and hub must also be removed. When removing drums, it may be necessary to back off the adjusting screw or release the shoe adjusting cams to provide ample lining-to-drum clearance. If this is not done, the brake drum may be very difficult to get off. ***Figure 45–44*** shows how the adjusting level can be lifted to allow the starwheel to be turned. If the wheel cylinder is being replaced on the rear wheel, remember to release the emergency brake.
5. With the brakes exposed, remove the springs that hold the brake shoes in place.

CAUTION: *Always use the correct tools to remove the return springs and holddown springs.*

Remove all other parts on the brake drum assembly. The exact parts will depend on the manufacturer and the type of vehicle. To aid in disassembly, always use wheel cylinder clamps as shown in ***Figure 45–45***. This clamp is used to hold the wheel cylinder piston in place during disassembly and reassembly.

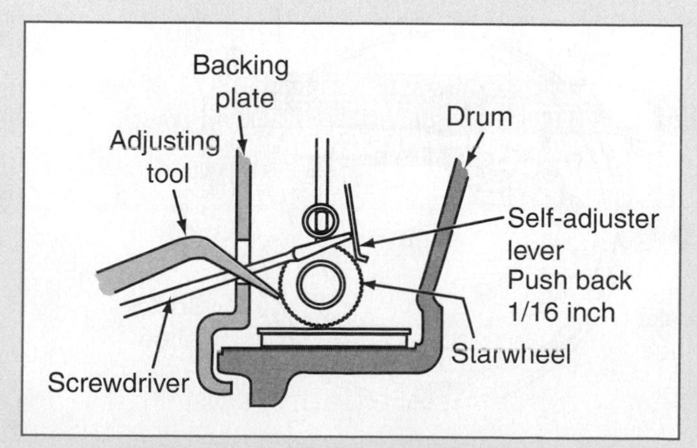

FIGURE 45–44 To back off the self-adjuster on most rear drum brakes, push in the adjusting lever and loosen the starwheel.

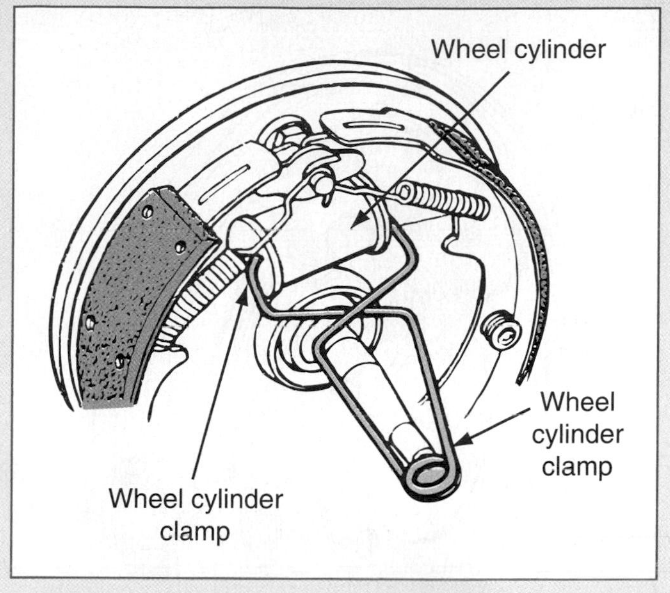

FIGURE 45–45 Use a wheel cylinder clamp to hold the wheel cylinder in place during repair. (Courtesy of Honeywell International, Inc.)

6. Check all springs and other parts for loss of tension and damage. Replace weak springs and other badly damaged parts.
7. Remove the link between the brake drums and the wheel cylinder.
8. When replacing or rebuilding a wheel cylinder, always purchase a wheel cylinder repair kit. Never use old parts on the wheel cylinder.
9. Remove the two rubber boots on either side of the wheel cylinder.
10. Pull out the pistons, cups, springs, and expanders from inside the wheel cylinder.
11. Flush and keep all wheel cylinder parts in clean brake fluid. Any dirt that gets into the hydraulic system may cause wear on the pistons and cups.
12. When rebuilding wheel cylinders, make sure there is no grease or oil on the parts. Grease can damage the rubber parts.
13. Inspect the wheel cylinder bore for scoring, pitting, and corrosion. An approved cylinder hone, ***Figure 45–46***, may be used to remove light roughness or deposits in the bore. Aluminum wheel cylinders and certain other wheel cylinders should not be honed. Check the manufacturer's service manual for your vehicle. Honing will help the cups seal better within the cylinder. Follow the manufacturer's recommended procedure for this operation.

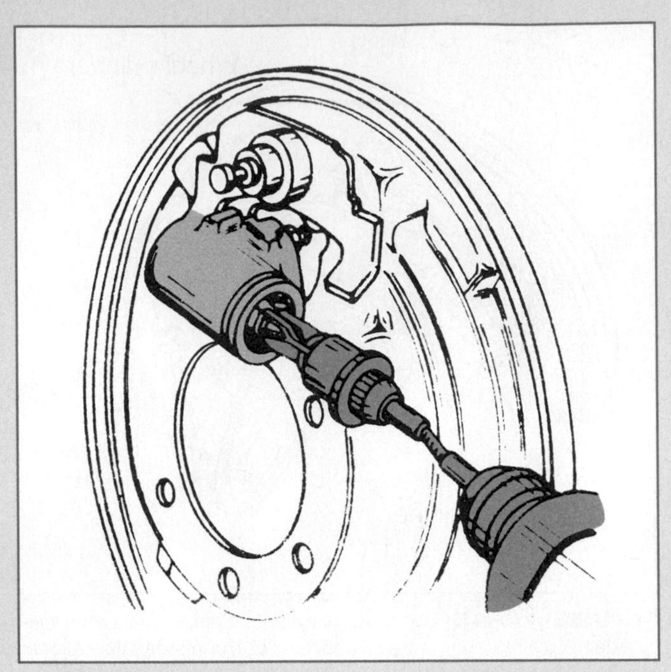

FIGURE 45-46 A hone can be used to clean up the internal surface of a steel cast wheel cylinder, but not an aluminum one. Lubricate the surface with brake fluid while honing. (Courtesy of Federal Mogul Products, Inc.)

14. Replace with the new cylinder kit and reassemble the drum brake assembly using the correct tools.
15. Adjust the brake linings by tightening the starwheel until there is a very slight drag between the drum and the brake lining.
16. Another method used to measure the inside of the brake drum is a suitable brake drum-to-shoe gauge. With the brake drum off the wheel, adjust the shoes to this dimension (***Figure 45–47***).
17. Once the brakes have been reassembled, it will be necessary to bleed the brakes. Use the general procedure described earlier.

PROBLEM: Worn Brake Pads or Linings

As the brakes are applied, there is a scraping sound of metal to metal. Also, for a considerable amount of time before this, there was a high-pitched squeaking sound when the brakes were applied.

DIAGNOSIS

Scraping sounds coming from the brakes are usually an indication of excessive wear on the brake pads or brake linings. If there is a metal scraping sound, it usually indicates that the brake pads or lining have been worn down to the metal, thus scraping the brake drum or disc. On many vehicles using disc brakes, there is a sensor spring to determine when the brake pads are worn and need replacement (***Figure 45–48***).

SERVICE

When the brake linings or pads need to be replaced, there are various checks to make. Use the following as a guideline when servicing brake components.

Drum Brakes

1. After safely placing the vehicle on a hoist or hydraulic jack, remove the wheels and expose the brake linings or pads.
2. Check brake drums for wear and distortion as shown in ***Figure 45–49***. If the drums are out of specifications, have them machined to correct specifications. Typically,

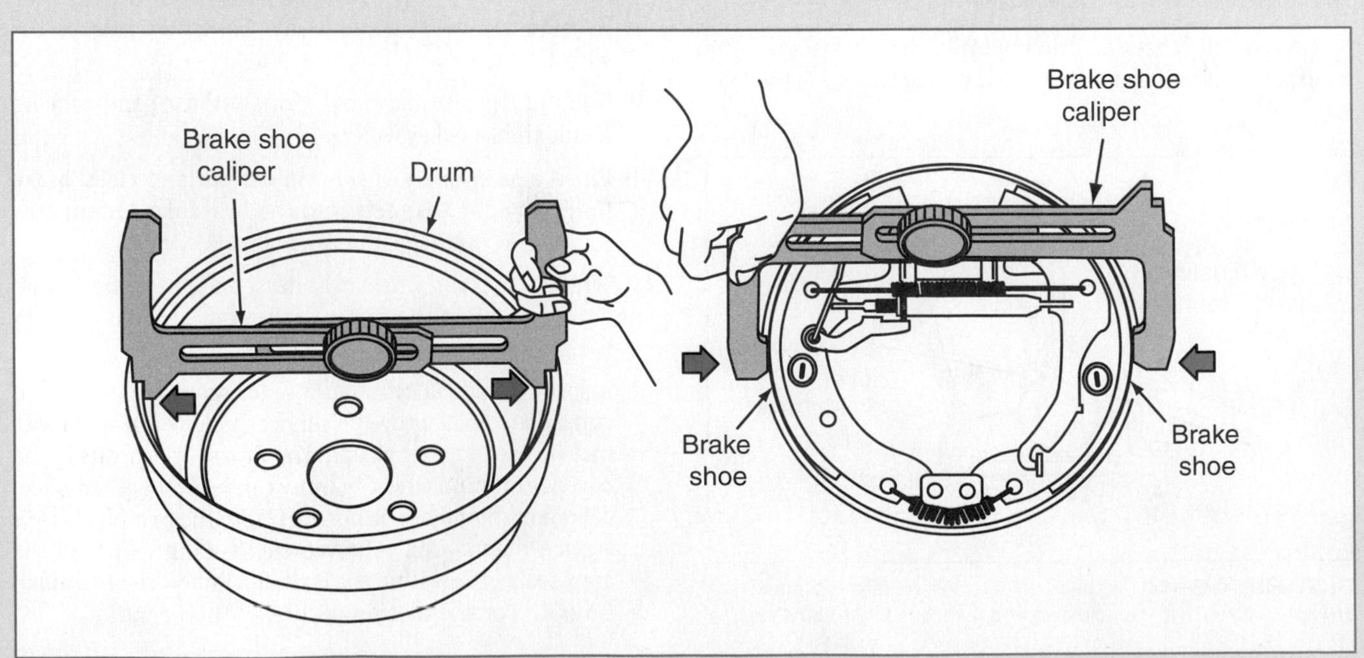

FIGURE 45-47 To adjust the brake shoes, measure the drum first, then set the brake shoes to this dimension.

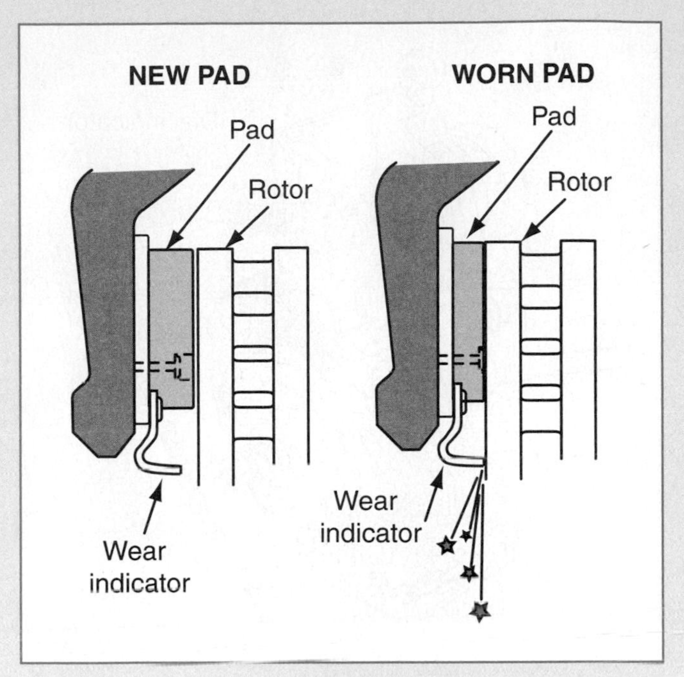

FIGURE 45–48 The sensor spring will squeak if the disc brake pads wear too far.

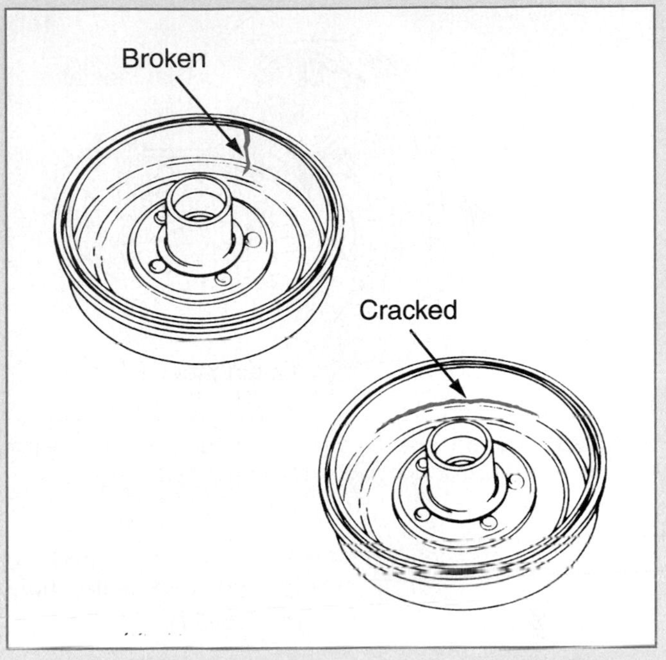

FIGURE 45–50 Inspect the drums for cracks. (Courtesy of Federal Mogul Products, Inc.)

the maximum amount to be taken off is about 0.060 inch.

3. Check the brake linings for excessive wear. New brake linings have approximately 0.250 inch of thickness to the width of the lining. When worn excessively, the lining has been worn down to the metal or rivets. Replace as necessary.
4. ***Figure 45–50*** shows areas where the drum on drum brakes could crack. Always check drums carefully for such cracks before reinstalling them in the vehicle.

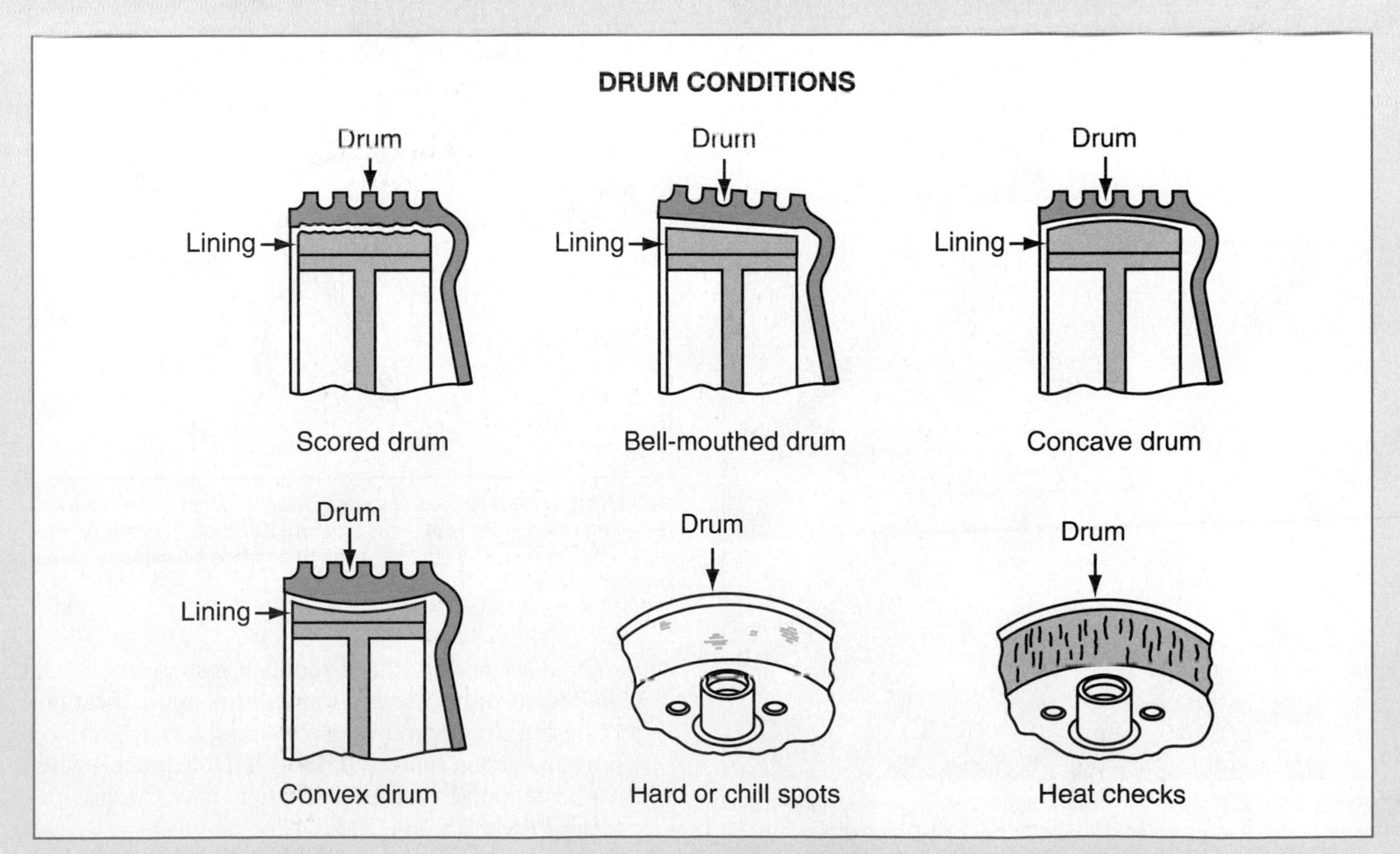

FIGURE 45–49 Check the drum for excessive wear and warping as shown. (Courtesy of Honeywell International, Inc.)

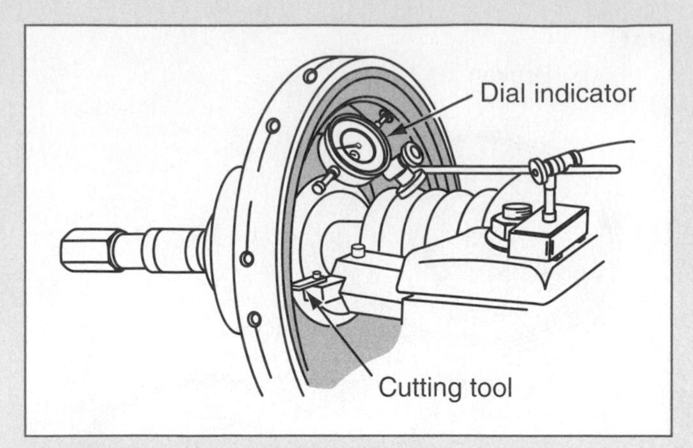

FIGURE 45–51 Use the correct procedure and equipment to recondition the internal surface of the drum brake.

5. Use only precision equipment to refinish or machine drums. Always follow the manufacturer's instructions for use of this equipment (***Figure 45–51***).

Disc Brakes

1. After the brake pads and disc have been exposed, check and inspect disc brakes for cracks or chips on the pistons, amount of wear on the pads, even wear on the brake pads, damage to the rotor, damaged seals, and cracks in the caliper housing.
2. Remove the caliper assembly and pads as shown in ***Figure 45–52***. Generally, the caliper is held in place by bolts or by steel locator pins. Follow the manufacturer's suggested procedure for removal.
3. Check the disc brake rotor for scoring, runout, parallelism, and thickness. Runout, for example, can be checked using a dial indicator as shown in ***Figure 45–53***.

FIGURE 45–52 To remove the caliper assembly, loosen and remove the locator pins holding the caliper assembly to the frame.

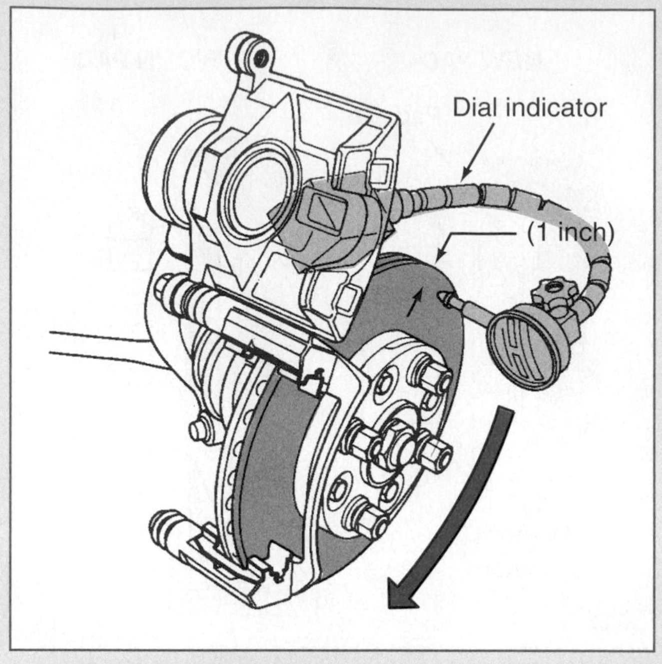

FIGURE 45–53 Runout on the disc of a disc brake system can be checked using a dial indicator.

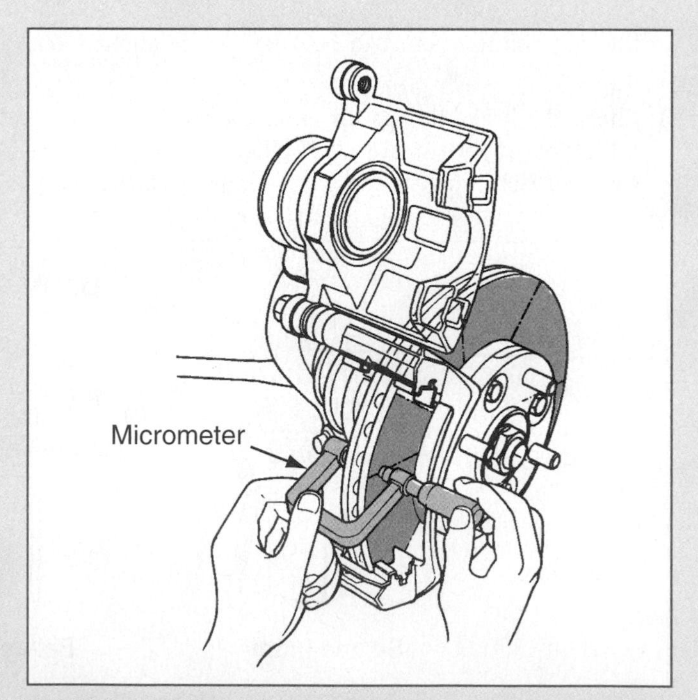

FIGURE 45–54 Use a suitable micrometer to check the thickness variation (parallelism) of the disc on a disc brake system. At least eight measurements should be taken and compared with the manufacturer's specifications.

4. Certain discs may be out-of-round or may vary in thickness. Measure the thickness at eight or more locations around the circumference of the lining contact surface. Use a suitable micrometer. If there is a difference greater than the manufacturer's specification, have the disc machined (***Figure 45–54***).

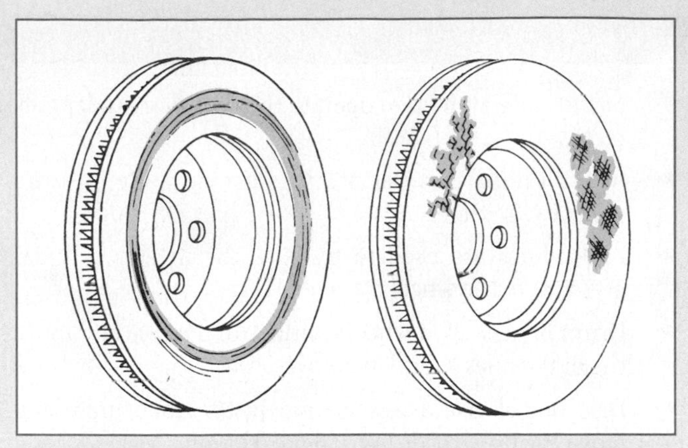

FIGURE 45–55 Check the rotors for signs of scoring or grooving and for signs of overheating. Overheating also causes a slight bluing on the rotor. (Courtesy of Federal Mogul Products, Inc.)

5. Scoring and grooving will sometimes occur on the disc as shown in ***Figure 45–55***. The rotors will require reconditioning for maximum performance to be developed.
6. If the caliper on a disc brake is leaking or sticking, it will need to be reconditioned. Follow the same general procedure described earlier for drum brake wheel cylinders.
7. Check to see if the piston inside the caliper moves freely with the caliper bore. Road dirt and rust can cause the piston to stick in the bore. When removing the seal, use a soft object such as a pencil so as not to scratch the cylinder walls (***Figure 45–56***).

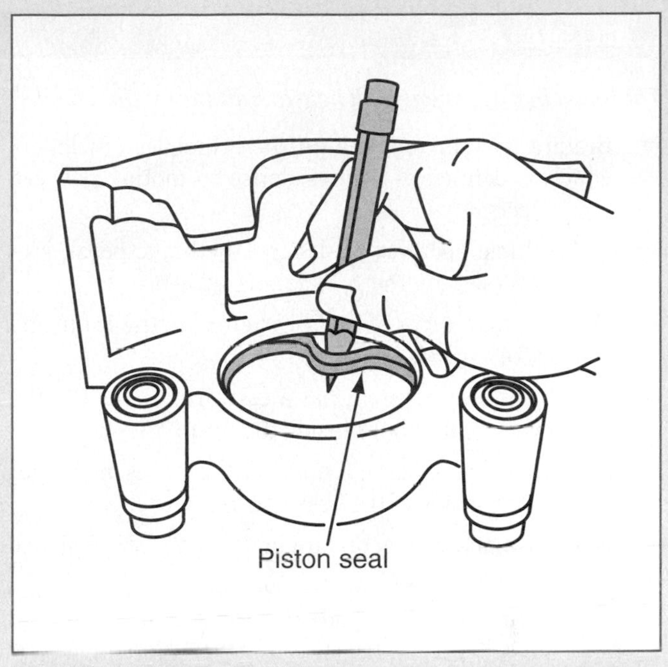

FIGURE 45–56 When removing the seal inside the cylinder on disc brakes, use a soft-pointed tool so that you do not scratch the cylinder surfaces.

Service Manual Connection

There are many important specifications to keep in mind when working with braking systems. To identify the specifications for your engine, you will need to know the VIN (vehicle identification number) of the vehicle, the type and year of the vehicle, and the type of engine. Although they may be titled differently, some of the more common braking system specifications (not all) found in service manuals are listed below. Note that these specifications are typical examples. Each vehicle may have different specifications.

Common Specification	Typical Example
Brake drum inside diameter	10.00 inches
Caliper bore diameter	2.59 inches
Caliper bore maximum cut	0.002 inch
Clearance between sensor face and flange	1.8 inches
Rotor lateral runout	0.003 inch
Rotor nominal thickness	0.870 inch
Rotor thickness variation and parallelism	0.0005 inch

In addition, service manuals provide specific directions for various service and testing procedures. Some of the more common procedures include:

- Brake pads replace
- Brake system bleeding
- Caliper overhaul
- Caliper replace
- Shoe and lining replacement

SUMMARY

The following statements will help to summarize this chapter:

- Braking systems operate on the principles of friction, which is defined as the resistance to motion between two objects.
- As a vehicle operator pushes on the brake pedal, friction increases and the vehicle slows down.
- Two common types of brake systems are the drum and disc brake systems.
- The drum brake system has a cast drum bolted to the wheel with brake shoes inside the drum.
- As the drum brakes are applied, the shoes move out against the inside of the brake drum.
- Both primary and secondary brake shoes are designed into the braking system.
- In a disc brake system, pads are forced against a rotor to stop the vehicle.
- A caliper is used to hold the two disc pads in place on a disc brake system.
- Brake systems also use hydraulic systems to transfer the operator's foot pressure to the brakes.
- A master cylinder is used to convert the mechanical force of the operator's foot to hydraulic pressure.
- Several ports, valves, and switches including the compensating port, proportioning valve, warning light switches, a residual pressure check valve, and the metering valve all help to operate the master cylinder correctly.
- A dual master cylinder provides two separate pressure changes for safety.
- Brake lines are used to transfer the master cylinder pressure to the wheel cylinders.
- Drum brakes use a wheel cylinder to operate and move the brake lines to produce friction.
- Disc brakes use a caliper and piston to operate and move the brake pads to produce friction.
- Disc brakes are used more extensively on the front wheels of a vehicle.
- Brake fluid must have the correct viscosity and boiling point, be noncorrosive and water-tolerant, and serve as a good lubricant, as well as have a low freezing point.
- Power brakes are designed to produce additional pressure that can be exerted on the hydraulic system.
- Power brakes use either a vacuum or hydraulic system to produce additional pressure.
- Several important service tips for servicing brakes include always using the correct tools and checking all brake system parts (disc, rotors, drums, shoes, springs, brake lines, etc).
- Brake systems should be bled and adjusted according to the manufacturer's specifications.

TERMS TO KNOW

Can you explain each of the following terms? Review the chapter until you can use each term correctly.

Backing plate
Bleeding
Brake shoes
Caliper
Compensating port
Computer-controlled brakes
Disc brakes
Fixed caliper
Floating caliper
Friction
Hydraulic system
Kinetic energy
Master cylinder
Primary shoe
Proportioning valve
Residual pressure
Secondary shoe
Self-energizing
Servo brakes
Tandem
Wheel cylinder

REVIEW QUESTIONS

Multiple Choice

1. Friction on the automobile brake linings depends on which of the following?
  a. Power steering pressure
  b. Tire size
  c. Speed of vehicle
  d. Pressure between the drum and linings
  e. Amount of spring pressure on the parking brake

2. Brake shoes are used on ________.
  a. Front disc brakes
  b. Drum-type brakes
  c. Brakes with a caliper
  d. The back of the master cylinder
  e. Piston-type brakes

3. When a brake shoe becomes tighter against the drum from rotation, this is called:
a. A master cylinder
b. Wheel cylinder
c. Hydraulic pressure
d. Self-energization
e. Compensating system

4. The ________ is located to the front of the wheel on vehicles with drum brakes.
a. Secondary shoe
b. Caliper
c. Master cylinder
d. Primary shoe
e. Relief valve

5. Which of the following uses a caliper?
a. Disc brakes
b. Drum brakes
c. Master cylinder
d. Star adjuster
e. Compensating port

6. The pads on a disc brake system move ________ the rotor.
a. Perpendicular to
b. Parallel to
c. Never touching
d. Attached to
e. None of the above

7. The ________ is/are directly attached to the steering assembly and frame on a disc brake system.
a. Floating caliper
b. Linings
c. Fixed caliper
d. Pads
e. Master cylinder

8. Hydraulic fluid used on a braking system:
a. Transfers the pressure from the master cylinder to the wheel cylinder
b. Cannot be compressed
c. Must withstand very high pressures
d. Is pressurized in the master cylinder
e. All of the above

9. The purpose of the master cylinder is to:
a. Reduce pressure on the brakes during braking
b. Produce the pressure on the brakes during braking
c. Produce the necessary friction during braking
d. Be used as a parking brake
e. Adjust the brakes

10. The ________ allow(s) fluid to flow from the reservoir to the master cylinder.
a. Forward stroke
b. Compensating port
c. Primary cup
d. Wheel cylinder
e. Brake linings

11. Brake fluid passes around the primary cup in the master cylinder:
a. During the return stroke
b. During the forward stroke
c. During acceleration
d. During parking brake action
e. During rapid stopping

12. A small amount of pressure is kept on the hydraulic fluid on certain brake systems:
a. To stop the car completely
b. By the compensating port
c. By the residual pressure check valve
d. By the parking brake
e. By the springs in the wheel cylinder

13. Two wheels braking from one chamber and two wheels braking from a second chamber:
a. Use a dual master cylinder
b. Use a diagonal brake system
c. Protect the system if one master cylinder fails
d. All of the above
e. None of the above

14. A ________ is used to reduce the pressure to the rear wheels during braking.
a. Proportioning valve
b. Compensating valve
c. Relief valve
d. Metering valve
e. None of the above

15. The primary shoe is ________.
a. Shorter than the secondary shoe
b. Longer than the secondary shoe
c. The same length as the secondary shoe
d. Placed to the rear on the wheel cylinder
e. Placed with the disc against the drum

16. Brake shoes can be adjusted ________.
a. Manually
b. Automatically
c. Using the starwheel
d. All of the above
e. None of the above

17. On a disc brake system, the pads are forced against the ________.
a. Rotor
b. Parking shoe
c. Bleeder screw
d. Piston seal
e. Piston

18. Which of the following is not a good characteristic of brake fluid?
a. Extremely low viscosity (thick)
b. High boiling point
c. Able to absorb and retain moisture
d. Able to lubricate
e. Low freezing point

19. Power brakes use ________ for increasing the pressure.
a. Vacuum
b. Lubricating oil
c. Cooling system fluid
d. The rpm of the engine
e. The operator's foot

20. Which of the following is a mode of operation on power braking systems?
a. Apply mode
b. Hold mode
c. Release mode
d. All of the above
e. None of the above

21. Hydro-boost brake systems use ________ to increase the pressure for the master cylinder.
a. Vacuum
b. Power steering fluid
c. Cooling system fluid
d. The rpm of the engine
e. The operator's foot

22. Which of the following is a good service tip when working on brakes?
a. Bleed the system of air
b. Replace linings that show excessive wear
c. Keep all wheel cylinder parts clean and coated with brake fluid
d. All of the above
e. None of the above

The following questions are similar in format to ASE (Automotive Service Excellence) test questions.

23. *Technician A* says that a spongy brake pedal is caused by air in the hydraulic system. *Technician B* says that a spongy brake pedal is caused by a weak return spring. Who is correct?
a. A only c. Both A and B
b. B only d. Neither A nor B

24. When a car is being stopped, it pulls to the right. *Technician A* says that the left wheel cylinder is leaking. *Technician B* says that the right wheel cylinder is leaking. Who is correct?
a. A only c. Both A and B
b. B only d. Neither A nor B

25. The left front wheel cylinder has just been replaced. *Technician A* says that the brake system should not be bled of air. *Technician B* says that the brakes must be bled to remove the air. Who is correct?
a. A only c. Both A and B
b. B only d. Neither A nor B

26. A wheel cylinder is leaking slightly. *Technician A* says that the wheel cylinder should be honed and the old parts replaced. *Technician B* says a new wheel cylinder kit should be put in after the cylinder is honed slightly. Who is correct?
a. A only c. Both A and B
b. B only d. Neither A nor B

27. During braking, a scraping sound is heard. *Technician A* says the problem is the bearings. *Technician B* says the brake pads are worn and touching the disc. Who is correct?
a. A only c. Both A and B
b. B only d. Neither A nor B

28. A vehicle is producing a squeaking sound during braking. *Technician A* says that the sound is produced by a sensor spring that squeaks if the brakes are worn too much. *Technician B* says the sound is being made from a bad wheel bearing. Who is correct?
a. A only c. Both A and B
b. B only d. Neither A nor B

Essay

29. Describe the process of bleeding the brakes.

30. What are the primary and the secondary shoes?

31. Describe the operation of a floating caliper.

32. What is a compensating port in the master cylinder?

33. Describe the purpose and operation of the proportioning valve.

34. Describe the operation of power brakes as compared with manual brakes.

Short Answer

35. In a hydraulic system, a force of 400 pounds was placed on the input piston, with an area of 1 square inch. Therefore, there would be a force of ________ pounds on a 1/2-square-inch output piston in the same hydraulic system.

36. When the right front wheel cylinder is leaking, it causes the car to pull to the ________.

37. In order to remove the air from the hydraulic circuit in a braking system the system must be ________.

38. The brake fluid level sensor can be located in the fluid reservoir or in the ________.

39. The parking brake switch ________ (closes/opens) when the parking brake is applied.

40. When a brake pad wear sensor is used, the sensor is located on the ________.

Applied Academics

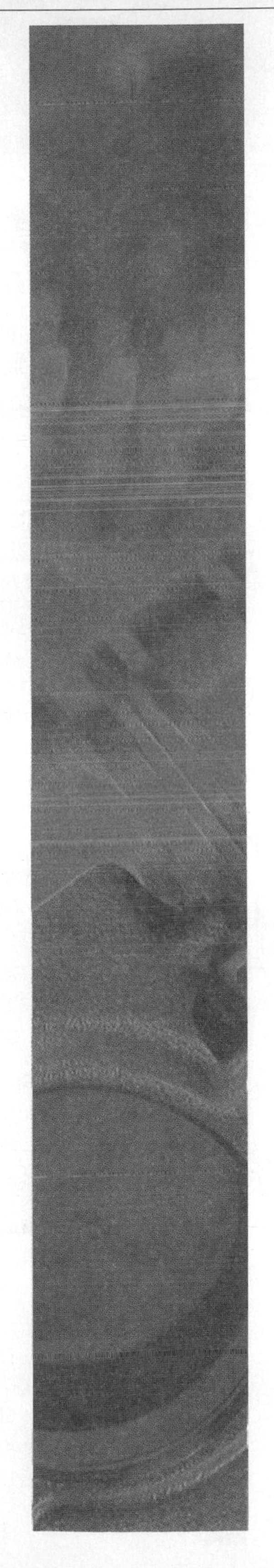

TITLE: Pascal's Law

ACADEMIC SKILLS AREA: Science

NATEF Automobile Technician Tasks:

The service technician can explain the dynamic control properties of a hydraulic system.

CONCEPT: PASCAL'S LAW

All fluid power circuits, including braking systems, follow certain predictable patterns. Pascal's Law helps to explain hydraulic principles of braking systems. This law states that a pressure applied to a confined fluid is transmitted undiminished to every portion of the surface of the containing vessel.

On the left in the diagram that follows, when a force is applied to the handle, the piston is forced downward. The pressure created in the fluid is equal and undiminished in all directions. Pascal's Law also states that pressure on a fluid is equal to the force applied, divided by the area. This can be shown as:

$$P = \frac{F}{A}$$

Where P = pressure in pounds per square inch
F = force applied in pounds
A = area to which the force is applied

Thus, 100 pounds of force on a 1-square-inch piston produces 100 pounds per square inch. When a force (driver's foot) presses down on the brake pedal, a pressure is built up in the master cylinder. This pressure is then transferred throughout the hydraulic lines to each wheel cylinder. The pressure is equal throughout the hydraulic brake system.

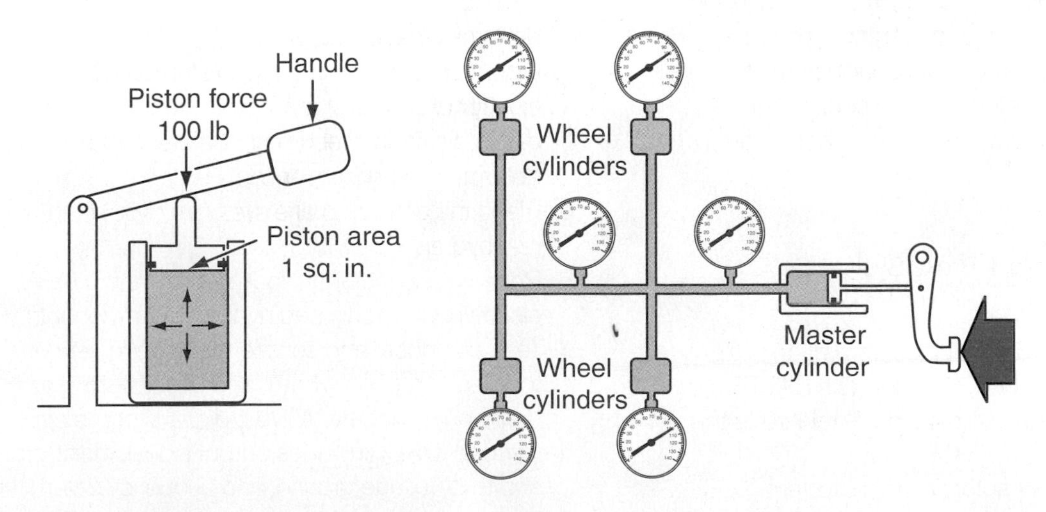

APPLICATION:

Considering Pascal's Law, as the brake pedal is pressed, why is there equal pressure on all of the wheel cylinders? What would happen to the pressure in a brake system if the master cylinder area or diameter were increased?

CHAPTER

46

Antilock Brake Systems

OBJECTIVES

After studying this chapter, you should be able to:

- Describe the purpose for using antilock brakes.
- Identify the basic components of antilock brakes.
- Define the basic principles and operation of antilock brakes.
- Analyze antilock brake driving characteristics.
- Identify, diagnose, and service various problems associated with antilocking brakes.

INTRODUCTION

Today, most vehicles use antilock braking systems. The service technician must know the basic operation of antilock brakes so that efficient service can be performed. This chapter introduces the basic operation of antilock brakes and their components, driving characteristics, and diagnosis and service.

46.1 PURPOSE OF ANTILOCK BRAKE SYSTEMS (ABS)

In a conventional brake system if the brake pedal is depressed excessively, the wheels can lock up before the vehicle comes to a stop. When this happens, driver control is reduced. Research has shown that the quickest stops occur when the wheels are prevented from lockup.

Certification Connection

ASE Connection: The information in this chapter can help you prepare for the National Institute for Automotive Service Excellence (ASE) certification tests. The test and content area most closely related to this chapter are:

Test A5—Brakes

- **Content Area**—Antilock Brake System (ABS) Diagnosis and Repair

NATEF Connection: Much of the information in this chapter is related to the NATEF tasks. The NATEF tasks and priority numbers most closely related to this chapter are:

1. Identify and inspect antilock brake system (ABS) components; determine necessary action. P-1
2. Diagnose poor stopping, wheel lockup, abnormal pedal feel or pulsation, and noise concerns caused by the antilock brake system (ABS); determine necessary action.
3. Diagnose an antilock brake system's (ABS) electronic control(s) and components using self-diagnosis and/or recommended test equipment; determine necessary action. P-1
4. Bleed the antilock brake system's (ABS) front and rear hydraulic circuits. P-2
5. Remove and install antilock brake system (ABS) electrical/electronic and hydraulic components.
6. Test, diagnose, and service ABS speed sensors, toothed ring (tone wheel), and circuits using a graphing multimeter (GMM)/digital storage oscilloscope (DSO) (includes output signal, resistance, shorts to voltage/ground, and frequency data). P-1
7. Diagnose antilock brake system (ABS) braking concerns caused by vehicle modifications (tire size, curb height, final drive ratio, etc.). P-3

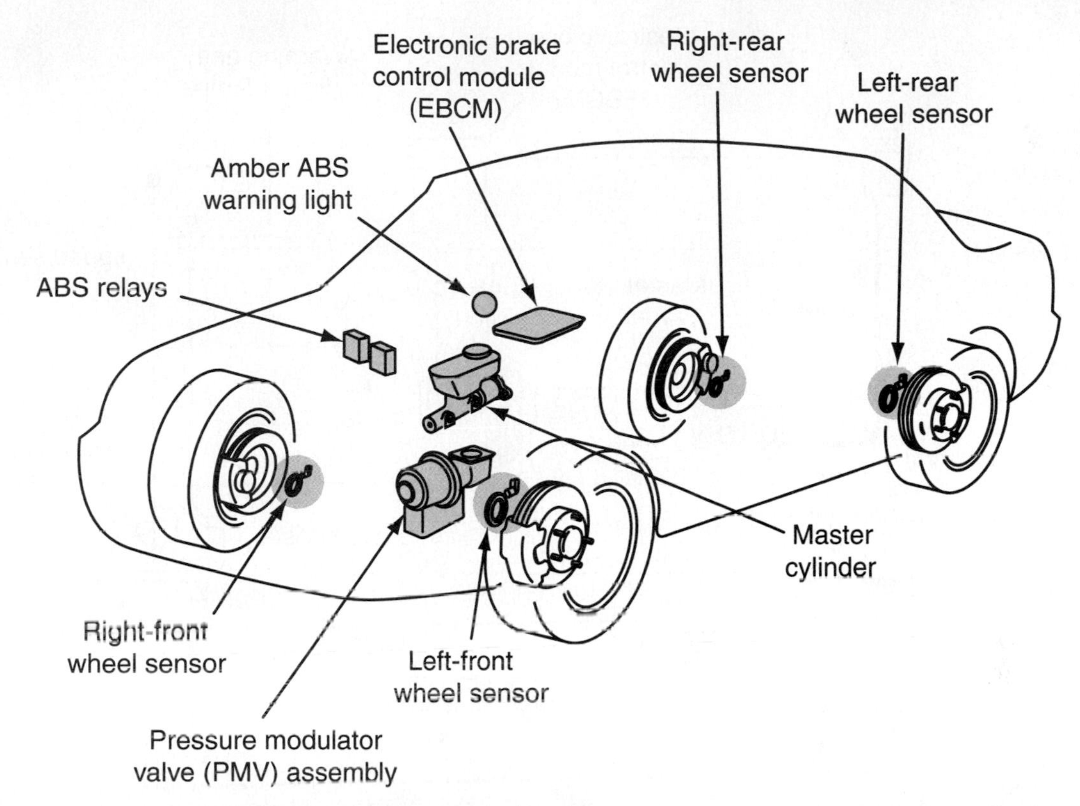

FIGURE 46–1 The components of a typical four-wheel antilock braking system.

The amount of lockup on a tire is referred to as **slip**. When a tire locks up completely during a stop, there is 100% slip. When a tire is rolling freely, there is 0% slip. Through research, it has been found that maximum braking force is generated when the tire slips about 10 to 30%. This means that some tire rotation is necessary to achieve maximum braking. Various factors affect slip on a tire. These include:

- Vehicle weight (loaded versus unloaded)
- Type of road surface (dirt, blacktop, concrete)
- Road condition (dry, wet, smooth, bumpy)
- Axle load distribution
- Sudden changes in the tire-road friction (e.g., puddles, loose dirt, ice)
- Braking in a turn

Antilock braking systems (ABS), as shown in ***Figure 46–1***, are designed to prevent wheel lockup under heavy braking conditions on any type of road condition. The result is that, during heavy braking, the driver:

- Retains directional stability (control of steering)
- Stops faster
- Retains maximum control of the vehicle

ABS GENERAL DESCRIPTION

Under normal conditions, the ABS on a vehicle operates in much the same way as standard brake systems with a split or dual master cylinder. If, during braking, wheel lockup begins to occur, the system will enter the antilock mode. When the braking system is in this mode, hydraulic pressure is released (in a pulsing fashion) on the wheel that is locking up. The amount of pressure released controls the wheel slip about 10 to 30%, thus preventing lockup.

Figure 46–2 shows the basic components (inputs and outputs) of a typical ABS. These include various hydraulic components and circuits and the electrical system, including the computer and relays.

46.2 ABS COMPONENTS

WHEEL SPEED SENSORS

A **wheel speed sensor** is located at each wheel. Its purpose is to sense the speed of each wheel. The speed is converted to an electrical signal and sent to the computer, often called the electronic brake control module (EBCM).

Figure 46–3 shows several examples of front and rear wheel sensor assemblies. The exact type will depend on the manufacturer and whether the vehicle is front- or rear-wheel drive.

The wheel speed sensor assemblies are made of a wheel speed sensor and a toothed wheel. As the toothed wheel passes by the sensor, a small ac voltage is produced. ***Figure 46–4*** shows that a permanent magnet is used as part of the wheel speed sensor to produce the voltage. Higher speeds

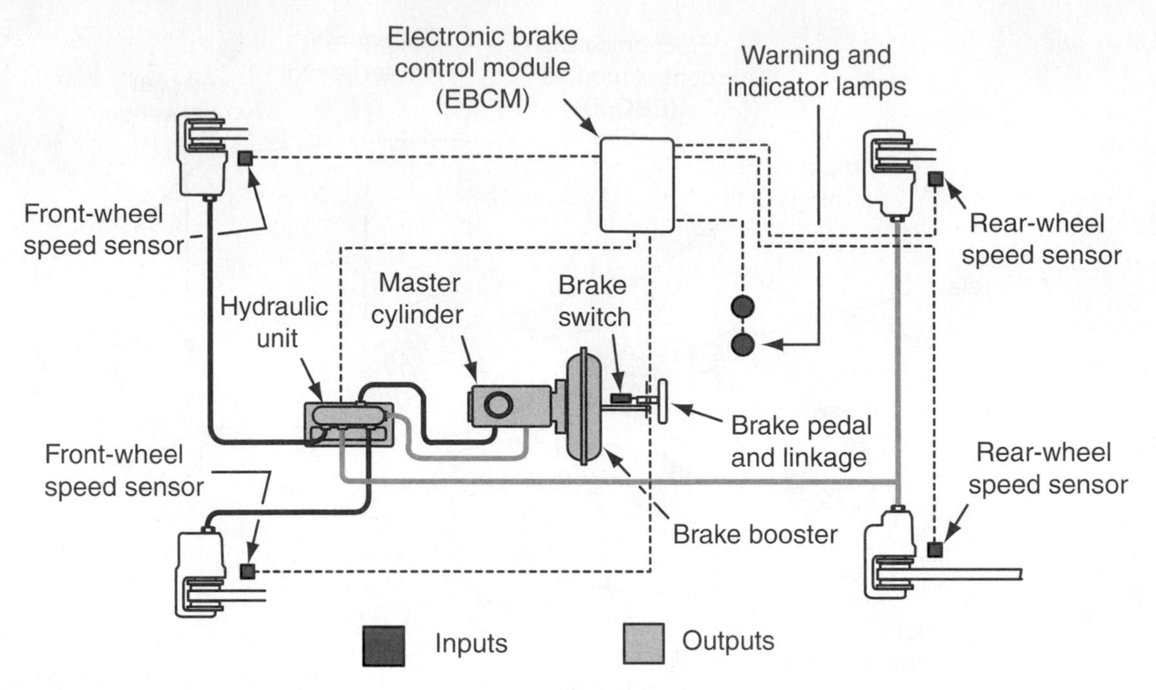

FIGURE 46-2 Antilock braking systems have various input and output components.

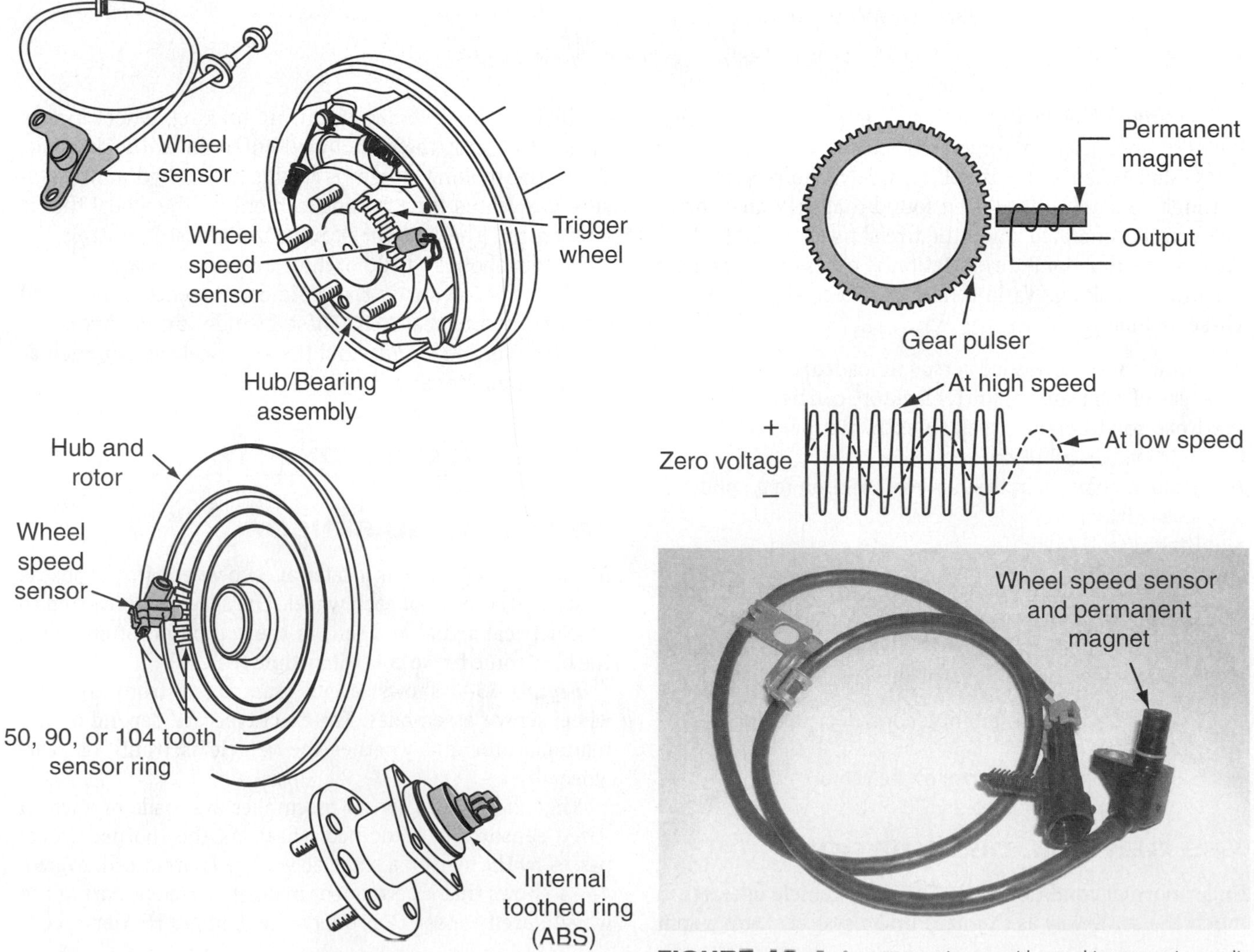

FIGURE 46-3 Wheel sensors have several configurations.

FIGURE 46-4 A permanent magnet is used to generate a voltage as the wheel turns, which in turn senses the wheel speed.

will produce a higher frequency. Lower speeds will produce a lower frequency.

PARTS LOCATOR

*Refer to photo #102 in the front of the textbook to see what an **ABS wheel speed sensor** looks like.*

PRESSURE MODULATOR ASSEMBLY

The **pressure modulator** assembly shown in ***Figure 46–5*** contains the wheel circuit modulator (solenoid) valves used to control or modulate the hydraulic pressure to the brake calipers. Inside hydraulic circuits are discussed later in this chapter.

PARTS LOCATOR

*Refer to photo #103 in the front of the textbook to see what a **pressure modular assembly** looks like.*

Pump Motor The pressure modulator assembly also contains a hydraulic pump motor. The pump motor takes low-pressure brake fluid from the fluid reservoir and pressurizes it for storage in an accumulator or for direct use in the antilock braking system. When the pressure on a caliper has been released because of an impending wheel lockup, pressure must be restored. The result is the production of a pulsing effect on the brakes. The pump motor helps to provide this extra buildup or increase in pressure.

MASTER CYLINDER

An ABS master cylinder works the same as that on a standard braking system (during normal braking). During antilock condition, it works with the pressure modulator assembly to control or modulate the hydraulic pressure to the calipers on each wheel.

ELECTRONIC BRAKE CONTROL MODULE (EBCM)

The **electronic brake control module (EBCM)** is a small control computer that receives wheel speed information. On the basis of the wheel speed information (if one wheel is decelerating too fast), the EBCM sends an electronic signal to the pressure solenoids. The solenoids then control the hydraulic pressure to the calipers.

Figure 46–6 shows an EBCM. In some cars, it is located in the trunk between the rear seat and the rear bulkhead trim panel. The primary functions of the EBCM are to detect wheel locking tendencies, control the brake system while in antilock mode, monitor the system for proper operation, and control the display of fault codes while in the diagnostic mode.

Concerning the fault codes, the EBCM has a self-diagnostic capability, including up to sixteen fault codes, depending on the vehicle, that may be stored in the EBCM. These can then be displayed for diagnostic purposes.

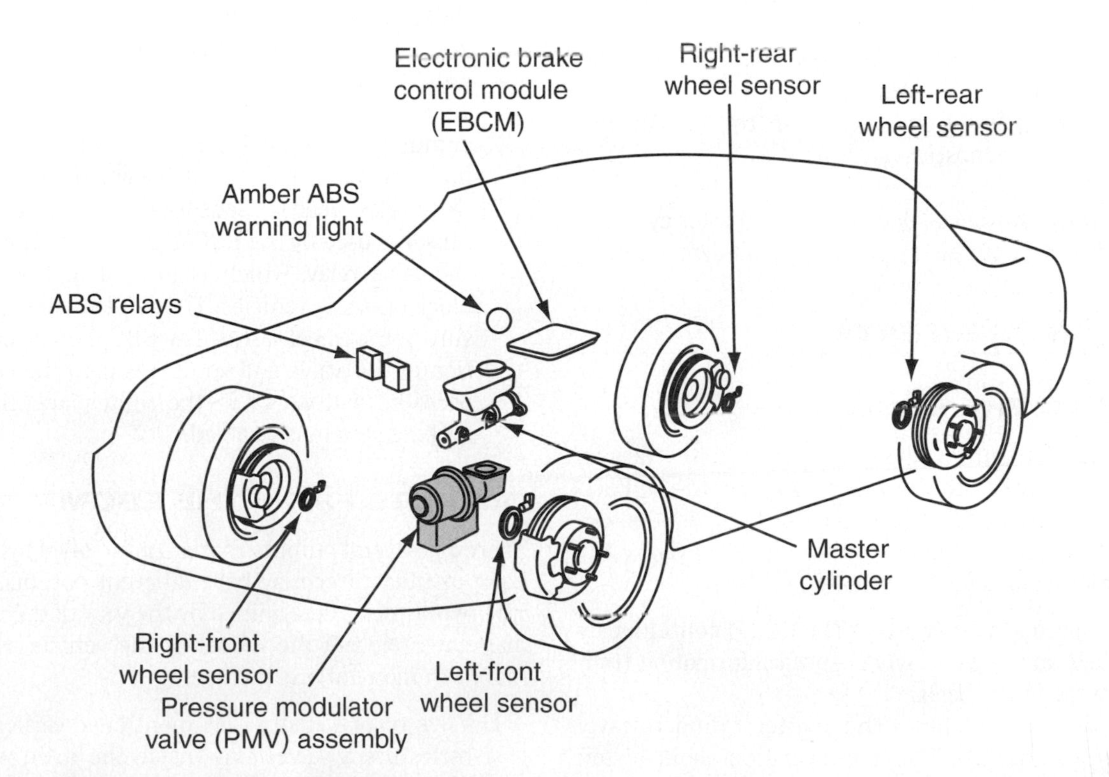

FIGURE 46–5 The pressure modulator valve is located below the hydraulic pump motor and is used to control the pressure to the calipers.

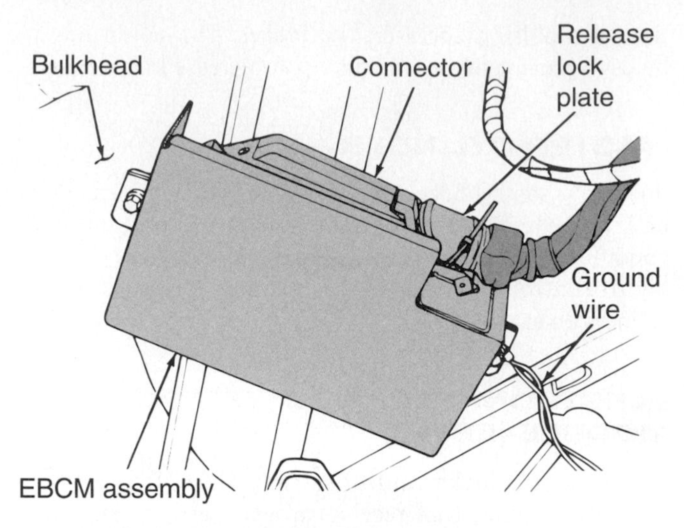

FIGURE 46-6 On some vehicles, the EBCM is located in the trunk area of the vehicle. (Courtesy of DaimlerChrysler Corporation)

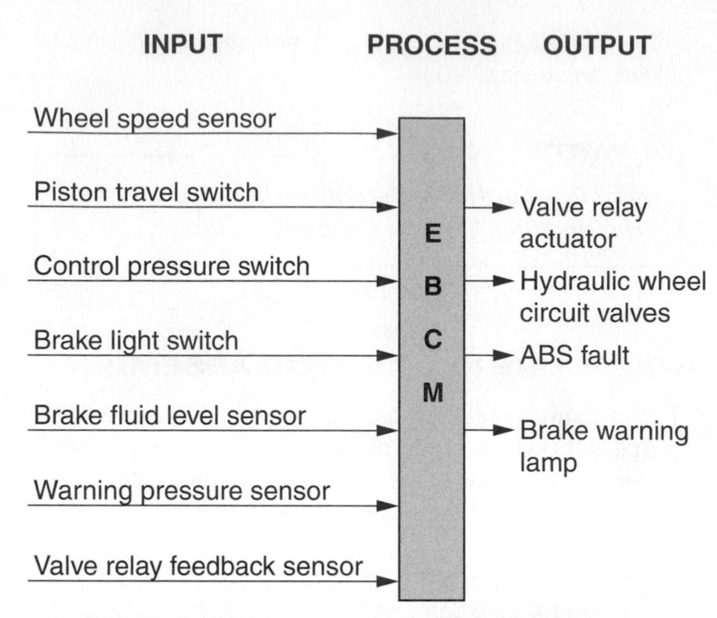

FIGURE 46-7 Various inputs and outputs are used in conjunction with the EBCM.

ADDITIONAL COMPONENTS ON ANTILOCK BRAKING SYSTEMS

Depending on the vehicle, manufacturer, and year, there are additional components of antilock braking systems. Some include:

- Antilock warning lamp—placed on the dashboard instrument cluster; used for monitoring ABS
- Fluid level sensor—located on the reservoir cap; senses the level of hydraulic brake fluid
- Pressure accumulator—located on the pressure modulator valve assembly; stores high pressure from the fluid pump
- Pressure switch—located on the accumulator; senses pressure in the accumulator
- Self-diagnostic function—located in the EBCM; provides various diagnostic checks on the antilock braking system
- Piston travel switches—located in the master cylinder; signal EBCM about piston travel for fault detection

46.3 ABS OPERATION

INPUTS TO THE EBCM

As with other computerized automotive systems, various sensors provide inputs to the EBCM (***Figure 46–7***). Each manufacturer may have more or less of these inputs, depending on the year and make of the vehicle. Seven of the more common input sensors include:

1. Wheel speed sensor (WSS)—The EBCM continuously monitors and receives wheel speed information from each of the four wheels.
2. Piston travel switches—The master cylinder travel switches signal the EBCM information about piston travel; mostly used for fault detection.
3. Control pressure switch—A control pressure switch is used as an input to the EBCM. This switch monitors the pressure in the system and is mostly used for fault detection.
4. Brake light switch—The brake light switch is used by the EBCM along with the travel switches for fault detection and to verify a proper brake pedal application.
5. Brake fluid level sensor—The brake fluid level sensor tells the EBCM when there is a lack of, or loss of, brake fluid. Knowing this information, the EBCM can disable the antilock function in the event of a loss of brake fluid.
6. Warning pressure sensor—The warning pressure sensor is used to tell the EBCM the pressure in the accumulator. Some antilock braking systems use an accumulator to hold high pressure in the system so high-pressure fluid is available when needed.
7. Valve relay feedback sensor—The valve relay feedback sensor is used by the EBCM to monitor the position of the valve relay, which is located in the main valve block on some vehicles. The valve relay provides pressure to the wheel valves. The EBCM monitors the position of this valve and compares it to the commanded position. If it is not in the right place, the antilock braking system is disabled.

OUTPUTS FROM THE EBCM

There are several outputs that are controlled by the EBCM. Each manufacturer may have different outputs. The exact type of outputs is determined by the year of the vehicle, the manufacturer, and the model of the vehicle. Four of the more common outputs include:

1. Valve relay actuator—As mentioned earlier, some vehicles use a valve relay inside the main valve block. The position of this valve is controlled by the EBCM.

When the EBCM energizes the valve relay, electrical power is provided to the four wheel valves.

2. Hydraulic wheel circuit valves—The EBCM controls each wheel circuit valve's position, which in effect controls the hydraulic brake pressure at each wheel.
3. ABS fault—The ABS fault line serves as a communication line from the EBCM to the antilock warning lamp on the dashboard. This electrical circuit is activated any time a condition is encountered that results in disabling of the antilock function.
4. Brake warning lamp—This output is used to turn on the red brake warning lamp if a condition is encountered by the EBCM that may result in a reduced braking ability. The brake warning lamp is also used to display ABS fault codes while in the diagnostic mode.

MODES OF OPERATION

There are several modes of operation on antilock braking systems. Each manufacturer may refer to these modes differently. The basic operation is still similar. The following typical modes for an antilock braking system are described in this section:

- Normal conditions
- Pressure hold
- Pressure drop
- Pressure increase

The normal condition mode operates when all four wheels are braking or slowing down equally. The other three conditions (pressure hold, drop, and increase) work when the computer senses an impending lockup. The cycle of pressure hold, drop, and increase occurs very rapidly, from 3 to 20 times per second.

PARTS LOCATOR

Refer to photo #104 in the front of the textbook to see what an ***electronic brake control module (EBCM)*** *looks like.*

Normal Conditions Refer to ***Figure 46–8***. Under normal conditions, when the brake pedal is depressed fluid pressure is sent out of the master cylinder to the modulator assembly. Inside the modulator assembly there are two modulator (solenoid) valves for each wheel caliper. One is an input, one an output modulator valve. Pressure from the master cylinder then goes through the normally open input modulator valves and into each caliper. The output modulator valve on the return line of

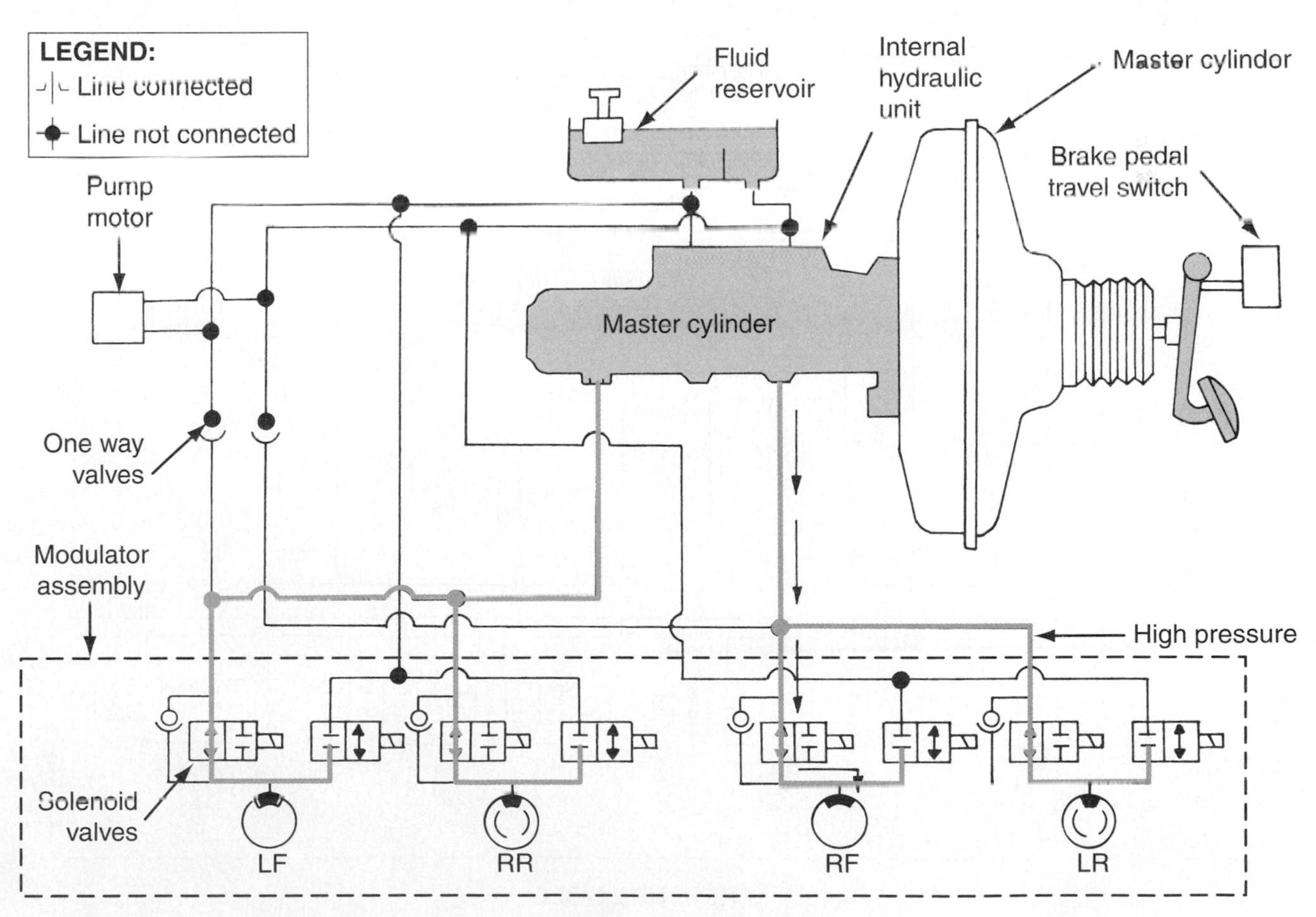

FIGURE 46–8 During normal conditions, the hydraulic pressure to each caliper is produced from the master cylinder.

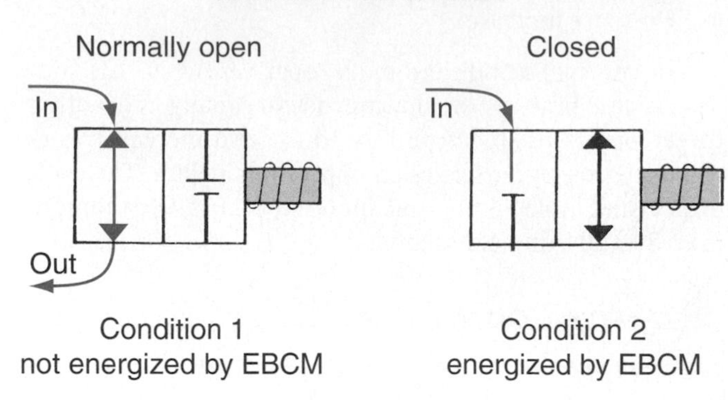

FIGURE 46–9 Inside the pressure modulator assembly there are small valves controlled by the computer. They are in either the normally open (not energized) or the closed (energized) position.

the caliper is closed. Thus, no fluid is sent back to the master cylinder.

When energized by the computer, the modulator valves are able to switch from a normally open to a closed position. ***Figure 46–9*** shows the two conditions for each of eight modulator valves. Each of the eight valves operates separately.

Pressure Hold When the EBCM determines that the wheel sensor shows an impending lockup, the first stage of the cycle is **pressure hold**. This means that no additional pressure can be produced from the master cylinder as the brake is applied further. The computer simply shuts off the pressure to the affected wheel. Refer to ***Figure 46–10***. This condition shows that the RF (right front) wheel is going to lock up. The EBCM energizes the input modulator valve so as to shut off further pressure from the driver. Note also that the output modulator valve is still closed, thus maintaining constant pressure in the caliper.

Pressure Drop Refer to ***Figure 46–11***. The next stage is called **pressure drop**. If the EBCM senses that the RF wheel will still lock up, it energizes the output modulator valve. This valve moves from a closed position to an open position. The result is that the internal pressure in the caliper is relieved. The hydraulic brake fluid is now sent back to the master cylinder.

Pressure Increase If, at this point, the EBCM senses that the lockup has been activated, the pressure needs to be increased again (***Figure 46–12***). Thus, the outlet modulator valve is de-energized, and the inlet modulator valve is energized. The modulator valve mode at this point is the same as normal conditions.

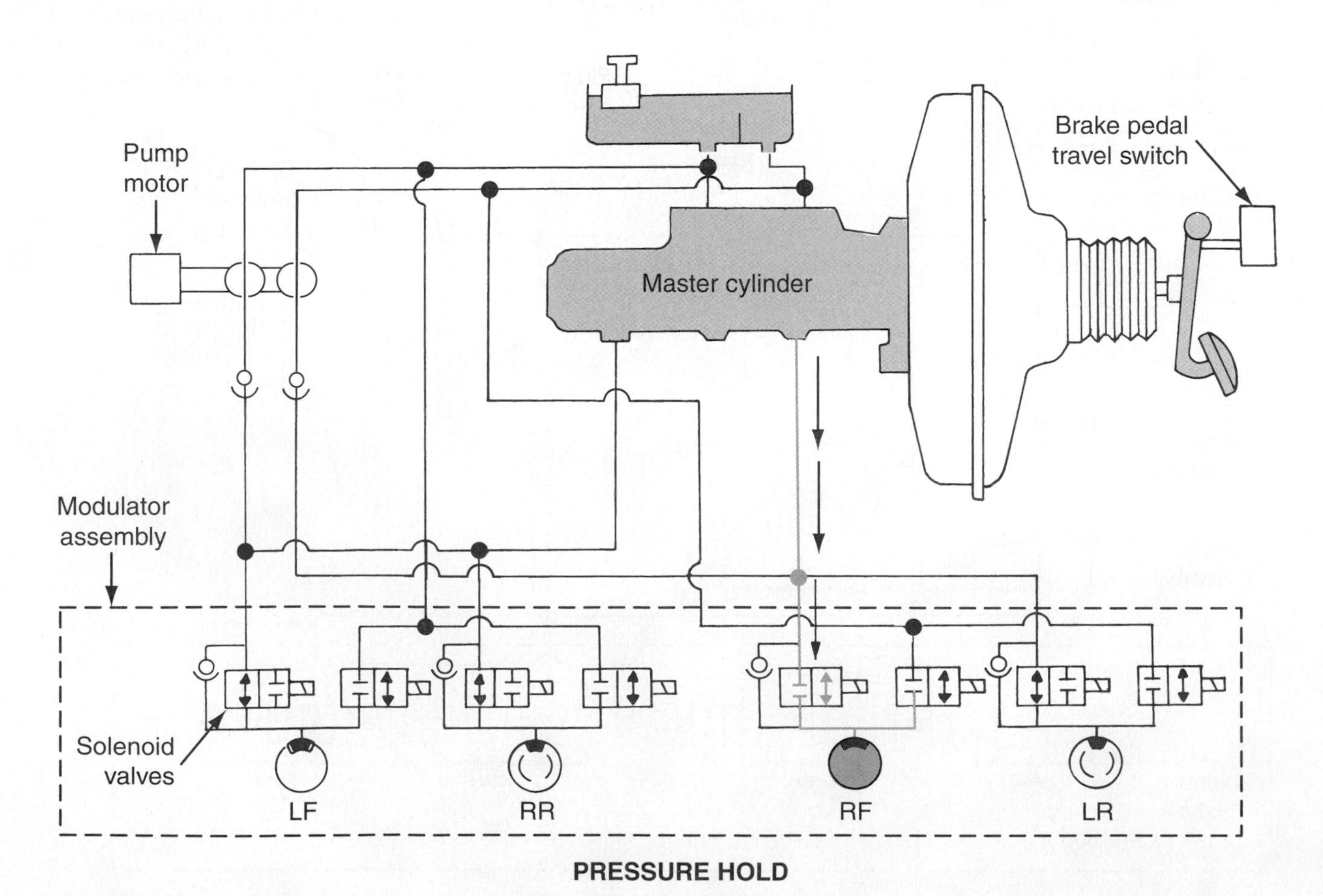

FIGURE 46–10 When an impending lockup of the right front (RF) wheel is sensed by the computer, the first mode of operation is the pressure hold. In this mode, hydraulic pressure is held constant at the RF caliper.

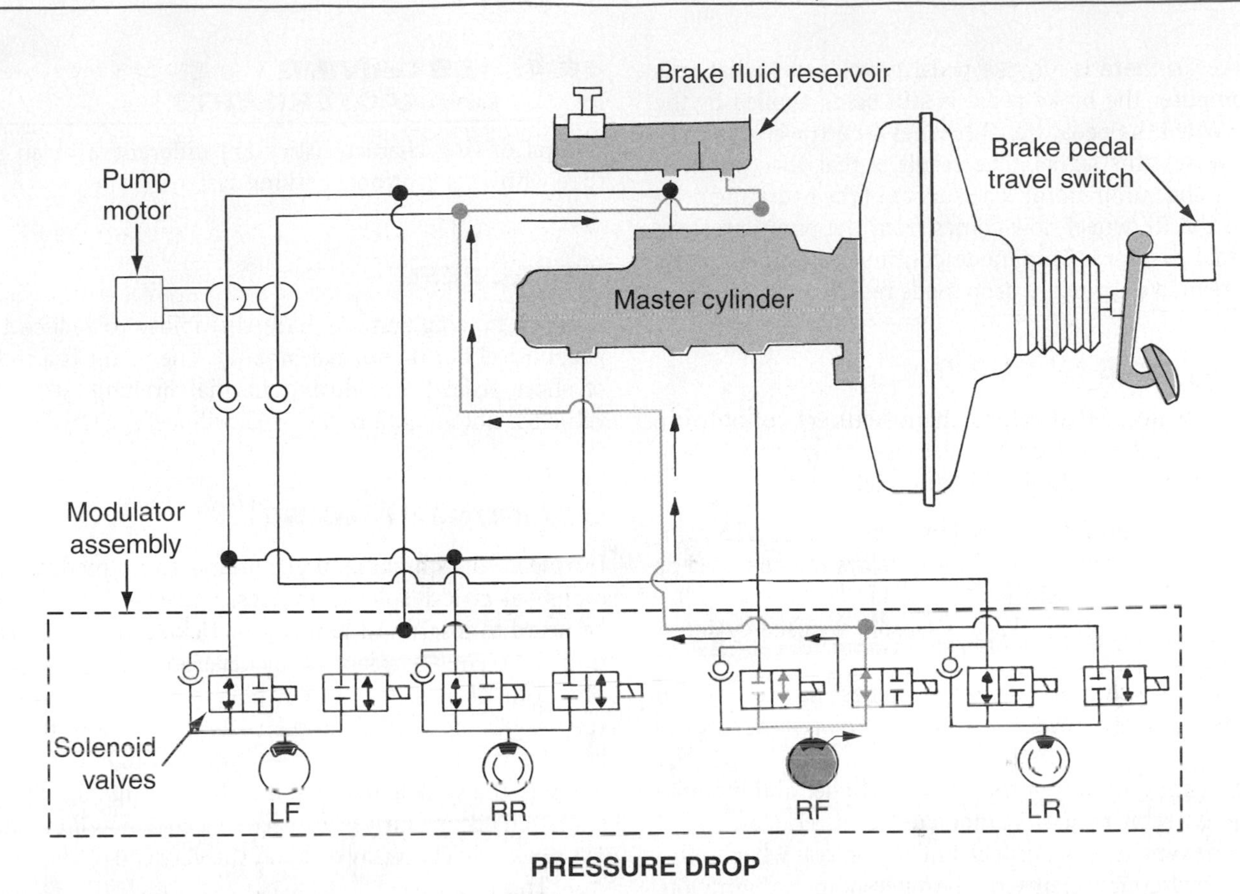

FIGURE 46–11 During the pressure drop mode, the RF caliper valves are energized so that hydraulic pressure is released at the RF caliper and returned to the brake fluid reservoir.

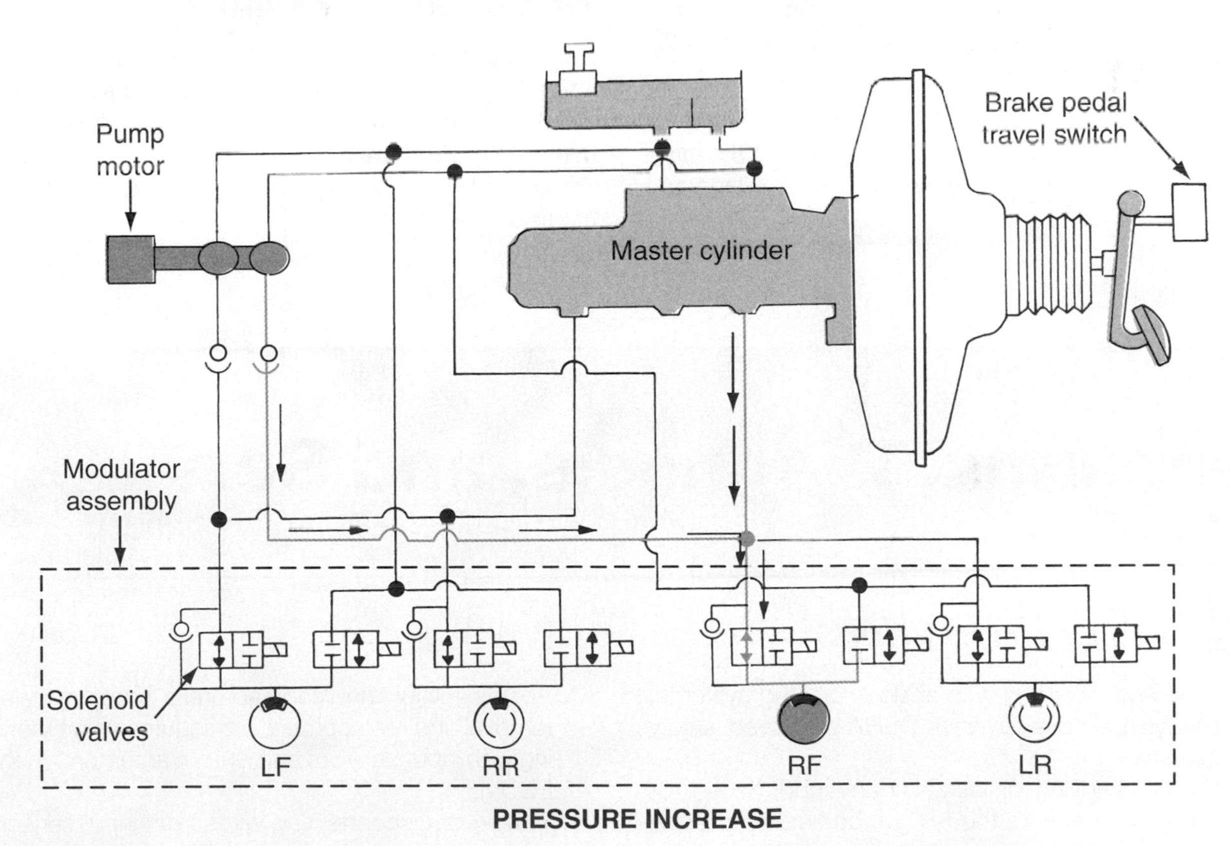

FIGURE 46–12 During the pressure increase mode, hydraulic pump motor pressure is used in addition to the master cylinder pressure to increase the hydraulic pressure again in the RF caliper.

However, there is a brake pedal travel switch that tells the computer the brake pedal is still being applied by the driver. When the pedal travel reaches 40% travel, the computer senses this signal. The result is that the computer turns on the pump motor. The source of the hydraulic pressure to the RF wheel now comes from the pump pressure and pedal pressure. This mode continues until either the pressure hold or pressure drop mode returns.

VARIATIONS IN DESIGN

It should be noted that vehicle manufacturers not only use the system just described but also use many variations to this design. Some of these variations include:

- Additional outputs are used on some vehicle ABS computers, for example, additional relays may be used as outputs from the EBCM.
- Several types of wheel speed sensors are used by different manufacturers.
- Motion sensors or mercury switches may be used to detect forward, reverse, or sideways motion during acceleration or deceleration.
- The configuration of the master cylinder and modulator assembly may vary among different vehicles.
- Some vehicles use antilock braking on rear wheels only.
- The hydraulic systems may be different in configuration.
- The type of solenoid valves may be shaped differently.
- Antilock braking systems can be either integrated (master cylinder and modular assembly in one unit) or nonintegrated (master cylinder and modular assembly built as separate units.

Even with these variations, the basic principles of antilock braking systems remain the same. Always refer to the manufacturer's specific service manual for correct design and operation.

46.4 ABS DRIVING CHARACTERISTICS

Several driving characteristics are different with an ABS than with a conventional braking system.

PEDAL FEEL

Antilock braking systems have the ability to reduce total pedal travel during normal braking. The result is a feeling of short pedal travel during normal braking. When the vehicle is not in motion, the pedal will feel springy.

COMPONENT NOISE

During antilock braking, brake pressures are modulated by cycling electrical (solenoid) valves. These cycling valves can be heard as a series of popping or ticking noises. In addition, the cycling may feel like a pulsation in the brake pedal. The cycle of pressure—build, hold, and drop—can occur very rapidly, up to 20 times per second. Generally, the pulsing does not cause pedal movement, although during hard braking, the vehicle may seem to be pulsing slightly. ABS operation occurs during all speeds except below 3 to 5 mph. Thus, wheel lockup may occur at the very end of an antilock stop. This is normal.

TIRE NOISE AND MARKS

During antilock braking, some wheel slip is desired (10 to 30%). Slip may result in some "skidding" of the wheel. This will depend on the road surface. Complete wheel lock normally leaves black tire marks on dry pavement. When in operation, antilock braking will leave noticeable light patch marks on the pavement.

Problems, Diagnosis, and Service

Safety Precautions

1. Whenever working with antilock braking systems, always be sure to wear OSHA-approved safety glasses.
2. When removing or installing any input sensor or output actuator to the EBCM, be sure to use the correct tools.
3. Always follow the manufacturer's recommended removal and replacement procedures when working with input sensors and output actuators on the EBCM.
4. Whenever disconnecting wires from input sensors or output actuators, always make sure to mark the

wires so that you know how they should be reattached.

5. Always use the recommended parts when replacing EBCM input sensors and output actuators.
6. If brake fluid spills while working on an antilock braking system, always clean up the spilled brake fluid as soon as possible to avoid slipping and causing injury.
7. Always be careful not to get brake fluid on the car's outside paint as it can damage the paint finish; it can even cause the paint to ripple or peel off.
8. Be careful not to breathe any of the dust particles produced from the brake pads or brake linings because these particles can cause permanent lung damage.

PROBLEM: Faulty Antilock Brakes

A vehicle with an antilock brake system has the antilock brake light on continuously. It should go off several seconds after the engine has started or the ignition switch is in the run position, but it stays on all the time.

DIAGNOSIS

Antilock braking systems have a built-in diagnostic system that helps the service technician identify and locate problems. Various diagnostic checks must be performed to determine the problem in antilock braking systems.

1. The first check is to visually inspect the brake system for leaks, low fluid, and disconnected wires. ***Figure 46–13*** shows one manufacturer's visual inspection chart.
2. Generally, each manufacturer has a functional check that must also be made to help diagnose the antilock braking system. A typical function check flowchart is shown for reference in ***Figure 46–14***.

VISUAL INSPECTION

ITEM	INSPECT FOR	CORRECTIVE ACTION
Brake fluid reservoir	Low fluid level	Add fluid as required. Determine cause of fluid loss and repair.
EBCM	Proper connector engagement External damage	Repair as required. Verify defect and repair as required.
Hydraulic assembly	Proper connector engagement External leaks Damaged wiring/connectors Control pressure switch Fluid level sensor defect	Repair as required. Repair leaks as required. Repair as required. Repair as required. Repair as required.
Pump/motor assembly and hoses	Proper assembly Damaged/leaking hoses Damaged wiring/connector	Install components properly. Repair hoses as required. Repair as required.
Parking brake	Full release	Operate manual release lever to verify operation. Adjust cable or repair release mechanism as required.
	Park brake switch	Verify correct operation. Repair as required.
Front- & rear-wheel speed sensors	Proper connector engagement Broken or damaged wires	Repair as required. Go to appropriate ABS code chart and verify fault. Correct as required.
OVPR & pump/motor relays	Proper connector engagement Loose wires or terminals	Repair as required. Repair as required.
Pump/motor ground	Corroded, broken, or loose eyelets	Repair as required.

VERIFY PROPER OPERATION AFTER REPAIRING ANY DEFECT. IF PROBLEM IS STILL PRESENT, PROCEED TO FUNCTIONAL CHECK.

FIGURE 46–13 Always visually inspect the antilock braking system before completing other diagnostic checks. (Courtesy of DaimlerChrysler Corporation)

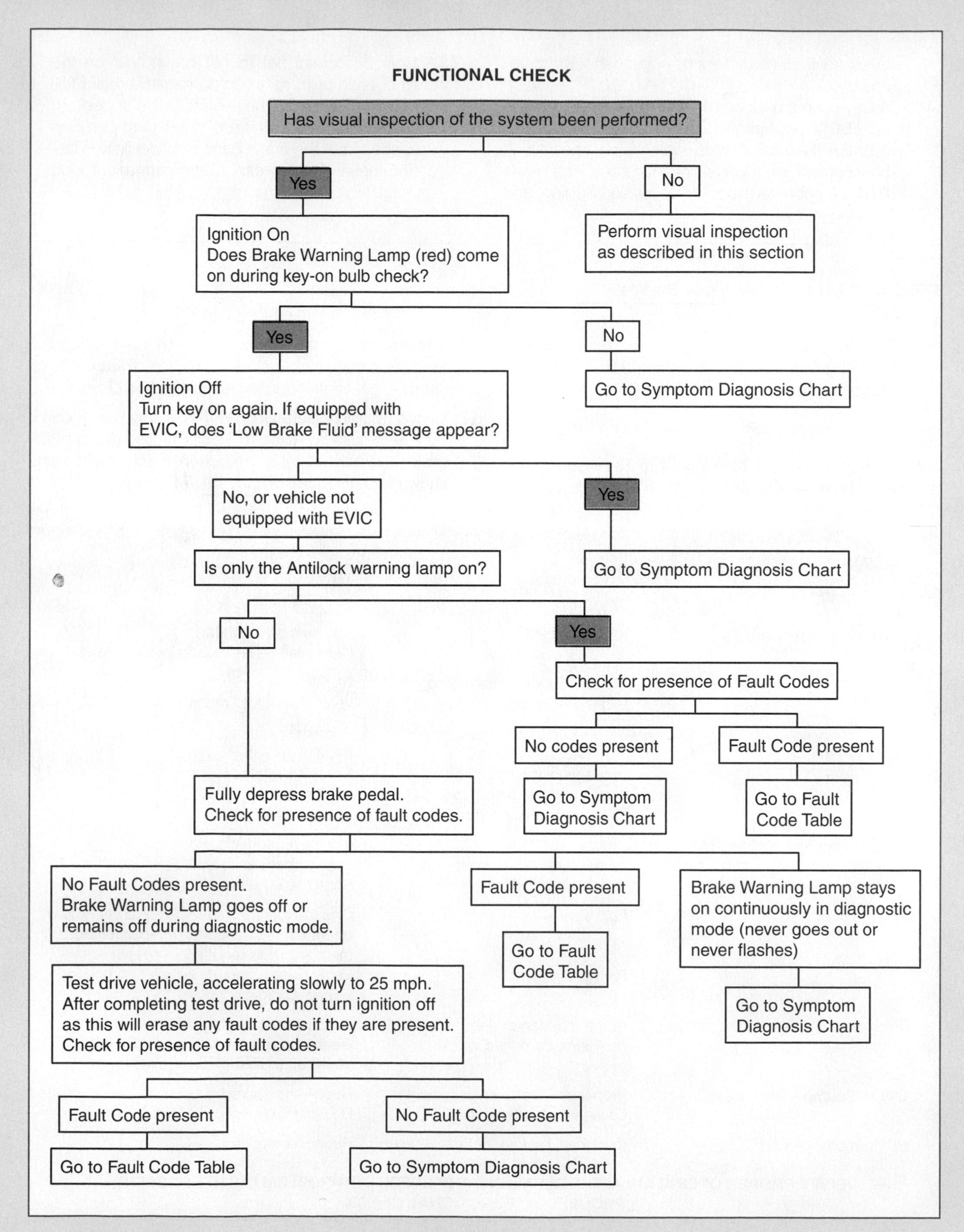

FIGURE 46-14 This flowchart shows how to perform a functional check to help diagnose the antilock braking system. (Courtesy of DaimlerChrysler Corporation)

FAULT CODE TABLE

FAULT CODE	DESCRIPTION
1	Left Front ABS Wheel Circuit Valve
2	Right Front ABS Wheel Circuit Valve
3	Right Rear ABS Wheel Circuit Valve
4	Left Rear ABS Wheel Circuit Valve
5	Left Front Wheel Speed Sensor
6	Right Front Wheel Speed Sensor
7	Right Rear Wheel Speed Sensor
8	Left Rear Wheel Speed Sensor
9	Left Front/Right Rear Wheel Speed Sensor
10	Right Front/Left Rear Wheel Speed Sensor
11	Replenishing Valve
12	Valve Relay
13	Excessive Displacement or Circuit Failure
14	Piston Travel Switches
15	Brake Light Switch
16	EBCM Error

FIGURE 46–15 During a functional check, the service technician checks for one or more of the trouble codes stored in the computer's memory. (Courtesy of DaimlerChrysler Corporation)

3. As part of a functional check, the service technician determines if there are any fault codes stored in the computer memory. ***Figure 46–15*** shows one manufacturer's trouble code table. The trouble codes are read in a manner similar to that used for engine trouble codes. Often the light used to check the trouble code is located directly on the computer in the trunk area.
4. On the basis of the results of this check, the service technician then goes to a flowchart dealing with the specific trouble code. ***Figure 46–16*** shows a typical flowchart relating to a faulty wheel speed sensor signal.

SERVICE

The service necessary on antilock braking systems is much the same as that for conventional brakes. However, often wheel sensors, wiring, the pressure modulating valves, and associated parts may need replacement. Refer to the manufacturer's service manual for correct service procedures on these parts.

In addition to the various components that can be checked and replaced, each manufacturer uses a set procedure to engage or enter the self-diagnostic fault code mode so that the fault codes stored on the EBCM can be determined. One

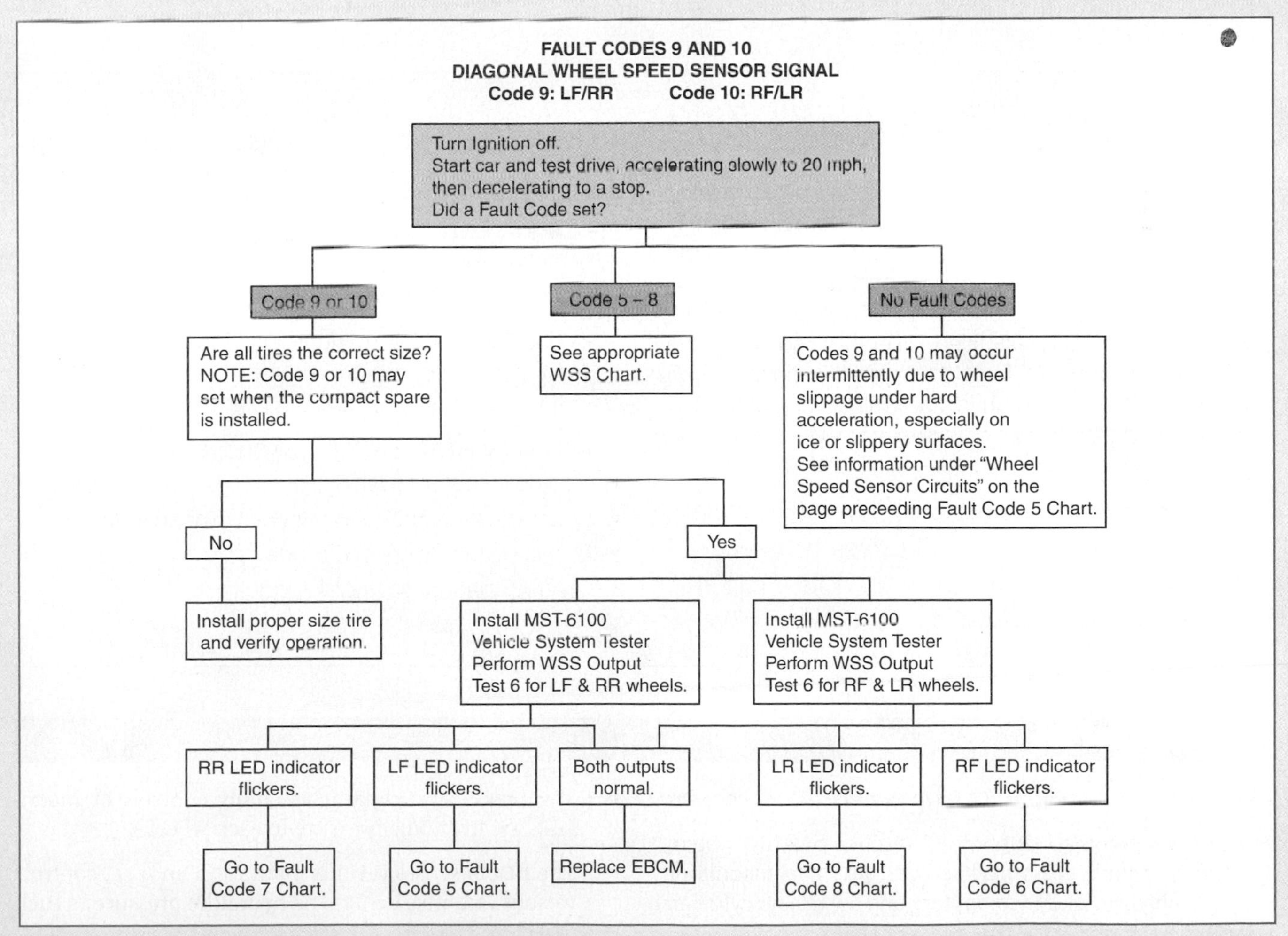

FIGURE 46–16 This flowchart shows the correct diagnostic procedure for checking the wheel speed sensors. (Courtesy of DaimlerChrysler Corporation)

manufacturer suggests that the service technician use the Tech II scanner. But not all service technicians have such a tool immediately available. So, to engage the self-diagnostic fault code mode and check the diagnostic fault codes, observe the blinking brake light on the dashboard by using the following general procedure:

1. First turn on the ignition. On many vehicles, when the ignition is turned off, the fault codes are erased. If the ignition has been turned off, the vehicle will need to be driven again to engage or store the fault codes. To test-drive the vehicle to set fault codes:
   a. Accelerate slowly to 25 miles per hour
   b. Repeat acceleration and stop two times
   c. After second stop, the fault code should be entered or set
2. Release the parking brake. This must be released so that the Brake light will not be turned on by the parking brake.
3. Check to see if a fault has been detected. Either the Brake warning lamp or the Antilock warning lamp should be illuminated. If not, verify the complaint by test-driving the vehicle. Proceed slowly, checking the operation of the service brakes.
4. Once a fault is detected, without turning off the ignition, depress the brake pedal fully with a firm pedal effort.
5. After about 5 seconds, if a fault code is present, the red Brake warning lamp will begin to flash. Count the number of times the Brake warning lamp flashes and compare the number to the fault code numbers. The number of times the lamp flashes indicates the number of the fault code.
6. When the lamp begins to flash, the brake pedal may be released. If the brake pedal is held down continuously, the fault code will repeat in about 10 seconds. Remember, the code is stored in memory as long as the ignition is on, but is erased as soon as the ignition is turned off.
7. If the lamp does not come on at all, crank the engine to perform a Brake warning lamp bulb check. If the lamp is bad, replace the lamp and redo the diagnostic fault check.
8. To exit ABS diagnostic service mode, simply release the brake pedal.

Service Manual Connection

There are many important specifications to keep in mind when working with antilock braking systems. To identify the specifications for your engine, you will need to know the VIN (vehicle identification number) of the vehicle, the type and year of the vehicle, and the type of engine. Although they may be titled differently, some of the common antilock braking system specifications (not all) found in the service manuals are listed here. Note that these specifications are typical examples. Each vehicle and engine may have different specifications.

Common Specification	Typical Example
Brake switch resistance	Less than 2 ohms
Pump motor voltage drop	12 volts
Brake travel switch resistance	5 ohms
Pressure switch in open position resistance	Infinite resistance
Wheel speed sensor resistance	800–1,800 ohms

In addition, service manuals provide specific directions for various service procedures. Some of the more common procedures include:

- Manual bleed for antilock braking systems
- Hydraulic system flushing
- Brake pressure modulator valve replacement
- Engaging the diagnostic mode
- Clearing diagnostic trouble codes

SUMMARY

The following statements will help to summarize this chapter:

- The basic components of an antilock braking system (ABS) include the wheel sensors, pressure modulator assembly, an EBCM (computer), and the master cylinder.
- The wheel sensors determine and monitor the speed of each wheel.
- The pressure modulator assembly controls or modulates the hydraulic pressure to each wheel.
- The EBCM senses various inputs and, in turn, controls various outputs such as the hydraulic pressure to each wheel.

- The master cylinder controls the fluid pressure to the wheel cylinders.
- Other ABS components include fluid pressure sensors, pressure accumulators, pressure switches, piston travel switches, warning lamps, and self-diagnostic function check modes in the EBCM.
- There are generally four modes of operation during antilock braking, including normal condition, pressure hold condition, pressure drop condition, and pressure release condition.
- A vehicle that has antilock braking may exhibit different driving characteristics from a conventional braking system, including pedal feel, pedal travel, and clicking noises, and, during hard braking, tire marks may be observed on the road surface.
- When diagnosing and servicing an ABS, always do a visual inspection first.
- After a visual inspection, a functional check should be made to determine the type of problem and the fault code.
- After the problem has been identified, continue with the manufacturer's suggested diagnostic checks on individual components by following the proper troubleshooting chart.

TERMS TO KNOW

Can you explain each of the following terms? Review the chapter until you can use each term correctly.

Antilock braking systems (ABS)
Electronic brake control module (EBCM)
Pressure drop
Pressure hold
Pressure modulator
Slip
Wheel speed sensor

REVIEW QUESTIONS

Multiple Choice

1. When a tire locks up completely during a stop, it is said that there is ________ slip.
a. Minimum
b. 100%
c. 50%
d. 25%
e. 10%

2. Which of the following factors affect slip on a tire?
a. Vehicle weight
b. Axle load distribution
c. Braking in a turn
d. All of the above
e. None of the above

3. Antilock braking systems are designed to prevent wheel lockup under:
a. Light braking conditions
b. Slippery road conditions
c. Heavy vehicle load conditions
d. Lightly loaded conditions
e. Maximum rpm conditions

4. Which of the following is not a component on an antilock braking system?
a. Wheel speed sensors
b. Pump and motor
c. Electronic brake control module (EBCM)
d. Front suspension springs
e. Hydraulic assembly

5. The ________ is used to sense the speed of each wheel.
a. Wheel speed sensor
b. Pump and motor
c. Electronic brake control module (EBCM)
d. Front suspension spring
e. Hydraulic assembly

6. Which of the following is a typical mode of operation for an antilock brake system?
a. Normal conditions
b. Pressure hold
c. Pressure drop
d. Pressure increase
e. All of the above

7. How many modular valves are there for each wheel caliper on an antilock brake system?
a. One
b. Two
c. Three
d. Four
e. Five

8. Which of the following is not a typical input sensor for the EBCM?
a. Wheel speed sensor
b. Piston travel switch
c. Suspension position switch
d. Brake light switch
e. Control pressure switch

9. Which of the following is an output actuator controlled by the EBCM?
a. Valve relay actuator
b. Hydraulic wheel circuit valves
c. ABS fault
d. Brake warning lamp
e. All of the above

10. The EBCM is often located in the ________ of the vehicle.
a. Trunk area
b. Wheels
c. Engine
d. Suspension
e. Front seat area

The following questions are similar in format to ASE (Automotive Service Excellence) test questions.

11. *Technician A* says that the first mode of operation when an impending lockup may occur is called the pressure hold. *Technician B* says that in the pressure hold, no additional hydraulic pressure can be produced. Who is correct?
a. A only c. Both A and B
b. B only d. Neither A nor B

12. *Technician A* says that one output from the EBCM is the hydraulic wheel circuit valves. *Technician B* says that one output from the EBCM is the brake warning lamp. Who is correct?
a. A only c. Both A and B
b. B only d. Neither A nor B

13. On a car with antilock brakes, the engine has been shut off but the ignition is still on. *Technician A* says that the diagnostic fault codes are still in the memory of the EBCM. *Technician B* says that a button must be pushed to identify the diagnostic fault code. Who is correct?
a. A only c. Both A and B
b. B only d. Neither A nor B

14. *Technician A* says to always perform a visual inspection before servicing any component on an antilock braking system. *Technician B* says to always perform a functional check when servicing the ABS. Who is correct?
a. A only c. Both A and B
b. B only d. Neither A nor B

15. *Technician A* says that on an ABS, the hydraulic system is being pulsed by the pressure modulator valves. *Technician B* says 1-4 speed sensors feed data into the EBCM to determine wheel speed. Who is correct?
a. A only c. Both A and B
b. B only d. Neither A nor B

Essay

16. Define the purpose of the functional check on an antilock braking system.

17. Define the term *slip* in relationship to antilock braking systems.

18. Explain the purposes of at least four inputs to the EBCM.

19. Define how a wheel speed sensor operates.

20. Explain hydraulically what happens in each of the four modes of operation on an antilock braking system.

Short Answer

21. The ________ takes low-pressure brake fluid and pressurizes it for storage in an accumulator.

22. On an antilock braking system, the ________ is located on the brake fluid reservoir cap.

23. The ________ located in the EBCM provides various diagnostic checks on the antilock braking system.

24. On an ABS, the ________ is measured at each wheel, and this information is sent to the computer.

25. Two driving characteristics that may be noticed with antilock braking systems are ________ and ________.

Applied Academics

TITLE: Uncontrolled Braking

ACADEMIC SKILLS AREA: Science

NATEF Automobile Technician Tasks:

The service technician demonstrates an understanding of friction and its effects on linear and rotational motion in the brake system.

CONCEPT:

Antilock brake systems have been designed to control vehicles in heavy braking conditions. The antilock brake system is designed to keep the vehicle straight and in control while braking on dangerous or slippery road surfaces. The diagram that follows shows some of the causes of uncontrolled braking. First, the road surface may be wet and slippery with ice or snow. In addition, slippery conditions may not be constant. Second, there may be changes in the road surface. For example, the road may be made of tar, then suddenly change to concrete. Third, the road conditions may also change—sand or gravel bumps, heaves, and potholes may be encountered. Fourth, the wheel load may be different depending on the weight of the vehicle. Fifth, the tires may be worn, thus causing a different friction between the tires and the road surface. Any or all of these conditions can combine to cause various degrees of uncontrolled braking.

CHANGES IN FRICTION

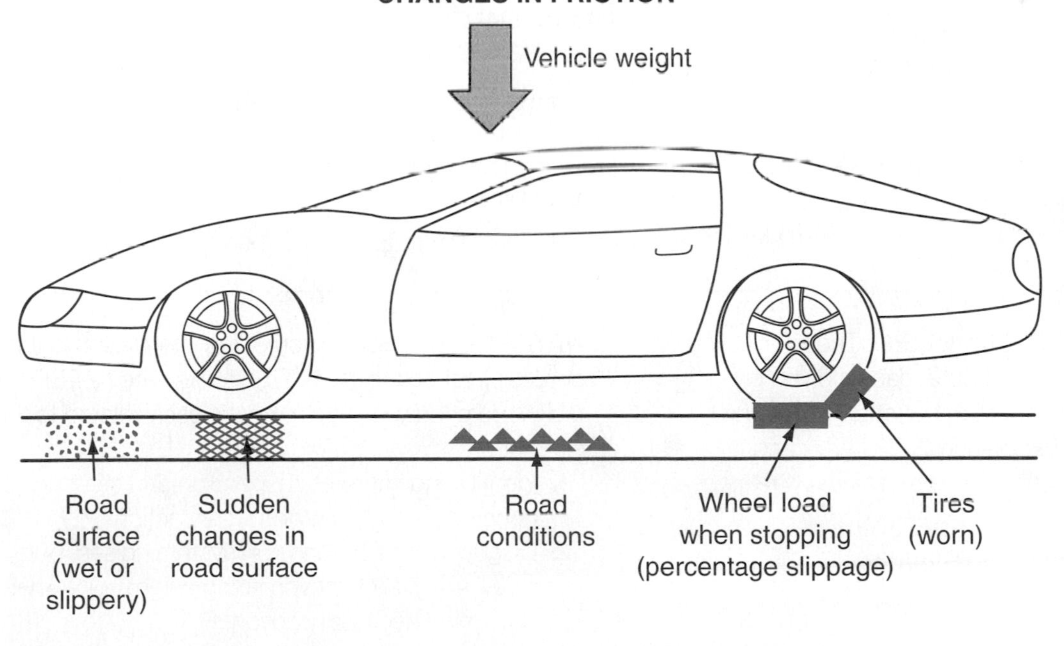

APPLICATION:

Can you explain, in terms of friction, how wheel load can change the friction between the tire and the road surface? What effect would there be on the friction between the wheel and road surface if new tires are placed on the vehicle? Why would wet or dry surfaces change the friction between the road and the tires?

CHAPTER 47 Suspension Systems

OBJECTIVES

After studying this chapter, you should be able to:

- Define the parts and operation of the front suspension system.
- Define the parts and operation of the rear suspension system.
- Analyze the purpose, parts, and operation of different types of shock absorbers.
- Compare the MacPherson strut suspension with other suspension systems, including parts and operation.
- Define the operation of computer-controlled suspension systems.
- Identify the purpose and operation of automatic level control and air suspension systems.
- State common problems, their diagnosis, and service suggestions concerning different types of suspension systems.

INTRODUCTION

The suspension system of a car is used to support its weight during varying road conditions. The suspension system is made of several parts and components. These include both the front and rear suspensions, the shock absorbers, and the MacPherson strut system. The objective of this chapter is to analyze the parts, operation, and diagnosis/service of different suspension systems.

47.1 FRONT SUSPENSION

PURPOSE OF THE FRONT SUSPENSION

The purpose of the front suspension is to support the weight of the vehicle. The suspension is also designed to provide a smooth passenger ride over varying road conditions and speeds.

Certification Connection

ASE Connection: The information in this chapter can help you prepare for the National Institute for Automotive Service Excellence (ASE) certification tests. The test and content area most closely related to this chapter are:

Test A4—Suspension and Steering

- **Content Area**—Suspension Systems Diagnosis and Repair

NATEF Connection: Much of the information in this chapter is related to the NATEF tasks. The NATEF tasks and priority numbers most closely related to this chapter are:

1. Identify and interpret suspension and steering concerns; determine necessary action. P-1
2. Diagnose strut suspension system noises, body sway, sway, and uneven riding height concerns; determine necessary action. P-1
3. Remove, inspect, and install upper and lower control arms, bushings, shafts, and rebound bumpers. P-3
4. Remove, inspect, and install upper and/or lower ball joints. P-2

5. Remove, inspect, and install steering knuckle assemblies. P-2
6. Lubricate suspension and steering systems. P-2
7. Remove, inspect, and install coil springs and spring insulators. P-2
8. Remove, inspect, and install transverse links, control arms, bushings, and mounts. P-2
9. Remove, inspect, and install leaf springs, leaf spring insulators (silencers), shackles, brackets, bushings, and mounts. P-3
10. Inspect, remove, and replace shock absorbers. P-1
11. Remove, inspect, and service or replace front and rear wheel bearings. P-1

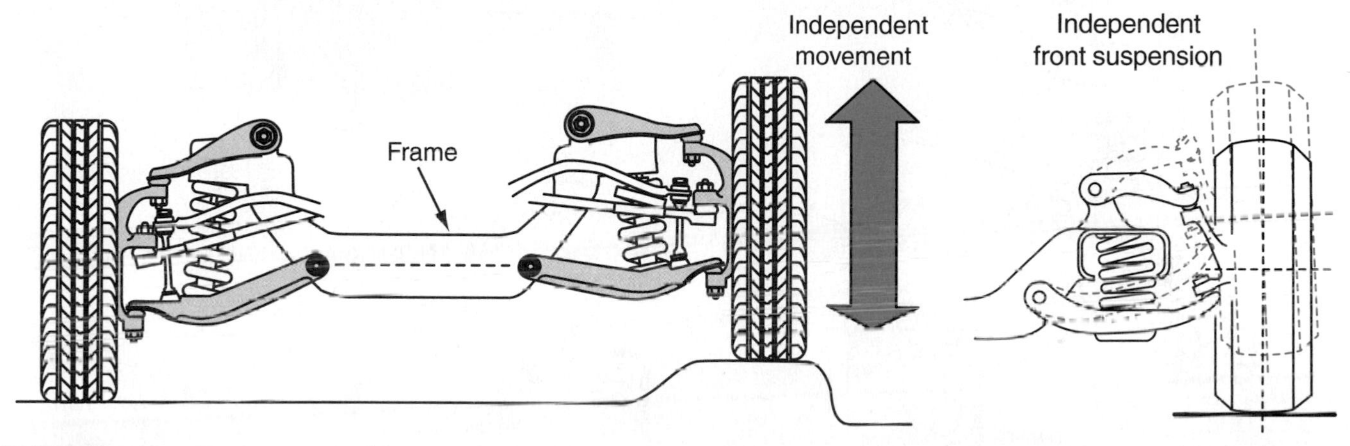

FIGURE 47–1 The front suspension is called independent front suspension. This means that each wheel acts independently when going over a bump.

There are several types of front-end suspension systems. Automobiles commonly use the independent front suspension system. This means that each wheel is independent from the other. For example, if the left wheel goes over a bump in the road, only the left wheel will move up and down (***Figure 47–1***). Certain types of trucks and other heavy-duty vehicles may use an I-beam suspension. This system has one main beam connecting each front wheel, so it is not an independent front suspension.

PARTS OF THE FRONT SUSPENSION SYSTEM

Although there are different types of front suspension systems, many of the parts are the same. ***Figure 47–2*** shows a common type of front suspension system and the related parts. These parts include:

- Ball joints (both upper and lower)
- Control arms, shaft bushings, and shims
- Sway bar, bushings
- Strut rod, bushings
- Coil springs
- Stabilizers
- Shock absorbers
- Steering knuckle and spindle

These parts are assembled to provide the entire front suspension. Each of these parts becomes a vital link in the front suspension operation. They must work properly to ensure driving safety and comfort.

Ball Joints The **ball joints** connect the spindle and steering knuckle to the upper and lower control arms. They are designed to do several things. The ball joints must carry the weight of the vehicle. They provide a pivot point for the wheel to turn. They also allow for vertical movement of the control arms when the vehicle goes over irregularities in

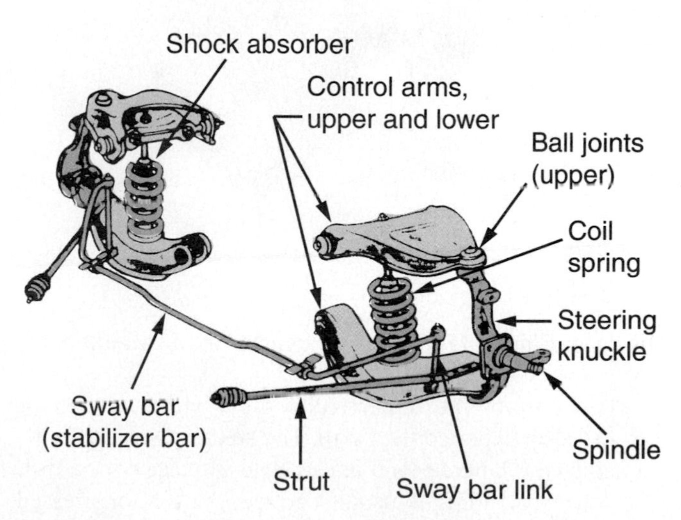

FIGURE 47–2 The parts of the front suspension system. (Courtesy of Dana Corporation)

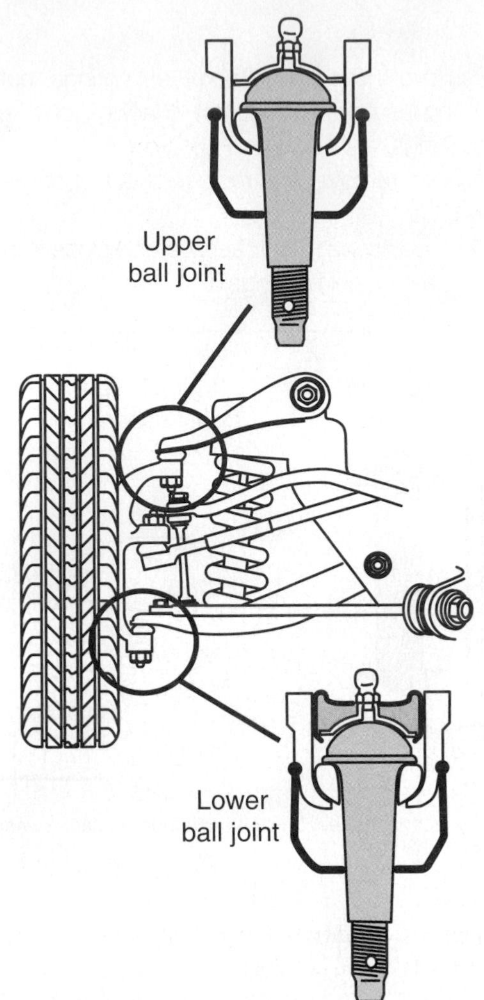

FIGURE 47–3 Ball joints connect the control arms to the steering knuckle.

the road. ***Figure 47–3*** shows a typical set of ball joints for upper and lower control arms.

The frame of the upper ball joint is either riveted or bolted to the upper control arm. The steering knuckle is attached to the tapered stud and is held in place with a nut. The lower ball joint is usually bolted, riveted, or pressed into the lower control arm (***Figure 47–4***). The steering knuckle is placed on the tapered stud and held in place with a nut. A rubber boot is placed around the assembly to keep grease in and dirt out.

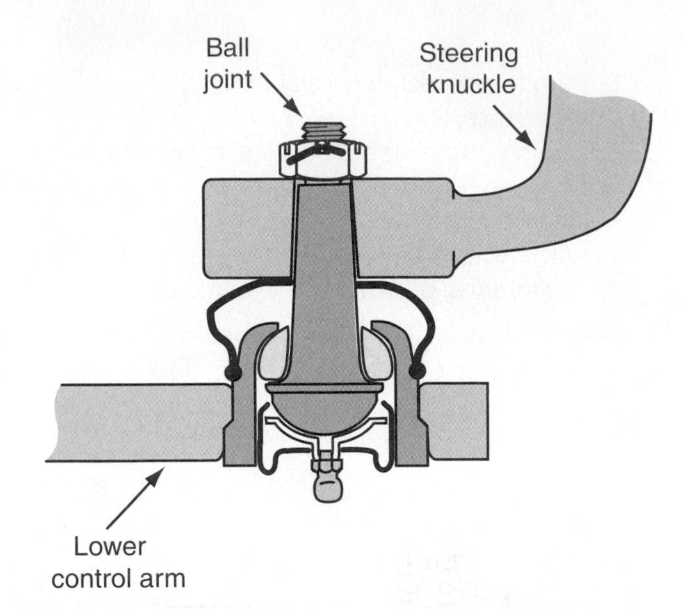

FIGURE 47–4 This ball joint connects the lower control arm to the steering knuckle.

Car Clinic: BALL JOINTS

CUSTOMER CONCERN:
Noise Heard in Front Suspension
The customer complains that there is a noise coming from the front end. The noise occurs only when entering or exiting the driveway at the customer's home. What is the problem?

SOLUTION: Noise such as this is often associated with the front ball joints. Test drive the vehicle, making slow turns, especially up and down a driveway ramp. The most likely cause of the problem is a worn lower ball joint on the left or the right side of the vehicle. Inspect the ball joints and replace where necessary.

Control Arms There are two **control arms**: an upper control arm and a lower control arm. Several arrangements are used for the control arms. There are single pivot control arms, double pivot control arms, and short and long control arms. ***Figure 47–5*** shows a comparison of control arms. The type of control arm depends on the year and manufacturer of the vehicle.

The other end of the control arm is attached to the frame of the vehicle. ***Figure 47–6*** shows how the upper control arm is attached to the frame. The upper control arm has a shaft that is bolted to the frame. The ends of the shaft carry bushings that are attached to the control arm. Shims are

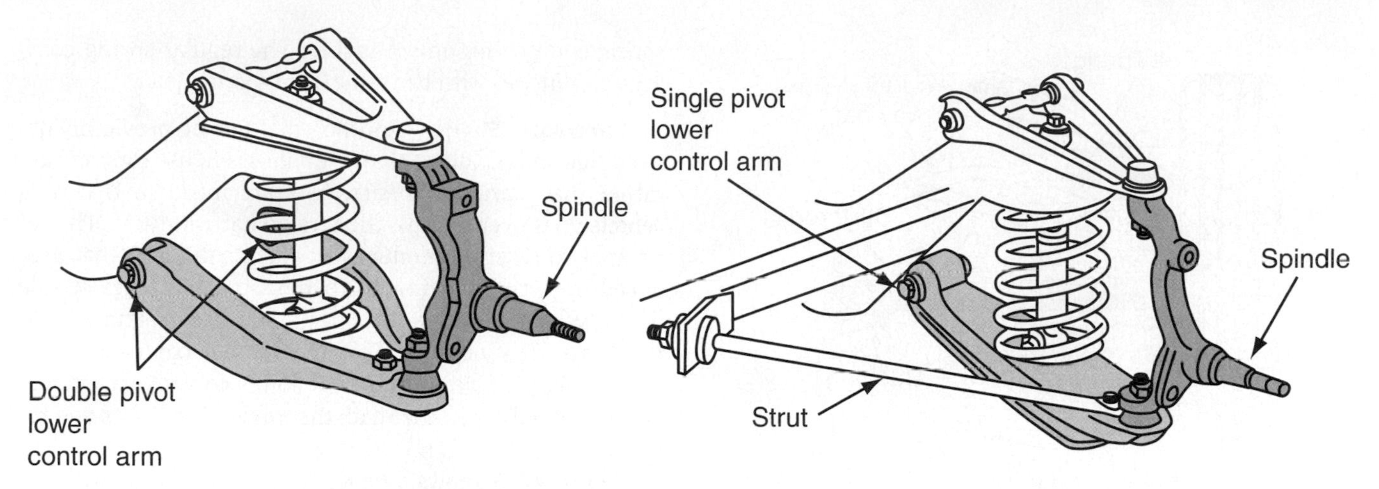

FIGURE 47–5 Either a single or a double lower control arm can be used. The single arm also uses a strut for support.

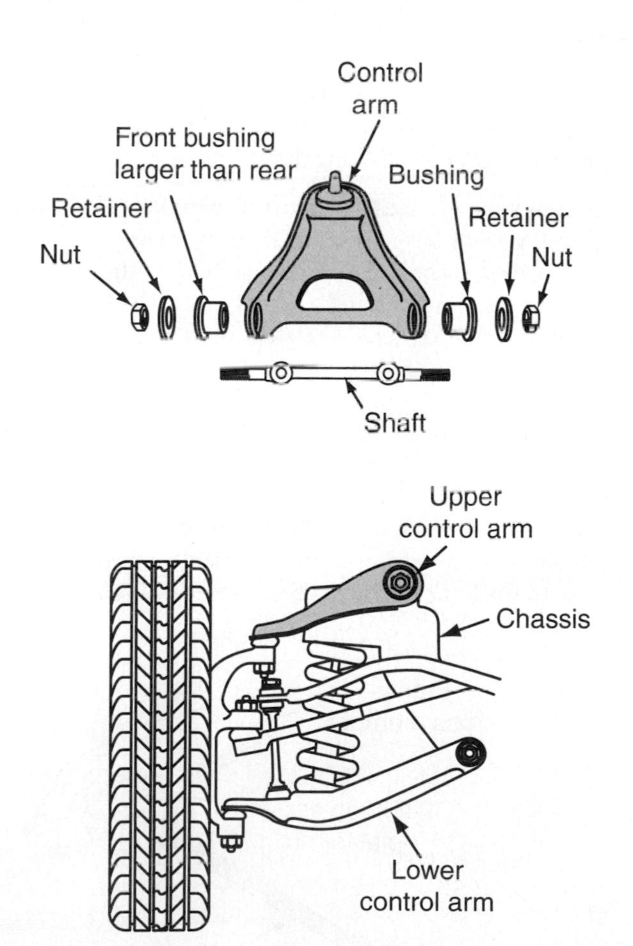

FIGURE 47–6 The parts of the upper control arm. The upper control arm is attached to the frame by bushings, shims, and the upper control arm shaft.

used to adjust the position of the shaft on the frame for alignment of the front suspension. The lower control arm is attached to the frame by bushings and bolts.

Car Clinic:
CONTROL ARM

CUSTOMER CONCERN:
Clunking Noise during Braking
The customer complains that when the vehicle's brakes are applied, there is a clunk from the front end. What could be the problem?

SOLUTION: Often a clunking noise is heard during braking if there is an excessively worn control arm bushing. If the bushing is worn, the control arm is able to move back and forth during braking or rapid acceleration. The clunking sound comes from the control arm shifting back and forth because of the braking momentum. Check the bushing in the control arms for wear. Look for shiny wear spots on the control arm as well. The solution is to replace the bushing in the control arms.

Sway Bar and Link (Stabilizer Bar and Link) The **sway bar** and sway bar link are also called the **stabilizer bar** and link. The sway bar link connects the lower control arm to the sway bar as shown in ***Figure 47–7***. The sway bar twists like a **torsion bar** during turns. It transmits cornering forces from one side of the vehicle to the other. This helps equalize the wheel loads and prevents excessive leaning of the car on turns. The sway bar link and sway bar are attached to the frame with rubber bushings and bolts.

Strut Rods The **strut rod** is used on vehicles that have single pivot lower control arms. They can be located either in front of or behind the control rod. They are designed to retain the lower control arms in their intended positions. They also provide a method of keeping the wheel in the right position for alignment. ***Figure 47–7*** shows the position of the strut rod. The sway bar and the sway bar link are also shown.

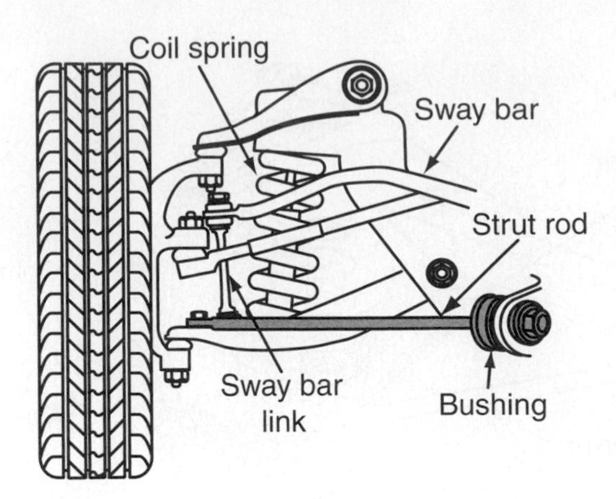

FIGURE 47-7 A strut rod is used to support the single pivot lower control arm.

Coil Springs The coil springs support the car's weight, maintain the car's **stance**, or height, and correctly position all the other suspension parts. Thus, if a spring sags a slight amount, the tires, shocks, ball joints, and control arms all work outside their normal positions. This condition can cause excessive or abnormal wear throughout the suspension systems.

Springs may be very flexible (regular) or very stiff (heavy-duty). The purpose of the springs is to absorb road shock and then return to their original position. A stiff spring produces a rough ride. A flexible spring may cause the vehicle to bounce too much. The best combination is to use a softer spring with a shock absorber.

Figure 47–8 shows the difference in compression of the springs when a 2,000-pound load is applied. The heavy-duty spring compresses only 4 inches. The regular spring compresses 5 inches with the same 2,000-pound load.

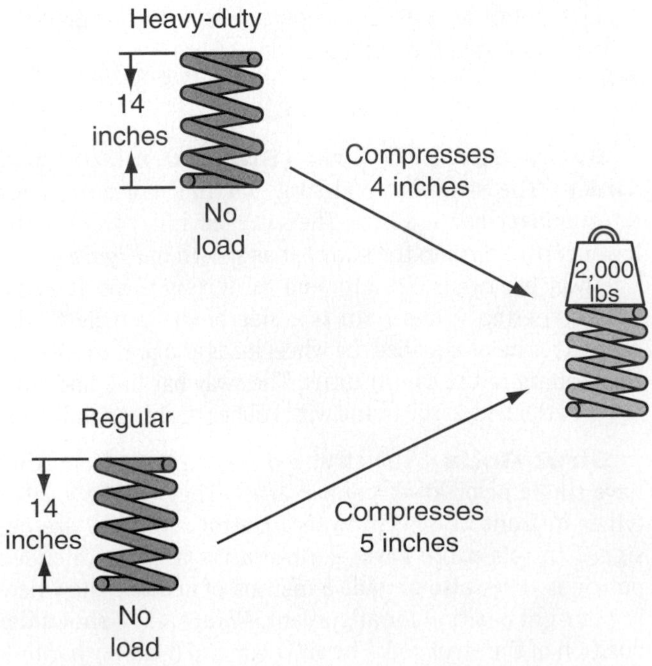

Heavy-duty
14 inches
No load
Compresses 4 inches
2,000 lbs
Regular
14 inches
No load
Compresses 5 inches

FIGURE 47-8 A heavy-duty coil spring will not compress as much under a load as the regular coil spring.

Torsion Bars Another method of providing desired ride and handling characteristics is to use torsion bars rather than springs. Torsion bars are made so that as a vehicle goes over bumps, the torsion bar will twist. The resistance to twisting produces an effect similar to that produced by springs. A torsion bar is attached to each side of the vehicle. One end of the torsion bar is attached to the frame. The other end of the torsion bar is attached to the lower control arm. As the lower control arm moves because of bumps in the road, the torsion bar twists and reduces the car's motion.

Figure 47–9 shows a typical torsion bar installation. In this case, the torsion bar is connected between the movement of the lower control arm and the nonmoving frame of the vehicle. As the lower control arm moves up and down over bumps in the road, the torsion bar continually twists, causing the up-and-down motion to be absorbed by the twisting motion. Note also that there is an adjuster nut on the frame end of the torsion bar. This adjustment can make the torsion bar change its twisting effort so that the vehicle can maintain proper ride height.

Steering Knuckle and Spindle Two other parts of the front suspension include the **steering knuckle** and the **wheel spindle**. The wheel spindle is the unit that carries the hub and bearing assembly with the help of the knuckle. In some vehicles, the steering knuckle and wheel spindle are one unit. ***Figure 47–10*** shows a steering knuckle and spindle. The steering knuckle is attached to the two control arms with ball joints. The wheel spindle carries the entire wheel load. Tapered roller bearings are used to reduce friction between the wheel and the spindle. The inner bearing on the spindle is usually larger than the outer bearing. It absorbs the greatest load because the wheel is placed as close to the knuckle as possible.

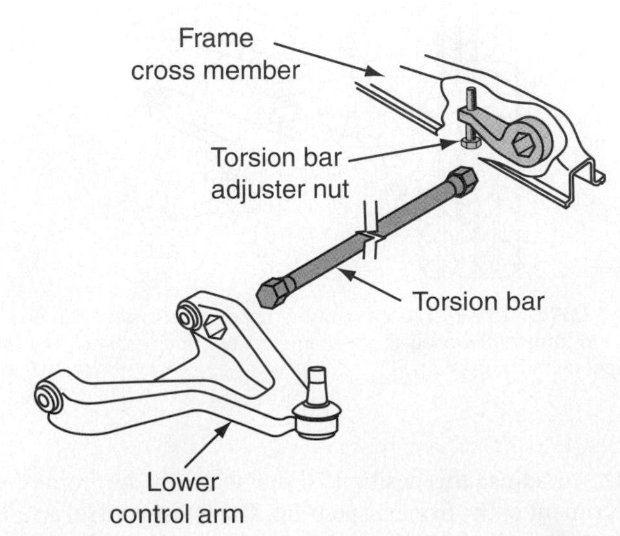

FIGURE 47-9 Torsion bars are used in place of springs on some vehicles.

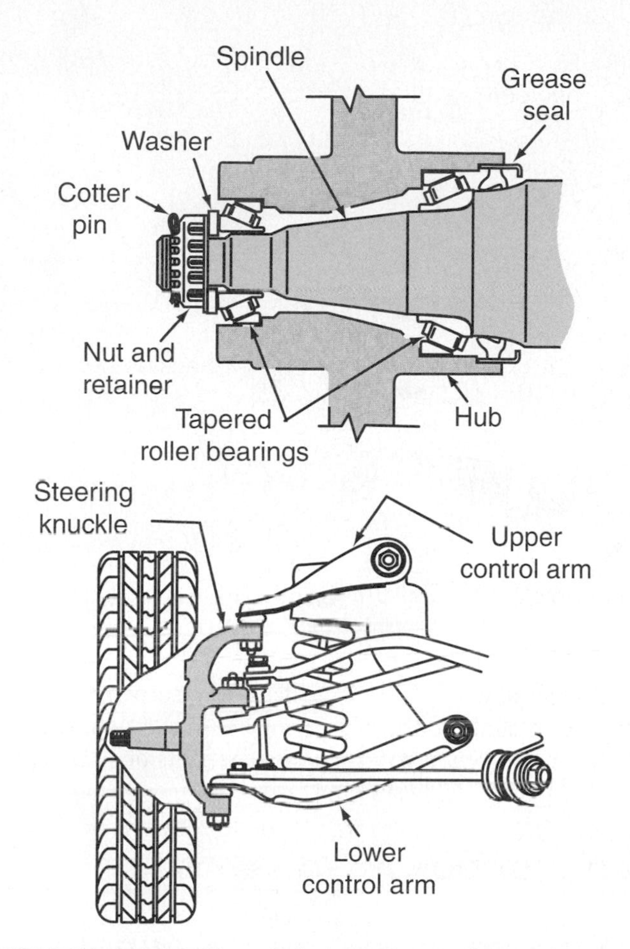

FIGURE 47–10 The spindle holds the wheel bearings and is attached to the steering knuckle.

FRONT WHEEL BEARINGS

The front wheel bearings are also considered part of the front suspension. There are two bearings on each front wheel spindle to support the wheel. Both bearings are called tapered roller bearings. On certain types of front-end suspension systems, ball bearings are used as well. The inner race of the bearing rides on the spindle. The outer race is lightly pressed into the wheel hub.

47.2 REAR SUSPENSION

The rear suspension system is an integral part of the total suspension system. There are typically two types of rear suspension systems: the solid axle type and the independent rear suspension type.

PURPOSE OF REAR SUSPENSION

All rear suspension systems serve the same purpose. They are designed to keep the rear axle and wheels in their proper position under the car body. The rear wheels must always track exactly straight ahead. The rear suspension axle allows each of the rear wheels to move up and down somewhat independently from the frame. This helps to maintain alignment and good vehicle control and provides passenger comfort. The spring assembly must also absorb a large amount of rear end torque from acceleration (on rear drive vehicles), side thrust from turning, and road shock from bumps.

LEAF SPRING

One common type of spring used on rear suspensions is the leaf spring. It consists of one or more leaves and usually has its end formed into eyes for connection to the vehicle frame. A U-bolt is used to hold the rear axle to the spring. This type of spring is called the semi-elliptical spring. The ends are higher than the center arch as shown in ***Figure 47–11***. One end of the spring is fixed to the frame. The other end of the spring is mounted to the frame by a spring **shackle** and bushing. The bushings are used to dampen noise and vibration from the road to the frame of the car. The spring shackle allows the spring to change length slightly during driving. During normal operation, the spring also bends because of acceleration, braking, or road conditions. The leaf spring supports the car frame, but it allows independent movement of the rear wheels. ***Figure 47–12*** shows the individual parts of a leaf spring rear suspension.

COIL SPRING REAR SUSPENSION

In a coil spring rear suspension, the spring is placed between a bracket mounted on the axle and the vehicle frame. The coil design is much the same as the front wheel coil spring. In addition to the coil springs, control arms and bushings are used. Control arms provide stability to the rear wheels during driving. The control arms are attached with bushings to the rear axle housing and the car frame. ***Figure 47–13*** shows a coil spring rear suspension system.

INDEPENDENT REAR SUSPENSION

Independent rear suspension is used on many front-wheel drive cars. Independent rear suspension means that each rear wheel is independent in its movement. This is much the same as the front suspension system. Although there

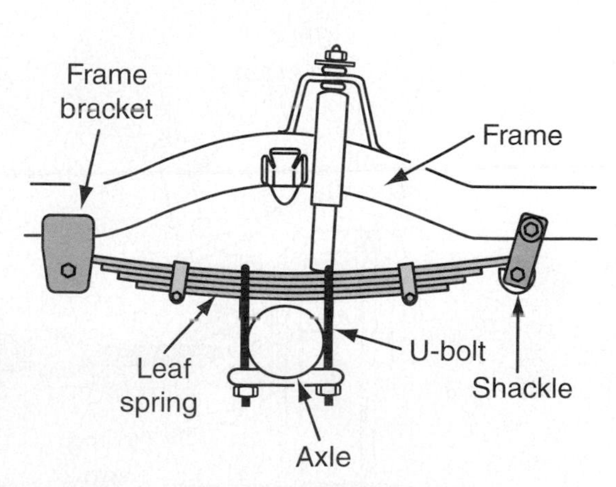

FIGURE 47–11 Both ends of the rear leaf spring are attached to the frame. A U-bolt attaches the axle to the rear spring.

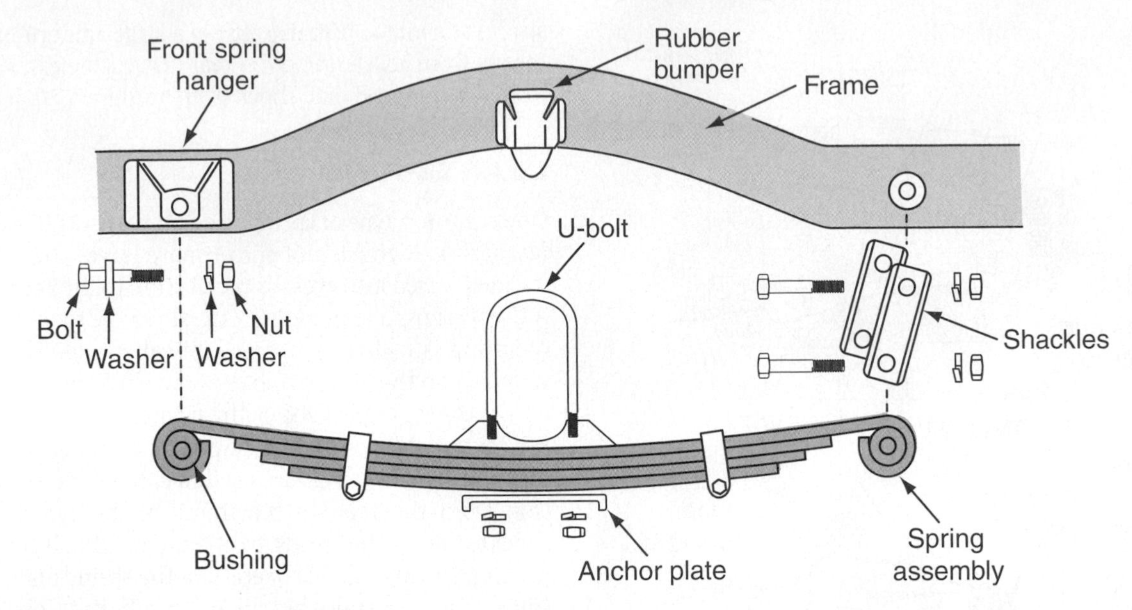

FIGURE 47-12 The parts of a leaf spring.

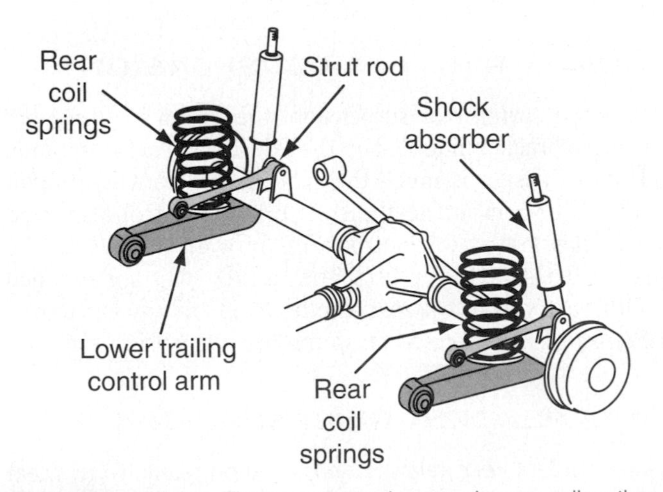

FIGURE 47-13 The rear suspension can also use coils rather than leaf springs.

are many designs for independent rear suspension, most systems include coil springs, control arms, struts, and stabilizer bars. ***Figure 47–14*** shows the parts of an independent rear suspension system.

47.3 SHOCK ABSORBERS

PURPOSE OF SHOCK ABSORBERS

Shock absorbers are hydraulic devices that help to control the up, down, and rolling motions of a car body. One shock absorber is used on each wheel. Each shock must control one wheel and axle motion. The car's springs support the body, but the shock absorbers work with the springs to control movements of the car body. A shock absorber can be considered a damper that controls energy stored in the springs under load. For this reason, shock absorbers are also called **oscillation** dampers.

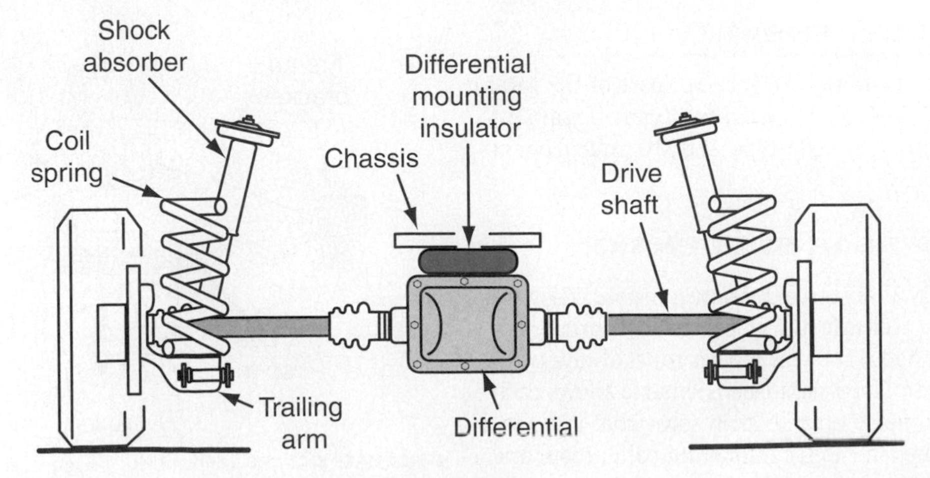

FIGURE 47-14 The parts of a typical independent rear suspension system.

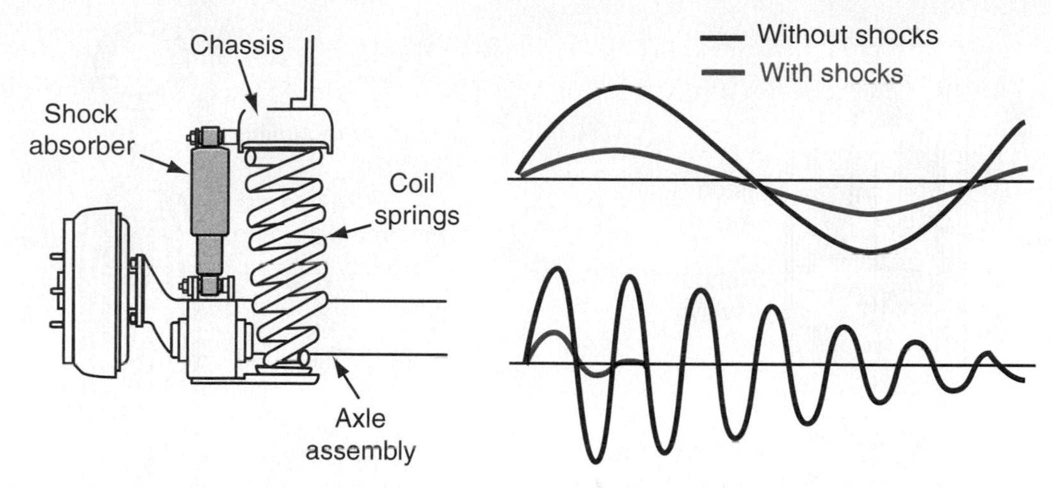

FIGURE 47–15 As the wheel and axle goes over a bump, the spring is compressed. The oscillations that are produced are shown with and without a shock absorber.

A shock absorber is placed parallel to the upward and downward motion of the car. It has two tasks:

1. To prevent excessive rolling and bouncing of the car body
2. To rapidly terminate the oscillation of the wheels and axle when they start moving up and down (called jounce and rebound)

These two factors are of major importance for driving comfort and safety.

Figure 47–15 shows how a shock absorber works. When there is a rise or bump in the road surface, the car axle immediately rises. Now the coil spring is compressed and starts to push the car body up. The impact acting on the vehicle is absorbed by the spring. The spring prevents the axle from touching the car body. After the springs have been compressed, they try to expand. This helps separate the car body from the axle. This entire action causes an oscillation motion to develop.

The oscillation motions are also shown. The shock absorber is placed between the axle and the car body, or frame. It is designed to reduce the number of oscillations produced after hitting a bump. For comparison, the oscillations are also shown when a shock absorber is placed between the car body and the axle.

PARTS LOCATOR

Refer to photo #105 in the front of the textbook to see what a ***shock absorber*** *looks like.*

SHOCK ABSORBER OPERATION

Shock absorbers are made to force a noncompressible liquid through small openings. ***Figure 47–16*** shows a schematic of how they work. When a compression force is produced from a road bump (called jounce), the piston rod is forced down. A pressure is produced in the oil below the piston in chamber B. The oil pressure forces oil outward and through the deflecting discs. Oil can only pass through these passageways at a certain speed. The pressure forces oil out into chamber C. The damping force originates from the resistance of oil flow at the narrow passages of the valve parts. In addition, oil passes into chamber A. Oil must flow into this chamber because it is getting larger as the piston is moved downward.

When the shock absorber rebounds, oil flows in the reverse direction. ***Figure 47–17*** shows the flow of oil when the shock absorber rebounds or returns to its original position. During rebounding, the piston rod is forced to extend out and upward. This action causes a vacuum to be produced inside chamber B. This vacuum draws oil from chamber A into chamber B through the rebound valve. Oil also flows

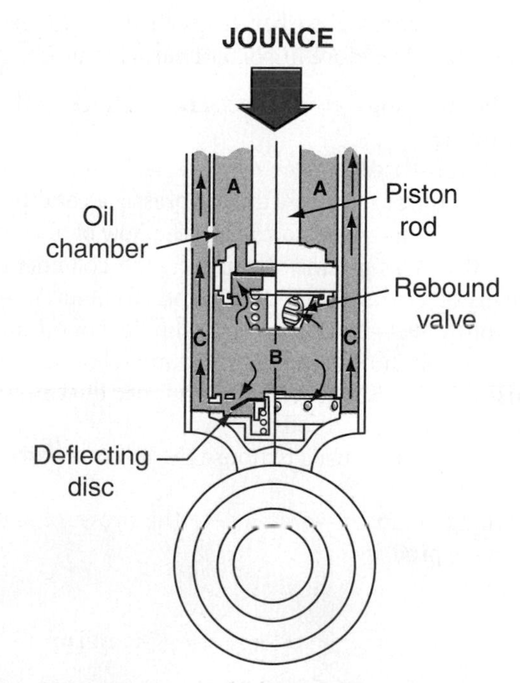

FIGURE 47–16 Under a compression force, or jounce, the oil is forced through small valves into reservoirs C and A.

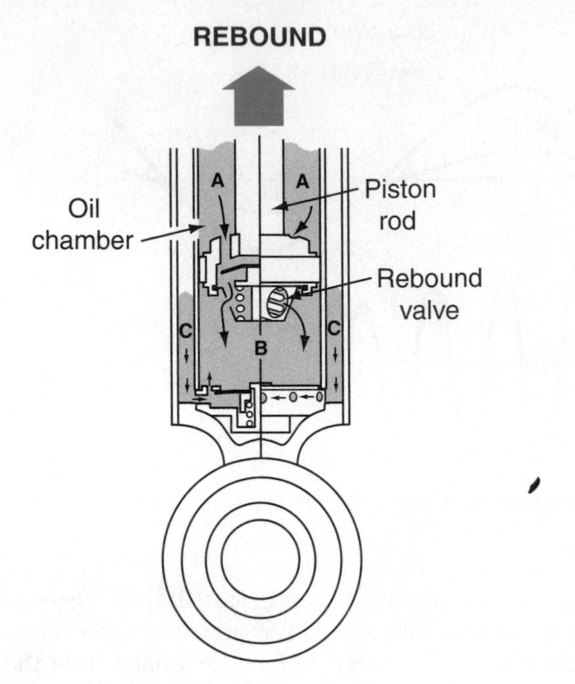

FIGURE 47-17 During rebound, oil is forced from reservoirs C and A into B.

from chamber C into chamber B. Air then expands to compensate for the loss of oil in chamber C. Since a part of the oscillating energy is converted to heat energy, shock absorbers that are working correctly get warm during operation.

PARTS OF A SHOCK ABSORBER

There are many types of shock absorbers, but the main principles of operation are the same. Although they may be different sizes, the parts are much the same. ***Figure 47–18*** shows a typical shock absorber. It is called the double-tube shock absorber. The more important parts include:

- Rubber bushing—used to attach the shock to the axle and frame
- Fluid seal—used to keep oil from leaking past the rod and into the atmosphere during pressure conditions
- Piston valving—used to control the flow of oil and produce the pressure and vacuum in the chamber above the piston during compression and rebound
- Bottom valve—used to control the flow of oil into the oil reservoir during compression and rebound
- Dust shield—used to keep dirt and road dust away from the seals and piston rod
- Container tube—used to house the internal parts of the shock absorber
- Working chamber—area where the pressure and vacuum are produced

TYPES OF SHOCK ABSORBERS

There are many types of shock absorbers. The following discussion describes some of the more common ones.

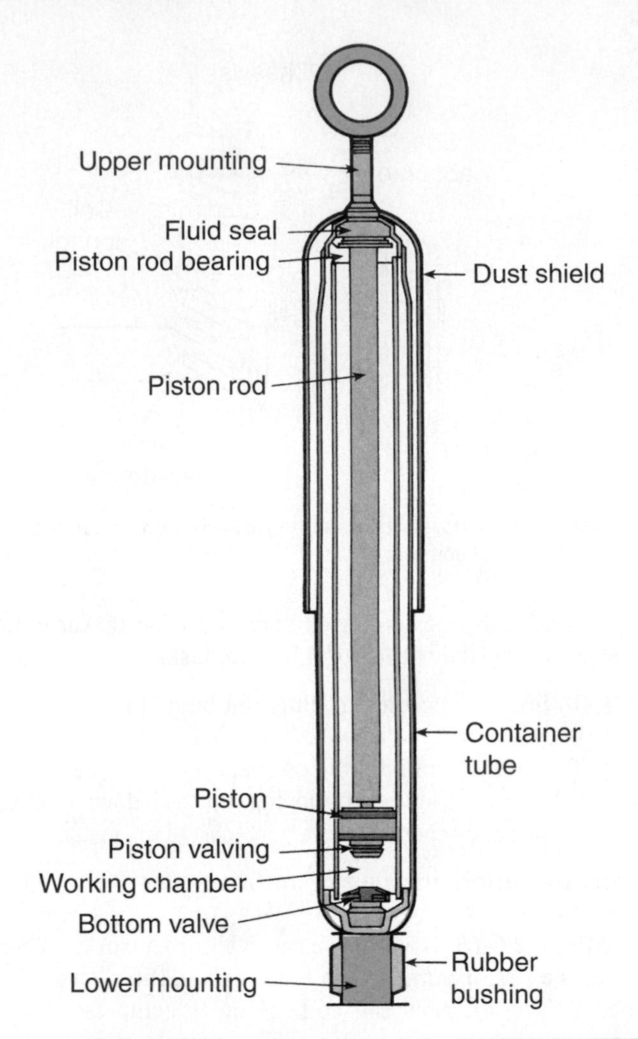

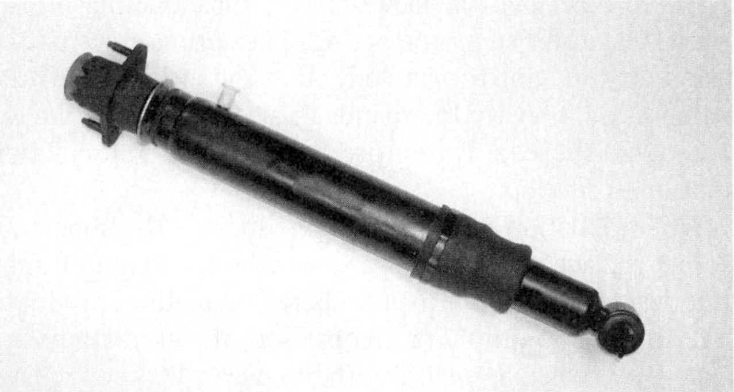

FIGURE 47-18 The parts of a shock absorber.

Spiral-Grooved Shock Absorbers When oil passes through the valves rapidly, some **aeration**, or foaming, is produced in the oil. Aeration is the mixing of air with the oils. When aeration occurs, the shock develops lag (piston moving through an air pocket that offers no resistance). This causes the shock absorber to work incorrectly and produces a poor ride.

One method used to reduce aeration is to use a spiral-grooved reservoir tube. The spiral grooves on the shock reservoir tend to break up the air bubbles. This action reduces lag.

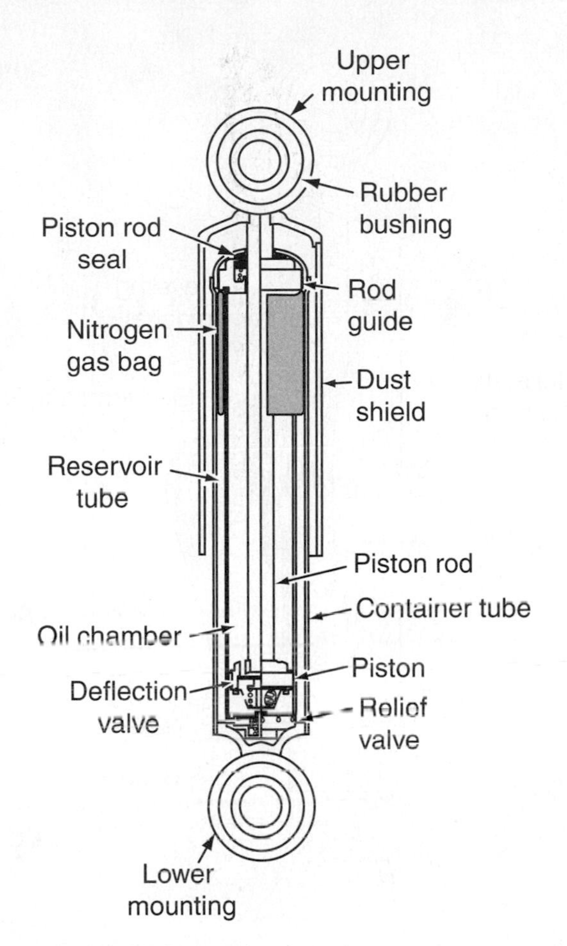

FIGURE 47–19 Gas-filled shock absorbers are used to reduce aeration in the shock absorber oil.

Gas-Filled Shock Absorbers Gas-filled shocks are also used to reduce aeration. If a pressure gas replaces the air in the shock absorber, air cannot mix with the oil to produce aeration. ***Figure 47–19*** shows one type of gas-filled shock absorber.

A nitrogen gas bag is used to help separate the oil and the gas. When oil is forced through the small holes in the deflection valve, high-pressure jets of oil are produced. These jets of oil are deflected by the deflection disc before they get to the gas bag. This action reduces foaming and aeration.

Air Shock Absorbers When an increased amount of load is placed in the car, the springs may not be able to keep the vehicle level (***Figure 47–20***). This condition can cause several problems. These include:

- Increased intensity of headlight beams at oncoming drivers even when the lights are on dim
- A change in steering geometry
- Reduction of comfort for the passengers
- Less steering control for the driver
- The possibility of bottoming out on bumps

One way to overcome these problems caused by heavy weight is to use air shock absorbers.

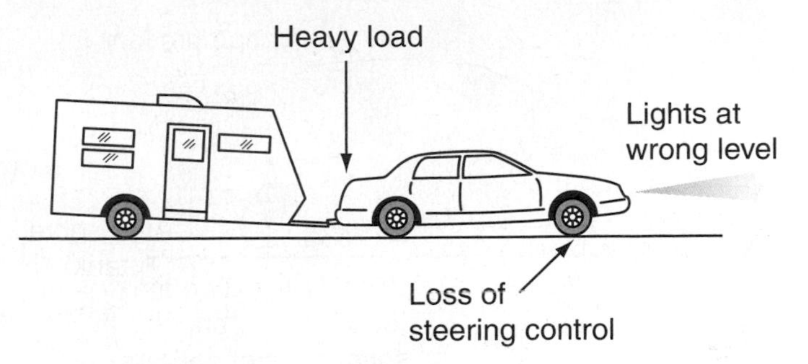

FIGURE 47–20 Heavy load at the rear of the vehicle can cause the steering geometry to change, reduce steering control, and increase the intensity of headlight beams at oncoming drivers.

Operation of Air Shock Absorbers ***Figure 47–21*** shows an air shock absorber. The unit is made by including an air chamber in the shock. A **bellows** is used to keep the air chamber sealed from the outside while the shock absorber is in different positions. The pressure inside the air chamber determines the amount of load that the vehicle can carry. The entire unit also uses the typical shock absorber system discussed previously.

The air is admitted to the air chamber by use of a standard tire valve. Pressure is produced by a small electric air pump called an air compressor. A height-sensing control

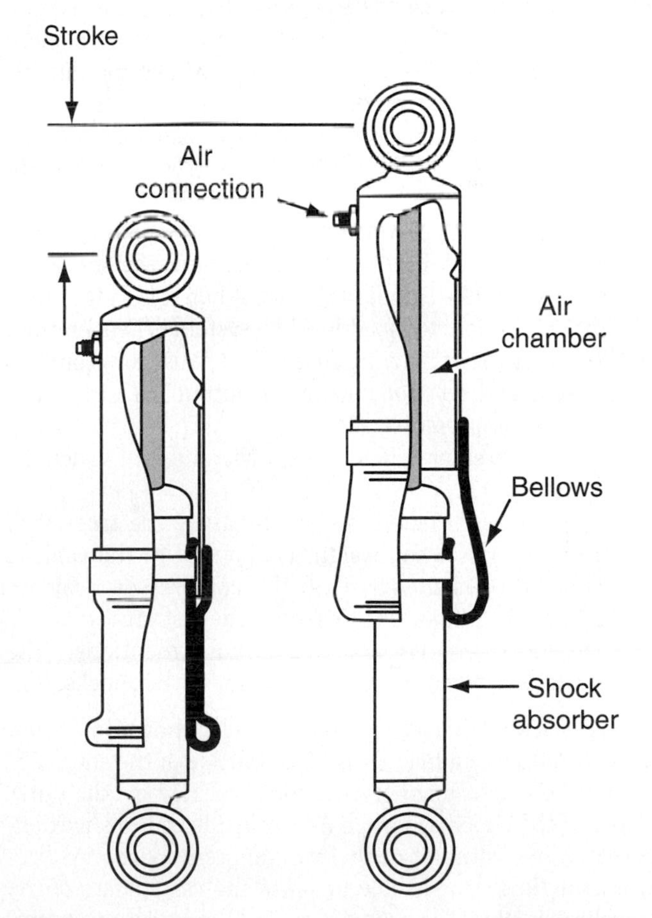

FIGURE 47–21 Air shocks are used to level the stance of the car. Compressed air is forced into a chamber to lift the shocks and level the car.

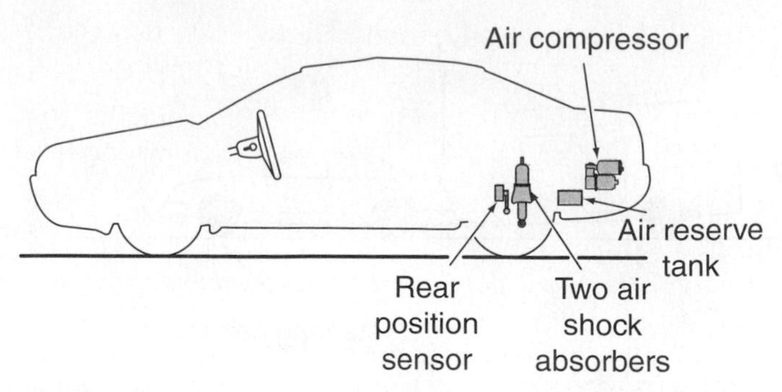

FIGURE 47–22 Air shocks use a small air compressor to produce air for extra support.

valve is also used. ***Figure 47–22*** shows a typical installation. Other common components include an air reserve tank, different types of air compressors, and a control valve.

AUTOMATIC LEVEL CONTROL

The automatic level control system adjusts the carrying load of the car when weight is added or removed from the vehicle. The system consists of several components:

- The air compressor is a positive-displacement, single-piston pump. It is powered by a 12-volt dc permanent magnet motor. The casting contains intake and exhaust valves for correct operation.
- The air dryer is used to dry the air by using a chemical. When air passes through this chemical, moisture is absorbed.
- A manual switch controls the compressor on certain systems. When it is in the off position, the shock absorbers act like standard shocks. When it is in the automatic position, the load-leveling system is in operation.
- The exhaust solenoid is used to exhaust air from the system and control maximum output pressure from the air compressor.
- The compressor relays control the different functions of the system.
- The electronic height sensors measure the amount of drop and rise when weight is changed in the vehicle. This signal is then sent to the compressor relays to change the amount of air sent to the system.
- The shocks are the same as previously mentioned. The air lines connect the air compressor to the shocks.

A shock absorber for such a system is shown in ***Figure 47–23***. A traveling magnet is placed inside of the shock absorber. An electromagnetic sensor is located on the outer frame of the shock absorber. As the inside of the shock absorber is moved up or down from different loads, the electromagnetic sensor is able to sense the magnetism of the traveling magnet. Electronic circuits then operate the compressor to inflate the shock absorbers or exhaust the air in order to maintain the desired pressure.

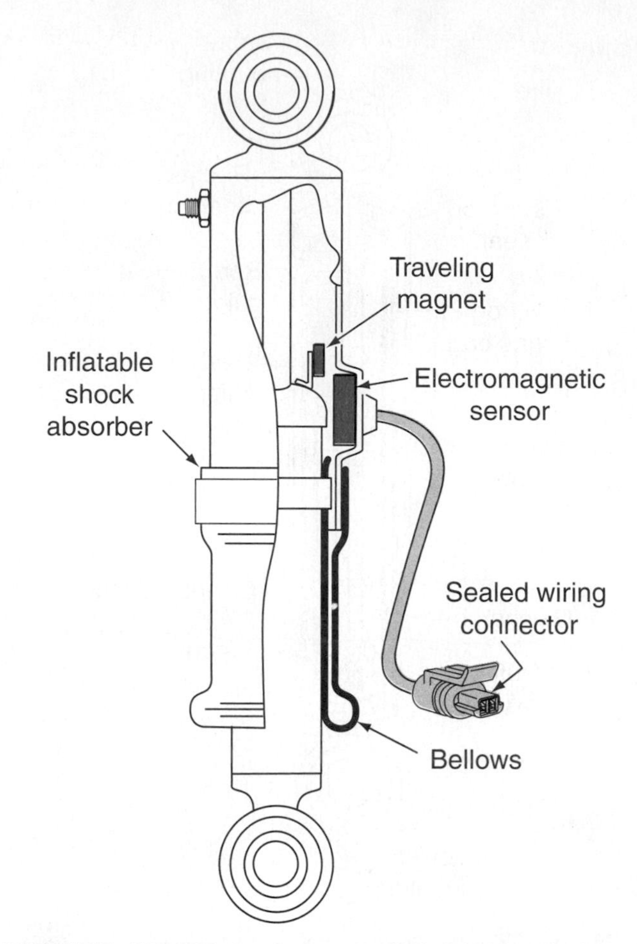

FIGURE 47–23 An electromagnetic height control sensor is used to measure the amount of load placed in the vehicle. The sensor tells the air compressor how much air should be added to the air shocks.

47.4 COMPUTER-CONTROLLED SUSPENSION SYSTEMS

PURPOSE OF COMPUTER-CONTROLLED SUSPENSION SYSTEMS

As with many vehicle systems, the suspension system can be controlled by computers and control modules. There are many advantages of using computers and control modules to control suspension. For example, most drivers like a smooth, soft ride while driving on a highway. Yet when the car is cornering or accelerating, a firmer ride is more comfortable. Suspension systems of years ago were designed for either a soft ride or a firm ride, but not both. Today, with the use of computers, both firm and soft rides can be produced, almost instantly. In addition, when vehicles have additional weight in the trunk or several extra passengers, a computer-controlled suspension system can keep the vehicle level continuously. The end result of using computer-controlled suspension systems is better control of the steering system, improved fuel mileage, a more attractive profile of the vehicle, improved vehicle stability and ride quality, and improved safety.

There are many designs for computer-controlled suspension systems. The following section will help the service technician understand many of the basic principles and operation. Note that there are many other systems; however, the basic principles are similar.

AIR SUSPENSION

One type of electronic suspension used on vehicles is called the air suspension system (***Figure 47–24***). Although there are shock absorbers, the system also uses four air springs, one on each wheel. As the front and rear height sensors feed information to the control module, the correct amount of air is sent to each air spring.

System operation is maintained by the addition or removal of air in the air springs. There is a predetermined height for both the front and rear sections of the car. The height sensors lengthen or shorten, depending on the amount of suspension travel. As weight is added, the vehicle body settles. As weight is removed, the vehicle body rises.

PARTS LOCATOR

Refer to photo #106 and photo #107 in the front of the textbook to see what ***air suspension springs*** *and* ***front height sensors*** *look like.*

The height sensors signal these changes to the control module. The control module then activates the air compressor through relays to change the amount of air in the air springs.

ACTIVE SUSPENSION SYSTEMS

Another design is called the **active suspension system.** Active suspension systems use a computer to control the position of the vehicle, including body roll, pitch, brake dive, acceleration squat, and ride height. The system can lower the entire car body for improved aerodynamics during highway driving. The vehicle can also be raised to increase ground clearance, which might be practical during heavy snow conditions.

In order to accomplish these adjustments, the shock absorbers have been replaced with hydraulic actuators and pressure sensors. The hydraulic actuators are placed on each wheel as part of the suspension system (***Figure 47–25***). The hydraulic actuator, along with a spring, support the weight of the vehicle at each wheel. The system is designed to react to varying road conditions and to make instant adjustments to keep the ride as smooth as possible.

Figure 47–26 shows a simplified schematic of an active suspension system and the hydraulic system. In operation, as the vehicle starts to go over a bump in the road, the wheel is forced upward. The upward movement causes the pressure in the hydraulic activator to increase. The increase in pressure is sensed by the pressure control valves. This

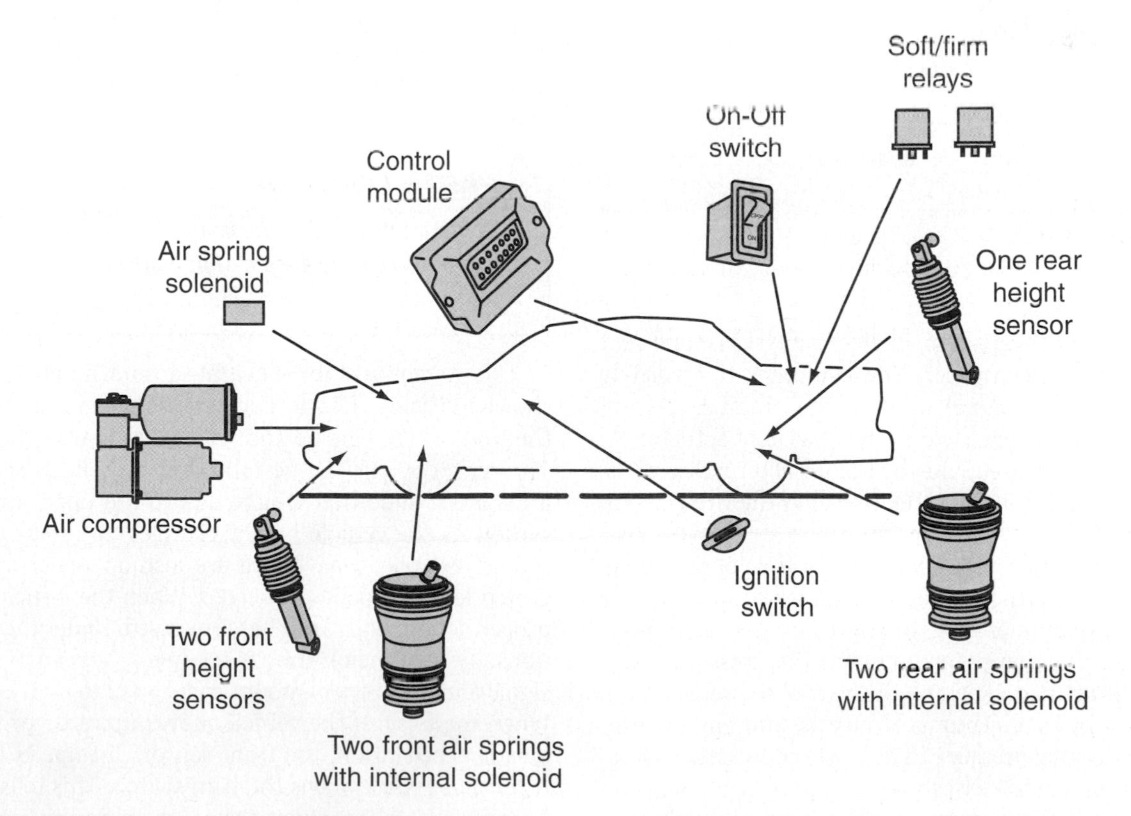

FIGURE 47–24 An air suspension system uses front height sensors, a rear height sensor, a control module, and air springs to keep the vehicle level under varying loads.

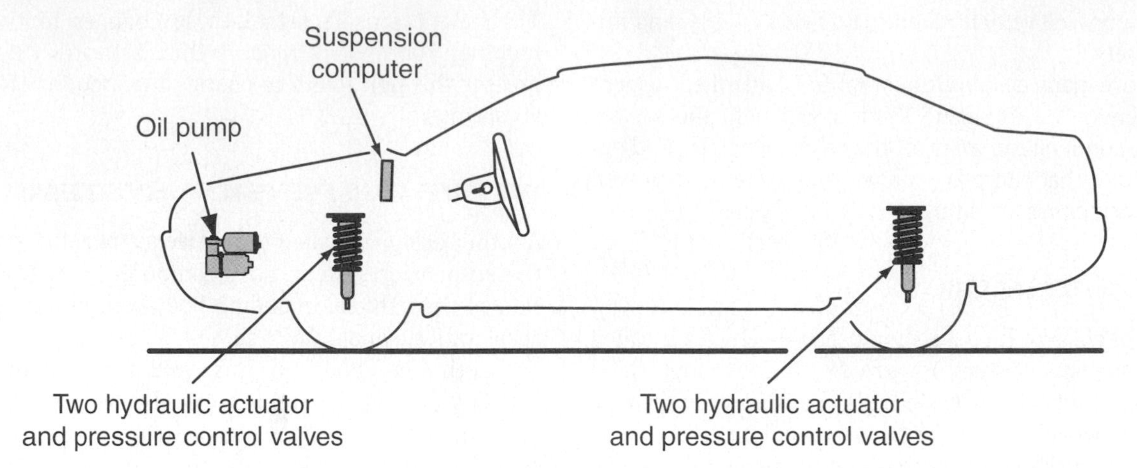

FIGURE 47-25 In an active suspension system, a hydraulic actuator is placed on each wheel to control the movement of the vehicle body over variations in road surface.

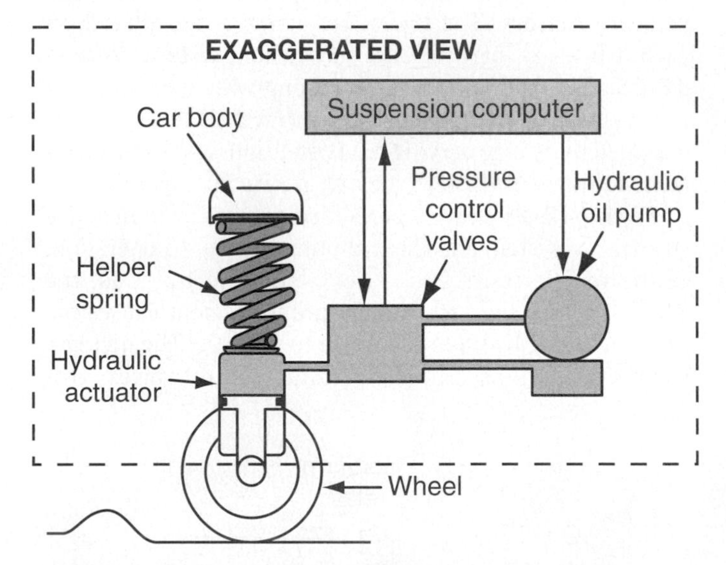

FIGURE 47-26 When the wheel goes over a bump, an increase in pressure is felt in the hydraulic actuator. The computer immediately alters the hydraulic pressure so the vehicle body remains at a constant position.

pressure is then converted to an electronic signal and sent to the suspension computer. The computer immediately sends an electronic signal back to the pressure control valves to release the pressure in the hydraulic actuator. As the pressure is released, the hydraulic actuator retracts. This results in keeping the vehicle level as the tire rides up over a bump.

As the vehicle then goes over the downside of the bump, the pressure inside the hydraulic actuator decreases. This decrease in pressure is sent to the computer as an electronic signal. The computer increases the pressure inside the hydraulic actuator, keeping the body of the vehicle from moving. The hydraulic pump, driven by the engine, produces the hydraulic pressure in the system. In actual operation, the pressure decreases and increases fast enough to keep the vehicle body in one position as the vehicle goes over a bump.

ELECTRONIC AIR SUSPENSION SYSTEMS

Another system is the electronic air suspension system. In this system, air is used to control the suspension rather than hydraulic fluid. ***Figure 47–27*** shows typical front and rear air springs. In this system one air spring is mounted on each wheel. The computer controls the amount of air going into each air spring. This system uses an air compressor with a compressor relay and various valves to operate or control the suspension system on the vehicle. The suspension control module (***Figure 47–28***) is located in the trunk of the vehicle. The control module output operates the compressor, a vent valve, and the air spring valves on each wheel. The control module controls the amount of air that is sent to the air springs. The result is that the vehicle height can be controlled under varying road conditions.

PARTS LOCATOR

Refer to photo #108 in the front of the textbook to see what an ***electronic suspension control module*** *looks like.*

This system uses three height sensors. A height sensor is shown in ***Figure 47–29***. Two of these sensors are located in the front of the vehicle and one in the rear of the vehicle. The sensor works in the following way. Each sensor has a magnetic slide that is attached to the upper end of the sensor. As the vehicle height changes, the magnetic slide moves up and down. There are also an upper and lower switch located inside the sensor. When the vehicle is at its proper height, both switches are closed. Under these conditions, the suspension control module receives a trim height signal and the proper amount of air is fed to each air spring. When the front of the vehicle moves upward, for example, during acceleration, the front sensors lift up. As the magnetic slide lifts, it opens the trim switch. This tells the suspension control module to exhaust air in the air springs, thus maintaining correct vehicle level. The opposite happens

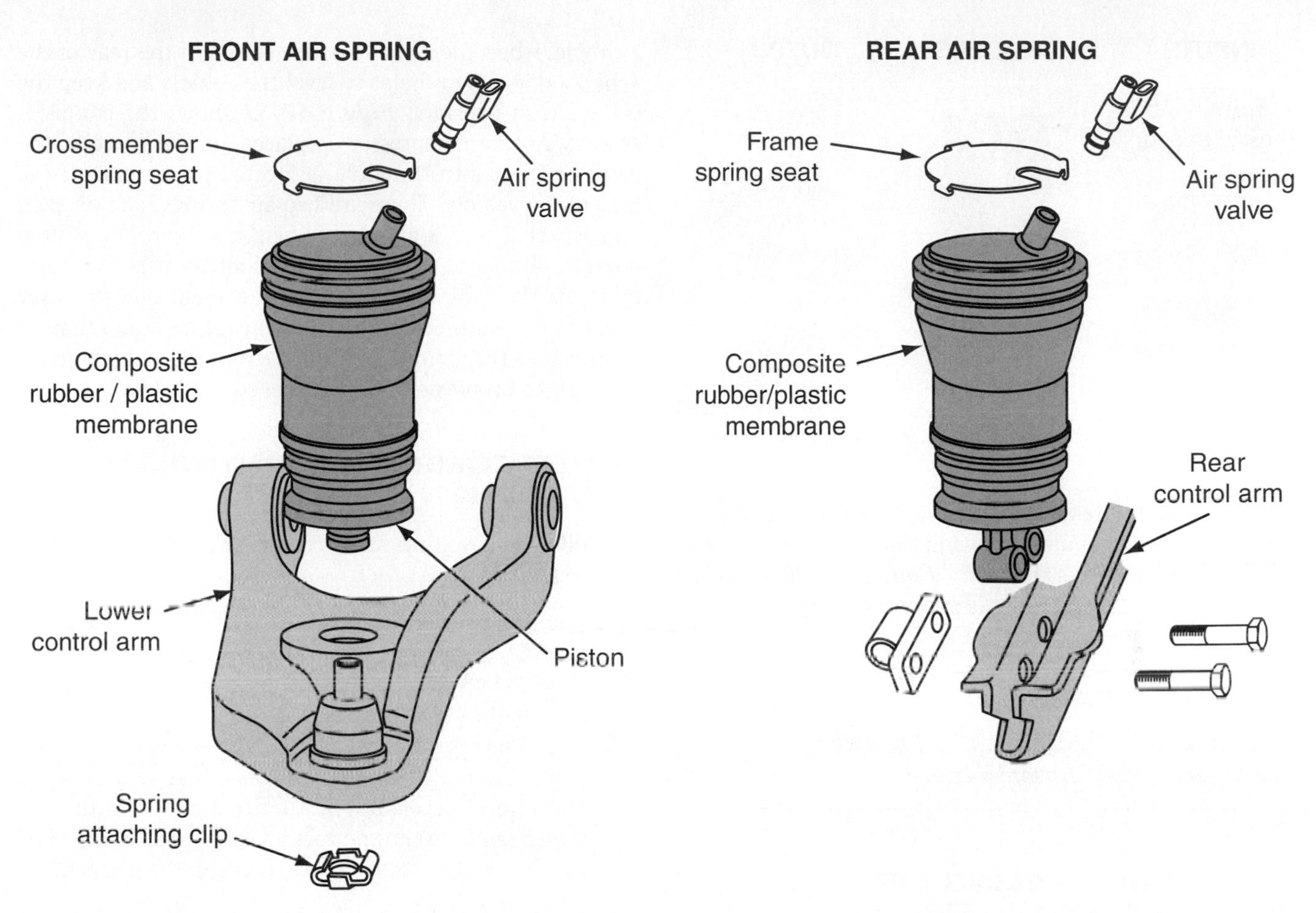

FIGURE 47–27 Front and rear air springs are used on an electronic air suspension system.

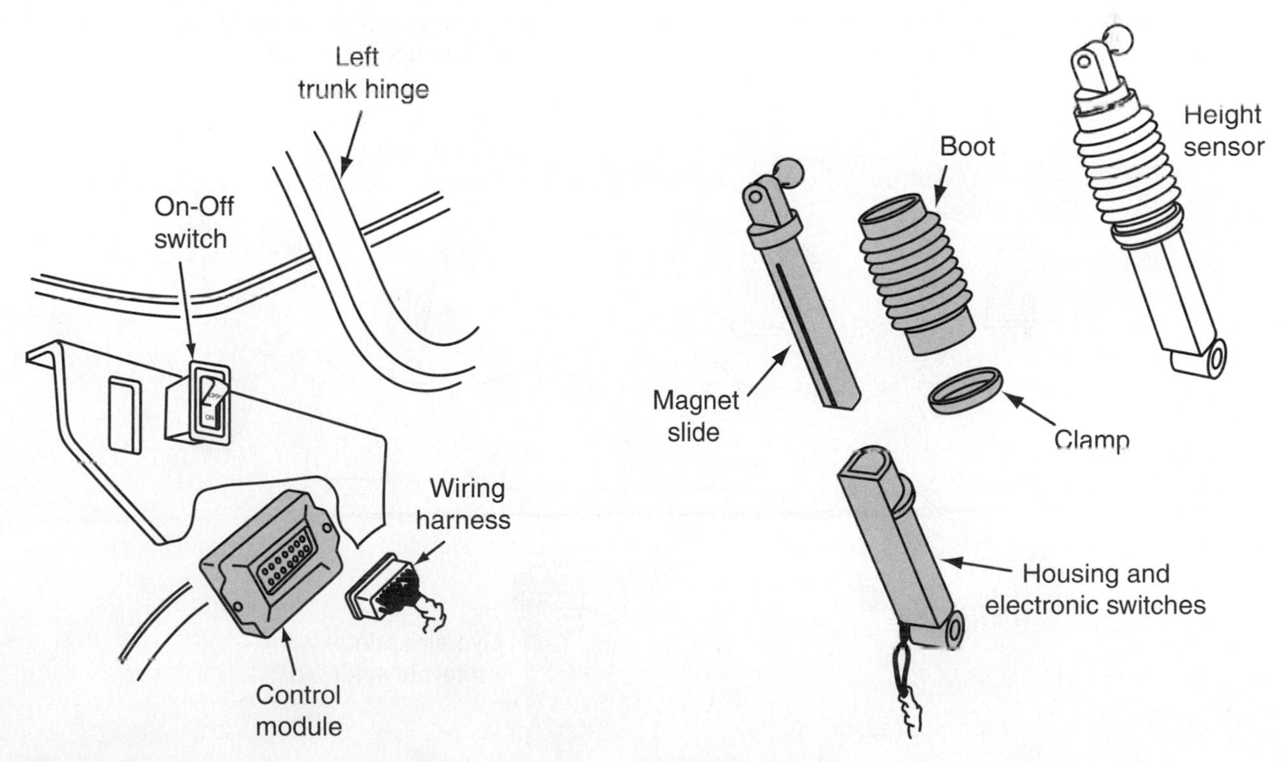

FIGURE 47–28 The suspension control module is often located in the trunk of the vehicle.

FIGURE 47–29 Height sensors are used to determine vehicle trim position.

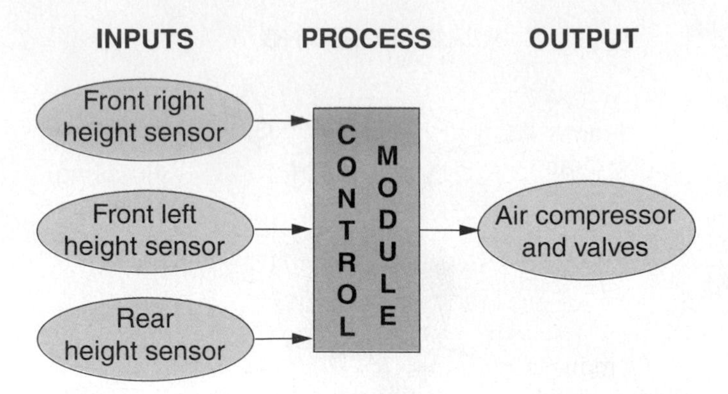

FIGURE 47–30 The typical electronic air suspension system uses various input sensors to control an air compressor and valves.

when the vehicle brakes and the front end drops. In this case, the control module sends additional air into the air springs to maintain correct trim. ***Figure 47–30*** shows a simple input, processing, and output schematic to illustrate electronic air suspension control.

PARTS LOCATOR

*Refer to photo #109 in the front of the textbook to see what a **rear height sensor** looks like.*

REAR LOAD-LEVELING AIR SUSPENSION SYSTEMS

Another computer-controlled suspension system is the rear load-leveling air suspension system. This system only controls the air and position of the rear of the vehicle. For example, when there is excessive weight in the rear of the vehicle, this system helps to level the vehicle and keep the vehicle trim constant. ***Figure 47–31*** shows the complete system. An air compressor is placed in the front of the vehicle. The control module is located in the rear, trunk area of the vehicle. There are two air springs, one on each rear wheel. There is also a rear height sensor. The system operates similar to other electronic control suspension systems. As the rear height sensor is moved, due to heavy weight or rapid acceleration, the control module exhausts or increases the amount of air in each of the air springs on the rear. An on-off switch is also placed in the trunk.

OTHER COMPUTER-CONTROLLED SUSPENSION SYSTEMS

The following are three other computer-controlled suspension systems.

1. The programmed ride control system uses a control module, a steering sensor, brake sensor, and several shock relays to determine firmness and smoothness to control the suspension system.
2. The computer command ride control system uses a control module, various inputs such as vehicle speed and rate of acceleration, a normal and firm driver's switch, and actuator motors on each wheel (controlled by the control module) to control the suspension system.
3. The automatic air suspension system uses several inputs and outputs to regulate and control the suspension system. ***Figure 47–32*** shows an example of the type of inputs and outputs for this system.

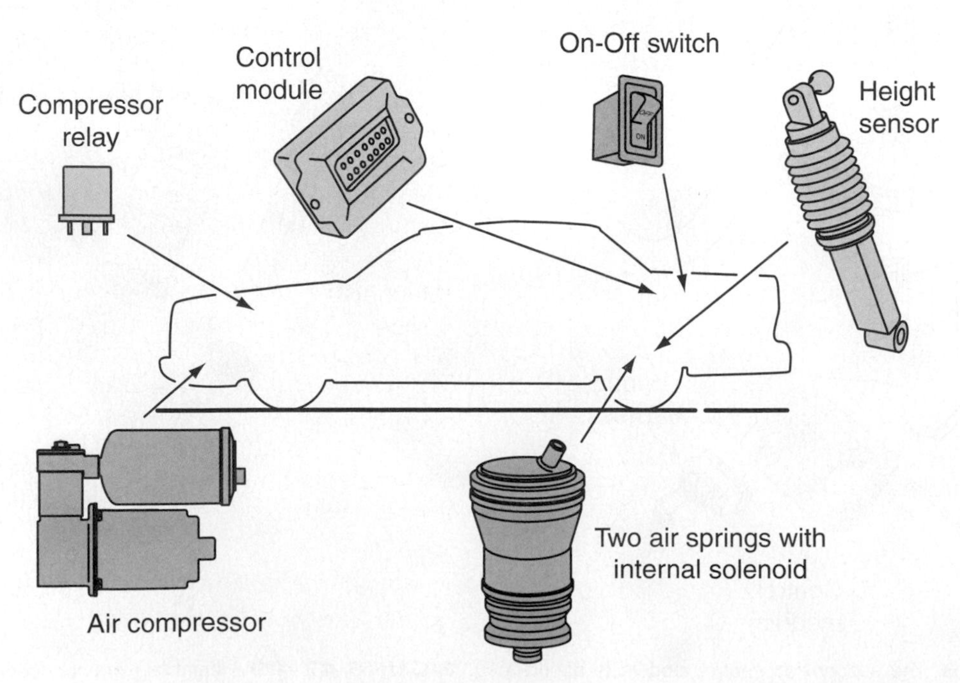

FIGURE 47–31 An automatic air suspension system uses two air springs, a rear height sensor, control module, compressor relay, air compressor, and an on-off switch for correct operation.

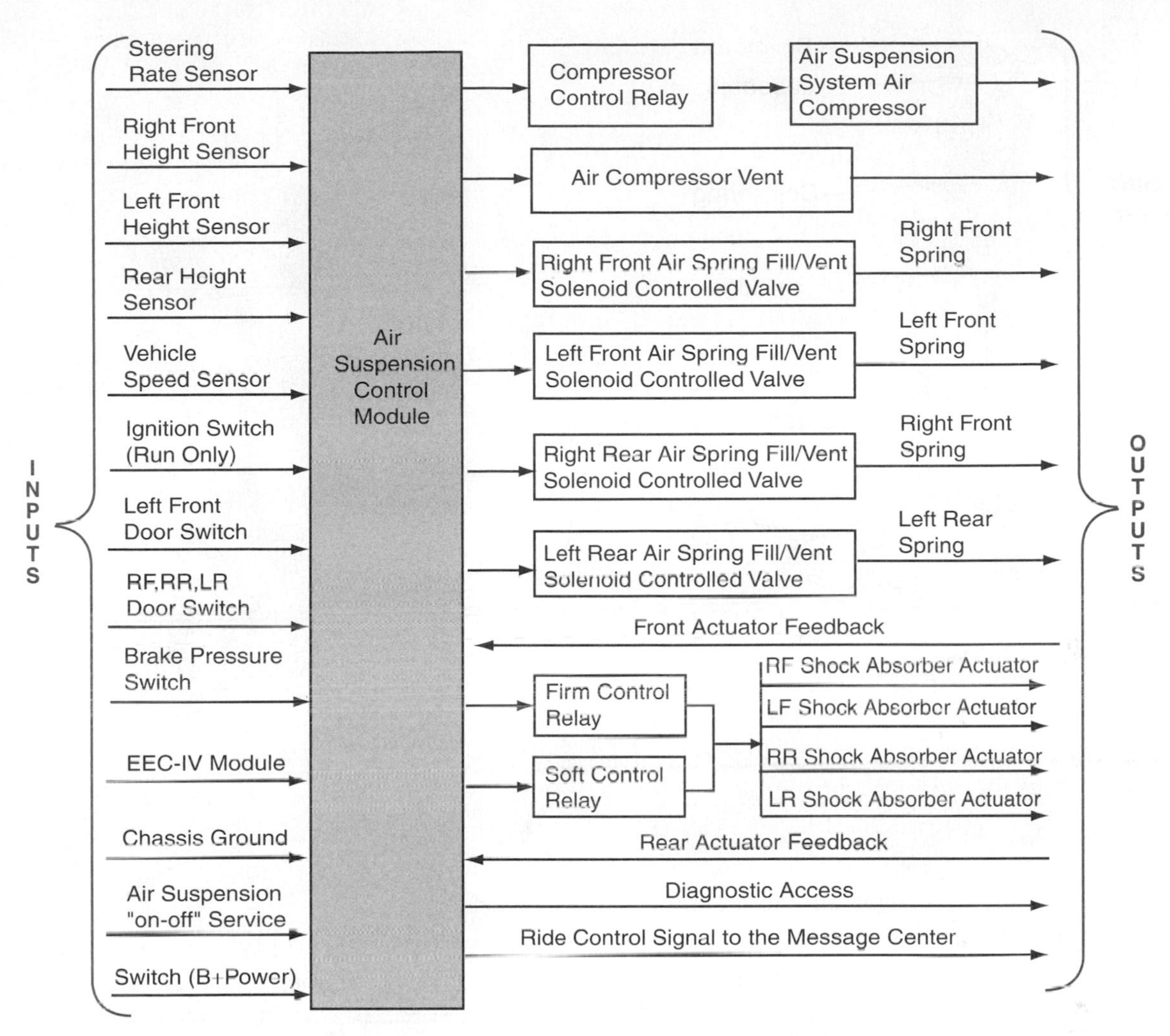

FIGURE 47–32 The automatic air suspension system uses various inputs and outputs for its complete operation.

47.5 MACPHERSON STRUT SUSPENSION

GENERAL DESCRIPTION OF THE MACPHERSON STRUT SUSPENSION

One other popular type of independent suspension system is the **MacPherson strut suspension**. Many imported and domestic vehicles utilize this system on front-wheel drive vehicles. Certain vehicles also use this system on the rear wheels. There is also a modified version of the MacPherson strut system. The MacPherson strut system is favored where space and weight savings are important. It is used by American, European, and Japanese auto manufacturers.

PARTS OF THE MACPHERSON STRUT SUSPENSION

Figure 47–33 shows a complete MacPherson strut suspension system as well as a strut cartridge. It is very much like a regular shock absorber and spring combined. The only difference is that the strut assembly is used as a structural part of the vehicle's suspension system. A more detailed drawing of a typical front strut is shown in ***Figure 47–34***.

The MacPherson strut suspension has eliminated the need for several common suspension parts. There is no upper control arm, and the upper ball joint is not needed. Vehicle weight is supported at the top of the strut assembly. The strut bearing is bolted directly to the shock tower. The shock tower is the part of the car body to which the MacPherson strut is attached. ***Figure 47–35*** shows the shock tower built into the car body. The lower part of the strut assembly is attached by bolts to the steering knuckle. The steering knuckle is attached to the lower control arm through a ball joint.

The lower control arm is bolted to the frame with conventional rubber bushings. The lower control arm ball joint is riveted to the lower control arm.

OPERATION OF THE MACPHERSON STRUT SUSPENSION

During turning, the entire strut assembly is turned. The strut assembly can turn because there is a bearing assembly on top of the strut assembly and a ball joint on the bottom

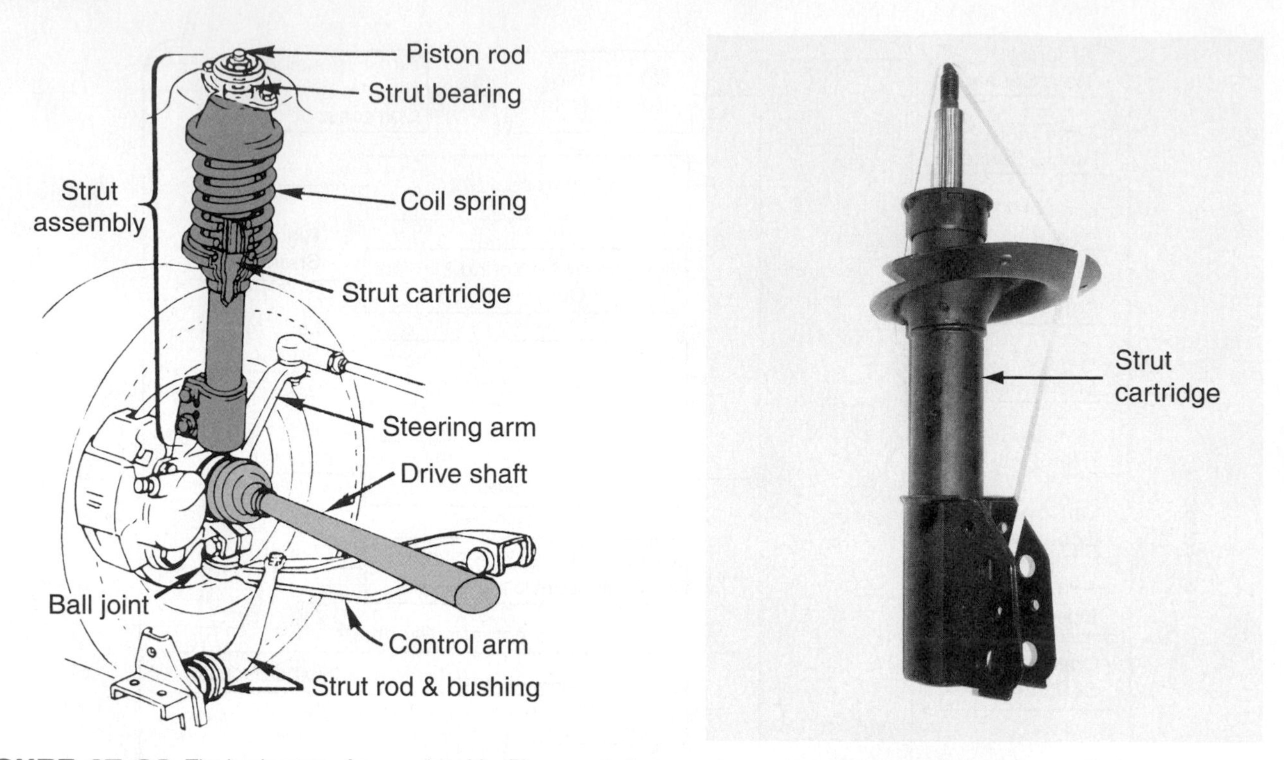

FIGURE 47-33 The basic parts of a complete MacPherson strut suspension system. (Courtesy of Dana Corporation)

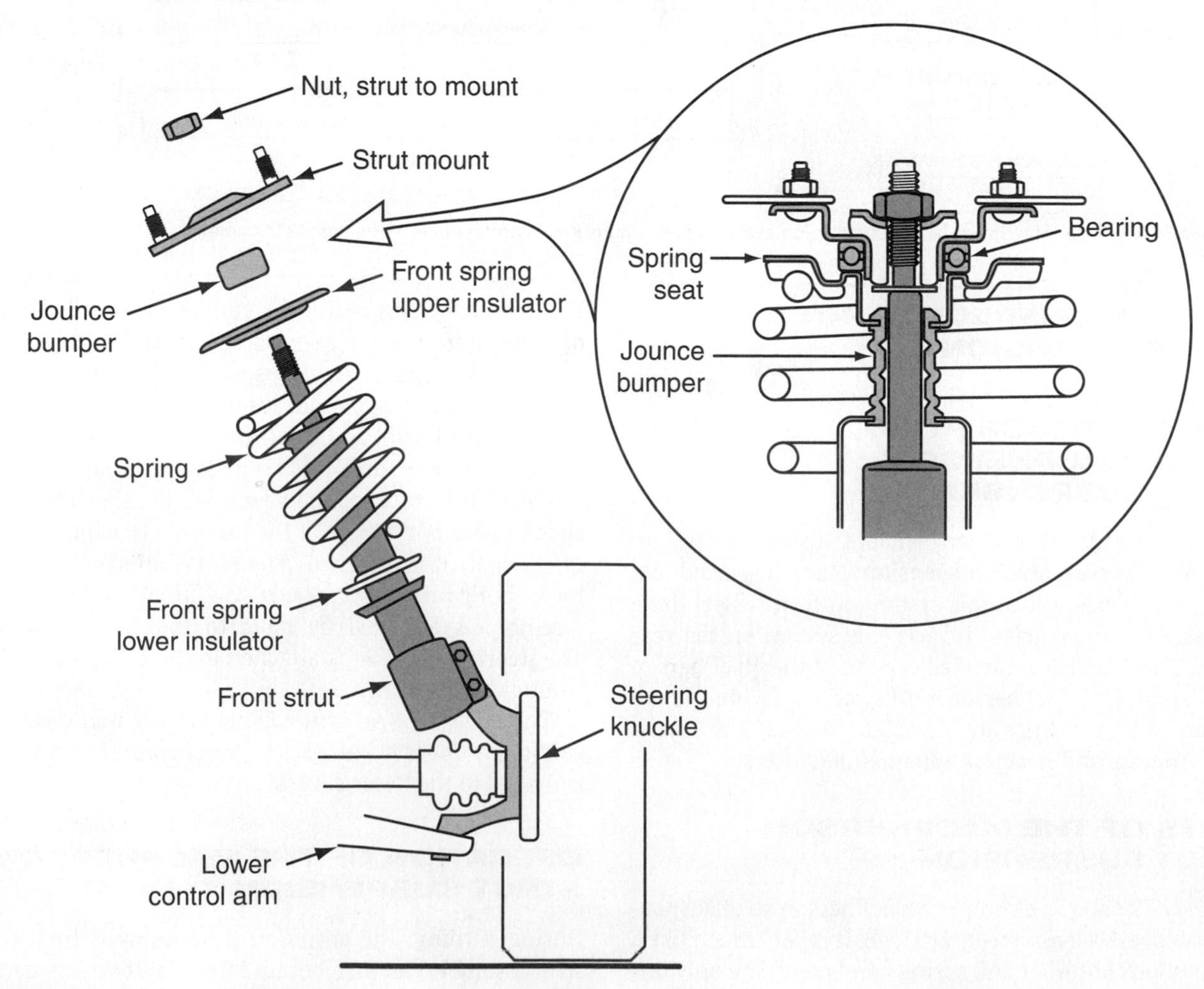

FIGURE 47-34 The detailed parts of the front strut suspension.

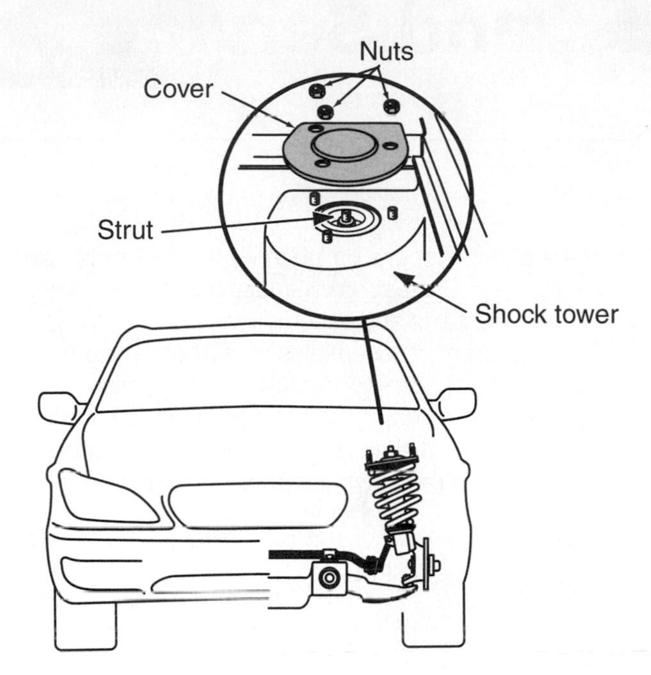

FIGURE 47-35 The MacPherson strut suspension is bolted to the shock tower as part of the wheel well.

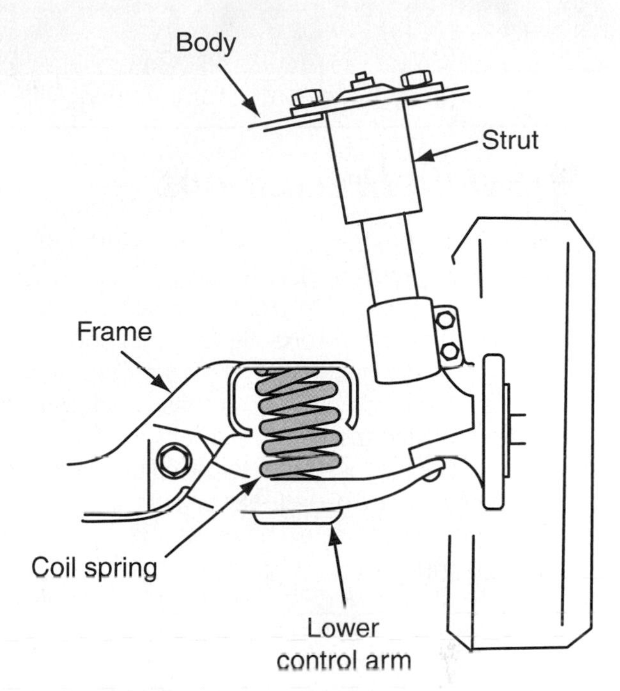

FIGURE 47-36 A modified MacPherson strut suspension has the strut assembly separated from the coil assembly.

of the assembly. The upper bearing and mount assembly takes the place of the upper control arm. The steering arm and linkage, disc brake caliper, and lower control arm ball joint are all attached to the steering knuckle. The drive shaft is connected directly to the wheel spindle through the steering knuckle. The spring is used for the same purpose as on other suspension systems. It supports the vehicle weight and maintains the car stance and height. The shock absorber, which is built into the system, helps to smooth out the oscillations from the spring.

ADVANTAGES OF USING MACPHERSON STRUT SUSPENSION

There are four advantages to the MacPherson strut suspension systems. These include:

1. They weigh less than the conventional two control arm system.
2. The system spreads the suspension load over a wider span of the car's chassis.
3. They take up less room in the engine compartment, which allows room for other components.
4. There are fewer moving parts than in the conventional two control arm system.

MODIFIED MACPHERSON STRUT SUSPENSION

Another type of MacPherson strut suspension is the modified system. ***Figure 47–36*** shows such a system. The system is basically the same, except the spring is placed between the frame and the lower control arm. With the spring located here, minor road vibrations are absorbed by the chassis rather than fed back to the driver through the steering system. A lower ball joint supports the vehicle weight. This system also eliminates the need for the upper control arm, bushings, and upper ball joints, which are used on the conventional suspension system.

Car Clinic: MACPHERSON STRUTS

CUSTOMER CONCERN:
Improper Alignment

A service technician has removed and replaced the MacPherson struts. After this procedure, the wheel alignment was off. What is the problem?

SOLUTION: When MacPherson struts are removed, the lower mounting bolts must be removed. This procedure disturbs the camber setting of the wheel assembly. Before removing the two lower mounting bolts, be sure to mark the position of the lower mounting bolt eccentric cam. If this is done, the front end can be reassembled, and the alignment can be put back to its original position by realigning the marks. However, it is still wise to check the camber after replacing the struts.

Problems, Diagnosis, and Service

Safety Precautions

1. The suspension system has many parts that are under high pressure and tension. These include the shocks, springs, and torsion bars. When these parts are removed incorrectly, they can pop out violently, causing serious injury. Make sure all tension has been removed from these components before removing them from the vehicle.
2. Never use high-pressure air to dry off bearings after they have been cleaned. Never spin the bearing with a high-pressure air hose, as the balls could dislodge and cause serious injury.
3. When removing MacPherson strut components, remember that some parts have high tension on them. Remove the tension or pressure before removing those components from the vehicle. To do this, extra support will be needed.
4. Parts of the suspension system are very heavy. Be careful not to crush your fingers or hands when moving the parts and assemblies.
5. Many parts and assemblies must be removed in the correct sequence. Incorrect disassembly may cause parts to drop or to spring out unexpectedly. Always follow the manufacturer's suggested procedure when removing suspension parts from the vehicle.
6. Always wear OSHA-approved safety glasses when working on a suspension system.
7. Always use the correct procedure for lifting the vehicle when working on suspension system components.

PROBLEM: Worn Suspension Components

As the vehicle goes down the road at higher speeds, there is a shimmy in the front end suspension.

DIAGNOSIS

Various components on the suspension system can cause a vehicle to shimmy. Probable causes include the ball joints, shock absorbers, and worn tie-rod ends. ***Figure 47–37*** shows two charts that will help you diagnose problems in the suspension system. If any of these components are worn excessively, the front suspension loosens up, and the vehicle may shimmy at different speeds. In addition, various noises may be heard from the front suspension if there is excessive wear on these parts.

Figure 47–38 shows a simple diagnostic flowchart to determine the condition of ball joints. ***Figure 47–39*** shows a similar troubleshooting chart for shock absorbers. Use these charts to help diagnose shimmy and noise problems in the front suspension system.

SERVICE

Use the following general procedures to service front suspension parts.

Ball Joints Ball joints can be checked for wear by inspecting the lower section of the ball joint. New ball joints will have a 0.050-inch clearance between the end of the grease **zerk** and the body of a ball joint. The grease zerk is a

FRONT WHEEL SHIMMY	
DIAGNOSIS	**SERVICE**
a. Tire and wheel out of balance	a. Balance tires
b. Worn or loose wheel bearings	b. Adjust wheel bearings
c. Worn tie-rod ends	c. Replace tie-rod ends
d. Worn ball joints	d. Replace ball joints
e. Incorrect front wheel alignment	e. Check and align front suspension
f. Shock absorber inoperative	f. Replace shock absorber

NOISE IN FRONT END	
DIAGNOSIS	**SERVICE**
a. Ball joints need lubrication	a. Lubricate ball joint
b. Shock absorber loose or bushings worn	b. Tighten bolts and/or replace bushings
c. Worn control arm bushings	c. Replace bushings
d. Worn tie-rod ends	d. Replace tie-rod ends
e. Worn or loose wheel bearings	e. Adjust or replace wheel bearings
f. Loose stabilizer bar	f. Tighten all stabilizer bar attachments
g. Loose wheel nuts	g. Tighten the wheel nuts to proper torque

FIGURE 47–37 Most front suspension problems produce either a shimmy or various noises. These problems can be diagnosed as shown on these charts.

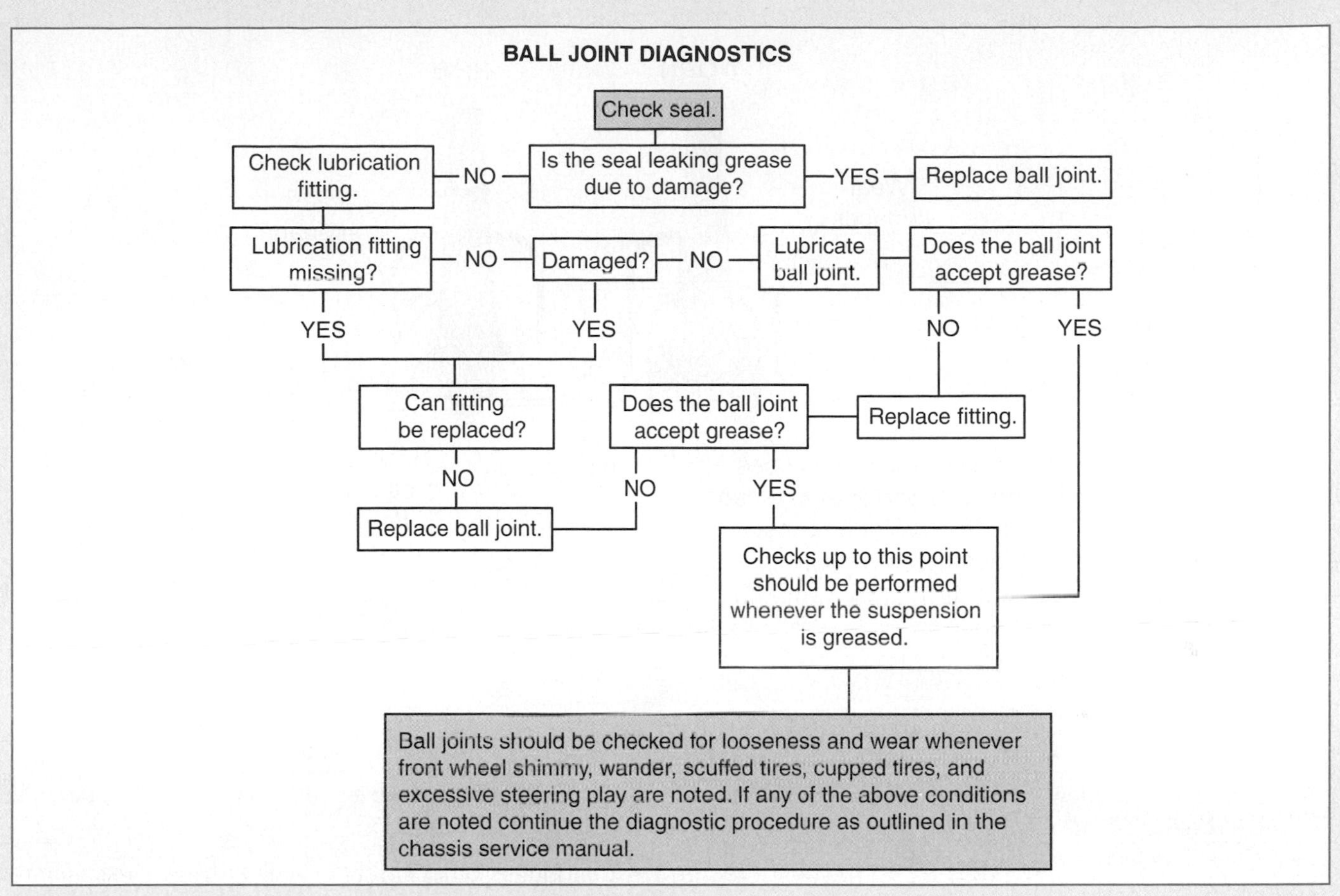

FIGURE 47-38 Follow this ball joint diagnostic procedure to determine problems with ball joints.

TROUBLESHOOTING CHART

PROBLEM	DIAGNOSIS	SERVICE
1. Shock absorber breaks down	Vehicle spring suspension travel limit stop defective.	Check rubber stop on the spring suspension travel; if necessary, replace it.
	Shock absorber performs improperly.	Replace shock absorber.
2. Shock absorber noises (rattling, rumbling)	Shock absorber mounting loose.	Fasten shock absorber properly.
	Protective tube loose.	Replace shock absorber.
	Protective tube grazes on cylinder tube.	Check offset between top and bottom shock absorber mountings.
	Shock absorber worn.	Exchange shock absorber.
3. Shock absorber inefficient	Oil loss due to defective seals or worn valves.	Exchange shock absorber.
4. Shock absorber leaky	Defective piston rod seal.	Exchange shock absorber.
5. Shock absorber works too hard	Wrong type shock absorber installed.	Install correct type according to vehicle specification.
	Valves not in order.	Exchange shock absorber.
6. Shock absorber works too smoothly	Wrong shock absorber installed.	Install correct type according to vehicle specification.
	Shock absorber worn out.	Install new shock absorber.
7. Bad driving quality	Damping efficiency fades.	Install new shock absorber.
8. Washing out (flattening) of tire profile	Damping efficiency has vanished or ceased to exist.	Install new shock absorber.

FIGURE 47-39 This troubleshooting chart can be used to solve problems with shock absorbers. (Courtesy of Sachs North America)

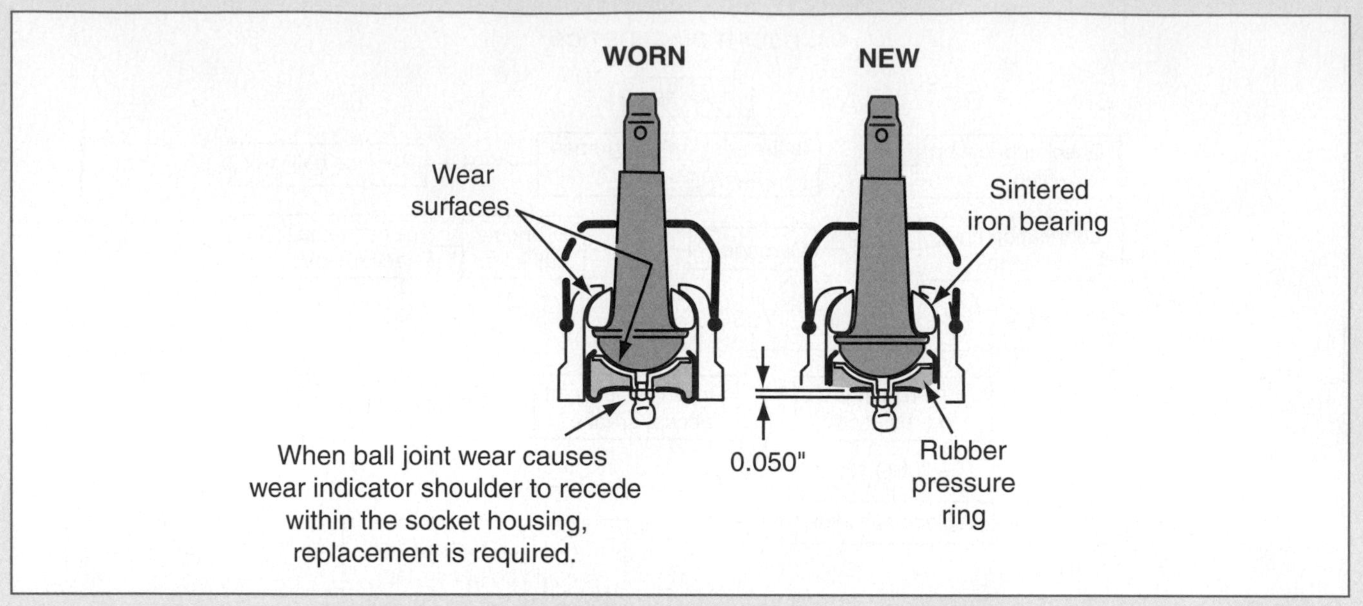

FIGURE 47–40 Ball joints can be checked for wear. There should be no less than 0.050-inch clearance between the grease zerk and the body of the ball joints.

small fitting used to allow grease to enter the bottom of the ball joint (***Figure 47–40***). Replace the ball joint if the clearance is less than 0.050 inch. Use the following general procedure to replace a ball joint.

1. Raise the vehicle and support it at the frame. Then remove the wheel and tire.
2. Position a suitable jack under the lower control arm spring seat, and raise the jack to compress the coil spring.

CAUTION: *The jack must remain in place when the ball joint is being replaced to hold the spring and control arm in position.*

3. Remove the cotter pins and nuts holding the ball joint stud to the steering knuckle. Now disconnect the joint from the knuckle using a pickle fork.
4. The ball joint must now be removed from the control arm. Remove the heads of the rivets that hold the ball joint to the control arms. Certain ball joints may have to be pressed out of the control arm. The ball joint should now be removable.
5. Place a new ball joint in the control arm. The new ball joint may have to be pressed or bolted in.
6. Install the ball joint stud into the steering knuckle and torque the nut to the manufacturer's specifications.
7. Install the cotter pin.
8. Grease the new ball joint.
9. Replace the wheel and tire.
10. Remove the vehicle from the jack and test it.

Shock Absorbers Bad shock absorbers have the following characteristics:

- Continuous bouncing of the body with every road bump
- Oscillation of the body with rough surface roads
- Lifting of the body when the car is accelerated

Bad shock absorbers cannot be repaired. In all cases, the shocks are replaced with new ones. Use the following general procedure to replace shock absorbers.

1. Raise and support the vehicle as needed.
2. Hold the shock absorber shaft with a suitable wrench.
3. Remove the upper retaining nut.
4. Remove the lower bolts that hold the shock absorber pivot arm to the control arm. Pull the shock absorber from the coil spring. Replace the shock absorber with a new one by reversing the removal procedures.
5. The rear shock absorbers are removed in much the same manner except the vehicle is supported in the rear. On some vehicles, the upper retaining nut may be located in the trunk area.

CAUTION: *On some vehicles, the rear springs may have to be supported by a jack to remove the shock absorber. Refer to the manufacturer's suggested procedure before removing rear shock absorbers.*

PROBLEM: Faulty Wheel Bearings

While driving down the road, the driver notices a slight growling noise coming from the front end of the vehicle. There is also a slight vibration.

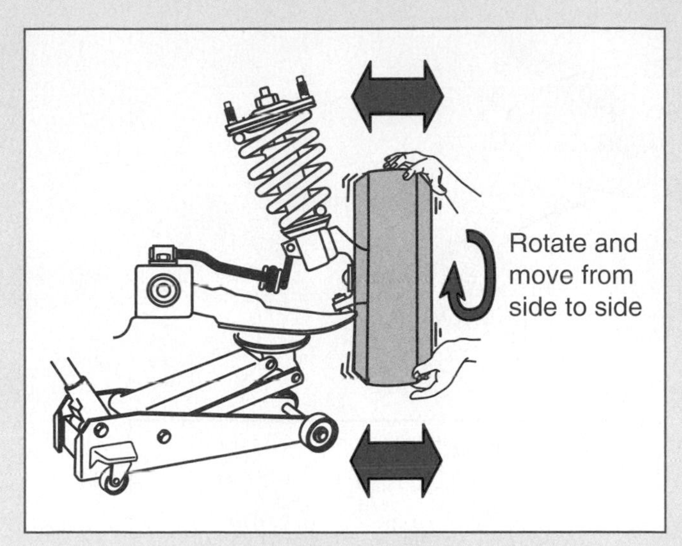

FIGURE 47–41 Check for loose or worn wheel bearings with the weight of the car off the wheel.

DIAGNOSIS

This type of problem is often diagnosed as a bad wheel bearing. When testing for bad wheel bearings, follow these general procedures:

1. Drive the car at low speed on a smooth road.
2. Turn the car to develop left and right motions, traffic permitting.
3. The noise should change because of the cornering loads being produced.
4. Jack up the wheels to verify the roughness at the wheels. Check for loose or worn wheel bearings with the weight of the car off the wheel. This procedure is shown in ***Figure 47–41***.
5. Rotate the wheel and verify roughness as the wheel turns.

SERVICE

If the wheel bearings seem rough or produce vibration in the front suspension, use the following general procedure to remove and replace the wheel bearings.

1. Remove the wheel from the hub.
2. Remove the small cover over the hub nut.
3. Remove the cotter pin.
4. Remove the spindle nut.
5. On disc brakes, pull off the caliper assembly.
6. Now pull off the drum or disc from the spindle assembly. The outside bearing will be pulled off as well.
7. Remove the inner seal that is pressed into the hub.
8. Using suitable tools, remove the inside bearing from the hub. Typically, a hammer and punch can be used. Remember to push on the part of the bearing that is pressed into the hub.
9. Clean the grease from the bearings and from the inside of the hub.
10. Check the condition of both the inner and outer bearings. The following conditions indicate a damaged bearing. Always check bearings carefully for each condition.
    - Galling—metal smears on roller ends caused by overheating, lubricant failure, or overload
    - Step wear—wear pattern on roller ends caused by fine abrasives
    - Indentations—surface depressions on race and rollers caused by hard particles of foreign materials
    - Etching—bearing surfaces appear gray or grayish black
    - Heat discoloration—bearing surfaces appear faint yellow to dark blue, resulting from overload and lubricant breakdown
    - Brinelling—surface indentations in the raceway caused by rollers either under impact loading or vibration while bearing is not rotating
11. Replace bearings that are damaged or show signs of wear.
12. Repack new bearings with grease. Make sure that grease has been repacked completely into the bearings before installation. When handling bearings, always use the guidelines listed in ***Figure 47–42***.
13. Install new bearings. The inner bearing will need to be pressed into the drum or disc hub. Be sure to put pressure only on the outer race of the bearing during installation.
14. Install the disc or drum hub onto the spindle.
15. Front wheel bearings must be adjusted correctly so that the right amount of load is placed on the bearings. Use the following procedure to adjust the wheel bearings.
    a. Rotate the wheel. While the wheel is rotating, torque the spindle nut to approximately 12 pounds. (Check the manufacturer's specifications for the correct torque.)
    b. Back off the nut until it is just loose, then retighten it by hand.
    c. Loosen the spindle nut until the cotter pin can be inserted. Do not, however, loosen the nut more than half flat on the nut. With the bearing properly adjusted, there should be about 0.001–0.005 inch end play.
    d. Check the front wheel rotation to see if the wheel rotates smoothly and easily, without excessive friction or noise.
    e. Insert the spindle nut cover and wheel.
    f. Remove the vehicle from the jack or hoist and test it.

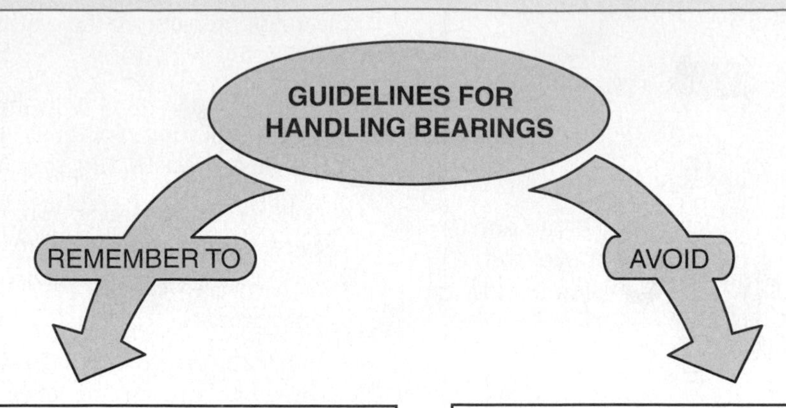

REMEMBER TO	AVOID
1. Remove all outside dirt from housing before exposing the bearing	1. Working in dirty area
2. Treat a used bearing as carefully as a new one	2. Using dirty, brittle, or chipped tools
3. Work in clean area with clean tools	3. Working on wooden benchtops
4. Handle with clean, dry hands or gloves	4. Handling with dirty, moist hands
5. Use solvents and flushing oils	5. Using gasoline (explosive)
6. Lay bearing on clean paper	6. Spinning uncleaned bearings
7. Protect disassembled bearing from rust and dirt	7. Spinning bearings with compressed air
8. Use clean rags to wipe bearing	8. Using dirty rags to wipe bearings
9. Keep bearings wrapped in oil-proof paper when not in use	9. Exposing bearings to rust or dirt
10. Clean inside of housing before replacing the bearing	10. Scratching or nicking the bearing surfaces

FIGURE 47-42 Bearings should be handled with care. Always follow these guidelines.

PROBLEM: Damaged Strut Assembly

A front-wheel drive vehicle produces a noise as the vehicle goes over a bump.

DIAGNOSIS

Often, if a front-wheel drive suspension system is damaged or worn, the problem is in the strut assembly.

1. A weak strut assembly on the MacPherson strut suspension can be checked by pushing downward, then quickly releasing near the fender over each strut. Any tendency to bounce more than once means the shock may be in poor condition and should be replaced.
2. Worn strut rod bushings may be checked by firmly grasping the strut rod and shaking it. Any noticeable play indicates excess wear, and replacement is needed.

SERVICE

Use the following general procedure to remove and replace the strut.

1. Raise and support the front of the vehicle, using all safety precautions mentioned earlier.
2. Remove the wheel and tire.
3. Support the lower control arm with a suitable jack stand.

4. Remove the brake hose bracket.
5. Mark the eccentric bolt position to the strut bracket to retain correct camber wheel alignment during reassembly.
6. Remove the strut-to-knuckle bolts.
7. Remove the cover from the upper end of the strut at the shock tower area.
8. Remove the nuts from the upper strut assembly.
9. Remove the strut assembly from the vehicle.
10. Reverse the procedure to install the strut.
11. Torque all bolts and nuts to the correct manufacturer's specifications.
12. It will probably be necessary to check the camber adjustment after the strut has been removed and replaced.

PROBLEM: Damaged Rear Suspension Components

While driving down the road, a customer hears noise from the rear suspension system.

DIAGNOSIS

Various problems can occur in the rear suspension system.

1. Visual inspection of the rear suspension system can reveal loose, worn, or broken parts. Leaf springs should bow upward at the ends. If the leaf springs are flat, they either have broken or have lost tension. Replace as necessary.
2. Check the coil springs for spots and cracks. Inspect the mounting plates for broken or missing pads. Make sure each coil is an equal distance from the coils above and below it. (Some springs are manufactured so that the spring coils are closer at the top.)
3. Check the vehicle for "dog" tracking. This means that the rear wheels track to the right or left of the front wheels. This condition can be checked by visually observing the alignment of the vehicle as it moves down the road. Possible causes for this condition and suggested service are shown in ***Figure 47–43***.

SERVICE

Depending on the damaged or worn part, replace as necessary. Refer to the manufacturer's service manual for correct procedures.

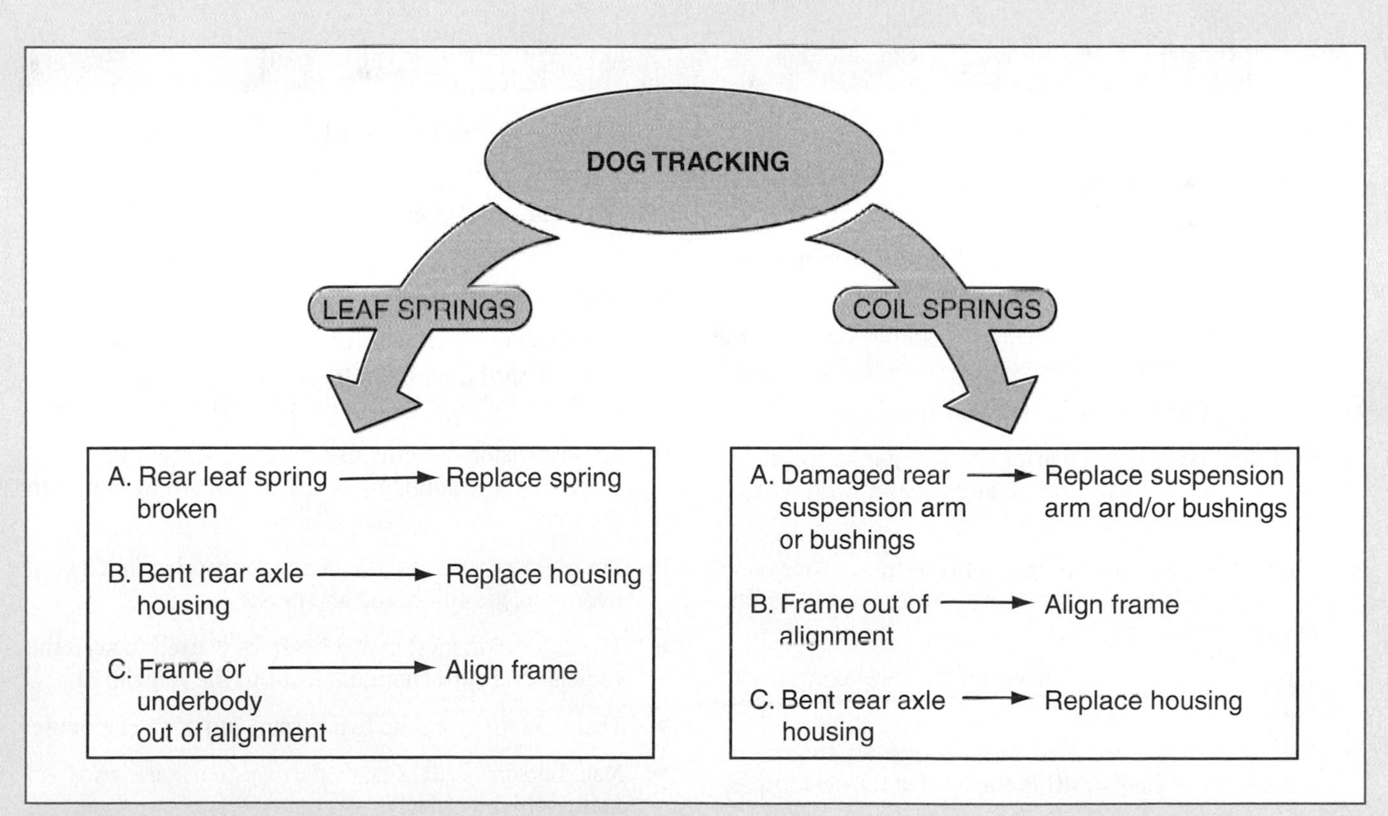

FIGURE 47–43 "Dog" tracking (rear wheels not tracking the same as the front wheels) can be caused by several problems in the rear suspension.

Service Manual Connection

There are many important specifications to keep in mind when working with suspension systems. To identify the specifications for your vehicle, you will need to know the VIN (vehicle identification number) of the vehicle, the type and year of the vehicle, and the type of engine. Although they may be titled differently, some of the more common suspension system specifications (not all) found in service manuals are listed below. Note that these specifcations are typical examples. Each vehicle may have different specifcations.

Common Specification	Typical Example
Ball joint nut torque	88 ft.-lb
Caliper-to-knuckle torque	58–72 ft.-lb
Maximum allowable movement at wheel outer rim diameter (14-inch wheel)	0.020 inch
Rear wheel bearing adjustment (end play)	0.004 inch
Starting torque on lower ball joint	26–87 ft.-lb
Tie-rod end torque	11–25 ft.-lb
Wheel hub end play	0.008 inch

In addition, service manuals provide specific directions for various service and testing procedures. Some of the more common procedures include:

- Ball joint inspection
- Coil spring replace
- Hub and bearing replace
- Rear wheel bearing adjust
- Shock absorber replace
- Spindle replace
- Stabilizer bar replace
- Steering knuckle replace
- Sway bar replace

SUMMARY

The following statements will help to summarize this chapter:

- The front suspension system supports the front of the vehicle.
- Most vehicles use an independent front suspension system.
- Ball joints are used to connect the spindle and steering knuckle to the control arms.
- There are upper and lower control arms.
- With the use of control arms and ball joints, the wheel can turn as well as move up and down during normal operation.
- Sway bars and links are used on the front suspension system to help transmit cornering forces from one side of the vehicle to the other.
- Strut rods are used to help retain the lower control arm in its position.
- Coil springs are used to support the weight of the vehicle and to help position the level of the vehicle.
- Torsion bars are used on some vehicles in place of springs.
- The spindle is used to carry the wheel bearings for the wheel.
- The rear suspension system supports the rear of the vehicle.
- The rear suspension must keep the rear wheels in line with the front wheels.
- Both coil springs and leaf springs support the rear of the vehicle.
- Leaf springs are made of spring steel.
- The shackle is a small link placed between the end of the leaf spring and the frame to allow the spring to lengthen and shorten slightly when going over bumps.
- All suspension systems use shock absorbers to absorb or dampen the bouncing of the car caused by bumps in the road.
- There are several designs of shock absorbers including hydraulic, gas-filled, and air shocks.
- The automatic level control system is used to keep the vehicle level with changing loads in the vehicle.
- There are many inputs to a suspension control module.
- MacPherson strut suspension systems are used on many vehicles today.
- The MacPherson strut system is much like a shock absorber and spring built into one unit.
- Diagnosis and service are important considerations when working on suspension systems.
- Ball joints should always be checked for wear.

- Front-wheel bearings should always be checked for damage and replaced accordingly.
- Shocks should be checked for leakage, excessive bouncing, and broken parts.
- Struts and sway bars and links should be checked for damaged bushings or broken parts.
- Many of the suspension system parts can be checked by visual inspection.
- Worn or damaged suspension parts can cause uneven tire wear, poor handling, or uncomfortable rides.

TERMS TO KNOW

Can you explain each of the following terms? Review the chapter until you can use each term correctly.

Active suspension system
Aeration
Ball joints
Bellows
Control arms
Independent rear suspension
MacPherson strut suspension
Oscillation
Shackle
Shock absorbers
Stabilizer bar
Stance
Steering knuckle
Strut rod
Sway bar
Torsion bar
Wheel spindle
Zerk

REVIEW QUESTIONS

Multiple Choioc

1. The front suspension system on vehicles is called:
a. Independent front suspension
b. Rigid suspension
c. I-beam suspension
d. Leaf spring suspension
e. Strut suspension

2. How many ball joints are used on each side of the standard front suspension system?
a. 1
b. 2
c. 3
d. 4
e. 5

3. The ball joints are attached to the:
a. Steering knuckle
b. Upper control arms
c. Lower control arms
d. All of the above
e. None of the above

4. Which of the following help to transmit cornering loads to the opposite wheel?
a. Strut rods
b. Control arms
c. Sway bars and links
d. Ball joints
e. Coil springs

5. To keep the single pivot lower control arm held in place, ________ are used.
a. Coil springs
b. Leaf springs
c. Ball joints
d. Strut rods
e. Stabilizer bars

6. Which of the following are used instead of coil springs on the front suspension?
a. Shock absorbers
b. Ball joints
c. Torsion bars
d. Stabilizer bars
e. Sway bar and link

7. Front-wheel bearings are of the ________ type.
a. Tapered-roller or ball-bearing
b. Needle-bearing
c. Bushing
d. Triple-ball
e. None of the above

8. The inner race of the wheel bearings rides on the:
a. Steering knuckle
b. Ball joint
c. Stabilizer bar
d. Control arm
e. Spindle

9. The rear suspension systems used on cars:
a. Never need service
b. Use a solid axle
c. Can be independent rear suspension
d. All of the above
e. B and C

10. One end of the leaf spring is attached to the frame. The other end of the leaf spring is attached to:
a. The shackle
b. The frame
c. The axle
d. The wheel
e. The bearing

11. Shock absorbers help to:
a. Reduce the number of oscillations of motion
b. Provide a smoother ride
c. Increase the stability of the car
d. All of the above
e. None of the above

12. Which of the following is/are not considered part of the shock absorber?
a. Piston rod and piston
b. Piston valves
c. Strut
d. Bottom valve
e. Container tube

13. Which of the following helps to reduce or dampen the motion of a shock absorber?
a. Position of the valves
b. Amount of oil that can pass through the valve
c. Size of the piston
d. Type of material on the valves
e. Addition of air or aeration of oil

14. Which of the following is considered a problem with shock absorbers?
a. They can heat up too much
b. They produce aeration inside, causing poor performance
c. They don't support the car properly
d. They leak transmission fluid
e. They lock up, causing a rough ride

15. Gas-filled shock absorbers are designed:
a. To reduce aeration
b. To improve safety
c. For automatic leveling
d. All of the above
e. None of the above

16. Which of the following systems uses an air pump and compressed air to level the car?
a. Oil shock absorbers
b. Automatic leveling systems
c. Leaf spring systems
d. MacPherson strut systems
e. Electrical suspension systems

17. The air suspension system uses which of the following?
a. Shock absorbers
b. Automatic height sensors
c. Control module
d. All of the above
e. None of the above

18. A bellows is used on the ________.
a. Springs
b. Air shock absorbers
c. MacPherson strut suspension
d. Standard shock absorbers
e. Torsion bars

19. The air suspension system forces air into the:
a. Shock absorbers
b. Air springs
c. Bellows
d. Stance
e. Coil springs

20. The upper portion of the MacPherson strut suspension:
a. Is attached to the shock tower
b. Uses a bearing
c. Turns as the car turns
d. All of the above
e. None of the above

21. Which is an advantage of a MacPherson strut suspension system?
a. It is lighter
b. It takes up more room
c. It is heavier
d. It requires increased maintenance
e. It has more moving parts

22. The MacPherson strut suspension system
a. Uses a standard shock absorber
b. Requires no maintenance
c. Uses extra control arms
d. Uses extra ball joints
e. Uses no bearings

23. To check the condition of the wear-indicating type ball joint, there should be a difference of ________ inch between the grease zerk and the body of the ball joint.
a. 0.010
b. 0.020
c. 0.030
d. 0.040
e. 0.050

24. To check for loose wheel bearings, the car must be:
a. Driven on a straight line at high speed
b. Placed on a jack with the weight removed
c. Placed on the road surface and shaken
d. Turned in one direction only
e. Loaded down with extra weight

25. A bad set of shocks causes which of the following?
   a. Car lifts when accelerated
   b. Car has more oscillations after a bump
   c. Car has less stability on the road
   d. All of the above
   e. None of the above

26. Front wheel shimmy can be caused by:
   a. Tires being in balance
   b. Worn tie-rods
   c. Worn ball joints
   d. All of the above
   e. B and C

The following questions are similar in format to ASE (Automotive Service Excellence) test questions.

27. *Technician A* says that the front-wheel bearings should be adjusted as tightly as possible by tightening the spindle nut to 100 ft.-lb. *Technician B* says that there should be approximately 0.500-inch play in the front-wheel bearing. Who is correct?
   a. A only   c. Both A and B
   b. B only   d. Neither A nor B

28. *Technician A* says that there is no check for testing the condition of ball joints. *Technician B* says that some ball joints can be checked for a clearance between the end of the grease zerk and the body of the ball joint. Who is correct?
   a. A only   c. Both A and B
   b. B only   d. Neither A nor B

29. There is a continuous bouncing of the vehicle body when the car goes over a bump. *Technician A* says the problem is bad shock absorbers. *Technician B* says the problem is a bad steering knuckle. Who is correct?
   a. A only   c. Both A and B
   b. B only   d. Neither A nor B

30. *Technician A* says there is no way to check the condition of a ball joint. *Technician B* says that a dial indicator can be used to check the movement of a ball joint. Who is correct?
   a. A only   c. Both A and B
   b. B only   d. Neither A nor B

31. A shimmy is felt in the front suspension of the vehicle as it drives down the road. *Technician A* says the problem is the shock absorbers. *Technician B* says the problem is the ball joints. Who is correct?
   a. A only   c. Both A and B
   b. B only   d. Neither A nor B

32. A car is "dog" tracking as it moves down the road. *Technician A* says the problem is the front shock absorbers. *Technician B* says the problem is the frame out of alignment with the wheels. Who is correct?
   a. A only   c. Both A and B
   b. B only   d. Neither A nor B

33. When handling bearings, *Technician A* says to keep bearings wrapped in oil-proof paper when not in use. *Technician B* says that when cleaning bearings, use clean solvent and flushing oils. Who is correct?
   a. A only   c. Both A and B
   b. B only   d. Neither A nor B

34. A noise is heard in the front suspension of the vehicle as it goes over a bump. *Technician A* says the problem is a worn shock absorber. *Technician B* says the problem is a worn tire. Who is correct?
   a. A only   c. Both A and B
   b. B only   d. Neither A nor B

Essay

35. What is the purpose of the sway bar?

36. Define the purpose and operation of a torsion bar.

37. Describe how to check the condition of shocks.

38. What is the purpose of gas-filled shocks?

39. What are several advantages of MacPherson strut suspension systems?

40. Describe the purpose of a wheel spindle.

41. What is the purpose of ball joints?

Short Answer

42. Active suspension systems are able to control body ________ and ________.

43. On an active suspension system, the computer is controlling ________.

44. New wear-indicating ball joints should have a clearance of about ________ inch between the end of the grease zerk and the body of the ball joint.

45. A car that has continuous bouncing of the body with every road bump has worn ________.

46. A damaged wheel bearing should be checked for several conditions. Two things to check for are ________ and ________.

Applied Academics

TITLE: Suspension Specifications

ACADEMIC SKILLS AREA: Science

NATEF Automobile Technician Tasks:

The service technician uses the information in service manual charts, tables, and graphs to determine manufacturers' specifications for steering and suspension system operation and the appropriate repair/replacement procedure.

CONCEPT:

There are many manufacturers' specifications that are important for the service technician to understand. The technician must be able to identify the specifications that are related to suspension systems and then determine what could happen if a vehicle is out of specifications. The following chart shows a series of common torque specifications for a suspension system. Each torque reading is shown in English and metric readings.

SUSPENSION TORQUE SPECIFICATIONS

COMPONENT	ENGLISH SPECIFICATIONS	METRIC SPECIFICATIONS
Axle spindle nut	181 ft.-lb	245 Nm
Ball joint castle nut	36 ft.-lb	49 Nm
Damper fork nut	47 ft.-lb	64 Nm
Damper fork pinch bolt	32 ft.-lb	44 Nm
Hub unit flange bolts	33 ft.-lb	44 Nm
Hydraulic fittings-to-the steering valve body	21–27 ft.-lb	28–37 Nm
Inner tie-rod end fastener	40 ft.-lb	54 Nm
Lower ball joint nuts/bolts	36–43 ft.-lb	49–59 Nm
Lower control arm front flange bolts	76 ft.-lb	103 Nm
Lower control arm rear bushing bolts	40 ft.-lb	54 Nm
Lower strut mount bolt	40 ft.-lb	55 Nm
Lug nuts	80 ft.-lb	108 Nm
Radius rod bushing bolt	17 ft.-lb	24 Nm
Radius rod-to-control arm flange bolts	76 ft.-lb	103 Nm
Strut flange nuts	47 ft.-lb	64 Nm
Strut mount nuts	28 ft.-lb	39 Nm
Sway bar bracket-to-body bolts	16 ft.-lb	22 Nm
Sway bar end link	14 ft.-lb	19 Nm
Sway bar link flange nut	22 ft.-lb	29 Nm
Tie-rod end	32 ft.-lb	43 Nm
Upper ball joint nuts/bolts	29–35 ft.-lb	39–47 Nm
Upper control arm nuts	47 ft.-lb	65 Nm
Upper strut mount nuts	28 ft.-lb	39 Nm
Wheel lug nuts	80 ft.-lb	108 Nm

APPLICATION:

Using the information in this chapter and the torque specifications shown in the figure, can you identify the torque for the "sway bar link flange nut"? What is the torque specification for the "ball joint castle nut"? What could happen if the torque specifications were not followed by a service technician and too much torque were applied to the lug nut?

Steering Systems

CHAPTER 48

OBJECTIVES

After studying this chapter, you should be able to:

- Identify the parts and operation of the standard steering system.
- Examine the operation of the steering gear.
- Define front end geometry including caster, camber, toe, steering axis inclination, turning radius, and four-wheel alignment.
- Describe the operation of power steering units and pumps.
- State common problems, their diagnosis, and service procedures on the steering system.

INTRODUCTION

The steering system is used to control the direction of the vehicle. The steering system is designed to control the direction of the front wheels over all types of road conditions, through turns, and at different speeds. It is made of a linkage system that is attached to the front wheels, the steering wheel, and the steering gear. Manual and power steering units are used. The purpose of this chapter is to analyze the parts and operation of the steering system components.

Certification Connection

ASE Connection: The information in this chapter can help you prepare for the National Institute for Automotive Service Excellence (ASE) certification tests. The tests and content areas most closely related to this chapter are:

Test A4—Suspension and Steering

- **Content Area**—Steering Systems Diagnosis and Repair
- **Content Area**—Wheel Alignment Diagnosis, Adjustment, and Repair

NATEF Connection: Much of the information in this chapter is related to the NATEF tasks. The NATEF tasks and priority numbers most closely related to this chapter are:

1. Identify and interpret suspension and steering concerns; determine necessary action. P-1
2. Inspect steering shaft universal joint(s), flexible coupling(s), collapsible column, lock cylinder mechanism, and steering wheel; perform necessary action. P-2
3. Remove and replace manual or power rack-and-pinion steering gear; inspect mounting bushings and brackets. P-1
4. Inspect power steering fluid levels and condition. P-1
5. Flush, fill, and bleed power steering system. P-2
6. Diagnose power steering fluid leakage; determine necessary action. P-2
7. Remove, inspect, and replace or adjust power steering pump belt. P-1
8. Remove and reinstall power steering pump. P-3
9. Remove and reinstall power steering pump pulley; check pulley and belt alignment. P-3.
10. Inspect and replace power steering hoses and fittings. P-2
11. Inspect and replace pitman arm, relay (centerline/intermediate) rod, idler arm and mountings, and steering linkage damper. P-2

continued

Certification Connection *continued*

12. Inspect, replace, and adjust tie-rod ends (sockets), tie-rod sleeves, and clamps. P-1
13. Differentiate between steering and suspension concerns using principles of steering geometry (caster, camber, toe, etc.). P-1
14. Diagnose vehicle wander, drift, pull, bad steering, bump steer, memory steer, torque steer, and steering return concerns; determine necessary action. P-1

48.1 STEERING SYSTEM PARTS AND OPERATION

PARTS OF A STEERING SYSTEM

The steering system is composed of three major subsystems (***Figure 48–1***). They are the steering column and wheel, the steering gear, and the steering linkage. As the driver turns the steering wheel, the steering gear transfers this motion to the steering linkage. The steering linkage turns the wheels to control the vehicle direction.

Although there are many variations of this system, these three major assemblies make up the steering system. Other variations include power steering and rack-and-pinion steering.

Steering Wheel and Column The purpose of the steering wheel and column is to produce the force necessary to turn the steering gear. The steering column has many parts. The exact type of steering wheel and column depends on the year of the car and the manufacturer. Major parts shown in ***Figure 48–2*** include:

- Steering wheel—produces the turning effort
- Upper and lower covers—conceal parts
- Universal joints—rotate at angles
- Coupling assembly—allows for a slight bending of the main and intermediate steering shaft
- Intermediate shaft—connects the coupling assembly to the universal joint
- Mounting bracket—holds the steering column in place

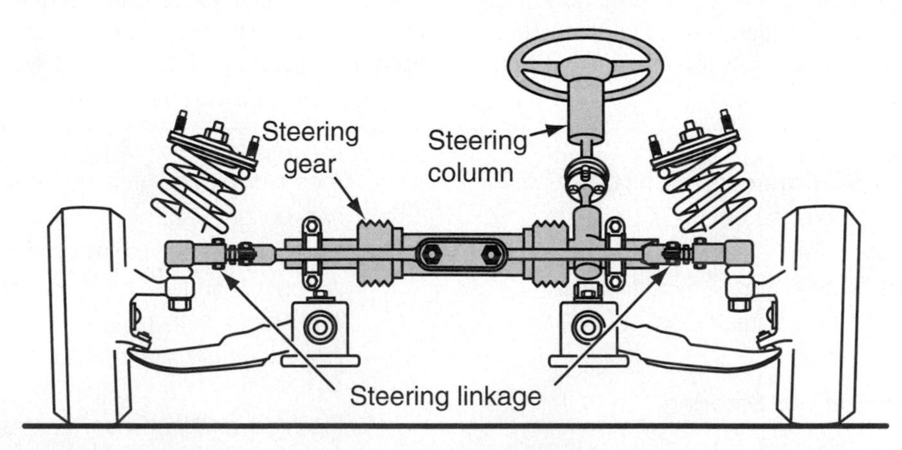

FIGURE 48–1 The three main parts of the steering system are the steering column, steering gear, and steering linkage.

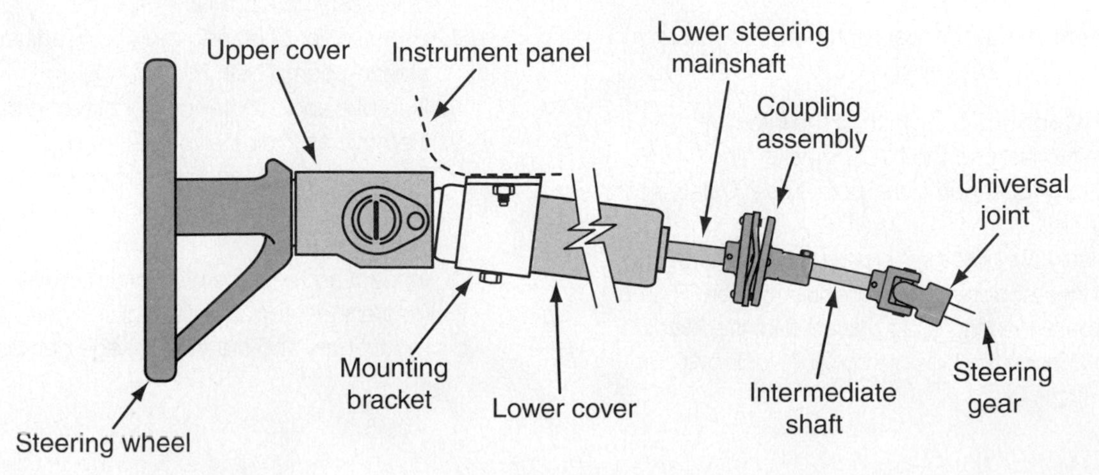

FIGURE 48–2 Steering columns have many parts and vary in design, depending on the manufacturer and the year of the vehicle.

Differences in the steering wheel and column include energy-absorbing or collapsible steering columns, tilt steering wheels, steering lock systems, and locations of turn signals and flasher controls.

Manual Steering Gear The purpose of the steering gear is to change the rotational motion of the steering wheel to reciprocating motion to move the steering linkage. Two styles are currently in use. These are the pitman arm, or recirculating ball steering gear, and the rack-and-pinion steering gear.

Pitman Arm Steering Gear One of the most common types of manual steering gears is called the **pitman arm** steering gear. Many manufacturers call this the recirculating ball and worm system. ***Figure 48–3*** shows such a system. In operation, as the steering shaft is turned, the wormshaft also turns. The wormshaft has spiral grooves on the outside diameter. The ball nut, which has mating spiral grooves inside, is placed over the wormshaft. Small steel balls circulate in the mating grooves and ball guides. As the balls move through the grooves and out, they return to the other side through the guides. This system provides a low-friction drive between the wormshaft and the ball nut.

Teeth on the ball nut mesh with the teeth on the sector shaft. The sector shaft is also known as the pitman shaft. As the wormshaft is rotated, the ball nut moves back and forth, to the left and right. As the ball nut moves back and forth, it causes the sector shaft, or pitman shaft, to rotate through a partial circle. The sector shaft is connected directly to the pitman arm, which controls the steering linkage.

Rack-and-Pinion Steering Gear The **rack-and-pinion system** has become the standard on most front-wheel drive cars sold in the United States. The rack-and-pinion system is used in conjunction with MacPherson struts and gives more engine compartment room for transverse-mounted engines.

Rack-and-pinion steering consists of a flat gear (the rack) and a mating gear called the pinion, ***Figure 48–4***. When the steering wheel and steering shaft turn, the pinion gear meshes with the teeth on the rack. This causes the rack to move left or right in the housing. This motion moves the remaining steering linkage to turn the front wheels. This system is very practical for small cars that require lighter steering capacity. It is a direct steering unit that is more positive in motion (less lost motion) than the standard

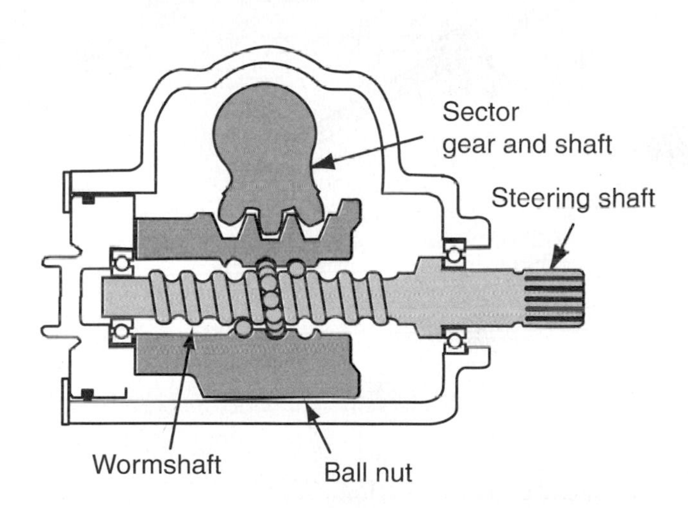

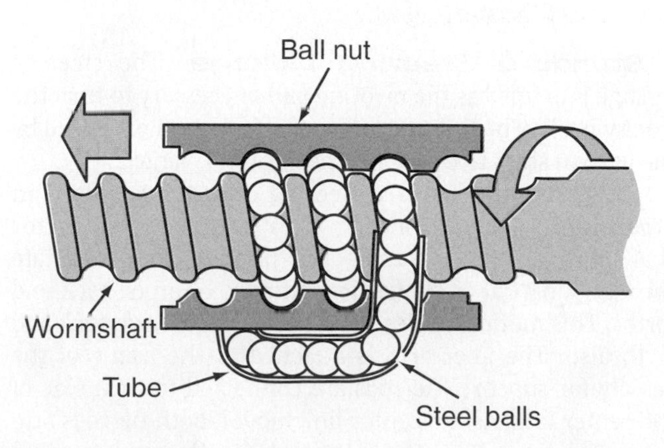

FIGURE 48–3 A manual steering gear uses a recirculating ball and worm system.

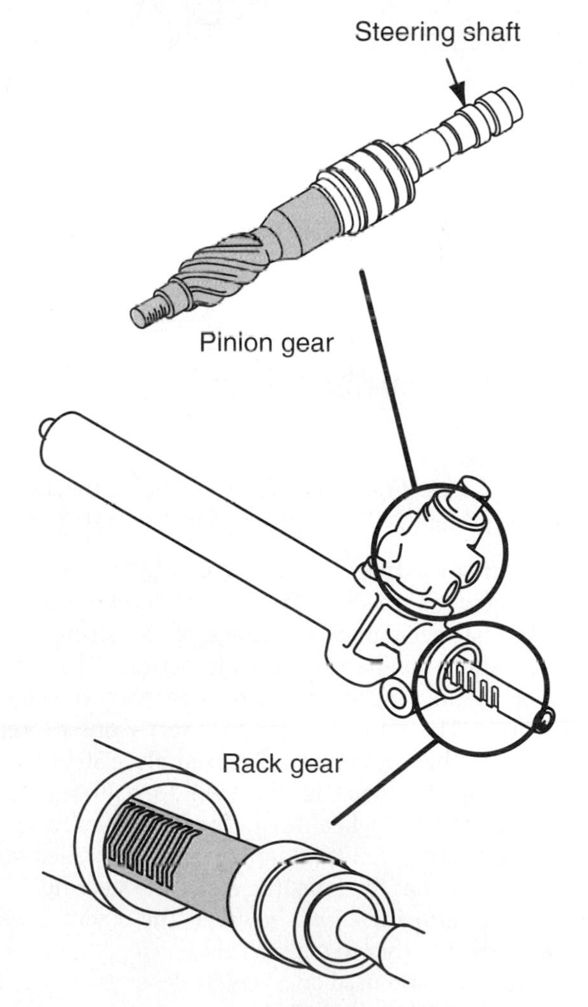

FIGURE 48–4 The rack and pinion uses a flat gear called the rack and a pinion gear, which is attached to the steering column.

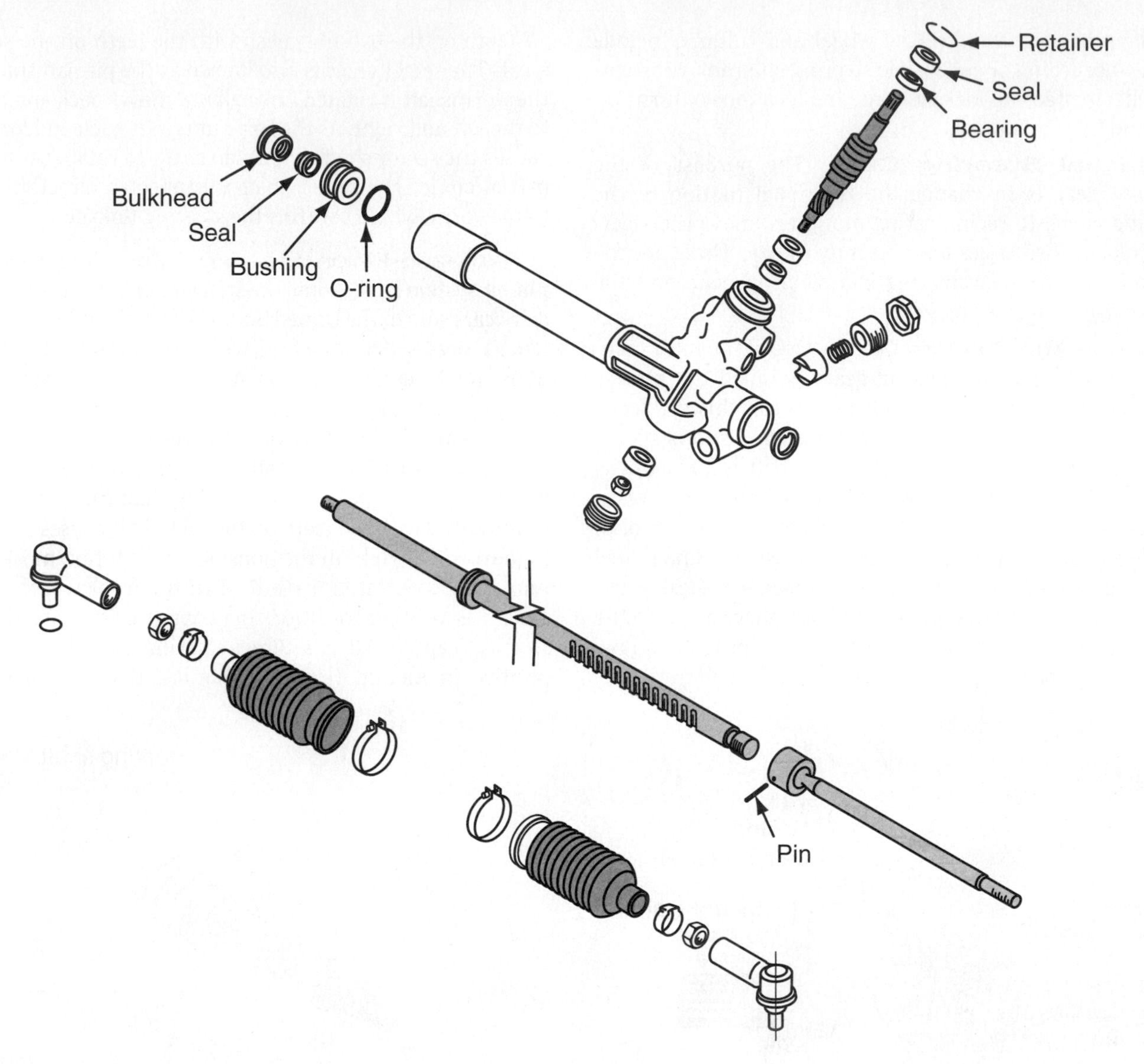

FIGURE 48–5 All parts of a typical rack-and-pinion system.

steering linkages. ***Figure 48–5*** shows the complete rack-and-pinion system with the housing and tie-rods.

Steering Ratio When the steering wheel is turned, a certain effort is needed. The amount of effort is determined by the mechanical advantage of the steering gear. Steering ratio is defined as the ratio between the degrees turned on the steering wheel and the degrees turned on the front wheels. The ratio is stated for exactly one degree of movement on the front wheels. For example, a 30 to 1 steering ratio means the steering wheel will turn 30 degrees for each degree of front wheel turn. The lower the ratio, the harder the steering. Lower steering ratios are called quick steering. The higher the ratio, the easier the steering. When the steering ratio increases, however, the steering wheel must be turned farther to make a turn.

The steering ratio used on a vehicle depends on four factors and differences in the steering system:

1. Manual or power steering
2. Weight and size of the vehicle
3. Type of steering gear
4. Size of the steering wheel

Standard Steering Linkage The steering linkage is defined as the pivoting parts necessary to turn the front wheels. The linkage connects the motion produced by the pitman shaft to the front wheels on the vehicle.

The parts of a standard steering linkage are shown in ***Figure 48–6***. The motion from the steering gear and sector shaft (pitman shaft) causes the pitman arm to rotate through a partial circle (reciprocating motion or back and forth). This motion causes the **center link** to move back and forth also. The idler arm is attached to the frame of the vehicle for support. Tie-rods are connected to each side of the center link. As the center link moves, both tie-rods also move. The tie-rods are then attached directly to the steering

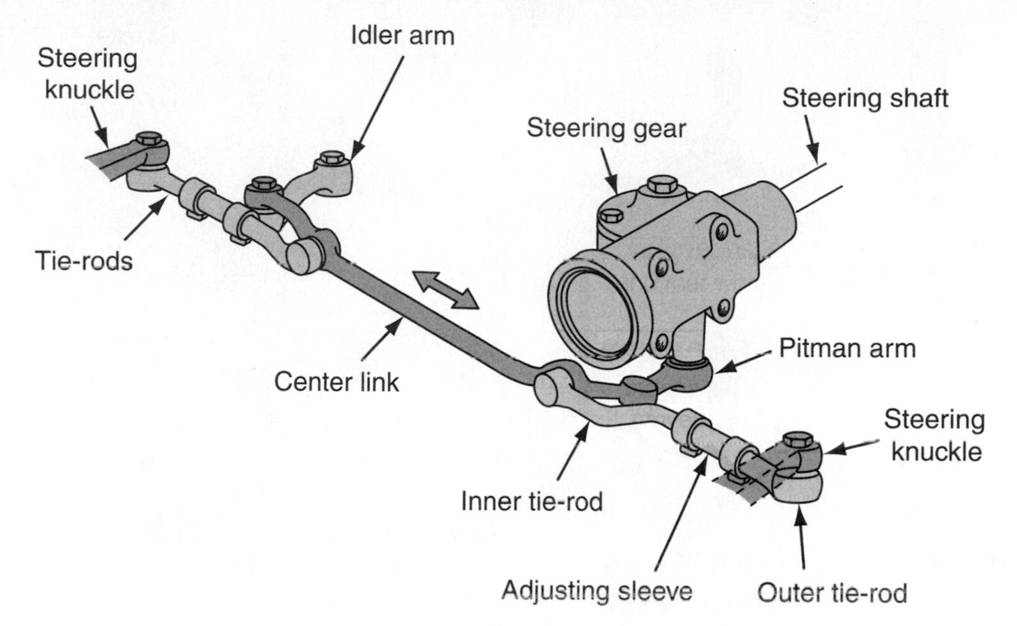

FIGURE 48–6 The steering linkage is made of tie-rods, an idler arm, a center link sleeve, and the pitman arm.

knuckle and wheel for turning. Adjusting sleeves are placed on each tie-rod for adjustment.

Pitman Arm Pitman arms can be of the wear or non-wear type. The wear-type pitman arm has a tapered ball stud that is connected to the center link. The other end is mounted on the steering gear sector shaft. The nonwear-style pitman arm has a tapered hole and seldom needs replacement. ***Figure 48–7*** shows the different types of pitman arms.

Center Link The center (drag) link can be designed in several ways. ***Figure 48–8*** shows several styles. The major difference is the method in which the other linkage is connected to the center link. The point of connection can be the pivot point, stud end, bushing end, or open taper end (nonwear).

Idler Arm **Idler arms** come in different designs as well. They differ mainly on the wear end of the arm. The different types include the bushing, taper, threaded, and constant

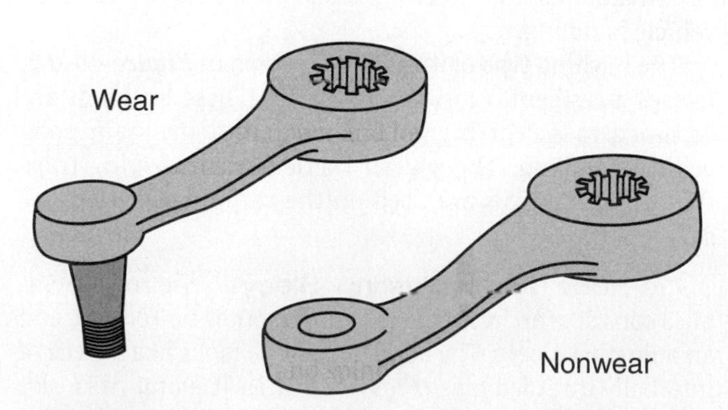

FIGURE 48–7 A vehicle can use one of two types of pitman arms: the wear or the nonwear type. (Courtesy of Dana Corporation)

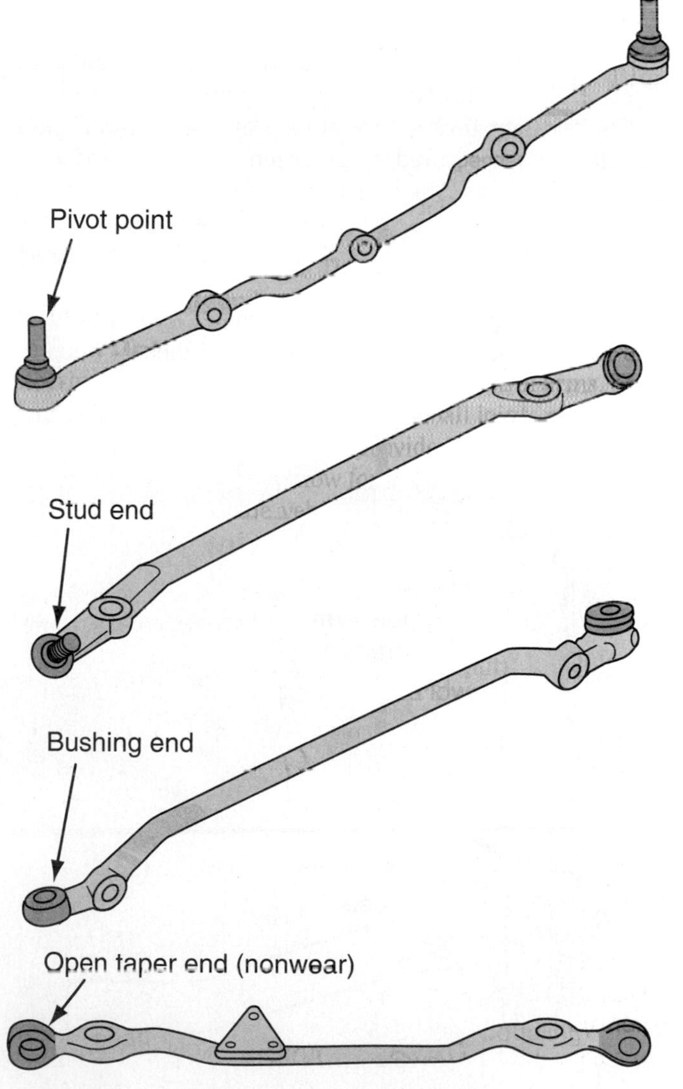

FIGURE 48–8 Different manufacturers use different types of center links. (Courtesy of Dana Corporation)

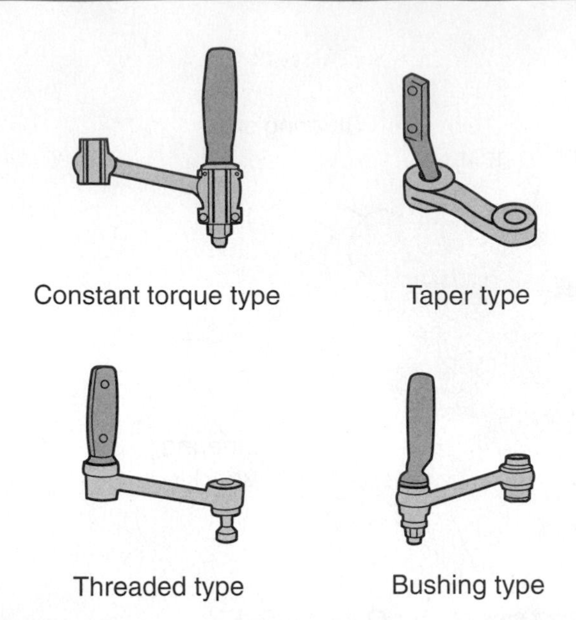

FIGURE 48–9 Different types of idler arms are used on the steering linkage. (Courtesy of Dana Corporation)

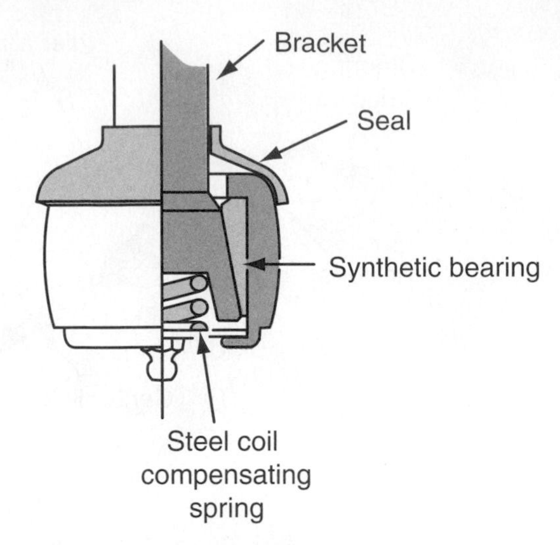

FIGURE 48–11 The taper-type idler arm contains a tapering bracket, synthetic bearings, and a compensating spring.

torque types of idler arm. ***Figure 48–9*** shows the different styles of the idler arms.

The constant torque type of idler arm shown in ***Figure 48–10*** is manufactured to precision tolerances and uses synthetic bearings. The bearings reduce friction and absorb road shock. The bearings are preloaded and preset at the factory. This type of idler arm has very low friction characteristics and is used on many vehicles.

The taper type of idler arm is shown in ***Figure 48–11***. It contains synthetic bearings, heat-treated tapered support brackets, and a compensating spring. The compensating spring takes up clearances produced by wear. The spring

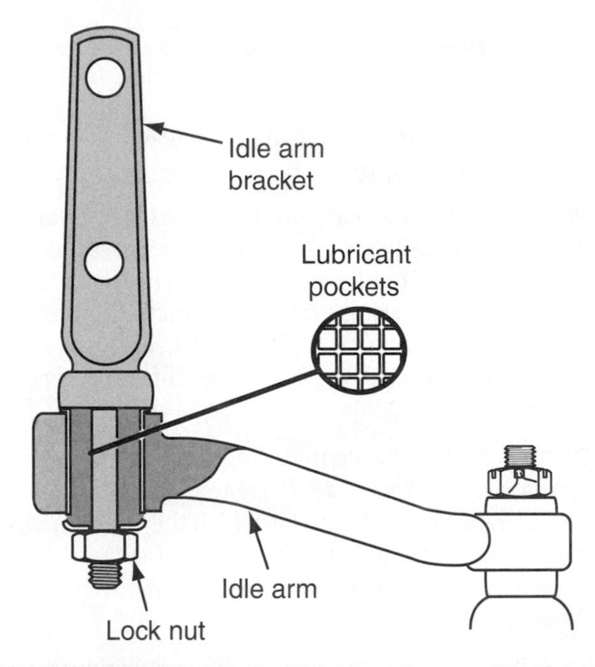

FIGURE 48–12 The bushing-type idler gear has lubricant pockets for lubrication.

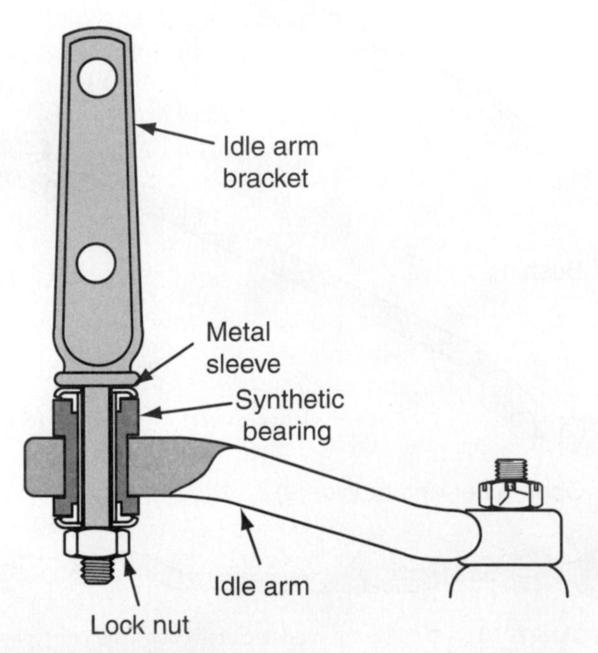

FIGURE 48–10 The constant torque type of idler arm uses precision tolerances and synthetic bearings.

also maintains the steering resistance desired for good vehicle handling.

The bushing type of idler arm is shown in ***Figure 48–12***. It uses a resilient, lubricated bushing. These bushings are designed to accept high shock loads and maintain good vehicle handling. The special waffle design bushing traps the lubricant. Seals are used on the end to make the unit self-contained.

Tie-Rods and Adjusting Sleeve Tie-rod assemblies consist of an inner tie-rod end, an outer tie-rod end, and an **adjusting sleeve**. The adjusting sleeve looks like a piece of internally threaded pipe (***Figure 48–13***). The unit has a slot that runs through the center. Adjusting sleeves also have two crimping, or squeezing, clamps. These are located at each end to lock the tie-rod together after adjustment.

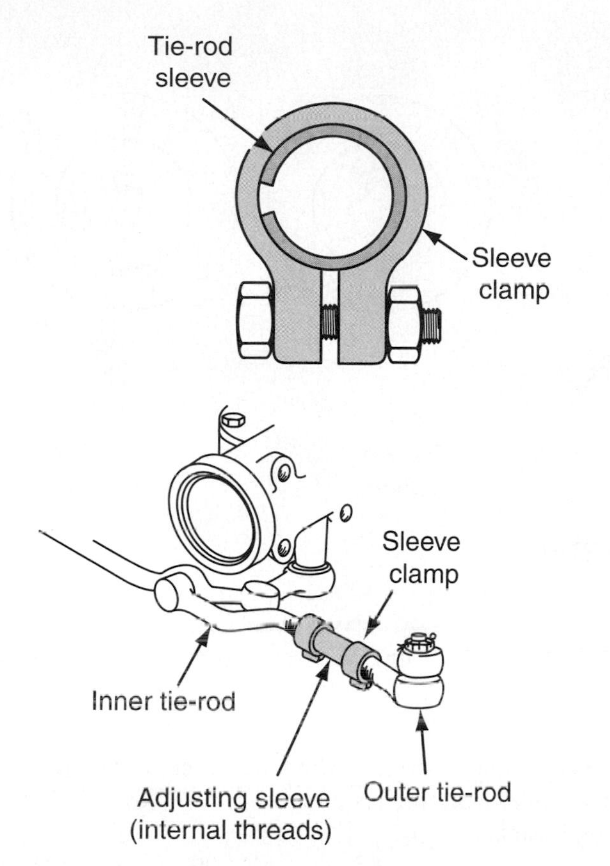

FIGURE 48–13 This sleeve is attached to the tie-rods to make adjustments for length.

The adjusting sleeve has threads inside. One end of the thread has a left-hand thread and the other end has a right-hand thread. This arrangement allows adjustment without disassembling the tie-rods.

The tie-rod ends have a rounded ball stud to allow both lateral and vertical movement. ***Figure 48–14*** shows an example of the rounded ball stud. The tension spring inside the tie-rod end is used to reduce road shock throughout the steering system. On some tie-rod ends, there is also a grease **zerk**. The zerk is the fitting through which grease is applied.

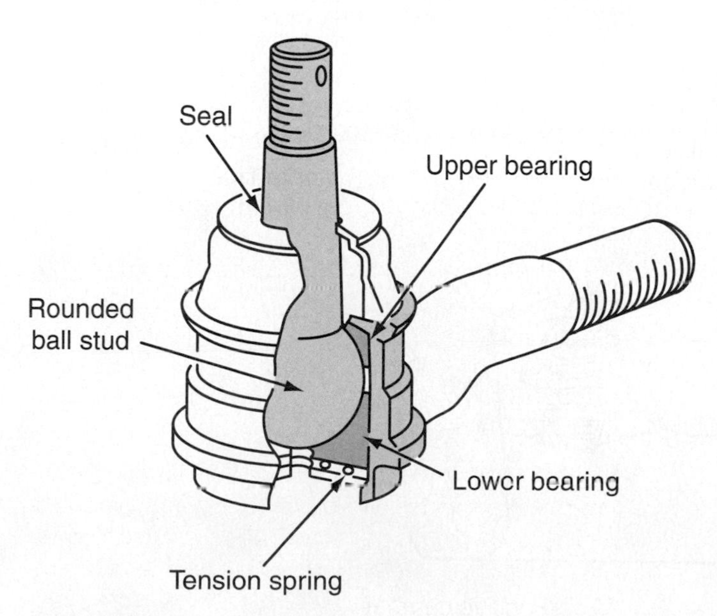

FIGURE 48–14 A rounded ball stud is used on the end of the tie-rods.

48.2 FRONT END GEOMETRY AND ALIGNMENT

PURPOSE OF WHEEL ALIGNMENT

Alignment is defined as the balancing of all forces created by friction, gravity, centrifugal force, and momentum while the vehicle is in motion. It is very important for the wheels of the vehicle to contact the road correctly. Wheel alignment is a check of how the wheels contact the pavement. The main purpose of wheel alignment is to allow the wheels to roll without scuffing, dragging, or slipping on the road. Good alignment results in:

- Better fuel economy
- Less strain on the front-end parts
- Directional stability
- Easier steering
- Longer tire life
- Increased safety

There are typically five angles that affect the steering alignment. These are caster, camber, toe, steering axis inclination, and turning radius.

Caster **Caster** is defined as the backward or forward tilt at the top of the spindle support arm. Backward tilt is called positive caster. Forward tilt is called negative caster. Caster angle is the distance between the centerline of the spindle support arm and the true vertical line. An example of caster is an ordinary bicycle or furniture caster (***Figure 48–15***). The bicycle is an example of positive caster. The furniture caster is an example of negative caster. Both examples tend to keep the rolling object going in a straight line. For example, a person can take his or her hands off the handlebars and still go in a straight line. The lead is defined as the distance at ground level between the two centerlines. ***Figure 48–16*** shows the lead.

Caster is designed into the front-end suspension of a car to do several things:

- Aids in the directional stability of the car by making the front wheels maintain a straight-ahead position
- Aids in returning the front wheels to a straight-ahead position when coming out of a turn
- Offsets the effect of road crown or curvature of the road

Too much caster causes hard steering, excessive road shock, and wheel shimmy. Too little caster causes wander, weave, and instability at high speeds. Unequal caster causes pulling to the side of least caster.

While positive caster does aid in directional stability, it also increases steering effort by the driver. This can be compensated for by power steering. Cars with manual steering

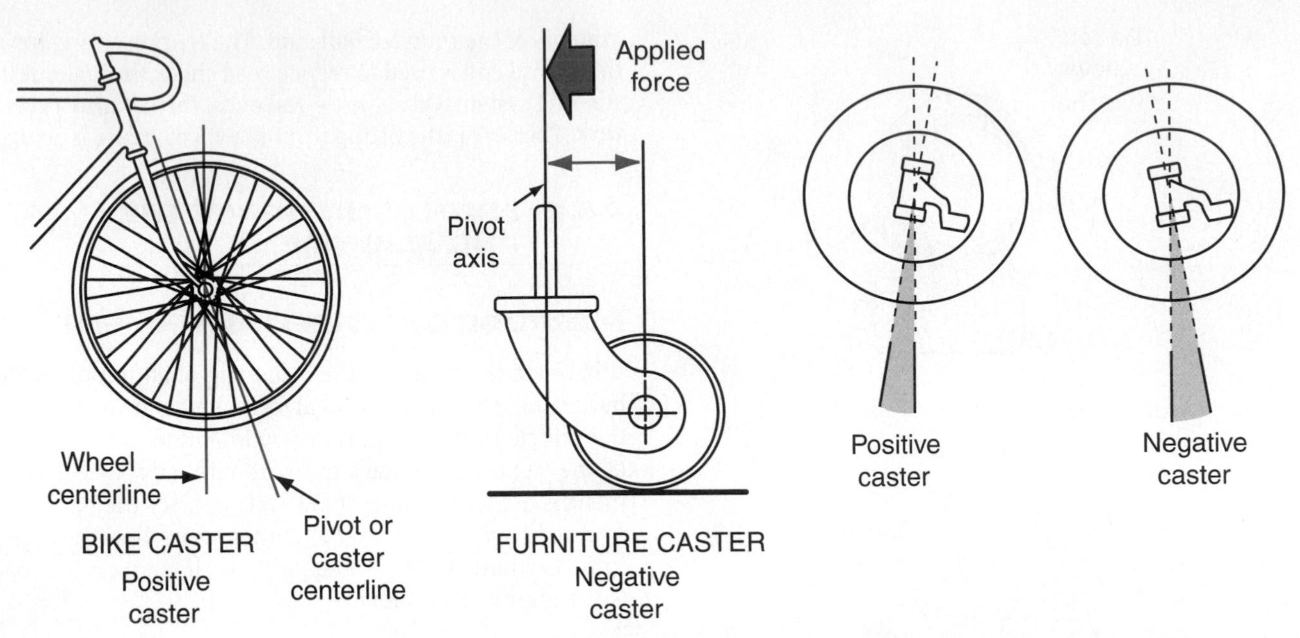

FIGURE 48-15 Caster is defined as the backward or forward tilt of the spindle support arm. It is much like a furniture caster.

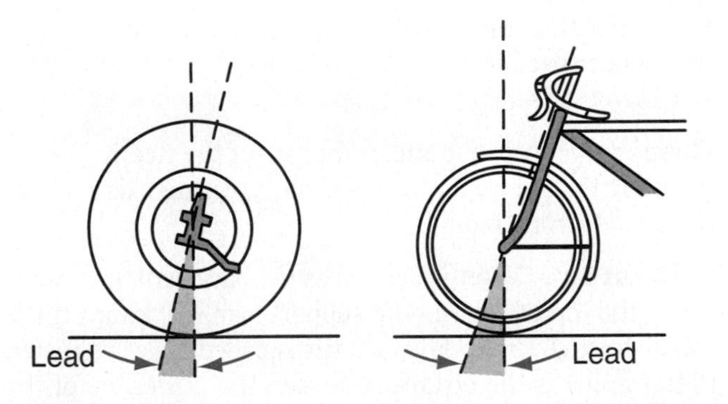

FIGURE 48-16 Lead is the distance at ground level between the two centerlines. A bicycle is shown as an example.

usually require a caster setting of near zero or even a negative angle. Negative caster settings are required on some newer cars.

Camber Camber is defined as the inward and outward tilt of the front wheels at the top. ***Figure 48–17*** shows examples of both negative and positive camber. Camber is measured as an angle in degrees from the centerline of the wheel to a true vertical line. The purpose of checking camber is to make sure the tire is vertical to the road. This position will make the tire tread uniform on both sides of the tire. This results in equal distribution of load and wear over the whole tire tread. When the camber setting is correct:

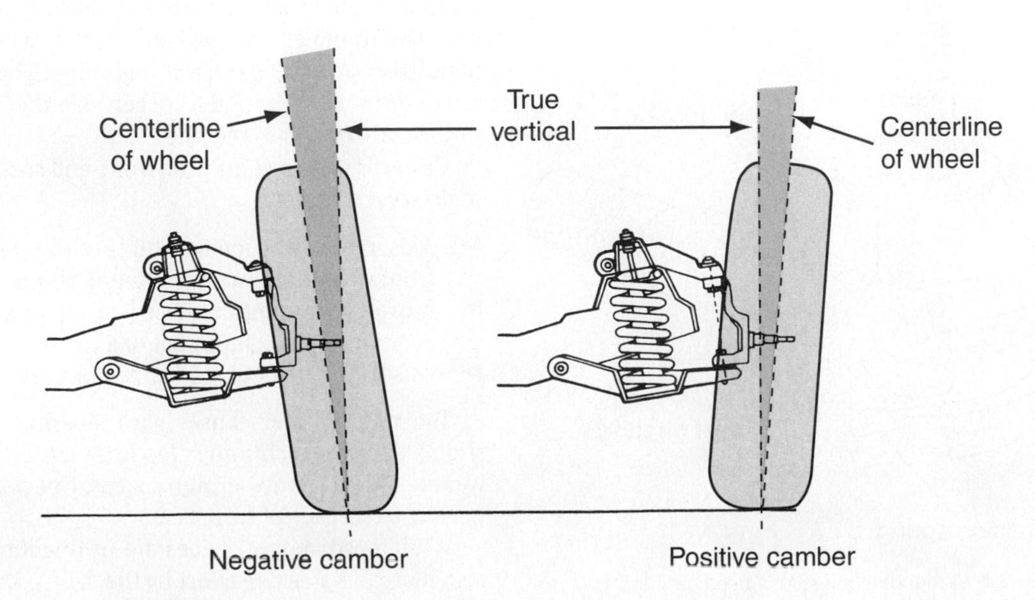

FIGURE 48-17 Negative and positive camber.

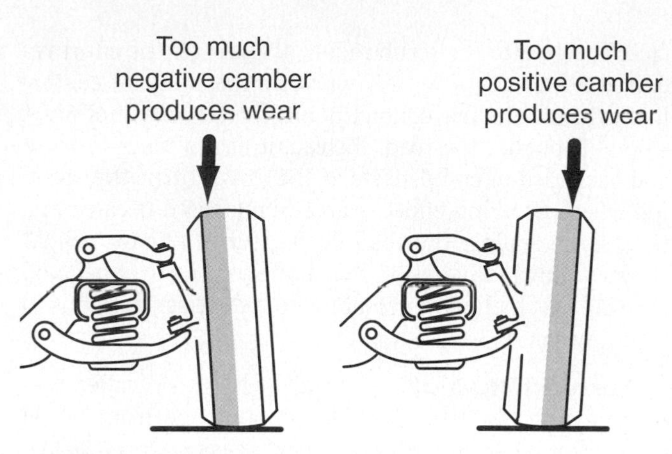

FIGURE 48-18 When the camber is incorrect, wear will increase on the sides of a tire.

- There will be maximum amount of tire tread in contact with the road surface.
- The road contact area of the tire will be directly under the point of load.
- The steering will be easier because the vehicle weight is placed on the inner wheel bearing and spindle.

When the camber is incorrect:

- There will be wear on the ball joint and wheel bearings.
- The steering will pull to one side.
- There will be excessive tire wear. This wear is shown in ***Figure 48–18***. Too much negative camber will cause wear on the inside of the tire. Too much positive camber will cause wear on the outside of the tire.

Wear on tires with incorrect camber is further explained in ***Figure 48–19***. Referring to this figure, notice the rolling radii at different parts of the wheel. At each point, the tire is rolling at different diameters. This causes the wheel to act like a cone. The cone has several rolling diameters and tends to want to roll in a circle. But since it is forced to roll in a straight line, the outer or smaller diameter tries to roll faster. This results in the outer part of the tread being ground off by slipping and scuffing.

Toe **Toe** is defined as the difference in the distance between the front and back of the front wheels. ***Figure 48–20*** shows toe. When dimension B is smaller than A, it is called a toe-in condition. When dimension B is greater than A, it is

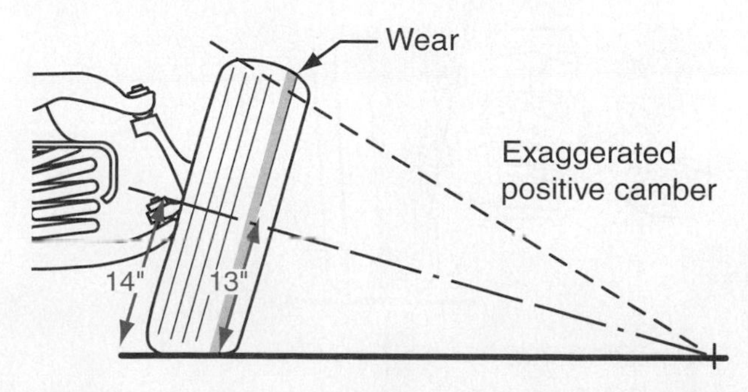

FIGURE 48-19 When camber is excessive, wear occurs because the tire is rolling at different radii.

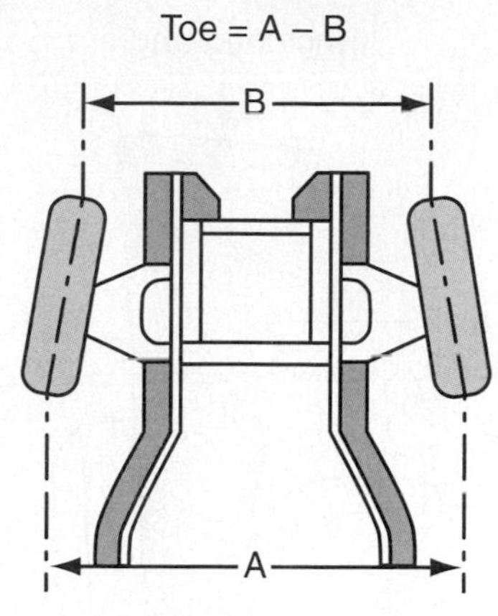

FIGURE 48-20 Toe is the difference in the distances between the front and back of the front tires.

called a toe-out condition. Toe is measured in inches or parts of an inch.

When a vehicle is moving forward, certain forces are developed. Braking and the rolling resistance of the tires force the front wheel outward in front. Vehicles are generally set with just a small amount of toe-in to help overcome these forces. Once in motion, clearances in the steering linkage allow the front of the tires to swing out. At this point, there should be zero amount of toe-in. In front-wheel drive vehicles, the tires may be purposely toed out to allow for other forces. Front-wheel drive vehicles tend to return the wheels to their proper straight-ahead position. Incorrect toe adjustment will cause the tires to wear excessively and will cause harder steering.

Steering Axis Inclination **Steering axis inclination** is defined as the inward tilt of the spindle support arm ball joints at the top. Steering axis inclination angle is the distance between the ball joint centerline and true vertical. This angle is not adjustable. It is shown in ***Figure 48–21***. The purposes of having a steering axis inclination angle are to:

- Reduce the need for excessive camber
- Provide a pivot point about which the wheel will turn, producing easy steering

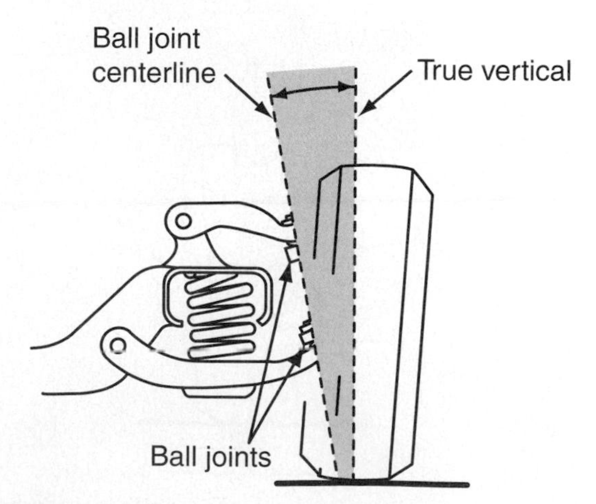

FIGURE 48-21 Steering axis inclination is the angle between the true vertical and the centerline of the ball joints.

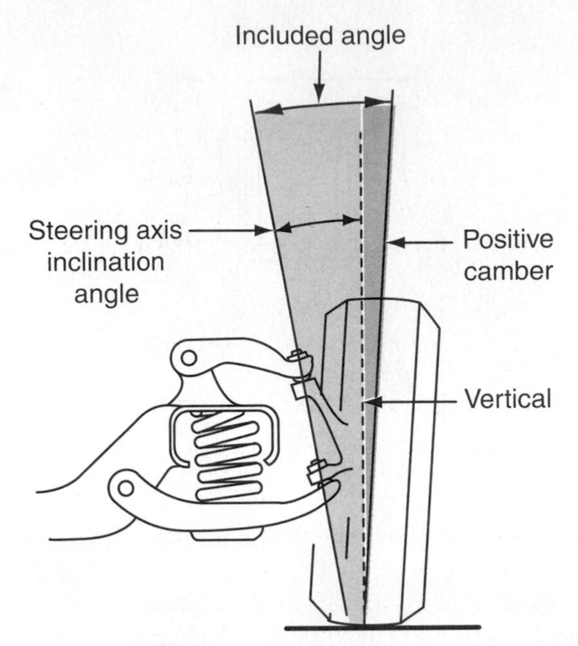

FIGURE 48-22 The included angle is the camber plus the steering axis inclination angle.

- Aid steering stability
- Lessen tire wear
- Provide directional stability
- Distribute the weight of the vehicle more nearly under the road contact area of the tire

Included Angle Certain manufacturers use the term *included angle* to illustrate information about steering axis inclination. The included angle is defined as the sum of the steering axis inclination angle and the camber. For example, ***Figure 48–22*** shows the included angle. Certain alignment charts specify the included angle instead of the steering axis inclination angle.

Scrub Radius **Scrub radius** is the distance between the centerline of the ball joints and the centerline of the tire at the point where the tire contacts the road surface. The greater the scrub radius, the greater the effort required to steer. During turning, when the ball joint centerline is inside the tire contact point, the tire does not pivot where it touches the road. Instead, it has to move forward and backward to compensate as the driver turns the steering wheel. Steering effort is greatly increased because the tires scrub against the road during turns. ***Figure 48–23*** shows scrub radius. Note that both positive camber and steering axis inclination combine to reduce scrub radius to a minimum.

Turning Radius **Turning radius**, also called toe-out on turns, is defined as the amount one front wheel turns more sharply than the other. It is measured in degrees. The major purpose for having the correct turning

CUSTOMER CONCERN:

Front End Shimmy

A car with 77,000 miles on it seems to have a bad shimmy. What are the causes of front end shimmy? What should be checked first?

SOLUTION: Shimmy can be caused by several problems, including:

- Not enough caster
- Toe-in out of specifications
- Loose steering linkage parts
- Too much play in the steering gear
- Bad shocks
- Bad suspension parts
- A combination of any of the above

These items should be checked carefully to determine the cause of shimmy on the front wheels.

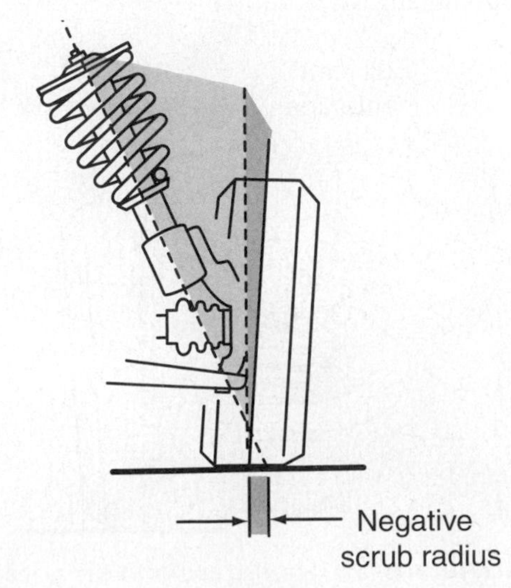

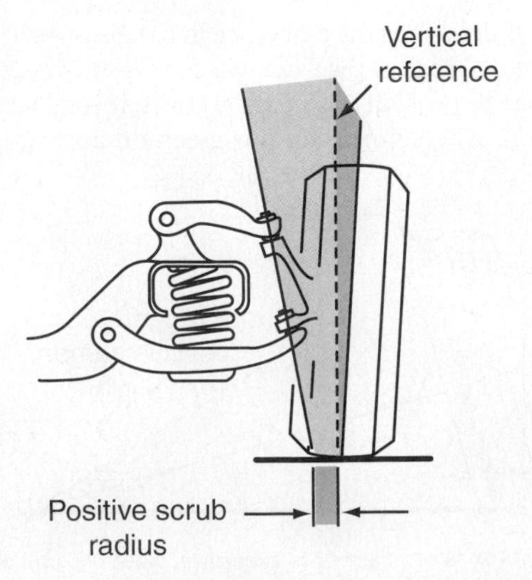

FIGURE 48-23 Scrub radius is the distance at the road surface between the centerline of the tire and the centerline of the ball joints.

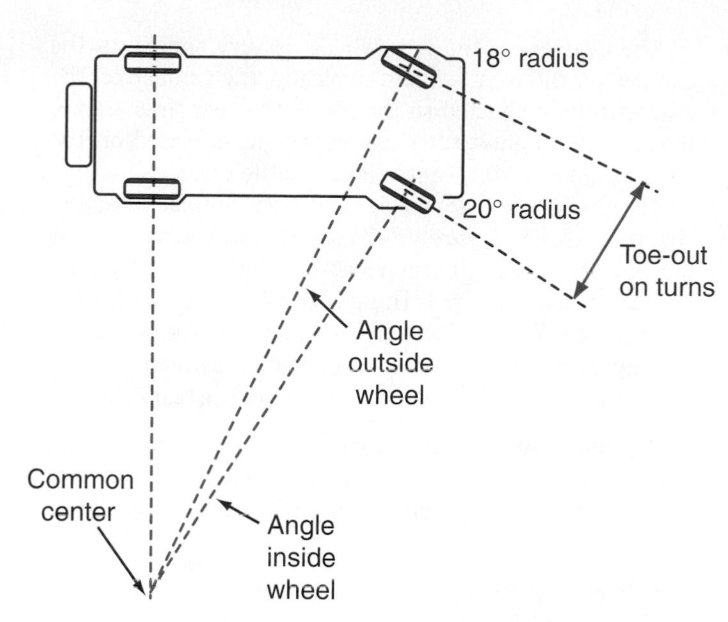

FIGURE 48-24 Turning radius is the amount one wheel turns more sharply than the other.

radius is to make the front wheels pivot around a common center. This is shown in ***Figure 48–24***. As the car turns around a corner, the outside wheel turns a radius of 18 degrees. The inside wheel turns a radius of 20 degrees. If the turning radius is incorrect, the front wheels will scrub against the road surface on turns.

Turning radius is usually not adjustable. On certain vehicles, however, it can be checked. Turning radius is checked after all other alignment checks have been made. A turning radius that is out of specifications usually indicates that some part of the steering linkage is bent or alignment is incorrect.

FOUR-WHEEL ALIGNMENT

In the past, the only type of alignment done on vehicles was front-wheel alignment. However, there has been an increased need to align both the front and rear wheels. In the automotive industry this is called all-wheel alignment, or **four-wheel alignment**. An example of the need for four-wheel alignment is shown in ***Figure 48–25***. The rear axle on this car is not perpendicular to the centerline. When the thrust line is different from the centerline, the car will veer off to one side. This type of problem can be diagnosed by alignment of all four wheels.

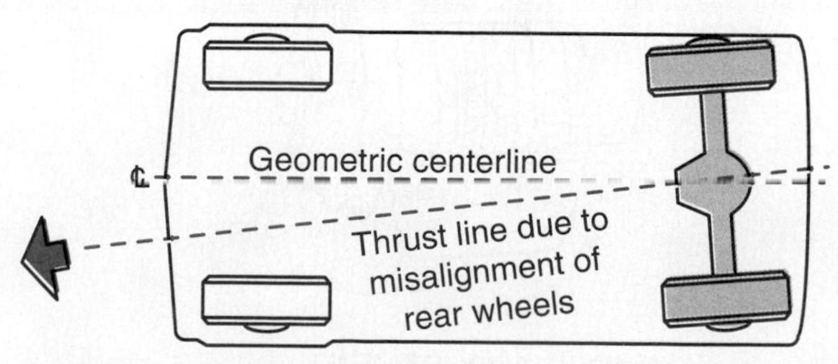

FIGURE 48-25 This vehicle needs four-wheel alignment because the rear wheels are not in line with the front wheels.

Four major vehicle design changes have caused the need for four-wheel alignment. They include the following:

1. Many vehicles today have front-wheel drive. Front-wheel drive vehicles are more sensitive to tracking problems than rear-wheel drive vehicles. Front-wheel drive vehicles demand that the rear wheels track directly behind the front wheels.
2. Modern tires are designed to allow for greater steering forces than older tires. The result is that there is less tolerance for misalignment between the front and rear wheels.
3. Modern tires are also more wear resistant. Thus, tires are more sensitive to misalignment.
4. Independent rear suspension makes it easier for the rear wheels to become misaligned because of bad road conditions.

The alignment machine shown in ***Figure 48–26*** is able to perform four-wheel alignment. These machines are extremely sophisticated and use computers to help the service technician align the wheels of the vehicle. Specific alignment procedures are stored in the computer for most vehicles. The computer also stores the alignment specifications for vehicles built over the past 25 years. ***Figure 48–27*** shows a close-up view of the specifications for vehicle wheel alignment.

Rear-Wheel Alignment Factors As part of four-wheel alignment, it is important to understand factors that may affect the alignment of the rear wheels. Many things can affect rear-wheel alignment angles. Some of the more important include:

- The rear subframes and axles can shift away from the centerline.
- The control arm bushings can wear out.

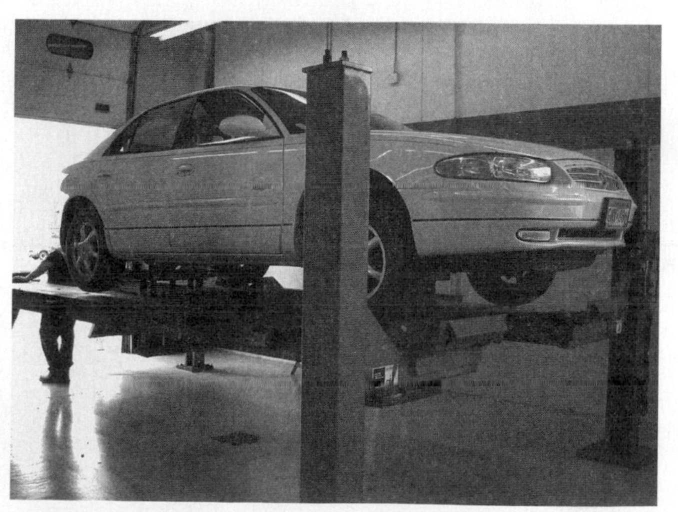

FIGURE 48-26 This computerized alignment machine gives directions and alignment procedures for most vehicles on the road today.

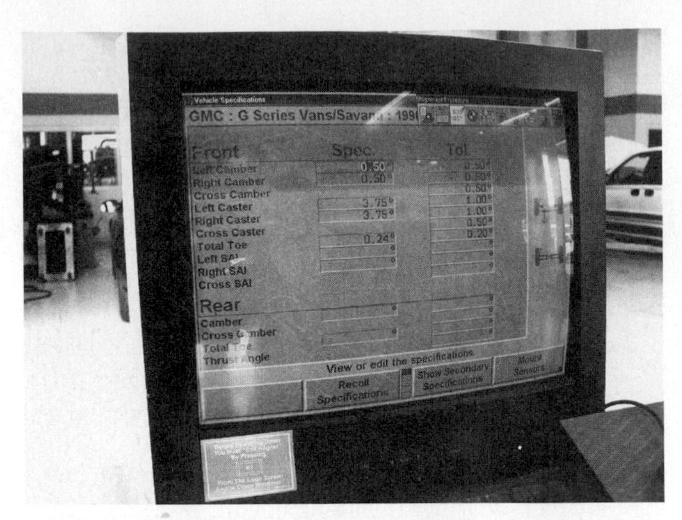

FIGURE 48-27 The computerized alignment machine stores alignment specifications for most vehicles on the road today.

- The springs can sag.
- Improper collision repairs or severe road shocks can bend rear suspension components out of specifications.

Because of these possible problems, it is always important to check the alignment of the rear wheels according to specifications recommended by the manufacturer.

Rear-Wheel Alignment Angles There are three angles that are commonly measured and adjusted when aligning the rear wheels. These angles are the thrust angle, the toe, and the camber.

The thrust angle is defined as the path that the rear wheels take. The thrust angle must be adjusted to be the same as the centerline of the vehicle. If not, the vehicle will "dog" track as it moves down the road.

The camber on the rear wheels is very similar to the camber on the front wheels. Typically, the camber on the rear wheels is adjusted so the top of the rear tires are adjusted slightly outward. Then, as weight is placed on the vehicle, the tires run perpendicular to the road.

The toe on the rear wheels is adjusted similar to that of the front wheels. ***Figure 48–28*** shows how the rear wheels are adjusted on an independent rear suspension system. The toe is adjusted by turning a tie-rod, just as with the front wheels. The camber is adjusted on this type of suspension by using a special tool to turn an adjustment nut. As the nut is turned, it adjusts the camber of the rear tires.

Four-Wheel Alignment Sequence of Checks The following is an example of the sequence of checks to be performed when doing a four-wheel alignment:

1. Perform a complete prealignment inspection.
2. Repair or replace any worn or damaged components identified during the prealignment inspection.
3. Position the vehicle on the alignment rack according to the manufacturer's instructions. Normalize the suspension by jouncing the front and rear wheels of the vehicle three times.
4. Measure and read the thrust angles at the rear axle.
5. If the rear alignment angles are adjustable, set the rear camber and align the thrust angle to the vehicle centerline by adjusting rear toe according to the manufacturer's specifications.
6. Measure and adjust the front angles in the following order: caster, camber, and toe.
7. If the vehicle is equipped with power steering, start the engine and rock the steering wheel back and forth 1/4 to 1/2 turn before adjusting toe. This step is not necessary if the vehicle has manual steering.

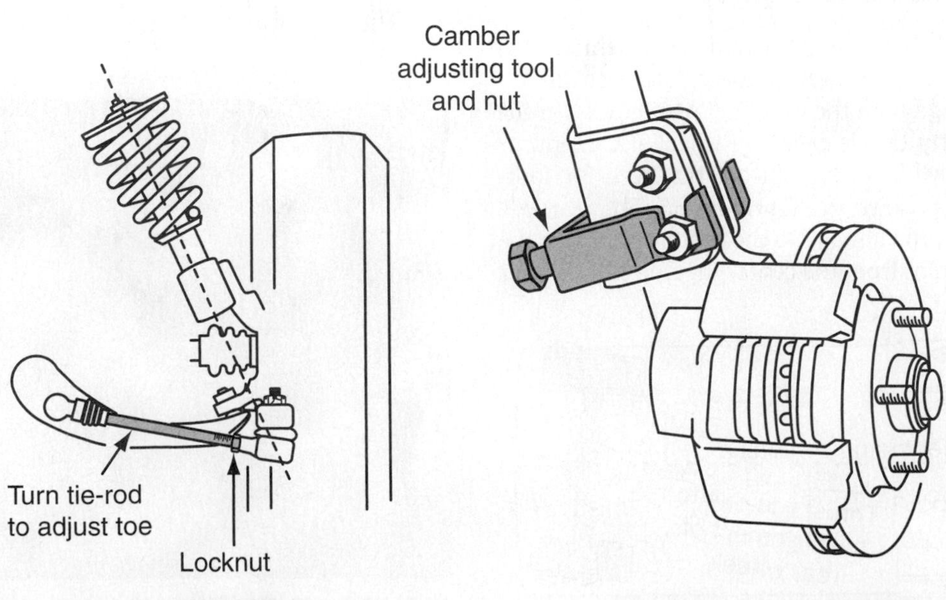

FIGURE 48-28 On vehicles with the rear wheels adjusted, both camber and toe are adjusted.

48.3 POWER STEERING

PURPOSE OF POWER STEERING

With manual steering, the driver is creating the forces needed to turn the steering gear. The only advantage that can be produced is by changing the steering ratio. Power steering is used on many vehicles to make steering easier for the driver, especially on heavier vehicles. The power steering system is designed to reduce the effort needed to turn the steering wheel. It reduces driver fatigue and increases safety during driving. Power steering systems are used both on the pitman arm steering gear system and the rack-and-pinion steering system.

MAJOR PARTS ON POWER STEERING SYSTEMS

Although there are many designs of power steering systems, there are two major parts in all power steering system designs. These are a hydraulic pump and the steering gear. These two units are connected by high-pressure hoses (*Figure 48–29*). The hydraulic pump produces fluid pressure. The pump is driven by a belt running from the crankshaft as shown in ***Figure 48–30***. It supplies the hydraulic pressure needed to operate the steering gear.

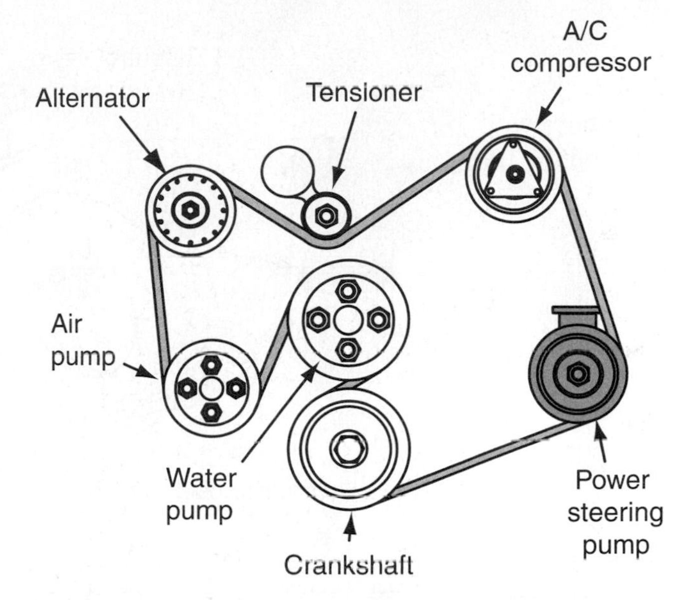

FIGURE 48–30 A belt drives the power steering pump from the crankshaft.

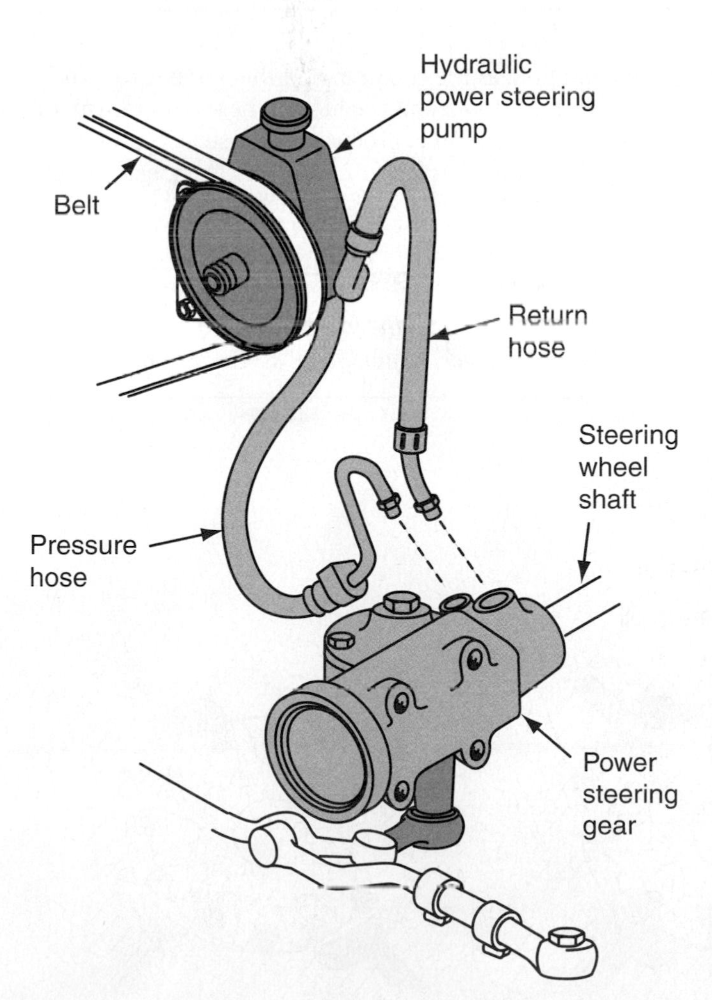

FIGURE 48–29 The power steering system uses a hydraulic pump and the steering unit to provide added pressure to turn the wheels.

The power steering unit is an integral part of either the steering gear or the rack-and-pinion arrangement. The hydraulic pressure is used to assist the motion of the steering gear or the rack-and-pinion gears. ***Figure 48–31*** shows a typical rack-and-pinion steering gear arrangement.

Power Steering Pump All power steering pumps are constant displacement or positive displacement pumps. They deliver different pressures, depending on the type and make of the vehicle. They use special power steering fluid that is recommended by the manufacturer. Automatic transmission fluid should not be used in power steering systems except in small quantities and then only to bring the fluid level up to the fill mark. If more transmission fluid is used in an emergency situation, the system should be drained, flushed, and refilled with power steering fluid as soon as possible.

The fluid is stored in a reservoir that is attached to the pump. There is usually a filter in the reservoir to prevent foreign matter from entering the system. A pressure relief valve is used to control excess pressure when the speed of the pump is increased. ***Figure 48–32*** shows a power steering pump circuit.

The pump can be of several designs. Three types of pumps are commonly used: the vane type, slipper type, and roller type. These are shown in ***Figure 48–33***. All three types work on the same principle. The center of the pump turns within an **eccentric** area. A suction is produced on one side of the pump housing. A pressure is produced on the other side of the housing.

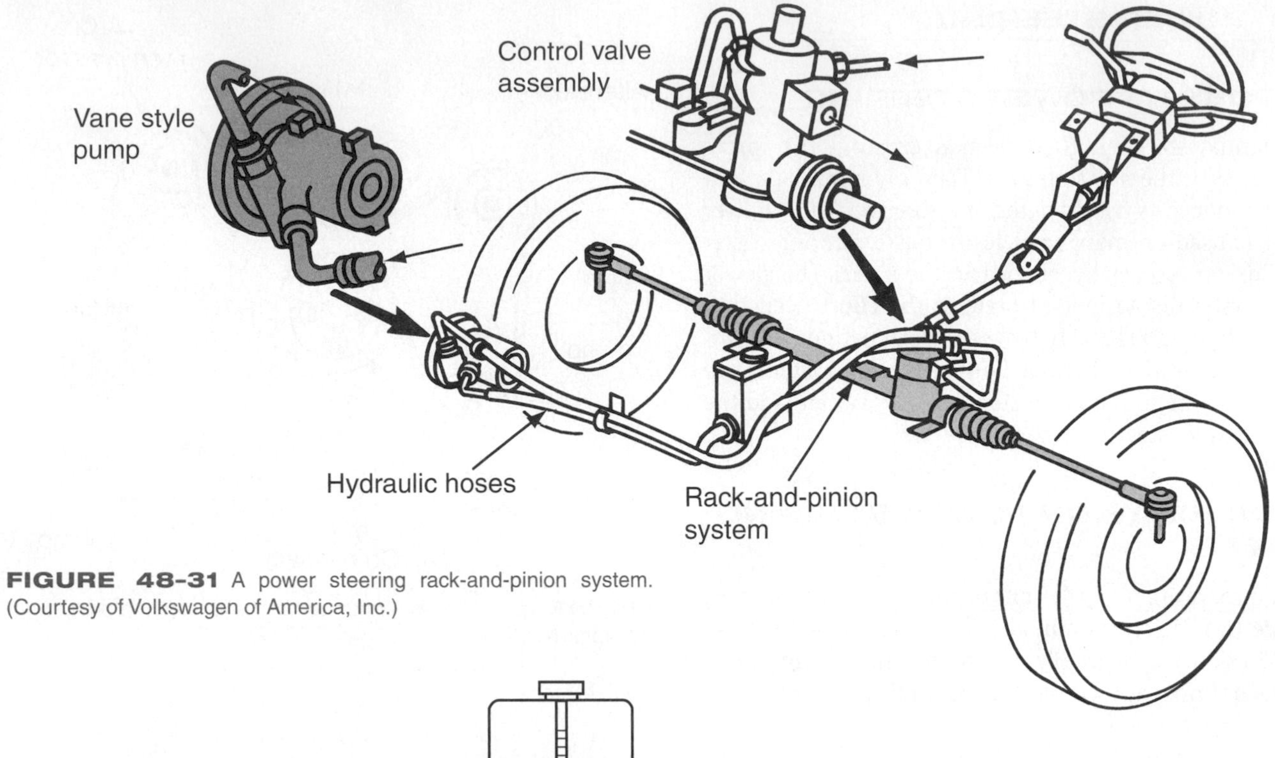

FIGURE 48-31 A power steering rack-and-pinion system. (Courtesy of Volkswagen of America, Inc.)

FIGURE 48-32 The parts of a power steering pump.

In operation, as the center part of the pump turns clockwise, there is a suction produced where the return oil comes in. The suction is produced because at this location the eccentric area gets larger and larger. Then, because the area gets smaller, a pressure is produced.

PARTS LOCATOR

*Refer to photo #110 in the front of the textbook to see what a **power steering pump and assembly** looks like.*

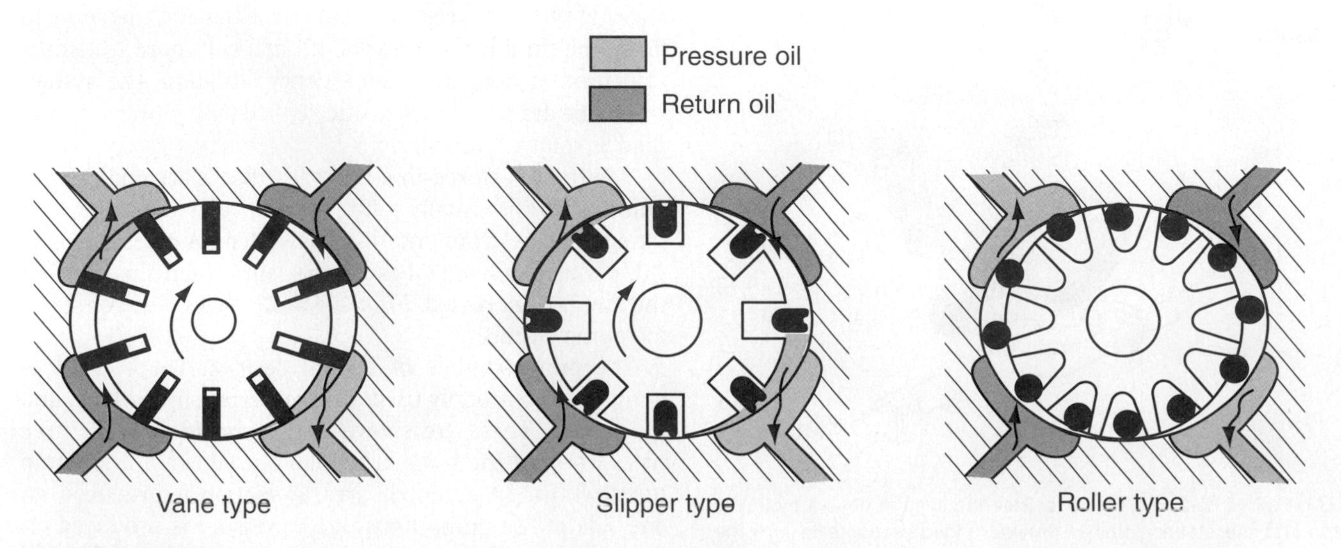

FIGURE 48-33 There are three types of power steering pumps: the vane type, slipper type, and roller type.

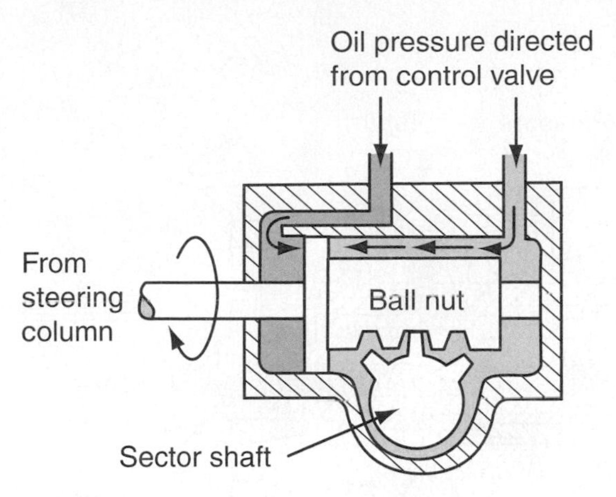

FIGURE 48–34 The parts of an integral power steering gear. Oil pressure pushes the ball nut to assist the driver when steering the vehicle.

INTEGRAL POWER STEERING GEAR

In ***Figure 48–34***, the sector shaft is turned by the piston and ball nut assembly. Normally, as the worm gear is turned by the steering wheel, the oil pressure is sent to the unit from the power steering pump. Oil is sent to both sides of the piston. This keeps the piston in a stable position. When the car is moving in a straight line, the pressures are equal on both sides of the piston. When the steering wheel is turned, higher oil pressure is directed to one side or the other to assist movement of the piston and ball nut assembly. Assisting the movement of this assembly makes it easier for the driver to turn the steering wheel.

CONTROL VALVES

Power steering control valves are built directly into the power steering gear assembly. The purpose of the control valve is to direct the oil pressure to one side or the other on the piston and ball nut assembly. When the steering wheel is turned, the control valve is positioned in such a way as to direct oil to the correct location. Two types of valves are commonly used: the sliding spool and rotary spool valves.

Sliding Spool Valve A sliding spool valve is shown in ***Figure 48–35***. As the steering wheel and the worm gear are turned, the sliding spool valve is moved slightly by linkage attached to the worm gear shaft. As this movement occurs, the internal spool opens a set of ports to allow high-pressure fluid to enter the correct side of the piston and ball nut assembly. Oil flows through internal passageways to get to the piston assembly. When the steering wheel is turned the other way, the oil is sent to the other side of the piston.

Rotary Spool Valve A rotary spool valve is also used on many vehicles to control the direction of oil through the steering gear. The rotary spool valve is shown in ***Figure 48–36***. When the steering wheel is turned, a twisting effort is produced through a torsion bar to rotate

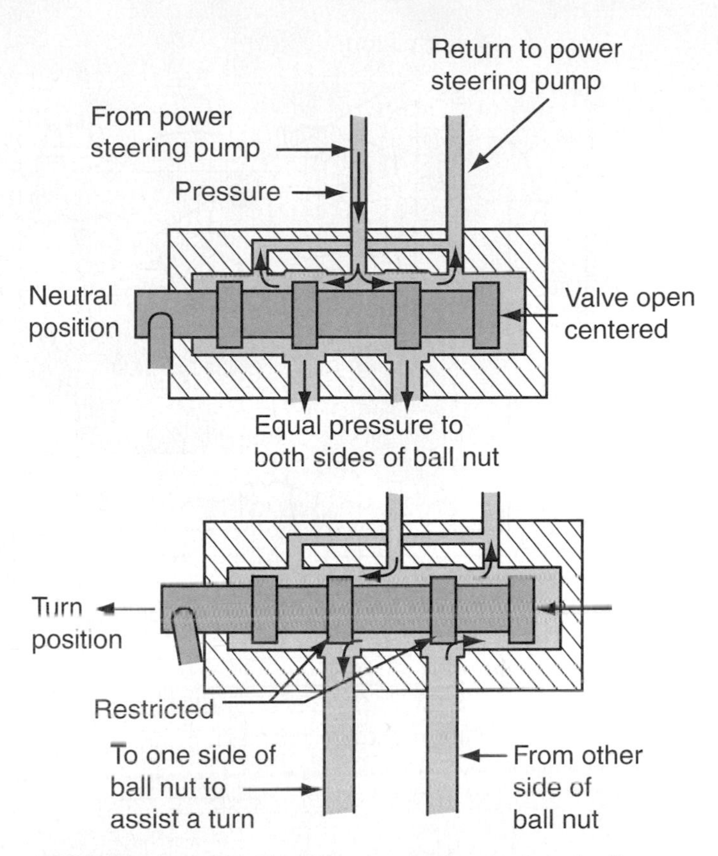

FIGURE 48–35 The sliding spool valve directs hydraulic pressure to the correct side of the ball nut for a left or right turn.

the internal spool slightly. This is done on an internal spline. As the spool rotates slightly, a different set of ports is opened and closed to allow oil pressure to flow to the correct side of the piston assembly. If the steering wheel is turned in the opposite direction, the oil will flow to the opposite side of the piston assembly. ***Figure 48–37*** shows a cutaway view of a typical power steering gear assembly.

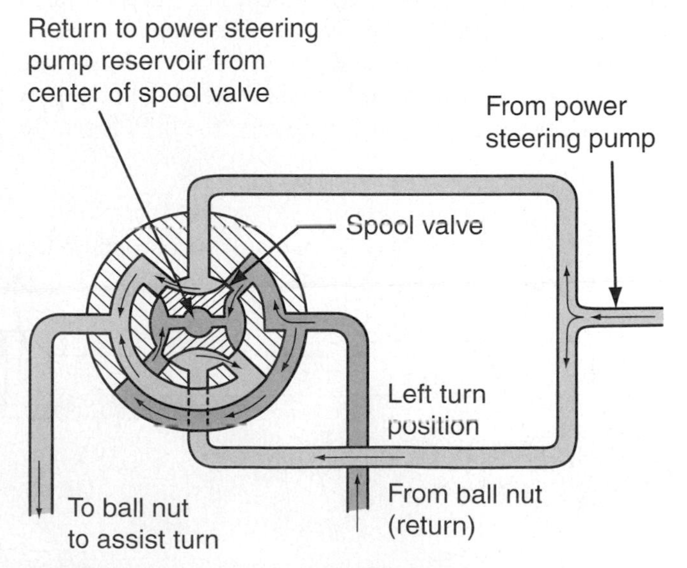

FIGURE 48–36 The rotary spool valve controls the direction of hydraulic pressure by twisting slightly to open or close different ports.

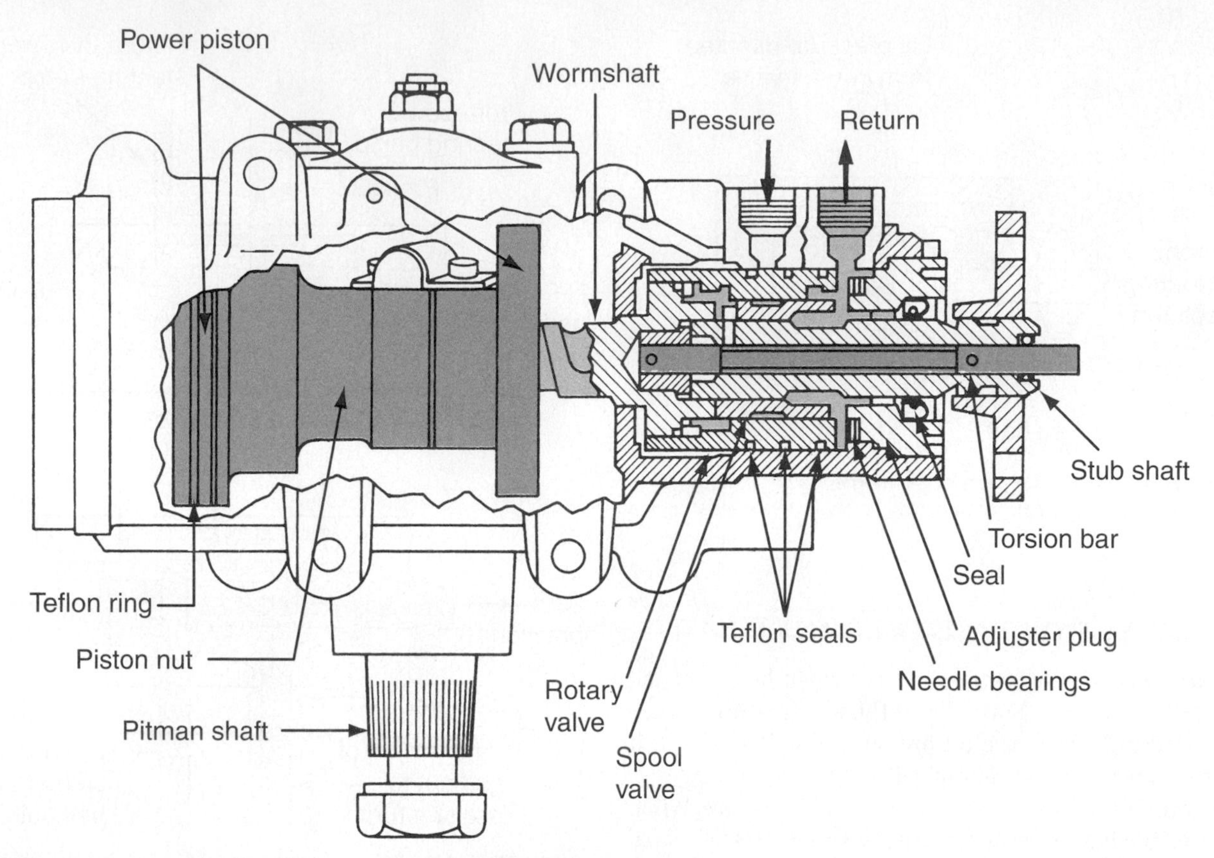

FIGURE 48–37 A complete power steering gear assembly with associated parts.

RACK-AND-PINION POWER STEERING

The rack-and-pinion power steering principles are much the same as the integral power steering principles. The major difference is that the pressure from the control valve operates the rack assembly. A power cylinder and piston assembly are placed on the rack (***Figure 48–38***). Oil pressure from the control valve then pushes or assists the movement of the rack. The control valve is attached to and is operated from the pinion gear as in other power steering systems.

A complete power rack-and-pinion system and parts are shown in ***Figure 48–39***. As the power steering pump produces the necessary hydraulic pressure, the fluid is sent to the rack-and-pinion steering gear. The pressure difference across the piston in the center assists the effort on the steering wheel.

*Refer to photo #111 in the front of the textbook to see what a **rack-and-pinion steering system** looks like.*

Electronic Power Steering Several manufacturers also use **electronic steering** systems to control the steering on a vehicle. In operation, an electrical motor is

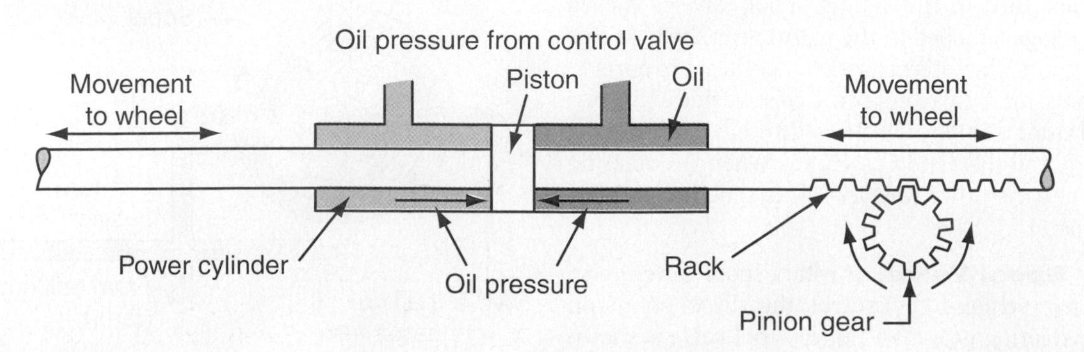

FIGURE 48–38 Power steering is accomplished on rack-and-pinion steering by a piston and power cylinder attached to the rack.

Car Clinic: RACK-AND-PINION GEARS

CUSTOMER CONCERN:
Rack-and-Pinion Steering Difficulty
A customer has complained that the steering system on the vehicle is becoming harder and harder to turn. The customer notices gear clunking when making sharp turns. What is the problem?

SOLUTION: This type of problem is most often produced by worn rack-and-pinion gears on the steering column. Over a period of time, the rack-and-pinion gears may wear, causing the pinion to slip or jump teeth on the rack. Normally, this can be easily identified by increasing amounts of slack in the steering wheel. The solution is to replace the rack-and-pinion steering gears.

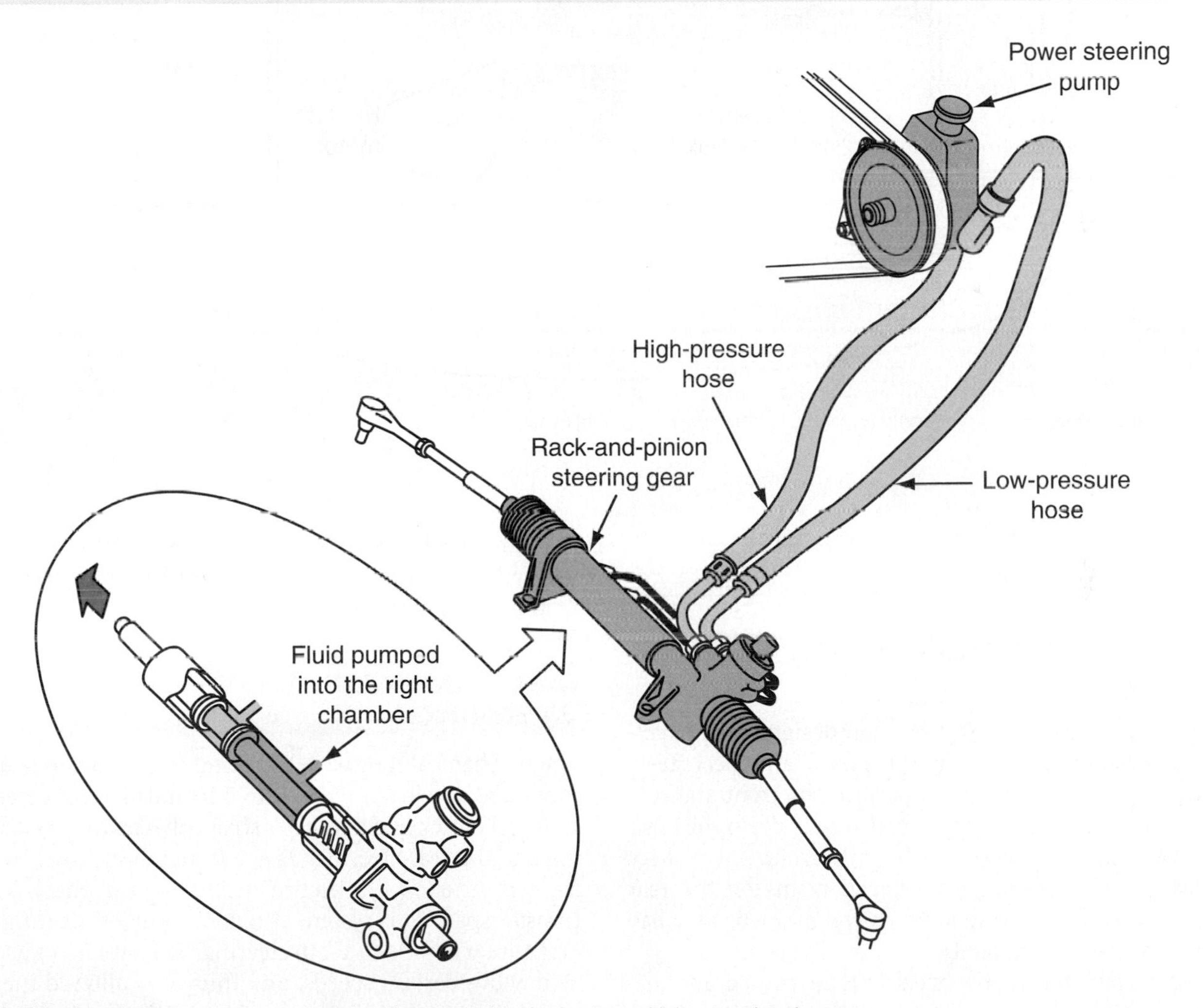

FIGURE 48–39 A complete rack-and-pinion power steering system with its associated parts.

placed inside the rack-and-pinion steering system. As the motor is turned, it assists the movement of the rack-and-pinion components. The result is that the steering effort is greatly reduced and is taken over by the electric motor.

The electric motor is able to rotate in both directions. The motor is controlled by a sensor placed on the rack-and-pinion housing, near the steering column. As the driver begins to turn the steering wheel, hydraulic pressure is built up. The increase in pressure is sensed by the steering effort sensor. The sensor then is able to select the direction of the motor movement and turn the motor off and on accordingly.

Figure 48–40 shows the typical components in an electronic power steering (EPS) system. As the steering wheel is turned, the steering sensor tells the EPS control unit to send hydraulic fluid and pressure to the steering gearbox. This action assists the steering effort when turning the front wheels.

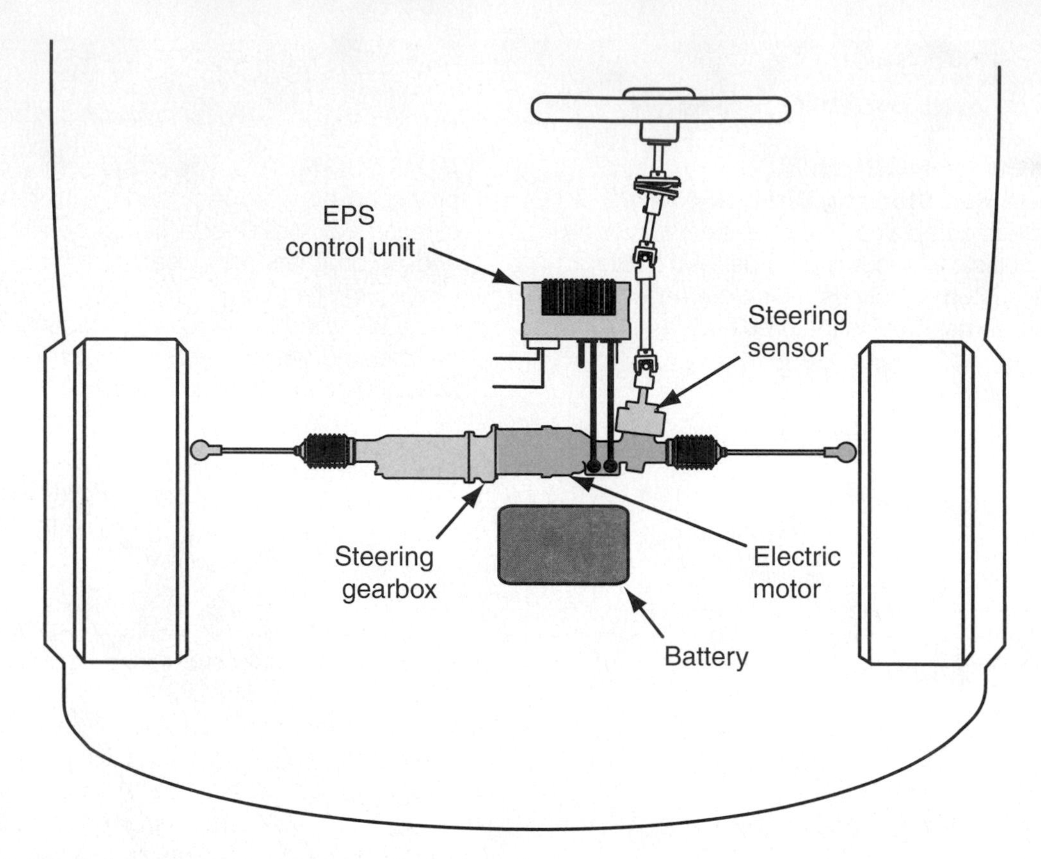

FIGURE 48-40 Typical components of an electronic power steering system.

48.4 FOUR-WHEEL STEERING SYSTEMS

PURPOSE OF FOUR-WHEEL STEERING

Several vehicle manufacturers are now designing and manufacturing four-wheel steering systems. Four-wheel steering systems are designed to allow both the front and rear wheels to turn or steer as the vehicle moves down the road. Four-wheel steering works well with front-wheel drive vehicles as well. Without drive mechanisms on the rear wheels, rear-wheel steering systems can easily be attached to the rear-wheel components.

In operation, the rear wheels can be turned approximately 5 degrees in either direction. Above a certain speed, for example, at 22 mph, the rear wheels steer in the same direction as the front wheels. The result is that the vehicle has very quick response for lane changes, sharp curves in the highway, and so on. Vehicle body motion during these types of turns and curves is also reduced, giving passengers a more comfortable ride. But at lower speeds, say below 22 mph, the rear wheels steer in the opposite direction of the front wheels. This results in improved maneuverability for U-turns, parallel parking, and so on.

Over the past few years, three types of four-wheel steering systems have evolved. These are the mechanical, hydraulic, and electronic four-wheel systems.

MECHANICAL FOUR-WHEEL STEERING

The mechanical four-wheel steering system was one of the first to be developed. This system included a front steering rack and pinion and a transfer shaft between the front steering and the rear steering. As the front wheels were turned, the turning effort was transmitted to the rear wheels via the transfer shaft. This system also used a second steering gear on the rear wheels to aid in steering. This system only operated above certain speeds, and thus only allowed the rear wheels to turn in the same direction as the front wheels.

HYDRAULIC FOUR-WHEEL STEERING

A second generation of four-wheel steering uses a hydraulic system to control the steering actions. This type of system turns the rear wheels only about 1 1/2 degrees and so only operates at speeds above 22 mph. ***Figure 48–41*** shows the typical parts and operation. To begin, the front-wheel steering movement is controlled by a standard rack-and-pinion

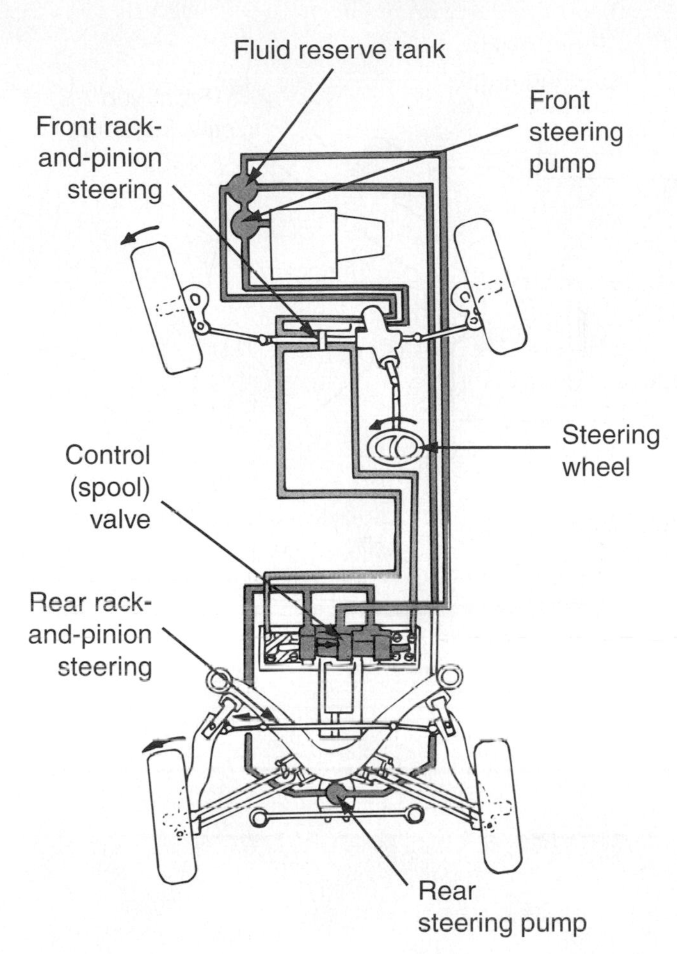

FIGURE 48–41 The hydraulic four-wheel system uses two rack-and-pinion gears (front and rear) to hydraulically control the turning of the front and rear wheels. (Courtesy of Mitsubishi Motor North America, Inc.)

gear system. In addition, some of the fluid is sent to the rear-wheel steering system to control the position of a control valve, also called a spool valve. As the front steering is turned one way, the spool valve moves in one direction. When the front steering is turned the other way, the spool valve moves in the other direction.

The spool valve is then used to control a second hydraulic circuit. This circuit uses fluid pressure developed by a second steering pump, driven from the differential. The fluid and pressure are again used to move a rack-and-pinion steering system similar to the front. However, the rack-and-pinion system is designed to allow only a small amount of movement. The rear wheels can turn no more than 1 1/2 degrees.

ELECTRONIC FOUR-WHEEL STEERING

Today, four-wheel steering systems are being controlled by electronics and a computer. Electronic four-wheel systems allow the rear wheels to turn either in the same direction as the front wheels (at higher speeds) or in the opposite direction as the front wheels (at lower speeds).

To accomplish this, a computer is used in conjunction with two sensors and two hydraulic actuators. ***Figure 48–42*** shows the inputs and outputs used for operation. First, a signal is sent to the computer from the vehicle speed sensor to let the computer know exactly how fast the vehicle is moving. This is needed to determine if the wheels will turn in the same or opposite direction. In addition, a sensor called the front steering angle sensor sends a signal to the computer, letting it know the number of degrees the vehicle is being turned. The four-wheel steering computer also knows the exact angle of the rear wheels from the rear-wheel sensors and the rear steering angle sensor. Based on these inputs, the computer tells the front steering gear and rear steering gear the exact amount for each to be turned. ***Figure 48–43*** shows the location of the major parts of such a four-wheel steering system.

There are many additional parts needed, such as hydraulic pumps (if hydraulic actuators are used rather than motors), solenoids, cutoff valves, and so on. These and other parts will continue to improve the efficiency and reliability of the system.

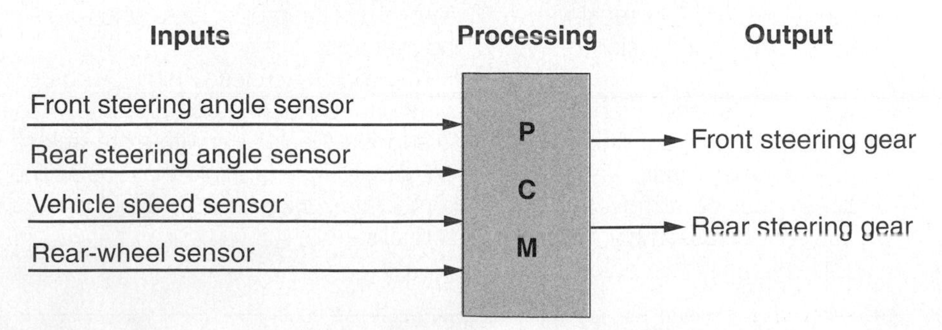

FIGURE 48–42 Electronic four-wheel steering uses several major inputs to control two outputs.

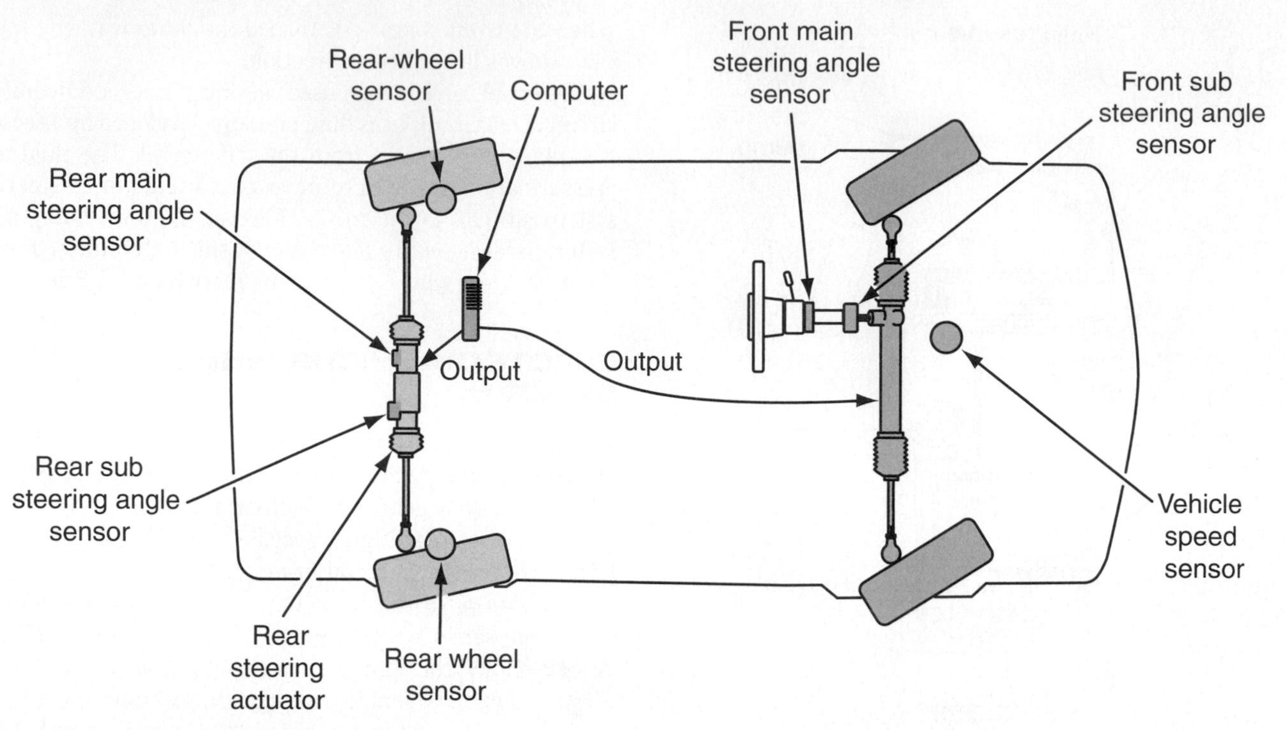

FIGURE 48-43 The major parts of an electronic four-wheel steering system.

Problems, Diagnosis, and Service

Safety Precautions

1. When removing the steering wheel, use the correct puller for safe removal.
2. The tie-rods may be stuck in place. Always use the correct tools to remove tie-rods (e.g., pickle fork).
3. Make sure the battery is disconnected to eliminate the possibility of the engine accidentally cranking over when you are checking the belt tension on the power steering unit.
4. When checking the tie-rod ends and steering mechanisms, be careful not to pinch your fingers between parts, as serious injury may result.
5. Many parts and assemblies must be removed in the correct sequence. Incorrect disassembly may cause parts to drop or spring out unexpectedly. Always follow the manufacturer's suggested procedure when removing steering system parts from the vehicle.
6. Always wear OSHA-approved safety glasses when working on steering systems.
7. Always use the proper tools when working on steering systems.
8. Follow all safety rules when jacking up a vehicle to work on the steering components. Also make sure to support the vehicle with extra jack stands.
9. Any power steering fluid that has spilled should be immediately wiped up to avoid injuries from slips and falls.

PROBLEM: Worn Steering Linkages

On a rear-wheel drive vehicle, the steering wheel has excessive play.

DIAGNOSIS

Problems such as this are generally in the steering linkages and gear or in the steering column. Other problems, such as poor alignment, can also cause excessive play. ***Figure 48–44*** shows a list of problems and their possible causes that can be used to diagnose similar steering system problems. Use this chart as a guideline for initial diagnosis of steering problems. When diagnosing the steering system, do the following preliminary checks:

1. Check all tires for proper inflation pressures and approximately the same tread wear.
2. Check the front-wheel bearings for proper adjustment.

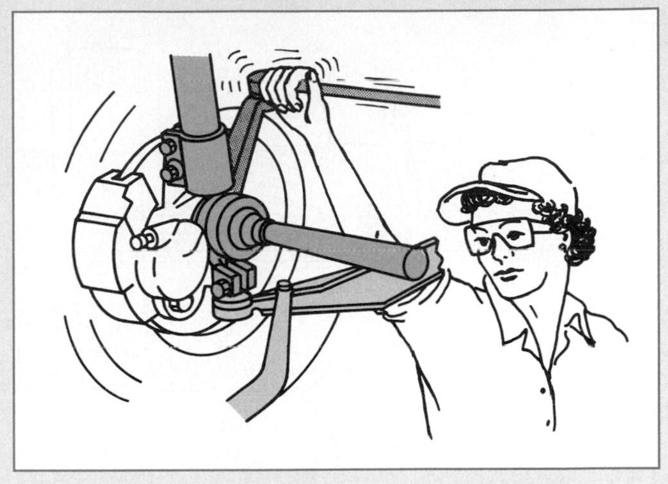

FIGURE 48–45 The tie-rod ends can be checked for wear by grasping the rod firmly and forcing it up and down. There should be no lost motion.

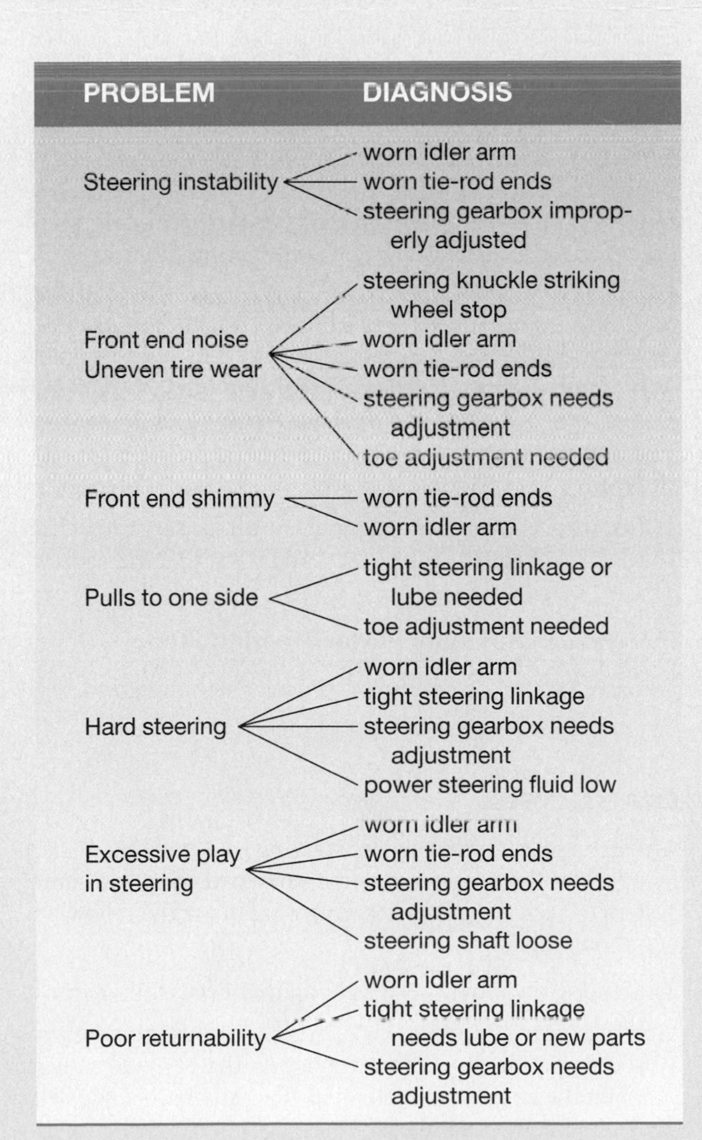

FIGURE 48–44 Each problem has several potential causes and diagnoses that need to be checked when troubleshooting a steering system. (Courtesy of Dana Corporation)

3. Check for loose or damaged ball joints, tie-rod ends, and control arms. Make sure all grease zerks have been lubricated.
4. Check for runout of wheels and tires.
5. Check to see if excess loads have been placed on the vehicle.
6. Consider the condition and type of equipment being used to check the alignment.
7. Check the tie-rod end for wear. This can be found by grasping the tie-rod end firmly and forcing it up and down or sideways to check for any lost motion. ***Figure 48–45*** shows a service technician performing this diagnostic check. If there is lost motion or sloppiness in movement, the tie-rods will need to be replaced.

SERVICE

Steering Gear and Linkages The following are common service procedures for steering gears and linkages.

1. When replacing tie-rods, first remove the retaining nut and cotter pin. Before removing the tie-rods, measure the distance from the adjusting sleeve to the tie-rod end center point. This allows a preliminary toe setting by placing the new tie-rod end in approximately the same position (***Figure 48–46***).
2. Loosen the tie-rod adjusting sleeve clamp nuts.
3. Remove the tie-rod end nut and cotter pin.
4. Use a pickle fork to remove the tie-rod from the steering knuckle. A pickle fork is shown in ***Figure 48–47***.
5. Some steering gears can be checked and adjusted for correct clearance or lash. Lash is the clearance between the sector shaft and the ball nut. Refer to the correct

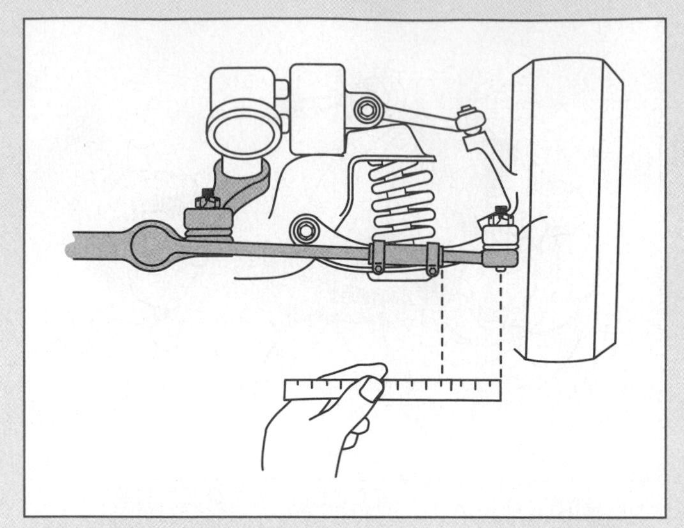

FIGURE 48–46 Before removing a tie-rod, check the distance from the adjusting sleeve to the ball joint grease zerk. This measurement gives a preliminary toe setting when you put the tie-rod back in the same position.

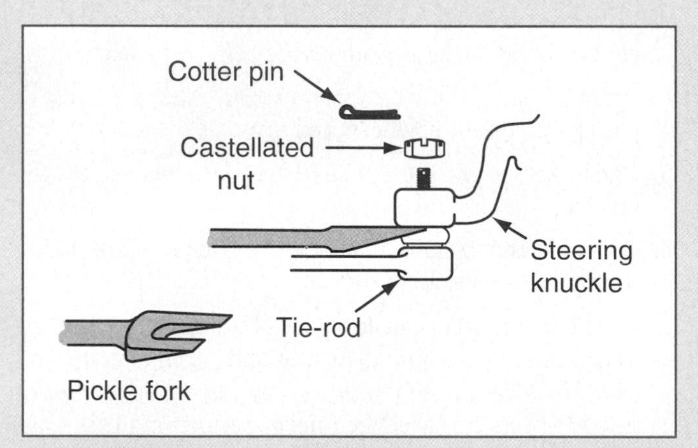

FIGURE 48–47 A pickle fork can be used to properly separate the tie-rod ball joints.

maintenance manual for the procedure. ***Figure 48–48*** shows the position of the lash adjuster screw located on top of the steering gear. The general procedure for adjusting the clearance is as follows:

a. If power steering is used, rotate the wormshaft through the complete range of travel. This is done to bleed air from the system. Then refill the reservoir to the top.
b. Place the steering gear in the center of its movement.
c. Loosen the locknut on the adjusting screw.
d. Tighten the adjusting screw until all backlash is removed; then tighten the locknut.
e. Operate the unit through its range of motion.
f. Loosen the locknut and adjusting screw again.
g. Tighten the adjusting screw again until all backlash is again removed.
h. Now tighten the adjusting screw an additional 3/8 of a turn. Now tighten the locknut.

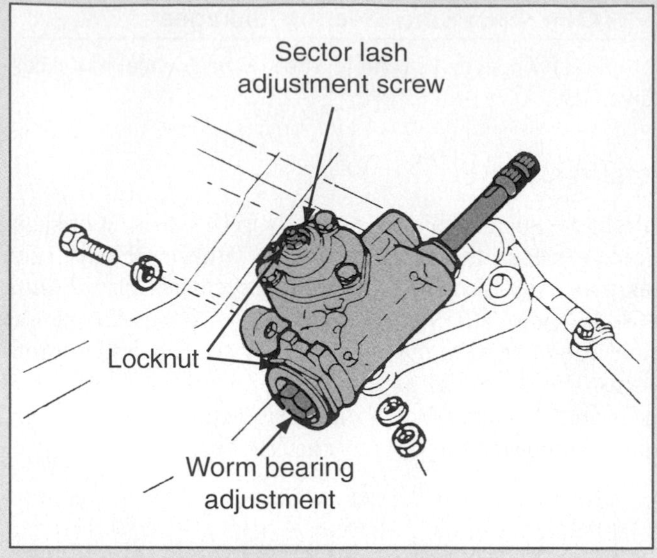

FIGURE 48–48 The lash adjuster screw is located on top of the steering gear assembly. Its purpose is to adjust the clearance between the sector shaft and the ball nut.

Steering Columns Many types of steering systems are used on vehicles. Each type has a different disassembly and assembly procedure. Refer to the manufacturer's service manual for the correct procedure. The following general service items can be done on steering columns:

1. First check the manufacturer's procedure for air bag safety.
2. Use a wheel puller to remove the steering wheel to get at the horn and turn signal mechanisms.
3. The universal joints can be replaced. However, the procedure and the parts required will depend on the manufacturer.
4. Electrical and mechanical problems can be serviced in the turn signal and flasher systems and the tilt mechanisms.

PROBLEM: Faulty Power Steering Pump

A buzzing sound comes from the power steering unit when the steering wheel is turned.

DIAGNOSIS

A buzzing sound from a power steering unit is caused either by a lack of fluid or by air in the power steering hydraulic system. Check the power steering system for the following common problems:

1. Steering column U-joint binding that needs lubrication.
2. Loose power steering belt, which causes momentary steering difficulty. When it is loose, the belt will squeal when the engine is accelerated. This shows that the belt is slipping on the pulleys.
3. Loose mounting bolts in the steering gear, which cause abnormal steering wheel kickback and poor control.

4. Leaky power steering lines. Check for loose fittings and damaged hoses.
5. A lack of hydraulic pressure, which can be caused by internal leaks past piston rings or valve body, worn seals, or misaligned housing bore.

SERVICE

To bleed the system of air, start the engine and turn the steering wheel throughout its range. Keep checking the level of the power steering fluid as needed. Turn the steering wheel throughout its range again and add fluid. Continue this procedure until all the air has been removed from the hydraulic system.

PROBLEM: Faulty Rack-and-Pinion Steering

A vehicle with a rack-and-pinion steering system has excessive play and looseness in the steering wheel.

DIAGNOSIS

Many components can cause similar problems on vehicles that have a rack-and-pinion steering system. Always check the rack-and-pinion steering system for leaks at the end seals, rack seals, and pinion shaft seals. ***Figure 48–49*** shows a diagnosis chart that can help the service technician determine the possible cause and correction. Two easy checks to diagnose rack-and-pinion steering are:

1. Check the rack housing bolts attached to the cross member. ***Figure 48–50*** shows the rack housing bolts. Look for movement between the rack housing and the cross member. This usually indicates worn mount bushings and can cause loose steering.
2. Check the inner tie-rod sockets. This can be done by having someone move the steering wheel back and forth slightly. Then place your hand on the flex-type bellows. Feel for looseness and wear in the inside connection (***Figure 48–51***). Repair as necessary.

RACK & PINION DIAGNOSIS

PROBLEM	DIAGNOSIS	SERVICE
Hard steering—excessive effort required at steering wheel	1. Low or uneven tire pressure. 2. Tight outer tie-rod end or ball joints. 3. Incorrect front-wheel alignment. 4. Bind or catch in gear.	1. Inflate to specified pressures. 2. Lube or replace as required. 3. Align to specifications. 4. Remove gear, disassemble, and inspect. a) Replace damaged or badly worn components (OPH).* b) If housing and tube assembly, rack, or pinion are damaged, replace with service assembly (THP).**
Poor returnability	1. Tight ball joints or end housing pivots. 2. Bent tie-rod(s). 3. Incorrect front-wheel alignment. 4. Bind or catch in gear.	1. Lube or replace as required. 2. Replace bent tie-rod(s). Align front end. 3. Align front end. 4. Remove gear, disassemble, and inspect. a) Replace damaged or badly worn components (OPH).* b) If housing and tube assembly, rack, or pinion are damaged, replace with service assembly (THP).**
Excessive play or looseness in steering system	1. Front-wheel bearings loosely adjusted. 2. Worn couplings or steering shaft U-joints. 3. Worn upper ball joints. 4. Loose steering wheel on shaft, tie-rods, steering arms, or steering linkage ball studs. 5. Worn outer tie-rod ends. 6. Loose frame to gear mounting bolts. 7. Deteriorated mounting grommets. 8. Excessive internal looseness in gear. 9. Worn rack bushing(s).	1. Adjust or replace as required. 2. Replace worn part(s). 3. Replace. 4. Tighten to specified torque. 5. Replace outer tie-rod ends. 6. Tighten to specified torque. 7. Replace mounting grommets. 8. Readjust gear. If still loose, disassemble and inspect. 9. Replace rack bushing(s).

* (OHP) One-piece housing only
** (TPH) Two-piece housing only

FIGURE 48–49 Use a diagnosis chart to help troubleshoot a rack-and-pinion power steering system.

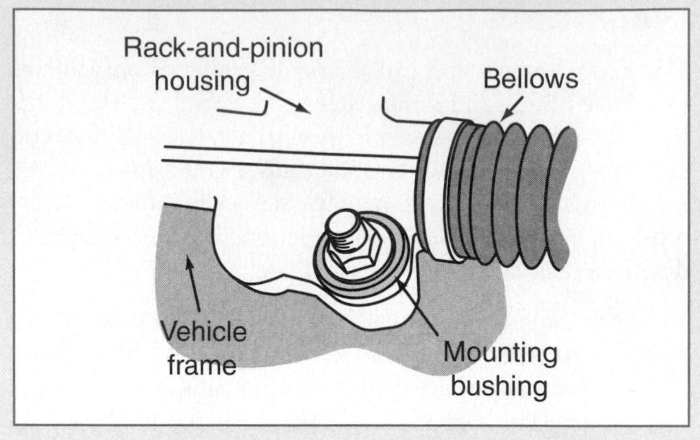

FIGURE 48-50 Check the rack-and-pinion housing bolt to the frame or cross member for looseness or worn bushings.

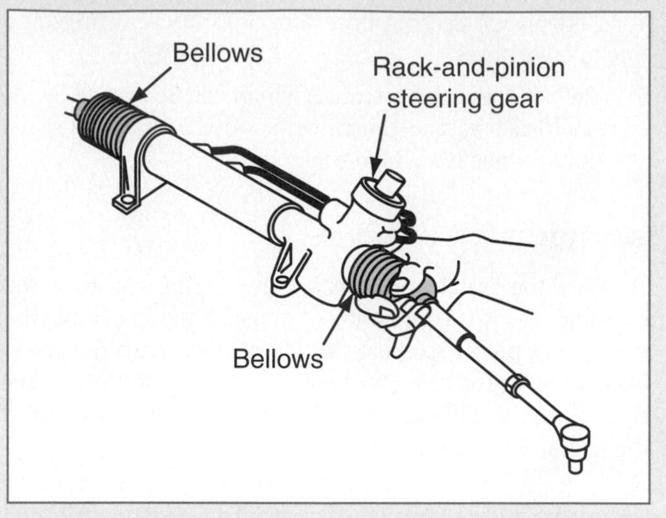

FIGURE 48-51 Check the rack-and-pinion tie-rods by pinching the bellows and feeling for looseness or wear.

SERVICE

Many service procedures are needed to repair the rack-and-pinion steering gear. The service procedures are different for each type and year of vehicle. For complete procedures for removal, disassembly, assembly, and overhaul, refer to the service manual.

PROBLEM: Incorrect Steering Alignment

The tires on a vehicle are wearing abnormally fast, and there is a definite pattern of wear on the tire tread.

DIAGNOSIS

One of the more common problems that produces rapid tire wear is incorrect wheel alignment. The alignment of both the front and rear tires needs to be checked. ***Figure 48–52*** shows a diagnosis chart for alignment. Match the type of tire wear on the tire to the symptoms. Then identify a probable cause by looking at the diagnosis chart.

ALIGNMENT DIAGNOSIS CHART

PROBLEM	DIAGNOSIS	PROBLEM	DIAGNOSIS
• Excessive tire wear on outside shoulder	• Excessive positive camber.	• Vehicle vibrates	• Defective tires. One or more of all 4 tires out-of-round. One or more of all 4 tires out-of-balance. Drive shaft bent. Drive shaft sprayed with undercoating.
• Excessive tire wear on inside shoulder	• Excessive negative camber.		
• Excessive tire wear on both shoulders	• Rounding curves at high speeds. Underinflated tires.		
• Sawtooth tire wear	• Too much toe-in or toe-out.		
• One tire wears more than the other	• Improper camber. Defective brakes. Defective shock absorber.	• Car tends to wander either to the right or left.	• Improper toe setting. Looseness in steering system or ball joints. Uneven caster. Tire pull.
• Tire treads cupped or dished	• Out-of-round tires. Out-of-balance condition. Defective shock absorber.	• Vehicle swerves or pulls to side when applying brakes.	• Uneven caster. Brakes need adjustment. Out-of-round brake drum. Defective brakes. Underinflated tire.
• Front wheels shimmy	• Defective idler arm bushing. Out-of-round tires. Out-of-balance condition. Excessive positive caster. Uneven caster.	• Car tends to pull either to the right or left when taking hands off steering wheel.	• Improper camber. Unequal caster. Tires worn unevenly. Tire pressure unequal.

FIGURE 48-52 Diagnosis charts can help the service technician troubleshoot alignment problems. (Courtesy of Hunter Engineering Co.)

ALIGNMENT DIAGNOSIS CHART

PROBLEM	DIAGNOSIS	PROBLEM	DIAGNOSIS
• Car is hard to steer	• Tires underinflated. Power steering defective. Too much positive caster. Steering system too tight or binding.	• Steering has excessive play or looseness.	• Loose wheel bearings. Loose ball joints or kingpins. Loose bushings. Loose idler arm. Loose steering gear assembly. Worn steering gear or steering gear bearings.

FIGURE 48–52 *continued*

SERVICE

It is beyond the scope of this textbook to describe complete alignment procedure. Specialized alignment equipment and computerized instruments should always be used to check and adjust for correct alignment. Computers are being used today to help determine exact procedures, measurements needed, and specifications for alignment. ***Figure 48–53*** shows a typical alignment machine.

Several additional service notes on alignment include:

1. All wheel alignment angles are interrelated. The adjustment order should be caster, camber, and toe.
2. The alignment angles can be adjusted by several methods, depending on the car. ***Figure 48–54*** shows how various adjustments can be used to change the alignment of the wheels.

FIGURE 48–53 Today, alignment is done with sophisticated computerized equipment on all four wheels.

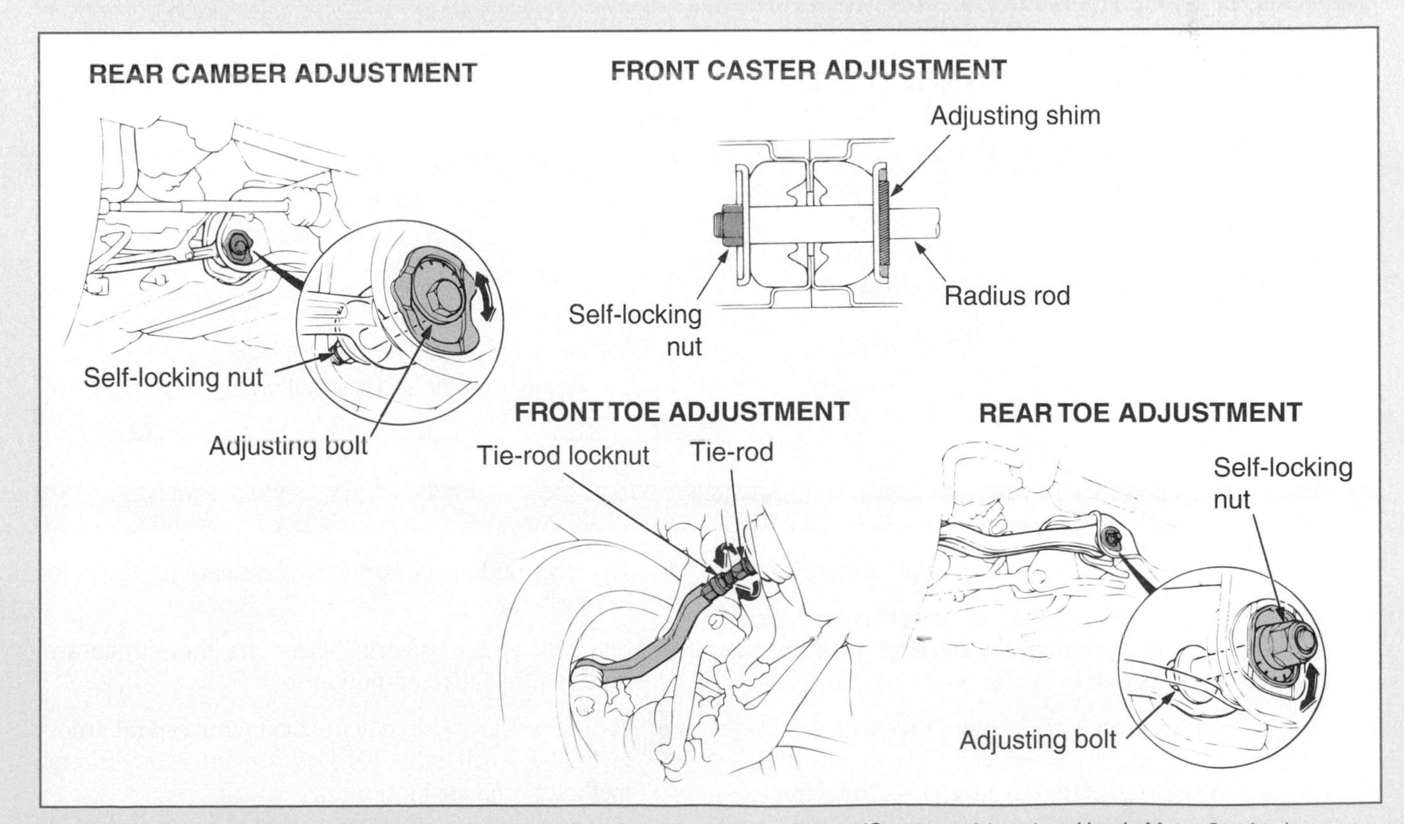

FIGURE 48–54 Various techniques are used to adjust the alignment on vehicles. (Courtesy of American Honda Motor Co., Inc.)

3. Alignment is somewhat different on the MacPherson strut and other front-wheel drive suspension systems. Depending on the manufacturer, certain adjustments cannot be made on the front suspension. Camber, for example, is sometimes built into the suspension and cannot be changed. On other vehicles, there is no caster adjustment. Refer to the vehicle manufacturer's specifications to determine exactly what adjustments can be done. ***Figure 48–55*** shows one method used to adjust camber on a knuckle-strut assembly. This method has an elongated bolt hole. A cam washer is placed on the end of the bolt to adjust for camber.

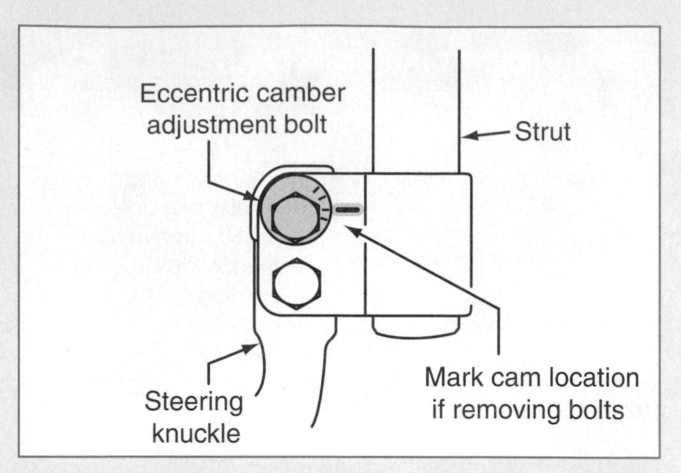

FIGURE 48-55 The cam washer (also called an eccentric washer) is used to adjust the camber on the MacPherson strut suspension.

Service Manual Connection

There are many important specifications to keep in mind when working with steering systems. To identify the specifications for your vehicle, you will need to know the VIN (vehicle identification number) of the vehicle, the type and year of the vehicle, and the type of engine. Although they may be titled differently, some of the more common steering system specifications (not all) found in service manuals are listed below. Note that these specifications are typical examples. Each vehicle may have different specifications.

Common Specification	Typical Example
Camber angle	
Left	+0.3 to +1.3 degrees
Right	+0.3 to +1.3 degrees
Caster angle	
Limits	+2.5 to +3.5 degrees
Desired	+3 degrees
Power steering belt tension (using gauge)	165 newtons

Common Specification	Typical Example
Power steering mounting bolts torque	45 ft.-lb
Toe-in	1/16–3/16 inch
Toe-out on turns	
Outer wheel	18.51 degrees
Inner wheel	20 degrees
Worm bearing preload	5–8 in.-lb
Yoke-to-rack clearance	0.005 inch

In addition, service manuals provide specific directions for various service and testing procedures. Some of the more common procedures include:

- Lower bearing shaft replace
- Piston and ball nut replace
- Power steering disassembly and assembly
- Rack yoke bearing preload adjustment
- Remove and install outer tie-rod
- Steering column remove and replace
- Steering gear repair and inspection

SUMMARY

The following statements will help to summarize this chapter:

- The steering system is made of three major components: the steering column, the steering gear, and the steering linkage.
- The steering column is used to produce the necessary force to turn the front wheels. The steering gear changes the motion of the steering wheel to reciprocating motion on the steering linkage.
- The steering linkage connects the steering gear to the front wheels.
- The two types of steering gears are the pitman arm style and the rack-and-pinion style.
- The pitman arm uses a wormshaft with a spiral groove in it and small steel balls to provide a low-friction method to change motion.

- As the driver turns the wormshaft, the pitman arm moves back and forth.
- The rack-and-pinion steering system uses a flat gear called a rack.
- The rack on a rack-and-pinion steering system meshes with the mating gear called a pinion gear.
- As the steering column is turned, the pinion gear turns, which causes the rack to move back and forth to control steering.
- The steering ratio is the ratio of the degrees turned on the steering wheel to the degrees turned on the front wheels.
- Steering linkage is typically composed of an idler arm, center link, tie-rods, and the pitman arm or rack.
- Wheel alignment is defined as the balancing of all forces created by friction, gravity, centrifugal force, and momentum while the wheels are in motion.
- There are five angles that are related to wheel alignment: caster, camber, toe, steering axis inclination, and turning radius.
- Caster is the forward or backward tilt of the spindle support.
- Camber is the inward or outward tilt of the front wheels.
- Steering axis inclination is the inward tilt of the spindle support arm ball joints at the top.
- Turning radius is the amount in degrees that one front wheel turns more sharply than the other front wheel.
- Toe is the inward or outward pointing of the front wheels.
- Power steering uses hydraulic pressure to aid or assist the driver in turning the steering wheel.
- A belt turns the power steering pump from the crankshaft. Power steering fluid is used as the hydraulic fluid in a power steering system.
- There are several types of power steering pumps including the vane, slipper, and roller types.
- The hydraulic pressure in a power steering unit is used to either move the rack or the ball nut in a pitman arm system.
- Four-wheel steering systems are designed to allow both the front and rear wheels to steer the vehicle.
- Various diagnostic and service procedures help service technicians properly service steering systems.

TERMS TO KNOW

Can you explain each of the following terms? Review the chapter until you can use each term correctly.

Adjusting sleeve
Alignment
Camber
Caster
Center link
Eccentric
Electronic steering
Four-wheel alignment
Idler arm
Pitman arms
Rack-and-pinion system
Scrub radius
Steering axis inclination
Toe
Turning radius
Zerk

REVIEW QUESTIONS

Multiple Choice

1. The three main components of the steering system are the steering column, steering gear, and
   a. Pitman arm
   b. Steering toe
   c. Steering linkage
   d. Rack and pinion
   e. Steering wheel

2. Universal joints are used in the ________ on a typical steering system.
   a. Steering wheel
   b. Steering column
   c. Pitman arm
   d. Tie-rods
   e. Steering linkage

3. The pitman arm steering gear uses:
   a. Recirculating balls
   b. A sector shaft
   c. A ball nut
   d. All of the above
   e. None of the above

4. Which of the following is/are not adjustable on a steering system?
   a. Tie-rods
   b. Camber
   c. Steering gear
   d. Toe-in
   e. Pitman arm

5. The purpose of the steering gear is to:
   a. Change rotary motion to reciprocating and back to rotary motion
   b. Change reciprocating motion to rotary motion
   c. Change rotary motion to reciprocating motion
   d. Adjust camber
   e. Adjust caster

6. Which type of steering system uses a flat gear?
   a. Tie-rod system
   b. Rack-and-pinion system
   c. Integral gear system
   d. Universal system
   e. Scrub system

7. When the steering wheel turns 60 degrees and the front wheel turns 3 degrees, the steering ratio is:
   a. 15 to 1
   b. 20 to 1
   c. 25 to 1
   d. 30 to 1
   e. 60 to 1

8. Which of the following is part of the steering linkage?
   a. Idler arm
   b. Center link
   c. Tie-rod
   d. All of the above
   e. None of the above

9. The adjusting sleeve is attached to the:
   a. Pitman arm
   b. Tie-rods
   c. Center link
   d. Idler arm
   e. U-joint

10. Which of the following is defined as the backward and forward tilt of the spindle support arm?
    a. Caster
    b. Camber
    c. Toe
    d. Steering axis inclination
    e. Turning radius

11. Which of the following is defined as the inward and outward tilt of the front wheels at the top?
    a. Caster
    b. Camber
    c. Toe
    d. Steering axis inclination
    e. Turning radius

12. Which of the following is defined as the distance between the front and back of the front wheels?
    a. Caster
    b. Camber
    c. Toe
    d. Steering axis inclination
    e. Turning radius

13. Which of the following is defined as the toe-out on turns or the amount one wheel turns more sharply on turns?
    a. Caster
    b. Camber
    c. Toe
    d. Steering axis inclination
    e. Turning radius

14. The included angle is a combination of the steering axis inclination and the ________.
    a. Toe-in
    b. Caster
    c. Camber
    d. Turning radius
    e. Toe-out

15. Power steering can be used:
    a. Only on pitman arm steering systems
    b. Only on rack-and-pinion steering systems
    c. Both A and B
    d. Neither A nor B
    e. Only when the vehicle is extremely light

16. The power steering pump is driven from the:
    a. Steering linkage
    b. Crankshaft by a belt
    c. Differential
    d. Drive shaft
    e. Steering gear

17. The control valves used in power steering are placed:
    a. On the external steering linkage
    b. In the integral gear and pitman arm assembly
    c. In the steering wheel
    d. On the rack and pinion
    e. All of the above

18. Which of the following spool valves changes passageways by rotating?
    a. The sliding spool valve
    b. The rotary spool valve
    c. The positive displacement spool valve
    d. The linkage spool valve
    e. The vane spool valve

19. Power steering on a rack-and-pinion steering system pushes which component?
    a. Pitman arm
    b. Rack
    c. Pinion
    d. U-joint
    e. Spool valve

20. Which of the following are methods used to adjust the alignment on a car?
    a. Shims
    b. Eccentric bolts
    c. Cams
    d. All of the above
    e. None of the above

21. Which of the following cannot be adjusted?
 a. Camber
 b. Steering axis inclination
 c. Steering gear lash
 d. Caster
 e. Toe

The following questions are similar in format to ASE (Automotive Service Excellence) test questions.

22. A buzzing sound is heard from the power steering unit. *Technician A* says that the fluid is low. *Technician B* says that there may be air in the power steering fluid. Who is correct?
 a. A only c. Both A and B
 b. B only d. Neither A nor B

23. *Technician A* says that one alignment adjustment will not affect the other alignment adjustments. *Technician B* says that one alignment adjustment is interrelated with all the other alignment adjustments and will affect the others. Who is correct?
 a. A only c. Both A and B
 b. B only d. Neither A nor B

24. When working on the steering system, several preliminary checks should be made. *Technician A* says that the tires should be properly inflated as they may affect the steering system. *Technician B* says that loose ball joints may affect the steering system. Who is correct?
 a. A only c. Both A and B
 b. B only d. Neither A nor B

25. A squealing sound is heard when the engine is accelerated. *Technician A* says the problem is a loose power steering belt. *Technician B* says the problem is loose tie-rods. Who is correct?
 a. A only c. Both A and B
 b. B only d. Neither A nor B

26. A vehicle has excessive play in the steering wheel. *Technician A* says the problem is power steering belts. *Technician B* says the problem is worn tie-rods. Who is correct?
 a. A only c. Both A and B
 b. B only d. Neither A nor B

27. *Technician A* says a cam washer is used to adjust the camber on MacPherson strut systems. *Technician B* says a cam washer is used to adjust the alignment angles on rear-wheel drive vehicles. Who is correct?
 a. A only c. Both A and B
 b. B only d. Neither A nor B

28. To remove a tie-rod, *Technician A* says to use a pickle fork. *Technician B* says tie-rods are attached to the bearings. Who is correct?
 a. A only c. Both A and B
 b. B only d. Neither A nor B

29. A car has hard steering. *Technician A* says the problem is a worn idler arm. *Technician B* says the problem is too much power steering fluid. Who is correct?
 a. A only c. Both A and B
 b. B only d. Neither A nor B

Essay

30. Describe how a rack-and-pinion steering system operates.

31. What is a pitman arm?

32. Define the term *steering ratio*.

33. List the parts and operation of the standard steering linkage.

34. What is the purpose of the idler arm?

35. What is the purpose of the adjusting sleeve on the tie-rods?

36. Define the term *caster.*

37. Define the term *camber*.

38. How does a sliding spool valve operate in a power steering system?

Short Answer

39. Front-wheel drive vehicles, better tires, and independent rear suspension vehicles are all reasons why vehicles use ________ alignment.

40. A/an ________ is placed inside the rack and pinion on an electronic steering system.

41. Excessive steering play can be caused by ________.

42. A buzzing sound in the power steering unit typically means that ________.

43. Abnormally fast tire wear is a sign of ________.

Applied Academics

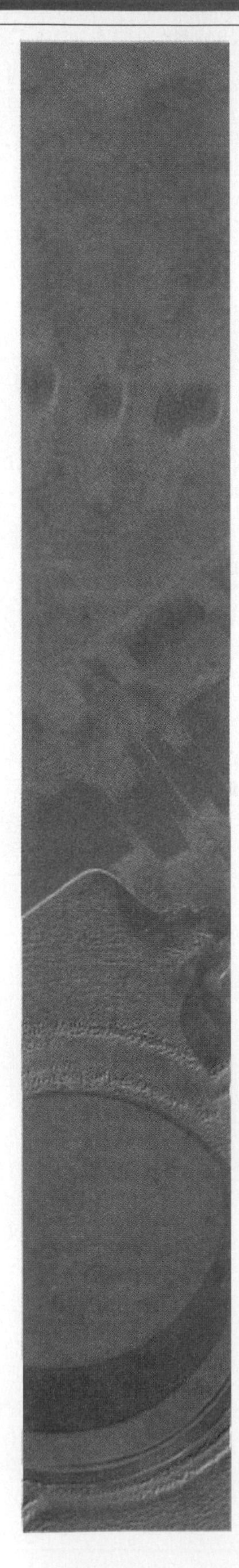

TITLE: Four-Link Steering Mechanism

ACADEMIC SKILLS AREA: Science

NATEF Automobile Technician Tasks:

The service technician can describe how steering and suspension geometry relates to how pulleys and levers are used to multiply force or transfer directions of force.

CONCEPT:

The steering system on all vehicles is theoretically constructed from the design and operation of a four-link mechanism as shown in the diagram that follows. It consists of two pivot points that are connected to the frame. The distance between these two points makes up one of the four links in the mechanism. In addition, there are three additional links that connect the two stationary points together. One is called the crank or driver, one is called the rocker or follower, and one is called the floating link. The exact length of each of these links, as well as the position of the two stationary pivot points, determine the exact motion that is produced by the four-link mechanism. All steering systems use the four-link mechanism as the basis for their design.

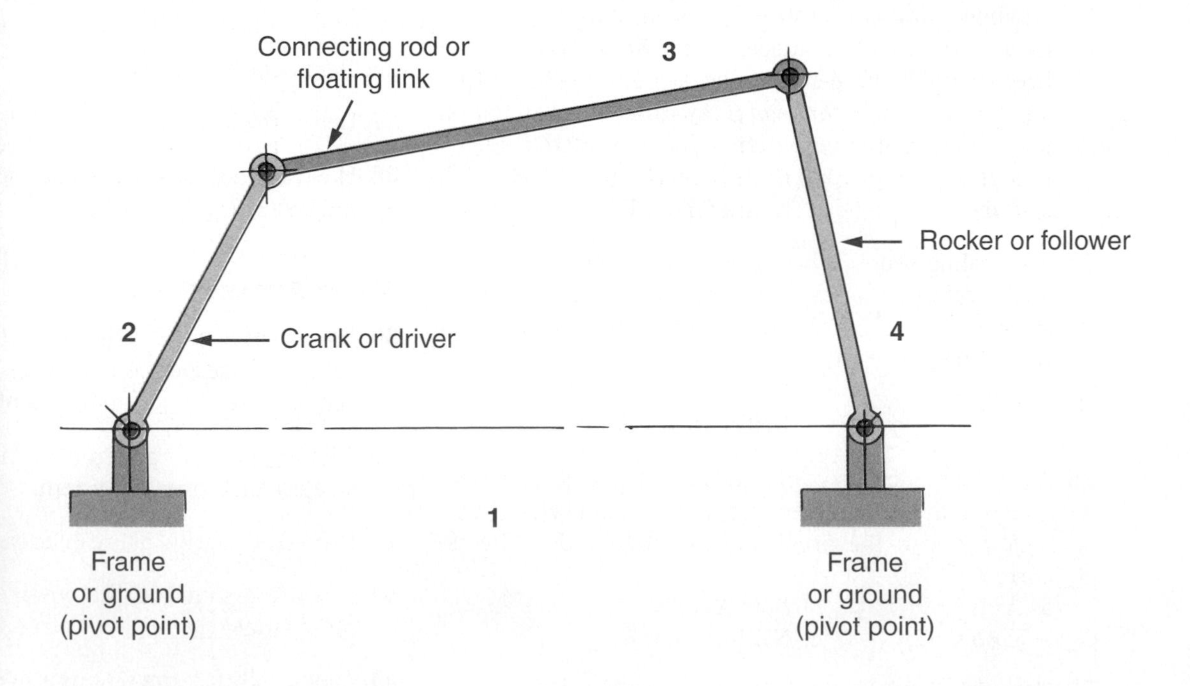

APPLICATION:

Can you identify how the four-link mechanics relates to the steering system geometry and linkages? Using the information in this chapter and after studying the four-link mechanism, what are the names of the parts of the steering system that represent each of the four links?

Tires and Wheels

CHAPTER 49

OBJECTIVES

After studying this chapter, you should be able to:

- Use tire terminology to define how tires are constructed.
- Identify different characteristics of tires.
- Compare different types of tires, including ply, radial, and spare tires.
- Identify how tires are sized.
- Analyze the purpose and operation of wheels and rims.
- Analyze tire problems, their diagnosis, and service procedures.

INTRODUCTION

Tires serve several important purposes. They are designed to carry the weight of the vehicle sufficiently, transfer braking and driving torque to the road, and withstand side thrust over varying speeds and conditions. This chapter explains how tires are designed, constructed, sized, and serviced.

49.1 TIRE CONSTRUCTION AND CHARACTERISTICS

DIFFERENCES IN TIRES

Different tire designs have been used on automobiles over the years to meet many demands. Originally, most vehicles used tube-type tires. However, tube tires were eventually

Certification Connection

ASE Connection: The information in this chapter can help you prepare for the National Institute for Automotive Service Excellence (ASE) certification tests. The test and content area most closely related to this chapter are:

Test A4—Suspension and Steering

- **Content Area**—Wheel and Tire Diagnosis and Repair

NATEF Connection: Much of the information in this chapter is related to the NATEF tasks. The NATEF tasks and priority numbers most closely related to this chapter are:

1. Diagnose tire wear patterns; determine necessary action. P-1
2. Inspect tires; check and adjust air pressure. P-1
3. Diagnose wheel/tire vibration, shimmy, and noise; determine necessary action. P- 2
4. Rotate tires according to manufacturer's recommendations. P-1
5. Measure wheel, tire, axle, and hub runout; determine necessary action. P2
6. Diagnose tire pull (lead) problem; determine necessary action. P-2
7. Balance wheel and tire assembly (static and dynamic). P-1
8. Dismount, inspect, repair, and remount tire on wheel. P-1
9. Reinstall wheel; torque lug nuts. P-1
10. Inspect and repair tire. P-2

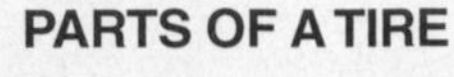

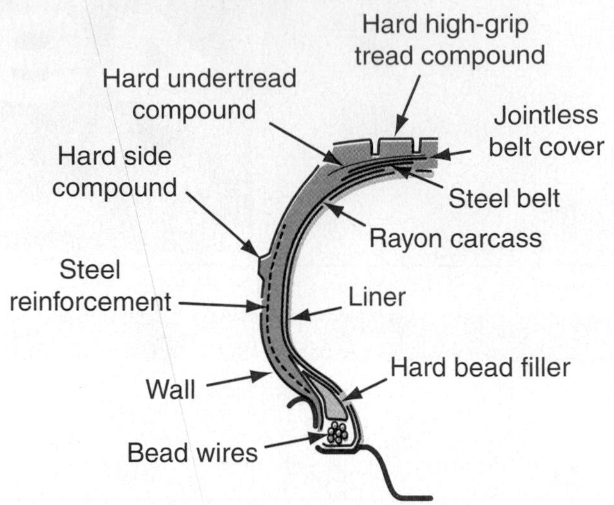

FIGURE 49–1 A tire has many parts. Each part plays an important part in tire operation.

replaced by tubeless-type tires. From that point on, different tread designs, internal construction, and belts and ply designs have been used on tires.

PARTS OF A TIRE

Although they seem simple, tires have several parts. ***Figure 49–1*** shows a cutaway view of a tubeless tire. The top or outside of the tire is called the **tread**. Its purpose is to produce friction for braking and torque for driving. The outside sides of the tire are called the wall. On some tires, the wall is made of a white material and is called the white wall. This is for the external looks of the tire. Note also that there is a steel reinforcement belt placed inside the wall. This steel reinforcement adds strength to the sides of the tire. The liner is the inside of the tire. A steel belt is also placed under the tread for strength and comfort in the vehicle.

Many tires also have additional layers, called **plies**, built or formed into the tire. The number of plies varies according to the use. For example, most automobile tires have two or four plies. Heavier vehicles such as vans and station wagons may use tires with up to eight plies for strength. The **carcass** is the strong, inner part of the tire that holds the air. The carcass is made of the layers or plies of fabric. It gives the tire its strength.

The wall and tread material is **vulcanized** into place. Vulcanizing is the process of heating rubber under pressure to mold the rubber into a desired shape. The **bead** wires, and remaining parts provide strength and durability. The tire is then attached to the wheel rim, and an air valve is used to admit the necessary air for pressure.

TUBE AND TUBELESS TIRES

The tube-type tire is mounted on the wheel rim with a rubber inner tube placed inside the casing. The inner tube is inflated with air. This causes the tire casing to resist change in shape. The tubeless tire is mounted directly on the rim. The air is retained between the rim and tire casing when inflated.

A tube tire cannot be used without a tube. Some service technicians will, however, use a tube in a tubeless tire in emergency situations. Tubes are sometimes installed by service technicians to eliminate hard-to-find slow leaks. Tubes are also used when imbedded dirt or rust prevents a tubeless tire from seating properly on the rim. However, using tubes for these reasons should be avoided if possible. Tubes are also useful with wire spoke wheels. The spokes tend to loosen in their sockets with long use, producing small leaks.

TIRE CHARACTERISTICS

Tires have four characteristics that are important in understanding their design. These include:

1. Tire traction—**Traction** is defined as a tire's ability to grip the road and to move or stop the vehicle.
2. Ride and handling—Tires are measured by their ride and handling ability. This is an indication of the degree of comfort a tire delivers to vehicle passengers. It is also a measure of the responsiveness to the driver's steering actions.
3. Rolling resistance—**Rolling resistance** is a term used to describe the pounds of force required to overcome the resistance of a tire to rotation. As rolling resistance of a tire decreases, fuel mileage typically increases.
4. Noise—All tires make a certain amount of noise. Tires can be made "quiet" by scrambling or changing the size, length, and shape of the tread elements. Scrambling prevents the sound frequency buildup that would develop if all of the patterns of the tread were spaced evenly.

CORDS

Within the tire, there are layers of plies or belts. These layers have **cords** running through them for strength. ***Figure 49–2*** shows several cords used inside the plies. ***Figure 49–3*** lists the advantages of different types of cords. Because newer synthetic fibers are constantly being developed, more advantages and better driving characteristics will continually occur in tire design.

RADIAL AND BIAS TIRES

There are typically three ways that the plies of the tire can be laid down. They can be positioned as a **bias**, belted bias, or belted **radial** tire. ***Figure 49–4*** shows the differences. The bias ply tire has layers of cord material running at an angle from bead to bead. Each cord runs opposite to the cord below it. This gives strength to the tire. However, the plies tend to work against each other during operation, which produces heat. In addition, bias ply tires tend to produce a certain amount of rolling resistance. These characteristics increase wear on the tread and shorten tire life.

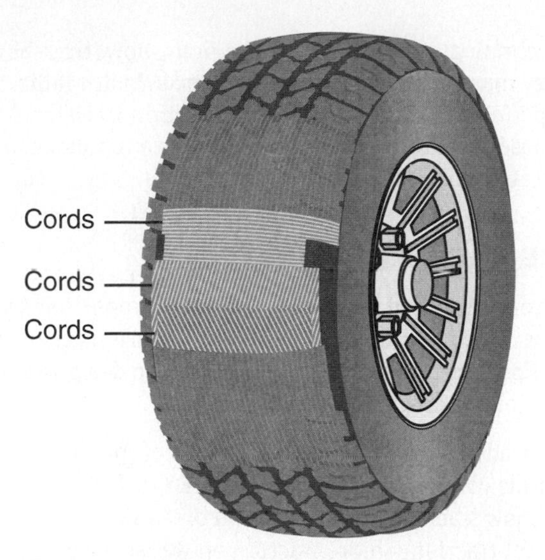

FIGURE 49-2 Cords are used inside the plies for strength.

TYPES OF TIRE CORDS

CORD TYPE	ADVANTAGES
Rayon	Soft ride, resilient, inexpensive
Polyester	Soft ride, more heat-resistant, inexpensive
Nylon	High heat resistance, excellent impact resistance, minimum flex
Steel	High heat resistance, excellent impact resistance, minimum flex

FIGURE 49-3 Different types of cords have different advantages.

The belted bias tire uses additional belts wrapped around the circumference of the body of the plies. The actual cords within the belts are manufactured at an angle. The design and addition of these belts add strength and stiffness to the tread. Because this tire typically has less rolling resistance than the bias ply tire, it lasts longer.

Car Clinic: TIRE TYPES

CUSTOMER CONCERN:
Shimmy Problems

A travel van with about 15,000 miles on it seems to have a severe shimmy when the vehicle hits a bump. This is especially noticeable on high-speed highway turns. The bias ply tires from the manufacturer are on the vehicle. The alignment has been checked and adjusted. The shimmy disappeared for about 500 miles, then returned.

SOLUTION: The problem could be caused by two things. First, is the vehicle used on bumpy roads? Severe bumps may cause the alignment to go out of specifications even after 500 miles. The second cause could be the tires. The weight distribution of large vehicles may cause the problem. It is extremely important for the tires to grip the road under all conditions. The better the grip, the less chance of shimmy. With a lot of weight in the rear of a van, the road grip on ply tires may not be enough. Try replacing the tires with a good set of radial tires. This will most likely solve the problem.

The cord material in the body of a belted radial or radial ply tire runs from bead to bead. It is not at an angle as in the other tires. Additional belts provide strength and durability. The belted radial tire holds more of the tread on the pavement during cornering and straight driving. In addition, the tire does not **squirm**. Squirm means that as the tire hits the road, it is moving or being pushed together. Less squirm results in better traction, less rolling resistance, less heat buildup, longer life, and better fuel mileage. In addition, newer radials have special materials in the cords to increase comfort and driveability.

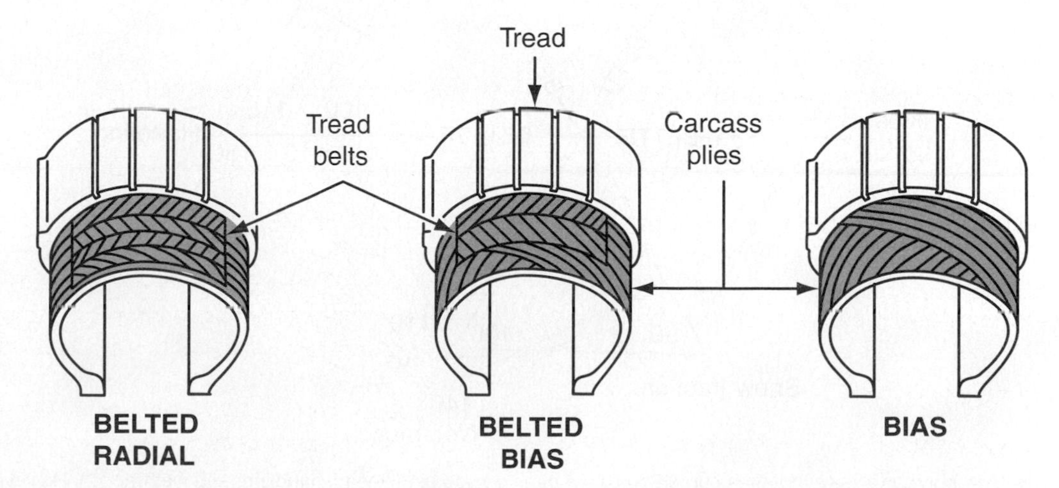

FIGURE 49-4 Tires can be bias ply, belted bias, or radial design. The difference is in how the plies are laid down on the tire.

TREAD DESIGN

The tread of a tire must work on all types of conditions. For example, vehicles operate on smooth pavement, gravel, wet pavement, and icy pavement. In addition, steering traction is different from rear-wheel drive traction. Today, tire designs are a compromise between these conditions and cost considerations.

Tire treads have been developed to provide better traction in wet conditions. Water must be squeezed from the contact area on the road surface. Treads have been developed to move the water away from the treads as efficiently as possible as the tire rolls. Of course, as the tire tread wears, this efficiency is reduced. ***Figure 49–5*** shows an example of different tread designs.

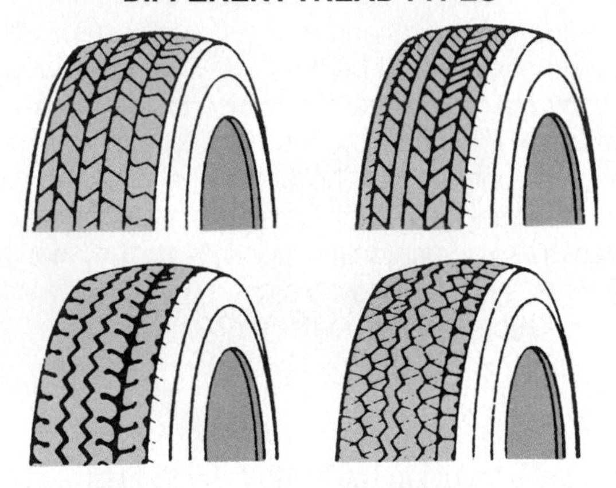

FIGURE 49-5 Tire treads designs vary and are based on the types of road conditions the vehicle will be driven on.

Technology is gradually changing snow tires as well. Some manufacturers have developed **hydrophilic** tread compounds. Hydrophilic means attraction to water. A tread composed of part of this compound has a tendency to stick to a wet or icy surface.

TIRE PERFORMANCE

Choosing the right tire is becoming an important task for the consumer. Tires can be designed for various characteristics. For example, one tire manufacturer designs tires for the following characteristics:

- Treadwear—more life on the tread of the tire
- Ride comfort—more comfortable ride for passengers
- Snow traction—more traction on snow
- Wet traction—more traction on wet surfaces
- Handling—improved handling during cornering
- High speed—more control at higher speeds

Many tires are sold as **all-season tires**. This type of tire gives equal importance to each of the characteristics just listed. However, an operator may need to improve particular characteristics. For example, during the winter months in certain parts of the country, a customer may still need to buy snow tires to get additional snow traction in severe driving conditions.

When one tire characteristic is improved, other characteristics have a tendency to be reduced. This relationship is shown in ***Figure 49–6***. This diagram is called a tire performance polygon. Four tires (four polygons) are shown for reference. The farther the points on the polygon get from

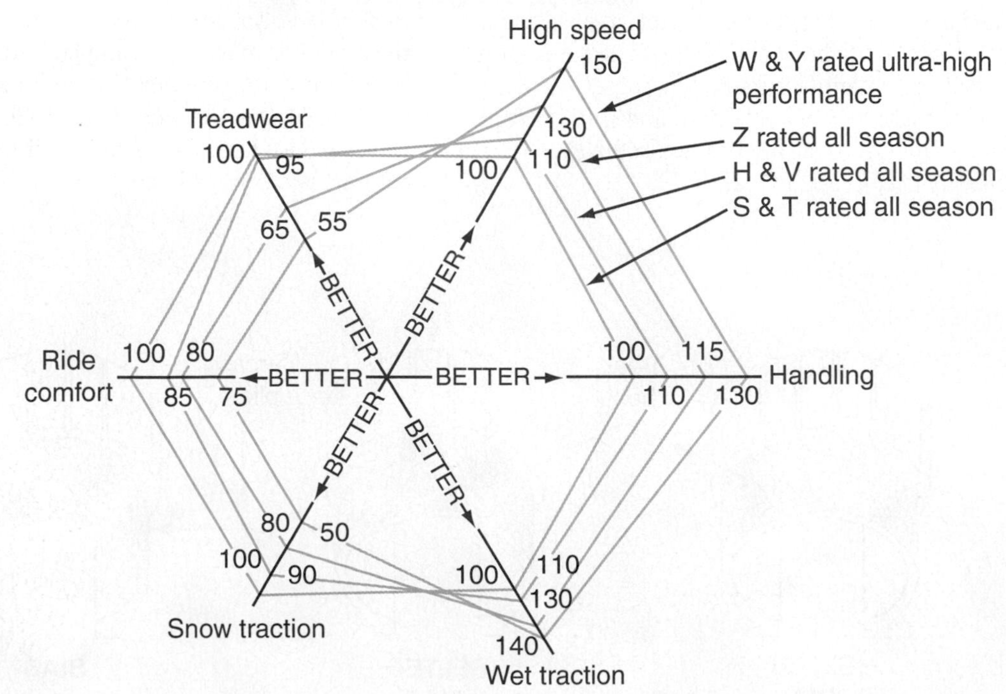

FIGURE 49-6 This polygon shows that tires can be designed for improved high speed, handling, and wet traction. However, snow traction, treadwear, and ride comfort characteristics are reduced. (Courtesy of Cooper Tire & Rubber Company)

the center, the better the characteristic. The all-season tire is designed for all-around use. Note that each characteristic point on this tire has been given a value of 100. The other numbers indicate the degree of improvement or reduction of the specific characteristic.

The other three tires are compared with the all-season tire. The other three tires, V-rated, H-rated, and S-rated are all designed for higher speeds, better handling, and improved wet traction. The result is that these tires lose comfort, treadwear, and snow traction characteristics. Knowing this, consumers can choose which performance characteristics are most important to their driving needs and which they are willing to sacrifice.

RUN-FLAT TIRE

Tire manufacturers have also designed tires that run more safely when the air has escaped through a puncture. ***Figure 49–7*** shows a run-flat type of tire. This type of tire has additional sidewall reinforcement. The additional sidewall has a thick material with several important characteristics. First, the material is flexible. Second, the material uses low-hysteresis rubber. Low-hysteresis material can return to its original shape after deflating. Also, the material has good thermal resistance. When the tire is running flat, the tire material has a tendency to heat up. The thermal resistance of a run-flat tire keeps it from being damaged by the additional heat buildup. The tire tread is designed to maintain comfort and handling similar to a standard tire, even when it is being run flat. A special bead design also is used to keep the bead from separating after pressure loss.

TIRE VALVES

The tire valve admits and exhausts the air into and out of the tire. ***Figure 49–8*** shows a typical tire valve. It has a central core that is spring loaded. This allows air to flow in only one direction, inward. When the small pin is depressed, air flows in the reverse, or outward, direction. When the valve core becomes defective, it can be unscrewed for removal and replaced. An airtight cap on the end of the valve produces an extra seal against valve leakage.

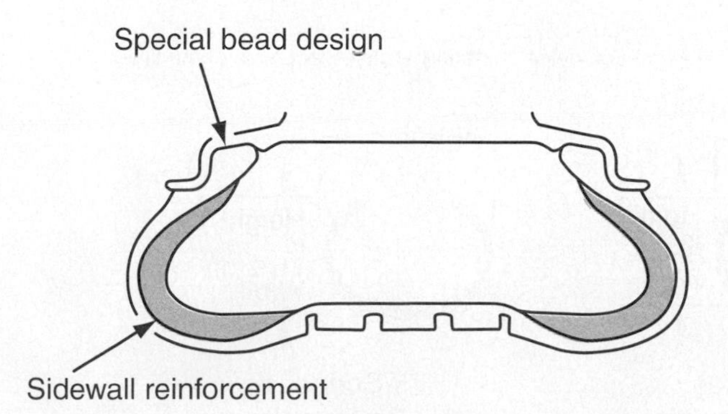

FIGURE 49–7 A run-flat tire has special sidewall reinforcements.

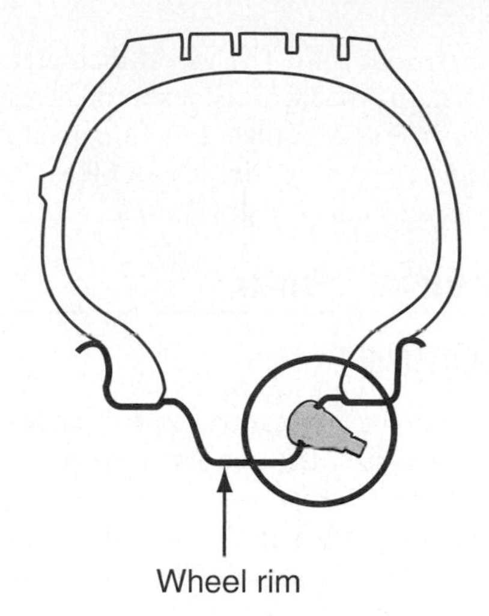

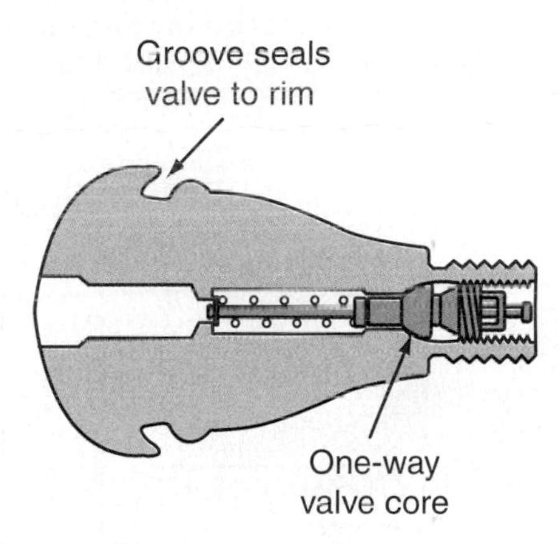

FIGURE 49–8 A tire valve is used to admit or exhaust air from the tire. A small one-way valve allows air to flow in only one direction unless the valve is opened.

COMPACT SPARE TIRE

Vehicles are built smaller than in the past and at times need more trunk space. Many cars use the compact spare tire to save space and weight. A compact spare tire is shown in ***Figure 49–9***. It is called a temporary-use-only spare tire. This

FIGURE 49–9 The compact spare tire is used to save weight and space in smaller vehicles.

tire has a narrow 4-inch-wide rim. The wheel diameter is usually 1 inch larger than the road wheels. It should be used only when one of the road tires has failed. Inflation pressure is much higher on the compact spare tire, about 60 psi, and the top-rated speed is lower than on regular tires.

49.2 IDENTIFYING TIRES

METRIC TIRE SIZES

Tires are sized according to the application in which they are to be used and their physical size. The size of the tire must be molded into the side of that tire. Most tires today are sized according to metric standards. In the past, however, there were other ways of identifying tire sizes. ***Figure 49–10*** shows how metric sizing works. The tire has several designations:

- The first letter of the size tells if the tire is used for passenger (P), temporary (T), light truck (LT) or commercial (C) use.
- The second designation identifies the section width of the tire. The section width is the distance in millimeters from one side of the tire to the other side of the tire when it is inflated normally. For example, the tire shown in ***Figure 49–10*** has a width of 215 mm.
- The third designation gives the **aspect ratio**. The aspect ratio is found by dividing the section height by the section width. The aspect ratio in ***Figure 49–10*** means the tire's height is 80% of the width. This is called the profile of the tire. Lower profile tires make the car closer to the road and reduces wind drag beneath the car body.

There are several common aspect ratios normally used on tires today. ***Figure 49–11*** shows four common aspect ratios. For example, the 75 series tire means that the height is 75% of the width of the tire. Notice that as the aspect ratio increases (60, 70, 75, 85) the vehicle sits higher from the ground. This causes both the comfort of the ride and the vehicle control to change. Generally, as the aspect ratio decreases, the comfort of the

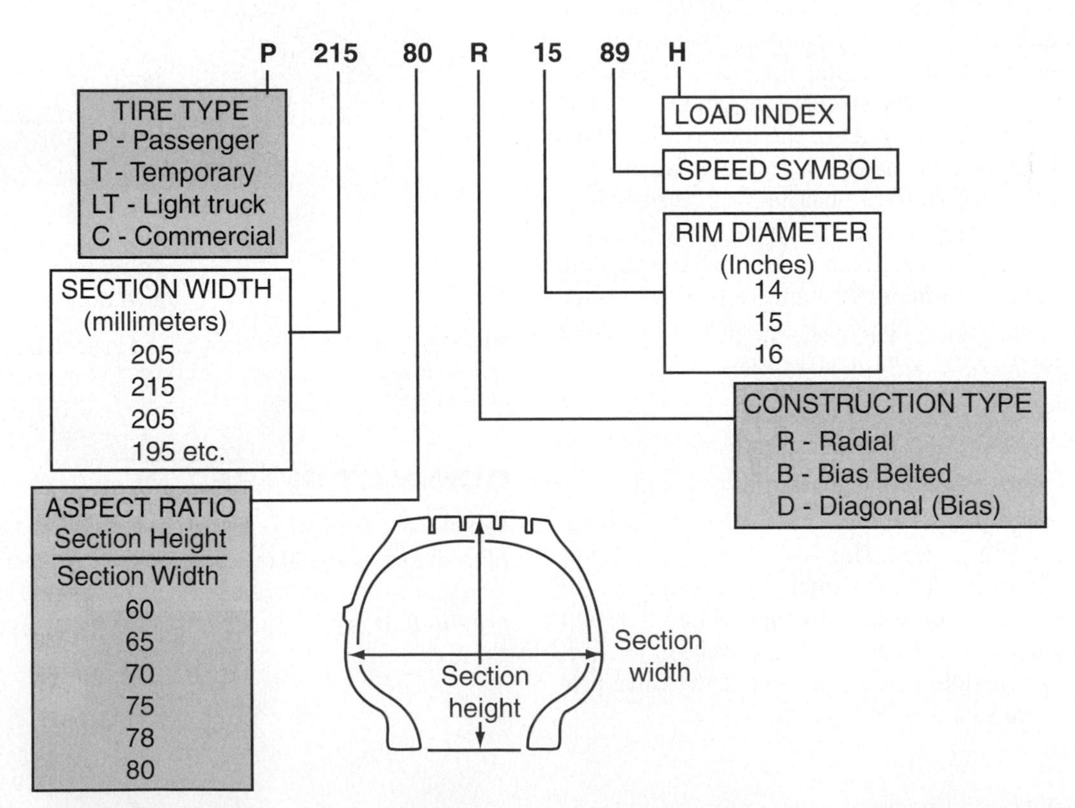

FIGURE 49–10 Tires are sized according to metric dimensions. Each designation helps to identify the type and size of the tire.

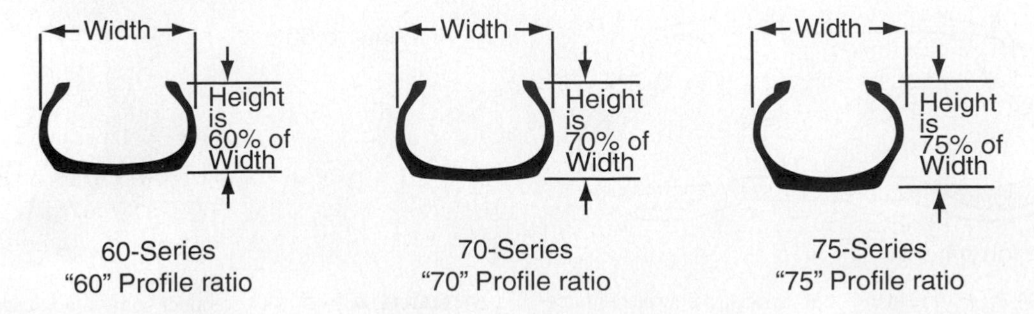

FIGURE 49–11 These are common tire aspect ratios. They affect comfort and vehicle control.

CUSTOMER CONCERN:

Car Pulls to One Side During Acceleration

The rear end of a vehicle with a limited slip differential seems to pull to one side during acceleration. Limited slip differentials are designed to provide equal push on both rear wheels. What could be the problem?

SOLUTION: The first thing to check is the rear tires. Make sure that the tires are both the same diameter. The tires have most likely been mismatched. Using a limited slip differential with two tire sizes will cause the car to pull to one side.

ride decreases, but the control of the vehicle increases. Also, as the aspect ratio increases, the ride or comfort increases, but the control of the vehicle decreases.

- The next designation gives the construction type. R means radial tire, B means belted, and D means diagonal (bias).
- The next designation gives the wheel rim in inches. The most common car rim sizes are 13, 14, 15, and 16 inches in diameter.
- The next two designations give the speed rating and the load index. These two designations help consumers compare speed and load ratings from one tire to another when purchasing tires. Load ratings are defined later is this chapter.

When a customer decides to buy a set of tires, there are many sizes to choose from. ***Figure 49–12*** shows an example of the variety of sizes that one manufacturer has available. Note that the variation in the aspect ratio ranges from 55 to 75. However, the aspect ratio may be even higher, such as 85. When selecting the correct tire size, always refer to the Owner's Manual so that the correct size can be determined.

WIDE TRACK RADIAL	WIDE TRACK BAJA
SIZE	SIZE
P185/70R14	P225/70R14
P195/70R14	P225/70R15
P205/70R14	P235/70R15
P215/70R14	P255/70R15
P215/70R15	P225/70R16
P215/65R15	P235/70R16
P215/60R14	P255/70R16
P235/60R14	P265/75R16
P235/60R15	P265/70R17
P255/60R15	P225/75R15
P275/60R15	P235/75R15
P185/60HR14	P235/75R15XL
P195/65HR15	P225/75R16
P205/65HR15	P235/75R16
P195/60HR15	P245/75R16
P205/60HR15	
P215/60HR15	
P225/60HR16	
P205/55HR16	

FIGURE 49-12 Several designations may be used to identify tire size.

TIRE PLACARD

A **tire placard** is permanently located on many vehicles. The placard is normally located on the rear edge of the driver's door. Refer to the placard for tire information. It lists the maximum vehicle load, tire size (including spare), and cold inflation pressure (including spare). ***Figure 49–13*** shows a tire placard.

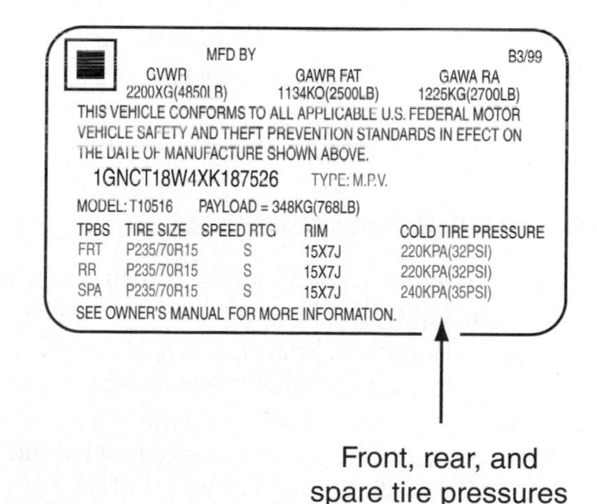

FIGURE 49-13 The tire placard lists the maximum vehicle load, tire size, and inflation data.

PARTS LOCATOR

*Refer to photo #112 in the front of the textbook to see what a **tire placard** looks like.*

UTQG DESIGNATION

Uniform tire quality grading (UTQG) symbols are required by law to be molded on the sidewall of each new tire sold in the United States. A typical grading may be 90 CB, 170 BC, or 140 AA. The number indicates the comparative tread life. A tire marked 140 should wear 40% longer than a tire marked 100. The first letter indicates comparative wet traction. A is best and C is worst. The second letter is a measure

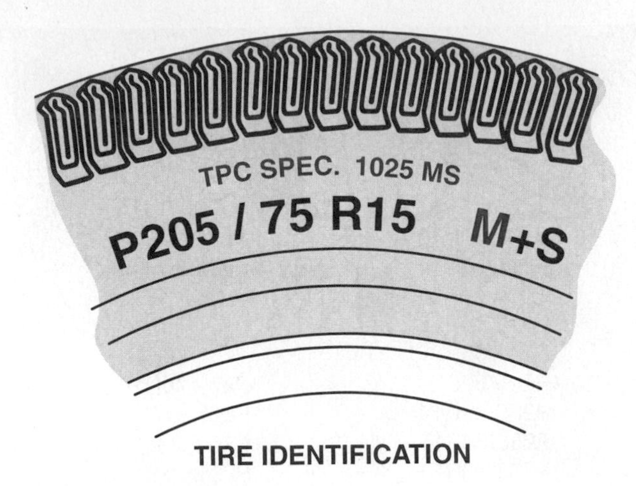

FIGURE 49–14 The TPC specification number is shown on this tire.

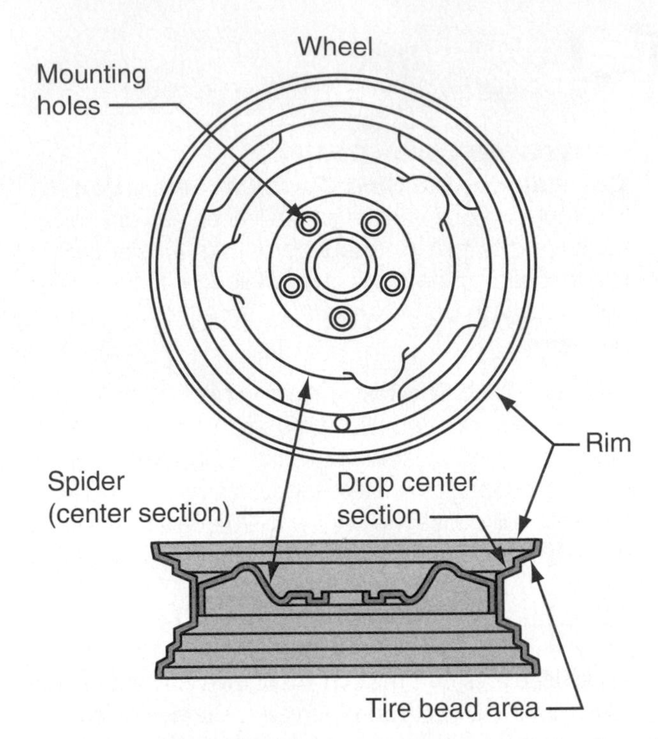

FIGURE 49–15 The wheel is used to support the tire. Several of its parts are identified here.

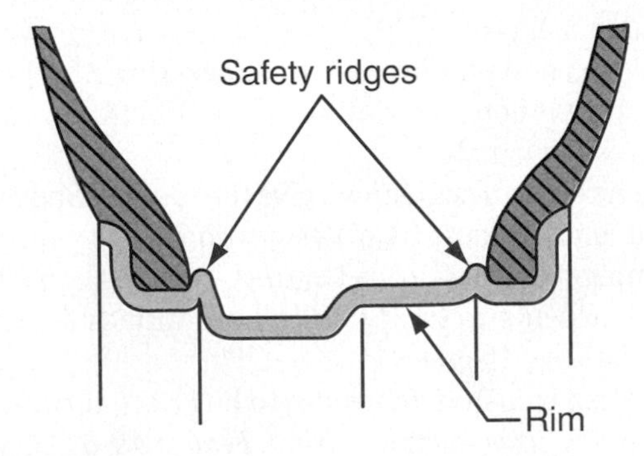

FIGURE 49–16 Safety rims have small ridges to stop the tire bead from getting into the dropped area during a blowout.

of the resistance to heat. Again, A is the best and C is the worst.

Some tire manufacturers use the **Department of Transportation (DOT)** designation. The DOT specification number indicates that the tire has met various tests of quality established by the Federal Department of Transportation.

TPC SPECIFICATION NUMBER

On most vehicles originally equipped with radial tires, a tire performance criteria (TPC) specification number is molded into the sidewall (***Figure 49–14***). This shows that the tire meets rigid size and performance standards developed for that particular automobile. It ensures a proper combination of endurance, handling, load capacity, ride, and traction on wet, dry, or snow-covered surfaces.

LOAD RANGE

Tires are measured for the amount of load they can withstand. They are typically identified by letter codes. The codes used are A, B, C, D, E, F, G, H, I, J, L, M, and N. As the size of the tire increases, the **load range** also increases. On some tires, the load range is stated in pounds for a specific tire pressure. For example, the load range molded into the tire may be, "Max Load, 585 kg (1,290 lb) at 240 kPa (35 psi)." ***Figure 49–14*** shows a tire with a load range of M.

49.3 WHEELS AND RIMS

DROP CENTER WHEELS

Tires are mounted on rims made of steel, aluminum, or other strong material. The parts of the wheel are shown in ***Figure 49–15***. Mounting holes are used to attach the wheel to the lugs on the axles. This wheel is called a **drop center wheel**, which means that the center of the wheel is made so that it has a smaller diameter than the rim. The wheel has a dropped center so that the tire can be easily removed. During removal and installation, the bead of the tire must be pushed into the dropped area. Only then can the other side of the tire be removed over the rim flange.

Safety rims are also being used on vehicles today. The safety rim has small ridges built into the rim (***Figure 49–16***). These ridges keep the tire from getting into the dropped center when a blowout occurs while driving and the tire becomes deflated. If the tire stays up on the rim, there is more driving stability than if the tire gets into the dropped center. Thus, safety is improved.

WHEEL SIZES

Common sizes for wheels are 13-, 14-, 15-, 16-, and 17-inch diameter rims. A 16-inch wheel is often used when a low-profile tire is placed on the vehicle. The low-profile tire is used for better handling. With the 16-inch wheel and the

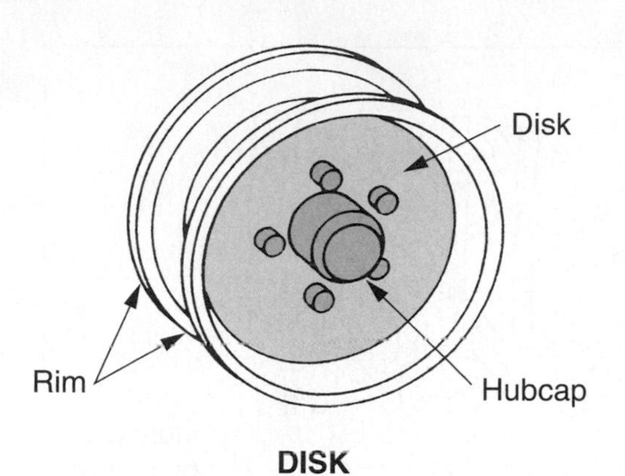

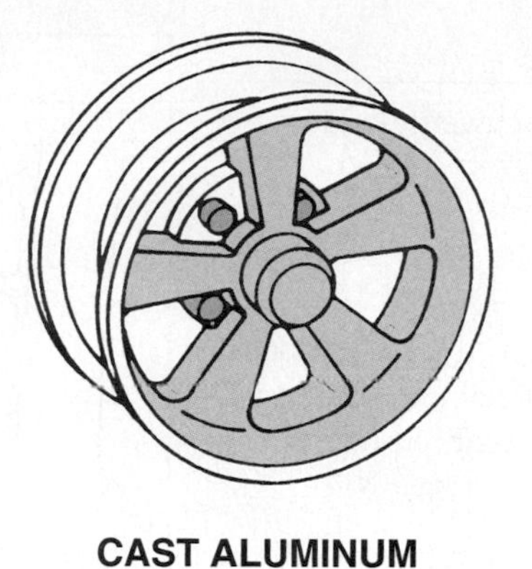

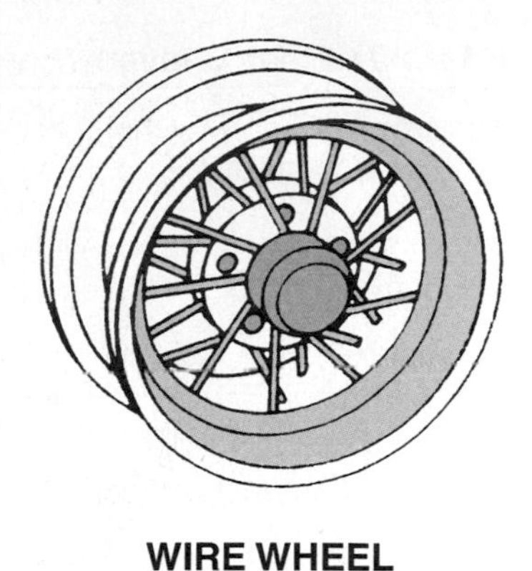

FIGURE 49–17 Generally, wheels are one of three types: the disk, cast aluminum, or wire wheel.

low-profile tire, the vehicle doesn't sit quite as low and in turn looks aesthetically pleasing. The 16- and 17-inch wheels are generally used on heavier duty and larger vehicles such as four-wheel drive and off-road vehicles. The exact wheel size used on a vehicle is determined by the automotive manufacturer. The width of the wheel rims is usually 4.5, 5, or 6 inches.

SPECIAL WHEELS

A variety of wheels are being manufactured today. They are normally made of steel or aluminum. Generally, wheels are one of three types: the disk wheel, the cast aluminum wheel, or the wire wheel (***Figure 49–17***). All three are currently being used by manufacturers. The consumer can also purchase a **mag wheel**. This type of wheel uses a light magnesium metal for the rim. The term *mag* is used today, however, to represent almost any type of wheel that uses special material and designs.

Problems, Diagnosis, and Service

Safety Precautions

1. To service tires and wheels, you must raise the vehicle off the floor. Always follow the correct procedure when using a hoist to lift a car. If a hydraulic jack is used, remember to use extra jack stands for support. Always block the other wheels to eliminate any possibility of the vehicle moving forward or backward.
2. When removing a tire from its rim, remember to use the proper tools and procedures. There is a great deal of pressure on the tire sides, and this pressure could cause serious injury during removal.
3. Always tighten the wheel nuts on the vehicle to the correct torque specifications and sequence. This will prevent the wheel nuts from coming loose and possibly causing serious injury during driving.
4. When a tire is being inflated after it has been replaced on the rim, the bead will often pop into place. Keep your fingers away from the bead during inflation. Also, never inflate a tire over 40 psi, because older rims may come apart and cause serious injury.
5. Tires that are underinflated will often get very hot when driven at high speeds. Always be careful not to touch a hot tire immediately after running at high speeds. It may burn your hands.
6. Be careful when handling a tire that has shredded at high speeds along a highway. There may be sharp metal wires protruding from inside the bead.
7. Never try to lift a tire without using proper posture and lifting technique. Keep the tire close to your body and lift with your legs.
8. Always wear OSHA-approved safety glasses when working on tires and rims.

PROBLEM: Uneven Wear on Tires

The front tires of a vehicle are wearing unevenly and now need to be replaced.

DIAGNOSIS

Uneven tire wear has several causes. Generally, it is caused by improper inflation, not rotating the tires, or misalignment. To determine the cause of uneven wear, perform the following diagnostic checks:

1. Before checking any problem with tires, make sure the front suspension, steering, and brake systems are working and adjusted correctly.
2. Incorrect front-end and rear-end tire geometry and operation can also cause excessive and uneven wear on the tires. Observe the tire wear patterns to determine the problem as shown in ***Figure 49–18***. If the problem appears to be in alignment, have the vehicle aligned according to the manufacturer's recommended procedure before proceeding.
3. Tires can be checked for excessive wear by observing the treadwear indicators or **tread bars** (***Figure 49–19***).

INCORRECT TIRE GEOMETRY			
Wear on one edge	Feathered edge	Bald spots	Scalloped wear
Excessive camber	Incorrect toe	Unbalanced wheel	Lack of rotation of tires or worn or out-of-alignment suspension
		Or tire defect	
Adjust camber to specs	Adjust toe to specs	Dynamic or static balance wheels	Rotate tires and inspect suspension

FIGURE 49–18 Different tire wear characteristics can be caused by incorrect alignment and front-end geometry.

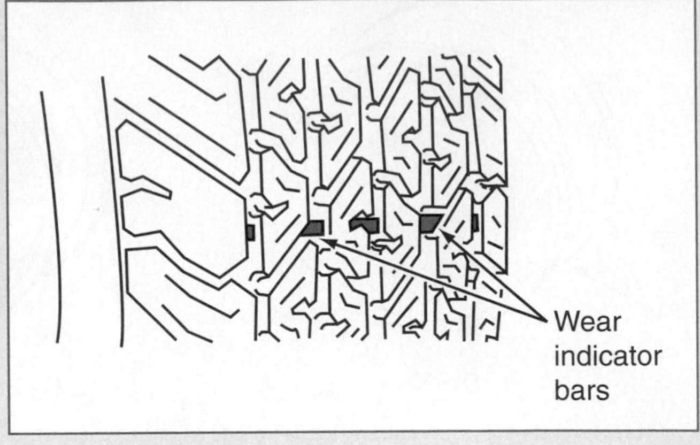

FIGURE 49–19 Check the tread bars on tires. If the tread bars show, the tire is ready to be replaced.

They look like narrow strips of smooth rubber across the tread. They will appear on a tire when it is worn down, either evenly or unevenly. When these tread bars are visible, the tire is worn and should be replaced.

4. Tires must be inflated correctly. The recommended tire pressure is determined by the type of tire, weight of the vehicle, and ride. In addition, tire inflation may change with temperatures. For example, cold weather reduces tire inflation pressure approximately 1 pound for every 10 degrees drop in temperature. Always check tire inflation in both summer and winter conditions.
5. Check the tires for overinflation. Overinflation of tires increases tire tension and prevents proper deflection of the sidewalls. These conditions result in wear in the center of the tread. The tire also loses its ability to absorb road shocks. Refer to ***Figure 49–20***.
6. Underinflation of tires distorts the normal contour of the tire body, causing the tire to bulge outward. It also increases internal heat. Heat can weaken the cords and cause the ply to separate. See ***Figure 49–20***.
7. Tire inflation can be measured in pounds per square inch (psi) or kilopascals (kPa). ***Figure 49–21*** shows the relationship between the two units.
8. Check whether the tires have been rotated according to the manufacturer's recommended time interval and procedure. There are several acceptable patterns for rotating tires (***Figure 49–22***). Always follow the manufacturer's recommended rotation procedure.
9. Check whether the weight of the vehicle has been increased with extra loads or objects. Always check the owner's manual for the maximum safe load limit the tires can handle as well as the tire inflation necessary.
10. A visual inspection of the tires can also tell if the inflation is correct or not. ***Figure 49–23*** shows how tires should look when they are properly inflated.

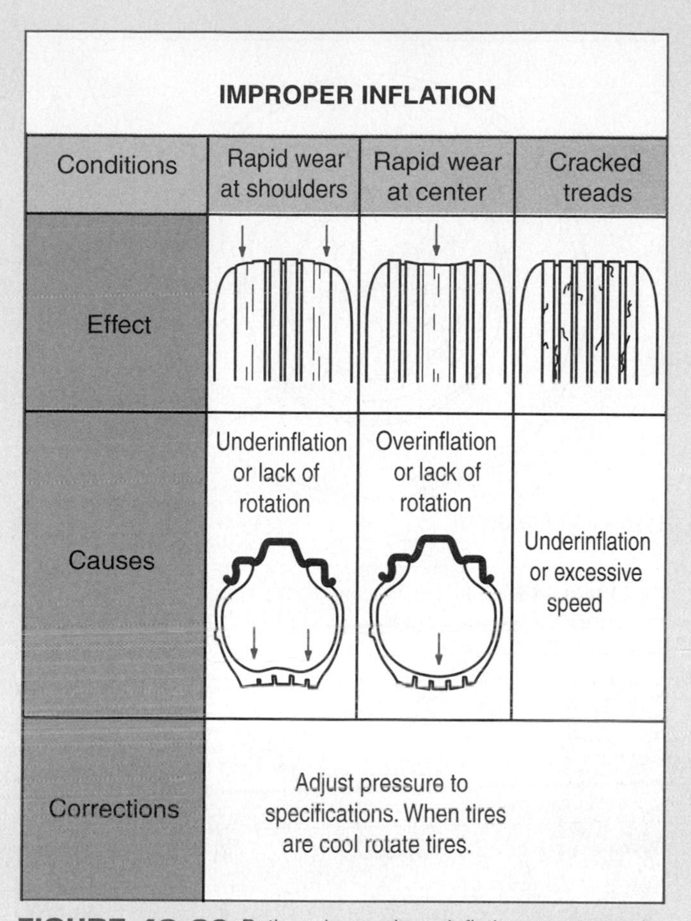

IMPROPER INFLATION

Conditions	Rapid wear at shoulders	Rapid wear at center	Cracked treads
Effect			
Causes	Underinflation or lack of rotation	Overinflation or lack of rotation	Underinflation or excessive speed
Corrections	Adjust pressure to specifications. When tires are cool rotate tires.		

FIGURE 49-20 Both under- and overinflation can cause tire damage and wear. Be sure that tires are inflated to the correct pressure recommended by the manufacturer.

TIRE INFLATION PRESSURE CONVERSION CHART

Inflation Pressure Conversion Chart (kilopascals to psi)

kPa	psi	kPa	psi
140	20	215	31
145	21	220	32
155	22	230	33
160	23	235	34
165	24	240	35
170	25	250	36
180	36	275	40
185	37	310	45
190	38	345	50
200	39	380	55
205	30	415	60

Conversion: 6.9 kPa=1 psi

FIGURE 49-21 Both psi (pounds per square inch) and kPa (kilopascals) can be used to measure the inflation pressure of tires.

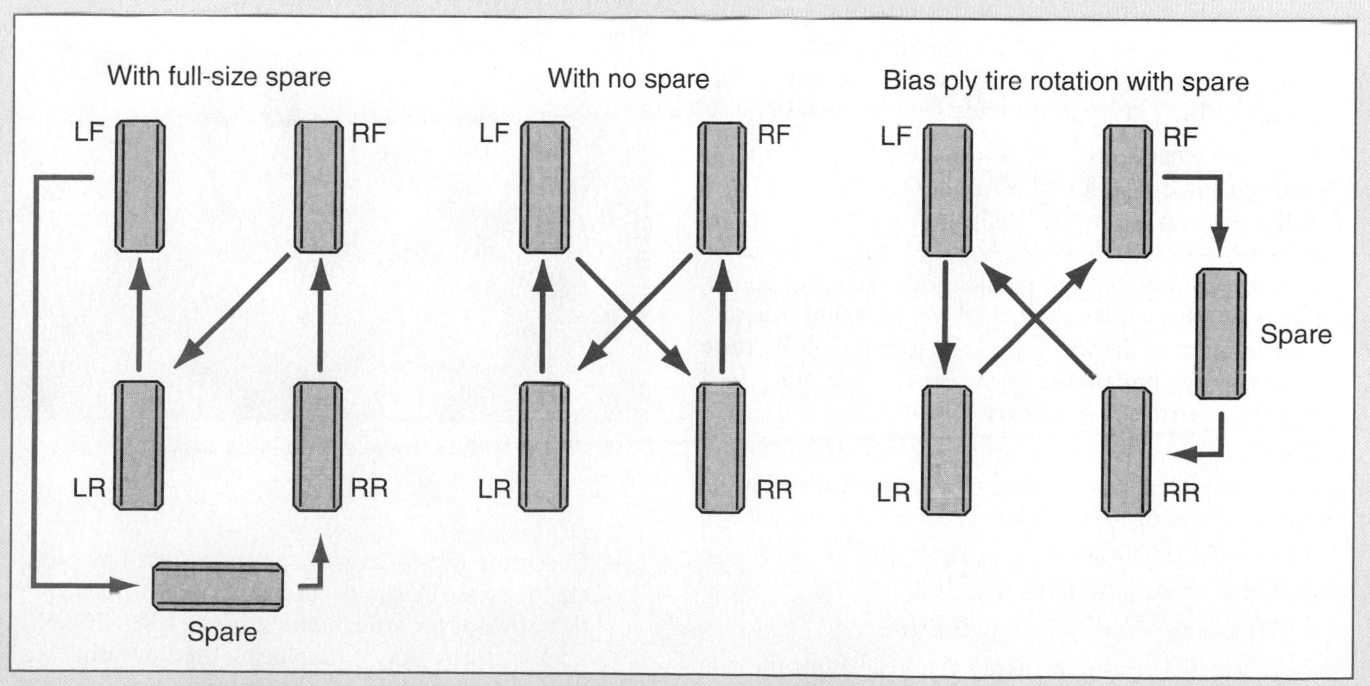

FIGURE 49-22 Tires can be rotated several ways to distribute the wear. Several common rotational patterns are shown. Follow the vehicle manufacturer's recommendations when rotating tires.

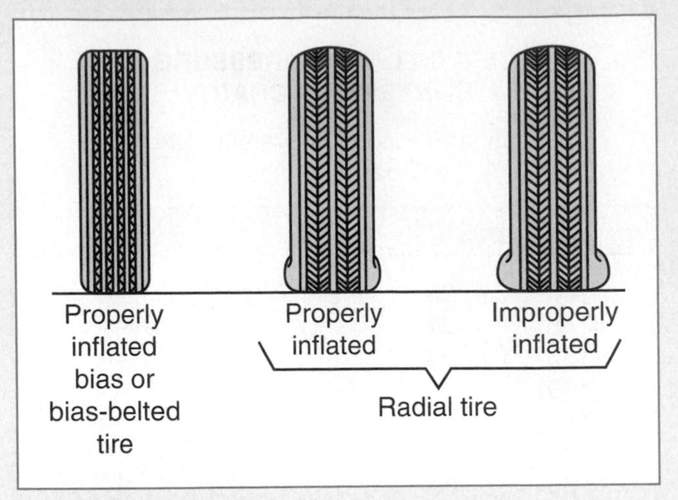

FIGURE 49–23 Various types of tires will have different appearances when properly inflated.

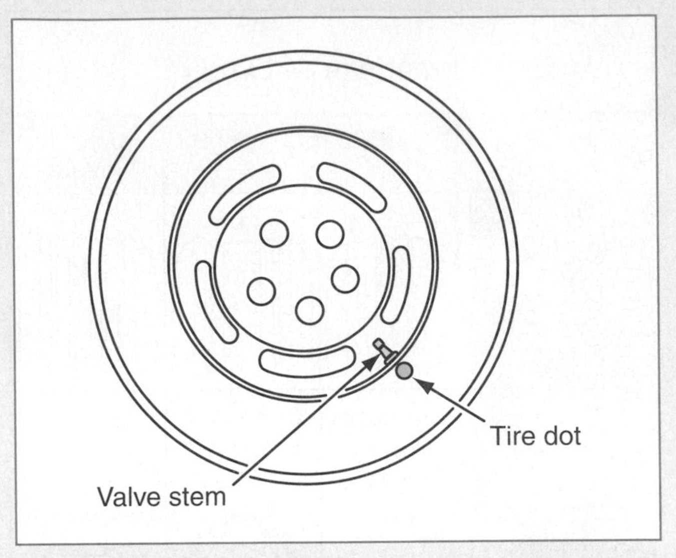

FIGURE 49–24 Align the dot on the tire with the valve stem to get the smoothest ride possible after dismounting a tire.

SERVICE

If it has been determined that the tires are producing uneven wear because of inflation, inflate the tires to the correct specification. Refer to the owner's manual. If the tires must be replaced, the following service items should be remembered.

1. Tires should be checked for inflation at least once a month and before long trips. Use an acceptable and accurate pressure gauge. Check tires when they are cold, and check the spare tire as well.
2. When replacing tires with those that do not have a TPC (tire performance criteria) specification number, use the same size, load range, and construction type (bias, belted bias, or radial) as the original tires on the car. A different type of tire may affect the ride, handling, speedometer and odometer readings, vehicle speed inputs to computers, and vehicle ground clearance.
3. Tires and wheels are match-mounted at the assembly plant on some vehicles. This means that the stiffest part of the tire radially, or the "high spot," is matched to the smallest radius, or "low spot," on the wheel. This is done to provide the smoothest possible ride. The high spot is marked with a yellow paint mark on the outer sidewall. The low spot of the wheel is at the location of the valve stem. Always mount the wheel so the tire and wheel markings are matched (***Figure 49–24***).
4. When removing and replacing tires, use a quality tire changer. Follow the procedures recommended by the manufacturer of the machine. Use the following general procedure to change a tire.
   a. Remove the air from the tire.
   b. Remove the valve stem from the tire.
   c. Unseat the tire by breaking the bead from the rim (***Figure 49–25***).
   d. Using the proper tools and pneumatic rams, remove the tire from the rim (***Figure 49–26***).

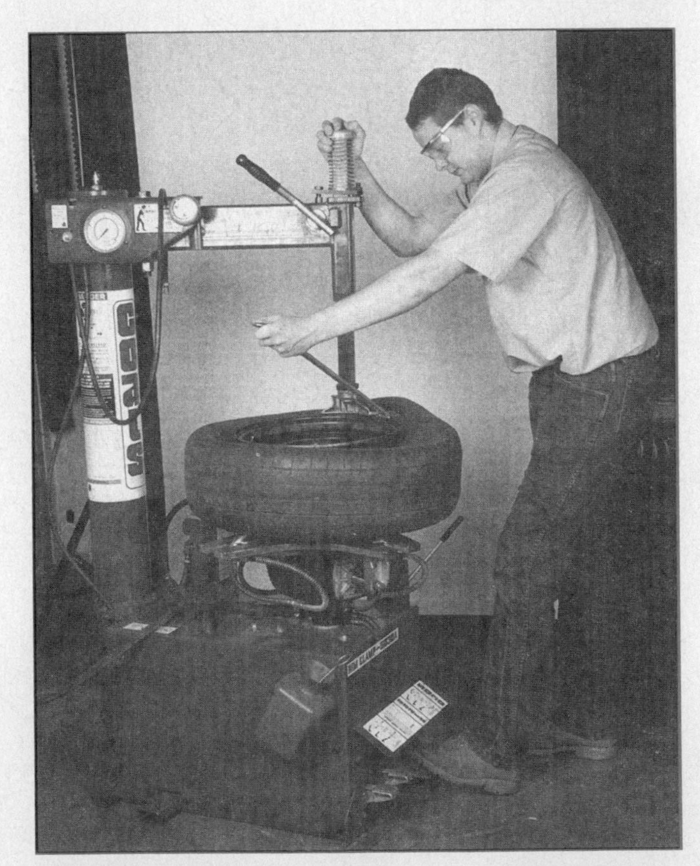

FIGURE 49–25 Use a tire changer to break the bead of the tire before removal.

   e. Check the rim for cracks or bends caused by hitting curbs and so on. It is dangerous to weld a cracked rim or to heat it for straightening (***Figure 49–27***). Most manufacturers suggest replacing the rim rather than trying to repair it.
   f. If the rim has no cracks and is not bent, clean the rim using a wire brush as shown in ***Figure 49–28***.

FIGURE 49-26 Use a tire changer to remove the tire from the rim.

FIGURE 49-28 Clean the rim with a wire brush before installing a new tire.

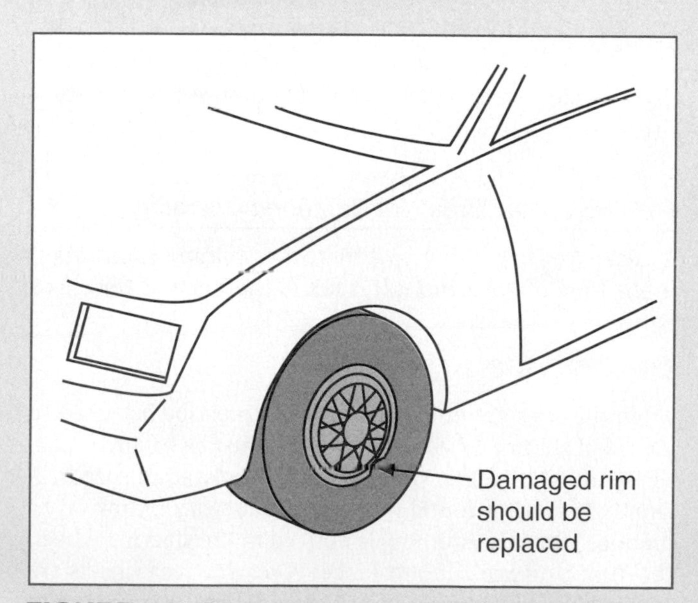

FIGURE 49-27 Never try to repair a rim that has been bent or broken. Always replace the rim.

g. Rubber compound should be liberally applied to the bead of the new tire (***Figure 49–29***).

h. Using the tire changer assembly, install the new tire onto the rim.

i. Reinstall the air valve.

j. Inflate to specifications, and check for leaks.

CAUTION: *Do not inflate the tire over 40 psi to set the bead. If the bead will not set, deflate the tire, push down the bead, and reinflate. Make sure rubber compound has been applied to the rim of the tire. If overinflated, the bead may jump over the rim and injure your hands.*

k. Install the tire and rim back on the vehicle. The tire lug nuts must be tightened in a specific sequence. Using an incorrect tightening sequence may cause the wheel, brake drum, or rotor to bind or bend. See ***Figure 49–30*** for the correct sequence for 4-, 5-, or 6-nut wheels.

FIGURE 49–29 Before installing the tire on the rim, apply rubber compound on the tire rim to aid in sealing.

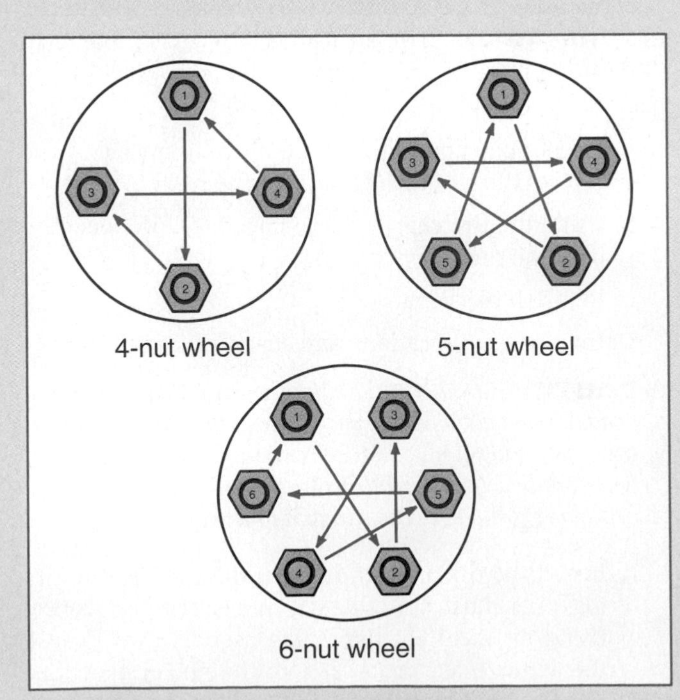

FIGURE 49–30 Wheel nuts should be tightened in the correct sequence and to proper torque specifications.

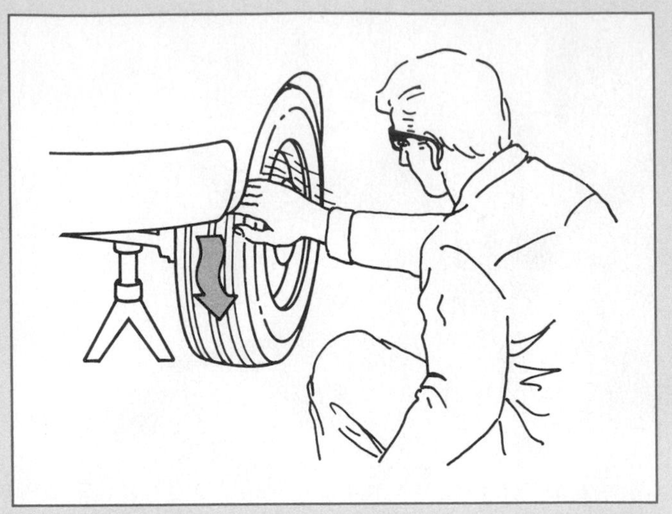

FIGURE 49–31 The tire can be checked for a defective tread by lifting the car and spinning the wheel. The tread will wobble from side to side if it is defective.

PROBLEM: Defective Radial Tire Belt

A vehicle is pulling to one side. The pulling begins at less than 10 miles per hour.

DIAGNOSIS

This condition may indicate a defective radial tire. Often, a belt may have shifted inside the tire body. To check this condition:

1. Make sure the suspension, steering, and brake systems are adjusted correctly.
2. Jack up the car, and spin the tire (***Figure 49–31***).
3. If the tread seems to wobble from side to side, replace the tire.

SERVICE

Use the same general procedure to remove and replace a tire as just described.

PROBLEM: Tires Not Balanced Correctly

As a car goes down the road, there is a shimmy or vibration in the front of the vehicle. It is felt in the steering wheel.

DIAGNOSIS

Often the tires get out of balance and must be balanced to get rid of shimmy or vibration. Tires must be balanced correctly in order to run smoothly and evenly at all speeds. A rim that is out-of-round may also produce a shimmy or vibration. When the shimmy is noticed in the steering wheel, the front tires are out of balance. When the rear wheels are out of balance, the shimmy is noticeable in the seat and floor of the vehicle.

Wheels get out of balance for two common reasons:

1. Tires wear unevenly, producing an unbalanced tire.
2. The car hits a bump, and the balance weights fall off.

There are two types of balancing: static and dynamic. **Static balance** is the equal distribution of weight around the axis of rotation. The tire is not spinning during the balancing process. Wheels that are unbalanced in a static condition cause a hopping or bouncing action. This is called wheel tramp. Severe wheel tramp can cause excessive wear and damage to the tire.

Dynamic balancing is the equal distribution of weight about the place of rotation. The tire is spinning during balancing. Tires are tested and checked in balancing machines as shown in ***Figure 49–32***. When a tire is spinning, it tends to move from side to side if it is out of balance. Wheels that are unbalanced during a dynamic or spinning condition cause the car to shimmy or vibrate when it is moving. Typically the vibration or shimmy will occur at a specific speed. Note that some tire-balancing models now are able to load the tire down during balancing to simulate actual road conditions.

A violent jolt, caused by hitting a curb or the like, may also cause a shimmy or vibration. In order to check for this condition, wheel (rim) runout must be measured. **Wheel runout** is a measure of the out-of-roundness of the rim. This can be measured by a dial indicator (***Figure 49–33***). Both inboard and outboard measurements should be taken on the rim. Radial and lateral (axial) runout on the rim can be taken with the tire on or off the rim. Compare the runout (radial and lateral) with the manufacturer's recommendations. Replace if necessary.

FIGURE 49–32 Tires are dynamically balanced by spinning them on this tire-balancing machine.

SERVICE

If the tires need to be balanced, follow the procedures listed by the manufacturer of the balancing machine. If the rim must be replaced because of runout, use the general procedure mentioned earlier.

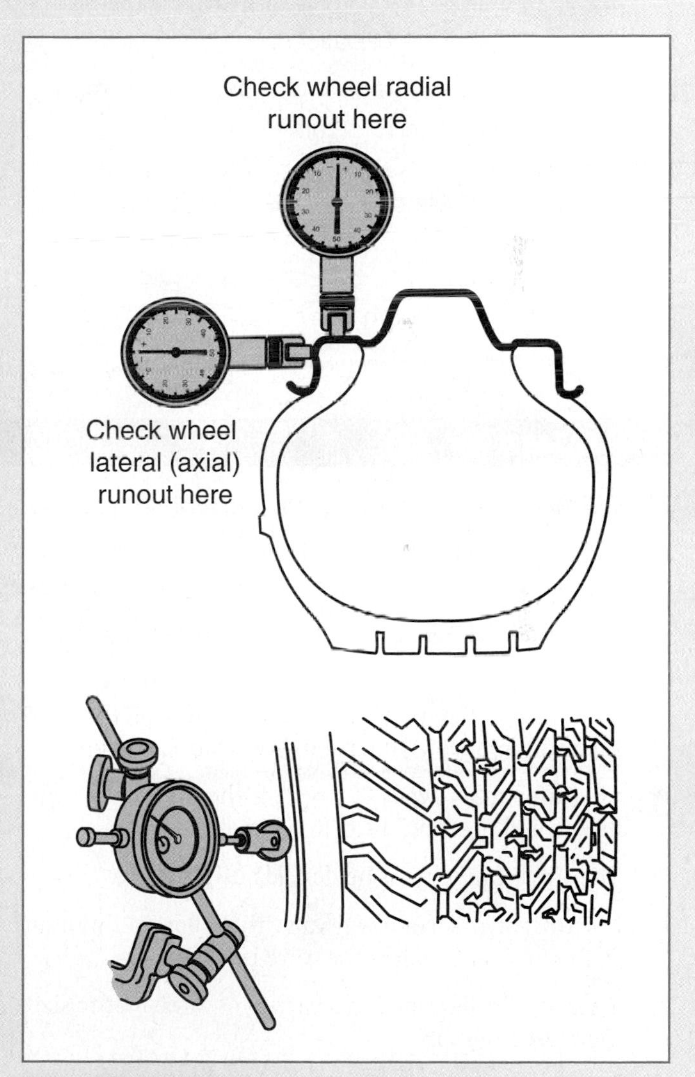

FIGURE 49–33 Wheel runout can be checked using a dial indicator. Both radial and lateral (axial) runout should be measured.

Service Manual Connection

There are many important specifications to keep in mind when working with tires and wheels. To identify the specifications for your vehicle, you will need to know the VIN (vehicle identification number) of the vehicle, the type and year of the vehicle, and the exact type of tire used. Although they may be titled differently, some of the more common tire and wheel specifications (not all) found in service manuals are listed below. Note that these specifications are typical examples. Each vehicle may have different specifications.

Common Specification	Typical Example
Front- and rear-wheel axial runout	
Steel wheels	0.039 inch
Aluminum wheels	0.045 inch
Inflation pressures	32 psi
Rim diameter	14 inches
Tire cord tensile strength	122,000 psi
Tire loading	1,100 lb
Tire sizing	P 185/70 R 14
Tire width	205 mm
Wheel nut torque	80 ft.-lb

In addition, service manuals provide specific directions for various service and testing procedures. Some of the more common procedures include:

- Balancing tires and wheels
- Inflation of tires
- Measuring wheel runout
- Selecting replacement tires
- Tire mounting and dismounting
- Tire rotation
- Wheel removal

SUMMARY

The following statements will help to summarize this chapter:

- Tires vary based on the tread design, the materials within the tire, and the number of belts, cords, plies, and beads.
- Both tube and tubeless tires are used in the automotive industry.
- Tires have different characteristics, such as tire traction, ride and handling, rolling resistance, and noise.
- The three main types of tires are the bias ply tire, the bias-belted tire, and the belted radial tire.
- The air within a tire is held inside using a tire valve.
- The tire valve is a one-way valve that allows air into the tire but not out, unless the valve is released.
- Tires can be identified in a variety of ways. Metric sizes are most common.
- Tires are identified by the tire type, section width, aspect ratio, construction type, rim diameter, speed rating, and load index.
- Many manufacturers use a tire placard to show maximum vehicle load, tire size, inflation pressure, and so on.
- Other designations for tires include the UTQG designation, the TPC specifications number, and load ranges.
- The design of the rim of a wheel is called the drop center design.
- Tires should always be inflated to their correct pressure for maximum life and comfort.
- When diagnosing tire problems, first make sure the steering, suspension, and brakes are in proper working order.
- Always replace tires when the tread bars are showing.
- Always balance tires according to the manufacturer's specifications.
- Always check for misaligned belts when diagnosing tire problems.

TERMS TO KNOW

Can you explain each of the following terms? Review the chapter until you can use each term correctly.

All-season tires
Aspect ratio
Bead
Bias
Carcass
Cords
Department of Transportation (DOT)
Drop center wheel
Dynamic balancing

Hydrophilic
Load range
Mag wheel
Plies
Radial
Rolling resistance
Safety rims
Squirm
Static balance
Tire placard
Traction
Tread
Tread bars
Vulcanized
Wheel runout

REVIEW QUESTIONS

Multiple Choice

1. Which part of the tire is used to produce the traction between the tire and road?
a. Cord
b. Carcass
c. Tread
d. Belts
e. Bead

2. Which part of the tire is used to hold the air inside the tire housing?
a. Cord
b. Carcass
c. Tread
d. Belts
e. Bead

3. As the rolling resistance of a tire increases, the fuel mileage of the vehicle:
a. Increases
b. Decreases
c. Remains the same
d. Increases then decreases as speed increases
e. None of the above

4. ________ are used within the tire plies or belts of a tire to increase the strength of the tire.
a. Rims
b. Cords
c. Tubes
d. Beads
e. Valves

5. Which type of tire uses plies that are crisscrossed over each other?
a. Radial
b. Radial belted
c. Bias
d. Bias radial
e. All of the above

6. Which type of tire has the lowest rolling resistance?
a. Bias
b. Bias ply
c. Bias-belted
d. Tube bias
e. Radial belted

7. The ________ tread tire has compounds used to attract water to the tread.
a. Tubeless
b. Hydrophilic
c. Belted
d. Corded
e. Tube

8. The compact spare tire should be used:
a. Whenever possible because it has low rolling resistance
b. Only on rough surfaces
c. Only in the city
d. Only when another tire is damaged and cannot be used
e. Only on wet surfaces

9. Which of the following is not molded into the side of a tire for sizing and identification on metric tire sizes?
a. Wheel size
b. Aspect ratio
c. Application
d. Tire speed
e. Construction type (radial, bias, etc.)

10. Most information about a car's tires can be found on the:
a. Steering wheel
b. Engine block
c. Tire placard
d. Undercarriage of the vehicle
e. Steering system

11. The maximum load a tire can safely withstand is called the:
a. Load range
b. Aspect ratio
c. Profile
d. Carcass
e. Construction type

12. Wheel rims are made with ________ to aid in removing and installing the tires on the wheels.
a. Cords
b. Vulcanized rubber
c. A dropped center
d. 16-inch rims
e. Flexible material

13. Which systems are important to check before diagnosing tire problems?
a. Steering systems
b. Brake systems
c. Suspension systems
d. All of the above
e. None of the above

14. Which of the following is critical for correct tire operation and wear?
a. Correct tire material
b. Correct tire inflation
c. Correct molding on the side of the tire
d. Correct casing material
e. None of the above

15. Which of the following can indicate excessive tread wear on a tire?
a. Tire inflation
b. Tread bars
c. Valve condition
d. Rim condition
e. Dropped wheel condition

16. Which of the following will affect tire inflation?
a. Hotter temperatures
b. Colder temperatures
c. Amount of pressure inside the tire
d. All of the above
e. None of the above

The following questions are similar in format to ASE (Automotive Service Excellence) test questions.

17. A vehicle is pulling to the right. *Technician A* says that the problem could be a defective radial tire. *Technician B* says that the problem could be in the dynamic balancing. Who is correct?
a. A only c. Both A and B
b. B only d. Neither A nor B

18. *Technician A* says that it is not important for the steering system to be correctly aligned before diagnosing tire problems. *Technician B* says that it is very important for the steering system to be correctly aligned before diagnosing tire problems. Who is correct?
a. A only c. Both A and B
b. B only d. Neither A nor B

19. *Technician A* says that the tread bars are used to show that the tire is in balance. *Technician B* says that the tread bars are used to shown out-of-round on the tire. Who is correct?
a. A only c. Both A and B
b. B only d. Neither A nor B

20. *Technician A* says that the tire pressure need not be changed for colder or warmer climates. *Technician B* says that the tire pressure should be changed for colder or warmer climates. Who is correct?
a. A only c. Both A and B
b. B only d. Neither A nor B

21. A vehicle has a shimmy in the front wheels. *Technician A* says the car needs to have the front wheel balanced. *Technician B* says that the rims should be checked for runout. Who is correct?
a. A only c. Both A and B
b. B only d. Neither A nor B

22. *Technician A* says that the tires need rotation after balancing. *Technician B* says that the tires need rotation after 100,000 miles. Who is correct?
a. A only c. Both A and B
b. B only d. Neither A nor B

23. *Technician A* says that the rims should be cleaned with a wire brush when the wheels are off. *Technician B* says that rubber compound should be applied to new tires before they are put on the rim. Who is correct?
a. A only c. Both A and B
b. B only d. Neither A nor B

24. A car has uneven wear on the tires. *Technician A* says to check the inflation pressure of the tires. *Technician B* says to check the power steering system. Who is correct?
a. A only c. Both A and B
b. B only d. Neither A nor B

Essay

25. List the parts of a tire and describe their purpose.

26. Describe at least three characteristics that tires are designed for.

27. Describe the difference between radial and bias tires.

28. Describe how tires are identified.

29. What is the purpose of a tire placard?

Short Answer

30. Tires are manufactured and designed for various characteristics including ________, ________, ________, ________, ________, and ________.

31. Which type of tire is designed for all of the preceding characteristics? ________

32. Three types of wheels made today are the ________, the ________, and the ________.

33. Uneven tire wear can be attributed to ________.

34. A shimmy in a moving vehicle is a sign of ________.

Applied Academics

TITLE: Radial Tire Design and Components

ACADEMIC SKILLS AREA: Language Arts

NATEF Automobile Technician Tasks:

The service technician supplies customers, associates, the parts supplier, and the supervisor with clarifying information about the steering and suspension problem.

CONCEPT:

Tires are manufactured for performance on the road. The figure that follows on page 934 shows the parts of a typical tire designed for maximum ride and performance.

1. Bead—provides the foundation of the tire; it is designed so the tire fits correctly on the rim, and it provides necessary strength to endure road stress from braking, acceleration, cornering, and so on.
2. Bead filler—stiffens the lower sidewall.
3. Flipper—strengthens the lower sidewall.
4. Liner—retains the air inside the tire.
5. Plies—provides the necessary carcass strength to contain the air volume in order to support the load.
6. Chafer—strengthens bead toe and reduces the possibility of "toe tear" during mounting and dismounting.
7. Sidewall—protects from weather and abrasion.
8. White sidewall (when applicable)—protects plies from weather; it is also aesthetically appealing.
9. Veneer (when applicable)—protects white sidewalls and allows for good adhesion between the belts and the sidewall.
10. Rim cushion—protects and aids in stiffening the lower sidewall.
11. Belts—carry the tire loads and maintain the radial tire profile.
12. Tread base—reduces heat buildup in the tread area and increases adhesion between the tread and top belt.
13. Tread—provides the traction designed to resist rapid wear.

APPLICATION:

Can you define the term *plies* to a customer who wants to buy tires? What is the purpose of the plies and how can it be explained in simple terms? How can you tell a customer the definition of belts? How do the belts differ from the plies?

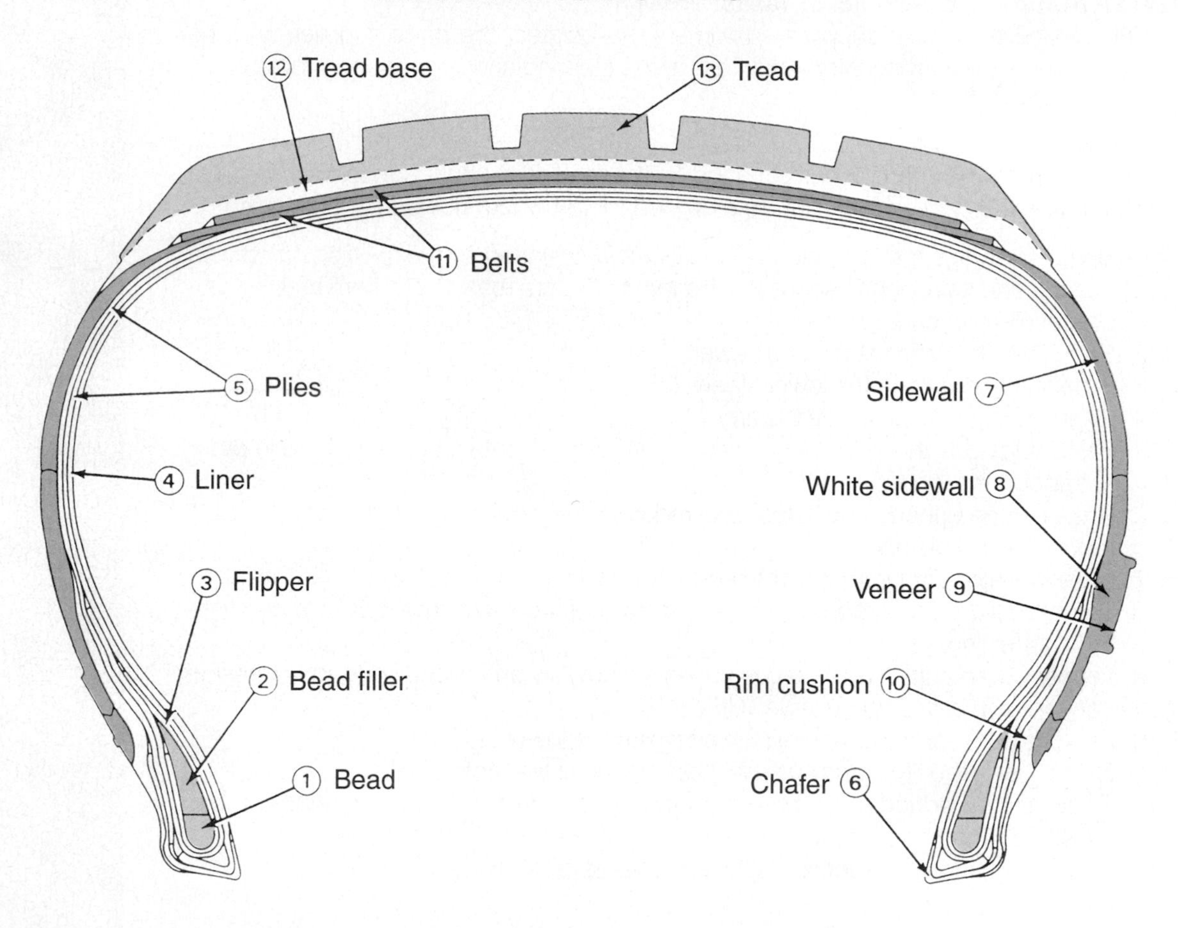
RADIAL TIRE COMPONENTS
12 Tread base
13 Tread
11 Belts
5 Plies
Sidewall 7
4 Liner
White sidewall 8
3 Flipper
Veneer 9
2 Bead filler
Rim cushion 10
1 Bead
Chafer 6

Vehicle Accessory Systems

SECTION

8

Scott Brown, 38

Owner

Connie and Dick's Service Center

Claremont, CA

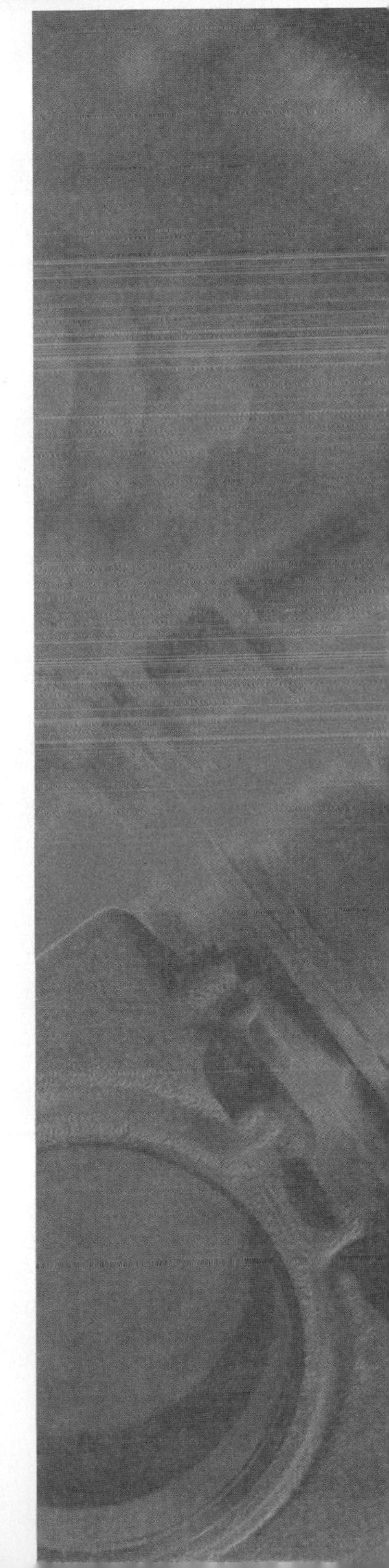

What's your educational background?

After high school, I went to technical trade school in Arizona for a year to get an automotive technology certificate. We attend training regularly to keep updated. I also participate in organizations like the International Automotive Technicians' Network, which allows me to collaborate with and learn from peers.

What career path did you take to become owner?

At first I was a line technician, doing heavy service like brakes, front ends, and engines. Then I noticed there was demand for tune up and fuel injection on vehicles. So I got trained in that, and I moved into a position at an aftermarket shop doing electronic diagnostics and managing that store and its technicians. I was in that position when I heard that the owners of Connie and Dick's were looking for someone to come to work for them and perhaps take over the business. In 1993, the owner wanted to retire, and I purchased his part of the business.

How did this path help you?

Having knowledge at the repair level benefits the shop as a whole because I can understand the whole process and the technicians' needs and values.

What's a typical day like for you?

I arrive around 6:30 A.M. I usually begin printing and reviewing the service history of customers scheduled to come in, so we're prepared with what's happened to the vehicles and what's due. Then I coordinate the distribution of repairs to technicians. I see if we need to order any parts. Throughout all that, customers bring in vehicles, and we'll analyze and estimate repairs. I may take customers home or for a ride in their vehicles to duplicate symptoms. I'll also pick up parts if necessary. And I'm constantly supervising my staff.

What's the best part of your job?

My favorite part is identifying and correcting difficult problems. I enjoy fixing things that are broken!

What advice do you have for a high school student interested in becoming an owner?

Really learn computers and how they function—electronics and mechanical systems are the future.

Air-Conditioning Systems

CHAPTER **50**

OBJECTIVES

After studying this chapter, you should be able to:

- State the principles of air conditioning.
- Examine the heat and refrigerant flow of an air-conditioning unit.
- Describe the purpose and operation of the common parts of an air-conditioning system.
- Examine the environmental effects of using R12 refrigerants.
- Describe the basic operation of computerized A/C systems.
- Identify various problems, their diagnosis, and service procedures used on A/C systems.

INTRODUCTION

Many automobiles today have air conditioning. Air conditioning removes the warm air inside the passenger compartment to the outside. This chapter is about air-conditioning principles and the parts necessary for air conditioning on vehicles.

50.1 AIR-CONDITIONING PRINCIPLES

There are several principles that help to explain air conditioning. These principles deal with heat flow, refrigerant, pressure, vaporization, and condensation.

Certification Connection

ASE Connection: The information in this chapter can help you prepare for the National Institute for Automotive Service Excellence (ASE) certification tests. The tests and content areas most closely related to this chapter are:

Test A7—Heating and Air Conditioning

- **Content Area**—A/C System Diagnosis and Repair
- **Content Area**—Refrigeration System Component Diagnosis and Repair
- **Content Area**—Refrigerant Recovery, Recycling, and Handling

NATEF Connection: Much of the information in this chapter is related to the NATEF tasks. The NATEF tasks and priority numbers most closely related to this chapter are:

1. Performance test A/C system; diagnose A/C system malfunctions using principles of refrigeration. P-1
2. Diagnose abnormal operating noises in the A/C system; determine necessary action. P-2
3. Inspect A/C compressor drive belts; determine necessary action. P-2
4. Remove and inspect A/C system mufflers, hoses, lines, fittings, O-rings, seals, and service valves; perform necessary action. P-2
5. Remove and reinstall receiver-drier or accumulator-drier; measure oil quantity; determine necessary action. P-1
6. Remove and install expansion valve or orifice (expansion) tube. P-2
7. Inspect evaporator housing water drain; perform necessary action. P-3

continued

Certification Connection *continued*

8. Diagnose malfunctions in the electrical controls of heating, ventilation, and A/C (HVAC) systems; determine necessary action. P-2
9. Inspect and test A/C-heater blower, motors, resistors, switches, relays, wiring, and protection devices; perform necessary action. P-1
10. Test and diagnose A/C compressor clutch control systems, determine necessary action. P-1
11. Diagnose malfunctions in the vacuum and mechanical components and controls of the heating, ventilation, and A/C (HVAC) system; determine necessary action. P-2
12. Check operation of automatic and semiautomatic heating, ventilation, and air-conditioning (HVAC) control systems; determine necessary action. P-3
13. Perform correct use and maintenance of refrigerant-handling equipment. P-1
14. Identify (by label application of use of a refrigerant identifier) and recover A/C system refrigerant. P-1
15. Recycle refrigerant. P-1
16. Label and store refrigerant. P-1
17. Evacuate and charge A/C system. P-1

PURPOSE OF AIR CONDITIONING

During normal summer driving conditions, a great amount of heat enters the passenger compartment. This heat comes from the engine and from the sun or outside air temperature. Air-conditioning (A/C) systems are designed to remove this excess heat so that passengers are comfortable. ***Figure 50–1*** shows heat flow in a vehicle. Heat is admitted into the passenger compartment from outside. The heat is removed by forcing it to flow through a heat exchanger in the passenger compartment. A fluid then absorbs the heat and transfers it to a second heat exchanger in the front of the vehicle.

Another reason for using air conditioning is that an air-conditioning system dehumidifies the air. Often in warm and damp driving conditions there is so much moisture inside the passenger compartment, the windows fog up. Although all cars have a defrost mode on their dashboard, when the air-conditioning system is turned on, the air inside the passenger compartment is dehumidified rapidly.

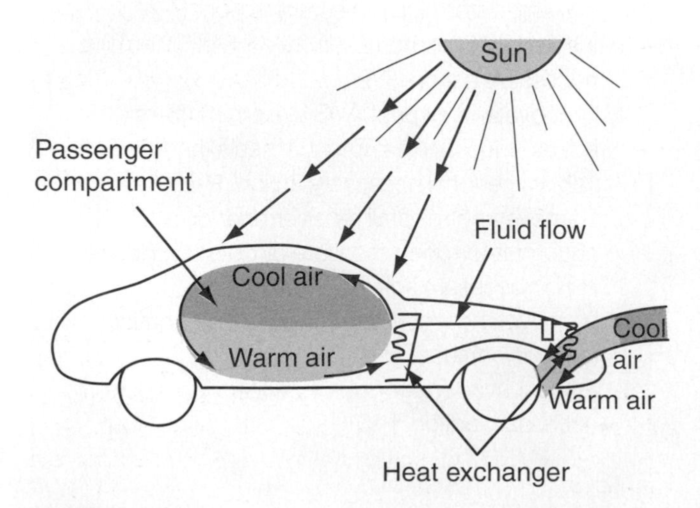

FIGURE 50–1 Heat inside the passenger compartment is sent through heat exchangers to the outside by use of a fluid refrigerant.

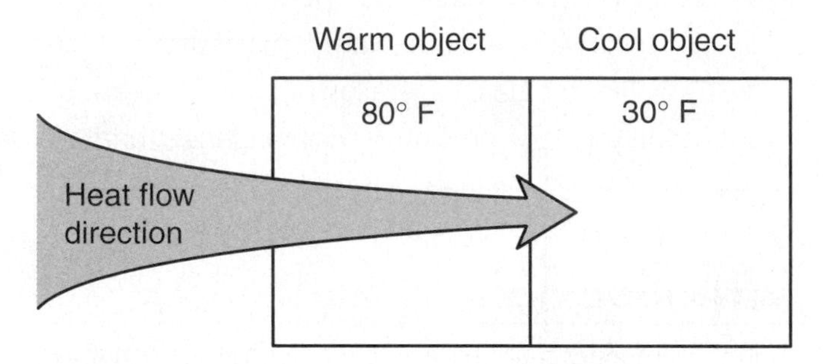

FIGURE 50–2 Heat always flows from a warmer to a colder object. Air-conditioning systems are designed to cause air to flow internally from warmer to colder areas.

HEAT FLOW

An air-conditioning system is designed to pump heat or Btus from one point to another. Heat can be measured in Btus (British thermal units), which was discussed in an earlier chapter. All materials or substances have heat in them down to −459 degrees Fahrenheit. At this temperature (absolute zero), there is no more heat. Also, heat always flows from a warmer to a colder object. For example, if one object were at 30 degrees Fahrenheit and another object were at 80 degrees Fahrenheit, heat would flow from the warmer object (80°F) to the colder object (30°F). The greater the temperature difference between the objects, the greater the amount of heat flowing, as shown in ***Figure 50–2***.

HEAT ABSORPTION

Objects exist in one of several forms. They can be in either a solid, a liquid, or a gas form. When objects change from one form to another, large amounts of heat can be absorbed. For example, an ice cube is in a solid form. When an ice cube melts, it absorbs a great amount of heat. In fact, all solids soak up huge amounts of heat without getting warmer when they change from a solid to a liquid.

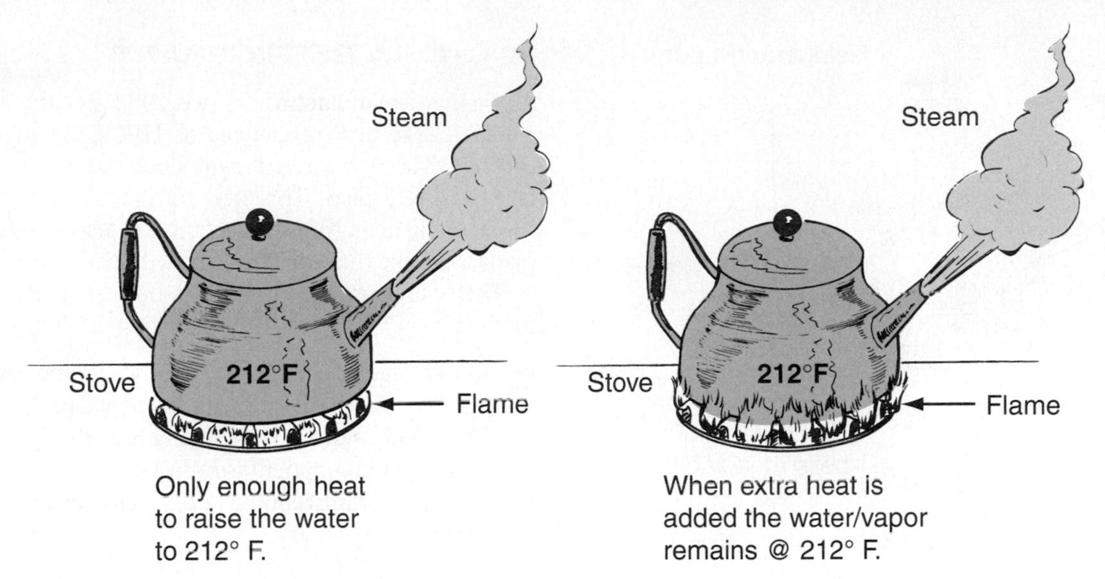

FIGURE 50-3 Heat is absorbed in large quantities when a liquid changes to a vapor as in the boiling kettle of water. The water in the tea kettle will stay at 212 degrees Fahrenheit even if more heat is applied below the kettle.

The same thing happens when a liquid changes to a vapor. Large amounts of heat are absorbed. For example, in ***Figure 50–3*** a tea kettle with water inside is warmed. As the burner heats up the water, the temperature of the water starts to rise. It continues to rise until it has reached 212 degrees Fahrenheit. At this point, the temperature will stay at 212 degrees Fahrenheit even when additional heat is applied. The water is changing to a vapor and is soaking or absorbing large quantities of heat. Although this heat does not appear on the thermometer, it is there. It is called **latent heat**, or hidden heat.

CONDENSATION

Condensation is the process of changing a vapor back to a liquid. Condensation is usually done by cooling the substance down, below its boiling point. All substances condense at the same point at which they boil. When a vapor is condensed, the heat removed from it is exactly equal to the amount of heat necessary to make it a vapor in the first place.

PRESSURE AND BOILING POINTS

Pressure also plays an important role in air conditioning. Pressure on a substance such as a liquid changes it boiling point. The greater the pressure on a liquid, the higher its boiling point. If pressure is placed on a vapor, it will condense at a higher temperature. In addition, as the pressure on a substance is reduced, the boiling point can also be reduced. For example, the boiling point of water is 212 degrees Fahrenheit. The boiling point can be increased by increasing the pressure on the fluid. It can also be decreased by reducing the pressure or placing the fluid in a vacuum.

Pressure on a fluid or vapor will also concentrate any heat in the substance. As the pressure on a fluid or vapor is increased, the temperature of the fluid or vapor tends to increase. For example, refer to ***Figure 50–4***. If pressure is exerted on a certain volume of vapor, the temperature of the vapor will be at a certain temperature, say 40 degrees Fahrenheit. If the volume of this vapor is reduced by increasing the pressure, the temperature of the vapor will be increased to 70 degrees Fahrenheit, without adding extra heat. This principle is used in air-conditioning systems.

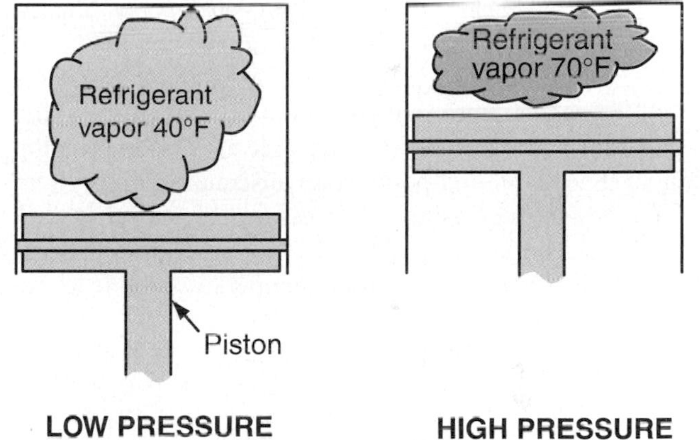

FIGURE 50-4 If a refrigerant vapor is compressed with pressure, its temperature will increase from 40 degrees Fahrenheit to 70 degrees Fahrenheit.

REFRIGERANTS

A **refrigerant** (abbreviated as "R") is used to transfer heat from inside the vehicle to the outside air. Today there are two major types of refrigerants used in automobiles. One is called R12 and the other is called R134a. These refrigerants have a very low boiling point (between –16 and –22 degrees Fahrenheit). This means that the refrigerant changes from a liquid to a vapor at these temperatures. If one were to place a flask of refrigerant inside a refrigerator, it would boil

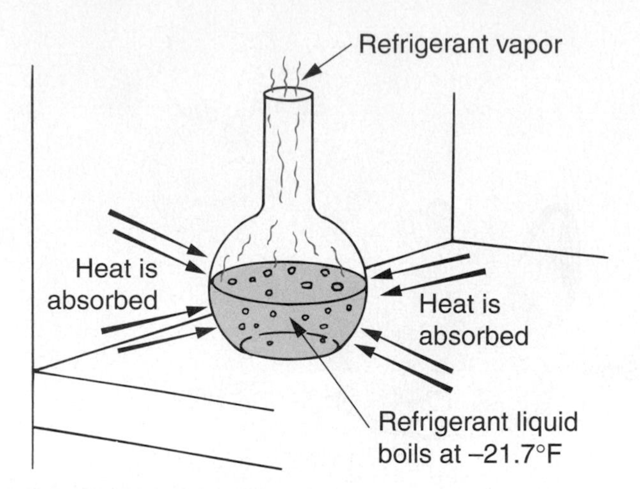

FIGURE 50–5 If refrigerant were placed in a flask inside a refrigerator, it would boil at around –20 degrees Fahrenheit, absorbing all heat surrounding it.

and draw heat away from everything around it (***Figure 50–5***). If the refrigerant were then pumped outside (along with the heat it absorbed), the inside of the refrigerator would be cooler. Note that the boiling temperature of the refrigerant is too low to operate inside an air-conditioning system. So the boiling point in an air-conditioning system is changed by increasing the pressure on the refrigerant until it has a boiling point of from 30 to 60 degrees Fahrenheit. This is the most appropriate temperature range for refrigerant to operate at.

Another important quality of refrigerant is that it dissolves in oil. This is necessary because the refrigerant circulates through the system with oil. Special oil is added to the air-conditioning refrigerant to lubricate the compressor.

REFRIGERANT AND THE ENVIRONMENT

Over the past few years research has shown that Freon R12 is one of the major contributors to reducing the **ozone layer** above the Earth. The ozone layer is a transparent shield of gases surrounding the Earth. It protects the Earth from harmful radiation from the sun. In addition, it helps keep the Earth at an average temperature of 15 degrees Celsius. When the ozone layer is reduced, increases of skin cancer result. Also, reducing the ozone layer causes global warming, known as the greenhouse effect.

The major greenhouse gases are carbon dioxide (CO_2), methane (CH_4), nitrous oxide (NO_x), and chlorofluorocarbons—**CFCs**. CFCs come from foams, aerosols, refrigerants, and solvents. Freon R12, a refrigerant, is classified as a chlorofluorocarbon, or CFC. In fact, Freon R12 is often referred to in the literature as CFC-12. Freon R12 is a major contributor to the greenhouse effect.

In 1987 many countries agreed to phase out, over a 10-year period, the use of products that use CFCs. Thus, new types of refrigerants have replaced R12.

HFC-134A REFRIGERANT

All vehicles manufactured since 1992 use the refrigerant called **R134a** also referenced as HFC-134a (hydrofluorocarbon 134a). This refrigerant does not harm the ozone layer like R12 does. This new refrigerant is also good for transferring heat. New types of oils are also being tested and made available that can be mixed with the refrigerant.

There are several drawbacks to using R134a. Some include:

- Certain design changes in air-conditioning systems have been necessary to work with the new type of refrigerant.
- The R134a system is generally less efficient than R12, and larger components are required.
- The refrigerant requires a different refrigerant oil and desiccant.

DIFFERENCES IN SYSTEM COMPONENTS

Today, service technicians work on vehicles that use either R12 or R134a as a refrigerant depending on the vehicles' year. It is important to read the manufacturer's manuals, specifications, and procedures to be aware of some of the differences in how the two refrigerants are handled, stored, and used.

R134a is not interchangeable with R12. This is true because the temperature and pressure characteristics of the two are different. ***Figure 50–6*** compares temperatures and pressures for the two types of refrigerants. For example, say that the temperature of R12 is 28 degrees Fahrenheit. At this temperature, its pressure is 28 pounds per square inch. However, at 28 degrees Fahrenheit, the pressure of R134a is 24 pounds per square inch. Because of the pressure and temperature differences, separate hoses, gauges, charging, and recycling equipment are necessary to service air-conditioned vehicles. This is done to avoid cross-contamination and damage to either system. Since R134a has different properties than R12, new vehicle air-conditioning systems are now designed to accommodate R134a. The pressure characteristics of R134a differ from those of R12. The boiling point of R134a is –16°F (–27°C) at sea level. This means that the R134a boils about 6°F higher than R12. This changes the temperature-pressure relationship curve. Because of this pressure difference, R134a system gauge pressure readings will differ slightly as compared to the previous R12 systems. Notable precautions and comparisons concerning R134a and R12 refrigerants include the following:

- Refrigerants R12 and R134a are not compatible.
- If R12 is used in a system that should use R134a, the R12 will not carry the lubricant throughout the system and the compressor will be damaged. The slightest trace of R12 in an R134a system will lead to compressor failure. When these two refrigerants are mixed, a sludge is formed that can damage system components.
- On some vehicles, the sight glass has been eliminated, therefore changing diagnostic procedures.

R134a Temperature - Pressure Chart

Evaporator range		Condenser range	
Temperature °C (°F)	Pressure kPa (PSI)	Temperature °C (°F)	Pressure kPa (PSI)
-9 (16)	106 (15)	38 (100)	857 (124)
-7 (20)	124 (18)	40 (104)	917 (133)
-4 (24)	144 (21)	42 (108)	980 (142)
-2 (28)	166 (24)	44 (112)	1045 (147)
0 (32)	188 (27)	47 (116)	1114 (162)
2 (36)	212 (31)	49 (120)	1185 (172)
4 (40)	238 (35)	51 (124)	1260 (183)
10 (50)	310 (45)	53 (128)	1337 (200)
16 (60)	392 (57)	57 (135)	1481 (215)
21 (70)	487 (71)	63 (145)	1704 (247)
27 (80)	609 (88)	68 (155)	1948 (283)
32 (90)	718 (104)	71 (160)	2079 (301)

R12 Temperature - Pressure Chart

Evaporator range		Condenser range	
Temperature °C (°F)	Pressure kPa (PSI)	Temperature °C (°F)	Pressure kPa (PSI)
-9 (16)	127 (18)	38 (100)	808 (117)
-7 (20)	145 (21)	40 (104)	859 (125)
-4 (24)	165 (24)	42 (108)	917 (133)
-2 (28)	190 (28)	44 (112)	969 (140)
0 (32)	207 (30)	47 (116)	1027 (149)
2 (36)	230 (33)	49 (120)	1087 (158)
4 (40)	255 (37)	51 (124)	1150 (167)
10 (50)	322 (47)	53 (128)	1215 (176)
16 (60)	398 (58)	57 (135)	1334 (194)
21 (70)	484 (70)	63 (145)	1519 (220)
27 (80)	580 (84)	68 (155)	1721 (250)
32 (90)	688 (100)	74 (165)	1940 (281)

FIGURE 50–6 These charts show the temperature and pressure differences between R12 and R134a refrigerants.

- The lubricant used with the R134a refrigerant is called polyalkylene glycol (PAG). It is a synthetic oil and is not compatible with the R12 refrigerant oil.
- Each refrigerant requires its own service equipment and recovery and recycling stations. Even though R134a does not contain chlorine, it must also be reclaimed.
- To avoid mixing R12 and R134a, SAE standards were established for the design of high- and low-side service ports used on R134a systems. SAE standard manifold gauges, hoses, and couplers are used to connect the gauge set to the refrigerant bottle. The threads have been changed and the service ports have been changed to require different service port couplers.
- These differences, along with the fact that R12 and R134a are not compatible, make it imperative that service technicians recognize which refrigerant is used in each A/C system that is being serviced.

FEDERAL CLEAN AIR ACT AMENDMENTS, SECTION 609

Because of the change from R12 to R134a, there have been various Clean Air Act amendments. The act states:

> Effective January 1, 1992, no person repairing or servicing motor vehicles for consideration may perform any service on a motor vehicle air conditioner involving the refrigerant for such air conditioner without properly using approved refrigerant recycling equipment and no such person may perform such service unless such a person has been properly trained and certified. The requirements of the previous sentence shall not apply until January 1, 1993, in the case of a person repairing or servicing motor vehicles for consideration at an entity which performed service on fewer than 100 motor vehicle air conditioners during calendar year 1990 and if such person so certifies . . . to the Administrator by January 1, 1992.

There were several parts that also identified how a person is to be certified. The important consideration is that now any service technician working with refrigerants must be trained and certified. Also, after November 15, 1995, R134a cannot be discharged into the air, and certification is required to purchase the old R12.

RECYCLING THE REFRIGERANT

All refrigerants must be prevented from escaping into the atmosphere; they can be recovered using various recycling tools. One such system is called the air-conditioning refrigerant recovery and recycling system—**ACR3**. During the recovery phase, the ACR3 system pulls refrigerant from the automotive air-conditioning system. During the recycling phase, the ACR3 system cleans and dries the refrigerant. After it has been recycled, the cleaned refrigerant can be stored for use at a later date. ***Figure 50–7*** shows a typical storage and recovery system. Note that the refillable storage

FIGURE 50–7 This machine is an air-conditioning refrigerant recovery and recycling (ACR3) system. It is used to recover and recycle used R12 and R134a. (Courtesy of SPX Corporation)

tank is located on the back. Follow the manufacturer's procedure when using similar machines to recover and recycle the refrigerant.

50.2 REFRIGERATION CYCLE OPERATION

BASIC HEAT TRANSFER

Heat exchangers are used to transfer heat in an air-conditioning unit. ***Figure 50–8*** shows how heat exchangers are used to pump heat from inside the car to the outside. When cool refrigerant is sent through the evaporator, which is located inside the passenger compartment, it cools the air passed over the evaporator fins. The refrigerant is boiling and vaporizes, absorbing the hot air from the car. The refrigerant vapor is then pumped outside the passenger compartment to a second heat exchanger. This one is called the condenser. Here, cooler outside air is forced through the fins on the condenser. The air is cooler than the refrigerant. Heat is now transferred from the refrigerant to the passing air, causing it to heat up. The heat in the car has essentially been pumped, via the refrigerant, from the inside to the outside of the car.

PURPOSE OF COMPRESSOR

All automotive air-conditioning systems have several major parts. These include the compressor, condenser, expansion tube or expansion valve, and evaporator. ***Figure 50–9*** shows the flow inside an air-conditioning system using all components.

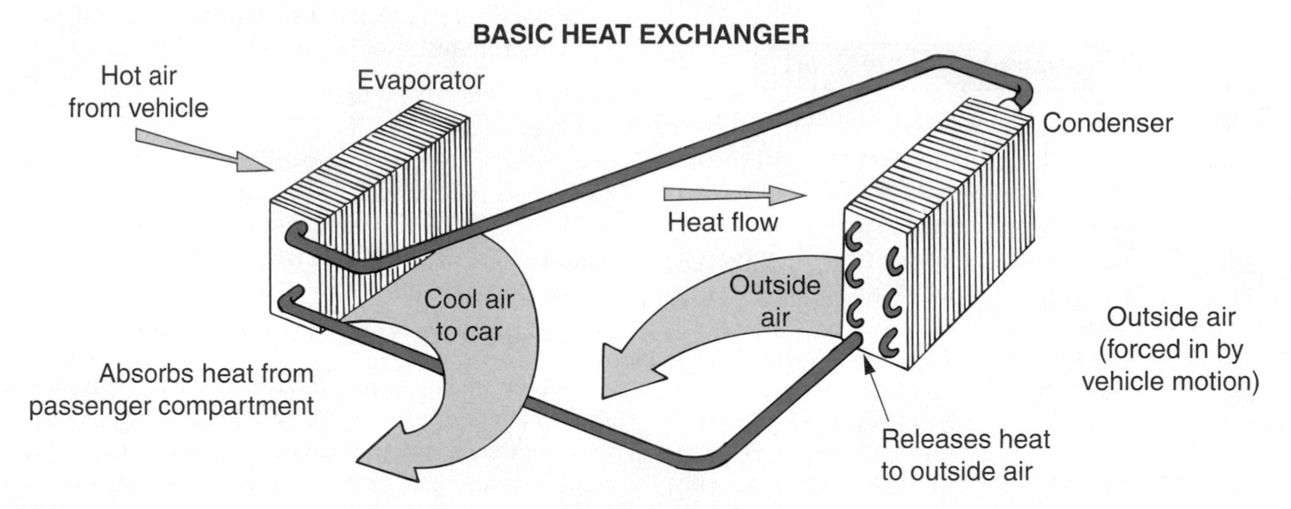

FIGURE 50–8 Heat transfer in an air-conditioning system is done by heat exchangers. The evaporator absorbs heat into a liquid refrigerant. The condenser gives off heat from the refrigerant.

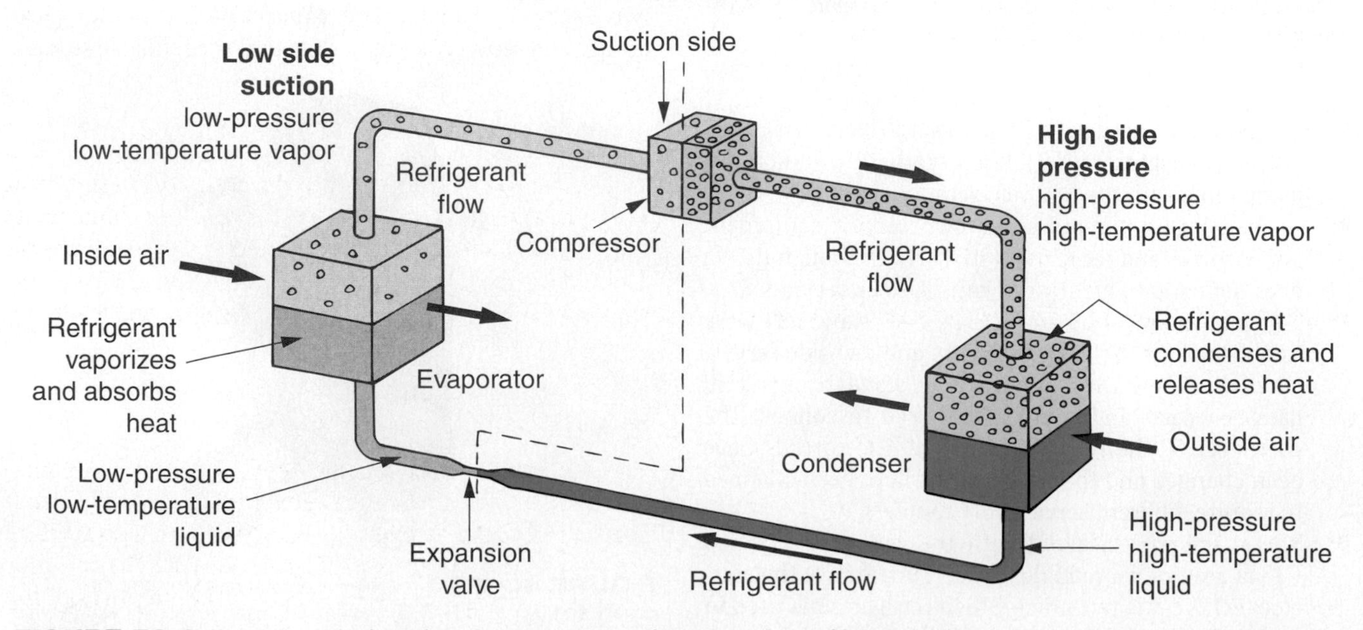

FIGURE 50–9 This schematic shows how the temperature and pressure of the refrigerant change throughout the air-conditioning cycle.

The compressor is the heart of the air-conditioning system. It has several purposes. First, the compressor moves the refrigerant. Second, the compressor compresses the refrigerant to change its boiling point. As the refrigerant flows through other components, it undergoes various changes in pressure and temperature. These changes are required for proper heat transfer to take place.

The compressor is powered by the vehicle engine through a clutch turned by a belt. It has an intake side (suction) and a discharge side (exhaust). When the refrigerant enters the compressor, it is a gas at low temperature and low pressure. When the refrigerant leaves the compressor, it is still a gas but it is at a very high pressure. Compressing the gas has also increased its temperature well above that of the outside air. The temperature of the gas must be higher than that of the outside air so that heat will flow in the correct direction, from inside the refrigerant to the outside air.

PURPOSE OF CONDENSER

The **condenser** is basically a heat exchanger. It consists of a series of tubes and fins. As the hot refrigerant gas flows through the tubes, it warms the fins. The fins provide enough surface area to transfer heat effectively. Heat transfer now takes place rapidly between the condenser and the outside air. During this rapid heat loss, the refrigerant is reduced in temperature below its boiling point. (Remember, the boiling point is increased because the vapor has been pressurized.) The refrigerant now condenses back into a liquid. As the refrigerant exits the condenser, it is a liquid at high pressure and high temperature.

PURPOSE OF EXPANSION VALVE

Next, the liquid refrigerant flows through an **expansion valve**. This valve is a restriction in the refrigerant flow. As the hot liquid refrigerant enters the valve, the restriction causes the pressure to build up behind it. As the liquid passes through the valve, there is a large pressure drop. This pressure drop changes the boiling point of the refrigerant. Now the refrigerant is in a condition in which it is ready to boil or evaporate just before it enters the evaporator.

PURPOSE OF EVAPORATOR

The **evaporator** is another version of a heat exchanger. In the evaporator, the heat is transferred from the passenger compartment air to the liquid refrigerant. Since the refrigerant entering the evaporator is at a lower temperature, it makes the evaporator fins "cold." Thus, warm car air circulating around the cold fins releases heat into the evaporator. As the heat is applied to the refrigerant, the refrigerant boils and changes to a vapor again. As it boils (just like the tea kettle earlier), it is able to absorb huge quantities of heat. The result is a heat loss in the inside passenger air. When the refrigerant exits the evaporator, it is once again a gas, at low temperature and low pressure. The system then continues to recycle back to the compressor, repeating the cycle.

Car Clinic:
EVAPORATOR

CUSTOMER CONCERN:
Musty Smell in Driver Compartment

A customer complains that the interior of the car smells musty when the air conditioning is turned on. Also, the windows fog up. What is the problem?

SOLUTION: In humid weather, a great deal of moisture forms in the air-conditioning evaporator. This moisture normally drains from the evaporator housing through tubes to the outside. If the water cannot drain, mold and mildew will form, causing an unpleasant odor. Make sure the evaporator housing drain is clear and water drips freely from the housing when the air conditioning is operating. Also make sure all hoses and tubes leading to the outside are clear of debris. The odor can be controlled by spraying a disinfectant or deodorant (available from most automotive parts stores) into the air intake in front of the windshield on most cars while the air conditioning is on.

HIGH SIDE AND LOW SIDE

Every air conditioner has two sides. There is a high side and a low side. The high side is that portion of the system where the refrigerant is at high pressure. The low side is that portion of the system where the refrigerant is at low pressure. A dotted line is used in ***Figure 50–9*** to show the dividing line between the high and low pressures. The high side includes the system between the discharge end of the compressor and the restriction of the expansion valve. It includes the condenser. The low side includes the portion from the restriction or expansion valve, through the evaporator, to the intake end of the compressor.

50.3 SYSTEM VARIATIONS

All air-conditioning systems use the basic components that were just explained. In addition, several other components are integrated into every system to keep it working properly. These components and their location make up the basic differences between one air-conditioning system and another.

ACCUMULATOR ORIFICE TUBE SYSTEM

The accumulator orifice tube system uses an additional **accumulator** and a tube for the expansion valve. The accumulator is located between the output of the evaporator and the input of the compressor (***Figure 50–10***). The accumulator performs three vital functions in the air-conditioning system:

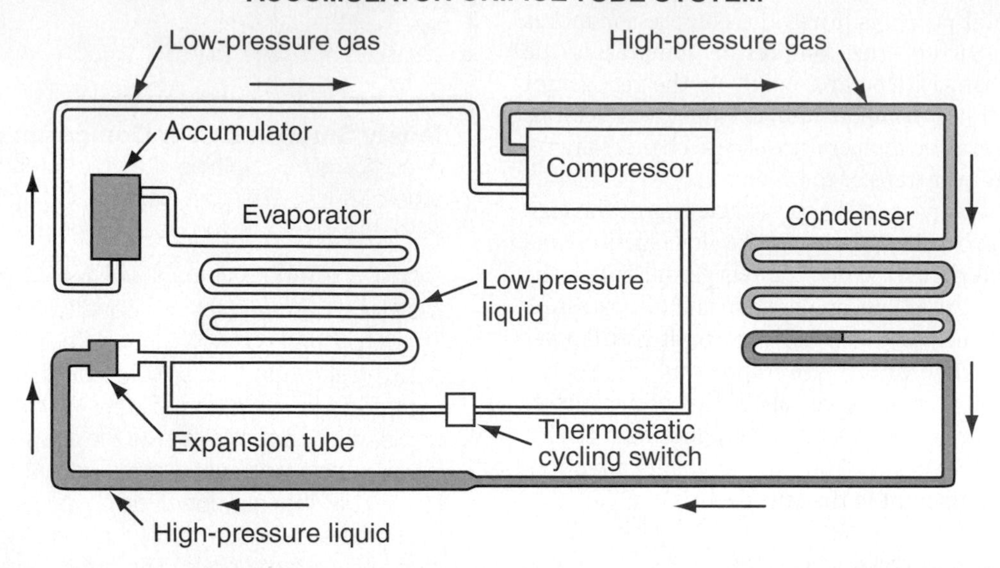

FIGURE 50–10 An accumulator is used on some air-conditioning systems to collect excess refrigerant and filter the refrigerant.

1. It collects excess refrigerant liquid, permitting only refrigerant gas to enter the compressor.
2. It contains a filtering element with a built-in **desiccant** (moisture-absorbing material) that helps to reduce moisture from the refrigerant.
3. It contains a filter screen that traps any foreign matter before it can reach the compressor.

In addition, this system uses an orifice tube as the expansion valve. It creates the pressure drop needed just before the refrigerant enters the evaporator. ***Figure 50–11A*** shows the accumulator on an air-conditioning system. It is most often located immediately after the evaporator in the vehicle passenger compartment. ***Figure 50–11B*** is a cutaway of a typical accumulator showing the desiccant and input-output lines.

CYCLING CLUTCH SYSTEM

Another type of air-conditioning system uses a receiver-filter-dryer along with a thermostatically controlled clutch. The receiver-filter-dryer is much the same as the accumulator in the preceding system. It is located on the high side of systems (***Figure 50–12***). In addition to performing filtra-

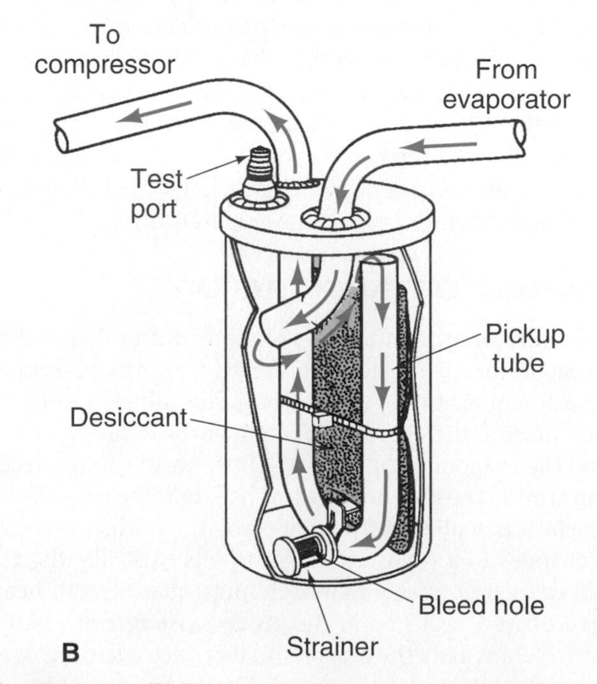

FIGURE 50–11 (A) The accumulator is located immediately after the evaporator. (B) This cutaway view shows the flow of refrigerant through the accumulator.

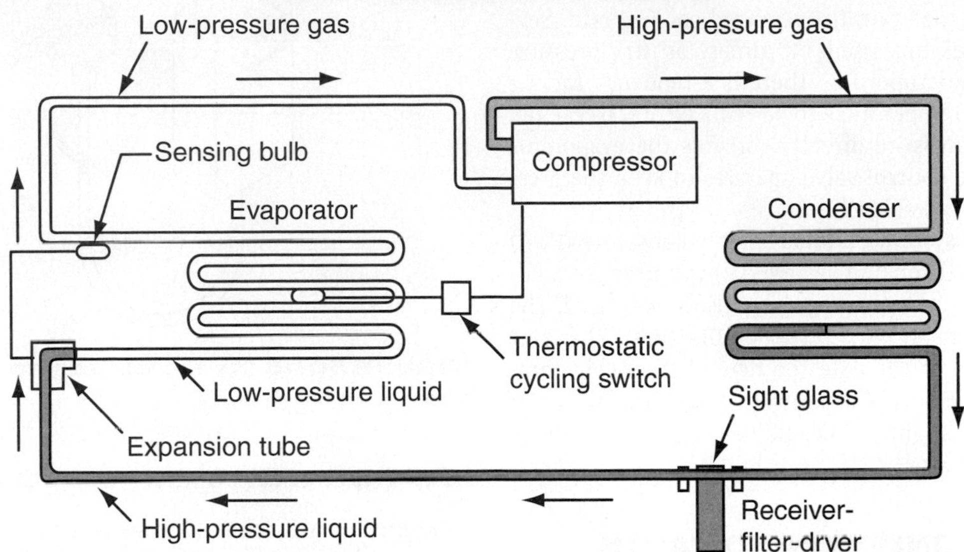

FIGURE 50-12 Certain air-conditioning systems use a receiver-filter-dryer. This device is placed on the high side of the system and is used to filter, remove moisture from, and act as a reservoir for the refrigerant. A cycling clutch is also used on this system to turn the compressor off and on.

tion and moisture removal, it acts as a reservoir, storing any excess refrigerant in the system.

This system also uses an expansion valve rather than an expansion tube to create the required pressure drop. In this system, the expansion valve controls the flow of refrigerant entering the evaporator. This metering function is achieved by a sensing bulb, which measures the temperature of the refrigerant leaving the evaporator. In turn, on the basis of this temperature, it changes the expansion valve opening to increase or decrease the flow rate.

A cycling clutch system also uses a compressor clutch to transmit power from the vehicle engine to the compressor. It is turned off and on by a magnetic coil. The magnetic coil is operated by another electrical circuit. The cycling clutch system engages or disengages the compressor according to the demands of the system. The cycle is controlled by a thermostatic switch that senses the temperature at the evaporator. A pressure-sensing switch is used on some models, rather than a temperature-sensing switch.

CONTROL VALVE SYSTEM

Certain older systems use valves to control the operation of the air-conditioning unit. ***Figure 50–13*** shows a typical control valve system. This system uses a noncycling clutch. The compressor operates continuously. Control devices are

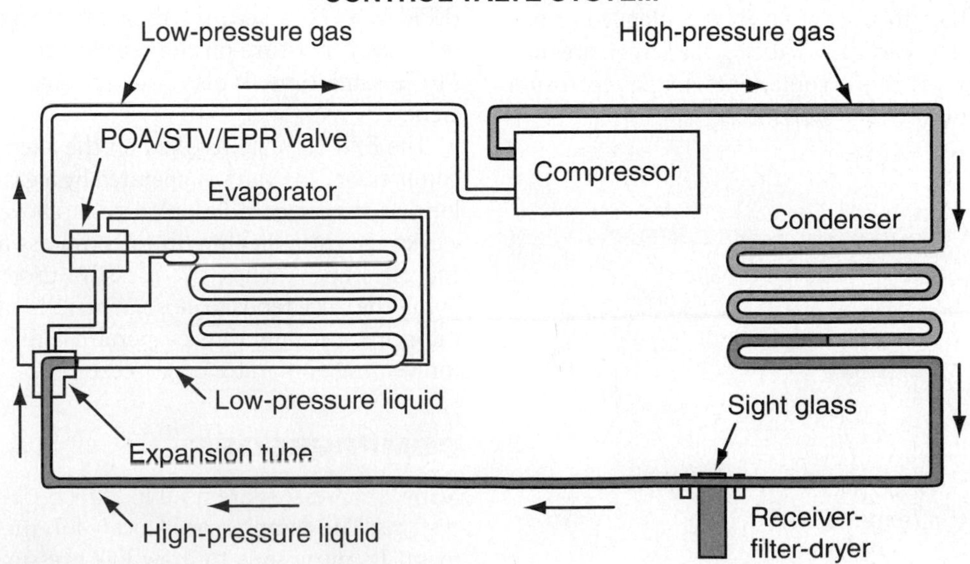

FIGURE 50-13 Older air-conditioning systems used several control valves to operate the system. Either a pilot-operated absolute (POA) valve or a suction throttling valve (STV) was used in this system.

used to adjust the temperature, pressure, and flow rate of the refrigerant to maintain the required cooling rate. Several control valves are used. At times, as the pressure changes inside the evaporator, there is a tendency for the evaporator inside temperature to be reduced and freeze up. (The evaporator pressure directly controls the evaporator temperature.) The control valve operates to keep the pressure at a predetermined level.

On some older systems, a pilot-operated absolute (POA) valve or a suction throttling valve (STV) was used. In addition, an evaporator pressure regulator (EPR) was used. The valve used depended on the vehicle manufacturer. The valve opened and closed to regulate the flow of the refrigerant. Controlling the pressure also controlled the temperature of the refrigerant, to slightly above 32 degrees Fahrenheit, as it exited the evaporator.

50.4 AIR-CONDITIONING PARTS

Once the operation cycle is understood, each component can be analyzed as to its design and function. This section analyzes the internal operation of several air-conditioning components.

ORIFICE TUBE

The orifice tube is one of the dividing points between the high- and low-pressure sides of the refrigerant system. The tube, with its mesh filter screen and orifice, provides a restriction to the high-pressure refrigerant in the liquid line. ***Figure 50–14*** shows an orifice tube. In addition to filtering, the orifice tube meters the flow of refrigerant into the evaporator as a low-pressure liquid. An orifice tube is normally located in the refrigerant line between the condenser outlet and the evaporator inlet.

Orifice tube failure is usually indicated by low discharge pressure and insufficient evaporator cooling. Orifice tube failure is usually due to restriction from a clogged orifice tube inlet screen. The particles that clog the screen are normally compressor particles, contamination, or corrosion particles that are loose in the system.

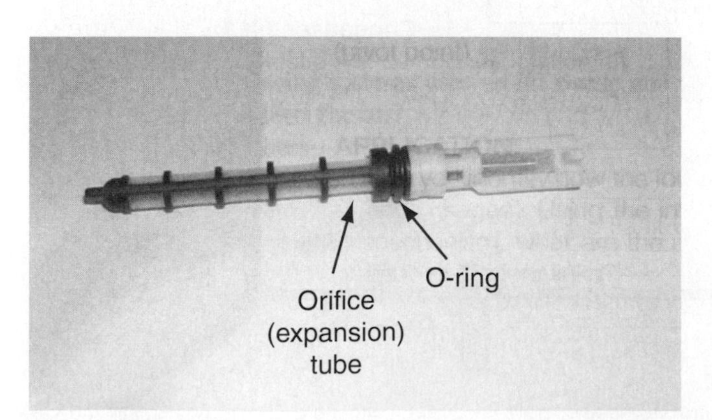

FIGURE 50–14 This orifice tube is used to meter the refrigerant.

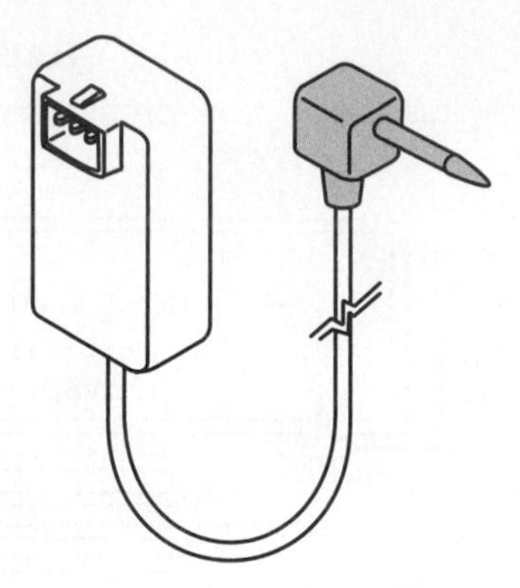

FIGURE 50–15 The evaporator thermistor is used to sense the evaporator temperature.

EVAPORATOR THERMISTOR

Thermistors are also used in some air-conditioning systems to control the evaporator temperatures. The thermistor is a sensor that converts the evaporator temperature into a resistance value. ***Figure 50–15*** shows a typical evaporator thermistor. The sensor is electrically connected to the system that controls the compressor. When the evaporator temperature drops below approximately 1 degree Centigrade (34 degrees Fahrenheit), the compressor shuts off, preventing formation of frost and ice on the evaporator fins.

PARTS LOCATOR

*Refer to photo #113 in the front of the textbook to see what an **evaporator sensor** looks like.*

EVAPORATOR PRESSURE REGULATOR (EPR) VALVE

The main function of the evaporator pressure regulator (EPR) valve is to keep the evaporator pressure low enough to prevent moisture on the evaporator core from freezing. At the same time, it also provides maximum cooling efficiency.

The EPR valve is installed on the suction passage of the compressor. The valve is operated by a gas-filled bellows. As long as the evaporator pressure is above a certain psi, it works against a diaphragm to compress a spring and hold the valve open. When the pressure drops below a certain point, the valve tends to close. When the valve closes, evaporator pressure, and thus evaporator temperature, increases (preventing evaporator core freeze-up).

COMPRESSORS

Many compressors are used in automotive air-conditioning systems. The compressor is located in the engine compartment. Its purpose is to draw low-pressure vapor from the evaporator and compress this vapor into a high-temperature, high-pressure vapor. It is also used to circulate the refrigerant throughout the system. The compressor is belt-driven by the engine crankshaft.

There are many types of compressors. One type of compressor works on a **swash plate** (or axial plate) pump arrangement. As shown in ***Figure 50–16***, the axial plate is attached to the center shaft at an angle. As the center shaft turns, the axial plate also wobbles as it turns. This action causes small double-ended pistons attached to the axial plate to move back and forth. The pistons are attached to the axial plate by large ball bearings. A suction and a pressure are created on the ends of the pistons. A valve is used to control the direction of the suction and pressure.

Figure 50–17 shows another schematic of the pistons and swash plate. Notice as the swash plate is turned, the pistons are forced to move in and out, creating a suction and pressure sequence.

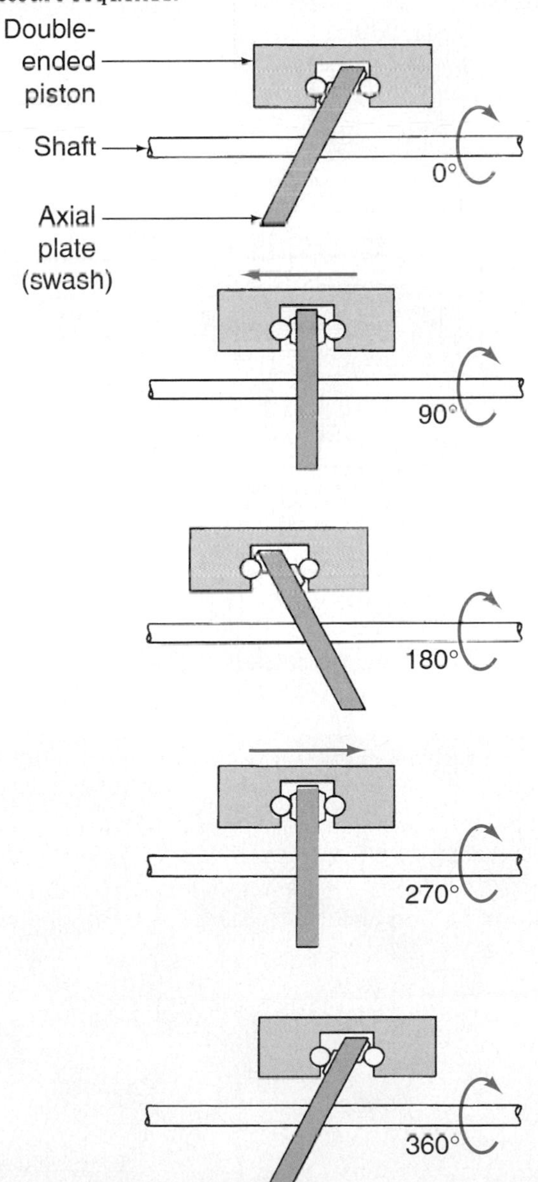

FIGURE 50–16 The compressor in an air-conditioning system uses a swash plate (axial plate) pump arrangement to move double-ended pistons back and forth. The piston movement produces suction and pressure to control the flow of refrigerant.

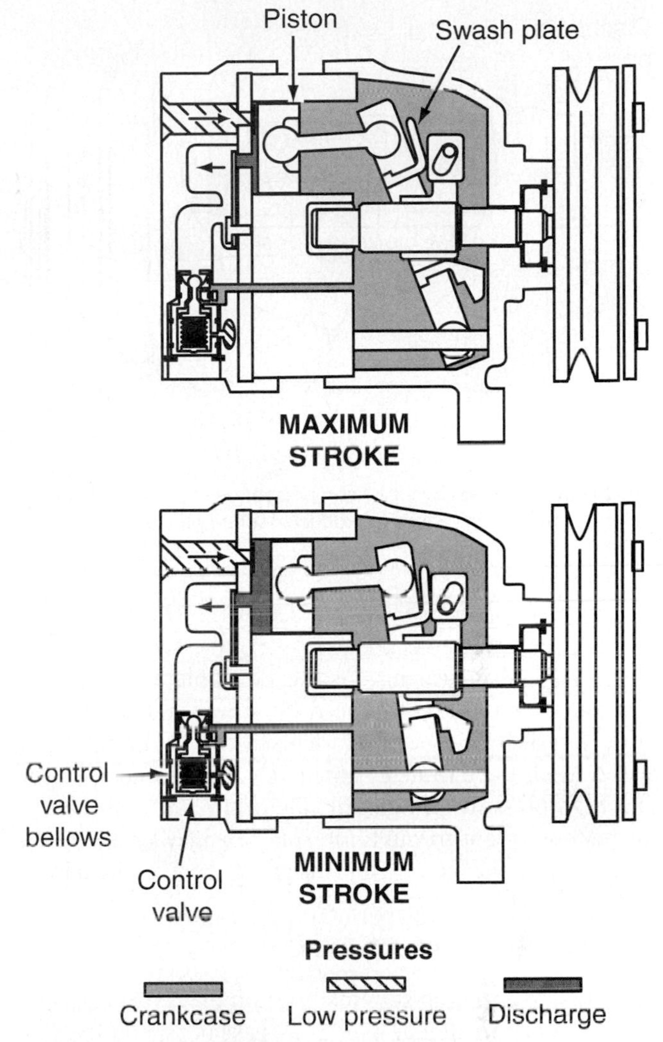

FIGURE 50–17 This compressor uses a swash plate and small pistons to produce the suction and pressure needed for air-conditioning systems.

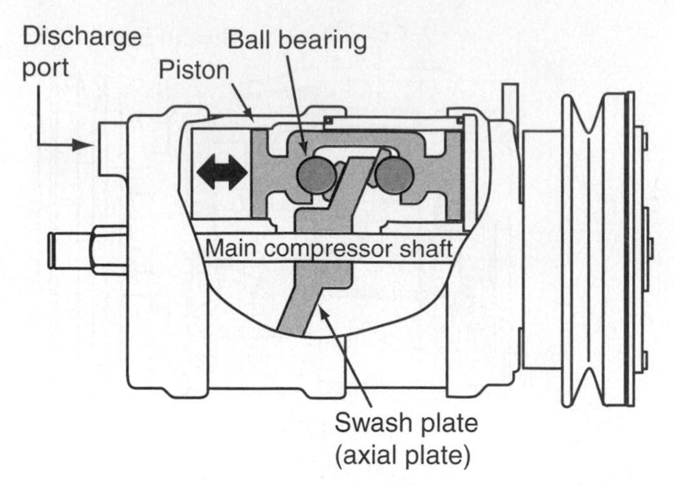

FIGURE 50–18 This A/C compressor uses a swash plate and sliding double-ended pistons to produce needed pressure.

A similar compressor design is shown in ***Figure 50–18***. In this drawing, pistons are attached to the swash plate and connected by ball bearings. As the swash plate rotates, suction and pressure are produced on either ends of the pistons. Note also that the pressure produced on the right-side piston needs to be routed to the left side of the compressor. The high-pressure refrigerant flow from the right side of the pistons is sent through internal passageways to get the refrigerant to the left side and discharge port on the pump.

COMPRESSOR CLUTCH

In all air-conditioning systems, the compressor is equipped with an electromagnetic clutch as part of the pulley. It is designed to engage and disengage the pulley to the compressor. ***Figure 50–19*** shows an electromagnetic clutch attached to the compressor. The clutch is engaged and disengaged by a magnetic field. When the controls call for compressor operation, the electrical circuit to the clutch is energized. The pulleys are connected to the compressor shaft. ***Figure 50–20*** shows the internal parts of an electromagnetic compressor clutch.

The compressor really does not need to be on all of the time. Therefore, on most vehicle A/C systems, the compres-

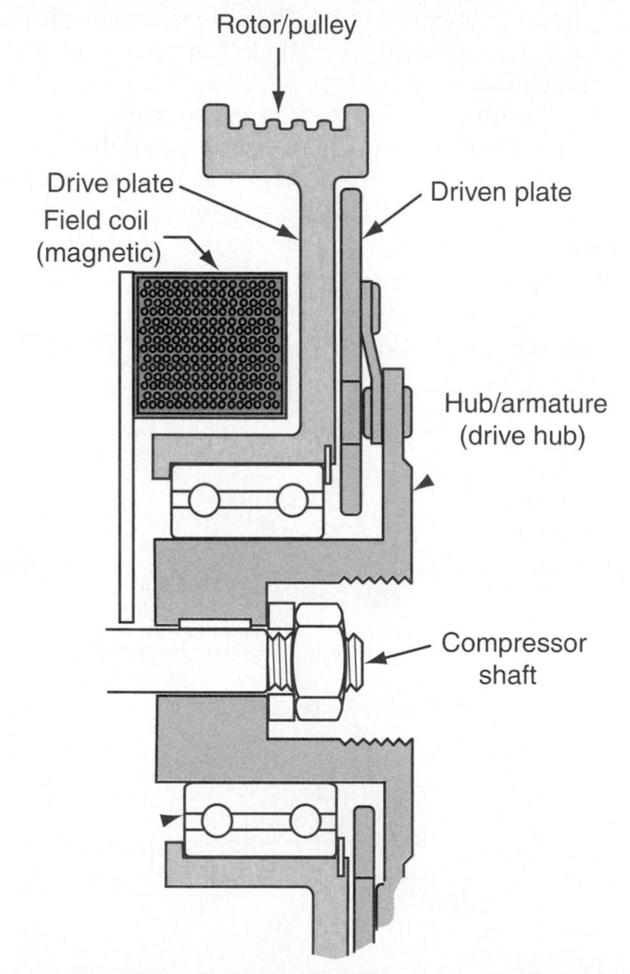

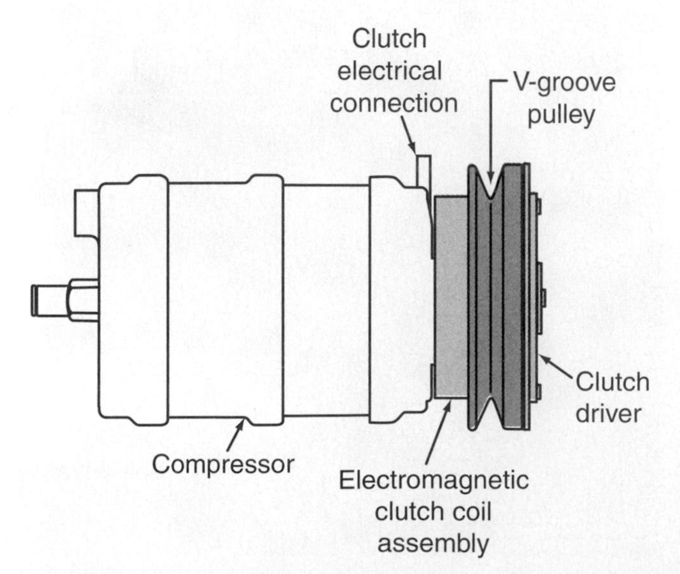

FIGURE 50–19 Electricity fed into the electrical connection causes the field windings in the clutch to energize, which locks up the V-groove pulley with the main compressor shaft.

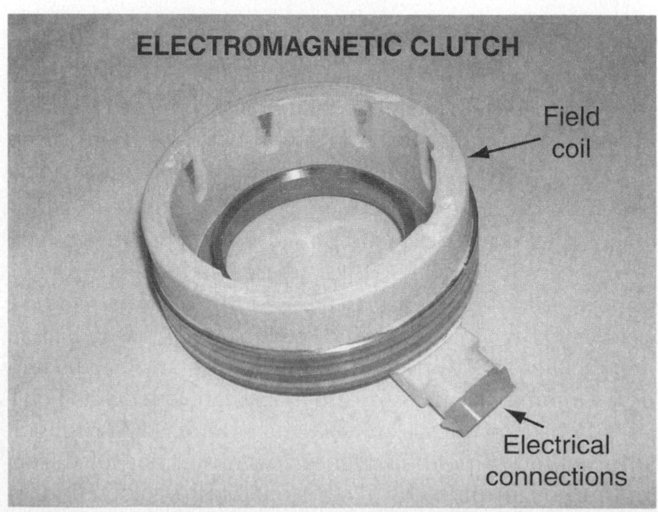

FIGURE 50–20 When engaging the electromagnetic compressor clutch, the drive plate and pulley assembly are forced by magnetism to work against the driven plate.

PARTS LOCATOR

*Refer to photo #114 in the front of the textbook to see what an **A/C compressor clutch** looks like.*

sor is cycled on and off according to the pressure on the low side. The clutches that cycle on and off are called cycling clutches. The cycling is determined by the pressure cycling switch. Refer to ***Figure 50–21***. The pressure cycling switch is located on the left bottom side of the circuit. It senses the pressure on the low side of the system. In operation, when the heater and A/C control assembly is placed in the norm, bi-lev, or HTG position, electricity flows from the fuse block, through the switch, through the A/C high-pressure cutout switch, and to the pressure cycling switch. When the switch closes, due to low pressure, voltage is sent to the PCM. The PCM is then signaled to close the A/C compressor control relay control (bottom of circuit). When this switch is closed, it energizes the A/C compressor control relay on the right side of the circuit. This causes the A/C compressor clutch to engage, causing the compressor to begin spinning, producing more pressure. If the pressure cycling switch is opened, the clutch will be released so the compressor swash plate stops turning. Note that the diode across the A/C compressor clutch is used to reduce voltage spikes as the clutch is cycled on and off.

PARTS LOCATOR

*Refer to photo #115 in the front of the textbook to see what an **A/C high-pressure cutout switch** looks like.*

RECEIVER-DEHYDRATOR

The receiver-dehydrator (dryer) shown in ***Figure 50–22*** is a storage tank for liquid refrigerant. The refrigerant flows from the condenser into the tank. Here a desiccant (moisture-absorbing material) removes moisture from the refrigerant. The refrigerant then flows through a filter screen into the

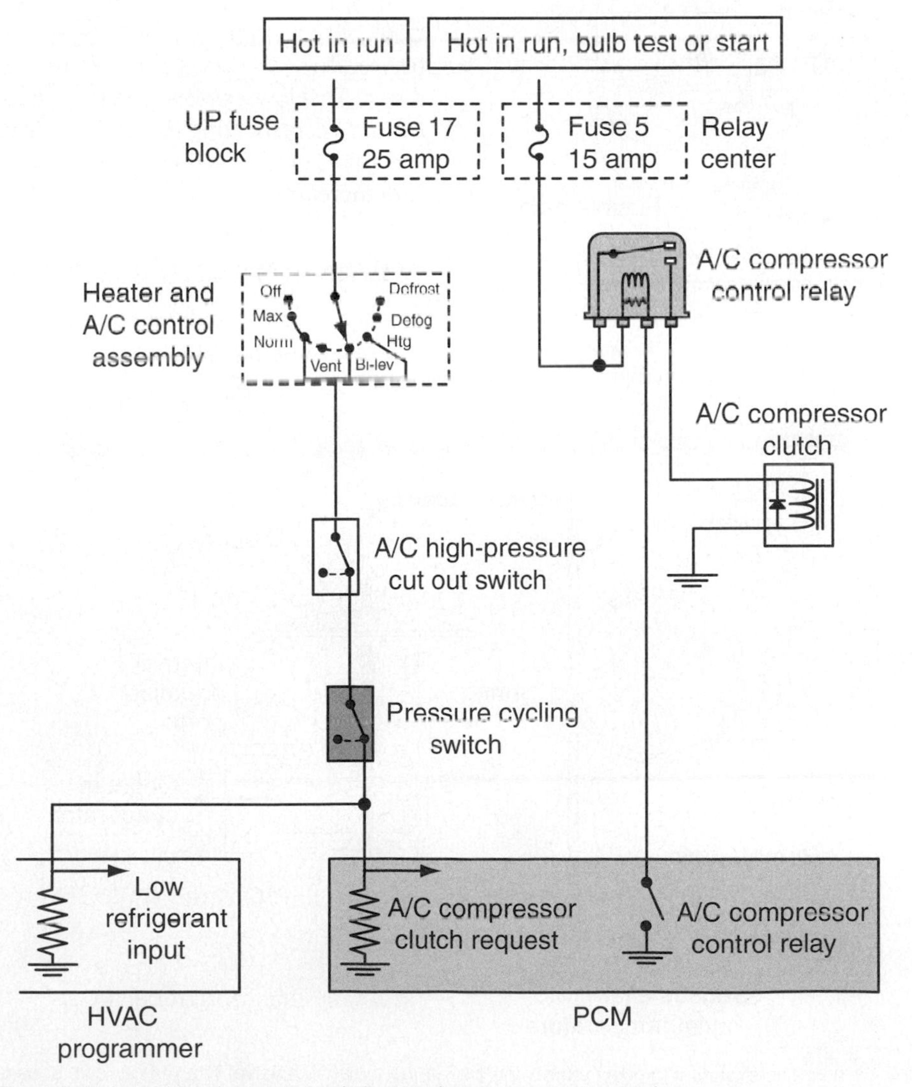

FIGURE 50–21 This circuit shows how the cycling switch controls the compressor clutch.

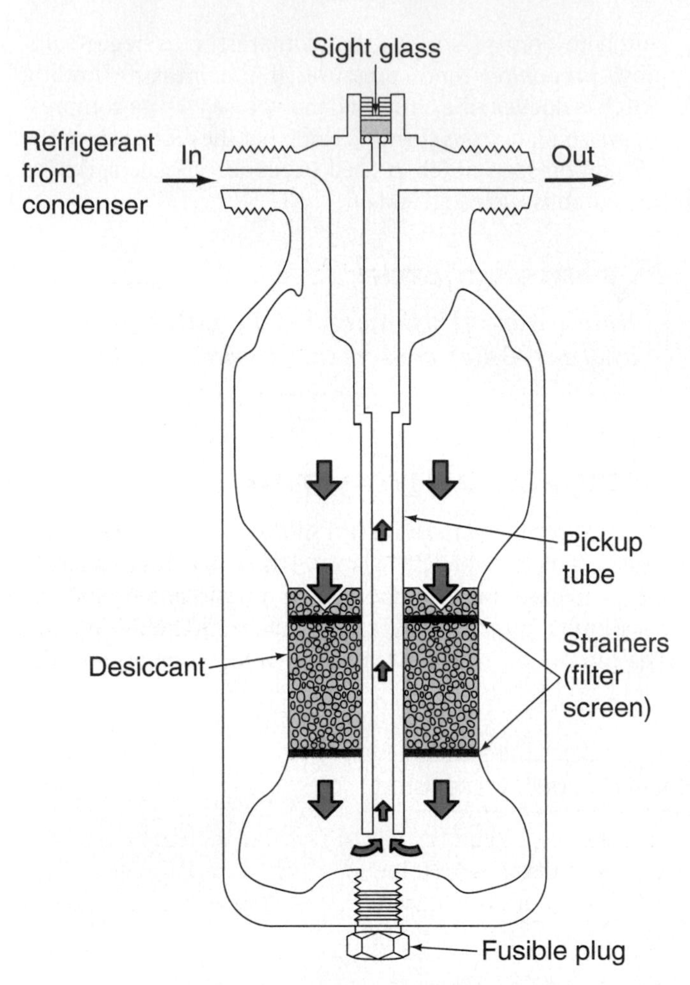

FIGURE 50-22 This receiver-dehydrator assembly is used to filter and remove moisture from the refrigerant. The desiccant is the moisture-absorbing material.

PARTS LOCATOR

*Refer to photo #116 in the front of the textbook to see what an **A/C sight glass** looks like.*

outlet. A sight glass is generally located on the top of the unit. The sight glass shows if there is enough refrigerant in the system. Some sight glasses are located directly in the line from the condenser.

THERMOSTATIC EXPANSION VALVE

The thermostatic expansion valve controls the supply of liquid refrigerant to the evaporator. The valve is considered a variable expansion valve. As shown in ***Figure 50–23***, it is controlled by two opposing forces. A spring pressure works against the power element pressure on a diaphragm. The balance between these two forces positions the seat and orifice to a certain size expansion valve. The power element senses the temperature of the evaporator outlet.

In operation, a decrease in the temperature of the evaporator outlet lowers the bulb temperature. When the bulb temperature decreases, the pressure in the diaphragm is reduced. This lower pressure causes the seat to move closer to the orifice, restricting the flow of refrigerant to the evaporator. The operation is reversed when the evaporator temperature increases.

THERMOSTATIC SWITCH

A **thermostatic switch** is used to cycle the electromagnetic clutch off and on. This switch senses evaporator tempera-

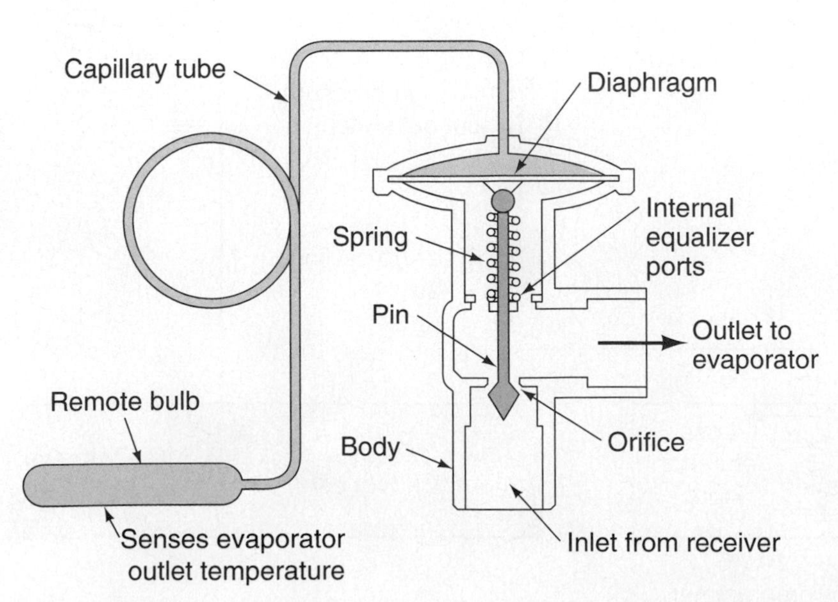

FIGURE 50-23 A thermostatic expansion valve controls the size of the expansion valve. The remote bulb senses the temperature of the evaporator outlet, which causes the diaphragm to move. The pressure from the diaphragm against the spring pressure sets the correct orifice size.

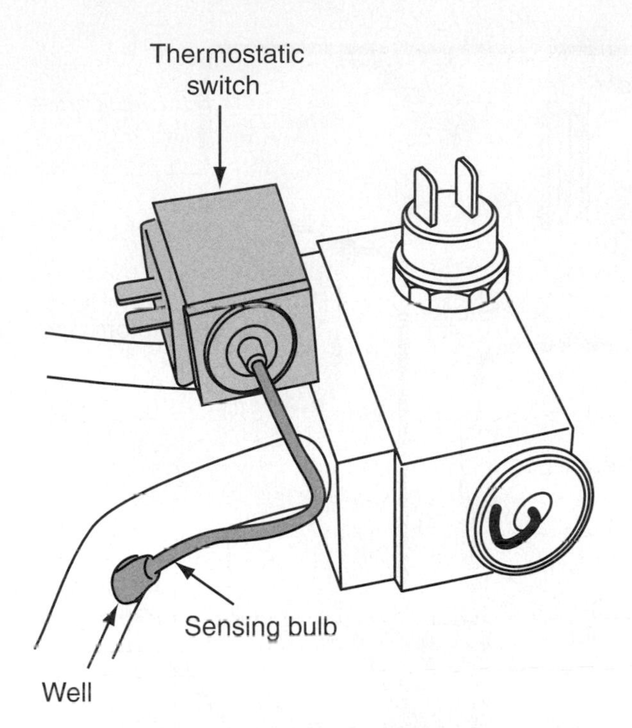

FIGURE 50-24 A thermostatic switch is used to cycle the clutch system on and off.

ture (***Figure 50–24***). The opening and closing of the internal contacts cycle the compressor. When the temperature of the evaporator approaches the freezing point, the thermostatic switch opens. This action disengages the compressor clutch. The compressor remains off until the evaporator temperature rises to a preset temperature. At this temperature, the switch closes, and the compressor resumes operation.

AMBIENT SWITCH

The **ambient** switch is used to sense outside air temperature. It is designed to prevent compressor clutch engagement when the air conditioning is not required. The switch is in series with the electromagnetic compressor clutch.

THERMAL LIMITER AND SUPERHEAT SWITCH

The thermal limiter and superheat switch is designed to protect the air-conditioning compressor against damage when the refrigerant is partially or totally lost. During this condition, the superheat switch heats up a resistor to melt a fuse in the thermal switch. The compressor ceases to operate and is protected from damage.

LOW-PRESSURE CUTOFF SWITCH

The low-pressure cutoff switch is located on the pressure side of the compressor. If low pressure is sensed, the switch opens a set of contact points. The electromagnetic clutch is now inoperative.

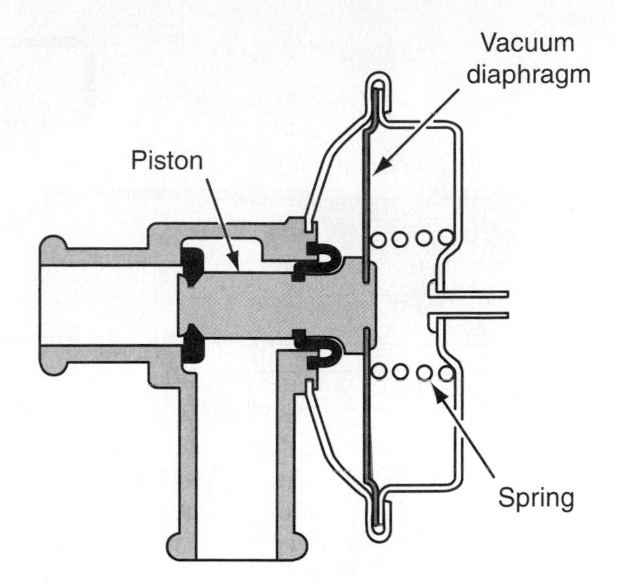

FIGURE 50-25 On some cars a water control valve is used to regulate the flow of engine coolant to the heater core during air conditioning.

WATER CONTROL VALVE

A water control valve is used in many air-conditioning systems (***Figure 50–25***). Its function is to regulate the flow of engine coolant to the heater core. On most vehicles, the water valve is closed when the air-conditioning controls are set at maximum cooling.

MUFFLER

A muffler is used to reduce the compressor noise (***Figure 50–26***). It is located on the discharge side of the compres-

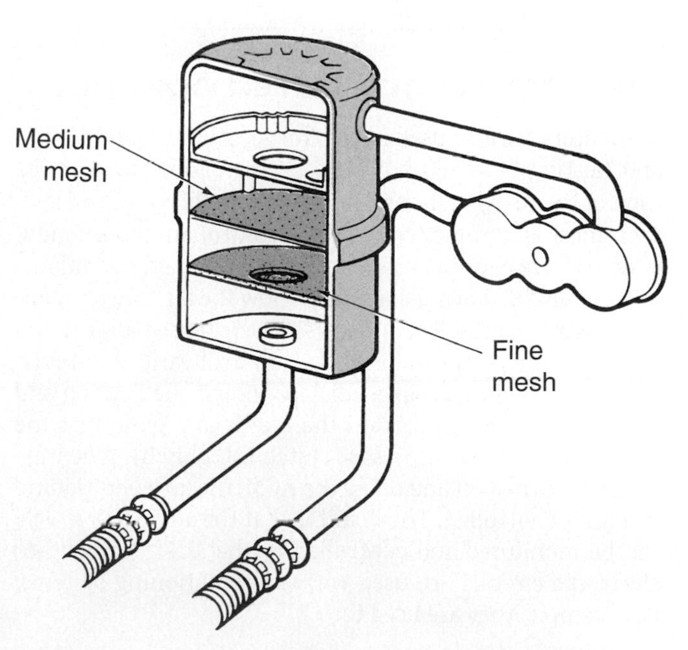

FIGURE 50-26 A muffler is used on air-conditioning systems to reduce the noise of the compressor.

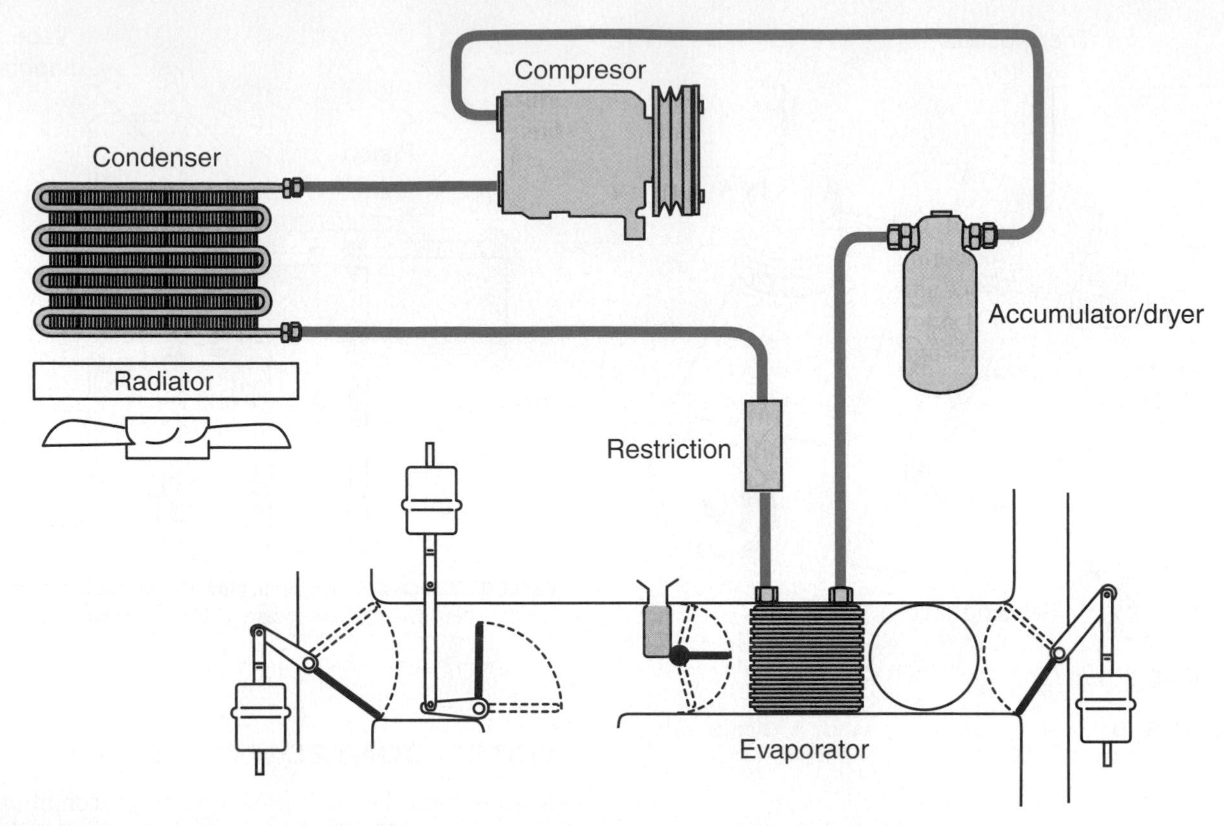

FIGURE 50–27 The major part of the air-conditioning and air distribution system and their locations.

sor. A complete air-conditioning system, with the location of several common parts highlighted, is shown in ***Figure 50–27***.

50.5 ELECTRONIC TEMPERATURE CONTROL AIR-CONDITIONING SYSTEMS

PURPOSE OF COMPUTER CONTROLS

Computers are also used to control vehicle air-conditioning and heating systems. It would be impossible to describe each of the many systems used on automobiles today. However, most electronic temperature control air-conditioning systems have certain common components and operations. ***Figure 50–28*** shows a diagram of how the air control module—**ACM**—works. The ACM is an electronic module in the vehicle that uses several inputs to control various outputs. It is generally located immediately behind the heating and air-conditioning controls on the dashboard. Note that the ACM controls the compressor clutch (AC clutch). When the clutch is turned off and on by the ACM, the air-conditioning system is controlled. The result is that the air in the vehicle can be monitored and controlled by the ACM. Many other electronic circuits are used with air-conditioning systems. For example, they are used to:

- Control blower speeds
- Operate blend doors to mix the cool and warm air
- Turn defrost systems off and on
- Operate floor and panel doors
- Operate cooling and engine coolant fans at different times and speeds

Figure 50–29 shows an electronic circuit for controlling a two-speed coolant fan on an air-conditioning system. For low-speed operation, the low-speed coolant fan relay is operated by the PCM. The PCM controls the low-speed fan control switch and it is turned on when the:

- Coolant temperature exceeds 213 degrees Fahrenheit
- Vehicle speed exceeds 47 mph
- A/C compressor is in operation

With the low-speed coolant fan relay energized, power comes through the low-speed coolant fan relay switch, through a coolant fan resistor, and into the coolant fan. The resistor drops the voltage, causing the coolant fan to run at low speed.

For high-speed operation, the high-speed coolant fan relay is operated by the PCM. The fan is turned on when the:

- Coolant temperature exceeds 226 degrees Fahrenheit
- A/C coolant fan pressure switch reads 275 psi or higher

During high-speed operation, the resistor is bypassed. The result is that higher voltage is available at the fan for high speed.

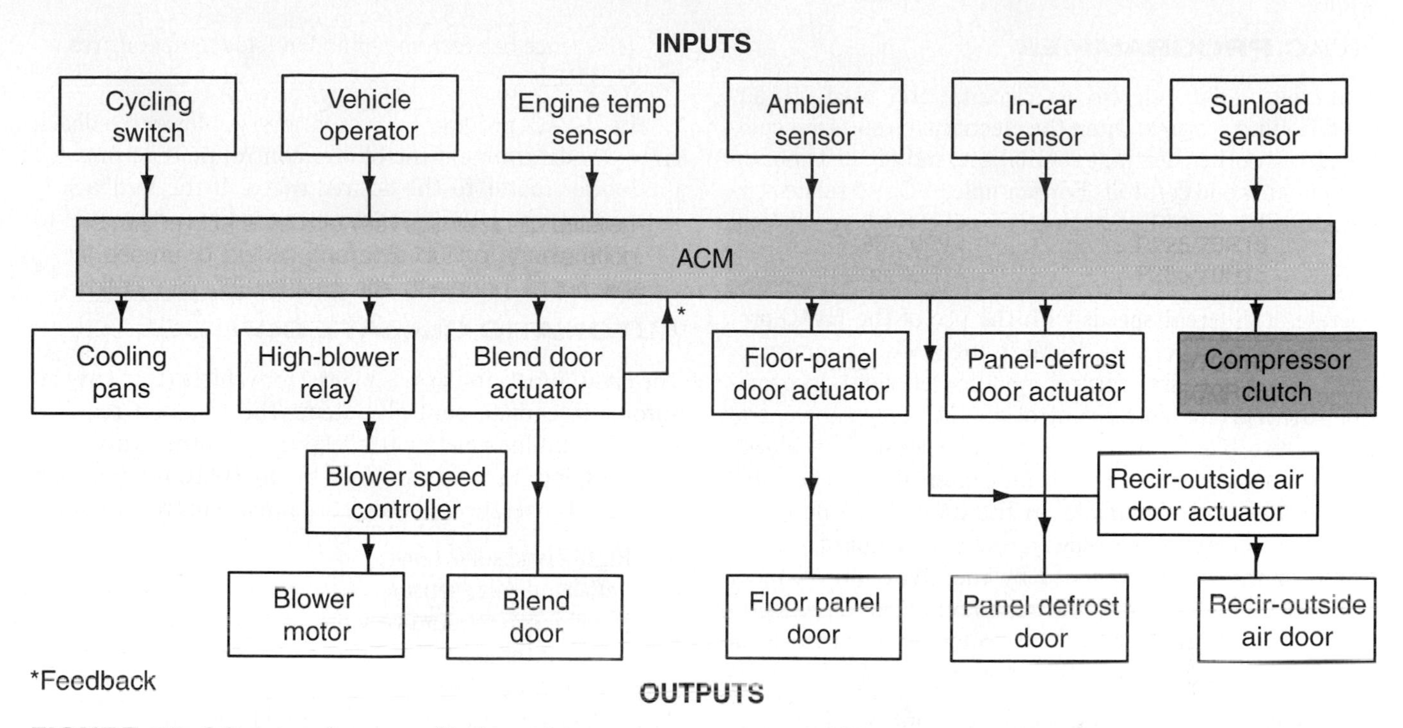

FIGURE 50-28 Automotive air-conditioning systems use computers for control. Note that one output is the compressor clutch turning off and on.

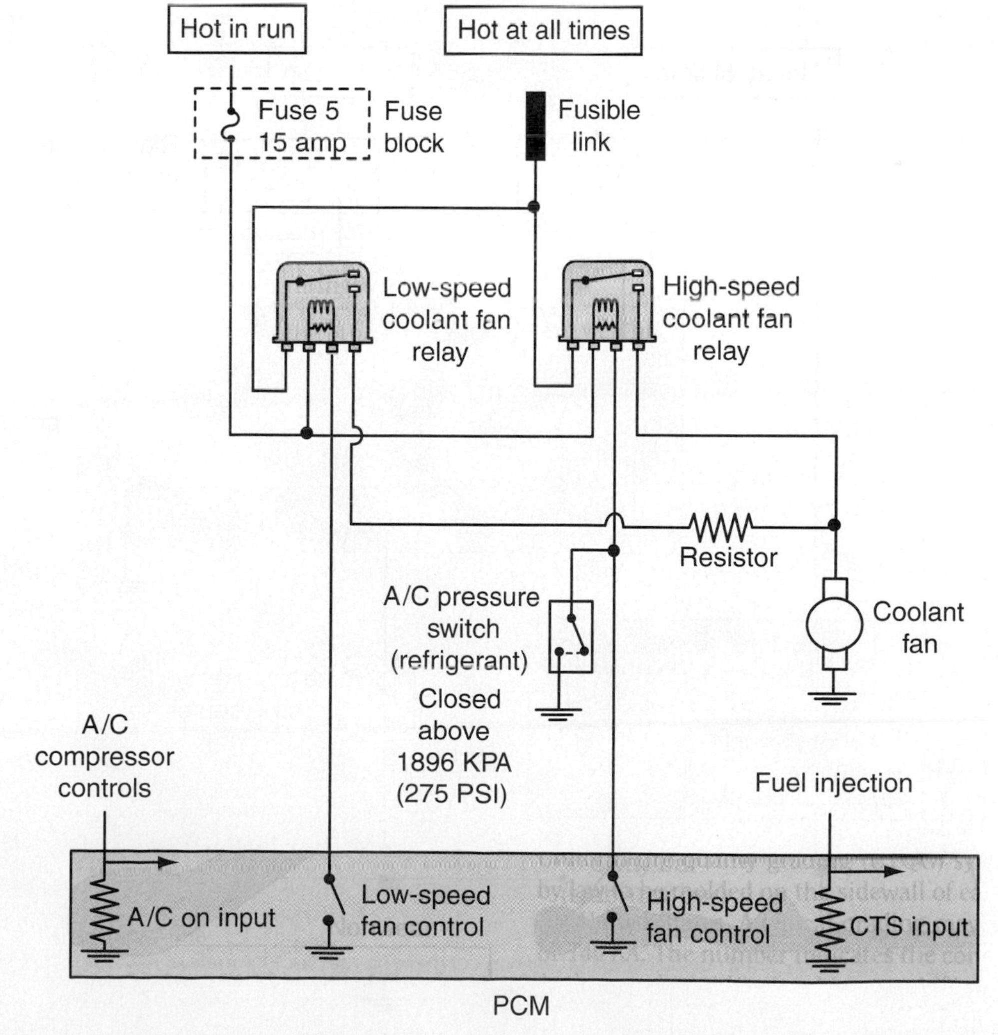

FIGURE 50-29 The PCM is used to control the low- and high-speed coolant fan on air-conditioning systems.

HVAC PROGRAMMER

Numerous other electronic circuits are used in air-conditioning systems. Often the electrical circuits are combined with other heating, ventilation, and air distribution components and controls. For example, A/C and related circuits can be controlled by an HVAC (heating, ventilation, and air conditioning) programmer. ***Figure 50–30*** shows a typical circuit used to turn the blower on and off and operate at different speeds with the use of the HVAC programmer. In this circuit, there is the HVAC programmer (center) and a blower control module (left side). In operation, the use of a blower control module allows the operation of the blower motor at various speeds. The blower motor is located near the bottom center of the schematic. The blower control module on the lower left side of the schematic operates the blower motor at a speed proportional to the signal produced by the HVAC programmer (located in the center of the schematic). The HVAC programmer provides the blower control module with a signal based on various inputs such as the:

- Blower mode selections on the A/C control assembly
- Temperature being set
- Difference between in-car and outside temperatures
- Sun load

The HVAC programmer monitors a blower feedback circuit to determine if the blower control module signaled the blower motor to the desired speed. If the feedback is not correct, the HVAC programmer modifies the signal to change the blower speed.

AUTOMATIC CLIMATE CONTROL

The HVAC programmer is also used for vehicles that have an automatic climate control system. The program requires various input information to effectively perform its task. The following input sensors are used by the HVAC programmer to correctly operate the automatic climate control circuits:

- Right-hand solar sensor
- Left-hand solar sensor
- In-vehicle temperature sensor
- Ambient temperature sensor
- Door position sensor
- Engine coolant temperature (ECT) sensor
- Vehicle speed sensor (VSS)

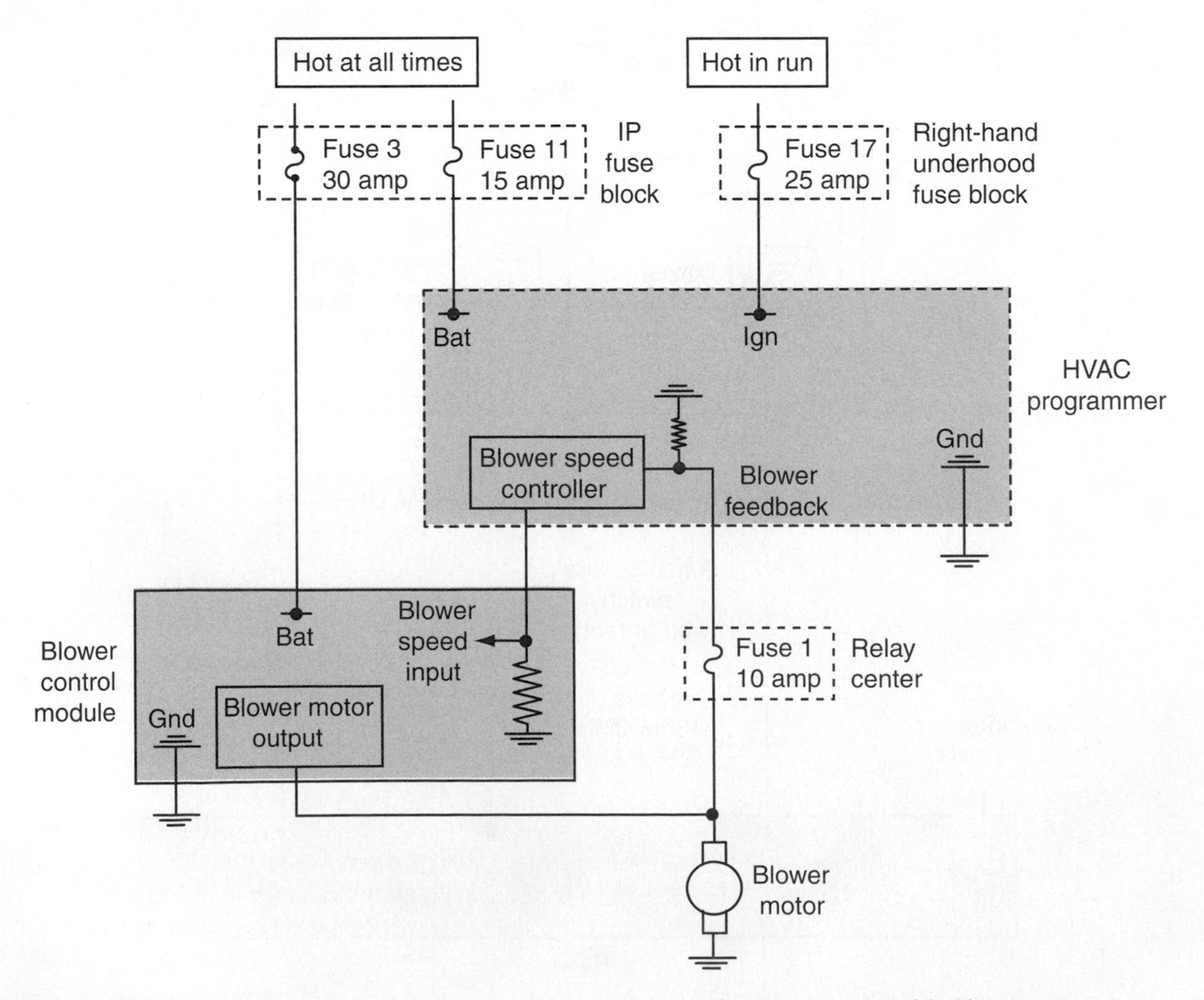

FIGURE 50-30 This circuit shows how the HVAC programmer is used electrically to vary the speed of the blower motor.

PARTS LOCATOR

*Refer to photo #117 in the front of the textbook to see what an **A/C solar sensor system** looks like.*

The right-hand and left-hand solar sensors provide the programmer with information on the heat load from the sun. They are used to modify the amount of cooling required in climate control. The in-vehicle temperature sensor provides the controller with information on how near it is to the occupant-desired temperature. The ambient temperate sensor provides the programmer with information on the outside air temperature. This information allows the programmer to modify its strategy of climate control.

The programmer determines the control strategy based on these inputs. The programmer then generates a number (0 to 225 or 0 to 100) that expresses the amount of heating or cooling required to meet the desired inside temperature. Zero represents maximum cooling, and 225 (or 100) represents maximum heating. Based on this number, the programmer determines blower and air door/valve control. By continuously monitoring the sensors, the programmer can achieve the desired in-vehicle temperature.

There are many other electrical circuits used to aid in climate control on the vehicle. The exact circuit, the type of programmer, the types and names of different control modules all will vary depending on the manufacturer, the vehicle model, and the year of the vehicle. Always refer to the manufacturer's specifications and procedures when servicing such circuits.

Car Clinic: A/C CONTROLS

CUSTOMER CONCERN:
A/C Blows Out Warm Air

A customer complains that the air conditioner works fine when the vehicle is first started in the morning. However, after about 5 to 10 minutes of driving, warm air begins to blow out of the vents.

SOLUTION: This type of problem could be produced by many defective parts. One easy component to check is the temperature control door. This door adjusts to control either warm air from the cooling system or cold air from the air-conditioning system. If this door is out of adjustment, it may be allowing warm air from the heater to be distributed in the passenger compartment. The warm air would come into the car after the engine was heated to operating temperature, after about 5 to 10 minutes. Check the service manual for the correct procedure to adjust the cable. Adjusting the cable may easily solve the problem.

On some vehicles, the door may be controlled by either vacuum or electric motors. These motors can also become defective and cause the same problem. Check the condition of the vacuum and electric motors, if they are used. If the problem is still not solved, further diagnostic procedures will have to be completed.

Problems, Diagnosis, and Service

Safety Precautions

1. Make sure the battery is disconnected when checking the belt tension on an air compressor. This will eliminate the possibility of the engine accidentally cranking over, causing serious injury.
2. On the high-pressure side of the compressor, the pipes carrying the refrigerant may become very hot. Be careful when touching these pipes to avoid burning your hands.
3. When using the propane leak detector, be sure the propane torch flame is away from other objects that could be burned.
4. Refrigerants are a chemical mixture that require special handling. Wrap a clean cloth around fittings, valves, and connections when performing work on the system. Never touch the refrigerant. If the refrigerant comes in contact with any part of your body, flush the exposed area with cold water and seek medical help immediately.
5. Refrigerant can cause severe frostbite and immediate freezing upon contact with skin or other tissue. If a refrigerant touches any part of the body, flush the exposed area with cold water and immediately seek medical help.
6. Never discharge any refrigerant into an enclosed area where there is an open flame.

continued

Safety Precautions *continued*

7. When removing an air compressor, be careful not to injure your hands and fingers.
8. Always wear OSHA-approved safety glasses to prevent any refrigerant from contacting your eyes.
9. Refrigerant containers purchased on the market are under high pressure. Always handle these containers with care and be careful not to drop them.

PROBLEM: Lack of Cold Air

A vehicle with an air-conditioning system is not producing cold air at the dashboard outlets.

DIAGNOSIS

Various general diagnostic checks can be performed on air-conditioning systems to solve similar problems.

CAUTION: *When servicing air-conditioning systems, remember that Freon R12 and Freon R134a have different service procedures. These differences were described earlier in this chapter.*

1. Check the outer surfaces of the radiator and condenser to make sure that airflow is not blocked by dirt, leaves, or other foreign material. Check between the condenser and radiator for foreign material and check the outer surfaces.
2. Check for restrictions or kinks in the evaporator core, condenser core, and refrigerant hoses and tubes.
3. Check for refrigerant leaks. It is customary to leak-check the refrigerant system whenever similar problems are encountered. A soapy water solution is often used to locate leaks in an air-conditioning system. Dye or trace solutions are also used to locate very difficult or slow leaks. A fluorescent tracer solution has become another popular method of leak detection on R134a systems (***Figure 50–31***).

FIGURE 50-31 Add fluorescent dye into a system, let it circulate. The dye escapes with the refrigerant or fluid and collects at all leak sites. Inspect with a UV or UV/blue light lamp and see all leaks glow brightly. (Courtesy of Tracer Products, Westbury, NY)

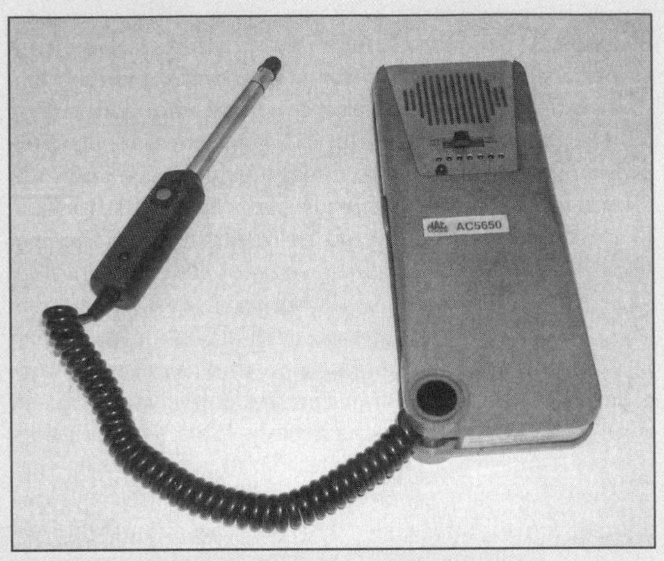

FIGURE 50-32 This electronic leak detector can be used in areas that are inaccessible or are hard to see.

4. ***Figure 50–32*** shows an electronic leak detector. This tester is used in areas that are hard to see or are inaccessible.

   CAUTION: *Be careful not to inhale the gas (phosgene gas) produced by this process. It is very dangerous.*

   Always follow the manufacturer's instructions regarding calibration, operation, and maintenance of this detector.
5. Check all air ducts for leaks or restrictions. Low airflow may indicate a restricted evaporator.
6. Check for proper drive belt tension.
7. Check the air discharge temperature and compare it with the manufacturer's specifications.
8. One common cause for lack of cold air is insufficient refrigerant. At temperatures higher than 70 degrees Fahrenheit, insufficient refrigerant can be observed through the sight glass. The sight glass is usually located in the top of the receiver-dehydrator. ***Figure 50–33*** shows the conditions viewed through the sight glass. A shortage of liquid refrigerant is indicated after about 5 minutes of compressor operation by the appearance of slow-moving bubbles (vapor) in the line. A broken column of refrigerant may appear in the glass, indicating insufficient charge. ***Figure 50–34*** shows a diagnosis chart for the sight glass check.

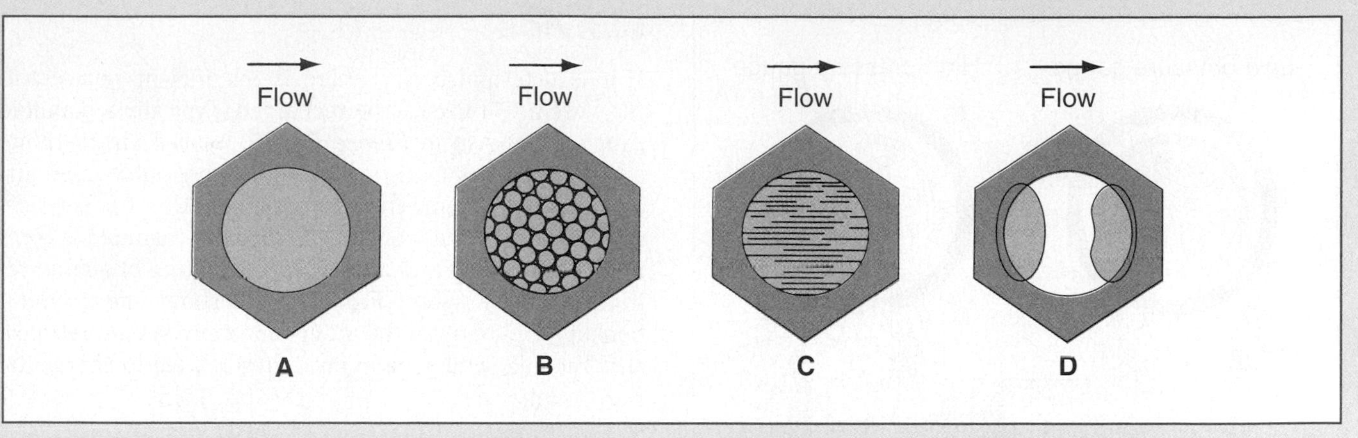

FIGURE 50-33 The sight glass on top of the dehydrator may show several conditions: (A) clear, (B) foamy or bubbly, (C) cloudy or streaked, or (D) broken column of refrigerant.

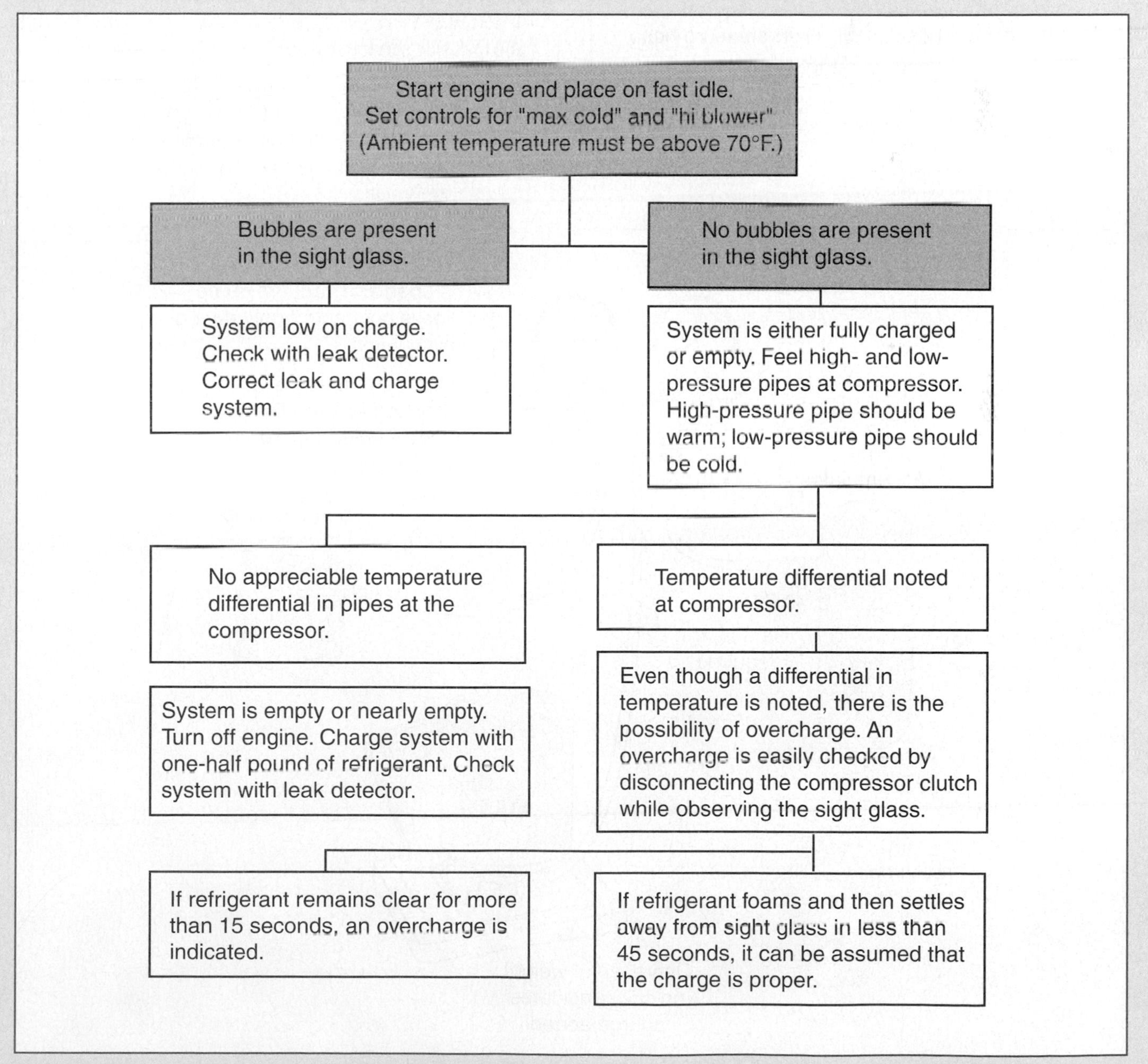

FIGURE 50-34 This diagnosis chart can be used to troubleshoot the air-conditioning system, starting with a look at the sight glass.

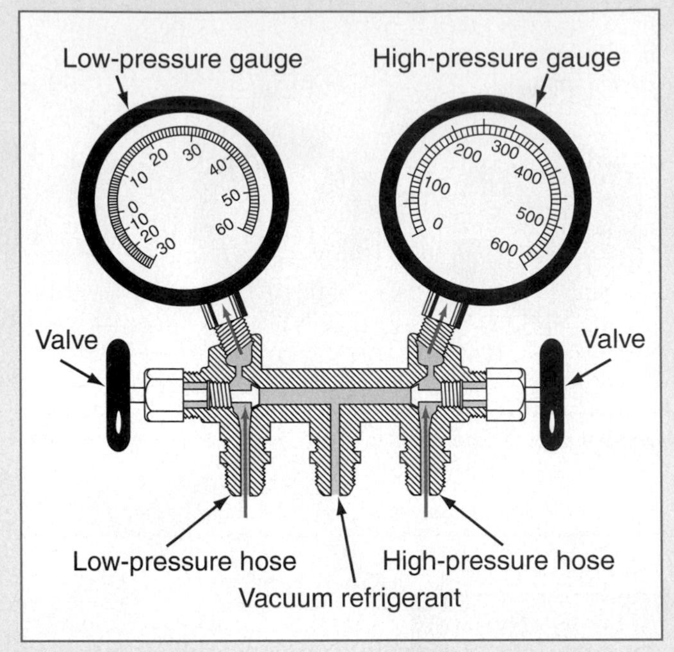

FIGURE 50-35 A manifold gauge is used to check air-conditioning system pressure, to help charge and discharge the system, and to evacuate the system.

SERVICE

If it is determined that there is insufficient refrigerant, the system will need to be recharged. Typically a manifold gauge set, shown in ***Figure 50–35***, is used for charging, discharging, evacuating, and diagnosing trouble in an air-conditioning system. The left gauge measures the low side, the right gauge the high side. The center manifold is common to both sides and is used for evacuating or adding refrigerant to the system. ***Figure 50–36*** shows one of several hookups for charging the air-conditioning system refrigerant. The following general procedure is used to charge the refrigerant:

1. With the A/C mode control lever Off, start the engine and bring it to operating temperature.
2. Invert the refrigerant drum, open the valve, and allow 1 pound of refrigerant to flow into the system through the low-side service fitting.
3. After the refrigerant has entered the system, immediately engage the compressor by setting the A/C control heat to Normal and the blower speed to High. The remainder of the refrigerant charge is now being brought into the system.

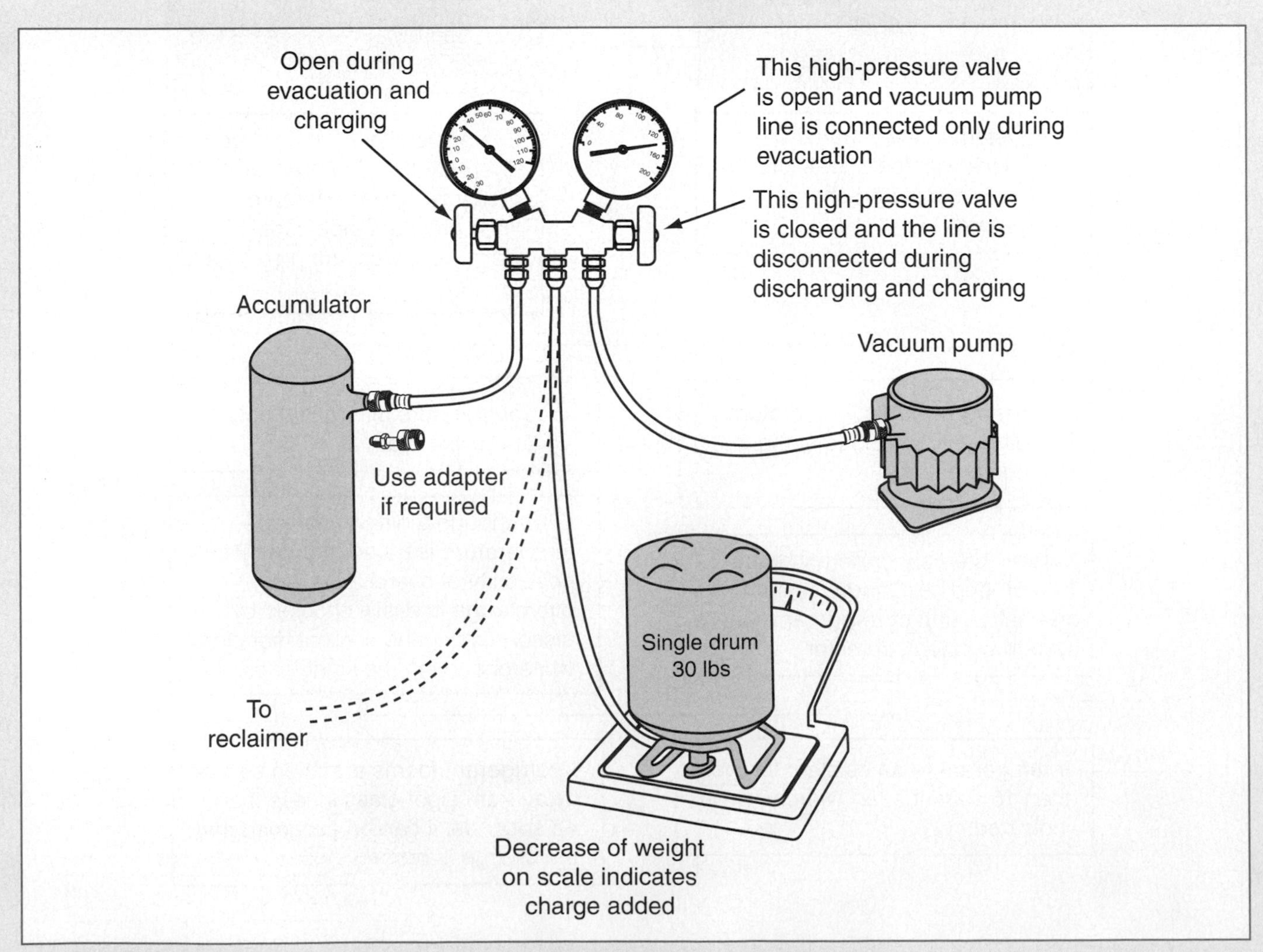

FIGURE 50-36 This schematic shows the hookup between the manifold gauge set and the A/C components for charging the system.

4. Turn off the refrigerant valve, and run the engine for 30 seconds to clear the lines and gauges.
5. With the engine running, remove the low-side charging hose adapter from the accumulator service fitting. Unscrew the adapter rapidly to avoid excessive refrigerant loss.
6. Replace the protective cap on the accumulator service fitting.
7. Turn the engine off.
8. Leak-check the system, and test the air temperature at the dash panel.

PROBLEM: Faulty Air-Conditioning Components

The air conditioner does not work.

DIAGNOSIS

Many components in an air-conditioning system can fail. Air-conditioning systems vary widely from vehicle to vehicle. Because of this variation, no universal or standard diagnostic procedure exists. However, the following diagnostic checks are common to most systems.

Compressor/Clutch Diagnosis

1. The most common reasons for a compressor failure are inadequate lubrication and contamination. Proper lubrication is critical to correct compressor operation. Contamination usually comes from having moisture in the refrigerant. Whenever an air-conditioning system is open to the atmosphere, it will absorb moisture from the air. All moisture must then be evacuated from the system, and the system must be recharged. Evacuation and recharging are done using the manifold gauge set discussed previously. Follow the procedure listed in the service manual for evacuating and charging the air-conditioning system.

CAUTION: *Be careful not to touch the refrigerant or get refrigerant in your eyes.*

Compressor malfunctions appear in one of four ways: noise, seizure, leakage, or low inlet and discharge pressure.

2. ***Figure 50–37*** shows an example of a diagnosis chart for troubleshooting a compressor that is not engaging or not operating. Many other diagnosis charts are available. Refer to the manufacturer's service manuals for more diagnosis charts.

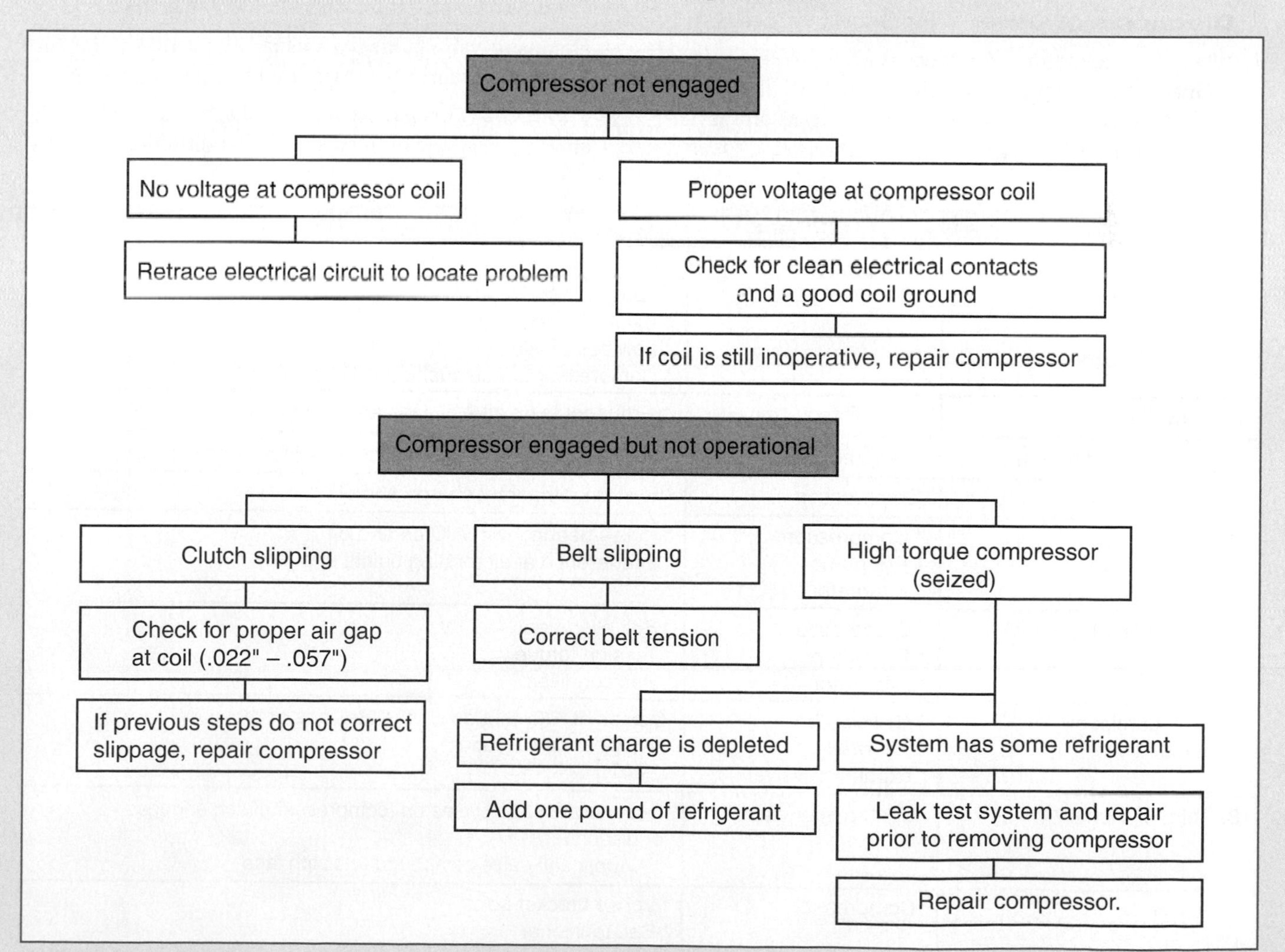

FIGURE 50-37 If the compressor is not operating or engaging, follow this diagnosis chart to determine the problem.

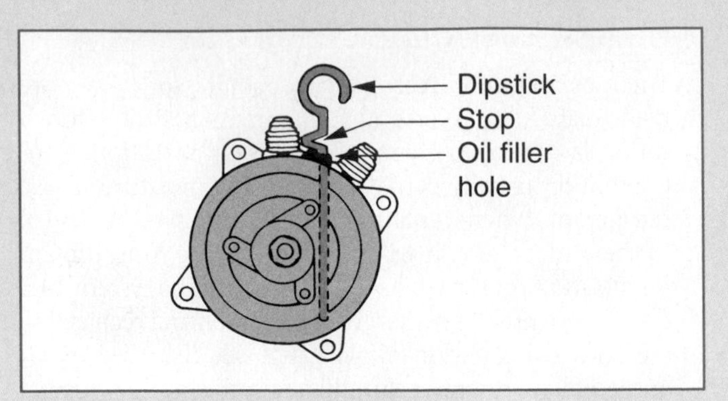

FIGURE 50-38 Some A/C compressors use a dipstick to check the oil inside the compressor.

3. On certain types of air-conditioning compressors, the oil of the compressor must be checked periodically. ***Figure 50–38*** shows how a dipstick is placed in the compressor to correctly check the oil.
4. Problems with the compressor clutch can usually be traced to the compressor. A worn or poorly lubricated compressor will put a strain on the clutch. Low voltage can also cause damage to the compressor clutch.

Diagnosis of Other A/C Components

1. When the evaporator is defective, the trouble shows up as an inadequate supply of cool air. A core partially plugged by dirt, a cracked case, a leaking seal, or leaks on the bottom, are generally the cause. Evaporator leaks are mostly the result of pinhole leaks that develop in the bottom section from corrosion. Corrosion will result if moisture enters the system and mixes with the refrigerant. External corrosion can be caused by atmospheric moisture, especially in areas that have high salt conditions.
2. The condenser may malfunction in two ways: it may leak refrigerant, or it may be restricted. Check as previously described.
3. The receiver-dehydrator (dryer) may fail because of a restriction inside the body. High pressures on the pressure side indicate a restriction at that point. Also, receiver-dehydrators frequently need to be replaced. These units become used up in normal service. The desiccant becomes saturated with absorbed moisture.
4. The thermostatic switch may fail and cause the compressor to run continuously or to shut off completely. Either may cause the evaporator to freeze up. Check the switch for correct adjustment according to the manufacturer's service manual.
5. Some older vehicles have a suction throttling valve (STV) that may need adjustment. It must keep the evaporator pressure above 29–30 psi. If the pressure falls below the specifications, the evaporator core may freeze up.
6. Expansion valve failures are usually indicated by low suction and discharge pressure and insufficient cooling.
7. ***Figure 50–39*** is a noise analysis diagnosis chart to help solve many of the problems in air-conditioning systems.

PROBLEM	DIAGNOSIS	SERVICE
1. Moan (occurs at a specific rpm)	Compressor	* Groundouts * Loose bracket bolts * Exhaust system grounds * Low belt tension (TRY:) * Compressor torque cushion (HD6 only)
2. Growl (idle)	Compressor	* Groundouts on all A/C lines and components
3. Growl (at all speeds)	Compressor	* All line clamps and anchor points (remove and check)
4. Chatter (idle)	Compressor	* System refrigerant charge and oil
5. Whine or grind (at all speeds)	* Compressor * P/S pump * Alternator	* Pulley bearing with A/C on and off * Interference at all rotating points of pulley
6. Squeal (at all speeds)	* Compressor * P/S pump * Alternator	* Belt alignment * Tension torque * Belt condition
7. Hiss, gurgle, or percolation	* Orifice tube * Thermal expansion valve	Occurs at A/C shut down (isolate lines or move O/T or TXV to new location)
8. Chirp or thump	Compressor clutch	Occurs when A/C cycles on (compressor clutch engages or disengages) *** Apply anti-seize compound to clutch face
9. Knock	Compressor	* Loose bracket bolts * Belt tensioner

FIGURE 50-39 Use this diagnosis chart to help determine the source and possible cause of A/C noise problems.

SERVICE

Most of the service for each of the preceding components is involved with cleaning or removal and replacement. The simplest and most effective cure for surface contamination is flushing the system with water. Flushing will remove debris and corrosive materials. Additional service includes discharging, evacuating, and recharging the system. Follow the manufacturer's service manual for each service needed.

Service Manual Connection

There are many important specifications to keep in mind when working with air-conditioning systems. To identify the specifications for your vehicle, you will need to know the VIN (vehicle identification number) of the vehicle, the type and year of the vehicle, and the type of engine used. Although they may be titled differently, some of the more common air-conditioning system specifications (not all) found in service manuals are listed below. Note that these are typical examples. Each vehicle may have different specifications.

Common Specification	Typical Example
A/C compressor belt tension (pounds on suitable belt tension gauge)	115 newtons
Compressor air gap	0.015–0.025 inch
Compressor discharge pressure	140–210 psig
Discharge air temperature on panel	35–46 degrees Fahrenheit
Evaporator suction (low-pressure side) pressure	18–21 psig
Refrigerant capacity	2.5 lb
Refrigerant oil viscosity	500
Total system capacity	8 ounces

In addition, service manuals provide specific directions for various service and testing procedures. Some of the more common procedures include:

- Accumulator replace
- Charging the air-conditioning system
- Checking compressor oil level
- Condenser replace
- Discharging and evacuating system
- Leak tests
- Oil charging
- Performance test

SUMMARY

The following statements will help to summarize this chapter:

- Heat always flows from a hotter to a colder direction.
- Heat can be absorbed in large quantities when a liquid turns to a gas.
- When a liquid is boiled, it turns into a vapor, and when it is cooled, it turns back to a liquid.
- Pressure has a direct effect on the boiling point of fluids.
- As a vapor is compressed under pressure, its temperature increases.
- Two types of refrigerant used on A/C systems are R12 and R134a. Which is better for the environment?
- A/C systems are designed to change the pressure of a refrigerant, causing it to boil and absorb large quantities of heat from the passenger compartment.
- Heat transfer in A/C systems is done by the use of heat exchangers.
- An evaporator is used to capture the heat inside the passenger compartment.
- A condenser is used to exhaust the captured heat in an A/C system.
- Heat is absorbed into the refrigerant in an A/C system.
- After the heat is in the refrigerant, the refrigerant is compressed.
- The expansion valve is used to reduce the pressure on the refrigerant.
- A swash plate compressor is used to compress the refrigerant at the right time.
- Computers are also used to control and operate A/C systems using many inputs to control blower speeds, blend doors, defrost systems, floor and panel doors, cooling and engine cooling fans, and compressor clutches.
- There are many diagnosis checks to make on A/C systems.

- The two biggest dangers to an A/C system are dirt and moisture in the system.
- Other problems often found in A/C systems include a damaged compressor, controls that are improperly adjusted, and accumulators and filters being used up.
- Always follow the manufacturer's suggested and recommended procedures when working on an A/C system.

TERMS TO KNOW

Can you explain each of the following terms? Review the chapter until you can use each term correctly.

Accumulator
ACM
ACR3
Ambient
CFCs
Condensation
Condenser
Desiccant
Evaporator
Expansion valve
Latent heat
Ozone layer
R134a
Refrigerant
Swash plate
Thermostatic switch
Vapor

REVIEW QUESTIONS

Multiple Choice

1. At what temperature is there a total lack of heat in any substance?
a. Boiling point
b. Absolute zero
c. Freezing point
d. 88 degrees Fahrenheit
e. –21.7 degrees Fahrenheit

2. Heat will always flow from a ________ substance.
a. Colder to a colder
b. Hotter to a hotter
c. Hotter to a colder
d. Colder to a hotter
e. None of the above

3. At what point can large quantities of heat be absorbed into a liquid or gas?
a. When a liquid boils to a vapor
b. When a vapor remains as a vapor
c. When a liquid freezes
d. All of the above
e. None of the above

4. Heat that is hidden is called:
a. Absolute heat
b. Warm heat
c. Specific heat
d. Latent heat
e. All of the above

5. When a vapor is changed back to a liquid, it is called:
a. Evaporation
b. Boiling point
c. Condensation
d. Accumulation
e. None of the above

6. As the pressure on a liquid increases, the boiling point of the liquid ________.
a. Decreases
b. Increases
c. Remains the same
d. Drops by only 2 degrees
e. None of the above

7. What is the boiling point of R134a that is not under pressure?
a. –2 degrees Fahrenheit
b. –22 degrees Fahrenheit
c. –42 degrees Fahrenheit
d. 30 degrees Fahrenheit
e. 65 degrees Fahrenheit

8. Heat transfer in an air-conditioning system is done in the:
a. Accumulator
b. Dehydrator
c. Compressor
d. Heat exchangers
e. Thermostatic switch

9. Which of the following processes is happening in the passenger compartment of an air-conditioning system?
a. The liquid refrigerant changes to a vapor.
b. Large amounts of heat are absorbed in the refrigerant.
c. The liquid refrigerant boils from the passenger compartment heat.
d. All of the above
e. None of the above

10. Condensation of the vapor refrigerant back to a liquid takes place at the:
a. Evaporator
b. Compressor
c. Condenser
d. Accumulator
e. Receiver

11. Which of the following components is on the high side of the air-conditioning circuit?
a. Evaporator
b. Condenser
c. Suction side of the compressor
d. Low-pressure side of the expansion valve
e. All of the above

12. Which of the following components is on the low side of the air-conditioning circuit?
a. Evaporator
b. Low-pressure side of the expansion valve
c. Suction side of the expansion valve
d. All of the above
e. None of the above

13. The ________ is used to store liquid refrigerant.
a. Evaporator
b. Condenser
c. Accumulator
d. Expansion valve
e. All of the above

14. The ________ is used to reduce the pressure and temperature of the refrigerant.
a. Expansion valve
b. Accumulator
c. Compressor
d. Receiver-dehydrator
e. Pressure regulator

15. The ________ freezes up if it is operated at the wrong pressures and temperatures.
a. Compressor
b. Condenser
c. Evaporator
d. Expansion valve
e. None of the above

16. To control evaporator temperature on the air-conditioning system, the ________ is controlled.
a. Pressure
b. Compressor speed
c. Engine speed
d. Condenser speed
e. Dc voltage

17. One common type of compressor uses:
a. A swash plate
b. A reed valve
c. Six pistons
d. All of the above
e. None of the above

18. What type of device is used to engage and disengage the compressor?
a. Friction clutch
b. Electromagnetic clutch
c. Centrifugal clutch
d. Roller clutch
e. None of the above

19. Which two elements are most dangerous to an air-conditioning system?
a. Dirt and moisture
b. Carbon and silicon
c. Dirt and dc voltage signals
d. Gasoline and oil
e. All of the above

20. Which of the following is an easy method used to check for refrigerant in the system?
a. Use a sight glass.
b. Loosen the system, and watch for fluid to flow out.
c. Apply an excess amount of pressure to get more cooling.
d. All of the above
e. None of the above

21. On an air-conditioning system, a propane torch is used to:
a. Create excess pressure
b. Check for pressure
c. Check for refrigerant leaks
d. Solder steel piping
e. Heat the evaporator during testing

22. A manifold gauge set is used on an air-conditioning system to:
a. Discharge the refrigerant
b. Charge the refrigerant
c. Evacuate the refrigerant
d. All of the above
e. None of the above

The following questions are similar in format to ASE (Automotive Service Excellence) test questions.

23. *Technician A* says that the most common problems with air compressors are contamination and lack of lubrication. *Technician B* says that the most common problem with air compressors is engine vibration. Who is correct?

a. A only
b. B only
c. Both A and B
d. Neither A nor B

24. *Technician A* says that when the air-conditioning system is open to the atmosphere, simply seal the hole and the system can be operated. *Technician B* says that an opening in the system means moisture will be absorbed into the system. Who is correct?

a. A only
b. B only
c. Both A and B
d. Neither A nor B

25. Small air bubbles are noticed in the sight glass on the air-conditioning system. *Technician A* says this is normal. *Technician B* says that the compressor is overcharging the Freon. Who is correct?

a. A only
b. B only
c. Both A and B
d. Neither A nor B

26. An air-conditioning system has insufficient cooling. *Technician A* says the problem is low refrigerant. *Technician B* says the condenser is clogged. Who is correct?

a. A only
b. B only
c. Both A and B
d. Neither A nor B

27. An air-conditioning system does not cool because the blower does not operate. *Technician A* says the problem is a blown fuse link. *Technician B* says the compressor clutch is faulty. Who is correct?

a. A only
b. B only
c. Both A and B
d. Neither A nor B

28. There are bubbles present in the sight glass. *Technician A* says there is insufficient refrigerant. *Technician B* says the system should be checked for refrigerant leaks. Who is correct?

a. A only
b. B only
c. Both A and B
d. Neither A nor B

29. When checking a malfunctioning air-conditioning system, *Technician A* says to make sure the condenser is not plugged with leaves or other foreign material. *Technician B* says to check for restrictions or kinks in the refrigerant lines. Who is correct?

a. A only
b. B only
c. Both A and B
d. Neither A nor B

Essay

30. Describe the process of condensation.

31. What happens to the boiling point of a fluid when the pressure increases or decreases?

32. Describe the purpose and operation of the expansion valve on an air-conditioning system.

33. What is the difference between the high and low sides of an air-conditioning system?

34. What is the purpose of an accumulator?

35. Describe how pressure is produced in the air-conditioning compressor.

Short Answer

36. R12 refrigerant is known to reduce the ________ above the Earth.

37. The major greenhouse gases are: ________, ________, ________, and ________.

38. To protect the environment, a refrigerant called ________ is now used in the design of A/C systems.

39. To prevent R12 from getting into the atmosphere during A/C servicing, a machine called ________ is used.

40. On an electronic A/C system, the ACM is used to control the ________.

Applied Academics

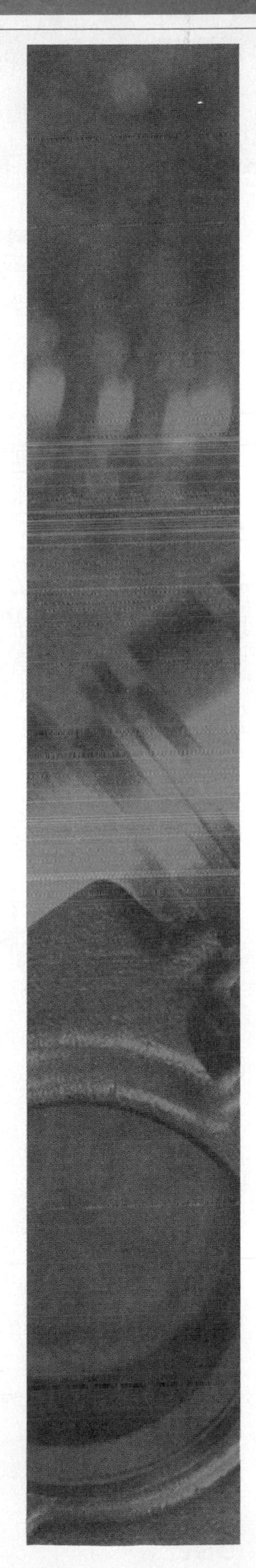

TITLE: Scientific Principles of Air Conditioning

ACADEMIC SKILLS AREA: Science

NATEF Automobile Technician Tasks:

The technician can describe the kinetic and potential energy relationships that occur in heating and air-conditioning systems.

CONCEPT:

Air-conditioning systems are designed around several scientific and physical principles about heat, gases, and pressures.

Heat

- Heat always flows from an area of higher temperature to an area of lower temperature.
- The greater the temperature differential, the faster heat flows.
- Heat continues to flow until the two temperatures are equal.

Liquids and Gases

- A specific amount of heat must be absorbed into the liquid to change liquid to a gas or vapor.

Pressure and Temperature Relationship of Heat and Gases

- If pressure acting on a liquid is increased, the boiling point of the liquid increases.
- If pressure acting on a liquid is decreased, the boiling point of the liquid decreases.

Pressure Temperature Relationship of Refrigerants

- For every pressure increase in a refrigerant, a corresponding temperature increase occurs.

Because of these laws, performance pressures in air-conditioning systems will change at different temperatures. In air-conditioning system service manuals, pressures are always given for at least five different ambient (outdoor) temperatures. This is shown in the following table.

PRESSURE/TEMPERATURE CHART

ENGLISH

Temp. F	Psig	Temp. F	Psig	Temp. F	Psig	Temp. F	Psig	Temp. F	Psig
65	74	75	87	85	102	95	118	105	136
66	75	76	88	86	103	96	120	106	138
67	76	77	90	87	105	97	122	107	140
68	78	78	92	88	107	98	124	108	142
69	79	79	94	89	108	99	125	109	144
70	80	80	96	90	110	100	127	110	146
71	82	81	98	91	111	101	129	111	148
72	83	82	99	92	113	102	130	112	150
73	84	83	100	93	115	103	132	113	152
74	86	84	101	94	116	104	134	114	154

APPLICATION:

What other application in the automobile engine uses the concept that as pressure increases, the boiling point of the fluid increases? Can you define the relationship between each of these scientific principles and the operation of an air-conditioning system? For example, how do each of these principles relate to air conditioning? How are these principles being practiced?

CHAPTER 51

Heating and Ventilation Systems

OBJECTIVES

After studying this chapter, you should be able to:

- State the purpose of the heating and ventilation systems used in automobiles.
- Identify common parts used on heating and ventilation systems.
- Analyze how the passenger ventilation system is designed in the automobile.
- Compare vacuum, mechanical, and computer controls on heating and ventilation systems in the automobile.
- Identify various problems, their diagnosis, and service steps used on heating and ventilation systems on automobiles.

INTRODUCTION

Several temperature controls are used to maintain a comfortable climate in the interior of an automobile. This chapter discusses the ventilation and heating systems that are used inside the passenger compartment.

51.1 HEATING AND VENTILATION SYSTEMS

PURPOSE OF HEATING AND VENTILATION SYSTEMS

Automobile heating and ventilation systems are used to keep the passenger compartment at a comfortable temperature. During the winter, cold outdoor temperatures may make the passenger compartment too cold. During the summer, high ambient temperatures may make the compartment too warm. Thus, both heating and ventilation are needed to keep vehicle occupants comfortable. In addition, as discussed in an earlier chapter, air conditioning may also be used.

FLOW-THROUGH VENTILATION

There are many systems used to heat and vent the passenger compartment. One type is called the flow-through ventilation system. A supply of outside air, which is called **ram air**, flows into the car when it is moving. When the car is not moving, a steady flow of outside air can be produced from the heater fan. ***Figure 51–1*** shows an example of the

Certification Connection

ASE Connection: The information in this chapter can help you prepare for the National Institute for Automotive Service Excellence (ASE) certification tests. The tests and content areas most closely related to this chapter are:

Test A7—Heating and Air Conditioning

- **Content Area**—Heating and Engine Cooling System Diagnosis and Repair
- **Content Area**—Operating Systems and Related Controls Diagnosis and Repair

NATEF Connection: Much of the information in this chapter is related to the NATEF tasks. The NATEF tasks and priority numbers most closely related to this chapter are:

1. Identify and interpret heating and air-conditioning concerns; determine necessary action. P-1
2. Diagnose temperature control problems in the heater/ventilation system; determine necessary action. P-2
3. Perform cooling system, cap, and recovery system tests (pressure, combustion leakage, and temperature); determine necessary action. P-1

4. Inspect engine cooling and heater system hoses and belts; perform necessary action. P-1
5. Inspect and test electric cooling fan, fan control system, and circuits; determine necessary action. P-1
6. Inspect and test heater control valve(s); perform necessary action. P-2
7. Remove and reinstall heater core. P-3

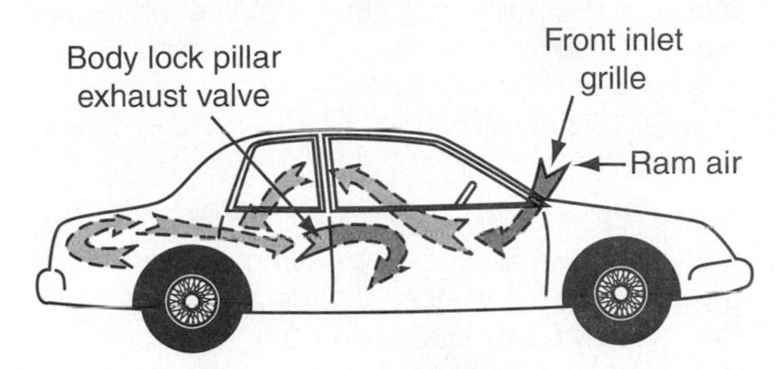

FIGURE 51-1 A flow-through ventilation system is used on some vehicles. Ram air is forced into the inlet grill and sent throughout the passenger and trunk compartments. The air then flows out of the vehicle through exhaust areas.

flow-through ventilation system. In operation, ram air is forced through an inlet grille. The pressurized air then circulates throughout the passenger and trunk compartment. From there, the air is forced outside the vehicle through an exhaust area.

On certain vehicles, air is admitted by opening or closing two vent knobs under the dashboard. The left knob controls air through the left inlet. The right knob controls air through the right inlet. The air is still considered ram air and is circulated through the passenger compartment.

FAN VENTILATION

Rather than using ram air (especially if the vehicle is stopped), a ventilation fan is used. The fan is located in the dashboard. It can be accessible from under the dashboard or from inside the engine compartment. ***Figure 51–2*** shows a typical ventilator assembly. A blower assembly is attached to the motor shaft. The entire unit is placed inside the blower housing. As the **squirrel cage blower** rotates, it produces a strong suction on the intake. A pressure is also created on the output. When the fan motor is energized by using the temperature controls on the dashboard, air is moved through the passenger compartment.

HEATER CORE

All ventilation and heating systems use a **heater core** to increase the temperature within the passenger compartment. ***Figure 51–3*** shows an example of a heater core. Hot fluid flowing through the cooling system is tapped off and sent to the heater core. The heater core is much like a small radiator. It is a liquid-to-air heat exchanger. As warm or hot engine coolant is circulated through the core, air can be heated as it flows through the core fins.

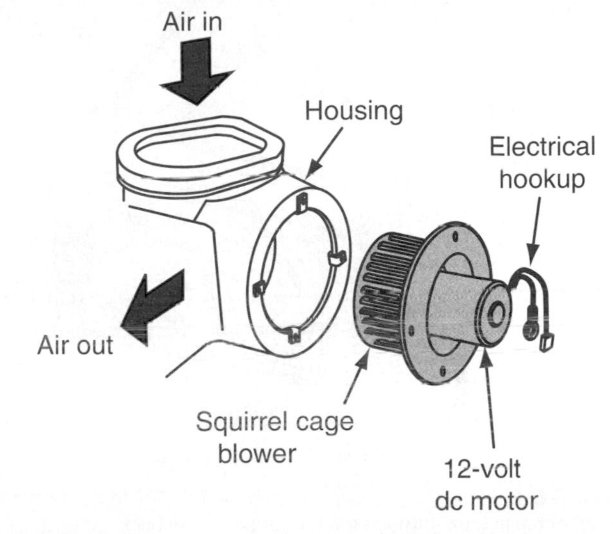

FIGURE 51-2 A fan is used to help move air throughout the passenger compartment. It is called a squirrel cage fan.

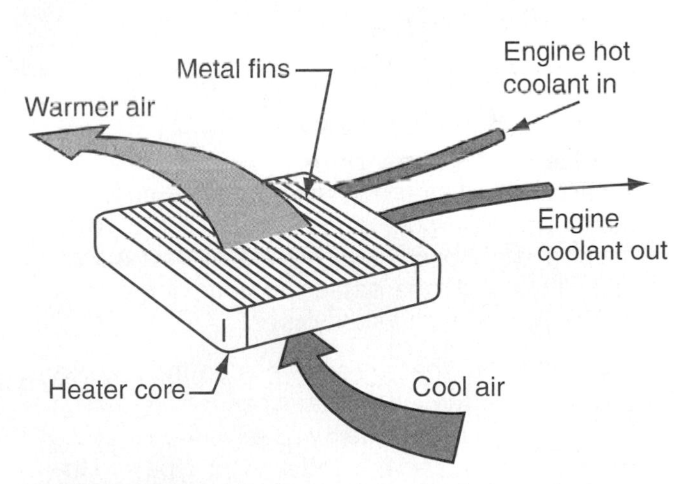

FIGURE 51-3 A heater core is used to increase the temperature of the air inside the passenger compartment.

Figure 51–4 shows the flow of coolant from the engine cooling system, through the heater core, and back to the cooling system. Near the top of the engine, just before the thermostat, a coolant line taps off coolant and sends it to the heater core. From the heater core, the coolant is sent back to the suction side of the cooling system. With this system, the thermostat does not have to be open to get heat to the heater core. Different manufacturers use other circuit connections.

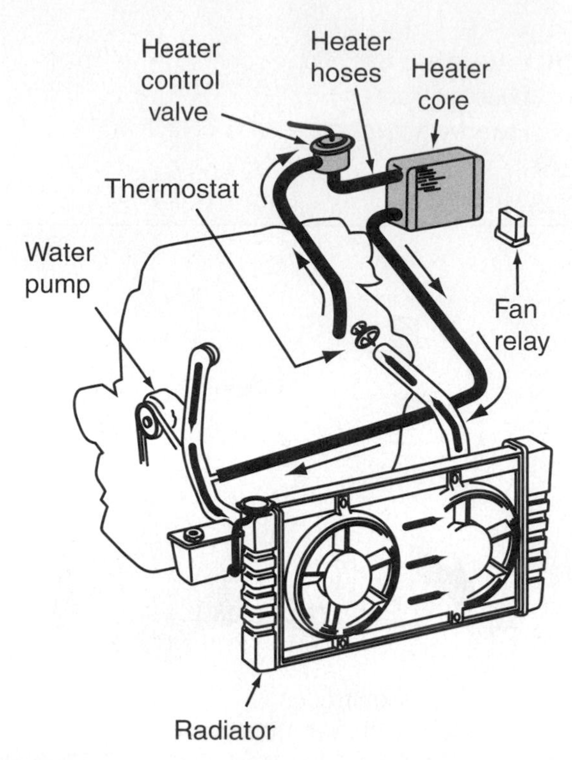

FIGURE 51–4 The coolant for the heater is sent from the upper portion of the engine, through the heater core, and back to the suction side of the engine cooling pump.

Car Clinic: HEATER CORE

CUSTOMER CONCERN:
Car Heater Doesn't Work
A car doesn't heat up in the passenger compartment, especially when it's very cold outside. When the engine is at operating temperature, heat is available. How could the passenger compartment get warmer, especially on short drives?

SOLUTION: The heater core is on the by-pass circuit around the thermostat when the thermostat is closed. First make sure that the thermostat is operating correctly and is the correct type for the vehicle. Then check to see if the heater core and the hoses to the core are hot before the radiator gets hot. They should be hot before the radiator. Also check to see if there are air bubbles in the coolant. On some vehicles, there may also be a shutoff valve to the heater core. Check the valve for correct operation.

Figure 51–5 shows a typical heater core. The engine coolant comes in through the tube on the top, and exits through the tube on the bottom. As air is passed through the fins, it is heated and sent to the passenger compartment.

Car Clinic: HEATER CORE

CUSTOMER CONCERN:
Moisture on the Inside Windows
A customer has complained that the windows always have moisture on the inside. It seems that the moisture is coming from the air vents when the blower is used for heating the inside compartment. The heating and ventilation ductwork was recently cleaned. The moisture also has a sweet smell. What could be the problem?

SOLUTION: When water vapor is found on the inside windows or vapor comes from the vents and the system is not plugged up, the problem is most likely a leaky heater core. Often, the heater core will leak (much like a radiator). If the heater core is leaking into the ductwork, it produces a vapor and moisture in the passenger compartment. The sweet smell comes from the antifreeze in the coolant. The heater core must be removed and either replaced or repaired.

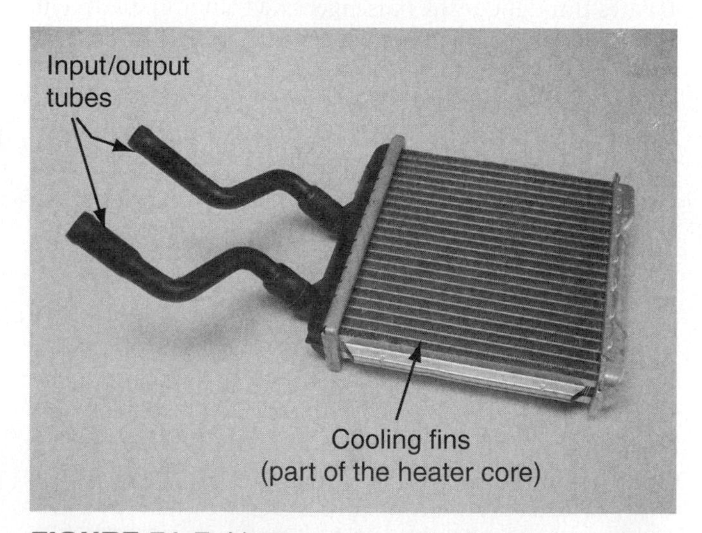

FIGURE 51–5 A heater core has cooling fins to help transfer the heat from the engine to the passenger compartment.

PARTS LOCATOR

*Refer to photo #118 in the front of the textbook to see what a **heater core** looks like.*

MECHANICAL DUCT CONTROLS

One method used to control ventilation and heating systems on older vehicles is by using **air ducts** with small doors that direct the flow of air. The doors can be controlled either mechanically or by vacuum. Mechanical control of the doors

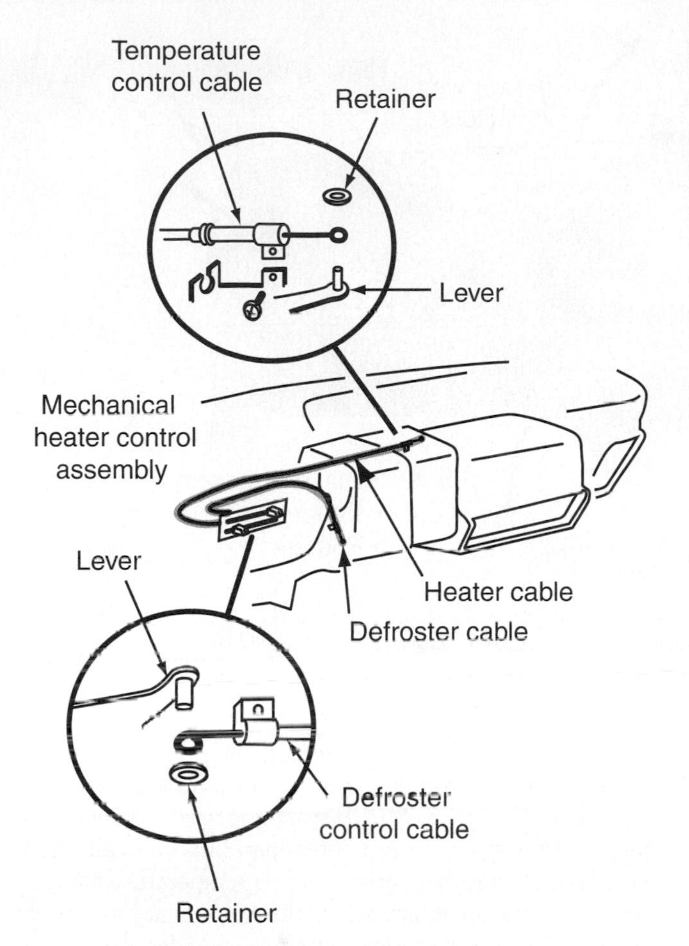

FIGURE 51-6 On older vehicles, opening and closing air duct doors controlled the ventilation and heating system. These doors were operated by mechanical cables. Here a temperature and defroster cable are used to control airflow. As the levers are moved on the control assembly, air duct doors open and close to direct air correctly.

is accomplished by moving the controls on the dashboard. As a control is moved, a cable attached to that control moves the duct door. ***Figure 51–6*** shows two control cables. One is used to operate an air duct door for the temperature control. One is used to operate an air duct door for the **defroster**. The cable is made of a strong steel wire that is wrapped within a flexible tube. As the control assembly knob is moved, the steel wire inside the flexible tube moves, which causes the position of the air duct door to change.

VACUUM DUCT CONTROLS

A second method used to operate the air duct doors is to use a vacuum-operated motor. ***Figure 51–7*** shows how a **vacuum motor** operates. The vacuum motor operates much like a solenoid. Instead of using electricity, vacuum is used. When vacuum is applied to the motor, it causes a small diaphragm inside the motor to move. This motion is transmitted by a rod to the air duct door. The air duct door is then moved from one position to another. In some vehicles, several vacuum motors may be used to operate air duct doors.

Car Clinic: BLOWER FAN

CUSTOMER CONCERN:
Low Volume of Air from Blower

A customer complains that the air volume from the heating and ventilation system is low. The fan motor is running, is able to change speeds, but has low air volume. What could be the problem?

SOLUTION: This type of problem is usually caused by some type of restriction in the air vents, in the intake of the ventilation system, or in the heater core. Check the suction side of the blower inlet. It may be restricted or blocked by leaves, tissues, or other debris that has been drawn into the system from the outside. Also check the outlet vents, making sure there are no restrictions. Check to make sure the venting ductwork has not become separated under the dashboard. Generally, if these items are checked and repaired, the problem should be eliminated.

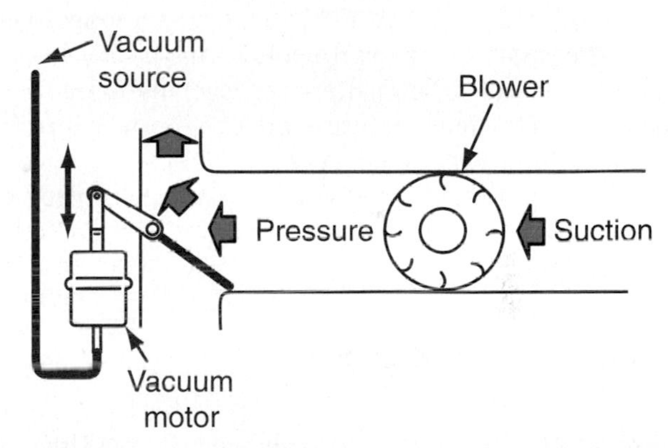

FIGURE 51-7 A vacuum motor can be used to control the operation of air duct doors. When a vacuum is applied to or removed from the vacuum motor, a diaphragm moves. This motion is transmitted by a small rod to the air duct door.

ELECTRICALLY CONTROLLED DUCTS

Another method to control the air duct doors is to use small electrical motors. On vehicles that use computers to control the heating and ventilation system, electrical motors can open and close duct passages. By using electrical control motors, the exact amount of cool and warm air can also be mixed. Depending on the manufacturer, this motor is referred to by different names. Common names include the function control motor, mode control motor, and blend-air door motor.

PARTS LOCATOR

*Refer to photo #119 in the front of the textbook to see what **electrically controlled ducts** look like.*

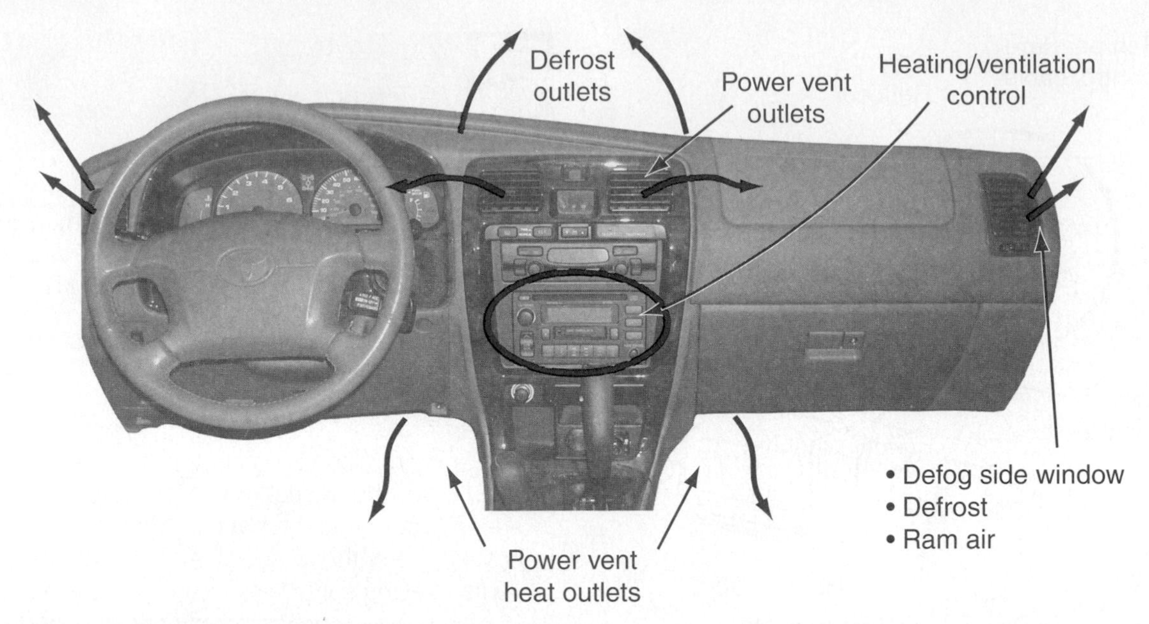

FIGURE 51-8 Air can be directed to many locations through the dashboard. Air duct doors, operated mechanically, by vacuum, or by electric motors open and close to direct air to the correct location.

AIR OUTLETS IN DASHBOARD

Vehicles today use a variety of ducts and passageways to get air into the passenger compartment. Each type of vehicle is different. A typical airflow pattern is shown in ***Figure 51–8***. Depending on the temperature controls, air can be directed to the feet, front windshield, center of the passenger compartment, or to the side windows. Temperature controls on the dashboard open and close doors that direct air to the correct location.

HEATER CONTROLS

There are numerous types and styles of heater controls. ***Figure 51–9*** shows a typical dashboard control. Although there are many types of controls, the functions that usually can be controlled include the fan speed, inside or outside air

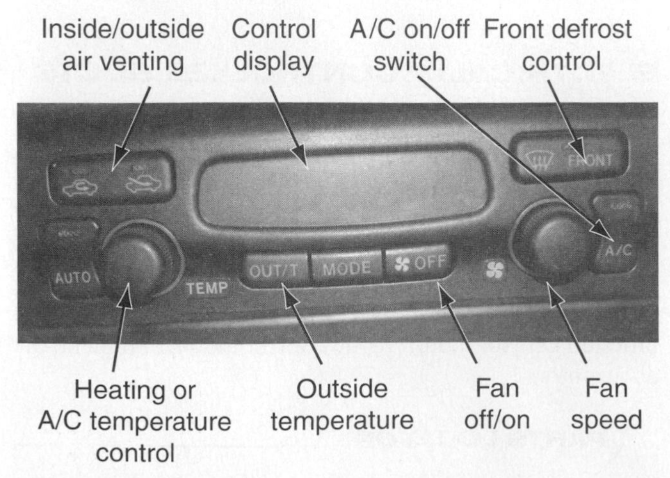

FIGURE 51-9 There are many types of heater controls. Typical controls include a fan control, a temperature control, and selector control (mode) adjustment.

venting, heating or air-conditioning temperature control, A/C on-off switch, front defrost control, and the control display. The mode button on this system is used to display the mode that the system is presently operating in. In addition, many vehicles now have an automatic temperature control. When this button is activated, the system automatically controls temperature. Also, on some vehicles, there is a **selector control lever** to change from heating, to venting, to defrosting, and so on.

Mechanical Control Heating Systems A mechanical heating system is shown in ***Figure 51–10***. Air is first pressurized by the blower. The air is then sent either through the heater core, bypassed, or mixed. The temperature door is positioned so that a mix of air can be obtained. In the uppermost position, airflow through the heater core is blocked. The air is not heated in this mode. When the temperature control on the dashboard is adjusted, it moves the temperature door so that more and more air flows through the heater core. The air temperature then increases accordingly. A second air duct door is used to stop all airflow into the passenger compartment. This occurs when the selector control lever is moved to the off position on the dashboard. The defroster door is also moved by the selector control lever. In the defrost position, the door is positioned so that air moves to the upper part of the dashboard against the windshield. Otherwise, the airflow is directed to the heat outlets in the center or lower portion of the dashboard.

Vacuum Control Heating Systems Another method used to control the heating and ventilation system is to use vacuum. Vacuum from the engine is used to open and close various vents and doors. When a vacuum is applied to the door, the door is either closed or opened. ***Figure 51–11*** illustrates the doors that can be controlled by the vacuum system. The vacuum doors are controlled by the

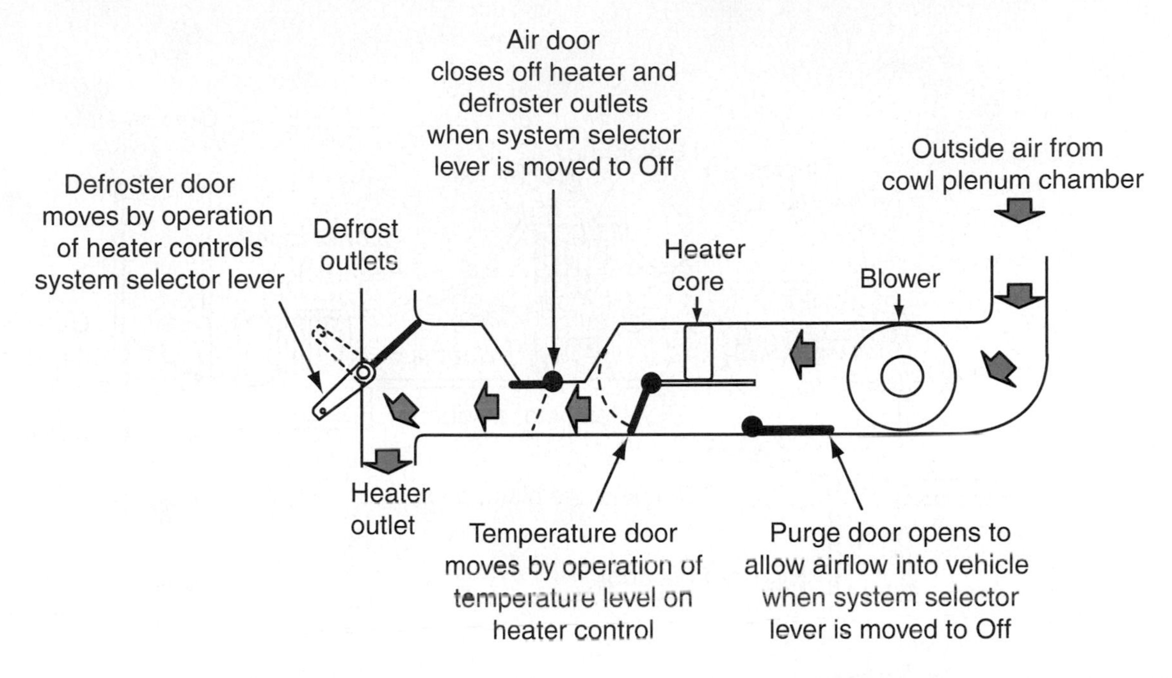

FIGURE 51-10 A typical heating system and airflow. Air from outside is pressurized by the blower. Depending on the position of the air duct doors, different temperatures and air flow can be obtained.

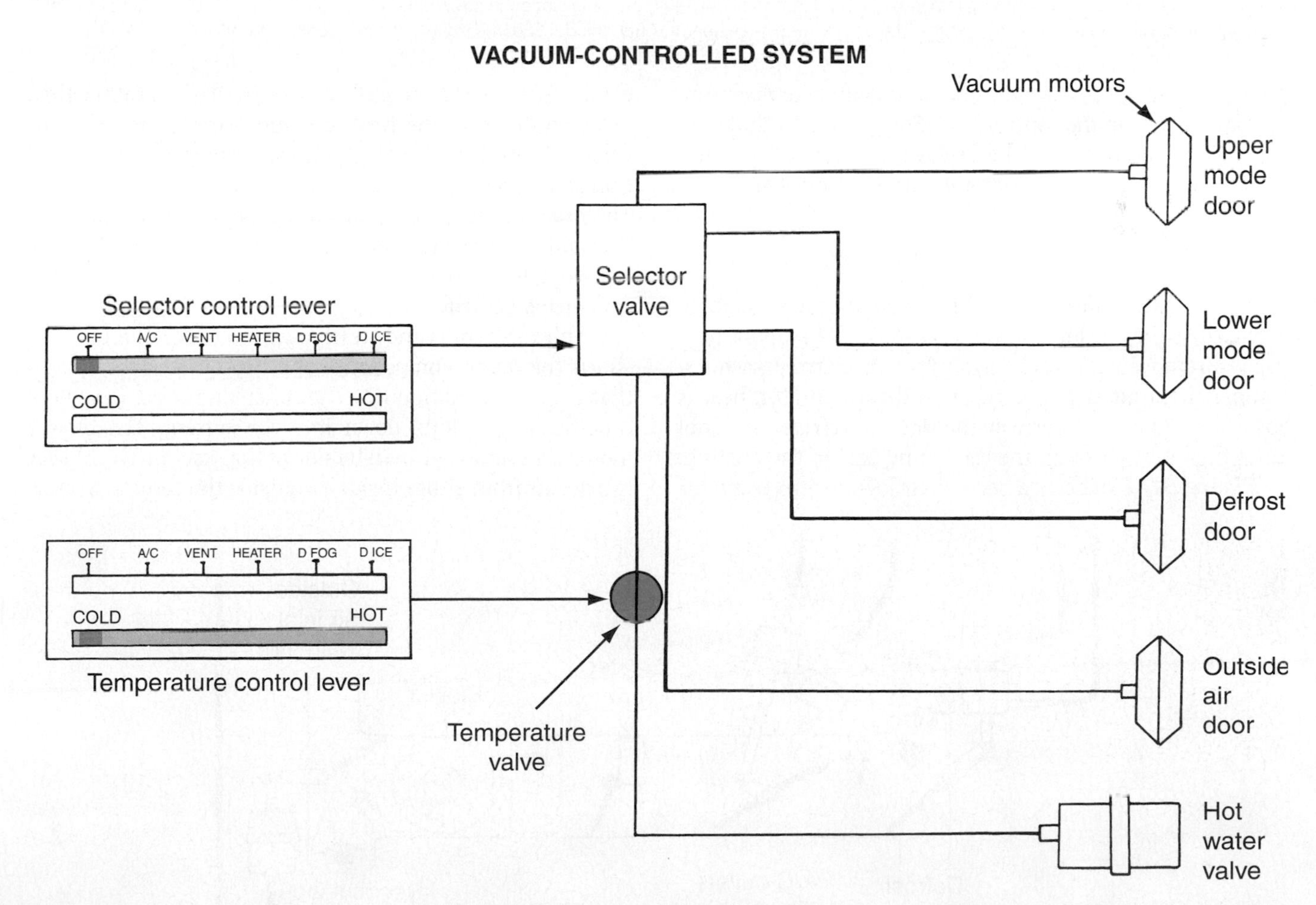

FIGURE 51-11 A vacuum system for controlling airflow is used on some vehicles. The selector control lever controls the selector valve. The temperature control lever controls the temperature valve. Depending on the position of the valve, different doors are moved to produce the required airflow and venting.

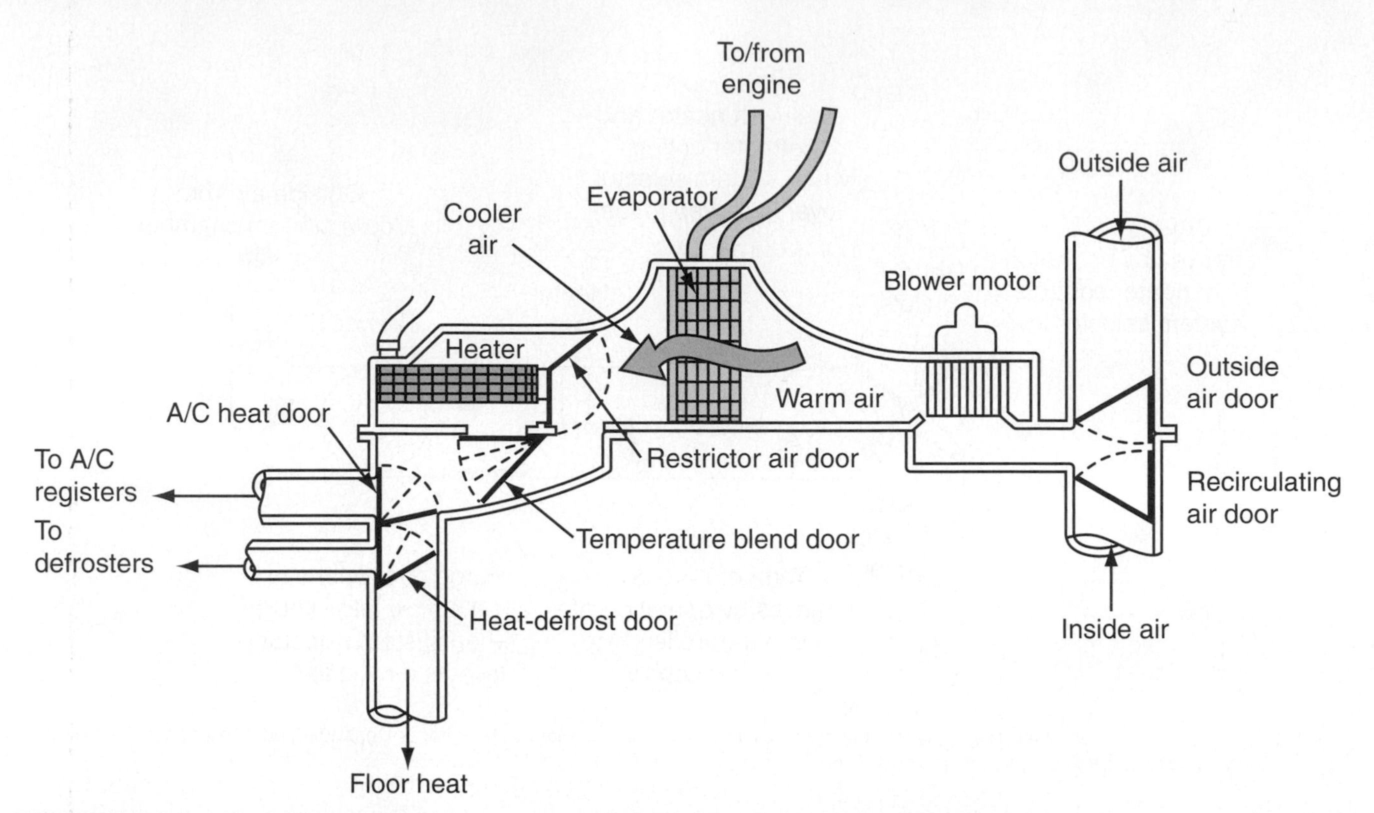

FIGURE 51–12 When an air conditioner is used, an additional evaporator is placed in the airflow. Warm air passing through the evaporator gives up its heat to the internal refrigerant. The cooler air is then sent through the ducting into the passenger compartment.

position of the selector switch and the amount of temperature that is adjusted on the control level. For example, if the selector control switch is set at the "defrost" position, the vacuum is directed to the defrost door for its correct operation.

AIRFLOW WITH AIR CONDITIONING

If the vehicle has air conditioning, an additional evaporator is placed in the air ducts (***Figure 51–12***). The evaporator is used to remove heat from the air flow. As warm passenger-compartment air is passed through the evaporator, heat is absorbed and carried away by the internal refrigerant. Cool air is then sent through the remaining part of the ducting.

Figure 51–13 shows a second airflow system using air conditioning. Here the airflow is recirculating, rather than using fresh air. If the fresh air duct were open, fresh air would be circulated throughout the system. Note the location of the evaporator, heater core, vacuum motors, and fan. The exact design of each airflow system is determined by the vehicle manufacturer, the year, and the model of the vehicle. These two designs show the variation of airflow and supporting controls.

A third system is shown in ***Figure 51–14***. The drawing shows the major components of a two-outlet door system that uses air conditioning. The schematic is a top view of a dashboard, looking down into the system. The blower motor is located on the right side of the system. The blower can get air from either inside or outside the vehicle. The air

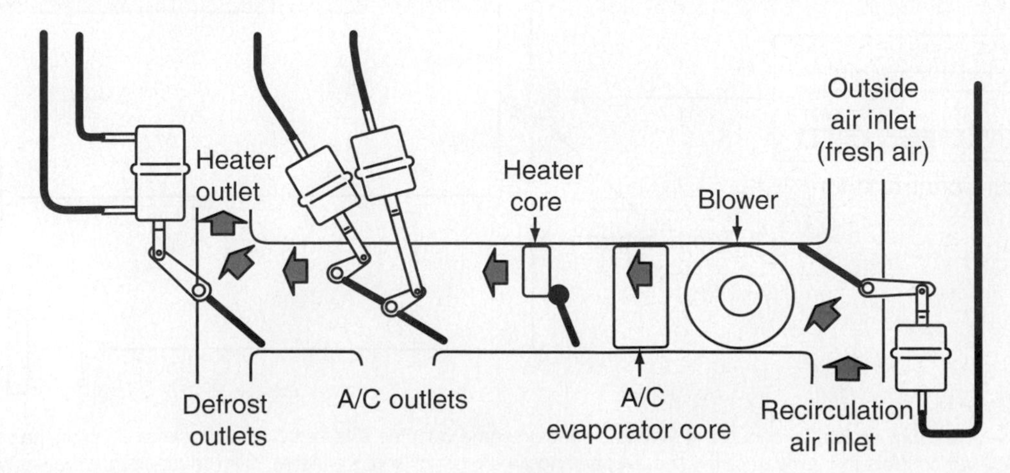

FIGURE 51–13 Typical airflow through a vehicle that has an evaporator used for air conditioning.

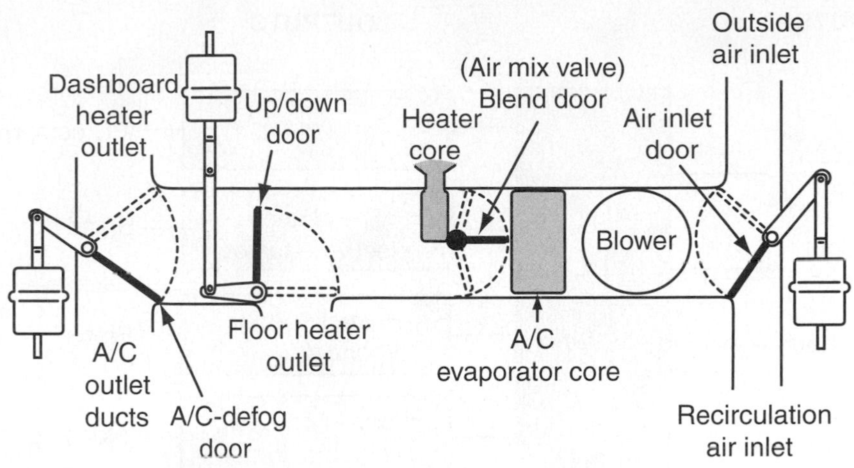

FIGURE 51–14 Another example of an air ducting system using the evaporator to provide air conditioning.

from the blower is then sent through the A/C evaporator core. If the air-conditioning system is operating, the air will be cooled. If the air-conditioning system is off, the air will not be cooled. The air then flows to the air mix valve. Depending upon its position, the air can either go through the heater core or bypass the heater core. The exact mix depends upon the position of the air mix valve. After the heater core, the air then flows to various valves to be exhausted either to the floor heater outlet, the A/C outlet ducts, or dashboard heater outlet. As with other systems, these valves can be controlled mechanically by using a vacuum or by electrical motors and actuators.

The fourth example shown in ***Figure 51–15*** shows the ducting for a rear heat/cool system. In this case, the diverter door determines if the A/C cool air or heating air goes to the rear left/right direction or the middle and upper right/left direction. This type of system is often used on some SUVs and similar vehicles.

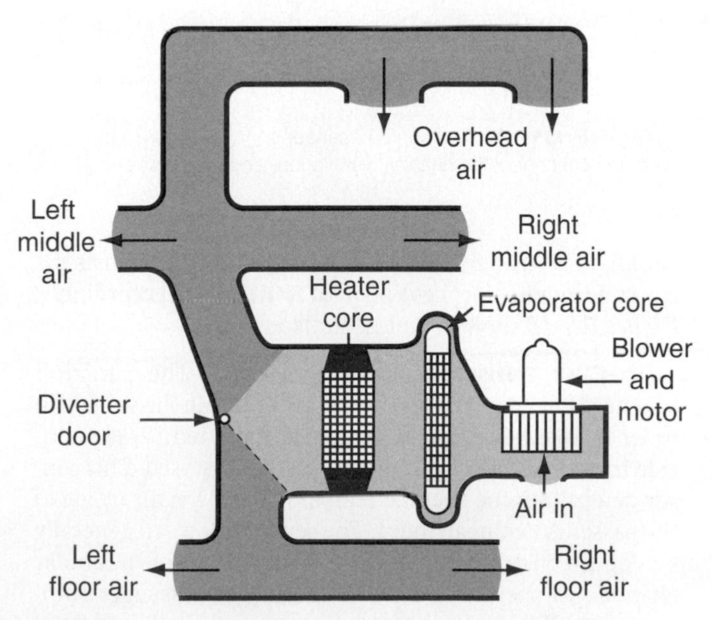

FIGURE 51–15 This illustration shows the airflow for a rear or overhead ducting system.

51.2 COMPUTERIZED AUTOMATIC TEMPERATURE CONTROL

PURPOSE OF COMPUTERIZED AUTOMATIC TEMPERATURE CONTROLS

Today many vehicle manufacturers are using computers to accurately control the air temperature in the passenger compartment. This type of system offers more choices on airflow and outlets than previous mechanical systems did. A computerized system operates using a microprocessor with a memory. The system is designed to regulate the temperature of air in the passenger compartment to the desired temperature level. The microprocessor measures the interior temperature and makes adjustments every few seconds. ***Figure 51–16*** shows a **computerized automatic temperature control**, including the major parts and their location.

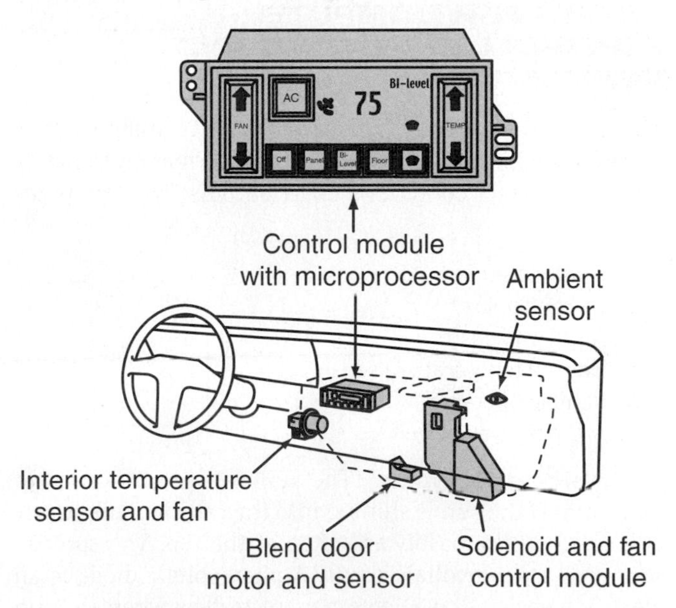

FIGURE 51–16 Many vehicles are using computers to control the ventilation and heating system. The computer senses the interior air temperature and adjusts the temperature every few seconds.

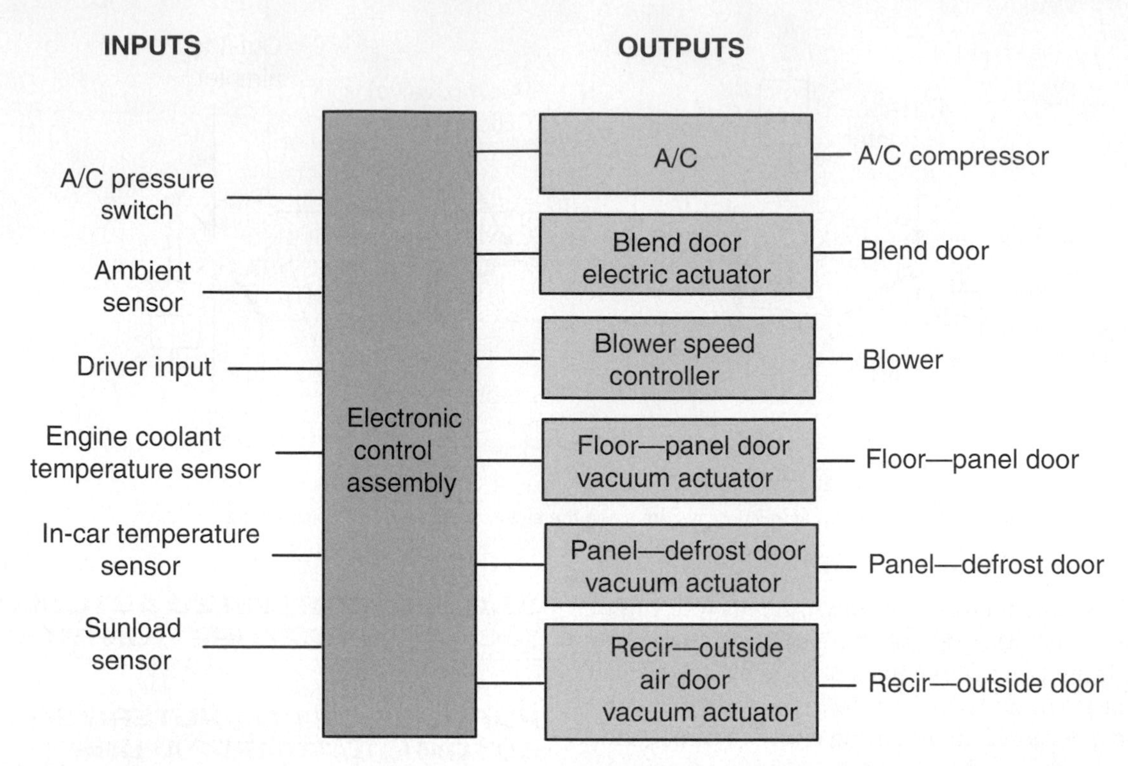

FIGURE 51–17 Many vehicles today use computers to control the heating and ventilation. The electronic control uses six or more inputs to control various outputs of the heating and ventilation system.

Figure 51–17 shows a typical computerized control system called the EATC (electronic automatic temperature control) system. This system uses an electronic control assembly to control various motors and switches. As with engine computers, not all computers are titled the same. For example, one manufacturer calls this computer the body control module (BCM).

INPUTS FOR COMPUTER-CONTROLLED HEATING AND VENTILATION

The EATC system uses a microcomputer to analyze six or more major inputs. It then uses this information to determine the correct conditions of six outputs. The inputs include:

1. Sunload sensor
2. In-car sensor
3. Ambient sensor
4. Engine temperature sensor
5. Vehicle operator
6. A/C pressure switch

Sunload Sensor The **sunload sensor** is used to determine if the sun is shining into the vehicle. This sensor is located on the dashboard often at the top. The sunload sensor is a photovoltaic diode. A photovoltaic diode is an electrical device that allows current to flow when the sun shines on its sensor. When the sun shines on the sensor, a voltage signal is sent to the computer to tell it the sun is shining. Based on this signal, the computer then adjusts the air conditioning or heating and ventilation accordingly. ***Figure 51–18*** shows a typical sunload sensor.

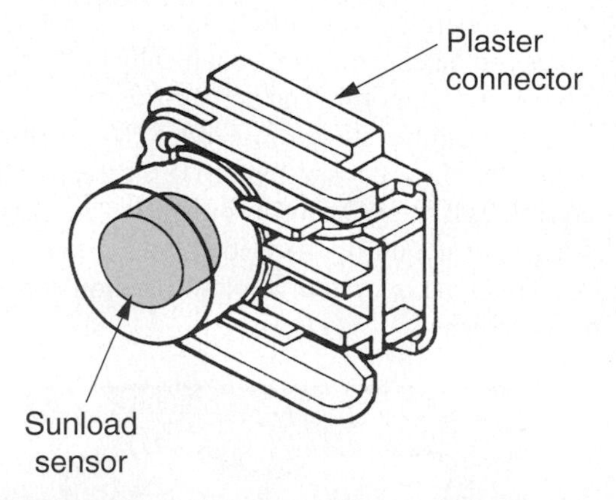

FIGURE 51–18 The sunload sensor sends a signal to the computer to help control the heating, ventilation, and air conditioning.

In-Car Temperature Sensor The computer also needs to know the air temperature inside the vehicle in order to operate correctly. To measure the temperature inside the car, an in-car temperature sensor is used. This sensor determines the average temperature of the air inside of the passenger compartment. The sensor is located generally behind the instrument panel or dashboard inside a tubular channel. In operation, a small amount of passenger compartment air is drawn through the channel. As the air flows past a small thermistor, its resistance changes. The change

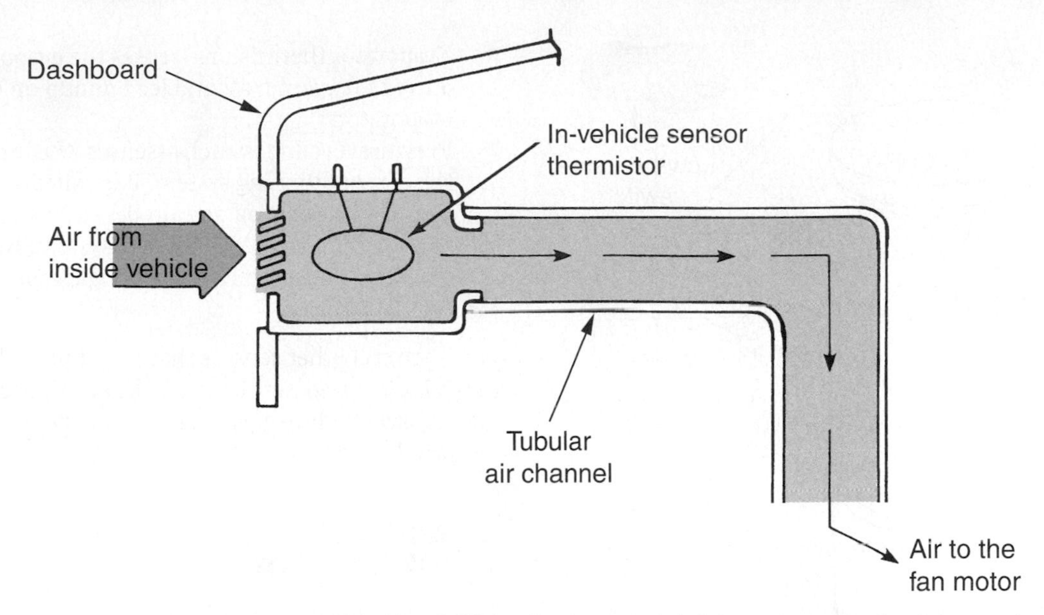

FIGURE 51–19 This in-car sensor is located behind the dashboard in a stream of air. It is used to sense the air temperature inside the passenger compartment.

in resistance is converted to a voltage signal for use as an input to the computer. ***Figure 51–19*** shows the airflow across the in-car temperature device.

Ambient Sensor The **ambient sensor** is a temperature sensing device that measures the air temperature outside the vehicle (***Figure 51–20***). The word *ambient* means surrounding, so the sensor measures the surrounding air temperature near the vehicle. Often the ambient temperature sensor is located in the front of the air-conditioning condenser or in front of the radiator. This sensor is a type of thermistor. However, it includes an electrical circuit and program to prevent false readings. For example, the sensor might get a false reading if it is stopped close behind another vehicle at a stop sign. The program measures the average temperature and sends this signal to the computer.

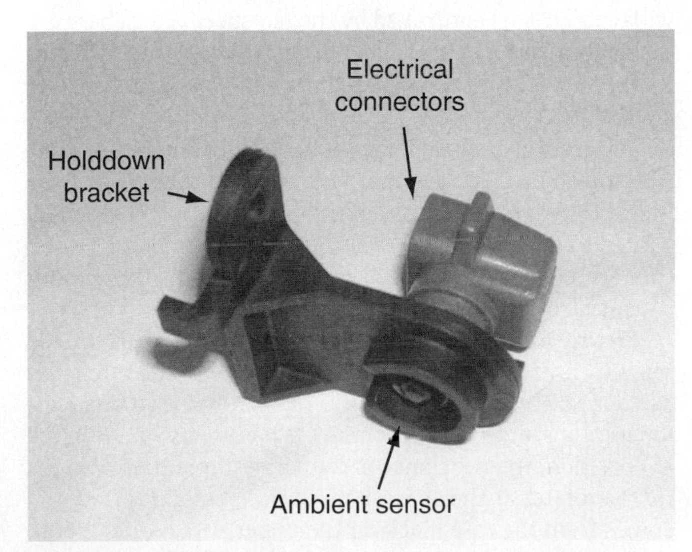

FIGURE 51–20 This ambient sensor is used to determine the air temperature outside the vehicle.

PARTS LOCATOR

*Refer to photo #120 in the front of the textbook to see what an **ambient sensor** looks like.*

Engine Temperature Sensor The engine temperature sensor is the same as the coolant temperature sensor. It is able to measure the temperature of the coolant by using a thermistor. Of course, this sensor also provides coolant temperature information to other computers used on the vehicle as well. This temperature sensor is needed because the computer needs to tell the heating system to send heated air to the passenger compartment when the engine is warm enough.

Vehicle Operator Sensor The vehicle operator sensor is a switch on the dashboard controls. This switch is used to determine the temperature mode that has been selected by the operator of the vehicle. For example, on some vehicles the temperature setting is set by the operator. On other vehicles, a simple switch is used to turn the air conditioning on and off. These switches and settings are considered inputs to the computer.

A/C Pressure Switch The A/C pressure switch is a low-pressure switch. It is able to measure the pressure inside the A/C system on the low side. The switch is generally located on the accumulator, as shown in ***Figure 51–21***. This is a normally closed, or NC, switch. It is designed to open when the low-side pressure drops below 2 to 8 psig. An

PARTS LOCATOR

*Refer to photo #121 in the front of the textbook to see what an **A/C pressure switch** looks like.*

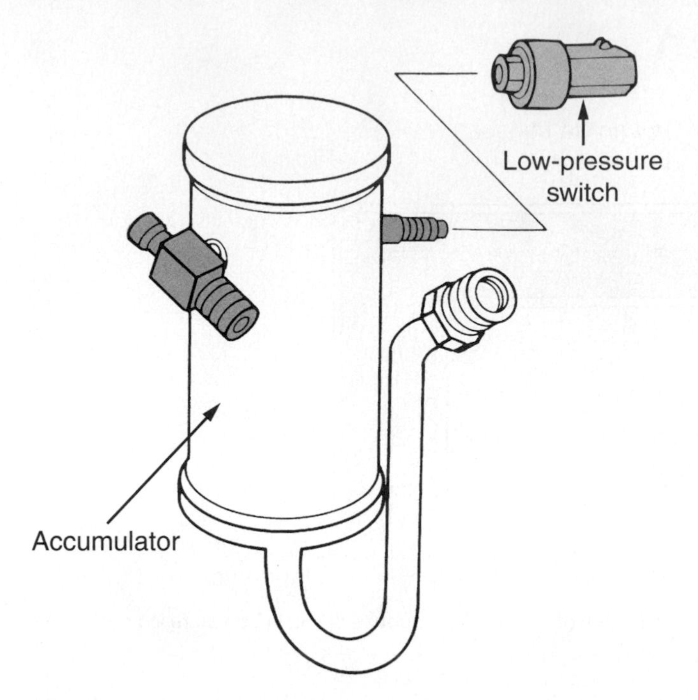

FIGURE 51-21 The low-pressure switch, which is located on the accumulator, senses the pressure on the low-pressure side of the A/C system and acts as an input to the computer.

open switch signals the computer to disengage the compressor clutch circuit during low-pressure operation such as when there is a loss of A/C refrigerant.

ADDITIONAL SENSORS

Other sensors may be used as well. The exact type and style of input sensor will be determined by the year, make, and model of the vehicle. Other input sensors include:

- A/C high-pressure switch—determines the pressure inside the air-conditioning system. This switch measures the high pressure being produced by the A/C system. It is a normally closed switch and opens if the air-conditioning system exceeds a preset pressure. It then closes when the pressure drops back to a normal range. For example, the pressure switch may open above 425 to 435 psig. It then will close when the pressure drops below 200 psig. This switch is a safety switch used to protect the compressor from too much pressure.
- Programmer—receives various electrical inputs from sensors and from the main control panel. Based on the inputs, the programmer provides output signals to the A/C system, to heater valves, blowers, and various mode doors.
- High-side temperature switch—measures the temperature of the high-pressure side of the air-conditioning system.
- Low-side temperature switch—measures the temperature of the low-pressure side of the air-conditioning system.
- Evaporator thermistor—senses the evaporator temperature to prevent frost and ice buildup on the fins of the evaporator.
- Pressure cycling switch—senses the pressure inside the air-conditioning system. It is called a cycling switch because it cycles off and on depending upon the pressure on the low-side of the A/C system. Its cycling controls the magnetic clutch to engage or disengage the A/C compressor.

There are still other sensors that might be included, such as the vehicle speed sensor, throttle position sensor, heater turn-on switch, brake booster vacuum switch, power steering switch, and so on.

COMPUTER-CONTROLLED OUTPUT COMPONENTS

There are many output devices used on computerized heating and ventilation systems. The output being controlled is based on the many inputs used.

On the basis of these inputs, the computer is able to make instant adjustments to the heating, A/C, and ventilation systems. The output components that are controlled include:

- Blower speed
- Blend door position
- Floor panel door position
- Defrost door position
- Recirculating outside door position
- A/C compressor clutch position

BLOWER MOTOR CONTROL

The blower motor is one of the output devices controlled by the electronic automatic temperature control module. The blower motor is responsible for creating an airflow through the air distribution system. The blower is an electric motor with a fan that is controlled by the fan speed dial. There are normally three types of blower controls used in vehicles today: the resistor pack, relay controlled system, and blower control module.

The resistor pack system will be discussed because of its widespread use. The resistor pack controls the speed of the motor by using a series of resistors. As the control switch is turned to different settings by the operator, different resistors are engaged to control the resistance of the circuit, thus controlling the motor speed.

Figure 51–22 shows a common resistor pack in the blower control circuit. In operation, a blower switch (located in the center of the circuit) has a switch that connects different resistors into the circuit. When the switch is in the LO position, the resistance is the highest and therefore the motor rotates at the slowest speed. In this case, electricity comes from the fuse block on the upper part of the circuit, through the mode selector switch, through all three resis-

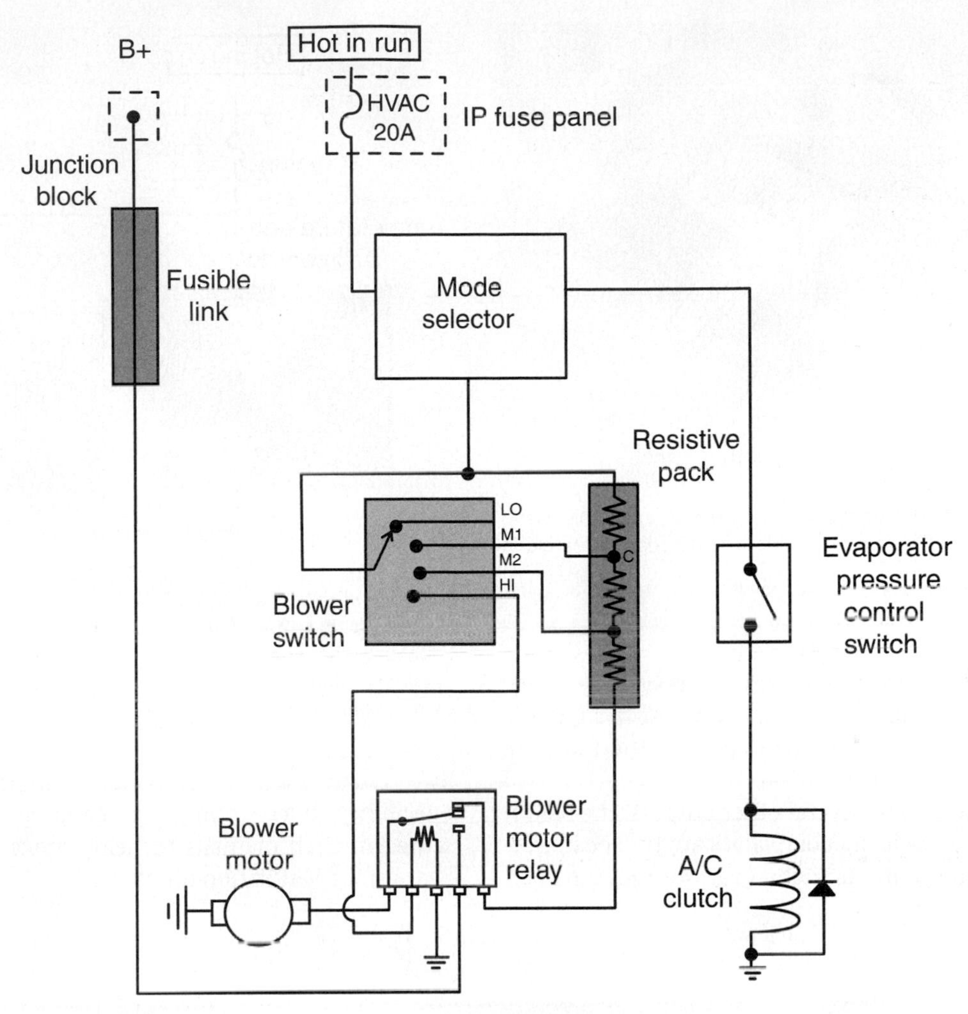

FIGURE 51-22 The resistor pack on the left side of this circuit determines the amount of current sent to the blower motor to achieve different speeds and, thus, to control airflow.

tors, down into the blower relay switch (bottom left of the circuit) and into the blower motor, then to ground.

When the blower switch is placed in the M1 position, the electricity flows from the mode selector switch, through the contacts inside the blower switch, to terminal C on the blower resistors, then through two bottom resistors, and again to the blower motor. In this case, there are only two resistors in the circuit.

When the blower switch is placed in the M2 position, only one resistor is used in the circuit. When the blower is placed in the HI position, electricity flows from the mode selector switch, through the blower switch, directly to the blower relay coil. When this coil is energized, it closes the contacts, causing electricity to flow from the fusible link through the blower relay switch, and directly to the blend door blower motor, without any resistances in the circuit. In this case, there is maximum current flow to the blower motor, thus the highest motor and fan speed.

Some vehicles use the relay control motor system. In this case, a series of relays and resistors is used to control the speed of the motor. The use of the blower control module allows the operation of the blower motor at various speeds. A blower control module is used to regulate the current sent to the blower. The blower control module receives a signal from the HVAC programmer, as discussed in the chapter on air-conditioning systems.

PARTS LOCATOR

Refer to photo #122 in the front of the textbook to see what an ***electronic blower relay*** *looks like.*

ELECTRONIC BLEND DOORS

On some vehicles, the blend doors and the air outlet doors are controlled electronically. This system uses electric motors to provide variable positioning of air doors and valves as shown in ***Figure 51–23***. The settings from the passenger controls determine the exact door or valve positioning. In many cases, the electric motors have electronic circuitry which positions the door and valve according to a controller signal. ***Figure 51–24*** shows such an example. In operation, first the temperature control is set by the passenger or

FIGURE 51-23 The computer controls this electronic blend door motor. The internal motor moves the level to position the blend door correctly.

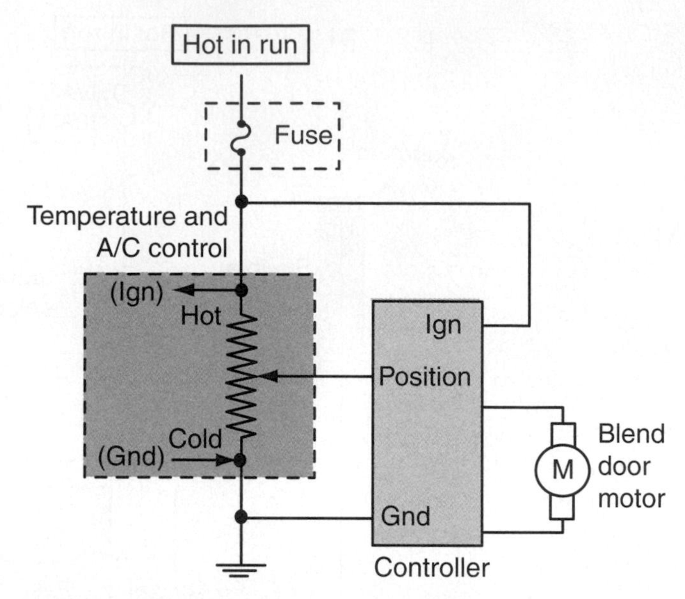

FIGURE 51-24 The setting of the temperature control acts as an input to the controller, which in turn sends an electronic signal to determine the direction of blower motor rotation and the amount of blower motor turning.

driver. The setting is basically changing the resistance, which sends a certain amount of voltage and current (a signal) to the controller. The controller then operates the motor to position the door correctly.

Although there may be several other outputs, the blower motor control and the blend door valves are the most used. To understand other outputs such as various vacuum operated actuators and valves, refer to vehicle service manuals. Specific troubleshooting and service procedures are presented in such manuals to help service technicians solve heating and ventilation problems.

Problems, Diagnosis, and Service

Safety Precautions

1. The fins on a heater core are very sharp. When working on or near a heater core, be careful not to scrape your hands against the core.
2. The heater core on heating and ventilation systems has coolant passing through it and can be very hot. Be careful not to touch these parts when they are hot.
3. The radiator is considered part of the heating system. Always keep hands away from the radiator fan. The fan may start automatically without warning for up to 30 minutes after the engine is turned off.
4. Many heating and ventilation systems use wire cables for control. Be careful when handling these cables, as they are very sharp and could puncture your skin.
5. Always wear OSHA-approved safety glasses when working with heating and ventilation systems.
6. Many electrical components are used on heating and ventilation systems. Always use the correct testing tools when diagnosing these systems.

PROBLEM: Leaking Heater Core

A car has liquid dripping on the front passenger compartment.

DIAGNOSIS

Coolant on the front floor mat or passenger compartment may indicate a leaking heater core. Complete the following diagnostic checks:

1. Check to see if antifreeze is mixed with the liquid. It will be oily if it contains antifreeze.
2. If it is oily, check all hose connections for leaks.
3. Check the drain passageways to make sure they are not clogged with foreign material.
4. Use a pressure checker to test the pressure in the cooling system. This procedure can detect a small leak in the heater core.

SERVICE

1. If the heater core is leaking, it must be removed and repaired or replaced. Follow the correct service manual for removal and replacement procedures.
2. During installation, make sure all hose connections are secure and tight.

PROBLEM: Faulty Electronic Control System

A vehicle with a computerized heating and ventilation system is not maintaining the interior air temperature at the correct level.

DIAGNOSIS

As do engine computers, electronic temperature controls have built-in diagnostics. There are many systems used in vehicles today. One system uses a BCM (body control module) along with the PCM (powertrain control module) to continually monitor the operating conditions for possible malfunctions. When a problem is detected, a two-digit numerical trouble code is stored within the computer. ***Figure 51–25*** shows an example of BCM diagnostic codes. These codes can be displayed when diagnosing the heating and ventilation system. A "Service Now" warning lamp will illuminate on the dashboard. When this occurs, it is necessary to follow the manufacturer's procedure to display the trouble codes. Each vehicle and year may be different, so the exact procedure will vary from vehicle to vehicle.

SERVICE

Once the exact trouble is detected, the faulty component will need to be replaced. Follow the manufacturer's procedure for removal and replacement of each item needing repair.

BODY CONTROL MODULE DIAGNOSTIC CODES

CODE	CIRCUIT AFFECTED
▼ F10	Outside Temp Sensor Ckt
▼ F11	A/C High Side Temp Sensor Ckt
▼ F12	A/C Low Side Temp Sensor Ckt
▼ F13	In-Car Temp Sensor Ckt
▼ F30	CCP to BCM Data Ckt
▼ F31	FDC to BCM Data Ckt
▼ F32	PCM-BCM Data Ckt's
▼ F40	Air Mix Door Problem
▼ F41	Cooling Fan Problem
□ F46	Low Refrigerant Warning
□ F47	Low Refrigerant Condition
□ F48	Low Refrigerant Pressure
▼ F49	High Temp Clutch Disengage
▼ F51	BCM Prom Error

□ Turns on "service air cond" light
▼ Does not turn on any light

Comments:
F11 turns on cooling fans when A/C clutch is engaged
F12 disengages A/C clutch
F32 turns on cooling fans
F30 turns on ft. defog at 75° F
F41 turns on "coolant temp/fans" light when fans should be on
F47 & F48 switches from "auto" to "econ"

FIGURE 51–25 These trouble codes are stored in the computer for diagnosis of the heating and ventilation system. (Courtesy of MOTOR Publications, Air Conditioning and Heater Manual, © The Hearst Corporation)

PROBLEM: Insufficient Heating

A vehicle has insufficient heating in the passenger compartment.

DIAGNOSIS

Many common diagnostic checks can be made to determine the cause of the problem. The following are common checks made on heating and ventilation systems:

1. A gurgle, whine, or "swish" in the heater may indicate that air is mixed with the coolant in the cooling system. Check the engine coolant level in the radiator. Also check for obstructions in the heater core or hoses. Repair as necessary.
2. On some vehicles, the temperature control cable between the dash and the temperature door may need adjustment. This can be done by adjusting the cable length shorter or longer. Follow the manufacturer's recommended procedure for adjusting the cable length. There

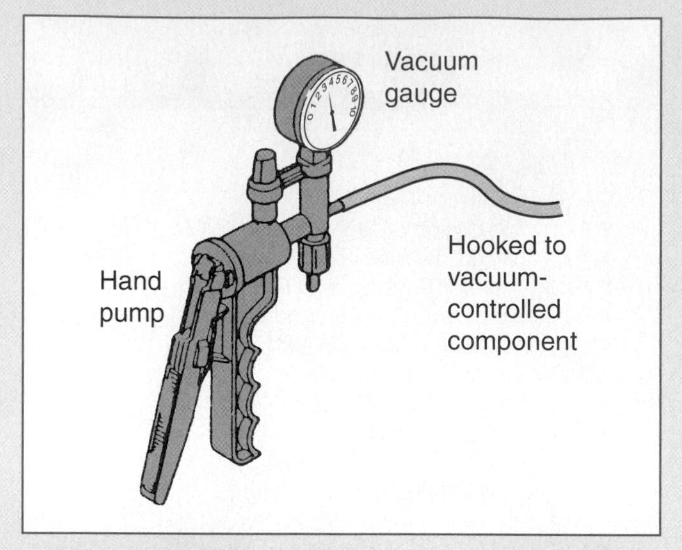

FIGURE 51–26 Use a vacuum pump and gauge to check the condition and operation of vacuum-controlled heating and ventilation system components.

should be a uniform effort from full cold to full hot. In addition, there should be an audible stop sound when the temperature door reaches the end position.

3. If the fan blower is inoperative, check the blower fuse. If the fuse is not bad, check for an open circuit between the ignition switch and the blower motor. If the circuitry appears to be correct, check the blower fan switch for damage. Also, check the blower motor resistance and compare with manufacturer's specifications.
4. Heating and ventilation systems that use vacuum hoses for control should be checked with a vacuum tester. ***Figure 51–26*** shows a typical vacuum tester. The vacuum tester is used to place a vacuum on the system to observe the sealing for leaks and to see if the components are operating correctly.
5. An engine that has a damaged cooling system with internal rust and other contaminants will usually develop insufficient heating problems. For example, if silicon and calcium deposits form in the radiator, they will form in the heater core as well. Such deposits cause poor heat transfer into the passenger compartment. The heater core will have to be removed and cleaned before it will operate correctly.
6. For a complete list of diagnostic checks concerning insufficient heating, refer to ***Figure 51–27*** to identify the diagnosis and service.

PROBLEM	DIAGNOSIS	SERVICE
Insufficient heating	Slow warming in car.	Incorrect operation of controls. Advise operator of proper operation of heater controls. Explain operation of vents and controls. Low coolant level; add coolant. Check control cable and blower operation.
	Objectionable engine or exhaust fumes in car.	Check for seal between engine compartment and plenum. Check for proper sealing between air inlet duct assembly and cowl. Locate and seal any other air leaks.
	Cold drafts on floor.	Check operation and adjustment of vent cables. Advise operator of proper operation of heater system. Advise operator to use blower to force air to rear seat area. Check to be sure front floor mat is under floor mat retainer at cowl.
	Insufficient heat to rear seat.	Obstruction on floor, possibly wrinkled or torn insulator material between front seat and floor. Advise operator to use high blower speed.
	Low engine coolant level—drop in heater air temperature at all blower speeds.	Check radiator and cooling system for leaks, correct and fill to proper level. Run engine to clear any air lock.
	Failure of engine cooling system to warm up.	Check engine thermostat; replace if required. Check coolant level.
	Kinked heater hoses.	Remove kink or replace hose.
	Foreign material obstructing water flow through heater core.	Remove foreign material if possible, otherwise, replace core—can usually be heard as squishing noise in core.
	Temperature door (valve) improperly adjusted. Air doors do not operate.	Adjust cable. Check installation and/or adjustment of air control or air-defrost cable.

FIGURE 51–27 Insufficient heating can be caused by many damaged or nonoperational components.

SERVICE

The service procedure will depend on the damaged or faulty part. Refer to the manufacturer's service manual for the correct procedure.

PROBLEM: Faulty Defrost System

There is an inadequate removal of fog and ice from the front windshield.

DIAGNOSIS

Several malfunctions can cause inadequate removal of fog and ice from the front windshield. ***Figure 51–28*** shows some of the common diagnosis and service suggestions. Each of these causes may have a certain procedure for diagnosing its condition. For example, assume that the blower motor is inoperative. The service manual will give a specific procedure to help diagnose the blower motor. ***Figure 51–29*** shows one of many charts used to help the service technician diagnose heating and ventilation components.

SERVICE

The exact service procedure will depend upon the problem or malfunction. Refer to the service manual for the correct service procedure to be used.

PROBLEM	DIAGNOSIS	SERVICE
Inadequate removal of fog or ice	Air door does not open. Defroster door does not open fully.	Check cable operation, lubricate if necessary, or replace if broken.
	Air door does not open.	Check installation and/or adjustment of air control or air-defrost cable.
	Temperature door does not open.	Check and adjust temperature control cable if necessary.
	Obstructions in defroster outlets at windshield.	Remove obstruction. Look for and repair loose instrument panel pad cover at defroster outlet.
	Damaged defroster outlets.	Reshape outlet flange with pliers. The outlet should have a uniform opening.
	Blower motor not connected.	Connect wire. Check ground.
	Inoperative blower motor.	Check heater fuse and wiring. Replace motor if necessary. See service manual for blower motor diagnosis.
	Inoperative blower motor switch.	Replace switch if necessary.

FIGURE 51–28 Inadequate removal of fog and ice from the windshield can be caused by several malfunctions.

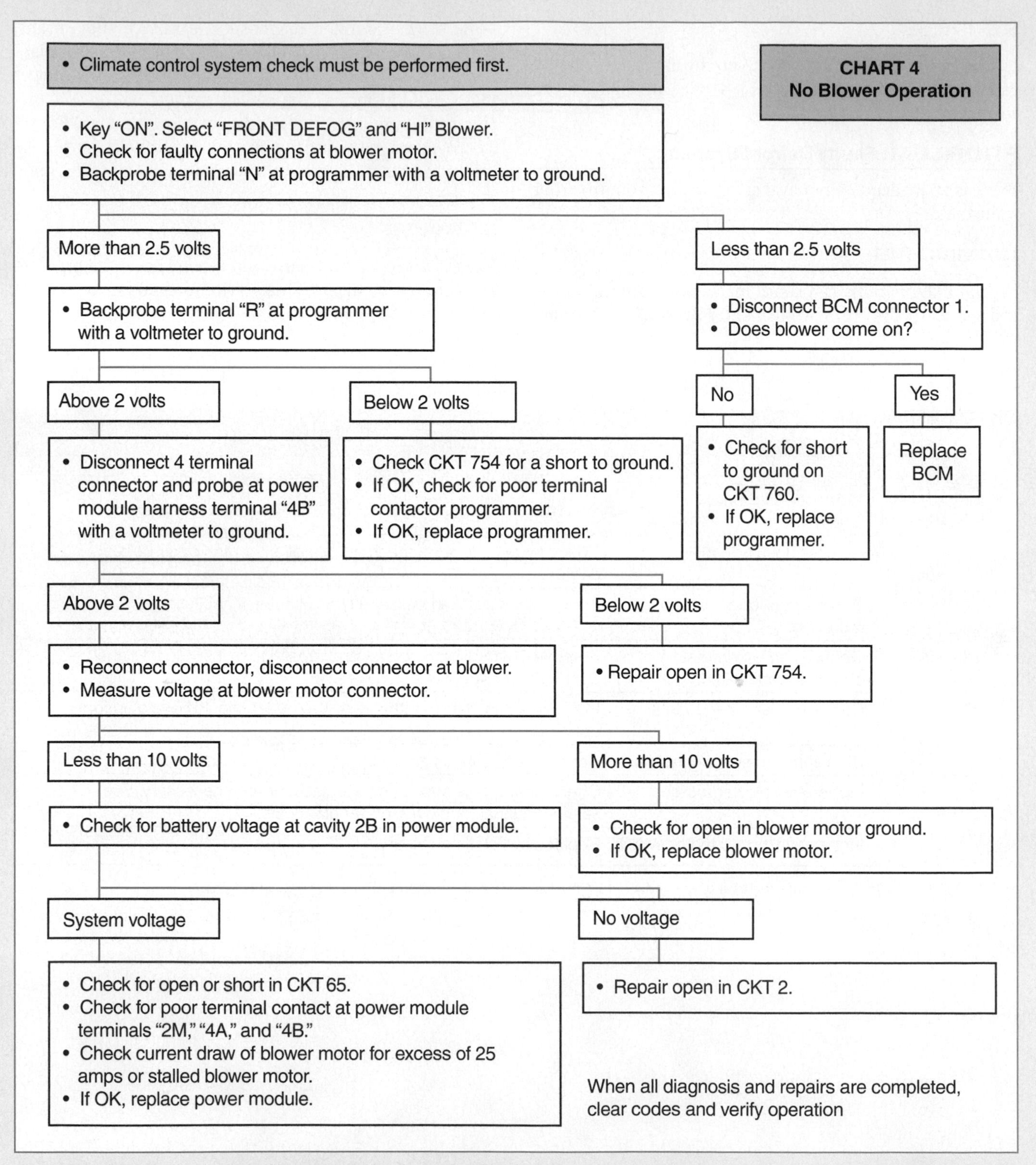

FIGURE 51-29 When a specific component has been identified through the trouble codes, various diagnosis charts are available to help the service technician further check the component. (Courtesy of MOTOR Publications, Air Conditioning and Heater Manual, © The Hearst Corporation)

Service Manual Connection

There are many important specifications to keep in mind when working with heating and ventilation systems. To identify the specifications for your vehicle, you will need to know the VIN (vehicle identification number) of the vehicle, the type and year of the vehicle, and the type of engine used. Although they may be titled differently, some of the more common heating and ventilation system specifications (not all) found in service manuals are listed below. Note that these are typical examples. Each vehicle may have different specifications.

Common Specification	Typical Example
Air floor temperature (floor outlet)	150 degrees Fahrenheit at 80 degrees Fahrenheit ambient
Automatic temperature control resistance	2,600 ohms
Blower motor resistance	1.79–2.06 ohms
Water temperature sensor resistance	5,073 ohms at 160 degrees Fahrenheit

In addition, service manuals provide specific directions for various service and testing procedures. Some of the more common procedures include:

- Blend door module replace
- Control display panel replace
- Damper control motor assembly replace
- Diagnosis and testing of automatic temperature controls
- Heater diagnosis
- Heater output performance test

SUMMARY

The following statements will help to summarize this chapter:

- Automobile heating and ventilation systems keep the passenger compartment comfortable during all weather conditions.
- The flow-through ventilation system uses ram air from the forward motion of the car to force air into the passenger area.
- Blower fans provide air movement when the vehicle is not moving.
- The blower fan is typically a squirrel cage design driven by a small dc motor.
- A heater core is used inside the heater and ventilation system to get heat into the vehicle.
- The heater core is a liquid-to-air heat exchanger.
- Mechanical duct controls use a steel wire attached to the dashboard controls.
- Vacuum duct controls are operated by engine vacuum.
- Air outlets are placed in various positions on the dashboard to direct air to the proper location inside the vehicle.
- Heater controls are used to adjust the temperature of the air entering the passenger compartment.
- To get different air temperatures, air duct doors open or close to allow more or less air through the heater core.
- The selector control lever is used to select the correct mode of operation for the heater and ventilation system such as defrost, heating, venting, and so on.
- With the increased use of computers on today's vehicles, more precise temperature control can be obtained.
- The computerized automatic temperature control monitors air temperatures inside and outside the vehicle.
- Several sensors are used to aid the computerized automatic temperature, including the sunload sensor, in-car temperature sensor, ambient sensor, engine temperature sensor, vehicle operator sensor, and the A/C pressure switch.
- The most common problems in heating and ventilation systems include leaky heater cores, misadjusted controls, and a faulty cooling system.

TERMS TO KNOW

Can you explain each of the following terms? Review the chapter until you can use each term correctly.

Air ducts
Ambient sensor
Computerized automatic temperature control
Defroster
Heater core
Ram air
Selector control lever
Squirrel cage blower
Sunload sensor
Vacuum motor

REVIEW QUESTIONS

Multiple Choice

1. Many vehicles are ventilated by using:
a. The heater core
b. Flow-through or ram ventilation
c. The evaporator
d. Engine coolant
e. None of the above

2. The fan motor and blower on a typical heating and ventilation system uses a/an:
a. Squirrel cage blower
b. Heater core motor
c. Evaporator to heat the air
d. Positive displacement cooling pump
e. Belt-driven pump

3. Heat is put into the air going into the passenger compartment at the:
a. Evaporator
b. Vacuum motor
c. Selector control lever
d. Heater core
e. None of the above

4. The actual thermal energy or heat put into the passenger compartment comes from the:
a. Electrical wires heating up
b. Evaporator
c. Coolant from the cooling system
d. Vacuum motor
e. Squirrel cage blower

5. To get different temperatures from the temperature control lever on the dashboard:
a. Cool air is mixed with warm air through the heater core
b. The evaporator pressure drops
c. The electricity on the coils decreases
d. The selector control lever is put in the off position
e. All of the above

6. The air duct doors can be operated from/by:
a. The vacuum motors
b. Mechanical cables
c. The control levers on the dashboard
d. All of the above
e. None of the above

7. The temperature control lever on the dashboard is used to:
a. Mix air to the correct temperature
b. Select venting or heating
c. Change the temperature of the engine coolant
d. All of the above
e. None of the above

8. What major component is used on the vacuum control heating system?
a. Vacuum motor
b. Thermostat
c. Vacuum coolant
d. All of the above
e. None of the above

9. When a car has air conditioning, which component is added into the airflow system?
a. Heater core
b. Evaporator
c. Heater blower
d. Computer control
e. Vacuum motor

10. Which type of system monitors air temperature, both inside and outside the vehicle, and makes an adjustment every few seconds to keep the temperature at the correct setting?
a. Evaporator system
b. Computerized automatic temperature control
c. Heater core automatic
d. Microprocessor control system
e. None of the above

11. Engine coolant leaking onto the front floor mats could be an indication of a:
a. Leaky vacuum motor
b. Poor sealing air duct door
c. Leaky heater core
d. Leaky control system
e. All of the above

12. What would need to be done if the heated air was not hot enough and the temperature control lever didn't seem to work smoothly?
a. Adjust the temperature control lever
b. Readjust the position of the evaporator
c. Readjust the position of the heater core
d. Replace all vacuum motors
e. Replace the fan motors and blower assembly

13. The ambient temperature sensor is used on the EATC ventilation system to sense the:
a. Pressure of the evaporator
b. Temperature of the inside of the passenger compartment
c. Temperature of the outside of the vehicle
d. Temperature of the sunload sensor
e. None of the above

14. On the resistor pack system for heating and ventilation, the resistors are used to:
a. Change the blower speed so that different amounts of air will be sent through the ventilation system
b. Allow the blower motor to spin at different speeds
c. Change the current being sent to the blower motor
d. All of the above
e. None of the above

The following questions are similar in format to the ASE (Automotive Service Excellence) test questions.

15. There is a gurgle or swishing sound coming from under the dashboard near the heating core. *Technician A* says that there is air in the cooling system. *Technician B* says that the dashboard controls are improperly adjusted. Who is correct?
a. A only c. Both A and B
b. B only d. Neither A nor B

16. Antifreeze and water are found on the front seat car mat. *Technician A* says that the heater core may be leaking and should be repaired. *Technician B* says the hose connections to the heater core may be leaking. Who is correct?
a. A only c. Both A and B
b. B only d. Neither A nor B

17. There is insufficient heating in the passenger compartment. *Technician A* says that because the cooling system has silicon and calcium built up in it, this may also damage the heater core. *Technician B* says that the dash controls may need to be adjusted. Who is correct?
a. A only c. Both A and B
b. B only d. Neither A nor B

18. The blower fan is inoperative. *Technician A* says to check the blower fan fuse. *Technician B* says to check the heater core for signs of being plugged. Who is correct?
a. A only c. Both A and B
b. B only d. Neither A nor B

19. A vehicle with a computerized heating and ventilation system has insufficient heating. *Technician A* says to check the trouble codes stored in the computer. *Technician B* says that the light-sensitive diode on the in-car sensor needs replacement. Who is correct?
a. A only c. Both A and B
b. B only d. Neither A nor B

20. *Technician A* says that a vacuum pump and gauge can be used to diagnose heating and ventilation systems. *Technician B* says that a compression gauge can be used to check the heating and ventilation systems. Who is correct?
a. A only c. Both A and B
b. B only d. Neither A nor B

21. *Technician A* says that air duct doors are controlled by vacuum. *Technician B* says that air duct doors are controlled by electrical motors. Who is correct?
a. A only c. Both A and B
b. B only d. Neither A nor B

22. A car has insufficient heating. *Technician A* says to check if the cooling system has warmed up. *Technician B* says to check if foreign material is blocking the water flow through the radiator core. Who is correct?
a. A only c. Both A and B
b. B only d. Neither A nor B

Essay

23. Describe the squirrel cage blower design.

24. What is the purpose of the vacuum motor on heating and ventilation systems?

25. What is the purpose of the reserve vacuum tank?

26. Describe the purpose and operation of the heater core.

27. Define the computerized automatic temperature control system.

Short Answer

28. Heating and ventilation ducts can be controlled by vacuum or by ________.

29. Name three of the six sensors used to monitor the EATC (electronic automatic temperature control) system. ________, ________, ________

30. Name three of the six components controlled by the computer on the EATC system. ________, ________, ________

31. If antifreeze is found in the front passenger compartment it is a sign that the ________ is leaking.

32. If there is air in the coolant, there will be a(n) ________, ________, or ________ sound in the heater core.

Applied Academics

TITLE: Heat Transfer on the Heater Core

ACADEMIC SKILLS AREA: Science

NATEF Automobile Technician Tasks:

The service technician is able to explain the concept of heat transfer in terms of conduction and/or convection in automotive heating and air-conditioning system failures.

CONCEPT:

Heat or thermal energy can transfer from one object to another in three ways. When heat is transferred by conduction, the transfer is between two objects that are touching each other. When heat is transferred by convection, heat transfer is from a solid object to a fluid substance such as water, coolant, or air. There is circulation of the heat in the liquid or gas. When heat is transferred by radiation, the heat is transferred from a solid to air as rays of heat. Heat is usually measured as wavelengths. The hotter the temperature of an object, the higher the frequency of radiation. The heater core shown here uses these three principles to transfer heat from the coolant to the passenger compartment when the heating system is being used on an automobile.

APPLICATION:

Using the three methods of heat transfer, can you identify where each type of transfer is occurring on the heater core and in the passenger compartment, and when heat is being transferred from the coolant to the passenger compartment? For example, what type of heat transfer occurs from the core tubes to the core fins? What type of heat transfer occurs from the fins to the surrounding air? What type of heat transfer occurs within the passenger compartment from the front seat to the back seat?

Cruise Control Systems

CHAPTER 52

OBJECTIVES

After studying this chapter, you should be able to:

- Identify the various parts of a standard cruise control system.
- State the basic operation of a standard cruise control system.
- Describe the parts and operation of electronic cruise control systems.
- Identify common problems, their diagnosis, and service procedures for cruise control systems.

INTRODUCTION

This chapter is about the cruise control systems used by many of today's vehicles. The cruise control system is used during highway driving to keep the vehicle at a constant speed. This chapter introduces the basic parts and operation of common cruise control systems. In addition, this chapter presents the parts and operation of electronic cruise control systems.

52.1 PURPOSE OF CRUISE CONTROL SYSTEMS

Cruise control systems are designed to allow the driver to maintain a constant speed without having to apply continual foot pressure to the accelerator pedal. Selected cruise speeds are easily maintained, and speed can be easily changed. Several override systems also allow the vehicle to be accelerated, slowed, or stopped.

When engaged, the cruise control components set the throttle position to the desired speed. The speed is maintained even with heavy loads and steep hills. Also, the cruise control is disengaged whenever the brake pedal is depressed. ***Figure 52–1*** shows the common components of a cruise control system and their location in the vehicle.

The operational controls of a cruise control system are usually located near the steering wheel so they can be easily operated by the driver. Usually the driver has the ability to turn the cruise control on and off as well as increase or

Certification Connection

ASE Connection: The information in this chapter can help you prepare for the National Institute for Automotive Service Excellence (ASE) certification tests. The test and content area most closely related to this chapter are:

Test A6—Electrical/Electronic Systems

- **Content Area**—Accessories Diagnosis and Repair

NATEF Connection: Much of the information in this chapter is related to the NATEF tasks. The NATEF tasks and priority numbers most closely related to this chapter are:

1. Inspect and test gauges and gauge sending units for cause of intermittent, high, low, or no gauge readings; determine necessary action. P-1
2. Inspect and test connectors, wires, and printed circuit boards of gauge circuits; determine necessary action. P-3
3. Inspect and test sensors, connectors, and wires of electronic instrument circuits; determine necessary action. P-2
4. Diagnose incorrect operation of cruise control systems; determine necessary action. P-3

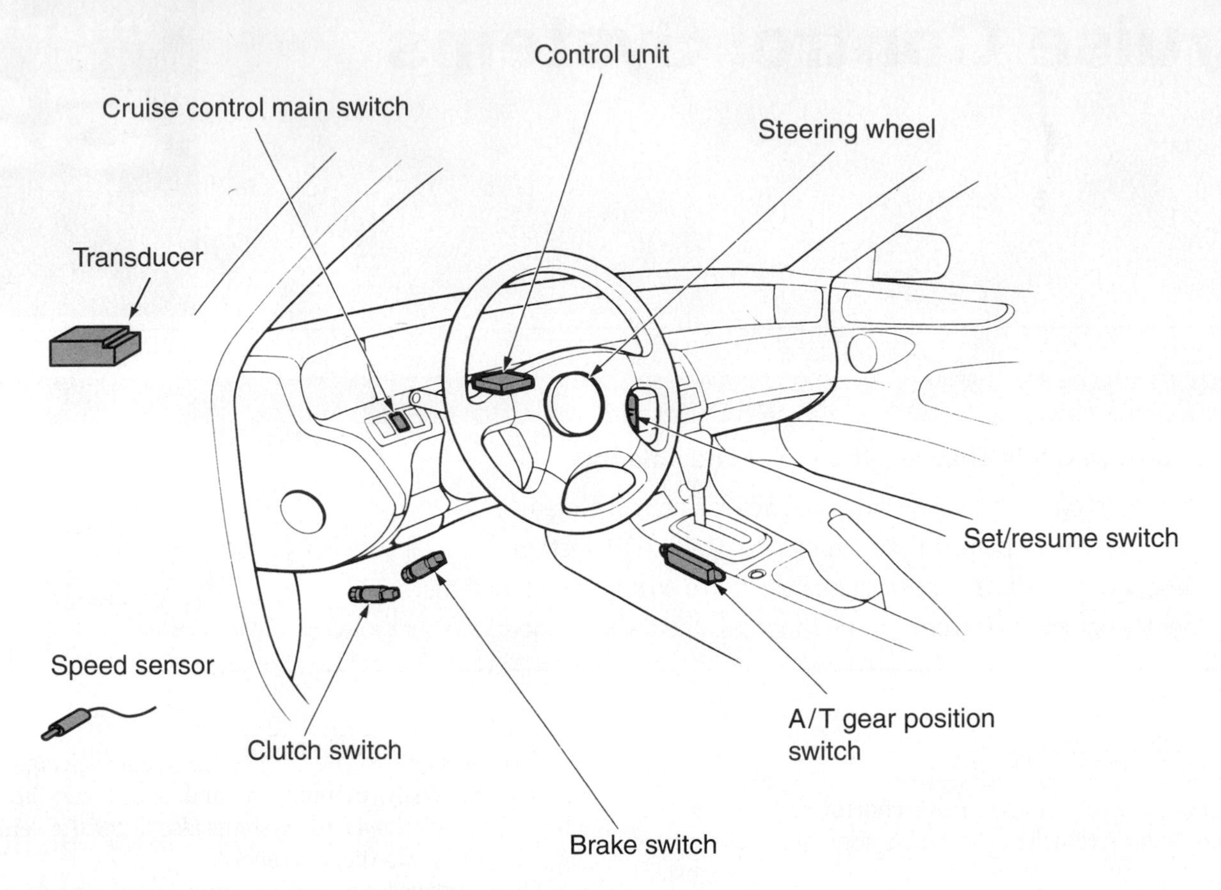

FIGURE 52–1 The parts and switches of a typical cruise control system.

decrease the speed. In addition, most cruise control systems have a coast button to slow the vehicle down in congested areas. Some of the common components of a cruise control system include the servo, push button controls, electric release switch, the transducer, brake/clutch switches, and the vacuum release valve.

There are many types of cruise control systems used on vehicles today. Because of the constant changes and improvements in technology, each cruise control system may have different components or slightly different operations. The exact type of system is determined by the year, make, and model of each vehicle. However, the most common types are the mechanical and the electronic cruise control systems.

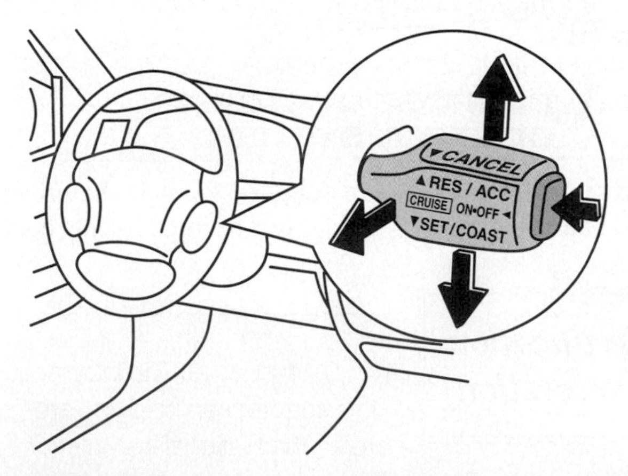

FIGURE 52–2 The cruise control switch is used to set or increase speed, resume speed, or turn the system on and off. (Courtesy of American Honda Motor Co., Inc.)

52.2 CRUISE CONTROL PARTS

CRUISE CONTROL SWITCH

The cruise control switch is located on the end of the turn signal or near the center or sides of the steering wheel. There are usually several functions on the switch, including off-on, resume, and engage buttons. The switch is different for resume and nonresume systems. ***Figure 52–2*** shows a cruise control switch. The switch has a main on/off control switch, a set/coast (cruise) switch, and a resume/accelerate switch.

TRANSDUCER

The **transducer** controls the speed of the vehicle when the cruise control is turned on. A transducer is a device that uses an electrical signal as the input. The transducer changes the electrical signal to a mechanical motion. When the cruise control is on and the transducer shown in ***Figure 52–3*** is operating, it is able to sense the vehicle speed based on electrical sensors and, in turn, control either the vacuum in a servo or the mechanical connection to the throttle.

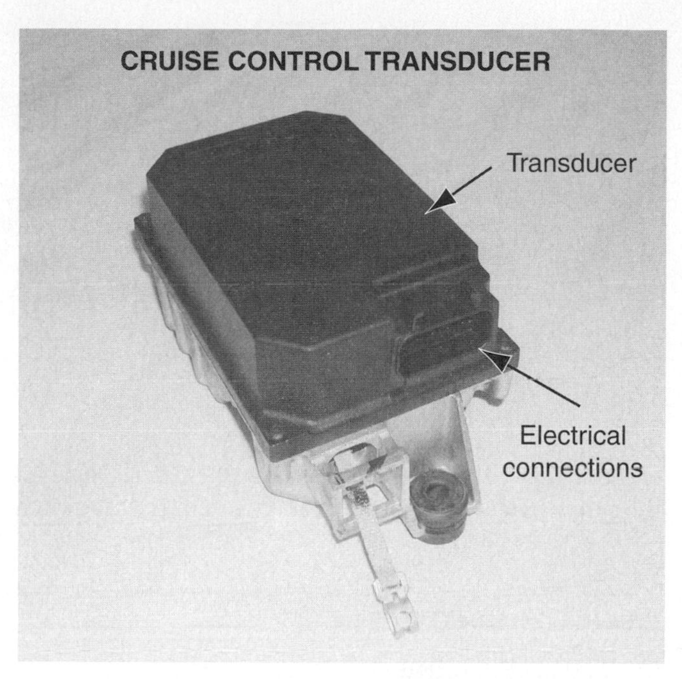

FIGURE 52–3 The transducer operates the servo or the mechanical connection used to control the position of the throttle.

Either the vacuum in the servo or the mechanical connection, in turn, controls the throttle position during cruise control operation. By controlling the throttle position, the speed of the vehicle is maintained according to the cruise control settings.

PARTS LOCATOR

Refer to photo #123 in the front of the textbook to see what a ***cruise control transducer*** *looks like.*

SERVO

The **servo** unit is connected to the throttle by a rod or linkage, a bead chain, or a **Bowden cable.** The servo unit positions the throttle to maintain the desired car speed by receiving a controlled amount of vacuum from the transducer. The variation in vacuum changes the position of the throttle.

An inside view of a computer-controlled servo is shown in ***Figure 52–4.*** The drawing at the top shows the servo being engaged. The drawing at the bottom shows the servo in a disengaged position. In operation, vacuum is always available from a line connected to the air intake, directly below the throttle plate. When the computer calls for the servo to be energized or turned on, a small solenoid is energized inside the servo. This is shown as the bottom solenoid. When the computer energizes this solenoid, it causes the vacuum valve to open on the bottom. When this happens, vacuum from the intake manifold is directed to the left-hand side of the servo diaphragm. The vacuum causes the spring and the rubber diaphragm to move to the left as shown by the arrows. As the spring and diaphragm move to

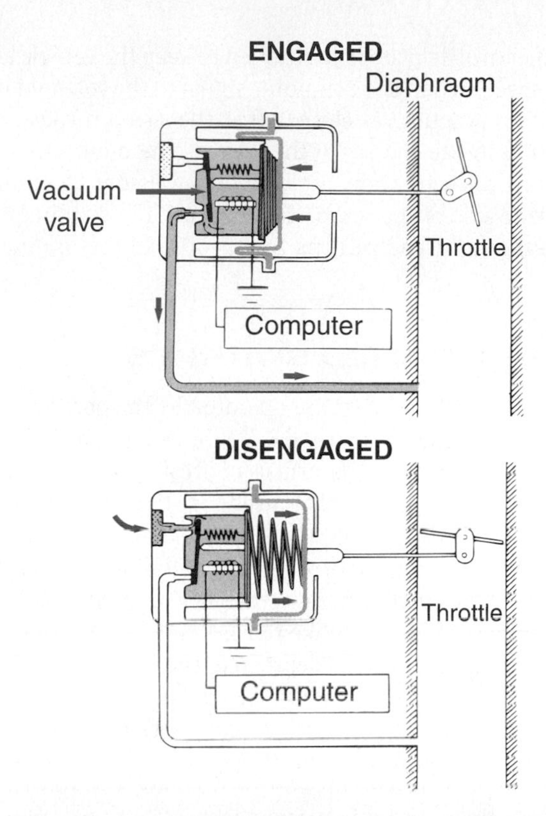

FIGURE 52–4 This servo is controlled by a computer signal, which in turn controls the amount of vacuum sent to the inside of the servo. Depending on the vacuum, the servo will either engage (top illustration) or disengage (bottom illustration).

Car Clinic:
VACUUM HOSES

CUSTOMER CONCERN:
Speed Varies while on Cruise Control
The operator of a vehicle notices that when the cruise control is engaged, the vehicle speed varies or fluctuates. The driver has indicated that there has been very little service on the vehicle and no service on the cruise control since the car was new.

SOLUTION: Several components may be damaged on the cruise control that cause its speed to fluctuate. One simple check is the condition of the vacuum hoses on the system. As a car gets older, the vacuum lines that are used to feed into and out of the vacuum servo have a tendency to leak. At times, the hoses develop cracks or cuts from weathering. This causes the vacuum in the servo to leak. Thus, the cruise control will vary its speed as it keeps trying to adjust the vacuum to its correct amount. Check all of the vacuum hoses for cracks and/or cuts. A vacuum gauge may be needed to check for leaks as well. Replace any hoses that are defective. Then check the operation of the cruise control to see if it continues to fluctuate.

the left, the throttle plate is positioned to keep the vehicle at a certain speed. When the computer signal to the solenoid is turned off, the vacuum is closed off at the vacuum valve. A small spring inside the servo then opens the other end of the vacuum valve to allow atmospheric pressure to enter the left-hand side of the servo. At this point, the diaphragm moves to the right and returns the throttle to the original position.

BRAKE-ACTIVATED SWITCHES

There are two switches that are operated by the position of the brake. When the brake pedal is depressed, the brake release switch disengages the cruise control system. A vacuum release valve is also used to disengage the system when the brake pedal is depressed.

On electronic cruise control systems, the brake release switch is used to signal other circuits as well. ***Figure 52–5*** shows examples of several brake release switches. In ***Figure 52–5A***, as the brake pedal is depressed, the lever moves. As the lever moves, it causes the cruise control circuit to disengage. The electrical connections complete the circuitry to the cruise control module. In ***Figure 52–5B***, examples of the simple cruise control brake and vacuum release switch are shown for reference.

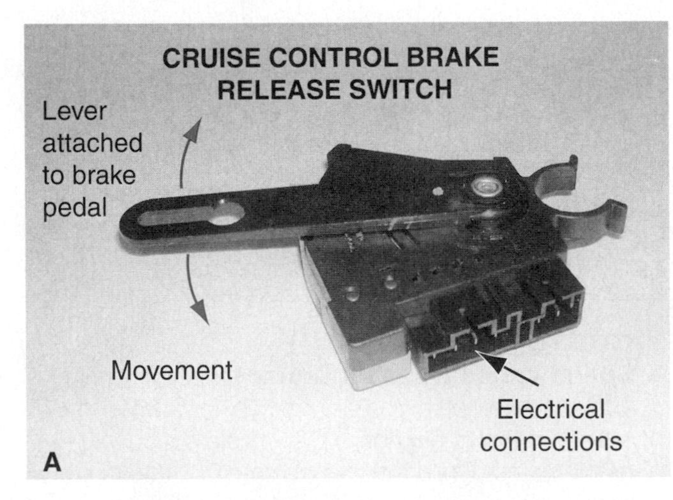

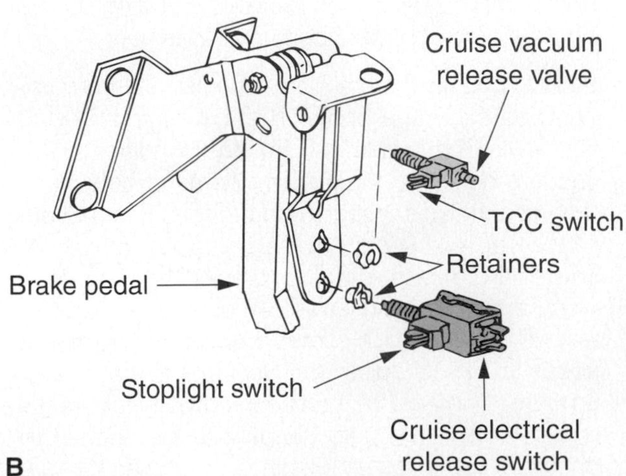

FIGURE 52–5 Release switches are used to disengage the cruise control system during braking. When the brake pedal is depressed, electrical circuits are opened to disengage the cruise control system.

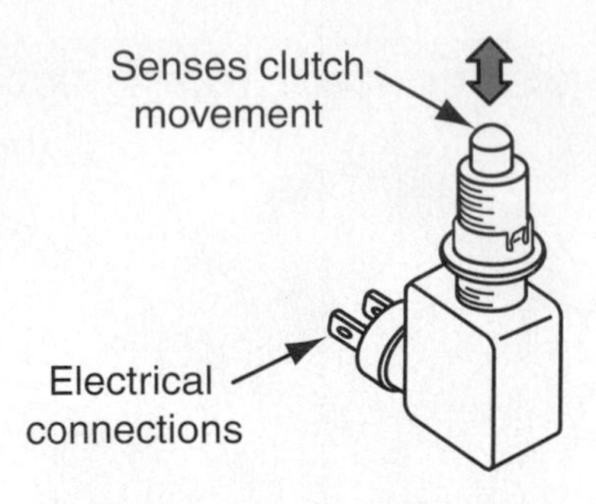

FIGURE 52–6 This clutch switch is used to disengage the cruise control on a vehicle with a manual transmission.

CLUTCH SWITCH

In addition to the brake switches, vehicles with manual transmissions also have a **clutch switch**. This switch disengages the cruise control the instant that the clutch is depressed. ***Figure 52–6*** shows an example of a clutch switch.

52.3 ELECTRONIC CRUISE CONTROL PARTS

BASIC PARTS AND OPERATION

Cruise control can also be obtained by using electronic components rather than mechanical components. Depending on the vehicle manufacturer and the type, year, and vehicle model, several additional components may be used. These include:

- Powertrain control module (PCM)—used to control the servo unit. The servo unit is again used to control the vacuum, which in turn controls the throttle.
- Vehicle speed sensor (VSS) buffer amplifier—used to monitor or sense vehicle speed. The signal created is sent to the electronic control module. A generator speed sensor may also be used in conjunction with the VSS.
- Clutch switch—used on vehicles with manual transmissions to disengage the cruise control when the clutch is depressed.
- Accumulator—used as a vacuum storage tank on vehicles that have low vacuum during heavy load and high road speeds.

Figure 52–7 shows the most common parts of a typical electronic cruise control system and their location. This particular type of system uses manifold vacuum or vacuum from a vacuum pump to drive the throttle servo. The servo controls the throttle position by changing the amount of vacuum in the servo. The amount of vacuum in the servo is controlled by the solenoid valve. The solenoid valve constantly modulates vacuum to the servo in response to com-

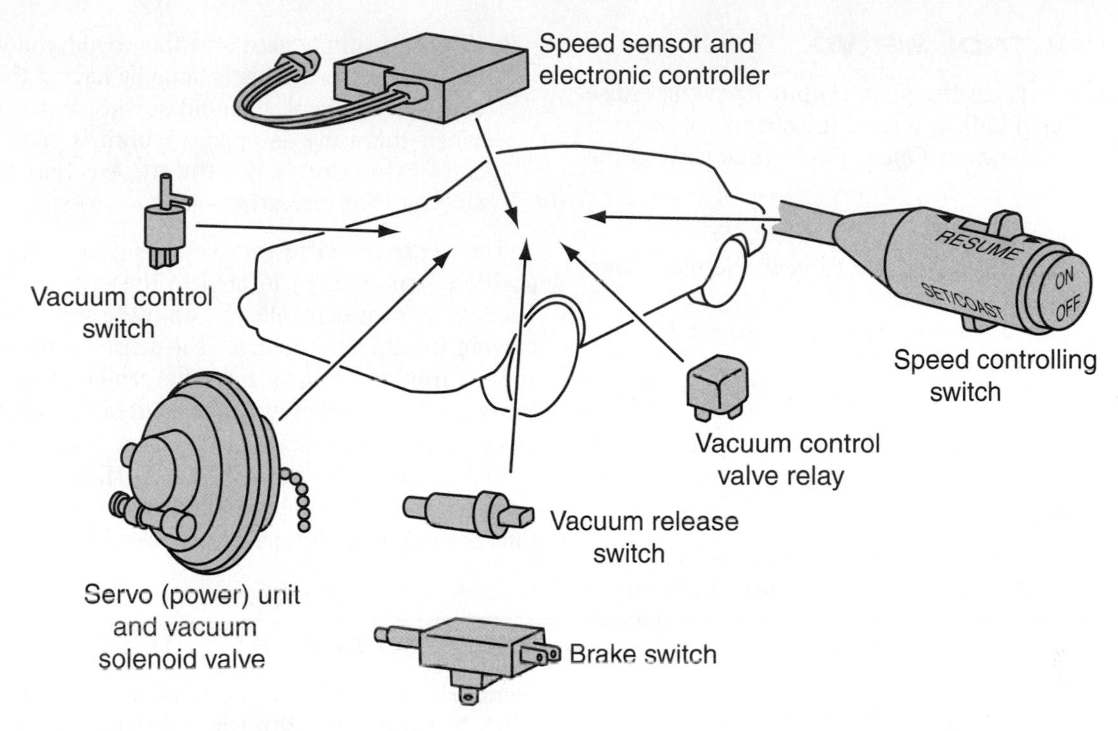

FIGURE 52-7 The major components of electronic cruise control.

mands from the electronic controller. The speed sensor is located on the back of the speedometer cluster. It provides a signal to the electronic controller indicating the vehicle speed. The electronic controller also receives signals from the selector switches and brake and clutch (manual transmission) switches. These inputs help to control the servo solenoid valve and, in turn, regulate the amount of vacuum in the servo, thus controlling the speed of the vehicle.

Figure 52–8 shows how electronic cruise control components work together. The throttle position is controlled by the servo unit. The servo unit uses a vacuum working against a spring pressure to operate an internal diaphragm. The servo unit vacuum circuit is controlled electronically by the controller. The controller has several inputs that help determine how it will affect the servo. Some common inputs include the:

- Brake release switch (clutch release switch)
- Speedometer, buffer amplifier, or generator speed sensor
- Turn signal mode switch (switch used on the turn signal to control the cruise control)

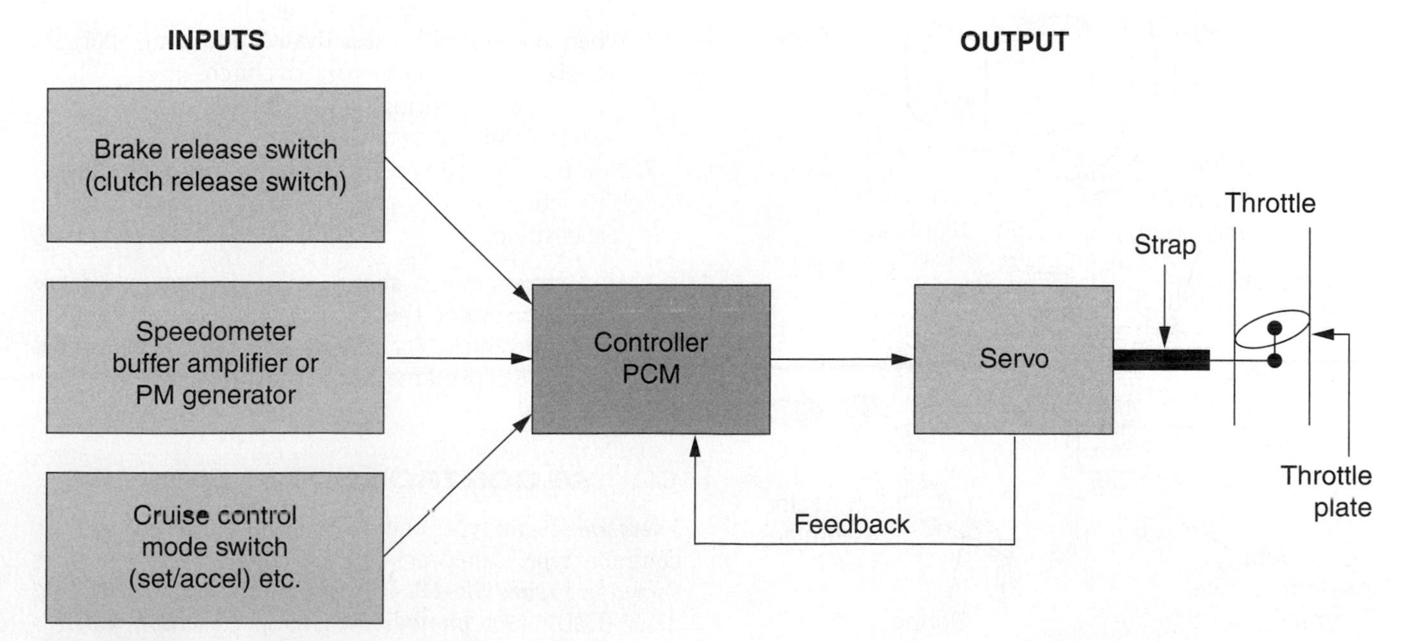

FIGURE 52-8 Electronic cruise control systems use a powertrain control module (controller) to operate a servo that controls the position of the throttle.

SPEED CONTROL SERVO

The servo is considered the main output from the cruise control computer (PCM). It is used to control the position of the throttle as shown in ***Figure 52–9***. It consists of the following:

- Vacuum-operated diaphragm
- Normally open solenoid valve to vent the diaphragm chamber to the atmosphere
- Normally closed solenoid valve to connect the diaphragm to the vacuum source
- Variable inductance position sensor

The servo operates the throttle position in response to signals from the electronic controller in three main modes. These modes include:

1. Steady-state cruise mode—In this mode, both the vacuum and vent valves are closed or sealed. The servo has a constant vacuum on the diaphragm and the vacuum works against the spring pressure to position the throttle at a steady point.
2. Vehicle losing speed—In this mode, the vehicle reduces its speed. This is done by having the PCM energize the vacuum solenoid valve. When this valve is open, vacuum from the vacuum source increases. This causes the throttle position to close off slightly, thus reducing speed.

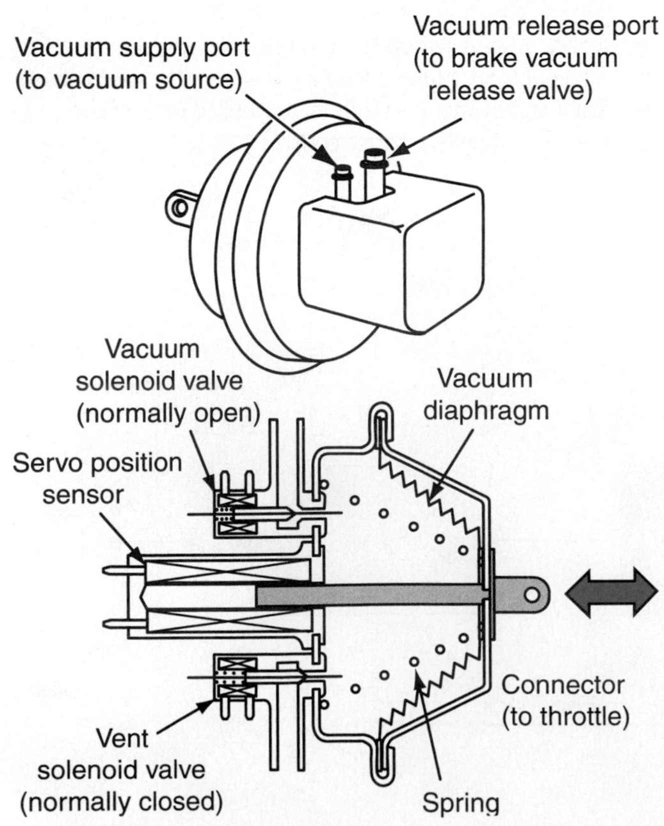

FIGURE 52–9 The cruise control servo controls the vacuum by operating two solenoid valves.

3. Vehicle gaining speed—In this mode, the vehicle increases its speed. This is done by having the PCM de-energize the vent solenoid so the vent valve opens. When this valve is open, vacuum is reduced in the servo. This causes the throttle position to increase slightly, thus increasing speed.

The controller also receives a signal from the **servo position sensor** (SPS) located in the center of the servo. The SPS tells the controller how fast the servo diaphragm is moving the throttle linkage. The sensor consists of a coil and an iron rod attached to the center of the servo diaphragm. As the rod moves in and out of the coil, the inductance of the coil changes. The change of inductance is sensed as a voltage signal by the PCM. As the rod moves within the coil, it changes the strength of its magnetic field, thus changing the voltage drop across the coil.

ELECTRO-MOTOR CRUISE CONTROL

Some vehicles that have cruise control do not use a vacuum servo to control the throttle position. Instead, these vehicles use an **electro-motor** as the output from the computer to control the position of the throttle. ***Figure 52–10*** shows the major parts of this system. The system eliminates the vacuum servo and incorporates the electronic controller and actuator into a single unit. Referring to this figure, the operation sequence is as follows:

1. When the vehicle speed is set or changed, the controller assembly adjusts throttle position by activating the stepper motor.
2. As the stepper motor turns, the pinion gear drives the drum gear assembly.
3. This causes the strap to wind up or release.
4. This results in a throttle position change.
5. When the system is deactivated, the controller de-energizes the solenoid-operated clutch.
6. This causes the spring on the clutch arm assembly to push the pinion gear away from the drum gear.
7. Now the spring on the drum gear unwinds the strap.
8. This action causes the throttle linkage to return to the idle position.

The electro-motor is shown in its position in ***Figure 52–11***. In this system, the electro-motor is located on the firewall of the vehicle. A cruise control cable is connected from the cruise control module to the throttle.

CRUISE CONTROL SPEED SENSORS

There are several types of cruise control speed sensors. One common type, called an optical vehicle speed sensor, is shown in ***Figure 52–12***. This system uses a light-emitting diode (LED) and a **photodiode**. This speed sensor, which is hooked to the speedometer, uses a light-sensitive semiconductor as a light-variable resistor. As it spins, it produces an

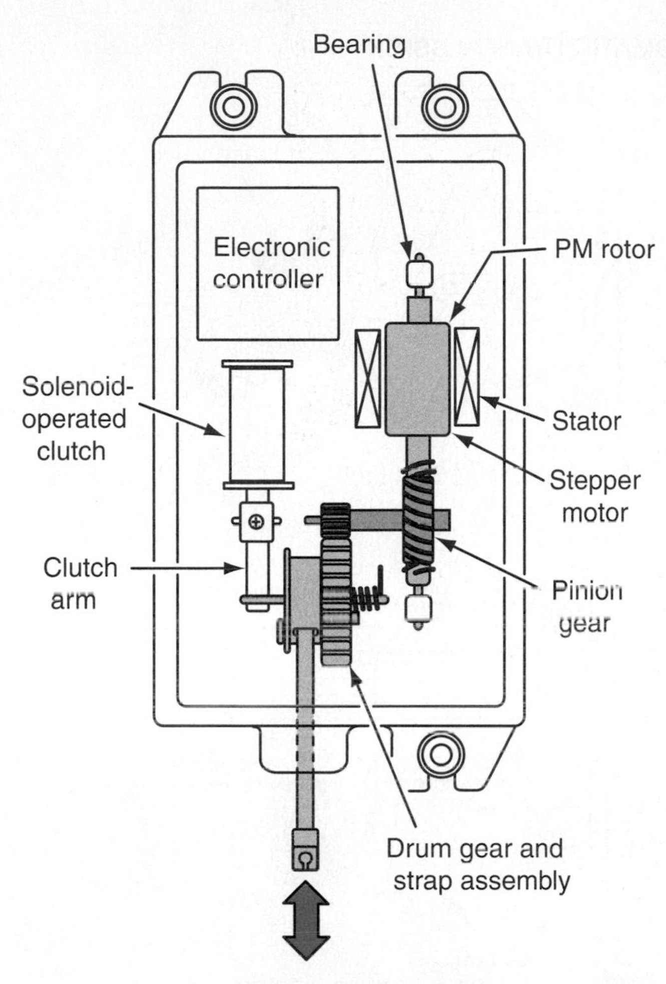

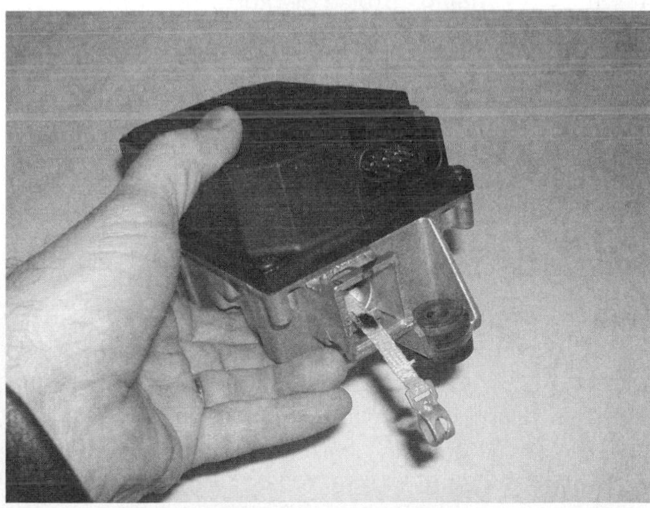

FIGURE 52–10 This electro-motor can be used in place of a servo to control the throttle position.

on-off voltage signal that is sent to the PCM and the cruise controller as an input.

A second type of speed sensor is called the PM—permanent magnetic—generator speed sensor. The sensor shown in ***Figure 52–13*** produces a voltage signal of about 4,000 pulses per mile of operation, which is sent to the cruise control PCM as an input signal.

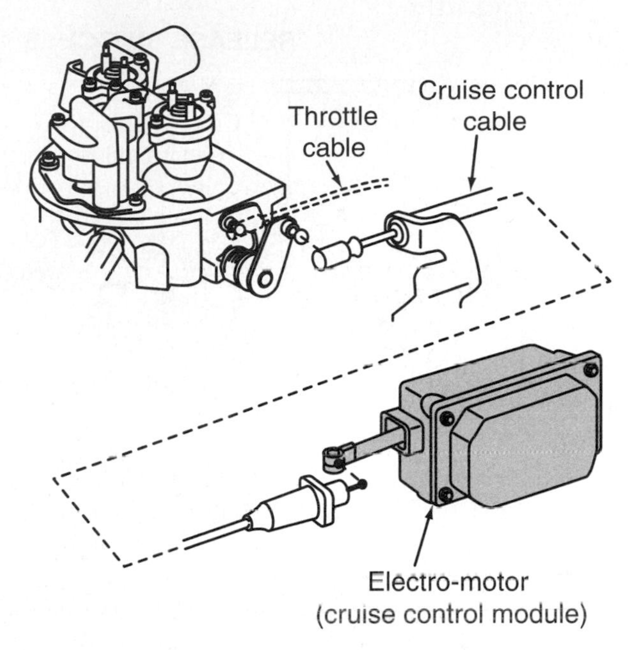

FIGURE 52–11 The electro-motor (cruise control module) is used to control the throttle position.

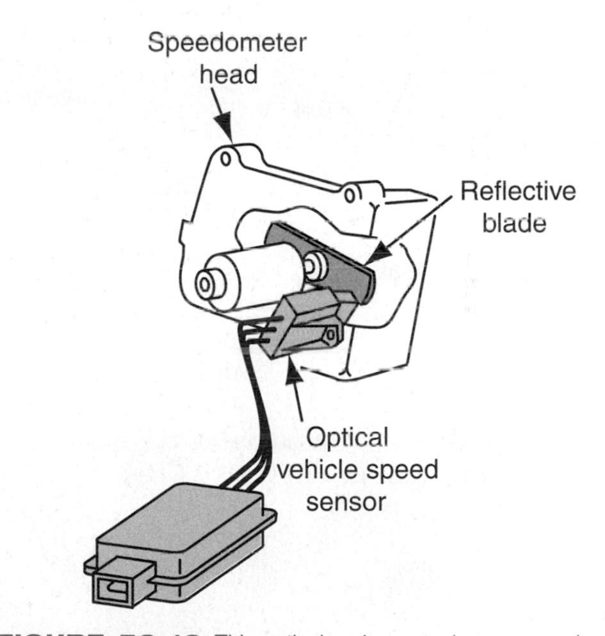

FIGURE 52–12 This optical cruise speed sensor produces a signal to be sent to the computer.

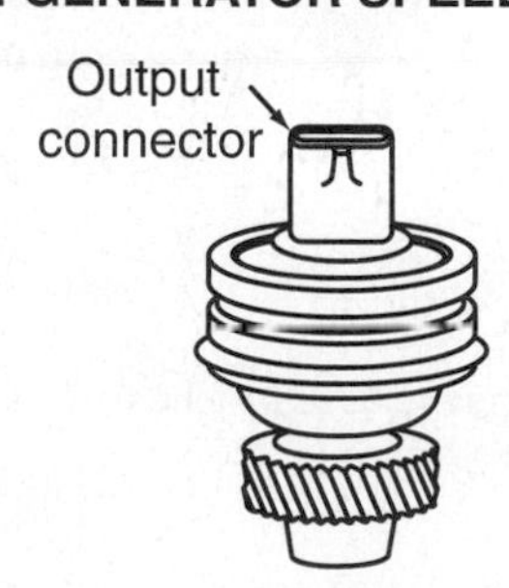

FIGURE 52–13 A permanent magnet (PM) generator speed sensor used for electronic cruise control systems.

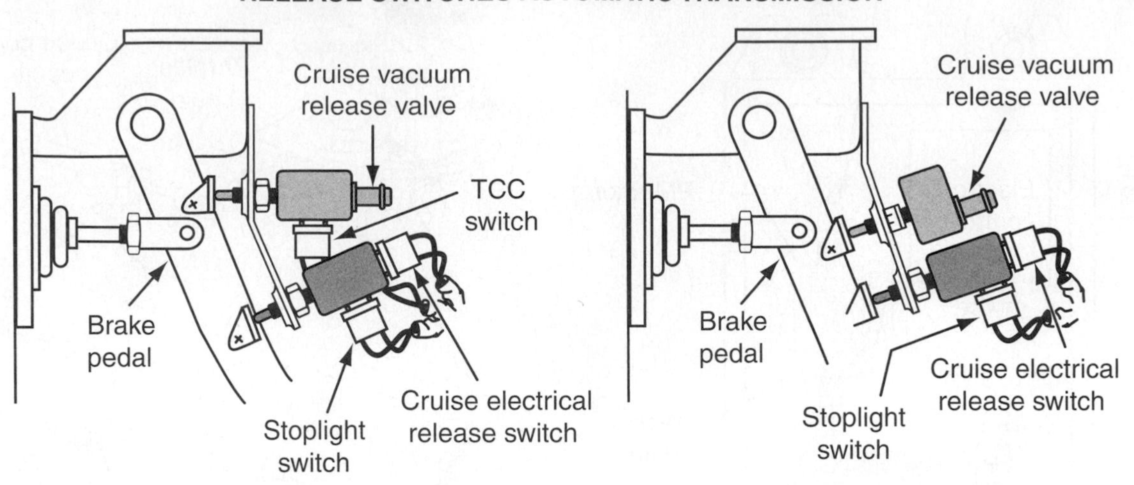

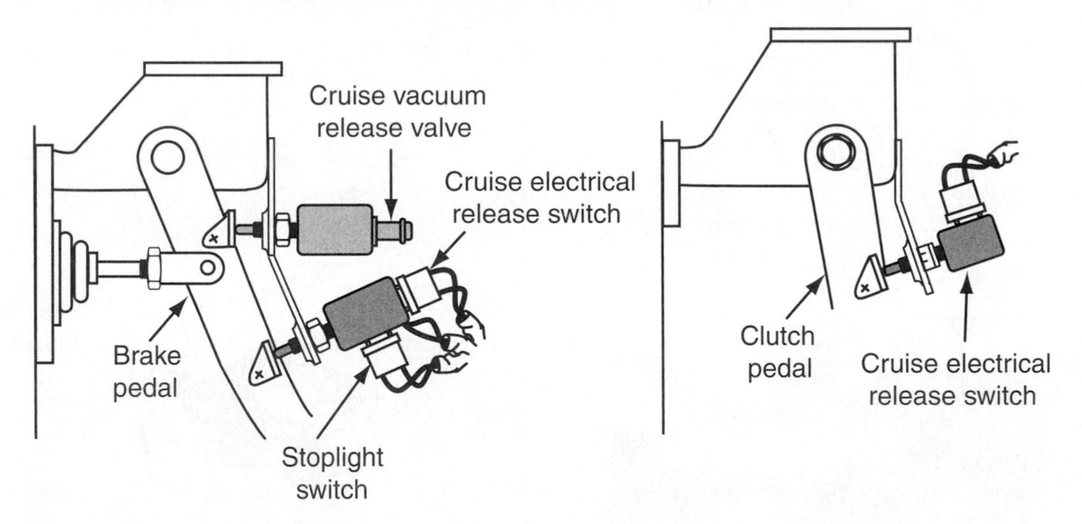

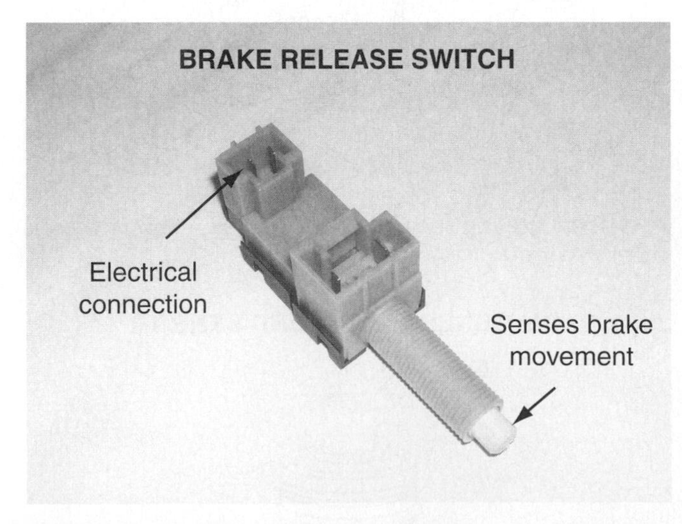

FIGURE 52–14 These release switches are used to shut off or disengage electronic cruise control systems.

RELEASE SWITCHES

There are several release switches that provide inputs to the cruise control PCM. The most common are the brake release switch, the clutch release switch used on manual transmissions, the automatic transmission release switch, and the vacuum release switch. ***Figure 52–14*** shows examples of several release switches. The brake pedal turns off the cruise control when the brake is depressed. The clutch release switch turns off the cruise control when the clutch is depressed. The automatic transmission switch (not shown) turns off the cruise control when the shift lever is moved to a mode other than drive. The vacuum release switch serves as a backup when the brake pedal is depressed. It opens a port that vents the servo unit to atmospheric pressure, causing the throttle to return to idle position.

52.4 ELECTRONIC CRUISE CONTROL OPERATION

To understand the complete operation of an electronic cruise control sytem, this section presents several electrical circuits of a typical electronic cruise control system. In effect, these circuits make up the entire cruise control circuit. In all of these circuits, the cruise control unit is considered the PCM, or powertrain control module, for the cruise control system. By following the operation of each circuit, the service technician can better understand the total cruise control system.

CRUISE CONTROL MAIN SWITCH CIRCUIT (INPUT)

The circuit shown in ***Figure 52–15*** shows how the electronic cruise control is turned on. In this circuit, when the operator switches the cruise control on, the cruise control main switch is turned on. This causes the indicator lamp on the dashboard to light up, telling the operator that the cruise control system is on. In addition, electricity is now sent directly to the cruise control unit and is considered one of the inputs to the computer. Also note that the brake pedal switch is in the normally closed position. When the brake pedal is depressed, this switch is opened.

SET/DECEL AND RESUME/ ACCEL CIRCUIT (INPUTS)

The circuit shown in ***Figure 52–16*** shows how the cruise control set and resume switch operates. Once again, these switches are considered inputs to the cruise control unit. When the Set/Decelerate (Set/Decel) switch is closed by the operator, a signal voltage is sent to the cruise control unit.

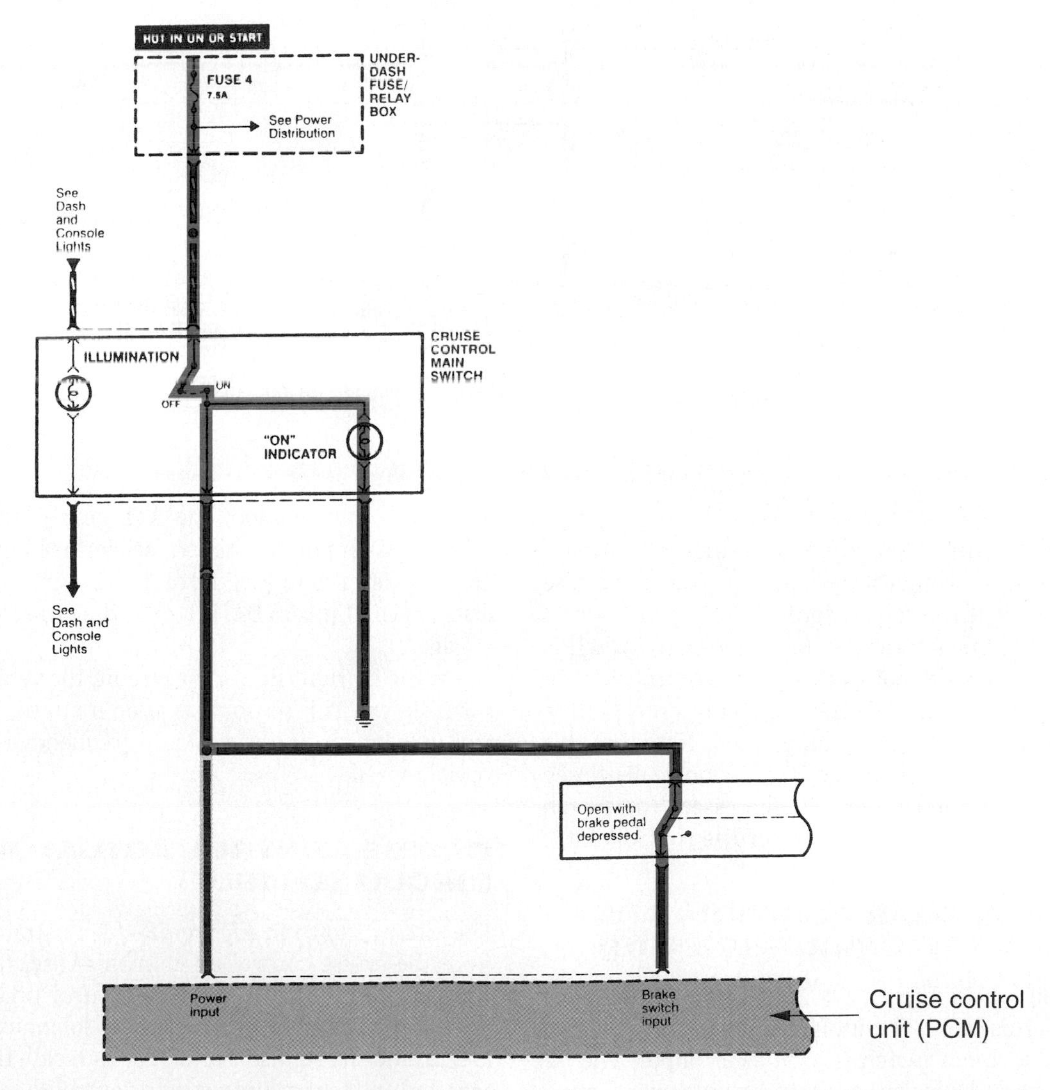

FIGURE 52–15 The main switch on the cruise control circuit is one of several inputs to the PCM. (Courtesy of American Honda Motor Co., Inc.)

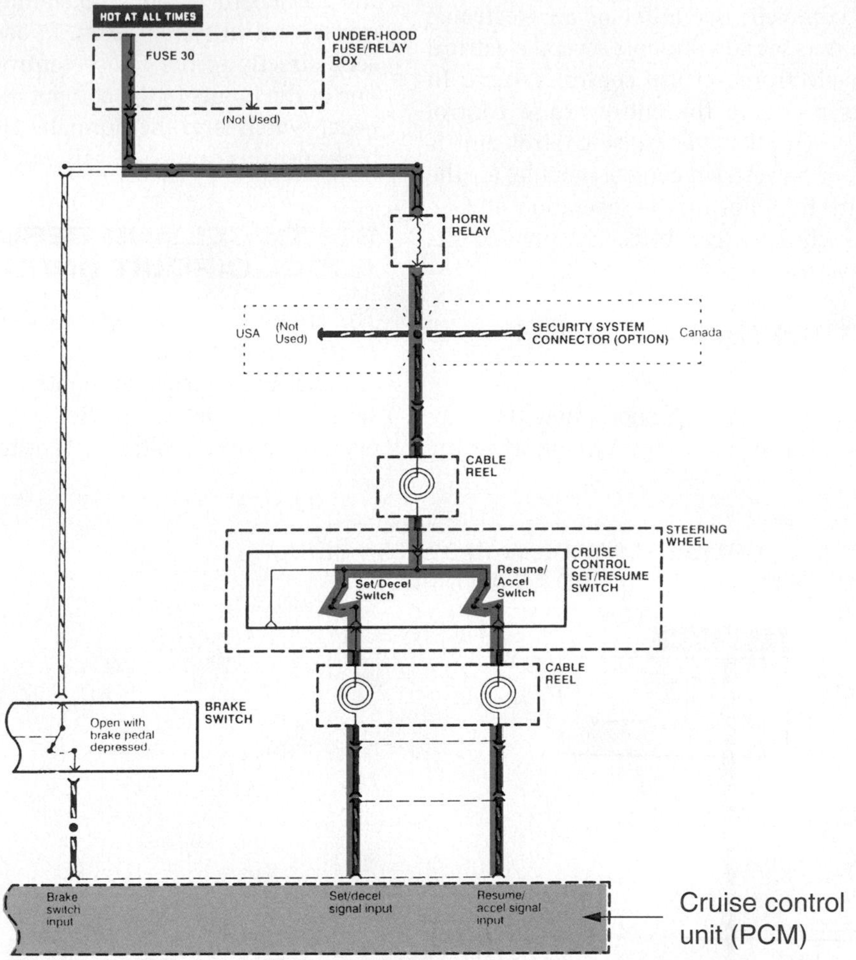

FIGURE 52–16 The cruise control Set/Decel and the Resume/Accel switch is used as an input to the PCM. (Courtesy of American Honda Motor Co., Inc.)

This tells the cruise control unit to set the throttle through the vacuum servo or by using the electro-motor. If the system is already set, closing this switch will cause the cruise control to decelerate the vehicle speed.

When the Resume/Accelerate (Resume/Accel) switch is closed, it also sends a signal to the cruise control unit. In this case, the control unit now sets the vacuum servo or the electro-motor to increase or resume speed. Note that this circuit also has a brake switch that is in the normally closed position. When the brake pedal is depressed, the switch opens and acts as another input to the cruise control unit.

CLUTCH, A/T GEAR POSITION, AND SPEED SENSOR CIRCUIT (INPUTS)

The circuit shown in ***Figure 52–17*** illustrates additional inputs to the cruise control unit. If the vehicle has a manual transmission, a clutch switch is used as an input. The instant the clutch pedal is depressed, the switch opens, causing the cruise control to be disengaged. If the vehicle has an automatic transmission, the A/T gear position switch is used as an input to the cruise control unit. When the gearshift lever is in P, R, N, or 1, the cruise control will be disengaged. If it is in D4, D3, or 2, the cruise control can be engaged.

On the far right side of the circuit, the vehicle speed sensor is shown. This sensor acts as an input to the cruise control unit by sending it a certain frequency signal to sense vehicle speed.

CRUISE CONTROL ACTUATOR CIRCUIT (OUTPUT)

The circuit shown in ***Figure 52–18*** illustrates the outputs from the cruise control unit. In this case, there are three wires coming out of the cruise control unit. These wires control the operation of the vacuum solenoid, the safety solenoid, and the vent solenoid. As you recall, these solenoids are used on the vacuum servo to control the vacuum entering and being vented from the vacuum source.

RELEASE SWITCHES (INPUTS)

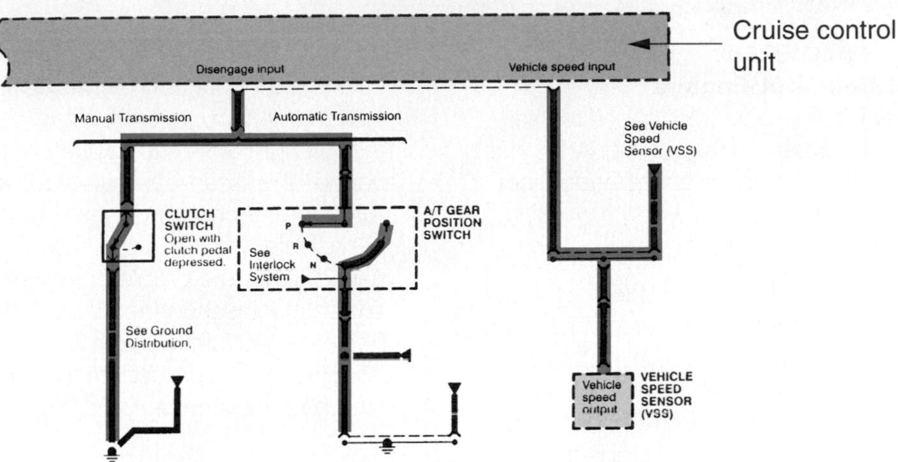

FIGURE 52-17 Release switches such as the clutch switch or the A/T gear position switch and the vehicle speed sensor are used as inputs to the computer. (Courtesy of American Honda Motor Co., Inc.)

CRUISE CONTROL ACTUATOR CIRCUIT (OUTPUT)

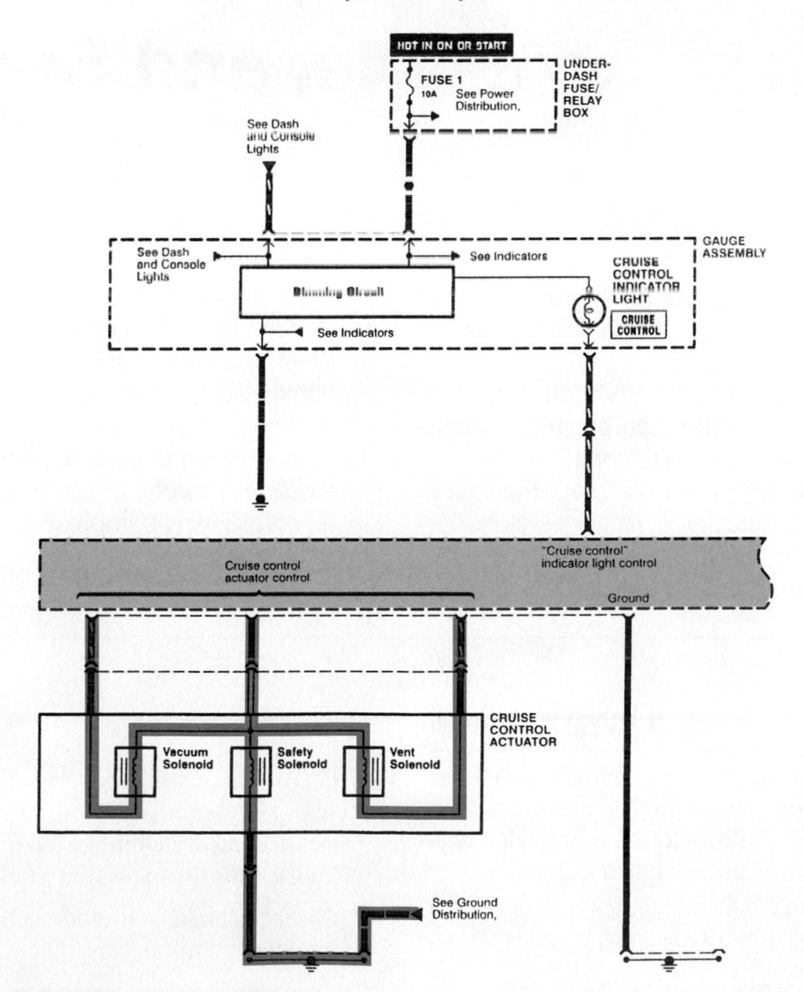

FIGURE 52-18 The main outputs on the electronic cruise control are the vacuum and vent solenoids and the safety solenoid. (Courtesy of American Honda Motor Co., Inc.)

Car Clinic: CLUTCH SWITCH

CUSTOMER CONCERN:

Cruise Control Does Not Engage

The owner of a vehicle has indicated that clutch service was recently performed on the vehicle. After the service was completed, the cruise control would not engage. The cruise control on-off switch is working, but the system cannot be engaged.

SOLUTION: The clue to this problem is that clutch work has recently been completed. The first check—and one of the easiest to make—is to see if the clutch release switch electrical connections have been replaced correctly. If the clutch release switch has been left disconnected, the PCM reads this the same as having the clutch pedal depressed. Make sure that the wire connectors have been completely inserted. Although there may be other problems that would cause the cruise control to stay disengaged, this is an easy and quick check to make. After double-checking these connections, road test the vehicle and see if the cruise control can be engaged. If so, the problem was in the clutch release switch electrical connections. If the system still will not engage, follow the manufacturer's suggested troubleshooting procedure to test the clutch release switch for resistance.

Problems, Diagnosis, and Service

Safety Precautions

1. Always use the correct tools when working on a cruise control system. This will eliminate the possibility of injury to the hands.
2. Many electrical wires on a cruise control system have sharp protrusions. When handling these wires, be careful not to puncture your hands.
3. Before removing of any cruise control part or electrical components, make sure you understand its operation and how it is hooked into the system. This precaution will eliminate many possible injuries.
4. Always wear OSHA-approved safety glasses when working on cruise control systems.
5. When working on any electrical parts of the cruise control system, always remove all jewelry (rings, watches, etc.).
6. When making checks while the engine is running, always be careful of spinning parts, including the air-conditioning compressor, various belts, and the cooling fan.

PROBLEM: Electrical and Mechanical Faults

Cruise control systems can have a variety of problems. Rather than specify a particular problem at this point, this section will present general diagnostic procedures for both electrical and mechanical problems with cruise control systems.

DIAGNOSIS

1. Always perform a visual inspection of all components.
2. Check the servo for leaks.
3. Check for blown fuses.
4. Bypass the low-speed switch, and check for operation.
5. With the car off and the ignition switch on, engage the cruise control switch. Listen for the solenoid to click or knock. This indicates that the solenoid is operative.
6. Check the brake-activated electric release switch. A small misadjustment can cause the system to be inoperative.
7. Check all electrical connectors for solid contact.
8. Check all electrical components for correct operation.

9. Check the vacuum circuit for correct operation. Check for correct vacuum at the transducer. Observe all vacuum lines for leaks and damaged rubber hoses. Use a vacuum pump if needed to check vacuum operation.
10. Check vacuum hoses that might be pinched.
11. Inspect the mechanical throttle linkage for ease of movement. The bead-chain type should have no slack in it with the engine at the correct idle speed.
12. If no external fault can be found, the transducer should be checked for vacuum leaks or incorrect operation. This is usually done at an authorized transducer repair facility.
13. On computerized speed control systems, a self-diagnostic system may also be used. Trouble codes are stored in the computer and can be accessed by the service technician. When a trouble code is identified, the service technician can then refer to a trouble code diagnostic chart. ***Figure 52–19*** is one of seven trouble code diagnostic charts available for one type of cruise control.

SERVICE

Each service procedure will be different for each component that might be faulty. Consult the maintenance or service manuals for the correct procedure to follow.

PROBLEM: Speed Will Not Set in System

A vehicle has a cruise control system that will not set.

DIAGNOSIS

There are many problems that can arise in any cruise control system. The exact problem will often be characteristic of a particular vehicle. To help the service technician troubleshoot and diagnose such a problem, various types of pinpoint tests are often necessary. ***Figure 52–20*** shows an example of the procedures recommended for the pinpoint test needed to diagnose this particular problem. There may be as many as sixty or more specific pinpoint tests possible for a particular vehicle manufacturer. The individual procedures lead

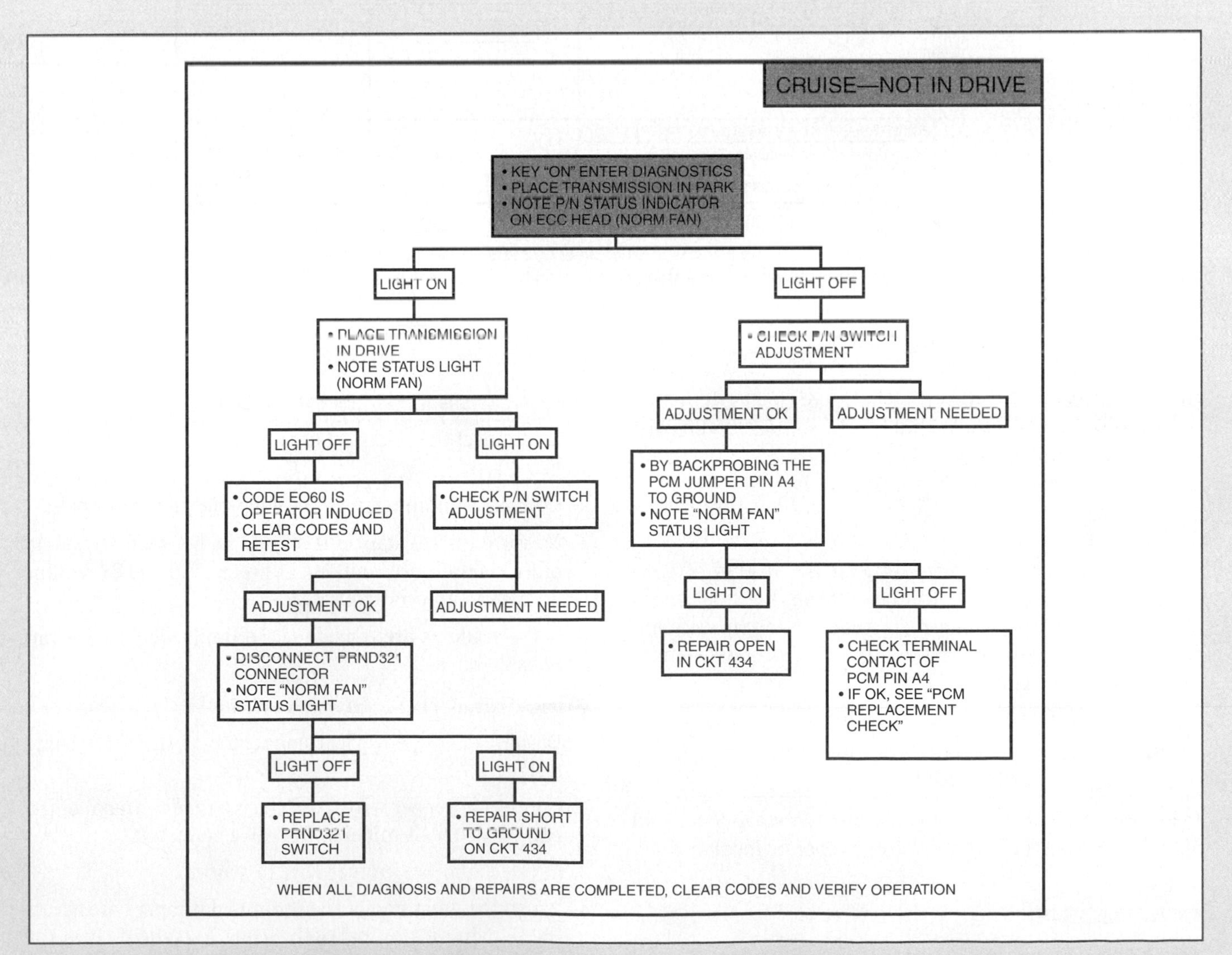

FIGURE 52–19 This type of diagnostic chart will help the service technician troubleshoot a specific component identified by a trouble code on the cruise control system. (Courtesy of MOTOR Publications, Auto Repair Manual, © The Hearst Corporation)

PINPOINT TEST
SPEED WILL NOT SET IN SYSTEM

	TEST STEP	RESULT →	ACTION TO TAKE
H1	CHECK THROTTLE LINKAGE		
	• Check the throttle linkage for proper operation and adjustment. • Does the throttle linkage operate properly?	Yes → No →	GO to H2. ADJUST or SERVICE as required.
H2	CHECK CONNECTIONS		
	• Check the system circuit connections. • Are the connections OK?	Yes → No →	GO to H3. SERVICE as required.
H3	PERFORM CONTROL SWITCH AND CIRCUIT TESTS		
	• Perform the Control Switch and Circuit Tests as described in this section. • Are the switches and/or circuits OK?	Yes → No →	GO to H4. SERVICE the switch circuit as required.
H4	CHECK SPEED CONTROL DUMP VALVE		
	• Perform the Speed Control Dump Valve Test as described in this section. • Is the speed control dump valve OK?	Yes → No →	GO to H5 for manual transaxle, H6 for automatic transaxles. ADJUST or SERVICE as required.
H5	CHECK CLUTCH PEDAL POSITION SWITCH (MTX)		
	• Check the clutch pedal position switch for proper operation. Perform the Clutch Pedal Position Switch Test as described in this section. • Does the clutch pedal position switch operate properly?	Yes → No →	GO to H6. SERVICE the switch as required.
H6	CHECK STOPLAMPS		
	• Apply the brakes and observe the stoplamps. Perform the Brake On/Off Switch and Circuit Test as described in this section. • Are the stoplamps working?	Yes → No →	GO to H7. SERVICE the lamps and circuit as required. VERIFY that the fuses are not open.
H7	PERFORM SPEED CONTROL ACTUATOR TESTS		
	• Perform the Speed Control Actuator Tests as described in this section. • Is the speed control amplifier OK?	Yes → No →	GO to H8. REPLACE the speed control servo.

FIGURE 52-20 This procedure is a typical pinpoint test used to diagnose and service a particular cruise control problem in a vehicle. (Courtesy of MOTOR Publications, Auto Repair Manual, © The Hearst Corporation)

the service technician through a series of checks. Based on each check, the service technician is routed to different points in the pinpoint test.

SERVICE

In this particular case, according to the results of each check and the action that has been taken, the final check shows that the speed control actuator was checked and found to be defective. Thus, the speed control servo component needs to be replaced.

PROBLEM: Defective Vacuum Servo and Solenoid Valve

It is suspected that a defective vacuum servo and solenoid valve has caused the cruise control to operate incorrectly.

DIAGNOSIS

There are several diagnostic checks to analyze the vacuum servo and solenoid valves.

Resistance Test (Engine Off)

1. Disconnect the electrical connector at the solenoid valve being tested.
2. Connect an ohmmeter to the solenoid valve terminals.
3. Measure the resistance. It should be between 35 and 48 ohms on the servo unit and between 23.5 and 27.5 ohms on the vacuum control valve.
4. If the readings are outside these limits, replace the applicable valve.

Functional Test (Engine Off)

1. Disconnect the electrical connector at the valve being tested.
2. Connect a jumper wire from the positive terminal of the battery to one terminal of the valve being tested.
3. Connect another jumper wire to ground.
4. Brush the other end of the grounded jumper wire across the remaining terminal of the valve. You should hear the vacuum control valve or servo unit solenoid valve open and close as you make and break contact with the termi-

nal. Both valves will click twice, indicating that each respective valve is functioning properly.

Functional Test (Engine Operational)

1. Prior to starting the engine, disconnect the bead chain, cable, or rod actuator at the servo unit.
2. Disconnect the electrical connectors at the vacuum control and servo unit valves.
3. Start the engine and let it idle. Install a jumper wire from the battery positive terminal to one terminal on both vacuum control valves and servo unit solenoid valve. Ground the second terminal on each valve. The servo diaphragm should fully retract, indicating both solenoid valves are functional.
4. Remove the battery jumper from the servo solenoid. The diaphragm should return to full extension, proving that no vacuum is reaching it. If it does not fully return, the servo unit solenoid valve is leaking and should be replaced.
5. Reinstall the battery jumper wire to the servo solenoid valve and remove the battery jumper from the vacuum control valve. The diaphragm should return to full extension. If it does not, the vacuum control valve is leaking and should be replaced.

SERVICE

Generally, since there are only a few parts that make up a standard cruise control system, most components can be removed and replaced with standard tools and procedures. Based on the results of the diagnostic procedures, replace the defective component as necessary.

Service Manual Connection

There are many important specifications to keep in mind when working with cruise control systems. To identify the specifications for your vehicle, you may need to know the VIN (Vehicle Identification Number), the type and year of the vehicle, and possibly the type of engine used. Although they may be titled differently, some of the more common cruise control system specifications (not all) found in service manuals are listed below. Note that these are typical examples. Each vehicle may have different specifications.

Common Specification	Typical Example
Cruise control switch contacts open	1/8–1/2 inch pedal travel
Disengagement switch brake pedal adjustment (extension from floor)	6.07 inches
Voltage at cruise control switch On	12 volts
SPS coil resistance	15–25 ohms
Vent valve solenoid resistance	35–55 ohms
Vacuum valve solenoid resistance	30–35 ohms

In addition, service manuals provide specific directions for various service and testing procedures associated with cruise control systems. Some of the more common procedures include:

- Speed control adjustment
- Speed control road test
- Removal of actuator cable
- Visual checks and vehicle preparation
- Instructions for using pinpoint tests
- Testing speed sensor
- Testing clutch release switch
- Testing A/T selector release switch
- Simulated road test

SUMMARY

The following statements will help to summarize this chapter:

- The cruise control system is used to keep the vehicle at a constant speed while driving on highways.
- Standard components in a cruise control system include the cruise control switch, transducer, vacuum servo, brake-activated switches, and various electrical and vacuum circuits.
- When in operation, the transducer senses the speed of the vehicle.
- To adjust the speed, the vacuum servo controls the position of the throttle.
- A variety of switch designs are used to engage and disengage the cruise control system.

- Common switches include the set, coast, accelerate, and resume controls.
- The most common release switches used on cruise control systems are the clutch, A/T selector, and the brake switch.
- Most cruise controls today are operated by electronic controls.
- An electronic module or computer controls electronic cruise control systems.
- There are various inputs to the electronic cruise control system (control module) including the vehicle speed sensor, brake switch, and the operator control switches.
- The output of an electronic cruise control module is the electro-motor, which is used to position the throttle correctly.
- Common problems with cruise control systems include fluctuating speeds, inability to turn the system on or off, and defective release switches.
- Many pinpoint tests are available from vehicle manufacturers to diagnose and troubleshoot cruise control systems.

TERMS TO KNOW

Can you explain each of the following terms? Review the chapter until you can use each term correctly.

Bowden cable
Clutch switch
Electro-motor
Photodiode
Servo
Servo position sensor
Transducer

REVIEW QUESTIONS

Multiple Choice

1. Which device on a cruise control system is used to sense speed and control the amount of vacuum?
a. Servo
b. Solenoid coil
c. Brake-activated switch
d. Transducer
e. None of the above

2. Which device on a cruise control system is used to adjust and control the position of the throttle?
a. Servo
b. Solenoid
c. Brake-activated switch
d. Transducer
e. Magnetic switch

3. Cruise control circuits operate by using:
a. Vacuum systems
b. Electrical systems
c. Mechanical systems
d. All of the above
e. None of the above

4. The servo unit uses vacuum forces working against ________ forces to operate an internal diaphragm.
a. Gravity
b. Electrical
c. Magnetic
d. Spring
e. All of the above

5. Which of the following is not considered a common check on a cruise control unit?
a. Check the servo for leaks
b. Check the brake-activated electric release switch
c. Check the metering pin and orifice
d. Check all electrical connections for solid contact
e. Check for solenoid operation

6. The operational controls for a cruise control system are usually located:
a. In the trunk of the car
b. On the steering wheel of the car
c. On the passenger side of the front seat
d. Inside the engine compartment
e. Near the tires

7. The vacuum servo can be connected to the throttle by a:
a. Chain
b. Bowden wire
c. Rod or linkage system
d. All of the above
e. None of the above

8. The accumulator on a cruise control system is used to:
a. Store vacuum
b. Store pressure
c. Sense the vehicle speed
d. Sense the transducer signal
e. Set the throttle position

9. The servo position solenoid (SPS) is used on an electronic cruise control system to:
a. Set the throttle position
b. Establish the amount of vacuum being used
c. Tell the controller how fast the servo linkage is moving
d. Set the correct cruise control speed
e. Signal the vehicle speed

10. The vacuum servo is replaced on some vehicles by a(n):
a. Solenoid valve
b. Electro-motor
c. Brake release switch
d. Clutch switch
e. A/T selector gear switch

The following questions are similar in format to ASE (Automotive Service Excellence) test questions.

11. *Technician A* says that a common check on the cruise control is to make sure all vacuum hoses are in good condition. *Technician B* says that a common check on the cruise control is to make sure that the bead chain on the throttle linkage has no slack in it. Who is correct?
a. A only c. Both A and B
b. B only d. Neither A nor B

12. *Technician A* says that electronic cruise controls do not have built-in diagnostics. *Technician B* says the electronic cruise controls keep trouble codes in the computer. Who is correct?
a. A only c. Both A and B
b. B only d. Neither A nor B

13. *Technician A* says that the brake release switch is used as an input to the cruise control PCM. *Technician B* says that the main on-off switch is used as an output from the cruise control PCM. Who is correct?
a. A only c. Both A and B
b. B only d. Neither A nor B

14. *Technician A* says that the SPS on a cruise control system is an input to the computer to determine brake position. *Technician B* says that the vehicle speed sensor is needed as an input to the cruise control PCM. Who is correct?
a. A only c. Both A and B
b. B only d. Neither A nor B

15. *Technician A* says that the vacuum valve on the vacuum servo is an output from the cruise control PCM. *Technician B* says that the vent valve on the vacuum servo is an output from the cruise control PCM. Who is correct?
a. A only c. Both A and B
b. B only d. Neither A nor B

Essay

16. Define the purpose and operation of the transducer on a cruise control system.

17. Explain how the vacuum servo is controlled by the computer on an electronic cruise control system.

18. Explain the purpose of the vacuum valve and the vent valve on the vacuum servo on a cruise control system.

19. What is the purpose of having an optical vehicle speed sensor or a PM generator speed sensor on an electronic cruise control system?

20. Explain the type and the purpose of using various release switches on the electronic cruise control system.

Short Answer

21. On computerized cruise control systems, ________ help the service technician identify the possible problem.

22. The vacuum servo can be replaced by a(n) ________ on some electronic cruise control systems.

23. Often, when troubleshooting an electronic cruise control system, it may be necessary to perform several ________ pinpoint tests.

24. The vacuum solenoid and the vent solenoid can be checked for ________ to determine their electrical condition.

25. If the clutch release switch was accidentally disconnected from the electronic cruise control circuit, the cruise control system would ________.

Applied Academics

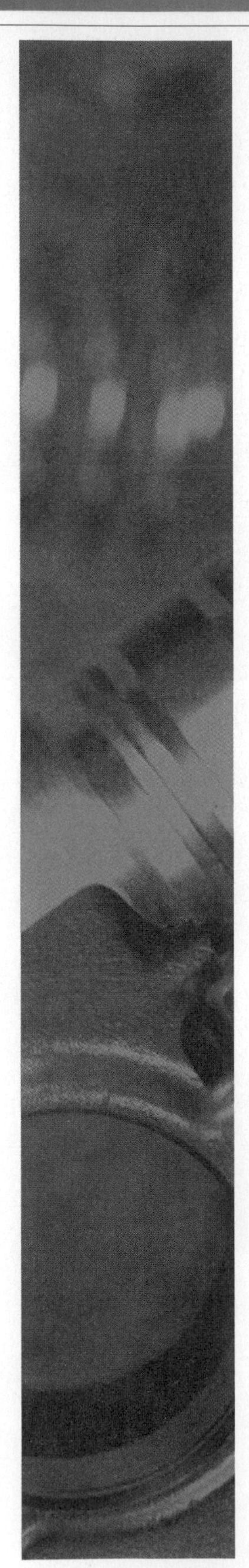

TITLE: Cruise Control and Mileage

ACADEMIC SKILLS AREA: Language Arts

NATEF Automobile Technician Tasks:

During discussions with customers, associates, and supervisors, the service technician makes inferences and predicts solutions to problems.

CONCEPT:

The cruise control system is designed to keep a vehicle at a constant speed. Keeping a vehicle at a constant speed has a tendency to improve fuel mileage and, in turn, decrease the amount of pollution produced by the vehicle. All vehicles have an optimum speed. In addition, fuel mileage decreases if the operator performs an erratic operation on the gas pedal, such as constantly accelerating and decelerating. The diagram that follows shows that on average most vehicles today get the best fuel mileage when the vehicle is traveling at a constant 50 miles per hour. If the vehicle goes slower than that, the fuel mileage drops. If the vehicle goes faster than that, the fuel mileage similarly drops. Although the highway speed limit is not set at 50 miles per hour, it is true that the more constant the vehicle speed, the better the fuel mileage and the less pollution from the engine.

MILEAGE AND SPEED

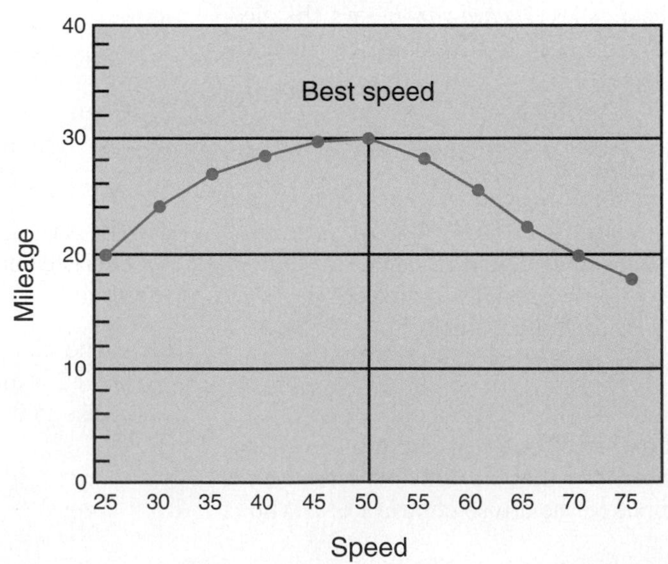

APPLICATION:

What would you say to convince a customer that if cruise control is used, it improves the fuel mileage? Why does a customer who has erratic driving habits get poorer fuel mileage? How does poorer fuel mileage cause more pollution?

Auxiliary and Electrical Systems

CHAPTER 53

OBJECTIVES

After studying this chapter, you should be able to:

- Identify various wiring circuits, schematics, and their symbols.
- Analyze several electrical circuits, including the headlights, defogger, power seats, and horn.
- Define the parts and operation of windshield wiper systems.
- Describe the basic operation of passive restraint systems.
- Identify common problems, their diagnosis, and service procedures for auxiliary and electrical systems.

INTRODUCTION

This chapter covers various auxiliary and electrical systems that are used on the automobile. These include systems such as windshield washers, headlights, horn circuits, and so on. These systems are designed in a variety of ways. Different vehicle manufacturers may or may not have similar systems. Many of these systems rely on electrical circuits for their operation and analysis. This chapter will help the service technician to analyze several auxiliary systems as well as to understand the basic circuitry of some systems.

53.1 READING ELECTRICAL CIRCUITS

SCHEMATICS

Electrical schematics are used to troubleshoot the electrical circuits on vehicles. Schematics subdivide vehicle electrical systems down into individual parts and circuits. Schematics show only the parts and how electrical current flows. Schematics do not show the actual position or physical appearance of the parts. For example, a 4-foot wire appears

Certification Connection

ASE Connection: The information in this chapter can help you prepare for the National Institute for Automotive Service Excellence (ASE) certification tests. The tests and content areas most closely related to this chapter are:

Test A6—Electrical/Electronics Systems

- **Content Area**—Lighting Systems Diagnosis and Repair
- **Content Area**—Gauges, Warning Devices, and Driver Information Systems Diagnoses and Repair
- **Content Area**—Horn and Wiper/Washer Diagnosis and Repair
- **Content Area**—Accessories Diagnosis and Repair

NATEF Connection: Much of the information in this chapter is related to the NATEF tasks. The NATEF tasks and priority numbers most closely related to this chapter are:

1. Identify and interpret electrical/electronic system concerns; determine necessary action. P-1
2. Diagnose electrical/electronic integrity for series, parallel, and series-parallel circuits using principles of electricity (Ohm's law). P-1
3. Use wiring diagrams during diagnosis of electrical and circuit problems. P-1
4. Measure source voltage and perform voltage drop tests in electrical/electronic circuits using a voltmeter; determine necessary action. P-1

continued

Certification Connection *continued*

5. Measure current flow in electrical/electronic circuits and components using an ammeter; determine necessary action. P-1
6. Check continuity and measure resistance in electrical/electronic circuits and components using an ohmmeter; determine necessary action. P-1
7. Locate shorts, grounds, opens, and resistance problems in electrical/electronic circuits; determine necessary action. P-1
8. Diagnose the cause of brighter than normal, intermittent, dim, or no light operation; determine necessary action. P-1
9. Diagnose incorrect horn operation; perform necessary action. P-2
10. Diagnose incorrect wiper operation; diagnose wiper speed control and park problems; perform necessary action. P-2
11. Diagnose incorrect washer operation; perform necessary action. P-2
12. Diagnose incorrect operation of motor-driven accessory circuits; determine necessary action. P-2
13. Diagnose incorrect heated glass operation; determine necessary action. P-3
14. Diagnose supplemental restraint system (SRS) concerns; determine necessary action. (Note: Follow the manufacturer's safety procedures to prevent accidental deployment.) P-2

no differently than a 2-inch wire. All parts are shown as simply as possible, with regard to function only.

ELECTRICAL SYMBOLS

The automotive industry uses **electrical symbols** on the schematics to help identify the circuit operation. Not all manufacturers use the same symbols, but they are generally similar in nature. ***Figure 53–1*** shows common symbols that are used by one manufacturer. Many of the symbols are designed to illustrate their meaning by the type of symbol. For example, connector symbols are shaped somewhat like actual connectors. Keep these symbols on hand to be able to read electrical schematics.

Several numbers and identifying characteristics are also shown on schematics. The color of the wire is represented by letters such as PNK (pink), YEL (yellow), BLU (blue), PPL (purple), ORN (orange), GRY (gray), DK GRN (dark green), and so on. The size of the wire, both metric and American Wire Gauge (AWG) sizes, and the component location (C 103) are also shown. When the service technician is ready to match the schematic parts to the actual hardware, this number is referenced to a **component location table**, along with the schematic. The table tells the technician exactly where to find the component. ***Figure 53–2*** shows an example of a component location table.

READING A SCHEMATIC

Figure 53–3 shows a typical schematic of the heater circuit. This schematic is used here only as an example of how to read it. The schematic is read from top to bottom. With the ignition switch in the Run position, voltage is applied to the fuse panel. From the fuse panel, electricity flows into the blower switch. The blower switch sets the blower speed by adding resistors in series with the blower motor. In LO, two resistors are connected to the blower motor. In MED, one resistor is connected to the blower motor. In HI, full voltage is applied directly to the blower motor.

FUSE BLOCK

Each circuit that is used on the automobile must be fused. Fuses are used to protect circuits. If a circuit becomes overloaded, the fuse melts and opens that particular circuit. Fuses are usually placed on a **fuse block**. The fuse block can be located in several areas. On some cars, the fuse block is located under the dashboard on the left side of the vehicle. On other vehicles, the fuse block is located under the hood.

CUSTOMER CONCERN:
Check Engine Light Is Erratic

It is noticed that, on the dashboard of a computer-controlled GM engine, the Check Engine light goes on intermittently or is erratic. What could be the problem?

SOLUTION: Most of the time, a component fault or light coming on intermittently indicates a faulty electrical connection. The diagnosis should include physical inspection of the wiring and connectors. Check all connections on the electrical circuits on the engine. Also, physically observe the connections for damaged or bent terminals.

SYMBOLS USED IN WIRING DIAGRAMS			
	Positive		Temperature switch
	Negative		Diode
	Ground		Zener diode
	Fuse		Motor
	Circuit breaker	C101	Connector 101
	Condenser		Male connector
	Ohms		Female connector
	Fixed value resistor		Splice
	Variable resistor	S101	Splice number
	Series resistors		Thermal element
	Coil		Multiple connectors
	Open contacts		Digital readout
	Closed contacts		Single filament bulb
	Closed switch		Dual filament bulb
	Open switch		Light-emitting diode
	Ganged switch (NO)		Thermistor
	Single pole double throw switch		PNP bipolar transistor
	Momentary contact switch		NPN bipolar transistor
	Pressure switch		Gauge

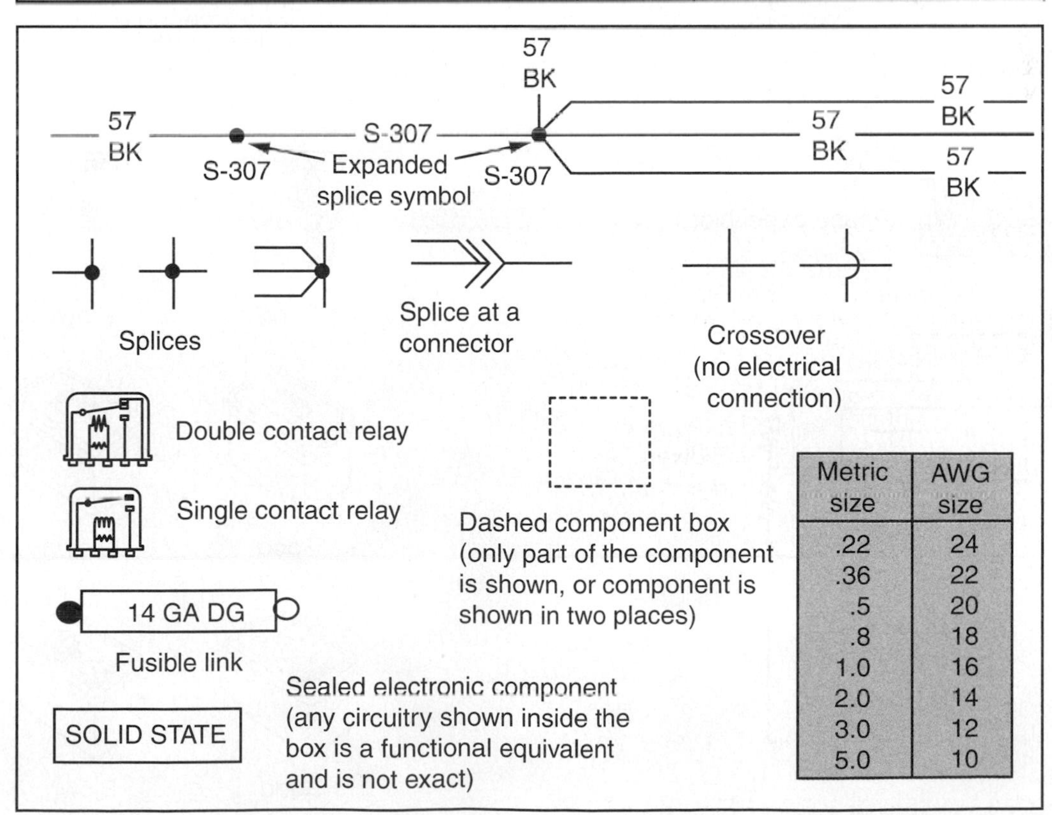

FIGURE 53–1 Various symbols are used to identify electrical circuits. Familiarity with these symbols will help the service technician identify problems in electrical systems.

ELECTRICAL PART	LOCATION	PAGE FIGURE
COMPONENTS		
Blower motor	RH rear of engine compartment	201-4-A
Blower resistors	On blower housing	201-4-E
Fuse block	Behind RH side of IP	201-4-A
Radio capacitor	Above blower motor	201-4-E
CONNECTORS		
C 118 (1 cavity)	Next to blower motor	201-8-B
C 209 (2 cavities)	Behind IP, near control head	201-10-C
C 220 (3 cavities)	Behind center of IP, near grommet	201-10-C
C 241 (2 cavities)	Behind IP, below control head	201-9-A
GROUNDS		
G 104	On EGR solenoid bracket	201-3-A
G 106	RH rear of engine compartment	201-4-A
SPLICES		
S 106	Engine harness, above water pump	201-18-B

FIGURE 53-2 Component location tables are available to help service technicians locate the specific parts on the vehicle that are labeled on the schematic.

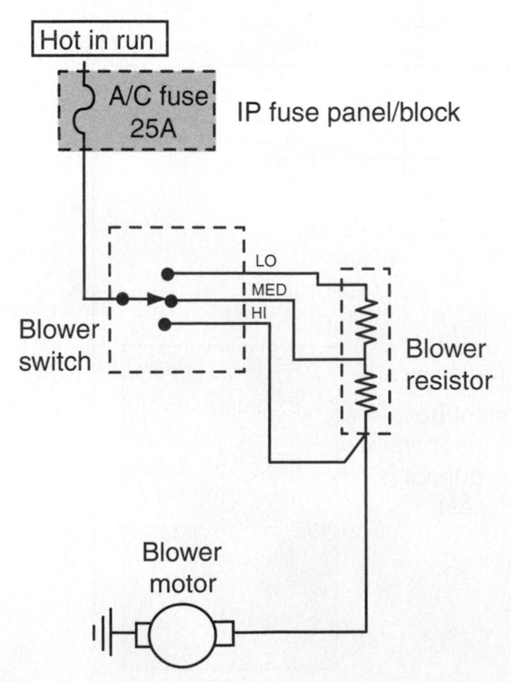

FIGURE 53-3 An example of a simple electrical schematic used to diagnose the heater circuit. Note how the symbols are used to help explain the circuit operation.

Figure 53–4 shows two common locations for the fuse block. The exact location depends on the vehicle manufacturer.

Several types of fuses are used in today's vehicles. ***Figure 53–5*** shows three types of commonly used fuses and how each fuse looks before and after it has blown. In most cases, the amperage capacity of the fuse is printed directly on top of the fuse. To locate the specific fuse that needs to be replaced, fuse boxes have a drawing on their cover that shows the name of each fuse and its location. ***Figure 53–6*** shows a typical fuse box and its cover. Note that on the cover each fuse is identified as to its circuit.

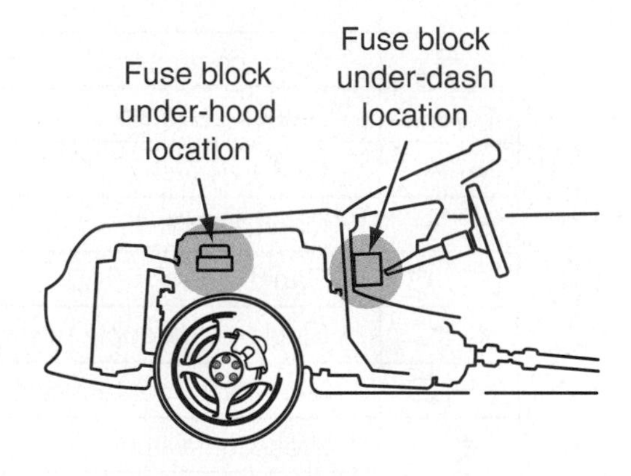

FIGURE 53-4 Fuse boxes are located in several places on the vehicle such as under the hood and under the dashboard.

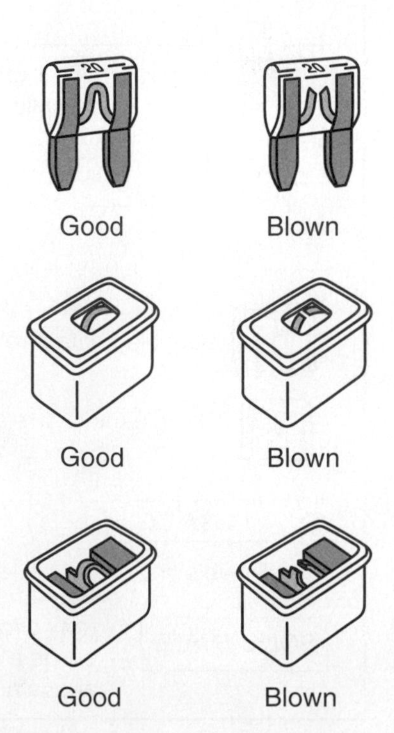

FIGURE 53-5 Three fuse types are shown here in good condition and when blown.

FUSE BOX

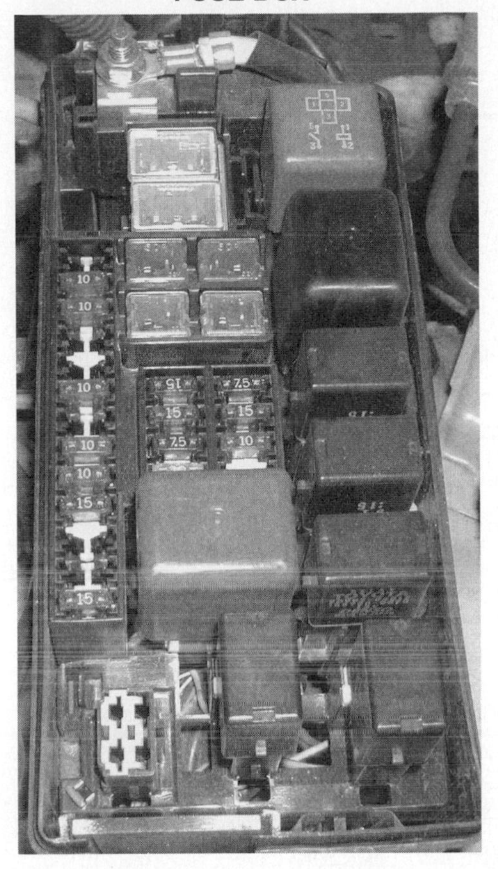

FUSE BOX COVER

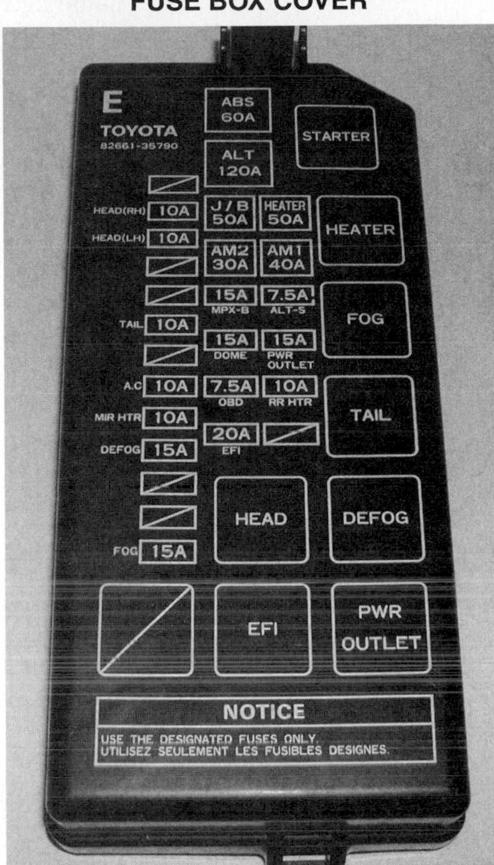

FIGURE 53–6 This fuse box, located under the hood, has both fuses and relays that can be replaced. The fuse box cover also has a chart that identifies each component.

Car Clinic:

FUSIBLE LINK

CUSTOMER CONCERN:

Multiple Electrical Problems

A new car with under 5,000 miles on the odometer has developed unusual electrical problems. For example, certain circuits do not work. These include the automatic door locks, the under-hood light, automatic antenna, and the radio. What could be the problem?

SOLUTION: The failure of multiple electrical circuits usually indicates the possibility of a burned electrical fusible link. Using the manufacturer's service manual, try to identify which fusible link is common to all of the circuits that are defective. This may take some time. Then check the appropriate fusible link for damage. It should be broken in the center, inside the wire insulation. A fusible link on new cars may be shorted out because the wiring harness rubs against the frame and shorts out. Replace the fusible link, but also identify what caused it to short out; for example, a shorted wiring harness.

FUSIBLE LINK

On many electrical circuits, there is a **fusible link**. A fusible link is a type of circuit protector in which a special wire melts to open the circuit when the current is excessive. The fusible link acts like an in-line fuse made of wire with a special nonflammable insulation.

53.2 ELECTRICAL CIRCUITS

As the automobile has developed over the years, more and more electrical circuits have been added to it. An example of some of the electrical circuits used include the clock, defogger, fog lamps, fuel injectors, gauges, headlights, ignition system, cigarette lighter, dome lights, power windows, horn, power seats, tachometer, and tailgate release, among others. All of these circuits can be identified and analyzed by observing their circuit operation on a schematic. This section looks at several electrical circuits that are commonly used on vehicles.

HEADLIGHTS

Figure 53–7 shows a headlight circuit. A complete analysis of the headlight circuit can be obtained by tracing the electricity through the schematic. Starting from the top:

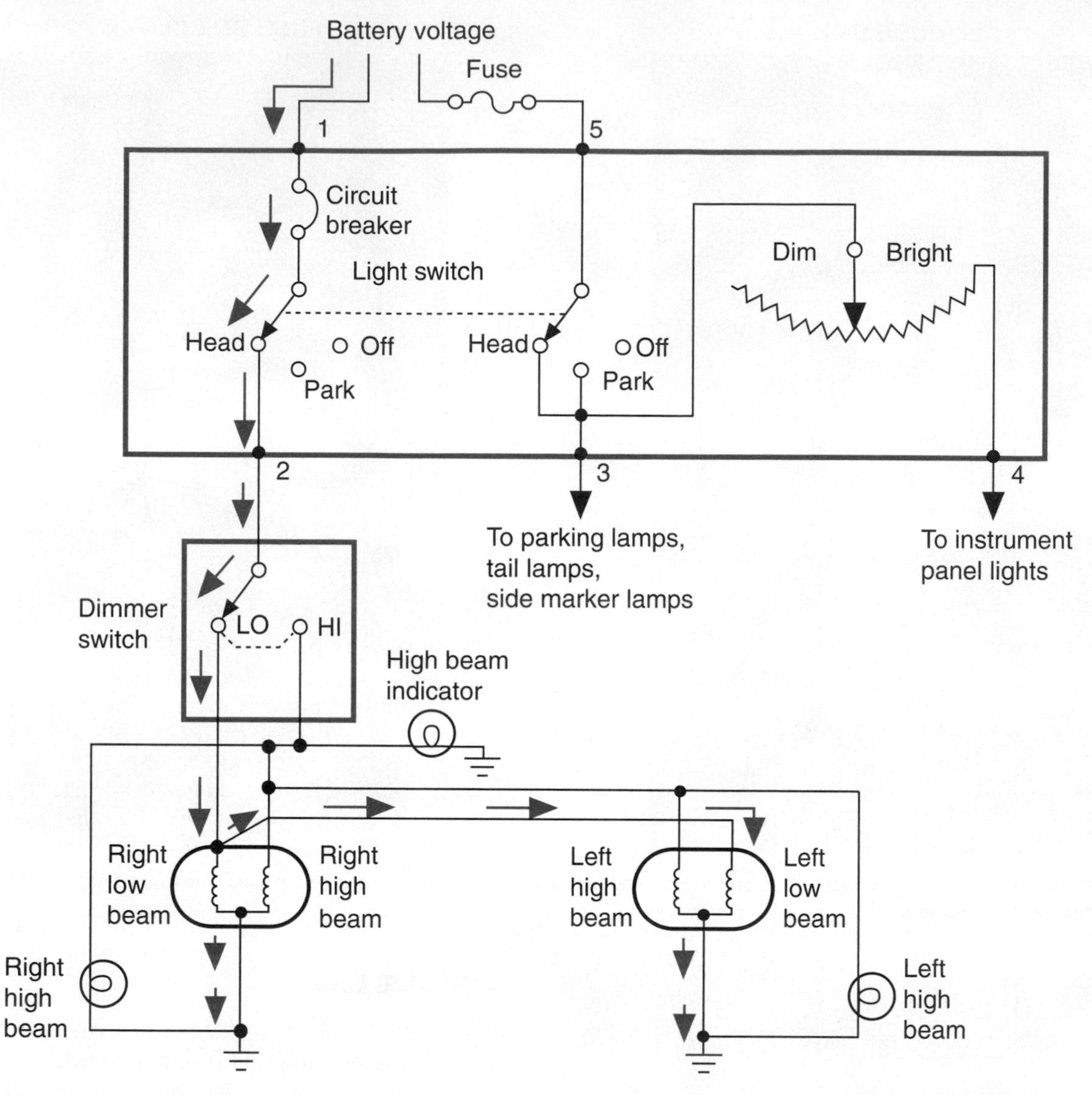

FIGURE 53–7 This vehicle schematic helps the service technician understand the headlight circuit and components.

1. Electricity is available to the light switch all of the time.
2. The electricity first flows through the circuit breaker.
3. The light switch has three positions. These include Off, Park, and Head.
4. When the light switch is in the HEAD position, current flows through the switch to the headlight **dimmer switch**.
5. The dimmer switch can be in one of two positions: LO or HI.
6. In the LO position, the current flows through the dimmer switch to each of the dual-beam headlights. One is for the left side, and one is for the right side of the vehicle.
7. There are two electrical circuits inside each dual-beam headlight. One circuit is for the high beam, and one is for the low beam.
8. Electricity flows through the low beam circuit to ground, completing the circuit.
9. When the dimmer switch is positioned to HI, several other circuits also operate.
10. One circuit sends electricity to the instrument panel's printed circuit to the HI beam indicator light.
11. From the HI connection on the dimmer switch, electricity flows directly to the HI beam circuit in each headlight.
12. After passing through the headlight, the electricity then returns to the battery through ground.

DEFOGGER

Figure 53–8 shows an example of a defogger electrical circuit. The defogger operates when voltage is applied to the rear window wires. The wires are on the inside surface of the glass. When current flows through them, the wires heat the window to remove fog and ice from the glass.

When the defogger control On-Off switch is moved to On (center of defogger control), the defogger control timer is

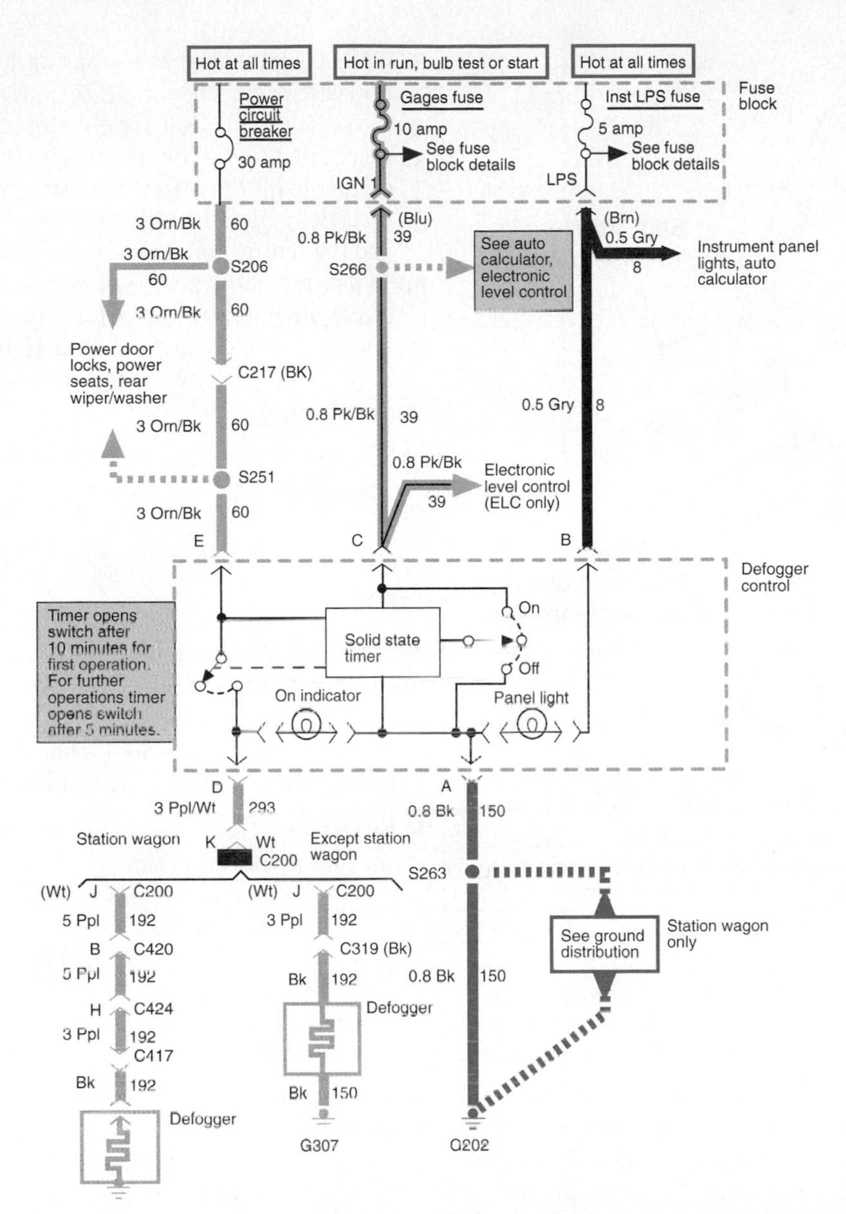

FIGURE 53–8 A defogger removes ice and fog from the rear windows. An electrical wire placed on the inside of the window heats up and removes the ice and fog.

turned on. The defogger control contacts close, and voltage is applied to the defogger and the On indicator. When the instrument panel light circuit provides power, voltage is also applied through the GRY wire to the defogger control panel light.

After the defogger control On-Off switch is released, the defogger control timer holds the defogger control contacts closed for 10 minutes for the first operation. The timer holds the contacts closed for 5 minutes for further operation. When the defogger control timer completes its cycle, the contacts open and voltage is removed from the defogger and the defogger control On indicator.

POWER SEATS

Many vehicles today have power seats that move the front seats upward, downward, forward, and backward. Many vehicles also use support mats within the seat to fit its shape to the driver or passengers. ***Figure 53–9*** shows the internal parts of a typical power seat. The lumbar (back curvature) supports are also shown. The electrical circuit for a standard power seat system can best be described by referring to ***Figure 53–10***. This circuit shows several important parts of the LH (left-hand) power seat switches circuit. The three motors in the middle are used to move the seat in a specific direction. The motors can run forward or backward, depending on the direction of current through the windings. The switches for the power seats are in the upper part of the schematic. The schematic can be traced from the top to the bottom as follows:

1. Voltage is available at the circuit breaker at all times.
2. Electricity flows to the upper wire in the LH power seat switch complex. The voltage is available at each switch, all of which are normally closed.

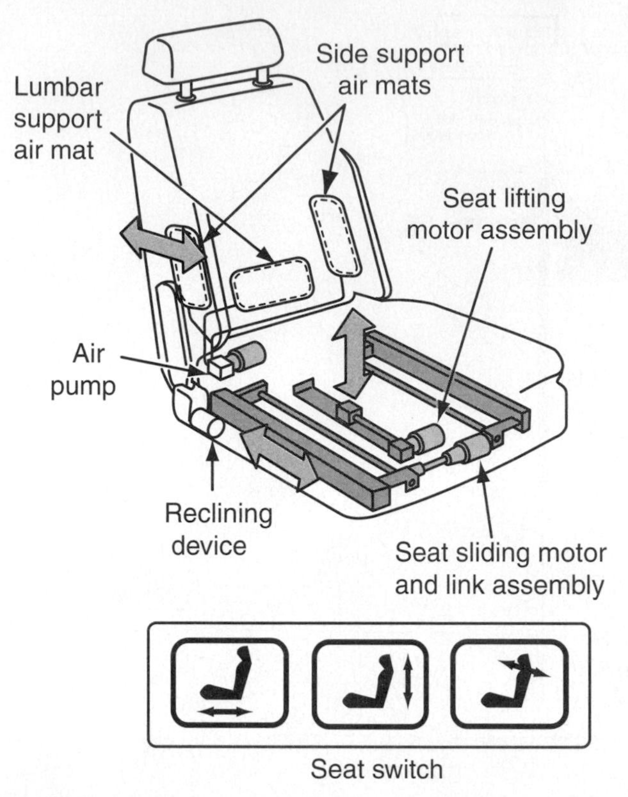

FIGURE 53-9 Power seats are used to adjust the seat and its cushions to best fit the driver and passengers for maximum comfort.

3. If the operator pushes the spring-loaded Rear Height Up switch (far left switch), electricity flows through the switch to the Rear Height Motor.
4. Electricity causes the motor to turn in the correct direction to lift the rear of the seat upward.
5. Electricity then flows through the "rear height down" and the "entire seat down" switch, back to ground.
6. To get the seat to go down, electricity flows in the opposite direction in the motor. The Rear Height Down switch must close and the Rear Height Up switch will open. This is done when the operator pushes the switch down.

When the other switches are closed or opened, the electricity flows to the correct motor to adjust the seat accordingly.

HORN

The horn is a simple circuit used by the operator of the vehicle to alert other drivers of danger. The horn circuit is shown in ***Figure 53–11***. It consists of several major components. These include two horns, a horn switch, a **horn brush/slip ring**, and a horn relay. The circuit operates as follows:

1. Electricity is sent from the battery directly through the fuse to the horn relay.

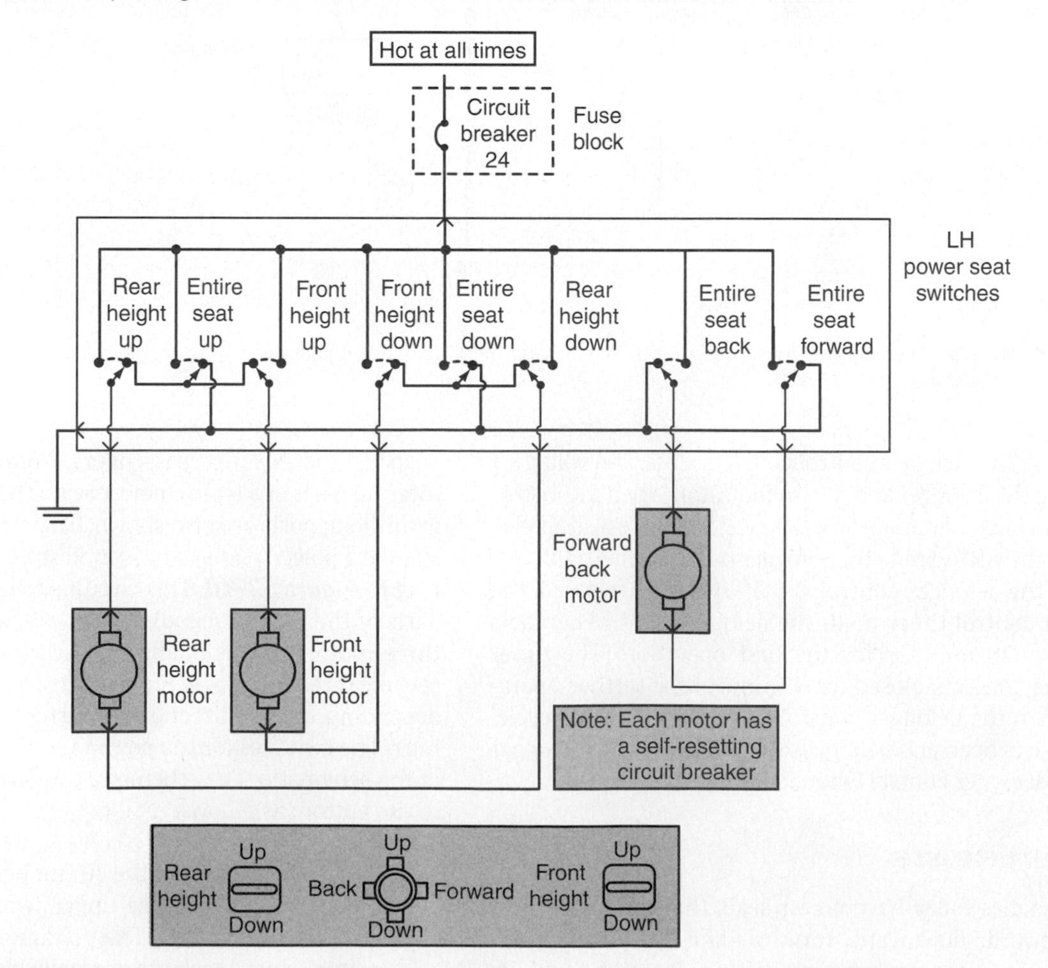

FIGURE 53-10 A typical power seat electrical circuit uses a series of switches to operate the reversible motors.

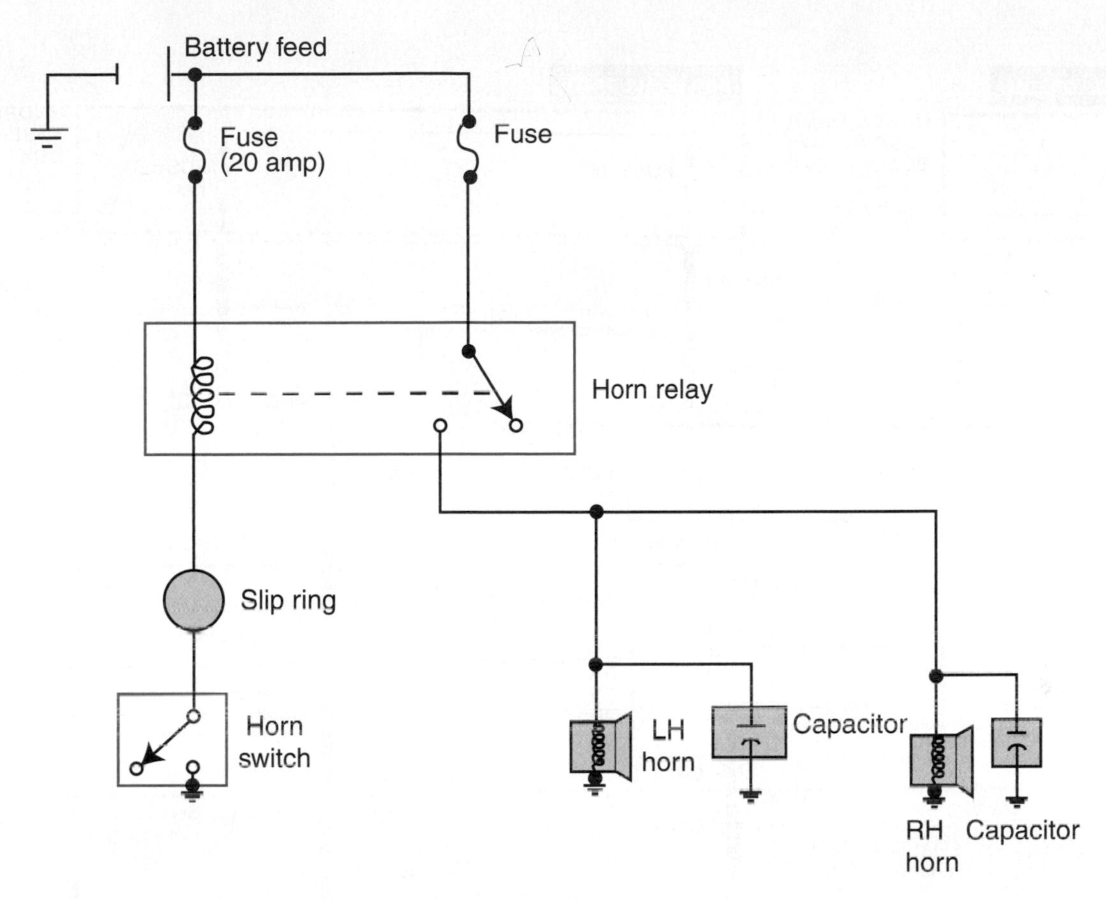

FIGURE 53–11 This horn schematic shows a horn relay used to operate the horns. Two horns, a horn switch, and a horn relay make up the circuit.

2. At this point, current wants to flow through the horn brush/slip-ring assembly. This assembly is used to keep the horn switch in contact with the horn relay when the steering wheel is turned.
3. When the horn switch is pressed, electrical current flows through the left side of the horn relay.
4. As the horn relay is energized, it closes the horn relay switch, causing the horns to operate.

POWER ANTENNA

Many vehicles today use a power antenna. The power antenna retracts the radio antenna when the vehicle ignition is shut off or if the radio is turned off. ***Figure 53–12*** shows a typical electrical circuit, illustrating how the power antenna operates. The electrical operation of the power antenna circuit works as follows:

1. When the ignition is turned on, the cigarette lighter relay is energized, closing the cigarette lighter switch.
2. With this switch closed, electricity can flow from Fuse 36 directly to the stereo radio/cassette player switch. When the radio is turned on, the switch is then closed and electricity now flows directly to the power antenna relay.
3. This action causes the power antenna relay coil to be energized, which closes the two switches (the switches move to the left).
4. Also note the position of the two limit switches. These switches are positioned by the position of the antenna. When the antenna is down, the limit switches are in the down position. When the antenna is up, the switches flip to the right and are now in the up position.
5. With the four switches correctly positioned, electricity flows from Fuse 37, through the right side power antenna relay switch, through the power antenna motor (in a downward direction), then up through the left limit switch, through the left power antenna relay switch, and finally to ground. In this case, the direction of the electricity flowing through the motor causes the antenna to go up.
6. If the antenna is up and the operator shuts off the stereo radio/cassette player switch, the power antenna relay is de-energized, causing the two power antenna relay switches to move to the right. Note that now the limit switches are in the Up position. Under these conditions, electricity flows from Fuse 37, through the right side power antenna relay switch, down through the right limit switch, upward through the power antenna motor, then through the left power antenna relay switch, and finally back to ground. Although many vehicles have similar circuits and switches, the principle is the same. Switches are used to cause electricity to flow through the power antenna motor in opposite directions to get the antenna to go up or down.

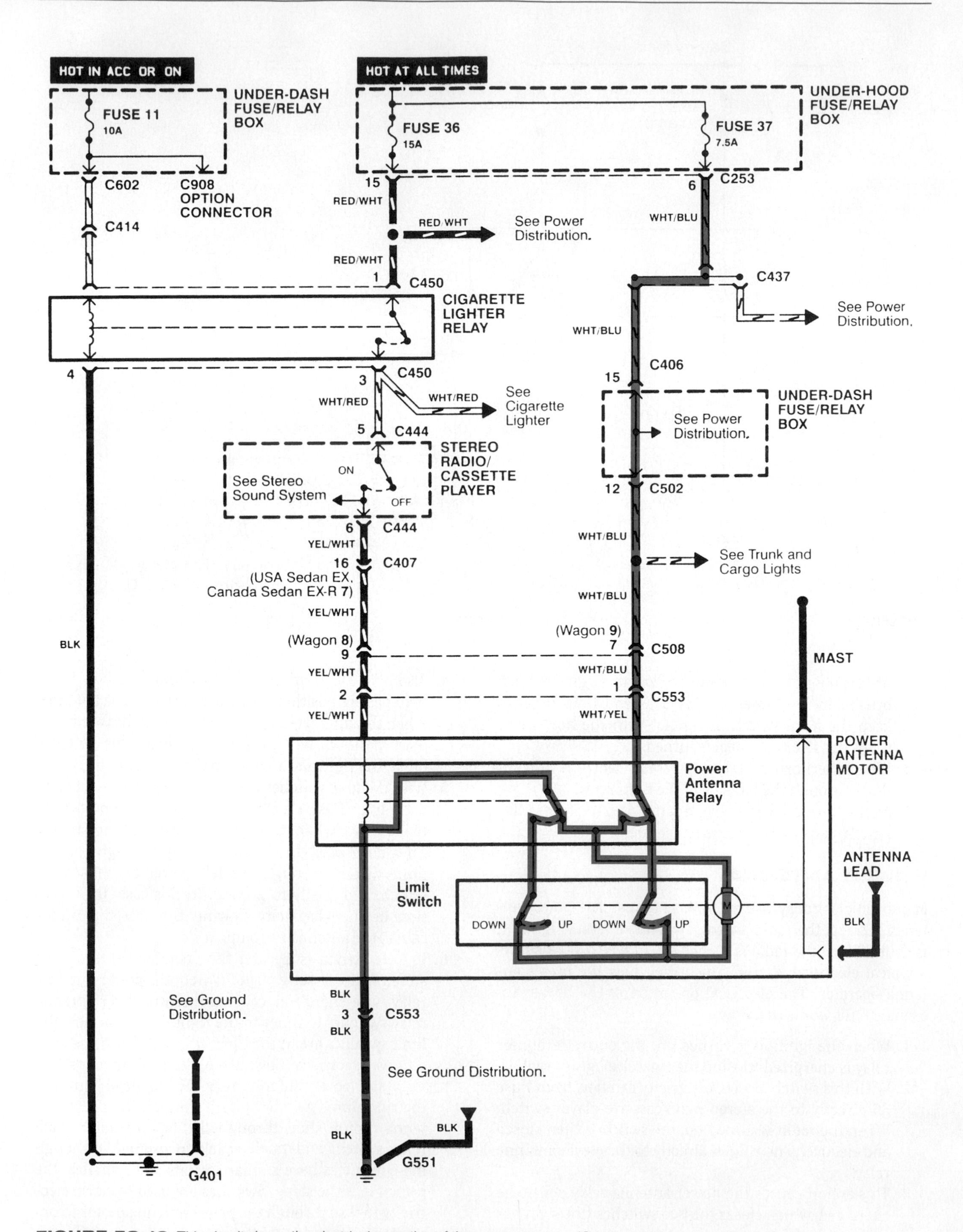

FIGURE 53-12 This circuit shows the electrical operation of the power antenna. (Courtesy of American Honda Motor Co., Inc.)

Note that this circuit is only an example. Many similar circuits are used for the power antenna. The type of schematic and the exact electrical flow is determined by the type, the year, and the model of the vehicle.

PARTS LOCATOR

*Refer to photo #124 in the front of the textbook to see what a **power antenna** looks like.*

POWER WINDOWS

Many vehicles today use electric power to raise and lower the vehicle windows. ***Figure 53–13*** is a typical electrical circuit, showing how power windows operate. The electrical operation of the LR power window circuit works as follows:

1. First locate the ignition switch, the left rear (LR) power window motor, and left rear power window switch. The position of the switches sends electricity through the motor in one of two directions. One direction raises the window, and the opposite direction lowers the window.
2. To operate, the master power window switch (located in the lower center of the circuit) must have the main switch turned on. The main switch is located on the driver's door. This allows the driver to control the operation of all windows in the vehicle.
3. When turned on, the LR window motor can be operated from either the front master switch (top left) or from the LR window switch (bottom left).
4. When the master switch LR window switch is pushed, electricity flows from the circuit breaker to the LR window switch in the master switch. The electricity then flows down through the LR window switch, through the motor, back through both LR window switches, and eventually through the main switch, to ground.
5. When the LR window switch is pushed, electricity flows from the circuit breaker, directly to the LR window switch, to the motor, and then back through the switches to ground.
6. So, depending upon which of the LR window switches are pushed, electricity flows through the motor in a specific direction to make the window go down.
7. To make the LR window go up, the up switch must be pushed. This causes the electricity to flow through the motor in the reverse direction.

Again, the two switches inside the left rear power window switch direct electricity through the motor in one of two directions. Also, note that the two switches located inside the master power window switch are those located at the driver's position. Therefore, the driver can also control the position of the rear left window from the driver's seat.

This shows just one part of the total electrical circuit that is used for the power window system. There is also a similar circuit for each of the other windows as well.

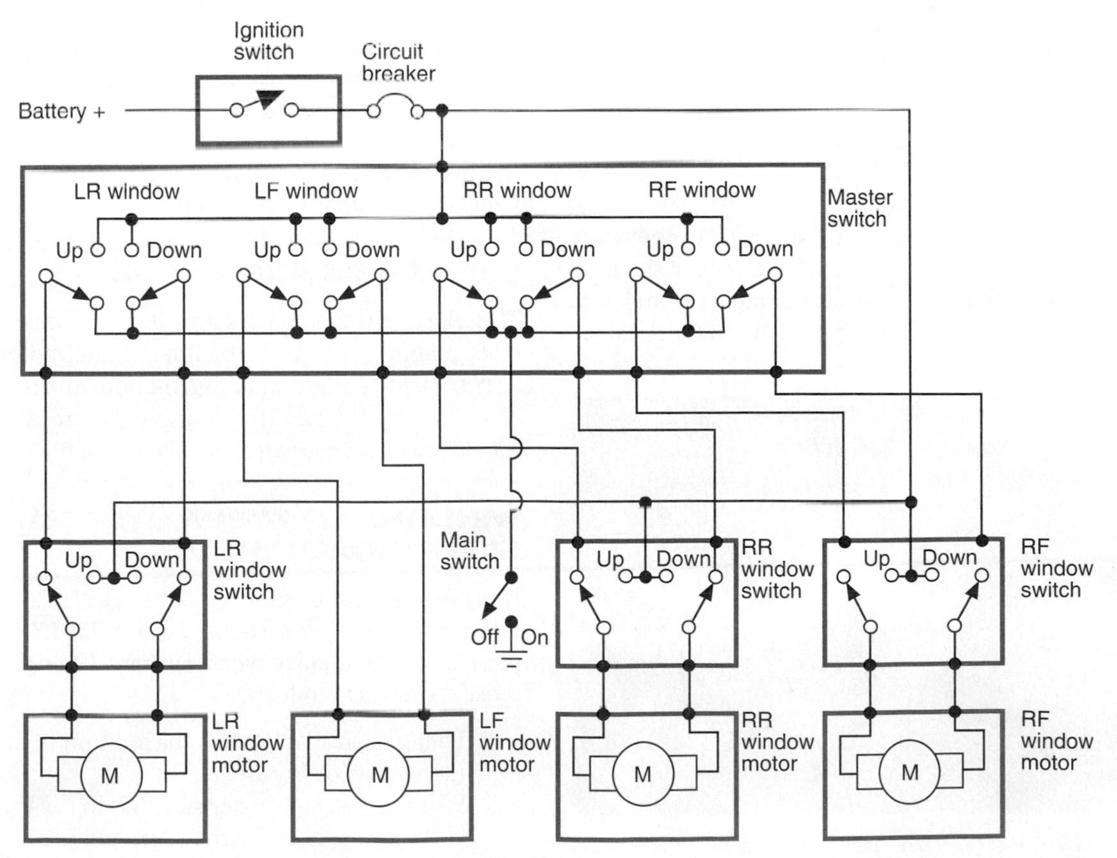

FIGURE 53–13 This circuit shows the electrical operation of the power windows.

53.3 WINDSHIELD WIPER SYSTEMS

There are several types of windshield wiper systems. Both rear and front window systems are commonly used. Common components usually include the motor, linkage mechanism, switch, electronic logic circuits, and a washer system.

WINDSHIELD WIPER MOTOR

The motor in a windshield wiper system uses a permanent magnet (PM) motor. Most front wiper motors have two speeds so that the wiper blade speed can be adjusted. In addition, some vehicles use a variable timer and speed control so the operator can control the wipers more precisely. ***Figure 53–14*** shows a typical motor assembly. It includes several components. The motor produces the rotational motion needed. The washer pump is located in the housing for pumping an alcohol-based spray onto the windshield. An internal mechanism is used to change the rotational motion of the motor to an oscillation motion, which is needed for the wiper blades. As the motor rotates, the output is an oscillation or a back-and-forth motion.

FLUID WASHER NOZZLE

Many cars have a fluid washer nozzle for the front or rear windshield washer system. The system consists of a fluid container, pump, fluid hoses and pipes, and nozzles or jets. ***Figure 53–15*** shows a washer system. When the operator turns on the washer, a small pump forces fluid through the hoses and pipes to the nozzle or jet by the windshield. The pump can be located either at the fluid container or wiper motor.

WINDSHIELD WIPER LINKAGE

Several arms and pivot shafts make up the linkage used to transmit the oscillating motion at the motor to the windshield wipers. ***Figure 53–16*** shows an example of the linkage. As the wiper motor oscillates, arm A moves from left to right. This moves arm B as well. As arm B moves, it causes the two pivot points to oscillate. The windshield wipers are connected to the two pivot points.

FIGURE 53–14 The motor assembly for a windshield wiper system includes a motor, washer, and oscillating mechanism.

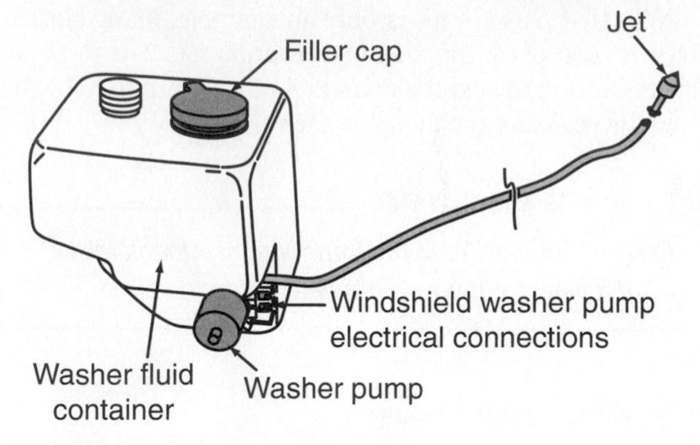

FIGURE 53–15 The windshield washer system uses a fluid container, a small washer pump, and hose and nozzles (jets) to deliver fluid to the windshield.

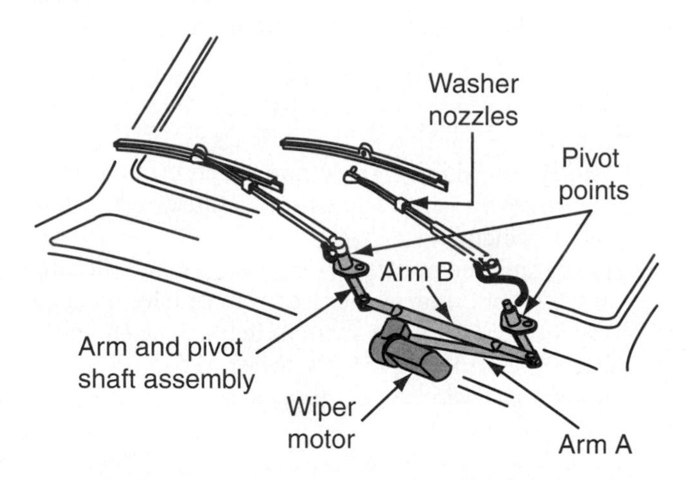

FIGURE 53–16 The oscillating motion from the wiper motor is transferred by a linkage system to the pivot points for the wipers.

WIPER ARM AND BLADE

The wiper arms and blades are attached directly to the two pivot points operated by the linkage and motor. The wiper arm transmits the oscillating motion to the wiper blade. The wiper blade wipes the windshield clear of water. ***Figure 53–17*** shows a wiper arm and blade assembly.

WINDSHIELD WIPER SYSTEM ELECTRICAL CIRCUIT

It is not practical to show all of the electrical circuits used on windshield wiper systems. ***Figure 53–18*** shows a complete circuit in a **pulse wiper system**. The operation of the pulse system is as follows:

1. Voltage is available when the ignition switch is on Accy (accessory) or Run.
2. When the wiper/washer switch is in the Pulse position, voltage is applied to the PNK and GRY wires on the wiper/washer motor module.

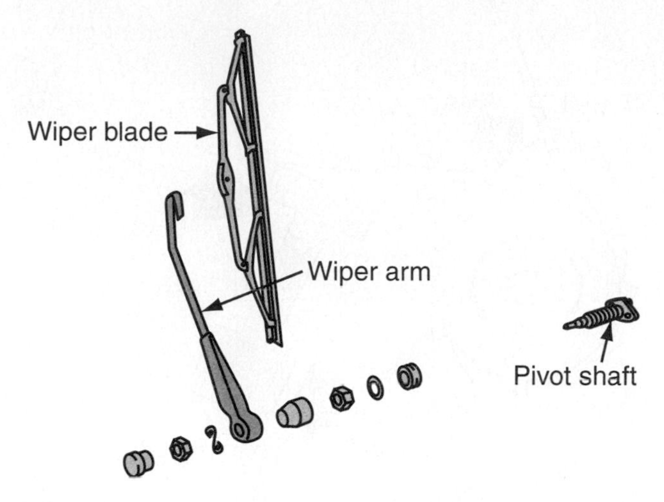

FIGURE 53-17 A wiper arm connects the oscillating motion from the linkage to the wiper blade. (Courtesy of Volkswagen of America, Inc.)

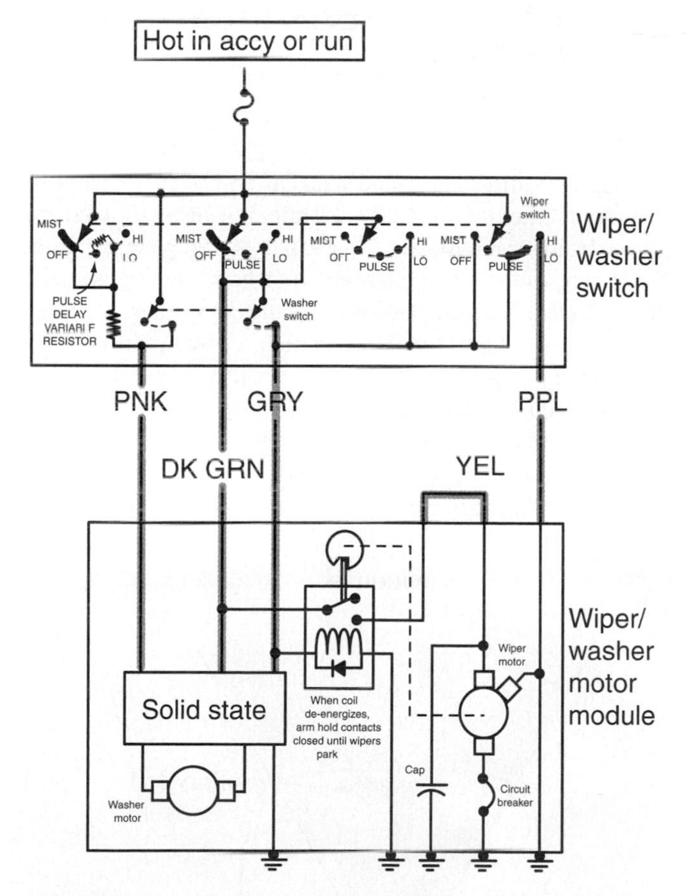

FIGURE 53-18 This circuit shows the complete operation of a pulse wiper system.

3. Voltage is now applied to the solid state control board. Voltage from the control board is sent out and to the coil inside the park relay. The coil pulls the switch down.
4. Another voltage from the control board is sent through the park relay switch, through the YEL wire, to run the wiper motor.
5. The park relay switch is held closed by the mechanical arm until the wipers have completed their sweep. The circuit is then opened, and the wipers remain parked until the control board again applies a pulse voltage to the park relay.
6. The length of delay time between sweeps is controlled by the 1.2 megohm **pulse delay variable resistor** in the wiper/washer switch. The time delay is adjustable from zero to 25 seconds on this circuit.

The LO speed operates as follows:

1. In the LO position, the wiper switch supplies voltage to the DK GRN wire as well as the PNK and GRY wires.
2. The park relay is again energized.
3. Battery voltage is applied continuously to the relay contacts and to the wiper motor. The wiper motor runs continuously at a low speed.

The HI speed operates as follows:

1. Battery voltage is applied directly to the wiper motor through the PPL wire.
2. Voltage is also applied to the DK GRN and the GRY wires to energize the park relay.
3. When turned OFF, the wipers complete the last sweep and park.

The washer operates as follows:

1. When the washer switch is held on for less than 1 second, voltage is applied to the motor module through the PNK and GRY wires.
2. The motor module turns on the washer motor for approximately 2 1/2 seconds.
3. The voltage on the GRY wire also operates the park relay.
4. The motor module also turns on the wiper motor for about 6 seconds.

53.4 SAFETY BELTS

PURPOSE OF RETRACTORS AND REELS

In the past, safety belts were cumbersome and binding. Today, retractors and reels rewind and loosen the belts as needed. The reel allows the safety belt wearer to move around inside the vehicle freely during normal conditions. This freedom makes some people skeptical. They feel the belt may not restrain them in a collision. However, the belts lock solidly when necessary.

Today's vehicles use both active and passive restraint systems. An active restraint system is one in which the driver and passengers manually put on their seat belts. A passive restraint system is one in which the seat belts automatically engage when the door is closed or when the ignition system is turned on. Depending on the year and manufacturer of the vehicle, either or both systems may be used on a particular vehicle.

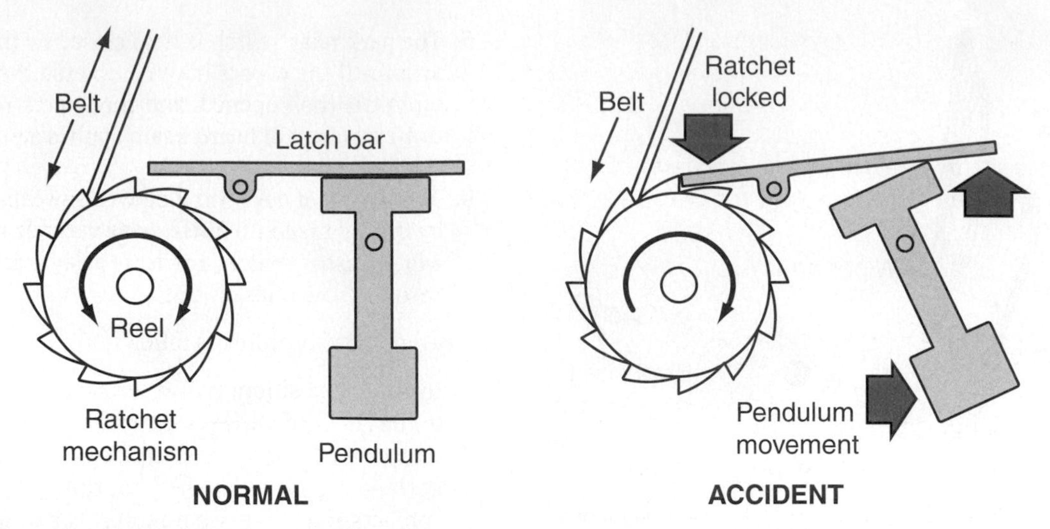

FIGURE 53-19 Some safety belts use a pendulum system to lock up when the vehicle is in a collision.

OPERATION OF SAFETY BELTS

The safety belt shown in ***Figure 53–19*** is called a car-sensitive belt. It uses a **pendulum** located in the car body and a ratchet mechanism. Under normal conditions, the pendulum and bar are in their resting position. The reel, which holds the belt, is free to rotate. As the occupant leans against the belt, it "gives" or unreels. Under accident conditions, such as a collision, the pendulum tilts toward the force of the impact. This causes the bar to engage the ratchet. The reel and seat belt now lock, restraining the occupant.

Many vehicles use a centrifugal seat belt retractor system. In this system, a seat belt that is pulled too quickly out of its retractor mechanism will lock up. However, if it is pulled out slowly, it will not lock up. This type of system allows the driver or passenger to move freely when the seat belt is on. However, when the driver or passenger is moved forward rapidly during an accident, the seat belt retractor immediately locks up.

Some vehicles also use a seat belt pretensioner system. This system consists of several components including the:

- Front air bag sensor
- SRS (supplemental restraint system) warning light on the dashboard
- Seat belt pretensioner assembly
- Air bag sensor assembly

In operation, when the air bag sensor on the front of the vehicle detects the shock of a severe frontal impact, the retractor automatically tightens the two front seat belts. This keeps the occupants safely restrained in the front seat during the impact.

53.5 AIR BAG (PASSIVE RESTRAINT) SYSTEMS

DESCRIPTION

Air bag systems, also called **passive restraint systems** or SIR (supplemental inflatable restraint) or SRS (supplemental restraint system), are now used on all new vehicles. Their purpose is to supplement seat belts during an accident. Driver's side air bags are designed to inflate through the steering column when the vehicle is involved in a front end accident of sufficient force. The result is that the driver is protected from serious injury. The restraint system is activated when an accident occurs up to 30% off the centerline of the vehicle (***Figure 53–20***).

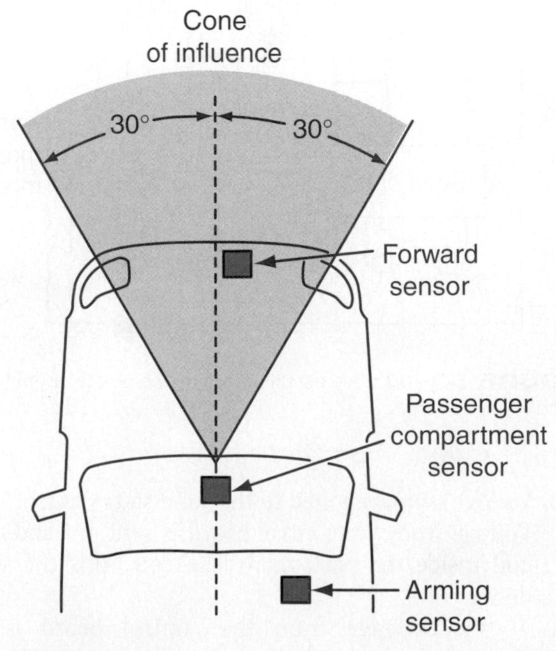

FIGURE 53-20 Sensors are located in several positions on a vehicle to sense an impending accident so the air bags can be inflated.

Most vehicles that are being manufactured today have air bags on both the driver's side and the passenger side. In fact, automotive manufacturers are now designing air bag systems for the sides of the vehicle (in the doors) to further protect passengers in serious accidents. ***Figure 53–21*** shows examples of three types of side impact air bags being used in vehicles today.

Note, however, that in recent tests, air bags have been shown to be dangerous to some passengers. Air bags inflate with such a powerful blast that they have severely injured small children and, in some cases, caused death. Because of such dangers, manufacturers are reducing the pressure used to inflate the air bag. In addition, it is a good idea to never let children sit in the front seat of a moving vehicle.

A Side impact head air bag

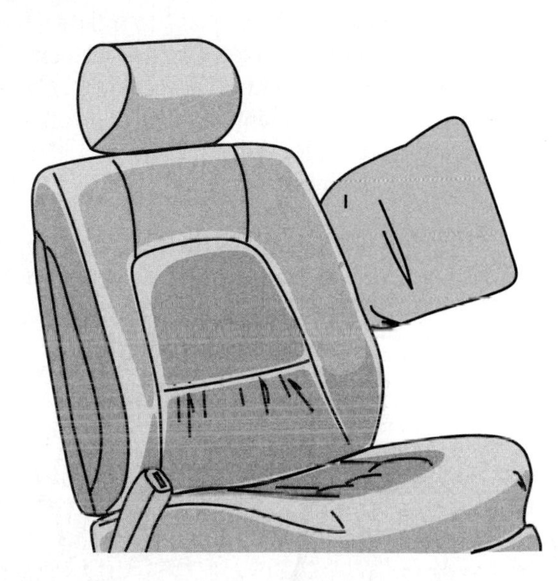

B Side impact seat air bag

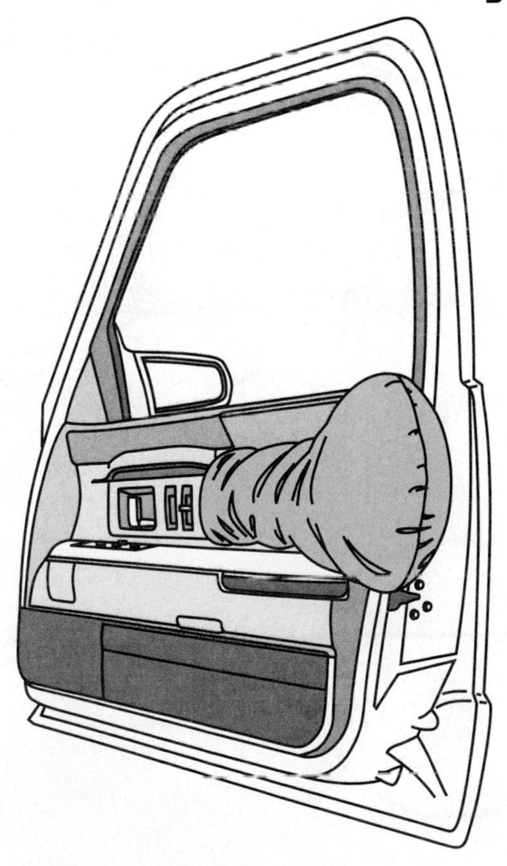

C Side impact door air bag

FIGURE 53–21 There are several variations in the design of side air bags.

COMPONENTS

Depending on the vehicle and the year, there are several designs. On one design, there are three sensors on the vehicle:

1. **Forward sensor**
2. Passenger compartment sensor
3. Arming sensor

These sensors cause the air bags to inflate when there is a rapid change in the vehicle's speed or velocity. Any rapid change means that an accident is occurring. ***Figure 53–22*** shows the major components and their location for a passive restraint system. Refer to the figure to identify the following thirteen components:

1. Control module—Acts as the passenger compartment sensor. The control module also contains diagnostic information for the restraint system. It monitors vehicle deceleration and combines this information with the other two sensors. On the basis of the information from these sensors, the control module signals the air bag to inflate or deploy.
2. Forward sensor—Located near the front grille of the vehicle. This sensor is able to detect a rapid drop in vehicle speed. It is normally opened electrically. When deceleration occurs, the switch is closed, signaling an impending accident.

The impact sensor can also be located directly within the SRS unit. Thus, there are several precautions that should be considered when working with the SRS unit. Some include:

a. Do not install any SRS system that shows signs of being dropped or improperly handled such as dents, cracks, or deformations.

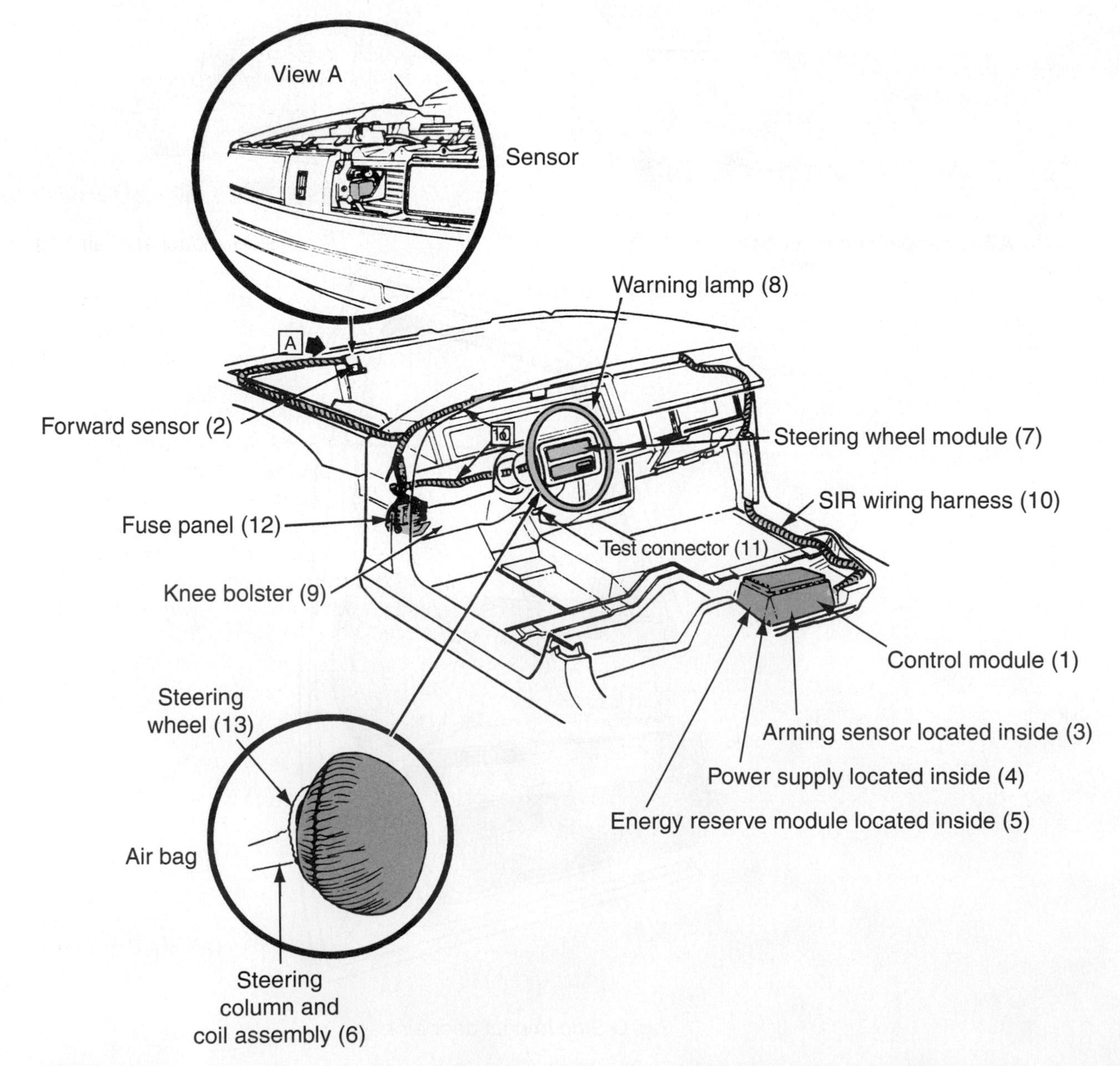

FIGURE 53–22 The major components of a passive restraint (air bag) system and their location. (Courtesy of DaimlerChrysler Corporation)

   b. Whenever an air bag has been activated, always replace the SRS unit as well.
   c. When installing an SRS unit, always avoid strong impacts (impact wrenches, hammers, etc.) near the SRS unit.
   d. Never disassemble the SRS unit.
   e. Store the SRS unit in a cool and dry place.
   f. Do not spill water on the SRS unit.
   g. Keep the SRS unit free from dust.
3. Arming sensor—A normally open mechanical sensor that closes during rapid deceleration. It is considered a backup to the other sensors.
4. Power supply—Provides an increased voltage to the air bag during low-voltage conditions to make sure it opens. The power supply also provides power for the diagnostics systems.
5. Energy reserve module—Provides extra electrical energy if the vehicle's battery is damaged or disconnected during the accident.
6. Steering column and coil assembly—A collapsible steering column and, inside the steering column, a coil. This assembly provides continuous electrical contact between the control module and the steering wheel module as the steering wheel is turned.
7. Steering wheel module—Made of the inflator assembly, which includes the electrical circuit and the inflatable bag. The steering wheel module is located inside the steering wheel hub area and is covered by a vinyl trim cover. When an accident occurs, the inflatable bag is blown out by high-pressure gas. The trim cover then opens or ruptures at the seams, allowing the bag to be inflated.
8. Warning lamp—Displays the words "Inflatable Restraint." It also helps the service technician diagnose the system.
9. Knee bolster—An energy-absorbing pad used to cushion the forward movement of the driver during an accident by restricting leg movement.
10. SIR wiring harness—Interconnects all of the system's components electrically.
11. Test connector—Used to start or initiate diagnosis of the air bag system.
12. Fuse panel—Most vehicles that have an air bag system have a specially designed fuse panel. It is located in the lower instrument panel trim pad and is hinged to swing out for easy access.
13. Steering wheel—Controls the direction of the vehicle during driving conditions.

CAUTION: *SRS Safety Precautions*

At times, it may be necessary to service components near and around the SRS. If this is the case, always follow the manufacturer's suggested service procedure for disabling or removing SRS system components. In addition, always follow these safety precautions:

1. *When carrying a live air bag module, point the bag and trim cushion away from your body. When placing a live air bag on a bench or other surface, always face the bag and trim cushion up and away from the surface.*
2. *Do not bump, strike, or drop any SRS component. Store SRS components away from any source of electricity (including static electricity), moisture, oil, grease, and extreme heat and humidity.*
3. *Do not cut, damage, or attempt to alter the SRS wiring harness or its yellow insulation.*
4. *Always disconnect both battery cables when working around SRS components or wiring.*
5. *Use only a digital multimeter when checking any part of the air bag system.*
6. *Do not install SRS components that have been recovered from wrecked or dismantled vehicles.*
7. *Always disable the air bag when working under the dashboard.*
8. *Always check the alignment of the air bag cable reel during steering-related service procedures.*
9. *Take extra care when working inside and around the dashboard. Avoid directly exposing the SRS unit or wiring to heat guns, welding, or spraying equipment.*
10. *If the vehicle is involved in a frontal impact or after a collision without air bag deployment, inspect the SRS unit for physical damage. If the SRS control unit is dented, cracked, or deformed, replace it.*
11. *Never disassemble the SRS control unit.*
12. *Avoid using impact wrenches, hammers, and so on in the area surrounding the SRS control unit.*
13. *Never reach through the steering wheel to start an air bag-equipped vehicle.*
14. *An SRS must be disarmed correctly before any of its components are disconnected or the air bag is removed.*

Failure to following any of these safety precautions may cause accidental deployment of the air bag, resulting in unnecessary SRS repairs and possible personal injury.

Problems, Diagnosis, and Service

Safety Precautions

1. Many electrical wires have very sharp protrusions. When handling wires, be careful not to puncture your hands.
2. Be careful when handling restraint system sensors. Never strike or jar a sensor in such a way that may cause deployment of the air bag. Before servicing any parts on a restraint system, wait 10 minutes after disconnecting the battery. This prevents accidental deployment and possible injury.
3. Before removing any auxiliary or electrical system component, make sure you understand its operation and how it is hooked into the system. This information will eliminate many possible injuries.
4. Make sure the battery is disconnected when working on electrical circuits to eliminate the possibility of accidentally starting the motor and experiencing an electrical shock.
5. Always wear OSHA-approved safety glasses when working on any auxiliary or electrical system.
6. Remove all metal jewelry (rings, watches, etc.) when working with electrical components.

PROBLEM: Electrical and Mechanical Faults

Many problems pertaining to auxiliary and electrical systems can occur. Rather than specify a particular problem, this section presents general diagnostic procedures for both electrical and mechanical problems and gives several examples.

DIAGNOSIS

Electrical Problems When diagnosing problems in electrical circuits, use the following general procedures:

1. Make sure you know what the exact problem is. Don't replace a component that is not faulty.
2. Refer to the electrical schematic. Read the schematic to make sure you know exactly how the electrical circuit and its components are supposed to operate.
3. Look for the possible cause. After viewing the schematic, consider possible causes. First check the basic components and those easiest to check, such as the fuse and ground.
4. Test for correct voltage during operation. Using a voltmeter, check for voltages at various parts of the schematic. On the basis of this information, again try to determine the cause of the problem.
5. Narrow down the problem to one point in the circuit. Try to isolate the problem. Make voltage checks and then try to narrow down the problem to only one component.
6. Find the cause and repair it. Use component location manuals to help identify where the part is positioned on the vehicle.
7. Check the circuit for correct operation. Make sure the cause has been corrected and replaced.
8. There are many diagnostic tools to help the service technician check electrical circuits and systems.
   a. To check electrical circuits, a fused **jumper wire** can be used to bypass a particular part of the circuit. ***Figure 53–23*** shows a fused jumper wire.
   b. Short finders are available to locate hidden shorts to ground.
   c. A fuse tester can be used to check for bad fuses.

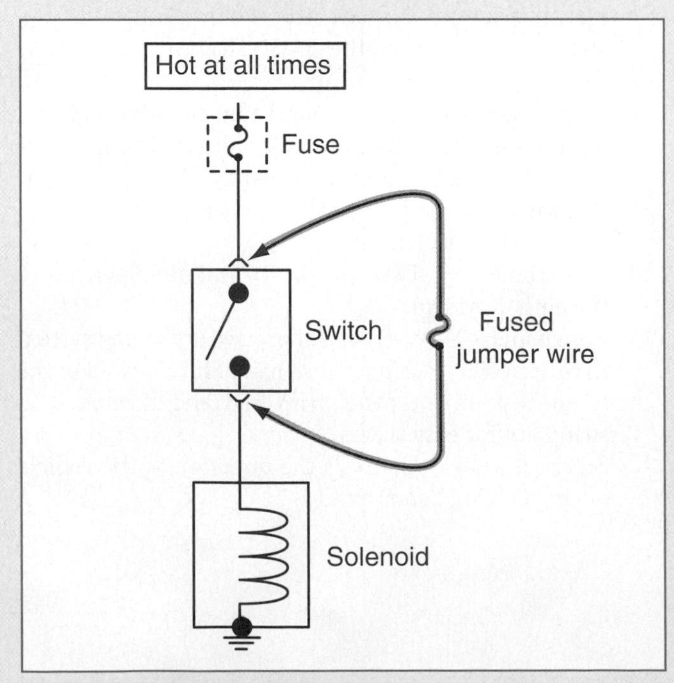

FIGURE 53–23 Jumper wires with a fuse can be used to bypass a component in an electrical circuit.

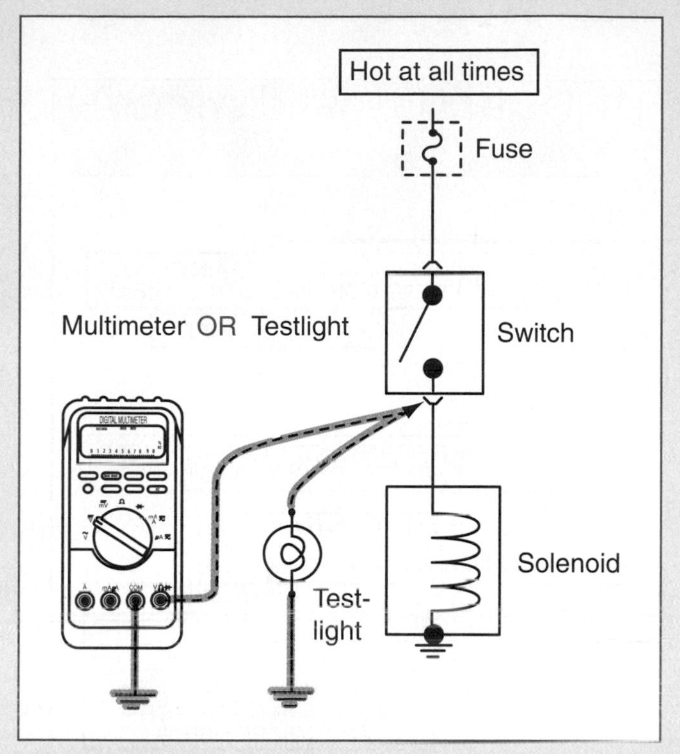

FIGURE 53–24 Connect the voltage meter leads across the component to check voltage drop on a component. In this case, the voltage drop of the solenoid would be read on the voltmeter.

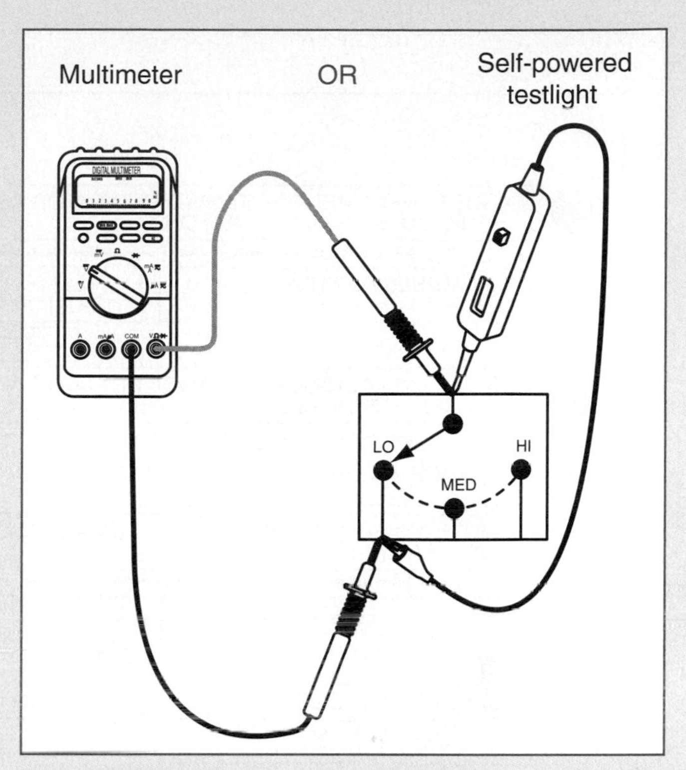

FIGURE 53–25 Use a self-powered test light or an ohmmeter to check continuity.

d. Testing for voltages can be done as shown in ***Figure 53–24***. One lead is grounded, and the other lead is connected to a specific test point. In this case, the voltage drop will be checked across the solenoid.

e. Test for continuity by using a self-powered test light or ohmmeter. ***Figure 53–25*** shows an example of testing the continuity of a blower switch in the low position.

Mechanical Problems Use the following general procedures to help diagnose mechanical problems related to auxiliary and electrical systems:

1. Operate the system through all conditions such as low, medium, high, off, and on.
2. Observe the malfunction or operational characteristics. Did the problem occur over a period of time or all at once? Ask the operator for information about its operation. Know the maintenance record of the system and vehicle.
3. Become completely familiar with how the system should operate. Use the operator's manual, electrical schematics, maintenance manuals, and component locator manual to aid the diagnostic procedure.
4. First, check the components that can be easily tested such as fuses, grounds, broken parts, and special noises.
5. Try to isolate the problem to a specific component within the system.
6. Check the suspected component for correct operation (through the total range of operation).
7. Replace the cause of the problem.
8. Check the system to make sure the problem has been corrected.
9. If there is still a problem, get a second opinion from another service technician.

Diagnostic Examples Numerous diagnostic checks can be performed on passive restraint systems. Each will depend on the year and type of vehicle and system. Most passive restraint systems are computer controlled. Thus, these systems have a built-in diagnostic system that monitors the electrical circuits. The diagnostic circuits, through the warning lamp, indicate the condition of the system.

Once it has been determined by the warning light that the system is inoperative, a diagnostic check can be made. ***Figure 53–26*** shows a typical diagnostic procedure to determine the cause of the problem.

Note that most items on a passive restraint system cannot be repaired. The following items are replaced, not repaired:

- Forward sensor
- Arming sensor
- Power supply

SIR DIAGNOSTIC SYSTEM CHECK

1
- NOTE "INFLATABLE RESTRAINT" INDICATOR AS IGNITION IS FIRST TURNED ON

"INFLATABLE RESTRAINT" INDICATOR BLINKS 7 TO 9 TIMES

"INFLATABLE RESTRAINT" INDICATOR COMES ON STEADY
- GO TO CHART C

"INFLATABLE RESTRAINT" INDICATOR DOES NOT COME ON
- GO TO CHART D

2
- NOTE "INFLATABLE RESTRAINT" INDICATOR AS ENGINE IS STARTED

"INFLATABLE RESTRAINT" INDICATOR COMES ON STEADY DURING CRANK
- NOTE "INFLATABLE RESTRAINT" INDICATOR WITH ENGINE AT IDLE AFTER CRANK

"INFLATABLE RESTRAINT" INDICATOR DOES NOT COME ON STEADY DURING CRANK
- GO TO CHART E

"INFLATABLE RESTRAINT" INDICATOR BLINKS 7 TO 9 TIMES AND THEN REMAINS ON.

"INFLATABLE RESTRAINT" INDICATOR BLINKS 7 TO 9 TIMES AND THEN FLASHES CODE 12.

"INFLATABLE RESTRAINT" INDICATOR BLINKS 7 TO 9 TIMES AND THEN GOES OUT.

3
- CURRENT SIR CODES EXIST
- WILL "FLASH CODES" OR A "SCAN TOOL" BE USED?

4
- GO TO CHART F

5
- NO CURRENT CODES EXIST
- WILL "FLASH CODES" OR A "SCAN TOOL" BE USED?

"SCAN TOOL"

"FLASH CODES"
- GO TO CHART B

"SCAN TOOL"

6
- CONNECT "SCAN TOOL" TO ALDL CONNECTOR, FOLLOW INSTRUCTIONS IN TOOL INSTRUCTION MANUAL
- IGNITION SWITCH "ON"
- REQUEST SIR CODE DISPLAY, RECORD ALL CURRENT AND HISTORY CODES ONTO REPAIR ORDER

CODES ARE DISPLAYED

NO CODES ARE DISPLAYED
- SIR SYSTEM IS FUNCTIONAL AND FAULT FREE
- NO FURTHER DIAGNOSIS IS NEEDED

"SCAN TOOL" INDICATES "NO DATA"
- IS VEHICLE EQUIPPED WIH A VIN E (LO3) ENGINE?

VERIFY PROPER SIR DIAGNOSTIC CARTRIDGE

YES
- CHECK SCAN TOOL CONNECTIONS
- IF SCAN TOOL IS OK, CHECK CKT 800 FOR AN OPEN, SHORT TO GROUND OR BATTERY
- REPLACE DERM IF CKT 800 IS OK

NO
- CAN ECM DATA BE READ USING THE SCAN TOOL?

YES
- CHECK CKT 800 FOR AN OPEN BETWEEN ALDL AND DERM CONNECTOR
- REPLACE DERM IF CKT 800 IS OK

NO
- CHECK SCAN TOOL CONNECTIONS
- CHECK CKT 800 FOR A SHORT TO BATTERY OR GROUND. CHECK FOR AN OPEN BETWEEN SPLICE AND ALDL CONNECTOR.

7
- IF A CODE 51 IS CURRENT, GO TO CODE 51 CHART
- IF A CODE 52 IS CURRENT, GO TO CODE 52 CHART
- IF A HISTORY OR CURRENT CODE 71 IS SET, GO TO CODE 71 CHART
- IF BOTH CODES 21 AND 21 ARE BOTH CURRENT, GO TO *CODE 21 AND 22 BOTH CURRENT* CHART
- IF ONLY HISTORY CODES EXIST GO TO CHART A1
- DIAGNOSE REMAINING *CURRENT CODES FROM LOWEST TO HIGHEST*

FIGURE 53-26 Passive restraint systems have a built-in diagnostic system to aid the service technician. Specific diagnostic procedures are available to help determine the exact cause of a problem. (Courtesy of MOTOR Publications, Auto Repair Manual, © The Hearst Corporation)

- Energy reserve module
- SIR control module
- Coil assembly
- Steering wheel module

SERVICE

Each service procedure will be different for each component that is faulty. Follow the maintenance or service manuals for the correct procedure to follow.

PROBLEM: Faulty Windshield Wiper System

The windshield wiper system is inoperative.

DIAGNOSIS

Many troubleshooting procedures are used to diagnose windshield wiper systems. The wiper may be inoperative, the motor may not shut off, it may be sluggish as it moves, or, as in this problem, it may be totally inoperative. ***Figure 53–27*** shows an example of a "symptom chart" for one

WIPER AND WASHER SYSTEMS

PROBLEM	DIAGNOSIS	SERVICE
Windshield washer inoperative	Blown fuse Windshield washer pump Multifunction switch Circuit	Go to Pinpoint Test A1
Windshield wipers inoperative in all switch positions	Blown fuse Multifunction switch Windshield wiper motor Circuit	Go to Pinpoint Test B1
Windshield wipers inoperative at high speed	Multifunction switch Windshield wiper motor Circuit	Go to Pinpoint Test C1
Windshield wipers inoperative at low speed	Multifunction switch Windshield wiper motor Circuit	Go to Pinpoint Test D1
Windshield wipers inoperative at interval setting	Multifunction switch Circuit	Go to Pinpoint Test E1
Windshield wipers continue to run when switch is turned Off	Multifunction switch Circuit	Go to Pinpoint Test F1
Rear window washer inoperative	Blown fuse Rear window washer pump Rear window switch Circuit	Go to Pinpoint Test G1
Rear window wiper inoperative	Blown fuse Rear window wiper motor Rear window switch Circuit	Go to Pinpoint Test H1
Rear window wiper will not turn Off	Rear window switch Circuit	Go to Pinpoint Test J1
Windshield wipers will not park at the proper position	Windshield wiper motor Multifunction switch Bent/cracked linkage	Go to Pinpoint Test K1
Rear window wiper will not park	Rear window wiper motor	REPLACE the rear window wiper motor

PINPOINT TEST - WIPER AND WASHER SYSTEMS

Pinpoint test A: Windshield washer inoperative

Test step		Result	Action to take
A1	Check washer switch		
	Perform the windshield wiper switch test component test in this section **Is the windshield wiper switch OK?**	Yes No	Go to A2 REPLACE the multifunction switch

FIGURE 53–27 This problem and diagnosis chart shows various problems, their diagnoses, and service steps to take to solve problems in a wiper and washer system.

manufacturer's wiper and washer system. In this particular case, various conditions are given along with a possible source of the problem. Then, based upon the possible source, various actions can be taken by the service technician. In each case, a specific pinpoint test is suggested.

For example, take the first condition, "Windshield Washer Inoperative." The source of the problem could be a blown fuse, the windshield washer pump, the multifunction switch, or the circuit. The service technician is instructed to go to pinpoint test A1. This particular test is shown directly below the symptom chart. Based upon the result of the pinpoint test, various actions are suggested to the service technician. In practice, there may be numerous pinpoint tests to complete to help solve a variety of problems.

SERVICE

Refer to the correct service manual, and replace the faulty part as necessary.

Service Manual Connection

There are many important specifications to keep in mind when working with auxiliary and electrical systems. To identify the specifications for your vehicle, you will need to know the VIN (vehicle identification number) of the vehicle, the type and year of the vehicle, and the type of engine used. Although they may be titled differently, some of the more common auxiliary and electrical system specifications (not all) found in service manuals are listed below. Note that these are typical examples. Each vehicle may have different specifications.

Common Specification	Typical Example
Rear wiper motor speed	33–44 rpm
Wiper washer switch (resistance)	Less than 0.05 ohm
Wiper motor current draw (no load)	3.5–4.0 amps
Wiper motor rated voltage	12 volts

In addition, service manuals provide specific directions for various service and testing procedures. Some of the more common procedures include:

- Air bag restraint system, disable
- Diagnosing passive restraint systems
- Fuse panel and flasher locations
- Horn sounder replace
- Wiper motor replace
- Windshield wiper switch replace

SUMMARY

The following statements will help to summarize this chapter:

- Various auxiliary systems and electrical circuits are used on the automobile.
- A schematic is a drawing of an electrical circuit that shows how it functions.
- Many electrical symbols are used on electrical schematics.
- The headlight circuit includes a dimmer switch, headlights with low and high beams, and a fuse or circuit breaker for protection.
- The defogger operates by using a defogger switch, a timer to hold the defogger on for a period of time, and a fuse panel.
- Power seats operate by using several reversible motors on the bottom of the seats.
- The horn circuit uses the horn button and a horn relay to operate correctly.
- The windshield wiper motor uses a permanent magnetic motor for operation.
- Each vehicle uses different mechanical linkage systems to connect the motor output to the wiper blades.
- Windshield wiper systems use an electrical circuit to time the wiper motion during the pulse mode of operation.
- Safety belts use retractors and reels that rewind and loosen for improved passenger comfort.

- All new vehicles now use passive restraint systems.
- Passive restraint systems use several sensors to determine if a vehicle crash is impending.
- Diagnosis and service of auxiliary and electrical systems use standard troubleshooting procedures to locate problems.
- Various troubleshooting tools are used on electrical circuits, including jumper wires, a short finder, a volt-ohmmeter, and a fuse tester.
- When troubleshooting an electrical circuit, always check the easiest components first.

TERMS TO KNOW

Can you explain each of the following terms? Review the chapter until you can use each term correctly.

Component location table
Dimmer switch
Electrical schematics
Electrical symbols
Forward sensor
Fuse block
Fusible link
Horn brush/slip ring
Jumper wire
Passive restraint systems
Pendulum
Pulse delay variable resistor
Pulse wiper system

REVIEW QUESTIONS

Multiple Choice

1. Electrical schematics are used to show the:
a. Parts of a circuit
b. Direction of electrical flow
c. Color of wires used in the circuits
d. All of the above
e. None of the above

2. To find the exact physical location of a part on the schematic, look at the:
a. Position on the schematic
b. Component location table
c. Size of the wire
d. Length of the wire
e. None of the above

3. To protect electrical circuits from overloaded conditions, ________ are used.
a. Fuses
b. Electrical connectors
c. Motors
d. ORN wires
e. PPL wires

4. Which of the following circuits uses a dimmer switch?
a. Horn
b. Headlight
c. Windshield wiper
d. All of the above
e. None of the above

5. A defogger system uses an electrical circuit that:
a. Only operates for a certain period of time
b. Uses an On-Off switch
c. Uses wires on the inside surface of the window
d. All of the above
e. None of the above

6. Electrical seats use motors that:
a. Have one electromagnet on each motor
b. Operate in only one direction
c. Operate in either direction, depending on the current
d. Are encased in oil
e. Operate on 1–2 volts

7. A horn circuit uses:
a. Seven switches to operate correctly
b. A solenoid and horn relay switch
c. Motors that operate in either direction
d. An electronic control module
e. All of the above

8. The mechanism in a windshield wiper system for changing rotational motion to oscillating motion is located:
a. On the windshield wipers
b. In the washer system
c. Inside the motor assembly
d. Under the dashboard
e. In the trunk

9. Which is not a part of a common windshield wiper system?
a. Fluid washer system
b. Motor
c. Wiper blades
d. Servo
e. Wiper linkage

10. Which component is used in the electrical circuit to adjust the time delay for pulsing the wipers?
a. Variable motor
b. Variable washer nozzle
c. Linkage
d. Wiper blades
e. Variable resistor

11. Safety belts are locked up during a collision by using a ________.
a. Magnetic circuit
b. Spur gear
c. Pendulum
d. Clutch
e. Magnetic clutch

12. Which of the following is not considered a good troubleshooting practice on electrical circuits?
a. Refer to an electrical schematic
b. Test for correct voltages
c. Narrow down the problem
d. Check the easiest and basic components first
e. Replace items and components until the problem is solved

The following questions are similar in format to ASE (Automotive Service Excellence) test questions.

13. *Technician A* says that when troubleshooting any electrical circuit, a jumper wire should never be used to bypass a part of the circuit. *Technician B* says that a voltmeter should be used to test electrical circuits. Who is correct?
a. A only c. Both A and B
b. B only d. Neither A nor B

14. When reading an electrical circuit for troubleshooting, *Technician A* says the flow of electricity normally goes from the left side of the page to the right. *Technician B* says the flow of electricity normally flows from the top downward through the circuit. Who is correct?
a. A only c. Both A and B
b. B only d. Neither A nor B

15. A vehicle has a passive restraint system. *Technician A* says that the diagnostic process is built into the system. *Technician B* says that the warning light needs to be off when diagnosing the system. Who is correct?
a. A only c. Both A and B
b. B only d. Neither A nor B

16. *Technician A* says that when diagnosing a mechanical system, always make visual checks first. *Technician B* says that when diagnosing a mechanical system, replace any components that you think might be faulty first. Who is correct?
a. A only c. Both A and B
b. B only d. Neither A nor B

17. A vehicle has a passive restraint system. *Technician A* says that the arming sensor can be repaired. *Technician B* says the forward sensor can be repaired. Who is correct?
a. A only c. Both A and B
b. B only d. Neither A nor B

Essay

18. Describe several electrical symbols used on electrical schematics.

19. What is the fuse block used for in an electrical circuit?

20. Describe the electrical operation of the defogger system used on vehicles.

21. What is the purpose of the component location table used along with certain electrical schematics?

22. Describe the purpose and operation of a fusible link.

Short Answer

23. On a passive restraint system, the ________ sensor is used to sense if a vehicle is involved in a crash.

24. On a passive restraint system, the ________ sensor is located in the front of the vehicle grille.

25. Three diagnostic tools used to check electrical circuits and systems include a(n) ________, a(n) ________, and a(n) ________.

Applied Academics

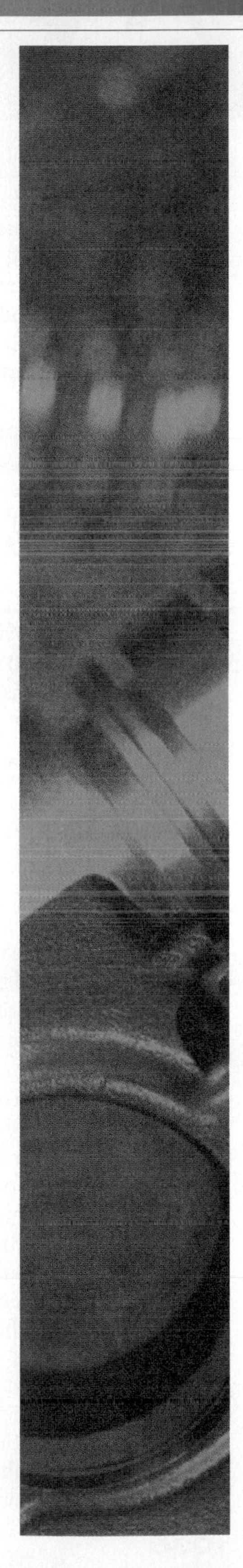

TITLE: Electrical Circuit Analysis

ACADEMIC SKILLS AREA: Science

NATEF Automobile Technician Tasks:

The service technician uses observations to identify electrical or electronic problem symptoms to develop a theory regarding the cause of the problem. The technician tests the hypothesis to determine the solution to the problem.

CONCEPT:

Many electrical circuits are detailed in the service manuals for each vehicle. Often, a technician has not seen the electrical circuit before. Thus, the technician needs to understand and analyze the circuit using all electrical knowledge. The circuit in the figure that follows shows the parking/marker lights circuits for a vehicle. When a customer is concerned that the right rear (RR) parking light is not working, the technician located this circuit for analysis.

APPLICATION:

Can you follow the electricity in the circuit to each of the components? Can you identify which component may be defective? What type of voltage, amperage, or resistance readings would need to be taken to determine the cause of the problem? What position would the headlight switch need to be in so that voltage drop readings could be taken? Can you determine the color of the electrical wires to each electrical component?

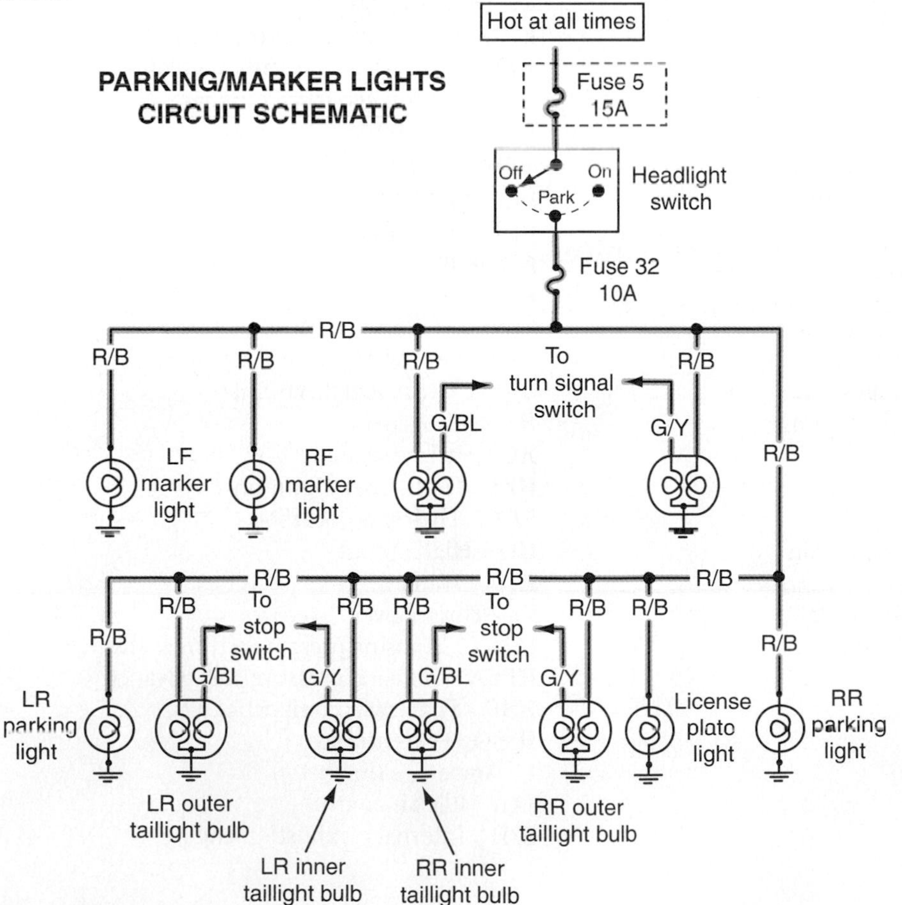

APPENDIX A Common Automotive Abbreviations

ABCM Antilock brake control module
ABS Antilock braking system
A/C Air conditioning
ac Alternating current
ACM Air control module
ACR3 Air-conditioning refrigerant recovery and recycling
ACT Air charge temperature
AFL Altitude fuel limiter
AIR Air injection reactor
ALCL Assembly line communication link
ALDL Assembly line diagnostic link
API American Petroleum Institute
ARB Air Resources Board
ASE Automotive Service Excellence, also called NIASE (National Institute for Automotive Service Excellence)
ASTM American Society for Testing and Materials
ATDC After top dead center
ATF Automatic transmission fluid
AWG American Wire Gauge
BAT Battery
BDC Bottom dead center
bhp Brake horsepower
BLU Blue wire
BMEP Brake mean effective pressure
BP Barometric pressure
BSFC Brake specific fuel consumption
BTDC Before top dead center
Btu British thermal unit
Btu/gal Btu per gallon
Btu/min Btu per minute
C Celsius
C^3I Computer-controlled coil ignition
CAFE Corporate Average Fuel Economy
cc Cubic centimeter (cm^3)
CCA Cold-cranking ampere
CCC Computer command control
CCC Computer-controlled combustion
CFC Chlorofluorocarbon
CID Cubic inch displacement
CIS Cylinder identification sensor
cm Centimeter
CO Carbon monoxide
CO_2 Carbon dioxide
CP Crankshaft position
CRT Cathode ray tube
CV Constant velocity
dB Decibel
dc Direct current
DECS Diesel electronic control system
DIS Direct ignition system
DIS Distributorless ignition system
DK GRN Dark green wire
DOHC Dual overhead camshaft
DOT Department of Transportation
E Voltage (electromotive force)
EBCM Electronic brake control module
ECC Electronic engine control
ECCS Electronic computer control system
ECCS Electronic constant engine control system
ECU Electronic control unit
EEC Electronic engine control
EFE Early fuel evaporation
EGR Exhaust gas recirculator
ELD Electric load detector
emf Electromotive force
EP Extreme pressure
EPA Environmental Protection Agency
EPR Evaporative pressure regulator
ESN Engine sequence number
EST Electronic spark timing
F Fahrenheit
fhp Frictional horsepower
ft.-lb Foot-pound
g Gram
g/mi Grams per mile
Gpm Gallons per minute
GRY Gray wire
GVW Gross vehicle weight
H Hydrogen
HC Hydrocarbons
HEI High-energy ignition
HFC Hydrofluorocarbon
HO High output
H_2O Water
hp Horsepower
HPAA Housing pressure altitude advance
HPCA Housing pressure cold advance
HSC High swirl combustion
H_2SO_4 Sulfuric acid
I Amperage (intensity)
IAC Idle air control
ICE Internal combustion engine

IEEGR Integrated electronic exhaust gas recirculating
ihp Indicated horsepower
in.-lb Inch-pound
k Kilo
KAM Keep-alive memory
km Kilometer
kPa Kilopascal
kW Kilowatt
L or **l** Liter
LOS Limited operational strategy
LPG Liquid petroleum gas
M Motor octane
MA Mechanical advantage
MAF Mass airflow
MAP Manifold absolute pressure
MAT Manifold air temperature
M/C Mixture control
MIL Military
mm Millimeter
mpg Miles per gallon
mph Miles per hour
ms Milliseconds
MSDS Material Safety Data Sheet
MVS Metering valve sensor
N North pole
N/A Not applicable
NAPA National Auto Parts Association
NC National Coarse threads
NF National Fine threads
NO_x Nitrogen oxide
O Oxygen
OASIS On-line Automotive Service Information System
OD Overdrive
OHC Overhead camshaft
OHV Overhead valves
ORN Orange wire
OSHA Occupational Safety and Health Act
P Power
PbO_2 Lead peroxide
PCM Powertrain control module
PCV Positive crankcase ventilation
PFI Port fuel injection
PIP Profile ignition pickup
PM Permanent magnet
PNK Pink wire
POA Pilot-operated absolute valve
PPL Purple wire
ppm Parts per million
PROM Programmable read-only memory
psia Pressure per square inch absolute
psig Pressure per square inch gauge
R134a Hydrofluorocarbon A/C refrigerant
R Research octane
R Resistance
R_1, R_2, etc. Resistor one, resistor two, etc.
RAM Random access memory
ROM Read-only memory
Rpm Revolutions per minute
RTV Room temperature vulcanizing
S South pole
SAE Society of Automotive Engineers
SEFI Sequential electronic fuel injection
SFI Sequential fuel injection
SIR Supplemental inflatable restraint
SO_4 Sulfate
SOHC Single overhead camshaft
STV Suction throttling valve
SUS Saybolt universal seconds
S/V ratio Surface to volume ratio
TAMP Timing, atomizing, metering, pressurizing
TBI Throttle body injection
TCC Torque converter clutch
TDC Top dead center
TEL Tetraethyl lead
TML Tetramethyl lead
TPC Tire performance criteria
TPI Tuned port injection
TPS Throttle position sensor
TRS Transmission regulated spark
TSCV Throttle solenoid control valve
TV Throttle valve
U-joint Universal joint
UNC Unified National Coarse threads
UNF Unified National Fine threads
USC U.S. Customary threads
UTQG Uniform Tire Quality Grading
VAT Voltage Amperage tester
VECI Vehicle emission control information
VI Viscosity index
VIN Vehicle identification number
VSS Vehicle speed sensor
W Watt
YEL Yellow wire

APPENDIX B Automotive Web Sites

PRODUCT	ADDRESS	COMPANY
A/C systems and technology	http://harrisonradiator.com	Harrison Radiator, also called Delphi
Alternative fuel information	http://afdc.doe.gov/questions.html	Alternative Fuels Data Center
Automotive batteries	http://www.exideworld.com	Exide Technologies
Automotive innovations, new batteries, etc.	http://www.extremetech.com	Extreme Tech
Automobile purchasing	http://www.autos.msn.com	MSN Autos
Bearings, automotive	http://www.crindustries.com	CR Industries
Bearings, pistons	http://www.federalmogul.com	Federal Mogul Corporation
Brake and clutch components	http://www.dana.com	Dana Corporation
Brake, friction, and sealing components	http://www.federalmogul.com	Federal Mogul Automotive
Braking components and systems	http://www.nucap.com	Nucap Industries
CAFE Standards	http://ita.doc.gov/td/auto/cafe.html	Office of Automotive Affairs
Certification and testing information	http://www.asecert.org	National Institute for Automotive Service Excellence
Clutch designs, materials, principles	http://www.ramclutches.com	Ram Clutches
Continuously variable transmission	http://www.edmunds.com	Edmunds Ownership
Cooling, lubrication, suspension, drive lines, hoses, sealing, bearing, aftermarket products	http://www.dana.com	Dana Corporation
Dynamometers	http://www.mustangdyne.com	MD Mustang Dynamometers
Dynamometers for automotive use	http://www.claytonindustries.com	Clayton Industries
Education about automobiles	http://www.howstuffworks.com	HowStuffWorks
Education, automotive information	http://www.sae.org	Society of Automotive Engineering
Electric vehicle information	http://www.evaa.org	Electric Vehicle Association of America
Electrical testing equipment for automobiles	http://www.tif.com	Advanced Testing Products
Electronic components	http://www.delphi.com	Delphi
	http://www.schneider-electric.com	Schneider Electric
Emission Standards	http://www.dieselnet.com/news	Miratech
Ethanol fuels	http://www.ethanolRFA.org	Renewable Fuels Association
Ethanol information	http://www.nwicc.com/etsp.htm	Ethanol—As a Fuel
Ethanol and reformulated fuels	http://www.ethanolRFA.org	Renewable Fuels Association
Ethanol vehicles	http://www.e85fuel.com	National Ethanol Vehicle Coalition
Fuel cell information center	http://www.fuelcells.org	Fuel Cells 2000
Fuels	http://www.aga.org	American Gas Association
Hybrid cars	http://www.hybridcars.com	Hybrid Cars
Instruments	http://www.ni.com/automotive	National Instruments
International Automotive Technician	http://www.iatn.net	IATN
Manuals	http://www.motor.com	Motor Information Systems
	http://www.mitchell.com	Mitchell International Inc.
Methanol fuels	http://www.methanol.org	American Methanol Institute
National Automotive Technician Education Foundation	http://natef.org	NATEF
Oil products	http://api-ec.api.org	American Petroleum Institute
Oxygenated gasoline information	http://www.chevron.com	Chevron
Performance, batteries, etc.	http://www.autospeed.com	Auto Speed

PRODUCT	ADDRESS	COMPANY
Pricing of vehicles	http://www.car-prices-usa.com	Car Prices-USA
Safety equipment	http://www.sellstrom.com	Sellstrom
Safety face shields	http://www.thermadyne.com	Thermadyne
School-to-work program	http://www.ayes.org	Automotive Youth Education System
Silicon and RTV sealants	http://www.loctite.com	Loctite Corporation
Sustainable transportation systems, solar	http://www.solectria.com	Solectria
Synthetic automotive parts	http://www.dowautomotive.com	Dow Automotive
Testing equipment—dynamometers	http://www.dynolab.com	Dynolab
Tire manufacturer	http://www.bfgoodrichtires.com	BF Goodrich
Tools, automotive	http://www.snapon.com	Snap-On
Transportation company	http://www.volvo.com/frameset.html	Volvo Corporation
Turbocharger manufacturer	http://www.atsturbo.com	Advance Turbo Systems
Turbochargers	http://www.egarrett.com/index.jsp	Garrett Engine Boosting Systems
Variable valve timing	http://www.bmwusa.com	BMW
Vehicle engineering	http://www.magnasteyr.com	Magna Steyr
Vehicle manufacturers	http://www.gm.com/flash_homepage	General Motors Corporation
	http://www.toyota.com/index.html	Toyota Incorporated
	http://www.honda.com	Honda Worldwide
	http://www.ford.com/en/default.htm	Ford Motor Company
	http://www.kia.com	KIA Motors America
	http://www.daimlerchrysler.com	DaimlerChrysler
	http://www.mitsubishi.com	Mitsubishi Companies
	http://www.audi.com	Audi Worldwide
	http://www.NissanUSA.com	Nissan USA
	http://www.mazdaUSA.com	Mazda North American Operations
42-volt battery system	http://www.autointell.com	Automotive Intelligence News

APPENDIX C

J1930 Terminology List

SAE J1930 standardizes automotive component names for all vehicle manufacturers.

OLD ACRONYM/TERM	NEW ACRONYM/ ABBREVIATION	NEW TERM
Accelerator	AP	Accelerator Pedal
Air Cleaner	ACL	Cleaner
A/C Air Conditioning	A/C	Air Conditioning
BP Barometric Pressure	BARO	Barometric Pressure
BATT+ Battery Positive	B+	Battery Positive Voltage
Camshaft Sensor	CMP	Camshaft Position
CARB Carburetor	CARB	Carburetor
Continuous Fuel Injection	CFI	Continuous Fuel Injection
After Cooler Intercooler	CAC	Charge Air Cooler
EEC	CL	Closed Loop
CTP Closed Throttle Position	CTP	Closed Throttle Position
CES CIS Clutch Engage Switch Clutch Interlock Switch	CPP	Clutch Pedal Position
CTO	CTOX	Continuous Trap Oxidizer
CPS VRS Variable Reluctance Sensor	CKP	Crankshaft Position
Self-Test Connector	DLC	Data Link Connector
Self-Test Mode	DTM	Diagnostic Test Mode
Self-Test Code	DTC	Diagnostic Trouble Code
CBD DS TFI Closed Bowl Distributor Duraspark Ignition Thick Film Ignition	DI	Distributor Ignition
EFE Early Fuel Evaporation	EFE	Early Fuel Evaporation
E2PROM	EEPROM	Electrically Erasable Programmable Read Only Memory
DIS EDIS Distributorless Ignition System Electronic Distributorless Ignition System	EI	Electronic Ignition

NEW ACRONYM/ OLD ACRONYM/TERM	ABBREVIATION	NEW TERM
Engine Coolant Level	ECL	Engine Coolant Level
ECT Engine Coolant Temperature	ECT	Engine Coolant Temperature
ECM Engine Control Module	ECM	Engine Control Module
RPM Revolutions Per Minute	RPM	Engine Speed
EPROM Erasable Programmable Read Only Memory	EPROM	Erasable Programmable Read Only Memory
EVP Sensor EVR Solenoid	EVAP	Evaporative Emission
EGR Exhaust Gas Recirculation	EGR	Exhaust Gas Recirculation
EDF Electro-Drive Fan	FC	Fan Control
FEEPROM Flash Electrically Erasable Programmable Read Only Memory	FEEPROM	Flash Electrically Erasable Programmable Read Only Memory
FEPROM Flash Erasable Programmable Read Only Memory	FEPROM	Flash Erasable Programmable Read Only Memory
FCS FFS FFV Fuel Compensation Sensor Flex Fuel Sensor	FF	Flexible Fuel
Fourth Gear	4GR	Fourth Gear
FP Fuel Pump	FP	Fuel Pump
ALT Alternator	GEN	Generator
GND Ground	GND	Ground
HEGO Heated Exhaust Gas Oxygen Sensor	HO_2S	Heated Oxygen Sensor
IAC Idle Air By-pass Control	IAC	Idle Air Control
Idle Speed Control	ISC	Idle Speed Control
DIS Module EDIS Module TFI Module	ICM	Ignition Control Module
IDFI Indirect Fuel Injection	IFI	Indirect Fuel Injection
Inertia Switch	IFS	Inertia Fuel Shutoff
ACT Air Charge Temperature	IAT	Intake Air Temperature
KS Knock Sensor	KS	Knock Sensor
CEL "Check Engine" Light "Service Engine Soon" Light	MIL	Malfunction Indicator Lamp
MAP Manifold Absolute Pressure	MAP	Manifold Absolute Pressure

Continued

OLD ACRONYM/TERM	NEW ACRONYM/ ABBREVIATION	NEW TERM
MDP Manifold Differential Pressure	MDP	Manifold Differential Pressure
MST Manifold Surface Temperature	MST	Manifold Surface Temperature
MVZ Manifold Vacuum Zone	MVZ	Manifold Vacuum Zone
MAF Mass Airflow	MAF	Mass Airflow
Mixture Control	MC	Mixture Control
EFI Electronic Fuel Injection	MFI	Multiport Fuel Injection
NVM Nonvolatile Memory	NVRAM	Nonvolatile Random Access Memory
Self-Test On-Board Diagnostic	OBD	On-Board Diagnostic
OL Open Loop	OL	Open Loop
COC Conventional Oxidation Catalyst	OC	Oxidation Catalytic Converter
EGO	O_2S	Oxygen Sensor
NDS NGS TSN Neutral Drive Switch Neutral Gear Switch Transmission Select Neutral	PNP	Park/Neutral Position
PTOX Periodic Trap Oxidizer	PTOX	Periodic Trap Oxidizer
PSPS Power Steering Pressure Switch	PSP	Power Steering Pressure
ECA ECM ECU EEC Processor Engine Control Assembly Engine Control Module Engine Control Unit	PCM	Powertrain Control Module
PROM Programmable Read Only Memory	PROM	Programmable Read Only Memory
MPA PA Thermactor II Managed Pulse Air Pulse Air	PAIR	Pulsed Secondary Air Injection
RAM Random Access Memory	RAM	Random Access Memory
ROM Read Only Memory	ROM	Read Only Memory
RM Relay Module RM	RM	Relay Module
GST NGS Generic Scan Tool New Generation STAR Tester Enhanced Scan Tool OBDII ST	ST	Scan Tool

OLD ACRONYM/TERM	NEW ACRONYM/ ABBREVIATION	NEW TERM
AM CT MTA Air Management Conventional Thermactor Managed Thermactor Air Thermactor	AIR	Secondary Air Injection
SEFI Sequential Electronic Fuel Injection	SFI	Sequential Multiport Fuel Injection
SRI Service Reminder Indicator	SRI	Service Reminder Indicator
SPL Smoke Puff Limiter	SPL	Smoke Puff Limiter
SC Supercharger	SC	Supercharger
SCB Supercharger Bypass	SCB	Supercharger Bypass
	SRT	System Readiness Test
Thermal Vacuum Switch	TVV	Thermal Vacuum Valve
Third Gear	3GR	Third Gear
TWC Three-Way Catalytic Converter	TWC	Three-Way Catalytic Converter
TWC & COC Dual Bed Three-Way Catalyst and Conventional Oxidation Catalyst	TWC+OC	Three-Way + Oxidation Catalytic Converter
TB Throttle Body	TB	Throttle Body
CFI Central Fuel Injection EFI	TBI	Throttle Body Fuel Injection
TP Throttle Position	TP	Throttle Position
CCC CCO MCCC Converter Clutch Control Converter Clutch Override Modulated Converter Clutch Control	TCC	Torque Converter Clutch
4EAT Module	TCM	Transmission Control Module
PRNDL	TR	Transmission Range
TC Turbocharger	TC	Turbocharger
VSS Vehicle Speed Sensor	VSS	Vehicle Speed Sensor
VR Voltage Regulator	VR	Voltage Regulator
VAF Volume Airflow	VAF	Volume Airflow
WV-OC Warm-up Oxidation Catalytic Converter	WU-OC	Warm-Up Oxidation Catalytic Converter
WU-TWC Warm-Up Three-Way Catalytic Converter	WU-TWC	Warm-Up Three-Way Catalytic Converter
Full Throttle WOT Wide Open Throttle	WOT	Wide Open Throttle

Bilingual Glossary

Note: Terms are highlighted in blue, followed by Spanish translation in bold. Numbers in parentheses indicate the chapter(s) in which each term is discussed.

4WD L (Lo) A four-wheel drive system often has both a low and high gear. The 4WD L (LO) gear is often used for off-road, rough terrain conditions for which higher speeds are not necessary. (44)

4RM B (BAJO) Sistema de cuatro ruedas motrices que suele tener una velocidad baja y alta. La velocidad 4RM B (BAJO) suele usarse fuera del camino, en los caminos disparejos en donde no se requieren las velocidades más altas. (44)

Absolute sensor A sensor used on the intake manifold to measure the intake manifold pressure in reference to a fixed pressure sealed inside the sensor. (38)

Sensor absoluto Sensor en el múltiple de admisión que sirve para medir la presión del múltiple de admisión en relación a una presión fija sellado dentro del sensor. (38)

Accumulator A device that cushions the motion of a clutch and servo action in an automatic transmission; a component used to store or hold liquid refrigerant in an air-conditioning system. (41) (50)

Acumulador Mecanismo que ablanda (acojina) el movimiento del embrague y servo-acción de una transmisión; un componente que sirve para almacenar o mantener un refrigerante líquido en un sistema de aire acondiciónado. (41) (50)

Acidity In lubrication, acidity denotes the presence of acid-type chemicals that are identified by the acid number. Acidity within oil causes corrosion, sludges, and varnish to increase. (20)

Acidez Referente a lubricación, acidez significa la presencia de químicos de tipo ácido que se identifican según el número del ácido. La acidez dentro del aceite causa un aumento de corrosión, residuos, y barniz. (20)

Acid rain A form of pollution produced when sulfur and nitrogen are emitted into the air. The mixture of these chemicals with water produces an acid solution that is found in rain. (23)

Lluvia ácida Una forma de contaminación producida cuando azufre y nitrógeno se emiten al aire. La mezcla de estos químicos con agua produce una solución líquida que se encuentra en lluvia. (23)

ACM A computer used to control air-conditioning system clutches; air control module. (50)

MCA Computadora (ordenador) que sirve para contolar embragues de sistemas de aire acondiciónado; módulo de control de aire. (50)

ACR3 A system called an air-conditioning refrigerant recovery and recycling system used to recover and recycle the refrigerant R12 from automobile air-conditioning systems. (50)

ACR3 Un sistema llamado "sistema para recuperar y reciclar refrigerantes de aire acondiciónado," el cual sirve para recuperar y reciclar el refrigerante R12 de los sistemas de aire acondiciónado de automóviles. (50)

Active suspension system A computerized suspension system able to control body roll, pitch, brake dive, acceleration squat, and ride height. (47)

Sistema de suspensión activa Sistema de suspensión computarizada capaz de controlar tambaleo de carrocería (movimientos angulares), inclinación, caída de pique al frenar, amarre sobre aceleración, y posición de altura correcta. (47)

Actuator A device that causes action or motion on another part. (28)

Activador Aparato que produce acción o movimiento sobre otra parte. (28)

Additives Chemical compounds added to a lubricant for the purpose of imparting new properties or improving those properties that the lubricant already has. (20)

Aditivos Elementos químicos añadidos a un lubricante para impartir nuevas propiedades o para mejorar las propiedades que ya tiene el lubricante. (20)

Adjusting sleeve An internally threaded sleeve located between the tie-rod ends. The sleeve is rotated to set toe-in/-out. (48)

Manga de ajuste Manga que tiene Rosca interior localizada entre los dos extremos de una barra de acoplamiento de las ruedas directrices. Se gira la manga para ajustar el ángulo de convergencia o divezrgencia de las ruedas delanteras. (48)

Aeration The process of mixing air with a liquid. Aeration occurs in a shock absorber from rapid fluctuations in movement. (47)

Aeración Proceso de mezclar aire con un líquido. Aeración ocurre dentro de un amortiguador cuando hay fluctuaciónes rápidas de movimiento. (47)

Aerodynamic The ease with which air can flow over the vehicle during higher-speed operation. An aerodynamically sound vehicle has very little wind resistance. (16)

Aerodinámico Facilidad con la que el aire puede fluir sobre un vehículo durante operación de alta velocidad. Un vehículo bien construido aerodinámicamente sufre muy poca resistencia al viento. (16)

Aftermarket Equipment sold to consumers after the vehicle has been manufactured. Aftermarket equipment and parts are sold locally by parts dealers. (27)

Postmercado Equipo vendido al cliente después de que se fabrica el vehículo. Equipo y refacciónes de postmercado se venden localmente por medio de comerciantes de partes. (27)

Airborne A term used to describe contaminants floating in air moving through the engine. The contaminants are light enough to be suspended in the airstream. (27)

Aereo Término que sirve para describir contaminates que flotan en el aire que pasa por el motor. Los contaminates son tan ligeros como para estar suspendidos en el chorro de aire. (27)

Air-cooled Removing heat from the engine by circulating air across the cylinder block and heads. (22)

Enfriado por aire El quitarle calor al motor por medio de circular aire a través del bloque de cilindros y las cabezas. (22)

Air ducts Tubes, channels, or other tubular structures used to carry air to a specific location. (51)
Conductos de aire Tubos, canales, u otras estructuras de tubería que sirven para llevar aire a un lugar específico. (51)

Air-fuel ratio The measure of the amount of air and fuel needed for proper combustion. The most efficient air-fuel ratio is 14.7 parts of air to 1 part of fuel. (13)
Proporción de aire y combustible Medida de la cantidad de aire y combustible que se requiere para combustión apropiada. La proporción correcta es 14.7 partes de aire por 1 parte de combustible. (13)

Air injection reactor (AIR) A type of emission control system that pumps fresh air into the exhaust. (37)
Reactor de inyección de aire Tipo de sistema de control de emisiones que inyecta aire limpio en el escape para gases. (37)

Air intake and exhaust systems The parts on the automobile engine used to get the air into the engine and the exhaust out of the engine, including air cleaner, muffler, tailpipe, and associated ducting. (3)
Sistema admision y escafe de aire Las partes de un motor automovilístico que sirven para hacer entrar aire en el motor y sacar los gases del motor, incluyendo el filtro de aire, silenciador, tubo de escape, y otra tubería relacionada. (3)

Airtight containers Containers used to hold waste and oily rags so that spontaneous combustion is eliminated. (4)
Contenedores herméticos Contenedores que sirven para guardar desperdicios y trapos engrasados para que se elimine la combustión espontánea. (4)

Alignment Fitting the crankcase to align with holes in the block. Also, the position of the vehicle wheels relative to each other and to the car body and frame. Caster, camber, and toe are typically adjusted. (17) (48)
Alineación El ajustarse el cigüeñal para que esté alineado con los agujeros del bloque. Tambíen, la posición de las ruedas del vehículo en relación a si mismas y a la carrocería y la armadura del coche. El caster camber y toe son ajustedos típicamente. (17) (48)

AllData A computerized information system for service technicians that includes selected service manuals and service bulletins. (11)
Todos datos Sistema de información computarizada para técnicos de servicio, el cual incluye manuales de servicio selecciónados y boletines de servicio. (11)

All-season tire A tire that is designed for all-around use in terms of treadwear, ride comfort, snow traction, wet traction, handling, and high speed. (49)
Llantas para todo clima Llanta que está diseñada para todas las ocasiones en el sentido de desgaste de huellas, tracción confortable, tracción sobre nieve, tracción sobre superficies mojadas, manejo, y alta velocidad. (49)

Ambient Surrounding area; for example, surrounding temperature or circulating air. (23) (50)
Ambiente Area ambiental; por ejemplo, temperatura ambiental o aire circulante. (23) (50)

Ambient sensor A sensor used on computerized automatic temperature control systems that senses the outside air temperature and uses this information as an input to the system. (51)
Sensor ambiental Sensor que se usa en sistemas automaticos de control computarizado automático de temperatura, el cual percibe la temperatura del aire de afuera y emplea está información como información de entrada al sistema. (51)

Amperage The number of electrons flowing past a given point in one second. (29)
Amperaje Número de electrones que pasan sobre un punto dado dentro de un segundo. (29)

Amperage capacity An indication of the length of time a battery can produce an amperage, or the amount of amperage that a battery can produce before being discharged. (31)
Capacidad de amperaje Indicación de la cantidad de tiempo que una batería puede producir un amperaje, o la cantidad de amperaje que una batería puede producir antes de descargarse. (31)

Anaerobic sealant A chemical sealant placed on a gasket in an engine to aid in sealing and to position the gasket during installation. (5)
Selladores anaerobios Sellador químico puesto en un empaque de un motor para ayudar a sellarlo y colocarlo durante su instalación. (5)

Analog signal A variable signal represented by a voltage within a given range. (30)
Señal análoga Señal variable representada por un voltaje dentro de un rango dado. (30)

Antifreeze A chemical solution added to the liquid coolant to protect against freezing and to raise the boiling point. (22)
Anticongelante Solución química añadida al refrigerante líquido que sirve para proteger contra congelación y para aumentar el punto de ebullición. (22)

Antifriction bearing A type of bearing that uses balls or rollers between the rotating shaft and the stationary part. (6)
Cojinete (Balero) antifricción Tipo de cojinete (balero) que utiliza bolas o rodillos (rolletes) entre un eje rotativo y una parte fija. (6)

Anti-icers Chemicals added to gasoline to eliminate the freezing of gasoline. (23)
Elementos de anti-hielo Químicos añadidos a la gasolina que sirven para eliminar la congelación de gasolina. (23)

Antilock braking system (ABS) A type of braking system that is able to sense the speed of each wheel; in conjunction with a computer, it controls the hydraulic braking pressure to eliminate wheel lockup. (46)
Frenos anti-inmovilizadores Tipo de sistema de frenar que puede percibir la velocidad de cada rueda; en combinación con la computadora (ordenador), controla la presión hidraúlica para evitar la inmovilización de las ruedas. (46)

Armature The iron or steel center of an electric motor or solenoid, which is located between two magnets; the movable or rotating part of a generator, which helps to cut the magnetic lines of force; the center moving part or rotational part of the starter. (24) (34) (35)
Armadura El centro férreo o de acero de un motor eléctrico o solenoide, el cual se encuentra entre dos campos magnéticos; la pieza móvil o rotativa de un generador, la cual ayuda a cortar las líneas de fuerza magnéticas; la parte móvil céntrica o parte rotativa del arrancador (Marcha). (24) (34) (35)

Asbestos A fibrous material used to fireproof objects. It may cause cancer. (4) (39)
Asbesto Material fibroso que sirve para hacer objetos anti-inflamables. Puede causar cancer. (4) (39)

ASE Abbreviation for the National Institute for Automotive Service Excellence. This organization is the corporation that certifies automotive service technicians. It also assists in automotive training programs. (2)
ESA Abreviatura para el Instituto Naciónal para la Excelencia en Servicio Automotriz. Está organización es la corporación que certifica a los técnicos de servicio automotriz. La organización ayuda en programas de entrenamiento automotriz. (2)

Aspect ratio The ratio of the height to width of the tire expressed as a percentage. (49)
Proporción de aspecto Proporción de la altura a la anchura de una llanta, expresada como porcentaje. (49)

Aspirator valve A device used to draw out fluids by suction. In this case, a pollution device used to draw fresh air by suction into the exhaust flow to reduce emissions. (37)
Válvula aspiradora Mecanismo que sirve para sacar flúidos por succión. En este caso, se emplea un aparato de (contaminación) polución para sacar aire fresco por succión en un flujo de escape para reducir partículas emitidas. (37)

Atom Part of a molecule that has protons, neutrons, and electrons. All things are made of atoms. (29)
Atomo Parte de una molécula que contiene protones, neutrones, y electrones. Todo material consta de átomos. (29)

Atomization The breaking down of a liquid into small particles (like a mist) by the use of pressure. (23)
Atomatización La descomposición de un líquido en partículas muy pequeñas (como neblina) por el uso de presión. (23)

Atomized A liquid is atomized when it is broken into tiny droplets of the liquid, much like a mist or spray form. (25)
Atomizado Se atomiza un líquido cuando se descompone en forma de gotitas, semejante a una neblina (llovizna) o a un "spray." (25)

Axial load A type of load placed on a bearing that is parallel to the axis of the rotating shaft. (6)
Carga axial (de eje) Tipo de carga colocada sobre un cojinete (balero) que se encuentra paralelo a la línea recta del eje rotativo. (6)

Axial motion Motion that occurs along the axis of a revolving shaft or parallel to the axis of a revolving shaft. (17)
Movimiento axial (de eje) Movimiento que ocurre a lo largo de la línea recta de un eje rotativo o paralelo a la línea recta de un eje rotativo. (17)

Babbitt A soft metal material on the inside of a bearing insert to allow for embedability (small dirt particles embedding into the metal). (6) (17)
"Babbitt" Material blando metálico puesto en el interior de un añadido de cojinete para permitir la empotrabilidad (partículas de mugre o tierra que se pueden empotrar (incrustar) en el metal). (6) (17)

Backdrive A process in a 4WD vehicle powertrain in which the front wheels are disconnected from the 4WD system, but as the front wheels turn, they turn the front differential, creating additional friction. (44)
Propulsión trasera Proceso en el tren de potencia de los vehículos de 4RM en el cual las ruedas delanteras son disconectadas del sistema 4RM, pero al girar las ruedas delanteras, hacen girar al diferencial delantero, creando fricción adiciónal. (44)

Backing plate A metal plate that serves as the foundation for the brake shoes and other drum brake hardware. (45)
Placa de respaldo Placa de respaldo (plato de anclate) metálica que sirve de fundación y soporte para zapatas (patines) de freno y otros elementos relaciónados al tambor del freno. (45)

Back injury Injury to the back usually cased by lifting heavy objects incorrectly. (4)
Daño de espalda Daño a la espalda causado típicamente por el levantar incorrectamente objetos pesados. (4)

Baffles A series of thin walls placed in a battery case to allow gases to condense during charging or discharging. (31)
Divisiones (Tabiques) Una serie de particiónes poco densas y colocadas dentro de un contenedor de batería para dejar que los gases se condensen durante el proceso de cargarse o descargarse. (31)

Ball joint A pivot point for turning a front wheel to right or left. Ball joints can be considered either nonloaded or loaded when carrying the car's weight. (47)
Articulación de rótula (esférica) Punto giratorio para mover (dirigir) una rueda delatera a la derecha o izquierda. Se pueden considerar las articulaciónes cargadas o no cargadas al llevar el peso del automóvil. (47)

Base of 10 The base unit in the metric system is 10. All units are increased or decreased in units of 10. One meter has 10 decimeters, 100 centimeters, and 1,000 millimeters. (9)
Base 10 La unidad básica del sistema métrico es 10. Todas las unidades se incrementan o se reducen en módulos de 10. Un metro contiene 10 decímetros, 100 centímetros, y 1,000 milímetros. (9)

Battery cell That part of a storage battery made from two dissimilar metals and an acid solution. A cell stores chemical energy for use later as electrical energy. (31)
Célda de batería (acumulador) Esa parte de una batería de carga, hecha de dos metales diferentes y una solución de ácido. Una célda almacena energía química para usarse más tarde como energía eléctrica. (31)

Battery signal An electrical signal that is read by the computer to determine the battery voltage. If the battery voltage is low, the computer will adjust the engine speed to compensate for low voltage. (25)
Señal de batería Señal eléctrica leida por la computadora (ordenador) para determinar el voltaje de la batería. Si el voltaje de la batería es bajo, la computador ajustará la velocidad del motor para compensar por el voltaje bajo. (25)

BDC Position of the piston; bottom dead center. (13)
PCI Posición del pistón; punto huerio inferior. (13)

Bead The edge of a tire's sidewall, usually made of steel wires wrapped in rubber, used to hold the tire to the wheel. (49)
Borde El borde del costado de una llanta, normalmente hecho de alambres de acero envueltos en hule, que sirve para mantener pegada al llanta a la rueda. (49)

Bearing A device used to eliminate friction between a rotating shaft and a stationary part. (6)
Cojinete Mecanismo que sirve para eliminar fricción entre un eje rotativo y una parte estática (estaciónaria). (6)

Bearing cap A device that retains the needle bearings that ride on the trunnion of a U-joint and is pressed into the yoke. (43)
Sombrerete de cojinete (Tapa de balero) Mecanismo que mantiene en posición los cojinetes de agujas que están asentados sobre el muñón de la junta cardán o universal y empotrados en el yugo. (43)

Bellows A flexible chamber that can be expanded to draw a fluid in and compressed to pressurize the fluid. (24) (47)
Fuelle Una cámara flexible que puede ser expandida para sorber un flúido y comprimida para presionar el flúido. (24) (47)

Belt tension gauge A gauge used to measure the tightness of a belt that is used to turn a generator, air compressor, and water pump on an automobile engine. (7)
Calibrador de tensión de banda Calibrador que sirve para medir la tensión de una banda que da vuelta a un generador, un comprimidor de aire, y una bomba de agua en el motor de un automóbil. (7)

Bias A diagonal line of direction. In relationship to tires, bias means that belts and plies are laid diagonally or crisscrossing each other. (49)

Sentido (Sesgo) Línea diagonal de dirección. Relaciónado a llantas, el sentido (sesgo) quiere decir que las correas y capas están puestas diagonal menté o entrecruzadas. (49)

Bimetallic strip Two pieces of metal, such as brass and steel, attached together. When heat is applied to the strip, the two metals expand at different rates. This causes the metal to bend. A bimetallic strip senses changes in temperature and causes a mechanical movement. (22)
Franja bimetálica Dos pedazos de metal, tal como latón y acero, conectados uno al otro. Cuando se les aplica calor, los dos metales se expanden a diferentes coeficientes de expansión. Está expansión hace que el metal se curve. Una franja bimetálica es sensible a cambios en temperatura produciendo un movimiento mecánico. (22)

Binary code A computer code made of combinations of the numbers 1 and 0. (30)
Códigos binarios Código computarizado hecho de combinaciónes de los números 1 y 0. (30)

Bleeding The act of removing air from the brake hydraulic system. (45)
Purga El acto de quitarle aire al sistema de frenos hidráulicos. (45)

Blow-by The gases that escape past the rings and into the crankcase area. These gases from the combustion process produce a positive crankcase pressure. (17) (37)
Compresión negativa Los gases que escapan alrededor de las anillos y entran en el área del cárter. Los gases del proceso de la combustión producen una presión de cárter positiva. (17) (37)

BMEP Brake mean effective pressure. A measure of pressure on top of the piston during the power stroke. (13)
PMEF Presión media efectiva de freno. Medida de presión arriba del pistón durante el ciclo de potencia. (13)

Body and frame The part of the automobile that supports all other components. The frame supports the engine, drive lines, differential, axles, and so on. The body houses the entire vehicle. (3)
Carrocería y bastidor (armadura) de coche La parte del automóvil que soporta todos los otros componentes. El bastidor soporta el motor, tren propulsor, diferencial, ejes etcétera. La carrocería incorpora todo el vehículo. (3)

Body control module The computer module used in a vehicle that controls various vehicle functions related to the body of the vehicle, rather than the engine in the vehicle. (30)
Módulo de control del armazón Módulo de una computadora (ordenador) que sirve para controlar las varias funciónes del vehículo relaciónado al armazón del vehículo, en contraste a las del motor del vehículo. (30)

Boiling point The temperature of a fluid when it changes from a liquid to a gas or vapor. (22) (23)
Punto de ebullición Temperatura de un flúido cuando cambia de líquido a un gas o vapor. (22) (23)

Bolt hardness Hardness of a bolt is determined by the number of lines on the head of the bolt. More lines mean a stronger bolt. (5)
Dureza de perno (tornillo) La dureza de un perno se determina según el número de líneas en la cabeza del perno. Más líneas significan un perno más fuerte o duro. (5)

Boost The increase in intake manifold pressure produced by a turbocharger. (28)
Aumento de fuerza Incremento en la presión del colector de admisión producido por un turboalimentador. (28)

Bore The diameter of the cylinder. (13)
Calibre Diámetro de un cilindro. (13)

Boss A cast or forged part of a piston that can be machined for accurate balance. (17)
Parte forjada de pistón Pieza fundida o forjada de un pistón, la cual puede ser trabajada (recortada) con máquina herramienta para lograr un balance preciso. (17)

Bourdon tube A curved tube that straightens as the pressure inside it is increased. The tube is attached to a needle on a gauge, which senses the movement of the tube and transmits it as a pressure reading. (22)
Tubo Bourdon Tubo curvado que se endereza mientras se incrementa la presión adentro. El tubo está conectado a una aguja de un indicador, que responde al movimiento del tubo y lo transmite como una indicación de presión. (22)

Bowden cable A small steel cable inside a flexible tube used to transmit mechanical motion from one point to another. (52)
Cable Bowden Pequeño cable de acero dentro de un tubo flexible usado para transmitir un movimiento mecánico de un punto a otro. (52)

Brake horsepower The horsepower available at the rear of the engine. (15)
Potencia de freno La potencia disponible atrás del motor. (15)

Brake shoes The curved metal parts faced with brake lining that are forced against the brake drum to produce the braking action. (45)
Zapatas (Patines) de freno Las partes metálicas curvadas junto a la guarnición del freno las cuales están forzadas contra al tambor del freno para producir la acción de frenar. (45)

Breaker points A set of contact points used in the ignition system to open and close the primary circuit. The points act like a switch operated from a distributor cam. (32)
Puntos de ruptura Juego de puntos de contacto usados en el sistema de encendido para abrir y cerrar el circuito principal. Los puntos funciónan como un interruptor puesto en marcha desde la leva distribuidora. (32)

Brinelling The process of producing small dents in a bearing surface due to heat and wear. (6)
Picaduras Proceso de producir abolladuras pequeñitas en la superficie de un cojinete (balero) debido al calor y fricción. (6)

Btu A unit of thermal or heat energy referred to as a British thermal unit. The amount of heat necessary to raise 1 pound of water 1 degree Fahrenheit. (23)
UTB Unidad de energía termal o calórica que lleva la nomenclatura de Unidad Termal Británica. La cantidad de calor necesaria para elevar una libra de agua un grado de temperatura F. (23)

Bushing A friction-type bearing usually identified as one piece. (6)
Cojinete (Balero) Un balero tipo de fricción usualmente identificado como una pieza. (6)

Bushing installers Tools used to provide correct alignment and applied forces when installing a bushing in a housing. (10)
Instaladores de cojinetes (baleros) Herramientas que sirven para proporcionar correcta alineación y fuerzas aplicadas al instalar un cojinete (balero) dentro de un hueco (Housing) estuche. (10)

Bushing pilot A guide that fits the inside of a circular bushing used to install a bushing correctly into a bore. The bushing pilot helps to keep the bushing aligned correctly during installation. (10)
Piloto del buje Guía que cabe en el interior de un buje circular que sirve para instalar correctamente un buje en el agujero. El piloto del buje mantiene alineado al buje correctamente durante la instalación. (10)

By-pass tube A tube directly in front of the thermostat. The coolant bypasses the radiator through this tube when cold. (22)

Tubo de desviación Tubo colocado delante del termóstato. El líquido refrigerante se desvia del radiador por el tubo cuando está frío. (22)

CAFE Corporate average fuel economy; the average fuel economy that manufacturers must meet each year as set by the Federal government. (36)

CAFE (Corporate average fuel economy); la economía media de combustible con la que los fabricantes deben cumplir según establecida por el gobierno federal. (36).

Calibrate To check or measure any instrument. (24)

Calibrar Revisar o medir cualquier instrumento. (24)

Caliper A C-shaped housing that fits over the rotor, holding the pads and containing the hydraulic components that force the pads against the rotors when braking. (45)

Calibrador Cubierta protectora en forma de una C que se coloca alrededor de un rotor, manteniendo así los cojinetes y conteniendo los componentes hidráulicos que empujan los conjinetes contra los rotores al frenar. (45)

Camber The inward or outward tilt of the wheel at the top. A wheel has a positive camber when the top is tilted out. (48)

Inclinación Inclinación de una rueda hacia dentro o fuera en la parte superior de la rueda. Una rueda muestra una inclinación positiva cuando la parte superior está inclinada hacia fuera. (48)

Cam ground piston Piston ground to a cam shape to aid in controlling expansion from heat. (17)

Pistón recortado por leva Pistón recortado a la forma de leva para ayudar a controlar la expansión debida al calor. (17)

Camshaft A shaft that is used to open and close the valves. (12)

Arbol de levas Arbol o eje que sirve para abrir y cerrar las válvulas. (12)

Camshaft position sensor A sensor used to measure the position of the camshaft, in regard to measuring the ignition timing of the engine. (33) (38)

Sensor de posición del árbol de levas Sensor que sirve para medir la posición del árbol de levas, en relacíon con la medida de la regulación del avance del encendido del motor. (33) (38)

Capacity specifications Specifications used to show quantity or amount of liquid in automobile components. (11)

Especificaciones de capacidad Especificaciones que sirven para indicar la cantidad de líquido encontrada en los componentes automovilísticos. (11)

Carbon canister A canister filled with carbon used to absorb and store fuel vapors that are normally exhausted into the air. (37)

Bote de carbón Bote lleno de carbón que sirve para absorber y almacenar los vapores de combustible que se escapan al aire. (37)

Carbon monoxide A pollutant in automotive exhaust that is produced when there is insufficient oxygen for combustion. It is a deadly, odorless, tasteless gas. (4) (36)

Monóxido de carbono Contaminante en los gases de escape automovilísticos que se produce cuando hay oxígeno insuficiente para hacer combustión. Es un gas (inodoro), (insipido) y mortal. (4) (36)

Carburetor A device used to mix air and fuel in the correct proportions. (12)

Carburador Aparato que sirve para mezclar aire y combustible según proporciones apropiadas. (12)

Carcass The inner part of the tire that holds the air for supporting the vehicle. (49)

Armadura La parte interior de una llanta que contiene el aire para soportar el vehículo. (49)

Cardan universal joint A universal joint sometimes known as the four-point or cross-and-bearing-type joint. This joint allows the transmission of power at an angle, but causes rhythmic variations in speed at the output yoke of the joint. (41)

Junta (Balero) de Cardán (junta universal) Junta (o balero) de Cardán a veces conocida como una junta de tipo cuatro puntos. Está junta permite la transmisión de potencia desde un ángulo pero produce variaciones rítmicas de velocidad en el punto de la brida (estribo o yugo) que hace la union. (41)

Case-hardened The outer surface of a metal that has been hardened to reduce the possibility of excessive wear. (17)

Endurecido por fuera, reforzado Superficie exterior de un metal que ha sido endurecida para reducir la posibilidad de fricción excesiva. (17)

Cast The process of shaping metal by heating to a liquid and pouring the hot metal into a sand mold. (17)

Fundir Proceso de formar metal por medio de calentamiento hasta producir un líquido metálico y vacían el metal supercaliente en un molde de arena. (17)

Caster The backward or forward tilt of the spindle support arm to the top. (48)

Divergencia Inclinación hacia atrás o hacia adelante de un huso de sopote hacia arriba. (48)

Catalyst A chemical that causes or speeds up a chemical reaction without changing its own composition. (20) (37)

Catalizador Químico que causa o produce una velocidad a lo largo de una reaccíon química sin cambiar su propia composición. (20) (37)

Catalytic converter A type of emission control device used to change the exhaust emission from the vehicle into harmless chemicals. (37)

Convertidor catalítico Aparato para controlar emisiones, el cual sirve para convertir las emisiones de escape del vehículo en químicos inofensivos. (37)

Cavitation A process in a cooling or water system that causes small vacuum bubbles to occur, which, upon implosion, damages the water pump blades. (22)

Cavitación Proceso dentro de un sistema de agua o un sistema refrigerante que hace que ocurran pequeñas burbujas de vacío, las cuales, al hacer implosión dañan las paletas de una bomba de agua. (22)

Cell density A measure of the amount of watts that can be discharged per hour, per pound of the battery. (31)

Densidad de célula Una medida de la cantidad de vatios que pueden ser descargados por hora, por libra de la batería (31).

Center link A steering linkage component connected between the pitman and idler arms. (48)

Conexión central Componente de conexión de manejo entre la barra de conexión y los brazos intermedios. (48)

Centrifugal pump A pump that draws coolant into its center and then uses centrifugal force to throw the coolant outward and into the cooling system. (22)

Bomba centrífuga Bomba que atrae el líquido refrigerante hacia su centro y luego emplea una fuerza centrífuga para aventar el refrigerante hace el sistema refrigerante. (22)

Cetane number The ignition quality of a diesel fuel. The time period or delay between injection and explosion of diesel fuel. (23)

Número cetano Calidad de encendido de un combustible diesel. El período o tardanza entre la inyección y la explosión del combustible diesel. (23)

CFC A chlorofluorocarbon or gas such as freon that damages the environment and ozone layer. (50)
CFC Un clorofluorocarburo o gas tal como el freón que produce daño al medio ambiente y la capa de ozono. (50)

Chamfer An angled cut on the end of a tooth of a gear. (35)
Bisel Corte diagonal en el extremo de un diente de una rueda dentada. (35)

Charging When the electrical flow of a battery is reversed, the battery is charged. The metals that are alike are converted to lead peroxide and sponge lead, and the electrolyte is now sulfuric acid and water. (31)
Cargando Cuando se pone al revés el flujo de la corriente eléctrica, se carga la batería. Los metales semejantes se convierten en peróxido de plomo y plomo esponjado, y el electrólito resulta en ácido sulfúrico y agua. (31)

Chassis dynamometer A dynamometer used to measure road horsepower. (15)
Dinamómetro de carrocería Un dinamómetro que sirve para calcular la potencia de caballos en una prueba de camino. (15)

Chemical hazards Hazards primarily from solvents, gasoline, asbestos, and antifreeze. (4)
Riesgos químicos Peligros provenientes principalmente de solventes, gasolina, asbestos, y anticongelantes. (4)

Closed loop A computer condition in which the air-fuel ratio is being controlled on the basis of various inputs to the computer. (25) (38)
Circuito cerrado Condicíon de computadora (ordenador) en la que la proporción de aire y combustible se controla a base de un juego de unas condiciónes pre-programadas. (25) (38)

Cloud point The temperature at which a diesel fuel begins to produce wax crystals due to cold temperatures. (23)
Punto de cristalización Temperatura en la que un combustible diesel empieza a producir cristales cerosos debido a temperaturas frías. (23)

Clutch The part of a manual transmission that is used to engage and disengage the transmission from the engine. (39)
Embrague Parte de una transmisión manual que sirve para engranar y desengranar la transmisión del motor. (39)

Clutch switch A switch connected to the clutch motion on a manual transmission vehicle with a cruise control system. When the clutch switch senses that the clutch has been depressed, it disengages the cruise control system. (52)
Conmutador del embrague Conmutador conectado al movimiento del embrague de un vehículo con transmisión manual y un sistema de control crucero. Cuando el conmutador del embrague percibe que se ha oprimido el embrague, desconecta al sistema de control crucero. (52)

Coefficient of drag A method used to check the air resistance of a moving vehicle. A measure of how much air is moved as the vehicle moves from one point to another. Generally the coefficient of drag ranges between 1.00 and 0.00. (15)
Coeficiente de resistencia Método que sirve para revisar la resistencia de aire de un vehículo en movimiento. Una medida de la cantidad de aire que está desplazado al pasar un vehículo de un punto a otro. En general, el coeficiente de resistencia tiene una distancia de medida de 1.00 a 0.00. (15)

Cogged timing belt A belt that has various cogs or raised notches on its inside, used to keep the camshaft and the crankshaft in time with each other. (7)
Correa de tiempo dentado Correa que tiene varias dientes o entalles en su interior, que sirve para sincronizar el árbol de levas con el cigüeñal. (7)

Cold soaking A method used to clean parts of an engine or fuel system in which the parts are placed in a container of specialized cleaning fluid for a certain amount of time. (10)
Remojo en frío Metodo de limpiar las partes de un motor o del sistema de combustible en el cual las partes se colocan en un recipiente de fluidos especiales para limpieza por un cierto periodo de tiempo. (10)

Color codes Various electrical or other components that are identified by giving them a color code to help identify their size or other characteristics. (21)
Códigos de color Componentes eléctricos u otros que se identifican dandoles un código de color para identificar su tamaño u otras características.

Combustion The process in which air and fuel are burned after being mixed to a correct ratio of 14.7 parts of air to 1 part of fuel. (12)
Combustión Proceso en el que el aire y el combustible queman después de mezclarse en una proporción correcta de 14.7 partes de aire a 1 parte de combustible. (12)

Combustion chamber The location inside the cylinder head or piston in which combustion of the air and fuel occurs. (12)
Cámara de combustión El hueco dentro de la cabeza del cilindro o pistón en el que ocurre la combustión del aire y combustible. (12)

Commutator A device that extracts the electrical energy from the rotating armature. When electrical energy is being extracted from the armature, ac voltage is converted to dc voltage; metal segments that are used on a starter motor to carry electricity to the armature. The commutator also reverses the current flow in the armature at the right time. (34) (35)
Commutador Aparato que extrae la energía eléctrica de un eje rotativo. Cuando la energía eléctrica se extrae de un eje, voltaje CA (corriente alterna) se convierte en un voltaje CC (corriente continua); segmentos metálicos que se usan en un motor de arranque para llevar electricidad al eje. El comutador también pone la marcha de la corriente en dirección opuesta en el eje en el momento apropiado. (34) (35)

Compensating port A passage for excess fluid to return to the reservoir when the brakes are released. (45)
Orificio compensador Pasaje para que el exceso de líquido regrese a un tanque o depósito cuando se sueltan los frenos. (45)

Component location table A table used with an electrical schematic that shows the actual location of the part being investigated. (53)
Tabla de localidad de componentes Una tabla (figura) usada con un esquemá electrico que indica el lugar exacto de la pieza que se investiga. (53)

Compression ratio A measure of how much the air has been compressed in a cylinder of an engine from TDC to BDC. Compression ratio will usually be from 8 to 1 to 25 to 1. (13)
Proporción de compresión Medida de la cantidad de aire que se ha comprimido en un cilindro de un motor de TDC a BDC. La proporción de compresión normalmente será de 8:1 a 25:1. (13)

Compression washer A washer used on an engine to reduce oil leakage when the threads of a bolt are in or near oil. (5)
Arandela de presión Arandela que sirve en un motor para reducir el escape o goteo de aceite cuando las roscas de un perno están en o cerca del aceite. (5)

Compressor A device on a gas turbine that compresses the air for combustion. (28)

Compresor Aparato de una turbina de gas que comprime al aire para combustión. (28)

Computer-controlled brakes A system that has a sensor on each wheel feeding electrical impulses into the on-board computer. As the vehicle is stopped, each wheel is stopped or slowed down at the same rate. This condition reduces skidding sideways during rapid braking. (45)
Frenos controlados por computadora (ordenador) Sistema que contiene un sensor en cada rueda, el cual manda impulsos eléctricos a una computadora a bordo. Al pararse el vehículo, cada rueda se detiene o se reduce en la velocidad de rodaje con la misma tasa de velocidad. Está condición reduce el deslizamiento de movimiento lateral (patinaje) durante el acto de frenar rápidamente. (45)

Computerized automatic temperature control A control system used to monitor and adjust the air temperature inside the passenger compartment. On the basis of several inputs, a small microprocessor adjusts air temperature about every few seconds. (51)
Control de temperatura autómatico computarizado Sistema de control que sirve para vigilar y ajustar la temperatura del aire dentro del compartimiento para pasajeros. A base de varios datos electrónicos enviados, un microprocesador ajusta la temperatura del aire cada rato de unos segundos. (51)

Condensation The process of reducing a gas to a more compact state such as a liquid. This is usually done by cooling the substance below its boiling point. For example, the moisture in the air in a fuel tank condenses to water. (24) (50)
Condensación Proceso de convertir un gas a un estado más condensado como es un líquido. Típicamente se hace el proceso por medio de enfriar la sustancia hasta un punto más bajo que su punto de ebullición. Por ejemplo, la humedad en el aire dentro de un tanque de combustible se condensa en agua. (24) (50)

Condense The process of cooling a vapor to below its boiling point. The vapor changes into a liquid. (23)
Condensar Proceso de enfriar un vapor hasta un punto más bajo que su punto de ebullición. El vapor se convierte en un líquido. (23)

Condenser A capacitor device used to protect the ignition points from corroding; a component in an air-conditioning system used to cool a refrigerant below its boiling point. (20) (50)
Condensador Un capacitador que sirve para proteger los puntos de ignición contra corrosión; un componente en un sistema de aire acondiciónado que sirve para enfriar un líquido refrigerante bajo de su punto de ebullición. (20) (50)

Conduction Transfer of heat between two solid objects. (22)
Conducción La transferencia de calor entre dos objetos sólidos. (22)

Conductor A substance with three or fewer electrons in the valence ring. (29)
Conductor Sustancia de tres electrones o menos en la banda de valencia. (29)

Configuration The figure, shape, or form of an engine. (16)
Configuración La figura, contorno o forma de un motor. (16)

Conical Having the shape or form of a cone. (26)
Cónico Tener la forma de un cono. (26)

Connecting rod The connecting link between the piston and crankshaft. (12)
Biela El vínculo de conexión entre el pistón y el cigüeñal. (12)

Constant velocity (CV) joint The CV joint consists of balls or tripods and yokes designed to allow the angular transfer of power without speed variations common to the Cardan universal joint. (43)
Junta (Balero) de velocidad constante (VC) La junta VC consiste en bolas o trípodes y balancines diseñados para permitir la transferencia de potencia angular sin variaciónes en velocidad comunes en una flecha universal Cardán. (43)

Contaminants Various chemicals in the oil that reduce its effectiveness, including water, fuel, carbon, acids, dust, and dirt particles; also impurities in fuel systems, including dirt, rust, water, and other materials; also chemicals that make the air impure, usually produced from the combustion process. (1) (20)
Contaminantes Varios químicos en el aceite, los cuales reducen la eficacia del aceite, incluyendo agua, combustible, carbón, ácidos, polvo, y partículas de tierra; también impurezas en los sistemas de combustible, como partículas de tierra, orín, agua, y otros materiales; también químicos que resultan en un aire contaminado, típicamente producido por el proceso de combustión. (1) (20)

Continuity An electrical circuit is said to have continuity when the circuit is complete and has no breaks. (10)
Continuidad Un circuito eléctrico se dice tener la continuidad cuando el circuito es completo y no tiene apertura. (10)

Continuous combustion Combustion of air and fuel that continues constantly. (12)
Combustión continua Combustión del aire y el combustible, la cual sigue constantemente. (12)

Continuously variable transmission A transmission that has an infinite number of gear ratios. Gear ratios are changed by changing the pulley size within the transmission. (14)
Transmisión continuamente variable Una transmisión con un número infinito de relaciones de engranaje. Se cambian las relaciones de engranaje cambiando el tamaño de la polea dentro de la transmisión. (14)

Control arm The main link between the vehicle frame and the wheels. The control arm acts as a hinge to allow the wheels to go up and down independently of the chassis. (47)
Brazo de control Vínculo principal entre el batidor de coche y las ruedas. El brazo de control funciona de gozne (bisagra) que permite que las ruedas tengan una acción amortiguadora de altibajos independiente del bastidor. (47)

Convection Transfer of heat by the circulation of the heated parts of a liquid or gas. (22)
Convección La transferencia de calor por circulación de las partículas de un líquido o gas. (22)

Conventional theory Electrons will flow from a positive to a negative point. (29)
Teoría convencional Los electrones flotarán de un punto positivo a uno negativo. (29)

Conversion factors Numbers used to convert between USC and the metric system. (9)
Factores de conversión Los números que sirven para convertir entre USC y el sistema métrico. (9)

Cooling system The subsystem on an engine used to keep the engine temperature at maximum efficiency. (3)
Sistema de refrigeración El subsistema de un motor que sirve para mantener la temperatura del motor a una eficiencia máxima. (3)

Cords The inner materials running through the plies that produce strength in the tire. Common cord materials are fiberglass and steel. (49)
Cuerdas Los materiales interiores que pasan por las capas que produce la fuerza de una llanta. El material de dos cuerdas son fibra de vidrio y acero. (49)

Core The center of the radiator, made of tubes and fins, used to transfer heat from the coolant to the air. (22)

Centro El centro de un radiador, hecho de tubos y aletas, que sirven para transferir el calor desde el líquido refrigerante hasta el aire. (22)

Core plugs Plugs inserted into the block that allow the sand core to be removed during casting. Also, at times these plugs will pop out and protect the block if the coolant freezes. (17)

Tapones céntricos Tapones insertados en el bloque que permiten que se quite la arena durante el proceso de fundición. También, a veces, tales tapones saltan y protegen el bloque si se congela el líquido refrigerante. (17)

Corrosive A substance that has the ability to corrode. (22)

Corrosive Substancia capaz de corroer. (22)

Counterweight Weight forged or cast into the crankshaft to reduce vibration. (17)

Contrapeso Un peso fundido en el cigüeñal para reducir vibraciónes. (17)

Crank signal A signal sent to the automotive computer to tell it that the engine is starting. The computer will then enrich the air-fuel ratio for easier starting. (35)

Señal de arranque Una señal enviada a la computadora del auto para comunicarle que el motor se arranca. La computadora en ese instante enriquece la proporción del aire y combustible para facilitar el encendido. (35)

Crankcase The area in the engine below the crankshaft. It contains oil and fumes from the combustion process. (17)

Cárter Area en el motor debajo del cigüeñal. Contiene aceite y gases gue resultan del proceso de combustión. (17)

Crankcase pressure The pressure produced in the crankcase from blow-by gases. (37)

Presión del cárter Presión producida en el cárter por los gases que se acumulan en el escape. (37)

Crankshaft A mechanical device that converts reciprocating motion to rotary motion. (12)

Cigüeñal Eje que convierte un movimiento recíproco a un movimiento rotativo. (12)

Cross The central component of the U-joint connecting the input and output yokes. (43)

Cruz El componente central de una junta universal que conecta las bridas de la entrada de potencia y las del rendimiento (o salida) de potencia. (43)

Crossfire injection A type of throttle body injection system that uses two injectors mounted on the manifold. Each injector feeds a cylinder on the opposite side by using a crossover port. (26)

Inyección de dos sentidos Tipo de sistema de inyección a la válvula de admisión, el cual emplea dos inyectores montados en el colector de admisión. Cada inyector suministra al cilindro situado al lado opuesto por medio del uso de un orificio de paso opuesto. (26)

Cubic centimeter A unit in the metric system to measure volume. There are 100 cubic centimeters in 1 liter. (9)

Centímetro cúbico Unidad del sistema metrico que sirve para medir volumen. Hay 100 centímetros cúbicos en 1 litro. (9)

Curtain area An engineering term that relates to the efficiency of the flow of air and fuel entering the combustion chamber. The curtain area represents how evenly the air and fuel are being admitted to the combustion chamber. The greater the curtain area, the more efficient the combustion. (18)

Area de velo Un término de ingeniería relacióndado a la eficiencia del flujo del aire y combustible, los cuales entran en la cámara de combustión. El área de velo representa que tan uniforme se admiten el aire y el combustible en la cámara de combustíon. Cuanto más grande sea el área del velo, tanto más eficiente será la combustión. (18)

Cycle The process of discharging and then recharging a battery. (31)

Ciclo Proceso de descargar y luego de cargar otra vez una batería. (31)

Cylinder Internal holes in the cylinder block. (12)

Cilindro Los huecos dentro del bloque de cilindros. (12)

Cylinder block Part of the engine that houses all components. The foundation of the engine. (12)

Bloque de cilindros Una parte del motor donde se encuentran todos los componentes. La fundación del motor. (12)

Cylinder head The top and cover for the cylinder; it houses the valves. (12)

Cabeza de cilindro La tapa para los cilindros; contiene las válvulas. (12)

Cylinder sleeve A round cylindrical tube that fits into the cylinder bore. Both wet and dry sleeves are used. (17)

Funda de cilindro Un tubo redondo y cilíndrico que cabe dentro de un hueco cilíndrico. Se emplean dos fundas, una mojada y otra seca. (17)

Cylinder taper The shape of the cylinder after it has been worn by the rings. (17)

Estrechamiento de cilindro La forma del cilindro después de haberse gastodo por los anillos. (17)

Data link A device in a computer system to send and receive digital signals. (30)

Vínculo de datos Aparato en el sistema de computadora que manda y recibe señales digitales. (30)

Data link connector A connector used on vehicles equipped with a computer for various controls (ignition, fuel injection, etc.) into which a scanner or other electrical device can be hooked so as to read out various fault codes. (30)

Conector de enlace de datos Conector en los vehículos equipados con computadora (ordenador) que controla varios aparatos (encendido, inyección de combustible, etcétera) al cual se puede conectar un explorador u otro aparato eléctrico para leer los códigos de averías. (30)

Dealership Privately owned service and sales organization that sells and services vehicles for the automobile manufacturer. (1)

Agencia automotriz Una organización de servicio y ventas con dueño particular que vende autos y servicio para un fabricante automotriz. (1)

Decibel A unit of sound measurement. Usually 90 to 100 decibels experienced for a long time can cause hearing damage. (4)

Decibelio Unidad de medida de sonido. Normalmente una cantidad de 90 a 100 decibelios a que se somete uno durante mucho tiempo puede resultar en un daño al oído. (4)

Deep cycle battery A battery that is designed to withstand continuous cycling, or discharging and charging. (31)

Batería de ciclo extenso Batería diseñada para aguantar un ciclismo continuo, o descargar y cargar. (31)

Deep well socket Socket used when a nut is located on a long stud. (8)

Manguito con cavidad profunda Manguito utilizando cuando la tuerca se encuentra en un perno prisionero largo. (8)

Deflection angle The angle at which the oil is deflected inside the torque converter during operation. The greater the angle of deflection, the greater the torque applied to the output shaft. (41)

Angulo de deflexión Angulo al cual se desvía el aceite dentro del convertidor de torsión durante la operación. Cuánto más sea la deflexión, tanto más será la torsión aplicada al eje de rendimiento. (41)

Defroster Part of the ventilation system on an automobile used to remove ice, frost, or moisture from the front windows. (51)

Descongelador Parte del sistema de ventilación del automóvil que sirve para quitar hielo, escarcha, o humedad de las ventanas delanteras. (51)

Department of Transportation (DOT) U.S. government agency that regulates transportation policy. (49)

Departamento de Transportes (DOT—Department of Transportation) Entidad gubernamental estadounidense que regula la política del transporte. (49)

Desiccant A material that absorbs moisture from a gas or liquid. A desiccant substance is used in an air-conditioning system to remove moisture from the refrigerant. (50)

Desecante Material que absorbe humedad de un gas o líquido. Una sustancia desecante se emplea en un sistema de aire acondiciónado para quitarle humedad al líquido refrigerante. (50)

Detent springs Small springs used in conjunction with a small ball. The springs push the ball against an indented part of a shaft, used to hold the shaft from sliding. (40)

Resortes de escape Resortes pequeños usados en combinación con una bolita. Los resortes empujan la bolita contra una parte dentada de un eje, los cuales sostienen en posición el eje para que no deslice. (40)

Detonation sensor A sensor used on computerized ignition systems that is able to measure if the engine is producing a "knock." The sensor is able to pick up small, high-frequency vibrations produced from engine knock. (33)

Detector de detonación Detector usado en sistemas de encendido computarizado, el cual puede percibir si el motor sufre golpes interiores. El detector puede sentir vibraciónes pequeñas de alta frecuencia producidas por un golpeteo en un motor. (33)

Diagnosis The process of finding and determining the cause of problems in the automobile. When the problem is identified, a solution is selected. (11)

Diagnosis Proceso de encontrar y determinar la causa de problemas de un automóvil. Al identificar un problema, se escoge una solución. (11)

Diagnostic circuit check A procedure used by a service technician to identify the type of fault codes that have been stored in a computer. (30)

Comprobación diagnóstico de circuitos Procedimiento usado po los tecnicos de servicio para identificar los códigos de averías que se han almacenado en una computadora (ordenador). (30)

Dial indicator A measuring tool used to check small clearances up to 0.001 inch. The clearance is read on a dial. (9)

Indicador de caratula Un instrumento de medición que sirve para verificar las aperturas pequeñas hasta una tolerancia de 0.001 de una pulgada. Se anota la cantidad de apertura en la caratula. (9)

Diaphragm A partition separating one cavity from another. A fuel pump uses a diaphragm to separate two cavities inside the pump. A diaphragm is usually made of a rubber and fiber material. (24)

Diafragma Una partición que separa una cavidad de otra. Una bomba de combustible emplea un diafragma para separar dos cavidades dentro de la bomba. Un diafragma normalmente es de un material de hule y fibra. (24)

Diaphragm spring A type of steel plate spring used in a pressure plate to engage or disengage the friction disc. (39)

Resorte de diafragma Tipo de resorte revestido de acero y usado en una plancha de presión para embragar o desembragar un disco de fricción. (39)

Die cast The process of shaping metal by forcing hot metal under pressure into a metal mold. (17)

Fundir a presión (troquelar) Proceso de formar metal a través de forzar metal super-caliente bajo presión en un molde metálico. (17)

Diesel engine An intermittent, internal combustion, reciprocating engine that uses the heat of compression to ignite the fuel. Fuel is supplied by a high-pressure fuel injector instead of a carburetor. (14)

Motor diesel Un motor intermitente, de combustión interna, con reciprocidad que emplea el calor de compresión para encender el combustible. En vez de un carburador, un inyector da alta presión suministra el combustible. (14)

Differential A gear assembly that transmits power from the drive shaft to the wheels. It also allows two opposite wheels to turn at different speeds for cornering and traction. (43)

Diferencial Un montaje de ruedas dentadas, que transmite potencia desde el eje de transmisión hasta las ruedas. También permite que dos ruedas opuestas giren a velocidades diferentes para doblar y tracción. (43)

Differential pressure Difference in pressure on an oil system between the input and the output of a filter. (21)

Presión diferencial Diferencia de presión en un sistema de aceite entre la entrada y la salida de un filtro. (21)

Differential sensor This type of sensor measures the intake manifold pressure and compares it to the atmospheric pressure. (38)

Sensor del diferencial Este tipo de sensor mide la presión del múltiple de admisión y lo compara con la presión atmosférica. (38)

Digital signal A computer signal that has either an on or off condition. (30)

Señal digital Una señal de computadora que tiene una condición de encendida o apagada. (30)

Dilution To make thinner or weaker. Oil is diluted by the addition of fuel and water droplets. (20)

Dilución Hacer un líquido más diluído o menos concentrado (fuerte). Se diluye el aceite por medio de la añadidura de gotitas de combustible y agua. (20)

Dimmer switch A switch in the headlight circuit used to switch electricity between LO and HI positions. (53)

Interruptor de regulador de voltaje Un interruptor en el circuito de los focos delanteros, el cual sirve para combiar la electricidad entre las posiciónes de luces bajas y luces altas. (53)

Diode Semiconductor in an electrical circuit to allow electrical current to flow in only one direction. (30)

Díodo Semiconductor en un circuito eléctrico que permite que una corriente eléctrica fluya en solo una dirección. (30)

Disc brakes Brakes in which the frictional forces act on the faces of a disc. (45)

Frenos de disco Frenos en los que las fuerzas de fricción rozan contra la supericie de un disco. (45)

Discharging When an electrical load is put on a battery, the battery is discharged. The internal parts of the battery are chemically changed to the same metals and water. (31)

Descargar Situación en la que una carga eléctrica se le pone a una batería, ésta descarga. Las partes internas de la batería están cambiadas químicamente a los mismos metales y la misma agua. (31)

Discrete sensor A sensor used on a computerized engine system that has only two modes or signals, such as open or closed. (38)
Sensor discreto Sensor que se usa en un sistema computerizado que solo tiene dos modos o señales, tal como abierto o cerrado. (38)

Disperse To scatter, spread, or diffuse. (27)
Dispersar Esparcir, disipar, o difundir. (27)

Dispersant A chemical added to motor oil to disperse or keep the particles of dirt from sticking together. (20)
Dispersante Químico añadido al aceite de motor para dispersar o prevenir que partículas de polvo o tierra se peguen. (20)

Displacement The volume the piston displaces from BDC to TDC. (13)
Desplazamiento El volumen que desplaza el pistón de BDC a TDC. (13)

Dissipate To become thinner and less concentrated. (1)
Disipar Llegar a estar más diluído y menos concentrado. (1)

Distillation The process at a refinery that separates the hydrocarbons into many products. Usually boiling a liquid to a vapor and condensing it are involved. (23)
Destilación Proceso de una refinería que separa los hidrocarburos en varios productos. Típicamente el hervir un líquido hasta vapor y el condensarlo son dos procesos. (23)

Distillation end point The temperature at which a fuel is completely vaporized. (23)
Punto final de la destilación Temperatura a la que un combustible está completamente vaporizado. (23)

Distributor cam Cams on the distributor shaft used to open and close the breaker points. There is one cam for each cylinder of the engine. (32)
Leva del distribuidor Levas en el eje del distribuidor, las cuales se emplean para abrir y cerrar los puntos de ruptura. Hay una leva para cada uno de los cilindros del motor. (32)

Distributorless ignition An ignition system that uses a computer to distribute the electrical spark, rather than a rotor and distributor cap. (33)
Ignición sin distribuidor Sistema de encendido que emplea una computadora para distribuir una chispa eléctrica, en vez de un rotor (rodete) y una tapa del distribuidor. (33)

Domains Small sections in a metal bar where atoms line up to produce a magnetic field. (29)
Dominios Secciónes pequeñas en una barra metal en la que los atomos se congregan para producir un campo magnético. (29)

Downshifting The process of shifting from a higher gear to a lower gear. (42)
Cambio descendente Proceso de cambiar de una velocidad alta a una velocidad más baja. (42)

Drive The end of a ratchet or other similar wrench used to hold the socket to the ratchet. (8)
Implemento de acciōna (o impulsión) El extremeo de una llave de cubo (trinquete, carraca, matraca) u otra llave semejante, que sirve para sostener firme la boquilla (casquilla, adaptador) a la llave de cubo. (8)

Drive lines Components on the vehicle that transmit the power from the engine to the wheels. (3)
Líneas (Flechas) de transmisión Componentes de un vehículo que transmite la potencia desde el motor a las ruedas. (3)

Driveline windup A condition in a four-wheel drive system when one of the four locked-up wheels tries to spin faster (for example, due to slippery conditions on that wheel). When this condition occurs, it produces excessive torque on the four-wheel drive system shafts and axles, possibly causing premature failure of some of the components and parts. (44)
Torsión de la flecha motríz Condición en un sistema de cuatro ruedas motrices en la cual una de las ruedas bloqueadas quiere girar más rapidamente (por ejemplo, debido a las condiciónes resbalosas en esa rueda). Cuando ocurre está condición, produce la torsión excesiva en las flechas y en los ejes del sistema de cuatro ruedas motrices, causando los fallos prematuros de algunos de los partes y componentes. (44)

Drive shaft A metal tube with I-joints or CV joints at each end, used to transmit power from a transmission to the wheels on differentials. (43)
Eje impulsor (Arbol de arrastre, Arbol motor, Flecha) Un tubo metálico con juntas I o juntas CV en cada extremo, que sirve para transmitir fuerza desde la transmisión a las ruedas que están en un diferencial. (43)

Drop center wheel A wheel that has its center dropped in with a smaller radius. A drop center wheel is used so that the tire can be easily removed. (49)
Rueda de centro reducido Una rueda que tiene su parte céntrica reducida con un radio más pequeño. Una rueda de centro reducido se emplea para que se le pueda quitar la llanta más fácilmente. (49)

Dual camshaft A type of engine that has two camshafts for opening and closing additional valves. (16)
Arbol de levas doble Tipo de motor que tiene dos árboles de levas para abriri y cerra válvulas adiciónales. (16)

Duty cyle The timing cycle of on and off of a particular engine component, such as the on and off sequence of fuel injectors. (25)
Ciclo de servicio Ciclo periódico de prendido o apagado de un componente particular de un motor, tal como la secuencia de prendido o apagado de los inyectores de combustible. (25)

DVOM A digital volt-ohm-ammeter, or a meter that reads voltage and resistance in ohms, that uses a digital readout rather than an analog readout or a pointer on a scale. (10)
DVOM Un voltiohmímetro digital, o un medidor que lee el voltaje y la resistencia en ohmios, que utiliza una lectura de salida digital en vez de una lectura análoga o una aguja indicadora en una gama. (10)

Dwell The length of time in degrees of distributor shaft rotation that the points remain closed. (32)
Duración Cantidad de tiempo en grados de la rotación del eje del distribuidor en que los puntos quendan cerrados. (32)

Dynamic balancing Equal distribution of weight on each side of a centerline of a wheel. Dynamic means moving or action, and dynamic balancing is done with the wheel moving or spinning. (49)
Equilibrio dinámico Distribución igual de peso a cada lado del centro de un rin. Dinámico significa en movimiento o acción, y equilibrio dinámico se hace con el rin en movimiento o giración. (49)

Dynamometer A device used to brake or absorb mechanical power produced from an engine for testing purposes in a laboratory situation. (15)
Dinamómetro Aparato que sirve para frenar o absorber la potencia mecánica producida por un motor con el propósito de hacer pruebas en una situación de laboratorio. (15)

Eccentric Circles having different center points. (21) (48)
Excéntrico Círculos que tienen diferentes puntos céntricos. (21) (48)

Eccentric journal A journal on a rotary engine crankshaft that is parallel to, but not on the same center point as, the other journals. (14)
Manga de flecha eccéntrica Cojinete del cigüeñal de un motor rotativo que es paralelo a, pero no ubicado en el mismo punto central que, los otros cojinetes. (14)

Eddy currents Small circular currents produced inside a metal core in the armature of a starter motor. Eddy currents produce heat and are reduced by using a laminated core. (35)
Remolino (corriente en remolino) Pequeñas corrientes circulares producidas dentro del centro de la armadura de un motor de arranque. Los remolinos producen calor y se reducen con un centro o núcleo laminado. (35)

EFE Early fuel evaporation. (37)
ECA Evaporación de combustible avanzado (o temprano). (37)

Efficiency A ratio of the amount of energy put into an engine to the amount of energy coming out of the engine. Gas engines are about 28% efficient. A measure of the quality of how well a particular machine works. (13) (15)
Eficiencia Proporción de la cantidad de energía que entra a un motro a la cantidad de energía que sale del motor. Los motores de gasolina tienen una eficiencia en el 28%. Es una medida de la calidad de funciónamiento de una máquina en particular. (13) (15)

EGR Exhaust gas recirculation. (37)
RGE Recirculación de gas de escape. (37)

Electric load detector An electric load sensor used to detect the exact amount of electrical load required by various components of an engine. (34)
Detector de carga eléctrica Un sensor de carga eléctrica que sirve para detectar la cantidad exacta de carga eléctrica requerida por varios componentes de un motor. (34)

Electrical schematic An electrical schematic system layout showing the parts, the wires, and the electrical flow of the circuit. (53)
Esquema eléctrico Un gráfica del sistema eléctrico que demuestra las partes, alambres, y el flujo de corriente del circuito. (53)

Electrical symbol A symbol used to identify an electrical part in an electrical schematic. (53)
Símbolo eléctrico Símbolo que sirve para identificar una parte eléctrica en un esquema eléctrico. (53)

Electricity The flow of electrons from a negative point to a more positive point. (29)
Electricidad El flujo de electrones desde un punto negativo a un punto más positivo. (29)

Electrochemical A process (as in a battery) that uses chemical action to produce and store electricity. (31)
Electroquímico Proceso (como en una batería) que utiliza una acción química para producir y almacenar electricidad. (31)

Electrolyte A solution of acid and water used as the acid in a battery. Many types of electrolyte are used, but the most common is sulfuric acid and water. (31)
Electrólito Solución de ácido y agua utilizada como el ácido en una pila. Se utilizan varios tipos de electrólitos, pero el más común es ácido sulfúrico y agua. (31)

Electromagnetic induction Producing electricity by passing a wire conductor through a magnetic field, causing the wire to cut the lines of force. (29)
Inducción electromagnética Producción de fuerza electromotriz por medio de pasar un conductor alámbrico por un campo magnético, dejando así que el alambre metálico corte las líneas de fuerza. (29)

Electromagnetism Producing magnetism by using electricity flowing through a wire. (29)
Electromagnetismo Producción de magnetismo por medio de electricidad que fluye a lo largo de un alambre metálico. (29)

Electro-motor An electric motor that is used on cruise control systems to control the position of the throttle. (52)
Electromotor Motor eléctrico que se usa en los sistemas de control crucero para controlar la posición del estrangulador. (52)

Electron The negative (–) part of the atom. (29)
Electrón Parte negativa (–) de un átomo. (29)

Electronic brake control module (EBCM) The computer used to control antilock braking systems. (28) (48)
EBCM (Electronic brake control module)—Módulo electrónico de control de frenos Computadora utilizada para controlar sistemas de frenos antibloqueo. (28)(48)

Electronic engine analyzer A testing instrument capable of measuring many readings such as rpm, spark voltages, timing, dwell, vacuum, and exhaust characteristics. (10)
Analizador electronico para el motor Instrumento de medición o prueba capaz de medir varias lecturas como rpm, voltajes de chispa, tiempo, duración, vacío, y características de gases de escape. (10)

Electron theory Electrons will flow from a negative to a positive point. (29)
Teoría de electrón Electrones fluyen de un punto negativo a un punto positivo. (29)

Electronic steering A steering system in which a small electric motor, in conjunction with a steering effort sensor, is used to move the rack-and-pinion components of a steering system. (48)
Manejo electrónico Sistema de manejo o conducción en el que se emplea un pequeño motor eléctrico, en combinación con un sensor del esfuerzo de manejo, para manipular los componentes de piñón y cremallera (barra dentada) del mismo sistema de manejo. (48)

Emission control Various devices placed on the vehicle and engine to reduce exhaust pollution. (36)
Control de emisión Varios aparatos montados en el vehículo y motor para reducir los contaminantes del escape. (36)

Emission standards The federal government has established certain emission and pollutant standards on all automobiles. Because of environmental damage in the past, car manufacturers must now meet strict emission standards. (26)
Estándares de emisión El gobierno federal ha establicedo ciertas normas de emisión y contaminantes para todo automóvil. Dado el daño medioambiental del pasado, los fabriciantes automotrices ahora tienen que cumplir con estándares de emisión muy estrictos. (26)

Energy The ability to do work. (12)
Energía Facultad de producir trabajo. (12)

Engine The power source that propels the vehicle forward or in reverse. The engine can be of several designs, including the standard gasoline piston engine, diesel engine, and rotary engine. (3)
Motor Fuente de poder que impulsa a un vehículo para adelante o para atrás. El motor puede ser de varios diseños, incluyendo el típico motor de gasolina y pistón, diesel, y rotativo. (3)

Engine dynamometer A dynamometer used to measure brake horsepower. (15)
Dinamómetro de motor Dinamómetro que sirve para medir la potencia de caballos frenada. (15)

Environmental Protection Agency (EPA) U.S. government agency that oversees environmental programs and issues. (36)

EPA (Environmental Protection Agency)—Agencia de Control del Medio Ambiente Entidad gubernamental estadounidense que supervisa programas y asuntos relacionados con el medio ambiente. (36)

Ergonomic hazards Conditions that relate to one's physical body or to its motion. (4)
Riesgos ergonómicos Condiciónes que se relaciónan al cuerpo físico o al movimiento. (4)

Ethanol A hydrocarbon produced from the distillation of corn, wheat, and so on, for use in making gasohol. (23)
Etanol Hidrocarburo producido de la destilación de maíz, trigo, *etcétera*, para producir "gasohol." (23)

Ether A highly volatile and flammable, colorless liquid used for starting diesel engines in cold weather. (23)
Eter Líquido altamente volátil y flamable, incoloro, que sirve para arrancar los motores diesel en condiciónes de mucho frío. (23)

Evaporator A component in an air-conditioning system used to heat a refrigerant above its boiling point. (50)
Evaporador (Vaporizador) Componente en un sistema de aire acondiciónado, que sirve para calentar un refrigerante más arriba de su punto de ebullición. (50)

Exhaust gas recirculation (EGR) An emission control that uses an EGR valve to recirculate exhaust gases back into the intake of the engine, thereby cooling the combustion slightly, which in turn reduces nitrogen oxide emissions. (37)
Recirculación de gas del escape (RGE) Control de emisión que usa una válvula de RGE para recircular los gases del escape a la entrada del motor, así enfriado ligeramente la combustión, lo cual por su turno disminuya las emisiones del óxido nitroso. (37)

Expansion valve A component in an air-conditioning system used to create a pressure on one side and reduce the pressure on the other side. (50)
Válvula de expansión Componente de un sistema de aire acondiciónado, que sirve para crear presión en un lado y reducir la presión del otro. (50)

Explosion-proof cabinet Cabinet used in the automotive shop to store gasoline and other flammable liquids. (4)
Gabinete a prueba de explosiones (antiexplosivo) Gabinete usado en talleres automotrices donde se guardan gasolina y otros líquidos inflamables. (4)

External combustion Combustion of air and fuel externally, or outside of the engine. (12)
Combustión externa Combustión del aire y combustible externamente, o sea, fuera del motor. (12)

Extreme pressure (EP) A term used to represent the consistency number of a particular type of grease. (20)
Presión extrema Término que sirve para representar el número de consistencia de un tipo grasa en particular. (20)

Eyewash fountain A water fountain that directs water to the eyes for flushing and cleaning. (4)
Fuente lavaojos Fuente de agua que dirige el agua a los ojos para aclarar y lavarlos. (4)

Fabric-reinforced rubber Rubber that has strands of fabric running through it so as to increase its strength. (7)
Hule reinforzado con fibra Hule que tiene incorporado las hebras de fibra para aumentar su fortaleza. (7)

Fastener Objects such as screws, bolts, splines, and so on, which hold together parts of the automobile. (5)
Sujetador Objectos tales como tornillos, pernos, estrías (chaveteros), *etcétera*, los cuales sujetan juntas las partes del automóvil. (5)

Fault codes Codes, such as numbers, stored in a computer that can be read out on the dashboard of a vehicle or by a scanner; used to identify a specific problem area for the service technician. (30)
Códigos de averías Códigos, tal como los números, almacenados en una computadora (ordenador) que se pueden aparecer en el tablero de instrumentos de un vehículo o en un explorador; que sirven para identificar una área específica de problemas para el técnico de servicio. (30)

Feeler gauge Small, thin metal blades or wires, each having a different thickness; used to measure small clearances such as valve clearances. (9)
Calibre de espesor (lámina calibradora, calibrador de separaciónes) Lengüetas o hijas metálicas degadas, cada una de las cuales tiene un grueso diferente; sirven para medir las separaciónes (huecos o distancias) como el espacio de válvulas. (9)

Field circuit An electrical circuit in a generator or alternator that causes the north and south poles to be energized. The field circuit can be either stationary, as in a generator, or rotating, as in an alternator. (34)
Circuito inductor Circuito eléctrico de un generador o alternador, que causa que los polos norte y sur sean energisados con energía. El circuito inductor puede ser fijo, como en un generador, o rotativo, como en un alternador. (34)

Filler neck restrictor A restriction plate located in the inlet of the fuel tank, used to prevent leaded fuel from being put into the gas tank of cars that require unleaded fuel. (24)
Restrictor del cuello de llenado (orificio de entrada) Placa restrictor situada en la boca de entrada de un tanque de combustible, que sirve para impedir que se le pongá combustible con plomo al tanque de autos que requieren un combustible sin plomo. (24)

Fillets Small, rounded corners machined on the crankshaft for strength. (17)
Tira o faja (listón, cinta) Pequeñas esquinas redon deadas de un cigüeñal maquinadas para darle más fuerza. (17)

First-aid box A kit made of various first-aid bandages, creams, and wraps for treating minor injuries. (4)
Caja de primeros auxilios Paquete de vendajes para primeros auxilios, cremas, y otros utensilios para tratar heridas menores. (4)

Fittings Fittings are used to connect together the ends of various hoses, pipes, and so on. (7)
Accesorios Los accesorios sirven para conectar las extremidades de varios mangueras, tubos, etcétera. (7)

Fixed caliper A disc brake caliper design where the caliper is rigidly mounted and creates braking force through opposing pistons. (45)
Calibrador fijo Diseño de calibrador para freno de disco, en el que el calibrador está montado rigidamente y produce una fuerza de frenada mediante pistones opuestos. (45)

Flare The end of a rigid pipe or fitting that is bent slightly outward, similar to a cone, so that it can be securely fitted to eliminate leakage. (7)
Extremo bocinado La extremidad de un tubo rígido o un accesorio que está ensanchado ligeramente, parecido a un cono, que se puede ajustar seguramente así eliminando las fugas. (7)

Flash point A diesel fuel's ignition point when being exposed to an open flame. (23)
Punto de inflamación Punto de encendido del combustible diesel al exponerse a una llama. (23)

Fleet service Service given to a fleet of vehicles owned by a particular company. (1)
Servicio de flotilla Servicio suministrado a una flotilla de vehículos cuyo dueño es una companía particular. (1)

Floating caliper A moving disc brake caliper with piston(s) on only one side of the rotor. (45)
Calibrador flotante Calibrador de freno de disco flotante con pistón(es) en un solo lado del rotor. (45)

Fluid coupling A fluid connection between the engine and transmission. The greater the speed, the better the fluid coupling between the two. (41)
Acoplamiento fluído Conexión para fluídos (líquidos) entre el motor y la transmisión. Cuanto más rápida la velocidad, tanto mejor sera el acoplamiento fluídico entre los dos. (41)

Fluidity A characteristic of a fluid such as oil, meaning the ease of flow. (20)
Fluidez Característica (atributo) de un fluído como aceite, que significa la facilidad de circulación (paso o movimiento). (20)

Flywheel A heavy circular device placed on the crankshaft. It keeps the crankshaft rotating when there is no power pulse. (12)
Rueda voladora Aparato circular y pesado montado en el cigüeñal. Mantiene al cigüeñal en movimiento rotativo cuando no hay potencia de pulsación. (12)

Forge The process of shaping metal by stamping it into a desired shape. (17)
Forjar Proceso de formar metal por medio de la acción de mazo (máquina de estampar) para estampar la forma deseada. (17)

Forward sensor The sensor used for air bag restraint systems. (53)
Sensor delantero Detector que sirve para sistemas protectores de bolas de aire. (53)

Four-gas analyzer An exhaust gas analyzer able to detect and measure exact amounts of hydrocarbons, carbon monoxide, carbon dioxide, and oxygen. (36)
Analizador de cuatro gases Analizador de gases de escape (emisiones), el cual es capaz de detectar y medir cantidades exactas de hidrocarburos, monóxido de carbono, bióxido de carbono, y oxígeno. (36)

Four-stroke engine An engine that has intake, compression, power, and exhaust strokes within two revolutions of the crankshaft. (13)
Motor de cuatro tiempos Motor que tiene entrada, compresión, encendido, y escape dentro de dos revoluciónes del cigüeñal. (13)

Four-valve head A cylinder head on an engine that has four valves. Two are used for intake and two are used for exhaust. (16)
Tapa (Cabeza) para cuatro válvulas Tapa de cilindro de un motor que tiene cuatro válvulas. Se emplean dos para entrada y otras dos para escape. (16)

Four-wheel alignment Alignment of both the front and rear wheels together. (48)
Alineación de cuatro ruedas Alineación de las dos ruedas delanteras y las dos traseras juntas. (48)

Franchised dealer A dealership that has a contract with the main car manufacturer to sell and service its automobiles. (1)
Agencia de franquicia Agencia automotriz que tiene contrato con la fabrica automotriz principal para vender y dar servicio a sus automóviles. (1)

Friction The resistance to motion between two bodies in contact with each other. (20) (45)
Fricción Resistencia a moción (movimiento) entre dos cuerpos que tiene contacto entre sí. (20) (45)

Frictional horsepower Horsepower lost to friction caused by bearings, road resistance, tire rolling resistance, and so on. (15)
Potencia en cabollos friccional Caballaje perdido a la fricción producida por los cojinetes (baleros), resistencia del camino, resistencia del rodaje de ruedas, *etcétera*. (15)

Friction bearing A type of bearing that uses oil between the rotating shaft and the stationary part. (6)
Cojinete (Balero) de fricción Tipo de cojinete (balero) que utiliza aceite entre el eje rotativo y la parte fija. (6)

Friction disc The part of a clutch system that is clamped between the flywheel and the pressure plate. The friction disc is the output of the clutch system. (39)
Disco de fricción Parte del sistema del embrague que se sujeta entre la rueda voladora y el plato de presión. El disco de fricción es la potencia útil del sistema de embrague. (39)

Friction modifier A chemical added to gear oil to enhance an oil's ability to reduce friction. (43)
Modificador de fricción Químico añadido al aceite para engranes para aumentar la capacidad del aceite de reducir la fricción. (43)

Fuel consumption The amount of fuel that is consumed or used by the vehicle. Four-, six-, and eight-cylinder engines all have different fuel consumption rates. (16)
Consumo de combustible Cantidad de combustible que consume o usa un vehículo. Motores de cuatro, sies, y ocho cilindros tienen diferentes tasas de consumo de combustible. (16)

Fuel injection Injecting fuel into the engine under pressure. (14)
Inyección de combustible El inyectar combustible en un motor bajo una presión. (14)

Fuel injector A carburetor mixes the air and fuel at a ratio of 14.7 to 1. However, today's vehicles are using fuel injectors. These injectors mix the fuel with the air just before the intake valve. The fuel is injected into the port under a low pressure. Computers control the amount of fuel injected. (24)
Inyector de combustible Ne un carburador mezcla el aire y el combustible en una proporción de 14.7 a 1. Sin embargo, los vehiculos modernos utilizan inyectores de combustible. Estos inyectores mezclan el combustible con el aire justamente antes de la válvula de entrada. El combustible es inyectado en el orificio bajo una presión baja. Computadoras (ordenadores) controlan la cantidad de combustible inyectado. (24)

Fuel system The subsystem on the engine used to mix the air and fuel correctly. (3)
Sistema de combustible El subsistema del motor que sirve para mezclar correctamente el aire y el combustible. (3)

Fuel vapors When gasoline heats up, it gives off vapors. The vapors take space and, at times, stop the fuel from flowing. Fuel vapors should be sent back to the fuel tank. (24)
Vapores de combustible Al calentarse la gasolina, emite vapores. Los vapores ocupan cierto espacio y a veces, impiden que pase el combustible. Los vapores de combustible deben ser enviados de vuelta al tanque de combustible. (24)

Fulcrum The support or point of rest of a lever; also called the pivot point. (40)
Eje fijo (Fulcro) Punto de apoyo sobre el cual descansa una palanca; llamado también punto de pivote. (40)

Full-time 4WD A type of four-wheel drive system in which the vehicle is always in 4WD. (44)
4RM continua Tipo de sistema de cuatro ruedas motrices en el cual el vehículo siempre está de 4RM. (44)

Full-wave rectification A process of rectifying a voltage from ac to dc by using diodes. The negative half cycle of the voltage is converted to a positive voltage. All cycles are used in full-wave rectification. (34)
Rectificación de onda completa Proceso de rectificar un voltaje de coriente alterna a corriente continua, utilizando díodos. El medio ciclo negativo del voltaje se convierte a un voltage positivo. Todos los ciclos se emplean en una rectificación de onda completa. (34)

Fuse block A small plastic block in a vehicle where all the electrical fuses are located. On some vehicles, the turn signal flashers and relays are also connected to the fuse block. (53)
Bloque portafusible (Placa de fusibles) Pequeña caja plástica en un vehículo donde se encuentran todos los fusibles. En ciertos vehículos, las luces intermitentes y conmutadores (relés) eléctricos también están conectados al mismo bloque (placa de fusibles). (53)

Fusible link A type of electrical circuit protector made from a special wire that melts when the current is excessive. (53)
Elemento fusible Tipo de protector de circuito eléctrico hecho de un hilo metálico especial que se funde cuando haya una corriente excesiva. (53)

Galling A displacement of metal, usually caused by a lack of lubrication, too loose fit, or capacity overloads on U-joints. (45)
"Galling" Desplazamiento de metal, normalmente producido por falta de lubricación, ajuste demasiado flojo, o sobrecargas más allá de la capacidad de la junta (el balero) universal. (45)

Gasket A rubber, felt, cork, steel, copper, or asbestos material placed between two parts to eliminate leakage of gases, greases, and other fluids. (5)
Empaque (Junta obturadora, Arandela, Guarnición de caucho o de metal) Material de caucho, corcho, acero, cobre, o asbesto colocado entre dos partes para eliminar la salida o escape de gases, lubricantes u otros líquidos. (5)

Gasket coolant A chemical sealant placed on a gasket in an engine to aid in sealing and to position the gasket during installation. (5)
Sellador de empaque Químico sellador puesto en un empaque de un motor para ayudar a sellar y colocarlo durante la instalación. (5)

Gasoline containers Specially OSHA-approved containers used to hold and store gasoline safely. (4)
Contenedores de gasolina Contenedores (cubos, tinajas) aprobados especialmente por "OSHA," los cuales sirven para guardar y almacenar gasolina con seguridad sin riesgos. (4)

Gauge pressure Pressure read on a gauge, at atmospheric pressure, starting with zero as a reference point. (26)
Presión manométrica Presión indicada en un manómetro empezando con cero como punto de referencia. (26)

General engine specifications Specifications used to identify a style and type of engine. (11)
Especificaciónes generales del motor Especificaciónes que sirven para identificar un estilo y clase de motor. (11)

Gram A metric unit of measure used to measure mass. (9)
Gramo Unidad métrica que sirve para medir masas o bultos. (9)

Grounded circuit A condition in an electrical circuit that causes the current to return back to the voltage source, such as the battery, thereby bypassing the intended load or resistor. (29)
Circuito en tierra Condición en un circuito eléctrico que causa que el corriente regresa al fuente del voltaje, tal como la batería, así desviandose de la carga o del resistor deseado. (29)

Hall-effect A device used in electronic and computerized ignition systems that provides a signal for the computer based on a rotating shaft. (33)
Efecto Hall Aparato usado en sistemas de encendido electrónicos y computarizados, el cual suministra una señal para la computadora (ordenador) basada en un eje rotativo. (33)

Harmonics When valve springs are opened and closed rapidly, they may vibrate. Periods of vibration are called harmonics. (18)
Armónicas Cuando se abren y se cierran rápidamente los resortes de válvula, puede que vibren. Se llaman armónicas a los períodos de vibración. (18)

Hazardous waste Any substance or waste that is dangerous to our environment or to the health and well being of persons. Also, any substance that is dangerous to other living plants and organisms. (4) (22)
Desechos tóxicos Cualquier sustancia o desecho que es peligroso al medio ambiente o al salud y bienestar de las personas. Tambien, cualquier sustancia que es peligrosa a las plantas y a los organismos. (4) (22)

Head The part of a bolt used to tighten it down with a wrench. (5)
Cabeza Parte de un perno (tornillo) que sirve para apretarlo con llave. (5)

Header A welded steel pipe used as an exhaust manifold. (27)
Colector (tubo colector) Tubo de acero soldado y usado como un colector múltiple de escape. (27)

Heat dam The narrow groove cut into the top of the piston. It restricts the flow of heat down into the piston. (17)
Contenedor de calor Ranura pequeña cortada en la cabeza de un pistón. Restringe el flujo de calor hacia abajo en el pistón. (17)

Heater core A small radiator-like heat exchanger. Hot coolant from the engine flows through the heater core. Airflow from a fan passes through the fins on the heater core, picking up heat to warm the passenger compartment. (51)
Tubería del calentador Intercambiador de calor como un pequeño radiador. Circula refrigerante caliente por los tubos de intercambio térmico. Pasa un flujo de aire por las aletas de los tubos, elevando así su temperatura para calentar el compartimiento de los pasajeros. (51)

Heat exchanger A device used to transfer heat from one medium to another. The radiator is a liquid-to-air heat exchanger. (22)
Intercambiador de calor Aparato que sirve para transferir calor de un cuerpo a otro. El radiador es un intercambiador de calor de líquido a aire. (22)

Heat stove An enclosed area made of thin sheet metal around the exhaust manifold. It preheats air passing over the exhaust manifold before the air enters the air cleaner snorkel. (37)
Estufa térmica Un área encerrada hecha de hojas delgadas metal de que cubren el colector múltiple de escape antes de que entre el aire al esnórkel de depurador del filtro aire. (37)

Heat transfer The process of moving heat from a warmer object to a colder object. (22)
Transferencia de calor Proceso de pasar calor de un objecto más caliente a uno más frío. (22)

Heat treated A process in which a metal is heated to a high temperature, then is quenched in a cool bath of water, oil, and brine (salt water). This process hardens the metal. (17)
Tratado térmicamente (termotratado) Proceso en que se calienta un metal a una temperatura alta, luego éste se baña en una solución de agua fresca, aceite, y agua salada. Este proceso endurece al metal. (17)

HEI High-energy ignition system; a type of ignition system used on vehicles in which the coil and distributor are physically combined into one unit. (33)
ESE Sistema de encendido de super-energia; tipo de sistema de encendido usado en vehículos en el cual la bobina eléctrica y el distribuidor se combinan físicamente en una sola unidad. (33)

Helicoil A device used to replace a set of damaged threads. (5)
"Helicoil" Aparato que sirve para reemplazar un juego de roscas dañadas. (5)

Hemispherical combustion chamber A type of combustion chamber that is shaped like a half circle. This combustion chamber has the valves on either side with the spark plug in the center. (18)
Cámara de combustión hemisférica Tipo de cámara de combustión formada en semicírculo. Esta cámara tiene válvulas en ambos lados con una bujía de encendido en el medio. (18)

High-test Gasoline that has an octane number near 90 to 95. (23)
Número alto de octano Gasolina que tiene un número de octano cercano a 90–95. (23)

Horn brush/slip ring An electrical contact ring used in the horn circuit. It is located in the steering wheel and is used to maintain electrical contact when the steering wheel is turned. (53)
Escobilla de bocina/anillo conductor Anillo de contacto eléctrico usado en el circuito de la bocina. Se sitúa en el volante de manejo y sirve mantener contacto eléctrico al girarse el volante. (53)

Horsepower A measure of work being done per time unit. One hp equals the work done when 33,000 pounds have been lifted 1 foot in 1 minute. (15)
Potencia en caballos (caballaje) Una medida de trabajo producido por unidad de tiempo. Un HP (caballo) equivale el trabajo producido cuando 33,000 libras (15,000 kg) se han levantado un pie (aproximadamente 33 centímetos) en un minuto. (15)

Housekeeping The type of safety in the shop that keeps floors, walls, and windows clean; lighting proper; containers correct; and tool storage correct. (4)
Administración intera Tipo de seguridad en un taller, la cual mantiene limpios los suelos, paredes, y ventanas; iluminación apropiada; contenedores apropiados y bien arregladas y guardadas las herramientas. (4)

Hybrid A vehicle that has two types of power sources such as an engine and a set of batteries to propel the vehicle. A sophisticated electronic controller determines exactly which or both of the power sources are to be used at specific loads and speed. (3) (14)
Híbrido Un vehículo con dos tipos de fuente de energía, tales como un motor y un conjunto de baterías para la propulsión del vehículo. Un sofisticado controlador electrónico determina exactamente cuál de las fuentes de energía se usará para cargas y velocidades específicas, o si se usarán las dos. (3)(14)

Hydraulic system A brake system in which brakes are operated and controlled by hydraulic fluid under pressure. (45)
Sistema hidráulico Sistema de frenas en que un líquido bajo presión hidráulica maneja y controla los frenos. (45)

Hydrocarbons A term used to describe the chemical combinations of hydrogen and carbon. A type of automotive pollution produced by incomplete combustion. Hydrocarbons are considered partly burned hydrogen and carbon molecules resulting from combustion. (23) (36)
Hidrocarburos Término que sirve para describir las combinaciones químicas de hidrógeno y carbón. Tipo de contaminación producida por combustión incompleta. Se consideran los hidrocarburos como moléculas de hidrógeno y carbón que resultan de una combustión incompleta. (23) (36)

Hydrometer An instrument used to measure the specific gravity of battery acid. (31)
Hidrómetro Instrumento que mide el peso específico (densidad) del ácido de batería. (31)

Hydrophilic Attraction to water. Some tires have a hydrophilic tread, which means the tread is attracted to water. (49)
Hidrófilo Atracción de agua. Algunas llantas tienen una rodadura (cara) hidrófila, que significa que la llanta se atrae al agua. (49)

Hypoid gear oil A thick oil used in differentials. (43)
Aceite hipoido para ruedas dentadas Aceite espeso usado en diferenciales. (43)

Idler arm A steering linkage component, fastened to the car frame, which supports the right end of the center link. (48)
Eje loco Componente de articulación de manejo, montado en el chasis automotriz, y que soporta el extremo derecho del eslabón central. (48)

Idler gear A third gear placed in a gear train, usually used to reverse the direction of rotation. (40)
Engranaje intermedio (loco) Tercera rueda dentada colocada en el tren de engranajes, usado normalmente para cambiar la rotación en dirección opuesta. (40)

Ignition system The subsystem on the engine used to ignite the air and fuel mixture efficiently. (3)
Sistema de encendido (ignición) Subsistema del motor que sirve para encender eficientemente la mezcla de aire y combustible. (3)

I-head A style of valve arrangement in an engine. I-head refers to the valves being placed directly above the piston in the cylinder head. (16)
Cabeza I Cierto arreglo de válvulas de un motor. Cabeza I se refiere a las válvulas que se encuentran justamente encima de la cabeza de cilindros. (16)

I/M 240 testing A specific type of vehicle emission test used to measure the emissions of the engine over a specific set of driving conditions. (36)
Prueba de I/M 240 Tipo de prueba específico de emisión que mide las emisiones del motor a través de un surtido específico de condiciónes de marcha. (36)

Impact socket Socket used for heavy-duty or high-torque applications. This type of socket is used with an impact wrench. (8)
Cubo (Tintero) de impacto Cubo que sirve para aplicaciones de alta torsión. Se emplea este tipo de cubo con una pistola de impacto. (8)

Impact wrench An air- or electric-operated power wrench that uses impacts during rotation to loosen or tighten bolts and nuts. (10)
Pistola de impacto (de choque, de golpe) Pistola que funciona a base del poder de aire o electricidad y que emplea impactos (golpes) durante la rotación para soltar o ajustar más fuerte los pernos y tuercas. (10)

Inclined surface A slope or slanted surface. An inclined surface is used in valve rotators. (18)
Superficie inclinada Superficie cuesta arriba o abajo. Se emplea una superficie inclinada en los rotadores de válvulas. (18)

Independent publishers Publishers that provide service information on automobiles. Examples are Motor manuals, Mitchell manuals, Chilton auto repair manual, and so on. (11)
Casa editorial independiente Casa editorial que suministra información de servicio acerca de automóviles. Unos ejemplos son manuales de

Motor, manuales Mitchell, manual de reparaciónes automovilísticas Chilton, *etcétera*. (11)

Independent rear suspension A rear suspension system composed of trailing arms (like control arms) and MacPherson struts or torsion bars. The system allows the rear wheels to move independently of each other. (47)
Suspensión trasera independiente Sistema de suspensión trasera compuesto de brazos de tracción (como brazos de control) y puntales de poste (codales, postes) MacPherson o barras de torsión. El sistema permite que las ruedas traseras se muevan independientemente, la una de la otra. (47)

Independent service Service provided by independent garages on all types and makes of vehicles. (1)
Servicio independiente Servicio suministrado por talleres independientes para todo tipo y marca de vehículos. (1)

Indicated horsepower Theoretical horsepower calculated by the manufacturer of the engine. (15)
Caballaje indicado Caballaje teórico calculado por el fabricante del motor. (15)

Inertia Objects in motion tend to remain in motion. Objects at rest tend to remain at rest. Inertia is the force keeping these objects in motion or at rest. (35)
Inercia Objetos en movimiento tienden a quedarse en movimiento. Objetos parados tienden a quedarse parados. Inercia es la fuerza que maintienen a los objetos en movimiento o parados. (35)

Inertia drive A drive system on a starter motor using inertia to turn a screw sleeve on a spline. (35)
Impulso inercial Sistema de propulsión de un motor de arranque que utiliza inercia para hacer girar una manguita de tornillo en un chavetero (estría o acanaladura). (35)

Inertia switch A switch used on fuel systems that opens the circuit whenever there is a sudden impact from a vehicle or crash. (24)
Interruptor inercial Interruptor usado en sistemas de combustible, el cual abre el circuito siempre que haya algún impacto de otro vehículo o choque. (24)

Infrared exhaust analyzer An instrument able to detect and measure hydrocarbons, carbon monoxide, carbon dioxide, and oxygen levels. (36)
Analizador infrarrojo de escape Instrumento capaz de detectar y medir los niveles de hidrocarburos, monóxido de carbono bióxido de carbono, y oxígeno. (36)

Inhibitor Any substance that slows down or prevents chemical reactions such as corrosion or oxidation. (20)
Retardador (Inhibidor) Cualquier sustancia que retarde o impida reacciones químicas tales como corrosión u oxidación. (20)

In-line Cylinders in an engine that are in one line or row, such as an in-line 4- or 6-cylinder engine. The cylinders are generally vertical as well. (16)
En línea (Alineado, En serie) Cilindros de un motro que se encuentran en una fila, tal como un motor de cuatro en línea o seis en línea. Los cilindros generalemente son verticales también son verticales. (16)

Insert guides Valve guides that are small cast cylinders pressed into the cylinder head. (18)
Guías interadas Guías de las válvulas son pequeños cilindros fundidos que son metidos a presión en la cabeza de los cilindros. (18)

Insulator A material with five or more electrons in the valence ring. (29)
Aislador Material con cinco electrones o más en la banda de valencia. (29)

Intake manifold vacuum The vacuum produced inside the intake manifold between the valve and the throttle plate. When the throttle plate is closed, there is high intake manifold vacuum. When it is open, there is low intake manifold vacuum. (26)
Vacío del colector múltiple de admisión Vacío producido dentro del colector de admisión entre la válvula y la mariposa (placa) de aceleración. Cuando está cerrada la mariposa, hay mucho alto vacío en el colector de admisión. Cuando está abierta la mariposa, hay poco vacío en el colector. (26)

Integral guides Valve guides that are manufactured and machined as part of the cylinder head. (18)
Guias integrales Guías de la válvula que se fabrican y maquilan como parte de la cabeza del cilindro. (18)

Integrated circuit A circuit board with many semiconductors forming a complex circuit. (30)
Circuito integrado (sólido) Microestructura integrada con muchos semiconductores que forman un circuito complejo. (30)

Intercooler A heat exchanger used on turbocharged engines; used to cool the intake air charge. (28)
Interenfriador (Radiador intermedio) Cambiador de calor usado en motores turbocargados; utilizado para enfriar la carga de aire que entra. (28)

Interference angle When the valve is ground at 45 degrees and the seat is ground at 44 degrees, the two angles will interfere with each other. Interference angles help valves seat faster. (18)
Angulo de interferencia Cuando se recorta (tornear) por máquina una válvula a 45 grados de ángulo y se recorta (tornear) el asiento a un ángulo de 44 grados, los dos ángulos se interfieren uno al otro. Los ángulos de interferencia ayudan a que las válvulas se asienten más rápidamente. (18)

Interference fit A fit between two parts that must be pressed together. Some piston pins have an interference fit with the piston. (17)
Conformidad (Ajuste) de interferencia El ajuste entre dos partes (o cuerpos) que tienen que presionar una contra la otra. Algunos ejes de pie de biela tienen una conformidad interferente con el pistón. (17)

Intermittent combustion Combustion of air and fuel that starts and stops. (12)
Combustión intermitente Combustión del aire y del combustible que comienza y se para. (12)

Internal combustion Combustion of air and fuel inside the engine. (12)
Combustión interna Combustión del aire y del combustible dentro del motor. (12)

Jack stand A stand used to support the vehicle when working under the car. Always support the car with such stands before working on the underside of the vehicle. (4)
Soporte de gato Soporte que sirve para sostener un vehículo al trabajar uno debajo de él. Siempre se debe soportar al vehículo con tales soportes antes de meterse debajo del vehículo. (4)

Jumper wire A wire used when troubleshooting an electrical circuit for bypassing or shorting out a specific component. (53)
Cables para diagnostico Alambre usado en la investigación de fallas o averías en un circuito eléctrico al desviar o cortar un componente específico. (53)

Keeper A small, circular, tapered metal piece that keeps the valve retainer attached to the valve. (18)
Retenedor (Fijador) Pequeña pieza metálica circular de forma cóncia que mantiene conectado el retén de válvula a la válvula. (18)

Keyway A machined slot on a shaft that holds a metal piece called the key. A slot and key used together attach a pulley or hub to a rotating shaft. (5) (19)

Ranura (cuña) de posición Ranura recortada en el eje que mantiene fija una pieza metálica llamada la clavija. Una ranura y clavija en combinación sirven para juntar una polea o núcleo de rodete a un eje rotativo. (5) (19)

Kinetic energy Energy in motion. (45)

Energía dinámica Energía en movimiento. (45)

Labyrinth seal A type of seal that uses centrifugal forces to eliminate leakage from a rotating shaft. The seal does not touch the rotating part. (5)

Sello de laberinto Tipo de sello que emplea fuerzas centrífugas para eliminar escapes o fugas de un eje rotativo. El sello no toca a la parte rotativa. (5)

Lag The time it takes for a turbocharger to increase the engine power after the throttle has been depressed. (28)

Retraso Tiempo que toma un turboalimentador para aumentar la potencia del motor después de deprimirse el pedal que opera el obturador de la gasolina. (28)

Laminated Describes a series of plates of metal placed together and used as a core in a magnetic circuit. (34)

Laminado Esto describe una serie de placas metálicos juntas y usadas como núcleo en un circuito magnético. (34)

Laminations Thin layers of soft metal used as the core for a magnetic field. (35)

Laminaciónes Capas delgadas de metal blando, usados como núcleo en un campo magnético. (35)

Latent heat Heat that is hidden or not readily observable. (50)

Calor latente Calor que está escondido o no se observa fácilmente. (50)

Lean mixture Too much air and not enough fuel for combustion. (13)

Mezcla pobre Demasiado aire e insuficiente combustible para hacer combustión. (13)

Lever A bar supported by a pivot point with a force applied to one end used to move a force exerted on the other. Force times distance moved on one side will always equal the force times distance moved on the other. Typically a lever is used to gain a force with a loss in distance. (40)

Palanca Barra sostenida por un punto de pivote con una fuerza aplicada en un extremo, la cual mueve una fuerza ejercida en el otro extremo. La fuerza por distancia movida por el otro lado. Típicamente se utiliza una palanca para ganar una fuerza a costo (expendas) de una distancia. (40)

L-head A valve arrangement that has the valves located in the block and not in the head. Engines that have L-head designs are commonly called flat-head engines. (16)

Cabeza L Arreglo de vávulas que coloca las válvulas en el bloque y no en la cabeza. Los motores que tienen un diseño de cabeza L se llaman comúnmente motores de cabeza plana. (16)

Lifter The small component that rides on the camshaft. The camshaft lobe lifts the lifter to aid in opening and closing the valves. (16)

Levantadores de válvulas Pequeño componente que se monta en el arbol de levas. El lóbulo del arbol de levas eleva el levantador para ayudar a abrir y cerrar las válvulas. (16)

Limited slip differential A differential that uses clutches or cones to lock up the side gears to the differential case. These components eliminate having one wheel spin faster, as on ice. (43)

Diferencial de desliz limitado Diferencial que utiliza embragues o conos para detener el engranaje lateral conectado a la caja del diferencial. Estos componentes eliminan que una rueda gire más rápido, como sobre hielo. (43)

Line contact The contact made between the cylinder and the torsional rings, usually on one side of the ring. The contact made between the valve and the valve seat. When an interference angle is used, only a small line contact is produced. (17) (18)

Contacto lineal Contacto hecho entre el cilindro y los anillos de torsión, normalmente en un lado del anillo. Contacto hecho entre la válvula y el asiento de la vávula. Al utilizarse un ángulo interferente, sólo se produce un contacto lineal pequeño. (17) (18)

Lines of force Invisible forces around a magnet. (29)

Líneas de fuerza Fuerza imaginaria o invisible alrededor de un campo magnético. (29)

Liquid-cooled Removing heat from the engine by circulating liquid coolant throughout the internal parts of the engine. (22)

Enfriado por líquido Reducción o eliminación de calor de un motor por un refrigerante circulante por todas las partes internas de un motor. (22)

Liter A unit in the metric system to measure volume. Equals 100 cubic centimeters. (9)

Litro Unidad en el sistema métrico que mide volúmen. Equivale a 100 centímetros cúbicos. (9)

Load The actual slowing down of the engine output shaft because of a brake applied to the shaft. The load could be that of driving up a hill. The load to the engine is increased in this case. (15)

Elemento disipador de potencia (carga) La reducción en velocidad del eje de salida de potencia del motor debido a la aplicación de un freno al eje. La carga podría ser la de ir cuesta arriba. En tal caso se aumenta al carga al motor. (15)

Load range The amount of load the tire is capable of supporting safely. The value is molded into the tire sidewall and is stated in pounds and/or kilograms. (49)

Margen de carga Cantidad de carga que la llanta puede soportar con seguridad. La cifra está estampada en el costado de la llanta y está expresada en libras o en kilogramos. (49)

Lobe The part of the camshaft that raises the lifter. (19)

Lóbulo Parte de un arbol de levas que eleva el levantador. (19)

Locking hubs Hubs that are located on the front wheels of some four-wheel drive vehicles, used to lock up the wheels with the spinning axles that are being driven by the four-wheel drive system. (44)

Cubos antideslizamientos Cubos ubicados en las ruedas delanteras de algunos vehículos de cuatro ruedas motrices, que enclavan las ruedas con los ejes rotativos arrastrados por el sistema de cuatro ruedas motrices. (44)

Lockup clutch A type of clutch in which a lockup system is able to lock up the torque converter turbine with the engine, eliminating slippage between the two. (41)

Embrague inmovilizador Tipo de embrague en el que un sistema de inmovilización puede inmovilizar la turbina del convertidor de torsión con el motor, eliminando así cualquier patinaje entre los dos. (41)

Loop scavenging A method used to remove the exhaust gases from the cylinder in a two-stroke engine. (14)

Vuelta de gases Método que sirve para eliminar los gases de escape del cilindro en un motor de dos tiempos. (14)

Lubrication system The subsystem on the engine that is used to keep all moving components lubricated. (3)

Sistema de lubricación Subsistema de un motor que sirve para mantener lubricado todo componente movible. (3)

MacPherson strut suspension An independent suspension system consisting of a coil spring, shock absorber, and upper bearing. (47)
Sistema de postes (codales, soportes) MacPherson Sistema independiente de suspensión que consiste en un resorte espiral, amortiguador contra choques, y cojinete (balero) superior. (47)

Magnetic pickup A sensor that is able to induce a charge of electricity as a magnet passes by. This sensor is usually used to measure crankshaft rotation or position. (26)
Detector magnético Sensor que puede producir una carga de electricidad al pasarse lateralmente un imán. Normalmente se utiliza este sensor para medir la rotación o posición de un cigüeñal. (26)

Magnetic pickup sensor A sensor that uses magnetism to generate a small electrical signal every time a rotating shaft turns past a specific point. This type of sensor is often used to sense or signal the rotational speed of a shaft or gear. (30) (38)
Sensor magnético captador Sensor que usa el magnetismo para generar una señal pequeña cada vez que una flecha rotativa gira por un punto específico. Este tipo de sensor suele usarse para percibir o señalar la velocidad giratoria de una flecha o un engranaje. (30) (38)

Mag wheel A type of wheel that uses magnesium metal for the rim, also any wheel that uses special materials and design. (49)
Rin de magnesio Tipo de rin que utiliza metal de magnesio en el borde. (49)

Main bearing clearance The clearance between the main bearing journal and the main bearings. (17)
Margen (holgura) del cojinete (balero) principal Espacio entre la manga del cojinete (balero) principal y el cojinete (balero) principal). (17)

Major thrust The thrust forces applied to the piston on the power stroke. (17)
Empuje principal Fuerzas de empuje que se aplican en el piston en la carrera de fuerza. (17)

Manual shifting mode One of several modes used on an electronic transmission in which the driver is able to shift manually from low gear to a higher gear, rather than having it done automatically. (41)
Modo de cambio de velocidad manual Uno de varios modos usado en una transmisión en el cual el conductor puede cambiar de una baja velocidad a una más alta manualmente, en vez de que ocurre automaticamente. (41)

Manual shift valve A valve in the automatic transmission that is controlled by the position of the gearshift lever. (43)
Válvula de cambio manual Valvula de una transmisión automática que es controlada por la posición de la palanca de cambio de velocidad. (43)

Manufacturer's service manual Service information and technical data supplied by the automotive manufacturer. (11)
Manual de servicio de la fabrica Información de servicio y datos técnicos suministrados por la fabrica automotriz. (11)

Master cylinder The main unit for displacing brake fluid under pressure in a hydraulic brake system. The master cylinder can be either single or dual design. (45)
Cilindro maestro Unidad principal que desplaza al líquido de freno bajo presión en un sistema hidráulico de freno. El cilindro maestro puede ser de un diseño sencillo o doble. (45)

Mechanical advantage A linkage or lever able to gain either distance or force. The rocker arm uses a mechanical advantage to gain distance to open the valve farther. (19)
Ventaja mecánica Articulación (sistema de enlace) o palanca que puede ganar distancia o fuerza. El balancín (palanca oscilante) utiliza una ventaja mecánica para ganar distancia para abrir una válvula aún más. (19)

Mechanical efficiency A measure of the mechanical operation of a machine. (15)
Eficiencia mecánica Medida de la operación mecánica de una máquina. (15)

Medium Usually referred to as a certain type of material acting as the environment. In this case, the area where dirt is captured in the fuel filter.
Medio Normalmente se le refiere al medio como cierto tipo de material en calidad del medioambiente. En este caso, el área donde se capturan partículas de tierra en el filtro para combustible.

Meter A device used to measure or control the flow of a liquid such as fuel. To control the amount of fuel passing into an injector, fuel is metered to obtain the correct measured quantity. (26)
Instrumento indicador Instrumento de medición que sirve para medir o controlar el flujo de un líquido tal como combustible. Para controlar la cantidad de combustible que entra al inyector, se mide el combustible para obtener la correcta cantidad medida. (26)

Methanol An alcohol product made from coal, wood, grain, or other biomass fuels used to increase oxygen in certain fuel. (23)
Metanol Un producto de alcohol hecho con carbón, madera, granos u otro combustible biomasa utilizado para aumentar el oxígeno en determinado combustible. (23)

Metric system A system of measurement based on the meter. All other units of length and volume are derived from the meter. These include centimeter, kilometer, millimeter, liter, cubic centimeter, and so on. (9)
Sistema métrico Sistema de medición basado en el metro. Todas las demás unidades de longitud y volumen se derivan del metro. Estos incluyen el centímetro, kilómetro, milímetro, litro, centímetro cúbico, *etcétera*. (9)

Metric thread A metric thread is measured by indicating the number of millimeters between each thread. (5)
Rosca métrica Se mide un rosca métrica indicando el número de milímetros entre cada rosca saliente. (5)

Micrometer A measuring tool used to accurately measure length to 0.001 of an inch. A distance measurement used to indicate the size of holes in a filter; 1 micrometer = 0.000039 inch; sometimes called a micron. (9) (24)
Micrómetro Instrumento de medición que sirve para medir con precisión hasta 0.001 de una pulgada de largo. Una medida de distancia utilizada para indicar el tamaño de agujeros en un filtro; 1 micrómetro = 0.000039 de pulgada; a veces se llama un micrón (micra). (9) (24)

Microprocessor A series of circuits using semiconductors and integrated circuits for computer applications. Microprocessors are capable of input, storage, and feeding out information to other circuits and systems on the automobile. (30)
Microprocesador Serie de circuitos que utilizan semiconductores y circuitos integrados para aplicaciones de computadoras (ordenadores). Los microprocesadores también son capaces de recibir información, almacenaje y envío de información a otros circuitos y sistemas en automóviles. (30)

Miles-per-gallon-per-person A term used to indicate how much energy is being used to transport people. It is calculated by multiplying the average miles per gallon of a vehicle, times the average number of people in the vehicle. (1)

Millas por galón por persona Expresión utilizada para indicar cuánta energía se utiliza para el transporte de personas. Se calcula multiplicando el promedio de millas por galón de un vehículo por el número promedio de personas en el vehículo. (1)

Millisecond One one-thousandth of a second, or 1×10^{-3} second, or 0.001 second. (30)
Milisegundo Una milésima de un segundo, o 1×10^{-3} de un segundo, o un 0.001 de un segundo. (30)

Minor thrust The thrust forces applied to the piston on the compression stroke. (17)
Empuje menor Fuerzas de empuje aplicados al piston en la carrera de compresión. (17)

Monolith A single body shaped like a pillar or long tubular structure used in some catalytic converters. (37)
Monolito Cuerpo sencillo en forma de una columna o estructura larga y tubular, usado como catalizador en un convertidor catalítico. (37)

Motor Any power device that imparts motion to an object. An engine is a motor because it causes the vehicle to move forward. An electric motor also causes motion to occur, such as the motor that moves the seat forward and backward. (3)
Motor Cualquier dispositivo de fuerza que imparta movimiento a un objeto. El motor hace que el vehículo se mueva hacia delante. El motor electrónico también provoca el movimiento, tal como el motor que mueve el asiento hacia delante y hacia atrás. (3)

MSDS An abbreviation for Material Safety Data Sheet, which contains hazardous ingredients, physical and chemical properties, fire and explosion data, reactivity data, health hazards, control measures, and precautions for using various chemicals in the automotive shop. (4)
HDSM Abreviadura de una Hoja de Datos sobre la Seguridad de Materiales, la cual contiene los ingredientes peligrosos, propiedades físicas y quimicas, datos sobre incendios y explosiones, datos sobre reactividad, riesgos contra la salud, y medidas de control, y precauciones para manipular varios químicos en un taller automotriz. (4)

Multiple-disc clutch A hydraulic clutch used in the automatic transmission. The clutch uses a series of discs, both friction and smooth metal, to lock up to rotating shafts. (41)
Embrague de discos múltiples Embrague hidráulico usado en una transmisión automática. El embrague utiliza una serie de discos, de ambos tipos: fricción y liso, para inmovilizar los ejes rotativos. (41)

National Lubricating Grease Institute (NLGI) One of several organizations that rate the consistency of greases. (20)
Instituto de Grasa Lubricante Nacional (IGLN) Una de varias organizaciones que evaluan la constancia (o consistencia) de grasas. (20)

Naturally aspirated An engine that uses the atmospheric pressure to force the air into the cylinders. (28)
Aspirado naturalmente Motor que utiliza la presión atmosférica para forzar el aire en los cilindros. (28)

NC National Coarse, an indicator of the number of threads per inch on a bolt. (5)
Grueso nacional Indicador del número de roscas por pulgada en un perno o tornillo. (5)

Needle valve A small valve used in the center of the fuel injector nozzle. The valve is shaped much like a thick needle and controls the opening and closing of ports for fuel injection. (26)
Válvula de aguja Pequeña válvula usada en el centro de la boquilla (pico o tober) del inyector del combustible. La forma de la válvula se parece a una aguja gruesa y controla la apertura y el cierre de los orificios de inyección del combustible. (26)

NF National Fine, an indicator of the number of threads per inch on a bolt. (5)
FN Fina Nacional, o indicador del número de roscas por pulgada de un perno o tornillo. (5)

Nitrogen oxide A type of automotive pollution produced when internal combustion temperatures reach 2,200–2,500 degrees Fahrenheit. (23) (36)
Oxido de nitrógeno Tipo de contaminante automotriz producido cuando las temperaturas de combustión interna alcanzan a 2.200–2.500 grados F. (23) (36)

OASIS A computerized information system for service technicians; stands for On-line Automotive Service Information System. (11)
"OASIS" Sistema de información computarizado para técnicos de servicio; significa Sistema de Información de Servicio en Línea Automotriz. (11)

Octane number A number used to identify the resistance to burning of gasoline. (23)
Número de octano Número que sirve para identificar la resistencia de combustión de gasolina. (23)

Off-road A term used to indicate the type of driving conditions generally encountered in rough and/or muddy terrain. (44)
Fuera del camino Término que se usa para indicar los tipos de condiciónes de conducir en que se encuentra terreno disparejo o lodoso. (44)

Ohm's law Voltage equals amperage times resistance ($E = I \times R$). (29)
Ley de Ohm Voltaje equivale a amperaje multiplicado por la resistencia ($E = I \times R$). (29)

Oil gallery The main or center passageway inside an engine block through which oil flows to feed all other components in the engine that need oil. (21)
Canalización de aceite Linea principal dentro del bloque motor por la cual fluye el aceite para suministrar todos los componentes del motor que necesitan el aceite. (21)

Oil relief A small machined area on the side of the lifter that allows oil to circle around the body of the lifter. (19)
Relieve de aceite Pequeña área en la parte lateral del (levantador de válvulas) alzaválvulas que permite que el aceite circule alrededor del cuerpo del alzaválvulas. (19)

On the fly A term applied to four-wheel drive systems that means the operator can shift from two-wheel drive to four-wheel drive while the vehicle is traveling (usually at any speed below 50 mph) down the road. (44)
En movimiento Termino aplicado a los sistemas de cuatro ruedas motrices que significa que el conductor puede cambiar de dos ruedas motrices a cuatro ruedas motrices mientras que el vehículo esté en movimiento (por lo general en cualquier elocidad menos de 50 mph) en el camino. (44)

Open circuit An electrical circuit that has a break or opening in a wire, producing infinite resistance. (29)
Circuito abierto Circuito eléctrico que tiene una apertura en un alambre, produciendo una resistencia infinita. (29)

Open loop A computer condition in which the air-fuel ratio is being controlled on the basis of a set of preprogrammed conditions. (25) (38)

Circuito abierto Condición en una computadora (ordenador) en la que la proporción del aire al combustible está controlada a base de unas condiciónes pre-programadas. (25) (38)

Operational specifications Specifications used to show how the vehicle operates, such as acceleration, tire inflation, and other general information. (11)
Especificaciónes operaciónales Especificaciones que sirven para indicar como funciona un vehículo, tal como aceleración, presión de llantas, y otra información en general. (11)

Opposed cylinder An engine that has two rows of pistons that are 180 degrees from each other. (16)
Cilindros opuestos Motor que tiene dos filas de pistones que están a 180 grados una del otro. (16)

Organic Pertaining to chemicals that are derived from living things. (20)
Orgánico Relacionado a químicos que se derivan de elementos vivientes. (20)

Organic material Another term for hydrocarbon pollution. (36)
Material orgánico Otro término que sirve para indicar contaminates de hidrocarburos. (36)

O-ring A type of static seal used to eliminate leakage between two stationary parts as fluid passes through them. (5)
Empaque O-ring Tipo de sello estático que sirve para eliminar escapes o goteras (fugas) entre dos partes fijas al pasar por ellas cualquier líquido. (5)

Oscillation Fluctuation or variation in motion or in electrical current. When the vehicle hits a bump in the road, the body oscillates up and down. Shocks are used to reduce oscillations. (47)
Oscilación Fluctuación o variación en movimiento o de corrientes eléctricas. Cuando el vehículo pasa por un bache o tope en el camino, el armázon de coche oscila en brincos. Se emplean amortiguadores para reducir los brincos. (47)

OSHA Occupational Safety and Health Act of 1970. This act provides safety regulations and rules for industry. (4)
"OSHA" Acta de Seguridad Industrial y Salud (ASIS) de 1970. Esta acta suministra los reglamentos y regals de seguridad para toda la industria. (4)

Overdrive A gear system used on a transmission to reduce the speed of the input. Normally, the highest gear ratio in a standard transmission is 1 to 1. Overdrive systems have a higher gear ratio of approximately 0.8 to 1. (40) (41)
Sobremarcha Sistema de engranaje usado en una transmisión para reducir la velocidad de la entrada de potencia. Normalmente la proporción de engranaje más alta en una transmisión estandard es 1 a 1. Sistemas de sobremarcha tienen una proporción de engranaje más alta de aproximadamente 0.8 a 1. (40) (41)

Overhaul and maintenance specifications Specifications used to service vehicle components such as pistons, crankshafts, rings, bearings, and so on. (11)
Especificaciones de reparaciones y mantenimiento Especificaciones que sirven para dar servicio a los componentes de vehículo tales como pistones, cigüeñales, anillos, cojinetes (baleros), *etcétera.* (11)

Overhead camshaft A camshaft located directly on top of the valves, used on I-head designs. (16)
Eje de leva superior (arriba) Eje de leva colocado directamente encima de las válvulas, el cual se emplea en motores con diseño de cabeza I. (16)

Overrunning clutch drive A type of drive on a starter motor that uses a series of rollers that lock up to cause the pinion gear to rotate. (35)
Sistema de mando de un embrague de rueda libre (de sobremarcha) Tipo de mando de un motor de arranque que emplea una serie de aparatos enrolladores que se inmovilizan para que gire la rueda dentada de piñón. (35)

Oxidation The process of combining oil molecules with oxygen. (20)
Oxidación Proceso de combinar moléculas de aceite y oxígeno. (20)

Oxidation inhibitors Additives to fuels and lubricants that reduce their reaction with oxygen. (20) (23)
Inhibidor de corrosión Aditivos en los combustibles y lubricantes que disminuyen su reacción con el oxígeno. (20) (23)

Oxides Chemicals that form when certain pollutants combine with oxygen. (36)
Oxidos Químicos que se forman cuando ciertos contaminantes se combinan con oxígeno. (36)

Ozone layer A transparent shield of gases surrounding the Earth. (50)
Capa de ozono Escudo protector de gases alrededor de la Tierra. (50)

Parallel circuit In this type of circuit, there is more than one path for the current to follow. (29)
Circuito paralelo En este tipo de circuito, hay más de un solo camino en el que la corriente puede circular. (29)

Park/neutral switch A switch that is connected to the mechanical motion of the shift lever so the computer can determine when the vehicle is shifted from neutral to a specific gear. (25)
Interruptor de park/neutral Interruptor que se conecta al movimiento mecánico de la palanca de cambiar velocidades para que la computadora (ordenador) puede determinar cuando el vehículo ha cambiado del neutro a una velocidad específica. (25)

Particulates A form of solid air pollution such as microscopic solid or liquid matter that floats in the air. (36)
Partículas Forma de contaminación de aire sólido tal como material microscópico que sólido o líquido flota en el aire. (36)

Parts distribution All service shops must have parts available. Parts distribution shops are retail businesses that sell parts for the automobile. (1)
Distribución de partes (piezas o refacciónes) Todo taller de servicio tiene que tener partes disponibles. Tiendas de distribución de partes son negocios de detallista que venden repuestos para el automóvil. (1)

Parts manager The person responsible for making sure the customer's parts are immediately available to the service technician. (2)
Gerente de partes y refacciones Persona encargada para asegurar de que las partes que pide el cliente estén al acceso inmediato del técnico (mecánico) de servicio. (2)

Parts specialist A person who sells automotive engine and vehicle parts. (2)
Especialista en partes y refacciones Persona que vende partes para el motor de automóvil y del vehículo, en general. (2)

Passive restraint system A supplemental safety system to safety belts using air bags that inflate immediately upon impact. (53)
Sistema de restricción pasiva Sistema suplemental de seguridad para los cinturones de seguridad utilizando cojines de aire que se inflan inmediatamente al chocar el auto. (53)

Passive seal A seal that has no extra springs or tension devices to help make the seal. O-ring seals on valves are called passive valve seals. (18)

Junta (Cierre) hermética pasiva Cierre que no tiene resortes extras o aparatos de tensión para ayudar a hacer la junta hermética. Junta tórcia (empaquetadura en O) en las válvulas se llaman cierres de válvula pasivos. (18)

Pendulum A swinging device with a weight on one end used to control the movement of a mechanism. Pendulums are used to lock safety belts during impact. (53)

Péndulo Aparato oscilante con un peso en un extremo que sirve para controlar el movimiento del aparato. Se utilizan los péndulos para cerrar (o inmovilizar) los cituromes de seguridad al momento de impacto. (53)

Performance chart A chart that has been produced from a dynamometer. It shows the horsepower, torque, and fuel consumption of an engine at various rpm. (15)

Esquema de funciónamiento (ejecución) Esquema que se ha hecho de una dinamométrico. Indica el caballaje (potencia en caballos), el torque, y consumo de combustible de un motor a varias rpm. (15)

Performance mode One of several modes used on an electronic transmission in which the shifting patterns are adjusted for rapid acceleration, rather than for high speed. (41)

Modo de ejecución Uno de varios modos usados en una transmisión automática en el cual los patrones de cambio de velocidad se ajustan para una aceleración rápida, en vez de para una alta velocidad. (41)

Periphery The external boundary of the torque converter. (41)

Periferia (Circunferencia) Borde externo de un convertidor de torque. (41)

Photochemical smog A type of smog produced when hydrocarbons and nitrogen oxides combine with sunlight. (1)

Contaminación de aire fotoquímico Tipo de contaminante de aire producido cuando se combinan los hidrocarburos y los óxidos de nitrógeno. (1)

Photodiode A type of light-sensitive semiconductor. In some cruise control systems, a photodiode uses light to create an electrical signal based upon the speed of a rotating shaft. (52)

Fotodiodo Tipo de semiconductor sensitivo a la luz. En algunos sistemas de control crucero, un fotodiodo usa la luz para crear una señal eléctrica basada en la velocidad de una flecha giratoria. (52)

Photoelectric A photoelectric device is one that uses light that is then converted to electricity. A photoelectric cell is used to take sunlight and convert it into electricity that is stored in batteries. (10)

Fotoeléctrico Un aparato fotoeléctrico es uno que usa la luz que luego se convierte en electricidad. Una célula fotoeléctrica se usa para convertir la luz del sol a la electricidad que se almacena en las baterías. (10)

Physical hazards Excessive levels of noise from vibration, temperature, and pressure factors. Also crushing hazards. (4)

Riesgos físicos Ruido de niveles excesivos que provienen de vibración, temperatura, y factores de presión. También riesgos aplastantes. (4)

Piezoelectric A material, crystalline in nature, that generates a small voltage when it is vibrated. (38)

Piezoeléctrico Material, de naturaleza cristalina, que genera un pequeño voltaje al ser vibrado. (38)

Pinging A sound heard in the automobile engine that is caused by two combustion fronts hitting each other inside the combustion chamber. (23)

Silbido (Zumbido) Sonido que se oye en un motor de automóvil, producido cuando se pegan dos frentes combustidos dentro de la cámara de combustión. (23)

Pinion gear The small gear attached to the armature shaft used to crank the flywheel ring gear. The pinion gear is also the smaller of two gears. (35)

Piñón diferencial Pequeño engrane conectado al eje que sirve para girar la corona dentada (engranaje) de la rueda voladora. (35)

Pinpoint testing A series of tests that are done on a vehicle's computer electrical system to help isolate problems, such as a broken wire, a bad wire connection, a defective part, and so on. Generally there are many pinpoint tests available for each system in an automobile. (30)

Pruebas diagnósticas precisas Una seria de pruebas efectuadas en el sistema eléctrico de la computadora (ordenador) de un vehículo para identificar los problemas, tal como un alambre roto, una conexión mala de alambre, una parte defectuosa, *etcétera*. Por lo general hay muchas pruebas diagnósticas precisas disponibles por cada sistema de un automóbil. (30)

Pintle The center pin used to control a fluid passing through a hole; a small pin or pointed shaft used to open or close a passageway. (26) (37)

Pivote central (Perno pinzote) Aguja central que sirve para controlar al flúido que pasa por un agujero; pequeña aguja (o perno afilado) que sirve para abrir o cerrar un pasadizo. (26) (37)

Piston A cylindrical object that slides in the cylinder. (12)

Pistón Objeto cilíndrico que se desliza dentro del cilindro. (12)

Piston slap The movement of the piston back and forth in the cylinder in a slapping motion. (17)

Golpe de pistón El movimiento del piston de un lado a otro en el cilindro. (17)

Pitch The angle of the valve spring twist. A variable pitch valve spring has unevenly spaced coils. (18)

Declive (Inclinación) Angulo de contorsión del resorte de válvula. Un resorte de un declive variable tiene espacios no uniformes entre las espirales. (18)

Pitman arm A steering linkage component that connects the steering gear to the linkage at the left end of the center link. (48)

Palanca (Brazo, Eslabón) Pitman Componente de articulación del manejo, el cual conecta el engranaje del volante a la articulación al extremo izquierdo del eslabón céntral. (48)

Pivot point The point on a rocker arm that is the center of its rotation. (19)

Punto de pivote El punto de un balancín que es el centro de su rotación. (19)

Planet carrier The part of a planetary gear system that connects the axis of the planet gears together. (41)

Portador planeta La parte del sistema del engranaje planetario, que conecta el eje del engranaje planetario. (41)

Planetary gear system A gear assembly that includes a sun gear, planet gears, and a ring gear. By locking up one gear, various gear ratios and speeds can be produced. (41)

Sistema de engranaje planetario Montaje de engranaje que incluye una rueda dentada de sol, ruedas dentadas de planeta, y una corona dentada. Cuando se inmoviliza una engrane, se producen varias proporciónes de engranaje y velocidades. (41)

Plastigage A small, thin, plastic strip that is used to help determine the clearance between the main and/or connection rod bearings and the crankshaft journals. (17)

Indicador plástico Pequeña y delgada cinta de plástico que sirve para ayudar a determinar el espacio entre los cojinetes principales y/o de la varilla de conexión y los muñones del cigüeñal. (17)

Platinum A silvery-white material that does not oxidize with air, but does act as a catalyst to change or affect other chemicals. (32) (38)

Platino Material de color plateado-blanco que no corroe en el aire, pero si funcióna como catalizador para cambiar o afectar otros químicos. (32) (38)

Pliers A tool used to grip or cut various objects when working on the automobile. (8)
Alicates (Pinzas) Herramienta que sirve para agarrar o cortar varios objetos al trabajar con un automóvil. (8)

Plies Layers of material that wrap around a tire. (49)
Envolturas Capas de material que envuelven a una llanta. (49)

Pneumatic jack A jack that uses air pressure to lift heavy objects. (10)
Gato neumático Gato que usa la presión de aire para levantar objetos pesados. (10)

Pole shoes Soft iron pieces that wire is wrapped around inside the starter motor. (35)
Expansión polar Piezas de metal blando alrededor de las cuales se envuelve un alambre metálico dentro del motor de arranque. (35)

Pollution Addition of harmful products to the environment. Types of pollution include air, water, noise, chemical, thermal, and nuclear. (1)
Contaminación Añadidura de productos peligrosos al medio-ambiente. Tipos de contaminación incluyen aire, agua, ruido, químico, térmico, y nuclear. (1)

Pollution control system The parts on an automobile engine used to reduce various emissions such as carbon monoxide, nitrogen oxide, and hydrocarbons. (3)
Sistema de control de contaminación Las partes de un motor de automóvil que sirven para reducir varias emisiones tales como monóxido de carbón, óxido de nitrógeno, e hidrocarburos. (3)

Poppet-type valve A type of valve, such as an automobile intake or exhaust valve, that operates from a crankshaft and opens and closes a port. (18)
Válvula de elevación Tipo de válvula, tal como una válvula de entrada o salida de un automóbil, operada por un cigüeñal que abre y cierra una puerta. (18)

Ported vacuum Vacuum taken from the carburetor slightly above the throttle plate. (25) (32)
Vacío enunorificio (Puerto) Vacío tomado desde el carburador un poco arriba de la mariposa (placa de aceleración. (25) (32)

Positive displacement In reference to a hydraulic or pneumatic pump, one that pumps an exact amount of fluid volume with each revolution of the pump. (21)
Desplazamiento positivo Referiendo a una bomba hidráulica o neumática, una que mueva una cantidad exacta de volumen de fluido con cada revolución de la bomba. (21)

Positive displacement pump A type of pressure pump that pumps an exact amount of fluid for each revolution. (37)
Bomba volumetrica (de desplazamiento) positiva Tipo de bomba de presión que bombea una cantidad exacta de líquido por cada revolución. (37)

Postignition Ignition that occurs after the engine ignition system is shut off due to carbon buildup in the combustion chamber. (23)
Post-ignición Encendido que ocurre depués de que se apaga (cierra) el sistema de encendido, debido a depositos de carbón dentro de la cámara de combustión. (23)

Potentiometer An electrical device that varies its resistance so as to change the voltage drop across the device, often used to show a difference in voltage drop as an input signal for the computer. (38)
Potenciómetro Aparato eléctrico que varía su resistencia para cambiar la caída del voltage por el aparato, suelen usarse para demostrar como una señal en la computadora (ordenador) una diferencia en caída del voltage. (38)

Pour point The temperature at which an oil ceases to flow because of being too cold. (20)
Punto de fluidez Temperatura a la que un aceite deja de fluír debido al frío y la congelación. (20)

Power A measure of work being done. (12)
Potencia Medición del trabajo que se produce. (12)

Power control module A module or computer used in an electronic transmission to aid in control of the shift solenoids. (42)
Módulo de control de potencia Módulo o computadora (ordenado) usado en una transmision electronica para ayudar a controlar los solenoides de cambio de velocidades. (42)

Powertrain control module (PCM) One of several computers used to control various components in a vehicle, often used for control of the electronic transmission. (41)
Módulo de control del tren de potencia (PCM) Una de varias computadoras (ordenadores) que sirve para controlar varios componentes en un vehículo, suele usarse para controlar la transmisión automática. (41)

Precombustion chamber A second combustion chamber placed directly off the main combustion chamber. The precombustion chamber is used to ignite a rich mixture of air and fuel. This mixture then ignites a lean mixture in the main combustion chamber. (18)
Cámara de pre-combustión Una segunda cámara de combustión montada justamente cerca de la cámara de combustion principal. Se utiliza la cámara de pre-combustión para encender una mezcla rica de aire y combustible. Está mezcla luego enciende una mezcla pobre en la cámara de combustión principal. (18)

Prefix A term used to indicate how may units the meter is increased or decreased. One thousand meters is equal to one kilometer. Kilo is the prefix. (9)
Prefijo Término que sirve para indicar cuantas unidades se aumenta o reduce el metro. Mil metros equivalen un kilómetro. Kilo- es el prefijo. (9)

Preignition The process of a glowing spark or deposit igniting the air-fuel mixture before the spark plug. (23)
Preignición Proceso de usar una chispa u otro material ardiente para encender una mezcla de aire y combustible antes que la bujía. (23)

Pressure The exertion of force on a body in contact with it. Pressure is developed within the cooling system and is measured in pounds per square inch on a gauge. (22)
Presión Aplicación de fuerza a un cuerpo, el cual tiene contacto con aquélla. Se aumenta la presión dentro del sistema refrigerante y se mide por libras por pulgada cuandrada en un aparato de medición. (22)

Pressure drop One condition in an antilock brake control module in which the hydraulic braking pressure is reduced to the locked-up brake caliper. (46)
Reducción de presión Una condición en un módulo de control de frenos anti-inmovilizador, dentro del cual la presión de los frenos hidráulicos está reducida al calibrador de freno inmovilizador. (46)

Pressure gauge A gauge used to read various pressures such as fuel pump, transmission oil, and fuel injection pressures. (9)
Manómetro (Indicador de presión) Instrumento de medición que sirve para medir (leer) varias presiones tales como bomba de combustible, aceite de transmisión, e inyección de combustible. (9)

Pressure hold One condition in an antilock brake control module in which no more braking pressure can be produced in the master cylinder. (46)
Retención de presión Una condición en un módulo de control de frenos anti-inmovilizador, en el que no se puede producir más presión de frenos en el cilindro maestro. (46)

Pressure modulator The assembly on an antilock braking system used to control the hydraulic pressure to the brake calipers. (46)
Modulador de presión Agrupación de piezas de un sistema de frenos anti-inmovilizador, la cual sirve para controlar la presión hidráulica a los calibradores de freno. (46)

Pressure plate The part in a clutch system used to squeeze or clamp the clutch disc between it and the flywheel. (39)
Platillo de presión La parte del sistema del embrague que sirve para agarrar o fijar al disco del embrague entre el embrague y la rueda voladora. (39)

Pressure regulator valve A valve used in an automatic transmission to regulate the pressure of the oil inside of the valve body. Also, a valve used in a fuel system to regulate the pressure of the fuel. (24) (41)
Válvula de presión del regulador Válvula en una transmisión automática que sirve para regular la presión del aceite dentro del cuerpo de la válvula. También, una válvula del sistema de combustible que sirve para regular la presión del combustible. (24) (41)

Pressure switch assembly (PSA) A switch that is used on electronic transmissions to tell the computer which gear has been selected by the driver or vehicle operator. (41)
Asamblea de interruptor de presión (PSA) Interruptor en una transmisión electrónica que sirve para informar a la computadora (ordenador) de cual velocidad ha sido selecciónada por el conductor o el operador del vehículo. (41)

Prevailing torque nuts Nuts designed to develop an interference fit between the nut and bolt threads. (5)
Tuercas de torque prevaleciente Tuercas diseñadas para desarrollar un ajuste interferente entre la tuerca y las roscas del perno. (5)

Preventive maintenance Maintenance performed on an engine to prevent potential repairs. (1)
Maintenimiento preventivo Mantenimiento dado al motor para prevenir algunas reparaciónes potenciales. (1)

Primary battery A type of battery that cannot be recharged after use. (31)
Batería primaria Tipo de batería que no puede recargarse después de usarse. (31)

Primary circuit A circuit in the ignition system that uses 12 volts to operate. It includes the ignition switch, ballast resistor or resistive wire, primary coil wires, condenser, and contact points. (32)
Circuito primario Circuito del sistema de encendido que utiliza 12 voltios para operarse. Se incluye el switch de encendido, resistencia reguladora o hilo resistor, sistema de bobinas primarias, condensador, y puntos de contacto. (32)

Primary shoe The forward shoe on a two-shoe drum brake system, often having shorter linings than the other. (45)
Zapata primaria Zapata delantera de un sistema de tambor de freno de dos zapatas. (45)

PROM Programmable read only memory, or permanent storage in a computer that can be easily accessed, removed, and/or replaced by installing a new chip. (30)
"PROM" Memoria programmable de lectura solamente, o almacenaje permanente en una computadora (ordenador) automotriz, la cual puede ser accesible fácilmente, eliminada y/o reemplazada con un nuevo (chip). (30)

Propane One of four gases found in natural gas. Methane, ethane, propane, and butane are in natural gas. Propane and butane have the highest amount of energy and can be made into a liquid by being put into a pressurized container. (23)
Propano Uno de cuatro gases encontrados en gas natural. Metano, etano, propano, y butano se encuentran en gas natural. Propano y butano tienen la cantidad de energía más alta y pueden convertirse en líquido al ponerse en un recipiente bajo presión. (23)

Property class A number stamped on the end of a metric bolt to indicate the hardness of the bolt. (5)
Clase de propiedad Número sellado (cuñado) en el cabo de un perno métrico para indicar la dureza del perno. (5)

Proportioning valve A valve in the brake hydraulic system that reduces pressure to the rear wheels to achieve better brake balance. (45)
Válvula proporcionadora Válvula de un sistema de freno hidráulico, la cual reduce la presión a las ruedas traseras para lograr un mejor balance de freno. (45)

Proton The positive (+) part of the atom. (29)
Protón Patre positiva (+) de un átomo. (29)

Psig A type of pressure scale read as pounds per square inch on a gauge. (22)
"Psig" Tipo de escala de presión expresada como libras por pulgada cuadrada en un indicador. (22)

Puller A tool attached to a shaft and gear, used to remove the gear from the shaft by applying certain pressures. (10)
Tirador Herramienta pegada a un eje y engrane, que sirve para quitar el engrane del eje por medio de aplicar ciertas presiones. (10)

Pulse delay variable resistor A resistor in the wiper system used to time delay the wiper motion from 0 to 25 seconds time delay. (53)
Resistencia variable de retardador de impulsos Resistencia en el sistema del limpiabrisas, el cual se emplea para retardar el movimiento del limpiabrisas desde cero a 25 segunods de retardo. (53)

Pulse width A term used to describe the length of time that a fuel injector or M/C solenoid is in the On position. (25) (26)
Impulso en anchura Termino para describir el tiempo que un inyector de combustible o un solenoid M/C está en posición prendida. (25) (26)

Pulse wiper system A wiper system using electronic circuits that cause the wipers to pulse or turn on one time, then off for a certain number of seconds. (53)
Sistema de limpiabrisas de pulso Sistema de limpiabrisas que utiliza circuitos electrónicos que hacen que los limpiabrisas pulsen u oscilen un rato, luego paren cierto número de segundos. (53)

Purge To separate or clean by carrying off gasoline fumes. The carbon canister has a purge line to remove impurities. (37)
Purga Separar o limpiar llevándose los gases de gasolina. El bote de carbón tiene una línea de purgas para eliminar impurezas. (37)

Pushrod Connector between the lifter and the rocker arm. (16)
Varilla de empuje Conector entre el levantaválvulas y el balancín. (16)

Quenching The cooling of the gases by pressing the gas volume out into a thin area. Quenching occurs inside the wedge-type combustion chamber in the quench area. (18)
Enfriamiento repentino Enfriamiento de gases por medio de apresionar el volumen de gas en una área reducida (angosta). El enfriamiento repentino ocurre dentro de la cámara de combustión de tipo cuña en el área de extincion. (18)

R134a A hydrofluorocarbon gas that is not damaging to the environment and ozone layer that can be used in air-conditioning systems. (50)
"R134a" Un gas de hidroflorocarburo que no daña ni al medioambiente ni a la capa de ozono y puede utilizarse en sistemas de aire acondicionado. (50)

Rack-and-pinion steering A steering system consisting of a flat gear (rack) and a mating gear (pinion). The pinion meshes with teeth on the rack causing the rack to move left or right. This motion moves tie-rods and the spindle arm to steer the front wheels. (48)
Dirección por piñón y cremallera Sistema de dirección consistente en un engrane plano (cremallera) y otra rueda dentada (piñón). El piñón se enlaca con dos dientes de la cremallera haciendo que la cremallera se mueve a la izquierda o a la derecha. Este movimiento hace que se muevan las barras tirantes (varillas de tensión) y el huso (mandril) para dirigir las ruedas delanteras. (48)

Radial Something that radiates from a center point. Radial tires have cord materials running in a direction from the center point of the tire, usually from bead to bead. (49)
Radial Algo que irradia de un punto central. Las llantas radiales contienen materiales de cuerda pasando en dirección desde el punto central de la llanta, normalmente de borde a borde. (49)

Radiation Transfer of heat by converting heat energy to radiant energy. (22)
Radiación Transferencia de calor por convirtiendo la energía calórifica a energía radiante. (22)

RAM Random access memory, or temporary storage of information in a computer. (30)
"RAM" Memoria de acceso al azar o almacenaje temporal de información en una computadora (ordenador). (30)

Ram air Air that is forced into the engine or passenger compartment by the force of the vehicle moving forward. (51)
Aire bajo presión dinámica Aire forzado en un motor o cabina de pasajero por la fuerza producida por el vehículo en movimiento hacia adelante. (51)

Reciprocating engine An engine in which the fuel energy moves parts up and down or back and forth. (12)
Motor de movimiento alternativo (de émbolos) Motro en el que la energía de combustible hace que se muevan las partes en movimiento altibajo o para atrás y hacia adenlante. (12)

Rectify To change one type of voltage to another. Usually ac voltage is rectified to dc voltage. (34)
Rectificar Cambiar un tipo de voltaje a otro tipo. Típicamente voltaje de corriente alterna se rectifica a voltaje de corriente directa. (34)

Refrigerant A liquid capable of vaporizing at low temperatures, such as ammonia or freon. (50)
Refrigerante Líquido capaz de vaporizar a temperaturas bajas, tal como amoníaco o freón. (50)

Regional offices and distributorships Offices owned and operated by the automobile company. They are considered to be the link between the automobile manufacturer and the dealerships. (1)
Oficinas regionales y destribuidoras Oficinas que pertenecen a los dueños y operadores de la compañía automotriz. Se consideran el vínculo entre la fabrica automotriz y las agencias. (1)

Regular gasoline Gasoline that has an octane number near 85 to 90. (23)
Gasolina regular Gasolina que tiene un número de octano cerca de 85-90. (23)

Relay An electromagnetic device by which the opening or closing of one circuit operates another device. A relay in a voltage regulator uses a set of points that are opened and closed by magnetic forces. The opening or closing of the points controls another circuit, commonly the field circuit. (30)
Relé (Conmutador) Aparato electromagnético por medio del cual el abrir y cerrar de un circuito opera a otro aparato. Un relé en un regulador de voltaje emplea un juego de puntos de encendido que se abren y cierran mediante fuerzas magnéticas. El abrir y cerrar de los platinos controla a otro circuito, comúnmente el circuito inductor. (30)

Reluctor In an electronic ignition system, a metal wheel with a series of tips used to produce the signal for the transistor. (33)
Reluctor En un sistema de encendido electrónico, una rueda metálica con una serie de puntas que sirven para producir la señal para el transistor. (33)

Residual pressure Remaining or leftover pressure. (45)
Presión residual Presión que sobra. (45)

Resistance The part in an electrical circuit that holds back the electrons; also called the load. (29)
Resistencia Parte del circuito eléctrico que retiene los electrones; también llamada la carga. (29)

Resonator A device used in an exhaust system to reduce noise, usually used in conjunction with a muffler. (27)
Resonador Aparato utilizado en el sistema de escape para reducir el rudio, normalmente usado en combinación con un silenciador (mofle). (27)

Revolutions per minute (rpm) The number of crankshaft revolutions occurring in an engine each minute. (15)
Revoluciones por minuto (rpm) Número de revoluciones por el cigüeñal que ocurre en un motor en cada minuto. (15)

Rich mixture Too much fuel and not enough air for efficient combustion. (13)
Mezcla rica Demasiado combustible e insuficiente aire para combustión. (13)

Road horsepower Horsepower available at the drive wheels of the vehicle. (15)
Caballaje de camino Caballaje disponible a las ruedas de transmisión del vehículo. (15)

Rocker arm An arm that has a pivot point in the center. One side is lifted by the camshaft movement and the other side moves down, opening the valves. (16)
Balancín Brazo que tiene un punto de pivote en el medio. El movimiento del arbol de levas levanta un extremo del balancín y el otro extremo va para abajo, abriendo así las válvulas. (16)

Rolling resistance A term used to describe the amount of resistance a tire has to rolling on the road. Tires that have a lower rolling resistance usually get better gas mileage. Typically, radial tires have lower rolling resistance. (49)
Resistencia al rodado Término que sirve para describir la cantidad de resistencia que una llanta tiene a la acción de rodar sobre la superficie de un camino. Llantas que tienen una resistencia más baja al rodamiento normalmente resultan en mejor economía de kilometraje. Típicamente las llantas radiales tienen una resistencia más baja al rodamiento. (49)

ROM Read-only memory, or permanent storage of information in a computer. (30)
"ROM" Memoria de lectura solamente, o almacenaje de información permanente de una computadora (ordenador). (30)

Rope seals A type of seal used on crankshafts shaped much like a small, thin rope. (5)
Cierres de cable Tipo de cierre (obturador) usado en los cigüeñales, el cual está formado como una pequeña soga (cuerda) delgada. (5)

Rotary engine An engine that uses a rotor rather than pistons to produce power. It is an intermittent, internal combustion engine, and motion is rotary (circular), not reciprocating. (12) (14)
Motor rotativo Motor que emplea un rotor (rodete) en vez de pistones para producir poder. Tiene un motor de combustión interna intermitente, y el movimiento es rotativo (circular), no recíproco. (12) (14)

Rotor The center of a rotary engine; the rotating component in a generator or alternator. Also, a three-pointed or -lobed object that replaces the pistons in a rotary engine. (14) (34)
Rotor Parte central de un motor rotativo; el componente giratorio en un generador o un alternador. También, un objeto de tres puntos o lóbulos que reemplaza los pistones en un motor rotativo. (14) (34)

Rpm Revolutions per minute on any rotating shaft. (15)
Rpm Revoluciónes por minuto de cualquier eje rotativo. (15)

RTV sealants A type of sealant that is able to cure in the presence of moisture and oxygen. (5)
Sellador "RTV" Tipo de sellador (empaque) capaz de curarse (enderezarse) en la presencia de humedad y oxígeno. (5)

Runner A cast tube on an intake or exhaust manifold used to carry air into or out of the engine. (27)
Burlete (Corredor) Tubo fundido de un colector de admisión o de escape, el cual se emplea para llevarse el aire para adentro o afuera del motor. (27)

Running gear Component on the automobile that is used to control the vehicle. This includes braking systems, wheels, and tires. (3)
Engrane (de funciónamiento) Componente de un automóvil que sirve para controlar al vehículo. Esto incluye el sistema de frenos, ruedas, y llantas. (3)

Rust inhibitors A type of chemical put into cooling and fuel systems to prevent rust from developing in the liquid. (20) (23)
Inhibidores de óxido (oxidación) Tipo de químico agredado a sistemas de aire acondiciónado o de combustible para prevenir oxidación en el líquido. (20) (23)

Safety glasses Glasses to be worn at all times when in the automotive shop. They should be designed with safety glass and side protectors, and they should be comfortable. (4)
Lentes protectores Lentes que se deben de usar todo el tiempo cuando una está presente en un taller automotriz. Deben de ser diseñados con cristal de seguridad y protectores laterales, y deben de ser cómodos. (4)

Safety rims A rim on a wheel that has inside ridges so that when a tire deflates, it stays on the rim. (49)
Aros (Rebordes protectores) de seguridad Aro de la rueda que tiene un borde interior para que, cuando se desinfla la llanta, está se quede en posición. (49)

Sales representative A person who sells new and used automobiles. (2)
Representante de ventas (Vendedor) Persona que vende automóviles nuevos y usados. (2)

Saybolt Universal Viscosimeter A meter used to measure the time in seconds required for 60 cubic centimeters of a fluid to flow through a hole on the meter at a given temperature under specified conditions. (20)
Viscosímetro Universal de Saybolt Medidor que sirve para medir el tiempo en segundos requerido para que 60 centímetros cúbicos de un líquido pasen por un agujero del medidor a cierta temperatura bajo condiciónes especificadas. (20)

Scanners Electronic analyzers used to analyze and diagnose engine performance. (10)
Exploradores Analizadores electrónicos que sirven para analizar y diagnosticar la eficiencia del motor. (10)

Screw-pitch gauge A gauge used to measure the number of threads per inch on a bolt. (5)
Calibre para determinar el paso de tornillo Calibre que sirve para medir el número de roscas por pulgada de un perno. (5)

Scrub radius The distance between the centerline of the ball joints and the centerline of the tire at the point when the tire contacts the road surface. (48)
Radio "scrub" Distancia entre la línea central de las juntas esféricas (o de rótulas) y la línea central de la llanta al punto donde haya contacto entre la llanta y la superficie del camino. (48)

Scuffing Scraping and heavy wear from the piston on the cylinder walls. (17)
Desgaste abrasivo Frotamiento y desgaste fuerte debido al movimiento de pistones contra las paredes interiores de los cilindros. (17)

Seal A device used on rotating shafts to keep oil or other fluid on one side of the seal, thus eliminating leakage. (5)
Cierre (Retenedor) Aparato usado en los ejes rotativos para impedir que el aceite u otro líquido pase de un lado al otro del cierre, y así eliminar la fuga o gotera del líquido. (5)

Sealant A thick liquid placed in engine parts to seal the parts from leakage. (5)
Sellador (Tapador) Líquido espeso puesto en partes del motor para ayudar a sellar y colocar con precisión el empaque durante el proceso de instalación. (5)

Seating When two metals must seal gases and liquids, they must be worked together to make a good seal. This process of getting two metal surfaces to seal is called seating. (18)
Asentamiento Cuando dos metales tienen que sellar gases y líquidos, tienen que ajustarse juntos para que haya un cierre preciso. El proceso de juntar las superficies de los dos metales para que ocluyan se llama asentamiento. (18)

Secondary circuit A circuit in the ignition system that uses 20,000 or more volts to operate. It includes the secondary coil windings, the rotor, distributor cap, coil and spark plug wires, and spark plugs. (32)
Circuito secundario Circuito del sistema de encendido que utiliza 20,000 voltios o más para operar. Incluye enrollados de bobina secundarios, rotor, tapa del distribuidor, bobina, cables de bujía, y bujías. (32)

Secondary shoe The rear shoe of a two-shoe drum brake system, often having a longer lining than the primary shoe. (45)
Zapata secundaria Zapata trasera de un sistema de tambor de freno de dos zapatas, el cual tiene, a veces, un forro más largo que el de la zapata primaria. (45)

Selector control lever A lever located on the dashboard used to select one of several heating and ventilation modes. (51)
Palanca de control del selector Palanca colocada en el panel de controles, que sirve para seleccionar uno de los varios modos de calefacción o ventilación. (51)

Self-energizing A drum brake arrangement where the braking action pulls the shoe lining tighter against the drum. (45)

Auto-activación (Auto-excitación) Sistema de tambor de freno en el que la acción de frenar aprieta al forro de zapata contra el tambor. (45)

Semiconductor A material with four electrons in the valence ring. (29)
Semiconductor Material de cuatro electrones en la banda de valencia. (29)

Series circuit A circuit in which there is only one path for the current to follow. (29)
Circuito en serie Circuito en que hay sólo una vía en la que circula la corriente. (29)

Serpentine belt A type of belt that snakes around various pulleys and other components that need to be turned or rotated. (7)
Serpentín correa Tipo de correa que culebrea por varias poleas u otros componentes que deben hacerse girar o dar vueltas. (7)

Service Bay Diagnosis System A computerized information network system that is connected to the manufacturer in Detroit, Michigan, used to answer service and diagnostic questions. (11)
Sistema de diagnóstico del área de servicio Red de información computerizada que está conectada con el fabricante, en Detroit, Michigan, la cual sirve para contestar preguntas acerca de servicio y diagnósticos. (11)

Service bulletins Technical service information provided by the manufacturer, used as updates for the service manuals. (11)
Boletín de servicio Información de servicio técnico suministrada por la fabrica, que se usa para mantener los manuales con información corriente. (11)

Service manager The person responsible for the entire service operation of the dealership. (2)
Gerente de servicio Persona responsable de toda la operación de servicio de una agencia automotriz. (2)

Service manual A manual provided by the manufacturer or other publisher that describes service procedures, troubleshooting and diagnosis, and specifications. (11)
Manual de servicio Manual suministrado por la fabrica u otra casa editorial que explica procedimientos de servicio, investigación diagnóstica de averías, y especificaciónes. (11)

Service procedures A set of listed steps used to disassemble, assemble, or repair an automotive component. (11)
Procedimientos de servicio Juego de pasos específicos que sirve para desmontar, motar, o reparar un componente automovilístico. (11)

Service representative A person who works in the area of providing service to the dealership from the car manufacturer. (2)
Representante de servicio Persona que trabaja en el área de proveer servicio de una fabrica a una agencia automotriz. (2)

Service technician A person who is actively involved in repair and maintenance of the total vehicle. (2)
Técnico de servicio Persona que está involucrada activamente en reparaciones y mantenimiento del vehículo entero. (2)

Servo A hydraulically operated component that operates or controls the operation of the transmission band on the automatic transmission; a device used on a cruise control system to maintain the speed of the vehicle. (41) (52)
Sistema servo Componente operado hidráulicamente, el cual controla la operación de la banda de transmisión que se encuentra en una transmisión automática; aparato usado en el sistemade control de crucero para mantener constante la velocidad del vehículo. (41) (52)

Servo brake A drum brake arrangement where the action of one shoe reinforces the action of the other shoe. (45)
Freno de servo Conjunto del tambor de freno en el que la acción de una zapata refuerza la acción de la otra. (45)

Servo position sensor A sensor used on an electronic transmission to determine the position of a particular servo, which in turn produces data to be sent to the computer for further transmission control. (52)
Sensor de posición del servo Sensor en una transmisión electrónica que sirve para determinar la posición de un servo en particular, que por su turno produce los datos que se env'an a la computadora (ordenador) por mayor control de la transmisión. (52)

Shackle The small arm between the frame and one end of the leaf spring. It is used to allow the spring to shorten and lengthen during normal driving conditions. (47)
Enganche Parte del enganche entre la armadura (armazón) del coche y un extemo del muelle (restore de hojas o láminas flexibles). Sirve para dejar que el muelle se acorte o se alargue (se estire) durante las condiciones de conducción normales. (47)

Shank The diameter of the bolt, usually measured in fractions of an inch or in millimeters. (5)
Fuste Diámetro de un perno o tornillo, normalmente medido en fracciones de una pulgada o milímetros. (5)

Shift solenoid A solenoid, controlled by an electronic signal, used in an electronic transmission to change the direction of a hydraulic circuit to shift the vehicle gears. (41)
Solenoide de cambio de velocidad Solenoide, controlado por una señal electrónica, en una transmisión electrónica que sirve para cambiar la dirección de un circuito hidraœlico para cambiar la velocidad en un vehículo. (41)

Shift valve A valve used in an automatic transmission that controls the oil flow to the clutches and transmission bands. (41)
Válvula de cambio Válvula usada en las transmisiones automáticas que controlan el pasaje de aceite al embrague y banda de transmisión. (41)

Shock absorber A device used on a suspension system to dampen the oscillations or jounce of the springs when the car goes over bumps. (47)
Amortiguador Aparato usado en un sistema de suspensión para absorber (disminuir) las oscilaciones o sacudidas de los resortes o muelles cuando pasa el vehículo sobre los topes y baches. (47)

Shop supervisor A supervisor in a dealership who is responsible for organizing work schedules and managing the service technicians. (2)
Supervisor de taller Un supervisor en una concesionaria, responsable por la organización de horarios de trabajo y administración de técnicos de servicio. (2)

Shroud An object that covers the area between the fan and the radiator. (22)
Tolva Objeto que cubre el área entre el ventilador y el radiador. (22)

Shrouding When a valve is placed close to the side of the combusion chamber, the air and fuel may be restricted by the side of the chamber. This restriction is referred to as shrouding. (18)
Tolvera (Restricción) Cuando se coloca una válvula cerca del lado de la cámara de combustión, el aire y el combustible pueden estar restringidos al lado de la cámara. Se refiere a está restricción como tolvera. (18)

Shunt More than one path for current to flow, such as a parallel part of a circuit. (35)
Desvío Más de un camino que la corriente puede seguir, tal como una parte paralela de un circuito. (35)

Siamese ports Intake or exhaust ports inside the cylinder head where two cylinders are feeding through the one port. (18)

Orificio de vaciado (Siamés) Orificio de entrada y escape dentro de la cabeza de cilindros donde dos cilindros utilizan el mismo orificio. (18)

Slant An in-line cylinder arrangement that has been placed at a slant. This arrangement makes the engine have a lower profile for aerodynamic design. (16)

Sesgo (Oblicuidad) Conjunto de cilindros en línea, los cuales están montados en un plano inclinado. Este arreglo hace que el motor tenga un perfil más bajo para un meson diseño aerodinámico. (16)

Slave cylinder A type of hydraulic cylinder used as a means to actuate the clutch mechanism. As the clutch pedal is pushed down, the hydraulic pressure produced in the slave cylinder is used to move the clutch mechanism. (39)

Cilindro esclavo Tipo de cilindro hidráulico usado como un medio de activar el mecanismo del embrague. Al deprimir el pedal para embrague, la presión hidráulica producida en el cilindro esclavo sirve para mover el mecanismo del embrague. (39)

Slip A term used to represent the amount of slippage between the tire and the road during a braking condition. When the tire locks up completely, 100% slip occurs. (46)

Resbalamiento (derrapar, patinar) Vocablo que sirve para representar la cantidad de deslizamiento entre la llanta y el camino durante una condición de frenar (perdida de agarre). Al inmovilizarse la rueda por completo, se ocurre un 100% de resbalamiento. (46)

Slip joint A splined shaft that can slide in a mating shaft to allow changes in drive shaft length. (43)

Junta corrediza Eje ranurado que se puede deslizar en otro eje compañero para permitir cambios en el largo del eje impulsor (flecha o eje motor). (43)

Slipper skirt A piston that has a cutaway skirt so that the piston can come closer to the counterweights. This makes the overall size of the engine smaller. (17)

"Slipper skirt" Pistón que tiene una falda o borde en corte para que el pistón pueda acercarse a los contrapesos. Esto hace que el tamaño, en general, del motor sea más pequeño. (17)

Slip ring A type of commutator used on an alternator made of two copper rings that are split in half. (34)

Anillo rozante Tipo de conmutador usado en un alternador y hecho de dos anillos de cobre que están divididos en dos mitades. (34)

Sludges Material formed as a result of oil in the presence of various acids. (20)

Fangos Material formado como resultado del aceite en la presencia de varios ácidos. (20)

Smog The combined effect of various chemicals put into the air by various forms of combustion. (1)

Esmog (Contaminación del aire) Efecto combinado de varios químicos puestos en el aire por varios modos de combustión. (1)

Smoking rules Only smoke in designated "smoking" areas. Dangerous explosive fuels in the shop may be ignited if this rule is not followed. (4)

Reglamentos de fumar Se permite fumar en áreas designadas para fumar. Combustibles explosivos en el taller pueden encenderse si no se obedece está regla. (4)

Snap rings Small rings, either external or internal, that are used to prevent gears and pulleys from sliding off the shaft. (5)

Anillos de seguridad (Snap rings) Pequeños anillos sujetadores, externos o internos, que sirven para prevenir que el engranaje y las poleas se deslicen afuera del eje. (5)

Snorkel tube A long, narrow tube attached to the air cleaner, used to direct air into the air filter. (37)

Esnórquel Tubo largo y angosto conectado al limpiador de aire, que sirve para dirigir aire en el filtro para aire. (37)

Socket points The number of points inside the socket head; 6, 8, and 12 points are most common. In applications where only a small amount of rotation of the ratchet is possible, use a 12-point socket. (8)

Puntos de dado Número de puntos dentro del dado; 6, 8, y 12 puntos son los números más comunes. Para aplicaciones en las que es posible una pequeña cantidad de rotacíon de llave de trinquete (matraca), use un dado de cubo de 12 puntos. (8)

Solenoid A device that converts electrical energy in a coil of wire to mechanical energy that moves back and forth. (30) (35)

Solenoide Aparato que convierte la energia eléctrica de una bobina de alambre a la energia mecánica que oscila. (30) (35)

Specialty shops Service shops that specialize in certain components of the automobile. Some include carburetor shops, body shops, transmission shops, muffler shops, and so on. (1)

Tiendas de especialidades Tiendas de servicio que se especializan en ciertos componentes de automóvil. Incluyen tiendas para carburadores, hojalaterías, transmisiones, silenciadores, *etcétera.* (1)

Specifications Any technical data, numbers, clearances, and measurements used to diagnose and adjust automobile components. They are also called specs. (11)

Especificaciones Cualquier dato técnico, números, distancias, y medidas, que sirven para diagnosticar y ajustar componentes automotrices. También se llaman "specs." (11)

Specific gravity The weight of a solution as related to water. Water has a specific gravity of 1.000. Sulfuric acid, being heavier than water, has a specific gravity of 1.835. (31)

Peso específico (Gravedad especifica) Peso de una solución con referencia al agua. El agua tiene un peso específico de 1.000. El ácido sulfúrico, al ser más pesado que el agua, tiene un peso específico de 1.835. (31)

Splines External or internal teeth cut into a shaft, used to keep a pulley or hub secured on a rotating shaft. (5)

Ranuras Estrías o acanaladuras, externas o internas cortadas en un eje, que sirven para mantener fija una polea o parte central en un eje rotativo. (5)

Spool valve A cylindrical rod with different-sized diameters. Usually the cylindrical body is placed inside a bore inside the transmission valve body. As the valve is moved in and out, different hydraulic circuits are operated. (41)

Válvula bobinada Cilíndrica con diferentes tamaños de diámetros. Normalmente la válvula cilíndrica se coloca dentro del hueco que está dentro del cuerpo de la válvula de transmisión. Al moverse la válvula, hacia adentro o afuera se operan diferentes circuitos hidráulicos. (41)

Sprag A pointed steel piece inside an overrunning clutch mechanism that allows rotation in one direction but locks up rotation in the opposite direction. (35)

"Sprag" Pieza de metal apuntada dentro del mecanismo del embrague de rueda libre (de sobremarcha), la cual permite la rotación en una dirección pero inmoviliza la rotación en la dirección opuesta. (35)

Squirm To wiggle or twist about a body. When applied to tires, squirm is the wiggle or movement of the tread against the road surface. Squirm increases tire wear. (49)

Retorcimiento Retorcerse alrededor de un cuerpo. Al referirse a las llantas, retorcimiento es un movimiento de las huellas contra la superficie del camino. Retorcimiento aumenta el desgaste de la llanta. (49)

Squirrel cage blower A type of air pressure fan shaped like a squirrel cage, used to move air throughout a system. The squirrel cage fan is run by a motor and placed inside a housing to improve its efficiency of operation. (51)
Ventilador de jaula de ardilla Tipo de ventilador de presión de aire formada como una jaula de ardilla, que sirve para mover el aire en el sistema. La jaula de ardilla funciona a base de un motor y está colocada dentro de una cubierta protectora para mejorar su eficiencia de operación. (51)

Stabilizer bar A reinforcement component on a suspension system that prevents the body from diving or leaning on turns. (47)
Barra estabilizadora Componente de reforzamiento de un sistema de suspensión que impide que el chasis se incline al dar vueltas. (47)

Stance The manner of standing or being placed. A vehicle's stance refers to the level or evenness of its position. (47)
Postura Una postura de estacionarse o de estar estacionado. La postura de un vehículo se refiere a la nivelación de su postura. (47)

Standard bolt and nut torque specifications A chart showing the standard torque for common sizes of bolts. (5)
Especificaciones de torque de perno (tornillos) y tuerca estándar Esquema que indica el torque estándar para los tamaños de pernos (tornillos) comunes. (5)

Starting and charging systems The subsystems on the engine used to start the engine and charge the battery. (3)
Sistemas de encendido (arranque) y carga Subsistemas de un motor, que sirve para arrancar el motor y cargar la batería. (3)

Static balance Equal distribution of weight around a center point. Static means stationary, and static balancing is done with the wheels stationary. (49)
Balanceo estático Distribución de peso igual alrededor de un punto central. Estático significa estacionario o fijo, y se logra hacer balanceo estático con las ruedas estacionarias. (49)

Stator The stationary part in an alternator that cuts the magnetic lines of force; the part of a torque converter that is stationary, used to direct the flow of fluid back to the rotary pump at the correct angle. (34) (41)
Estator Parte estacionaria (fija) en un alternador, la cual interrumpe las líneas de fuerza magnéticas; la parte de un convertidor torque, que está estacionaria, que sirve para dirigir el flujo de líquido hacia la bomba rotativa a un ángulo correcto. (34) (41)

Steering axis inclination The inward tilt of the spindle support arm ball joints at the top. (48)
Inclinación del eje de la dirección Inclinación hacia el interior de la pieza del extremo del eje delantero que soporta en los más alto de la palanca de las juntas esféricas. (48)

Steering knuckle A part of the front suspension that connects the wheel to the suspension system and the tie-rod ends for steering. The wheel spindle is also attached to the steering knuckle. (47)
Nudo de dirección (de mando) Parte de la suspensión delantera que conecta la rueda con el sistema de suspensión y con las barras de acoplamiento que sirven para dirigir. (47)

Stellite A very hard metal made from cobalt, chromium, and tungsten, used for insert-type valve seats. (18)
Estelita Un metal muy duro hecho de cobalto, cromo, y tungsteno (volframio), usado en los asientos de tipo-insertado de las válvulas. (18)

Stoichiometric ratio A 14.7 to 1 air-fuel ratio. This is the best ratio to operate an internal combustion engine. (13) (26)
Relación estoiquiometricia (Óptima) Proporción de 14.7 a 1 de aire y combustible. Está es la mejor proporción para operar un motor de combustión interna. (13) (26)

Stratified To layer or have in layers. (16)
Estratificado (en capas o láminas) Láminar o poner en capas. (16)

Stratified charged engine An engine that has an additional small combustion chamber. The air-fuel mixture in this chamber is very rich. The air-fuel in the regular chamber is leaner. The small chamber ignites the larger chamber mixture, reducing emissions. (16)
Motor de cargo estratificado Motor que tiene un pequeña cámara de combustión adiciónal. La mezcla de aire y combustible en es cámara es muy rica. La mezcla de aire y combustible en la cámara normal es más pobre. La cámara más pequeña enciende la mezcla de la cámara normal, reduciendo así emisiones. (16)

Stroke The distance from TDC to BDC of piston travel. (13)
Ciclo Distancia de "TDC" a "BDC" del movimiento del pistón. (13)

Strut rod A rod on the suspension system located ahead of or behind a lower control arm to retain the arm in its intended position. (47)
Puntal Barra del sistema de suspensión situada delante de o detrás de un brazo de control más bajo para mantener al brazo en su posición normal. (47)

Sulfur A chemical in diesel fuel that produces pollution. When mixed with oxygen and water, sulfur produces a strong acid. (23)
Azufre Un químico en combustible diesel que produce contamina-ción. Al mezclarse con oxígeno y agua, el azufre produce un acido fuerte. (23)

Sump A pit or well where a fluid is collected. Oil is collected in the oil sump. (21)
Sumidero Del carter pozo donde se deposita un flúido. Aceite es depositado en el sumidero (colector) de aceite. (21)

Sunload sensor A sensor placed on the dashboard to determine the amount of sun coming into the vehicle. (51)
Sensor de luz solar Sensor situado en el panel de controles para determinar la cantidad de luz solar que entra en el vehículo. (51)

Supercharger A device placed on a vehicle to increase the amount of air, and therefore, the amount of fuel that is sent into the engine. (16)
Supercargador (Sobrealimentador, Turboalimentador) Aparato situado en el vehículo para aumentar al cantidad de aire, y por tanto, la cantidad de combustible que entra en el motor. (16)

Supercharging The process of forcing air into an engine cylinder with an air pump. The forced air can come from a blower or turbocharger. (28)
Sobrealimentación Proceso de forzar aire en el cilindro del motor con una bomba para aire. El aire forzado puede provenir de soplador o turbocargador. (28)

Supporting careers The automotive industry supports careers in the following areas, claims adjusting, vocational teaching, auto body repairing, frame and alignment repair, specialty shops, and others. (2)
Carreras de soporte La industria automotriz soporta carreras en las siguientes áreas; ajustes de reclamos, enseñanza vocaciónal, reparaciónes en hojalatería, reparaciones en armadura y alineación, talleres en especialidades, y otros. (2)

Surging A sudden rushing of water from the water pump. (22)
Oleada (Oleaje o Pulsación) Pulsación repentina de agua desde la bomba de agua. (22)

Suspendability The ability of a fluid to suspend heavier dirt particles within the oil, rather than letting them fall to the bottom. (20)
Suspendibilidad Habilidad de un flúido de suspender las partículas de tierra más pesadas en el aceite, en vez de dejar que se hundan al fondo. (20)

Suspension system Components that support the total vehicle, including springs, shock absorbers, torsion bars, axles, and connecting linkages. (3)
Sistema de suspensión Componentes que suportan al vehículo entero, incluyendo los resortes, amortiguadores, barras de torsión, ejes, y palancas de vínculo. (3)

Swash plate An angular plate attached to the bottom of the four pistons on a Stirling engine. As the pistons move downward, the swash plate is turned. Also, a mechanical system that is used for pumping. An angled plate is attached to a center shaft, and pistons are attached to the plate along the axis of the shaft. As the shaft rotates, the pistons move in and out of a cylinder, producing a suction and pressure. (50)
Placa "swash" Placa angular conectada a la parte inferior de los cuatro pistones de un motor Stirling. Al moverse para abajo los pistones, se gira una placa "Swash". Sistema mecánico que sirve para bombear. Una placa angulada está conectada a un eje central, y se juntan pistones a la placa a lo largo de la línea recta del eje. Al girarse el eje, los pistones se mueven adentro y afuera de un cilindro, produciendo así una succión y presión. (50)

Sway bar A bar on the suspension system that connects the two sides together. It is designed so that during cornering, forces on one wheel are shared by the other. (47)
Barra de balanceo Barra del sistema de suspensión que junta los dos lados. Está diseñda para que al tomar una curva, las fuerzas sobre una rueda están distribuidas en parte sobre la otra. (47)

Synchronizer An assembly in a manual transmission used to make both gears rotate at the same speed before meshing. (40)
Sincronizador Conjunto de dispositivos dentro de una transmisión automática, que sirve para hacer que los dos engranajes roten a la misma velocidad antes de o entrealazarse. (40)

Synthetic A product made by combining various chemical elements (a manmade product). (20)
Sintetico Producto hecho por la combinación de varios elementos químicos (producto hecho por seres humanos). (20)

Tachometer A meter used to measure speed of any rotating shaft. (10)
Tacómetro Aparato que sirve para medir la velocidad de cualquier eje rotativo. (10)

Tandem Meaning one object behind the other. (45)
Tándem Significa que un objeto está detrás de otro. (45)

Tang A projecting piece of metal placed on the end of the torque converter on automatic transmissions. The tangs are used to rotate the oil pump. (41)
Cola (Rabo) Pieza de metal saliente (proyectante) colocada en el extremo de un convertidor de torque en las transmisiones automáticas. Los rabos sirven para rotar las bombas de aceite. (41)

Tappets Another term for valve lifters. (19)
Botador (Taquete, Levantaválvulas) Otro término para alzaválvulas. (19)

TDC Position of the piston; top dead center. (13)
"TDC" Posición del pistón; punto muerto en la parte superior. (13)

Tensile strength The amount of pressure per square inch the bolt can withstand just before breaking when being pulled apart. (5)
Resistencia a la tracción (a la tensión) Cantidad de presión por pulgada cuadrada que un perno puede aguantar antes de romperse en dos partes. (5)

Testlight A small light attached to a wire (inside a sharp, pointed terminal) used to determine if electricity is at a certain point in an electrical circuit. (10)
Luz de prueba Pequeña luz conectada a un alambre a (dentro de una terminal puntiaguda) que sirve para determinar si electricidad pasa por un punto determinado en un circuito electrico. (10)

Tetraethyl lead A chemical formerly added to gasoline to increase the octane and aid in lubrication of the valves. (23)
Plomo tetraetilo Químico agregado a gasolina para incrementar el octano y ayudar a lubricar las válvulas. (23)

Thermal efficiency A measure of how effectively an engine converts heat energy in fuel into mechanical energy at the rear of the engine. (15)
Eficiencia térmica Medición de la eficacia con que un motor convierte a energía mecánica en un punto detrás del motor. (15)

Thermistor A sensor that is able to change electrical resistance on the basis of a change in temperature. (26)
Termistor Sensor que puede cambiar resistencia eléctrica a base de un cambio en temperatura. (26)

Thermostat The part on a cooling system that controls the engine coolant to its highest operating temperature. (22)
Termostato (Termóstato) Parte del sistema refrigerante que controla al flúido refrigerante hasta su temperatura de operación más alta. (22)

Thermostatic switch A heat-sensitive switch used to turn on and off an air-conditioning compressor. (50)
Conmutador (Interruptor) termostático Interruptor sensible al calor que sirve para encender y apagar un compresor de aire acondiciónado. (50)

Threaded fasteners A type of fastener such as bolts, studs, setscrews, cap screws, machine screws, and self-tapping screws that has a thread on it. (5)
Sujetadores roscados Tipo de sujetador tales como pernos, espigas, tornillos fijadores de posición, tornillos de capa, tornillos para metales, tornillos autoroscantes que contienen roscas. (5)

Threads per inch A number used to identify bolts, showing the number of threads per inch on the bolt. (5)
Roscas por pulgada Número que sirve para identificar pernos, indicando el número de roscas por pulgada en el perno. (5)

Three-phase Voltages produced from a generator or alternator can be either single-phase or three-phase. Three-phase voltages are electrically 120 degrees apart from each other. (34)
Trifásico (de tres fases) Voltajes producidos en un generador (dínamo) o alternador pueden ser o de fase sencilla o de tres fases. Voltajes trifásicos están separados eléctricamente a 120 grados uno del otro. (34)

Throttle body The part of a fuel system where fuel is injected into the air stream. The throttle body injector is located above the throttle plate. (26)
Cuerpo del regulador Parte del sistema de combustible en la que el combustible es inyectado en el chorro de aire. El cuerpo inyector del regulador está situado sobre la placa del regulador. (26)

Throttle plate The plate or circular disk that controls the amount of air going into an engine. It is usually controlled by the position of the operator's foot. (26)

Placa del regulador (Mariposa) Placa o disco circular que controla la cantidad de aire que entra en el motor. Normalmente es controlada por la posición del pie del conductor. (26)

Throttle valve (TV) A valve used in the automatic transmission that changes oil flow on the basis of throttle position of the engine. (41)
Válvula reguladora (de mariposa) Válvula usada en una transmisón automática que controla el flujo del aceite en base a la posición del regulador del motor. (41)

Throw The distance from the center point of the crankshaft to the center point of the connecting rod. (13)
Carrera (Juego) Distancia desde el punto central del cigüeñal hasta el punto central de la barra de conexión. (13)

Thrust bearing An antifriction bearing designed to absorb any thrust along the axis of the rotating shaft. (6)
Cojinete de empuje (Quiciónera) Cojinete de antifricción diseñado para absorber cualquier empuje a lo largo de la línea recta de un eje rotativo. (6)

Thrust load Another name for axial load. (6)
Tracción axial Otro término para carga axial. (6)

Thrust plate The plate used to bolt the camshaft to the block, which absorbs camshaft thrust. (19)
Placa de empuje Placa que sirve para sujetar con pernos al arbol de levas al bloque la cual absorbe el empuje del arbol de levas. (19)

Timing The process of identifying when air, fuel, and ignition occur in relation to the crankshaft rotation. (13)
Sincronización Proceso de identificar el momento cuando aire, combustible y encendido ocurren en relación con la rotación del cigüeñal. (13)

Timing diagram A graphic method used to identify the time at which all of the events of the four-stroke engine operate. (13)
Esquema de reglaje (de puestas a punto) Método gráfico que sirve para identificar el tiempo en que funciónan todos los eventos de un motor de cuatro tiempos. (13)

Timing light A strobe light placed on the number 1 cylinder spark plug wire, which flashes when electricity is available at the spark plug. (10)
Luz de tiempo Luz intermitente montada en el hilo de la bujía del número 1, la cual destella cuando la electricidad llega a la bujía. (10)

Tire placard A permanently located sticker on the vehicle that gives tire information such as load, pressure, and so on. (49)
Anuncio de llanta Calcomanía situada permanentemente en el vehículo la cual da información sobre las llantas, tal como carga, presión, *etcétera.* (49)

Tire scuffing The process that occurs when a vehicle turns a corner, when both the inner and outer wheels are locked together as in some four-wheel drive vehicles. Because of the different turning radiuses, one tire will have a tendency to scuff or slip on the road surface, causing excessive wear. (44)
Desgaste de llanta Proceso que ocurre al dar la vuelta un vehículo, cuando ambas ruedas interiores e exteriores se enclavan como en algunos vehículos de cuatro ruedas motrices. A causa de la diferencia en el radio de virage, una llanta tiene la tendencia de gastarse o patinar en la superficie del camino, causando el desgaste. (44)

Toe (in, out) The inward or outward pointing of the front wheels as measured in inches or millimeters. (48)
Convergencia o divergencia de las ruedas delanteras Dirección hacia adentro o hacia afuera de las ruedas delanteras medida en pulgadas o milímetros. (48)

Torque A twisting force applied to a shaft or bolt. (5) (15)
Torsión, Torque Fuerza de torsión aplicada a un eje o perno. (5) (15)

Torque converter The coupling between the engine and transmission on an automatic transmission. It is also used to multiply torque at lower speeds. (41)
Convertidor de torsión La union entre el motor y la transmisión de una transmisión automática. Sirve también para multiplicar la torsión a velocidades más bajas. (41)

Torque converter clutch (TCC) valve A valve used in conjunction with the lockup clutch mechanism on an automatic transmission. (41)
Válvula de embrague del convertidor de torsión Válvula que sirve en conjunto con el mecanismo del embrague de inmovilización de una transmisión automática. (41)

Torque specifications Specifications used to tell the service technician the exact torque that should be applied to bolts and nuts. (11)
Especificaciones de torsión Especificaciónes que sirven para informar al técnico de servicio la torsión fuerzas exacta que se debe aplicar a pernos y tuercas. (11)

Torque-to-yield bolt A bolt that has been tightened at the manufacturer to a preset yield or stretch point. (5)
Pernos apretados a torsión antes de ceder (estinanse) Perno que se ha apretado en la fabrica a un punto de deformacion o rendimiento pre-determinado. (5)

Torque wrench A wrench used to measure the amount of torque or twisting force applied to a bolt or nut. (9)
Llave indicadora de torsión Llave que sirve para medir la cantidad de fuerza de torsión aplicada a un perno o tuerca. (9)

Torsional rings Rings that have a slight twist when placed within the cylinder wall. These are made by adding a chamfer or counterbore on the ring. (17)
Anillos (Aros) torsionales Anillos que tienen una pequeña distorsión cuando se colocan en la pared del cilindro. Estos se hacen con la añadidura de un bisel o contrataladro en el anillo. (17)

Torsional vibration A vibration produced in a spinning shaft, caused by torque applied to the shaft. (17) (39)
Vibración torsional Vibración producida en un eje giratorio, que resulta por un par de torsión aplicado al eje. (17) (39)

Traction A tire's ability to hold or grip the road surface. (49)
Tracción Habilidad que tiene una llanta de pegarse o agarrarse a la superficie del camino. (49)

Transaxle A type of transmission used on front-wheel drive vehicles where the engine is crosswise. The transmission is designed so the differential is built in the transmission and the output goes directly to the front wheels. (40)
Ejetransversal Tipo de transmisión utilizado en vehículo de tracción delantera en la que el motor está en posición transversal. Se diseña la transmisión para que el diferencial esté construido en la transmisión y la potencia que se produce va directamente a las ruedas delanteras. (40)

Transducer A device on many electrical computer circuits that uses one type of energy as the input and another type of energy as the output, such as a piezoelectric crystal that uses motion as the input and electricity as its output. (37) (38) (52)
Transductor Aparato en muchos circuitos eléctricos de una computadora (ordenador) que usa un tipo de energía como entrada y otro tipo de energía de salida, tal como un cristal piezoeléctrico que utiliza el movimiento de entrada y la electricidad como salida. (37) (38) (52)

Transfer case An additional component added to a four-wheel drive vehicle in which the power from the engine is divided between both the front and rear wheels equally. (44)

Caja de transferencia Componente adiciónal añadido en un vehículo de cuatro ruedas motrices en el cual la potencia del motor se divide igualmente entre las ruedas delanteras y las traseras. (44)

Transistor A semiconductor used in circuits to turn off or on a second circuit; also used for amplification of signals. (30)
Transistor Semiconductor empleado en circuitos para apagar o encender un segundo circuito; también sirve para amplificación de señales. (30)

Transmission band A type of hydraulic clutch that uses a metal band fitted around a clutch housing. As the band is tightened, the housing rotation is stopped. (41)
Banda de transmisión libre Tipo de embrague hidraúlico que emplea una banda metálica ajustada alrededor de la columna del embrague. Al aprestarse la banda, se para la rotación de la columna. (41)

Transmission regulated spark (TRS) An emission control system that regulates the spark advance depending upon what gear the vehicle is in, which in turn reduces vehicle emissions. (37)
Chispa regulada por la transmisión (TRS) Sistema de control de emisión que regula el tiempo electrónico de chispa según la velocidad en la que marcha el vehículo, que por su turno disminuya las emisiones del vehículo. (37)

Tread The outer surface of a tire used to produce friction with the road for starting and stopping. (49)
Huella Superfice exterior de una llanta que sirve para producir fricción con el camino para impulsarse o pararse. (49)

Tread bars Narrow strips of rubber molded into the tread. When the tread bars show, the tire is worn enough to be replaced. (49)
Bandas de rodadura Franjas angostas de goma moldeadas en las huellas. Cuando aparecen las bandas de rodadura, se ha desgastado suficiente como para reponerse la llanta. (49)

Tripod The central part of certain CV joints. It has three arms or trunnions with needle bearings and rollers running in grooves or races in the assembly. (43)
Trípode Parte central de ciertas juntas CV. Tiene tres muñones (soportes giratorios) con cojinetes de agua y rodillos colocados en surcos en el montaje. (43)

Trouble code A code stored inside a vehicle's computer that tells the service technician the exact component that is faulty. (25)
Código de averías Código almacenado dentro de la computadora (ordenador) del vehículo que informa al tecnico de servicio exactamente cual componente está defectuoso. (25)

Troubleshooting Another term for diagnosis. (11)
Investigación de problemas Otro término para diagnóstico. (11)

TRS Transmission-regulated spark. (39)
"TRS" Chispa controlada por la transmission (CCT). (39)

Trunnion The arm or arms of the four-point U-joint, which serves as the inner bearing surface or race. (43)
Muñón (Gorrón, espiga, soporte giratorio) Brazo o brazos de una junta universal de cuatro puntos, el cual sirve como la superficie de cojinete interior o anillo-guía. (43)

Tuned ports Intake ports used on fuel injection engines, designed to produce equal and minimum restriction to the airflow. (26)
Puertos de admisión sintonizados Orificios de admisión utilizados en motores de inyección de combustible, diseñados para producir restricción igual y mínima en el flujo de aire. (26)

Tune-up specifications Specifications primarily used during a tune-up on an automobile. (11)
Especificaciones de afinación Especificaciónes que sirven principalmente cuando se afina el motor. (11)

Turbine A component in a gas turbine engine that changes the energy in the gases into rotary motion for power. A vaned type of wheel being turned by a fluid such as exhaust gases passing over it. The rotary part or vaned wheel inside a torque converter, used to turn the transmission. (28) (41)
Turbina Componente en un motor de turbina de gas, el cual cambia la energía de los gases en movimiento rotativo para crear potencia. Un tipo de rueda o volante con veletas que se deja girar cuando un flúido tal como gases de escape pasan sobre las veletas. La parte rotativa o rueda con veletas dentro de un convertidor de par de torsión, que sirve para hacer girar la transmisión. (28) (41)

Turbocharged An engine that uses the exhaust gases to turn a turbine. The turning turbine forces in extra fresh air for more performance. (27)
Turbocargado Motor que utiliza los gases de escape para propulsar una turbina. La turbina giratoria hace que entre más aire fresco para más rendimiento. (27)

Turbocharger A device that uses the energy in the engine exhaust gases to turn a compressor, which pumps more air into the engine. (16)
Turbocargador Aparato que utiliza la energía en los gases de salida del motor para dar vueltas a un comprimidor, lo cual inyecta más aire en el motor. (16)

Turbulence A term used to describe combustion chambers. It means rapid movement and mixing of air and fuel inside the combustion chamber. (18)
Turbulencia Término que sirve para describir las cámaras de combustión. Significa movimiento rápido y combinación de aire y combustible dentro de la cámara de combustión. (18)

Turning radius The amount (in degrees) that one front wheel turns more sharply than the other front wheel during a turn. (48)
Radio de giro (vuelta) Cantidad (en grados) que una rueda delantera gira más agudamente que la otra rueda delantera durante el giro. (48)

Two-stroke engine A type of engine that requires only two strokes (or one crankshaft revolution) to produce the intake, compression, power, and exhaust sequence. (14)
Motor de dos tiempos Tipo de motor que requiere sólo dos tiempos (o una revolución del cigüeñal) para producir la sequencia de entrada de aire, compresión, potencia, y escape. (14)

Type A fire A fire resulting from the burning of wood, paper, textiles, and clothing. (4)
Incendio de tipo A Fuego que resulta de la combustión de madera,papel, textiles, y ropa. (4)

Type B fire A fire resulting from the burning of gasoline, greases, oils, and other flammable liquids. (4)
Incendio de tipo B Fuego que resulta de la combustión de gasolina, grasas, aceites, y otros líquidos inflamables. (4)

Type C fire A fire resulting from the burning of electrical equipment, motors, and switches. (4)
Incendio de tipo C Fuego que resulta de la combustión de equipo eléctrico, motores, y conmutadores (interruptores). (4)

Type D fire A fire resulting from the burning of various types of metals, such as sodium, titanium, and zirconium. (4)
Fuegos de tipo D Fuego que resulta del incendio de varios tipos de metales, tal como el sodio, el titanio, y el zirconio. (4)

Upshifting The process of shifting from a lower gear to a higher gear. (41)
Cambio ascendente Proceso de cambiar la velocidad de una velocidad baja a una más alta. (41)

USC measurements U.S. Customary (standard English) measurements, including feet, inches, miles, pounds, ounces, and so on. (9)
Medidas de EE.UU acostumbradas (inglés estándar) Medidas, incluyendo pies, pulgadas, millas, libras, onzas, *etcétera.* (9)

Vacuum An enclosed space in which the pressure is below zero psig. (22) (25)
Vacío Espacio encerrado en que la presión está más abajo de cero libras por pulgada cuadrada ("Psig"). (22) (25)

Vacuum gauge A gauge designed to read various vacuum readings on an engine, the most common being intake manifold vacuum and pollution control equipment. (9)
Vacuómetro (Manómetro de vacío) Indicador que sirve para leer varias indicaciones de vacío de un motor, las más comunes son las del vacío del múltiple de entrada y equipo para control de contaminación. (9)

Vacuum modulator A diaphragm used on automatic transmissions that controls the throttle valve on the basis of engine intake manifold vacuum. (41)
Modulador de vacío Diafragma utilizado en transmisiones automáticas, el cual controla a la válvula de la mariposa a base del vacío del múltiple de entrada del motor. (41)

Vacuum motor A small diaphragm inside a housing operated by vacuum working against a spring pressure. When vacuum is applied, the diaphragm moves. This movement is then used to open or close small air doors or other apparatus. (51)
Motor de vacío Pequeño diafragma dentro de una cubierta protectora, el cual es operado por un vacío que funciona contra una presión de resortes. Al aplicarse el vacío, se mueve el diafragma. Este movimiento sirve para abrir y cerrar pequeños orificios de aire u otros aparatos. (51)

Valence ring The outer orbit of electrons in an atom. (29)
Banda de valencia Orbita exterior de electrones de un átomo. (29)

Valve A device used to open and close a port to let intake and exhaust gases in and out of the engine. (12)
Válvula Aparato que sirve para abrir y cerrar un orificio (puerto) para dejar de los gases de entrada y de escape pasen adentro o afuera del motor. (12)

Valve body The part of an automatic transmission used to direct the oil flow to different parts of the transmission. The valve body also houses most of the valve for control. (41)
Cuerpo de válvula Parte de la tranmisión automática que sirve para dirigir el flujo del aceite a diferentes partes de la transmisión. El cuerpo de la válvula también acomoda a la mayoría de la válvula para control. (41)

Valve bounce When a valve is forced to close because of spring pressure, the valve may bounce when it closes. This action can damage the seats or break the valve in two. (18)
Rebotre de la válvula Cuando se cierra la válvula a fuerza debido a la presión de resorte, puede que se reborte la valvula al cerrarse. Está acción puede dañar los asientos o causar que se rompa la válvula. (18)

Valve clearance The clearance or space between the valve and the rocker arm. As the parts heat up, the clearance is reduced because of expansion. This keeps the valves from remaining open when the engine is hot. (16)
Holgura de la válvula Distancia o espacio entre la válvula y el balancín. Al calentarse las partes, se reduce la distancia debido a la expansión. Esto previene que queden abiertas las válvulas cuando esté caliente el motor. (16)

Valve face The part of a poppet valve that actually touches the seat for sealing in the cylinder head. (18)
Cara (Sombrete) de la válvula La parte de una válvula de elevacíon que realmente toca al asiento para sellar la cabeza del cilindro. (18)

Valve float If a valve spring is not strong enough to close the valve, the valve may float or stay open slightly longer than designed. This condition will limit the maximum rpm an engine can develop. (18)
Válvula flotante Si el resorte de la válvula no tiene suficiente fuerza para cerrar la válvula, está puede flotar o quedar abierta un poco más tiempo que lo anticipado por su diseño. Está condición limitará las rpm máximas que puede producir un motor. (18)

Valve guide The part in the cylinder head that holds the stem of the valve. (18)
Guía de la válvula Parte de al cabeza de la válvula que sostiene el vástago (varilla) de válvula. (18)

Valve train clearance The clearance between the lifters, rocker arms, and valves. This clearance is necessary because as the parts heat up, they will expand. The valve train clearance allows for this expansion. (19)
Holgura del tren de la válvula Distancia o espacio entro los levantaválvulas, balancines y las válvulas. Esta distancia es necesaria porque al calentarse las partes, se expanden. La holgura del tren de la válvula acomoda esta expansión. (19)

Vapor A substance in a gaseous state. Liquids become vapor when they are brought above their boiling point. (50)
Vapor Substancia en un estado gaseoso. Los líquidos se convierten en vapor cuando superan su punto de ebullición. (50)

Vaporize The process of passing from a liquid to a gas. Fuel is vaporized when it is heated. (23) (25)
Vaporizar Proceso de pasar de un líquido a un gas. Combustible se vaporiza al calentarse. (23) (25)

Vapor lock Vapor buildup that restricts the flow of gasoline through the fuel system. Vapor lock occurs from heating the fuel, causing it to turn to a vapor. (23)
Tapón de vapor Obstructión por vapores que restringe el flujo de gasolina por el sistema. Tapón de vapor ocurre porque se ha calentado el combustible, produciendo así un vapor. (23)

Variable displacement Type of engine that is able to change its displacement by using either four or eight cylinders. (16)
Motor de desplazamiento (cilindrado) variable Motor que puede cambiar su cilindrada por medio de usar o cuatro u ocho cilindros. (16)

Variable resistor A resistor that is able to change its resistance by moving a lever or handle. Often a variable resistor is used to control some electric circuit that needs a variable resistance, such as a variable speed fan motor used for heating the passenger compartment of a vehicle. (21) (38)
Resistor variable Resistor capaz de cambiar su resistencia al mover una palanca o manívela. Muchas veces un resistor variable sirve para controlar algun circuito eléctrico que requiere una resistencia variable, tal como un motor de ventilador de velocidad variable que sirve para calentar el compartimento de pasajero del vehículo. (21) (38)

Variable valve timing A mechanical-hydraulic-electrical system able to change the point at which the valve opens and closes. (19)
Reglaje de las válvulas varible Sistema mecánico-hidráulico-eléctrico que puede cambiar el punto al cual la válvula abre y cierra. (19)

Varnish A deposit in an engine lubrication system resulting from oxidation of the motor oil. Varnish is similar to, but softer than, lacquer. (20)
Barniz Depósito en el sistema de lubricación del motor que resulta de la oxidación del aceite de motor. El barniz es similar a, pero más suave que la laca. (20)

VAT A voltage amperage tester used for checking the output of a charging system. (34)
"VAT" Analizador de amperaje y voltaje (AAV) que sirve para comprobar el rendimineto de un sistema de carga. (34)

V configuration A style of engine that has two rows of cylinders that are approximately 90 degrees apart and in a V shape. (16)
Configuración V Diseño del motor que tiene dos filas de cilindros que están separados aproximadamente a 90 grados y tienen una forma de V. (16)

Vehicle Emission Control Information (VECI) Emission control information shown directly on a label on each vehicle. (36)
Información de Control de Emisiones de Vehículo (ICEV) Información de control de emisiones indicada directamente en la calcomanía de cada vehículo. (36)

Vehicle speed sensor A sensor that is used in conjunction with a computer to measure the speed of the vehicle, used as an input to the computer. (41)
Sensor de velocidad del vehículo Sensor que sirve junto con una computadora (ordenador) para medir la velocidad del vehículo, usado como una entrada para la computadora (ordenador). (41)

Venturi A restriction in a tube where air or liquid is flowing. A venturi always causes a vacuum to be created at the point of greatest restriction. (25)
"Venturi" Restricción en un tubo donde circula aire o líquido. Un venturi siempre causa la producción de un vacío al punto de la restricción más grande. (25)

Vernier caliper A measuring tool used to accurately measure length to 0.001 inch. (9)
Calibre de Nonio (de Vernier) Herramienta de medición que sirve para medir con precision hasta 0.001 de pulgada. (9)

Vernier scale A scale for measuring in which two lines are adjusted to line up vertically with each other. (9)
Escala Vernier (Nonio) Escala que sirve para medir, en la cual dos líneas se ajustan para alinearse verticalmente una con la otra. (9)

VIN The vehicle identification number, located on the left front of the dashboard, which represents various data such as the model of the vehicle, year, body, style, engine type, and serial number. (11)
"VIN" Número de identificación de vehículo (NIV), situado delante y a la izquierda del panel de controles, el cual representa varios datos tales como el modelo del vehículo, año, carrocería, estilo, tipo de motor, y número de serie. (11)

Viscosity A fluid property that causes resistance to flow. The higher the viscosity, the greater the resistance to flow. The lower the viscosity, the easier for the fluid to flow. (20)
Viscosidad Propiedad de fluído que produce resistencia al flujo. Cuanto más alta la viscosidad, tanto más resistencia haya en el flujo. Cuanto meonos sea la viscosidad, tano más facil es que fluya el fluido. (20)

Viscosity index A common term used to measure a fluid's change of viscosity with a change in temperature. The higher the viscosity index, the smaller the relative change in viscosity with temperature. (20)
Indice de viscosidad Término común que sirve para medir el cambio en la viscosidad de un flúido con un cambio en la temperatura. Cuanto más alto el índice de viscosidad, tanto más pequeño sea el cambio relativo en viscosidad con temperatura. (20)

Viscous coupling A coupling with a thick fluid place inside its housing, used on all-wheel drive vehicles to eliminate tire scuffing and excessive wear. (44)
Acoplamiento viscoso Acoplamiento con un compartimento para fluido espeso dentro de su cárter, usado en los vehículos de todas ruedas motrices para eliminar el desgaste de las llantas y el gasto excesivo. (44)

Volatility The ease with which a fuel is able to ignite. (23)
Volatilidad Facilidad con la que un combustible puede encenderse. (23)

Voltage The push or pressure used to move electrons along a wire. (29)
Voltaje Fuerza o presión que sirve para empujar a los electrones a lo largo de un alambre. (29)

Voltage drop Voltage lost at each resistor, usually defined as $I \times R$ drop. (29)
Perdida de voltaje Voltaje perdido en cada resistencia, normalmente definido como caida $I \times R$. (29)

Volt-ohm-ammeter Also called a multimeter; a testing instrument able to read voltage, resistance, and amperage in an electrical circuit. (10)
Voltiohmímetro-amperímetro Llamado también un multímetro; intsrumento de medición para comprobar que haya voltaje, resistencia, y amperaje en un circuito eléctrico. (10)

Volumetric efficiency A measure of how well air flows in and out of an engine. (15)
Rendimiento volumétrico Medición de la eficiencia del flujo del aire hacia adentro y afuera de un motor. (15)

Vulcanized A process of heating rubber under pressure to mold it into a special shape. (49)
Vulcanizado Proceso de calentar goma bajo presión para moldearla de una forma especial. (49)

Waste spark A spark occurring during the exhaust stroke on a computerized ignition system. (33)
Chispa de desecho Chispa que ocurre en el recorrido de escape en un sistema de encendido computarizado. (33)

Wattage A measure of the total power of an electrical circuit, calculated by multiplying the voltage in the circuit times the amperage in the circuit. (29)
Wataje (Vataje, Vatiaje) Medición de la potencia total de un circuito eléctrico, calculada al multiplicar el voltaje en el circuito por el amperaje en el circuito. (29)

Watt's Law Power equals voltage times amperage ($P = E \times I$). (29)
Ley de Watt Potencia equivale al voltaje multiplicado por el amperaje; ($P = E \times I$). (29)

Waveform A graphical representation of the voltage output on a charging system. (34)
Forma de onda Representación gráficia del rendimiento (salida o capacidad) de voltaje en un sistema de carga. (34)

Wedge-shaped combustion chamber A type of combustion chamber that is shaped similar to a wedge or V. This chamber is designed to increase the movement of air and fuel to aid in mixing. (18)
Camára de combustión cuneiforme Tipo de cámara de combustión que tiene una forma similar a la de una cuña o V. Esta cámara es diseñada para aumentar el movimiento de aire y combustible para ayundar a mezclarlos. (18)

Wheel cylinder A device used to convert hydraulic fluid pressure to mechanical force for brake applications. (45)
Cilindro de rueda Un aparato que sirve para convertir presión de flúido hidráulico en una fuerza mecánica para aplicaciones de frenos. (45)

Wheel runout A measure of the out-of-roundness of a wheel or tire. (49)
Falta de redondez de llanta o rueda Medición de la falta de la redondez verdadera de una llanta o una rueda. (49)

Wheel speed sensor A sensor on each wheel used to monitor speed in an antilock braking system. (46)
Sensor de velocidad de la rueda Un sensor ubicado en cada rueda, utilizado para monitorear la velocidad en el sistema de frenos antibloqueo. (46)

Wheel spindle The short shaft on the front wheel upon which the wheel bearings ride and to which the wheel is attached. (47)
Muñon de rueda Eje corto en la rueda delantera sobre el cual los cojinetes de rueda se sientan y al cual está conectada la rueda. (47)

Winter mode One of several modes used on an electronic transmission in which the shifting patterns change for winter driving. (41)
Modo de invierno Uno de varios modos usados en una transmisión electrónica en el cual los patrones de cambio de velocidad cambia para el manejo en el invierno. (41)

Work Work is defined as the result of a force applied to a mass, moved a certain distance. Work = Force × Distance. (15)
Trabajo Se define el trabajo como el resultado de una fuerza aplicada a una masa, movida a cierta distancia. Trabajo = Fuerza × Distancia. (15)

Work-based learning Learning that is oriented toward some aspect of the working world. Often work-based learning occurs in a classroom in order to prepare for a specific career or job after the learning has been completed. (2)
Aprendizaje en mano de obra El aprendizaje dirigido hacia algún aspecto del mundo del trabajo. El aprendizaje en mano de obra ocurre en una sala de clase como preparación para una carrera específica o un trabajo después de que se haya completado el aprendizaje.(2)

Yoke The Y-shaped metal device that is attached to the drive shaft. (43)
Yugo Aparato metálico en forma de una Y, el cual está conectado al eje impulsor (motor). (43)

Zener diode A type of diode that requires a certain amount of voltage before it will conduct electricity. This voltage is used to control transistors in voltage regulators. (34)
Díodo Zener Tipo de díodo que requiere una cantidad de voltaje determinada antes de que conduzca electricidad. Este voltaje sirve para controlar transistores en reguladores de voltaje. (34)

Zerk A lubrication fitting through which grease is applied to a steering joint with a grease gun. (47) (48)
"Zerk" Conector de lubricación por el cual se aplica grasa a una junta de dirección con un pistola (jeringa) para engrasar. (47) (48)

Zirconia A chemical used in various ceramics; often used as an abrasive. (38)
Zirconio Química usado en various ceramicas; suele usarse como un abrasivo. (38)

Index

Note: Page numbers in bold type refer to figures.